2010 Twenty-Fifth Annual IEEE Applied Power Electronics Conference and Exposition

(APEC 2010)

Palm Springs, California, USA
21 – 25 February 2010

Pages 768-1541

IEEE Catalog Number:	CFP10APE-PRT
ISBN:	978-1-4244-4782-4

Copyright © 2010 by the Institute of Electrical and Electronic Engineers, Inc
All Rights Reserved

Copyright and Reprint Permissions: Abstracting is permitted with credit to the source. Libraries are permitted to photocopy beyond the limit of U.S. copyright law for private use of patrons those articles in this volume that carry a code at the bottom of the first page, provided the per-copy fee indicated in the code is paid through Copyright Clearance Center, 222 Rosewood Drive, Danvers, MA 01923.

For other copying, reprint or republication permission, write to IEEE Copyrights Manager, IEEE Service Center, 445 Hoes Lane, Piscataway, NJ 08854. All rights reserved.

***This publication is a representation of what appears in the IEEE Digital Libraries. Some format issues inherent in the e-media version may also appear in this print version.**

IEEE Catalog Number: CFP10APE-PRT
ISBN 13:   978-1-4244-4782-4
Library of Congress No.: 90-643607
ISSN: 1048-2334

Additional Copies of This Publication Are Available From:

Curran Associates, Inc
57 Morehouse Lane
Red Hook, NY 12571 USA
Phone: (845) 758-0400
Fax: (845) 758-2633
E-mail: curran@proceedings.com
Web: www.proceedings.com

TABLE OF CONTENTS

Session A1L-A: DC-DC Converter I
Tuesday, February 23, 8:30 - 10:10
Session Chairs: Van Niemela, *Fairchild Semiconductor*
Haidong Yu, *Phoenix International*

Minimum Deviation Digital Controller IC for Single and Two Phase DC-DC Switch-Mode Power Supplies 1
Aleksandar Radić, *University of Toronto, Canada*
Zdravko Lukić, *University of Toronto, Canada*
Aleksandar Prodić, *University of Toronto, Canada*
Robert de Nie, *NXP Semiconductors, Netherlands*

Modeling and Design Considerations of Coupled Inductor Converters 7
Guangyong Zhu, *Auscom Engineering, Inc., United States*
Kunrong Wang, *Dell, Inc., United States*

Design Procedure for High Frequency Operation of the Modified Series Resonant APWM Converter with Improved Efficiency and Reduced Size 14
Darryl J. Tschirhart, *Queen's University, Canada*
Praveen K. Jain, *Queen's University, Canada*

Expiremantal Results and Study of a Modified Adaptive Bus Voltage Controller 19
Jaber A. Abu Qahouq, *University of Alabama, United States*
Gautam Muralidhar, *University of Alabama, United States*

Session A1L-B: AC-DC Power Factor Correction Topologies I
Tuesday, February 23, 8:30 - 10:10
Session Chairs: Gerry Moschopoulos, *University of Western Ontario*
Omer Onar, *Illinois Institute of Technology*

Bridgeless Buck PFC Rectifier 23
Yungtaek Jang, *Delta Products Corporation, United States*
Milan M. Jovanović, *Delta Products Corporation, United States*

An Active-Clamped Full-Wave Zero-Current-Switched Quasi-Resonant Boost Converter in Power Factor Correction Application 30
E. Firmansyah, *Kyushu University, Japan*
S. Abe, *Kyushu University, Japan*
M. Shoyama, *Kyushu University, Japan*
S. Tomioka, *TDK-Lambda Corporation, Japan*
T. Ninomiya, *Nagasaki University, Japan*

Novel Adaptive Master-Slave Method for Interleaved Boundary Conduction Mode (BCM) PFC Converters 36
Hangseok Choi, *Fairchild Semiconductor, United States*

A Novel Bridgeless Single-Stage Half-Bridge AC-DC Converter 42
Woo-Young Choi, *Virginia Polytechnic Institute and State University, United States*
Wen-Song Yu, *Virginia Polytechnic Institute and State University, United States*
Jih-Sheng Lai, *Virginia Polytechnic Institute and State University, United States*

Session A1L-C: Power Electronics for Utility Interface I
Tuesday, February 23, 8:30 - 10:10
Session Chairs: Zareh Soghomonian, *BMT Syntek Technologies*
Jin Wang, *Ohio State University*

Power Quality Improvement at Medium-Voltage Grids Using Hexagram Active Power Filter 47
Jun Wen, *University of California, Irvine, United States*
Liang Zhou, *University of California, Irvine, United States*
Keyue Smedley, *University of California, Irvine, United States*

A Generalized Capacitor Voltage Balancing Scheme for Flying Capacitor Multilevel Converters 58
Mostafa Khazraei, *Missouri University of Science and Technology, United States*
Hossein Sepahvand, *Missouri University of Science and Technology, United States*
Keith Corzine, *Missouri University of Science and Technology, United States*
Mehdi Ferdowsi, *Missouri University of Science and Technology, United States*

An Active Damping Technique for a Current Source Inverter Employing a Virtual Negative Inductance 63
Ahmed Salah Morsy, *Texas A&M University at Qatar, Qatar*
Shehab Ahmed, *Texas A&M University at Qatar, Qatar*
Prasad Enjeti, *Texas A&M University at Qatar, Qatar*
Ahmed Massoud, *Qatar University, Qatar*

Maximum Solar Power Transfer in Multi-Port Power Electronic Interface 68
Wei Jiang, *University of Texas at Arlington, United States*
Babak Fahimi, *University of Texas at Arlington, United States*

Session A1L-D: Passive Devices I
Tuesday, February 23, 8:30 - 10:10
Session Chairs: Laura Lyle, *Wright Patterson Air Force Base*
Mike Schutten, *General Electric*

SMD Inductors Based on Soft-Magnetic Powder Compacts 74
Etsuo Otsuki, *Toho Zinc Co., Ltd., Japan*
Kenichiro Ishii, *Toho Zinc Co., Ltd., Japan*
Shinya Nakano, *Toho Zinc Co., Ltd., Japan*

High Density Low Profile Coupled Inductor Design for Integrated Point-of-Load Converter 79
Qiang Li, *Virginia Polytechnic Institute and State University, United States*
Yan Dong, *Virginia Polytechnic Institute and State University, United States*
Fred C. Lee, *Virginia Polytechnic Institute and State University, United States*

Relationship of Quality Factor and Hollow Winding Structure of Coreless Printed Spiral Winding (CPSW) Inductor 86
Y.P. Su, *Virginia Polytechnic Institute and State University, United States*
Xun Liu, *ConvenientPower HK Ltd., China*
C.K. Lee, *Hong Kong Polytechnic University, China*
S.Y.R. Hui, *City University of Hong Kong, China*

Modeling of Adaptable-Diameter Burners Formed by Concentric Planar Windings for Domestic Induction Heating Applications 92
Jesus Acero, *University of Zaragoza, Spain*
Claudio Carretero, *University of Zaragoza, Spain*
Ignacio Millan, *University of Zaragoza, Spain*
Oscar Lucía, *University of Zaragoza, Spain*
Jose-Miguel Burdío, *University of Zaragoza, Spain*
Rafael Alonso, *University of Zaragoza, Spain*

Session A1L-E: Controls in Motor Drives I
Tuesday, February 23, 8:30 - 10:10
Session Chairs: Jonathan Kimball, *Missouri S&T*

Flux Concentration and Pole Shaping in a Single Phase Hybrid Switched Reluctance Motor Drive 98
Uffe Jakobsen, *Aalborg University, Denmark*
Kaiyuan Lu, *Aalborg University, Denmark*

Parameter Independent Maximum Torque Per Ampere (MTPA) Control of IPM Machine Based on Signal Injection 103
Sungmin Kim, *Seoul National University, Korea, South*
Young-Doo Yoon, *Seoul National University, Korea, South*
Seung-Ki Sul, *Seoul National University, Korea, South*
Kozo Ide, *Yaskawa Electric Corporation, Japan*
Koji Tomita, *Yaskawa Electric Corporation, Japan*

Performance Analysis of Three-Phase Capacitor Motor in Frequency Control System 109
Zheng-Feng Ming, *Xidian University, China*
Guang-Zheng Ni, *Zhejiang University, China*
Bing-Zhong Yang, *Chongqing University, China*

Efficiency Improvement by Changeover of Phase Windings of Multiphase Permanent Magnet Synchronous Motor with Outer-Rotor Type 112
Young-Gook Kim, *Pusan National University, Korea, South*
Chae-Bong Bae, *Pusan National University, Korea, South*
Jang-Mok Kim, *Pusan National University, Korea, South*
Hyun-Cheol Kim, *Agency for Defense Development, Korea, South*

Session A1L-F: Digital Controls in DC-DC Converters I
Tuesday, February 23, 8:30 - 10:10
Session Chairs: Dragan Maksimović, *University of Colorado at Boulder*
Jason Neely, *Purdue University*

Digital Power Controller with Non-Linear Variable Switching Frequency 120
Jaber A. Abu Qahouq, *University of Alabama, United States*

**Digital Charge Balance Controller with an Auxiliary Circuit for Superior Unloading
Transient Performance of Buck Converters** 124
Eric Meyer, *Queen's University, Canada*
Dong Wang, *Queen's University, Canada*
Liang Jia, *Queen's University, Canada*
Yan-Fei Liu, *Queen's University, Canada*

One-Step Digital Dead-Time Correction for DC-DC Converters 132
April Zhao, *University of Toronto, Canada*
Armin Akhavan Fomani, *University of Toronto, Canada*
Wai Tung Ng, *University of Toronto, Canada*

**The Practical Aspects of Utilizing Digital Power Controller for
Monitoring of Power Supply Operation** 138
Oleg Volfson, *Intersil Corporation, United States*

Session A1L-G: Wind Power
Tuesday, February 23, 8:30 - 10:10
Session Chairs: Morgan Kiani, *University of Texas at Arlington*

**A Unity Power Factor, Maximum Power Point Tracking Battery Charger for
Low Power Wind Turbines** 143
Gustavo Gamboa, *University of Central Florida, United States*
John Elmes, *University of Central Florida, United States*
Christopher Hamilton, *University of Central Florida, United States*
Jonathan Baker, *University of Central Florida, United States*
Michael Pepper, *University of Central Florida, United States*
Issa Batarseh, *University of Central Florida, United States*

**Maximum Power Point Tracking of a Wind Energy Conversion System Using
Adaptive Nonlinear Approach** 149
Majid Pahlevaninezhad, *Queen's University, Canada*
Suzan Eren, *Queen's University, Canada*
Alireza Bakhshai, *Queen's University, Canada*
Praveen Jain, *Queen's University, Canada*

A Hybrid Wind-Solar Energy System: a New Rectifier Stage Topology 155
Joanne Hui, *Queen's University, Canada*
Alireza Bakhshai, *Queen's University, Canada*
Praveen Jain, *Queen's University, Canada*

Dynamic Operation and Control of a Hybrid Wind-Diesel Stand Alone Power Systems 162
A.M.O. Haruni, *University of Tasmania, Australia*
A. Gargoom, *University of Tasmania, Australia*
M.E. Haque, *University of Tasmania, Australia*
M. Negnevitsky, *University of Tasmania, Australia*

Session A2L-A: DC-DC Converter II
Tuesday, February 23, 10:40 - 11:55
Session Chairs: Van Niemela, *Fairchild Semiconductor*
　　　　　　　　　　Haidong Yu, *Phoenix International*

SystemC-AMS Modeling and Simulation of Digitally Controlled DC-DC Converters 170
Matteo Agostinelli, *University of Klagenfurt, Austria*
Robert Priewasser, *University of Klagenfurt, Austria*
Mario Huemer, *University of Klagenfurt, Austria*
Stefano Marsili, *Infineon Technologies Austria AG, Austria*
Dietmar Straeussnigg, *Infineon Technologies Austria AG, Austria*

Modeling of Digitally Controlled Voltage Regulator Modules 176
Yi Sun, *Linear Technology, United States*
Fred C. Lee, *Virginia Polytechnic Institute and State University, United States*
Jian Li, *Linear Technology, United States*

**Design and Comparison of Digital Control Loops Analytical Models, Laboratory
Measurements, and Simulation Results** 183
Philip Cooke, *Infineon Technologies, United States*
Thomas G. Wilson, Jr., *SIMPLIS Technologies, United States*
Rohan Samsi, *Primarion, United States*

Session A2L-B: AC-DC Power Factor Correction Topologies II
Tuesday, February 23, 10:40 - 11:55
Session Chairs: Gerry Moschopoulos, *University of Western Ontario*
　　　　　　　　　　Omer Onar, *Illinois Institute of Technology*

Digital Control for Efficiency Improvements in Interleaved Boost PFC Rectifiers 188
Fu-Zen Chen, *University of Colorado at Boulder, United States*
Dragan Maksimović, *University of Colorado at Boulder, United States*

**Reduction of the Output Capacitor in Power Factor Correctors by
Distorting the Line Input Current** 196
Diego G. Lamar, *Universidad de Oviedo, Spain*
Javier Sebastián, *Universidad de Oviedo, Spain*
Manuel Arias, *Universidad de Oviedo, Spain*
Arturo Fernández, *Universidad de Oviedo, Spain*

**Universal-Input Single-Stage PFC Flyback with Variable Boost Inductance for
High-Brightness LED Applications** 203
Yuequan Hu, *Delta Products Corporation, United States*
Laszlo Huber, *Delta Products Corporation, United States*
Milan M. Jovanović, *Delta Products Corporation, United States*

Session A2L-C: Power Electronics for Utility Interface II
Tuesday, February 23, 10:40 - 11:55
Session Chairs: Zareh Soghomonian, *BMT Syntek Technologies*
Jin Wang, *Ohio State University*

**High Frequency High Efficiency Bidirectional DC-DC Converter Module Design for
10 kVA Solid State Transformer** 210
Haifeng Fan, *Florida State University, United States*
Hui Li, *Florida State University, United States*

**Synchronization of Three-Phase Converters and Virtual Microgrid Implementation
Utilizing the Power-Hardware-in-the-Loop Concept** 216
O. Vodyakho, *Florida State University, United States*
C.S. Edrington, *Florida State University, United States*
M. Steurer, *Florida State University, United States*
S. Azongha, *Florida State University, United States*
F. Fleming, *Florida State University, United States*

**A Single-Stage Grid-Connected Inverter with Wide Range Reactive Power
Compensation Using Energy Storage System (Ess)** 223
Liming Liu, *Florida State University, United States*
Zhichao Wu, *Florida State University, United States*
Hui Li, *Florida State University, United States*

Session A2L-D: Passive Devices II
Tuesday, February 23, 10:40 - 11:55
Session Chairs: Laura Lyle, *Wright Patterson Air Force Base*
Mike Schutten, *General Electric*

Polymer Bonded Soft Magnetics for EMI Filter Applications in Power Electronics 231
S. Egelkraut, *University of Erlangen-Nürnberg, Germany*
L. Frey, *University of Erlangen-Nürnberg, Germany*
M. Rauch, *Fraunhofer Institute for Integrated Systems and Device Technology, Germany*
A. Schletz, *Fraunhofer Institute for Integrated Systems and Device Technology, Germany*
M. März, *Fraunhofer Institute for Integrated Systems and Device Technology, Germany*

Lead-Acid Battery Modeling and State of Charge Monitoring 239
J.F. Araujo Leão, *Universidade Federal de Campina Grande, Brazil*
L.V. Hartmann, *Universidade Federal de Campina Grande, Brazil*
M.B.R. Corrêa, *Universidade Federal de Campina Grande, Brazil*
A.M.N. Lima, *Universidade Federal de Campina Grande, Brazil*

**Voltage and Current Ripple Considerations for Improving Lifetime of Ultra-Capacitors
Used for Energy Buffer Applications at Converter Inputs** 244
Supratim Basu, *Bose Research Pvt. Ltd., India*
Tore M. Undeland, *Norwegian University of Science and Technology, Norway*

Session A2L-E: Controls in Motor Drives II
Tuesday, February 23, 10:40 - 11:55
Session Chairs: Jonathan Kimball, *Missouri S&T*

Implementation and Operational Investigations of Bipolar Gate Drivers 248
Jean-Christophe Crebier, *Grenoble Institute of Technology, France*
Manh Hung Tran, *Grenoble Institute of Technology, France*
Jean Barbaroux, *Grenoble Institute of Technology, France*
Pierre-Olivier Jeannin, *Grenoble Institute of Technology, France*

**A Method for Impact Assessment of Faults on the Performance of Field-Oriented
Control Drives: a First Step to Reliability Modeling** 256
Ali M. Bazzi, *University of Illinois at Urbana-Champaign, United States*
Alejandro Dominguez-Garcia, *University of Illinois at Urbana-Champaign, United States*
Philip T. Krein, *University of Illinois at Urbana-Champaign, United States*

A Fault Tolerant Control System for Hexagram Inverter Motor Drive 264
Liang Zhou, *University of California, Irvine, United States*
Keyue Smedley, *University of California, Irvine, United States*

Session A2L-F: Digital Controls in DC-DC Converters II
Tuesday, February 23, 10:40 - 11:55
Session Chairs: Dragan Maksimović, *University of Colorado at Boulder*
　　　　　　　　　　Jason Neely, *Purdue University*

Power Analog to Digital Converter for Voltage Scaling Applications 271
M.C. Gonzalez, *Universidad Politécnica de Madrid, Spain*
M. Vasić, *Universidad Politécnica de Madrid, Spain*
P. Alou, *Universidad Politécnica de Madrid, Spain*
O. Garcia, *Universidad Politécnica de Madrid, Spain*
J.A. Oliver, *Universidad Politécnica de Madrid, Spain*
J.A. Cobos, *Universidad Politécnica de Madrid, Spain*
H. Visairo, *Intel Corporation, Mexico*

**A Digital Pulse-Width Modulator for Phase-Shift Operation of
Full-Bridge Isolated DC-DC Converters** 277
L. Corradini, *University of Colorado at Boulder, United States*
D. Maksimović, *University of Colorado at Boulder, United States*

**Digitally Controlled Integrated Pseudo-CCM SIMO Converter with
Adaptive Freewheel Current Modulation** 284
Yi Zhang, *University of Arizona, United States*
Dongsheng Ma, *University of Arizona, United States*

Session A2L-G: Fuel Cells
Tuesday, February 23, 10:40 - 11:55
Session Chairs: Morgan Kiani, *University of Texas at Arlington*

Analysis of Pulse-Link DC-AC Converter for Fuel Cells Applications Operated in Zero-Current-Slope Mode 289
Kentaro Fukushima, *Kyushu University, Japan*
Isami Norigoe, *I.N. Laboratory, Japan*
Masahito Shoyama, *Kyushu University, Japan*
Tamotsu Ninomiya, *Nagasaki University, Japan*
Yosuke Harada, *Ebara Densan Ltd., Japan*
Kenta Tsukakoshi, *Ebara Densan Ltd., Japan*

A Minimum Power-Processing Stage Fuel Cell Energy System Based on a Boost-Inverter with a Bi-Directional Back-Up Battery Storage 295
Minsoo Jang, *University of Sydney, Australia*
Vassilios G. Agelidis, *University of Sydney, Australia*

Power Conditioning System for Fuel Cell with 2-Stage DC-DC Converter 303
Byung M. Han, *Myongji University, Korea, South*
Jun-Young Lee, *Myongji University, Korea, South*
Yu-Seok Jeong, *Myongji University, Korea, South*

Session B1L-A: DC-DC Converter III
Wednesday, February 24, 8:30 - 10:10
Session Chairs: Alireza Khaligh, *Illinois Institute of Technology*
Sheldon Williamson, *Concordia University*

Real-Time FPGA-Based Hardware-in-the-Loop Development Test-Bench for Multiple Output Power Converters 309
O. Lucía, *University of Zaragoza, Spain*
O. Jiménez, *University of Zaragoza, Spain*
L.A. Barragán, *University of Zaragoza, Spain*
I. Urriza, *University of Zaragoza, Spain*
J.M. Burdío, *University of Zaragoza, Spain*
D. Navarro, *University of Zaragoza, Spain*

Oversampled Digital Controller IC Based on Successive Load-Change Estimation for DC-DC Converters 315
Zdravko Lukić, *University of Toronto, Canada*
Aleksandar Radić, *University of Toronto, Canada*
Aleksandar Prodić, *University of Toronto, Canada*
Simon Effler, *University of Limerick, Ireland*

Novel Nonlinear Control of Dual Active Bridge Using Simplified Converter Model 321
Diogenes D. Molina Cardozo, *University of Arkansas, United States*
Juan Carlos Balda, *University of Arkansas, United States*
Derik Trowler, *University of Arkansas, United States*
H. Alan Mantooth, *University of Arkansas, United States*

A Novel Digital Single-Wire Quasi-Democratic Stress Share Scheme for Paralleled Switching Converters 328

Karl Rinne, *Powervation Ltd., Ireland*
Anthony Kelly, *Powervation Ltd., Ireland*
Eamon O'Malley, *Powervation Ltd., Ireland*

Session B1L-B: AC-DC Conversion Control Strategies
Wednesday, February 24, 8:30 - 10:10
Session Chairs: Alireza Khaligh, *Illinois Institute of Technology*
Omer Onar, *Illinois Institute of Technology*

Minimum-Sensing Current Control of Three-Phase PFC Converters 336

Zhonghui Bing, *Rensselaer Polytechnic Institute, United States*
Jian Sun, *Rensselaer Polytechnic Institute, United States*

Direct Power Control of a Dual Converter Operating As Synchronous Rectifier 343

José Restrepo, *Universidad Simón Bolívar, Venezuela*
José M. Aller, *Universidad Simón Bolívar, Venezuela*
Alexander Bueno, *Universidad Simón Bolívar, Venezuela*
Julio C. Viola, *Universidad Simón Bolívar, Venezuela*
Alberto Berzoy, *Universidad Simón Bolívar, Venezuela*
Thomas Habetler, *Georgia Institute of Technology, United States*

A Low-Cost Adaptive Multi-Mode Digital Control Solution Maximizing AC-DC Power Supply Efficiency 349

Yong Li, *iWatt Inc., United States*
Jerry Zheng, *iWatt Inc., United States*

Average Modeling and Control for Three-Phase Three-Level Non-Regenerate Rectifier with Unbalanced DC Loads 355

Rixin Lai, *GE Global Research Center, United States*
Fred Wang, *University of Tennessee - Knoxville and Oak Ridge National Laboratory, United States*
Rolando Burgos, *ABB Inc., United States*
Dushan Boroyevich, *Virginia Polytechnic Institute and State University, United States*

Session B1L-C: Active Power Filter
Wednesday, February 24, 8:30 - 10:10
Session Chairs: Jingjun Liu, *Xi'an Jiaotong Univ.*
Jin Wang, *Ohio State University*

A Waveform Control Technique for High Power Shunt Active Power Filter Based on Repetitive Control Algorithm 361

Zhiqiang Wang, *Zhejiang University, China*
Chuan Xie, *Zhejiang University, China*
Chao He, *Zhejiang University, China*
Guozhu Chen, *Zhejiang University, China*

A Combined Series-Parallel Active Filter System Implementation Using Generalized Non-Active Power Theory 367

Mehmet Ucar, *Kocaeli University, Turkey*
Sule Ozdemir, *Kocaeli University, Turkey*
Engin Ozdemir, *Kocaeli University, Turkey*

A Novel Control Method for Unified Power Quality Conditioner (UPQC) Under Non-Ideal Mains Voltage and Unbalanced Load Conditions 374

Metin Kesler, *Kocaeli University, Turkey*
Engin Ozdemir, *Kocaeli University, Turkey*

Resonant Current Regulation for Transformerless Hybrid Active Filter to Suppress Harmonic Resonances in Industrial Power Systems 380

Tzung-Lin Lee, *National Sun Yat-sen University, Taiwan*
Yen-Ching Wang, *National Sun Yat-sen University, Taiwan*
Josep M. Guerrero, *Technical University of Catalonia, Spain*

Session B1L-D: Semiconductor Devices
Wednesday, February 24, 8:30 - 10:10
Session Chairs: Carl Blake, *Transphorm*
Chuck Mullett, *ON Semiconductor*

Performance Evaluation of High Voltage Super Junction MOSFETs for Zero-Voltage Soft-Switching Inverter Applications 387

Sung-Yeul Park, *University of Connecticut, United States*
Pengwei Sun, *Virginia Polytechnic Institute and State University, United States*
Wensong Yu, *Virginia Polytechnic Institute and State University, United States*
Jih-Sheng Lai, *Virginia Polytechnic Institute and State University, United States*

New 1.7kV IGBT Chip with Fine Pattern and Optimized Buffer Layer 392

John F. Donlon, *Powerex, Inc., United States*
Eric R. Motto, *Powerex, Inc., United States*
K. Satoh, *Mitsubishi Electric Corp, Japan*
K. Suzuki, *Mitsubishi Electric Corp, Japan*
Y. Yoshihiura, *Mitsubishi Electric Corp, Japan*
T. Takahashi, *Mitsubishi Electric Corp, Japan*

Novel Thermally Enhanced Power Package 398

Juan A. Herbsommer, *Texas Instruments, United States*
Jonathan Noquil, *Texas Instruments, Philippines*
Chris Bull, *Texas Instruments, United States*
Osvaldo Lopez, *Texas Instruments, United States*

Recent Advances in Silicon Carbide MOSFET Power Devices 401

Ljubisa D. Stevanovic, *GE Global Research, United States*
Kevin S. Matocha, *GE Global Research, United States*
Peter A. Losee, *GE Global Research, United States*
John S. Glaser, *GE Global Research, United States*
Jeffrey J. Nasadoski, *GE Global Research, United States*
Stephen D. Arthur, *GE Global Research, United States*

Session B1L-E: Sensorless Techniques in Motor Drives
Wednesday, February 24, 8:30 - 10:10
Session Chairs: Patrick Chapman, *University of Illinois*

Start-Up Transient Improvement for Sensorless Control Approach of PM Motor 408
Dong Jiang, *Virginia Polytechnic Institute and State University, United States*
Rixin Lai, *Virginia Polytechnic Institute and State University, United States*
Fred Wang, *University of Tennessee - Knoxville, United States*
Rolando Burgos, *Virginia Polytechnic Institute and State University, United States*
Dushan Boroyevich, *Virginia Polytechnic Institute and State University, United States*

**Sensorless Position Control of Skewed Rotor Induction Machines Based on
Multi Saliency Extraction** 414
T.M. Wolbank, *Vienna University of Technology, Austria*
M.K. Metwally, *Vienna University of Technology, Austria*

**Fuzzy Gain Scheduling PI Controller for a Sensorless Four Switch Three
Phase BLDC Motor** 420
Chung-Wen Hung, *National Yunlin University of Science and Technology, Taiwan*
Jen-Ta Su, *National Taiwan University, Taiwan*
Chih-Wen Liu, *National Taiwan University, Taiwan*
Cheng-Tsung Lin, *DynaPack Co., Ltd., Taiwan*
Jhih-Han Chen, *National Yunlin University of Science and Technology, Taiwan*

Equivalent EMF Based Position Observers for Sensorless Synchronous Machines 425
Jingbo Liu, *Rockwell Automation, United States*
Thomas Nondahl, *Rockwell Automation, United States*
Peter Schmidt, *Rockwell Automation, United States*
Semyon Royak, *Rockwell Automation, United States*
Mark Harbaugh, *Rockwell Automation, United States*

Session B1L-F: Modeling, Simulation & Control I
Wednesday, February 24, 8:30 - 10:10
Session Chairs: Mahesh Krishnamurthy, *Illinois Institue of Technology*

An Improved Winding Loss Analytical Model of Flyback Transformer 433
Wei Yuan, *Zhejiang University, China*
Xiucheng Huang, *Zhejiang University, China*
Peipei Meng, *Zhejiang University, China*
Guoxing Zhang, *Zhejiang University, China*
Junming Zhang, *Zhejiang University, China*

**Identification of the Material Properties Used in Domestic Induction Heating
Appliances for System-Level Simulation and Design Purposes** 439
Jesus Acero, *University of Zaragoza, Spain*
Oscar Lucía, *University of Zaragoza, Spain*
Ignacio Millan, *University of Zaragoza, Spain*
Luis Angel Barragán, *University of Zaragoza, Spain*
Jose-Miguel Burdío, *University of Zaragoza, Spain*
Rafael Alonso, *University of Zaragoza, Spain*

A Retrofit 60 Hz Current Sensor for Non-Intrusive Power Monitoring at the Circuit Breaker 444

Zachary Clifford, *Massachusetts Institute of Technology, United States*
John J. Cooley, *Massachusetts Institute of Technology, United States*
Al-Thaddeus Avestruz, *Massachusetts Institute of Technology, United States*
Zack Remscrim, *Massachusetts Institute of Technology, United States*
Dan Vickery, *Massachusetts Institute of Technology, United States*
Steven B. Leeb, *Massachusetts Institute of Technology, United States*

Session B1L-G: Vehicle Electronics I
Wednesday, February 24, 8:30 - 10:10
Session Chairs: Ali Emadi, *Illinois Institute of Technology*

Feasibility of Capacitor Voltage Regulation and Output Voltage Harmonic Minimization in Cascaded H-Bridge Converters 452

Hossein Sepahvand, *Missouri University of Science and Technology, United States*
Mostafa Khazarei, *Missouri University of Science and Technology, United States*
Mehdi Ferdowsi, *Missouri University of Science and Technology, United States*
Keith Corzine, *Missouri University of Science and Technology, United States*

Examination of a PHEV Bidirectional Charger System for V2G Reactive Power Compensation 458

Mithat C. Kisacikoglu, *University of Tennessee, United States*
Burak Ozpineci, *Oak Ridge National Laboratory, United States*
Leon M. Tolbert, *University of Tennessee and Oak Ridge National Laboratory, United States*

Optimal Selection and Design of the Supercapacitor Module for Fuel Cell Vehicles 466

Sang-Hyun Kim, *Soongsil University, Korea, South*
Tae-Hoon Kim, *Soongsil University, Korea, South*
Wook Kim, *Soongsil University, Korea, South*
Jong-Hak Lee, *Soongsil University, Korea, South*
Woojin Choi, *Soongsil University, Korea, South*

Efficiency Evaluation of a 55kW Soft-Switching Module Based Inverter for High Temperature Hybrid Electric Vehicle Drives Application 474

Pengwei Sun, *Virginia Polytechnic Institute and State University, United States*
Jih-Sheng Lai, *Virginia Polytechnic Institute and State University, United States*
Hao Qian, *Virginia Polytechnic Institute and State University, United States*
Wensong Yu, *Virginia Polytechnic Institute and State University, United States*
Chris Smith, *Azure Dynamics Inc., United States*
John Bates, *Azure Dynamics Inc., United States*
Beat Arnet, *Azure Dynamics Inc., United States*
Alexander Litvinov, *Powerex Inc., United States*
Scott Leslie, *Powerex Inc., United States*

Session B2L-A: DC-DC Converter IV
Wednesday, February 24, 14:00 - 15:40
Session Chairs: Jin Wang, *Ohio State University*
Wayne Weaver, *Michigan Technological University*

Real-Time Hybrid Model Predictive Control of a Boost Converter with Constant Power Load 480
Jason Neely, *Purdue University, United States*
Steve Pekarek, *Purdue University, United States*
Ray DeCarlo, *Purdue University, United States*
Nir Vaks, *Purdue University, United States*

Predictive Control of Buck Converter Using Nonlinear Output Capacitor Current Programming 491
Victor Sui-pung Cheung, *City University of Hong Kong, China*
Henry Shu-hung Chung, *City University of Hong Kong, China*
Huai Wang, *City University of Hong Kong, China*

Analysis of a High Performance Voltage Regulator with Non-Linear Multi-Mode Control: Bandwidth and Large Transient Response 499
S. Pan, *Queen's University, Canada*
P.K. Jain, *Queen's University, Canada*

Multi-Output Synchronously-Rectified Forward Converter with Load Transient Considered ... 507
K.I. Hwu, *National Taipei University of Technology, Taiwan*
Y.T. Yau, *National Taipei University of Technology, Taiwan*

Session B2L-B: System Integration I
Wednesday, February 24, 14:00 - 15:40
Session Chairs: Shamala Chickamenahalli, *Intel*

Symmetric Current Balancing Circuit for Multiple DC Loads 512
Sungjin Choi, *Samsung Electronics Co., Ltd., Korea, South*
Pankaj Agarwal, *Samsung Electronics Co., Ltd., Korea, South*
Teahoon Kim, *Samsung Electronics Co., Ltd., Korea, South*
Joonhyun Yang, *Samsung Electronics Co., Ltd., Korea, South*
Baikhee Han, *Samsung Electronics Co., Ltd., Korea, South*

A Simple Method for Configuring Multi-PWM Channels for Multi-Level Converter Applications Based on PWM IP Core 519
Haibing Hu, *Nanjing University of Aeronautics and Astronautics, China*
Xiaodong Ding, *Nanjing Guojun Electric Co., Ltd., China*
Tao Xue, *Nanjing Sute Electric Co., Ltd., China*
Wenxi Yao, *Zhejiang University, China*
Zhengyu Lu, *Zhejiang University, China*

Technology Roadmapping for Power Supply in Package (PSiP) and Power Supply on Chip (PwrSoC) 525
Raymond Foley, *University College Cork, Ireland*
Finbarr Waldron, *Tyndall National Institute, Ireland*
John Slowey, *University College Cork, Ireland*
Arnold Alderman, *Anagenesis Inc., United States*
Brian Narveson, *Texas Instruments, United States*
Cian Ó'Mathúna, *Tyndall National Institute, Ireland*

Technology Road Map for High Frequency Integrated DC-DC Converter 533
Qiang Li, *Virginia Polytechnic Institute and State University, United States*
Michele Lim, *Virginia Polytechnic Institute and State University, United States*
Julu Sun, *Virginia Polytechnic Institute and State University, United States*
Arthur Ball, *Virginia Polytechnic Institute and State University, United States*
Yucheng Ying, *Virginia Polytechnic Institute and State University, United States*
Fred C. Lee, *Virginia Polytechnic Institute and State University, United States*
K.D.T. Ngo, *Virginia Polytechnic Institute and State University, United States*

Session B2L-C: Resonant DC-DC Converters I
Wednesday, February 24, 14:00 - 15:40
Session Chairs: Dustin Becker, *Emerson Network Power*
Russell Spyker, *USAF*

A New Valley-Detection Method for the Quasi-Resonance Switching 540
Gwan-Bon Koo, *Fairchild Semiconductor, Korea, South*
Sang-Cheol Moon, *Fairchild Semiconductor, Korea, South*
Jin-Tae Kim, *Fairchild Semiconductor, Korea, South*

Secondary-Side Control of a Constant Frequency Series Resonant Converter Using Dual-Edge PWM 544
Darryl J. Tschirhart, *Queen's University, Canada*
Praveen K. Jain, *Queen's University, Canada*

A Non-Insulated Resonant Boost Converter 550
Peng Shuai, *RWTH Aachen University, Germany*
Yales R. De Novaes, *ABB, Switzerland*
Francisco Canales, *ABB, Switzerland*
Ivo Barbi, *Federal University of Santa Catarina, Brazil*

Analysis and Design of a Low-Profile Resonant LCC Converter 557
A. Pawellek, *University of Erlangen-Nürnberg, Germany*
A. Bucher, *University of Erlangen-Nürnberg, Germany*
T. Duerbaum, *University of Erlangen-Nürnberg, Germany*

Session B2L-D: Miscellaneous Applications
Wednesday, February 24, 14:00 - 15:40
Session Chairs: Alejandro Dominguez-Garcia, *University of Illinois*

ZVS and ZCS DC-DC PWM Full-Bridge Fuel Cell Converters 564
Ahmad Mousavi, *University of Western Ontario, Canada*
Pritam Das, *University of Western Ontario, Canada*
Gerry Moschopoulos, *University of Western Ontario, Canada*

**Effective Switching Mode Power Supplies Common Mode Noise Cancellation
Technique with Zero Equipotential Transformer Models** 571
Yick Po Chan, *University of Hong Kong, China*
Man Hay Pong, *University of Hong Kong, China*
Ngai Kit Poon, *University of Hong Kong, China*
Chui Pong Liu, *University of Hong Kong, China*

**50W Power Device (PD) Power in Power Over Ethernet (PoE) System with Input
Current Balance in Four-Pair Architecture with Two DC-DC Converters** 575
Haimeng Wu, *Zhejiang University, China*
Zhengshi Wang, *Zhejiang University, China*
Jiande Wu, *Zhejiang University, China*
Xiangning He, *Zhejiang University, China*
Yan Deng, *Zhejiang University, China*

High-Resolution Physically-Windowed Sensors for Power Electronics Applications 580
Warit Wichakool, *Massachusetts Institute of Technology, United States*
James Paris, *Massachusetts Institute of Technology, United States*
Al-Thaddeus Avestruz, *Massachusetts Institute of Technology, United States*
Steven B. Leeb, *Massachusetts Institute of Technology, United States*

Session B2L-E: LED Lighting I
Wednesday, February 24, 14:00 - 15:40
Session Chairs: Regan Zane, *University of Colorado*

**Edison Revisited: Impact of DC Distribution on the Cost of
LED Lighting and Distributed Generation** 588
Brinda A. Thomas, *Carnegie Mellon University, United States*

A Novel Passive Off-Line Light-Emitting Diode (LED) Driver with Long Lifetime 594
S.Y.R. Hui, *City University of Hong Kong, China*
S.N. Li, *City University of Hong Kong, China*
X.H. Tao, *City University of Hong Kong, China*
W. Chen, *City University of Hong Kong, China*
W.M. Ng, *City University of Hong Kong, China*

Improving Current Regulation for Offline LED Driver 601
Jianwen Shao, *STMicroelectronics, United States*

LED Driver Circuit with Inherent PFC 605
D. Aguilar, *University of Minnesota, United States*
C.P. Henze, *Analog Power Design Inc., United States*

Session B2L-F: Power Electronics for Utility Interface III
Wednesday, February 24, 14:00 - 15:40
Session Chairs: Hui Li, *Florida State University*
Miaosen Shen, *United Technologies Research Center*

A New Circuit Design and Control to Reduce Input Harmonic Current for a
Three-Phase AC Machine Drive System Having a Very Small DC-Link Capacitor 611
Hyunjae Yoo, *Samsung Heavy Industries Co., Ltd., Korea, South*
Seung-Ki Sul, *Seoul National University, Korea, South*

State-Space Modeling, Analysis, and Implementation of
Parallel Inverters for Microgrid Applications 619
Chien Liang Chen, *Virginia Polytechnic Institute and State University, United States*
Jih-Sheng Lai, *Virginia Polytechnic Institute and State University, United States*
Daniel Martin, *Virginia Polytechnic Institute and State University, United States*
Yuang-Shung Lee, *Fu-Jen Catholic University, Taiwan*

Efficiency Improvement of Grid-Tied Inverters at Low Input Power Using
Pulse Skipping Control Strategy 627
Haibing Hu, *University of Central Florida, United States*
Wisam Al-Hoor, *University of Central Florida, United States*
Nasser Kutkut, *University of Central Florida, United States*
Issa Batarseh, *University of Central Florida, United States*
John Shen, *University of Central Florida, United States*

Phase Locked Loop for Unbalanced Utility Conditions 634
Carlos D. Rodríguez-Valdez, *Rockwell Automation, United States*
Russ J. Kerkman, *Rockwell Automation, United States*

Session B2L-G: Isolated DC-DC Converters I
Wednesday, February 24, 14:00 - 15:40
Session Chairs: Alexis Kwasinski, *The University of Texas at Austin*
Sheldon Williamson, *Concordia University*

Analysis and Design Considerations for EMI and Losses of
RCD Snubber in Flyback Converter 642
Peipei Meng, *Zhejiang University, China*
Xinke Wu, *Zhejiang University, China*
Jianyou Yang, *Zhejiang University, China*
Henglin Chen, *Zhejiang University, China*
Zhaoming Qian, *Zhejiang University, China*

A High Output Power Density 400/400V Isolated DC-DC Converter with
Hybrid Pair of SJ-MOSFET and SiC-SBD for Power Supply of Data Center 648
Rejeki Simanjorang, *National Institute of Advanced Industrial Science and Technology, Japan*
Hiroshi Yamaguchi, *National Institute of Advanced Industrial Science and Technology, Japan*
Hiromichi Ohashi, *National Institute of Advanced Industrial Science and Technology, Japan*
Takashi Takeda, *NTT Facilities Inc., Japan*
Mikio Yamazaki, *NTT Facilities Inc., Japan*
H. Murai, *NTT Facilities Inc., Japan*

A 500 W Push-Pull DC-DC Power Converter with a 30 MHz Switching Frequency 654
John S. Glaser, *GE Global Research, United States*
Juan M. Rivas, *GE Global Research, United States*

Input-Series Connnected High Frequency DC-DC Converters with One Transformer 662
Deshang Sha, *Beijing Institute of Technology, China*
Zhiqiang Guo, *Beijing Institute of Technology, China*
XiaoZhong Liao, *Beijing Institute of Technology, China*

Session B3L-A: Renewable Energy
Wednesday, February 24, 16:10 - 17:25
Session Chairs: Chris Edrington, *Florida State University*
Alex Huang, *North Carolina State University*

**Simple Photovoltaic Solar Cell Dynamic Sliding Mode Controlled Maximum Power Point
Tracker for Battery Charging Applications** 666
Emil A. Jimenez-Brea, *University of Puerto Rico-Mayaguez, Puerto Rico*
Eduardo I. Ortiz-Rivera, *University of Puerto Rico-Mayaguez, Puerto Rico*
Andres Salazar-Llinas, *University of Puerto Rico-Mayaguez, Puerto Rico*
Jesus Gonzalez-Llorente, *University of Puerto Rico-Mayaguez, Puerto Rico*

An Enhanced Circuit-Based Model for Single-Cell Battery 672
Jiucai Zhang, *University of Nebraska-Lincoln, United States*
Song Ci, *University of Nebraska-Lincoln, United States*
Hamid Sharif, *University of Nebraska-Lincoln, United States*
Mahmoud Alahmad, *University of Nebraska-Lincoln, United States*

A High Frequency Battery Model for Current Ripple Analysis 676
Jin Wang, *Ohio State University, United States*
Ke Zou, *Ohio State University, United States*
Chingchi Chen, *Ford Motor Company, United States*
Lihua Chen, *Ford Motor Company, United States*

Session B3L-B: System Integration II
Wednesday, February 24, 16:10 - 17:25
Session Chairs: Shamala Chickamenahalli, *Intel*

**A Novel Power Line Communication Technique Based on
Power Electronics Circuit Topology** 681
Jiande Wu, *Zhejiang University, China*
Chushan Li, *Zhejiang University, China*
Xiangning He, *Zhejiang University, China*

**Compact Temperature Compensation of Inductive Fly-back Clamps for Integrated
Power Switches Using a High-Voltage Base-Current-Compensated V_{be} Multiplier** 686
Timothy P. Duryea, *Texas Instruments, United States*
Hoi Lee, *University of Texas at Dallas, United States*

Optimal Design for the Damping Resistor in RCD-R Snubber to Suppress Common-Mode Noise 691

Peipei Meng, *Zhejiang University, China*
Henglin Chen, *Zhejiang University, China*
Sheng Zheng, *Zhejiang University, China*
Xinke Wu, *Zhejiang University, China*
Zhaoming Qian, *Zhejiang University, China*

Session B3L-C: Resonant DC-DC Converters II
Wednesday, February 24, 16:10 - 17:25
Session Chairs: Dustin Becker, *Emerson Network Power*
Russell Spyker, *USAF*

A High-Efficient LLCC Series-Parallel Resonant Converter 696

Christian P. Dick, *RWTH Aachen University, Germany*
Furkan Kaan Titiz, *RWTH Aachen University, Germany*
Rik De Doncker, *RWTH Aachen University, Germany*

Accurate Switching Loss Model and Optimal Design of a Current Source Driver Considering the Current Diversion Problem 702

Jizhen Fu, *Queen's University, Canada*
Zhiliang Zhang, *Nanjing University of Aeronautics and Astronautics, China*
Andrew Dickson, *Queen's University, Canada*
Yan-Fei Liu, *Queen's University, Canada*
P.C. Sen, *Queen's University, Canada*

Bidirectional Operation of Resonant Voltage Divider 710

K.I. Hwu, *National Taipei University of Technology, Taiwan*
Y.T. Yau, *National Taipei University of Technology, Taiwan*

Session B3L-D: RF Applications
Wednesday, February 24, 16:10 - 17:25
Session Chairs: Alejandro Dominguez-Garcia, *University of Illinois*

Multiple-Input Buck Converter Optimized for Accurate Envelope Tracking in RF Power Amplifiers 715

M. Rodríguez, *University of Oviedo, Spain*
P.F. Miaja, *University of Oviedo, Spain*
A. Rodríguez, *University of Oviedo, Spain*
J. Sebastián, *University of Oviedo, Spain*

Switching Capacities Based Envelope Amplifier for High Efficiency RF Amplifiers 723

M. Vasić, *Universidad Politécnica de Madrid, Spain*
O. García, *Universidad Politécnica de Madrid, Spain*
J.A. Oliver, *Universidad Politécnica de Madrid, Spain*
P. Alou, *Universidad Politécnica de Madrid, Spain*
D. Diaz, *Universidad Politécnica de Madrid, Spain*
J.A. Cobos, *Universidad Politécnica de Madrid, Spain*

High Efficiency Power Amplifier for High Frequency Radio Transmitters 729

M. Vasić, *Universidad Politécnica de Madrid, Spain*
O. García, *Universidad Politécnica de Madrid, Spain*
J.A. Oliver, *Universidad Politécnica de Madrid, Spain*
P. Alou, *Universidad Politécnica de Madrid, Spain*
D. Diaz, *Universidad Politécnica de Madrid, Spain*
J.A. Cobos, *Universidad Politécnica de Madrid, Spain*
A. Gimeno, *Universidad Politecnica de Madrid, Spain*
J.M. Pardo, *Universidad Politécnica de Madrid, Spain*
C. Benavente, *Universidad Politécnica de Madrid, Spain*
F.J. Ortega, *Universidad Politécnica de Madrid, Spain*

Session B3L-E: LED Lighting II
Wednesday, February 24, 16:10 - 17:25
Session Chairs: Regan Zane, *University of Colorado*

Applying One-Comparator Counter-Based Sampling to Current Sharing Control of Multi-Channel LED Strings 737

K.I. Hwu, *National Taipei University of Technology, Taiwan*
Y.T. Yau, *National Taipei University of Technology, Taiwan*

High Frequency PWM Dimming Technique for High Power Factor Converters in LED Lighting 743

D. Gacio, *University of Oviedo, Spain*
J.M. Alonso, *University of Oviedo, Spain*
J. Garcia, *University of Oviedo, Spain*
L. Campa, *University of Oviedo, Spain*
M. Crespo, *University of Oviedo, Spain*
M. Rico-Secades, *University of Oviedo, Spain*

A RGB-Driver for LED Display Panels 750

Jaber Hasan, *University of Arkansas, United States*
Do Hung Nguyen, *University of Arkansas, United States*
Simon S. Ang, *University of Arkansas, United States*

Session B3L-F: Power Electronics for Utility Interface IIII
Wednesday, February 24, 16:10 - 17:25
Session Chairs: Hui Li, *Florida State University*
                  Miaosen Shen, *United Technologies Research Center*

A Low Investment Single-Phase to Three-Phase Converter Operating with Reduced Losses 755

José A.A. Dias, *Instituto Federal de Educação, Ciência e Tecnologia da Paraíba, Brazil*
Euzeli C. dos Santos, *Universidade Federal de Campina Grande, Brazil*
Cursino B. Jacobina, *Universidade Federal de Campina Grande, Brazil*

Voltage and Power Balance Control for a Cascaded Multilevel Solid State Transformer 761
Tiefu Zhao, *North Carolina State University, United States*
Gangyao Wang, *North Carolina State University, United States*
Jie Zeng, *North Carolina State University, United States*
Sumit Dutta, *North Carolina State University, United States*
Subhashish Bhattacharya, *North Carolina State University, United States*
Alex Q. Huang, *North Carolina State University, United States*

**Grid-Connected Voltage Source Inverter for Renewable Energy Conversion System
with Sensorless Current Control** 768
Suzan Eren, *Queen's University, Canada*
Majid Pahlevaninezhad, *Queen's University, Canada*
Alireza Bakhshai, *Queen's University, Canada*
Praveen Jain, *Queen's University, Canada*

Session B3L-G: Isolated DC-DC Converters II
Wednesday, February 24, 16:10 - 17:25
Session Chairs: Alexis Kwasinski, *The University of Texas at Austin*
Sheldon Williamson, *Concordia University*

**Design of an 99%-Efficient, 5kW, Phase-Shift PWM DC-DC Converter for
Telecom Applications** 773
U. Badstuebner, *ETH Zurich, Switzerland*
J. Biela, *ETH Zurich, Switzerland*
J.W. Kolar, *ETH Zurich, Switzerland*

**DC-DC Transformer Multiphase Converter with Transformer Coupling for
Two-Stage Architecture** 781
M.C. Gonzalez, *Universidad Politécnica de Madrid, Spain*
P. Alou, *Universidad Politécnica de Madrid, Spain*
O. Garcia, *Universidad Politécnica de Madrid, Spain*
J.A. Oliver, *Universidad Politécnica de Madrid, Spain*
J.A. Cobos, *Universidad Politécnica de Madrid, Spain*
H. Visairo, *Intel Corporation, Mexico*

**A Comparison of Classical Two Phase (2L) and Transformer – Coupled (XL)
Interleaved Boost Converters for Fuel Cell Applications** 787
Kevin J. Hartnett, *University College Cork, Ireland*
Marek S. Rylko, *University College Cork, Ireland*
John G. Hayes, *University College Cork, Ireland*
Michael G. Egan, *University College Cork, Ireland*

Session C1L-A: Applications of DC-DC Converter I
Thursday, February 25, 8:30 - 10:10
Session Chairs: Chuck Mullett, *ON Semiconductor*
Kevin Parmenter, *Freescale*

**Design Considerations for Narrow Vdc Based Power Delivery Architecture in
Mobile Computing System** 794
Xiaoguo Liang, *Intel Asia-Pacific Research & Development Ltd., China*
Gnanavel Jayakanthan, *Intel Asia-Pacific Research & Development Ltd., China*
Meng Wang, *Intel Asia-Pacific Research & Development Ltd., China*

Active Clamp Boost Converter with Switched Capacitor and Coupled Inductor 801
Yi Zhao, *Zhejiang University, China*
Wuhua Li, *Zhejiang University, China*
Bo Yang, *Zhejiang University, China*
Xiangning He, *Zhejiang University, China*

Unified Modulation for Three-Phase Current-Fed Bidirectional DC-DC Converter
Under Varied Input Voltage 807
Zhan Wang, *Florida State University, United States*
Hui Li, *Florida State University, United States*

Integrated Switched-Capacitor Voltage Doubler with Clock Transition Periods
Boosting and Transfer Blocking Techniques 813
Phong Ngo, *University of Arizona, United States*
Dongsheng Ma, *University of Arizona, United States*

Session C1L-B: AC-DC Conversion Misc. Topics I
Thursday, February 25, 8:30 - 10:10
Session Chairs: Frank Cirolia, *Emerson Network Power*
Alireza Khaligh, *Illinois Institute of Technology*

A Novel Class of Multipulse Converters Based on High-Frequency-Operated Transformers ... 818
Sheng Zheng, *Zhejiang University, China*
Dong Chen, *Zhejiang University, China*
Hai Lin, *Zhejiang University, China*
Yousheng Wang, *Zhejiang University, China*
Zhaoming Qian, *Zhejiang University, China*
Fang Z. Peng, *Michigan State University, United States*

A High Efficiency Flyback Converter with New Active Clamp Technique 823
Xiucheng Huang, *Zhejiang University, China*
Weijing Du, *Zhejiang University, China*
Wei Yuan, *Zhejiang University, China*
Junming Zhang, *Zhejiang University, China*
Zhaoming Qian, *Zhejiang University, China*

Analysis and Design of a Novel Integrated Three-Phase Single-Stage
AC-DC PWM Full-Bridge Converter 829
Dunisha Wijeratne, *University of Western Ontario, Canada*
Gerry Moschopoulos, *University of Western Ontario, Canada*

Three-Phase Voltage Doubler Rectifier Based on Three-State Switching Cell for
Uninterruptible Power Supply Applications Using FPGA 837
Raphael A. da Câmara, *Universidade Federal do Ceará, Brazil*
P.P. Praça, *Universidade Federal do Ceará, Brazil*
C.M.T. Cruz, *Universidade Federal do Ceará, Brazil*
R.P. Torrico-Bascopé, *Universidade Federal do Ceará, Brazil*
C.E.A. Silva, *Universidade Federal do Ceará, Brazil*
D.S. Oliveira, Jr., *Universidade Federal do Ceará, Brazil*
L.H.S.C. Barreto, *Universidade Federal do Ceará, Brazil*

Session C1L-C: Grid Interconnection I
Thursday, February 25, 8:30 - 10:10
Session Chairs: Ali Bazzi, *University of Illinois*
Patrick Chapman, *University of Illinois*

Multi-Loop Control Algorithms for Seamless Transition of Grid-Connected Inverter 844
Qin Lei, *Michigan State University, United States*
Shuitao Yang, *Michigan State University & Zhe Jiang University, United States*
Fang Z. Peng, *Michigan state University, United States*

Digital Controller Development for Grid-Tied Photovoltaic Inverter with Model Based Technique 849
Zhigang Liang, *North Carolina State University, United States*
Larry Alesi, *MegaWatt Solar Inc., United States*
Xiaohu Zhou, *North Carolina State University, United States*
Alex Q. Huang, *North Carolina State University, United States*

High-Performance and Cost-Effective Multiple Feedback Control Strategy for Standalone Operation of Grid-Connected Inverter 854
Qin Lei, *Michigan State University, United States*
Shuitao Yang, *Zhejiang University, United States*
Fang Z. Peng, *Michigan State University, United States*

Current Control Optimization for Grid-Tied Inverters with Grid Impedance Estimation 861
Guoqiao Shen, *Zhejiang University, China*
Jun Zhang, *Zhejiang University, China*
Xiao Li, *Zhejiang University, China*
Chengrui Du, *Zhejiang University, China*
Dehong Xu, *Zhejiang University, China*

Session C1L-D: Inverter I
Thursday, February 25, 8:30 - 10:10
Session Chairs: Russell Spyker, *USAF*
Haidong Yu, *Phoenix International*

A New Direct Peak DC-Link Voltage Control Strategy of Z-Source Inverters 867
Yu Tang, *Nanjing University of Aeronautics and Astronautics, China*
Jukui Wei, *Nanjing University of Aeronautics and Astronautics, China*
Shaojun Xie, *Nanjing University of Aeronautics and Astronautics, China*

High Performance Voltage Regulation of Current Source Inverters 873
S.A.S. Grogan, *Monash University, Australia*
D.G. Holmes, *Monash University, Australia*
B.P. McGrath, *Monash University, Australia*

Development of a New Voltage Source Inverter (VSI) Average Model Including Low Frequency Harmonics 881
S. Ahmed, *Virginia Polytechnic Institute and State University, United States*
D. Boroyevich, *Virginia Polytechnic Institute and State University, United States*
F. Wang, *University of Tennessee - Knoxville, United States*
R. Burgos, *ABB US Corporate Research Center, United States*

Realization and Improvement of Repetitive Control in Rotating Frame for Active Power Filter System 887

Baifeng Chen, *Wuhan University, China*
Xiaoming Zha, *Wuhan University, China*
Jinwu Gong, *Wuhan University, China*
Suxuan Guo, *Wuhan University, China*
Jianjun Sun, *Wuhan University, China*

Session C1L-E: PWM in Motor Drives I
Thursday, February 25, 8:30 - 10:10
Session Chairs: Dionysios Aliprantis, *Iowa State University*

Current Constraints of PWM Rectifier Under Unbalanced Voltage Supply 895

Miroslav Chomat, *Institute of Thermomechanics, Czech Rep.*
Ludek Schreier, *Institute of Thermomechanics, Czech Rep.*
Jiri Bendl, *Institute of Thermomechanics, Czech Rep.*

Space Vector PWM for a Direct Matrix Converter Based Open-End Winding AC Drives with Enhanced Capabilities 901

Ranjan K. Gupta, *University of Minnesota, United States*
Apurva Somani, *University of Minnesota, United States*
Krushna K. Mohapatra, *University of Minnesota, United States*
Ned Mohan, *University of Minnesota, United States*

Evaluation of the Hybrid Four-Level Converter Employing Half-Bridge Modules for Two Different Modulation Schemes 909

Alessandro L. Batschauer, *Santa Catarina State University, Brazil*
Arnaldo J. Perin, *Federal University of Santa Catarina, Brazil*
Samir A. Mussa, *Federal University of Santa Catarina, Brazil*
Marcelo L. Heldwein, *Federal University of Santa Catarina, Brazil*

A Comparative Study of Space Vector PWM Strategy for Dual Three-Phase Permanent-Magnet Synchronous Motor Drives 915

Yanhui He, *Xi'an Jiaotong University, China*
Yue Wang, *Xi'an Jiaotong University, China*
Jinlong Wu, *Xi'an Jiaotong University, China*
Yupeng Feng, *Xi'an Jiaotong University, China*
Jinjun Liu, *Xi'an Jiaotong University, China*

Session C1L-F: Magnetics in DC-DC Converters
Thursday, February 25, 8:30 - 10:10
Session Chairs: Arnold Alderman, *PSMA*

A Novel Coupled Inductor for Interleaved Converters 920

Qianhong Chen, *Nanjing University of Aeronautics and Astronautics, China*
Ligang Xu, *Nanjing university of aeronautics and astronautics, China*
Xiaoyong Ren, *Nanjing University of Aeronautics and Astronautics, China*
Lingling Cao, *Nanjing University of Aeronautics and Astronautics, China*
Xinbo Ruan, *Nanjing University of Aeronautics and Astronautics, China*

Transformer's Capacitance Effect on the Operation of Triangular-Current Shaped Soft-Switched Converters 928
Ilya Zeltser, *Ben-Gurion University of the Negev, Israel*
Sam Ben-Yaakov, *Ben-Gurion University of the Negev, Israel*

An Input and Output Ripple Free Converter with a Four-Winding Coupled Inductor 935
Zhuomin Feng, *Zhejiang University, China*
Zhe Zhang, *Zhejiang university, China*
Duo Li, *Zhejiang University, China*
Min Chen, *Zhejiang University, China*
Zhaoming Qian, *Zhejiang University, China*

Investigation on Transformer Design of High Frequency High Efficiency DC-DC Converters 940
Dianbo Fu, *Virginia Polytechnic Institute and State University, United States*
Fred C. Lee, *Virginia Polytechnic Institute and State University, United States*
Shuo Wang, *Virginia Polytechnic Institute and State University, United States*

Session C1L-G: Photovoltaics I
Thursday, February 25, 8:30 - 10:10
Session Chairs: Robert Balog, *Texas A&M University*

A DSP-Based Single-Stage Maximum Power Point Tracking PV Inverter 948
Wen Long Yu, *National Taiwan University of Science and Technology, Taiwan*
Ting-Peng Lee, *National Taiwan University of Science and Technology, Taiwan*
Guan-Hong Wu, *National Taiwan University of Science and Technology, Taiwan*
Qing Su Chen, *National Taiwan University of Science and Technology, Taiwan*
Huang-Jen Chiu, *National Taiwan University of Science and Technology, Taiwan*
Yu-Kang Lo, *National Taiwan University of Science and Technology, Taiwan*
Frank Shih, *Macroblock Inc., Taiwan*

A Simple Mixed-Signal MPPT Circuit for Photovoltaic Applications 953
P. Mattavelli, *University of Padova, Italy*
S. Saggini, *University of Udine, Italy*
E. Orietti, *University of Padova, Italy*
G. Spiazzi, *University of Padova, Italy*

Low-Power Maximum Power Point Tracker with Digital Control for Thermophotovoltaic Generators 961
Robert C.N. Pilawa-Podgurski, *Massachusetts Institute of Technology, United States*
Nathan A. Pallo, *Massachusetts Institute of Technology, United States*
Walker R. Chan, *Massachusetts Institute of Technology, United States*
David J. Perreault, *Massachusetts Institute of Technology, United States*
Ivan L. Celanovic, *Massachusetts Institute of Technology, United States*

11-Level Cascaded H-Bridge Grid-Tied Inverter Interface with Solar Panels 968
Faete Filho, *University of Tennessee, United States*
Yue Cao, *University of Tennessee, United States*
Leon M. Tolbert, *University of Tennessee, United States*

Session C2L-A: Applications of DC-DC Converter II
Thursday, February 25, 10:40 - 11:30
Session Chairs: Chuck Mullett, *ON Semiconductor*
Kevin Parmenter, *Freescale*

**A Life Prediction Scheme for Electrolytic Capacitors in
Power Converters without Current Sensor** 973
H.M. Pang, *University of Hong Kong, China*
M.H. Bryan Pong, *University of Hong Kong, China*

**Load-Interactive Steered-Inductor DC-DC Converter with
Minimized Output Filter Capacitance** 980
S.M. Ahsanuzzaman, *University of Toronto, Canada*
Amir Parayandeh, *University of Toronto, Canada*
Aleksandar Prodić, *University of Toronto, Canada*
Dragan Maksimović, *University of Colorado at Boulder, United States*

Session C2L-B: AC-DC Conversion Misc. Topics II
Thursday, February 25, 10:40 - 11:30
Session Chairs: Frank Cirolia, *Emerson Network Power*
Alireza Khaligh, *Illinois Institute of Technology*

**EMI Filter Design for High Switching Frequency Three-Phase/Level
PWM Rectifier Systems** 986
M. Hartmann, *ETH Zurich, Switzerland*
H. Ertl, *Vienna University of Technology, Austria*
J.W. Kolar, *ETH Zurich, Switzerland*

**Self-Driven AC-DC Synchronous Rectifier for Power Applications – A Direct
Energy-Efficient Replacement for Traditional Diode Rectifier** 994
W.X. Zhong, *City University of Hong Kong, China*
W.C. Ho, *ConvenientPower HK Ltd., China*
X. Liu, *ConvenientPower HK Ltd., China*
S.Y.R. Hui, *City University of Hong Kong, China*

Session C2L-C: Grid Interconnection II
Thursday, February 25, 10:40 - 11:30
Session Chairs: Ali Bazzi, *University of Illinois*

A Robust Control Scheme for Grid-Connected Voltage Source Inverters 1002
Shuitao Yang, *Zhejiang University and Michigan State University, China*
Qin Lei, *Michigan State University, United States*
Fang Z. Peng, *Michigan State University, United States*
Zhaoming Qian, *Zhejiang University, China*

**Application of Active NPC Converter on Generator Side for
MW Direct-Driven Wind Turbine** 1010
Jun Li, *North Carolina State University, United States*
Alex Q. Huang, *North Carolina State University, United States*
Subhashish Bhattacharya, *North Carolina State University, United States*
Wei Jing, *China University of Mining and Technology, United States*

Session C2L-D: Inverter II
Thursday, February 25, 10:40 - 11:30
Session Chairs: Russell Spyker, *USAF*
Haidong Yu, *Phoenix International*

Nonlinear Modeling of Switched Reluctance Motor Using Different Methods 1018
Jun Cai, *Nanjing University of Aeronautics and Astronautics, China*
Zhiquan Deng, *Nanjing University of Aeronautics and Astronautics, China*
Zeyuan Liu, *Nanjing University of Aeronautics and Astronautics, China*

Simplified Synchronous Reference Frame Control of the
Three Phase Grid Connected Inverter 1026
Abad Lorduy, *Carlos III University of Madrid, Spain*
Antonio Lázaro, *Carlos III University of Madrid, Spain*
Andrés Barrado, *Carlos III University of Madrid, Spain*
Cristina Fernández, *Carlos III University of Madrid, Spain*
Isabel Quesada, *Carlos III University of Madrid, Spain*
Carlos Lucena, *Carlos III University of Madrid, Spain*

Session C2L-E: PWM in Motor Drives II
Thursday, February 25, 10:40 - 11:30
Session Chairs: Dionysios Aliprantis, *Iowa State University*

A Novel Direct Digital SPWM Method for Multilevel Voltage Source Inverters 1034
Wanmin Fei, *Nanjing Normal University, China*
Yanli Zhang, *Nanjing Normal University, China*
Bin Wu, *Ryerson University, Canada*

Weight Oriented Optimal PWM in Low Modulation Indexes for
Multilevel Inverters with Unbalanced DC Sources 1038
Damoun Ahmadi, *Ohio State University, United States*
Ke Zou, *Ohio State University, United States*
Jin Wang, *Ohio State University, United States*

Session C2L-F: Measurement and Testing
Thursday, February 25, 10:40 - 11:30
Session Chairs: Patrick Chapman, *University of Illinois*

Oscillation-Test Technique for Buck Voltage Regulator 1043
Jing-Yi Huang, *National Cheng-Kung University, Taiwan*
Chun-Hsun Wu, *National Cheng-Kung University, Taiwan*
Le-Ren Chang-Chien, *National Cheng-Kung University, Taiwan*
Soon-Jyh Chang, *National Cheng-Kung University, Taiwan*

Core Loss Predictions for General PWM Waveforms from a
Simplified Set of Measured Data 1048
Charles R. Sullivan, *Thayer School of Engineering at Dartmouth, United States*
John H. Harris, *Thayer School of Engineering at Dartmouth, United States*
Edward Herbert, *FMTT, Inc., United States*

Session C2L-G: Photovoltaics II
Thursday, February 25, 10:40 - 11:30
Session Chairs: Robert Balog, *Texas A&M University*

**High-Efficiency Inverter with H6-Type Configuration for Photovoltaic
Non-Isolated AC Module Applications** 1056
Wensong Yu, *Virginia Polytechnic Institute and State University, United States*
Jih-Sheng Lai, *Virginia Polytechnic Institute and State University, United States*
Hao Qian, *Virginia Polytechnic Institute and State University, United States*
Chris Hutchens, *Virginia Polytechnic Institute and State University, United States*
Jianhui Zhang, *National Semiconductor Corporation, United States*
Gianpaolo Lisi, *National Semiconductor Corporation, United States*
Ali Djabbari, *National Semiconductor Corporation, United States*
Greg Smith, *National Semiconductor Corporation, United States*
Tim Hegarty, *National Semiconductor Corporation, United States*

**Analyzing the Optimal Matching of DC Motors to
Photovoltaic Modules via DC-DC Converters** 1062
Jesus Gonzalez-Llorente, *University of Puerto Rico-Mayaguez, Puerto Rico*
Eduardo I. Ortiz-Rivera, *University of Puerto Rico-Mayaguez, Puerto Rico*
Andres Salazar-Llinas, *University of Puerto Rico-Mayaguez, Puerto Rico*
Emil Jimenez-Brea, *University of Puerto Rico-Mayaguez, Puerto Rico*

Session C3L-A: Load Management Interface I
Thursday, February 25, 14:00 - 15:40
Session Chairs: Siamak Abedinpour, *Freescale*
Jonathan Kimball, *Missouri S&T*

**Performance Analysis of an Interleaved High Step-Up Converter with
Voltage Multiplier Cell** 1069
Wuhua Li, *Zhejiang University, China*
Yi Zhao, *Zhejiang University, China*
Yan Deng, *Zhejiang University, China*
Xiangning He, *Zhejiang University, China*

FPGA-Based Multi-Phase Digital Pulse Width Modulator with Dual-Edge Modulation 1075
Martin Scharrer, *University of Limerick, Ireland*
Mark Halton, *University of Limerick, Ireland*
Tony Scanlan, *University of Limerick, Ireland*
Karl Rinne, *University of Limerick, Ireland*

Phase Doubler for High Power Voltage Regulators 1081
Chun Cheung, *Intersil Corporation, United States*
Weihong Qiu, *Intersil Corporation, United States*
Emil Chen, *Intersil Corporation, United States*
Greg Miller, *Intersil Corporation, United States*

Automatic Multi-Phase Digital Pulse Width Modulator 1087
Simon Effler, *University of Limerick, Ireland*
Mark Halton, *University of Limerick, Ireland*
Karl Rinne, *University of Limerick, Ireland*

Session C3L-B: Power Electronics in Motor Drives I
Thursday, February 25, 14:00 - 15:40
Session Chairs: Chris Edrington, *Florida State University*
Patrick Chapman, *University of Illinois*

A Simple Current Sharing Scheme for Dual Three-Phase Permanent-Magnet Synchronous Motor Drives 1093
Yanhui He, *Xi'an Jiaotong University, China*
Yue Wang, *Xi'an Jiaotong University, China*
Jinlong Wu, *Xi'an Jiaotong University, China*
Yupeng Feng, *Xi'an Jiaotong University, China*
Jinjun Liu, *Xi'an Jiaotong University, China*

Multilevel Current Source Inverter Topologies Based on the Duality Principle 1097
Jianyu Bao, *Ningbo Institute of Technology, Zhejiang University, China*
Weibing Bao, *Zhejiang University of Science and Technology, China*
Siran Wang, *Zhejiang University, China*
Zhongchao Zhang, *Zhejiang University, China*

3-Level Power Converter with High-Voltage SiC-PiN Diode and Hard-Gate-Driving of IEGT for Future High-Voltage Power Conversion Systems 1101
Kazuto Takao, *Toshiba Corporation, Japan*
Yasunori Tanaka, *National Institute of Advanced Industrial Science and Technology, Japan*
Kyungmin Sung, *Ibaraki National College of Technology, Japan*
Keiji Wada, *Tokyo Metoropolitan University, Japan*
Takashi Shinohe, *Toshiba Corporation, Japan*
Takeo Kanai, *Toshiba Mitsubishi-Electric Industrial Systems Corporation, Japan*
Hiromichi Ohashi, *National Institute of Advanced Industrial Science and Technology, Japan*

18 kW Three Phase Inverter System Using Hermetically Sealed SiC Phase-Leg Power Modules 1108
Hui Zhang, *Tuskegee University, United States*
Leon M. Tolbert, *University of Tennessee, United States*
Jung Hee Han, *Global Power Electronics, United States*
Madhu S. Chinthavali, *Oak Ridge National Laboratory, United States*
Fred Barlow, *University of Idaho, United States*

Session C3L-C: DC-DC Converter V
Thursday, February 25, 14:00 - 15:40
Session Chairs: Frank Ciriola, *Emerson*
Arnold Alderman, *PSMA*

Multiphase Optimal Response Mixed-Signal Current-Programmed Mode Controller 1113
Jurgen Alico, *University of Toronto, Canada*
Aleksandar Prodić, *University of Toronto, Canada*

Switching Loss Analysis of Closed-Loop Gate Drive 1119
Lihua Chen, *Michigan State University, United States*
Fang Z. Peng, *Michigan State University, United States*

Modeling and Analysis of Closed-Loop Gate Drive 1124
Lihua Chen, *Michigan State University, United States*
Baoming Ge, *Michigan State University, United States*
Fang Z. Peng, *Michigan State University, United States*

**Black-Box Modeling of DC-DC Converters Based on Transient Response
Analysis and Parametric Identification Methods** 1131
V. Valdivia, *Carlos III University of Madrid, Spain*
A. Barrado, *Carlos III University of Madrid, Spain*
A. Lázaro, *Carlos III University of Madrid, Spain*
C. Fernández, *Carlos III University of Madrid, Spain*
P. Zumel, *Carlos III University of Madrid, Spain*

Session C3L-D: Transportation
Thursday, February 25, 14:00 - 15:40
Session Chairs: Jaber Abu Qahouq, *University of Alabama*
Dionysios Aliprantis, *Iowa State University*

**Harmonic and Balance Compensation Using Instantaneous Active and
Reactive Power Control on Electric Railway Systems** 1139
A. Bueno, *Universidad Simón Bolívar, Venezuela*
J.M. Aller, *Universidad Simón Bolívar, Venezuela*
J. Restrepo, *Universidad Simón Bolívar, Venezuela*
T. Habetler, *Georgia Institute of Technology, United States*

**Review of Non-Isolated Bi-Directional DC-DC Converters for Plug-in Hybrid Electric
Vehicle Charge Station Application at Municipal Parking Decks** 1145
Yu Du, *North Carolina State University, United States*
Xiaohu Zhou, *North Carolina State University, United States*
Sanzhong Bai, *North Carolina State University, United States*
Srdjan Lukic, *North Carolina State University, United States*
Alex Huang, *North Carolina State University, United States*

**Control of Plug-in Hybrid Electric Vehicles for Mobile Power Generation and
Grid Support Applications** 1152
Gui-Jia Su, *Oak Ridge National Laboratory, United States*
Lixin Tang, *Oak Ridge National Laboratory, United States*

Interface Issues of Mining Haul Trucks Operating on Trolley Systems 1158
Joy Mazumdar, *Siemens Industry Inc., United States*
Walter Koellner, *Siemens Industry Inc., United States*
Rohit Moghe, *Georgia Institute of Technology, United States*

Session C3L-E: Power Converter Applications I
Thursday, February 25, 14:00 - 15:40
Session Chairs: Vajapeyam Sukumar, *Maxim Integrated Products*

Regenerative AC Electronic Load with One-Cycle Control 1166
In Wha Jeong, *University of California, Irvine, United States*
Mikhail Slepchenkov, *University of California, Irvine, United States*
Keyue Smedley, *University of California, Irvine, United States*
Franco Maddaleno, *University of California, Irvine, United States*

A High Efficiency Regulated Charge Pump Over Wide Input and Load Range 1172
Rong Guo, *North Carolina State University, United States*
Liyu Yang, *North Carolina State University, United States*
Alex Huang, *North Carolina State University, United States*
John Endredy, *RF Micro Devices, United States*

High Performance, High-Power Capacitor Charging:
Focus on Pulse-to-Pulse Repeatability 1177
A. Pokryvailo, *Spellman High Voltage Electronics Corporation, United States*
C. Carp, *Spellman High Voltage Electronics Corporation, United States*
C. Scapellati, *Spellman High Voltage Electronics Corporation, United States*

Generalized AC-DC Single-Phase Boost Rectifier 1183
C.B. Jacobina, *Universidade Federal de Campina Grande, Brazil*
Euzeli dos Santos, *Universidade Federal de Campina Grande, Brazil*
Nady Rocha, *Universidade Federal de Campina Grande, Brazil*

Session C3L-F: Utility Interface Applications
Thursday, February 25, 14:00 - 15:40
Session Chairs: Ali Davoudi, *University of Illinois*

Parallel Connection of Two Shunt Active Power Filters with Losses Optimization 1191
E.C. dos Santos, Jr., *Universidade Federal de Campina Grande, Brazil*
C.B. Jacobina, *Universidade Federal de Campina Grande, Brazil*
A.M. Maciel, *Universidade Federal de Campina Grande, Brazil*

Design and Implementation of an Improved Controller for Parallel-Connected
400 Hz Frequency Converters 1197
B. Tamyurek, *Eskisehir Osmangazi University, Turkey*
E. Birdane, *Kaynak Electronic Machine Industry and Trade Co. Ltd., Turkey*
Adil Ceyhan, *Kaynak Electronic Machine Industry and Trade Co. Ltd., Turkey*

Study on the Impact of the Complex Impedance on the Droop Control Method
for the Parallel Inverters 1204
Wei Yao, *Zhejiang university, China*
Mingzhi Gao, *Zhejiang University, China*
Zheng Ren, *Zhejiang university, China*
Min Chen, *Zhejiang University, China*
Zhaoming Qian, *Zhejiang University, China*

A Three-Phase Adaptive Approach to Extract Harmonic and Reactive Currents 1209
D. Yazdani, *Queen's University, Canada*
A. Bakhshai, *Queen's University, Canada*
P.K. Jain, *Queen's University, Canada*

Session C3L-G: Soft Switching Techniques I
Thursday, February 25, 14:00 - 15:40
Session Chairs: Jason Neely, *Purdue University*
Wayne Weaver, *Michigan Technological University*

Analysis, Optimized Design and Adaptive Control of a ZCS Full-Bridge Converter Without Voltage Over-Stress on the Switches 1214
Xin Zhang, *Nanjing University of Aeronautics and Astronautics, China*
Henry Shu-hung Chung, *City University of Hong Kong, China*
Xinbo Ruan, *Huazhong University of Science and Technology, China*
Adrian Ioinovici, *Holon Institute of Technology, Israel*

Analysis and Design of a Novel ZVS-PWM DC-DC Converter for Bidirectional Applications with Steep Conversion Ratio 1222
Pritam Das, *University of Western Ontario, Canada*
Ahmad Mousavi, *University of Western Ontario, Canada*
Gerry Moschopoulos, *University of Western Ontario, Canada*

Three-Level Phase-Shift ZVS-PWM DC-DC Converter with High Frequency Transformer for High Performance Arc Welding Machines 1230
Tomokazu Mishima, *Kure National College of Technology, Japan*
Hisayuki Sugimura, *Daihen Corporation, Japan*
Khairy Fathy Sayed, *Kyungnam University, Korea, South*
Soon Kurl Kwon, *Kyungnam University, Korea, South*
Mutsuo Nakaoka, *Kyungnam University and Yamaguchi University, Japan*

Fully Soft-Switched Bidirectional Resonant DC-DC Converter with a New CLLC Tank 1238
Wei Chen, *Zhejiang University, China*
Siran Wang, *Zhejiang University, China*
Xiaoyuan Hong, *Zhejiang University, China*
Zhengyu Lu, *Zhejiang University, China*
Shaoshi Ye, *Delta Electronics (Shanghai) Co., LTD., China*

Session C4L-A: Load Management Interface II
Thursday, February 25, 16:10 - 17:25
Session Chairs: Siamak Abedinpour, *Freescale*
Jonathan Kimball, *Missouri S&T*

Optimal Phase Changing Frequency Determination for Multiphase Voltage Regulator Modules 1243
Anand Ramamurthy, *North Carolina State University, United States*
Subhashish Bhattacharya, *North Carolina State University, United States*
Chris Thompson, *Intersil Corporation, United States*
Jon Day, *Intersil Corporation, United States*

A New Digital Adaptive Voltage Positioning Technique with Dynamically Varying Voltage and Current References 1248
S. Pan, *Queen's University, Canada*
P.K. Jain, *Queen's University, Canada*

**A Three-Level Buck Converter and Digital Controller for
Improving Load Transient Response** 1256
Zhenyu Zhao, *Exar Corp, Canada*
Aleksandar Prodić, *University of Toronto, Canada*

Session C4L-B: Power Electronics in Motor Drives II
Thursday, February 25, 16:10 - 17:25
Session Chairs: Chris Edrington, *Florida State University*
Patrick Chapman, *University of Illinois*

Trends in MW-Rated VSI Technology and Reliability for Adjustable Speed Drives 1261
Hiromi Hosoda, *Toshiba Mitsubishi-Electric Industrial Systems Corporation, Japan*
Mostafa Al Mamun, *Toshiba Mitsubishi-Electric Industrial Systems Corporation, Japan*
Teruo Yoshino, *Toshiba Mitsubishi-Electric Industrial Systems Corporation, Japan*

**Development of a Compact 750KVA Three-Phase NPC Three-Level Universal
Inverter Module with Specifically Designed Busbar** 1266
Jun Wang, *Zhejiang University, China*
Binjian Yang, *Zhejiang University, China*
Jing Zhao, *Zhejiang University, China*
Yan Deng, *Zhejiang University, China*
Xiangning He, *Zhejiang University, China*
Xu Zhixin, *Zhejiang University of Science and Technology, China*

Common Mode Voltage in DC-Fed Motor Drive System and its Impact on the EMI Filter 1272
Fang Luo, *Virginia Polytechnic Institute and State University &
Huazhong University of Science and Technology, United States*
Shuo Wang, *GE Aviation Systems, United States*
Fred Wang, *University of Tennessee - Knoxville and Oak Ridge National Laboratory, United States*
Dushan Boroyevich, *Virginia Polytechnic Institute and State University, United States*
Nicolas Gazel, *Virginia Polytechnic Institute and State University, United States*
Yong Kang, *Huazhong University of Science and Technology, China*

Session C4L-C: DC-DC Converter VI
Thursday, February 25, 16:10 - 17:25
Session Chairs: Arnold Alderman, *PSMA*
Frank Cirolia, *Emerson Network Power*

**Black-Box Modeling of Three Phase Voltage Source Inverters Based on
Transient Response Analysis** 1279
V. Valdivia, *Carlos III University of Madrid, Spain*
A. Lázaro, *Carlos III University of Madrid, Spain*
A. Barrado, *Carlos III University of Madrid, Spain*
P. Zumel, *Carlos III University of Madrid, Spain*
C. Fernández, *Carlos III University of Madrid, Spain*
M. Sanz, *Carlos III University of Madrid, Spain*

Digital Autotuning of DC-DC Converters Based on Model Reference Impulse Response ... 1287
A. Costabeber, *University of Padova, Italy*
P. Mattavelli, *University of Padova, Italy*
S. Saggini, *University of Udine, Italy*
A. Bianco, *STMicroelectronics, Italy*

High-Fidelity and High-Speed Modeling and Simulation for Power Conversion Systems ... 1295
Chunchun Xu, *GE Global Research, United States*
Luis Garces, *GE Global Research, United States*
Paul Szczesny, *GE Global Research, United States*

Session C4L-D: Aerospace
Thursday, February 25, 16:10 - 17:25
Session Chairs: Jaber Abu Qahouq, *University of Alabama*
Dionysios Aliprantis, *Iowa State University*

Electrical Power Distribution System (HV270DC), for Application in More Electric Aircraft .. 1300
D. Izquierdo, *EADS, Spain*
R. Azcona, *EADS, Spain*
F.J. López del Cerro, *EADS, Spain*
Carlos Fernández, *EADS, Spain*
Bernardo Delicado, *EADS, Spain*

Supercapacitor-Based Energy Management for Future Aircraft Systems 1306
R. Todd, *University of Manchester, United Kingdom*
D. Wu, *University of Manchester, United Kingdom*
J.A. dos Santos Girio, *University of Manchester, United Kingdom*
M. Poucand, *University of Manchester, United Kingdom*
A.J. Forsyth, *University of Manchester, United Kingdom*

Buck Boost Regulator (B²R) for Spacecraft Solar Array Power Conversion 1313
Olivier Mourra, *European Space Agency, Netherlands*
Arturo Fernandez, *European Space Agency, Netherlands*
Ferdinando Tonicello, *European Space Agency, Netherlands*

Session C4L-E: Power Converter Applications II
Thursday, February 25, 16:10 - 17:25
Session Chairs: Vajapeyam Sukumar, *Maxim Integrated Products*

Quadratic Power Conversion for Industrial Applications 1320
Gerry Moschopoulos, *University of Western Ontario, Canada*

Multiple-Output Resonant Inverter Topology for Multi-Inductor Loads 1328
O. Lucía, *University of Zaragoza, Spain*
J.M. Burdío, *University of Zaragoza, Spain*
I. Millán, *University of Zaragoza, Spain*
J. Acero, *University of Zaragoza, Spain*

Variable Frequency Pulse Density Modulation[1] for Efficient High Frequency
Operation of Series Resonant Converters Operating As Voltage Regulators 1334
Darryl J. Tschirhart, *Queen's University, Canada*
Praveen K. Jain, *Queen's University, Canada*

Session C4L-F: General Lighting
Thursday, February 25, 16:10 - 17:25
Session Chairs: Ali Davoudi, *University of Illinois*

Flexible-Controlled High Power-Density Automotive HID Electronic Ballast Using Full-Digital Control Mode 1340
Xinyi Yang, *Zhejiang University, China*
Biwen Xu, *Zhejiang University, China*
Chongguang Ma, *Zhejiang University, China*
Min Chen, *Zhejiang University, China*
Zhaoming Qian, *Zhejiang University, China*

A "Class-A2" Ultra-Low-Loss Magnetic Ballast for T5 Fluorescent Lamps 1346
S.Y.R. Hui, *City University of Hong Kong, China*
D.Y. Lin, *City University of Hong Kong, China*
W.M. Ng, *City University of Hong Kong, China*
W. Yan, *City University of Hong Kong, China*

Simple Triac Dimmable Compact Fluorescent Lamp Ballast and Light Emitting Diode Driver 1352
Andre Tjokrorahardjo, *International Rectifier, United States*

Session C4L-G: Soft Switching Techniques II
Thursday, February 25, 16:10 - 17:25
Session Chairs: Jason Neely, *Purdue University*
Wayne Weaver, *Michigan Technological University*

High Efficiency Soft-Switched Step-Up DC-DC Converter with Hybrid Mode LLC+C Resonant Tank 1358
Wei Chen, *Zhejiang University and Delta Electronics (Shanghai) Co., LTD., China*
Xiaoyuan Hong, *Zhejiang University, China*
Siran Wang, *Zhejiang University, China*
Zhengyu Lu, *Zhejiang University, China*
Shaoshi Ye, *Delta Electronics (Shanghai) Co., LTD., China*

A Family of Zero Current Switching Switched-Capacitor DC-DC Converters 1365
Dong Cao, *Michigan State University, United States*
Fang Zheng Peng, *Michigan State University, United States*

Analysis and Design of the Half Bridge Magnetizing Inductor Resonant(L_mC) DC-DC Converter 1373
B.-C. Hyeon, *Seoul National University, Korea, South*
B.-H. Cho, *Seoul National University, Korea, South*

Session B4P-H: AC-DC Conversion
Thursday, February 25, 11:30 - 13:30

A High Power Density Single Phase PWM Rectifier with Active Ripple Energy Storage 1378
Ruxi Wang, *Virginia Polytechnic Institute and State University, United States*
Fred Wang, *University of Tennessee - Knoxville and Oak Ridge National Laboratory, United States*
Dushan Boroyevich, *Virginia Polytechnic Institute and State University, United States*
Puqi Ning, *Virginia Polytechnic Institute and State University, United States*

Design Considerations for High Efficiency Buck PFC with Half-Bridge Regulation Stage .. 1384
Bernard Keogh, *Texas Instruments (Cork) Ltd., Ireland*
George Young, *Texas Instruments (Cork) Ltd., Ireland*
Hagen Wegner, *Texas Instruments (Cork) Ltd., Ireland*
Colin Gillmor, *Texas Instruments (Cork) Ltd., Ireland*

A Novel Variable Frequency Soft Switching Method for Flyback Converter with Synchronous Rectifier 1392
Xiucheng Huang, *Zhejiang University, China*
Weijing Du, *Zhejiang University, China*
Wei Yuan, *Zhejiang University, China*
Junming Zhang, *Zhejiang University, China*
Zhaoming Qian, *Zhejiang University, China*

Optimal Design of a Compact 99.3% Efficient Single-Phase PFC Rectifier 1397
J. Biela, *ETH Zurich, Switzerland*
J.W. Kolar, *ETH Zurich, Switzerland*
G. Deboy, *Infineon Technologies Austria AG, Austria*

DCM Boost PFC Converter with High Input PF 1405
Kai Yao, *Nanjing university of aeronautics and astronautics, China*
Xinbo Ruan, *Nanjing University of Aeronautics and Astronautics, China*
Xiaojing Mao, *Nanjing University of Aeronautics and Astronautics, China*
Zhihong Ye, *Lite-on Technology Corp., China*

Interleaved Forward Converter with Ripple-Free Circuit for Humane Killer Poultry Applications 1413
S.-Y. Tseng, *Chang Gung University, Taiwan*
T.-Y. Chiang, *Chang Gung University, Taiwan*
K.-C. Wang, *Chang Gung University, Taiwan*
S.-A. Chuang, *Chang Gung University, Taiwan*

The Optimal Control Strategy for Rectifier Side of Low Switching Frequency Back-to-Back Converter 1419
Kai Tan, *Chinese Academy of Sciences, China*
Qiongxuan Ge, *Chinese Academy of Sciences, China*
Zhenggang Yin, *Chinese Academy of Sciences, China*
Congwei Liu, *Chinese Academy of Sciences, China*
Yaohua Li, *Chinese Academy of Sciences, China*

Transformer Structure and its Effects on Common Mode EMI Noise in Isolated Power Converters 1424
Pengju Kong, *Virginia Polytechnic Institute and State University, United States*
Fred C. Lee, *Virginia Polytechnic Institute and State University, United States*

A Single-Stage Single-Phase Bi-Directional Grid Interface Circuit with Digital Lookup Table Based Control 1430
Evan Reutzel, *University of California, Berkeley, United States*
Seth Sanders, *University of California, Berkeley, United States*

Session B4P-J: DC-DC Converter VII
Thursday, February 25, 11:30 - 13:30

A High Performance Dual Output DC-DC Converter Combined the Phase Shift Full Bridge and LLC Resonant Half Bridge with the Shared Lagging Leg 1435
Yu Chen, *Huazhong University of Science and Technology, China*
Xuejun Pei, *Huazhong University of Science and Technology, China*
Li Peng, *Huazhong University of Science and Technology, China*
Yong Kang, *Huazhong University of Science and Technology, China*

Dual Output DC-DC Converter with Shared ZCS Lagging Leg 1441
Yu Chen, *Huazhong University of Science and Technology, China*
Li Peng, *Huazhong University of Science and Technology, China*
Xuejun Pei, *Huazhong University of Science and Technology, China*
Yong Kang, *Huazhong University of Science and Technology, China*

A Novel ZVS Full-Bridge Converter with Auxiliary Circuit 1448
Zhong Chen, *Nanjing University of Aeronautics and Astronautics, China*
Biao Ji, *Nanjing University of Aeronautics and Astronautics, China*
Feng Ji, *Nanjing University of Aeronautics and Astronautics, China*
Lei Shi, *Nanjing University of Aeronautics and Astronautics, China*

An Active Clamp ZVT Converter with Input-Parallel and Output-Series Configuration 1454
Yi Zhao, *Zhejiang University, China*
Wuhua Li, *Zhejiang University, China*
Weichen Li, *Zhejiang University, China*
Xiangning He, *Zhejiang University, China*

A Parallel Front-End LCL Resonant Push-Pull Converter with a Coupled Inductor for Automotive Applications 1460
Yuan Yisheng, *East China Jiaotong University, China*
Chen Min, *Zhejiang University, China*
Qian Zhaoming, *Zhejiang University, China*

A Novel Full Bridge Dual Output DC-DC Converter with Complementary Pulse Widths and Frequency Modulation 1464
Yu Chen, *Huazhong University of Science and Technology, China*
Xuejun Pei, *Huazhong University of Science and Technology, China*
Li Peng, *Huazhong University of Science and Technology, China*
Yong Kang, *Huazhong University of Science and Technology, China*

Analysis and Design Considerations of an Improved ZVS Full-Bridge DC-DC Converter ... 1471
Zhong Chen, *Nanjing University of Aeronautics and Astronautics, China*
Biao Ji, *Nanjing University of Aeronautics and Astronautics, China*
Feng Ji, *Nanjing University of Aeronautics and Astronautics, China*
Lei Shi, *Nanjing University of Aeronautics and Astronautics, China*

A New Resonant Gate Driver for Switching Loss Reduction of High Side Switch in Buck Converter 1477

Xin Zhou, *North Carolina State University, United States*
Zhigang Liang, *North Carolina State University, United States*
Alex Huang, *North Carolina State University, United States*

Switching Loss Analysis Considering Parasitic Loop Inductance with Current Source Drivers for Buck Converters 1482

Zhiliang Zhang, *Nanjing University of Aeronautics and Astronautics, China*
Jizhen Fu, *Queen's University, Canada*
Yan-Fei Liu, *Queen's University, Canada*
P.C. Sen, *Queen's University, Canada*

Improved Asymmetric Space Vector Modulation for Voltage Source Converters with Low Carrier Ratio 1487

Di Zhang, *Virginia Polytechnic Institute and State University, United States*
Fred Wang, *University of Tennessee - Knoxville and Oak Ridge National Laboratory, United States*
Said El-Barbari, *GE Global Research, Germany*
Juan Sabate, *GE Global Research, United States*
Dushan Boroyevich, *Virginia Polytechnic Institute and State University, United States*

A Hybrid Switching Scheme for LLC Series-Resonant Half-Bridge DC-DC Converter in a Wide Load Range 1494

Woo-Young Choi, *Virginia Polytechnic Institute and State University, United States*
Bong-Hwan Kwon, *Pohang University of Science and Technology, Korea, South*
Jih-Sheng Lai, *Virginia Polytechnic Institute and State University, United States*

Session B4P-K: Motor Drives & Inverters I
Thursday, February 25, 11:30 - 13:30

Industrial Servo Applications of Linear Induction Motors Based on Dynamic Maximum Force Control 1498

Haidong Yu, *Phoenix International, United States*
Babak Fahimi, *University of Texas at Arlington, United States*

A Soft-Switching Interleaved Three-Level Inverter 1503

Yuan Yisheng, *East China Jiaotong University, China*
Chen Min, *Zhejiang University, China*
Qian Zhaoming, *Zhejiang University, China*

Reducing Common-Mode Voltage in Three-Phase Sine-Triangle PWM with Interleaved Carriers 1508

Jonathan W. Kimball, *Missouri University of Science and Technology, United States*
Maciej Zawodniok, *Missouri University of Science and Technology, United States*

Dynamic DC-Bus Voltage Control Strategies for a Three-Phase High Power Shunt Active Power Filter 1514

Zhiqiang Wang, *Zhejiang University, China*
Chuan Xie, *Zhejiang University, China*
Jing Zhang, *Zhejiang University, China*
Guozhu Chen, *Zhejiang University, China*

A Simplified Three Phase Three-Level Zero-Current-Transition Active Neutral-Point-Clamped Converter with Three Auxiliary Switches 1521

Jin Li, *Xi'an Jiaotong University and Virginia Polytechnic Institute and State University, China*
Jinjun Liu, *Xi'an Jiaotong University, China*
Dushan Boroyevich, *Virginia Polytechnic Institute and State University, China*

Comparison and Implementation of a 3-Level NPC Voltage Link Back-to-Back Converter with SiC and Si Diodes 1527

Mario Schweizer, *ETH Zurich, Switzerland*
Thomas Friedli, *ETH Zurich, Switzerland*
Johann W. Kolar, *ETH Zurich, Switzerland*

A Novel PWM Control Method to Eliminate the Effect of Dead Time on the Output Waveform for Hybrid Clamped Multilevel Inverters 1534

Jing Zhao, *Zhejiang University, China*
Xiangning He, *Zhejiang University, China*
Yunlong Han, *Zhejiang University, China*
Yan Chen, *Zhejiang University, China*
Rongxiang Zhao, *Zhejiang University, China*

Study on Wide Range Robust Speed Sensorless Control of Medium Voltage Induction Motor 1542

Siran Wang, *Zhejiang University, China*
Zhengyu Lu, *Zhejiang University, China*

Fault Detection and Diagnostics for Non-Intrusive Monitoring Using Motor Harmonics 1547

Uzoma A. Orji, *Massachusetts Institute of Technology, United States*
Zachary Remscrim, *Massachusetts Institute of Technology, United States*
Christopher Laughman, *Massachusetts Institute of Technology, United States*
Steven B. Leeb, *Massachusetts Institute of Technology, United States*
Warit Wichakool, *Massachusetts Institute of Technology, United States*
Christopher Schantz, *Massachusetts Institute of Technology, United States*
Robert Cox, *Massachusetts Institute of Technology, United States*
James Paris, *Massachusetts Institute of Technology, United States*
James L. Kirtley, Jr., *Massachusetts Institute of Technology, United States*
Les K. Norford, *Massachusetts Institute of Technology, United States*

Reliability Evaluation of Three-Level Inverters 1555

Yi Ding, *Nanyang Technological University, Singapore*
Poh Chiang Loh, *Nanyang Technological University, Singapore*
Kuan Khoon Tan, *Nanyang Technological University, Singapore*
Peng Wang, *Nanyang Technological University, Singapore*
Feng Gao, *Nanyang Technological University, Singapore*

Parallel Operation of PWM Inverters for High Speed Motor Drive System 1561

Un-Kwan Cho, *Seoul National University, Korea, South*
Jung-Sik Yim, *Seoul National University, Korea, South*
Seung-Ki Sul, *Seoul National University, Korea, South*

Session B4P-L: Active Components
Thursday, February 25, 11:30 - 13:30

Reverse Conduction of a 100 a SiC DMOSFET Module in High-Power Applications 1568
R.A. Wood, *US Army Research Lab, United States*
D.P. Urciuoli, *US Army Research Lab, United States*
T.E. Salem, *US Naval Academy, United States*
R. Green, *US Army Research Lab, United States*

Investigation of 1.2 kV SiC MOSFET for High Frequency High Power Applications 1572
Honggang Sheng, *Monolithic Power Systems, United States*
Zheng Chen, *Virginia Polytechnic Institute and State University, United States*
Fred Wang, *University of Tennessee - Knoxville, United States*
Alan Millner, *MKS Instruments, United States*

Comparative Analysis of Power Stage Losses for Synchronous Buck Converter in Diode Emulation Mode Vs. Continuous Conduction Mode at Light Load Condition 1578
Yang Chen, *International Rectifier Corp., United States*
Peyman Asadi, *International Rectifier Corp., United States*
Parviz Parto, *International Rectifier Corp., United States*

Controllable dv/dt Behaviour of the SiC MOSFET/JFET Cascode an Alternative Hard Commutated Switch for Telecom Applications 1584
Daniel Aggeler, *ETH Zurich, Switzerland*
Juergen Biela, *ETH Zurich, Switzerland*
Johann W. Kolar, *ETH Zurich, Switzerland*

Integral Micro-Channel Liquid Cooling for Power Electronics 1591
Ljubisa D. Stevanovic, *GE Global Research, United States*
Richard A. Beaupre, *GE Global Research, United States*
Arun V. Gowda, *GE Global Research, United States*
Adam G. Pautsch, *GE Global Research, United States*
Stephen A. Solovitz, *Washington State University Vancouver, United States*

3000V, 25A Pulse Power Asymmetrical Highly Interdigitated SiC Thyristors 1598
Ahmed Elasser, *GE Global Research, United States*
Peter Losee, *GE Global Research, United States*
Stephen Arthur, *GE Global Research, United States*
Zachary Stum, *GE Global Research, United States*
Jerome Garrett, *GE Global Research, United States*
Michael Schutten, *GE Global Research, United States*

Session B4P-M: System Integration III
Thursday, February 25, 11:30 - 13:30

Low Inductance Power Module with Blade Connector 1603
Ljubisa D. Stevanovic, *GE Global Research, United States*
Richard A. Beaupre, *GE Global Research, United States*
Eladio C. Delgado, *GE Global Research, United States*
Arun V. Gowda, *GE Global Research, United States*

Design of Multi-Turn LTCC Inductors for High Frequency DC-DC Converters 1610
Laili Wang, *Xi'an Jiaotong University, China*
Yunqing Pei, *Xi'an Jiaotong University, China*
Xu Yang, *Xi'an Jiaotong University, China*
Xizhi Cui, *Xi'an Jiaotong University, China*
Zhaoan Wang, *Xi'an Jiaotong University, China*
Guopeng Zhao, *Xi'an Jiaotong University, China*

Session B4P-N: Utility Interface
Thursday, February 25, 11:30 - 13:30

Topological Research and Comparison of Low Harmonic Input Three-Phase Rectifier with Passive Auxiliary Circuit 1616
Zhong Chen, *Nanjing University of Aeronautics and Astronautics, China*
Yingpeng Luo, *Nanjing University of Aeronautics and Astronautics, China*
Yinyu Zhu, *Nanjing University of Aeronautics and Astronautics, China*

The Reactive Power Compensation and Harmonic Filtering and the Over-Voltage Analysis of the ITER Power Supply System 1622
L. Xu, *Chinese Academy of Sciences, China*
Z. Sheng, *Chinese Academy of Sciences, China*
P. Fu, *Chinese Academy of Sciences, China*
G. Gao, *Chinese Academy of Sciences, China*
I. Benfatto, *Iter Organization, France*
A.D. Mankani, *Iter Organization, France*
J. Tao, *Iter Organization, France*

Optimal Design Method of Three-Phase Rectifier with Near-Sinusoidal Input Currents 1627
Zhong Chen, *Nanjing University of Aeronautics and Astronautics, China*
Yingpeng Luo, *Nanjing University of Aeronautics and Astronautics, China*
Yinyu Zhu, *Nanjing University of Aeronautics and Astronautics, China*
Shunqing Wang, *Nanjing University of Aeronautics and Astronautics, China*

An Analysis on the Influence of Interface Inductor to STATCOM System with Phase and Amplitude Control and Corresponding Design Considerations 1633
Guopeng Zhao, *Xi'an Jiaotong University, China*
Jinjun Liu, *Xi'an Jiaotong University, China*

Vector Oriented Control of Voltage Source PWM Inverter As a Dynamic VAR Compensator for Wind Energy Conversion System Connected to Utility Grid 1640
Mahmoud M.N. Amin, *Florida International University, United States*
O.A. Mohammed, *Florida International University, United States*

Control System Design for Bi-Directional Power Transfer in Single-Phase Back-to-Back Converter Based on the Linear Operating Region 1651
Janeth Alcalá, *Universidad Autónoma de San Luis Potosi, Mexico*
Víctor Cárdenas, *Universidad Autónoma de San Luis Potosi, Mexico*
Emanuel Rosas, *Universidad Autónoma de San Luis Potosi, Mexico*
Ciro Núñez, *Universidad Autónoma de San Luis Potosi, Mexico*

Comparative Analysis of Low-Pass Output Filter for Single-Phase Grid-Connected Photovoltaic Inverter 1659
Hanju Cha, *Chungnam National University, Korea, South*
Trung-Kien Vu, *Chungnam National University, Korea, South*

Design and Development of Generation-I Silicon based Solid State Transformer 1666
Subhashish Bhattacharya, *North Carolina State University, United States*
Tiefu Zhao, *North Carolina State University, United States*
Gangyao Wang, *North Carolina State University, United States*
Sumit Dutta, *North Carolina State University, United Kingdom*
Seunghun Baek, *North Carolina State University, United States*
Yu Du, *North Carolina State University, United States*
Babak Parkhideh, *North Carolina State University, United States*
Xiaohu Zhou, *North Carolina State University, United States*
Alex Q. Huang, *North Carolina State University, United States*

Power Calculation Method Used in Wireless Parallel Inverters Under Nonlinear Load Conditions 1674
Zheng Ren, *Zhejiang University, China*
Mingzhi Gao, *Zhejiang University, China*
Qiong Mo, *Zhejiang University, China*
Kun Liu, *Zhejiang University, China*
Wei Yao, *Zhejiang University, China*
Min Chen, *Zhejiang University, China*
Zhaomin Qian, *Zhejiang University, China*

A Real-Time Fault Diagnosis System for UPS Based on FFT Frequency Analysis 1678
Won-Sul Shim, *Kangwon National University, Korea, South*
Gi-Taek Kim, *Kangwon National University, Korea, South*
Ha-Jin Jung, *Powertron Engineering Co. Ltd., Korea, South*
Deuk-Soo Kim, *Powertron Engineering Co. Ltd., Korea, South*

Control Strategy for a Buck-Boost Type Direct Interface Converter Using an Indirect Matrix Converter with an Active Snubber 1684
Koji Kato, *Nagaoka University of Technology, Japan*
Jun-Ichi Itoh, *Nagaoka University of Technology, Japan*

A PI Control Algorithm of Three-Level APF with Little Static Misadjustment for Tracking Harmonic Current 1692
Yingjie He, *Xi'an Jiaotong University, China*
Jinjun Liu, *Xi'an Jiaotong University, China*
Zhaoan Wang, *Xi'an Jiaotong University, China*
Yunping Zou, *Huazhong University of Science and Technology, China*

A Novel Topology of LLC Resonant Inverter with Two Resonant Tanks for Power Conditioning System 1698
Eun-Soo Kim, *Jeonju University, Korea, South*
Kwang-Ho Lee, *Jeonju University, Korea, South*
Bong-Gun Chung, *Jeonju University, Korea, South*
Joo-Hoon Kim, *Jeonju University, Korea, South*
Moon-Ho Kye, *Powerplaza, United States*

Analysis and Realization of a Fast Repetitive Controller in Active Power Filter System 1704
Jinwu Gong, *Wuhan University, China*
Xiaoming Zha, *Wuhan University, China*
Suxuan Guo, *Wuhan University, China*
Baifeng Chen, *Wuhan University, China*
Jianjun Sun, *Wuhan University, China*

Session B4P-P: Modeling, Simulation & Control II
Thursday, February 25, 11:30 - 13:30
Session Chairs: Jonathan Kimball, *Missouri S&T*
Omer Onar, *Illinois Institute of Technology*

**On Extended Kalman Filters with Augmented State Vectors for the
Stator Flux Estimation in SPMSMs** 1711
T.J. Vyncke, *Ghent University, Belgium*
R.K. Boel, *Ghent University, Belgium*
J.A.A. Melkebeek, *Ghent University, Belgium*

State Equations Based Resonant Converters Modeling Technique 1719
Yingqi Zhang, *GE Global Research, China*
P.C. Sen, *Queen's University, Canada*

**Design Considerations and Expiremantal Results of an Adaptive Frequency
Controller Under Variable Line and Load Conditions** 1723
Jaber A. Abu Qahouq, *University of Alabama, United States*
Wisam Al-Hoor, *University of Central Florida, United States*
Issa Batarseh, *University of Central Florida, United States*

**Modeling and Mitigation of Dynamic Load Beat-Frequency Oscillation in
Multiphase Voltage Regulators with High-Gain Peak Current Control Scheme** 1727
Chen-Hua Chiu, *National Taiwan University, Taiwan*
Dan Chen, *National Taiwan University, Taiwan*
Ching-Jan Chen, *National Taiwan University, Taiwan*
Wei-Hsu Chang, *RichTek Technology Corporation, Taiwan*

Half-Wave Symmetry SHE-PWM Method for Multilevel Voltage Inverters 1732
Wanmin Fei, *Nanjing Normal University, China*
Xiaoli Du, *Nanjing Normal University, China*
Bin Wu, *Ryerson University, Canada*

PI Type Dynamic Decoupling Control Scheme for PMSM High Speed Operation 1736
Hao Zhu, *Tsinghua University, China*
Xi Xiao, *Tsinghua University, China*
Yongdong Li, *Tsinghua University, China*

**High Performance Positive and Negative Sequence Filters in
Stationary Frame Based on Complex Transfer Function** 1740
Jingxin Mao, *Beijing Jiaotong University, China*
Fei Lin, *Beijing Jiaotong University, China*
Hong Li, *Beijing Jiaotong University, China*
Xiaojie You, *Beijing Jiaotong University, China*
Trillion Q. Zheng, *Beijing Jiaotong University, China*

Simulation Study of Parameter Influence on Dynamic Voltage Rise Control 1745
Ming Li, *Xi'an Jiaotong University, China*
Xiong Fang, *Xi'an Jiaotong University, China*
Yue Wang, *Xi'an Jiaotong University, China*
Leqiang Zhang, *Xi'an Jiaotong University, China*
Ke Wang, *Xi'an Jiaotong University, China*
Guopeng Zhao, *Xi'an Jiaotong University, China*

Shaping of the Noise Spectrum in Power Electronic Converters 1749
Cristian Lascu, *University of Nevada, Reno, United States*
Andrzej M. Trzynadlowski, *University of Nevada, Reno, United States*
R. Lynn Kirlin, *University of Victoria, Canada*

**Grid Interactions and Stability Analysis of Distribution Power Network with
High Penetration of Plug-in Hybrid Electric Vehicles** 1755
Omer C. Onar, *Illinois Institute of Technology, United States*
Alireza Khaligh, *Illinois Institute of Technology, United States*

**Rapid Simulation of Fourth-Order Multi-Resonant LLCC Converters with
Capacitive Output Filter** 1763
A. Bucher, *University of Erlangen-Nürnberg, Germany*
T. Duerbaum, *University of Erlangen-Nürnberg, Germany*

**FHA-Based Voltage Gain Function with Harmonic Compensation for
LLC Resonant Converter** 1770
Hong Huang, *Texas Instruments, United States*

Session B4P-Q: Aerospace & Transportation
Thursday, February 25, 11:30 - 13:30

**Analysis and Design of LCC Resonant Inverter for the
Tranportation Systems Applications** 1778
Mohamed Youssef, *Bombardier Transportation Inc., Canada*
Jaber A. Abu Qahouq, *University of Alabama, United States*
Mohamed Orabi, *South Valley University, Egypt*

**A Multi-Resolution Control Strategy for DSP Controlled 400Hz Shunt
Active Power Filter in an Aircraft Power System** 1785
Haibing Hu, *Nanjing University of Aeronautics and Astronautics, China*
Wei Shi, *Nanjing University of Aeronautics and Astronautics, China*
Jianren Xue, *Nanjing Sute Electric Co., Ltd., China*
Ying Lu, *Nanjing University of Aeronautics and Astronautics, China*
Yan Xing, *Nanjing University of Aeronautics and Astronautics, China*

Battery Discharge Regulator for Space Applications Based on the Boost Converter 1792
A. Fernandez, *European Space Agency, Netherlands*
F. Tonicello, *European Space Agency, Netherlands*
J. Aroca, *European Space Agency, Netherlands*
O. Mourra, *European Space Agency, Netherlands*

Electromagnetic Compatibility Results for an LCC Resonant Inverter for the Tranportation Systems 1800

Mohamed Youssef, *Bombardier Transportation Inc., Canada*
Jaber A. Abu Qahouq, *University of Alabama, United States*
Mohamed Orabi, *South Valley University, Egypt*

Torque Impulse for Experimental Modal Analysis in Transmitted Vibration Study of Engine-Generators 1804

Elias Ayana, *Cummins Power Generation & University of Minnesota, United States*
Steve Seidlitz, *Cummins Power Generation, United States*
Sze Kwan Cheah, *Cummins Power Generation, United States*
Ned Mohan, *University of Minnesota, United States*

Session B4P-R: Power Converters & Applications
Thursday, February 25, 11:30 - 13:30

Review and Analysis of the AC-DC Converter of ITER Coil Power Supply 1810

P. Fu, *Institute of Plasma Physics, China*
G. Gao, *Institute of Plasma Physics, China*
L.W. Xu, *Institute of Plasma Physics, China*
Z.Q. Song, *Institute of Plasma Physics, China*
Z.C. Sheng, *Institute of Plasma Physics, China*
I. Benfatto, *ITER Organization, France*
J. Tao, *ITER Organization, France*
A.D. Mankani, *ITER Organization, France*
J.S. Oh, *National Fusion Research Institute, Korea, South*
C. Neumeyer, *Princeton Plasma Physics Laboratory, United States*

Fault Tolerance on Interleaved Inverter with Magnetic Couplers 1817

K. Guépratte, *Grenoble Electrical Engineering Laboratory, France*
D. Frey, *Grenoble Electrical Engineering Laboratory, France*
P.-O. Jeannin, *Grenoble Electrical Engineering Laboratory, France*
H. Stephan, *Grenoble Electrical Engineering Laboratory, France*
J.-P. Ferrieux, *Grenoble Electrical Engineering Laboratory, France*

Latest Practical Developments of Triplex Series Load Resonant Frequency-Operated High Frequency Inverter for Induction-Heated Low Resistivity Metallic Appliances in Consumer Built-in Cooktops 1825

Hideki Sadakata, *Panasonic Corporation, Japan*
Atsushi Fujita, *Panasonic Corporation, Japan*
Shinichiro Sumiyoshi, *Panasonic Corporation, Japan*
Hideki Omori, *Panasonic Corporation, Japan*
Bishwajit Saha, *Kyungnam University / Yamaguchi University, Korea, South*
Tarek Ahmed, *Kyungnam University / Yamaguchi University, Korea, South*
Mutsuo Nakaoka, *Kyungnam University / Yamaguchi University, Korea, South*

A Study of Novel Flyback Converter with Very Low Power Consumption at the Standby Operating Mode 1833

Eun-Soo Kim, *Jeonju University, Korea, South*
Bong-Gun Chung, *Jeonju University, Korea, South*
Sang-Ho Jang, *Jeonju University, Korea, South*
Mun-Gi Choi, *LG Innotek, Korea, South*
Moon-Ho Kye, *Powerplaza, United States*

Improved Two-Stage DC-Coupled Gate Driver for Enhancement-Mode SiC JFET 1838
Robin Kelley, *SemiSouth Laboratories Inc., United States*
Andrew Ritenour, *SemiSouth Laboratories Inc., United States*
David Sheridan, *SemiSouth Laboratories Inc., United States*
Jeff Casady, *SemiSouth Laboratories Inc., United States*

Design and Implementation of Multi-Channel Land Fowls Stunner with
Current Sharing Controller 1842
S.-Y. Fan, *Wufeng Institute of Technology, Taiwan*
S.-Y. Tseng, *Chang Gung University, Taiwan*
Y.-H. Su, *Chang Gung University, Taiwan*
W.-C. Wu, *Wufeng Institute of Technology, Taiwan*

High Voltage Generator Using Boost/Flyback Hybrid Converter for Stun Gun Applications .. 1849
S.-Y. Tseng, *Chang Gung University, Taiwan*
C.-M. Yang, *Chang Gung University, Taiwan*
K.-C. Wang, *Chang Gung University, Taiwan*
G.-W. Hsu, *Chang-Gung University, Taiwan*

Session C5P-H: DC-DC Converter VIII
Thursday, February 25, 11:30 - 13:30

A Method to Analysis and Design for Long Life Power Converter 1857
H.M. Pang, *University of Hong Kong, China*
M.H. Bryan Pong, *University of Hong Kong, China*

DC-DC Converter for Gate Power Supplies with an Optimal Air Transformer 1865
Christoph Marxgut, *ETH Zurich, Switzerland*
Jürgen Biela, *ETH Zurich, Switzerland*
Johann W. Kolar, *ETH Zurich, Switzerland*
Reto Steiner, *ABB, Switzerland*
Peter K. Steimer, *ABB, Switzerland*

A Digitally Controlled DC-DC Buck Converter Using Frequency Domain ADCs 1871
Hani Ahmad, *Arizona State University, United States*
Bertan Bakkaloglu, *Arizona State University, United States*

Low-Dropout (LDO) Regulator Output Impedance Analysis and
Transient Performance Enhancement Circuit 1875
Sungkeun Lim, *North Carolina State University, United States*
Alex Q. Huang, *North Carolina State University, United States*

A Design for Small Time-Delay Control Circuit for DPWM- POL 1879
Yoichi Ishizuka, *Nagasaki University, Japan*
Yusuke Yamada, *Nagasaki University, Japan*
Fumitoshi Hirose, *Nagasaki University, Japan*
Mariko Nishi, *Nagasaki University, Japan*
Hirofumi Matsuo, *Nagasaki University, Japan*

Low Profile LLC Series Resonant Converter with Two Transformers 1885
Eun-Soo Kim, *Jeonju University, Korea, South*
Joo-Hoon Kim, *Jeonju University, Korea, South*
Sung-In Kang, *LG Innotek, Korea, South*
Jun-Ho Park, *LG Innotek, Korea, South*
Jae-Sam Lee, *LG Innotek, Korea, South*
Dong-Young Huh, *LG Innotek, Korea, South*
Yong-Chae Jung, *Namseoul University, Korea, South*

Adaptive Frequency Control for ZVS Synchronous Boost Converters Operated in Average Current Mode 1890
Ben York, *Virginia Polytechnic Institute and State University, United States*
Rae-Young Kim, *Virginia Polytechnic Institute and State University, United States*
Jih-Sheng Lai, *Virginia Polytechnic Institute and State University, United States*

Power Saving Control Strategies and Their Implementation in DC-DC Converter for Data and Telecommunication Power Supply 1897
Rais Miftakhutdinov, *Texas Instruments Inc., United States*

Session C5P-J: DC-DC Converter IX
Thursday, February 25, 11:30 - 13:30

Analysis and Optimized Design of an Efficient High-Voltage Converter with High Output Capacity 1904
Huai Wang, *City University of Hong Kong, China*
Henry Shu-hung Chung, *City University of Hong Kong, China*
Adrian Ioinovici, *Holon Institute of Technology, Israel*

A Novel Three-Phase Three-Level ZVS PWM DC-DC Converter 1911
Eloi Agostini Junior, *Federal University of Santa Catarina, Brazil*
Ivo Barbi, *Federal University of Santa Catarina, Brazil*

Optimize the Synchronous Rectifier for LCC Converters 1919
Feng Zheng, *Xidian University, China*
Zhengfeng Ming, *Xidian University, China*

Digital Control Scheme for Robust Clock Tuning and PWM Phase Synchronization in Digitally Controlled Multi-POL Applications 1922
Eamon O'Malley, *Powervation Ltd., Ireland*
Karl Rinne, *Powervation Ltd., Ireland*
Anthony Kelly, *Powervation Ltd., Ireland*
Basil Almukhtar, *Powervation Ltd., Ireland*
Paul Kelleher, *Powervation Ltd., Ireland*

Control Scheme and Transient Performance of Sigma VR 1927
Pengjie Lai, *Virginia Polytechnic Institute and State University, United States*
Julu Sun, *Virginia Polytechnic Institute and State University, United States*
Fred C. Lee, *Virginia Polytechnic Institute and State University, United States*

A Three-Phase Current-Fed Push-Pull DC-DC Converter with Active Clamp for Fuel Cell Applications 1934
Sangwon Lee, *Seoul National University of Technology, Korea, South*
Sewan Choi, *Seoul National University of Technology, Korea, South*

Resonant Voltage Divider with Startup Considered 1942
K.I. Hwu, *National Taipei University of Technology, Taiwan*
Y.T. Yau, *National Taipei University of Technology, Taiwan*

LLC Resonant Converter with Two Resonant Tanks 1949
Eun-Soo Kim, *Jeonju University, Korea, South*
Joo-Hoon Kim, *Jeonju University, Korea, South*
Kwang-Ho Lee, *Jeonju University, Korea, South*
Yong-Seog Jeon, *Jeonju University, Korea, South*
Jae-Sam Lee, *LG Innotek, Korea, South*
Dong-Young Huh, *LG Innotek, Korea, South*

Session C5P-K: Motor Drives & Inverters II
Thursday, February 25, 11:30 - 13:30

A Digital Control Strategy for Brushless DC Generators 1957
Nikola Milivojevic, *Illinois Institute of Technology, United States*
Igor Stamenkovic, *Illinois Institute of Technology, United States*
Mahesh Krishnamurthy, *Illinois Institute of Technology, United States*
Ali Emadi, *Illinois Institute of Technology, United States*

Space Vector Based PWM Scheme Without Sector Identification for a 4-Level
Dual Inverter Fed Induction Motor Drive with Asymmetrical DC Link Voltages 1963
G. Shiny, *College of Engineering Trivandrum, India*
M.R. Baiju, *College of Engineering Trivandrum, India*

Control Method for a Novel Converter Topology for Permanent Magnet Drives 1970
Philip Brockerhoff, *Universität der Bundeswehr München, Germany*
Martin Schulz, *Universität der Bundeswehr München, Germany*

A Voltage Controlled Adjustable Speed PMBLDCM Drive Using a Single-Stage
PFC Half-Bridge Converter 1976
Sanjeev Singh, *Indian Institute of Technology Delhi, India*
Bhim Singh, *Indian Institute of Technology Delhi, India*

Comparison of HF Signal Injection Methods for Sensorless Control of
PM Synchronous Motors 1984
Eisenhawer de M. Fernandes, *Universidade Federal de Campina Grande, Brazil*
Alexandre C. Oliveira, *Universidade Federal de Campina Grande, Brazil*
Cursino B. Jacobina, *Universidade Federal de Campina Grande, Brazil*
Antonio M.N. Lima, *Universidade Federal de Campina Grande, Brazil*

A Robust Sensorless Fault Diagnosis Algorithm for Low Cost Motor Drives 1990
Seung-deog Choi, *Texas A&M University, United States*
Bilal Akin, *Texas Instruments Inc., United States*
Mina M. Rahimian, *Texas A&M University, United States*
Hamid A. Toliyat, *Texas A&M University, United States*

High Dynamic Performance Constrained Optimal Control of Induction Motors 1995
Sébastien Mariéthoz, *ETH Zurich, Switzerland*
Alexander Domahidi, *ETH Zurich, Switzerland*
Manfred Morari, *ETH Zurich, Switzerland*

PMSM Control Based on Edge Field Measurements by Hall Sensors 2002
Sungyoon Jung, *Pohang University of Science and Technology, Korea, South*
Beomseok Lee, *Pohang University of Science and Technology, Korea, South*
Kwanghee Nam, *Pohang University of Science and Technology, Korea, South*

Bridged-T Speed Controller for High Performance Switched Reluctance Motor Drives 2007
Gregory Pasquesoone, *University of Akron, United States*
Iqbal Husain, *University of Akron, United States*
Robert J. Veillette, *University of Akron, United States*

**Reducing Losses in Multilevel Coupled Inductor Inverters Using
Interleaved Discontinuous SVPWM** 2013
Behzad Vafakhah, *University of Alberta, Canada*
Andy Knight, *University of Alberta, Canada*
John Salmon, *University of Alberta, Canada*

A Novel Elevator Load Torque Identification Method Based on Friction Mode 2021
Xiaoyuan Hong, *Zhejiang University, China*
Zhe Deng, *Zhejiang University, China*
Siran Wang, *Zhejiang University, China*
Lijun Hang, *Zhejiang University, China*
Wuhua Li, *Zhejiang University, China*
Zhengyu Lu, *Zhejiang University, China*

**A Novel Digital Current Control Strategy for Torque Ripple Reduction in
Permanent Magnet Synchronous Motor Drives** 2025
Haidong Yu, *Phoenix International, United States*

Session C5P-L: Passive Components
Thursday, February 25, 11:30 - 13:30

**Evaluation of a SiC Power Module Using Low-on-Resistance IEMOSFET and
JBS for High Power Density Power Converters** 2030
Kazuto Takao, *Toshiba Corporation, Japan*
Takashi Shinohe, *Toshiba Corporation, Japan*
Shinsuke Harada, *National Institute of Advanced Industrial Science and Technology, Japan*
Kenji Fukuda, *National Institute of Advanced Industrial Science and Technology, Japan*
Hiromichi Ohashi, *National Institute of Advanced Industrial Science and Technology, Japan*

**A Novel Integrated Power Inductor in Silicon Substrate for
Ultra-Compact Power Supplies** 2036
Mingliang Wang, *University of Florida, United States*
Jiping Li, *University of Florida, United States*
Khai D.T. Ngo, *Virginia Polytechnic Institute and State University, United States*
Huikai Xie, *University of Florida, United States*

A Class of Coupled Inductors Based on LTCC Technology 2042
Laili Wang, *Xi'an Jiaotong University, China*
Yunqing Pei, *Xi'an Jiaotong University, China*
Xu Yang, *Xi'an Jiaotong University, China*
Xizhi Cui, *Xi'an Jiaotong University, China*
Zhaoan Wang, *Xi'an Jiaotong University, China*
Guopeng Zhao, *Xi'an Jiaotong University, China*

Optimising the High Frequency Bandwidth and Immuntity to Interference of Rogowski Coils in Measurement Applications with Large local dV/dt 2050
Christopher R. Hewson, *Power Electronic Measurements Ltd., United Kingdom*
William F. Ray, *Power Electronic Measurements Ltd., United Kingdom*

PFC Inductor Selection Made Easy by "PL Product" 2057
Welly Chou, *Precision Incorporated, United States*

Evaluation of LTCC Capacitors and Inductors in DC-DC Converters 2060
Laili Wang, *Xi'an Jiaotong University, China*
Yunqing Pei, *Xi'an Jiaotong University, China*
Xu Yang, *Xi'an Jiaotong University, China*
Bo Song, *Xi'an Jiaotong University, China*
Zhaoan Wang, *Xi'an Jiaotong University, China*
Guopeng Zhao, *Xi'an Jiaotong University, China*

Session C5P-M: Vehicle Electronics II
Thursday, February 25, 11:30 - 13:30

Bi-Directional Charging Topologies for Plug-in Hybrid Electric Vehicles 2066
Dylan C. Erb, *Illinois Institute of Technology, United States*
Omer C. Onar, *Illinois Institute of Technology, United States*
Alireza Khaligh, *Illinois Institute of Technology, United States*

Session C5P-N: Renewable Energy Systems
Thursday, February 25, 11:30 - 13:30
Session Chairs: Robert Balog, *Texas A&M University*

Multi-Channel Three-Port DC-DC Converters As Maximum Power Tracker, Battery Charger and Bus Regulator 2073
Zhijun Qian, *University of Central Florida, United States*
Osama Abdel-Rahman, *ApECOR, United States*
Haibing Hu, *University of Central Florida, United States*
Issa Batarseh, *University of Central Florida, United States*

A Smart and Simple PV Charger for Portable Applications 2080
Weichen Li, *Zhejiang University, China*
Yuzhen Zheng, *Zhejiang University of Science and Technology, China*
Wuhua Li, *Zhejiang University, China*
Yi Zhao, *Zhejiang University, China*
Xiangning He, *Zhejiang University, China*

RTDS-Based Real Time Simulations of Grid-Connected Wind Turbine Generator Systems .. 2085
Gyeong-Hun Kim, *Changwon National University, Korea, South*
Young-Ju Kim, *Changwon National University, Korea, South*
Minwon Park, *Changwon National University, Korea, South*
In-Keun Yu, *Changwon National University, Korea, South*
Byeong-Mun Song, *Baylor University, United States*

Investigation of Fully Digital Controlled Li-Ion Battery Power Recovery System 2091
Siran Wang, *Zhejiang University, China*
Xia Zhou, *Zhejiang University, China*
Jifeng Chen, *Zhejiang University, China*
Wenxi Yao, *Zhejiang University, China*
Zhengyu Lu, *Zhejiang University, China*

A Novel Control System for Harmonic Compensation by Using Wind Energy
Conversion Based on DFIG Technology 2096
Grazia Todeschini, *Worcester Polytechnic Institute, United States*
Alexander E. Emanuel, *Worcester Polytechnic Institute, United States*

A Transformerless Modular Permanent Magnet Wind Generator System with
Minimum Generator Coils 2104
Xibo Yuan, *Tsinghua University, China*
Yongdong Li, *Tsinghua University, China*
Jianyun Chai, *Tsinghua University, China*

Small-Signal Modeling and Analysis of the Double-Input Buckboost Converter 2111
Deepak Somayajula, *Missouri University of Science and Technology, United States*
Mehdi Ferdowsi, *Missouri University of Science and Technology, United States*

A Novel Power Distribution Strategy for Parallel Inverters in Islanded Mode Microgrid 2116
Xuan Zhang, *Xi'an Jiaotong University, China*
Jinjun Liu, *Xi'an Jiaotong University, China*
Ting Liu, *Xi'an Jiaotong University, China*
Linyuan Zhou, *Xi'an Jiaotong University, China*

Direct Power Control of Doubly-Fed Generator Based Wind Turbine Converters to
Improve Low Voltage Ride-Through During System Imbalance 2121
Murali M. Baggu, *Missouri University of Science and Technology, United States*
Luke D. Watson, *Missouri University of Science and Technology, United States*
Jonathan W. Kimball, *Missouri University of Science and Technology, United States*
Badrul H. Chowdhury, *Missouri University of Science and Technology, United States*

Active Damping for Torsional Vibrations in PMSG Based WECS 2126
Hua Geng, *Ryerson University, Canada*
Dewei Xu, *Ryerson University, Canada*
Bin Wu, *Ryerson University, Canada*
Geng Yang, *Tsinghua University, China*

Voltage and Frequency Stabilization Using PI-Like Fuzzy Controller for the
Load Side Converters of the Stand Alone Wind Energy Systems 2132
Ameen Gargoom, *University of Tasmania, Australia*
Abu Mohammad Osman Haruni, *University of Tasmania, Australia*
Md. Enamul Haque, *University of Tasmania, Australia*
Michael Negnevitsky, *University of Tasmania, Australia*

Dual-Stage Converter to Improve Transfer Efficiency and Maximum Power Point
Tracking Feasibility in Photovoltaic Energy-Conversion Systems 2138
Sairaj V. Dhople, *University of Illinois at Urbana-Champaign, United States*
Ali Davoudi, *University of Illinois at Urbana-Champaign, United States*
Patrick L. Chapman, *University of Illinois at Urbana-Champaign, United States*

A Novel Approach of Maximizing Energy Harvesting in Photovoltaic Systems Based on Bisection Search Theorem 2143

Peng Wang, *Nanyang Technological University, Singapore*
Haipeng Zhu, *Nanyang Technological University, Singapore*
Weixiang Shen, *Nanyang Technological University, Singapore*
Fook Hoong Choo, *Nanyang Technological University, Singapore*
Poh Chiang Loh, *Nanyang Technological University, Singapore*
Kuan Khoon Tan, *Nanyang Technological University, Singapore*

Simple Control Design for a Three-Port DC-DC Converter Based PV System with Energy Storage 2149

Sixifo Falcones, *Arizona State University, United States*
Raja Ayyanar, *Arizona State University, United States*

A Self-Powered Power Management Circuit for Energy Harvested by a Piezoelectric Cantilever 2154

Na Kong, *Virginia Polytechnic Institute and State University, United States*
Travis Cochran, *Virginia Polytechnic Institute and State University, United States*
Dong Sam Ha, *Virginia Polytechnic Institute and State University, United States*
Hung-Chih Lin, *National Tsing Hua University, Taiwan*
Daniel J. Inman, *Virginia Polytechnic Institute and State University, United States*

A Maximum Power Point Tracker Implementation for Photovoltaic Cells Using Dynamic Optimal Voltage Tracking 2161

Emil Jimenez-Brea, *University of Puerto Rico-Mayaguez, Puerto Rico*
Andres Salazar-Llinas, *University of Puerto Rico-Mayaguez, Puerto Rico*
Eduardo Ortiz-Rivera, *University of Puerto Rico-Mayaguez, Puerto Rico*
Jesus Gonzalez-Llorente, *University of Puerto Rico-Mayaguez, Puerto Rico*

Development of the Novel Control Algorithm for the Small Proton Exchange Membrane Fuel Cell Stack Without External Humidification 2166

Tae-Hoon Kim, *Soongsil University, Korea, South*
Sang-Hyun Kim, *Soongsil University, Korea, South*
Wook Kim, *Soongsil University, Korea, South*
Jong-Hak Lee, *Soongsil University, Korea, South*
Woojin Choi, *Soongsil University, Korea, South*

Session C5P-P: Modeling, Simulation & Control III
Thursday, February 25, 11:30 - 13:30
Session Chairs: Jonathan Kimball, *Missouri S&T*
Omer Onar, *Illinois Institute of Technology*

Stabilization of Constant-Power Loads by Passive Impedance Damping 2174

Mauricio Céspedes, *Rensselaer Polytechnic Institute, United States*
Troy Beechner, *Rensselaer Polytechnic Institute, United States*
Lei Xing, *Rensselaer Polytechnic Institute, United States*
Jian Sun, *Rensselaer Polytechnic Institute, United States*

An Adaptive External Ramp Control of the Peak Current Controlled Buck Converters for High Control Bandwidth and Wide Operation Range 2181

Liyu Yang, *North Carolina State University, United States*
Jinseok Park, *North Carolina State University, United States*
Alex Q. Huang, *North Carolina State University, United States*

Masterless Multirate Control of Parallel DC-DC Converters 2189
Anthony Kelly, *Powervation Ltd., Ireland*
Karl Rinne, *Powervation Ltd., Ireland*
Eamon O'Malley, *Powervation Ltd., Ireland*

FPGA-Based Spectral Envelope Preprocessor for Power Monitoring and Control 2194
Zachary Remscrim, *Massachusetts Institute of Technology, United States*
James Paris, *Massachusetts Institute of Technology, United States*
Steven B. Leeb, *Massachusetts Institute of Technology, United States*
Steven R. Shaw, *Montana State University, United States*
Sabrina Neuman, *Massachusetts Institute of Technology, United States*
Christopher Schantz, *Massachusetts Institute of Technology, United States*
Sean Muller, *Massachusetts Institute of Technology, United States*
Sarah Page, *Massachusetts Institute of Technology, United States*

Sigma-Delta Modulation of Multi-Phase High Frequency Converters 2202
Jonathan W. Kimball, *Missouri University of Science and Technology, United States*
Kyle Roger Eckler, *Missouri University of Science and Technology, United States*
Luke Watson, *Missouri University of Science and Technology, United States*

**Specialized Digital Signal Processor for Control of Multi-Rail/Multi-Phase
High Switching Frequency Power Converters** 2207
James Mooney, *University of Limerick, Ireland*
Mark Halton, *University of Limerick, Ireland*
Abdulhussain E. Mahdi, *University of Limerick, Ireland*

**Computer-Aided Design for Class-E Switching Circuits Taking into
Account Optimized Inductor Designs** 2212
Natsumi Sagawa, *Chiba University, Japan*
Hiroo Sekiya, *Chiba University and Wright State University, Japan*
Marian K. Kazimierczuk, *Wright State University, United States*

Characterization of IGBT Modules for System EMI Simulation 2220
Tao Qi, *Rensselaer Polytechnic Institute, United States*
Jeff Graham, *Fairchild Controls Corporation, United States*
Jian Sun, *Rensselaer Polytechnic Institute, United States*

**A Mathematical Model for Online Electrical Characterization of Thermoelectric
Generators Using the P-I Curves at Different Temperatures** 2226
Eduardo I. Ortiz-Rivera, *University of Puerto Rico-Mayaguez, Puerto Rico*
Andres Salazar-Llinas, *University of Puerto Rico-Mayaguez, Puerto Rico*
Jesus Gonzalez-Llorente, *University of Puerto Rico-Mayaguez, Puerto Rico*

**A Novel Method for Permanent Magnet Demagnetization Fault Detection and
Treatment in Permanent Magnet Synchronous Machines** 2231
Amir Khoobroo, *University of Texas at Arlington, United States*
Babak Fahimi, *University of Texas at Arlington, United States*

Session C5P-Q: Alternative Energy Applications
Thursday, February 25, 11:30 - 13:30

Series Connection of IGBT 2238
The-Van Nguyen, *Grenoble Institute of Technology, France*
Pierre-Olivier Jeannin, *Grenoble Institute of Technology, France*
Eric Vagnon, *Grenoble Institute of Technology, France*
David Frey, *Grenoble Institute of Technology, France*
Jean-Christophe Crebier, *Grenoble Institute of Technology, France*

**Three Phase Linear Permanent Magnet Energy Scavenger Based on
Foot Horizontal Motion** 2245
Igor Stamenkovic, *Illinois Institute of Technology, United States*
Nikola Milivojevic, *Illinois Institute of Technology, United States*
Cong Zheng, *Illinois Institute of Technology, United States*
Alireza Khaligh, *Illinois Institute of Technology, United States*

**Bidirectional Communication Techniques for Wireless Battery Charging
Systems and Portable Consumer Electronics** 2251
W.P. Choi, *ConvenientPower HK Ltd. and City University of Hong Kong, China*
W.C. Ho, *ConvenientPower HK Ltd., China*
X. Liu, *ConvenientPower HK Ltd., China*
S.Y.R. Hui, *City University of Hong Kong, China*

**Proposal of a DC-DC Converter with Wide Conversion Range Used in
Photovoltaic Systems and Utility Power Grid for the Universal Voltage Range** 2258
Jonas Reginaldo de Britto, *Universidade Federal de Uberlândia, Brazil*
Fábio Vincenzi Romualdo da Silva, *Universidade Federal de Uberlândia, Brazil*
Enane Antônio Alves Coelho, *Universidade Federal de Uberlândia, Brazil*
Luiz Carlos de Freitas, *Universidade Federal de Uberlândia, Brazil*
Valdeir José Farias, *Universidade Federal de Uberlândia, Brazil*
João Batista Vieira, Jr., *Universidade Federal de Uberlândia, Brazil*

Characterization of a 5 kW Solid Oxide Fuel Cell Stack Using Power Electronic Excitation .. 2264
John J. Cooley, *Massachusetts Institute of Technology, United States*
Eric Seger, *Montana State University, United States*
Steven Leeb, *Massachusetts Institute of Technology, United States*
Steven R. Shaw, *Montana State University, United States*

**Photovoltaic Parallel Resonant DC-Link Soft Switching Inverter Using
Hysteresis Current Control** 2275
Young-Ho Kim, *Sungkyunkwan University, Korea, South*
Jun-Gu Kim, *Sungkyunkwan University, Korea, South*
Young-Hyok Ji, *Sungkyunkwan University, Korea, South*
Chung-Yuen Won, *Sungkyunkwan University, Korea, South*
Yong-Chae Jung, *Namseoul University, Korea, South*

**Supercapacitor-Based Hybrid Storage Systems for Energy Harvesting in
Wireless Sensor Networks** 2281
S. Saggini, *University of Udine, Italy*
F. Ongaro, *University of Udine, Italy*
C. Galperti, *Politecnico di Milano, Italy*
P. Mattavelli, *University of Padova, Italy*

The Faulty Module Bypass for Thermoelectric Generation 2288
Wei Qian, *Michigan State University, United States*
Fang Z. Peng, *Michigan State University, United States*
Sangmin Han, *Michigan State University, United States*

Maximum Power Point Tracking Feasibility in Photovoltaic Energy-Conversion Systems . 2294
Sairaj V. Dhople, *University of Illinois at Urbana-Champaign, United States*
Ali Davoudi, *University of Illinois at Urbana-Champaign, United States*
Gerald Nilles, *University of Illinois at Urbana-Champaign, United States*
Patrick L. Chapman, *University of Illinois at Urbana-Champaign, United States*

Session C5P-R: Lighting Applications
Thursday, February 25, 11:30 - 13:30

**Realization of a General LED Lighting System Based on a Novel Power Line
Communication Technology** 2300
Chushan Li, *Zhejiang University, China*
Jiande Wu, *Zhejiang University, China*
Xiangning He, *Zhejiang University, China*

Solid-State Lamp with Integral Occupancy Sensor 2305
John J. Cooley, *Massachusetts Institute of Technology, United States*
Dan Vickery, *Massachusetts Institute of Technology, United States*
Al-Thaddeus Avestruz, *Massachusetts Institute of Technology, United States*
Amy Englehart, *Massachusetts Institute of Technology, United States*
James Paris, *Massachusetts Institute of Technology, United States*
Steven B. Leeb, *Massachusetts Institute of Technology, United States*

**A 0.9 PF LED Driver with Small LED Current Ripple Based on
Series-Input Digitally-Controlled Converter Modules** 2314
Qingcong Hu, *University of Colorado at Boulder, United States*
Regan Zane, *University of Colorado at Boulder, United States*

**A Novel Dimmable Electronic Ballast for Compact Fluorescent Lamps Using
Phase-Cut Incandescent Lamp Dimmers with Wide Dimming Range and
Low Dimming Level Lamp Ignition Capability** 2321
John Lam, *Queen's University, Canada*
Praveen K. Jain, *Queen's University, Canada*

Author Index

Grid-Connected Voltage Source Inverter for Renewable Energy Conversion System with Sensorless Current Control

Suzan Eren, Majid Pahlevaninezhad, Alireza Bakhshai, Praveen Jain

Department of Electrical and Computer Engineering
Queen's University
Kingston, Canada
suzan.eren@queensu.ca, majid.pahlevaninezhad@queensu.ca, alireza.bakhshai@queensu.ca, praveen.jain@queensu.ca

Abstract—This paper introduces a novel sensorless control approach for a three-phase grid-connected voltage source inverter utilized in a renewable energy conversion system. Renewable energy conversion applications are beginning to require a faster response to the changing power demands of the grid. Therefore, the methods that are based on conventional power theory can no longer comply with the speed requirements because they require low-pass filters in order to generate the power feedback. The closed loop control is based on an instantaneous power approach using real-time values to calculate the reference currents. Thus, the control system has fast tracking of power references compared to conventional methods. In addition, the proposed approach uses a sliding-mode observer to estimate the inverter output current as well as the grid current. This sensorless approach makes the control system robust against measurement noise and phase error, and it reduces system costs. The state-of-the-art PR controller is employed in order to provide zero steady state error for the inverter currents as well as disturbance rejection. Theoretical analysis and simulation results are provided in order to demonstrate the validity and effectiveness of the proposed control method.

I. INTRODUCTION

Renewable energy systems have recently received a lot of attention due to many factors. Diminishing fossil-fuel reserves, energy security concerns, and increased global warming all contribute to the rise of renewable energy system usage. Using renewable energy systems can eliminate harmful emissions from polluting the environment while also offering inexhaustible resources of primary energy. Cost and uncontrollability are the main challenges preventing renewable energy systems from becoming primary energy sources.

Most renewable energy systems are connected to the utility grid through a voltage source inverter. The voltage source inverter is controlled by a grid-side controller, which is responsible for meeting the power demands of the grid and controlling the quality of the injected current [1]-[3].The control methods based on the conventional power theory require a low-pass filter in the control loop in order to provide the power feedback. Therefore, a delay is inevitably introduced into the control loop. Also, conventional control methods mostly use PI controllers, which have the problems of limited bandwidth and very poor disturbance rejection.

The PR controller has been proposed in many recent works, [7]-[9], as an effective grid-side controller. It does not introduce any phase delay into the control loop at the resonant frequency. Also, it has infinite gain at the resonant frequency, which ensures that it has perfect reference tracking. In [4]-[6], a sliding-mode controller is used in the control loop and a discrete proportional resonant (PR) controller is used to remove harmonics. This method is robust and has fast dynamics compared to conventional control methods using the PI controller. However, the control loop can only be used for islanded mode and is not designed for grid-connected operation. In addition, this method does not address the problem of power feedback delay.

Sensorless control methods have been introduced recently for grid-side control. The basic number of sensors required in a three-phase grid-connected renewable energy system is five (two ac currents, two ac voltages, and one dc voltage) [10]. Using sensors can introduce phase delay and noise distortion into the measured signal. To avoid these problems, sensorless control techniques can effectively be employed to estimate the signal, while reducing the number of sensors being used to save costs [11]-[13]. Different sensorless techniques have been reported in the literature to achieve sensorless control, most commonly the Kalman filter [11] and the sliding mode observer.The sliding mode observer has two main advantages compared to other methods. The first is that it has a very high gain and fast dynamics. The second is that it is robust against parameter variations [14]-[19].

This paper introduces a novel sensorless control approach for a three-phase grid-connected voltage source inverter utilized in a renewable energy conversion system. The first feature of the proposed approach is an instantaneous power-based power feedback. Thus, the control system has fast tracking of power references compared to conventional methods. The second feature of the proposed approach is that it employs a sliding-mode observer to estimate both the inverter output current and grid current, which eliminates the phase-error, noise distortion, and cost introduced by using sensors. The third feature of the proposed approach is the utilization of a PR controller to provide perfect tracking and disturbance rejection. Finally, this approach uses an inclusive system model that considers local loads at the point-of-common-coupling, unlike other sensorless control approaches [10]-[12]. In Section II, a theoretical analysis is provided to show how the structure of the proposed method is derived. In Section III, simulation results are provided which demonstrate the feasibility of the proposed method. Finally, Section IV provides a conclusion.

II. THEORETICAL ANALYSIS

A. System Model

There are two main models used in the literature to represent the interface between a renewable energy system and the utility grid. The first model only considers an inductive filter between the voltage source inverter and the utility grid [2]-[3]. The second model considers this inductive filter as well as the grid-side impedance and the filter capacitance at the point of common coupling [20]. Finally, there is a third model, which has all of the features of the second model and it also includes the local loads at the point-of-common-coupling (PCC). This paper considers the third model to simulate the system and design the sliding-mode observer (Fig. 1).

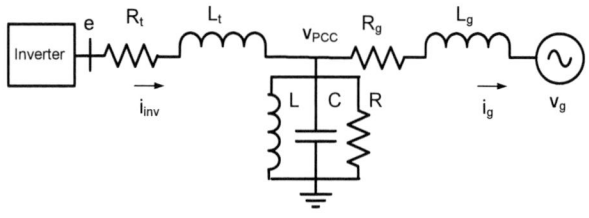

Fig.1 .System Model

According to Fig. 1, the system model is given by (1).

$$
\begin{aligned}
\frac{di_{inv}}{dt} &= -\frac{R_t}{L_t}i_{inv} - \frac{1}{L_t}v_p + \frac{1}{L_t}e \\
\frac{dv_p}{dt} &= \frac{1}{C}i_{inv} - \frac{1}{C}i_g - \frac{1}{CR}v_p - \frac{1}{C}i_L \\
\frac{di_g}{dt} &= -\frac{R_g}{L_g}i_g + \frac{1}{L_g}v_p - \frac{1}{L_g}v_g \\
\frac{di_L}{dt} &= \frac{1}{L}v_p
\end{aligned}
\tag{1}
$$

where e is the output voltage of the inverter, vg is the grid voltage, v_p is the voltage at the point of common coupling, i_{inv} is the inverter output current, i_g is the grid current, and i_L is the current through the inductor.

B. Proposed Closed Loop Control System

In Fig. 2, the proposed closed loop control system is illustrated. In this figure, the reference values for active and reactive power are inserted to the current reference generation unit. This unit generates the reference values for the inverter currents. Then, the inverter currents are controlled through three PR controllers. The current feedback into the control loop is provided by the sliding-mode observer unit, which estimates both the inverter output currents as well as the grid currents. Each feature of the proposed approach will be explained in detail in the following subsections.

1) Current Reference Generator

Fig. 3 shows the current reference generation unit. This unit calculates the reference current values for the voltage source inverter based on the instantaneous value of the power delivered to the PCC. According to Fig. 3, the reference current is given by (2) and (3). Since the CRG uses instantaneous power for its calculations, it provides a very fast prediction of the inverter output power.

$$
i_{inv,ref} = \frac{p^*(t)}{v_{PCC}(t)}
\tag{2}
$$

$$
p(t) = V\sin(\omega t)I\sin(\omega t + \theta) = P[1 - \cos(2\omega t)] + Q\sin(2\omega t)
\tag{3}
$$

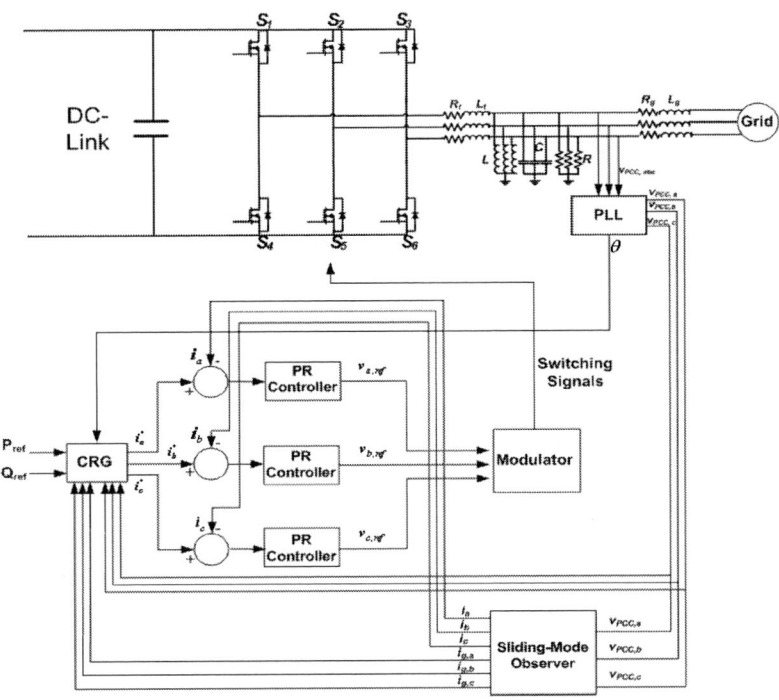

Fig. 2. Proposed Closed-Loop Control System

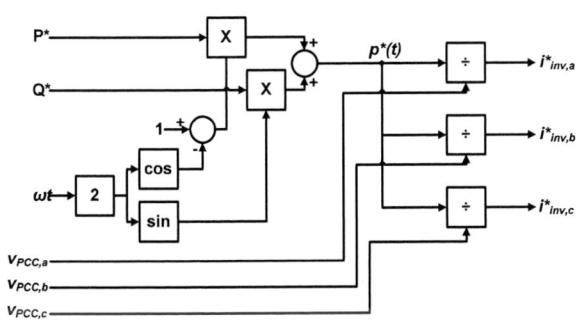

Fig.3.Current Reference Generation Unit

2) Proportional-Resonant (PR) Controller

The proportional resonant controller has successfully been applied in previous works, and is considered to be a state-of-the-art control method for grid-side renewable energy system applications [7]-[9].

The PR controller is given by (4).

$$\frac{u}{e} = \frac{\alpha_1 s + \alpha_2}{s^2 + \omega_n^2} \qquad (4)$$

where u is the desired output voltage of the voltage source inverter, and e is the inverter current error.

3) Sliding-Mode Observer

The sliding-mode observer is employed to estimate the inverter output currents and the grid currents using the inverter voltage, the grid voltage and the voltage at PCC. Due to high gain, the sliding-mode observer approach is robust to disturbances and system parameter uncertainties [15]. Using the reduced order sliding mode observer design, the PCC voltage estimator is given by the following equations (5).

$$\frac{d\hat{i}_{inv}}{dt} = -\frac{R_t}{L_t}\hat{i}_{inv} - \frac{1}{L_t}\hat{v}_p + \frac{1}{L_t}e + L_1 \operatorname{sgn}(v_p - \hat{v}_p)$$

$$\frac{d\hat{v}_p}{dt} = \frac{1}{C}\hat{i}_{inv} - \frac{1}{C}\hat{i}_g - \frac{1}{CR}\hat{v}_p - \frac{1}{C}\hat{i}_L + L_2 \operatorname{sgn}(v_p - \hat{v}_p)$$

$$\frac{d\hat{i}_g}{dt} = -\frac{R_g}{L_g}\hat{i}_g + \frac{1}{L_g}\hat{v}_p - \frac{1}{L_g}v_g + L_3 \operatorname{sgn}(v_p - \hat{v}_p)$$

$$\frac{d\hat{i}_L}{dt} = \frac{1}{L}\hat{v}_p + L_4 \operatorname{sgn}(v_p - \hat{v}_p)$$

$$(5)$$

v_p, e, and v_g are the measurable states of the system, while i_{inv} and i_g have to be estimated using (5). By appropriately choosing L_i, the sliding mode occurs along the manifold v_p-$v_{p,estimated}$=0. Using the sliding mode observer to estimate the inverter output currents and grid currents allows the distortion

978-1-4244-4782-4/10 $26.00 © 2010 IEEE

and delay caused by using a sensor to be avoided, while also reducing costs.

III. SIMULATION RESULTS

The feasibility of the proposed closed loop system was verified using MATLAB/Simulink and PSIM. In Fig. 4 and Fig. 5, the performance of the PR controller is shown. Fig. 4 illustrates the inverter reference current and the actual inverter current produced by the PR controller. It can be seen that the implemented PR controller is fast and has zero steady state error. Fig. 5 illustrates the inverter estimated current and the inverter actual current. According to this figure, the sliding observer is very fast. Fig. 6 and Fig. 7 show the system performance in terms of power flow control. In these figures a step change is applied to the active power reference value at t=0.02s and a step change is applied to the reactive power reference value at t= 0.15s. It can be seen that the active and reactive power track their references. Finally, Fig. 8 illustrates the reference and actual inverter currents.

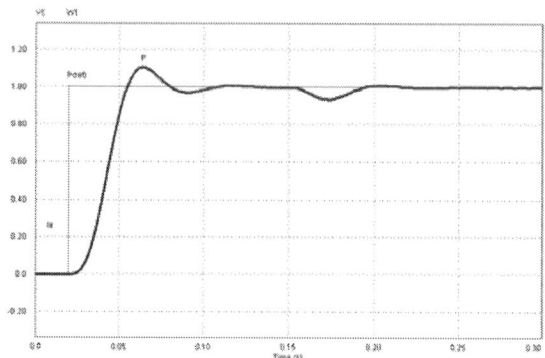

Fig.6 . Active Power Tracking

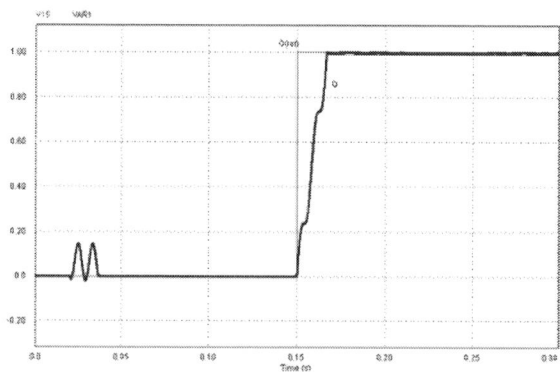

Fig.7 .Reactive Power Tracking

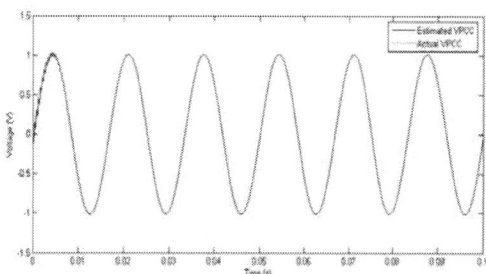

Fig.4 . Actual V_{PCC} and Estimated V_{PCC} provided by sliding mode observer

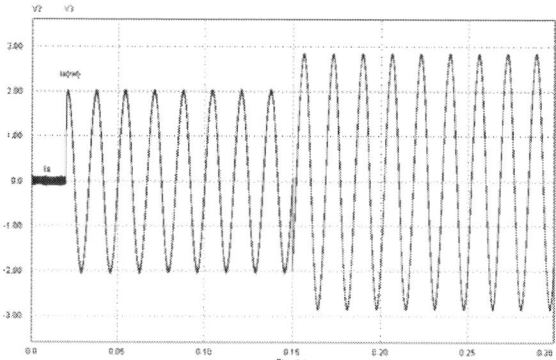

Fig.8 .Reference and Actual Inverter Output Currents

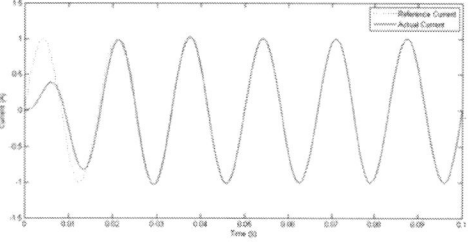

Fig. 5. Reference Inverter Current and Actual Inverter Current produced by PR controller

IV. CONCLUSION

This paper presented a new approach to control the active and reactive power flow from the voltage source inverter to the utility grid. In this approach, the current reference for the voltage source inverter was generated based on the instantaneous power at the input. This provided a fast response for power flow control. In addition, a PR controller was used to control the voltage source inverter output current. This controller can guarantee perfect tracking and a fast response for different load conditions. In the closed loop structure a sliding mode observer was used to estimate the inverter output currents and grid currents. Due to the fact that the sliding mode observer has a very high gain, it was able to estimate the actual inverter output currents and grid currents very quickly and with no error, as seen in the simulation results. In conclusion, the proposed structure is well-suited for three-phase grid-connected renewable energy system applications.

REFERENCES

[1] Blaabjerg, F.; Teodorescu, R.; Liserre, M.; Timbus, A.V., "Overview of Control and Grid Synchronization for Distributed Power Generation Systems," *Industrial Electronics, IEEE Transactions on* , vol.53, no.5, pp.1398-1409, Oct. 2006

[2] Timbus, A.V.; Teodorescu, R.; Blaabjerg, F.; Liserre, M.; Dell'Aquila, A., "Independent synchronization and control of three phase grid converters," *Power Electronics, Electrical Drives, Automation and Motion, 2006. SPEEDAM 2006. International Symposium on* , vol., no., pp.1246-1251, 23-26 May 2006

[3] Chen, Z.; Hu, Y; Blaabjerg, F., "Control of Distributed Power Systems," *Power Electronics and Motion Control Conference, 2006. IPEMC 2006. CES/IEEE 5th International* , vol.3, no., pp.1-6, 14-16, Aug. 2006

[4] Min Dai; Marwali, M.N.; Jin-Woo Jung; Keyhani, A., "Power Flow Control of a Single Distributed Generation Unit," *Power Electronics, IEEE Transactions on* , vol.23, no.1, pp.343-352, Jan. 2008

[5] Marwali, M.N.; Keyhani, A., "Control of distributed generation systems-Part I: Voltages and currents control," *Power Electronics, IEEE Transactions on* , vol.19, no.6, pp. 1541-1550, Nov. 2004

[6] Min Dai; Marwali, M.N.; Jin-Woo Jung; Keyhani, A., "A Three-Phase Four-Wire Inverter Control Technique for a Single Distributed Generation Unit in Island Mode," *Power Electronics, IEEE Transactions on* , vol.23, no.1, pp.322-331, Jan. 2008

[7] Liserre, M.; Teodorescu, R.; Blaabjerg, F., "Multiple harmonics control for three-phase grid converter systems with the use of PIRES current controller in a rotating frame," *Power Electronics, IEEE Transactions on* , vol.21, no.3, pp.836-841, May 2006

[8] Timbus, A.V.; Ciobotaru, M.; Teodorescu, R.; Blaabjerg, F., "Adaptive resonant controller for grid-connected converters in distributed power generation systems," *Applied Power Electronics Conference and Exposition, 2006. APEC '06. Twenty-First Annual IEEE* , vol., no., pp. 6 pp.-, 19-23 March 2006

[9] Teodorescu, R.; Blaabjerg, F.; Liserre, M.; Loh, P.C., "Proportional resonant controllers and filters for grid-connected voltage-source converters," *Electric Power Applications, IEE Proceedings -* , vol.153, no.5, pp.750-762, September 2006

[10] Liserre, M.; Pigazo, A.; Dell'Aquila, A.; Moreno, V.M., "An Anti-Islanding Method for Single-Phase Inverters Based on a Grid Voltage Sensorless Control," *Industrial Electronics, IEEE Transactions on* , vol.53, no.5, pp.1418-1426, Oct. 2006

[11] [11] Ahmed, K.H.; Massoud, A.M.; Finney, S.J.; Williams, B.W., "Sensorless Current Control of Three-Phase Inverter-Based Distributed Generation," *Power Delivery, IEEE Transactions on* , vol.24, no.2, pp.919-929, April 2009

[12] Zitao Wang; Liuchen Chang; Meiqin Mao, "DC voltage sensorless control strategy for three-phase grid-connected inverter," *Power Electronics Specialists Conference, 2008. PESC 2008. IEEE* , vol., no., pp.323-329, 15-19 June 2008

[13] Ahmed, K.H.; Massoud, A.M.; Finney, S.J.; Williams, B.W., "An autonomous adaptive voltage sensorless controller of inverter-based islanded distributed generation system," *Industrial Electronics, 2008. ISIE 2008. IEEE International Symposium on* , vol., no., pp.1586-1591, June 30 2008-July 2 2008

[14] Utkin, V.I., "Sliding mode control design principles and applications to electric drives," *Industrial Electronics, IEEE Transactions on* , vol.40, no.1, pp.23-36, Feb 1993

[15] Haskara, I.; Ozguner, U.; Utkin, V., "On variable structure observers," *Variable Structure Systems, 1996. VSS '96. Proceedings., 1996 IEEE International Workshop on* , vol., no., pp.193-198, 5-6 Dec 1996

[16] Derdiyok, A.; Guven, M.K.; Rehman, H.; Inanc, N.; Longya Xu, "Design and implementation of a new sliding-mode observer for speed-sensorless control of induction machine," *Industrial Electronics, IEEE Transactions on* , vol.49, no.5, pp. 1177-1182, Oct 2002

[17] Davila, J.; Fridman, L.; Levant, A., "Second-order sliding-mode observer for mechanical systems," *Automatic Control, IEEE Transactions on* , vol.50, no.11, pp. 1785-1789, Nov. 2005

[18] Drakunov, S.; Utkin, V., "Sliding mode observers. Tutorial," *Decision and Control, 1995., Proceedings of the 34th IEEE Conference on* , vol.4, no., pp.3376-3378 vol.4, 13-15 Dec 1995

[19] Young, K.D.; Utkin, V.I.; Ozguner, U., "A control engineer's guide to sliding mode control," *Control Systems Technology, IEEE Transactions on* , vol.7, no.3, pp.328-342, May 1999

[20] [A. Yazdani; R. Iravani, "A unified dynamic model and control for the voltage-sourced converter under unbalanced grid conditions," *Power Delivery, IEEE Transactions on* , vol.21, no.3, pp.1620-1629, July 2006.

Design of an 99%-Efficient, 5 kW, Phase-Shift PWM DC-DC Converter for Telecom Applications

U. Badstuebner, J. Biela, and J.W. Kolar

Power Electronic Systems Laboratory, ETH Zurich
Email: badstuebner@lem.ee.ethz.ch; www.pes.ee.ethz.ch

Abstract—In the last decade power electronic research focused on the power density maximization mainly to reduce initial systems costs [1]. In the field of data centers and telecom applications, the costs for powering and cooling exceed the purchasing cost in less than 2 years [2]. That causes the changing driving forces in the development of new power supplies to efficiency, while the power density should stay on a high level.

The commonly used DC-DC converter in the power supply unit (PSU) for data centers and telecom applications are full bridge phase-shift converters since they meet the demands of high power and efficient power conversion, a compact design and the constant operation frequency allows a simple control and EMI design.

The development of the converter with respect to high efficiency has a lot of degrees of freedom. An optimization procedure based on comprehensive analytical models leads to the optimal parameters (e.g. switching frequency, switching devices in parallel and transformer design) for the most efficient design.

In this paper a 5 kW, 400 V - 48..56 V phase-shift PWM converter with LC-output filter is designed for highest efficiency ($\eta \geq 99\%$) with a volume limitation and the consideration of the part-load efficiency. The components dependency as well as the optimal design will be explained. The realized prototype design reaches a calculated efficiency of $\eta = 99.2\%$ under full load condition and a power density of $\rho = 36\,\text{W/in}^3$ (2.2 kW/liter).

I. INTRODUCTION

Since 1970, the power density of power electronic converters roughly doubles every 10 years, mainly caused by the increasing switching frequency due to the continuous improvements of power switching devices and the decreasing volume of magnetic components [3]. In the area of power supply units (PSU) for data center and telecom applications, this evolution led to the main developing focus on compact design and a capital expenditure (CapEx) measured by the square feet occupied, rather than power consumption [2]. However, the demand for data centers is continuously increasing and the rising energy prices result in powering and cooling costs, which are higher as the purchase cost in less than two years [2]. This causes a change of the driving force in power converter system development towards high efficient power conversion. However, high power density is still required, which leads to multi-objective targets for the system development.

In [4], the prototype of a 5 kW, 400 V to 48..56 V phase-shift PWM DC-DC converter for telecom application is presented, which was optimized with respect to highest achievable power density. There, the design process was based on comprehensive analytical models of the converter system with an automatic optimization procedure, which results in a power density of $\rho = 147\,\text{W/in}^3$ and a measured efficiency of $\eta =$

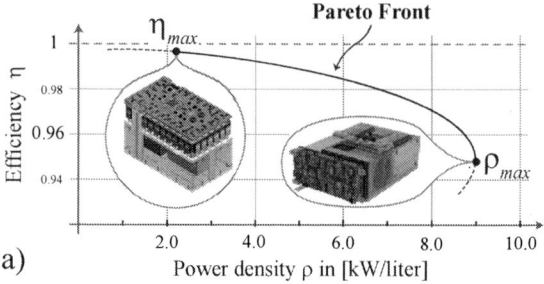

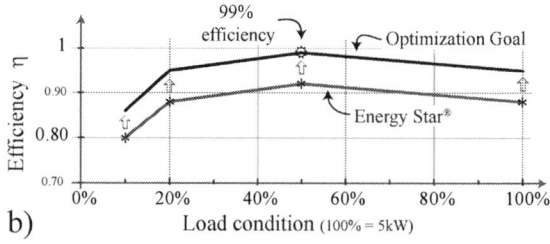

Figure 1. a) Efficiency η vs. power density ρ. The best possible solution of a multi-objective optimization according to the weights w_η for efficiency and w_ρ for power density for a defined topology is presented by the Pareto front. b) Optimization goal the efficiency curve through the point $\eta = 0.99$ raised from the values proposed in [5].

94.75 %. These points determine the point of highest power density ρ_{max} in the η-ρ-plane in Fig. 1 a).

In this paper, a design process is presented to reach the point of highest efficiency ($\eta_{max} \geq 99\%$) for a phase-shift PWM, 5 kW DC-DC converter with LC output filter as shown in Fig. 2, which is plotted in the diagram Fig. 1 a) as well. The curve between these two optimal points is called Pareto front, which present the optimal points for varying weights w_η of the efficiency and w_ρ of the power density in the optimization procedure. This allows the OEMs a classification of present system and builds the basis for road maps, as well as the identification of unachievable designs.

In the optimization procedure, the part load efficiency of the converter system is considered as well. The reference efficiency curve for the optimization is related to the efficiency points proposed in the Energy Star® requirements of computer serves [5], which have been moved to the point, where the maximum efficiency is 99 %, as shown in Fig. 1 b).

The optimization procedure and the comprehensive analytical electrical and magnetic models are described in **Section II**. After the presentation of the optimization algorithm, the calculation of the topology specific operation point, i.e. all relevant current and voltage waveforms, is described. Within

978-1-4244-4782-4/10 $26.00 © 2010 IEEE

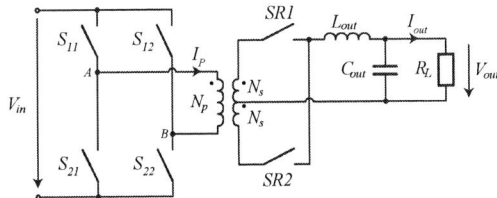

Figure 2. Schematic of the selected phase-shift PWM converter for 99 % efficiency. (V_{in}=400 V, V_{out}=48..54 V, P_{out}=5 kW).

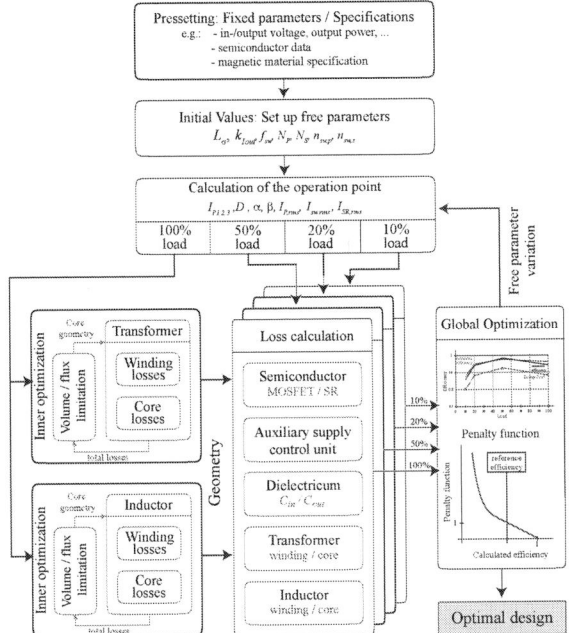

Figure 3. Automatic efficiency optimization procedure for the phase-Shift PWM DC-DC converter considering the part load efficiency.

inner optimization procedures for the transformer and output inductor, the optimal geometry parameters for the cores are determined, considering the core losses and HF-winding losses in the magnetic devices.

The optimization procedure results in the optimal design parameters for the converter given in **Section III**. A prototype design, which reaches a calculated full load efficiency of η = 99.2 % and a power density of ρ = 36 W/in^3 (2.2 kW/liter) is presented in section **Section IV**.

II. EFFICIENCY OPTIMIZATION

For the design of the selected phase-shift converter and the given specifications, the component values must be chosen so that the efficiency becomes maximal. However, this task is challenging because the components interdepend to some degree from each other. In this section the design process is described based on an optimization procedure, which automatically finds the optimal component values of the converter system. The underlying analytical models are described after the presentation of the procedure in section II-A.

The starting point of the optimization procedure in Fig. 3 are the fixed parameters for the converter design process, i.e. the electrical specifications (e.g. output power P_{out}) and magnetic constraints (e.g. maximum flux density in the transformer or output inductor core, respectively).

With the initial set of the free parameters (the switching frequency f_{sw}, the primary and secondary turns number of the transformer N_p and N_s, the leakage inductance L_σ, the allowed output inductor current ripple characterized by k_{Iout}, as well as the number of parallel primary $n_{sw,p}$ and synchronous $n_{sw,s}$ rectifier MOSFET's) the optimization loop is launched.

The first step in the loop is the calculation of the operation point based on a coupled inductance model of the center tapped transformer. There, the output inductance L_{out} and output capacitance C_{out} are determined for full-load conditions and so, all component values are defined in order to determine all relevant current and voltage waveforms (cf. Fig. 5). In addition, the characteristic waveforms for the part load conditions (cf. [5] and Fig. 1) of the converter have to be calculated as well as explained in section I.

The core of the optimization procedure is the calculation of the losses in the converter system. For the magnetic components, i.e. the center-tapped transformer and the output inductor, the geometry set-up is found in two inner optimization loops. There, the geometry parameters (cf. Fig. 4), are varied systematically, while the volume and flux density stay in the preset limit until the optimum parameter are found, which results in the minimum component losses (considering HF-winding and core loses) under full load condition. The actual loss calculation for the four load conditions (i.e. 10 %, 20 %, 50 %, and 100 %) are running in a for-next loop or in parallel, alternatively. For the given amount of the power switches in parallel, which are optimizided during the optimization process as well, the conduction losses are determined. For the full bridge MOSFETs, the switching losses are considered as well, which is especially important for low load conditions. Besides that, the gate drive losses are added in order to calculate the total semiconductor losses. For magnetic components, the core and winding losses are determined. The HF-losses in the windings are considered for the transformer as well, whereas the HF-losses in the output inductor have only a minor influence. The dielectric losses in the output capacitor are determined with the loss factor. Additional losses in the control unit and auxiliary supply are considered as constant over the whole load range.

The resulting part-load losses, which define the actual efficiency curve η_{act} as depicted in Fig. 1 b) are now compared with the reference efficiency curve η_{ref} (labeled as Optimization goal in Fig. 1 b)). The deviations $\Delta\eta$ between reference and actual efficiency values are the input for a penalty function, which is defined as:

$$penalty(\Delta\eta) = \begin{cases} \dfrac{1}{1-\eta_{ref}} \cdot (1-\eta_{act}), & \text{for } \Delta_\eta < 0 \\ \dfrac{1}{(1-\eta_{ref}+\eta_{act})^{20}}, & \text{for } \Delta_\eta \geq 0. \end{cases} \quad (1)$$

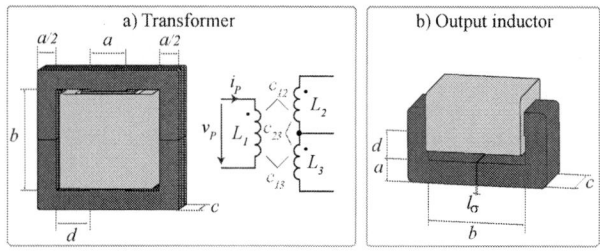

Figure 4. Geometry model of transformer and output inductor. In addition, the coupled inductance model of the transformer is presented.

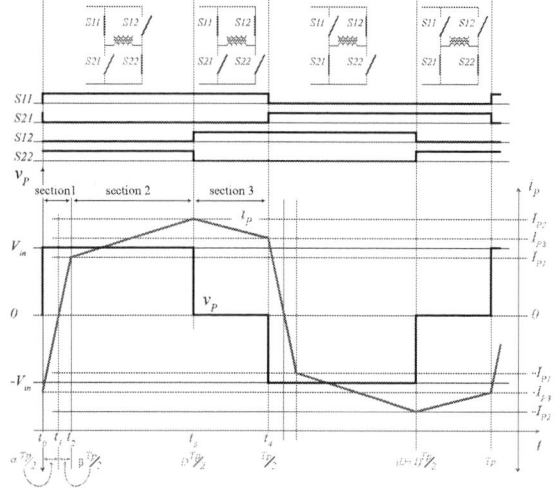

Figure 5. Principle primary side current / voltage waveforms with the characteristic points I_{P1}, I_{P2}, I_{P3} and the switching states.

The sum of the penalty values forms the optimization criteria. The global optimization algorithm changes the free converter parameters systematically, until the minimum penalty value is found. The outputs of the optimization procedure are the converter design parameters which directly enables a prototype assembly.

A. Analytical Models

The analytical models can be divided into 4 parts: First, the formulas for calculating the operation point are explained. After that, the equations for the losses in the power semiconductor devices are derived. In the third subsection, the calculation of the magnetic components is presented and finally the losses in the capacitors are given.

1) Operation point: The models for determining the operation point are based on a coupled inductors equivalent circuit of the center-tapped transformer, as presented in fig:MagGeometry a) on the right hand side. With the set inductance L_1 and the free parameter L_σ, the couple factor c_{12}, the inductance L_2 (which is considered to be equal to L_3)

and the transmission ratio ktr can be calculated as:

$$
\begin{aligned}
c_{12} &= 1 - L_\sigma/L_1 \\
L_2 &= L_1 \cdot N_s/N_p \cdot c_{12} \\
ktr &= \sqrt{L_1/L_2} \cdot c_{12},
\end{aligned}
\tag{2}
$$

whereas N_P is the number of primary turns and N_S the number of one secondary turns.

In Fig. 5, there are three different piecewise linear current sections, whose current slope are determined by the center-tapped transformer (cf. Fig. 4) and the output inductance, as described in the following.

a) Section 1 ($t = t_0..t_2$): The input voltage is applied to the primary side of the transformer and the current slope is limited by the leakage inductance. With the assumption, that the inductances L_2 and L_3 and whose coupling factors in respect to L_1 (c_{12} and c_{13}) are equal, the formula for calculating the current slope in this section is simplified to:

$$
\frac{\Delta i_p}{\Delta(t_0..t_2)} = V_{in} \cdot \frac{1 + c_{23}}{L_1 \cdot (1 + c_{23} - 2 \cdot (c_{12})^2)}
\tag{3}
$$

b) Section 2 ($t = t_2..t_3$): In this section, power is transferred form the input to the output stage. With the applied input voltage V_{in} the resulting output voltage V_{out} and the calculated output inductance L_{out}, the current slope of section 2 is given by:

$$
\frac{\Delta i_p}{\Delta(t_2..t_3)} = \frac{V_{in} \cdot (L_{out} + L_2) - c_{12} \cdot \sqrt{L_1} \cdot \sqrt{L_2} \cdot V_{out}}{L_1 \cdot (L_{out} + L_2 \cdot (1 - (c_{12})^2))}
\tag{4}
$$

The output inductance L_{out} is determined by the allowed output current ripple k_{Iout} and the output current I_{out} under full load condition:

$$
L_{out} = \frac{V_{in}/ktr - V_{out}}{\frac{P_{out}}{V_{out} \cdot k_{Iout}}} \cdot ktr \cdot V_{out}/V_{in} \cdot Tp/2
\tag{5}
$$

c) Section 3 ($t = t_3..t_4$): The converter is in the freewheeling phase, where the transformer is short circuited by the high-side or low-side MOSFETs, respectively. The current slope is determined by the output side and can be calculated with:

$$
\frac{\Delta i_p}{\Delta(t_3..t_4)} = \frac{V_{out} \cdot c_{12} \cdot \sqrt{L_2}}{\sqrt{L_1} \cdot (L_{out} + L_2 \cdot (1 - (c_{12})^2))}
\tag{6}
$$

With the current slopes, the piecewise linear current sections confer Fig. 5 can be expressed as:

$$
\begin{aligned}
\frac{\Delta i_p}{\Delta(t_0..t_2)} &= \frac{I_{P1} + I_{P3}}{(\alpha + \beta) \cdot Tp/2} \\
\frac{\Delta i_p}{\Delta(t_2..t_3)} &= \frac{I_{P2} - I_{P1}}{(D - \alpha - \beta) \cdot Tp/2} \\
\frac{\Delta i_p}{\Delta(t_3..t_4)} &= \frac{I_{P2} - I_{P3}}{(1 - D) \cdot Tp/2}
\end{aligned}
\tag{7}
$$

where Tp is the switching period and D is the duty cycle, which is defined as:

$$
D = ktr \cdot V_{out}/V_{in} + \alpha + \beta
\tag{8}
$$

By inserting (3)-(6) and (8) in (7) and solving for I_{P1}, I_{P2} and I_{P3}, the result solutions are only depending on α and β.

To eliminate α and β, two additional equations have to be found. The first expression is the periodicity of the primary current:

$$0 = \frac{\Delta i_p}{\Delta(t_0..t_2)} \cdot \beta \cdot Tp/2 + \frac{\Delta i_p}{\Delta(t_2..t_3)} \cdot (D - \alpha - \beta) \cdot Tp/2$$
$$- \frac{\Delta i_p}{\Delta(t_3..t_4)} \cdot (1 - D) \cdot Tp/2 - \frac{\Delta i_p}{\Delta(t_0..t_2)} \cdot \alpha \cdot Tp/2 \qquad (9)$$

The second expression is the equation of the average output power:

$$P_{out} = V_{out} \cdot \frac{2}{Tp} \cdot ktr \cdot \left[\quad \dots \right.$$

$$\int_{\alpha \cdot Tp/2}^{(\alpha+\beta)Tp/2} \frac{\Delta i_p}{\Delta(t_0..t_2)} \cdot (t - \alpha \cdot Tp/2) \cdot dt$$

$$+ \int_{(\alpha+\beta) \cdot Tp/2}^{D \cdot Tp/2} I_1 + \frac{\Delta i_p}{\Delta(t_2..t_3)} \cdot (t - (\alpha+\beta) \cdot Tp/2) \cdot dt \qquad (10)$$

$$+ \int_{D \cdot Tp/2}^{Tp/2} I_2 - \frac{\Delta i_p}{\Delta(t_3..t_4)} \cdot (t - D \cdot Tp/2) \cdot dt$$

$$\left. + \int_{Tp/2}^{(1+\alpha) \cdot Tp/2} I_3 - \frac{\Delta i_p}{\Delta(t_0..t_2)} \cdot (t - Tp/2) \cdot dt \quad \right]$$

with

$$I_1 = \frac{\Delta i_p}{\Delta(t_0..t_2)} \cdot \alpha \cdot Tp/2$$

$$I_2 = I_1 + \frac{\Delta i_p}{\Delta(t_2..t_3)} \cdot (D - \alpha - \beta) \cdot Tp/2 \qquad (11)$$

$$I_3 = I_2 - \frac{\Delta i_p}{\Delta(t_3..t_4)} \cdot (1 - D) \cdot Tp/2.$$

Equations (9) and (10) can be solved for α and β by inserting the characteristic current points from the solutions of (7). (The solutions for I_{P1}, I_{P2} and I_{P3}, as well as for α and β are omitted for the sake of brevity.) The definition of the current and voltage waveforms allows the calculation of the rms-values, the derivations of the rms-values and harmonics for transformer, inductor, capacitor and semiconductor currents for the following loss calculations.

B. Semiconductor Losses

The losses in the full bridge switches are derived with the $R_{DS,on}$, the gate charge Q_G, the energy equivalent output capacitance $C_{oss,eq}$ at 400 V and the energy in the output capacitance $E(V)$ as function of the applied voltage for a preselected MOSFET (cf. section III-A). The conduction losses are calculated with:

$$P_{cond} = \frac{R_{DS,on} \cdot I_{sw,rms}^2}{n_{sw,p}} \qquad (12)$$

with the rms-value $I_{sw,rms}$ of the current through the MOSFET and $n_{sw,p}$, the number of parallel connected MOSFETs.

Since the converter topology offers zero voltage switching (ZVS) by inserting an interlock delay between the switching states, switching losses are almost zero in principle. However, especially at part load conditions, the interlock delay might not be sufficient for a complete resonant dis-/charge of the MOSFETs in one bridge leg, i.e. a residual voltage V_{res} is remaining, which has to be discharged by the the MOSFET. The residual voltage is calculated based on a RLC-resonant circuit consisting of the leakage inductance L_σ, the energy equivalent output capacitance $C_{oss,eq}$ and the $R_{DS,on}$ of the parallel connected MOSFETs. With the characteristic energy curve of the used MOSFET, which is described by piecewise polynomial function dependent on the applied residual voltage, the switching losses P_{sw} of one bridge leg can be determined as:

$$P_{sw} = 2 \cdot n_{sw,P} \cdot E(V_{res}) \cdot f_{sw}, \qquad (13)$$

where $n_{sw,P}$ is the number of parallel full bridge MOSFETs and f_{sw} is the switching frequency. Since the switch-off current are different for the two bridge legs, the switching losses are calculated separately.

In addition the losses in the gate driver P_{drive} are considered as well and are calculated for each switch with:

$$P_{drive} = n_{sw,P} \cdot (V_{GS,on} \cdot Q_G \cdot f_{sw} + P_{driver}), \qquad (14)$$

taking the number of parallel connected MOSFETs $n_{sw,P}$ and losses of the driver P_{driver} into account. Note, that the applied on-gate-source voltage $V_{GS,on}$ has an positive value and the off-gate-source voltage $V_{GS,off}$ is considered to be zero in the assembly and thus omitted in (14).

The conduction and gate drive losses in the synchronous rectifier switches are calculated in the same manner as the full bridge switches with (12) and (14). Since the rectifier MOSFETs are turned-on during the free-wheeling phase, where the voltage is approximately zero over the switches, the switching losses are negligible. Recovery losses in the body diode are avoided because the MOSFETs are turned off at the point, where the current is completely commuted from SR1 to SR2 cf. Fig. 2 or vice versa, respectively.

C. Losses in the Magnetic Components

In the inner optimization procedure, the geometry parameters of the magnetic components cf. Fig. 4 are varied systematically in order to obtain the minimum losses, with the maximum flux density B_{max} and component volume (bounding box) as constraint, since the losses are decreasing continuously for higher volumes as explained in section III-B. For the assembly, foil windings are considered, whose optimal foil thickness can be calculated with [6]:

$$d_{opt} = \frac{1}{\Psi^{\frac{1}{4}}} \cdot \sqrt{\frac{2 \cdot \pi \cdot f_{sw} \cdot I_{rms}}{I'_{rms}} \cdot \delta_0} \qquad (15)$$

with the rms-values of the currents I_{rms} in the windings and whose derivations I'_{rms}, respectively. Ψ is defined as:

$$\Psi = \frac{5 \cdot N^2 - 1}{15}, \qquad (16)$$

the skin depth δ_0, the resistivity ρ of copper and the permeability of free space μ_0. The peak-to-peak flux density ΔB in the transformer is approximately defined via:

$$\Delta B = \frac{V_{in} \cdot D \cdot Tp/2}{N_P \cdot A_{core}} \qquad (17)$$

with number of primary turns N_P and the cores cross section area A_{core}. The maximum flux density B_{max} is half of the peak-to-peak flux density. With the extended Steinmetz formula in [7], the core losses can be determined:

$$P_{co,tr} = \frac{k_i \cdot \Delta B^{(\alpha_S - \beta_S)}}{Tp} \cdot 2 \cdot \left(\frac{V_{in}}{N_P \cdot A_{core}} \cdot D \cdot Tp/2 \right) \cdot Vol_{co} \qquad (18)$$

with the core volume Vol_{co} and

$$k_i = \frac{k_S}{2^{(\beta_S+1)} \cdot \pi^{(\alpha_S-1)} \cdot \left(0.2761 \cdot \frac{1.7061}{\alpha_S + 1.354} \right)}. \qquad (19)$$

For the calculation of the winding losses, the HF-losses due to the skin and proximity effect are considered. The underlying model is based on a one-dimensional approach [8]. With the magnitudes of the current harmonics $I_{P,h,n}$ (with $n = 0..n_h$ and n_h the number of calculated harmonics) the primary winding losses due to the skin effect in the windings are:

$$P_{skin,P} = R_{DC,P} \cdot (I_{P,h,0})^2 + \dots$$
$$+ \sum_{n=1}^{n_h} R_{DC,P} \cdot \frac{\nu_P}{4} \cdot \frac{\sinh \nu_P + \sin \nu_P}{\cosh \nu_P - \cos \nu_P} \cdot (I_{P,h,n})^2 \qquad (20)$$

with the dc-resistance $R_{DC,P}$ of the primary winding:

$$R_{DC,P} = \frac{l_{w,P} \cdot \rho}{d_{opt,P} \cdot b_{foil}} \qquad (21)$$

whereas $l_{w,P}$ is the length of the primary winding, $d_{opt,P}$ is the optimal foil thickness, b_{foil} is the winding width and ν_P is defined as:

$$\nu_p = \frac{d_{opt,P}}{\sqrt{\frac{\rho}{\pi \cdot n \cdot f_{sw} \cdot \mu_0}}} \qquad (22)$$

The losses due to the proximity effect in the primary winding are defined as:

$$P_{proxP} = \sum_{n=1}^{n_h} \sum_{m=1}^{N_P} \frac{b_{foil} \cdot \rho}{\sqrt{\frac{\rho}{\pi n f_{sw} \mu_0}}} \cdot \frac{\sinh \nu_P + \sin \nu_P}{\cosh \nu_P - \cos \nu_P} \dots$$
$$\cdot \left(\frac{1}{2 \cdot b} \cdot I_{P,h,n} \cdot (2 \cdot m - 1) \right)^2 \cdot l_{w,P} \qquad (23)$$

The winding losses in the secondary windings are calculated with the same approach as for the primary windings, however, since the winding order is: primary - secondary 1 - secondary 2, two cases have to be considered for calculating the proximity effect: Current is flowing in secondary 1 leads only to losses this winding. If the current is flowing in secondary 2, the H-field in secondary 1 is not zero which results in additional losses in secondary 1.

The losses in the output inductor can be calculated with the same approach as for the transformer, however, HF-losses only have minor influence to the winding losses, since the

output current has only a small ac-component compared to the dc-component. Because of the negligible ac-component, the foil thickness with (15) would result in large values and thus, its limited to $300 \, \mu m$. The peak-to-peak flux ΔB_{Lout} can be calculated with:

$$\Delta B_{Lout} = \frac{L_{out} \cdot k_{Iout} \cdot I_{out}}{N_L \cdot A_{core,L}} \qquad (24)$$

with the number of output inductor turns N_L and the cross-section area $A_{core,L}$ of the inductor core. The maximum flux density is defined as

$$B_{max,L} = \frac{L_{out} \cdot (I_{out} + 1/2 \cdot k_{Iout} \cdot I_{out})}{N_L \cdot A_{core,L}} \qquad (25)$$

D. Dielectric losses in the output capacitor

With the applied output capacitor from muRata [9] ($2.2 \, \mu F$ / $100 \, V$ / X7R / 1210 housing) and the given loss factor $\tan \delta$ the losses are determined with:

$$P_{Cout} = \frac{I_{Cout,rms}^2 \cdot \tan \delta}{2 \cdot \pi \cdot f_{sw} \cdot C_{out}} \qquad (26)$$

with the rms-value of the capacitor ripple current $I_{Cout,rms}$. There, it is assumed, that the output current I_{out} has only a dc-component and the entire ripple current is flowing in the output capacitance. The capacitance can be calculated with the allowed voltage ripple V_{pp} by solving the equation with respect to C_{out}:

$$V_{pp} = \frac{1}{C_{out}} \cdot \left[\int_0^{1/2 \cdot D \cdot Tp/2} \frac{I_{out} \cdot k_{Iout}}{D \cdot Tp/2} \cdot t \cdot dt + \dots \right.$$
$$\left. + \int_0^{1/2 \cdot (1-D) \cdot Tp/2} \frac{I_{out} \cdot k_{Iout}}{(1-D) \cdot Tp/2} \cdot t \cdot dt \right] \qquad (27)$$

The losses in the auxiliary supply and control unit are considered to be constant over the entire load range and set to $3 \, W$.

III. OPTIMIZATION RESULTS

In this section the results of the optimization procedure are presented and discussed. The results have been proved with simulations.

A. Optimum Number of Parallel Switches

Before the first start of the optimization procedure, several switches have been compared with respect to the resulting losses. For preselecting the MOSFETs, a figure of merit (FOM) could be defined based on the on-conductance $G_{DS,on}$, which should be high for small conduction losses, and the energy-equivalent output capacitance of the device $C_{oss,eq}$, which should be small in order in order to obtain small/no switching losses over almost the whole power range: $FOM = G_{DS,on}/C_{oss,eq}$ [10]. This FOM is usually chosen for hard-switching devices topology and might not result in best choice for soft-switching topologies, as it will be presented.

In Fig. 6 the losses of the full bridge are presented for a fixed operation point in the optimization procedure, which may not result in the optimal design. There, Infineon

CoolMOS™IPW60R045CP [11] are applied, as an example. It can be seen, that the conduction losses P_{cond} decrease with the increasing number of the parallel switches $1/n_{sw}$, whereas the driver losses P_{drive} increases linearly with n_{sw}. The switching losses P_{sw} are approximately zero until the interlock delay between the switching states is not sufficient for the total charge transfer of the output capacitors of the two MOSFETs in a bridge leg, where the losses are increase drastically. If the number of parallel switches and the respectively capacitance reaches a value, where the capacitors could almost not be dis-/charged and the MOSFETs are completely switched hard off, the switching losses increase linearly with n_{sw}.

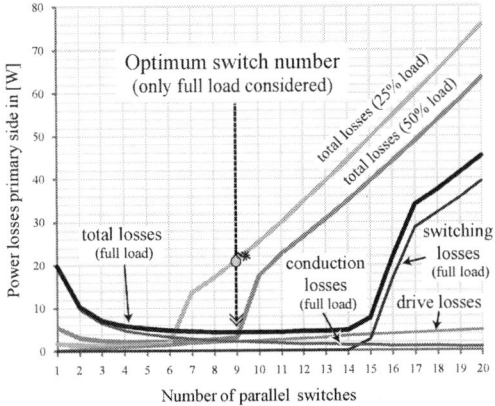

Figure 6. Dependency of parallel H-bridge switches number on device losses. (Infineon CoolMOS™IPW60R045CP [11] considered.)

If the number of parallel switches would be chosen for full load, the respective optimal point in the example of Fig. 6 would be $n_{sw,P} = 9$ switches in parallel, where the total losses on the primary side are calculated to be 4.4 W. Due to the small losses, the temperature in the devices is small and thus the on-resistance of 25 °C was considered for the comparison of several switching devices. However, at part load (e.g. 25 % output power) the total full-bridge losses increases drastically by an factor of almost 5 to 21.1 W.

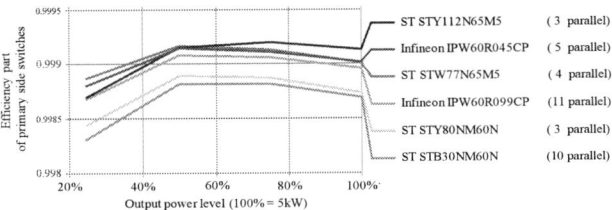

Figure 7. Efficiency contribution of full bridge switches (Examples of the considered MOSFETs). The free-parameters are fixed for the calculation, which may not present the parameters resulting in the optimal design. The curves show the optimum of the semiconductor losses with the respective optimal number of parallel switches.

In Fig. 7 the efficiency contribution of analyzed MOSFETs in the range of 600..700 V is presented. There, the free parameters have been fixed and the respective optimal number of switches which leads to the minimum losses is plotted. For the prototype, the STY112N65M5 from STMicroelectronics

[12] has been chosen, since they offer the best conditions in terms of the optimization criteria as described in section II and because of the small $R_{DS,on} = 19\,\text{m}\Omega$ (typ), only 2 MOSFETs are needed for the final prototype design.

The same procedure is performed for the secondary side switches. However, the switching losses are approximately zero over the entire load range since the synchronous rectifiers are turned on with zero voltage condition and the current-capacitance-ratio is high enough to ensure ZVS. The selected MOSFETs for the synchronous rectifiers are IRFP4668PbF from International Rectifier [13] ($R_{DS,on} = 8\,\text{m}\Omega$ (typ.) / 200 V). The optimum number of parallel switches are in this case $n_{sw,s}=11$, which results in 9.3 W total losses in the synchronous rectifier.

B. Losses in the Magnetic Components

In the inner optimization procedure of the magnetic device, the geometry parameters are varied until minimum possible losses result. In Fig. 8, the allowed maximum transformer volume is varied as constraint in the inner optimization procedure with respect to the transformer losses. For each allowed volume, the optimal values for the transformer geometry are chosen so that the overall losses became minimal, considering the maximum allowed flux density in the core. The limited flux density leads to a minimum core area and/or number of primary turns N_P, respectively. The resulting optimized losses are higher because of a higher flux and smaller windings. The more volume allowed, the higher the foil width and/or the smaller the number of primary turns, and thus the smaller are the resulting losses. However, the curve becomes flat, if even a higher volume is used as the constraint in the procedure. There, an increase of the core area would lead to a reduction of the core losses but the skin/proximity losses are increasing due to a larger winding length, which have to be balanced with a larger winding in order to reduce the dc-resistance.

The optimized power losses for the maximum allowed transformer volume are presented in Fig. 8, where the geometry parameters confer Fig. 4 are not limited. It is shown, that the volume in the optimization procedure must be limited in order to obtain realizable geometry parameters and volumes.

C. Optimized Design Parameters

For the practical converter design, the volume point of $18.3\ in^3$ (0.3 liter) was chosen as limit for both, transformer and output inductor. First runs with the optimization procedure, where the geometry parameters of transformer and output inductor have not been limited, result in geometry parameters for the magnetic components, which are close to available standard cores. These results have been considered in order to limit the geometry parameters a and b (cf. Fig. 4) with respect to the standard core parameters in the inner optimization procedures. The best results have been reached with parameter limits of the EPCOS E70/33/32 E-Core [14] for the transformer and the Metglas AMCC 320 [15] for the inductor, which are presented in the following.

In table I the optimization results of the free parameters are presented. Conferring the described optimization criteria, the

978-1-4244-4782-4/10 $26.00 © 2010 IEEE

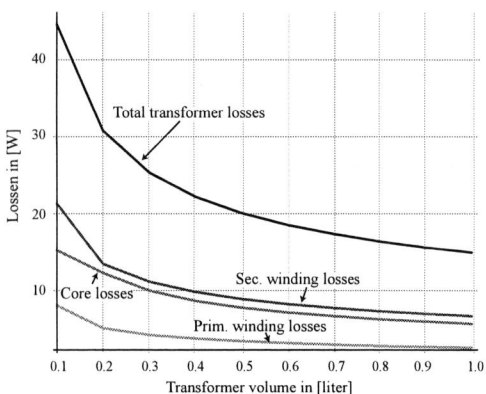

Figure 8. Transformer losses in dependency of transformer volume limit (bounding box). The curves represent the minimal losses resulting from the optimized geometry parameters with the respectively volume limits.

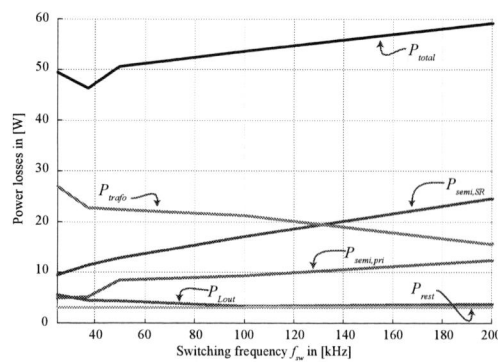

Figure 9. Losses of the optimized Phase-Shift PWM converter as a function of frequency. (P_{total} = total losses, $P_{semi,pri}$ = primary side semiconductor losses, $P_{semi,SR}$ = synchronous rectifier losses, P_{trafo} = transformer losses, P_{Lout} = output inductor losses, P_{rest} = losses in aux. supply, control unit, output capacitors).

Table I
FREE PARAMETER OPTIMIZATION RESULTS WITH A VOLUME LIMIT OF
$18.3\, in^3$ (0.3 LITER). (f_{sw} = SWITCHING FREQUENCY, N_P / N_S / N_L =
PRIMARY / SECONDARY / OUTPUT INDUCTOR TURNS NUMBER, L_σ =
TRANSFORMER LEAKAGE INDUCTANCE, L_{out} = OUTPUT INDUCTANCE).

f_{sw}	N_P	N_S	N_L	$n_{sw,P}$	$n_{sw,S}$	L_σ [μF]	L_{out} [μF]
16.0 kHz	22	3	5	2	12	2.0	48.8
25.0 kHz	22	3	4	2	11	1.8	43.0
37.5 kHz	15	2	3	2	9	1.0	34.6
50.0 kHz	15	2	3	1	8	1.3	34.0
100.0 kHz	15	2	2	1	6	1.8	23.1
200.0 kHz	7	1	2	1	4	1.8	1.8

optimal design with the best efficiency characteristic can be found at a switching frequency of 25 kHz.

The resulting optimized efficiency values η are given below:

Frequency	$\eta_{10\%}$	$\eta_{20\%}$	$\eta_{50\%}$	$\eta_{100\%}$
Energy star®	80.00 %	88.00 %	92.00 %	88.00 %
Goal	86.09 %	94.70 %	99.00 %	94.70 %
16.0 kHz	96.34 %	98.06 %	98.93 %	98.93 %
25.0 kHz	97.42 %	98.60 %	99.13 %	99.01 %
37.5 kHz	96.98 %	98.39 %	99.09 %	99.07 %
50.0 kHz	97.24 %	98.51 %	99.09 %	98.99 %
100.0 kHz	97.31 %	98.53 %	99.06 %	98.93 %
200.0 kHz	96.52 %	98.12 %	98.88 %	98.82 %

The power losses of the components as function of the switching frequency are presented in Fig. 9. As shown in Fig. 8 and Fig. 9, respectively, the losses in the lower frequency range are mainly determined by the magnetic components because of the volume limitation. At higher switching frequencies, the number of parallel full bridge switches decreases because the switching losses, due to not completely dis-/charge of the drain-source capacitor especially in the low-load condition, increase with the switching frequency. Additionally, the smaller number of parallel MOSETs results in higher conduction losses, as well. The pend in the total losses curve after 37.5 kHz is mainly caused by this effect, since the number of parallel switches in the full bridge changes from two to one. The most significant impact in the decrease of the total

power losses with increasing switching frequency have the losses in the synchronous rectifier, which is caused by the continuously decrease of parallel switching devices (cf. table I). Since the synchronous rectifier MOSFETs operate with ZVS, switching losses does not force the optimization procedure to reduce the switches. The decrease is caused by the increasing gate-driver losses with higher switching frequencies, which lead to a significant efficiency drop at lower load-conditions. That is why the number of synchronous rectifier switches is reduced by the optimization algorithm. However, this leads to higher conduction losses at full load.

The transformer losses are higher for lower frequency because of the volume limitation as described before. Due to the flux density limitation, the minimum number of turns rises, which leads to higher winding losses. Because of the decreasing turns numbers and the smaller required cross section areas for higher switching frequencies, the losses in the magnetic components are decreasing as well. For frequencies higher than 200 kHz, the winding losses (due to the proximity and skin effect), as well as the core losses are increasing again. For the output inductor, the losses are mainly caused by the conduction losses, which are decreasing for higher frequency due to the decreasing turns number. However, for higher switching frequencies, HF-losses in the winding and higher core losses are increasing, like for the transformer. (Note, that switching frequencies above 200 kHz have been omitted in this paper, since for higher frequencies, additional losses have especially to be considered in the cores of the magnetic components due to the non-uniform flux distribution. The accuracy of the applied models in the optimization procedure would decrease.)

The residual losses (control unit, auxiliary supply, output capacitors) stay approximately constant over the entire frequency range.

IV. PROTOTYPE

In Fig. 10 the prototype design for the proposed converter is presented. Standard components have been applied:

Transformer	EPCOS E70/33/32, 2 in parallel
Output inductor	Metglas AMCC320 (cut legs)
Full bridge MOSFETs ..	ST STY112N65M5, 2 in parallel
Sync. rectifier MOSFETs	IR IRFP4668PbF, 11 in parallel
Gate driver	IXYS IXDD414
Digital control	TI TMS320F2808

In order to utilize the winding window with a high copper fill factor, Litz wires have been used instead of foil wires (primary winding: 175/0.2 mm; secondary winding: 600/0.2 mm; inductor winding: 1200/0.2 mm) .

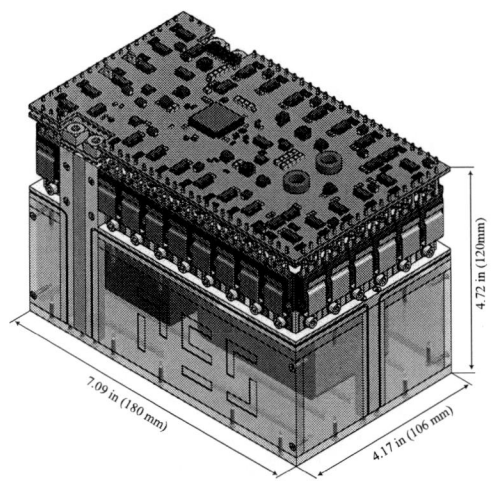

Figure 10. Prototype design of the proposed high efficient DC-DC converter. ($\eta = 99.2\%$ at full load, $\rho = 36$ W/in^3 (2.2 kW/liter)).

Because of the selected components, the volumes are higher as the volume limit of $18.3\,in^3$ (0.3 liter), used in the optimization procedure. For the transformer, the volume results in $29.2\,in^3$ (0.48 liter) and for the output inductor $23.1\,in^3$ (0.39 liter). This leads to further improvement for the efficiency as presented below. The total losses decrease to 37.0 W (full load) and the part load efficiency result in:

Load condition	10 %	20 %	50 %	100 %
Efficiency	98.0 %	99.0 %	99.3 %	99.2 %

In Fig. 11 the loss distribution is depicted for full load conditions. The major loss distribution has the transformer together with the output inductor (55 %), followed by the power semiconductor losses (37 %).

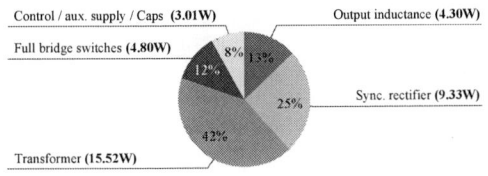

Figure 11. Loss distribution at full load (5 kW) for the prototype design.

The resulting power density of this design is $\rho = 36$ W/in^3 (2.2 kW/liter). Since the simulations have validated the analytical models and based on the experiences with the power density optimization (e.g. [10] and [4]), the metrological

validation with the prototype seems very promising. However, variances could occur mainly because of the not considered contact resistances, which are strongly depended on the design and assembly.

V. CONCLUSION

In this paper, the design process of an ultra high efficient 400V/48..50V DC-DC converter for data center and telecom application is presented. An optimization procedure based on comprehensive analytical models, considering the part-load efficiency, was applied to find the optimal design. The final prototype design results in a calculated efficiency of $\eta = 99.2\%$ at full load and offers a flat efficiency curve over the entire load range. At 10 % load, i.e. 500 W, the converter system exhibits still an efficiency of 98.0 %. The power density of the realized prototype is $\rho = 36$ W/in^3 (2.2 kW/liter).

ACKNOWLEDGMENT

The authors would like to thank Mr. Harder as representative of the European Center for Power Electronics (ECPE) [16] for funding this project. In addition, the authors would like to acknowledge the support of Dr. Kirchenberger and Dr. Rauscher (ST Microelectronics) for providing the full bridge MOSFET samples, Mr. Spring (Rotima AG) for providing the Litz wires and Mr. Rappel (EPCOS AG) for providing E-core samples.

REFERENCES

[1] J. W. Kolar, U. Drofenik, J. Biela, M. L. Heldwein, H. Ertl, T. Friedli, and S. D. Round, "PWM Converter Power Density Barriers," in *Proceedings of the 4th Power Conversion Conference (PCC'07)*, apr 2007, pp. 9–29.
[2] K. G. Brill, "Moore's law economic meltdown," jun 2008. [Online]. Available: www.forbes.com
[3] J. W. Kolar, J. Biela, and U. Badstubner, "Impact of power density maximization on efficiency of dc-dc converter systems," in *7th Internatonal Conference on Power Electronics (ICPE'07)*, oct 2007, pp. 23–32.
[4] U. Badstuebner, J. Biela, B. Faessler, D. Hoesli, and J. W. Kolar, "An Optimized 5 kW, 147 W/in^3 Telecom Phase-Shift DC-DC Converter with Magnetically Integrated Current Doubler," in *Twenty-Fourth Annual IEEE Applied Power Electronics Conference and Exposition (APEC'09)*, vol. 24, feb 2009, pp. 21 – 27.
[5] ENERGY STAR®, "ENERGY STAR® Program Requirements for Computer Servers," aug 2008. [Online]. Available: www.eu-energystar.org/
[6] W. G. Hurley, E. Gath, and J. D. Breslin, "Optimizing the ac resistance of multilayer transformer windings with arbitrary current waveforms," in *IEEE Transaction on Power Electronics*, vol. 15, no. 2, mar 2000, pp. 369–376.
[7] K. Venkatachalam, C. R. Sullivan, T. Abdallah, and H. Tacca, "Accurate Prediction of Ferrite Core Loss with Nonsinusoidal Waveforms Using Only Steinmetz Parameters," in *Proceedings of the 8th IEEE Workshop on Computers in Power Electronics (COMPEL'02)*, jun 2002, pp. 36–41.
[8] J. A. Ferreira, "Improved analytical modeling of conductive losses in magnetic components," in *IEEE Transactions on Power Electronics*, vol. 9, jan 1994, pp. 127–131.
[9] "Homepage of muRata Manufacturing Co., Ltd." [Online]. Available: www.murata.com
[10] J. W. Kolar, J. Biela, and J. Miniboeck, "Exploring the Pareto Front of Multi-Objective Single-Phase PFC Rectifier Design Optimization - 99.2 % Efficiency vs. 7 kW/dm^3 Power Density," in *14th International Power Electronics and Motion Control Conference (IPEMC'09)*, may 2009.
[11] "Homepage of Infineon." [Online]. Available: www.infineon.com
[12] "Homepage of ST Microelectronics." [Online]. Available: www.st.com
[13] "Homepage of International Rectifier." [Online]. Available: www.irf.com
[14] "Homepage of EPCOS AG." [Online]. Available: www.epcos.com
[15] "Homepage of Metglas Inc." [Online]. Available: www.metglas.com
[16] "Hompage of the European Center for Power Electronics e.V." [Online]. Available: www.ecpe.org

DC-DC Transformer Multiphase Converter with Transformer Coupling for Two-Stage Architecture

M.C.Gonzalez, P.Alou, O.Garcia,J.A. Oliver and
J.A.Cobos
Centro de Electrónica Industrial
Universidad Politécnica de Madrid
Madrid, España
carmen.gsanchez@upm.es

H.Visairo
Systems Research Center, Mexico
Intel Corporation
Guadalajara, Mexico
horacio.visairo-cruz@intel.com

Abstract— **This paper presents an analysis of a very high efficiency topology based on a multiphase converter where the coupling among the phases is done by means of transformers. Since energy is not stored in the transformers, it is transferred directly from the input to the output. This topology along with its control strategy, were previously reported in state of the art. Some advantages of the proposed concept are that, thanks to the minimum storage of energy, the transfer rate of the energy is decoupled from switching frequency and can be chosen in order to increase efficiency. The regulation capability of this topology is limited and the topology acts as a dc-dc transformer. In this paper, the proposed topology is analyzed in detail and design guidelines are presented in order to optimize the design of the converter. These guidelines are applied to a specific design in order to build an optimized prototype.**

I. INTRODUCTION

Voltage Regulator Modules are converters dedicated to supply large amounts of energy with very fast slew rates at very low voltages (1.1 V). These converters generally operate under very small duty cycle conditions because the available input voltage is usually 12 V. When running under these conditions the efficiency and overall performance of the converter is degraded. In state of the art, the use of a two-stage architecture [1,2] has been proposed as a way to solve this problem since it enables the use of a lower intermediate voltage. This principle is illustrated in Figure 1. In this way, the converter which is adjacent to the load (Point Of Load converter) is allowed to run under better duty cycle conditions and therefore achieve better efficiency so it can be operated at higher switching frequencies, as in [2] where the POL is operated in the MHz range. When the system is adequately configured, improvements regarding size and efficiency can be achieved [1]. One of the requirements for the adequate configuration of the system is to choose a first stage or pre-regulator with very high efficiency. In this kind of applications, dc-dc transformers become an interesting choice. A dc-dc transformer can be described as a converter whose output voltage is for every instant of time, proportional to the

input voltage; the proportion is given by the 'transformation ratio' of the dc-dc converter, which is usually operated under fixed duty cycle conditions; hence, if the input voltage varies, the dc-dc transformer will be unable to adjust the duty cycle in order to compensate for the input voltage variation. In a two-stage architecture, this fine regulation would be accomplished by the POL converter. Two voltage dividers (used as first-stage topologies) based on different technologies are found in the literature. These topologies operate as 2:1 dc-dc transformers. The application of a voltage divider based on switching capacitor technology can be found in [4], while a resonant voltage divider is reported in [3] as a pre-regulator candidate in a two-stage architecture.

In this paper, a dc-dc transformer with different 'transformation ratios' and based on PWM technology is presented. The main characteristic of this multiphase converter is that the magnetic coupling among the phases is done by tight coupled transformers. Hence it is called multiphase converter with transformer coupling. Its operating principle, which was presented in [5] is reviewed in Section II along with a brief review of inductor-coupled multiphase converters. Design guidelines for this kind of converters are given in Section III; these design guidelines are based on a losses model which is also briefly explained. This methodology is applied and validated with two different prototypes. Section IV presents and compares these prototypes. Finally, conclusions are presented in Section V.

II. PROPOSED CONCEPT AND STATE OF THE ART

The present work focuses on proposing a control strategy that can be applied to a coupled inductor converter. This topology is illustrated in Figure 2b. First of all, Figure 2a. illustrates the operating principle of a coupled inductor converter when d<50%. From this figure, it can be seen that the magnetic structure (represented by an ideal transformer) acts as an energy adder, adding the voltages v_A and v_B. The combined waveform resulting at the node v_C is a pulsating voltage that needs to be filtered by the output inductor

---

This work was supported by Intel Corporation

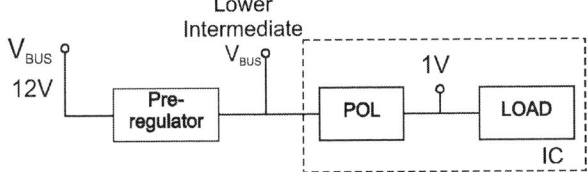

Figure 1. Two-stage architecture enables a lower bus voltage which could allow Point Of Load converter (POL) to operate at higher frequency and reduce its size.

(L_{FILTER}) and capacitor. One of the main advantages of coupled-inductor converter is that the inductance seen by the converter during steady state is different form that seen under dynamic conditions [8]. It is desirable that inductance seen under steady state is greater, so smaller phase current ripple can be achieved. However, the values of L_{FILTER} and output capacitor should be adequately chosen, along with the operating frequency and magnetizing inductance in order to filter the pulsating voltage and the current ripple at v_C and also provide the required dynamic response and efficiency.

The main voltage waveforms of the proposed operating principle is illustrated in Figure 2.c; it is based on keeping the sum of the voltages v_A and v_B (inputs to the magnetic structure) constant for every instant of time, the voltage in the magnetic structure is given by:

$$v_A - v_C = v_C - v_A \qquad (1)$$

if the sum of v_A and v_B is constant for every instant of time, the voltage at the node v_C will be also constant., and equal to:

$$v_C = \frac{v_A + v_B}{2} \qquad (2)$$

seen that $v_C = v_O$ at any instant of time, hence the output inductance (L_{FILTER}) can be reduced or theoretically eliminated from the converter, minimizing the energy storage in the converter. In this way, tightly coupled transformers can be used instead of coupled inductors. This implies that a converter working under the proposed concept can only operate at certain duty cycles; as in the example, the operating waveforms shown are those corresponding to d=50%. If the value of L_{FILTER} can be represented only by the leakage inductance inherent to the construction of a very tight coupled transformer, the energy will not be stored as it is in coupled inductors, but it will be transferred directly from the input to the output through the transformers that couple the phases. Designing a converter with a minimum value of L_{FILTER} can provide certain advantages regarding dynamic response and efficiency.

For instance, if L_{FILTER} is minimized, the predominant dynamic under a load step is that of the transformer. The response of the transformer to a load step is also illustrated in Figure 2.b. When a load step occurs at the common point of the transformer, it is immediately seen at the input, and equally distributed among the phases carrying i_1 and i_2. A step in the load will be seen almost immediately at the input, since the element that opposes to the current change (filter inductance) has been reduced to its minimum. This can be done by using an adequate interleaving technique while placing the windings in the transformer. With the proposed control technique and the converter designed to operate with minimum storage of energy, the transfer of the energy under a load step is independent from the switching frequency, and f_{SW} can be chosen to minimize losses instead of the accomplishment of a specific dynamic response. It has been said before, that this converter has limited regulation capability and it behaves as a DC-DC transformer. However, if more phases are added to the magnetic structure, more duty cycles where the proposed control strategy is achieved become

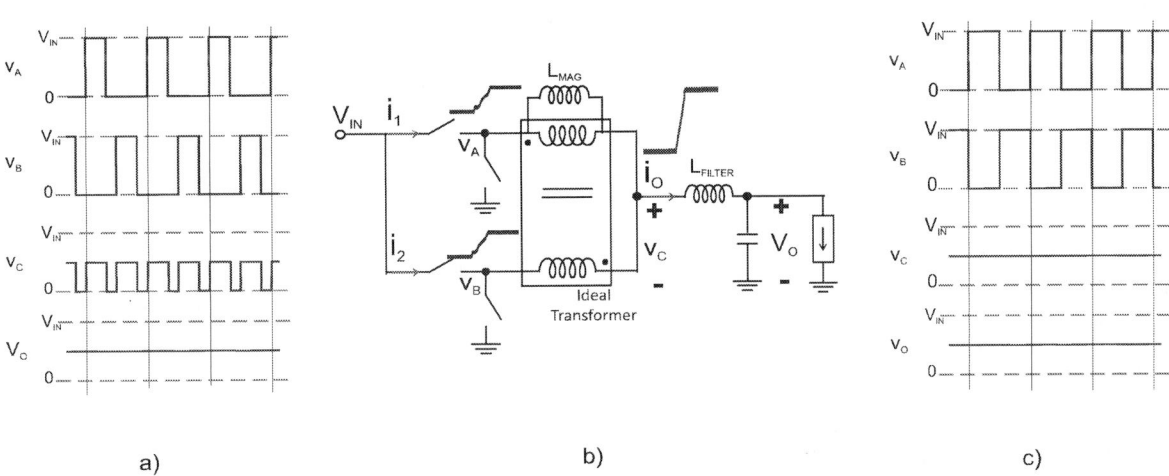

Figure 2. a) Operating waveforms for a coupled-inductor multiphase converter for d<50%. b) Magnetic structure in a coupled multiphase converter can be represented by an ideal transformer (plus L_{MAG} and leakage inductance); in an ideal transformer a change in the output current is seen immediately at the input of the transformer and distributed equally among the input phases. c) Proposed control strategy achieves that $v_C = v_O$, v_C is constant for every instant of time. Since there is no pulsating voltage to filter at v_C, L_{FILTER} is no longer necessary and can be minimized to its minimum (L_{LK} inherent to the construction of the transformer). Coupling between the phases is done by a tightly coupled transformer.

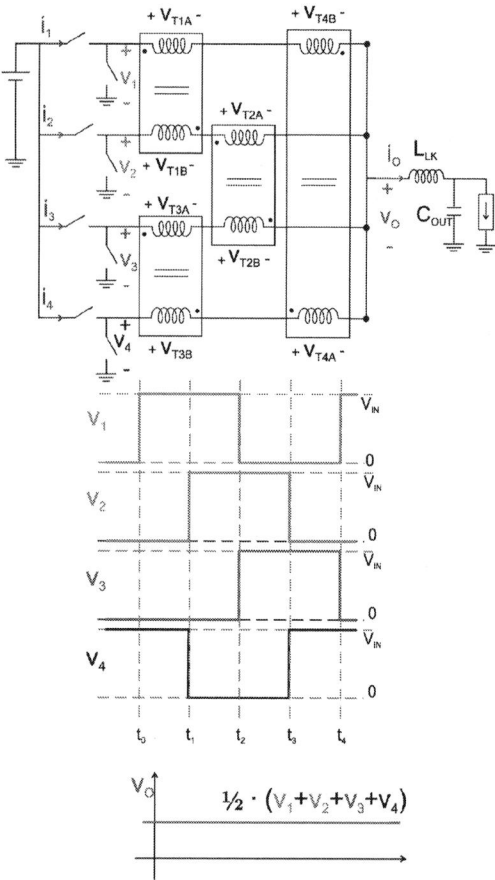

Figure 3. Schematic of proposed topology and main operating waveforms. Coupling between phases is done by means of tightly coupled transformers

can be seen in Figure 3. The voltages of the transformers can then be found by the following equations:

$$v_{T1A} = \frac{3}{4}v_1 + \frac{1}{4}v_3 + \frac{1}{2}v_4 - \frac{3}{2}v_O \qquad (3)$$

$$v_{T2A} = -\frac{1}{4}v_1 - \frac{3}{4}v_3 - \frac{1}{2}v_4 + \frac{3}{2}v_O \qquad (4)$$

$$v_{T3A} = -\frac{1}{4}v_1 + \frac{1}{4}v_3 + \frac{1}{2}v_4 + \frac{1}{2}v_O \qquad (5)$$

$$v_{T4A} = -\frac{1}{4}v_1 + \frac{1}{4}v_3 - \frac{1}{2}v_4 + \frac{1}{2}v_O \qquad (6)$$

and the output voltage equals:

$$v_O = \frac{v_1 + v_2 + v_3 + v_4}{4} \qquad (7)$$

for every instant of time.

Based on the assumption of the constant input – constant output voltage, the proposed topology can be represented with the model shown in Figure 4, where the magnetic structure is represented by dependent current and voltage sources. The input and output impedance are represented in this model since they have a significative impact on the dynamic response. It is important to point out, that the $L_{LEAKAGE}$ stands for the equivalent leakage inductance of the transformers.

There are many strategies to couple the phases using discrete transformers, the one shown in Figure 3 is used in order to build the prototype covered in section IV.

III. DESIGN GUIDELINES AND LOSSES MODEL

In a converter operating with minimum storage of energy, the dynamic response is decoupled from the switching frequency, so the criteria for choosing f_{SW} is independent from the dynamic response. The values of L_{MAG} (magnetic core) and f_{SW} can be chosen in order to minimize the losses. However, size optimization should also be taken into account. For a fixed value of L_{MAG}, a f_{SW} value at which the losses are minimized exists; aiming to find this point, along with the optimum MOSFETs, a design methodology was developed. This methodology helps to find the appropriate losses and size. The process starts with some given specifications (V_{IN}, V_{OUT}, I_{OUT}, current ripple, etc.) and the objective is to find the optimum combination of size and losses by evaluating different combinations of transformer configurations (turns, core size and material), MOSFETs and switching frequencies. The proposed methodology is based on a losses model that was verified using a preliminary prototype.

This losses model was implemented taking into account conduction losses, switching losses (6) and circulating energy losses. The validation of this model is shown in Figure 6, where measurements for different loads at a frequency of 100 kHz are compared with the theoretical calculation of the losses; It is desirable that this model works in a wide range of frequencies and loads.

available; then it is possible to achieve different values of output voltage and the topology behaves as a DC-DC transformer with multiple conversion ratios.

When the concept is extended to multiple phases, the duty cycles where the control strategy is achieved are related to the number of phases and are given by: $d = k/n$, where n is the number of phases; k is an integer which represents the number of cells that are simultaneously transferring energy from the input and the range of its values is comprised from 0 to $n-1$. For example, in a four-phase converter ($n = 4$), three different levels of output voltage would be available, those corresponding to duty cycles of 25%, 50% and 75%, when k equals 1, 2 and 3 respectively. A schematic of a four-phase topology is illustrated in Figure 3, along with the operating waveforms that corresponds to d=50%. It can be seen that the switching cycle has been divided into four periods (t0 to t4). For every instant of time, there are two phases which are simultaneously connected to V_{IN}.

At t0-t1, the value of the voltages v1 and v4 is equal to V_{IN}, while the other phases are connected to ground. At t1-t2, v1 and v2 equal V_{IN}. In the same way, the other phases connect consecutively to VIN for the other instants of time, as

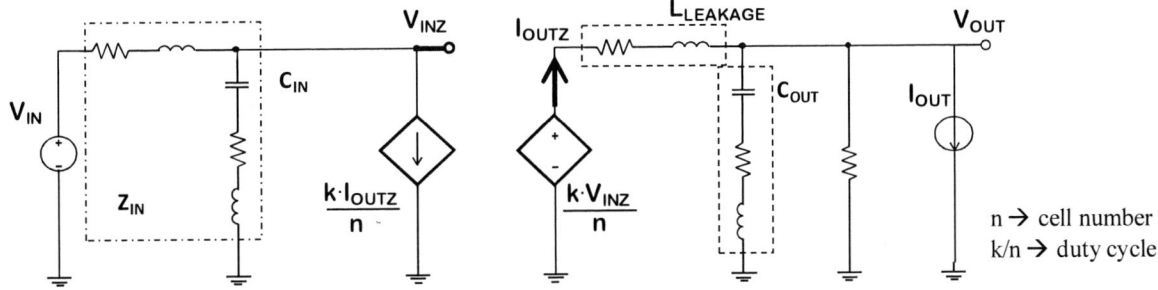

Figure 4. Model of the proposed topology. The magnetic structure is modeled as a current and voltage dependent sources.

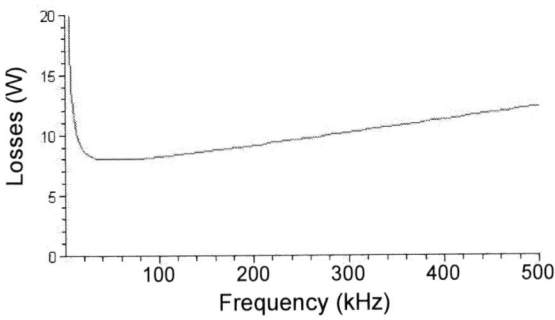

Figure 5. Losses for maximum load (I_{OUT}=30A) over a wide range of frequencies.

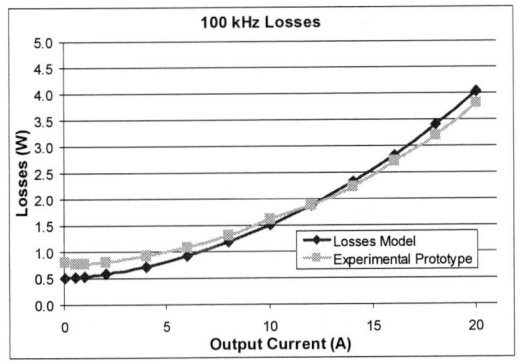

Figure 6. Losses measurements and model prediction for different loads at 100kHz frequency

Figure 7. Picture of the implemented prototype. The magnetic structure is comprised of four pairs of E/14/3.5/5 cores

The losses taken into account are the following:

1) Conduction losses. These power losses include losses due to average current and circulating energy losses. Conduction losses account for losses in the R_{DSon} of MOSFETs, in the DC windings of the transformer and an estimation of the parasitic resistance of the PCB:

$$PR_{DSon} = \text{n} \cdot (RonHI) \cdot \text{d} \cdot \text{f} \int_{0}^{\frac{1}{f}} i_n^2 \, dt + \text{n} \cdot (RonRS) \cdot (1 - d) \cdot \text{f} \int_{0}^{\frac{1}{f}} i_n^2 \, dt$$

$$PR_{PCB} = \left(R_{PCB} \right) \cdot Iout^2$$

$$PR_{DC\_TRANSFORMER} = \text{n} \cdot RDC \cdot \left(\frac{Iout}{n} \right)^2$$

2) Switching losses. These losses are calculated by implementing the model proposed in [6] where, given a MOSFET model which includes the parasitic inductance of the PCB, apart from the parasite capacitances (C_{ISS}, C_{OSS}, C_{RSS}); in this paper, expressions for the V_{DS} and I_{DS} of the

MOSFET are derived, making it possible to calculate the losses due to ON and OFF MOSFET transitions.

3) Circulating energy losses. These include losses in the magnetic core and losses due to the AC resistance in the windings of the transformer. The core losses are calculated by Steinmetz equation:

$$Pv = k \cdot f_{SW}{}^{\alpha} \cdot B^{\beta}$$

in order to calculate losses due to AC resistance of the transformer windings, (which gains importance when operating at high frequency), an AC resistance model has been obtained from PEmag. With this model, the AC resistances at different frequencies can be calculated, and use them in the formula:

$$PR_{AC} = \text{n} \cdot \left(RMS_1^2 \cdot RAC_1 + RMS_2^2 \cdot RAC_2 + \ldots + RMS_n^2 \cdot RAC_n \right)$$

Figure 5 shows the prediction of the losses using this model for an output current of 30 A within a wide range of operating frequencies. From this curve, the frequency for less losses for a given L_{MAG} at I_{OUT}=30 A can be chosen.

978-1-4244-4782-4/10 $26.00 © 2010 IEEE

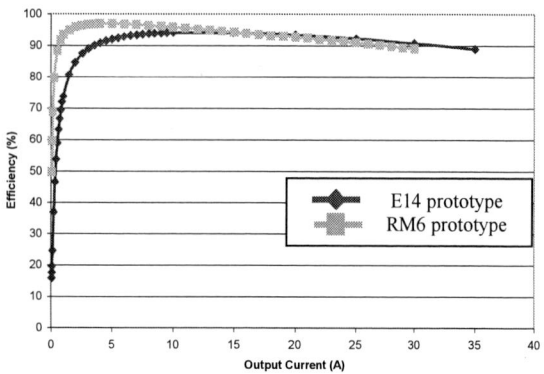

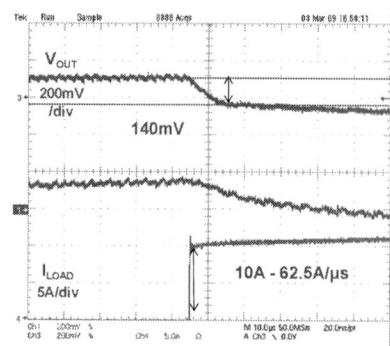

Figure 9. For a 10 A load step (60 A/us) only twenty 22uF capacitors are needed to maintain the voltage deviation within 5% of V_{OUT}.

Figure 8. At frequency of 100 kHz, the efficiency of the prototype is higher than 90% for a wide load range: from 3 A to 30 A.

IV. DESIGN OF A FOUR-PHASE PROTOTYPE APPLYING PROPOSED METHODOLOGY

Based on the proposed minimum energy storage concept, two four-phase converters were implemented in order to verify the concept. The design specifications for both converters are: V_{IN}=12V, V_{OUT}=3V, I_{OUT}=30A and I_{RIPPLE}<5%·I_{OUT}. The main difference between these two prototypes is that they were designed with different criteria. The first of them was designed prioritizing the reduction of the switching losses, this is mainly reflected in an increased efficiency at medium and light load. Operating frequency of this prototype is around 40 kHz and the size of the transformers employed to couple the phases are RM6 and the magnetizing inductance is 100µH. In the design of the second prototype, the proposed design methodology is applied with the aim of optimizing the size and power losses. The size of the cores, resulting from the application of the methodology is E14 (L_{MAG}=10uH). This prototype is operated at 100 kHz. In order to develop this solution, the following design guidelines, based on proposed design methodology, were followed:

1. The number of phases was selected in agreement with the required conversion ratio. For this case V_{IN}/V_{OUT}=4. Any multiple of 4 can achieve the desired conversion ratio. Four would be the minimum number of phases at which the desired conversion ratio can be achieved, when $k = 1$ (since $d = k/n$, $d = 25\%$); for simplicity, four phases have been selected.

2. When size of the prototype is a main constraint, the core can be chosen in function of its size as the next step. Assuming that the turns of the transformer are integrated into the PCB, maximum number of turns fitting into the window area of the core is determined when a PCB technology is chosen, hence determining the maximum magnetizing inductance and the minimum operating frequency that accomplishes the specified current ripple. In this case, planar E14 cores are chosen.

3. At this point (with initial L_{MAG}, f_{SW}, and current ripple determined), a preliminary losses evaluation can be done taking into account the losses in the transformer (DC and circulating energy losses). Since full load output power is known, acceptable losses in the transformer can be considered

to be less than 5% of total output power at full load. In order to achieve this losses level, turns number should be decreased and more windings can be paralleled, thus reducing DC losses. After this, it is interesting to return to step 2 in order to choose a different core size and evaluate different configurations for the transformer.

4. When a set of acceptable transformer designs has been done, the next step is to evaluate different MOSFETs with this set of transformers. The evaluation of different MOSFETs has to be carried out based on a losses model explained above. After selecting the less power losses combination for L_{MAG}, f_{SW} different set of MOSFETs can be evaluated. After choosing the right MOSFETs for the selected operating frequency, a prototype can be build. The prototype with E14 cores is shown in Figure 7. The turns of the transformer are integrated into de PCB and in this case, a 12 layer PCB was chosen for the implementation of the prototype. Its efficiency is shown in Figure 8 along with the efficiency of the prototype built with RM6 cores. The size of RM6 prototype is greater than that of the optimized E14 prototype (effective volume for RM6 is 1090mm³ while for E14 is 300 mm³). Although at heavy load, the efficiency for both prototypes is very similar, the peak efficiency for RM6 cores is higher than that for E14: 96.9% and 94 % respectively.

For both prototypes, efficiency is high in a wide load of ranges, for E14 prototype efficiency is greater than 90% for 3A to 30 A, while for RM6, this range goes from 1 A to 25 A. When using the RM6 prototype with V_{IN}=12 V and V_{OUT}=6 V, the measured efficiency is 98% from 4 A to 6 A (24 W to 36 W). Figure 9 shows output voltage deviation (when a load step occurs) for the topology built with E14 cores: the output voltage deviation for V_{OUT}=3 V is shown in this figure (ΔV_{OUT}=140 mV). This is due to the drop in the equivalent series resistance of the converter at 10 A (for this prototype, measured R_{SERIES} equals 10 mΩ approx.). It is important to point out, that this voltage drop cannot be compensated since the prototype can only operate with duty cycles of 25%, 50% and 75%. The slew rate of the applied load step is 60A/µs. This response is achieved with only 20 MLC Capacitors of 22uF at the output and a 470uF-Oscon at the input.

978-1-4244-4782-4/10 $26.00 © 2010 IEEE

V. Conclusions

The proposed topology, which can be considered as multiphase converter with transformer coupling, is controlled with a constant-input, constant-output control strategy, which allows a direct transfer of energy from the input to the output. Since stored energy is minimized in the converter, the dynamic response of the topology is decoupled from the operating frequency and this parameter can be chosen in order to optimize losses or size of the converter. It is important to point out, that due to the lack of energy storage in the proposed topology, there are only certain duty cycles where the converter can be operated and these points correspond to the duty cycles where the proposed control strategy is achieved.

The proposed design methodology is based on a losses model. With this model, different transformer configurations (size, turns) can be evaluated in order to determine the less power losses operating frequency. This methodology is applied to optimize the design of a prototype, and its optimum frequency is around 100 kHz with small cores (E14). Since switching frequency is kept in a low value (100 kHz) switching losses can be reduced and the obtained efficiency is high in a wide load range (>90% for 3 A to 30 A) and a peak efficiency of 94% was measured. When using bigger cores (RM6) higher efficiency can be achieved: peak efficiency of ≈97% and the range where η>90% is from 1A to 25 A. This efficiency can be achieved while maintaining a fast dynamic response thus allowing a small size of the output capacitor. The high efficiency presented by this topology enables its use as a first stage, in two-stage architectures, where the converter in the second stage (the converter placed close to the load), can benefit from a lower input voltage. Regulation capability of this topology is limited, but if multiple-phases are considered, output voltage can be changed among different values and the topology can be considered as a DC-DC transformer with multiple ratios.

References

[1] Julu Sun, Ming Xu, Yucheng Ying and Fred C. Lee. "High Power Density, High Efficiency system Two-stage Power Architecture for Laptop Computers". Power Electronics Specialists Conference, 2006. 37th IEEE pages 1-7. June 2006

[2] Yuancheng Ren, Ming Xu, Kaiwei Yao, Yu Meng and F.C. Lee. "Two-stage approach for 12 V VR". Power Electronics, IEEE Transactions on, Nov. 2004

[3] K. I. Hwu, and Y. T. Yau, "A Simple Resonant Voltage Divider", Applied Power Electronics Conference and Exposition, 2009. APEC 2009

[4] Ming Xu, Julu Sun, and Fred C. Lee, "Voltage Divider and its Application in the Two-stage Power Architecture", Applied Power Electronics Conference and Exposition, 2006. APEC 2006

[5] M.C.Gonzalez, L.Laguna, P.Alou, O.Garcia,J.A.Cobos and H.Visairo, "New control strategy for energy conversion based on coupled magnetic structures", Power Electronics Specialists Conference, 2008. PESC 2008. IEEE

[6] Yuancheng Ren, Ming Xu, Jinghai Zhou, and Fred C. Lee, "Analytical Loss Model of Power MOSFET", IEEE Transactions on Power Electronics, Vol. 21, No. 2, March 2006

[7] P.Zumel, O.Garcia, J.A.Cobos, J. Uceda, "Tight magnetic coupling in multiphase interleaved converters based on simple transformers", Applied Power Electronics Conference and Exposition, 2005. APEC 2005. Twentieth Annual IEEE pp: 385- 391

[8] Pit-Leong, Peng Xu, Bo Yang, F.C. Lee "Performance Improvements of Interleaving VRMs with Coupling Inductors" IEEE Transactions on Power Electronics, Vol.16 July 2001

A Comparison of Classical Two Phase (2L) and Transformer – Coupled (XL) Interleaved Boost Converters for Fuel Cell Applications

Kevin. J. Hartnett (kevinh@rennes.ucc.ie), Marek. S. Rylko, John. G. Hayes (john.hayes@ucc.ie), Michael G. Egan

University College Cork, Ireland

Abstract – **This paper investigates power interfaces for a PEM fuel cell. The main focus of the investigation is to analyze and test the effects of part-load operation on component selection and stresses with an emphasis on the magnetic components. The standard two-phase interleaved boost with a discrete inductor per phase is compared with the transformer-coupled two-phase interleaved boost, consisting of a single input inductor in series with a phase-coupling transformer. The converter characteristics are investigated for the experimental V-I inputs derived from the polarization curve of an industrial PEM fuel cell. Experimental validation is presented for a 3 kW design. Magnetic sizing of air-cooled components for power converters up to 45 kW are additionally investigated.**

Keywords – interleaved boost converter; fuel cell; magnetic coupling

I. INTRODUCTION

If fuel cell electric vehicles (FCEV) are to become an alternative option to internal combustion engine (ICE) vehicles then they will have to exhibit the following characteristics: high-power-density, high efficiency, fast transient response and quick start up along with a comparable size and weight to today's ICE based vehicles. Fig.1 shows a typical FCEV architecture which combines the high energy density of the fuel cell with the high power density of the battery pack. Efficient, robust and price-competitive power conversion in the FCEV powertrain is critical for optimal fuel economy, low maintenance cost and customer satisfaction.

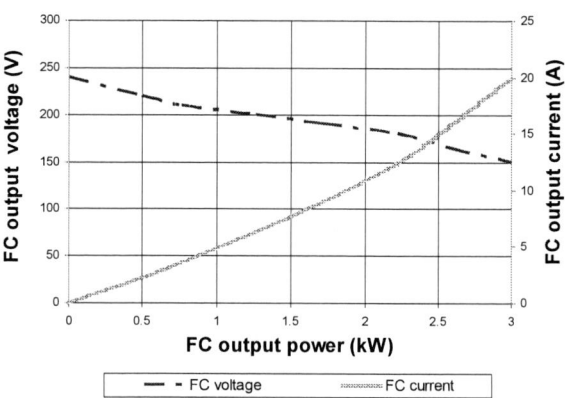

Fig. 2. Typical fuel cell V-I characteristic.

Various design requirements such as efficiency, boost converter/fuel cell characteristics and cost have to be considered. Magnetic components in dc-dc converters can account for between 20 to 30% of the converters volume [1].

An industrial PEM fuel cell polarization curve [2, 3] is presented in Fig. 2. This experimental curve follows the typical theoretical curve with significant voltage reduction with load due to the activation, ohmic and oxidation voltage drops from the no-load voltage defined by the Nernst equation. The curve shows that the PEM fuel cell voltage characteristic has a significant voltage drop over the output power range. In this case, the PEM fuel cell voltage varies from 250 V at no load to 150 V at full load.

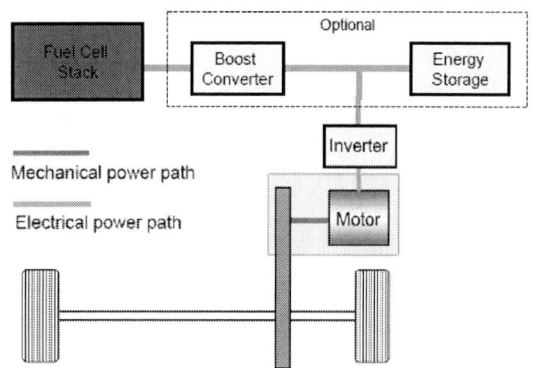

Fig. 1. Typical FCEV powertrain architecture.

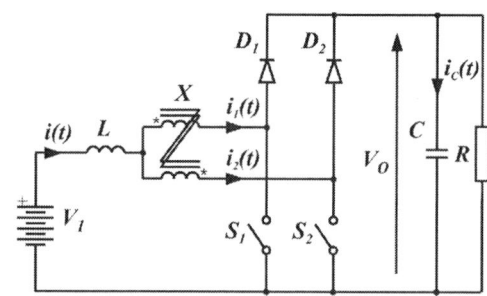

Fig. 3. Transformer coupled boost topology.

978-1-4244-4782-4/10 $26.00 © 2010 IEEE

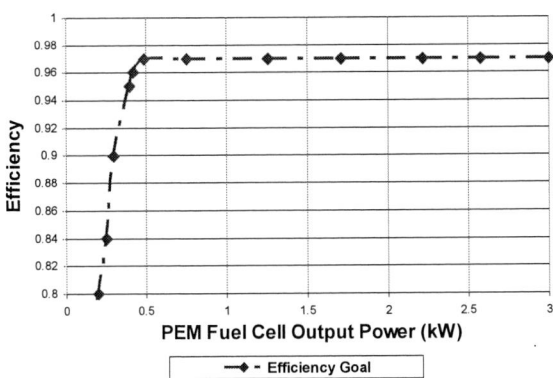

Fig. 4. Specified efficiency requirements.

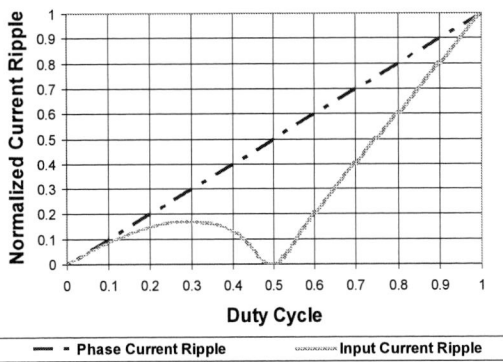

Fig. 5. Normalized input and phase current ripple for two-phase boost.

This paper presents two non-isolated dc-dc boost converters that can be interfaced with the PEM fuel cell: a classical interleaved two-phase boost with discrete inductors in each phase and a transformer-coupled boost consisting of a single input inductor in series with a phase-coupling transformer. The transformer-coupled boost topology is presented in Fig. 3.

Multiphase interleaved boost converters allow for input and output current ripple cancellation which results in a size reduction of the input filter capacitors. The XL design extends this current ripple cancellation to the phase current and allows for the reduction of magnetic and ohmic losses along with an overall size reduction. The coupled-transformer topology is similar to what is presented in [4, 5]. This paper will investigate the performance of the two-phase and XL topologies over the operating range of the PEM fuel cell presented in Fig. 2. The efficiencies of both converter topologies will also be analyzed. Fig. 4 illustrates the typical efficiency requirements for the boost topologies. The efficiency goal corresponds to the efficiency target presented in [6]. This calls for efficiency requirements of greater than 97% over a wide load current range corresponding to a PEM fuel cell output current of 2.5 A to 20 A.

The topology presented in [5] has high-phase current ripple due to the low magnetizing inductance of the transformer. However, input current ripple is not affected by the magnetizing inductance. Thus, it is possible to use different magnetizing inductance values to alternate between continuous conduction mode (CCM) and discontinuous conduction mode (DCM) operation per phase as is shown in the high-power 45 kW paper design.

Section II gives a description of the two boost topologies that are to be compared in this paper. Section III presents a medium-to-high power paper design of the magnetic components for both topologies. Section IV presents a 3 kW prototype design and experimental validation of both converters.

II. BOOST TOPOLOGIES

The two boost topologies to be compared in this paper are presented in this section: a classical interleaved two-phase boost with a single inductor per phase (2L) and a transformer-coupled boost consisting of a single input inductor in series with a transformer that couples each phase (XL). An alternative topology to the XL that is not considered in this paper is an integrated magnetic approach similar to that presented in [1, 7-13]. The XL topology is a compromise between the classical and integrated magnetic approaches and benefits from the use of a high saturation magnetic material for the inductor design. The transformer is constructed from low power loss ferrite.

A. Two-Phase-Boost Topology

The classical two-phase design uses two discrete inductors. A multiphase converter design allows for phase disabling which permits phases to operate at a corresponding fraction of the total converter power as shown in [6]. When operating in two-phase mode, phase interleaving causes portions of the inductor phase current to overlap as a function of duty cycle, D. This results in reduced current ripple at the input and output and full current ripple cancellation is achieved at $D = 0.5$ for two-phase operation. Hence, the input and output capacitors can be sized accordingly which results in converter weight and size reduction. The input and phase currents of the two phase topology are presented in Fig. 5. The phase current ripple is greater than the input current ripple and this will result in higher ac power loss in the inductor and also require the semiconductors to be sized accordingly. The phase and input current ripple in Fig. 5 are normalized to the peak value of the input current ripple when $D = 1$.

B. XL Boost Topology

The two-phase transformer-coupled boost converter is presented in Fig. 3. This topology is realized by using a single input inductor along with a transformer that couples between phases.

$$\Delta I_{phase} = \begin{cases} \dfrac{V_1}{4L} \dfrac{1-2D}{1-D} DT & for \ 0 \le D \le 0.5 \\[2ex] \dfrac{V_1}{4L}(1-2D)T & for \ 0.5 \le D \le 1 \end{cases} \quad (1)$$

$$\Delta I_{input} = \begin{cases} \dfrac{V_1}{2L} \dfrac{1-2D}{1-D} DT & for \ 0 \le D \le 0.5 \\[2ex] \dfrac{V_1}{2L}(2D-1)T & for \ 0.5 \le D \le 1 \end{cases} \quad (2)$$

The input and phase current ripple is triangular with a frequency equal to twice the switching frequency. The phase current is a combination of half of the input ripple and the magnetizing current ripple. Note that the magnetizing current is cumulatively coupled within the transformer. Equation (1) and (2) show the XL phase and input current ripple respectively. Fig. 6 depicts the converter waveforms for $D \in [0; 0.5]$ and Fig. 7. shows the converter waveforms for $D \in [0.5; 1)$.

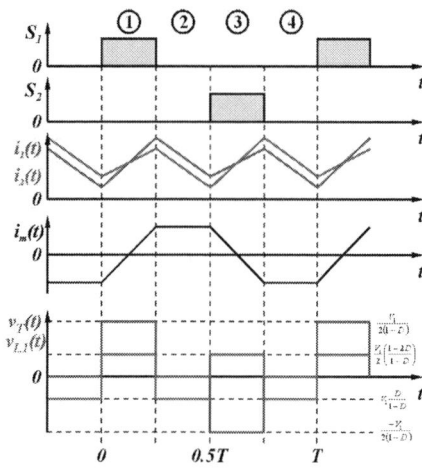

Fig. 6. XL switching waveforms for $D \in [0;0.5]$.

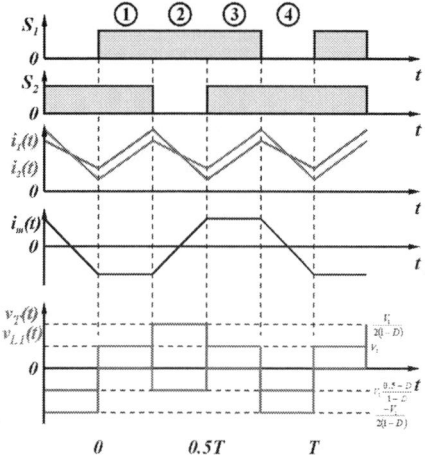

Fig. 7. XL switching waveforms for $D \in [0.5;1)$.

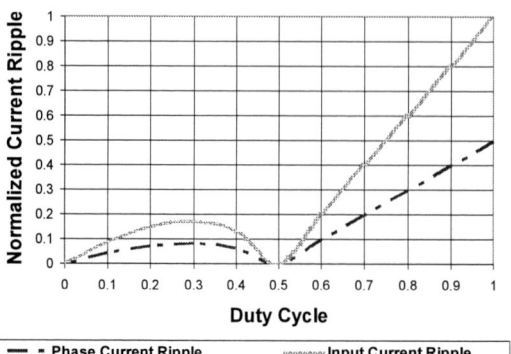

| — · Phase Current Ripple | 〜〜〜〜 Input Current Ripple |

Fig. 8. Normalized input and phase current ripple for XL boost converter.

In the case of $D \in [0.5; 1)$, it can be seen that during sub-interval 1, when switch S_1 and S_2 are turned on, that there is no voltage across the transformer. Over this sub-interval the magnetizing current is a constant value. Conversely, all the input voltage, V_1 is across the inductor and so the phase currents increase. During sub-interval 2 the voltage across the transformer is equal to $V_1/2(1-D)$ and so the magnetizing current is increasing. At this point, there is a negative voltage across the input inductor and so the phase current starts to decrease. This is repeated for the sub-intervals 3 and 4.

Fig. 8 shows how the normalized input and phase current ripple vary with D. The currents in Fig. 8 are normalized against the maximum input current ripple which occurs when $D = 1$. Similar to the two-phase design there is current ripple cancellation at $D = 0.5$ but due to the coupling effect of the transformer there is the added benefit of phase current ripple cancellation. In this design the phase-current ripple is less than the input current ripple if the magnetizing inductance is high enough. This allows for reduced output capacitors as well as downsizing of the switching components and reduction of core losses in the magnetic component. However, high magnetizing inductances may be impractical at low frequencies due to the high number of turns needed.

For $D \in [0; 0.5]$ the highest current ripple condition occurs at $D = 0.3$. The elevated current ripple at $D = 0.3$ may significantly reduce part load efficiency as it will induce increased core and ac losses in the inductor.

It should be noted, that the desired inductor must be able to deal with the full dc input current as well as input current ripple over the operating range of the converter. Hence, the inductor is constructed from high dc-bias material such as amorphous metal or silicon steel. These materials have a high material power loss. The transformer, on the other hand, is designed with a bi-filar winding. This eliminates the dc magnetization and thus a low dc-bias material like ferrite can be used for its construction.

III. MEDIUM-TO-HIGH POWER MAGNETIC COMPONENTS

In this section, magnetic components for high-power converters are investigated. A 45 kW paper design is used to compare and contrast both converter topologies. Both converters are compared for the same input current ripple and input power. Design curves are presented for area product vs. normalized input current ripple. The magnetic components are designed and optimized using a modification of the inductor design flowchart presented in [14]. For simplicity, all the magnetic components are designed and optimized for natural convection cooling. Both designs are temperature limited by the core material with 99 % converter efficiency. The magnetic core material is Metglas 2605SA1. The converters are designed to operate at input voltage, $V_{in} = 150$ V and output voltage, $V_{out} = 375$ V. This corresponds to a duty ratio, $D = 0.6$. The input inductor and transformer of the XL were designed at full power, $P_{in} = 45$ kW. The phase inductors of the 2L converter were designed for $P_{in} = 22.5$ kW each.

The phase inductor of the 2L and the input inductor of the XL were designed and optimized for total minimum area product vs. normalized input current ripple. The 2L inductors were designed per phase, and so the phase current ripple of the 2L is 3 times greater than the input current ripple, as can be seen in Fig. 5. Thus, it was necessary to determine a normalized input current ripple ratio so the 2L and XL could be compared like-for-like.

Both converters were analysed over the 20 to 75 kHz frequency range. The results of the 25 kHz analysis are presented in Fig. 9. The minimum area product for the XL converter is at 15 % input current ripple which corresponds to an actual current ripple of 45 A and a phase current ripple of 22.5 A. The minimum area product for the two phase inductor is at approximately 5 % input current ripple which corresponds to an input current ripple of approximately 15 A or a phase current ripple of approximately 45 A in the 2L converter. The transformer of the XL was designed for minimum area product vs. phase magnetizing current ripple.

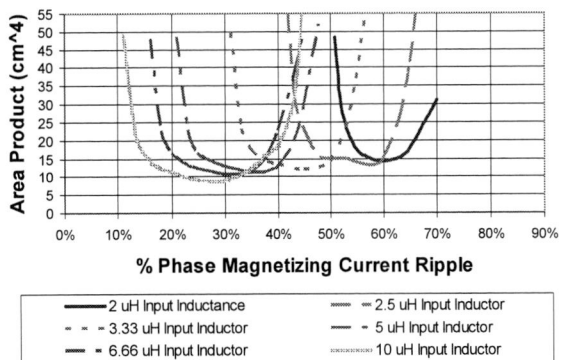

Fig. 10. XL transformer area product vs. phase magnetizing current ripple.

The transformer design was evaluated by setting the input inductance to a constant value. This input inductance value was determined by the minimum area product from the input inductor design flowchart. The phase current ripple, which is a combination of half the input ripple and the cumulative magnetizing current ripple, was varied by changing the magnetizing inductance of the transformer. Fig. 10 illustrates the transformer area product vs. phase magnetizing current ripple for varying input inductances at 25 kHz switching frequency. Table I presents a breakdown of the design results at 25 kHz. This shows that the XL magnetic components have a 65 % lower area product then the 2L magnetic components.

Thus, it is possible to design the XL for different transformer magnetizing inductances. A large magnetizing inductance results in a low magnetizing current and vice versa. As presented in [5], it is therefore possible to operate the transformer of the XL in DCM or CCM while not affecting the input and output current ripple. The high-ripple design allows for zero-voltage switching (ZVS) while reducing the total amount of components traditionally required for ZVS implementation.

Another factor for choosing a high-ripple design is due to the bi-filar wound transformer. This inherently contains a large parasitic capacitance and may introduce unwanted ringing when the converter is switching. Designing the transformer to have a low magnetizing inductance will reduce the number of turns required and reduce the parasitic capacitance.

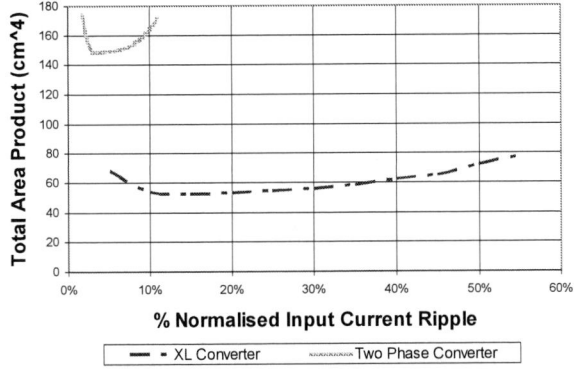

Fig. 9. Total inductor area product of XL and 2L vs. normalised input current ripple.

Table I. Comparison of 2L and XL area product at 25 kHz

			Two Phase Design	Coupled Design	
			2L	XL	
			Single Phase Inductor	Input Inductor	Transformer
Switching Frequency	fs	kHz	25	25	25
Inductance	L	uH	100	7	22
Input Current Ripple	ΔI_in	A	15	45	-
Phase Current Ripple	ΔI_phase	A	45	-	50
Area Product	AP	cm^4	75	41	11
Total Area Product	AP_total	cm^4	150	52	
Power Loss	P	W	225	170	35
Total Power Loss	P_total	W	450	205	

978-1-4244-4782-4/10 $26.00 © 2010 IEEE

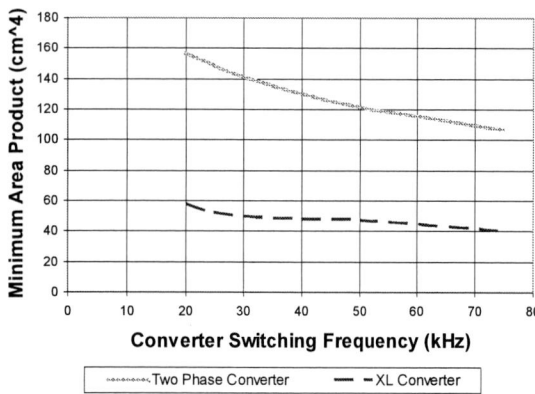

Fig. 11. Minimum area product vs. switching frequency.

Fig. 11 presents the minimum area product for each topology over the converter switching frequency. This illustrates that for each switching frequency the area product for the XL converter is approximately 65 % lower than that of the two phase converter. Intuitively, the area product decreases as the frequency increases. This is due to the decrease in required inductance as the frequency increases for constant current ripple.

Fig. 12 illustrates how the total area product of the XL and 2L converters vary with input power. At each power level the input current ripple of the XL and 2L converters are 15 % and 5 % respectively. It can be seen that as the input power is increased the area product of both converters increases and is due to increased input current and higher dc bias. If liquid cooling was employed then it would be possible to shrink the size of the magnetic components by a factor of three or more [15].

In the XL converter the input and phase currents are operating at twice the switching frequency due to interphase coupling. Thus, this factor gives the XL a significant size saving advantage over the 2L design.

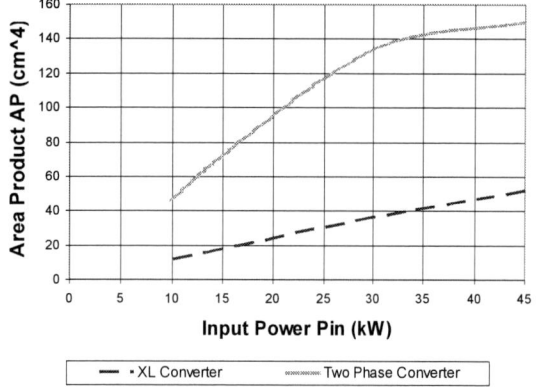

Fig. 12. Total area product vs. input power.

IV. 3 kW PROTOTYPE DESIGN AND EXPERIMENTAL RESULTS

This section details a 3 kW prototype design used to compare and validate both converter topologies. The experimental results are also presented. The predicted ohmic and magnetic core losses were determined from the design flowchart. The flowchart is supported by 2D FEA, which is used to estimate the eddy current loss in the winding. Each topology is modeled in Pspice and the phase current is decomposed into harmonics, which are used as the inputs to the model in the FEA solver. The ac power loss is the total of the power losses generated by all harmonics. The flowchart then predicts the estimated temperature rise of the magnetic component and this is compared with the actual experimental temperature rise.

A. 3 kW Prototype Design

Two 3 kW experimental converters were built in order to validate both topologies. In order to compare like-for-like the input inductor in the XL topology was designed to have the same input current ripple as the two phase converter. The converters are designed to operate at various input voltages according to the fuel cell specification given in Fig. 2, V_{in} = 150 to 250 V and fixed output voltage, V_{out} = 375 V. The two-phase inductors are 390 uH each while the XL inductor is 195 uH and the transformer magnetizing inductance is 3.4 mH. Both converters operate at 20 kHz. All three inductors were constructed using Metglas 2605SA1 cores with the same geometry. The transformer of the XL was constructed from a 3C93 ferrite core. The transformer has a bi-filar winding configuration that allows for the elimination of dc flux. Two switching cell poles were constructed using Infineon IPW60R045CP CoolMOS power transistors and Infineon STD12S60 Silicon Carbide diodes as the switching devices. A UCC28070 controller IC from Texas Instruments was used to generate interleaved two-phase gate signals. The converters were placed in a thermal chamber where they operated until steady state was reached. The inductors temperature rise was measured to validate the design.

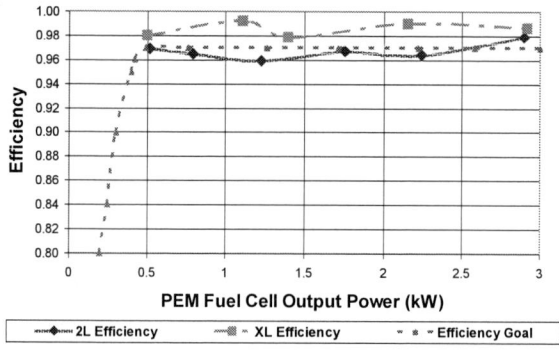

Fig. 13. Boost converter efficiency vs. PEM fuel cell output power.

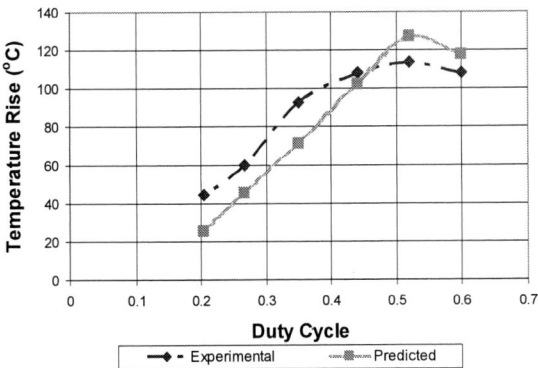

Fig. 14. Temperature rise of classical boost inductor.

B. Experimental Validation

The following section presents the results of the experimental validation of a 3 kW 2L and XL boost converter. Fig. 13 shows the efficiency of the classical two-phase and XL boost converters vs. duty cycle. The XL boost converter shows greater efficiency than the two-phase converter in meeting the required efficiency targets. The two-phase discrete inductors are exposed to a much higher flux swing than the single input inductor of the XL. Thus, the higher phase-current ripple in the two-phase design results in increased core power loss along with increased ac power loss in the windings, which reduces its efficiency. The input inductor of the XL converter is exposed to a lower flux swing resulting in a low core power loss.

Fig. 14 illustrates the temperature rise in the two-phase boost inductors which correlate well with the predicted temperature rise.

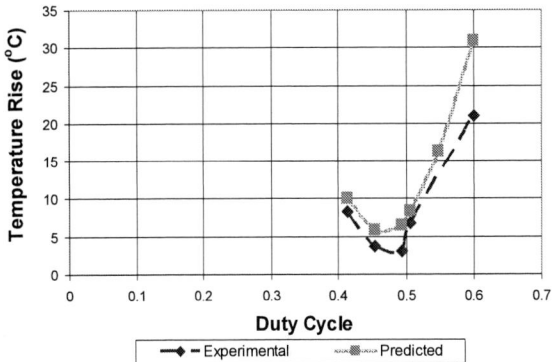

Fig. 16. Temperature rise of XL inductor.

Fig. 15 presents the input and phase current ripple vs. duty cycle for the two-phase converter. Input current ripple cancellation is visible at $D = 0.5$. The experimental results correspond well to the theoretical results predicted in Fig. 5.

Fig. 16 illustrates the temperature rise in the inductor of the XL converter, as expected the minimum temperature rise is visible at approximately $D = 0.5$ which corresponds to input and phase current ripple cancellation. The inductor power loss is significantly lower than in the two phase inductor. The temperature increases for $D < 0.5$ and this corresponds to the increased current ripple at $D = 0.3$ shown in Fig. 8. The input and phase current ripple vs. duty cycle, D, for the XL boost is shown in Fig. 17. It is clear the input current ripple does not go to zero at $D = 0.5$. This is due to current imbalance between the phases because the controller is not optimized. Further work is required here. The two-phase design operates in DCM while the XL design operates in CCM.

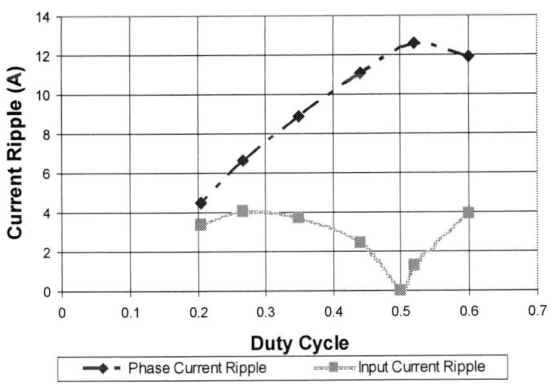

Fig. 15. Phase and input current ripple of two-phase converter.

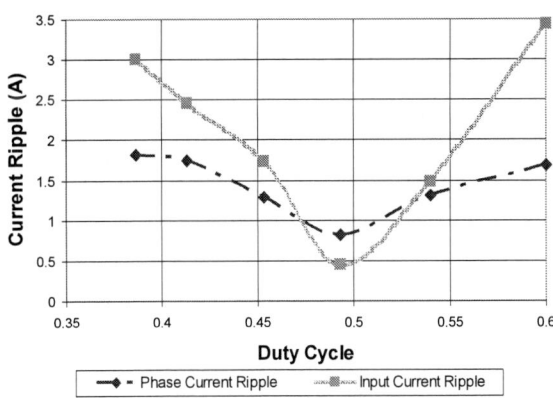

Fig. 17. Phase and input current ripple of XL converter.

978-1-4244-4782-4/10 $26.00 © 2010 IEEE

V. CONCLUSIONS

In this paper two boost converters for an automotive fuel cell application are compared. The classical two-phase boost is compared with a transformer-coupled boost over the wide load range. A high-power design of the magnetic components of both converters is investigated. Design curves were developed to show the difference in the size for both magnetic designs over the input current ripple range. The area products of the XL magnetic components are significantly smaller than that of the 2L design.

It is also presented that the phase current ripple could be changed by varying the magnetizing inductance of the transformer. A lower magnetizing inductance results in higher phase current ripple, thus, ZVS may be facilitated. A lower magnetizing inductance also reduces the amount of turns required and this helps to minimize the capacitance associated the bi-filar winding.

Two prototype converters were designed for the same input current ripple at rated power of 3 kW. The XL temperature rise and current ripple vs. duty cycle waveforms are presented and compared to the two-phase boost converter. The XL converter shows superior efficiency over the same size classical two-phase converter.

REFERENCES

[1] S. Chandrasekaran, L. U. Gokdere, "Integrated Magnetics for Interleaved Dc-Dc Boost Converter for Fuel Cell Powered Vehicles", *IEEE Power Electronics Specialists Conference*, 2004, pp356-361.

[2] Kirubakaran A, et al., "A review on fuel cell technologies and power electronic interface", *Renewable Sustainable Energy Review* (2009).

[3] Ehsani M, et el, "Modern Electric, Hybrid Electric, and Fuel Cell Vehicles – Fundamentals, Theory, and Design", *CRC Press*, 2005.

[4] G. Calderon-Lopez, A.J. Forsyth, D.R. Nuttall, "Design and performance evaluation of a 10-kW interleaved boost converter for a fuel cell electric vehicle," *5th International Power Electronics and Motion Control Conference*, vol. 2, pp. 1328-1332, 2006.

[5] G. Calderon-Lopez, A.J. Forsyth, "High-Power Dual-Interleaved ZVS Boost Converter with Interphase Transformer for Electric Vehicles", *Applied Power Electronics Conference and Exposition*, APEC 2009, pp. 1073-1077, 2009.

[6] B. Eckardt, M. Marz, "A 100kW automotive powertrain DC/DC converter with 25 kW/dm3 by using SiC", *International Conference Power Electronics, Intelligent Motion*, PCIM Europe 2006, pp. 185-190.

[7] A. Fratta, P. Casasso, G. Guglielmi, S. Nieddu, G.M. Pellegrino, "New design concepts and realization of hybrid dc-dc coupling reactors for light EVS", *IEEE* , 2003, pp. 2877-2882.

[8] A. Fratta, G. Griffero, S. Nieddu, G.M. Pellegrino, F. Villata, "New hybrid iron-ferrite-core couplings upgrade effectiveness of H-bridge-based power conversion structures," *IEEE* , 2002, pp. 884-889.

[9] A. Fratta, P. Guglielmi, F. Villata, A. Vagati, "Efficiency and cost-effectiveness of ac drivers for electric vehicles improved by a novel boost dc-dc conversion structure," *Power Electronics in Transportation*, 1998, pp. 11-19, 1998.

[10] Wei Wen, Yim-Shu Lee, "A Two-Channel Interleaved Boost Converter with Reduced Core Loss and Copper Loss", *2004 35th Annual IEEE Power Electronics Specialists Conference*, Aachen, Germany, 2004.

[11] Jieli Li, T. Abdallah, and C. R. Sullivan, "Improved calculation of Core loss with nonsinusoidal waveforms", in *Conference Record of the 2001 IEEE Industry Applications Conference. 36th IAS Annual Meeting*, 2001, pp. 2203–2210.

[12] Jens Czogalla, Jieli Li, Charles R. Sullivan, "Automotive Application of Multi-phase Coupled-Inductor DC-DC Converter", *IEEE Industry Applications Society Annual Meeting*, vol.3, pp 1524-1529, October 2003.

[13] Jieli Li, Anthony Stratakos, Aaron Schultz, Charles R. Sullivan, "Using Coupled Inductors to Enhance Transient Performance of Multi-Phase Buck Converters", *Applied Power Electronics Conference and Exposition*, APEC 2004, vol.2, pp. 1289-1293, 2004.

[14] Marek S. Rylko, Brendan J. Lyons, John G. Hayes, Michael G. Egan, "Magnetic Material Comparisons for High-Current Gapped and Gapless Foil Would Inductors in High Frequency DC-DC Converters", *European Power Electronics and Drives Association*, EPE-PEMC, 2008.

[15] Marek S. Rylko, Kevin J. Hartnett, John G. Hayes, Michael G. Egan, "Magnetic Material Selection for High Power High Frequency Inductors in DC-DC Converters", *Applied Power Electronics Conference and Exposition*, APEC 2009, pp. 2043 -2049, 2009.

Design Considerations for Narrow Vdc based Power Delivery Architecture in Mobile Computing System

Xiaoguo Liang, Gnanavel Jayakanthan, Meng Wang

Mobile Platform Group
Intel Asia-Pacific Research& Development Ltd.
Shanghai, 200241, China
E-mail: xiaoguo.liang@intel.com

Abstract- Narrow Vdc power architecture is one of initiatives to extend battery life(EBL) in mobile computing system. Advantages of it is analyzed in detail. All design considerations for such possible concerns as charger controller selection, pre-charge issue, thermal stress, Energy Star compliant related, is illustrated and solutions to some concerns are also provided in this paper. Finally, test is conducted in a typical Notebook system to verify all theoretical analysis and proposed solutions to those concerns.

I. INTRODUCTION

Extending Battery Life(EBL) has been increasingly a strong demand from end users for mobile computing devices such as Notebook, Netbook and Mobile Internet Device(MID). Statistic data shows 2/3 of users consider battery life as top or second priority of all factors in selecting a new notebook and above 50% users are willing to pay $30~$40 for one additional hour battery life [1]. Narrow Vdc(NVDC) based power delivery architecture, proposed by Intel, is one of many initiatives to extend battery life for mobile system. This technology has evolved for two generations, first generation NVDC is called NVDC-I, and second generation NVDC is called Adaptive Mobile Power System(AMPS) [2]. However, for many reasons, NVDC-I has been adopted by some Notebook/Netbook OEMs in industry while AMPS still not adopted yet. In this paper, we only focus on NVDC-I architecture and all terminology 'NVDC' will refer to NVDC-I architecture in following section.

A typical conventional power delivery architecture in Notebook is shown in Fig.1. All on-board VRs are actually buck converter in today's design and it can be seen on-board VRs can get power from AC adaptor or battery and then regulate this voltage down to required voltage rails for processor, DDR memory, chipsets I/O and other devices. Battery charger IC will control a buck converter(consists of S_4, S_5, L_1, and C_2) to charge battery and manage Power Path Switches(consists of S_1, S_2 and S_3) to be ON/OFF in a right way to ensure all on-board VRs either get power from AC adaptor or battery. Since AC adaptor output is 19V and battery output ranges from 8.7V to 12.6V with a most popularly used 3S2P(3 cells in series, 2 cells in parallel) li-ion cell configuration, bus voltage V_{DC} will range from 8.7V to 19V for all on-board VRs. By contrast, NVDC power delivery architecture is shown in Fig.2, and the key difference from conventional counterpart is that charger's output is used to both power on-board VRs and charge battery, and AC adaptor's

output will be not directly used to power on-board VRs any more. It indicates that bus voltage range of V_{DC} in NVDC architecture is identical to voltage range of battery, i.e. only ranging from 8.7V to 12.6V for a battery with typical 3S2P cell configuration. It could be seen bus voltage range of on-board VRs in NVDC architecture is dramatically reduced over conventional architecture in Notebook system.

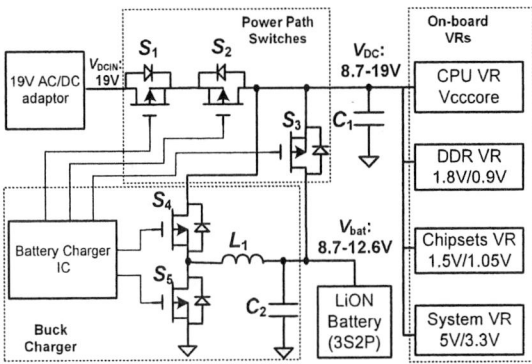

Fig. 1. Conventional Power delivery Arch. in Notebook System

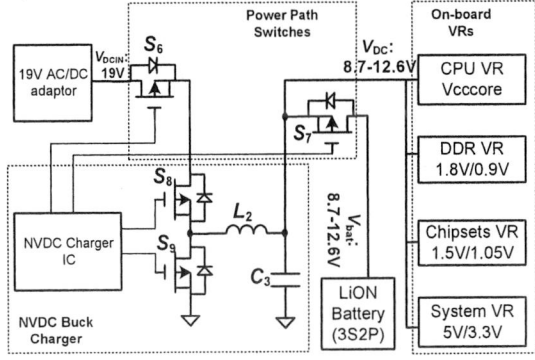

Fig. 2. NVDC power delivery Arch. in Notebook System

It's well known that a wider range of input voltage for a bulk regulator will not benefit its efficiency optimization design if output voltage/current is fixed. Main reason is switch loss is a bottle-neck for efficiency improvement and many papers have very detailed analysis on this before, so this paper will not explain this in detail[3] [4] [5]. As mentioned above, NVDC power architecture can narrow input voltage range of on-board VRs, then of course can improve conversion efficiency

978-1-4244-4782-4/10 $26.00 © 2010 IEEE

of these VRs to extend battery life for a mobile computing system.

In this paper, advantages of NVDC architecture is analyzed firstly, and then design considerations for this technology in practice are analyzed in detail. Finally, test in a typical Notebook system is conducted and theoretical analysis is verified by the test result.

II. ADVANTAGES of PROPOSED NVDC POWER ARCHITECTURE

A. Extending battery life

In NVDC architecture, as mentioned above, a narrow input voltage range can benefit optimization of key power components design for on-board VRs(buck regulator) to achieve high efficiency to extend battery life. This can be explained from several aspects:

Firstly, for switching MOSFET, lower input voltage can reduce switching loss especially for upper-side MOSFET, which has been well explained in many previous papers[3] [4] [5]. The other thing is lower input voltage make it possible to use some lower voltage rating MOSFET like 20V or 25V FET with better performance to further reduce switching loss. Table 1 is one example to illustrate performance improvement of 25V MOSFET over 30V counterpart. It can be seen that 25V FET is much better than 30V FET especially on switching loss associated parameters.

Table 1. Comparison of 30V and 25V MOSFET

	IRFH7932Pbf	CSD16403Q5A
Vds/Vgs	30V/±20V	25V/+16V,-12V
Package	Power SO8	Power SO8
Rdson/Qg@Vgs=4.5V	3.3mΩ/34nC	2.9mΩ/13.3nC
Crss	420pf	115pf
Vendor	IR	TI

Secondly, for inductor, inductance value is calculated by equation(1) in buck regulators:

$$L_f = \frac{V_o \cdot (1 - V_o/V_{in})}{\Delta I \cdot f_s} \tag{1}$$

where L_f is output inductance, V_{in} is input voltage, V_o is output voltage, ΔI is allowable ripple current across inductor and f_s is switching frequency. And for on-board VRs, V_o is generally fixed, f_s typically is designed at 300KHz so that it's a constant, and ΔI typically is designed at 30% of full load current so that it's also a constant.

It can be seen in equation (1) that higher V_{in} will lead to larger inductance and vice versa. With a specified input voltage range, inductance is designed in accordance with maximum input voltage. It's obvious maximum input voltage is much lower in NVDC architecture than that in conventional architecture, then required inductance for on-board VRs is much smaller in NVDC arch. It's a common sense that DC resistance(DCR) of one inductor is smaller if inductance is smaller with same package. Then DCR loss of inductor for all on-board VRs in NVDC architecture is reduced than that in conventional architecture.

Totally, reduction of MOSFET switching loss and DCR loss of inductor both contribute a high conversion efficiency for on-board VRs in NVDC architecture to extend battery life.

B. Power delivery cost down chances

As shown in Fig.1 and Fig.2, at first glance, it's easy to consider NVDC architecture much more expensive on cost that conventional architecture. After looking into details from whole system standpoint, the story will be different. A relatively detailed comparison on cost is shown in Table 2 and it can be seen NVDC architecture even save around $0.5, compared with its counterpart. It is mainly due to:

Firstly, even though NVDC architecture introduces high current-rating MOSFET and inductor for charger converter itself, it removes one P-ch power MOSFET on power path from AC adaptor to input of charger. That P-ch FET is required in conventional arch. to block AC adaptor power into system under battery learning mode for Notebook system, while this requirement can be realized just by disabling charger in NVDC arch. So overall offset effect make cost of inductors and MOSFETs used in NVDC architecture only little bit higher that its counterpart

Secondly, for bus decoupling capacitors both for charger output and on-board VRs' input in NVDC architecture, 16V low voltage rating ceramic capacitor could be adopted instead of 25V capacitor in conventional architecture. With the fact that many pieces of this kind of cap is required in a whole system and 16V capacitor is cheaper than 25V capacitor, overall cost saving is significant on this in NVDC architecture.

Then above two factors together will make NVDC architecture not cost higher but provide cost down opportunities instead. It should be noted cost information in Table 2 is just some typical marketing price and not supposed to be 100% accurate since cost is affected by many factors in practice. The data here mainly focus on cost gap instead of absolute price information. The charger controller price is not shown since it's still sensitive at this moment, but from a long term standpoint charger IC price should be at same level since complexity and technology used for both kinds of controllers is similar.

III. DESIGN CONSIDERATIONS FOR NVDC

POWER ARCHITECTURE

A. NVDC charger controller selection

Most of known pulse width modulation(PWM) control schemes like voltage mode , current mode, constant on-time and constant-off time mode, are still applicable to today's Notebook charger controller , especially voltage mode control is very widely used. However, some special care needs to be taken when these control schemes are applied to charger in NVDC power architecture. Firstly, it

978-1-4244-4782-4/10 $26.00 © 2010 IEEE

	Conventional Arch.		NVDC Arch.		NVDC over conventional Arch
Items	Parts description	Cost($)	Parts description	Cost($)	Cost Delta($)
Charger IC	available from most of VR IC vendors	N/A	available from 2 vendors now	N/A	0.1
Mosfets	S_1, S_2, S_3- 30V PFET, handle full current rating by system	0.45	S_6, S_7- 30V PFET, handle full current rating by system	0.3	-0.15
	S_4, S_5- 30V NFET, handle battery charging current	0.15	S_8, S_9-30V NFET, handle full current rating by system	0.3	0.15
Inductor	L_1- 6.5x7x3mm package, handle battery charging current	0.13	L_2- 10x11x4mm package, handle whole system current	0.16	0.03
Caps	C_1, C_2- typically 20 pcs 25V/10uF ceramic used	1.4	C_3-typically 20pcs 16V/10uF ceramic used	0.7	-0.7
				Total	-0.57

Table. 2. cost comparison for NVDC arch. and conventional arch.

can be seen that battery is only load for charger in conventional power architecture and battery is passive component and then charger will not face fast load transient during operation, however both battery and on-board VRs are loads for charger in NVDC power architecture which means charger has to face frequent load transient during operation since on-board VRs are dynamic load seen from NVDC charger in a computing system. Secondly, to meet the high light load efficiency requirement by Notebook, a voltage mode or current mode PWM charger controller will generally adopt the pulse frequency modulation(PFM) to reduce switching related loss, gate driving loss of charger converter when load enters discontinuous mode(DCM), and a transition between PFM and PWM always occurs when charger load current switches between light load and heavy load dynamically. And it has been verified, in a voltage or current mode charger controller, this transition is always not smooth since circuit inside controller need time to complete this mode transition and the time delay is always inevitable[6]. Based on above analysis, voltage mode and current mode-based control mechanism is not suitable to implement NVDC charger controller. Fortunately, constant on-time control has seamless transition from PFM in light load condition to PWM mode in heavy load condition because this control method uses output ripple of charger as a PWM ramp signal to compare with a reference voltage to regulate output voltage[6]. Therefore

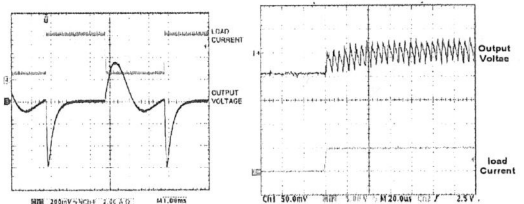

(a) voltage mode controller   (b) Constant on time controller

Fig. 3 Charger Load transient with different kind of controller.

constant on-time control could be a good candidate for NVDC charger controller.

Test waveforms are shown in Fig.3 to compare load transient between different kinds of charger controllers. Fig.3(a) test is from one vendor's voltage mode NVDC charger controller, where big output voltage undershoot/overshoot is observed during load transient from light load to heavy load. By contrast, Fig.3(b) test is from the other vendor's constant on-time NVDC charger controller, and transition from light load to heavy load is very smooth and no obvious undershoot is observed for output voltage.

It should be noted that big overshoot and undershoot of charger output voltage in NVDC architecture will cause potential risk for battery in practical operation. Considering such a scenario, battery is being charged and close to terminated voltage(12.6V for 3 cells in series configuration), and suddenly a big overshoot across bus voltage V_{DC} comes because of load current change from heavy load to light load, then the voltage seen at battery may exceed cell allowable maximum voltage which is risky for battery in two areas: firstly, some surge energy into battery will trigger battery internal protection to cut off battery from Notebook system; secondly, repeat of this over voltage across battery cell may shorten battery life cycle since battery cell life cycle is very sensitive to charge voltage after it's above nominal terminated voltage level(4.2V/cell typically for Li-ion battery). Given load transient is a frequent event for charger in NVDC architecture, it's very important to choose a right charger controller to ensure a good load transient performance during PFM to PWM mode transition.

Moreover, driver capability of charger controller should be enough to drive high current rating MOSFET in NVDC charger.

B. Pre-charge Design for NVDC charger

For a typical Li-on battery, pre-charge process is required for battery cells before going into a normal constant current charge process when battery is over-discharged or over depleted and voltage per cell is below one certain value(typically 3V/cell for Li-ion battery). As shown in Fig.1 and Fig.2 above, one key difference between NVDC architecture and its counterpart is that bus voltage V_{DC} in NVDC architecture is common input of battery and on-board VRs during charging process(S_7 is ON), and it can cause one serious problem that computing system cannot start up if pre-charge process is required when battery is extremely discharged or depleted and total battery voltage is below 5V. Reason behind is that input voltage V_{DC} is identical to battery charging voltage and must be also clamped to such a low voltage below 5V which is too low to start up system VR(5V rail), and then whole system cannot start up without 5V rail ramping up normally. This problem more easily happen in a Netbook system than a Notebook system since Notebook typically use 3S2P cell configuration while Netbook uses a 2 cells in series(2S2P or 2S1P) configuration, and 2 cells in series configuration is more liable to drop below 5V if battery is not charged and uncared for some months in practical application. However, it doesn't mean Notebook will never meet such problem. For some extreme cases, battery even could be so depleted that its voltage drops below 2V or even worse if the battery is not cared for too long time.

One conventional solution to this pre-charge problem is shown in Fig.4 and NVDC power architecture is simplified here. Under pre-charge operation mode, NVDC charger still regulate bus voltage V_{DC} to terminated voltage(say, 12.6V for 3 cells in series configuration of battery), and sense the charge current I_{bat} and control it within required value of pre-charge current $I_{pre-chg}$ by controlling S_7 operate in a linear mode. In this way, pre-charge current can be controlled just by adjusting gate driving voltage of S_7. Although this ideas seems perfect, it may face two concerns as following:

♦ Power loss of S_7 during pre-charge process

Power loss across S_7 during pre-charge process can be represented by following equation:

$$P_{loss\_S_7} = (V_{DC} - V_{bat}) * I_{pre-chg} \qquad (1)$$

Where P_{loss_S7} is power dissipation of S_7, V_{DC} is charger output voltage, V_{bat} is battery voltage and $I_{pre-chg}$ is pre-charge current.

It can be seen power loss of S_7 is significant during pre-charge process, especially when battery is deeply depleted and V_{bat} approaches zero. And this will cause not only more power consumption from wall power but thermal stress for S_7 which means a high performance switch with higher cost may be needed here to meet this critical requirement

♦ Accuracy of pre-charge current control

Today's Notebook system typically monitors charge current by a 20mohm sensing resistor, and under pre-charge mode typically 200mA charging current is allowed so that only total 4mV voltage is sensed to charger controller as charging current information, then only 30% or even worse accuracy can be achieved since charger controller has some inherent offset error from amplifier and other factors[7] [8]. Under such extreme condition as that battery could deeply depleted so that its voltage is very low, say, around 2V, a even smaller

pre-charge current like tens of mA should be used to charge battery until battery voltage rises up to one certain threshold level and can be recognized by embedded controller(EC) in Notebook system. So accuracy of pre-charge current control is even worse for this extreme condition. It will make pre-charge current easily out of programmed value.

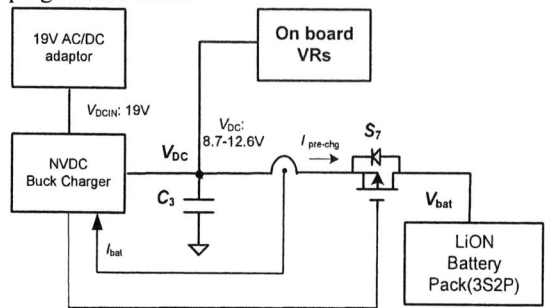

Fig.4 conventional linear pre-charge for NVDC charger

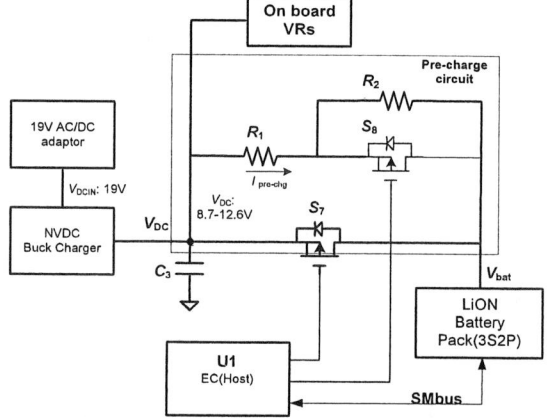

Fig.5 proposed pre-charge scheme for NVDC charger

One robust pre-charge method is proposed in Fig.5 [9], and it consists of two resistor R_1,R_2, one small switch S_8 ,one power switch S_7 and embedded control(EC), and EC is used to communicate with battery via SMbus to get battery voltage, current information and also control ON/OFF of S_7 and S_8 . it has three working modes from pre-charge to normal charge, and two assumptions are made to simply the analysis of its working principle: firstly, S_7, S_8 are ideal switches; secondly, 3 cells in series configuration for battery is used which means battery cut off voltage is around 9V. And its working principle is described as below:

♦ *Deep pre-charge MODE($V_{bat}<V_{th}$)*

This is a default MODE, under this mode, battery voltage is lower than one certain threshold value V_{th}, S_7 is OFF, S_8 is also OFF. Pre-charge current can be represented by following equation:

$$I_{pre-chg1} = \frac{(V_{DC} - V_{bat})}{R_1 + R_2} \qquad (2)$$

♦ *Normal pre-charge MODE($V_{th}<V_{bat}<9V$)*

Under this mode, battery voltage is higher than V_{th} and lower than cut off voltage. S_7 still OFF, S_8 is ON. Pre-charge current can be represented by following equation:

$$I_{pre-chg2} = \frac{(V_{DC} - V_{bat})}{R_1} \qquad (3)$$

- *Normal charge MODE(9V<V_{bat}<12.6V)*

Under this mode, battery voltage exceeds cut off voltage and enter normal charge process, and S_7 is ON, S_8 is OFF. NVDC charger will take over the control and complete battery charging process.

It should be noted that V_{th} is programmed by EC and ON/OFF of S_7, S_8 is also controlled by EC since EC can monitor battery voltage information via SMbus and decide which mode to enter to charge battery. However, under such condition as battery voltage is ultra low so that SMbus circuit stop working inside battery, EC cannot detect related battery information via SMbus. For this case, default mode(Deep pre-charge MODE) will ensure no problems since S_7, S_8 is OFF by default and battery is pre-charged with a very small current even without EC's involvement.

Power dissipation of R_1, R_2 can be represented by equation (4) and (5) below:

Under deep pre-charge mode,

$$P_{R1} = \frac{(V_{DC} - V_{bat})^2 \cdot R_1}{(R_1 + R_2)^2}, P_{R2} = \frac{(V_{DC} - V_{bat})^2 \cdot R_2}{(R_1 + R_2)^2} \qquad (4)$$

Under normal pre-charge mode,

$$P_{R1} = \frac{(V_{DC} - V_{bat})^2}{R_1}, P_{R2} = 0 \qquad (5)$$

Where P_{R1}, P_{R2} represents power dissipation of R_1 and R_2 respectively. In practice, since $R_1 << R_2$ should be satisfied to meet requirement of $I_{pre-chg1} << I_{pre-chg2}$, let $V_{bat}=0$ in equation(4), maximum power dissipation of R_2 can be represented by following:

$$P_{R2\_max} = \frac{V_{DC}^2 \cdot R_2}{(R_1 + R_2)^2} \qquad (6)$$

let $V_{bat}=V_{th}$ in equation(5), maximum power dissipation of R_1 can be represented by following equation:

$$P_{R1\_max} = \frac{(V_{DC} - V_{th})^2}{R_1} \qquad (7)$$

Generally speaking, for R_1, using several 0805 sized normal resistors in series or in parallel is enough to handle required power across it; while for R_2, one 0603 normal resistor is enough to handle the power. Also it's a good choice to program V_{th} within 4~6V range to minimize P_{R1_max}.

Detailed Analysis in [9] also shows that pre-charge current accuracy is below 2% for this proposed method just by using such normal external parts as 1% resistor of R_1 and R_2, 0.5% V_{DC} regulation accuracy which is very

popularly used in today's Notebook charger design.

From above analysis ,it can be seen that proposed pre-charge method has three advantages over conventional way: higher pre-charge current accuracy; lower charging power dissipation; and low cost adder since R_1, R_2 is normal 0805,0603 size resistor, S_8 is a small FET and S_7 and EC is always needed regardless of pre-charge method and shouldn't be counted into cost adder factors.

C. Thermal design for NVDC power Arch.

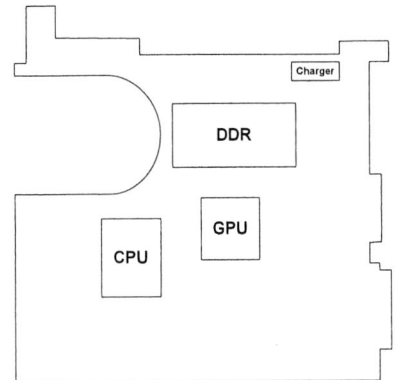

Fig.6 one motherboard placement example

Since NVDC charger will convert all power required by whole Notebook system instead of just charging battery in conventional architecture, the power loss of NVDC charger itself should be higher than conventional charger when system is running under heavy load, then one intuitional concern for NVDC architecture is whether it will face some thermal challenge so that Notebook chassis skin temperature is too high to pass thermal test. For this, an optimized components placement and layout routing is helpful to minimize this size effect, and following is some general guide:

Place NVDC charger far from other key heat-generating parts such as CPU, GPU(graphic processor unit)DDR memory on motherboard in case thermal aggregation effect;

Not place NVDC charger underneath some sensitive location like palm rest where user always put their hands over there during daily usage of Notebook. For this location, both a maximum temperature and a maximum temperature difference between palm rest left side and right side need to be controlled within specification;

Route as large as possible for charger output power plane on external or internal layer of PCB and also put more vias from charger GND to ground plane(one or two internal complete layers), and this will help an even thermal distribution across PCB.

Fig.6 is one good illustration of key parts placement on one Notebook system motherboard.

D. Energy Star requirement

Current Energy Star regulation for PC was effective in

978-1-4244-4782-4/10 $26.00 © 2010 IEEE

	In NVDC Arch.	In Conventional Arch.
MOSFET of On-boards VRs	High side: CSD16410Q5A(Rdson/Qg:9.6mΩ/3.9nC), Vds:25V Low side: CSD16413Q5A(Rdson/Qg:4.1mΩ/9nC), Vds:25V	High side: IRF7821(Rdson/Qg:9.1mΩ/9.3nC), Vds:30V Low side: IRF7862(Rdson/Qg:3.7mΩ/30nC), Vds:30V
Inductor of on-board VRs	using same package as in conventional arch., and reduce inductance by around 1.5X for each VR	Two kind of package used: 10mmx10mmx4mm, 6mmx7mmx3mm, and inductance designed to keep inductor ripple current within around 30% of full load current

Table. 4 Key parts design criteria for on-board VRs to compare battery life effect.

July 2007 and it specifies that wall power of Notebook PC with internal graphics and WOL(Wake-on-Lan) enabled should below 14W in S0(idle), 2.4W in S3(Standby) and 1.7W in S5(off) in order to get certified for a Energy Star logo[10]. As explained above, NVDC charger itself will introduce extra power loss and then it's concern whether total power dissipation seen from input of AC adaptor will be much more difficult to pass above Energy Star regulation than conventional power architecture.

Several factors need to be taken into account to get a qualitative understanding into this concern. Firstly, In NVDC architecture, charger power loss is higher than that in conventional power architecture, but input power of on-board VRs is lower due to a higher conversion efficiency , and this will bring about an offset effect; secondly, system required power associated with Energy Star test condition is relatively small and NVDC charger conversion efficiency is also very high even under light load condition with today's charger controller, so power dissipation of NVDC charger is generally small. Then overall effect will result in power seen from output of AC adaptor only little bit higher in NVDC architecture than that in conventional architecture, so does it for power seen from wall power or input of AC adaptor given same adaptor is selected for comparison here.

IV. EXPERIMENTAL RESULTS

A Montevina platform based 14inch Notebook system with following basic configuration is used for a series of test to verify above analysis: CPU: Intel®Core™2 Duo P8400@2.26GHz; Chipsets: Intel®Cantiga Express Chipset Family; DDR: KVR-DDR2-667x512MBx2; HDD: Fujitsu MHV2060BH-PL-60GB; LCD: CHI MEI N141I1-L02, 14.1 inch, CCFL backlight; ODD: TSSTcorp CDDVDW TS-L633A; Battery: Dynapak Li-ion battery 11.1V/4.8Ah; I/O: 4xUSB, 1xVGA, 1xRJ11, 1xRJ45, 1xHDMI. OS: Windows Xp SP2

A. Extending battery life analysis

For battery life analysis, using following configuration: benchmark: MobileMark 2005 , productivity mode; on-boards VR MOSFETs: choose 30V part for Non-NVDC architecture, and 25V parts for NVDC architecture; on-board VRs inductor: switching frequency fixed, inductor package fixed and inductance fine-tuned to keep same current ripple requirement both in NVDC

architecture and conventional architecture; power test point: from battery output(AC adaptor absent). Table 4 shows key parts design criteria details.

Test shows NVDC architecture can save around 0.15W under battery running mode over conventional architecture. Total power consumption of system is around 14.5W under the specified benchmark above, then total battery life can be extended by around 1% by using NVDC arch. it seems the effect is not so significant mainly because total on-board VRs power loss accounts for a small percentage of total system power consumption with running above benchmark. However, when total system power reduces significantly by adopting LED backlight display panel in the future, the effect of VR power loss reduction with NVDC arch. will become more obvious.

B. Pre-charge test

According to Fig.5, one design is implemented by using following parameters: R_1: 40Ω(two 20Ω in series, 0805, 1/8W,1%); R_2: 1kΩ(0603,1%); S_7:FDS6679AZ; S_8:FDN360P. Battery is deeply discharged to around 3V then charged again and whole charge process is plotted in Fig.7. V_{th} is programmed at 6V, and transition voltage from normal pre-charge to normal charge is programmed at 9.2V by EC software. Fig.7 shows actual test for V_{th} is at 6.02V and transition voltage from normal pre-charge to normal charge is at 9.19V, very close to programmed value. Also it can be seen pre-charge current is not sensitive to charging voltage under deep pre-charge mode since it's too small, while pre-charge current decreases with increase of charging voltage under normal pre-charge mode. It verifies above theoretical analysis.

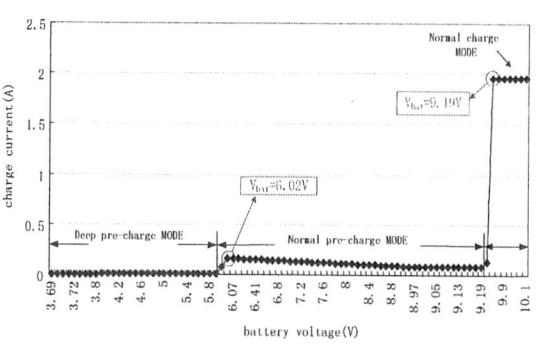

Fig.7 voltage vs. current during whole charge process

In Table 3, test result shows effect of V_{th} on charging power dissipation, charging time, temperature of R_1. it's obvious that increasing V_{th} from 3.2V to 6V can make power dissipation decrease from 0.45W to 0.26W(42% reduction), and temperature of R_1 drop by almost 10℃. By the way, power dissipation of conventional linear pre-charge method is also shown there, and verify that proposed method has significant lower charging power dissipation over linear pre-charge method.

Table 3 Effect of Vth on pre-charge

Condition[1]	Proposed pre-charge method			Conventional solution
	Power loss(W)	Max. R1 temp(℃)	Pre-charge time(s)	Power loss(W)
Vth=3.2V	0.454	56.5	8.8	0.95
Vth=6.0V	0.263	47.2	8.5	0.8
1.Test under room temperature Ta=25℃				

C. Thermal test result

Fig.8 shows temperature scan result of C cover of Notebook chassis at 25℃ ambient temperature by using an infrared camera with system running a popularly used thermal analysis tool TAT by Intel(100% loading) and the battery is being charged concurrently. It can be seen maximum temperature is 34.6℃, well within allowable specification 37℃(12℃ above ambient temperature for high touch area, and 23℃ above ambient temperature for low touch area). It indicates NVDC architecture faces no problems to pass thermal specification by Notebook.

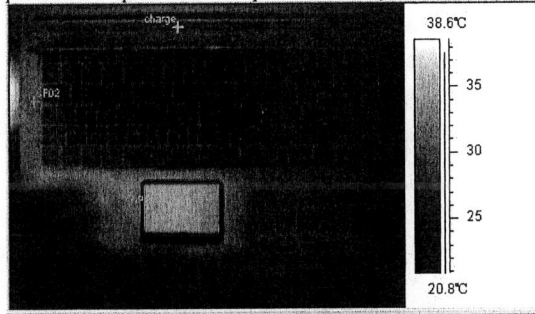

Label	Value [℃]	Width	Min	Max	Max - Min	Avg
Image			20.5	38.0	17.6	
charge	26.2					
SP02	32.0					
SP03	34.6					

Fig.8 C cover temperature scan

D. Energy Star requirement test

Due to test equipment limited, power dissipation from AC adaptor's output instead of from wall power is tested to see effect of NVDC architecture on total Notebook system power dissipation. Test result is shown in Table 5 and it can be seen, compared with conventional architecture, NVDC architecture increases total system power dissipation by 0.25W,0.08W under system idle and S3 state respectively, while reduces total system power dissipation by 0.08W surprisingly under S5 state.

Reduction of power under S5 state just indicates that advantage of NVDC architecture on switching related loss reduction dominates its disadvantage when load is quite low.

Table 5 Energy requirement test

	idle	S3 state	S5 state
NVDC Arch.	11.07W	1.28W	0.56W
Conventional Arch.	10.82W	1.2W	0.64W

Then it can be concluded it's not a challenge for NVDC architecture to pass Energy Star test requirement since its adder of power dissipation is very minor under idle and S3 state, and moreover its power dissipation is even reduced under S5 state, compared with conventional architecture.

V. Conclusions

The advantage of NVDC power delivery architecture in mobile computing system is analyzed and design considerations for all concerns for this technology is also analyzed and related solutions for these concerns are provided as well. Test is conducted in a typical Notebook system to verify the theoretical analysis and proves the effectiveness of those proposed solutions.

Reference

[1] Intel, "User perception of mobile PC battery life", *EBLS005*, Intel Developer Forum(((IDF), 2008.

[2] Intel, "Intel Adaptive Mobile Power System Application Note", Revision 1.0, Nov. 2006

[3] Jia Wei and Fred C. Lee, "Two-Stage Voltage Regulator for Laptop Computer CPUs and Corresponding Advanced Control Schemes to Improve Light-load Performance" , in *Proc. IEEE Appl. Power Electron. Conf. (APEC'04)*, 2004, pp. 1294-1300.

[4] Yuancheng Ren, Ming Xu, Kaiwei Yao, Yu Meng, Fred C. Lee and Jinghong Guo, "Two-stage Approach for 12V VR" in *Proc. IEEE Appl. Power Electron. Conf. (APEC'04)*, 2004, pp. 1306-1312.

[5] Jia Wei, "High Frequency High-Efficiency Voltage Regulators for Future Microprocessors", Ph.D. dissertation of Virginia Polytechnic Institute & State University, Sep.15,2004

[6] Chuan Ni and Tateishi Tetsuo , "Adaptive Constant On-Time(D-CAP[TM]) Control Study in Notebook Applications", TI application note SLVA281B, Jul. 2007.

[7] Maxim, datasheet of Max8724, http://www.maxim-ic.com/ quick_view2.cfm/qv_pk/3767

[8] TI, datasheet of bq24740, http://focus.ti.com/docs/prod /folders/print/bq24740.html

[9] Xiaoguo Liang, Gnanavel Jayakanthan, "One Robust Pre-charge Method for Battery in NVDC Based Mobile System," in *Intel Design & Test Technology Conference (DTTC'09)*, 2009.

[10] EPA, "ENERGY STAR® program requirements for computers", http://www.energystar.gov.

Active Clamp Boost Converter with Switched Capacitor and Coupled Inductor

Yi Zhao, Wuhua Li, Bo Yang, Xiangning He
College of Electrical Engineering, Zhejiang University
Hangzhou, 310027, P.R. China
Email: woohualee@zju.edu.cn

Abstract—A high step-up converter with a coupled inductor and a switched capacitor is proposed in this paper. By using the coupled inductor, the voltage gain of the converter is extended and the voltage stress of the main switch is decreased to reduce the conduction losses. Moreover, the reverse-recovery problem of the output diode is alleviated by the leakage inductance. By using the switched capacitor, the resonance between the leakage inductance and the stray capacitor of the output diode is eliminated. Additionally, an active clamp circuit is induced here to depress the voltage spikes of the main switch and to recycle the energy stored in the leakage inductance. Both the main switch and the clamp switch work with zero-voltage-switching (ZVS) performance, which reduces the switching losses significantly. At last, a 500W prototype is built to verify the analysis. The highest efficiency of the prototype is 96% and the efficiency is over 93% in a wide load range.

I. INTRODUCTION

Usually, high efficiency and high step-up DC-DC converters are necessary components in renewable generation system and uninterruptable power supply (UPS), because the relatively low voltages of the sources like photovoltaic arrays and acid batteries is required to be converted to high voltages for inverters [1, 2]. However, the classic boost converter is not suitable for high step-up conversion, since low on-resistance switches can not be adopted and the reverse-recovery problem of the output diode is severe due to the extremely high duty-cycle [3, 4].

Coupled inductors are induced to solve the above-motioned problems. A buck-boost type converter was presented in [5], and a boost type converter was presented in [5, 6]. In these converters, by using the coupled inductors, the voltage stress of the main switch is reduced, so low on-resistance switches can be adopted to reduce the conduction losses. The current falling rate through the output diode is limited by the leakage inductance of the coupled inductor. Thus the reverse-recovery problem is partly solved. At the same time, an active clamp circuit is employed for three reasons: First of all, the voltage spikes across the main switch caused by the leakage inductance are eliminated. Next, the energy stored in the leakage inductance is recycled. The last

This work is sponsored by the National Nature Science Foundation of China (50907058), the Power Electronics S&E Development Program of Delta Environmental & Education Foundation (DREM2009001) and the China Postdoctoral Science Foundation (200902625).

but not least, the main switch and the auxiliary switch are all switched with ZVS performance, which reduces the switching loss significantly. So the converters with coupled inductors are good choices for high step-up applications.

Unfortunately, the voltage stress of the output diodes in [6] is higher than the output voltage. Also the resonance between the leakage inductance and the stray capacitor of the output diode increases the losses and causes the EMI problem, which increases the voltage stress of the output diode. Therefore a RCD snubber is employed in [7] and a set of clamp circuit is employed in [8] to mitigate the problems.

In this paper, a novel converter with a coupled inductor and a switched capacitor is proposed to fulfill the high step-up requirements and to solve the above-mentioned problems. Compared with the converter in [6], the voltage stress of the output diode is reduced and the resonance between the leakage inductance and the stray capacitor of the output diode is eliminated. Furthermore a higher conversion ratio is obtained by the added switched capacitor. Moreover, when the main switch is in the on-state, the coupled inductor transfers the energy like a forward converter, and when the main switch is in the off-state, the coupled inductor transfers the energy as a flyback converter. Therefore, the volume of the magnetic core is reduced due to the fully utilization of the coupled inductor.

II. PROPOSED CONVERTER AND OPERATION ANALYSIS

The proposed converter is illustrated in Fig.1 (a). A classic boost converter can be obtained by the components out of the markers. The clamp circuit marked in the rectangular loop is composed of a switch S_c and a capacitor C_c. The circuit marked in the elliptical loop is composed of a winding of a coupled inductor, a switched capacitor and a diode, which is to extend the voltage gain and to reduce the voltage stress of the switch. The inductor L_a is coupled to the inductor L_b. The coupling reference is marked by "*", and the turns ratio is expressed by $N=n_2/n_1$.

The coupled inductor can be considered as an ideal transformer, a parallel magnetizing inductance and a series leakage inductance [9]. The equivalent circuit of the proposed converter is shown in Fig.1 (b). L_a and L_b represent the primary and the secondary winding respectively. C_p is the parallel capacitor to achieve ZVS, which includes the parasitic capacitors of the main switch and the clamp switch. C_{s1} and C_{s2} are the equivalent variable capacitors when the diodes begin to suffer reverse voltages, otherwise in the rest time, C_{s1} and C_{s2} are too small to affect the circuit performance.

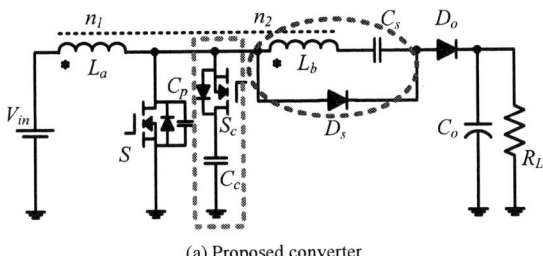

(a) Proposed converter

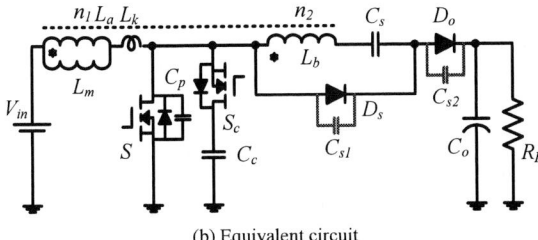

(b) Equivalent circuit

Figure 1.   Proposed converter and the equivalent circuit

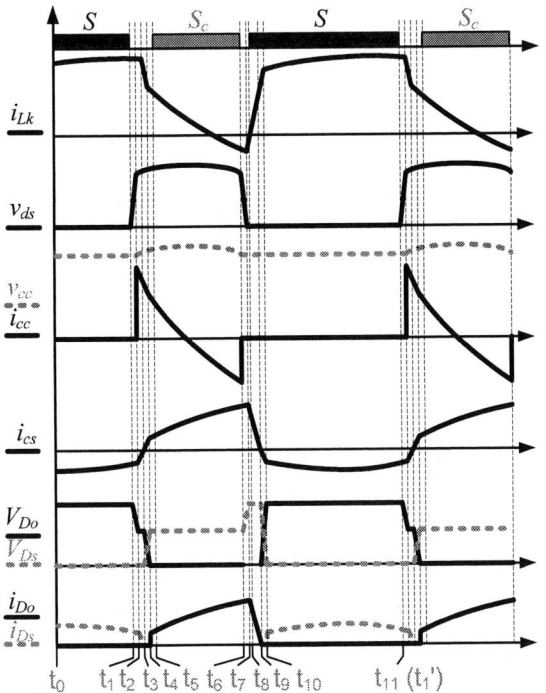

Figure 2.   Key waveforms of the converter

There are ten main subintervals during one switching cycle. The key waveforms are shown in Fig.2, and the explanation of each waveform is the following: i_{Lk} is the current through the leakage inductance L_k, also it's the input current of the converter; v_{ds} is the voltage across the main switch S; v_{cc} is the voltage across the clamp capacitor C_c; i_{cc} is the current though the clamp circuit S_c and C_c; i_{cs} is the current through the switched capacitor C_s; v_{Do}, v_{Ds}, i_{Do} and i_{Ds} are the voltages and currents of the diodes D_o and D_s. The equivalent circuits for each subinterval are shown in Fig.3.

Subinterval 1 [$t_0 \sim t_1$]: Before t_1, the main switch S is in the on-state while the clamp switch S_c is in the off-state. During this period the output diode D_o is anti-biased. The magnetizing inductance is charged by the input voltage, thus the magnetizing current increases gradually almost in a linear way. At the same time, the energy is transferred to the switched capacitor C_s by the coupled inductor, as the coupled inductor works in forward mode. The current through the secondary winding is limited by the leakage inductance L_k.

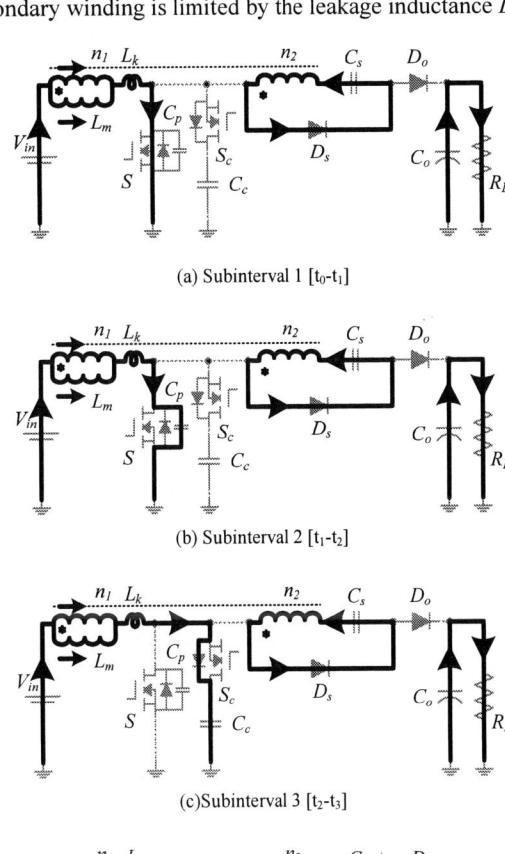

(a) Subinterval 1 [t_0-t_1]

(b) Subinterval 2 [t_1-t_2]

(c) Subinterval 3 [t_2-t_3]

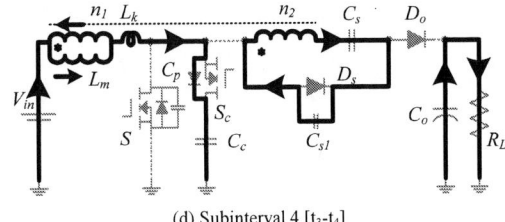

(d) Subinterval 4 [t_3-t_4]

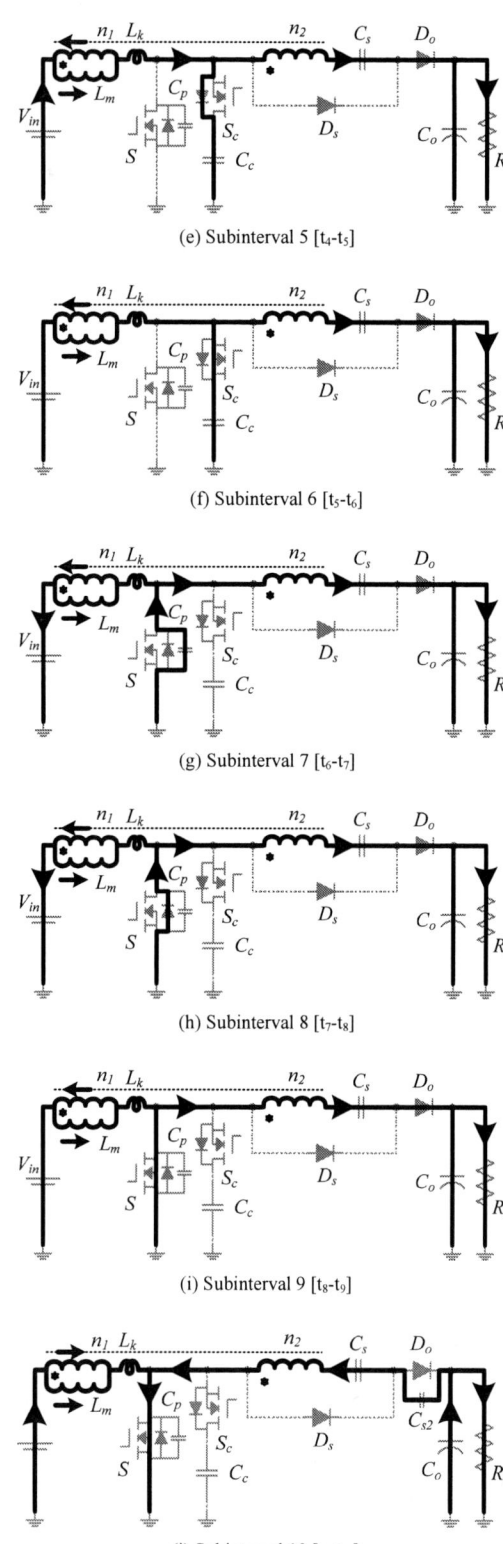

(e) Subinterval 5 [t_4-t_5]

(f) Subinterval 6 [t_5-t_6]

(g) Subinterval 7 [t_6-t_7]

(h) Subinterval 8 [t_7-t_8]

(i) Subinterval 9 [t_8-t_9]

(j) Subinterval 10 [t_9-t_{10}]

Figure 3.   Operation processes of the proposed converter

Subinterval 2 [t_1~t_2]: The main switch is turned off at t_1, then the parallel capacitor C_p and the leakage inductance L_k begin to resonate, and C_p is charged by the current of the leakage inductance from zero. So the main switch is turned off with ZVS duo to the existence of C_p.

Subinterval 3 [t_2~t_3]: At t_2, the switching voltage of the main switch reaches the clamp capacitor voltage, so the body diode of the clamp switch is forced to conduct. Then v_{ds} is clamped to v_{cc} by the anti-parallel diode of the clamp switch S_c. Since the clamp capacitor C_c is much larger than C_p, Almost all the current flow through C_c. After t_2, the leakage inductance is discharged by the voltage of C_c. The current through L_k decreases almost linearly as well as the current through the secondary winding.

Subinterval 4 [t_3~t_4]: At t_3, the current through the secondary winding falls to zero, and the equivalent capacitor C_{s1} begins to resonate with the leakage inductance L_k. The voltage across C_{s1} rises while the current through L_k decreases. The current falling rate of the diode D_s is limited by the leakage inductance. Therefore, the reverse-recovery problem of D_s is alleviated.

Subinterval 5 [t_4~t_5]:   The voltage of C_{s1} is high enough to cause D_o to conduct at t_4. Since the output capacitor C_o is much larger than C_s, most of the current flows through C_o. During this period the energy stored in the magnetizing inductance and the switched capacitor is transferred to the load. The coupled inductor works in flyback mode. The leakage inductance and the clamp capacitor begin to resonate.

Subinterval 6 [t_5~t_6]: The turn-off signal is applied to the clamp switch at t_5. The clamp switch S_c is turned off when its body diode is conducting, thus ZVS turned off is achieved. This movement does not change the working state of the converter.

Subinterval 7 [t_6~t_7]: The clamp switch is turned off at t_6. After that the leakage inductance and the clamp capacitor stop resonating and new resonance is formed between the leakage inductance L_k and the parallel capacitor C_p. The voltage across the clamp switch S_c increases from zero while the voltage of C_p decreases at the same rate. The energy stored in C_p is transferred to L_k. C_p causes the clamp switch to be turned off with ZVS.

Subinterval 8 [t_7~t_8]: The voltage of the parallel capacitor C_p falls to zero at t_7, and then the body diode of the main switch begins to conduct. C_p and L_k stop resonating. The increasing rate of the current through L_k is controlled by the output voltage, as well as the decreasing rate of the current through D_o

Subinterval 9 [t_8~t_9]: The turn-on signal is applied to the main switch when its anti-parallel diode is in the on-state. So the main switch is turned on with ZVS. Other part of the circuits work as in subinterval 8.

Subinterval 10 [t_9~t_{10}]: At t_9, the current through the output diode decreases to zero. The equivalent capacitor of the output diode C_{s2} begins to resonate with the leakage inductance. The voltage of C_{s2} increases while the voltage of D_s decreases at the same rate. At t_{10}, the voltage of D_s drops to zero. After that the magnetizing inductance is charged by the input source and the energy is transfer to C_s by the coupled inductor again.

III. PERFORMANCE ANALYSIS

To simplify the analysis, the voltages across the clamp capacitor C_c and the switched capacitor C_s are considered to be constant, and the same assumption is made to the current through the magnetizing inductance.

A. Voltage gain

The parasitic parameters of are not considered in this analysis, including the leakage inductance, the parallel capacitor and the equivalent capacitors of the diodes. The effect caused by the dead times between the main switch and the clamp switch is neglected. A whole switching period can be separated into two pieces. During the former period, the main switch is in the on-state and the clamp switch is in the off-state. The magnetizing inductance is charged by the input voltage.

$$V_{Lm\_charge} = V_{in} \tag{1}$$

And the voltage of the switched capacitor can be expressed by:

$$V_{cc} = N \cdot V_{in} \tag{2}$$

In the latter period, the main switch is in the off-state and the clamp switch is in the on-state. The voltage on the magnetizing inductance can be derived by:

$$V_{Lm\_discharge} = \frac{V_{out}}{N+1} - V_{in} \tag{3}$$

According to the inductor volt-second balance principle, the voltage gain can be obtained by:

$$M = \frac{V_{out}}{V_{in}} = \frac{N+1}{1-D} \tag{4}$$

The voltage gain of the proposed converter is extended greatly compared with the classic boost converter, even higher than that of the converter in [6]. The relationship among the conversion ratio, the duty-cycle and the turns ratio are sketched in Fig.4.

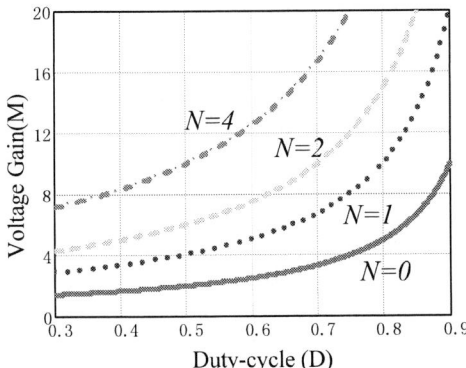

Figure 4. Voltage gain of the proposed converter

As shown in Fig.4, when the turns ratio is zero, the voltage gain of this converter is the same as that of the classic boost converter. As the turns ratio increases, the voltage gain increases significantly.

B. Voltage stress of the switch

By neglecting the voltage ripple of the clamp capacitor, the voltage stress of the main switch is given by:

$$V_{stress\_switch} = \frac{V_{in}}{1-D} = \frac{V_{out}}{N+1} \tag{5}$$

The voltage stress of the clamp switch is the same as that of the main switch. The proportion of the voltage stress and the output voltage with different turns ratios is sketched in Fig.5. It can be derived that the voltage stress decreases sharply with the increasing of the turns ratio. So high performance switches can be used here to improve the efficiency.

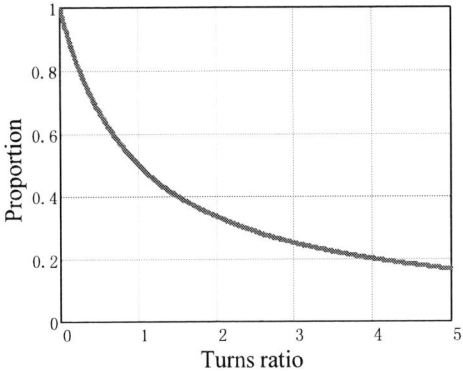

Figure 5. Voltage stress reduction effect

C. Effect of the leakage inductance

The leakage inductance causes both the main switch and the clamp switch to work with ZVS. Meanwhile, the current variation rate of the output diode current is limited by the leakage inductance. When the energy is transferred to the switched capacitor, the current impulsion is depressed by the leakage inductance effectively. Especially to the diodes D_o and D_s, the reverse-recovery problem is partly solved by the series leakage inductance. The relationship between the leakage inductance and the current falling rate of D_o and D_s is expressed by:

$$\frac{di_D(t)}{dt} = \frac{V_{out}}{N^2 \cdot L_k} \tag{6}$$

IV. EXPERIMENTAL RESULTS AND ANALYSIS

To demonstrate the theoretical analysis, a 500W prototype was built, and the parameters of the converter are described as follow: P_{out}: 500W; V_{out}: 380V; V_{in}: 40~56V; f_s: 100 kHz; n_2/n_1: 14/28; L_m: 95μH; L_k: 1.9μH; C_c: 2.2μF; C_s: 4.7μF; C_o: 4.7μF: S: STW90NF20; S_c: IRF250N; D_s and D_o: MUR1560. The waveforms are obtained at full load, and the input voltage is 40V.

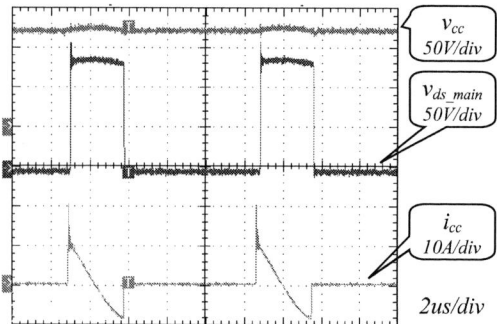

Figure 6. Performance of the clamp circuit

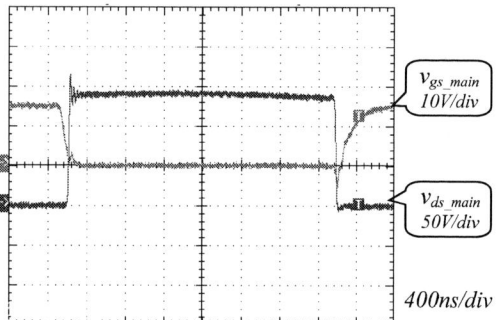

Figure 7. ZVT performance of the main switch

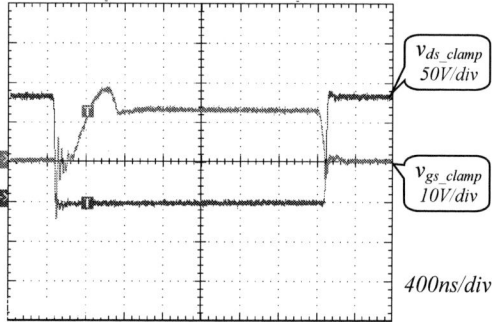

Figure 8. ZVT performance of the clamp switch

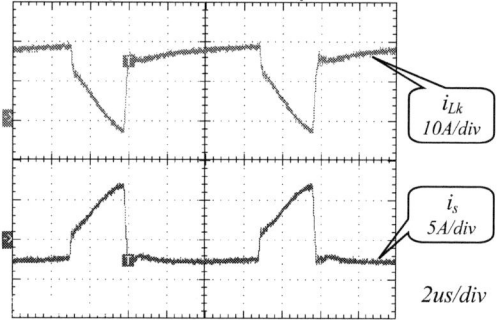

Figure 9. Currents through the windings of the coupled inductor

through the clamp capacitor. The voltage spikes caused by the leakage inductance are depressed greatly.

ZVT performances of the main switch and the clamp switch are illustrated in Fig.7 and Fig.8 respectively. It is shown that the ZVT of the main switch and the auxiliary switch are achieved during the whole switching transitions, which reduces the switching losses greatly.

The currents through the windings of the coupled inductor are shown in Fig.9. The current through the primary winding is also the input current, including the current through the magnetizing inductance and the current through the secondary winding reflected to the primary side.

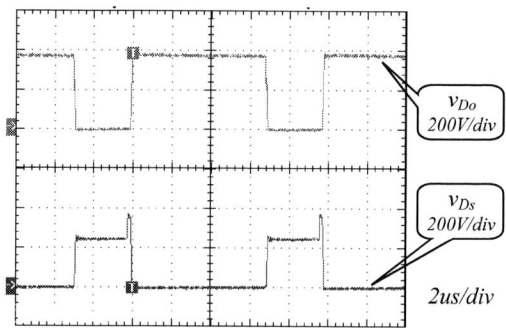

Figure 10. Voltages across the diodes

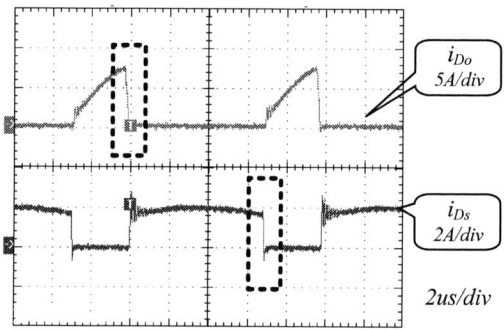

Figure 11. Current through the diodes

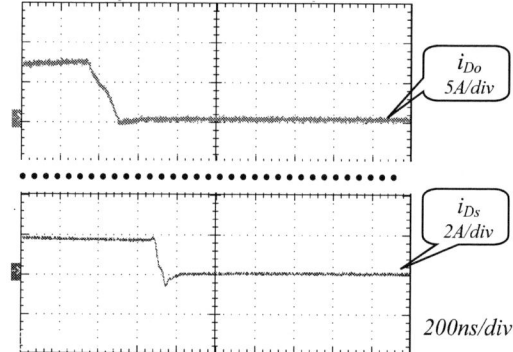

Figure 12. Detail current of the diodes

The performance of the clamp circuit is shown in Fig.6. When the main switch is turned off, the voltage of the main switch is clamped to that of the clamp capacitor by the clamp switch. Then the current through the leakage inductance flows

The voltages across the diodes are shown in Fig.10. The voltage stresses of both the diodes are equal to the output voltage, and no voltage spikes are added to the waveforms.

978-1-4244-4782-4/10 $26.00 © 2010 IEEE

The current though the diode is shown in Fig.11 and the details of the reverse recovery current marked in Fig.11 are shown in Fig.12. The leakage inductance limits the current falling rate and the reverse recovery problem is partly solved.

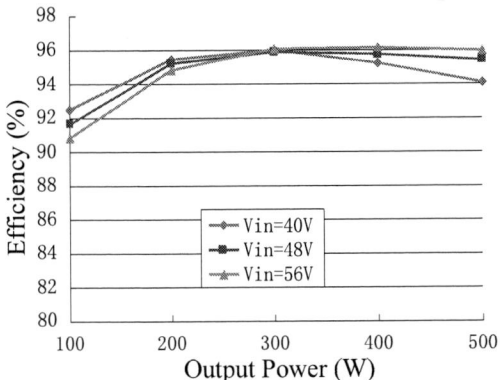

Figure 13. Efficiency of the converter

The efficiency of the prototype at different loads under different input voltages is sketched in Fig.13. The highest efficiency of the converter is 96.1%, and the efficiency is higher than 93% over a wide load range.

CONCLUSION

A novel converter for high step-up applications is presented in this paper. A coupled inductor and a switched capacitor are employed in the converter to extend the voltage gain and depress the voltage stress of the switches. Meanwhile, the problems caused by the leakage inductance are solved by the active clamp circuit. The ZVS condition is achieved for all the active switches, which reduces the switching losses greatly. The reverse-recovery problem is overcome due to the leakage inductance, which is also helpful to solve the EMI problem. The experimental results show that the proposed converter is a good candidate for high step-up applications.

REFERENCES

[1] C. E. A. Silva, R. P. T. Bascope and D. S. Oliveira, "Proposal of a New High Step-Up Converter for UPS Applications", In IEEE 2006 ISIE Conf., 2006, pp.1288-1292.

[2] S. B. Kjaer, J. K. Pedersen and F. Blaabjerg, "Power Inverter Topologies for Photovoltaic Modules-A Review", In Proc. IEEE 2002 IAS Conf., 2002, vol.2, pp.272-288.

[3] W. Li and X. He, "ZVT interleaved boost converters for high-efficiency, high step-up DC-DC conversion", IEE Proceedings Electric Power Applications, Volume.2, Issue.2, pp.284-290, Mar. 2007.

[4] A. A. Fardoun and E. H. Ismail, "Ultra step-up dc-dc converter with reduced switch", In IEEE 2008 ICSET., 2008, pp.425-430.

[5] Qun Zhao, Fengfeng Tao, Yongxuan Hu and Lee, F.C., "Active-clamp DC/DC converters using magnetic switches", In IEEE 2001 Applied Power Electronics Conference and Exposition, 2001, Volume.2, pp.946-952.

[6] T. F. Wu, Y. S. Lai, J. C. Hung and Y. M. Chen, "Boost Converter with Coupled Inductors and Buck-Boost Type of Active Clamp", In IEEE 2005 Power Electronics Specialists Conference, 2005, pp.309-405.

[7] Qun Zhao and F. C. Lee, "High-efficiency, high step-up DC-DC converters", IEEE Transactions on Power Electronics, Volume.18, Issue.1, part.1, pp.65-73, Jan. 2003.

[8] T. F. Wu, Y. S. Lai, J. C Hung, and Y. M. Chen, "An improved boost converter with coupled inductors and buck-boost type of active clamp", In IEEE 2005 IAS, 2005, Volume.1, pp.639-644.

[9] Dong Wang, Xiangning He and Rongxiang Zhao, "ZVT Interleaved Boost Converters with Built-In Voltage Doubler and Current Auto-Balance Characteristic", IEEE Transactions on Power Electronics, Volume.23, Issue.6, pp.2847-2854, Nov. 2008.

Unified Modulation for Three-phase Current-fed Bidirectional DC-DC Converter Under Varied Input Voltage

Zhan Wang，Hui Li

Center for Advanced Power System
Florida State University
Tallahassee, FL, 32310, USA
zwang@caps.fsu.edu, hli@caps.fsu.edu

Abstract -- In this paper, a new three-phase current-fed bidirectional dc-dc converter is proposed. The topology integrates a three-phase boost converter with a three-phase galvanically isolated bidirectional dc-dc converter. Keeping dc bus voltage constant allows high efficient energy conversion over a wide input voltage and maintains ZVS conditions in the whole operation range. By using compact small inductance dc inductors the ZVS conditions can be improved in low power range and the switching losses of the lower switches are reduced. At the same time the input current ripple can be alleviated by three-phase interleaved structure. Finally, a 6-kW hardware prototype is constructed and tested. The experimental results verify it can maintain high efficiency over wide input voltage and power range.

I. INTRODUCTION

The multiphase voltage-fed bidirectional isolated DC-DC converters have been reported in [1, 2]. They have the advantages of high efficiency and high power density for high current applications. However, they cannot achieve high efficiency at varied input voltages, therefore they are not suitable for energy storage elements such as ultracapacitors.

This paper proposes a three-phase current-fed bidirectional dc-dc converter. Compared to its voltage-fed counterparts, it can achieve zero-voltage switching (ZVS) under a varied input voltage with 50% variation. In addition, the interleaved structure can reduce input current ripple and output voltage ripple resulting in improved efficiency and life cycles for energy storage elements especially for batteries. The interleaved structure also reduces the size of components including inductors and capacitors to improve power density. Furthermore, the analysis of three-phase converter can be extended to n-phase (n >= 3) dc-dc converters.

In order to achieve ZVS operation under varied input voltage, duty cycle must be introduced to work with phase shift angle to regulate the power flow. The duty cycle and phase shift control leads to two operating modes with

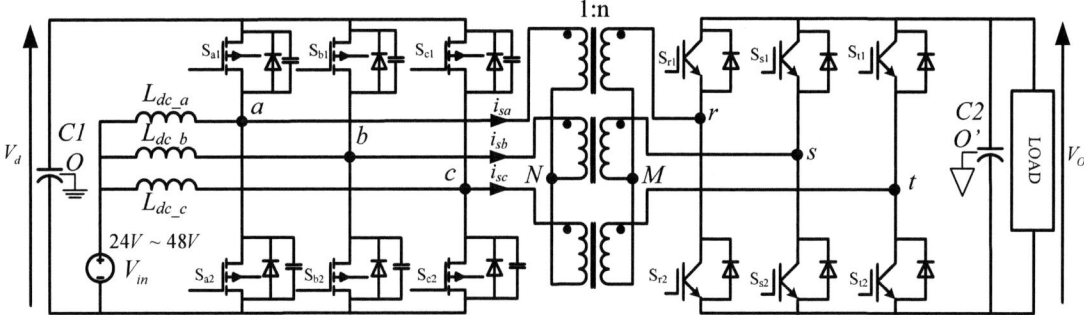

Fig.1. Proposed three-phase current-fed ZVS bidirectional dc/dc converter

---

*This work was supported by the National Science Foundation under Award Number ECCS-0641972.

different (ϕ, D) for the proposed dc-dc converter, which increases the power flow control complexity. The analysis shows there is small error by using one operating mode instead of two modes. Therefore a unified modulation is proposed to simplify the power flow control with enough accuracy. The design guidelines of key components of this converter are provided. Finally, the experimental results are presented to show the high efficiency operation over a varied input voltage.

II. TOPOLOGY AND OPERATION PRINCIPLE

A. Topology description

The topology of proposed converter is shown in Fig. 1, which integrates a three-phase boost converter with a three-phase galvanically isolated bidirectional dc-dc converter. The boost stage can maintain the dc bus voltage constant on primary side, so as to allow high efficiency conversion over wide varied input voltage source. A low voltage energy storage element such as battery or ultracapacitor can be connected in the input side. The high voltage side can be connected to the DC bus to provide power output to inverter for vehicle applications. Three single-phase high frequency transformers are connected in a Y/Y type to reduce circulating current, and the leakage inductance of each transformer is implemented by adjusting the distance between the primary and secondary windings. The three-phase DC inductors are implemented by three single planar inductors with low inductances.

B. Operating modes

Considering the input voltage with 50% variation, the duty cycle D is selected in the range of [1/3, 2/3]. In order to maintain high efficiency [3, 4], the phase shift angle ϕ should be no more than 15°. The operation principle with fixed duty cycle D = 1/3 have been analysed in [2]. Using the same method, the low voltage side output voltages, v_{aO}, v_{bO}, v_{cO} can be expressed by (1), (2) and (3), respectively:

$$v_{aO} = \begin{cases} V_d/2, & 0 < \omega t \le 2D\pi \\ -V_d/2, & 2D\pi < \omega t \le 2\pi \end{cases} \quad (1)$$

$$v_{bO} = \begin{cases} -V_d/2, & 0 < \omega t \le 2\pi/3 \\ V_d/2, & 2\pi/3 < \omega t \le 2(D+1/3)\pi \\ -V_d/2, & 2(D+1/3)\pi < \omega t \le 2\pi \end{cases} \quad (2)$$

$$v_{cO} = \begin{cases} V_d/2, & 0 < \omega t \le 2(D-1)\pi \\ -V_d/2, & 2(D-1)\pi < \omega t \le 4\pi/3 \\ V_d/2, & 4\pi/3 < \omega t \le 2\pi \end{cases} \quad (3)$$

Where V_d is primary side dc link voltage, and $\omega = 2\pi f$. Since

$$\begin{cases} V_{aO} = V_{aN} + V_{NO} \\ V_{bO} = V_{bN} + V_{NO} \\ V_{cO} = V_{cN} + V_{NO} \\ V_{aN} + V_{bN} + V_{cN} = 0 \end{cases} \quad (4)$$

the neutral point voltage and transformer voltage on primary side of phase a can be derived in (5) and (6). v_{bN}, v_{cN} are lagging v_{aN} 120°, 240° respectively. The voltages on the transformer primary side are derived in (1-6). The voltages on secondary side are derived considering having a ϕ lag in phase.

$$v_{NO} = \begin{cases} V_d/6, & 0 < \omega t \le 2(D-1/3)\pi \\ -V_d/6, & 2(D-1/3)\pi < \omega t \le 2\pi/3 \\ V_d/6, & 2\pi/3 < \omega t \le 2D\pi \\ -V_d/6, & 2D\pi < \omega t \le 4\pi/3 \\ V_d/6, & 4\pi/3 < \omega t \le 2(D+1/3)\pi \\ -V_d/6, & 2(D+1/3)\pi < \omega t \le 2\pi \end{cases} \quad (5)$$

$$v_{aN} = \begin{cases} V_d/3, & 0 < \omega t \le 2(D-1/3)\pi \\ V_d/3, & 2(D-1/3)\pi < \omega t \le 2\pi/3 \\ V_d/3, & 2\pi/3 < \omega t \le 2D\pi \\ -V_d/3, & 2D\pi < \omega t \le 4\pi/3 \\ -2V_d/3, & 4\pi/3 < \omega t \le 2(D+1/3)\pi \\ -V_d/3, & 2(D+1/3)\pi < \omega t \le 2\pi \end{cases} \quad (6)$$

In order to get minimum peak current [5], set $d = V_o'/V_d = 1$, where V_o' is the output voltage referred to the primary side. Considering the selected operating range D = [1/3, 2/3] and ϕ = [-15°, 15°], the converter is operated in two different modes, which are shown in Fig. 2. The key waveforms of each mode are shown in Fig. 3, where v_{aN} is the transformer primary side voltage of phase a and v_{rN} is transformer secondary side voltage of phase a. Specially, when D = 1/3 or 2/3, v_{NO} will become a dc voltage with value $-\pi/6$ or $\pi/6$ respectively, and the transformer phase voltages will change from 3 level voltages to 2 level voltages which are shown in (7) and (8):

$$v_{aN} = \begin{cases} 2V_d/3, & 0 < \omega t \le 2\pi/3 \\ -V_d/3, & 2\pi/3 < \omega t \le 2\pi \end{cases}, D = 1/3 \quad (7)$$

$$v_{aN} = \begin{cases} V_d/3, & 0 < \omega t \le 4\pi/3 \\ -2V_d/3, & 4\pi/3 < \omega t \le 2\pi \end{cases}, D = 2/3 \quad (8)$$

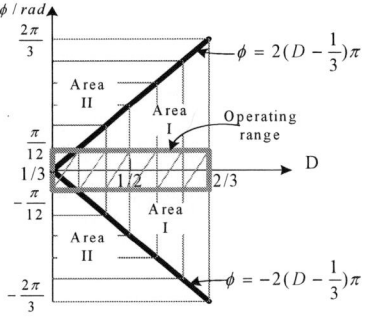

Fig. 2 Two Operating modes with different (ϕ,D)

The output power can be derived in (7) and shown in Fig. 4 using the methods discussed in [1-2, 6]. Since average current of leakage inductance i_{sa} in one switching period is zero:

$$\int_0^{2\pi} i_{sa}(\theta)d\omega t = 0 \tag{9}$$

$i_{sa}(0)$ can be derived by (9):

$$i_{sa}(0) = -\frac{V_{in}\phi}{3\omega L_s} \tag{10}$$

$i_{sa}(\theta)$ of each time interval can be obtained in terms of (10). According to the power equation

$$P_o = \frac{3}{2\pi}\int_0^{2\pi} v(\theta)i_{sa}(\theta)d\omega t \tag{11}$$

The total output power can be calculated in (12)

The operating range is the pink part shown in Fig.2. It is mainly located in area I and cover less than 1/10 of area II. In order to simplify the power flow control, P_1 can be approximately equal to P_2 under the defined operating range. Fig.5 shows the power flow difference between P_1 and P_2 using P_1/P percentage as a measuement where the maximum error is no more than 6% using P_1 to replace P_2 at the valid operating range.

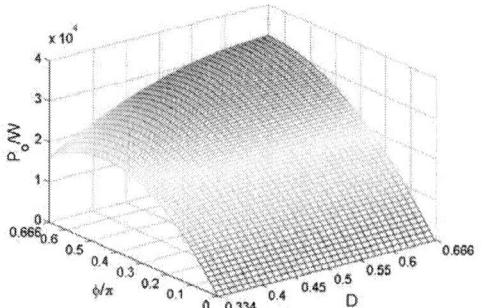

Fig. 4 The output power P_o versus (D,Φ)

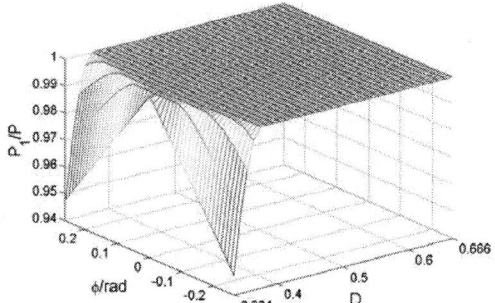

Fig. 5 The P_1/P at selected oprerating range

$$P_o = \begin{cases} P_1 = V_{in}^2\phi(4\pi-3\phi)/6\pi D^3\omega L_s & |\phi|\le 2\left(D-\frac{1}{3}\right)\pi, \frac{1}{3}\le D\le\frac{2}{3} \\ P_2 = V_{in}^2(-27\phi^3+12\pi\phi-36D^3\pi^2+24D\pi^2+36D\pi\phi-4\pi^2)/36\pi D^3\omega L_s & |\phi|> 2\left(D-\frac{1}{3}\right)\pi, \frac{1}{3}\le D\le\frac{2}{3} \end{cases} \tag{12}$$

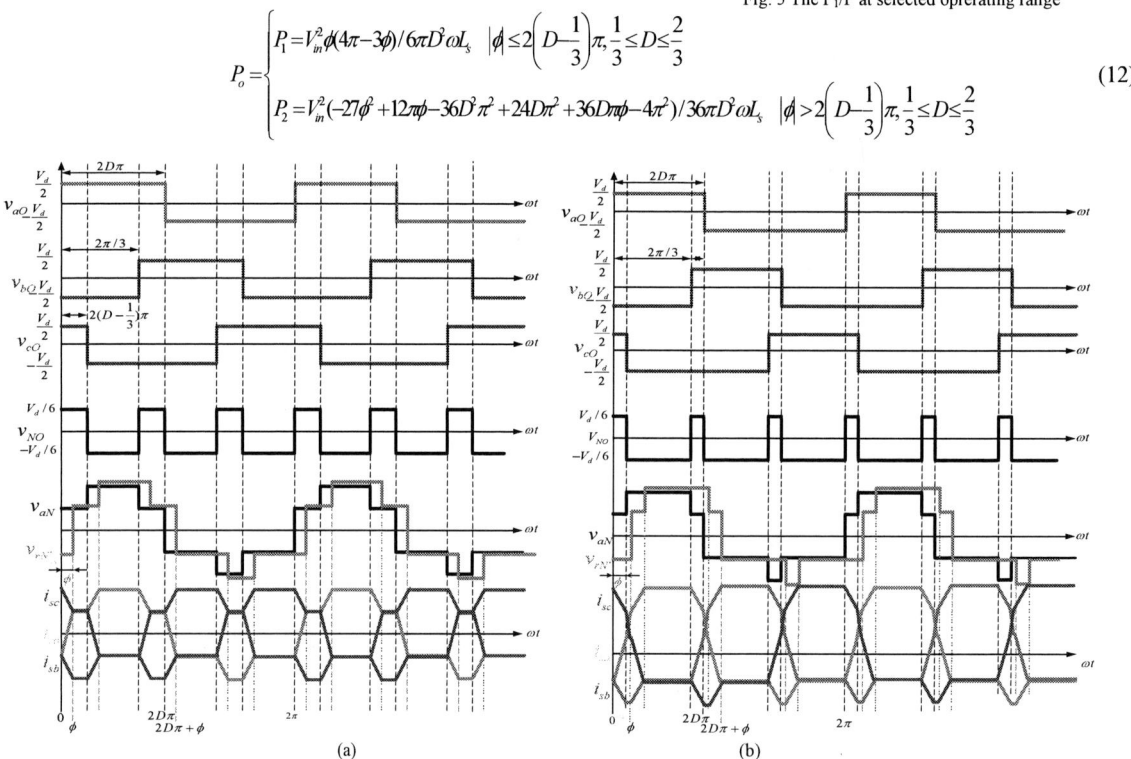

Fig. 3 The key waveforms of two modes: (a) Mode I where $\phi < 2(D-1/3)\pi$, and (b) Mode II where $\phi > 2(D-1/3)\pi$

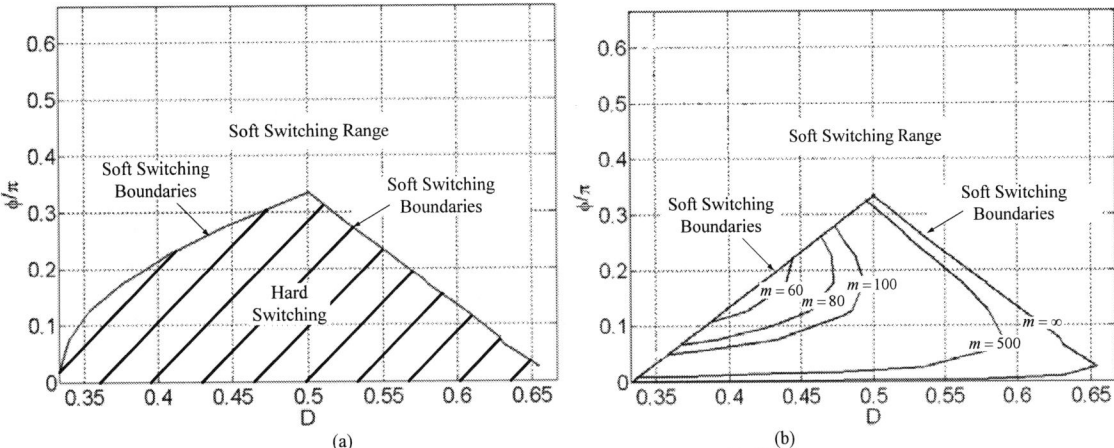

(a)　　　　　　　　(b)

Fig. 6 ZVS range versus (D,Φ) (a) with $m = \infty$, and (b) with different m

C. ZVS analysis

For the boost mode ($\phi > 0$), the ZVS conditions in phase a are:

$$\begin{cases} S_{a1} : I_{dc}(0) - I_{sa}(0) > 0 \\ S_{r1} : I_{sa}(\phi) > 0 \\ S_{a2} : I_{sa}(2D\pi) - I_{dc}(2D\pi) > 0 \\ S_{r2} : I_{sa}(2D\pi + \phi) < 0 \end{cases} \qquad (13)$$

In (13), the ZVS condition of primary lower switches S_{a2} is the most critical [6]. It can be rewritten in:

$$I_{sa}(2D\pi) \geq I_{dc\_avg} - \frac{1}{2}\Delta I_{dc} \qquad (14)$$

Where

$$\begin{cases} I_{sa}(2D\pi) = \frac{1}{9}V_{in}\frac{3\phi}{D\omega L_s} \\ I_{dc\_avg} = \frac{P_1}{3V_{in}} \qquad\qquad Mode\ I \\ \Delta I_{dc} = \frac{2\pi(1-D)V_{in}}{D\omega L_{dc}} \end{cases} \qquad (15)$$

$$\begin{cases} I_{sa}(2D\pi) = \frac{1}{9}V_{in}\frac{3\phi}{D\omega L_s} \\ I_{dc\_avg} = \frac{P_1}{3V_{in}} \qquad\qquad Mode\ II \\ \Delta I_{dc} = \frac{2\pi(1-D)V_{in}}{D\omega L_{dc}} \end{cases} \qquad (16)$$

It shows that ZVS condition also depends on the dc current ripple ΔI_{dc}. Define $m = L_{dc}/L_s$. Fig. 6(a) shows the ZVS cannot be maintained in the selected operation range if current ripple is zero, i.e. $m = \infty$. The ZVS range will increase as m increases. Fig. 6(b) gives the relationship between soft switching boundary and m. It can be seen that when $m < 55$, ZVS condition can be maintained in the whole operation range.

III. CONVERTER DESIGN GUIDELINES

A. Transformer design

A 6kW converter is designed with varying low input voltage (24V - 48V) and rated high output voltage 288V. The low side dc bus voltage $V_d = V_{in}/D$ is keeping constant. Therefore the transformer turns ratio is determined by $n = V_o/V_d = DV_o/V_{in} = 4$. The leakage inductance is integrated into the transformer. Comparing to traditional transformers, PCB planar transformers provide better consistency and flexibility. The transformer consists of 2 E64-3C92 planar cores with separate primary and secondary PCBs, and the desired leakage inductance can be controlled by adjusting the distance between the two PCBs. The single phase transformer prototype is shown in Fig. 7. The primary and secondary PCBs are separated by a thickness h_Δ and the leakage inductance for the non-interleaved structure is given in (17) [6]:

$$L_s = \mu_0 N^2 \frac{l_w}{b_w}\left(\frac{h_1 + h_2}{3} + h_\Delta\right) \qquad (17)$$

Where μ_0 is the permeability of air (H/m), N is turns ratio, l_w is the mean length of traces (m), and b_w is the width of primary trace (m). h_1 and h_2 are the primary and secondary side copper weight, and h_Δ is the distance between them.

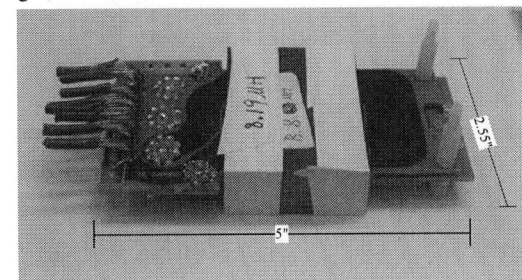

Fig. 7 Photo of planar transformer prototype

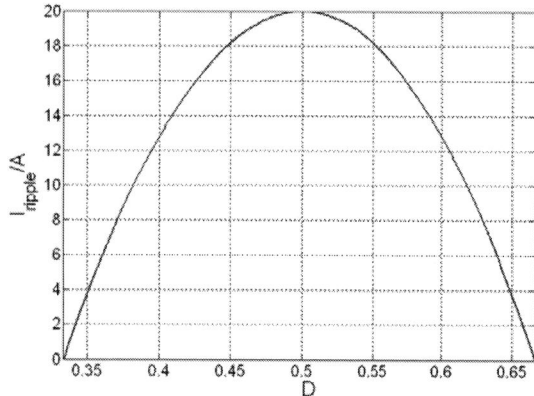

Fig. 8 Input current ripple vs duty cycle

B. DC inductor design

The key requirements of DC inductor design are to maintain the ZVS and achieve small size. The DC inductance should meet the inequation $L_{dc} < 55L_s$. In the experiment, the DC inductor is designed as $L_{dc} = 15L_s = 7.5uH$, so the phase current ripple ΔI_{dc} is as large as $2I_{dc}$ at rated power, and at the same time ZVS can be guaranteed. Thanks to the three-phase interleaved converter, the current ripples can be alleviated significantly [7]. The total current ripple is given as:

$$I_{ripple} = \left| \frac{3V_d(D-1/3)(D-2/3)}{fL_{dc}} \right| \qquad (17)$$

Fig. 8 shows the total input current ripple will reach zero at D = 1/3 and 2/3 but maximum at D = 1/2.

IV. EXPERIMENTAL VERIFICATION

A 6kW three-phase DC-DC converter was built in the laboratory. The input voltage varies from 24V to 48V, and output voltage is 288V. Fig. 9 shows a photo of 6-kW prototype with a liquid-cooled heat sink. The size of this converter is 13.7″ by 7.8″. The DC inductors and transformers are implemented with PCBs and 6 E64-3C92 planar cores. The DC inductor is 7.5uH and the leakage inductance of each transformer is 510nH referred to primary side.

Fig. 10 shows the transformer current with different duty cycle. The currents become 2 level waveforms in D=1/3, 2/3. Fig. 11 shows the DC inductor currents and total input current with D = 1/3, 1/2 and 2/3, respectively. The large dc inductor current ripple of each phase is benefit for device ZVS operation and inductor lower volume size but a total small current ripple can be achieved by interleaving three-phase currents. It can be seen that when D = 1/3 and 2/3, the input current ripple is minimum, and when D= 1/2, the current ripple reaches maximum value. When input voltage is fixed, duty cycle is selected as D = V_d/V_{in}. Fig. 12 shows the efficiency from 450W

to 4.5kW with different input voltage. The efficiencies in different input voltage keep stable, and the highest efficiency is 96.4%, which happens at V_{in} = 36V and P_o = 2.3kW.

Fig. 9 Photo of 6kW experimental prototype

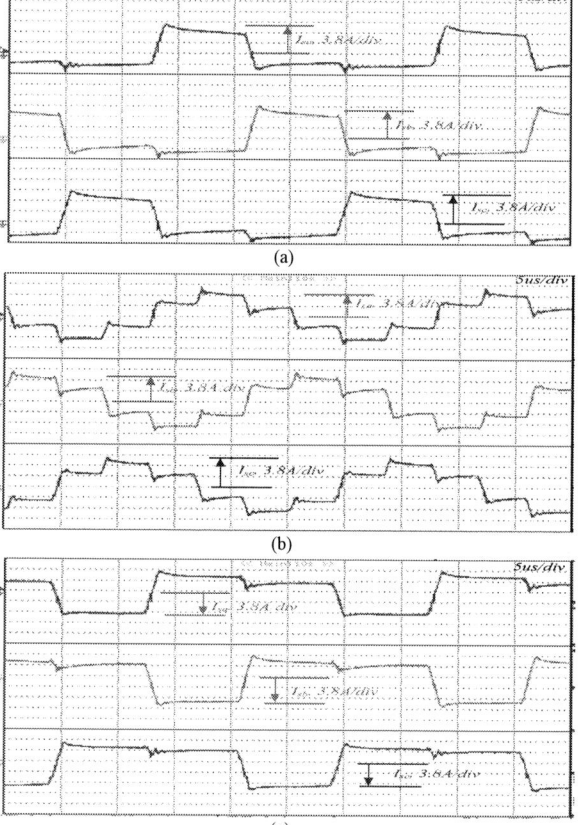

Fig.10 Transformer current waveforms on secondary side in different D, P_o = 3.2kW: (a) D = 1/3; (b) D = 1/2; (c) D = 2/3

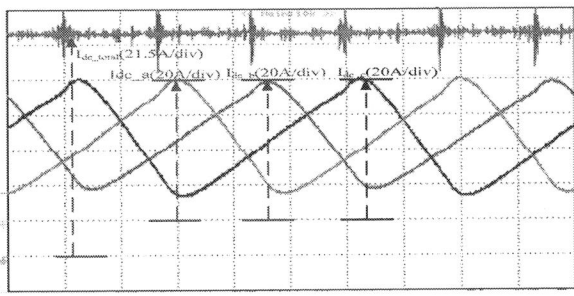

(a)

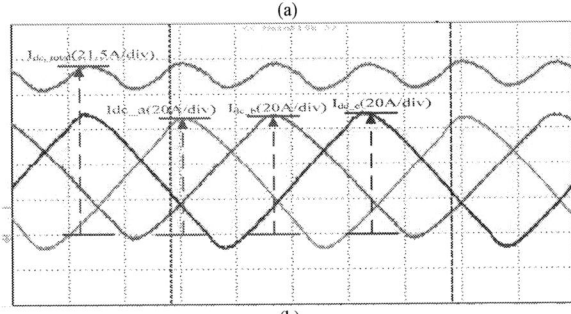

(b)

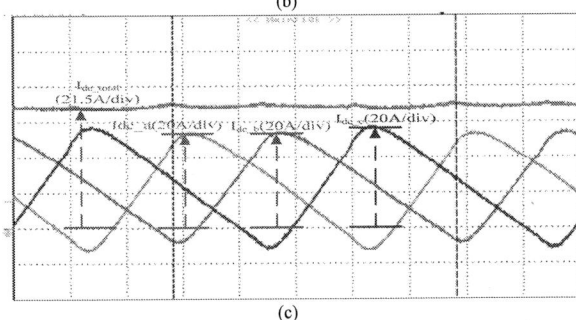

(c)

Fig. 11 Three phase inductor current and total input current at $P_o = 3.2kW$: (a) I_{dc} when D = 1/3, (b) I_{dc} when D = 1/2, (c) I_{dc} when D = 2/3

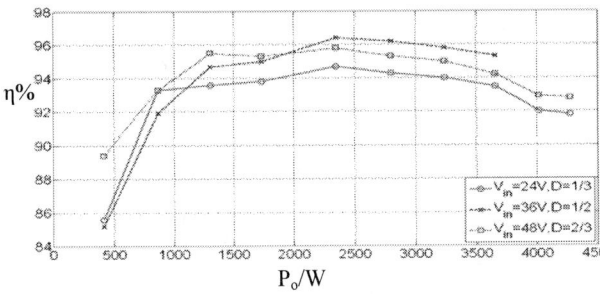

Fig. 12 Measured efficiency of proposed converter with V_{in}=24V, 36V and 48V

V. CONCLUSION

This paper presented a three-phase bidirectional isolated dc-dc converter therefore it can be used for energy storage elements. The converter operates in two modes when input voltage has 50% variation. Considering the specified (ϕ, D) range of the application, the power equation in mode I can be used to simplify the power flow analysis. Theoretical analysis shows that the maximum error between actual power and estimated power is no more than 6%. ZVS condition is maintained by generating large dc current ripple using small dc inductances and three-phase interleaved structure can make the three-phase current ripples cancelled each other. Three single PCB planar transformers achieve similar integrated leakage inductance and alleviate imbalance issue. Duty cycle regulation is integrated with phase shift control to keep the primary side dc voltage to be constant resulting in converter high efficiency in different input voltage. The 6-kW experimental prototype was built in the laboratory. The preliminary experimental results were derived to verify the high efficiency operation of proposed converter for energy storage element application.

REFERENCES

[1] Rik W. A. A. De Doncker, Deepakraj M. Divan, and Mustansir H. Kheraluwala, "A three-phase soft-switched high-power-density dc/dc converter for high-power applications," IEEE Transactions on Industry Applications, vol. 27, no. 1, pp. 63-73, Janurary/Februrary 1991.

[2] Gui-Jia Su, and Lixin Tang, "A three-phase bidirectional dc-dc converter for automotive applications," in Proc. of 2008 IEEE IAS Annual meeting (IAS'08), Oct. 5-9, 2008, Edmonton, Alberta, Canada, pp. 1-7.

[3] M. N. Gitaua, G. Ebersohna, and J. G. Kettleboroughb, "Power processor for interfacing battery storage system to 725v dc bus," Energy Conversion and Management, vol. 48, pp. 871-881, March 2007.

[4] N. Schibli, "Symmetrical multilevel converters with two quadrant dc-dc feeding," EPFL, Theses, 2000.

[5] Haimin Tao, A. Kotsopoulos, J. L. Duarte, and M. A. M. Hendrix, "Triple-half-bridge bidirectional converter controlled by phase shift and PWM", in Proc. of 2006 IEEE APEC Annual meeting, (APEC'2006) , March 7-9, Dallas, USA, pp. 1256-1262

[6] Fang Z. Peng, Hui Li, Gui-Jia Su and Jack S. Lawler, "A New ZVS Bidirectional DC-DC Converter for Fuel Cell and Battery Application," IEEE Transactions on Power Electronics, vol. 19, no. 1, pp. 54-65, Janurary 2004.

[7] Jeremy Ferrell, Jih-Sheng Lai, Troy Nergaard, Xudong Huang, Lizhi Zhu and Roy Davis, "The role of parasitic inductance in high-power planar transformer design and converter integration," in Proc. of 2004 IEEE APEC Annual meeting, (APEC'2004), February 22-26, 2004, Anaheim, California, pp. 510- 515 Vol.1.

[8] M. Gerber, J. A. Ferreira, I. W. Hofsajer, and N. Seliger, "Interleaving optimization in synchronous rectified dc/dc converter," in Proc. IEEE 35th Annual Power Electronics Specialists Conf. (PESC'2004), Aachen, Germany, pp. 4655-4661

Integrated Switched-Capacitor Voltage Doubler with Clock Transition Periods Boosting and Transfer Blocking Techniques

Phong Ngo and Dongsheng Ma

Integrated System Design Laboratory
Department of Electrical and Computer Engineering
The University of Arizona
Tucson, AZ 85721, USA
Email: {phong, dma}@email.arizona.edu

Abstract—**In this paper, a CMOS switched-capacitor voltage doubler is proposed. It employs the techniques of clock synchronization and charge transfer blocking to minimize the reversion loss. The clock transition period detection and boosting circuit modules allow continuous charge action to the output node, which significantly improves operation performances. The proposed voltage doubler was designed using IBM 180nm CMOS process, with a 1.2V supply voltage. Under no-load condition, it achieves 99.92% of ideal voltage level with 8mV voltage ripple, while consuming only 9.6μW of quiescent power. With a load ranging from 20kΩ to 200kΩ, the up-conversion ratio performs 45% better than the prior arts.**

I. INTRODUCTION

With the proliferation of battery-powered portable devices and the fast development of semiconductor industry, integrated power electronics attracts ever-increasing attentions in low power VLSI system designs. Recent trend in integrated power electronics has been focusing on the framework of deep sub-micrometer technologies and low power applications, where supply voltages are reduced significantly. Hence, a well-regulated on-chip power supply is critical to system stability and performance as well as battery run-time.

Nowadays, most of integrated power supplies are implemented with three different types of DC-DC conversion circuits: switching converter, switched-capacitor (**SC**) power converter (or charge pump for open-loop design) and linear regulator. For those applications that require on-chip voltage(s) higher than the supply voltage, such as LCD drivers, EEPROMs and electrostatic actuator drivers, boost or non-inverting flyback switching converter and SC power converter are usually the best solutions. However, due to large EMI noise introduced by the inductive components, switching converter is not preferred in noise-sensitive

designs. SC power converters become more desirable under these circumstances.

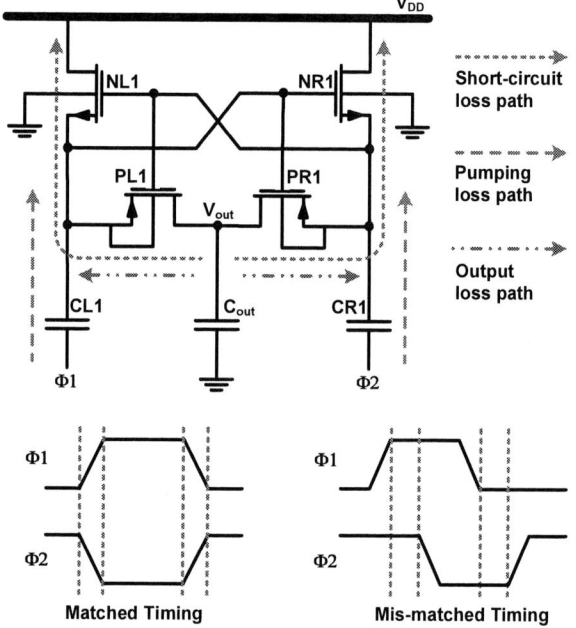

Figure 1. Conventional cross-coupled voltage doubler with the illustruation of reversion power loss.

Among different structures of SC power converters, cross-coupled voltage doublers are among the most commonly-used [1-6]. As illustrated in Fig. 1, it usually utilizes a pair of complementary control clock signals, Φ1 and Φ2, to simultaneously charge the output capacitor Cout while pre-charging the pumping capacitors, CL1 or CR1. This allows the doubler to operate at a doubled equivalent switching frequency compared to a Dickson charge pump [7]. Accordingly, output voltage ripples and filtering capacitor can be reduced. In addition, because the voltage drop across each turned-on power switch is equal to the drain-to-source voltage V_{ds} (<100 mV) rather than a threshold voltage V_{th} in a Dickson charge pump, the design is determined to be more efficient. Today, almost all on-chip charge pump designs are based on the cross-coupled voltage doubler topology.

However, this topology suffers from substantial reversion loss during clock transition period due to timing mismatch, which occurs between high voltage nodes and low voltage nodes in the power stage. The reversion loss usually consists of a pumping loss, an output loss and a short-circuit loss [8]. As depicted in Fig. 1, The pumping loss is caused by a reverse charge transfer from a boosting node to the input power source. The output loss is induced by a reverse charge transfer from the output of the converter to a boosting node. The short-circuit loss is invoked by a reverse charge transferred from the output of the converter to the input power source. In addition, an analysis of PMOS structure in Fig. 2, reveals that an adaption of using PMOS charge transfer transistors in Fig. 1 would suffer from significant leakage loss if the bulk potential is too low. As an effort to corresponding loss reduction, Favrat proposed a solution to maintain transistor bulk at higher voltage potential than drain and source terminals to avoid leakage loss [1]. The work in [2] reported an attempt to prevent V_{out} from being shorted to V_{DD} by using so-called "break-before-make" mechanism. Other topologies, such as in [6], propose separating pre-charge operation from charge transfer operation to avoid reversion loss.

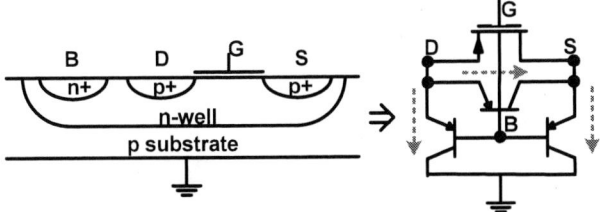

Figure 2.   Cross section and loss analysis of a PMOS transitor.

Motivated by above researches, this paper proposes clock synchronization and charge transfer blocking techniques to minimize the reversion power loss while introducing clock transition period detection and boosting circuit modules to improve up-conversion performance. The rest of this paper is organized as follows. Section II introduces the proposed voltage doubler circuit and elaborates its operation scheme. Design issues on key circuit blocks are also addressed in this section. Then fully transistor-based simulation results are provided in Section III to verify the design ideas. Finally, Section IV concludes this research effort.

II.   CIRCUIT STRUCTURE & OPERATION SCHEME

The transfer blocking technique was initially introduced in [8]. As illustrated in Fig. 3, two transfer blocking transistors, NL2 and NR2, gated by a locally boosted control signal TCO, are added in series with charge transfer transistors, NL3 and NR3. Pre-charge transistors, NL1 and NR1, are separately driven through auxiliary pre-charge circuit NL0 and NR0. During circuit operation, the control signal TCO would be activated when an overlapping of main boosting or auxiliary pre-charge clock signals is detected. TCO will then turns OFF NL2 and NR2 and blocks reverse charge transfer, resulting in no reversion loss.

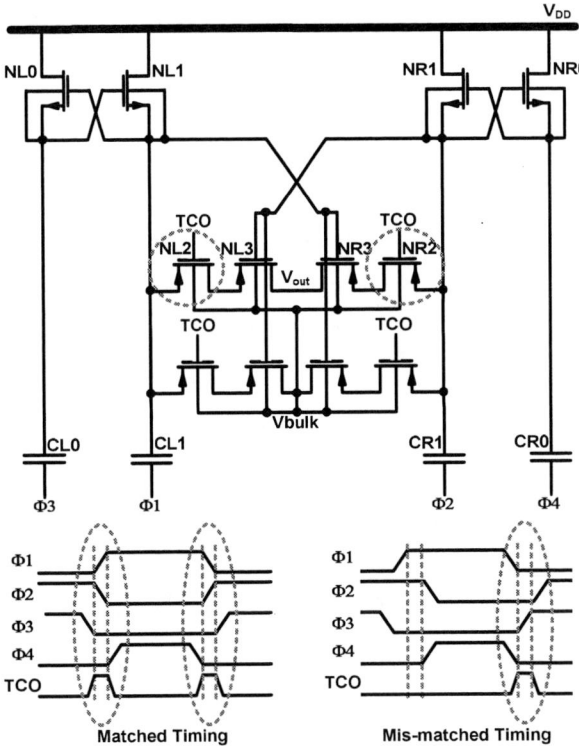

Figure 3.   Circuit schematic and timing diagram of the design in [8].

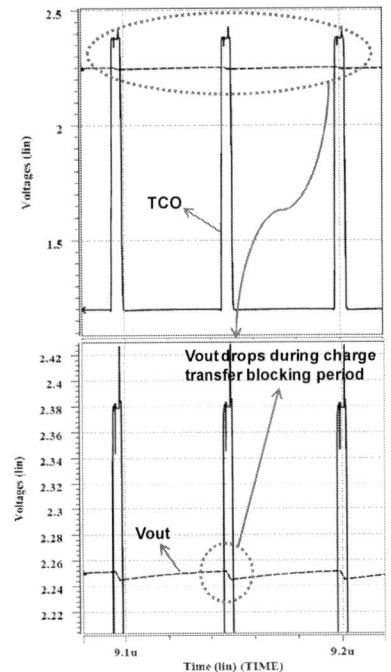

Figure 4.   Simulated V_{out} with reference to TCO in [8].

978-1-4244-4782-4/10 $26.00 © 2010 IEEE   814

Although the design was significantly improved compared to conventional works, it requires rather complicated control circuitry to monitor all clock signal transitions and relative mismatch conditions in order to produce transfer blocking control signal TCO. In addition, added auxiliary pre-charge circuit in order to allow relaxed timing restriction increase the complexity of overall circuit profile. Finally and most critical draw-back of [6] structure, during charge transfer blocking period, the disconnection between V_{out} and charge paths will cause significant voltage drop at V_{out}, as illustrated in Fig. 4.

In order to overcome these drawbacks, we proposed a new voltage doubler with circuit schematic and timing diagram in Fig. 5. Transfer blocking control signal TCO detect and activated during the transition periods of Φ1 and Φ2, as shown in Fig. 5. TCO1 is a duplicate of TCO which locally boosted to V_{DD}~$2V_{DD}$ in order to efficiently switch ON and OFF transfer blocking transistors NL2 and NR2. Similarly, TCO2 is complementary with TCO which is boosted to $2V_{DD}$ ~$3V_{DD}$ to effectively control the clock transition period boosting transistor PB1.

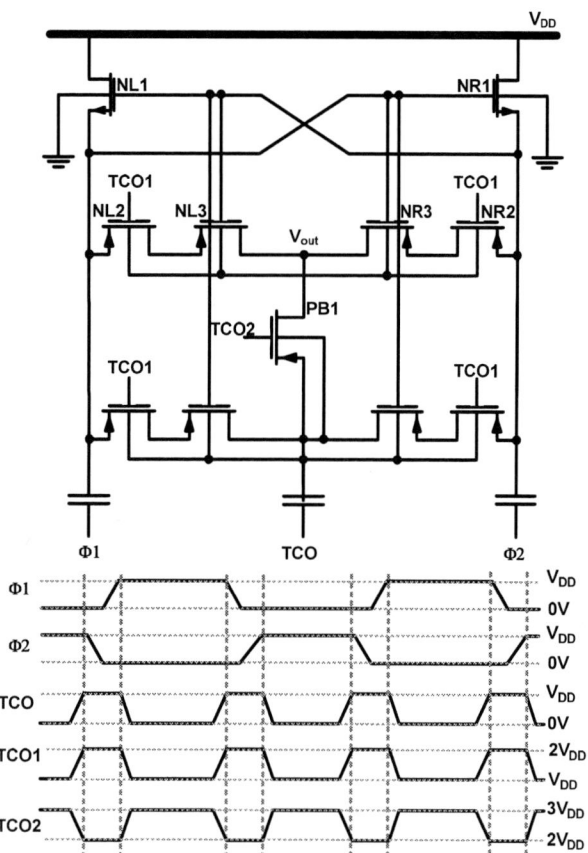

Figure 5. Circuit schematic and timing diagram of the proposed voltage doubler.

The proposed CMOS voltage doubler does not suffer from pumping loss due to charge flow through NL1 or NR1. Since Φ1 and Φ2 are ensured to be non-overlapping, NL1 or NR1, will not turn ON until NR3 or NL3 is completely OFF and vice versa. In addition, it has no output loss or short-

circuit loss since TCO1 would be activated during clock transition periods of Φ1 and Φ2, which turns OFF NL2 or NR2 to eliminate any reverse charge transfer.

Simple level shifter structure in Fig. 6 was utilized to generate desired boosted level of TCO correspondents. Since timing mismatch is critical to overall performance of voltage doubler, we proposed a power supply independent clock generator in Fig. 7. Integrated with independent clock generator, detection structure utilizes reference clock signal to provide a pair of synchronized and non-overlapping complementary clock control signals Φ1 and Φ2. Meanwhile, an integrated clock transition period detection provides a simple but effective generation of control signal TCO.

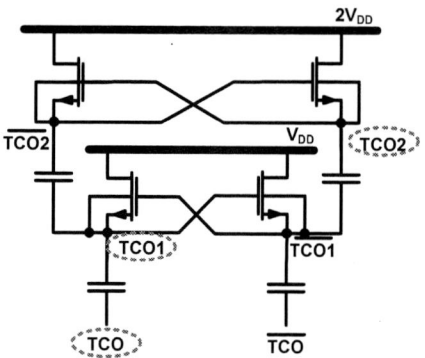

Figure 6. Circuit schematic of a level-shifter to generate boosted **TCO** correspondents.

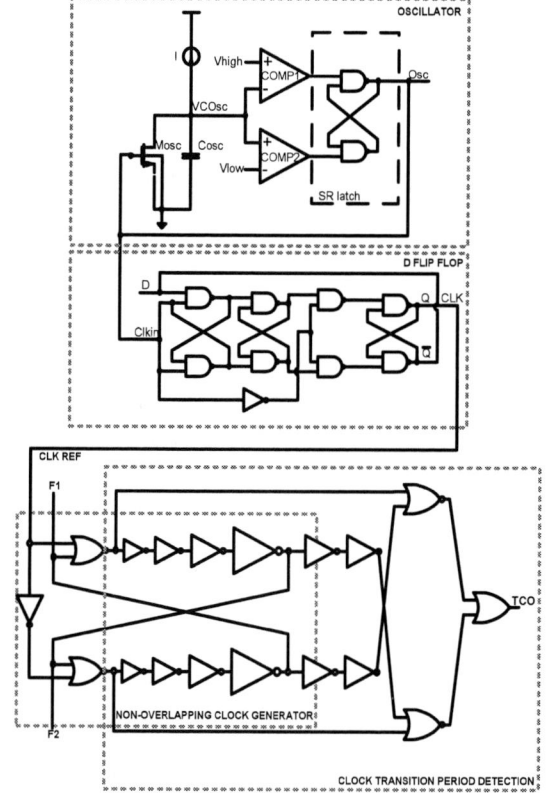

Figure 7. Circuit schematic of a proposed supply-independent control signal generator.

978-1-4244-4782-4/10 $26.00 © 2010 IEEE

In order to maintain the highest voltage potential of PMOS bulk to avoid leakage loss of PMOS structure, as shown in Fig. 5, the PMOS bulks are separated from V_{out}, but connected to the auxiliary boosting node, which drives no load and hence operates at higher voltage.

As shown in Fig. 8, during clocks transition period, TCO activated and raises Vbulk potential to 3VDD, meanwhile TCO2 turns ON boosting transistor PB1 and allows charging action from Vbulk to output node. In addition, TCO1 turns OFF NL2 and NR2 to block reverse charge transfer. The proposed voltage doubler control scheme allows continuous charge action at V_{out} throughout entire clock period with no reversion loss and hence significantly improves performance over prior arts.

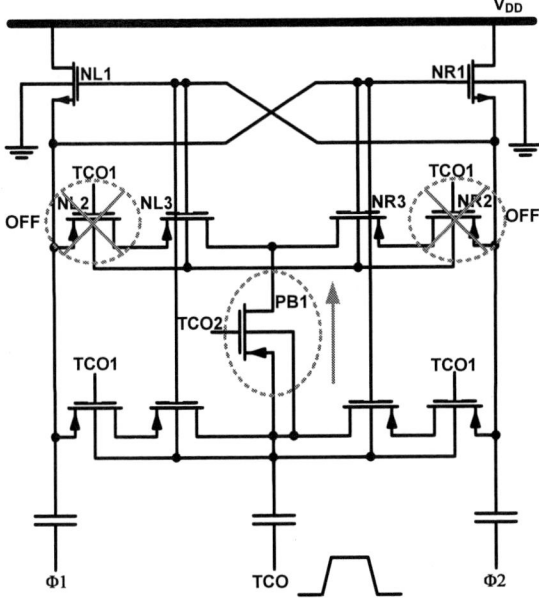

Figure 8.   Operation of the proposed design during clock transition periods.

III. SIMULATION RESULTS

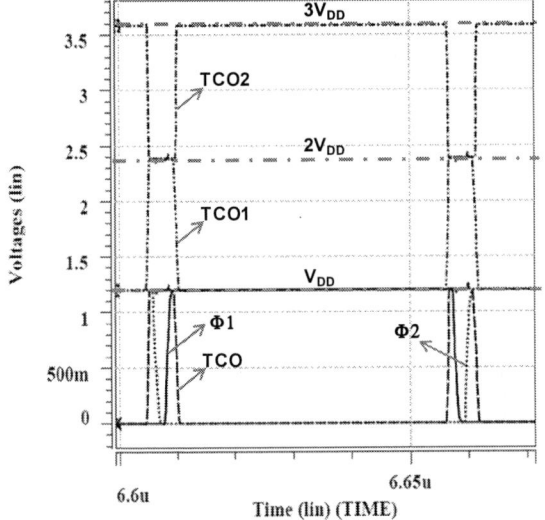

Figure 9.   Simulated non-overlapping clock & control signals.

In order to verify the performance and effectiveness of the proposed design, a series of CMOS voltage doublers have been designed and compared, using IBM 180nm CMOS process. These include a conventional voltage doubler [1], a CMOS charge pump with charge transfer blocking for no reversion loss [8] and the proposed work. All the designs are operated under a 1.2-V power supply and simulated in fully-transistor based HSPICE environment. Each voltage doubler is designed with main pumping capacitance of 40pF and auxiliary pumping capacitance of 10pF. Reservoir capacitor C_{out} is set to be 80pF at the output node V_{out}.

Fig. 9 demonstrates the successful generation of control signal TCO to effectively envelope the non-overlapping clocks Φ1 and Φ2 and its boosted correspondent TCO1 and TCO2.

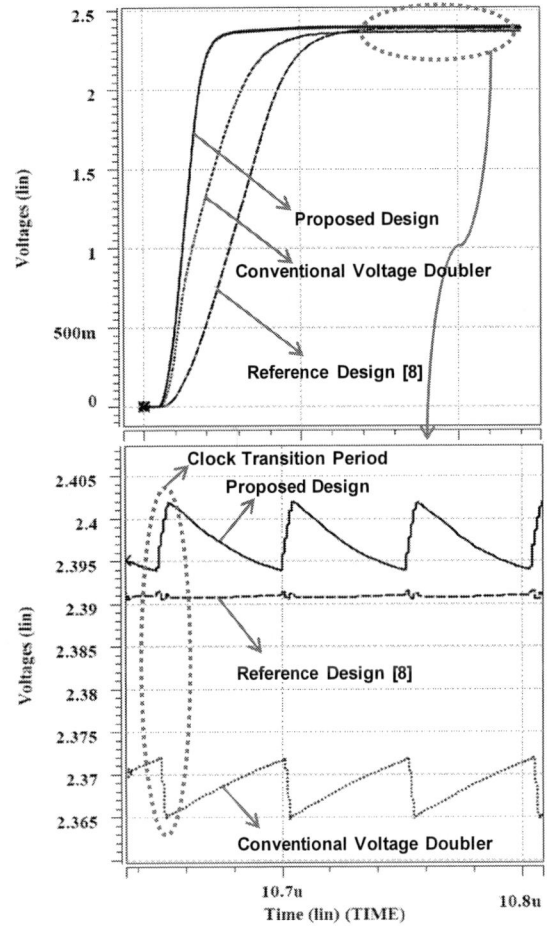

Figure 10. Performance comparison: (upper) in the case of start-up transient and (lower) in the case of the steady state.

Simulated output waveforms under no load condition are shown in Fig. 10. In the upper trace, the simulation proves that the proposed design achieves start-up time transient of less than 2μs, which is much faster than the prior arts. This is a result of clock transition detection and boosting mechanism. This scheme, as explained in Section 2, allows continuous charging action at output node throughout entire clock period. The lower trace of Fig. 10 demonstrates the effective employment of the charge path PB1, which the proposed design, preserves Vout more accurately than the other works. The proposed CMOS voltage doubler achieves highest output voltage level at approximately at 2.398V or 99.92% of ideal output level with only 8mV of voltage ripple.

Quiescent power consumption of proposed design was also examined for validation of efficiency. Under no-load condition, switched-capacitor output stage only consumes approximately 500nA. Meanwhile, control & clock signals generator circuitry utilizes approximately 7.5µA. Overall, proposed voltage doubler only consumes 8µA or 9.6µW of quiescent power.

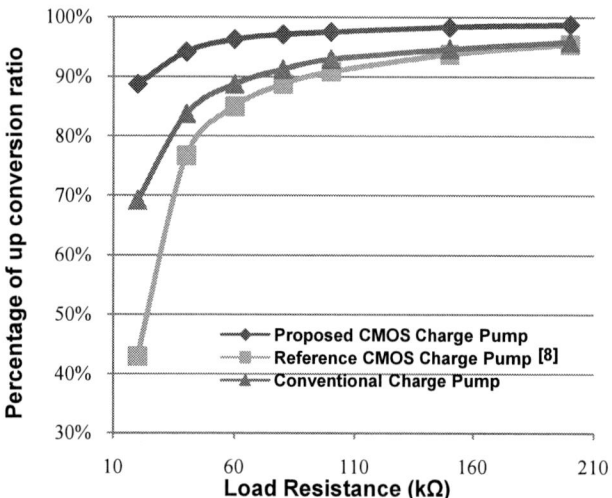

Figure 11. Performance comparison on conversion ratio versus load resistance .

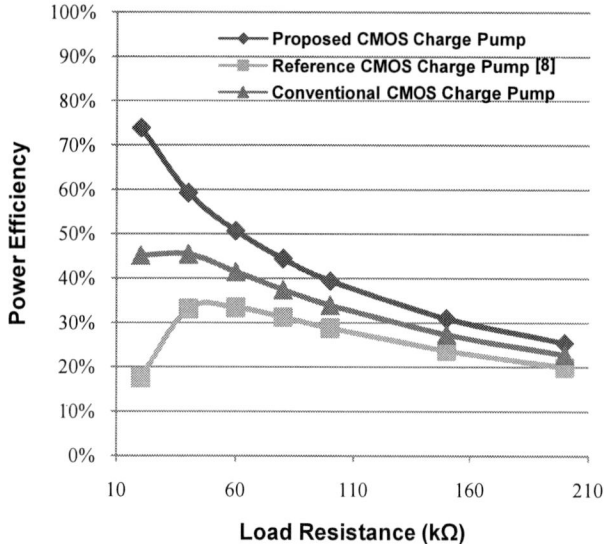

Figure 12. Performance comparison on efficiency.

As shown in Fig. 11 and Fig.12, as the output ressistive load varies from 20kΩ to 200kΩ, the proposed design maintains the highest up-conversion ratio and power efficiency performance. Under no-load condition, proposed design achieves 99.92% of ideal voltage level with 8mV voltage ripple, while consuming only 9.6µW of quiescent power. With a load ranging from 20kΩ to 200kΩ, the up-conversion ratio performs 45% better than that of the prior arts. These improvements of proposed CMOS voltage doubler over prior arts are due to adaption of charge transfer blocking technique to minimize reversion loss while clock transition period detection and boosting technique maintains output voltage at higher level.

IV. CONCLUSION

The paper introduced an integrated voltage doubler which utilizes clock synchronization and charge transfer blocking techniques to minimize reversion loss. Clock transition period detection and boosting circuit modules significantly improve the performance. The design exhibits less reversion loss, faster start-up time and maintain accurate conversion ratio.

REFERENCES

[1] P. Favrat, P. Deval, and M. J. Declercq, "A high-efficiency CMOS voltage doubler," *IEEE J. Solid-State Circuits*, Vol. 33, No. 3, pp.410–416, Mar. 1998.

[2] F. Su, W.-H. Ki, and C.-Y. Tsui, "High efficiency cross-coupled doubler with no reversion loss," *IEEE International Symposium on Circuit and System*, pp. 2761-2764, May 2006.

[3] N. S. Mohan, D. Ma, "Low-Ripple CMOS Switched-Capacitor Power Converter With Closed-Loop Interleaving Regulation", *IEEE Custom Integrated Circuits Conferences (CICC)*, pp. 245-248, Sept. 2006.

[4] L. Su, D. Ma, "Design and Optimization of Integrated Low-Voltage Low-Power Monolithic CMOS Charge Pumps," *IEEE International Symposium on Power Electronics, Electrical Drives, Automation and Motions (SPEEDAM)*, pp. 43-48, Ichita, Italy, June 2008.

[5] D. Ma, L. Su, M. Somasundaram, "Integrated Interleaving SC Power Converters with Analog and Digital Control Schemes for Energy-Efficient Microsystems," *Journal of Analog Integrated Circuits and Signal Processing*, accepted for publication.

[6] C. Falconi, G. Savone, and A. D'Amico, "High light-load efficiency charge pumps," *IEEE International Symposium on Circuit and* System, vol.2, pp. 1887–1890, May 2005.

[7] J. F. Dickson, "On-chip high-voltage generation in MNOS integrated circuits using an improved voltage multiplier technique," *IEEE J. Solid-State Circuits*, Vol. 11, No. 3, pp. 374–378, Jun. 1976.

[8] J-Y Kim, Y-H Jun, and B-S Kong; "CMOS Charge Pump With Transfer Blocking Technique for No Reversion Loss and Relaxed Clock Timing Restriction," *IEEE J. Circuit and System II*, vol. 56, no. 1, pp. 11-15, Jan. 2009.

[9] Y. Nakagome *et al.*, "An experimental 1.5 V 64 Mb dram," *IEEE J. Solid-State Circuits,"* vol. 26, pp. 465–472, Apr. 1991.

[10] T. B. Cho and P. R. Gray, "A 10-bit, 20 MS/s, 35 mW pipeline A/D converter," in *IEEE Custom Integrated Circuits Conf.*, 1994, pp. 499-502.

[11] P. E. Allen and D. R. Holberg, *CMOS Analog Circuit Design*. Philadelphia, PA: Saunders, 1987.

[12] G. van Steenwijk, K. Hoen, and H. Wallinga, "Analysis and design of a charge pump circuit for high output current applications," in *European Solid-State Circuits Conf.*, 1993, pp. 118–121.

[13] S. Sakiyama, J. Kajiwara, M. Kinoshita, K. Satomi, K. Ohtani, and A. Matsuzawa, "An on-chip high-efficiency and low-noise dc–dc converter using divided switches with current-control technique," in *IEEE Int. Solid-State Circuits Conf.*, Feb. 1999, pp. 156–157.

[14] S. J. Jou, S. H. Kuo, J. T. Chiu, and T. H. Lin, "Low switching noise and load-adaptive output buffer design techniques," *IEEE J. Solid-State Circuits*, vol. 36, no. 8, pp. 1239–1249, Aug. 2001.

A Novel Class of Multipulse Converters Based on High-frequency-operated Transformers

Sheng Zheng[1], Dong Chen[1], Hai Lin[1], Yousheng Wang[1], Zhaoming Qian[1], Fang Z. Peng[2]

1. College of Electrical Engineering, Zhejiang University, Hangzhou 31007, China
2. Dept. of Electrical and Computer Engineering, Michigan State University, East Lansing, MI 48824, USA
Email: zhengowen@zju.edu.cn

Abstract—This paper presents a novel class of multipulse converter based on high-frequency-operated multipulse transformers. Utility line currents are 12-pulse in nature with the cancellation of 5th and 7th harmonic components. The method demonstrates a considerable reduction in size and weight of the multipulse transformer, due to higher frequency operation. Furthermore, the magnetic components on the rectifier dc sides are also exposed to higher frequency and can be reduced in size and weight. The absence of bidirectional switches is also achieved by employing a modified modulation strategy. The proposed converter is suitable for medium-voltage rectifier systems, which demand clean input current. And the proposed converter employs modularity in design, which offers several benefits, such as reduction in cost and flexibility in design. This paper presents construction design, analysis, simulation and experiment results for a 3300 V, 20 kVA prototype system.

I. INTRODUCTION

High-power and high voltage rectifiers are very popular in industrial applications such as in traction substations, LV AC drives, MV drive etc[1]. Diode rectifiers, in virtue of being inherent ruggedness and simplicity, are attractive in these applications. However, they include several large lower harmonic component in the input currents and do not meet harmonic current content restrictions in several international standards. Subsequently multipulse converters are primarily employed to cancel several lower harmonic components in the input line current.

A 12-pulse configuration[2] is able to cancel the harmonic components such as the fifth and seventh harmonics, but the (12m+1)th harmonics still remain in the input currents. To further reduce the (12m+1)th harmonics, several strategies [3]–[5] were proposed for 12-pulse diode rectifiers. These methods, employing phase-shifting transformer operated at low frequency, cause several crucial problems, such as huge size and weight and great magnetic loss.

Other contribution over the years improved multipulse converters. Introducing an autoconnected transformer can significantly reduce the voltampere rating of transformers used in multipulse converters.[6][7] Later, a new class of multipulse rectifiers formed by autoconnected electronic transformers(ACET) was improved so that the proposed concept could enable ACET to be operated at higher frequency. Further reduction in size and weight could also be achieved.[8][9] But the appearance of bidirectional switches detracts from its simplicity and reliability.

This paper presents a novel class of multipulse converter with high-frequency-operated transformers, which use a modulation strategy to avoid the presence of bidirectional switches. The paper also presents theoretical investigations on the principle of the reduction in the 5th and 7th harmonic components, and clarifies the limitations and advantages of this scheme. Furthermore, the feasibility of these multipulse converters is verified by simulation and experimental results. Simulation and experimental results are also presented. The advantages of the proposed converter can be summarized as follows:

- Significant reduction in both size and weight of magnetic components due to higher frequency operation;

- Utility input current can be near sinusoidal due to the cancellation of 5th and 7th harmonic by the 12-pulse high-frequency-operated multipulse transformers, further cancellation of 11th and 13th harmonic can be achieved by the 24-pulse high-frequency-operated multipulse transformers;

- Low EMI can be achieved by the absence of switching frequency components in the utility input current;

- The controller design can be simplified due to the absence of bidirectional switches;

- The output DC voltage regulation can be achieved by changing the pulse-width;

- Modularity in design makes the process automation and total cost reduction being possible in the system.

II. PROPOSED 12-PULSE APPROACH

The proposed converter topology for a 3300V industry application is shown in Fig.1, it employs a high-frequency-operated multipulse transformer. This series scheme makes the system applicable to medium-voltage industry, and the number of power cells can easily be decided by voltage rating of both input grid and semiconductor devices.

High-frequency voltages, modulated by these power cells, are applied to primary windings independently. At the DC side, the 12-pulse series bridge rectifier circuit is equipped with a LC output filter in order to eliminate the higher frequency component (ripple frequency is 12 times the main frequency, 600Hz). Fig.2 shows the core construction of the phase-shifting transformer.

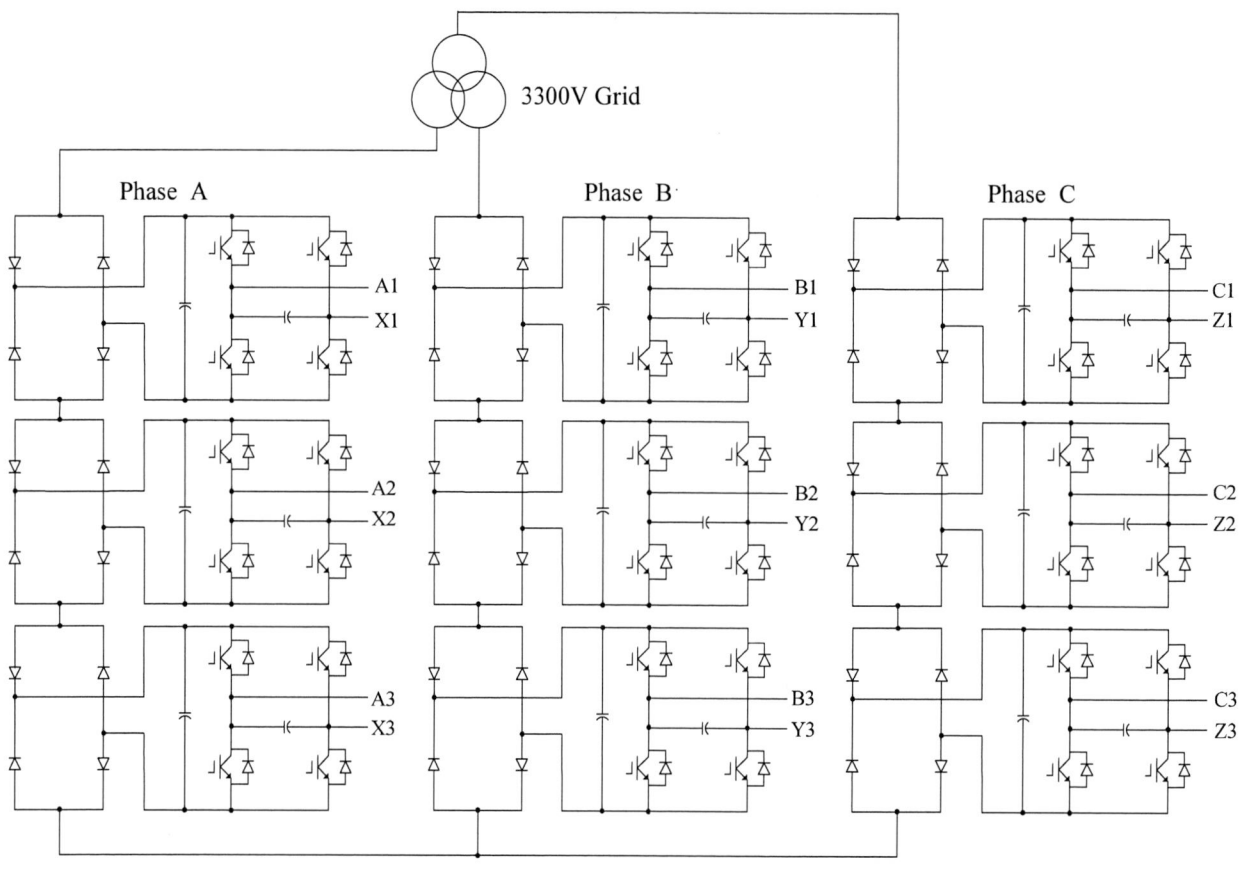

Figure 1. Circuit diagram of proposed multipulse converter based on high-frequency-operated transformer

A. Power Circuit

Three power cells are connected in series in each phase, and the group of power cells are WYE connected, with a floating neutral. Due to the same number of turns and the symmetrical arrangement in every single-phase transformer, all the output voltages of power cells can be naturally equal. Consequently equal input voltage sharing of power cells will be achieved. Instead of building a single power cell in each phase at medium voltage, the proposed approach can produce the required voltage ratings by combining several low-voltage power cells. Single power cells are consisted of a rectifier bridge and a full-bridge inverter, in addition, all these power devices can be rated at 1200V or 1700V, which would take advantage of the high manufacturing volume of low-voltage devices. If the voltage rating is elevated to 6600V, Fig.1 would be extended to have six power cells in series in each phase.

Neutral point is not needed to connect to the neutral line of the grid, due to the absence of zero-sequence current. Neglecting deadtime and synchronous operation of these power cells, the utility input current harmonics at the switching frequency can be cancelled by placing small line-to-line capacitors at the input, and EMI will be reduced to practical levels at the same time.

Fig.3 shows the vector diagram of Yy and Yz winding connections of the 12-pulse series bridge rectifier, in this transformer prototype, instead of connect as Δ connection, secondary windings connected as a $30°$ zigzag connection. Due to this $30°$ phase-shifting, the number of turns of both a_1x_1 and a_2x_2, a_3x_3 differ from each other by a factor of $\sqrt{3}$.

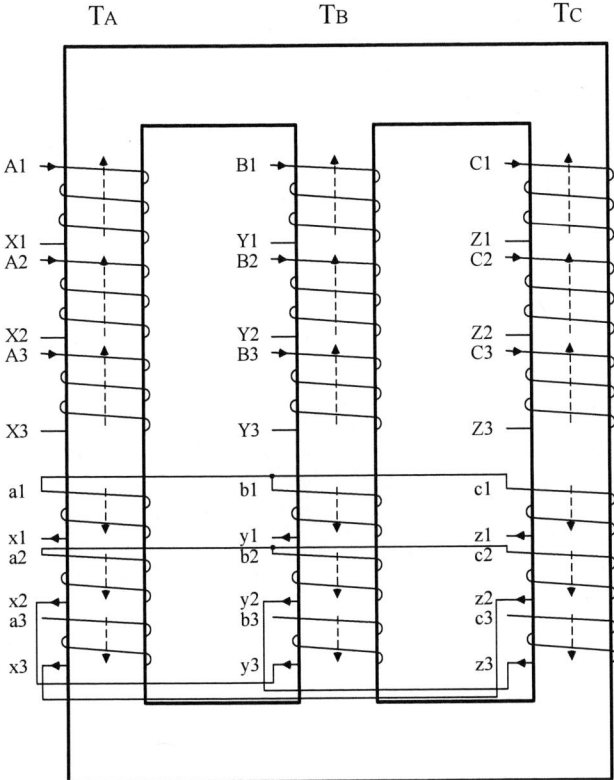

Figure 2. The core construction of the phase-shifting transformer

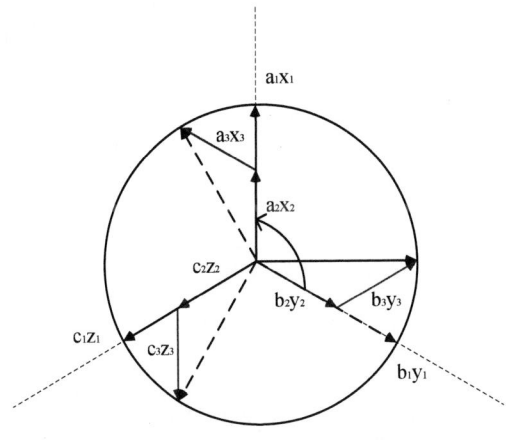

Figure 3. Vector diagram of Yy and Yz transformer winging connection

B. Modulation Strategy

The power cells all receive commands from one central controller. These commands are passed to the cells over fiber-optic cables in order to maintain the 3300V class isolation.

The input phase voltage needs to be specially modulated to feed the high-frequency transformer. If inverting the phase voltage directly, the utility of bidirectional switches can not be avoided. In the other way, phase voltages must be rectified

and then modulated by switching action, shown in Fig.4, the output waveform is also shown. The voltages applied to the transformer primary windings are essentially consisted of higher frequency components at the switching frequency and consequently induce higher frequency voltages across the secondary windings.

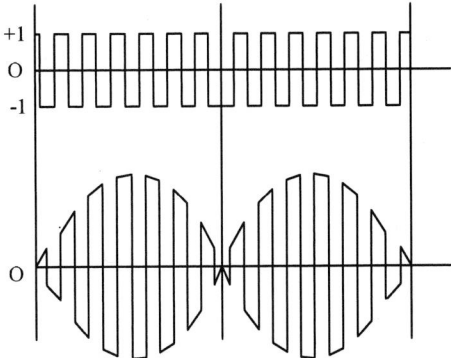

Figure 4. Switching action for high-frequency modulation circuit and primary winding voltage

III. ANALYSIS OF 12-PULSE HIGH-FREQUENCY-OPERATED RECTIFIER SYSTEM

The operation of the 12-pulse high-frequency-operated rectifier system is explained in the previous section. As section II has explained, secondary windings voltages of 12-pulse transformer which feed into the rectifier bridges, consist of a 100-Hz component along with superimposed higher frequency terms. So the rectifier input voltages, generated by Yy and Yz winding connections, are given by

$$
\begin{cases}
V_{a\_\sec} = \dfrac{N_2}{N_1} \cdot S_a(d) \cdot \sqrt{2} V_s \sin(\omega_s t) \\[2mm]
V_{b\_\sec} = \dfrac{N_2}{N_1} \cdot S_b(d) \cdot \sqrt{2} V_s \sin(\omega_s t - \dfrac{2\pi}{3}) \\[2mm]
V_{c\_\sec} = \dfrac{N_2}{N_1} \cdot S_c(d) \cdot \sqrt{2} V_s \sin(\omega_s t + \dfrac{2\pi}{3})
\end{cases}
\tag{1}
$$

And

$$
\begin{cases}
V_{a\_ter} = \dfrac{N_3}{N_1} \cdot S_a(d) \cdot \sqrt{2} V_s \sin(\omega_s t - \dfrac{\pi}{6}) \\[2mm]
V_{b\_ter} = \dfrac{N_3}{N_1} \cdot S_b(d) \cdot \sqrt{2} V_s \sin(\omega_s t - \dfrac{2\pi}{3} - \dfrac{\pi}{6}) \\[2mm]
V_{c\_ter} = \dfrac{N_3}{N_1} \cdot S_c(d) \cdot \sqrt{2} V_s \sin(\omega_s t + \dfrac{2\pi}{3} - \dfrac{\pi}{6})
\end{cases}
\tag{2}
$$

Similarly, the voltage of primary windings can be expressed as follows:

$$\begin{cases} V_{a\_pri} = S_a(d) \cdot \sqrt{2}V_s \sin(\omega_s t) \\ V_{b\_pri} = S_b(d) \cdot \sqrt{2}V_s \sin(\omega_s t - \dfrac{2\pi}{3}) \\ V_{c\_pri} = S_c(d) \cdot \sqrt{2}V_s \sin(\omega_s t + \dfrac{2\pi}{3}) \end{cases} \quad (3)$$

The switching function of power cells $S_a(d)$, $S_b(d)$, $S_c(d)$ assume an instantaneous voltage of ± 1, and is Quasi-PWM square wave in shape(Fig.4). When one phase voltage cross zero volt, the switching function of power cells in corresponding phase should be inverted. Therefore, the switching function of each phase can be expressed as

$$S_b(d) = S_a(d, \omega_s t - \frac{2\pi}{3}), S_c(d) = S_a(d, \omega_s t + \frac{2\pi}{3}) \quad (4)$$

Consequently, the superimposed higher frequency components in the rectifier input voltage modulate the rectifier bridges input currents into a Quasi-PWM shape. Suppose the inductor at the DC-side is sufficiently large in value, the inductor current is constant dc. To provide a path for the constant inductor to flow, at least two of the rectifier diodes must conduct at any given instant in time. So each rectifier input current is nonzero for 120 degrees out of each line half cycle and zero for the remaining 60 degrees. In this case, the rectifier input current will not be constant dc during the nonzero period, but modulated as an instantaneous current of $\pm Io$ per switching period.

IV. SIMULATION AND EXPERIMENTAL RESULTS

A 3300 V, 20 kVA multipulse converter based on high-frequency-operated transformers has been built up in the laboratory, and the switching frequency is set at 20 KHz. The simulation will be presented and selected results are also provided using the recorder Tek TDS3014B.

Fig.5(a) shows voltages applied to primary windings of the three single-phase transformers, from this figure, it is clear that these voltages consist of not only a 100-Hz component but also a higher frequency switching action component(20KHz).

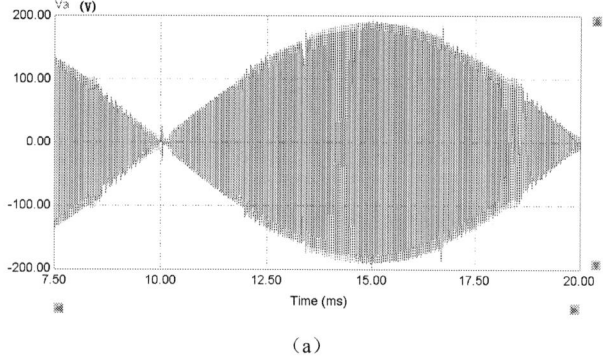

(a)

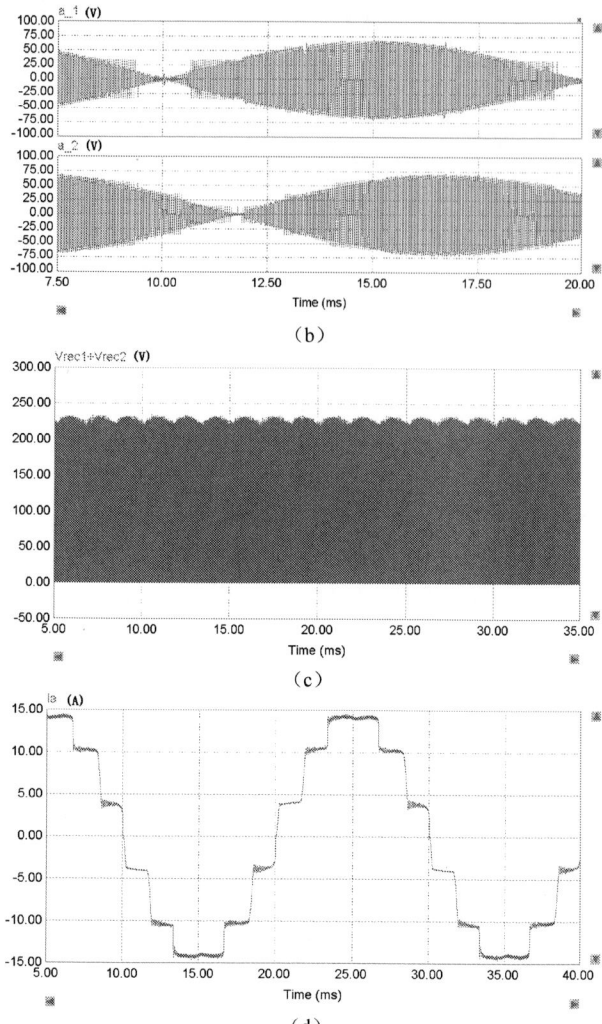

(b)

(c)

(d)

Figure 5. Simulation results for the proposed converter. (a). Single-phase inverters output waveform. (b). The Yy And Yz winding connection waveforms.(c).12-Pulse birdge rectifier output waveform. (d)Utility input current waveforms.

Following transformer winding connection shown in Fig.5, Y-Y and Y-Z winding connection generate the same voltage but with a 30° phase-shifting(Fig.5(b)). Fig.5(c) shows 12-pulse rectifier waveform, consisting of both 100-Hz component and higher frequency component. Consequently, a LC filter seems to be necessary. The most important simulation, shown in Fig.5(d), is utility input current waveform, a established 12-pulse input current.

Fig.6 shows the experimental results at rated load condition. The voltages induced in the secondary windings consist of a fundamental frequency component(100 Hz) and a higher frequency switching action component(20 kHz), the 30° phase-shifting has also been demonstrated in this figure. Another curve is presented as the utility input current. Fig.7 is the prototype of one power cell and the multipulse transformer.

978-1-4244-4782-4/10 $26.00 © 2010 IEEE   821

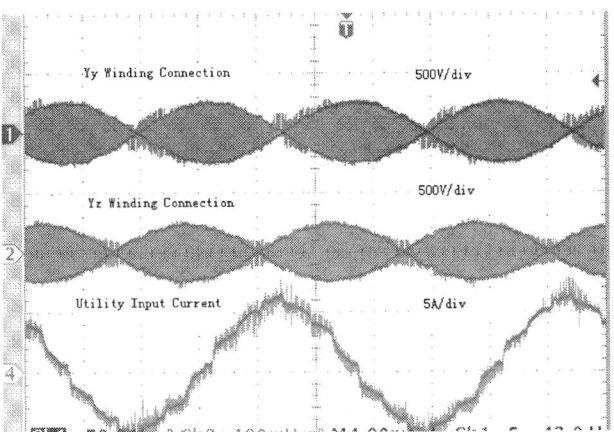

Figure 6. Experimental results of the proposed converter

V. CONCLUSION

In this paper, a novel class of multipulse converter based on high-frequency-operated multipulse transformers has been proposed. The absence of bidirectional switches is achieved by the proposed modulation strategy and the phase voltages are modulated to feed the high-frequency transformer. Due to the higher frequency operation, the prototype system verifies considerable reduction in size and weight of the multipulse transformer and the magnetic components on the rectifier dc sides. This paper also theoretically investigates the principle of the harmonic reduction in the utility input current, and how to realize modularity in design. Simulation and experimental results verify the clean input current characteristics of the 12-pulse rectifier system, with the cancellation of 5th and 7th harmonic components.

REFERENCES

[1] A. Siebert, A. Troedson, and S. Ebner, "AC to dc power conversion now and in the future," IEEE Trans. Ind. Appl., vol. IA-38, no. 4, pp.934–940, Jul./Aug. 2002.

[2] J. R. Rodriguez et al., "Large current rectifiers: state of the art and future trends," IEEE Trans. Ind. Electron., vol. 52, no. 3, pp. 738–746,Jun. 2005.

[3] S. Choi, P. Enjeti, H. Lee, and I. Pitel, "A new active interphase reactor for 12-pulse rectifiers provides clean power utility interface," IEEE Trans. Ind. Appl., vol. 32, no. 6, pp. 1304–1311, Nov./Dec. 1996.

[4] M.Villablanca, J. Nadal, and M. Bravo, "A12-pulse ac-dc rectifier with high-quality input/output waveforms," IEEE Trans. Power Electron.,vol. 22, no. 5, pp. 1875–1881, Sep. 2007.

[5] Shoji Fukuda, Masaaki Ohta, "An Auxiliary-Supply-Assisted Harmonic Reduction Scheme for 12-Pulse Diode Rectifiers," IEEE Trans. Power Electron.,vol. 23, no. 3, pp. 1270–1277, May. 2008.

[6] S. Choi, B. Lee, and P. Enjeti, "New 24-pulse diode rectifier systems for utility interface of high-power AC motor drives," IEEE Trans. Ind. Appl., vol. 33, pp. 531–541, Mar./Apr. 1997.

[7] S. Choi, P. Enjeti, and I. Pitel, "New polyphase transformer arrangements with reduced KVA capacities for harmonic current reduction in rectifier type utility interface," IEEE Trans. Power Electron., vol. 11, pp. 680–690, Sept. 1996.

[8] K. Harada, H. Sakamoto, and M. Syoyama, "Phase controlled DC-AC converter with high-frequency switching," IEEE Trans. Power Electron., vol. 3, pp. 406–411, July 1988.

[9] Moonshik Kang, B. O. Woo, P. Enjeti, and Ira J. Pitel, "Autoconnected-Electronic-Transformer-Based Multipulse Rectifiers for Utility Interface of Power Electronic Systems," IEEE Trans. Ind. Appl., vol. 35, no. 3, pp. 646–656, May./June. 1999.

(a)

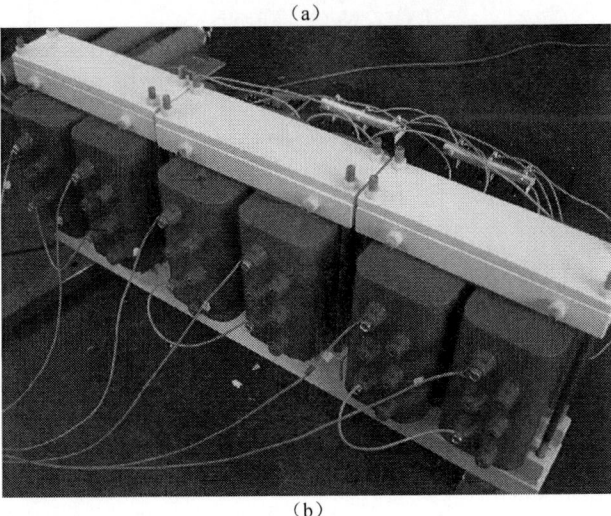

(b)

Figure 7. The prototype of one power cell module and the multipulse transformer.

A High Efficiency Flyback Converter with New Active Clamp Technique

Xiucheng Huang, Weijing Du, Wei Yuan, Junming Zhang, Zhaoming Qian

College of Electrical Engineering, Zhejiang University
38 Zheda Road, Hangzhou, China 310027
E-mail: zhangjm@zju.edu.cn

Abstract-This paper proposes a Flyback converter with a non-complementary active clamp method. With the proposed control method, the energy in the leakage inductance can be fully recycled and the soft switching is achieved for the main switch. Compared to the conventional active clamp technique, the proposed method features high efficiency on both heavy load and light load conditions and the efficiency is almost not affected by the leakage inductance. The detailed operation principle and design considerations are presented. Performance of the proposed method is validated by the experimental results from a 16V/4A prototype.

I. INTRODUCTION

With more and more emphasis on the environment protection and energy saving, the efficiency and standby power loss of the power supply are much concerned. The average efficiency instead of the full load efficiency is more important for external power supply. Therefore, both the light load efficiency and full load efficiency need to be carefully considered, which creates a new challenge for the power supply design.

Flyback converters are widely adopted for low power applications due to its simplicity and low cost. The leakage inductance of the power transformer causes high power loss and voltage spikes due to the parasitic resonance between the leakage inductance and the parasitic capacitances of the primary switch. Usually, a dissipative clamp circuit is necessary to dissipate the leakage energy and suppress the voltage spike. How to further improve the efficiency of a Flyback converter still remains as a challenge.

The lossless snubber for single ended converter was proposed to recycle the leakage energy, but the difficulties in parameters optimization and the large circulating energy limit the efficiency improvement [1]. The active clamp Flyback converter can recycle the leakage energy and achieve soft switching for both main and auxiliary switches [2~8]. Although the efficiency is high at full load condition, it is sensitive to the leakage inductance and clamp capacitance variation. Furthermore, the conventional complementary gate signal and constant frequency control methods result in poor efficiency at light load condition due to the conduction loss of its resonant circulation current, which also leads to lower average efficiency. Other topologies use half-bridge structure to absorb the leakage energy, such as AHB (asymmetrical half-bridge), asymmetrical Flyback or LLC converter, but most of them are not suitable for wide input range application [9~12].

This paper proposes a Flyback converter with a non-complementary active clamp circuit to achieve soft switching and high efficiency in whole load range. The power stage shown in Fig. 1 is the same as the conventional active-clamp circuit, but the gate signals are thoroughly different. In the proposed control method, the auxiliary switch is turned on for a short time before the main switch is turned on, which can be adopted in variable-frequency control to greatly improve efficiency at light load condition.

The detailed operation principle will be illustrated in section II. The design considerations are given in section III. Section IV will present the detailed experimental results from a 64W (16V/4A) prototype with universal AC input. A comparison of the experimental results with the other control scheme is also provided to validate the advantages of the Flyback converter with proposed non-complementary control signal.

II. OPERATION PRINCIPLES

Fig.1 shows the circuit configuration of the Flyback converter with a non-complementary active clamp circuit. L_m is the transformer magnetizing inductance. L_k is the transformer leakage inductance and C_C is the clamp capacitor. S_W is the primary main switch, and D_R is the output rectifier diode. Auxiliary switch S_a can be a NMOS or PMOS as shown in Fig 1. C_{oss} is the equivalent parasitic capacitance of S_W, S_a and the parasitic winding capacitance of the transformer. The transformer turns ratio is N.

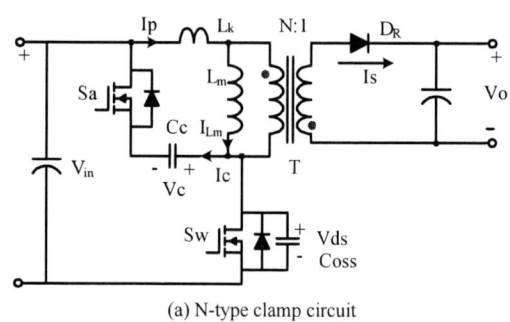

(a) N-type clamp circuit

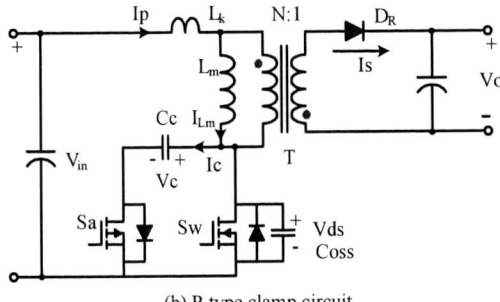

(b) P-type clamp circuit

Fig.1. Topology of the proposed active clamp Flyback converter

To simplify analysis of the steady state circuit operation, the clamp voltage V_C in either type of clamp circuit is assumed to be constant. The N-type in DCM operation is used as an example, and the steady state waveform and equivalent circuit are shown in Fig.2 and Fig.3 respectively. Each operation mode is simply described below.

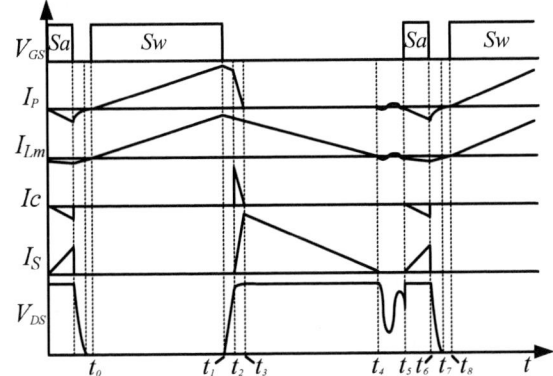

Fig. 2. Steady state waveforms with proposed control method

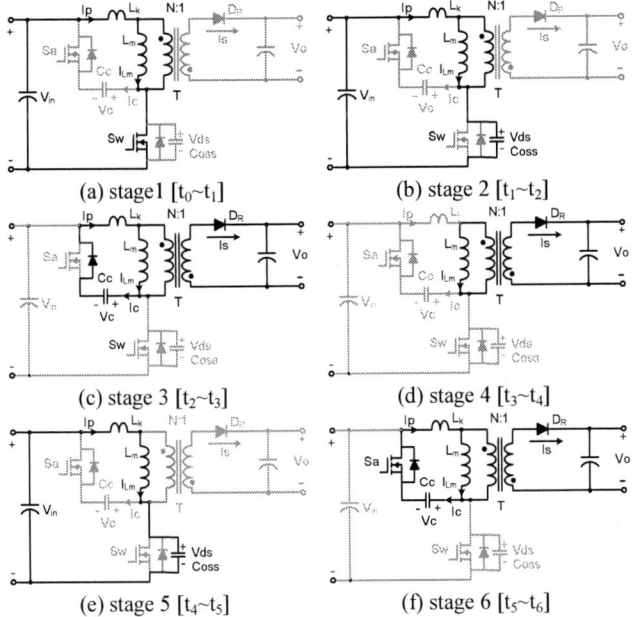

(a) stage1 [t_0~t_1]

(b) stage 2 [t_1~t_2]

(c) stage 3 [t_2~t_3]

(d) stage 4 [t_3~t_4]

(e) stage 5 [t_4~t_5]

(f) stage 6 [t_5~t_6]

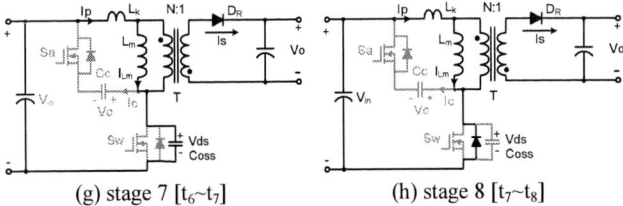

(g) stage 7 [t_6~t_7]

(h) stage 8 [t_7~t_8]

Fig. 3. Equivalent circuits in steady state operation

a) Mode 1 [t_0-t_1]

In this mode, primary side switch S_W is on and the auxiliary switch S_a is off. The energy is stored to the magnetizing inductor and the primary side current I_p increases linearly, which is the same as the conventional Flyback converter.

b) Mode 2 [t_1-t_2]

At t_1, when S_W turns off, C_{oss} is charged up by the magnetizing current. Due to relative large magnetizing inductance, the voltage V_{ds} is increased linearly.

c) Mode3 [t_2-t_3]

There are two possible situations in this mode. Suppose m is the ratio of the leakage inductance and the magnetizing inductance. If clamp voltage V_C is larger than $(1+m)NV_O$, then the secondary side rectifier D_R is turned on at t_2 when the voltage V_{ds} reaches $V_{in}+(1+m)NV_O$. The anti-paralleled diode of S_a turns on later until the voltage across the capacitor C_{oss} reaches to $V_{in}+V_C$ which is charged by L_k. If V_C is smaller than $(1+m)NV_O$, then anti-paralleled diode of S_a turns on first, and clamp voltage is charged by both L_m and L_k. D_R turns on once the voltage across L_m reaches $-NV_O$. During this mode, the difference between the magnetizing current and primary current is delivered to secondary side. And the energy in the leakage inductance is absorbed by the clamp capacitor. This mode can be treated as a primary to secondary commutation period. As soon as the current in the leakage inductance reaches zero, this mode is finished. Due to the large clamp capacitor, there is no parasitic ring or voltage spike, which will reduce the EMI noise and the voltage rating of S_W.

d) Mode 4 [t_3-t_4]

At t_3, the current through leakage inductance is zero and the anti-paralleled diode of S_a is off. The magnetizing energy is delivered to the load as conventional Flyback converter and the magnetizing current decreases linearly.

e) Mode 5 [t_4-t_5]

At t_4, magnetizing current decreased to zero, and D_R turns off. A parasitic resonance occurs between L_m and C_{oss} as conventional Flyback at DCM condition.

f) Mode 6 [t_5-t_6]

At t_5, auxiliary switch S_a is turned on. The voltage across the L_m and L_k is clamped to $-V_C$, and secondary winding is forward biased, so D_R is on. The current through L_k increases reversely. The magnetizing current increases reversely too, but the magnitude may be smaller than the leakage current. These negative current is used to achieve ZVS of main switch S_W. The absorbed leakage energy in Mode 3 is transferred to the output side and the leakage inductance again.

g) Mode 7 [t_6-t_7]

At t_6, the auxiliary switch S_a turns off. The negative current I_p discharges the parasitic capacitor C_{oss}. As soon as the leakage inductor current I_p reaches I_{Lm}, the secondary D_R is off, and both L_m and L_k discharge C_{oss}.

h) Mode 8 [t_7-t_8]

At t_7, the output capacitor C_{oss} voltage decreased to zero and the anti-parallel diode of main switch S_W turns on. The primary side switch S_W should be turned on before the primary current I_p changes the polarity.

Based on the description above, the proposed circuit can be applied to any control scheme to recycle the leakage energy, such as constant frequency or variable frequency.

III. DESIGN CONSIDERATIONS

We simply assume that the clamp voltage almost constant in the steady state operation, which is usually true with a relative large clamp capacitor. Also, the resonant period for the leakage inductance L_k and clamp capacitor C_C is much longer than the auxiliary switch S_a on time T_a, which means the charge and discharge of leakage current can be simply assumed to be linearly.

The clamp voltage is self-balanced based on the charge balance or energy balance, which means that the charge into the clamp capacitor during Mode 3 should be equal to the charge out of the clamp capacitor during Mode 6 for balancing. Thus, the primary current I_p should be satisfied:

$$i_p(t_2) = i_p(t_6) = I_{p\text{-}neg} = I_{pk} \qquad (1)$$

I_{pk} and $I_{p\text{-}neg}$ represents the positive and negative peak value of the primary side current.

And the current commutation time T_{com} also equals to the auxiliary switch on time T_a. Thus, the clamp voltage can be expressed as

$$V_c = N \cdot V_o + \frac{L_k \cdot I_{pk}}{T_a} \qquad (2)$$

Based on the previous analysis and assumption, we can discuss the parameters and control method selection for performance optimization.

1) ZVS Operation Range

During the switching transient, i.e. Mode 7, more detailed waveforms is shown in Fig 4. In this mode, the leakage inductor current I_p increases reversely. If the leakage energy E_{Lk} is larger than the parasitic capacitor energy E_{Coss}, the ZVS operation of primary switch S_W can be achieved easily.

$$E_{Lk} = \frac{1}{2} \cdot L_k \cdot I_{p\text{-}neg}^2 \geq E_{Coss} = \frac{1}{2} \cdot C_{oss} \cdot (V_{in} + N \cdot V_o)^2 \qquad (3)$$

I_{p_neg} is given in (1) and the value is related to the load condition. If $E_{Lk} > E_{Coss}$. Once V_{ds} reaches zero, the leakage current drops fast due to the voltage applied to L_k becomes $V_{in} + NV_O$. When leakage current I_p reaches the magnetizing current I_{Lm}, both of them increase linearly because the voltage is clamped to V_{in} as shown in Mode 8 of Fig. 3.

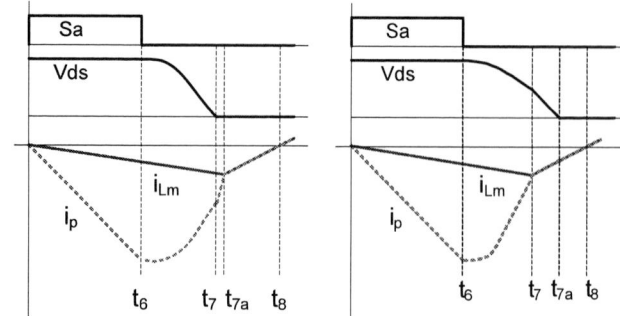

(a) Large leakage energy   (b) Small leakage energy

Fig. 4 Switching transient with large and small leakage energy for ZVS

Under DCM operation or critical DCM operation (available for almost all VF control schemes), the magnetizing current increases reversely (secondary side rectifier diode D_R is on) as shown in Fig. 4. The ZVS of S_W can be maintained even E_{Lk} is smaller than E_{Coss} due to the magnetizing energy E_{Lm}, which helps to realize ZVS of S_W as shown in Fig.4 (b). The expression is shown below:

$$\frac{1}{2}(L_k \cdot I_{p\text{-}neg}^2 + L_m \cdot I_{Lm\text{-}neg}^2) \geq \frac{1}{2} \cdot C_{oss}(V_{in} + N \cdot V_o)^2 \qquad (4)$$

$$I_{Lm\text{-}neg} = N \cdot V_o \cdot L_m^{-1} \cdot T_a \qquad (5)$$

Where I_{Lm_neg} is negative peak value of magnetizing current.

Due to relative large magnetizing inductance, even a very small negative magnetizing current can achieve ZVS of S_W if the dead time is sufficient. However, in the real application, the larger dead time will result in low equivalent switching frequency, which will deteriorate the overall efficiency. Increasing the auxiliary switch's on-time also helps to increase the negative magnetizing current to help ZVS operation, but it causes extra conduction loss, which may offset the saved switching loss.

2) Optimized Control Scheme

From (6), it is preferred that the negative peak current keeps constant with load variation. In most VF control scheme, the converter usually operates in CRM condition, such as QR control and off-time control. For QR operation, the peak current is adjusted according to the load condition, which means the ZVS condition may not be satisfied at light load condition. The constant frequency control scheme is similar to QR operation. For the off-time controlled method, the peak magnetizing current and the frequency are used to regulate the output power. Therefore, off-time control method is preferred for the proposed control scheme.

3) Auxiliary switch on-time T_a

Auxiliary switch on-time decides the circulating energy and the clamp voltage as given in (2). The on-time T_a can be designed by the desired maximum clamp voltage at the full load condition. It is an important parameter, which also affects the selection of clamp capacitance as analyzed below.

4) Leakage inductance L_k and dead time

For simplicity, we need to select the leakage inductance to

satisfy equation (3) for ZVS operation. With preferred off-time control, the I_{p_neg} in the equation is determined by the maximum switching frequency at full load condition. Therefore, the required dead time T_d should be smaller than the quarter the resonant period:

$$T_d \leq \frac{2 \cdot \pi}{4} \cdot \sqrt{L_k \cdot C_{oss}} \qquad (6)$$

As explained above, the magnetizing current also helps to achieve the ZVS under CRM or DCM operation, which may need a larger dead time. Thus, the dead time may slightly larger than the calculated value given by (6). But too large dead time will decrease the equivalent switching frequency, which results in higher peak current for full load condition, so that the conduction loss will be higher. In practical design, 200~500ns is a suitable value for proper operation.

5) Clamp Capacitance C_C

The previous analysis is based on the assumption that the clamp voltage is almost constant and the resonant period formed by the leakage inductor and clamp capacitor should be much longer than the auxiliary switch on time T_a. Usually, the voltage ripple should be within 5% or 10% to treat it as constant. The clamp capacitance can be expressed as

$$C_c \geq \frac{I_{pk} \cdot T_a}{2 \cdot V_c \cdot ripple\%} \qquad (7)$$

Where V_C is given in (2) and ripple% is the maximum allowed clamp voltage ripple percentage, such as 5% or 10%. It is clear that smaller T_a helps to reduce the clamp capacitance.

Also, the resonant period should be several times larger than T_a to treat it as a linear charge/discharge as given in (8). Usually, k should be above 5.

$$2\pi \cdot \sqrt{L_k \cdot C_c} \geq k \cdot T_a \qquad (8)$$

Based on these two equations, we can determine the clamp capacitance. Though large capacitance makes the operation more close to the ideal condition, it doesn't help much for the clamping performance and usually causes high cost and bulky volume. For practical application, the clamp capacitance is usually ranges from 47nF to 220 nF.

6) Auxiliary switch S_a

For the auxiliary switch S_a, the current rating is quite small, the RMS current ratio of the auxiliary switch S_a to main switch S_W is given in (9). Theoretically a very small current rating MOSFET can be used. But the on resistance will affect the real negative peak current I_{p-neg}.

$$\frac{I_{rms-Sa}}{I_{rms-Sw}} = \sqrt{\frac{T_a}{T_{on}}} \qquad (9)$$

Where I_{rms-Sw} and I_{rms-Sa} are the RMS current of the main switch S_W and auxiliary switch S_a respectively.

Compared to the conventional active clamp Flyback converter with complementary gate drive signal, the ZVS operation of auxiliary switch is lost. Due to the small device rating, the turning on switching loss is small.

IV. EXPERIMENTAL RESULTS

The experimental results from a 16V/4A prototype with universal AC input are shown from Fig. 5 to Fig. 8. The key parameters are listed in Table. 1. To achieve high average efficiency and light load efficiency, the off time control is adopted in the prototype, which is usually impossible for conventional active-clamp Flyback control method. The maximum switching frequency is set to around 65 kHz. The transformer core is PQ26/20 with PC40 equivalent material. A PMOS is used as auxiliary switch for simple gate drive.

Table.1. Key parameters of the experimental prototype

Parameters	Symbol	Value
AC input voltage	V_{in}	AC 90-265V (RMS)
Output	V_o/I_o	16V/4A
Primary side switch	S_W	SPA11N60C3
Rectifier diode	D_R	V40100C
Auxiliary switch	S_a	FQP3P50 (Ron=4.9 Ohm)
Bridge rectifier		KBJ406
Transformer turns ratio	$n=N_p:N_{s1}:N_{s2}$	24:4:3
Magnetizing inductance	L_m	260uH
Leakage inductance	L_k	1.5 uH
Clamp capacitor	C_c	150nF
Controller		NCP1351B
Auxiliary switch on time	T_a	400ns
Dead time	T_d	400ns

Fig. 5 shows the gate drive signal, drain to source voltage (V_{ds}) of S_W and the primary side current I_p at different loads and input conditions. It is clear that the soft switching for the main switch S_W is achieved at any load and input condition due to off time control. The parasitic ring is eliminated by the clamp capacitor. Thus, a lower voltage rating MOSFET can be used as primary switch for lower conduction loss and better efficiency.

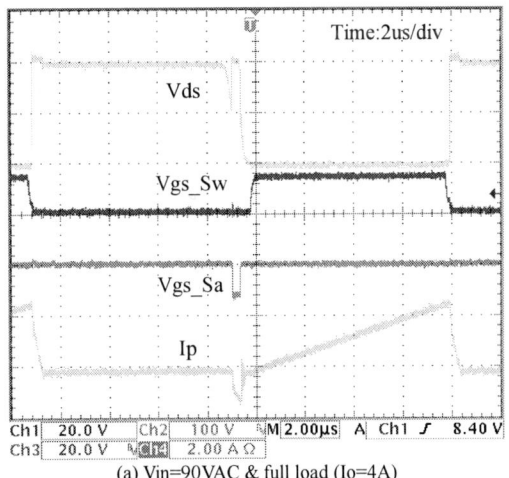

(a) Vin=90VAC & full load (Io=4A)

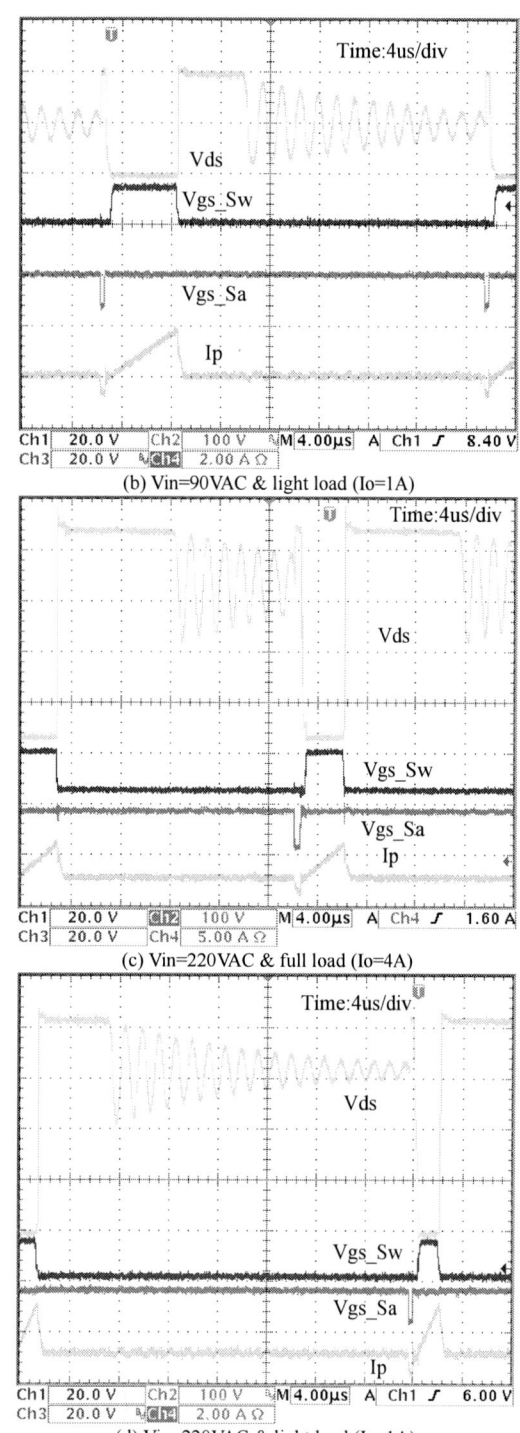

(b) Vin=90VAC & light load (Io=1A)

(c) Vin=220VAC & full load (Io=4A)

(d) Vin=220VAC & light load (Io=1A)

Fig. 5.   Switching waveforms and primary side current at different inputs
and load conditions

Fig.6 shows the waveforms of the primary current I_p, clamp current I_c and secondary current I_s at different load conditions. The leakage energy is absorbed and transferred to the output side and input side when auxiliary switch S_a turns on. The peak value of the clamp current I_c is smaller than peak value of the primary side current I_p shown in Fig. 6. Because

the clamp voltage is larger than $(1+m)NV_O$ which is discussed in section II, part of the leakage energy will be absorbed by the capacitor C_{oss} to raise the voltage V_{ds} to $V_{in}+V_C$ and then turns on the anti-paralleled diode of S_a. Therefore, the charger absorbed by the clamp capacitor during Mode 3 is slightly reduced, which will leads to a reduced negative peak current at ideal condition. Also the extra damping effect introduced by the circuit power loss reduces the negative peak value. The more detailed waveform at main switch turns off and auxiliary switch turns on is shown in Fig.7 and Fig.8 respectively. The commutation period shown in Fig 8 determines the circulating energy.

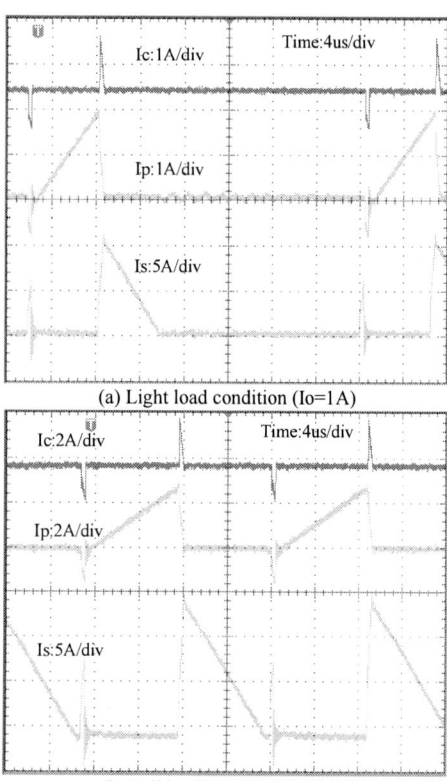

(a) Light load condition (Io=1A)

(b) Full load condition (Io=4A)

Fig.6 Primary side current Ip, secondary side current Is and clamp circuit
current Ic at different load

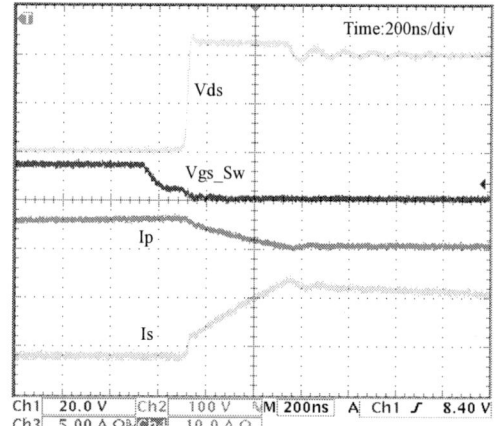

Fig.7 The transition between Ip and Is at Sw turns off

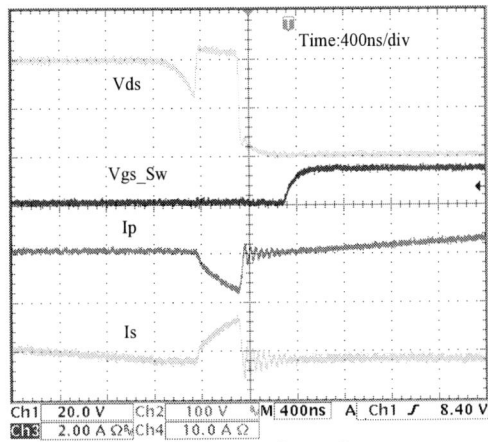

Fig.8 Vds, Ip and Is waveform at Sa turns on

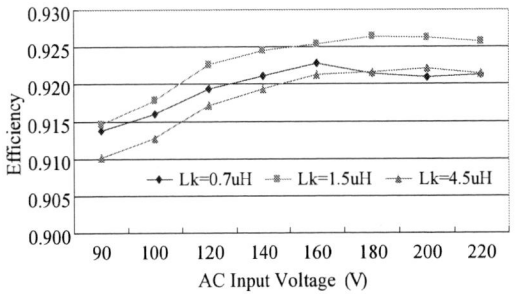

Fig. 10 Average efficiency vs. leakage inductance

The measured efficiency of the prototype is shown in Fig. 9. The average efficiency is measured and calculated at 25%/50%/75%/100% load conditions. Also, the efficiency with conventional RCD clamp circuit (R=100K, C=2.2nF) at the same condition is presented for comparison. The proposed converter efficiency at high line and light load conditions is much higher than conventional one.

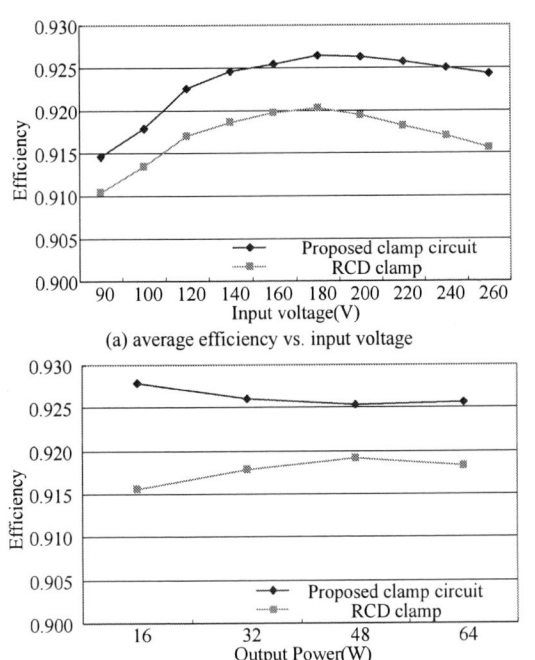

(a) average efficiency vs. input voltage

(b) efficiency vs. output power at Vin=220V

Fig. 9 Efficiency comparison between Flyback with proposed active-clamp method and conventional RCD clamp

Fig.10 shows the average efficiency versus the leakage inductance. It is clear that the efficiency is not sensitive for the leakage inductance variation, which makes it very attractive for applications with low cost transformers.

V CONCLUSION

This paper proposes a soft-switching and high efficiency Flyback converter with new active-clamp technique. The proposed circuit has very attractive features, such as low device stress, soft-switching, high efficiency on both full load and light load conditions and can be adopted to various control schemes. Also, it is not sensitive for leakage inductance variation. All the advantages make it suitable for low power offline application with strict efficiency and standby power requirement. A 16V/4A prototype with off time control scheme is built and compared to conventional Flyback converter with dissipative clamp circuit. The experimental results confirm the theoretical analysis and the advantages mentioned above.

Acknowledgment

This work is supported by China National Science Fund, No. 50907061.

REFERENCE

[1] Tamotsu Ninomiya, et al, "Analysis and Optimization of a Nondissipative LC Turn-Off Snubber", *IEEE Trans. Power Electronics*, vol. 3, No.2, April 1988.

[2] Choi, C.T., Li, C.K., and Kok, S.K., "Control of an Active Clamp Discontinuous Conduction Mode Flyback Converter", in *Proc. 1999 IEEE Power Electronics and Drive Systems Conf.*, vol. 2, pp. 1120–1123.

[3] Robert Watson, Fred C. Lee, Guichao C. Hua, "Utilization of an Active-Clamp Circuit to Achieve Soft Switching in Flyback Converters", *IEEE Trans. Power Electronics*, vol. 11, pp. 162-169, January 1996.

[4] Yu-Kang Lo, and Jing-Yuan Lin, "Active-Clamping ZVS Flyback Converter Employing Two Transformers," *IEEE Trans. Power Electronics*, vol. 22, No. 6, November 2007.

[5] Gwan-Bon Koo, Myung-Joong Youn, "A New Zero Voltage Switching Active Clamp Flyback Converter", in *Proc. 2004 IEEE PESC Conf.*, pp. 508-510.

[6] P. Alou, et al, "A Low Power Topology Derived from Flyback with Active Clamp Based on a very Simple Transformer", in *Proc. 2006 IEEE APEC Conf.*, pp. 627-632

[7] E.H. Wittenbreder, "Zero Voltage Switching Pulse with Modulated Power Converters", U.S. Patent 5 402 329, March 1995.

[8] David A. Cross, "Clamped Continuous Flyback Power Converter", U.S. Patent 5 570 278, Oct. 29, 1996.

[9] T. M. Chen, C.-L. Chen., "Analysis and Design of Asymmetrical Half Bridge Flyback Converter", in *Proc. 2002 IEE Electr. Power Appl. Conf.*, vol. 149, No. 6.

[10] Bor-Ren Lin, Cheng-Chang Yang. "Analysis, Design and Implementation of an Asymmetrical Half-Bridge Converter", in *Proc. 2005 IEEE ICIT Conf.*, pp. 1209-1214.

[11] Bing Lu, et al, "Optimum Design Methodology for LLC Resonant Converter", in *Proc. 2006 IEEE APEC Conf.*, pp. 6~11.

[12] D. Fu, F. C. Lee, Y. Qiu, F. Wang, "A Novel High-Power-Density Three-Level LCC Resonant Converter with Constant-Power- Factor-Control for Charging Applications", *IEEE Transaction on Power Electronics*, 2008.

978-1-4244-4782-4/10 $26.00 © 2010 IEEE

Analysis and Design of a Novel Integrated Three-Phase Single-Stage AC-DC PWM Full-Bridge Converter

Dunisha Wijeratne
University of Western Ontario
Department of Electrical and Computer Engineering
London, Ontario, Canada
dwijerat@uwo.ca

Gerry Moschopoulos
University of Western Ontario
Department of Electrical and Computer Engineering
London, Ontario, Canada
gmoschopoulos@eng.uwo.ca

Abstract-A new three-phase ac-dc single-stage PWM full-bridge converter is proposed in the paper. The converter can simultaneously perform input power factor correction and isolated dc-dc conversion, operate using standard phase-shift PWM control, and is simple and low cost. In the paper, the operation of the new converter is explained and analyzed, its design is discussed, and experimental results that confirm its feasibility are presented.

I. INTRODUCTION

Single-switch ac-dc rectifiers, regardless of whether they are buck or boost, are much simpler than six-switch rectifiers and do not need the sophisticated control methods that the six-switch rectifiers need to perform input PFC. In fact, single-switch rectifiers can naturally perform input PFC without the need for any input current sensing or online calculation methods. Two separate switch-mode converters are still needed, however, to perform three-phase ac-dc power conversion with transformer (transformer) isolation so there is still a need to reduce cost and complexity. Moreover, the switch in the front-end ac-dc converter is subjected to considerable stress, which limits the power level that this approach can be used.

Several three-phase ac-dc converters that integrate the functions of PFC and isolated dc-dc conversion in a single power converter have been proposed to reduce cost and complexity. Previously proposed three-phase single-stage ac-dc full-bridge converters, however, have at least one of the following drawbacks, which have prevented their wide-spread use:

- They are implemented with three separate ac-dc boost converter modules, one for each input phase [4].

- The input currents must be discontinuous with large ripples and peak for input power factor correction to be achieved.

- The input currents are distorted and contain a significant amount of low frequency harmonics.

- The transformer primary side dc bus voltage may become excessive (> 900-1000V), which makes large, high-voltage bulk capacitors necessary (which may exceed 900-1000V) [6]- [9].

- The converter must be controlled using very sophisticated techniques.

A new three-phase ac-dc single-stage PWM full-bridge converter that has none of the above drawbacks is proposed in the paper and is shown in Fig. 1. In the paper, the operation of the new converter is explained and analyzed, its design is discussed, and experimental results that confirm its feasibility are presented.

II. MODES OF OPERATION

Fig. 2 shows a typical input capacitor voltage waveform. A near sinusoidal current can be ensured if this voltage is discontinuous. Figs 3 and 4 respectively show typical converter waveforms and equivalent circuit diagrams that illustrate the modes of operation that the proposed converter

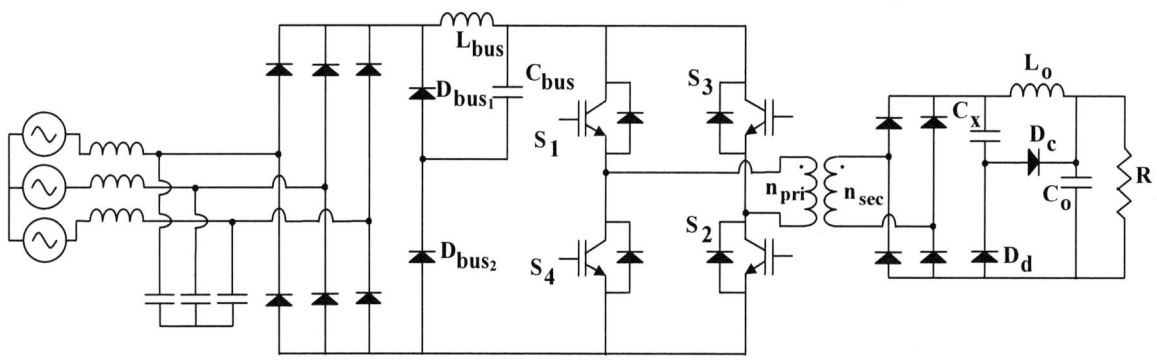

Fig. 1. Proposed three-phase single-stage ac-dc converter.

978-1-4244-4782-4/10 $26.00 © 2010 IEEE

goes through during a half-switching-cycle. In Fig.. 4, the output capacitor and load are shown as an equivalent voltage source V_o and the transformer magnetizing current is neglected. An arbitrary k^{th} half-switch cycle where $V_{a,k}=V_{pk}$, $V_{b,k}=V_{c,k}=-V_{pk}/2$ is considered, in which $V_{a,k}$, $V_{b,k}$, $V_{c,k}$ refer to respective instantaneous phase voltages at the particular cycle and V_{pk} is the peak phase voltage.

The significant modes of operation the converter goes through are as follows:

Mode 0 (t<t₀) Fig. 4(a): Before $t=t_0$, switch S_1 is on and the other switches in the full-bridge are off. There is very little current flowing in the transformer primary and secondary windings. The secondary side auxiliary capacitor C_x is discharging with a constant current I_{Lo}. Current in the dc bus is freewheeling through diode D_{bus1}, inductor L_{bus}, and capacitor C_{bus}. Diode D_{bus2} is off because it is reverse biased. The input capacitors charge while the dc bus current freewheels.

At $t=t_0$, $v_{Ca,k}$ is at its peak. If $I_{a,k}$ is the line current for the kth cycle and $V_{Ca,k(pk)}$ is the peak voltage of C_a, then $V_{Ca,k(pk)}$ can be expressed as

$$v_{Ca,k}(t=t_0)=V_{Ca,k(t_0)}=V_{Ca,k(pk)}=\frac{I_{a,k}}{C_a}(1-D)\frac{T_s}{2} \quad (1)$$

where D is the duty ratio, T_s is the switching period and C_a is the value of input capacitor C_a. $V_{Cb,k(t0)}$ is the voltage of input capacitor C_b at $t=t_0$ and can be expressed as

$$V_{Cb,k(t_0)}=\frac{I_{b,k}}{C_b}(1-D)\frac{T_s}{2}=-\frac{V_{Ca,k(t_0)}}{2}=-\frac{I_{a,k}}{2C_a}(1-D)\frac{T_s}{2} \quad (2)$$

The current of C_b in Mode 0 is half of that of C_a for the kth cycle. The resulting rectifier voltage, $V_{rec,k(t0)}$ can be given as

$$V_{rec,k(t_0)}=V_{Ca,k(t_0)}-V_{Cb,k(t_0)}=\frac{3I_{a,k}}{2C_a}(1-D)\frac{T_s}{2} \quad (3)$$

Mode 1 (t₀<t<t₁) Fig. 4(b): At $t=t_0$, switch S_2 is turned on with ZCS and current starts to flow into the full-bridge from the dc bus. $i_{Lbus,k}$ is made up of line current $I_{a,k}$ and the discharging current $i_{Ca,k}$ of C_a. Since diagonally opposite switches S_1 and S_2 are on, power is transmitted to the load from the input side. Primary current is reflected to the secondary side of the transformer and this current feeds the load while charging C_x through the auxiliary circuit diode

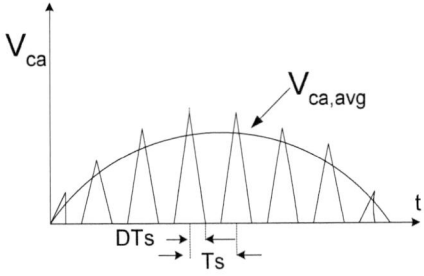

Fig. 2. Input capacitor voltage for a half-line-cycle.

D_c. C_x reaches its peak voltage at the end of Mode 1.

The current in L_{bus}, $i_{Lbus,k}$ can be written as

$$i_{Lbus,k}=I_{a,k}+i_{Ca,k} \quad (4).$$

Applying Kirchoff's voltage law to the dc bus circuit results in

$$v_{rec,k}=1.5v_{Ca,k}=v_{Lbus,k}+V_{Cbus} \quad (5)$$

where $v_{Lbus,k}$ and $V_{Cbus,k}$ are dc bus inductor and capacitor voltages. Replacing $v_{Lbus,k}$ by voltage-current relationships of L_{bus} and C_a gives the following modal equation,

$$L_{bus}C_a\frac{d^2(v_{Ca,k})}{dt^2}+1.5v_{Ca,k}-V_{Cbus}=0 \quad (6)$$

At $t=t_1$, $V_{Ca,k(1)}=V_{Cbus}/1.5$.

Mode 2 (t₁<t<t₂) Fig. 4(c): At $t=t_1$, C_x reaches its peak voltage and stops being charged. The input capacitors continue to discharge into the full-bridge. $v_{Ca,k}$ decreases and so does the voltage impressed across the rectifier by capacitor C_a. As the voltage of $C_{bus}(V_{Cbus})$ is constant throughout the cycle, the reduction in the rectifier voltage makes the voltage across L_{bus} decrease. The voltage across L_{bus}, $v_{Lbus,k}$, is negative and the inductor current decreases. The voltage across C_{bus} is applied across the primary winding and primary current is equal to the reflected load current. Mode 2 ends when C_a is completely discharged.

Since the transformer primary current $i_{Llk,k}$ is equal to $i_{Lbus,k}$ and $i_{Llk,k}$ is also the reflected load current, therefore, $i_{Llk,k}$ can be written as

$$i_{Llk,k}=i_{Lbus,k}=I_{a,k}+i_{Ca,k}=nI_{Lo} \quad (7)$$

Mode 3 (t₂<t<t₃) Fig. 4(d): During this mode, C_a is clamped to zero voltage; therefore the voltage applied across diode bridge rectifier is also zero. When the rectifier voltage becomes zero, D_{bus1} and D_{bus2} in the dc bus circuit become forward biased and they begin to conduct at $t=t_2$. L_{bus} discharges into C_{bus} through D_{bus1}. C_{bus} continues to discharge into the full-bridge though D_{bus2} so that V_{Cbus} is still applied across the transformer primary. The current in the transformer is same as in Mode 2. At $t=t_3$ the mode comes to an end with S_1 being turned off.

Mode 4 (t₃<t<t₄) Fig. 4(e): At $t=t_3$, when S_1 is turned off, the output switch capacitors, C_1 and C_4 of S_1 and S_4 start to charge and discharge respectively. The input capacitors begin to be charged by the line currents, and the voltage across the rectifier is no longer zero. This voltage is placed across D_{bus2} and the diode is reversed biased during Mode 4. L_{bus} discharges through D_{bus1} into C_{bus}. Once C_4 across switch S_4 is completely discharged, S_4's body diode conducts and S_4 can be turned on with ZVS.

The charging of C_a during Mode 4 can be expressed as

$$I_{a,k}=C_a\frac{dv_{Ca,k}}{dt} \quad (8)$$

978-1-4244-4782-4/10 $26.00 © 2010 IEEE

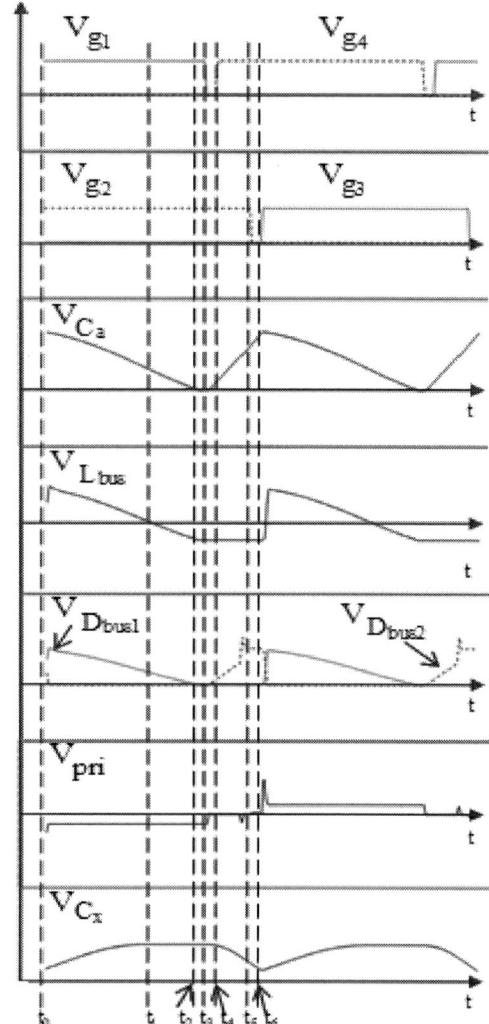

Fig. 3. Typical converter waveforms.

which can be rewritten in terms of $v_{Ca,k}$ at $t=t_4$

$$v_{Ca,k}(t=t_4) = V_{Ca,k(t_4)} = \frac{I_{a,k}}{C_a}(t_4 - t_3)$$ (9)

Mode 5 (t₄<t<t₅) Fig. 4(f): The converter enters a freewheeling mode at that start of this mode. Transformer primary current freewheels through S_4, S_2 and the primary winding. The input capacitors continue to charge and D_{bus2} is reversed biased due to the rectifier output voltage across it. Current also freewheels through L_{bus}, C_{bus}, and D_{bus1}, and no current enters the full-bridge during this mode. The voltage input to the full-bridge is zero, but a counter voltage is impressed across L_{lk} due to the presence of C_x on the transformer secondary side. V_{Cx} keeps the secondary side rectifier voltage above zero and forces voltages across the transformers secondary and primary.

This counter voltage decreases the primary current. As a result, secondary current also falls below I_{Lo}. In order to keep

I_{Lo} constant, C_x discharges though D_d into L_o. At $t=t_5$, the primary current is completely reset by the auxiliary capacitor C_x. Practically, there can be a small amount of magnetizing current in the primary winding, and, therefore some snubbing would be required for switches S_2 and S_3 when they turn off.

Considering the voltage-current relationship of L_{lk}, (10) is derived to show the ZCS operation of the converter where $v_{Cx,k}$ is the voltage of C_x,

$$-\frac{v_{Cx,k}}{n} = L_{lk}\frac{d(i_{L_{lk},k})}{dt}$$ (10).

As $i_{Cx,k}$ is the difference of the I_{Lo} and the reflected primary current to the secondary side,

$$i_{Cx,k} = -(I_{Lo} - \frac{i_{L_{lk},k}}{n})$$ (11).

By means of the voltage-current relationship of C_x, (10) and (11) the following modal equation can be derived

$$n^2 C_x L_{lk}\frac{d^2(v_{Cx,k})}{dt^2} + v_{Cx,k} = 0$$ (12).

<u>Mode 6 (t₅<t<t₆):</u> At $t=t_5$, S_2 is turned off with ZCS. The input capacitors continue to charge and current freewheels in the dc bus. There are no currents in the full-bridge and in the transformer. C_x continues to discharge until switch S_3 is turned on with ZCS at $t=t_6$, as current through the leakage inductor cannot change suddenly. This is the beginning of the next half cycle and the converter enters Mode 1, but with S_2 and S_3 on instead of S_1 and S_4.

The proposed converter can be operated with standard phase-shift PWM control with input currents that are continuous. It can be designed with a large dc bus capacitor that can store sufficient energy to feed the dc output and meet hold-up time requirements when the input ac supply is absent. The converter is simple and inexpensive as it only has four active switches.

III. CONVERTER ANALYSIS

In order to develop a procedure that can be used to design the proposed converter, the steady-state operation of the converter must be analyzed to determine its behavior for any given set of specifications (line-to-line input voltage $V_{ll,rms}$, output voltage V_{out}, output current I_{out}, and switching frequency f_s) and any given set of component values (input capacitors $C_a=C_b=C_c$, input inductors $L_a=L_b=L_c$, duty ratio D, transformer turns ratio n $=n_{sec}/n_{pri}$, output inductor L_o). After the analysis has been performed, important converter characteristics can be determined and then used to develop the design procedure. The key parameters that need to be determined for the design of the converter are the dc bus capacitor voltage V_{Cbus}, the instantaneous input capacitor voltage $v_{Ca,k}$, and the peak switch voltage stress $V_{sw,pk}$.

Given the interdependency of key variables such as duty ratio D, turns ratio n and input capacitance C_a, the analysis of the converter cannot be performed using equations with

978-1-4244-4782-4/10 $26.00 © 2010 IEEE

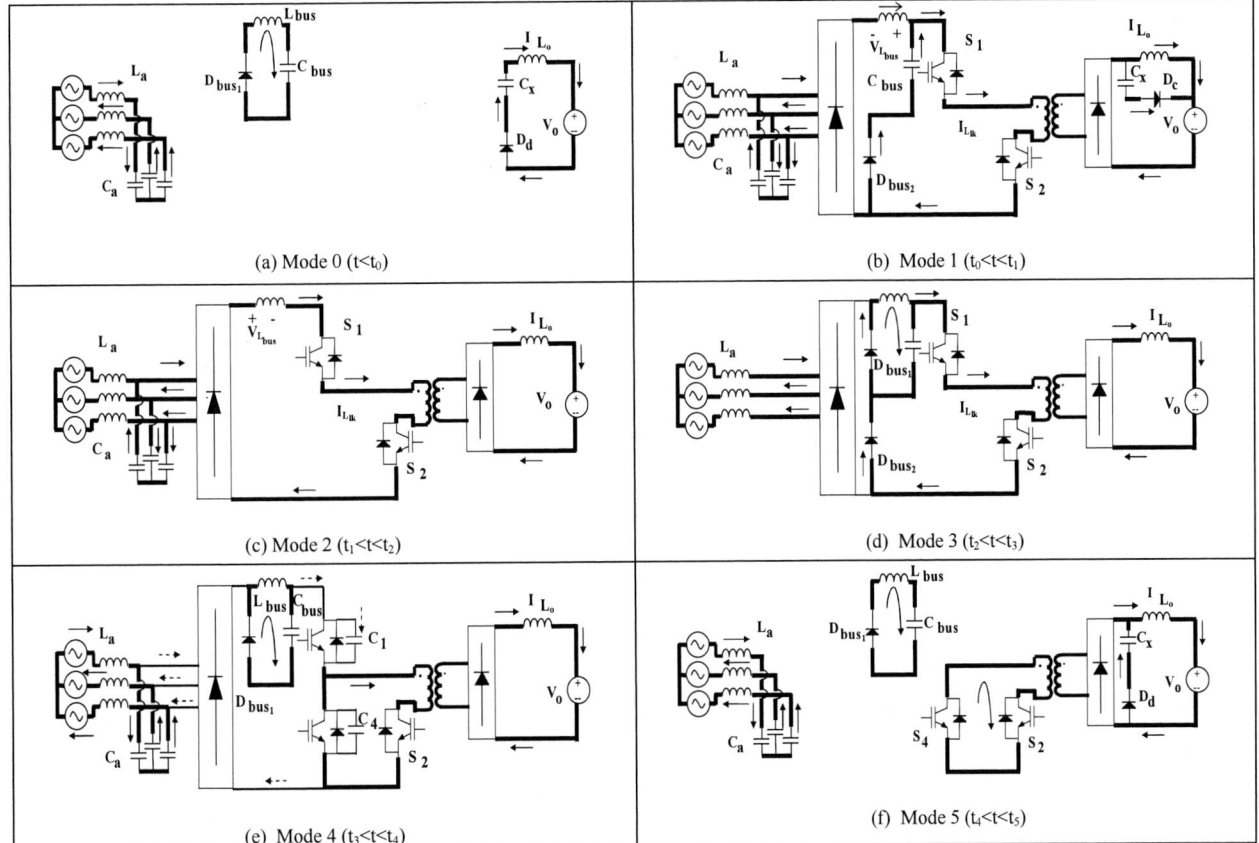

(a) Mode 0 (t<t₀) (b) Mode 1 (t₀<t<t₁)

(c) Mode 2 (t₁<t<t₂) (d) Mode 3 (t₂<t<t₃)

(e) Mode 4 (t₃<t<t₄) (f) Mode 5 (t₄<t<t₅)

Fig. 4. Modes of converter operation.

closed form solutions, and some sort of computer program must be used. There are many ways to write such a program; one way is suggested here.

A valid steady-state operating point for the converter for a given set of specifications and component values can be found by selecting a value of D, determining a value of dc bus voltage V_{Cbus}, and then checking to see if this value of V_{Cbus} leads to an average voltage of zero across dc bus inductor L_{bus} over a period of one-sixth of the line cycle. If a non-zero average voltage results, then the computer program can adjust D so that a new value of V_{Cbus} that results in zero average voltage across L_{bus} is found. Once a valid operating point has been determined, the process can be repeated with different sets of component values to find a range of valid operating points. Once such a range has been determined, important converter characteristics can be identified and a design procedure can be established based on these characteristics.

The equations that were derived in the previous section of this paper can be used as part of the program. There are $n_{sw}=T_L$ (line period)/6T_s (switching period) for each 60° section of the line cycle (one-sixth the 60Hz line cycle). The program can be made to sweep through any switching cycle k, by starting with some set of initial conditions for key converter voltages and currents, determining their final values at the end of a mode, and then using these "final" values as initial conditions for the next mode. This process can be continued until the end

of the kth cycle, when a new switching cycle starts, then repeated until the end of the n_{sw}^{th} cycle. A check of the average voltage across L_{bus} can then be made.

The following assumptions can be made to simplify the analysis and the computer program:

- The input line currents and source voltages can be considered to be constant during a switching cycle as the switching period T_s is much smaller than the line period T_L.

- The low 360Hz frequency ripple that exists in the dc bus capacitor voltage and current and the dc bus inductor current can be neglected. The dc bus capacitor voltage is constant.

- The input line currents coming out of the source can be assumed to be sinusoidal with no ripple.

- The effects of the auxiliary circuit and the leakage inductance of the transformer can be neglected until after a range of valid steady-state operating points has been determined. The analysis of the dc-dc full-bridge section of the converter can be performed once a set of component values for the input section, the transformer turns ratio, and the output inductor has been chosen. Since this analysis is very similar to that of a ZVZCS-PWM dc-dc full-bridge converter, it will not be presented here.

- The worst-case operating condition occurs when the converter is operating with maximum load.

The computer program can be implemented to perform the following steps:

Step 0: Select the set of specifications and components values to be considered. Select a value of D (any value) to start the process of determining a corresponding dc bus capacitor voltage V_{Cbus}.

Step 1: Determine the instantaneous line currents. This can be done using the following set of equations

$$
\begin{aligned}
i_{a,k} &= I_{a,pk} \sin\left(2\pi f_s t + \frac{\pi}{2}\right) \\
i_{b,k} &= I_{b,pk} \sin\left(2\pi f_s t + \frac{\pi}{2} + \frac{2\pi}{3}\right) \\
i_{b,k} &= I_{c,pk} \sin\left(2\pi f_s t + \frac{\pi}{2} + \frac{4\pi}{3}\right)
\end{aligned}
\tag{13}
$$

where

$$
I_{a,pk} = I_{b,pk} = I_{c,pk} = \sqrt{2}\,\frac{(V_{out}I_{out})}{\sqrt{3}V_{ll,rms}}
\tag{14}.
$$

Step 2: Determine the dc bus capacitor voltage V_{Cbus}. Do so by assuming that the current through the output inductor L_o, I_{Lo}, is continuous so that

$$
V_{Cbus} = \frac{V_{out}}{nD}
\tag{15}.
$$

Step 3: Verify that i_{Lo} is actually continuous. This can be done by using the following equation to see if the average current component of i_{Lo}, which equals I_{out}, is less than the peak current ripple (Δ_i_{Lo}).

$$
\Delta\_i_{Lo} = 0.5\left(\frac{nV_{Cbus} - V_{out}}{L_o}\right)\left(\frac{D}{2f_s}\right) < I_{out}
\tag{16}.
$$

If (16) is not satisfied, then the value of L_o needs to be changed to make i_{Lo} continuous.

Step 4: Determine $I_{Lbus,k}$, the current flowing through dc bus inductor L_{bus}. This can be done by considering that the net dc current of C_{bus} should be zero over a 60° angle when the converter is operating in steady-state. If the current ripple in $i_{Cbus,k}$ is ignored, then the average value of the current flowing into C_{bus} for k^{th} cycle can be written as

$$
I_{Cbus,k,in\_ave} = I_{Cbus,k,out\_ave} = D(nI_{out})
\tag{17}
$$

using the reflected load current, nI_{out}.

If the 360Hz ripple of $i_{Lbus,k}$ is neglected, then the dc component of $i_{Lbus,k}$, out of L_{bus}, $I_{Lbus,k}$, should be equal to nDI_{out}.

Step 5: Track the instantaneous input capacitor voltages (e.g. $v_{Ca,k}$). The shape of these voltage waveforms needs to be determined so that the peak value of these voltages (and thus the peak switch stress) can be found. The shape is also needed for the next step, which involves using the voltage across inductor L_{bus}, $v_{Lbus,k}$, to confirm that the operating point under consideration is a valid one. The average value of $v_{Lbus,k}$ over a 60° portion of the line cycle, $V_{Lbus(ave)}$, must be zero if the converter is operating in steady-state. The value of $V_{Lbus,(ave)}$ can be found from $v_{Lbus,k}$, which can be found from the shape of the diode bridge rectifier output voltage $v_{rec,k}$. The shape of $v_{rec,k}$ is just a reflection of the input capacitor line-to-line voltage waveforms.

With the values of D and $C_a = C_b = C_c$ set, the peak voltage of C_a for the kth switching cycle is

$$
V_{Ca,k(pk)} = \frac{I_{a,k}}{C_a}(1-D)\frac{T_s}{2}
\tag{18}
$$

where $I_{a,k}$ is the instantaneous current of phase a for k^{th} cycle. The values $V_{Cb,k(pk)}$ and $V_{Cc,k(pk)}$ are the same as that for $V_{Ca,k(pk)}$. Since the input section of the converter is disconnected from the dc bus when the switches in the full-bridge are off, the input line current for each phase is the only current available to charge an input capacitor.

To determine the shape of the input capacitor voltage waveforms when the switch is on and the capacitors are discharging, the current that is available to discharge these capacitors needs to be known. The discharging current of C_a is the difference of the $I_{Lbus,k}$ and $I_{a,k}$ and is

$$
i_{Ca,k} = I_{Lbus,k} - I_{a,k} \forall t \in [t_0, t_2]
\tag{19}.
$$

At $t=t_2$, $v_{Ca,k}$ is zero; time $t = t_2$ can be found by

$$
t_2 = t_0 + \frac{C_a\left|V_{Ca,k(pk)}\right|}{I_{Lbus,k} - \left|I_{a,k}\right|}
\tag{20}.
$$

To prepare for the next step, the rectifier voltage $v_{rec,k}$, can be found as follows

$$
\begin{aligned}
v_{Ca,k} > v_{Cb,k} > v_{Cc,k} &\Rightarrow v_{rec,k} = \left|v_{Ca,k}\right| + \left|v_{Cc,k}\right| \\
v_{Cb,k} > v_{Cc,k} > v_{Ca,k} &\Rightarrow v_{rec,k} = \left|v_{Cb,k}\right| + \left|v_{Ca,k}\right| \\
v_{Cc,k} > v_{Ca,k} > v_{Cb,k} &\Rightarrow v_{rec,k} = \left|v_{Cc,k}\right| + \left|v_{Cb,k}\right|
\end{aligned}
\tag{21}
$$

where $v_{rec,k}$ depends on the maximum and the minimum phase capacitor voltages. With $v_{rec,k}$ known, the voltage across inductor L_{bus} during a switching cycle k, $v_{Lbus,k}$, can be determined. When the voltage across L_{bus} is positive and energy is placed into L_{bus} during Modes 1 and 2, $v_{Lbus,k}$ is

$$v_{Lbus,k} = v_{rec,k} - V_{Cbus} \; \forall t \in [t_0 - t_2] \qquad (22)$$

When current in L_{bus} freewheels through C_{bus} and dc bus diode D_{bus1}, $v_{Lbus,k}$ is

$$v_{Lbus,k} = -V_{Cbus} \; \forall t \in [t_2 - t_6] \qquad (23)$$

Step 6: Confirm that the average value of the voltage across inductor L_{bus}, $V_{Lbus(ave)}$ is zero, as it should if the converter is operating in steady-state. This can be done by summing the positive voltage across L_{bus} with the negative voltage as follows

$$\sum_{t_0}^{t_1}(v_{rec,k} - V_{Cbus}) = \sum_{t_2}^{t_6} V_{Cbus} \qquad (24).$$

If $V_{Lbus(ave)}$ is not zero, then the operating point under consideration is invalid and D, n and/or C_a should be changed to find a valid operating point.

Step 7: Confirm that the input capacitor voltages are discontinuous and that the input line currents meet the desired harmonic standards. Once a valid, steady-state operating point has been determined, it should be verified that the input capacitors voltages are discontinuous to confirm the assumption of sinusoidal input line currents. The input currents will be distorted if the input capacitor voltages are not fully discontinuous.

The preceding steps of the analysis are done with the converter operating with maximum load. The performance of the converter needs to be checked under light load conditions, however, because this is the worst-case condition for input current distortion. This is because there is less available current to discharge the input capacitors, which increases the likelihood of partially continuous capacitor voltages, which, in turn, lead to current distortion. For example, this verification process can be done with the converter operating at 600W and seeing if IEC 61000-3-2 Class A standards are met.

Both of the above "checks" can be performed by having the program sweep through an entire line cycle and looking at the input capacitor voltages and line currents.

Step 8: Repeat the previous steps to determine values for D and V_{Cbus} for other sets of component values.

IV. CONVERTER DESIGN

Once a range of valid steady-state operating points has been determined, these points can then be plotted as shown in Fig. 5, so that key converter characteristics can be seen. The main characteristics of the converter are the ones related to dc bus capacitor voltage V_{Cbus}, input capacitor peak voltage $V_{Ca,pk}$, switch stress $V_{sw,pk}$. These characteristics depend on the selection of duty ratio D, turns ratio n, and input capacitor C_a.

A procedure for the selection of components can be developed using the design curves shown in Fig. 5. The procedure will be demonstrated below with an example. For the example, the converter will be designed for the following specifications: line-to-line input voltage $V_{ll,rms}$=208V, output voltage V_{out}=48V, maximum output current I_{out}=40A, and switching frequency f_s=50kHz.

The following should be noted about the design procedure/example and the characteristic curves shown in Fig. 5.

- The procedure is based on the computer program described in Section III of this paper.

- The procedure is iterative given the interdependency of key parameters. Only the final iteration is presented in the design example.

- The operating points in the graphs shown in Fig. 5 have been derived for maximum load conditions.

- The operating points in the graphs shown in Fig. 5 have been derived, with the assumption that the input line currents are sinusoidal and the input capacitor voltages are discontinuous.

- The upper limit for $V_{sw,pk}$ has been set to 1200V, as 1200V devices are the highest rated devices that are commonly available.

- The design of the dc bus inductor L_{bus} and bus capacitor C_{bus} is similar to that of most ac-dc converters as they should be designed like a low pass filter to reduce the 360Hz component that is present in the diode bridge rectifier voltage. The design of these components, therefore, is not shown here.

A. Selection of transformer turns ratio n

The value of n should be such that the converter is able to provide the required output dc voltage; this is not possible if n $=n_{sec}/n_{pri}$ is too low. If n is too big, however, then the transformer primary current will not be extinguished before S_2 or S_3 are turned off. This is something that cannot be allowed to happen as there is no path for current after S_2 or S_3 are turned off. A large value of n will increase the amount of time needed for the primary current to be extinguished since the larger n is, the larger the current will be. As a result, the time that can be allowed for this current to extinguish will not be sufficient unless the duration of power transfer mode (and thus the duty ratio D) is reduced.

The value of n that is selected, therefore, should be the smallest value that allows the converter to deliver the required output voltage and that allows for the selection of the other parameters. For this iteration, a value of n=0.4 is selected.

B. Determination of Duty Ratio D and Input Capacitors C_a, C_b, C_c

With the value of n=0.4 selected in the previous step, graphs such as the ones shown in Fig. 5 can be drawn. The key graph in this step is the one shown in Fig. 5(b), which is a graph of $V_{sw,pk}$ vs. D for various values of C_a. The values for D and $C_a=C_b=C_c$ are interrelated.

If D is too small, then the input capacitors may not have enough time to fully discharge, which would result in input

978-1-4244-4782-4/10 $26.00 © 2010 IEEE

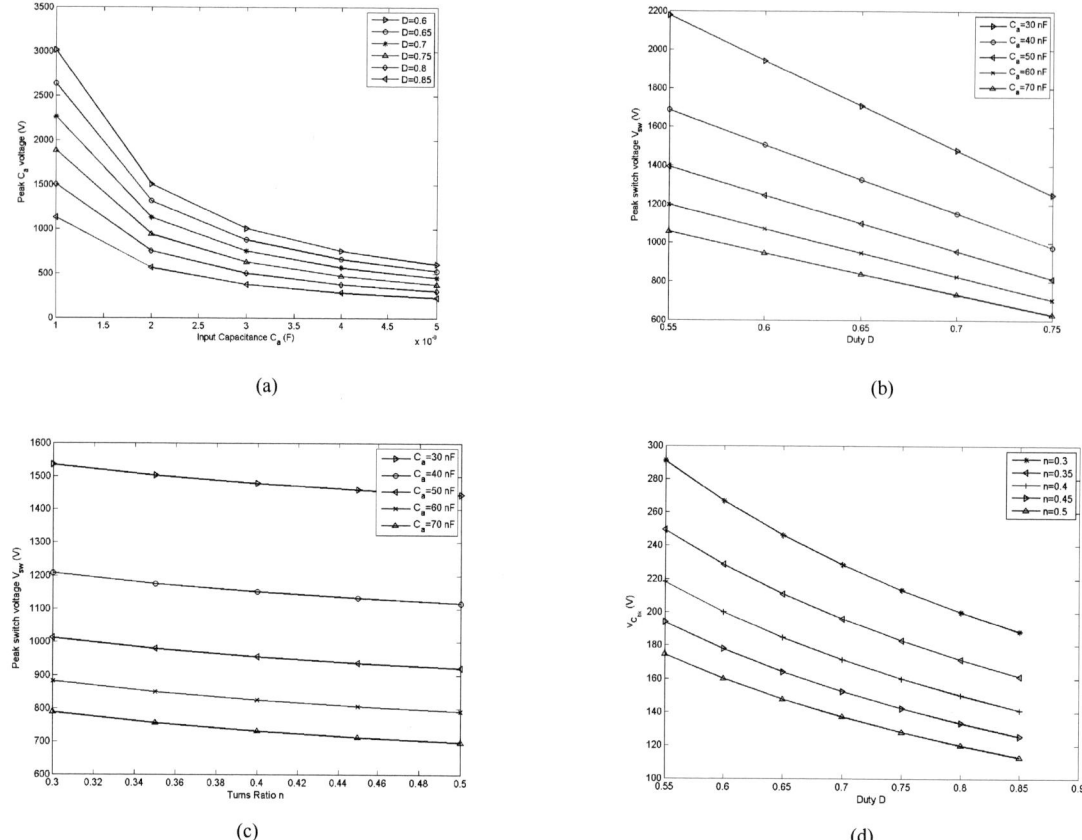

(a)

(b)

(c)

(d)

Fig. 5. Steady-state characteristic curves ($V_{ll,rms}$=208 V, V_{out}=48 V, f_s=50 kHz) (a) Effect of varying D & C_a on $V_{Ca,pk}$
(n=0.4), (b) Effect of varying D & C_a on $V_{sw,pk}$ (n=0.4), (c) Effect of varying n & C_a on $V_{sw,pk}$ (D=0.7), (d) Effect of varying D and n on V_{Cbk} (C_a=48 nF).

capacitor voltages that are not fully discontinuous and distorted input currents. This can be avoided by making C_a very small, but doing so would result in very high values of $V_{sw,pk}$. If D is too large, then an insufficient amount of time will be given to charge the input capacitors. Since it is energy from these capacitors that is eventually transferred to the output through the dc bus, an insufficient amount of energy would result in the converter being unable to provide the required output dc voltage.

Another consideration that needs to be taken into account is the amount of time needed for the primary current to be extinguished when the converter is in a freewheeling mode of operation. If D is very large, then $(1-D)T_s/2$ will not be enough time for this to happen, and the converter will not work properly as a result.

Taking the above considerations into account, D=0.7 and C_a=50nF are chosen. This will result in $V_{Ca,pk}$, which provides sufficient margin for 1200V devices to be used.

C. Design of the ZVZCS Auxiliary Circuit

With the values of n=0.4 and D=0.7 that have been a determined from previous steps, the dc bus voltage can be determined from (15) to be V_{Cbus}=178V. With values for n, D, and V_{Cbus} known, a value for capacitor C_x, which is the critical component needed to be extinguished the primary current when the converter is operating in freewheeling mode, can be

determined if the leakage inductance of the transformer L_{lk} is known.

Assuming a typical value of L_{lk}=7µH, a value of C_x=23µF can be found using the procedure described in [10-11], which is for a dc-dc ZVZCS full-bridge converter. Since the proposed converter has the same ZVZCS topology, the procedure used in [10-11] is suitable for the selection of C_x.

D. Determination of Input Inductors L_a, L_b, L_c

The design of the converter in the previous steps is based on the assumption that the input currents are perfectly sinusoidal. In reality, this is not true as these currents will have high-frequency ripple and low-frequency harmonics. As the load is increased, the low-frequency harmonics will be reduced as there is more current available to discharge the input capacitors. This makes it more likely that the input capacitor voltages will be fully discontinuous and thus more likely that the input currents will be sinusoidal. As a result, it is the high-frequency ripple that is more dominant when the converter is operating with heavy load.

The amount of high-frequency ripple in the input currents is dependent on the value of the input inductors, L_a=L_b=L_c. Since values of n, D, and V_{Cbus} are known, the dc bus inductor current can be determined to be $I_{Lbus,out-ave}$=16A, using standard buck converter equations. With $I_{Lbus,out-ave}$ and D known, the harmonic content of the current flowing out of the rectifier

978-1-4244-4782-4/10 $26.00 © 2010 IEEE   835

can be determined by taking the Fast Fourier Transform (FFT($i_{rec,k}$)). Based on this harmonic content, a value for $L_a=L_b=L_c$ can be determined by considering the ac source voltage as a harmonic ground and the input phase inductors and phase capacitors as L-C filters. If this is done, a value of $L_a=1.3$mH can be found.

E. Check for Compliance of Harmonic Standard

The final step of the procedure is to confirm that the converter's input line currents comply with the required harmonic standards. One such standard for three-phase converters is the IEC 61000-3-2 Class A standard.

Since the converter's input line currents are most likely to be distorted when operating under light load conditions, the converter's operation should be confirmed for these conditions. The method for doing so, however, will not be presented here due to limitations in space.

V. EXPERIMENTAL RESULTS

An experimental prototype of the proposed converter was built to confirm its feasibility. It was built for an input line-line voltage of $V_{ll,rms}=208$V, an output voltage of $V_{out}=48$V, a maximum output power of $P_{out,max}=2$kW and a switching frequency of $f_s=50$kHz. Typical converter waveforms are shown in Figs. 6 and 7. It can be seen from Fig. 6 that the converter can operate with near sinusoidal input currents as a result of the input capacitor voltages being discontinuous. Fig. 7(a) shows the ZVS switching of a switch in one leg. It can be seen negative current through the body diode of the switch drops the voltage across it to zero so that it can turn on with ZVS. Fig. 7(b) shows how the primary transformer current is extinguished and the current in the switch drops to zero before the switch is turned off.

ACKNOWLEDGMENT

The author would like to acknowledge funding from the National Sciences and Research Council of Canada (NSERC) for this work.

VI. CONCLUSION

A new three-phase, single-stage, ac-dc full-bridge converter was presented in this paper. The proposed converter is simple and can operate with an excellent input power factor with all of the advantages of a ZVZCS full-bridge converter using standard phase shift PWM control. In the paper, the operation of the new converter was explained, a method of analyzing its steady-state operation was presented, and a procedure for the design of its key components was demonstrated with an example. The feasibility of the converter was confirmed with results obtained from an experimental prototype.

REFERENCES

[1] E. H. Ismail and R. Erickson, "Single-switch 3Φ PWM low harmonic rectifiers," in *IEEE Trans. on Power Elec.*, vol. 11, no. 2, Mar. 1996, pp. 338-346.

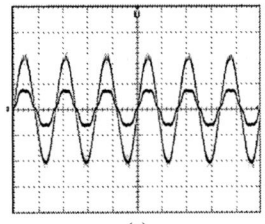 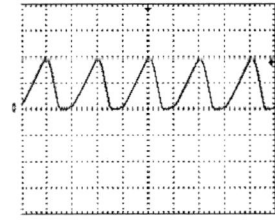

(a)   (b)

Fig. 6. (a) Input phase current and phase voltage (V: 85V/div, I: 10A/div, t: 8ms/div), (b) Voltage across capacitor C_a (V: 200V/div, t: 5µs/div).

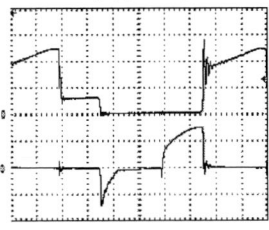 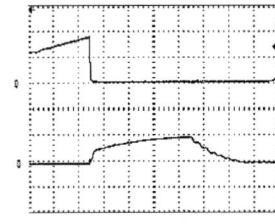

(a)   (b)

Fig. 7. (a) Switch voltage (top) and switch current (bottom) of ZVS leg (V: 400V/div, I: 10A/div, t: 2.5µs/div), (b) Switch current (top) and switch voltage (bottom) of ZCS leg (V: 400V/div, I: 20A/div, t: 1µs/div).

[2] Y. Jang and M. M. Jovanović, "Design considerations and performance evaluation of a 6-kW, single-switch, three-phase, high-power factor, multi-resonant, zero-current-switching buck rectifier," in *IEEE INTELEC Conf. Rec.* 1997, pp. 715-722.

[3] H. M. Suraywanshi, M. R. Ramteke. K. L. Thakre, and V. B. Borghate, "Unity-power-factor operation of three-phase ac-dc soft switched converter based on boost active clamp topology in modular approach," in *IEEE Trans. on Power Elec.*, vol. 23, no. 1., Jan. 2008, pp. 229-236.

[4] J. Contreras, and I. Barbi, "A three-phase high power factor PWM ZVS power supply with a single power stage," in *IEEE PESC Conf. Rec.* 1994, pp. 356-362.

[5] F. S. Hamdad and A. K. S. Bhat, "A novel soft-switching high-frequency transformer isolated three-phase AC-to-DC converter with low harmonic distortion," in *IEEE Trans. on Power Elec.* Vol. 19, no. 1, Jan. 2004, pp. 35 - 45.

[6] P. M. Barbosa, J. M. Burdio, and F. C. Lee, "A three-level converter and its application to power factor correction," in *IEEE Trans. on Power Elec.*, vol. 20, no. 6., Nov. 2005, pp. 1319-1327.

[7] Y. Xie, Y. Fang, and H. Li, " Zero-voltage-switching three-level three-phase high-power-factor rectifier," in *IEEE IECON Conf. Rec.* 2007, pp. 1962-1967.

[8] F. Cannales, P. Barbosa, C. Aguilar, and F. C. Lee, "A quasi-integrated AC/DC three-phase dual-bridge converter," in *IEEE PESC Conf. Rec.* 2001, pp. 1893-1898.

[9] J. G. Cho, J. W. Beak, C. Y. Jeong, D. W. Yoo, H. S. Lee, and G. H. Rim," Novel zero-voltage and zero-current-switching (ZVZCS) full bridge PWM converter with low output current ripple," in *IEEE INTELEC Conf. Rec.* 1997, pp. 257-262.

[10] J. G. Cho, J. W. Beak, C. Y. Jeong, D. W. Yoo, H. S. Lee, and G. H. Rim, "Novel zero-voltage and zero-current-switching (ZVZCS) full bridge PWM converter using a simple auxiliary circuit," in *IEEE APEC Conf. Rec.*,1998 pp. 834-839.

[11] Y. Jang, and D. L. Dillman, and M. M. Jovanović, "Three-phase isolated high power factor rectifier using soft-switching two-switched forward converter," in *IEEE APEC Conf. Rec.*, 2007, pp. 809-815.

Three-phase Voltage Doubler Rectifier Based on Three-state Switching Cell for Uninterruptible Power Supply Applications Using FPGA

Raphael A. da Câmara, P.P. Praça, C.M.T. Cruz, R.P. Torrico-Bascopé, C.E.A. Silva, D.S.Oliveira Jr., L.H.S.C. Barreto

Energy Processsing and Control Group, Electrical Engineering Department
Universidade Federal do Ceará
Fortaleza-CE, Brazil
raphaelpur@gmail.com

Abstract— **This paper presents a three-phase voltage doubler rectifier based on three-state switching cells for Uninterruptible Power Supply (UPS) applications using FPGA. Its main features are: high power factor, reduced conduction losses, weight and volume, simple control strategy based on One-cycle Control (OCC), and connection between input and output enabling the use of inverter and bypass. A theoretical analysis, simulation results and preliminaries experimental results from a 9kW development stage lab model are presented.**

I. INTRODUCTION

An equipment that has been highlighting in the power electronics on its ability to supply clean and reliable power to critical loads such as industrial processes, computers, network servers, telecommunications systems, medical systems, even in situations of power outages or anomalies of the mains is the Uninterruptible Power Supply (UPS). An UPS can be classified into three types: On-line, Line-interactive and Stand-by [1]. Among the different types of UPS systems, the on-line UPS system is widely recognized as the superior topology in performance, power conditioning and load protection [2].

On-line UPS systems consist of a rectifier, a battery set, an inverter, and a bypass. A typical single-phase on-line UPS system based on full-bridge converters is shown in Fig. 1 (a). In this configuration it is normally required an isolating

transformer for proper operation of the bypass circuit. This isolation transformer, when operating at the grid frequency, both size and cost are considerable.

Others topologies were proposed in literature to overcome this problem, using the isolation transformer in a high frequency DC link [3-5]. Although this UPS topology incorporating a high frequency transformer reduces weight of the system, it has increased the number of active switches and power stages, compromising the system's overall efficiency and reliability.

Transformerless UPS incorporating a common neutral bus line using a half-bridge converter and inverter has attracted special interest for applications in computer and telecommunication systems. A typical single-phase on-line UPS system is shown in Fig. 1 (b). This type of system is highly cost-effective and acceptable due to its total power conversion efficiency improvement, volume and weight reduction [6-9]. However, some disadvantages are found as: unbalance between the upper and lower side DC link capacitors, AC-DC and DC-AC converters switches are exposed to total DC link voltage [10].

The single-phase three-level rectifier with a half-bridge inverter can be advantageous for many applications [11-12]. In this converter only half of the DC link voltage is applied across the rectifier switches and the current flows through only

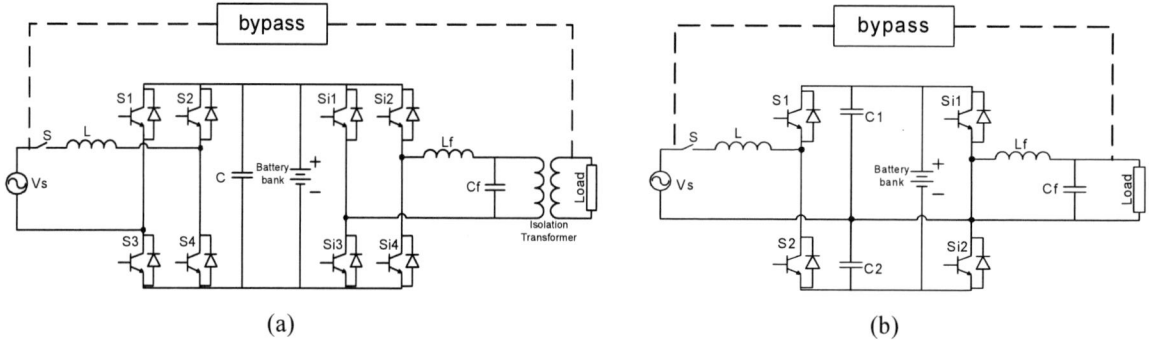

Figure 1. A typical single-phase on-line UPS system: (a) based on full-bridge converters; (b) based on half-bridge converters.

This work was supported by CAPES – Higher Education Improvement Coordination two or three power semiconductors simultaneously. Therefore, this converter presents less conduction losses, and a common neutral bus line is connected between the middle point of DC capacitors link, input and load, making it possible to realize the bypass operation without an isolating transformer. However, with the intention to work with output power over few kilowatts, this converter has high weight, volume, current stress on the semiconductors devices presenting a high cost of components and low efficiency. These drawbacks are solved by using of three-state switching cell [13] in the single-phase three-level rectifier presented in Fig. 2 [14]. But if the output power arise more than few kilowatts, it's necessary to work with three-phase topologies. Thus, the single-phase converter presented in Fig. 2 is updated to three-phase topology presented in Fig. 3.

Within this context, this paper presents a three-phase voltage doubler rectifier based on three-state switching cells for UPS application that presents power factor correction (PFC), reduced current stress on semiconductors devices, reduced volume and weight of the magnetic components, simple control strategy based on OCC using FPGA and connection between the middle point of DC link, input and load, enabling the bypass operation without an isolating transformer. The theoretical analysis, simulation results and preliminaries experimental results of a development stage lab model of 9kW output power are presented to validate the proposal.

II. PROPOSED CONVERTER CIRCUIT

A. Circuit Description

The proposed converter is shown in Fig. 3. It basically

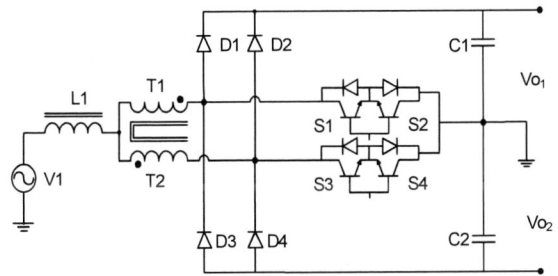

Figure 2. Single-phase rectifier topology.

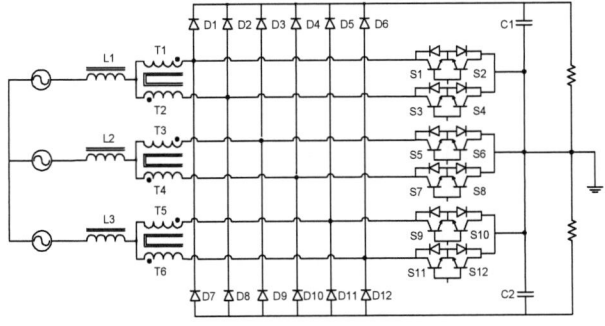

Figure 3. Proposed 3-phase rectifier topology.

consists of three single-phase rectifier, presented in Fig. 2, star-connected like VIENNA [15] but with three-state switching cell. Every single-phase rectifier has one inductor, one autotransformer with windings T1 and T2, four diodes, two bi-directional controlled switches. The converter operates only in continuous conduction mode (CCM). The converter operation modes are defined by comparison between input voltage and output voltage in function of controlled switches duty cycle of each phase. While the input phase voltage is less than the half of output voltage, the converter operates in overlapping mode (duty cycle > 0.5) and, while the input phase voltage is greater than the half of output voltage, the converter operates in non-overlapping mode (duty cycle < 0.5).

B. Converter Operation Analysis

The operation of proposed three-phase rectifier is similar to single-phase rectifier. Thus, to simplify the analysis, single-

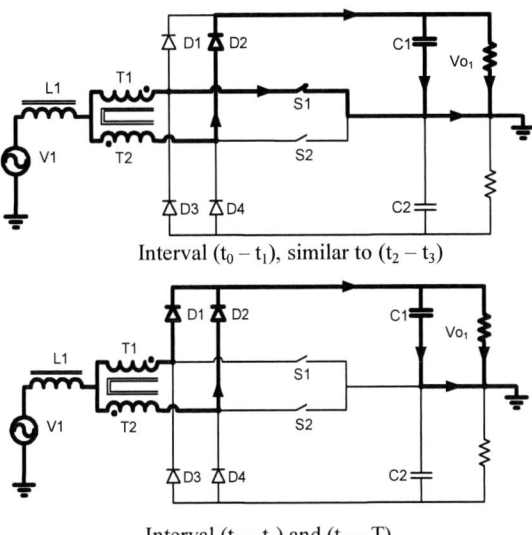

Figure 4. Operating stages in non-overlapping mode.

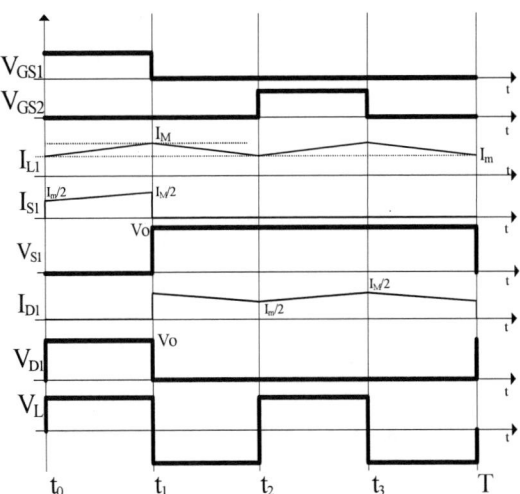

Figure 5. Main theoretical waveforms in overlapping mode.

978-1-4244-4782-4/10 $26.00 © 2010 IEEE

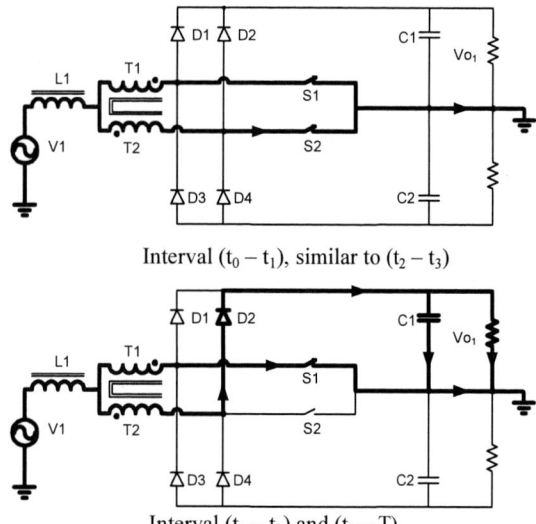

Interval ($t_0 - t_1$), similar to ($t_2 - t_3$)

Interval ($t_1 - t_2$) and ($t_3 - T$)

Figure 6. Operating stages in overlapping mode.

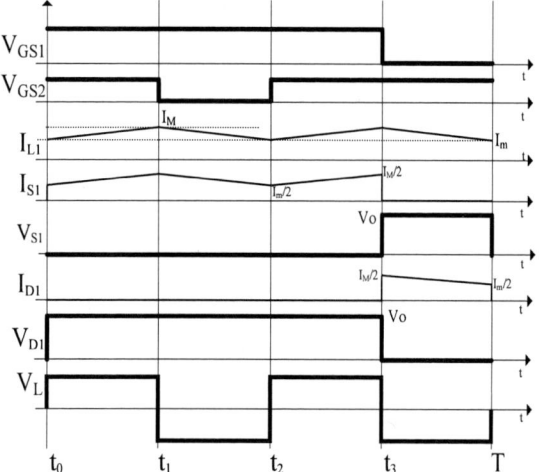

Figure 7. Main theoretical waveforms in overlapping mode.

phase rectifier operation is presented. The operating stages in non-overlapping mode are shown in Fig. 4 according to the main waveforms in Fig. 5. During the time intervals ($t0 - t1$) and ($t2 - t3$) occurs the energy storage in the inductor L1 and the current increases linearly. Half of the load current flows through diode D2 in time interval ($t0 - t1$) or D1 in time interval ($t2 - t3$) and half flows through switch S1 in time interval ($t0 - t1$) or S2 in time interval ($t2 - t3$). This way, the current stresses of switches are reduced. The transfer of the energy stored in the inductors to the load occurs in time intervals ($t1 - t2$) and ($t3 - T$). The operation of the topology in the negative semi-cycle of the input voltage is analogous to the positive one.

The operating stages in the overlapping mode are shown in Fig. 6 according to the main waveforms in Fig. 7. During the time intervals ($t0 - t1$) and ($t2 - t3$) occurs the energy storage in the inductor L1 and there is no power transfer from the

mains to the load. The transfer of the energy stored to the load occurs in time intervals ($t1 - t2$) and ($t3 - T$). The operation of the topology in the negative semi-cycle of the input voltage is analogous to the positive one.

Note that, for both operating modes, the inductor current frequency is the double of switching frequency enabling the size and volume reduction of magnetic core.

The duty cycle in each phase must vary on each commutation period for constant switch frequency in order to control the average value of the output voltages. Departing of the validation relation for the classic boost converter static gain, output voltage by input voltage, it has:

$$\frac{V_o}{V_{in}} = \frac{1}{1-D} . \tag{1}$$

Replacing adequately, it arrives the three-phase variation duty cycle term. The Fig. 8 represents the duty cycle variation expressed by (2).

$$\begin{cases} D_a(\omega t) = 1 - \dfrac{1}{\beta} \cdot \sin(\omega t) \\[2mm] D_b(\omega t) = 1 - \dfrac{1}{\beta} \cdot \sin(\omega t + 120°) \\[2mm] D_c(\omega t) = 1 - \dfrac{1}{\beta} \cdot \sin(\omega t - 120°) \end{cases} \tag{2}$$

Being β defined as the relation between output voltage and input peak voltage, by the following expression:

$$\beta = \frac{V_o}{V_{inpk}} . \tag{3}$$

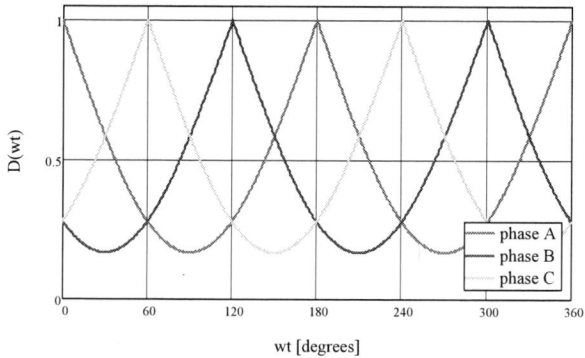

Figure 8. Three-phase duty cycle variation.

III. CONTROL STRATEGY

The control goal is to maximize the power factor, adjusting the input current waveform until it is equal to the input voltage waveform. The modulation strategy used is based on the appropriate variation of duty cycle at a constant frequency.

978-1-4244-4782-4/10 $26.00 © 2010 IEEE

The proposed control strategy is based on One-Cycle Control (OCC) technique for VIENNA rectifier [16]. Its main features arc:

- Constant switching frequency.

- Simple and reliable. This controller is composed of one or two integrators with reset along with some flip-flops, comparators, and some logic and linear components.

- No need for multipliers that are required to scale the

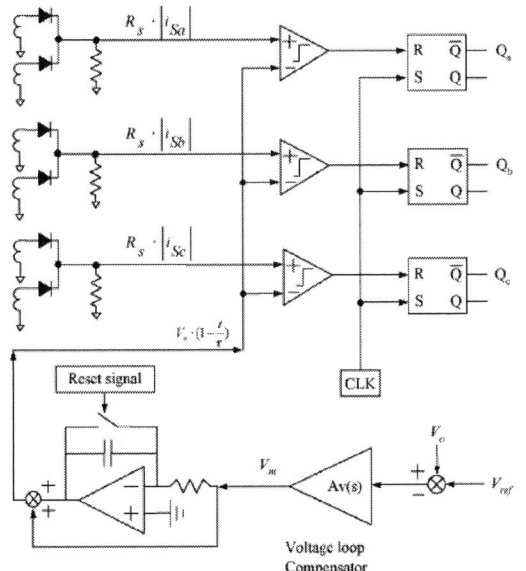

Figure 9. Schematic of three-PFC controller for VIENNA rectifier[16].

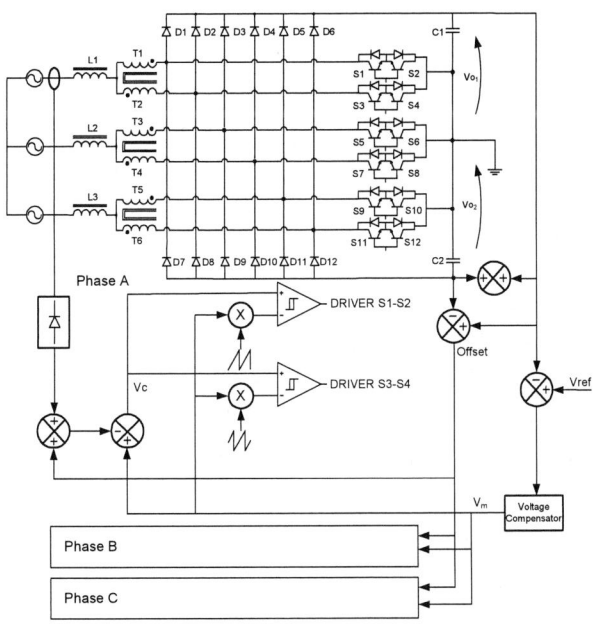

Figure 10. Proposed control block diagram of the 3-phase rectifier topology.

current reference according to the load level as used in many other control approaches.

- No three-phase input ac voltage sensors are required.

The schematic of three-PFC controller for the VIENNA rectifier is shown in Fig. 9. However, this controller is very complex for digital implementation. A simple way to achieve the same OCC response is multiplying a traditional saw-tooth carrier with signal v_m from voltage compensator. Throughout v_m is possible changing the peak value of the saw-tooth carrier without changing your period. To this converter is used two saw-tooth carrier with half phase shifted. The power stage and new control circuit schematic diagram is show in Fig. 10.

IV. DIGITAL IMPLEMENTATION

The digital control is implemented using the Cyclone® II EP2C20F484C7 FPGA, programmed with the software Quartus II®. It has the following features [17]:

- 18,752 LEs;

- 52 M4K RAM blocks;

- 315 I/O pins;

- 4 digital PLLs.

The development board has the following hardware features:

- Altera Cyclone® II EP2C20 FPGA device;

- 50 MHz, 27MHz and 24MHz oscillators;

- Two 40-pin expansion headers;

- 8-MB SDRAM memory, 512-kB SRAM, and 4-MB flash memory.

The controller was implemented using VHDL language and the block schematic diagram at Quartus® II software. In Fig. 11 is shown the top-level entity, which represents all controller hardware and its submodules for each phase.

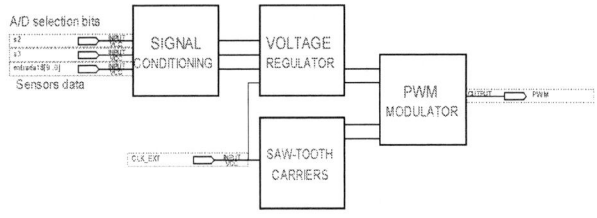

Figure 11. The block schematic diagram in Quartus® II.

A. Signal Conditioning

This block, shown in Fig. 12, is the entity that provides the values of voltage and current, using the A/D samples, generating an error signal, and offset signal between the output voltages to control the unbalanced output voltage in DC link. The "somador" block and the "lpm_add_subX" blocks are megafunctions provided by ALTERA® in Quartus® II software. The "demux" block were made using VHDL language.

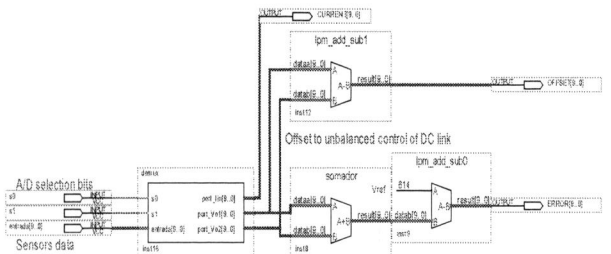

Figure 12. The signal conditioning block in Quartus® II.

B. Voltage Regulator

In the Fig. 13 is the voltage regulator block, which receives the error signal, offset signal and current signal from the previously block. This error is applied as an input signal, and the control signal is calculated using the following expression:

$$U_k = K_4 \cdot U_{k-1} - K_3 \cdot U_{k-2} + K_2 \cdot e_k + K_1 \cdot e_{k-1} - K_0 \cdot e_{k-2}. \quad (4)$$

U_k represents the current control signal, U_{k-1} represents the previously control signal, U_{k-2} represents the last control signal, e_k is the actual input error, e_{k-1} represents the previously input error and e_{k-2} represents the last input error. Constants K_4 to K_0 are the gains in the controller.

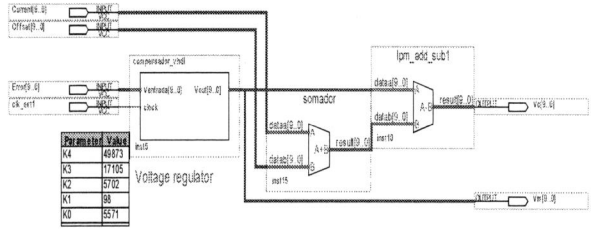

Figure 13. The voltage regulator block in Quartus® II.

C. Saw-tooth Carriers

This block is responsible to generate two saw-tooth carriers with half phase shifted and multiplying these carriers with the vm signal from voltage regulator block and it is shown in Fig. 14. The carriers are made by counters and comparators and the multiplier is made using VHDL.

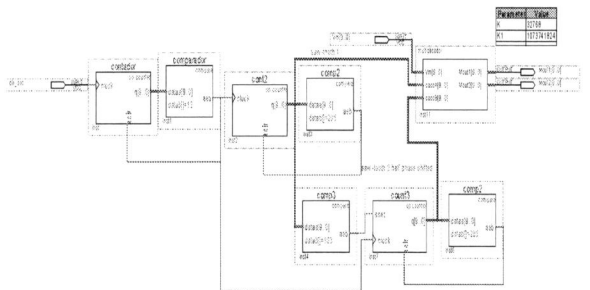

Figure 14. The saw-tooth carriers block in Quartus® II.

D. PWM Modulator

The block responsible to do the PWM modulation is shown in Fig. 15 and it is composed by two comparators. A voltage control v_c from voltage regulator block is compared with two saw-tooth carriers half phase shifted multiplied with v_m from saw-tooth carriers block. The comparator output is PWM signal used to control the switches at each phase in converter.

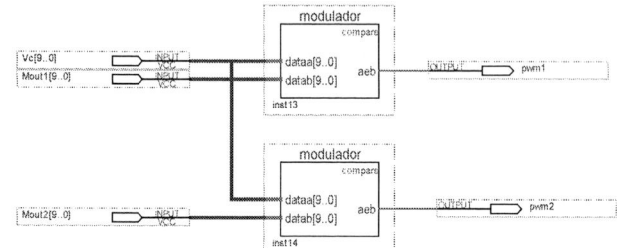

Figure 15. The PWM modulator block in Quartus® II.

V. EXPERIMENTAL RESULTS

A. Converter Specifications

The proposed three-phase voltage doubler boost rectifier based on three-state switching cells design specifications are shown in Table I. The converter switching frequency was assumed fs = 20kHz. All following design are made for each phase.

TABLE I.   DESIGN SPECIFICATIONS

Output power	Po = 9kW
Mains input voltage	V1 = 110Vac
Output voltage	Vo=200+200Vdc
Mains frequency	f_r = 60Hz
Output voltage ripple	ΔVo = 5%.Vo
Input current ripple	ΔI1 = 20%.I1
Theoretical efficiency	η = 0,97

1) Inductor: The inductance is calculated by (5), considering Vo = 200V:

$$L1 = \frac{V_o}{16 \cdot \Delta I1 \cdot f_s} = 62,87 \mu H. \quad (5)$$

To the project is adopted L1 = 60μH. The rms inductor current and the inductor peak current is determined using (6) and (7), respectively:

$$I_{rmsL1} = \frac{Po}{3 \cdot \eta \cdot V1} = 28,12A. \quad (6)$$

$$I_{pL1} = \sqrt{2} \cdot I_{rmsL1} = 39,76A. \quad (7)$$

2) Autotransformer: The rms winding current and the peak current is calculated by (8) and (9):

$$I_{rmsT1} = \frac{I_{rmsL1}}{2} = 14,06A \; . \tag{8}$$

$$I_{pT1} = \frac{I_{pL1}}{2} = 19,88A \; . \tag{9}$$

3) Controlled switches: The switch average current and rms current is given by (10) and (11), respectively:

$$I_{avgS1} = \frac{(4 \cdot \beta - \pi) \cdot Io}{3 \cdot \pi \cdot \eta} = 4,93A \; , \tag{10}$$

$$I_{rmsS1} = \frac{2 \cdot Io}{3 \cdot \eta} \cdot \sqrt{\frac{\beta \cdot (3 \cdot \pi \cdot \beta - 8)}{6 \cdot \pi}} = 8,19A \; , \tag{11}$$

4) Diodes: The diode average current is given by (12):

$$I_{avgD1} = \frac{Io}{4} = 11,25A \; . \tag{12}$$

5) Filter capacitors: The capacitance value of C1 = C2 is defined by (13):

$$C1 \geq \frac{I_{pL1}}{2 \cdot \pi \cdot f_r \cdot 12 \cdot \beta \cdot \Delta V_o} \cong 1360\mu F \; . \tag{13}$$

The converter parameters are presented in Table II.

TABLE II.   PARAMETERS OF CONVERTER

Inductor L1	L = 60µH NEE – 65/33/26 (Thornton Ipec) N_L = 16 turns (37 x 22AWG)
Autotransformer	NEE – 55/28/21 (Thornton Ipec) N_p = N_s = 12 turns (15 x 22AWG)
Switches S1 – S4	IRGP50B60PD1
Diodes D1 - D4	30EPH06
Capacitors C1 and C2	C = 1360µF (2x680µF/ 350V)

B. Simulation Results

Fig. 16 shows the input phase voltage (Va) and input

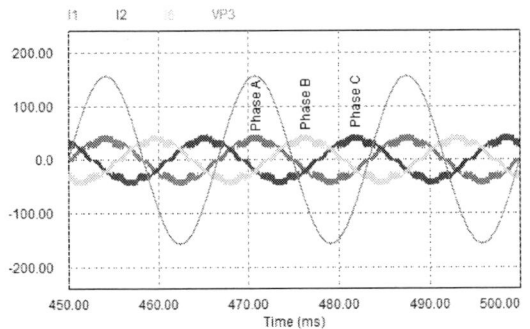

Figure 16. Input phase voltage (V_a) and input currents in each phase waveforms.

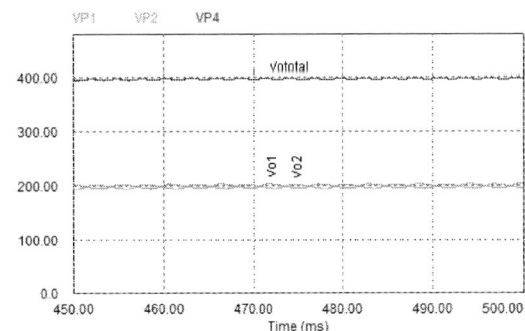

Figure 17. Ouput voltage in each capacitor (V_{01} and V_{02}) and total output voltage (V_{ototal}).

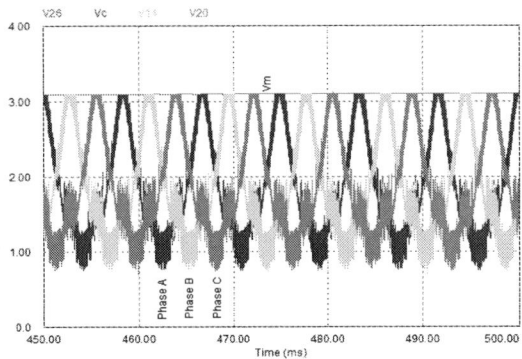

Figure 18. Control voltages in each phase and voltage v_m waveforms.

currents in each phase waveforms where power factor correction achieved is 99.25%. Fig. 17 shows the output voltages in each capacitor and total output voltage waveforms. Fig. 18 shows the control voltages v_c in each phase and the output of voltage feedback regulator v_m waveforms.

C. Experimental Results

A lab model of the proposed converter is actually in development stage. Its picture is shown in Fig. 19. Thus,

Figure 19. Lab model picture.

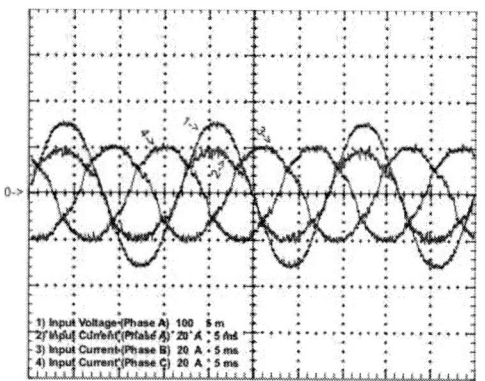

Figure 20. Input phase voltage and input currents in each phase (100V/div, 20A/div, 5ms/div).

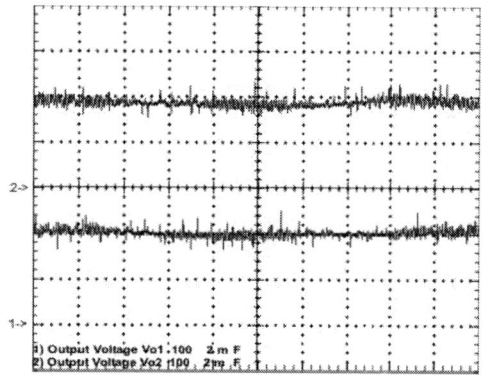

Figure 21. Output voltage in each capacitor V_{01} and V_{02} (100V/div, 2ms/div).

preliminaries experimental results are presented. Fig. 20 shows the input phase voltage and input currents in each phase waveforms where power factor correction is observed. Fig. 21 shows the output voltage in each output capacitor.

VI. CONCLUSIONS

A three-phase rectifier with PFC characteristics based on three-state switching cell and its theoretical analysis, digital implementation, simulation results, and preliminaries experimental results is presented in this paper. The important features observed in the proposed converter are: high input power factor, new simple control scheme based on OCC using digital control with FPGA, reduced conduction losses, reduced volume and weight of the magnetic components (about 53% if compared with VIENNA topology) and connection between input and output enabling by-pass without isolating transformer.

ACKNOWLEDGMENT

The authors would like to thank the Energy Processing and Control Group – GPEC and the Electrical Engineering Department for all material, physical and mental support.

REFERENCES

[1] S. B. Bekiarov, A. Emadi, "Uninterruptible Power Supplies:Classification, Operation, Dynamics, and Control," *in IEEE 2002 Applied Power Electronics Conference*, 2002, pp. 597-604.

[2] J. M. Guerrero, L. G. Vicuna, J. Uceda, "Uninterruptible power supply systems provide protection", *IEEE Transactions on Industrial Electronics Mag.*, vol. 1, no. 1, pp. 28-38, 2007.

[3] K. Hirachi, et al., "Switched-mode PFC rectifier with high-frequency transformer link for high-power density single phase UPS", *in Proc. of PESC'97*, vol. 01, pp. 290-296, 1997.

[4] R. Krishnan, "Design and development of a high frequency on-line uninterruptible power supply", *in Proc. of IECON'95*, vol. 01, pp. 578-583, 1995.

[5] R. P. Torrico-Bascopé, E. M. Sá Jr, C. G. C. Branco, F. L. M. Antunes, "PFC pre-regulator with high frequency isolation using full-bridge chopper for UPS applications", *in Proc. of INDUSCON'04*, vol. 01, 2004.

[6] K. Hirachi, M. Sakane, S. Niwa, T. Matsui, "Development of UPS using new type of circuits", *in IEEE 1994 INTELEC*, 1994, pp. 635-642.

[7] N. Hirao, T. Satonaga, T. Uematsu, T. Kohama, T. Ninomiya, M. Shoyama, "Analytical considerations on power loss in a three-arm-type Uninterruptible Power Supply", *in IEEE 1998 PESC*, 1998, vol. 2, pp. 1886-1891.

[8] K. Hirachi, A. Kajiyama, T. Mii, M. Nakaoka, "Cost-effective bidirectional chopper-based battery link UPS with common input-output bus line and its control scheme", *in IEEE 1996 IECON*, 1996, vol. 3, pp.1681-1686.

[9] S. B. Bekiarov, A. Nasiri, A. Emadi , "A new reduced parts on-line single phase UPS system", *in IEEE 2003 IECON*, 2003, vol. 1, pp. 688-693.

[10] C. G. C. Branco, C. M. T. Cruz, R. P. Torrico-Bascope, F. L. M. Antunes, L. H. S. C. Barreto, "A transformerless single phase on-line UPS with 110V/220V input output voltage", *in IEEE 2006 Applied Power Electronics Conference*, 2006, pp. 348-354.

[11] J. C. Salmon, "Circuit topologies for single-phase voltage doubler boost rectifiers", *IEEE Transactions on Power Electronincs*, vol. 8, no. 4, pp. 521-529, Oct. 1993.

[12] G. J. Su, T. Ohno, "A new topology for single phase UPS systems", *in 1997 PCC*, 1997, vol. 2, pp. 913-918.

[13] G. V. T. Bascopé, I. Barbi, "A single phase PFC 3kW converter using a three-state switching cell," *in IEEE 2004 Power Electronics Specialist Conference*, 2004, vol. 5, pp. 4037-4042.

[14] R. A. da Câmara, P. P. Praça, C. M. T. Cruz, R. P. Torrico-Bascopé, "Voltage doubler boost rectifier based on three-state switching cell for UPS applications", *in IEEE 2009 IECON*, 2009, vol. 1, pp. 947-952.

[15] J. C. Salmon, "Circuit topologies for PWM boost rectifiers operated from 1-phase and 3-phase ac supplies and using either single or split dc rail voltage outputs", *in IEEE 1995 Applied Power Electronics Conference*, 1995, pp. 473-479.

[16] C. Qiao and K. M. Smedley, "Three-phase unity-power-factor star-connected switch (VIENNA) rectifier with unified constant-frequency integration control", *IEEE Transactions on Power Electronics*, vol. 18, no. 4, pp. 952-957, 2003.

[17] ALTERA, "Cyclone II FPGA Starter Development Board – Reference Manual", 2006.

Multi-loop control algorithms for Seamless Transition of Grid-connected Inverter

Qin Lei
ECE department,
Michigan State University
East Lansing, MI, USA
leiqin@msu.edu

Shuitao Yang
ECE department,
Zhengjiang University
Hangzhou, Zhejiang, PRC
styang@msu.edu

Fang Z. Peng
ECE department,
Michigan State University
East Lansing, MI, USA
fzpeng@egr.msu.edu

Abstract-The grid-connected inverter works as a controlled current source in grid-connected mode, while operates as a controlled voltage source in standalone mode. So in case of utility faults or intentional islanding, the inverter has to change its control strategy from current control to voltage control. This paper first proposed a multi-loop voltage controller with capacitor differential voltage feedback inner loop and voltage reference feedforward for standlone system especially designed to maintain the voltage continuity and decrease the dynamic response time in transition. However, the turn-off characteristics of the SSR which is used as switch here makes the transition last for a long time up to half a cycle. So in order to force the grid currents through the SSR switches to decrease to zero at much less time and make the voltage fluctuates within permissible levels during SSRs turn-off period, the voltage control based voltage amplitude regulation, instantaneous voltage regulation algorithms and current control based zero current regulation algorithms have been adopted in transition. After disconnection from the grid, the inverter will recover its voltage to a rated level. Simulation and experiments are carried out to verigy the proposed controllers and algorithms.

I. INTRODUCTION

In the transition from grid-connected to standalone operation, Solid State Relay (SSR) is used in this paper as the switch between DG and grid which can not turn off right after the driving signal has been removed until the current drops to zero. In order to keep the output voltage less distorted, it is preferred to force the current falling down to zero as soon as possible. Generally, the inductor current could be driven to zero by establishing a negative voltage [1] on it or by controlling the current to be zero directly. Fig. 1 shows the system configuration for transition. A multi-loop voltage controller with capacitor differential voltage feedback inner loop and voltage reference feedforward has been adopted for standalone system and a current controller with grid voltage feedforward has been chosen for grid-connected system. Based on these, some new control strategies for transition have been analyzed in this paper. Simulations and experiments have been carried out to verify the operation principle and features.

II. CONTROL STRATEGY FOR STANDALONE AND GRID CONNECTED SYSTEM

A. Multi-loop controller for standalone system

For standalone system, a multi-loop voltage controller with capacitor current feedback as inner loop can eliminate the output LC filter resonant peak to increase stability in load disturbance and enhance dynamic performance [2]. Compared to inductor current feedback, it has lower gain for distortion load current hence has better disturbance rejection capability. However, it has high requirement for the accuracy and dynamic performance of the current sensor because capacitor current is small-scale. So capacitor voltage differential feedback is utilized in this paper [3] to replace capacitor current feedback as shown in Fig.2, which saves a high quality required current sensor. In addition, considering the grid-connected to standalone transition performance, only a PI controller would make the initial output voltage be zero, hence make it discontinuous to the previous voltage. In order to overcome this limitation, a reference voltage feed-forward synchronized with grid is proposed to maintain the waveform continuity and also boost dynamic response. Fig. 3 shows the step response for the closed-loop transfer function with and without voltage reference feedforward. It can be verified that adding a feedforward can boost the dynamic response which is beneficial in transition. Also, a single proportional controller is selected to replace PI controller in order to reduce the harmonics in the waveform which will be analyzed as follows. The output voltage-to-reference voltage and output voltage-to-load current transfer function can be expressed as follows:

$$
\begin{aligned}
v_o(s) = {}& G_{vo\_vref}(s) \cdot v_{ref} - G_{vo\_io}(s) \cdot i_o = \\
& \frac{(1+K_v)e^{-sT_s}}{L_f C_f s^2 + (R_f C_f + K_d e^{-sT_s})s + K_v e^{-sT_s} + 1} v_{ref} \\
& - \frac{L_f s + R_f}{L_f C_f s^2 + (R_f C_f + K_d e^{-sT_s})s + K_v e^{-sT_s} + 1} i_o
\end{aligned} \tag{1}
$$

Fig. 4 compares the bode diagrams for voltage closed-loop transfer function with P and PI controller. By only using a P controller can damp the amplitude peak rise in PI controller between 60HZ and 900 HZ to zero hence reduces the

978-1-4244-4782-4/10 $26.00 © 2010 IEEE

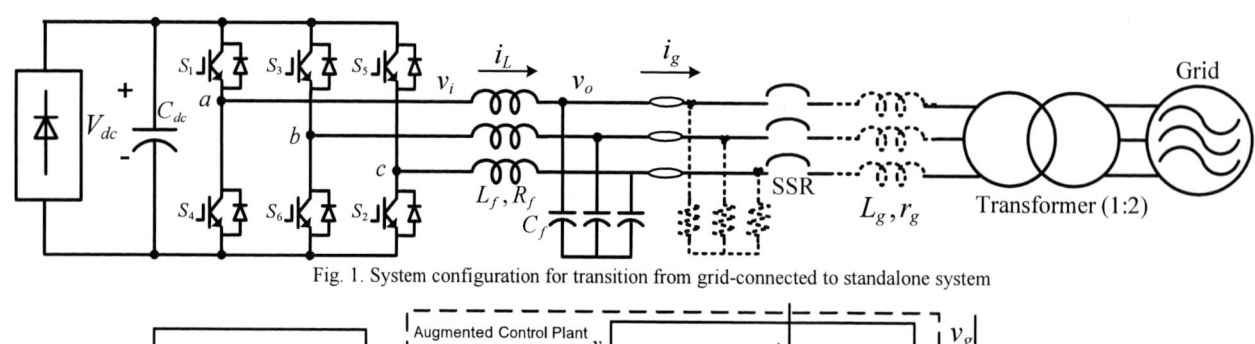

Fig. 1. System configuration for transition from grid-connected to standalone system

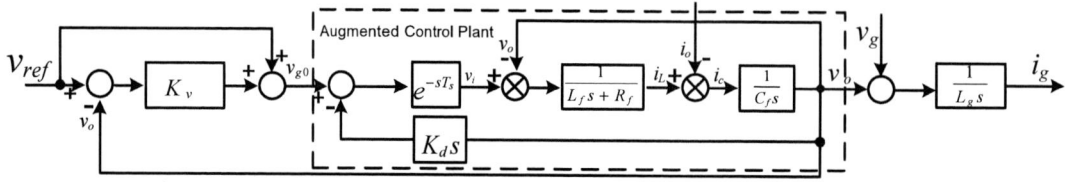

Fig. 2. Voltage control block diagram for standalone system

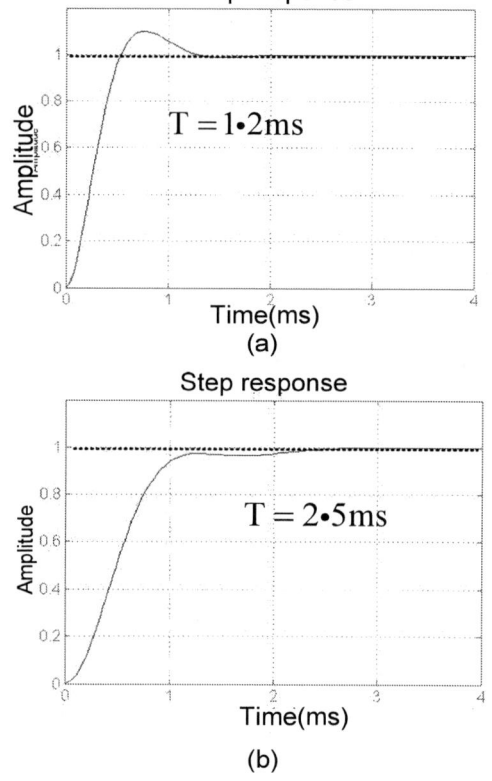

Fig. 3 Step response of the closed loop transfer function

(a)with voltage feedforward (b)without voltage feedforward

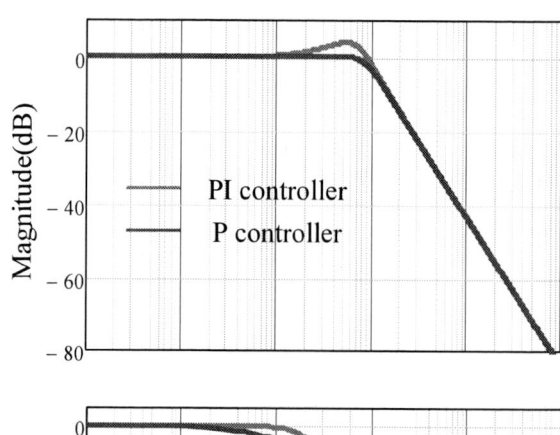

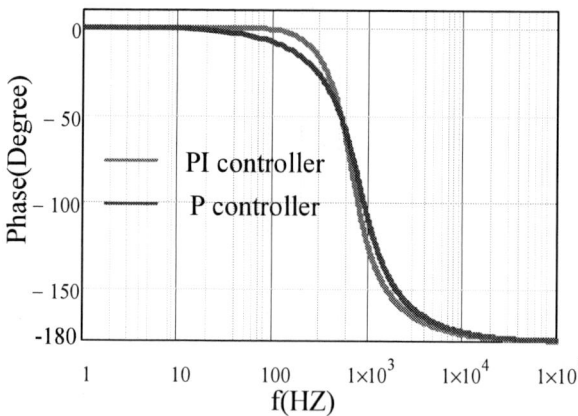

Fig. 4. Bode diagrams of $G_{vo\_vref}(s)$ with P or PI controller

harmonic components. Also with feedforward, when K_v=0.3, the steady state error is within 0.07%, but without feedforward loop, the steady state is 10% even when K_v rises up to 9. Obviously that feedforward path enhances the steady-state performance and system stability.

B. Current controller for grid-connected system

A convertional proporitional-integrator (PI) controller along with a capacitor voltage v_o feed-forward compensator has been used here for grid-connected operation[4-5], as shown in Fig.5 (a). Capacitor voltage feed-forward is employed to reduce the effect of the grid voltage and to increase dynamic response. Fig.5 (b) shows the simplified control block diagram. Without considering the grid

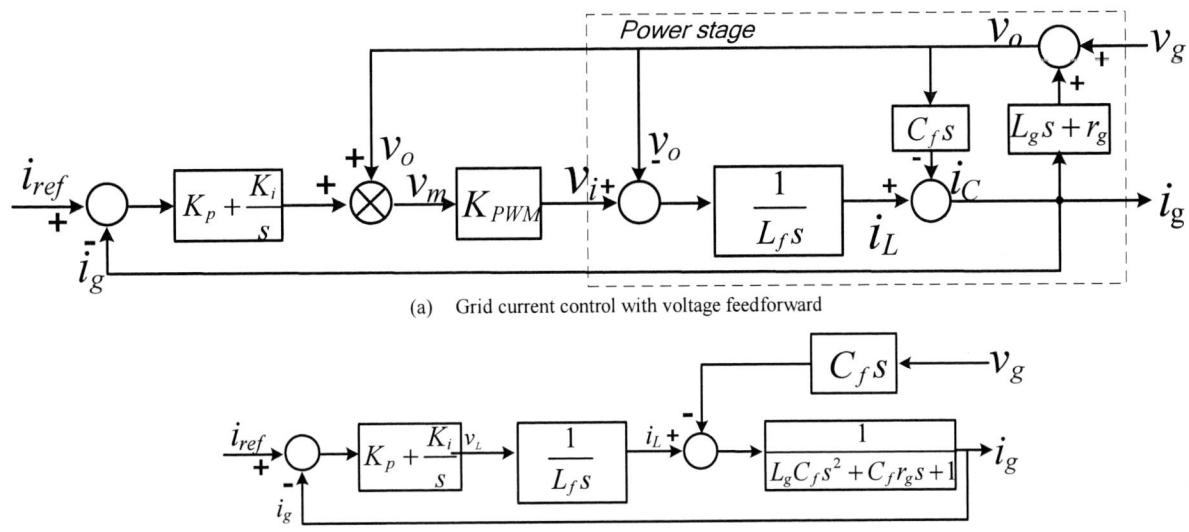

(a) Grid current control with voltage feedforward

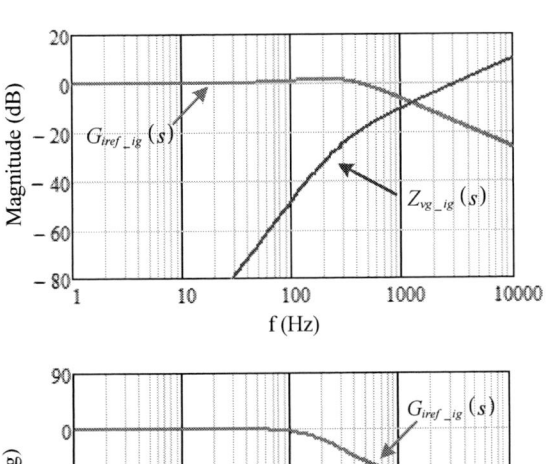

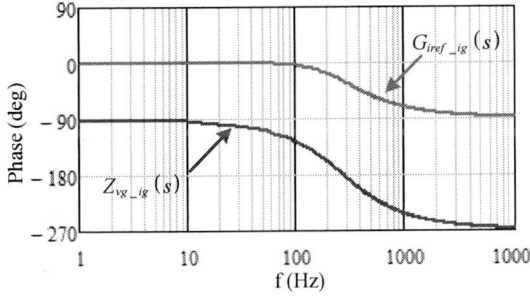

(b) Simplified control block diagram

Fig. 5. Current control block diagram for grid-connected system

impedance, the closed-loop transfer function of the grid current can be obtained as:

$$I_g(s) = G_{iref\_ig}(s)I_{ref}(s) - Z_{vg\_ig}(s)V_g(s)$$
$$= \frac{K_p s + K_i}{L_f s^2 + K_p s + K_i}I_{ref}(s) - \frac{L_f C_f s^3}{L_f s^2 + K_p s + K_i}V_g(s) \quad (2)$$

The bode plot of closed loop transfer function is shown in Fig. 6. It has small tracking error and high distortion rejection. With considering Z_g, the closed-loop transfer function of the grid current can be rewritten as:

$$I_g(s) = G'_{iref\_ig}(s)I_{ref}(s) - Z'_{vg\_ig}(s)V_g(s) =$$
$$\frac{K_p s + K_i}{L_f L_g C_f s^4 + L_f C_f r_g s^3 + L_f s^2 + K_p s + K_i}I_{ref}(s)$$
$$-\frac{L_f C_f s^3}{L_f L_g C_f s^4 + L_f C_f r_g s^3 + L_f s^2 + K_p s + K_i}V_g(s) \quad (3)$$

Compared to previous case, there will be an additional second-order transfer function $G_a(s)$ in the control loop,

$$G_a(s) = \frac{1}{L_g C_f s^2 + C_f r_g s + 1} = \frac{\omega_n^2}{s^2 + 2\xi_a \omega_n s + \omega_n^2} \quad (4)$$

,where $\omega_n = 1/\sqrt{L_g C_f}, \xi_a = r_g \sqrt{C_f/L_g}$. If the line impedance is high inductive, r_g is relatively small, then the $G_a(s)$ in (4) will be "under-damping", which may cause the instability. So the compensator parameter has been chosen accordingly to adjust the steady-state error and system stability margin.

III. PRINCIPLE AND ANALYSIS OF TRANSFER STRATEGIES

A. Voltage control based transfer strategies

The voltage amplitude regulation, and instantaneous voltage regulation [6] are both based on applying voltage control strategy in transition time. Fig. 7 shows the voltage and current phasor diagram for above two strategies. Voltage amplitude regulation method represses the current by increasing or decreasing voltage amplitude while keep phase the same in the transition. Once the grid current is force to decrease to zero, the SSR are turned off and the reference output voltage is recovered to the rated value. It takes less time to complete the transfer process hence minimize the voltage distortion. Instantaneous voltage regulation is to generate a constant voltage difference holding a fixed ratio to the initial grid current at transition start, which can settle the transition time to a fixed value. The detailed equations between transition time Δt and other parameters are shown

Fig.6. Bode diagrams of $G_{iref\_ig}(s)$ and $Z_{vg\_ig}(s)$

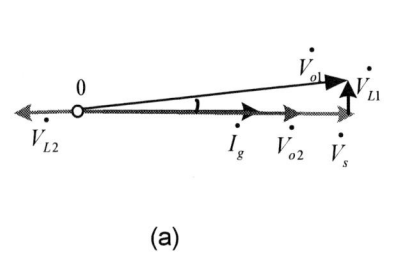
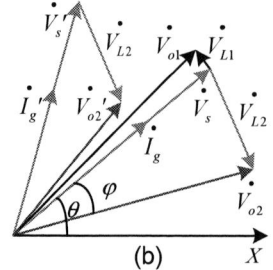

(a)　　　　　　　　**(b)**

Fig. 7.Phasor diagram at (a)Voltage amplitude regulation (b) Instantaneous voltage regulation

Table I. Equations for voltage, current and transition time

	i_g	v_s	v_{o2}	v_L	Δt
Voltage Amplitude Regulation	$I_g \sin(\omega t)$	$V_{sm} \sin(\omega t)$	$V_{02m} \sin(\omega t)$	$v_L = v_{o2} - v_s = L_g \dfrac{di_g}{dt}$	$\Delta t = \dfrac{-i_g}{di_g/dt} = \dfrac{L_g I_{gm}}{V_{sm} - V_{o2m}}$
Instantanous Voltage Regulation	$I_g \sin(\omega t + \theta)$	$V_{sm} \sin(\omega t + \theta)$	$v_s + v_b$	$v_L = V_b = kv_L(t_0)$	$\Delta t = \dfrac{L_g i_g(t_0)}{V_b} = \dfrac{1}{k\omega} = Cons \tan t$

in Table.1. The variables used are defined as: v_s -- grid phase voltage; v_{o1} -- initial inverter output phase voltage; v_{o2} -- regulated inverter output phase voltage; i_g --grid side phase current; t_0 --the moment that the drive signal of SSR is given; v_L --grid side inductor regulated voltage; V_b --pre-set voltage drop on the inductor.

B. Current control based transfer strategies

The principle of zero current regulation is to retain the current control mode in transition but change the current reference to zero. After the current drops to zero, SSR will turn off and the system shifts to voltage control. Due to the delay of zero current sensing, there is a blank time between disconnection and control mode shift in which voltage is out of control. However, the zero current regulation dynamic response depends on the step response time of current control loop, which is shorter than the time for voltage control loop, thus it gets better dynamic performance. However, this method doesn't rely on the grid side voltage and also don't need to sense the voltage accurately. So it is preferred in grid voltage short circuit or highly disturbed case.

IV. SIMULATION AND EXPERIMENTAL RESULT

The system parameter in the simulation and experiment is: $\omega = 377 rad/s$, $f_{sw} = 10kHZ$, $V_{sm} = 85V$, $V_{dc} = 200V$, $L_f = 1mH$, $C_f = 50uF$, $L_g = 0.1mH$, $r_g = 0.1\Omega$, $I_{ref} = 20A$

Assume grid absorbs current from inverter. Fig. 8 (a) (b) (c) show the simulation results for the three strategies respectively. The current reference in grid-connected mode is set to be 20A. In Fig.8 (a), the inverter output voltage reference is set to be 0.8 times of the grid voltage in the transition. According to the equations in table I, the transition time is calculated to be 0.2ms. The total transition time is 0.8ms by adding the caculated value and voltage loop step response rising time 0.6ms, which is coincident with the simulation results.

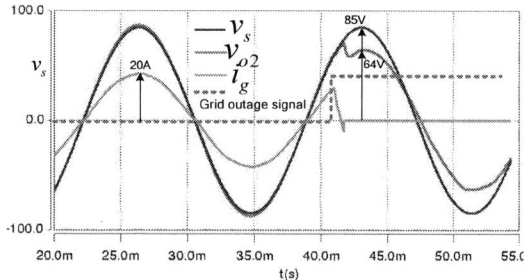

(a)

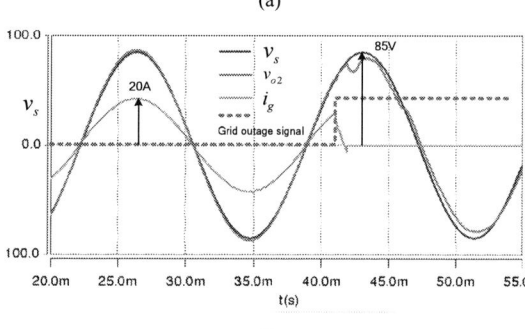

(b)

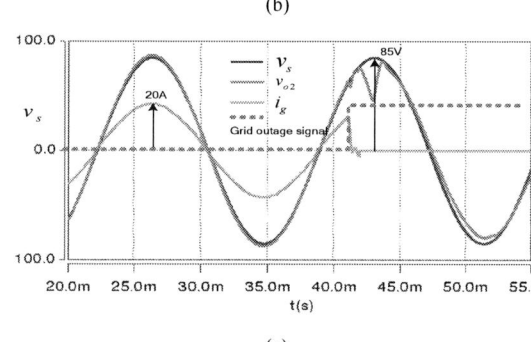

(c)

Fig. 8. Simulation results for grid voltage V_s, inverter output voltage V_{o2}, grid side current i_g in transition using different strategy (a) voltage amplitude regulation (b) voltage instantaneous value regulation (c) zero current regulation

978-1-4244-4782-4/10 $26.00 © 2010 IEEE　　　　847

In Fig.8 (b), the constant voltage difference is 20V, so the transition time is fixed at 1ms as calculated, which is the same as simulation result. In Fig. 8 (c), as mentioned before current regulation method has relatively small transition time but bigger voltage distortion. Fig. 9 shows the experimental results for voltage instantaneous value regulation and zero current regulation and Fig. 10 shows the picture of experiment hardware. The three upper sinusoidal waveforms are the inverter output voltage and the three lower ones are the current before the load (load are connected in parallel with capacitor on the grid side). So the current after the transition is equal to the load current which is not zero as simulated one. The green one is the grid outage signal, which also indicates the time that the SSR off signal is sent out. From the experimental waveforms, it can be seen that the voltage distortion and transition time are both in a reasonable range. Voltage control based regulation will last for longer time than current control based strategy, while the current

Fig. 10. Experiment hardware prototype

control one may cause a relatively big voltage amplitude and phase change. Voltage based algorithms are used when the grid maintain its voltage after transition and current based algorithm is used when grid voltage is highly distorted.

V. CONCLUSION

This paper proposes the multi-loop voltage controller with reference feedforward strategy for standalone system in order to obtain good stiffness and dynamic performance in transition. Also voltage based and current control based algorithm are adopted in transition to force the current to decrease to zero at a short time. The simulation and experimental results show that the proposed controller and algorithms can provide seamless transfers between the two operating modes for the inverter, avoiding the temporarily uncontrolled output voltage. It can be valuable for grid-connected inverters such as PV and fuel cell generation system.

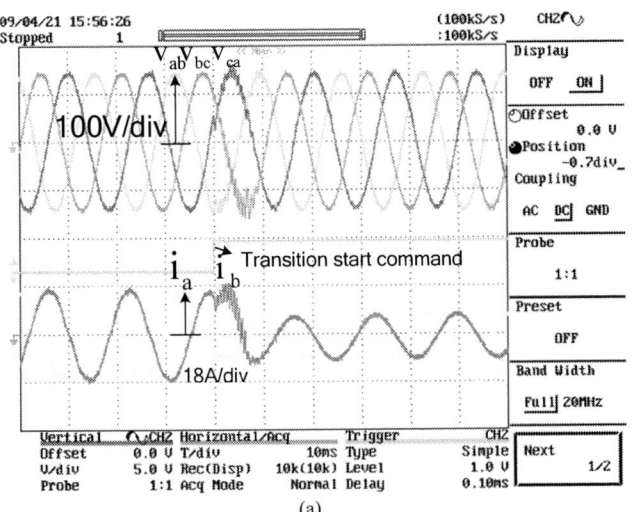

(a)

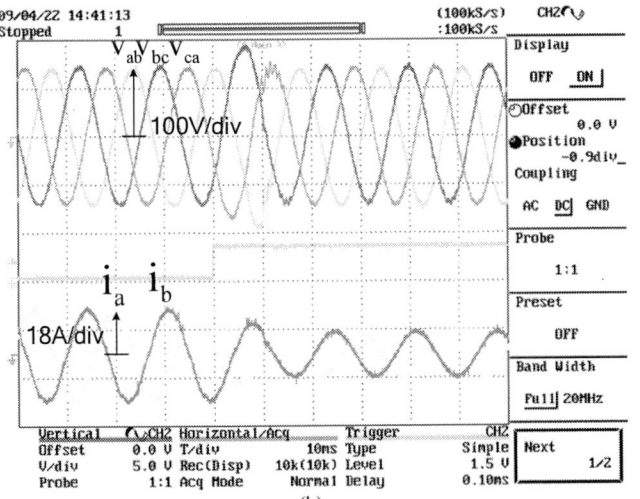

(b)

Fig. 9. Experiment results for inverter output line to line voltage and phase current in transition using two strategies (a) Instantaneous voltage regulation (b)Zero Current regulation

REFERENCE

[1] Guoqiao Shen, Dehong Xu and Danji Xi, "Novel seamless transfer strategies for fuel cell inverters from grid-tied mode to off-grid mode," Applied Power Electronics Conference and Exposition, 2005. APEC 2005. Twentieth Annual IEEE, Volume 1, 6-10 March 2005 Page(s):109 - 113 Vol. 1.

[2] Abdel-Rahim, N.M. and Quaicoe, J.E., "Analysis and design of a multiple feedback loop control strategy for single-phase voltage-source UPS inverters," Power Electronics, IEEE Transactions, on Volume 11, Issue 4 July 1996 Page(s):532 – 541.

[3] Shuitao Yang, Xinping Ding, Jinyun Liu and Zhaoming Qian, "Analysis and Design of a Cost-Effective Voltage Feedback Control Strategy for EPS Inverters," Power Electronics Specialists Conference, 2007. PESC 2007. IEEE 17-21 June 2007 Page(s):477 – 482.

[4] D. M. Brod and D. W. Novotny, "Current control of VSI-PWM inverters," IEEE Trans. Ind. Appl., vol. IA-21, no.4, pp. 562–570, Jul./Aug. 1985.

[5] M. P. Kazmierkowski and L. Malesani, "Current control technique for three-phase voltage-source PWM converters: A survey," IEEE Trans. Ind. Electron., vol. 45, no. 5, pp.691–703, Oct. 1998.

[6] Guoqiao Shen, Dehong Xu and Xiaoming Yuan, "Instantaneous Voltage Regulated Seamless Transfer Control Strategy for Utility-interconnected Fuel cell Inverters with an LCL-filter,"Power Electronics and Motion Control Conference, 2006.IPEMC 2006. CES/IEEE 5th International, Volume 3,14-16 Aug. 2006 Page(s):1- 5

Digital Controller Development for Grid-Tied Photovoltaic Inverter with Model Based Technique

Zhigang Liang[1], *Student Member, IEEE*, Larry Alesi[2],

Xiaohu Zhou[1], *Student Member, IEEE*, Alex Q. Huang[1], *Fellow, IEEE*

[1]FREEDM Systems Center
North Carolina State University
Raleigh, NC, 27695 USA
Email: zliang2@ncsu.edu

[2]MegaWatt Solar Inc
437 Dimmocks Mill Road, Suite 39
Hillsborough, NC 27278, USA
Email: Larry.Alesi@megawattsolar.com

Abstract— the main objective for a grid-tied photovoltaic (PV) inverter is to feed the harvested energy from PV panels to the grid with high efficiency and high power quality. A digital controller is the "heart" of the PV system: it calculates the maximum power point (MPP) and regulates the output current to meet the utility inter-connection standards. This paper introduced a new approach "model based design" to develop current controller for PV inverter. This design methodology filled the gap between computer simulation and hardware implementation of the digital controller and made an easy way for the implementation of complex and high level algorithm in a digital signal processor (DSP) [1]. The inverter's small signal model in D-Q rotating frame is derived and a double loop current controller is designed. The performance of the developed controller has been verified by both of the simulation in MATLAB and experimental results from a 2kW single phase PV inverter prototype.

I. INTRODUCTION

A grid-tied PV inverter must meet requirements from the utility companies. Detailed requirements can be found in IEEE1547 [2], EN61000-3-2 (applied in Europe) [3] and certain parts in U.S. National Electrical Code (NEC) [4]. Table 1 [5] gives a summary of the requirements. Basically, the inverter's controller should regulate the output current into sinusoidal and it should make response to islanding situation. Moreover, the inverter should make the PV panel work at its MPP and the DC bus voltage usually needs to be regulated in a two stage PV inverter. For these tough control tasks, a sophisticated DSP is usually required. The conventional digital controller development needs a lot of effort to translate control algorithm to executable code in DSP which is time consuming. To some extent, this design procedure impeded the verification of control algorithm. In this paper, the model based method has been adopted to design the controller for PV inverter. Firstly, the system control block diagram is described and small signal model in DQ rotating frame for DC/AC converter with LCL filter is derived. Then the building blocks of the current controller, such as Phase Locker Loop (PLL)

TABLE I.   A SUMMERY OF STANDARDS FOR GRID-TIED PV APPILCATIONS [5]

ISSUE	IEC61727 [3]	IEEE1547 [5]	EN61000-3-2 [4]
Nominal power	10 kW	30 kW	16 A × 230 V ≈ 3.7 kW
Harmonic currents	(3-9) 4.0%	(2-10) 4.0%	(3) 2.30 A
(Order – h) Limits	(11-15) 2.0%	(11-16) 2.0%	(5) 1.14 A
	(17-21) 1.5%	(17-22) 1.5%	(7) 0.77 A
	(23-33) 0.6%	(23-34) 0.6%	(9) 0.40 A
		(> 35) 0.3%	(11) 0.33 A
			(13) 0.21 A
			(15-39) 2.25/h
	Even harmonics in these ranges shall be less than 25% of the odd harmonic limits listed.		Approximately 30% of the odd harmonics -see standard.
Maximum current THD	5.0%		-
Power factor at 50% of rated power	0.90	-	
DC current injection	Less than 1.0% of rated output current.	Less than 0.5% of rated output current.	< 0.22 A -corresponds to a 50 W half-wave rectifier.
Voltage range for normal operation	85% - 110%	88% - 110%	-
	(196 V – 253 V)	(97 V – 121 V)	
Frequency range for normal operation	50 ± 1 Hz	59.3 Hz to 60.5 Hz	-

block, single phase D-Q transformation blocks, anti- windup PID control block and other computation blocks have been designed and verified by simulation in MATLAB. Finally, these blocks are optimized for code generation in the model based environment and experimental results on a PV inverter prototype verified the performance of designed controller.

II. SYSTEM STRUCTURE AND SMALL SIGNAL MODEL FOR DC/AC INVERTER WITH LCL FILTER

Fig. 1 depicted the system structure. Basically this is a two stage system and the DC/AC stage with its control is the main concern of this paper. Since LCL filter can attenuate the harmonic current into the grid and it's not sensitive to grid side impedance variation, it makes the inverter interface with

---

This work is sponsored by MegaWatt Solar Inc and made use of ERC shared facilities supported by the National Science Foundation under Award Number EEC-08212121.

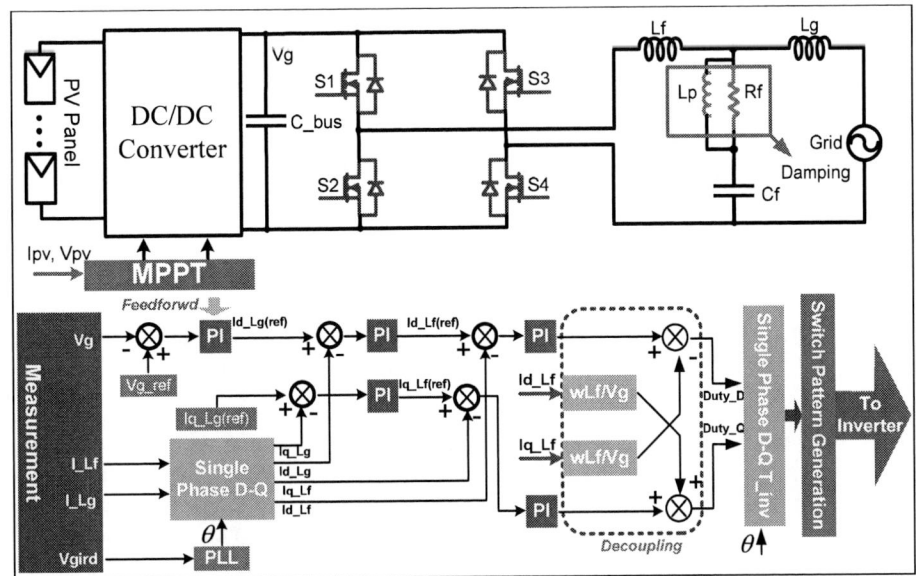

Figure1. System diagram and the controller architecture for DC/AC stage

the grid easier [6] [7].Two more components Lp and Rf are added to prevent the resonant of the LCL filter and also maintain a lower loss of the damping components [6]. The currents through inductors Lf and Lg are sampled and a double current loop controller is proposed to regulate the output current. This controller will sense both I_Lg and I_Lf. The power balance will be achieved once the DC bus voltage is regulated to a constant value, if the second order ripple is ignored. Since the controller can acquire high frequency current information from I_Lf, the regulation performance of the current controller is improved. Also it has a function of cycle by cycle current limitation.

In order to implement the control loop in a rotating D-Q frame, an Imaginary Orthogonal Circuit of the DC/AC inverter has been built with the concept in [8]. Equation (1) (2) [8] give the definition of the rotating transformation matrix for transformation from stationary frame to DQ rotating frame. Equation (3) (4) [8] give the reversed transformation matrix.

$$T = \begin{bmatrix} \cos(\omega t) & \sin(\omega t) \\ -\sin(\omega t) & \cos(\omega t) \end{bmatrix} \tag{1}$$

$$\begin{bmatrix} X_D \\ X_Q \end{bmatrix} = T \begin{bmatrix} X_R \\ X_I \end{bmatrix} = T \begin{bmatrix} X_M \cos(\omega t + \varphi) \\ X_M \sin(\omega t + \varphi) \end{bmatrix} = X_M \begin{bmatrix} \cos \varphi \\ \sin \varphi \end{bmatrix} \tag{2}$$

$$T_{inv} = T^T \begin{bmatrix} \cos(\omega t) & -\sin(\omega t) \\ \sin(\omega t) & \cos(\omega t) \end{bmatrix} \tag{3}$$

$$\begin{bmatrix} X_R \\ X_I \end{bmatrix} = T_{inv} \begin{bmatrix} X_D \\ X_Q \end{bmatrix} = \begin{bmatrix} X_M \cos(\omega t + \varphi) \\ X_M \sin(\omega t + \varphi) \end{bmatrix} \tag{4}$$

Also the equations describing the average model for inverter with LCL filter can be written as:

$$\frac{d}{dt} \begin{bmatrix} I_R \\ I_I \end{bmatrix} = \frac{V_g}{L_f} \begin{bmatrix} D_R \\ D_I \end{bmatrix} - \frac{1}{L_f} \begin{bmatrix} V_R \\ V_I \end{bmatrix} \tag{5}$$

$$\frac{d}{dt} \begin{bmatrix} V_R \\ V_I \end{bmatrix} = \frac{1}{C_{eq}} \begin{bmatrix} I_R \\ I_I \end{bmatrix} - \frac{1}{Z_{eq} \cdot C_{eq}} \begin{bmatrix} V_R \\ V_I \end{bmatrix} \tag{6}$$

$$(Z_{eq} = Z_l + L_g \cdot S \ ; \ C_{eq} = \frac{R_f \cdot C_f + L_p \cdot C_f \cdot S}{L_p \cdot R_f \cdot C_f \cdot S + R_f})$$

Where Vg is DC bus voltage; V_R, V_I, I_R, I_I: they are circuit parameters shown in Fig. 2 (a); Z_l is the grid side impedance;

Fig. 2 (a) shows the derived average model in stationary frame. Applying DQ transformation equation to Fig. 2(a), the average model in DQ frame can be derived, as shown in Fig. 2 (b).

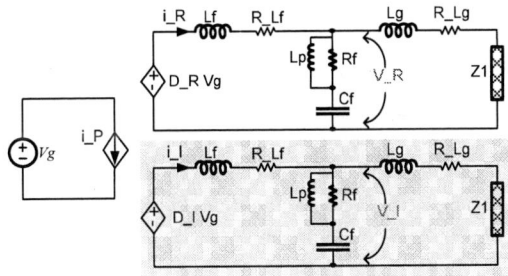

(a) Inverter model in stationary frame

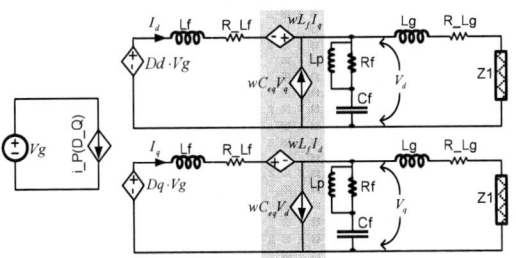

(b) Inverter model in D_Q rotating frame

Figure2. Averaged inverter models in stationary frame and rotating frame

978-1-4244-4782-4/10 $26.00 © 2010 IEEE

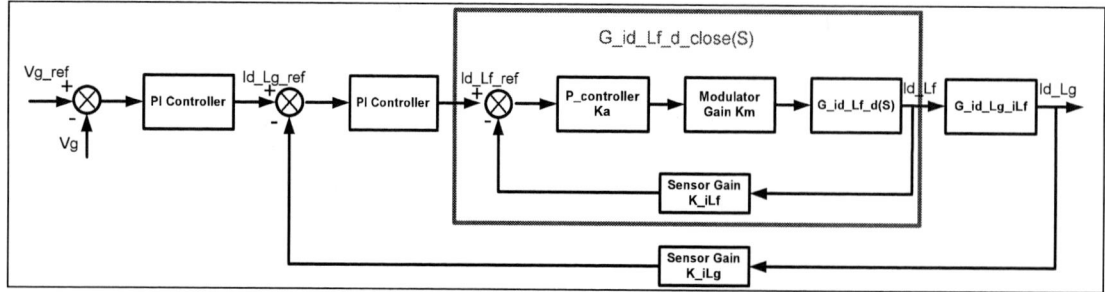

Figure3. Proposed double current loop controller block diagram (take D-axis as an example)

Applying perturbation to the average circuit model, the small signal transfer functions for compensator design can be derived. As shown in Fig. 3, three transfer functions (TFs) need to be derived:

① TF of control to inductor current i_Lf: G_i_Lf(S)

$$G_{i\_Lf\_d}(S) = \frac{V_g}{L_f \cdot S + Z_c} \ , \ \text{where} \begin{cases} Z_c = Z_{eq} \, // \, Z_{rf} \\ Z_{rf} = R_f \, // \, L_p \end{cases} \tag{7}$$

② TF of control to inductor current i_Lg: G_i_Lg(S)

$$G_{i\_Lg\_d}(S) = G_{i\_Lf\_d\_close}(S) \frac{1 + Z_{rf} \cdot C_f \cdot S}{L_g \cdot C_f \cdot S^2 + (Z_l + Z_{rf}) \cdot C_f \cdot S + 1} \tag{8}$$

where $G_{i\_Lf\_d\_close}(S) = \dfrac{k_a \cdot k_m \cdot G_{i\_Lf\_d}(S)}{1 + k_a \cdot k_m \cdot k_{iL} \cdot G_{i\_Lf\_d}(S)}$

③ TF of control to DC bus voltage Vg: G_Vg(S)

$$G_{Vg\_d}(S) = (-2) \cdot V_g \cdot D \cdot \frac{Z_{in}}{D^2 \cdot Z_{in} + Z_c} \tag{9}$$

where $\begin{cases} Z_{in} = \text{Input Impedance of Inverter} \\ Z_c = Z_{eq} \, // \, Z_{rf} \end{cases}$

The system is a 4th order system but some of the zeros and poles are close to each other. Therefore, it behaves like a 2nd order system which can be compensated well with a conventional PI controller. Also this is a coupled system: any changes in D/Q axis will cause change in Q/D axis since there are two coupling factors αI_q, βV_q for D-axis and αI_d, βV_d for Q-axis. For many cases, the decoupling method shown in Figure.1 can be used to realize the decoupling of the D and Q control loops since β is usually very small in a practical system. This is also true for inverters with a normal LC filter since the perturbation of grid side voltage does not exist if assuming the grid is an ideal voltage source with zero internal impedance.

III. DESIGN OF CURRENT CONTROLLER BUILDING BLOCKS WITH MODEL BASED TECHNIQUE

Fig. 5 depicted the developed digital current controller model in Simulink for TMS320F28335 DSP. The "EPWM" and "ADC" block is related to specific hardware in DSP and their parameters can be set accordingly in these blocks. A target support block "EZDSP" is necessary to set parameters for specific DSP. The other blocks in Fig. 5 are "pure" computation blocks which can be verified by simulation in MATLAB. The single phase PLL block and "AB- DQ"/"DQ-

AB" blocks are doing the job of tracking grid voltage phase and convert signals between stationary and rotating D-Q frame, respectively. In order to generate 1/4 cycle delayed signal from I_Lg and I_Lf for single phase D-Q transformation [8], a "real to image conversion" block has been built as shown in Fig. 5. The input of this block is ADC sampling results for I_Lg/I_Lf and the grid frequency information from PLL. Choosing the ADC sampling frequency f_sampling (usually equals to the switching frequency fs, as shown in Fig. 4), then the amount of samples n in a 1/4 cycle can be calculated by equation (10). Feed this value n to an "adaptive digital delay" block in MATLAB will generate a signal which is exactly 1/4 cycle delayed of original signal no matter what the signal's frequency is.

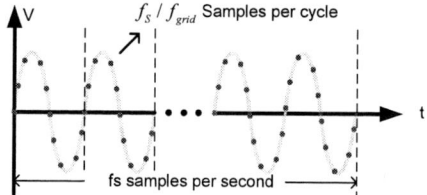

Figure4. Relationship between fs and fgrid

$$n = \frac{f_{Sampling}}{f_{grid}} \times \frac{1}{4} \tag{10}$$

In the model based design, more attention should be paid for model verification. Sometimes the block which functions well in MATLAB simulation may get poor performance when it's running on DSP. This is partly because of the limited "real-time" computation capability of DSP. In Fig. 5, the PLL block and "real-image" block are already optimized to reduce the computation cycles in DSP. At the meanwhile, all of these blocks are in the discrete time domain and they are sensitive to their discretization frequency f_d. A too large f_d will increase the computation burden of DSP and a too small f_d may lead to an inaccurate control result. Careful choosing f_d is critical to the system design.

IV. SIMULATION AND EXPERIMENTAL RESULTS

The developed digital current controller has been tested on a 2kW PV inverter prototype. Table 2 gives the circuit parameters for the DC/AC inverter. This PV inverter utilized an improved resonant DC/DC converter topology which can

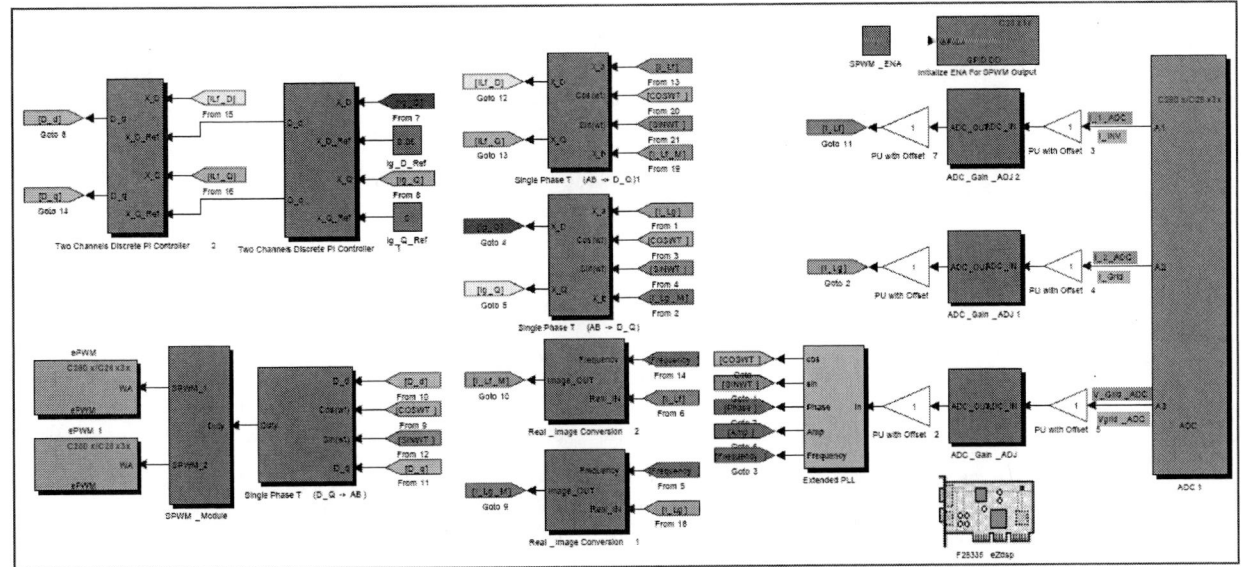

Figure5. Developed digital current controller model for TMS320F28335 with model based technique

maintain high efficiency in a wide input range from 125V to 550V DC. The capability for DC side grounding makes it possible to operate with thin-film panels [9]. Also the inverter can be a string inverter or multi-string inverter [10] which shares a common DC/AC stage. In the experiment, the clock of TMS320F28335 is set to 150MHz and the clock for ADC block is set to 75MHz which is fast enough for this application. The acquire window size of ADC is 2 ADC clock cycles. The PLL block's f_d is two times of switching frequency to ensure the phase tracking accuracy. The "real-image" block's f_d is 33.33 kHz and a 150 data point buffer in DSP RAM is needed to store 1/4 cycle data of 60Hz signal. In order to verify the operation of PLL block before closing the control loop, the signal "phase" (0 - 2 π) from PLL is fed to one of the DPWM blocks to do a high frequency modulation and then it is recovered by a 4th order analog filter. Fig. 6 shows that the recovered phase signal drops from 2 π to almost zero at the zero-crossing point of the reference signal. The filter bandwidth limited the phase waveform can not drops to zero ideally. Fig.7 and Fig. 8 shows the Bode plots for inner and outer current loop after compensation.

TABLE II: KEY PARAMETERS FOR INVERTER PROTOTYPE

Input Voltage	125V-550V
Grid voltage	208-277 RMS 50/60Hz
Input Capacitor	0.5 mF
DC Bus Capacitor	1 mF
Lf	2 mH
Lg	0.5 mH
Lp	0.25 mH
Rp	2.5 Ω
Cf	10 µF

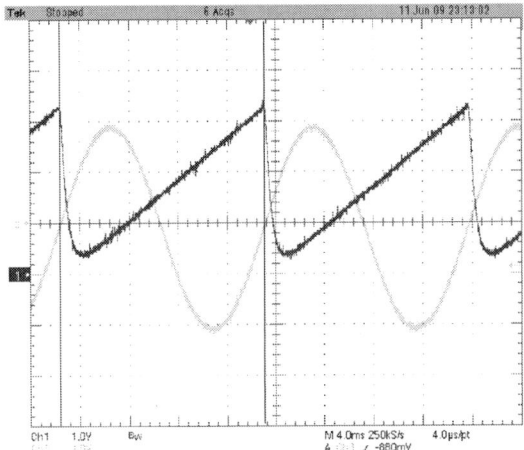

Figure 6 phase signal (blue) from PLL and reference (green)

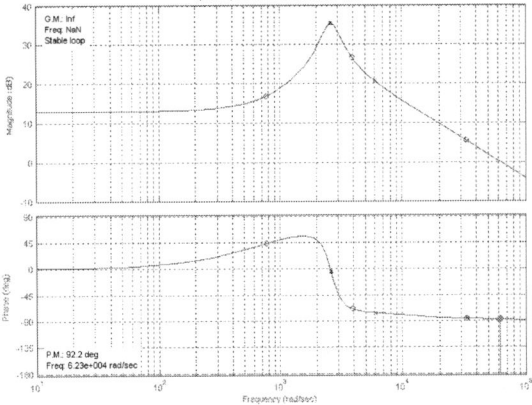

Figure7. Inner current loop Bode diagram after compensation

978-1-4244-4782-4/10 $26.00 © 2010 IEEE

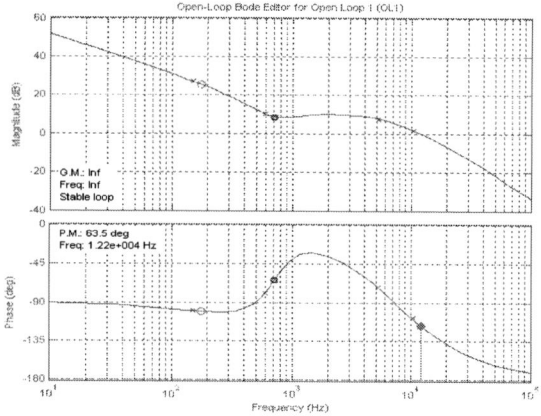

Figure8. Outer current loop Bode diagram after compensation (full load)

Fig. 9 shows the inverter output current I_Lg which is in phase with the grid voltage and most part of the high frequency components have been filtered out with the help of LCL filter. Fig. 10 shows the THD for I_Lg is 1.6%.

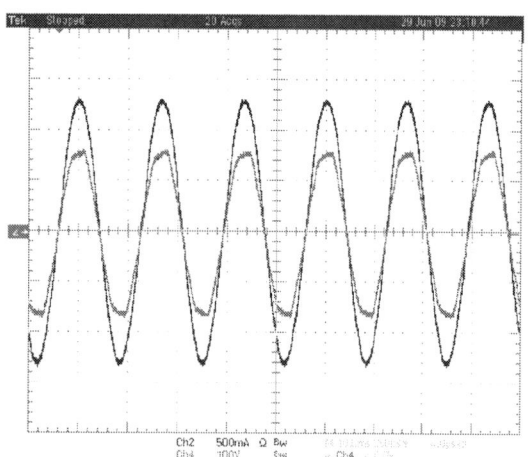

Figure9. Current I_Lg (blue) and grid voltage (purple)

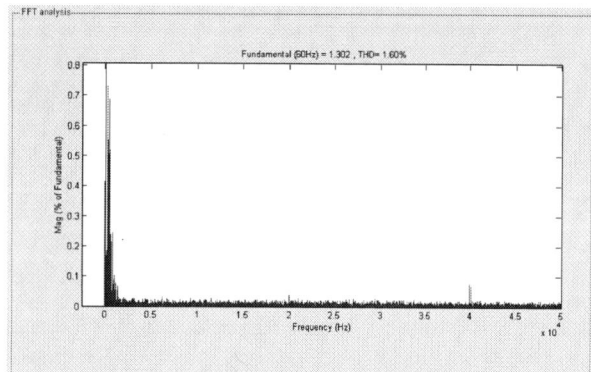

Figure10. FFT analysis results for current I_Lg

V. CONCLUSIONS

A current controller for the grid-tied single phase inverter is developed with model based technique. No coding is required

but model verification and optimization are crucial to the system performance. The average model for single phase inverter with LCL filter in D-Q rotating frame is derived. Also a double current loop controller with decoupling block is proposed and implemented in a DSP. The experimental results show that the current controller developed with this technique has a good performance. The MPPT controller has also been developed by this approach and system test with "solar tree" [11] from MegaWatt Solar is on the way.

REFERENCES

[1] Available:
http://www.mathworks.com/products/matlab/userstories.html?file=198
93

[2] IEEE Standard for Interconnecting Distributed Resources with Electric Power Systems, IEEE Std. 1547, 2003.

[3] Limits for Harmonic Current Emission (Equipment Input Current<16A per Phase), EN61000-3-2, 1995

[4] 2005 National Electrical Code, National Fire Protection Association, Inc., Quincy, MA, 2005

[5] Soeren Baekhoej Kjaer, John K. Pedersen, Frede Blaabjerg, "A review of single-phase grid-connected inverters for photovoltaic modules", IEEE Transactions on Industry Applications, Volume. 41, NO.5, September/October, 2005

[6] Araujo, Samuel Vasconcelos and etc, "LCL filter design for grid-connected NPC inverters in offshore wind turbines", Power Electronics, 2007. ICPE '07. 7th Internatonal Conference on, Oct. 2007 Page(s):1133 – 1138

[7] Teodorescu, R.; Blaabjerg, F.; Borup, U.; Liserre, M.; "A new control structure for grid-connected LCL PV inverters with zero steady-state error and selective harmonic compensation", Applied Power Electronics Conference and Exposition, 2004. APEC '04. Nineteenth Annual IEEE, Volume 1, 2004 Page(s):580 - 586 Vol.1

[8] Zhang.R., Cardinal, M. and etc, "A grid simulator with control of single-phase power converters in D-Q rotating frame", Power Electronics Specialists Conference, 2002., Volume: 3, On pages(s): 1431-1436 vol.3

[9] Zhigang Liang, Larry Alesi and etc, "A novel wide input range photovoltaic inverter with a dual mode DC/DC stage", Proceeding of FREEDM System Center First Annual Conference.

[10] Soeren Baekhoej Kjaer, John K. Pedersen, Frede Blaabjerg, "A review of single-phase grid-connected inverters for photovoltaic modules", IEEE Transactions on Industry Applications, Volume. 41, NO.5, September/October, 2005

[11] Available: http://www.megawattsolar.com/technology

High-performance and cost-effective multiple feedback control strategy for standalone operation of grid-connected inverter

Qin Lei
ECE department,
Michigan State University
East Lansing, MI, USA
leiqin@msu.edu

Shuitao Yang
ECE department,
Zhengjiang University
Hangzhou, Zhejiang, PRC
styang@msu.edu

Fang Z. Peng
ECE department,
Michigan State University
East Lansing, MI, USA
fzpeng@egr.msu.edu

Abstract-This paper presents a multi-loop control strategy with capacitor voltage differential and low pass filter feedback as inner loop, with P controller and voltage reference feedforward as outer loop for voltage-source grid-connected inverters standalone operation. A multi-loop controller with capacitor current feedback can effectively eliminate the inherent high-resonant peak of the output LC-filter to increase system stability margin. It also has better disturbance rejection capability compared to inductor current feedback. Capacitor voltage differential feedback has the same function as capacitor current feedback while also has its own merits that it can save a highly accurate and fast responded current sensor. However, the differential function will enlarge some small disturbance in the output voltage, especially in the high frequency range so as to cause stability problem. Hence a low pass filter (LPF) is introduced to use with differential feedback to reject the high frequency disturbance. In addition, in order to improve the dynamic response, eliminate the tracking phase error caused by PI controller and also maintain a continuous output voltage waveform in the transition from grid-connected to standalone, a voltage reference feedforward has been added to the outer loop. Simulation and experimental results are given in this paper to verify that the system, with proposed control strategy, possesses very high stability, low THD and fast dynamic response when supplying linear and nonlinear loads.

I. INTRODUCTION

Voltage source inverters (VSIs) are now widely used in many grid applications to interface distributed generation (DG) systems (for instance photovoltaics, wind, fuel cells and microturbines) to the utility system. The traditional average voltage feedback has very slow dynamic response and poor performance under nonlinear loads. A dead-beat control [1][2] can make the capacitor voltage exactly tracks the reference voltage while its control algorithm depends on the load parameters. The repetitive control can obtain low THD on output voltage [3][4]. Better performance with non-linear load can be achieved by instantaneous voltage feedback control or multiple feedback loop control

with inductor [5] or the capacitor [6] current feedback as inner loop. In this paper, a multi-loop control strategy with capacitor voltage differential and low pass filter as inner loop, with PI and voltage reference feedforward as outer loop has been proposed to achieve better stability and dynamic performance. The control strategy is verified with simulation and experiment results and proves to be a effective solution for standalone operation of grid-connected inverter.

II. COMPARISON OF CAPACITOR CURRENT FEEDBACK AND INDUCTOR CURRENT FEEDBACK

A. Inner Current loop and outer voltage loop design with considering the control delay

Fig. 1 shows the configuration of three-phase grid-connected VSI with LC filters and local load. It will operate in grid-connected or standalone mode by controlling the switch SSR. This paper concentrates on the control strategy for standalone mode. The system parameters used in this paper is: switching frequency: 10kHz; output frequency: 60Hz; IGBT dead time: 1.5us; DC-link voltage: 200V; output phase voltage(rms): 60V; output capacity: 3KVA.

Fig.2 shows the control block diagram for traditional multi-loop voltage controller with capacitor current feedback or inductor current feedback. The inner open loop transfer function has a high resonant peak which will cause instability which is shown in Fig. 3.It can be eliminated by capacitor current feedback. The closed-loop transfer function of inner current loop is:

$$
i_c = \frac{K_{p2}e^{-sT_s}C_f s}{L_f C_f s^2 + \left(K_{p2}e^{-sT_s} + R_f\right)C_f s + 1} i_{cref}
$$
$$
- \frac{L_f C_f s^2 + C_f R_f s}{L_f C_f s^2 + \left(K_{p2}e^{-sT_s} + R_f\right)C_f s + 1} i_o
\tag{1}
$$

where i_{cref} is current reference generated by voltage loop, i_C is measured capacitor current and i_o is the load current. And the bode diagrams of this closed-loop transfer function with different K_{p2} without considering the control delay is shown in Fig. 3 (a). The bandwidth of $I_C/I_C{}^*$ can be widen by using a larger K_{p2}, to achieve perfect reference tracking at all input frequencies, a

978-1-4244-4782-4/10 $26.00 © 2010 IEEE

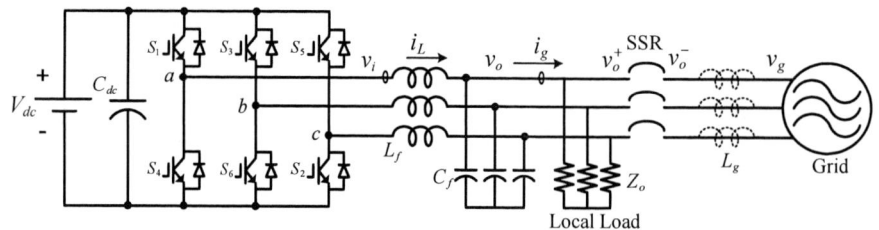

Fig.1 Configuration of three-phase grid-connected VSI with LC filters and local load

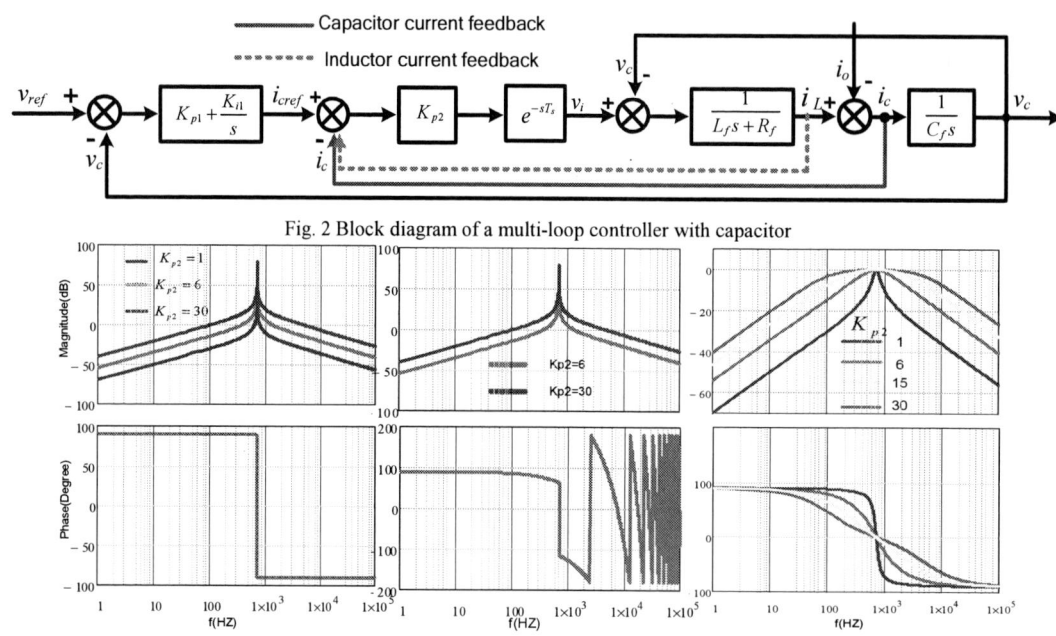

Fig. 2 Block diagram of a multi-loop controller with capacitor

Fig. 3 Bode diagrams of inner loop transfer function (a) open loop without delay (b) open loop with delay (c) closed-loop without delay at different K_{p2}

faster dynamic response and the complete blocking of disturbance. In order to obtain unity gain from output frequency to half of the switching frequency [6], the gain value should be designed to be 30. However, in practice, a digital control implementation for above multiple-loop control method introduces a time delay, usually equal to one switching period T_s which strongly limits the system bandwidth and affects the dynamic performance [7][8]. Thus the high gain would degrade the control loop stability. This can be verified by the root-locus of the inner current closed-loop which is shown in Fig.4. The arrows show the moving trend of the root locus when K_{p2} increases. Without considering the delay, the complex roots will move towards the real axis. But with the delay, the complex roots will move away from real axis, which will cause the system oscillation. This can be further verified by the bode diagrams of the inner current open loop with delay in Fig. 3 (b). When K_{p2}=30, the phase angle at crossover frequency enters into the oscillation region and the system become unstable. Based on this gain selection criterion, K_{p2} is set as 6, which will give a reasonable bandwidth and enough stability margin. For the outer loop PI controller design, the crossover frequency f_{CI} is set to be 300HZ which guarantees enough attenuation

at switching frequency and enough phase margin at crossover frequency. K_{P1} and K_{i1} are selected to obtain unit gain at this crossover frequency. Also, the zero of the PI compensator is set the same as the corner frequency of LC filter [6], Which leads to $K_{P1} = 0.05, K_{i1} = 300$.

B. Capacitor current feedback and inductor current feedback comparison

The close-loop transfer function for aforementioned multi-loop controller with capacitor current feedback is

$$v_o = \frac{(K_{p1}s + K_{i1})K_{p2}e^{-sT_s}}{L_fC_fs^3 + (K_{p2}e^{-sT_s} + R_f)C_fs^2 + (K_{p1}K_{p2}e^{-sT_s} + 1)s + K_{i1}K_{p2}e^{-sT_s}}v_{ref}$$
$$- \frac{L_fs^2 + R_fs}{L_fC_fs^3 + (K_{p2}e^{-sT_s} + R_f)C_fs^2 + (K_{p1}K_{p2}e^{-sT_s} + 1)s + K_{i1}K_{p2}e^{-sT_s}}i_o \quad (2)$$

where V_{cref} is the given voltage reference and v_o is the output voltage.

If inductor currents are used as inner current feedback variables, the closed-loop transfer function is:

$$v_o = \frac{(K_{p1}s + K_{i1})K_{p2}e^{-sT_s}}{L_fC_fs^3 + (K_{p2}e^{-sT_s} + R_f)C_fs^2 + (K_{p1}K_{p2}e^{-sT_s} + 1)s + K_{i1}K_{p2}e^{-sT_s}}v_{ref}$$
$$- \frac{L_fs^2 + (K_{p2}e^{-sT_s} + R_f)s}{L_fC_fs^3 + (K_{p2}e^{-sT_s} + R_f)C_fs^2 + (K_{p1}K_{p2}e^{-sT_s} + 1)s + K_{i1}K_{p2}e^{-sT_s}}i_o \quad (3)$$

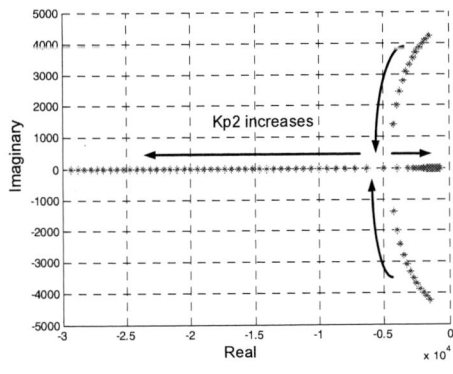
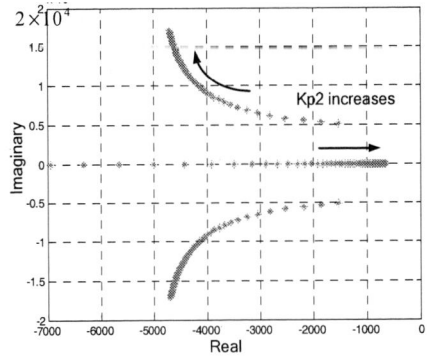

(a) Without considering control delay   (b) With considering control delay

Fig.4 Root locus of inner current closed-loop without and with control delay when K_{p2} increases

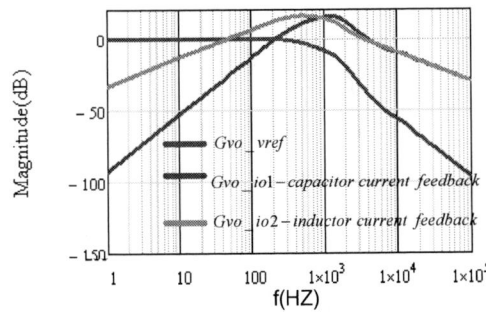
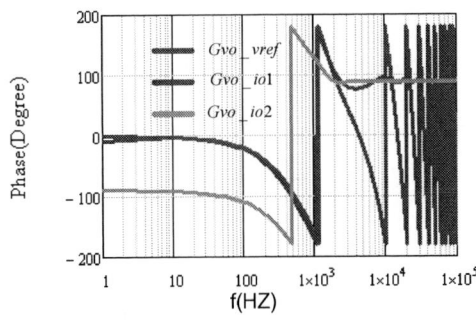

Fig. 5 Bode diagrams of the outer voltage closed-loop transfer function with capacitor current or inductor current feedback as inner loop

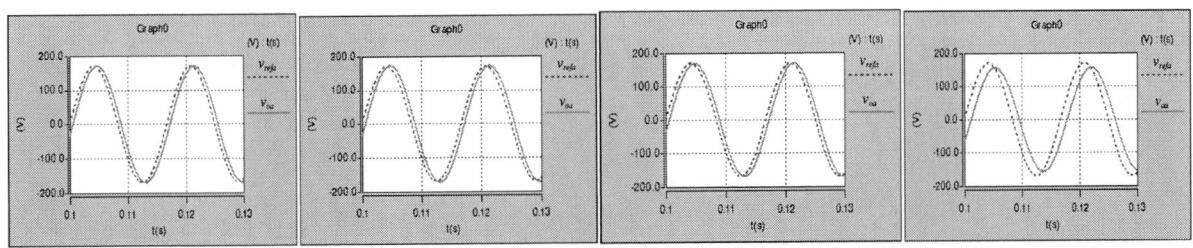

(a) $R = 100\,\Omega$ (Y con.) with CVDF   (b) $R = 5\,\Omega$ (Y con.) with CVDF   (c) $R = 100\,\Omega$ (Y con.) with ICF   (d) $R = 5\,\Omega$ (Y con.) with ICF

Fig. 6 Simulation results for output voltage (pink one) and reference voltage(blue one) at (a-b) capacitor voltage differential feedback (CVDF) (c-d) inductor current feedback (ICF)

It is notable that in the capacitor current feedback and inductor current feedback, the output voltage-to-reference voltage transfer function $G_{vo\_vref}(s)$ is exactly the same, while the voltage-to-output current transfer function $G_{vo\_io}(s)$ is different from each other. Fig.5 shows the bode diagrams of closed-loop voltage transfer function using two control schemes respectively with aforementioned designed parameters. Obviously, capacitor current feedback strategy has lower output voltage to load current gain at low frequency range than inductor current feedback strategy, which means it has much better disturbance rejection capability. This can be further verified by the simulation results in Fig. 6. It compares capacitor voltage differential feedback control scheme and inductor current feedback control scheme with respect to the output voltage at various load condition. When the load changes from 100 to 100, with capacitor

current feedback, the output voltage has no much change but with inductor current feedback, the tracking error increases obviously. It shows that the former scheme has better stiffness to the load disturbance.

III. MULTI-LOOP CONTROLLER WITH CAPACITOR VOTLAGE DIFFERENTIAL FEEDBACK

A. Control block diagram and design of feedback coefficient

In capacitor current feedback control, a highly accurate current sensor is needed because the capacitor current is in small scale. It also requires the sensor has good dynamic performance to guarantee fast dynamic response for inner loop. Since capacitor current can be calculated from the voltage, capacitor voltage differential feedback can be used to replace capacitor current feedback so that the current sensor can be saved [9]. The control block diagram of this strategy is shown in Fig. 7. Like the capacitor current feedback, the

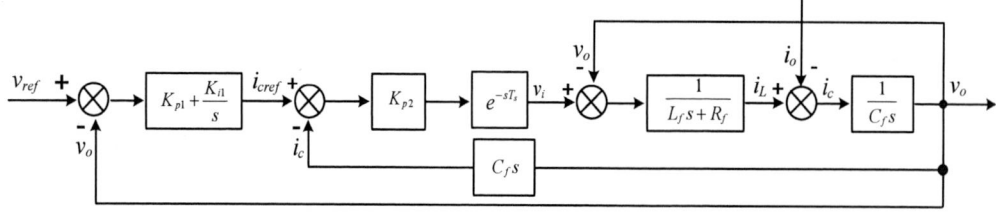

Fig. 7 Control block diagram of capacitor voltage differential feedback

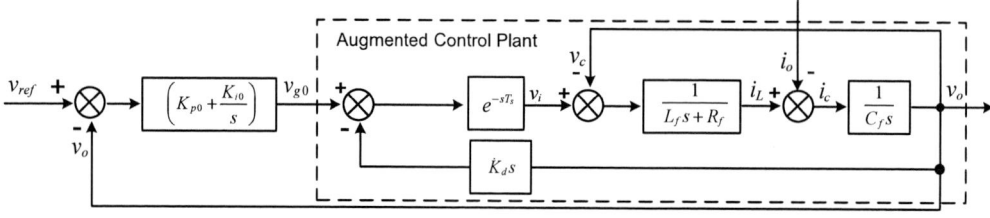

Fig. 8 Simplified diagrams of capacitor voltage differential feedback

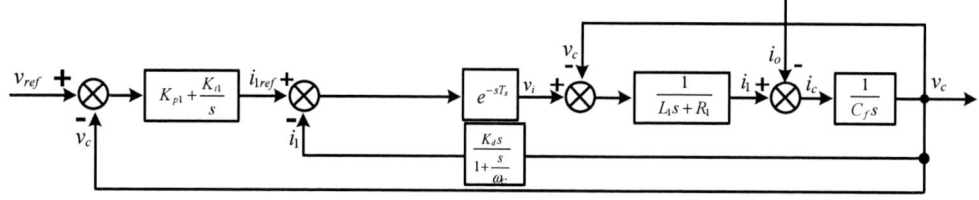

Fig.9 Bode diagrams of capacitor voltage differential feedback with LPF

capacitor voltage differential feedback can also eliminate the resonant peak of L-C filter. The control diagram can be simplified into Fig. 8, from which we can design the K_d and K_{p0} in another view, as given in [9]. The closed-loop transfer function of the inner loop :

$$G_{vo\_vgo}(s) = \frac{v_o}{v_{g0}} = \frac{1}{L_1 C_f s^2 + R_1 C_f s + K_d s + 1} = \frac{\omega_n^2}{s^2 + 2\xi\omega_n s + \omega_n^2} \quad (4)$$

,which can also be expressed as a normalized second-order transfer function. So the damping ratio can be calculated as:

$$\xi = \frac{R_f C_f + K_d}{2\sqrt{L_f C_f}} \quad (5)$$

From equation (4) and (5), it is known that the LC filter transfer function has a high-resonant peak because the damping ratio ξ is close to zero because R_f is very small; after introducing the output voltage differential feedback, the damping ratio ξ changes from zero to a variable value, which can be adjusted by the differential coefficient K_d. When K_d is equal to the designed value 0.3m, the damping ratio is about 0.67, which is very close to the technical optimum (TO) value 0.707. Thus this control method is coincident with aforementioned capacitor current feedback.

B. Capacitor voltage differential feedback with low pass filter

However, the differential function will enlarge some small disturbance in the output voltage, especially in

the high frequency range so as to cause stability problem. Hence a low pass filter (LPF) is introduced to use with differential feedback to reduce the high frequency disturbance. With a low pass filter, the feedback transfer function becomes:

$$F = C_f s / (1 + s / \omega_C) \quad (6)$$

, where the cut off frequency is designed to be $f_c = \omega_C / 2\pi = 2kHz$. It has the same amplitude and phase as differential feedback at low frequency, while has an additional -20dB/dec drop in amplitude after cut

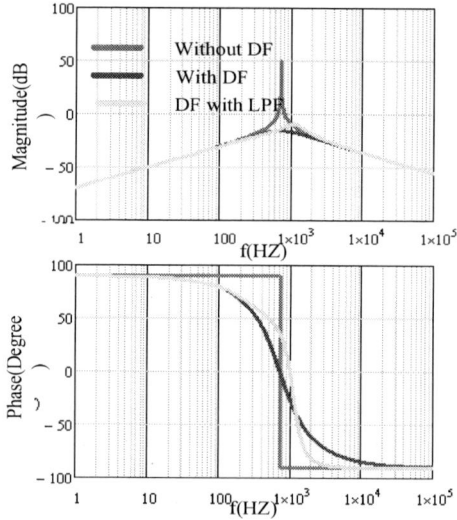

Fig. 10. Bode diagrams of inner open loop transfer function

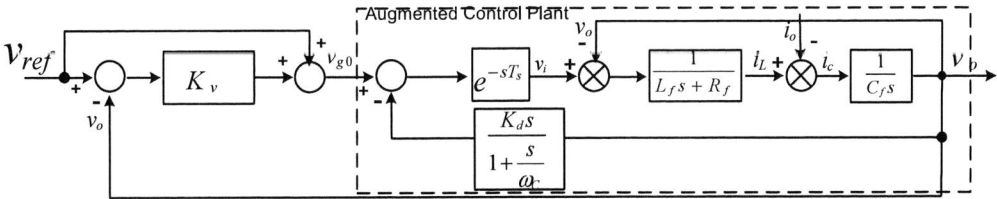

Fig.11. Proportional compensator plus feed-forward control strategy

off frequency thus can suppress the high frequency distortion.

C. Proportional control with voltage reference feedforward

Considering the transient performance when the inverter changes from grid-connected operation mode to standalone mode, with proposed outer voltage PI controller, the initial output voltage is zero which causes the voltage fall to zero immediately after the mode switching. So the voltage waveform becomes discontinuous in transition. In order to overcome this disadvantage, a voltage reference feed-forward path is added to the outer loop, which is shown in Fig.11. The reference voltage is generated by multiplication of voltage amplitude and phase angle which is synchronized with grid voltage in grid-connected operation mode. So the initial output voltage of this standalone system which is equal to the grid voltage at the moment of control mode switching. Hence it maintains the continuity of voltage waveform.

In addition, the feedforward path can also improve the steady-state performance. Without feedforward path, the outer loop gain needs to be designed to be a large value to obtain small steady-state error which will make the system face stability problem. Moreover, feedforward path can enhance the dynamic response which is beneficial in transition process or nonlinear load case.

Also, a single proportional controller is used instead of PI controller to reduce the harmonics which is analyzed as follows. With considering the control delay, the output voltage-to-reference voltage and output voltage-to-load current transfer function under P controller can be expressed as follows:

$$
\begin{aligned}
v_o(s) &= G'_{vo\_vref}(s) \cdot v_{ref} - G'_{vo\_io}(s) \cdot i_o \\
&= \frac{(1+K_v)e^{-sT_s}}{L_f C_f s^2 + (R_f C_f + K_d e^{-sT_s})s + K_v e^{-sT_s} + 1} v_{ref} \\
&\quad - \frac{L_f s + R_f}{L_f C_f s^2 + (R_f C_f + K_d e^{-sT_s})s + K_v e^{-sT_s} + 1} i_o
\end{aligned}
\tag{7}
$$

Fig. 12 compares the bode diagrams of output voltage to voltage reference closed loop transfer function under two controllers; it shows that with PI controller in which $K_{p0} = 0.3$, $K_{i0} = 1800$, there is a peak at frequency between 60HZ to 900HZ. It raises the portion of the 5th, 7th and 11th harmonics in the output voltage. The peak magnitude decreases as K_i decreases and it decreases to zero when $K_i = 0$. But with only P controller, the magnitude gain keeps around 0 dB thus the steady-state error stays in a reasonable range. So using a single proportional controller can reduce the harmonic components in the output voltage and still track the reference voltage very well. In addition, PI controller will introduce a phase error between output voltage and reference voltage which will cause a large line current in the transition. In conclusion, a propotional controller with voltage reference feedforward, can maintain the voltage continuity, lower the voltage THD and eliminate the phase error between output voltage and referernce voltage while still keep the magnitude error in a small range.

IV. EXPERIMENTAL RESULTS

The multi-loop controller with capacitor voltage differential feedback and the proposed new controller are implemented in the laboratory. For capacitor voltage differential feedback strategy, Fig. 12 (a) (b) shows the steady-state load voltage and DC voltage waveforms of the inverter at different load conditions. These figures indicate that the proposed control strategy is capable of producing a nearly perfect sinusoidal load voltage with small steady-state error. Also shown in those figures is that the load current has no appreciable influence on the voltage loop performance. Fig. 13 shows the output voltage with low pass filter added. The total harmonic distortion is measured to be 3% at no load (worst case), which is better than Fig. 12 (a) (measured as 4%). That verifies

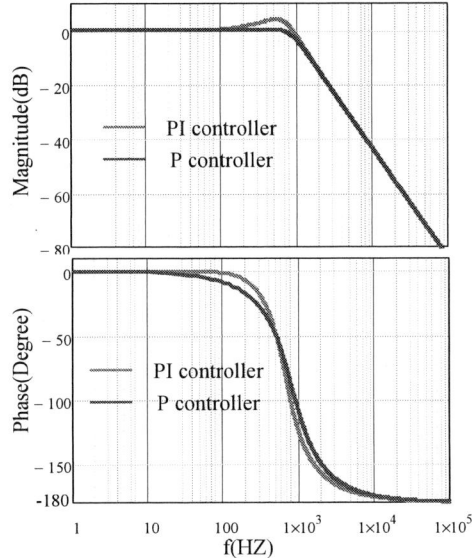

Fig. 12 Bode diagrams of $G'_{vo\_vref}(s)$ with P and PI controller

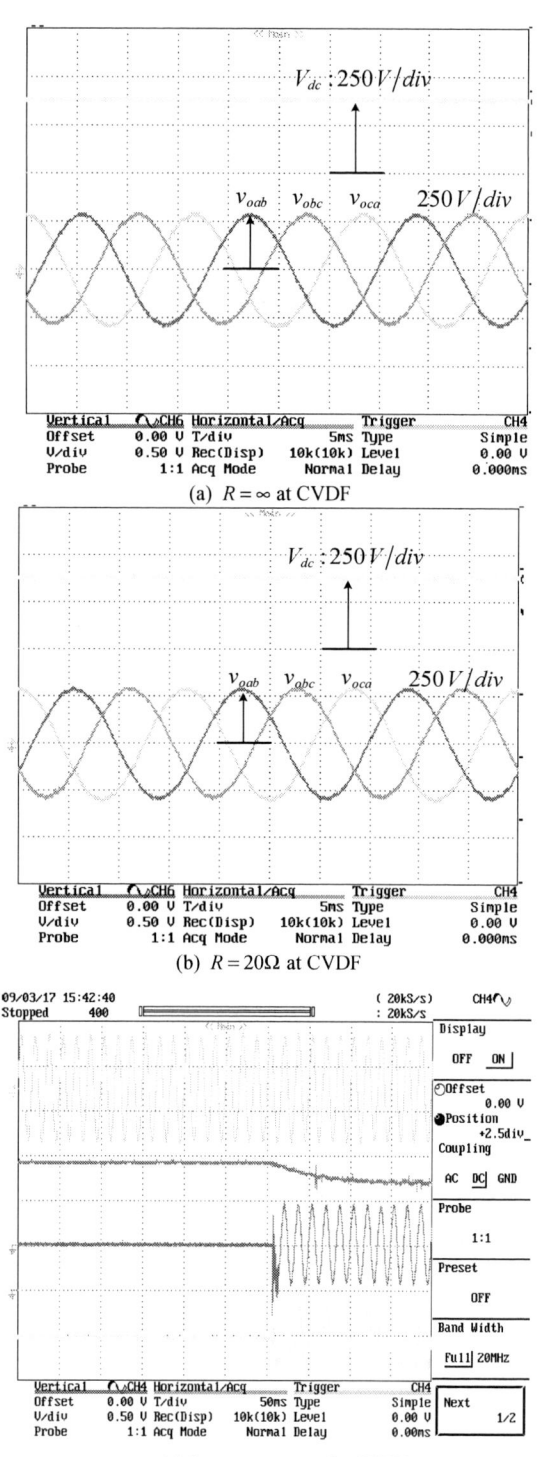

(a) $R = \infty$ at CVDF

(b) $R = 20\Omega$ at CVDF

(c) Start-up process for CVDF

Fig. 12 Experiment results for output line to line voltage for capacitor voltage differential feedback strategy

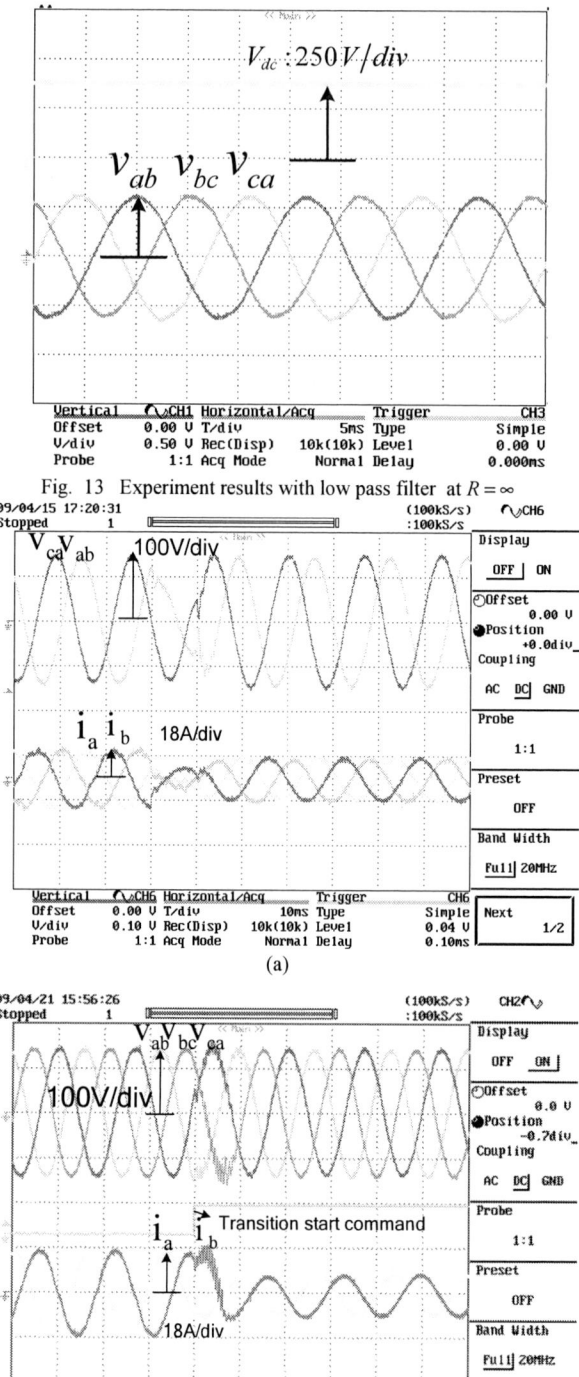

Fig. 13 Experiment results with low pass filter at $R = \infty$

(a)

(b)

Fig. 14. Transition from grid-connected to standalone (a)with PI and no feedforward (b) with P and voltage feedforward

that it reduces the harmonics components in the output voltage. Fig. 14 (a) shows the output line to line voltage and phase current waveforms in the transition from grid-connected to standalone under two control strategies, with PI controller and with P controller plus voltage feedforward. In the former strategy, the voltage

dips to zero at transition start and recovers to normal value in a long time. With voltage reference feedforward and P controller, the voltage becomes continuous and the transition time has been reduced which supresses the output voltage distortion in advance.

978-1-4244-4782-4/10 $26.00 © 2010 IEEE

V. CONCLUSION

The utilized voltage differential inner loop can eliminate the inherent high-resonant peak of the output LC-filter that may cause instability. The low pass filter can reduce the high frequency disturbance caused by differential function. A P control strategy with voltage reference feedforward can improve the system dynamic performance and lower the output THD, also make better performance in transition. Both the simulation and the experiment results verify that the system with proposed control strategy can produce perfect sinusoidal output voltage with small steady-state error. It possesses fast dynamic response and good stability.

REFERENCE

[1] A. Kawamura, T. Haneyoshi, R.G. Hoft, "Deadbeat controlled PWM inverter with parameter estimation using only voltage sensor," IEEE Transaction on Power Electronics, 1988, 3(2):118-125.

[2] L. Mihalache, "Dsp control method of single-phase inverters for UPS application," Proceedings of the 2002 Applied Power Electronics Conference, 2002: 590-596.

[3] Y.Y. Tzou, R.S. Ou, S.L. Jung, M.Y. Chang, "High-performance programmable AC power source with low harmonic distortion using DSP-based repetitive control technique," IEEE Transaction on Power Electronics, 1997, 12(2): 715-725.

[4] Y.Y. Tzou, S.L. Jung, H.C. Yeh, "Adaptive repetitive control of PWM inverters for very low THD AC-voltage regulation with unknown loads," IEEE Transaction on Power Electronics, 1999,14(5),973-981.

[5] H.Y. Wu, D. Lin, D. Zhang, et al, "A current-mode control technique with instantaneous inductor-current feedback for UPS inverter," Proceedings of the 1999 Applied Power Electronics Conference, 1999: 951-957.

[6] N.M. Abdel-Rahim, J.E. Quaicoe, "Analysis and Design of a multiple feedback loop control strategy for single-phase voltage-source UPS inverter," IEEE Trans. on Power Electronics, 1996, 11(4):532-541.

[7] P. Mattavelli, F. Polo, S. Sattin, F. D. Lago, "Dynamic Improvement in UPS by means of control delay minimization," Proceedings of the 2004 Industry Applications Society Annual, 2004, 2(1):843-849.

[8] H. Deng, R. Qruganti, D. Sprinivasan, "PWM methods to handle time delay in digital control of a UPS inverter," IEEE Power Electronics Letters, 2005, 3(1):1-6.

[9] Shuitao Yang, Xinping Ding, Jinyun Liu and Zhaoming Qian, "Analysis and Design of a Cost-Effective Voltage Feedback Control Strategy for EPS Inverters," Power Electronics Specialists Conference, 2007. PESC 2007. IEEE 17-21 June 2007 Page(s):477 – 482.

Current Control Optimization for Grid-tied Inverters with Grid Impedance Estimation

Guoqiao Shen，Jun Zhang, Xiao Li, Chengrui Du, Dehong Xu

Power Electronics Institute, Department of Electrical Engineering,
Zhejiang University, Zheda Road 38#
Hangzhou, Zhejiang province, China, 310027

Abstract—**The control of a PWM inverter is affected by variable inductance present in the power line of the grid. This paper presents a new implementation of grid-tied inverter with weighted-average-current feedback control (WAC) and an embedded online technique to determine the grid line impedance. By adding an inter-harmonic voltage periodically in the modulation of the PWM and measuring the current response, it is possible to estimate the load impedance connected to the inverter, and the grid impedance can be obtained by subtracting the low-pass filter from the load impedance. Consequently, current control optimization can be carried out such as rearrangement of poles or zeroes and adjustment of the current loop gains to obtain the desired dynamic response even under changing line conditions. The presented technique can be implemented with the existing sensors and the digital processor of the inverter, provides a low cost and adaptive approach for current control of grid-tied inverters. Simulation on a 5kW grid-tied inverter demonstrates the control method and the impedance measurement.**

I. INTRODUCTION

The steady increase in renewable power generations, such as photovoltaic systems, wind turbines, fuel-cells power systems, calls for new and better control methods in response to the utility grid connection. Many current control methods have been developed for the grid-tie inverters to improve the performance of the system.

In most case, the utility grid is considered to be stiff with low impedance that is usually neglected. In fact, many distributed generations are installed under weak grid conditions due to low power transformers and long power lines [1], where the grid impedance may be of a wide range. Therefore, the control of these PWM inverters will be affected by variable inductance present in the power line. The solution to this problem leads to the estimation of the grid impedance and an adaptive current control.

Various methods have been used to measure power line impedance. One solution is to attach a large separate hardware device such as capacitors or resistors for the measuring purpose[2], but this solution is more expensive and difficult to integrate with the existing inverter control. Another solution is to use the existing sensors and the digital processor of the

grid-tie inverter to implement the measuring method. Numerous publications exist in this field [3-6]. Typically, a disturbance is introduced in the inverter output and then the voltage and current responses at Point of Common Coupling (PCC) are extracted to calculate the grid impedance.

This paper presents a new implementation of grid-tied inverter with WAC control and an embedded online technique to determine the grid line impedance. By adding an inter-harmonic voltage periodically in the modulation of the PWM and measuring the current response, the transfer function from the modulation voltage to the grid current is investigated, and then the grid impedance can be obtained without a precise knowledge of the grid voltage response which is easy to be disturbed by grid background distortions. Current control optimization can be carried out to obtain the desired dynamic response under changing line conditions. Besides, the knowledge of the grid impedance at the fundamental frequency can also be used to detect a utility failure and prevent the inverter from islanding operation. The presented technique can be implemented with the existing sensors and the MCU (Micro-Controller Unit) of the inverter. Simulation on a 5kW grid-tied inverter demonstrates the control method and the impedance measurement.

II. CURRENT CONTROL UNDER WEAK GRID CONDITION

Fig.1 shows a system topology for a 5kW grid-connected VSI inverter for fuel cell generation. The topology has a low-pass LCL-filter connecting the inverter output to the grid.

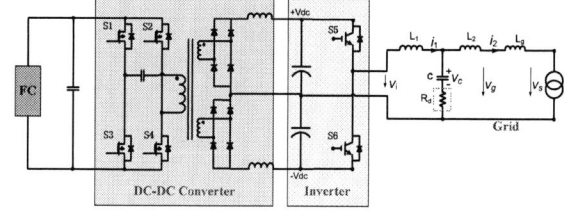

Fig.1. System topology for the grid-connect fuel cell inverter

The LCL-filter is mainly used to achieve decreased switching ripple with only a small increase in filter hardware compared with the L-filter. The transfer function from inverter output voltage $V_i(s)$ to inverter current $I_1(s)$, $G_{Vi\text{-}I1}(s)$, and the transfer function from inverter output voltage to grid current $I_2(s)$, $G_{Vi\text{-}I2}(s)$, can be expressed as (1) and (2).

$$G_{Vi-I1}(s) = \frac{I_1(s)}{V_i(s)} = \frac{(1-\alpha)LCs^2 + R_dCs + 1}{\alpha(1-\alpha)L^2Cs^3 + R_dLCs^2 + Ls} \qquad (1)$$

$$G_{Vi-I2}(s) = \frac{I_2(s)}{V_i(s)} = \frac{R_dCs + 1}{\alpha(1-\alpha)L^2Cs^3 + R_dLCs^2 + Ls}$$

(2)

Where, $L=L_1+L_2+L_g$, $\alpha=L_1/L$, and neglect ESR of the inductor. R_d is the resistor connected in series with C of the LCL-filter, which may be used for the purpose of resonance damping.

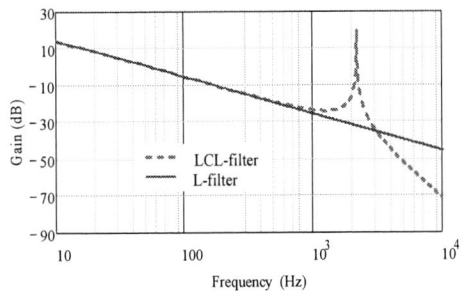

Fig.2. Bode plots of the transfer function of the filters $G_{Vi-I2}(s)$

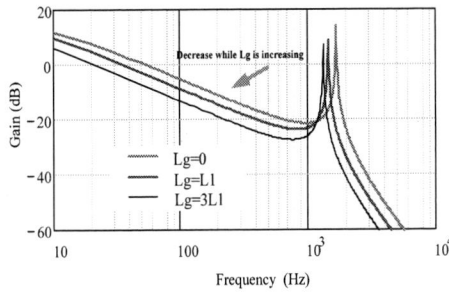

Fig.3. Bode plots for the V-I gains $G_{Vi-I2}(s)$ with different grid impedances

Fig.2 shows the bode plots of the transfer function $G_{Vi-I2}(s)$ of an LCL-filter and an L-filter with same total value of inductance (L=3mH). The LCL-filter has more attenuation to the high frequency switching ripple compared with the L-filter. In low frequency region, the amplitude gains of these two filters are almost same. When the inverter is connected with a weak grid, the grid line impedance will be in series with the grid side inductance of the LCL-filter, and it may change in wide range. Hence, the V-I gain of the filter may be affected both in amplitude and the position of poles, as shown in Fig. 3.

The block diagram of the fuel cell inverter with a current controller is shown in Fig.4. The open loop transfer function can be expressed as (3),

$$G_o(s) = \frac{I_f(s)}{E_i(s)} = G_C(s) \cdot G_{INV}(s) \cdot G_{V-I}(s) \cdot H(s) \qquad (3)$$

Whatever the current regulator is used, adaptive parameters are desirable to match the changing line conditions associated with $G_{V-I}(s)$ in (3), i.e., $G_{Vi-I1}(s)$ in (1) or $G_{Vi-I2}(s)$ in (2), in order to obtain the desired system stability and dynamic response.

Therefore, the grid impedance estimation is necessary for the inverter.

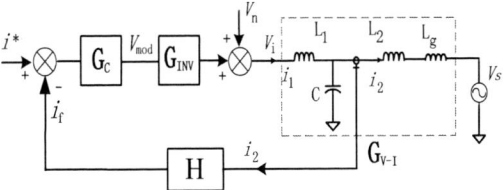

Fig.4. Block diagram of the current controlled grid-tied inverter

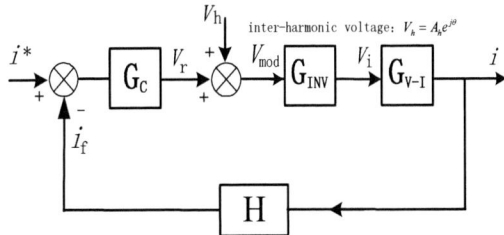

Fig.5. Block diagrams of the inverter controller with inter-harmonic disturbance

III. PROPOSED METHOD FOR GRID IMPEDANCE ESTIMATION

A. Principle of the method

The proposed grid impedance estimation method is based on inter-harmonic injection which is employed in the literature [5], and a new impedance calculation approach is presented in this paper. Generally, the existing methods calculate the grid impedance from the measured voltage and current responses at the PCC to the inter-harmonic disturbance. Because the voltage response to the inter-harmonic disturbance is quite small compared with the grid voltage and it is easy to be interfered by the background grid voltage noises, a delicate voltage sensor is required. For the proposed method, the inter-harmonic disturbance within the digital processor of the inverter is used instead of the grid voltage for the grid impedance estimation. The principle can be explained as following.

Fig.5 shows the block diagrams of the inverter controller. Impedance estimation is realized by adding a proper disturbance V_h to the reference voltage for the PWM modulator. The current response to the disturbance can be derived as (4)

$$\frac{I(s)}{V_h(s)} = \frac{G_{INV}(s) \cdot G_{V-I}(s)}{1 + G_{INV}(s) \cdot G_{V-I}(s) \cdot H(s) \cdot G_C(s)} \qquad (4)$$

If the frequency of the inter-harmonic voltage is selected to be 75Hz, i.e., $\omega_h = 150\pi$, the current response to the disturbance is

$$\frac{I(\omega_h)}{V_h(\omega_h)} = \frac{G_{INV}(j\omega_h) \cdot G_{V-I}(j\omega_h)}{1 + G_C(j\omega_h) \cdot G_{INV}(j\omega_h) \cdot G_{V-I}(j\omega_h) \cdot H(j\omega_h)} \qquad (5)$$

978-1-4244-4782-4/10 $26.00 © 2010 IEEE

Because the inter-harmonic frequency is quite low compared with the resonant frequency of the LCL-filter as shown in Fig. 2, $G_{V-I}(j\omega_h)$ can be simplified as (6).

$$G_{V-I}(j\omega_h) = \frac{1}{Z_1(\omega_h) + Z_2(\omega_h) + Z_3(\omega_h)} \quad (6)$$

Here, $Z_1(\omega_h) = R_1 + j\omega_h L_1$, $Z_2(\omega_h) = R_2 + j\omega_h L_2$, $Z_g(\omega_h) = R_g + j\omega_h L_g$. Then, the grid impedance can be derived as (7)

$$Z_g(\omega_h) = \frac{1}{G_{V-I}(j\omega_h)} - Z_1(\omega_h) - Z_2(\omega_h) \quad (7)$$

From (5) and (7), we can obtain the estimation of the grid impedance by (8).

$$Z_g(\omega_h) = \left[\frac{V_h(\omega_h)}{I(\omega_h)} - G_C(j\omega_h) \cdot H(j\omega_h) \right] \cdot G_{INV}(j\omega_h) - Z_1(\omega_h) - Z_2(\omega_h) \quad (8)$$

In (8), $G_C(j\omega_h)$, $G_{INV}(j\omega_h)$, $H(j\omega_h)$ and the LCL-filter are designed beforehand, and $V_h(\omega_h)$ is given by MCU of the inverter, therefore it is possible to calculate the grid impedance through the current response to a given inter-harmonic disturbance. The inter-harmonic current response can be extracted with Discrete Fourier Transformation (DFT) analysis as [5] and [6].

If the initial grid impedance is known as $Z_{ini}(\omega_h)$ and the inverter system is not changed after the grid condition variation, the grid impedance can be updated by an increment as (9),

$$Z_g(\omega_h) = Z_{ini}(\omega_h) + \Delta Z_g(\omega_h)$$

$$\Delta Z_g(\omega_h) = \left[\frac{V_h(\omega_h)}{I(\omega_h)} - \frac{V_h(\omega_h)}{I_{ini}(\omega_h)} \right] \cdot G_{INV}(j\omega_h) \quad (9)$$

Here, $I_{ini}(\omega_h)$ is the current response to $V_h(\omega_h)$ for the initial grid impedance.

B. Extension of the method

Grid voltage feed-forward is widely used in current controls for the grid-connected inverters. Fig.6 shows the system diagram of the fuel cell inverter with grid voltage feed-forward. Then, the block diagram of the inverter control can be described as Fig.7.

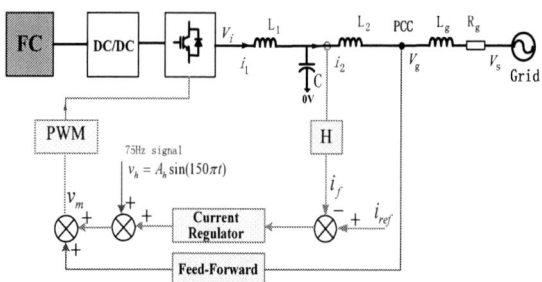

Fig.6 System diagram with grid voltage feed-forward

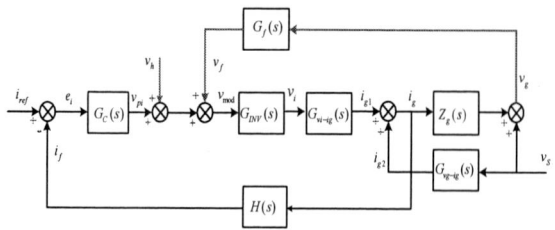

Fig.7 Block diagram of the inverter with grid voltage feed-forward

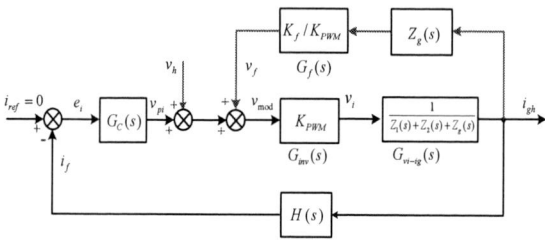

Fig.8 Block diagrams for the response to the inter-harmonic disturbance

In Fig.7, $G_f(s)$ is the feed-forward gain, defined to be the inverse of the PWM gain of the inverter multiplied with a coefficient K_f. For low frequency inter-harmonic disturbance, Fig.7 can be simplified as Fig.8, where K_{PWM} represents the inverter voltage gain which is close to the dc voltage of the inverter. Hence, the close-loop gain from $V_h(\omega_h)$ to grid current is derived as (10)

$$\frac{I(\omega_h)}{V_h(\omega_h)} = \frac{K_{PWM}}{Z_1(\omega_h) + Z_2(\omega_h) + (1-K_f)Z_g(\omega_h) + K_{PWM}G_C(j\omega_h)H(j\omega_h)} \quad (10)$$

Then, on the condition that $K_f \neq 1$, we can obtain $Z_g(\omega_h)$ by (11),

$$(1-K_f)Z_g(\omega_h) = \left(\frac{V_h(\omega_h)}{I(\omega_h)} - G_C(j\omega_h)H(j\omega_h) \right) K_{PWM} - Z_1(\omega_h) - Z_2(\omega_h) \quad (11)$$

When $K_f = 0$, that means no grid voltage feed-forward, (11) is same to (8).

With an initial grid impedance known as $Z_{ini}(\omega_h)$, $Z_g(\omega_h)$ can be obtained by (12),

$$Z_g(\omega_h) = Z_{ini}(\omega_h) + \Delta Z_g(\omega_h)$$

$$\Delta Z_g(\omega_h) = \left[\frac{V_h(\omega_h)}{I(\omega_h)} - \frac{V_h(\omega_h)}{I_{ini}(\omega_h)} \right] \cdot \frac{K_{PWM}}{1-K_f} \quad (12)$$

From (11), for different feed-forward coefficients, e.g. K_{f1} and K_{f2}, their corresponding inter-harmonic currents, I_{h1} and I_{h2}, will satisfy (13),

$$Z_g(\omega_h) = \left[\frac{V_h(\omega_h)}{I_{h2}} - \frac{V_h(\omega_h)}{I_{h1}} \right] \cdot \frac{K_{PWM}}{K_{f1} - K_{f2}} \quad (13)$$

Equation (13) means that the grid impedance can be obtained by changing of the feed-forward coefficient. When $K_{f1} = 1$, the grid impedance will disappear in (10), and the current response

I_{h1} is equivalent to $I_{ini}(\omega_h)$ with zero initial grid impedance, as shown in (12). Therefore, we have,

$$Z_g(\omega_h) = \left[\frac{V_h(\omega_h)}{I_{h2}} - \frac{V_h(\omega_h)}{I_{h1}} \right] \cdot \frac{K_{PWM}}{1 - K_{f2}} \qquad (14)$$

IV. CURRENT CONTROL OPTIMIZATION

In this paper, a new implementation of grid-tied inverter with proposed grid impedance estimation and current control optimization is presented, as shown in Fig.9. The weighted-average-current feedback control (WAC) is employed in order to degrade the control system with LCL-filter from 3rd order to 1st order and suppress the resonant peak without significant passive damping of filter [7]. Current control is implemented through a proportional plus resonant (P+R) regulator [8], providing the modulation voltage for the PWM inverter. A 75Hz inter-harmonic disturbance is periodically added to the modulation voltage for a short duration (4 cycles every 5 seconds) for the sake of grid impedance estimation. The control parameters are optimized according the grid impedance in following procedure.

During the start up procedure of the inverter, because of the unknown grid impedance, the weight value for the WAC feedback control is designed according to the inductances of the LCL-filter, $\beta = 1 - L_1/(L_1 + L_2)$, the proportional gain of the PR controller is small to ensure the system stable operation.

After several times grid impedance estimations, the average of the grid impedance is obtained, and the weight value for the WAC should be adjusted by (15)

$$\beta = 1 - L_1/(L_1 + L_2 + L_g) \qquad (15)$$

The proportional coefficient of the PR controller K_P can also be enlarged within the limit of system stability to increase the low frequency gains and bandwidth of the control loop, which will allow more resonant compensators to be plugged into the current regulator for current harmonics attenuation.

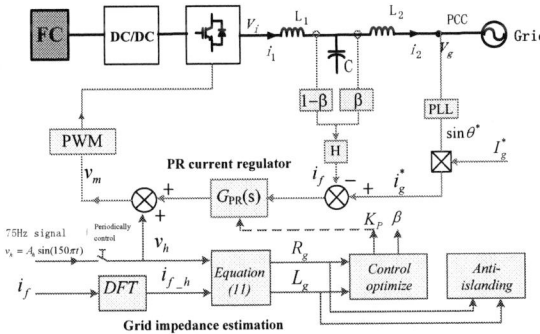

Fig. 9 Control diagram for the grid-tied inverter

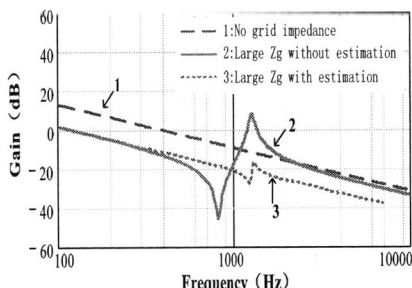

Fig.10 Vi-If transfer function in different cases

Fig.10 shows the V_i-I_f transfer function in the case of 1) no grid impedance, hence no resonant peak because of WAC; 2) with large grid impedance and no estimation, resulting in a significant mismatch of WAC weight value, current regulator (including K_P) designed according to no grid impedance condition may be unstable; 3) with large grid impedance and control parameters optimized according to approximately impedance estimation, the suppression of the resonant peak allows the increase of the K_P of the current regulator.

V. SIMULATION RESULTS

A 5kW grid-connected inverter model is built on Matlab. The grid voltage is 220Vac/50Hz, switching frequency is 16 kHz, and the dead time is 2.5us. Parameters of the LCL-filter are selected as $L_1=L_2=1.5mH$, C=7uF with a damping resister 0.5Ω in series. The ESR of the inductor is 0.1Ω.

Table 1 shows the simulation results of the grid impedance estimation for five cases of grid line conditions, without grid voltage feed-forward in the inverter current control. The inter-harmonic disturbance is 90 degree leading to the grid voltage, 75Hz in frequency and 0.1pu in amplitude. From the results in Table1, it is obvious that the estimation error for Rg is significant, and the error for Lg is relatively small.

Table.1. Simulation results of the grid impedance estimation (No voltage feed-forward)

Real grid impedance		Estimation of the grid impedance		Error for Rg	Error for Lg
Rg	Lg	Rge	Lge	(%)	(%)
0.19	6mH	0.199	6.068mH	4.74	1.13
0.19	3mH	0.29	3.015mH	31.58	0.50
0.19	1.5mH	0.238	1.171mH	25.26	-8.93
1.19	1.5mH	1.324	1.316mH	14.54	-4.40
1.19	4.5mH	1.408	4.188mH	18.32	-6.93

Fig.11, Fig.12 and Fig.13 are the simulation results of the grid impedance on the inverter with grid voltage feed-forward. In Fig.11, the line resistance is set fixed to be 0.2Ω, and the real grid inductance varies from 1mH to 6mH. The maximum estimation error (Percentage) appears in the low inductance region, it is no more than 10% in other value. In Fig.12, the line inductance is set fixed to be 1.5mH, and the grid line resistance varies from 0.1Ω to 0.7Ω. The estimation error in average is bigger than that of inductance. Fig.13 shows the estimation results when the grid line resistance and inductance are increasing synchronously. The average of horizontal errors is bigger too.

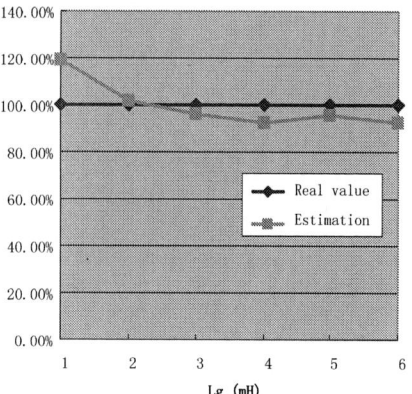

Fig.11 Estimation results when the grid inductance varies from 1mH to 6mH

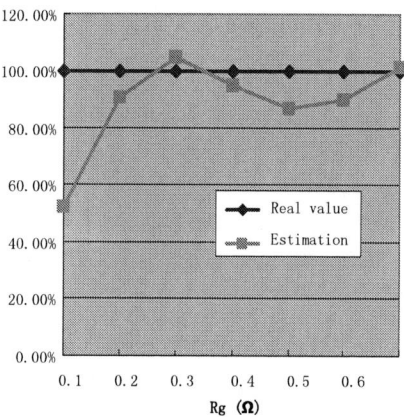

Fig.12 Estimation results when the grid line resistance varies from 0.1Ω to 0.7Ω

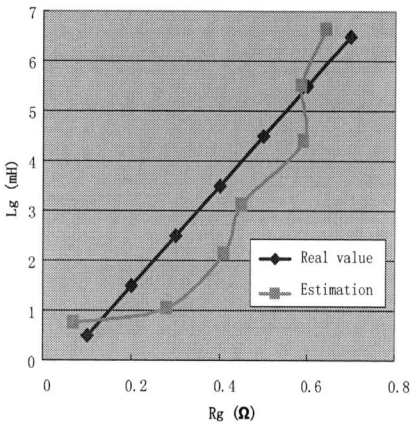

Fig.13 Estimation results when the grid line resistance and inductance are increasing synchronously.

VI. EXPERIMENTAL RESULTS

A 10kW grid-connected inverter prototype is built. The grid voltage is 220Vac/50Hz, switching frequency is 16 kHz, and the dead time is 2.5us. Parameters of the LCL-filter are selected as L_1=0.732mH, L_2=0.745mH, C=20uF without damping resistor in series. The PI regulator with WAC control is applied to the inverter with $K_P = 2.4 = 2.4$, $K_i = 676$. The dc voltage is 720V totally for the half-bridge inverter, and it is well regulated by a prepositive dc to dc converter.

Fig.14 shows the waveforms of the grid voltage and grid current of the inverter with WAC control. The weight value for the WAC feedback control is designed according the inductances of the LCL-filter, i.e., $\beta = 0.5$ as shown in Fig.9. The inverter is steadily operating in this condition.

Fig.15 shows the waveforms of the grid voltage and grid current when a 1mH inductor is inserted in the grid line for the foregoing inverter. Due to the mismatch of the weight value for the WAC feedback control, the system is not stable, and oscillation appears in the waveform of the current.

Fig.16 shows the waveforms of the grid voltage and grid current after a retuned weight value for the WAC feedback control according to the total grid side inductance with the inserted 1mH inductor. The inverter becomes stable again and has good performance with large grid impedance. This proves the necessity of the grid impedance estimation and the adaptive current control for the grid-connected inverters using in a weak grid with a wide range of impedance.

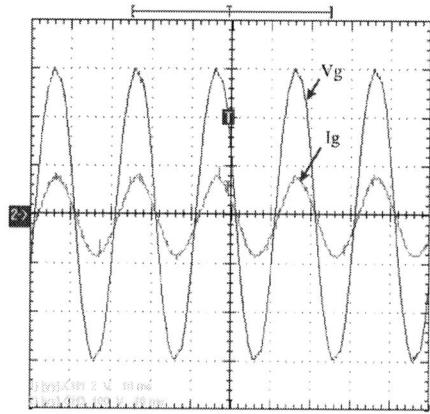

Vg:100V/div, Ig: 20A/div, Time: 10ms/div

Fig.14 Waveforms of the grid voltage and grid current of the inverter before insertion of the 1mH inductor.

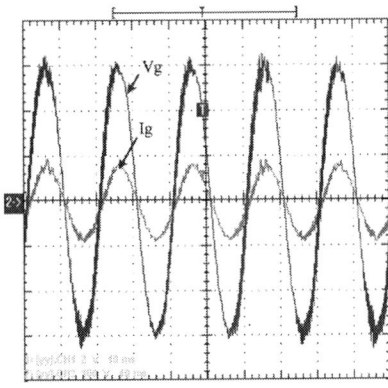

Vg:100V/div, Ig: 20A/div, Time: 10ms/div

Fig.15 Waveforms of the inverter output after insertion of the 1mH inductor (without adaptive control).

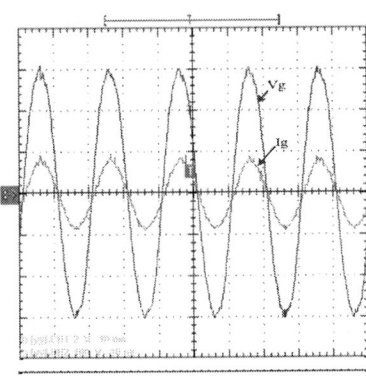

Vg:100V/div, Ig: 20A/div, Time: 10ms/div

Fig.16 Waveforms of the inverter output with the insertion of the 1mH inductor (with adaptive control)

VII. CONCLUSION

This paper presents a new implementation of grid-tied inverter with weighted-average-current feedback control

(WAC) and an embedded online technique to determine the grid line impedance. The presented technique can be implemented with the existing sensors and the digital processor of the inverter, provides a low cost and adaptive approach for current control of grid-tied inverters, especially for the weak grid which always has a wide range of line impedance. Simulation and experimental results are given to validate the control method and the impedance measurement.

ACKNOWLEDGMENT

This work was supported by National Natural Science Foundation of China (50877072), the Specialized Research fund for the Doctoral Program of Higher Education of China (J20070715) and the National High Technology Research and Development of China 863 Program (2007AA05Z243). It is sponsored by grants from the Power Electronics Science and Education Development Program of Delta Environmental & Educational Foundation.

REFERENCES

[1] H. Bindner and P. Lundsager, "Integration of wind power in the power system," in Proc. IECON '02, vol. 4, Nov. 2002, pp. 3309–3316.
[2] M. Harris, A. Kelley: Instrumentation for Measurement of Line Impedance, Proc. On APEC'94, Vol. 2, pp 887-893, 1994
[3] Ciobotaru, M.; Teodorescu, R.; Rodriguez, P.; Timbus, A.; Blaabjerg, F.; "Online grid impedance estimation for single-phase grid-connected systems using PQ variations", Power Electronics Specialists Conference, 2007. PESC 2007. IEEE, 17-21 June 2007, Page(s):2306 - 2312
[4] Liserre, M.; Blaabjerg, F.; Teodorescu, R.; "Grid Impedance Estimation via Excitation of LCL -Filter Resonance", IEEE Transactions on Industry Applications, Volume 43, Issue 5, Sept.-Oct. 2007, Page(s):1401 - 1407
[5] L. Asiminoei, R. Teodorescu, F. Blaabjerg, and U. Borup, "A Digital Controlled PV-Inverter With Grid Impedance Estimation for ENS Detection", IEEE Trans. on Power Electronics, Vol. 20, No.6, pp. 1480-1490, November 2005.
[6] Braca, V.; Petrella, R.; "An improved method for grid impedance estimation by digital controlled PV inverters suitable for ENS detection", in Proc. PEMD 2008. 2-4 April 2008, Page(s):386 - 390
[7] Guoqiao Shen, Dehong Xu, Luping Cao, Xuancai Zhu, "An Improved Control Strategy for Grid-Connected Voltage Source Inverters with an LCL Filter", IEEE Trans. on Power Electronic, vol23, no.4, pp. 1899-1906, July 2008.
[8] D.N. Zmood, D.G. Holmes, "Stationary frame current regulation of PWM inverters with zero steady-state error," IEEE Trans. Power Electron., vol.18-3, pp.814-822, May 2003

A New Direct Peak DC-link Voltage Control Strategy of Z-source Inverters

Yu Tang, Jukui Wei, Shaojun Xie
College of Automation Engineering
Nanjing University of Aeronautics and Astronautics
tangyu@nuaa.edu.cn

Abstract- This paper proposes a new direct peak dc-link voltage control strategy of Z-source inverters. For Z-source inverters, the peak dc-link voltage is the actual voltage fed to the inversion stage. Controlling the peak dc-link voltage at the constant level can effectively decrease the voltage stress and simplify the design process of the inversion stage. A direct peak dc-link voltage detection method is developed with a simple sampling circuit, and then controlled directly. Therefore, the constant peak dc-link voltage can be achieved in different input and load conditions. The peak ac output is calculated in real-time by sampling the ac phase voltage, and it is controlled to the desired value without steady-state error. With the proposed direct peak dc-link voltage control and peak ac output control, both the peak dc-link voltage and ac output is controlled to a desired level under different input and load condition. Experimental results are presented to demonstrate the feasibility of the proposed control strategy.

I. INTRODUCTION

The Z-source inverter (ZSI) is a recently proposed inverter topology to perform the buck-boost conversion in a single-stage [1-3]. It can get a desired output even higher than available dc bus voltage by utilizing the shoot-through state. Because the shoot-through state becomes a unique operation state, the ZSI operates without dead-time, so the output distortion is reduced and the reliability is also enhanced. The traditional ZSI suffers high voltage across the Z-source capacitors with discontinuous input current. In recently, two types of modified ZSI topologies, called as the embedded ZSI and the series ZSI in this paper, have been developed with several advantages [4-6].

Each of these topologies features a different structure and characteristic. The embedded ZSI presents the unsymmetrical Z-source network, and features low voltage stress across one of the two Z-source capacitors with continuous input current. The series ZSI (the inverter bridge is in series with the input source) also shows a symmetrical Z-source network, and features low voltage stress across both Z-source capacitors with discontinuous input current. With the same modulation schemes developed in traditional topology, the modified topologies can perform the same voltage boost ability with reduced system cost, as discussed carefully in published papers [4-6].

The control strategy of ZSIs is an important issue and several feedback control strategies have been investigated in recent

publications. For ZSIs, the control of dc-link stage is a key point while the control of ac side is similar to that in voltage-source inverter, therefore the controller design is focused on how to control the dc-link voltage. Existing control strategies of ZSIs including the following two types: the Z-source capacitor voltage (ZCV) control [7] and the peak dc-link voltage (PDV) control [8-9]. For ZCV control strategy, the capacitor voltage is controlled to a constant value. In this condition, the PDV fed to the inversion stage is varied under different input voltage, which increases the voltage stress across the switches and introduces difficulty in ac side controller design. The merits of PDV control strategy are obvious. It is more effective with simplified ac side controller design and lower voltage stress across the switches. The dc-link voltage in ZSIs is a high frequency pulse waveform: it reaches its peak value in non-shoot-through state and keeps zero in shoot-through state. Such a pulse voltage introduces difficulty in detecting its peak value directly. A peak value sensing circuit is introduced in [8], and then a direct PDV controller is designed based on it. The additional sensing circuit remains to be carefully designed and may increase the complexity of the design. An indirect PDV control strategy is proposed in [9]. By sampling the ZCV and then introducing a divider algorithm to the controller, the peak value is calculated based on the steady state equation and controlled indirectly.

In this paper, a PDV sensing method is developed and then controlled directly. The PDV is controlled to the desired value under different input and load condition with fast transient performance. The peak ac voltage feedback control is adopted to obtain the desired output voltage without steady-state error. The proposed control strategy is simple in realization without introducing complex sensing circuit and the divider algorithm in calculation. Experimental results will be presented to verify the analysis.

II. CHARACTERISTIC OF DC-LINK VOLTAGE IN Z-SOURCE INVERTERS

For traditional ZSI, as shown in Fig.1, the inverter bridge is paralleled with the Z-source network. There are two states in one switching period: shoot-through and non-shoot-through state. Fig.2 shows the equivalent circuits. When in the shoot-through state as shown in Fig.2 (a), the inductors charge and the capacitors discharge, there is no energy exchange between the

978-1-4244-4782-4/10 $26.00 © 2010 IEEE

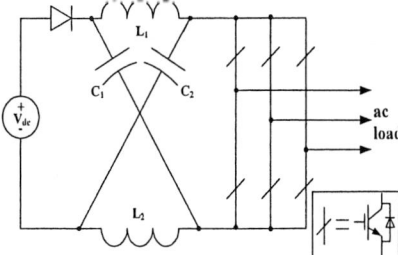

Fig.1 Traditional ZSI

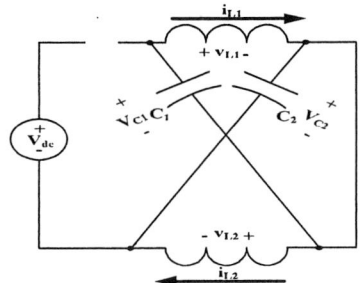

(a)　Shoot-through state

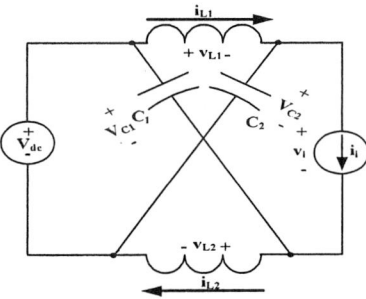

(b)　non-shoot-through state

Fig.2 Equivalent circuits of Traditional ZSI

power source and load; when in the non-shoot-through state as shown in Fig.2 (b), the inductors discharge and the capacitors charge, the energy is transferred from the power source to the load.

As analyzed above, in shoot-through state, the dc-link voltage fed to the inversion stage (the voltage across the positive and negative rail of bridge arms) is zero; while in non-shoot-through state, the dc-link voltage reaches the peak value. Fig.3 shows the waveforms of the shoot-through signal and dc-link voltage of traditional ZSI, the PDV can be expressed as

$$\hat{v}_i = 2V_C - V_{dc} = \frac{1}{1-2D_0}V_{dc} \qquad (1)$$

The embedded ZSI is shown in Fig.4. The power source is embedded to the Z-source network. Fig.5 shows the equivalent circuits. When in the shoot-through state as shown in Fig.5 (a), the inductors charge and the capacitors discharge, the energy is transferred from the power source to the inductor; when in the non-shoot-through state as shown in Fig.5 (b), the inductors discharge and the capacitors charge, the energy is transferred from the power source to the load.

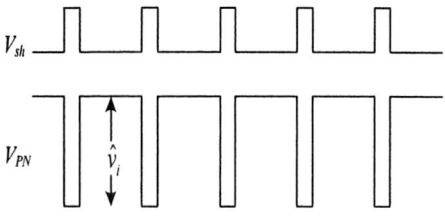

V_{sh} : the shoot-through signal，V_{PN} : the dc-link voltage

Fig.3 The shoot-through signal and dc-link voltage of ZSI

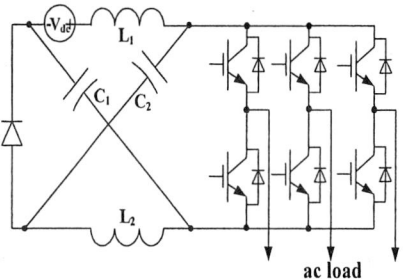

Fig.4　Embedded ZSI

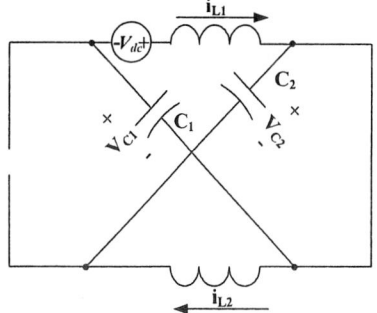

(a)　Shoot-through state

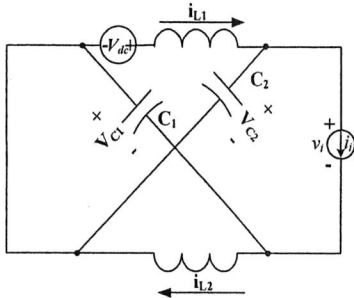

(b) non-shoot-through state

Fig.5 Equivalent circuits of Embedded ZSI

For embedded ZSI, the characteristic of the dc-link voltage is the same as the traditional ZSI, and can also be described in Fig.3. The PDV of embedded ZSI is

$$\hat{v}_i = V_{C1} + V_{C2} = \frac{1}{1 - 2D_0} V_{dc} \qquad (2)$$

Fig.6 shows the series ZSI. The Z-source network is in series with the power source and inverter bridge. Fig.7 shows the equivalent circuits. When in the shoot-through state as shown in Fig.7 (a), the inductors are charged and the capacitors are discharged, the energy is transferred from the power source to the inductor; when in the non-shoot-through state as shown in Fig.7 (b), the inductors are discharged and the capacitors are charged, the energy is transferred from the power source to the load.

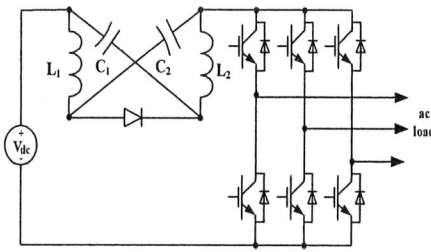

Fig.6   SeriesZSI

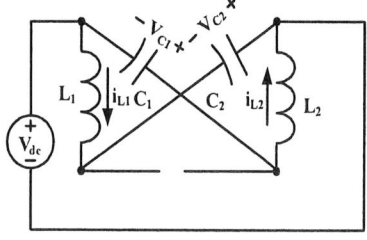

(a) Shoot-through state

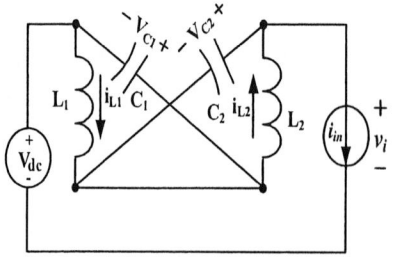

(b) non-shoot-through state

Fig.7 Equivalent circuits of Series ZSI

For series ZSI, the characteristic of the dc-link voltage is also the same shown in Fig.3. The PDV of series ZSI is

$$\hat{v}_i = 2V_C + V_{dc} = \frac{1}{1 - 2D_0} V_{dc} \qquad (3)$$

Therefore the PDV and boost ability of the three topologies is exactly the same. The boost factor B is determined by shoot-through duty ratio D_0 and can be expressed as

$$B = \frac{1}{1 - 2D_0} \qquad (4)$$

The output voltage of the three topologies is also the same, and the peak phase voltage is

$$\hat{v}_p = M \frac{\hat{v}_i}{2} = MB \frac{V_{dc}}{2} \qquad (5)$$

The modulation schemes of these ZSIs are the same, and all the present modulation schemes developed for traditional ZSI, such as the simple boost control, the maximum boost control, the maximum constant boost control and the SVPWM scheme, can be utilized in modified topologies directly.

Fig.8 shows the SVPWM scheme of voltage-source inverter (VSI) and ZSIs. By inserting the shoot-through state at the transit of the switching state in VSI while keeping the same active intervals, the output voltage is boost without any distortion. In addition, the switching frequency keeps the same as in VSI and the equivalent frequency of Z-source network is greatly enhanced, because there are six shoot-through intervals in one switching cycle.

III. DIRECT PEAK DC-LINK VOLTAGE SAMPLING METHOD

A direct PDV sampling method is proposed in this paper. For series ZSI, the PDV sampling circuit is shown in Fig.9. Three sampling resistance R_1, R_2, R_3 and a LEM is utilized to sample the PDV. The voltage V_{AB} is sampled by a LEM. The parasitical resistance in primary side of the LEM is much smaller than R_3 and can be omitted. To simplify the analysis, the equivalent sampling circuit is given in Fig.10.

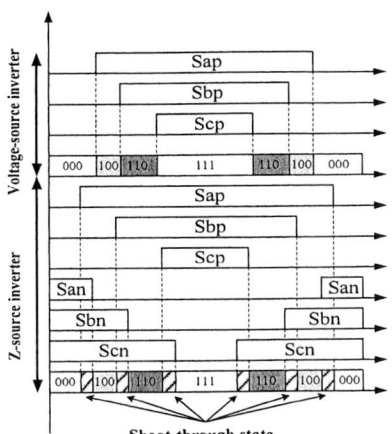

Fig.8 SVPWM Schem of ZSIs

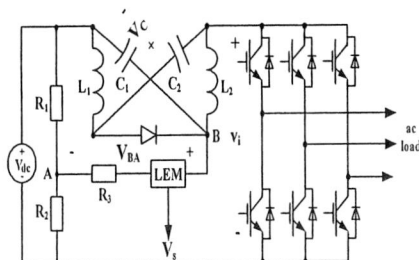

Fig.9 Direct PDV Sampling circuit of Series ZSI

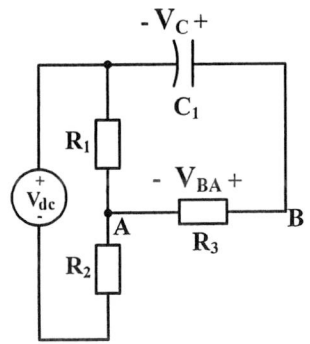

Fig.10 Simplified sampling circuit of Series ZSI

According to the KCL rule of point A, we can get

$$\frac{V_{BA}-V_C}{R_1}+\frac{V_{BA}}{R_3}=\frac{V_C+V_{dc}-V_{BA}}{R_2} \qquad (6)$$

Selecting $R_1=R_2=R$, we can get

$$V_{BA}=\frac{R_3}{R+2R_3}(V_{dc}+2V_C)=\frac{R_3}{R+2R_3}\hat{v}_i \qquad (7)$$

That means V_{BA} is proportional to PDV. Assumption the sampling ratio of the LEM is k, the voltage Vs get from the secondary side of the LEM is

$$V_s=kV_{BA}=\frac{kR_3}{R+2R_3}\hat{v}_i \qquad (8)$$

V_s is also proportional to the PDV, therefore the PDV can be detected easily without delay and fast transient response is desired. For traditional ZSI and embedded ZSI, the PDV sampling circuit is similar, and is given in Fig.11 and Fig.12, respectively. The principles are similar to that in series ZSI. The PDV can be sampled directly by the proposed sampling method for all the ZSI topologies.

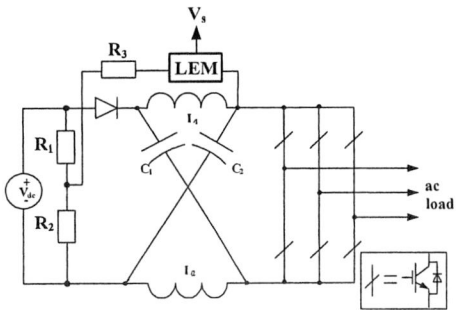

Fig.11 Direct PDV Sampling circuit of Traditional ZSI

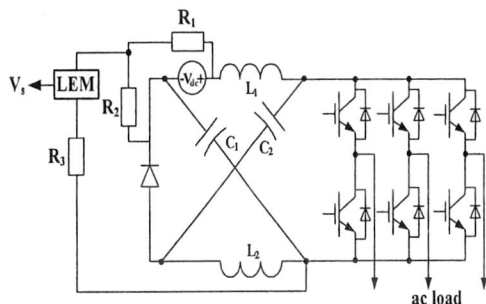

Fig.12 Direct PDV Sampling circuit of Embedded ZSI

IV. CONTROL STRATEGY OF Z-SOURCE INVERTER

Taking the series ZSI as an example, the control strategy is shown in Fig.13. For traditional and embedded ZSIs, the control strategy is similar and can be derived easily from Fig.13. The shoot-through duty ratio D_0 is utilized to control the PDV adopting the direct PDV sampling circuit discussed above. The peak ac output is calculated and controlled to the desired value by the modulation index M, then the SVPWM signals are produced and send to the switches.

As can be derived from (3), the PDV is determined by shoot-through duty ratio D_0, therefore by controlling D_0, we can get the required PDV under different input voltage and load condition. In the voltage-boost operation mode, the input voltage V_{dc} is less than the required PDV, then D_0 is larger than 0 and shoot-through state is utilized to perform voltage boost conversion; while in voltage-buck operation, V_{dc} is high enough to produce the output, then D_0 is set to 0 and shoot-through state is not used, the ZSI performs the voltage buck conversion the same as the VSI.

The three-phase output voltage is detected to calculate the peak value of the phase voltage by

$$V_p = \sqrt{\frac{2}{3}(v_a{}^2 + v_b{}^2 + v_c{}^2)} \qquad (9)$$

V_p is a constant DC value, therefore there is no steady-state error in the reference peak value and the peak output voltage with PI controller. By using the proposed control strategy, the PDV is controlled directly to its required value, and the output voltage can be controlled without steady-state error.

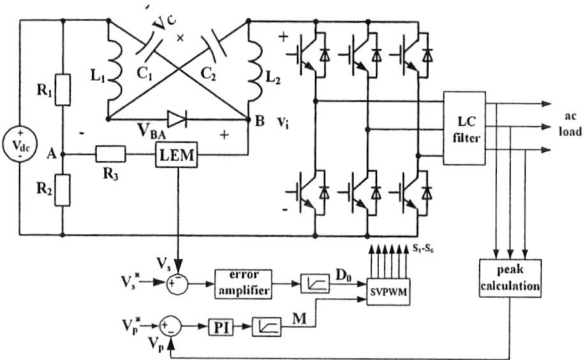

Fig.13 Control strategy of series ZSI

IV. EXPERIMENTAL VERIFICATION

A prototype is built up in the lab to verify the sampling method and control strategy. The parameters used are:

1) Input voltage V_{dc}=200-400V;
2) Output phase voltage: 110V/50Hz;
3) Load: three-phase resistance load R=24 Ω ;
4) Z-source network: L_1=L_2=500 μ H, C_1=C_2=2000 μ F;
5) Output filters: L_f=1000 μ H, C_f=15 μ F;
6) Switching frequency: 10 KHz.

In the control strategy, the reference value of PDV is 370V, that means when V_{dc} is less than 370V, shoot-through state is introduced, while V_{dc} is larger than 370V, shoot-through state is not inserted into the switching state.

Fig.14 shows the experimental results when input voltage is 200V. The Z-source capacitor voltage is 85V and PDV is boost to 370V to produce the desired output. In shoot-through state, the dc-link voltage is zero and the inductor current increases, while in non shoot-through state, the dc-link voltage is kept at 370V and the inductor current decreases. By using the SVPWM strategy as shown in Fig.8, there are six shoot-through states in one switching cycle. Fig.15 shows the experimental results when input voltage is 300V. With the increase of input voltage, the Z-source capacitor voltage is decreased to 35V and the PDV is also controlled at 370V well. Fig 16 shows the experimental results when input voltage is 400V. The input voltage is higher than the desired PDV, therefore the shoot-through state is not introduced and the inverter performs as a voltage-source inverter. The dc-link voltage is equal to 400V and there is no current ripple across the inductor.

V. CONCLUSION

A direct PDV control strategy for ZSIs is proposed in this paper. The PDV can be detected directly by utilizing a simple sampling circuit, and is controlled to the desired under varied input voltage by regulating the shoot-through duty ratio. The peak value of ac output is calculated instantaneously and controlled precisely without steady-state error. Experimental results have been given to verify the analysis and proposed control strategy.

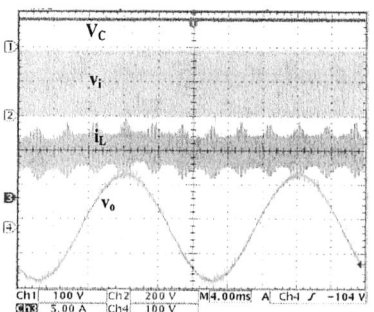

(a) Waveforms in line frequency

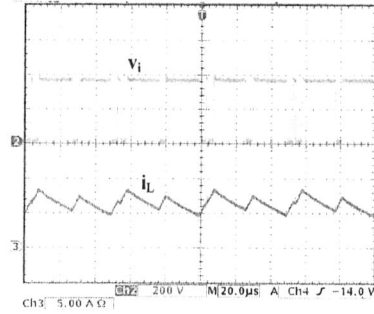

(b) Waveforms in switching frequency

Fig.13 Experimental results of series ZSI under V_{dc}=200V

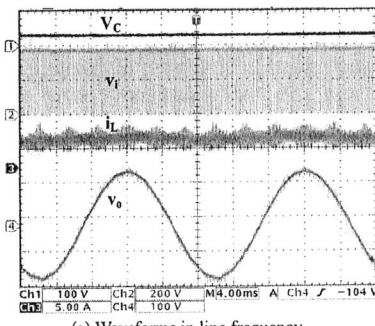

(a) Waveforms in line frequency

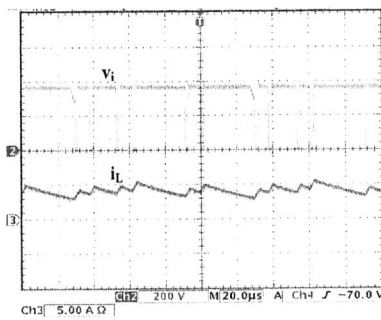

(b) Waveforms in switching frequency

Fig.14 Experimental results of series ZSI under V_{dc}=300V

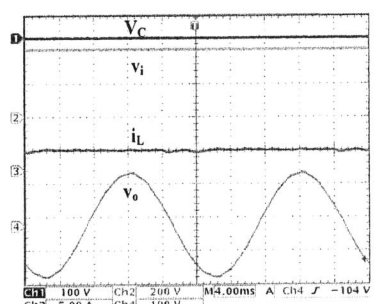

Fig.15 Experimental results of series ZSI under V_{dc}=400V

ACKNOWLEDGMENT

This work was supported by Delta Environmental and Educational Foundation under the Power Electronics Science and Education Development Program (DREG2008002) , the Natural Science Foundation of Jiangsu Province of China (BK2008391) and Foundation of NUAA. The authors thanks for the financial support.

REFERENCES

[1] F. Z. Peng, Z-source inverter, IEEE Trans. on Industry Applications, vol. 39, no. 2, 2003: 504-510.

[2] M. S. Shen, J. Wang, A. Joseph, et al, Constant boost control of the Z-source inverter to minimize current ripple and voltage stress, IEEE Trans. on Industry Applications, 2006, 42(3):770-777

[3] P. C. Loh, D. M. Vilathgamuwa, Y. S. Lai , et al, Pulse-width modulation of Z-source inverters, IEEE Trans. on Power Electronics, 2005, 20(6): 1346-1355

[4] Yu Tang, Shaojun Xie, Chaohua Zhang and Zegang Xu, Improved Z-source inverter with reduced Z-source capacitor voltage stress and soft-start capability, IEEE Trans. on Power Electronics, 2009, 24 (2):409-415.

[5] J. Anderson, F. Z. Peng. Four quasi-Z-Source inverters, IEEE PESC, 2008: 2743~2749.

[6] F. Gao, P. C. Loh, F. Blaabjerg, C. J. Gajanayake, Operational analysis and comparative evaluation of Embedded Z-source inverters, IEEE PESC, 2008: 2757-2763.

[7] Q. Tran, T. Chun, J. Ahn and H. Lee, Algorithms for controlling both the DC boost and AC output voltage of Z-source inverter, IEEE Trans. on Industrial Electronics, vol. 54, no. 5, 2007: 2745-2750.

[8] X. P. Ding, Z. M. Qian, S. T. Yang, et al. A direct peak DC-link boost voltage control strategy in Z-source inverter, IEEE APEC, 2007: 648~653.

[9] C. J. Gajanayake, D. M. Vilathgamuwa, P. C. Loh, Development of a comprehensive model and a multiloop controller for Z-source inverter DG systems, IEEE Trans. on Industrial Electronics, vol. 54, no. 4, 2007: 2352-2359.

High Performance Voltage Regulation of Current Source Inverters

S.A.S. Grogan, D.G. Holmes, B.P McGrath

Department of Electrical and Computer Systems Engineering
Monash University
Melbourne, Australia

Abstract—Current source inverters offer advantages of voltage boost, short circuit protection, reduced EMI and direct regeneration. While CSI control strategies are less developed than for a VSI, the topologies are functional duals and have much in common in a control sense. In particular, since CSI voltage regulation is the dual of VSI current regulation, current regulation control strategies for a VSI can be readily implemented as equivalent voltage control strategies for a CSI. This paper shows how a high performance PI stationary frame and P+Resonant CSI voltage regulator can be analytically designed and optimised, while in particular taking into account the second order response of a CSI filter/load combination.

I.   INTRODUCTION

While voltage source inverters (VSI) are by far the dominant topology for power electronic conversion, current source inverters (CSI) do offer some specific advantages in particular applications. These include voltage boost from a lower magnitude input source, implicit output short circuit protection, significantly reduced EMI because of the capacitive voltage filters located directly on the inverter output, and an ability to directly support regeneration when the DC current is provided by an input SCR rectifier [1]. Hence CSI's can be useful in applications where their advantages outweigh the obvious fundamental CSI disadvantage of DC series inductor losses [2][3].

In general, modulation and control strategies for a CSI are less developed than they are for a VSI, primarily because of more limited usage of CSI's in practice. However, while the two topologies are not exact duals, they have much in common in a space vector sense [4], and hence leading edge VSI modulation and control strategies can often be applied to a CSI without significant modification to achieve a similar performance to a VSI [5]. In particular, the dual of VSI current regulation is CSI voltage regulation. Thus it could be expected that the advanced current regulation strategies that have been developed for a VSI, such as hysteresis, synchronous frame DQ, stationary frame P+Resonant [6] and predictive control, can be directly applied to create equivalent CSI voltage regulation strategies, as has been reported [5].

However, from a control perspective, a VSI is a first order system, while a CSI is a second order system with a relatively low frequency resonant peak caused by the output filter/load network. Hence the strategies that are used to set controller gains for a VSI are not necessarily directly applicable to a CSI. Furthermore, it is commonly suggested in the literature [7] that the resonant effect of a second order system should be

mitigated using compensation techniques such as active damping, to ensure system stability under all operating conditions. Optimising the gains of a system with such an additional feedback path is more complex than for a simple first order VSI closed loop regulator, and needs additional investigation.

The primary approach presented in this paper is to use a simple stationary frame Proportional-Integral (PI) AC voltage regulation strategy for a CSI, since recent work has shown how this approach can be very effective for VSI current regulation once sampling/transport delays and back EMF disturbance effects are taken into account [8]. Using the principles developed in this work, a strategy is proposed in this paper to optimize controller PI gains for a CSI, and the performance of the resulting system is then compared against the more advanced P+Resonant regulator to determine its range of effectiveness. Finally, the concepts are applied to a system where active damping has been included as a means of further improving system stability. The results of this analysis show that when controller gains are set to their maximum possible values, active damping offers little or no additional benefit to either controller performance or stability margins.

All analytical development presented in this paper is supported by detailed time domain switching simulations and matching experimental results.

II.   MODELLING A THREE PHASE CURRENT SOURCE INVERTER CONTROLLED BY A SIMPLE PI CONTROLLER

Figure 1 shows the structure of a voltage controlled CSI, where the switching converter system feeds into an RL load with a back EMF, connected in parallel with the (mandatory) output filter capacitor C. This structure provides a generic representation of most types of three phase current fed AC load systems, unless there is significant unbalanced cross coupling between the phase load elements.

The CSI voltage regulation control system operates by comparing the measured output voltages against target reference commands. These errors are then used to generate modulation signals which are transformed into VSI space vector switching commands using standard PWM strategies [9]. Finally, the space vectors are mapped to equivalent CSI space vectors and thence to device switching signals, as described in [4]. (This approach avoids having to develop a separate CSI PWM strategy.) Fig. 2 shows the experimental CSI switched phase currents and filtered output voltages achieved using this control/modulation strategy.

978-1-4244-4782-4/10 $26.00 © 2010 IEEE

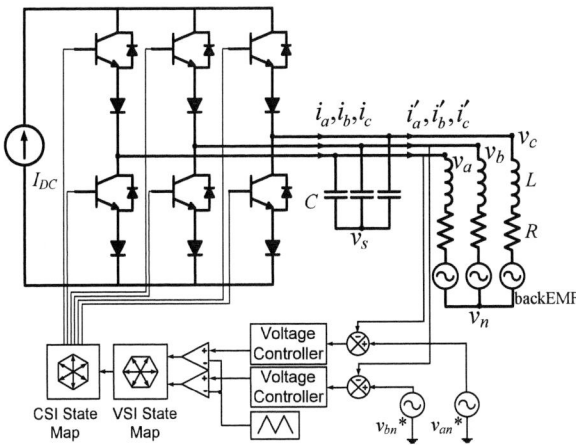

Fig. 1: Three Phase Voltage Controlled CSI

Since the focus of this paper is the voltage regulation response, which is significantly slower than the converter switching frequency, the PWM modulator can be modeled as a simple linear amplifier with a gain of I_{DC}. Essentially, this means that over any particular PWM half carrier cycle, the CSI modulator can command any required current from each phase leg, provided that the sum of the three phase leg positive peak current magnitudes is less than I_{DC}. It can readily be shown that only two independent voltage regulation loops are required for this control system, as follows:

Defining the load star point voltage as $v_n(t)$, the relationship between the RL load phase currents $i_a'(t)$, $i_b'(t)$, $i_c'(t)$ and the converter output voltages is given by

$$v_{an}(t) = Ri_a'(t) + L\frac{di_a'(t)}{dt} + emf_a(t) \tag{1a}$$

$$v_{bn}(t) = Ri_b'(t) + L\frac{di_b'(t)}{dt} + emf_b(t) \tag{1b}$$

$$v_{cn}(t) = Ri_c'(t) + L\frac{di_c'(t)}{dt} + emf_c(t) \tag{1c}$$

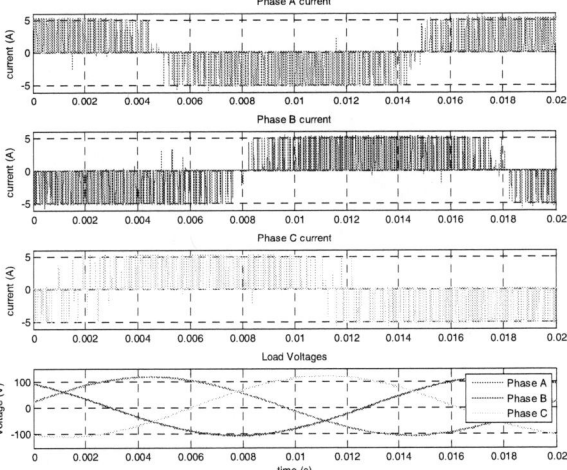

Fig. 2: Experimental CSI switched phase currents and load voltages with two independent closed loop regulators

KCL constrains the sum of the load currents to be zero, i.e.

$$i_a'(t) + i_b'(t) + i_c'(t) = 0 \tag{2}$$

Summing (1), substituting from (2), and recognizing that the sum of the three backEMF's must be zero since they are a balanced three phase system, identifies that the sum of the converter output voltages must also be zero, i.e.

$$v_{an}(t) + v_{bn}(t) + v_{cn}(t) = 0 \tag{3}$$

Thus the converter output voltages constitute a balanced three phase set, as could be expected, and hence the load star point $v_n(t)$ is in fact a true zero voltage neutral point.

The CSI output currents for each phase are given by

$$i_a(t) = i_a'(t) + C\frac{dv_{as}(t)}{dt} = i_a'(t) + C\frac{d\{v_{an}(t) + v_{ns}(t)\}}{dt} \tag{4a}$$

$$i_b(t) = i_b'(t) + C\frac{dv_{bs}(t)}{dt} = i_b'(t) + C\frac{d\{v_{bn}(t) + v_{ns}(t)\}}{dt} \tag{4b}$$

$$i_c(t) = i_c'(t) + C\frac{dv_{cs}(t)}{dt} = i_c'(t) + C\frac{d\{v_{cn}(t) + v_{ns}(t)\}}{dt} \tag{4c}$$

where $v_s(t)$ is the star point voltage of the capacitor filter bank. Once again, KCL constrains the sum of the converter currents to be zero, so that

$$i_a(t) + i_b(t) + i_c(t) = 0 \tag{5}$$

Substituting (2), (3) and (5) into (4) identifies that $v_{ns}(t) = 0$, which means that the capacitor network star point $v_s(t)$ must also be a true zero voltage neutral point. From this analysis, it then immediately follows that each load phase voltage can be separately and independently controlled by simply regulating the converter current injected into that load/filter phase leg.

Furthermore, since the CSI current reference modulation commands are constrained according to (5), it also follows that only two independent voltage regulation loops are required because the third phase leg current can be directly calculated as the simple inverse sum of the first two phase leg current references (this is the same concept as the need for only two control loops to control currents in a three phase VSI, as described in [8]). The third phase leg voltage must automatically follow on to make up a three phase voltage set because of the current that is forced to flow through it.

The CSI voltage regulation requirement therefore reduces to creating two phase current modulation reference signals, determined so as to make their associated load phase voltages track a commanded sinusoidal target. In addition, since each of the two regulated phase legs are independent of each other, their controllers will also be independent, and can be considered simply as two separate single phase systems.

Fig. 3 shows the system "average model" closed loop control block diagram in the Laplace domain, based on these concepts. $G(s)$ is the load transfer function block, which is the parallel combination of the series RL load and the filter capacitor C, while $H(s)$ is the controller regulator. From Fig. 1, and taking linearised frequency dependent impedances for each of the load elements, $G(s)$ for each phase load element (without back EMF) can be developed as:

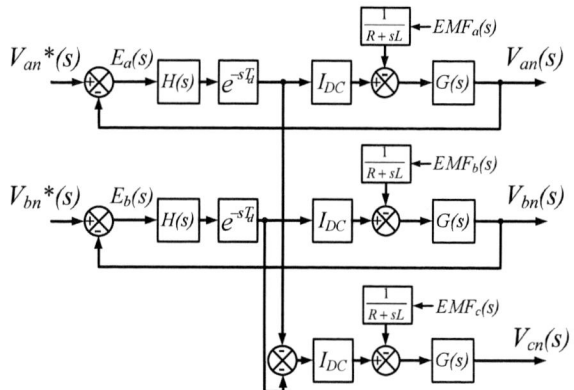

Fig. 3: Linearised "average" CSI Voltage Regulation System

$$G(s) = \frac{1}{sC} \| (R+sL) \Rightarrow \frac{R+sL}{s^2 LC + sRC + 1} \qquad (6)$$

which unlike a VSI is clearly a second order system with a possible underdamped oscillatory response at $\omega_r \approx 1/\sqrt{LC}$, depending on the damping ratio of the denominator. Such a system is well known to potentially produce a resonant transfer function gain peak, as is shown in Fig. 4 for the base test CSI system used in this paper, which has the load and operating parameters listed in TABLE I.

The transfer function $H(s)$ of the two single phase stationary frame PI controllers is well known, and can be immediately expressed as:

$$H(s) = K_P \left(1 + \frac{1}{s\tau_r} \right) = \frac{K_P}{s\tau_r} (1 + s\tau_r) \qquad (7)$$

where K_P = proportional gain and τ_r = integral time constant.

As discussed in [8], the effect of PWM on any digital modulator is to introduce a "quarter" carrier transport delay in the forward path gain, since the modulation command is "locked in" by the controller at the start of each half carrier period (note that if symmetrical modulation is used, this transport delay doubles). Also, a further effect of digital control is to introduce an additional "sampling" delay, because the controller must measure the voltage to be controlled in sufficient time before the start of the next PWM half carrier period to complete the control loop calculation. In practice, this delay is typically another "half" carrier period, since the measurement of the controlled voltage must be done synchronously to avoid including switching ripple in the measurement, which can adversely effect the control stability. Fig 3 shows how these delays have been included into the system by adding e^{-sT_d} into the forward path, as has been shown in [8] to be an effective method of modeling the dynamic performance of this type of sampled control system.

TABLE I.   TEST SYSTEM PARAMETERS

Parameter	Value	Parameter	Value
f_0	50Hz	f_c	5000Hz
R	6Ω	L	500 mH
C	320uF	I_{DC}	10A

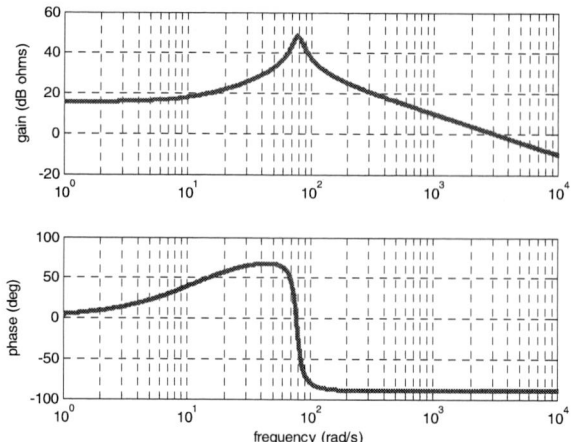

Fig. 4: CSI load plant frequency response, no controller, no digitization delays, $\omega_r = 79.2$ rad/s

Using standard control system concepts, the open loop transfer function for each phase of the system shown in Fig. 3 can be found by combining (6) and (7) with the Laplace expression developed for transport delay and including a forward linear gain representation of the CSI of I_{DC}, to give

$$H(s)e^{-sT_d} I_{DC} G(s) = \frac{K_p(1+s\tau_r)}{s\tau_r} e^{-sT_d} \frac{I_{DC} R(1+sT_p)}{s^2 LC + sRC + 1} \qquad (8)$$

$$= \frac{V_x(s)}{E_x(s)}, x \in \{a,b\}, \text{ with } T_p = L/R.$$

III.   DETERMINING THE MAXIMUM PI CONTROLLER GAINS

The maximum possible controller gains for the system described by (8) can now be analytically determined using the concepts developed in [8], whereby the proportional gain K_P is set to achieve a required phase margin ϕ_m at the cross over frequency ω_c (i.e. the frequency at which the loop gain becomes unity). This means that the phase angle of the open loop transfer function (8) at ω_c should be set to

$$\angle \left\{ H(j\omega_c)e^{-j\omega_c T_d} I_{DC} G(j\omega_c) \right\} = -\pi + \phi_m \qquad (9)$$

Standard control theory suggests that a suitable phase margin for satisfactory control performance with adequate transient damping is 40^0 [10], although the method is suitable for designing for any required phase margin in any given system.

The phase angle of (8) is given by

$$\angle \frac{V_x(s)}{E_x(s)} = \angle \left\{ \frac{K_p I_{DC} R}{\tau_r} \frac{(1+j\omega_c \tau_r)e^{-j\omega_c T_d}(1+j\omega_c T_p)}{j\omega_c (1 - \omega_c^2 LC + j\omega_c RC)} \right\} \qquad (10)$$

$$= \tan^{-1}(\omega_c \tau_r) - \omega_c T_d + \tan^{-1}(\omega_c T_p)$$
$$- \pi/2 - \tan^{-1}\left(\omega_c RC/(1 - \omega_c^2 LC) \right) \qquad (11)$$

For most systems, ω_c will be well above the plant pole frequency, and hence the angular contribution of

$\tan^{-1}\left(\omega_c T_p\right) - \tan^{-1}\left(\omega_c RC \big/ \left(1 - \omega_c^2 LC\right)\right)$ is approximately $-\pi/2$. Substituting this approximation into (9) and (11) gives

$$\phi_m \approx \tan^{-1}\left(\omega_c \tau_r\right) - \omega_c T_d \qquad (12)$$

$$\Rightarrow \quad \omega_c = \frac{\tan^{-1}\left(\omega_c \tau_r\right) - \phi_m}{T_d} \approx \frac{\pi/2 - \phi_m}{T_d} \qquad (13)$$

if $\tau_r \approx 10/\omega_c$. Note that setting τ_r in this way allows the integrator contribution to be completed approximately a decade below the crossover frequency as suggested in [8]. K_P can now be found by using (8) to determine the value that achieves unity open loop gain at ω_c. This gives the maximum possible PI gain values for a given phase margin ϕ_m as

$$K_p = \frac{\tau_r}{RI_{DC}}\omega_c \sqrt{\frac{\left(1-\omega_c^2 LC\right)^2 + \left(\omega_c RC\right)^2}{\left(1+\omega_c^2 \tau_r^2\right)\left(1+\omega_c^2 T_p^2\right)}} \approx \frac{\omega_c C}{I_{DC}} \qquad (14)$$

and $\quad \tau_r \approx 10/\omega_c \qquad (15)$

These gains are the maximum possible that can be achieved while still maintaining an adequate stability margin. Fig. 5(a) shows the open loop frequency response for the test system with these gains, where it can be seen that at the target frequency of 50Hz (314 rad/s) the system still has a substantial open loop gain of 32dB. Fig. 5(b) shows the theoretical, simulated and matching experimental time domain transient response to a step change voltage command for this system, which has minimal steady state magnitude and phase error despite the very simple controller strategy that has been used.

IV. "GAIN" BENEFITS OF A RESONANT 2ND ORDER SYSTEM

A common proposition in the literature is that the resonance of a 2nd order system such as occurs with a CSI load/filter is a stability hazard that requires compensation. However, the work presented in this paper shows that this resonance can actually enhance the system closed loop performance once the controller gains are properly optimized.

Looking firstly at the stability issue, it can be seen from Fig 4 that the phase contribution of a second order plant begins at zero, and asymptotes to -90° as frequency increases, with a positive phase excursion around the resonant frequency. From (7), it can also be seen that the phase contribution of the PI controller begins at -90°, and returns to zero as frequency increases. Hence the open loop phase of the overall system (excluding the effects of transport delay) can never go below -180°, and so this 2nd order system is still always stable irrespective of the controller gain settings. Of course this simple analysis breaks down at higher frequencies once the phase shift introduced by controller transport delay is taken into account, but it is exactly this effect which limits the controller gains as has been discussed in Section III, and hence there is no stability issue from this delay contribution if the gains are set correctly as described.

In fact, the resonant effect of a 2nd order load system is quite similar to the resonant gain of more advanced controllers such as a P+Resonant system, except that the resonance is not specifically tuned to the target reference frequency. Hence it

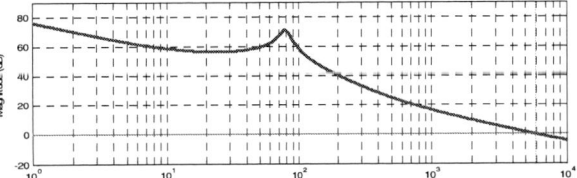

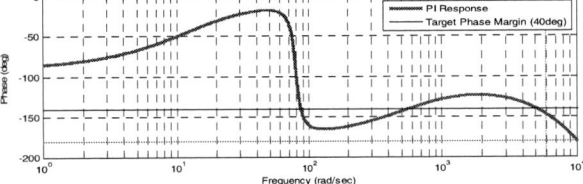

(a)

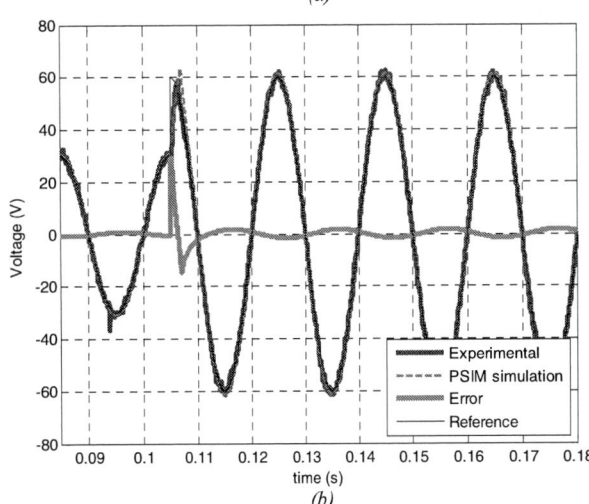

(b)

Fig. 5: CSI system (a) frequency and (b) transient response with maximum possible PI gains
$K_p= 0.185, \ \tau_r= 0.00172s, \ \omega_r = 79.2 \ rad/s$

could be expected that if the load resonance does match the target frequency, the system closed loop performance will actually be enhanced rather than degraded by the load resonant response. Fig. 6 confirms this expectation, showing the frequency response and time domain transient response for the CSI system with parameters as described in TABLE II, where the load parameters have been adjusted to place the load resonance almost at the target reference frequency of 50 Hz (314 rad/s). Both the steady state and transient performance are clearly improved under this resonant match condition, compared to the response previously shown in Fig. 5. Hence instead of attempting to damp out a 2nd order system resonant response, it is actually advantageous if this resonance occurs in the frequency region of the reference target signal.

TABLE II. TEST SYSTEM PARAMETERS: LOAD RESONANCE AT 51 HZ

Parameter	Value	Parameter	Value
f_0	50Hz	f_c	5000Hz
R	11Ω	L	**160mH**
C	60uF	I_{DC}	10A

$$V_{xn}(s) = (R + sL)(I_x(s) - sCV_{xn}(s)) + EMF_x(s) \quad (18)$$

which rearranges to

$$
\begin{aligned}
V_{xn}(s) &= \frac{(R + sL)I_x(s) + EMF_x(s)}{1 + sRC + s^2 LC}, \quad x \in \{a,b,c\} \\
&= G(s)I_x(s) + G(s)\frac{EMF_x(s)}{R + sL}
\end{aligned}
\quad (19)
$$

From Fig 3, the current $I_x(s)$ injected into each phase leg by the CSI is given by the PI controller forward transfer function multiplied by the measured voltage error (excluding the effects of transport delay for simplicity since it has minimal effect at the target reference frequency), i.e.

$$I_x(s) = I_{DC}H(s)\{V_{xn}^*(s) - V_{xn}(s)\} = I_{DC}H(s)E_x(s) \quad (20)$$

where $E_x(s)$ is the error between the target phase voltage and the measured phase voltage, $x \in \{a,b,c\}$.

Finally, substituting (20) into (19) and rearranging, gives:

$$V_x^*(s) - E_x(s) = I_{DC} \cdot G(s)H(s)E_x(s) + G(s)\frac{EMF_x(s)}{R + sL} \quad (21)$$

$$\rightarrow E_x(s) = \frac{V_x^*(s) - G(s)EMF_x(s)/(R + sL)}{1 + I_{DC}G(s)H(s)}, x \in \{a,b,c\} \quad (22)$$

Eqn. (22) can be separated into two sections, to define a closed loop reference tracking error term $E_I(s)$ and a disturbance rejection error term $E_D(s)$, given by

$$E_I(s) = \frac{E_x(s)}{V_x^*(s)}\bigg|_{EMF_x(s)=0} = \frac{1}{1 + I_{DC}G(s)H(s)} \quad (23)$$

$$E_D(s) = \frac{E_x(s)}{EMF_x(s)}\bigg|_{V_x^*(s)=0} = \frac{-G(s)/(R + sL)}{1 + I_{DC}G(s)H(s)} \quad (24)$$

Fig. 7 shows the variations of these error terms with frequency for the system described in TABLE III, where the load parameters have been adjusted to match the 3 kW induction motor that was used for experimental verification of this part of the work. From this plot, it can be seen that unlike the results for a VSI reported in [8], the tracking error and disturbance rejection error are much the same until the frequency gets close to the load resonance. This suggests that there is little value in attempting to separately compensate for the disturbance injection, especially since the magnitude of the reference command will be very similar to that of the disturbance injection signal. Hence if the simple PI regulator can adequately track the required output voltage command, it most likely will be able to also accomodate a backEMF disturbance without requiring a separate feedforward

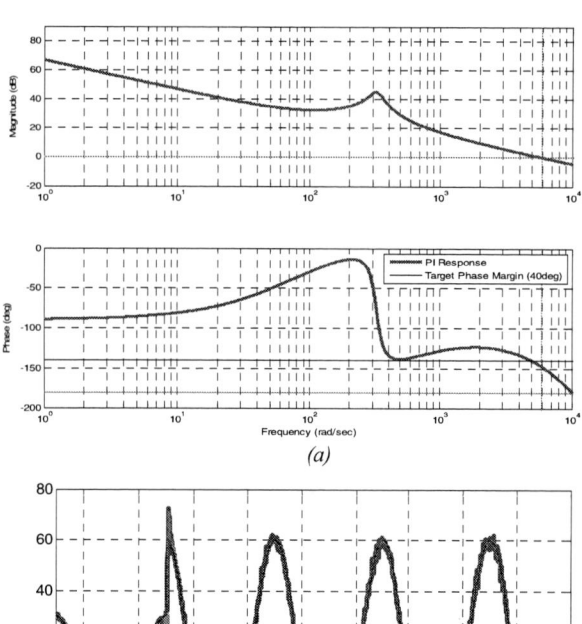

Fig. 6: CSI system (a) frequency and (b) transient response with load resonance at the target fundamental frequency
$K_p = 0.0346, \tau_r = 0.00172s, \omega_r = 321.7\ rad/s$

V. CSI REJECTION OF BACKEMF DISTURBANCE

The analysis so far has concentrated on the performance of a CSI feeding into a passive RL load with a resonant frequency that is near to or even less than the target reference frequency, to explore possible stability issues caused by this resonance. For this case of course, the load voltage depends only on the injected current from the CSI since there is no backEMF contribution. In contrast, loads such as an induction motor must always include a backEMF whose magnitude varies with motor speed independently of the current injected by the converter, as shown in Fig. 1. The implications of this load model extension will now be considered.

In the Laplace domain, (1) and (4) can be written as

$$V_{xn}(s) = (R + sL)I_x'(s) + EMF_x(s), x \in \{a,b,c\} \quad (16)$$

$$I_x(s) = I_x'(s) + sCV_{xn}(s), \quad x \in \{a,b,c\} \quad (17)$$

Rearranging (17), and substituting into (16) gives

TABLE III. TEST SYSTEM PARAMETERS – 3KW INDUCTION MOTOR

Parameter	Value	Parameter	Value
f_0	50Hz	f_c	5000 Hz
R	5.4Ω	L	43 mH
C	60uF	I_{DC}	10A

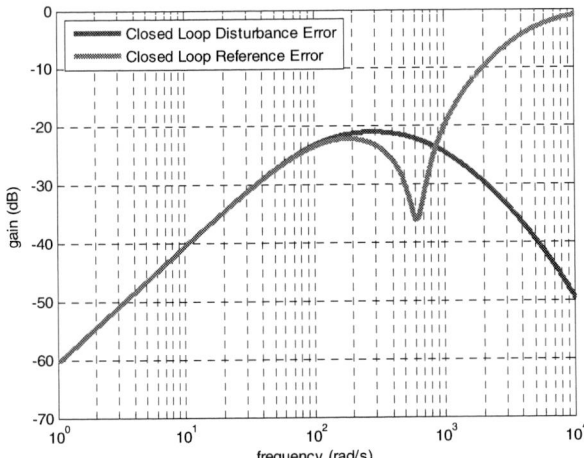

Fig. 7: Reference Tracking Error verses Disturbance Rejection Error for a three phase voltage regulated CSI feeding into a 3 kW induction motor, $K_p = 0.0344$, $\tau_r = 0.00172s$, $\omega_r = 622$ rad/s

compensation signal. Fig. 8 confirms this expectation, by showing the simulated and matching experimental transient and steady state responses for the controller feeding into the 3 kW motor load described in TABLE III. Despite the apparently excellent voltage regulation achieved by this system, the tracking error has a relatively large peak magnitude of approximately 8V (-22 dB), as also predicted by Fig. 7. Partially, this is also because the phase of the error signal is nearly in quadrature compared to the target reference, so that the degradation of the controller's performance is actually quite small despite the relatively large tracking error. This result is typical of a voltage regulated CSI under these conditions, and further confirms the effectiveness of a properly tuned simple PI regulation system in this application.

VI. CSI VOLTAGE REGULATION USING A P+RESONANT CONTROLLER

The analysis presented in Section III allows the maximum possible gains to be analytically determined for a PI regulator to maintain a required stability margin while still achieving the

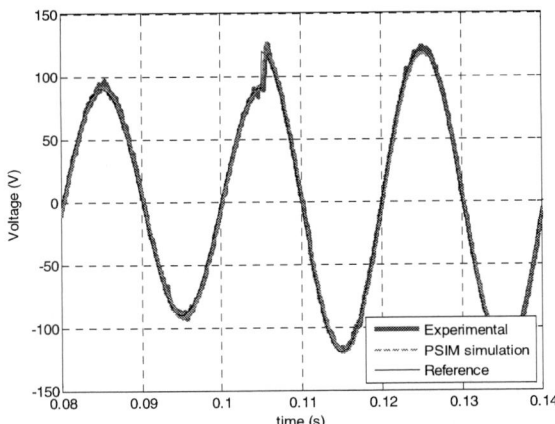

Fig. 8: CSI system transient response for a three phase voltage regulated CSI feeding into a 3 kW induction motor $K_p = 0.0344$, $\tau_r = 0.00172s$, $\omega_r = 622$ rad/s

best possible dynamic performance. However, (13) and (14) also show that the overriding limitation on gain is the controller sampling/transport delay T_d, which is in turn directly dependent on the PWM switching frequency. Hence for a higher power converter with a lower switching frequency, it may just not be possible to get an acceptable (and stable) dynamic performance even when the controller gains are set to their maximum possible values. In such situations, it is necessary to use a controller that can provide significantly increase gain at the target reference frequency without increased higher frequency gain (which would compromise stability), such as a synchronous frame or a P+Resonant controller [8]. For this paper, a P+Resonant controller using a rapidly implementable and robust Second Order Generalised Integration (SOGI) algorithm [11] has been implemented to illustrate the issue.

The general form of a P+Resonant controller is given by

$$H_{P+R}(s) = K_P\left(1 + \frac{1}{\tau_r}\frac{s\omega_o}{s^2 + \omega_o^2}\right) \tag{25}$$

from which it can be seen how the resonant term in the denominator dramatically increases the controller gain at the target frequency ω_o without particularly affecting the high frequency proportional gain. As identified in [8], the gains of this controller are set to the same values as have already been determined for the simple stationary frame PI controller.

For this part of the investigation, the PWM switching frequency was reduced to a low 600 Hz, to greatly exacerbate the phase roll-off of the forward path gain, because of the very large transport delay that occurs at this switching frequency. TABLE IV lists the main parameters for this system. The resulting frequency and transient step responses for both a simple PI and a P+Resonant regulator, with the gains set to their maximum values for both cases, are shown in Fig. 9. These results clearly show the benefit that can be achieved using the more sophisticated P+Resonant controller in these low switching frequency conditions.

However, the conclusions of Section IV suggest an alternative strategy that may be useful in some situations. If the load resonant frequency is set to the target reference frequency, the system provides "free gain" that will improve the steady state and transient performance even for a simple PI regulator. Fig. 10 illustrates this effect by setting the load resonance to 51 Hz (321.7 rad/s), almost directly at the target reference frequency of 50 Hz. While the resulting performance of the simple PI regulator is still not as good as the P+Resonant system, it is quite acceptable, and is most certainly completely stable without any need for active damping. However, the strategy will of course only work for applications requiring a fixed reference frequency, since it requires specific load/filter resonance tuning.

TABLE IV. TEST SYSTEM PARAMETERS – LOW SWITCHING FREQ

Parameter	Value	Parameter	Value
f_0	50Hz	f_c	**600 Hz**
R	6Ω	L	500 mH
C	320uF	I_{DC}	10A

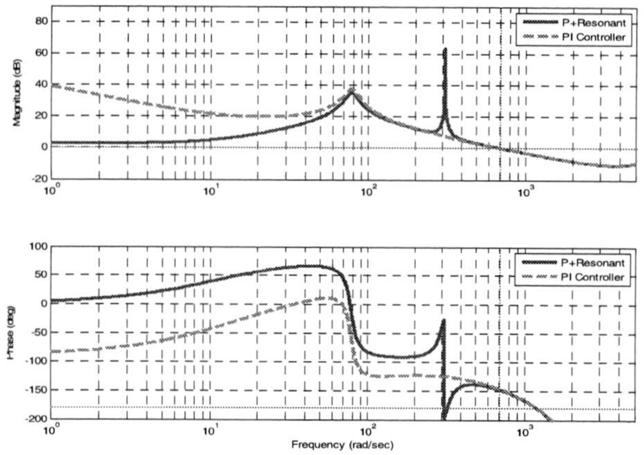

Fig. 9: CSI system (a) frequency and (b) transient response with 600 Hz switching frequency, controlled by simple PI and P+Resonant regulators with maximum possible gains, $K_p = 0.022$, $\tau_r = 0.0143s$, $\omega_r = 79\ rad/s$

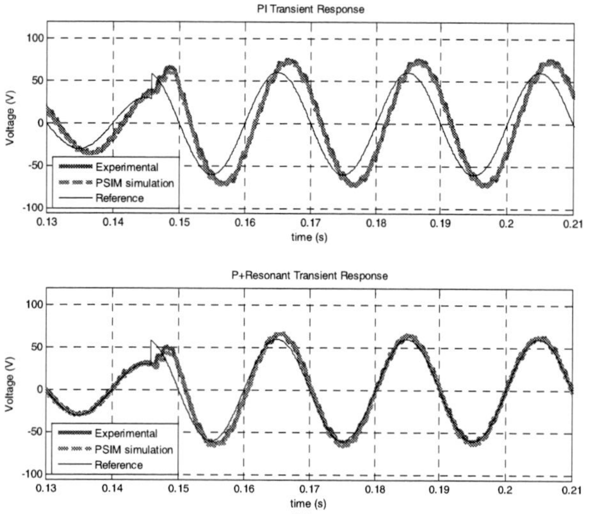

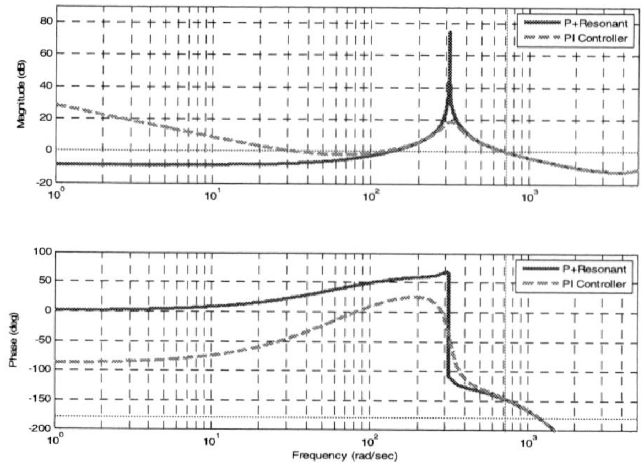

Fig. 10: CSI system (a) frequency and (b) transient response with 600 Hz switching frequency, controlled by simple PI and P+Resonant regulators with maximum possible gains, plant resonant frequency at target reference frequency $K_p = 0.022$, $\tau_r = 0.0143s$, $\omega_r = 321.7\ rad/s$

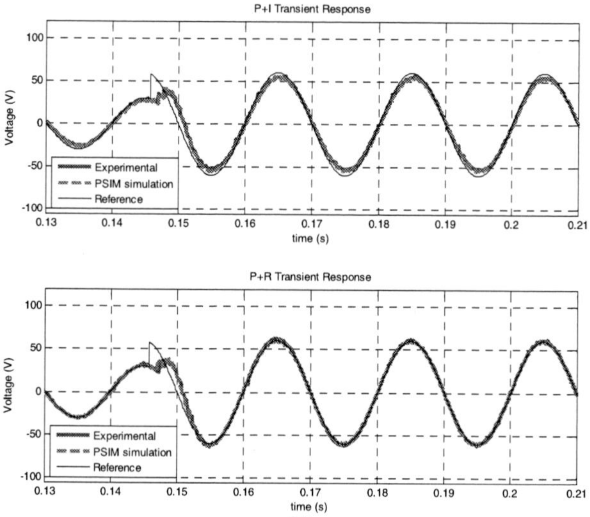

VII. THE ROLE OF ACTIVE DAMPING

Active damping is commonly reported in the literature as critical to maintaining the stability of system with a second order load/filter resonance [7]. A typical approach is shown in Fig. 11, where the additional feedback path is used to smooth out the system resonance. The modified closed loop response for this system is given in (26). However, as the active damping is more aggressively applied (R_D becomes smaller), the overall gain of the system reduces, and the controller gains must be re-tuned to maintain optimal system performance.

$$\frac{V(s)}{V^*(s)} = \frac{G(s)I_{DC}H(s)}{1+G(s)I_{DC}\left(H(s)+1/I_{DC}R_D\right)} \quad (26)$$

If the active damping loop is incorporated into system, the expression for the open loop transfer function phase now becomes

$$\phi = \tan^{-1}\left(\omega_c T_p\right) - \omega_c T_d - \tan^{-1}\left(\frac{\left(RC + L/R_D\right)\omega_c}{1 + R/R_D - \omega_c^2 LC}\right) \quad (27)$$

if the integral time constant remains at the same approximation of $\tau_r \approx 10/\omega_c$. Hence the expression for maximum K_P with active damping can be expressed as (28).

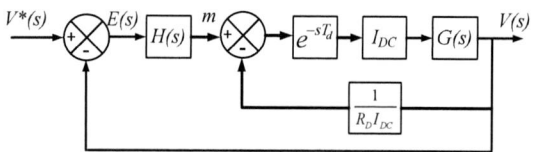

Fig. 11: Single Phase CSI Voltage Regulator with active damping

$$K_p = \frac{\omega_c \tau_r}{I_{dc}\sqrt{1+(\omega_o \tau_r)^2}} * \sqrt{\frac{\left(\omega_c RC + \frac{L}{R_D}\right)^2 + \left(1 + \frac{R}{R_D} - LC\omega_c^2\right)^2}{\left(\omega_c^2 L^2 + R^2\right)}}$$

(28)

There is no doubt that active damping achieves its goal of reducing the system resonant peak, as shown in Fig 12 for the test system with an increasing active damping ratio. However, it is equally clear from Fig. 12 that the overall loop gain also reduces with increasing active damping, so that K_p for the system must be increased again to achieve the same target phase margin as before. Fig. 13 shows the result with and without active damping once K_p has been retuned to its maximum values using the principles presented in this paper, and it can be seen that the introduction of active damping has contributed nothing to either the high frequency gain response, or to the stability margin. In fact, its only contribution appears to be to reduce the system load/filter resonance, which has been shown in this paper to actually be potentially advantageous in certain applications. Hence the benefits of including active damping into a correctly tuned PI control system (or a P+Resonant system since the gains are the same) seem at best uncertain, and at worst counterproductive.

VIII. SUMMARY

The work presented in this paper shows how PWM transport and sampling delays are a significant factor that must be considered when designing a high performance stationary frame PI or P+Resonant voltage controller for a Current Source Inverter. From this understanding, a deterministic analytical method is presented which calculates the maximum possible PI and P+Resonant gains that can be achieved for this system. The controller does not need active damping to maintain stability or improve performance despite the highly resonant nature of the plant. The analytical development is supported by detailed simulation and experimental results.

REFERENCES

[1] E. Wiechmann, P. Aqueveque, R. Burgos, J. Rodriguez "On the Efficiency of Voltage Source and Current Source Inverters for High-Power Drives", IEEE Transactions on Industrial Electronics, 2008, pp. 1771 – 1782.

[2] V.D. Colli, P. Cancelliere, F. Marignetti, R. Di Stefano, "Voltage control of current source inverters", IEEE Transactions of Energy Conversion, Vol. 21, No. 2, June 2006, pp. 451 – 458.

[3] L. Tan, Y.Li, C. Liu, P. Wang , X. Lv, Z. Li "High Performance Controller for Voltage-Controlled Current Source Inverter with Nonlinear Loads", Energy Conversion Congress and Exposition, 2009 pp. 2942 – 2948.

[4] D.N. Zmood and D.G. Holmes, "A Generalised Approach to the Modulation of Current Source Inverters", Conference Record of the Power Electronics Specialist Conference 1998, PESC'98, pp. 739-746, 1998.

[5] D.N. Zmood, D.G. Holmes, "Improved Voltage Regulation for Current Source Inverters", IEEE Transactions on Industry Applications, Vol. 37, No. 4, July-Aug. 2001 pp. 1028 - 1036

[6] D. N. Zmood and D.G. Holmes, "Frequency Domain Analysis of Linear Current Regulators", IEEE Transactions on Industry Applications, Vol. 37, No. 2, March-April 2001, pp. 601 - 610.

[7] J. Ma, B. Wu, S. Rizzo, "Active Damping Control of PWM CSI High Power Induction Motor Drives", in IEEE 2000 Power Electronics Specialists Conference , 2000, pp. 61-66.

[8] D.G. Holmes, T.A. Lipo, B.P. McGrath, W.Y. Kong, "Optimized Design of Stationary Frame Three Phase AC Current Regulators", IEEE Transactions on Power Electronics, Vol. 24, No. 11, Nov. 2009, pp. 2417 – 2426.

[9] D.G. Holmes & T.A. Lipo, Pulse Width Modulation for Power Converters, New York: IEEE Press, 2003.

[10] G. F. Franklin, J. D. Powell, and M. L. Workman, Digital control of dynamic systems, 3rd ed. Menlo Park, Calif.: Addison-Wesley, 1998.

[11] E. Bueno et. al. F.J. Rodriguez, "Discrete-time implementation of second order generalized integrators for grid converters", IEEE IECON, pp. 176-181, 2008.

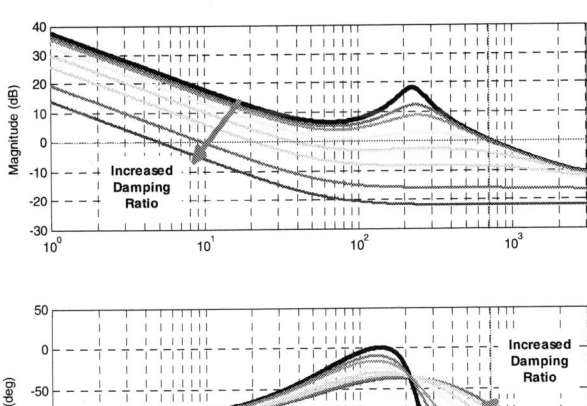

Fig. 12 – CSI system open loop frequency response for differing Damping Ratios. The black trace is optimally tuned with no active damping.

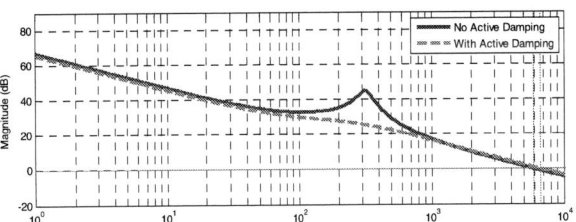

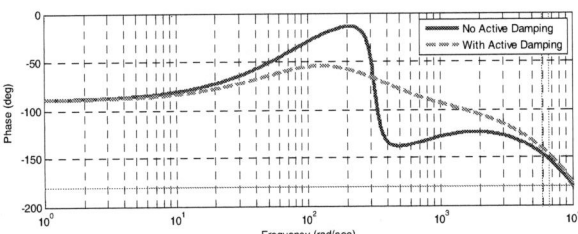

Fig. 13: CSI system open loop frequency response with optimally tuned PI controller, with and without active damping

Development of a New Voltage Source Inverter (VSI) Average Model Including Low Frequency Harmonics

S. Ahmed[1], D. Boroyevich[1], F. Wang[2], and R. Burgos[3]

[1]Center for Power Electronic Systems
(CPES)
The Bradley Department of Electrical
and Computer Engineering
Virginia Polytechnic Institute and State
University
Blacksburg, VA
sahassan@vt.edu

[2]Oak Ridge National Lab
University of Tennessee
Knoxville,TN
fred.wang@utk.edu

[3]ABB U.S. Corporate
Research Center
Raleigh, NC
rbugos@ieee.org

Abstract— This paper presents an enhanced average model of a voltage source inverter (VSI) that can accurately predict some of the low frequency phenomena only seen by the switching models. These phenomena include the dead time, voltage drop across switching devices (switch & diode), different modulation and minimum pulse width. Simulation and experimental results with a 2 kW prototype is used for validation purposes. The paper shows the simulation results of both complete switching and proposed average model depicting a great match between the two. The enhanced average model showed very good matching with the switching model, about 2% difference in time domain and also harmonic spectrum analysis.

I. INTRODUCTION

Analysis of power electronics systems is complicated when using complete switching models. The switching models are detailed accurate models that can predict the actual and transient behavior. However, power electronics systems can have hundreds of power electronics converters and motor drive systems which if presented with detailed component models will result in practical constraints in the size of the simulation system that can be achieved. In addition, it will place constraints on the computing time and will also increase the computer storage [1]. Therefore, the demand for the development of the accurate, time-efficient models becomes very strong.

This paper presents accurate average model of a voltage source inverter (VSI) that can predict some of the phenomena only seen by the switching models. The paper first presents the ideal average model of the VSI model which has been presented in literature and then adds some of the low frequency phenomena that is usually only seen with the switching model in order to reach a more realistic model. From these phenomena are the various types of modulation [2-

3], dead time that is inevitable to prevent the shoot-through phenomenon [4-8] analyzing the non-linearities, voltage and current distortion due to that, minimum pulse-width constraint [9-11] and other phenomena due to the inherent characteristics of the switching devices as the voltage drop, and the turn on/off time of the switches. Research concentrated on developing compensation methods for these types of phenomena in a switching model but none actually implemented them in an average model. In this paper, some of the methods developed in the research are used in a reverse way to introduce the low frequency harmonics into the proposed average model to become more realistic as the switching model and then compensate for that. The paper then shows the simulation results of both complete switching and proposed average model depicting a great match between the two. Then finally, it verifies the results experimentally with a prototype that matches the simulation model implemented. The paper also discusses the saving in simulation time introduced by this enhanced average model.

II. PROPOSED VOLTAGE SOURCE INVERTER AVERAGE MODEL

Fig. 1 shows the circuit schematic of the VSI model considered in this paper, comprised of a power stage, controller, modulation and passive elements. The power stage is modeled as an ideal average model with the set of equations (1) in the *abc*-coordinates. However, a diode bridge is added in parallel as shown in Fig. 1 to reveal the converter real phase-leg operation. The controller is in the *dq*-frame.

$$\begin{bmatrix} v_a \\ v_b \\ v_c \end{bmatrix} = \begin{bmatrix} d_a \\ d_b \\ d_c \end{bmatrix} v_{dc} \ \& \ i_{dc} = i_a d_a + i_b d_b + i_c d_c \quad (1)$$

978-1-4244-4782-4/10 $26.00 © 2010 IEEE

The inverter parameters is as follows, it is a 2 kW prototype implemented in Saber running at about 1kW, the dc link voltage V_{dc}=300 V, the output phase RMS voltage is 60 V_{rms}, phase current is 4 A_{rms} and the load resistor is 15 Ω. Fig. 2 shows some of the simulation results for the ideal VSI average model defined here. It shows the output voltage with an RMS voltage of 59.897 V, RMS current of phase A of 3.9932 A and duty cycle of the three phases respectively.

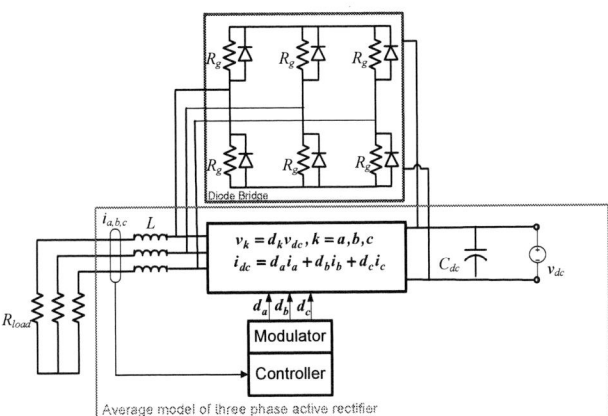

Figure 1 Circuit schematic of VSI topology considered in this paper.

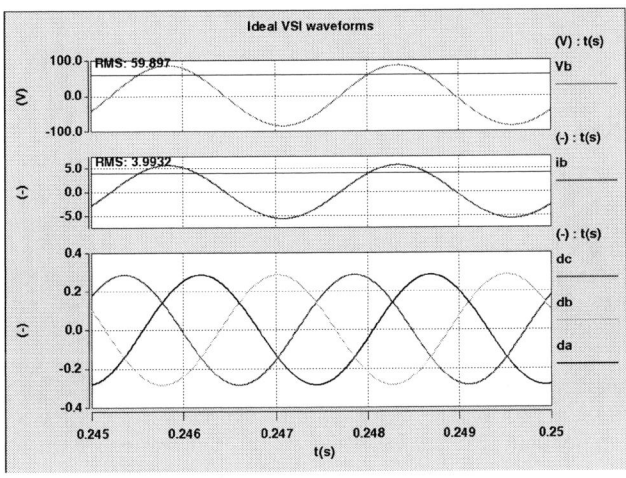

Figure 2 Simulation results for ideal VSI average model proposed.

III. DIFFERENT LOW FREQUENCY PHENOMENA MODELED

A. Modulation Control Methods

The ideal inverter simulation in the previous section shows pure sinusoidal PWM waveforms where it just represents the fundamental signals with no injected harmonics. The first modification for the model to be more realistic is to include the zero-sequence into the PWM waveforms. This signal depends on the distribution of the zero vector and the type of modulation control used [2]. In [2] a unified representation for the zero-sequence signal $e_i(t)$ is given in (2):

$$ e_i = k_o(1 - u_{max}) + (1 - k_o)(-1 - u_{min}), \ 0 \le k_o \le 1 \quad (2) $$

and u_i is the fundamental sinusoidal waveforms, u_{min} and u_{max} is the minimum and maximum of the three phases respectively.

For continuous symmetrical PWM (SYPWM), $k_o = 0.5$ and this means the two zero vectors are equally distributed. Fig. 3 shows the modulation signals for one phase, with the injected zero sequence. For discontinuous PWM, there were different ones implemented as the two extremes, the DPWMMAX where the zero vector (111) is used for the six sectors and the DPWMMIN where the zero vector (000) is used in this case for the six vectors. In addition to those two, the 2-phase right aligned (2-Φ RA) SVM presented in Fig. 4 is also implemented.

TABLE I shows the zero-sequence representations and space voltage vectors used for all six sectors to achieve this discontinuous 2-Φ RA. In addition, to the zero sequence injection the model is capable of discrete sampling and for this case it is 20 kHz. The effect of the sampling can be seen in the staircase steps effect on the modulation signal as shown in Fig. 3. Fig 5 compares the continuous and discrete sampling effect on phase A current.

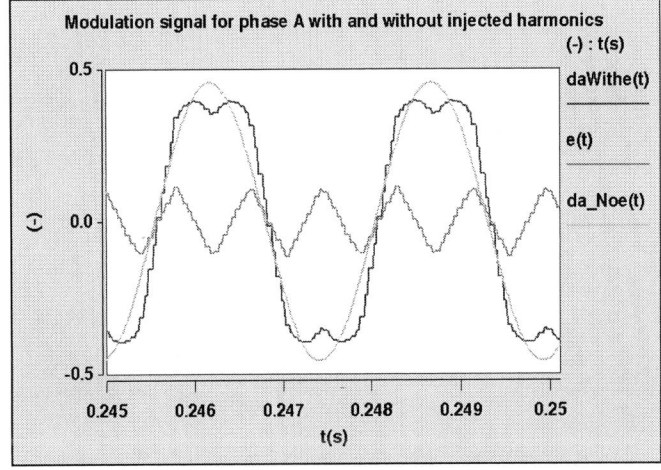

Figure 3 Modulation signals for continuous symmetrical PWM.

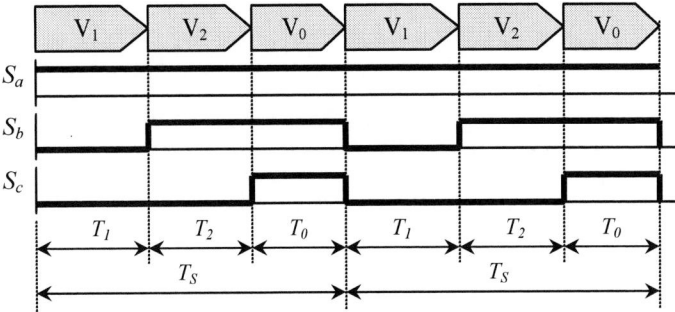

Figure 4 Discontinuous 2-Φ RA modulation in sector 1.

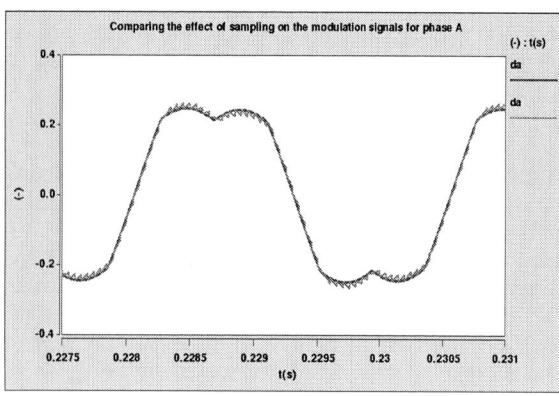

Figure 5 Comparing the phase A duty cycle with continous and discrete sampling of 20kHz.

TABLE I.   THE ZERO SEQUENCE CALCULATED FOR EACH SECTOR FOR 2 2-Φ RA MODULATION

Sector NO.	Space voltage vectors	Zero sequence $e_i(t)$
I	$T_s=T_1(100)+T_2$ $(110)+T_0(111)$	$e_i(t) = \dfrac{T_0}{Ts} + \dfrac{1}{3}\dfrac{T_2}{Ts} - \dfrac{1}{3}\dfrac{T_1}{Ts}$
II	$T_s=T_1(110)+T_2$ $(010)+T_0(000)$	$e_i(t) = \dfrac{-T_0}{Ts} - \dfrac{1}{3}\dfrac{T_2}{Ts} + \dfrac{1}{3}\dfrac{T_1}{Ts}$
III	$T_s=T_1(010)+T_2$ $(011)+T_0(111)$	$e_i(t) = \dfrac{T_0}{Ts} + \dfrac{1}{3}\dfrac{T_2}{Ts} - \dfrac{1}{3}\dfrac{T_1}{Ts}$
IV	$T_s=T_1(011)+T_2$ $(001)+T_0(000)$	$e_i(t) = \dfrac{-T_0}{Ts} - \dfrac{1}{3}\dfrac{T_2}{Ts} + \dfrac{1}{3}\dfrac{T_1}{Ts}$
V	$T_s=T_1(001)+T_2$ $(101)+T_0(111)$	$e_i(t) = \dfrac{T_0}{Ts} + \dfrac{1}{3}\dfrac{T_2}{Ts} - \dfrac{1}{3}\dfrac{T_1}{Ts}$
VI	$T_s=T_1(101)+T_2$ $(100)+T_0(000)$	$e_i(t) = \dfrac{-T_0}{Ts} - \dfrac{1}{3}\dfrac{T_2}{Ts} + \dfrac{1}{3}\dfrac{T_1}{Ts}$

B. Dead time and Non-linearities due to Turn On/off and Voltage Drop on Switching Devices

The second important phenomenon implemented is the dead time. The dead time is a blanking time to avoid the so-called shoot-through of the dc link. This time guarantee that both switches in an inverter leg never conduct simultaneously [4]. For switching model, dead time is generated by delaying the switching time as shown in Fig. 6. From Fig. 6, it can be seen that time error in the on time duration resulting from dead time is given by (3).

$$T_{err} = \left[\left(-\frac{T_{off}}{2} + \frac{T_{on}}{2} + T_d\right)\Big/T_s\right] sign\,(i)$$

$$\&\quad sign\,(i) = \begin{bmatrix} 1: when\ i > 0 \\ -1: when\ i < 0 \end{bmatrix} \quad (3)$$

For the ideal average model, the error time is added to the commanded time duration T_{com} so that the effective time duration T_{eff} is the summation of both as given in (4):

$$T_{eff} = T_{com} + \left[\left(-\frac{T_{off}}{2} + \frac{T_{on}}{2} + T_d\right)\Big/T_s\right] sign\,(i) \quad (4)$$

In addition to the effect of the dead time and the turn on and off time of the switches, there are some non-linearities caused by the voltage drop across the switch (V_{ce}) & the on voltage of the diode (V_d). This can be realized in the model by adding the error caused by the voltage drop to the error time in (3) giving the new effective duty cycle as in (5):

$$d_{eff} = d_{com} + \left[\left(\left(-\frac{T_{off}}{2} + \frac{T_{on}}{2} + T_d\right)\Big/T_s\right) + d_a(V_{ce}/V_{dc}) + (1-d_a)(V_d/V_{dc})\right] sign\,(i$$

$$(5)$$

Fig 7 shows the phase A current with and without the effect of a dead time of $5\mu s$.

C. Minimum Pulse-width

Another important source of distortion is the minimum pulse-width (MPW) limitations that result when small on-time duration of the switches can not be achieved. Such limitations may have to be enforced to prevent damage of the semiconductors switches as this MPW can be smaller than the on and off time of the switch [10]. In a switching model, this can be implemented easily by comparing the width of the pulse of the gating signal to the minimum pulse allowed duration and if it is smaller then it gets deleted. However, in the proposed average model, a limit is enforced on the duty cycle achievable. The minimum allowed duty cycle is given in (6) and if the duty cycle is smaller then it is clipped to the minimum one.

$$d_{min} = \left(1/T_s\right)\left(T_{deadtime} + T_{MPW}\right) \quad (6)$$

Fig. 8 depicts the difference between phase A current when the MPW phenomenon is added to the model.

978-1-4244-4782-4/10 $26.00 © 2010 IEEE   883

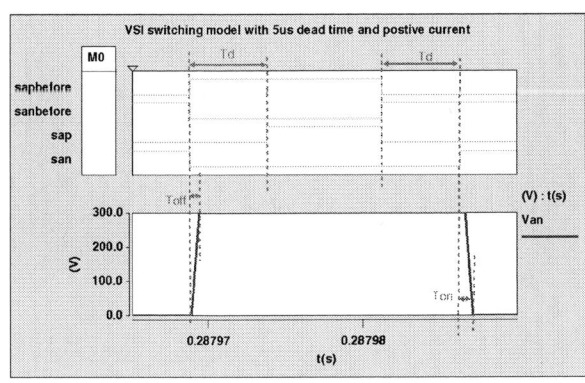

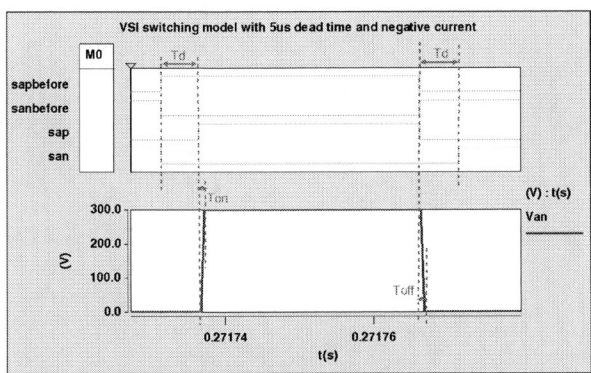

Figure 6 PWM voltage waveforms for positive current (top) and negative current (bottom). From the top for each one: phase A positive gate signal without dead time, negative gate signal without dead time, positive gate signal with dead time, negative gate signal with dead time and line to neutral phase A voltage.

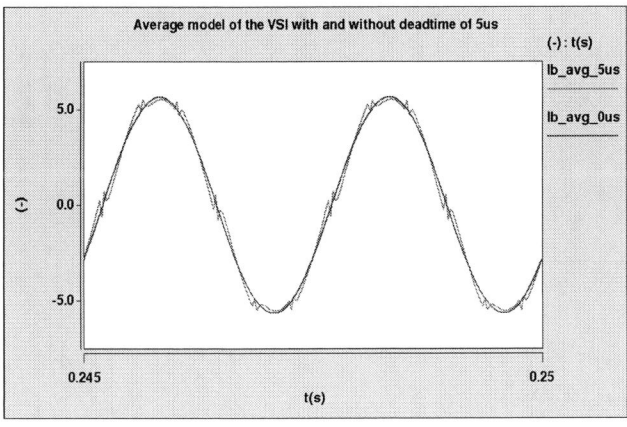

Figure 7 Phase A current for proposed model with and without dead time.

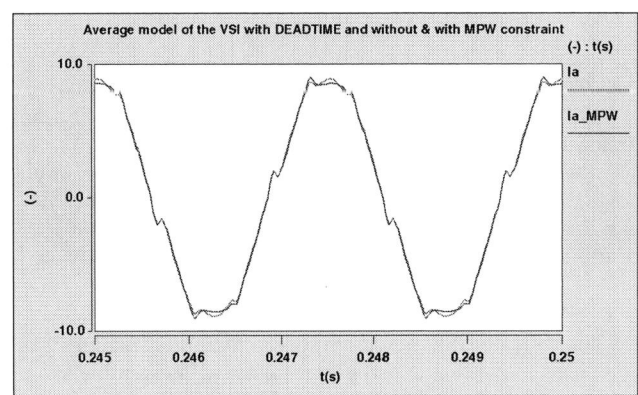

Figure 8 Phase A current with and without minimum pulse width.

IV. COMPARING THE PERORFMANCE OF SWITCHING AND AVERAGE MODEL

The average model proposed in the previous section with all the different low frequency harmonics phenomena discussed is compared to a switching model with the same exact parameters and the results are compared and shown in Fig. 9 and 10 respectively. Fig. 9 shows the comparison between the real switching and average model when no distortion is enforced. It compares phase B current. Fig. 10 shows the same current with distortion added: $5\mu s$ dead time, on and off time of $1\mu s$, voltage drop across the switch of 0.2V, diode on voltage of 0.7V and $1\mu s$ minimum pulse width. Fig. 10 compares the average switching current waveform (averaged over one switching cycle) versus the average model current. From looking at the time domain waveform, it can be seen that the two waveforms matches very good. However, a more accurate way of verifying the results is to calculate the deviation percentage between both waveforms. This percentage is calculated using equation (7),

$$Deviation = \frac{1}{X_{RMS}} \sqrt{\frac{1}{N} \sum_{i=1}^{N} \left(x_{i_{swi}} - x_{i_{avg}} \right)^2} \quad (7)$$

Using the above equation, each point is subtracted from the corresponding one and the relative error is calculated and normalized using RMS current. The percentage error between the two models is 2.32%.

In addition to comparison above, the models simulation time was compared for a 0.2s window. The ideal average model with no distortion has the least simulation time (1.7s) followed by the new enhanced average model with distortion (2.38min) and finally the switching model almost 4 times more than the enhanced average model time (8.1s). So although there is a penalty paid when using the enhanced average models on the simulation time, the benefit of low frequency modeling is more significant than the time in this case. This is because the enhanced average model still has much lower simulation time than the switching model almost halved and could predict well the performance as shown.

V. EXPERIMENTAL SETUP & RESULTS VERIFICATION

A. Setup

A 2 kW prototype experiment is conducted to verify the performance of the new developed model. Fig. 11 shows the construction. A 6-pack IGBT IPMs from Fuji (6MBP20RH060) (600V, 20A) is used. The output inductance per phase is 600μH. The control board is DSP-FPGA digital controller.

B. Verification

To validate the new average model, the model was compared to the real experimental model with exact the same parameters as mentioned above. The validation tests are divided into three sets, power quality verification, impedance measurement and EMI verification. However, this section will concentrate on the first one and the later two are beyond the scope of this paper. The power quality verification will include normal steady state verification where the ac output current waveforms are measured and compared versus the average simulation model as done in the previous section. The error percentage is again calculated the same way using equation (7) where now the waveforms compared is the averaged experimental versus the new developed average model. The second part of the power quality verification is the frequency characteristics. This part will include comparison between individual harmonics and the total percentage error is also calculated.

Fig. 12 compares both the experimental phase B current averaged over one cycle with the average simulation model current. The percentage error is given as 8.74% calculated with equation (7).

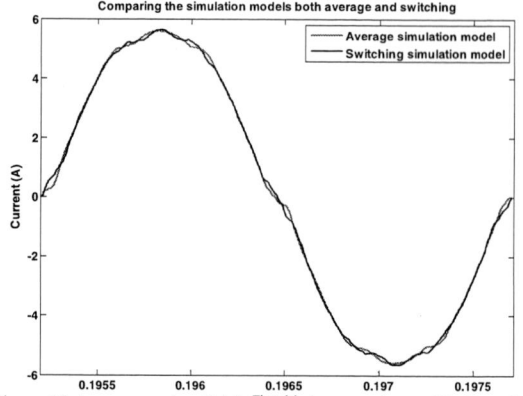

Figure 10 Average and switching model comparison with distortion.

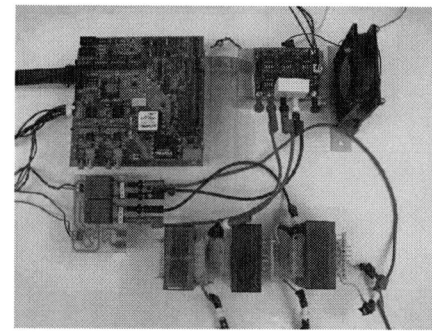

Figure 11 Experimental setup.

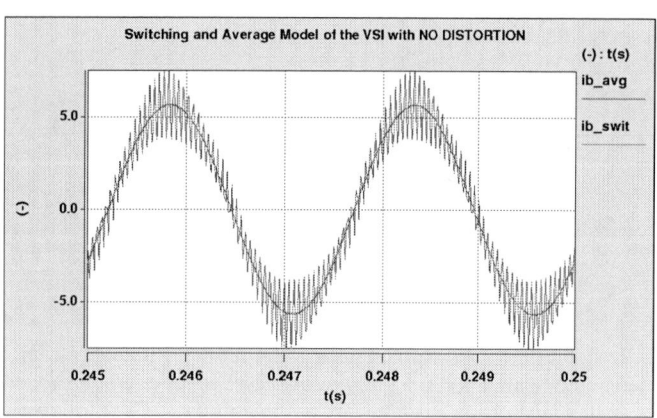

Figure 9 Average and switching model comparison without distortion.

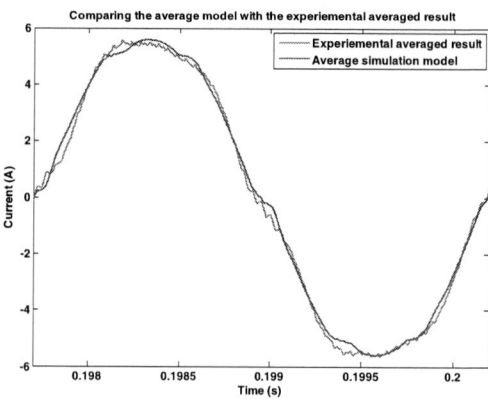

Figure 12 Average and switching model comparison with distortion.

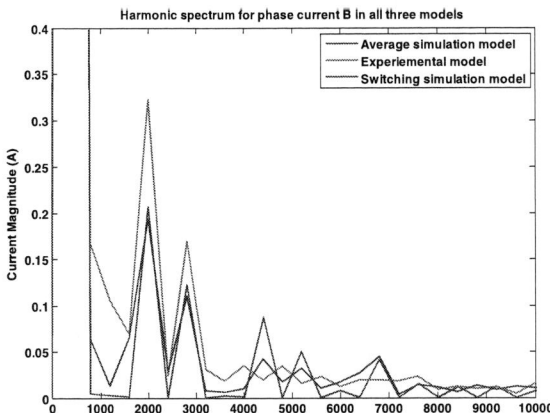

Figure 13 Average and switching model comparison with distortion.

Fig. 13 compares the harmonic spectrum of phase B current for the three models (real switching model, enhanced average model and experimental prototype). The fundamental harmonic is about 5.72A (magnitude) and the figure zooms on the sub harmonics that are high enough to be investigated, therefore, it concentrates from the 5th harmonic till half the switching frequency (10kHz) which is around the 23rd harmonic. It can be seen that the simulation models (average and switching) matches very well in most points while the experimental is little higher. Calculating the percentage of deviation or error using equation (7) and over the region given in Fig 13, it is found to be 0.82% between the experimental prototype and the enhanced average model and 1.85% between the real switching simulation model and the average model.

VI. CONCLUSIONS

This paper has presented a new average model for Voltage Source Inverter (VSI) model. This average model can capture low frequency harmonics that are usually phenomena modeled only with switching models. The paper discussed the process of implementing these different phenomena and compared the results of the proposed average model to a complete switching model. Experimental results obtained with a 2 kW Voltage Source Inverter prototype verified the proposed average model.

ACKNOWLEDGMENT

This work was conducted under the CPES project "Stability of AC Power Systems-Phase II" sponsored by The Boeing Company.

This work made use of Engineering Research Center Shared Facilities supported by the National Science Foundation under NSF Award Number EEC-9731677.

REFERENCES

[1] B. Needham, P. Eckerling, and K. Siri, "Simulation of large distributed dc power systems using averaged modeling techniques and the Saber Simulator," in Proceed. *IEEE APEC'94*, vol. 2, pp. 801–807, Feb. 1994.

[2] K. Zhou, and D. Wang, "Relationship between space-vector modulation and three-phase carrier-based PWM: a comprehensive analysis [three-phase inverters]," *IEEE Trans on Indus Elect*, vol.49, no.1, pp.186–196, Feb 2002.

[3] V. Blasko, "Analysis of a hybrid PWM based on modified space-vector and triangle-comparison methods," *IEEE Trans on Indus Appl.*, vol.33, no.3, pp.756–764, May/Jun 1997.

[4] A. Munoz-Garcia, and T.A. Lipo, "On-line dead time compensation technique for open-loop PWM-VSI drives," *IEEE APE '98*, vol.1,pp.95–100 vol.1, 15–19 Feb 1998.

[5] J. Choi; S. Sul, "Inverter output voltage synthesis using novel dead time compensation," *IEEE Trans. On Pow Elect* , vol.11, no.2, pp.221–227, Mar 1996.

[6] J. Choi; S. Sul, "A new compensation strategy reducing voltage/current distortion in PWM VSI systems operating with low output voltages," *IEEE Trans on Indus Appl.*, vol.31, no.5, pp.1001–1008, Sep/Oct 1995.

[7] C. Attaianese, V. Nardi, and G. Tomasso, "A novel SVM strategy for VSI dead-time-effect reduction," *IEEE Trans on Indus Appl.*, vol.41, no.6, pp. 1667–1674, Nov.-Dec. 2005.

[8] C. Attaianese, D. Capraro, and G. Tomasso, "Hardware dead time compensation for VSI based electrical drives," *IEEE ISIE 2001.*, vol.2, pp.759–764 vol.2, 2001.

[9] B.A. Welchko, S.E. Schulz, S. Hiti, "Effects and Compensation of Dead-Time and Minimum Pulse-Width Limitations in Two-Level PWM Voltage Source Inverters," *IEEE IAS'06* , vol.2, pp.889–896, 8–12 Oct. 2006.

[10] D.A. Grant, "Technique for pulse dropping in pulse-width modulated inverters," *IEEE Proceed. Elect Power Appl.* , vol.128, no.1, pp.67–72, January 1981.

[11] D. Grant, R. Seidner, "Technique for pulse elimination in pulsewidth-modulation inverters with no waveform discontinuity," *IEEE Proceed. Elect Power Appl.*, vol.129, no.4, pp.205–210, July 1982.

978-1-4244-4782-4/10 $26.00 © 2010 IEEE

Realization and Improvement of Repetitive Control in Rotating Frame for Active Power Filter System

Baifeng Chen, Xiaoming Zha, *Member, IEEE,* Jinwu Gong, Suxuan Guo, and Jianjun Sun

School of Electrical Engineering
Wuhan University
Wuhan, China
Cbaifeng@yahoo.cn, Xmzha@whu.edu.cn

Abstract—The repetitive control, which is based on a delay of fundamental period, is widely used in grid-connected voltage-source converters, such as uninterruptible power supply (UPS), photovoltaic (PV) inverter and active power filter (APF). This paper presents four repetitive controllers in rotating frame for APF system, whose repetitive period is T, T/2, T/3, and T/6 respectively. Compared with traditional repetitive control, these repetitive controllers have fast dynamic response and also have similar structure, which could also be easily implemented in digital processor for APF system. Regeneration spectrum and sensitivity function based analysis methods provide a way for parameters design and performance comparison between each controller. Simulation and experiment results demonstrate the validity and superiority of each controller.

Index Terms--Repetitive control; active power filter; rotating frame; regeneration spectrum; sensitivity function.

I. INTRODUCTION

Nonlinear loads, such as diode/thyristor rectifier, generate significant harmonic problems in distribution power system. These harmonic problems would bring energy loss, poor power factor, and harmful disturbance to other appliances. Active power filter (APF) has been widely used to deal with power quality issues [1]. However, implementing a fast and accurate current controller is essential for APF applications. Traditional linear proportional-integral (PI) controllers have the fast tracking speed, but it also has the drawback of steady-state error in the stationary frame. To achieve high compensation effect and fast dynamic response, various instantaneous feedback control schemes, such as deadbeat control, multi-loop control, have been proposed. Which could achieve both high-compensation and fast dynamic response. But more variables also need to be sensed.

The proportional-resonant (PR) control and repetitive control, which is originated from internal-model principle of control theory [2-4], is widely used in UPS, DVR, and APF systems. The Transfer function of PR control is

$$H_{PR} = K_p + \frac{K_i s}{s^2 + \omega_0^2} \tag{1}$$

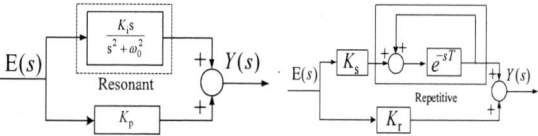

Fig. 1. PR controller and Repetitive controller

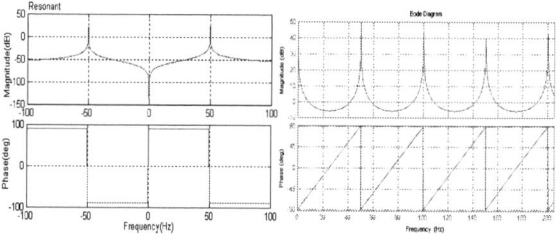

Fig. 2. Bode plots of resonant loop and repetitive loop

The block diagram of PR control and theoretical Bode plots of resonant loop are shown in Fig. 1 and 2, which shows that the PR controller has a theoretically infinite resonant peak at special frequency and could eliminate the steady-state error to zero at that frequency, which is therefore similar to a generalized AC integrator, as proven in [5]. So, by using multiple PR controllers, the APF could eliminate selectively order harmonics [6-7]. This control strategy also has been used in UPS and single-phase PV inverters [8]. But a bank of resonant controllers in digital processor would undoubtedly complicate the controller implementation [9].

The repetitive control is usually described as

$$H_{re} = K_r + \frac{K_s}{e^{sT} - 1} \tag{2}$$

Fig. 2 shows that repetitive control has numerous infinite gains and could eliminate all harmonic signals. But the repetitive control is based on one fundamental period, which would influence the dynamic characteristic.

To avoid the disadvantages of the PI controller, PR controller and conventional repetitive controller, this paper propose a new method for implementation of repetitive control. That is implementation in rotating frame. Part III presents half fundamental period (T/2) based repetitive controller, which could eliminate all odd harmonics. Part IV presents T/3 and T/6 based repetitive controllers, which could eliminate all $(3k \pm 1)^{th}$ or $(6k \pm 1)^{th}$ harmonics in stationary frame. Based on regeneration spectrum theory and sensitivity function, part V presents analysis and design results of proposed controllers in rotating frame, which demonstrates the advantages and disadvantages of each controller. Part VI presents the simulation and experiment results, which verify the validity and superiority of proposed repetitive controllers.

II. APF SYSTEM CONFIGURATION

The simple three-phase shunt APF system configuration is presented in Fig. 3. The closed-loop control system is given out in the Fig. 4. $I_{Lh}(s)$ and $I_{Sh}(s)$ represent the harmonic current at load side and system side respectively. $G_{AF}(s)$ stands for the transfer function of PWM inverter and Second-order Low-pass filter with 3 kHz cut-off frequency (w_q) with 0.707 quality factor $(1/\xi_q)$. APF parameters and system parameters are presented in Table I. PWM inverter model is equal to a small delay loop; the switching frequency $(1/T_{PWM})$ is 6 kHz, the transfer function of $G_{AF}(s)$ is

$$G_{AF}(s) = \frac{\omega_q^2}{s^2 + 2\xi_q \omega_q s + \omega_q^2} \cdot \frac{1}{0.5 T_{PWM} \cdot s + 1} \quad (3)$$

The transfer function of interface inductor is

$$G_{sp}(s) = \frac{1}{sL_f + R} \quad (4)$$

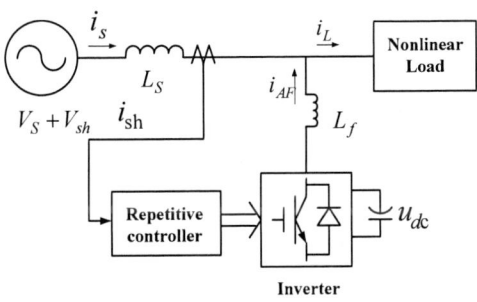

Fig. 3. Block diagram of APF configuration system

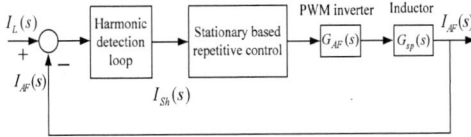

Fig. 4. Block diagram of closed-loop control system

TABLE I. APF SYSTEM PARAMETERS

System	Value	APF	Value
System voltage	220V	L_f	0.25 mH
L_s	0.005 mH	R	0.01
F	50Hz	U_{dc}	1000V

III. T AND T/2 BASED REPETITIVE CONTROLLER

A. T Based Repetitive Controller

Traditional repetitive control is designed in stationary frame, which needs three repetitive controllers. When they are implemented in rotating frame, it would only need two controllers. Another issue for traditional repetitive controller is that three controller should work synchronously. Otherwise, the sum of three-phase output currents may not be zero, which would bring unpredictable problems. However, even if the two controllers in rotating frame are out of synchronization, the output currents would still be balanced. The block diagram of rotating frame based control system is presented in Fig. 5. The one fundamental period based repetitive control is presented in Fig. 6. The transfer function of T based repetitive controller is

$$H_{RE1} = \frac{Y(S)}{E(S)} = K_r + K_s \cdot \frac{K_f \cdot e^{-sT}}{1 - K_f \cdot e^{-sT}} \quad (5)$$

K_f is near to 1 in practical, such as $K_f=0.95$. In order to facilitate the analysis and design, let $K_f=1$ at first.

$$G_{re1} = \frac{e^{-sT}}{1 - e^{-sT}} = -\frac{1}{2} + \frac{1}{T}\left[\frac{1}{s} + \frac{2s}{s^2 + \omega_0^2} + \frac{2s}{s^2 + (2\omega_0)^2} + \cdots \right] \quad (6)$$

Equation (6) shows T based repetitive control contains all resonant controllers and could eliminate all harmonics under rotating frame, through d-q reverse transformation, this controller could still eliminate all harmonics in stationary frame and would have no restriction for industry application.

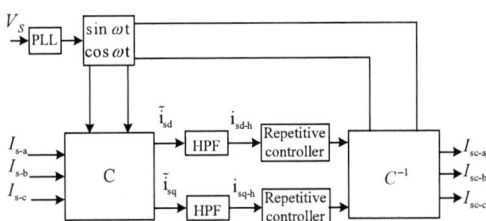

Fig. 5. Rotating frame based control system

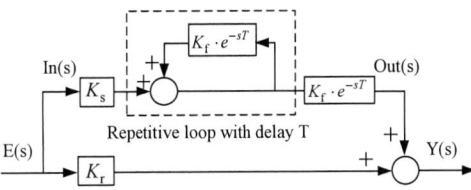

Fig. 6. Block diagram of T based repetitive control

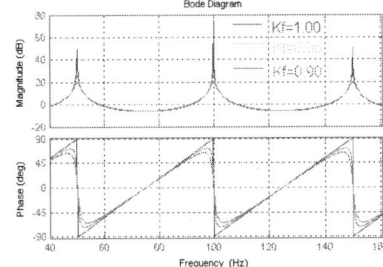

Fig. 7. Bode plots of T based repetitive loop

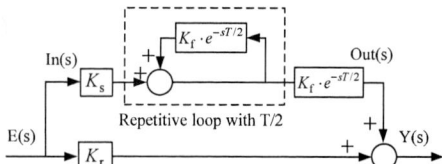

Fig. 8. Block diagram of T/2 based repetitive control

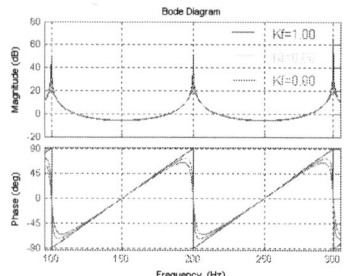

Fig. 9. Bode plots of T/2 based repetitive loop

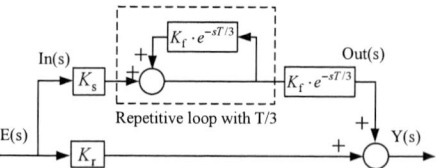

Fig. 10. Block diagram of T/3 based repetitive control

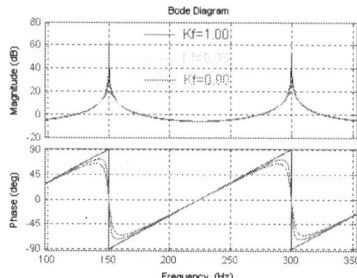

Fig. 11. Bode plots of T/3 based repetitive loop

B. T/2 Based Repetitive Controller

In practical power distribution system, only odd harmonics exist. So, considering the even harmonics, the T based repetitive control system is a redundant control system.

One significant characteristic of d-q transformation is the frequency shifting, which means the odd harmonics would appear as even harmonics (2nd, 4th, 6th...) in rotation frame. So the T/2 based repetitive controller could be implemented in rotating frame, which is shown in Fig.8. The transfer function of T/2 based repetitive controller is:

$$H_{RE2} = \frac{Y(S)}{E(S)} = K_r + K_s \cdot \frac{K_f \cdot e^{-sT/2}}{1 - K_f \cdot e^{-sT/2}} \qquad (7)$$

$$G_{re2} = \frac{e^{-sT/2}}{1 - e^{-sT/2}} = -\frac{1}{2} + \frac{2}{T}\left[\frac{1}{s} + \frac{2s}{s^2 + (2\omega_0)^2} + \frac{2s}{s^2 + (4\omega_0)^2} + \cdots\right] \quad (8)$$

Equation (8) shows T/2 based repetitive control could eliminate all even harmonics under rotating frame. Through d-q reverse transformation, this controller could eliminate all odd harmonics in stationary frame and would have no restriction for industry application.

IV. EASE OF T/3 AND T/6 BASED REPETITIVE CONTROL

The harmonic current of three-phase rectifier is $(6k \pm 1)^{th}$ harmonics, which is presented as

$$i = I_1 \sin \omega t + \sum_{\substack{n=6k\pm1 \\ k=1,2,3\dots}}^{\infty} (-1)^k I_n \sin \omega t \qquad (9)$$

Through d-q transformation, $(6k \pm 1)^{th}$ harmonics will be presented as $6k^{th}$ harmonics in rotating frame. So in addition to the T and T/2 based repetitive controllers, T/3 and T/6 based repetitive controllers could also be implemented in rotating frame.

A. T/3 Based Repetitive Controller

The T/3 based repetitive control is presented in Fig. 10. The transfer function is

$$H_{RE3} = \frac{Y(S)}{E(S)} = K_r + K_s \cdot \frac{K_f \cdot e^{-sT/3}}{1 - K_f \cdot e^{-sT/3}} \qquad (10)$$

$$G_{re3} = \frac{e^{-sT/3}}{1 - e^{-sT/3}} = -\frac{1}{2} + \frac{3}{T}\left[\frac{1}{s} + \frac{2s}{s^2 + (3\omega_0)^2} + \frac{2s}{s^2 + (6\omega_0)^2} + \cdots\right] \quad (11)$$

Equation (11) and Bode plots in Fig. 11 show T/3 based controller could eliminate all $3k^{th}$ harmonics in rotating frame. But this controller could not eliminate all odd harmonics in stationary frame, which is one restriction.

B. T/6 Based Repetitive Controller

T/6 is the minimum period of harmonic current under rotating frame. The T/6 based repetitive control is presented in Fig. 12. The transfer function is

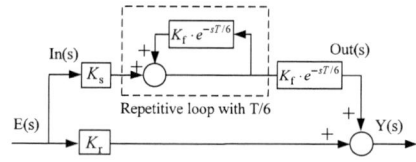

Fig. 12. Block diagram of T/3 based repetitive control

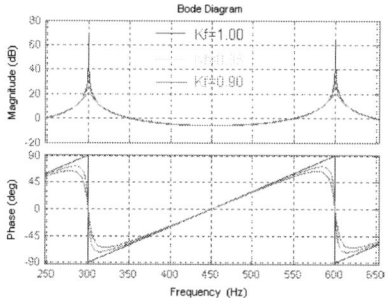

Fig. 13. Bode plots of T/6 based repetitive control

978-1-4244-4782-4/10 $26.00 © 2010 IEEE

$$H_{RE6} = \frac{Y(S)}{E(S)} = K_r + K_s \cdot \frac{K_f \cdot e^{-sT/6}}{1 - K_f \cdot e^{-sT/6}} \qquad (12)$$

$$G_{re6} = \frac{e^{-sT/6}}{1 - e^{-sT/6}} = -\frac{1}{2} + \frac{6}{T}\left[\frac{1}{s} + \frac{2s}{s^2 + (6\omega_0)^2} + \frac{2s}{s^2 + (12\omega_0)^2} + \cdots\right] \qquad (13)$$

Equation (13) and Bode plots in Fig. 13 show T/6 based controller could eliminate all $6k^{th}$ harmonics in rotating frame, in other words, all $(6k \pm 1)^{th}$ harmonics in stationary frame.

V. ANALYSIS AND DESIGN

Many literatures, such as [10-11], have studied the design and analysis of the repetitive control system. The state-space based methods proposed by Hara (1985) could be applied in multivariable repetitive systems and discrete-time repetitive controllers. However, the analysis and synthesis procedures are lack of measure of relative stability for repetitive control system. Since the tradeoff between stability and accuracy is an important aspect of repetitive control system performance. In the absence of a relative stability measure for repetitive control system, controller design relies excessively on computer simulation or experimental evaluation [12].

Srinivasan and Shaw introduced a so-called regeneration spectrum to estimate the locus of the dominant closed-loop poles of the repetitive control system [13]. The characteristic equation of a continuous-time, time-invariant, time-delayed system with a single time delay T_D is given by

$$P(s) + Q(s)e^{-sT_D} = 0 \qquad (14)$$

The regeneration spectrum for such a system is defined as a plot of $R(w)$ versus the frequency w.

$$R(\omega) = \left|\frac{Q(s)}{P(s)}\right| \qquad (15)$$

The regeneration spectrum based analysis and design process for repetitive control are as following:

(1) The control system should be stable in the absence of the repetitive control action. That is the feedback control system with K_r should be stable first.

(2) The time delay T_D should be larger than the dominant time constant of the system in the absence of repetitive control. (This condition is ensured by the high-frequency inverter)

(3) $R(w)<1$, for all frequency. The control system would be stable.

Notice: when condition (2) is satisfied, the regeneration spectrum bore a very simple relationship to the dominant characteristic roots of (14):

$$\alpha_{i,-i} \cong \frac{\ln(R(\omega_i))}{T_D} \qquad (16)$$

Where $\alpha_{i,-i}$ are the real parts of complex conjugate roots with their imaginary parts equal to w_i in magnitude. The (16) is that the continuum of complex numbers $\ln(R(w))/T_D \pm jw$

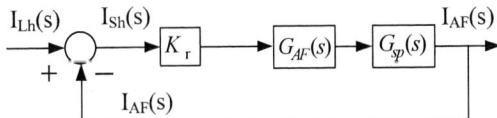

Fig. 14. Feedback control system without repetitive action

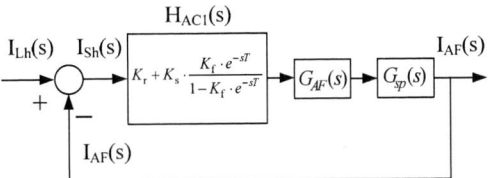

Fig. 15. Feedback closed-loop control system in stationary frame

for w varying from 0 to ∞ includes the characteristic root locations of the system (14) closest to the imaginary axis. When $R(w)<1$, the closed-loop roots would be in the left half of the complex plane, which ensure the system stability.

Except the stability analysis with regeneration spectrum, this paper also combines the sensitivity function for steady-state error and convergence analysis. For four controllers are presented in this paper, the analysis and design of T based repetitive controller would be elaborated; the results of other three controllers would be given out.

A. Control System Stability Analysis

The control system analysis would be carried out in stationary frame. The frame transformation proposed by Zmood at 2001 is

$$H_{AC}(s) = \frac{1}{2}\left[H_{DC}(s + j\omega_0) + H_{DC}(s - j\omega_0)\right] \qquad (17)$$

The condition (1) provides the way for analysis and design of K_r. The value of K_r would not change through frame transformation in (17). The design feedback system of K_r could be realized through Bode plots analysis. Bode plots analysis shows that the feedback system without repetitive loop would have sufficient margin when $K_r = 1.5$.

When the repetitive loop is put into use, the transfer function of T based repetitive controller is

$$H_{AC1}(s) = \frac{1}{2}\left[H_{RE1}(s + j\omega_0) + H_{RE1}(s - j\omega_0)\right]$$

$$= K_r + K_s \cdot \frac{K_f \cdot e^{-sT}}{1 - K_f \cdot e^{-sT}} \qquad (18)$$

Firstly, defining $G_0(s) = G_{AF}(s)G_{sp}(s))$. The closed-loop transfer function is:

$$G_{AC1} = \frac{(K_r - K_f K_r e^{-sT} + K_f K_s e^{-sT}) \cdot G_0(s)}{1 + K_r G_0(s) + K_f \cdot (K_s G_0(s) - 1 - K_r G_0(s)) \cdot e^{-sT}} \qquad (19)$$

The characteristic equation is

$$1 + K_r G_0(s) + K_f \cdot (K_s G_0(s) - 1 - K_f G_0(s)) \cdot e^{-sT} = 0 \qquad (20)$$

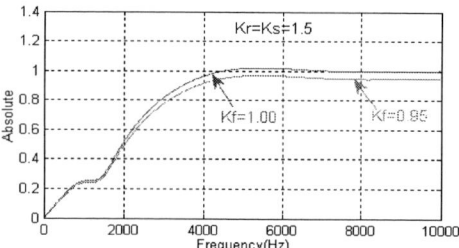

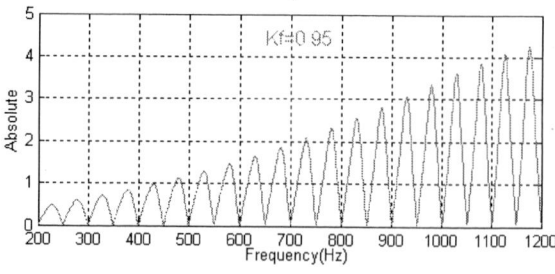

Fig. 16. Block diagram and bode plots of PR controller

The regeneration spectrum is:

$$R_1(\omega) = K_f \cdot \left| 1 - \frac{K_s G_0(s)}{1 + K_r G_0(s)} \right| \qquad (21)$$

The K_f would influence the stability directly. Considering the tradeoff between system stability and other aspects of control system performance, the parameters design should take into steady-state error and convergence characteristic.

B. Steady-state Error Analysis

In industry application, $K_f=1$ would bring unstable problems, but too small K_f would cause significant steady-state error. This paper chooses $K_s=K_r=1.5$, the reason for K_s will be presented in convergence analysis. When $K_f=0.95$, the regeneration spectrum would exceed 1 in high frequency band. when $K_f=0.95$, the value of $R_1(w)$ is shown in Fig. 16, which indicates the $R_1(w)$ is almost equal to 0.95 and less than 1 in high frequency band, which meets the stability requirement. So this paper chooses $K_f=0.95$.

When defining $T(s)=G_0(s)/(1+K_r G_0(s))$. The sensitivity function would be

$$S_1(s) = \frac{E(s)}{R(s)} = \frac{I_{Sh}(s)}{I_{Lh}(s)}$$

$$= \frac{1}{1 + (K_r + K_s \cdot \dfrac{K_f e^{-sT}}{1 - K_f e^{-sT}}) \cdot G_0(s)}$$

$$= \frac{1}{1 + K_r G_0(s)} \cdot \frac{1 - K_f e^{-sT}}{1 - K_f \cdot (1 - K_s T(s)) \cdot e^{-sT}} \quad (22)$$

The sensitivity function with out repetitive loop is:

$$S_r(s) = \frac{E_r(s)}{R(s)} = \frac{1}{1 + K_r G_0(s)} \qquad (23)$$

The expression brought by repetitive loop is:

$$Y_{RE1}(s) = \frac{1 - K_f e^{-sT}}{1 - K_f \cdot (1 - K_s T(s) \cdot e^{-sT}} \qquad (24)$$

The sensitivity function implies the steady-state error or compensation effect. Sensitivity function should have small value at 5th, 7th low-frequency harmonics, and 23rd, 25th high-frequency harmonics. The overall plot of sensitivity function is presented in Fig. 17(a), which indicates that the T based

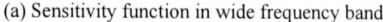

(a) Sensitivity function in wide frequency band

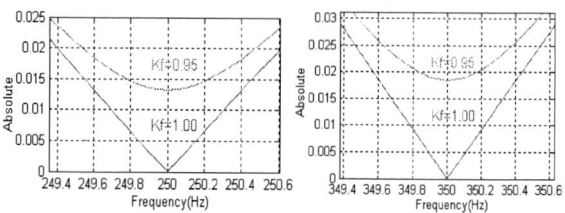

(b) Value at 5th harmonic   (c) Value at 7th harmonic

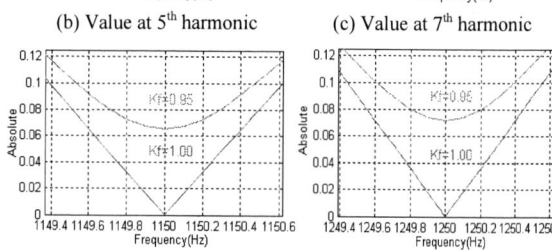

(d) Value at 23rd harmonic   (e) Value at 25th harmonic

Fig. 17. Sensitivity function of T based repetitive controller

repetitive controller have a very small sensitivity function for all harmonics. The value at 5th, 7th, 23rd, 25th are presented in Fig. 17(b), (c), (d), (e) respectively, which indicate that system would have no steady-state error when $K_f=1$.

The introduction of K_f improves the system stability on one hand, but it brings the steady-state error on another hand. The selection of 0.95 in this paper meets both the stability requirement and steady-state error. Under the system stability, the low frequency steady-state error is about 1.5%; the high frequency error is about 6.5%.

C. Convergence Analysis

When considering

$$\left| K_f \cdot (1 - K_s T(s)) \cdot e^{-sT} \right| = R_1(\omega) \cdot e^{-sT} < 1 \qquad (25)$$

The Maclaurin expansion of $S_1(s)$ and error would be

$$S_1(s) = \frac{1 - K_f e^{-sT}}{1 + K_r G_0(s)} \cdot \prod_{k=0}^{\infty} \left[K_f (1 - K_s T(s)) \cdot e^{-sT} \right]^k \qquad (26)$$

$$E_1(s) = S_1(s) R(s) = \frac{R(s)}{1 + K_r G_0(s)} \cdot [1 -$$

$$\sum_{k=1}^{\infty} K_f K_s T(s) \cdot \left[K_f (1 - K_s T(s)) \right]^{k-1} \cdot (e^{-sT})^k$$

$$= E_r(s) \cdot [1 - \sum_{k=1}^{\infty} K_f K_s T(s) \cdot \left[K_f (1 - K_s T(s)) \right]^{k-1} \cdot (e^{-sT})^k \quad (27)$$

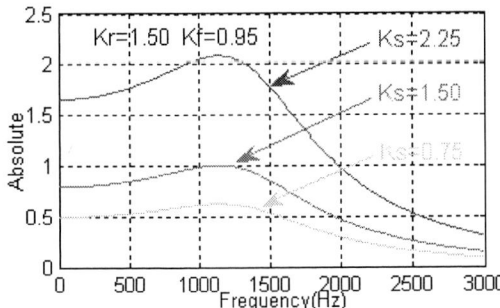

Fig. 18. Plots of convergence expression

The $E_r(s)$ in (27) presents the error caused by feedback control of K_r. The inverse Laplace transforms result of $E_1(s)$ is

$$e(t) = e_r(t) - e_r(t-T) \cdot L^{-1}\{K_f K_s T(s)\}$$

$$-\cdots$$

$$-e_r(t-kT) \cdot L^{-1}\{K_f K_s T(s) \cdot [K_f(1-K_s T(s))]^{k-1}\}$$

$$-\cdots]$$

$$= e_1(t) + e_2(t-T) + e_3(t-3T) + e_4(t-4T) + \ldots \quad (28)$$

Equation (28) indicates that the error in the first period is independent of repetitive loop, that in the third period is equal to the error corresponding to e_3 superimposed over the combined steady-state errors of e_1 and e_2 etc. When the K_s is chosen such that $K_s T(s)$ is nearly unity within interesting frequency range, the convergence rate would be fast. For the importance role to convergence rate, the $K_s T(s)$ is defined as convergence expression $Q_1(s)$:

$$Q_1(s) = K_s T(s) = \frac{K_s G_0(s)}{1 + K_r G_0(s)} \quad (29)$$

The plot of $Q_1(s)$ is shown in Fig. 18, which indicates that 1.5 is more preferable for K_s.

D. T/2, T/3 and T/6 Based Repetitive Controllers

For T/2 based repetitive controller, the equivalent controller in stationary frame is

$$H_{AC2}(s) = K_r - K_s \cdot \frac{K_f \cdot e^{-sT/2}}{1 + K_f \cdot e^{-sT/2}} \quad (30)$$

The regeneration spectrum is:

$$R_2(\omega) = R_1(\omega) = K_f \cdot \left| 1 - \frac{K_s G_0(s)}{1 + K_r G_0(s)} \right| \quad (31)$$

Convergence expression $Q_2(s)$ is:

$$Q_2(s) = K_s T(s) = \frac{K_s G_0(s)}{1 + K_r G_0(s)} \quad (32)$$

The sensitivity function is:

$$S_2(s) = \frac{1}{1 + K_r G_0(s)} \cdot \frac{1 + K_f e^{-sT/2}}{1 - K_f \cdot (K_s T(s) - 1) \cdot e^{-sT/2}} \quad (33)$$

The T/3 and T/6 based repetitive controllers also have the same regeneration spectrum and convergence expression with T and T/2 based repetitive controllers. The only difference is the equivalent controller in stationary frame and the sensitivity function. So these four types controllers should choose the same control parameters, that is $K_f=0.95$, $K_s=K_r=1.5$. The difference in sensitivity function means the difference in steady-state error, which would be discussed with dynamic response in simulations and experiments.

Sensitivity function of T/3 based repetitive controller is:

$$S_3(s) = \frac{1}{1 + K_r G_0(s)} \cdot \frac{(K_f^2 e^{-2sT/3}) + K_f e^{-sT/3} + 1}{1 - K_f \cdot [K_s T(s)(K_f e^{-sT/3} + 0.5) - K_f e^{-sT/3} - 1)] \cdot e^{-sT/3}} \quad (34)$$

Sensitivity function of T/6 based repetitive controller is:

$$S_6(s) = \frac{1}{1 + K_r G_0(s)} \cdot \frac{(K_f^2 e^{-sT/3}) - K_f e^{-sT/6} + 1}{1 - K_f \cdot [K_s T(s)(K_f e^{-sT/6} - 0.5) - K_f e^{-sT/6} + 1)] \cdot e^{-sT/6}} \quad (35)$$

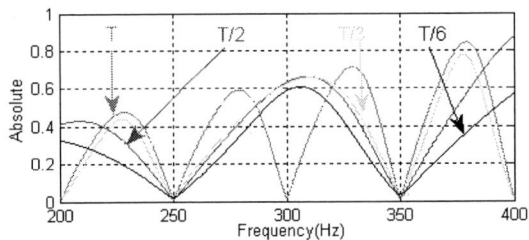

(a) Sensitivity function in low frequency band

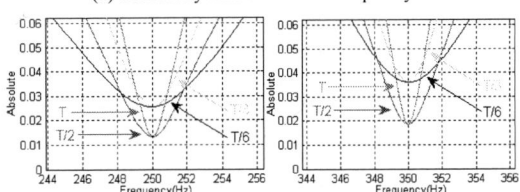

(b) Value at 5th harmonic (c) Value at 7th harmonic

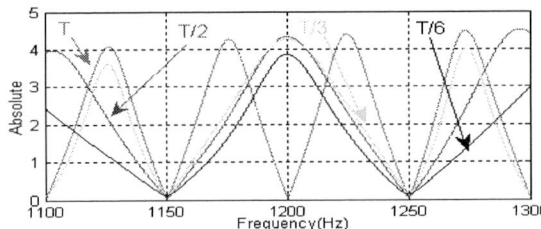

(d) Sensitivity function in high frequency band

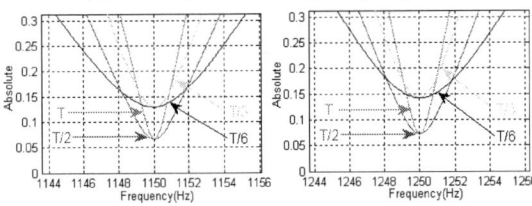

(e) Value at 23rd harmonic (f) Value at 25th harmonic

Fig. 19. Sensitivity function comparison of four controllers

The comparison of four different Sensitivity functions in Fig. 19 shows that:

(1) All sensitivity functions have little value at $(6k\pm1)^{th}$ harmonics. But at frequency point of $(6k\pm1)$ harmonics, the sensitivity function of T based repetitive controller is similar with T/2 based repetitive controller's; The sensitivity function of T/3 based repetitive controller is similar with T/6 based repetitive controller's.

(2) At point of $(6k\pm1)^{th}$ harmonics, compared with T/3 and T/6 based repetitive controllers, the T and T/2 based repetitive controllers have higher compensation efficiency, which is presented in the TABLE II.

(3) The sensitivity function of T/6 based repetitive have wider opening, which means good robustness for frequency variations at $(6k\pm1)^{th}$ harmonics.

(4) T based repetitive controller could eliminate all harmonics; T/2 based could eliminate all odd harmonics; T/3 based could eliminate all $(3k\pm1)^{th}$ harmonics; T/6 based could only eliminate all $(6k\pm1)^{th}$ harmonics.

The TABLE III presents the performance comparison of response time, compensation efficiency, compensation objects, and robustness for frequency variations.

TABLE II.   STEADY-STATE ERROR

Error (%) VS frequency(Hz)	Time delay based repetitive controllers			
	T	T/2	T/3	T/6
250	1.31%	1.31%	2.56%	2.56%
350	1.85%	1.85%	3.61%	3.61%
550	2.94%	2.94%	5.75%	5.75%
650	3.50%	3.50%	6.85%	6.85%
1150	6.54%	6.54%	12.87%	12.87%
1250	7.21%	7.21%	14.18%	14.18%

TABLE III.   PERFORMANCE COMPARISON

Performance evaluations	Time delay based repetitive controllers			
	T	T/2	T/3	T/6
Response time	0.02s	0.01s	0.0067s	0.0033s
Compensation efficiency	Good	Good	Ordinary	Ordinary
Compensation object	All harmonics	Odd harmonics	$(3\pm1)^{th}$ harmonics	$(6k\pm1)^{th}$ harmonics
Frequency variations	Poor	Good	Ordinary	Best

VI.   SIMULATION AND EXPERIMENT

Four Simulations on MATLAB/Simulink are carried out to verify the validity of proposed four repetitive controllers in rotating frame. The system parameters are already given out in part II, the control parameters are $K_r=K_s=1.5$, $K_f=0.95$.

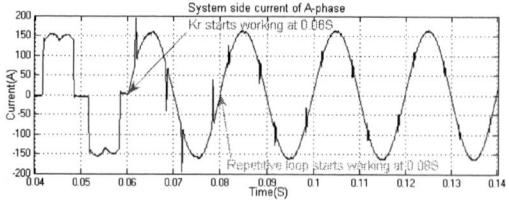

(a) T based repetitive controller's performance

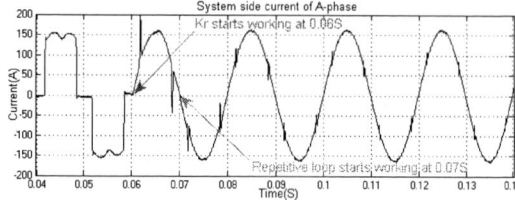

(b) T/2 based repetitive controller's performance

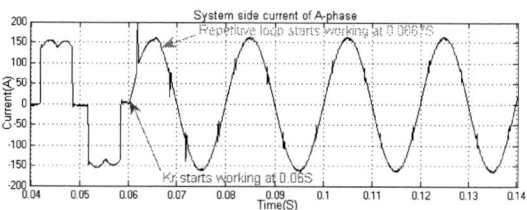

(c) T/3 based repetitive controller's performance

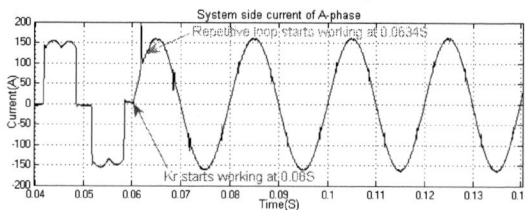

(d) T based repetitive controller's performance

Fig. 20.   Simulation results

Simulation results are presented in Fig. 20. In the simulation, the APF starts to work at 0.06s. The repetitive loop of T based repetitive controller starts working at 0.080s; T/2 based repetitive controller starts working at 0.070s; T/3 based repetitive controller starts working at 0.067s; T/6 based repetitive controller starts working at 0.063s. The Total Harmonic Distortion (THD) under steady-state condition are 2.14%, 2.17%, 2.34%, 2.31%.

To test the validity in industry application, these four repetitive controllers are used in the control system of a 100-kVA shunt active filter, which is shown in Fig. 22. The control circuit is fully digitized and based on FPGA and DSP. FPGA mainly performs harmonic detection, repetitive control realization, PWM generation and DC voltage controlling. DSP mainly performs man-machine interface function, data analysis and device protection. The experiments results are shown in the Fig.21. The CH1 channel represents 80A/div (current); the CH2 channel represents 100V/div (voltage).

Without compensation, system has 45 A (RMS) harmonic current and the THD is 33.2%. With the compensation of APF, THD is reduced to 5.17% for T based repetitive controller, 5.25% for T/2 based controller, 5.44% for T/3 based controller, and 5.62% for T/6 based controller.

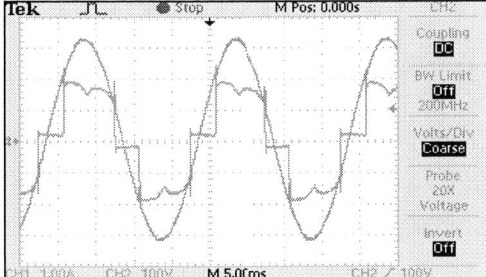

(a) System voltage and current without compensation

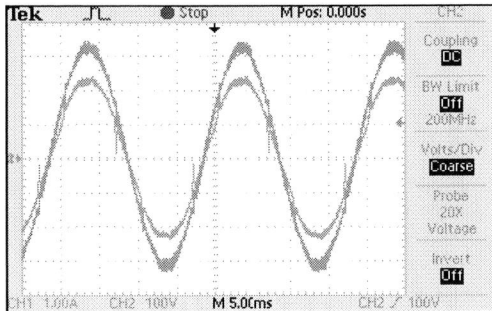

(b) T based repetitive controller's performance

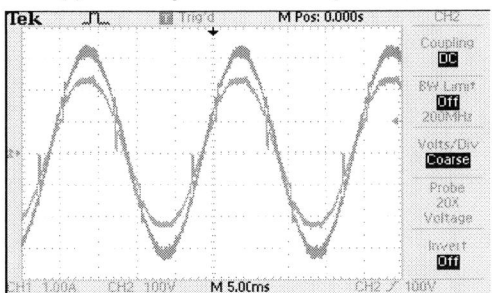

(c) T/2 based repetitive controller's performance

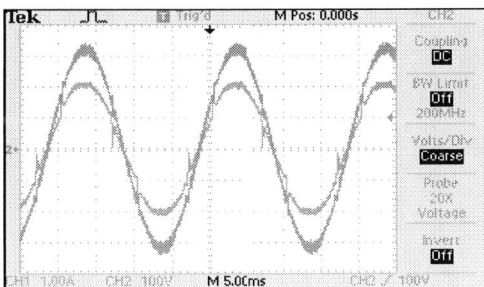

(d) T/3 based repetitive controller's performance

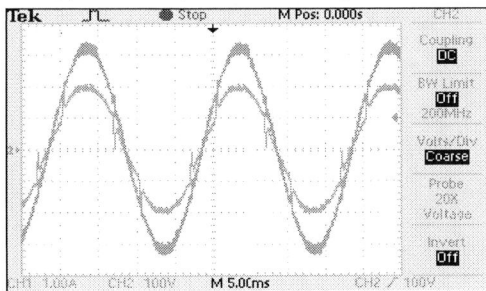

(e) T/6 based repetitive controller's performance

Fig. 21. Experiment results

Fig. 22. Experiment Platform

VII. CONCLUSION

This paper has presented four repetitive controllers in rotating frame for APF system, whose repetitive period is T, T/2, T/3, and T/6 respectively. These four repetitive controllers have different dynamic response time, compensation efficiency. They also suitable for different compensation object and have different robustness for frequency variations. All this performance characteristics have been discussed in the paper and verified by the simulation and experiments, which provides certain reference values for further application in DVR, UPS, PV, and other grid-connected converters.

REFERENCES

[1] J. Shlabbach, D. Blume and T. Stephanblome, "Voltage quality in electrical power systems," IEE power series No. 36, England, 2001.

[2] M Santolo, A Perfetto, "Comparison of different control techniques for active filter applications," Fourth IEEE International Caracas Conference on Devices, Circuits and Systems, Aruba, 2002.17-19.

[3] D. N. Zmood, D. G. Holmes and Bode, "Frequency domain analysis of three-phase linear current regulator", IEEE Trans. Ind. Appl., 2001, 37, pp. 601–610

[4] D. N. Zmood and D. G Holmes. "Stationary frame current regulation of PWM inverters with zero steady-state error", IEEE Trans. Power Electron., 2003, 18, pp. 814–822

[5] W. X. Yuan, and J. Allmeling. "Stationary frame generalized integrators for current control of active power filters with zero steady-state error for current harmonics of concern under unbalanced and distorted operating conditions", IEEE Trans. Ind. Appl., 2002, 38, pp. 523–532

[6] P. Mattavelli, "A closed-loop selective harmonic compensation for active filters", IEEE Trans. Ind. Appl., 2001, 37, pp. 81–89

[7] P. Mattavelli, "Synchronous-frame harmonic control for highperformance AC power supplies", IEEE Trans. Ind. Appl., 2001, 37, pp. 864–872

[8] R. Teodorescu, F. Blaabjerg, "A new control structure for grid-connected PV inverters with zero steady-state error and selective harmonic compensation" .

[9] J. Kauraniemi, T. I. Laakso and I. Hartimo, "Delta operator realizations of direct-form IIR filters," IEEE Trans. Circuits Syst., vol. 45, pp. 41–51, Jan. 1998.

[10] S. Hara, Y. Yamammoto, T. Omata, and M. Nakano, "Repetitive control system: A new type servo system for periodic exogenous signals," IEEE Trans. Automat. Contr., vol. 33, pp. 659–666, July 1988.

[11] M. Tomizuka, T. C. Tsao, and K. K. Chew, "Discrete-time domain analysis and synthesis of repetitive controllers," in Proc. Amer. Contr. Conf, vol. 2, Atlanta, GA, June 1988, pp. 860–886.

[12] S. HARA, T. OMATA, and M. NAKANO "Synthesis of repetitive control systems and its application,"Proc 24th Conf Decision Contr .1985.1384-1392.

[13] K. Srinivasan and F. R. Shaw, "Analysis and design of repetitive control systems using the regeneration spectrum," Trans. ASME, vol. 113, pp. 113–217, 1991.

Current Constraints of PWM Rectifier under Unbalanced Voltage Supply

Miroslav Chomat, Ludek Schreier, Jiri Bendl
Department of Electrical Machines, Drives, and Power Electronics
Institute of Thermomechanics ASCR, v.v.i.
Prague 8, Czech Republic
E-mail: chomat@it.cas.cz

Abstract—**Non-standard conditions in the power network such as voltage unbalance negatively affect operation of electric drives. The unbalance can be caused by a failure in the network or by an unbalanced load in the vicinity of the affected drive. Unsymmetrical voltages at the input of a voltage source inverter cause pulsations in the DC-link voltage when not properly taken care of. This may result in a significant reduction of power capabilities and, therefore, limited controllability of the drive. The paper deals with the analysis of the constraints caused especially by the current stress of the converter's parts when operated from an unbalanced voltage supply. The influence of circuit parameters on the operating regions and their constraints are particularly dealt with.**

I. INTRODUCTION

Certain non-standard operating conditions in the power grid such as voltage unbalance may negatively affect operation of electrical devices including electric drives. Such an unbalance can be caused by a failure in the network or by an unbalanced load in the vicinity of the affected drive. An unbalanced system of supply voltages at the input of a voltage source inverter causes pulsations in the DC-link voltage when not properly taken care of. This may result in significantly

reduced power capabilities and, therefore, limited controllability of the drive.

Numerous methods dealing with such a situation in electric drives have been developed and introduced into practice. The main purpose of this paper is to identify and describe the additional consequences for operation of an electric drive fed by an unbalanced system of supply voltages. The main attention is paid to the restrictions arising from the current constraints given by the rating of electronic components in the front-end rectifier of the drive.

II. CONTROL METHOD

A simplified scheme of the drive under investigation is shown in Fig. 1. Suitable control of the front-end AC/DC converter can be employed in order to draw constant input power from the power network even at unbalanced voltage supply [1-5]. The switching functions for the front-end AC/DC converter are generated so that a constant voltage across the DC bus can be maintained. Series combinations of inductance and resistance are considered at the input terminals of the inverter. The proper choice of the switching functions may optimize the power quality effect of the inverter on the

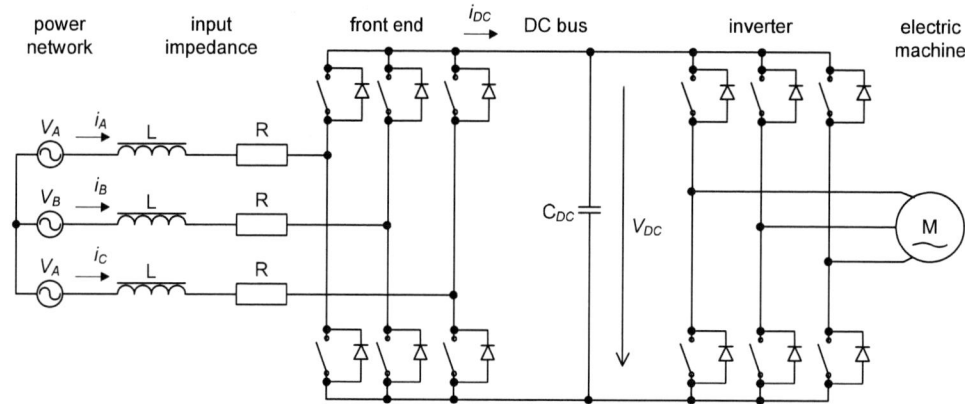

Figure 1. Scheme of system under investigation.

This work was supported by the Grant Agency of the Czech Republic under research grant No. 102/09/1273 and by the Institutional Research Plan AV0Z20570509.

power grid. Detailed description of the control algorithm referenced throughout this paper is in [6, 7]. The analysis of the operating regions has been presented in [8, 9].

III. LIMITATION OF CONTROL DUE TO UNBALANCED VOLTAGE SUPPLY

The necessity to generate the negative sequence component of the switching functions in order to eliminate the effect of the supply-voltage unbalance on the DC-link voltage pulsations reduces the control range for the positive sequence component of the switching functions. This is due to the fact that in individual phases the maximum of the switching function can only reach one at most at any given time. Another constraint results from the current rating of the converter. The resulting constraints depend on the value and type of the unbalance.

Analysis of the limitation corresponding to various types of unbalanced supply voltages has been carried out. The reference parameters of the input impedance were chosen to be $R = 0.1\ \Omega$ and $L = 10$ mH. The input phase voltages had nominal voltage amplitudes of 230 V, nominal frequency of 50 Hz, and mutual phase shifts of $120°$ to form a three-phase voltage system in the case of the symmetrical system. The DC-link voltage was set to 400 V.

The choice of the positive sequence component of the switching functions from the available control range affects both the magnitude of the DC-link current and the currents in individual input phases. Fig. 2 shows what magnitudes of the DC-link current correspond to the coordinates from the available control range. The unbalance was formed by setting the magnitude of the voltage in phase A to 0.75 p.u. The corresponding maximal input phase current magnitude, calculated as the maximum of all the phase currents, is shown in Fig. 3. It can be seen from Fig. 2 that the resulting DC-link current decreases in the vertical direction of the operating region, whereas the maximal input current in Fig. 3 decreases in the horizontal direction. The corresponding measure of the current unbalance is depicted in Fig. 4 and the average power factor of all the three input phases is depicted in Fig. 5.

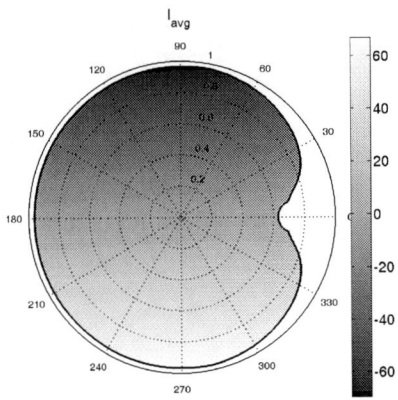

Figure 2.   DC-link current under unbalanced voltage supply ($L = 10$ mH, $R = 0.1\ \Omega$, $V_{dc} = 400$ V).

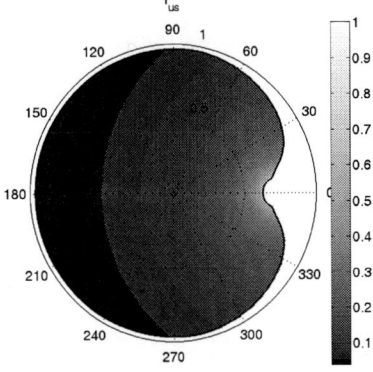

Figure 4.   Input current unbalance under unbalanced voltage supply ($L = 10$ mH, $R = 0.1\ \Omega$, $V_{dc} = 400$ V).

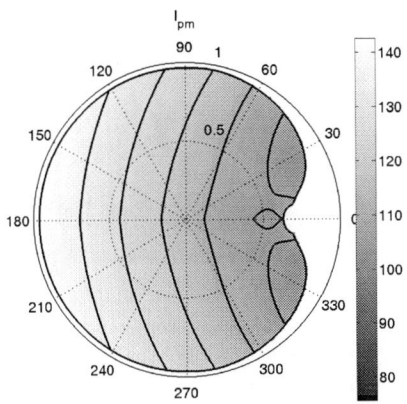

Figure 3.   Maximal input phase current under unbalanced voltage supply ($L = 10$ mH, $R = 0.1\ \Omega$, $V_{dc} = 400$ V).

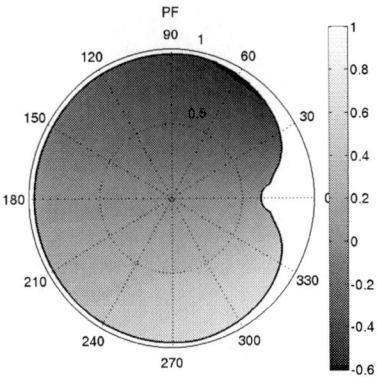

Figure 5.   Power factor under unbalanced voltage supply ($L = 10$ mH, $R = 0.1\ \Omega$, $V_{dc} = 400$ V).

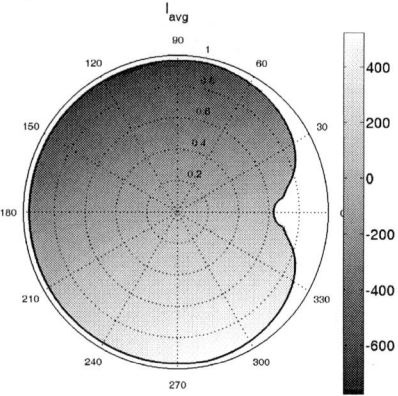

Figure 6. DC-link current under unbalanced voltage supply (L = 1 mH, R = 0.1 Ω, Vdc = 400 V).

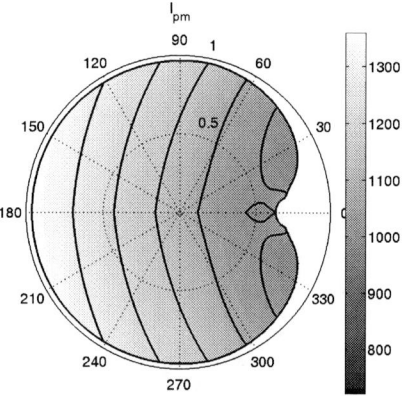

Figure 7. Maximal input phase current under unbalanced voltage supply (L = 1 mH, R = 0.1 Ω, Vdc = 400 V).

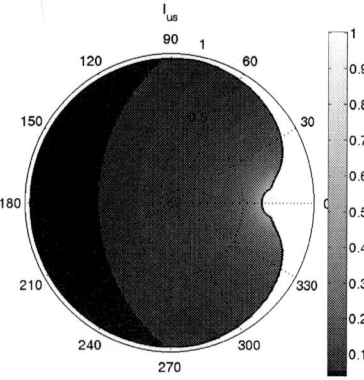

Figure 8. Input current unbalance under unbalanced voltage supply (L = 1 mH, R = 0.1 Ω, Vdc = 400 V).

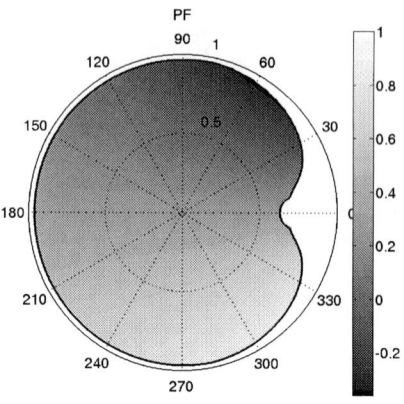

Figure 9. Power factor under unbalanced voltage supply (L = 1 mH, R = 0.1 Ω, Vdc = 400 V).

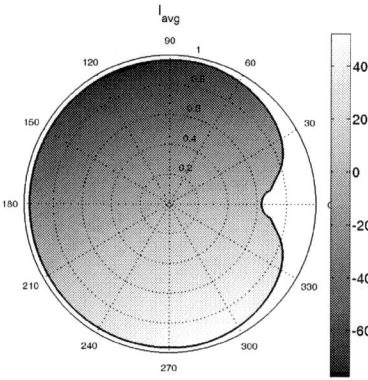

Figure 10. DC-link current under unbalanced voltage supply (L = 1 mH, R = 1 Ω, Vdc = 400 V).

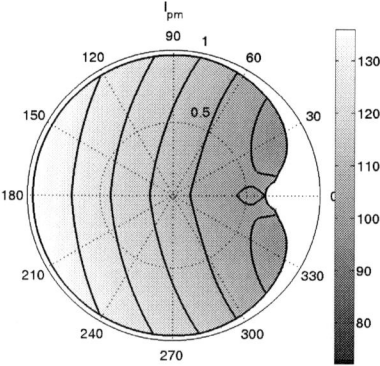

Figure 11. Maximal input phase current under unbalanced voltage supply (L = 1 mH, R = 1 Ω, Vdc = 400 V).

978-1-4244-4782-4/10 $26.00 © 2010 IEEE

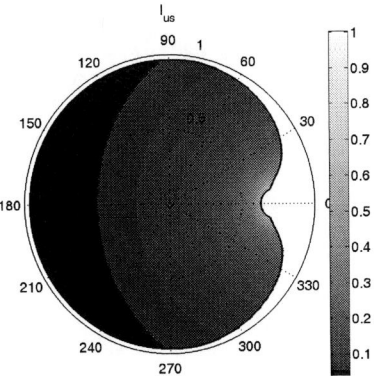

Figure 12. Input current unbalance under unbalanced voltage supply (L = 1 mH, R = 1 Ω, Vdc = 400 V).

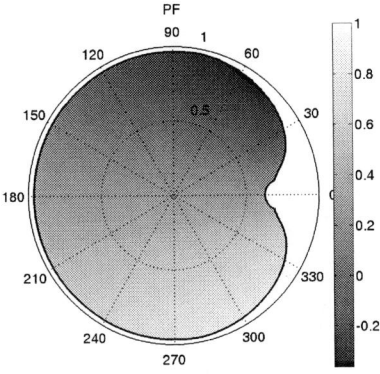

Figure 13. Power factor under unbalanced voltage supply (L = 1 mH, R = 1 Ω, Vdc = 400 V).

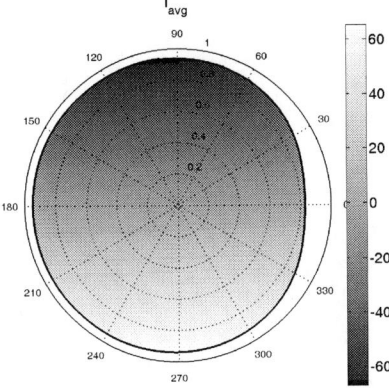

Figure 14. DC-link current under unbalanced voltage supply (L = 10 mH, R = 0.1 Ω, Vdc = 200 V).

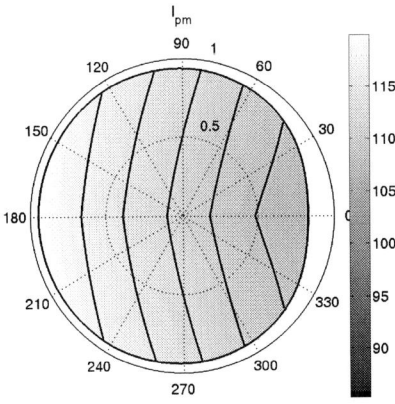

Figure 15. Maximal input phase current under unbalanced voltage supply (L = 10 mH, R = 0.1 Ω, Vdc = 200 V).

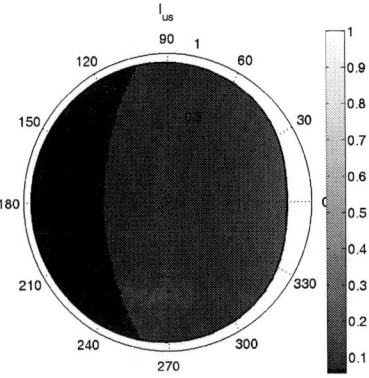

Figure 16. Input current unbalance under unbalanced voltage supply (L = 10 mH, R = 0.1 Ω, Vdc = 200 V).

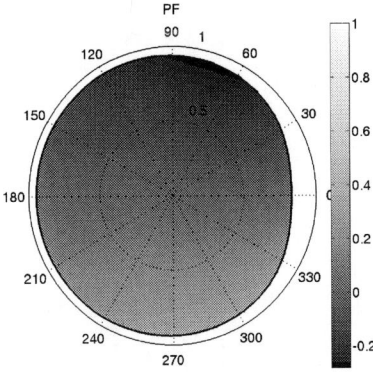

Figure 17. Power factor under unbalanced voltage supply (L = 10 mH, R = 0.1 Ω, Vdc = 200 V).

978-1-4244-4782-4/10 $26.00 © 2010 IEEE

If we change the value of the input inductance from 10 mH to 1 mH, the constraints caused by the switching functions remain the same as can be seen from Figs. 6 through 9. However, both the DC-link current and the input current increased nearly ten times as the input reactance represents the main limiting factor for the currents entering the rectifier. The excessive values of the currents would, in a case of a real rectifier, impose additional restrictions to the operating regions resulting from current stress of electronic components in the bridge. This can also be considered in the shape of new borders of operating regions.

A different situation arises when the input resistance is increased ten times to 1 Ω. The corresponding electrical quantities are shown in Fig. 10 through 13. The increase in the DC-link and input phase currents is not as dramatic as the resistance plays less significant role in limiting the currents than the inductance. The values of the currents are similar to the ones in the first case.

A change in the DC-link voltage introduces, on the other hand, a noticeable change in the shape of constraints caused by the limitation of the switching functions. Figs. 14 through 17 show the situation for the decrease in the DC-link voltage from 400 V to 200 V and Figs. 18 through 21 show the situation for the increase to 600 V. In the latter case, a rise of an isolated restricted area in the right hand side of the figure completely surrounded by available control space can be noticed.

Measurements on an experimental system identical to the simulated one have been carried out in order to verify the investigated method. The scope traces in Fig. 22 show the measured current in phase A and the DC link current when the negative-sequence in the supply voltage is not compensated for by the control method and the DC link current, therefore, contains significant component pulsating with a frequency of 100 Hz, twice the fundamental network frequency. The case when unbalanced voltage system is compensated by the investigated control method is illustrated in Fig. 23. It can be seen that the pulsating component of the DC link current has been effectively eliminated by the investigated method.

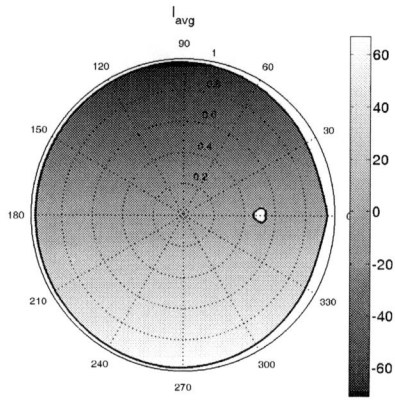

Figure 18. DC-link current under unbalanced voltage supply (L = 10 mH, R = 0.1 Ω, Vdc = 600 V).

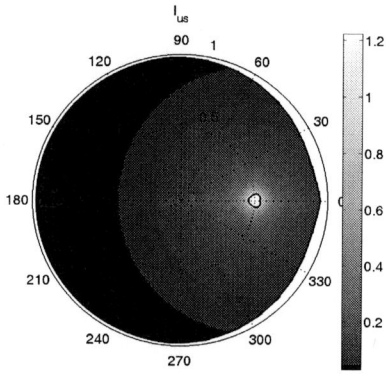

Figure 20. Input current unbalance under unbalanced voltage supply (L = 10 mH, R = 0.1 Ω, Vdc = 600 V).

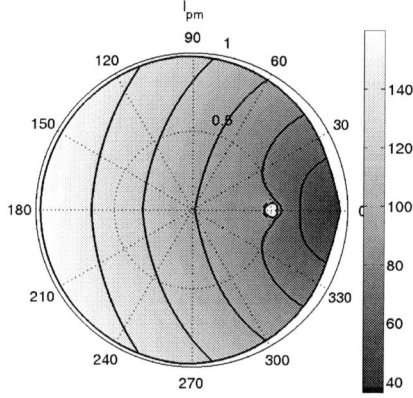

Figure 19. Maximal input phase current under unbalanced voltage supply (L = 10 mH, R = 0.1 Ω, Vdc = 600 V).

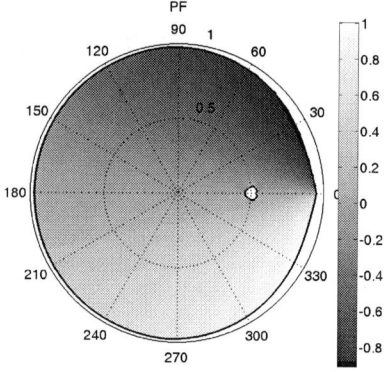

Figure 21. Power factor under unbalanced voltage supply (L = 10 mH, R = 0.1 Ω, Vdc = 600 V).

978-1-4244-4782-4/10 $26.00 © 2010 IEEE

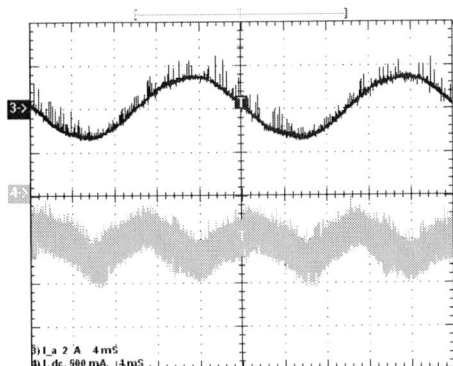

Figure 22. Phase A current and DC-link current under unbalanced voltage supply without elimination of pulsating component.

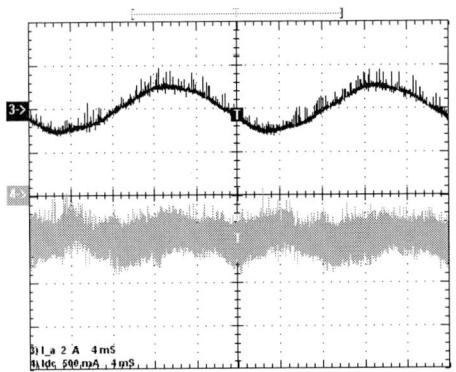

Figure 23. Phase A current and DC-link current under unbalanced voltage supply with elimination of pulsating component.

IV. CONCLUSIONS

The need to compensate for the negative sequence component in the supply voltages at the input of active bridge rectifier brings along significant restriction in the available operating region for the switching functions. This restriction can be analytically evaluated and particular constraints may be identified for different types and levels of the unbalance. The shape of the constraints and the resulting electrical quantities depend on some circuit parameters and electrical quantities in the system. The influence of selected parameters has been investigated using mathematical and numerical models.

ACKNOWLEDGMENT

This work was supported by the Grant Agency of the Czech Republic under research grant No. 102/09/1273 and by the Institutional Research Plan AV0Z20570509.

REFERENCES

[1] A.V. Stankovic, T.A. Lipo: "A Novel Control Method for Input Output Harmonic Elimination of the PWM Boost Type Rectifier Under Unbalanced Operating Conditions," *IEEE Trans. on Power Electronics*, 16, 2001, pp. 603-611.

[2] A.V. Stankovic, T.A. Lipo, "A Generalized Control Method for Input-Output Harmonic Elimination of the PWM Boost Type Rectifier Under Simultaneous Unbalanced Input Voltages and Input Impedances," *Power Electronics Specialists Conference*, 2001, pp. 1309-1314.

[3] K. Lee, T.M. Jahns, W.E. Berkopec, T.A. Lipo, T.A.; "Closed-form analysis of adjustable-speed drive performance under input-voltage unbalance and sag conditions," *IEEE Trans. on Industry Applications*, 2006, vol. 42, no. 3, pp. 733-741.

[4] A.M. Cross, P.D. Evans, A.J. Forsyth, "DC Link Current in PWM Inverters with Unbalanced and Non Linear Loads," *IEE Proc.-Electr. Power Appl.*, 1999, vol. 146, no. 6, pp. 620-626.

[5] H. Song, K. Nam: "Dual Current Control Scheme for PWM Converter Under Unbalanced Input Voltage Conditions," *IEEE Trans. on Industrial Electronics*, 46, 1999, pp. 953-959.

[6] M. Chomat, L. Schreier, "Control Method for DC-Link Voltage Ripple Cancellation in Voltage Source Inverter under Unbalanced Three-Phase Voltage Supply," *IEE Proceedings on Electric Power Applications*, vol. 152, no. 3, 2005, pp. 494 – 500.

[7] M. Chomat, L. Schreier, J. Bendl, "Operation of Adjustable Speed Drives under Non Standard Supply Conditions," *IEEE Industry Applications Conference/42th IAS Annual Meeting*, New Orleans, USA, 2007.

[8] M. Chomat, L. Schreier, J. Bendl, "Influence of Circuit Parameters on Operating Regions of PWM Rectifier Under Unbalanced Voltage Supply," *IEEE International Electric Machines* and *Drives Conference*, Miami, USA, 2009, pp. 416-421.

[9] M. Chomat, L. Schreier, J. Bendl, "Operating Regions of PWM Rectifier under Unbalanced Voltage Supply," *International Conference on Industrial Technology*, Gippsland, Australia, 2009, pp. 510 - 515.

Space Vector PWM for a Direct Matrix Converter Based Open-End Winding AC Drives with Enhanced Capabilities

Ranjan K. Gupta, Apurva Somani, Krushna K. Mohapatra and Ned Mohan
Department of Electrical and Computer Engineering, University of Minnesota
200 Union Street SE, 4-174 EE/CSci Bldg, Minneapolis, MN 55455 USA
Phone: (612) 625-3362
Email: mohan@umn.edu

Abstract—**This paper introduces a space vector PWM technique for a direct matrix converter based three-phase open-end winding ac machine drive. With the proposed PWM technique, the following simultaneous capabilities of the open-end winding drive system are achieved: 1) Machine phase voltage up to 1.5 times the input phase voltage in the linear modulation, 2) Controllable grid power factor, and 3) Elimination of the high-frequency common-mode voltage at the machine terminals. Elimination of the high-frequency common-mode voltage results in elimination of the high-frequency bearing current and reduced conducted EMI. The proposed modulation scheme is implemented on a dSPACE and FPGA based control platform. Experimental results on a hardware prototype verify the abovementioned capabilities of the drive system.**

I. INTRODUCTION

The future trend is a high efficiency, high power-density power electronic system which may be operated at a high temperature. These requirements are important for naval and aviation applications where size and weight are critical parameters [1]. Presently, voltage source based back-to-back converter is widely used in electric drives which uses bulky storage capacitors for power conversion. The storage capacitors are prone to poor performance at high temperatures and susceptible to hazardous failures. Therefore, the use of storage capacitors not only increases the system size and weight but also causes reliability degradation. Matrix converters [2]–[4] are well known for direct ac-ac power conversion which eliminate the use of storage capacitors. This leads to reduced system size and weight resulting in a higher power-density converter system.

This paper describes how matrix converters, one at each side of an open-end winding ac machine (shown in Fig. 1) [5] are modulated synchronously, to achieve the following simultaneous advanced features:

- Machine phase voltage up to 1.5 times the input phase voltage in linear-modulation, which means that the rated torque capability region of the machine can be extended to 150% of the rated speed,
- The peak voltage-stress across the slot-insulation within the motor is equal to the peak of input phase voltage

i.e. a factor of at least $\sqrt{3}$ lower as compared to the conventional back-to-back converter
- Controllable grid power factor, and
- Elimination of the high-frequency common-mode voltage at the machine terminals.

The high-frequency common-mode voltage cause bearing current and conducted EMI in electric drive systems [6], [7]. Reduction in the detrimental effects of high-frequency common-mode voltage can be achieved by modifying the PWM strategy [8]–[10] and/or using the common-mode filters at the output of drive system [7], [11], [12]. In the proposed space vector PWM of the matrix converter (modulated with only rotating vectors), at any instant, sum of the phase voltages at the machine terminals is zero [5], [13]. Therefore, the common-mode voltage is not present at the output terminals of the converter. Elimination of the common-mode voltage leads to significant savings in the common-mode filter which further leads to increase in power-density of the converter system.

With the abovementioned capabilities along with the assumption that the cost of copper will increase and the cost of semiconductor will decrease in future, the open-end drive shown in Fig. 1 offers a more silicon solution to a high performance drive with high power-density and advanced features. The topology shown in Fig. 1 is a special case of multi-input multi-output direct frequency changers reported in the past [2], [14]. However, common-mode voltage elimination in the context of electric drive applications was not discussed. Open-end winding ac machines supplied through dc-link based systems that need storage capacitors have been proposed [15]–[18]. A direct-link drive, without storage capacitor, for open-end winding ac machine has also been reported recently [19]. However, the grid power factor is not controllable in the direct-link drive when the converter is modulated for common-mode voltage elimination. Simulation results of the open-end winding drive shown in Fig. 1 was presented in [5]. However, PWM strategy and corresponding modulation equations were not described. In this paper, a new space vector PWM strategy is introduced and related modulation equations are derived. To generate the gate signals for 18 four-quadrant switches

978-1-4244-4782-4/10 $26.00 © 2010 IEEE

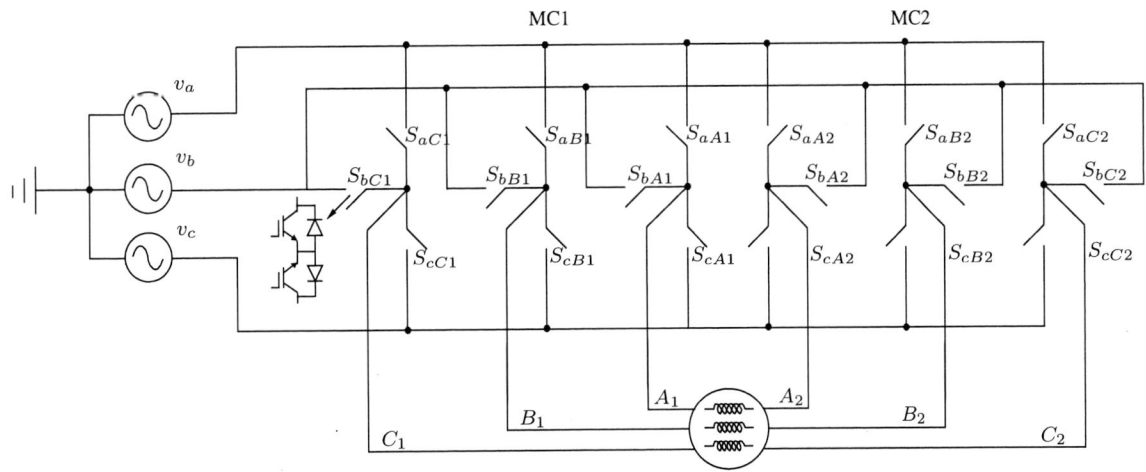

Fig. 1. Direct matrix converter based open-end winding ac machine drive

TABLE I
ROTATING VOLTAGE VECTORS WITH ZERO COMMON-MODE VOLTAGE AT THE MACHINE TERMINALS (A_1,B_1,C_1)

Group	$\vec{v}_{o1}(t)$	$v_{A1}(t)$	$v_{B1}(t)$	$v_{C1}(t)$	$v_{cm1}(t)$
CCW-rotating	$\vec{v}_{abc}(t) = \frac{3}{2}\hat{V}_i e^{j\omega_i t}$	$v_a(t)$	$v_b(t)$	$v_c(t)$	0
	$\vec{v}_{bca}(t) = \frac{3}{2}\hat{V}_i e^{j(\omega_i t - 2\pi/3)}$	$v_b(t)$	$v_c(t)$	$v_a(t)$	0
	$\vec{v}_{cab}(t) = \frac{3}{2}\hat{V}_i e^{j(\omega_i t + 2\pi/3)}$	$v_c(t)$	$v_a(t)$	$v_b(t)$	0
CW-rotating	$\vec{v}_{acb}(t) = \frac{3}{2}\hat{V}_i e^{-j\omega_i t}$	$v_a(t)$	$v_c(t)$	$v_b(t)$	0
	$\vec{v}_{cba}(t) = \frac{3}{2}\hat{V}_i e^{-j(\omega_i t - 2\pi/3)}$	$v_c(t)$	$v_b(t)$	$v_a(t)$	0
	$\vec{v}_{bac}(t) = \frac{3}{2}\hat{V}_i e^{-j(\omega_i t + 2\pi/3)}$	$v_b(t)$	$v_a(t)$	$v_c(t)$	0

five triangle-carrier comparisons along with sector information are required. This new PWM strategy is implemented on a dSPACE and FPGA based control platform. Experimental results on a hardware prototype verify the operation and the advance features of the drive system.

II. SPACE VECTOR PWM FOR OPEN-END WINDING DRIVE

The open-end winding ac machine driven by two matrix converters MC1 and MC2 is shown in Fig. 1. Machine terminals (A_1, B_1, C_1) and (A_2, B_2, C_2) are connected to the output terminals of the matrix converters MC1 and MC2 respectively. MC1 and MC2 are connected to the same three-phase grid defined by: $v_a(t) = \hat{V}_i \cos(\omega_i t)$, $v_b(t) = \hat{V}_i \cos(\omega_i t - 2\pi/3)$ and $v_c(t) = \hat{V}_i \cos(\omega_i t + 2\pi/3)$, where ω_i is the grid frequency and $\hat{V}_i$ is the amplitude of the grid phase voltage. Voltage space vectors at the two sets of three-phase machine terminals are defined by (1) and (2) respectively where terminal voltages are referenced to the system ground. The instantaneous common-mode voltage $v_{cm1}(t)$ and $v_{cm2}(t)$ at the machine terminals (A_1, B_1, C_1) and (A_2, B_2, C_2) are defined by (3) and (4) respectively.

$$\vec{v}_{o1}(t) = v_{A1}(t) + v_{B1}(t)e^{j2\pi/3} + v_{C1}(t)e^{j4\pi/3} \quad (1)$$

$$\vec{v}_{o2}(t) = -(v_{A2}(t) + v_{B2}(t)e^{j2\pi/3} + v_{C2}(t)e^{j4\pi/3}) \quad (2)$$

$$v_{cm1}(t) = \frac{v_{A1}(t) + v_{B1}(t) + v_{C1}(t)}{3} \quad (3)$$

$$v_{cm2}(t) = \frac{v_{A2}(t) + v_{B2}(t) + v_{C2}(t)}{3} \quad (4)$$

In a 3x3 matrix converter drive, machine terminals (A_1, B_1, C_1) can be connected to the grid terminals (a, b, c) in 27 different switching combinations. Switching combinations that results in zero common-mode voltage at the machine terminals (A_1, B_1, C_1) are shown in Table I [5], [13]. These six switching combinations correspond to six instantaneous rotating space vectors used in the modulation of MC1. Using (1), expressions for rotating space vectors are given in Table I. As an example, when the machine terminals (A_1, B_1, C_1) are connected to the grid terminals (b, c, a), the output voltage space vector $\vec{v}_{o1}(t)$ is given by: $\vec{v}_{o1}(t) = \vec{v}_{bca}(t) = v_b(t) + v_c(t)e^{j2\pi/3} + v_a(t)e^{j4\pi/3} = \frac{3}{2}\hat{V}_i e^{j(\omega_i t - 2\pi/3)}$. Based on the direction of rotation in the complex plane the rotating space vectors are grouped as counter-clockwise (CCW)-rotating and clockwise (CW)-rotating vectors. Similar observations can be made for zero common-mode voltage at the machine terminals (A_2, B_2, C_2). Combining $\vec{v}_{o1}(t)$ and $\vec{v}_{o2}(t)$, the instantaneous voltage space vector $\vec{v}_o(t)$ across the open-end winding can be written as: $\vec{v}_o(t) = \vec{v}_{o1}(t) + \vec{v}_{o2}(t)$.

At any instant, when the matrix converters are modulated using CCW-rotating vectors, $\vec{v}_{o1}(t)$ can be equal to any one of the CCW-rotating vectors $\vec{v}_{abc}(t)$, $\vec{v}_{bca}(t)$ and $\vec{v}_{cab}(t)$. Similarly, at any instant, $\vec{v}_{o2}(t)$ can be equal to any one of the voltage space vector $-\vec{v}_{abc}(t)$, $-\vec{v}_{bca}(t)$ and $-\vec{v}_{cab}(t)$. Based on the combinations of $\vec{v}_{o1}(t)$ and $\vec{v}_{o2}(t)$, Fig. 2 shows the resultant voltage space vector across the open-end winding. There are six active vectors $\vec{v}_{p1}(t),..,\vec{v}_{p6}(t)$ rotating in CCW

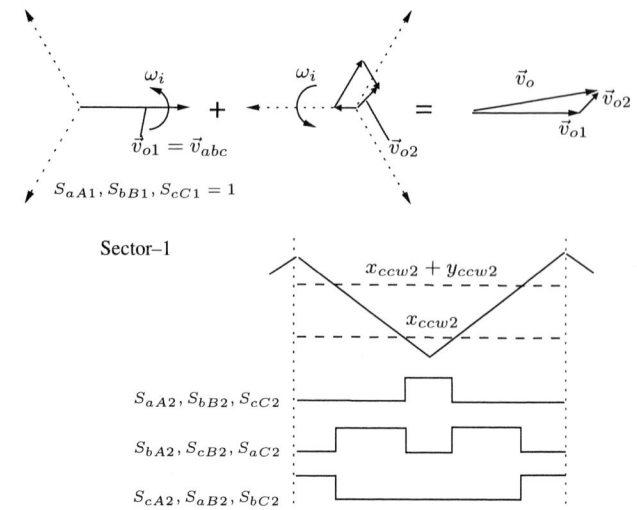

Fig. 4. PWM in sector–1 using only CCW-rotating vectors

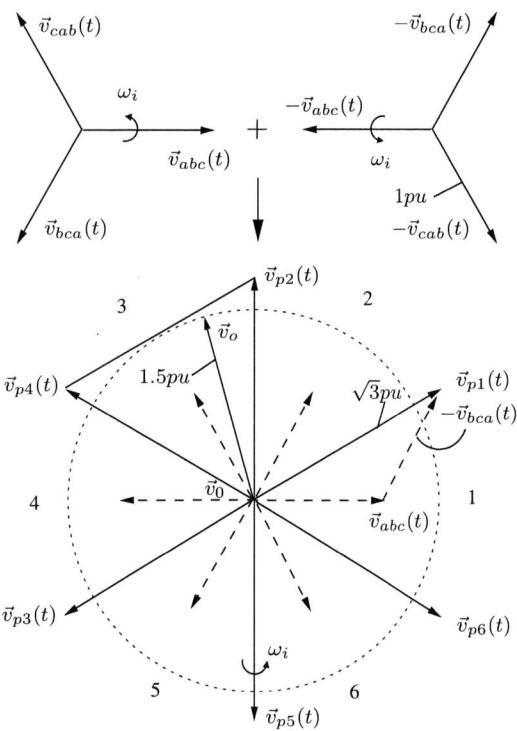

Fig. 2. CCW rotating vectors across the open-end winding

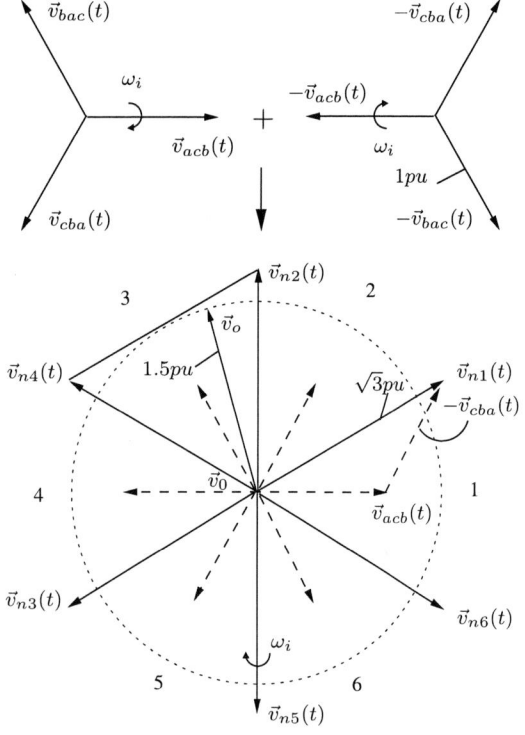

Fig. 3. CW rotating vectors across the open-end winding

direction and three zero vectors indicated as $\vec{v}_0$ [5], [20]. In each sector, output voltage is synthesized using two adjacent active vectors and a zero vector. Likewise, as shown in Fig. 3, if the matrix converters are modulated using CW-rotating vectors, six active vectors $\vec{v}_{n1}(t),..,\vec{v}_{n6}(t)$ rotating in CW direction and three zero vectors result across the open-end winding. From Fig. 2, it can be observed that the magnitude of $\vec{v}_o$ is limited to 1.5 times the magnitude of $\vec{v}_{abc}(t)$ and therefore, the maximum value of converter voltage gain q is $q_{\max} = 1.5$. Similar conclusion can be drawn from Fig. 3. Hence, the matrix converters in the open-end winding drive can be modulated using either CCW-rotating or CW-rotating vectors, for common-mode voltage elimination, with $q_{\max} = 1.5$.

A. Matrix converters modulated with CCW rotating vector

In a switching time period T_s, the average voltage space vector across the open-end winding can be written as: $\vec{v}_o = \vec{v}_{o1} + \vec{v}_{o2}$, where $\vec{v}_{o1}$ and $\vec{v}_{o2}$ are the average values of $\vec{v}_{o1}(t)$ and $\vec{v}_{o2}(t)$ in T_s respectively. Vectors $\vec{v}_{o1}$ and $\vec{v}_{o2}$ are synthesized using (5) and (6) respectively.

$$\vec{v}_{o1} = x_{ccw1}\vec{v}_{abc} + y_{ccw1}\vec{v}_{bca} + z_{ccw1}\vec{v}_{cab} \tag{5}$$

$$\vec{v}_{o2} = -x_{ccw2}\vec{v}_{abc} - y_{ccw2}\vec{v}_{bca} - z_{ccw2}\vec{v}_{cab} \tag{6}$$

Fig. 4 shows the synthesis of output voltage vector $\vec{v}_o$ in sector–1. In this sector, the references for $\vec{v}_{o1}$ and $\vec{v}_{o2}$ are chosen as $\vec{v}_{abc}(t)$ and $\vec{v}_o(t) - \vec{v}_{abc}(t)$ respectively. Therefore, in sector–1, MC1 is clamped to a fixed switching state $S_{aA1} = S_{bB1} = S_{cC1} = 1$ to generate $\vec{v}_{o1} = \vec{v}_{abc}(t)$ and MC2 is modulated using (6) to generate $\vec{v}_{o2} = \vec{v}_o(t) - \vec{v}_{abc}(t)$. On the other hand, in sector–2, MC2 is clamped to a fixed switching state to generate $\vec{v}_{o2} = -\vec{v}_{bca}(t)$ whereas MC1 is modulated using (5) to generate $\vec{v}_{o1} = \vec{v}_o(t) - (-\vec{v}_{bca}(t))$. Table II shows the references for $\vec{v}_{o1}$ and $\vec{v}_{o2}$ in each sector. Referring to Fig. 1 and 4, the four-quadrant switch S_{aA1} is ON for $S_{aA1} = 1$ and OFF for $S_{aA1} = 0$. Similarly, status of other four-quadrant

TABLE II
REFERENCE VOLTAGE VECTORS IN SPACE VECTOR PWM WITH ONLY CCW ROTATING VECTORS

sector	$\vec{v}_{o1}$	$\vec{v}_{o2}$	ϕ_v
1	$\vec{v}_{abc}(t)$	$\vec{v}_o(t) - \vec{v}_{abc}(t)$	0
2	$\vec{v}_o(t) - (-\vec{v}_{bca}(t))$	$-\vec{v}_{bca}(t)$	$-2\pi/3$
3	$\vec{v}_{cab}(t)$	$\vec{v}_o(t) - \vec{v}_{cab}(t)$	$2\pi/3$
4	$\vec{v}_o(t) - (-\vec{v}_{abc}(t))$	$-\vec{v}_{abc}(t)$	0
5	$\vec{v}_{bca}(t)$	$\vec{v}_o(t) - \vec{v}_{bca}(t)$	$-2\pi/3$
6	$\vec{v}_o(t) - (-\vec{v}_{cab}(t))$	$-\vec{v}_{cab}(t)$	$2\pi/3$

TABLE III
REFERENCE VOLTAGE VECTORS IN SPACE VECTOR PWM WITH ONLY CW ROTATING VECTORS

sector	$\vec{v}_{o1}$	$\vec{v}_{o2}$	ϕ_v
1	$\vec{v}_{acb}(t)$	$\vec{v}_o(t) - \vec{v}_{acb}(t)$	0
2	$\vec{v}_o(t) - (-\vec{v}_{cba}(t))$	$-\vec{v}_{cba}(t)$	$-2\pi/3$
3	$\vec{v}_{bac}(t)$	$\vec{v}_o(t) - \vec{v}_{bac}(t)$	$2\pi/3$
4	$\vec{v}_o(t) - (-\vec{v}_{acb}(t))$	$-\vec{v}_{acb}(t)$	0
5	$\vec{v}_{cba}(t)$	$\vec{v}_o(t) - \vec{v}_{cba}(t)$	$-2\pi/3$
6	$\vec{v}_o(t) - (-\vec{v}_{bac}(t))$	$-\vec{v}_{bac}(t)$	$2\pi/3$

switches are defined.

$$x_{ccw1} = \frac{1}{3} + \frac{2}{3}\cos(\phi_v) + \frac{2}{3}m\cos(\omega_o - \omega_i)t$$
$$y_{ccw1} = \frac{1}{3} + \frac{2}{3}\cos(\phi_v + \frac{2\pi}{3}) + \frac{2}{3}m\cos((\omega_o - \omega_i)t + \frac{2\pi}{3})$$
$$z_{ccw1} = \frac{1}{3} + \frac{2}{3}\cos(\phi_v - \frac{2\pi}{3}) + \frac{2}{3}m\cos((\omega_o - \omega_i)t - \frac{2\pi}{3})$$
$$\rightarrow \text{for sector} - 2, 4, 6 \tag{7}$$

$$x_{ccw2} = \frac{1}{3} + \frac{2}{3}\cos(\phi_v) - \frac{2}{3}m\cos(\omega_o - \omega_i)t$$
$$y_{ccw2} = \frac{1}{3} + \frac{2}{3}\cos(\phi_v + \frac{2\pi}{3}) - \frac{2}{3}m\cos((\omega_o - \omega_i)t + \frac{2\pi}{3})$$
$$z_{ccw2} = \frac{1}{3} + \frac{2}{3}\cos(\phi_v - \frac{2\pi}{3}) - \frac{2}{3}m\cos((\omega_o - \omega_i)t - \frac{2\pi}{3})$$
$$\rightarrow \text{for sector} - 1, 3, 5 \tag{8}$$

The variables x_{ccw1}, y_{ccw1} and z_{ccw1} are the fraction of time T_s for which vectors $\vec{v}_{abc}(t)$, $\vec{v}_{bca}(t)$ and $\vec{v}_{cab}(t)$ are applied, and the variables x_{ccw2}, y_{ccw2} and z_{ccw2} are the fraction of time T_s for which vectors $-\vec{v}_{abc}(t)$, $-\vec{v}_{bca}(t)$ and $-\vec{v}_{cab}(t)$ are applied at the machine terminals. Solving (5) and (6), expressions for these variables generalized for any sector are given by (7) and (8) where $0 \leq m \leq 1.5$. Based on the sector information, the value of ϕ_v is taken from Table II. The sector information is used to clamp one of the matrix converters to a fixed switching state whereas triangle carrier comparison (as shown in Fig. 4) is used to generate the modulation signals for the other matrix converter modulated using (5) or (6). The range of $(\omega_o - \omega_i)t$ is different in each sector. For example in sector–1 it is equal to $-\pi/6 \leq (\omega_o - \omega_i)t \leq \pi/6$. It can be verified by applying $-\pi/6 \leq (\omega_o - \omega_i)t \leq \pi/6$ and $0 \leq m \leq 1.5$ that x_{ccw2}, y_{ccw2}, z_{ccw2} etc. are bounded in the region $0 \leq x_{ccw2}, y_{ccw2}, z_{ccw2} \leq 1$. Similarly verification in other sectors can be shown.

B. Matrix converters modulated with CW rotating vectors

Analysis similar to presented above can be performed in the modulation of matrix converters using only CW-rotating vectors. The modulation functions are given by (9)–(12) where $0 \leq m \leq 1.5$. Based on the location of output voltage vector $\vec{v}_o(t)$, the references for $\vec{v}_{o1}$ and $\vec{v}_{o2}$ are given in Table III.

$$\vec{v}_{o1} = x_{cw1}\vec{v}_{acb} + y_{cw1}\vec{v}_{bac} + z_{cw1}\vec{v}_{cba} \tag{9}$$
$$\vec{v}_{o2} = -x_{cw2}\vec{v}_{acb} - y_{cw2}\vec{v}_{bac} - z_{cw2}\vec{v}_{cba} \tag{10}$$
$$x_{cw1} = \frac{1}{3} + \frac{2}{3}\cos(\phi_v) + \frac{2}{3}m\cos(\omega_o + \omega_i)t$$
$$y_{cw1} = \frac{1}{3} + \frac{2}{3}\cos(\phi_v - \frac{2\pi}{3}) + \frac{2}{3}m\cos((\omega_o + \omega_i)t - \frac{2\pi}{3})$$
$$z_{cw1} = \frac{1}{3} + \frac{2}{3}\cos(\phi_v + \frac{2\pi}{3}) + \frac{2}{3}m\cos((\omega_o + \omega_i)t + \frac{2\pi}{3})$$
$$\rightarrow \text{for sector} - 2, 4, 6 \tag{11}$$
$$x_{cw2} = \frac{1}{3} + \frac{2}{3}\cos(\phi_v) - \frac{2}{3}m\cos(\omega_o + \omega_i)t$$
$$y_{cw2} = \frac{1}{3} + \frac{2}{3}\cos(\phi_v - \frac{2\pi}{3}) - \frac{2}{3}m\cos((\omega_o + \omega_i)t - \frac{2\pi}{3})$$
$$z_{cw2} = \frac{1}{3} + \frac{2}{3}\cos(\phi_v + \frac{2\pi}{3}) - \frac{2}{3}m\cos((\omega_o + \omega_i)t + \frac{2\pi}{3})$$
$$\rightarrow \text{for sector} - 1, 3, 5 \tag{12}$$

From Table II and III it is observed that in any sector only one matrix converter is switching. Therefore, the number of switching transitions and mapping requirement are minimized. It is also observed that in the expressions (7–8), (11-12), many trigonometric identities are common. Hence, the overall computational requirement in implementation of theses expressions is minimized.

III. ANALYSIS OF GRID CURRENT

The three-phase open-end winding machine under balanced loading condition can be considered as a three-phase balanced current source. Therefore, corresponding to six rotating voltage space vectors at the machine terminals (Table. I), there are six rotating current space vectors at the input of the matrix converter [5], [21]. As an example, when the machine terminals (A_1, B_1, C_1) are connected to the grid terminals (a, b, c), the input current space vector $\vec{i}_{s1}(t)$ for MC1 is given by: $\vec{i}_{s1}(t) = \vec{i}_{ABC}(t) = i_A(t) + i_B(t)e^{j2\pi/3} + i_C(t)e^{j4\pi/3} = \frac{3}{2}\hat{I}_o e^{j(\omega_o t - \rho)}$, where $i_A(t)$, $i_B(t)$ and $i_C(t)$ are the machine phase currents with an amplitude $\hat{I}_o$ and frequency ω_o; ρ is the machine power factor angle. Combining MC1 and MC2, at any instant the grid current space vector can be written as: $\vec{i}_s(t) = \vec{i}_{s1}(t) + \vec{i}_{s2}(t)$ where $\vec{i}_{s2}(t)$ is the input current space vector for MC2. The average grid current space vector in a switching time period is expressed as: $\vec{i}_s = \vec{i}_{s1} + \vec{i}_{s2}$ where $\vec{i}_{s1}$ and $\vec{i}_{s2}$ are the average values of $\vec{i}_{s1}(t)$ and $\vec{i}_{s2}(t)$ in a switching time period T_s.

Referring to Fig. 2, if the matrix converters are modulated using only CCW-rotating vectors then for example in sector–1, $\vec{v}_{o1} = \vec{v}_{abc}$ and $\vec{v}_{o2} = \vec{v}_o - \vec{v}_{abc} = -x_{ccw2}\vec{v}_{abc} - y_{ccw2}\vec{v}_{bca} - z_{ccw2}\vec{v}_{cab}$. Therefore, the input current space vectors to the matrix converters in sector–1 can be expressed as: $\vec{i}_{s1} = \vec{i}_{ABC}$ and $\vec{i}_{s2} = -x_{ccw2}\vec{i}_{ABC} - y_{ccw2}\vec{i}_{CAB} - z_{ccw2}\vec{i}_{BCA}$. Thus, the resultant grid current space vector $\vec{i}_s$ is given by (13) where $0 \leq m \leq 1.5$.

$$
\begin{aligned}
\vec{i}_s &= \vec{i}_{s1} + \vec{i}_{s2} \\
&= (1 - x_{ccw2})\vec{i}_{ABC} - y_{ccw2}\vec{i}_{CAB} - z_{ccw2}\vec{i}_{BCA} \\
&= \vdots \\
\vec{i}_s &= \frac{3}{2}m\hat{I}_o e^{j(\omega_i t - \rho)}
\end{aligned}
\tag{13}
$$

Similarly, the expressions for grid current space vector in other sectors can be derived and it is verified that $\vec{i}_s$ is the same in all sector as given by (13). It is observed from (13) that the grid power factor is lagging and it is equal to the machine power factor $\cos\rho$. Hence, the grid power factor is fixed and it can not be controlled to any other value. Similarly, if the matrix converters are modulated using only CW-rotating vectors it is observed that the grid power factor is leading and it is equal to $-\cos\rho$. The expression for grid current space vector in sector–1 (Fig. 3) is derived and given by (14). It can be observed from (14) as well that the grid power factor is not controllable to any other value.

$$
\begin{aligned}
\vec{i}_s &= \vec{i}_{s1} + \vec{i}_{s2} \\
&= (1 - x_{cw2})\vec{i}_{ACB} - y_{cw2}\vec{i}_{BAC} - z_{cw2}\vec{i}_{CBA} \\
&= \vdots \\
\vec{i}_s &= \frac{3}{2}m\hat{I}_o e^{j(\omega_i t + \rho)}
\end{aligned}
\tag{14}
$$

IV. GRID POWER FACTOR CONTROL

In order to control the grid power factor both CCW and CW-rotating vectors are used in a switching time period to generate the resultant output vector $\vec{v}_o$. This is achieved by applying CCW-rotating vectors at both ends of the open-end winding followed by CW-rotating vectors. If $\vec{v}_{ccw}$ and $\vec{v}_{cw}$ are the average voltage vectors across the open-end winding in T_s then the $\vec{v}_o$ can be expressed as 15. Similarly, the resultant grid current space vector can be expressed as (16) where, $\vec{i}_{ccw}$ and $\vec{i}_{cw}$ are the average grid current space vectors resulting from the application of CCW and CW-rotating voltage space vectors across the open-end winding.

$$
\vec{v}_o = \vec{v}_{ccw} + \vec{v}_{cw} \tag{15}
$$
$$
\vec{i}_s = \vec{i}_{ccw} + \vec{i}_{cw} \tag{16}
$$

Let us assume $\vec{v}_{ccw} = x\frac{3}{2}m\hat{V}_i e^{j\omega_o t}$ and $\vec{v}_{cw} = (1 - x)\frac{3}{2}m\hat{V}_i e^{j\omega_o t}$ where $0 \leq x \leq 1$ and $0 \leq m \leq 1.5$. Substituting $\vec{v}_{ccw}$ and $\vec{v}_{cw}$ in (15), the resultant voltage space vector across the open-end winding is given by (17). Using (13) and (14) the expression for grid current vector is given by (18) where $r = \sqrt{\cos^2\rho + (1 - 2x)^2\sin^2\rho}$ and $\phi = \tan^{-1}((1 - 2x)\tan\rho)$.

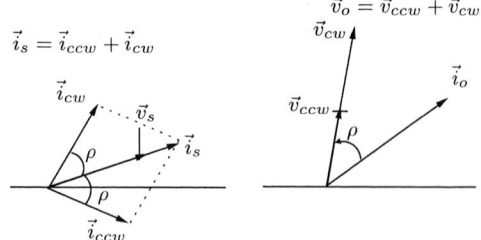

Fig. 5. Grid power factor control

From (18) it is observed that by varying x in the range $0 \leq x \leq 1$, the grid power power factor can be controlled in the range $|\cos\rho| \leq |\cos\phi| \leq 1$. Also, it can be observed from (17) that the location of $\vec{v}_o$ in the complex plane as well as its magnitude is independent of x. Hence, the converter voltage gain is independent of the grid power factor $\cos\phi$. Therefore, in the grid power factor range $|\cos\rho| \leq |\cos\phi| \leq 1$, the converter voltage gain up to 1.5 is possible. Comparing with a 3x3 matrix converter drive modulated with conventional space vector PWM technique [22], the converter voltage gain is limited to $0.866 \times \cos\phi$ for a grid power factor $\cos\phi$. It should be noted that the grid power factor control range is limited to $|\cos\rho| \leq |\cos\phi| \leq 1$ in the proposed PWM scheme. However, control in the range $0 \leq |\cos\phi| \leq |\cos\rho|$ at the expense of decrease in converter voltage gain can be achieved as described in [5], [20]. The proposed PWM technique is similar to reported in [23] for a 3x3 matrix converter drive. However in [23], elimination of common-mode voltage was not considered.

$$
\begin{aligned}
\vec{v}_o &= \vec{v}_{ccw} + \vec{v}_{cw} \\
&= x\frac{3}{2}m\hat{V}_i e^{j\omega_o t} + (1 - x)\frac{3}{2}m\hat{V}_i e^{j\omega_o t} \\
&= \frac{3}{2}m\hat{V}_i e^{j\omega_o t} \\
\vec{i}_s &= \vec{i}_{ccw} + \vec{i}_{cw} \\
&= x\frac{3}{2}m\hat{I}_o e^{j(\omega_i t - \rho)} + (1 - x)\frac{3}{2}m\hat{I}_o e^{j(\omega_i t + \rho)} \\
&= \frac{3}{2}mr\hat{I}_o e^{j(\omega_i t + \phi)}
\end{aligned}
\tag{17}
$$
$$
\tag{18}
$$

With the help of space vectors, Fig. 5 shows the output voltage synthesis for unity grid power factor (at $x = 0.5$). At $x = 0.5$, half of the desired voltage vector $\vec{v}_o$ is synthesized by CCW-rotating vectors and the remaining half is synthesized by CW-rotating vectors. The grid current vector $\vec{i}_{ccw}$ corresponding to $\vec{v}_{ccw}$ lags the grid voltage space vector $\vec{v}_s = \vec{v}_{abc}$ by machine power factor angle ρ, whereas $\vec{i}_{cw}$ corresponding to $\vec{v}_{cw}$ leads $\vec{v}_s$ by angle ρ. Hence, the resultant grid current space vector $\vec{i}_s$ is in phase with $\vec{v}_s$. Fig. 6 shows the output voltage synthesis at $t = 0$ when grid phase-a voltage $v_a(t)$ is peaking. At $t = 0$, $\vec{v}_o$ is assumed to be in sector–2 with respect to both CCW and CW-rotating vectors. The voltage space vectors shown by solid lines are the average voltage space vector in T_s. Modulation signals corresponding to the output voltage synthesis described

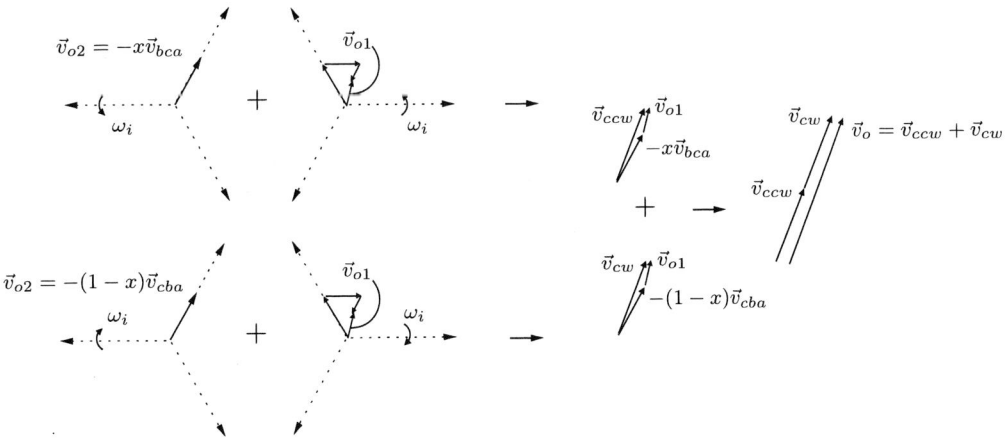

Fig. 6. Example of output voltage synthesis

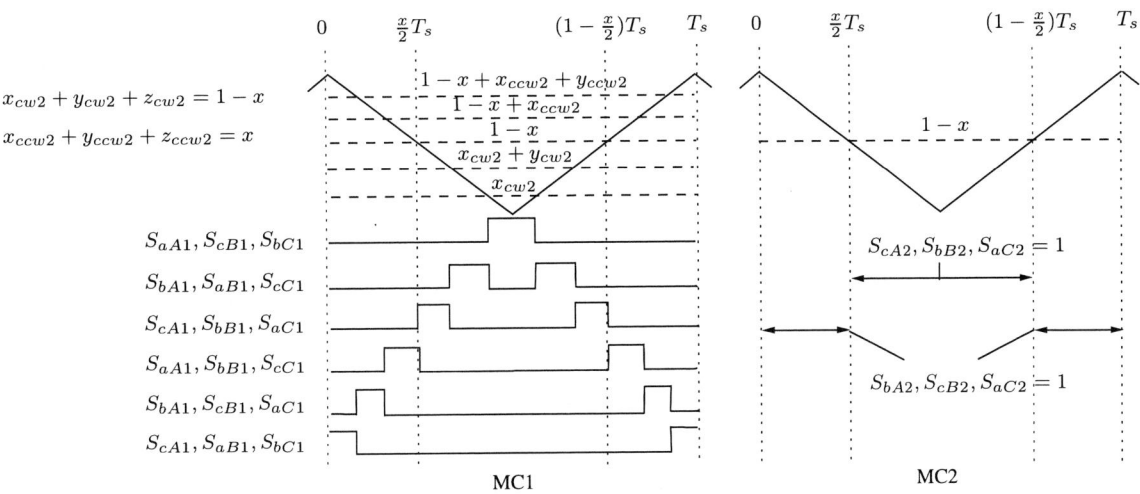

Fig. 7. Modulation signals for the matrix converters corresponding to Fig. 6

in Fig. 6 are shown in Fig. 7. CCW-rotating vectors are applied during $0 \leq t \leq \frac{x}{2}T_s$ and $(1-\frac{x}{2})T_s \leq t \leq T_s$ and CW-rotating vectors are applied during $\frac{x}{2}T_s \leq t \leq (1-\frac{x}{2})T_s$. Hence, the matrix converters are modulated with CW and CCW-rotating vectors for time $(1-x)T_s$ and xT_s in a switching time period T_s. Referring to Fig. 6 and Fig. 7,

- $0 \leq t \leq \frac{x}{2}T_s$ and $(1-\frac{x}{2})T_s \leq t \leq T_s$: From Table II, the references for $\vec{v}_{o1}$ and $\vec{v}_{o2}$ in sector–2 are $\frac{3}{2}m\hat{V}_ie^{j\omega_ot} + \vec{v}_{bca}(t)$ and $-\vec{v}_{bca}(t)$ respectively. The matrix converter MC2 is clamped to a fixed state $S_{bA2}, S_{cB2}, S_{aC2} = 1$ and the matrix converter MC1 is modulated using (7).

- $\frac{x}{2}T_s \leq t \leq (1-\frac{x}{2})T_s$: From Table III, the references for $\vec{v}_{o1}$ and $\vec{v}_{o2}$ in sector–2 are $\frac{3}{2}m\hat{V}_ie^{j\omega_ot} + \vec{v}_{cba}(t)$ and $-\vec{v}_{cba}(t)$ respectively. The matrix converter MC2 is clamped to a fixed state $S_{cA2}, S_{bB2}, S_{aC2} = 1$ and the matrix converter MC1 is modulated using (11).

Modulation equations (7) and (11) are multiplied by x and $(1-x)$ respectively such that in T_s, $x_{cw} + y_{cw} + z_{cw} + x_{ccw} + y_{ccw} + z_{ccw} = 1$. Hence, to generate the gate signals for 18 four quadrant switches five triangle-carrier comparisons are

required along with the location of $\vec{v}_o$ with respect to CCW and CW-rotating vectors.

V. EXPERIMENTAL RESULTS

The experimental setup of a laboratory prototype of the open-end winding drive system is shown in Fig. 8. The matrix converters are constructed using six *Microsemi* IGBT module APTGT50TDU60PG. The proposed space vector PWM strategy is implemented on a ds1103 dSPACE control platform which communicates in real-time with control desk for the desired control task. The dSPACE system sends multiplexed gate signals and sector information to a Xilinx Spartan-3 FPGA board. Demultiplex logic to generate gate signals for 18 four-quadrant switches used in the matrix converters MC1 and MC2 is implemented on the FPGA system. The commutation and protection logic of the matrix converters are also implemented on the FPGA system. For a simplified commutation control, the commutation strategy uses a dead-time of $1.6\mu s$ between each switching transition. Therefore, during the dead-time machine terminals are connected to

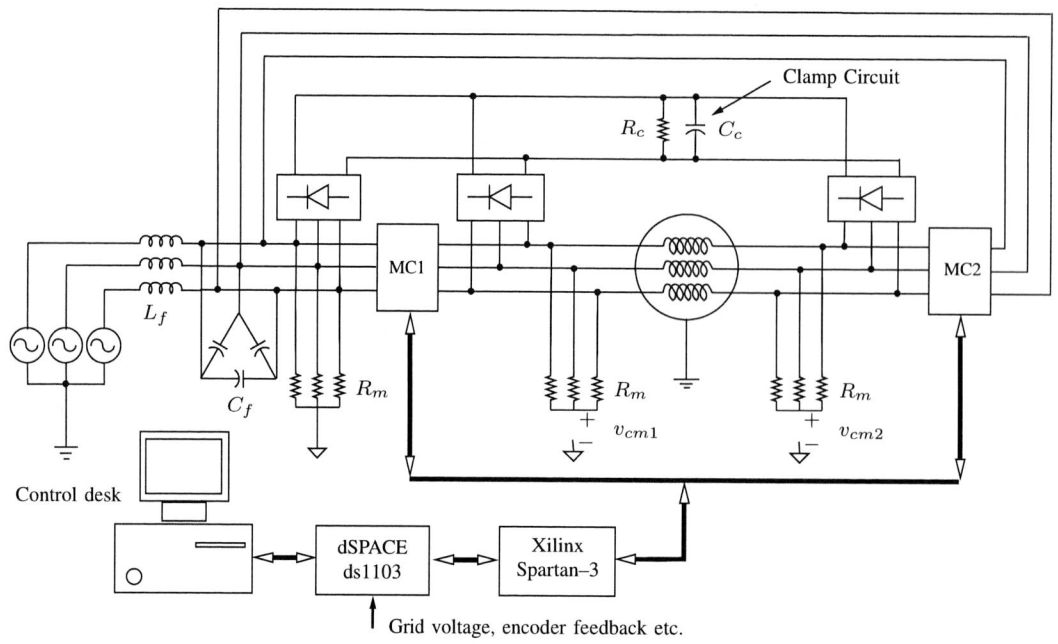

Fig. 8. Experimental Setup

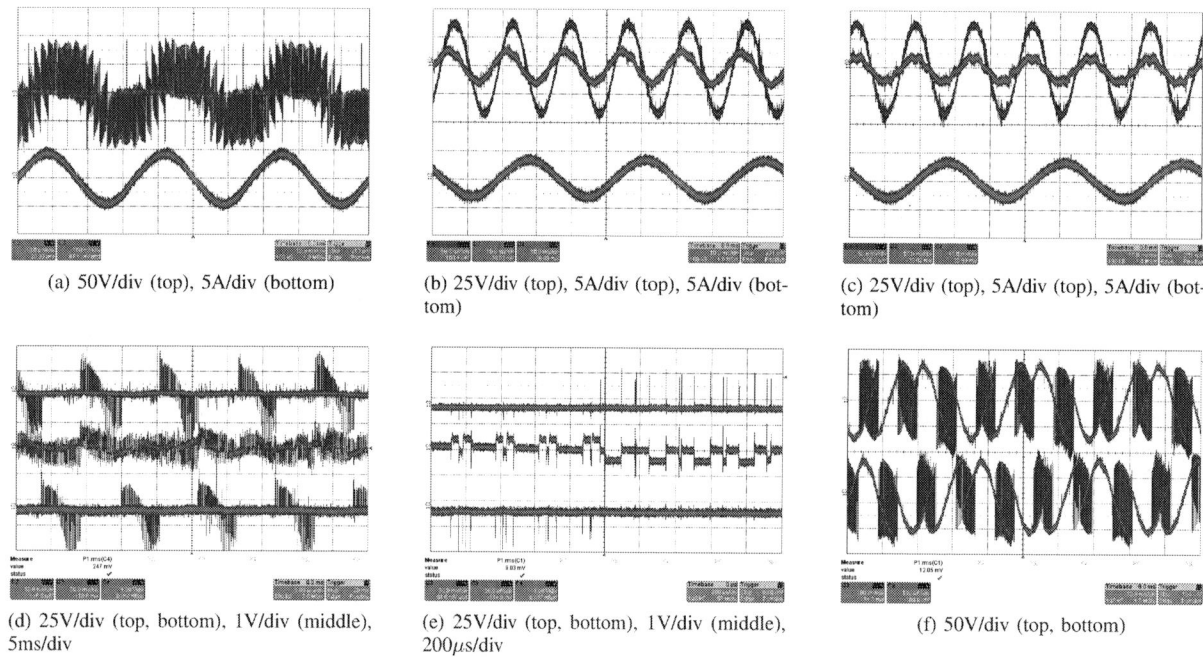

(a) 50V/div (top), 5A/div (bottom)

(b) 25V/div (top), 5A/div (top), 5A/div (bottom)

(c) 25V/div (top), 5A/div (top), 5A/div (bottom)

(d) 25V/div (top, bottom), 1V/div (middle), 5ms/div

(e) 25V/div (top, bottom), 1V/div (middle), 200μs/div

(f) 50V/div (top, bottom)

Fig. 9. Experimental Results: (a) Output phase–A voltage (top) and phase–A current (bottom) for leading grid power factor, (b) Leading grid power factor, grid phase–a voltage and phase–a current (top), load phase–A current (bottom), (c) Lagging grid power factor and (d) Unity grid power factor

the clamp circuit. The value of passive components used in the experimental setup are as follows: $L_f = 1.1\ mH$, $C_f = 40\ \mu F$, $f_s(1/T_s) = 4\ kHz$, $R_c = 410\ \Omega$, $C_c = 37\ \mu F$ and $R_m = 18\ k\Omega$. To measure the common-mode voltage at the machine terminals, star terminals are created using resistors R_m as shown in Fig. 8. For the demonstration of grid power factor control, the open-end winding drive is loaded with an RL load ($R = 6\Omega, L = 45mH$) of fixed $\rho = 54.7°$ at 30Hz. The grid phase-phase voltage is fixed at a low voltage of $65V$ (peak). The experimental waveforms at $m = 1.1$ and $x = 0$ for 30Hz output are shown in Fig. 9(a)–(b). As expected at $x = 0$, leading grid power factor can be observed from Fig. 9(b). For the same loading condition (at $m = 1.1$), by changing x, the waveforms for unity grid power factor are shown in Fig. 9(c).

To demonstrate the elimination of common-mode voltage a three-phase open-end winding induction motor is used. A three-phase dual voltage 230 V/480 V, 12 leads, 60 Hz induction motor is configured to use as an open-end winding machine. The RMS value of phase-phase 60 Hz grid voltage is fixed at 90 V. The matrix converters are modulated for $x = 1$ to drive the induction motor at no-load operating condition with constant volt-hertz control. The RMS value of machine phase-phase voltage at 30 Hz is required to be $\cong 115$ V which is achieved by setting $m = 1.3$. The common-mode voltages v_{cm1} and v_{cm2} at both ends of the motor windings are shown in Fig. 9(d)(top,bottom). The shaft voltage measured between the rotor-shaft and motor frame is also shown in Fig. 9(d)(middle). Fig. 9(d) is expanded to observe the waveforms in switching time period range as shown in Fig. 9(e). It can be observed that the common-mode voltage is zero in the entire switching time period except at the switching transitions. Also, the transitions in the shaft voltage and common-mode voltages happen at the same instant. These transition correspond to the commutation of the matrix converters from one switching state to another switching state during which the machine terminals are connected to the clamp circuit. Therefore, the sum of the terminal voltages is not zero. The motor terminal voltages v_{A1} and v_{A2} measured with respect to system ground are shown in Fig. 9(f). The peak values of terminal voltages are limited to the amplitude of grid phase voltage. However, during the switching transitions higher terminal voltages are observed due to the connection of machine terminals to the clamp circuit. Unlike the dead-time based commutation scheme used in the present experimental setup, if the current-direction based commutation schemes such as those summarized in [4] are used then during switching transitions, the machine terminals will not be connected to the clamp circuit. However, due to practical differences in the switching characteristic of the IGBTs finite common-mode voltage during switching transitions is expected.

VI. Conclusion

A space vector PWM technique for an open-end winding ac machine drive driven by matrix converters is described in this paper. Several simultaneous benefits are illustrated such as the elimination of switching common-mode voltage at the machine terminals, and the grid power factor control. Implementation of the proposed PWM strategy is practical which is verified by its implementation on a standard control platform ds1103 dSPACE. Experimental results presented verify the operation and advanced features of the drive.

References

[1] T. Ericsen, N. Hingorani, and Y. Khersonsky, "Power electronics and future marine electrical systems," *Industry Applications, IEEE Transactions on*, vol. 42, no. 1, pp. 155–163, Jan.-Feb. 2006.

[2] L. Gyugyi and B. Pelly, *Static power frequency changers: theory, performance, and application*. John Wiley & Sons, 1976.

[3] A. Alesina and M. Venturini, "Analysis and design of optimum amplitude nine-switch direct ac-ac converters," *IEEE Trans. Power Electron.*, vol. 4, pp. 101–112, Jan. 1989.

[4] P. Wheeler, J. Rodriguez, J. Clare, L. Empringham, and A. Weinstein, "Matrix converters: a technology review," *IEEE Transactions on industrial electronics*, vol. 49, no. 2, pp. 276–288, 2002.

[5] K. K. Mohapatra and N. Mohan, "Open-end winding induction motor driven with matrix converter for common-mode elimination," in *Proc. IEEE International Conference on Power Electronics, Drives and Energy Systems (PEDES06)*, New Delhi, India, Dec. 2006, pp. 1–6.

[6] S. Chen, T. Lipo, and D. Fitzgerald, "Source of induction motor bearing currents caused by PWM inverters," *Energy conversion, IEEE transactions on*, vol. 11, no. 1, pp. 25–32, Mar 1996.

[7] G. Skibinski, R. Kerkman, and D. Schlegel, "EMI emissions of modern PWM AC drives," *Industry Applications Magazine, IEEE*, vol. 5, no. 6, pp. 47–80, Nov/Dec 1999.

[8] H. J. Cha and P. Enjeti, "An approach to reduce common-mode voltage in matrix converter," *Industry Applications, IEEE Transactions on*, vol. 39, no. 4, pp. 1151–1159, July-Aug. 2003.

[9] M. Cacciato, A. Consoli, G. Scarcella, and A. Testa, "Reduction of common-mode currents in PWM inverter motor drives," *Industry Applications, IEEE Transactions on*, vol. 35, no. 2, pp. 469–476, Mar/Apr 1999.

[10] H. Zhang, A. Von Jouanne, S. Dai, A. Wallace, and F. Wang, "Multilevel inverter modulation schemes to eliminate common-mode voltages," *Industry Applications, IEEE Transactions on*, vol. 36, no. 6, pp. 1645–1653, Nov/Dec 2000.

[11] A. Muetze, "Scaling Issues for Common-Mode Chokes to Mitigate Ground Currents in Inverter-Based Drive Systems," *IEEE Transactions on Industry Applications*, vol. 45, no. 1, pp. 286–294, 2009.

[12] M. Swamy, K. Yamada, T. Kume, Y. America, and I. Waukegan, "Common mode current attenuation techniques for use with PWM drives," *IEEE Transactions on Power Electronics*, vol. 16, no. 2, pp. 248–255, 2001.

[13] R. K. Gupta, K. K. Mohapatra, and N. Mohan, "A Novel Three-Phase, Switched Multi-Winding Power Electronic Transformer," in *Proc. IEEE Energy Conversion Congress and Exposition (ECCE) 2009*, San Jose, CA, Sep. 2009.

[14] M. Braun and K. Hasse, "A direct frequency changer with control of input reactive power," in *IFAC Control in Power Electronics and Electrical Drives*, Lausanne, Switzerland, 1983, pp. 187–194.

[15] M. Baiju, K. Mohapatra, R. Kanchan, and K. Gopakumar, "A dual two-level inverter scheme with common mode voltage elimination for an induction motor drive," *Power Electronics, IEEE Transactions on*, vol. 19, no. 3, pp. 794–805, May 2004.

[16] H. Stemmler and P. Guggenbach, "Configuration of high power voltage source inverter drives," in *Proc. EPE'93 conference*, Brighton, U.K., 1993, pp. 7–12.

[17] Y. Kawabata, M. Nasu, T. Nomoto, E. Ejiogu, and T. Kawabata, "High-efficiency and low acoustic noise drive system using open-winding ac motor and two space-vector-modulated inverters," *Industrial Electronics, IEEE Transactions on*, vol. 49, no. 4, pp. 783–789, Aug 2002.

[18] V. Somasekhar, K. Gopakumar, M. Baiju, K. Mohapatra, and L. Umanand, "A multilevel inverter system for an induction motor with open-end windings," *Industrial Electronics, IEEE Transactions on*, vol. 52, no. 3, pp. 824–836, June 2005.

[19] A. Somani, R. Gupta, K. Satish, K. Mohapatra, and N. Mohan, "A PEBB-based direct-link drive for open-ended AC machines," in *Proceedings of the SCSC 2007*. Society for Computer Simulation International San Diego, CA, USA, 2007, pp. 118–123.

[20] R. K. Gupta, K. K. Mohapatra, and N. Mohan, "Open-End Winding AC Drives with Enhanced Reactive Power Support to the Grid While Eliminating Switching Common-Mode Voltage," in *Proc. Industrial Electronics Society, 2009. IECON'09. The 35th Annual Conference of the IEEE*, Porto, Portugal, Nov. 2009.

[21] K. Mohapatra, R. Gupta, S. Thuta, A. Somani, A. Umarikar, K. Basu, and N. Mohan, "New Research on AC-AC Converters without Intermediate Storage and Their Applications in Power-Electronic Transformers and AC Drives," *IEEJ Transactions on Electrical and Electronic Engineering*, vol. 4, no. 5, 2009.

[22] L. Huber and D. Borojevic, "Space vector modulated three-phase to three-phase matrix converter with input power factor correction," *IEEE Transactions on industry applications*, vol. 31, no. 6, pp. 1234–1246, 1995.

[23] M. Milanovic and B. Dobaj, "Unity input displacement factor correction principle for direct ac to ac matrix converters based on modulation strategy," *IEEE Trans. Circuits Syst. I*, vol. 47, pp. 221–230, Feb. 2000.

Evaluation of the Hybrid Four-level Converter Employing Half-Bridge Modules for Two Different Modulation Schemes

Alessandro L. Batschauer[1], Arnaldo J. Perin[2], Samir A. Mussa[2], and Marcelo L. Heldwein[2],

Santa Catarina State University (UDESC)
Electrical Engineering Department — NPEE
Campus Universitário Prof. Avelino Marcante s/n
Joinville, BRAZIL, 89223-100
batschauer@joinville.udesc.br

Federal University of Santa Catarina (UFSC)
Electrical Engineering Department — INEP
Campus Universitário Trindade, PO box 5119
Florianópolis, BRAZIL, 88040-970
arnaldo.perin; samir; heldwein; @inep.ufsc.br

Abstract—**A novel hybrid three-phase multilevel converter is proposed for medium-voltage applications. The converter employs a conventional three-phase voltage source inverter (VSI) linking series connected half-bridge modules at each phase. With the proposed connection, a large portion of energy can be processed by the VSI without the need for insulation or by employing a single transformer, while smaller power shares are processed within the half-bridge modules. Thus, the requirements for galvanically insulated dc sources are reduced. Modularity is naturally achieved. A modulation scheme for a four-level version is proposed and analyzed in detail. This scheme allows unidirectional power flow in all dc sources and, consequently enables diode bridges to be employed in the rectification input stage for unidirectional applications.**

I. INTRODUCTION

Multilevel topologies are a major part of medium voltage power electronics equipment and research in this field has led to various solutions [2]–[9]. Hybrid multilevel topologies based on the series connection of three-phase VSI or NPC with full-bridge modules (H3phCFB) have been proposed [5], [8], [10]–[12] as alternative to the Cascaded Full-Bridge (CFB) converters [4], [13]. Following the hybridization approach a multilevel converter topology has been introduced in [1]. This hybrid topology (H3phCHB) utilizes a three-phase inverter (cf. Fig. 1) with its output terminals connected in series to a pair, or multiple pairs (cascade), of half-bridge converters connected with inverse polarity.

The demand for medium voltage converters grows based on the lowering costs for the technology and the necessity of areas such as high power drives, renewable energy generation, power quality and naval propulsion. In this context, the search for solutions in this field becomes important as does the need for a careful evaluation of the proposed topologies. This work aims at a thorough evaluation of the four-level converter shown in (cf. Fig. 1) [1] regarding two possible modulation strategies. The first strategy is based on the switching of all turn-off devices at the switching frequency, while in the second modulation strategy only the devices of the half-bridge converters operate at the switching frequency. In the second strategy, the semiconductors of the VSI switch at the fundamental output frequency or do not switch at all for low modulation indexes. These modulation strategies are explained in section II. The computation of current stresses, conduction and switching losses is introduced in section III. Finally a comparison of both schemes regarding current efforts and

losses is performed and experimental results are presented.

II. MODULATION STRATEGIES

Two modulation schemes for the four-level H3phCHB are presented in this section. The first scheme is based on conventional carrier based PWM signals for all switches, here named CONVENTIONAL modulation. The PWM signals generation logic for this scheme is depicted in Fig. 2(b). In the second modulation scheme, named HYBRID modulation (described in details in [1]), the semiconductors of the three-phase VSI switch at low frequency in order to reduce the VSI switching losses. Thus, only the half-bridge modules switch at the switching frequency. Switches S_{jo} and $S_{jo'}$, with $o = A, B, C$ and $j = 1, 2, 3$, are switched in a complementary way. This modulation scheme is further divided into two operation modes *LM* and *HM* depending on the modulation index M and described in the following:

- **$0 \leq M \leq 0.5$ (*LM*):** the VSI has all switches either clamped to the positive or the negative rail. The clamping can be changed at every modulation cycle in order to balance the losses at all semiconductors. The half-bridge modules process all the active power transferred to the load.

- **$M > 0.5$ (*HM*):** each VSI leg switches a single time per modulation period (cf. Fig. 2(c)) and the half-bridge modules handle a smaller power share.

III. SEMICONDUCTOR CURRENT EFFORTS

Based on the modulation strategy presented in section II the methodology for the computation of current efforts in all semiconductors is introduced in the following. The considerations are made for a phase-leg comprising two half-bridge modules and a VSI leg, exemplarily for phase A. Furthermore, the modulation index is limited to $M > 2/(3\sqrt{3})$ and only the HYBRID modulation derivation is shown for the sake of brevity. The same methodology can be applied for the other modulation schemes.

Neglecting the output current ripple, possible high frequency harmonic contents and assuming balanced loads, the phase current is defined as $i_A(\varphi) = I_p \sin(\varphi - \Phi)$, where Φ is the load phase displacement angle limited from 0 to $+\pi/2$ in the analysis (inductive loads). The angles where the modulating function has its values on the limit between two different carriers are defined with $\theta_M = \arcsin\left(\frac{1}{3M}\right)$.

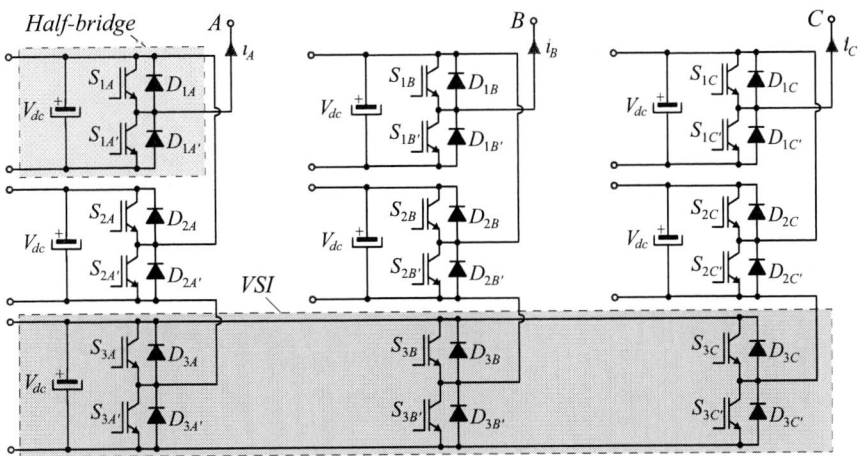

Fig. 1. Circuit schematic of the hybrid four-level converter employing half-bridge modules and a three-phase inverter [1].

Given the modulation pattern *HM*, the duty-cycle for switch S_{1A} is exemplarily given by

$$
d_{S1A}(\varphi) = \begin{cases} \frac{1}{2} + \frac{3}{2}M\sin(\varphi) & ; 0 \leq \varphi \leq \theta_M \\ 1 & ; \theta_M \leq \varphi \leq \pi - \theta_M \\ \frac{1}{2} + \frac{3}{2}M\sin(\varphi) & ; \pi - \theta_M \leq \varphi \leq \pi \\ 1 & ; \pi \leq \varphi \leq \pi + \theta_M \\ \frac{3}{2} + \frac{3}{2}M\sin(\varphi) & ; \pi + \theta_M \leq \varphi \leq 2\pi - \theta_M \\ 1 & ; 2\pi - \theta_M \leq \varphi \leq 2\pi \end{cases}
$$

(1)

for switch S_{2A} by

$$
d_{S2A}(\varphi) = \begin{cases} 0 & ; 0 \leq \varphi \leq \theta_M \\ -\frac{1}{2} + \frac{3}{2}M\sin(\varphi) & ; \theta_M \leq \varphi \leq \pi - \theta_M \\ 0 & ; \pi - \theta_M \leq \varphi \leq \pi \\ \frac{1}{2} + \frac{3}{2}M\sin(\varphi) & ; \pi \leq \varphi \leq \pi + \theta_M \\ 0 & ; \pi + \theta_M \leq \varphi \leq 2\pi - \theta_M \\ \frac{1}{2} + \frac{3}{2}M\sin(\varphi) & ; 2\pi - \theta_M \leq \varphi \leq 2\pi \end{cases}
$$

(2)

and for the VSI switch switch S_{3B} by

$$
d_{S3A}(\varphi) = \begin{cases} 1 & ; 0 \leq \varphi \leq \pi \\ 0 & ; \pi \leq \varphi \leq 2\pi \end{cases}
$$

(3)

For the complementary switches, the duty-cycle is generally defined as $d_{SjA'}(\varphi) = 1 - d_{SjA}(\varphi)$, with $j = 1, 2, 3$.

Thus, the average values for the semiconductor currents are computed with

$$
I_{SjA,avg} = \frac{1}{2\pi} \int_0^{2\pi} i_o(\varphi) \cdot d_{SjA}(\varphi) d\varphi
$$

(4)

$$
I_{DjA',avg} = \frac{1}{2\pi} \int_0^{2\pi} i_o(\varphi) \cdot d_{SjA'}(\varphi) d\varphi,
$$

(5)

and the RMS values with

$$
I_{SjA,rms}^2 = \frac{1}{2\pi} \int_0^{2\pi} d_{SjA}(\varphi) \cdot (i_o(\varphi))^2 d\varphi
$$

(6)

$$
I_{DjA',rms}^2 = \frac{1}{2\pi} \int_0^{2\pi} d_{SjA'}(\varphi) \cdot (i_o(\varphi))^2 d\varphi.
$$

(7)

The integration limits for switches S_{1A} and S_{2A} can be divided into the intervals given in (1). However, a further subdivision is necessary due to the load phase displacement Φ. Therefore, two cases must be analyzed, namely: Case I : $0 < \Phi < \theta_M$ and Case II : $\theta_M < \Phi < \frac{\pi}{2}$. These two cases are irrelevant for the derivations regarding S_{3A} as long as the VSI switches at low frequency. In this case, the average currents across the semiconductors of the three-phase VSI are

$$
I_{S3,avg} = I_{S3',avg} = \frac{Ip}{2\pi}[1 + \cos(\Phi)]
$$

(8)

$$
I_{D3,avg} = I_{D3',avg} = \frac{Ip}{2\pi}[1 - \cos(\Phi)],
$$

(9)

and the RMS values are given by

$$
I_{S3,rms} = I_{S3',rms} = \frac{Ip}{2}\sqrt{\frac{\sin(2\Phi)}{2\pi} + 1 - \frac{\Phi}{\pi}}
$$

(10)

$$
I_{D3,rms} = I_{D3',rms} = \frac{Ip}{2}\sqrt{\frac{\Phi - \sin(2\Phi)}{\pi}}.
$$

(11)

The integration of the local average/rms values of the semiconductors current lead to involved expressions for the devices of the half-bridge modules. However, normalizing the current values by dividing them by the current peak value I_p and plotting against the variations of modulation index M and phase angle ϕ gives insight on the behavior of the currents. This is done in Fig. 3, where Fig. 3(a) and (b) show the current efforts for the switches and Fig. 3(c) and (d) for the anti-parallel diodes. It is noticed that switches S_{jo}

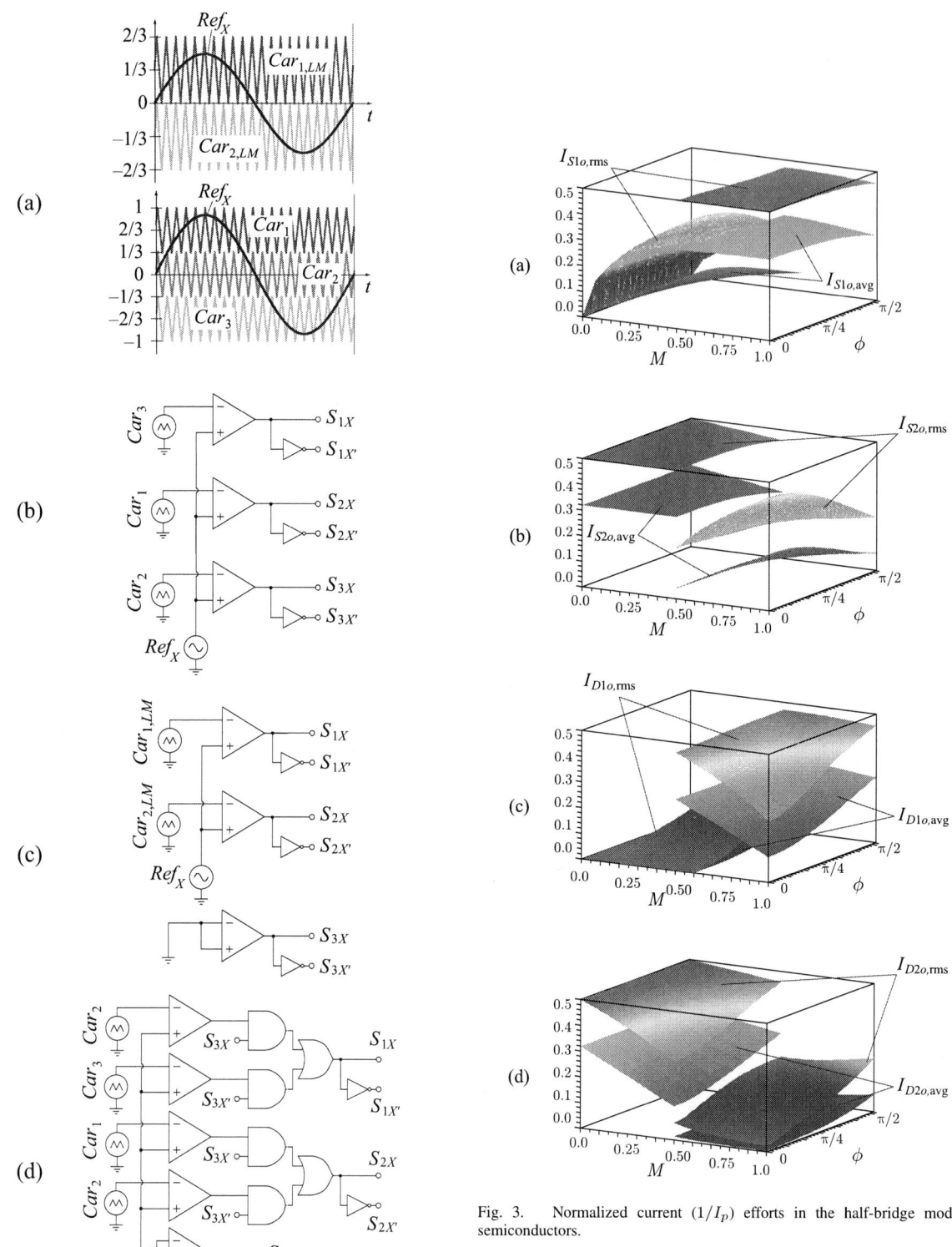

Fig. 2. PWM generation logic for modulation strategies: (a) carrier signals; (b) CONVENTIONAL; (C) HYBRID modulation with LM ($M \leq 1/2$), and; (d) HYBRID modulation with HM ($M > 1/2$).

Fig. 3. Normalized current ($1/I_p$) efforts in the half-bridge modules semiconductors.

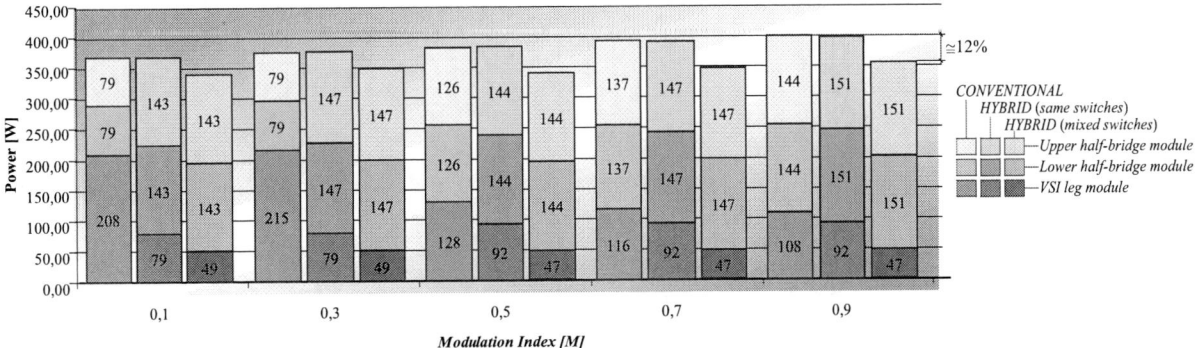

Fig. 4. Comparison of estimated losses and loss distribution for both modulation strategies for a single converter phase-leg comprising two half-bridge modules and a VSI leg. First (yellows) and second (blues) bars are computed with the complete converter based on SKM75GB063D. Third (greens) bar refers to a converter employng low speed IGBTs SKM145GB066D on the VSI module. Conditions: $I_p = 70$ A, $V_{dc} = 400$ V, $f_o = 50$ Hz, $f_s = 20$ kHz and $\Phi = 0°$.

($o = A, B, C$) present higher currents than switches $S_{jo'}$ for higher modulation indexes. On the other hand, diodes D_{jo} present lower currents than diodes $D_{jo'}$ for $M \geq 0.5$. The opposite behaviour is observed for low modulation indexes. Thus, losses are reasonably balanced for a switch comprising an IGBT and its diode, even though the IGBTs, or the diodes, for a half-bridge module can present distinct current demands. The calculations results have been validated for multiple operation points through computational simulation results obtained from the software PSIM. The observed errors have been smaller than 0.5% for all simulation conditions.

A. Losses Comparison

Conduction losses P_{con} are computed with the derived RMS and average current values, so that

$$P_{con} = V_{con}I_{S/D,avg} + r_{con}I_{S/D,rms}^2, \qquad (12)$$

where V_{con} is the forward voltage drop of the respective switch S or diode D and r_{con} its resistance.

Following the approach presented in reference [14], the switching loss energy is approximated with a second order polynomial. A further simplification adds the switching loss energy contributions for all switches into only three coefficients κ_0, κ_1 and κ_2, which represent the sum of the coefficients that model the switch turn-on and turn-off losses and the diodes' reverse recovery. The switching loss energy as a function of the switched current is given by

$$W_{sw}(I_{sw}) = \kappa_0 + \kappa_1 I_{sw} + \kappa_2 I_{sw}^2. \qquad (13)$$

The total switching losses are

$$P_{sw,tot} = \frac{1}{2\pi} \int_0^{2\pi} f_s W_{sw}(i_g)\, d\omega t. \qquad (14)$$

In order to properly compare both modulation strategies the inverter conduction and switching losses are computed for an exemplary design. The output current peak is set at 70 A, with a dc-link voltage of $V_{dc} = 400$ V, $f_o = 50$ Hz, $f_s = 20$ kHz

and a null load displacement angle. In this context, the output power varies with the modulation index. Fig. 4 shows the comparison results for a four-level converter phase-leg. In this graph, the two first bars for each modulation index represent the losses for the converter when employing IGBT model SKM75GB063D for the switches in, both, half-bridge and VSI modules. It is observed that the total losses are very similar for all modulation indexes. However, the loss distribution among the half-bridge and VSI modules present large variation. With the CONVENTIONAL modulation the loss distribution for high modulation indexes is more favorable, while it is seen that for $M < 1/3$ the VSI semiconductors are highly overloaded presenting a loss peak much higher than for any other operating condition with the HYBRID modulation. Since the VSI semiconductors switch at low frequency, low speed IGBTs are a better choice for the converter employing the HYBRID modulation. With this, a further design has been performed utilizing IGBT model SKM145GB066D, which is a trench device with much lower forward conduction voltage drop. The losses for this design are shown in the third bar of Fig. 4 and indicate that a total loss reduction around 10% is achieved at all modulation indexes. Thus, reducing cooling efforts and increasing overall efficiency.

IV. Experimental Verification

Experimental verification is carried out in lab prototype based on nine IGBT (Semikron SKM75GB063D) half-bridge modules (cf. Fig. 1). The dc sources are insulated through three-phase transformers with the secondaries connected to diode bridges and electrolytic capacitors. The average dc supply value is around 400 V. The switching frequency is set to 4.08 kHz and the output fundamental voltage is 60 Hz. The hardware has been built to offer safe operation margins and flexibility and, thus, is not optimized for specific operation points. The employed RL load presents $R = 740$ Ω and $L = 111$ mH connected in delta, leading to a current displacement angle around 4°@60 Hz. Both modulation strategies are implemented in a DSP (TMS320F2812) in

978-1-4244-4782-4/10 $26.00 © 2010 IEEE

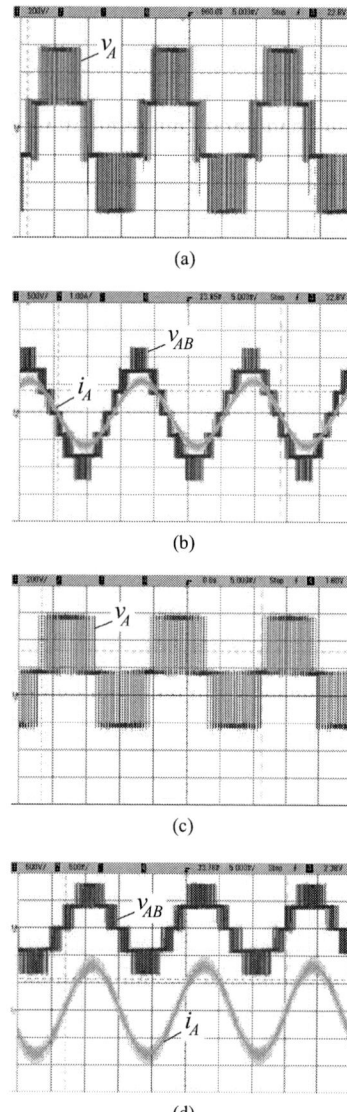

(a)

(b)

(c)

(d)

Fig. 5. Experimental results for $f_o \cong 60Hz$ and $f_s \cong 4kHz$: (a) phase voltage v_A for $M = 0.9$ (HM); (b) line voltage v_{AB} and phase current i_A for $M = 0.9$ (HM)); (C) phase voltage v_A for $M = 0.5$ (LM); (b) line voltage v_{AB} and phase current i_A for $M = 0.5$ (LM)).

an open-loop scheme. The phase and line voltages for low and high modulation indexes are shown in Fig. 5 for the HYBRID modulation scheme. The line voltage waveforms for the CONVENTIONAL modulation are strictly the same as the shown in Fig. 5.

Measurements displaying different operating conditions, i.e. modulation index and load displacement angle, are displayed in Fig. 6. This figure shows the measured switch and diode currents RMS and average values for the given operating conditions along with the respective waveform.

Fig. 7 shows a comparison between the theoretical analysis based current efforts and experimentally measured values. The

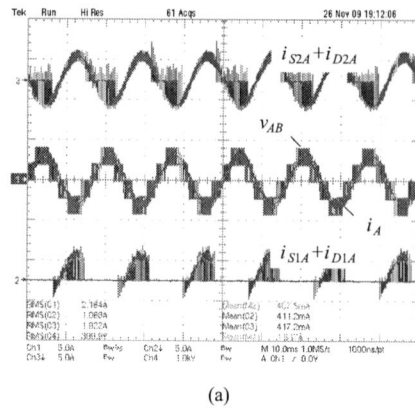

(a)

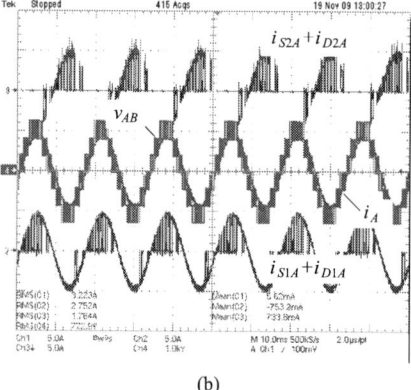

(b)

Fig. 6. Experimental results for $f_o \cong 60Hz$ and $f_s \cong 4kHz$: (a) phase voltage v_A for $M = 0.9$ (HM); (b) line voltage v_{AB} and phase current i_A for $M = 0.9$ (HM)); (C) phase voltage v_A for $M = 0.5$ (LM); (b) line voltage v_{AB} and phase current i_A for $M = 0.5$ (LM)).

current values are given as fuctions of the load RMS value (100%) for a modulation index $M = 1.0$. The modulation index is varied from $M = 0.1$ up to $M = 1.0$ and three different loads have been tested with phase displacement angles of $\Phi = 10°$, $45°$ and $85°$. The currents have been measured at the power module's input, so that IGBT and diode current are combined accordingly. It is seen that a very good agreement is achieved for all tested conditions, thus, certifying the validity of the performed theoretical analysis.

V. CONCLUSIONS

The analysis of the semiconductor efforts and losses oriented to the design of the newly proposed four-level hybrid converter has been performed. The computation of current efforts has been presented, where experimental results based on a built prototype have been shown certifying the validity of the theoretical analysis. It was shown that the proposed hybrid modulation is able to provide better losses distribution among the power semiconductors and to limit the maximum device loss to a lower level when compared to a fully high frequency switched converter. Furthermore, it was shown that

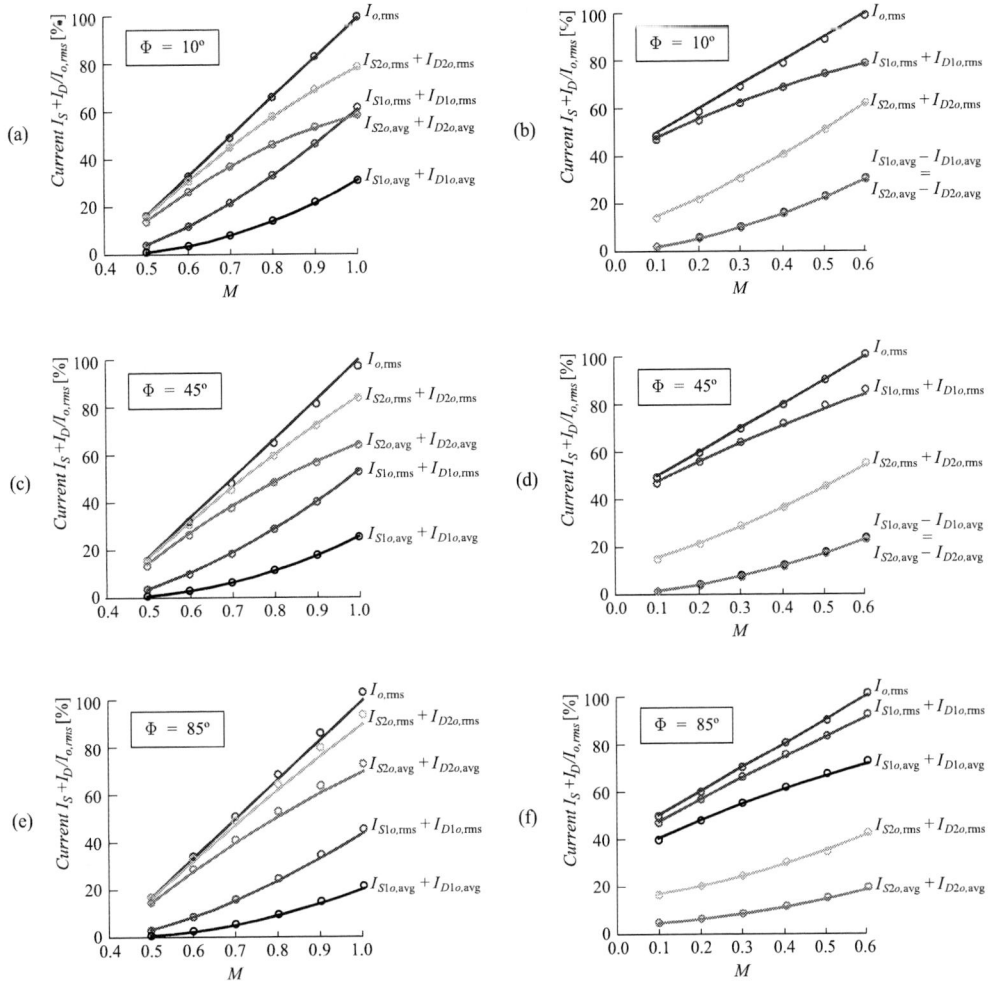

Fig. 7. Comparison between measured and theoretical values for the current efforts across the power semiconductors of the half-bridge module. Current values for: (a) and (b) load displacement angle $\Phi = 10°$; (c) and (d) load displacement angle $\Phi = 45°$; (e) and (f) load displacement angle $\Phi = 85°$.

for the case where the VSI devices are replaced with lower speed and lower forward voltage drop IGBTs, the four-level hybrid converter is able to achieve higher efficiency figures.

REFERENCES

[1] A. L. Batschauer, S. A. Mussa, A. J. Perin, and L. M. Heldwein, "Hybrid multilevel converter employing half-bridge modules," in *Brazilian Power Electronics Conference*, Bonito, Brazil, 2009, pp. CD–ROM.

[2] R. H. Baker, "Bridge converter circuit," Patent U.S. 4,270,163, 1981.

[3] A. Nabae, I. Takahashi, and H. Akagi, "A new neutral-point-clamped pwm inverter," *IEEE Transactions on Industry Applications*, vol. IA-17, no. 5, pp. 518–523, 1981, 0093-9994.

[4] F. Peng and J. S. Lai, "Multilevel cascade voltage source inverter with separate dc sources," Patent U.S. 5,642,275, 1997.

[5] B.-S. Suh, G. Sinha, M. D. Manjrekar, and T. A. Lipo, "Multilevel power conversion - an overview of topologies and modulation strategies," in *International Conference on Optimization of Electrical and Electronic Equipments*, vol. 2, 1998, pp. AD–11–AD–24.

[6] T. A. Meynard, H. Foch, F. Forest, C. Turpin, F. Richardeau, L. Delmas, G. Gateau, and E. Lefeuvre, "Multicell converters: derived topologies," *IEEE Transactions on Industrial Electronics*, vol. 49, no. 5, pp. 978–987, 2002, 0278-0046.

[7] J. Rodriguez, L. Jih-Sheng, and P. Fang Zheng, "Multilevel inverters: a survey of topologies, controls, and applications," *IEEE Transactions on Industrial Electronics*, vol. 49, no. 4, pp. 724–738, 2002, 0278-0046.

[8] S. Mariethoz and A. Rufer, "New configurations for the three-phase asymmetrical multilevel inverter," in *Industry Applications Conference*, vol. 2, 2004, pp. 828–835.

[9] J. Rodriguez, S. Bernet, W. Bin, J. O. Pontt, and S. Kouro, "Multilevel voltage-source-converter topologies for industrial medium-voltage drives," *IEEE Transactions on Industrial Electronics*, vol. 54, no. 6, pp. 2930–2945, 2007, 0278-0046.

[10] T. A. Lipo and M. D. Manjrekar, "Hybrid topology for multilevel power conversion," Patent W.O. 99/41 828, 1999.

[11] P. K. Steimer and M. D. Manjrekar, "Practical medium voltage converter topologies for high power applications," in *Industry Applications Conference*, vol. 3, 2001, pp. 1723–1730 vol.3.

[12] K. A. Corzine, "Cascaded multi-level h-bridge drive," Patent U.S. 6,697,271, 2001.

[13] M. Marchesoni, M. Mazzucchelli, and S. Tenconi, "A nonconventional power converter for plasma stabilization," *IEEE Transactions on Power Electronics*, vol. 5, no. 2, pp. 212–219, 1990.

[14] U. Drofenik and J. W. Kolar, "A general scheme for calculating switching- and conduction-losses of power semiconductors in numerical circuit simulations of power electronic systems," in *International Power Electronics Conference (IPEC'05)*, Niigata, Japan, 2005, pp. CD–ROM.

A Comparative Study of Space Vector PWM Strategy for Dual Three-Phase Permanent-Magnet Synchronous Motor Drives

Yanhui He, Yue Wang, Jinlong Wu, Yupeng Feng, Jinjun Liu
School of Electrical Engineering of Xi'an Jiao Tong University
Xi'an Jiao Tong University
Xi'an, China
huihuigui@yahoo.com.cn

Abstract—The main goal of this paper is to perform a comparative study of different space vector pulse width modulation (SVPWM) techniques for dual three-phase permanent-magnet synchronous motor (DTP-PMSM) drives. Simulation results are provided to emphasize the advantages and disadvantages of each method. Experimental tests have been carried out to validate the most promising strategy which gives satisfactory results in terms of current harmonic minimization and low implementation complexity.

I. INTRODUCTION

During last 30 years the interest in multi-phase AC motor drives has considerably increased, especially for high-power applications. The controlled power can be divided among more inverter legs to reduce the single static switches current stress without the need for parallel techniques or multilevel converter. The multi-phase AC motor drives also show several advantages over the conventional three-phase ones such as reducing the amplitude of torque pulsation, lowering the dc link current harmonics and higher reliability[1-3].

The DTP-PMSM has two three-phase windings spatially shifted by 30 electrical degrees with isolated neutrals. The stator windings are fed by a current controlled PWM six-phase voltage source inverter. A DTP-PMSM and the inverter arrangement are illustrated in Fig.1. The one three-phase system is composed of stator windings A, C and E. Another one is B, D and F. The two sets of windings have spatially shifted by 30 electrical degrees with isolated neutrals.

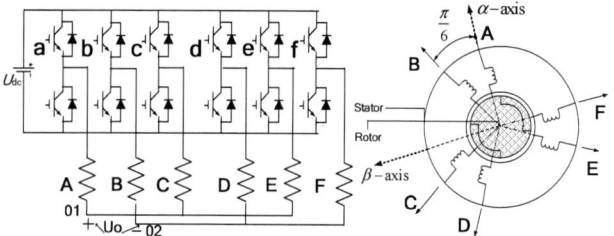
Fig. 1 Voltage source inverter fed DTP-PMSM

During the last years, many authors have presented several SVPWM techniques [4-7]. The aim of the space vector PWM techniques is to reduce the stator current harmonics which have been observed in the machine stator current spectrum. The main goal of this paper is to provide a comprehensive comparison between the performances of these modulation techniques based on several criteria: current harmonic minimization, utilization of DC bus and implementation complexity with low cost fixed-point DSP platforms.

II. SVPWM CONTROL OF A DTP-PMSM

A combinatorial analysis of the inverter switch states shows 64 switching modes. Each voltage vector is represented by a decimal number corresponding to the binary number ($S_a, S_b, S_c, S_d, S_e, S_f$). According to vector space decomposition theory [6], the original six-dimensional machine system can be decomposed into three orthogonal subspaces, (α-β), (z_1-z_2) and ($0_1, 0_2$) subspaces. The voltage vectors projected in ($0_1, 0_2$) subspace will be null since the neutrals of the two windings sets are isolated. Hence, each voltage vector can be decomposed into (α-β) and (z_1-z_2) subspaces shown in Fig.2. The space voltage vectors can be divided into four groups based on the amplitude of the voltage vectors, shown in below.

$$V_{\min} = 0.2989 U_{dc}, V_{\mathrm{mi}d2} = 0.5774 U_{dc},$$

$$V_{\mathrm{mi}d1} = 0.8164 U_{dc}, V_{\max} = 1.1154 U_{dc}$$

where U_{dc} is the DC bus voltage.

The main goal of any PWM control strategy is to impose the (α-β) reference stator voltage vector generated by the control system. In addition, the average voltage vectors generated in the other two subspaces should be zero to avoid large current harmonics, due to the fact that these current harmonics are limited only by the stator leakage inductance and stator resistance[6,7]. According to the selection of the voltage space voltage vector which are used to synthesize reference voltage vectors, there are mainly three kinds of SVPWM.

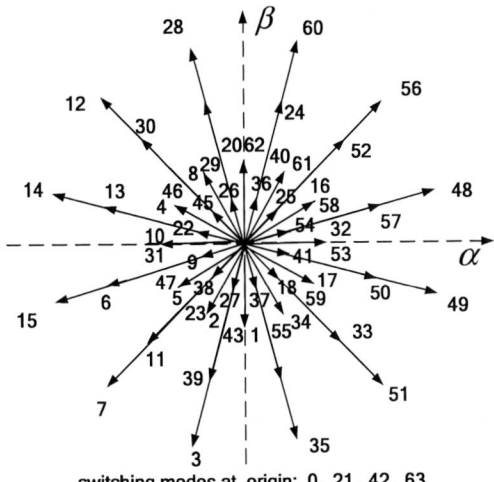

switching modes at origin: 0, 21, 42, 63

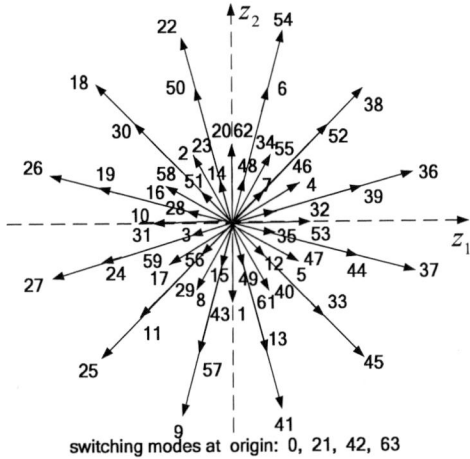

switching modes at origin: 0, 21, 42, 63

Fig.2 Presentation of the inverter voltage vectors in $(\alpha\text{-}\beta)$ and $(z_1\text{-}z_2)$ subspaces

A. Conventional SVPWM Technique

In conventional SVPWM technique[4], the reference voltage vector is synthesized according to the switching state vector maps on the $(\alpha\text{-}\beta)$ subspace only as shown in Fig. 3. It is can be seen from Fig.3 that the average volt–seconds of these switching vectors on the $(z_1\text{-}z_2)$ subspace will not become zero during one sampling interval. In other words, harmonics of the order k =6m $\pm$ 1(m =1, 3, 5...) are generated in the drive system due to the elimination of some switching state vectors that contribute in harmonic reduction.

When the six-phase VSI operates with 12-step output voltages, the fundamental voltage amplitude reaches the maximum value, shown in (1).

$$U_{max} = \frac{2}{\pi} U_{dc} \qquad (1)$$

Define the modulation index as:

$$m = \frac{U}{U_{max}}, 0 \le m \le 1 \qquad (2)$$

where U is the amplitude of the output voltage.

The linear modulation region of the conventional SVPWM is the inscribed circle of the dodecagon with maximum amplitude. According to (2), the maximum modulation index can be abstained, shown in (3), which increases 7.73% compared to three-phase one.

$$m_{max} = \frac{V_{max}\cos 15°}{\sqrt{3}U_{max}} = 0.9771 \qquad (3)$$

When a voltage vector switches to adjacent one, only one power device need to switch on or off, as shown obviously in Fig.4. Hence, it is able to efficiently reduce switching loss.

However, the conventional SVPWM technique has heavy computation burden since it is necessary to calculate the three dwell times of the space voltage vectors in 12 sectors. Hence, it requires hardware with fast calculation ability in practical application.

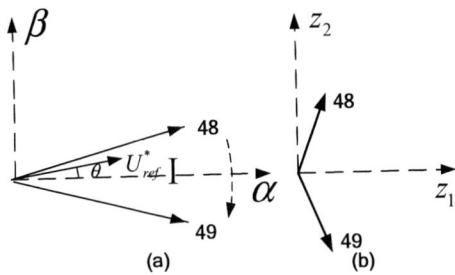

Fig.3 Conventional SVPWM Technique

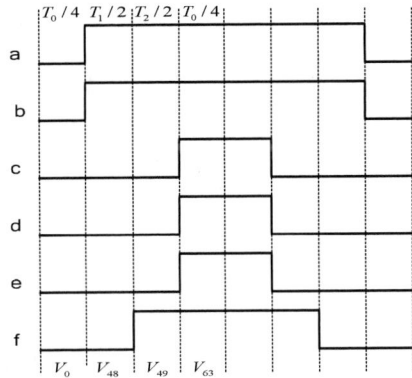

Fig.4 The SVPWM waveform in sector Ⅰ using conventional SVPWM

B. SVPWM Based on a Vector Decomposition Technique

In vector decomposition technique[5], four adjacent non-zero vectors and one zero vector are used. The goal is 1) to make the average volt–seconds of the switching state vectors equal the reference vector on the $(\alpha\text{-}\beta)$ subspace, and 2) to maintain the average volt–seconds of the switching vectors on the $(z_1\text{-}z_2)$ subspace to be zero. The vectorial pictures of switching vectors are depicted in Fig.5. Due to the minimisation of the mean voltage in $(z_1\text{-}z_2)$ subspace, this modulation strategy offers great advantages concerning the reduction of the extra-losses produced by the current harmonics.

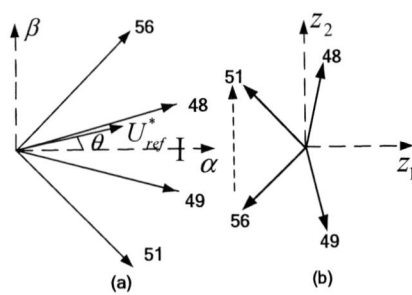

Fig.5 Vector Decomposition Technique

When $tV_{zero}=0$, the SVPWM technique based on vector decomposition has maximum modulation index. According to (2), we can get $m_{max}=0.907$, same as three-phase SVPWM.

During a switching period one power device is always necessary to switch three times using in vector decomposition technique. For example, the transistor "e" switches three times shown in Fig.6. Hence, the switch device losses are more than conventional technique.

Since the SVPWM based on vector decomposition technique need to calculate five dwell times of the space voltage vectors in 12 sectors, it is more complex than the conventional SVPWM.

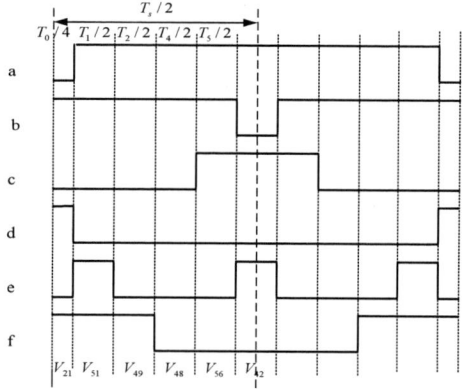

Fig.6 The SVPWM waveform in sector I based on vector decomposition

C. SVPWM Based on Vector Classification Technique

This technique is to operate each of the inverters independently with space vector PWM strategy[6]. Each inverter will then generate the hexagon voltage vector diagram. This can be achieved by means of two different techniques as shown in Fig.7.

In the first approach shown in Fig.7 (a), the reference voltage is rotated while space vector modulators are not. The second approach is shown in Fig.7 (b). The two space voltage vector diagrams will be separated by 30 electrical degrees but the same reference voltage is applied. In both methods, the

harmonic currents such as 5th, 7th and 17th, etc. are inherently eliminated without extensive computations.

The vector classification is simpler than the vector space decomposition technique from the point of view of the implementation complexity, because the switching state vectors are mapped into the inverter basis frame rather than the machine basis frame. And the maximum modulation index increases by 3.59% compared to vector decomposition technique.

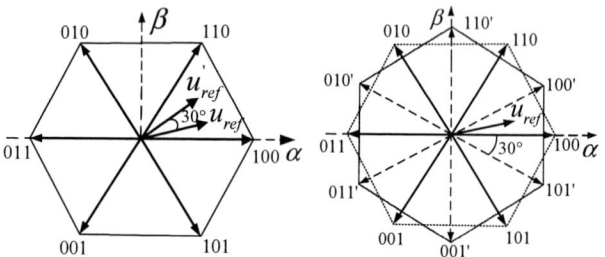

(a) The reference voltage vector is rotated by 30 electrical degrees

(b) Space voltage vector diagrams separated by 30 electrical degrees

Fig.7 Vector classification technique

III. SIMULATION AND EXPERIMENTAL RESULTS

The simulation and experimental tests have been performed with a 3kW, 380V DTP-PMSM. The motor parameters are shown in Table 1.

Table 1 Motor specification

Rated speed	300 rpm
Stator resistance	4.177Ω
d axis synchronous reactance	9.451Ω
q axis synchronous reactance	9.451Ω
magnet flux linkage	0.928Wb
Number of pole pairs	10

A. Simulation Results

The simulations were set up such that a reference speed command is rated speed with 100% load. The simulation results of three techniques are shown in Fig.8-Fig.10 respectively.

From the simulation results, it can be seen that large current harmonic appear in motor phase currents for the conventional SVPWM method. The amplitude of the 5th order current harmonic has already exceeded the fundament current. However, by contrast, these harmonics are practically eliminated for the vector decomposition method. The vector classification have almost the same results as the vector decomposition technique which are satisfactory from the point of view of harmonic minimization since it can inherently eliminate low order harmonics without extensive computations.

Therefore, it can be verified that the SVPWM based on vector classification technique have more advantage in term of stator harmonic current minimization and can be considered as a low-cost solution.

978-1-4244-4782-4/10 $26.00 © 2010 IEEE

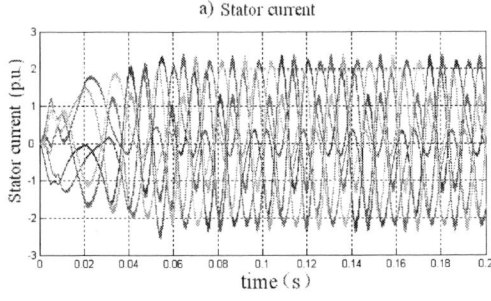

a) Stator current

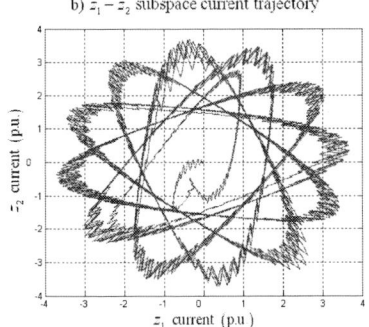

b) $z_1 - z_2$ subspace current trajectory

Fig.8 Simulation results of conventional SVPWM technique

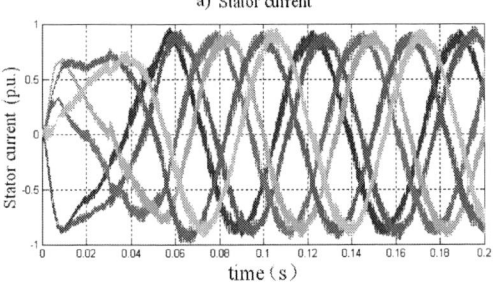

a) Stator current

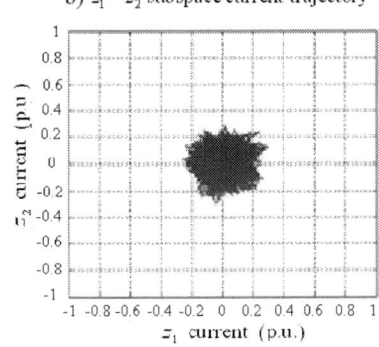

b) $z_1 - z_2$ subspace current trajectory

Fig.9 Simulation results of SVPWM based on vector decomposition technique

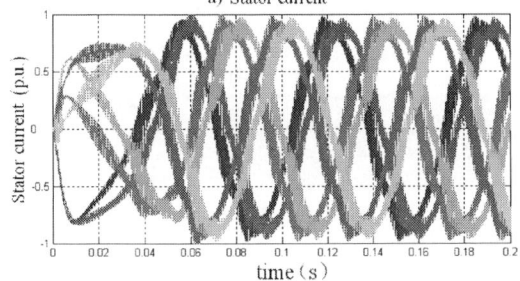

a) Stator current

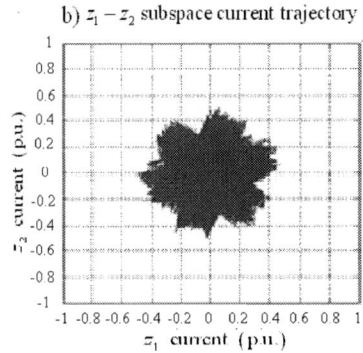

b) $z_1 - z_2$ subspace current trajectory

Fig.10 Simulation results of SVPWM based on vector classification technique

B. Experimental Results

The experimental results using vector classification technique is shown in Fig.11-14.

The waveforms of the motor stator current and speed during motor starting with load are shown in Fig.11. We can see that the control system has very good speed tracking performance.

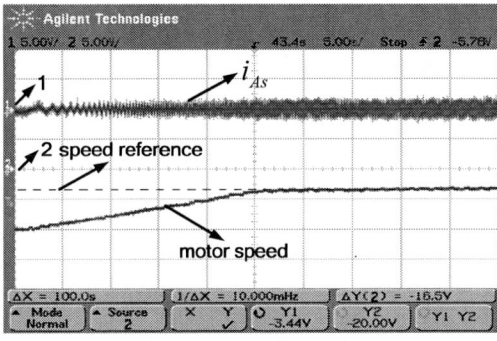

Fig.11 Stator current and motor speed during motor starting with load

Fig.12 shows the waveforms of the motor stator current and speed with step load. The speed reference is set to 30rpm and the load is immediately changed from 11.3Nm to 26.4Nm. It can be seen that the motor speed tracks speed command quite well.

The A and B phase stator current of the motor with 11.3Nm load in steady state are shown in Fig.13. The phase displacement between them is 29 electrical degrees and the amplitude difference is very small. It is suggested that the stator current of the DTP-PMSM have good symmetry.

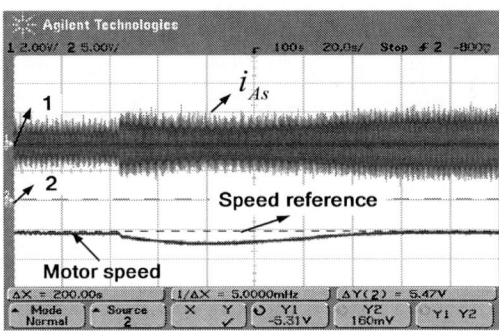

Fig.12 Stator current and motor speed response with step load

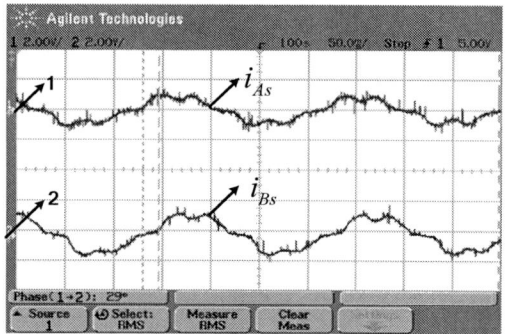

Fig.13 Stator current i_{As} and i_{Bs} in steady state with load

Harmonic analysis of the A phase stator current is shown in Fig.14. The motor speed is 30rpm and the load is 11.3Nm. The fundamental current is 0.63rms. The content percent of the 5th order harmonic current is 7.04%, and 7th order harmonic current is 2.54%. It had proved that the content of the 5th and 7th order harmonic current are almost 2-5 times than fundamental current with no load using conventional SVPWM technique in [4] and [6]. Moreover, with rated load the 5th order harmonic current is nearly 70% of the fundamental current and 7th order harmonic current is about 30% illustrated in [5] and [7].

As expected from the simulation results, SVPWM control based on vector classification technique gives satisfactory results with regard to harmonic current minimization and implementation complexity.

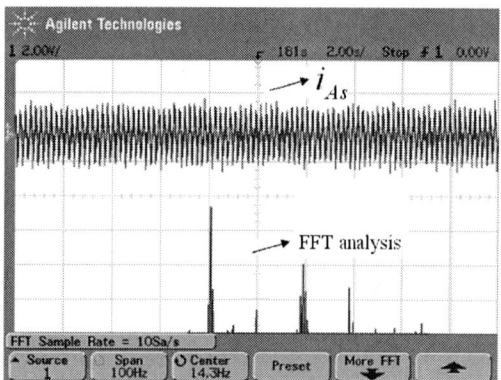

Fig.14 Stator current and current spectrum

IV. CONCLUSIONS

The main goal of the proposed paper is to perform a comprehensive comparative study based on simulation and experimental results of SVPWM strategies for DTP-PMSM drives.

The conventional SVPWM is simpler to implement than vector decomposition technique, because only three dwell times must be calculated. Moreover the simple selection of switching state vectors grants a low switching frequency and provides a good utilization of the inverter. However, a lot of low order harmonic currents will flow in the machine phases.

The vector space decomposition technique gives much better results from the point of view of harmonic minimization. The modulation strategy offers great advantages concerning the reduction of the extra-losses produced by the current harmonics. However, the computational requirements imposed by this strategy are quite large.

The vector classification techniques offers similar good results with vector decomposition technique from the point of view of harmonic minimization. Furthermore, the vector classification is simpler than the vector space decomposition technique and conventional technique from the point of view of the implementation complexity, because the switching state vectors are mapped into the inverter basis frame rather than the machine basis frame. And the maximum modulation index increases by 3.59% compared to vector decomposition technique.

Consequently, the vector classification techniques can be easily implemented with low cost fixed-point DSP controllers.

REFERENCES

[1] E.A. Klingshirn. High Phase Order Induction Motors Part I—Description and Theoretical Considerations[J]. IEEE Transactions on Power Apparatus Systems,1983,102(1):47-59.

[2] N.A. Ai-Nuaim, H.A. Toliyat. A novel method for modeling dynamic air-gap eccentricity in synchronous machines based on modified winding function theory [J]. IEEE Trans. Energy Convers,1998,13(2):156-162.

[3] R.H. Nelson, P.C. Kranse. Induction Machine Analysis For Arbitrary Displacement Between Multiple Winding Sets[J].IEEE Transactions on Power Apparatus Systems,1994,9(3):841-848.

[4] K. Gopakumar, V.T. Ranganathan. Split-Phase Induction Motor Operation From Pwm Voltage Source Inverter[J].IEEE Transactions on Industry Applications,1993,29(5):927-932.

[5] Y.F. Zhao, T.A. Lipo. Space Vector PWM Control Of Dual Three Phase Machine Using Vector Decomposition[J]. IEEE Transactions on Industry Applications,1995,31(5):1100-1108.

[6] A.R Bakhshai, G. Joos.Space Vector Pwm Control of a Split-phase Induction Machine Using The Vector Classification Technique[J]. Conf.Rec.IEEE PESC, 1998,(2):802-807.

[7] D. Hadiouche,L. Baghli,A. Rezzoug.Space-vector PWM techniques for dual three-phase AC machine analysis,performance evaluation,and DSP implementation. IEEE Transactions on Industry Applications,2006,42(4):1112-1122.

[8] R. Bojoi, A. Tenconi.Complete Analysis and Comparative Study of Digital Modulation Techniques for Dual Three-phase AC Motor Drives[J].Conf.Rec.IEEE PESC,2002,(2):851-857.

A Novel Coupled Inductor for Interleaved Converters*

Qianhong Chen, Ligang Xu, Xiaoyong Ren, Lingling Cao, Xinbo Ruan
College of Automation Engineering
Nanjing University of Aeronautics & Astronautics
Nanjing, China
Chenqh@nuaa.edu.cn

Abstract—Aiming to reduce the volume and weight further, a novel coupled inductor is proposed in this paper. The implementation approaches and the operation principles are analyzed in detail along with the comparison analysis with three-leg coupled inductor. Two prototypes of two-phase 1kW interleaved Boost converters are built to verify the theoretical analysis using the proposed or traditional three-leg coupled inductors respectively. In prototype, the volume and weight are reduced to about 40% of the traditional three-leg coupled inductor with the proposed coupled inductor.

I. INTRODUCTION

The shortage of energy and resources gives rise to the requirement of reducing the spillage of material and consumption of energy, as well as decreasing the consumption of scarce resources. For power electronics, that means to increase the efficiency and power density, as well as to reduce the consumption of copper and iron material. Interleaving technology has been widely used due to the advantages of better dynamic response, better thermal management, smaller filters, less EMI content and better packaging [1]. It is also an effective approach to adjust the consumption ratio of material with more "silicon" instead of "copper" and "iron" [2]. Integrated Magnetics (IM) has been widely applied in the interleaved converters so as to reduce the volume of the magnetic components, and to attain higher power density further [2]. The magnetic integration method applied in the interleaved converters can be divided into coupled and decoupled approach [1], [2]. Due to the difficulty in mass production, decoupled inductors have limited usage in multi-phase converters [2]. For two-phase converters, the inductors can be directly coupled or inversely coupled. Reference [2] has made a detailed analysis and comparison between the two coupling way, which have been the basis of a lot of researches[3-4]. Reference [2] discloses that inverse coupling can increase the steady equivalent inductance and decrease the transient equivalent inductance, improving both the steady-state and dynamic performances of the converters. In order to apply the inversely coupled inductor in multiphase converter, different approaches of coupled inductors (CI) using one core or multi-core are proposed in [3], [4] respectively. Aiming to further reduce the weight and volume of CI, [5] proposed the

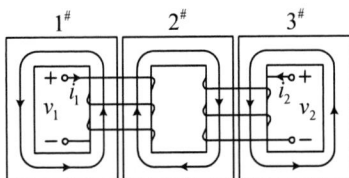

Figure 1. Coupled inductor proposed in [5]

CI as shown in Fig.1, where v_1, v_2, i_1, i_2 are the voltage and current of two inductors respectively. Due to the ampere-turn counteraction of two inductors, there is only ac flux in core 2#. The similar idea also be proposed in [6], along with the several application methods of this idea in multi-phase converters.

To further reduce the weight and volume of CI, a novel CI is proposed in this paper, in which one phase winding is separated into several parts having different function and flux characteristics, providing much convenience for optimization. The implementation approaches and the operation principles are analyzed in detail along with the comparison analysis with three-leg CI. Two prototypes of two-phase 1kW interleaved Boost converters are built to verify the theoretical analysis using two types of coupled inductors respectively.

II. PROPOSED COUPLED INDUCTOR

A. Description of the proposed CI

The proposed CI of two phase converter is as shown in Fig. 2(a), where T_{12} is the closed coupled transformer of channel 1 and 2 with the same turns N_T for two windings; L is the independent inductor with turns of N_L. Node a is the common point; c_1, c_2 are the output points, b_1, b_2 are the internal points, of the CI respectively. Neglecting the leakage inductance of T_{12}, the transformer model of the proposed CI is shown in Fig. 2(b), whose parameters satisfies

$$\begin{cases} L_m = N_T^2 / R_T \\ L = N_L^2 / R_L \end{cases} \quad (1)$$

Where R_T, R_L are the reluctance of T_{12} or L respectively.
Similar with [6], there are several implementation methods for multiphase converters, such as centralized structure, closed

---

* Project Supported by National Natural Science Foundation of China (50507009) and Natural Science Foundation of Jiangsu (SBK200922464).

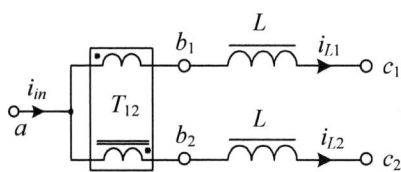

(a) Proposed coupled inductor of two-phase converter   (b)The model of the coupled inductor

Figure 2. Proposed coupled inductor of two-phase converter and it's model.

(a)Centralized   (b) Closed chain

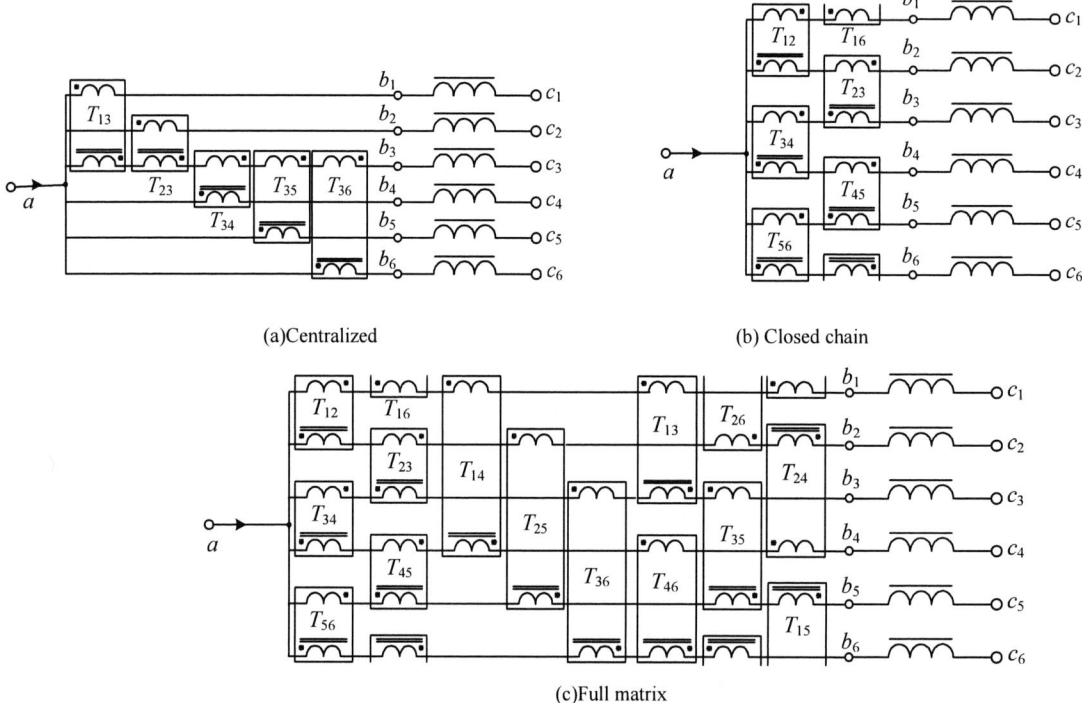

(c)Full matrix

Figure 3. Different structures of the coupled inductor for 6 phases

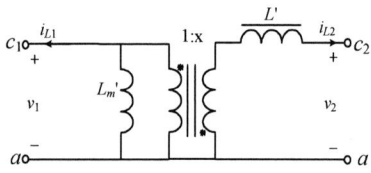

Figure 4. Simplified implementation approach for 2- phase coupled inductor

chain structure and so on. Fig.3 illustrates the details of the CI for 6 phases.

B. Simplified implementation approach for the proposed CI.

Neglecting the nonlinearity, the 2-phase CI as shown in Fig.2(b) can be considered as a two-port linear network, with the port equation as (2).

$$
\begin{bmatrix} \dot{V}_1 \\ \dot{V}_2 \end{bmatrix} = j\omega \begin{bmatrix} L+L_m & -L_m \\ -L_m & L_m+L \end{bmatrix} \begin{bmatrix} \dot{I}_{L1} \\ \dot{I}_{L2} \end{bmatrix}
\tag{2}
$$

From linear circuit theory, only three parameters are necessary and sufficient to represent the output characteristics of a two-port linear network. Using this concept, another simpler implementation approach for 2-phase CI can be derived as Fig.4 shows, whose port equation satisfies

$$
\begin{bmatrix} \dot{V}_1 \\ \dot{V}_2 \end{bmatrix} = j\omega \begin{bmatrix} L_m' & -xL_m' \\ -xL_m' & x^2 L_m'+L' \end{bmatrix} \begin{bmatrix} \dot{I}_1 \\ \dot{I}_2 \end{bmatrix}
\tag{3}
$$

Comparing (2) and (3) yields

$$
L_m' = L+L_m \quad x = \frac{L_m}{L+L_m} \quad L' = \frac{L^2+2L \cdot L_m}{L+L_m}
\tag{4}
$$

It is obviously that this method can reduce component counts and avoid the difference of the independent inductance. This method can also be applied to multi-phase CI.

For example, Fig.5 (a) gives the basic transformer part of the multiphase CI with centralized structure, and Fig. 5(b) is

978-1-4244-4782-4/10 $26.00 © 2010 IEEE   921

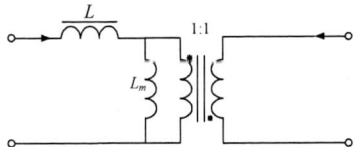

(a) Basic transformer part of multiphase CI with centralized structure

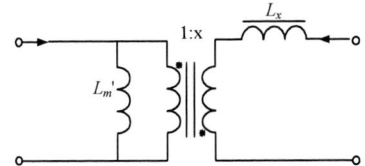

(b) An equavalent circuit of that in Fig.5(a)

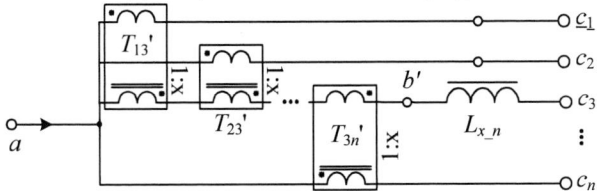

(c) Simplified implementation approach

Figure 5. Simplified implementation approach for multiphase CI

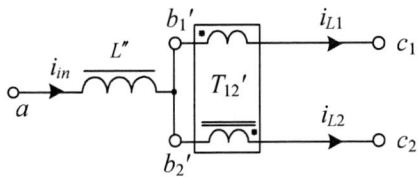

Figure.6 Equivalent circuit with the same current ripple at point a

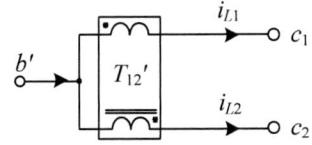

(a) Center tapped transformer

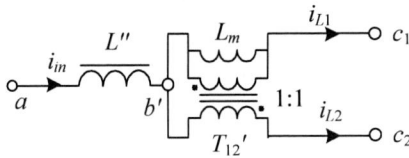

(b) Equivalent circuit of 2-phase CI with the same input current ripple

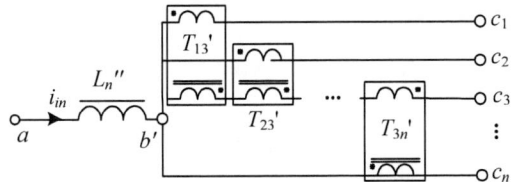

(c) Equivalent circuit of multiphase CI with the same input current ripple

Figure.7 Equivalent circuit of proposed CI with the same input current ripple

one of the Fig.5(a)'s equivalent circuit. By comparing their port equations, the parameter relation of the two models can

be attained. And then, the simplified multiphase CI is achieved, as shown in Fig.5(c), whose parameters satisfies

$$L_m{}' = L + L_m, \ x = \frac{L_m}{L + L_m}, \ L_x = \frac{L \cdot L_m}{L + L_m}, \ L_{x\_n} = \frac{L^2 + nL \cdot L_m}{L + L_m} \quad (5)$$

C. Equivalent circuit with the same current ripple at the common point.

For the proposed 2-phase CI as shown in Fig.2, the input ripple current is determined by,

$$\frac{di_{in}}{dt} = \frac{di_{L1}}{dt} + \frac{di_{L2}}{dt} = \frac{v_{b1} - v_{c1}}{L} + \frac{v_{b2} - v_{c2}}{L} \quad (6)$$

The winding voltage of T_{12} satisfies

$$v_a - v_{b1} = -(v_a - v_{b2}) \quad (7)$$

Substituting (7) into (6) yields

$$\frac{di_{in}}{dt} = \frac{v_{ac1} + v_{ac2}}{L} = \frac{2v_a - (v_{c1} + v_{c2})}{L} = \frac{v_a - \dfrac{v_{c1} + v_{c2}}{2}}{\dfrac{L}{2}} \quad (8)$$

The voltage of the center tap of the transformer as shown in Fig.7 (a) is given by

$$v_{b'} = (v_{c1} + v_{c2})/2 \quad (9)$$

Combing (8), (9) and Fig.7(a), the equivalent circuit of the proposed CI with the same current ripple at the common point can be achieved, as shown in Fig.7(b). By combing (8) and Fig.7(b), it can be easily seen that,

$$L'' = L/2 \quad (10)$$

Similarly, the equivalent circuit of multiphase CI with the same input current ripple can be attained, as shown in Fig.7(c).

And thus

$$L_n{}'' = L/n \quad (11)$$

III.   OPERATION PRINCIPLES OF THE COUPLED INDUCTOR

A. Principle of Operation.

For simplicity, only the coupled inductor shown in Fig. 2 is discussed here. From Fig. 2(b), i_{L1}, i_{L2} is given by

$$\begin{cases} \dfrac{di_{L1}}{dt} = \dfrac{v_{ac1} - v_{ab1}}{L} = \dfrac{di_{L2}}{dt} + \dfrac{v_{ab1}}{L_m} \\ \dfrac{di_{L2}}{dt} = \dfrac{v_{ac2} - v_{ab2}}{L} \end{cases} \quad (12)$$

Since the winding ratio of transformer T_{12} is 1, hence

$$v_{ab1} = -v_{ab2} \quad (13)$$

Defining

$$m = L/L_m \quad (14)$$

Combining (4)-(6) yields

$$\begin{cases} \dfrac{di_{L1}}{dt} = \dfrac{v_{ac1}(m+1)+v_{ac2}}{L(2+m)} \\ \dfrac{di_{L2}}{dt} = \dfrac{v_{ac1}+v_{ac2}(m+1)}{L(2+m)} \end{cases} \quad (15)$$

$$\begin{cases} \dfrac{d\phi_{L1}}{dt} = \dfrac{v_{ac1}(m+1)+v_{ac2}}{N_L(2+m)} \\ \dfrac{d\phi_{L2}}{dt} = \dfrac{v_{ac1}+v_{ac2}(m+1)}{N_L(2+m)} \end{cases} \quad (16)$$

$$d\phi_T/dt = (v_{ac1}-v_{ac2})/N_T(2+m) \quad (17)$$

$$v_{ab1} = \frac{d\phi_T}{dt}N_T(m+1) = (v_{ac1}-v_{ac2})(m+1)/(2+m) \quad (18)$$

Equation (18) indicates that the voltage across the transformer winding is proportional to the voltage difference of v_{ac1} and v_{ac2}. The voltage drop across the transformer winding leads to the decrease of the voltage across the inductor and its ripple current as well.

In steady state, the volt-second integral of v_{ac1} and v_{ac2} equal to zero, thus

$$V_1 D + V_2(1-D) = 0 \quad (19)$$

Fig. 8 illustrates the waveforms of the voltage across the

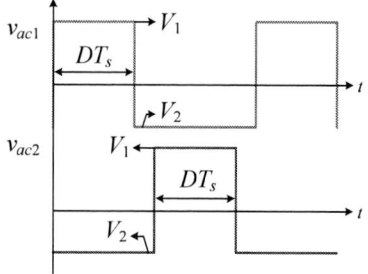

Figure 8 (a) Waveforms of v_{ac1} and v_{ac2} at $D \leq 0.5$

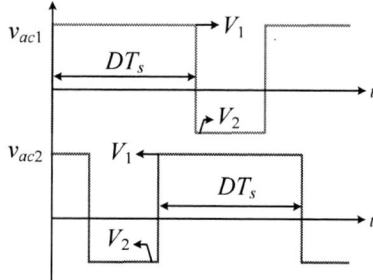

Figure 8 (b) Waveforms of v_{ac1} and v_{ac2} at $D \geq 0.5$

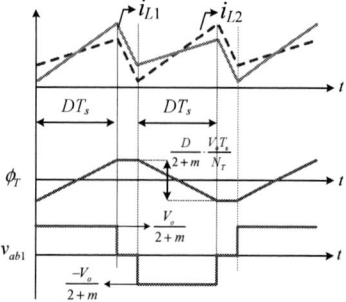

Figure 9 (a) Waveforms of i_L, v_{ab1} and ϕ_T at $D \leq 0.5$

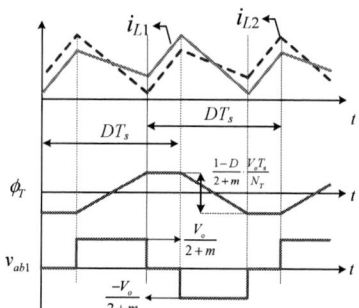

Figure 9 (b) Waveforms of i_L, v_{ab1} and ϕ_T at $D \geq 0.5$

TABLE. I Characteristics in different operation stages

	Stage 1 $[0, DT_s]$	Stage 2 $[DT_s, T_s/2]$	Stage 3 $[T_s/2, (1+2D)T_s/2]$	Stage 4 $[(1+2D)T_s/2, T_s]$
$D \leq 0.5$ ($V_1+V_2>0$)	$v_{ac1}=V_1, v_{ac2}=V_2$	$v_{ac1}=V_2, v_{ac2}=V_2$	$v_{ac1}=V_2, v_{ac2}=V_1$	$v_{ac1}=V_2, v_{ac2}=V_2$
	$\begin{cases}\dfrac{di_{L1}}{dt}=\dfrac{V_1(m+1)+V_2}{L(2+m)}>0 \\ \dfrac{di_{L2}}{dt}=\dfrac{V_1+V_2(m+1)}{L(2+m)} \\ v_{ab1}=\dfrac{V_1-V_2}{2+m}>0\end{cases}$	$\begin{cases}\dfrac{di_{L1}}{dt}=\dfrac{V_2}{L}<0 \\ \dfrac{di_{L2}}{dt}=\dfrac{V_2}{L}<0 \\ v_{ab1}=0\end{cases}$	$\begin{cases}\dfrac{di_{L1}}{dt}=\dfrac{V_1+V_2(m+1)}{L(2+m)} \\ \dfrac{di_{L2}}{dt}=\dfrac{V_1(m+1)+V_2}{L(2+m)}>0 \\ v_{ab1}=-\dfrac{V_1-V_2}{2+m}<0\end{cases}$	$\begin{cases}\dfrac{di_{L1}}{dt}=\dfrac{V_2}{L}<0 \\ \dfrac{di_{L2}}{dt}=\dfrac{V_2}{L}<0 \\ v_{ab1}=0\end{cases}$
	Stage 1 $[0, (2D-1)T_s/2]$	Stage 2 $[(2D-1)T_s/2, T_s/2]$	Stage 3 $[T_s/2, DT_s]$	Stage 4 $[DT_s, T_s]$
$D \geq 0.5$ ($V_1+V_2 \leq 0$)	$v_{ac1}=V_1, v_{ac2}=V_1$	$v_{ac1}=V_1, v_{ac2}=V_2$	$v_{ac1}=V_1, v_{ac2}=V_1$	$v_{ac1}=V_2, v_{ac2}=V_1$
	$\begin{cases}\dfrac{di_{L1}}{dt}=\dfrac{V_1}{L}>0 \\ \dfrac{di_{L2}}{dt}=\dfrac{V_1}{L}>0 \\ v_{ab1}=0\end{cases}$	$\begin{cases}\dfrac{di_{L1}}{dt}=\dfrac{V_1(m+1)+V_2}{L(2+m)} \\ \dfrac{di_{L2}}{dt}=\dfrac{V_1+V_2(m+1)}{L(2+m)}<0 \\ v_{ab1}=\dfrac{V_1-V_2}{2+m}>0\end{cases}$	$\begin{cases}\dfrac{di_{L1}}{dt}=\dfrac{V_1}{L}>0 \\ \dfrac{di_{L2}}{dt}=\dfrac{V_1}{L}>0 \\ v_{ab1}=0\end{cases}$	$\begin{cases}\dfrac{di_{L1}}{dt}=\dfrac{V_1+V_2(m+1)}{L(2+m)}<0 \\ \dfrac{di_{L2}}{dt}=\dfrac{V_1(m+1)+V_2}{L(2+m)} \\ v_{ab1}=\dfrac{V_2-V_1}{2+m}<0\end{cases}$

coupled inductor v_{ac1}, v_{ac2}. It can be seen from Fig. 8 that the operation of the CI in one cycle includes four stages. When D ≤0.5, it can be divided into four stages of $[0, DT_s]$, $[DT_s, T_s/2]$, $[T_s/2, (1+2D)T_s/2]$ and $[(1+2D)T_s/2, T_s]$. When D≥0.5, it can be divided into four stages of $[0, (2D-1)T_s/2]$, $[DT_s, T_s/2]$, $[T_s/2, (1+2D)T_s/2]$ and $[(1+2D)T_s/2, T_s]$. Substituting the value of v_{ac1}, v_{ac2} in different stages into (15)-(18), the equation of i_{L1}, i_{L2}, v_{ab1} and flux of T_{12} (ϕ_T) can be obtained, as well as their waveforms. Tab.1 illustrates the analysis results in different operation stages and Fig. 9 gives the operation waveforms of i_{L1}, i_{L2}, v_{ab1} and flux of T_{12} (ϕ_T).

B. Current ripple.

- Input current ripple Δi_{in}

Substituting the value of v_{ac1}, v_{ac2} in different stages into (8), the input current ripple in different stages can be estimated as shown in Tab. II.

If output voltage V_o of the converter is constant, the expression of the Δi_{in} can be changed correspondingly. For Buck and Buck-Boost converter, V_2 equals to $-V_o$, and for Boost converter, V_1-V_2 equals to V_o. Thus the expression of Δi_{in} can be further changed using (19), as stated in Tab. II. The maximum input current ripple Δi_{in_max} can then be calculated, as Tab. III shows.

- Inductor current ripple Δi_L

When D≤0.5, i_{L1} keeps increasing during stage 1 and keeps decreasing during stage 2, stage 4. So, the inductor current ripple is the maximum value between the current ripple during stage 1 and stage 3. Combining $m>0$, $V_1>0$, $V_2<0$ with Tab. I,

TABLE.II Input current ripple in different stages

D≤0.5				
Stage	$[0, DT_s]$, $[T_s/2, (1+2D)T_s/2]$		$[DT_s, T_s/2]$, $[(1+2D)T_s/2, T_s]$	
Δi_{in}	$\frac{V_1+V_2}{L}DT_s$		$\frac{V_2}{L}(1-2D)T_s$	
	Boost	$\frac{V_1-V_2}{L}D(1-2D)T_s$	Boost	$\frac{V_1-V_2}{L}(1-2D)(-D)T_s$
	Buck & Buck-Boost	$\frac{-V_2}{L}(1-2D)T_s$	Buck & Buck-Boost	$\frac{V_2}{L}(1-2D)T_s$
D≥0.5				
Stage	$[0, (2D-1)T_s/2]$, $[T_s/2, DT_s]$		$[(2D-1)T_s/2, T_s/2]$, $[DT_s, T_s]$	
Δi_{in}	$\frac{V_1}{L}(2D-1)T_s$		$\frac{V_1+V_2}{L}(1-D)T_s$	
	Boost	$\frac{V_1-V_2}{L}(2D-1)(1-D)T_s$	Boost	$\frac{V_1-V_2}{L}(1-2D)(1-D)T_s$
	Buck & Buck-Boost	$\frac{-V_2}{L}\cdot\frac{(1-D)(2D-1)}{D}T_s$	Buck & Buck-Boost	$\frac{V_2}{L}\cdot\frac{(1-D)(2D-1)}{D}T_s$

Tab. III The maximum input current ripple Δi_{in_max}

	D≤0.5	D≥0.5
Boost	$\Delta i_{in\_max}=\frac{V_1-V_2}{8L}T_s\Big\vert D=0.25$	$\Delta i_{in\_max}=\frac{V_1-V_2}{8L}T_s\Big\vert D=0.75$
Buck & Buck-Boost	$\Delta i_{in\_max}=\frac{-V_2}{L}(1-2D_{min})T_s\Big\vert D=D_{min}$	$\Delta i_{in\_max}=0.172\cdot\frac{-V_2}{L}T_s\Big\vert D=\frac{\sqrt2}{2}$

it can be find that the changing ratio of i_{L1} in stage 1 is larger than that in stage 3. Therefore, Δi_{L1} equals to the current ripple during stage 1 when D≤0.5. Similarly, It can be concluded that Δi_{L1} equaling to the current ripple during stage 4 when D≥0.5.

Thus, Δi_{L1} of Boost converter with V_o constant is given by

$$\Delta i_{L1}=\begin{cases} D\left(\frac{1+m}{2+m}-D\right)\cdot\frac{(V_1-V_2)T_s}{L} & D\le 0.5 \\ (1-D)\left(D-\frac{1}{2+m}\right)\frac{(V_1-V_2)T_s}{L} & D\ge 0.5 \end{cases} \tag{20}$$

Δi_{L1} of Buck and Buck-Boost converter is determined by

$$\Delta i_{L1}=\begin{cases} \left(\frac{1+m}{2+m}-D\right)\cdot\frac{(-V_2)T_s}{L} & D\le 0.5 \\ \frac{1-D}{D}\left(D-\frac{1}{2+m}\right)\frac{(-V_2)T_s}{L} & D\ge 0.5 \end{cases} \tag{21}$$

Considering $\Delta i_L=\Delta i_{L1}=\Delta i_{L2}$, to get the maximum value of (20) yields

$$\begin{cases} \Delta i_{L\_max}=\left(\frac{1+m}{4+2m}\right)^2\frac{(V_1-V_2)T_s}{L} \\ D=\frac{1+m}{4+2m},or\quad\frac{3+m}{4+2m} \end{cases} \tag{22}$$

To Get the maximum value of (21) yields

$$\Delta i_{L\_max}=\begin{cases} \frac{1+m}{2+m}\cdot\frac{(-V_2)T_s}{L}, & D\to 0, \\ \frac{\left(\sqrt{2+m}-1\right)^2}{2+m}\frac{(-V_2)T_s}{L}, & D=\frac{1}{\sqrt{2+m}}\ge 0.5 \end{cases} \tag{23}$$

C. Characteristics of the magnetic components.

- Function of the magnetic components

It can be seen from (8) and Tab. II that, the main function of inductor L is to smooth the i_{in} which is independent of T_{12}. Seen from (12), (18), Fig.(9), the transformer T_{12} is to act as a voltage source proportional to the voltage difference of v_{ac1} and v_{ac2}, resulting in the reduction of the inductor's current ripple. In other words, T_{12} is to decrease the current ripple by reducing the circulating current among different channels.

- Flux density equations of inductor L

Assuming the maximum DC current of each phase is $I_{L\_dc\_max}$, according to the Ohm's law of reluctance circuit and (23), the maximum flux density of L is estimated to be as following,

$$\begin{aligned} B_{L\_max} &=\frac{\phi_{L\_dc\_max}}{Ae_L}+\frac{\Delta\phi_{L\_max}}{2Ae_L} \\ &=\frac{N_L I_{L\_dc\_max}}{R_L Ae_L}+\frac{(1+m)^2}{(4+2m)^2}\frac{(V_1-V_2)T_s}{2N_L} \end{aligned} \tag{24}$$

Where R_L is the reluctance of L, Ae_L is the cross-sectional area of the inductor core.

- Flux density equations of transformer T_{12}

According to the Faraday's law and the value of v_{ab1} in different stages, the ac flux of T_{12} is given by

$$\Delta\phi_T = \begin{cases} \dfrac{D}{2+m}\cdot\dfrac{(V_1-V_2)T_s}{N_T} & D\le 0.5 \\[2ex] \dfrac{1-D}{2+m}\cdot\dfrac{(V_1-V_2)T_s}{N_T} & D\ge 0.5 \end{cases} \quad (25)$$

$\Delta\phi_T$ arrives the maximum value when D=0.5, that is,

$$\Delta\phi_{T\_\max} = \frac{1}{2(2+m)}\cdot\frac{(V_1-V_2)T_s}{N_T} \quad (26)$$

Assuming the maximum DC current difference between two phases is ΔI_{L_DC}, according to the Ohm's law of reluctance circuit and (26), the maximum flux density of T_{12} can be determined by

$$B_{T\_\max} = \frac{N_T\Delta I_{L\_DC}}{R_T Ae_T} + \frac{1}{4(2+m)}\cdot\frac{(V_1-V_2)T_s}{N_T Ae_T} \quad (27)$$

Where R_T is the reluctance of T_{12}, Ae_T is the cross-sectional area of the transformer core.

IV. COMPARISON OF TWO TYPES OF COUPLED INDUCTORS IN TWO-PHASE INTERLEAVED BOOST CONVERTER

A. Selection of m.

For two-phase interleaved Boost converter, V_1, V_2 satisfy,

$$\begin{cases} V_1 = V_{in} \\ V_2 = V_{in} - V_o \end{cases} \quad (28)$$

Substituting (1), (14), (28) into (22) and (27) gives

$$\begin{cases} \Delta i_{L\_\max} = \left(\dfrac{1+m}{2+m}\right)^2\dfrac{V_o T_s}{4L} = f(m)\dfrac{V_o T_s}{4L} \\[2ex] B_{T\_\max} = \dfrac{N_L^2\Delta I_{L\_DC}}{mN_T R_L Ae_T} + \dfrac{1}{4(2+m)}\cdot\dfrac{V_o T_s}{N_T Ae_T} \end{cases} \quad (29)$$

It can be seen from (29) that reducing m can decrease the Δi_L but increase the DC bias of the flux density in the transformer due to the current unbalance at the same time.

Fig.10 gives the curves of f(m) defined by (29). Fig.10 (a) is with the large variation range for m, and Fig.10 (b) is with the small variation range for m. Normally the ratio of the unbalanced current to the averaged DC current is controlled to within 5%, so the minimum m is suggested to be around 0.05. Otherwise, the DC bias of the flux density in transformer is about the similar level with that of inductor, which will require a larger core and the increased winding turns. To ensure the small current ripple, it is suggested that m is less than 0.15. In conclusion, m is suggested to be chosen ranging from 0.05 to 0.15.

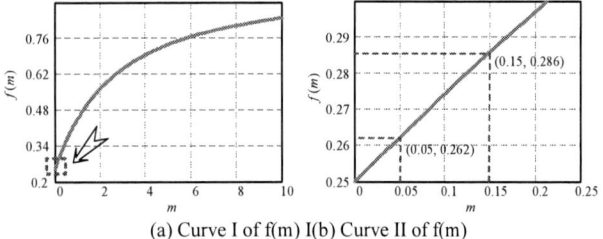

(a) Curve I of f(m) I(b) Curve II of f(m)

Figure 10. Curves of f(m)

B. Flux density characteristics of two types of CI

- **Proposed CI**

Defining

$$a = \Delta i_{L\_\max}\big/I_{L\_DC\_\max} \quad (30)$$

Substituting (22), (28), (30) into (24) yields

$$B_{L\_\max} = \frac{(1+m)^2}{2(4+2m)^2}\frac{V_o\cdot T_s}{N_L\cdot Ae_L}\left(1+\frac{2}{a}\right) \quad (31)$$

Due to the effect of T_{12}, a is small meaning the ac flux component of the inductor is relatively small.

Neglecting the unbalance of the average current in two channels, there is only AC flux in transformer core. The maximum flux density of the transformer can be calculated from (29)

$$B_{T\_\max} = \frac{V_o\cdot T_s}{4(2+m)N_T Ae_T} \quad (32)$$

- **Three-leg inversely coupled CI**

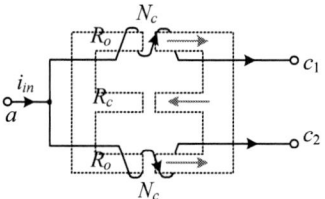

Figure 11 The traditional three-leg inversely coupled inductor

Fig. 11 illustrates the traditional three-leg inversely coupled inductor, with the equivalent circuit as shown in Fig. 2 (b). The transfer equation of the two circuit parameters is as shown in (33).

$$\begin{cases} L_m = R_c N_c^2\big/R_o(R_o+2R_c) \\ L = N_c^2\big/(R_o+2R_c) \\ m = R_o/R_c \end{cases} \quad (33)$$

Where R_o, R_c are the reluctance of the core's out leg and center leg respectively.

Using Faraday's law and Ohm's law of the reluctance circuit, the flux density of the core can be derived,

$$\begin{cases} \Delta B_{o\_\max} = V_o T_s/4N_c Ae_o \\ \Delta B_{c\_\max} = V_o T_s/8N_c Ae_c \\ B_{o\_dc\_\max} = N_c I_{L\_dc\_\max}\big/Ae_o(R_o+2R_c) \\ B_{c\_dc\_\max} = 2N_c I_{L\_dc\_\max}\big/Ae_c(R_o+2R_c) \end{cases} \quad (34)$$

Where Ae_o, Ae_c are the cross-sectional area of the core's out leg and center leg respectively.

Substituting $Ae_c=2Ae_o$, (30), (33) into (34) yields

$$\begin{cases} B_{o\_max} = \dfrac{(1+m)^2}{2(4+2m)^2}\dfrac{T_s \cdot V_o}{N_c A e_o}[\dfrac{2}{a}+\dfrac{(2+m)^2}{(1+m)^2}] \\[4mm] B_{c\_max} = \dfrac{(1+m)^2}{2(4+2m)^2}\dfrac{T_s \cdot V_o}{N_c A e_o}(\dfrac{2}{a}+\dfrac{(2+m)^2}{4(1+m)^2}) \end{cases} \quad (35)$$

C. Comparisions of two types of CI

The main difference of the two types of CI lies in two: the maximum DC current capability, the volume and weight.

- Comparison of the maximum DC current capability

It can be seen from (31) and (32) that the maximum DC current only affect the flux density of the independent inductor. Equation (35) indicates that the flux density of the out leg of the three-leg CI is larger than that of center leg. So, the flux density of the independent inductor and the out leg of the three-leg CI, with different maximum DC current, the same ripple current, winding turns and cross-sectional area, will reach the same maximum flux density allowable by the core material. Combining (30), (31) and (35) gives

$$1+\frac{2I_{L\_dc\_max}}{\Delta i_{L\_max}} = \frac{2I_{o\_dc\_max}}{\Delta i_{L\_max}}+\frac{(2+m)^2}{(1+m)^2} \quad (36)$$

Where $I_{L\_dc\_max}$, $I_{o\_dc\_max}$ are the allowable maximum DC current of the two types of CI respectively. Due to the separation of the winding of the proposed CI, the AC flux component of inductor is less than that in out leg of the Three-leg CI. That is the reason of $1 < (2+m)^2/(1+m)^2$ stated in (36). Therefore, the proposed CI have larger DC current capability.

Considering $m \in (0.05, 0.15)$, the ratio of $I_{L\_dc\_max}$ to $I_{o\_dc\_max}$ satisfies

$$\frac{I_{L\_dc\_max}}{I_{o\_dc\_max}} = 1+(1.2475,\quad 1.406)\frac{\Delta i_{L\_max}}{I_{o\_dc\_max}} > 1 \quad (37)$$

- Comparison of the volume and weight

Defining the cross section area of winding wire as S, window filling ratio as Q, the volume of the inductor can be approximately estimated as following

$$V_L = \frac{N_L S}{Q}\cdot Ae_L \quad (38)$$

Substituting (31) into (38) gives

$$V_L = \frac{S}{Q}\cdot\frac{(1+m)^2}{2(4+2m)^2}\frac{V_o \cdot T_s}{B_{L\_max}}(1+\frac{2}{a}) \quad (39)$$

Similarly, the volume of T_{12} can be approximately estimated

$$V_T = \frac{S}{Q}\frac{V_o \cdot T_s}{4(2+m)B_{T\_max}} \quad (40)$$

The total volume of the proposed CI is determined by

$$V_{all} = V_T + 2V_L$$

$$= \frac{S}{Q}\cdot\frac{(1+m)^2 V_o \cdot T_s}{2(4+2m)^2}[\frac{4+2m}{(1+m)^2 B_{T\_max}}+\frac{(2+\frac{4}{a})}{B_{L\_max}}] \quad (41)$$

For the three leg CI, the volume should be estimated according to the out leg which has higher flux density than center leg,

$$V_{thr} = \frac{S}{Q}\frac{(1+m)^2}{2(4+2m)^2}\frac{V_o T_s}{B_{o\_max}}(\frac{4}{a}+\frac{(4+2m)^2}{2(1+m)^2}) \quad (42)$$

Using the cores of the same magnetic material gives

$$B_{T\_max} = B_{L\_max} = B_{o\_max} \quad (43)$$

Combining (41), (42), (43) yields

$$\frac{V_{thr}}{V_{all}} = 1+\frac{2+2m}{4+2m+(2+4/a)(1+m)^2} > 1 \quad (44)$$

For instance, the volume ratio of the three leg CI to the proposed CI is 1.12 with a=0.4, m=0.1.

In addition, since the main flux component of the independent inductor is DC flux and of transformer is AC flux, we can choose different core material correspondingly, which will provide much more space for optimization.

It is obviously that the separation of windings and cores of the proposed CI offers much more flexibility for optimization compared with the three-leg CI.

V. EXPERIMENTAL RESULTS

Two prototypes of two-phase 1kW interleaved Boost converters are built to verify the theoretical analysis using two types of coupled inductors respectively.

The Spec. are as follows: V_{in_min}=80Vdc, V_{in_max}=375Vdc, f_s=100kHz, V_o=390Vdc, P_o=1kW, f_s=100kHz.

Fig. 12(a) and (b) shows the waveforms of inductor current of two channels at full load and 310V input. The current ripple agrees well with (20). According to the analysis, the current ripple arrives at the maximum value of 1.65A when V_{in}=102 or 287Vdc, in accordance with the experimental results as shown in Fig. 13(a) and (b).

Fig. 14 and Fig. 15 are the photos of the converter and the coupled inductors. The three-leg coupled inductor is made of EE55/28/21 with volume of 44 cm^3 and weight of 216g. The coupled inductor in Fig. 15(a) is made of three ferrite toroids: TX36/23/15*2 and TX39/20/13*1, with volume of 26.98 cm^3 and weight of 125g. The coupled inductor in Fig. 15(b) is made of one ferrite toroids TX39/20/13 and two Kool Mu toroids 77934, with volume of 17.8 cm^3 and weight of 95g. The volume and weight are reduced to about 40% of the three-leg coupled inductor.

VI. CONCLUSIONS

A novel coupled inductor is proposed in this paper to reduce the volume and weight of magnetic components by wingding and core separation. The implementation approaches along with the operation principles are analyzed in detail. Comparisons of analysis and experimental results verify the effectiveness of the proposed approach. The volume and weight of the proposed coupled inductor for a 1kW boost converter has been reduced to about 40% of the traditional three-leg coupled inductor, both having the same equivalent electrical circuit.

REFERENCES

[1] Zumel P, Garcia O, Cobos J.A, Uceda J. Magnetic Integration for Interleaved Converters. Proc. IEEE APEC, 2003: 1143 – 1149.

[2] Pit-Leong Wong, Peng Xu, P. Yang and F.C. Lee. Performance Improvements of Interleaving VRMs with Coupling Inductors.

IEEE Trans. on Power Electronics, 2001, 16(4): 499–507.

[3] Jieli Li, Stratakos, A., Schultz, A, Sullivan, C.R. Using coupled inductors to enhance transient performance of multi-phase buck converters. Proc. IEEE APEC, 2004: 1289- 1293.

[4] Ming Xu, Yucheng Ying, Qiang Li and Fred C. Lee. Novel coupled-inductor multi-phase VRs. Proc. IEEE APEC, 2007: 113-119.

[5] Yang Yu-gang, Yu Qing-guang, Li Hong-zhu, Liu Chun-xi, "Research on planar integrated magnetics whose DC-bias can be eliminated in 4 phase interleaving VRM. Proceedings of CSEE, 2006, 26(24): 179-185.

[6] Zumel, P., Garcia, O., Cobos, J.A., Uceda, J. Tight magnetic coupling in multiphase interleaved converters based on simple transformers. Proc. IEEE APEC, 2005, 1(6-10): 385 - 391.

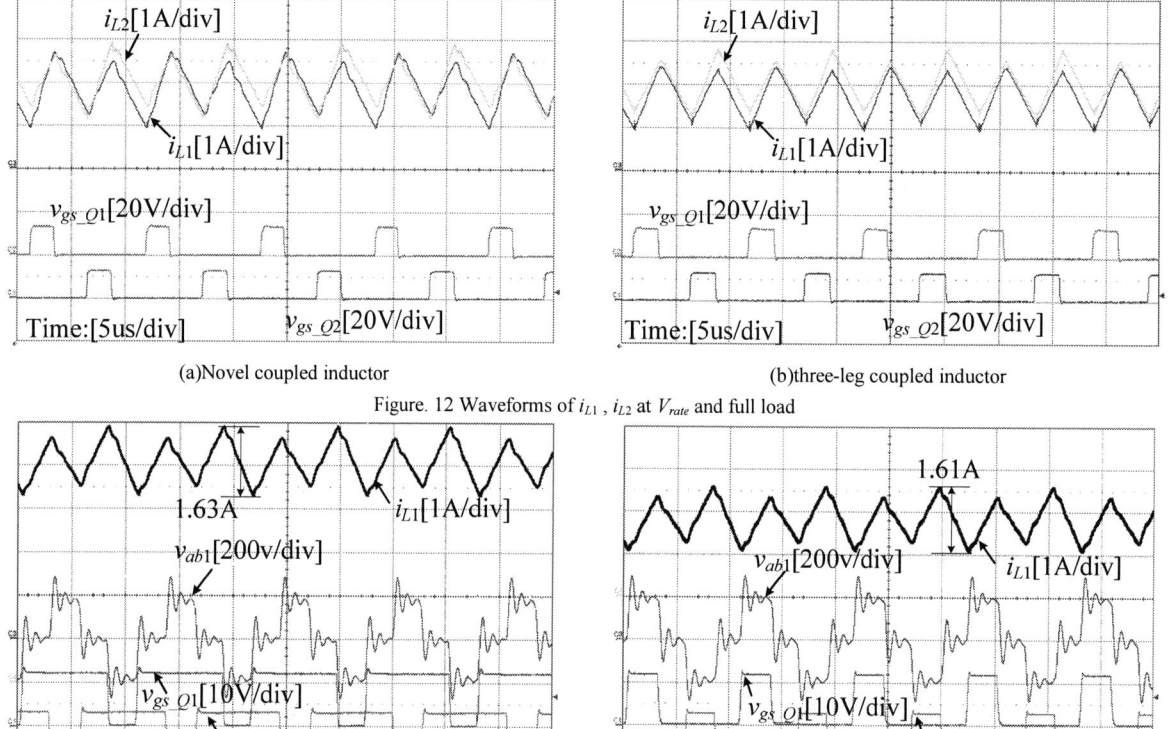

(a)Novel coupled inductor   (b)three-leg coupled inductor

Figure. 12 Waveforms of i_{L1}, i_{L2} at V_{rate} and full load

(a)V_{in}=102VDC   (b)V_{in}=287VDC

Figure. 13 The waveform of i_L and v_{ab1} at full load

Figure. 14 The photo of converter

Figure. 15(a) Photo of magnetics I

Figure. 15(b) Photo of magnetics II

Transformer's Capacitance Effect on the Operation of Triangular-Current Shaped Soft-Switched Converters

Ilya Zeltser, *Student Member, IEEE*, and Sam Ben-Yaakov, *Fellow, IEEE*
Power Electronics Laboratory, Department of Electrical and Computer Engineering
Ben-Gurion University of the Negev, P.O. Box 653, Beer-Sheva 84105, ISRAEL.
Phone: +972-8-646-1561; Fax: +972-8-647-2949;
Emails: ilyaz@ee.bgu.ac.il, sby@ee.bgu.ac.il; Website: www.ee.bgu.ac.il/~pel

Abstract - This paper addresses the effect of the isolating transformer parasitic capacitor in soft switched converter topologies that apply triangular inductor current. When a transformer is included in the circuit, it will not only add (leakage) inductance to the converter but will also add capacitance. This capacitance will cause a partial resonance behavior and therefore may alter the voltage transfer ratio of the converter. The objective of this study was to analyze the effect of this capacitance on the behavior of Triangular Current Shaped (TCS) soft-switched converters.

The isolated TCS converter was analyzed and the theoretical results were verified by simulations and experimentally at a 1kW power level. The analysis show that the transformer's capacitance boosts the voltage gain of the converter, changes the shape of the inductor current and hence modifies the RMS value of the current. It was found though, that by proper design the RMS current (for same power level) may in fact be reduced and consequently, will result in lower conduction losses as compared to the case with no capacitor.

I. INTRODUCTION

In classical soft switched converters, a resonant network is normally applied to achieve soft switching [1]. Another approach for achieving soft switching, that is similar to resonant load converters but without applying a resonant capacitor, was reported in [2-6]. In the latter cases the current waveforms are non sinusoidal but rather triangular. Due to the fact that the resonant capacitor can be omitted, these topologies are especially suitable for high current applications [2, 5].

The Triangular Current Shaped (TCS) converter can be built without [6] or with transformer isolation. If a high input to output voltage ratio is required, as for example in fuel cells applications, or laser drivers, a step up isolation transformer will be normally introduced between the input and output sides of the converter. In some applications, an isolation transformer may be required for safety reasons even in low output voltage applications. When a transformer is included, it will not only add (leakage) inductance to

the converter but will also add capacitance. This capacitance will cause a partial resonance behavior and therefore may alter the voltage transfer ratio of the converter. The objective of this study was to analyze the effect of this capacitance on the behavior of isolated TCS converter.

II. THE BASIC TRANSFORMER ISOLATED TCS CONVERTER

As background, we first consider the isolated topology shown in Fig. 1 with no winding capacitance. We assume that the isolated converter (Fig. 1) is fed from an input voltage Vin, runs at a switching frequency f_s, and the output voltage is V_{out}. The current is rectified at the secondary side by means of a voltage doubler. The latter is beneficial when extra gain is needed and may simplify the design of the isolating transformer. To reduce the RMS value of the inductor current, it is advantageous to operate this converter by the phase shifted PWM control scheme. In this operational mode, the circulating current is reduced by locking the primary current within the bridge switches rather than injecting it back to the input [6]. This is achieved by shorting the current via the two upper switches (Q1, Q2 - ON) or the two lower switches (Q3, Q4 - ON) of the input bridge (Fig. 1).

The output voltage of the bridge (V_{MR}-V_{ML}) and the inductor current for this mode of operation are sketched in Fig. 2. It is further assumed that the isolation transformer has a transformer ratio n and the voltage reflected to its primary side (V_{refl}) is lower than the input voltage.

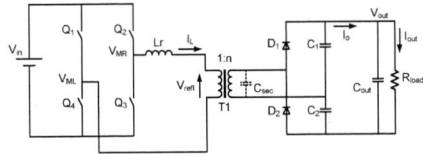

Figure 1.   Isolated TCS converter with voltage doubler at the output side.

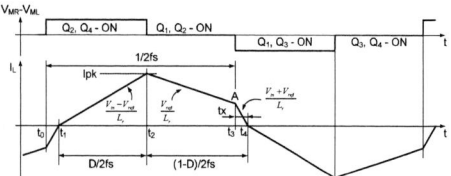

Figure 2. Inductor current of an isolated TCS converter.

Based on the inductor current shape (Fig. 2), three operational stages can be identified:

a: time interval t1- t2

During this time interval, the current through the inductor rises linearly with a slope determined by the difference between the input voltage and the voltage reflected to the primary of the transformer (V_{refl}). This interval ends when transistor Q4 turns off and Q1 turns on (after a small dead time). The peak current of the inductor will thus be calculated as:

$$I_{pk} = \frac{V_{in}}{L_r}\frac{D}{2f_s}(1-k)$$ (1)

where $k = V_{refl}/V_{in}$; $V_{refl} = V_{out}/2n$; L_r - main inductor; f_s - switching frequency; $D = \frac{t_2 - t_1}{2f_s}$.

b: time interval t2-t3

During this interval the voltage applied to the inductor is the voltage reflected to the primary of the isolation transformer (V_{refl}). The current through the inductor decreases with a slope V_{refl}/L_r. This interval ends when transistor Q2 turns off and Q3 turns on.

c: time interval t3-t4

The voltage across the inductor is equal to the sum of the input and the reflected voltages. The inductor current drops to zero.

The switching at point A at t_3 (Fig. 2) may occur at non-zero current and consequently the current during the time interval tx (t_3-t_4) will be fed back to the input, increasing the RMS value of the input current. To prevent the energy return to the input during this time interval (t_3-t_4), the frequency (f_s) and duty cycle (D) can be adjusted such that the residual current is close to zero (tx=0) as shown in Fig. 3. In this case, the switching will be under zero current switching conditions and if some residual current is allowed, it will maintain zero voltage switching at turn on.

From Fig. 3, the voltage applied to the inductor during t_1-t_2 is $V_{in}(1-k)$, and during t_2-t_3 the inductors voltage is $V_{in}k$. Since the average voltage across the inductor must be kept zero, one can write:

$$V_{in}(1-k)\frac{D}{2f_s} = V_{in}k\frac{1-D}{2f_s}$$ (2)

From (2), by canceling out the similar terms and rearranging:

$$k = D$$ (3)

That is, in this operation mode, the duty cycle (D) sets the ratio of the reflected to primary output voltage (V_{refl}) to the input voltage. Considering the transformer ratio and the gain contribution of the voltage doubler at the secondary side of the transformer, the output to the input ratio will be given as follows:

$$\frac{V_{out}}{V_{in}} = 2nD$$ (4)

III. TRANSFORMER ISOLATED TCS CONVERTER ANALYSIS (WITH WINDING CAPACITANCE)

Fig. 4 depicts the shape of the inductor current of the isolated converter when the parasitic winding capacitor is taken into consideration. Based on Fig. 4, four operational stages can be identified:

a: time period t_0-t_1 (resonant phase).

At the beginning of this time interval (Q2 and Q4 are on) the voltage across the capacitor is negative and its absolute value equals to the half the output voltage. During this period the current through the main inductor is sinusoidal due to the resonance between L_r and the equivalent parasitic capacitor at the secondary winding of the transformer (C_{sec}). The inductor current and the capacitor voltage are given as follows:

$$I_{Lr}(t) = V_{in}\frac{1+k}{Z_r}\sin(2\pi \cdot f_r \cdot t)$$ (5a)

$$V_{Csec}(t) = V_{in}[1-(1+k)\cos(2\pi \cdot f_r \cdot t)]$$ (5b)

where $k = V_{refl}/V_{in}$; $V_{refl} = V_{out}/2n$; $Z_r = \frac{1}{n}\sqrt{\frac{L_r}{C_{sec}}}$;

$f_r = \frac{1}{2\pi \cdot n\sqrt{L_r \cdot C_{sec}}}$.

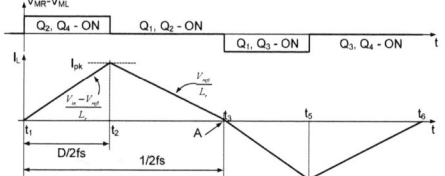

Figure 3. Inductor current of an isolated TCS converter operating under ZCS and ZVS at turn-on.

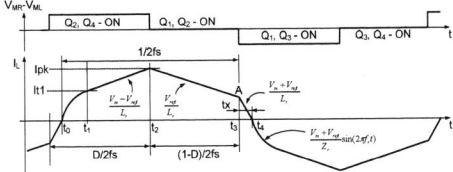

Figure 4. Inductor current of an isolated TCS converter with parasitic capacitance.

This time interval ends when the voltage across the capacitor reaches half of the output voltage (in the voltage doubler configuration). At this instance, the relevant output diode starts to conduct clamping the voltage across the capacitor C_{sec} to $V_{out}/2$.

The time instance t_1 can be calculated by substituting $V_{Csec}(t_1) = V_{out}/2 = k \cdot n \cdot V_{in}$ into (5b):

$$t_1 = \frac{1}{2\pi f_r} \arccos \frac{1-k}{1+k} \qquad (6)$$

Substituting (6) into (5a) the current at the end of this time period will be found as follows:

$$I(t_1) = 2\sqrt{k}\,\frac{V_{in}}{Z_r} \qquad (7)$$

b: time interval t_1- t_2

During this time interval, the current through the inductor rises linearly with a slope determined by the difference between the input voltage and the voltage reflected to the primary of the transformer. This interval ends when transistor Q4 turns off and Q1 turns on (after a small dead time). The peak current of the inductor will thus be calculated as:

$$I_{pk} = 2\sqrt{k}\,\frac{V_{in}}{Z_r} + \frac{V_{in}}{Z_r} 2\pi f_r (1-k)(t_2 - t_1) \qquad (8)$$

c: time interval t_2-t_3

The output of the input bridge is shorted via Q1, Q2, so the voltage applied to the inductor is equal to the voltage reflected to the primary of the isolation transformer ($V_{refl} = V_{out}/2n$). The current through the inductor decreases with a slope V_{refl}/L_r. This interval ends when transistor Q2 turns off and Q3 turns on. The inductor current at the end of this time interval is found to be:

$$I(t_3) = I_{pk} - \frac{V_{in}}{Z_r} k 2\pi f_r (t_3 - t_2) \qquad (9)$$

d: time interval t_3-t_4

The voltage applied to the inductor is equal to the sum of the input and the reflected voltages. The inductor current drops to zero.

The duration of this time interval is found to be:

$$t_x = \frac{I(t_3)}{V_{in}/Z_r} \frac{1}{2\pi f_r (1+k)} \qquad (10)$$

As discussed below, this time interval can be adjusted for better performance.

Similarly to the case with no capacitance, to prevent the energy return to the input during the time interval t_3-t_4, and to maintain zero current switching at turn off of Q2 and zero current and voltage switching at turn on of Q3, the frequency (f_s) and duty cycle (D) should be adjusted such that the residual current is close to zero (t_x=0). This is demonstrated in Fig. 5. This operation mode is denoted here as Borderline Conduction Mode (BCM). The duty cycle in BCM is referenced as D_{BL}

In BCM, the current increase during the time interval t_0-t_2 is equal to the drop in the current during t_2-t_3. From (8), expressing the time interval (t_2-t_0) as $D_{BL}/2f_s$ we get:

$$\Delta I_{t0-t2} = I_{pk} = 2\sqrt{k}\,\frac{V_{in}}{Z_r} + \frac{V_{in}}{Z_r} 2\pi f_r (1-k)\left(\frac{D_{BL}}{2f_s} - t_1\right) (11)$$

where: t_1 is given by (6).

Similarly,

$$\Delta I_{t2-t3} = I_{pk} = \frac{V_{in}k}{Z_r} 2\pi f_r \frac{1 - D_{BL}}{2f_s} \qquad (12)$$

By equating (11) and (12) the duty cycle for the BCM was found to be:

$$D_{BL} = k - \frac{m}{\pi}\left[2\sqrt{k} - (1-k)\arccos\left(\frac{1-k}{1+k}\right)\right] \qquad (13)$$

The normalized average output current, delivered to the load in BCM was found from (5a), (6), (7), (11), (12) and (13) to be:

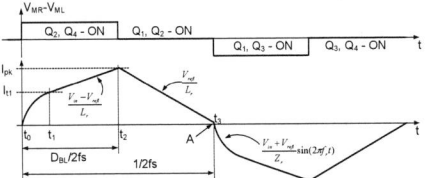

Figure 5. Inductor current of an isolated TCS converter operating in BCM (with parasitic capacitance).

$$I_{out\,av}^{*} = \frac{k}{2n}\begin{bmatrix} \frac{m}{\pi}(1-k)\arccos^2\left(\frac{1-k}{1+k}\right)+ \\ \left(\frac{4m}{\pi}\sqrt{k}+2(1-k)\right)\arccos\left(\frac{1-k}{1+k}\right)+ \\ 4\sqrt{k}+\frac{\pi}{m}(1-k)-\frac{4m}{\pi} \end{bmatrix} \quad (14)$$

where: $I_{out\,av}^{*} = \dfrac{I_{out\,av}}{V_{refl}/Z_r}$, $I_{out\,av}$ - average current

delivered to the load and $m = f_s/f_r$.

The same procedure was used to derive the expression for the normalized rms value of the inductor current:

$$I_{rms}^{*} = \sqrt{\begin{bmatrix} (1+k)^2\frac{m}{2\pi}\left[\arccos\left(\frac{1-k}{1+k}\right)-\frac{2\sqrt{k}(1-k)}{(1+k)^2}\right]+ \\ \left[\frac{\pi^2k^2}{3m^2}(1-D_{BL})^3+\frac{1}{3}\left(D_{BL}-\frac{m}{\pi}\arccos\left(\frac{1-k}{1+k}\right)\right)\right]\times \\ \left(4k+\frac{\pi^2k^2}{m^2}(1-D_{BL})^2+\frac{2k\sqrt{k}\pi}{m}(1-D_{BL})\right) \end{bmatrix}} \quad (15)$$

where: $I_{rms}^{*} = \dfrac{I_{rms}}{V_{refl}/Z_r}$, I_{rms} – rms value of inductor current.

In this operational mode, the power delivered to the load will be adjusted by controlling both the switching frequency {m} and the duty cycle (D_{BL}). The latter needs to be set according to (13) for each operating conditions in terms of input to output voltage ratio and power level.

For a duty cycles lower then D_{BL} the inductor current will cross the zero before the relevant transistor of the leading lag is switched off. For example, in the half switching cycle described above (Fig. 5), if the inductor current becomes negative prior to the time instance t_3, the body diode of Q4 will start conducting. This will result in hard switching and a current spike due to the reverse recovery of this diode when the complimentary switch Q1 turns on. Consequently, for safe operation, the duty cycle must be always higher than D_{BL} for every switching frequency. This will assure not only near zero current switching but also current lag and hence zero voltage switching at turn on (e.g. of Q1 in above example).

IV. THE EFFECT OF THE TRANSFORMER'S CAPACITANCE IN BCM OPERATION

Comparing the duty cycle D_{BL} found for the case when the transformer's capacitance is taken into account (13), to the no capacitance case (3) one finds that for the same voltage conversion ratio ($2nk$), the

steady state duty cycles will be different. This difference is due to the second term in (13). Fig. 6 plots the value in the brackets of the second term of (13) for k below unity.

It follows from this plot that the second term of (13) is always positive for the range of interest ($0 < k < 1$). Consequently, the duty cycle D_{BL} in the case with the winding capacitance is lower then the reflected to primary output to input ratio (k). This is unlike the case with no winding capacitor where the ratio k is set by the duty cycle (3). Putting this in other words - for the same duty cycle, the ratio k in the case of the converter with the capacitor will always be higher. It means that the capacitance adds some extra voltage gain to the converter. On the other hand, the reactive current charging the capacitor at the beginning of every cycle (time interval t_0-t_1 in Fig. 4&5) will result in a change of the RMS value of the inductor current. Interestingly, this will not necessarily lead to higher conduction losses. In fact, for some operating conditions, the ratio of the RMS current to the average current (which is proportional to the output power for a fixed reflected voltage) can even be better (lower) than for the case with no capacitance at all. This is demonstrated in Figs. 7-8 where the RMS-to-average current ratio is plotted vs. the output power for different reflected voltages (different k values and hence output voltages). These figures show that when the resonant (parasitic) capacitor is relatively small, the ratio is lower as compared to the case with no capacitance, for a quite wide operating range. These curves were calculated for L=0.55µH, Vin=24V, and for a switching frequency range of 75kHz down to 25kHz. The horizontal line at about 1.55 is the theoretical value ($2/\sqrt{3}$) of the pure triangular case.

V. SIMULATION AND EXPERIMENTAL

The behavior of the isolated TCS soft switched converter was verified by running a full circuit (cycle-by-cycle) simulation and conducting measurements on an experimental converter. Fig 9 shows the cycle-by-cycle simulation model of the converter.

The simulation model comprises of an input bridge implemented by switches S1-S4 and body diodes D1-D4, main inductor L1, isolation transformer (L3, L4), and voltage doubler (D12, D13, C2, C3).

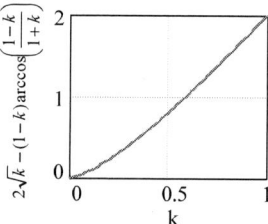

Figure 6. Plot of the value in the brackets of the second term of (13).

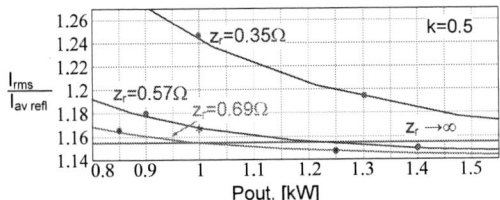

Figure 7. Ratio of input rms current to output average current reflected to the primary for different parasitic capacitors and primary-to-input voltage ratio (k) 0.5. Vin=24V. Solid line: theoretical predictions; Dots: cycle-by-cycle simulation results. Asterisk : Experimental converter

The circuit is fed by a dc voltage source V8. The output voltage is set by the DC voltage source V10 according to a chosen k ratio. The winding capacitance is modeled by capacitor C_{sec}. Simulation results for the ratio I_{rms}/I_{av_refl} at different operating conditions are shown as dots in Fig. 7 and 8.

Fig. 10 shows the experimental results measured on an isolated 1kW TCS converter (Fig. 1) for Vin=24V. The output voltage was about 4kV. The circuit was operated at BCM. The input bridge comprised of IRFP4368 (International Rectifiers, USA) MOSFETs with typical Rds_{on} of $1.46m\Omega@25^0C$. The main inductor was about 0.55μH (leakage of the isolation transformer and stray inductances in the circuit), output capacitors were 0.3μF each, and load was about 15kΩ. The transformer was built on an E65 core and comprised of two primary turns and 338 secondary. The converter was assembled on 4 layers PCB. The outer layers that were used for carrying the power were of 3oz copper. The equivalent parasitic capacitor of the secondary winding was about 60pF. The switching frequency was set to 35kHz and the output power was about 1kW. The efficiency was measured to be 81.7%. The peak inductor current measured experimentally (Fig. 9) was about 170A. The peak inductor current calculated theoretically at the operating conditions stated above is 173A. The current at the end of the capacitor charging interval (I_{t1}) calculated from (7) is 60A. I_{t1} found from the Fig. 9 is about 60A as well.

Fig. 11 depicts the estimated losses distribution for the above described conditions.

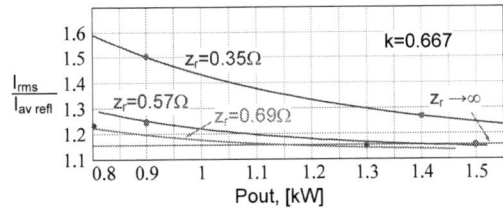

Figure 8. Ratio of input rms current to output average current reflected to the primary for different parasitic capacitors and primary-to-input voltage ratio (k) 0.667. Vin=24V. Solid line: theoretical predictions; Dots: cycle-by-cycle simulation results.

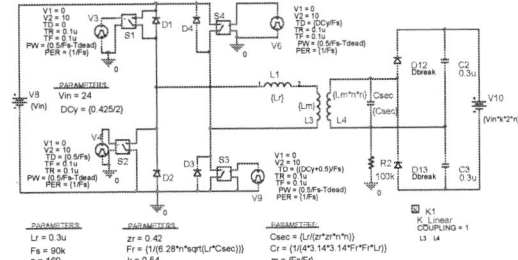

Figure 9. PSPICE cycle-by-cycle simulation model.

The conduction losses estimation is based on a primary rms current of 100A. The switching losses were calculated assuming linear change of the drain current and voltage at turn off, and switching time of 300ns. The transformer cupper losses were measured by first mapping the temperature of the transformer winding for different power dissipation levels. It was accomplished by injecting a DC current into the secondary winding of the transformer. The power was measured by monitoring the current and the voltage of the secondary winding and the surface temperature of the transformer was measured for different injected power levels. The copper losses in the experimental setup were then determined by measuring the surface temperature of the transformer while operating the converter and comparing it to the temperature versus power curve.

The loss estimation chart presented in Fig. 11 suggests that that the conduction losses (MOSFET, wiring and PCB) are dominant. The wiring and PCB losses were estimated by subtracting from the total power loss (180W) the estimated losses of the active and passive devices. The balance, 60W, is assumed to be dissipated by inter-wiring and PCB of the experimental breadboard. It should be noted that with an rms current of 100A, 6mΩ of stray resistance will dissipate the 60W. The PCB losses were relatively high due to the fact that the copper gauge was 3Oz while using only two layers to carry the power.

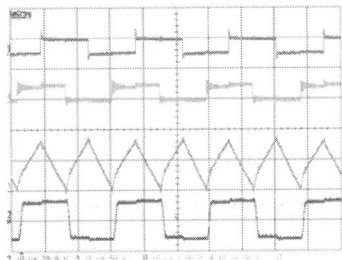

Figure 10. Experimental results.

Vin=24.24V; Iin=51.7A; Vout≈4kV; Iout=254mA; Vrefl=11.9V.

Upper trace: lead lag mid. Point, 50V/div; Second trace from top: lagging lag mid. point 50V/div; Third trace from top: rectified inductors current 100A/div; Lower trace: primary voltage 50V/div; Horizontal scale: 10us/div.

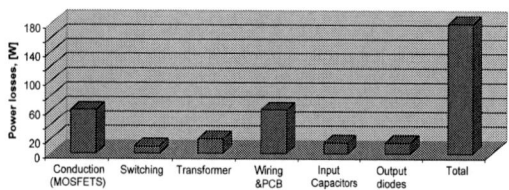

Figure 11. Estimated power losses distribution of experimental breadboard.

A second reason for the high PCB conduction losses is the small package of the MOSFETs (TO247) that has rather small and closely spaced pins. This dictates a layout with sections of narrow traces. Thermal imaging revealed that the temperature of some parts of the PCB were as high as 70^0C when the system was operated under room temperature conditions. Fig. 12 shows the measurement results for the converter when 100pF capacitor was added to the secondary of the isolation transformer. Together with the parasitic winding capacitance the equivalent secondary capacitance was about 160pF. The measurements were conducted for the same output voltage (4kV). The input voltage was somewhat lower as compared to the case with no additional capacitance, as predicted by theoretical analysis. The efficiency was about 80.6%. The peak current in this case was about the same but the current at the end of the resonant phase was higher and the resonant phase lasted longer. According to Fig. 10, the $I_{t1} \approx 100A$ (I_{t1} calculated from (7) is 98A) and $t_1 \approx 2\mu s$ (the value calculated from (6) is $1.9\mu s$). The higher rms value of the inductor current, due to the higher reactive current charging the capacitor, explains the lower efficiency (as compared to the case with no additional capacitor) measured in this experiment.

VI. DISCUSSION AND CONCLUSIONS

This study presented an isolated TCS converter and investigated the effect of the transformer's parasitic capacitance on its behavior.

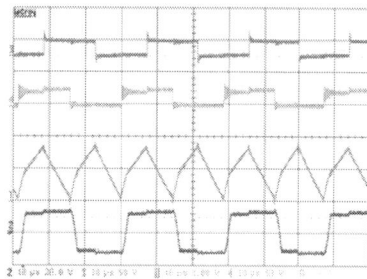

Figure 12. Experimental results. $C_{sec} \approx 160$pF.

Vin=23 V; Iin=55A; Vout≈4kV; Iout=253mA; Vrefl=11.9V.

Upper trace: lead lag mid. point 50V/div; Second trace from top: lagging lag mid. point 50V/div; Third trace from top: Rectified inductors current 100A/div; Lower trace: primary voltage 50V/div; Horizontal scale: 10us/div.

The proposed one stage topology features high voltage gain, current sourcing behavior at the output and ability to operate with high currents at the input. The latter stems from the fact that no series capacitor is applied at the input side. These features makes this topology beneficial for high power, low input, and high output application like, for example a microwave oven fed from a 24V battery or fuel cell converters where high input-to-output gain is needed.

The results of the analysis show that the capacitance of the transformer's winding boosts the voltage gain of the converter and shapes the inductor current, changing thereby its RMS value. It was found though, that this change can lead to higher or lower RMS currents for same power level. This is because the resonant part of the current that is caused by the transformer's capacitance has two opposing effects, as demonstrated in Figs. 13&14. These figures show the inductor and the output current for two different capacitors. On the one hand , the resonant portion of the current adds a reactive current to the primary side since during the resonant part no current is outputted to the secondary (area A in Figs. 13&14). On the other hand, the shape of the current that passes through the transformer to the output, has a lower RMS to average current ratio. This is because the first part of the current (area B in Figs. 13&14) has a trapezoidal shape while in the non resonant case the shape is triangular [6]. In the triangular case, the RMS to average current is $2/\sqrt{3}$ while the ratio calculated for the trapezoidal shape is always lower. When the transformer's capacitance is large (Fig. 14), the reactive current effect is dominant and the overall RMS to average current is higher (Figs. 7-8). However, when the capacitance is small (Fig. 13) the reactive current is relatively low and current is outputted to the load during the majority of the switching cycle. In this case the mitigating effect of the lower RMS value of the trapezoidal shape prevails and the overall conduction losses will be lower. The normalized analysis and curves presented in this paper may help designers to select the optimal operating conditions, such as the optimal reflected voltage, for maximum efficiency.

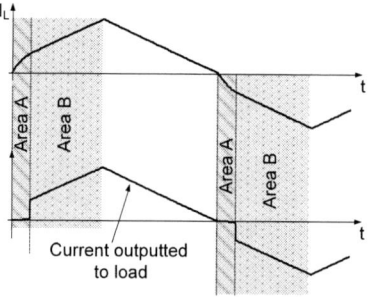

Figure 13. Effect of parasitic capacitance. Small capacitance.

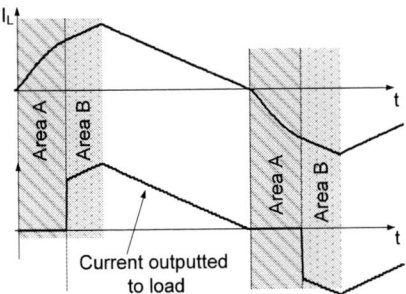

Figure 14. Effect of parasitic capacitance. Large capacitance.

The experimental converter was not designed to operate at the optimal area of operation (the I_{rms}/I_{av_refl} ratio was 1.162 (Fig. 7), which is slightly above $2/\sqrt{3}$ ratio, found when the current is purely triangular). The major reason for this was the requirement to operate at nominal power for quite wide input voltage range (18V-30V). As a result, the reflected voltage had to be kept at relatively low value. This is because according to the analysis developed in this study, for the low input voltage (18V), and the given leakage inductance, the reflected voltage needed to be as low as 12V to allow transferring the nominal power without lowering the switching frequencies down to the audible range. In addition, due to the lack of space a relatively small core for the main transformer had to be chosen. Notwithstanding these deficiencies, the efficiency of the experimental converter was found to be on par with the case of, say, two serially connected stages of lower gain each. This is due to the fact that the converter under study was operated under zero current and zero voltage at turn on, so the switching losses were relatively small. Moreover, it follows from the losses analysis (Fig. 11) that the efficiency can even be improved by utilizing thicker PCB layers (e.g. 4-6oz) and a bigger transformer.

REFERENCES

[1] R. W. Erickson, "Resonant Conversion," chap. 19 in *Fundamentals of Power Electronics*, Chapman & Hall, N.Y., 1997.

[2] S. Sato, S. Moisseev, and M. Nakaoka, "A novel synchronous rectifiers based ZVS-PWM DC-DC power converter with extended soft-switching operational range," *IEEE The 25th International Telecommunications Energy Conference, INTELEC'03, pp. 268-273, 19-23 October 2003.*

[3] I. D. Jitaru, "A 3 kW soft switching DC-DC converter," *IEEE* Fifteenth Annual Applied Power Electronics Conference and Exposition, APEC 2000, vol. 1, pp. 86 – 92, 6-10 Feb. 2000.

[4] Y. Zhang, P. C. Sen, "A new ZVS phase-shifted PWM DC-DC converter with push-pull type synchronous rectifier," IEEE Canadian Conference on Electrical and Computer Engineering, CCECE 2003, vol. 1, pp. 395-398, 4-7 May 2003.

[5] F. Krismer, J. Biela, and J. W. Kolar, "A Comparative Evaluation of Isolated Bi-directional DC/DC Converters with Wide Input and Output Voltage Range," *IEEE Industry Applications Conference, Fourteenth IAS Annual Meeting, vol. 1, pp. 599 - 606, 2-6 October 2005.*

[6] I. Zeltser, and S. Ben-Yaakov, "Modeling, analysis and simulation of "AC inductor" based converters," *IEEE Power Electronics Specialists Conference, PESC-2007, pp. 2128-2134, 2007.*

An Input and Output Ripple Free Converter with a Four-winding Coupled Inductor

Zhuomin Feng, Zhe Zhang, Duo Li, Min Chen, Zhaoming Qian

Electrical Engineering College, Zhejiang University, Hangzhou(310027), China

Abstract - **An input and output ripple free Step-down converter is proposed in this paper. The two phase interleaved converter has only one inductor with four coupled windings, which decreases both the input and output ripple dramatically. Based on the analyses, the coupled inductor can also improve the transient response performance, and reduce the size and weight of the converter. Detailed analyses and simulations are performed. Experimental results are also presented for validation.**

I. INTRODUCTION

The Step-down converter has been widely used now in VRM, LED driver, solar energy and so on. Traditional buck converter is one of the most popular topology, because it needs few components with simple control strategy. However, the input of buck converter is a pulse current and a LC filter is needed at the input to reduce the current pulse and the EMI problem, which increases the volume of the converter and may also bring instability problems [1, 2].

For the applications which have special requirements for the input and EMI, a zero ripple Step down topology is proposed [3, 4, 5]. This topology belongs to the Cuk converter family and has very low input ripple current. It can be used in battery equalizer, PFC, and aerospace applications [6, 7, 8]. Furthermore, when using the battery as the source, low input ripple current is desired for extending the life time of the battery and also the battery's output voltage stability.

Small ripple and fast transient response are also the essential targets of step down converter. For steady state operation of DC-DC converter, small current ripple is much preferred. It can provide less conduction losses both in switches and filter capacitors. Therefore in traditional applications, bigger inductor is always used to reduce the current ripple. But large inductance

also reduces the current slew rate which results in slow transient response [9, 10].

In this paper, a novel interleave step-down converter is proposed. Four windings are coupled in this converter, so only one magnetic component is needed. The coupled inductor can reduce the input and output ripple dramatically. Hence, smaller LC input filter can be used. Based on analyses, the coupled inductor can improve the transient response of the converter. This paper is organized as follows: in Section II detail theoretical analyses will be given; the simulation and experiment results are presented in Section III and Section IV; finally some conclusions are given in Section V.

II. THEORETICAL ANALYSES

The propose converter is shown in Fig. 1. Some basic assumptions are given first to ease the analyses. The four winding inductors are the same ($L_1=L_2=L_3=L_4$); and the coupling inductors between the four windings are equal ($M=M_{ij}$, i=1,2,3,4; j=1,2,3,4; i≠j); the input and output voltage (V_i, V_o) are almost constant during switching period; the switching of Q_1 and Q_2 are fast enough.

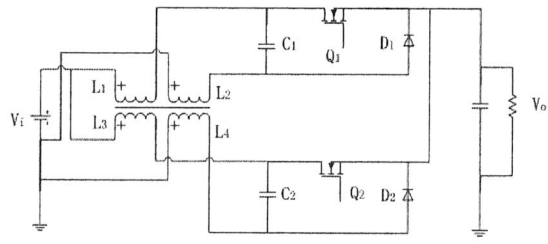

Fig. 1 The proposed step down converter

This work is sponsored by Science and Technology Planning Project of Zhejiang Province, China. Project No: 2009C31005

978-1-4244-4782-4/10 $26.00 © 2010 IEEE

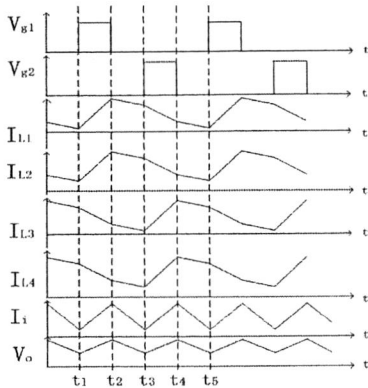

Fig. 2 The typical waveforms

Fig. 2 shows some typical waveforms of the proposed converter. The operating principles and steady-state analysis are presented in detail as follows.

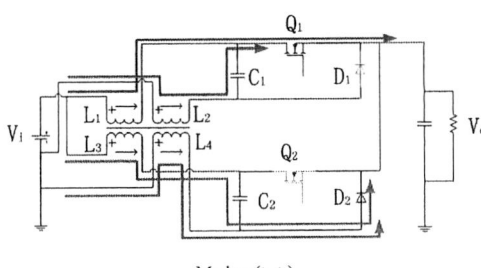

Mode a (t₁-t₂)

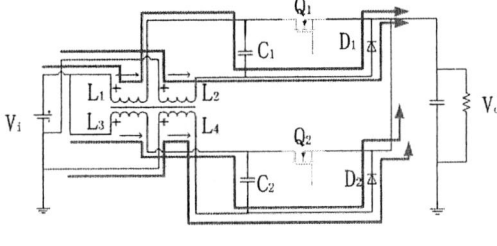

Mode b (t₂-t₃)

Fig. 3 Operating modes（Red: increase; Blue: decrease）

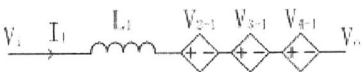

Fig. 4 Coupled model of L_1

Fig. 3 illustrates two of the operating modes of the proposed converter. In Mode a (t₁-t₂), when Q_1 and D_2 turn on, Q_2 and D_1 are off. I_{L1} and I_{L2} increase (red line) while I_{L3} and I_{L4} decrease (blue line). As above assumptions, the winding inductors are the same and the coupling inductors between the windings are equal. The coupled model of L_1 is shown in Fig. 4. V_{2-1}, V_{3-1} and V_{4-1} are the voltage coupled from L_2, L_3 and L_4 respectively. According to Kirchhoff's voltage law, the voltage across L_1 and L_3 can be expressed as

$$L\frac{di_1}{dt} + M\frac{di_2}{dt} + M\frac{di_3}{dt} + M\frac{di_4}{dt} = V_i - V_o \tag{1a}$$

$$L\frac{di_3}{dt} + M\frac{di_1}{dt} + M\frac{di_2}{dt} + M\frac{di_4}{dt} = -V_o \tag{1b}$$

Since the average voltage across every winding must be zero, in one duty cycle, we have

$$\frac{di_1}{dt} = \frac{di_2}{dt} = \frac{(L+M)V_i - (L-M)V_o}{(L+3M)(L-M)} \tag{2a}$$

$$\frac{di_3}{dt} = \frac{di_4}{dt} = \frac{2MV_i + (L-M)V_o}{(L+3M)(M-L)} \tag{2b}$$

For input and output current,

$$\frac{di_i}{dt} = \frac{V_i - 2V_o}{L+3M} \tag{3a}$$

$$\frac{di_o}{dt} = 2\frac{V_i - 2V_o}{L+3M} \tag{3b}$$

In Mode b (t₂-t₃), Q_1 and Q_2 turn off while D_1 and D_2 conduct. The currents I_{L1}, I_{L2}, I_{L3}, I_{L4} will decrease. Similar to above analyses,

$$\frac{di_1}{dt} = \frac{di_2}{dt} = \frac{di_3}{dt} = \frac{di_4}{dt} = -\frac{V_o}{L+3M} \tag{4a}$$

$$\frac{di_i}{dt} = \frac{di_1}{dt} + \frac{di_3}{dt} = -\frac{2V_o}{L+3M} \tag{4b}$$

$$\frac{di_o}{dt} = \frac{di_1}{dt} + \frac{di_2}{dt} + \frac{di_3}{dt} + \frac{di_4}{dt} = -\frac{4V_o}{L+3M} \tag{4c}$$

According equation (2b) and equation (4a), the falling rate of I_{L3} and I_{L4} is smaller in Mode b than that in Mode a, as shown in Fig. 2.

In Mode c (t3-t4), Q_1 and D_2 turn on and Q_2 and D_1 are off, which is similar as Mode a. The falling rate of I_{L1} and I_{L2} is expressed same as equation (2b) and the raising rate of I_{L3} and I_{L4} is expressed as equation (2a).

In Mode d (t4-t5), both of Q_1 and Q_2 are off and D_1 and D_2 conduct; the current I_{L1}, I_{L2}, I_{L3}, I_{L4} in the coupled inductors will fall same as equation (4a).

Based on above analyses, the input and output ripple of the interleaved converter can be expressed as:

$$D = \frac{V_o}{V_i} \qquad (5)$$

$$\Delta I_i = \frac{(V_i - 2V_o)DT}{L + 3M} \qquad (6)$$

$$\Delta V_o = \frac{(V_i - 2V_o)DT^2}{8(L + 3M)C_o} \qquad (7)$$

$$\frac{\Delta I_o}{\Delta D} = \frac{2(V_i - 2V_o)}{(L + 3M)f} \qquad (8)$$

Equation (6) is the expression of the input current ripple of the proposed converter. Small ripple current can be achieved by proper design of the coupled inductor. And continuous input current can be obtained even without any LC filter.

Equation (7) shows the output ripple which is smaller than the traditional bulk converter ($\Delta V_o = \frac{(V_i - V_o)DT^2}{8LC_o}$), when the coupling inductor is large enough, as shown in Fig. 5. The transient response speed is expressed as (8). Therefore, both the steady-state and dynamic performances can be improved by adjusting corresponding equivalent inductances.

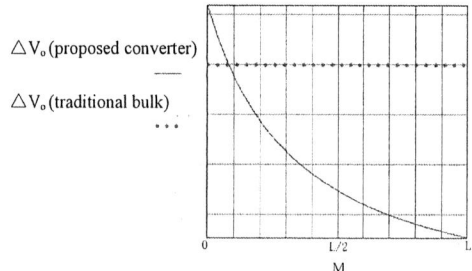

Fig. 5 Comparison of output ripple

III. SIMULATION RESULTS

Simulations of the proposed converter were conducted with SABER. The proposed converter system parameters are listed as in Table 1.

Table 1 Converter's parameters

Input voltage:	12V
Output voltage:	3.3V
Load current:	20A
Winding inductor:	20u
Switching frequency:	100 kHz
C_1 and C_2:	20u
Output filter capacitor:	1500u
Coupling coefficient (k):	0.9

For comparison, the simulation of traditional buck with the same parameters is shown in Fig. 6. It is clear that the input current is pulse and the peak-to-peak current is more than 20A. It will give the input source or the front-end circuit in cascade system great challenges in the application. In this case, LC filter is need, which increases the volume of the converter and may also bring instability problems.

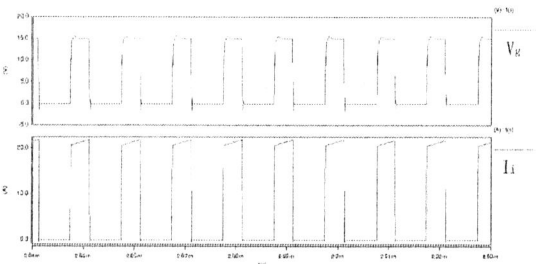

Fig. 6 Gate voltage and input current of traditional buck

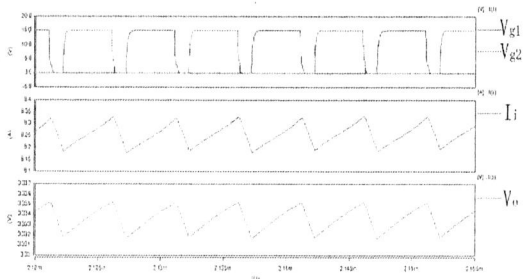

Fig. 7 Input and output ripple

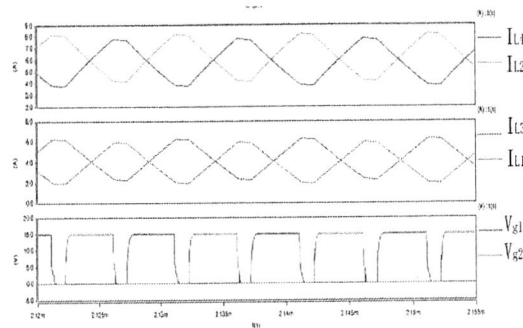

Fig. 8 Coupled inductor current

Fig. 6 shows the gate voltage of Q_1 and Q_2, the input current I_i and the output voltage V_o. It is clear that the input current of the proposed converter is continuous, and the input current ripple ($0.3A_{p-p}$) and the output voltage ripple ($5mV_{p-p}$) are very small. Fig. 7 illustrates the current (I_{L1}, I_{L2}, I_{L3}, I_{L4}) in the four windings and the gate voltage of Q_1 and Q_2. The current in the four windings work in interleaving modes and mach the analyses well.

IV. EXPERIMENT RESULT

In order to verify the theoretical analysis and simulation, a prototype was built according to the parameters in table 1. The four winding inductors were coupled in one core (Magnetics cool mu core). And the prototype was test without any input LC filter.

Fig. 9 shows the input current (I_i). It is continuous with 7.08A Rms value and the peak-to-peak current is measured as around 0.7A. The output voltage ripple ($\triangle V_o$) is shown in Fig. 10. The peak-to-peak voltage ripple is about 50mV.

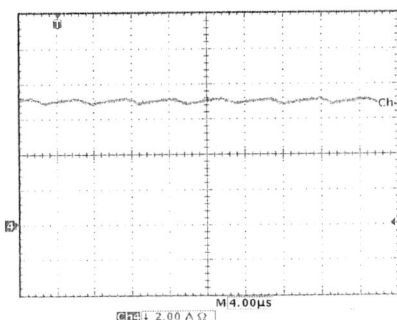

Fig. 9 Input current

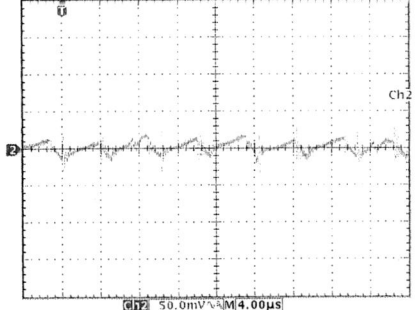

Fig. 10 Output voltage ripple

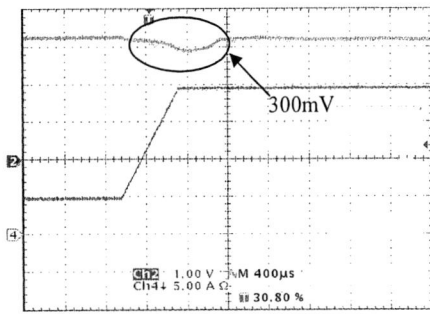

Fig. 11 Output current change from 5A to 20A

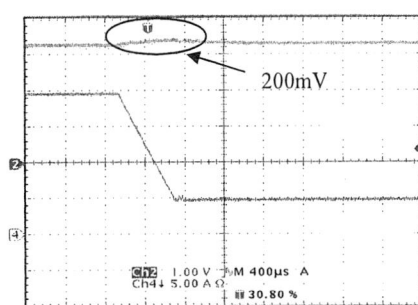

Fig. 12 Output current change from 20A to 5A

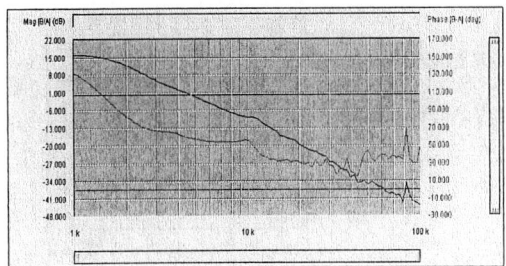

Fig. 13 Loop gain at 20A

978-1-4244-4782-4/10 $26.00 © 2010 IEEE   938

Fig. 11 and Fig. 12 show the transient response of the proposed converter (Ch4 is the output current. Ch2 is the output voltage). When the output changes from 5A to 20A in about 500uS, the output voltage has a dip about 300mV. When the output changes from 20A to 5A in about 500uS, the output has a surge about 200mV.

Fig.13 shows the close loop gain of the proposed converter. There is about 60 deg phase margin. And the gain margin is more than 15 dB while the crossing frequency is about 5 kHz. Stable system could be designed with simple control.

V. CONCLUSIONS

This paper proposed a four-winding coupled interleave converter. Both analysis and experiment have shown that the coupled inductor can reduced the input and output ripple dramatically. Zero input and output ripple can be achieved, together with fast transient response. The proposed converter is suitable to be used in the application which has special requirement for the input.

REFERENCES

[1] Y. Ren. "High Frequency, High Efficiency Two-Stage Approach for Future Microprocessors". Ph. D. Dissertation, Virginia Tech 2005.

[2] D. Fu, P. Kong, S. Wang, F.C. Lee, M. Xu, "Analysis and Suppression of Conducted EMI Emissions for Front-end LLC Resonant DC/DC Converters", in Proc. IEEE PESC, 2008, P: 1144-1150

[3] J. L. White, W. J. Muldoon. "Two-inductor Boost and Buck converter", Proc. Of the IEEE Power Electronics Specialists Conference, pp. 387-392, 1987.

[4] A. Capel, H. Spruyt, A. Weinberg, D. O'Sullivan, A. Crausaz, JC. Marpinard, "A versatile zero ripple topology", Proc, of the IEEE Power Electronics Specialists Conference, pp. 133-141, 1988.

[5] R. Martinelli, C. Ashley. "Coupled inductor boost converter with input and output ripple cancellation", Proc, of the IEEE Applied Power Electronics Conference, pp. 567-572, 1991.

[6] Z. Zhang, S. Cuk, "A high efficiency 1.8kW battery equalizer", Proc. Of the IEEE applied Power Electronics Conference, pp. 221-227, 1993.

[7] V. Grigore, J. Kyyra, "A Step-down Converter with Low Ripple Input Current for Power Factor Correction" APEC 2000 Volume 1, 6-10 Feb. 2000 Page(s):188-194 vol.1

[8] P. Perol, "An Efficient Low Cost Modular Power System for Fully Regulated Bus in Low Earth Orbit Appications", Proc. Of six European Space Power Conference, May 2002, pp.375-381

[9] Pit-Leong Wong, Peng Xu, Bo Yang, fred C. Lee. "Performance Improvements of interleaving VRMs with Coupling inductors", Power Electronics, IEEE Transactions on. Volume 16, issue 4, July 2001 Page(s): 499-507

[10] Xiangcheng Wang, Qingshui li and Issa Batarseh, "Large Signal Compensation Network Design for VRM Transient Response Improvement", IEEE IECON2005, pp.145-150

Investigation on Transformer Design of High Frequency High Efficiency DC-DC Converters

Dianbo Fu, Fred C. Lee and Shuo Wang
Center for Power Electronics Systems
Virginia Polytechnic Institute and State University
Blacksburg, VA 24061 USA

Abstract- **This paper studies the high-frequency high-efficiency transformer design. Several novel concepts are proposed to reveal the essence of the transformer design. In order to minimize the winding loss, several winding structure are proposed and compared. The planar transformer winding with PCB or spiral windings are discussed. The key factors to achieve the low winding losses are presented. The transformer design of the LLC resonant converter is investigated. The impacts on integration of magnetizing inductance and leakage inductance are studied.**

I. INTRODUCTION

Performance per watt requirements are driving higher energy efficiency and density metrics in power conversion technologies. Hence there is a large incentive to increase the overall efficiency and density of power delivery systems [1-16].

Hard switching PWM converters suffer high switching losses. Low efficiency and low power density are the major drawbacks. Soft switching PWM circuits, such as full bridge phase shift PWM converters are widely applied for dc-dc power conversions. Soft switching PWM converters can achieve zero-voltage-switching (ZVS). Therefore, lower switching loss and higher frequency can be accomplished. Nevertheless, issues of PWM converters still exist. First of all, ZVS may or may not be maintained for the whole load range. Secondly, the secondary side rectifiers snap off drastically and hence suffer high switching loss and high voltage stress. For computing and consumer electronics applications, such as server, desktop, flat panel TV, etc, holdup time operation is required. Conventional PWM converters have to sacrifice normal operation efficiency to extend its operation range [1-8].

Shown in Fig. 1, LLC resonant converters are studied and draw a lot of attention since 2002 [1, 2, 6–11]. The advantages of LLC resonant converters are: (i) ZVS capability for zero to full load range; (ii) low turn off current for primary side switches, so switching loss is low; (iii) Zero-current-switching (ZCS) is achieved for secondary side rectifiers; (iv) voltage gain boost capability without deterioration of efficiency at normal condition, suitable for hold-up time operation. In comparison with soft switching PWM converters, LLC resonant converters can achieve higher frequency and high power density with better efficiency. Therefore, LLC resonant converters are good candidates for next generation high-power-density high-efficiency dc-dc converters.

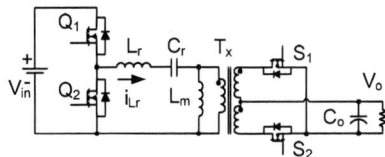

Fig. 1. LLC resonant converters.

For LLC resonant converters, topology investigation and optimization have been studied extensively [1, 2, 6–11]. Finally, the key issues for achieving high efficiency and high power density turn to the magnetic design. Normally, magnetizing inductance L_m acts as one of the resonant component and can be integrated with the transformer. At high frequency, the series resonant inductor L_r can also be integrated with the transformer. With the integration, the size of the magnetic component of LLC resonant converters can be substantially reduced. However, how to design transformer with low losses is not very clear. What is the impact of the magnetic integration is still concealed. On the other hand, magnetic component design is among the most important aspects to achieve high efficiency for dc-dc converters. For low-voltage high-current applications, center-tapped structure with multiple parallel windings is widely applied. How to reduce the losses of center-tapped windings becomes crucial.

This paper studies the high-frequency high-efficiency transformer design. Several concepts are proposed to reveal the essence of the transformer design. The planar transformer winding with PCB or spiral windings are discussed. The termination effect at high frequency is presented. In order to minimize the winding loss, several winding structures are proposed and compared. The impacts of magnetizing inductance and leakage inductance are studied. All the analysis is verified by the finite element analysis (FEA). According to the investigation, the optimal transformer structure is proposed. Generally speaking, the presented design methodologies can be also extended to PWM and other resonant converters. Finally, the theoretical analysis and the proposed optimal transformer

978-1-4244-4782-4/10 $26.00 © 2010 IEEE

design are verified and demonstrated on a 1 kW, 1 MHz, 390 /12 V LLC resonant converter prototype.

II. TRANSFORMER CURRENT EXCITATION ANALYSIS AND TRANSFORMATION

The center-tapped transformer structure is normally chosen for low-voltage high-current applications. The winding current of LLC transformer is plotted in Fig. 2. Each secondary side winding only conducts half cycle. To obtain the winding loss, normally, Fourier analysis is performed under the assumption of linear magnetic system. The total winding loss is the sum of each harmonic. Obviously, winding loss analysis of center tapped structure is very complicated.

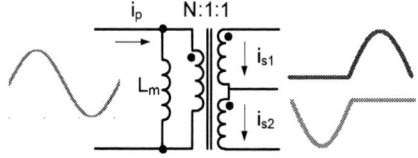

Fig. 2. Center-tapped transformer configuration.

Normally, to avoid circuit unbalanced operation, the center-tapped windings are designed as symmetrically as possible. Thus, physically speaking, the operation mode of the winding in first half cycle should exactly equal the winding of the second half cycle provided the good symmetry is achieved. In this case, from the winding loss point of view, equivalently, one winding carries all the secondary side current and another winding carries none. Following this equivalent operation concept, the current of one secondary winding can be transferred to another secondary winding and the current distribution will also reflect faithfully. Hence, it becomes very convenient to study the current distribution and winding losses. The equivalent transformation results are shown in Fig. 3-Fig. 7. Fig. 3 illustrates an example of a planar transformer with sandwich interleave structure. The outer side PCB windings are primary side windings. The inner PCB windings are interleaved secondary side windings. Both primary and secondary side windings are connected in series, for instance. The current excitation of original circuit is shown in Fig. 4. The transformed current excitation is shown in Fig. 5. The instantaneous winding losses of these two excitations are depicted in Fig. 6 and Fig. 7 respectively. It can be observed the two results are identical. Hence, with the proposed transformation method, complicated Fourier analysis and associated FEA can be avoided. It greatly simplifies the winding loss analysis and design. With the same concept, the current excitation transformation can be extended to other resonant converters and PWM converters with symmetrical operation and winding designs.

Fig. 3. Planar transformer structure example.

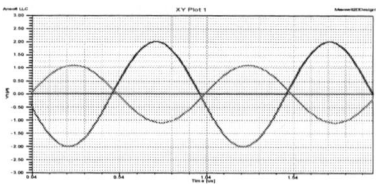

Fig. 4. Original winding current of LLC transformer.

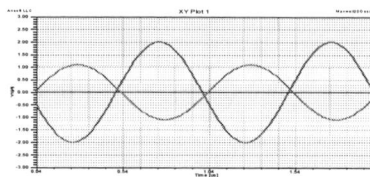

Fig. 5. Transformed winding current of LLC transformer.

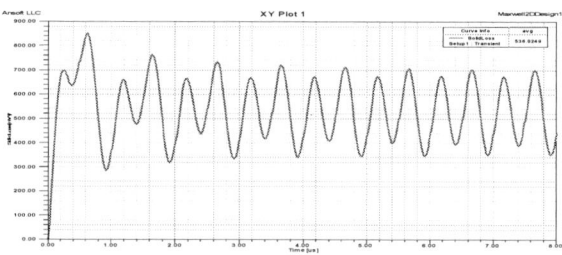

Fig. 6. Instantaneous total winding loss of LLC transformer with original current excitation.

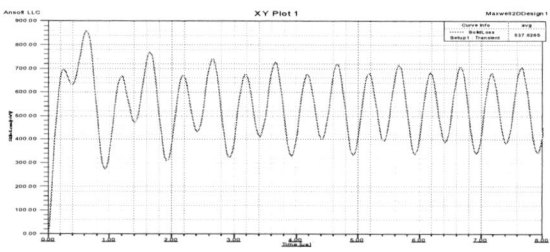

Fig. 7. Instantaneous total winding loss of LLC transformer with transformed current excitation.

LLC resonant converters are normally operate at the resonant frequency to achieve the highest efficiency. Thus, the resonant tank current is sinusoidal waveform. Although the primary side winding current is sinusoidal current, the secondary side winding currents are distorted owing to the effect of magnetizing current. The simulated current waveforms and their fundamental components are drawn in Fig. 8. Here, 1 kW, 1 MHz, 390 V/12 V LLC resonant converter is simulated as an example. The transformer structure is shown in Fig. 9. The transformer turns ratio is 16:1:1. Red bars represent the 16 primary side windings in series. Blue and green bars represent the center-tapped secondary side windings, which are 1 turn for each winding and in parallel. The Fourier analysis of the secondary side current and their related winding losses are illustrated in Fig. 10. The FEA results indicate more than 99% of the total loss is contributed by the fundamental component. As a result, the higher order harmonics can be ignored.

978-1-4244-4782-4/10 $26.00 © 2010 IEEE

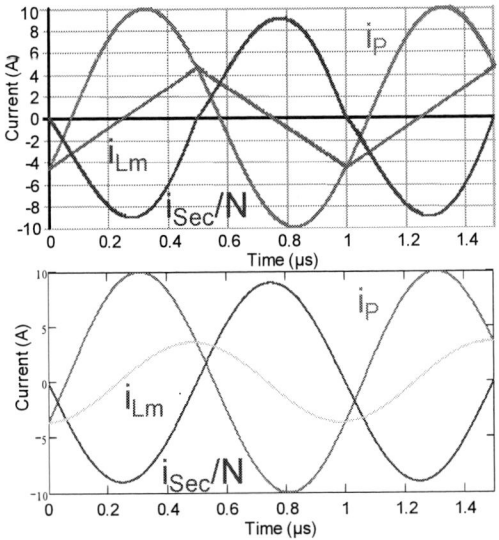

Fig. 8. Typical waveforms of LLC resonant converters and their fundamental components.

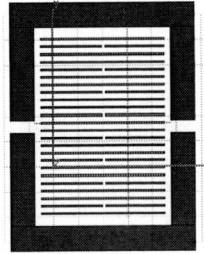

Fig. 9. Transformer winding structure for 1kW 1MHz 390V/12V LLC.

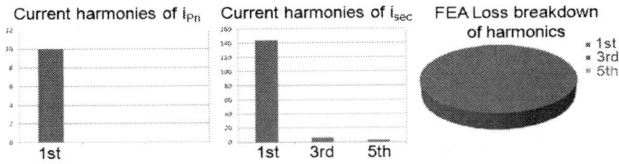

Fig. 10. Harmonic analysis and the loss breakdown of each harmonic.

Furthermore, the transformer winding current can be decomposed into ideal transformer current excitation and magnetizing current excitation, which are illustrated in Fig. 11.

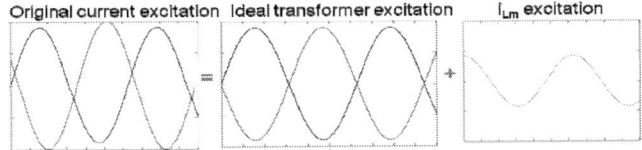

Fig. 11. Transformer winding current decomposition.

According to the orthogonal principle, if the sinusoidal currents are 90 degree phase shift (orthogonal), the sum of their excited losses equals the total loss. This is because the product of orthogonal components equals zero. Thus, it can be concluded if the decomposed ideal transformer current excitation and magnetizing current excitation are orthogonal, the total losses can be directly divided into two parts, which

excited by the decomposed parts respectively. According to the calculation, the phase shift between the decomposed ideal transformer current excitation and the magnetizing current excitation is not exactly 90 degree. Nevertheless, the phase shift is only a few degree deviations. For the waveforms shown in Fig. 8, the phase shift is 85 degree. Thus, it almost meets the orthogonal assumption and the pseudo-orthogonal results are very close to the original results. The FEA losses are shown in Fig. 12 - Fig. 14, respectively. The exact total winding loss is 13.1W. The sum of decomposed loss is 13W. The error is 0.1%.

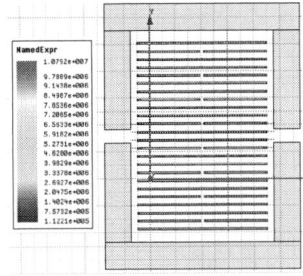

Fig. 12. Current distribution with original excitation.

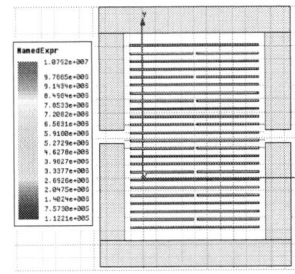

Fig. 13. Current distribution of decomposed ideal transformer excitation.

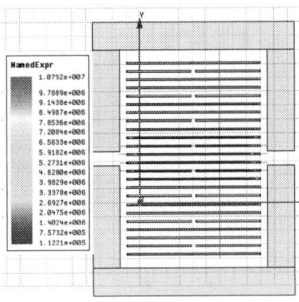

Fig. 14. Current distribution for decomposed magnetizing current excitation.

According to the current decomposition strategy, it is very convenient to reveal the essence of the winding losses. FEA shows that only 3.2W is generated by the ideal transformer excitation. The interleaving structure helps to reduce the ac winding loss effectively. However, the magnetizing current induces 9.8W winding loss, which can be divided into two parts. One is induced eddy current loss on the secondary side windings. Another one is the eddy current loss owing to the fringing flux. Flux lines of the transformer are plotted in Fig. 15. The fringing flux is very strong close to the gap. Thus, high eddy currents are induced on the copper plates near the gap.

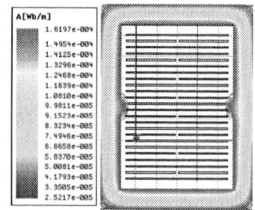

Fig. 15. Flux distribution of the transformer.

The relationship of winding loss and gap distance is illustrated in Fig. 16. If the gap distance is reduced, the fringing effect is reduced. Also, small gap leads to higher magnetizing inductance. Higher magnetizing inductance means less magnetizing current. Thus, both the current magnitude and induced eddy currents on the secondary side are reduced.

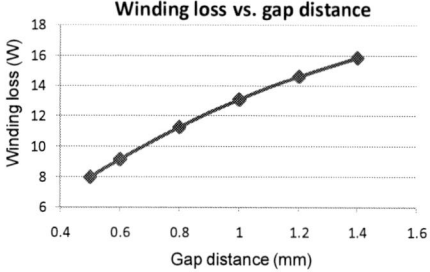

Fig. 16. Winding loss comparison for different gaps.

The magnetizing current leads to extra winding losses. To reduce the winding loss, it is preferred to achieve large magnetizing inductance with small gap. However, because the magnetizing inductance is normally determined by the whole system design [1, 2, 6–11], it is not convenient to change the magnetizing inductance in some circumstances. Other methods should be investigated for loss reduction.

III. TERMINATION LOSS ANALYSIS AND THE IMPROVED TRANSFORMER STRUCTURE

For conventional transformer design, vertical winding structure is normally adopted. The conventional vertical transformer structure is drawn in Fig. 17. For low frequency and low cost design, solid or multi-strand wires are applied for the primary side winding. At high frequency, litz wires are chosen. For off-line dc-dc converters, the secondary side current is much higher than the primary side. Therefore, copper foils are selected for the secondary side winding. To alleviate ac winding loss, the primary side and the secondary side windings are interleaved as a sandwich structure.

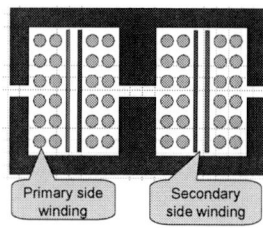

Fig. 17. Conventional vertical transformer structures.

For conventional vertical transformer structure, the secondary side windings are connected with the motherboard via copper poles, which are shown in Fig. 18. However, this structure suffers high termination losses. Due to proximity effect, the current surges in adjacent sides of terminals. Meanwhile, very little current goes through the center of the terminals because of the skin effect. Thus, very high losses and hot spot are generated. According to the FEA result, 8 W loss dissipates at winding terminals for 1 kW, 1 MHz LLC resonant converters. Termination losses sacrifice efficiency significantly.

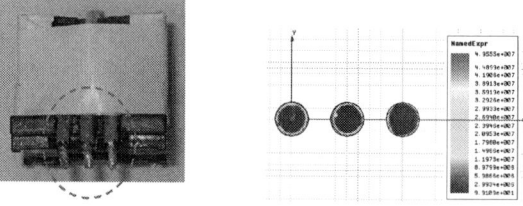

Fig. 18. Transformer secondary side terminals and associated current distribution of conventional vertical transformer structures.

The conventional planar transformer and the secondary side structure are depicted in Fig. 19. For low-voltage high-current applications, synchronous rectifiers (SR) are popular to achieve lower conduction loss. The circuit schematic of the transformer and SR structure is shown in Fig. 19. For high current applications, paralleled winding structure is a common practice. However, this structure suffers high termination losses. The current distribution of the termination is shown in Fig. 20. Similar to copper poles situation, the current surges in adjacent terminals. Thus, very high losses and hot spot are generated. According to the FEA result, 6.5 W loss dissipates at winding terminals for 1 kW, 1 MHz, 390 V/12 V LLC resonant converters. This termination loss deteriorates efficiency significantly. High termination loss is a barrier to achieve high efficiency.

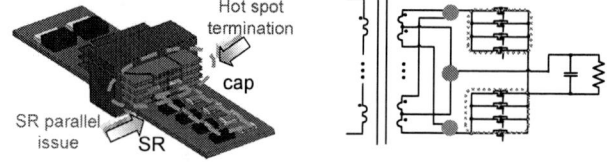

Fig. 19. Conventional transformer and the secondary side structure.

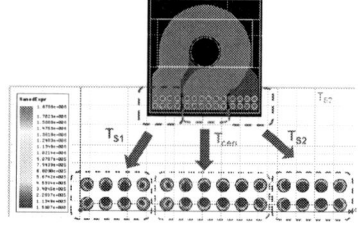

Fig. 20. Current distribution of termination for the conventional transformer design.

Alternatively, in [13], the interleaved termination approach was proposed to reduce the ac termination loss. With non-interleaved termination configuration, the multiple pins do not contribute current conduction effectively owing to strong proximity effect. With interleaved termination approach, the current can be distributed more evenly among the termination pins. Thus, lower conduction loss can be attained. However, the design of the interleaved termination is very complicated. High level interleaving is required to effectively reduce the proximity effect. Taking the clearance distance between interleaved pins into account, the effective conduction copper area before reaching the termination pins will be reduced considerably. This will increase the conduction loss. More importantly, for center-tapped transformer structure, three terminals are adopted. Interleaved termination design becomes quite difficult and may be impractical in some cases, especially in planar PCB transformer design.

On the other hand, to reduce the high current conduction loss, high number of parallel SR is a must. For the conventional design, all SRs are placed on the mother board. Physically speaking, large loop of the secondary side rectifiers is inevitable. Although multi-layer printed circuit board (PCB) with heavy coppers can reduce dc conduction loss, there is little effect to reduce the high frequency ac losses. Consequently, large distribution loss will be generated. In addition, for large number of SR devices, it is extremely hard to achieve symmetrical layout for each SR. Hence, current sharing of SR is a severe problem owing to different path impedance of each SR.

In [14], a PCB winding concept is proposed. The secondary side winding is integrated with the device PCB board. Virtually, termination is achieved at dc output. Thus, the termination loss can be reduced considerably. However, there are some limitations for transformer structure in [14]. The primary side windings are interleaved with the secondary side windings as one. For each cell, the primary side winding is connected in series. To finish the whole transformer, multiple cells are connected in parallel. As a result, this structure simplifies the secondary side connection but the primary side connection becomes very complicated. In addition, since all primary side and secondary side windings are paralleled, current sharing between windings might be a potential issue for some circumstances, especially for high current applications.

Thus, improved transformer and the secondary side structure are critical to improve the system efficiency and need further investigation.

To achieve a simple transformer without ac termination problems, an improved transformer structure is proposed and depicted in Fig. 21.

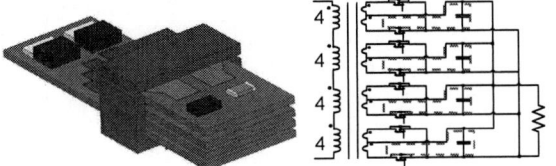

Fig. 21. An improved simple transformer structure with low termination loss.

The proposed improved transformer winding structure owns several merits.

(1) Because SR devices and output capacitor filter are directly placed on each winding plate, the ac termination issues can be solved.

(2) SR is integrated with the transformer winding. Each device winding carry one conducting SR at any time. Thus, there is no SR parallel issue.

(3) The primary side winding are placed in series and can be interleaved with the secondary side windings. Interleaving structure is beneficial to reduction of ac winding loss. The primary side winding also facilitates even current distribution among the secondary side windings. More importantly, no complicated series-parallel connection exits. Simple structure contributes easy manufactory.

IV. PLANAR TRANSFORMER DESIGN WITH MIXED LITZ WIRES AND PCB WINDINGS

It is well known the litz wire can reduce the ac losses. Due to the split and twisted structure, the litz wire is not sensitive to the internal and external flux. To reduce the fringing flux losses, litz wires are applied for the primary side windings.

Due to high output current, the secondary side winding is designed with one turn. The optimal number of SR can be obtained with trade off of conduction loss and switching loss. High number of SR leads to excessive driving loss. Thus, eight SRs are applied and are grouped into four sets. Then the four secondary PCB windings are chosen to carry four sets of SR.

There are several possible transformer structures. The different structures of the combined litz & PCB windings are depicted in Fig. 22. The detail examination is studied to find the most suitable transformer winding structure. 4oz (2 skin depth) copper PCB is used for FEA. PCB width is 9.6mm.

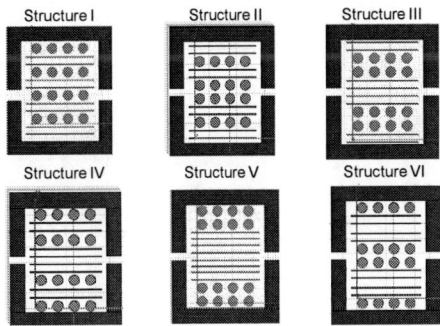

Fig. 22. Possible transformer structure with combined litz & PCB windings.

Current, flux and winding loss distribution of Structure I are illustrated in Fig. 23. Normally, this structure is widely applied for conventional transformer design due to its easy interleaving structure. However, because the interleaving structure is not symmetrical, the current distribution on the secondary side windings is very different. The bottom PCB winding is only interleaved by one set of primary side winding, while the other three are interleaved with two sets of primary side windings. Consequently, the bottom PCB winding conducts much less current than the other three. The winding loss on the secondary side is very different.

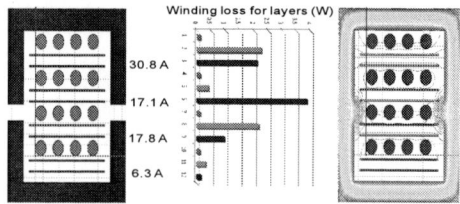

Fig. 23. Winding configuration, net current among conducting secondary windings, winding loss, flux and current distribution of Structure I.

Compared with structure I, PCB windings of structure II are farther from the gap, which is shown in Fig. 24. Thus, the fringing effect loss can be further reduced. However, more PCB windings are not complete interleaved with the primary side windings. The current distribution on PCB windings is even worse. As a result, the total winding loss cannot be reduced.

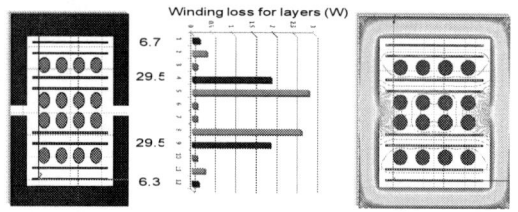

Fig. 24. Winding configuration, net current among conducting secondary windings, winding loss, flux and current distribution of Structure II.

Depicted in Fig. 25, Structure III achieves good symmetrical winding placement. Thus, the current distribution among windings is more even. However, the drawbacks are high fringing effect losses. It also should be noted, the fringing effect also affects the current sharing between the PCB windings. This structure is still not attractive.

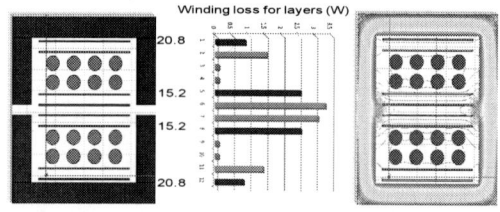

Fig. 25. Winding configuration, net current among conducting secondary windings, winding loss, flux and current distribution of Structure III.

Structure IV and structure V owns the same problems as structure III. They are not good structures at all. PCB windings are close to the gap, which induced high fringing effect losses. The whole winding layout is not symmetrical from the interleaving point of view. Thus, high loss is inevitable. These structures are not practical and only exist for completing the possible structure construction.

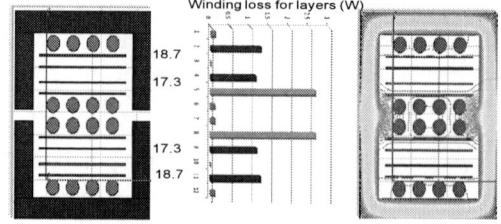

Fig. 26. Winding configuration, net current among conducting secondary windings, winding loss, flux and current distribution of structure VI.

Drawn in Fig. 26, The PCB windings of Structure VI are placed far from the gap. Hence, it leads less fringing effect loss. The primary and the secondary side windings are interleaved symmetrically. The current distribution among different PCB windings is very good. Therefore, this structure is acceptable with consideration of avoiding fringing flux and creating even current sharing.

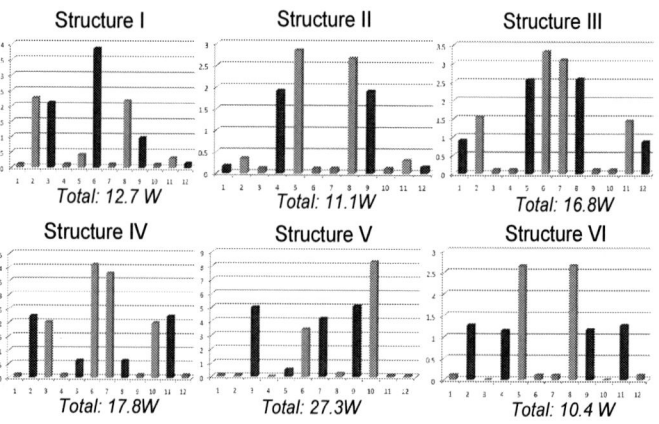

Fig. 27. Winding loss comparison for different structures.

Winding loss comparisons of different structures are illustrated in Fig. 27. Due to the advantages of low fringing flux loss and even current distribution, structure VI is chosen as the preferable transformer winding structure.

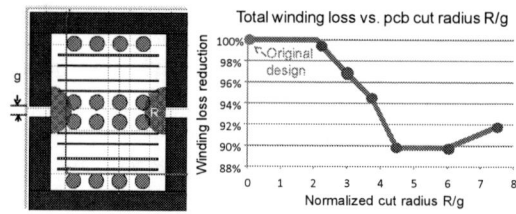

Fig. 28. Winding reshape concept to reduce the fringing effect losses.

Because majority of the fringing flux has been avoid by place the PCB winding far from the gap, it is easy to further reduce the fringing effect loss by reshaping the PCB winding. Shown in Fig. 28, the winding loss can be reduced by slightly cutting the copper adjacent to the gap. With more cutting area of copper, the high surge current can be avoided. Therefore, lower winding loss can be achieved. However, excessive copper cut leads to higher overall winding resistance. Therefore, there is

978-1-4244-4782-4/10 $26.00 © 2010 IEEE

an optimal cutting area range. Within this range, the winding loss can be reduced effectively.

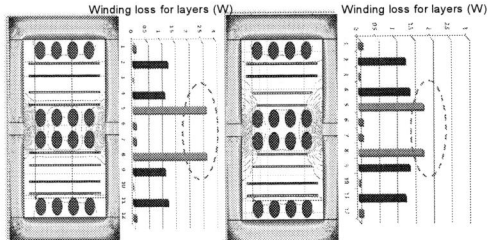

Fig. 29. Winding loss comparison for different gaps.

Fig. 29 illustrates the detail FEA result. Apparently, less fringing effect loss benefits the winding loss reduction. Shown in Fig. 28, 10% winding loss reduction can be accomplished. It should be noted the virtual 'cutting' can be easily achieved in initial PCB winding layout. Thus, there is no manufactory problem at all. Distributed air gap is another effective way to reduce winding loss. However, manufactory might be difficult. Meanwhile, at 1 MHz, the suitable magnetic material with low permeability and low core loss is very rare.

V. IMPACTS OF MAGNETIC COMPONENTS INTEGRATION

For LLC resonant converter, there are three magnetic components. To meet the power density requirement, it is preferable to apply magnetic components integration. Thus, fewer magnetic components, less passive components size, lower core loss and simple connection can be achieved.

The magnetizing inductance can be chosen as the parallel resonant inductor of LLC resonant converters. The impacts of magnetizing inductance have been discussed. To reduce the winding loss, it is preferred to achieve large magnetizing inductance with small gap. Litz wires and PCB winding reshaping techniques can reduce the adverse effect related to the gap.

It is preferable to integrate the leakage inductance as the series resonant inductor. However, the natural leakage inductance of the transformer might not be large enough for this purpose. Normally, it is not convenient to adjust the leakage inductance with increasing the space between the primary side and the secondary side windings. It leads to poor winding window utilization.

Insertion of magnetic shunt is an effective way to create the required leakage inductance. The concept of the utilization of magnetic shunt is shown in Fig. 30. A magnetic material can be inserted between the primary side and the secondary side windings. Thus, the induced leakage energy can be increased if the high permeability materials are applied. The leakage inductance can be adjusted with the variation of the magnetic shunt thickness. In this application, low permeability material FPC is applied. The permeability is 9. Thus, it is high enough to create the needed leakage inductance. Compared with the normal ferrite materials with much higher permeability, it is not very sensitive to the thickness of the shunt layer. Thus, very

precise leakage inductance can be obtained. For 1 kW, 1 MHz, 390 V/12 V LLC converter design, 1 μH leakage inductance is desired. According to FEA result, 0.4mm magnetic shut is inserted.

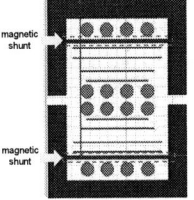

Fig. 30. Winding loss comparison for different gaps.

Fig. 31 illustrates the flux lines and the H field distribution for the winding structures with and without magnetic shunt. The major flux lines and H field do not change significantly. Compared with $\mu_r>1000$ for the main core, the permeability of magnetic shunt is so low that it can hardly affect the main flux. Nevertheless, magnetic shunt layers do attract more flux and induce a slight fringing effect close to the edge of the shunt layers. Because magnetic shunt is not an excellent magnetic conductor, the induced extra flux bends slightly. The curved leakage flux generates a little bit more winding loss. The detail loss breakdown of each winding layer is depicted in Fig. 32. According to the curved leakage flux, 0.5W extra winding loss will be induced. Nevertheless, it is still acceptable for 1 kW converters. Easy design and convenient implementation make this integration method attractive. FEA simulation results validate the analysis.

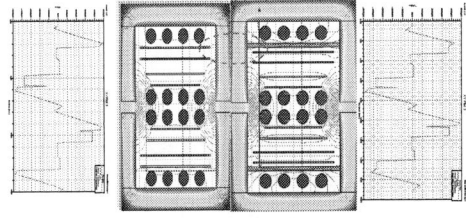

Fig. 31. H field comparison with and without magnetic shunt.

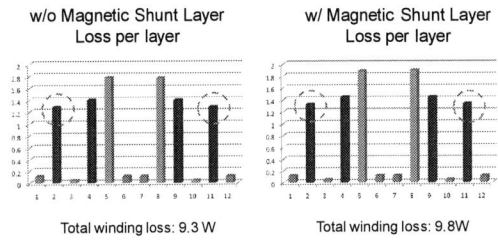

Fig. 32. Winding loss comparison with and without magnetic shunt.

Another factor affects the winding loss is the thickness of the winding copper. The winding loss is heavily affected by the winding copper thickness at the high frequency operation. The relationship of winding loss versus copper thickness is given in Fig. 33. It can be observed that thicker copper does not lead to lower winding loss at the high frequency, such as 1 MHz case.

On the contrary, too heavy copper increases the total winding loss. This can be explained by the eddy current effects. With thicker conductor, more eddy current can be induced. Thus, the loss is even higher. In addition, more copper thickness means higher cost. Thus, optimal copper thickness should be chosen. In 1 MHz case, 2 oz. and 3 oz. copper achieve the lowest winding loss. With consideration of cost, 2 oz. copper is chosen for the improved design. As a result, the optimal copper thickness is chosen as 1 skin depth of the operating frequency.

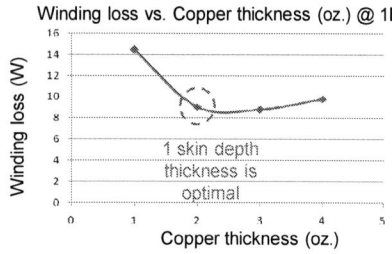

Fig. 33. Winding loss comparison of winding loss for different inserted magnetic shunt structures.

VI. EXPERIMENTAL RESULTS

Shown in Fig. 34, a 1 MHz 1 kW 390 V/12 V prototype is built to verify the proposed transformer design and analysis for LLC resonant converters. The L_m is chosen as 14 μH. The resonant inductor L_r is 1 μH and the resonant capacitor is chosen as 20 nF. The primary side device is chosen as STW45NM50FD. BSC016N04LS is chosen as the secondary side device.

Fig. 34. The prototype of 1 kW, 1 MHz, 390 V/12 V LLC resonant converter.

High frequency operation and magnetic integration contribute size reduction of passive components. All magnetic elements have been integrated into one component. 200W/in³ power density is achieved. 95.5% efficiency is achieved at 1kW condition. Further improvement might be achieved with improved high current side arrangement, such as winding and SR, etc.

VII. CONCLUSION

This paper describes some novel concepts to achieve the high efficiency of the transformer at high frequency. To simplify the analysis, the current transformation of center-tapped structure is proposed. Complicated Fourier analysis and associated FEA

can be avoided. The proposed transformer current decomposition reveals the essence and difficulties of the transformer design. Great attention should be paid to alleviate the extra winding loss owing to magnetizing current. High termination loss is another stringent challenge of transformer design. An improved transformer structure is proposed to reduce the termination loss and simplify the transformer winding configuration. Optimal transformer structure is proposed and verified with the FEA and hardware. Low loss is achieved for transformer with integration of extra two resonant inductors. The proposed concepts might be extended to other PWM and resonant dc-dc converters.

Finally, a 1M Hz, 1k W LLC resonant converter with SR is built to verify the theoretical analysis. High efficiency and high power density are accomplished.

REFERENCES

[1] B. Yang, F.C. Lee, A.J. Zhang, G. Huang, "LLC resonant converter for front end DC/DC conversion," in *Proc. IEEE APEC*, 2002, pp. 1108–1112.

[2] B. Lu, W. Liu, Y. Liang, F.C. Lee, J.D. van Wyk, "Optimal design methodology for LLC resonant converter," in *Proc. IEEE APEC*, 2006, , pp. 533–538.

[3] Y. Xie, J. Sun, and J. S. Freudenberg, "Power Flow Characterization of a Bidirectional Galvanically Isolated High Power DC/DC Converter Over a Wide Operating Range", *IEEE Trans. on Power Electron.*, 2009.

[4] Y. Xie, J. Sun, and J. S. Freudenberg, "Integrated Voltage Regulation and Current-sensorless Power Tracking Control for A DC Hybrid Power System", in *Proc. American Control Conference*, 2008, pp. 591-596.

[5] Y. Xie, R. Ghaemi, J. Sun, and J. S. Freudenberg, "Implicit Model Predictive Control of a Full Bridge DC/DC Converter", *IEEE Trans. On Power Electron.*, 2009.

[6] Y. Gu, Z. Lu, L. Hang, Z. Qian, G. Huang, "Three-level LLC series resonant DC/DC converter," IEEE Transactions on PE, Volume 20, Issue 4, July 2005 Page(s):781 – 789

[7] D. Fu, F. C. Lee, Y. Qiu, F. Wang, "A Novel High-Power-Density Three-Level LCC Resonant Converter with Constant-Power-Factor-Control for Charging Applications", *IEEE Trans. on Power Electron.*, 2008.

[8] D. Fu, Y. Liu, F.C Lee, M. Xu, "A Novel Driving Scheme for Synchronous Rectifiers in LLC Resonant Converters", *IEEE Trans. on Power Electron.*, 2009.

[9] D. Fu, B. Lu, F.C. Lee, "1MHz High Efficiency LLC Resonant Converters with Synchronous Rectifier," in *Proc. IEEE PESC*, 2007, pp. 2404-2410.

[10] D. Fu, F.C. Lee, Y. Liu, M. Xu, "Novel Multi-Element Resonant Converters for Front-end DC/DC Converters" in *Proc. IEEE PESC*, 2008, pp. 250-256.

[11] D. Fu, P. Kong, S. Wang, F.C. Lee, M. Xu, "Analysis and Suppression of Conducted EMI Emissions for Front-end LLC Resonant DC/DC Converters", in Proc. IEEE PESC, 2008, pp. 1144-1150.

[12] J.P. Vandelac, P.D. Ziogas, "A novel approach for minimizing high-frequency transformer copper losses," *IEEE Trans. on Power Electron.*, Vol. 3, Iss. 3, pp: 266-277, Jul. 1988.

[13] Ph.D dissertation, Modeling, Analysis, Design of High-Frequency High-Density Low-Profile Power Transformers, Ning Dai, 1996.

[14] C. Yan, F. Li, J. Zeng, T. Liu, J. Ying, "A Novel Transformer Structure for High power, High Frequency converter," in *Proc. IEEE PESC*, 2007, pp. 214- 218.

[15] R. Lai, F. Wang, Y. Pei, R. Burgos, D. Boroyevich, "Minimizing Passive Components in High-frequency High-density AC Active Voltage Source Converters", in *Proc. IEEE PESC*, 2007, pp: 672-677.

[16] R. Lai, F. Wang, R. Burgos, Y. Pei, D. Boroyevich, B. Wang, T.A. Lipo, V.D. Immanuel, K.J.A. Karimi, "Systematic Topology Evaluation Methodology for High-Density Three-Phase PWM AC-AC Converters," *IEEE Trans. on Power Electron.*, Vol. 23, Iss. 6, pp:2665-2680, Nov. 2008.

A DSP-Based Single-Stage Maximum Power Point Tracking PV Inverter

Wen Long Yu, Ting-Peng Lee, Guan-Hong Wu,
Qing Su Chen, Huang-Jen Chiu, Yu-Kang Lo
Dept. of Electronic Engineering
National Taiwan University of Science and Technology
Taipei, Taiwan
hjchiu@mail.ntust.edu.tw

Frank Shih
Technical Marketing, Macroblock Inc.
Hsinchu, Taiwan
Frank.shih@mblock.com.tw

Abstract—**This paper presents the design and implementation of a DSP-based single-stage photovoltaic (PV) inverter system which can extract maximum power from solar panel. A perturbation and observation (P&O) method is realized and cooperated with a digital unipolar voltage switching sinusoidal pulse width modulation. The studied single-stage topology provides the benefits of low cost, simple configuration and good overall efficiency compared with the conventional two-stage structure. Moreover, this paper also presents a new current sensing technique for maximum power point tracking (MPPT) to enhance the system flexibility and MPPT accuracy. A prototype system is implemented and tested. The experimental results are shown to verify the feasibility of the proposed scheme.**

I. INTRODUCTION

While fossil fuels exhaustion and greenhouse effects are widely concerned around the world, one of the most important issues toward to these problems is to find alternative energy for long-term solutions. Green energy offering the promise of clean and abundant energy gathered from self-renewing sources such as solar energy, geothermal energy and wind source are broadly developed. Solar cells are unique in that they directly convert the incident solar radiation into electricity. Photovoltaic (PV) power management concepts are essential to extract as much power as possible from the solar energy. PV energy systems are being extensively studied because of its benefits of environmental friendly and renewable characteristics. Unlike oil, gas and coals, solar energy does not emit greenhouse gases, or cause pollutions, and is expected to enhance the feasibility of lowering cost and increasing conversion efficiency [1]. While the studies for the improvements of the manufacturing process and materials remain as future targets, this paper focuses on the research of energy conversion system and the developments of high performance single-stage PV inverter system with maximum power point tracking (MPPT). Figure 1 shows the characteristic curves of PV cell. The maximum power point is changed nonlinearly. As shown in Figure 2, it depends on the amount of irradiation and the environmental temperature.

This work was supported in part by the Macroblock, Inc. and National Science Council of Taiwan through grant number NSC 96-2628-E-011-115-MY3

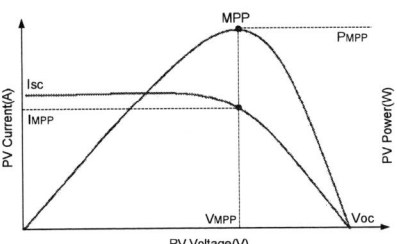

Figure 1 Characteristic Curves of PV Cell

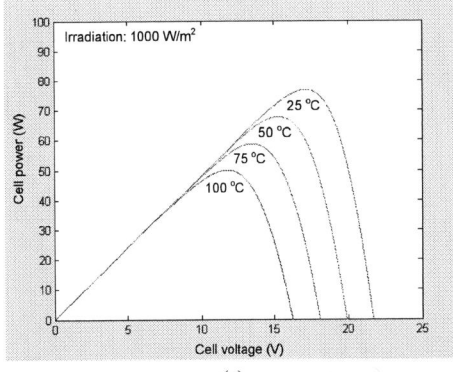

(a)

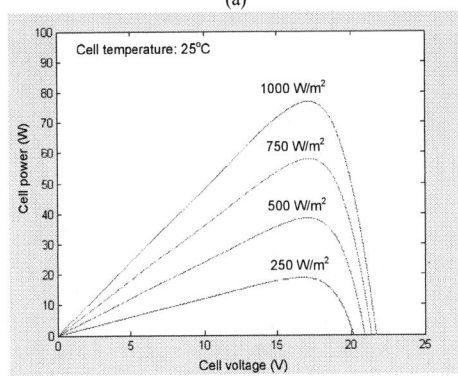

(b)

Figure 2 P/V Curves under Different (a) Temperature and (b) Irradiation Conditions

A conventional photovoltaic system shown in Figure 3(a) consists of a DC/DC converter for tracking the MPP of the photovoltaic array and a DC/AC inverter for supply AC voltage output [2-4]. An interesting alternative solution is the application of a single-stage topology shown in Figure 3(b). The advantages of the single-stage topology are a high power density, simple configuration and low circuit cost compared with the conventional two-stage structure [5, 6]. This paper proposes a PV inverter system which converts solar energy into an AC sinusoidal voltage output and tracks the maximum power point with a single-stage topology.

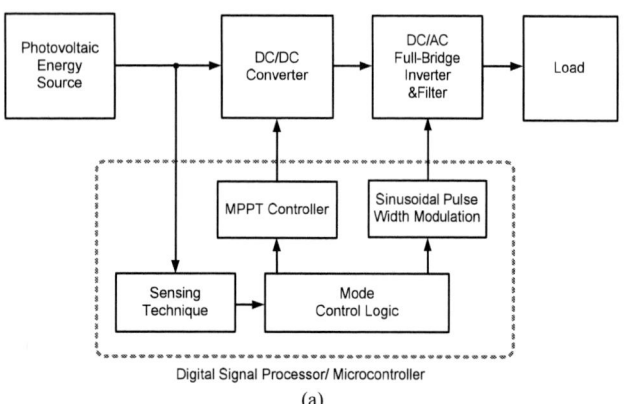

(a)

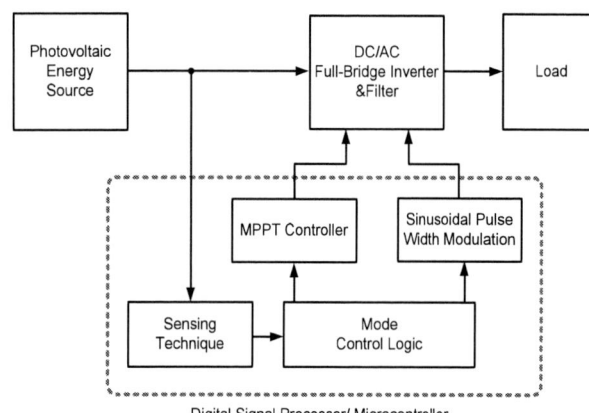

(b)

Figure 3(a) Conventional Two-Stage and (b) Single-Stage PV Inverter Systems

II. SYSTEM CONFIGURATION

Figure 4(a) shows the system configuration of the studied single-stage PV inverter with MPPT function. The current sensing point is moved from PV panel side to inverter side for raising system flexibility and increasing MPPT accuracy [7, 8]. However, the sensed current is discontinuous. As the sampling waveforms shown in Figure 4(b), the PWM interruption of upper switch T_{A+} is enable for A/D conversion when the event time-base counter is incrementing for the lower switch T_{A-}. Therefore, it can avoid extracting switching spike while the A/D converter is just activated and provide more options for sampling point by appropriately controlling the dead-time.

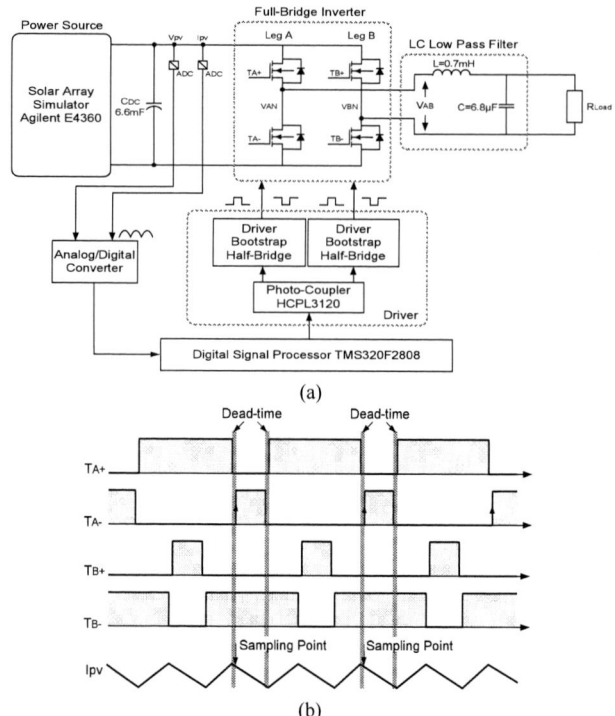

(a)

(b)

Figure 4(a) System Configuration of the Studied Single-Stage MPPT Inverter (b) Current Sampling Waveforms

In this study, the unipolar voltage switching scheme is used to realize the sinusoidal waveform output. As shown in Figure 5, the two sinusoidal reference signals $V_{control}$ and -$V_{control}$ are compared with the triangular waveform V_{tri}. The output voltage with unipolar voltage switching scheme changes between V_{DC} to $-V_{DC}$ input voltage levels. Furthermore, the frequency of the output voltage is twice as that of the triangular waveform, due to the control signals used for the two legs. It is clear that using sinusoidal PWM with uniploar voltage switching scheme results in a smaller ripple in the current on the DC side of the inverter [9, 10].

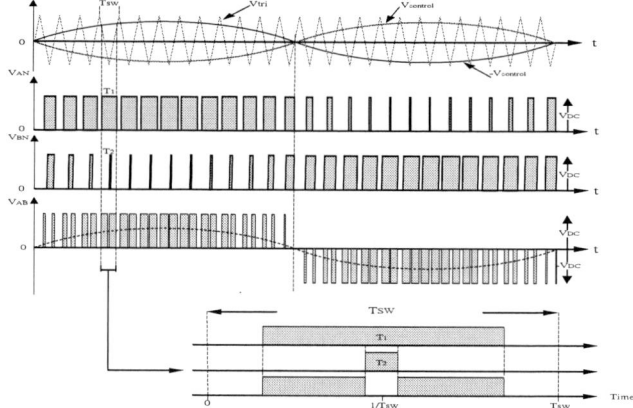

Figure 5 Sinusoidal PWM with Unipolar Voltage Switching

The relationship between the output voltage and the switching at leg A and leg B is given in Equation (1). As shown in Figure 6, there are four switching states.

$$\begin{bmatrix} V_{AB} \\ V_{BA} \end{bmatrix} = V_{DC} \cdot \begin{bmatrix} 1 & -1 \\ -1 & 1 \end{bmatrix} \cdot \begin{bmatrix} T_{A+} \\ T_{B1} \end{bmatrix} \quad (1)$$

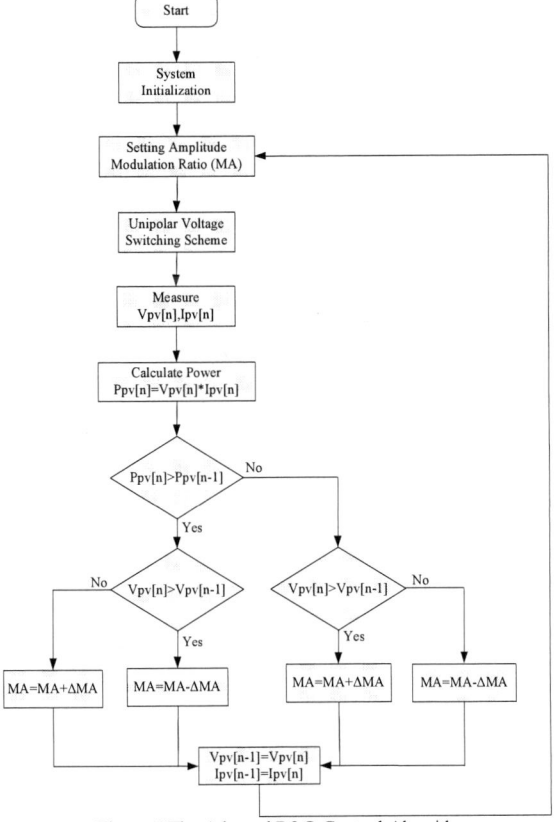

Figure 6 Switching States of Full-bridge Inverter

The perturbation and observation algorithm has been widely used in maximum power point tracking, because of its easy implementation and the fewer measured parameters which are required. Figure 7 shows the flow chart of the adopted P&Q algorithm of this paper. The MPP tracker operates by periodically perturbing the solar array terminal voltage and comparing its output power with that of the previous perturbation cycle. If the solar array voltage changes and power increases, the control system moves the solar array operating point in the same direction; otherwise, the operating point is moved in the opposite direction. After that the subsequent perturbed cycle is generated in the same way.

Figure 7 The Adopted P&O Control Algorithm

III. EXPERIMENTAL VERIFICATIONS

A prototype system is implemented and tested to verify the feasibility of the studied scheme. Circuit specifications of the laboratory prototype are shown in Table 1. The measured unipolar voltage switching PWM waveforms are shown in Figure 8. Figure 9 shows the measured MPPT waveforms of the studied single-stage PV system. High MPPT accuracy can be achieved. The output voltage and current waveforms of the studied PV inverter are measured and shown in Figure 10. It can be observed that the output waveforms are presented with low THD features. Figure 11 shows the performance comparisons of the conventional and proposed schemes. According to the experimental results, the MPPT accuracy and overall efficiency of the studied PV system can achieve **98.7%** and **88.8%** compared with the conventional topology with 97.8% MPPT accuracy and 87.4% overall efficiency.

Table 1 Circuit Specifications of the Laboratory Prototype

Parameter	Value
Maximum Input Power ($P_{in,max}$)	50 W
Switching Frequency (f_{sw})	120 kHz
Input Voltage/Current (V_{in}/I_{in})	17 V/2.941 A
Output Voltage/Frequency (V_{out}/f_{out})	11 Vrms/60 Hz
Full-Bridge Switching (T_{A+}, T_{A-}, T_{B+}, T_{B-})	IRF2807(40V/75A/2mΩ)
Filter Inductance (L)	700 µH
Filter Capacitance (C)	6.8 µF

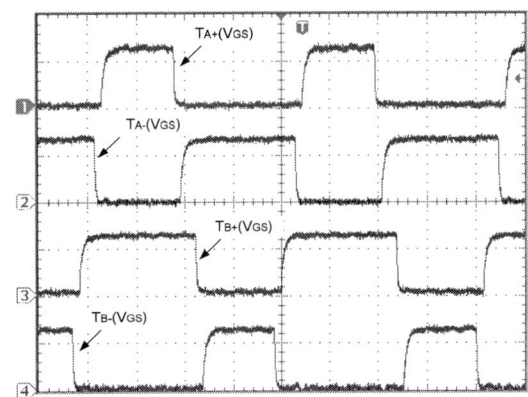

Figure 8 Unipolar Voltage Switching Waveform (Time: 2µs/div、CH1/TA+: 10V/div、CH2/TA-: 10V/div、CH3/TB+: 10V/div、CH4/TB-: 10V/div)

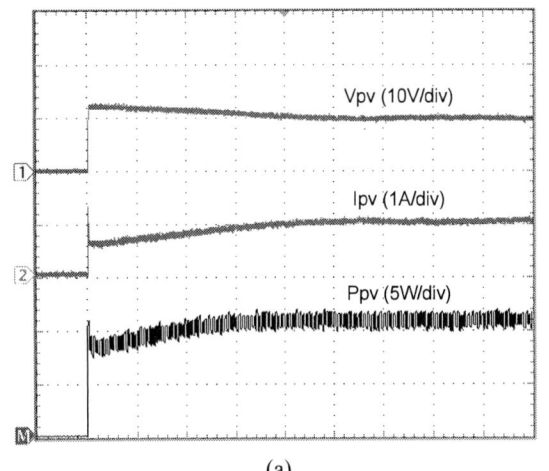

(a)

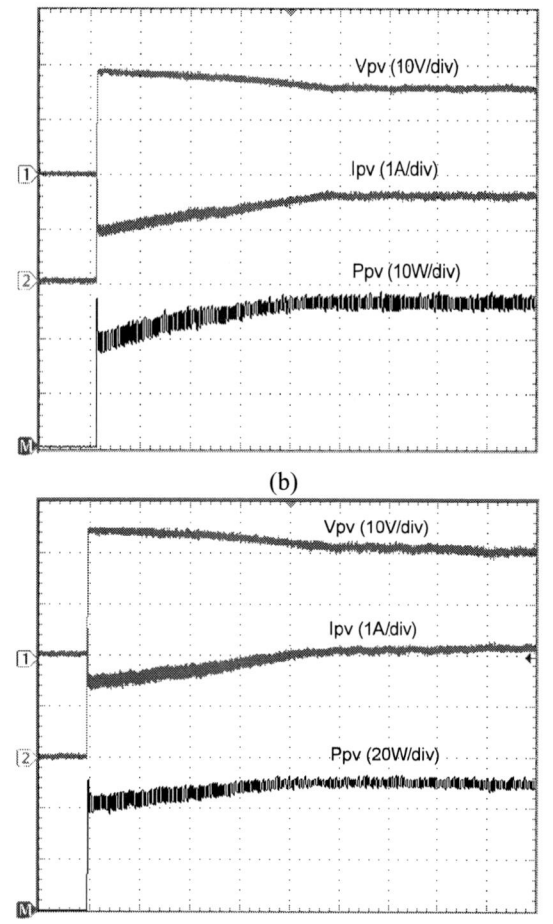

(b)

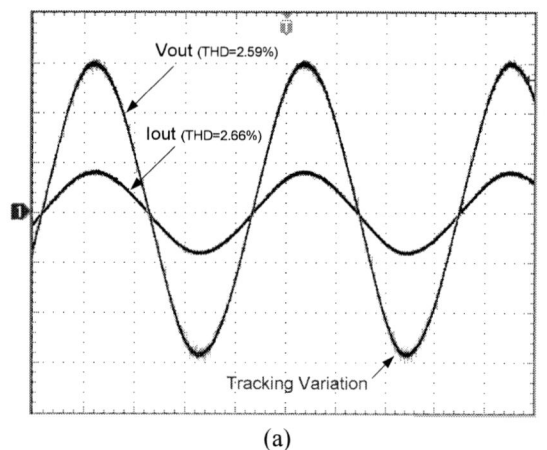

(c)

Figure 9 Measured MPPT Waveforms at (a) 10W, (b) 25W and (c) 50W Input.

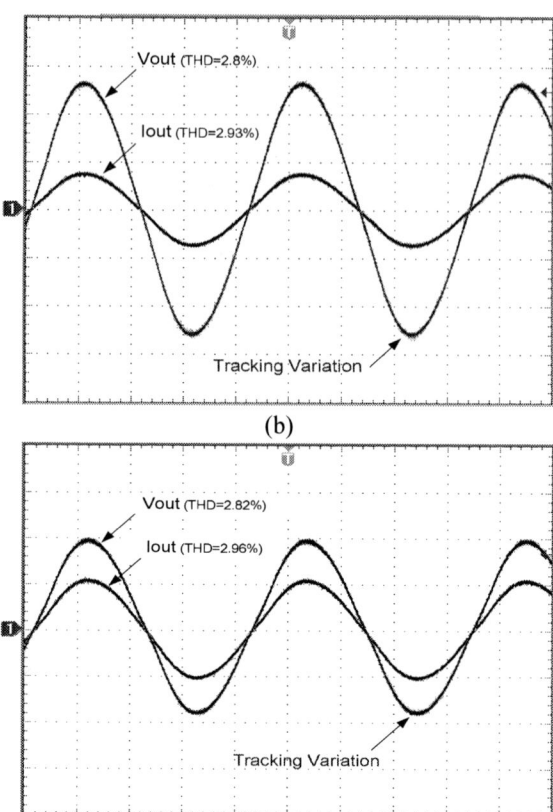

(b)

(c)

Figure 10 Measured Output Waveforms at (a) 10W, (b) 25W and (c) 50W Input.

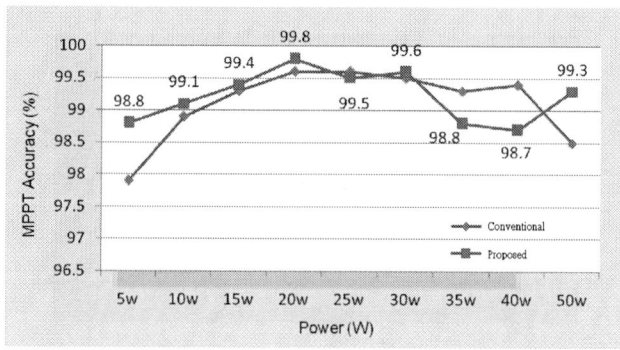

(a)

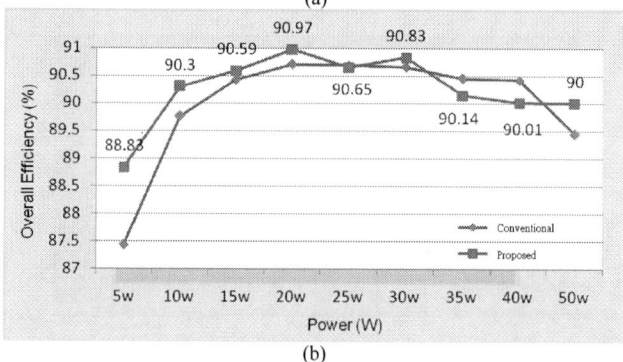

(b)

Figure 11 Performance Comparisons of the Conventional and Proposed Schemes (a) MPPT Accuracy and (b) Overall Efficiency.

978-1-4244-4782-4/10 $26.00 © 2010 IEEE

IV. CONCLUSIONS

This paper presents a DSP based single-stage PV inverter with maximum power point tracking function. High MPPT accuracy and high conversion efficiency can be achieved by using a simple single-stage configuration. A prototype system is implemented and tested to verify the proposed scheme. In the future, a module series technique with synchronization design will provide the further application of grid-tied photovoltaic system with a single-stage MPPT inverter.

REFERENCES

[1] T. Esram and P. L. Chapman, "Comparison of Photovoltaic Array Maximum Power Point Tracking Techniques," IEEE Trans. on Energy Conversion, vol. 22, no. 2, pp. 439-449, 2007.

[2] T. Brekken, N. Bhiwapurkar, M. Rathi, N. Mohan, C. Henze and L. R. Moumneh, "Utility-connected power converter for maximizing power transfer from a photovoltaic source while drawing ripple-free current," IEEE Power Electronics Specialists Conf., PESC, vol.3, pp. 1518-1522, 2002.

[3] B. Yu, Y. Jung, J. So, H. Hwang and G. Yu, "A Robust Anti-islanding Method for Grid-Connected Photovoltaic Inverter," IEEE Photovoltaic Energy Conversion Conf., vol. 2, pp. 2242-2245, 2006.

[4] H. Shan, Y. Kang, S. Duan, Y. Zhang, M. Yu, Y. Liu, G. Chen and F. Luo, "Research on Novel Parallel Current Sharing Control Technique of the Stand-Alone Photovoltaic Inverter," IEEE Industrial Electronics Society Conf., IECON, pp. 1645-1649, 2007.

[5] M. Ciobotaru, R. Teodorescu and F. Blabjerg, "Control of single-stage single-phase PV inverter," Power Electronics and Applications, pp. 1-10, 2005.

[6] T. J. Liang, Y. C. Kuo and J. F. Chen, "Single-stage photovoltaic energy conversion system," IEE Proceedings-Electric Power Applications, vol. 148, Issue 4, pp.339-344, 2001.

[7] A. Barnett, C. Honsberg, D. Kirkpatrick, S. Kurtz, D Moore, D. Salzman, R. Schwartz, J Gray, S. Bowden, K. Goossen, M. Haney, D. Aiken, M. Wanlass and K. Emery, "50% Efficient Solar Cell Architectures and Designs," IEEE Photovoltaic Energy Conversion Conf., pp. 2560-2564, 2006.

[8] L. A. Dobrzanski, L. Wosinska, B. Dolzanska and A. Drygala, "Comparison of Electrical Characteristics of Silicon Solar Cells," Journal of Achievements in Mechanical and Materials Engineering, vol. 8, pp. 215-218, 2006.

[9] G. J. Yu, Y. S. Jung, J. Y. Choi and G. S. Kim, "A Novel Two-Mode MPPT Control Algorithm Based on Comparative Study of Existing Algorithms," Solar Energy, vol. 76, no.4, pp. 455-463, 2004.

[10] C. Hua, J. Lin and C. Shen. "Implementation of a DSP-controlled photovoltaic system with peak power tracking," IEEE Trans. on Industrial Electronics, vol. 45, no. 1, pp. 99-107, 1998.

A Simple Mixed-Signal MPPT Circuit
for Photovoltaic Applications

P. Mattavelli[*], S. Saggini [**], E. Orietti[***], G. Spiazzi[***]

[*] DTG, University of Padova, Vicenza, Italy. e-mail: paolo.mattavelli@unipd.it
[**] DIEGM, University of Udine, Udine, Italy. e-mail: stefano.saggini@uniud.it
[***] DEI, University of Padova, Padova, Italy. e-mail: spiazzi@dei.unipd.it

Abstract— The paper proposes a mixed-signal circuit approach for the Maximum Power Point Tracking (MPPT) for interfacing photovoltaic (PV) panels with batteries or inverters connected to the utility grid. The proposed circuit is aimed to a low cost single-chip IC implementation and it is based on the incremental conductance method, widely used in the literature when more complex microcontrollers or DSPs are used. Analog-to-Digital Converters (ADC) or Digital to Analog Converters (DAC) are avoided and the digital logic is reduced to a few elementary operations (a XOR port, a flip-flop and a counter). The analog part of the circuit implements most of the MPPT algorithm by exploiting the linear region of a field-effect transistor to measure and to store the absolute conductance of the operating point, and by using few analog switches, operational amplifiers and a comparator, to elaborate, after the duty-cycle perturbation, the value of the incremental conductance and to compare it with the absolute one. The digital logic imposes the timing of the different phases and the direction of the search algorithm. Simulation and experimental results on a 35W PV panel, when a high-frequency dc-dc boost converter is employed as a MPP tracker, verify the validity and effectiveness of the proposed solution.

I. INTRODUCTION

As a renewable energy source, the photovoltaic (PV) generation is gaining increasing interest and importance at different power levels. In order to maintain the output power of the PV array at the optimal operating point when environmental conditions (i.e. solar irradiation, temperature, etc.) are changing, real-time Maximum Power Point Tracking (MPPT) techniques are usually needed.

In the literature [1-11], there is a large number of techniques available, using perturb and observe (P&O), hill climbing and incremental conductance methods. In the kW range, the advancements of low cost microcontrollers (μCs) and Digital Signal Processors (DSPs) make the digital implementation of MPPT algorithms very effective. Instead, in low power applications (from a few watts up to a few hundreds of watt), μCs or DSPs and the related analog conditioning circuit may be not cost-effective and there is the need for low cost, possibly single-chip, MPPT circuits.

Within these applications, fractional open-circuit voltage or short-circuit current [2] strategies are a simple way to get close to the maximum power. The approximation involved in these solutions and the periodical disconnection of the PV panel, needed to measure the open-circuit voltage or the short-circuit current, make these approaches less attractive. An alternative solution for low power applications is the Ripple Correlation Control (RCC), which has proved to be a suitable technique for analog implementation [3-11]. This control technique is based on the correlation existing between PV panel power and voltage or current ripples, and it exploits the inherent AC signals caused by the switching action of the converter used to process the generated power. In general terms, RCC features: a) no need of prior PV panel characterization; b) asymptotical convergence toward the Maximum Power Point (MPP); c) switching ripple exploitation and d) high speed of convergence. In RCC, however, the inductive and capacitive parasitic components may have a significant impact on the ability of RCC to drive the system toward the true MPP, especially when the switching frequency of the dc-dc converter is pushed toward hundreds of kHz or more [12].

The aim of the paper is to develop a simple MPPT circuit, with a complexity similar or less than RCC, where most of the MPPT operations are obtained by an analog circuit, while the digital logic is used to handle the search algorithm. The proposed solution is based on the incremental conductance method, using a field-effect transistor operating in its linear region to store the absolute conductance, and some other analog components (op-amps, comparators, switches) to calculate the incremental conductance and to compare it with the absolute one. The proposed solution avoids the use of Analog-to-Digital and Digital-to-Analog Converters and it requires a simple digital logic.

The paper is organized as follows: section II describes the incremental conductance method and the circuit used to handle the different phases of the search algorithm. Section III reports the performances using a simulation model and Section IV reports the experimental results which confirm the theoretical expectations.

978-1-4244-4782-4/10 $26.00 © 2010 IEEE

II. MPPT Circuit Based on Incremental Conductance

The representation of a dc-dc converter connected to a PV panel, is shown in Fig. 1, where, for the sake of explanation, a boost converter is used. The output voltage generator V_o that appears in Fig. 1 represents either a battery, in case of a stand alone application, or a grid-connected inverter, for example in micro-inverter applications, controlled so as to keep its input voltage constant. Being the output voltage V_o constant, the average input voltage $\bar{v}_p$ and thus, the Maximum Power operating point can be controlled by varying the converter duty-cycle, using an MPPT algorithm. In the case an input voltage (or current) feedback control is needed, the MPPT circuit determines the required voltage (or current) reference for the inner loop.

A. MPPT based on Incremental Conductance

Figure 2 shows the typical $i_p - v_p$ and $p_p - v_p$ curves of a PV panel, where the Maximum Power Point is highlighted. The incremental conductance method is based on the consideration that, at the condition of maximum power i.e. $dp_p/dv_p = 0$, we can write:

$$\frac{dp_p}{dv_p}\bigg/_{MPP} = \frac{d(i_p v_p)}{dv_p}\bigg/_{MPP} = \frac{di_p}{dv_p}\bigg/_{MPP} V_{MPP} + I_{MPP} = 0$$

$$-\frac{di_p}{dv_p}\bigg/_{MPP} = \frac{I_{MPP}}{V_{MPP}} \qquad (1)$$

This condition says that, at the MPP, the incremental conductance $-\Delta G$ is equal to the absolute conductance G. Thus, a search algorithm is used based on the consideration that:

$$\begin{cases} G > -\Delta G & for \quad v_p < V_{MPP} \\ G < -\Delta G & for \quad v_p > V_{MPP} \end{cases} \qquad (2)$$

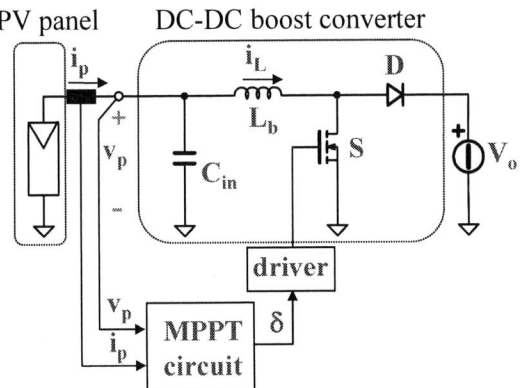

Figure 1. Dc-dc boost converter used to interface a PV panel with a battery (stand-alone applications) or a grid-connected inverter

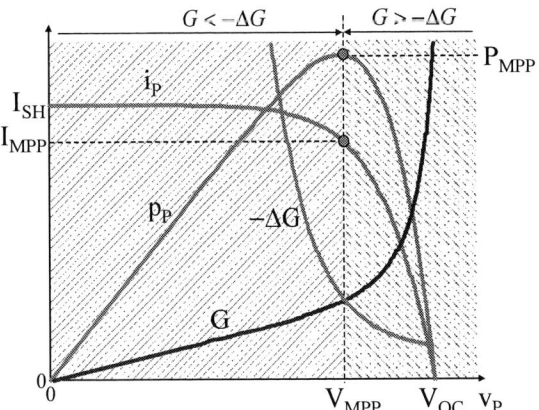

Figure 2. Typical static curves of a PV panel: $i_p - v_p$, $p_p - v_p$, (absolute conductance) $G - v_p$, (incremental conductance) $-\Delta G - v_p$

B. Absolute conductance phase: $\varphi = 1$

For a given operating point at instant k, the absolute conductance $G(k)$ is determined imposing the PV voltage $v_p(k)$ and current $i_p(k)$ to a field-effect transistor (the MOSFET M in Fig. 3). In Fig. 3 the PV current i_p and voltage v_p are the *sensed* PV current and voltage; moreover, the current generator i_p represents the voltage-to-current converter from the sensed PV current.

This phase is denoted as *absolute conductance phase* and, referring to Fig. 3, it corresponds to $\varphi = 1$. In fact, through the feedback loop, the operational amplifier forces the MOSFET's channel conductance, to equal the PV panel absolute conductance $G(k) = i_p(k)/v_p(k)$. The two resistors R shown in Fig. 3 are used to eliminate the quadratic term of the drain-source voltage from the drain current in the MOSFET linear region. At the end of this phase, the logic signal φ changes to zero, thus freezing the MOSFET's channel conductance value by means of the holder capacitor C_{hold1}.

C. Incremental conductance and compare phase: $\varphi = 0$

At instant $k+1$, a perturbation on the duty-cycle is applied, forcing a perturbation on the PV operating point. Referring to Fig. 4, the drain voltage $v_d(k+1)$ is given by:

$$v_d(k+1) = v_p(k) + \Delta i_p(k)/ G(k) , \qquad (3)$$

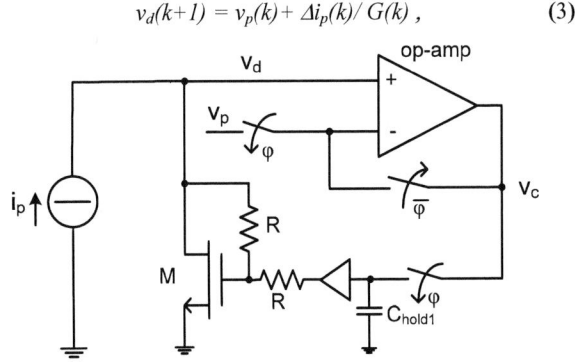

Figure 3. Circuit implementation of the absolute conductance phase

978-1-4244-4782-4/10 $26.00 © 2010 IEEE

where $\Delta i_p(k)$ is the current variation of the PV panel due to the perturbation and $G(k+1)=G(k)$ since the absolute conductance, stored as MOSFET's channel conductance, is kept the same of the previous phase.

Following the circuit of Fig. 4, the positive terminal of the comparator is the average between the drain voltage $v_d(k+1)$ and the PV voltage $v_p(k+1)$, i.e.:

$$v_+(k+1) = \frac{v_d(k+1)+v_p(k+1)}{2} =$$
$$= v_p(k) + \frac{v_p(k)}{i_p(k)}\frac{\Delta i_p(k)}{2} + \frac{\Delta v_p(k)}{2} \qquad (4)$$

In order to compare the absolute conductance $G(k)$ with the differential conductance $\Delta G(k)$, defined as:

$$\Delta G(k+1) = \frac{i_p(k+1)-i_p(k)}{v_p(k+1)-v_p(k)}, \qquad (5)$$

the PV voltage, that was sampled and stored at instant k (v_{ps} in Fig. 4), is imposed at the negative terminal of the comparator ($v_-(k+1) = v_p(k)$), as shown in Fig. 4. Thus, the output of the comparator is high if:

$$\frac{v_p(k)}{i_p(k)}\Delta i_p(k) > -\Delta v_p(k) \qquad (6)$$

Thus, in the case of a duty-cycle increase, $\Delta v_p(k) < 0$ and $\Delta i_p(k) > 0$, the comparator is high if:

$$\frac{v_p(k)}{i_p(k)} > -\frac{\Delta v_p(k)}{\Delta i_p(k)} \Rightarrow G(k) < -\Delta G(k) \qquad (7)$$

and viceversa. Instead, in the case of a duty-cycle decrease, $\Delta v_p(k) > 0$ and $\Delta i_p(k) < 0$, the comparator is high if:

$$\frac{v_p(k)}{i_p(k)} < -\frac{\Delta v_p(k)}{\Delta i_p(k)} \Rightarrow G(k) > -\Delta G(k) \qquad (8)$$

and viceversa. Thus, taking into account the sign of the perturbation and using the status of the comparator output, it is possible to implement the incremental conductance method (2).

D. Control logic

The control logic implements the typical search algorithm using (2), (7) and (8). Taking into account that the comparator output (COMP) has two states (high=1, low=0) and the perturbation (P) is either increasing the PV voltage (P=1) or decreasing the PV voltage (P=0), the implementation of (7), (8) and (2) is represented by four cases:

1. If COMP=1 and P=1, using (8) and (2) $v_p<V_{MPP}$, thus the next perturbation (denoted as S) must increase the PV voltage, i.e. S=1.

2. If COMP=0 and P=1, using (8) and (2) $v_p>V_{MPP}$, thus the next perturbation must decrease the PV voltage, i.e. S=0.

3. If COMP=1 and P=0, using (7) and (2) $v_p>V_{MPP}$, thus the next perturbation (denoted as S) must decrease the PV voltage, i.e. S=0.

4. If COMP=0 and P=0, using (7) and (2) $v_p<V_{MPP}$, thus the next perturbation (denoted as S) must increase the PV voltage, i.e. S=1.

This control logic is summarized in Table I. Taking into account that a positive PV voltage perturbation (P=1) is obtained by decreasing the duty-cycle (and vice-versa), the output S represents the direction (up/down) of the counter, whose output is the complement of the duty-cycle, as shown in Fig. 5. It is worth noting that the circuit requires only a XOR port, a flip-flop and a counter. The counter increases or decreases a digital number, representing the converter complement of the duty-cycle, which is then applied to a Digital-Pulse Width Modulator.

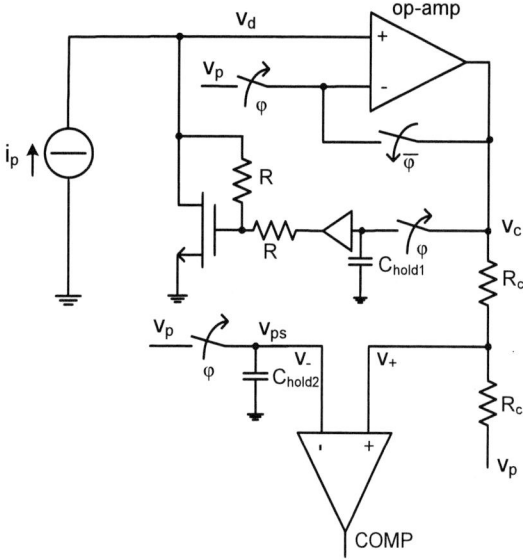

Figure 4. Circuit implementation of the incremental conductance and compare phase

Table I – True Table of Incremental Method

COMP	P	S
1	1	1
1	0	0
0	1	0
0	0	1

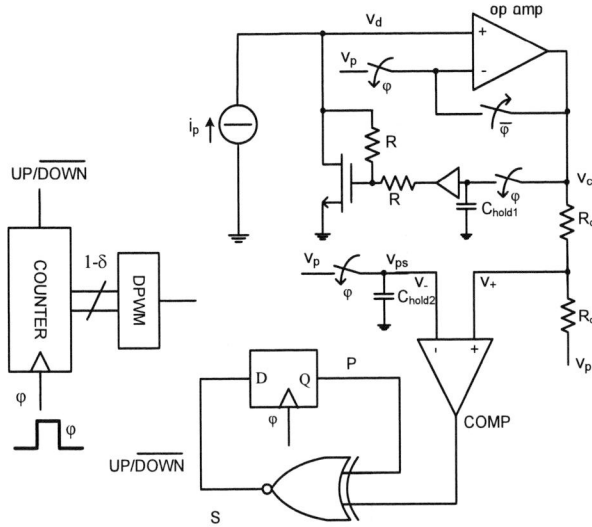

Figure 5. Proposed mixed-signal MPPT circuit

An example of the search is reported in Fig. 6, where the initial condition at instant k is $v_p(k) > V_{MPP}$. Following a negative input voltage perturbation ($v_p(k+1) < v_p(k)$ or P=0) and moving along the PV curve, it is evident that $v_+(k+1) > v_-(k+1)$; thus COMP=1 and the next perturbation (S) needs to decrease the PV voltage (S=0), as confirmed by the second row of Table I.

III.   LIMIT CYCLE OSCILLATIONS

In order to evaluate expression (2), two PV panel operating points must be used since the incremental conductance has to be calculated. Moreover, as in any MPPT algorithm which introduces perturbations on the operating point, the PV current and voltage continuously move around the MPP, thus causing unavoidable Limit Cycle Oscillations (LCOs), whose amplitude depends on the quantization level of the duty-cycle perturbation. This section briefly describes the foreseen number of LCOs.

Fig. 7 shows an example of such typical oscillation with reference to the absolute conductance and incremental

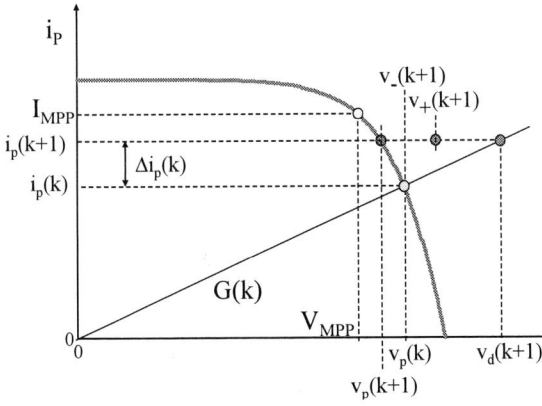

Figure 6. Signals at instant *k* and *k+1* when $v_p > V_{MPP}$.

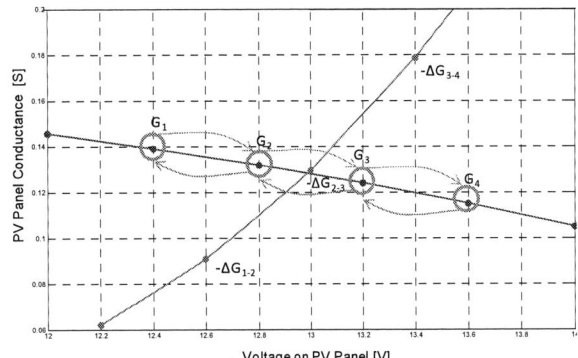

Figure 7. MATLAB evaluation of conductance and incremental conductance showing a 4-level LCO caused by the proposed algorithm.

conductance comparison performed by the proposed MPPT algorithm. The dots in the incremental conductance (for example the dot $-\Delta G_{2-3}$) have been plotted with a shift on x axis in order to visually indicate that the same incremental conductance value can be found by perturbing the system from point 2 to point 3 and viceversa.

Let's consider the PV panel to be in the operating point 2 (i.e. the absolute conductance G_2 is stored). During the compare phase ($\varphi=0$), let's suppose to perturb the system to a higher PV voltage. The panel is therefore forced to operate at point 3. During this phase the incremental conductance ΔG_{2-3} is evaluated and compared to the hold conductance G_2. While the system is already operating at a PV voltage higher than V_{MPP} ($G_3 < -\Delta G_{2-3}$), the algorithm is not able to correctly evaluate such situation since the comparison is performed using the old conductance value (G_2) leading to additionally increment the panel voltage since $G_2 > -\Delta G_{2-3}$. Thus, the system goes to operating point 4. The new absolute conductance G_3 is now stored and then compared to the incremental conductance ΔG_{3-4} leading the algorithm to reduce the PV panel voltage. As a conclusion, two operating points (G_3 and G_4) are needed after the theoretical MPP. Following the same reasoning, decreasing the output voltage two operating points are needed after the theoretical MPP (G_2 and G_1). In the example of Fig. 7, the number of levels of LCOs introduced by the proposed algorithm is four. The condition for the 4-level LCO is that the incremental conductance value at MPP is between the two adjacent absolute conductance values stored by the algorithm (in the case of Fig. 7: $G_2 > -\Delta G_{2-3} > G_3$).

Fig. 8 depicts two opposite situations that occur when the incremental conductance value at MPP is higher or conversely lower of both the adjacent absolute conductance values. In these situations, following the reasoning of the previous paragraph and Fig.8, it can be seen that the proposed MPPT approach forces the system to a three-level LCO.

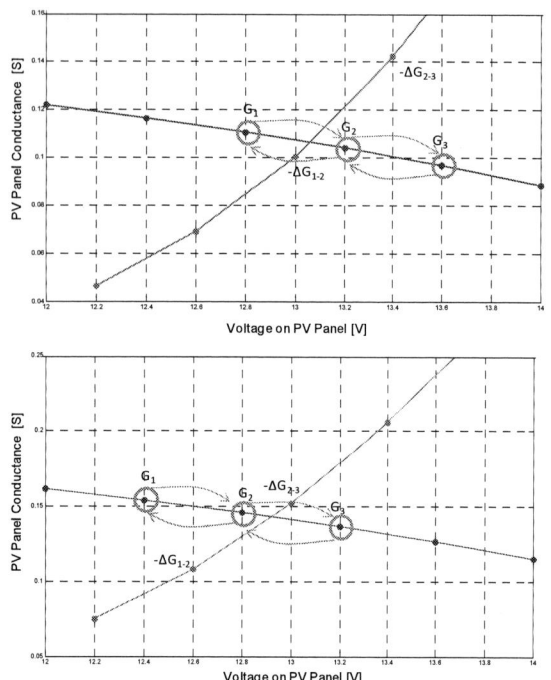

Figure 8. MATLAB evaluation of conductance and incremental conductance showing two conditions for 3-level LCO caused by the proposed algorithm.

IV. SIMULATION RESULTS

In order to verify the performance of the proposed circuit, a Matlab/Simulink model has been developed. The PV parameters are: N_{cell} = 36, I_{SH} = 2.5 A, V_{oc} = 17.24 V, I_{MPP} = 2.30 A, V_{MPP} = 12.73 V. A dc-dc boost converter has been interfaced between the PV panel and the battery, having the following parameters: C_{in} = 100 μF, L_b = 19 μH, V_o = 24 V, f_s = 200 kHz. The PV panel output current was sensed with a resistor R_{sense} = 0.11 Ω. Fig. 9 shows the performances of the proposed circuit under steady-steady, while Fig. 10 depicts the system evolution during a step variation for the PV short circuit current (I_{SH}=2.5A→0.8A): note the ability to correctly track the MPP with an inherent limit cycle oscillation around the MPP and with a good dynamic response. In this case the selected duty-cycle increment caused a variation of the PV voltage of around 100 mV each step, ensuring a small effect on the generated power.

V. EXPERIMENTAL RESULTS

An experimental prototype with the parameters outlined in the previous section has been realized. The quantization level of the duty-cycle is 1.78%, leading to a quantization of the PV voltage of around 430mV. The control circuit has been implemented with Component of the Shelf (COS) and the digital logic and DPWM implemented in FPGA. The frequency of the MPPT perturbation was set at 1 kHz.

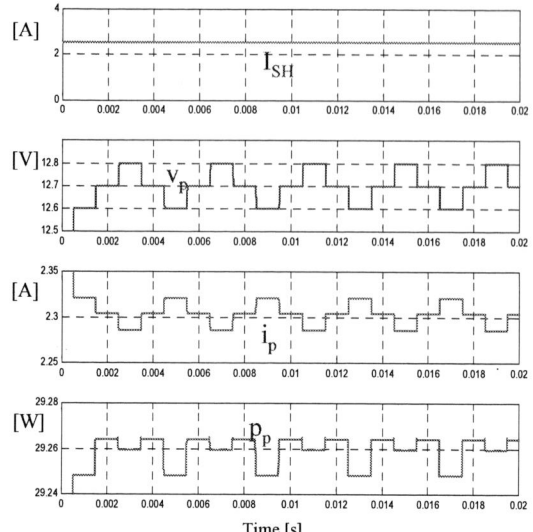

Figure 9. Typical PV waveforms in steady-state with the proposed search algorithm

In order to solve some noise issues due to the fact that, during the compare phase (φ=0), voltages v_+ and v_- at the input of the comparator are very closed to each other, an additional filtered differential amplifier with a high-gain (40) was added in the circuit. The RC filter has been designed with a single pole around 10 kHz to allow the suppression of boost switching frequency noise and not to influence the MPPT performances. The final circuit is depicted in Fig 11.

The circuit has been tested both with a 35W PV panel and with an PV panel emulation, realized with a DC voltage source and a series resistance. The results reported hereafter are based on the PV emulation so as to verify the circuit performance under controlled conditions. The typical steady-state waveforms are reported in Fig. 12 (a) and (b), highlighting the two possible Limit-Cycle Oscillations of the MPP search algorithm, the 4-level LCO and 3-level LCO, respectively. Fig. 12 shows also that the variation of the operating point introduces an inherent oscillation on the generated power, thus reducing the effectiveness of the MPPT.

Since a PV panel emulation is used, it was possible to measure the effectiveness of the proposed MPPT, defined as the ratio of output power of the PV panel and the theoretical output power at the MPP. The measurements of MPPT effectiveness under different MPPT conditions are reported in table II, together with the simulation results obtained under the same operating conditions. The results are very good in spite of some noise issues in the analog circuit.

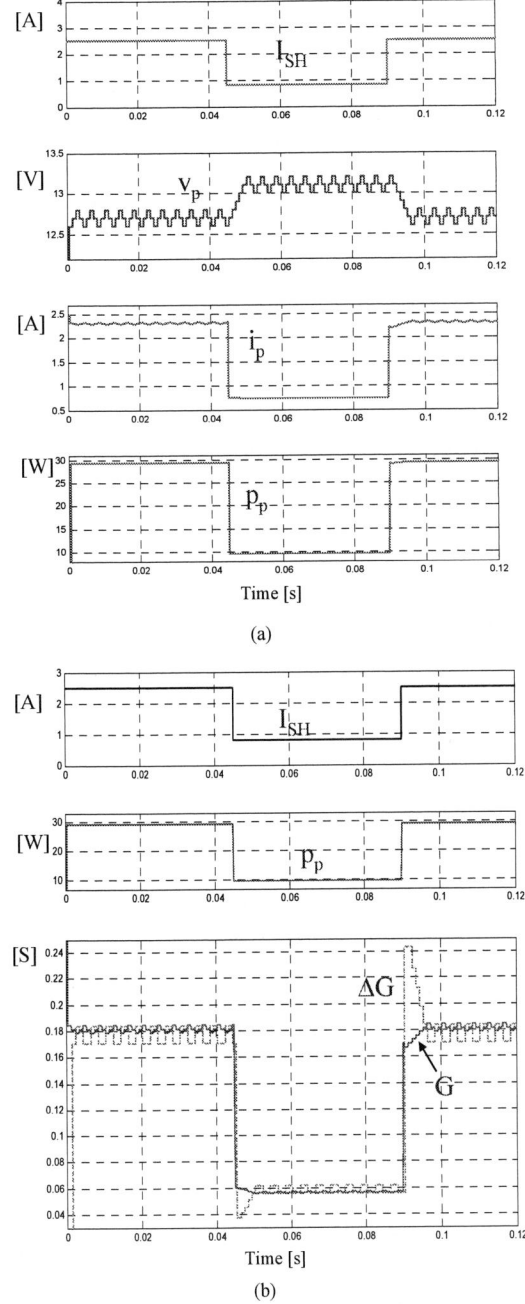

(a)

(b)

Figure 10. Step variation of the PV short-circuit current a) PV panel voltage and current waveforms; b) absolute and incremental conductance waveforms

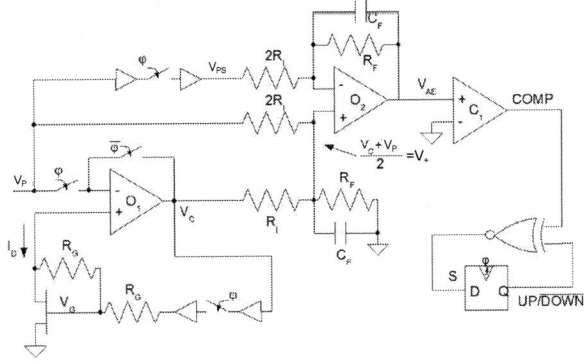

Figure 11. Final implementation of the proposed mixed-signal MPPT circuit.

Using the same PV emulation circuit, the transient response of the system has been verified by means of a voltage step of the DC voltage source from zero to 14V. This test allows to indirectly check the MPPT response to PV short-circuit current variations. The results are shown in Fig. 13, illustrating both the PV current and PV voltage evolutions superimposed to the emulator DC voltage. It can be noted that the MPPT algorithm is able to quickly sense the step as

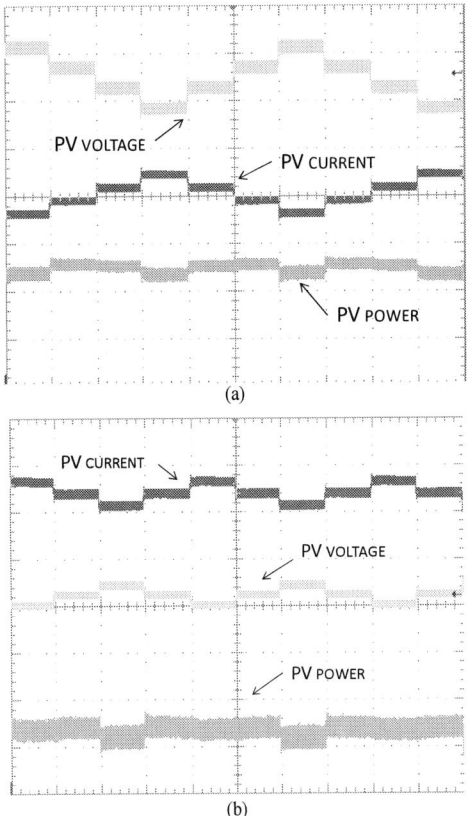

Figure 12. Circuit waveforms under steady-state conditions using PV panel emulation (power-supply dc voltage V_{dc} (a): V_{dc}=15V, PV$_{VOLTAGE}$ (1V/div), PV$_{CURRENT}$ (200mA/div), PV$_{POWER}$ (250 mW/div) and (b): V_{dc}=20.53V): PV$_{VOLTAGE}$ (2V/div), PV$_{CURRENT}$ (200mA/div), PV$_{POWER}$ (200 mW/div), (time scale: 1ms/div).

highlighted by the PV voltage and to recover to MPP in approximately 20 ms.

Finally, Fig. 14 and Fig. 15 show the waveforms for the main signals used during phase φ=1 and phase φ=0, respectively. The clock signal φ has been added to both pictures to improve readability.

As depicted in Fig. 14, the FET gate voltage is updated after φ rising edge and kept constant by the sample&hold circuit during the clock period in order to store the panel absolute conductance at φ=1. Meanwhile also the panel voltage is stored as depicted by V_{PS} waveform. The FET gate voltage is driven by O_1 op-amp output, therefore the V+ node exhibits a large negative voltage during φ=1, while during the compare phase it is very close to V_{PS} voltage at V- node). For this reason a filtered amplifier was needed to increase the circuit robustness. The O_2 amplifier output is depicted in Fig.15 together with the comparator threshold (gnd voltage) and the comparator output voltage itself that is used by MPPT circuit to decide for next perturbation point.

Table II – MPPT effectiveness

V_{OPT} [V] (=V_{EMUL}/2)	Simulation Results		Measurement results	
	V_{MPPT} [V] (error [%])	η [%]	V_{MPPT} [V] (error [%])	η [%]
6 V	5.88 V (-1.93 %)	99.46	6.08 (1.4 %)	99.40
7 V	6,98 (-0.33 %)	99.66	6.97 (-0.4 %)	99.71
8 V	7,80 (-2.51 %)	99.67	8.12 (1.5 %)	99.68
9 V	8,76 (-2.68 %)	99.72	9.05 (0.6 %)	99.80
10 V	9,78 (-2.18 %)	99.77	10.01 (0.1 %)	99.84
11 V	10,82 (-1.62 %)	99.81	11.13 (1.2 %)	99.93
12 V	11,81 (-1.59 %)	99.84	12.21 (1.8 %)	99.89
13 V	13,00 (0.01 %)	99.93	13.42 (3.2 %)	99.84

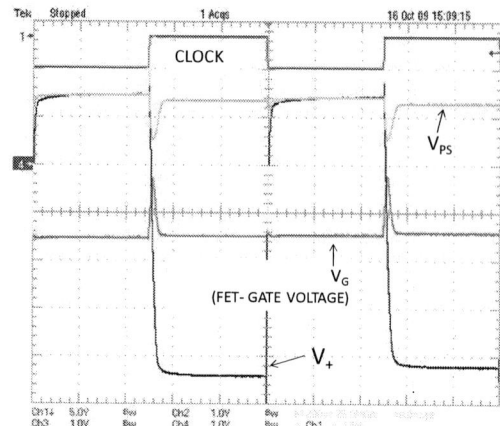

Figure 14. Measurement on the prototype showing the circuit behavior during absolute conductance phase (time scale: 200 μs/div; V_G, V_+ and V_{PS} are 1V/div).

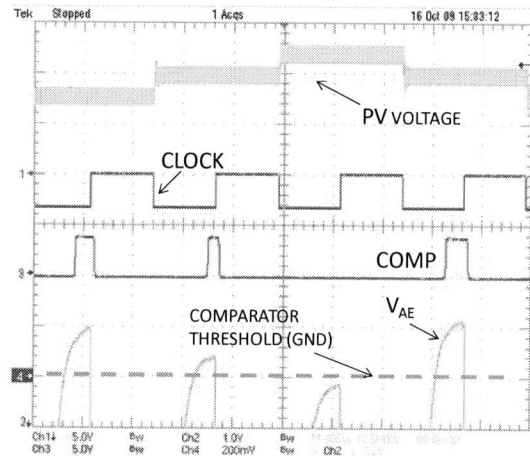

Figure 15. Measurement on the prototype showing the circuit behavior during compare phase (time scale: 400 μs/div; PV voltage 1V/div, amplified error V_{AE} is 200 mV/div).

VI. CONCLUSION

The paper has proposed a simple mixed-signal circuit approach for the Maximum Power Point Tracking for photovoltaic applications. The circuit is aimed to a low cost single-chip IC implementation and it is based on the incremental conductance method. Compared to existing methods, ADCs or DACs are avoided and the digital logic is reduced to a few elementary operations (a XOR port, a flip-flop and a counter). The analog part of the circuit implements most of the MPPT algorithm by exploiting the linear region of a field-effect transistor and by using few analog components (op-amp, comparator and some analog switches). Simulation and experimental results have verified the performance of the proposed solution.

ACKNOWLEDGMENT

The authors would like to thank Eng. Francesco Sanguin for his contribution in the experimental setup and measurements.

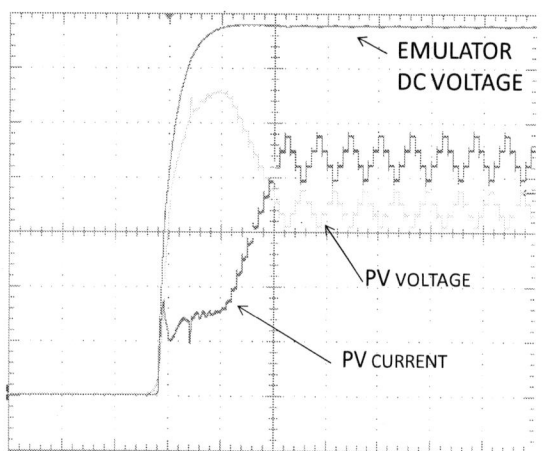

Figure 13. Photovoltaic emulator voltage step 0→14 V: PV_{VOLTAGE} (2 V/div), PV_{CURRENT} (200 mA/div), V_{DC-EMULATOR} (2V/div) (time scale: 10 ms/div)

REFERENCES

[1] F. Liu, S.Duan, F. Liu, B. Liu, Y. Kang "A Variable Step Size INC MPPT Method for PV Systems", IEEE Transactions on Industrial Electronics, Vol. 55, N°7, July 2008, pp.2622-2628.

[2] M.soum, H. Dehbonei, E. F. Fuchs, "Theoretical and experimental analyses of photovoltaic systems with voltage and current-based maximum power-point tracking", IEEE Transactions on Energy Conversion, Vol. 17, N°4, Dec. 2002, pp. 514-522.

[3] P. T. Krein, "Ripple correlation control, with some applications," in Proc. 1999 IEEE Int. Symp. Circuits Syst., 1999, pp. 283–286.

[4] P. Midya, P. T. Krein, R. J. Turnbull, R. Reppa, and J. Kimball, "Dynamic maximum power point tracker for photovoltaic applications," in Proc. 27th Annu. IEEE Power Electron. Spec. Conf., 1996, pp. 1710–1716.

[5] D. L. Logue and P. T. Krein, "Observer-based techniques in ripple correlation control applied to power electronic systems," in Proc. 32nd Annu. IEEE Power Electron. Spec. Conf., 2001, pp. 2014–2018.

[6] D. L. Logue and P. T. Krein, "Optimization of power electronic systems using ripple correlation control: a dynamic programming approach," in Proc. 32nd Annu. IEEE Power Electron. Spec. Conf., 2001, pp. 459–464.

[7] Y. H. Lim and D. C. Hamill, "Simple maximum power point tracker for photovoltaic arrays," Electron. Lett., vol. 36, pp. 997–999, May 2000.

[8] Y. H. Lim and D. C. Hamill, "Synthesis, simulation and experimental verification of a maximum power point tracker from nonlinear dynamics," in Proc. 32nd Annu. IEEE Power Electron. Spec. Conf., 2001, pp. 199–204.

[9] J. R. Wells, P. L. Chapman, and P. T. Krein, "Fundamental aspects of ripple correlation control of electric machinery," in Proc. 34th Annu. IEEE Power Electron. Spec. Conf., 2003, pp. 1659–1662.

[10] T. Esram, J. W. Kimball, P. T. Krein, P. L. Chapman, P. Midya, "Dynamic Maximum Power Point Tracking of Photovoltaic Arrays Using Ripple Correlation Control", IEEE Transactions on Power Electronics, Vol.21, N°5, September 2006, pp.1282-1291.

[11] H. Sugimoto and H. Dong, "A new scheme for maximum photovoltaic power tracking control," in Proc. Power Conv. Conf., 1997, pp. 691–696.

[12] S. Spiazzi, S. Buso, P. Mattavelli, "Analysis of mppt algorithms for photovoltaic panels based on ripple correlation techniques in presence of parasitic components" 10th Brazilian Power Electronics Conference (COBEP), October 2009, Bonito (Brazil).

Low-Power Maximum Power Point Tracker with Digital Control for Thermophotovoltaic Generators

Robert C.N. Pilawa-Podgurski, Nathan A. Pallo, Walker R. Chan, David J. Perreault, Ivan L. Celanovic

MASSACHUSETTS INSTITUTE OF TECHNOLOGY

CAMBRIDGE, MASSACHUSETTS 02139

EMAIL: pilawa@mit.edu

Abstract—This paper describes the design, optimization, and evaluation of the power electronics circuitry for a low-power portable thermophotovotaic (TPV) generator system. TPV system is based on a silicon micro-reactor design and low-bandgap photovoltaic (PV) diodes. We outline critical system-level challenges associated with TPV power generation, and propose a power electronics architecture that addresses these challenges. We present experimental data from a compact, highly efficient peak power tracker and show how the proposed architecture enables increased energy extraction compared to conventional methods. The operation of the power tracker is verified with low-bandgap PV cells illuminated by a quartz halogen lamp producing a PV diode output power of 0.5 W, and above 99% tracking efficiency is demonstrated. Additionally, the complete system operation is verified with the power tracker connected to GaInAsSb PV diodes and a silicon micro-reactor, producing 150 mW of electrical power.

Index Terms—maximum power point tracker, thermo-photovoltaic, TPV, MPPT, lossless current sensing, digital control, micro burner, portable power

I. INTRODUCTION

THE possibility of statically converting heat into electricity–without moving parts–has captured the imagination of scientists and engineers for nearly two centuries. Since the discovery of the thermoelectric, photovoltaic and thermionic effects, there have been significant efforts towards developing devices that can perform this conversion with good efficiencies. One of the promising technologies to convert heat (more precisely radiant heat) into electricity is thermophotovoltaics (TPV). TPV converts heat into thermal radiation photons that are in turn converted into electron current via the photovoltaic effect, as shown in the inset of Figure 1. While TPV power conversion is in many aspects similar to solar photovoltaics (PV), there are several key differences. The TPV emitter typically operates at temperatures between 1100K-1500K, and hence the peak of the radiated spectrum is shifted towards longer wavelengths. This is illustrated in Figure 1 which shows spectral irradiance of a blackbody emitter at 1100K that peaks around 2.6 μm; in stark contrast with solar spectrum that peaks around 480 nm. Indeed, TPV requires low-bandgap PV diodes such that the bandgap is better matched to the peak infrared (IR) radiation since only

This work was sponsored in part by the U.S. Army Research Office through the Institute for Soldier Nanotechnologies.

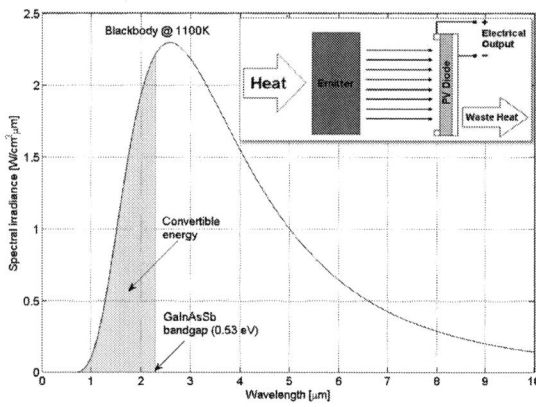

Fig. 1. Radiated spectral power distribution of a blackbody emitter at 1100K. Inset shows a block diagram of a thermophotovoltaic energy conversion process.

photons with energies above the PV diode bandgap can generate electron-hole pairs; as represented by shaded area under the blackbody curve in Figure 1. Furthermore, a TPV thermal emitter and TPV diode are in close proximity, thereby enabling photon recycling; a process were photons reflected from the TPV diode can be reabsorbed by the emitter. Due to the close proximity, TPV cells operate at more than two orders of magnitude higher energy densities than solar PV (as shown in Figure 1), but are exposed to spatially non-uniform incident photon flux, which can be challenging from a system design perspective.

The TPV concept was first proposed in the 1950s [1]. However, high-efficiency operation has only recently been enabled through scientific and technological advancements in two critical areas: low-bandgap semiconductor materials, and photonic crystals. High-performance low-bandgap semiconductor diodes such as GaInAsSb enable quantum efficiencies approaching unity for a wavelength range between 1 and 2.3 μm. The addition of photonic crystals (PhC) allows for spectral shaping of the thermal radiation so that its spectrum is almost perfectly matched to the diode electronic bandgap [2], [3]. These two technologies combined have brought TPV

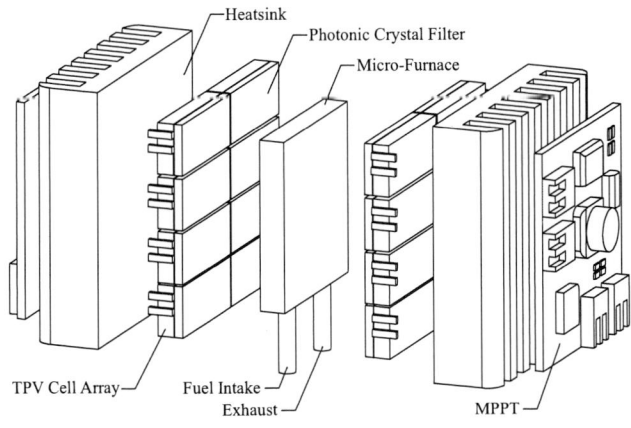

Fig. 2. Illustrative drawing of burner and TPV cells for portable power generation.

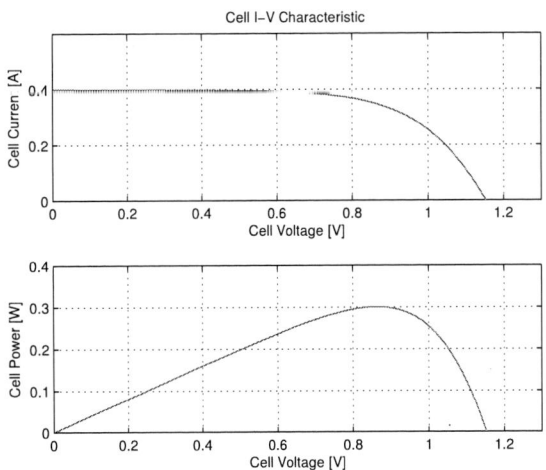

Fig. 3. I-V (top) and P-V (bottom) characteristic of TPV cell used in this work for a typical operating point.

to the forefront of portable power generation, demonstrating above 20% efficiency in converting radiative heat into electricity [4]. With new PhC designs and optimized TPV diodes 30% conversion efficiency is within reach.

In this paper we focus on the low-power, micro-fabricated, butane powered TPV generator, as shown in Figure 2. It comprises a silicon micro-fabricated fuel reactor that acts as a radiant heat source [5], low-bandgap GaInAsSb PV diodes [6], and a low-power power electronics module. The key advantages of the TPV technology for micro-scale power generation are: high energy density, no moving parts, robust multi-fuel operation, and high efficiency. High energy density stems from the energy density of butane, which is almost two orders of magnitude higher than current Li-ion batteries.

Although significant headway has been made on the device level there have been very few attempts at complete TPV system level demonstrations. One of the critical components in a fully integrated micro-TPV system is the low-power power electronics converter. This paper, to the best of our knowledge, presents the first systematic and rigorous treatment of the design, optimization, and testing of a low-power maximum power point tracking (MPPT) converter for a TPV power generator system. To this end, we describe the power electronics subsystem for the TPV system of Figure 2, address some unique challenges associated with this application, and outline the solutions implemented to achieve a high performance overall system. Although our focus is on a micro-fabricated TPV generator, this approach is applicable to other TPV systems such as radioisotope powered TPV, and solar-TPV.

In this paper, section II gives an overview of the electrical characteristics of the low-bandgap TPV diodes, along with the associated system challenges from a power electronics point of view. Section III contains a detailed description of the proposed power electronics system, followed by experimental results presented in section IV. Finally, section V concludes the paper.

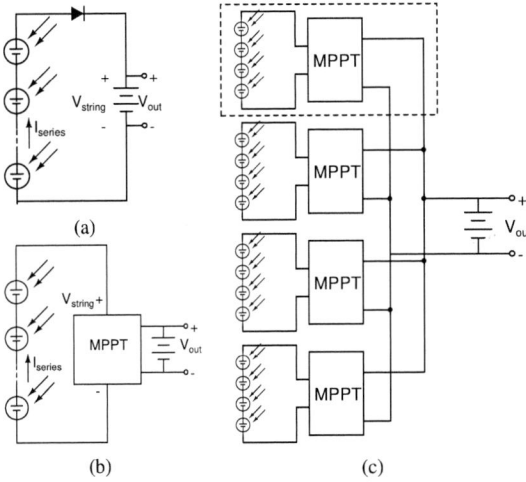

Fig. 4. (a) Simple cell connection, which does not extract the maximum power from the cell. (b) Conventional method with series-connected cells attached to MPPT. (c) Multi-MPPT method employed in this work.

II. TPV CELL CHARACTERISTICS

Shown in Figure 3 is the I-V characteristic for one TPV module, which consists of four series-connected GaInAsSb PV diodes [6]. The bottom graph of the figure shows the corresponding power versus voltage graph, which clearly shows a maximum power point (MPP). This point typically changes with operating conditions such as incident irradiation and cell junction temperature, and must therefore be continuously tracked to ensure that the maximum power is extracted from the cell. Figures 4a and 4b illustrate two common methods to connect photovoltaic cells to their loads. In Figure 4a all the cells are connected in series, and are directly connected to the load, a battery in this example. A diode is typically placed in series with the cells to prevent the battery from discharging through the cells during low light conditions. This approach, while simple, is typically very inefficient. Ignoring the small

978-1-4244-4782-4/10 $26.00 © 2010 IEEE   962

voltage drop across the diode, the string voltage V_{string} is restricted to be equal to the battery voltage V_{out} at all times, which is typically not the same as the MPP voltage (V_{MPP}). For a particular operating irradiation level and temperature, the series-connected cells' V_{MPP} may coincide with V_{out}, but at all other times, less than the maximum power is extracted from the cells. Figure 4b shows a method which is typically used to circumvent this limitation. By placing a dc-dc converter between the series-connected cells and the load, the string voltage V_{string} can be controlled to equal V_{MPP} at all times. The dc-dc converter, acting as a maximum power point tracker (MPPT), continuously tracks V_{MPP} by adjusting its conversion ratio in response to changes in operating conditions.

The method of Figure 4b is often adequate for solar photovoltaic applications, where the solar irradiation is a plane-wave, ensuring uniform illumination of all cells in the series string. Provided the cells are properly matched in terms of their electrical characteristics, they will then produce equal currents. The situation is different in the TPV application considered here. Since the burner is positioned close to the TPV diode (2-3 millimeters), the irradiation is non-uniform and depends on the relative position of the diode with respect to the burner. In addition, the temperature distribution across the burner surface is non-uniform. This leads to mismatched cell photocurrents, with the cell receiving the most irradiation producing the most current. If a method similar to that of Figure 4b is employed in this situation, the string current I_{string} is limited to the value of the least irradiated cell. Thus, all other cells are operating at a cell current that is below their peak current, resulting in a total output power that can be substantially lower than the maximum achievable. The result is similar to that observed in solar panels with partial shading, as discussed in [7], [8]. The non-uniform irradiation in this application prevents efficient energy extraction with the stacking of many cells in series to achieve a high output voltage.

Figure 4c shows the architecture we propose to ameliorate these concerns. In this architecture, four diodes are connected in series and form a module. Each module is then connected to its own individual MPPT, and the outputs of all MPPTs are connected in parallel. The choice of four cells per module was made to provide a large enough working voltage (approximately 1 V) for the MPPTs to ensure efficient power conversion by the electronics. Using this architecture, current mismatch is limited to only four cells, all of which are placed in close proximity to each other, thereby minimizing the negative effects of non-uniform irradiation. The boxed area of Figure 4c highlights the system components that are considered in this work, which constitute four series-connected cells and one MPPT.

III. PEAK POWER TRACKING IMPLEMENTATION

Figure 5 shows a schematic drawing of the implementation of one of the MPPTs of Figure 4c. The low-voltage PV module (with the I-V characteristics of Figure 3) is used to charge a lithium-ion battery, which acts as intermediate energy storage for the system. The power electronic circuit is implemented as

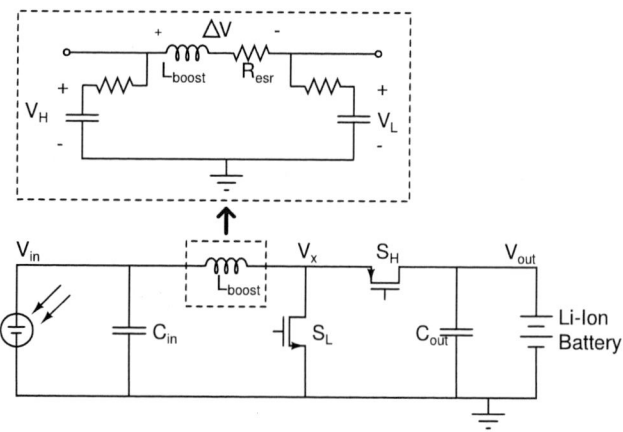

Fig. 5. Schematic drawing of MPPT, and an illustration of the lossless current sensing scheme used.

a boost converter, which performs both the requisite voltage transformation and the peak power tracking.

A. Control Algorithm and Implementation

The boost converter shown in Figure 5 has an input/output voltage relationship given by:

$$V_{out} = \frac{V_{in}}{1 - D} \qquad (1)$$

where D is the duty cycle of the bottom switch (S_L). In this synchronous rectification implementation, the top switch (S_H) is turned on when the bottom switch is off. The boost converter can be controlled to achieve peak power tracking by perturbing the duty cycle in a certain direction (increase or decrease), and observe whether the delivered power increased or decreased due to this perturbation. If the power increased, the controller continues to perturb the duty cycle in the same direction, but if the power decreased, the direction of the perturbation is changed. With this method, the controller eventually settles on the peak power point of Figure 3, where it oscillates to within the finest resolutions of the duty cycle command and sensors. This method, often called hill climbing, or perturb and observe, [9] is one of the most common MPPT algorithms used to date. Figure 6 shows a flow chart of the MPPT algorithm. The initial starting point for the duty cycle is determined by performing a coarse sweep of the duty cycle at startup, and recording the duty cycle corresponding to the maximum output power observed. This approach ensures that peak power tracker can quickly lock in on the maximum power point.

The algorithm described above is well-suited for an implementation in digital form, and we have chosen to use a microcontroller for our implementation. In addition to keeping state and running the tracking algorithm, the microcontroller can be used to perform analog to digital conversion, generate the PWM signals, perform temperature measurements, and handle communication. The ability of the microcontroller to handle a variety of functions is very beneficial in this low-power application, where the power loss of the auxiliary components must be kept to a minimum. An additional benefit

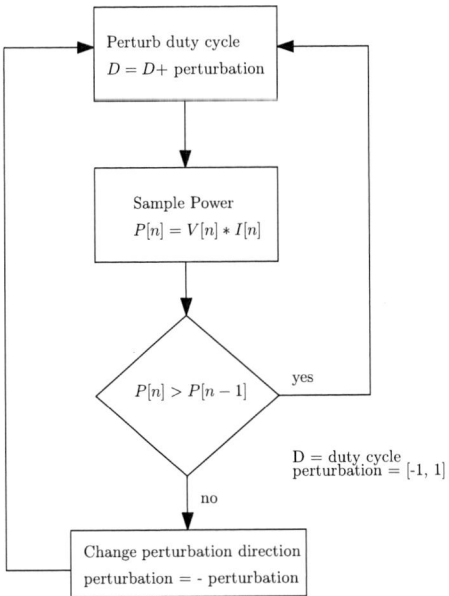

Fig. 6. Flow chart illustrating the operation of perturb and observe.

of a multi-function chip such as the micro-controller is the significant space savings that can be realized compared to an implementation with discrete devices for each function.

B. Voltage and Current Measurement

In the general case, both current and voltage must be measured to find the maximum power point (see [7] for a discussion of cases where only one of the two needs to be measured). Typically, only the average values need to be measured, which reduces bandwidth requirements and enables the use of low-power analog to digital converter (ADC) architectures. Furthermore, the absolute value of current and voltage is not required, since the minimum or maximum power points are found relative to the other possible operating points. The ADC thus needs high resolution, but not high absolute precision, a characteristic that can be leveraged to obtain high performance while maintaining low power consumption.

The microcontroller used, the 8-bit ATtiny861 from Atmel, provides a multiplexed 10-bit ADC, along with an internal bandgap reference. The 10-bit precision can be further extended in the digital domain by oversampling and decimation [10]. The input and output voltages can thus easily be measured with this built-in ADC with sufficient resolution.

A more difficult challenge is that of current sensing, which is typically done with a current-sense resistor. The addition of a current-sense resistor in the current path introduces an undesired power loss, which decreases overall converter efficiency. For this reason, the current-sense resistor is typically made small, and the subsequently small voltage drop is sensed with a low-noise, high gain amplifier. In this application, with a total output power of less than 500 mW, the additional power consumption and area of a low-noise amplifier for current sensing, together with the added power loss of a current-sense resistor, was deemed too high, so alternative implementations were investigated.

Another current-sensing option is that of a hall-effect sensor, which measures the magnetic field associated with a current. With no added resistor in the current path, the only power loss is that of the magnetic sensing circuitry, which can unfortunately be quite large. Indeed, in this application it was found that the static power consumption of this method was much too large for acceptable system efficiency.

Figure 5 illustrates the current-sensing technique used in the power tracker. To maximize overall system efficiency, lossless current sensing [11] is used, where the average voltage drop across the inductor is measured. The relationship between inductor current I_L and sensed voltage ΔV is given by:

$$\langle I_L \rangle R_{esr} = \langle \Delta V \rangle = \langle V_H \rangle - \langle V_L \rangle, \qquad (2)$$

where R_{esr} is the parasitic resistance of the inductor. The average voltages, $\langle V_H \rangle$ and $\langle V_L \rangle$ are produced by first-order RC low-pass filters. These two voltages are then sampled by the differential ADC of the micro-controller with a built-in gain of 32, which gives a reading directly proportional to the inductor current. It should be noted that the common concern with this current sensing method, the tolerance and temperature coefficient of R_{esr}, is not a problem in this application. Since, for our tracking algorithm, we are only concerned with relative changes of the current, any static offset of R_{esr} has no effect on the peak power tracking. Furthermore, the time constant of any temperature-induced variation of the R_{esr} value is much larger than the chosen sampling time, so the tracking can be made insensitive to this variation as well. In our converter implementation, a relative change in current of less than 1 mA can be resolved using this method, as confirmed by experimental measurements. It should be noted that this current sensing is achieved without the need for a power-consuming series-sense resistors, and that the amplifier and ADC are built-in to the microcontroller, and thus consume negligible additional power and take up no additional area.

It should be emphasized that a key enabler to the use of this current-sensing technique is the fact that the application does neither require absolute accuracy of the current, nor instantaneous current values. Thus, the tolerance of the inductor resistance is not critical, and the low-pass filters can be designed to provide significant averaging over a relatively long time.

IV. PROTOTYPE AND EXPERIMENTAL RESULTS

A. Converter prototype

An experimental prototype of the MPPT converter has been developed and characterized. Figure 7 shows a photograph of the peak power tracker, and Table I lists the converter specifications; converter efficiency includes all control and gate driver losses. The tracking efficiency is a measurement of how close the tracking algorithm operates to the true maximum power point, and is given by:

$$\eta_{tracking} = \frac{\langle P_{in} \rangle}{P_{MPP}}, \qquad (3)$$

Fig. 7. Photograph of the peak power tracker.

TABLE I
CONVERTER SPECIFICATIONS

Input Voltage	0.3-1.1 V
Output Voltage	1.5-4.2 V
Output Power	500 mW
Switching Frequency	250 kHz
Converter Efficiency	90%
Tracking Efficiency	>99%

TABLE II
COMPONENT LISTING

Device	Model	Value	Manufacturer
S_1	BSO300N03S		Infineon Tech.
S_2	SI2351DS		Vishay Siliconix
L	MSS5131-822ML	8.2 μH	Coilcraft
R_H, R_L	0603	100 kΩ	Panasonic
C_H, C_L	0603	10 μF	Murata
C_{IN}	0805	3 μF	Murata
C_{OUT}	0805	50 μF	Murata
Microcontroller	ATtiny861		Atmel
Gate Driver	LM5111		National Semi.

where P_{in} corresponds to the converter input power, and P_{MPP} is the output power of the TPV module at the maximum power point. Due to the low operating point of the TPV module (~ 1 V), it is difficult to make a high precision input power measurement of the converter without also perturbing the actual operating point of the converter. An easier, but strictly speaking less accurate, approximation of the tracking efficiency can be found by calculating the ratio:

$$\eta_{tracking,approx.} = \frac{\langle P_{out} \rangle}{P_{out,max}}, \qquad (4)$$

where $P_{out,max}$ corresponds to the maximum output power from the converter. This is only an approximation, and will over-estimate the tracking efficiency because $P_{out,max}$ will not correspond to the exact peak power point, owing to the finite resolution of the digital PWM implementation. However, with proper knowledge of the cell I-V curve (Figure 3) and the tracking algorithm step-size (PWM resolution is this implementation), one can find an upper bound on the error in the approximation given by 4, and from there calculate a minimum tracking efficiency. Using this technique, the tracking efficiency of the converter considered here was found to be above 99%.

The converter design was guided by the desire to achieve small system size and weight, while maintaining high efficiency. As can be seen in Figure 7, the majority of the circuit board area is taken up by connectors, while the converter core (switching devices, micro-controller, and passive components) take up a relatively small area. Figure 8 provides a detailed schematic drawing of the converter. As shown, the converter can be powered either from the Li-Ion battery output, or from an external power supply. Table II lists the components used in the experimental prototype.

B. Converter experimental verification

To evaluate the performance of the peak power tracker, the converter was initially connected to a PV diode illuminated by a quartz halogen lamp. The lamp brightness and distance from the cell was adjusted to match the expected power output from the cell when illuminated by the micro-reactor (500 mW).

This enabled initial characterization of the converter without the added complexity of the micro-reactor dynamics. Figure 9 (top) shows the output power of the converter over time, and illustrates the MPPT startup algorithm for this experimental setup. Initially, the converter steps its duty cycle through a coarse sweep to find the approximate point of the MPP. The duty cycle corresponding to the maximum power observed is recorded, and once the sweep is concluded, the duty cycle is set to this value. At this point, the converter enters the hill-climbing phase (perturb and observe), and uses a fine step-size to reach the MPP. Note that the step-size of the hill-climbing algorithm is too small to be visible in the top plot.

The steady-state behavior of the hill-climbing algorithm is shown in the bottom of the figure, which shows the converter output power versus time in steady-state. This is a zoomed-in version of the top plot, and shows the discrete steps in power corresponding to a 1-bit change in duty cycle. The total PWM resolution of the micro-controller is 10 bits. The converter oscillates around the MPP to within the resolution of the PWM signal and the current and voltage sensors. Because the sensing and duty cycle control have similar resolution, the hill-climbing algorithm is limited by sensing noise, and occasionally takes one extra step in the wrong direction. It should be noted that the sampling interval for the MPPT algorithm has been set to several seconds, as seen in Figure 9 (bottom). This was done to enable high accuracy power measurements by the external instruments used to characterize the converter, and is not a fundamental limit of the converter itself. If desired, the MPPT algorithm can be set to sampling frequencies considerably higher (on the order of several kHz) without a noticeable impact on tracking efficiency. In this application, however, the system time constant of any change in maximum power point is long enough such that the sampling frequency of Figure 9 is sufficient to allow efficient energy extraction from the PV

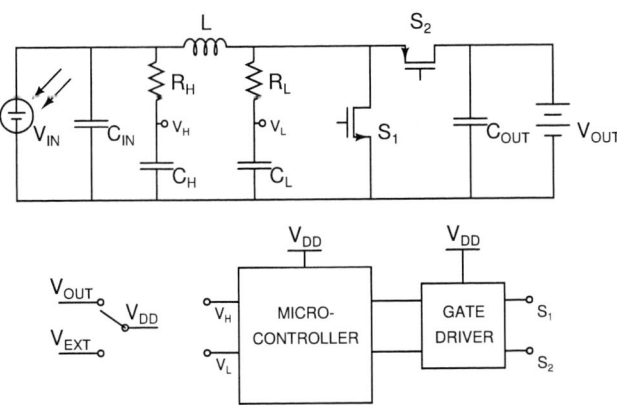

Fig. 8.   Schematic drawing of converter.

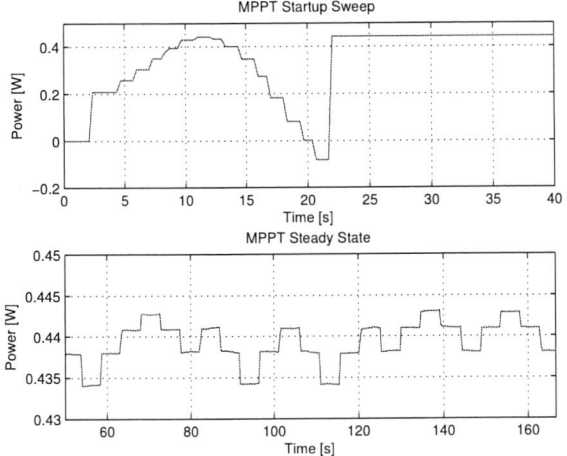

Fig. 9.   Experimental data showing startup behavior of power tracker (top), and steady-state performance (bottom).

module.

C. Micro-reactor experimental results

In order to fully evaluate the MPPT converter performance we tested it with an experimental system setup similar to the one depicted in Figure 2. The PV cells were illuminated with the micro-reactor, shown in the photo of Figure 10. The reactor is a 10 mm by 10 mm by 1 mm silicon slab with a serpentine, platinum catalyst-loaded channel running through it [5]. A mixture of butane and oxygen is fed into one end of the channel; carbon dioxide and water vapor are exhausted from the other end. With a butane flow of 8 sccm (standard cubic centimeters per minute) and 80 sccm of oxygen, the average surface temperature is 850°C. For reference, an ordinary pocket lighter burns 15 sccm of butane.

In the experimental setup, the two GaInAsSb PV cells are located directly above the burner and another two cells are located below the burner as shown in Figure 10. These four PV cells are connected in series and their output is connected to the MPPT converter. Experimental data from the complete system setup is shown in Figure 11, which shows converter

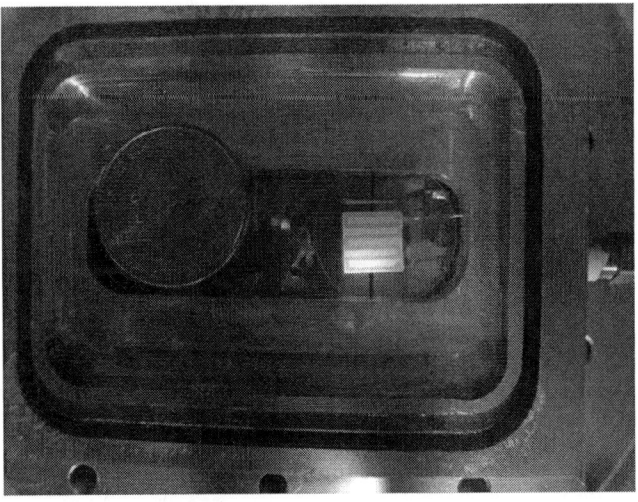

Fig. 10.   Photograph of the experimental setup with the top two PV cells removed and a US quarter for scale. The MEMS burner and the bottom two PV cells are visible.

output power versus time. As expected, this plot looks similar to Figure 9, but there are some notable differences. This first generation micro-reactor assembly has a typical output power of 150 mW, due to the cell being placed at a distance from the burner that is too far for optimum power transfer. Despite this, the demonstrated system output power is more than two orders of magnitude higher than what has previously been achieved [12]. The measured energy density of this micro-TPV system is 75 mW/cm^2. For comparison, the best power densities reported for micro scale direct methanol fuel cell (DMFC), with comparable size to this TPV system, are in the range from 4 to 30 mW/cm^2 [13]. It should be noted that while this early burner prototype has a lower efficiency than the fuel cell presented in [13], previous TPV results [4] show that a comparable efficiency to that of a fuel cell system is achievable. With better system packaging and by further optimizing the system design we are targeting a micro-TPV system power density of 250-300 mW/cm^2.

One of the difficulties encountered during system testing was that the burner experiences occasional temperature fluctuations due to condensed butane entering the fuel supply. Butane is delivered to the burner as a gas but occasional droplets, representing additional fuel, can enter the inlet stream. When a droplet enters the burner, there is a sudden increase in temperature as it burns. Figure 11 captures such an event, which occurs slightly before time t=45 seconds, with a correspondingly large increase in output power, followed by an exponential decay back to steady-state. The time constant associated with this event is such that the MPPT algorithm may take one or two steps in the wrong direction during the increasing power phase, followed by a continuous change of direction during the exponential decay, since the output power at each sample time is lower than the previous sample. The result is that while the converter may operate slightly off of the peak power point during this transient event, it is guaranteed

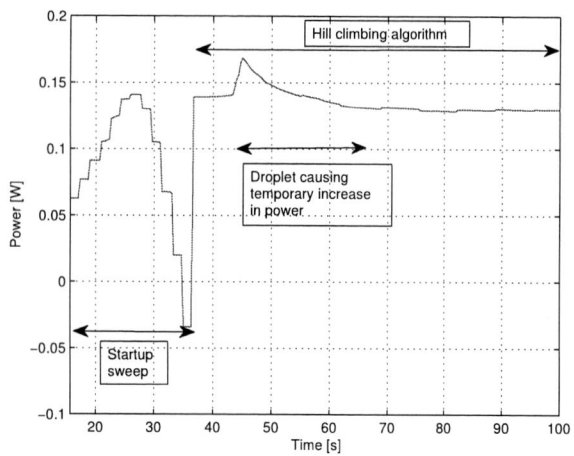

Fig. 11. Experimental data showing the output power of the MPPT as a function of time. The temporary increase in output power around time t=45 seconds is due to a butane droplet forming and causing an increase in burner temperature.

not to move more than a few steps in the wrong direction, ensuring a quick return to the maximum power point once the burner has returned to equilibrium.

V. CONCLUSION

We have demonstrated a power electronics architecture suitable for TPV power generation. The proposed architecture addresses challenges that are unique to the TPV application, and enables a substantial increase in energy capture compared to conventional methods. We incorporate low-power sensing techniques and achieve high power conversion efficiency and small size. A tracking efficiency above above 99% is demonstrated. In addition, we have demonstrated a micro-TPV system generating 150 mW with the MPPT converter. These results pave the way towards a fully integrated, fully functional micro-scale TPV power generator.

REFERENCES

[1] H. Kolm, "Solar-battery power source," quarterly progress report, solid state research, group 35, MIT-Lincoln laboratory, 1956.

[2] I. Celanovic, N. Jovanovic, and J. Kassakian, "Two-dimensional tungsten photonic crystals as selective thermal emitters," *Applied Physics Letters*, vol. 92, pp. 193101–+, May 2008.

[3] I. Celanovic, F. O'Sullivan, M. Ilak, J. Kassakian, and D. Perreault, "Design and optimization of one-dimensional photonic crystals for thermophotovoltaic applications," *Optics Letters*, vol. 29, p. 863, 2004.

[4] B. Wernsman, R. Siergiej, S. Link, R. Mahorter, M. Palmisiano, R. Wehrer, R. Schultz, G. Schmuck, R. Messham, S. Murray, C. Murray, F. Newman, D. Taylor, D. DePoy, and T. Rahmlow, "Greater than 20 percent radiant heat conversion efficiency of a thermophotovoltaic radiator/module system using reflective spectral control," *IEEE Transactions on Electron Devices*, vol. 51, no. 3, pp. 512– 515, 2004.

[5] B. Blackwell, *Design, fabrication, and characterization of a micro fuel processor*. PhD thesis, Massachusetts Institute of Technology, 2008.

[6] M. W. Dashiell, J. F. Beausang, H. Ehsani, G. J. Nichols, D. M. Depoy, L. R. Danielson, P. Talamo, K. D. Rahner, E. J. Brown, S. R. Burger, P. M. Fourspring, W. F. TopperJr., P. F. Baldasaro, C. A. Wang, R. K. Huang, M. K. Connors, G. W. Turner, Z. A. Shellenbarger, G. Taylor, J. Li, R. Martinelli, D. Donetski, S. Anikeev, G. L. Belenky, and S. Luryi, "Quaternary InGaAsSb thermophotovoltaic diodes," *IEEE Transactions on Electron Devices*, vol. 53, pp. 2879–2891, Dec. 2006.

[7] C. Sullivan and M. Powers, "A high-efficiency maximum power point tracker for photovoltaic arrays in a solar-powered race vehicle," in *Proc. 24th Annual IEEE Power Electronics Specialists Conference PESC '93 Record*, pp. 574–580, 1993.

[8] A. Woyte, J. Nijs, and R. Belmans, "Partial shadowing of photovoltaic arrays with different system configurations: literature review and field test results.," *Solar Energy*, vol. 74, pp. 217–233, 2003.

[9] T. Esram and P. Chapman, "Comparison of photovoltaic array maximum power point tracking techniques," *IEEE Transaction on Energy Conversion*, vol. 22, no. 2, pp. 439–449, 2007.

[10] Atmel, "AVR121: Enhancing ADC resolution by oversampling," application note, 2005.

[11] X. Zhou, P. Xu, and F. Lee, "A novel current-sharing control technique for low-voltage high-current voltage regulator module applications," *IEEE Transactions on Power Electronics*, vol. 15, no. 6, pp. 1153–1162, 2000.

[12] O. Nielsen, L. Arana, C. Baertsch, K. Jensen, and M. Schmidt, "A thermophotovoltaic micro-generator for portable power applications," in *TRANSDUCERS, Solid-State Sensors, Actuators and Microsystems, 12th International Conference on, 2003*, vol. 1, pp. 714–717 vol.1, June 2003.

[13] A. Kamitani, S. Morishita, H. Kotaki, and S. Arscott, "Miniaturized microdmfc using silicon microsystems techniques: performances at low fuel flow rates," *J. Micromech. Microeng.*, vol. 18, p. 125019, 2008.

978-1-4244-4782-4/10 $26.00 © 2010 IEEE

11-level Cascaded H-bridge Grid-tied Inverter Interface with Solar Panels

Faete Filho, Yue Cao, Leon M. Tolbert
Electrical Engineering and Computer Science Department
The University of Tennessee
Knoxville, TN 37996-2100, USA

Abstract—**This paper presents a single-phase 11-level (5 H-bridges) cascade multilevel DC-AC grid-tied inverter. Each inverter bridge is connected to a 200 W solar panel. OPAL-RT lab was used as the hardware in the loop (HIL) real-time control system platform where a Maximum Power Point Tracking (MPPT) algorithm was implemented based on the inverter output power to assure optimal operation of the inverter when connected to the power grid as well as a Phase Locked Loop (PLL) for phase and frequency match. A novel SPWM scheme is proposed in this paper to be used with the solar panels that can account for voltage profile fluctuations among the panels during the day. Simulation and experimental results are shown for voltage and current during synchronization mode and power transferring mode to validate the methodology for grid connection of renewable resources.**

Keywords – **multilevel converter, cascaded H-bridges, solar panel, photovoltaic, MPPT, PWM**

I. INTRODUCTION

Because energy resources and their utilization will be a prominent issue of this century, the problems of natural resource depletion, environmental impacts, and the rising demand for new energy resources have been discussed fervently in recent years. Several forms of renewable zero-pollution energy resources, including wind, solar, bio, geothermal and so forth, have gained more prominence and are being researched by many scientists and engineers [1-2].

Solar cell installations involve the use of multiple solar panels or modules, which can be connected in series or in parallel to provide the desired voltage level to the inverter. The cascaded H-bridge multilevel inverter topology requires a separate DC source for each H-bridge so that high power and/or high voltage that can result from the combination of the multiple modules in a multilevel inverter would favor this topology [3-7]. To maximize the energy harvested from each string, a maximum power point tracking (MPPT) strategy is needed. The task of finding the optimum operation point might increase the complexity and component count as the number of isolated DC sources increase. The approach chosen to deal with the number of input sources was to monitor AC output power parameters instead of DC input measurements [8].

Traditional multilevel inverters include cascaded H-bridge inverter, diode clamped inverter, and flying capacitors inverter. This paper focuses on the single-phase 11-level (5 H-bridges) cascade multilevel inverter.

II. MULTILEVEL INVERTER AND PV INTERFACE

An overview of the system is shown in Fig. 1. The core component of this inverter design is the four-switch combination shown in Fig. 1. By connecting the DC source to the AC output by different combinations of the four switches, Q_{11}, Q_{12}, Q_{13}, and Q_{14}, three different voltage output levels can be generated for each DC source, +Vdc, 0, and -Vdc. A cascade inverter with N input sources will provide (2N+1) levels to synthesize the AC output waveform. The DC source in the inverter comes from the PV arrays, and the switching signals come from the multicarrier sinusoidal pulse width modulation (SPWM) controller. The 11-level inverter connects five H-bridges in series and is controlled by five sets of different SPWM signals to generate a near sinusoidal waveform [9-11].

The connection to the grid is done through a variable transformer to assure that at any time the number of H-Bridges used can be controlled, the grid voltage generated by the inverter is met and also to give more flexibility to the experiment since irradiance levels might not be enough. For that reason, an additional fixed 10mH inductance was added as the connection inductance for power transferring mode.

The individual solar panel output power is proportional to solar irradiance variations that occur during the day. The MPPT algorithm will work sensing the output power so no feedback from the individual panels is provided to reduce the number of sensors [12].

As can be seen in Fig. 1, the lower panels, in terms of control signals, will deliver more energy than the upper panels. In order to avoid uneven power to be drawn from the panels by the inverter, a different inverter control approach for the SPWM scheme is proposed here to be used with the solar panels that can account for the voltage profile variation of the panels that occurs during the day. The MPPT and grid synchronization algorithm are fed by output and voltage current signals to generate the gate driver signals as shown in Fig. 2. In Fig. 3 are shown the inverter and its cycle by cycle SPWM control methodology. The irradiance profile over a day changes a few orders of magnitude than a 60 Hz system [13]. That means that a control change action over the modulation index can be taken over a few cycles of the 60 Hz control system. It is desired to get the same amount of power from each string, which cannot be achieved using a conventional SPWM approach. For example, the lower panels

978-1-4244-4782-4/10 $26.00 © 2010 IEEE

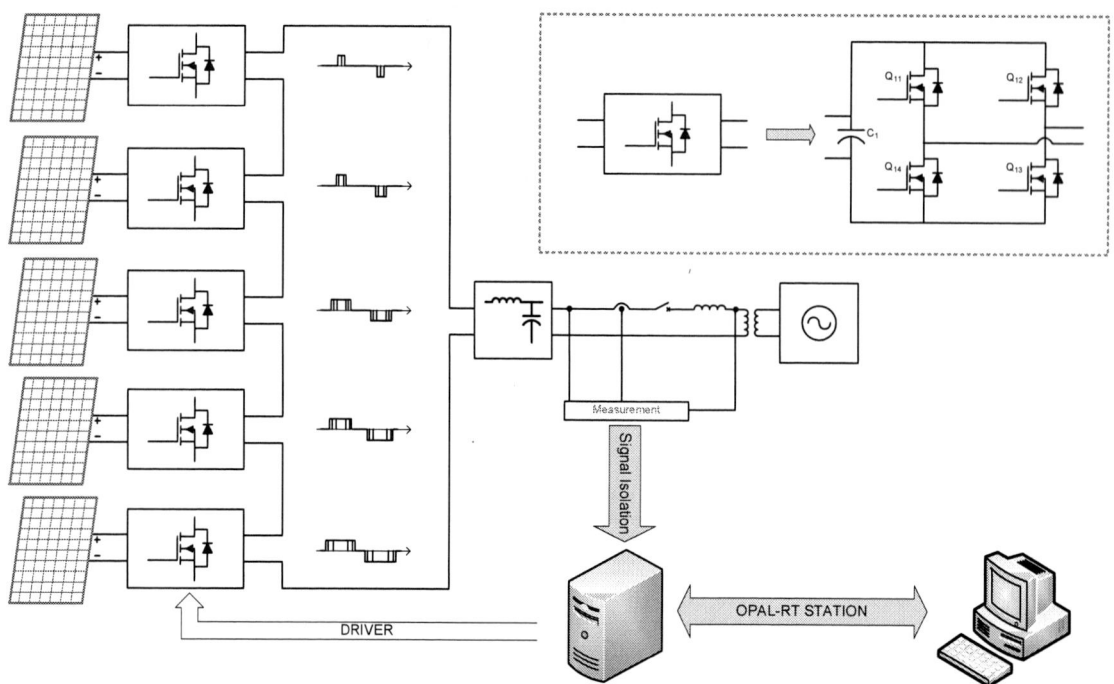

Fig. 1. Multilevel inverter system overview.

in Fig. 1 would send more power than the upper panels as they are switching for a longer time. The sinusoidal nature of the current comes as another factor that makes the power drawn from different panels uneven. The multilevel cascade topology does not require any of the H-bridges to be switched in a determined sequence as would be the case for a diode clamped multilevel (DCM) converter. This gives freedom to switch the H-bridges in the circuit in any order, which can be used as strategy to equalize the power transferred from individual panels. The control strategy implemented shifts the carrier signal over N cycles in the case of a (2N+1) level inverter to make it possible to draw the same amount of

power from each string. Shifting the carrier down for each cycle is the same as physically changing the position of the H-bridges shown in Fig 3(a) for an 11-level inverter. In that figure five cycles of the fundamental frequency are needed to have each panel switch position with the other four. The energy stored in the capacitor will come to help in this process in a cycle-by-cycle basis to avoid a considerable voltage drop due to its considerable large capacitance (1000 uF).

III. SYNCHRONIZATION AND TRACKING CONTROL SYSTEM

Synchronization between inverter and grid means that both will have the same phase angle, frequency and amplitude. This can be done noise proof with respect to the grid by sensing the grid voltage in a Phase Locked Loop (PLL) [14-17]. Typical PLL algorithms include inverse Park-based PLL, Hilbert transformer-based PLL, and transport delay-based PLL. The one to be included in this design is the transport delay-based PLL. Fig. 4(a) shows the block diagram of the PLL algorithm. Notice that the delayed angle can be directly controlled at the computer station during the experiment to provide the signal in quadrature with the grid, which is the input to the Park transform block.

The PLL output is the actual angle position of the grid voltage. This signal is used to generate the sine wave that is used as the reference signal to the control system, which will generate the SPWM signals to drive the switches. The time required for synchronization will be dependent on the PI block parameters. Fig. 4(b) shows the PLL synchronization simulation. In that figure, the PLL starts its synchronization at 0.03 second, and it is in synchronization after about 0.13 second. Since the angle is now known, it is possible to control

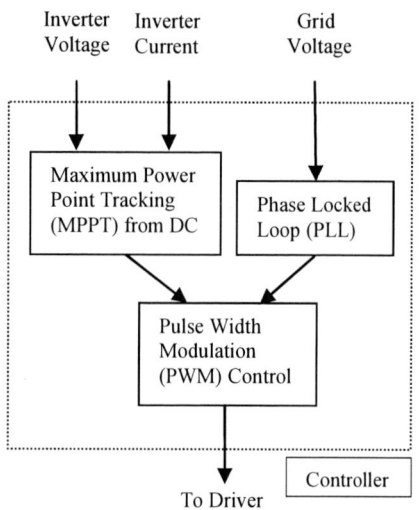

Fig. 2. Control block diagram.

978-1-4244-4782-4/10 $26.00 © 2010 IEEE

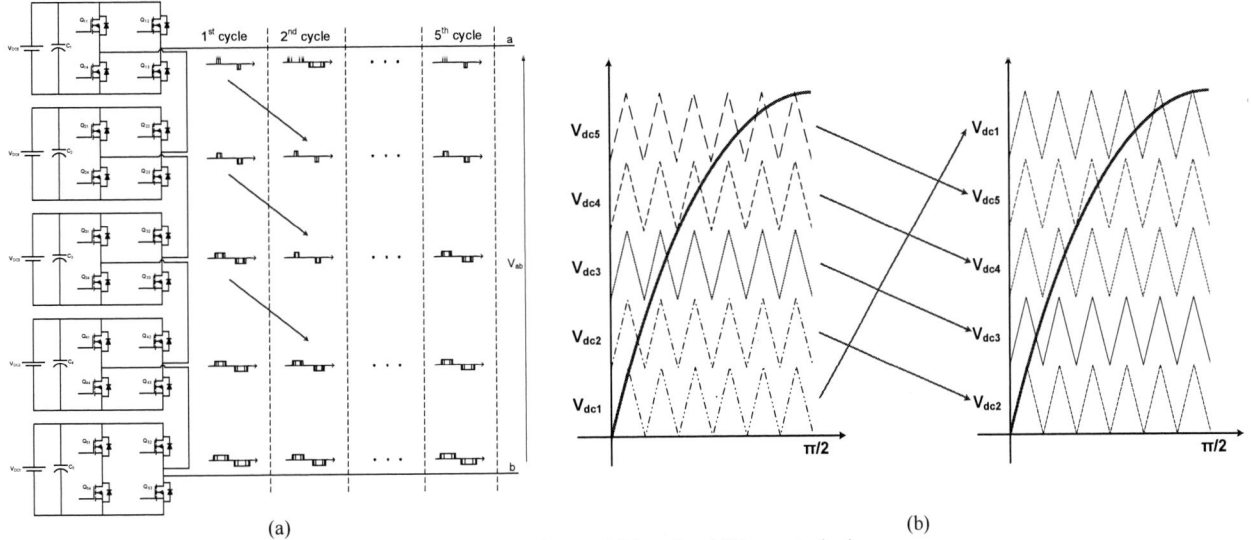

Fig. 3. (a) Inverter topology and (b) carrier shifting control scheme.

the phase difference between inverter and grid by controlling δ. This allows the power flow to be controlled according to (1).

$$P = \frac{V_{inv} \cdot V_{grid}}{X_L} \sin \delta \quad (1)$$

where V_{inv} is the inverter voltage, V_{grid} is the grid voltage, X_L is the connection impedance, and δ is the angle between grid and inverter.

Instead of sensing the individual panel voltages, the maximum power point tracking (MPPT) algorithm determines the optimal point of operation of the panel by calculating the output power and phase angle variation [18]. It monitors output voltage and current parameters by making small changes on the phase angle and looking at the power variation, as in a hill climbing optimization method to track the maximum power point.

IV. EXPERIMENTAL DEMONSTRATION

Each one of the five H-bridges has its own 200 W PV panel connected as an independent source. The panels' specification can be seen in Table 1. The control signals to the bridges are sent by the OPAL-RT workstation where software

and I/O boards are installed. The system acquires grid voltage and inverter output current and voltage to the control block (OPAL-RT workstation) to generate the driver signals to the inverter. Fig. 5 shows the experimental solar panel and the multilevel inverter setup.

TABLE I. SOLAR PANEL SPECIFICATIONS

Sanyo HIP-200DA3	
Rated Power	200 W
Maximum power voltage	56.2 V
Maximum power current	3.56 A
Open circuit voltage	68.8 V
Module efficiency	16.5 %

The RT-Lab control platform, which connects the software (PWM, PLL, MPPT) with the hardware (solar panel, grid, 11-level cascaded H-bridge inverter), to create a real time platform, is the main tool to perform the experiments. Due to hardware limitation, the maximum achievable frequency for the SPWM signals is 2 kHz which requires bulk filtering components as shown in Table 2. A 2 kHz discrete-time RT-Lab drives the inverter with five solar panels as DC inputs and

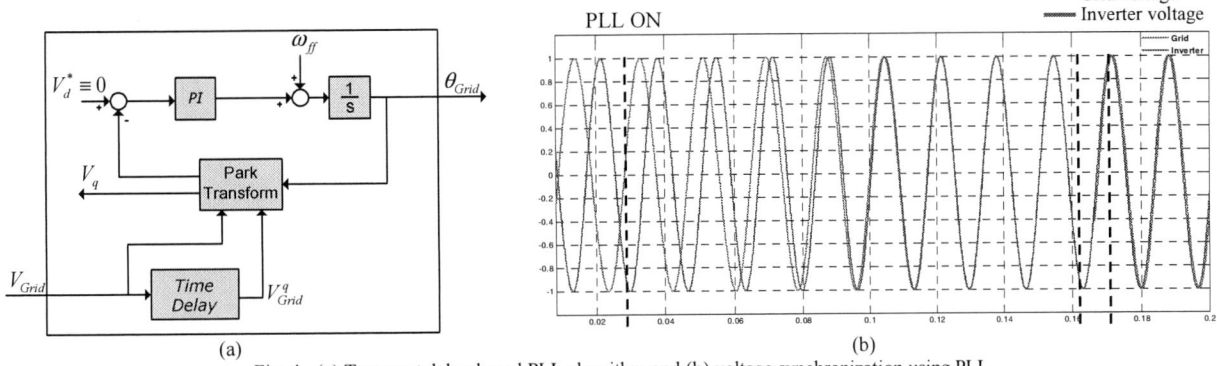

Fig. 4. (a) Transport delay-based PLL algorithm and (b) voltage synchronization using PLL.

(a)

(b)

Fig. 5. (a) Solar panel arrangement and (b) experimental setup.

provides a waveform to interface with the 60 Hz AC grid. Fig. 6(a) shows unfiltered inverter and filtered sinusoidal grid voltages with all five levels of H-bridge inverter under SPWM control and in synchronization. The inverter voltage was stepped up using a transformer to give more freedom during the experiments, for example, to run with all the panels working and to assure that the same amplitude was kept under cloudy conditions when the voltage drops under load condition. In Fig. 6(b), a phase shift is applied by the MPPT algorithm, and power is transmitted from the solar panels to the grid. Another case is shown in Fig. 7 with inverter output current THD shown in Fig. 7(b).

TABLE II. PASSIVE COMPONENTS SPECIFICATION

Filter Inductance	1 mH
Filter Capacitance	92 uF

Connection inductance	10 mH

V. CONCLUSIONS

This paper presented an eleven-level cascade H-bridge inverter, which uses PLL and MPPT with separate solar panels as DC sources to interact with the power grid. A SPWM approach was presented to deal with the uneven power transferring characteristics of the conventional SPWM modulation technique. This technique proved to be successful due to the irradiance profile and the use of capacitors to smooth the voltage fluctuation. The system was driven at 2 kHz because of speed constrains of the control platform, which required bulk filter components.

Grid connection results were shown using the proposed MPPT algorithm. Future work includes the use of a DSP platform to increase switching frequency and reduce filter

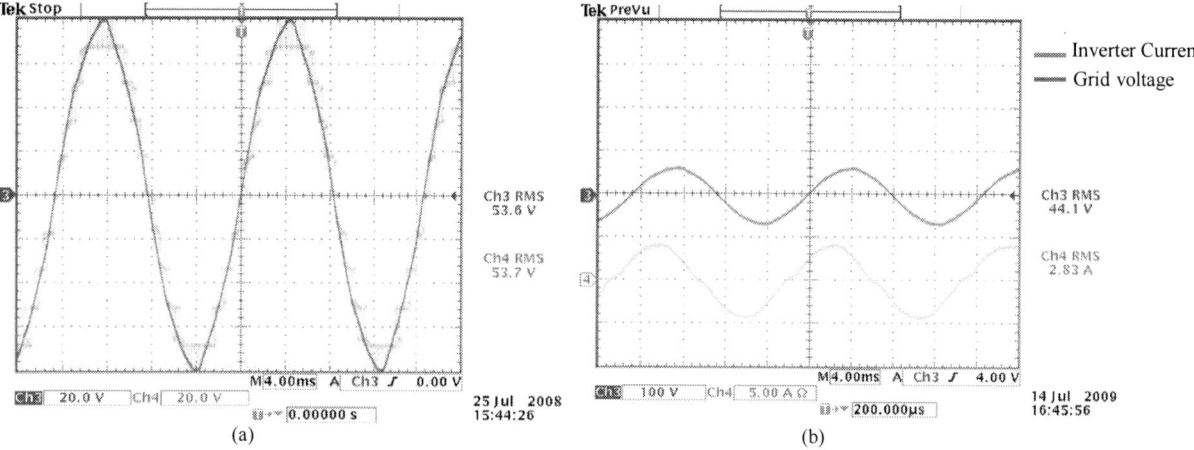

(a) (b)

Fig. 6. Experimental voltage and current waveforms. (a) Utility voltage (purple) and unfiltered output voltage (green) and (b) voltage and current waveforms for grid connected multilevel inverter.

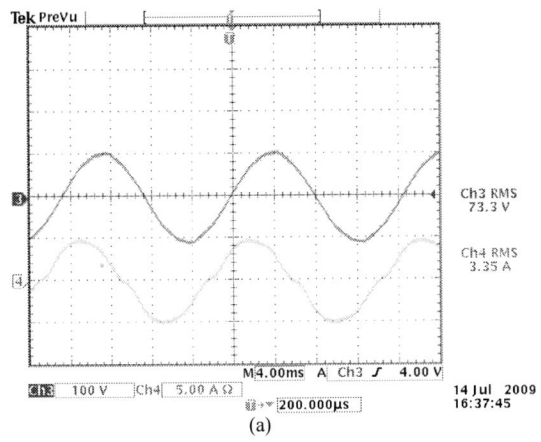 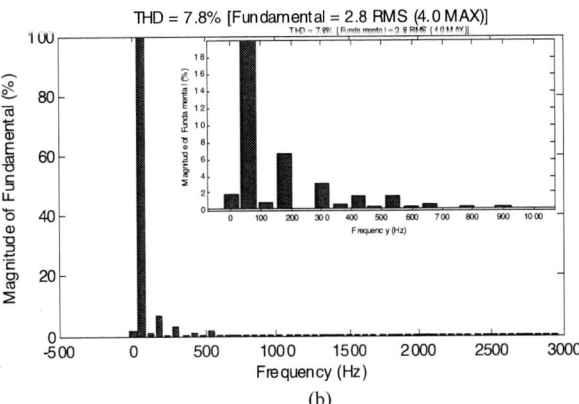

(a)　　　　　　　　　　　　　　　　(b)

Fig. 7. (a) Grid voltage (purple) and current (green) and (b) output current THD.

requirements. The entire PV system structure and its interaction with the grid through PLL and MPPT algorithms were shown by the simulation and experimental results.

REFERENCES

[1] J. M. Carrasco, L. G. Franquelo, J. T. Bialasiewicz, E. Galvan, R. C. P. Guisado, Ma. A. M. Prats, J. I. Leon, N. Moreno-Alfonso, "Power-Electronic Systems for the Grid Integration of Renewable Energy Sources: A Survey," *IEEE Transactions on Industrial Electronics*, vol. 53, no. 4, pp. 1002-1016, June 2006.

[2] A. J. Morrison, "Global Demand Projections for Renewable Energy Resources," *IEEE Canada Electrical Power Conference*, 25-26 Oct. 2007, pp 537-542.

[3] J. Rodriguez, S. Bernet, Bin Wu, J. O. Pontt, S. Kouro, "Multilevel Voltage-Source-Converter Topologies for Industrial Medium-Voltage Drives, " *IEEE Transactions on Industrial Electronics*, vol. 54, no. 6, pp. 2930-2945, Dec. 2007.

[4] L. M. Tolbert, F. Z. Peng, "Multilevel Converters as a Utility Interface for Renewable Energy Systems," *IEEE Power Engineering Society Summer Meeting*, Seattle, Washington, July 15-20, 2000, pp. 1271-1274.

[5] S. Khomfoi, L. M. Tolbert, "Multilevel Power Converters," *Power Electronics Handbook*, 2nd Edition Elsevier, 2007, ISBN 978-0-12-088479-7, Chapter 17, pp. 451-482.

[6] S. Busquets-Monge, J. Rocabert, P. Rodriguez, S. Alepuz, J. Bordonau, "Multilevel Diode-clamped Converter for Photovoltaic Generators with Independent Voltage Control of Each Solar Array," *IEEE Transactions on Industrial Electronics*, vol. 55, July 2008, pp. 2713-2723.

[7] E. Ozdemir, S. Ozdemir, L. M. Tolbert, B. Ozpineci, "Fundamental Frequency Modulated Multilevel Inverter for Three-phase Stand-alone Photovoltaic Application," *IEEE Applied Power Electronics Conference and Exposition*, Feb. 24-28, 2008, pp. 148-153.

[8] S. A. Khajehoddin, A. Bakhshai, P. Jain, "The Application of the Cascaded Multilevel Converters in Grid Connected Photovoltaic Systems," *IEEE Canada Electrical Power Conference*, 25-26 Oct. 2007, pp. 296-301.

[9] S. Ozdemir, E. Ozdemir, L. M. Tolbert, S. Khomfoi, "Elimination of Harmonics in a Five-level Diode-clamped Multilevel Inverter Using

Fundamental Modulation," *International Conference on Power Electronics and Drive Systems*, Nov. 27-30, 2007, pp. 850-854.

[10] J. S. Lai, F. Z. Peng, "Multilevel Converters - A New Breed of Power Converters," *IEEE Transactions on Industry Applications*, vol. 32, no. 3, May/Jun. 1996, pp. 509-517.

[11] B. Kavidha, K. Rajambal, "Transformerless Cascaded Inverter Topology for Photovoltaic Applications," *India International Conference on Power Electronics*, Chennai, India, Dec. 19-21, 2006, pp. 328-331.

[12] O. Alonso, P. Sanchis, E. Gubia, L. Marroyo, "Cascaded H-bridge Multilevel Converter for Grid Connected Photovoltaic Generators with Independent Maximum Power Point Tracking of each Solar Array," *IEEE Power Electronics Specialist Conference*, 15-19 June 2003, pp. 731-735.

[13] A. Abete, R. Napoli, F. Spertino, "A Simulation Procedure to Predict the Monthly Energy Supplied by Grid Connected PV Systems," *Photovoltaic Energy Conversion, 2003. Proceedings of 3rd World Conference on* , vol. 3, 12-16 May 2003, pp. 2427-2430.

[14] E. Villanueva, P. Correa, J. Rodriguez, "Control of a Single Phase H-bridge Multilevel Inverter for Grid-connected PV Applications," *Power Electronics and Motion Control Conference*, Poznan, Poland, Sept. 1-3, 2008, pp. 451-455.

[15] R. B. Godoy, H. Z. Maia, F. J. T. Filho, L. G. Junior, J. O. P. Pinto, G. S. Tatibana, "Design and Implementation of a Utility Interactive Converter for Small Distributed Generation," *IEEE Industry Applications Conference*, Oct. 8-12, 2006, pp. 1032-1038.

[16] S. M. Silva, B. M. Lopes, B. J. C. Filho, R. P. Campana, W. C. Boaventura, "Performance Evaluation of PLL Algorithms for Single-phase Grid-connected Systems," *IEEE Industry Applications Society Annual Meeting*, Seattle, Washington, October 3-7, 2004, pp. 2259-2263.

[17] A. Pandey, N. Dasgupta, A. K. Mukerjee, "A Simple Single-sensor MPPT Solution," *IEEE Transactions on Power Electronics*, vol. 22, no. 2, March 2007, pp. 698-700.

[18] T. Esram, P. L. Chapman, "Comparison of Photovoltaic Array Maximum Power Point Tracking Techniques," *IEEE Transactions on Energy Conversion*, vol. 22, no. 2, June 2007, pp. 439-449.

A life prediction scheme for electrolytic capacitors in power converters without current sensor

Pang H. M.
Department of Electrical and Electronic Engineering
The University of Hong Kong
Hong Kong SAR, China

Pong M. H. Bryan, Senior Member IEEE
Department of Electrical and Electronic Engineering
The University of Hong Kong
Hong Kong SAR, China

Abstract—**Predicting the expected life of switching power supply is essential since unexpected failure of the subsystem can produce enormous loss. Electrolytic capacitor is the weakest among various power components in a power converter. Monitoring the Equivalent Series Resistance (ESR) variation of the electrolytic capacitor, achieving by voltage and current ripple, can estimate the converter life. Currently, Hall Effect sensor or others are current sensing options but all of them add series impedance to the capacitor and deteriorate capacitor voltage waveform. A sensor-less current waveform prediction method is proposed. Popular current mode control with the switch current signal is used. Repetitive sampling on the switch current allows capacitors current waveforms prediction without any current sensor at capacitor nodes. Together with the voltage waveform acquired, the ESR value can be calculated.**

I. INTRODUCTION

Power converter is an essential subsystem in various electronic equipments. Failures of power converters can lead to imminent or stoppage of the whole system. Early and accurate prediction of faults would allow preventive maintenance to be performed, reducing the costs of outage-time and repairs [7]. A better utilization of the converter, achieving by knowing the life of the device, favors green environment as well.

Currently, most power supplies only have their life estimation done in the design stage but this is not sufficient [1, 2]. Electrolytic capacitor is often the weakest component so it represents the converter life. Fig. 1 shows the failure distribution of different components in a static converter [13, 14]. Electrolytic capacitor accounts for largest portion of failures of most power converters. However, the useful life of electrolytic capacitor is strongly affected by the operating conditions [3-9]. There are numerous reasons but mainly the dry-out of electrolyte leads to evident short lifespan of electrolytic capacitor. In order to estimate power converter life accurately it is essential to monitor the operating conditions of the electrolytic capacitor and make appropriate life-span compensation.

Many researchers use various methods to predict electrolytic life [3-9]. High reliability, high power and cost insensitive applications like Uninterruptable Power Supply and DC Bus Capacitor Bank favor monitoring of the capacitor pressure and power devices. A. Riz et al. implemented inner gas pressure measurement approach with an industrial-level equipment setup [3]. M. L. Gasperi suggested a compromised model for ESR estimation from inner vapor pressure that no pressure sensing was required [4]. V. A. Sankaran et al. examined the life model by Gasperi. The experiment showed the model over-predicted the life of capacitor, suggesting the vapor pressure data alone is far from enough [5]. S. K. Maddula presented a capacitor model from Arrhenius' rule of thumb and used it in the dc bus of regenerative IM drives but accurate estimation of the core temperature was critical but difficult as self-heating effect is included [6]. E. C. Aeloiza suggested a real time ESR deterioration approach. ESR calculation is based on the assumption that under steady state power loss only comes from ESR and thus ESR is power loss produced by RMS current [7]. Direct monitoring approach is more accurate but current sensor introduces Equivalent Series Inductance (ESL) that totally changes the voltage waveform of the capacitor, as shown by fig. 2. Y. M. Chen et al. proposed a processor-free online failure prediction method for choke

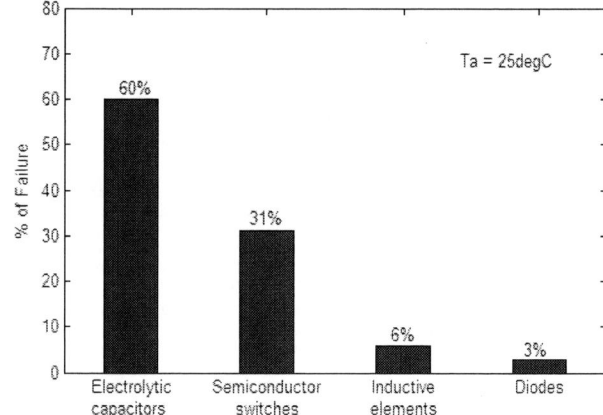

Figure 1. Distribution of failure of power components

978-1-4244-4782-4/10 $26.00 © 2010 IEEE

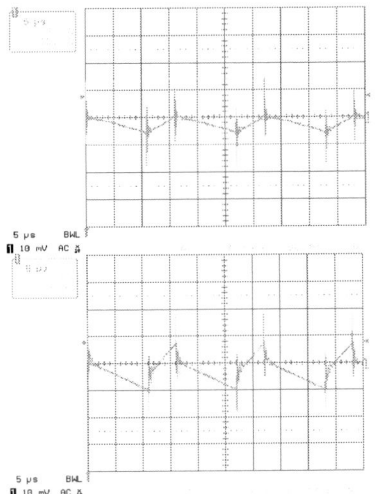

Figure 2. Capacitor voltage ripple waveform without (Upper) and with (Lower) Hall Effect current sensor

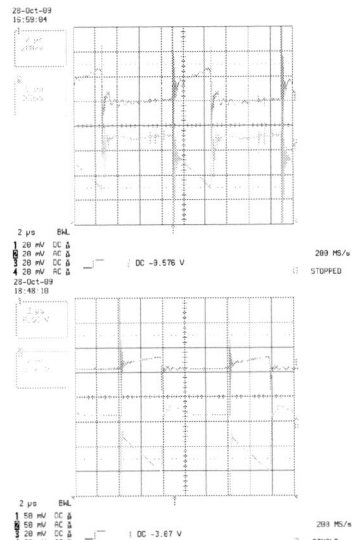

Figure 3. Capacitor voltage ripple waveform (Upper) and current ripple waveform (Lower).

capacitor [8]. Voltage ripple variation could be detected analogically but voltage ripple change due to load variation was not considered. Hao Ma et al. proposed ESR identification system by capturing inductor current and capacitor voltage. This cannot totally eliminate impact of current sensor on the waveforms. Industrial PC was required for complicated calculation as well [9].

This paper proposes a new method to predict electrolytic capacitor life in a power converter. ESR deterioration provides direct and accurate estimation of capacitor life [7, 8]. Capacitor current information is essential for ESR monitoring but it is not desirable to put in any current sensor. Therefore, this paper aims at introducing capacitor current prediction incorporated into voltage ripple measurement to determine the state of the capacitor. Repetitive sampling on switch current, which is readily available in current mode control, allows low-speed waveform acquisition. Capacitor current can then be obtained accurately from the switch current without deteriorating the capacitor voltage waveform. Capacitor voltage can also be captured by repetitive sampling with few modifications on the Sample-and-hold circuit (S/H). With the current and voltage waveforms, ESR is equal to ratio of capacitor voltage ripple to inductor current ripple. The assumption is that ESL is negligible comparing with ESR. This is valid when the capacitor leads are cut to be short and no hall sensor is installed [8]. Fig. 3 shows the experimental captured voltage and current ripple of input capacitor, which closely follow the characteristic by (1).

$$V_{ac} = ESR \times I_{L,ac} \qquad (1)$$

The dry-out of electrolyte is the evident for short lifespan of electrolytic capacitor. The liquid electrolyte has rather conspicuous temperature characteristics and so does the thermal stress have a decisive effect on the capacitor's life expectancy [5, 10-17]. The heat dissipation generated by the ripple current on ESR is an important factor affecting the useful life. An increase in ambient temperature or in internal temperature rise caused by ripple current accelerates

evaporation of the electrolyte [5, 15-17]. By monitoring temperature and ripple current, and with the appropriate model, the life estimation of capacitor can be easily done and so can the power converter life be predicted.

Low cost Analog-to-digital conversion (ADC) and microprocessor can achieve the acquisition and calculation, enabling mass production. Fig. 4 shows the proposed system setup on forward converter. It is well known that forward input capacitor life is critical and ESR monitoring is therefore applied to the input capacitor.

II. CURRENT CAPTURING TECHNIQUE

Sensor-less approach to predict the capacitors' information is more preferable as any senor attached to electrolytic capacitor will distort the voltage waveform across it. Applying sensor also means heavy cost barrier to mass production. Typical current mode control scheme detect the switch current peak to control the switching duty. The current through the switching semiconductor is already available. If more information can be obtained from this current waveform, waveforms of input and output capacitor can be predicted.

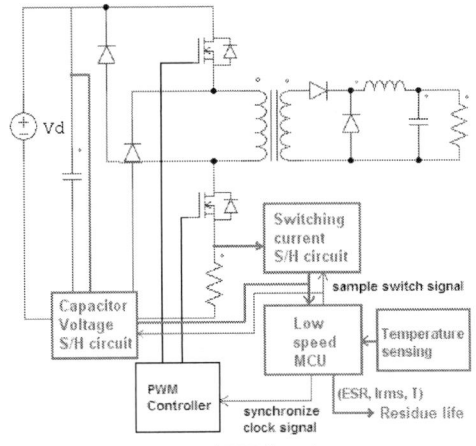

Figure 4. Proposed ESR detection system.

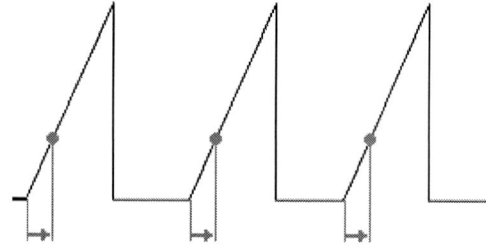

Figure 5. Repetitively sampling a point for a number of cycles before A/D conversion is completed

An essential point is the acquisition of the switching current into the digital microcontroller platform. Direct high-speed ADC of switch current to digital platform is achievable but it is neither economical nor essential. Sample-and-hold circuit and repetitive sampling technique can achieve the same purpose with a slow ADC. This is making use of the repetitive nature of the switching current waveform. The current waveform is sampled at a point for a number of cycles, and waits for the A/D conversion to complete before moving onto the next point. Repeating the sampling throughout the switching cycle, the current waveform can be acquired with low-cost ADC and S/H.

Acquiring the switch current waveform, ESR and the ripple currents RMS[2] of the input (Idrms2) and output capacitors (Iorms2) can be calculated. This can be applied to all the basic power converter topologies, namely the buck, boost and buck-boost converters and their isolation counterparts Forward and Flyback converters. Equations for Forward and Flyback capacitors' ripple current calculations in CCM and DCM are listed. Fig. 6 and fig. 7 shows the ideal capacitor current waveforms and experimental current waveforms respectively.

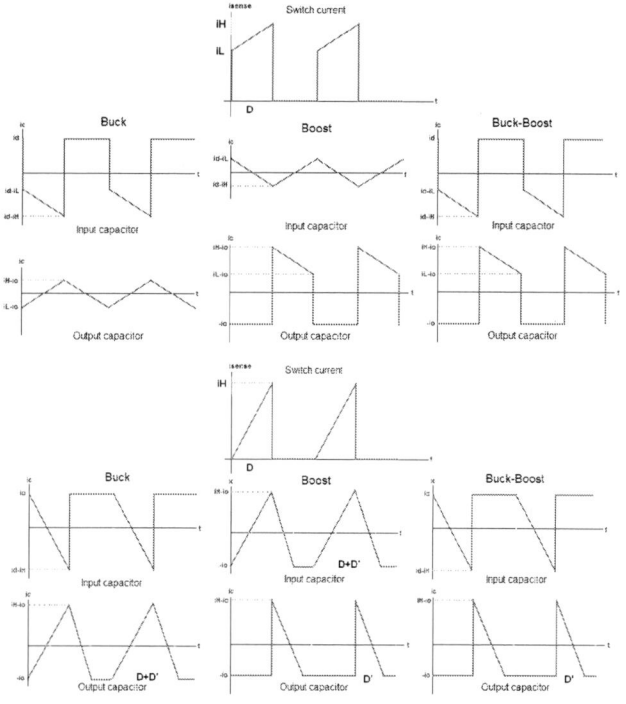

Figure 6. Switch current and capacitors current for CCM and DCM cases

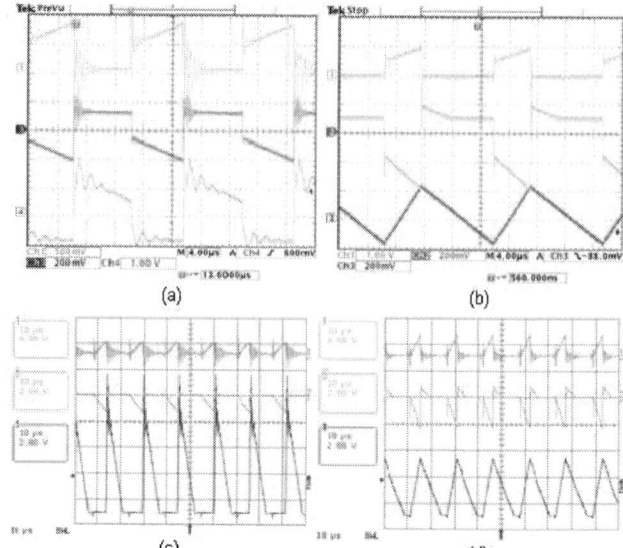

Figure 7. Experimental waveforms for a)CCM Flyback, b)CCM Forward, c)DCM Flyback, d)DCM Forward (upper: switch current; middle: input capacitor current; lower: output capacitor current)

a) CCM Forward converter:

$$I_{d\,rms}{}^2 = \frac{1}{12}[(i_L{}^2 + i_H{}^2)(-3D^2 + 4D) + i_L i_H(-6D^2 + 4D) + i_m{}^2 D_r] \tag{2}$$

$$I_{o\,rms}{}^2 = \frac{1}{12}(n(i_H - i_m) - ni_L)^2 \tag{3}$$

where iL: lower current peak; iH: higher peak; D: switch duty; im: transformer magnetizing current; Dr: reset duty; n: transformer turn ratio.

Note that the magnetizing current should be much smaller than that of switch load current for equation (2) to be accurate. Otherwise higher order equation is required.

b) CCM Flyback converter:

$$I_{d\,rms}{}^2 = \frac{1}{12}[(i_L{}^2 + i_H{}^2)(-3D^2 + 4D) + i_L i_H(-6D^2 + 4D)] \tag{4}$$

$$I_{o\,rms}{}^2 = \frac{1}{12}[(i_{L\sec}{}^2 + i_{H\sec}{}^2)(-3D^2 + 2D + 1) + i_{L\sec} i_{H\sec}(-6D^2 + 8D - 2)] \tag{5}$$

where iLsec=n*iL; iHsec=n*iH.

c) DCM Forward converter:

$$I_{d\,rms}{}^2 = \frac{1}{12}i_H{}^2(-3D^2 + 4D) \tag{6}$$

978-1-4244-4782-4/10 $26.00 © 2010 IEEE

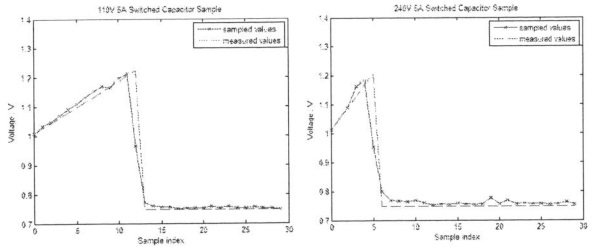

Figure 9. Comparison between switch current waveforms captured by repetitive sampling (solid line) and measured waveforms from scope (dotted line).

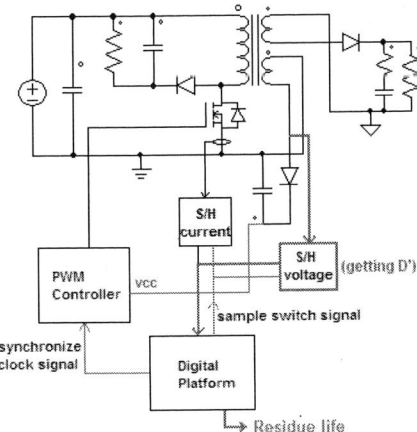

Figure 8. Current capturing system for DCM Flyback

$$I_{orms}^2 = \frac{1}{12} n(i_H - i_m)^2 [-3(D+D')^2 + 4(D+D')] \quad (7)$$

d) DCM Flyback converter:

$$I_{drms}^2 = \frac{1}{12} i_H^2 (-3D^2 + 4D) \quad (8)$$

$$I_{orms}^2 = \frac{1}{12} i_{Hsec}^2 (-3D'^2 + 4D') \quad (9)$$

Flyback converter does not require a filter choke. This is an advantage of the topology from device view point. It brings problem when predicting the output capacitor current since the secondary current fall-time and D' then highly depend on output leakage inductance. This gives rise to large calculation error. One additional voltage sense at auxiliary IC supply winding is implemented to obtain the unknown directly. Fig. 8 shows the system configuration with additional voltage feedback. It does not require any extra winding or special-designed transformer but a simple S/H circuit to sense D' effectively.

III. EXPERIMENTAL RESULTS

A. Capacitor current prediction

The performance of the S/H circuit is verified in fig. 9. Small discrepancy is found between actual waveform (dotted line) and sampled one (solid line). This is mainly caused by the switching noise and under-sampling at current peak. Some minor effects include sample switch gate discharging current and capacitor leakage. Both analog and digital filter design can help reducing noise. Freescale microcontroller MC68HC908MR32 was written to perform the bit-shifting and current ripple calculation at a clock frequency of 4 MHz. Higher sampling frequency can improve the accuracy of peak value detection.

Two 120W CCM converters (Forward and Flyback) and low power DCM converter were built to verify the accuracy of the current calculation system. The parameters extracted from the current waveforms and the calculated current ripple square RMS values at different input voltages and loads, obtained from the microprocessor, are shown in tables in the appendix with their corresponding readings from the measurement equipment by current probes. The Error rows give the percentage errors between the calculated output and input capacitor current ripple square RMS values (Iorms and Idrms) and the measured values.

Calculated capacitor current values for CCM Forward converter have errors range from 1.36% to 7.25%. Sources of errors mainly come from calculation and data acquisition. Acquisition error mainly involves sampling error mentioned previously, magnetizing current peak and reset duty estimation error (in Forward case) and errors from high frequency oscillation by parasitic elements. Calculation error includes truncation error in calculating process and over-simplification of equations. Even so, errors are less than 10%, which is well acceptable. Likewise, the errors of calculated capacitor currents are low for other operation modes and topology. Some gives error as low as 0.45%, which is comparable to measurement errors. These verify the accuracy of the capacitor current prediction method.

B. Online ESR monitoring

The proposed online ESR monitoring method can be applied to different switch mode power converters. A conventional 2-FET Forward converter, as shown in fig. 4, was built to test the performance of the method. Specifications of the Forward converter are listed as follow.

Input voltage Vin = 360 V

Output voltage Vout = 24 V

Maximum output current Iout = 8 A

Switching frequency fs = 100 kHz

Input capacitor Cout = 180 μF, 450 V, 105 C,

Rubycon MXG series

Measured ESR = 306mΩ (25℃, 100 kHz)

Transformer Lm = 3.3mH

Transformer turn ratio n = 5

Output inductor Lout = 60 μH

Fig. 10 shows the experimental captured voltage and current ripple of input capacitor at full load (8A) condition. The measured results are listed in the table. Note that the

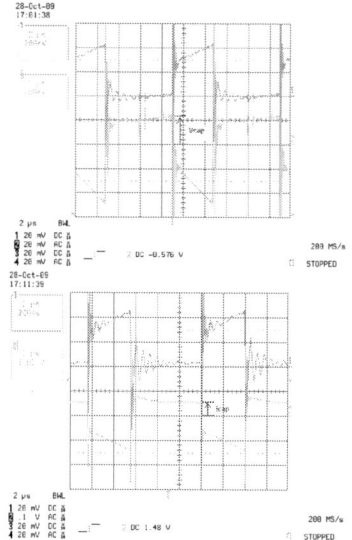

Figure 10. Forward input capacitor voltage ripple waveform (Upper) and current ripple waveform (Lower)

reading for captured ripple voltage (Vcap) and ripple current (Icap) were taken when either voltage probe or current probe was inserted solely. Both voltage and current probes were removed while reading the ESR value. This is essential as these probes will interrupt the sampling system, by means of inserting impedance or providing leakage path. The calculated results' errors range from 2% to 8%, which is well within the acceptable range. Experimental results show that the proposed online monitoring can be applied to the switching-mode power converter successfully.

TABLE I.   *RESULTS FOR ESR MONITORING*

	Half load: 4A		Full load: 8A	
	Captured	Measured	Captured	Measured
IL	0.517	0.510	1.244	1.230
IH	1.177	1.170	1.928	1.920
D	0.388	0.381	0.388	0.381
Icap	0.282	0.277	0.563	0.565
Vcap	0.099	0.095	0.189	0.196
Irms2	0.186	0.194	0.613	0.625
Irms Error	4.12%		1.92%	
ESR	0.329	0.306	0.330	0.306
ESR Error	7.52%		7.84%	

IV. RESIDUE LIFE REPORTING

With the calculated real time capacitor current ripple and ESR, capacitor self-heating loss can be monitored. And the remaining capacitor life can be worked out with an appropriate life model. Several electrolytic capacitor manufacturers provide their life equations [10-11]. It is generally agreed that the effect of temperature on capacitor life is dictated by the Law of Arrhenius [5, 9-12]. The temperature-dependent life model is established to the familiar "life doubles in every 10°C" rule in electrolytic capacitor industry. Table below shows one of the proposed life models [4, 9, 11]. Note that the accuracy of life prediction heavily depends on the life equation itself. But with the real-time monitoring, the deviation due to changing operational

conditions can be eliminated and the prediction accuracy can be improved accordingly. This is also what this paper aims for.

TABLE II.   *RESULTS FOR ESR MONITORING*

	Equations
Arrhenius law	$\dfrac{L_p}{L_r} \approx 2^{\frac{T_r - T_p}{10}}$
Capacitor core temperature	$T_{core} = T_A + \alpha \Delta T$
Temperature rise by self heating	$\Delta T = \dfrac{I^2 \times ESR}{\beta \times S}$
Thermal resistance	$\beta = 2.3 \times 10^{-3} \times S^{-0.2}$
Capacitor surface area	$S = 2\pi(r^2 + rh)$

Adding additional or build-in temperature sensing equipment for ambient temperature, the microprocessor can work out the core temperature of the capacitor and thus the life degradation by the model. The program flowchart is shown in fig. 11. Programmed shifting subroutine generates sampling gate signals for sampling circuit and performs ADC for captured samples (1). Temperature is also captured next (2). When all data inputs are finished, the program analyzes the stored sample data to obtain useful parameters like duty and peak values. Current RMS values are then calculated from programmed equations (3). ESR is also calculated (4). With all necessary data is ready, capacitor life degradation in the predefined period is then calculated from the model (5) and is used to renew the residue life of the capacitor (6).

How to obtain the life degradation? The capacitor life under certain working condition is repeatedly calculated by the microcontroller. The obtained values are then used to modify the remaining life of the capacitor as shown in fig. 12. The calculated life is the reciprocal of slope. This can be easily proved by assuming the capacitor is working under the

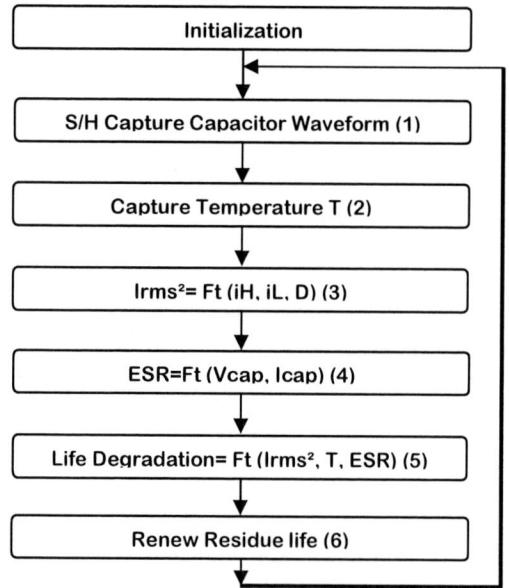

Figure 11. Main program flowchart

978-1-4244-4782-4/10 $26.00 © 2010 IEEE

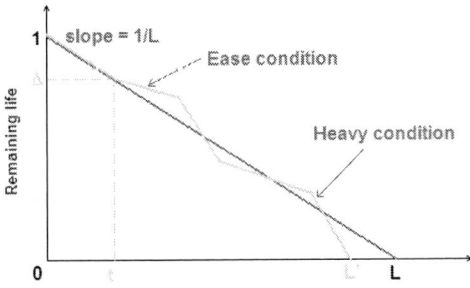

Figure 12. Graph of remaining life against time

rated condition throughout its life, as shown by grey straight line. Then the total operating time is L as remaining life portion drops from 1 to 0 and the slope is $m = 1/L$. Time t is the predefined data renew period which means the new capacitor life result is available every t second. The heavier the working condition over the period gives sleeper slope and the life degradation is faster. With known time t and slope m, the new remaining life portion Δ is projected on Y-axis. Storing up the value Δ, the remaining life of the capacitor can be known. This value can be sent out as PWM duty percentage for computer monitoring.

V. CONCLUSIONS

A new method to predict power converter life through estimation of electrolytic capacitor ESR and ripple current is presented. This method employs no current sensor to measure capacitor current. The popular current mode control current signal is taken. The input and output capacitor currents are calculated. Together with capacitor voltage ripple monitoring the capacitor ESR can be estimated. Power loss on ripple current can be worked out and the life can accurately be estimated. This method is geared towards low cost mass produced power converter. The platform employs a simple microprocessor and waveform digitization technique. Implementation to two power converters verifies the current prediction accuracy. A completed prototype that can tell the ESR is built. One life model employing core temperature estimation derived from ESR deterioration and operating conditions is shown with detailed instructions for implementation.

APPENDIX

TABLE III.   *CURRENT CAPTURE RESULT FOR CCM FORWARD*

I:120V O:5A	iL	iH	D	Iorms	Idrms
Calculated	0.649	0.201	0.310	1.250	1.130
Measured	0.648	0.208	0.310	1.300	1.190
Error				3.85%	5.04%
I:120V O:10A	iL	iH	D	Iorms	Idrms
Calculated	1.020	0.558	0.341	1.350	3.740
Measured	1.000	0.552	0.341	1.370	3.690
Error				1.46%	1.36%
I:240V O:5A	iL	iH	D	Iorms	Idrms
Calculated	0.643	0.160	0.154	2.270	0.609
Measured	0.648	0.162	0.155	2.130	0.619
Error				6.57%	1.62%
I:240V O:10A	iL	iH	D	Iorms	Idrms
Calculated	1.040	0.504	0.170	2.890	2.220
Measured	1.040	0.504	0.165	2.820	2.070
Error				2.48%	7.25%

TABLE IV.   *CURRENT CAPTURE RESULT FOR DCM FORWARD*

I:120V O:2A	iH	D	Dd	Iorms	Idrms
Calculated	0.434	0.280	0.640	1.813	0.511
Measured	0.430	0.281	0.627	1.866	0.494
Error				2.84%	3.44%
I:240V O:2A	iH	D	Dd	Iorms	Idrms
Calculated	0.438	0.143	0.711	2.404	0.246
Measured	0.430	0.138	0.700	2.344	0.256
Error				2.56%	3.91%

TABLE V.   *CURRENT CAPTURE RESULT FOR CCM FLYBACK*

I:120V O:5A	iL	iH	D	Iorms	Idrms
Calculated	0.132	0.493	0.448	25.7	0.505
Measured	0.134	0.49	0.449	25.3	0.530
Error				1.58%	4.72%
I:120V O:10A	iL	iH	D	Iorms	Idrms
Calculated	0.452	0.809	0.465	89.1	1.810
Measured	0.451	0.797	0.470	89.7	1.850
Error				0.67%	2.16%
I:240V O:5A	iL	iH	D	Iorms	Idrms
Calculated	0.030	0.484	0.279	21.9	0.314
Measured	0.040	0.494	0.284	22.0	0.345
Error				0.45%	8.99%
I:240V O:10A	iL	iH	D	Iorms	Idrms
Calculated	0.271	0.727	0.296	54.5	0.988
Measured	0.269	0.730	0.294	54.8	1.020
Error				0.55%	3.14%

TABLE VI.   *CURRENT CAPTURE RESULT FOR DCM FLYBACK*

I:120V O:2A	iH	D	Dd	Iorms	Idrms
Calculated	0.318	0.371	0.472	7.306	0.157
Measured	0.310	0.374	0.475	7.150	0.159
Error				2.18%	1.26%
I:240V O:2A	iH	D	Dd	Iorms	Idrms
Calculated	0.315	0.205	0.472	7.131	0.099
Measured	0.310	0.213	0.472	7.258	0.095
Error				1.75%	4.21%

REFERENCES

[1] Xijin Tian, "Design-for-reliability and implementation on power converters", Reliability and Maintainability Symposium 2005, 24-27 Jan. 2005, pp. 89- 95

[2] A. D. Dominguez-Garcia and P.T. Krein, "Integrating reliability into the design of fault-tolerant power electronics systems", Power Electronics Specialists Conference, PESC 2008, 15-19 June 2008, pp. 2665-2671

[3] A. Riz, D. Fodor, O. Klug and Z. Karaffy, "Inner gas pressure measurement based life-span estimation of electrolytic capacitors", Power Electronics and Motion Control Conference, 2008. EPE-PEMC 2008. 13th, 1-3 Sept. 2008, pp. 2096-2101

[4] M. L. Gasperi, "Life prediction modeling of bus capacitors in AC variable-frequency drives", IEEE Transactions on Industry Applications, Volume 41, Issue 6, Nov.-Dec. 2005, pp.1430-1435

[5] V. A. Sankaran, F. L. Rees and C. S. Avant, "Electrolytic capacitor life testing and prediction", Industry Applications Conference, 1997. Thirty-Second IAS Annual Meeting, IAS '97, Volume 2, 5-9 Oct. 1997, pp.1058-1065

[6] S. K. Maddula and J. C. Balda, "Lifetime of Electrolytic Capacitors in Regenerative Induction Motor Drives", Power Electronics Specialists Conference, 2005. PESC '05. IEEE 36th, 16-16 June 2005 pp.153-159

[7] E. C. Aeloiza, J. H. Kim, P. Ruminot and P. N. Enjeti, "A Real Time Method to Estimate Electrolytic Capacitor Condition in PWM Adjustable Speed Drives and Uninterruptible Power Supplies", Power Electronics Specialists Conference, 2005. PESC '05. IEEE 36th, 16 June 2005, pp.2867-2872

[8] Yaow-Ming Chen, Hsu-Chin Wu, Ming-Wei Chou and Kung-Yen Lee, "Online Failure Prediction of the Electrolytic Capacitor for LC Filter of Switching-Mode Power Converters", IEEE Transactions on Industrial Electronics, Volume 55, Issue 1, Jan. 2008, pp.400-406

[9] Hao Ma and Linguo Wang, "Fault diagnosis and failure prediction of aluminum electrolytic capacitors in power electronic converters", IEEE Industrial Electronics Society 31st Annual Conference, IECON 2005, 6-6 Nov. 2005, pp. 6 pp.-

[10] CDE Cornell Dubilier, "Application Guide Aluminum Electrolytic Capacitors" www.cornell-dubilier.com

[11] Rubycon Corporation, "Technical Notes for Electrolytic Capacitor" www.rubycon.com

[12] Maniktala, Sanjaya., "Switching power supply design & optimization", McGraw-Hill Professional, 2005, pp. 5-7, 324, 361-369

[13] P. Venet, F. Perisse, M. H. El-Husseini and G. Rojat, "Realization of a smart electrolytic capacitor circuit", IEEE Industry Applications Magazine, Volume 8, Issue 1, pp. 16-20, Jan/Feb 2002

[14] Reliability Prediction of Electronic Equipment, Military Handbook 217 F, 1995.

[15] K. Harada, A. Katsuki and M. Fujiwara, "Use of ESR for deterioration diagnosis of electrolytic capacitor", IEEE Transactions on Power Electronics, Volume 8, Issue 4, pp. 355-361, Oct 1993

[16] Afroz M. Imam, "Condition monitoring of electrolytic capacitors for power electronics applications", PhD. Dissertation, Georgia Institute of Technology, 2007

[17] M. L. Gasperi, "A method for predicting the expected life of bus capacitors", Industry Applications Conference, 1997. Thirty-Second IAS Annual Meeting, IAS '97, Volume 2, 5-9 Oct 1997, pp. 1042-1047

Load-Interactive Steered-Inductor DC-DC Converter with Minimized Output Filter Capacitance

S. M. Ahsanuzzaman[1], Amir Parayandeh[1], Aleksandar Prodić[1], Dragan Maksimović[2]

[1]. ECE Department, University of Toronto, Toronto, CANADA
[2]. ECEE Department, University of Colorado at Boulder, USA

Abstract— The paper introduces a concept of load-interactive digitally controlled point-of-load (POL) dc-dc converters where the size of the output filter capacitor can be minimized based on the enhanced load-related communication between the digital load and the converter controller. Based on the information about upcoming load changes, the inductor current of the dc-dc converter is adjusted such that the converter output current matches the new load, thus minimizing the energy storage requirements for the output filter capacitor. The effectiveness of this approach is verified on a digitally controlled 3.3 V/ 30 W experimental prototype, where the dc-dc converter is a steered-inductor non-inverting buck-boost. Transient response results show that, compared to standard converters with near-time optimal control, the new system has about three times smaller voltage deviation, allowing for a proportionally significant reduction in the output capacitor size.

I. INTRODUCTION

In recent years, advanced "power aware" digital chipsets have been introduced in devices such as cell phones, computers, and other electronic equipment having a large portion of digital hardware. These digital loads communicate with power supplies and, to a certain extent, regulate their operation to reduce the overall power consumption [1-4]. Examples include dynamic/adaptive voltage scaling (DVS/AVS) and adaptive body biasing (ABB) techniques, where the supply voltage of the digital signal processors is dynamically changed, depending on the processing load, to minimize dynamic and static power consumption [1-6]. Also, the latest Intel chipsets, provide the switched-mode power supply (SMPS) with an indication of the current demand it expects over a period of time [7], so that this information can be used for phase shading that improves efficiency of multi-phase SMPS, as shown in [8]. Methods that dynamically perform efficiency optimization of a segmented power stage based on the prediction of the load current from a digital signal stream processed by the electronic load have also been shown [9]. So far, communication between the digital load and the SMPS, has mostly been motivated by power savings.

The objective of this paper is to show how this interaction can further be utilized to minimize the overall size of the SMPS, by dramatically reducing bulky and costly energy storage output filter capacitors. The novel system of Fig.1 consists of a modified non-inverting buck-boost converter, i.e. steered-inductor buck-boost, a digital controller, and a communication block providing load current information.

The modified buck-boost converter has an additional switch SW_5, for current steering [12], which, as it will be shown here, provides dramatic improvements of transient response in both buck and boost modes. It should be noted that the non-inverting buck boost is a common stage in numerous battery powered applications (e.g. [15]), so compared to those systems, no additional switches or conduction losses are introduced.

To provide the load current information, i.e. *pre-load* command, a block similar to those implemented in the latest generation of Intel processors [7] is used. Based on the *pre-load* command, during heavy-to-light load transients, the power stage is reconfigured, to steer the inductor current away from the output capacitor into the source. Similarly, the load information is used during the light-to-heavy transient to ramp up the inductor current to the upcoming load value before the transient occurs.

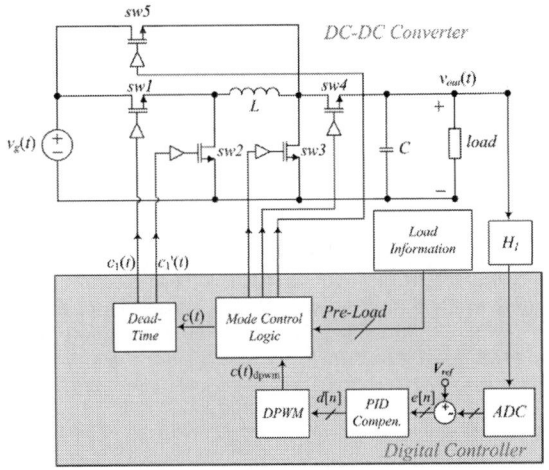

Fig.1: Digital Controller with modified buck-boost converter

This work has been supported by the Natural Science and Research Council of Canada (NSERC)

II. PRINCIPLE OF OPERATION

In a point-of-load (POL) dc-dc converter, the bulky output filter capacitor is sized based on the energy storage required to supply current during large step-load transients. This capacitor size is usually significantly larger compared to the size required to meet the output voltage ripple specification. The latency in the inductor current change, i.e. its limited slew rate, compared to that of the load is the major factor in determining the value of the output filter capacitance.

The converter of Fig. 1 is a modification of the steered-inductor topology proposed in [12]. This topology also resembles the conventional non-inverting buck boost converter [13], except for an addition of SW_5. This switch allows flexible and bidirectional control over the slope of the inductor current, which, in conjunction with the load information communicated by the load, as described in the following subsections, is a key element in providing dramatic improvements in the dynamic response. In the buck mode, SW_4 is on, SW_3 is off at all times while SW_1 and SW_2 change their states at the switching rate. When operating as a boost, SW_1 is on, SW_2 is off at all times and switches SW_3 and SW_4 are active.

It should be noted that constant operation in buck-boost mode (instantaneous conduction of SW_1 and SW_3 followed by the conduction of SW_2 and SW_4), is also possible. In this case, it is avoided in steady state operation, to minimize the number of active switching components operating at switching frequency and, consequently, to reduce the switching losses. Furthermore, in the buck mode, the rms current and consequently, the conduction losses, are also reduced.

A. Light-to-Heavy Load Transient Improvement

When operating as a conventional buck, the rate of change of the inductor current during a light-to-heavy load transient is limited by:

$$\frac{di_L}{dt} = \frac{V_g - V_{out}}{L}, \tag{1}$$

where V_g is the input voltage, V_{out} is the output voltage and L is the converter inductance. As a result, during a sudden light-to-heavy load change, the output capacitor initially discharges providing the difference between the load and output currents, until the inductor current reaches the load value.

In the boost mode, when the input voltage is relatively low (compared to that of the output), the slope of the inductor current during a light-to-heavy load transient is limited by

$$\frac{di_L}{dt} = \frac{V_g}{L}, \tag{2}$$

which causes a latency in the following of the load current. Moreover, to respond to a load transient, a fast acting controller initially disconnects the inductor from the load, leaving the output capacitor to provide all the current. In a properly designed SMPS, this capacitor is usually selected to

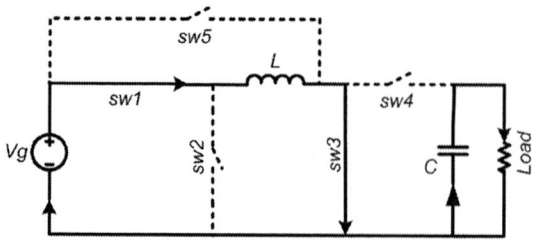

Fig.2 Light-to-Heavy Steer Inductor Operation

be large enough, so that the charge lost during transients does not significantly affect the output voltage.

In the system introduced here, the converter switches in the steering/boost mode of operation, based on a pre-load command. In this mode SW_1 and SW_3 are conducting to increase the inductor current. In some cases, this action is periodically replaced with the conduction of switches SW_1 and SW_4 to maintain the output voltage regulation, as described in the following section. During the steering/boost mode, the controller increases the inductor current to the load level prior to the load change, such that, ideally, at the transient instant, the two currents are matched. This results in a minimum amount of charge supplied by the capacitor. As shown in Fig.2, during the pre-charge phase, the inductor current is charged at the maximum instantaneous rate equal to V_g/L.

B. Heavy-to-Light Load Transient Improvement

During a heavy-to-light load transient the inductor current slew rate of the buck converter is limited by:

$$\frac{di_L}{dt} = \frac{-V_{out}}{L}, \tag{3}$$

In modern low power devices, where the output voltage of the converter is often very low (0.8-1.2V) [14], this limit severely slows down recovery from heavy-to-light load transients causing the output voltage overshoots. To minimize the overshoots, in a conventional converter, the output filter capacitor must be oversized.

In a traditional boost the situation is similar; the slope of the inductor current, limited by

$$\frac{di_L}{dt} = \frac{V_g - V_{out}}{L}, \tag{4}$$

causes the capacitor overcharge. This is a significant problem when the difference between the input and output voltage of the converter is small, which is often the case in numerous portable applications, where a non-inverting buck-boost is used to slightly boost the battery voltage [16].

Unlike the light-to-heavy load transient operation that requires charging up the inductor current prior to the load transient, the heavy-to-light load operation is performed at the point when the actual load transition occurs. At that instant, the controller closes SW_2 and SW_5, as shown in Fig. 3, to discharge the inductor current to the new steady state value while driving the current back to the source. It can be seen that this action not only disconnects the inductor from the capacitor and prevents overcharging but, in the buck mode,

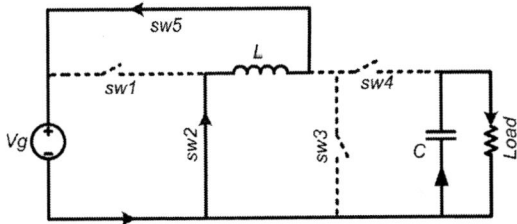

Fig.3 Heavy-to-Light Steer Inductor Operation

also changes the current slew rate to a significantly higher value:

$$\frac{di_L}{dt} = \frac{-V_g}{L} \qquad (5)$$

In this case, the preload command supplies the controller with the information about the new steady state current value, which is used to determine the point when the regular buck or boost operation could resume.

III. PRACTICAL IMPLEMENTATION

The controller architecture, shown in Fig.1, incorporates PID compensator that is used for steady-state voltage regulation with a block named *mode control logic*. The mode control logic is a finite state machine (FSM), which chooses different operating modes for the converter, depending on the pre-load information and the conditions in the circuit.

The state transitions of the FSM are shown in Fig. 4. During the steady state, a modified PID controller operates the converter as a buck or a boost, depending on the input and output voltages (as in [17]). Once the 'Pre-Load' command is received, the mode control logic determines the required steer-inductor current action. In the case of a light-to-heavy load transient, *steering/boost* is activated prior to the occurrence of the load step, to achieve the required inductor current. Once the desired current is reached, the FSM generates the 'Done Signal' to inform the load that the power consumption can be increased. In the case of the heavy-to-light load transient, SW_2 and SW_5 are turned on to steer the inductor current back to the source V_g.

Fig.5 demonstrates the operation of the mode control logic. The logic governs the inductor current steering after receiving the 'Pre-Load' command during steady state. Once the operation is completed the "Done Signal" is created to inform the load that the inductor current has reached the new requested value. It can also be seen that for the changes of load from a very light to a heavy value (as shown in Fig. 5) the steer operation reduces to a single switching action, where only SW_1 and SW_3 are turned on until the desired current is reached. Such simplified operation is possible as long as the inductor current can be charged to the new steady state value before a significant change of the capacitor voltage due to the load current occurs. Otherwise, a more advanced control action is required, as described in the following subsection.

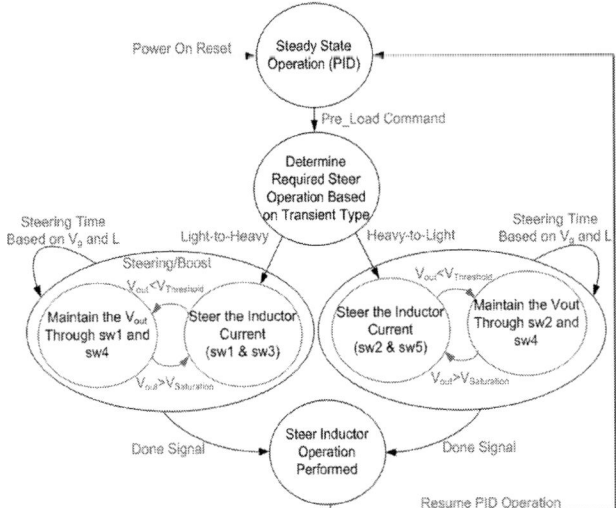

Fig.4. State flow diagram of the mode control logic

A. Preventing Significant Capacitor Discharge During Inductor Current Steering

As shown in Figs. 2 and 3, during the inductor current steering, the output capacitor is not connected to the inductor. As a result, it discharges through the load that causes a voltage drop. When the load change is significant, this voltage drop cannot be neglected. In order to minimize this variation, a threshold for the maximum allowable voltage dip is set, the output is monitored, and a periodic switching between *steering and boost* operation is activated accordingly.

The operation in this mode is illustrated in Fig.6. In the case of light-to-heavy load transient, if the output voltage drops below the threshold ($V_{out} < V_{Threshold}$), the mode control logic momentarily stops the steer operation to charge up the output capacitor back to the desired level by turning on SW_1 and SW_4. After the voltage is recovered, the logic reestablishes the steering through SW_1 and SW_3. This allows increasing the inductor current while maintaining the output voltage. Similarly, during a heavy-to-light load transient in order to prevent capacitor discharging beyond the threshold the steer operation is periodically paused and the output capacitor is charged by the inductor by turning off SW_5 and turning on SW_4 (Fig. 4).

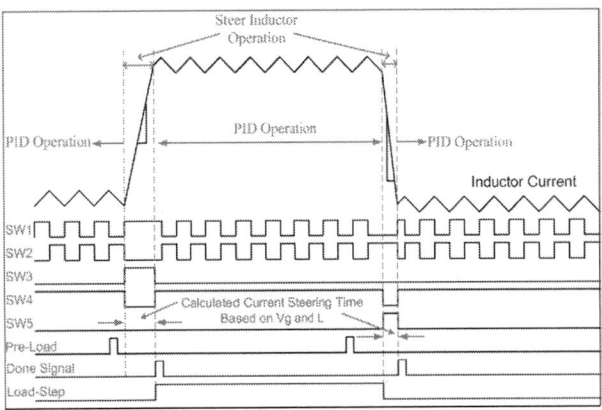

Fig.5. Gating signals for different modes of operation.

978-1-4244-4782-4/10 $26.00 © 2010 IEEE

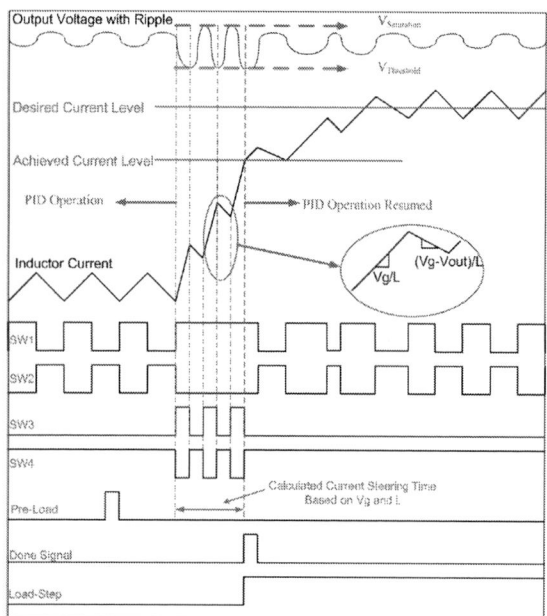

Fig.6. The converter operation as a boost converter during light-to-heavy load transient.

IV. EXPERIMENTAL SYSTEM AND RESULTS

To verify the operation of the system shown in Fig. 1 an experimental system was built. The 30-W, 3.3-V, 500 kHz power stage is designed for the input voltage ranging between 1.5V and 12V. To compare the performance, a time optimal controller capable of recovering output voltage through a single on-off action, similar to [10], was developed as well.

The experimental results presented in Figs. 7 to 10 show comparisons of the transient responses of both converters and

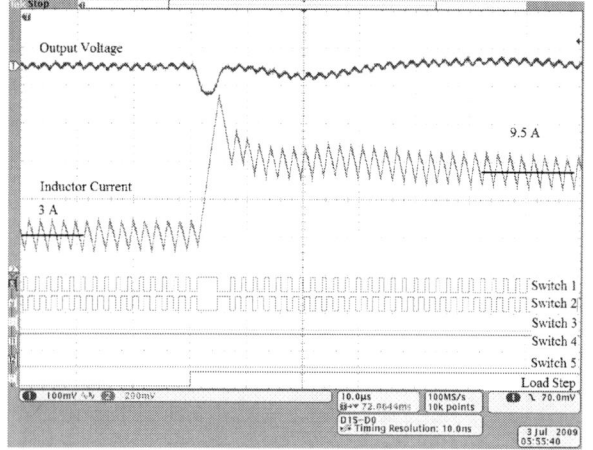

Fig.7. The light-to-heavy load transient of the time-optimal controller– Ch1: output converter voltage (100mV/div); Ch2: inductor current (3.5A/div).

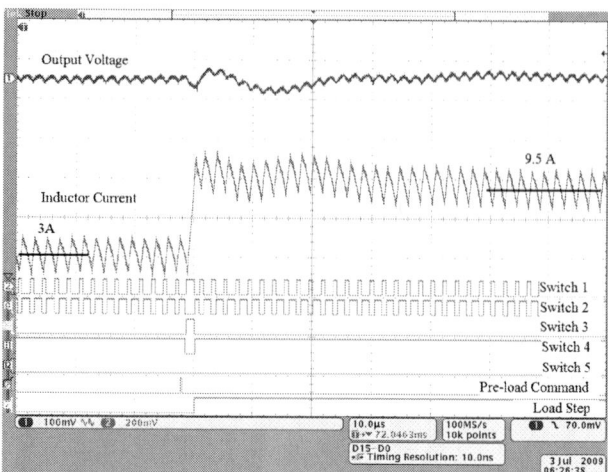

Fig.8. The light-to-heavy load transient of the proposed load-interactive controller– Ch1: output converter voltage (100mV/div); Ch2: inductor current (3.5A/div).

verify superior performance of the load-interactive steered-inductor topology. The waveforms show the comparison for the case when the converter is operating as a buck with a 10V input and 3.3 V output. Fig. 11 and 12 show the converter switching in *steer and boost* operation mode.

Fig. 7 shows the transient response of the time-optimal controller to a 6.5A light-to-heavy load transient. The gating signals of SW_1 and SW_2 depict the action of the time-optimal control during the transient. Fig. 8 shows the response of the load-interactive system. It can be seen that the new system has less than one half of the voltage undershoot for the same load step. Moreover, unlike the time optimal controller, the load-interactive system exposes the inductor to a much smaller peak current preventing possible saturation. The pre-load command that informs the converter prior to the occurrence of the load step is also shown. Based on the pre-load command SW_1 and SW_3 are turned on, to charge up the inductor to the desired level. Once the desired current is achieved the regular

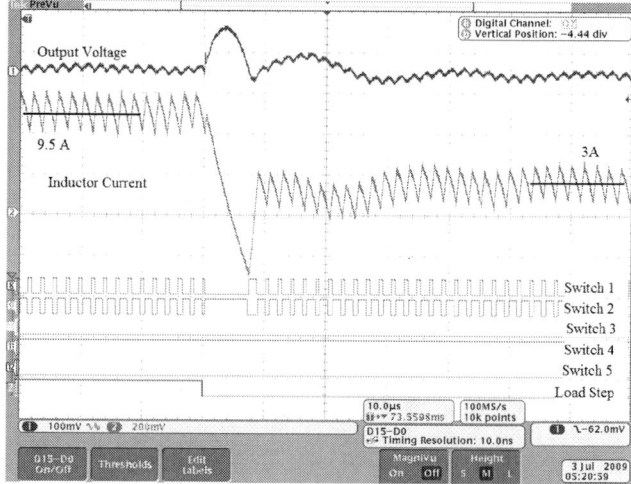

Fig.9. The heavy-to-light load transient of the time-optimal controller– Ch1: output converter voltage (100mV/div); Ch2: inductor current (3.5A/div).

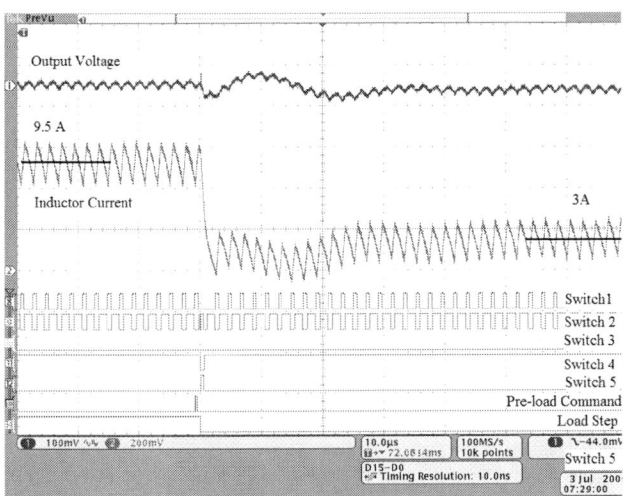

Fig.10. The heavy-to-light load transient of the proposed load-interactive controller– Ch1: output converter voltage (100mV/div); Ch2: inductor current (3.5A/div).

PID operation is resumed and it continues until the next pre-load command is received.

Figs. 9 and 10 show the transient responses of the time-optimal controller and that of the proposed load-interactive converter to a 6.5A heavy-to-light load step, respectively. A similar improvement to that for the light-to heavy transient case can be observed. The waveforms of Fig.10 also experimentally verify the operation described in Section II. It can be seen that the steer inductor operation during the heavy-to-light load transient is performed at the point of the actual load step. The digital gating signals of SW_2 and SW_5 indicate that the inductor current is driven back to the source until it reaches the desired value.

Figs. 11 and 12 show the converter operation in *steer and boost* mode, due to the output voltage drop below threshold voltage. In Fig. 11 it is shown that the inductor current level

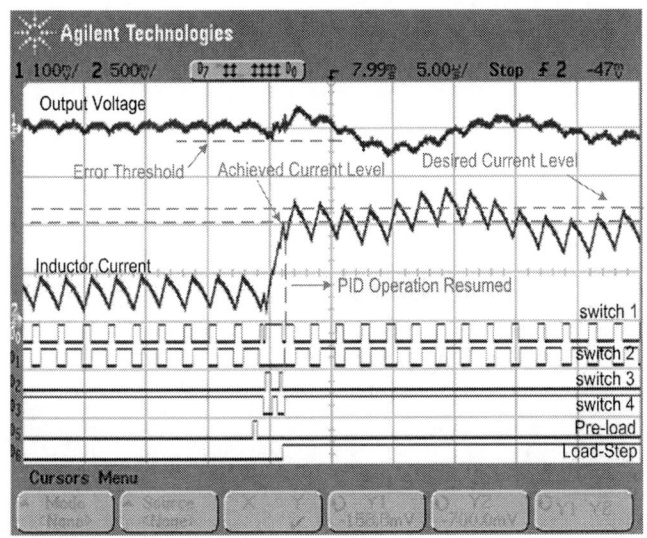

Fig.11.Steer and Boost mode of operation– Ch1: output converter voltage (100mV/div); Ch2: inductor current (5A/div).

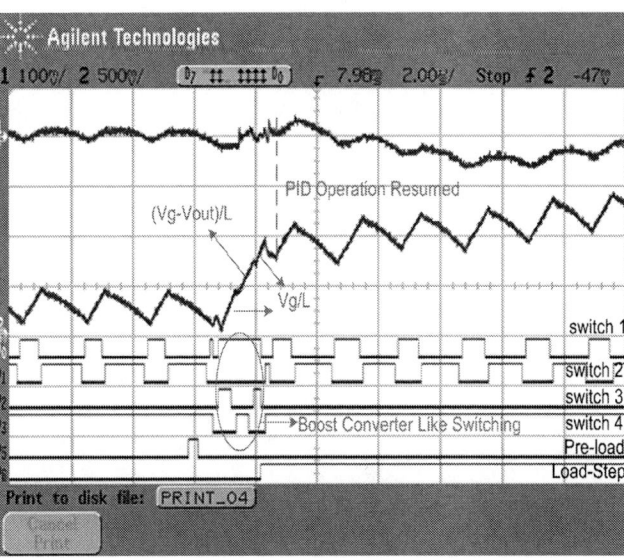

Fig.12. Boost converter like switching during steer and boost operation– Ch1: output converter voltage (100mV/div); Ch2: inductor current (5A/div).

achieved is below the desired load current level due this operation. Nevertheless, this operation brings the inductor current close to the desired level, while maintaining the output voltage in the acceptable range. Fig. 12 zooms into this operation to show the boost converter like switching. As this operation is taking place during buck mode, when SW_1 and SW_4 conduct to restore the output voltage, the inductor current increases with a slope (V_g-V_{out}/L) which is lower than the slope during the steer operation (V_g/L).

CONCLUSIONS

A load interactive flexible digitally controlled dc-dc converter with minimized output filter capacitor is introduced. It is shown that by using a modified buck-boost converter, named steered inductor buck-boost, and by improving communication between the converter and its digital load the size of the output capacitor can be drastically reduced. Based on the information about the load current, the inductor current is steered into or out of the capacitor, at a much higher slew rate compared to the conventional designs, minimizing the current provided by the capacitor during load transients. Theoretically, the newly proposed system opens a possibility of designing dc-dc converters such that the output filter capacitor is selected only based on the maximum allowable switching ripple, thus drastically reducing the overall size of the dc-dc converters.

REFERENCES

[1] T. Burd, , T. Pering, A.Stratakos and R.Brodersen. " A Dynamic Voltage Scaled Microprosseor System," In *IEEE ISSCC* pp.294-295, Feb 2000.

[2] R.. Gonzalez,; B.M Gordon, , M.A Horowitz, "Supply and threshold voltage scaling for low power CMOS," *IEEE Jour. Solid-State Circ.*, vol, 32, no. 8, pp 1210-1216, August 1997.

[3] M. Miyazaki, M. Ono, G. Hattori, T. Shiozawa, K. Uchiyama, K. Ishibashi, "A 1000-MIPS/W microprocessor using speed adaptive threshold-voltage CMOS with forward bias," In *IEEE ISSCC* pp.420-425, Feb 2000.

[4] Kuroda, T. Fujita, T. Mita, S. Nagamatsu, T. Yoshioka, S. Suzuki, K. Sano, F. Norishima, M. Murota, M. Kako, M. Kinugawa, M. Kakumu, M. Sakurai, "A 0.9-V, 150-MHz, 10-mW, 4 mm2, 2-D discrete cosine transform core processor with variable threshold-voltage (VT) scheme," *IEEE Jour. Solid-State Circ.* Vol.31 , no.11, Nov 1996, pp 1770-1779.

[5] A.P. Chandrakasan, R.W. Brodersen , "Minimizing Power consumption in Digital CMOS circuits," *Proc. IEEE* vol. 83, NO 4, April 1995, pp. 498 – 523.

[6] Kao, J.T. , Miyazaki, M., and A.P. Chandrakasan ," A 175-mV Multiply-Accumulate Unit Using an Adaptive Supply Voltage and Body Bias Architecture," *IEEE Jour. Solid-State Circ.* vol. 37, Nov.2002, pp. 1545 – 1554.

[7] "Intel VRM 11.1 Design Guideline," Available online at http://www.intel.com/technology/itj/index.htm.

[8] Z. Lukić, Z. Zhao, A.Prodić, and D. Goder "Digital Controller for Multi-Phase DC-DC Converters with Logarithmic Current Sharing," in *Proc. IEEE PESC*, 2007, June 2007.

[9] O. Trescases, G. Wei, A. Prodić, W.T. Ng, K. Takasuka, T. Sugimoto, and H. Nishio "A Digital Predictive On-Line Energy Optimization Scheme for DC-DC Converters," in *Proc. IEEE APEC Conf*, March 2007.

[10] Z. Zhao and A. Prodić, "Continuous-time digital controller for high frequency dc–dc converters," *IEEE Trans. Power Electron.*, vol. 23, no.2, pp. 564–573, Mar. 2008.

[11] Z.Yousefzadeh, A. Babazadeh, B. Ramachandran , E.Alarcón, L.Pao,D. Maksimovic " Proximate Time-Optimal Digital Control for Synchronous Buck DC–DC Converters," *IEEE Trans. Power Electron.*, vol. 23, no.2, pp. 564–573, July. 2008.

[12] A. Stupar, Z. Lukić, and A. Prodić, "Digitally-Controlled Steered-Inductor Buck Converter for Improving Heavy-to-Light Load Transient Response," in *Proc. IEEE PESC conf*, 2008.

[13] Robert W. Erickson and Dragan Maksimović, "*Fundamentals of Power Electronics,*" Second Edition, New York: Springer Science+Business Media, 2001.

[14] "International Technology Roadmap for Semiconductors 2005 Edition Executive Summary," pp. 81-82. Available online at http://www.itrs.net/Links/2005ITRS/ExecSum2005.pdf

[15] P. Henry , "New Advances in Portable Electronics," in *APEC Plenary Session* , March 2009.

[16] R. Paul, L. Corradini, D. Maksimović, "Σ-Δ Modulated Digitally Controlled Non-Inverting Buck-Boost Converter for WCDMA RF Power Amplifiers," in *Proc. IEEE APEC Conf*, March 2009.

[17] A. Prodić and D. Maksimović, "Digital PWM Controller and Current] ator for a Low-Power Switching Converter," *in Proc., IEEE Computers in Power Electronics Conference (IEEE COMPEL)*, July 2000, pp. 123-128.

EMI Filter Design for High Switching Frequency Three-Phase/Level PWM Rectifier Systems

M. Hartmann[*], H. Ertl[†] and J. W. Kolar[*]

[*] Power Electronic Systems Laboratory, Swiss Federal Institute of Technology
Zurich, Switzerland; Email: hartmann@lem.ee.ethz.ch
[†] Institute of Electrical Drives and Machines, Power Electronics Section
University of Technology Vienna, Austria; Email: j.ertl@tuwien.ac.at

Abstract— **The actual attenuation characteristic of EMI filters in practice often differs from theoretical predictions and minor changes could result in a significant performance improvement. Whereas the performance of the differential-mode (DM) filter stage usually can be well predicted, the common-mode (CM) behavior is more difficult to handle. This is especially true for three-phase PWM rectifier systems, which show a large high-frequency CM voltage at the rectifier output. In this work the possible CM noise current paths of a three-phase/level PWM rectifier are analyzed where parasitic capacitances to the heat sink and to earth are considered. Additionally, a concept to significantly reduce CM emissions is discussed in detail. Based on the proposed models an EMI filter design for a system with 1 MHz switching frequency is shown. Experimental verification of the designed EMI filter is presented by impedance and conducted emission (CE) measurements taken from a 10 kW hardware prototype. Several practical aspects of filter realization like component arrangement, shielding layers, magnetic coupling etc. are discussed and verified by measurements.**

I. INTRODUCTION

For new application fields (e.g. power electronics in aircraft [1]) modern rectifier systems have to meet high requirements concerning efficiency, weight and compactness. Active three-phase rectifiers offer the possibility to comply with rigorous low-frequency current harmonics limits but show a large high frequency noise level. Passive low-pass filters employing inductors and capacitors in connection with resistors providing passive damping can be used to attenuate resulting conducted emissions (CE) of the systems [2]-[3]. These passive filter elements take up a relatively large portion of overall system volume and can only be reduced in volume by increasing the switching frequency.

It is common and very helpful to split the generated EMI emissions into a common-mode (CM) and differential-mode (DM) component. Whereas DM noise currents flow in and out through the phases, CM currents return via earth. Hence, different filter strategies and filter elements have to be applied to handle the two emission types. However, as will be shown in section II, asymmetrical currents to earth caused by asymmetrical impedances of the rectifier system also generate DM noise. These type of emissions are called "mixed-mode noise" (MM) and their origin was analyzed in [4]-[5] for single-phase flyback converters. MM noise in three-phase diode front-end converters was discussed in [6]-[7].

The performance of the DM filter can be well predicted. Commonly (dependent on the required attenuation) multi-stage LC-filters are applied [8]. Also "zero-ripple" DM filter concepts have been proposed [9]. On the contrary, CM emissions are mainly determined by parasitic elements like capacitances of semiconductors to the heat sink, capacitances between heat sink and earth, magnetic couplings of inductors etc. and are therefore difficult to identify and quantify.

In this work an EMI filter design and realization for an ultra-compact three-phase/level PWM rectifier (cf., **Fig. 1** where

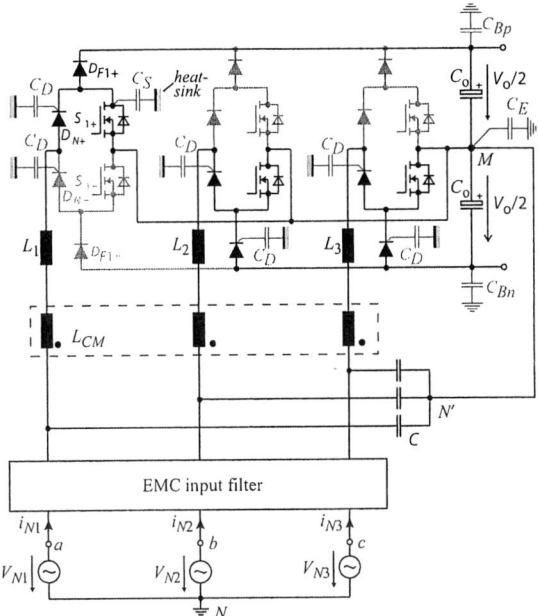

Fig. 1: Schematic of the three-phase/level PWM rectifier including relevant parasitic capacitances from semiconductors to heat sink (C_S, C_D) and from the DC output rails to earth (C_{Bp}, C_{Bn} and C_E) for $i_{N1} > 0$, $i_{N2}, i_{N3} < 0$. Additionally, the proposed CM voltage reduction concept is shown, where the output voltage midpoint M is connected to an artificial mains star point N'. High frequency CM currents are limited by a three-phase CM inductor L_{CM}.

the parasitic capacitances relevant for $i_{N1} > 0$, $i_{N2}, i_{N3} < 0$ are shown) [10] will be discussed. The realized filter should show minimum volume in order to achieve a highest possible rectifier power density. Accordingly, a switching frequency of 1 MHz is chosen [11].

In [12] the formation of the CM voltage was analyzed in detail but for sake of simplicity only a single lumped capacitor from the output voltage midpoint M to earth was used to model the CM current paths. This is a reasonable approach to get an overview of the CM behavior of a system but turns out to be not sufficiently accurate for designing the EMI filter. Hence, a more detailed CM model considering parasitic capacitances of the semiconductors to the heat sink and from the output voltage rails to earth is developed in section II.

In [12] also a concept for minimizing the high-frequency CM emissions was proposed (further concepts can be found in [13]). There, the output voltage midpoint M is connected to an artificial mains star-point N' formed by capacitors C. Whereas the low-frequency CM voltage used to increase the input voltage range of

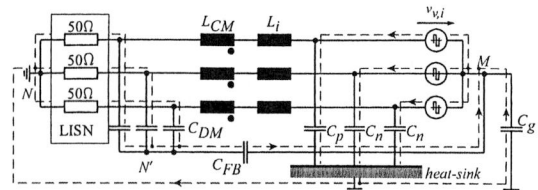

Fig. 2: Detailed noise model valid for $i_{N1} > 0$, $i_{N2}, i_{N3} < 0$ if the heat sink is connected to earth. Only current paths involving the parasitic capacitors C_p and C_n are shown.

the rectifier drops across the capacitors C, the high frequency CM output voltage is attenuated by a low-pass filter action of the boost inductors L_i and the capacitors C. Unfortunately, this concept results in a considerably increased ripple of the boost inductor currents and/or in higher copper and core losses. Therefore, the basic concept is extended here by placing a three-phase CM inductor L_{CM} in series to the boost inductors L_i which considerably reduces the additional high-frequency current ripple.

The DM filter design as well as the design of the proposed CM filter inductor will be discussed in section III where also a simplified method for determining the CM noise level is proposed. Section IV deals with the practical realization of the filter considering available magnetic materials. Measurements taken from a 10 kW rectifier prototype verify the effectiveness of the proposed EMI filter in section V. In this section also several practical aspects of filter realization like the physical components arrangement, shielding layers, and magnetic couplings etc. are discussed with reference to measurement results.

II. CONVERTER NOISE MODEL

The semiconductors of a power electronics converter are typically mounted on a common heat sink which usually is connected to earth. Therefore, parasitic capacitances to earth exist which are not included the basic circuit diagram. In order to fully understand the propagation of the resulting CM noise currents these capacitances have to be considered and/or a detailed CM noise model has to be derived. Accordingly, in the following the modeling approach given in [19]-[20] for single-phase PFC will be extended to three-phase systems.

In **Fig. 1**, the relevant parasitic capacitors between semiconductors and heat sink are drawn for $i_{N1} > 0$, $i_{N2}, i_{N3} < 0$. The capacitors C_S and C_D represent the stray capacitance of a MOSFET's drain and a diode's cathode to the heat sink which is approximately 60 pF for the applied TO220 package. Actually, these capacitances are present for all semiconductors of the rectifier system, but capacitors which are not carrying current in $-30° < \varphi_N < 30°$ are not shown. The capacitors C_{Bp} and C_{Bn} model the stray capacitances of the positive and negative output voltage rail to earth. The capacitor C_E models the parasitic capacitance of the output voltage midpoint to earth but also includes possible parasitic capacitances of the load. Therefore, the capacitance of C_E can be comparably large, i.e. several nF. In order to develop a high-frequency CM model of the circuit the MOSFETs are replaced by voltage sources $v_{v,i}$, which impress the switched voltage waveforms. Similarly, the diodes D_{F1+}, D_{F2-} and D_{F3-} are replaced by current sources showing the same pulsed current waveform. The impedance of the output capacitor C_o is typically very small at switching frequency (1 MHz) and therefore modeled as a short-circuit. As can be verified easily, the current sources do not contribute to the CM noise (cf., [20]). The mains diodes are permanently on during a half mains period and are, hence

replaced by a short-circuit. The resulting three-phase noise model of the rectifier is depicted in **Fig. 2**. The phase leg with positive input current (i_{N1}) shows a total capacitance of $C_p = 2C_D + C_S$ to the heat sink and the phase-legs with negative input currents (i_{N2} and i_{N3}) show a total capacitance of $C_n = 2C_D$ which is different to C_p. Note that this model is only valid for $i_{N1} > 0$, $i_{N2}, i_{N3} < 0$, i.e. $-30° < \varphi_N < 30°$, and that the capacitance of C_p and C_n is changing if one of the input phase currents changes sign, i.e. every 60°. The capacitor C_g models all capacitances from the output voltage rails and the output voltage midpoint to earth ($C_g = C_E + C_{Bp} + C_{Bn}$). The detailed function of the CM filtering based on a connection of M with N' will be discussed later. Note, that the voltage sources $v_{v,i}$ include DM emissions as well as CM emissions.

Now, different possibilities for defining the heat sink potential exist where the most important ones are:

1) heat sink connected to earth
2) heat sink connected to the output voltage midpoint M
3) heat sink floating.

All three connection types result in different CM behavior. If the heat sink is connected to M the noise currents through C_p and C_n are directly guided back to the noise source and thus no additional external CM noise occurs. The largest impact on the noise emissions can be observed if the heat sink is connected to earth and this scenario will be further discussed.

Unfortunately, the three phases show different capacitive couplings to earth and therefore a separation into CM and DM emissions is not directly possible. Due to asymmetry the currents of the phases towards earth also generate DM noise. This noise can be called "mixed-mode noise" (cf., [4]-[7]). For the sake of brevity this is not further analyzed in this work but is subject for further research on this topic.

III. FILTER DESIGN

In order to design a proper EMI filter the CM and DM noise levels of the three-phase rectifier system are required. A computer simulation is used to determine the DM and CM voltage waveforms (cf., **Fig. 3(a)** and **Fig. 3(b)**) generated by the rectifier system. The assumed system specifications are listed in TABLE I. Note, that different modulation strategies result in different CM and DM voltage waveforms which would finally lead to different EMI filter requirements. However, for the sake of brevity this is not further discussed in this paper. In three-phase PWM rectifier systems, a third harmonic signal is added to the sinusoidal phase voltage reference values in order to increase the input voltage modulation range for a given output DC voltage. A triangular shaped signal

$$v_{h3} = \frac{\hat{V}_N}{6}\text{tri}(3\varphi) \tag{1}$$

is employed in the implemented modulator which results in a triangular low-frequency component ($v_{CM,avg}$) of the CM voltage (cf., **Fig. 3(b)**).

According to the calculation scheme given in [8] the quasi-peak (QP) or peak (PK) weighted DM and CM spectrum can be calculated. The calculated spectra (using peak-detection) of the voltages shown in **Fig. 3(a)-(b)** are depicted in **Fig. 3(c)-(d)**, together with the limit defined by CISPR11 class A [14]. Note, that the result of this calculation are only the spectra of the simulated voltage waveforms and that, contrary to [8], the influence of the LISN is not considered. The EMI filter has now to be designed such that the generated emissions do not exceed the

TABLE I: Specifications of the analyzed PWM rectifier system.

V_{Ni}	230 V
f_N	50/60 Hz
f_s	1 MHz
V_o	800 V_{DC}
P_o	10 kW

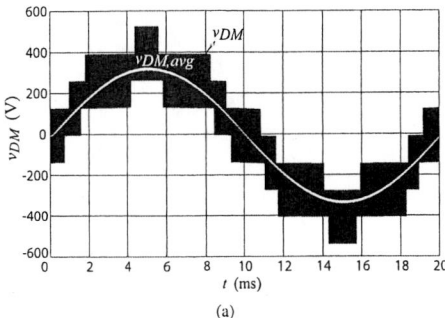

(a)

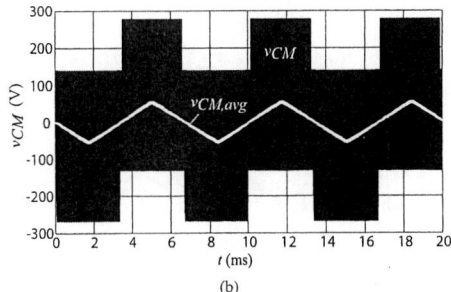

(b)

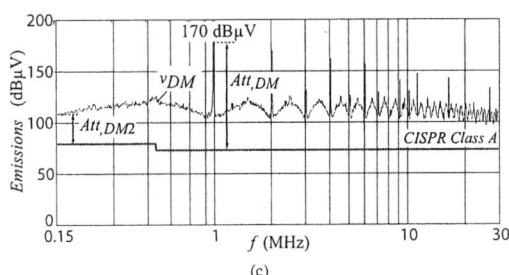

(c)

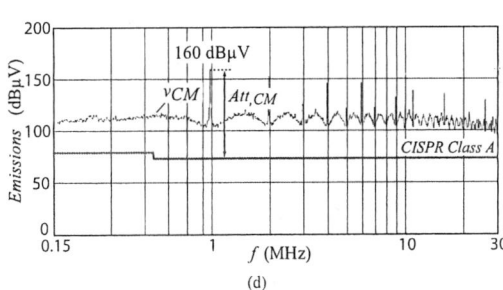

(d)

Fig. 3: Simulated voltage waveforms of the rectifier system operated at an output power of $P_o = 10$ kW and corresponding predicted spectra using peak-detection; a) DM voltage; (b) CM voltage; (c) Predicted DM emission and (d) predicted CM emission.

CISPR11 class A. The influence of the LISN has to be considered later in the EMI filter design process.

According to **Fig. 3**, the main amplitudes of harmonics occur at multiples of the switching frequency. Hence, the required attenuation of the DM and CM filter can be calculated by comparing the simulated emissions with the limit specified in CISPR11. For the DM filter this results in a required attenuation of

$$Att_{DM}[\text{dB}] = v_{DM}(f_s)[\text{dB}\mu\text{V}] - Limit[\text{dB}\mu\text{V}] + margin[\text{dB}] \cong 103\,\text{dB} , \qquad (2)$$

where a margin of 6 dB is included. The required attenuation of the CM filter can be calculated to

$$Att_{CM}[\text{dB}] = v_{CM}(f_s)[\text{dB}\mu\text{V}] - Limit[\text{dB}\mu\text{V}] + margin[dB] \cong 93\,\text{dB} . \qquad (3)$$

The employed process for calculating the noise spectra is a time consuming task. The result of this calculation are noise amplitudes over the whole frequency range but as shown before only the amplitude of the emissions generated at the switching frequency fundamental is used for EMI filter design (in case of $f_s \geq 150$ kHz). In [15] an approximation method for determining the EMI emission has been presented, where only the fundamental DM component at the switching frequency is considered. However, this method can also be applied to determine the CM filter requirements. Therefore, the rms value of the CM voltage $V_{CM,rms}$ which comprises a low-frequency component $V_{CM,h3,rms}$ and a high-frequency component $V_{CM,noise,rms}$ containing all switching frequency harmonics has to be calculated. This could be done purely analytically but for sake of simplicity the rms value of the total CM voltage is calculated using the simulated CM voltage shown in **Fig. 3(b)**. The rms value of the high-frequency CM noise $V_{CM,noise,rms}$ is therefore given by

$$V_{CM,noise,rms}^2 = V_{CM,rms}^2 - V_{CM,h3,rms}^2 . \qquad (4)$$

Substituting (1) in (4) yields

$$V_{CM,noise,rms}^2 = V_{CM,rms}^2 - \frac{\left(\frac{\hat{V}_N}{6}\right)^2}{3} \qquad (5)$$

which results in an estimated noise level of

$$V_{CM,noise} = 164\,\text{dB}\mu\text{V} . \qquad (6)$$

The difference to the result given in **Fig. 3(d)** is only 4 dBμV and therefore, the proposed procedure is a very simple method to estimate the EMI filter requirements.

According to **Fig. 3(c)-(d)** a relatively large noise floor of $\approx$ 110 dB is generated by the rectifier system. The reason for this can be found in the time behavior of the DM and CM voltages. As reported in [16] and [17] carrier sideband harmonics are present in the spectrum of a PWM signal with low-frequency local average. This leads to an increased noise floor which has to be considered in the DM filter design. Therefore, the DM filter has to realize an attenuation of at least

$$Att_{DM2}[\text{dB}] = v_{DM}(150\,\text{kHz})[\text{dB}\mu\text{V}] - Limit[\text{dB}\mu\text{V}] + margin[\text{dB}] \cong 37\,\text{dB} \qquad (7)$$

at the lower frequency limit for conducted emissions measurements, i.e. at 150 kHz. Accordingly, at least one filter stage has to be designed such that the required attenuation at 150 kHz is realized, that means the volume of this filter stage can not be reduced by a high switching frequency. Among other limitations like the lack of available high frequency magnetic materials this is a main limitation of EMI filter volume reduction by increasing the switching frequency.

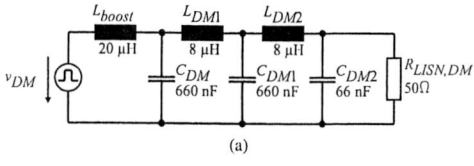

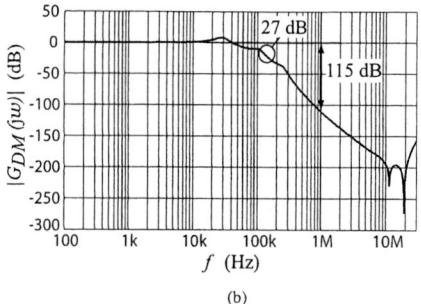

(b)

Fig. 4: (a) Equivalent single-phase DM model and (b) calculated transfer function $G_{DM}(j\omega)$ of the designed DM filter.

A. DM Filter Design

For DM filter design, in addition to the required filter attenuation also the phase displacement of the mains currents resulting from the currents drawn by the filter capacitors has to be considered. If a maximum phase displacement should be limited to $10°$ at an output power of $0.1 P_o$, the DM filter capacitors are limited to a total capacitance of $C_{DM} = 3.5\,\mu F$ per phase. According to the specifications given in (2) and (7) an LC-filter with three stages is used for realization of the DM filter. The single-phase equivalent circuit of the DM filter is shown in **Fig. 4(a)**. The boost inductor of the rectifier can be used to realize the first filter stage. The inductance value of the boost inductor is defined by the maximum allowed current ripple. For the realization at hand a maximum current ripple of $\Delta i_{L,pp} = 0.2 \hat{I}_{N,i}$ shall be allowed which results in an inductance of $L_i = 20\,\mu H$. In [18] it was shown that a maximum attenuation for a multi-stage LC-filter can be achieved if all inductance values and all capacitance values are equal, which also implies that the cut-off frequencies of all filter stages are identical. Unfortunately, (2) as well as (7) have to be satisfied which is not possible with regard to a minimum filter volume by application of this criteria. Additionally, this concept shows the problem of multiple equal filter resonance frequencies. Hence, the cut-off frequencies of the filter stages are selected in a distributed manner and the resulting filter components are chosen also considering aspects of practical realization, which will be further discussed in section IV. The calculated transfer function $G_{DM}(j\omega)$ of the designed filter is depicted in **Fig. 4(b)**.

B. CM Filter Design

For a first step design of the CM filter the influence of the parasitic capacitors C_p and C_n (cf., **Fig. 2**) is neglected. For selecting the CM capacitors safety regulations have to be considered where the leakage earth current is limited to several mA. This results in a limited total capacitance connected to earth which fundamentally influences CM filter design. If a conventional multistage LC low-pass filter would be realized at least three stages would be necessary. Furthermore, the large CM component of V_o would still be present. However, by connecting the output-voltage midpoint M to an artificial star-point N' as shown in **Fig. 1**, the CM component of the output voltage can reduced significantly without violating the safety regulations. For the

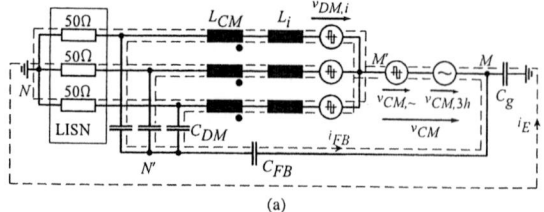

(a)

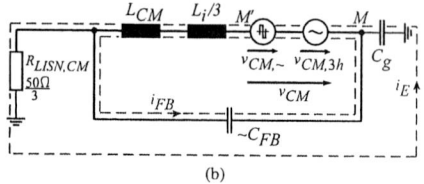

(b)

Fig. 5: (a) Conducted noise equivalent circuit of the three-phase/level PWM rectifier system for the proposed CM filter concept if C_p and C_n are neglected; (b) simplified CM equivalent circuit.

formation of N' the capacitors C_{DM} of the first DM filter stage can be used advantageously. An equivalent circuit of the proposed CM filter concept is shown in **Fig. 5(a)**. There, the converter voltage $v_{v,i}$ is split into a DM voltage $v_{DM,i}$ and a CM voltage v_{CM}

$$v_{v,i} = v_{DM,i} + v_{CM} \qquad (8)$$

where the CM voltage

$$v_{CM} = v_{CM,\sim} + v_{CM,3h} \qquad (9)$$

comprises a high-frequency component $v_{CM,\sim}$ and a low-frequency component $v_{CM,3h}$ which represents the mentioned third harmonic injection. The CM filter path is realized with a capacitor C_{FB} in series to the DM capacitors forming N'. Additionally, the capacitance C_g is shown which represents the lumped capacitance between M and earth which is significantly influenced by the load. If this parasitic capacitance is neglected for a first analysis, the generated CM voltage appears across circuit L_i, L_{CM} and the series connection of the capacitors C_{FB} and C_{DM}. The capacitors C_{FB} and C_{DM} are not connected to earth and are therefore not limited in capacitance by equipment safety regulations. The realization of a CM inductor which is able to handle the third harmonic voltage component $v_{CM,3h}$ without saturation is possible but would result in a very large and bulky element. If the feedback capacitor C_{FB} is dimensioned such that it represents a short-circuit for the high-frequency CM signals ($v_{CM,\sim}$) but a high impedance element for the third harmonic component ($v_{CM,3h}$), the low-frequency component will drop across the feedback capacitor and only $v_{CM,\sim}$ has to be handled by the inductors. Unfortunately, three-phase CM inductors typically allow only a very small zero-sequence current without saturation because of their very high permeability. Because of

$$i_{FB,3h} \approx C \frac{dv_{CM,3h}}{dt} . \qquad (10)$$

the feedback capacitor has to be as small as possible to prevent saturation of the CM inductor. Hence, the relatively large DM capacitors C_{DM} (660 nF per phase) solely can not be used for realization and a low capacitance feedback capacitor C_{FB} is connected in series to the star-point N'.

978-1-4244-4782-4/10 $26.00 © 2010 IEEE

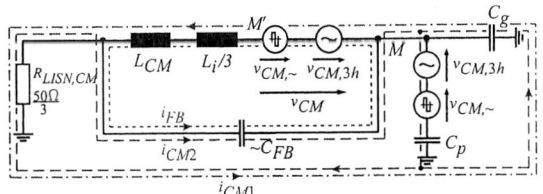

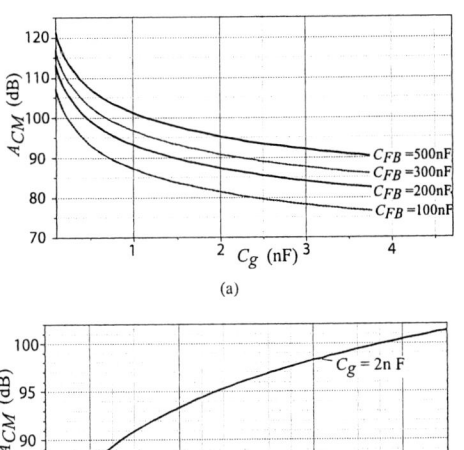

Fig. 7: CM equivalent circuit of the three-phase/level PWM rectifier with heat sink connected to earth if C_p and C_n are assumed to be equal.

value given in (3). According to **Fig. 6(b)**, a higher capacitance of C_{FB} would increase the attenuation but this would also increase the low-frequency current $i_{CM,h3}$ and is therefore no option. For the realization presented in this paper a compromise of $C_{FB} = 200$ nF is used. Hence, an additional CM filter stage is required which has to realize the missing attenuation of ≈ 20 dB.

Up to this point, the capacitors C_p and C_n of the extended noise model given in **Fig. 2** have been neglected for CM filter design. According to (8) the total converter noise $v_{v,i}$ can be divided into DM and CM emissions, if the two lumped parasitic capacitors are assumed to be equal ($C_p = C_n$). Hence, a simplified CM model can be drawn which is shown in **Fig. 7**. It is obvious, that even if the proposed CM concept operates ideally ($i_{CM1} = 0$) a CM current i_{CM2} through the LISN exists (marked as blue dashed line). This current is caused by the parasitic capacitances of the semiconductors to the heat sink (if the heat sink is connected to earth) and can only be reduced by insertion of an additional CM filter stage. However, as already discussed, this additional filter stage is needed anyway in order to realize the required attenuation given in (3). It has to be stated here, that the proposed CM filter concept (connection of M with N') supports the propagation of this type of CM emissions and that a large capacitance C_g would help to reduce the emissions. However, the advantage of an output voltage without high-frequency CM component clearly dominates this drawback.

IV. Filter Realization

After determination of the EMI filter topology and calculation of the filter performance the realization of the passive components is an important design step. Here, a proper magnetic material has to be chosen for realization of the inductors. In **Fig. 8(a)** the complex permeability $\mu = \mu' - j\mu''$ of the widely used ferrite materials N97 and N49 from EPCOS Inc. are plotted. According to **Fig. 8(a)**, the real part of the permeability μ' of material N97 is only constant up to ≈ 1 MHz and drops quickly for higher frequencies. Moreover, also the imaginary part μ'' which is related to core losses rises steeply. Consequently, the magnetic material N97 can not be applied for realizing the boost inductor and is also not a good option for DM filter inductor realization. A realization using material N49 would be possible for $f_s = 1$ MHz but unfortunately, no suitable core size is commercially available. The permeability of the powder core-material -8 from Micrometals Inc. stays constant up to 100 MHz (cf., **Fig. 8(b)**) and shows acceptable losses and is therefore used for implementation of the DM and boost inductors. Single layer winding toroids T90-8 with $N = 16$ turns are used for realizing the DM inductors with very small parasitic capacitance. This results in a high self-resonance frequency of 25.9 MHz. For the realization of the DM capacitors C_{DM1} and C_{DM2} three 220 nF/630V X7R ceramic capacitors in parallel are used. Unfortunately, the capacitance of these ceramic capacitors is strongly dependent on the applied voltage which results in a much smaller effective capacitance. Hence, for the

Fig. 6: (a) Attenuation A_{CM} of the proposed CM filter concept as a function of earth capacitance C_g and (b) achieved attenuation as a function of C_{FB} for $C_g = 2$ nF ($f = 1$ MHz).

The CM inductor on the other hand has to hold the high-frequency CM voltage and core saturation is avoided if

$$B_{sat} > B_{CM,max} = \frac{\int_0^{T_p} v_{CM,\sim} dt}{N\,A_{fe}} = \frac{\frac{V_o}{3} T_s}{N\,A_{fe}} \quad (11)$$

where $T_s = 1/f_s$ and T_p denotes the maximal length of a CM pulse which was set to $T_p = T_s$.

Since the CM inductors are placed in series to the boost inductors L_i the full phase current including the high-frequency ripple flows through their windings. The high-frequency DM current ripple does not cause core losses because of the mutual compensation of the magnetomotive forces but copper losses caused by skin and proximity effect have to be considered.

In **Fig. 5(b)** the CM equivalent circuit is shown. If the lumped capacitance C_g is not neglected, part of the total CM current doesn't return through the feedback path i_{FB}. Depending on the size of C_g this results in a notable earth current through the line impedance stabilization network (LISN). In order to limit this current the feedback capacitor should be large which, however, is in discrepancy with the design criteria given in (10). Depending on the parasitic capacitances C_g and C_{FB} the attenuation A_{CM} can be calculated by

$$
\begin{aligned}
A_{CM} &= 20 \log_{10}\left(\left|\frac{1 + A\,s + \underline{Z}_L\,C_g\,C_{FB}\,R_{LISN,CM}\,s^2}{R_{LISN,CM}\,C_g\,s}\right|\right) \\
A &= C_g\,(\underline{Z}_L + R_{LISN,CM}) + \underline{Z}_L\,C_{FB}
\end{aligned}
\quad (12)
$$

where $\underline{Z}_L$ denotes the total impedance of the CM inductor and $R_{LISN,CM}$ is the equivalent high-frequency CM LISN impedance ($R_{LISN,CM} = 16.7\,\Omega$). In **Fig. 6(a)** the resulting attenuation of the realized filter is plotted as a function of the earth capacitance C_g. It is obvious that the required attenuation of 93 dB can only be realized with the proposed concept if either $C_{FB} > 500$ nF or $C_g < 1$ nF. A higher amount of capacitance to earth would result in an attenuation lower than the required

978-1-4244-4782-4/10 $26.00 © 2010 IEEE

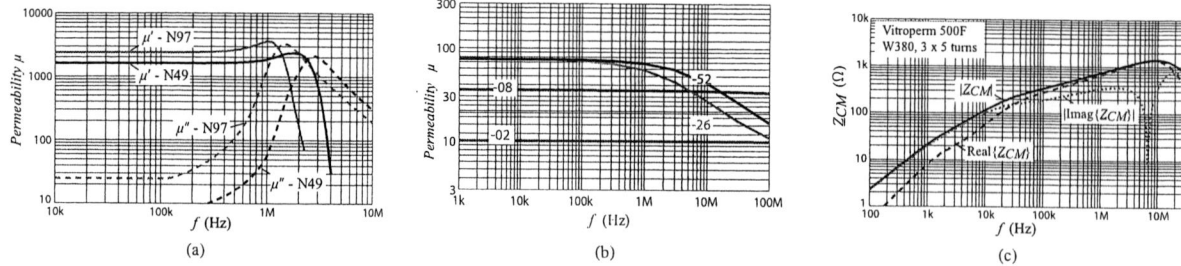

Fig. 8: Real and imaginary part of complex permeability $\mu = \mu' - j\mu''$ for (a) ferrite materials N97 and N49 of EPCOS and (b) initial permeability for powder core materials from Micrometals. (c) Measured impedance of the realized three-phase CM inductor ($N = 5$ turns) employing Vitroperm 500F from VAC.

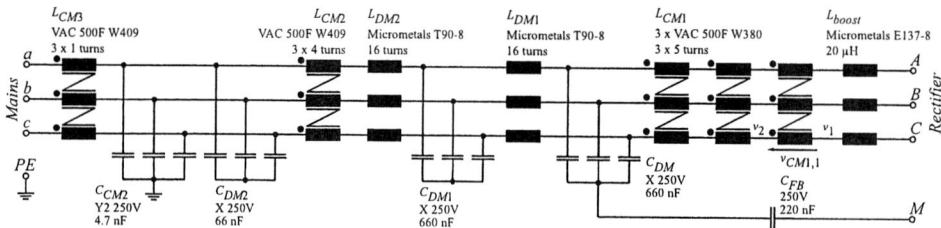

Fig. 9: Complete schematic of the realized EMI filter including information on the used materials.

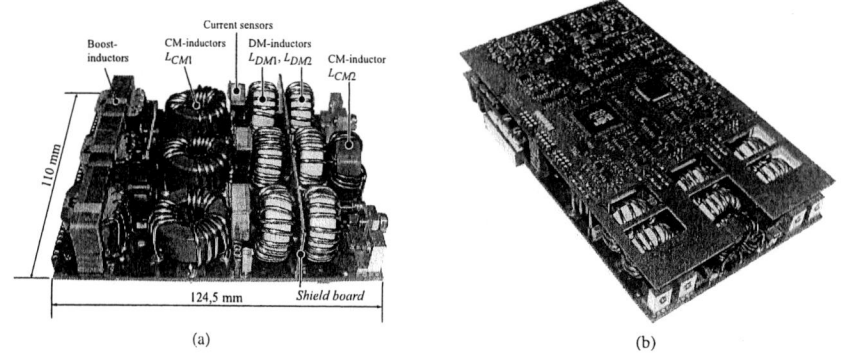

Fig. 10: (a) Realized prototype of EMI filter and (b) 10 kW laboratory prototype of the ultra-compact rectifier system (Dimensions: 195 mm x 110 mm x 33 mm).

realization at hand five capacitors have to be used. This can be avoided by application of foil capacitors resulting however in a larger volume.

For realization of the CM inductors the nanocrystalline material Vitroperm 500F of Vacuumschmelze Inc. is used. The measured complex insertion impedance $\underline{Z}_{CM}$ of the CM inductor is plotted in **Fig. 8(c)**. It can be seen that the inductor exhibits a substantial real part of $\underline{Z}_{CM}$ at $f = 1\,\text{MHz}$ which has to be considered for the design of the CM filter stage. Three inductors are connected in series for realization of L_{CM1} in order to limit the core losses. For the second CM filter stage a core W409 (also utilizing Vitroperm 500F) is used in conjunction with 4.7 nF Y2-rated ceramic capacitors which show a very small volume.

The complete schematic of the realized filter including detailed information on the used materials is given in **Fig. 9**. For arrangement of the different filter stages the impedance mismatch concept described in [21] has to be used. According to this concept the impedance of the last DM filter stage (C_{DM2}) should be much smaller than the impedance of the LISN ($R_{LISN,DM} = 50\,\Omega$) which is given for the arrangement shown in **Fig. 9**. This should also be considered for the last CM filter stage (C_{CM2}). However,

the LISN shows a reduced CM impedance $R_{LISN,CM} = 16.7\,\Omega$ and in addition the impedance reduction of the last CM filter stage is limited by equipment safety regulation (C_{CM2}). Therefore, an additional CM inductor (L_{CM3}) is needed to fulfill the impedance mismatch constraint. This CM inductor is realized with a ferrite core which is clamped on the power cable.

The CM inductors are placed in series to the DM inductors because the stray inductance of the CM inductors then can be used advantageously to increase the DM attenuation. The realized EMI filter prototype is shown in **Fig. 10(a)** together with a picture of the 10 kW laboratory prototype of the rectifier (**Fig. 10(b)**). The overall dimensions of the EMI filter board are 124.5 mm x 110 mm x 33 mm which results in a power density of 22.1 kW/dm^3 for the EMI filter. The overall rectifier system with the dimensions 195 mm x 110 mm x 33 mm shows a power density of 14.1 kW/dm^3.

V. EXPERIMENTAL RESULTS

In **Fig. 11** the results of an EMI measurement according to CISPR 11 (frequency range 150 kHz...30 MHz) are shown together with the limits of CISPR 11 class A. A three-phase DM/CM noise

978-1-4244-4782-4/10 $26.00 © 2010 IEEE

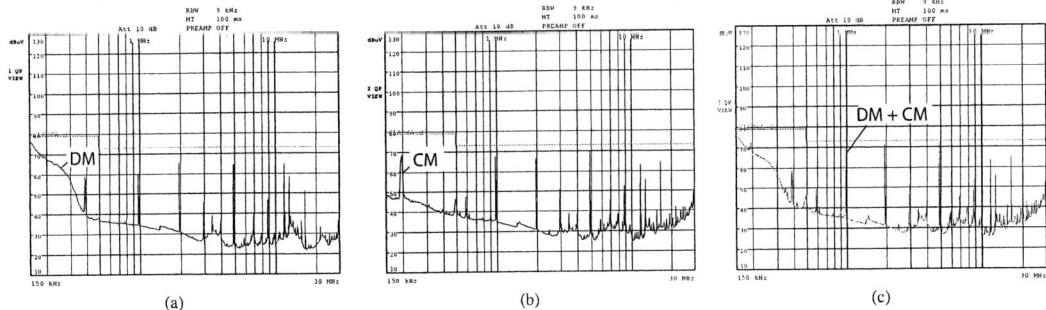

Fig. 11: Final CE measurements of the realized rectifier system; (a) DM emissions, (b) CM emissions and (c) total conducted emissions.

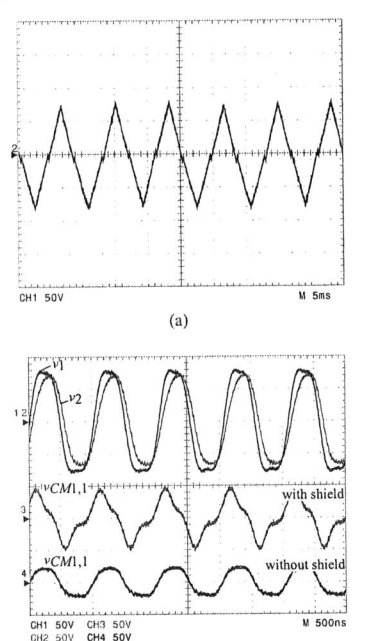

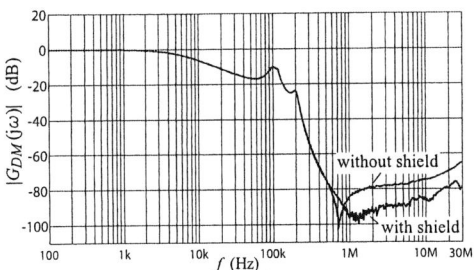

Fig. 13: Measured transfer function $G(j\omega)$ of the DM filter with and without shield.

Fig. 12: (a) Measured CM output voltage component form (from M to earth) employing the proposed CM filter concept and (b) measured voltage of CM inductor $L_{CM1,1}$ with and without PCB shield layer connected to M.

separator was applied to measure the DM and CM noise separately [22]. As expected, the main peaks in the spectra occur at f_s and $2f_s$ and are well below the limit. The peak at 5 MHz results from a shielding layer in the printed circuit board and will be discussed later. The peak in the DM spectrum at 400 kHz has its source in the auxiliary power supply and the peak of the CM emissions at 200 kHz is caused by the auxiliary supplies of the gate drives.

Fig. 12(a) shows the voltage of the output midpoint against earth. No high-frequency CM voltage is present and only the third harmonic triangular signal $v_{CM,h3}$, used for increasing the rectifier modulation range is measured. This verifies the proper operation of the proposed CM filter concept. Deviations from the triangle signal can be observed at the zero-crossings. There, the CM cokes are able to hold the total CM voltage v_{CM} without saturation and the voltage stays zero. After reaching light saturation C_{FB} is charged and the system operates as intended.

As can be seen in **Fig. 10(a)**, the DM inductors C_{DM1} and C_{DM2} are arranged face to face and are therefore magnetically coupled. This parasitic effect reduces the attenuation of the DM

filter stage and hence a magnetic shield (0.1 mm thick Mu-metal foil glued on a PCB with solid copper layer) is inserted between the two stages to reduce the coupling (also shown in **Fig. 10(a)**). In **Fig. 13** the measured transfer function $G_{DM}(j\omega)$ of the filter with and without the shield is plotted. Whereas for frequencies below 800 kHz no difference occurs, an improvement of $\approx 10\,\text{dB}$ can be measured for frequencies above 1 MHz. It has to be stated here that the dynamic range of the used network analyzer Bode100 [23] is 100 dB and that the measurement results are therefore limited to this value.

Next, the influence of the arrangement of the first DM and CM filter stages (impedance mismatch) is examined. For this purpose, the DM filter capacitors C_{DM2} are moved behind the CM choke L_{CM2} (cf., **Fig. 14(a)**). As nothing is changed for the CM path the three Y2-capacitors (4.7 nF) now constitute an additional DM filter stage with the leakage inductance of the CM choke. Unfortunately, the impedance of the Y2-capacitors $|Z_{C,CM2}|_{1\,\text{MHz}} = 1/\omega C_{CM2} = 33.8\,\Omega$ is in the same range as the DM impedance of the LISN ($R_{LISN,DM} = 50\,\Omega$) and therefore doen't fulfill the impedance mismatch criteria. This results in 10 dB higher noise level compared to a realization satisfying the impedance mismatch criteria (cf., **Fig. 14(b)**). Therefore, the arrangement of filter stages has to be handled with care in order to achieve the desired attenuation.

In the following the question if a solid copper layer in the printed circuit board (PCB) covering the whole power and EMI arrangement could act as an advantageous shielding layer is discussed. There, the intention is to connect this copper layer to M in order to catch high-frequency noise currents similar to the proposed CM filter concept. Unfortunately, this shield layer also introduces a (capacitive) coupling path from the interconnections of the three CM chokes forming L_{CM1} to M (cf., **Fig. 9**). Because of these capacitive couplings, a uniform voltage distribution between the three CM chokes is inhibited. According to the measurement given in **Fig. 12(b)** a phase-shift of the voltage

978-1-4244-4782-4/10 $26.00 © 2010 IEEE

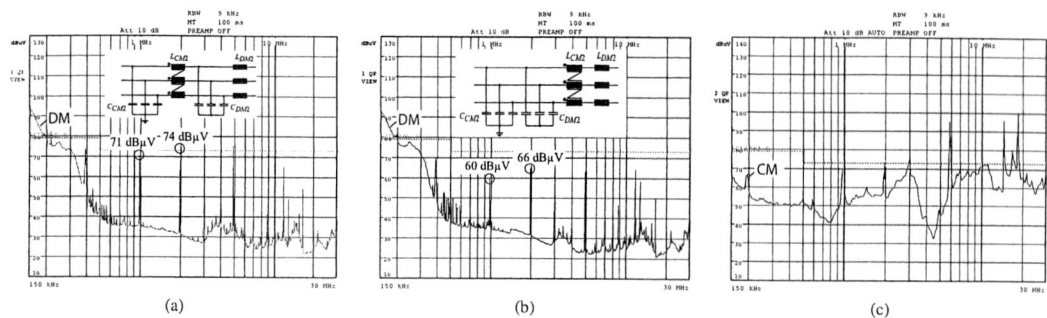

Fig. 14: (a) Measured DM emissions if the CM filter stage arrangement violates the impedance mismatch criteria for the DM filter stage and (b) measured DM emissions if filter satisfies the impedance mismatch criteria. Measured CM emissions with unfavorable copper shielding layer in the printed circuit board. The CM filter is short circuited by the shielding layer for frequencies above 5 MHz.

v_2 (after the first CM inductor $L_{CM1,1}$) drives this inductor into saturation ($v_{CM1,1}$ with shield). However, if the shielding layer is not connected to M a uniform voltage distribution occurs which can be verified by the measured voltage ($v_{CM1,1}$ without shield) which amplitude corresponds to $v_{CM,\sim}/3$.

If the shielding layer is left open, another effect can be observed: The copper layer covers the whole EMI filter and causes a capacitive coupling which forms a low-impedance path short circuiting the EMI filter at higher frequencies. An EMI measurement only considering CM emissions with a copper layer beneath the whole EMI filter is given in **Fig. 14(c)** and verifies increased emissions. For the final realization this copper layer was cut after the CM inductors L_{CM1} but the remaining part still causes a noise peak at 5 MHz in **Fig. 11**. Hence, shielding layers have to be handled with special care in order not to degrade the filter performance.

VI. CONCLUSION

This paper presented the design of an EMI filter for an ultra-compact 10 kW three-phase/level PWM rectifier. Filter requirements have been derived using computer simulations. A specific CM filter strategy has been proposed, where the output of the rectifier shows no high-frequency CM voltage. The performance of the novel filter concept has been analyzed and the CM inductor L_{CM} turned out to be the key element for a successful implementation of the proposed concept. For a better understanding of CM behavior an extended CM model of the rectifier system has been derived which demonstrated also the occurrence of "mixed-mode emissions". A 10 kW laboratory prototype with a power density of 14.1 kW/dm³ and a switching frequency of 1 MHz has been realized. The performance of the designed EMI filter has been verified by measurements taken from this prototype. Modifications on several filter parts have shown that a careful component selection as well as a proper arrangement and layout is essential for achieving a satisfactory EMI filter performance.

REFERENCES

[1] J.A. Rosero, J.A. Ortega, E. Aldabas, and L. Romeral, "Moving towards a more electric aircraft," *IEEE Magazine on Aerospace and Electronic Systems*, Vol.22, No.3, March 2007, pp.3-9.

[2] R.W. Erickson, and D. Maksimovic, "Fundamentals of Power Electronics (2nd ed.)," *Springer Science+Business Media, LLC*, 2001.

[3] A. Nagel, and R.W. De Doncker, "Systematic design of EMI-filters for power converters," *Proc. of the IEEE IAS 2000*, Oct 2000, Vol.4, pp.2523-2525.

[4] S. Qu, and D. Chen, "Mixed-mode EMI noise and its implications to filter design in offline switching power supplies," *IEEE Trans. Power Electron.* Vol.17, No.4, Jul 2002, pp.502-507.

[5] J. Meng, and M. Weiming, "A new technique for modeling and analysis of mixed-mode conducted EMI noise," *IEEE Trans. Power Electron.*, Vol.19, No.6, Nov. 2004, pp. 1679-1687.

[6] W. Shen, F. Wang, and D. Boroyevich, "Conducted EMI characteristic and its implications to filter design in 3-phase diode front-end converters," *Proc. of the IAS 2004*, Vol.3, 3-7 Oct. 2004, pp. 1840-1846.

[7] W. Shen, F. Wang, D. Boroyevich, and Y. Liu, "Definition and acquisition of CM and DM EMI noise for general-purpose adjustable speed motor drives," *Proc. of the PESC'04*, Vol.2, 20-25 June 2004, pp. 1028-1033.

[8] T. Nussbaumer, M.L. Heldwein, and J.W. Kolar, "Differential Mode Input Filter Design for a Three-Phase Buck-Type PWM Rectifier Based on Modeling of the EMC Test Receiver," *IEEE Trans. Ind. Electron.*, Vol.53, No.5, Oct. 2006, pp.1649-1661.

[9] G. Laimer, and J.W. Kolar, "'Zero'-ripple EMI input filter concepts for application in a 1-U 500 kHz Si/SiC three-phase PWM rectifier."*Proc. of the INTELEC'03*, 19-23 Oct. 2003, pp. 750-756.

[10] Yifan Zhao, Yue Li, and T.A. Lipo, "Force commutated three level boost type rectifier," *IEEE Trans. Ind. Appl.*, Vol.31, No.1, Jan/Feb 1995, pp.155-161.

[11] M.L. Heldwein, and J. W. Kolar, "Impact of EMC Filters on the Power Density of Modern Three-Phase PWM Converters," *IEEE Trans. Power Electron.*, Vol.24, No.6, June 2009, pp.1577-1588.

[12] J.W. Kolar, U. Drofenik, J. Minibock, and H. Ertl, "A new concept for minimizing high-frequency common-mode EMI of three-phase PWM rectifier systems keeping high utilization of the output voltage," *Proc. of the APEC 2000*, Vol.1, pp.519-527.

[13] H. Joergensen, S. Guttowski, and K. Heumann, "Comparison of Methods to Reduce the Common Mode Noise Emission of PWM Voltage-Fed Inverters," *Proc. of the 37th Int. Power Conf. Conf.*, Nuremberg, Germany, May 26-28 1998, pp. 273-280.

[14] IEC Internationl Special Committee on Ratio Interference - C.I.S.P.R.(2004), " Specification for Industrial, scientific and medical (ISM) radio-frequency equipment - Electromagnetic disturbance characteristics - Limits and methods of measurement - Publication 11," Geneve, Switzerland.

[15] K. Raggl, T. Nussbaumer, and J.W. Kolar, "Guideline for a Simplified Differential Mode EMI Filter Design," Accepted for future publication in the *IEEE Trans. Ind. Electron.* (2009).

[16] D.G. Holmes, "A general analytical method for determining the theoretical harmonic components of carrier based PWM strategies," *Proc. of the 33rd IAS Industry Appl. Conf.*, vol.2, 12-15 Oct 1998, pp.1207-1214.

[17] H. Deng,L. Helle, Yin Bo, and K.B. Larsen, "A General Solution for Theoretical Harmonic Components of Carrier Based PWM Schemes," *Proc. of the APEC 2009*, 15-19 Feb. 2009, pp.1698-1703.

[18] M.L. Heldwein, and J.W. Kolar, "Design of Minimum Volume Input Filters for an Ultra Compact Three-Phase PWM Rectifier," *Proc. of the 9th Brazilian Power Electron. Conf. (COBEP'07)*, 2007, Vol.1, pp.454-461.

[19] H. Ye et al., "Common mode noise modeling and analysis of dual boost PFC circuit," *Proc. of the INTELEC 2004*, 19-23 Sept. 2004, pp. 575-582.

[20] S. Wang, P. Kong, and F.C. Lee, "Common Mode Noise Reduction for Boost Converters Using General Balance Technique." *IEEE Trans. Power Electron.*, Vol.22, No.4, July 2007

[21] S. Shuo Wang; F.C. Lee, W.G. Odendaal, "Improving the performance of boost PFC EMI filters," *Proc of the APEC '03*, 9-13 Feb. 2003, Vol.1, pp. 368-374.

[22] M.L. Heldwein, T. Nussbaumer, F. Beck, and J.W. Kolar, "Novel three-phase CM/DM conducted emissions separator," *Proc. of the APEC '05*, Vol.2, 6-10 March 2005, pp.797-802.

[23] Omicron electronics. (2008). Handbook of the network vector analyzer: Bode100 [Online]. Available: www.omicron-labs.com

[24] IEC International Special Committee on Ratio Interference - C.I.S.P.R.(1977), "C.I.S.P.R. Specification for Radio Interference Measuring Apparatur and Measurement Methods - Publication 16," Geneve, Switzerland.

Self-Driven AC-DC Synchronous Rectifier for Power Applications

- A Direct Energy-Efficient Replacement for Traditional Diode Rectifier

W.X. Zhong[1], W.C. Ho[2], *Member IEEE*, X. Liu[2] and S.Y.R. Hui[1], *Fellow IEEE*

[1]Center for Power Electronics
City University of Hong Kong
Hong Kong, China

[2]ConvenientPower HK Ltd.
Hong Kong, China

Abstract—**Synchronous rectification has previously been adopted in switched mode circuits for reducing the conduction losses particularly in low-voltage and high-current applications. This paper presents a generalized "self-driven" AC-DC synchronous rectification technique that can be used to develop AC-DC synchronous rectifier that behaves like a diode bridge, but with much reduced conduction losses and without control integrated circuits. This generalized concept can be extended from single-phase to multi-phase systems. Experiments based on 1kW and 2kW single-phase systems have been successfully conducted for capacitive, inductive and resistive loads. Very significant power loss reduction (over 50%) has been achieved in the rectification stage at both 110V and 220V AC mains operations. This patent-pending circuit can be regarded as a direct replacement of general-purpose diode rectifier. Due to the reduction of power loss, further reduction in the size and cost of the heatsink or thermal management for the power circuit becomes possible.**

I. INTRODUCTION

As early as 1990 [1], synchronous rectifiers based on the use of power MOSFETs to replace diodes for reducing the conduction losses has been widely used in low-voltage high-current applications [1]-[16]. Synchronous rectifier techniques are primarily applied to various versions of DC-DC converters such as buck converters [2], [3], flyback converter and boost-buck converters [4], [5], half-bridge converters [6], [7] and LCC resonant converters [8], [9]. To reduce the cost of the gate drive circuits, self-driven techniques have been an active research topic in synchronous rectifiers [2], [7], [9], [10], although gate control integrated circuit for driving synchronous rectifiers is also commercially available [11]. Other research aspects include the use of soft-switching technique [6], [9], [12]. Besides DC-DC converters, synchronous rectification techniques have been applied to 3-phase full-bridge AC-DC converter based on a 3-phase fully-

controlled bridge [13] and even to 5-level converter [14]. While the self-driven technique uses the changing voltage polarity of the coupled windings to control the switching of the power MOSFETs, other techniques tend to use control integrated circuits to provide the gating signals. The first attempt to replace a general-purpose diode bridge with synchronous rectifier for low power and low voltage (3V-5V) applications appears in [15] in which the synchronous rectification technique is applied to a centre-tap rectifier topology. Customized charge pump circuit is however needed in the proposal of [15] in order to provide a suitable dc power supply for the gate drive. As such proposal aims at low-voltage applications, it is not suitable for mains voltage operations.

In this paper, a new AC-DC synchronous rectifier circuit including self-driven control circuit for high-voltage power applications is presented. This new circuit is designed to behave like a traditional diode bridge, except that its conduction loss is much smaller than that of a diode bridge. This principle has been successfully demonstrated in two single-phase synchronous rectifiers of power up to 2kW for capacitive, inductive and resistive loads with significant loss reduction exceeding 50% when compared with a diode bridge at both 110V and 220V mains. Consequently, the thermal management and heatsink requirements can be reduced and a more compact rectifier can be achieved for general-purpose AC-DC applications.

II. SINGLE-PHASE SYNCHRONOUS RECTIFIER FOR MAINS VOLTAGE OPERATIONS

A. Self-driven Synchronous Rectifier Circuit

Fig.1 shows a traditional single-phase diode bridge with the two conducting paths highlighted. Since a diode will only turn off after the current reverse-recovery process, there should be at least one current-controlled MOSFET in each current path of the equivalent diode-bridge circuit. This diode circuit in Fig.1 is to be replaced by the proposed circuit concept in Fig.2. This self-driven AC-DC synchronous rectifier is expected to be a single circuit that can directly replace a traditional power diode bridge in 110V, 60Hz and 220V, 50Hz mains. In Fig.2, M1 and M2 form a pair of switches of one conducting path while M3 and M4 for the other pair in this single-phase system. To simulate the turn-on and turn-off conditions of a diode, the switches should be turned on when the voltage across them is forward-biased and turned off when the switch current is reversed (i.e. like diode current reverse recovery). Therefore, there should be at least one current-controlled MOSFET in each current path of the equivalent diode-bridge circuit.

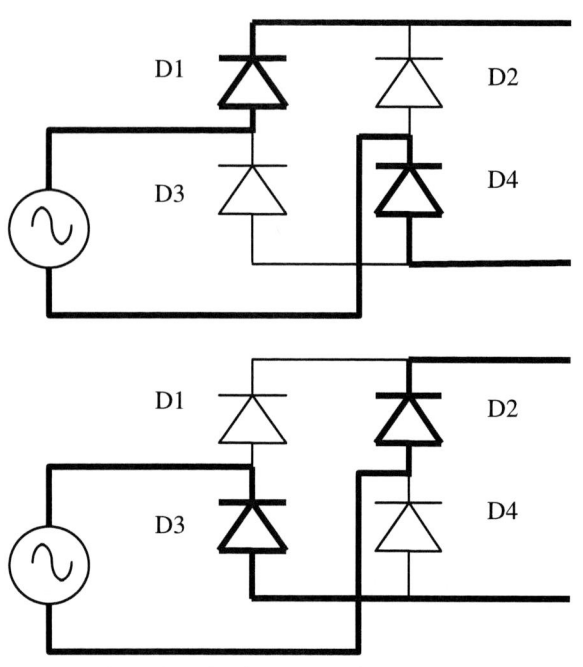

Fig.1. Conducting paths of a diode bridge

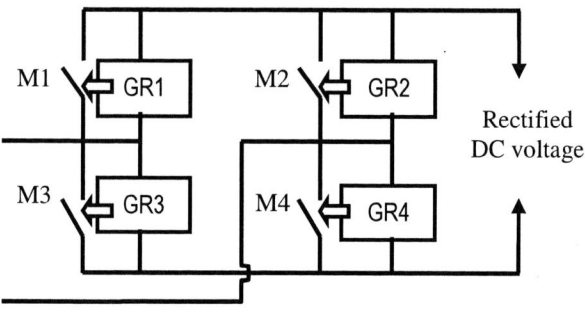

Fig.2. Concept of the proposed AC-DC synchronous rectifier

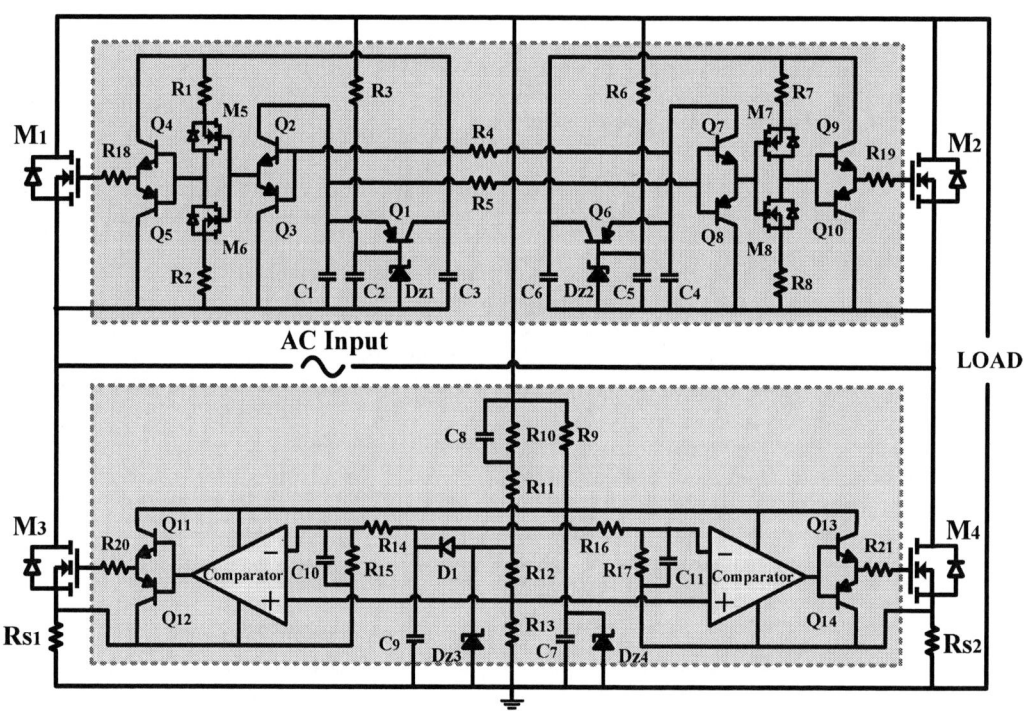

Fig.3. Proposed Self-Driven Synchronous Rectifier

978-1-4244-4782-4/10 $26.00 © 2010 IEEE

The actual circuit implementation of the concept in Fig.2 is shown in Fig.3. It should be noted that only the four power MOSFETs M1 to M4 are power devices. The self-driven gate-drive circuits for the four MOSFETs, highlighted in the shaded boxes, are of very low power, and can in principle be integrated. The whole circuit can be divided into two parts: high-side and low-side circuits. Both high-side and low-side parts are symmetrical. It is important to note that the four body diodes of power MOSFETs M1 to M4 form a standard diode bridge. This means that even if the gate-drive circuits are not ready for operation immediately at the start-up of the circuit, standard diode bridge is inherent in the proposed circuit in Fig.3.

B. Operating Principle - Initial Gate-drive Start-up

In this circuit, the power diodes D_1 and D_2 of Fig.1 are replaced by n-type enhancement power MOSFETs (Fig.3) which are controlled by sensing the voltage across the power MOSFETs. They are known as "voltage-controlled self-driven" (VCSD) switches and will be turned on when the switches are forward-biased. The diodes D_3 and D_4 of Fig.1 are replaced by power MOSFETs which are controlled by sensing the current through the power MOSFETs. These are called current-controlled self-driven (CCSD) switches, because they are turned off whenever the current is reversed. In general, each current path should contain at least one CCSD switch to ensure automatic turn-off (as a diode turns off after current reverse recovery). In this study, it is assumed that the two lower MOSFET switches M_3 and M_4 have a current-controlled gate drive circuit (i.e. the turn-off state of which is determined by sensing the current in the switch). Self-driven gate drive circuits for M_3 and M_4 are grounded with the power circuits and their power supplies can thus be derived from the DC voltage of the synchronous rectifier.

As shown in Fig. 3, there are 3 capacitors in each of the two high-side driving circuits (C_1, C_2 and C_3 for M_1; C_4, C_5 and C_6 for M_2). Each upper gate drive has three driving stages. Taking M_1 as an example, Q_2 & Q_3; M_5 & M_6; Q_4 & Q_5 form the three driving stages. Q_2 and Q_3 are for signal amplifying and providing a charging path for the power supplies of the driving circuit of M_2; M_5 and M_6 form an inverter; and Q_4 and Q_5 are for fast driving of the power MOSFET M1. C_1 and C_4 will be charged up as the power supplies of the first-stage driving pair. Before C_1 and C_4 have been charged up to a certain threshold voltage, say 10V, the driving logic in the circuit will not be ready. C_3 and C_6 are to be charged as the power supplies for driving the M_1 and M_2 respectively. In start-up stage, C_2 and C_5 are designed to be charged up faster than C_1 and C_4, until they reach a certain voltage which is decided by the zener diodes D_{Z1} and D_{Z2}. Bipolar transistors Q_1 and Q_6 are used to ensure that C_3 and C_6 will not be charged before C_1 and C_4 have been charged up to a voltage higher than the voltage of C_2 and C_5. Therefore the MOSFETs M1 and M2 will not switch before the driving logic has been set up.

C. Operation Principle of the High-Side Gate Drives

C_1 and C_4 will be charged and discharged at line frequency. Assume that C_1 and C_4 will be charged up to V_1 in every charging period and will be discharged to V_2 which is a little lower that V_1 after every discharging period. Here we use v_{BC3} and v_{BC8} to represent the base-collector voltages of BJT Q_3 and Q_8, respectively. With the aid of the timing diagram in Fig.4, the operation of the high-side gate drives can be explained as follows:

a) Before t_1 : v_{in} is higher than V_1. Both the current directions in R_4 and R_5 are from left to right as illustrated in Fig. 4(a). v_{BC3} is clamped to zero by the p-n junction between the collector and base of Q_3. So the p-channel MOSFET M_5 is turned on and the n-channel MOSFET M_6 is turned off, and such condition keeps the gate-source voltage of M_1 high and thus turning M_1 on. Meanwhile, current in R_5 flows through the p-n junction between the base and collector of Q_7, which clamps v_{BC8} to V_1. So M_7 is turned off and M_8 is turned on to keep the gate-source voltage of M_2 low.

b) $t_1 - t_2$: As illustrated in Fig. 4(b), when v_{in} becomes lower than V_1 after t_1, current in R_4 changes its direction. Q_2 is turned on and Q_3 becomes off. v_{BC3} starts to increase from zero. At the end of this interval, v_{BC3} reaches the gate threshold voltage of M_6.

c) $t_2 - t_3$: M_6 begins to conduct at t_2. v_{GS1} starts to decrease from high. v_{BC3} continues increasing. Before it reaches the threshold voltage for M_5 to switch-off at t_3, M_5 and M_6 will conduct simultaneously as illustrated in Fig. 4(c). Within this interval, v_{GS1} will fall down to the gate threshold voltage of M_1, the time of which can be slightly controlled by the ratio of R_1 and R_2. The interval ends when v_{GS1} falls to zero and M_1 is turned off.

d) $t_3 - t_4$: As shown in Fig. 4(d), M_5 is turned off at t_3 and M_6 keeps on to keep M_1 off. Current flows through the source drain diode of M_1.

e) $t_4 - t_0$: As illustrated in Fig. 4(e), when v_{in} becomes lower than $V_1 - V_2$ after t_4, current in R_5 changes its direction. v_{BC8} begins to decrease from V_1, indicating that Q_7 is turned off and Q_8 is turned on. While at t_4, v_{BC3} reaches V_2 and will be clamped to the voltage of C_3 by the p-n junction between the base and collector of Q_2.

f) $t_0 - t_5$: Because $V_1 - V_2$ is a positive value, v_{in} commutates at t_0 which is after t_4. At t_0, the source drain diode of M_1 begins off and the source drain diode of M_2 begins to conduct naturally as illustrated in Fig. 4(f). The interval ends when v_{BC8} falls down to the threshold voltage for M_7 to switch on.

g) $t_5 - t_6$: As shown in Fig. 4(g), M_7 begins to conduct at t_5, and v_{BC8} continues decreasing. Before v_{BC8} falls down to the gate threshold voltage of M_8 at t_6, M_7 and M_8 conduct simultaneously. At the end of the interval, v_{GS2} reaches high and M_2 is turned on.

h) $t_6 - t_7$: At t_6, M_8 is turned off and v_{BC8} continues decreasing as illustrated in Fig. 4(h). The interval ends when v_{BC8} reaches zero.

i) After t_7 : v_{in} is lower than $-V_2$. v_{BC8} is clamped to zero by the p-n junction between the collector and base of Q_8 as illustrated in Fig. 4(i).

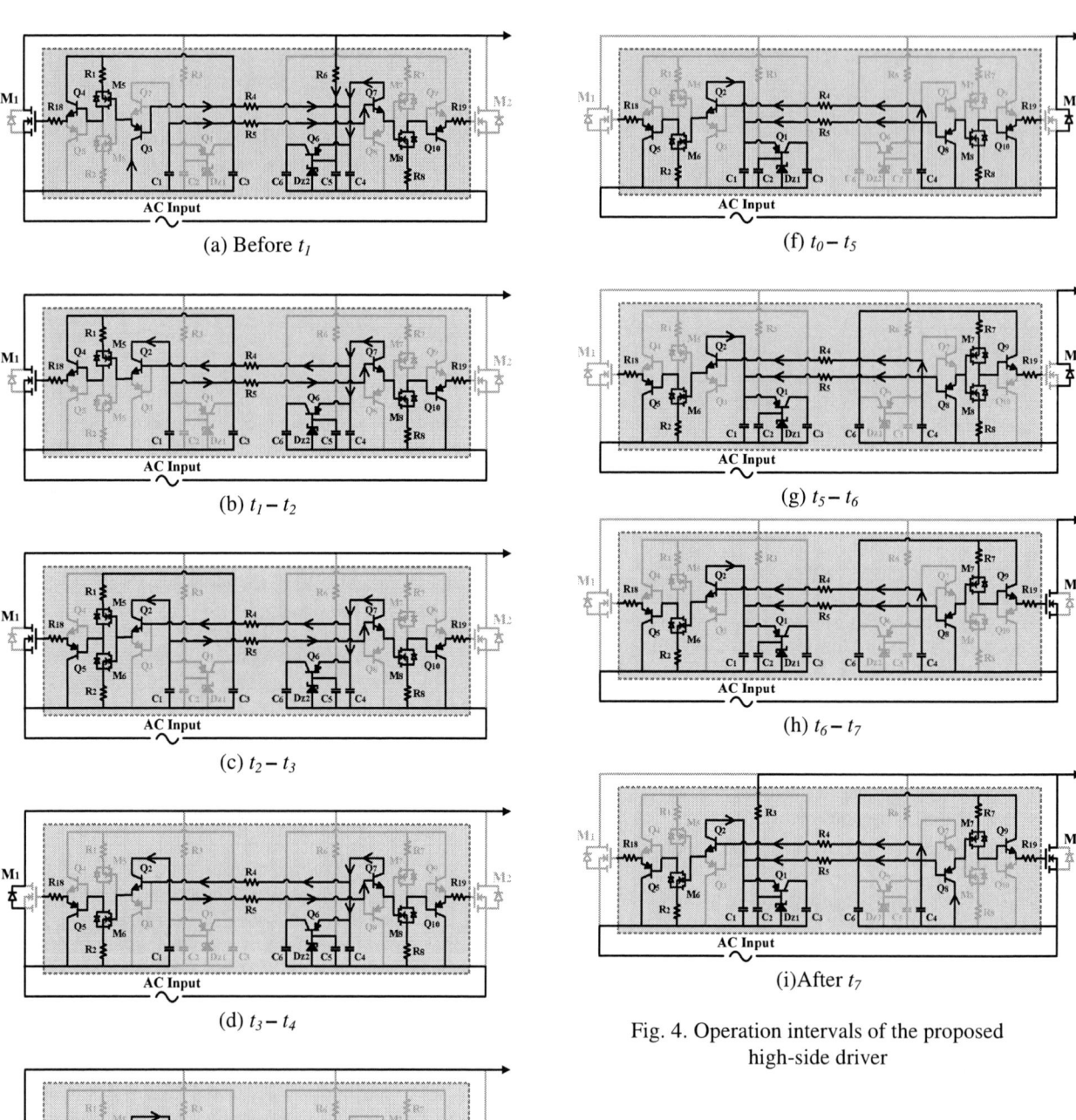

(a) Before t_1

(b) $t_1 - t_2$

(c) $t_2 - t_3$

(d) $t_3 - t_4$

(e) $t_4 - t_0$

(f) $t_0 - t_5$

(g) $t_5 - t_6$

(h) $t_6 - t_7$

(i) After t_7

Fig. 4. Operation intervals of the proposed high-side driver

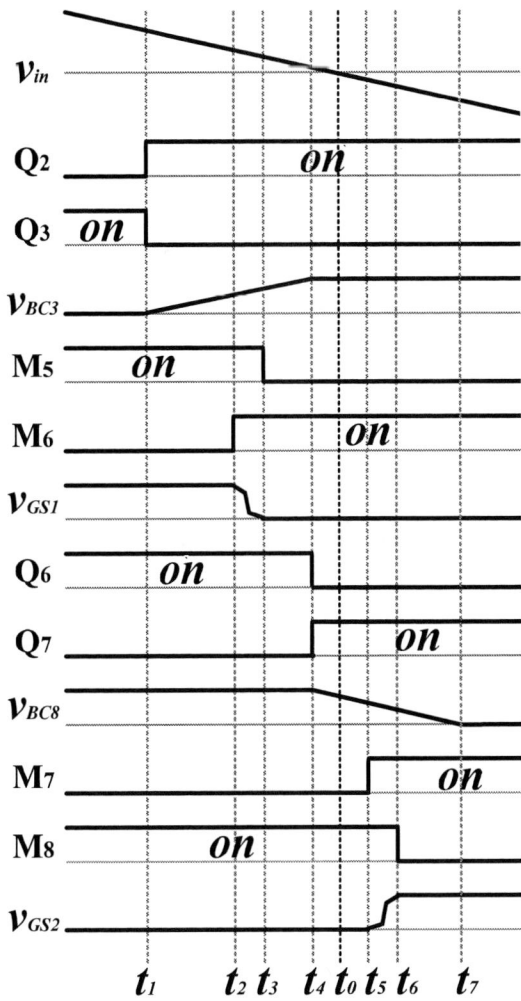

Fig. 5. Timing diagram for the
high-side gate drive circuits

D. Operation Principle of the Low-Side Gate Drives

The low-side MOSFETs are controlled by "current-controlled" gate drive circuits. Current sensing resistors R_{S1} and R_{S2} and comparators are used to detect the MOSFET currents and drive the MOSFETs. In principle, the on-state resistors of the MOSFETs can also be used to replace the sensing resistors if desired. As illustrated in Fig. 6, C_9 is used to give a positive voltage for the inverting inputs of the comparators which can provide a safe margin set by the potential divider comprising resistors R_{14} to R_{17}. The voltage of C_7, stabilized by the zener diode D_{Z4}, is the power supply for the comparators. C_9 will be charged up to the designated voltage before the voltage of C_7 reaches a voltage high enough for the comparators to work. This arrangement ensures that the low-side MOSFETs M_3 and M_4 to be switched only when the proper logic control is ready.

E. Additional Circuit to ensure turning off at zero-crossing points

The self-driven synchronous rectifier is designed to cope with resistive, capacitive, and inductive loads. Typical driving waveforms are illustrated in Fig.7. When the self-driven synchronous rectifier is connected to inductive load, input current will commutate very quickly after every half cycle. The comparators may not respond quickly enough to turn off the MOSFETs M_3 and M_4 under such a fast current change, which may cause fatal short circuit situation. Therefore, an extra circuit (Fig. 8) is added to provide a small positive signal to the non-inverting inputs of the two comparators. The signal is generated from the rectified voltage as illustrated in Fig. 1, and it can guarantee the comparators to turn off the MOSFETs in time. In Fig. 8, D_{z3} is a zener diode of about $10V$. When the rectified voltage begins to rise form zero, the voltage of Dz3 will rise until it reaches its rating voltage. And then it will keep at this voltage with very small fluctuation. And R_{12} and R_{13} here form a voltage divider for scaled-down voltage signal with reduced fluctuation. The rising period and falling period of the signal can be slightly shaped by the value of C_8 and the ratio of R_{10} and R_{11}.

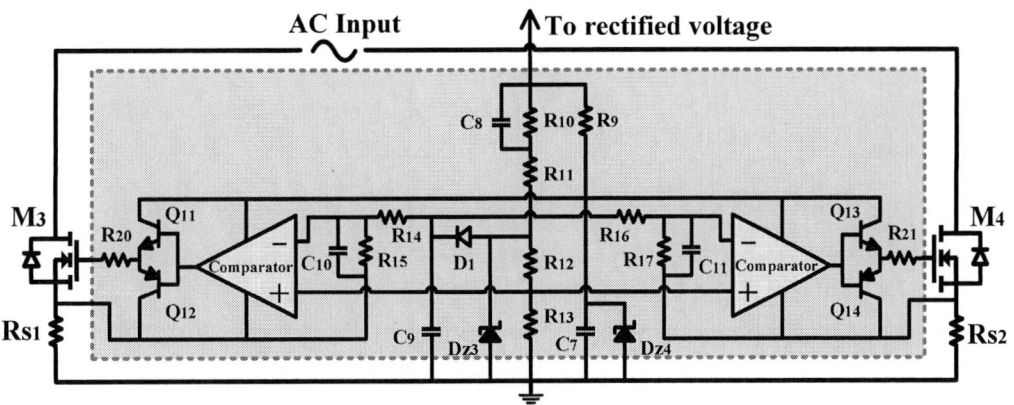

Fig. 6. Low-side circuit of the proposed SR

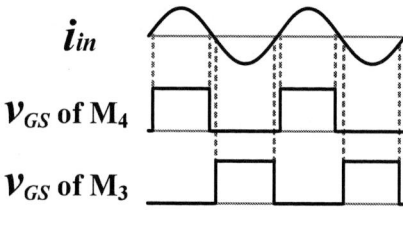

(a) Resistive load

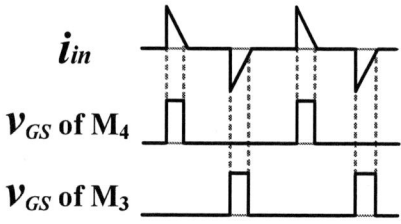

(b) Capacitive load

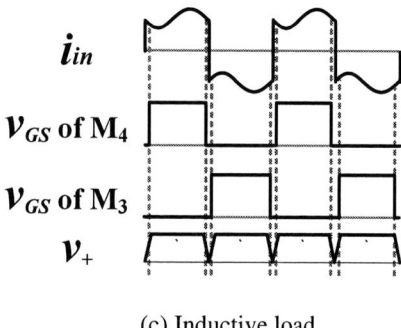

(c) Inductive load

Fig. 7. Driving waveforms of the low-side circuit

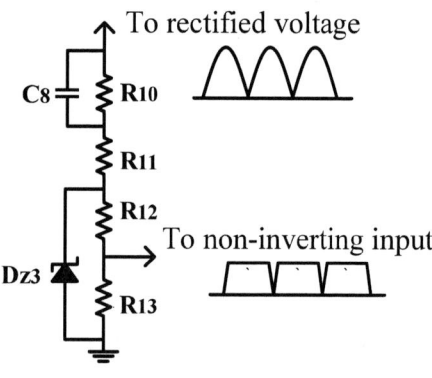

Fig. 8. Additional circuit to guarantee the comparators to turn off

III. EXPERIMENTAL VERIFICATION BASED ON A SINGLE-PHASE SYNCHRONOUS RECTIFIER

A. *Resistive Load*

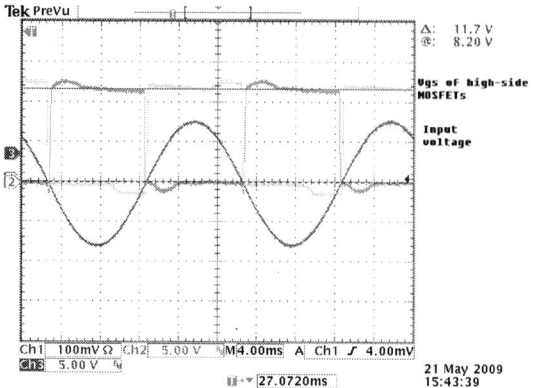

Fig. 9. Gate-source voltages of the high-side (VCSD) MOSFETS & Input voltage

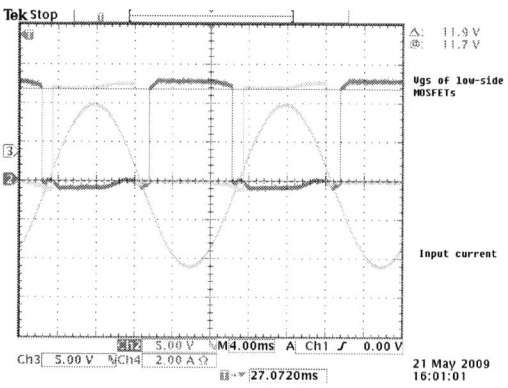

Fig. 10. Gate-source voltages of the low-side (CCSD) MOSFETs & Input Current

B. *Capacitive-Resistive Load*

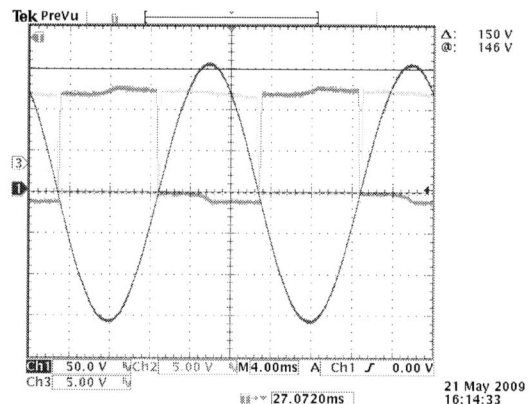

Fig. 11 Gate-source voltages of the high-side (VCSD) MOSFETs & Input voltage

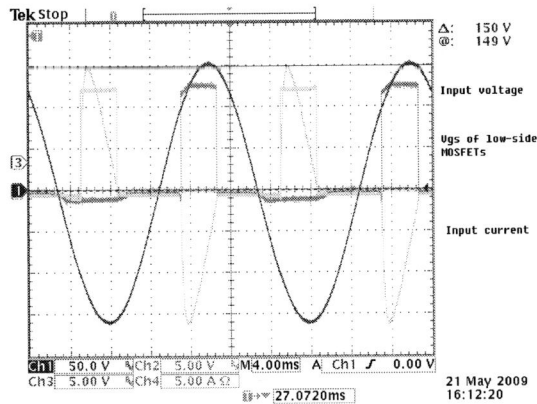

Fig. 12 Gate-source voltages of the low-side (CCSD) MOSFETs & Input voltage & Input Current

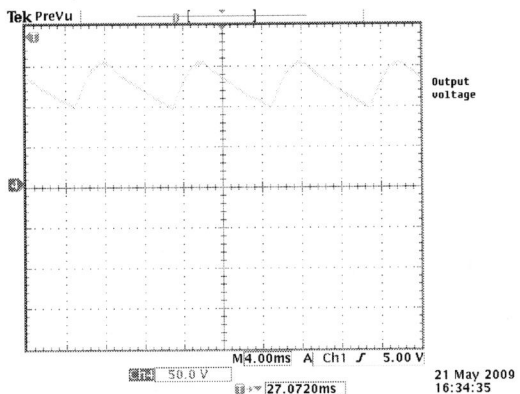

Fig. 13 Output capacitor voltage

C. Inductive-Resistive Load

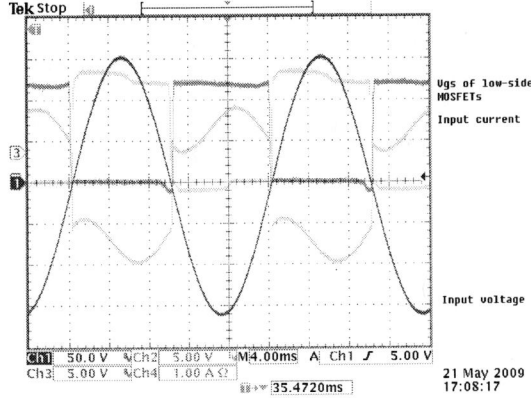

Fig. 14 Gate voltage of VCSD MOSFETs (light blue & pink) & Input voltage (dark blue) & Input current (green)

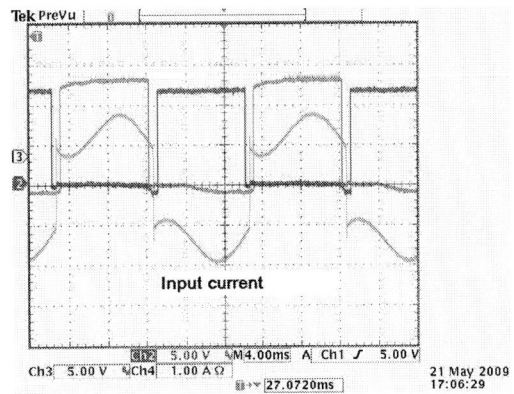

Fig. 15 Gate voltage of CCSD MOSFETS (pink and dark green) & Input current (green)

Fig. 9 shows the gate voltage signals of the top (VCSD) MOSFETs (pink and green color) and the input voltage (blue color) for a resistive load. Fig. 10 shows the corresponding gate signals for the bottom MOSFETs and the input current (green). For the capacitive-resistive load, the results are included in Fig. 11-Fig. 13. It is noted that the input current (green waveform in Fig. 13) looks exactly like a typical current for a diode bridge with an output capacitor. The capacitor voltage is shown in Fig. 13. For the inductive-resistive load, the measurements are included in Fig. 14 and 15. Again, they are typical waveforms as obtained from a diode bridge with LR load.

D. Power Loss Comparison of diode bridge and proposed self-driven synchronous rectifier

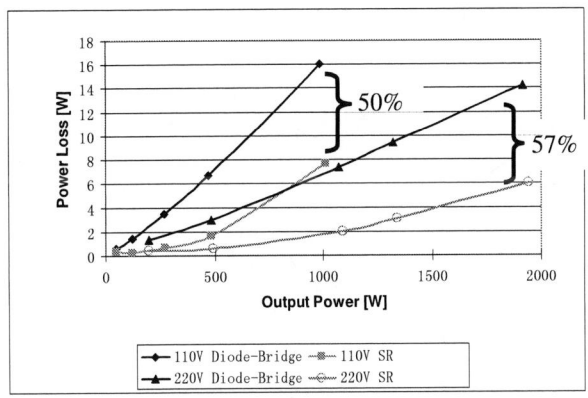

Fig. 16 Power loss comparison of diode rectifier and self-driven rectifier with resistive load

Fig. 16 shows power loss of the synchronous rectifier and the diode rectifier under the mains operation of 110V (up to 1kW output power) and 220V (up to 2kWoutput power). At 1kW and 110V operation, a 50% power loss reduction has been achieved. At 2kW and 220V operation, the power loss reduction is 57%.

IV. CONCLUSIONS

This paper presents a self-driven AC-DC synchronous rectifier that can replace a diode bridge as a general-purpose AC-DC rectifier. The proposal relies on simple circuits and can provide the normal rectification functions for resistive, capacitive and inductive loads as a traditional diode rectifier would do, except that the power loss of the proposed self-driven synchronous rectifier is at least 50% less than that of a diode bridge. This circuit has been practically and successfully demonstrated in a 1kW system under 110V mains operation (with 50% power loss reduction) and a 2kW system under 220 mains operation (with 57% power loss reduction). Costs reduction in the size of the heatsink and the energy-saving can compensate for the minor increase in the component counts of the inexpensive and low-power components. High system compactness due to reduced heat dissipation can also be achieved. This proposal can in principle be extended to multiphase rectification systems [17], [18].

ACKNOWLEDGEMENT

The authors are grateful to the Centre for Power Electronics for its support for this project. The permission to publish this work by Convenientpower HK Ltd. is also acknowledged.

REFERENCES

[1] Tabisz, W.A.; Lee, F.C.; Chen, D.Y.; "A MOSFET resonant synchronous rectifier for high-frequency DC/DC converters", 21st Annual IEEE Power Electronics Specialists Conference, 1990. PESC '90 Record., 11-14 June 1990 Page(s):769 – 779

[2] Fukushima, K.; Hashimoto, T.; Ninomiya, T.; Segawa, T.; "Analysis of Abnormal Phenomenon in Common-Source-type Forward Converter with Self-driven Synchronous Rectifier", CES/IEEE 5th International Power Electronics and Motion Control Conference, 2006. IPEMC 2006. Volume 2, 14-16 Aug. 2006 Page(s):1 – 5

[3] Murata, K.; Harada, K.; Harimoto, T.; "Self turn-on loss of MOSFET as synchronous rectifier in DC/DC buck converter - in case of a low driving impedance" IEEE Power Electronics Specialists Conference, 15-19 June 2008 Page(s):3348 – 3353

[4] Bong-Suck Kim; Ho-Seon Ryu; Man-Su Shin; Joo-Hyun Lee; Ik-Hun Lim; "Design of Flyback Converter with Voltage Driven Synchronous Rectifier", 12th International Power Electronics and Motion Control Conference, 2006. EPE-PEMC 2006. Aug. 2006 Page(s):631 – 635

[5] Jaschke, R.; "Conduction Losses in DC/DC-Converters as buckboost/boostbuck synchronous rectifier types" CPE '07

Compatibility in Power Electronics, 2007. May 29 2007-June 1 2007 Page(s):1 - 10

[6] Garcia, O.; Cobos, J.A.; Uceda, J.; Sebastian, J.; "Zero voltage switching in the PWM half bridge topology with complementary control and synchronous rectification", 26th Annual IEEE Power Electronics Specialists Conference, 1995. Volume 1, 18-22 June 1995 Page(s):286 - 291

[7] Jeong, G.-Y.; "High efficiency asymmetrical half-bridge converter using a self-driven synchronous rectifier" IET Proceedings of Power Electronics, Volume 1, Issue 1, March 2008 Page(s):62 – 71

[8] Yu Ma; Xiaogao Xie; Zhaoming Qian; "Frequency-Controlled LCC Resonant Converter with Synchronous Rectifier", 7th International Conference on Power Electronics and Drive Systems, 2007. PEDS '07. 27-30 Nov. 2007 Page(s):1442 – 1445

[9] Dianbo Fu; Ya Liu; Lee, F.C.; Ming Xu; "A Novel Driving Scheme for Synchronous Rectifiers in LLC Resonant Converters" IEEE Transactions on Power Electronics, Volume 24, Issue 5, May 2009 Page(s):1321 - 1329

[10] Cobos, J.A.; Alou, P.; Garcia, O.; Uceda, J.; Rascon, M.; "New driving scheme for self driven synchronous rectifiers" IEEE APEC '99. Fourteenth Annual Applied Power Electronics Conference and Exposition, 1999. Volume 2, 14-18 March 1999 Page(s):840 - 846

[11] Janssen, Eric; "GreenChip SR: Synchronous Rectifier controller IC", IEEE International Symposium on Industrial Electronics, 2007. ISIE 2007. 4-7 June 2007 Page(s):2319 – 2325

[12] Hong Mao; Abdel Rahman, O.; Batarseh, I.; "Zero-Voltage-Switching DC–DC Converters With Synchronous Rectifiers", IEEE Transactions on Power Electronics, Volume 23, Issue 1, Jan. 2008 Page(s):369 - 378

[13] Bong-Hwan Kwon; Jang-Hyoun Youm; Jee-Woo Lim; "A line-voltage-sensorless synchronous rectifier", IEEE Transactions on Power Electronics, Volume 14, Issue 5, Sept. 1999 Page(s):966 – 972

[14] Portillo, R.; Carrasco, J.M.; Leon, J.I.; Galvan, E.; Prats, M.M.; "Modeling of Five-Level Converter Used in a Synchronous Rectifier Application" IEEE 36th Power Electronics Specialists Conference, 2005. PESC '05. 16-16 June 2005 Page(s):1396 – 1401

[15] Mihaiu, M.I.; "Toward the 'ideal diode' using power MOSFET in full wave synchronous rectifiers for low voltage power supplies" International Symposium on Power Electronics, Electrical Drives, Automation and Motion, 2008. SPEEDAM 2008. 11-13 June 2008 Page(s):1384 – 1387

[16] Po-Yuan Chen; Jinno, M.; Yu-Min Shie; "Research on the Reverse Conduction of Synchronous Rectifiers" IEEE Transactions on Industrial Electronics, Volume 55, Issue 4, April 2008 Page(s):1570 – 1575

[17] W. C. Ho, S. Y. R. Hui and X. Liu, "Single-phase self-driven full-bridge synchronous rectification", US patent application: US12/194,921, Aug. 20, 2008.

[18] W. C. Ho, S. Y. R. Hui, X. Liu and W. P. Choi, "Generalized AC-DC synchronous rectification techniques for single- and multi- phase systems", US patent application: US 12/274,469, Nov. 20, 2008.

A Robust Control Scheme for Grid-Connected Voltage Source Inverters

Shuitao Yang[1,2], Qin Lei[2], Fang Z. Peng[2], and Zhaoming Qian[1]

1. College of Electrical Engineering, Zhejiang University, Hangzhou 31007, China
2. Dept. of Electrical and Computer Engineering, Michigan State University, East Lansing, MI 48824, USA
Email: shuitaoyang@zju.edu.cn

Abstract—This paper analyzes the stability problem of the grid-connected voltage source inverter (VSI) with LC filters, which demonstrates that the possible grid-impedance variations have a significant influence on the system stability when conventional proportional-integrator (PI) controller is used for grid current control. As the grid inductive impedance increasing, the low-frequency gain and bandwidth of the PI controller have to be decreased to keep the system stable, thus degrading the tracking performance and disturbance rejection capability. To deal with this stability problem, an H∞ controller with the explicit robustness in terms of grid-impedance variations is proposed to incorporate the desired tracking performance and the stability margin. By properly selecting the weighting functions, the synthesized H∞ controller exhibits high gains at the vicinity of the line frequency, similar to the traditional proportional-resonant (PR) controller, meanwhile it has enough high frequency attenuation to keep the control loop stable. An inner inverter-output-current loop with high bandwidth is also designed to get better disturbance rejection capability. The selection of weighting functions, inner inverter-output-current loop design, and system disturbance rejection capability are discussed in detail in this paper. Both simulation and experimental results of the proposed H∞ controller as well as the conventional PI controller are given and compared, which validates the performance of the proposed control scheme.

I. INTRODUCTION

As a new means of power generation, distributed generation (DG) is experiencing a rapid development. The DG systems based on renewable energy source or micro-sources such as fuel cells, photovoltaic (PV) cells, wind turbines, and micro-turbines can reduce greenhouse gas emissions, improve power system efficiency and reliability, and relieve today's stress on power transmission and distribution infrastructure [1], [2]. In grid-connected DG systems, single-phase or three-phase pulse-width modulation (PWM) voltage-source inverters (VSIs) are often used for interfacing the renewable energy source to the utility grid, and the current control of the grid-connected inverters plays a predominant role in feeding a grid with high-quality power.

The stationary reference frame proportional-integrator (PI) controller, also called the ramp comparison current controller, is commonly used for current controlled inverters because of its simplicity and easy implementation [3], [4]. Nevertheless, it is regarded as a unsatisfactory solution for ac current regulation because of large steady-state tracking error. Through this tracking error can be reduced by increasing the PI gain and bandwidth, unfortunately, it will also push the systems towards their stabilitiy limits. In contrast, the synchronous frame PI controller can theoretically achieve zero steady-state tracking error by shifting the base-band information back to dc, however, it requires significant computation arising from the need for multiple reference frames [4]-[6], [12]. Also it's difficult to be applied to single-phase inverters. The newly developed proportional-resonant (PR) can achieve virtually the same steady-state and transient performance as a synchronous frame PI controller for both single-phase and three-phase inverters, which is a potential candidate for the grid-connected inverters [7]-[10].

When connecting the inverter to the utility grid, either a pure inductor (L) or a LCL filter can be used as the inverter output stage. The LCL filter instead of L filter is more attractive because it can not only provide higher high-frequency harmonics attenuation with the same inductance value, but also allow the inverter to operate in both standalone and grid-connected modes, which makes it a universal inverter for DG applications [14]. However, the system incorporating LCL filters is of third order, and it has an inherent high-resonant peak at the resonant frequency of the LCL filter, which will make the current control instable if the controller is not suitably designed. To avoid this stability problem, the passive or active damping methods are usually needed [11]-[13].

In [14], [15], an admittance compensator along with a quasi-resonant-proportional controller was proposed. Using the inverter output current instead of the grid current as the feedback signal, the control system can be simplified to a first-order system thus it is possible to keep the control loop stable with high loop gain and bandwidth. However, from the whole system view, the filter capacitor and the grid-side inductor form a parallel resonant circuit, and harmonic current from inverter output in the vicinity of resonant frequency can be amplified excessively and may cause the resonance of the grid current. Reference [16] proposed a new control strategy with feedback of grid current plus part of the capacitor current. In this way, the inverter control system can also be degraded from third-order to first-order due to the counteraction between zeros and poles. The main drawbacks of this method are that the grid current is not directly controlled and will be affected by capacitor current, moreover, the zeros counteract poles only when the values of both inverter-side inductor and grid-side inductor are well known, which is difficult in practice since the grid impedance is different depending on the grid stiffness, and changes with time, furthermore, it is hard to estimate. Alternatively adding a large inductor in grid side will make the variations of grid impedance relatively smaller, but it's not a cost-effective solution since manufactures tend to minimize the number and

978-1-4244-4782-4/10 $26.00 © 2010 IEEE

volume of the magnetic components. In fact, large grid impedance variation will seriously affect the performance of the current control, challenging the control of grid-connected inverter and the grid filter design in terms of stability. A theoretical analysis of the grid stiffness influence on current control is given in [17].

Taking the grid impedance into consideration, in the condition that the grid impedance is big enough when compared to the filter capacitor impedance at the switching frequency, the grid-side filter inductor can be eliminated, in other words, we can use LC filter instead of LCL filter to reduce one filter inductor. By optimally selecting the value of filter capacitor, this condition is easy to be satisfied since the total grid-impedance includes the line impedance and the internal impedance of the grid, in some applications, a transformer is used to couple the inverter to the grid which will equivalently increase the grid impedance.

To address all the aforementioned issues, this paper proposes a robust control scheme for grid-connected voltage source inverters with LC filters. The objective is to achieve small steady-state tracking error and low total harmonic distortion (THD) of the grid current while keep the system stable in the predefined variational ranges of the grid impedance.

II. STABILITY PROBLEM OF THE PI CONTROLLED SYSTEM CAUSED BY GRID-IMPEDANCE VARIATIONS

Fig.1 shows the configuration of the three-phase grid-connected VSI with LC filters, where $Z_g = r_g + j\omega L_g$ is the total grid impedance, including solid state relay (SSR) resistance, line impedance and the grid internal impedance. The equivalent series resistance (ESR) of the filter inductor L_f and capacitor C_f are ignored here. Taking the grid impedance into consideration, in the condition that the grid impedance Z_g is big enough when compared to the capacitor impedance C_f (e. g. C_f and Z_g with the predefined values as given in Table I), the grid-side filter inductor can be eliminated, which means we can use LC filter instead of LCL filter to reduce one filter inductor.

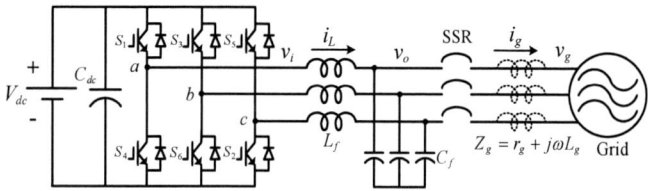

Fig.1. Configuration of three-phase grid-connected VSI with LC-filter.

The influence of the grid-impedance variations on the current control is investigated and disscussed based on the stationary reference frame PI controller along with a capacitor voltage v_o feed-forward compensator, as shown in Fig.2 (a), where K_{PWM} is the inverter gain, which can be regarded as unity by measuring the dc-link voltage, e^{-sT_d} represents the control delay, T_d is equal to one switching period in this paper, and v_m is the modulation signal. Here feed-forward controller is employed to reduce the effect of the grid voltage and to increase dynamic response [4], [18]. Fig.2 (b) shows the simplified control blcok diagram corresponding to Fig.2 (a) without considering the control delay.

TABLE I
SYSTEM PARAMETERS

	Parameter	Value
Grid	Grid phase voltage	$120\ V_{RMS}$
	Frequency	$60\ Hz$
	Grid resistance r_g	$\in [0.1, 0.5]\,\Omega$, with nominal value $0.2\ \Omega$
	Grid inductance L_g	$\in [0.05, 0.3]\,mH$, with nominal value $0.15\ mH$
Inverter	Rated Power	$5\ kVA$
	Switching Frequency	$10\ kHz$
	Sampling frequency	$10\ kHz$
	L_f	$1\ mH$ (4.4% p. u.)
	C_f	$50\ \mu F$ (6.1 p. u.)

Before embarking on studying the influence of grid impedance on the current control loop, let us consider the grid impedance Z_g is small enough and can be ignored first. Without considering the grid impedance, the output voltage v_o is equal to the grid voltage v_g, the current-loop transfer function is the same as an L-filter based inverter. Assuming $K_{PWM}=1$, the following closed-loop transfer function of the grid current can be obtained as:

$$I_g(s) = G_{iref\_ig}(s)I_{ref}(s) - Z_{vg\_ig}(s)V_g(s)$$
$$= \frac{K_p s + K_i}{L_f s^2 + K_p s + K_i}I_{ref}(s) - \frac{L_f C_f s^3}{L_f s^2 + K_p s + K_i}V_g(s). \quad (1)$$

where $G_{iref\_ig}(s)$ and $Z_{vg\_ig}(s)$ represent reference-to-grid current and grid voltage-to-grid current transfer functions, respectively. With $K_p=3$, $K_i=3000$, the crossover frequency of the open loop is set at 500 Hz, and the bode diagrams of $G_{iref\_ig}(s)$ and $Z_{vg\_ig}(s)$ are shown in Fig.3.

When considering the grid impedance Z_g, there will be an additional second-order transfer function $G_a(s)$ in the control loop as shown in the shadow area of Fig. 2 (b),

$$G_a(s) = \frac{1}{L_g C_f s^2 + C_f r_g s + 1}. \quad (2)$$

Equation (2) can be written as the following normalized form:

$$G_a(s) = \frac{\omega_n^2}{s^2 + 2\xi_a \omega_n s + \omega_n^2} \quad (3)$$

where $\omega_n = 1/\sqrt{L_g C_f}$ is natural angular frequency, and $\xi_a = 0.5 r_g \sqrt{C_f / L_g}$ is the damping ratio. As is well known, for ξ_a less than one, the unit-step response of (3) will exhibit over-shooting and ringing, while the frequency response of (3) will have a resonant peak, so called "under damping". Specifically, if the line impedance Z_g is highly inductive, ξ_a will be close to zero, thus leading to high resonant peak and may causing the system instable. The stability problem will be more serious when L_g is large, resulting ω_n close to line frequency. With considering Z_g, the closed-loop transfer function of the grid current can be rewritten as

$$I_g(s) = G'_{iref\_ig}(s)I_{ref}(s) - Z'_{vg\_ig}(s)V_g(s)$$
$$= \frac{K_p s + K_i}{L_f L_g C_f s^4 + L_f C_f r_g s^3 + L_f s^2 + K_p s + K_i}I_{ref}(s) \quad (4)$$
$$- \frac{L_f C_f s^3}{L_f L_g C_f s^4 + L_f C_f r_g s^3 + L_f s^2 + K_p s + K_i}V_g(s).$$

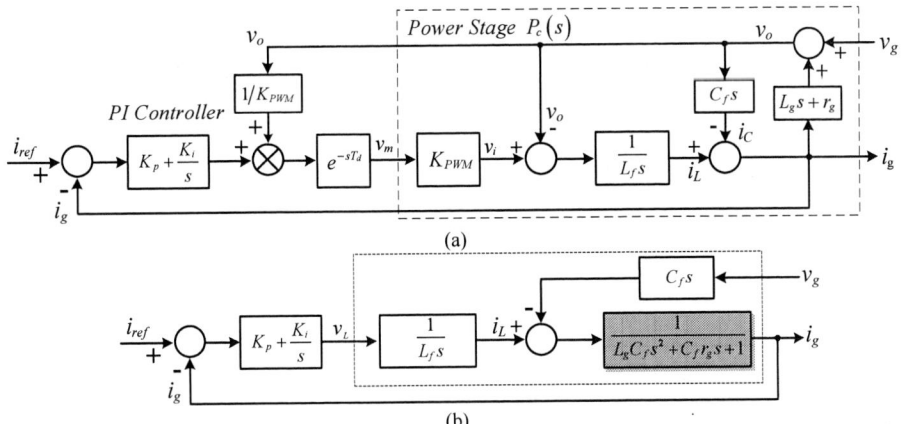

(a)

(b)

Fig.2. (a) block diagram of the PI controller along with the capacitor voltage feed-forward compensator, (b) simplified control block diagram.

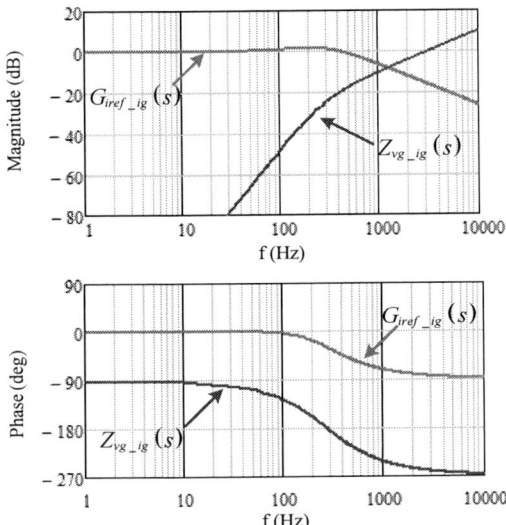

Fig.3. Bode diagrams of $G_{iref\_ig}(s)$ and $Z_{vg\_ig}(s)$.

Fig.4 shows the root loci of the control system with (a) r_g=0.1 Ω and L_g changing from 1 μH to 100 μH, (b) L_g= 10 μH, 100 μH separately and r_g changing from1 mΩ to 1 Ω, where K_p=3, K_i=3000 and the arrow direction represents the corresponding value increasing. Fig.4 (a) reveals that with a constant r_g, as L_g increasing, the dominant roots of the characteristic equation move toward the right half-plane, while Fig.4 (b) depicts that with a constant L_g, as r_g increasing, the roots move toward the left half-plane and away the imaginary axis. Therefore, when grid impedance Z_g increases and keeps highly inductive, resulting a less damped system, the poles are attracted to imaginary or even right half-plane, which may cause the system oscillatory or even instable. In addition, even with the enough damping ratio ξ_a (ξ_a=1), as both L_g and r_g increasing together, the poles are still attracted to the imaginary axis, resulting less stability margin, as shown in Fig.5. Therefore, even with enough damping ratio, large value of L_g will still limit the system bandwidth and affect the system stability. In conclusion, the low-frequency gain and bandwidth will be seriously limited by large value of L_g, which will consequently decrease the tracking performance and disturbance rejection capability.

Hence, large grid impedance variation is challenging the control of grid-connected inverter and the grid filter design in terms of stability.

It is notable that if we decrease the capacitance of C_f, with the same value of L_g, the natural frequency ω_n will increase and the system stability problem will be alleviated. Unfortunately, smaller C_f will increase the inverter output impedance in the stand-alone mode, thus increasing the THD of output voltage. In addition, the control delay between the sampling instant and duty-cycle update instant also decreases the stability margin of the system and strongly limits the bandwidth of the control loop [19]. Here we focus on analyzing the stability problem caused by the grid impedance variation, and we won't discuss the details of the control delay's impact on the system stability.

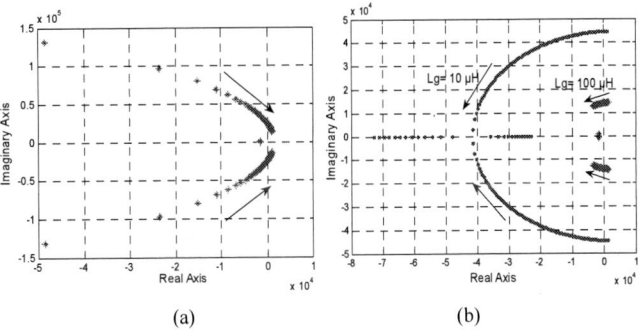

(a)   (b)

Fig.4. Root loci of the control system with (a) r_g=0.1 Ω and L_g changing from 1 μH to 100 μH, (b) L_g=10 μH, 100 μH separately and r_g changing from 1 $m\Omega$ to 1 Ω.

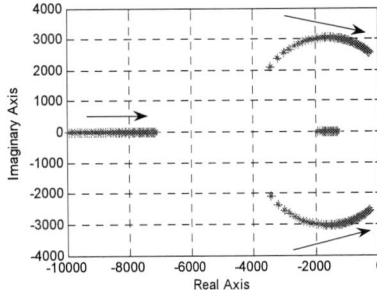

Fig.5. Root locus of the control system with L_g changing from 100 μH to 1 mH, and ξ_a=1.

III. DESIGN OF THE ROBUST CONTROLLER FOR GRID-CONNECTED INVERTERS

To deal with the stability problem caused by the uncertainty of the grid impedance, an H∞ controller is proposed in this paper. The design goal is to get desired tracking performance, such as small steady-state tracking error and low THD of the grid current, while keep the system stable in the predefined variational ranges of the grid impedance as given in the Table. I. The H∞ control theory has been introduced in the early 1980s and opened a new direction in robust control design. Recently, this approach has been applied to the control of active power filters (APF), uninterrupted power supply (UPS), dc-dc boost converters, and dynamic voltage restorer (DVR) to effectively mitigate the effects of the uncertain parameters [20]–[24].

Fig.6 shows (a) the proposed control block and (b) the standard H∞ control configuration with weighting functions. It's notable that $1/L_f s$ as shown in Fig. 2 (b) is merged into the desired H∞ controller $K(s)$, then the H∞ controller only needs to deal with the uncertain second-order control plant, otherwise third-order control plant needs to be compensated, which will be more difficult. In Fig. 6(a), i_d is the equivalent current disturbance, and the design goal of $K(s)$ is to get both desired tracing performance and robustness in the predefined variational ranges of L_g, r_g. In Fig.6 (b), $G_{aN}(s)$ is the nominal plant; z, y, w and u are the controlled output, the measured output, the exogenous input, and the control input, respectively, W_1, W_2, and W_3 are the weighting functions for tracking error performance, the weight on the controller transfer function, and robust performance, respectively. The H∞ controller synthesis is conducted by the singular value loop shaping using mixed-sensitivity approach. The objective is to synthesize the stabilizing controller $K(s)$ so that the H∞ gain from w to z is less than 1, i.e.,

$$\left\| T_{wz} \right\|_\infty < 1 \ \textit{or equivalently} \ \left\| \begin{matrix} W_1 S \\ W_3 T \end{matrix} \right\|_\infty < 1 \tag{5}$$

where $S(s) = 1/(1 + G_{aN}(s)K(s))$ is the sensitivity transfer function, and $T(s) = G_{aN}(s)K(s)/(1 + G_{aN}(s)K(s))$ is called the complementary sensitivity transfer function since they satisfy

$$T(s) + S(s) = 1. \tag{6}$$

It can be seen from (5) that the mixed-sensitivity approach is simply the shaping of $T(s)$ (transfer function from reference to output, or closed-loop transfer function) and $S(s)$ (transfer function from reference to error), by properly selecting their respective weighting function W_1 and W_3, respectively. Typically, we would choose W_1 to have high gain inside the desired control bandwidth to achieve good disturbance attenuation (i.e., tracking performance), and choose W_3 to have high gain outside the control bandwidth, which helps to ensure good stability margin (i.e., robustness). W_2 is the weight on the controller transfer function. A small value (0.1) is assigned to W_2 to ensure the D_{12} matrix of the augmented plant is of full rank [23].

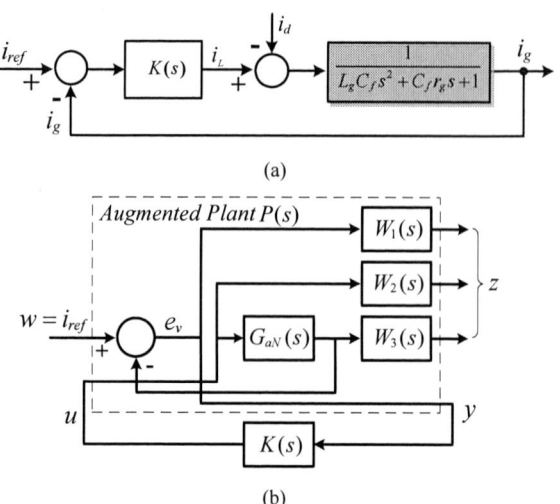

(a)

(b)

Fig.6. (a) proposed robust control block, (b) standard H∞ control configuration with weighting functions

The H∞ loop-shaping design may be processed as following steps:

1) Make the proper selection of weighting functions

2) Perform a standard H∞ synthesis to obtain an adequate controller which makes the closed-loop system having the desired loop shapes.

3) Reduce the order of $K(s)$ for practical implementation while keep almost the same performance as the original controller.

A. Weighting Function Selection for Tracking Error Performance

According to the procedure described above, we consider the determination of weighting function W_1 first. Since the sensitivity transfer function $S(s)$ is the gain from reference to error, for the sinusoidal reference voltage, small tracking error means that the gain of $S(s)$ must be small at the line frequency or in a small neighborhood around the line frequency. From (5), $S(s)$ is shaped in frequency according to the profile specified by $1/W_1$. Hence the weighting function W_1 should exhibits high gains at only the vicinity of the line frequency while providing smaller gains at all the other frequencies. Here, W_1 is selected as a standard second-order weighting function similar to [22], [23],

$$W_1 = \frac{k_1 \omega_0^2}{s^2 + 2\xi\omega_0 s + \omega_0^2} \tag{7}$$

where ω_0 is set as the line frequency, $\omega_0 = 2\pi \cdot 60$. The k_1 in the numerator gives a freedom for adjusting the tracking error over the whole frequency range, and the damping ratio ξ provides another degree of freedom for specifically regulating the tracking error performance at the line frequency ω_0, a smaller ξ gives a larger resonant peak but narrower bandwidth around the frequency ω_0. When $\xi \to 0$, the resulting H∞ controller will act like an ideal PR controller, which can theoretically achieve zero steady-state error at the line frequency. In this paper, $k_1 = 2$, $\xi = 0.01$.

B. Weighting Function Selection for Robust performance

From Fig.6, the nominal plant $G_{aN}(s)$ is expressed as

$$G_{aN}(s) = \frac{1}{L_{gN}C_f s^2 + C_f r_{gN} s + 1}. \qquad (8)$$

where L_{gN}, r_{gN} is the nominal value of the grid equivalent inductance and resistance, respectively. The nominal value L_{gN}=0.15 mH, r_{gN}=0.2 Ω as given in Table I. This parameter uncertainty is transformed to multiplicative output uncertainty, and the resulting relative plant uncertainty with respect to the nominal plant, expressed as

$$\Delta(s) = \sigma \left(\frac{G_a - G_{aN}}{G_{aN}} \right). \qquad (9)$$

$\Delta(s)$ is the plant uncertainty, G_a is the disturbed plant as given in (2), corresponding to the parameter variations $L_g \in [0.05, 0.3] \, mH$, $r_g \in [0.1, 0.5] \, \Omega$, and $\sigma(H)$ stands for the singular values of transfer function H. To achieve required robustness, the condition $\|W_3 T\|_\infty < 1$ must be satisfied. The weighting function W_3 is determined by the worst profile of $\Delta(s)$. One can choose the weighting function W_3 that just bounds the worst case uncertainty spectrum from above. The worst case of $\Delta(s)$ can be obtained when L_g is maximum (L_g=0.3 mH) and r_g (r_g=0.1 Ω) is minimum.

An important aspect that must not be overlooked is the fact that the sensitivity $S(s)$ and complementary sensitivity $T(s)$ must meet the constraint (6), which leads to a basic criterion that the 0 dB crossover frequency of W_1 must be sufficiently below the 0 dB crossover frequency of $1/W_3$. Hence W_3 is chosen as

$$W_3 = k_3(2.066 \cdot 10^{-7} s^2 + 0.909 \cdot 10^{-3} s + 1). \qquad (10)$$

where k_3=0.8. Fig.7 shows relative plant uncertainty $\Delta(s)$ with different parameter variations (the solid line indicating the worst case of the plant uncertainty) and the weighting function W_3. In addition, the weighting functions $W_1 \sim W_3$ need to be proper transfer functions (i.e. the degree of the numerator does not exceed the degree of the denominator) as required by Matlab *mixsyn* function. Therefore, two poles far away with the line frequency are added in the denominator of W_3, leading to

$$W_3' = \frac{1.058 \cdot 10^5 s^2 + 4.655 \cdot 10^8 s + 5.12 \cdot 10^{11}}{s^2 + 1.6 \cdot 10^6 s + 6.4 \cdot 10^{11}}. \qquad (11)$$

C. Mixed-Sensitivity H∞ Controller Synthesis

After selecting the necessary weighting functions, mixed-sensitivity optimization control design can be conducted by the Robust Control Toolbox in Matlab software to synthesize an H∞ controller $K(s)$ such that the H∞ norm of the weighted mixed sensitivity is minimized. For the previous version of Matlab software, we need to create a state-space model of an augmented plant $P(s)$ at first, then synthesize the $K(s)$. While for the latest version, the H∞ controller can be directly synthesized by using *mixsyn* function with the defined $G_{aN}(s)$, $W_1(s)$, $W_2(s)$ and $W_3(s)$, thus resulting a sixth-order controller, which is the same order as the augmented plant model. By using the model reduction function *reduce* in Matlab, the original sixth-order $K(s)$ is then reduced to following third-order for easy implementation with almost the same performance as the original controller

$$K(s) = \frac{608.4 s^2 + 2.825 \cdot 10^6 s + 3.65 \cdot 10^8}{s^3 + 2122 s^2 + 1.581 \cdot 10^5 s + 3.005 \cdot 10^8}. \qquad (12)$$

Fig. 8 shows the singular values of the original sixth-order $K(s)$ and the reduced third-order $K(s)$. The third-order $K(s)$ has almost the same singular values at low frequencies, but less high-frequency attenuation at the vicinity of natural angular frequency ω_n. The singular values of $W_1(s)$, $W_3(s)$, and the resulting $S(s)$, $T(s)$ are shown in Fig.9. As expected, the designed H∞ controller in Fig.8 exhibits significant gain at the line frequency to ensure nearly zero steady-state error. Also, it can be seen that the singular value of H∞ controller falls quickly at high frequencies, making the control system immune to the resonant peak of $G_a(s)$ in (3) as well as high-frequency switching or measurement noises.

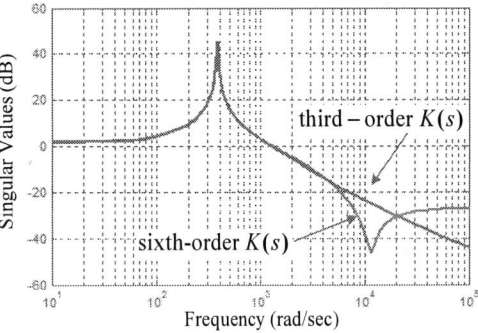

Fig.8. Singular values of the original sixth-order $K(s)$ and the reduced third-order $K(s)$.

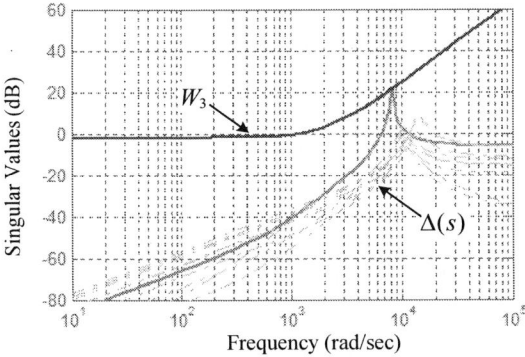

Fig.7. Singular values of $\Delta(s)$ and weighting function W_3.

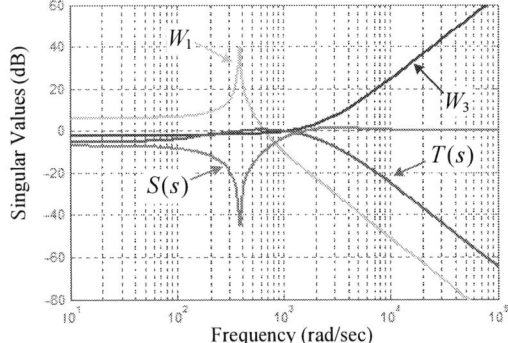

Fig.9. Singular values of W_1, W_3, $S(s)$, and $T(s)$.

D. The Design of the Inner Inverter-Output Current Feedback Loop

As aforementioned, the $1/L_f s$ as shown in Fig. 2 (b) is merged into the designed H∞ controller $K(s)$ in Fig.6 (a). The $1/L_f s$ is a physical transfer function, and needs to be restored from $K(s)$. There are two options to restore it:

1) Split the designed $K(s)$ into two parts, the physical transfer function $1/L_f s$ and the controller $K'(s) = K(s) \cdot L_f s$.

2) Remain the designed $K(s)$, but add an inner inverter-output current feedback loop. Ideally, the closed-loop gain of this inner current loop, including the transfer function $1/L_f s$, can be regarded as unity.

From the control block shown in Fig.2, method 2 employs a closed-loop control for inverter-output current, unlike the method 1 that uses an open-loop control. With properly designed high bandwidth, method 2 will have much better disturbance (e.g. dead time effect, model deviation and so on) rejection capability than method 1. Finally, Fig.10 shows the overall control bock diagram: H∞ controller combined with an inner inverter-output current control loop, and the three-phase phase-locked loop (PLL) for grid synchronization [1]. In Fig.10, for each phase, the three-phase grid-current reference

$$i_{refa} = \hat{I}_{ref} \cdot \sin \theta$$
$$i_{refb} = \hat{I}_{ref} \cdot \sin \left(\theta - 120° \right) \qquad (13)$$
$$i_{refc} = \hat{I}_{ref} \cdot \sin \left(\theta + 120° \right),$$

where θ is the output of the PLL as shown in Fig.10.

A simple proportional compensator K_{pc} is used in the inner inverter-output current feedback loop to get better control performance. Ideally, the loop gain and bandwidth of the inner loop should be maximized by using a higher value of K_{pc}, to achieve perfect reference tracking at all input frequencies, afaster dynamic response and the complete disturbance rejection. A high gain of K_{pc} gives the desired performance, and meanwhile stability problem would arise during physical implementation due to the control delay, measurement noises, and so on. In this paper, $K_{pc} = 5$ is chosen.

IV. SIMULATION AND EXPERIMENTAL RESULTS

The performance of the proposed H∞ controller was first verified via computer simulation, and then realized and tested in the laboratory. The simulation and experimental results based on conventional PI controller were also given as a comparison. In our simulation and experiment, a transformer with turn ratio 1: 2 is used to couple the 3-phase inverter to the grid. Therefore, the inverter output voltage v_o is about half of the grid voltage v_g. To be consistent, all the parameters in both simulation and experiment are the same as given in Table I.

A. Simulation Results

The simulations have been carried out under continuous domain with a switching frequency of 10 kHz and one switching period control delay. To investigate the tracking performance as well as the system stability, simulations have been developed under different (weak or stiff) grid conditions.

Fig.11 shows the simulation results of the PI controller with grid impedance $[r_g, L_g]$ changing from $[0.1\ \Omega, 0.05\ mH]$ to $[0.2\ \Omega, 0.2\ mH]$ at the instant $t=0.2s$, where $K_p=4$, $K_i=4000$. From Fig.11, it can be seen that with L_g increasing, the system exhibits oscillations and tends to instability, which is a good agreement with the analysis in section II. Through decreasing the loop gain (e.g. $K_p=3$, $K_i=3000$) can sustain the system stable, however, it will significantly decrease the tracking performance and the disturbance rejection capability, thus resulting higher THD of the grid current. Fig.12 shows the simulation results of proposed H∞ controller with grid impedance $[r_g, L_g]$ changing from from $[0.1\ \Omega, 0.05\ mH]$ to $[0.2\ \Omega, 0.3\ mH]$ at the instant $t=0.2s$, which reveals that the system can keep stable with the grid impedance variations in a wide range. In addition, the simulation results of H∞ controller with amplitude of grid-current reference (denoted as $\hat{I}_{ref}$) step change from 10 A to 20A at the instant $t = 0.2\ s$ are also given in Fig.13. This figure states that with the grid current reference step change, the output grid current can faithfully follow the reference after a short regulation time.

B. Experimental Results

In the simulation, the continuous-time controller is used, while in the experiment, the third-order H∞ controller in (12) need to be discretized as shown in the following expression with the coefficients calculated by the bilinear transformation method,

$$K(z) = \frac{b_0 + b_1 \cdot z^{-1} + b_2 \cdot z^{-2} + b_3 \cdot z^{-3}}{1 + a_1 \cdot z^{-1} + a_2 \cdot z^{-2} + a_3 \cdot z^{-3}}. \qquad (14)$$

The coefficients $a_1 \sim a_3$ and $b_1 \sim b_3$ in (14) should keep enough decimal digits to avoid loss of accuracy. The discrete H∞ controller in (14) was implemented with a 16-bit fixed-point

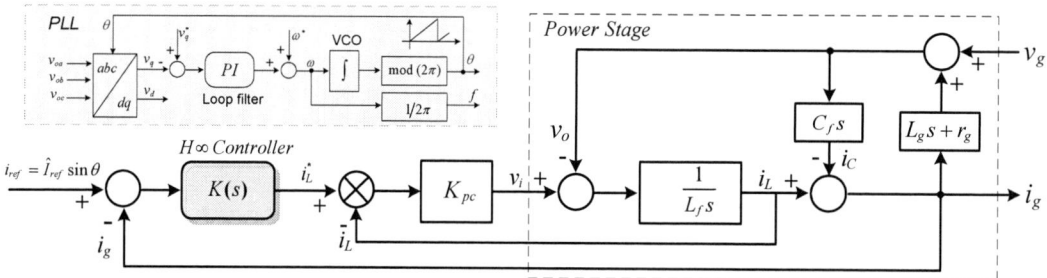

Fig.10. The overall control block diagram: H∞ controller combined with an inner inverter-output current control loop, and the three-phase PLL for grid synchronization.

digital signal processor (DSP), TMS320LF2407 in our experiments, which can operate at 40 million instructions per second (MIPS). Nevertheless, 32-bit fixed-point or floating-point DSP is strongly recommended for better calculation speed and avoiding the numerical error during the calculation. To verify the advantages of the proposed control scheme, the experiments based on the conventional PI controller and the H∞ controller are developed with two different grid impedance. The tuned parameters of the PI controller in the experiment are K_p=3, K_i=3000. For PI control, this combination of parameters gives the best performance in terms of both THD of the grid current and system stability

In the first case, the measured grid impedance $[r_g, L_g]$ =[0.1 Ω, 0.15 mH]. Fig.14 (a) and (b) show the experimental results of the conventional PI controller with different amplitude of grid-current reference, $\hat{I}_{ref}$ =10A, 20A respectively, while Fig.15 (a) and (b) show the corresponding experimental results of the proposed H∞ controller. The start-up processes for H∞ controller is also given in Fig.15 (c), respectively. From Fig.14-Fig.15, we see that in both control methods, the grid voltage and the corresponding grid current are in phase and near-unity power factor is achieved. However, the grid-current waveforms of the H∞ controller are much better than that of PI controller.

In the second case, with an intentionally added inductor, the grid impedance $[r_g, L_g]$ =[0.1 Ω, 0.3 mH]. Due to the page limit, the experimental waveforms are not shown, but the measured THD of the grid current for all cases are listed in Table II. It can be seen that the grid current THD of H∞ controller are always lower than that of PI controller and almost keep unchanged with the grid-impedance variations, which satisfy the THD requirement of IEEE Std. 1547-2003 (i.e. 5%).

From the measured waveforms and THD of grid current, it is evident that proposed H∞ controller has satisfactory performance while applying to grid-connected inverters to deal with the possible grid impedance variations. In the simulation and experiments, the grid impedance variations are within the predefined range as given in Table. I. In the condition that the grid impedance variations are beyond this range, we can use the same design method but re-select the weighting functions and synthesize a new H∞ controller to get the satisfactory control performance as well.

V. CONCLUSION

In the grid-connected voltage source inverter with LC filters, the possible wide range of grid-impedance variations can challenge the design of the controller, especially when the grid-impedance is highly inductive. This paper proposes an H∞ controller with the explicit robustness in terms of grid impedance variations to incorporate the desired tracking performance and stability margin. By properly selecting the weighting functions, the synthesized H∞ controller exhibits high gains at the vicinity of the line frequency, similar to the traditional PR controller, meanwhile it has enough high frequency attenuation to keep the control loop stable. An inner inverter-output-current loop with high bandwidth is also designed to get better disturbance rejection capability. The selection of weighting functions, inner inverter-output-current

loop design, and system disturbance rejection capability are discussed in detail in this paper. Both simulation and experimental results of the proposed H∞ controller, with comparison to the conventional PI controller, are given to validate the performance of the proposed control scheme. It should be noted again that the proposed H∞ controller can be easily applied to single-phase grid-connected inverter as well since it is developed in the stationary reference frame.

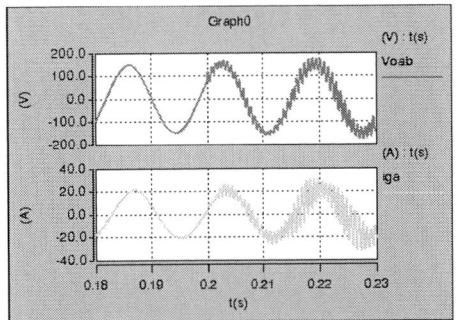

Fig.11. Simulation results of PI controller with grid impedance $[r_g, L_g]$ changing from [0.1 Ω, 0.05 mH] to [0.2 Ω, 0.2 mH] at the instant t=0.2s.

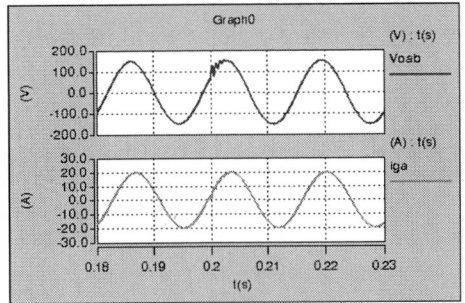

Fig.12. Simulation results of H∞ controller with grid impedance $[r_g, L_g]$ changing from [0.1 Ω, 0.05 mH] to [0.2 Ω, 0.3 mH] at the instant t=0.2s.

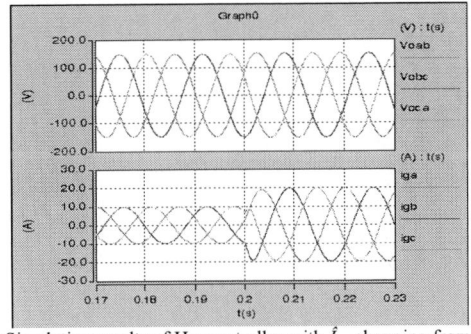

Fig.13. Simulation results of H∞ controller with $\hat{I}_{ref}$ changing from 10 A to 20A at the instant t=0.2s.

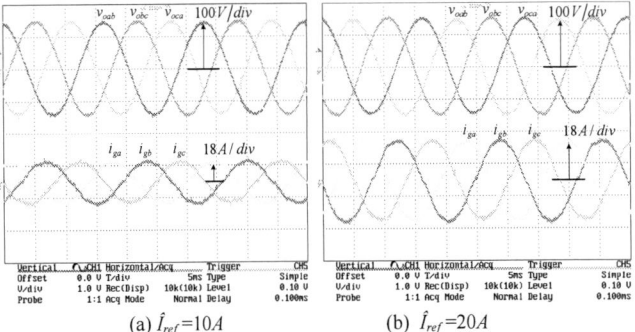

(a) $\hat{I}_{ref}$=10A   (b) $\hat{I}_{ref}$=20A

Fig.14. Experimental results of the conventional PI controller with different $\hat{I}_{ref}$.

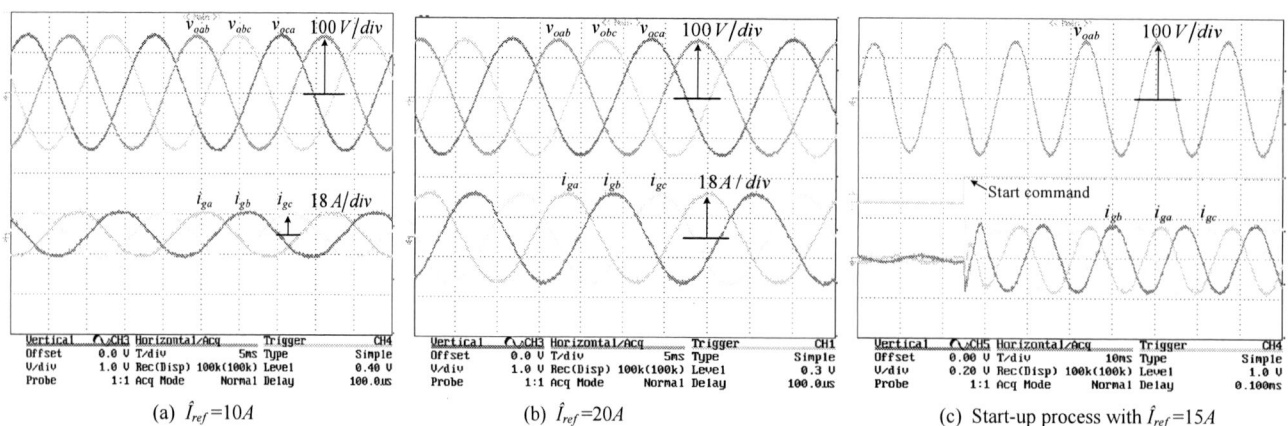

(a) $\hat{I}_{ref}=10A$ (b) $\hat{I}_{ref}=20A$ (c) Start-up process with $\hat{I}_{ref}=15A$

Fig.15. Experimental results of the proposed H∞ controller in steady-state and the start-up process.

TABLE II
THD OF GRID CURRENT (%)

Grid Impedance	Ctrl. Method	$\hat{I}_{ref}=10A$	$\hat{I}_{ref}=15A$	$\hat{I}_{ref}=20A$
$[r_g, L_g]=$ [0.2 Ω, 0.15 mH]	PI Controller	4.327%	3.681%	3.184%
	H∞ Controller	3.398%	2.441%	2.385%
$[r_g, L_g]=$ [0.2 Ω, 0.3 mH]	PI Controller	5.277%	4.859%	4.553%
	H∞ Controller	3.709%	2.401%	2.206%

REFERENCES

[1] Frede Blaabjerg, Remus Teodorescu, and Marco Liserre, "Overview of control and grid synchronization for distributed power generation systems," *IEEE Trans. Ind. Electron.*, vol. 53, no. 5, pp. 1398–1409, Oct. 2006.

[2] J. M. Carrasco, L. G. Franquelo, J. T. Bialasiewicz, E. Galvan, R. C. P. Guisodo, M. A. M. Prats, J. I. Leon, and N. Moreno-Alfonso, "Power-electronics systems for the grid integration of renewable energy source: a survey," *IEEE Trans. Ind. Electron.*, vol. 53, no. 4, pp. 1002–1016, Aug. 2006.

[3] J. Selvaraj, and N. A. Rahim, "Multilevel inverter for grid-connected PV system employing digital PI controller," *IEEE Trans. Ind. Electron.*, vol. 56, no. 1, pp. 149–158, Jan. 2009.

[4] M. P. Kazmierkowski and L. Malesani, "Current control technique for three-phase voltage-source PWM converters: A survey," *IEEE Trans. Ind. Electron.*, vol. 45, no. 5, pp.691–703, Oct. 1998.

[5] Zitao Wang and Liuchen Chang, "A dc voltage monitoring and control method for three-phase grid-connected wind turbine inverters," *IEEE Trans. Power Electron.*, vol. 23, no. 3, pp. 1118–1125, May 2008.

[6] E. Twining and D. G. Holmes, "Grid current regulation of a three-phase voltage source inverter with an LCL input filter," *IEEE Trans. Power Electron.*, vol. 18, no. 3, pp. 888–895, May 2003.

[7] M. Castilla, J. Miret, J. Matas, L. G. Vicuna, and J. M. Guerrero, "Linear current control scheme with series resonant harmonic compensator for single-phase grid-connected photovoltaic inverters," *IEEE Trans. Ind. Electron.*, vol. 55, no. 7, pp. 2724–2733, Jul. 2008.

[8] D. N. Zmood and D. G. Holmes, "Stationary frame current regulation of PWM inverters with zero steady-state error," *IEEE Trans. Power Electron.*, vol. 18, no. 3, pp. 814–822, May 2003.

[9] Xiaoming Yuan, Willi Merk, Herbert Stemmler, and Jost Allmeling, "Stationary-frame generalized integrators for current control of active power filters with zero steady-state error for current harmonics of concern under unbalanced and distorted operating conditions," *IEEE Trans. Ind. Appl.*, vol. 38, no.2, pp. 523–532, Mar./Apr. 2002.

[10] Fei Wang, M. C. Benhabib, J. L. Duarte, and Marcel A. M. Hendrix, "Sequence-decoupled resonant controller for three-phase grid-connected inverters," in *Proc. IEEE APEC*, Washington DC, USA, Feb. 2009, pp. 1–7.

[11] M. Liserre, F. Blaabjerg, and S. Hansen, "Design and control of an LCL-filter based active rectifier," *IEEE Trans. Ind. Appl.*, vol. 41, no.5, pp. 1281–1291, Sept./Oct. 2005.

[12] V. Blasko and V. Kaura, "A novel control to actively damp resonance in input LC filter of a three-phase voltage source inverter," *IEEE Trans. Ind. Appl.* vol. 33, no. 2, pp. 542–550, Mar./Apr. 1997.

[13] N. Abdel-Rahim and J. E. Quaicoe, "Modeling and analysis of a feedback control strategy for three-phase voltage-source utility interface systems," *in Proc. 29th IAS Annu. Meeting*, 1994, pp.895–902.

[14] S. Park, C. Chen, J. Lai, and S. Moon, "Admittance compensation in current loop control for a grid-tie LCL fuel cell inverter," *IEEE Trans. Power Electron.*, vol. 23, no. 4, pp. 1716–1723, July 2008.

[15] C. Chen, J. Lai, Y. Wang, S. Park, and H. Miwa, "Design and control for LCL-based inverters with both grid-tie and standalone parallel operations," in *Proc. IEEE Ind. Appl. Conf.*, Alberta, Canada, Oct. 2008, pp. 1–7.

[16] Guoqiao Shen, Dehong Xu, Luping Cao, and Xuancai Zhu, "An improved control strategy for grid-connected voltage source inverters with an LCL Filter," *IEEE Trans. Power Electron.*, vol. 23, no. 4, pp. 1899–1906, Jul. 2008.

[17] M. Liserre, R. Teodorescu, and F. Blaabjerg, "Stability of photovoltaic and wind turbine grid-connected inverters for large set of grid impedance values," *IEEE Trans. Power Electron.*, vol. 21, no. 1, pp. 263–272, Jan. 2006.

[18] Yunwei Li, D. M. Vilathgamuwa, P. C. Loh, "Design, analysis, and real-time testing of a controller for multibus microgrid system," *IEEE Trans. Power Electron.*, vol. 19, no. 5, pp. 1195–1204, Sept. 2004.

[19] Shuitao Yang, Xinping Ding, Jinyun Liu, and Zhaoming Qian, "Analysis and Design of a Cost-Effective Voltage Feedback Control Strategy for EPS Inverters," in *Proc. IEEE PESC*, Florida, USA, Jun. 2007, pp.477–482.

[20] G. Willmann, D. F. Coutinho, L. F. A. Pereira, and F. B. Libano, "Multiple-loop H-infinity control design for uninterruptible power supplies," *IEEE Trans. Ind. Electron.*, vol. 54, no. 3, pp. 1591–1602, Jun. 2007.

[21] Rami Naim, George Weiss, and Shmuel Ben-Yaakov, "H∞ control applied to boost power converters," *IEEE Trans. Power Electron.*, vol. 12, no. 4, pp. 677–683, Jul. 1997.

[22] Tzann-Shin Lee, S.-J. Chiang, and Jhy-Ming Chang, "H∞ loop-shaping controller designs for the single-phase UPS inverters," *IEEE Trans. Power Electron.*, vol. 16, no. 4, pp. 473–481, Jul. 2001.

[23] Yun Wei Li, D. M. Vilathgamuwa, and Poh C. Loh, "Robust control scheme for a microgrid with PFC capacitor connected," *IEEE Trans. Ind. Appl.* Vol. 43, no. 5, pp. 1172–1182, Sept. / Oct. 2007.

[24] Yun Wei Li, D. M. Vilathgamuwa, F. Blaabjerg, and Poh C. Loh, "A robust control scheme for medium-voltage-level DVR implementation," *IEEE Trans. Ind. Electron.*, vol. 54, no. 4, pp. 2249–2261, Aug. 2007.

Application of Active NPC Converter on Generator Side for MW Direct-driven Wind Turbine

Jun Li, Alex Q.Huang, Subhashish Bhattacharya
FREEDM systems Center
North Carolina State University
Raleigh, NC, USA

Wei Jing
School of Information and Electrical Engineering
China University of Mining and Technology
Xuzhou, Jiangsu, China

Abstract—**3L-NPC topology is usually used in MW wind turbine (WT) systems with full-scale converter configuration. However, due to its drawback of unequal device loss distribution, the converter rated power, and thus the WT unit capacity is limited. Moreover, in cased of device failure in generator converter, in order to protect the WT system, the converter has to shut down to disconnect the WT. This paper presents the application of active NPC (ANPC) converter on generator side. Loss-balancing schemes are discussed, and thermal performance of NPC and ANPC generator converters are compared. Also, the control scheme of generator converter under single device failure condition is proposed to maintain the WT in service and continue to provide real power, which brings benefits in the reliable and economic aspects to the WT system. Simulation results are provided to validate the proposed control methods.**

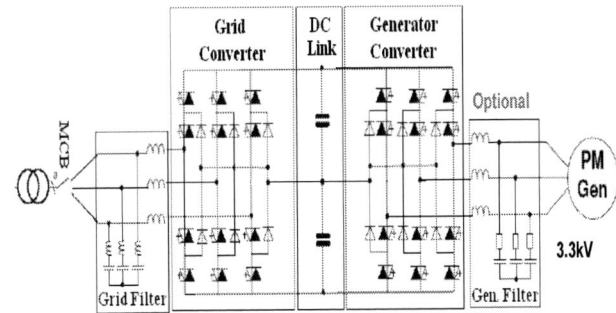

Figure.1. Circuit diagram of 4-quadrant B2B 3L-NPC wind power converter

I. INTRODUCTION

MW Wind turbine (WT) system with direct-driven PMSG and full-scale converter configuration is becoming more attractive for offshore wind power generation. The main advantages include: low mechanical maintenance requirement, robust gearless construction, high efficiency of PM machine, better grid-code fulfillment, and etc [1] [2].

Back-to-back 3L-NPC converters are commonly used in multi-MW WT system for power conditioning [3]. The basic circuit diagram of a 4-quadrant 3L-NPC wind power converter is shown in Fig.1. However, due to the drawback of unequal loss distribution among the semiconductor devices in 3L-NPC topology, the output power of the converter is limited. Since the trend of WT manufacture is toward higher unit capacity, it is a practical solution to increase the maximum output power of the converter by optimizing devices utilization, so as to match the large WT capacity. Moreover, during device failure condition (open or short circuit failure) in generator converter, 3L-NPC converter can not operate properly anymore, therefore the converter and WT have to shut down to protect the system, which will reduce the system reliability, and cause economic losses.

In [4], the active NPC (ANPC) topology was introduced to overcome the loss balancing problem of NPC converter. By equalizing the loss distribution, a substantial increase in the

output power or switching frequency of the converter can be obtained.

This paper presents the application of 3L-ANPC converter on generator side of MW direct-driven wind turbine system. Different loss-balancing schemes are discussed, and the thermal characteristics and performance of 3L-NPC and 3L-ANPC generator converters are compared. Also, the control scheme of generator converter under single device failure condition is proposed, which is effective for both open-circuit and short-circuit failure cases. The fault tolerant operation of ANPC converter enables the WT to maintain in service and continue providing real power. This unique feature is very meaningful especially for offshore wind farm. Since the period between two service visits is long for offshore wind site, the disconnection of WT due to device failure will be very costly. Therefore, the proposed solution enables to bring benefits in terms of reliable and economic aspects. Simulation results are provided to validate the proposed methods.

II. ACTIVE NPC (ANPC) CONVERTER

The topology of 3L-ANPC converter is shown in Fig. 2. As seen, it derives from standard 3L-NPC topology by replacing the clamping diodes with active switches.

Moreover, through proper combination of the 3L-ANPC converter and other two-level or multilevel converters, such as flying capacitor or H-bridge topology, n-level (n≥3) ANPC converter can be derived [5] [6].

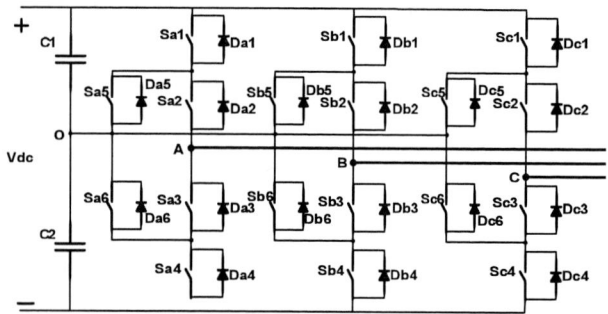

Figure.2. 3L-ANPC converter topology

TABLE I
SWITCHING STATES OF 3L-ANPC CONVERTER

Switching States	Switching Sequence						Output Voltage
	Sa1	Sa2	Sa3	Sa4	Sa5	Sa6	
+	1	1	0	0	0	1	+Vdc/2
0U2	0	1	0	0	1	0	0
0U1	0	1	0	1	1	0	0
0L1	1	0	1	0	0	1	0
0L2	0	0	1	0	0	1	0
-	0	0	1	1	1	0	-Vdc/2

For 3L-ANPC converter, the switching states, switching sequence and the output voltage of one phase are shown in Table I. Compared with 3L-NPC topology, it has four '0' switching states, named '0U1', '0U2', '0L1' and '0L2'. When output voltage is '0', the conduction losses distribution among the inner and clamping switches and diodes can be actively controlled by appropriately choosing one of the '0' switching states to construct the load current flow path. During the switching commutations of '+'↔'0' or '-'↔'0', the switching losses of the devices can also achieve optimal distribution by selecting a proper '0' switching state based on loss-balancing scheme.

III. THERMAL ANALYSIS OF ANPC GENERATOR CONVERTERS FOR MW WT SYSTEM

A. Loss-Blancing Schemes of ANPC Converter

The PWM strategy and loss-balancing method have a great impact on the loss distribution of semiconductor devices, and thus affect the thermal performance of the ANPC converter. Some typical loss-balancing schemes are presented below:

1) Feedback-controlled loss-balancing scheme[4] [7]

This scheme sends the feedback signals of load current, dc-link voltage and cooling water temperature to the built-in loss and thermal model. Based on these information, the losses and junction temperatures of the devices are calculated on-line, which is then used to select the proper commutations to keep the hottest of all the devices as cool as possible. However, the real-time junction temperature calculation needs fast digital controller such as FPGA, and it also increases the control complexity.

2) Feedforward loss-balancing scheme[8]

Different from the previous feedback-controlled scheme, feedforward loss-balancing control calculates the device loss

and junction temperature for all relevant operating points off-line by computer simulation instead of real-time calculation. The optimal ratios between different types of commutations, which are functions of modulation index and power factor, are stored in a lookup table. Finally the predefined ratio or sequence of commutation is implemented by controller. This method is very simple, but the loss-balancing effect is not good for fast transient and abnormal load conditions, and this method also requires the behavior of the modulation methods at all operating points to be predictable.

3) Natural doubling-frequency loss-balancing scheme[9]

In this scheme, the reference voltage Sr is compared with two carrier signals with 180° phase shift on the horizontal axis. Fig. 3 (a) and (b) show the switching states and output voltage for the positive and negative half cycle of reference voltage respectively. These commutation sequences lead to a natural doubling of the apparent switching frequency. Each switch commutes at the switching frequency f_s and the output voltage has an apparent switching frequency equal to $2f_s$.

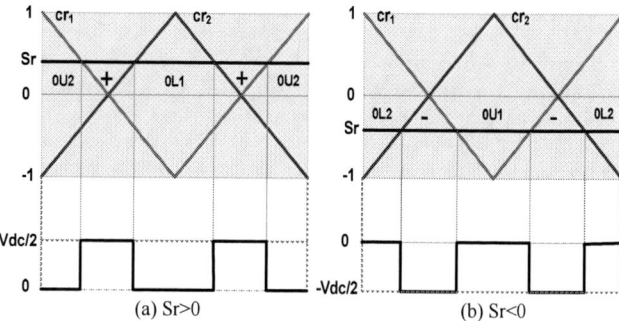

Figure.3. Switching states and output voltage of natural doubling-frequency PWM scheme of 3L-ANPC converter

The loss balancing effect of this method may not be as good as the other two schemes, since it uses completely natural commutation rather than that based on converter operating points and feedback sensor signals. However, it is easy to implement, and more convenient to derive the analytical equations for converter power loss and thermal calculations. Therefore, natural doubling-frequency scheme is used for the thermal analysis of 3L-ANPC generator converter in the next part.

B. Thermal Calculus Methodology of 3L-ANPC Converter

In this paper, the 3L-ANPC generator converter is connected to a PMSG with 3.3 kV rated stator voltage, so the dc-link voltage is chosen to be 5 kV [10].

The 4500V/4000A Gen-4 Emitter Turn-off (ETO) thyristors, shown in Fig. 4, are used as the main switches of the 3L-ANPC converter. For anti-parallel diodes and clamping diodes, ABB's fast recovery diodes 5SDF 10H4502 are used.

Figure.4. Picture of 4500V/4000A Gen-4 ETO

The PWM modulation strategy for the converter is shown in Fig. 3. The reference voltage with 3rd harmonic injection and load current expressions are given by:

$$V = m \cdot \cos wt - \frac{1}{6}m \cdot \cos 3wt \qquad (1)$$

$$I = I_{amp} \cdot \cos(\alpha) = I_{amp} \cdot \cos(wt - \phi) \qquad (2)$$

Here, m is modulation index, I_{amp} is the load current amplitude, and Φ is the load impedance angle.

The device power loss consists of switching loss and conduction loss, while switching loss includes turn-on loss and turn-off loss. Due to the use of di/dt inductor in ETO-based 3L-ANPC generator converter, the ETO turn-on loss is very small and can be neglected.

The ETO's turn-off loss at 125°C junction temperature under 2.5 kV dc-bus voltage is expressed by:

$$E_{off}(I) = 0.0041 \times I + 0.21 \qquad (3)$$

Here, E_{off} (I) is the turn-off energy in Joule, and I is the load current in Ampere.

The ETO on-state voltage equation at 125°C junction temperature is expressed by:

$$V_{on}(I) = 0.0014 \times I + 1.1 \qquad (4)$$

Here, V_{on} (I) is the on-state voltage in Volts.

The switching loss P_{sw} and conduction loss P_{cond} are expressed by:

$$P_{sw} = f_{sw} \cdot \frac{1}{2\pi} \cdot \int_{0}^{2\pi} E_{off}(I) \cdot d\alpha \qquad (5)$$

$$P_{cond} = \frac{1}{2\pi} \cdot \int_{0}^{2\pi} I \cdot V_{on}(I) \cdot D(\alpha) \cdot d\alpha \qquad (6)$$

The device total loss P_{loss} is:

$$P_{loss} = P_{sw} + P_{cond} \qquad (7)$$

Here, f_{sw} is the device switching frequency, α is the load current phase angle, ranging between 0 to 2π, and $D(\alpha)$ is the device switching duty cycle.

The duty cycles of S1, S2 and S6 are given in (8) ~ (10) respectively.

$$d_{S1} = \begin{cases} m \cdot \cos wt - \frac{1}{6}m \cdot \cos 3wt, & wt \in [0, \frac{\pi}{2}]and[\frac{3\pi}{2}+\phi, 2\pi] \\ 0, & wt \in [\frac{\pi}{2}, \frac{3\pi}{2}+\phi] \end{cases} \qquad (8)$$

$$d_{S2} = \begin{cases} \frac{1}{2}\left(1+m \cdot \cos wt - \frac{1}{6}m \cdot \cos 3wt\right), & wt \in [0, \frac{\pi}{2}+\phi]and[\frac{3\pi}{2}+\phi, 2\pi] \\ 0, & wt \in [\frac{\pi}{2}+\phi, \frac{3\pi}{2}+\phi] \end{cases} \qquad (9)$$

$$d_{S6} = \begin{cases} \frac{1}{2}\left(1-m \cdot \cos wt + \frac{1}{6}m \cdot \cos 3wt\right), & wt \in [0, \frac{\pi}{2}]and[\frac{3\pi}{2}+\phi, 2\pi] \\ \frac{1}{2}\left(1+m \cdot \cos wt - \frac{1}{6}m \cdot \cos 3wt\right), & wt \in [\frac{\pi}{2}, \frac{\pi}{2}+\phi] \\ 0, & wt \in [\frac{\pi}{2}+\phi, \frac{3\pi}{2}+\phi] \end{cases} \qquad (10)$$

Similarly, the power loss of anti-parallel diodes and clamping diodes can be calculated. The tested loss characteristics of the diode 5SDF 10H4502 at 125°C junction temperature are:

$$E_{switching}(I) = 0.00018 \times I + 0.8 \qquad (11)$$

$$V_{on}(I) = 5.674 \cdot 10^{-4} \times I + 2.307 \qquad (12)$$

Here, $E_{switching}$ (I) is the diode switching energy in Joule. The switching losses of the diode include its turn-off loss and reverse recovery loss. V_{on} (I) in (12) is the diode on-state voltage in Volts, and I is the load current in Ampere.

The duty cycles of D1, D2 and D5 are given in (13)~(15) respectively.

$$d_{D1} = \begin{cases} m \cdot \cos wt - \frac{1}{6}m \cdot \cos 3wt, & wt \in [\frac{3\pi}{2}, \frac{3\pi}{2}+\phi] \\ 0, & wt \in [0, \frac{3\pi}{2}]and[\frac{3\pi}{2}+\phi, 2\pi] \end{cases} \qquad (13)$$

$$d_{D2} = \begin{cases} \frac{1}{2}\left(1+m \cdot \cos wt - \frac{1}{6}m \cdot \cos 3wt\right), & wt \in [\frac{\pi}{2}+\phi, \frac{3\pi}{2}+\phi] \\ 0, & wt \in [0, \frac{\pi}{2}+\phi]and[\frac{3\pi}{2}+\phi, 2\pi] \end{cases} \qquad (14)$$

$$d_{D5} = \begin{cases} \frac{1}{2}\left(1-m \cdot \cos wt + \frac{1}{6}m \cdot \cos 3wt\right), & wt \in [0, \frac{\pi}{2}]and[\frac{3\pi}{2}+\phi, 2\pi] \\ \frac{1}{2}\left(1+m \cdot \cos wt - \frac{1}{6}m \cdot \cos 3wt\right), & wt \in [\frac{\pi}{2}, \frac{\pi}{2}+\phi] \\ 0, & wt \in [\frac{\pi}{2}+\phi, \frac{3\pi}{2}+\phi] \end{cases} \qquad (15)$$

Due to the symmetric structure of the 3L-ANPC converter, the power loss of S1 & S4 are the same. Others with the same power losses are: S2 & S3, S5 & S6, D1 &D4, D2 &D3 and D5 & D6.

Considering the cost, reliability and compact structure, heat pipes based air cooling system is used for ETO-based ANPC generator converter. The physical arrangement for a single-pole stack of the converter is shown in Fig. 5. Both ETOs and diodes are double-side cooled using the heat pipes with 2.2 kW heat removal capability.

For ETO, the sum of the thermal resistances from junction to Anode and that from case to heatsink is 26 K/kW, and the sum of thermal resistance from junction to Cathode and that from case to heatsink is 35 K/kW; for diodes, the thermal resistance from junction to heatsink is 15 K/kW for both sides; for heat pipe, the thermal resistance from contact point to the air is 18 K/kW with 5 m/s flow rate of cooling air.

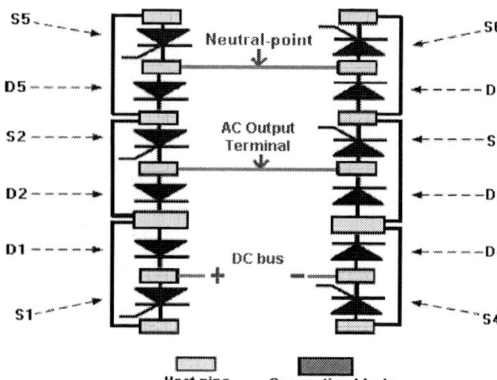

Figure.5. Physical arrangement of a single-pole stack of ETO-based 3L-ANPC converter

Using the analytical equations (1) ~ (15), all the devices power loss can be calculated. Then with the information of power losses and thermal resistances, the device steady-state operating junction temperatures can be estimated.

C. Thermal Performance Comparison of NPC and ANPC Generator Converters for MW WT System

According to the calculus methodology above, the thermal analysis of ANPC generator converter is implemented. In the calculation, the apparent switching frequency is selected to be 1 kHz, so the actual device switching frequency of ANPC converter is 500 Hz due to the natural doubling-frequency PWM strategy. The ambient temperature is 45 °C.

To find the converter peak output power rating, two thermal limitations must be satisfied: one is that the steady-state operating junction temperatures of all the devices are below 125 °C; the other is that the power dissipation on all the devices must be lower than the heat removable capability of the heat pipes.

For thermal performance comparison, the thermal analysis of ETO-based NPC generator converter is also implemented [10]. The physical arrangement for a single-pole stack of ETO-based NPC generator converter is shown in Fig. 6.

Thermal analysis parameters and results are shown in Table II. The device power loss distribution and steady-state junction temperature of ANPC and NPC generator converters under peak output power condition are shown in Fig. 7 and Fig.8.

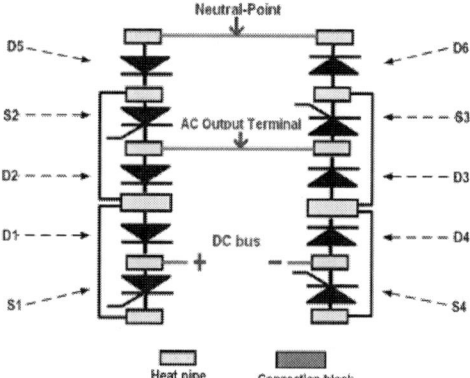

Figure.6. Physical arrangement of a single-pole stack of ETO-based 3L-NPC converter

TABLE II
THERMAL ANALYSIS AND COMPARISON OF ANPC AND NPC GENERATOR CONVERTERS

DC voltage	5 kVDC	
Switching frequency	1 kHz (apparent f_{sw})	
Line-line voltage	3.3 kVAC	
Power factor	$\cos(\Phi) = -0.95$	
	NPC	Active NPC
PWM modulation	SPWM with 3rd harmonic injection	Doubling-frequency PWM with 3rd harmonic injection
Modulation index	1.078	1.078
Peak RMS current	1.346 kA	1.539 kA
Peak output power	7.69 MVA	8.74 MVA
Total device losses	13.78 kW	15.87 kW
Maximum power loss on individual heat pipe	2.2 kW	2.2 kW

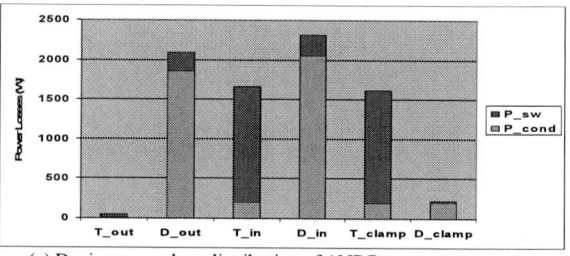

(a) Device power loss distribution of ANPC generator converter

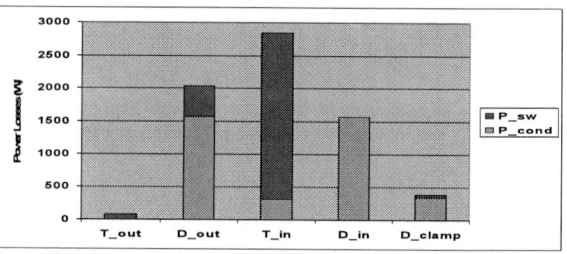

(b) Device power loss distribution of NPC generator converter
Figure.7. Devices power loss distribution comparison

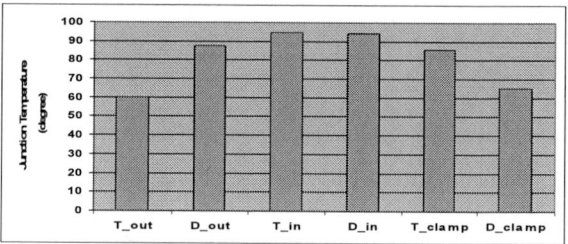

(a) Device junction temperatures of ANPC generator converter

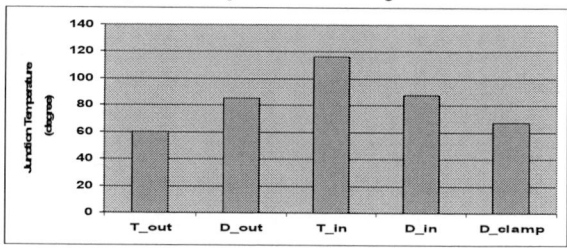

(b) Device junction temperatures of NPC generator converter
Figure.8. Devices junction temperatures comparison

From the results, it is observed that the maximum output power of NPC and ANPC generator converters are 7.69 MVA and 8.74 MVA respectively. The ANPC converter enables to obtain an increased power of 14% higher than NPC converter, and it can be applied to support up to 8 MW wind turbines.

The maximum heat removal capability of the heat pipe is the thermal limitation in this study. The maximum power loss on individual heat pipe reaches 2.2 kW limit before the steady-state junction temperature of the hottest device reaches 125°C. As seen in Fig. 8, the hottest device junction temperature of ANPC generator converter is much lower than that in NPC converter. Therefore, if heat pipes with sufficient cooling capability are equipped, the peak power rating of ANPC converter will be much higher, and its power level is expected to be 20% higher than that of NPC converter.

IV. WIND TURBINE OPERATION DURING DEVICE FAILURE OF 3L-ANPC GENERATOR CONVERTER

In Fig.1, under device failure condition on generator converter (open or short-circuit failure), the converter can not continue to work properly anymore, therefore the WT must be disconnected from the system, which will cause economic losses. Since today's WTs power ratings are multi-MW level, and the next generation unit is expected to even higher, it is beneficial to maintain the WT tied to the grid to continue supporting real power under certain devices failure condition, This is also meaningful for the operation stability of the power system with high percentage penetration of wind power.

To achieve this goal, the WT operation under single device failure condition on ANPC generator converter is analyzed first. And then WT fault tolerant control schemes are proposed in this section.

A. Device Failure Analysis of ANPC Generator Converter

In ANPC generator converter, the switches and diodes may fail in either open-circuit fault or short-circuit fault. Some examples are shown in Fig.9. The direction of current Ia is defined as positive when it flows out of the converter. It is observed that in Fig.9 (a) and (b), when D5 or S2 fails open, if $Ia>0$ and switching state is 0U2/0U1, the output voltage is actually '-Vdc/2' rather than '0'. In Fig. 8(c), when D2 fails in open, if $Ia<0$ and switching state is +/0U2/0U1, the current path will be cut-off, which may cause over-voltage on the terminals of PMSG and generator converter, and destroy them. Fig.9 (d)~(f) show device short-circuit failure cases for S2/D2 and S5/D5. As seen, the upper or lower dc-link capacitor will be discharged through a short-circuit path, which is formed by

the failed device, thus some devices may be destroyed due to over-current, and other devices may have to stand the full dc-link voltage.

B. Fault Tolerant Capability of ANPC Converter

The fault tolerant capability of ANPC converter was introduced in [11]. Through switching states reconfiguration and reference voltage modification, the ANPC converter is still able to generating three-phase symmetrical current to the load under single device failure condition. The fault tolerant principles are summarized as below:

1) Sa5(D5) or Sa6(D6) open-circuit failure case (*Type a*): ANPC converter is degraded to NPC converter by choosing a proper switching state, thus voltage and current waveforms are not effected;

2) Other device failure cases (*Type b*): the phase containing the failed device will be connected to dc neutral-point throught a proper '0' switching state, and meanwhile the reference voltage is modified to maintain three-phase line-line voltage balance. The maximum modulation index will be reduced to 0.5774, thus the available maximum output voltage of the ANPC converter is limited.

C. Generator Converter Control

The rotor aerodynamics is expressed by:

$$P_T = \frac{1}{2}\rho A_r C_P(\beta,\lambda)V_w^{\;3} \qquad (16)$$

where P_T is the extracted wind power, ρ is the air density, A_r is the area swept by the rotor, C_p is the power coefficient, $\lambda=\omega_r\cdot R/V_w$ is the tip speed ratio, ω_r is the rotor angular speed, R is the blade radius, V_w is the wind speed, β is the pitch angle of the rotor blade.

The typical wind power curve is shown in Fig.10.

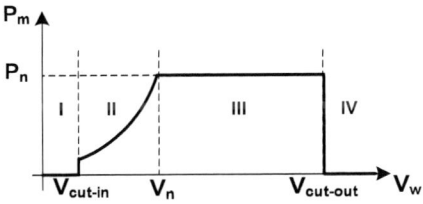

Figure.10. Typical wind power curve

1) Generator control under normal condition

Under normal operation (no device failure), depending on the wind speed, different control methods are used for generator converter [1].

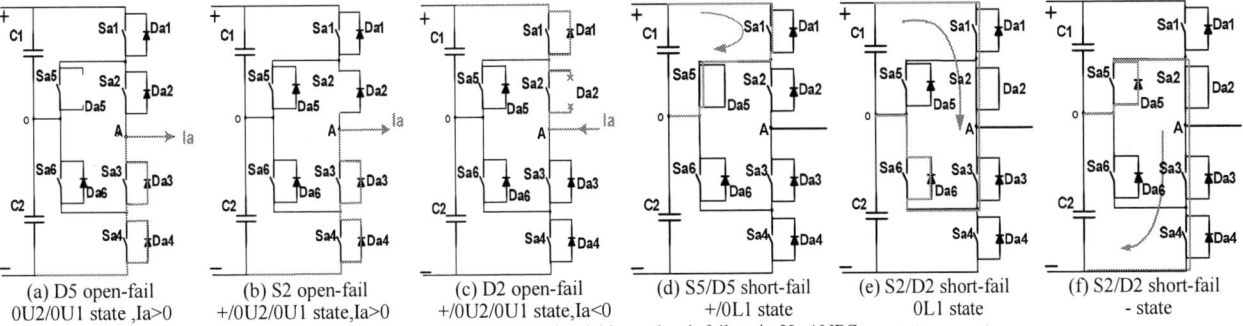

| (a) D5 open-fail 0U2/0U1 state ,Ia>0 | (b) S2 open-fail +/0U2/0U1 state,Ia>0 | (c) D2 open-fail +/0U2/0U1 state,Ia<0 | (d) S5/D5 short-fail +/0L1 state | (e) S2/D2 short-fail 0L1 state | (f) S2/D2 short-fail - state |

Figure.9. Examples of device open-circuit/short-circuit failure in 3L-ANPC generate converter

a) When the wind speed is between cut-in speed V_{cut_in} and rated speed V_n (Region II), the controller of generator converter implements the MPPT algorithm to extract the maximum power from wind by adjusting generator speed to keep the constant optimal tip speed ratio λ_{opt}. During this period, pitch angle is always fixed to zero.

b) When the wind speed is between rated speed V_n and cut-out speed V_{cut_out} (Region III), the controller of generator converter needs to limit the output power at the rated level by adjusting pitch angle.

In this paper, we focus on the control with wind speed in Region II. The grid converter controls the dc-link voltage, neutral-point voltage and grid-side power factor. For generator converter, the control diagram is shown in Fig.11, in which field oriented control (FOC) is used to perform speed control of PMSG by decoupling the flux linkage and torque control. Outer loop sets the torque reference with MPPT algorithm.

The optimal torque reference is given by [12]:

$$T_{em\_ref} = K \cdot \omega_r^2 \qquad (17)$$

$$K = \frac{1}{2}\rho \cdot A_r \cdot R^3 \frac{C_{p\max}}{\lambda_{opt}^3} \qquad (18)$$

$$C_{p\max} = C_p\left(\beta\big|_{\beta=0}, \lambda_{opt}\right) \qquad (19)$$

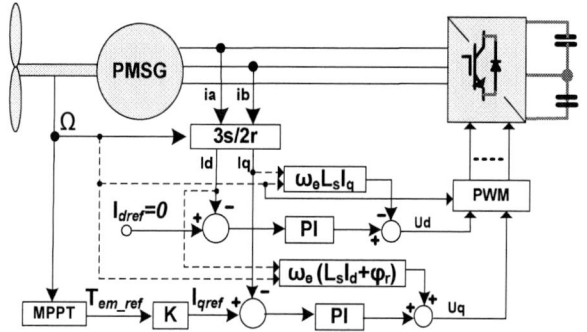

Figure.11. Control diagram of generator converter

2) Generaotor control under deivce failure condition

Under *Type (a)* device failure condtion, as described in section IV (B), since the maximum modulation index and output voltage are not influenced by the fault tolerant control scheme, the control of generator converter can still extract the maximum power for the whole wind speed range in Region II. The rated modulation index is 1.078. The WT operation is almost the same as the that in normal condition.

Under *Type (b)* device failure condition, the maximum modulation index is reduced to 0.5774, so the adjustable speed range of the PMSG will be limited. Table III lists the main parameters of the 5MW WT and PMSG. With the fault tolerant control dealing with type (b) device failure, the maximum achievable rotor speed, which corresponds to the modulation index of 0.5774 is calculated by:

$$\omega_r = \frac{0.5774}{1.078} \times 14.8 \; rpm = 7.9272 \; rpm \qquad (20)$$

TABLE III
THE MAIN PARAMETER OF 5 MW WT AND PMSG

Wind Turbine Parameters		PMSG Parameters	
Rate power P_n	5 MW	Rate power P_n	5 MW
Rated rotor speed	14.8 rpm	Rated L-L voltage	3.3 kV
Rotor diameter	116 m	Power factor	0.95
Cut in wind speed	2.5 m/s	Phase resistor R_s	50 mΩ
Cut out wind speed	25 m/s	d-aix inductance	3.5218 mH
Rated wind speed	11.8 m/s	q-aix inductance	3.5218 mH
Optimum tip speed ratio	7.6179	PM flux	10.1486Wb
Maximum power coefficient	0.4746	Number of pole	118

When constant λ_{opt} is kept, the wind speed is:

$$V_w = \frac{0.5774}{1.078} \times 11.8 \, m/s = 6.3203 \; m/s \qquad (21)$$

In Region II, when wind speed is below 6.3203 m/s, the MPPT can still be obtained by adjusting rotor speed. When wind speed is above 6.3203 m/s, the MPPT is not available anymore due to the maximum rotor speed limit. Under this condition, we can transit the controller of generator converter from MPPT mode to constant speed mode, where the rotor speed is controlled to be constant at 7.9272 rpm.

The power curve under constant rotor speed mode is shown in Fig.12. It is observed that at different wind speed, although at rotor speed ω_r=7.9272 rpm (0.83 rad/s), it can not extract the same maximum power as that in normal operation, it still generates the output power which is the allowable maximum value under the device failure condition, therefore we named it "Sub-MPPT" control. If rotor speed is lower than 0.83 rad/s, as seen in Fig.12, the output power will become smaller.

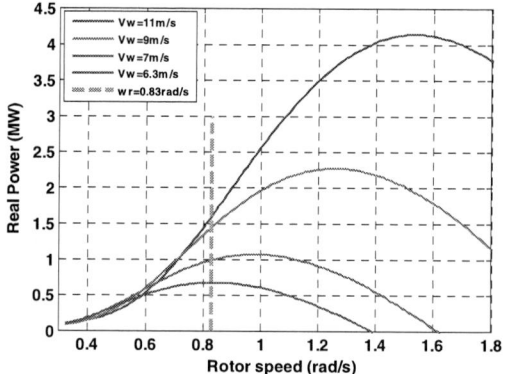

Figure.12. Power curve with constant rotor speed control

Based on the analysis of WT and generator converter control and operation under *type (a)* and *type (b)* device failure conditions, Fig.13 shows the control diagram of ANPC generator converter during fault tolerant operation. With the information of position and failure-type of the failed device, and wind speed, one proper control mode will be selected to generate Iq reference. The block output I_{q_ref} is then sent to the inner current loop of the controller in Fig. 11.

With the presented control scheme, the ANPC generator converter allows to continue extracting power from wind under device failure condition.

978-1-4244-4782-4/10 $26.00 © 2010 IEEE   1015

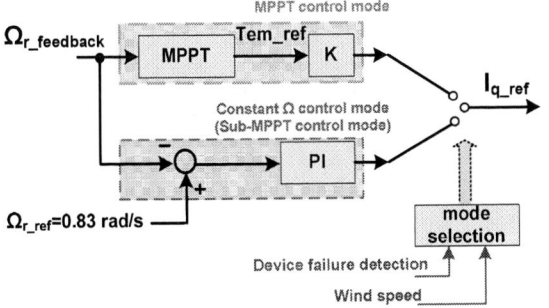

Figure.13. Mode selection diagram under fault tolerant operation

V. SIMULATION RESULTS

To verify the proposed fault tolerant control scheme of 3L-ANPC generator converter, the simulation of a 5MW direct-driven WT system is implemented in MATLAB. The main parameters of WT system are given in Table II and Table III. In the simulation, the device failure occurs at 1.5s in phase A. The waveforms of the extracted real power, generator current, generator torque, rotor speed, dc-link voltage, neutral-point voltage are plotted for each test case.

Fig. 14 shows generator side waveforms at V_w=11.5m/s with D5 open-circuit failure at 1.5s. This device failure condition belongs to type (a) failure, so the fault tolerant control will degrade the ANPC converter operation to NPC converter operation, and thus the output voltage is not reduced. Therefore it can be observed from the waveforms that the operation of WT and generator almost have no change after D5 fails, and maximum power extraction is still effective.

Fig. 15 shows generator side waveforms at V_w=6m/s with D2 short-circuit failure at 1.5s. This failure belongs to type (b), so the maximum modulation index is limited to 0.5774 with fault tolerant control. However, since wind speed V_w is below 6.3203 m/s, the MPPT can still be achieved. The waveforms are almost the same as normal operation. Because the fault tolerant control connects one-phase of the generator converter to neutral-point after D2 fails, the oscillation of the neutral-point voltage becomes larger, which increases the torque ripple slightly.

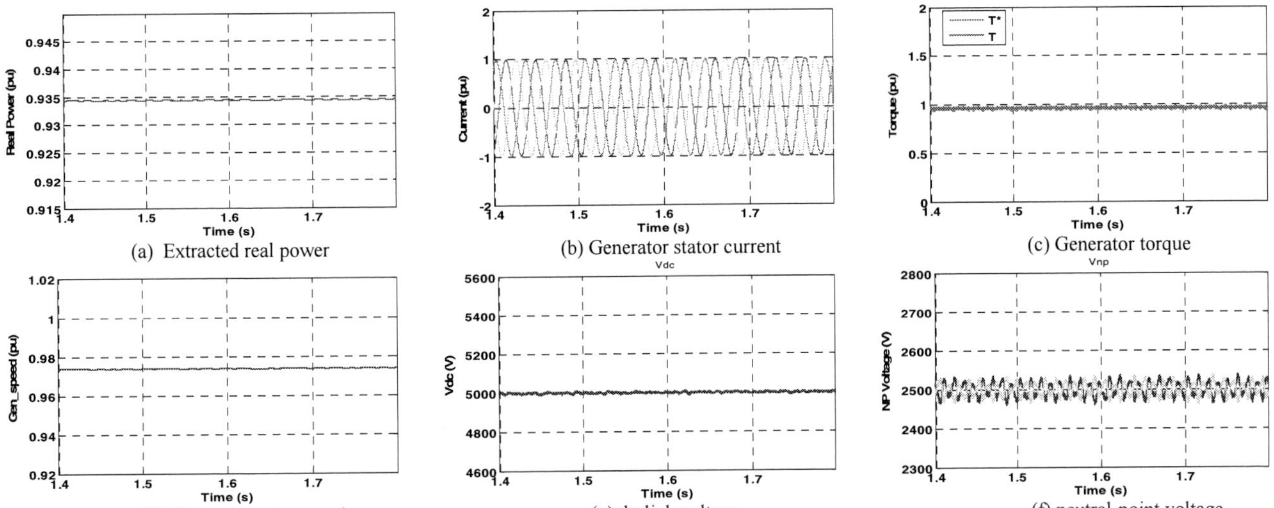

(a) Extracted real power

(b) Generator stator current

(c) Generator torque

(d) Generator rotor speed

(e) dc-link voltage

(f) neutral-point voltage

Figure.14. Generator side waveforms under D5 open-circuit failure for 11.5 m/s wind speed with fault tolerant control

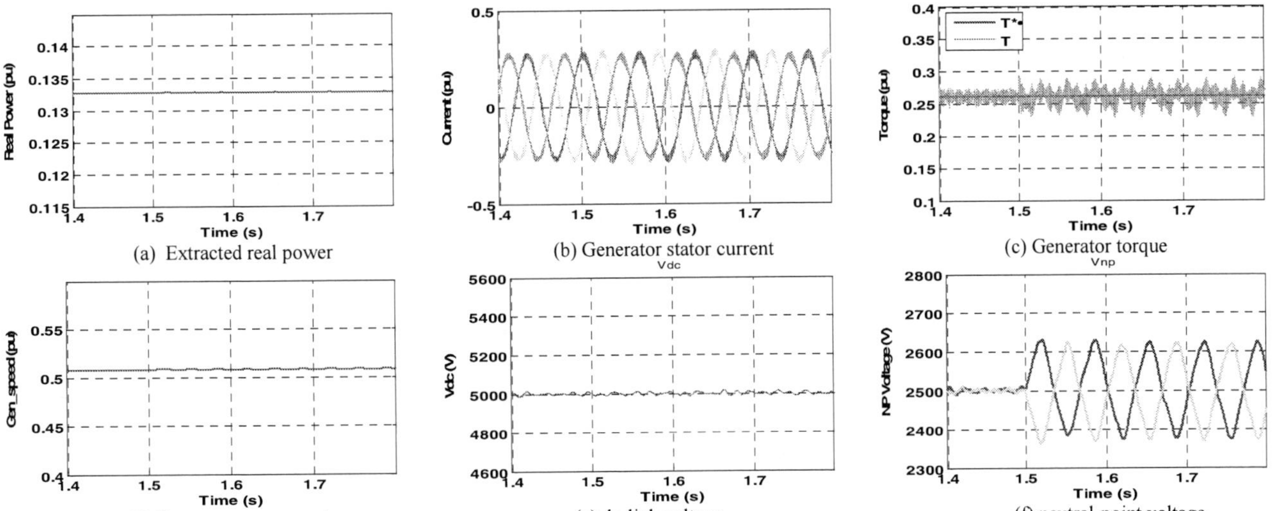

(a) Extracted real power

(b) Generator stator current

(c) Generator torque

(d) Generator rotor speed

(e) dc-link voltage

(f) neutral-point voltage

Figure.15. Generator side waveforms under D2 short-circuit failure for 6 m/s wind speed with fault tolerant control

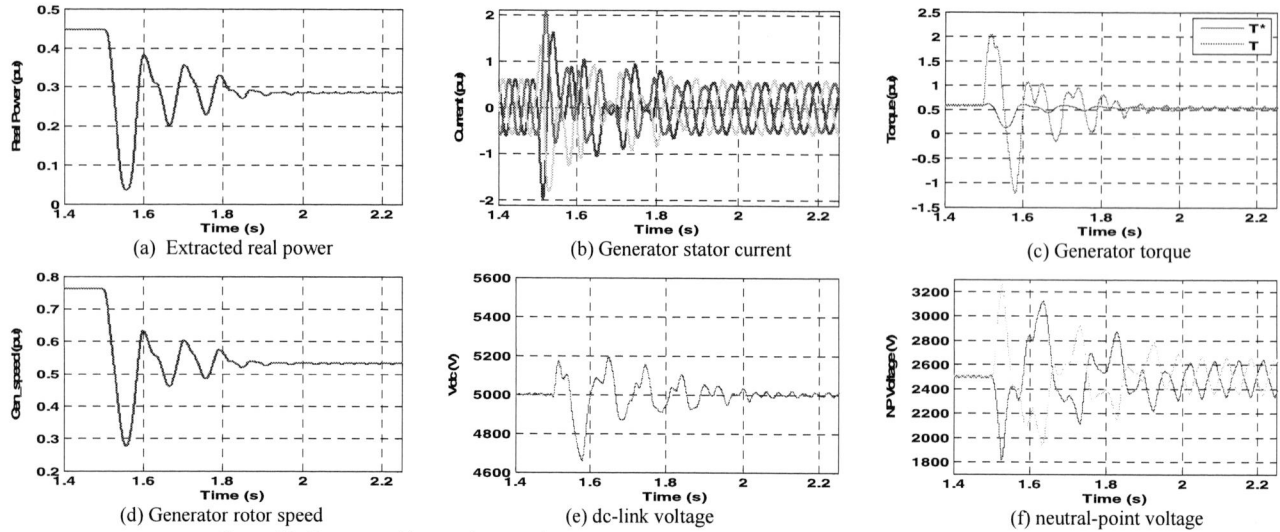

(a) Extracted real power   (b) Generator stator current   (c) Generator torque

(d) Generator rotor speed   (e) dc-link voltage   (f) neutral-point voltage

Figure.16. Generator side waveforms under D2 open-circuit failure for 9 m/s wind speed with fault tolerant control

Fig. 16 shows generator side waveforms at V_w=9m/s with D2 open-circuit failure at 1.5s. This failure belongs to type (b), so the maximum modulation index is limited to 0.5774 with fault tolerant control. Since wind speed V_w is above 6.3203 m/s, controller selects constant rotor speed control mode instead of MPPT control mode. The reduction of the extracted real power and rotor speed can be observed in the waveform. The transient time is around 0.4 s before the WT and generator reach the new steady-state operating point after D2 failure.

VI. CONCLUSION

This paper presents the application of 3L-ANPC converter on generator side in MW direct-driven WT system. The loss-balancing schemes are discussed and the thermal calculus methodology of ANPC converter is derived. The thermal performance is compared between ETO-based NPC and ANPC generator converter. It is shown that the loss distribution control allows a substantial increase of the ANPC generator converter power rating, and thus it can be applied for larger WTs. Moreover, the fault tolerant control of the ANPC generator converter under single device failure condition is proposed, which enables the WT to maintain in service and continue providing real power. This feature is able to improve the system reliability and economics benefits. Simulation results show the correctness of the proposed control methods.

ACKNOWLEDGMENT

It made use of ERC shared facilities supported by the National Science Foundation under Award Number EEC-08212121.

REFERENCES

[1] Z.Chen, J.M.Guerrero, F.Blaabjerg, "A review of the state of the art of power electronics for wind turbines," *IEEE Trans. Power Electron.*, vol. 24, no.8, pp.1859-1875, Aug.2009.

[2] H.Li, Z.Chen, "Overview of different wind generator systems and their comparison," *IET Proc. Renewable Power Generation*, vol.2, no. 2, pp.123-138, Jun.2008.

[3] A.Faulstich, J.Steinke, F.Wittwer, "Medium voltage converter for permanent magnet wind power generators up to 5MW," *2005 European Conference on Power Electronics*, pp.11-14, Sep.2005.

[4] T.Bruckner, S.Bernet, "Loss balancing in three-level voltage source inverters applying active NPC switches," in *Proc. IEEE PESC 2001*, vol.2, pp.1135-1140, Jun.2001.

[5] P.Barbosa, P.K.Steimer, L.Meysenc, J.Steinke, M.Winkelnkemper, N.Celanovic, "Active neutral-point-clamped multilevel converters," in *Proc. IEEE PESC 2005*, pp.2296-2301, Jun.2005.

[6] T.Chaudhuri, P.Steimer, A.Rufer, "Introducing the Common Cross Connected Stage (C3S) for the 5L ANPC multilevel inverter," in *Proc. IEEE PESC 2008*, pp.167-173, Jun.2008.

[7] T. Bruckner, S. Bernet, H. Guldner, "The active NPC converter and its loss-balancing control," *IEEE Trans. Ind. Electron.*, vol. 54, no. 6, pp.855-868, Jun.2005.

[8] T. Bruckner, S. Bernet, P.K.Steimer, "Feedforward Loss Control of Three-Level Active NPC Converters," *IEEE Trans. Ind. Appl.*, vol.43, no.6, pp.1588-1596, Nov-Dec. 2007.

[9] D.Floricau, E.Floricau, G.Gateau, 'Three-level active NPC converter: PWM strategies and loss distribution," in *Proc. IEEE IECON 2008*, pp.3333-3338, Nov.2008.

[10] J.Li, A.Q.Huang, W.Jing, "7MVA ETO Light NPC converter for multi-MW direct-driven wind turbine application," in *Proc. IEEE PEMWA 2009*, pp.1-7, Jun.2009.

[11] J. Li, A.Q.Huang, S.Bhattacharya, G.Tan, "Three-level active neutral-point-clamped (ANPC) converter with fault tolerant ability," *in Proc. IEEE APEC 2009*, pp.840-845, Feb. 2009.

[12] K.Johnson, L.Fingersh, M.Balas, L.Pao, "Methods for increasing region2 power capture on a variable speed wind turbine," *J. Solar Energy Eng.*, vol.126, no.4, pp.1092-1100, 2004.

978-1-4244-4782-4/10 $26.00 © 2010 IEEE

Nonlinear Modeling of Switched Reluctance Motor Using Different Methods

Jun Cai, Zhiquan Deng, Zeyuan Liu

Aero-Power Sci-Tech Center, College of Automation and Engineering, Nanjing University of
Aeronautics and Astronautics, Nanjing 210016, China
E-mail: jcai_yjs05@126.com

Abstract—Accurate modeling of flux-linkage characteristics is the basis of design and control of switched reluctance motor(SRM). The flux-linkage characteristic of SRM is a function of both the excitation current and rotor position. But due to the highly nonlinear characteristics of SRM, modeling is cumbersome. In this paper, three effective algorithms for modeling of SRM are investigated, which are based on sub-regional mathematical model, neural network and adaptive neural fuzzy inference systems(ANFIS) respectively. The performance of each method is validated and compared via matlab tools. Simulation results show that all these models can accurately and successfully modeling of SRM, while RBF neural network is the best. Moreover, a new SRM dynamic simulation model based on RBF is developed for verification. Simulation and experimental results validated the accuracy of RBF modeling algorithm.

I. INTRODUCTION

Switched reluctance motor(SRM) are drawing great attention for variable speed drive applications, which have several advantages such as simple and robust construction, high torque to weight ratio, simple power converter structure, and its good fault-tolerant performance etc. Modeling of SRM is motivated by the requirement of accurate performance estimation to implement the optimal excitation and control tactics. But due to the motor's double salient structure, its magnetic characteristics are highly nonlinear. The flux-linkage appears to be a complex nonlinear function of both the phase current and rotor position. Though accurately modeling of SRM is cumbersome, it can be grouped as function approximation problems.

Recently, several attempts have been made to model the SRM using the data obtained by experimental results or finite element analysis. Some are based on mathematical models, and others employ the intelligent techniques. The former always suffer from lack of accuracy and its complexity. While the latter are model free algorithm, though a little time consuming, are well suit for modeling the systems with highly nonlinearity [1]. A Fourier cosine series based analytical expression of flux was proposed by Terrey [2]. A B-spline based flux expression was developed in [3]. A parametric piecewise Hermite cubic spline function with a combination of

Fourier series is presented in [4]. In [5]-[6] the use of fuzzy logic for modeling estimation and prediction of SRM are investigated. In [7]-[9], approaches of SRM modeling based on neural networks with different structures are discussed.

In this paper, three different algorithms for flux-linkage modeling, which are based on a sub-regional mathematical model, neural network and adaptive neural fuzzy inference system(ANFIS) respectively, are investigated systematically. All these developed models are validated via Matlab/simulink tools. For different application circumstance, each method has its own merits and demerits. Simulation results are presented to demonstrate the effectiveness of these modeling schemes, and provide the theory reference to select a modeling algorithm. Finally, a new dynamic simulation method based on RBF model is developed for analyzing the dynamical performance of SRM. Simulation and experimental results validated the accuracy of RBF modeling algorithm.

II. FINITE ELEMENT ANALYSIS OF SRM

A. Prototype Chosen

In this section, a 3-phase 12/8 poles rated 7.5kw prototype motor are chosen as the study object. The specification of the prototype is given in Table.1.

TABLE I. SPECIFICATION OF THE SRM

parameter	value
Numbers of phases	3
Stator/rotor poles	12/8
Air gap	0.03(mm)
Rated power	7.5kw
Max rotor speed	60000rpm
Rated voltage	270vDC

B. Finite Element Analysis of the SRM

Accurately calculation of flux linkage characteristics is the basis of modeling of SRM. Two dimensional (2-D) FEM based on magnetic vector potential method (MVP) is an effective method to obtain this characteristic. In this paper, the

---

Programs for New Century Excellent Talents in University

978-1-4244-4782-4/10 $26.00 © 2010 IEEE

finite element model(FEM) of the prototype is established using ANSYS software. While neglecting the magnetic effect of end region of the motor, the 2-D FEA Boundary value conditions can be expressed as (1), and the finite element meshed model of SRM is shown in Fig.1.

$$\begin{cases} \dfrac{\partial}{\partial x}[\dfrac{1}{\mu}\dfrac{\partial A}{\partial x}]+\dfrac{\partial}{\partial y}[\dfrac{1}{\mu}\dfrac{\partial A}{\partial y}]=-J_c \\ A\big|_0 =0 \end{cases} \quad (1)$$

Where A is magnetic potential vector, μ is material magnetic permeability and J_c is the winding surface current density.

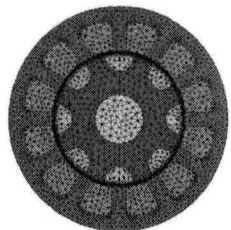

Figure 1.　FE Meshed model of the 12/8 SRM

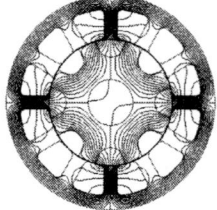

(a) unaligned position

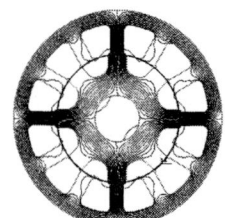

(b) aligned position

Figure 2.　Distributions of magnetic field in different rotor positions

The flux linkage $\psi(\theta,i)$ is a function of phase current and rotor position, and the magnetic vector potential A in each node and flux density B in each element of the flux field can be calculated by FEA, so the flux and flux linkage can be represented as

$$\phi(i,\theta) = l_{Fe}\int\vec{B}\bullet d\vec{S} = l_{Fe}\int\nabla\times\vec{A}\,dl \quad (2)$$

$$\psi(i,\theta) = N\phi(i,\theta) \quad (3)$$

Where l_{Fe} is the length of the motor in axis direction; N is the number of stator coils.

In this way, the flux linkage characteristics (ψ-θ-i) can be calculated using the 2-D FEM as discussed above, with the analysis conditions given in Table.2.

The results of FEA are shown in Fig.3 and Fig.4. The distributions of magnetic field in the aligned and unaligned positions are shown in Fig.3 and the calculated (ψ-θ-i) relationship curve is shown in Fig.4. As can be seen from the figures, the inner magnetic field appears great nonlinearity and the flux-linkage varied with both the current and rotor position.

TABLE II.　CONDITION OF FEM ANALYSIS

Excitation phase	one phase only
Excitation current	0-80 (A) every 4 (A)
Analysis range	0-22.5(deg) every 1.5 deg

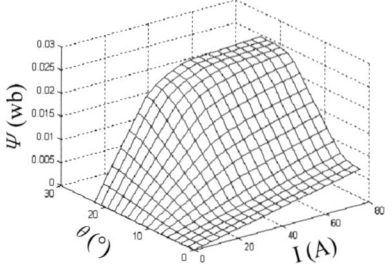

Figure 3.　Calculated (ψ-θ-i) relationship curve

III.　SUBREGIONAL MATHEMATICAL MODEL

In the operation process, there are two kinds of saturation (general saturation and local saturation) exist. The general saturation is caused by the phase current, which is the main saturation in SRM. When the phase current is lower than a certain limit, the flux linkage varies linearly with the current, but as the current increases further, the influence of saturation will become severely. In traditional mathematical models, the saturation is always ignored. In this part, a new sub-regional mathematical model cooperating with LM optimization method for modeling is developed. In this model, the modeling region is distributed according to the saturation current value and a periodical Fourier series is used as the estimated model.

A.　Subregional Model

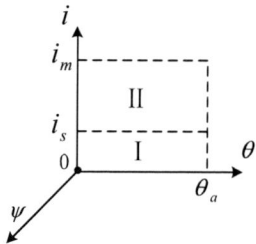

Figure 4.　Division of the flux linkage characteristitics

The flux-current characteristics can be divided into two regions as shown in Fig.4. The reference current is i_s. Thus, the proposed model for flux-linkage is characterized by (4)

$$\psi(\theta,i) = \sum_{n=0}^{k}\sum_{m=0}^{l}N_{1,n,m}i^{m+1}\cos(nN_r\theta) \quad (0\leq i\leq i_s)$$

$$= \sum_{n=0}^{k}\sum_{m=0}^{l}N_{2,n,m}i^{m+1}\cos(nN_r\theta) \quad (i\geq i_s) \quad (4)$$

978-1-4244-4782-4/10 $26.00 © 2010 IEEE　　1019

Where $N_{a,n,m}$. is the coefficients of Fourier series, when $0 \leq i \leq i_s$, a=1, for else a=2.

B. Levenberg-Marqurdt Optimization Method

Levenberg-Marqurdt(LM) optimization method is the integration of gradient descent algorithm and Guasi-Newton algorithm. The search direction of LM method is

$$S(X^{(k)}) = -(H^{(k)} + \lambda^{(k)} I)^{-1} J^T e \qquad (5)$$

$$X^{(k+1)} = X^{(k)} + S(X^{(k)}) \qquad (6)$$

At the beginning, λ get a big value, the training process is similar as the gradient descent algorithm. When the optimal point is closer, λ is descent to zero and $S(X^{(k)})$ is changed into Newton algorithm direction. Generally, when $f(X^{(k+1)}) < f(X^{(k)})$, λ is descent to some extent. In this algorithm, the Hessian matrix can be computed approximately as

$$H = J^T J \qquad (7)$$

Where J is the Jacobian matrix that contains first derivations of the estimated errors, e is error vector of the estimated model, λ is a non negative parameter and can be adjusted, I is the standard matrix.

Without computing the Hessian matrix directly, the LM algorithm possesses of very fast convergence capability, so it is quite useful in solving the nonlinear least squares curve fitting problem. .

C. Modeling Results And Analysis

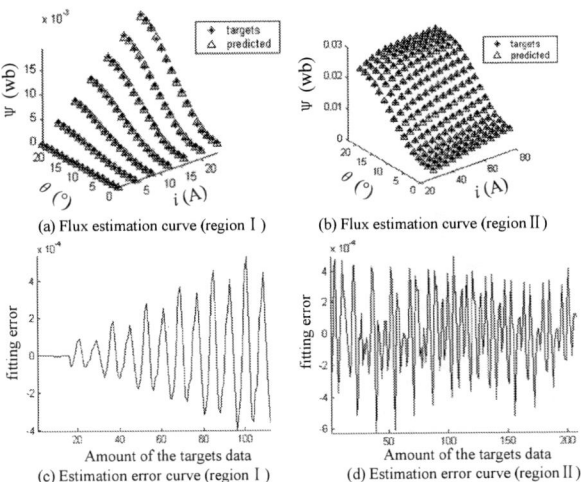

(a) Flux estimation curve (region I)   (b) Flux estimation curve (region II)

(c) Estimation error curve (region I)   (d) Estimation error curve (region II)

Figure 5. The modeling results

The validation results of the proposed mathematical model are shown in Fig.5. As can be seen from Fig.5(a) and (b), the flux linkage characteristic surface obtained by the proposed model can approach the target value smoothly and continuously. Considering the general saturation, the max estimation error in each estimation region is lower than 5e-4,

which is represented in Fig.5(c) and (d). Appropriately adjusting the k and l can make the flux model more accurately.

As is discussed above, the lower order Fourier series based flux linkage model is simple enough, cooperating with LM algorithm, the model can approximate real characteristics rapidly. Moreover, other dynamic parameters such as incremental inductance, BEMF factor and instantaneous torque can also be obtained from this analytical flux model, which make it more convenient to analyze the dynamic performance of SRM. But it still has some disadvantages. In order to enhance the accuracy, one should adjust the order of the Fourier series repetitiously to choose a better one for each estimation region, it is not self-adaptive but very time consume. Once the modeling task is finished, the coefficients of Fourier series are fixed and can not be adjusted adaptively.

IV. MODELING BASED ON ANFIS

A. Adaptive Neural Fuzzy Inference System

The adaptive neural fuzzy inference system(ANFIS) is implemented in the framework of adaptive networks using a hybrid learning procedure, whose membership function parameters are tuned using a back-propagation algorithm combined with a least square method, so ANFIS combines the benefits of BPNN and fuzzy logic algorithm. ANFIS is capable of dealing with uncertainty and imprecision of human knowledge, which has self-organized ability and inductive inference function to learn from the data, thus, it is no doubt an excellent function approximation tool.

ANFIS is a multilayer feed-forward network. Each node of the network performs a particular function on incoming signals as well as a set of parameters pertaining to this node. A simple multilayer architecture of ANFIS is shown in Fig.6.

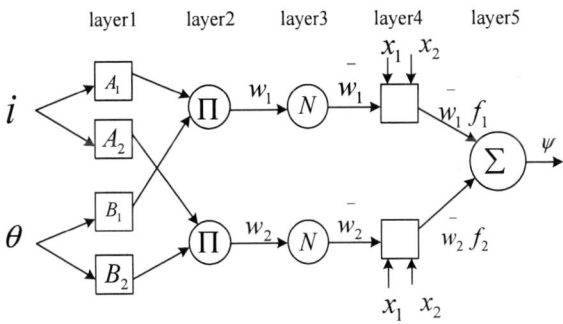

Figure 6. A simple multilayer architecture of ANFIS

ANFIS implements a first-order Sugeno-style fuzzy system, which can be expressed as

Rule i: If x is A_i and y is B_i, then $f_i = p_i x + q_i y + r_i$.

Where A_i and B_i are the fuzzy sets in the antecedent, and p_i, q_i and r_i are the design parameters that are determined by the training process.

As shown in Fig.6, the ANFIS is formed with five layers. The detailed explanation of each layer is as follows.

Layer 1: Every node i in this layer is a square node with a node function

$$O_{1,i} = \mu_{Ai}(x_1), i = 1,2 \tag{8}$$

$$O_{1,i} = \mu_{B(i-2)}(x_2), i = 3,4 \tag{9}$$

Where $x_i(i=1,2)$ is the input of the node i; A_i or $B_{(i-2)}$ are the linguistic label correlated with the node function. While $O_{1,i}$ is the degree of membership function of the fuzzy set A. Usually we choose $\mu_{Ai}(x)$ to be bell-shaped with maximum equal to 1 and minimum equal to 0, such as

$$\mu_{Ai}(x) = \exp\left\{ -\left(\frac{x - c_i}{a_i} \right)^2 \right\} \tag{10}$$

Where (a_i, c_i) is the parameter set. As the values of these parameters change, the bell shape functions vary accordingly. In this paper, function (10) is chosen as the input membership function.

Layer 2: Every node of this layer calculates the firing strength of a rule by multiplying the degree of membership function. For instance

$$O_{2,i} = \mu_{Ai}(x_1) \times \mu_{Bi}(x_2), i = 1,2 \tag{11}$$

Layer 3: Every node in this layer is a circle node labeled N. The ith node calculates the ratio of the ith rule's firing strength to the sum of all rules' firing strengths:

$$O_{3,i} = \bar{w}_i f_i = \frac{w_i}{w_1 + w_2}, i = 1,2 \tag{12}$$

Layer 4: Every node i of this layer is the adaptive node, the output is

$$O_{4,i} = \bar{w}_i f_i = \bar{w}_i(p_i x_1 + q_i x_2 + r_i), i = 1,2 \tag{13}$$

Layer 5: The node of this layer computes the overall output as the summation of all the incoming signals.

$$O_{5,i} = \sum_i \bar{w}_i f_i = \frac{\sum_i w_i f_i}{\sum_i w_i}, i = 1,2 \tag{14}$$

B. ANFIS Learning Algorithm

In order to minimize the error between the ANFIS output and the targets value, in this paper, a hybrid algorithm is employed, which is used for identification of the premise and consequent parameters. In the forward pass of learning algorithm, functional signals go forward till layer 4 and the consequent parameters are identified by the least-squares(LS) estimator. In the backward pass of the hybrid learning algorithm, the error rates propagate backward and the premise parameters are updated by the gradient descent.

C. Flux Modeling Based On ANFIS

ANFIS provided a simple way of modeling the flux linkage which can take into consideration of the static and dynamic effects of the motor. In this paper, the ANFIS under consideration has two-input rotor position (θ), current (i), and one-output flux-linkage (ψ). The input-output training data is chosen from the static magnetization characteristics calculated by FEA. The corresponding multilayer architecture of ANFIS is shown in Fig. 6, which is composed of the parameters as listed in table.3. By using the hybrid learning algorithm, the error between the desired output ψ and the actual output of ANFIS are evaluated until the estimation goal is reached. After training, the mapping of the estimated flux linkage characteristics can be obtained.

TABLE III. ARCHITECTURE PARAMETERS OF THE ANFIS MODEL

Number of membership functions for each input	7
Number of fuzzy rules	49
Number of nodes	131
Number of linear parameters	147
Number of nonlinear parameters	42
Total number of parameters	189
Number of training data pairs	336

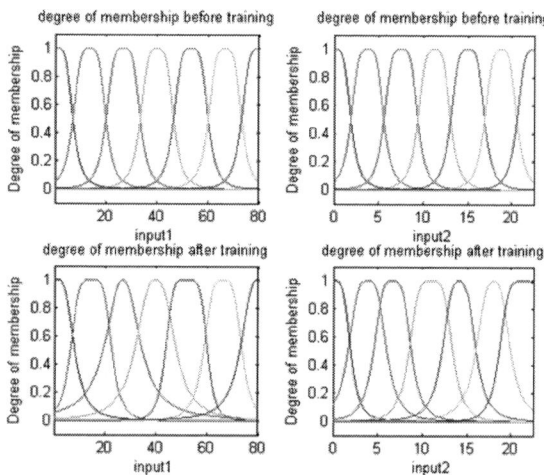

Figure 7. Membership function adjustment of the input variables

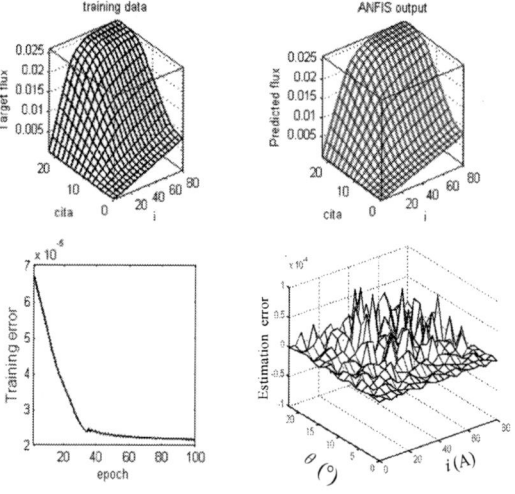

Figure 8. The training results based on ANFIS

978-1-4244-4782-4/10 $26.00 © 2010 IEEE

The ANFIS training results are shown in Fig.7 and Fig.8. Fig. 7 illustrates the self adjustment process of the inputs membership functions. As shown in Fig. 8, the root mean square error is descent to about 2.1e-5 after 100 epochs. The estimated flux-linkage mapping surface is compared to the static magnetization characteristics surface, and the estimation error is shown Fig.8. It is clearly that the ANFIS model can approximate the real flux characteristics with great accuracy.

V. MODELING BASED ON NEURAL NETWORK

As is discussed before, modeling of the flux linkage characteristics can be grouped as function approximation problems. Artificial neural networks(ANNs) have proven to be a powerful tool in various applications, including adaptive dynamic modeling, adaptive control, optimization and expert systems, which make them idea candidates to modeling the flux-linkage characteristics. Back-Propagated Networks(BP) and Radial Basis Function Networks (RBF) are both well-known developments of the Delta rule for single layer networks. Both can learn arbitrary mappings or classifications. In this part, the modeling performance of these two networks is investigated.

A. Modeling Based On BPNN

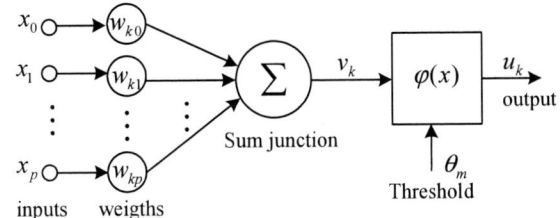

Figure 9. Structure of a neuron of BPNN

BPNN is composed of three layers: the input layer, the hidden layer and the output layer. As illustrated in Fig.9 and Fig 10, each neuron receives signals from the neurons in the previous layer, and each of those signals is multiplied by a separate weight value. The weighted inputs are summed, and passed through a limiting function which scales the output to a fixed range of values. The output of the limiter is then broadcast to all of the neurons in the next layer.

In this paper, a three-layer BPNN with two hidden layers is chosen for modeling. The architecture and training parameters of the proposed network is given in Table4, and the topology is shown in Fig10. The static magnetization characteristics calculated by FEA is also used as the training data. The detailed analysis of the input-output relationship of each layer and the training algorithm are as follows

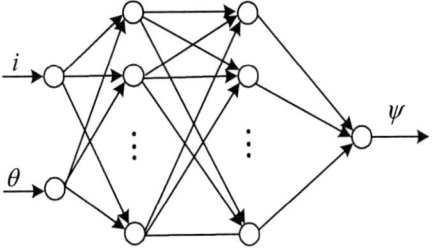

Figure 10. Topology of the proposed BPNN

TABLE IV. THE ARCHITECTURE AND TRAINING PARAMETERS

Number of hidden layer	2
Number of neuron in the hidden layer 1	8
Number of neuron in the hidden layer 2	8
Hidden layer neuron activation function	sigmoid
Output layer neuron activation function	linear
Training goal	2e-5
Learning algorithm	LM

The output of the first hidden layer is

$$
\begin{bmatrix} u_1 \\ u_2 \\ \vdots \\ u_r \end{bmatrix} = \varphi \left\{ \begin{bmatrix} w_{11}^{(1)} & w_{12}^{(1)} \\ w_{21}^{(1)} & w_{22}^{(1)} \\ \vdots & \vdots \\ w_{r1}^{(1)} & w_{r2}^{(1)} \end{bmatrix} \begin{bmatrix} k_1 x_1 \\ k_2 x_2 \end{bmatrix} + \begin{bmatrix} b_1^{(1)} \\ b_2^{(1)} \\ \vdots \\ b_r^{(1)} \end{bmatrix} \right\} \quad (15)
$$

The output of the second hidden layer is

$$
\begin{bmatrix} v_1 \\ v_2 \\ \vdots \\ v_l \end{bmatrix} = \varphi \left\{ \begin{bmatrix} w_{11}^{(2)} & w_{12}^{(2)} & \cdots & w_{1r}^{(2)} \\ w_{21}^{(2)} & w_{22}^{(2)} & \cdots & w_{2r}^{(2)} \\ \vdots & \vdots & \vdots & \vdots \\ w_{l1}^{(2)} & w_{l2}^{(2)} & \cdots\vdots & w_{lr}^{(2)} \end{bmatrix} \begin{bmatrix} u_1 \\ u_2 \\ \vdots \\ u_l \end{bmatrix} + \begin{bmatrix} b_1^{(2)} \\ b_2^{(2)} \\ \vdots \\ b_l^{(2)} \end{bmatrix} \right\}
$$

(16)

The network output ψ is

$$
\psi = f \left(\begin{bmatrix} w_1^{(3)} & w_2^{(3)} & \cdots & w_l^{(2)} \end{bmatrix} \begin{bmatrix} v_1 \\ v_2 \\ \vdots \\ v_l \end{bmatrix} + \begin{bmatrix} b^{(3)} \end{bmatrix} \right) \quad (17)
$$

Where, x_1, x_2 are the input variables, k_1, k_2 are the unitary coefficients, $w_{rk}^{(1)}$ are the input weights, $b_r^{(1)}$, $b_l^{(2)}$ is the bias. φ is the sigmoid function. $w_{lr}^{(2)}$ are the weights in the second layer. $w_l^{(3)}$ are the output weights, $b^{(3)}$ is the output bias, f is a linear function.

The error back propagation(BP) algorithm is used in this application, which achieves minimization between the target function and the actual output of the ANN under training. The output layer error is computed first, and then it is back propagated through ANN to compute the equivalent error of each hidden node. The ANN weights and bias are adjusted by LM back propagation algorithm, since its convergence is faster than other conventional back-propagation methods. Corresponding simulations are done via Matlab/simulink. The simulation results are shown in Fig.11. The mean square error(MSE) vs. training epochs for LM algorithm is shown in Fig.11(a), which shows that the training goal 2e-5 is met after 200 epochs. The BPNN mapping surface and the estimation results is shown in Fig.11(b) and (c), which indicated that the ANN model can smoothly approximate the flux-linkage characteristics with only little estimation error as shown in Fig11(d).

978-1-4244-4782-4/10 $26.00 © 2010 IEEE

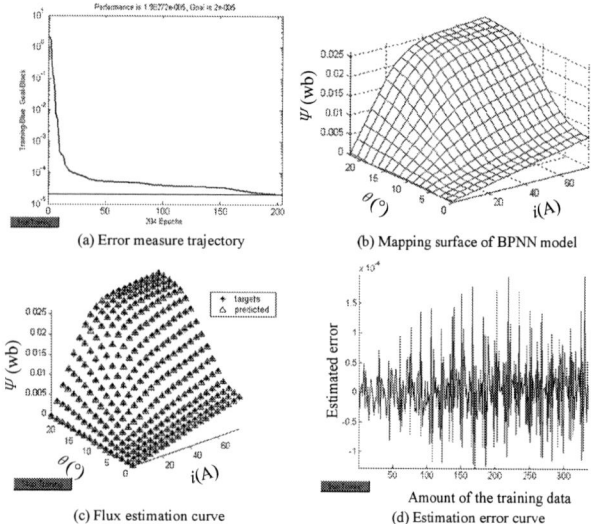

(a) Error measure trajectory

(b) Mapping surface of BPNN model

(c) Flux estimation curve

(d) Estimation error curve

Figure 11. The modeling results of BPNN

B. Modeling Based On RBF

Radial basis functions (RBF) is embedded in a two layer neural network, where each hidden unit implements a radial activated function. The output units implement a weighted sum of hidden unit outputs. Their excellent approximation capabilities have been studied in many literatures. Like BPNN, RBF can learn arbitrary mappings. The primary difference is in the hidden layer. RBF hidden layer units have a receptive field which has a centre, that is, a particular input value at which they have a maximal output. Their output tails off as the input moves away from this point. The structure of a unit of the RBF layer is shown in Fig.12(b). Generally, the hidden unit function is a Gaussian

$$\varphi_k(X_j) = \exp(-\frac{\left\|X_j - c_j\right\|^2}{2\sigma_k^2}) \qquad (18)$$

The output layer implements a weighted sum of hidden-unit outputs

$$f(X_j) = \sum_{k=1}^{n} \omega_k \varphi_k(X_j) \qquad (19)$$

Where X_j is the ith input vector; c_k is center of the kth hidden layer node; σ_k is the width of the kth hidden layer node.

The training of the network consist of two parts: the first is training the weights in RBF layer using unsupervised learning method; the second is training the weights of output layer using supervised learning method. The weights of RBF layer is trained by adjust the weighs vector w_{ij} to convergent with the input vector p_j, to let the RBF output approach **1** in each $w_{ij} \rightarrow p_j$ point. When the network is running, each nerve cell will output the training input data according to the closer extent between the input vectors and the weights vectors. In this way, if the input vector far from the weights vectors, the output of the RBF layer is close to zero, so the effect of these

output can be neglected in linear layer. While other outputs will approach to 1, and the weighted sum of these values is the output of the whole RBF network. The input-output relationship in RBF layer is represented in Fig.13. The network output is compared to the target output, and a mean-squared error signal is calculated. By employ the least square method, the weights of the output layer can be modified.

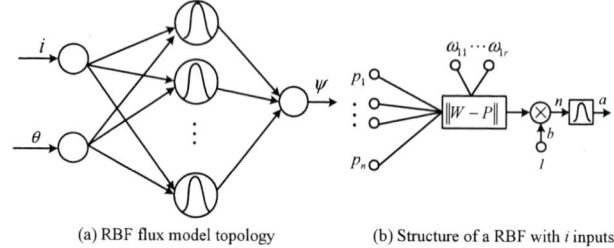

(a) RBF flux model topology

(b) Structure of a RBF with i inputs

Figure 12. The topology of RBF flux linkage model

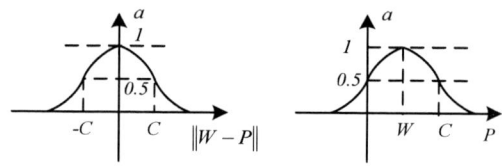

Figure 13. The input/output relationship in RBF layer

As is discussed above, theoretically, if the RBF layer has enough nerve cells, the RBF network can approximate any nonlinear functions with any expected precision. In this section, a flux linkage model based on RBF is investigated. The topology of the network is shown in Fig.12(a), with the phase current and rotor position as inputs, the flux linkage as output. The training data is obtained from the static magnetization characteristics, without any pretreatment of the sample data, the RBF can be trained immediately. Observing from Fig.14, the RBF predicted flux-linkage can well approximate the targets value, while the estimation errors is lower than 4e-16 as shown in Fig14 (b), which is much better than any other modeling algorithm investigated in above sections.

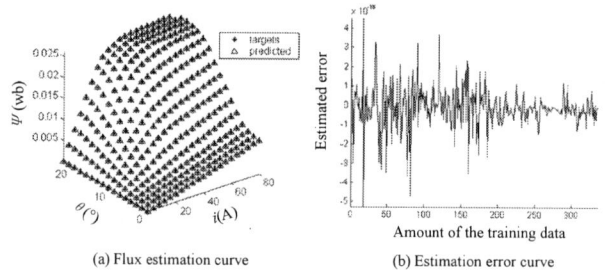

(a) Flux estimation curve

(b) Estimation error curve

Figure 14. The modeling results of RBF

VI. ANALYSIS OF THE MODELING PERFORMANCE

A. Comparison of the Modeling Methods

The simulation results obtained in above sections show that all of the modeling algorithms can accurately modeling of the flux-linkage characteristics. But for different application circumstance, each method has its own merits and demerits. A comprehensive comparison of them is summarized in the following table. Based on the modeling results, it is clearly that, though the proposed sub regional model is simple enough compared to traditional analytical models, it is still suffered from its convergent rate, too much time consumed in this process making it not suitable for real-time control. While the others belong to the category of intelligent control methods, in approximation applications, they get some commonness. ANFIS combines the benefits of BPNN and fuzzy logic algorithm, the convergence rate and generalization capabilities are all better than normal BPNN. However, BPNN always suffer from a long training time and easily getting into local value. While RBF overcomes these problems, the network not only possesses locally approximating capability, but also the optimal approximating performance. The training process is simple, while the convergence rate is quite fast, it is no doubt that RBF is the best modeling algorithm for SRM modeling.

TABLE V.   COMPARISON OF THE PROPOSED MODELING ALGORITHMS

Contents	Subregional	ANFIS	BPNN	RBF
Model free	not	yes	yes	yes
Precision	good	better	better	Best
Modeling complexity	low	lower	lower	lowest
Memory	more	less	more	more
Time consume	263.254s	0.297s	3.078s	0.125s
Robustness	low	high	high	high
Adaptability	not	yes	yes	yes
Convergence mode	whole	locally	whole	locally
Generalization	poor	better	good	better

B. Dynamic Simulation

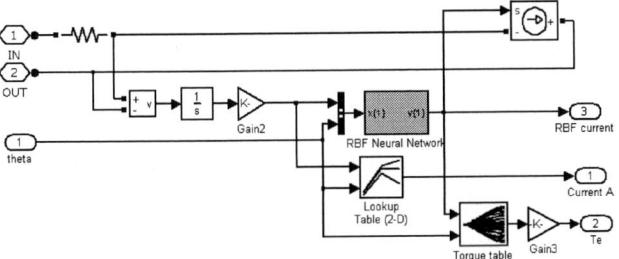

Figure 15   One phase simulation model with RBF

As is concluded in part A, RBF is considered to be the best modeling method for SRM modeling. In the following sections, the dynamic performance of RBF model is investigated. For SRM dynamic simulation, a converse flux model $i(\psi,\theta)$ based on RBF network is established. The training data is also obtained from the static magnetization characteristics. As the training is over, the estimation error is shown in Fig.16 (a). As can be seen from which, the max estimation error is lower than 5×10^{-12} that can be neglected. Fig.16(b) shows the RBF current mapping surface. By replacing the traditional current table module, a new SRM dynamic simulation system based on RBF current model is obtained, which is composed of power converter module, phase winding module, current hysteresis module, RBF current module and position estimation module etc. The one phase simulation model is shown in Fig.15 and the dynamic simulation is implemented in Matlab/Simulink environment. Fig.17 depicts the simulated phase current during the steady

state operation around the rotational speed of 10000rpm, with a DC voltage of 90V. As shown in which, the peak phase current reached about 5A. With the proposed RBF current model the phase current can be smoothly estimated in real time simulation process.

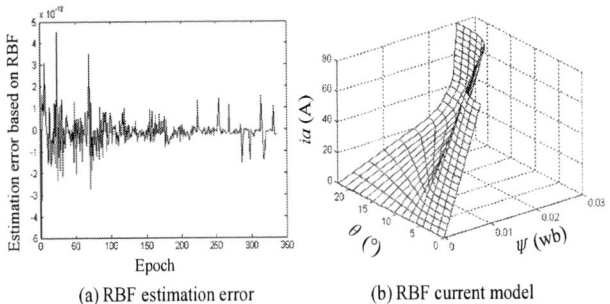

(a) RBF estimation error   (b) RBF current model

Figure 16.   The modeling results of RBF

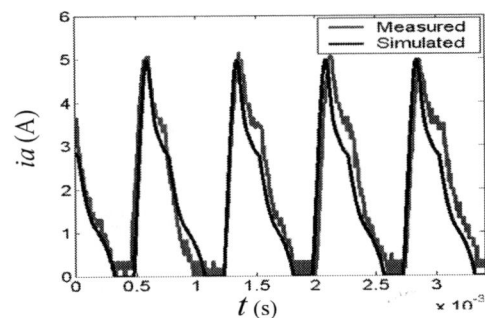

Figure 17   Simulated and measured phase current during the steady state operation around n=10000 rpm, I_{chop}=6A, Vdc=90V

C. Experiment Validation

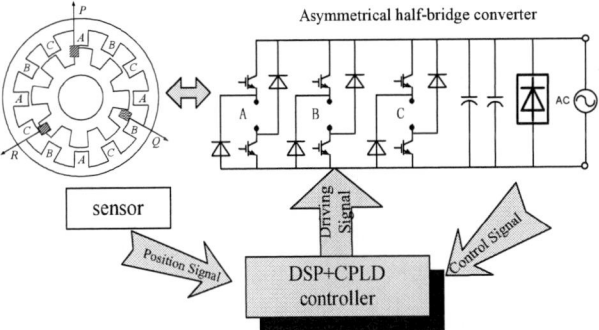

Figure 18.   The composition of the SRM drive system

In this section, the accuracy of the new dynamic simulation model is verified by comparison of simulated and the measured phase current waveforms. The block diagram of the SRM drive system is shown in Fig.18, which consists of SRM motor、DSP+CPLD hybrid controller and the power converters etc. In this paper, the prototype is a three phases 12/8-structure Double-salient motor, the specification of which is given in table 1; the power convert is an three-phase asymmetrical half bridge, with two IGBTs and two fast diodes per phase; and the position signal can be obtained by three optical electrical position sensors.

978-1-4244-4782-4/10 $26.00 © 2010 IEEE

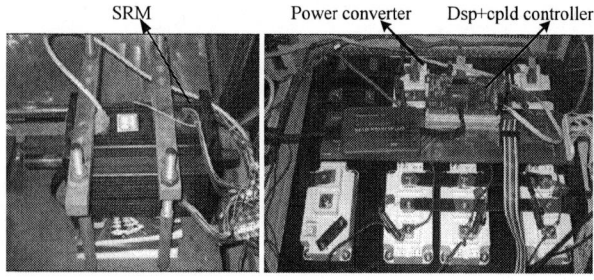

Figure 19   Experimental system

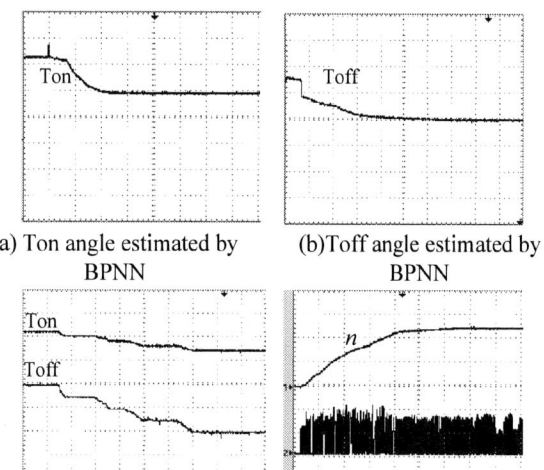

(a) Ton angle estimated by BPNN   (b)Toff angle estimated by BPNN

(c) Ton and Toff angles estimated by 1-D look up table   (d) rotor speed waveform

Figure 20   The waveforms of ton and toff angles

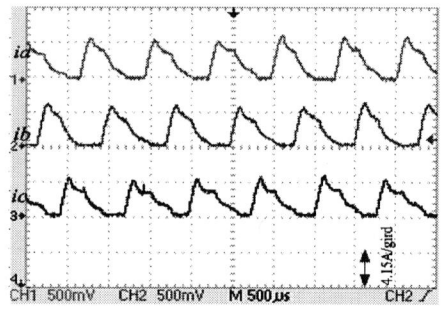

Figure 21   Measured phase current with
n=10000 rpm, I_{chop}=6A ,Vdc=90V

The experiment is implemented in the hardware system as shown in Fig.19, with a DC voltage of 90V and the rotor speed of 10000rpm. In this study, a hybrid control strategy is employed which combines the current chopping control(CCC) and angle position control(APC) strategies. The chopping reference current is fixed in 6A, while the turn on and turn off angles are adjusted respectively by using a single hidden layer BPNN with one input and one output, which was trained using the optimal turn on-off angle vs. rotor speed data pairs, where the traditional one-dimension look-up table method is replaced. The online-adjustment waveform of ton-off angle based on BPNN and 1-D loop-up table are shown in Fig.20. As can be seen, the 1-D look-up table obtains the ton-off angle step by

step that is not continuously, which may result in rotor speed fluctuation severely, however, the BPNN algorithm overcomes the problem, which can estimate the ton-off angle smoothly and continuously. The rotational speed response waveform is shown in Fig.20(d). Fig.21 shows the measured phase current during the steady state operation around the rotor speed of 10000rpm, where the peak phase current is about 5.05A. As compared with the simulated phase current, in Fig.17 the measured waveform is in well agreement with the simulation results, which fully validated the accuracy of the new dynamic simulation model based on RBF.

VII. CONCLUSIONS

Three different algorithms for flux-linkage modeling are investigated and compared, which provides a theory reference for selecting a modeling algorithm. As discussed in this paper, RBF is no doubt the best method for SRM modeling. Moreover, a new dynamic simulation strategy based on RBF model is developed for analyzing the dynamical performance of SRM. Simulation and experimental results verified that the new simulation model can reflect the practical operation characteristics of SRM more accurately than traditional dynamic simulation methods. The nonlinear modeling methods presented will be quite useful for analyzing and optimizing the control strategy of SRM drive systems.

REFERENCES

[1]   K.M. Rahman, S. Gopalakrisnan, B. Fahimi, A.V. Rajarathnam, M. Ehsani. "Optimized torque control of switched reluctance motor at all operational regimes using neural network." *IEEE Trans on industry applications*, No.3. vol. 37, pp. 904-913, May 2001

[2]   D. A.Terry and J.H.Lang. "Modeling a nonlinear variable-reluctance machine model implemented in the saber circuit simulator", electric machine and power systems,1996,vol.24,pp.199-209

[3]   D. G. Manzer, V. Matthew, J.S. Thorp. "Variable reluctance motor characterization", IEEE Transaction on Industrial electronics, vol.36, No.1, Feb.1989,PP.56-63

[4]   S. Cao; Tseng K.J; "a new method for accurate analytical modeling of switched reluctance motor", Power electronic drives and energy systems for industrial growth, 1998. proceedings ,vol 2,1-3 Dec.1998,pp.540-545

[5]   D. A. Cheok, N. Ertugrul, "Use of fuzzy logical for modeling, estimation, and prediction in switched reluctance motor drives". IEEE Trans. on industrial electronics. Vol. 46. No.6, Dec 1999.pp 1207-1224

[6]   D. A. Cheok, N. Ertugrul, "High robustness and reliability of fuzzy logic based position estimation for sensorless switched reluctance motor drives." *IEEE Trans. on power electronics*, vol. 15, No 2, pp: 319-334. Mar 2000.

[7]   E.Mese, and D.A. Torry, "An approach for sensorless position estimation for switched reluctance motor using artificial neural networks," *IEEE Trans. on power electronics*. No.1. pp:66-75. Jan 2002

[8]   J. M. Lian, Y.G. Lee, S. D. Sudhoff, S.H. Zak, "Self-organizing radial basis function network for real time approximation of continuous-time dynamical systems", IEEE Trans. on neural networks. Vol 19, issue 3, pp:460-474, March 2008.

[9]   S. Paramasivam, S. Vijayan, M. Vasudevan, R. Arumugam, R. Krishnan, "Real-Time Verification of AI Based Rotor Position Estimation Techniques for a 6/4 Pole Switched Reluctance Motor Drive." *Magnetics, IEEE Trans.on*, Vol 43, Issue 7, pp:3029-3222, July 2007

Simplified Synchronous Reference Frame Control of the Three Phase Grid Connected Inverter

Abad Lorduy, Antonio Lázaro, Andrés Barrado, Cristina Fernández, Isabel Quesada and Carlos Lucena.
Carlos III de Madrid University, Electronic Power Department.
Power Electronic Systems Group.
Avda. Universidad, 30; 28911, Leganés, Madrid, SPAIN
E-mail: alorduy@ing.uc3m.es

Abstract— **This paper presents the optimized control of a three-phase grid connected inverter, considering as basis of analysis the observation of the dynamics of the system with two inputs and two outputs (TITO SYSTEM).This model allows the design of the inverter control in a direct way, by means of two PI controllers to control the instantaneous active and reactive power. By means of the control of the instantaneous active power p, the inverter duty cycle, *dd*, is controlled in a direct way. By means of the control of the instantaneous reactive power q, the inverter duty cycle, *dq*, is controlled in a direct way. Thus, the implementation of the power control becomes easier, with DC signals, obtained with PI controllers and with a very simple algorithm regarding calculations. Also, a phase locked PLL is used for achieve the synchronization of the system with the grid frequency. The proposed control is compared with the conventional Synchronous Reference Frame Control (SRFC), showing the advantages of this new simplified control.**

I. Introduction

Nowadays there are applications that require a very fast dynamic response in order to improve the system efficiency, like in the bidirectional control of three-phase inverters, regenerative break of motors, power active filters, etc [1] [2].

In general, an optimum control of the three-phase inverter must have the following characteristics: assure the stability of the system for most working points, fast and non-oscillating dynamic response, low computing cost algorithm, ability to adapt to different applications [1] [2].

In order to manipulate the system plant according the required specifications, it is necessary to represent it by means of the appropriate mathematical models. The three-phase inverter can be modelled like a TITO SYSTEM (two inputs two outputs) [8] [9] [10] [11].

The transient behaviour of the current injected into the grid of the open-loop response of a three-phase grid-connected inverter, corresponds typically to a second order underdamped system. To control effectively the closed-loop three phase inverter, the key point is to eliminate as quick as possible the existing underdamped oscillations of the transient response, keeping of course the specified reference.

The proposed control has a structure similar to a direct power control DPC [6] [7], but with different blocks in the implementation; it is possible to say, that the proposed control is a linear DPC , achieving very fast dynamic responses due to the almost exact cancellation of the transient states.

The proposed control is simpler, with two proportional-integral PI and has a mathematics base for generating a new Simplified Synchronous Reference Frame Control SSRFC, without use of the cross coupling feedforward terms.

II. Synchronous Reference Frame Control

The SRFC of the three phase grid connected inverter is based on the transformation of the AC signals of the abc coordinates system into a synchronous reference frame. Thus, the observing of the system is done in DC signals. In the case of grid connected inverters, the reference frequency to which the frame will be synchronized is that of the grid.

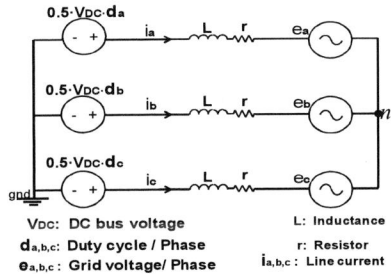

Figure 1. Average model in abc coordinates for the Three Phase Grid Connected Inverter.

In Figure 1 it has been represented the model of the grid connected three phase inverter in abc coordinates. The dynamic expressions that describe the system behavior in SRFC are obtained from the previous model [3] [4] [5]. The voltage equations for each phase are included in (1):

$$L \cdot \frac{di_a}{dt} = 0.5 V_{DC} \cdot d_a - r \cdot i_a - e_a$$

$$L \cdot \frac{di_b}{dt} = 0.5 V_{DC} \cdot d_b - r \cdot i_b - e_b \qquad (1)$$

$$L \cdot \frac{di_c}{dt} = 0.5 V_{DC} \cdot d_c - r \cdot i_c - e_c$$

Now, expressed in matrix form:

$$\frac{d}{dt}\left([I_{abc}]\right) = \frac{V_G}{L}\cdot[D_{abc}] - \frac{r}{L}\cdot[I_{abc}] - \frac{1}{L}\cdot[E_{abc}] \quad (2)$$

The transformation of the dynamic equations into the synchronous reference frame is achieved by means of the abc/dq and dq/abc transformation matrixes:

$$[I_{abc}] = T^{-1}\cdot[I_{dqo}] \qquad [D_{abc}] = T^{-1}\cdot[D_{dqo}]$$

$$[E_{abc}] = T^{-1}\cdot[E_{dqo}] \qquad T\cdot\frac{d}{dt}\left(T^{-1}\right) = -\begin{bmatrix} 0 & -\omega & 0 \\ \omega & 0 & 0 \\ 0 & 0 & 0 \end{bmatrix}$$

Where T is the abc/dq transformation matrix.

Applying this transformation to (2) and after some algebraic manipulation, the equations (3) and (4) are obtained.

$$\frac{d}{dt}\left[I_{dqo}\right] = -T\cdot\frac{d}{dt}\left(T^{-1}\right)\cdot\left[I_{dqo}\right] + \frac{0.5V_{DC}}{L}\cdot\left[D_{dqo}\right] - \frac{r}{L}\cdot\left[I_{dqo}\right] - \frac{1}{L}\cdot\left[E_{dqo}\right] \quad (3)$$

$$\frac{d}{dt}\begin{bmatrix} i_d \\ i_q \\ i_0 \end{bmatrix} = -\begin{bmatrix} 0 & -\omega & 0 \\ \omega & 0 & 0 \\ 0 & 0 & 0 \end{bmatrix}\cdot\begin{bmatrix} i_d \\ i_q \\ i_0 \end{bmatrix} + \frac{0.5V_{DC}}{L}\cdot\begin{bmatrix} d_d \\ d_q \\ d_0 \end{bmatrix} - \frac{r}{L}\cdot\begin{bmatrix} i_d \\ i_q \\ i_0 \end{bmatrix} - \frac{1}{L}\cdot\begin{bmatrix} e_d \\ e_q \\ e_0 \end{bmatrix} \quad (4)$$

From (4), the two usual synchronous d-q frame equations are derived, see (5) and (6)

$$L\cdot\frac{di_d}{dt} = L\omega\cdot i_q + 0.5V_{DC}\cdot d_d - r\cdot i_d - e_d \quad (5)$$

$$L\cdot\frac{di_q}{dt} = -L\omega\cdot i_d + 0.5V_{DC}\cdot d_q - r\cdot i_q - e_q \quad (6)$$

Considering that the d-axes is in phase with the grid voltage vector, thus, ed=Vg and eq=0. Where, Vg is the magnitude of grid voltage. Then, assuming small signal and a constant DC bus voltage, the equations (7) and (8) are obtained.

$$(L\cdot s + r)\cdot\hat{i}_d = 0.5V_{DC}\cdot\hat{d}_d + L\omega\cdot\hat{i}_q - \hat{V}_g \quad (7)$$

$$(L\cdot s + r)\cdot\hat{i}_q = 0.5V_{DC}\cdot\hat{d}_q - L\omega\cdot\hat{i}_d \quad (8)$$

The sets of equations (5), (6) and (7), (8) correspond, respectively, to the average and small signal model of the grid connected three phase inverter in synchronized d-q axes. On these equations, it can be observed that i_d and i_q are coupled through the terms Lωi_q and Lωi_d, [3] [4] [5]. Traditionally, in the control design, current feedforwards loops are used to compensate this coupling. However, it implies a significant increase of the computational cost of the control algorithm. Additionally, the physical implementation of the feedforward loops implies other disadvantages. Some of them are listed below:

- The Lω value used within the control algorithm is an estimated value. So it can differ from the actual one, since ω can vary slightly and the L value is not accurately known.

- Assumed that the estimated value of Lω is close to the actual value, then, an accurate compensation of the coupled terms will be achieved. However, the compensation of the coupled terms during the transient response could not be so good due to the fact that the system frequency varies.

- The determination of the switching of the inverter, due to its digital nature, will provide an average voltage which is only an approximation to that needed for a perfect compensation.

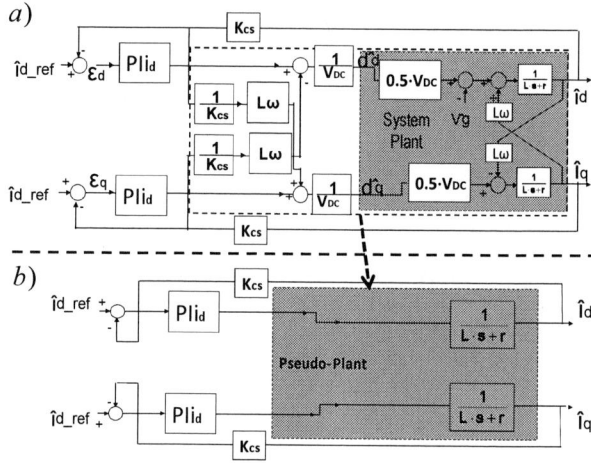

Figure 2. Model of the dq conventional control.

In Figure 2a the conventional model is given, including the crossed compensation loops needed to develop the algorithm of the d-q control. In Figure 2b, perfect coupling compensation *is assumed*, therefore, is generated a corresponding Pseudo-plant (nominal plant of design). This model in Figure 2b is used in d-q control to determine the PI controllers. *The good implementation hardware* of the control model of Figure 2a, should behave as the model of Figure 2b.

III. TITO SYSTEM MODEL THREE PHASE INVERTER

The simplified d-q control proposed in this paper is based on redefining the model of the three phase grid connected inverter. The mathematical expressions of the proposed model can be generated from equations (7) and (8) with the following algebraic manipulations: First, (7) must be multiplied by -Lω and (8) by (L·s+r), adding the results, is obtained the equation (9). Second, (7) must be multiplied by (L·s+r), and (8) by Lω, adding the results, is obtained the equation (10).

Isolating variables i_d and i_q in (9) and (10), the matrix expression (11) can be obtained, which represents a new way of observing the system. In this model, the cross coupling terms of current do not appear.

978-1-4244-4782-4/10 $26.00 © 2010 IEEE   1027

$$0.5 \cdot V_{dc} \cdot \hat{d}_q \cdot (L \cdot s + r) - 0.5 \cdot V_{dc} \cdot \hat{d}_d \cdot L\omega + \hat{V}_g \cdot L\omega = [(Ls + r)^2 + (L\omega)^2] \hat{i}_q \quad (9)$$

$$0.5 \cdot V_{dc} \cdot \hat{d}_q \cdot L\omega + 0.5 \cdot V_{dc} \cdot \hat{d}_d \cdot (L \cdot s + r) - \hat{V}_g (L \cdot s + r) = [(L \cdot s + r)^2 + (L\omega)^2] \hat{i}_d \quad (10)$$

$$\begin{bmatrix} \hat{i}_d \\ \hat{i}_q \end{bmatrix} = \begin{bmatrix} \dfrac{0.5 V_{dc}(L \cdot s + r)}{(L \cdot s + r)^2 + (L\omega)^2} & \dfrac{0.5 V_{dc} \cdot L\omega}{(L \cdot s + r)^2 + (L\omega)^2} \\ \dfrac{-0.5 V_{dc} \cdot L\omega}{(L \cdot s + r)^2 + (L\omega)^2} & \dfrac{0.5 V_{dc}(L \cdot s + r)}{(L \cdot s + r)^2 + (L\omega)^2} \end{bmatrix} \begin{bmatrix} \hat{d}_d \\ \hat{d}_q \end{bmatrix} + \begin{bmatrix} \dfrac{-(L \cdot s + r)}{(L \cdot s + r)^2 + (L\omega)^2} \\ \dfrac{L\omega}{(L \cdot s + r)^2 + (L\omega)^2} \end{bmatrix} \hat{V}_g \quad (11)$$

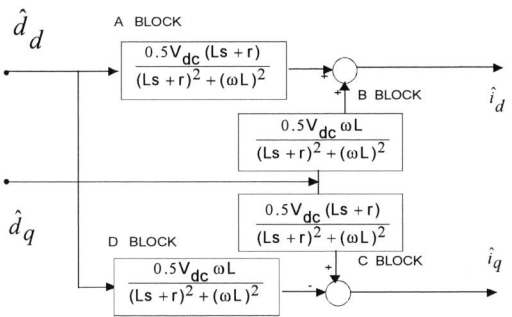

Figure 3.   Small signal model (TITO SYSTEM).

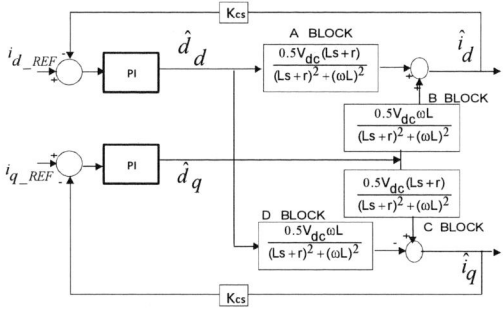

Figure 4.   Closed-loop of the proposed control.

The analysis of the proposed control is carried out assuming constant grid voltage and constan DC bus voltage. Thus, the inverter model corresponds to that represented in Figure 3 and equation (12).

$$\begin{bmatrix} \hat{i}_d \\ \hat{i}_q \end{bmatrix} = \begin{bmatrix} \dfrac{0.5 V_{dc}(L \cdot s + r)}{(L \cdot s + r)^2 + (L\omega)^2} & \dfrac{0.5 V_{dc} \cdot L\omega}{(L \cdot s + r)^2 + (L\omega)^2} \\ \dfrac{-0.5 V_{dc} \cdot L\omega}{(L \cdot s + r)^2 + (L\omega)^2} & \dfrac{0.5 V_{dc}(L \cdot S + r)}{(L \cdot s + r)^2 + (L\omega)^2} \end{bmatrix} \begin{bmatrix} \hat{d}_d \\ \hat{d}_q \end{bmatrix} (12)$$

This model does not present coupling between i_d and i_q in the system output, but each of the current, depend of the two duty cycles (*dd* and *dq*). This small signal model is a classical two input two output model, known as TITO SYSTEM.

It is reported in literature the difficulty of the control design in a direct way for this kind of structure TITO System [8] [9] [10] [11]. The solutions proposed in the literature, are usually specific for a given application.

However, given the small signal model of the inverter, proposed in this paper, the form of the internal transfer functions and the ortogonality of the duty cycles *dd* and *dq*, provides an easier way for the control design. The detailed process for the control design is given along the following section.

IV.   SIMPLIFIED SYNCHRONOUS REFERENCE FRAME CONTROL

The closed loop block diagram of the simplified control proposed is given in Figure 4. It can be seen that the regulation is carried out directly by means of two PI controllers that provides the signals *dd* and *dq* form the regulation of i_d and i_q, respectively.

Although each of the currents (i_d and i_q), depend of both duty cycles (see (9) and (10)), the control is designed directly respect of an alone duty cycle, due to the fact that *dd* and *dq* are orthogonal to each other. So, *dd* and *dq* are related through the tangent of the phase shift δ angle, between the grid voltage and the inverter voltage (13). This relationship can be observed in Figure 5, where the vectors of the grid voltage, inverter output voltage and impedance voltage are represented.

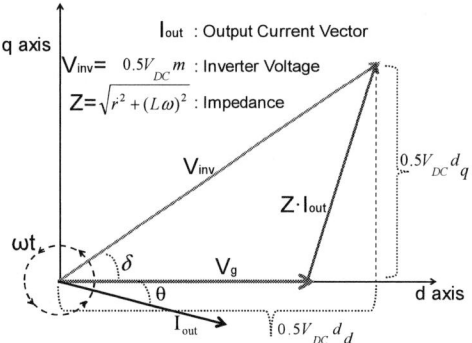

Figure 5.   Synchronous Refence Frame.

978-1-4244-4782-4/10 $26.00 © 2010 IEEE   1028

$$\tan \delta = \frac{d_q}{d_d} \quad (13)$$

Both duty cycles, *dd* and *dq*, can be isolated from (13). Their mathematical expressions are given in (14) and (15):

$$d_q = d_d \tan \delta \quad (14) \qquad d_d = d_q \cot \delta \quad (15)$$

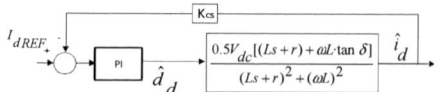

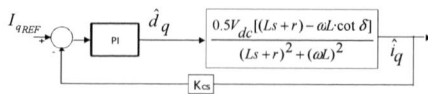

Figure 7. Simplified model for the control design. Two SISO system.

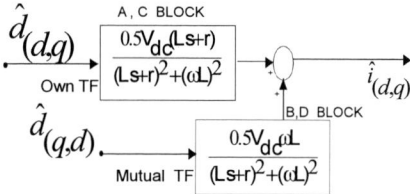

Figure 6. Small signal model. Mutual and Own transfer function.

In the model represented in Figure 3, the A-block and C-blocks have the same structure and can be called own transfer functions and the B-block and D-block can be called mutual transfer functions. The simplified representation of these transfer functions is shown in Figure 6. In general, the influence of the mutual transfer function on the final trajectory of the current is variable, and its weight depends of the value of the duty cycle for a given operation point.

With (14) and (15) can be quantified the influence of mutual transfer functions. i_d and i_q current can be expressed as a function of the δ angle and one of the duty cycles.

Thus, applying (14) in the row corresponding to the i_d current in (12), i_d can be expressed as a function of duty cycle *dd*, given in (16). Similarly, applying (15) in the row corresponding to the i_q current in (12), the i_q current is expressed as a function of duty cycle *dq* and the resulting transfer function is given in (17).

$$\frac{\hat{i}_d}{\hat{d}_d} = \frac{0.5 V_{dc} [(L \cdot S + r) + L\omega \tan \delta]}{(L \cdot S + r)^2 + (L\omega)^2} \quad (16)$$

$$\frac{\hat{i}_q}{\hat{d}_q} = \frac{0.5 V_{dc} [(L \cdot S + r) - L\omega \cot \delta]}{(L \cdot S + r)^2 + (L\omega)^2} \quad (17)$$

Therefore, there are two trajectories with a single input and a single output (SISO). In Figure 7 the simplified closed loop model is given, including the transfer functions (16) and (17).

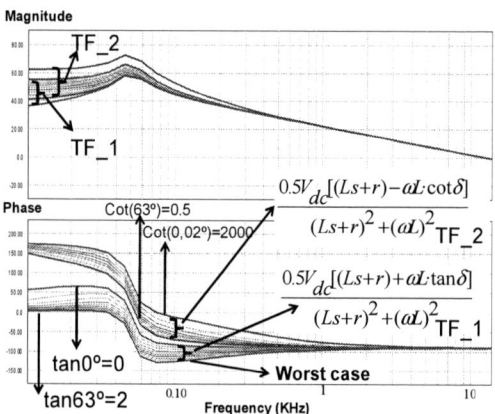

Figure 8. Bode diagrams of the equation (16) and (17).

The Bode diagrams of both transfer functions (16) and (17) are represented in Figure 8. As an example, a family of functions from $\delta=0°$ to $\delta=63°$ are given. The plant of the grid connected inverter varies from an operation point to another, when the system is observed in a Synchronous Reference Frame, independently from the model or the control type used. For example, the shaded area model in Figure 2a and the model in Figure 3, both vary according to the operating point.

The fact that the inverter plant varies with the operation point is not so important. On the contrary, it will provide accurate information in order to determine the worst case (nominal plant of design) for the design of the PI controllers.

In order to optimize the design of the PI controllers, it is necessary to select the worst case between (16) and (17). In Figure 8, it can be observed that the frequency response for (16) and (17) are very similar in magnitude. However, regarding the phase are very different. It can be also noted that the (16) is always the nearest phase to -180°. Additionally, when the δ angle increases, the phase of the transfer functions (16), corresponding to the i_d current will be nearer to -180°. Therefore, the worst case will correspond to (16) evaluated for the biggest δ angle. The biggest δ angle, depends on the active power p and reactive power q, generated in the system and is expressed by the equation (18) [12].

978-1-4244-4782-4/10 $26.00 © 2010 IEEE   1029

$$\tan\delta = \frac{L\omega\,p + r\cdot q}{V_g^2 - L\omega q + r\cdot p} \qquad (18)$$

Theoretically, there is a limit of stability of steady state and occurs when the δ angle reaches 90 ° [16]. When the angle approaches 90 °, the tangent of the angle will tend to infinity; therefore, in practice it is preferable to do an estimation of a maximum angle with the equation (18).

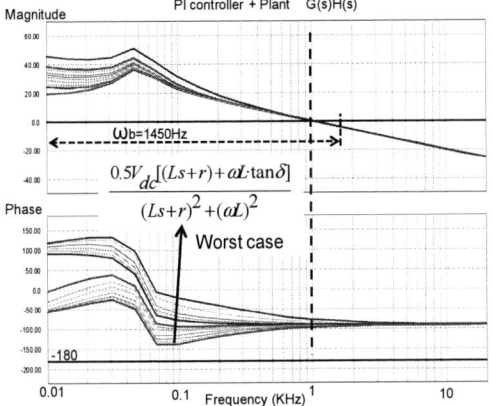

Figure 9. Bode diagrams of the G(s)H(s). PI controller design.

The bode diagrams of the open loop transfer functions are represented in Figure 9. The worst case transfer function for which the PI controllers have been design has been highlighted. It should be noted that only one PI controlled have been designed, and it will be the same for both trajectories. The designed PI controllers will have to guarantee stability for any operation point, enough attenuation of high frequency, good gain in low frequency and fast dynamic response.

The mathematical expression of the closed-loop transfer function is given in (19):

$$\frac{i_{d\_ref}}{\hat{i}_d} = \frac{(K_p + \dfrac{K_i}{S})\dfrac{0.5V_{dc}[(L\cdot S + r) + L\omega\tan\delta]}{(L\cdot S + r)^2 + (L\omega)^2}}{1 + K_{cs}(K_p + \dfrac{K_i}{S})\dfrac{0.5V_{dc}[(L\cdot S + r) + L\omega\tan\delta]}{(L\cdot S + r)^2 + (L\omega)^2}} \quad (19)$$

The schematic of the proposed control in terms of output current is given in Figure 10, where the computational cost of the control algorithm has been reduced by means of eliminating the coupling loops. Nevertheless, the system admits a further reduction of the computational cost through the redefinition of the model included in Figure 7 in terms of active and reactive power. Details are given along the next section.

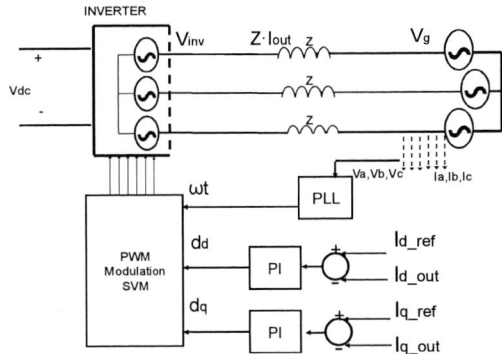

Figure 10. Schematic of the proposed control in term of output current

V. DIRECT POWER CONTROL WITH LINEAR CONTROLLER

In order to achieve a simplified control system in terms of active and reactive power, the model given in Figure 7 must be scaled by the magnitude of grid voltage.

The active and reactive power can be expressed as:

$$p = \frac{3}{2}(V_d i_d + V_q i_q) \qquad q = \frac{3}{2}(V_q i_d - V_d i_q)$$

Due to the synchronization, it can be said that Vd=Vg and Vq=0. Therefore, the active and reactive power expressions are simplified and included in (20).

$$p = \frac{3}{2}V_g i_d \qquad q = -\frac{3}{2}V_g i_q \qquad (20)$$

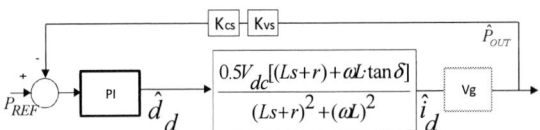

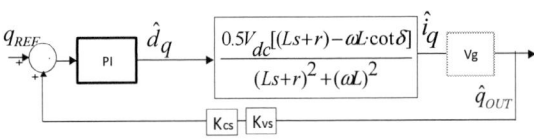

Figure 11. Proposed Direct Power Control with linear controller.

Now, if the two SISO systems of Figure 7 are multiplied by Vg, then a simplified model in terms of active and reactive power is obtained. The block diagram of this model is represented in Figure 11.

A new PI controller must be designed taking into account that the loop gain has been multiplied by the grid voltage Vg, as it can be noted in (21)

$$\frac{p_{ref}}{P_{out}} = \frac{(K_p + \dfrac{K_i}{S})\dfrac{0.5V_{dc}V_g[(L \cdot S + r) + L\omega\tan\delta]}{(L \cdot S + r)^2 + (L\omega)^2}}{1 + K_{cs}K_{vs}(K_p + \dfrac{K_i}{S})\dfrac{0.5V_{dc}V_g[(L \cdot S + r) + L\omega\tan\delta]}{(L \cdot S + r)^2 + (L\omega)^2}} \quad (21)$$

In Figure 12, the scheme of the proposed control in terms of the active and reactive power is given.

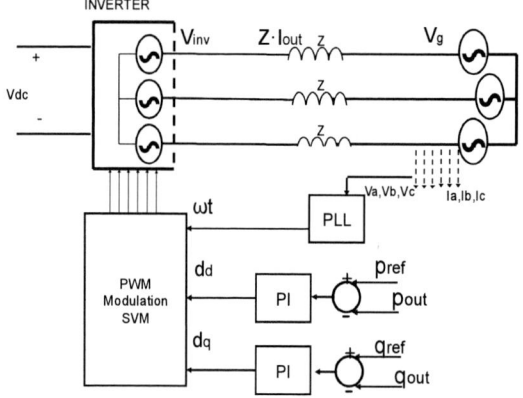

Figure 12. Schematic of the proposed control in term of the active power and reactive power.

It should be remarked that the scheme of Figure 12 is very similar to that of a direct power control (DPC) [6] [7], but with the following differences: the use of a modulator block, a phase tracker (PLL) and the PI controllers.

Therefore, the proposed control can be seen as a DPC with linear controllers. In order to make the comparison, in Figure 13 is showed the schematic of a traditional DPC.

The DPC allows controlling both the active and the reactive powers; however it is a non linear control based on hysteresis controllers.

The major drawback of these hysteresis controllers is that "their operation is somewhat rough because of the inherent randomness caused by the limit cycle; therefore, protection of the converter is difficult" [15].

Thus, the proposed simplified control DPC with PI controllers is an interesting alternative approach.

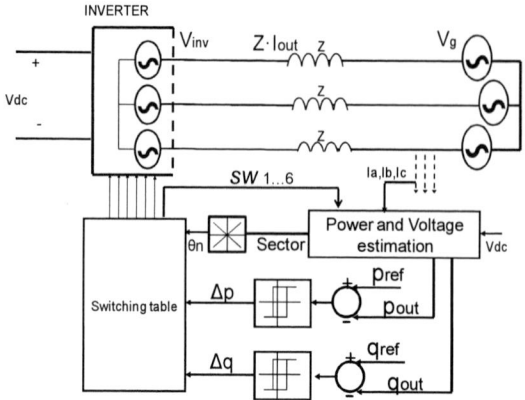

Figure 13. Direct power control DPC of the inverter.

VI. VALIDATION

The physical implementation of the system given in Figure 12 is formed by the following blocks: Space Vector Modulation Subroutine [13], grid frequency detector [14], PI controller subroutine and the subroutine in charge of the calculation of the instantaneous active and reactive power. The mathematical expressions used to calculate both the active and reactive power are (22) and (23).

$$p = V_a i_a + V_b i_b + V_c i_c \quad (22)$$

$$q = 0.577(V_{ab}i_c + V_{bc}i_a + V_{ca}i_b) \quad (23)$$

The characteristic values of the system are: DC bus voltage 800V, Inductance 5mH, Resistor 0.2Ω, grid voltage 220V 50Hz.

Assuming a maximum value in the system of δ=60° and with the characteristic values aforementioned, the transfer function given in (24) is the one used to design the PI controllers.

$$G(s)H(s) = \frac{K_{cvs} \cdot (K_p + \dfrac{K_i}{S})400 \cdot 220 \cdot [(0.005 \cdot S + 0.2) + 1.57 \cdot \tan 60°]}{(0.005 \cdot S + 0.2)^2 + (1.57)^2} \quad (24)$$

The resulting PI is defined by Kp=0.0004 and Ki=0.01. This controller is able to provide a bandwidth of 1450Hz with a phase margin higher than 70°. A good dynamic response is achieved. The Bode diagram under these circumstances is represented in figure 9.

In Figure 14 and 15 the waveforms obtained from the closed loop simulation have been represented. It can be observed the oscillations of the dynamic behavior between the A-block and B-blocks, and the C-block and D-blocks, which cancel each other out.

978-1-4244-4782-4/10 $26.00 © 2010 IEEE

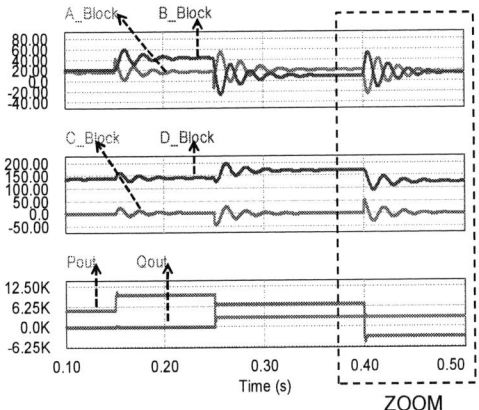

Figure 14. Internal dynamic of the proposed model in close loop.

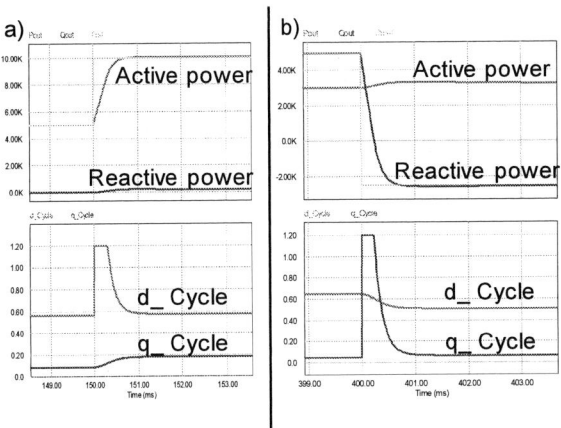

Figure 16. Step active power and step reactive power.

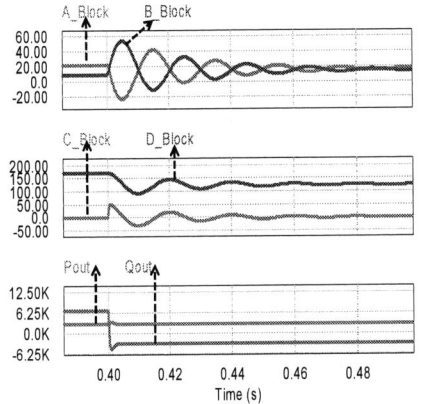

Figure 15. Zoom of internal dynamic. Figure (14).

Taking into account that the Laplace inverse transform of A, B, C and D blocks are damped sines and cosines, the use of compensated loops with the same PI controller design, will lead to an almost exact internal cancelation of the disturbances.

In a transient state, the internal dynamics does not disappear, but waveform generated by A-block is opposite to B-block, and waveform corresponding to C-block is in phase with D-block, achieving the effective cancelations and obtaining the very fast dynamic response. This fact leads to very fast transient states in the output power signals (see Figure 14 and 15).

In the case of a transient due to an active power step, it provokes a very fast change in the duty cycle dd, however the steady state is achieved due to the change of the duty cycle dq, (see Figure 16a). Similarly, a reactive power step provokes a fast change in the duty cycle dq but the steady state is achieved though the variation of the duty cycle dd.

A good design of the PI controller will ensure that most of the transient states will be solved properly as described before.

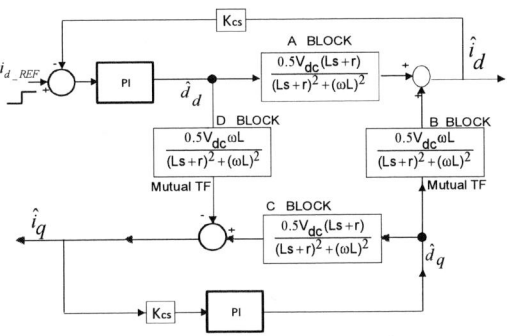

Figure 17. Model of the system for a step of the reference id.

In figure 17 is displayed the system model when a step of the i_d reference current occurs. Qualitatively, it can be said that, due to the block symmetry, the closed loops of each of the output currents are coupled by the mutual transfer functions. This fact allows the exchange of information between the loops and so achieves a perfect synchronization which leads to the equilibrium of the system. The effect described above is similar to Physics classical problem of the Coupled Oscillators.

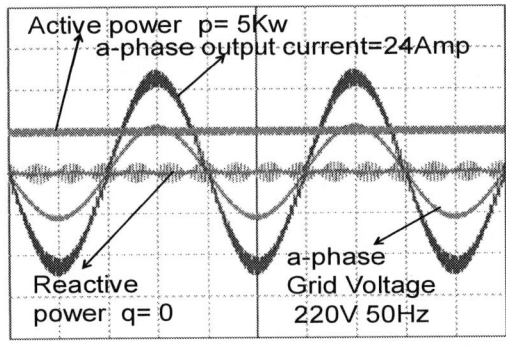

Figure 18. Waveforms. The current and the voltage are in phase.

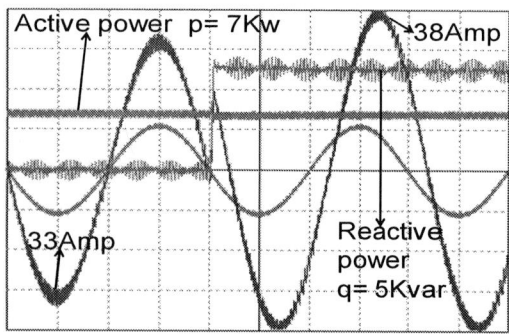

Figure 19. Waveforms. Step Reactive power 5Kvar.

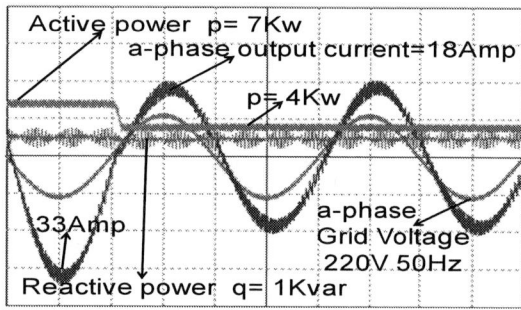

Figure 20. Waveforms. Step active power 7Kw ; 4Kw.

In figure 18, the reactive power, q=0, therefore, the current and the voltage are in phase.

In figure 19 and 20, the fast dynamic response obtained by the proposed control is observed comparing with a fundamental period of the output current.

VII. CONCLUSIONS

Along the present paper, a new model of a three phase grid connected inverter has been presented, which allows the generation of a simplified Synchronous Reference Frame control SSRFC.

The control algorithm proposed is much simpler and reduces significantly the computational cost than the conventional SRFC. It is due to the fact that the feedforward loops are no longer needed in the proposed simplified control.

The results of the waveforms show a fast dynamic response and a good track of the reference signals. Additionally, the hardware implementation is simple, with two PI controllers to control the instantaneous active and the instantaneous reactive power.

The proposed control also can be considered as a Direct Power Control with Linear Controller.

REFERENCES

[1] Giovanni Petrone, Giovanni Spagnuolo, Remus Teodorescu, Mummadi Veerachary, and Massimo Vitelli. "Reliability Issues in Photovoltaic Power Processing Systems " IEEE Transaction on Industrial Electronic , Vol. 55, No. 7, july 2008 2569

[2] Frede Blaabjerg, Remus Teodorescu, Marco Liserre, and Adrian V. Timbus . "Overview of Control and Grid Synchronization for Distributed Power Generation Systems" IEEE Transaction on Industrial Electronic, Vol. 53, No. 5, october 2006.

[3] Jong-Woo Choi and Seung-Ki Sul, "Fast Current Controller in Three-Phase AC/DC Boost Converter Using - Axis Crosscoupling" IEEE Transaction on Power Electronic, Vol. 13, No. 1, ene 1998.

[4] Hong-seok Song and Kwanghee Nam. " Dual Current Control Scheme for PWM Converter Under Unbalanced Input Voltage Conditions" IEEE Transaction on Industrial Electronic, Vol. 46, No. 5, october 1999.

[5] Zitao Wang and Liuchen Chang. "A DC Voltage Monitoring and Control Method for Three-Phase Grid-Connected Wind Turbine Inverters" IEEE Transaction on Power Electronic, Vol. 23, No. 3, may 2008.

[6] Toshihiko Noguchi, Hiroaki Tomiki, Seiji Kondo and Isao Takahashi. "Direct Power Control of PWM Converter Without Power-Source Voltage Sensors" IEEE Transaction on Industry Applications, Vol. 34, No. 3, may/june 1998 473.

[7] L eonardo Augusto Serpa, Srinivas Ponnaluri, Peter Mantovanelli Barbosa , and Johann Walter Kolar. "A Modified Direct Power Control Strategy Allowing the Connection of Three-Phase Inverter to the Grid through LCL Filters" IEEE Transaction on Industry Applications, VOL. 43, NO. 5, SEP/OCT 2007.

[8] M. Zhuang D.P. Atherton, "PID controller design for a TITO system" IEE Proc.-Control Theory Appl., Vol. 141, No. 2, March 1994

[9] Arjin Numsomrarr, Theerachai Wongkhurrr, Tianchai Suksri, Phongchai Nilas and Jutarut Chaoraingem "Design of Decoupled Controller for TITO System using Characteristic Ratio Assignment" International Conference on Control, Automation and Systems 2007 Oct. 17-20, in COEX, Seoul, Korea.

[10] Y.Y.Liu,W.D.Zhang and L.L. Ou , "Analytical decoupling PI/PID controller design for two-by-two processes with time delays" IET Control Theory Appl., Vol. 1, No. 1, January 2007.

[11] Qiang Xiong and Guangyu Jin, "Iterative decentralized PID tuning based on gain and phase margins for TITO SYSTEM" International Journal of Innovative Computing, Information and Control ICIC International °c 2007 ISSN 1349-4198 Volume 3, Number 2, April 2007 pp. 385—396.

[12] A. Lorduy, A. Lazaro, C. Fernandez, I. Quesada, A. Barrado, "Novel Simplified Controller for Three Phase Grid Connected Inverter Based on Instantaneous Complex Power" Applied Power Electronics Conference and Exposition, 2009. APEC 2009. Twenty-Fourth Annual IEEE , vol., no., pp.1306-1312, 15-19 Feb. 2009.

[13] D. Grahame Holmes, T.Lipo, "Pulse Width Modulation For Power Converters" Principles and Practice. IEEE Press. Series on Power Engineering.

[14] Francisco D. Freijedo, Jesu´s Doval-Gandoy, Óscar López ,and Enrique Acha. "A Generic Open-Loop Algorithm for Three-Phase Grid Voltage/Current Synchronization With Particular Reference to Phase, Frequency, and Amplitude Estimation" IEEE Transaction on Power Electronic, Vol. 24, No. 1, January 2009.

[15] Marian P. Kazmierkowski, R. Krishnan, Frede Blaaabjerg " Control In Power Electronics - Selected Problems" Academic Press. Series In Engineering.

[16] Jhon J. Grainger, William Stevenson Jr. "Book: Power System Analysis"

978-1-4244-4782-4/10 $26.00 © 2010 IEEE

A Novel Direct Digital SPWM Method for Multilevel Voltage Source Inverters

Wanmin Fei, Yanli Zhang
School of Electrical and Automation Engineering
Nanjing Normal University, Nanjing, China

Bin Wu
Department of Electrical and Computer Engineering
Ryerson University, Toronto, Ontario, Canada

Abstract—A novel direct digital Sine Pulse Width Modulation (SPWM) method based on equal-area theorem for multilevel inverters is proposed in this paper. The principle of the proposed modulation schemes is elaborated, and the best scheme with the lowest THD is identified. The THD and fundamental voltage produced by the proposed method with modulation index M from zero to its maximum value are calculated and compared with carrier-based SPWM method. It is demonstrated that the proposed direct digital SPWM method can provide far lower THD while its fundamental voltages are on par with the carrier-based SPWM method. Other advantages of the proposed modulation scheme include simplicity, digital oriented, and generalized for multilevel inverter with any levels and any topologies. Simulations and experiments are carried out. The experimental results agree well with simulations, which prove the validity of the proposed digital SPWM method.

I. INTRODUCTION

PWM method is one of the most important techniques of multilevel inverters, which is employed mainly for control of fundamental and elimination of undesired harmonics. So far various multilevel PWM techniques have been studied and plenty of results are published such as multiple-carrier based modulation, selective harmonic elimination modulation (SHE-PWM), space vector PWM (SVPWM) and some combination PWM [1] [2]. Among these, SVPWM is the most popular one due to its simplicity both in hardware and software, and its relatively good performance at low modulation ratio. But the SVPWM becomes very difficult to achieve when the level increases. SHE-PWM method can eliminate selected low order harmonics and has the advantages of better output waveform, lower switching frequency and easy filter design. However, complicated computation and large number of storage space is needed. Therefore, SVPWM and SHE-PWM can not be employed in situations when low cost, high simplicity and short trial-manufacture time are required.

Natural sampling multi-carrier SPWM based on analog circuit is much easier for realization and has the advantages of high quality waveform and high response speed. However it has a poor flexibility and is difficult to revise. The carriers are very difficult and complicated to define and realize when the frequency of the inverter is to be varied in a wide range. Digital based regular sampling SPWM is flexible and easy to

revise, but it has higher harmonics contents and lower fundamental control accuracy. Several optimized regular SPWM methods for two-level inverters are proposed [3]; however, they are not applicable to multilevel inverters. So a digitally implemented natural sampling SPWM method is proposed in [4]. The number of intersection points of reference signal with the multiple carriers of multilevel inverters varies with the changing of frequency and modulation depth, and asymmetry problems and even order harmonics especially DC component which are very harmful to motors and other loads can be produced. Stair-case modulation method [5] is a direct digital method, but the inverter works at the fundamental frequency and the switching capability of nowadays power semiconductor devices can not be fully utilized. With the rapid development of DSP, the capability and speed of computation with lower cost are becoming higher and higher, which made the direct digital SPWM method based on equal-area theorem much more attractive.

II. EQUAL-AREA THEOREM BASED DIRECT DIGITAL SPWM METHOD

The basic principle of equal-area theorem based SPWM method with quarter-wave symmetry is shown in Fig. 1. In order that the output voltage at 90° should be the maximum value, the first quarter cycle is divided into J+0.5 sections. Where J is a positive integer; T_f is the length of each section; E is the voltage of a level, T_{gi} is the cg of the reference sine signal in the ith section. Substitute the reference signal in each section with a rectangular or a protruding PWM waveform and let the area of the PWM waveform equal to that of the reference signal, the width of the top rectangle T_{wi} can be expressed as:

$$T_{wi} = \frac{U_m}{\omega E}\left[\cos \omega(i-1)T_f - \cos \omega i T_f\right] - (N-1)T_f \qquad (1)$$

where U_m is the peak value of the reference signal, ω is the radian frequency of the output voltage.

The equal-area theorem says that responses tend to be identical when input signals have same area and time durations of input impulses become very small. In other words, the condition for responses to be identical under input

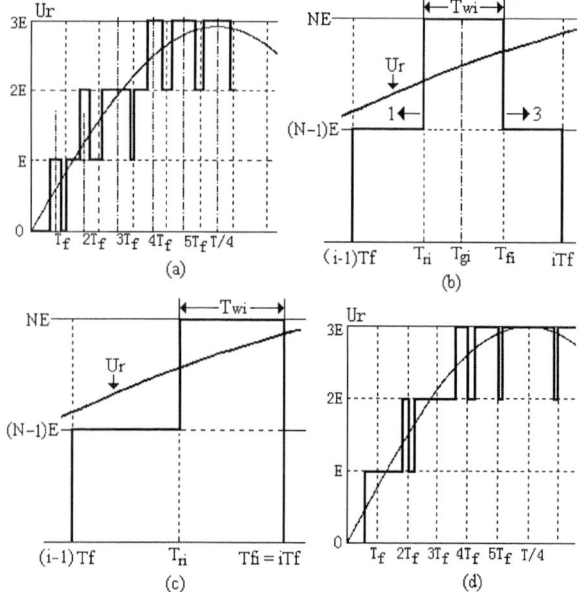

Fig .1 Principle of equal-area theorem based SPWM method

$$T_{ri} = \frac{2T_{gi}[T_{wi} + (N-1)T_f] - (N-1)(2i-1)T_f^2 - T_{wi}^2}{2T_{wi}} \tag{5}$$

$$T_{fi} = \frac{2T_{gi}[T_{wi} + (N-1)T_f] - (N-1)(2i-1)T_f^2 + T_{wi}^2}{2T_{wi}} \tag{6}$$

Sometimes the falling edge T_{fi} obtained according to Eq. (6) may be greater than iT_f, the upper limit of the ith section, and then the rising edge and falling edge should be revised according to the following equations:

$$T_{ri} = iT_f - T_{wi} \tag{7}$$

$$T_{fi} = iT_f \tag{8}$$

For Cg Centered and Cg Superposed SPWM schemes, if the area of the reference sine signal is between $(N–1)ET_f$ and NET_f in the ith section and between NET_f and $(N+1)ET_f$ in the $(i+1)^{th}$ section, then iT_f is another rising edge. For a multilevel inverter with L levels, there are $(L-1)/2-1$ such rising edges at most in the first quarter cycle.

C. Flush Right SPWM Scheme

Flush Right SPWM scheme means to flush right the top rectangular in each section, as shown in Fig.1 (c) and (d), and thus the rising edge T_{ri} and falling edge T_{fi} in the ith section are expressed by Eq. (7) and (8). In this scheme, the first quarter cycle can be divided into J (integer) sections, and the SPWM waveform extended according to the symmetry characteristics satisfies the requirement that the voltage at 90 degree is maximum.

For Flush Right SPWM scheme, if the area of the reference sine signal is between $(N–1)ET_f$ and NET_f in the ith section and between NET_f and $(N+1)ET_f$ in the $(i+1)^{th}$ section, then $(i-1)T_f$ is the falling edge of $(i–1)^{th}$ section and the another rising edge besides the one expressed by Eq. (7) of ith section and as a result the PWM waveform remains unchanged at this point. For a multilevel inverter with L levels, there are $(L-1)/2-1$ such points at most in the first quarter cycle.

III. ANALYSIS AND EVALUATION OF HARMONIC PERFORMANCE

The performance of a SPWM scheme can be evaluated with two parameters, THD to specify the quality of waveform and amplitude of fundamental voltage to illustrate the specification in fundamental control. A five-level inverter is taken as an example. In order that the number of switching points in a quarter cycle under each of the three SPWM schemes is the same, such as 10, the first quarter cycle in Cg Centered SPWM and Cg Superposed SPWM is divided equally into four and a half sections and in Flush Right SPWM scheme into six sections for M greater than 0.5. The first quarter cycle is divided equally into five sections for M equal or less than 0.5, then 10 switching points for Cg Centered and Cg Superposed schemes and 9 switching points for Flush Right scheme in a quarter-cycle can be obtained.

Up to 50th harmonics are included in the calculation, and the two parameters are defined as follows:

signals with the same area and different shapes is that the duration must be short enough or the switching frequency of the inverter must be sufficiently high. In general, multilevel inverters are great power equipment and work at low frequencies. From Fig.1 it can be seen that the shape of PWM waveform varies with the changing of the location of the top rectangular in each section. Responses under signals with same area and different shapes could be different when the switching frequency is not high enough. Based on this assumption and take a five-level inverter as an example, three typical SPWM schemes are investigated and compared with each other according to THD and amplitude of fundamental voltage. The methods for obtaining all the rising and falling edges in the first quarter cycle, which are sufficient to express the SPWM result, are described as follows.

A. Cg Centered SPWM Scheme

Simply set the top rectangular at the center of each section, and the rising edge T_{ri} and falling edge T_{fi} in the ith section can be expressed as:

$$T_{ri} = (i-0.5)T_f - 0.5T_{wi} \tag{2}$$

$$T_{fi} = (i-0.5)T_f + 0.5T_{wi} \tag{3}$$

B. Cg Superposed SPWM Scheme

The cg of the reference sine signal in the ith section is:

$$T_{gi} = \frac{AT_f + B}{C} \begin{cases} A = (i-1)\cos(\omega(i-1)T_f) - i\cos(i\omega T_f) \\ B = \frac{1}{\omega}(\sin(\omega i T_f) - \sin(\omega(i-1)T_f)) \\ C = \cos(\omega(i-1)T_f) - \cos(\omega i T_f) \end{cases} \tag{4}$$

Let the cg of the PWM waveform to be equal to that of the reference signal expressed by Eq. (4), we can obtain the rising edges T_{ri} and falling edge T_{fi} in the ith section as follows:

$$THD = \frac{1}{V_{1m}} \sqrt{\sum_{i=1}^{24} V_{(2i+1)m}^2} \qquad (9)$$

where M is modulation index, L is the output level of line-to-neutral voltage, V_{im} is the peak value i^{th} harmonics and can be expressed by:

$$V_{nm} = \frac{4E}{n\pi} \sum_{k=1}^{N} p_k \cos(n\omega t_k), \text{ n=1, 3, 5, ..., 49,}$$

$$\text{and } p_k = \begin{cases} 1, \alpha_k & \text{for rising edge} \\ -1, \alpha_k & \text{for falling edge} \end{cases} \qquad (10)$$

where t_k is the kth switching points, including rising and falling edges.

Let E=1V, THD and amplitude of fundamental voltage versus M are calculated according to Eq. (1) – (10), and are illustrated in curves shown in Fig.2.

From Fig. 2 we can see clearly that for quarter-wave symmetry SPWM waveform, with the moving of cg of PWM waveform in each section from left to right, the THD decreases distinctly and the fundamental voltage increases slightly. With almost the same relationship of fundamental voltage with modulation index, Flush Right digital SPWM method has the minimum THD and is the optimum direct digital SPWM scheme except for M less than 0.07 where Cg Centered scheme gets better. Therefore, we choose the Flush Right direct digital SPWM method when M is greater than 0.07 (this point could vary when the parameters such as level, M and frequency change) and Cg Centered scheme when M is less than 0.07 as the direct digital SPWM method.

It can be seen from Equ. (1) – (3), (7) and (8) which are used that the computation is very simple.

IV. COMPARISON WITH LEVEL-SHIFTED CARRIED-BASED SPWM SCHEME

In this section, we try to demonstrate how the new method behaves as compared with the well known carrier-based SPWM method. Because the level-shift carrier-based SPWM method can provide smaller THD than the phase-shift scheme [3], the former is taken as an example.

In order that all the THDs and fundamental voltages are calculated with the same preconditions and at frequencies as close as possible to the new direct digital SPWM method, the frequency and the initial phase angle of the carriers are 1050 Hz and 90.72 degree, In-Phase-Disposition(IPD) [1][2] scheme is adopted. PWM waveforms obtained from the carrier-based SPWM scheme described above are quarter-wave symmetric. Therefore, the THD and the fundamental voltages can be calculated according to Equ. (9) and (10). The calculation results are depicted in Fig. 2 in curves with signs of "*". The number of switching points of carrier-based SPWM method within a quarter cycle varies with the change of modulation index from ten to twelve when modulation index M is greater than 0.5 and remains constant at eleven when M is smaller than 0.5. It can be seen form Fig. 2 that THD obtained with the new optimum direct digital SPWM method is distinctly

smaller than that obtained with the carrier-based SPWM method.

From Fig. 2 we can see that the Cg Superposed direct digital SPWM method has almost the same performance with level-shift multi-carrier SPWM method. By adopting Flush Right scheme in most range of M and Cg Centered scheme in a small area when M is very small, we can obtain a direct digital SPWM scheme which is much better than the traditional level-shift carrier-based SPWM method. The novel direct digital SPWM method can provide a smaller THD with a lower frequency. Most importantly, it can bring great flexibility in practice especially when the frequency is to be varied in a large scale.

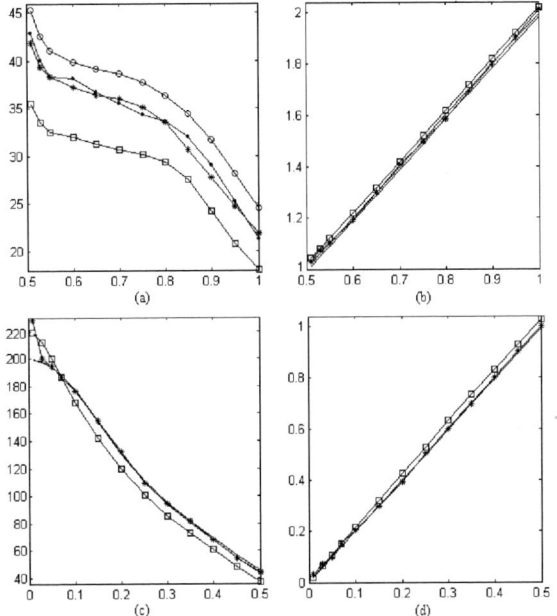

Fig. 2. THD and amplitude of fundamental voltage versus M under different SPWM schemes. (a) THD versus M for $0.5 < M \leq 1$. (b) Fundamental voltage versus M for $0.5 < M \leq 1$. (c) THD versus M for $0 < M \leq 0.5$. (b) Curves of fundamental voltage versus M for $0 < M \leq 0.5$.
□: Flush right; *: Multi-carrier based SPWM method; o: cg centered; •: cg superposed.

V. SIMULATION AND EXPERIMENT

Take a five-level cascade inverter as an example, let E=100V, M=0.9, J=6 for proposed digital SPWM method and carrier-based SPWM, simulations based on Psim 6 are carried out. The simulation waveforms and their frequency spectrums are shown in Fig. 3, where (a), (b) are that of the proposed digital SPWM method, and (c), (d) are that of carrier-based SPWM method. It can be seen that the switching times in a quarter-cycle for novel digital SPWM method is ten which is smaller than twelve for carrier-based SPWM method. The THD of the waveforms based on the proposed digital SPWM method and carrier-based SPWM method is 24.5% and 27.8%, respectively.

A cascade five-level inverter is constructed and experimentally researched based on the novel digital SPWM method with the same parameters for simulation. The

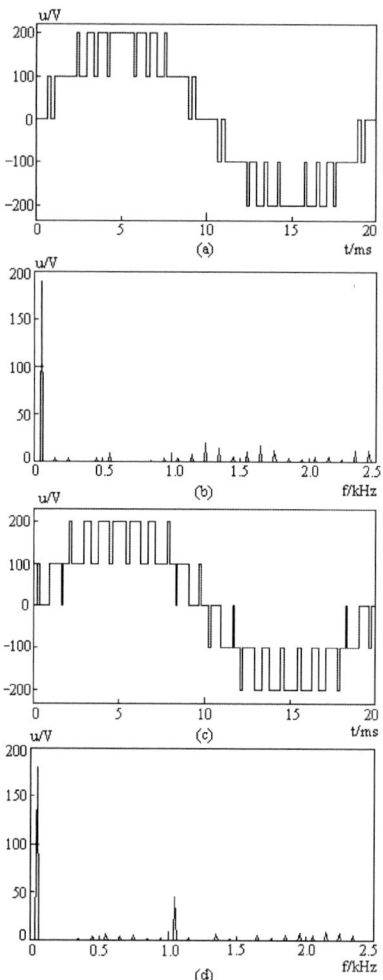

Fig. 3. Simulated waveform and its frequency spectrum. (a), (b) Phase-voltage and its frequency spectrum with proposed digital SPWM method. (c), (d) Phase-voltage and its frequency spectrum with carrier-based level-shift SPWM method.

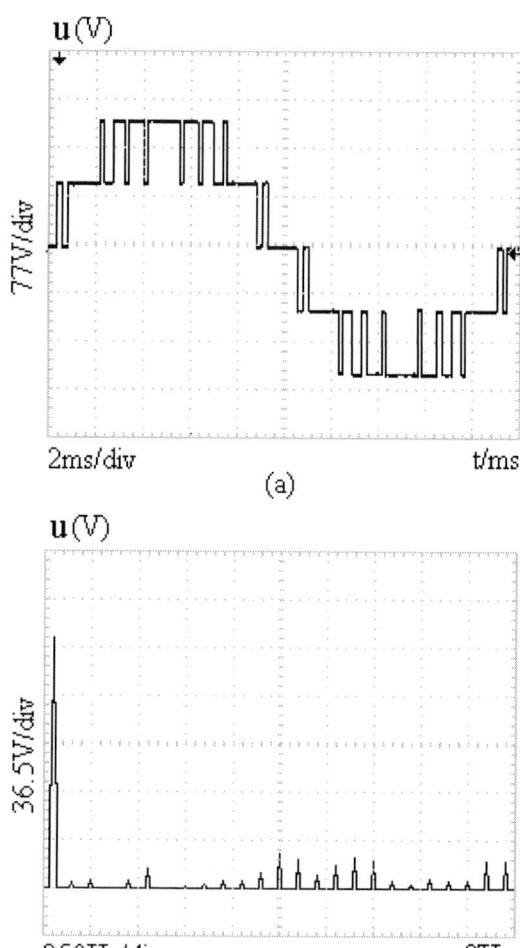

Fig. 4. Experimental waveform with the proposed digital SPWM method. (a) Phase-voltage. (b) Frequency spectrum.

waveform and its frequency spectrum are shown in Fig. 4. It can be seen from Fig. 3 (a), (b) and Fig. 4 that the experiments agree well with simulation.

VI. CONCLUSION

It can be demonstrated clearly that for quarter-wave symmetry SPWM waveform, the fundamental voltage increases and the *THD* decreases with the moving of cg of PWM waveform in each section from left to right. With almost the same relationship of fundamental voltage with modulation index, Flush Right digital SPWM method has the minimum *THD* and is the optimum direct digital SPWM scheme except for *M* less than 0.07 where Cg Centered scheme becomes better. By combining the Flush Right scheme with Cg Centered scheme, we can obtain a direct digital SPWM scheme that is much better than the traditional level-shift carrier-based SPWM method. Other important advantages of the new direct digital SPWM method include simplicity, digital oriented, and generalized for multilevel inverter with any levels and any topologies including Cascade

H-Bridge inverters, Diode-Clamped inverters, Flying Capacitor inverters and any other type of inverters. Simulations and experiments are carried out. Experimental results agree well with the computation and simulations, which proves the validity of the proposed SPWM method.

REFERENCES

[1] G. Carrara, S. Gardella, et al, "A New PWM Method: A Theoretical Analysis," *IEEE Trans. Power Electon.*, vol. 7, no. 3, pp. 497-505, July 1992.

[2] Bin Wu, "High Power Converter Systems", Wiley - IEEE Press & John Wiley, March 2006, ISBN: 0–4717–3171–4.

[3] Sidney R. Bowes and Derrick Holliday, "Optimal Regular-Sampled PWM Inverter Control Techniques," IEEE Trans. Ind. Electron., vol. 54, no. 3, pp. 1547–1559, June 2007.

[4] Geoffrey R. Walker, "Digitally-Implemented Naturally Sampled PWM Suitable for Multilevel Converter Control", *IEEE Trans. Power Electron.*, vol. 18, no. 6, pp. 1322-1329, Nov. 2003.

[5] Mouzhi Dong, Bin Wu, N. Zargari, J. Rodriguez, "A novel digital modulation scheme for multilevel cascaded H-bridge inverters", *Proc. IEEE Power Electronics Specialists Conference*, 15-19 June 2008, pp.1675 – 1680.

Weight Oriented Optimal PWM in Low Modulation Indexes for Multilevel Inverters with Unbalanced DC Sources

Damoun Ahmadi, Ke Zou, and Jin Wang
The Ohio State University
205 Dreese Lab, 2015 Neil Avenue
Columbus, OH 43210, USA
damoun@ece.osu.edu

Abstract—Selective harmonics elimination for the staircase voltage waveform generated by multilevel inverters has been widely studied for medium and high voltage applications. Most published strategies are for multilevel inverters with balanced dc levels and single switching per level. This paper proposes an Optimal Pulse Width Modulation (OPWM) method that utilizes equal area criteria and harmonic injection. In particular, for low modulation indexes, the weight oriented OPWM is proposed to determine switching angles to achieve accurate harmonic elimination.

Keywords: Equal Area Criteria, Harmonics Injection, Optimal PWM, Weight Oriented

I. INTRODUCTION

Because of the recent development of flexible ac transmission devices, medium voltage drives, and distributed generations, design and implementation of multilevel inverters circuits have been widely studied [1]-[3]. For these medium and high power applications, selective harmonic elimination based optimal pulse width modulation can be utilized to reduce the switching frequency while simultaneously keeping the total harmonics distortion (THD) under the numbers specified in relevant regulations and standards [4]-[6].

In 1964, the basic idea for selective harmonic elimination was proposed by Turnbull based on the Fourier series [4]. In this method, harmonic components are described as functions of the switching angles in trigonometric terms. Then a multiple variable equation group is formed, with one equation to guarantee the amplitude of the fundamental component, and the other equations ensuring elimination of the harmonics. If N is the total number of switching transitions, then, by calculating the switching angles, N-1 harmonics can be eliminated [6]-[10].

Some earlier algorithms use the Newton-Raphson method or linearization for real time calculation and harmonic elimination [11]-[12]. Some recent research on this topic utilizes advanced algorithms and control theories such as fuzzy logic solution, resultant theory, and sliding mode

control [13]. In [14] and [15], online calculations of the switching angels are introduced. However, all these methods are still based on solving complex groups of equations. Therefore, when the number of switching transients is high, it is quite difficult or time consuming to solve with current computation methods [16], [17].

To eliminate harmonic components in multilevel inverters, four simple equations based on equal area criteria and the harmonic injection method have been proposed recently [7]. In [18], this method was proposed for optimal PWM switching angles on inverters. Then in [19], this method was utilized for unbalanced dc levels in multilevel topologies. The main problem associated with the basic four-equation method is the amplitude difference between the desired and resulted fundamental voltage.

To solve this problem, in [20] and [21], modulation indexes are categorized in two groups: low modulation indexes with an additional switching angle when an extra dc level is available; and high modulation indexes with an adjusted switching angle, where no extra voltage level is available, as shown in Figure 1.

In lower modulation indexes, fewer dc levels are utilized to synthesize the staircase waveform. Therefore, an extra voltage level is available and can be used for fundamental voltage compensation. Based on this idea, an additional switching angle in the extra voltage level is calculated to achieve desired fundamental voltage. For unbalanced dc sources, the following two equations are utilized:

- First, the fundamental voltage based on switching angles from θ_1 to θ_m is calculated with the following equation:

$$V_{1m} = \sum_{i=1}^{m} \frac{4V_{dc(i)}}{\pi} \cos(\theta_i), \qquad m < N \qquad (1)$$

Then, an additional switching angle is calculated to

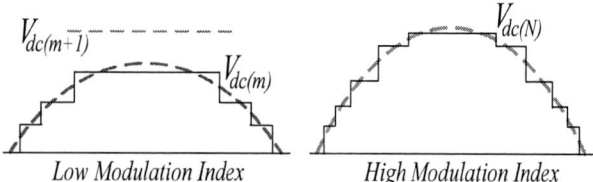

Fig. 1: Two groups of modulation indexes in multilevel inverters with unbalanced dc levels.

compensate for the difference between V_{1m} and fundamental voltage:

$$\theta_{m+1} = a\cos(\frac{\pi}{4V_{dc(m+1)}}(V_F - V_{1m})) \qquad (2)$$

Thus, based on the four-equation based method, and through (1) and (2), harmonic components are eliminated by "m" switching angles and the reference fundamental voltage is compensated adequately by θ_{m+1}.

For the adjustment switching angle in higher modulation indexes, the fundamental voltage is corrected by the switching angle modification performed in the last DC level. This adjustment for unbalanced dc sources will be defined by the following equations:

- For fundamental voltage correction, an "adjustment" switching angle is calculated to by the following equation:

$$\theta_N^* = a\cos(\frac{\pi}{4V_{dc(N)}}(V_F - V_{1N})) \qquad (3)$$

where V_{1N} is the total fundamental voltage generated by switching angles from θ_1 to θ_N.

- This "adjustment" angle is used to modify the switching angle for the last voltage level:

$$\theta_{N(\text{modified})} = a\cos(\cos(\theta_N) + \cos(\theta_N^*)) \qquad (4)$$

Based on the switching angle adjustment for the last dc level, the desired voltage magnitude in the fundamental frequency can be achieved. However, the "adjustment" switching angle also causes new harmonic components used in the final modulation waveform in equal area criteria.

In this paper, the four-equation method is further developed for a multilevel inverter with unbalanced dc levels and multiple switching angles per level.

II. OPTIMAL PWM COMBINED WITH FOUR EQUATION BASED METHOD

In the basic four-equation based method in multilevel inverters, the harmonics selected for elimination are limited by the number of available dc levels. To overcome this problem, in each dc level, the number of switching angles can be increased to eliminate more harmonic components, as shown in Fig. 2. This is very helpful especially for low modulation indexes where limited dc levels are available.

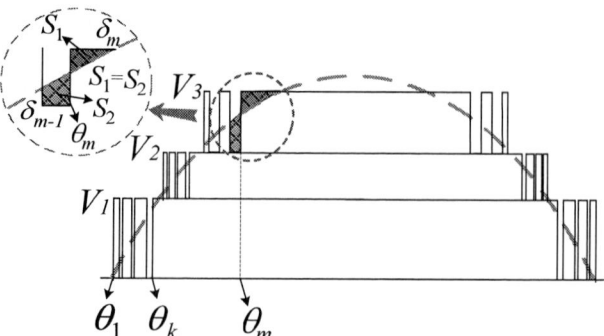

Fig. 2. Optimal PWM with four-equation based method on multilevel inverters with unbalanced dc sources.

In theory, there is no limitation for the number of switching angles used for each level. Multiple carrier based PWM has been the most popular method. But the multiple carrier approach will not solve the issue with unbalanced dc levels. Furthermore, if the number of switching angles is limited by the switching loss, the multiple carrier PWM will not be able to realize effective harmonic elimination efficiently. So in this paper, the optimal PWM method is combined with the four-equation method to achieve selected harmonic elimination for multilevel inverters. In the basic four-equation method, switching angles are determined with equal area criteria by utilizing two adjacent junction points of the reference waveform and the dc voltage levels. When optimal PWM is used to increase the number of the angles for each level, switching angles cannot be found by just utilizing two major junction points. Thus, the area between the two junction points should be distributed in a reasonable pattern to achieve sub-junction points for switching angle calculation. With two sub-junction points, δ_{k-1} and δ_k, the switching angles are determined through the following equation:

$$\theta_k = 1/V_{dc(k)} * (\sum_{i=1}^{k} V_{dc(i)}\delta_k - \sum_{i=1}^{k-1} V_{dc(i)}\delta_{k-1} + V_F(\cos(\delta_k) - \cos(\delta_{k-1}))$$
$$- \frac{h_5}{5}(\cos(5\delta_k) - \cos(5\delta_{k-1}))... - \frac{h_m}{m}(\cos(m\delta_k) - \cos(m\delta_{k-1}))) \qquad (5)$$

For medium and high modulation indexes, in each voltage level, the available area can be equally divided for corresponded switching angles, as shown in Fig. 3 (a). However, for low modulation indexes, this strategy does not work well. This is simply because that with less voltage levels, larger "area" is needed to compensates selected harmonic components precisely. So the distribution of the δ_{k-1} and δ_k, becomes very crucial. Therefore, to overcome this problem in low modulation index, a weight oriented solution is proposed in this paper to decide the δ_{k-1} and δ_k.

III. WEIGHT ORIENTED SOLUTION

In this four-equation and OPWM combined method for multilevel inverters with unbalanced dc sources, the harmonics that are injected into the reference waveform are calculated by the following equations through Fourier series:

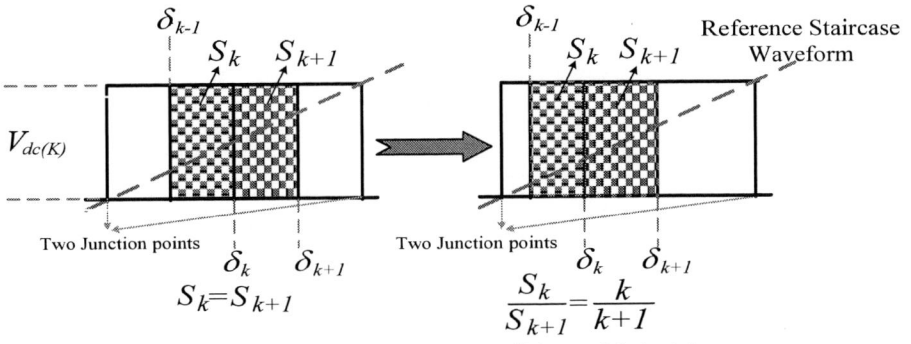

(a) Medium and high modulation index;   (b) low modulation index

Fig. 3. Two methodologies for area division in OPWM and four-equation combined method.

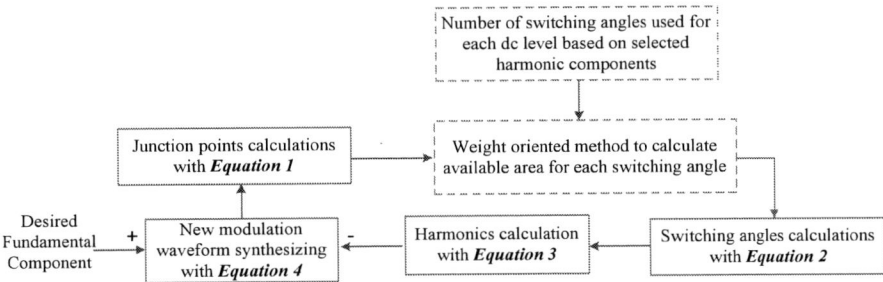

Fig. 4. Block diagram for weight oriented solution in low modulation indexes.

$$h_m = \sum_{k=1,2,..,N}^{N} \frac{4V_{dc(k)}}{m\pi}(\cos(m\theta_k) - \cos(m\delta_k)) \qquad (6)$$

It is clear that the magnitude of harmonic content decreases as the order of the harmonics increases. Therefore, the corresponded area necessary to eliminate higher order harmonics is decreased as well. Based on this observation, the area division for low modulation indexes is determined by the weight of the harmonics, which is shown in Fig. 3(b).

In this weight orientated solution, there is more area available for lower harmonic components with higher magnitude. Therefore, select harmonics can be eliminated in low modulation indexes with better accuracy. The procedure for this method is illustrated in Fig. 4.

If λ_k is defined by the difference between two sub-junction points, δ_k and δ_{k-1}; then the following equations can be used for λ_k:

$$\lambda_k = \delta_k - \delta_{k-1} \qquad (7)$$

$$\frac{\lambda_{k+1}}{\lambda_k} = \frac{k+1}{k} \qquad (8)$$

In this case, for a symmetric waveform, sub junction points are determined until $\pi/2$. Therefore for first level, if we assume that, switching angles are between 0 and $\pi/2$, then:

$$\pi/2 = \sum_{k=1}^{m} \lambda_k = \lambda_1 * \sum_{k=1}^{m} k = \lambda_1 * \frac{m*(m+1)}{2} \qquad (9)$$

So:

$$\lambda_1 = \frac{\pi}{m*(m+1)} \qquad (10)$$

Then, through (7) and (8), other sub-junction points can be determined easily.

A similar procedure can be used if the number of levels is more than one. In those cases, λ_k is calculated between two main junction points for each dc level. Thus, based on this procedure, switching angles are defined more effectively for lower modulation indexes. This idea is verified in simulations and experimental tests in next part.

To test the proposed method, it is applied to two cascade multilevel inverter based cases: 1) one H-bridge and 2) two H-bridges with unbalanced dc levels. Five switching angles are used in both cases for harmonic elimination. The results in Table 1, show that five harmonic components can be eliminated effectively, thus verifying the advantage of the proposed method in low modulation indexes.

IV. SIMULATION RESULTS

To verify the switching angles shown in Table 1, a multilevel inverter with unbalanced dc levels is simulated for low modulation indexes. Table 2 shows the selected harmonic components for the simulated voltage waveform shown in Fig. 5(a) and (b). It can be seen that these harmonics are minimized successfully.

Table 1. Sample points based on the four-equation and OPWM combined method for low modulation indexes

# levels	Modulation Index	Switching Angles (rad.)					Harmonics (% based on fundamental output)					
		θ_1	θ_2	θ_3	θ_4	θ_5	1st	5th	7th	11th	13th	17th
1	0.1348	0.2438	0.5220	0.7486	1.0185	1.3050	100.00	0.00	0.8949	0.00	1.6620	1.1135
	0.1414	0.2450	0.5070	0.7461	1.0091	1.2887	100.00	0.00	0.5404	0.00	1.0035	1.1360
2	0.1901	0.3546	0.5497	1.0294	1.1829	1.3403	100.00	0.00	0.00	0.00	0.00	0.00
	0.2278	0.1159	0.4645	0.7975	1.2422	1.4205	100.00	0.00	0.00	0.00	0.00	0.00
	0.2545	0.1763	0.4578	0.8411	1.1378	1.3779	100.00	0.00	0.00	0.00	0.00	0.00

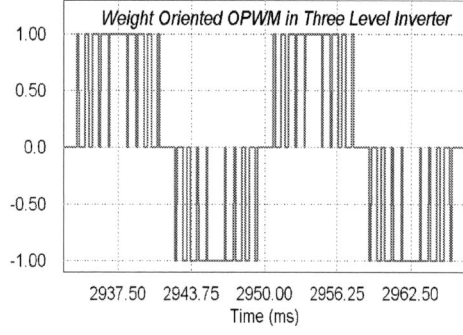

(a) Three level inverters

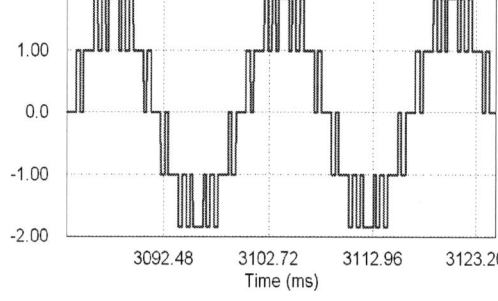

(b) Five level inverters with unbalanced DC sources

Fig. 5. Simulated voltage based on four-equation and OPWM combined method for multilevel inverters with unbalanced DC links.

Table 2. Simulation results for output voltage and selected harmonic components on different modulation indexes.

Modulation Index	Switching Angles (rad.)					Harmonics (% based on fundamental output)					
	θ_1	θ_2	θ_3	θ_4	θ_5	1st	5th	7th	11th	13th	17th
0.1414	0.2450	0.5070	0.7461	1.0091	1.2887	100.00	0.1532	0.5542	0.1673	0.9094	0.9700
0.2545	0.1763	0.4578	0.8411	1.1378	1.3779	100.00	1.1723	0.1256	0.8080	0.8237	0.3047

V. CASE STUDY

To verify the accuracy of the proposed method, 5 calculated switching angles, the same used for simulation, are applied to multilevel inverters. The experimental setup is shown in Fig. 6. In Fig. 7 (a) and (b), the output voltages for single inverter and a multilevel inverter with unbalanced dc voltage are shown respectively. In the unbalanced case, the second dc level is 80% of first level. From the harmonics analysis in Table 3, it can be seen that the weight oriented method can be used successfully in low modulation indexes.

VI. CONCLUSION

Most published methods for harmonic elimination in multilevel inverters are proposed for balanced dc sources with a single switching angle for each dc level. In this paper, a modified four-equation method is proposed for unbalanced dc levels with multiple switching angles per level. A weight oriented solution is used to identify the junction points needed in the four-equation method for low modulation

Fig. 6. Experimental setup for multilevel inverter with unbalanced dc sources.

indexes. Simulation and experimental results are shown to verify the proposed method.

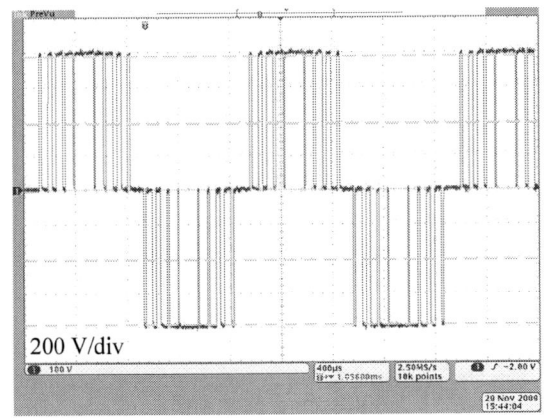

(a) Three level inverters

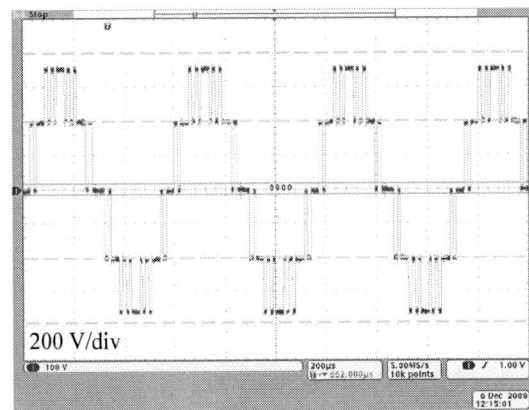

(b) Five level inverters with unbalanced DC sources

Fig. 7. Experimental results for single and multilevel inverter with unbalanced dc sources.

Table 7. Selected harmonics analysis for two case studies.

Modulation Index	Switching Angles (rad.)					Harmonics (% based on fundamental output)					
	θ_1	θ_2	θ_3	θ_4	θ_5	1st	5th	7th	11th	13th	17th
0.1414	0.2450	0.5070	0.7461	1.0091	1.2887	100.00	2.2365	1.8975	1.2392	0.8451	0.3652
0.2545	0.1763	0.4578	0.8411	1.1378	1.3779	100.00	3.0636	1.4018	0.8134	0.7811	0.5384

REFERENCES

[1]. Tolbert, L.M, F. Z. Peng, T. G. Habetler, "Multilevel Converters for Large Electric Drives", *IEEE Trans on Industry Applications*, Vol. 35, No. 1, Jan 1999, pp-36-44.

[2]. F. Z. Peng, "A generalized multilevel inverter topology with self voltage balancing," *IEEE Trans. Ind. Applicat.* vol. 37, pp. 611–618, Apr. 2001.

[3]. K. A. Corzine, S. D. Sudhoff, E. A. Lewis, D. H. Schmucker, R. A. Youngs, and H. J. Hegner, "Use of multi-level converters in ship propulsion drives," in *Proc. All Electric Ship Conf.* vol. 1, London, U.K , pp. 155–163, Sep. 1998.

[4]. F. G. Turnbull, "*Selected harmonic reduction in static DC-AC inverters,*" IEEE Trans. Communication and Electronics, vol. 83, July 1964, pp. 374-378.

[5]. J. Sun, S. Beineke, and H. Grotstollen, "*Optimal PWM based on real-time solution of harmonics elimination equations,*" *IEEE Trans. Power Electron.*, vol. 11, no. 4, pt. 2, pp. 612–621, July. 1996.

[6]. J. Pontt, J. Rodriguez, and R. Huerta, "*Mitigation of non-eliminated harmonics of SHEPWM three-level multi-pulse three-phase active front end converters with low switching frequency for meeting standard IEEE-519-92,*" IEEE Trans. Power Electron., vol. 19, no. 6, pp. 1594–1600, Nov. 2004.

[7]. J. Wang, Y. Huang, and F.Z. Peng, " A practical harmonics elimination method for multilevel inverters" *in Proc. IEEE on Ind. Applicat.* vol. 3, pp. 1665-1670, Oct. 2005.

[8]. L. M. Tolbert, F. Z. Peng, and T. G. Habetler, " Multilevel PWM Methods at Low Modulation Indices", *IEEE Trans. Power Elect.* vol. 15, no. 4, pp. 719-725, Jul 2000.

[9]. J. N. Chiasson, L. M. Tolbert, K. J. McKenzie, and Z. Du, "A unified approach to solving the harmonic elimination equations in multilevel converters," *IEEE Trans. on Power Elect.* vol. 19, no. 2, pp. 478–490, Mar. 2004.

[10]. J. Vassallo, J. C. Clare, P. W. Wheeler, "Power-equalized harmonic-elimination scheme for utility-connected cascaded H-bridge multilevel convertes," *Ind Elect. Society, 2003. IECON '03. The 29th Annual Conf. of the IEEE*, vol. 2, pp. 1185 – 1190, Nov. 2003.

[11]. H. S.patel and R. G.Hoft," Generalized technique of harmonics elimination and voltage vontrol in Thyristor inverts: Part I HarmonicElimination", *IEEE Trans. Ind. Applicat.* Vol., IA-9, N0.3, JUNE, 1973, pp. 310 -317.

[12]. G. Carrara, S. Gardella, M. Marchesoni, R. Salutari, and G. Sciutto, "A new multilevel PWM method: A theoretical analysis," *IEEE Trans. Power Elect.* vol. 7, no. 4, pp. 497–505, Jul. 1992.

[13]. J. N. Chiasson, L. M. Tolbert, K. J. McKenzie, and Z. Du, "Control of a multilevel converter using resultant theory," *IEEE Trans. on Control System Tech.* vol. 11, no. 3, pp. 345–354, May 2003.

[14]. J. Sun, S. Beineke, and H. Grotstollen, "Optimal PWM based on real-time Solution of harmonic elimination equations," *IEEE Trans Power Elect.* vol. 11, no. 4, pp. 612-621, Jul. 1996.

[15]. Y. Liu, H. Hong, and A. Q. Huang, "Real-time calculation of switching angles minimizing THD for multilevel inverters with Step modulation", *IEEE Trans on Ind. Elect.* vol. 56, no. 2, Feb 2009, pp. 285-293.

[16]. J. N. Chiasson, L. M. Tolbert, Z. Du, and K. J. McKenzie, "The use of power sums to solve the harmonic elimination equations for multilevel converters," *Euro. Power Elect. Drives*, vol. 15, no. 1, pp. 19–27, Feb. 2005.

[17]. J. Pontt, J. Rodriguez, and R. Huerta, "Mitigation of non-eliminated harmonics of SHEPWM three-level multi-pulse three-phase active front end converters with low switching frequency for meeting standard IEEE-519-92," *IEEE Trans. Power Elect.*, vol. 19, no. 6, pp. 1594–1600, Nov. 2004.

[18]. Jin Wang, Damoun Ahmadi and Renxiang Wang, "Optimal PWM Method based on Harmonics Injection and Equal Area Criteria", *IEEE Energy Conversion Congress and Exposition (ECCE)*, Sep. 2009.

[19]. Damoun Ahmadi and Jin Wang, "Selective Harmonic Elimination for Multilevel Inverters with Unbalanced DC Inputs", *IEEE Vehicle Power and Propulsion Conference (VPPC)*. Sep. 2009.

[20]. Damoun Ahmadi, Jin Wang, "Full Study of a Precise and Practical Harmonic Elimination Method for Multilevel Inverters" *IEEE Applied Power Electronics Conference and Exposition (APEC)*, pp. 871 – 876, Feb. 2009.

[21]. J. Wang, D. Ahmadi, "A precise and practical harmonic elimination method for multilevel inverters," IEEE Trans. Industry Application, in press

Oscillation-Test Technique for Buck Voltage Regulator

Jing-Yi Huang, Chun-Hsun Wu, Le-Ren Chang-Chien, Soon-Jyh Chang

Department of Electrical Engineering, National Cheng-Kung University, Tainan, Taiwan

Abstract—**In this paper, a vector-less simple technique based on the oscillation-test strategy (OTS) is proposed for the test of the Buck regulator. In the test mode, the circuit under test (CUT) of the buck regulator could be easily transformed into an oscillator for generating the oscillating signal. By observing the frequency and amplitude of the oscillating signal, we could detect the CUT to see if it is faulty or fault-free. Simulation results show that the fault coverage for hard faults could reach up to 99%. Based on the facts of the simple structure and high fault coverage, OTS is considered as a useful technique in the testing of the Buck regulator.**

I. INTRODUCTION

In recent years, portable devices such as cell phones, PDAs and notebooks have become very popular due to their mobile convenience. From the viewpoint of system on a chip (SoC), multi-voltage levels are generally required in these portable devices. With battery that can only provide a single voltage, the role of a voltage regulator is to convert the battery voltage to any specified voltage for the integrated circuits.

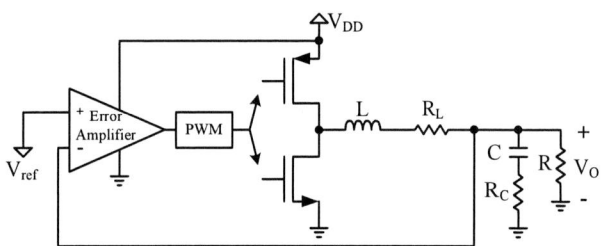

Figure 1. Simplified Buck regulator with voltage mode control

One of the most popular DC-DC voltage regulators for integrated circuits is Buck regulator. Fig. 1 shows a simplified Buck regulator with voltage mode control. Due to the high conversion efficiency, Buck regulators are widely used to supply power to the integrated circuits. Due to the trend of SoC design, we see that the Buck regulator is being integrated into the chip system. It naturally shows that the embedded Buck regulator becomes part of the integrated circuit.

With the increasing complexity of integrated circuit, yield of the final product is getting harder to be guaranteed. Hence, testing of the integrated circuits plays an important role in the design flow. Unlike digital circuits, test of analog circuits is more complex and expensive due to the long testing times, poor fault coverage, and the necessity for dedicated test equipment [1]. Therefore, enhancing the fault coverage is one of the important issues on the testing of the integrated circuit.

Fault coverage is defined as the ratio of the number of detectable faults to the total number of possible faults. As far as test of final product is concerned, a good test should indicate high fault coverage. One of the most effective methods to test the fault coverage of the analog circuit is the oscillation-test strategy (OTS). In 1996, the OTS methodology based on the design-for-testability (DFT) skill was presented to reduce the complexity of analog or mix-signal testing [2]. The result was promising with very high fault coverage. As Buck regulators become part of the analog system, OTS is a good technique to check whether the regulator is faulty or not.

The presentation of this paper is organized as follows. Section II briefly describes an overview of OTS. Section III introduces the OTS application to the testing of Buck regulators. Fault model and simulation results are represented in Section IV. Conclusion is drawn in Section V.

II. OSCILLATION TEST STRATRGY OVERVIEW

OTS is a conceptual vector-less test method which eliminates the complexity of test generation (TG). The principle of OTS is described as follows: A large circuit under test (CUT) is divided into smaller functional blocks, and each of these blocks is converted into an oscillator. The oscillation frequency could reflect the function of CUT. If some of these blocks are faulty, the oscillation frequency will deviate from that of the fault-free circuit.

The general way to perform OTS in CUTs is to create adequate feedback loop within the circuit for transforming CUTs into oscillators. In this way, the oscillation signal could be measured at circuit output without any input test vector. The oscillation frequencies of the output signal can be predicted using Barkhausen criterions [1] or Routh-Hurwitz rules [3]. In addition to the oscillation frequency, oscillation amplitude and supply current could be taken into account to improve the fault coverage [4]. Since OTS is suitable to perform built-in self test (BIST) in a chip system, OTS has been applied to test OPamps [5], filters [1], [6]-[7] and analog to digital converters (ADCs) [8] with high fault coverage. In next section, detailed illustration of OTS employed in the chip-scaled Buck regulators will be presented.

III. APPLY OTS TO TEST BUCK REGULATOR

In the hardware configuration, a Buck regulator is converted into an oscillator for further test in its oscillation frequency and amplitude. In this case, we regard the error amplifier, the two power MOSFETs and the output LC filter as a CUT which is illustrated in Fig. 2. The two power MOSFETs form an inverter to drive the LC filter. The structure of the error amplifier is shown in Fig. 3 [9].

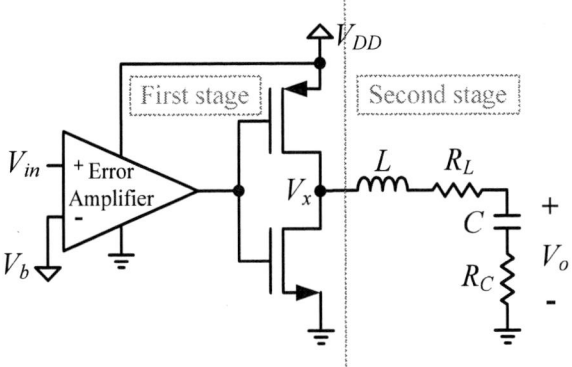

Figure 2. CUT based on Buck regulator

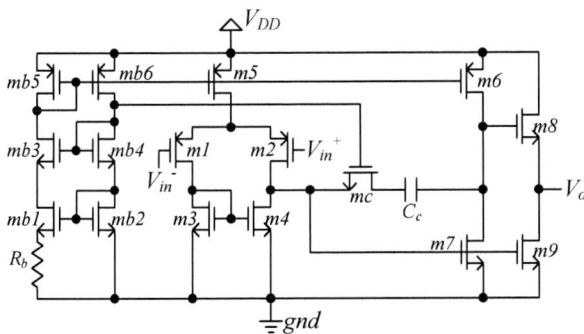

Figure 3. The structure of error amplifier

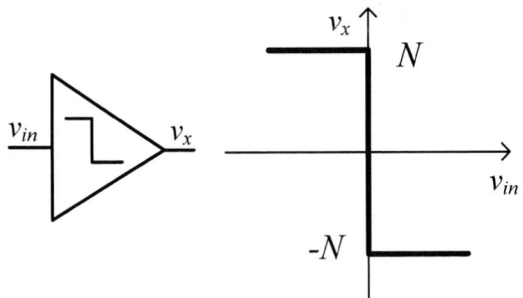

Figure 4. Small-signal model and transfer curve of the first stage in Fig. 2

The CUT shown in Fig. 2 can be divided into two stages. The first stage consists of the open-loop error amplifier with a bias voltage V_b as well as the inverter. Due to the high gain and inversion feature, once the voltage difference between V_{in} and V_b is not equal to zero, the output of the first stage V_x would saturate to V_{DD} or gnd like an inverting relay. The bias voltage V_b is set to be $0.5V_{DD}$ such that the first stage would act as a symmetric inverting relay. Fig. 4 shows the equivalent small-signal model and transfer curve of the first stage. In Fig. 4, v_{in} represents the voltage difference between V_{in} and V_b in Fig. 2. N represents the saturated voltage and it is equal to $1.65V$ in this case.

The nonlinear transfer function of the v_{in}-to-v_x transfer characteristic in Fig. 4 could be approximated by using the describing function method [10]. It could be expressed as

$$H_1(A) = \frac{4 \cdot N}{\pi \cdot A} \angle - \pi. \tag{1}$$

In (1), A is the amplitude of the sinusoidal input voltage and the $-\pi$ in the phase represents the inversion feature of the first stage. Therefore, the first stage contributes a $-180°$ phase shift.

The second stage in Fig. 2 includes the output LC filter. The small-signal transfer function could be expressed as follows.

$$H_2(j\omega) = \frac{Z_C(j\omega)}{Z_L(j\omega) + Z_C(j\omega)} \tag{2}$$

where

$$Z_C(j\omega) = \frac{1}{j\omega C} + R_C \tag{3}$$

$$Z_L(j\omega) = j\omega L + R_L \tag{4}$$

$Z_L(j\omega)$ and $Z_C(j\omega)$ are the impedance of output inductor and capacitor including the effective series resistors, R_L and R_C, respectively.

The second stage circuit is a low pass filter. It transforms the square wave from the first stage's output to a sinusoidal wave. Noted that the phase delay induced by the second stage is less than $180°$ because an ESR zero may boost up the phase from the phase drop caused by the complex poles. As a result, the total phase shift of the two stages in Fig. 2 could not achieve $360°$ such that it fails to meet the phase conditions from Barkhausen criterions.

In order to sustain an oscillation for this CUT, additional phase delay must be provided. As a result, the CUT in Fig. 2 is reconfigured by adding a voltage divider into the feedback path of the first stage. The new CUT is shown in Fig. 5. The function of the voltage divider is to provide a hysteretic band which results in extra delay. The voltage division ratio, β, could be expressed as

$$\beta = \frac{R_y}{R_x + R_y}. \tag{5}$$

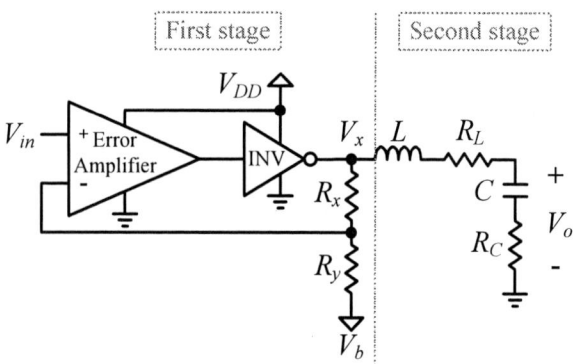

Figure 5.   Reconfigured CUT

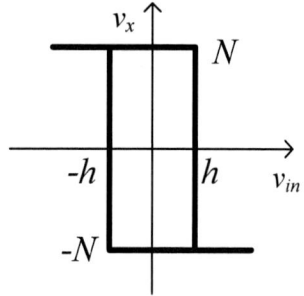

Figure 6.   Small-signal transfer curve of the first stage in Fig. 5

TABLE I.   PARAMETERS OF THE CUT IN FIG. 5

V_{DD}	3.3V
L	10µH
R_L	100mΩ
C	10µH
R_C	100mΩ
V_b	1.65V
N	1.65V
β	0.1 (Rx=9kΩ, Ry=1kΩ)
h	0.165V

Fig. 6 shows the small-signal transfer curve of this reconfigured CUT, where h is the hysteretic band controlled by β and N by the following relation,

$$h = \beta \cdot N . \quad (6)$$

The transfer characteristic in Fig. 6 is just like one of an inverting Schmitt trigger, which is equivalent to

$$H_1'(A) = \frac{4 \cdot N}{\pi \cdot A} \angle -\pi - \sin^{-1} \frac{h}{A} . \quad (7)$$

From (7), we could find the hysteretic band h induces extra delay. All parameters of the CUT in Fig. 5 are listed in Table I.

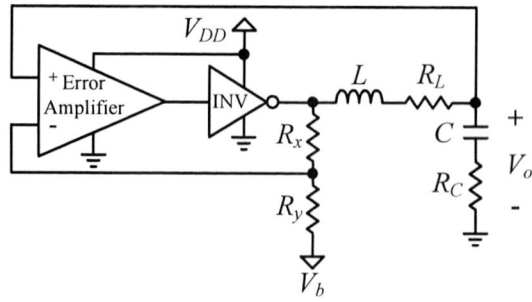

Figure 7.   Oscillator based on Buck regulator

In this paper, we use a direct feedback from V_o to the "+" input of error amplifier to transform the CUT into an oscillator. The schematic diagram of the circuit is illustrated in Fig. 7. The loop gain could be written from (2) and (7) as

$$L(A, j\omega) = H_1'(A)H_2(j\omega) \quad (8)$$

where A is the amplitude of V_o.

From Barkhausen criterions, the oscillation conditions for the oscillator in Fig. 7 should be

$$\left| L(A_{osc}, j\omega_{osc}) \right| = 1 \quad (9)$$

$$\angle L(A_{osc}, j\omega_{osc}) = 360° . \quad (10)$$

We could solve (9) and (10) by using the parameters from Table I. The oscillation frequency and amplitude are predicted as

$$f_{osc} = \frac{\omega_{osc}}{2\pi} = 36.43(kHz) \quad (11)$$

$$A_{osc} = 0.51(V) \quad (12)$$

$$A_{osc,pp} = 2 \cdot A_{osc} = 1.02(V) . \quad (13)$$

Following that, HSPICE is employed to verify the predicted frequency and amplitude of the oscillating signal. The simulated waveform of fault-free CUT is shown in Fig. 8 and the results are summarized in Table II. It shows that the simulated results of the oscillation frequency and peak-to-peak amplitude are *37.64kHz* and *0.9V,* respectively. These values are close to those we predicted from the mathematical calculation from (11) and (13).

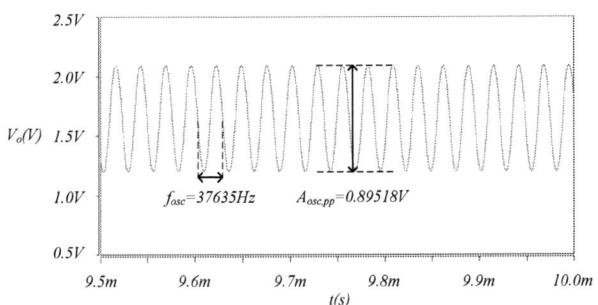

Figure 8. Simulation waveform of fault-free CUT

TABLE II. OSCILLATION FEATURES OF FAULT-FREE CUT

Frequency(f_{osc})	37.64kHz
Peak-to-peak amplitude ($A_{osc,pp}$)	0.90V
Average	1.65V

Although the analytical results are close to the simulated ones, little difference could still exist from the following reasons. First, the describing function is just an approximation of the nonlinear characteristic, and it is assumed that the input to the nonlinear block is a purely sinusoidal wave. In reality, the output waveform V_o may consist of some higher-order harmonics. Second, the error amplifier and the inverter might have some unexpected effects such as propagation delay or offset which cause unpredictable phase shift and gain variation. These unexpected effects would also affect the simulation results.

In the following section, we will use the results in the Table II as the nominal values for the reference of fault detection.

IV. FAULT MODEL AND FAULT SIMULATION

Conventionally, analog faults can be classified into two types: hard faults and soft faults [2]. Hard faults, also known as catastrophic faults, could be open circuit, short circuit or a large deviation from the designed values of the CUT parameters. They are usually caused by random defects like dust particles. If there are some hard faults in the CUT, the original circuit may not work correctly and may cause system failure. On the other hand, soft faults are caused by the fluctuation of manufacturing process, and usually result in a small deviation from their normal performance.

From [2], it shows that most of analog faults are hard faults such as shorts and opens in transistors. It also states that if a test method can cover 100% of hard faults, it can also indicate the majority of soft faults. For this reason, this paper only considers the opens and shorts of transistors as hard faults.

In this paper, an open fault is modeled as a large resistor in series with the faulty node. On the other hand, a short fault is modeled as a small resistor between the two short nodes. During fault simulation, the open and short faults are modeled using the resistors 10MΩ and 10Ω, respectively. It is noted

that the gate opens of transistors (floating gates) cannot be directly modeled as a large resistor in series with the gates because the input resistance of gates is up to tens of MΩ. So we employ another model as described in [4] to model open fault in gates. This model utilizes a zero-current source to reflect the small current through the open net and the voltage discontinuity due to the open. The fault models mentioned above are shown as Fig. 9 and Fig. 10.

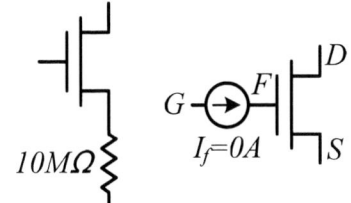

Figure 9. Fault models for open faults

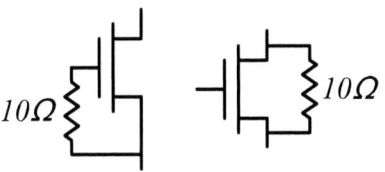

Figure 10. Fault models for short faults

In Fig. 9, the left model is equivalent to a drain open or a source open. On the other hand, the right model is equivalent to a gate open where F is the floating terminal and G is the driving node. The left side model in Fig. 10 represents a gate-drain short or a gate-source short. The right side model is a drain-to-source short fault.

TABLE III. HARD FAULT COVERAGE FROM SIMULATION RESULTS

Oscillation features / Fault types	Oscillation frequency	Oscillation amplitude	Both features
Open faults	98.44%	98.44%	98.44%
Short faults	100%	100%	100%
Hard faults	99.07%	99.07%	99.07%

The fault simulation is performed with single fault injection, where the faults are extracted from the schematic as shown in Fig. 2 and Fig. 3. Only the opens of the three terminals (drain, source and gate) and the shorts between the three terminals of a transistor are simulated. There are 64 open faults and 43 short faults including the power MOSFETs totally in this case. In order to judge the CUT is fault-free or faulty, we have to define a tolerance band. The tolerance band is defined as ±5% around the normal oscillation frequency and amplitude of output voltage in this paper. If either of the oscillation frequency or amplitude is out of the tolerance band, we regard the CUT is faulty.

From the simulation results, we find that most of the faulty circuits with hard faults lose oscillation and their output voltage V_o would saturate to V_{DD} or gnd. The fault coverage of hard faults is summarized in Table III. The table shows that the fault coverage of hard faults is up to *99%*.

TABLE IV.   SOFT FAULT COVERAGE FROM SIMULATION RESULTS

Oscillation features / Fault types	Oscillation frequency	Oscillation amplitude	Both features
L	75%	50%	75%
C	100%	100%	100%
R_L	100%	25%	100%
R_C	100%	75%	100%
Soft faults	93.75%	53.13%	93.75%

In addition to the hard fault detection, we also inject soft faults into this CUT. The variation of device parameters in the output filter is a common problem which would cause stability issues or degrade the performance of regulators. Thus, we take the variation of device parameters in the output filter into consideration. The inductance, capacitance and ESRs of output filter are individually changed every *5%* from *-20%* to *+20%* from its nominal values. Simulation results are summarized in Table IV. From Table IV, we could find that the oscillation amplitude to the variation of output filter is less sensitive compared to the oscillation frequency.

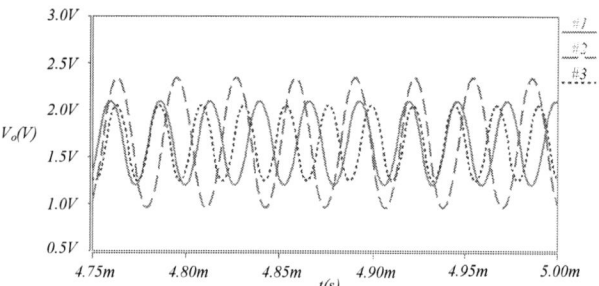

Figure 11. Simulation waveforms comparison

There are three examples of simulation waveforms named *#1*, *#2* and *#3*. The three waveforms are shown in Fig. 11. *#1* is the fault-free waveform. #2 is the waveform with an open-drain fault in the transistor mb3 in Fig. 3. The oscillation frequency and amplitude deviations of *#2* from the fault-free ones are *-16.34%* and *54.44%* respectively, so both of them are larger than the tolerance band. Thus, we could detect this hard fault and regard the CUT as faulty. On the other hand, #3

is the waveform with a *-20%* capacitance variation. The oscillation frequency and amplitude deviations from the fault-free ones are *23.09%* and *3.33%* respectively. The amplitude variation is less than the tolerance band, but the oscillation variation is larger than the tolerance band. Consequently, this CUT is still regarded as faulty. So, the comparison in Fig. 11 validates the principle of OTS.

From these simulation results, we find that the high fault coverage, especially in hard fault detection, is very promising.

V.   CONCLUSION

This paper describes the OTS application to the test on a chip-scaled Buck regulator. Without any input test vector, OTS reduces the complexity of test generation and improves the test efficiency. Just adding a few components to form a simple feedback loop, high fault coverage could be achieved for the test of Buck regulator, which is very promising in the test of IC power manufacturing.

ACKNOWLEDGMENT

The authors would like to thank the National Science Council for the support of research project under grant (NSC 96-2221-E-006-316-MY2) and the National Chip Implementation Center (CIC) for chip fabrication (D35-97E-07).

REFERENCES

[1] K. Arabi and B. Kaminska, "Oscillation-Test Methodology for Low-Cost Testing of Active Analog Filters," IEEE Trans. Instrumentation and Measurement, Vol. 48, no. 4, pp. 798-806, Aug. 1999.

[2] K. Arabi and B. Kaminska, "Oscillation-Test Strategy for Analog and Mixed-Signal Integrated Circuits," IEEE Conf. VLSI Test Symposium, pp. 476-482, 1996.

[3] J. Roh and J. A. Abraham, "A Comprehensive Signature Analysis Scheme for Oscillation-Test," IEEE Trans. CAD of ICs and Sys., Vol. 22, no. 10, pp. 1409-1423, Oct. 2003.

[4] J. Font, J. Ginard, E. Roca, J. Segura and E. Garcia, "Oscillation-Test Technique for CMOS Operational Amplifiers by Monitoring Supply Current," Analog Integrated Circuits and Signal Processing, Vol. 33, no.2, pp. 213-224, Nov. 2002.

[5] K. Arabi and B. Kaminska, "Design for Testability of Embedded Integrated Operational Amplifiers ," IEEE J. Solid-State Circuits, Vol. 33, no. 4, pp. 573-581, April 1998.

[6] G. Huertas, D. Vazquez, E. J. Peralias, A. Rueda and J. L. Huertas, "Practical oscillation-based test of integrated filters," IEEE J. Design and Test of Computers, Vol. 19, no.6, pp. 64-72, Nov.-Dec. 2002.

[7] M. W. T. Wong, "On the Issue of Oscillation Test Methodology," IEEE Trans. Instrumentation and Measurement, Vol. 49, no. 2, pp. 240-245, April 2000.

[8] K. Arabi and B. Kaminska, "Testing Analog and Mixed-Signal Integrated Circuits Using Oscillation-Test Method," IEEE Trans. CAD of ICs and Sys., Vol. 16, no. 7, pp. 745-753, July 1997.

[9] D. Johns and K. Martin, "Analog Integrated Circuit Design," John Wiley & Sons, 1997.

[10] G. F. franklin, J.D. Powell and A. Emami-Naeini, "Feeback Control of Dynamic Systems," 5th edition, Pearson Education, 2006.

Core Loss Predictions for General PWM Waveforms from a Simplified Set of Measured Data

Charles R. Sullivan, John H. Harris
Thayer School of Engineering at Dartmouth
charles.r.sullivan@dartmouth.edu
http://engineering.dartmouth.edu/inductor
8000 Cummings Hall, Hanover, NH 03755, USA

Edward Herbert
closs@eherbert.com
http://fmtt.com/
1 Dyer Cemetery Rd., Canton, CT 06019, USA

Abstract—A method to generalize square-wave core-loss data to predict core loss with any common rectangular voltage waveform is proposed. An automated measurement system was used to collect the required square-wave core characterization data for ferrite and powdered-iron cores, and to collect additional data to assess the accuracy of the method for other voltage waveforms. Measurement data is presented and the application of the method in power-electronics design is discussed.

I. INTRODUCTION

Core-loss data published by core manufacturers is based on sinusoidal excitation, whereas most applications in switching power supplies and other types of power electronics circuits use rectangular voltage waveforms. Rectangular waveforms can be described by the voltage, period, and duty cycles of the positive and negative portions of the waveform. This leads to a wide diversity of different possible test conditions, and it is not practical for manufacturers to test all possible waveforms that might be used by customers. Approximate methods to estimate expected core loss with rectangular waveforms based on sinusoidal data [1]–[8] exist, but are difficult to use, are inherently limited in accuracy, and are not in wide use in industry.

In this paper, we introduce a new approach that uses a simplified set of square-wave measurements to produce easy-to-use data that can be applied to calculate loss for any rectangular-voltage waveform. This approach is expected to provide higher accuracy than is possible starting from data based on sinusoidal waveforms, and is expected to be easier to use than existing methods for nonsinusoidal waveforms. The method can be applied to computerized optimizations or in hand calculations using graphical data. Although the data required is different from conventional sinusoidal measurements, the amount of data needed is no more than the amount of data collected in traditional loss characterization.

In order to implement and evaluate the new method, we have developed an automated excitation and data collection system under computer control. This allows rapidly gathering the proposed square-wave characterization data set, and also facilitates scanning through other rectangular waveform

sets in order to assess the accuracy of the generalization from the characterization data.

Previous methods for predicting core loss with rectangular waveforms based on sinusoidal data are reviewed in Section II. The new calculation method is described in Section III and the measurement system in Section IV. Measurement results are presented and used to assess the accuracy of the method in Section V. A guide to applying the method in practical design is provided in Section VI. Section VII further discusses the future application and improvement of this approach.

II. PREVIOUS METHODS FOR CORE LOSS WITH NONSINUSOIDAL WAVEFORMS

For sinusoidal waveforms, loss is often estimated by a power law equation [9], [10]

$$\overline{P_v} = k f^\alpha \hat{B}^\beta \tag{1}$$

where $\hat{B}$ is the peak flux amplitude, $\overline{P_v}$ is the time-average power loss per unit volume, and f is the frequency of sinusoidal excitation, and k, α, and β are constants found by curve fitting. A similar equation, but without the frequency dependence, was proposed by Steinmetz in 1892 [11], and so (1) is often referred to as the Steinmetz equation. Unfortunately, the Steinmetz equation, as well as the data provided by manufacturers of magnetic materials, is based only on sinusoidal excitation, and nonsinusoidal waveforms result in different losses [1], [2], [5]. DC bias can also significantly affect loss [12], [13], [14].

More detailed models, based on physical phenomena producing loss, have been studied [15]–[18]. However, especially for ferrites, there is not yet a clear consensus on a practical physically-based model that properly includes dynamic and nonlinear effects [5].

Initial attempts to make use of Steinmetz-equation parameters and extend the calculation to address arbitrary waveforms allowed improved loss estimates, but have significant limitations. The "modified Steinmetz equation" (MSE) [1], [2], [3] works well for waveforms with small harmonic content, but exhibits anomalies with large harmonic content [5], as does the model introduced in [4],

978-1-4244-4782-4/10 $26.00 © 2010 IEEE

as discussed in [6]. The "generalized Steinmetz equation" (GSE) was introduced in [5] to overcome anomalies in the MSE, and although it overcomes the problems with the MSE, it has poor accuracy in some regions [6].

A satisfactory method of using Steinmetz-equation parameters to roughly estimate loss with non-sinusoidal waveforms, the "improved generalized Steinmetz equation" (iGSE) was introduced in [6]. The same equation was independently discovered in [7], [8], where it was called the natural Steinmetz extension (NSE). Comparisons of different approaches in [24] confirm results in [6], [7], [8] showing that this method can work well in many situations.

An additional refinement introduced in [6] is to decompose a waveform that includes minor loops in the hysteresis curve, and separately analyze the loss in each minor loop. This was shown to be essential for accurately modeling such cases. An automated algorithm is described in [6] to perform this decomposition, but is unnecessary for waveforms without minor loops.

Despite these improvements, the iGSE remains an approximate prediction method, and, in particular, is dependent on the accuracy of the underlying Steinmetz model for sinusoidal loss. Unfortunately, the best-fit Steinmetz parameters are known to vary with frequency [25], [5]. For waveforms with a harmonic content over a wide frequency range, choosing the appropriate parameters can be problematic [5]. Some solutions to this problem that work for sinusoidal waveforms (e.g., [25], [26]) are not applicable with the iGSE. Summing several power-law terms is one option that can be used to better capture the wide-range frequency behavior while retaining compatibility with the general approach of the iGSE, at the price of additional complexity [7].

The approach in this paper is to directly measure loss with square waveforms, rather than trying to extend data from sinusoidal loss measurements to square waveforms. The advantages relative to the iGSE and related methods are both simplicity and accuracy. The challenge to developing such a method is to be able to take data for a reasonably constrained set of parameters, and be able to use the results to predict loss for a wider range of practical waveforms. This is discussed in the next section.

III. CALCULATING CORE LOSS FROM A SIMPLIFIED DATA SET

Consider a core with voltage waveforms such as those shown in Fig. 1, typical of power electronics applications, applied to a winding. The flux in the core will ramp up or down during each positive or negative voltage pulse, respectively. We hypothesize that the energy loss incurred during each of these flux transitions depends only on the amplitude and duration of the pulse, and that there is no loss during periods of zero applied voltage (constant flux). If this is the case, we can decompose any of the rectangular waveform types shown in Fig. 1(b) into a set of two pulses,

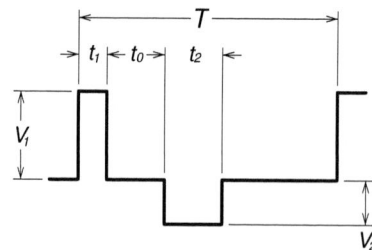

(a) Waveform parameter definitions

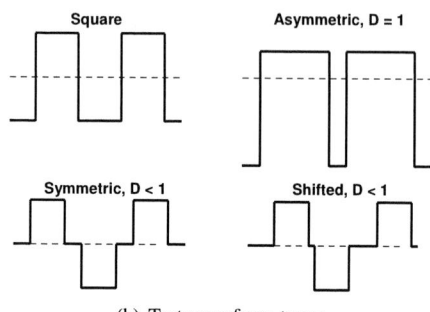

(b) Test waveform types

Fig. 1: Waveforms (voltage vs. time): parameters and test waveform types. Square waves are used for characterizing materials; the other test waveforms are used to test the validity of the composite waveform hypothesis.

calculate the energy loss associated with each pulse, and sum them to find the total energy loss per cycle. We call this hypothesis the *composite waveform hypothesis*.

If the composite waveform hypothesis proves to be a good approximation, we can predict core loss for any of the waveforms in Fig. 1(b) if we know the loss for a square pulse as a function of its amplitude and duration. While we might estimate that loss from sinusoidal data using one of the methods describe in Section II ([1]–[8]), a more accurate approach is to collect measured test data with square voltage waveforms, for which we can assume that the loss associated with each pulse is one half of the per-cycle energy loss. This requires data as a function of two parameters, such as flux amplitude and frequency, as used in conventional sinusoidal loss characterization. The parameters may also be described in terms of applied voltage per turn (corresponding to flux ramp rate) and on-time t_1 (one half the period for square waves).

The method we propose starts with characterizing a core material by measuring loss data for square waveforms. One half of the measured energy loss per cycle is the energy lost for a single pulse of the applied amplitude and on-time. If the composite waveform hypothesis is accurate, the same loss per cycle will be incurred for that applied voltage and on-time in any composite waveform. For the waveforms we consider here (Fig. 1), the waveform comprises two pulses. To find the total energy loss per cycle one sums the per-

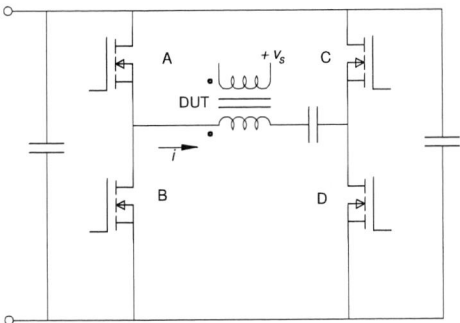

Fig. 2: Full-bridge excitation circuit. The device under test, a magnetic core, is labeled DUT.

TABLE I: Sample cores.

Manufacturer	Part	Material	Turns used
Magnetics	42206-TC	R ferrite	5
Micrometals Inc.	T80-52	–52 iron powder	21

pulse loss data for each of the two sets of pulse parameters (amplitude and on-time) from the corresponding square-wave measurements.

In Sections IV and V we report on experimental measurements conducted for two purposes: 1) to collect square-wave data for sample cores as is necessary for this method, and 2) to assess the accuracy of the method and of the composite waveform hypothesis. We note that all of the previous methods for predicting loss with non-sinusoidal waveforms discussed in Section II depend on some version of the composite waveform hypothesis, even though this assumption is rarely stated. Thus, tests of this hypothesis are important for other approaches to predicting non-sinusoidal losses as well as for validating the approach proposed here.

Predictive core loss models built up from fundamental physical principals are not available for most core loss mechanisms, and so theoretical analysis of the composite waveform hypothesis is not possible. However, for core loss produced purely by classical eddy-current effects, physical models are well established, and analytical solutions exist for some geometries. It can be shown that, for classical eddy-current loss in core material layers thicker than or comparable to an electromagnetic skin depth, the composite waveform does not always hold exactly. However, it may still be a useful approximation, and may hold exactly for other types of losses. Thus experimental evaluation is needed to assess its utility.

IV. Measurement System

We use a two-winding loss measurement [8] on toroidal core samples with 5 or 21 turns to match core characteristics to our drive system capabilities. The drive winding is connected to a full-bridge switching network through a blocking capacitor (Fig. 2). The gates of the four MOSFETs are controlled by an arbitrary function generator through a logic circuit and optically isolated gate drivers. Both the arbitrary function generator and the dc power supply feeding the bridge circuit are automatically controlled by a computer to allow synthesis of a sequence of different rectangular voltage waveforms.

Current in the drive winding is sensed with a Tektronix P 6021 wide-band passive ac current probe. This avoids the delay inherent in an active current probe; delay measurements verify that the delay is negligible. Flux is calculated from the voltage v_s across a sense winding. These signals are acquired by a digital storage oscilloscope under computer control to automatically collect waveforms from a sequence of measurements. After the waveforms are allowed to reach steady-state, 512 periods are averaged, and the average is recorded with 1000 points per period. Core loss and other parameters are calculated off line from the acquired data.

The core temperature can be controlled by immersion in a heated oil bath. All the results reported here are for an oil-bath temperature of 80 °C. The automated data collection allows acquiring data for a single excitation in less than two seconds; a pause of about two seconds precedes the next excitation. Even without the oil bath, this results in very little temperature rise; with the oil bath, temperature deviations are negligible.

For verification of the test system, a large air-core toroidal transformer was constructed. This would, in principle, provide a zero-core-loss reference. However, large stray capacitance in the transformer led to excessive ringing in the waveforms and negative power dissipation numbers. Future work will develop a better reference design to allow a useful air-core test.

V. Experimental Results

Two sample cores were tested: one ferrite core and one powdered-iron core, as listed in Table I.

A. Characterization

Fig. 3 presents square-wave loss data in two different formats for the ferrite core. Fig. 3(a) uses a format similar to that used for sinusoidal loss data on many datasheets. Fig. 3(b) shows a Herbert curve in which core loss is plotted as a function of on-time (t_1 in Fig. 1(a)), parameterized by the voltage per turn during that on-time [27]. The Herbert curve is convenient for use in design as discussed in Section VI; in addition, it directly illustrates the effect of period on power loss and can help guide the choice of switching frequency. Starting at a low switching frequency, increasing frequency (and thus reducing the pulse width) decreases losses, but beyond a certain frequency, further increases not only result in diminishing returns, but actually increase core losses. This point corresponds to pulse widths

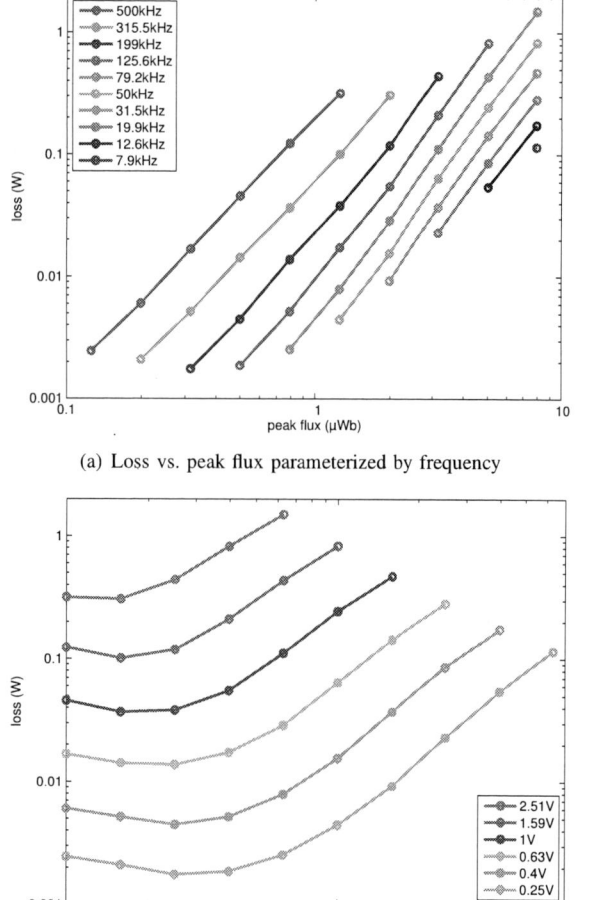

(a) Loss vs. peak flux parameterized by frequency

(b) Herbert curves: loss vs. on-time parameterized by voltage per turn

Fig. 3: Square-wave loss data for the ferrite core in Table I presented in two different formats.

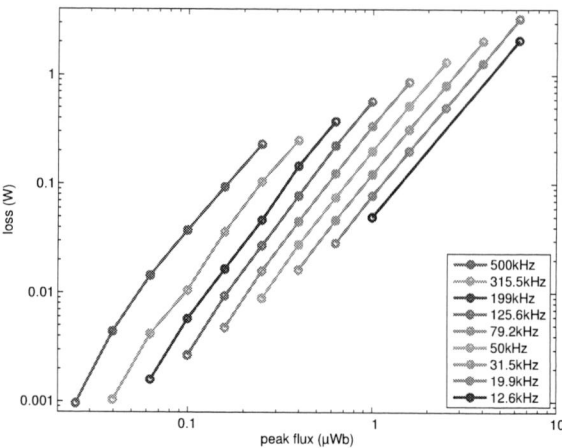

(a) Loss vs. peak flux parameterized by frequency

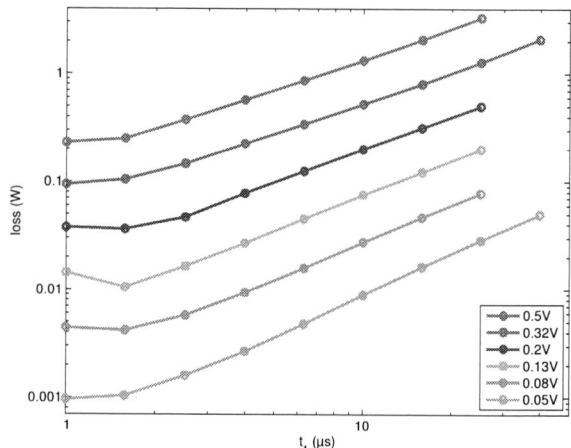

(b) Herbert curves: loss vs. on-time parameterized by voltage per turn

Fig. 4: Square-wave loss data for the powdered-iron core in Table I presented in two different formats.

of 1.5 to 3 μs for the ferrite material tested, and thus periods of 3 to 6 μs, and square-wave frequencies of 167 to 333 kHz. This is generally consistent with the behavior seen in plots of "performance factor" $B \cdot f$ for fixed power loss provided by some manufacturers [28], [29], [30]. However, the frequency beyond which performance degrades is lower in our data than in plots of performance factor for the same material (about 600 kHz [28]), presumably because of the harmonic energy content of the square-wave excitation.

Square-wave loss data for the powdered-iron core is shown in the same two formats in Fig. 4, and shows similar trends though the values are different.

B. Verification

Additional data was collected to assess the accuracy of the method described in Section III for predicting loss for other waveforms using only data from square-wave measurements. The first category of these tests is experiments

using asymmetric waveforms as shown in Fig. 1(b) (upper right). The results of these experiments are plotted in Fig. 5. Each curve is for fixed width and amplitude of the first pulse (fixed V_1 and t_1 as defined in Fig. 1(a)) with the width of the second pulse (t_2) varying. The amplitude of the second pulse was adjusted for zero average voltage. The widths and amplitudes were all chosen to match data in the original characterization data set such that no interpolation was needed to predict loss, and the energy loss per cycle could be predicted from two points in the characterization data: the energy loss per cycle for a square wave of amplitude V_1 and half-period t_1 ($E_{\mathrm{sqr}}(V_1, t_1)$) and the energy loss per cycle for a square wave of amplitude V_2 and half-period t_2, as

$$E_c = \frac{1}{2}\left(E_{\mathrm{sqr}}(V_1, t_1) + E_{\mathrm{sqr}}(V_2, t_2)\right) \qquad (2)$$

978-1-4244-4782-4/10 $26.00 © 2010 IEEE

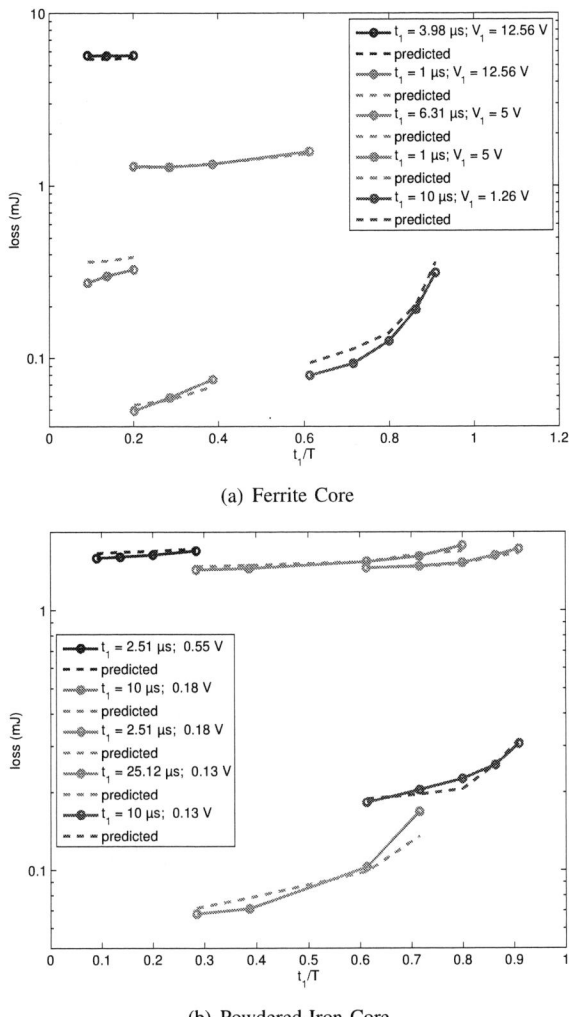

(a) Ferrite Core

(b) Powdered-Iron Core

Fig. 5: Losses with asymmetric waveforms as predicted by the proposed approach compared to measured results.

The measured asymmetric waveform loss is compared to loss predicted from (2) in Fig. 5, showing excellent matching over a wide range of asymmetry ratios (t_1/T) and amplitudes, for both the ferrite core and the powdered iron core. This confirms that the composite waveform hypothesis is a good approximation for asymmetric waveforms.

Test results for waveforms with a zero-voltage time t_0 between pulses are shown in Figs. 6 and 7. In each test, the positive and negative voltage pulses have constant amplitude and duration (as listed in the figure legend), but the zero-voltage time between pulses, t_0, is varied. In Figs. 6(a) and 7(a), the waveform is always symmetric and the overall period is expanded as t_0 is increased (marked "Symmetric, $D < 1$" in Fig. 1(b)). In Figs. 6(b) and 7(b), the period remains constant but the waveform is skewed with one of the two zero-voltage periods shrinking as the

other expands (marked "Shifted, $D < 1$" in Fig. 1(b)).

Based on the composite waveform hypothesis (see Section III), we would expect Figs. 6(a) and 7(a) to show constant energy loss per cycle as the zero-voltage time, and thus the period, increases, with no loss occurring during the zero-voltage time. The data in Fig. 6(a), for the ferrite core, show significant variations as t_0 increases, as much as about 40%, in some cases increasing and in others decreasing. The data in Fig. 7(a), for the powdered-iron core, show much less variation, matching the expectation from the composite waveform hypothesis very closely, with only slight increases in loss for long off-times, which may be a result of measurement artifacts.

The results for the shifted pulse waveforms, in Figs. 6(b) and 7(b), show little variation in loss as the pulse position is shifted (increasing one off-time while decreasing another), as would be expected from the composite waveform hypothesis, but the ferrite-core loss is slightly different from that predicted from the square-wave characterization data using (2), whereas the powdered-iron core matches the predicted loss more closely. This is consistent with the results shown in Figs. 6(a) and 7(a). For the ferrite core, the relatively low variation in loss as the pulse position shifts, compared to that shown in Figs. 6(a) and 7(a), could be explained by the effects of one off-time increasing offsetting the effect of the other decreasing for a net zero effect on loss. Alternatively, if the trends shown in Figs. 6(a) and 7(a) are due to measurement artifacts associated with the expanding period, this could also explain the relatively flat behavior seen in Figs. 6(b) and 7(b), because the shifted pulse experiments are immune to any errors associated with waveform period. However, the difference in behavior between the two cores seen in Fig. 6(a) and 7(a) indicates that the trends seen in the data are in fact due to the true behavior of the cores, not to any unexpected measurement artifacts.

VI. DESIGN

The loss data derived from square-wave measurements can be provided to a magnetics designer in various forms, including tabulated data or curve-fit functions for use in software, and various types of graphical presentation. The loss data can be presented as loss per unit volume, or as total loss for a specific core, to simplify calculations for the designer. Here we discuss working from graphical data in the "Herbert curve" format of Fig. 3(b). These curves are parameterized by voltage per turn applied to a winding. It's also possible to provide curves like this for a specific *component* with a given number of turns, parameterized by voltage.

In general, for waveforms as shown in Fig. 1(a), based on (2), one can calculate loss from a Herbert plot as

$$P = \frac{1}{T}\left(P_{\text{sqr}}(V_1/N, t_1) \cdot t_1 + P_{\text{sqr}}(V_2/N, t_2) \cdot t_2\right) \quad (3)$$

978-1-4244-4782-4/10 $26.00 © 2010 IEEE

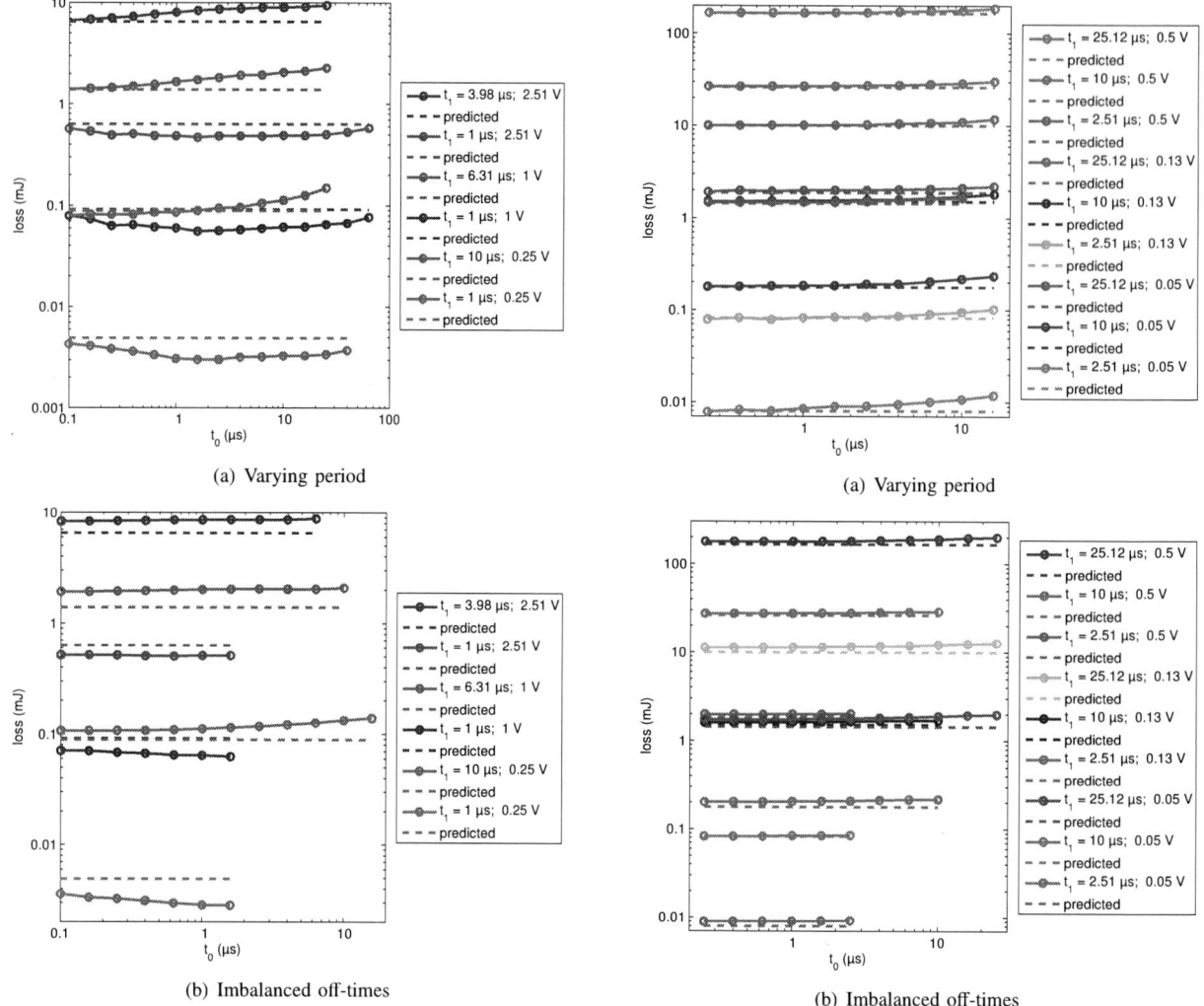

(a) Varying period

(b) Imbalanced off-times

Fig. 6: Experimental data testing the extension of ferrite-core square-wave data to waveforms incorporating zero-voltage time t_0. This is done by varying the period and extending off-time t_0, or by shrinking one zero-voltage time while expanding the other to maintain a constant period. The legend gives the on-time t_1 and the per-turn pulse voltage for each curve.

(a) Varying period

(b) Imbalanced off-times

Fig. 7: Experimental data testing the extension of powdered-iron-core square-wave data to waveforms incorporating zero-voltage time t_0. This is done by varying the period and extending on-time t_1, or by shrinking one zero-voltage time while expanding the other to maintain a constant period.

In the case of symmetric waveforms, $P_{\text{sqr}}(V_1/N, t_1) = P_{\text{sqr}}(V_2/N, t_2)$, and the loss calculation reduces to

$$P = \frac{2t_1}{T} P_{\text{sqr}}(V_1/N, t_1). \qquad (4)$$

Consider, for example, a component operating at 50 kHz with a symmetric waveform with a duty cycle of 63%, 12 turns, and a pulse amplitude of 4.8 V. The period is 20 μs, and the positive and negative pulse widths are $t_1 = t_2 = 0.63 \cdot 10\ \mu s = 6.3\ \mu s$. To find the correct curve to examine in Fig. 3(b), we find the voltage per turn which is 4.8/12 = 0.4 V. As shown in Fig. 8, we can read off the power loss for this pulse width and voltage per turn as $P_{\text{sqr}} = 7.9$ mW. According to (4) we scale this result by the ratio

$2t_1/T = 12.6\ \mu s/20\ \mu s$, to get a predicted power loss of $7.9 \cdot 0.63 = 5$ mW.

An interesting design space to explore is to maintain constant frequency and average voltage, but to vary the pulse width and period. The relevant data is along a curve of constant volt-seconds on the Herbert plot—the dashed line in Fig. 9. To get power loss from these points, assuming constant frequency of 50 kHz, we then scale these points down by the ratio $2 \cdot t_1/20\ \mu s$ to get the solid line in Fig. 9. This rise of this curve to the left illustrated the disadvantage of using shorter duty cycles for a given average voltage or volt-second requirement.

As an asymmetric example, consider the same 12-turn winding, with a 12 V, 10 μs pulse applied in one direction,

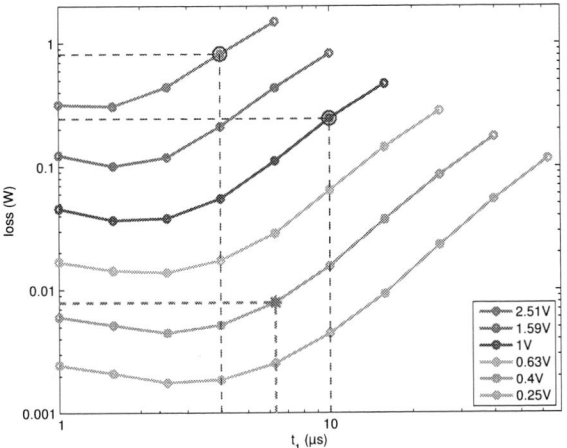

Fig. 8: Reading data from a Herbert plot for the examples discussed in the text. The lower red point marked with * is for the first, symmetric example. The upper blue circles are for the second, asymmetric example.

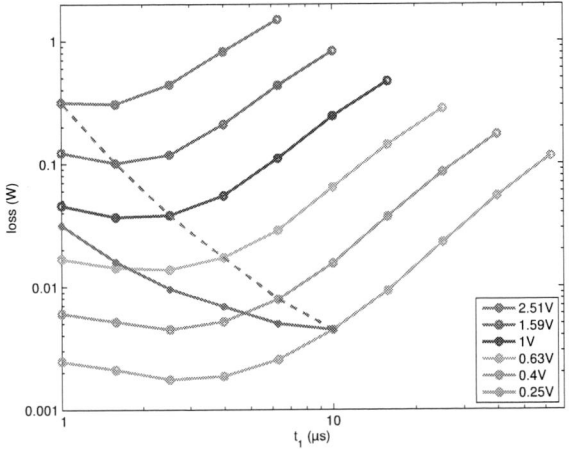

Fig. 9: Example exploring different duty cycle options for a constant average voltage (constant volt-seconds). The dashed line is the raw data points used from the plot; the solid line shows the result scaled by the ratio of on-time to period to calculate actual loss, as discussed in the text.

and a 30 V, 4 μs pulse applied in the other direction, with a 20 μs overall period (50 kHz) as before (the waveform includes a total of 6 μs of zero-voltage time). $P_{\text{sqr}}(V_1/N, t_1)$ and $P_{\text{sqr}}(V_2/N, t_2)$ are read off Fig. 8 as 244 mW and 818 mW. The overall power loss can then be found from (3) as

$$P = 50 \text{ kHz} \,(244 \text{ mW} \cdot 10 \,\mu\text{s} + 818 \text{ mW} \cdot 4 \,\mu\text{s}) = 286 \text{ mW}. \tag{5}$$

VII. Discussion

The results in Figs. 5 and 5(b) show that the composite waveform hypothesis holds well for asymmetric waveforms, and the method provides excellent accuracy. Fig. 7 shows that, for the powdered-iron core tested, it also holds very well for waveforms with zero-voltage periods, although zero voltage times can cause significant deviation in the ferrite core tested. This variation is not predicted by the composite-waveform hypothesis; nor is it predicted by any of the methods discussed in Section II. Additional work to better characterize and model this behavior could lead to more accurate loss predictions. However, even with this error, the approach described here is expected to be more accurate than other methods which are subject to the same error, and additionally entail error due to using sinusoidal data to predict square-wave loss.

In addition to being more accurate than other methods, the new approach is also easier to use than methods like the iGSE. Thus, we believe that it would be beneficial for core manufacturers to characterize square-wave loss and provide that data graphically, electronically or both, either on a per-unit-volume basis or on a per-core basis.

As presented here, the method is only applicable to waveforms with one positive voltage pulse and one negative pulse. However, it could also be easily applied to waveforms with minor loops by separating the minor loops following the approach in [6], as long as each constituent loop comprises only one positive pulse and one negative pulse. Adapting the method to waveforms with a series of voltage pulses of the same polarity but differing amplitudes is less straightforward. The corresponding analysis in the iGSE (eq. (13) in [6]) includes a factor that depends on the total flux excursion as well as the flux change for a given pulse, and it may be necessary to introduce similar factors to accurately model losses in such cases using square-wave loss data. However, most power electronics applications use waveforms with only one positive voltage pulse and one negative pulse, such that that the analysis here applies directly.

VIII. Conclusion

The proposed measurement and loss calculation approach allows generalizing square-wave core-loss data to predict core loss with any common rectangular voltage waveform. An automated measurement system has been used to collect the required square-wave core characterization data for ferrite and powdered-iron cores, and to collect additional data to assess the accuracy of the method for other voltage waveforms. Measurements show good correlation, but also exhibit behavior not yet explained by published models, which may lead to new insights and more accurate models. Despite the minor discrepancies, the loss prediction method yields higher accuracy, and is easier to use, than other methods for non-sinusoidal waveforms.

IX. ACKNOWLEDGEMENTS

Thanks to the Power Sources Manufacturers' Association and the National Institute for Standards and Technology (NIST) for support of this work.

REFERENCES

[1] M. Albach, T. Durbaum, and A. Brockmeyer, "Calculating core losses in transformers for arbitrary magnetizing currents a comparison of different approaches." in *27th Annual IEEE Power Electronics Specialists Conference*, vol. 2, Jun. 1996, pp. 1463–8.

[2] J. Reinert, A. Brockmeyer, and R. De Doncker, "Calculation of losses in ferro- and ferrimagnetic materials based on the modified Steinmetz equation," in *Proceedings of 34th Annual Meeting of the IEEE Industry Applications Society*, 1999, pp. 2087–92 vol.3.

[3] A. Brockmeyer, "Dimensionierungswekzeug für magnetische bauelemente in stromrichteranwendungen," Ph.D. dissertation, Aachen University of Technology, 1997.

[4] J. Liu, T.G. Wilson, Jr., R. Wong, R. Wunderlich, and F. Lee, "A method for inductor core loss estimation in power factor correction applications," in *Proceedings of APEC 2002 - Applied Power Electronics Conference and Exposition*, 2002, p. 439.

[5] Jieli Li, T. Abdallah, and C. R. Sullivan, "Improved calculation of core loss with nonsinusoidal waveforms," in *IEEE Industry Applications Society Annual Meeting*, 2001, pp. 2203–2210.

[6] K. Venkatachalam, C. R. Sullivan, T. Abdallah, and H. Tacca, "Accurate prediction of ferrite core loss with nonsinusoidal waveforms using only Steinmetz parameters," in *IEEE Workshop on Computers in Power Electronics*, 2002.

[7] A. Van den Bossche, D. Van de Sype, and V. Valchev, "Ferrite loss measurement and models in half bridge and full bridge waveforms," in *IEEE Power Electronics Specialists Conference*, June 2005, pp. 1535–1539.

[8] A. van den Bossche and V. Valchev, *Inductors and Transformers for Power Electronics*. Taylor and Francis Group, 2005.

[9] S. Mulder, "Power ferrite loss formulas for transformer design," *Power Conversion & Intelligent Motion*, vol. 21, no. 7, pp. 22–31, Jul. 1995.

[10] E. C. Snelling, *Soft Ferrites, Properties and Applications*, 2nd ed. Butterworths, 1988.

[11] C. P. Steinmetz, "On the law of hysteresis," *AIEE Transactions*, vol. 9, pp. 3–64, 1892, reprinted under the title "A Steinmetz contribution to the ac power revolution", introduction by J. E. Brittain, in *Proceedings of the IEEE* 72(2) 1984, pp. 196-221.

[12] A. Brockmeyer, "Experimental evaluation of the influence of dc-premagnetization on the properties of power electronic ferrites," in *APEC '96. Eleventh Annual Applied Power Electronics Conference*, 1996, pp. 454–60.

[13] A. Brockmeyer and J. Paulus-Neues, "Frequency dependence of the ferrite-loss increase caused by premagnetization," in *Twelfth Annual Applied Power Electronics Conference and Exposition*, 1997, pp. 375–80.

[14] W. K. Mo, D. Cheng, and Y. Lee, "Simple approximations of the dc flux influence on the core loss power electronic ferrites and their use in design of magnetic components," *IEEE Transactions on Industrial Electronics*, vol. 44, no. 6, pp. 788–99, 1997.

[15] G. Bertotti, "General properties of power losses in soft ferromagnetic materials," *IEEE Transactions on Magnetics*, vol. 24, no. 1, pp. 621–630, 1988.

[16] ——, *Hysteresis in magnetism: for physicists, materials scientists, and engineers*. Academic Press, 1998.

[17] K. H. Carpenter, "Simple models for dynamic hysteresis which add frequency-dependent losses to static models," *IEEE Transactions on Magnetics*, vol. 34, no. 3, pp. 619–22, 1998.

[18] J.-T. Hsu and K. Ngo, "A Hammerstein-based dynamic model for hysteresis phenomenon," *IEEE Transactions on Power Electronics*, vol. 12, no. 3, pp. 406–413, 1997.

[19] H. Saotome and Y. Sakaki, "Iron loss analysis of Mn-Zn ferrite cores," *IEEE Transactions on Magnetics*, vol. 33, no. 1, pp. 728–34, 1997.

[20] W. Roshen, "Ferrite core loss for power magnetic components design," *IEEE Transactions on Magnetics*, vol. 27, no. 6, pp. 4407–15, 1991.

[21] P. Tenant and J. J. Rousseau, "Dynamic model of magnetic materials applied on soft ferrites," *IEEE Transactions on Power Electronics*, vol. 13, no. 2, pp. 372–9, 1998.

[22] F. Fiorillo and A. Novikov, "An improved approach to power losses in magnetic laminations under nonsinusoidal induction waveform," *IEEE Transactions on Magnetics*, vol. 26, no. 5, pp. 2904–10, 1990.

[23] D. Jiles, "Frequency dependence of hysteresis curves in 'non-conducting' magnetic materials," *IEEE Transactions on Magnetics*, vol. 29, no. 6, pp. 3490–2, 1993.

[24] W. Shen, F. F. Wang, D. Boroyevich, and C. W. Tipton, "Loss characterization and calculation of nanocrystalline cores for high-frequency magnetics applications," *IEEE Transactions on Power Electronics*, vol. 23, no. 1, 2008.

[25] R. Ridley and A. Nace, "Modeling ferrite core losses," *Switching Power Magazine*, vol. 3, no. 1, pp. 6–13, 2002.

[26] C. Oliver, "A new core loss model," *Switching Power Magazine*, vol. 3, no. 2, 2002.

[27] E. Herbert, "User-friendly data for magnetic core loss calculations," Aug. 2008, http://fmtt.com/Coreloss2009.pdf.

[28] "Ferrite material selection guide," 2000, magnetics Division of Spang & Company, Bulletin No. FC-S1.

[29] "Soft ferrites and accessories, data handbook," 2009, Ferroxcube, Eindhoven, The Netherlands.

[30] "Ferrites and accessories: SIFERRIT materials," 2006, EPCOS AG, Munich, Germany.

[31] K. Carpenter and S. Warren, "A wide bandwidth, dynamic hysteresis model for magnetization in soft ferrites," *IEEE Transactions on Magnetics*, vol. 28, no. 5, pp. 2037–41, 1992.

[32] I. Mayergoyz, *Mathematical models of hysteresis*. Springer-Verlag, 1991.

[33] P. Han, G. R. Skutt, J. Zhang, and F. C. Lee, "Finite element method for ferrite core loss calculation," in *Applied Power Electronics Conference and Exposition*, vol. 1, 1995, pp. 348–353.

High-Efficiency Inverter with H6-Type Configuration for Photovoltaic Non-Isolated AC Module Applications

Wensong Yu, Jih-Sheng Lai, Hao Qian, and Chris Hutchens

Bradley Department of Electrical and Computer Engineering
Virginia Polytechnic Institute and State University
Blacksburg, VA, USA
wensong@vt.edu, jslai@vt.edu, hqian@vt.edu, Hutchens@vt.edu

Jianhui Zhang, Gianpaolo Lisi, Ali Djabbari, Greg Smith, and Tim Hegarty

National Semiconductor Corporation
Santa Clara, CA, USA
Jianhui.Zhang@nsc.com, Gianpaolo.Lisi@nsc.com, Ali.Djabbari@nsc.com, Greg.Smith@nsc.com,Tim.Hegarty@nsc.com

Abstract—**A novel, high-efficiency inverter using MOSFETs for all active switches is presented for photovoltaic, non-isolated, ac module applications. The proposed H6-type configuration features high efficiency over a wide load range, low ground leakage current, no need for split capacitors, and low output ac-current distortion. The detailed power stage operating principles, PWM scheme, and novel bootstrap power supply for the proposed inverter are described. Experimental results of a 300 W hardware prototype show that not only are MOSFET body diode reverse-recovery and ground leakage current issues alleviated in the proposed inverter, but also that 98.3% maximum efficiency and 98.1% European Union efficiency are achieved.**

I. INTRODUCTION

Photovoltaic (PV) ac modules may become a trend for future PV systems because of their greater flexibility in distributed system expansion, easier installation due to their "plug and play" nature, lower manufacturing cost from modular and scalable production, and higher system-level energy harnessing capabilities under shaded or PV manufacturing mismatch conditions as compared to the single- or multi- string inverters [1-4]. A number of inverter topologies for PV ac module applications have been reported so far with respect to the number of power stages, location of power-decoupling capacitors, use of transformers, and types of grid interface [5-15]. Unfortunately, these solutions suffer from one or more of the following major drawbacks: (1) the limited-lifetime issue of electrolytic capacitors for power decoupling [5-9]; (2) limited input voltage range for the available panels in the market [10-12]; (3) high ground leakage current when the unipolar pulse-width-modulation (PWM) scheme is used in a transformer-less PV system [13]; (4) low system efficiency if an additional high-frequency bidirectional converter is employed [14-16]; and (5) increased cost and complexity of the circuit if energy in the transformer leakage inductance is recycled by either an active snubber or soft-switching circuit [17-19].

Since galvanic insulation in an ac module for PV application is not required by code, a non-isolated ac module

combining a non-isolated high step-up converter and a high-efficiency inverter with H6-type configuration, as shown in Fig. 1, can be used to solve the above issues. Reference [20] reported a dc/dc converter with a single active switch – combining boost, flyback, and charge-pump circuits to achieve wide input range, high voltage gain, high efficiency, and low cost simultaneously as a part of a 20-70 V input, 180-200 V output, and 97.4% peak efficiency PV integrated ac module. This paper concentrates on the second power stage – the inverter circuit.

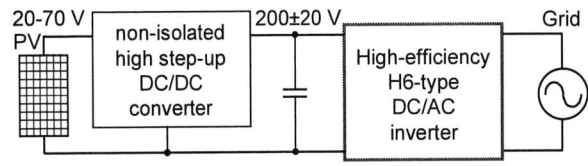

Fig. 1 PV ac module application of the H6-type inverter

The simplest inverter using hybrid MOSFETs and IGBTs with unipolar PWM to achieve high efficiency is shown in Fig.2. The high side IGBTs serve as line frequency polarity selection switches and low side MOSFETs operate in high frequency sinusoidal PWM to control the output voltage or current. The high efficiency of the hybrid 4-swich inverter can be achieved over wide load range because the MOSFETs can avoid the fixed voltage-drop losses and significantly reduce the turn-off losses without tail current as compared to the case with IGBTs. However, the hybrid 4-swich inverter with unipolar PWM is not suitable for non-isolated AC module application because the high ground leakage current is generated through the parasitic capacitance of the PV panel due to high frequency voltage swing at the PV terminals [21, 22]. Reference [23] presented the 5-switch high efficiency inverter with unipolar PWM, as shown in Fig. 3, to solve the high ground leakage current issue. The MOSFET S_5 and S_1 or S_2 operate in SPWM and share half the DC bus voltage, high frequency voltage swing at the PV terminals is eliminated. Note that MOSFETs cannot be used as S_3 or S_4 in the inverter because of the MOSFET body diode slow reverse-recovery issue. Moreover, the cost-effective solution using bootstrap

978-1-4244-4782-4/10 $26.00 © 2010 IEEE

technology with the integrated chips to drive the high-side and mid-side active switches has not been presented.

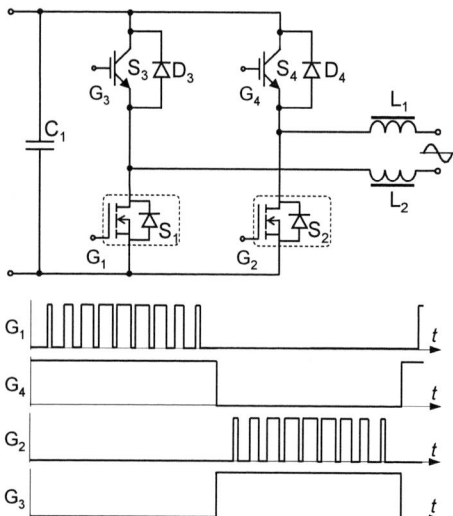

Fig. 2 Simplest inverter using hybrid MOSFETs and IGBTs with unipolar PWM to achieve high efficiency.

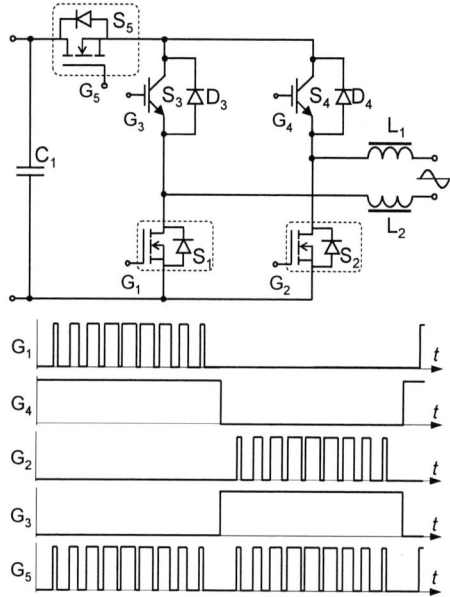

Fig. 3 High efficiency 5-switch inverter without high ground leakage current issue.

In this paper, a novel, high-efficiency inverter using MOSFETs for all active switches is proposed for photovoltaic, non-isolated, ac module applications. The presented H6-type configuration features high efficiency over a wide load range, low ground leakage current, no need for split capacitors, and low output ac-current distortion. Detailed power stage operating principles, PWM scheme, and novel bootstrap power supply for the proposed inverter are described. To verify the validity of the circuit and the improved performance of the proposed inverter, a 300 W hardware prototype – targeted at PV non-isolated ac module applications – has been designed, fabricated and tested. Experimental results show

that not only are MOSFET body diode reverse-recovery and ground leakage current issues alleviated in the proposed inverter, but also that 98.3% maximum efficiency at about half of rated output power and 98.1% European Union (EU) efficiency are achieved.

II. THE PROPOSED INVERTER TOPOLOGY AND OPERATION ANALYSIS

Fig.4 shows the circuit diagram of the proposed inverter with H6-type configuration, which is composed of six power MOSFETs (S_1-S_6), two freewheeling diodes (D_1 and D_2), and two split inductors (L_1 and L_2) as a low-pass filter. This circuit is well suited for non-isolated ac module applications thanks to the following advantages: (1) high efficiency over a wide load range by using MOSFETs for all active switches since their intrinsic body diodes are naturally inactive; (2) low ground leakage current because the voltage applied to the parasitic ground-loop capacitance contains only low-frequency components; (3) no need for limited lifetime electrolytic capacitors for a split dc link similar to that of the half-bridge family of inverters with two or three levels; (4) smaller output inductance as compared to that of the common full-bridge inverter with bipolar PWM switching; and (5) low output ac current distortion because there is no need to have dead time for the proposed circuit since the same phase-leg active switches never all turn on during the same PWM cycle.

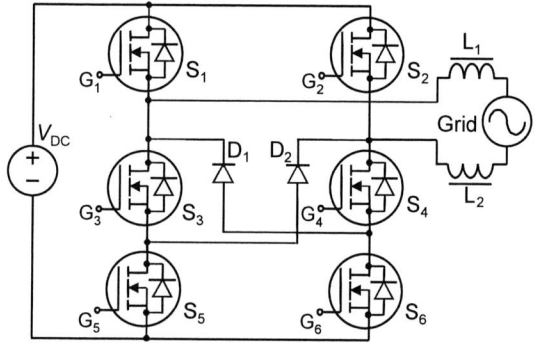

Fig. 4 Circuit diagram of the proposed inverter with H6-type configuration.

Fig. 5 illustrates the PWM scheme for the proposed inverter. As shown in Fig. 5 (a), the top device in one leg and the bottom device in the other leg are switched simultaneously in the PWM cycle and the middle device operates as a polarity selection switch in the grid cycle. As shown in Fig. 5 (b), if the sinusoidal control voltage $v_{control}$, which is synchronized with grid voltage, is higher than the triangular carrier voltage $v_{carrier}$, the gating voltage G_1 and G_6 are active; otherwise, G_1 and G_6 are inactive. And if $v_{control}$ is higher than zero, the gating voltage G_4 is active; otherwise, G_4 is inactive. Similarly, the comparison of ($-v_{control}$) with $v_{carrier}$ or zero results in the logical signals to control G_2, G_5 and G_3, respectively.

Fig.6 shows the four topological stages in one grid cycle for the proposed inverter. Note that the point N is the dc link negative terminal, and the point E is the grid negative terminal. The four operation modes are briefly described as follows. During the grid positive half cycle, switch S_4 remains

978-1-4244-4782-4/10 $26.00 © 2010 IEEE

on, whereas S_1, S_6 and D_1 commutate at the PWM switching frequency. When S_1, S_6 and S_4 are on and the other switches and diodes are off, the inductor current is charging, as shown in Fig. 6 (a). Under the condition that the inductance values of L_1 and L_2 are identical, the inductor voltage can be found as

$$v_{L1} = v_{L2} = 0.5 \cdot (v_{DC} - v_G) \qquad (1)$$

And the grid voltage v_G is calculated by

$$v_G = v_{DC} \cdot M \cdot \sin(\omega t) \qquad (2)$$

where v_{DC} is the dc link voltage; M is the modulation index; and ω is the angular frequency of the grid.

From (1) and (2), the ground potential v_{EN1} in the charging interval during positive grid half cycle can be expressed

$$v_{EN1} = 0.5 \cdot v_{DC} \cdot \left[1 - M \cdot \sin(\omega t)\right] \qquad (3)$$

In the freewheeling interval during the positive grid half cycle shown in Fig. 6 (b), the S_1 and S_6 simultaneously turn off and S_4 and D_1 are on. The voltages of the inductor L_1 and L_2 are given as

$$v_{L1} = v_{L2} = -0.5 \cdot v_G \qquad (4)$$

Under the condition that the S_1 and S_6 share the dc link voltage when they are simultaneously turned off, the voltage stress of the S_6 can be found as

$$v_{S6} = 0.5 \cdot v_{DC} \qquad (5)$$

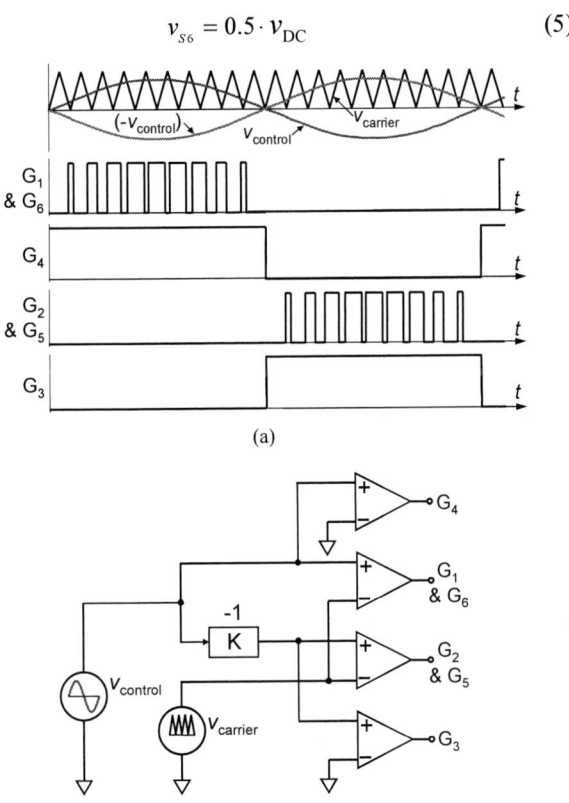

(a)

(b)

Fig. 5 PWM scheme for the proposed inverter: (a) signals in time domain; and (b) implemented circuit.

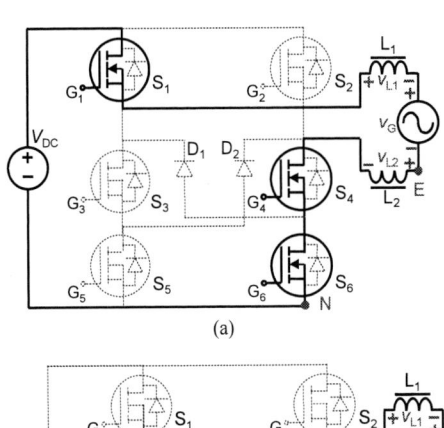

(a)

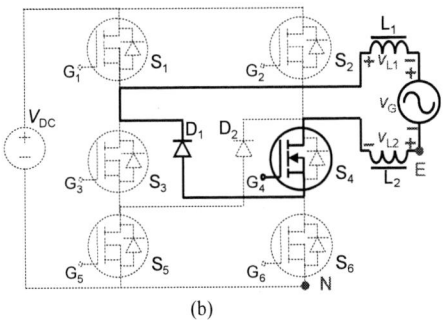

(b)

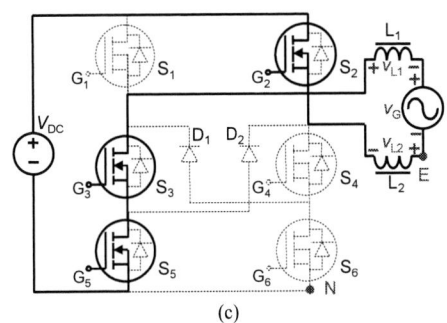

(c)

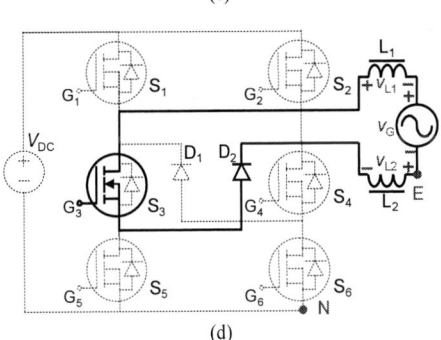

(d)

Fig. 6 Topological stages of the proposed inverter: (a) charging interval and (b) freewheeling interval during positive grid half cycle, (c) charging interval and (d) freewheeling interval during negative grid half cycle.

From (2), (4) and (5), the ground potential v_{EN2} in the freewheeling interval during positive grid half cycle can be expressed

$$v_{EN2} = 0.5 \cdot v_{DC} \cdot \left[1 - M \cdot \sin(\omega t)\right] \qquad (6)$$

Based on the fact that (6) is identical to (3), the PWM switching frequency voltage of the ground potential is avoided. The operation modes similarly change during the grid negative half cycle. From Fig. 6 (a)-(d), it can be seen that

978-1-4244-4782-4/10 $26.00 © 2010 IEEE

the body diodes of the MOSFETs are naturally inactive and the high-frequency voltage of the ground potential is avoided during the whole grid cycle. As a result, MOSFETs can be employed as all the active switches and high ground leakage current can be avoided.

The output inductance can be calculated based on the design criterion that the maximum magnitude of the peak-to-peak current ripple is less than (10~20) % of the rated output current I_{rated}. The peak-to-peak inductor current ripple can be derived as

$$\Delta i_{pk} = (v_{DC} - v_G) \cdot D \cdot T_S / (L_1 + L_2) \qquad (7)$$

And the duty cycle D in the proposed inverter is calculated by

$$D = M \cdot \sin(\omega t) \qquad (8)$$

From (2), (7) and (8), the peak-to-peak ripple of the inductor current can be derived as

$$\Delta i_{pk} = \frac{0.25 \cdot v_{DC} \cdot T_S}{L_1 + L_2} \cdot [1 - (1 - 2M \sin \omega t)^2] \qquad (9)$$

where T_S is the PWM switching period.

when

$$1 - 2M \sin(\omega t) = 0 \qquad (10)$$

The maximum peak-to-peak ripple of the inductor current in the whole grid cycle is calculated by

$$\Delta i_{pk,\max} = \frac{0.25 \cdot v_{DC} \cdot T_S}{L_1 + L_2} \leq (10 \sim 20)\% \cdot I_{rated} \qquad (11)$$

In the proposed inverter, the output inductance then can be calculated as

$$(L_1 + L_2) \geq \frac{0.25 \cdot v_{DC} \cdot T_S}{(10 \sim 20)\% \cdot I_{rated}} \qquad (12)$$

In the conventional full-bridge inverter using bipolar PWM scheme, the output inductance can be calculated as

$$L_{out} \geq \frac{0.5 \cdot v_{DC} \cdot T_S}{(10 \sim 20)\% \cdot I_{rated}} \qquad (13)$$

Thus, the output inductance in the proposed inverter is half that of the conventional full-bridge inverter using bipolar PWM scheme.

III. NOVEL BOOTSTRAP POWER SUPPLY AND INTEGRATED GATE DRIVERS FOR THE PROPOSED INVERTER

Although the proposed inverter has the distinctive advantages over the conventional full-bridge inverter, there are also some shortcomings associated with this method: two more active switches and their gate drivers and the individual power supplies. A cost-effective solution to power the high-side and mid-side gate drives for the proposed inverter using the bootstrap power supply technique is shown in Fig. 7. Four small capacitors C_{a1}-C_{a4} and diodes D_{a1}-D_{a4} are employed to

transfer energy to the high-side and mid-side switches from a single non-isolated power supply E_a. Note that the energy of the C_{a1} can not be transferred from C_{a3} which in the same leg because the middle device S_3 and its body diode are never on when S_1 operates at high-frequency, such that inter-leg connections of the bootstrap circuit for the proposed inverter are necessary. The novel bootstrap power supply is preferred over isolated auxiliary supply for each gate drive thanks to the compact size and compatibility with integrated chips. It makes the proposed inverter more appealing for PV ac module applications.

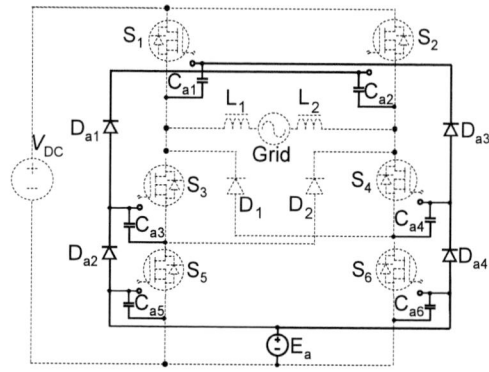

Fig. 7 Circuit diagram of the bootstrap power supply for the proposed inverter.

The high-side and middle-side MOSFETs in the same phase leg can be driven by one piece of the integrated chip FAN7385, as shown in Fig. 8.

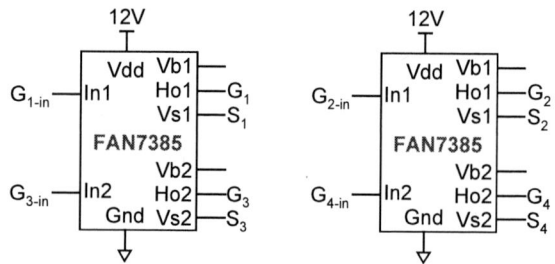

Fig. 8 High-side and middle-side gate drivers using integrated chips FAN7385 for the proposed inverter.

IV. EXPERIMENTAL VERIFICATIONS

A 300 W hardware prototype has been designed, fabricated and tested to verify the validity of the proposed inverter targeted at PV non-isolated ac module application. The main devices $S_1 \sim S_6$ are 250 V, 42.5 mΩ MOSFETs (FDB2710), the freewheeling diodes D_1 and D_2 are ultrafast diodes CMR5U-04 (400 V/5 A), the auxiliary diodes $D_{a1} \sim D_{a4}$ are ultrafast diodes MURA160 (600 V/1 A), and the bootstrap capacitors $C_{a1} \sim C_{a4}$ are 1 μF, 25 V X7R. A low-pass filter ($L_1 = L_2 = 0.8$ mH, $C_f = 0.68$ μF) is used as the output filter. The three pieces of dual-channel high-side gate driver integrated chips FAN7385 with the proposed bootstrap power supply are designed to produce the matched gating signals for the six power MOSFETs. A digital control board with Spartan-3E FPGA is used as the sinusoidal output current controller.

978-1-4244-4782-4/10 $26.00 © 2010 IEEE

Specifications of the inverter are as follows: dc link voltage V_{DC} = 180-200 V; output power P_o = 300 W; output voltage $V_{o,RMS}$ = 120 V; switching frequency f_{sw} = 30 kHz.

The experimental gating signals in the grid cycle and in PWM cycle are shown in Fig. 9(a) and (b), respectively. It can be seen that the experimental gating signals v_{G1}, v_{G6} and v_{G4} agree with the analysis results of the PWM scheme and the proposed bootstrap circuit works well by observing that the gate drive voltage level of the middle and top switches are kept constant during the grid cycle. The gate drive signal v_{G1} of the top switch in one leg and the v_{G6} of the bottom switch in the other leg are matched with each other.

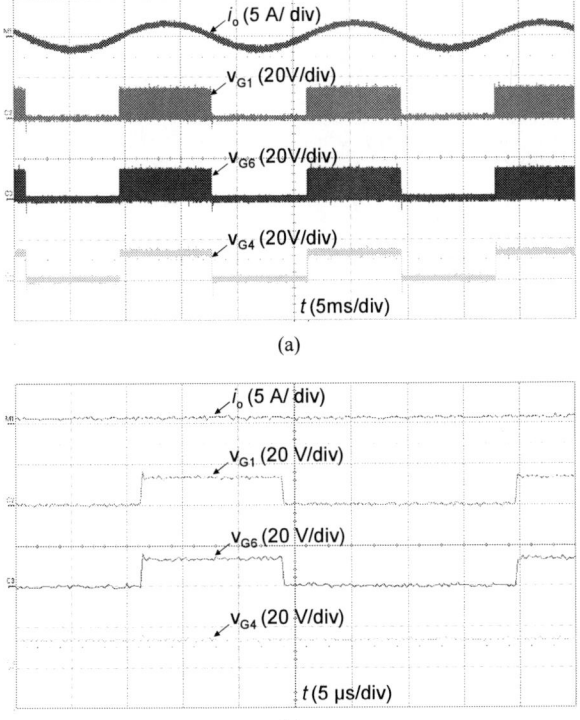

(a)

(b)

Fig. 9 Experimental gating signals: (a) in the grid cycle and (b) in the PWM cycle.

The experimental switches voltage waveforms under 200 V dc link conditions are shown in Fig. 10. The voltage stress of the top switch S_1 and the middle switch S_4 is 200V, which is the same as the dc link voltage. The voltage stress of the bottom switch S_6 is about half of the dc link voltage, as shown in Fig. 10 (a). It can be seen from Fig. 10 (b) that the switches S_1 and S_6 almost evenly share the dc link voltage when they switch off simultaneously.

Fig. 11 shows the experimental waveforms of the ground potential under full-load conditions. The testing results of the ground potential v_{EN} agree with the (6). The high ground leakage current is avoided by observing that the high-frequency voltage of the ground potential is less than 10 V_{RMS} under full-load conditions.

The experimental waveforms of the output current under the 120 V_{RMS} ac voltage and full-load conditions are described in Fig. 12. This figure shows that the proposed inverter presents high power-factor and low harmonic distortion.

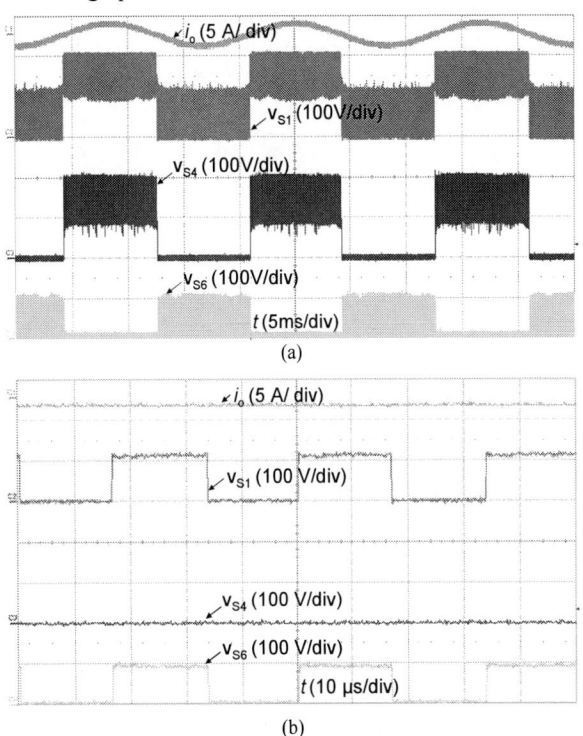

(a)

(b)

Fig. 10 Experimental switches voltage waveforms: (a) in the grid cycle and (b) in the PWM cycle.

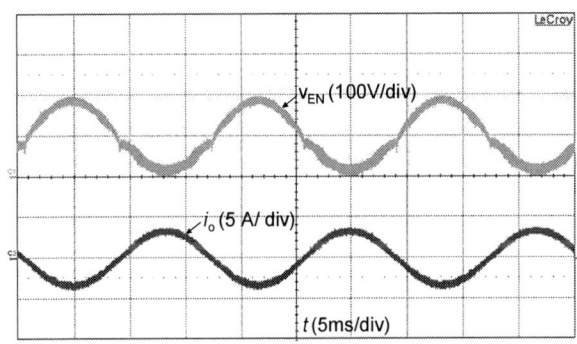

Fig. 11 Experimental waveforms of the ground potential under full-load conditions.

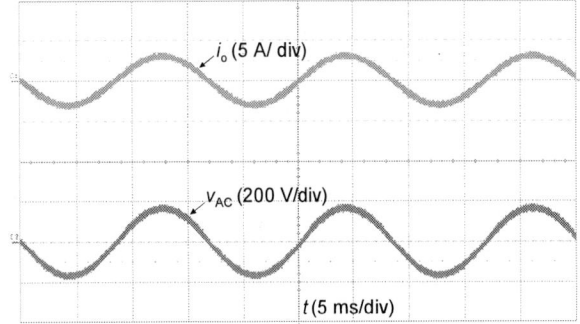

Fig. 12 Experimental waveforms of the output current under the 120 V_{RMS} ac voltage and full-load conditions.

Fig.13 illustrates the experimental results for efficiency under different dc link voltages and output loads with the 30 kHz switching frequency and 120 V_{RMS} ac voltage conditions. The maximum experimental efficiency of the prototype is 98.3% at about half of rated output power and its European Union efficiency is 98.1%.

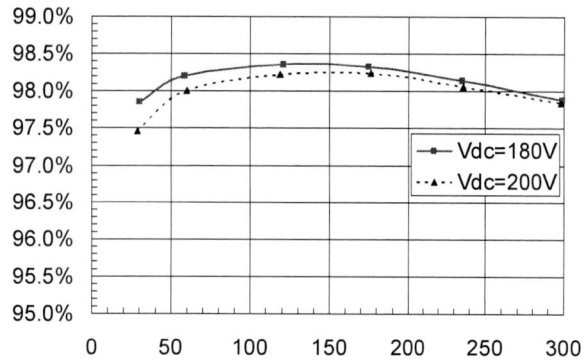

Fig. 13 Experimental results of efficiency as a function of the input voltage and the output power.

V. CONCLUSION

This paper proposes a novel single-phase inverter with H6-type configuration as a part of a wide input range, high-efficiency, and long lifetime PV non-isolated 300 W ac module. Key features of the proposed circuit are as follows: (1) high efficiency over a wide load range by using MOSFETs for all active switches since their body diodes are naturally inactive; (2) low ground leakage current even if no transformer is used; (3) no need for bulky capacitors for a split dc link like what three-level inverters do; (4) 50% reduction of the output inductance as compared to that of the common full-bridge inverter with bipolar PWM switching; (5) low output ac current distortion because the PWM dead time for the proposed circuit is eliminated; and (6) simple bootstrap power supply and integrated gate drivers for the inverter. A 300 W hardware prototype has been designed, fabricated and tested. Experimental results verify the validity of the novel circuit and show 98.1% European efficiency.

REFERENCES

[1] M. Calais, J. Myrzik, T. Spooner, and V. G. Agelidis, "Inverters for single-phase grid connected photovoltaic systems -- An overview," in *Proc. IEEE PESC'02*, vol. 2, 2002, pp. 1995-2000.

[2] F. Blaabjerg, Z. Chen, and S. B. Kjaer, "Power electronics as efficient interface in dispersed power generation systems," *IEEE Trans. Power Electron.*, vol. 19, no. 5, pp. 1184-1194, Sep. 2004.

[3] S. B. Kjaer , J. K. Pedersen and F. Blaabjerg "A review of single-phase grid-connected inverters for photovoltaic modules," *IEEE Trans. Ind. Appl.*, vol. 41, pp. 1292, Sep./Oct. 2005.

[4] Quan Li; and P. Wolfs, "A review of the single phase photovoltaic module integrated converter topologies with three different dc link configurations," *IEEE Trans. Power Electron.*, vol. 23, no. 3, pp. 1320-1333, May 2008.

[5] M. Fornage, "Method and apparatus for converting direct current to alternating current," U.S. Patent applications 2007 0 221 267 A1, Sep. 27, 2007.

[6] S. Saha and V. P. Sundarsingh, "Novel grid-connected photovoltaic inverter," *Proc. Inst. Elect. Eng.*, vol. 143, pp. 219-224, Mar. 1996.

[7] A. Lohner, T. Meyer, and A. Nagel, "A new panel-integratable inverter concept for grid-connected photovoltaic systems," in *Proc. IEEE ISIE'96*, vol. 2, 1996, pp. 827-831.

[8] S. B. Kjaer and F. Blaabjerg, "Design optimization of a single phase inverter for photovoltaic applications," in *Proc. IEEE PESC'03*, vol. 3, 2003, pp. 1183-1190.

[9] A. C. Kyritsis, E. C. Tatakis, and N. P. Papanikolaou, "Optimum design of the current-source flyback inverter for decentralized grid-connected photovoltaic systems," *IEEE Trans. Energy Conversion*, vol. 23, no.1 pp. 281-293, Mar. 2008.

[10] B. Sahan, A. N. Vergara, N. Henze, A. Engler and P. Zacharias, "A single-stage PV module integrated converter based on a low-power current-source inverter," *IEEE Trans. Ind. Electron.*, vol. 55,no. 7, pp. 2602-2609, Jul. 2008.

[11] H. Patel and V. Agarwal, "A single-stage single-phase transformer-less doubly grounded grid-connected PV interface," *IEEE Trans. Energy Conversion*, vol. 24, no. 1 Mar. 2009.

[12] S. Funabiki , T. Tanaka and T. Nishi "A new buck-boost-operation-based sinusoidal inverter circuit," in *Proc. IEEE PESC* ,2002, pp. 1624-1629.

[13] J. M. Chang, W. N. Chang and S. J. Chiang, "Single-phase grid-connected PV system using three-arm rectifier-inverter," *IEEE Trans. Aerosp. and Electron. Syst.*, vol. 42, no. 1, pp. 211-219, Jan. 2006.

[14] A. C. Kyritsis, N. P. Papanikolaou and E.C. Tatakis, "Enhanced current pulsation smoothing parallel active filter for single stage grid-connected AC-PV modules," in *Proc. IEEE EPE-PEMC'2008*, 1-3 Sept. 2008, pp. 1287 - 1292.

[15] T. Shimizu , K. Wada and N. Nakamura, "Flyback-type single-phase utility interactive inverter with power pulsation decoupling on the dc input for an ac photovoltaic module system," *IEEE Trans. Power Electron.*, vol. 21, pp. 1264-1272, Sep. 2006.

[16] P.T. Krein and R.S. Balog, "Cost-effective hundred-year life for single-phase inverters and rectifiers in solar and LED lighting applications based on minimum capacitance requirements and a ripple power port," in *Proc. IEEE APEC 2009*, Washington, DC, 15-19 Feb. 2009, pp. 620 - 625.

[17] Q. Li and P. Wolfs, "A current fed two-inductor boost converter with an integrated magnetic structure and passive lossless snubbers for photovoltaic module integrated converter applications," *IEEE Trans. Power Electron.*, vol. 22, no. 1, pp. 309-320, Jan. 2007.

[18] M. Andersen and B. Alvsten, "200 W low cost module integrated utility interface for modular photovoltaic energy systems," in *Proc. IEEE IECON*, 1995, pp. 572-577.

[19] C. Rodriguez and G. Amaratunga, "Long-lifetime power inverter for photovoltaic AC modules," *IEEE Trans. Ind. Electron.*, vol. 55,no. 7, pp. 2593-2601, Jul. 2008.

[20] W. Yu, C. Hutchens, J.-S. Lai, J. Zhang, G. Lisi, A. Djabbari, G. Smith, and T. Hegarty, "High efficiency converter with charge pump and coupled inductor for wide input photovoltaic ac module applications," in *Proc. IEEE ECCE 2009*, San Jose, CA, 20-24 Sep. 2009, pp. 3895 - 3900.

[21] S.V. Araujo, P. Zacharias, and B. Sahan, "Novel grid-connected non-isolated converters for photovoltaic systems with grounded generator," in Proc. IEEE PESC 2008, 15-19 Jun. 2008, pp.58-65.

[22] R. Gonzalez, E. Gubia, J. Lopez, and L. Marroyo, "Transformerless Single-Phase Multilevel-Based Photovoltaic Inverter," IEEE Trans. Ind. Electron., vol.55, no.7, pp.2694-2702, July 2008.

[23] M. Victor, F. Greizer, S. Bremicker, U. Hubler "Method of converting a direct current voltage of a source of direct current voltage, more specifically of a photovoltaic source of direct current voltage, into an alternating current voltage," U.S. Patent applications 2005 0 286 281 A1, Dec. 29, 2005.

Analyzing the Optimal Matching of DC Motors to Photovoltaic Modules via DC-DC converters

Jesus Gonzalez-Llorente[*], Eduardo I. Ortiz-Rivera[†], Andres Salazar-Llinas[‡] and Emil Jimenez-Brea[§]

Department of Electrical and Computer Engineering

University of Puerto Rico at Mayaguez

Mayaguez, PR 00681-9042

jesus.gonzalez@ece.uprm.edu[*], eduardo.ortiz@ece.uprm.edu[†], andres.salazar@ece.uprm.edu[‡], emil.jimenez@ece.uprm.edu[§]

Abstract—Because to the nonlinear behavior of the photovoltaic (PV) cells, dc-dc power converters are added for matching the load to the photovoltaic modules (PVM). In this paper, we use mathematical models in order to examine the behavior of the off-grid photovoltaic system composed by: PV generator, dc-dc converter and dc motor. We compare different converter topologies (step-up, step-down and step-down/step-up) and evaluate the feasibility of being used as interface to attain operation around the maximum power point (MPP). Our analysis found the relationships between the optimal duty ratio and the maximum power, and between the optimal duty ratio and the motor speed; using these relationships, the simplest topology to meet the requirement can be selected as interface. Moreover, a simple but reliable maximum power point tracking (MPPT) method and a controller are implemented on a microcontroller and tested in real weather conditions. The MPPT provides an approximation to the optimal voltage or to the optimal current in a straightforward way, and the controller adjusts the duty ratio of the power converter, improving the matching of the PVM supplying a dc motor, when operation around MPP is obtained.

I. INTRODUCTION

Since the photovoltaic (PV) generators are dc sources, these generators are very useful to supply dc motors. There are many applications of PV systems where the load is a dc motor, such as: refrigeration, telecommunication and water pumping, among others applications [1].

When a dc motor is directly connected to a photovoltaic module, the operating point of the PVM is very far from its maximum power point (MPP) [2], [3]. In order to improve the performance, the operating point must be closer to MPP; for this purpose is needed matching of the DC motor to the PVM. The matching could be reached by selecting carefully the dc motor according to motor I-V curve, mechanical load characteristics and PVM parameters [4]–[6], and by including an maximum power point tracker (MPPT) [2], [3], [7].

The interfacing circuit consists of dc-dc power converters, which can vary the current coming from the PV array; thus its duty ratio is adjusted to a value until optimal matching is achieved [8]. Step-down and step-up topologies such as nonisolated buck and boost dc-dc converters are widely used as photovoltaic interfaces due to their advantages of simplicity and efficiency [9].

Step-up dc-dc power converters were used as circuit interface between a PVM and a DC motor in [2], [3], [10]–[12], whereas in [13]–[17] step-down dc-dc converters were employed. Furthermore, in [18], [19] step-down/step-up converter can be found. All of them utilized varied ways to adjust the duty ratio of power converter, but none of them specified the necessary conditions for optimal matching with the used topology. Thus, we studied different types of dc-dc converters, including step-up, step-down and step-down/step-up topologies in order to determine the conditions for attaining the matching.

We used an analytical method for comparing the different power converters, obtaining expression for the optimal duty ratio. Besides, this paper presents a schema with MPPT for the system consisted of PV array, dc-dc converter and dc motor, which can track the maximum power point without dependence on temperature, irradiance, or the kind of mechanical load. Finally, experimental results under real weather conditions are presented.

II. INTERFACING THE PHOTOVOLTAIC MODULE TO THE DC MOTOR

Without any interfacing circuitry between the PVM and the dc motor, the operating point depends on the temperature, the irradiance, PVM specifications and the dc motor parameters. Then, if the dc motor characteristic I-V is superimposed on a set of photovoltaic I-V curves, the operating point is given by intersection between this curves [2]. The characteristics curves can be obtained by using the PV array and dc motor models, which are described below.

A. Mathematical models

The relationship of the current (I) with respect to the voltage (V), for any PV array is given in (1). This model takes into consideration the short-circuit current (I_x) and the open-circuit voltage (V_x) at any given irradiance level (E_i) and temperature (T), the PVM characteristic constant (b), and the numbers of in series and in parallels modules with the same electrical characteristics (s and p, respectively). The PVM exponential model is fully described in [20], [21].

$$I(V) = \frac{p \cdot Ix}{1 - exp\left(\frac{-1}{b}\right)} \cdot \left[1 - exp\left(\frac{V}{b \cdot s \cdot Vx} - \frac{1}{b}\right)\right] \quad (1)$$

Electrical side of a dc motor with constant field flux can be described for (2), and the torque balance equation is given by (3) [22].

$$V_a = R_a \cdot i_a + L_a \frac{di_a}{dt} + K_e \cdot \omega_m \qquad (2)$$

$$T_e = J \cdot \frac{d\omega_m}{dt} + B_m \cdot \omega_m + T_L \qquad (3)$$

In the above equations R_a, L_a, K_e, V_a and i_a are armature resistance, armature inductance, back emf constant, armature voltage and current respectively. J, B_m, and T_L are the moment of inertia of the motor and connected load, constant viscous friction coefficient and load torque, respectively. The electromagnetic torque, T_e, is proportional to the current through the armature winding and can be written as $T_e = K_e \cdot i_a$.

The PVM and DC motor $I - V$ curves are illustrated in Fig. 1(a), where different operating points are shown according to the irradiance conditions and to the kind of dc motor load. Moreover, The dc motor $P - V$ curves for different load characteristics, are superimposed on a set of PVM $P - V$ curves for different irradiance in Fig. 1(b) [12]. It shows that for some load (T_L) the motor voltage is always lower than PVM optimal voltage. The torque-speed characteristics of the load is given by $T_L = c_1 \cdot \omega_m + c_2$, where the constants c_1 and c_2 depend on the chosen position of the braking magnet of an eddy current brake.

As can be seen from Fig. 1, there is only few conditions where the operating point is near to the maximum power point in direct coupling. The operating point of the PV array can be moved to the maximum power point using a dc-dc converter as interface. This is possible, because the input resistance (R_i) of a power converter in continuous conduction mode (CCM) depends on the duty ratio (D) and the load resistance (R_o) [23], [24]. Fig 2 shows the relationship between the normalized input resistance (R_i/R_o) and the duty ratio D of power converter for CCM in steady state.

III. OPTIMAL MATCHING WITH MPPT

This section describes the derivation of the optimal duty ratio for each one of the basic dc-dc power converter: step-down, step-up, and step-down/step-up. Since the optimal duty ratio depends on the voltage at maximum power of the PVM (V_{op}), a simple mathematical method to approximate this method is shown.

A. Derivation of the optimal duty ratio for each topology

For the buck chopper converter, shown in Fig. 3(a), the duty ratio is expressed by $D = V_o/V_i$; the power converter output voltage V_o is equal to the motor armature voltage, V_a. The power converter input voltage V_i is equal to PVM voltage V, or equal to V_{op} if the terminal voltage of the PVM is operating in the maximum power point.

The optimal duty ratio can be obtained by substitution of V_a in $D = V_a/V_{op}$, where V_{op}, is the terminal voltage of the PVM corresponding to the maximum power point and V_a, in steady condition, is given by $V_a = R_a \cdot I_a + K_e \cdot \omega_m$ [14]. Then the optimal duty ratio is given by (4):

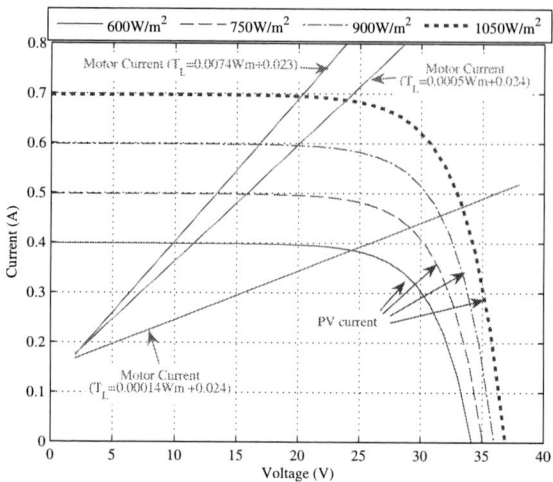

(a) I-V curves DC motor and PVM

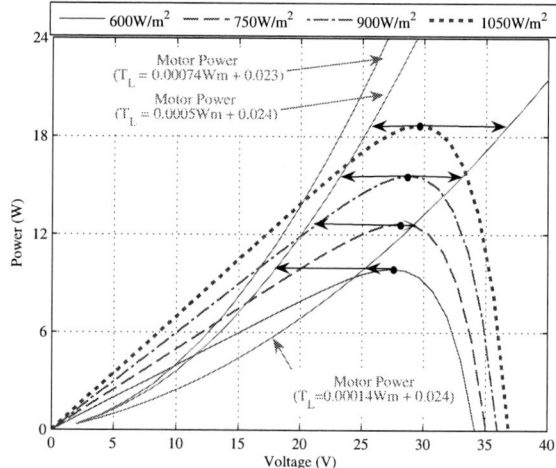

(b) Power curves DC motor and PVM

Fig. 1.   PV and DC motor Curves

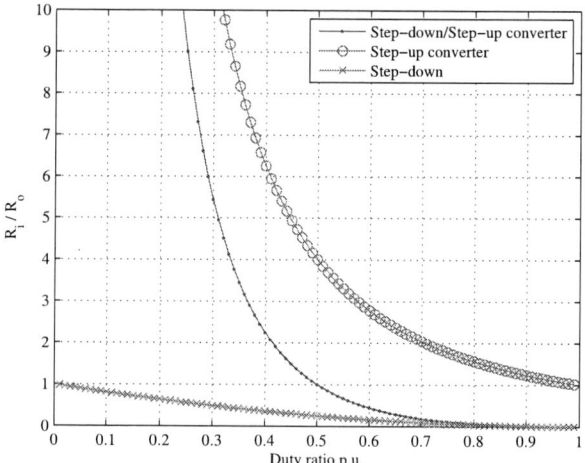

Fig. 2.   Normalized input resistance in CCM of power converters

$$D = \frac{V_a}{V_{op}} = \frac{1}{V_{op}}\left[K_e{\cdot}\omega_m + R_a\left(\frac{B_m{\cdot}\omega_m + T_L}{K_e}\right)\right] \quad (4)$$

For the Boost converter, shown in Fig. 3(b), whose output-to-input voltage conversion ratio, is given by

$$\frac{V_o}{V_{in}} = \frac{1}{1-D} \quad (5)$$

similar procedure is done, and from (6) optimal duty cycle is derived and shown in (7).

$$\frac{1}{1-D} = \frac{V_a}{V_{op}} = \frac{1}{V_{op}}\left[K_e{\cdot}\omega_m + R_a\left(\frac{B_m{\cdot}\omega_m + T_L}{K_e}\right)\right] \quad (6)$$

$$D = 1 - \frac{K_e{\cdot}V_{op}}{K_e^2{\cdot}\omega_m + R_a{\cdot}B_m{\cdot}\omega_m + T_L R_a} \quad (7)$$

The relationship of the input and the output voltage of the Buck-Boost, Cuk and SEPIC converter is given by

$$\frac{V_o}{V_{in}} = \frac{D}{1-D} \quad (8)$$

Fig. 3(c) shows a Buck-Boost converter as interface. If $V_o = V_a$ and $V_{in} = V_{op}$, then, the optimal duty ratio can be derived from (9), and is given by (10)

$$\frac{D}{1-D} = \frac{V_a}{V_{op}} = \frac{1}{V_{op}}\left[K_e{\cdot}\omega_m + R_a\left(\frac{B_m{\cdot}\omega_m + T_L}{K_e}\right)\right] \quad (9)$$

$$D = \frac{R_a{\cdot}B_m{\cdot}\omega_m + K_e^2{\cdot}\omega_m + R_a{\cdot}T_L}{R_a{\cdot}B_m{\cdot}\omega_m + K_e^2{\cdot}\omega_m + V_{op}{\cdot}K_e + R_a{\cdot}T_L} \quad (10)$$

Table I summarize the relationship between optimal the duty ratio and the motor speed for each topology. The optimal voltage V_{op} can be calculated by Linear Reoriented Coordinates Method (LRCM) [25].

B. Maximum power point tracking method

The Linear Reoriented Coordinates Method (LRCM) was proposed in [25], but the verification was based only in simulation results. This paper presents experimental verification applying a simpler equation to approximate the optimal voltage V_{op}, using the measurement of open-circuit voltage (V_x) and the characteristic constant of PV array (b).

The equations of the approximated optimal voltage (V_{ap}) and the approximated optimal current (I_{ap}) are given by (11) and (12) respectively; however V_{ap} is preferable to track the MPP because the PVM voltage has a slower dynamics than the PVM current [9]. The MPPT is completed with the selected dc-dc converter and a PI controller that adjusts the power converter duty ratio in order to maintain the PVM voltage at its setpoint, which is given by (V_{ap}).

$$V_{ap} = V_x + b{\cdot}V_x{\cdot}ln\left(b - b{\cdot}exp(-\tfrac{1}{b})\right) \le V_{op} \quad (11)$$

$$I_{ap} = I_x{\cdot}\frac{1 - b + b{\cdot}exp\left(\frac{-1}{b}\right)}{1 - exp\left(\frac{-1}{b}\right)} \ge I_{op} \quad (12)$$

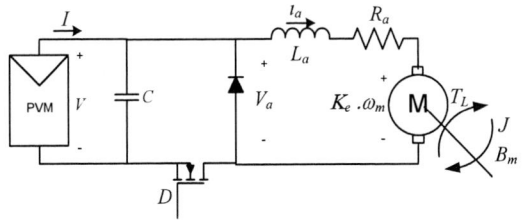

(a) Step-down converter

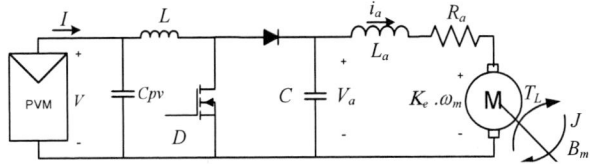

(b) Step-up converter

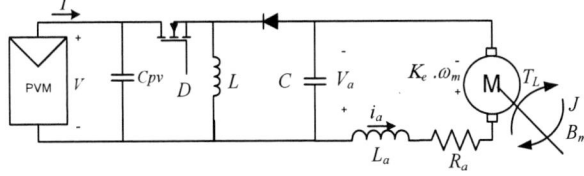

(c) Step-down/Step-up converter

Fig. 3. Different power converters as interface

TABLE I
SUMMARY OF OPTIMAL DUTY RATIO

Power converter	Optimal Duty Ratio
Step-down	$D = \frac{1}{V_{op}}\left[K_e{\cdot}\omega_m + R_a\left(\frac{B_m{\cdot}\omega_m + T_L}{K_e}\right)\right]$
Step-up	$D = 1 - \frac{K_e{\cdot}V_{op}}{K_e^2{\cdot}\omega_m + R_a{\cdot}B_m{\cdot}\omega_m + T_L R_a}$
Step-down/Step-up	$D = \frac{R_a{\cdot}B_m{\cdot}\omega_m + K_e^2{\cdot}\omega_m + R_a{\cdot}T_L}{R_a{\cdot}B_m{\cdot}\omega_m + K_e^2\omega_m + V_{op}{\cdot}K_e + R_a{\cdot}T_L}$

C. Implementation of MPPT

Fig. 4 represents the proposed system using LRCM. The control system consisted of: ATMEL's ATmega88 8-bit microcontroller; interface circuits which comprise of sensors and signal conditioners connected to the analog-to-digital converter (ADC) of the microcontroller, and a driver for the power MOSFET.

The transistor Qv disconnects the PVM each time the LRCM is executed, allowing to measure the open-circuit voltage V_x. The voltage divider composed of R_1 and R_2 produces a fraction of the PVM voltage, which can be connected to the microcontroller ADC through a signal conditioner. This voltage is used to calculate the error with respect to the desired voltage V_{ap}.

The main program of the microcontroller only initialize its internal modules and wait for the timer overflow interruption. The interrupt routine executes the LRCM to update the setpoint, by disconnecting of the PVM at regular intervals and updating V_{ap}. Moreover, this routine uses a interrupt counter to obtain a constant sampling time for the controller.

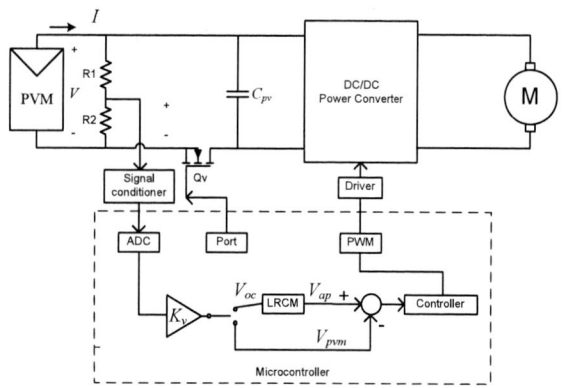

Fig. 4. Digital MPPT with LRCM

The control variable of a PI can be expressed by (13) in ideal form.

$$u(t) = K_p \left[e(t) + \frac{1}{T_i} \int_0^\infty e(t)dt \right] \quad (13)$$

where $u(t)$ is the output of the controller and $e(t) = r(t) - y(t)$, in which $r(t)$ is the optimal voltage and $y(t)$ is the actual voltage of the PVM. K_p and T_i are known as proportional gain and integral time respectively. The trapezoidal approximation for the integral term, which is also known as Bilinear transform or Tustin's method is used for discretization of (13), thus this becomes:

$$u(t_k) = K_p \cdot e(t_k) + K_i \left(\frac{e(t_k) + e(t_{k-1})}{2} \right) T_S + u(t_{k-1}) \quad (14)$$

IV. SIMULATION AND EXPERIMENTAL RESULTS

The systems is tested with a separately excited dc motor with constant field; therefore, the induced emf is proportional to the rotor speed, likewise in a permanent magnet dc motor. The motor parameters are listed in table II.

The motor is loaded through an eddy current brake, where the position of the braking magnet define the load torque (T_L) applied. In this case the load torque is proportional to speed and it is given by $T_L = c_1 \cdot \omega_m + c_2$; the constants c_1 and c_2 are calculated from real data and are shown in table III.

The PV array used for this study consist of two PVMs BP SX10M connected in series. The electrical characteristics of PVM BP SX10M are presented in table IV. The PVM characteristic constant is $b = 0.084$ [12].

TABLE II
DC MOTOR PARAMETERS

Symbol	Parameter	Value	Units
R_a	Armature resistance	8.57	Ω
L_a	Armature inductance	58.7	mH
K_e	Back emf constant	0.1485	$\frac{V}{rad/sec}$
J	Moment of inertia	45.5×10^{-6}	$Kg \cdot m^2$
B_m	Viscous friction coefficient	94.8×10^{-6}	$\frac{N \cdot m}{rad/seg}$

TABLE III
LOAD TORQUE CHARACTERISTIC FOR DIFFERENT POSITIONS,
$T_L = c_1 \cdot \omega_m + c_2$

Position	Constant c_1	Constant c_2
3	0.00014	0.024
4	0.00038	0.023
5	0.00055	0.024
6	0.00074	0.023

TABLE IV
PVM BP SX10M SPECIFICATIONS AT STC

Symbol	Parameter	Value	Units
I_{sc}	Short-circuit Current	0.65	A
V_{oc}	Open-circuit Voltage	21.0	V
P_{max}	Maximum Power	10.0	W
V_{op}	Voltage at P_{max}	16.8	V
I_{op}	Current at P_{max}	0.59	A
TCi	Temperature coeff. of I_{sc}	(0.065 ±0.015)	$\%/C$
TCv	Temperature coeff. of V_{oc}	-(80 ±10)	mV/C

A. Comparison of circuit interfacing

Using the derived expressions for the optimal duty ratio, which are shown in Table I, the relationship between optimal duty ratio and maximum power can be calculated for each studied topology. Therefore, it is possible evaluate this relationship at different loads in order to determine if the topology is suitable to extract the maximum power.

Fig. 5 shows the relationship between optimal duty ratio and maximum power at different irradiance values for step-down, step-up and step-up/step-down converters, likewise Fig. 6 shows the relationship between optimal duty ratio and maximum speed.

Since the duty ratio must be between 0.0 and 1.0 p.u. carefully selection of the topology should be done according to the kind of load; by example, a step-down converter is suitable to match the PVM and the DC motor if the load is set up to the position 4, where the load torque is $T_L = 0.00038 \cdot \omega_m + 0.023$.

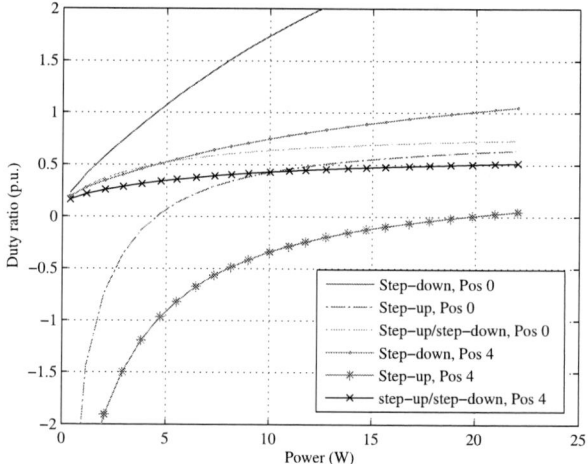

Fig. 5. Duty ratio vs Maximum power

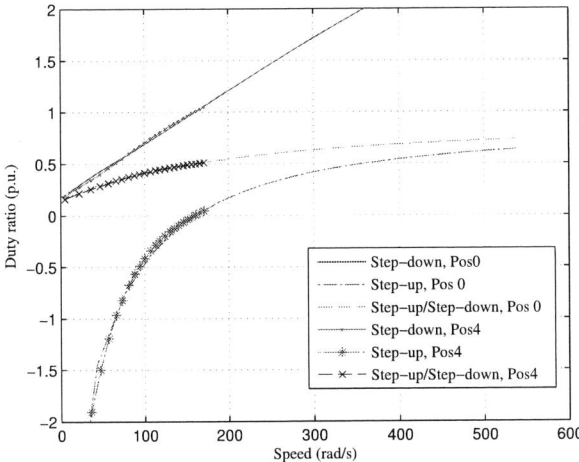

Fig. 6. Comparisons of optimal duty ratio for each topology

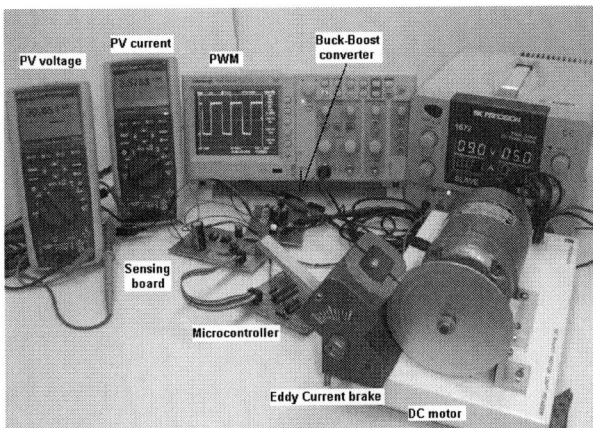

(a) System operation

(b) PV array

Fig. 7. Experiment setup

B. Experimental Results with Step-down/step-up topology

Fig. 7 shows the experimental setup for the system, when the dc motor is used in shunt configuration; the motor started succesfully while in direct coupling the dc motor did not start at the same environmental conditions.

The realized system was also tested with constant field for different weather conditions and constant torque load. Fig. 8 shows the PVM power, comparing the measured value with the theoretical value for different values of short-circuit current (I_x) and open-circuit voltage (V_x), which are shown in the top of each bar. Variations of irradiance and temperature produces the different values of I_x and V_x.

The method was also tested in real environmental conditions under load variations. The position of the braking magnet was changed in order to modify the load characteristics. The systems started with the braking magnet in the position 3; the position was modified each 25 seconds approximately, following the sequence: Position 3, position 5, position 4 and position 6.

Fig. 9 (top) shows the irradiance values and the output power of the PVM for the test condition. The average value of irradiance is equals to $1134W/m^2$. In spite of the variations of the load, the PV array kept providing the same power (15.6W), shown in Fig. 9 (bottom). However, the disconnection of the PVM produces output power reduction, in "in press" [26] is shown how to estimate V_x without disconnection of the PVM.

The controller adjusted the duty ratio, which is shown in Fig. 10, keeping the PV array operating around of the maximum power point. The motor speed, shown in Fig.10 changed according to the load characteristics.

The PV array voltage and the PV array current are shown in Fig. 11. The load variations did not affect the performance of the MPPT method, the PV array voltage had a variation only of $0.4V$ around of the approximated optimal voltage.

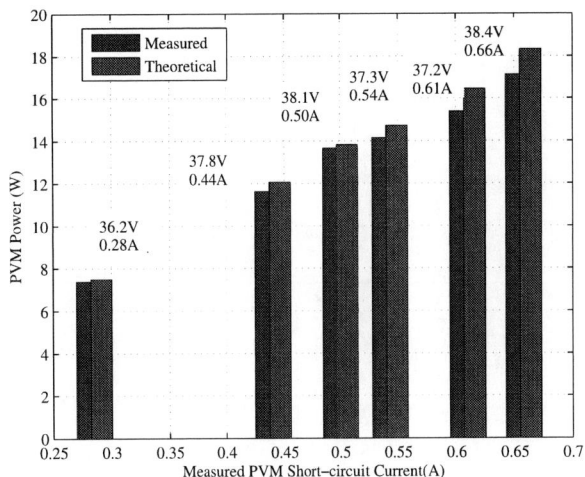

Fig. 8. PVM power at different environment conditions

978-1-4244-4782-4/10 $26.00 © 2010 IEEE

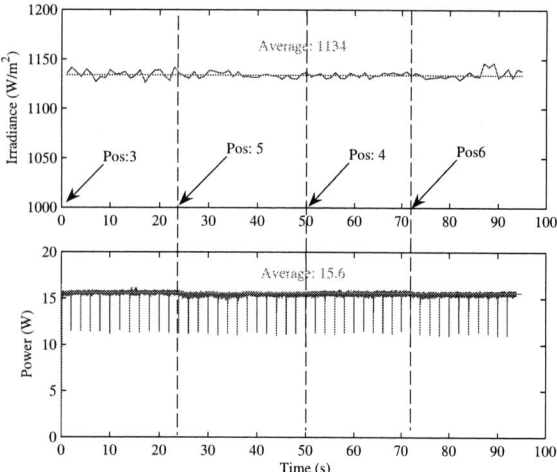

Fig. 9.   Irradiance and PVM Power

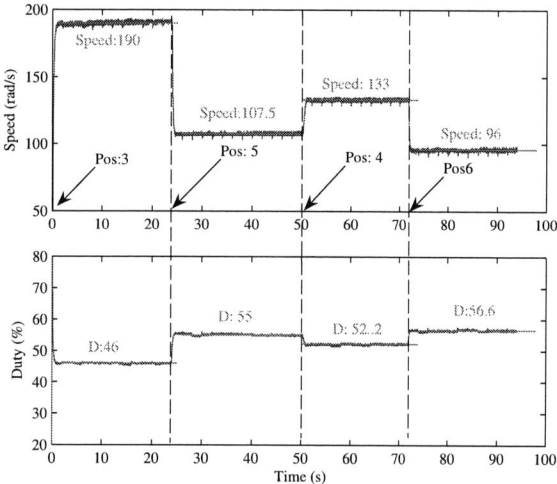

Fig. 10.   Motor Speed and Duty ratio

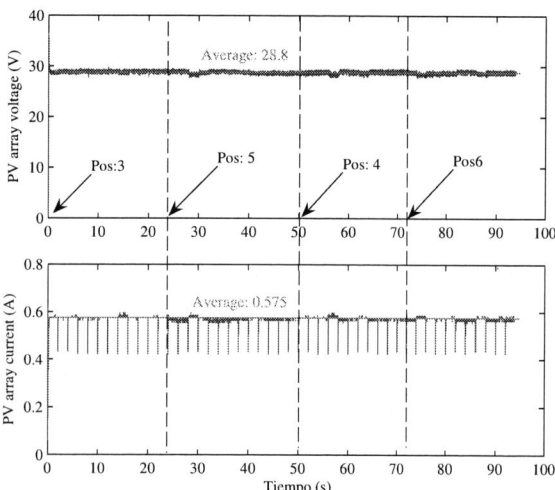

Fig. 11.   PVM voltage and PVM current

V. CONCLUSIONS

This paper presented a photovoltaic array feeding a dc motor; step-up, step-down and step-down/step-up dc/dc power converters were investigated as interface between PVM and the dc motor. Based on the mathematical model an expression for optimal duty ratio was derived for each dc/dc converter. Then, it showed the relationship between the optimal duty ratio and the maximum power point; as well as the relationship between the optimal duty ratio and the maximum speed. This comparative study allows to choose the suitable dc/dc power converter to supply a dc motor through a PVM. The analysis of the input resistance in steady state for the power converter was helpful to set up the action of the controller.

This paper also showed that it is possible to track the maximum power at different irradiance levels using the Linear Reoriented Coordinated Method (LRCM), only measurement of open-circuit voltage is required. When the load is changed, the LRCM keeps the PVM operating around the maximum power point in spite of variation of irradiance and temperature. A simplified method to track the maximum power point without iteration was shown. This method was implemented with the advantage of low cost and simple configuration. If the disconnection of the PV array were a unsuitable for the application. The LRCM could be implemented using irradiance measurement or a pilot PVM to estimate the open-circuit voltage.

ACKNOWLEDGMENT

The authors gratefully acknowledge the contributions of all the members that belong to the Mathematical Modeling and Control of Renewable Energies for Advance Technology & Education (M_{inds}^2 CREATE) Research Team at UPRM.

REFERENCES

[1] G. Masters, *Renewable and Efficient Electric Power Systems*.   John Wiley & sons, Inc., 2004.

[2] S. Alghuwainem, "Steady-state performance of dc motors supplied from photovoltaic generators with step-up converter," *Energy Conversion, IEEE Transaction on*, vol. 7, no. 2, pp. 267–272, Jun 1992.

[3] ——, "Matching of a dc motor to a photovoltaic generator using a step-up converter with a current-locked loop," *Energy Conversion, IEEE Transaction on*, vol. 9, no. 1, pp. 192–198, Mar 1994.

[4] W. Fam and M. Balachander, "Dynamic performance of a dc shunt motor connected to a photovoltaic array," *Energy Conversion, IEEE Transaction on*, vol. 3, no. 3, pp. 613–617, Sep 1988.

[5] M. Saied, "Matching of dc motors to photovoltaic generators for maximum daily gross mechanical energy," *Energy Conversion, IEEE Transaction on*, vol. 3, no. 3, pp. 465–472, Sep 1988.

[6] ——, "A study on the matching of dc motors to photovoltaic solar arrays," *Electrical Machines and Systems, 2001. ICEMS 2001. Proceedings of the Fifth International Conference on*, vol. 1, pp. 652–655 vol.1, 2001.

[7] S. Singer and J. Appelbaum, "Starting characteristics of direct current motors powered by solar cells," *Energy Conversion, IEEE Transaction on*, vol. 8, no. 1, pp. 47–53, Mar 1993.

[8] T. Esram and P. Chapman, "Comparison of photovoltaic array maximum power point tracking techniques," *Energy Conversion, IEEE Transaction on*, vol. 22, no. 2, pp. 439–449, June 2007.

[9] W. Xiao, W. G. Dunford, P. R. Palmer, and A. Capel, "Regulation of photovoltaic voltage," *Industrial Electronics, IEEE Transactions on*, vol. 54, no. 3, pp. 1365–1374, June 2007.

[10] K. Hussein, I. Muta, T. Hoshino, and M. Osakada, "Field and armature control for separately excited dc motors powered by photovoltaic cells," *Photovoltaic Energy Conversion, 1994., Conference Record of the Twenty Fourth. IEEE Photovoltaic Specialists Conference - 1994, 1994 IEEE First World Conference on*, vol. 1, pp. 1169–1172 vol.1, Dec 1994.

[11] M. Akbaba, "Optimization of the performance of solar powered permanent magnet dc motor drives," *Electrical Machines and Power Electronics, 2007. ACEMP '07. International Aegean Conference on*, pp. 725–729, Sept. 2007.

[12] J. Gonzalez-Llorente, E. I. Ortiz-Rivera, and A. Diaz, "A maximum power point tracker using positive feedforward control based on the dc motor dynamics and pvm mathematical model," in *Electric Machines and Drives Conference, 2009. IEMDC '09. IEEE International*, May 2009, pp. 259–264.

[13] S. Shokrolla, N. Twieg, and A. Sharaf, "A photovoltaic powered separately excited dc motor drive for rural/desert pump irrigation," *Electrical Machines and Drives, 1993. 6th International Conf. on*, pp. 406–411, Sep 1993.

[14] K. Venkatesan and D. Cheverez-Gonzalez, "Matching dc motors to photovoltaic generators for maximum power tracking," *Applied Power Electronics Conf. and Exp. APEC '97 Conf. Proc. 1997., 12th Annual*, vol. 1, pp. 514–519 vol.1, Feb 1997.

[15] A. Tariq and M. Asghar, "Matching of a separately excited dc motor to a photovoltaic panel using an analog maximum power point tracker," *Industrial Technology, 2006. ICIT 2006. IEEE International Conf. on*, pp. 1020–1025, Dec. 2006.

[16] A. Sharaf and L. Yang, "A novel maximum power trecking controller for a stand-alone photovoltaic dc motor drive," *Electrical and Computer Engineering, 2006. CCECE '06. Canadian Conference on*, pp. 450–453, May 2006.

[17] A. Sharaf, E. Elbakush, and I. Altas, "An error driven pid controller for maximum utilization of photovoltaic powered pmdc motor drives," *Electrical and Computer Eng., 2007. CCECE 2007. Canadian Conf. on*, pp. 129–132, April 2007.

[18] A. Hussein, K. Hirasawa, J. Hu, and J. Murata, "The dynamic performance of photovoltaic supplied dc motor fed from dc-dc converter and controlled by neural networks," *Neural Networks, 2002. IJCNN '02. Proceedings of the 2002 International Joint Conference on*, vol. 1, pp. 607–612, 2002.

[19] E. E. Jimenez-Toribio, A. A. Labour-Castro, F. Muniz-Rodriguez, H. R. Perez-Hernandez, and E. I. Ortiz-Rivera, "Sensorless control of sepic and cuk converters for dc motors using solar panels," in *Electric Machines and Drives Conference, 2009. IEMDC '09. IEEE International*, May 2009, pp. 1503–1510.

[20] O. Gil-Arias and E. I. Ortiz-Rivera, "A general purpose tool for simulating the behavior of pv solar cells, modules and arrays," *Control and Modeling for Power Electronics, 2008. COMPEL 2008. 11th Workshop on*, pp. 1–5, Aug. 2008.

[21] E. Ortiz-Rivera and F. Peng, "Analytical model for a photovoltaic module using the electrical characteristics provided by the manufacturer data sheet," *Power Electronics Specialists Conf., 2005. PESC '05. IEEE 36th*, pp. 2087–2091, Sept. 2005.

[22] R. Krishnan, *Electric Motor Drives: Modeling, Analysis and Control.* Prentice-Hall, 2001.

[23] E. Ortiz-Rivera, "Maximum power point tracking using the optimal duty ratio for dc-dc converters and load matching in photovoltaic applications," in *Applied Power Electronics Conference and Exposition, 2008. APEC 2008. Twenty-Third Annual IEEE*, Feb. 2008, pp. 987–991.

[24] E. Ortiz-Rivera, M. Sidrach-de Cardona, J. Galan, and J. Andujar, "Comparative analysis of buck-boost converters used to obtain iv characteristic curves of photovoltaic modules," in *Power Electronics Specialists Conference, 2008. PESC 2008. IEEE*, June 2008, pp. 2036–2042.

[25] E. rtiz Rivera and F. Peng, "A novel method to estimate the maximum power for a photovoltaic inverter system," *Power Electronics Specialists Conf., 2004. PESC 04. IEEE 35th Annual*, vol. 3, pp. 2065–2069 Vol.3, 20-25 June 2004.

[26] E. Jimenez-Brea, A. Salazar-Llinas, E. Ortiz-Rivera, and J. Gonzalez-Llorente, "A maximum power point tracker implementation for photovoltaic cells using dynamic optimal voltage tracking," in *Applied Power Electronics Conference and Exposition, 2010. APEC 2010. Twenty-Fifth Annual IEEE*, in press.

978-1-4244-4782-4/10 $26.00 © 2010 IEEE

Performance Analysis of an Interleaved High Step-Up Converter with Voltage Multiplier Cell

Wuhua Li, Yi Zhao, Yan Deng, Xiangning He
College of Electrical Engineering, Zhejiang University
Hangzhou, 310027, P.R. China
Email: woohualee@zju.edu.cn

Abstract—**A novel interleaved high step-up converter with voltage multiplier cell is proposed in this paper to avoid the narrow turn-off period and to reduce the large power device current ripple, which is existed in the conventional boost converters in high step-up applications. The voltage multiplier cell is composed of the secondary windings of the coupled inductors, the series capacitor and two diodes. Furthermore, the switch voltage stress is reduced due to the transformer function of the coupled inductors, which makes low voltage rated MOSFETs available to reduce the conduction losses. In addition, Zero current switching (ZCS) turn-on soft switching performance is realized to reduce the switching losses. And the output diode turn-off current falling rate is controlled by the leakage inductance of the coupled inductors, which alleviates the diode reverse-recovery problem. Finally, a 1kW 40V-input 380V-output prototype is built and tested to verify the effectiveness of the presented converter.**

I. INTRODUCTION

Nowadays, the renewable energy grid-connected system with photovoltaic and fuel cells calls for high step-up and high efficiency DC/DC converters because the low voltage generated by the photovoltaic and the fuel cells should be boosted to a high voltage for the grid-connected applications [1-4].

The conventional interleaved boost converter is widely employed for the front-end applications [5-7], which is given in Fig.1. Unfortunately, it is not suitable for high step-up system due to its disadvantages of extreme duty cycle, large current ripple, high switch voltage stress and severe output diode reverse-recovery problem.

An interleaved high step-up DC/DC converter is derived by inserting a voltage multiplier cell into the conventional interleaved boost converter in this paper. The voltage multiplier cell is used to extend the voltage gain and to minimize the current ripple without extreme duty cycle. Moreover, the switch voltage stress is reduced as the turns ratio of the coupled inductors increases. The diode reverse-recovery problem can be alleviated by the leakage inductance

This work is sponsored by the National Nature Science Foundation of China (50907058), the Power Electronics S&E Development Program of Delta Environmental & Education Foundation (DREM2009001) and the China Postdoctoral Science Foundation (20080440197).

of the coupled inductors. Furthermore, ZCS turn-on soft switching performance for the switches is realized. At last, a 1kW 40V-to-380V prototype with f_s=100 kHz is built to verify the effectiveness of the proposed converter.

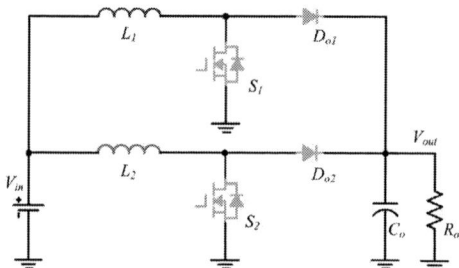

Fig.1. Conventional interleaved boost converter.

II. PROPOSED CONVERTER AND OPERATION ANALYSIS

The proposed interleaved high step-up converter with voltage multiplier cell is shown in Fig.2 (a). It is a conventional interleaved boost converter in the left dashed circle and the voltage multiplier cell is given in the right dashed circle. The primary windings of the two coupled inductors are n_1 turns and the secondary windings are n_2 turns. The coupling reference is pointed by " ° " and "*". The voltage multiplier cell is composed of the secondary windings L_{1b} and L_{2b} of the two coupled inductors, the series capacitor C_m, the regenerative diode D_r and the output diode D_o.

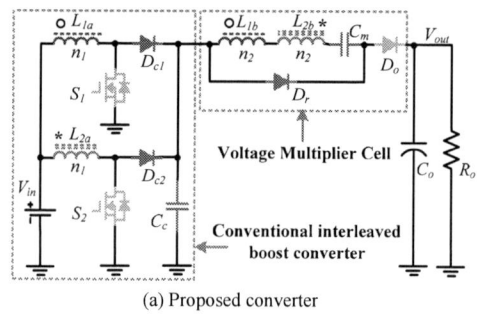

(a) Proposed converter

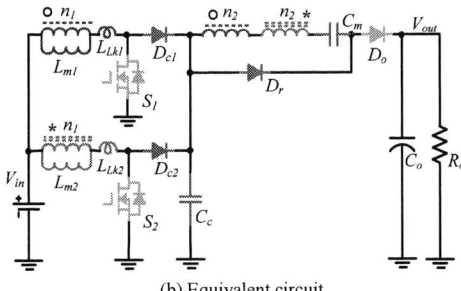

(b) Equivalent circuit

Fig.2. Proposed converter and its equivalent circuit.

The equivalent circuit of the proposed converter is introduced in Fig.2 (b), where L_{m1}, L_{m2} are the magnetizing inductors; L_{Lk1}, L_{Lk2} are the leakage inductances; S_1 and S_2 are the power switches; D_{c1} and D_{c2} are the clamp diodes; C_c is the clamp capacitor; D_r is the regenerative diode; C_m is the series capacitor; D_o is the output diode; C_o is the output capacitor; V_{in} and V_{out} are the input and output voltages, respectively. N is defined as the turns ratio n_2/n_1.

The key steady waveforms of the proposed converter are shown in Fig.3. There are eight stages in one switching period. The corresponding equivalent circuits for each operational stage are shown in Fig.4.

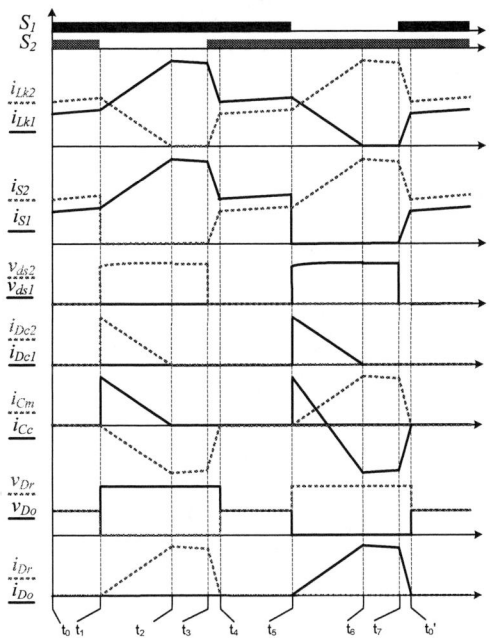

Fig.3. Key waveforms

Stage 1 [t_0, t_1]: S_1 and S_2 are both in turn-on state before t_1. D_{c1} and D_{c2}, D_r and D_o are all in turn-off state. L_{m1}, L_{m2} and L_{Lk1}, L_{Lk2} are charged linearly by the input voltage. The voltage stress of D_{c1}, D_{c2} is the voltage on C_c. The voltage stress of D_r is equivalent to the voltage on C_m. And the voltage stress of D_o is the output voltage minus the voltages on C_m and C_c.

Stage 2 [t_1, t_2]: At t_1, S_2 turns off, which makes D_{c2} and D_r turn on. The energy stored in L_{m2} and L_{Lk2} begins to transfer

to C_c. The current through D_r and C_m is determined by L_{Lk1} and L_{Lk2}. The current through S_1 increases and the current through D_{c2} decreases linearly as the current through D_r increases in a linear way.

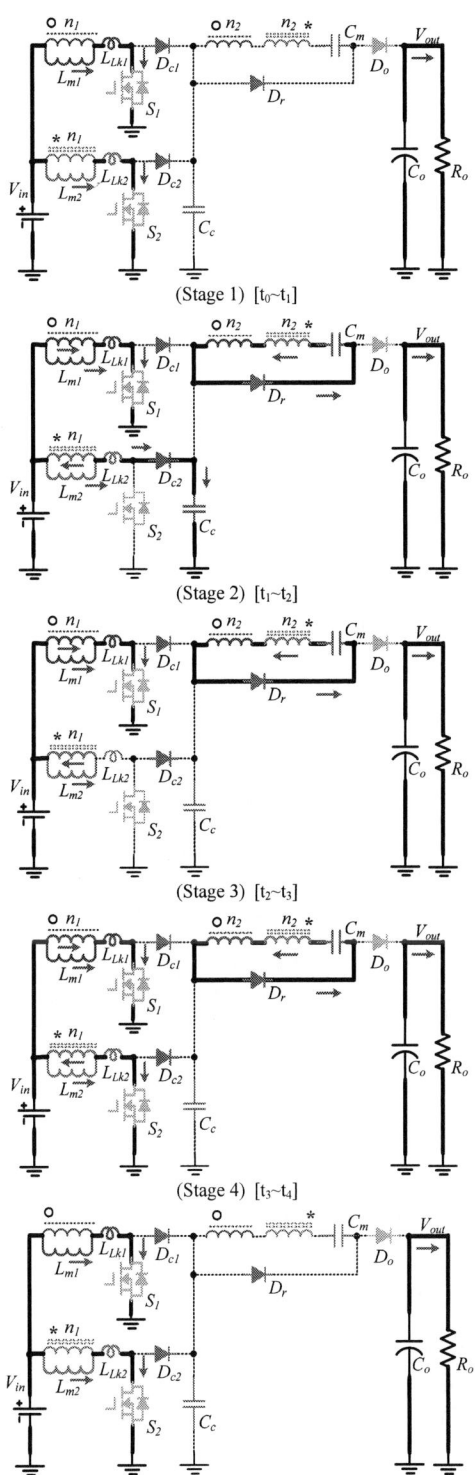

(Stage 1) [t_0~t_1]

(Stage 2) [t_1~t_2]

(Stage 3) [t_2~t_3]

(Stage 4) [t_3~t_4]

(Stage 5) [t_4~t_5]

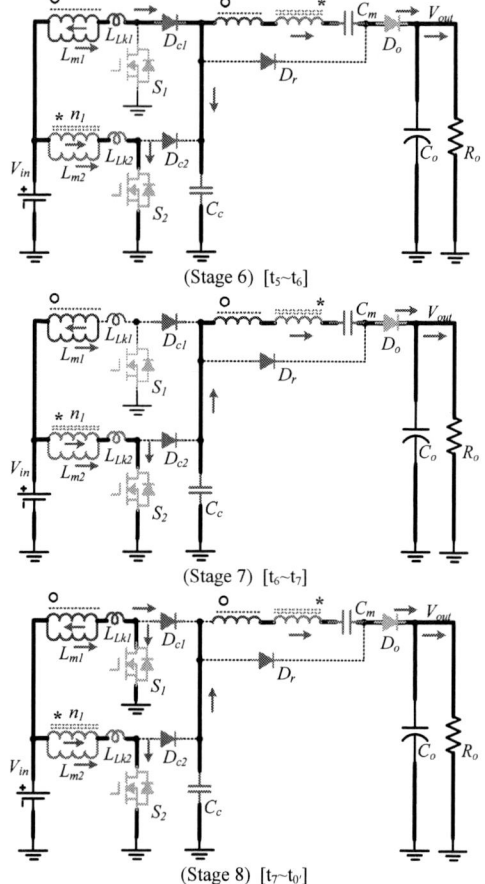

(Stage 6) [t_5~t_6]

(Stage 7) [t_6~t_7]

(Stage 8) [t_7~t_0']

Fig.4. Operation stages of the proposed converter.

Stage 3 [t_2, t_3]: At t_2, the current through L_{Lk2} decreases to zero and D_{c2} turns off naturally. There is no reverse-recovery problem for the clamp diodes. The current of S_1 is equal to the summation of the current of L_{m1} and L_{m2}.

Stage 4 [t_3, t_4]: At t_3, S_2 turns on. Due to L_{Lk2}, S_2 turns on with ZCS soft switching condition. The current falling rate through D_r is controlled by L_{Lk1} and L_{Lk2}, which alleviates the diode reverse-recovery problem. This stage ends when the current through D_r decreases to zero.

Stage 5 [t_4, t_5]: At t_4, D_r turns off. The operation of this stage is similar to that of stage 1.

Stage 6 [t_5, t_6]: At t_5, S_1 turns off and D_{c1} turns on. The energy begins to transfer to the load through D_o. The second windings of the coupled inductors and the capacitor C_m serve as voltage sources to extend the voltage gain and reduce the switch voltage stress. The current increasing rate through D_o is controlled by L_{Lk1} and L_{Lk2}.

Stage 7 [t_6, t_7]: At t_2, the current through L_{Lk1} decreases to zero and D_{c1} turns off naturally. The current through S_2 is the summation of the current of L_{m1} and L_{m2}.

Stage 8 [t_7, t_8]: At t_7, S_1 turns on with ZCS soft switching performance. The current falling rate through D_o is limited by L_{Lk1} and L_{Lk2}, which alleviates the output diode reverse-

recovery problem. When the current through D_o turns off, a new switching period begins.

III. STEADY CIRCUIT PERFORMANCE ANALYSIS

A. Voltage gain expression:

It can be seen that a conventional interleaved boost converter is existed in the left dashed block as shown in Fig.2(a). The voltage on the clamp capacitor C_c can be derived by

$$V_{Cc} = \frac{1}{1-D} \cdot V_{in} \qquad (1)$$

where V_{in} is the input voltage and D is duty cycle of the switches.

During S_1 turns on and S_2 turns off, the voltage on C_m can be derived by

$$V_{Cm} = \frac{N}{1-D} \cdot V_{in} \qquad (2)$$

When S_2 is in turn-on state and S_1 is in turn-off state, the energy is transferred to the load through D_o. The voltage gain of the proposed converter can be derived by

$$M = \frac{V_{out}}{V_{in}} = \frac{2 \cdot N + 1}{1 - D} \qquad (3)$$

From (3), it can be seen that a high voltage gain can be obtained without an extreme duty cycle as the turns ratio of the coupled inductors increases, which makes the proposed converter suitable for high step-up and high power applications.

B. Voltage stress analysis:

The voltage ripple on the capacitors is ignored to simplify the voltage stress analysis on the components of the proposed converter. The voltage stresses of S_1, S_2, D_{c1}, D_{c2} are equal to that of C_c, which are given by

$$V_S = V_{Dc} = V_{Cc} = \frac{V_{out}}{2 \cdot N + 1} \qquad (4)$$

The voltage stress of D_r and D_o can be derived by

$$V_{Dr} = V_{Do} = \frac{2 \cdot N}{2 \cdot N + 1} \cdot V_{out} \qquad (5)$$

It can be seen that the voltage stress of the regenerative diode is the same as that of the output diode. The diode voltage stress increases as the turns ratio increases, but it is always lower than the output voltage.

C. Soft switching performance and diode reverse-recovery problem alleviation:

Due to the leakage inductance, ZCS turn-on soft switching performance is realized for S_1 and S_2, which reduces the switching losses. When the switch turns off, the leakage

978-1-4244-4782-4/10 $26.00 © 2010 IEEE

energy is transferred to the clamp capacitor through the clamp diode, which absorbs the voltage spikes on the MOSFETs and recycles the leakage energy. Furthermore, the clamp diodes turn off naturally and there is no reverse-recovery problem for the clamp diodes. Moreover, the diode current falling rate of D_r and D_o is controlled by the leakage inductances of the coupled inductors, which alleviates the output diode reverse-recovery problem and reduces the reverse-recovery losses.

From above analysis, it can be concluded that the proposed interleaved high step-up converter has the following significant advantages:

1. The voltage gain is extended and the switch voltage stress is reduced due to the voltage multiplier cell;

2. ZCS turn-on soft switching performance is achieved and the diode reverse-recovery problem is alleviated due to the leakage inductances of the coupled inductors;

3. The leakage energy is recycled losslessly by the passive clamp scheme.

So the proposed converter is suitable for high step-up, high efficiency and high current conversion.

IV. DESIGN CONSIDERATIONS

A. Turns ratio design:

The turns ration selection is the key design parameter for the proposed converter because it determines the power device voltage stress, the switch duty cycle, which is given by

$$N = \frac{1}{2} \cdot \left[\frac{V_{out}}{V_{in}} \cdot (1-D) - 1 \right] \qquad (6)$$

One the duty cycle of the switches is selected, the turns ratio of the coupled inductors can be designed easily.

B. Leakage inductance design:

The leakage inductance of the coupled inductors is designed to limit the diode current falling rate and alleviate the diode reverse-recovery problem.

$$\frac{di_{Dr}}{dt} = \frac{di_{Do}}{dt} = \frac{V_{out}}{N \cdot (2 \cdot N + 1) \cdot L_{Lk}} \qquad (7)$$

where L_{Lk} is the summation of the leakage inductances L_{Lk1} and L_{Lk2}. The detailed leakage inductance can be designed according to the turns ratio and the diode current falling rate selection.

C. Capacitors design:

How to suppress the voltage ripple on the clamp capacitor C_c and the series capacitor C_m to an acceptable value is the main consideration for the capacitor design. The relationship between the voltage ripple and the output power can be derived by

$$C = \frac{P_{out}}{V_{out} \cdot \Delta V_c \cdot f_s} \qquad (8)$$

where P_{out} is the output power, V_{out} is the output voltage, ΔV_c is the voltage ripple on the capacitor C_c or C_m and f_s is the switching frequency. A large capacitor can reduce the voltage ripple. Unfortunately, the large capacitor is bulky and costly. So a compromise for the capacitor design should be made between the voltage ripple cancellation and the cost.

V. SIMULATED AND EXPERIMENTAL VERIFICATIONS

In order to verify the effectiveness of the proposed converter, a 1kW 40V-input 380V-output prototype is built and tested. The specifications of the converter are shown as follows:

V_{in}: 40V; V_{out}: 380V; P_{out}: 1kW; f_s: 100 kHz; n_2/n_1: 22/22;

L_m: 95μH; C_o: 470 μF; C_c: 47μF; C_m: 4.7μF;

S_1, S_2: Two pieces of IRF250N;

D_{c1}, D_{c2}: RURG1540; D_r, D_o: MUR1560.

The following experimental results from Fig.5 to Fig. 10 are given under 1kW full load condition.

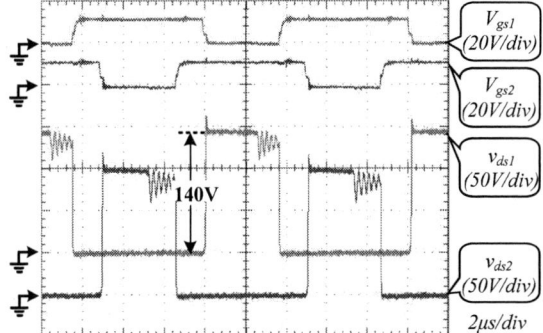

Fig.5. Experimental results of v_{gs1}, v_{gs2}, v_{ds1} and v_{ds2}.

The experimental results of the gate signals v_{gs1}, v_{gs2} and the MOSFET drain-source voltages v_{ds1}, v_{ds2} are shown in Fig. 5. It can be seen that the switch turn-off period of the presented converter is extended greatly, which can minimize the current ripple and optimize the dynamic response. And the switch voltage stress of the proposed converter is only about 140V, which is far lower than the high output voltage. So low voltage rated MOSFETs with low R_{DS_ON} are available to reduce the conduction losses and improve the circuit performance.

The experimental results of the leakage inductance current i_{Lk1}, the switch current i_{S1}, the clamp capacitor voltage v_{Cc} and the switch drain-source voltage v_{ds1} of the presented converter are given in Fig. 6. It can be seen that the drain-source voltage is clamped to a low level and the leakage energy is absorbed effectively. The voltage on the clamp capacitor is nearly constant. The current reflected effect is plotted in the dashed block, which is caused by the transformer effect of the coupled inductors.

978-1-4244-4782-4/10 $26.00 © 2010 IEEE

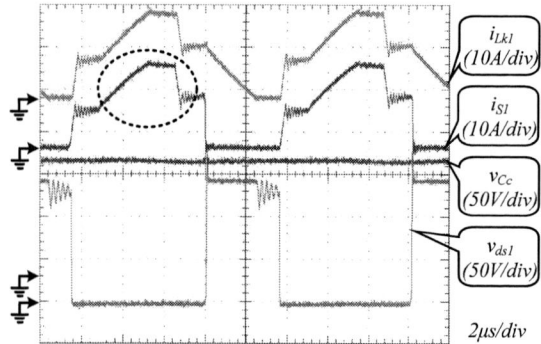

Fig.6. Experimental results of i_{Lk1}, i_{S1}, v_{Cc} and v_{ds1}.

The experimental results of the voltage v_{Dc1} and the current i_{Dc1} on the clamp diode D_{c1} is illustrated in Fig.7. The voltage stress of the clamp diode is the same as that of the switches, which is about 140V in the prototype. The current through the clamp diode deceases to zero before it turns off, which means that there is no reverse-recovery problem for the clamp diode.

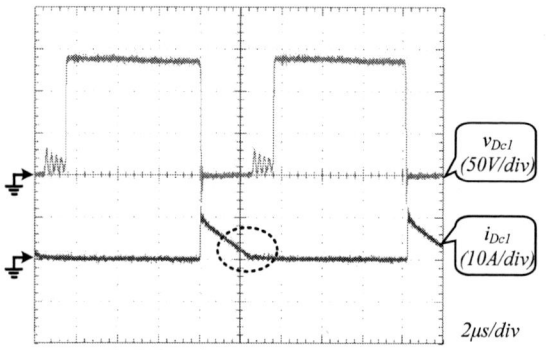

Fig.7. Experimental results of v_{Dc1} and i_{Dc1}.

The experimental waveforms of the voltage v_{Cm} and the current i_{Cm} on the series capacitor C_m are shown in Fig.8. The voltage stress of C_m is about 110V and the voltage ripple on C_m is small, which means that the series capacitor design is reasonable. The current through C_m is the same as that through the secondary windings of the coupled inductors.

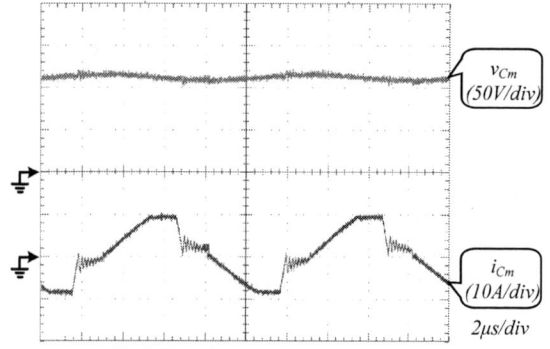

Fig.8. Experimental results of v_{Cm} and i_{Cm}.

The experimental results of the current i_{Do}, the voltage v_{Do} on the output diode D_o and the current i_{Dr}, the voltage v_{Dr} on the regenerative diode D_r are given in Fig. 9. The voltage stress of the output diode D_o and the regenerative diode D_r is

approximately 250V, which is consistent with the theoretical analysis in previous section. The two diodes operate in the interleaved mode and the reverse-recovery current is small.

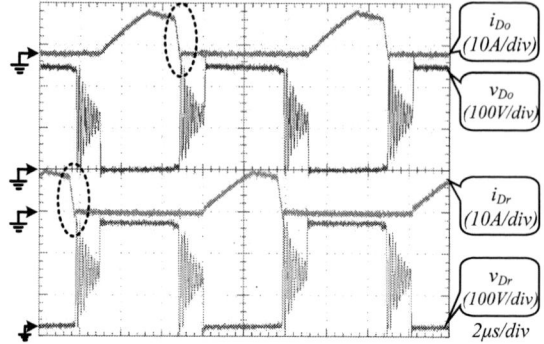

Fig.9. Experimental results of i_{Do}, v_{Do}, i_{Dr} and v_{Dr}.

The detailed experimental waveforms of the current i_{S1} and the voltage v_{ds1} on the switch S_1 is plotted in Fig. 10. It can be seen that ZCS turn-on soft switching performance is achieved for the power switches, which reduces the switching losses and the EMI noise.

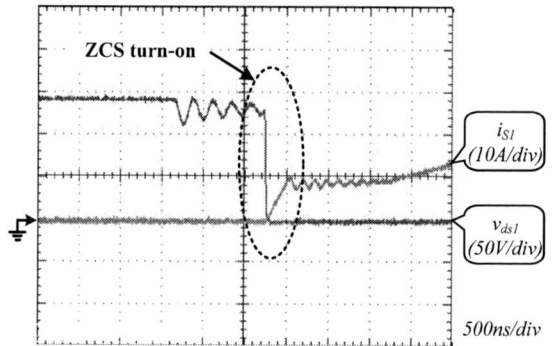

Fig.10. Detailed experimental results of i_{S1} and v_{ds1}.

VI. CONCLUSION

An interleaved high step-up boost converter with voltage multiplier cell is introduced and developed in this paper. A new design parameter of the turns ratio selection of the coupled inductors is employed to extend the voltage gain, which greatly reduced the peak current and the voltage stress of the switches compared with the conventional boost topology. This means significant reduction of the conduction losses. Furthermore, the switching losses are reduced due to the ZCS turn-on soft switching performance. Moreover, the diode turn-off current falling rate is controlled by the leakage inductances of the coupled inductors, which minimizes the diode reverse-recovery current and greatly reduce the reverse-recovery losses compared with the conventional boost converter. In additional, the leakage energy is absorbed and the voltage ringing on the switches is suppressed without adding any other active power devices. Experimental results have proved that the proposed interleaved converter with voltage multiplier cell is a suitable candidate for high step-up and high efficiency DC/DC conversion.

REFERENCES

[1] Q. Zhao, F. C. Lee, "High-efficiency, high step-up DC-DC converters," *IEEE Trans. Power Electron.*, vol. 18, no.1, pp. 65-73, Jan. 2003.

[2] X. Huang, X. Wang, T. Nergaard, J. S. Lai, X. Xu, L. Zhu, "Parasitic ringing and design issues of digitally controlled high power interleaved boost converters," *IEEE Trans. Power Electron.*, vol.19, no.5, pp.1341-1352, Sept. 2004.

[3] W. Li, X. He, "An Interleaved Winding-Coupled Boost Converter with Passive Lossless Clamp Circuits," *IEEE Trans. Power Electron.*, vol. 22, no.4, pp. 1499-1507, Jul. 2007.

[4] W. Li, X. He, "A Family of Interleaved DC/DC Converters Deduced from a Basic Cell with Winding-Cross-Coupled Inductors (WCCIs) for High Step-Up or Step-Down Conversions," *IEEE Trans. Power Electron.*, vol. 22, no.4, pp. 1499-1507, Jul. 2008.

[5] P. W. Lee, Y.S. Lee, D. K. Cheng, X. C. Liu, "Steady-state analysis of an interleaved boost converter with coupled inductors," *IEEE Trans. Ind. Electron.*, vol. 47, no.4, pp. 787-795, Aug. 2000.

[6] R. W. Erickson, D. Maksimovic, *Fundamentals of power electronics*, 2nd ed. Kluwer Academic Publishers, 2001.

[7] W. Li, X. He, "ZVT Interleaved Boost Converters for High-Efficiency, High-Step-Up DC/DC Conversion," in *Proc. IET-Elect. Power Applicat.*, vol. 1, no. 2, pp. 284–290, March. 2007.

FPGA-Based Multi-Phase Digital Pulse Width Modulator with Dual-Edge Modulation

Martin Scharrer, Mark Halton, Tony Scanlan and Karl Rinne

Dept. Electronic and Computer Engineering,
University of Limerick
Limerick, Ireland
Email: martin.scharrer@ul.ie

Abstract—This paper proposes a new FPGA-based architecture for a multi-phase digital pulse width modulator (MP-DPWM). A novel fine-leading/coarse-trailing edge modulation is applied to allow the sharing of a single fine resolution block for all phases. Specifically, the architecture takes advantage of Digital Clock Manager (DCM) blocks available in modern FPGAs to produce four clock phases from a single clock input to increase the resolution by two bit. An optimized counter/shift-register block is detailed which reduces the size and increases the maximum clock frequency of the architecture for certain numbers of phases. The design was successfully implemented on a low-cost Xilinx Spartan-3 FPGA 9-bit resolution with a switching frequency of 1 MHz and 2–16 phases.

I. INTRODUCTION

In recent years an increasing research interest in digital control of switching mode power supplies (SMPS) is clearly evident [1, 2]. A key reason for this is that digital control has a number of advantages over analog control: such as programmability, reduced sensitivity to external influences, the use of adaptive or other advanced control algorithms along with simpler implementation and prototyping.

A digital controller uses a digital pulse width modulator (DPWM) to generate the control signals for the power supply switches. A sufficiently high resolution of the DPWM is critical for the stability of the output voltage. Conversely, a DPWM resolution that is lower than an ADC resolution leads to limit cycling [3, 4]. To avoid high clock frequencies (i.e. $2^{\#\text{Bit Resolution}} \times f_{\text{switching}}$) needed for simple counter-comparator DPWMs, a range of different hybrid architectures have been proposed and implemented in ICs [5–7] and in field programmable gate arrays (FPGAs) [8–14].

FPGAs are widely used by practicing engineers to prototype and validate their designs. FPGA implementations of digital systems allow for easy and fast prototyping without or before an implementation as an application specific integrated circuit (ASIC).

Modern high-current switched mode power supplies use multiple inductors which are driven in phases in order to split the inductor current and to increase the control loop performance and efficiency. Each phase requires its own DPWM driver signal. This signals are generated by a multi-phase digital pulse width modulator (MP-DPWM), also referred to as digital multiphase modulators (MPM). Different architectures have been proposed in the literature which either require the

replication of timing blocks for each phase [15–18] or share the fine resolution block between the phases [19–22].

Furthermore, some multi-phase architectures only support a single common duty cycle for all phases. This limits the controller performance and only permits passive current sharing. Other architectures support independent duty cycles for each phase. This allows the update of the duty cycle with a rate of $P \cdot f_{\text{sw}}$ instead of f_{sw} where P is the number of phases and f_{sw} is the switching frequency. Also variances in the power supply inductors can be compensated by adjusting the duty cycle for the corresponding phases accordingly.

In this paper a new FPGA-based architecture for multi-phase digital pulse width modulators supporting separate duty-cycles for each phase is proposed. It allows the sharing of the fine resolution block across all phases to reduce the implementation size. The architecture is explained in detail, followed by its implementation and verification on an FPGA.

II. PROPOSED MULTI-PHASE DPWM ARCHITECTURE

The proposed architecture is shown in Fig. 1. The course resolution N_c is achieved by a trailing-edge counter/comparator style DPWM stage. A synchronous fine resolution block detailed in [10] for a single-phase trailing-edge DPWM is adapted to support leading-edge modulation. This block increases the DPWM resolution (N) by two bit and requires four out-of-phase clock signals ($0°, 90°, 180°, 270°$) which are produced using dedicated digital clock managers (DCM) available in modern FPGAs. These PLL-type blocks can multiple the frequency and shift the phase of existing clock signals. The rest of the architecture is clocked by the $0°$ clock.

A. Counter/Comparator Block

This block contains a single counter with $N_C = N - 2$ bit which is shared across all phases. An array of N_C-bit adders subtracts the phase difference of each phase from the counter value. The resulting values, "phase values", drive two arrays of comparators. With reference to Fig. 1, each comparator in the top array tests if the corresponding phase value is all-ones. This indicates that the value will be zero after the next clock edge. Their output signals are named "counter full". Because some parts of the DPWM need to be triggered when any of the phase values is all-ones, these signals are or-ed together into a single signal name "any counter full".

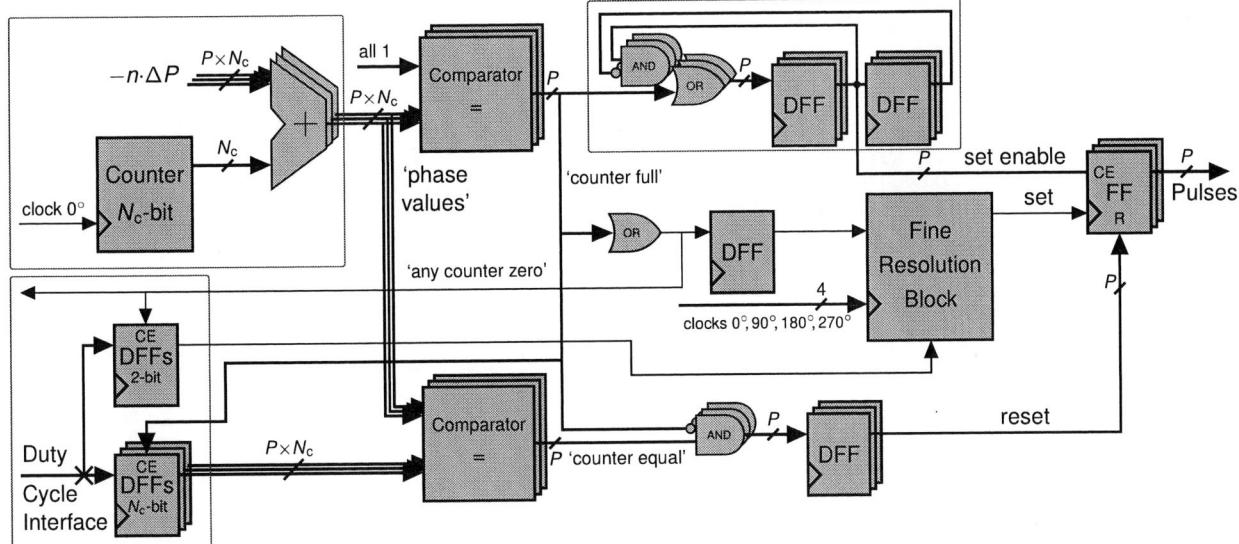

Fig. 1: Schematic of proposed DPWM Architecture.

The comparators in the bottom array test if the corresponding phase value is equal to the most significant N_C bit of the duty cycle of this phase. Their output signals are named "counter-equal".

B. Duty Cycle Interface

The duty cycle values arrive from a DPWM-external source, e.g. a PID controller, and are registered one clock cycle before the begin of the switching period by using the "counter full" signals as clock-enable. The duty cycle interface shown in the lower left corner of Fig. 1 uses a single serial N_c-bit wire to receive all duty cycles. The duty cycle of the next phase must be written on this wire and held until it is registered. The lower two LSBs are only needed by the fine resolution block for a few clock cycles directly after they are registered. Therefore a single 2-bit register can be shared by all phases which is controlled by the "any counter full" signal. This signal is also provided to the external block as handshake signal which indicates that the current duty cycle was registered and it is safe to write the next one onto the bus. The "counter full" signals can also be provided to indicate which phase was processed.

If required by the surrounding system a different interface can be used. Alternatively a parallel interface which provides each duty cycle on an N_c-bit wire or a chip-internal bus like AMBA APB which allows addressing of each duty cycle input register separately. To support such interfaces only the two LSB registers must be changed. They cannot be shared and therefore a P-to-1 multiplexer (MUX) is needed to supply the current phase LSBs to the fine resolution block. This MUX uses the "counter full" signals as an one-hot select signal.

C. Output Pulse Registers

The DPWM pulses are produced using D-Flip-Flops (DFFs) due to the absence of Set/Reset-FFs in FPGAs. The output

signal of the fine resolution block drives the clock inputs of all pulse DFFs and acts as a shared 'set' signal. The clock enable (CE) inputs of the DFFs are used only to set the pulse register of the current phase. In order to avoid setup and hold time violations of the pulse register, these enable signals must be active one clock cycle before any possible 'set' signal and stay active during the following clock cycle. This is achieved by registering and extending the corresponding "counter-full" signals using the circuit shown in Fig. 2. Because a simple circuit as 2(a) could produce a glitch at the clock transition the circuit 2(b) is proposed which produces a stable, registered clock-enable signals.

The pulse registers are reset by feeding the registered version of the corresponding "counter-equal" signal to their asynchronous reset input. This makes the reset effectively synchronous to the $0°$ clock.

For very high duty cycles (i.e. the last four fine resolution steps, $D = \{2^N-4, 2^N-1\}$), the reset signal will overlap with the set signal of the next switching cycle. Therefore to counteract this, the reset signal is disabled by an AND gate when the "counter-full" signal is high. The resulting early saturation to 100% duty cycle has no practical influence for use in power supplies.

D. Leading-Edge Fine Resolution Block

The proposed synchronous fine resolution block (FRB) for leading-edge modulation is an adaption of the FRB detailed in [10]. Signal timing and a schematic for this block are shown in Fig. 3 and Fig. 4 respectively. The proposed block contains four DFFs and a 4-to-1 multiplexer (MUX). A pulse generated by a DFF outside the block is used as an input for the four DFFs. This input pulse is produced in the last clock cycle before the start of the next switching period and acts as a timing reference. Each DFF is clocked using one of the

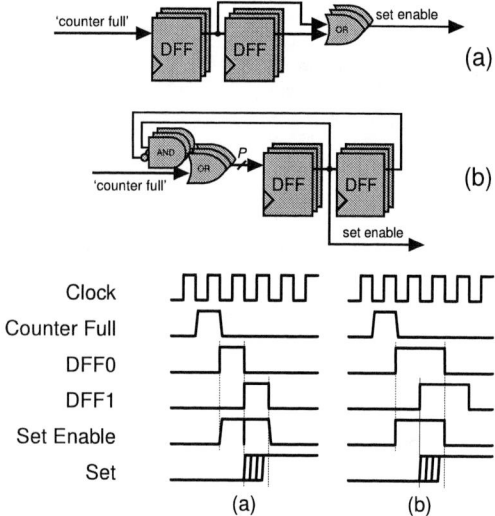

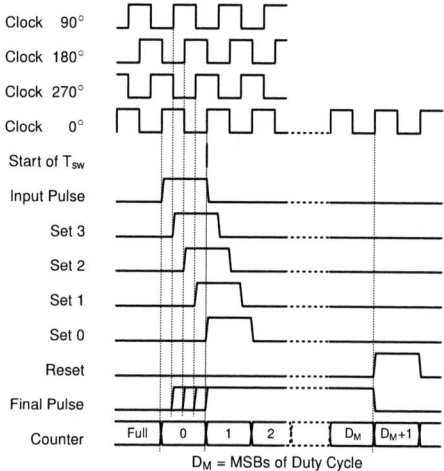

Fig. 2: Schematic and timing diagram of `set enable` logic: (a) simple circuit with potential glitch, (b) proposed glitch-free circuit.

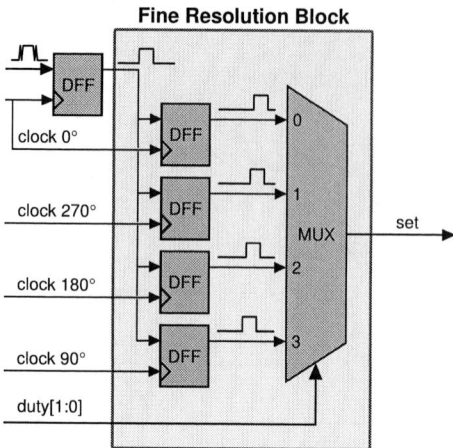

Fig. 4: Schematic of Fine Resolution Block.

Fig. 3: Timing diagram showing fine leading and coarse trailing pulse modulation.

four clock phases $0°(\hat{=}360°)$, $270°$, $180°$, $90°$ and therefore delays the input pulse by a full, ¾, ½ and ¼ clock period, respectively. The DFF outputs are then multiplexed using the two LSBs of the duty cycle as a select signal.

If the duty cycle LSBs are both zero (b00), the full cycle delayed pulse ($0°$ clock) is selected and the block output signal goes high at the start of the switching period. If the LSBs are b01 (=1), b10 (=2) or b11 (=3), the leading edge of the output pulse starts 1, 2 or 3 quarters of a clock cycle earlier, resulting in a longer DPWM pulse.

With the move of the fine-resolution modulation from the trailing to the leading edge, it can be ensured that only one fine-resolution modulation is required at the same time. This

allows the sharing of one single fine-resolution block among all phases. Note that if coarse and fine-resolution would be performed at the same edge, such time multiplexing would not be possible, as two different phases might require a fine-resolution modulation at the same time.

E. Counter/Shift-Register Block

This paper also proposes a new counter/shift-register block which can replace the larger counter and all adders (Fig. 5a), leading to a smaller circuit size and to a larger maximal clock frequency due to a reduced logic path. The proposed block can be applied when the number of phases is a power of two. The block consists of a $(N_C - N_p)$ bit counter and a N_p-bit, $P/2$-stage shift-register, where P is the number of phases and $N_p = \log_2 P$. The counter holds the common LSBs which are identical for all phases as long as $P = 2^n$. The stages of the shift-register represent the phase-specific MSBs and are loaded, i.e. reset, with the phase-differences at start-up avoiding the requirement of any adders (Fig. 5b).

The cyclic shift-register is clocked every time the counter is full. The required control signal can be reused for one of the comparators and therefore does not require additional logic gates. The number of required shift-register stages is reduced from P to $P/2$ by utilizing the symmetry of binary numbers. The values of a phase n and a phase $n + P/2$ differ only by the MSB while the other $N_P - 1$ bit are identical. The last $P/2$ stages can therefore be replaced by the values of the first $P/2$ stages where the MSB is inverted (Fig. 5c).

III. EXPERIMENTAL VERIFICATION

The FPGA implementation of the proposed architecture has been verified by simulation and lab measurements.

A. Implementation

The proposed architecture has been implemented using the hardware description language Verilog. A behaviour/RTL-level coding style has been used to make the implementation vendor independent. The only exception is the clock generating

978-1-4244-4782-4/10 $26.00 © 2010 IEEE

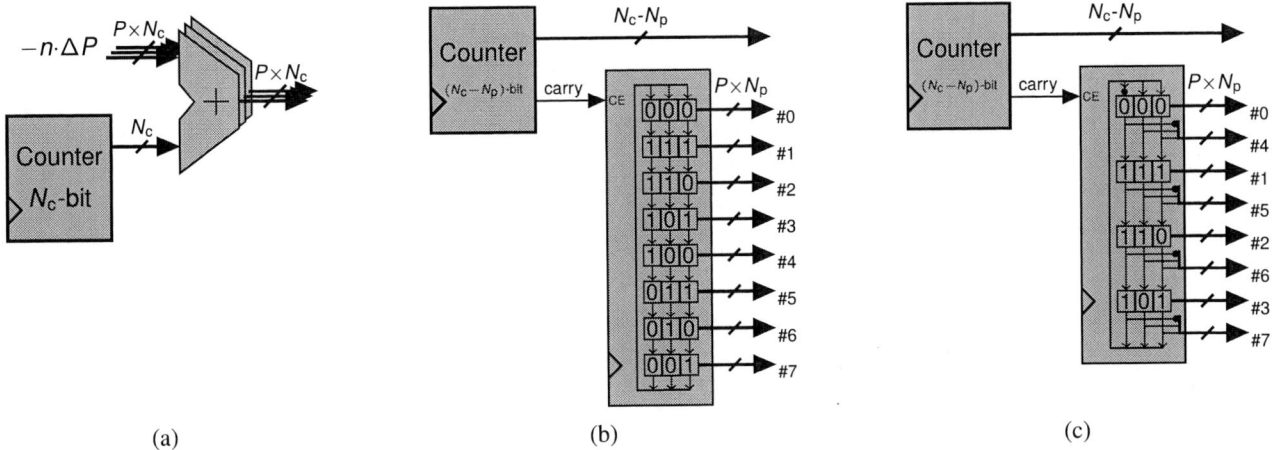

Fig. 5: General Counter/Adders Block (a) versus non-optimized (b) and optimized Counter/Shift-Register Block (c) (shown for $P = 8$, $N_p = 3$).

module (DCM) which must be instantiated using a vendor specific template. However, this module can be encapsulated in an abstract Verilog module which itself instantiates the corresponding clock generator of the used vendor.

A low-cost Xilinx Spartan-3 FPGA has been chosen to verify the architecture. This FPGA provides four DCMs where only one is needed to produce the required four phases of the clock. Additional DCMs can be used to multiply an external clock source to the required clock frequency. For a switching frequency of $f_{sw} = 1$ MHz, the on-board clock of 50 MHz has been multiplied in two steps to 128 MHz, that is required for a DPWM with 9-bit ($2^{9-2} = 128$) resolution. This results in the use of three DCMs blocks, leaving one for additional tasks. If needed this number can be reduced by matching the different clock frequencies to a common divider. This can be done either by applying an external clock oscillator with a more suitable clock frequency, e.g. 64 MHz, or by using a clock frequency which is a multiple of the FPGA clock, e.g. 100 MHz, which would lead to an odd switching frequency, e.g. $100 \, \text{MHz}/2^{9-2} = 0.781$ MHz.

B. Simulation Results

The implemented Verilog code has been first simulated using a behavioural simulation model also written in Verilog HDL to verify the correct functional behaviour of the circuit. The synchronous nature of the fine resolution block simplifies this step. Attention must be paid to the correct timing of the comparator and phase enable signals to ensure proper multi-phase operation.

A subsequent post-synthesis gate-level simulation has been performed to include the effect of logic and routing delays into account. Routing delays in the used Spartan-3 FPGA can be around 1ns which is in the range of one LSB ($\approx$2ns). Largely different routing delays in critical paths – between the four DFFs and the MUX of the FRB – can lead to non-linear, even non-monotonic behaviour. For this reason manual placing

of the components of the FRB and at least the final output FFs is strongly recommended. The implementation techniques proposed in [10] have been applied to ensure monotony and reduce the offset and non-linearity error. This has been verified by the post-synthesis simulation results.

C. Synthesis Results

The proposed architecture can be implemented by using only a small number of DFFs and Look-Up Tables (LUTs), while the exact size depends on the resolution, the number of phases and the synthesis constraints. A fixed number of DCMs are required as mentioned earlier. An additional Input/Output Block (IOB) which contains the pulse register is required for every phase. The synthesis results for a Xilinx Spartan-3 FPGA (speed grade 4) are shown in Table I. The area results are accurate for all FPGAs with 4-input LUTs. Higher frequencies can be achieved with faster FPGAs, e.g. Xilinx Virtex.

D. Measurement Results

The output pulses of a 16-phase DPWM implemented on a Xilinx Spartan-3 FPGA have been measured and are shown in Fig. 6. The switching frequency is 1 MHz. Fig. 6a displays

TABLE I: Number of Phases vs. Area and Speed

# Phases	FFs	LUTs	Max. f [MHz]
2	24	29	148.7
4	47	41	147.5
8	95	66	148.2
10	131	88	147.2
12	161	105	142.7
14	189	122	138.6
16	215	138	136.4

Pulse Register	P	IOBs
Clock Generation	$1 \ldots 3$	DCMs
Clock Buffers	$5 \ldots 7$	GBUFs

the coarse trailing-edge modulation by superposition of three pulses with duty cycles of 6.25% ($1/16$), 25% and 37.5%. The plot also shows the phase difference of T_{sw}/P between the pulses. The leading-edge modulation of the synchronous fine resolution block is shown in Fig. 6b. It compares the four possible fine steps together with the internal reference input pulse of the fine resolution block. This verifies the pulse timing shown in Fig. 3.

The achieved linearity is shown in Fig. 7. The behaviour is strictly monotonic. The absolute error, i.e. the difference between the ideal and measured pulse width, is around 0.1 LSB ($\approx$200ps). This error is far below the critical level of a 0.5 LSB which allows the implemented DPWM to fully use its 9-bit resolution effectively. The fine resolution block was manually placed as described by [10] to reduce the routing delay differences and to correct the offset error.

IV. CONCLUSIONS

In this paper an architecture suitable for implementation of high-speed, high-resolution multi-phase DPWM with high linearity on low-cost FPGAs has been proposed. It supports different duty cycles for every phase to allow active current sharing. The outlined dual-edge modulation allows sharing of the fine resolution block between all phases without the risk of sharing violations. The achieved linearity is approximately 0.1 LSB. The architecture has been implemented and verified on a Xilinx Spartan-3 FPGA with 9-bit resolution and up to 16 phases operating at a switching frequency of 1 MHz.

V. ACKNOWLEDGMENT

This work was funded under Enterprise Ireland Technology Development fund (CFTD/219/05).

REFERENCES

[1] C. Mullett, "A 5-year power technology roadmap," in *Applied Power Electronics Conference and Exposition, 2004. APEC '04. Nineteenth Annual IEEE*, vol. 1, 2004, pp. 11–17 Vol.1.

[2] Y.-F. Liu and P. Sen, "Digital control of switching power converters," in *Control Applications, 2005. CCA 2005. Proceedings of 2005 IEEE Conference on*, 2005, pp. 635–640.

[3] A. Peterchev and S. Sanders, "Quantization resolution and limit cycling in digitally controlled pwm converters," *Power Electronics, IEEE Transactions on*, vol. 18, pp. 301–308, 2003.

[4] H. Peng, A. Prodic, E. Alarcon, and D. Maksimovic, "Modeling of quantization effects in digitally controlled DC-DC converters," *Power Electronics, IEEE Transactions on*, vol. 22, pp. 208–215, 2007.

[5] A. Dancy and A. Chandrakasan, "Ultra low power control circuits for pwm converters," in *Power Electronics Specialists Conference, 1997. PESC '97 Record., 28th Annual IEEE*, vol. 1, 1997, pp. 21–27 vol.1.

[6] E. O'Malley and K. Rinne, "A programmable digital pulse width modulator providing versatile pulse patterns and supporting switching frequencies beyond 15 mhz," in *Applied Power Electronics Conference and Exposition, 2004. APEC '04. Nineteenth Annual IEEE*, vol. 1, 2004, pp. 53–59 Vol.1.

[7] K. Wang, N. Rahman, Z. Lukic, and A. Prodic, "All-digital dpwm/dpfm controller for low-power DC-DC converters," in *Applied Power Electronics Conference and Exposition, 2006. APEC '06. Twenty-First Annual IEEE*, 2006, p. 5 pp.

[8] J. Quintero, M. Sanz, A. Barrado, and A. Lazaro, "FPGA based digital control with High-Resolution synchronous DPWM and High-Speed embedded A/D converter," in *Applied Power Electronics Conference and Exposition, 2009. APEC 2009. Twenty-Fourth Annual IEEE*, 2009, pp. 1360–1366.

[9] Y. Gao, S. Guo, Y. Xu, S. X. Lin, and B. Allard, "FPGA-based DPWM for digitally controlled high-frequency DC-DC SMPS," in *Power Electronics Systems and Applications, 2009. PESA 2009. 3rd International Conference on*, 2009, pp. 1–7.

[10] M. Scharrer, M. Halton, and T. Scanlan, "FPGA-based digital pulse width modulator with optimized linearity," in *Applied Power Electronics Conference and Exposition, 2009. APEC 2009. Twenty-Fourth Annual IEEE*, 2009, pp. 1220–1225.

[11] A. de Castro and E. Todorovich, "DPWM based on FPGA clock phase shifting with time resolution under 100 ps," in *Power Electronics Specialists Conference, 2008. PESC 08. 2008 IEEE 39th Annual*, 2008, pp. 3054–3059.

[12] M. G. Batarseh, W. Al-Hoor, L. Huang, C. Iannello, and I. Batarseh, "Window-Masked segmented digital clock manager–FPGA based digital pulse width modulator technique," *Power Electronics, IEEE Transactions on*, vol. PP, no. 99, p. 1, 2009.

[13] S. C. Huerta, A. de Castro, O. Garcia, and J. Cobos, "FPGA based digital pulse width modulator with time resolution under 2 ns," in *Applied Power Electronics Conference, APEC 2007 - Twenty Second Annual IEEE*, 2007, pp. 877–881.

[14] R. Foley, R. Kavanagh, W. Marnane, and M. Egan, "A versatile digital pulsewidth modulation architecture with area-efficient FPGA implementation," in *Power Electronics Specialists Conference, 2005. PESC '05. IEEE 36th*, 2005, pp. 2609–2615.

[15] A. Peterchev, J. Xiao, and S. Sanders, "Architecture and IC implementation of a digital VRM controller," *Power Electronics, IEEE Transactions on*, vol. 18, no. 1, pp. 356–364, 2003.

[16] A. Wu, J. Xiao, D. Markovic, and S. Sanders, "Digital PWM control: application in voltage regulation modules," in *Power Electronics Specialists Conference, 1999. PESC 99. 30th Annual IEEE*, vol. 1, 1999, pp. 77–83 vol.1.

[17] A. Dancy, R. Amirtharajah, and A. Chandrakasan, "High-efficiency multiple-output DC-DC conversion for low-voltage systems," *Very Large Scale Integration (VLSI) Systems, IEEE Transactions on*, vol. 8, no. 3, pp. 252–263, 2000.

[18] Z. Lukic, C. Blake, S. C. Huerta, and A. Prodic, "Universal and Fault-Tolerant multiphase digital PWM controller IC for High-Frequency DC-DC converters," in *Applied Power Electronics Conference, APEC 2007 - Twenty Second Annual IEEE*, 2007, pp. 42–47.

[19] T. Carosa, R. Zane, and D. Maksimovic, "Scalable digital multiphase modulator," *Power Electronics, IEEE Transactions on*, vol. 23, no. 4, pp. 2201–2205, 2008.

[20] T. Carosa, R. Zane, and D. Maksimovic, "Digital multiphase modulator a power D/A perspective," in *Power Electronics Specialists Conference, 2006. PESC '06. 37th IEEE*, 2006, pp. 1–6.

[21] T. Carosa, R. Zane, and D. Maksimovic, "Implementation of a 16 phase digital modulator in a 0.35 μm process," in *Computers in Power Electronics, 2006. COMPEL '06. IEEE Workshops on*, 2006, pp. 159–165.

[22] R. Foley, R. Kavanagh, W. Marnane, and M. Egan, "Multiphase digital pulsewidth modulator," *Power Electronics, IEEE Transactions on*, vol. 21, no. 3, pp. 842–846, 2006.

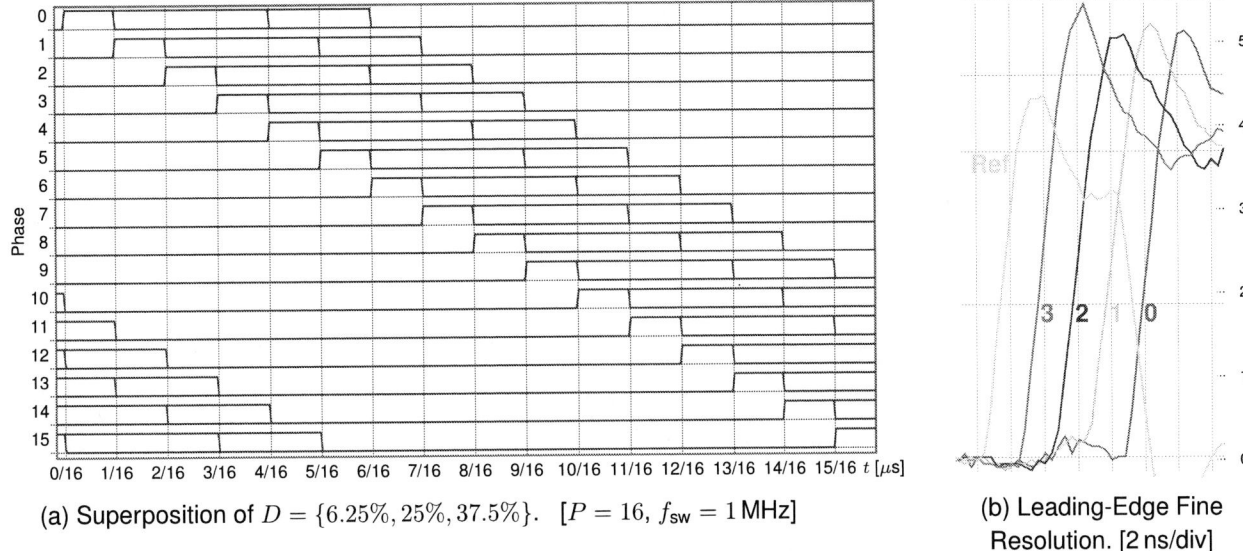

(a) Superposition of $D = \{6.25\%, 25\%, 37.5\%\}$. [$P = 16$, $f_{sw} = 1\,\text{MHz}$]

(b) Leading-Edge Fine Resolution. [2 ns/div]

Fig. 6: Measured Multi-Phase DPWM Pulses.

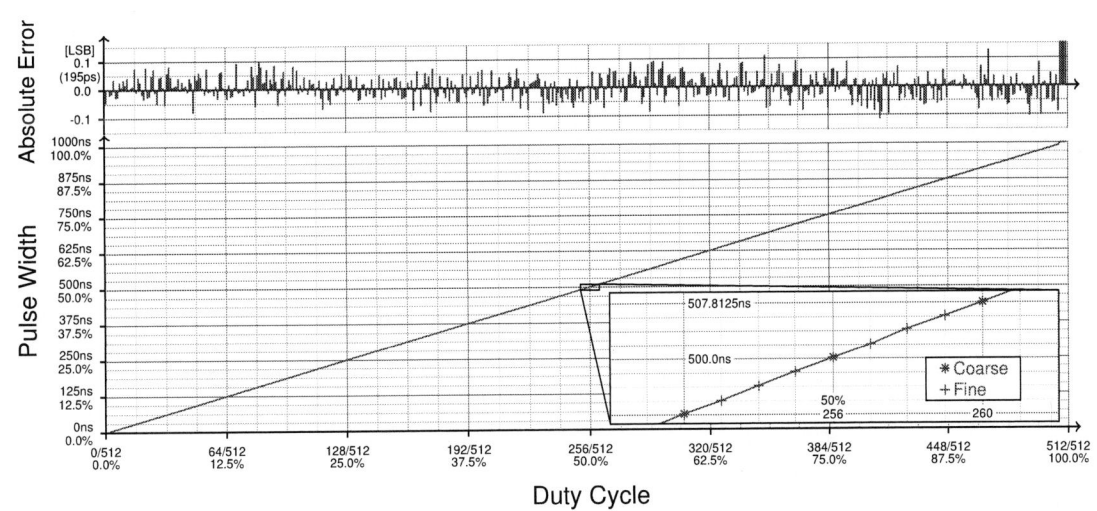

Fig. 7: Measured Linearity of DPWM Pulses

Phase Doubler for High Power Voltage Regulators

Chun Cheung[1]   Weihong Qiu[2]   Emil Chen[2]   Greg Miller[3]

Intersil Corporation[1]
65 Readington Road
North Branch, NJ, 08876

Intersil Corporation[2]
1001 Murphy Ranch Road
Milpitas, CA 95035

Intersil Corporation[3]
4020 Stirrup Creek Drive
Durham, NC 27703

Abstract— **As the power consumption of microprocessors increases, a high phase count voltage regulator is required to meet the power hunger (>180A) and high efficiency requirements. In general, the higher the phase number count, the more PWM outputs and current sensing inputs are required for the PWM controller. Developing a very high phase count PWM controller becomes very challenging and requires a high pin count package, resulting in high cost and complex layout design. In this paper, a Phase Doubler, using a single PWM input to drive two interleaved phases, is proposed to overcome these problems. The Phase Doubler can be further expanded to quadruple the phase count, i.e, control 4 phases with only a single PWM input. Some experimental data of a 12-phase voltage regulator are included to verify the concept.**

I. INTRODUCTION

Multiphase Buck converters have been a staple of microprocessor power regulation for many generations of computing platforms. Standard mobile solutions typically use 2-phase or 3-phase controllers and typical desktop solutions employ anywhere from 3 to 6 phases. However, as the power consumption of the modern microprocessor continues to increase, especially in over-clocking applications, higher phase count multiphase voltage regulators are being utilized to simultaneously provide high output current (>180A), high efficiency and low ripple current [1-9].

One key complication in designing a high phase count controller is the routing of current sense information for each phase. Each additional phase requires at least 3 pins: one for the PWM output, and two for differential current sensing signals. For a 12+ phase design, the total pin count on the package becomes rather prohibitive. A 24-phase controller needs as many as 102 pins, as shown in Figure 1.

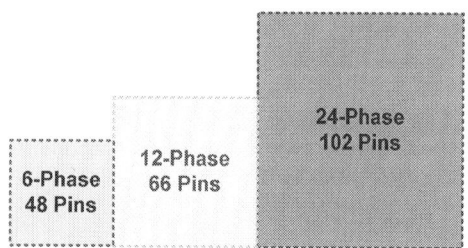

Figure 1. Phase Number vs Package Size & Pin Count

The inner circuit design of high phase count controller is also very complex and challenging. Due to cost constraints

and the complexity of designing a high phase count PWM controller, few controllers with greater than 8 output phases exist in today's market.

To overcome these challenges, a phase doubling scheme is proposed in this paper. This scheme will accept one PWM signal from the PWM controller and send two interleaved PWM signals to two phase power stages. The Phase Doubler can help double or quadruple the phase count of the existing 4- or 6-phase controller [7-8] to achieve an 8 to 24-phase voltage regulator solution. The detailed description about this scheme is included with typical application circuits. Some experimental results are provided to verify the proposed scheme.

II. BENEFITS OF HIGH PHASE COUNT SYSTEMS

In typical applications for microprocessor power, the optimal current level per phase is typically between 15-30A; thus, a multiphase voltage regulator is required to meet high current demand. Some key benefits of high phase count systems in high power applications are highlighted.

A. Improved Efficiency

At high load current, efficiency is improved by spreading the load out across many phases [1-5]. This is primarily because resistive losses become a very large component of the total loss budget at high current levels. Figure 2 is a chart showing the efficiency improvement with additional phases with increasing load.

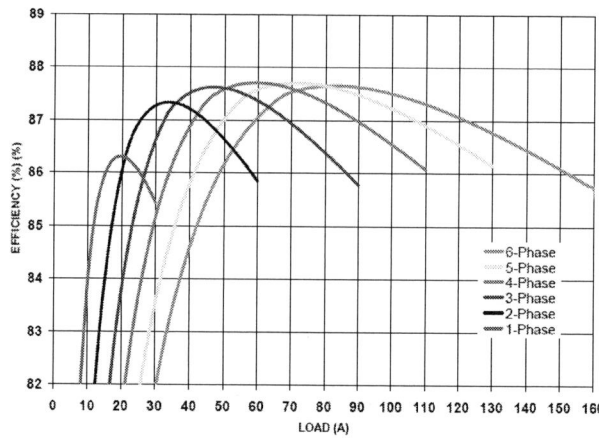

Figure 2. Efficiency improvement with increased phase count at heavy load

B. Better Thermal Dissipation

Since the load current is carried by more phases, the power devices in each phase will handle less current and can be lower cost. In addition, the power devices are likely to be spread over a larger area on the board. Both factors result in improved heat dissipation for higher phase count systems. By reducing the system's operating temperature, component reliability is improved.

C. Reduced Ripple Current

Increasing the interleaving phase count also reduces the magnitude of the ripple on both the input and output current. Figure 3 shows the ripple value for a 24-phase voltage regulator with the following parameters:

- Input voltage: 12V
- Output voltage: 1.6V
- Duty cycle: 13.3%
- Load current: 200A
- Output Phase inductor: 500nH
- Phase switching frequency: 200kHz

In this example, the 24-phase voltage regulator can run in 6-phase, 8-phase, 12-phase or 24-phase interleaving mode. In 6-phase interleaving mode, every 4 phases runs synchronously. In this case it is equivalent to a 6-phase voltage regulator, which yields 18.73A and 12.93A input and output ripple current, respectively. The 24-phase interleaving regulator significantly drops these values to 4.05A and 0.78A, respectively. As shown in Table 1, both input and output ripple currents are reduced when more phases are running in interleaving mode.

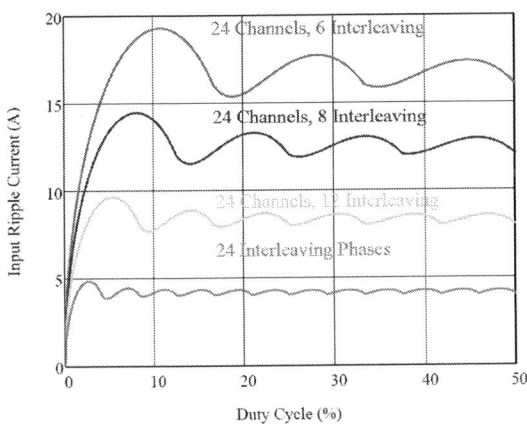

a. Input current ripple

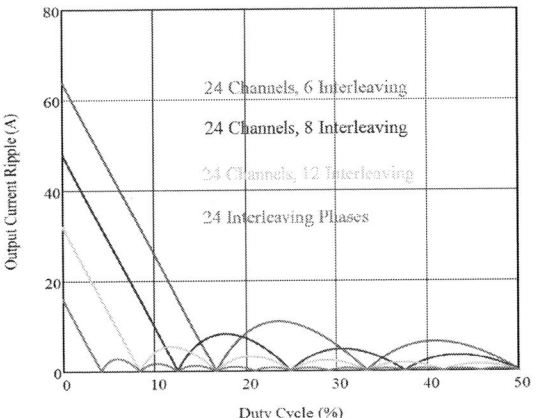

b. Output current ripple

Figure 3. Current ripple improvement

Table 1: Input Ripple Current (Unit: A)

# of Interleaved phases	6	8	12	24
Input Ripple Current	18.73	11.64	8.79	4.05
Output Ripple Current	12.93	2.70	4.83	0.78

III. BENEFITS OF HIGH PHASE COUNT SYSTEMS

It is clear that high-phase count has benefits for high power applications. However, high phase count systems incur complex design and high cost for the controller design. To mitigate these issues, phase doubling technique is proposed here. The basic concept of phase doubling scheme is to double the phase count of the power stage without increasing the complexity to the PWM controller and system layout design.

Figure 4(a) shows the block diagram of the Phase Doubler. It receives one PWM signal from the controller, and sends two PWM signals to two power stages (Phase A and Phase B). There are two ways to distribute the PWM signals:

- Interleaving mode: alternate the input PWM signal between two power stages as shown in Figure 4(b). This ping-pong approach halves the switching frequency of the output power stage, but also provides the benefit of 180 degrees phase shift between two power stages. It gains all typical benefits of interleaved multiphase design.

- Synchronous mode: just copy the input PWM signal to two power stages as shown in Figure 4(c). Therefore two phase power stages run in synchronous mode with switching at the same time. It can reduce the required power rating on the power stage devices. However it cannot achieve other benefits compared to the interleaved design.

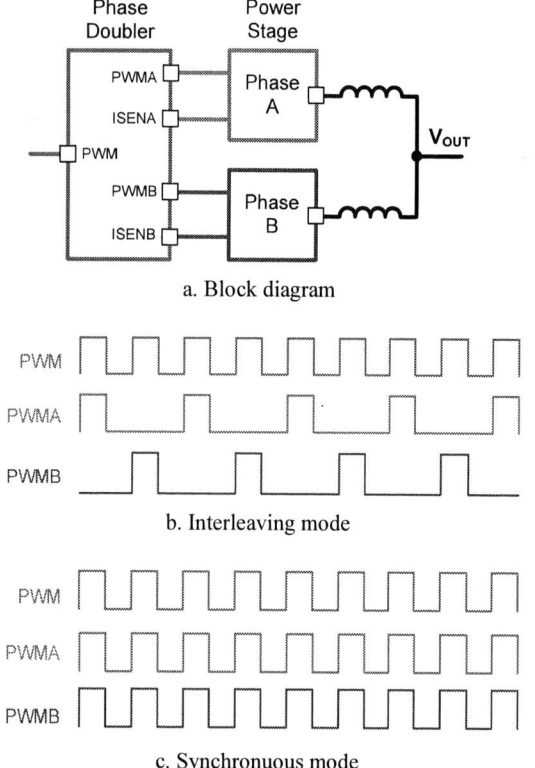

a. Block diagram

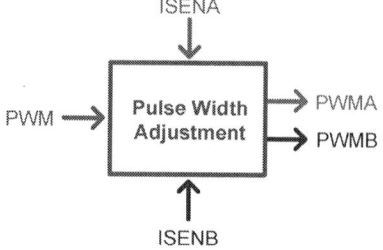

b. Interleaving mode

c. Synchronuous mode

Figure 4. Phase Doubler block diagram and operational waveforms

The preferred operational scheme is to run in interleaving mode and will be the focus of this paper. One obvious drawback of this ping-pong technique is that it limits the maximum duty cycle to 50%. However, this is not an issue for low duty cycle applications such as microprocessor power regulation.

Alternating the PWM pulses between two power stages ideally creates two phases from a single PWM input. In a real system, the duty cycle of the input PWM signal may not be the same for two continuous pulses. Additionally, the two power stages will not be identical, even though they are designed with the same components. They may have different propagation delays, and slightly mismatched power MOSFETs, output inductors, and PCB layout. Therefore, even if they are driven by identical input pulses, the upper MOSFETs will be turned on for different periods of time. It will cause unbalanced currents between the two power stages. Therefore it is necessary to implement a current balancing function inside the Phase Doubler to compensate for the mismatches in the system and avoid potential system failure.

Like conventional phase current balance technology, the phase currents are monitored by the Phase Doubler. Based on the phase current difference, the duty cycle of the output PWM signal can be adjusted in a way to zero the difference. The duty cycle of the PWM signal to the phase with less output current is increased slightly, while the duty cycle of the PWM signal to the other phase is slightly reduced. It will eventually

keep the two phase currents close to each other. Figure 5 shows the simple block diagram of the current balance function.

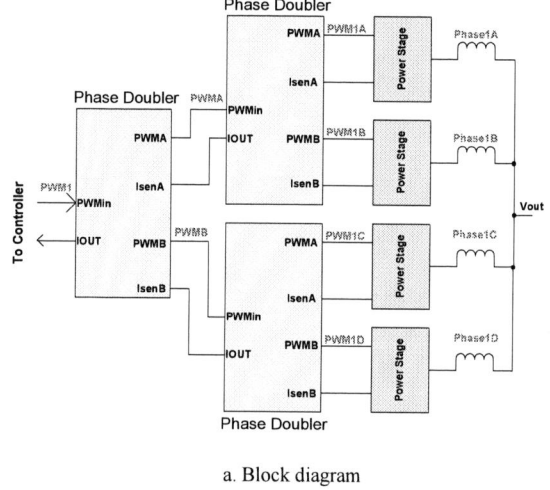

Figure 5. Pusle Width Modulator for current balance

The technique of doubling phases can be further expanded to quadruple the phase count by cascading the Phase Doubler. Figure 6 illustrates how this can be accomplished. In the first stage, one Phase Doubler will convert the input PWM1 signal into two PWM signals, i.e. PWMA and PWMB. Then PWMA signal is converted into PWM1A and PWM1B by another Phase Doubler, while PWMB is converted into PWM1C and PWM1D by the third Phase Doubler. All four PWM outputs are fed to four power stages. This effectively utilizes one PWM signal to control interleaved 4-phase power stages.

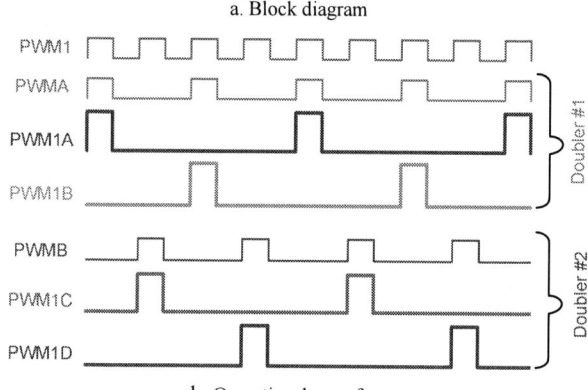

a. Block diagram

b. Operational waveforms

Figure 6. Cascaded doubler technique

978-1-4244-4782-4/10 $26.00 © 2010 IEEE   1083

Since the phase currents are sensed by the Phase Doubler for current balance purposes, the sensed phase current can be summed and fed back to the previous stage to further simplify the system design. Therefore in cascaded applications, the second-stage phase doublers will keep the phase currents of its two power stages balanced and send the summed phase current signal to the first-stage phase doubler. The first-stage phase doubler will maintain the current balance between two second-stage phase doublers based on the current signals from those phase doublers. Additionally, the first-stage phase doublers can feed the summed sensed current information to the PWM controller.

The switching frequency of the PWM controller in this cascaded phase doubler architecture must be set at four times the desired phase switching frequency, and the maximum effective duty cycle will be limited to 25%.

IV. APPLICATION CIRCUITS AND EXPERIMENTAL RESULTS

A. Phase Doubler Application Circuit

An example application circuit with the Phase Doubler is provided in Figure 7. ISL6611A is a Phase Doubler with a dual-driver and integrated current sensing and balancing circuitry. The chip monitors the voltage drop across the lower MOSFETs to obtain phase current information. It can drive two interleaved power stages with one PWM input signal.

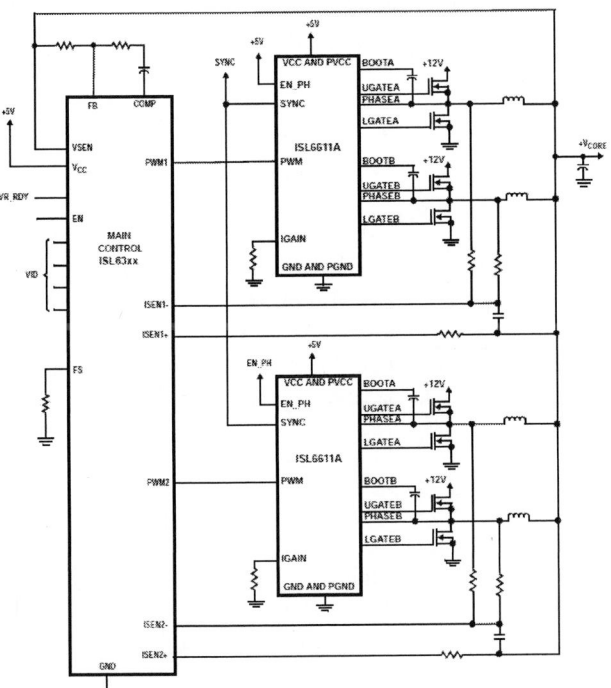

Figure 7.   Application circuit of 2-to-4 phase case

This approach can be used to simply scale up a single phase regulator to 2 phases to improve ripple cancellation, efficiency and system performance. It can also be used with

the ISL6334 or ISL6336 family, Intersil's 4 and 6-phase controllers for VR11.1 generation platforms, to provide 8 or 12 phase interleaved operation.

The ISL6611A also has the sync pin to run in the synchronous mode shown in Figure 4(c). This function can also be utilized to improve the transient response by turning on both phases at the same time during step-up load transient events.

B. Achieving 24-Phase Operation (Cascaded Doublers)

As discussed above, the technique of doubling phases can be further expanded to quadruple the phase count by cascading the phase doublers. Specifically for cascading, the ISL6617, a driverless Phase Doubler, is introduced into the market. The ISL6617 integrates the current sensing and balancing features to maintain the current balance between two secondary phases. The ISL6617 can work with the ISL6611A or Dr. MOS for higher phase count system design. To simplify the system complexity and layout design, the ISL6617 includes an IOUT pin to feed the average sensed current signal of the paired phases to the PWM controller. A typical application circuit with ISL6617 is shown in Figure 8.

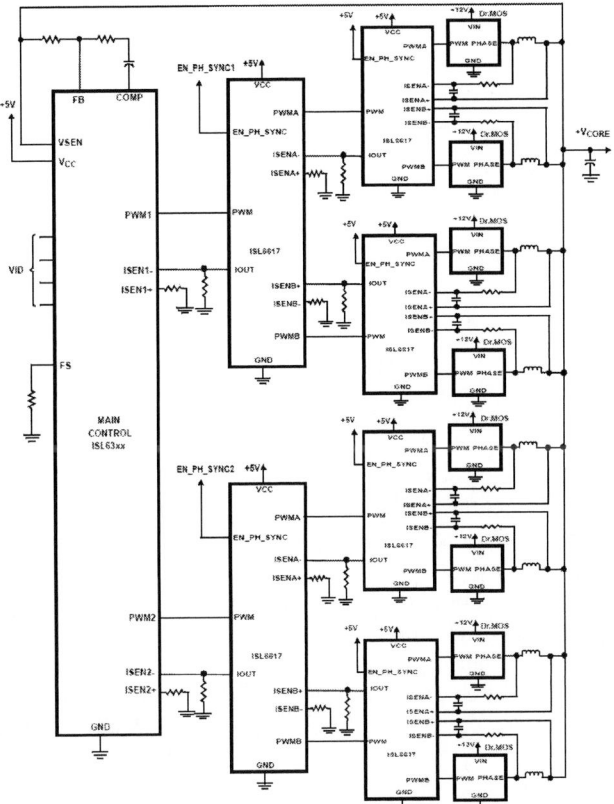

Figure 8.   Application circuit of 2-to-8 phase case

Figure 9 shows the 24-phase voltage regulator developed by Gigabyte to support the Intel® Core™ i7/i5 series processor using the phase doublers.

Figure 9. Gigabyte P55-UD6 motherboard featuring 24-phase voltage regulator with Phase Doubler

Figure 10 is a scope capture of four Phases signals that are generated from the same input PWM in a 24-phase system. By looking at the duty cycle measurement, one can see they are well matched, i.e., the phase currents are well balanced.

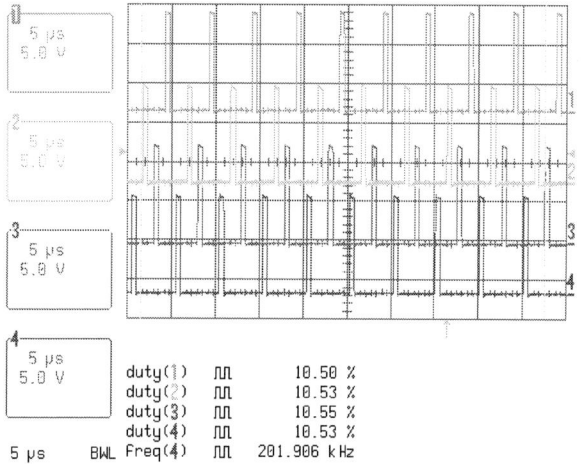

Figure 10. Measured duty cycle of 4 Phases generated by a single PWM

C. Experimental Data for 12-Phase System

The phase shift, the current balance and the efficiency are among the critical performance data in a multiphase system with the Phase Doubler. One 12-phase system based on ISL6336 and ISL6611A was built and tested to verify the performance.

Figure 11 provides bench data demonstrating the current balancing during a load apply and release. Very good current balance is achieved in both cases.

The measured efficiency of the 12-phase system is compared to that with other schemes in Figure 12. The green trace and pink trace in Figure 12 are plots of the measured efficiency of a 12-phase regulator. The green trace data is for

the system with the Phase Doubler in interleaving mode, while the pink trace data is for the same regulator in synchronous mode. As expected, the efficiency for interleaving mode is slightly better because of reduced input current ripple. The light blue trace shows a lower efficiency of an interleaved 6-phase regulator with same amount of MOSFETs and inductors. From this data, it is evident that a 12-phase system in interleaving mode provides the highest efficiency.

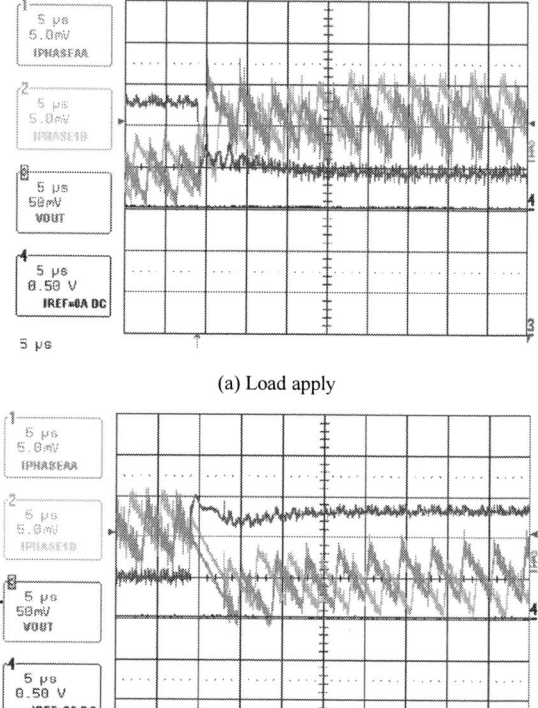

(a) Load apply

(a) Load release

Figure 11. Phase currents during load transient events

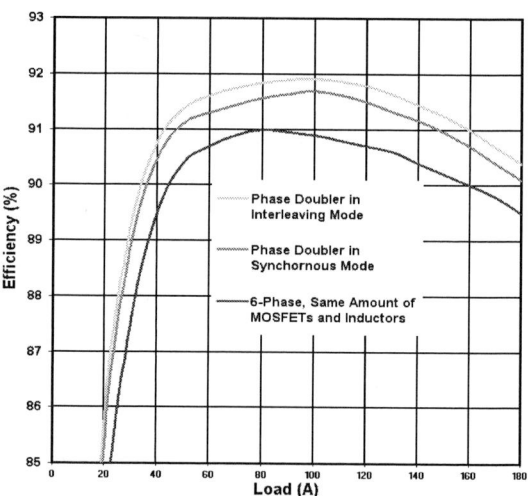

Figure 12. Measured efficiency for a 12-phase design

V. CONCLUSION

Creating a high phase count regulator can be accomplished by using a phase doubling technique with current balancing. This approach is a cost-effective way to increase phase count while maintaining good current sharing. The techniques described here have been used by Intersil and Gigabyte Technology to develop 12+ phase Vcore solutions.

VI. ACKNOWLEDGEMENTS

The authors would like to thanks Bob Isham, Paul Sferrazza, and Bogdan Duduman at Intersil for their input and discussions on the proposed scheme. We would also like to acknowledge Gigabyte Technology for providing us with valuable data and feedback.

[1] Qahouq, J.A.A.; Abdel-Rahman, O.; Huang, L.; Batarseh, I.; "On Load Adaptive Control of Voltage Regulators for Power Managed Loads: Control Schemes to Improve Converter Efficiency and Performance", Power Electronics, IEEE Transactions on, Volume 22, Issue 5, 2007 Page(s):1806 – 1819.

[2] Jia Wei; Lee, F.C.; "Two-stage voltage regulator for laptop computer CPUs and the corresponding advanced control schemes to improve light-load performance", IEEE APEC 2004 Page(s):1294 – 1300.

[3] P. Zumel, C. Fernandez, A. de Castro, O. Garcia, "Efficiency improvement in multiphase converter by changing dynamically the number of phases," IEEE PESC 2006, Page(s)1-6.

[4] Lukic, Z.; Zhenyu Zhao; Prodic, A.; Goder, D.; "Digital Controller for Multi-Phase DC-DC Converters with Logarithmic Current Sharing" IEEE PESC 2007, Page(s):119 – 123.

[5] Qiu, Weihong; Cheung, Chun; Xiao, Shangyang; Greg, Miller; "Power Loss Analyses for Dynamic Phase Number Control in Multiphase Voltage Regulators", IEEE APEC 2008 Page(s):102–108.

[6] "ISL6611A: Phase Doubler with Integrated Drivers and Phase Shedding Function", Intersil datasheet, available at www.intersil.com

[7] "ISL6617: PWM Doubler", Intersil datasheet, available at www.intersil.com

[8] "ISL6334: VR11.1, 4-Phase PWM Controller with Light Load Efficiency Enhancement and Load Current Monitoring Features", Intersil datasheet, available at www.intersil.com

[9] "ISL6336: VR11.1, 6-Phase PWM Controller with Light Load Efficiency Enhancement and Load Current Monitoring Features", Intersil datasheet, available at www.intersil.com

Automatic Multi-Phase Digital Pulse Width Modulator

Simon Effler, Mark Halton and Karl Rinne
Department of Electronic & Computer Engineering
University of Limerick
Limerick, Ireland
Email: simon.effler@ul.ie

Abstract—**Current demands on switched-mode power supplies to deliver higher output power with improved efficiency are leading to an increased use of multi-phase power converters. With an increasing number of phases, special multi-phase digital pulse width modulators (DPWMs) prove advantageous over the parallel use of conventional DPWMs. In this paper a "smart" multi-phase DPWM is presented which incorporates a duty cycle distribution algorithm. This algorithm is based on the fastest execution of the duty cycle input command with respect to the number of switching actions per phase and switching cycle. The system provides good dynamic current sharing during transients and enables the use of "faster" digital loop compensators. Intrinsic support of a variable number of active phases (phase-shedding operation) and improved scalability over conventional designs complete the feature set. The proposed system has been implemented on an FPGA system and tested with a four-phase buck converter.**

I. INTRODUCTION

With switched-mode power supplies (SMPSs) moving to higher output power and smaller output voltages, the need for multi-phase power converters is continuously growing. With an increasing number of phases, the benefits of a control loop bandwidth closer to or even higher than the switching frequency of the individual phases are well-acknowledged. As simple digital pulse width modulators (DPWMs) are still used in current multi-phase designs, these tend to be one of the limiting factors of the loop bandwidth, and hence the need for improved "smart" DPWMs is clearly evident. Additionally, most of the existing architectures are not able to drive a varying number of phases which is required in systems with phase shedding operation, e.g. [1], [2].

Standard multi-phase power converter, e.g. [3], comprising of N phases, use N conventional DPWM modulators to generate the control signals for the power stage switches [4]–[6]. Synchronisation between the individual modulators ensures proper phase shift. Naturally sampled DPWMs update the duty cycle once per switching cycle, which allows the update of the inputs of the modulator up to N-times the phase switching frequency. However, each individual DPWM is only updated once per cycle leading to additional problems such as current mismatch during load transients. Most designs overcome these issues by limiting the loop bandwidth, and thereby compromising the system performance.

For single phase applications, this issue has been addressed using several different concepts, such as charged-balanced control [7], [8], linear–non–linear control [9], [10] and multi-sampled DPWMs [11]–[14]. However to date, most of these concepts have not been applied to multi-phase applications. As a consequence, multi-phase converters are still driven by standard multi-phase modulators.

Additionally, multi-phase DPWMs comprising of conventional single-phase DPWMs require synchronisation between the individual phases in order to achieve an optimal phase shift between the individual power stages. This is mandatory to gain full advantage of the parallelization of multiple phases. The phase shift is subject to the number of phases, i.e. when the number of phases changes during run-time (phase shedding), a resynchronisation is required which can led to additional implementation requirements.

For multi-phase converters, special scalable solutions have been presented in the literature [15]–[18], which typically share area consuming hardware resources across the phases. By doing so, the design detailed in [15], [16] duplicates the same duty cycle value for all phases, and therefore restricts the implementation of current sharing techniques.

In [17], [18], an approach based on a digital-to-analogue converter architecture (DAC) is presented. The duty cycle input command is considered as an input of a DAC representing the delivered output power, which is then distributed over the individual phases. This scheme accommodates the update of the duty cycle with frequencies greater than the actual switching frequency and scales favourably with the number of phases. However, it does not support phase shedding and disrespects the number of switching actions per switching cycle and phase which can lead to a undesired increase in effective switching frequency.

In this paper, a new multi-phase DPWM scheme is presented addressing the issues outlined. It provides good hardware utilization, limits the number of switching actions per phase and cycle and supports phase shedding. The system design level is presented, followed by the proposal of the new duty cycle distribution scheme. This is followed by its implementation and verification on an FPGA.

978-1-4244-4782-4/10 $26.00 © 2010 IEEE

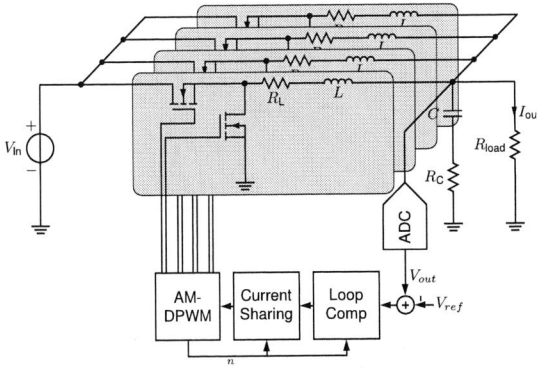

Figure 1. AM-DPWM with a four-phase buck converter.

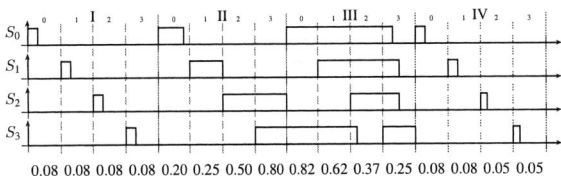

0.08 0.08 0.08 0.08 0.20 0.25 0.50 0.80 0.82 0.62 0.37 0.25 0.08 0.08 0.05 0.05

Figure 2. Standard DPWM modulation scheme.

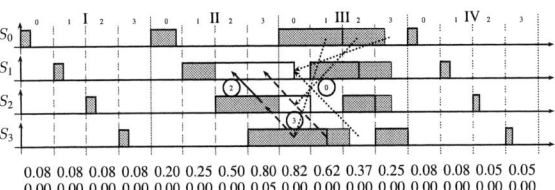

0.08 0.08 0.08 0.08 0.20 0.25 0.50 0.80 0.82 0.62 0.37 0.25 0.08 0.08 0.05 0.05
0.00 0.00 0.00 0.00 0.00 0.00 0.00 0.05 0.00 0.00 0.00 0.00 0.00 0.00 0.00 0.00

Figure 3. Duty cycle distribution with the new modulation scheme.

II. SYSTEM OVERVIEW

The proposed system (Fig. 1) consists of an N-phase power stage, an analog-digital-converter (ADC), a digital loop compensator (Loop Comp), current sharing, and a "smart" DPWM modulator (AM-DPWM), incorporating a new duty cycle distribution algorithm. This algorithm is based on the fastest possible execution of the duty cycle commands while still ensuring that each phase switches only twice (on/off) per switching cycle. A dynamic number of active phases (phase shedding) is incorporated by design together with an optional sigma-delta functionality to improve the effective resolution.

III. MULTI-PHASE DPWM

A. Distribution Scheme

Before detailing the new modulation scheme, it is first necessary to understand the modulation output of conventional multi-phase DPWMs. As an example, a typical modulation output of a four-phase system is shown in Fig. 2. For each "subcycle" (latin numbering), the duty cycle (shown below the waveform) is applied to the currently active DPWM. While the system restricts the number of switching operations per cycle intrinsically, the delay in the application of the duty cycle and the resulting distribution of the output signal are not optimal. For the given example, the phases zero (S_0) and three (S_3) take most of the transient current leading to a large current mismatch immediately after the transient. Also the delay between the reception of the duty cycle command and the application to the power stage can be up to one full DPWM cycle.

The concept of the proposed distribution scheme is based on the fastest possible execution of the duty cycle command respecting the number of allowed switching actions per phase and switching cycle. This can be quantified in the following criteria:

- Each phase is allowed to switch up to two-times per cycle (on once, off once).

- Only the next phase in the cycle is additionally turned on at the start of a subcycle.
- If a phase is still on at the end of a subcycle, it can be kept on for "longer".
- Phase shedding (number of available phases) and phase alignment (phase shift) are respected.

Adhering to these criteria leads to a distribution of the duty cycle as shown in Fig. 3 where the standard duty cycle distribution (Fig. 2) is included for comparison purposes (shaded). As before, the duty cycle input command for each subcycle is listed in the first line below the waveform, while the second line represents the residue forwarded to the next subcycle. The arrows illustrate the redistribution of the duty cycles among the phases compared with a standard modulation scheme where the encircled number indicates the respective subcycle; only the first three redistribution steps are highlighted (in order: solid, dashed, dotted). At the start of each subcycle, the duty cycle is distributed over the currently active phases with the priority given to the phase turned on last. Only if this phase is required to be on for the entire subcycle and if the previous phase is still on, will the latter's duty cycle be extended by the remaining duty cycle. This procedure is continued for subsequent phases.

This distribution algorithm can be expressed mathematically as

$$D_{\mathrm{n,k}} = \begin{cases} 0, & \text{if } D_{\mathrm{n,k-1}} \neq \frac{1}{N} \text{ and } i \neq 0, \\ \min(\frac{1}{N}, \max(0, D_{\mathrm{k}} - \frac{i}{N})) & \text{otherwise,} \end{cases} \quad (1)$$

where $D_{\mathrm{n,k}}$ is the duty cycle for each phase n and time instant k, D_{k} is the duty cycle input command and $i = (k + n) \bmod N$. N is the number of phases, which is passed to the DPWM as a parameter depending on the current operation conditions (phase shedding).

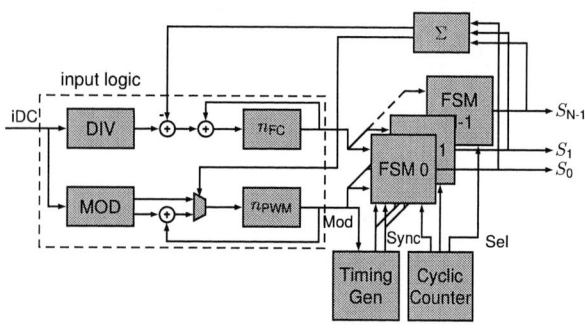

Figure 4.   DPWM Implementation Diagram.

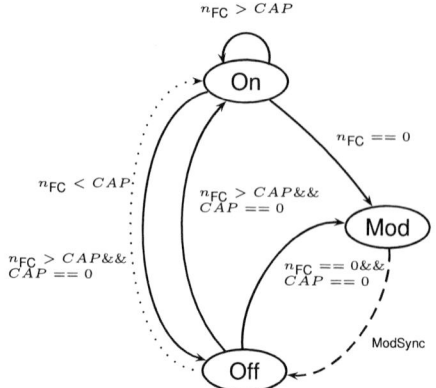

Figure 5.   Phase State Machine.

Table I
AREA VS. NUMBER OF PHASES.

# of phases	3	4	6	8
Standard DPWMs	204	242	336	433
AM-DPWM	155	240	379	503

B. Implementation

The overall implementation of the proposed DPWM is based on the block diagram shown in Fig. 4. The system comprises of an input logic, one finite-state-machine (FSM) for each controlled phase, a cyclic-counter unit and a shared high-resolution DPWM module.

The input logic normalises the duty cycle input command so that the following logic blocks are independent of the current number of phases and can be implemented more efficiently. Along with a serial integer division and a modulo operation used for the normalisation, internal integral action ensures the proper application of the entire incoming duty cycle signal.

One standard high-resolution DPWM block provides the synchronization signal for the subcycle timing depending on the number of available phases and the inter-subcycle modulation for one phase if required. The cyclic-counter provides information about the current subcycle for the individual FSMs. The FSMs control the output modulation of the phases where one FSM is required per phase. Their implementation is relatively simple and independent of the number of available phases as this is handled by the input logic.

With reference to Fig. 3 and Fig. 5, each phase can be in one of three possible states:

- Off: The phase is switched off.
- On: The phase is switched on for the entire subcycle.
- Mod: The phase is modulated during the subcycle.

At the start of each subcycle, each individual FSM evaluates the current situation. Dependent on the current operation conditions, the output of the respective phase is set. It is either turned on, turned off or modulated via the auxiliary high-resolution DPWM module. This is done by a comparison of the number of currently required phases, n_{FC}, with a cyclic counter value, CAP, which allows a hardware effective implementation of (1).

One additional advantages of the proposed architecture is the scalability with the number of phases (Table I). Parallel single-phase DPWM (with common resynchronization logic) scale about linear with the number of phases. However, in order to achieve sufficient resolution most of today's designs will use a hybrid architecture using analog and digital components. This increases the required area and usually involves manual layout. The proposed architecture requires digital blocks only, despite one high-resolution DPWM module, and hence is fully synthesisable. The auxiliary units require a larger initial area compared to standard DPWMs with a break-even around four phases.

IV. LOOP COMPENSATION

In order to test the performance of the proposed DPWM in closed loop operation, some additional digital blocks are required in a digital control loop (Fig. 1). Namely, a loop compensator and, due to the nature of multi-phase power converters, current sharing functionality. In the following sections, these two blocks with their system specific implementation details and requirements are described.

A. Compensator Design

Loop compensation techniques for power converters, both in the analog and the digital domain, are well-developed. A lot of research has been focused on several different techniques. For the system presented in this paper, one specific requirement has to be taken into account, i.e. the dynamic change in the number of phases. The compensator must be able to control all plants arising from a variying number of phases which results in a modification of the power converter's transfer function.

978-1-4244-4782-4/10 $26.00 © 2010 IEEE

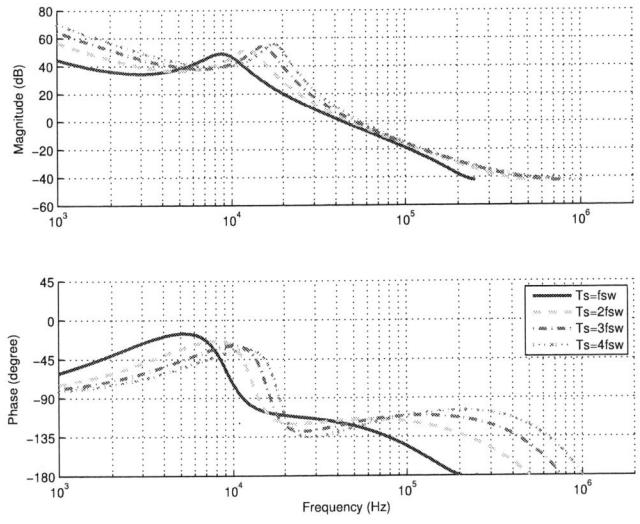

Figure 6. Bode plots of the open loop for different sampling times.

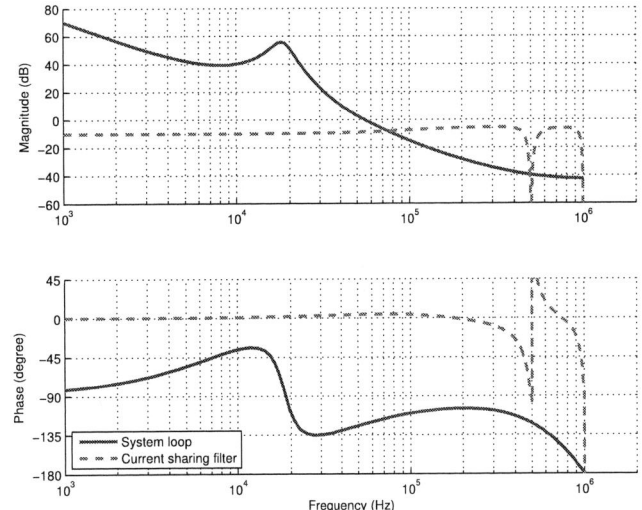

Figure 7. Bode plot for control loop and additional filter.

This is caused firstly by a change of the equivalent circuit model and secondly by a change in sampling time.

For demonstration purposes, a PID-based compensator, designed using standard design techniques proves sufficient. This compensator is "retimed" for different sampling frequencies by changing its clock frequency without a change in its coefficients. This can be viewed as a modification of the compensator in the time/frequency-domain due to a change of its discretization time/frequency. The compensator remains constant in the z-domain as the coefficients are not altered. Generally speaking, a change in the sampling frequency towards lower frequencies "slows" down the compensator and moves the corner frequency to the left. Bode plots of the resulting open loop systems are shown in Fig. 6. Note that the described technique does not provide the best possible loop compensation, but proves sufficient for testing the proposed modulator. It is expected that more advanced compensation schemes, e.g. adaptive control techniques, will provide better performance.

B. Current Sharing Technique

In multi-phase power converters current sharing is a mandatory need, due to tolerances and mismatches between the individual phases. Different strategies have been investigated in the past and can be broadly categorized into two groups, i.e. active and passive techniques. Active techniques require current sensing of the individual inductor currents and distribute the current equally among the phases. Passive techniques do not require current sensing and use alternative measures to estimate the phase current. The technique employed in this paper is based on the concept detailed in [19], [20] where current sharing as a result of duty cycle matching is proposed. This balances the current among the phases based on loss minimization instead of current equalization.

To ensure duty cycle matching among the phases, several different techniques are available. However, all of them introduce additional delays into the control loop and hence degrade its performance. In this paper, a technique originally proposed for oversampled, single-phase power converters [12] is adapted to achieve passive current sharing without phase delay. A comb-filter is inserted into the system loop which rejects signals (harmonics) at multiples of the switching frequency without delaying the control loop. The bode plot of the filters transfer function is shown in Fig. 7. One advantage of the proposed method is the relatively good decoupling of the design procedures of the current sharing filter and the loop compensator.

Additionally, the proposed comb-filter (Fig. 8) is simple to implement and does not required any dedicated multipliers. Adders and shifters are sufficient in practice. Like the loop compensator, the comb-filter is retimed when the number of phases is changed. Unlike the loop compensator, the comb-filter requires some additional action during retiming.

Note when the filter is retimed with any modifications, its rejection frequencies are shifted as they are relative to the sampling frequency, e.g. $\frac{1}{4}f_{sa}$ and $\frac{1}{2}f_{sa}$. However, in order to ensure proper current distribution, the rejection bands need to match the switching frequency and its harmonics. While four, two and one phase operation can be covered by one filter for harmonics at $\frac{1}{4}f_{sa}$ and $\frac{1}{2}f_{sa}$, different rejection bands are required for three phase operation. For this case, the harmonics are at $\frac{1}{3}f_{sa}$ and $\frac{2}{3}f_{sa}$. Subsequently, a small modification of the comb-filter from fourth-order into a third-order system is required to move the rejection bands into the correct positions. With reference to Fig. 8, this is achieved with five multiplexers and one additional gain stage (1/3). When the multiplexer control signal, T, is set to one, the filter is in third-order mode, otherwise it is in fourth-order mode.

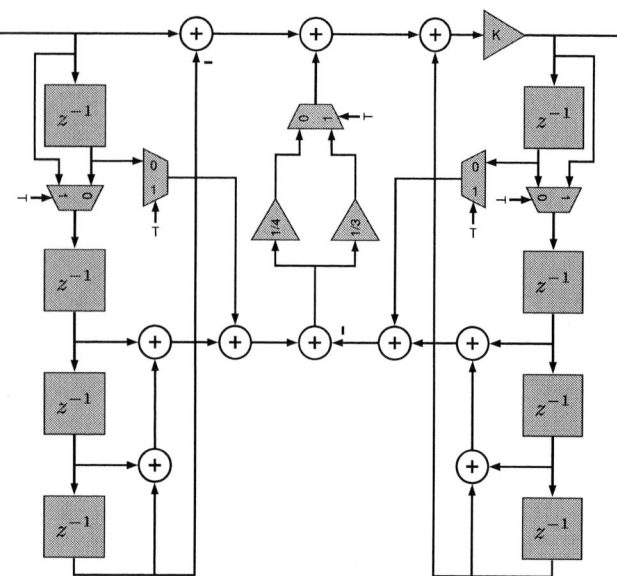

Figure 8. Implementation of a comb-filter with selectable order.

Table II
TECHNICAL DETAILS OF THE PROTOTYPE SYSTEM.

Input voltage	12 V
Output voltage	1.5 V
Number of phases	1-4
Phase switching frequency	500 kHz
Phase inductance	680 nH
Total output capacitance	450 μF
Total output power	120 W

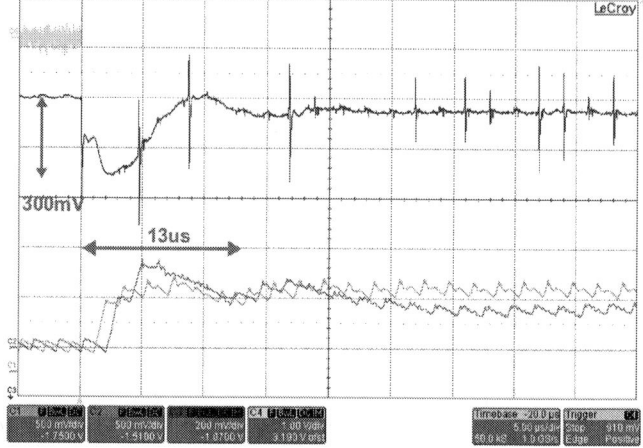

Figure 9. System response of the new DPWM to a 70 A load step.

V. VERIFICATION

A four-phase buck converter with 500 kHz switching frequency per phase has been built to prove the concept in practice. The full technical details are given in Table II.

In Fig. 9 and Fig. 10, the system's voltage response to a 70 A load transient is shown. Fig. 11 shows the response for a system using a standard DPWM for comparison. In Fig. 11 and Fig. 9, the respective output voltage is shown (blue), together with two of the four phase currents (yellow and magenta). Note that the large noise spikes are due to pure HF-coupling of the measurement probe. In Fig. 10, the respective digital control signals (DPWMs outputs) are shown, together with the output voltage as reference signal. Also the outputs of the standard DPWM (D_{12} to D_{15}) and the new modulator (D_8 to D_{11}) are both shown in the same diagram, where only the latter set is used as switch control signals.

Using the new distribution algorithm, the transient response is improved by almost 40% compared to regular DPWMs. The maximum deviation is significantly reduced which is also due to an increase in the maximum loop gain. In this prototype, the gain of the standard DPWM loop is reduced by a factor of two compared with the new modulator, as a higher gain would cause a significant output voltage overshoot.

In Fig. 12, the behaviour of the DPWM during a phase shedding operation is shown. The number of phases is changed from three to four via an external control signal (D_7). The respective phase (D_{15}) is switched from tri-state mode into standard operation. At the same time, the phases are realigned with a phase shift of 90°. The phase shedding procedure does not cause any perturbation of the output voltage due to the internal distribution scheme. Note that in Fig. 12 (output voltage resolution 20mV/DIV), the unexpected increase in ripple voltage during the change in the number of phases is caused by the switching noise of the just-enabled phase as this is the nearest to the measurement probe.

VI. CONCLUSIONS

A new multi-phase DPWM modulation scheme has been presented based on the fastest possible distribution of the duty cycle input command to the power stage with respect to the number of allowed switching actions per cycle. Utilizing a new distribution algorithm, the system provides superior transient performance and inductor current distribution over systems using conventional modulators. The proposed modulator has been implemented in an FPGA prototype and its performance assessed for a four-phase buck converter. Results presented show clear advantages of the proposed system over standard DPWM modulators with a maximum performance increase of approx. 40%.

ACKNOWLEDGMENT

This work is partly funded by the Irish Research Council for Science, Engineering and Technology: funded by the National Development Plan.

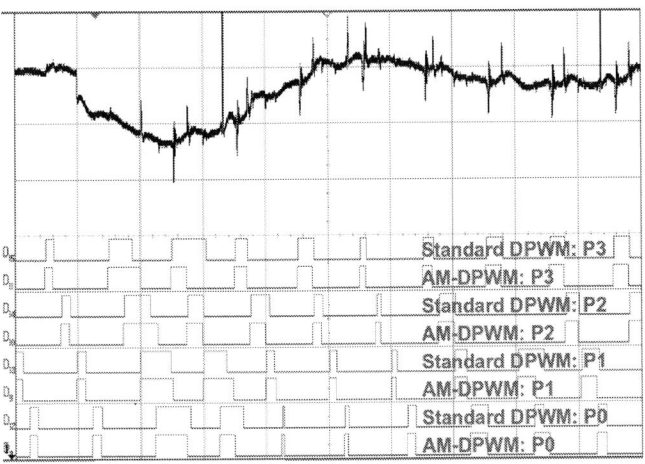

Figure 10. System response of the new DPWM to a 70 A load step with DPWM output signals.

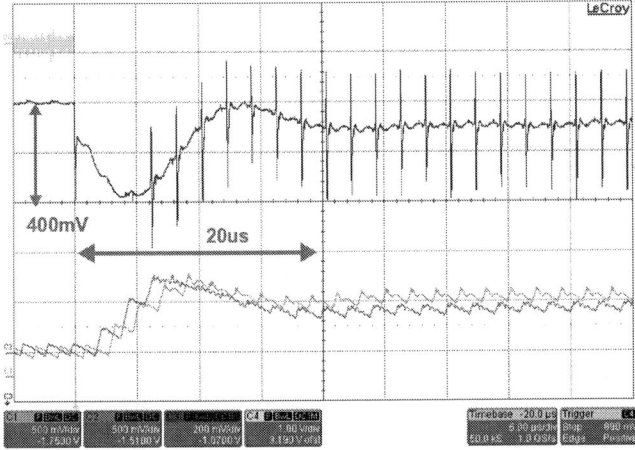

Figure 11. System response of a standard DPWM to a 70 A load step.

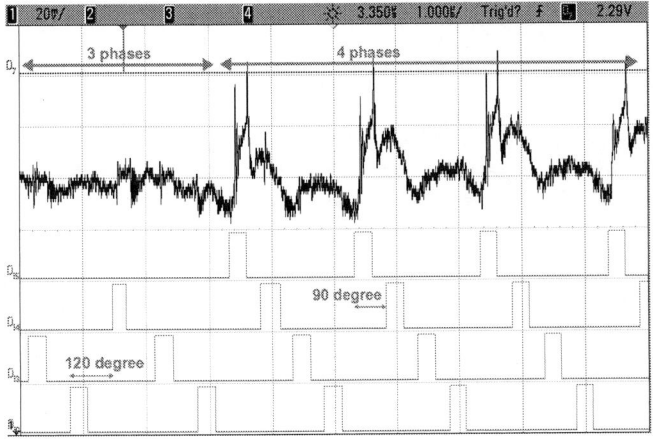

Figure 12. Phase shedding operation with the new modulation scheme.

REFERENCES

[1] P. Zumel, C. Fernandez, A. de Castro, and O. Garcia, "Efficiency improvement in multiphase converter by changing dynamically the number of phases," in *Power Electronics Specialists Conference, 2006. PESC '06. 37th IEEE*, 6 2006, pp. 1–6.

[2] L. T. Jakobsen, O. Garcia, J. A. Oliver, P. Alou, J. A. Cobos, and M. A. E. Andersen, "Interleaved buck converter with variable number of active phases and a predictive current sharing scheme," in *Proc. IEEE Power Electronics Specialists Conference PESC 2008*, 2008.

[3] Z. Lukic, C. Blake, S. C. Huerta, and A. Prodic, "Universal and fault-tolerant multiphase digital PWM controller IC for high-frequency DC-DC converters," in *Proc. APEC 2007 - Twenty Second Annual IEEE Applied Power Electronics Conference*, 2 2007, pp. 42–47.

[4] A. Kelly and K. Rinne, "High resolution DPWM in a DC-DC converter application using digital sigma-delta techniques," in *Power Electronics Specialists, 2005 IEEE 36th Conference on*, 9 2005, pp. 1458–1463.

[5] Z. Lukic, N. Rahman, and A. Prodic, "Multibit sigma-delta PWM digital controller ic for dc/dc converters operating at switching frequencies beyond 10 mhz," *IEEE Trans. Power Electron.*, vol. 22, no. 5, pp. 1693–1707, 2007.

[6] M. Scharrer, M. Halton, and T. Scanlan, "FPGA-based digital pulse width modulator with optimized linearity," in *Proc. Twenty-Fourth Annual IEEE Applied Power Electronics Conference and Exposition APEC 2009*, 2 2009, pp. 1220–1225.

[7] E. Meyer, Z. Zhang, and Y.-F. Liu, "An optimal control method for buck convertersusing a practical capacitor chargebalance technique," *IEEE Trans. Power Electron.*, vol. 23, no. 4, pp. 1802–1812, 7 2008.

[8] ——, "Digital charge balance controller with low gate count to improve the transient response of buck converters," in *Proc. IEEE Energy Conversion Congress and Exposition*, 2009.

[9] J. Quintero, A. Barrado, M. Sanz, and A. Lazaro, "Digital control with asynchronous linear-non-linear compensator," in *Proc. Twenty-Third Annual IEEE Applied Power Electronics Conference and Exposition APEC 2008*, 2 2008, pp. 491–497.

[10] A. Barrado, J. Quintero, A. Lazaro, C. Fernandez, P. Zumel, and E. Olias, "Linear-non-linear control applied in multiphase VRM," in *Power Electronics Specialists, 2005 IEEE 36th Conference on*, 9 2005, pp. 904–909.

[11] L. Corradini and P. Mattavelli, "Modeling of multisampled pulse width modulators for digitally controlled DC-DC converters," *IEEE Trans. Power Electron.*, vol. 23, no. 4, pp. 1839–1847, 2008.

[12] L. Corradini, P. Mattavelli, E. Tedeschi, and D. Trevisan, "High-bandwidth multisampled digitally controlled DC-DC converters using ripple compensation," *IEEE Trans. Ind. Electron.*, vol. 55, no. 4, pp. 1501–1508, 2008.

[13] L. Corradini and P. Mattavelli, "Analysis of multiple sampling technique for digitally controlled DC-DC converters," in *Power Electronics Specialists Conference, 2006. PESC '06. 37th IEEE*, 6 2006, pp. 1–6.

[14] S. Effler, Z. Lukic, and A. Prodic, "Oversampled digital power controller with bumpless transition between sampling frequencies," in *Proc. IEEE Energy Conversion Congress and Exposition*, 2009.

[15] R. F. Foley, R. C. Kavanagh, W. P. Marnane, and M. G. Egan, "An area-efficient digital pulsewidth modulation architecture suitable for FPGA implementation," in *Proc. Twentieth Annual IEEE Applied Power Electronics Conference and Exposition APEC 2005*, vol. 3, 3 2005, pp. 1412–1418.

[16] R. Foley, R. Kavanagh, W. Marnane, and M. Egan, "Multiphase digital pulsewidth modulator," *IEEE Trans. Power Electron.*, vol. 21, no. 3, pp. 842–846, 5 2006.

[17] T. Carosa, R. Zane, and D. Maksimovic, "Digital multiphase modulator – a power D/A perspective," in *Power Electronics Specialists Conference, 2006. PESC '06. 37th IEEE*, 6 2006, pp. 1–6.

[18] ——, "Scalable digital multiphase modulator," *IEEE Trans. Power Electron.*, vol. 23, no. 4, pp. 2201–2205, 7 2008.

[19] A. Kelly, "Current share in multiphase DC-DC converters using digital filtering techniques," *IEEE Trans. Power Electron.*, vol. 24, no. 1, pp. 212–220, 1 2009.

[20] ——, "Current share in multiphase SMPCs by digital filtering," in *Proc. IEEE Power Electronics Specialists Conference PESC 2008*, 2008.

A Simple Current Sharing Scheme for Dual Three-Phase Permanent-Magnet Synchronous Motor Drives

Yanhui He, Yue Wang, Jinlong Wu, Yupeng Feng, Jinjun Liu
School of Electrical Engineering of Xi'an Jiao Tong University
Xi'an Jiao Tong University
Xi'an, China
huihuigui@yahoo.com.cn

Abstract—**Stator current sharing will be unbalance in dual three-phase permanent-magnet synchronous motor (DTP-PMSM) drives since two three-phase systems have inherent asymmetries. In this paper, a simple current sharing scheme is proposed to eliminate the inherent asymmetries of the two three-phase systems. It will enable the motor to operate with balance current sharing between the two three-phase winding sets and has low implementation complexity. The control scheme is validated by simulation and experiment with a 3kW DTP-PMSM drive prototype.**

I. INTRODUCTION

During last 30 years the interest in multi-phase AC motor drives has considerably increased, especially for high-power applications. The controlled power can be divided among more inverter legs to reduce the single static switches current stress without the need for parallel techniques or multilevel converter. The multi-phase AC motor drives also show several advantages over the conventional three-phase ones such as reducing the amplitude of torque pulsation, lowering the dc link current harmonics and higher reliability [1-3].

Over the years various topics pertinent to this motor drive system have been covered in literature. These mainly include four key issues: machine modeling, space vector PWM techniques, current control solutions and fault-tolerant control. Machine modeling has two different paths. The first approach had been discussed in [3, 4] in which the machine is represented with two pairs of (d, q) windings. The second is on the basis of vector space decomposition theory [5, 6]. The aim of the space vector PWM techniques [6-10] is to reduce the stator current harmonics which have been observed in the machine stator current spectrum. The current control solutions that have been discussed in literature are to restraint the stator current sharing unbalance due to asymmetries between the two three-phase stator windings sets. Bojoi investigated the current sharing control of dual-three-phase induction motor in synchronous and stationary reference frame [11-13]. In [14], the current unbalance is settled by six current regulators, which has increased the system complexity and control difficulties. To ensure current sharing and low current harmonics, Oguchi

presented a novel six-phase inverter system in which 60-step output voltages were generated [15]. Several authors have addressed the issue of fault tolerance, focusing for the large part on the behavior and compensation strategies of machines in which a single phase is open circuited and on for open-circuit faults [16-18].

Dual three-phase permanent-magnet synchronous motors (DTP-PMSMs) combine high efficiency and power density characteristics, which are primarily responsible for their wide appeal. However, because two three-phase winding sets have inherent asymmetries, stator current sharing will be unbalance if there doesn't take some measures. Furthermore, in high power 4-quadrant dual-three-phase motor drives, separate dc power supplies that is used commonly to avoid the circulating current is a potential unbalance current sharing when the separated dc links have some difference. It is necessary to eliminate unbalance current sharing of the two stator winding sets because it would introduce sixth order torque ripple while dual-three-phase PMSM should not.

In this paper, we briefly discuss the modeling of a DTP-PMSM based on vector space decomposition to set forth the current control fundamentals. Then, a simple current sharing scheme is proposed to eliminate current sharing unbalance based on the mathematic model. Finally, implementations on a 3kW dual three-phase PMSM prototype are presented to validate the proposed scheme.

II. MACHINE MODEL

The DTP-PMSM has two three-phase windings spatially shifted by 30 electrical degrees with isolated neutrals. The stator windings are fed by a current controlled PWM six-phase voltage source inverter. A DTP-PMSM and the inverter arrangement are illustrated in Fig.1. The one three-phase system is composed of stator windings A, C, E. Another one is B, D, F. The two sets of windings spatially have shifted by 30 electrical degrees with isolated neutrals.

According to vector space decomposition theory [6], the original six-dimensional machine system can be decomposed into three orthogonal subspaces, i.e. (α, β) , (μ_1, μ_2)

978-1-4244-4782-4/10 $26.00 © 2010 IEEE

and (z_1, z_2) subspaces by using the transformation matrix $[T]_6$, shown in (1).

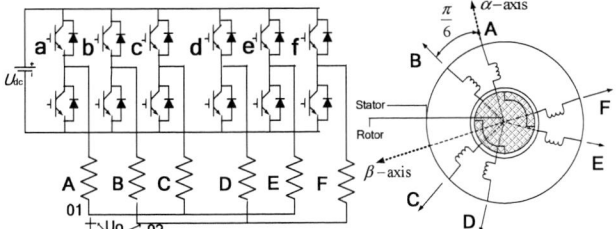

Fig. 1 Voltage source inverter fed DTP-PMSM

$$[T]_6 = \sqrt{\frac{1}{3}}\begin{bmatrix} 1 & \sqrt{3}/2 & -1/2 & -\sqrt{3}/2 & -1/2 & 0 \\ 0 & 1/2 & \sqrt{3}/2 & 1/2 & -\sqrt{3}/2 & -1 \\ 1 & -\sqrt{3}/2 & -1/2 & \sqrt{3}/2 & -1/2 & 0 \\ 0 & 1/2 & -\sqrt{3}/2 & 1/2 & \sqrt{3}/2 & -1 \\ 1 & 0 & 1 & 0 & 1 & 0 \\ 0 & 1 & 0 & 1 & 0 & 1 \end{bmatrix} \quad (1)$$

Because the neutrals of the two windings sets are isolated, the voltage vectors projected in (z_1, z_2) subspace will be null. Only these components which are mapped in the (α, β) subspaces are transformed to the synchronous (d, q) reference frame, using the transformation matrix $[T]_{s/r}$, shown in (2).

$$[T]_{s/r} = \begin{bmatrix} \sin\theta & \cos\theta & 0 & 0 & 0 & 0 \\ -\cos\theta & \sin\theta & 0 & 0 & 0 & 0 \\ 0 & 0 & 1 & 0 & 0 & 0 \\ 0 & 0 & 0 & 1 & 0 & 0 \\ 0 & 0 & 0 & 0 & 1 & 0 \\ 0 & 0 & 0 & 0 & 0 & 1 \end{bmatrix} \quad (2)$$

Then the dual-three-phase PMSM mathematic model can be obtain in the synchronous (d, q) reference frame as follows:

A. *Machine Model in (d,q) subspace*

$$\frac{d}{dt}\begin{bmatrix} i_d \\ i_q \end{bmatrix} = \begin{bmatrix} -\dfrac{R_s}{L_d} & \omega_s \dfrac{L_q}{L_d} \\ -\dfrac{L_d}{L_q}\omega_s & -\dfrac{R_s}{L_q} \end{bmatrix}\begin{bmatrix} i_d \\ i_q \end{bmatrix} + \begin{bmatrix} \dfrac{1}{L_d} & 0 \\ 0 & \dfrac{1}{L_q} \end{bmatrix}\begin{bmatrix} u_d \\ u_q \end{bmatrix} - \frac{\sqrt{3}\omega_s}{L_q}\begin{bmatrix} 0 \\ \psi_f \end{bmatrix} \quad (3)$$

B. *Machine Model in (μ_1, μ_2) subspace*

$$\frac{d}{dt}\begin{bmatrix} i_{\mu 1} \\ i_{\mu 2} \end{bmatrix} = \begin{bmatrix} -\dfrac{r_s}{L_{ls}} & 0 \\ 0 & -\dfrac{r_s}{L_{ls}} \end{bmatrix}\begin{bmatrix} i_{\mu 1} \\ i_{\mu 2} \end{bmatrix} + \begin{bmatrix} -\dfrac{1}{L_{ls}} & 0 \\ 0 & -\dfrac{1}{L_{ls}} \end{bmatrix}\begin{bmatrix} u_{\mu 1} \\ u_{\mu 2} \end{bmatrix} \quad (4)$$

where r_s is the stator resistance; ω_s is synchronous angular velocity; L_d is d axis synchronous inductance; L_q is q axis synchronous inductance; L_{ls} is stator leakage inductance and

ψ_f is the permanent magnet flux linkage; i_d, i_q and u_d, u_q are stator currents and voltages respectively in (d,q) subspace; $i_{\mu 1}$, $i_{\mu 2}$ and $u_{\mu 1}$, $u_{\mu 2}$ are in (μ_1, μ_2) subspace.

It is observed immediately from the above equations that all the electromechanical energy conversion related variable components are mapped into the $(d\text{-}q)$ subspace and the non-electromechanical energy conversion related variable components are transformed to the (μ_1, μ_2) and (z_1, z_2) subspaces. Hence, the dynamic equations of the machine are totally decoupled. As a result, the analysis and control of the machine can be greatly simplified.

III. CURRENT CONTROL SCHEME

A. *Conventional control scheme for DTP-PMSM*

From the modelling of the DTP-PMSM, we can see that the current control of the three-phase motors can be extended to the six-phase drives, called conventional control scheme in this paper, as depicted in Fig.2.

Since conventional control scheme views six phase current of DTP-PMSM as an integral system, the stator currents in the (d,q) subspace represent only mean current of two three-phase winding sets. Moreover, it is observed that the six-phase voltage source inverter has only one modulation ratio, i.e. the two three phase system have the same pole voltage. Therefore, the current difference of two three-phase winding sets can not be reflected. If two three phase systems have asymmetries, the unbalance current sharing will appear clearly.

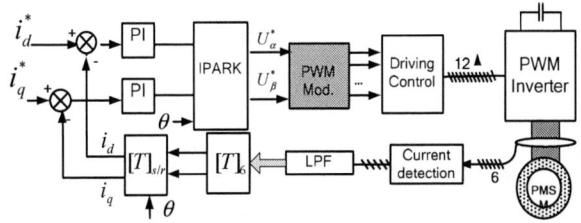

Fig.2 Conventional control scheme

The simulations were set up such that fundamental current amplitude of one three-phase winding set is 1.2A and the other set is 0.8A. The simulation results are shown in Fig.3.

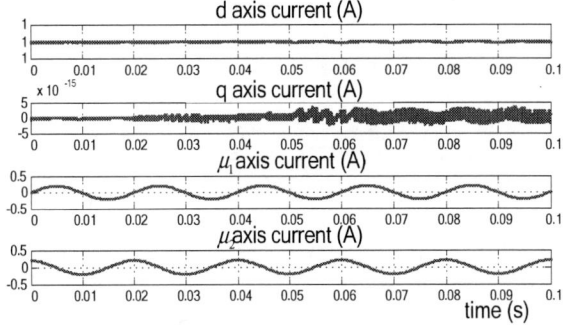

Fig.3 Stator current after coordinate transformation using $[T]_6$ and $[T]_{s/r}$

It is observed that (d,q) subspace represent only mean value of two three-phase winding sets current. It is the key point that using conventional current control is not able to eliminate the unbalance current sharing between the two three-phase winding sets. Furthermore, it can be see that there has a lot of current component in (μ_1, μ_2) subspace which will introduce sixth order torque ripple. Because of the neutrals of the two windings sets are isolated (Fig.1), the current in (z_1, z_2) subspace has no components.

B. Proposed current sharing scheme for DTP-PMSM

The control method depicted in Fig.2 is just simply extended from three-phase system which has only one d-q synchronous frame current control. It is not able to restraint the current unbalance between the two three-phase winding sets when the two three-phase systems have asymmetries. In this paper, a simple current sharing scheme for solving the unbalance current sharing problem is shown in Fig.4. The transformation matrix $[T]_3$ is three-phase Clark transformation matrix.

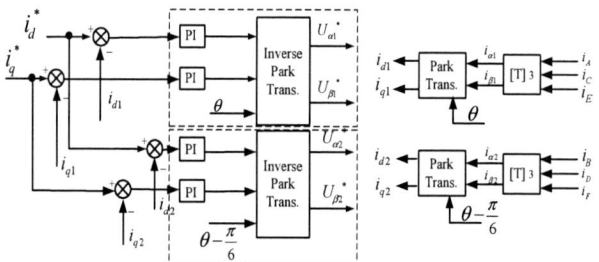

Fig.4 Proposed current sharing scheme

The proposed scheme views the DTP-PMSM as two independent three-phase systems. The proposed strategy operates each of the three-phase inverter independently with two current regulators which have just four PI current regulators. Therefore, the system complexity and control difficulties can be simplified entirely compared with the strategy presented in [14]. The reason for the current unbalance elimination is that the two three-phase systems have different pole voltages waveforms. Hence, the proposed scheme may have two modulation ratios which would reflect the differences of the two three-phase systems, such as different dc links and asymmetry of two stator winding sets.

IV. SIMULATION AND EXPERIMENTAL RESULTS

The simulation and experimental tests have been performed with a 3kW, 380V DTP-PMSM. The motor parameters are shown in Table 1.

Table 1 Motor specification

Rated speed	300 rpm
Stator resistance	4.177Ω
d axis synchronous reactance	9.451Ω
q axis synchronous reactance	9.451Ω
magnet flux linkage	0.928Wb
Number of pole pairs	10

The direct rotor field oriented control block diagram for DTP-PMSM using the proposed current sharing control scheme is shown in Fig.5. Maximum torque per ampere

method has been introduced to enhance torque production and reduce input current. The rotor position from an incremental photoelectric encoder has been used to calculate motor speed.

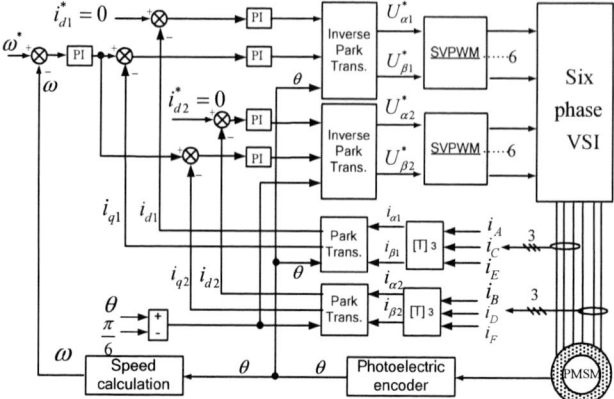

Fig.5 The proposed DRFOC for dual-three-phase PMSM

Different PWM modulation techniques are compared in terms of motor stator current harmonics content and implementation complexity in [8]. The PWM modulator has adopted the strategy which had discussed in [10] which can considerably decrease the instruction execution time and reduce the software complexity of the multi-dimension SVPWM algorithm. Moreover, the harmonic currents are inherently eliminated without extensive computations.

A. Simulation Results

The simulations were set up such that a reference speed command is nearly 25% of rated speed with 100% load. The simulation results of the conventional control scheme are shown in Fig.6. The simulation results using the proposed current sharing scheme are shown In Fig.7. It suggests that the unbalance current sharing between the two three-phase sets has been eliminated using the proposed current sharing scheme.

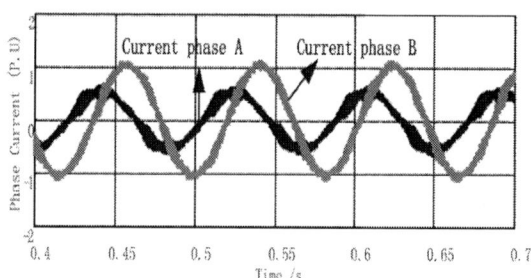

Fig.6 Conventional control scheme

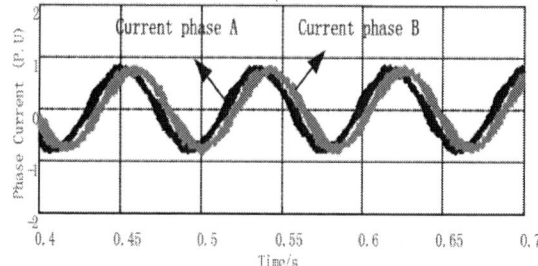

Fig.7 Proposed current sharing scheme

B. Experimental Results

The experiments have been performed for the verification of the proposed control algorithm. A TMS320F2812 DSP system is used for the digital processing of the proposed algorithm. The reference speed command is nearly 50% of rated speed with 70% load. The experimental results are shown in Table.2. From the experimental results we can see that the stator current unbalance in the two three-phase stator windings sets can be eliminated using the proposed current sharing scheme.

Table.2 Experiment results comparisons

Control method	Winding I (Phase A)	Winding II (Phase B)
Conventional control scheme	1.405 A (RMS)	2.719 A (RMS)
Proposed current sharing scheme	1.983 A (RMS)	2.006 A (RMS)

The pole voltages of the conventional method and the proposed method are shown in Fig.8 and Fig.9. From the experimental results we can see that the stator current unbalance in the two three-phase stator windings sets can be eliminated and the current harmonic content can be also substantially reduced. Furthermore, it can be seen that the proposed scheme have two modulation ratios according to different pole voltages waveforms. It is the key point that it is able to eliminate the unbalance current sharing between the two three-phase systems.

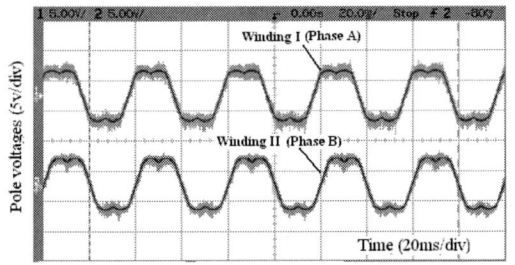

Fig.8 Pole voltages using conventional control scheme

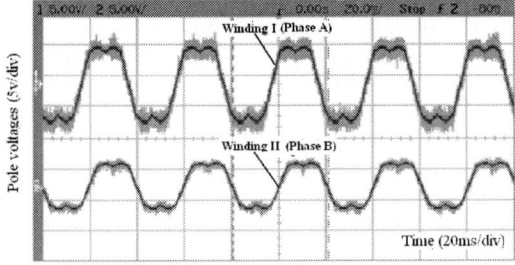

Fig.9 Pole voltages using proposed control scheme

V. CONCLUSIONS

A simply current sharing strategy for DTP-PMSM is presented in this paper. The proposed current sharing control has two sets of current regulators in synchronous frame. The stator current unbalance in two three-phase stator windings sets can be eliminated since it have two different pole voltages. The experimental results, presented for a direct rotor field oriented control of a 3kW motor prototype, have confirmed the correctness of the proposed current sharing control scheme.

REFERENCES

[1] E.A. Klingshirn. High Phase Order Induction Motors Part I—Description and Theoretical Considerations[J]. IEEE Transactions on Power Apparatus Systems,1983,102(1):47-59.

[2] N.A. Ai-Nuaim, H.A. Toliyat. A novel method for modeling dynamic air-gap eccentricity in synchronous machines based on modified winding function theory [J]. IEEE Trans. Energy Convers,1998,13(2):156-162.

[3] R.H. Nelson, P.C. Kranse. Induction Machine Analysis For Arbitrary Displacement Between Multiple Winding Sets[J].IEEE Transactions on Power Apparatus Systems,1994,9(3):841-848.

[4] T.A Lipo, A d-q model for Six-Phase Induction machines. Proc.Int. Conf. on Elec.Machines ICEM, Athens, Greece, 1980, pp. PEE4/4 860-867.

[5] D. Hadiouche, R. Abderrezak. Study and simulation of space vector PWM control of double-star induction motors[J].CIPE in Acapulco,Mexico,October 15-19,2000:42-47.

[6] Y.F. Zhao, T.A. Lipo. Space Vector PWM Control Of Dual Three Phase Machine Using Vector Decomposition[J]. IEEE Transactions on Industry Applications,1995,31(5):1100-1108.

[7] K. Gopakumar, V.T. Ranganathan. Split-Phase Induction Motor Operation From Pwm Voltage Source Inverter[J].IEEE Transactions on Industry Applications,1993,29(5):927-932.

[8] R. Bojoi, A. Tenconi.Complete Analysis and Comparative Study of Digital Modulation Techniques for Dual Three-phase AC Motor Drives[J].Conf.Rec.IEEE PESC,2002,(2):851-857.

[9] D. Hadiouche,L. Baghli,A. Rezzoug.Space-vector PWM techniques for dual three-phase AC machine analysis,performance evaluation,and DSP implementation. IEEE Transactions on Industry Applications,2006,42(4):1112-1122.

[10] A.R Bakhshai, G. Joos.Space Vector Pwm Control of a Split-phase Induction Machine Using The Vector Classification Technique[J]. Conf.Rec.IEEE PESC, 1998,(2):802-807.

[11] R. Bojoi, M. Lazzari.Digital Field Oriented Control for Dual-three-phase Induction Motor Drives[J]. IEEE Transactions on Industry Applications,2003,39(3):752-759.

[12] R. Bojoi, E. Levi, F. Farina, A. Tenconi and F. Profumo.Dual three phase induction motor drive with digital current control in the stationary reference frame.IEE proc.Electr.Powe Appl., Vol. 153, No.1, January 2006.

[13] G.K. Singh, K. Nam and S.K. Lim. A Simple Indirect Field-Oriented Control Scheme for Multiphase Induction Machine. IEEE Transactions on Industrial Electronics, Vol.52, No.4, August 2005, pp. 1177-1184.

[14] R.O.C. Lyra and T.A. Lipo. Torque density improvement in a six-phase induction motor with third harmonic current injection. IEEE Transactions on Industrial Application, Vol.38, No.5, September/Octorber 2002, pp. 1351-1360.

[15] K. Oguchi, A. Kawaguchi, T. Kubota and N. Hoshi, "A Novel Six-Phase Inverter System with 60-Step Output Voltages for High-Power Motor Drives", IEEE Trans. Ind. Applicat., Vol. 35, No. 5, pp. 1141-1149, Sept./Oct. 1999.

[16] G. K. Singh and V. Pant, "Analysis of a multiphase induction machine under fault condition in a phase-redundant AC drive system," Electr. Mach. Power Syst., vol. 28, no. 6, pp. 577–590, Jun. 2000.

[17] H. A. Toliyat, "Analysis and simulation of five-phase variable-speed induction motor drives under asymmetrical connections," IEEE Trans.Power Electron., vol. 13, no. 4, pp. 748–756, Jul. 1998.

[18] Y. Zhao and T. A. Lipo, "Modeling and control of a multi-phase induction machine with structural unbalance. Part I. Machine modeling and multi-dimensional current regulation," IEEE Trans. Energy Convers., vol. 11,no. 3, pp. 570–577, Sep. 1996.

Multilevel Current Source Inverter Topologies Based on the Duality Principle

Jianyu Bao
Ningbo Institute of Technology
Zhejiang University
Ningbo, P.R.China
baojy@nit.zju.edu.cn

Weibing Bao
Zhijiang College
Zhejiang University of Technology
Hangzhou, P.R.China
weibing-bao@163.com

Siran Wang and Zhongchao Zhang
Department of Electrical Engineering
Zhejiang University
Hangzhou, P.R.China
Siran.Wang09@gmail.com

Abstract—This paper explores the issue of constructing multi-level current source inverter (MCSI) topologies using the duality principle. In order to extend the application of duality to the three-phase power electronic circuits which are non-planar, two methods for planar treatment with three-phase voltage source inverter (VSI) are presented. Based on such equivalent transformation, an improved topology of three-phase flying-capacitor multi-level voltage source inverter (MVSI) is obtained, then a new kind of three-phase MCSI topology is derived from the application of the duality on such MVSI topology. This allows the wealth of existing knowledge relating to modulation of multilevel VSI to be immediately mapped to the multilevel CSI. Analytical results show that the derived MCSI and its corresponding MVSI present dual features such as operations, modulations and control strategies, etc. Some simulation results are given to verify the proposed MCSI structure.

I. Introduction

Multilevel converters [1] can offer substantial benefits for higher power applications compared to two level converters, including reduced harmonics, and increased power ratings because of reduced switching device voltage and current stresses. Nowadays, with the development of SMES technology [2], there is an emerging interest in research to explore how MCSI's can be topologically constructed. In [3], the idea "current multilevel" was mentioned and a kind of MCSI topology was derived by using the generalized current multilevel cell. Based on the current multilevel cell, some extended single-phase MCSI topological constructions [4][5] are derived. To meet with the requirement in higher power applications, it is meaningful to explore how a three-phase MCSI topology can be arranged. Recently, there emerged some published papers about multilevel CSI which were focused on three-phase topological constructions (not including multiple configurations) [6]-[10]. These proposed three-phase topologies can be classified into two circuit types, one is direct type [6], but it's PWM control is complex and should be preset in advance; the other is combination type

[7]-[10], which is often composed by many multilevel cells in parallel and the whole system was carrier phase-shifted (CPS) SPWM controlled. It seems difficult to develop MCSI's by using the present knowledge about MVSI's through direct duality because most of the common MVSI's, especially their three-phase structures are all non-planar circuits. However, it seems logical to find an equivalent MVSI topology which has planar circuit firstly, and then it is possible to use the duality principle to conform three-phase MCSI's.

This paper aims to explore how to construct the MCSI topologies using the duality principle on the existent MVSI topologies. Although the traditional three-phase MVSI's are non-planar, their equivalent circuits can be planar through some modifications. This study begins by introducing how a traditional non-planar VSI is transformed to a planar VSI circuit with the same electrical characteristics in Section II. In Section III, a three-phase MVSI circuit with planarity is obtained by modifying the traditional three-phase flying-capacitor MVSI topology, and then a kind of three-phase MCSI structure is dually derived from such planar three-phase MVSI. In Section IV, the operational principle of this new three-phase MCSI is analyzed in detail. In Section V, a simulation model has been built up to verify the proposed MCSI structure.

II. The Planar Methods to Treat with VSI Circuit

The duality principle has proven to be a useful tool in the field of power electronics [11][12]. However, the duality transformation can only be applied to circuits that are in planar fashion, namely those that can be drawn with no "wires" crossing each other without a connection [11]. In order to investigate MCSI's by using the present knowledge about MVSI's, it is important to find the original topologies of three-phase VSI's with planarity before using the duality theory.

A. The Method of Adding an Extra Leg

In the traditional topology of 3-phase 6-switch 2-level VSI, by adding an extra leg circuit (S_1^* in series with S_4^*), and assigning S_1 and S_1^*, S_4 and S_4^* with the same driving pulses,

978-1-4244-4782-4/10 $26.00 © 2010 IEEE

a new 3-phase 2-level VSI topology is obtained as shown in Fig. 1a, which is in a planar fashion. Therefore, the duality transformation can be directly applied to it, Fig. 1b gives the corresponding dual.

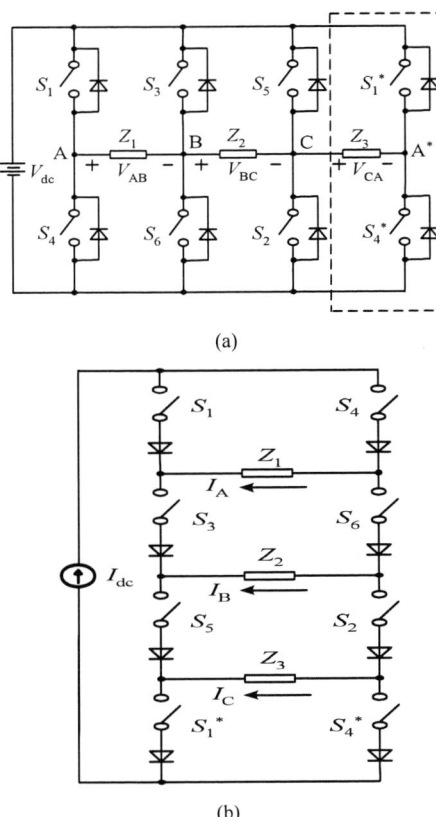

(a)

(b)

Fig. 1.   (a) Two-level VSI with four legs, (b) dual structure

B. The Method of Adding an Extra Voltage-Source

Another way to get the planar circuit from the traditional topology of 3-phase 6-switch 2-level VSI, is that, adding an extra voltage-source V_{dc2} to the third leg circuit (S_5 in series with S_2), another modified VSI topology is obtained as shown in Fig. 2a, which is also in a planar fashion. Similarly, it is easy to obtain its corresponding dual topology by directly applying the dual transformation, Fig. 2b shows its dual structure.

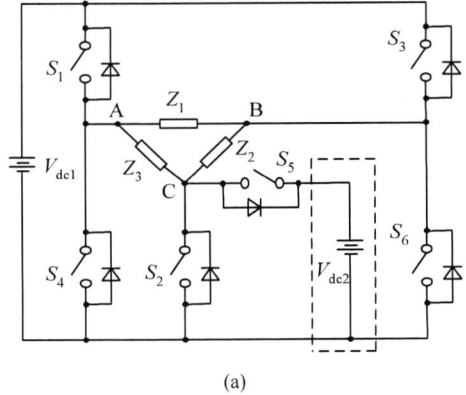

(a)

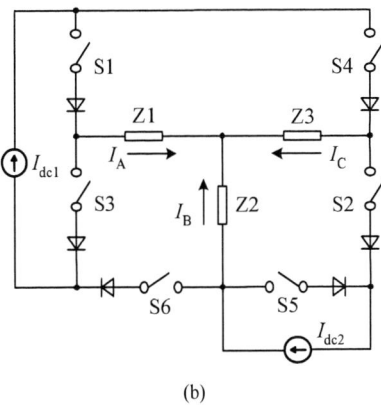

(b)

Fig. 2.   (a) Two-level VSI with 2 voltage-sources, (b) dual structure

C. Comparison with the Two Methods

By using the two planar methods described above, the equivalent three-phase VSI with planarity can be derived from the traditional topology of 3-phase 6-switch 2-level VSI, this allows the wealth of existing knowledge relating to modulation of VSI to be immediately applied to the CSI. However, from the view point of practicability, three-phase load connection mode of the dual structure in Fig. 1b is more uncommon compared to that of the dual structure in Fig. 2b. Therefore the method of adding an extra voltage-source is more attractive and will be extended to deal with the three-phase MVSI circuit.

III.   CONFORMATION OF THE THREE-PHASE MCSI

In this section, how to construct the MCSI topology by using the present knowledge about MVSI's [1] will be discussed. For cascaded H-bridge MVSI and diode-clamp MVSI, due to their circuit structures, it is hard to get the equivalent topology whether by adding an extra leg or by adding an extra voltage-source. However, for flying-capacitors MVSI, by adding an extra voltage-source to the third leg, a three-phase equivalent circuit with planar graph can be obtained. Fig. 3a shows the improved three-phase flying-capacitors 5-level VSI, obviously, it presents the planar property, so the duality transformation can be directly applied to it.

Fig. 3b gives the corresponding dual, namely a new three-phase 5-level CSI, in which the bidirectional conducting switches are transformed into unidirectional switches by duality, and the same assignations show the dual elements in two dual structures. According to the duality principle, in each phase there are two complementary switch pairs: (S_{x1}, S_{x4}), (S_{x2}, S_{x3}) (x=a, b, c), and the dc current source $2I_{dc}$ is equally divided by inductors (or sharing-inductors) L_y (y=a, b, c), so the currents through every inductor are all I_{dc}. Furthermore, the circuit connections in each phase are very similar to the single-phase 5-level CSI described in [3]. Therefore, according to the extension method of the single-phase MCSI, the three-phase MCSI with more output current levels can be easily derived from this 5-level structure.

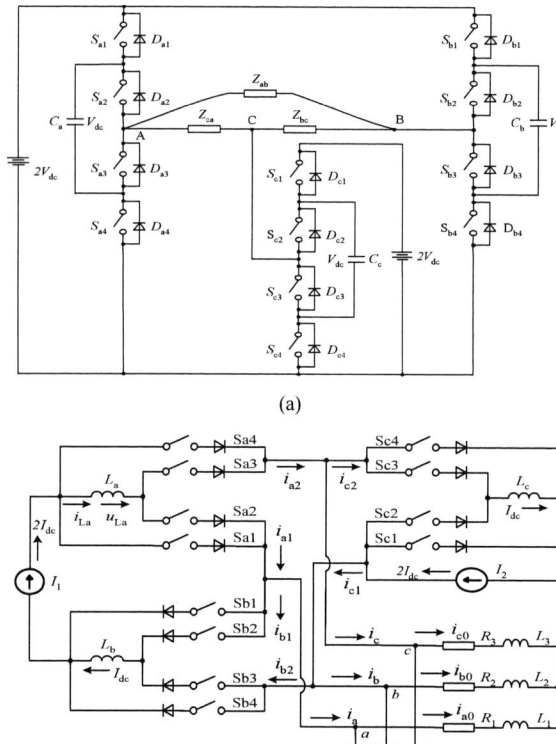

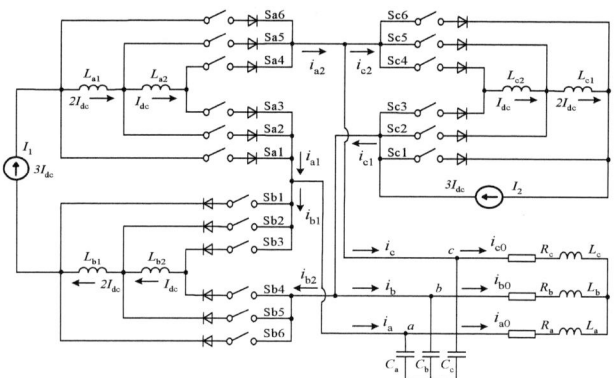

Fig. 3.   (a) The improved three-phase flying-capacitors 5-level VSI
(b) dual structure

Fig.4 shows the 7-level CSI structure, which is made up of 18 switches and 6 sharing-inductors. According to the duality principle, in each phase there are three complementary switch pairs: (S_{x1}, S_{x6}), (S_{x2}, S_{x5}), (S_{x3}, S_{x4}) (x=a, b, c).

Fig. 4.   The new three-phase 7-level CSI

IV.   OPERATION ANALYSIS

A.   Mechanism of 5-level Current Generation

Taking the 5-level circuit shown in Fig. 3b as an example, the operational modes for generating 3-level current of i_{a1} can be obtained as follows:

- $i_{a1}=+2I_{dc}$, when S_{a1}, S_{a2} are on at the same time;
- $i_{a1}=+I_{dc}$, when S_{a1}, S_{a3} or S_{a2}, S_{a4} are on simultaneously;
- $i_{a1}=0$, when S_{a3}, S_{a4} are on at the same time.

In similar way, the 3-level current of i_{b1} can be generated as follows:

- $i_{b1}=+2I_{dc}$, when S_{b1}, S_{b2} are on at the same time;
- $i_{b1}=+I_{dc}$, when S_{b1}, S_{b3} or S_{b2}, S_{b4} are on simultaneously;
- $i_{b1}=0$, when S_{b3}, S_{b4} are on at the same time.

While the total output current of phase-a is i_a, and $i_a= i_{a1}-i_{b1}$. Because i_{a1} and i_{b1} are all 3-level currents, through subtraction with each other, the 5-level current is generated. On the viewpoint of duality, just like that of 5-level VSI, the 5-level line voltage is obtained by the subtraction between two phase-voltages (3-level). On the same way, the generation modes for generating 5-level current of i_b, i_c can be deduced.

B.   Five-Level CSI PWM

It is well established that the carrier disposition PD PWM strategy creates the lowest line-to-line harmonic distortion for a multilevel VSI [13]. According to the dual relationship, there is a correspondence among their available switching states and that these switching states produce the same effect in the output. The 5-level CSI is PD PWM controlled, that is, when the reference waveforms are compared against the upper triangular carrier, the phase leg outputs switch between $+2I_{dc}$ and $+I_{dc}$, while when they are compared against the lower triangular carrier, the phase leg outputs switch between $+I_{dc}$ and 0.

V.   SIMULATION RESULTS

The new three-phase 5-level CSI shown in Fig. 3b has been verified in a detailed SIMULINK simulation that fully defines all component elements of the circuit. DC current source $2I_{dc}$ was 100A; L_k(k=1,2,3)=5mH, R_x(x=a,b,c)=4Ω, L_x(x=a,b,c)=20mH, C_x(x=a,b,c)=25μF; the CSI was operated with a modulation index of 0.9, an output frequency of 50 Hz, and a carrier switching frequency of 1050 Hz (pulse ratio = 21).

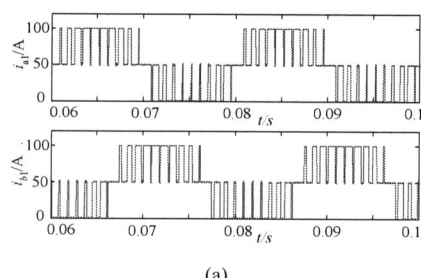

(a)

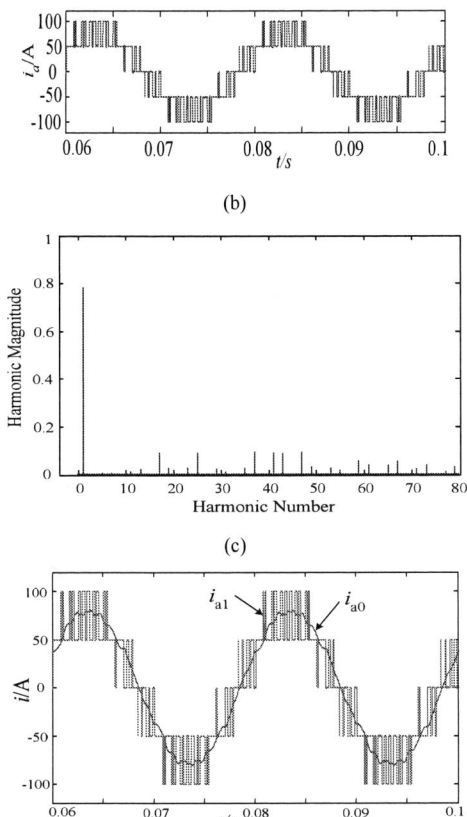

(b)

(c)

(d)

Fig. 5.   (a) 3-level switched current outputs for i_{a1} and i_{b1}, (b) 5-level switched current output for i_a, (c) current spectrum of i_a, (d) output currents (before and after filtering)

Fig. 5a shows the 3-level switched current outputs for i_{a1} and i_{b1} with the VSI modulator operating under PD PWM with simple sinusoidal fundamental references. Similar to the generation of the 5-level line-line voltage in VSI, the 5-level switched current i_a is obtained by subtraction between i_{a1} and i_{b1}, which is shown in Fig. 5b, and its associated harmonic spectrum is illustrated in Fig. 5c. The close match achieved between the spectrums for this modulation strategy, which confirms the effectiveness of the modulation strategy in the 5-level CSI. In Fig. 5d, i_{a0} is the filtered output current, which is almost sinusoidal.

VI. CONCLUSION

This paper has presented a dual approach to the conformation of a three-phase multilevel CSI. Two planar methods for three-phase 2-level VSI circuit are established by using the equivalent transformation. An improved three-phase 5-level flying capacitors VSI with planarity is obtained by applying one of the two methods, then a new three-phase 5-level CSI is derived from applying the direct duality to the improved three-phase 5-level flying capacitors VSI. Because

the derived MCSI and their corresponding MVSI share some analogous properties, the switching states selected by the 3-level VSI modulator can be mapped to the 5-level CSI through duality. The result is a PWM modulator for a 5-level CSI which achieves exactly the same harmonic performance as its equivalent 3-level VSI controller.

The concept has been verified by a detailed SIMULINK simulation to give confidence as the validity of the dual approach. Based on the duality, the established knowledge such as operations, modulations and control strategies, etc., about the MVSI's can also apply to their corresponding MCSI's, and vice versa. So the research on the two multilevel inverters, MVSI's and MCSI's, can be used for reference each other.

REFERENCES

[1] J. Rodriguez, J. Lai, F.Z. Peng, "Multilevel inverters: A Survey of Topologies, Controls and Applications," *IEEE Trans. on Ind. Eletron.*, 2002. 49(4): 724-738.

[2] M. Steurer, C.A. Luongo, P.R. Ribeiro, S. Eckroad, " Interaction Between a Superconducting Coil and the Power Electronics Interface on a 100 MJ SMES System," *IEEE Trans. on Applied Superconductivity*, 2003, 32(2):1806-1809.

[3] F.L.M. Antunes, H.A.C. Braga, Ivo Barbi, "Appication of a Generalized Current Multilevel Cell to Current-Source Inverters," *IEEE Trans. on Industrial Electronics*, 1999, 46(1): 31-38.

[4] Y. Xiong, D.J. Chen, S.Q. Deng, Z.C. Zhang, "A New Single-Phase Multilevel Current-Source Inverter," *IEEE Conference on APEC2004*, 2004, pp.1682-1685.

[5] J.Y. Bao, D.G. Holmes, Z.H. Bai, Z.C. Zhang, D.H. Xu, " PWM Control of a 5-level Single-Phase Current-Source Inverter with Controlled Intermediate DC-link Current," *IEEE Power Electron. Specialists Conf.*, 2006, pp.1633-1638.

[6] Y. Xiong, Y.L. Li, X. Yang, et al., "A New Three-Phase Five-Level Current-Source Inverter," *IEEE Conference on APEC05*, 2005, Vol.1, pp.424-427.

[7] J.Y. Bao, Z.H. Bai, Q.S. Wang, Z.C. Zhang, "A New Three-Phase 5-Level Current-Source Inverter," *Journal of Zhejiang University SCIENCE A*, 2006, 7(12): 1973-1978.

[8] S.S. Kwak, H.A. Toliyat, "Multilevel Converter Topology Using Two Types of Current-Source Inverters," *IEEE Trans. on Industry Applications*, 2006, 42(6):1558-1564.

[9] Z.H. Bai, Z.C. Zhang, Y. Zhang, "A Generalized Three-Phase Multilevel Current Source Inverter with Carrier Phase-Shifted SPWM," *IEEE Conference on PESC07*, 2007, pp.2055-2060.

[10] Y.L. Familiant, D.G. Holmes, T.A. Lipo, B.P. McGrath, "A General Modulation Strategy for a Five-Level Three-Phase Current Source Inverter with Regulated Intermediate DC Link Currents," *IEEE Conference on IAS07*, 2007, pp.581-588.

[11] S.D. Freeland, "Techniques for the Practical Application of Duality to Power Circuits," *IEEE Trans. on Power Electronics*, 1992, 7(2):374-384.

[12] V.G. Agelidis, G. Joos, " On Applying Graph Theory Toward a Unified Analysis of Three-Phase PWM Inverter Topologies," *IEEE Conference on PESC93*,1993, pp.408-415.

[13] B.P. McGrath, D.G. Holmes, "Multicarrier PWM Strategies for Multilevel Inverters," *IEEE Trans. on Industrial Electronics*, 2002, 49(4):858-867.

3-Level Power Converter with High-Voltage SiC-PiN diode and Hard-Gate-Driving of IEGT for future high-voltage power conversion systems

[1]Kazuto Takao, [2]Yasunori Tanaka, [3]Kyungmin Sung, [4]Keiji Wada,
[1]Takashi Shinohe, [5]Takeo Kanai, and [2]Hiromichi Ohashi

[1]Corporative Research & Development Center, Toshiba Corporation, 1. Komukai-Toshiba-cho, Saiwaiku, Kawasaki, Japan
[2]National Institute of Advanced Industrial Science and Technology, 1-1-1, Umezono, Tsukuba, Ibaraki, Japan
[3]Ibaraki National College of Technology, 866 Nakane, Hitachinaka, Ibaraki, Japan
[4]Tokyo Metropolitan University, 1-1 Minami Osawa, Hachioji, Tokyo, Japan.
[5]Toshiba Mitsubishi-Electric Industrial System Corporation, 1.Toshiba-cho, Fuchu, Tokyo, Japan

Abstract—Reductions in the size and weight of medium-voltage power converters are essential for saving space of power conversion systems and cutting their cost. Volumes of magnetic components such as transformers and LC filters are significant in medium-voltage power converters. High-switching-frequency operation is essential for reducing the volume of magnetic components. In this work, hybrid pairs of 6 kV SiC-PiN diodes and 4.5 kV Si-IEGTs have been applied to realize the high-switching-frequency operation of medium-voltage power converters. For low switching losses and series operation of power devices, a gate-driving technique with an extremely low gate resistance called hard gate driving is employed. Switching characteristics of the hybrid pair are measured experimentally. It has been demonstrated that the total switching loss can be reduced up to 50% with the hybrid pair. In order to demonstrate a 2 kHz switching frequency operation of the hybrid pair, which is about 4 times higher than that of conventional medium-voltage power converters, a 378 kVA prototype 3-level inverter has been designed and constructed.

I. INTRODUCTION

Power conversion technology applying high-power semiconductors supports power systems, industrial systems, railway systems and other social infrastructures [1]. In the context of efforts to realize a low-carbon society, attention worldwide has recently focused on smart grids, mega-solar projects, and wind power generation. Power converters for these high-power applications often need bulky and heavy transformers and LC filters. Reductions in their size and weight are essential for saving space occupied by power conversion systems and cutting their cost. In order to dispense with transformers, high-voltage 3-level power converters have been utilized for practical purposes [2]. On the other hand, high switching frequency operation is required to reduce the size and weight of LC filters.

Power semiconductor devices should operate with low switching loss for the high switching frequency operation. High-voltage silicon IGBTs / IEGTs (Si-IGBTs / Si-IEGTs) and PiN diodes (Si-PiN diodes) are widely utilized in medium-voltage power converters. Switching frequency of these high-voltage power devices has been restricted to around 1 kHz owing to their large switching losses.

Silicon carbide (SiC) power devices are expected to be applied as next-generation power devices offering performance superior to that of silicon (Si) power devices in terms of low power loss, high-speed switching, and high-temperature operation. Power loss reduction effects attributable to hybrid pairs of SiC PiN diodes and Si-IGBTs / Si-IEGTs have recently been reported [3]-[6]. SiC-PiN diodes have extremely low reverse recovery current and low switching loss characteristics compared to those of conventional Si-PiN diodes. The use of SiC-PiN also reduces turn-on losses of Si-IGBTs. In addition, because the turn-off peak power of SiC-PiN diodes is much lower than that of Si-PiN diodes, an extremely low gate resistance operation [7], [8] can be applied for the further reduction of the turn-on losses. For the above-mentioned reasons, hybrid pairs of Si-IGBTs / Si-IEGTs and Si-PiN diodes have the potential to increase switching frequency of medium-voltage power converters.

In this work, hybrid pairs of 6 kV SiC-PiN diodes and 4.5 kV Si-IEGTs have been applied to realize the high switching frequency operation of medium-voltage power converters. Switching characteristics of the hybrid pair are measured experimentally to investigate the lower limitation of the external gate resistance and estimate the switching losses. In order to demonstrate a 2 kHz switching frequency operation of the hybrid-pair, which is about 4 times higher than that of conventional medium-voltage power converters, a 378 kVA prototype 3-level inverter has been designed and constructed.

978-1-4244-4782-4/10 $26.00 © 2010 IEEE

II. 6 kV SiC-PiN DIODE

6 kV class SiC-PiN diodes have been fabricated by AIST. Fig. 1 shows the fabricated SiC-PiN diodes on a 2-inch wafer. The active area of the SiC-PiN diodes is 4 mm x 4mm. The cross-sectional view of the SiC-PiN diodes is illustrated in Fig.2. The drift layer is doped 1 x 10^{15} cm^{-3}, and has a thickness of 70 μm.

Fig. 3 shows the forward *IV* characteristics of two parallel-connected SiC-PiN diodes. The on-voltage decreases with increasing the temperature. The-on voltages at 50 A (156 A/cm2) are 3.52 V at 25 °C and 3.33 V at 150 °C. Fig. 4 shows the typical reverse *IV* characteristics of the SiC-PiN diode. As shown in Fig. 4, the breakdown voltage of 7 kV is achieved.

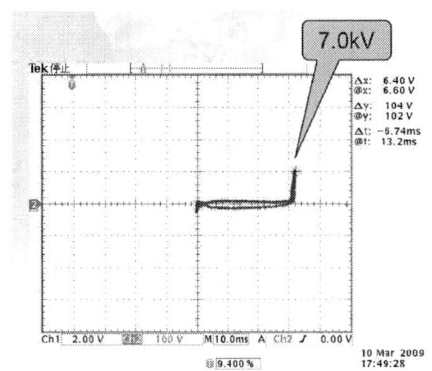

Fig. 4. Reverse *IV* characteristics of the SiC-PiN diode

III. CHARACTERISTICS OF HARD DRIVING HYBRID PAIR

A. Experimantal setup

Switching characteristics of the SiC-PiN diode and Si-IEGT hybrid-pair are measured in an inductive load chopper circuit illustrated in Fig. 5, and the experimental setup is shown in Fig. 6. Switching tests are implemented by a

Fig. 1. 6 kV SiC-PiN diodes

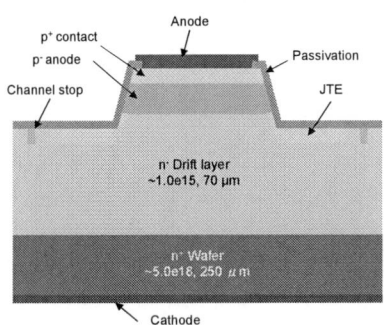

Fig. 2. Cross-sectional view of the 6 kV SiC-diode

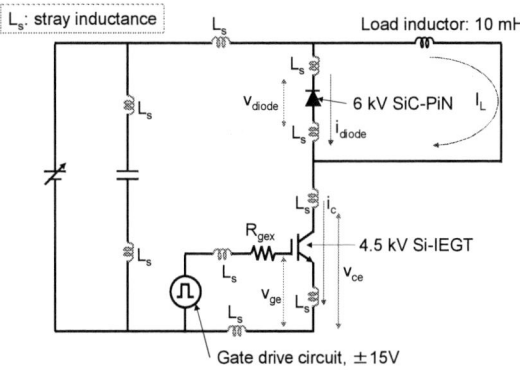

Fig. 5. Chopper circuit for switching tests

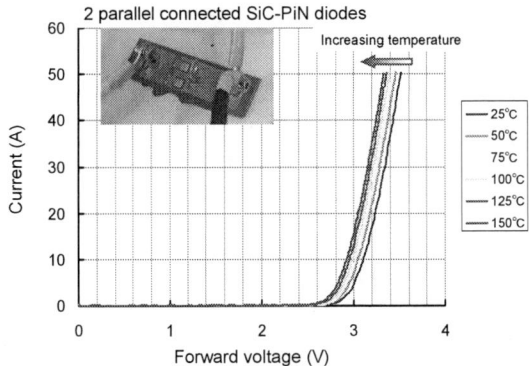

Fig. 3. Forward *IV* characteristic of 2-paralell SiC-PiN diodes.

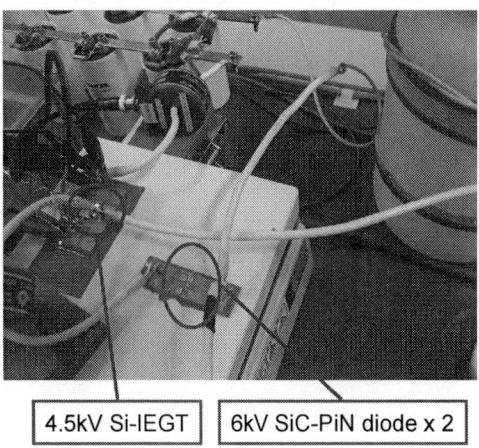

Fig. 6. Setup for switching tests

double-pulses method. The experimental conditions are summarized in Table I. In the experiments, parallel-connected SiC-PiN diodes and one Si-IEGT are utilized. The external gate resistance of the Si-IEGT (R_{gex}) is varied from 27 Ω to 117.5 Ω to evaluate the dependence of the switching characteristics on the switching speed. From the evaluation results, the lower limitation of the external gate resistance is extracted.

TABLE I. EXPERIMENTAL CONDITIONS FOR SWITCHING CHARACTERISTICS MEASUREMENTS

DC voltage : V_{dc}	2.5 kV
Load current : I_L	10, 20, 30, 40, 50 A
Gate applied voltage: V_{gs}	± 15 V
External gate resistance: R_{gex}	378 kVA
Device temperature : T_j	25, 125 °C
Stray inductance : L_s	4.2 μH

B. Switchng waveforms

Fig. 7 shows recovery waveforms of the SiC-PiN diodes. As seen in Fig. 7, current waveforms have no oscillation. In contrast, oscillations are found in voltage waveforms. The peak recovery current increases with increasing device temperature because minority carriers in the SiC-PiN diodes increase with the device temperature (T_j). The surge voltages due to the inductive voltage by the circuit stray inductance are confirmed in the voltage waveforms. In the case of $R_{gex} = 27$ Ω, the peak surge voltages (V_{dpeak}) are 2960 V at $T_j = 25$ °C and 3300 V at $T_j = 125$ °C, and these values exceed the input DC voltage (V_{dc}). On the other hand, in the case of $R_{gex} = 47$ Ω, V_{dpeak} is less than V_{dc}.

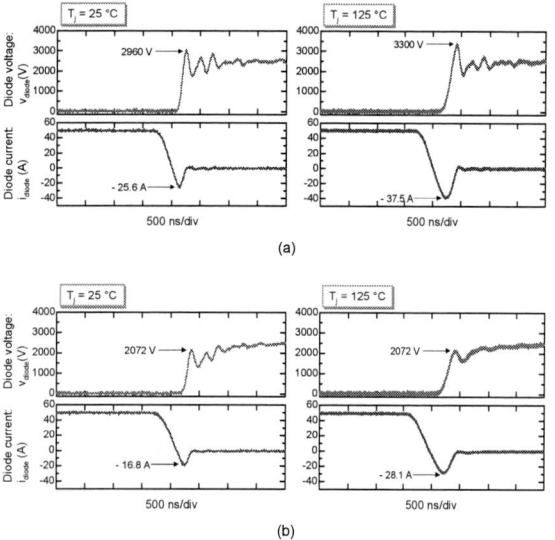

Fig. 7. Recovery waveforms of the SiC-PiN diode. (a) $R_{gex} = 27$ Ω, (b) $R_{gex} = 47$ Ω

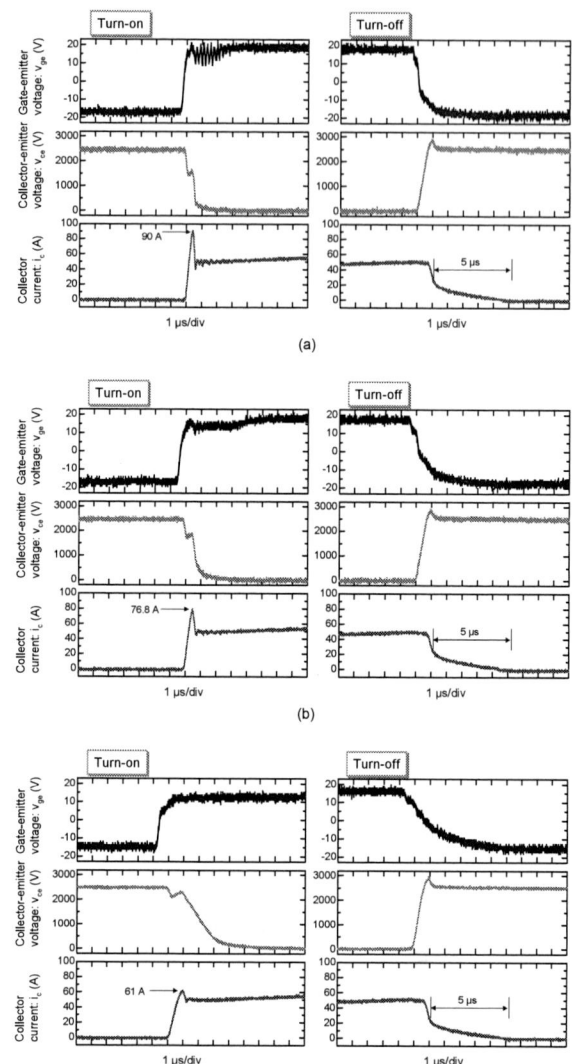

Fig. 8. Switching waveforms of the Si-IEGT. (a) $R_{gex} = 27$ Ω, (b) $R_{gex} = 47$ Ω, (c) $R_{gex} = 117.5$ Ω

Fig. 8 shows switching waveforms of the Si-IEGT. Turn-on speed increases with decreasing R_{gex}. In contrast, turn-off speed is independent of R_{gex}. The peak current at turn-on increases with decreasing R_{gex}.

From the switching waveforms shown in Fig. 7 and Fig. 8, it is confirmed that decrease of R_{gex} is effective for reducing the turn-on energy of the Si-IEGT. However, V_{dpeak} of the SiC-PiN diode increase with decreasing R_{gex}. This means that V_{dpeak} would exceeds breakdown voltage of the SiC-PiN diodes or Si-IEGT at very low R_{gex}. Therefore, the lower limitation of R_{gex} should be extracted to prevent the breakdown.

C. Extraction of the lower limitation of the external gate resistance

Fig. 9 shows the dependence of V_{dpeak} on the load current (I_L) obtained at V_{dc} = 2500 V. V_{dpeak} at T_j = 125 °C is larger than that at T_j = 25 °C. In the case of T_j = 25 °C, V_{depak} is the maximum around I_L = 20 A. In the case of T_j = 125 °C, V_{dpeak} is the maximum around I_L = 20 A at R_{gex} = 27 Ω however V_{dpeak} increases with decreasing I_L in the range of $R_{gex} \geq$ 39 Ω.

In this work, it is defined that the maximum value of V_{dpeak} is the static breakdown voltage of the Si-IEGT (4500 V) and R_{gex} in this condition is determined as the lower limitation. In order to find the lower limitation of R_{gex}, the relationship between R_{gex} and V_{dpeak} at the condition of T_j = 125 °C and I_L = 20 A is investigated because V_{dpeak} is the maximum around I_L = 20 A. Fig. 10 shows the relationship between R_{gex} and V_{dpeak}. From the approximated curve, V_{dpeak} exceeds the static breakdown voltage of the Si-IEGT at the condition of $R_{gex} \leq$ 26 Ω. Therefore, the lower limitation

of R_{gex} can be determined as 26 Ω. As R_{gex} for the 4.5 kV Si-IEGT is around 100 Ω / chip~200 Ω / chip in present-day power converter systems, R_{gex} can be reduced to less than 1/4 by using the SiC-PiN diodes.

D. Evaluation of switching energy

Fig. 11 shows the dependence of turn-on energy (E_{on}) and turn-off energy (E_{off}) of the Si-IEGT on I_L. E_{on} decreases with decreasing R_{gex}. In contrast, E_{off} is independent of R_{gex}. Fig. 12 shows the dependence of recovery loss (E_{dsw}) of the SiC-PiN diode on I_L. E_{dsw} increases with decreasing R_{gex} because the recovery charge increases with the switching speed.

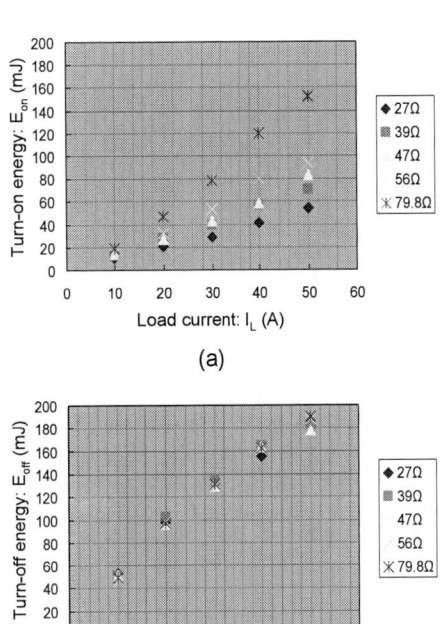

(a)

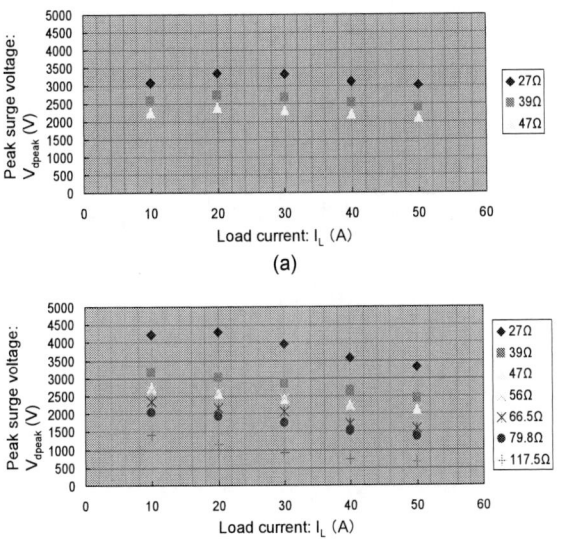

(a)

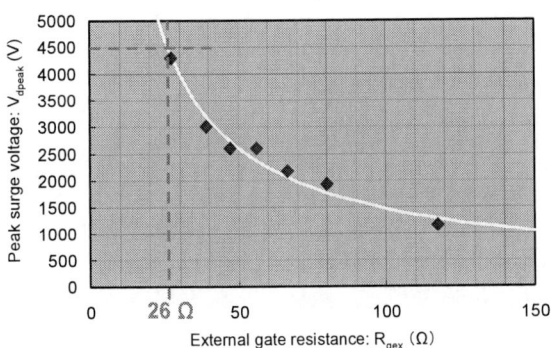

(b)

Fig. 9 Load current vs. peak surge voltage of the SiC-Pin diode (a) T_j = 25 °C, (b) T_j = 125 °C

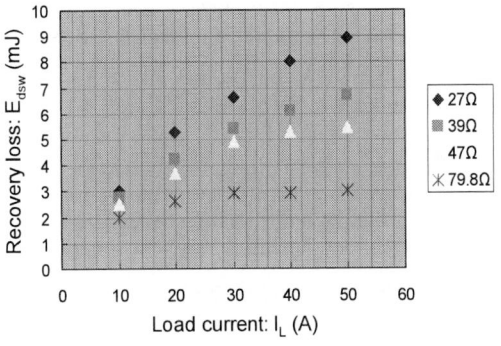

(b)

Fig. 11 Load current vs. turn on / off energies of the Si-IEGT (a) turn on, (b) turn off

Fig. 10 External gate resistance vs. peak surge voltage of the SiC-PiN diode at I_L = 20 A and T_j = 125 °C

Fig. 12 Load current vs. recovery loss energy of the SiC-PiN diode

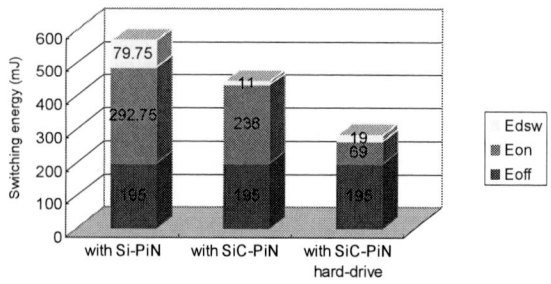

Fig. 13 Comparison of switching energies at I_L = 50 A

Fig. 13 shows the comparison of the total switching energy $(E_{on} + E_{off} + E_{dsw})$ for (a) a conventional Si-PiN diode and Si-IEGT pair with normal gate driving (R_{gex} = 117.5 Ω), (b) SiC-PiN diode and Si-IEGT hybrid pair with normal gate driving, and (c) SiC-PiN diode and Si-IEGT hybrid pair with hard gate driving (R_{gex} = 27 Ω). In the normal gate driving condition, 86% and 18% reductions of E_{dsw} and E_{on} are confirmed by using the SiC-PiN diode. In this case, the reduction effect of the total switching energy is 22%. By applying the hard gate driving to the hybrid pair, 76% reduction of E_{on} and 76% reduction of E_{dsw} can be obtained compared with the conventional Si-PiN diode and Si-IEGT pair, and a large reduction effect of the total switching energy of 50% can be achieved. In the case of the hybrid pair with the hard gate driving, the percentage of the E_{off} in the total switching energy is 69%. This percentage is larger than for the conventional Si-PiN diode and Si-IEGT pair. This indicates that the reduction of E_{off} is the key issue from the viewpoint of achieving further reduction of the power loss of the hybrid-pair.

IV. PROTOTYPE 3-LEVEL POWER CONVERTER FOR DEMONSTRATION OF 2 KH OPERATION

A. Specifications of prototype 3-level power converter

In order to demonstrate a 2 kHz switching frequency operation of the SiC-PiN diode and Si-IEGT hybrid pair with hard gate driving, a 378 kVA prototype 3-level power converter has been designed. The overview and equivalent circuit of the prototype 3-level power converter are shown in Fig. 14 and the specifications are listed in Table II.

TABLE II.   SPECIFICATIONS OF THE PROTOTYPE 3-LEVEL POWER CONVERTER

Input DC voltage	± 5 kV
Output AC voltage	6.3 kV$_{rms}$
Output AC current	60 A$_{rms}$
Rated power	378 kVA

The prototype 3-level power converter employs 100 A class hybrid pair and clamp diode modules. The modules are connected in 2 series to control 5 kV dc voltages. Fig. 15

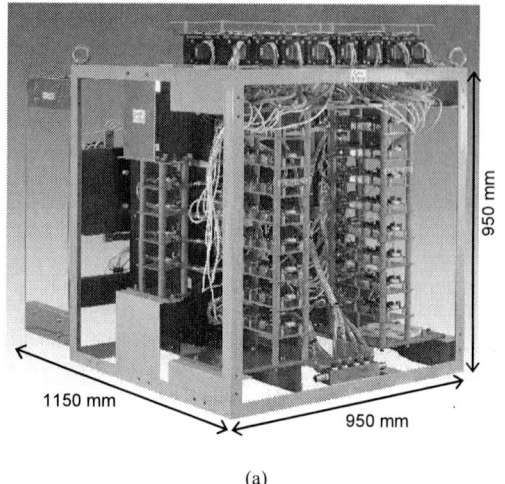

(a)

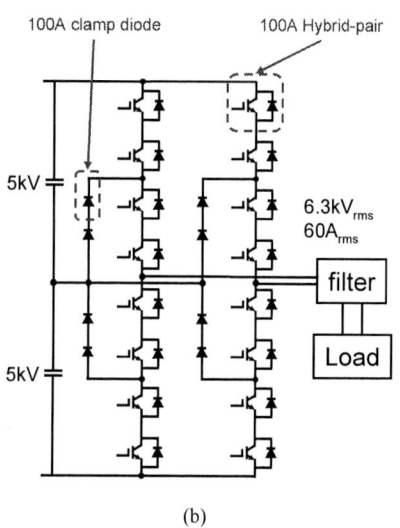

(b)

Fig. 14 Prototype 3-level power converter using 6 kV SiC-PiN diode and 4.5 kV Si-IEGT hybrid-pairs (a) overview, (b) equivalent circuit of the converter

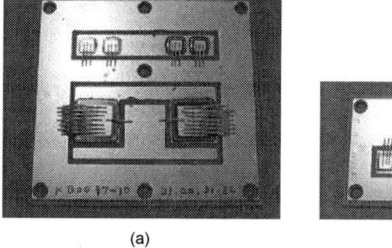
(a)

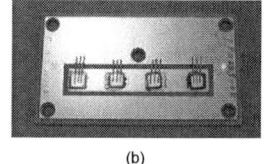
(b)

Fig. 15 DBC substrates of (a) hybrid-pair and (b)clamp diode modules

978-1-4244-4782-4/10 $26.00 © 2010 IEEE   1105

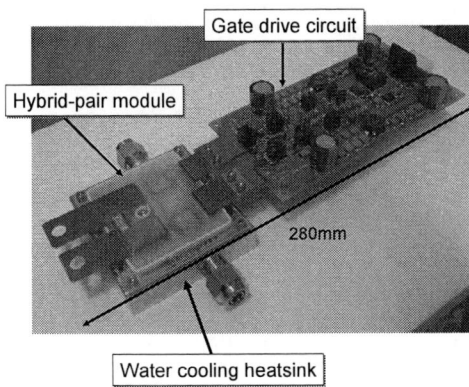

Fig. 16 Overview of the 100 A hybrid-pair switching unit

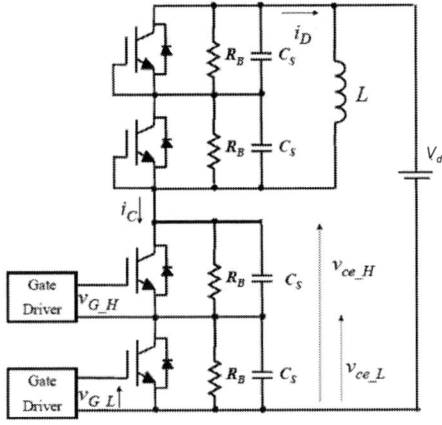

Fig. 17 Test circuit of 2-series connection

shows DBC substrates in the modules. Hybrid pair modules use two 4.5 kV IEGT chips and four 6 kV SiC-PiN diode chips, and clamp diode modules use four 6 kV SiC-PiN diode chips.

Fig. 16 shows the 100 A hybrid-pair switching units used in the prototype power converter. The switching unit comprises the DBC substrate, water-cooling-type heatsink, and gate drive circuit. The heatsink is designed based on the power loss data of the hybrid pair.

B. Switchng characteristics of 2-series connected hybrid-pair modules

Switching characteristics of the 2 series connected 100 A hybrid-pair switching units are measured by the doubl-pulse tests. The equivalent circuit of the test setup is shown in Fig.17. Snubber capacitors (4.7 nF) are connected in parallel to the switching units to improve the voltage sharing at switching instants. The switching waveforms of the high-side and low-side units are shown in Fig. 18. As seen in the figure, both units successfully operate under the conditions of $V_{dc} = 5$ kV, $I_L = 100$ A, and $R_{gex} = 50$ Ω.

C. 2 kHz operation tests rerults

The preliminary experiment of the prototype power converter has been implemented under the condition that the DC voltage is ± 1.2 kV. Fig. 19 shows the collector-emitter voltages of high-side and low-side hybrid-pair switching units. The voltage balance between 2 series-connected switching units and the 2 kHz switching frequency operation can be confirmed.

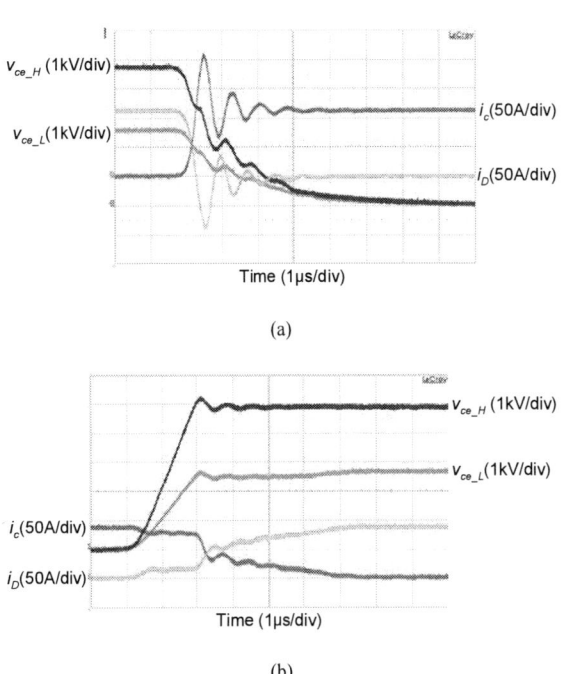

(a)

(b)

Fig. 18 Switching waveforms of the high-side and low-side hybrid-pair switching units (a) turn on, (b) turn off

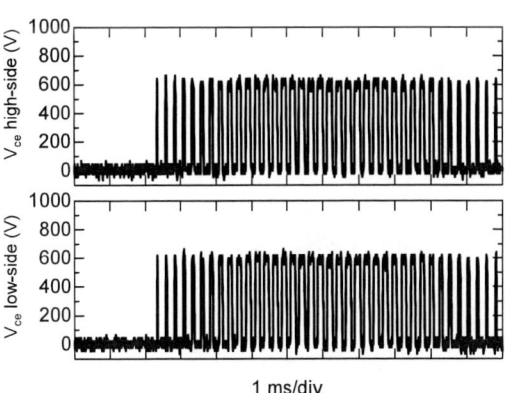

Fig. 19. Collector-emitter voltages of high-side and low-side hybrid-pair switching units

V. CONCLUSION

Hybrid pairs of 6 kV SiC-PiN diodes and 4.5 kV Si-IEGTs are evaluated in terms of their ability to increase switching frequency of medium-voltage power converters. By the use of the hybrid pair, an external gate resistance of a Si-IEGT can be reduced to less than 1/4 compared to that of a conventional Si-PiN diode and Si-IEGT pair. By applying a hard gate driving technique, switching losses can be reduced up to 50%. This result demonstrates that the hard gate driving is attractive for increasing the switching frequency. In order to demonstrate a 2 kHz switching frequency operation of the SiC-PiN diode and Si-IEGT hybrid pair with hard gate driving, a 378 kVA prototype 3-level power converter was designed. In preliminary experiments of the prototype power converter, series voltage balance and 2 kHz operation of switching units has been confirmed.

REFERENCES

[1] T. Matsumoto, H. Tai, and T. Shinohe, "Power Conversion Technology Appliying High-Power Semiconductors," Toshiba review, vol. 63, no. 11, pp. 2–8, 2008. (*in Japanease*)

[2] R. Jakob, C. Keller, G. Mohlenkamp, and B. Gollentz, "3-Level High Power Converter with Press Pack IGBT," in *Proc. CD-ROM, 12th European Conference on Power Electronics and Applications (EPE 2007)*, Aalborg, Denmark, September, 2007.

[3] K. Sung, K. Suzuki, Y. Tanaka, T. Ogura, and H. Ohashi, "A Study on Switching Frequency Limitation of High Voltage Power Converters in Combination of Si-IEGT and SiC-PiN Diode," in *IEEE 2006 Applied Power Electronics Conference*, 2006, pp. 455-459.

[4] W. Bartsch, S. Gediga, H. Koehler, R. Sommer and G. Zaiser "Comparison of Si- and SiC-Powerdiodes in 100A-Modules," in *Proc. CD-ROM, 12th European Conference on Power Electronics and Applications (EPE 2007)*, Aalborg, Denmark, September, 2007.

[5] T. Kinjo, K. Takao, K. Sung, Y. Tanaka, T. Ogura and H. Ohashi, "Exact Power Loss Estimation Method for High Voltage Power Converters with 5 kV SiC-PiN Diode," in *IEEE 2008 Applied Power Electronics Conference*, 2008, pp. 1247-1251.

[6] T. Kinjo, K. Takao, Y. Tanaka, K. Sung, and H. Ohashi, "Quantitative Study on Operation Frequency Limitation of Multi-level High Voltage Power Converter Equipped with Si-IEGT and SiC-PiN Diode," in *IEEE 2008 Power Electronics Specialists Conference*, 2008, pp. 2909-2913.

[7] M. Tsukuda, I. Omura, T. Domon, W. Saito, and T. Ogura, "Demonstration of High Output Power Density (30 W/cc) Converter using 600 V SiC-SBD and Low Impedance Gate Driver," in Proc. *Int. Power Electronics Conf. (IPEC-Niigata)*, 2005, pp.1184-1189

[8] K. Takao, C. Ota, J. Nishio, T. Shinohe and H. Ohashi, "Design Consideration of High Power Density Inverter with Low-on-voltage SiC-JBS and High-speed Gate Driving of Si-IGBT," in *IEEE 2009 Applied Power Electronics Conference*, 2009, pp. 397-400.

18 kW Three Phase Inverter System Using Hermetically Sealed SiC Phase-Leg Power Modules

Hui Zhang[1], Leon M. Tolbert[2, 4], Jung Hee Han[3], Madhu S. Chinthavali[4], Fred Barlow[5]

[1]Electrical Engineering Department
Tuskegee University
Tuskegee, Al 36088

[2]Electrical Engineering and Computer Science
The University of Tennessee
Knoxville, TN 37996-2100

[3]Global Power Electronics
27 Mauchly Ste 206
Irvine, CA 92618

[4]Power Electronics and Electric Machinery Research Center
Oak Ridge National Laboratory
Knoxville, TN 37932

[5]Electrical and Computer Engineering
University of Idaho
Moscow, Idaho 83844-1023

Abstract—**Power electronics play an important role in electricity utilization from generation to end customers. Thus, high-efficiency power electronics help to save energy and conserve energy resources. Research on silicon carbide (SiC) power electronics has shown their better efficiency compared to Si power electronics due to the significant reduction in both conduction and switching losses. Combined with their high-temperature capability, SiC power electronics are more reliable and compact. This paper focuses on the development of such a high efficiency, high temperature inverter based on SiC JFET and diode modules. It involves the work on high temperature packaging (>200 °C), inverter design and prototype development, device characterization, and inverter testing. A SiC inverter prototype with a power rating of 18 kW is developed and demonstrated. When tested at moderate load levels compared to the inverter rating, an efficiency of 98.2% is achieved by the initial prototype without optimization, which is higher than most Si inverters.**

Keywords—**Silicon Carbide (SiC), SiC JFET, Efficiency, High temperature, Inverter, Packaging.**

I. Introduction

Power electronics is the technology that enables the efficient generation, transmission, distribution, utilization, and storage of electricity. Today 40% of energy consumption is electrical energy, more than 50% of which is controlled by power electronics and will only increase in the future [1]. For example, in electricity generation, especially for renewable technologies such as solar, wind, etc, power electronics are inevitable to work as power interface with the grid or connected loads, or as the interface for energy storage[2][3]. Another example, today's motor drives consume 50%-60% of all electricity consumption [4]. With application of power electronics as controller to motor drives, tremendous energy savings can be achieved [5-6]. Thus, high efficiency and high

reliability power electronics are needed in every steps of electricity utilization from generation to end customers.

Research on SiC power electronics has revealed their better efficiency compared to Si power electronics due to the significant reduction in both conduction and switching losses [7-10]. Presently, Si IGBTs are commonly used switching component in a converter because of the low conduction and well controllable gate. These bipolar Si devices are supposed to be substituted by SiC unipolar devices such as JFETs and MOSFETs which have low switching losses and can switch fast due to the absence of minority carriers. Similarly, SiC diodes can also take the place of Si PiN diodes as freewheeling diodes because of the negligible reverse recovery current.

Furthermore, with a high-temperature package, the high temperature capability of SiC power electronics can be utilized [11]. Consequently, SiC based systems are more reliable and compact in high temperature applications such as aircraft, automobile, space exploration, deep gas/oil extraction, geothermal, etc [12-13]. This work will focus on the development and demonstration of such a high efficiency, high temperature inverter. The inverter performance will also be illustrated and confirmed by experiments.

II. Design and Development of SiC Module

Towards high efficiency and high temperature SiC inverter, a SiC JFET-based phase leg module with hermetic metal housing is developed as shown in Fig. 1(a). The package is designed to work at a temperature of at least 200 °C ambient. More details will be given in the Section II.A. Each module is for a single phase leg and is composed by six 1200 V SiC JFETs (normally-on) and two 1200 V Schottky diodes from SiCED as well as a thermistor to detect the temperature inside the module (see Fig. 1(b)). Three SiC JFETs are in parallel in

(a) CTE-matched metal lid and housing.

(b) Six SiC JFETs and two SiC Schottky diode attached to substrate and are wire bonded with aluminum alloy wire.

Fig. 1. SiC JFET phase leg module with hermetic metal housing.

order to achieve a higher current rating (~30A). To verify their performance, high temperature testing is conducted, and the details are shown in the Section II.B.

A. Package Developement

The packages used in this inverter were built based on hermetic metallic housings and direct bond copper (DBC) substrates as shown in Fig. 2. Metal housings were selected due to the high reliability associated with these packages even at temperatures well above 200 °C. The use of coefficient of expansion (CTE) matched metals, and glass isolated electrical connections ensures thermal and mechanical stability far superior to traditional plastic modules.

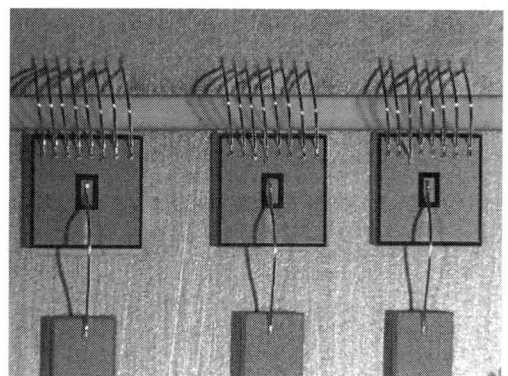

Fig. 2. SiC JFETs bonded to a DBC BeO substrate within the power modules housing.

Each individual module contains a single DBC substrate that is used to interconnect the SiC based diodes and JFETs. Beryllium oxide (BeO) was chosen for the substrate in this application due to its superior thermal performance, but versions of this module could also be fabricated with Alumina (Al_2O_3) or Aluminum Nitride (AlN) substrates, with slightly lower thermal performance [14-17]. The SiC devices were attached to these substrates through the use of a polyimide based conductive adhesive. This adhesive is stable to well above 200 °C and therefore can be used when many of the traditional solders would melt.

The SiC devices were then wire bonded with aluminum alloy wire to form the connections between the die, substrate traces, and external electrical connections. In modules of this nature the metallization must be carefully considered to avoid failures caused by inter-metallic alloy formation at the die wire bond interface, or the substrate wire bond interface. If selected inappropriately inter-diffusion, which is accelerated at high temperatures, can lead to void formation and a reduction in the strength of the bond welds. The most famous case is the Al / Au inter-metallic [18] but similar effects have been observed in the Al / Cu system [19]. In this case a surface finish of Nickel was selected since it has been established that the Ni / Al system is far more stable at high temperatures than many of the alternative systems [20-22].

After assembly of the modules, a low modulus encapsulation was used to suppress arcing in and around the die and wire bonds within the metal housings. This material is applied in a liquid state and therefore flows around the structures within the module. Once cured it becomes a soft rubber like material that is stable to temperatures greater than 200 °C. A detailed analysis of this material indicates no change in the breakdown voltage even after more than 1000 hours of exposure to high temperatures. Finally, a CTE matched metal lid was sealed in place.

B. Characterization of the SiC JFET Modules

The SiC JFET Modules are tested at different ambient temperatures from 25°C to 200 °C with an increment of 25 °C. As expected, at static state, the on-state resistances of both JFETs and diodes in the modules increase with temperature, as shown as Fig. 3(a) and 3(b). Compared to those comparable Si devices, not only the resistance values but also the variation with temperature is smaller. Thus, the SiC devices are more efficient in terms of conduction loss. The transfer characteristics of the JFETs in Fig. 3 (c) are nearly constant for tested temperatures. This means that the switching losses of the JFETs will not significantly change with temperature. This is also confirmed by the switching tests.

The two switches in the modules are also tested individually using the circuit shown in Fig. 4 at the temperature range from 25°C to 200 °C with the increment of 50 °C. With a pure inductive load, the current in the switches can be controlled by adjusting the duty ratio of the first pulse when applying a double-pulse control signal. Commercial gate driver IC HCNW3120 is selected to drive the SiC JFETs. With proper design of power supplies voltages, the gate driver IC can generate -20V to turn off the JFETs, and 0 V for turn on.

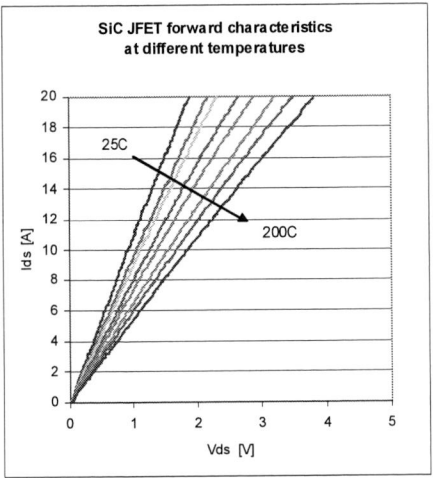

(a) SiC JFETs forward characteristics

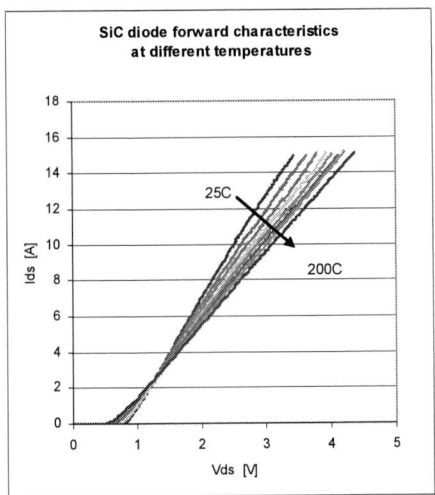

(b) SiC diodes forward characteristics

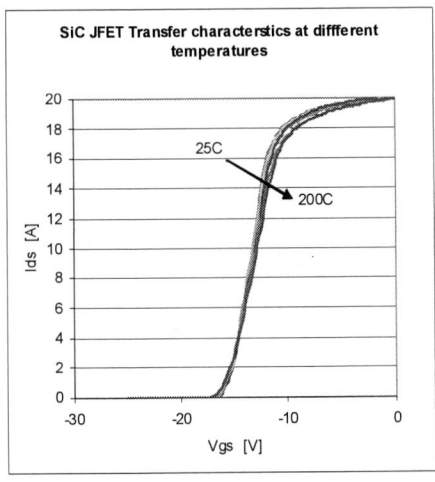

(c) SiC JFETs transfer characteristics

Fig. 3.   Static characteristics of the SiC module.

The gate current of JFETs for the worse case at the temperature of 200 °C and the drain current of 8 A is shown in Fig. 4. The peak gate current is about 1.6 A when using a gate resistance of 5 Ω.

With the aforementioned gate drive design, the real switching waveform of the JFETs without any snubber at a test condition of 200V (dc bus voltage), 10A (drain current) and 200°C (ambient temperature) is shown in Figure 5. The switching losses including both turn-on loss and turn-off loss are calculated at each test condition. They are plotted versus the drain current at each tested temperature in Fig. 6. The switching losses in the SiC devices are almost constant (increase only slightly) when temperature increases, while the increase of the switching losses of Si devices is much more obvious. Thus, the substitution of SiC devices for Si devices will improve system efficiency, and the higher temperature and the higher frequency, the more benefits that the system will gain.

III.   EXPERIMENTS AND SiC INVERTER EFFICIENCY

A SiC inverter is built using three of the SiC modules shown in Fig. 1(a). The dimension of the inverter is 25 cm × 10 cm × 11 cm including the extruded heatsink and control boards (see Fig. 6).

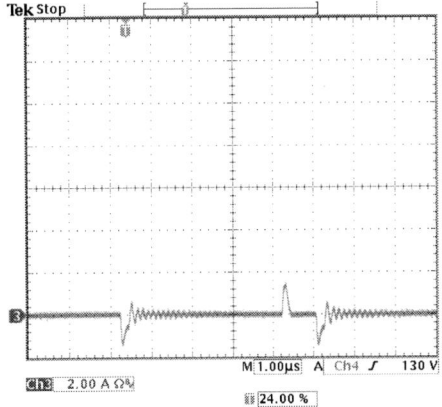

Fig. 4. Gate current waveform @200V/8A/200°C.

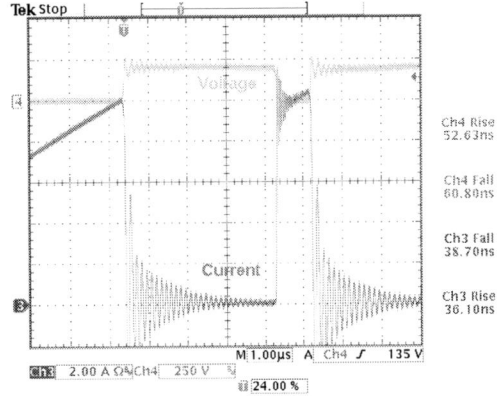

Fig 5. Switching waveform of the SiC JFETs @200V/10A/200°C

978-1-4244-4782-4/10 $26.00 © 2010 IEEE

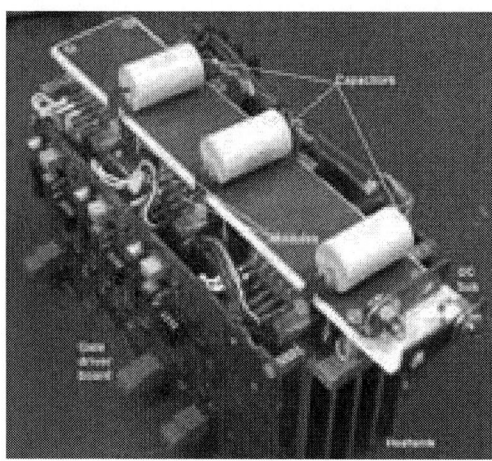

Fig. 6. Prototype SiC JFET inverter

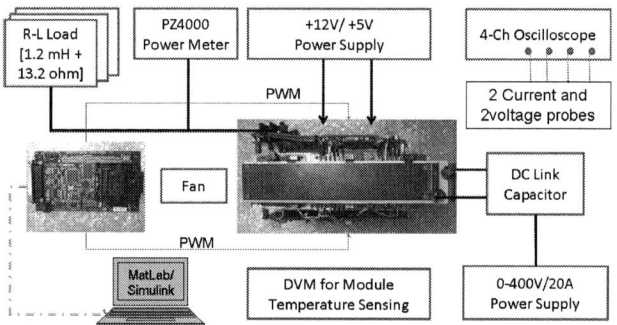

Fig. 7. Experimental system setup

The SiC inverter is tested with an RL load. As shown in Fig. 7, the inverter is connected to a DC power supply and feeds ac power to a 3-phase RL load, where R=10 Ω and L=2 mH.

The switches in the inverter are controlled by SVPWM signals generated by the DSP board. The control program is developed in Matlab Simulink, and controllable parameters can be modified on line. The input DC voltage, current and 3-phase output voltage, current are monitored and measured by oscilloscope and PZ4000. Some of experimental waveforms of input voltage, output voltage and output current are shown in Fig. 8. The input voltage in the figure is 500 V. The frequency of the output voltage is 60 Hz, and its magnitude is set by a modulation index of 0.85. As shown from Fig. 8(a)-(c), the output current on the ac side is a clean sinusoidal waveform, and closer to a fine sinusoidal waveform at increasing switching frequency.

The values of input and output current/voltage are recorded using PZ4000. The input power and output power are also calculated by PZ4000. Then, the power loss and efficiency of the inverter can be calculated based on this information. Fig. 9 shows the inverter efficiencies at 60 Hz fundamental output frequency with a modulation index of 0.85 and three different switching frequencies (10, 15, 20 kHz). The maximum efficiency, 98.2%, is achieved at a switching

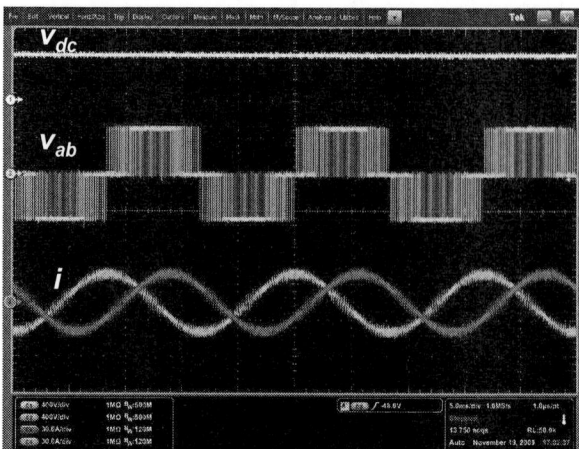

(a) at 5 kHz

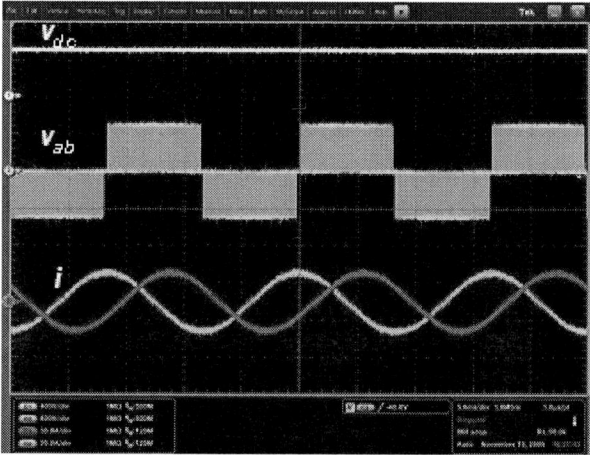

(b) at 10 kHz

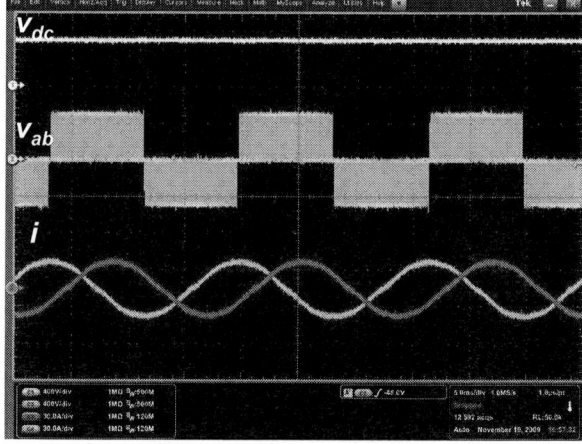

(c) at 20 kHz

Fig. 8. Experimental waveforms of the SiC inverter at 500 Vdc.

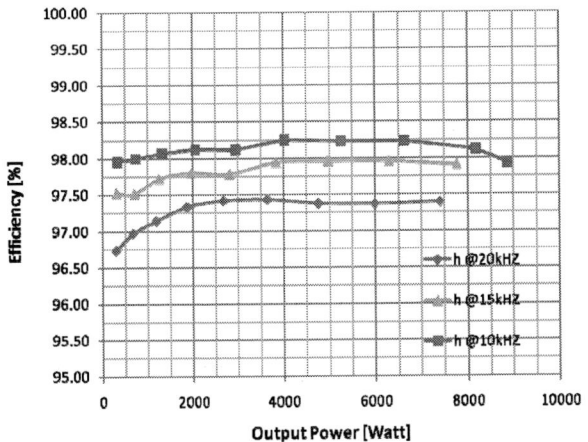

Fig. 9. Efficiency of the SiC inverter at different output power levels and switching frequencies.

frequency of 10 kHz at 4 kW output power range. The temperature inside the modules is less than 100 °C at this operating condition. If further increase the output power over about 8 kW, the module temperatures will increase quickly as monitored by the thermistors inside the package. This indicates that the cooling capability of the present heasink design is not enough for the power-level over 8 kW.

IV. CONCLUSIONS AND FUTURE WORK

SiC phase-leg modules with high-temperature package (200 °C) are designed and demonstrated. Each switch inside the modules is composed of three SiC JFETs (1200V/10V) in parallel. A SiC inverter composed of three such modules is developed and tested. It achieves efficiency as high as 98.2%. To fully utilize the device ratings and achieve better efficiency, the cooling capacity of present heatsink needs to be improved. This may involve the work of further improving the packaging technique, optimizing the thermal management, and selecting high temperature passive components.

REFERENCES

[1] European Center for Power Electronics (ECPE), "Position paper on energy efficiency – the Role of power electronics," http://www.ecpe.org/download/power_electronic/Position_Paper_Energy_Efficiency.pdf, March 2007.

[2] F. Blaabjerg, Z. Chen, "Power electronics as an enabling technology for renewable energy integration," *Journal of Power Electronics*, vol. 3, no. 2, pp. 81-89, Apr. 2003.

[3] D. Mahoney, L. Casey, "Presentation on high-megawatt converter technology workshop," *Proceedings of the High Megawatt Converter Workshop*, Gaithersburg, MD, January 24, 2007.

[4] D. Peters, "SiC power devices for industrial inverters," http://siced.com/download/CY5063ab4aX1190ed29602XY353c/Abstract_Peters.pdf?ITServ=_CY11cb9072X1225bfb7e84X30c6.

[5] H. R. Chang, E. Hanna, A. V. Radun, "Demonstration of silicon carbide (SiC)-based motor drive," *Annual Conference of the IEEE Industrial Electronics Society*, 2003, pp. 1116-1121.

[6] J. M. Hornberger, E. Cilio, B. McPherson, R. M. Schupbach, A. B. Lostetter, "A fully integrated 300°C, 4 kW, 3-Phase, SiC motor drive module," *IEEE Power Electronics Specialists Conference*, 2007, pp. 1048-1053.

[7] A. Elasser, T.P. Chow, "Silicon carbide benefits and advantages for power electronics circuits and systems," *Proceeding of IEEE*, vol. 90, no. 6, pp. 969-986, June 2002.

[8] A. R. Hefner, R. Singh, J. S. Lai, D. W. Berning, S. Bouche, and C. Chapuy, "SiC power diodes provide breakthrough performance for a wide range of applications," *IEEE Tranactions on Power Electronics*, vol. 16, pp. 273-280, Mar. 2001.

[9] L. M. Tolbert, H. Zhang, M. S. Chinthavali, B. Ozpineci, "SiC-based power converters for high temperature applications," *Materials Science Forum*, vols. 556-557, 2007, pp. 965-970.

[10] H. Zhang, L. M. Tolbert, "Efficiency of SiC JFET-based inverters," *IEEE Conference on Industrial Electronics and Applications*, Xi'an, China, May 25-27, 2009, pp. 2056-2059.

[11] D. Bergogne, H. Morel, D. Planson, D. Tournier, P. Bevilacqua, B. Allard, R. Meuret, S. Viellard, S. Rael, F. MeibodyTabar, "Towards an airborne high temperature SiC inverter," *IEEE Power Electronics Specialists Conference*, 2008, pp. 3178-3183.

[12] H. Zhang, L. M. Tolbert, B. Ozpineci, "Impact of SiC Devices in Plug-in Hybrid Electric Vehicles," *IEEE Industry Applications Society Annual Meeting*, Edmonton, Canada, October 5-9, 2008.

[13] C. Buttay, D. Planson, B, Allard, D, Bergogne, " State of the art of high temperature power electronics," *Microthrem*, Lodz, Poland, 2009.

[14] M. A. Occhionero, R.W. Adams, and K. P. Fennessy, "A New Substrate for Electronic Packaging: Aluminum Silicon Carbide (AlSiC) Composites," *Proceedings of the Forth Annual Portable by Design Conference*, Electronics Design, March 24–27, 1997, pp. 398-403.

[15] E. S. Dettmer, B. M. Romenesko, H. K. Charles, Jr., B. G. Carkhuff, and D. J. Merrill, "Steady-State Thermal Conductivity Measurements of AlN and SiC Substrate Materials," *IEEE Transactions on Components, Hybrids, and Manufacturing Technology*, 12(4), December 1989, pp. 543-547.

[16] T. R. Bloom, "The Reliability of AlN Power Hybrids Using Cu Thick Film Conductive," *Proceedings of the 40th Electronic Components and Technology Conference*, May 20–23, 1990, pp. 111-115.

[17] F. Barlow and A. Elshabini, "High-Temperature High-Power Packaging Techniques for HEV Traction Applications", ORNL/TM-2006/515, published by UT-Battelle, LLC, Oak Ridge National Laboratory, Oak Ridge, Tennessee, Nov. 2006.

[18] G. Harman, Wire Bonding in Microelectronics: Materials, Processes, Reliability, and Yield, Second Edition, McGraw-Hill Professional, June 1, 1997.

[19] D. Olsen, K. James, "Effects of Ambient Atmosphere on Aluminum — Copper Wirebond Reliability," *IEEE Transactions on Hybrids and Manufacturing Technology*, vol. 7, no. 4, December 1984, pp. 357-362.

[20] D. Palmer, "Hybrid Microcircuitry for 300°C Operation," *IEEE Transactions on Parts, Hybrids, and Packaging*, vol. 13, no. 3, September 1977, pp. 252-257.

[21] J. T. Benoit, S. Chin, R. R. Grzybowski, L. Shun-Tien, R. Jain, P. McCluskey, T. Bloom, "Wire Bond Metallurgy for High Temperature Electronics," pp. 109–113 in High Temperature Electronics Fourth International Conference, HITEC'98, June 14–18, 1998.

[22] J. T. Benoit, R. R. Grzybowski, D. B. Kerwin, "Evaluation of Aluminum Wire Bonds for High Temperature (2WC) Electronic Packaging," *Transactions of the Third International High Temperature Electronics Conference*, 1996, pp. III-17-23.

Multiphase Optimal Response Mixed-Signal Current-Programmed Mode Controller

Jurgen Alico, Aleksandar Prodic

Laboratory for Power Management and Integrated SMPS
Dept. of Electrical and Computer Engineering
University of Toronto, Toronto, M5S 3G4, ON, Canada
jurgen.alico@utoronto.ca, prodic@ele.utoronto.ca

Abstract— **This paper presents a simple and practical implementation of a proximity optimal-time (OT) response controller for multiphase interleaved dc-dc switch-mode power supplies (SMPS). This novel solution enables equal current sharing between phases not only in steady state, but also during load transients. It also achieves a bump-less transition between the transient and steady state, where the interleaved operation is resumed without any delay. To minimize calculation burden and hardware complexity, a single digital voltage loop and multiple analog current loops are combined to implement a capacitor-charge balance based optimal-time recovery algorithm. The interface between the loops is provided through a structure consisting of a sample-and-hold circuit (S&H) and a relatively slow successive approximation digital-to-analog converter (DAC) providing control signals for all the current loops. The effectiveness of the controller is demonstrated on a 2-phase, 5V-to-1.8V, 20W, interleaved buck converter operating at a 1MHz switching frequency. The experimental results verify equal current sharing under all operating conditions, bump-less transition between the modes, and demonstrate that upon a transient converter reaches new steady state in the virtually fastest possible time.**

I. INTRODUCTION

In low-power dc-dc SMPS used for supplying various portable and consumer electronic devices, the speed of the recovery time after a load transient is of a key importance. By reducing the time the SMPS needs to reach the new steady state after a transient, the size of the output filter of this cost-sensitive device can be significantly reduced. With the ultimate goal of achieving the fastest possible response for a given power stage, i.e. recovery through a single on-off action of the power switches, numerous methods have been developed. Those include trajectory path methods [1, 2], non-linear/ linear controllers [3]-[7], and methods based on the capacitor charge balance [8]-[12]. Most of these optimal-time response methods utilize recent advances in digital control of low-power SMPS to implement relatively sophisticated algorithms. While the digital optimal controllers have proven superior dynamic performance compared to analog solutions, their operation in multiphase interleaved systems has not been demonstrated in the listed previous art. Even though, in the multiphase interleaved systems [13]-[18] the speed of the transient response is one of the most important parameters.

Two problems can be correlated to the absence of the digital optimal control solutions for the multiphase systems: hardware complexity required for implementation and a lack of solution for equal current sharing during load transients. The optimal controllers usually rely on the instantaneous matching of the output load and inductor current, which, in the interleaved multiphase system, is challenging to achieve. Due to the time-shift between the phases, a large mismatch between instantaneous inductor current values usually exists.

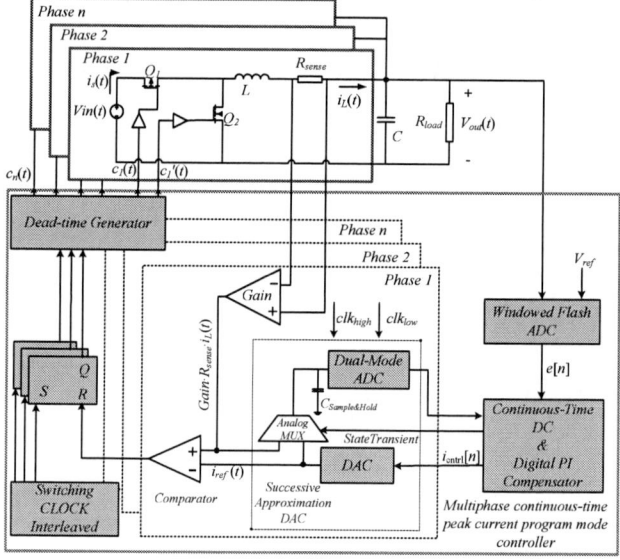

Fig. 1. Multiphase time-optimal controller regulating operation of an interleaved buck converter

---

This work of Laboratory for Power Management and Integrated SMPS is supported by Toshiba Corporation

This causes difficulties in matching the sum of inductor currents to that of the load and, equally importantly, in providing equal current sharing during transients. Furthermore, mode transition problems, causing possible instability, which have been noticed in numerous single-phase implementations [1]-[12], have not been addressed for the multiphase system.

The main goal of this paper is to introduce a mixed-signal solution for the optimal control of multiphase interleaved dc-dc SMPS. The novel system, shown in Fig. 1, provides not only recovery from load transients in virtually fastest possible time but also proper dynamic current sharing during transients and bump-less transition between the modes. Furthermore, as it will be described in later sections, due to its mixed-signal realization, at light loads, the controller allows automatic transition to more efficient pulse frequency mode (PFM) of voltage regulation.

II. SYSTEM OVERVIEW

Fig.1 shows the block diagram of the novel multiphase current program mode controller architecture. The voltage feedback loop is built as a digital block and the current loops are analog. Fig. 2 shows a more detailed diagram of the voltage loop that sets reference values for the current loops, either through a successive approximation DAC or through the sample and hold circuits, depending on the load conditions.

The controller has three distinctive modes of operation. In steady state, at medium and heavy loads, it works as a conventional mixed-signal peak current-programmed mode (CPM) system, similar to the one proposed in [19]. During transients, the continuous-time digital controller (CT-DC) similar to [10] is active, to achieve virtually fastest possible response. At light loads it behaves as a pulse frequency modulation (PFM) regulator.

A. Steady state

In steady state, the output voltage is sampled continuously by a windowed flash analog-to-digital converter (ADC), which produces an error signal $e[n]$. For small errors, i.e. $|e[n]| \leq 1$, the system is in steady state, operating as a CPM regulator. In an n-phase system, the analog references for the current loops are formed by n successive approximation DACs.

In each switching cycle, the new current reference is formed as:

$$i_{cntrl}[n] = K \cdot e[n] + i_{cntrl}[n-1], \qquad (1)$$

where $e[n]$ is the error signal, K is an integrating coefficient, and $i_{cntrl}[n-1]$ the value of the control signal of one cycle before. The analog equivalent of $i_{cntrl}[n]$ is also captured by the sample and hold circuit (S&H) of Fig.2. In the following switching cycle the sample is used to form new "one cycle before" value after being converted to its digital equivalent, with the *Dual-Mode ADC* that, in this mode, takes samples at the switching rate. The use of sample and hold capacitor makes the implementation of the DAC relatively simple and, as it will be described later, significantly reduces the complexity of the time-optimal response algorithms compared to all-digital implementation [5]-[11]. More importantly, the S&Hs allow practical implementation of the time-optimal algorithm for multiphase systems.

After being converted to its analog domain, the current control reference is compared with the representation of the inductor current. The output of the analog comparator resets the SR latch to produce duty pulses and complete the current loop [20].

B. Load transient mode

The transient is sensed by the *mode control logic* block, which triggers the optimal control mode at the time instant when the absolute value of the output voltage error exceeds 1, i.e. $|e[n]| \geq 2$. At that moment the main switch is immediately turned on or off, depending on the type of the transient. Simultaneously, a direct connection between the current sensing circuit of Fig.1 and S&H is established, through analog multiplexer, and the dual-mode ADC is set to operate in asynchronous mode reassembling operation of a continuous time processor [21]-[24]. This is performed to capture the peak/valley point of the inductor current during transients and consequently transfer it into its digital equivalent. As described in the following section, the captured peak/valley value is used by the *Optimal Δi* calculator, which sets two current references, to achieve a bump-less time-optimal response.

C. Pulse Frequency Modulation

When the current drops below the peak inductor current ripple value, the controller automatically switches to PFM mode of regulation. In this mode, the clock generator from Fig.1 is suspended, the current reference $i_{cntrl}[n]$ is set to a fixed value $i_{PFM}[n]$ (smaller than the equivalent current ripple), and the SR latch is clocked by the windowed ADC.

Each switching cycle is initiated by the ADC. It begins when the error signal, $e[n]$, becomes larger than 1 and ends when the inductor current exceeds the analog equivalent of $i_{PFM}[n]$.

It should be noted that, unlike in voltage mode digital pulse frequency regulators [25], [26], this implementation virtually does not introduce any hardware overhead, minimizing the overall cost of the controller implementation.

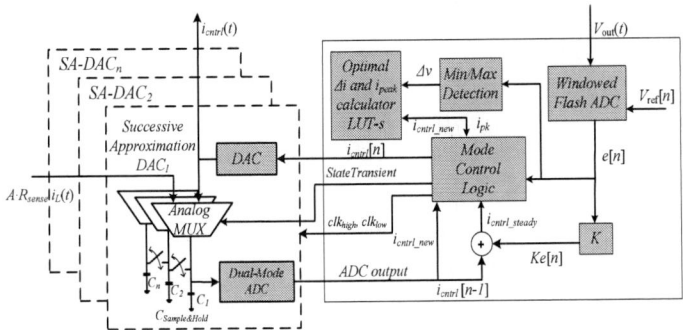

Fig. 2. Block diagram of the voltage loop

III. FAST TRANSIENT RESPONSE IN CPM BASED ON CAPACITOR CHARGE BALANCE ALGORITHM

In this section, the operation of a single-phase optimal response CPM that utilizes capacitor-charge balance algorithm is first explained. Then, it is shown how the concept can be extended to multiphase interleaved systems.

A. Single-phase operation

Compared to the voltage mode control [10], implementation of the continuous-time digital controller for CPM is significantly simpler. The reason is that only two current control reference values are necessary. Those are the peak/valley value during a transient and the new steady-state after the transient, where, as shown here, the first one does not need any calculations since it can be directly captured by the S&H of Fig.2.

The operation of the introduced system can be explained by looking at the diagrams of Fig.3. They show the key waveforms of a single-phase converter during a light-to-heavy load transient. During steady state, the fast recovery system monitors the output voltage continuously. As soon as the load transient is detected, the analog current reference i_{ref} (Fig.1) is set to its maximum allowable value, to prevent inductor saturation. Simultaneously, the block named *Optimal Δi* generator is activated. This block sets two reference values shown in Fig.3: i_{peak} - the maximum current and the new steady state value, labeled as i_{cntrl_new}. To simplify implementation, only i_{peak} is calculated while i_{cntrl_new}, which is equal to the current at voltage valley point [10], is captured with the S&H. This process is followed by the calculation of Δi, also shown in Fig. 3,

The calculated Δi is used to set the digital peak current reference as:

$$i_{peak}[n] = i_{cntrl\_new}[n] + \Delta i[n]. \qquad (2)$$

During this calculation period, the speed of the DAC is increased such that the new reference value is be set before $i_L(t)$ actually reaches it.

The actual calculation of the current difference is performed using a simple capacitor charge balance based algorithm, starting from the equation stating that the lost of the capacitor charge due to voltage variation needs to be equal to the extra charge brought by the inductor, i.e.

$$Q = C\Delta v = \frac{1}{2}\Delta i(t_{on} + t_{off}) \qquad (3)$$

For an ideal buck converter the expression for Δi becomes:

$$\Delta i = \sqrt{\frac{2C(1-D)V_{ref}\Delta v}{L}} \qquad (4)$$

It should be noted that this system completely eliminates stability problems related to transition between optimal and steady state mode. Furthermore, even when the voltage recovery is not optimal, the current reaches its proper steady

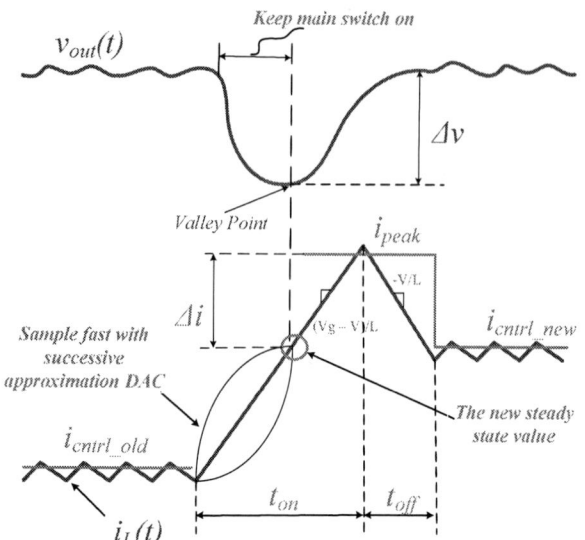

Fig. 3. The key wavforms of a single-phase power stage during a light-to-heavy load transient. Top: output voltage; Bottom: the inductor current.

state at the end of the on-off the cycle. This is because the new steady state is not determined through calculations but by capturing the actual new current value of the inductor current at peak/valley point.

B. Multiphase opertion

Compared to the voltage mode of regulation, parallel operation in CPM is simpler, since the inductor would behave as a voltage-controlled current source. Therefore, a multiphase converter behaves as a set of current sources charging a parallel combination of the output capacitor and the load resistor. However, the main obstacle in implementing an optimal-time controller for multiphase systems is maintaining balanced current sharing between the phases right after the transient. As it can be seen from Fig.4, showing waveforms of

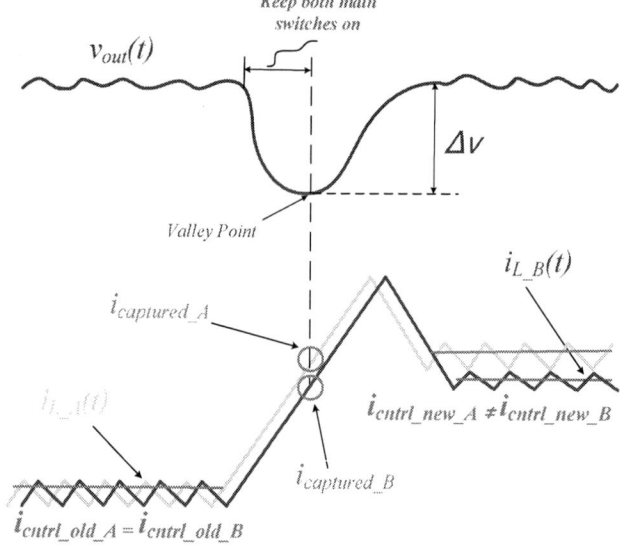

Fig. 4. The key wavforms of a two-phase interleaved buck converter during a light-to-heavy load transient. Top: output voltage of the converter; Bottom: inductor phase currents.

a two-phase interleaved system, unlike in the single-phase case, at the valley point the instantaneous inductor currents are usually not equal to their steady state values. This is caused by the time-shift of the phase currents.

To resolve this, the sample-and-hold circuits of Fig.1 are used, as well as the fact that, at the valley/peak point, the sum of phase currents is equal to the output current. When the valley/peak voltage point is detected, the inductor current of each phase is captured and saved in the respective sample and hold capacitor. Then, the outputs of all sample-and-hold capacitors are briefly short-connected such that the average value of all the sample circuits is obtained in all phases. As results, the new steady state references of all phases are the same and their sum is equal to the output current.

Averaging can be obtained in the digital domain as well, after the new control values have been digitized and divided by the number of phases. For 2^k (k =1, 2, 3..) phases, this operation is simple, since the division can be reduced to an arithmetical shift. However, doing digital averaging for other numbers is not trivial and does require a considerable time.

IV. SIMULATIONS AND EXPERIMENTAL VERIFICATION

The operation of the new controller is verified both through experiments and simulations. The system of Fig.1 was verified experimentally on a two-phase 5V-to-1.8V, 20W buck converter operating at 1MHz. The digital part of the continuous-time controller feedback loop was implemented with an Altera DE-2 FPGA board and off-shelf ADCs. Simulations are performed to verify the system operation for a larger number of phases, i.e. for three and four phases.

A. Experimental Results

Figs. 5 to 9 demonstrate load transient operation. From Fig. 5 showing response for 8 A light-to-heavy load changes, it can be seen that the two phase system recovers to a new steady state through a single on-off action of each main switch and that equal current sharing is maintained before and

after the transient. The waveforms also verify seamless, i.e. bump-less, transition between transient and steady state modes as well as operation without stability problems.

Figs. 6 and 7 show how $i_{ref}(t)$ (the analog comparator reference signal of Fig.1) changes during transient and inherent current protection feature of the system. As described in Section III, in Fig.6, it can be seen that the reference is set first to the peak value, then to the the new reference value. The waveforms of Fig.7 show the case when the peak value is limitted by the maximum allowable reference produced by the controller, to prevent inductor saturation.

Fig. 8 shows that the developed control system has the ability to recover in optimal time to successive load transients occurring during on/off sequence calculation. The first transient is a light-to-heavy load change of 2.2 A and it is followed by another 2 A step.

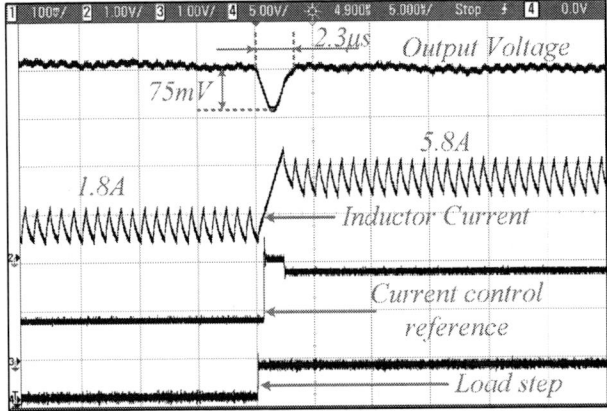

Fig.6. Key controller and power stage wavforms during a light-to-heavy load transient. Top: Output voltage (100 mV/div); Ch.2: Signal from a current sense circuit; Ch.3: Current control reference $i_{ref}(t)$; Ch.4: Load step command. The time scale is 5µs/div.

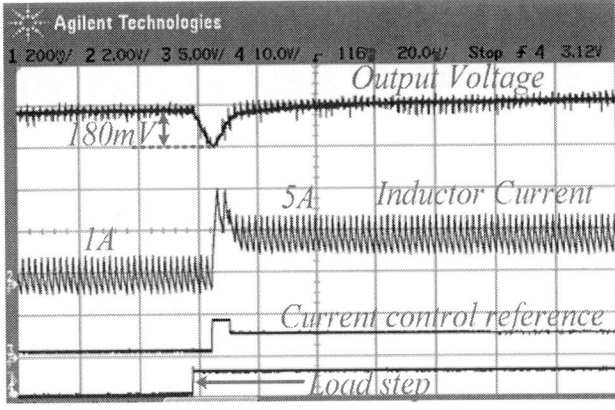

Fig. 7. Operation of current protection. The waveforms from top to bottom are: Output voltage (200mV/div); Amplified inductor current; Current control reference; Load step signal. The time scale is 20µs/div.

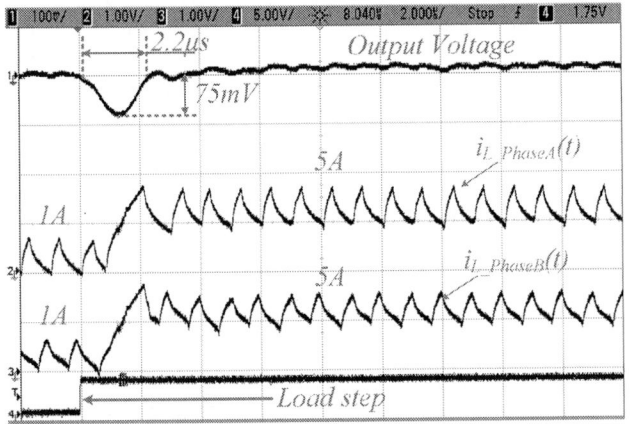

Fig.5. Transient response on a two-phase interleaved buck converter for an 8A light-to-heavy load step change. Ch.1: Output voltage (100 mV/div); Ch2 and 3: Signals from current sense circuit; Ch.4: Load step command. The time scale is 2µ/div.

978-1-4244-4782-4/10 $26.00 © 2010 IEEE   1116

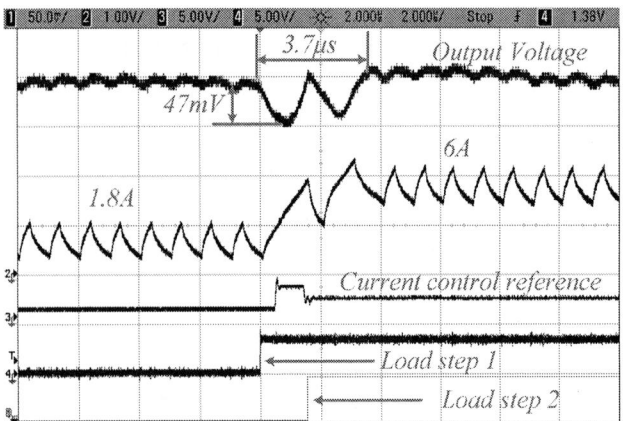

Fig. 8. Two consecutive transients using CT-DC. Top to bottom: Output voltage (50mV/div); Sensed inductor current; Current control reference; First load change command ; Second load change command. The time scale is 2µs/div.

Fig. 9 shows automatic transition between the continous current CPM mode pulse-frequency regulation. It can be seen that, upon a quick recovery from the initial change, controller "recognizes" very light load conditions and automatically switches to PFM regulation, as described in Section II.

B. Simulations

Based on the 2-phase experimental prototype a model of the converter is made and extra phases with the same current rating are added. The obtained model is then tested through simulations using the mixed-signal NCSIM simulator of the Cadence [27]. All the analog parts of the system were developed using Verilog AMS and the digital parts were in Verilog HDL. Simulation results are shown in Figs. 10 and 11. In these simulations "voltage probes" are set at the capacitors of sample and hold circuits and their numerical values are produced on the plots. It can be seen that both in the 3-phase cases and the 4-phase case equal current sharing is maintained before and after the transient and that bumpless transition is achieved in all cases.

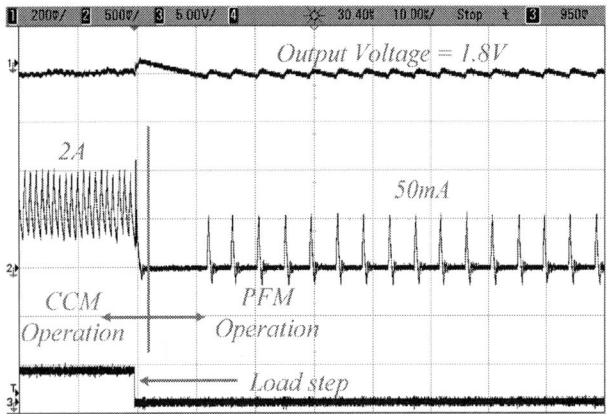

Fig.9. Transition between CPM and PFM modes for a heavy-to-light load transient. Ch.1: Output voltage (200mV/div); Ch.2: Signal from current sense circuit of a phase; Ch.3: Load step signal. The time scale is 10µs/div.

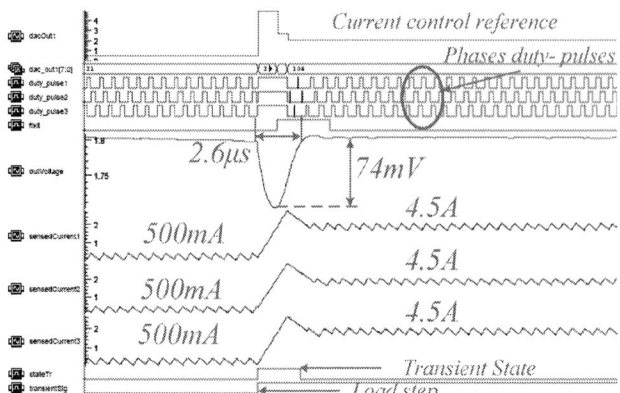

Fig. 10. Load transient rtesponse of a 3-phase interleaved system (simulation results).

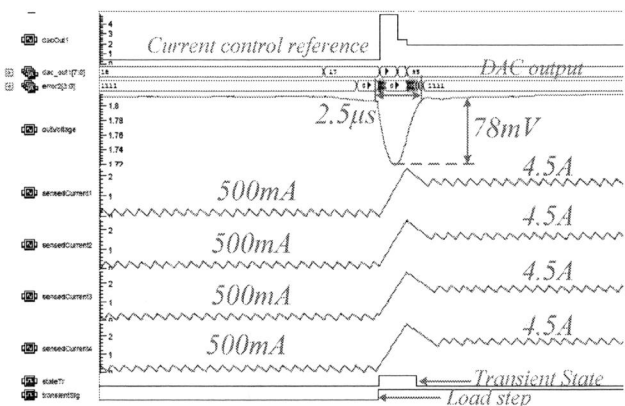

Fig. 11. Load transient rtesponse of a 3-phase interleaved system (simulation results).

V. CONCLUSION

An optimal-time control method and system for multiphase interleaved converters is introduced. The controller is implemented as a mixed-signal system, where the voltage loop is digital and current loops are implemented in analog manner. The optimal response is obtained utilizing simplified version of capacitor charge balance algorithm. To ensure proper dynamic current sharing during transients and bumpless transition between the modes of the controller, a simple solution based on short-connecting of the capacitors of sample-and-hold circuits is presented. The effectiveness of the controller is demonstrated on a two-phase experimental prototype, where both virtually the fastest possible transient response for a given power stage and equal current sharing in all operating conditions, including transients, are demonstrated.

REFERENCES

[1] R. D. Middlebrook, "Small-signal modeling of pulse-width modulated switched-mode power converters," *Proc. IEEE*, vol. 76, no. 4, pp. 343–354, Apr. 1988

[2] G. E. Pitel and P. T. Krein, "Trajectory paths for DC-DC converters and limits to performance," in *Proc. IEEE Workshop Computers in Power Electronics*, 2006, pp. 40–47.

978-1-4244-4782-4/10 $26.00 © 2010 IEEE

[3] A. Soto, P. Alou, J. A. Oliver, J. A. Cobos, and J. Uceda, "Optimum control design of PWM-buck topologies to minimize output impedance," in *Proc. IEEE APEC Conf. 2000*, Mar. 10–14, 2002, vol. 1, pp. 426–432.

[4] S. Saggini, P. Mattavelli, M. Ghioni, M. Redaelli, "Mixed-signal voltage-mode control for dc-dc converters with inherent analog derivative action," *IEEE Trans. Power Electron.*, vol. 23, Issue. 3, pp. 1485–1493, May 2008.

[5] A. Costabeber, L. Corradini, P. Mattavelli, S. Saggini, "Time Optimal, Parameters-Insensitive Digital Controller for DC-DC Buck Converters", *IEEE Power Electronics Specialists Conference*, 2008, pp. 1243-1249

[6] L. Corradini, A. Costabeber, P. Mattavelli, S. Saggini, "Time Optimal, Parameters-Insensitive Digital Controller for VRM Applications with Adaptive Voltage Positioning", *IEEE Workshop on Control and Modeling for Power Electronics (COMPEL)*, 2008, pp. 1-8

[7] A. Soto, A. P. Alou, and J. A. Cobos, "Nonlinear digital control breaks bandwidth limitations," in *Proc. IEEE APEC Conf.*, 2006, pp. 42–47.

[8] G. Feng, E. Meyer, Y. F. Liu, "A new digital control algorithm to achieve optimal dynamic performance in DC-to-DC converters," *IEEE Trans. Power Electron.*, vol. 22, Issue. 4, pp. 1489–1498, July 2007.

[9] G. Feng, W. Eberle, and Y. Liu, "High performance digital control algorithms for DC-DC converters based on the principle of capacitor charge balance," in *Proc. IEEE PESC Conf. 2006*, Jun. 18–22, 2006, pp. 1740–1743.

[10] Z. Zhao, A. Prodic, "Continuous-Time Digital Controller for High-Frequency DC-DC Converters", *IEEE Transactions on Power Electronics*, vol. 23, no. 2 March 2008, pp. 564-573

[11] E. Meyer, G. Feng, and Y. F. Liu, "Novel digital controller improves dynamic response and simplifies design process of voltage regulator module," in *Proc. IEEE APEC Conf.*, 2007, pp. 1447–1453.

[12] E. Meyer, Z. Zhang, Y.-F. Liu, "An Optimal Control Method for Buck Converters Using a Practical Capacitor Charge Balance Technique", *IEEE Transactions on Power Electronics*, Volume 23, Issue 4, July 2008, pp. 1802-1812

[13] A. Peterchev, J. Xiao, and S. Sanders, "Architecture and IC implementation of a digital VRM controller," *IEEE Transactions on Power Electronic Circuits*, vol. 18, Jan 2003.

[14] Z. Lukic, C. Blake, S. C. Huerta, A. Prodic, "Universal and Fault-Tolerant DPWM Controller IC for High-Frequency DC-DC Converters", in *Proceedings, IEEE APEC*, pp. 42–47, 2007.

[15] T. Carosa, R. Zane, D. Maksimovic, "Scalable Digital Multiphase Modulator," in *IEEE Transactions on Power Electronics*, Volume 23, Issue 4, July 2008, pp. 2201-2205

[16] Y. Zhang, R. Zane, D. Maksimovic, "System Modeling and Digital Control in Modular Masterless Multi-phase DC-DC Converters," *IEEE Power Electronics Specialists Conference(PESC)*, 2006, pp. 1-7

[17] Z. Lukic, S. M. Ahsanuzzaman, A. Prodic, Z. Zhao, "Self-Tuning Sensorless Digital Current-Mode Controller with Accurate Current Sharing for Multi-Phase DC-DC Converters", in *Proceedings, IEEE APEC*, pp. 264–268, 2009.

[18] Jaber A. Abu Qahouq, Lilly Huang, Doug Huard and Allan Hallberg, "Novel Current Sharing Schemes for Multiphase Converters with Digital Controller Implementation," in *Proc. IEEE Applied Power Electronics Conf.*, Feb. 2007, pp. 148 – 156.

[19] O. Trescases, Z. Lukic, W. Ng, and A. Prodic, "A low power mixed-signal current-mode DC-DC converter using a one-bit delta sigma DAC," in *Proceedings, IEEE Applied*, pp. 700–704, 2006.

[20] R. Erickson and D. Maksimovic, *Fundamentals of Power Electronics*. Kluwer Academic Publishers, 2001.

[21] Y. Tsividis, Mixed-Domain Systems and Signal Processing Based on Input Decomposition", *IEEE Trans. Circ. Syst.I: Regular Papers*, Vol. 53, pp.2145 –2156, Oct. 2006..

[22] Y. Tsividis. "Continuous-time digital signal processing," in *Proc. Electronics Letters*, vol. 39, pp. 1551 – 1552, Oct. 2003.

[23] Y. Tsividis, G. Cowan, Y. W. Li, and K. Shepard, "Continuous-Time DSPs, Analog/Digital Computers and Other Mixed-Domain Circuits", *Proc. of ESSCIRC Conf.* 2005. pp. 113-116.

[24] Y. Tsividis, "Digital signal processing in continuous time: a possibility for avoiding aliasing and reducing quantization error," in *Proc. IEEE ICASSP Conf.*, 2004, Vol. 2, 17-21 May 2004, pp. 589-92.

[25] K. Wang, N. Rahman, Z. Lukić, and A. Prodić , "All-digital DPWM/DPFM controller for low-power DC-DC converters" in *Proc. IEEE Applied Power Electronics Conference (APEC)*, 2006, March 2006, pp. 719-724.

[26] N. Rahman and A. Prodić, "Digital Pulse-Frequency/Pulse-Amplitude Modulator for Improving Efficiency of SMPS Operating Under Light Loads," in *Proc., IEEE Computers in Power Electronics Conference (COMPEL'06)*, July 2006, pp.149-153.

[27] Cadence design Systems available online at http://www.cadence.com

Switching Loss Analysis of Closed-Loop Gate Drive

Lihua Chen, and Fang Z. Peng
Dept. of Electric and Computer Engineering
Michigan State University
East Lansing, MI USA
Chenlih2@msu.edu

Abstract—**In this work, the IGBT turn-on and turn-off waveforms are normalized in order to analyze the closed-loop gate drive switching losses. The relationships of energy losses and controlled switching speed, voltage overshoot, current overshoot are derived. Theoretical analysis results show good agreement with experimental results. The proposed analysis methodology and results can provide guidelines for the gate drive design in real applications.**

I. Introduction

To overcome the drawbacks of the conventional gate drive, a novel closed-loop gate drive method featuring active control of switching speed for high power MOSFET and IGBTs has been reported recently [1-2]. In the proposed new method, the switching speed, specifically di/dt, is derived from a measurement of the voltage across the parasitic inductance of the power module package and a closed-loop control is employed to adaptively adjust the switching speed according to preset references. As a result, both the voltage overshoot and current overshoot can be effectively controlled in the turn-off and turn-on transients, respectively. As new features, the closed-loop gate drive fully utilizes gate drive signal information of amplitude and duty cycle (two control freedoms), and provides capabilities for the system modulator to effectively control both the power electronic circuit's steady-state and transient behaviors simultaneously.

In practice, the switching energy loss is also a major concern in applications utilizing semiconductor devices. In this work, the relationships between the switching losses and controlled switching speed, voltage overshoot, current overshoot are derived and can provide guidelines for practical gate drive designs. For example, during fault protection, what gate drive trader-off design should be made to effectively limit fault current, control voltage overshoot, and reduce excessive energy loss to avoid power device thermal breakdown.

II. Turn-off Switching Loss Analysis

The control of the IGBT turn-off speed aims to control the voltage overshoot. So, the analysis of the turn-off energy loss will focus on finding the relationship between the energy loss and voltage overshoot. To analyze the IGBT turn-off energy loss with the closed-loop gate control, normalized ideal turn-off waveforms shown in figure 1 will be used for theoretical analysis.

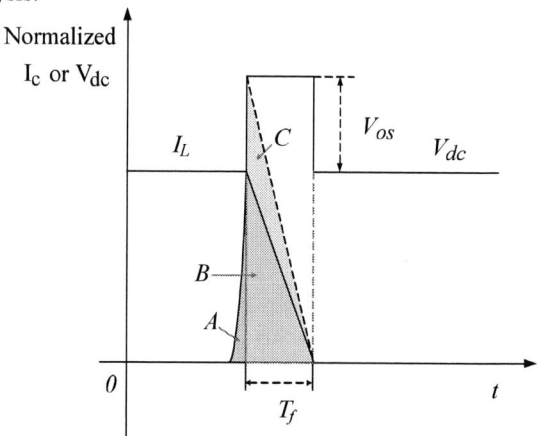

Figure 1. Ideal turn-off waveforms for energy loss analysis

With the closed-loop gate control method, the initial rate of change of collector-emitter voltage was not reduced and the energy loss associated with this, which is the area A shown in figure 1, is very small compared to the energy loss caused by the IGBT current decay. Therefore, in the following energy loss analysis, only the energy loss during the current decay is considered. The energy loss during the current decay, which is the combination of area B and C in figure 1, is expressed as:

$$E_F = \frac{1}{2}(V_{dc} + V_{os}) \times I_L \times T_f \tag{1}$$

where, T_f is the duration of collector current decay and can be expressed as:

$$T_f = \frac{I_L}{di_c / dt} \tag{2}$$

Actually area B is the energy loss if there is no voltage overshoot during IGBT turn-off, and area C is the additional energy loss caused by the voltage overshoot. The voltage overshoot is expressed as:

$$V_{OS} = L_S \times di_C / dt \qquad (3)$$

where L_s is the total parasitic inductance of the main circuit. Combining and rearranging (1) through (3), the following relation is found:

$$E_F = \frac{1}{2} L_S I_L^2 + \frac{1}{2} I_L^2 \frac{V_{dc}}{di_C / dt} \qquad (4)$$

For ideal switching, *di/dt* is infinite and, from (4), the turn-off energy loss for each turn-off of ideal switching is simplified as:

$$E_0 = \frac{1}{2} L_s I_L^2 \qquad (5)$$

Here E_0 is defined as the base energy loss, which is actually the energy stored in the parasitic inductance at the moment the IGBT is switched off.

The voltage overshoot ratio, σ, is defined as follows:

$$\sigma = \frac{V_{OS}}{V_{dc}} = \frac{L_s \times di_C / dt}{V_{dc}} \qquad (6)$$

Combing and rearranging (4) through (6) gives:

$$E_F = \frac{1}{2} L_s I_L^2 + \frac{1}{2} L_s I_L^2 \frac{1}{\sigma} = E_0 (1 + \frac{1}{\sigma}) \qquad (7)$$

The ratio of turn-off energy loss EF over base energy loss E_0 can be expressed as a function of the voltage overshoot ratio σ:

$$\frac{E_F}{E_0} = (1 + \frac{1}{\sigma}) \qquad (8)$$

Relationship (8) was plotted using Matlab software and is shown in figure 2.

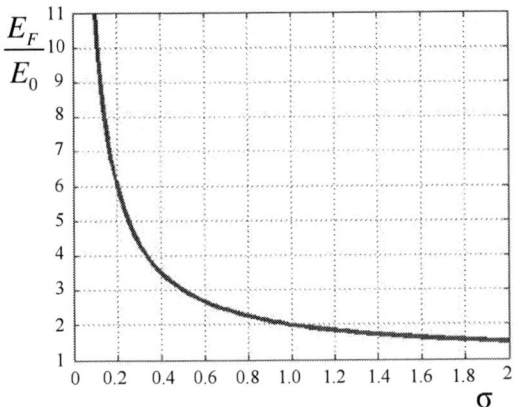

Figure 2.   Theoretical energy loss vs. voltage overshoot

The curve in figure 2 implies that the IGBT suffers from much higher turn-off energy loss if the voltage overshoot is controlled to be very small, while the overshoot is very high if the switching energy loss is controlled to be very small. In real applications, according to the operating constraints, a trade-off can be made, according to (8), in the design between the energy loss and the voltage overshoot according to this relationship.

To validate the switching loss analysis for the closed-loop gate control for IGBTs, a gate drive board has been built and tested on a chopper circuit with a laminated bus-bar structure. Detailed circuit design of the gate drive board has been described in previous work [1]. Two single IGBT modules, CM1000HA-24H from Powerex, rated at 1 kA, 1200 V, were chosen to build this circuit. Figure 3 shows the test circuit configuration and signal monitoring. A double-pulse method has been used for the test of the closed-loop gate drive. The first long pulse is used to raise the load current to a desired value, e.g., 1 kA. The second short pulse, 10μs, is used for characterizing the IGBT switching losses. The switching transients are captured by the oscilloscope with a small time scale and high resolution to verify the proposed switching loss analysis.

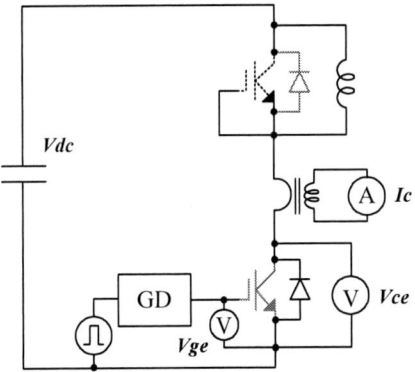

Figure 3.   Test circuit configuration and signal monitoring.

During the test, the actual turn-off energy losses, E_{Toff}, of different controlled voltage overshoots have been measured using the math function of the oscilloscope. Table I lists the voltage overshoots and turn-off energy losses with different control V_{ref}. During the tests, the voltage V_{dc} was set to 300V and the initial IGBT turn off current is conditioned to 1 kA. Since ideal switching does not exits in real applications, here the turn-off energy loss of the conventional gate drive control, E_{Coff}, was used as a base. The voltage overshoot ratio $\sigma = V_{os}/V_{dc}$, and the turn-off energy loss ratio E_{Toff}/E_{Coff} are also listed in Table I. While the voltage overshoot was controlled to be approximately equal to the dc voltage, V_{dc}, the increased turn-off energy loss is less than 10% of that of the conventional gate drive (CGD) method. However, if the voltage overshoot is controlled to be one third of V_{dc}, the turn-off energy loss was twice that of the CGD method. A plot of E_{Toff}/E_{Coff} as a function of σ is shown in figure 4. The measurement results demonstrate good agreement with the theoretical analysis shown in figure 2. Therefore, the analysis of turn-off energy loss in this work can provide guidelines for the voltage overshoot control under hard-switching operation.

TABLE I. VOLTAGE OVERSHOOT AND TURN-OFF ENERGY LOSS

V_{ref}(V)	V_{os}(V)	V_{os}/V_{dc}	E_{Toff}(J)	E_{Toff}/E_{Coff}
-2.5	100	0.333	0.355	2.09
-3	120	0.4	0.325	1.91
-4	150	0.5	0.27	1.59
-5	180	0.6	0.235	1.38
-6	200	0.667	0.215	1.26
-7	220	0.733	0.205	1.21
-8	240	0.8	0.195	1.15
-9	280	0.933	0.185	1.09
-10	310	1.03	0.18	1.06
CGD	480	1.6	0.17	1

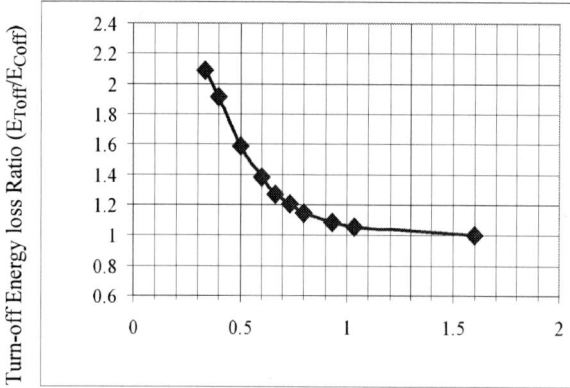

Figure 4.   Experimental energy loss vs. voltage overshoot

III.   TURN-ON SWITCHING LOSS ANALYSIS

In a similar manner, the IGBT turn-on energy loss can be analyzed. Management of the current overshoot and the associated ringing problem is the main objective in controlling the IGBT turn-on speed. However, because the diode's reverse recovery process is a very complicated phenomenon, the diode's peak reverse recovery current does not solely depend on the rate of change of the diode conduction current [3]. It is also heavily dependent on the diode's physical structure and initial conduction current. Here the current ratio, σ_1, is defined as (9), and I_{rr} is diode reverse recovery current, and I_L is initial load current right before switching on.

$$\sigma_1 = \frac{I_{rr}}{I_L} \qquad (9)$$

To analyze the IGBT turn-on energy loss with closed-loop gate control, the ideal turn-on waveforms have been normalized and sketched in figure 5 and these will be used for theoretical analysis.

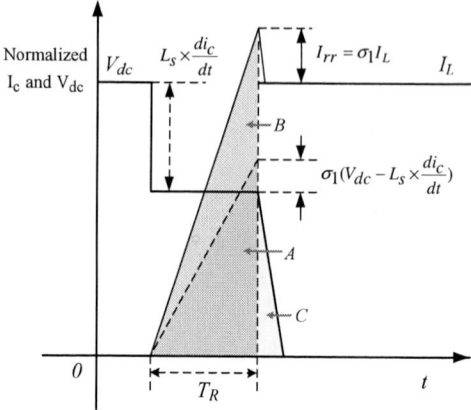

Figure 5.   Ideal IGBT turn-on waveforms for energy loss analysis

As shown in figure 5, area A represents the energy loss during the IGBT collector current rise, and area B is actually the reduced energy loss by the parasitic inductance. Area C can be used to approximately represent the energy loss during the IGBT collector voltage decay since the diode snap-off is a very fast transient. In the following energy loss analysis, only the energy loss caused by the current rise is considered. The energy loss during the current rise, which is area A in figure 5, is expressed as:

$$E_R = \frac{1}{2}(1+\sigma_1)I_L \times (1+\sigma_1)(V_{dc} - L_s \times \frac{di}{dt}) \times T_R \qquad (10)$$

where, T_R is the duration of collector current increase and can be expressed as:

$$T_R = \frac{(1+\sigma_1)I_L}{di_c/dt} \qquad (11)$$

Combining and rearranging (10) and (11), the following relation is found:

$$E_R = [\frac{1}{2}I_L^2 \frac{V_{dc}}{di_c/dt} - \frac{1}{2}L_sI_L^2] \times (1+\sigma_1)^3 \qquad (12)$$

For ideal switching, di/dt is only limited by the parasitic inductance and this is the maximum di/dt that can be achieved.

$$\left(\frac{di}{dt}\right)_{max} = \frac{V_{dc}}{L_s} \qquad (13)$$

During the IGBT turn-on, the switching speed ratio, α, is defined as:

$$\alpha = \frac{di_c/dt}{(di_c/dt)_{max}} \qquad (14)$$

Combing and rearranging (12) through (14) gives:

$$E_R = \frac{1}{2} I_L{}^2 \times L_S (\frac{1}{\alpha} - 1) \times (1 + \sigma_1)^3 \qquad (15)$$

Here E_0 is defined as the base energy loss, which is actually the energy stored in the parasitic inductance at the moment the IGBT is fully switched on. If the load current has very limited change during each switching cycle and the parasitic inductances in each switching loop has the same value, then this base energy is the same as that defined as in (5) in subsection II for IGBT turn-off analysis. However, this condition is not necessary true in real system, since the equivalent parasitic inductance for IGBT turn-on is not the same as that of the IGBT turn-off, in most cases.

$$E_0 = \frac{1}{2} L_s I_L{}^2 \qquad (16)$$

The ratio of turn-off energy loss E_R over base energy loss E_0 can be expressed as a function of the current overshoot ratio σ_1 and turn-on speed ratio α:

$$\frac{E_R}{E_0} = (\frac{1}{\alpha} - 1) \times (1 + \sigma_1)^3 \qquad (17)$$

Relationship (17) was also plotted using Matlab software and is shown in figure 6. In this plot, the range of switching speed ratio α varies from 0.1 to 1, and the curves, from bottom to top, represent the current overshoot ratio, $\sigma1$, of 0, 0.05, 0.1, 0.2, 0.3, and 0.5, respectively. The curves imply that the IGBT suffers from much higher turn-on energy loss if the switching speed is controlled to be slow. Therefore, in real applications, according to the operating constraints, a trade-off should be made in the design between the switching speed and the current overshoot in term of the additional energy loss.

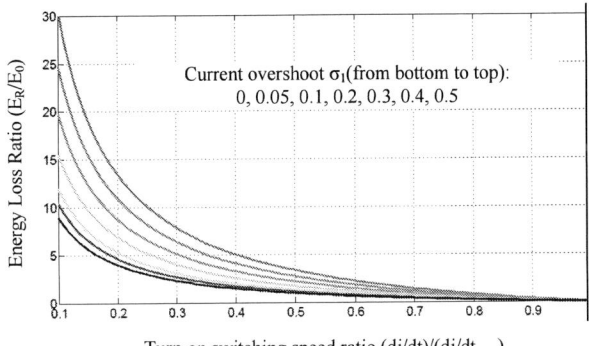

Figure 6.   Theoretical energy loss vs. switching speed

To validate this turn-on energy loss analysis, the actual turn-on energy losses, E_{Ton}, of different controlled switching speed have been measured using the math function of the oscilloscope. Table II lists the current overshoots and turn-off energy losses with different control references V_{ref}. During the tests, the turn-on current is 1 kA for all conditions. It should be pointed out here that the measured turn-on energy loss includes both that associated with the IGBT collector current increase and collector voltage decrease. Since ideal switching does not exits in real applications, here the turn-on energy loss

of conventional gate drive control, E_{Con}, was used as a base energy loss. The current overshoot ratio $\sigma_1 = I_{rr}/I_{rr(CGD)}$, and the turn-on energy loss ratio E_{Ton}/E_{Con} are also listed in Table II. The test results indicate that when the current overshoot is controlled to be low, e.g. less than 50% of that of the CGD method, the turn-off energy loss was significantly increased.

TABLE II. CURRENT OVERSHOOT AND TURN-ON ENERGY LOSS

V_{ref} (V)	di/dt (A/μs)	[di/dt]/ [di/dt(CGD)]	I_{rr} (A)	$I_{rr}/$ $I_{rr(CGD)}$	E_{Ton} (J)	$E_{Ton}/$ E_{Con}
3	340	0.213	20	0.125	620	21.38
4	590	0.369	45	0.281	350	12.07
5	730	0.456	60	0.375	260	8.97
6	940	0.588	70	0.438	200	6.90
7	1140	0.713	95	0.594	140	4.83
8	1230	0.769	110	0.688	110	3.80
10	1333	0.833	115	0.719	110	3.80
CGD	1600	1	160	1	29	1

The following two relations are plotted in figure 7 for fully understanding the IGBT turn-on control.

- E_{Ton}/E_{Con} vs. α : measured turn-on energy loss ratio vs. switching speed ratio;

- σ_1 vs. α: current overshoot ratio vs. switching speed ratio.

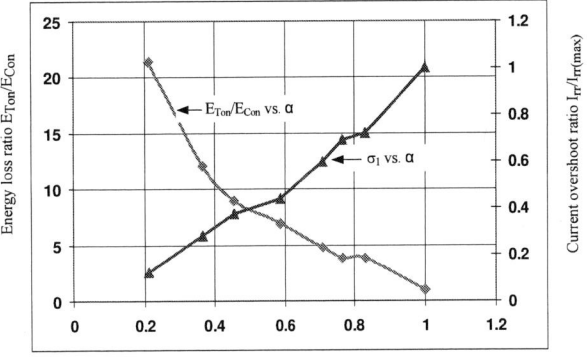

Figure 7.   Experimental *energy loss vs. switching speed*

The measurement results shown in figure 7 demonstrate that as the switching speed is controlled to be slow, the current overshoot can be effectively reduced. However, even with the effect of reduced current overshoot, the overall turn-on energy loss is monotonously increased as the switching speed decreases except that the switching speed ratio changes from 0.83 to 0.87. It should be pointed out here that in real applications [1], it found that the energy loss associated with the IGBT collector voltage V_{ce} decay increases quite a lot, since the dv/dt is very low, e.g. 500 V/μs, when compared to that of the IGBT turn-off transient, and this can cause IGBT turn-on switching loss estimation inaccuracy.

IV. CONCLUSION AND DISCUSSION

In this work, the IGBT turn-on and turn-off waveforms haven been normalized in order to analyze the closed-loop gate drive, and the relationships of energy losses and controlled switching speed, voltage overshoot, current overshoot are derived. Theoretical analysis results show good agreement with experimental results. The proposed analysis methodology and results can provide guidelines for the gate drive design in real applications.

Compared to that of IGBT turn-off energy loss vs. voltage overshoot, the analysis results of turn-on energy loss in this work imply that the IGBT sacrifices much higher turn-on energy loss if the current overshoot is controlled to be small. In practice, the parasitic inductances are inevitable and relatively large because of the large-scale size and complicated structure of high power devices and conversion systems. Fortunately the parasitic inductances serve as self-snubbers during the IGBT turn-on, and limit the switching speed. Therefore the current overshoot and associated problems are not as critical as the voltage overshoot that are caused by the fast snap-off during IGBT turn-off. However, in lower power conversion systems, due to their simple and compact structure, the parasitic inductances can be reduced to a very small value to limited voltage overshoot during turn-off. Therefore, during turn-on, a trade-off design of turn-on switching speed and energy loss with the closed-loop gate control might be more practical and meaningful.

REFERENCES

[1] L. Chen, and F. Z. Peng, "Closed-Loop Gate Drive for High Power IGBTs," in Proc. IEEE APEC'09, Feb. 15-19, 2009, Washington, DC. pp. 1331–1337.

[2] L. Chen, and F. Z. Peng, "Active Fault Protection for High Power IGBTs," in Proc. IEEE APEC'09, Feb. 15-19, 2009, Washington, DC. pp. 2050–2054.

[3] H. R. Muhammad, "Power Electronics-Circuits, Devices, and Applications," 3rd edition, Upper Saddle River, NJ, Pearson Prentice Hall, 2004.

[4] A. R. Hefner, Jr., "An Investigation of the Drive Circuit Requirements for the Power Insulated Gate Bipolar Transistor (IGBT)," IEEE Trans. on Power Electronics, vol. 6, issue 2, April 1991 pp. 208 – 219.

[5] F. Blaabjerg and J. K. Pederson, "An optimum drive and clamp circuit design with controlled switching for a snubberless PWM-VSI-IGBT inverter leg," in Proc. IEEE PESC'92, 1992, pp. 289–297.

[6] H. G. Eckel and L. Sack, "Optimization of the turn-off performance of IGBT at over-current and short-circuit current," in Proc. EPE'93, 1993, pp. 317–322.

[7] A. N. Githiari, R. J. Leedham, and P. R. Palmer, "High performance gate drives utilizing the IGBT in the active region," in Proc. IEEE PESC'96, 1996, pp. 1754–1759.

[8] S. Igarashi, S. Takizawa, M. Tabata, M. Takei, and K. Kuroki, "An Active Gate Drive Circuit for IGBTs to Realize Low-noise and Snubberless System," ISPSD, 1997, pp. 69-72.

[9] S. Musumeci, A. Raciti, A. Testa, A. Galluzzo, and M. Melito, "A new adaptive driving technique for high current gate controlled devices," in Proc. IEEE APEC, 1994, pp. 480–486.

[10] V. John, B. Suh and T.A. Lipo, "High performance active gate drive for high power IGBTs," IEEE Trans. Industry Appl., VOL. 35, NO. 5, Sept./Oct. 1999.pp.1108-1117.

[11] C. Gerster and P. Hofer, "Gate-controlled dv/dt and di/dt limitation in high power IGBT converters," EPE Journal, Vol. 5, no 3/4, January 1996, page 11-16.

[12] P. R. Palmer and H. S. Rajamani, "Active voltage control of IGBTs for high power applications," IEEE Trans. Power Electron., vol. 19, no. 4, pp. 894–901, Jul. 2004.

[13] Y. Wang, P. R. Palmer T. C. Lim, S. J. Finney, A. T. Bryant, "Real-time Optimization of IGBT/Diode Cell Switching under Active Voltage Control,", IEEE IAS conference, 8-12 Oct. 2006.

Modeling and Analysis of Closed-Loop Gate Drive

Lihua Chen, Baoming Ge, and Fang Z. Peng
Dept. of Electric and Computer Engineering
Michigan State University
East Lansing, MI USA
Chenlih2@msu.edu

Abstract—**In this work, the closed-loop gate drive is modeled and mathematically analyzed. The simulated results show good agreement with experimental results. The reported analysis method and tool can provide guidelines for the closed-loop gate drive design in real applications. This work proposed a new mathematical method for IGBT switching transient behaviors modeling and gate drive control analysis. In the proposed method, equivalent circuits are developed according different power device operating conditions and circuit behaviors are mathematically described with state equations.**

I. INTRODUCTION

A novel closed-loop gate drive method featuring active control of switching speed for high power MOSFETs and IGBTs has been recently reported [1-2]. In comparison with conventional gate drive, this new method features many advantages including closed-loop control the switching speed, specifically *di/dt*, according to preset control references; control the voltage overshoot during IGBT turn-off transient, and control the current overshoot during turn-on transient. This new gate drive method fully utilizes gate drive signal information of amplitude and duty cycle (two control freedoms), and provides capabilities for the system modulator to effectively modulate both the power electronic circuit's steady-state and transient behaviors simultaneously, which has not been found in previous literatures.

This work focuses on modeling and analysis of the proposed closed-loop gate drive. The IGBT, gate drive circuit, and main test circuit are modeled and combined; circuit parameters are extracted; the IGBT turn-on and turn-off transients are divided into several modes according operating conditions, and equivalent circuits are developed to mathematically express the circuit behaviors. The state equations for closed-lop control are derived and simulated. The analytical results show very good agreement with experimental results which have been reported in [1].

II. CIRCUIT MODELING AND PARAMETER EXTRACTION

A. Proposed Closed-loop Gate Drive and Test Circuit

Figure 1 shows the proposed closed-loop gate drive schematic circuit. For the on-state and off-state control of the IGBT, the positive inputs of the Op Amps, OP_1 and OP_2, are zero volts, and the control input signal is amplified by OP_1, OP_2 and a power stage. The IGBT gate voltage is firmly clamped to either turn-on voltage V_{CC}, in most applications +15V, or turn-off voltage V_{EE}, usually -8V to -15V. This leads to stable on and off states for the IGBT.

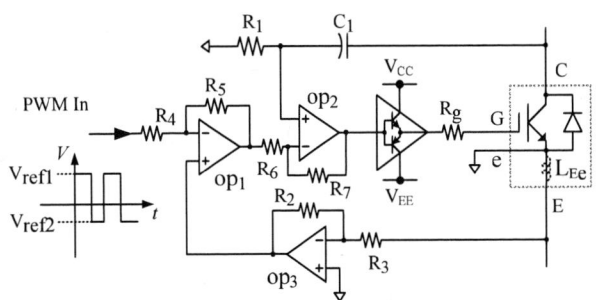

Figure 1.   Proposed cloase-loop gate drive circuit

During the IGBT turn-on and turn-off transients, a substantial voltage appears across the parasitic inductance, L_{Ee} (shown with a dotted line in Fig. 1), of the main emitter and Kelvin emitter of the IGBT module. This voltage is linearly proportional to the rate of change of the IGBT collector current, di_c/dt,. The voltage is measured and adjusted by Op Amp OP_3, and fed to the positive input of OP_1. During the switching transient, the error between the control input voltage and the measured voltage overshoot is amplified and used to adaptively adjust the IGBT gate drive voltage, V_{GG}. As a result, the IGBT switching speed can be actively manipulated by varying the control input voltage reference, V_{ref1} and V_{ref2}. Specifically, change the amplitude of V_{ref1} to control the turn-on switching speed and change the amplitude of V_{ref2} to control the turn-off switching speed [1].

A closed-loop gate drive board has been built and tested on a chopper circuit with a laminated bus-bar structure. Figure 2 shows the test circuit configuration and signal monitoring. Two single IGBT modules, CM1000HA-24H from Powerex, rated at 1kA/1200V, were chosen to build this circuit. Detailed circuit design and experimental results have been reported in previous work [1]. A double-pulse method has been used for the test of the closed-loop gate drive. The first long pulse is used to raise the load current to a desired value, e.g., 1 kA.

978-1-4244-4782-4/10 $26.00 © 2010 IEEE   1124

The second short pulse, 10μs, is used for characterizing the IGBT switching behaviors.

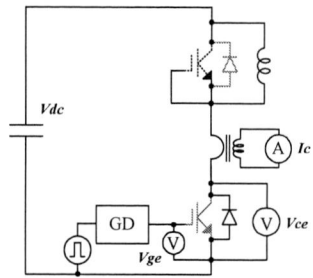

Figure 2. Test circuit configuration and signal monitoring

B. Circuit Modeling

Figure 3 shows the modeling of the proposed closed-loop gate control for IGBT with inductive load, specifically, the combination of figure 1 and figure 2. where:

- The IGBT is modeled in its active operating region, simply speaking, a gate voltage controlled current source with parasitic components considered.
- The opposite anti-parallel power diode is modeled as an ideal diode D_I, in-series with resistor R_D and parasitic inductor L_D.
- The proposed closed-gate drive is modeled as K_I through K_4, which reflect the decomposed gains of different signal paths.
- V_{ref}: PWM control signal amplitudes.
- L_s: lumped main circuit parasitic inductance.
- R_g: gate resistance.
- L_g: gate drive loop parasitic inductance.
- C_{ge}: gate-emitter parasitic capacitance.
- C_{gc}: gate-collector parasitic capacitance.
- L_e: IGBT chip to Kelvin emitter parasitic inductance.
- L_E: Kelvin emitter to IGBT emitter terminal parasitic inductance.
- L_C: IGBT chip to main collector parasitic inductance.
- R_0: Early effect equivalent resistance.
- C_0: output parasitic capacitance including that of anti-parallel diode.

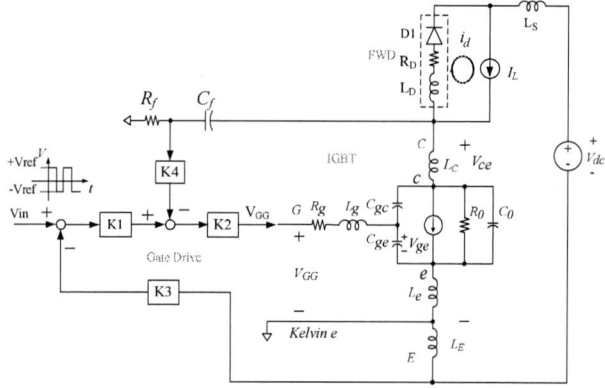

Figure 3. Modeling of the proposed closed-loop gate control

C. Circuit Parameter Extraction

During the IGBT operates in its active region, its collector current is actively controlled by its gate voltage. The transconductance, g_m, can be extracted from the manufacture power device datasheet. The output characteristics of IGBT CM1000HA-24H can be downloaded from *Powrex Inc.* website [3]. From which, the output curve can be extracted and drawn in blue line in figure 4. By using MATLAB's polynomial curve fitting function, this output characteristics can be fitted as the equation (1).

$$I_C = g_{m1}(V_{ge}-V_T) + g_{m2}(V_{ge}-V_T)^2 \qquad (1)$$

Where: $g_{m1} = 69$, $g_{m2} = 66.7$, and $V_T = 7$ V.

The fitting curve is shown in pink line in figure 4. This equation fitting curve shows good agreement with that extracted from the IGBT datasheet, and equation (1) will be used for following analysis.

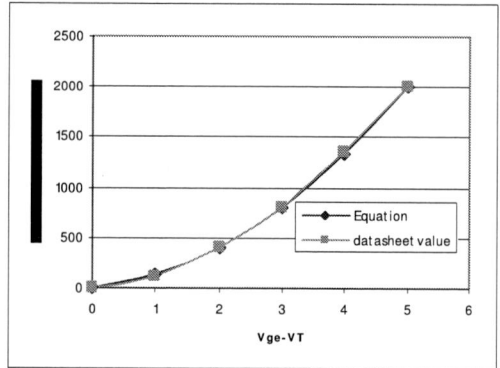

Figure 4. Polynomial curve fitting of IGBT output characteristics

Other extracted circuit and power device parameters to be used for following analysis are summarized in table I.

TABLE I. EXTRACTED CIRCUIT AND IGBT PARAMETERS

R_g	L_g	C_{ge}	C_{gc1}	C_{gc2}
4.05 Ω	100nH	100nF	100nF	75nF
C_{gc3}	L_e	L_E	V_T	R_f
0.2nF	2nH	9nH	7V	375 Ω
C_f	L_s	K_1	K_2	K_3
22pF	150nH	-5	-2	1
K_4	g_{m1}	g_{m2}		
-1	69	66.7		

III. TURN-ON CLOSED-LOOP GATE CONTROL ANALYSIS

The IGBT turn-on process starts when a turn on command is received. The initial condition can be assumed as:

978-1-4244-4782-4/10 $26.00 © 2010 IEEE   1125

- IGBT collector current I_c is zero.
- V_{ce} is equal to the dc bus voltage V_{dc}, e.g. 300V used for analysis.
- V_{ge} is clamped to turn-off voltage V_{EE}, e.g. -8V used for analysis.
- The load current freewheels through the opposite anti-parallel diode.

From the equivalent circuit shown in figure 3, the following relationships can be derived. .

$$V_{GG} = L_g \frac{di_g}{dt} + R_g i_g + v_{ge} + L_e \frac{d(i_g + i_c)}{dt} \qquad (2)$$

$$V_{GG} = L_g \frac{di_g}{dt} + R_g i_g + v_{ge} + L_e \frac{d(i_g)}{dt} + L_e \frac{d(i_c)}{dt} \qquad (3)$$

During the initial IGBT gate charge duration ($V_{ge} \leq V_T$), the equivalent circuit shown in figure 3 can be simplified and illustrated as figure 5.

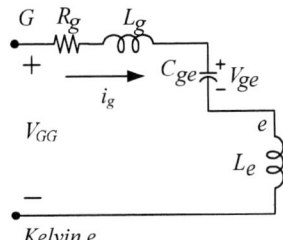

Figure 5. Equivalent circuit during the initial turn-on gate charge ($V_{ge} \leq V_T$)

when $V_{ge} \leq V_T$, $I_c = 0$, then (3) can be rewritten as (4).

$$V_{GG} = (L_g + L_e) \frac{di_g}{dt} + R_g i_g + v_{ge} \qquad (4)$$

The relationship between the gate voltage and gate current can be expressed as (5) through (7).

$$V_{ge} = V_{ge}(0) + \frac{1}{C_{ge}} \int i_g dt \qquad (5)$$

$$i_g = C_{ge} \frac{dv_{ge}}{dt} \qquad (6)$$

$$\frac{di_g}{dt} = C_{ge} \frac{d^2 v_{ge}}{d^2 t} \qquad (7)$$

Combine (4), (6) and (7), the (4) can be rewritten as (8).

$$V_{GG} = (L_g + L_e) C_{ge} \frac{d^2 v_{ge}}{d^2 t} + R_g C_{ge} \frac{dv_{ge}}{dt} + v_{ge} \qquad (8)$$

Here define:

$$Z_1 = v_{ge} \qquad (9)$$

$$Z_2 = \overset{\bullet}{Z_1} = \frac{dv_{ge}}{dt} \qquad (10)$$

$$\overset{\bullet}{Z_2} = \frac{d^2 v_{ge}}{d^2 t} \qquad (11)$$

Therefore, the equation (8) can be rewritten as (12).

$$V_{GG} = (L_g + L_e) C_{ge} \overset{\bullet}{Z_2} + R_g C_{ge} Z_2 + Z_1 \qquad (12)$$

Then the state equations can be expressed as (13) and (14).

$$\overset{\bullet}{Z_1} = Z_2 \qquad (13)$$

$$\overset{\bullet}{Z_2} = -\frac{1}{(L_g + L_e) C_{ge}} Z_1 - \frac{R_g C_{ge}}{(L_g + L_e) C_{ge}} Z_2 + \frac{-1}{(L_g + L_e) C_{ge}} V_{GG} \qquad (14)$$

This state ends when the gate voltage V_{ge} reaches the threshold voltage V_T, e.g. 7 V for the IGBT CM1000HA-24H.

When V_{ge} is greater than the V_T, equation (1) applies. Equation (1) also can be rewritten as:

$$v_{ge} = \sqrt{\frac{i_c}{g_{m2}} + \frac{g^2_{m1}}{4g^2_{m2}}} - \frac{g_{m1}}{2g_{m2}} + V_T \qquad (15)$$

From (15), the following two equations can be derived.

$$\frac{dv_{ge}}{dt} = \frac{1}{2g_{m2}} \frac{1}{\sqrt{\frac{i_c}{g_{m2}} + \frac{g^2_{m1}}{4g^2_{m2}}}} \frac{di_c}{dt} \qquad (16)$$

$$\frac{d^2 v_{ge}}{d^2 t} = \frac{1}{2g_{m2}} \frac{1}{\sqrt{\frac{i_c}{g_{m2}} + \frac{g^2_{m1}}{4g^2_{m2}}}} \frac{d^2 i_c}{d^2 t} - \frac{1}{4g^2_{m2}} \frac{1}{(\frac{i_c}{g_{m2}} + \frac{g^2_{m1}}{4g^2_{m2}})^{\frac{3}{2}}} (\frac{di_c}{dt})^2 \qquad (17)$$

Combining (3), (6) and (7), equation (3) can be rewritten as (18).

$$V_{GG} = (L_g + L_e) C_{ge} \frac{d^2 v_{ge}}{d^2 t} + R_g C_{ge} \frac{dv_{ge}}{dt} + v_{ge} + L_e \frac{di_c}{dt} \qquad (18)$$

Here define $X_1 = i_c$ and $X_2 = \overset{\bullet}{X_1} = \frac{di_c}{dt}$, combining equations (16), (17) and (18), equation (18) can be rewritten as:

$$V_{GG} = (L_g + L_e)C_{ge}[\frac{1}{2g_{m2}} \frac{1}{\sqrt{\frac{X_1}{g_{m2}} + \frac{g^2_{m1}}{4g^2_{m2}}}} \overset{\bullet}{X_2} - \frac{1}{4g^2_{m2}} \frac{1}{(\frac{X_1}{g_{m2}} + \frac{g^2_{m1}}{4g^2_{m2}})^{\frac{3}{2}}} X_2^2] +$$

$$R_g C_{ge} \frac{1}{2g_{m2}} \frac{1}{\sqrt{\frac{X_1}{g_{m2}} + \frac{g^2_{m1}}{4g^2_{m2}}}} X_2 + \sqrt{\frac{X_1}{g_{m2}} + \frac{g^2_{m1}}{4g^2_{m2}}} - \frac{g_{m1}}{2g_{m2}} + V_T + L_e X_2$$

(19)

Equation (19) can also be rearranged as (20).

$$V_{GG} - V_T = (L_g + L_e)C_{ge} \frac{1}{2g_{m2}} \frac{1}{\sqrt{\frac{X_1}{g_{m2}} + \frac{g^2_{m1}}{4g^2_{m2}}}} \overset{\bullet}{X_2} - \frac{1}{4g^2_{m2}} \frac{(L_g + L_e)C_{ge}}{(\frac{X_1}{g_{m2}} + \frac{g^2_{m1}}{4g^2_{m2}})^{\frac{3}{2}}} X_2^2 +$$

$$\left(R_g C_{ge} \frac{1}{2g_{m2}} \frac{1}{\sqrt{\frac{X_1}{g_{m2}} + \frac{g^2_{m1}}{4g^2_{m2}}}} + L_e \right) X_2 + \sqrt{\frac{X_1}{g_{m2}} + \frac{g^2_{m1}}{4g^2_{m2}}} - \frac{g_{m1}}{2g_{m2}}$$

(20)

Then the state equations can be expressed as:

$$\overset{\bullet}{X_1} = X_2 \tag{21}$$

$$\overset{\bullet}{X_2} = \frac{2g_{m2}\sqrt{\frac{X_1}{g_{m2}} + \frac{g^2_{m1}}{4g^2_{m2}}}}{(L_g + L_e)C_{ge}} \left\{ \frac{1}{4g^2_{m2}} \frac{(L_g + L_e)C_{ge}}{(\frac{X_1}{g_{m2}} + \frac{g^2_{m1}}{4g^2_{m2}})^{\frac{3}{2}}} X_2^2 - \left(R_g C_{ge} \frac{1}{2g_{m2}} \frac{1}{\sqrt{\frac{X_1}{g_{m2}} + \frac{g^2_{m1}}{4g^2_{m2}}}} + L_e \right) X_2 \right.$$

$$\left. - \sqrt{\frac{X_1}{g_{m2}} + \frac{g^2_{m1}}{4g^2_{m2}}} + \frac{g_{m1}}{2g_{m2}} + V_{GG} - V_T \right\}$$

(22)

For the closed-loop gate control, V_{GG} is governed by the following equation:

$$V_{GG} = K_1 K_2 V_{in} - K_1 K_2 K_3 L_E \frac{di_c}{dt} + K_2 K_4 R_f C_f \frac{dv_{ce}}{dt} \tag{23}$$

Where:

$$v_{ce} = V_{dc} - L_s \frac{di_c}{dt} \tag{24}$$

Therefore,

$$\frac{dv_{ce}}{dt} = -L_s \frac{d^2 i_c}{d^2 t} \tag{25}$$

Combining (23) and (25), equation (23) can be rewritten as:

$$V_{GG} = K_1 K_2 V_{in} - K_1 K_2 K_3 L_E \frac{di_c}{dt} - K_2 K_4 R_f C_f L_s \frac{d^2 i_c}{d^2 t} \tag{26}$$

Using MATLAB Simulink software, with the parameters and initial conditions derived above, the state equations (13) and (14) can be solved. Then the ending state is used as the initial condition for equations (21), (22) and (26). During which, the ending condition of (13) and (14) is transferred to the initial condition of (21), (22) and (26) by the following equation, which is derived from (1).

$$\frac{di_c}{dt} = [g_{m1} + 2g_{m2}(v_{ge} - V_{th})] \frac{dv_{ge}}{dt} \tag{27}$$

Figure 6 shows the plotted IGBT turn-on collector current waveforms with different control references of 3V, 5V, 7V, and 10V, respectively. These calculated waveforms are compatible with the test waveforms with similar control references, which have been reported in [1]. Figure 7 shows the IGBT turn-on di/dt waveforms with different control references. For comparison, the simulated turn on di/dt and that of the experimental result are summarized in table II.

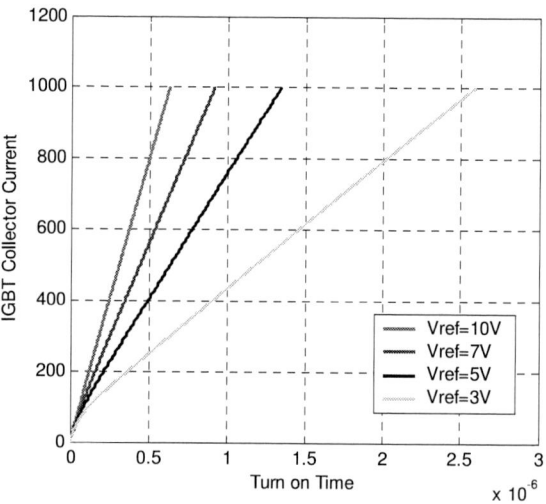

Figure 6. IGBT turn on current for different turn on control voltage references (from left to right: Vref = 10V, 7V, 5V, 3V, respectively).

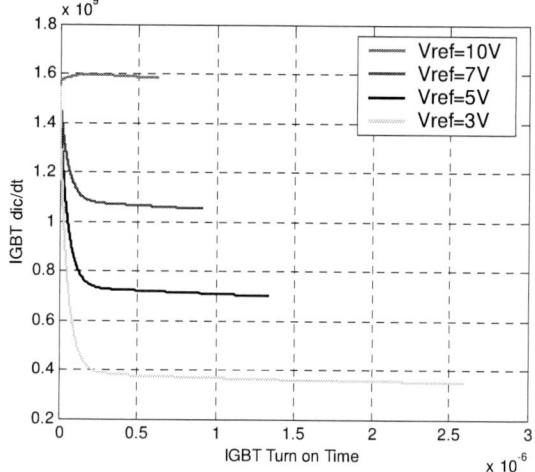

Figure 7. IGBT turn on di/dt for different turn on control voltage references (from top to bottom: Vref = 10V, 7V, 5V, 3V, respectively).

TABLE II. COMPARISON OF THE TURN-ON DI/DT RESULTS

V_{ref}(V)	3	5	7	10
Tested di/dt (A/µs)	340	730	1140	1333
Simulated di/dt (A/µs)	350	710	1060	1580

IV. TURN-OFF CLOSED-LOOP GATE CONTROL ANALYSIS

Using similar methods described above, the IGBT turn-off closed-loop control can be analyzed. The IGBT turn-off process starts when a turn off command is received. The initial condition can be assumed as:

- IGBT collector current I_c is equal to the load current I_L.
- V_{ce} is equal to the saturation voltage $V_{ce(sat)}$, e.g., 2.5V when the collector current is 1000 A for CM1000HA-24H.
- V_{ge} is clamped at the turn-on voltage V_{CC}, e.g. 15V.
- The opposite anti-parallel diode current is zero.

Three modes are defined in the following analysis:

A. Mode 1:

During the initial discharge of the IGBT gate parasitic capacitance, the equivalent circuit can be shown in figure 8. The turn on saturation voltage V_{ce} is approximately 2.5 V for the IGBT CM1000HA-24A when it conducts a current of 1000 A.

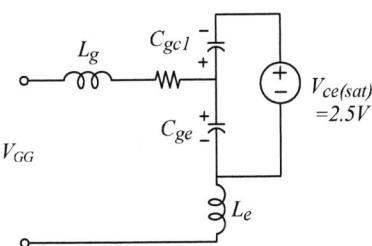

Figure 8.   Equivalent circuit during the initial turn-off gate discharge (Vge≥10.39V)

From the equivalent circuit shown in figure 8, the following equation can be derived.

$$V_{GG} = (L_g + L_e)\frac{di_g}{dt} + R_g i_g + V_{ge}(0) + \frac{1}{C_{ge}+C_{gc1}} \int i_g dt \quad (28)$$

During mode 1, V_{GG} is clamped to V_{EE}, e.g. -8V. C_{gc} is simply linearized to 100 nF for the IGBT CM1000HA-24H. This mode ends when V_{ge} decreases to 10.39V, at which time the IGBT collector current is 1000A. Then V_{ge} maintains the same value and the gate drive starts to reversely charge C_{gc}. The equivalent circuit can be modified to figure 9.

B. Mode 2:

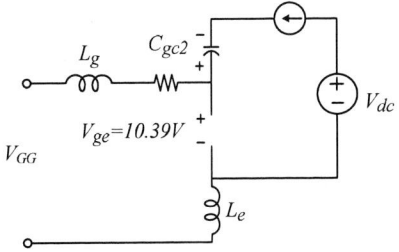

Figure 9.   Equivalent circuit during the initial turn-off gate discharge (Vge = 10.39V, Vce≤Vge)

From the equivalent circuit shown in figure 9, the following two equations can be derived.

$$V_{GG} = (L_g + L_e)\frac{di_g}{dt} + R_g i_g + V_{ge} \quad (29)$$

$$v_{gc} = V_{gc}(1) + \frac{1}{C_{gc2}} \int i_g dt \quad (30)$$

where $V_{gc(1)}$ is the ending voltage across C_{gc} of mode 1 and V_{GG} is governed by the following equation.

$$V_{GG} = -8V - K_2 K_4 V_f = -8V + -K_2 K_4 \frac{R_f C_f}{R_f C_f S + 1}(-v_{gc} + V_{ge}) \quad (31)$$

During mode 2, C_{gc} is also linearized to 75 nF for the IGBT CM1000HA-24H. This mode ends when V_{ce} rises to the same voltage of Vge, which is 10.39V and V_{gc} is 0V. Then the capacitance C_{gc} dramatically decreases. This is mainly attributed to the rapid expansion of the depletion region. When V_{ce} is greater than V_{ge}, C_{gc} can be finally linearized as 0.2 nF for the IGBT CM1000HA-24H. The equivalent circuit can be modified to figure 10.

C. Mode 3:

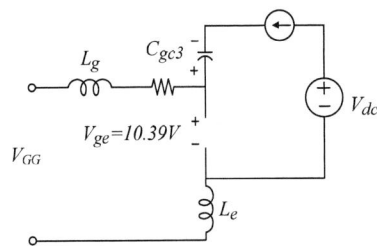

Figure 10. Equivalent circuit during the initial turn-off gate discharge (Vge=10.39V, Vce≥Vge)

During mode 3, the equations (29) and (31) still apply. Equation (30) needs to be slightly modified to (31), in which $V_{gc(2)}$ is the final voltage across C_{gc} in mode 2.

$$v_{gc} = V_{gc}(2) + \frac{1}{C_{gc2}} \int i_g dt \quad (32)$$

Mode 3 ends when V_{ce} is charged up to the dc bus voltage V_{dc}. Then the IGBT collector current starts to decay and the equations (21), (22), (26) apply until the IGBT is turned off. It should be pointed out here that the IGBT tail current phenomenon has not been considered in this analysis.

The MATLAB Simulink software is also used for the IGBT turn off analysis. Modes 1, 2, and 3 are simulated one after another. The previous ending condition is used as the initial condition of the next. Also the continuity of the current is applied to change from one mode to another. The ending condition of mode 3 is used as the initial condition of the state equations (21), (22) and (26) to solve the IGBT turn off control. During which, equations (6) and (27) are used for the transformation of these two conditions.

Figure 11 shows the IGBT turn-off collector current waveforms with different control references of -2.5V, -5V, -7V, and -10V, respectively. These waveforms are compatible with the test results, which have been reported in [1] with similar control references. Figure 12 shows the IGBT turn-off di/dt waveforms with different control references. For comparison, the simulated turn-off di/dt and that of experimental result are summarized in table III.

TABLE III. COMPARISON OF THE TURN-OFF DI/DT RESULTS

$V_{ref}(V)$	-2.5	-5	-7	-10
Tested di/dt (A/μs)	-580	-1020	-1290	-2150
Simulated di/dt (A/μs)	-600	-1050	-1300	-1900

Examining tables II and III, it can be found the simulated turn-on and turn-off di/dt is comparable with that of the tested result. This confirms that the model of the power device, parameter extraction, and the control analysis method are correct. The slight discrepancy could be attributed to measurement errors and modeling parameter inaccuracy.

Figures 7 and 12 also show that the initial di/dt is very high in both the turn-on and turn-off control. In the simulation, this high di/dt is a direct result of the initial fast charge and discharge of the gate parasitic capacitance. In real applications, this high di/dt does not show in the test waveforms [1], and this is due to the small transconductance of the IGBT when Vge is slightly greater than V_T. Also the parasitic inductance within the IGBT module can help to limit this initial high di/dt. Appropriately modifying the circuit modeling can reduce this initial di/dt discrepancy between the simulated results and those of experiments. However, this is out of the major concerns of this work and ignored here.

V. CONCLUSION AND DISCUSSION

This work proposed a new mathematical method for IGBT switching transient behaviors modeling and gate drive control analysis. In the proposed method, equivalent circuits are developed according different power device operating conditions and circuit behaviors are mathematically described with state equations and simulated one after another, during which, the previous ending condition is used as the initial condition of the next, also the IGBT current continuity is applied to move from one mode to another. The simulated results show good agreement with that of experimental results. Therefore, the reported modeling and analysis tool can provide guidelines for the closed-loop gate drive design in real applications.

Finally, it should be pointed out here that equation (1) is fitted according to the static output characteristics of the power device. The IGBT switching transients are very complicated phenomena and really difficult to be accurately described, even though this type of power device has been widely used in power conversion systems for many years. A more accurate IGBT model may be developed later and used for control analysis as the IGBT becomes more fully understood.

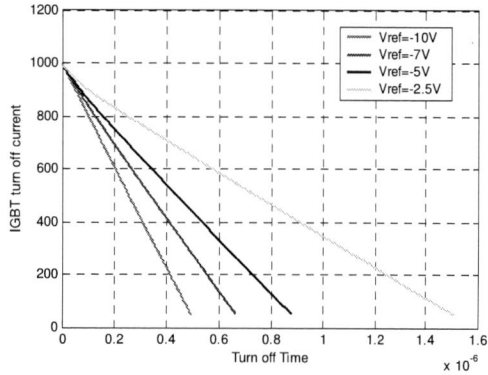

Figure 11. IGBT turn off current for different turn off control voltage references (from left to right: Vref = -10V, -7V, -5V, -2.5V, respectively).

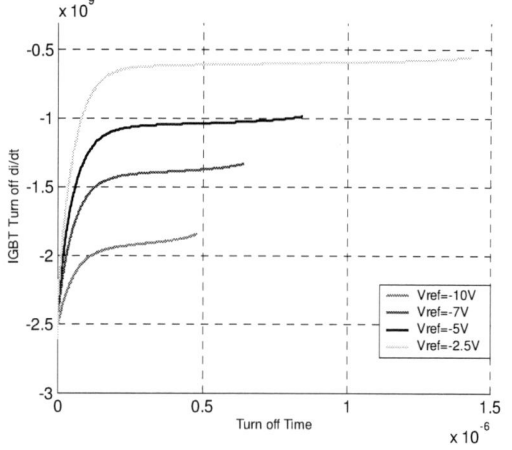

Figure 12. IGBT turn off di/dt for different turn off control voltage references (from bottom to top Vref = -10V, -7V, -5V, -2.5V, respectively).

REFERENCES

[1] L. Chen, and F. Z. Peng, "Closed-Loop Gate Drive for High Power IGBTs," in Proc. IEEE APEC'09, Feb. 15-19, 2009, Washington, DC. pp. 1331–1337.

[2] L. Chen, and F. Z. Peng, "Active Fault Protection for High Power IGBTs," in Proc. IEEE APEC'09, Feb. 15-19, 2009, Washington, DC. pp. 2050–2054.

[3] IGBT CM1000HA-24H datasheet: *http://www.pwrx.com/pwrx/docs/ cm1000ha24h.pdf.*

[4] A. R. Hefner, Jr., "An Investigation of the Drive Circuit Requirements for the Power Insulated Gate Bipolar Transistor (IGBT)," IEEE Trans. on Power Electronics, vol. 6, issue 2, April 1991 pp. 208 – 219.

[5] F. Frisina, M. Melito, S. Musurneci, R. Pagano, A. Raciti, "Transient Behavior of IGBTs Submitted to Fault Under Load Conditions," IEEE IAS Annual Meeting, October 2002, vol. 3, pp. 2182-2189.

[6] H. G. Eckel and L. Sack, "Optimization of the turn-off performance of IGBT at over-current and short-circuit current," in Proc. EPE'93, 1993, pp. 317–322.

[7] P. R. Palmer and H. S. Rajamani, "Active voltage control of IGBTs for high power applications," IEEE Trans. Power Electron., vol. 19, no. 4, pp. 894–901, Jul. 2004.

[8] A. T. Bryant, Y. Wang, S. J. Finney, T. C. Lim, and P. R. Palmer, "Numerical Optimization of an Active Voltage Controller for High-Power IGBT Converters," IEEE Tran. On Power Electronics. VOL. 22, NO. 2, MARCH 2007.

[9] H. R. Muhammad, "Power Electronics-Circuits, Devices, and Applications," 3rd edition, Upper Saddle River, NJ, Pearson Prentice Hall, 2004.

[10] V. K. Khanna, "IGBT Theory and Design," IEEE Press, Wiley-Interscience, 2003.

Black-Box Modeling of DC-DC Converters Based on Transient Response Analysis and Parametric Identification Methods

V. Valdivia, A. Barrado, A. Lázaro, C. Fernández, P. Zumel
Carlos III University of Madrid
Electronics Technology Department
Power Electronics Systems Group
Leganés, Madrid, Spain
virgilio.valdivia@uc3m.es

Abstract— Today, "black-box" behavioral models of power converters are becoming interesting for system-level analysis. These models can be used to evaluate the response of power electronics systems which are composed of commercial converters, since they can be fully parameterized by analyzing the actual converter response. In this paper a new identification method of "black box" models for DC-DC converters is presented. This method is based on the analysis of the step transient response of the converter, which is obtained by means of simple experiments. The identification of the model is carried out using well established fitting algorithms, which can be applied using commercial tools.

I. INTRODUCTION

Nowadays, power electronics systems composed of multiple converters are growing; e.g. the system of the "more electric aircraft" [1]-[5]. Modeling and simulation are essential tools for the system-level analysis. Conventional models are based on the internal structure of the converter (topology, control, etc.). However, power electronics systems are usually composed of commercial converters and because of confidentiality of manufacturers, the system designer does not have access to all the necessary data to parameterize these models. Hence behavioral models, which are based on the external behavior of the converter are required, as well as identification procedures of them (a black-box approach). These models only compute the signals of interest regarding system level analysis and can be fully parameterized by analyzing the actual converter response. First approaches to behavioral models of power converters and their identification procedures have been recently proposed [6]-[13].

References [8]-[11] propose the model shown in Fig. 1. It is a two-port g-parameters network composed of four dynamic systems: the output impedance Z_o, the audio-susceptibility G_o, the back current gain H_i and the input admittance Y_i (V_{ss} is the steady-state output voltage in case of regulated converters). These dynamic systems can exhibit either linear or nonlinear behavior. If they can be considered linear into the operating

range of the converter, they can be modeled using a single linear time invariant model "LTI", defined by a single transfer function. If nonlinear effects are found to be significant, Wiener, Hammerstein and Polytopic structures have been proposed to model them [9], [11]. These structures are composed of LTI models combined with nonlinear static functions (Fig. 2). This model computes the average behavior of the converter (below half the switching frequency, approximately) that may be accurate enough regarding system-level dynamic analysis. Thus, the large-signal behavior of the converters can be accurately modeled by means of it.

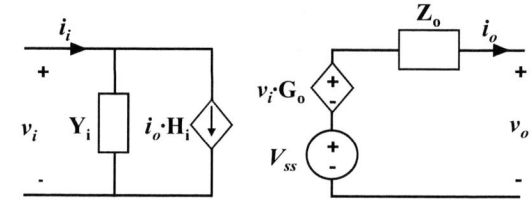

Figure 1. Black-box g-parameters model of a DC-DC converter

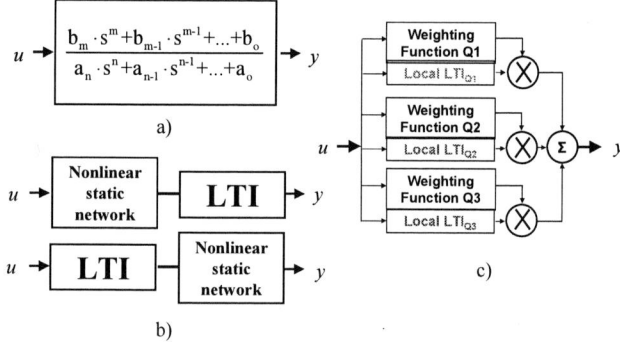

Figure 2. Structures of the dynamic systems that compose a black-box model. u ≡ input, y ≡ output. a) Linear Time Invariant model. b) Hammerstein model (up) and Wiener model (down) c) Polytopic model (example in which the working operating range is split in three partitions and the weighting functions are driven by the input signal)

Although the first approach to this structure was reported two decades ago [14], [15], the first identification procedure was recently proposed [8]. This procedure is based on the frequency response of the converter, obtained using a network analyzer. Good accuracy can be achieved by means of this approach. However, relatively expensive equipment is required; furthermore measuring the frequency response of a power converter is not exempt of complexity, being particularly complex in high power applications [16].

An alternative identification method was proposed in [12], which is based on the transient response data. The response is split as a sum of first and second order subsystems responses, and each subsystem is identified by simple equations. Nevertheless, [12] focuses on the identification of the output network of the model and the method presents some intrinsic limitations (sometimes it is difficult to split the response as a sum of subsystems responses).

In this paper an easy identification procedure of the whole model (i.e. both the input network and the output network) based on the step transient response analysis, is proposed. On the one hand, two kinds of experiments are proposed, which can be performed straightforwardly using inexpensive equipment. On the other hand, parametric identification algorithms are proposed to fit the models from the transient response data. These algorithms are well established; they can be applied using commercial tools and allow the identification of the models in a more general way. The performance of the proposed methodology (step tests and parametric identification algorithms) is discussed both through simulation and experimental results obtained on step-up and step-down PWM converters.

Along the paper, *uppercase* signal names mean steady-state values and *lowercase* signal names mean varying values.

II. PROPOSED EXPERIMENTS

The two-port-network model shown in Fig. 1 is an un-terminated model; therefore the dynamic systems composing it should only model the internal converter dynamics. In order to parameterize the model, the converter response has to be analyzed. The experiments should be chosen in such a way that the influences of the external equipments (source and load) are minimized. Furthermore the experiments have to be "informative enough", meaning that the obtained data allow an accurate fitting of the LTI model in the frequency bandwidth of interest. As a consequence, the choice of a proper input signal is a critical task. In this paper step tests are proposed, since they can be easily carried out in power electronics laboratories with low cost equipment and the measured transient waveforms can be used to accurately fit LTI models, as shown in [12], [13]. The step slew rate limits the identified frequency bandwidth.

A. Output Current Step Test

This test is performed to identify both the output impedance Z_o and the back current gain H_i. The output current step has to be carried out while the input voltage is kept constant (1). This step can be carried out either by using resistor or an electronic load (programming a current step), as

shown in Fig. 3. If an electronic load is used, it should provide a high enough slew rate, in order to allow a proper identification of the parameters of the model.

$$H_i(s) = \left. \frac{i_i(s)}{i_o(s)} \right|_{v_i(s)=0} \qquad Z_o(s) = -\left. \frac{v_o(s)}{i_o(s)} \right|_{v_i(s)=0} \qquad (1)$$

In practice, it is not possible to keep the input voltage absolutely constant, since it will be perturbed because of the output impedance of the input voltage source. Therefore, a source with low enough output impedance should be selected, in order to avoid both the audio-susceptibility influence on the v_o transient response and the input-admittance influence on the i_i transient response. If this perturbation is significant, the output impedance of the source should be characterized (e.g., by means of its step response) and after that its effect can be subtracted from the identified models, as illustrated in [10].

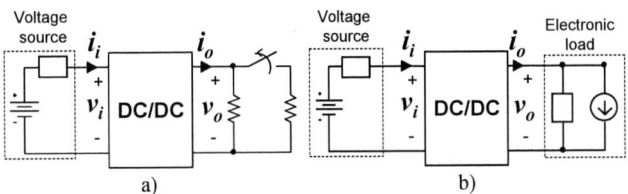

Figure 3.   Output current step tests. a) Resistive load based test. b) Electronic load based test

B. Input Voltage Step Test

This test is applied to identify both the input admittance and the audio-susceptibility. The input voltage step has to be applied while the output current is kept constant (2). Two implementations are proposed.

$$Y_i(s) = \left. \frac{i_i(s)}{v_i(s)} \right|_{i_o(s)=0} \qquad G_o(s) = \left. \frac{v_o(s)}{v_i(s)} \right|_{i_o(s)=0} \qquad (2)$$

1)   Array of diodes based test: The test shown in Fig. 4a is based on the interconnection of an array of diodes in parallel with a switch. When the switch is off, there is a voltage drop in the diodes $V_d = n \cdot V_f$, where n is the number of diodes and V_f is their threshold voltage. When the switch is turned on, a voltage step with magnitude $\approx V_d$ is generated at the input of the converter (neglecting switch conduction voltage drop). If the internal resistance of the switch is low enough, a step with a high slew rate can be performed.

2)   Two DC voltage sources based test: The test shown in Fig. 4b is based on the use of two voltage sources ($V_{dc2} > V_{dc1}$). When the switch is turned on, a voltage step with magnitude $\approx V_{dc2} - V_{dc1}$ is generated at the input of the converter. This test may be useful if the converter is fed from a high voltage bus and, as a consequence, high magnitude steps are needed.

Both tests can be carried out using either an electronic load or a passive load. If an electronic load is used, output currents perturbations may be negligible. If a passive load is selected, i_o perturbations could be significant and affect the measurements, in a similar way to that previously explained. If these perturbations are found to be significant, the load impedance should be characterized and after that its effect can be subtracted from the measurements in a similar way to explained before.

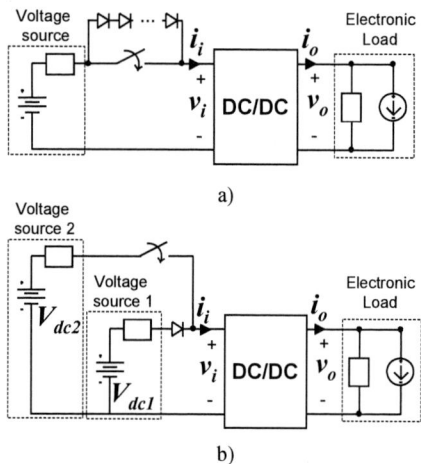

Figure 4. Input voltage step tests. a) Array of diodes based test b) Two DC voltage sources based test

C. Linear and nonlinear dynamic systems identification:

As indicated before, LTI models are used to model both linear and nonlinear dynamic systems. If the dynamic system can be supposed linear, the LTI model can be identified from a single step response. If the dynamic system is nonlinear, the nonlinear structures can be identified as follows:

1) Polytopic structure: This structure is composed of a set of local "small-signal" LTI models which are parameterized for a certain operating condition [11]. Their responses are combined by means of weighting functions. To identify each LTI model, small step tests applied on different operating conditions are performed (Fig. 5).

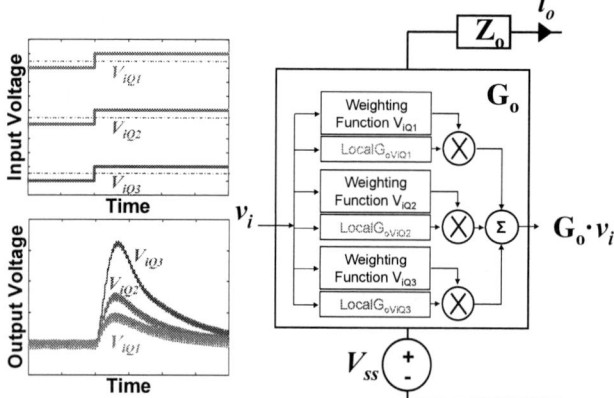

Figure 5. Audiosusceptibility modeling by means of Polytopic structure

2) Wiener and Hammerstein structures: These structures are composed of a nonlinear static function cascaded with an LTI model [7], [9], [17], [18]. First, the static nonlinear model is identified by analyzing the steady-state response of the converter at different operating points. Next, the effect of the static network is subtracted from the measured response and finally the LTI model is identified.

III. PARAMETRIC IDENTIFICATION

There are two main methods for LTI models identification: non-parametric methods and parametric methods [17]. Parametric methods are applied to identify models which are defined by a set of parameters. They are based on the least-squares method and the extensions of it (for instance, the prediction error method), and they can be applied to fit models from transient response data and frequency response data [17]-[19]. Transfer function models and state-space models are derived. Since these models are suitable for simulation, parametric methods are the proper ones. In fact, parametric methods are used to identify state-space model from frequency response data in [8]-[11].

The present paper focuses on parametric methods applied to identify transfer function models from transient response data. There are several structures of transfer function models, and they differ on how they model the dynamic relation between input, output and disturbance. The common ones are shown in Fig. 6 and discussed below.

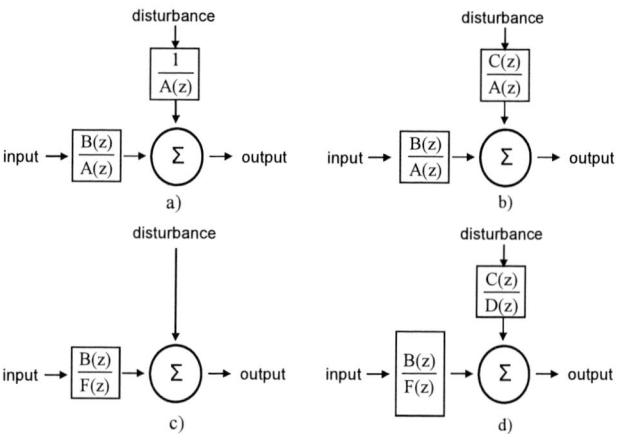

Figure 6. Common transfer function models. a) ARX model. b) ARMAX model c) OE model d) BJ model

a) Autoregressive with Exogenous Input Model ARX: This structure is the simplest one and can be idenfied by means of a linear regression. White noise characteristics are assumed for the disturbance signal. However, both the disturbance to output dynamics and the input to output dynamics cannot be separately described.

b) Autoregressive Moving Average with Exogenous Input Model ARMAX: This structure adds flexibility with respect to ARX, due to the the addition of a moving average polynomial in the disturbance to output transfer function.

978-1-4244-4782-4/10 $26.00 © 2010 IEEE

c) Output Error Model OE: ARX and ARMAX have the same denominator in both input to output and disturbance to output transfer functions. The OE model is characterized by a disturbance model independent of the input to output transfer function, what may be more natural from a physical point of view. White noise characteristics for the disturbance are assumed.

d) Bob-Jenkins Model: This structure is an enhanced version of the OE model, in terms of flexibility, since the disturbance is filtered through an autoregressive moving average filter. Therefore "colored noise" characteristics of the disturbance can be considered. Generally the identification of this structure is the most complex.

Selecting a proper model structure is the first task during the identification procedure, as well as the number of parameters of the model to be identified. A trade-off in terms of simplicity and performance (i.e. fitting accuracy) may be considered. Moreover, the order of the model to be identified has to be selected (i.e. the number of parameters which define the LTI model).

The Output Error model is suggested in this paper. Although white noise characteristics are assumed for the disturbance, just two polynomial orders (*B(z)* and *F(z)*) have to be defined before identification. Moreover the fitting accuracy is good in practice, as will be shown along the following sections. Identification algorithms of OE models have been widely discussed in literature [17]-[19] and there are commercial tools that allow the identification of them (e.g. MATLAB System Identification Toolbox [20]). An OE model can be identified using MATLAB by means of *"oe"* function, which performs the fitting using a prediction-error method PEM. Nevertheless, if the OE structure does not provide good identification results, other structures should be tried.

Notice that discrete models are derived by applying these algorithms (Fig. 6), since they handle discretized transient data. The models can be transformed into the continuous domain, looking for a simpler implementation in circuit simulators, for example, using the *"d2c"* function of MATLAB. The sampling frequency of the transient waveforms should be high enough to avoid accuracy problems when transforming the models.

IV. VALIDATION THROUGH SIMULATION

The scope of this section is to evaluate the performance of the proposed identification procedure by means of simulation. A voltage mode controlled (VMC) boost converter has been modeled (switching model) and simulated using PSIM [21] (Fig. 7). The objective is to identify the dynamic systems that compose the black-box model and to carry out a comparison of the small-signal frequency response of the identified dynamic systems with the real ones. Throughout this example only one operating point for each kind of experiment is considered. Large-signal black-box modeling issues will be discussed in the next section.

The input voltage is 10V±10%, the output voltage is 16.8V and the rated output power is 10W. The VMC boost converter exhibits a nonlinear behavior, since its dynamic response depends not only on the input voltage but also on the output current. Therefore, output current and input voltage steps with low magnitude, applied on different operating points within the operating range of the converter, should be performed to identify local-linear models for each operating point. If a narrow enough operating range is going to be considered, just one linear model may be sufficient to model each dynamic system.

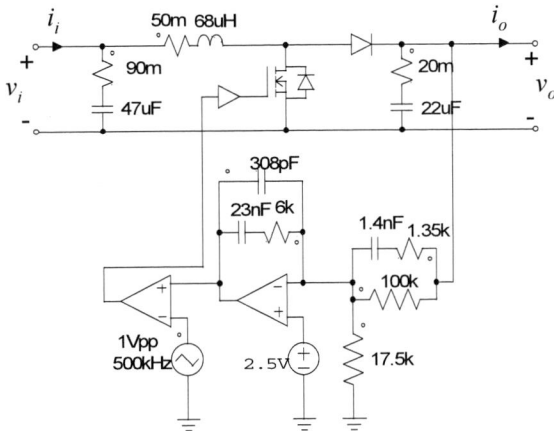

Figure 7. Voltage mode controlled boost converter simulated using PSIM

Fig. 8 shows the response of the converter when a resistive load step is carried out while v_i is kept constant. Before applying the identification algorithms, the simulated transient waveforms were filtered in order to attenuate their switching ripple content and the steady-state value of the signals (the value before the step is performed) was subtracted. After that, Z_o and H_i were identified using the *"oe"* function of MATLAB and converted to continuous time domain using the *"d2c"* function (the sampling time was high enough to avoid inaccuracy due to discretization). Fig. 9 shows the resulting fits. As can be seen, highly accurate fits are achieved.

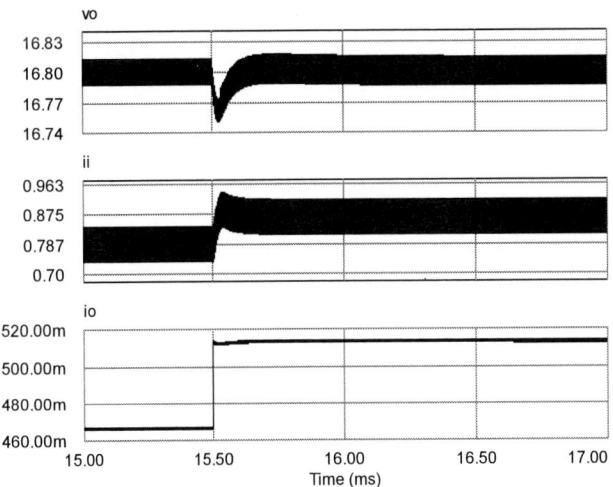

Figure 8. Resistive load step from 7.8W to 8.6W. V_i=10V

The frequency response of the identified models is compared with the small-signal frequency response of the switching model (measured using ac-sweep capabilities of PSIM) in Fig. 10. It can be seen that the frequency responses of the fitted LTI models and the switching model match properly up to half of switching frequency f_{sw}.

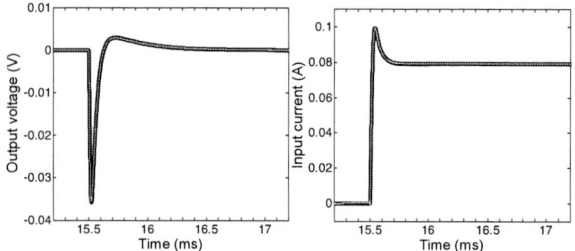

Figure 9. Resistive load step transient response fitting using MATLAB. Black lines, switching model response. Grey lines, fitted model response. The swithcing model response corresponds to Fig. 8 after filtering and substracing the steady-state value before the step is performed)

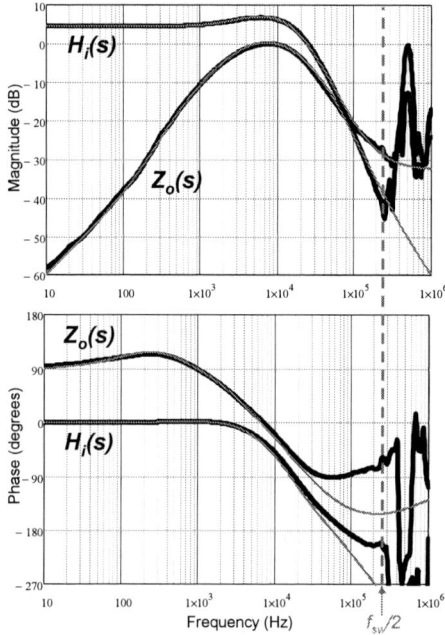

Figure 10. Comparison of the frequency response of fitted $Z_o(s)$ and $H_i(s)$ (grey lines) with the small-signal frequency response of the switching model simulated using PSIM (black lines). V_i=10V, I_o=0.49A

Fig. 11 shows the response of the converter when an input voltage step is carried out while i_o is kept constant. The corresponding fits are shown in Fig. 12 and the comparison of the frequency response of the fitted LTI models with the response of the switching model is shown in Fig. 13. It can be seen that the step responses were fitted with high accuracy and the frequency responses closely match up to half of f_{sw}. In this case, no filtering was necessary to obtain a proper fit.

Therefore, it is concluded that the proposed identification procedure allows identifying the linear time invariant systems that compose a black-box model accurately.

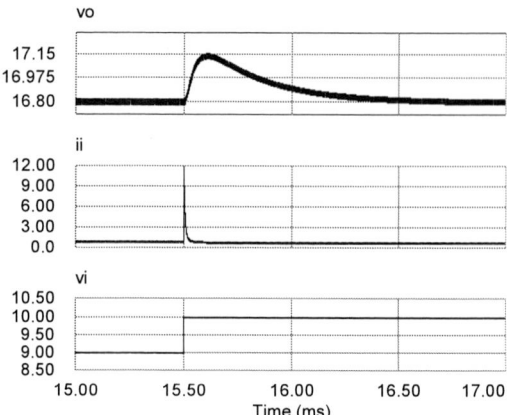

Figure 11. Input voltage step from 9V to 10V. I_o=0.51A

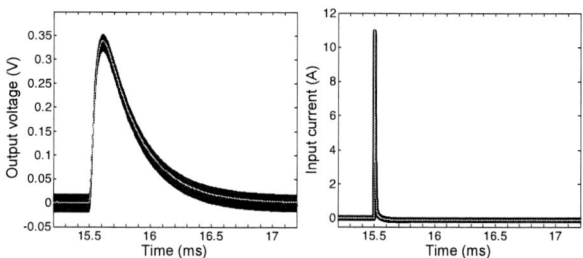

Figure 12. Input voltage step transient response fitting using MATLAB. Black lines, switching model response. Grey lines, fitted model response. The swithcing model response corresponds to Fig. 11 after substracing the steady-state value before the step is performed)

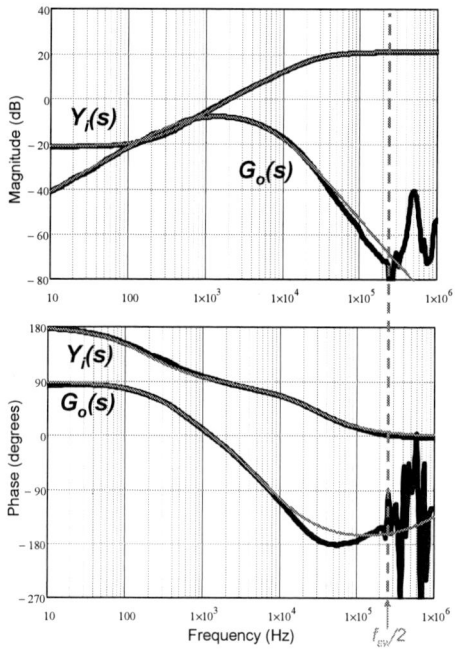

Figure 13. Comparison of the frequency response of the fitted $G_o(s)$ and $Y_i(s)$ (grey lines) and the small-signal frequency response of the switching model simulated using PSIM (black lines). V_i=9.5V, I_o=0.51A

978-1-4244-4782-4/10 $26.00 © 2010 IEEE

The fitted small-signal models are given in (3) (notice that they correspond to one operating point). Some high frequency terms (above f_{sw}) were removed from some of the identified models in order to reduce their orders. Notice that the resulting orders are relatively low (third and fourth order). Low order models are suitable when modeling a power converter for system level simulation, looking for computational cost optimization [10], [12], [14].

$$Z_o(s) = \frac{2.4 \cdot 10^4 \cdot s^3 + 1.3 \cdot 10^{11} \cdot s^2 + 1.3 \cdot 10^{14} \cdot s + 3.2 \cdot 10^{14}}{5 \cdot s^4 + 1.7 \cdot 10^6 \cdot s^3 + 1.4 \cdot 10^{11} \cdot s^2 + 2.9 \cdot 10^{15} \cdot s + 8.1 \cdot 10^{18}}$$

$$H_i(s) = \frac{2.3 \cdot 10^{-2} \cdot s^3 - 5.3 \cdot 10^4 \cdot s^2 + 3.0 \cdot 10^{10} \cdot s + 4.9 \cdot 10^{14}}{1.0 \cdot s^3 + 2.2 \cdot 10^5 \cdot s^2 + 1.6 \cdot 10^{10} \cdot s + 2.9 \cdot 10^{14}} \quad (3)$$

$$Y_i(s) = \frac{4.4 \cdot 10^9 \cdot s^2 + 1.7 \cdot 10^{14} \cdot s - 2.0 \cdot 10^{17}}{1.0 \cdot s^3 + 4.0 \cdot 10^8 \cdot s^2 + 8.1 \cdot 10^{13} \cdot s + 2.2 \cdot 10^{18}}$$

$$G_o(s) = \frac{1.3 \cdot 10^2 \cdot s^2 + 8.4 \cdot 10^8 \cdot s + 1.8 \cdot 10^9}{1.0 \cdot s^3 + 1.0 \cdot 10^5 \cdot s^2 + 2.1 \cdot 10^9 \cdot s + 6.1 \cdot 10^{12}}$$

V. EXPERIMENTAL VALIDATION

A model of Texas Instruments PTN78020WAZ step-down switching converter [22] has been identified by means of experimental measurements using the proposed procedure. The model has been implemented in PSIM and its simulated response has been compared with actual measurements. The converter has been set to provide a regulated output voltage V_o = 2.5V, and the operating ranges have been V_i = [8V-15V] and P_o = [4W-13W]. A voltage source HP6012B has been used.

Every test has been carried out setting several operating points, in order to evaluate the dynamic systems dependence on both V_i and I_o operating points.

A. Output current steps

The output current steps have been carried out using resistors. No significant Z_o dependence on neither I_o nor V_i was found. With regard to H_i, no significant dependence on I_o was found, but significant dependence on V_i has been measured. This dependence comes from the "constant power load" behavior of regulated converters (the input current decreases when the input voltage increases if the output power is kept constant). Fig. 14 shows this fact. In both diagrams, the output current step magnitude is the same, but the upper one corresponds to V_i= 8V and the lower one corresponds to V_i= 15V. The output current transient response is practically the same in both cases, but amplitude of the input current variation is inversely proportional to the input voltage level.

Therefore, Z_o has been modeled using a single LTI model. Wiener structure was selected to model H_i, in order to consider the "constant power source" behavior of the converter as well as the efficiency. The efficiency was modeled by means of a 2-D look-up-table, in order to compute it as a function of both I_o and V_i.

B. Input voltage steps

The test shown in Fig. 4a has been performed using a single diode, a MOSFET and the electronic load HP 6050A. A set of input voltage steps with small amplitude has been tested setting different I_o and V_i operating points.

No significant Y_i dependence on neither I_o nor V_i was found. Regarding G_o, no significant dependence on I_o was found, but significant dependence on V_i has been encountered (the higher V_i, the lower G_o). Fig. 15 shows the input voltage dependence analysis. In both diagrams, the input voltage step magnitude is the same, but the upper one corresponds to a 7.7V to 8.4V input voltage step and the lower one corresponds to an 11.0V to 11.7V voltage step. The input current transient response is approximately the same in both cases, but the output voltage exhibits much less overshoot in the second case.

Therefore, a single LTI model was selected to model Y_i, and a Polytopic structure was chosen to model G_o. When using a Polytopic structure, the range partitioning and the weighting functions have to be properly selected [11]. Generally, more partitions should mean higher model accuracy. Moreover, it has to be taken into account that the sum of the weighting functions has to be always equal to one. As a simple example, equally spaced partition of the input voltage range in two regions and triangular weighting functions were chosen. The input voltage steps have been applied around each weighting function maximum point. Since the magnitude of the steps is low enough, the identification of each small signal LTI model is accurate.

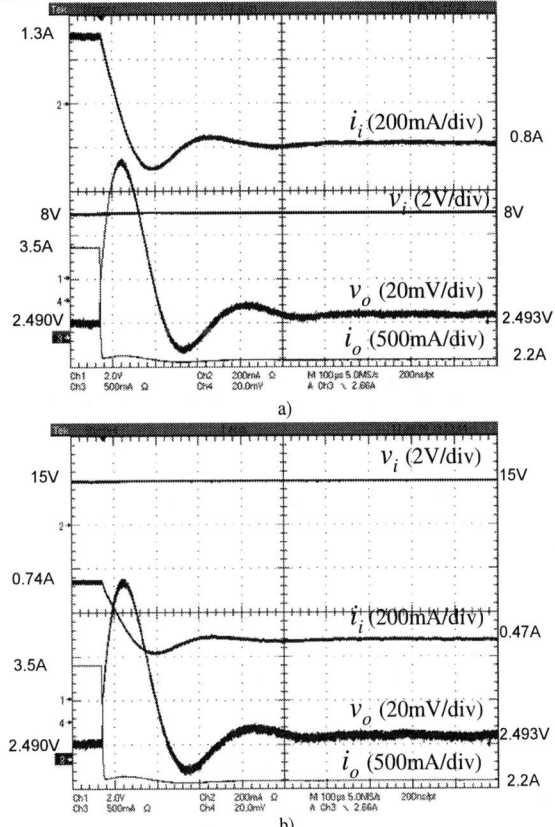

Figure 14. Output current steps transient response of Texas Instruments PTN78020WAZ from 3.5A to 2.2A. a) V_i = 8V. b) V_i = 15V

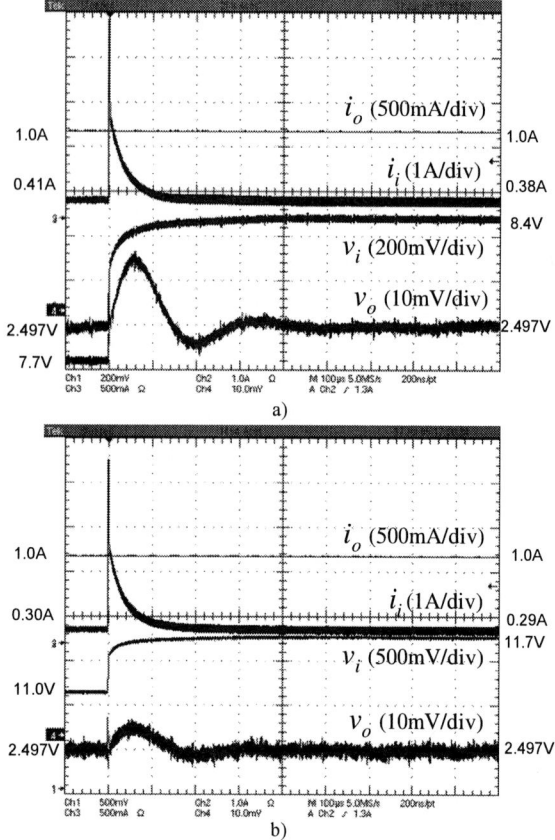

a)

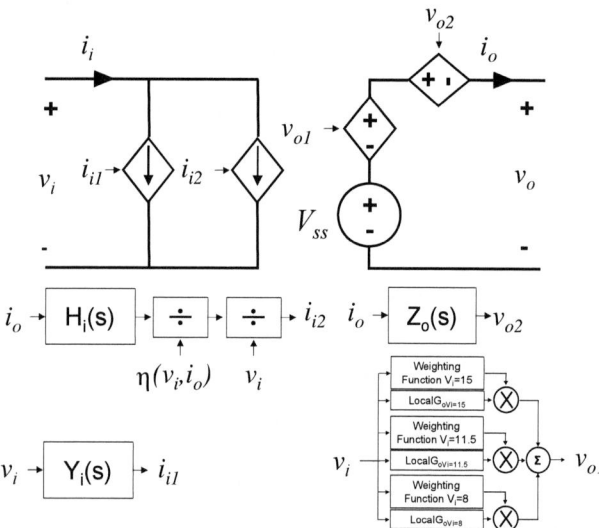

b)

Figure 15. Input voltage steps transient response of Texas Instruments PTN78020WAZ $I_o = 1A$. a) From 7.7V to 8.4V. b) From 11.0V to 11.7V

The resulting model after applying the identification procedures exposed in this paper is shown in Fig. 16. Notice that, since a Wiener model is used to model H_i, the DC gain of Y_i has to be subtracted from the identified model because the i_i static dependence on v_i is already modeled by H_i.

Figure 16. Black-Box model of Texas Instruments PTN78020WAZ

C. Model validation

In order to validate the proper behavior of the black-box model, a set of experiments have been carried out on the actual converter in the laboratory and after that these experiments have been reproduced by simulation using the identified black-box model.

Fig. 17 shows a comparison between both the actual and the simulated response when a load step is applied. A comparison between the actual response and the simulated response when an input voltage step is applied is shown in Fig. 18 (corresponding to Fig. 14a and Fig. 15a, respectively). These responses were used for identification. It can be seen that the response of the black-box model closely matches the response of the actual converter in both cases.

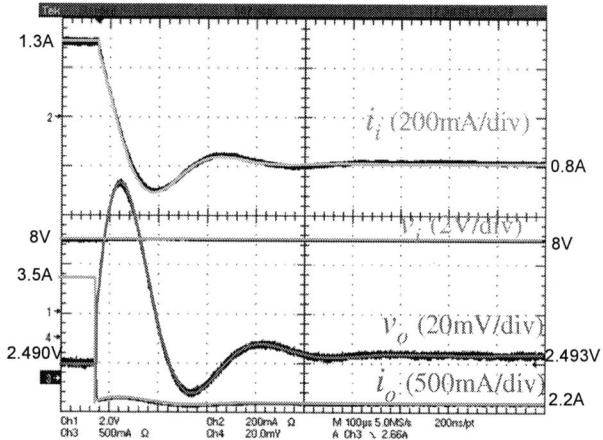

Figure 17. Comparison of the actual converter response with the black-box model simulated response when a load step is applied. Black waveforms: experimental response. Color waveforms: simulated response.

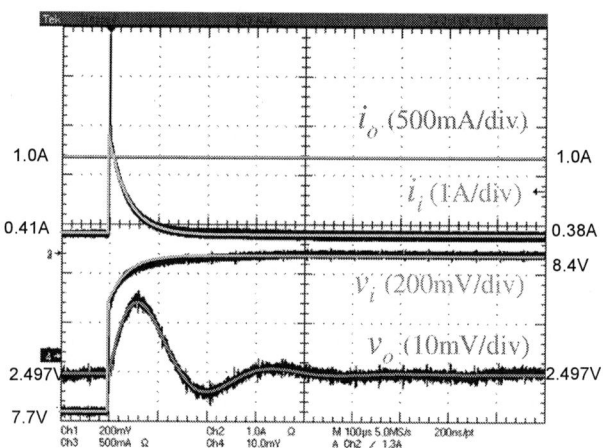

Figure 18. Comparison of the actual converter response with the black-box model simulated response when an input voltage step is applied. Black waveforms: experimental response. Color waveforms: simulated response.

Finally, Fig. 19 shows the comparison when an exponential current evolution is applied while the input voltage is perturbed. The aim of this test was to apply large-signal perturbations simultaneously on both the input voltage and the output current. These perturbations are different from those used for identification. An electronic load was placed at the output in order to generate the exponential current waveform. Additionally, a 1Ω resistor was placed in series between the DC voltage source and the input of the converter, in order to perturb the input voltage. It can be seen that both the model response and the actual converter response closely match also in this case. Therefore, both the model and the identification procedures have been experimentally validated.

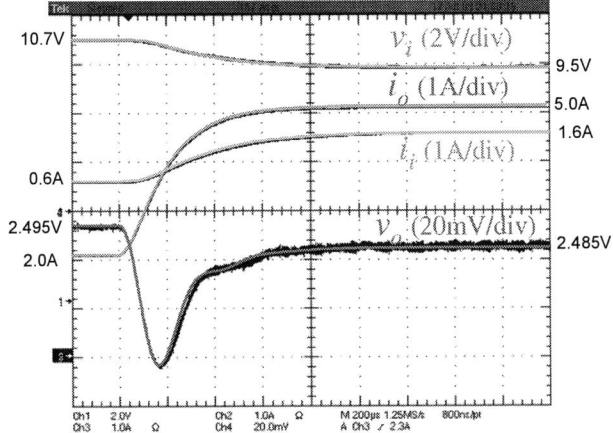

Figure 19. Comparison of the actual converter response with the black-box model simulated response when an exponential current is applied and simultaneously the input voltage is perturbed. Black waveforms: experimental response. Color waveforms: simulated response.

VI. CONCLUSIONS

In this paper a black-box modeling procedure of dc-dc converters, based on transient response analysis, has been proposed. The model is identified by means of simple step tests carried out on the converter and the step response is fitted by applying easily usable and well established fitting algorithms. Firstly, the performance of the identification procedure has been evaluated by means of simulation using a step-up switching converter. Finally, a commercial step-down switching converter has been modeled using the proposed method. The responses of the identified models and the converters have been compared in terms of both the small-signal and the large-signal response. They have matched properly in all cases. Therefore, it has been demonstrated that the proposed methodology works properly.

ACKNOWLEDGMENT

This work has been supported by the Spanish Ministry of Education and Science, through the research projects "Diseño y Modelado de Sistemas Aeroespaciales, Nivel Subsistema" (Code: DPI2006-14866) and "Sistemas de Alimentación Para Aplicaciones Embarcadas y Portátiles Basadas en Fuentes y Dispositivos de Almacenamiento de Energía Emergentes" (Code: DPI2009-12501).

REFERENCES

[1] A. Emadi, M. Ehsani, "Aircraft power systems: technology, state of the art and future trends", IEEE Aero. Electron. Systems Maganize, vol. 15, no. 1, pp. 28-32, Jan. 2000.

[2] J.A. Rosero, J.A. Ortega, E. Aldabas, and L. Romeral, "Moving towards a more electric aircraft", IEEE Aero. Electron. Systems Maganize, vol. 22, no. 3, pp. 3-9, Mar. 2007.

[3] R. Jayabalan, B. Fahimi, A. Koenig, S. Pekarek, "Applications of power electronics-based systems in vehicular technology: state of the art and future trends", in Proc. of IEEE Power Electron. Specialists Conf. (PESC), pp. 1887-1894, 20-25 Jun. 2004.

[4] D. Boroyevich, R. Burgos, R., L. Arnedo, Fei Wang; "Synthesis and Integration of Future Electronic Power Distribution Systems", in Proc. of Power Conv. Conf. - Nagoya, (PCC), pp: 1-8, 2-5 Apr. 2007.

[5] J.A. Oliver, R. Prieto, L. Laguna, J.A. Cobos, "Modeling and simulation requirements for the analysis and design of DC distributed power electronic systems", in Proc. of IEEE Power Electronics Congress (CIEP), pp. 204 - 209, 24-27 Aug. 2008.

[6] J.A. Oliver, R. Prieto, V. Romero, J.A. Cobos, "Behavioral modeling of dc-dc converters for large-signal simulation of distributed power systems", in Proc. of IEEE Appl. Power Electron. Conf. (APEC), pp. 1054 - 1060, 19-23 Mar. 2006.

[7] J. Oliver, R. Prieto, J. Cobos, P. Alou, and O. Garcia, "Hybrid Wiener-Hammerstein Structure for Grey-Box Modeling of DC DC Converters," in Proc. of IEEE Appl. Power Electron. Conf. (APEC), Page(s):280-285, 15-19 Feb. 2009.

[8] L. Arnedo, R. Burgos, F. Wang, D. Boroyevich, "Black-Box Terminal Characterization Modeling of DC-to-DC Converters", in Proc. of IEEE Appl. Power Electron. Conf. (APEC), pp. 457– 463, 25 Feb.-1 Mar. 2007.

[9] L. Arnedo, R. Burgos, D. Boroyevich, F. Wang, "System-Level Black-Box Dc to Dc Converter models", in Proc. of IEEE Appl. Power Electron. Conf. (APEC), Page(s):1476 – 1481, 15-19 Feb. 2009.

[10] L. Arnedo, D. Boroyevich, R. Burgos, F. Wang, "Un-terminated frequency response measurements and model order reduction for black-box terminal characterization models", in Proc. of IEEE Appl. Power Electron. Conf. (APEC), pp. 1054 - 1060, 24-28 Feb. 2008.

[11] L. Arnedo, D. Boroyevich, R. Burgos, F. Wang, "Polytopic Black-Box modeling of DC-DC converters", in Proc. of IEEE Power Electron. Specialists Conf. (PESC), pp. 1015 - 1021, 15-19 Jun. 2008.

[12] Valdivia, V.; Barrado, A.; Lazaro, A.; Zumel, P.; Raga, C., "Easy Modeling and Identification Procedure for "Black Box" Behavioral Models of Power Electronics Converters with Reduced Order Based on Transient Response Analysis", in Proc. of IEEE Appl. Power Electron. Conf. (APEC), pp:318–324, 15-19 Feb. 2009.

[13] Valdivia, V.; Barrado, A.; Lazaro, A.; Zumel, P.; Raga, C., "New Nonlinear Dynamic "Grey Box" Behavioral Modeling and Identification of Voltage Mode Controlled Buck Derived DC-DC Converters", in Proc. of IEEE Appl. Power Electron. Conf. (APEC), Page(s):312 – 317, 15-19 Feb. 2009.

[14] B.H. Cho, and F.C. Y. Lee, "Modeling and analysis of spacecraft power systems," IEEE Trans. Power Electron., vol. 3, no. 1, pp. 44–54, May. 1988.

[15] P.G. Maranesi, V. Tavazzi, and V. Varoli, "Two-port characterization of PWM voltage regulators at low frequencies," IEEE Trans. Ind. Electron., vol. 35, no. 3, pp. 444-450, Aug. 1988.

[16] Y. Panov, M. Jovanovic, "Practical issues of input/output impedance measurements in switching power supplies and application of measured data to stability analysis", in Proc. of IEEE Appl. Power Electron. Conf. (APEC), Page(s):1339 – 1345, Vol. 2, 6-10 March. 2005.

[17] L. Ljung, "System Identification: theory for the user, 2nd ed.", Prentice Hall, 1999.

[18] O. Nelles, "Nonlinear System Identification", Springer 2001.

[19] Y. Zhu, "Multivariable system identification for process control", Elsevier, 2001.

[20] MATLAB R2008b, http://www.mathworks.com

[21] PSIM 8.0 software by PowerSim, http://www.powersimtech.com/

[22] TIPTN78020WAZ, http://focus.ti.com/lit/ds/symlink/ptn78020w.pdf

Harmonic and Balance Compensation using Instantaneous Active and Reactive Power Control on Electric Railway Systems

A. Bueno, J. M. Aller and J. Restrepo
Grupo de Sistemas Industriales de Electrónica de Potencia
Universidad Simón Bolívar
Caracas 1080A, Venezuela

T. Habetler
School of Electrical and Computer Engineering
Georgia Institute of Tecnology
Atlanta, Georgia

Abstract—**This work presents a general filtering and unbalance compensation scheme for electric traction systems. The proposed method uses an active filter controlled with the instantaneous active and reactive power, to reduce the harmonic current distortion and the negative sequence obtained by the system under unbalanced operation in steady state. The proposed filter is evaluated using open delta (V-V) and Scott transformers in the power substation. The scheme has been simulated and experimentally validated. Experimental and simulation results show the controller advantages and the applicability of the proposed method in railway systems .**

Index Terms—**Harmonics, Active filter, Transformer, Locomotive, Traction application.**

I. INTRODUCTION

Electric traction systems for passengers and goods use different power transformer configurations, in order to feed single phase systems from the three phase supply. In general, three-phase to two single phase conversion schemes use transformers connected in open delta $(V - V)$, Scott or Le Blanc configurations [1]. In a practical application, the load associated with each single-phase circuit does not compensate each other, due to the variable demands in the transport system and railroad line profile. Also, the use of uncontrolled rectification to feed the traction load contribute to the total unbalance seen from the three phase supply. This unbalance is due mainly to the injection of current harmonics to the main three-phase system depending on the transformer connection and harmonic order [2].

It is then required the use of filters and unbalance compensators to ensure proper system operation and to raise the power quality [3].

These problems are usually addressed, in practice, with the use of passive power quality compensators such as reactive power compensation capacitors and passive filters, and they are single-phase equipment installed in each feeder of the traction substation. Usually, the coupling factor between two feeders is negligible due to the independent operation of each passive compensator. Moreover, passive equipment does not have the dynamic capability to adjust to changes in load, where over and under compensation happen frequently as a result of continuous change in load conditions.

Different active power quality compensators have been proposed in [4]–[6] to solve the unbalance problem. All of them employ two single-phase converters that have a common DC bus and the simultaneous compensation of harmonic content and unbalance can not be achieved with these schemes. Also, when the compensation is made from the single phase side, the instantaneous active and reactive power definition is difficult to use in the compensation of harmonics and power unbalance [7] [8].

In this work a compensation scheme is proposed to provide simultaneous correction of harmonic content and load unbalance for railroad systems using open delta or Scott transformers in the power substation. This scheme is based on the instantaneous active and reactive power description of the system [9], using space vector representation of the state variables, and the application of direct power control (DPC) to attain the required correction by minimizing a cost function obtained from the instantaneous active and reactive mismatch [10]–[12].

The control strategies presented in this work are both, simulated using a state variables model representation and experimentally validated using a DSP based modular power electronic system able to emulate the electric traction system operating conditions, the open delta, the Scott transformer, the filtering and the load balancing converters [13].

The generality of the proposed filtering technique using instantaneous active and reactive power can be extended to any other transformer configuration in the power substation. Multilevel converter technology can facilitate the industrial implementation because reduces the specifications of the power electronics switches and the voltage stress ($\frac{dv}{dt}$) on the magnetic components like coupling transformers and/or inductors [14].

II. HARMONIC AND UNBALANCE COMPENSATION SYSTEM

Figure 1 shows the proposed control scheme. A shunt active filter is used, directly connected to the power system using a voltage rising transformer. The active filter uses a power converter configured as an active three-phase PWM rectifier, connected to the three-phase side.

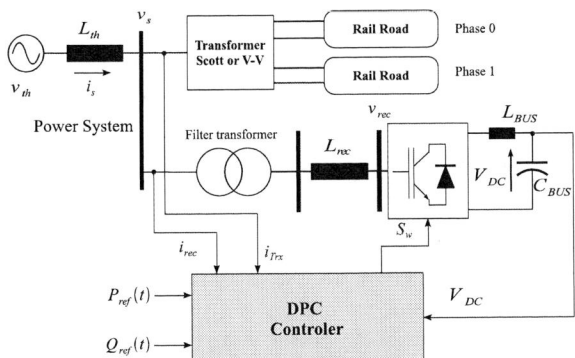

Figure 1: Proposed compensation scheme

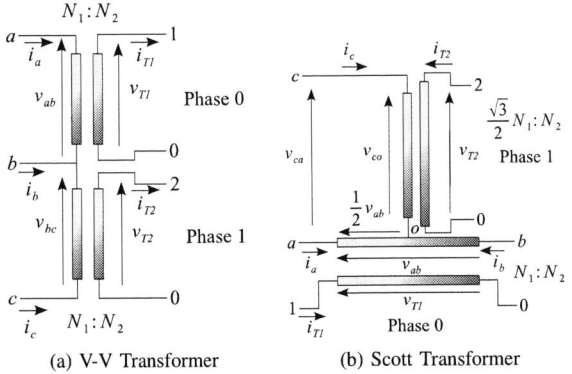

(a) V-V Transformer   (b) Scott Transformer

Figure 2: Proposed compensation scheme.

Figure 2 shows the open delta $(V - V)$ and Scott transformers used frequently to connect a traction substation to the electric grid. These connection schemes generate two single-phase networks from the three-phase power system. Each single-phase circuit is used to feed a 60 to 100 km rail track.

The simulation of the steady state and dynamic behavior for the traction system under unbalance conditions and with harmonic current injection uses a space vector model of the open delta and Scott transformer, uncoupling the differential equations in the transformer model [7]. Additionally, the filter and its control have been modeled using a space vector representation [15].

The power invariant space vector transformation is defined as,

$$\vec{x} = \sqrt{\frac{2}{3}} \left(x_a(t) + \alpha x_b(t) + \alpha^2 x_c(t) \right) \quad \alpha = e^{j\frac{2\pi}{3}} \quad (1)$$

A. V-V Transformer space vector model

For the ideal $V - V$ transformer configuration shown in Fig. 2a, its model can be obtained considering the transformer ratio and using Ampere and Faraday Laws [1]:

$$v_{ab} = \frac{N_1}{N_2} v_{T1}; \; v_{bc} = \frac{N_1}{N_2} v_{T2}; \; i_a = \frac{N_2}{N_1} i_{T1}; \; i_c = \frac{N_2}{N_1} i_{T2} \quad (2)$$

The voltage and current space vectors calculated in the transformer's primary winding as function of the secondary winding voltages and currents are:

$$\vec{v}_s = \sqrt{\frac{2}{3}} \frac{N_1}{N_2} \left(v_{T1} - \alpha^2 v_{T2} \right)$$
$$\vec{i}_s = \sqrt{\frac{2}{3}} \frac{N_2}{N_1} \left[(1 - \alpha) i_{T1} + \left(\alpha - \alpha^2 \right) i_{T2} \right] \quad (3)$$

B. Scott Transformer space vector model

For the ideal Scott transformer shown in figure 2b, its model can be obtained considering the transformer ratio and using Ampere and Faraday Laws [1]:

$$v_{ab} = \frac{N_1}{N_2} v_{T1}; \; v_{co} = \frac{\sqrt{3}}{2} \frac{N_1}{N_2} v_{T2};$$
$$\frac{\sqrt{3}}{2} \frac{N_1}{N_2} i_c = i_{T2}; \; \frac{1}{2} \frac{N_1}{N_2} (i_a - i_b) = i_{T1} \quad (4)$$

The voltage and current space vectors calculated in the transformer's primary winding as function of the secondary winding voltages and currents are:

$$\vec{v}_s = \sqrt{\frac{3}{2}} \frac{N_1}{N_2} \frac{1}{1-\alpha^2} \left(v_{T1} - j v_{T2} \right)$$
$$\vec{i}_s = \sqrt{\frac{2}{3}} \frac{N_2}{N_1} \left((1 - \alpha) i_{T1} + \sqrt{3} \alpha^2 i_{T2} \right) \quad (5)$$

C. Active and reactive power control

The DPC controller is based in the instantaneous apparent power from the current and voltage space vectors definitions [7]:

$$\vec{s} = \vec{v}_s \cdot \vec{i}_s^* = (v_{s\alpha} + j v_{s\beta}) \cdot (i_{s\alpha} + j i_{s\beta})^* = p + jq \quad (6)$$

From Fig. 1, the active filter can be modeled as,

$$\vec{v}_s = \vec{v}_r + R\vec{i}_s + L\frac{d\vec{i}_s}{dt} \quad (7)$$

A discrete time version of this equation is obtained by replacing the derivative with a first order Euler approximation, and the estimated supply current for the next control cycle becomes

$$\widehat{\vec{i}_s}(k + 1) = \vec{i}_s(k) + \Delta\widehat{\vec{i}_s}(k) \quad (8)$$

where

$$\Delta\widehat{\vec{i}_s}(k) = \frac{T_s}{\hat{L}} \left[\widehat{\vec{v}}_s(k) - \widehat{\vec{v}}_r(k) - \hat{R}\vec{i}_s(k) \right] \quad (9)$$

From (6), the estimated active and reactive power for the next sampling period can be written as,

$$\vec{s}(k + 1) = \vec{s}(k) + \Delta\vec{s}(k) = \vec{s}(k) + \Delta p(k) + j\Delta q(k) =$$
$$= \vec{v}_s(k) \cdot \vec{i}_s(k)^* + \Delta\widehat{\vec{v}}_s(k) \cdot \vec{i}_s(k)^* + \widehat{\vec{v}}_s(k+1) \cdot \Delta\widehat{\vec{i}_s}(k)^* \quad (10)$$

Replacing (9) in (10), the change in apparent power is:

$$\Delta\vec{s}(k) = \Delta\widehat{\vec{v}}_s(k) \cdot \vec{i}_s(k)^* + \cdots$$
$$\cdots + \widehat{\vec{v}}_s(k+1) \cdot \frac{T_s}{\hat{L}} \left[\widehat{\vec{v}}_s(k) - \widehat{\vec{v}}_r(k) - \hat{R}\vec{i}_s(k) \right]^* \quad (11)$$

For a sinusoidal voltage source power supply, the estimated $\widehat{\vec{v}}_s(k + 1)$ is obtained by rotating in $\Delta\theta = \omega T_s$ rads.

978-1-4244-4782-4/10 $26.00 © 2010 IEEE

$$\widehat{\vec{v}}_s(k+1) = \vec{v}_s(k) \cdot e^{j\omega T_s} \qquad (12)$$

$$\Delta\widehat{\vec{v}}_s(k) = \vec{v}_s(k)\left(e^{j\omega T_s} - 1\right) \qquad (13)$$

The complex apparent power $\vec{s}(k)$ is a function of the supply voltage, and also changes with the rectifier voltage $\widehat{\vec{v}}_r(k)$ that can be manipulated to obtain the commanded value $p_{ref} + j\,q_{ref}$. Defining $\Delta\vec{s}_0(k)$ as an independent voltage vector term in $\widehat{\vec{v}}_r(k)$, the following change in active and reactive power is obtained

$$\Delta\vec{s}_0(k) = \frac{T_s}{\hat{L}}|\vec{v}_s(k)|^2 \cdot e^{j\omega T_s} + \cdots \qquad (14)$$

$$\cdots + \vec{v}_s(k) \cdot \vec{i}_s(k)^* \left[\left(1 - \frac{T_s\hat{R}}{\hat{L}}\right)e^{j\omega T_s} - 1\right]$$

For a given reference in active and reactive power the change in power for proper compensation becomes a function of the converter voltage $\widehat{\vec{v}}_r(k)$. The apparent power variation needed to change from the actual to the demanded value in the following sampling period, p_{ref} and q_{ref}, are given by the following expressions

$$\Delta\vec{s}(k) = \Delta\vec{s}_0(k) - \frac{T_s}{\hat{L}}\left[\widehat{\vec{v}}_s(k+1) \cdot \widehat{\vec{v}}_r(k)^*\right] \qquad (15)$$

$$\vec{\epsilon}_s(k) = \underbrace{p_{ref} - \Re\{\vec{s}(k)\}}_{\epsilon_p(k)} + j\underbrace{q_{ref} - \Im m\{\vec{s}(k)\}}_{\epsilon_q(k)} \qquad (16)$$

In the OVSS algorithm, also known as predictive direct power control, a cost function is evaluated for a set of the converter voltages $\vec{v}_r$, and the value of this voltage providing the minimum value for the cost function is employed in the next control cycle [16]–[18]. In this case the cost function is

$$J(k) = \eta_p\left(\epsilon_p(k) - \Re|\Delta\vec{s}(k)|\right)^2 + \eta_q\left(\epsilon_q(k) - \Im m|\Delta\vec{s}(k)|\right)^2 \qquad (17)$$

where η_p and η_q control the relative importance of the active and reactive parts in the system.

The proposed control technique is based in the selection of the voltage vector that minimized the cost function (17) expressed by the active and reactive power errors. However, since zero is the global minimum for the cost function, instead of testing among several candidate vectors for the best choice, the proposed techniques computes with a closed form formula, the voltage vector for this minimum. Forcing to zero the cost function (17), $J(k) = 0$,

$$\vec{\epsilon}_s(k) - \Delta\vec{s}(k) = \mathbf{0} \qquad (18)$$

Replacing (15) and (16) in (18)

$$\Delta\vec{s}_0(k) - \frac{T_s}{\hat{L}}\left[\widehat{\vec{v}}_s(k+1) \cdot \widehat{\vec{v}}_r(k)^*\right] = \epsilon_p(k) + j\epsilon_q(k) \qquad (19)$$

Finally, replacing (12) into (19), the absolute optimum converter voltage required to attain the commanded active and reactive power becomes

$$\widehat{\vec{v}}_r(k) = \widehat{v}_{r\alpha}(k) + j\,\widehat{v}_{r\beta}(k) = \frac{\hat{L}}{T_s}\left[\frac{\Delta\vec{s}_0(k) - \vec{\epsilon}_s(k)}{\vec{v}_s(k) \cdot e^{j\omega T_s}}\right]^* \qquad (20)$$

This voltage is synthesized in the converter using standard space vector modulation (SVM) [19]. As with other DPC algorithms, the reactor parameters are required for computing the estimated value of the power system voltages, the active and reactive power and the update value for the converter voltage indicated in (20).

The proposed algorithm has many advantages over existing methods, among them it provides an instantaneous correction of the active and reactive power flowing into the converter, reduces the ripple in the instantaneous power and currents, resulting in a low harmonic distortion and have low computational demands.

III. SIMULATION RESULTS

The scheme shown in Fig. 1 has been modeled using the space vector representation of the state variables [7]. Both, the $V - V$ and Scott transformers have been included in these simulations. The rail road system is represented using the measured harmonic currents distribution, injected to the power system in the secondary side of each transformer [20]. The three phase power system is modeled using a space vector Thevenin equivalent. Also, space vector representations of the power transformer ($V - V$ or Scott), IGBT converter and the filter inductor are used in the simulation. The per unit parameter used in simulations are shown in Table I.

Table I: Parameters of the filter scheme model

L_{th}	R_{Trx}	L_{Trx}	R_{trx}	L_{trx}	R_{rec}	L_{rec}	V_{DC}
0.037	0.01	0.1	0.01	0.1	0.005	0.05	3

Table II shows the current Total Harmonic Distortion (THD) and the unbalance relation between positive and negative sequences (I_2/I_1) [21], for uncompensated and compensated cases using $V - V$ and Scott transformer. The simulation uses maximum unbalance by operating on one single phase circuit with the other under no load, which is the most demanding operating condition. The active filter injects the harmonic content used by the single phase rail road load. The proposed control scheme reduces by more than 50% the THD for both transformer connections.

Table II: Simulated total harmonic distortion and unbalance

Simulated Cases	Uncompesated		Compensated	
	THD	I_2/I_1	THD	I_2/I_1
$V - V$ rectifier	.4065	.9238	.1804	.0835
$V - V$ rail road	.2054	.9216	.1110	.0732
Scott rectifier	.4015	.9235	.1228	.0838
Scott rail road	.2264	.9384	.0747	.0813

Figure 3 shows the simulated instantaneous currents flowing into the power system without compensation and with the proposed active and reactive compensation. The simulations show the balancing effect on the power system current as well as the THD reduction obtained with the active filter controlled by the instantaneous active and reactive power. Both transformers ($V - V$ and Scott) have a similar current behavior when a single phase rectifier load is connected in one secondary.

IV. EXPERIMENTAL RESULTS

For the experimental test, the proposed algorithm was implemented on a custom build floating point DSP (ADSP-21061-40 MHz) based test-rig. The power stage uses six 50A, 1200V, IGBTs with two 2200 μF 400 V series connected capacitors in the DC link. The input inductors have an 7mH; the PWM signals are provided by a motion co-processor ADMC-201 operating at 10 kHz. The rail road load was implemented in only one single phase circuit using a single phase rectifier bridge with an R-L (50-200 Ω, 40 mH) load in the DC side. The sampling frequency is synchronized by the motion co-processor at the beginning of each PWM cycle. Fig. 4 shows the power module and the DSP based processing unit. The electrical parameters for the power circuit in the experimental tests are the same to those used in the simulations, and shown in Table III. The $V - V$ and Scott transformer connections were built using two single-phase – 480 : 240 – 120 V, 1 kVA transformers. The Scott transformer was built with two additional single phase variable transformers.

Table III: Parameter test-rig

R_{rec}	L_{rec}	C	V_{DC}	T_s	V
$20\,m\Omega$	$7.0\,mH$	$1100\,\mu F$	$200 \sim 600V$	$100\,\mu s$	$208\,V$

	f	f_s	R_{LOAD}	C_{LOAD}	L_{LOAD}
	$60\,Hz$	$100\,kHz$	$50\,\Omega$	$2200\,\mu F$	$17\,mH$

A. $V - V$ Transformer

Figures 5a to 5d show the current waveforms and spectrums measured on a three phase $V - V$ transformer test bench feeding a non-linear load, with and without the proposed compensation scheme. The measurements were obtained using a power quality analyzer type "B" [22]. Comparing the compensated and the uncompensated results, it can be observed that the compensator reduces unbalance and harmonic distortion in the system. The unbalance is reduced from 94.7% (uncompensated value) to 16.8% (compensated value), and the system's current THD to values below 18.4% in all phases, with a significant reduction in the third and seventh harmonics which are the more significant components present in the harmonic spectra generated by vector controlled converters used in locomotives.

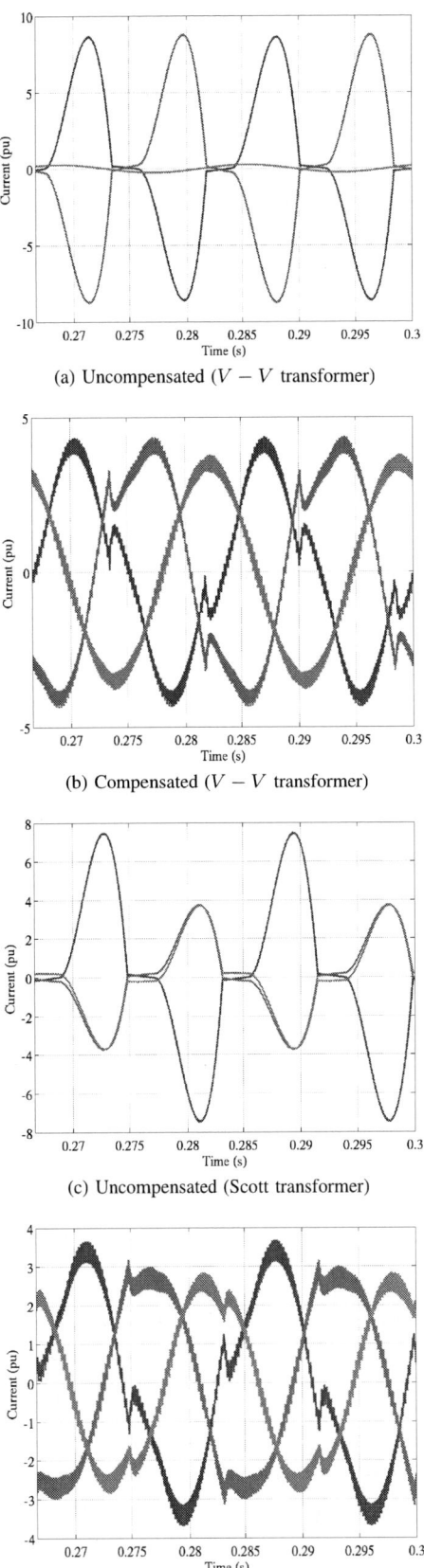

(a) Uncompensated ($V - V$ transformer)

(b) Compensated ($V - V$ transformer)

(c) Uncompensated (Scott transformer)

(d) Compensated (Scott transformer)

Figure 3: Simulated active filter effect on power system currents feeding a single phase rectifier load

Figure 4: Experimental rig

B. Scott Transformer

Figures 5e to 5h show the current waveforms and the harmonic spectrum, measured on a three phase Scott transformer test bench feeding a non-linear load, with and without the proposed compensation scheme. The non-linear load was the same used in the $V - V$ transformer case. Comparing the compensated and the uncompensated results, it can be observed that the compensator reduces harmonic content and balance the three phase load. The compensator reduces the current unbalance from 94.7% (uncompensated value) to 16.8% (compensated value), and the system's current THD to values below 18.4% in all phases, with a significant reduction in the third and seventh harmonics which are the more significant components present in the harmonic spectra generated by vector controlled converters used in locomotives.

V. CONCLUSIONS

The proposed compensation scheme reduces negative sequence currents that circulate in the uncompensated system feeding an electric traction system using a power system transformer connected in $V - V$ or Scott configuration. The scheme reduces the current THD to values allowed by international regulations, and regulates the power factor observed in the common coupling point between the traction substation and the power system. The proposed compensation scheme implementation using an instantaneous power control algorithm with direct space vector representation, reduces the system's current THD to allowable ranges ($< 20\%$) and reduces the overall unbalance from 97% to 18% for worse case operation. The compensation algorithm is able to control the power factor measured at the coupling point under all considered conditions. From the simulation and experimental results it is found that there is a compromise between the amount of unbalance correction and harmonic reduction that can be achieved. This is due to the finite amount of energy stored by the active filter in its input inductance and dc-link capacitor.

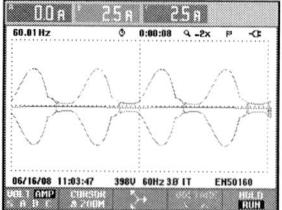

(a) Uncompensated ($V - V$)

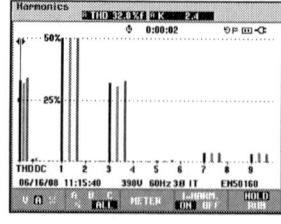

(b) Harmonics uncompensated ($V - V$)

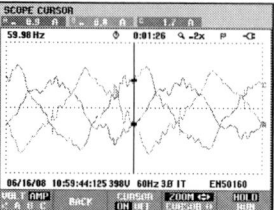

(c) Compensated ($V - V$)

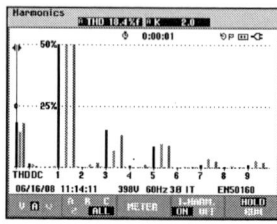

(d) Harmonics compensated ($V - V$)

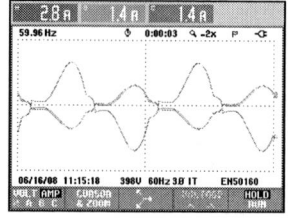

(e) Uncompensated (Scott)

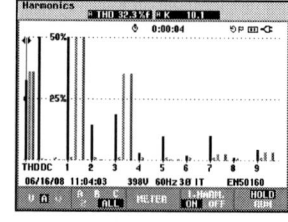

(f) Harmonics uncompensated (Scott)

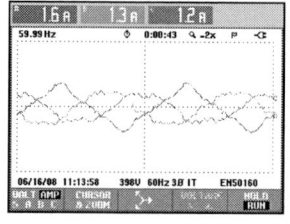

(g) Compensated (Scott)

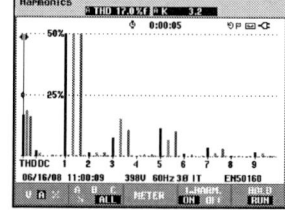

(h) Harmonics compensated (Scott)

Figure 5: Experimental active filter effects on power system currents feeding a single phase rectifier load

ACKNOWLEDGMENT

The authors want to express their gratitude to the Dean of Research and Development Bureau (DID) of the Simón Bolívar University, for the annual financial support provided to the GSIEP (registered as GID-04 in the DID) to perform this work.

REFERENCES

[1] B.-K. Chen and B.-S. Guo, "Three phase models of specially connected transformers," *Power Delivery, IEEE Transactions on*, vol. 11, no. 1, pp. 323–330, 1996.

[2] G. W. Chang, H.-W. Lin, and S.-K. Chen, "Modeling characteristics of harmonic currents generated by high-speed railway traction drive converters," *IEEE Transactions on Power Delivery*, vol. 19, pp. 766–733, apr 2004.

978-1-4244-4782-4/10 $26.00 © 2010 IEEE

[3] R. E. Morrison, "Power quality issues on ac traction systems," in *Proceedings Ninth International Conference on Harmonics and Quality of Power, 2000*, vol. 2, pp. 709–714, 2000.

[4] S. T. Senini and P. J. Wolfs, "Novel topology for correction of unbalanced load in single phase electric traction systems," in *IEEE 33rd Annual Power Electronics Specialists Conference, 2002. pesc 02. 2002*, vol. 3, pp. 1208–1212, 2002.

[5] M. Goto, T. Nakamura, Y. Mochinaga, and Y. Ishii, "Static negative-phase-sequence current compensator for railwaypower supply system," in *International Conference on Electric Railways in a United Europe, 1995.*, pp. 78–82, mar 1995.

[6] R. Cutri and L. M. Jr, "Reference currents determination techniques for load unbalance compensation," Tech. Rep. 11, 2005.

[7] J. M. Aller, A. Bueno, and T. Pagá, "Power system analysis using space vector transformation," *IEEE Transaction on Power System*, vol. 17, no. 4, pp. 957–965, 2002.

[8] H. Li, F. Zhuo, L. Wu, W. Lei, J. Liu, and Z. Wang, "A novel current detection algorithm for shunt active power filters in harmonic elimination, reactive power compensation and three-phase balancing," in *IEEE 35th Annual Power Electronics Specialists Conference, 2004. PESC 04. 2004*, vol. 2, pp. 1017–1023, jun 2004.

[9] J. M. Aller, A. Bueno, J. Restrepo, M. I. Giménez, and V. Guzmán, "Advantages of instantaneous reactive power definition in three phase system meausrements," *IEEE Power Engineering Review*, vol. 19, no. 6, pp. 49–50, 1999.

[10] S. A. Larrinaga, M. A. R. Vidal, E. Oyarbide, and J. R. T. Apraiz, "Predictive control strategy for DC/AC converters based on direct power control," *IEEE TRANSACTIONS ON INDUSTRIAL ELECTRONICS*, vol. 54, pp. 1261–1271, June 2007.

[11] P. Antoniewicz, M. P. Kazmierkowski, S. Aurtenechea, and M. A. Rodriguez, "Comparative study of two predictive direct power control algorithms for three-phase AC/DC converters," in *Power Electronics and Applications, 2007 European Conference on*, (Aalborg), pp. 1–10, Sept. 2007.

[12] P. Cortés, J. Rodríguez, P. Antoniewicz, and M. Kazmierkowski, "Direct power control of an AFE using predictive control," *IEEE Transactions on Power Electronics*, vol. 23, pp. 2516–2523, Sept. 2008.

[13] M. Giménez, V. Guzmán, J. Restrepo, J. Aller, A. Bueno, J. Viola, A. Millán, and A. Cabello, "Plataforma: Development of an integrated dynamic test system for power electronics systems performance analysis," *Revista de la Facultad de Ingeniería – UCV*, vol. 23, no. 3, pp. 91–92, 2008.

[14] Y. Cheng, C. Qian, M. L. Crow, S. Pekarek, and S. Atcitty, "A comparison of diode-clamped and cascaded multilevel converters for a statcom with energy storage," *IEEE TRANSACTIONS ON INDUSTRIAL ELECTRONICS*, vol. 53, pp. 1512–1521, Oct. 2006.

[15] J. A. Restrepo, J. M. Aller, J. C. Viola, A. Bueno, and T. G. Habetler, "Optimum space vector computation technique for direct power control," *IEEE Transactions on Power Electronics*, vol. 24, pp. 1637–1645, June 2009.

[16] J. Restrepo, J. Viola, J. M. Aller, and A. Bueno, "A simple switch selection state for svm direct power control," in *Proceedings of the 2006 IEEE International Symposium on Industrial Electronics, ISIE 2006*, pp. 1112–1116, July 2006.

[17] S. Larrinaga, M. Rodríguez, E. Oyarbide, and J. Torrealday, "Predictive control strategy for DC/AC converters based on direct power control," *IEEE Trans. Ind. Electron.*, vol. 54, pp. 1261–1271, June 2007.

[18] P. Antoniewicz, M. P. Kazmierkowski, S. Aurtenechea, and M. A. Rodriguez, "Comparative study of two predictive direct power control algorithms for three-phase AC/DC converters," in *Proceedings of the 2007 European Conference on Power Electronics and Applications, EPE 2007*, pp. P1–P10, Sept. 2007.

[19] M. Malinowski, "Adaptive modulator for three-phase pwm rectifier/inverter," in *Proceedings of the 9th International conference on power electronics and motion control EPE-PEMC 2000*, vol. 1, (Kosice, Slovak Republic), pp. 35–41, Sept. 2000.

[20] S. R. Huang and B. N. Chen, "Harmonic study of the le blanc transformer for taiwan railway's electrif ication system - power delivery, ieee transactions on," *IEEE TRANSACTIONS ON POWER DELIVERY*, vol. 17, no. 2, pp. 495–499, 2002.

[21] IEEE, *IEEE 519-1992 Recommended practices and requirements for harmonic control in electrical power systems*. pub-IEEE-STD:adr: IEEE Standards Coordinating Committee, 1992.

[22] IEC, *IEC 61000-4-15 Testing and measurement techniques – Flickermeter Functional and Desing Specifications*. International Electrotechnical Commission Standard, 1997.

978-1-4244-4782-4/10 $26.00 © 2010 IEEE

Review of Non-isolated Bi-directional DC-DC Converters for Plug-in Hybrid Electric Vehicle Charge Station Application at Municipal Parking Decks

Yu Du, Xiaohu Zhou, Sanzhong Bai, Srdjan Lukic and Alex Huang

FREEDM Systems Center

Department of Electrical and Computer Engineering, North Carolina State University

Raleigh, NC, 27695, U.S.A.

Email: ydu4@ncsu.edu

Abstract — **There is a growing interest on plug-in hybrid electric vehicles (PHEV's) due to energy security and green house gas emission issues, as well as the low electricity fuel cost. As battery capacity and all-electric range of PHEV's are improved, and potentially some PHEV's or EV's need fast charging, there is increased demand to build high power off-board charging infrastructures. A charge station architecture for municipal parking decks has been proposed, which has a DC microgrid to interface with multiple DC-DC chargers, distributed renewable power generations and energy storage, and provides functionalities of normal and rapid charging, grid support such as reactive and real power injection (including V2G), current harmonic filtering and load balance. Several non-isolated bi-directional DC-DC converters suited for charge station applications have been reviewed and compared, as the major focus of this paper. Half bridge converter is a good candidate but it is difficult to maintain high efficiency in wide battery pack voltage range. A variable frequency pulse width modulation (VFPWM) scheme is proposed to mitigate this issue. Finally three-level bi-directional DC-DC converter is suggested to be employed in this application. A 10kW prototype verifies that 95.1-97.9% full load efficiency can be achieved in charging mode with 180-360V battery pack voltage. In addition, the inductor size is only one third of the half bridge counterpart, which is a great advantage for high power converters.**

I. INTRODUCTION

Nowadays about 62% of crude oil used in United States is refined into gasoline for transportation. The associated energy security and green house gas emission problems are well known [1]. Hybrid electric vehicles (HEV's) is one of the solutions to address these issues, because the fuel economy has been improved by optimizing internal combustion engine (ICE) efficiency, regenerating brake energy and shutting down ICE during the idle time. After more than one million HEV's are driven on the road today, there is a growing interest on plug-in hybrid electric vehicles (PHEV's), which is defined by IEEE-USA's Energy Policy Committee as (1) a battery storage system of 4kWh or more, used to power the motion of

the vehicle, (2) a means of recharging that battery system from an external source of electricity, and (3) an ability to drive at least 10 miles in all-electric mode consuming no gasoline [2]. PHEV's can be power by electricity from various sources, including emerging renewable power generations, and benefit from lower fuel (electricity) cost. Green house gases such as CO_2 emission is expected to be greatly reduced due to much less petroleum consumption for daily commuters who drive PHEV's mainly in all-electric mode.

Major automakers are preparing to launch the first models in a new generation of PHEV's in 2010. Several global and U.S. market research reports [3-4] indicate there will be a rapid growth on PHEV's sales. Reference [4] forecasts that PHEV's will follow a similar adoption pattern as that of HEV's over the past few years, and by 2015, a total of 1.7 million PHEV's will be on the road worldwide.

Currently major PHEV's are designed for an all-electric range of several tens miles to meet daily commute requirement, due to the high cost of on-board battery energy storage system. An on-board charger is usually employed for slow overnight charging in home garage. However, as battery capacity and all-electric range of PHEV's are improved, and potentially some PHEV's or EV's in the future need fast charging to extend all-electric drive range, there is increased requirement to build off-board charge station infrastructures. In Part II the motivations to build municipal parking deck charging stations are discussed and a charge station architecture has been proposed with the functionalities of normal and rapid charging, grid support such as reactive and real power (V2G mode) injection, current harmonic filtering and load balance. The proposed charging station has a DC power distribution bus, which can be considered as a microgrid, to interface with DC-DC chargers, distributed renewable power generations and energy storage system. In Part III, several non-isolated bi-directional DC-DC converters suited for charge station application have been reviewed and compared. Half bridge converter is considered to be a good

---

This work made use of ERC shared facilities supported by the National Science Foundation under Award Number EEC-08212121.

candidate but it is difficult to maintain high efficiency in wide battery pack voltage range. In Part IV, a variable frequency pulse width modulation (VFPWM) scheme is proposed to improve the power efficiency of half bridge converter at lower battery pack voltage, because major portion of energy is delivered to battery in low pack voltage range. The efficiency improvement is verified by experimental test. In Part V, a three-level bi-directional DC-DC converter is proposed to be employed in charging station applications. A 10kW prototype verifies that 95.1-97.9% full load efficiency can be achieved in charging mode with 180-360V battery voltage. In addition, the inductor size is only one third of the half bridge counterpart.

II. ACHITECTURE OF MUNICIPAL PARKING DECK CHARGING STATION

A. Motivations to Build Charge Station Infrastructure

Currently most PHEV's use single-phase on-board charger to refuel their batteries, which is the common practice for both converted PHEV's and several ones that will be commercialized soon. On-board charger can either use independent power converter, or leverage the power stage of drive train and motor inductance [5]. The power rating of on-board charger is low, limited to the current rating of wall plug. For example, 120V/12A (Level I) or 240V/32A (Level II) single-phase AC input according to SAE-J1772, and so it is suited for slow overnight recharging. However, with advancement of battery technology, the energy density and power density of battery packs are improved and battery cost decreases, the desire of more on-board battery capacity for more all-electric range and less gasoline consumption is possible for future PHEV's or EV's. Therefore there will be a trend for battery chargers to shift from compact low power on-board installation to shared large high-power off-board charging station in the future.

The second motivation to build this charge station infrastructure is to use a DC link to interface with distributed renewable power generations, which can be considered as a microgrid. With high penetration of PHEV's, it is necessary to install new power generation capacity. For municipal parking deck applications, it is quite possible that PHEV's will be charged during daytime and refueled with peak power electricity. The distributed renewable power generations can potentially provide a solution for this issue because they will not only reduce the power demand from the grid during peak time, or even inject real power back to support the grid, but also recharge PHEV battery packs with green energy source and make transportation industry cleaner. The charge station can install its own solar power generation, fuel cell power generation and wind turbines [6-7]. For example, PV panels generate power during daytime and can reduce the power demand from grid during peak load time. In addition, there is less safety or noise concern to install PV in municipal parking decks. On the other hand, vehicle to grid (V2G) operation provides a potential solution of using PHEV batteries as energy storage devices for high penetration of distributed wind and solar power generations and smoothes their intermittent power [8].

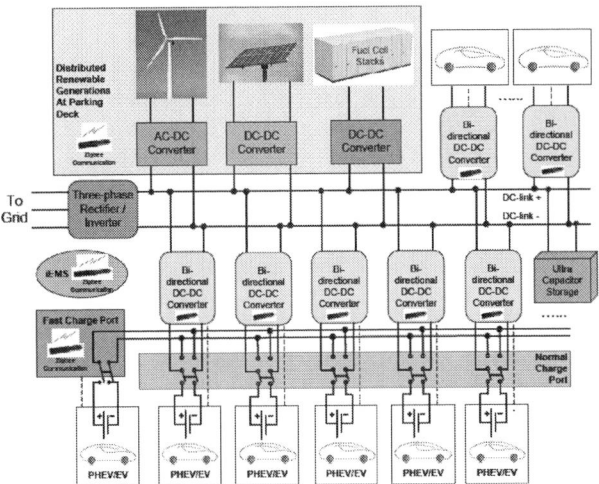

Figure 1: Proposed Charging Station Architecture with DC Power Distribution and Renewable Generations

The third motivation is to fast charge a PHEV/EV battery pack. In order to improve the all-electric range of PHEV's and EV's, one will either increase the on-board battery capacity or recharge the battery pack in very short time at a fast charging station. However, all-electric range will be eventually limited by battery capacity due to the size and weight as well as the cost of batteries. Therefore, high power fast charging will potentially solve EV range problem by recharging its battery in 10 to 30 minutes, like what people do in today's gasoline station, as long as the battery packs are capable of accepting high rate charging current.

B. Charging Station with DC Power Distribution

The proposed architecture of municipal charge station at parking decks is shown in Fig.1. Compared to discrete AC-DC and DC-DC chargers, the proposed charge station uses 750V or higher DC distribution bus with one high power three-phase AC-DC converter as grid interface. This architecture has several advantages. The specific cost of high power AC-DC stage is lower than that of discrete low power AC-DC converters if AC power distribution bus is used. The three-phase rectifier is rated for average power rather than peak power if ultra capacitor energy storage is installed to filter the ripple power. DC distribution bus is easy to integrate distributed renewable power generations such as wind, PV, fuel cells and other energy storage devices. The power of each DC-DC channel is rated for normal slow charge rating to minimize cost. On the other hand, the parallel of several DC-DC stages provides a high power fast charging channel, assuming only small portion of PHEV's will require this service. With bi-directional DC-DC converters, energy stored in PHEV batteries can be fed back to grid, which is called V2G operation. An intelligent energy management system (iEMS) with wireless Zigbee communication platform can coordinate system operation [9]. The charge station can provide several grid support functions such as reactive power injection, peak power generation, harmonic current filter, and load balance [10].

The bi-directional DC-DC converters are basic building blocks for municipal charging stations and they are the interface between PHEV battery system and the DC distribution grid. Bi-directional DC-DC converters with low cost, high efficiency and high reliability are important for the charging stations. The non-isolated bi-directional DC-DC converters can be considered for this application and their performance is compared in the following sections.

III. NON-ISOLATED BI-DIRECTIONAL DC-DC CONVERTERS

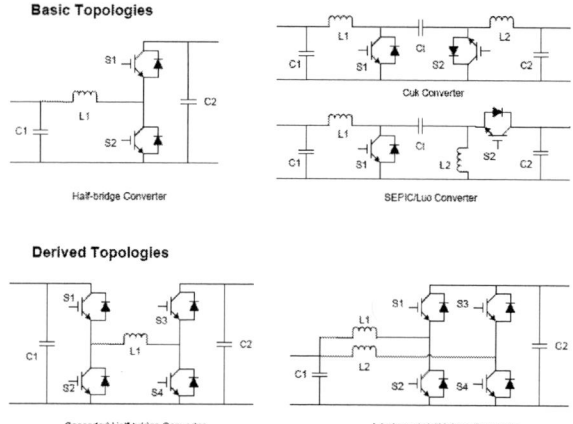

Figure 2: Several Non-isolated Bi-directional DC-DC Converters Reported in Literatures

TABLE I.   SPECIFICATIONS FOR BI-DIRECTIONAL DC-DC CONVERTERS IN MUNICIPAL PARKING DECK CHARGE STATION

Rated Power		10kW	
DC-link Voltage		750V	
Battery Pack Voltage		180V-360V	
Maximal Inductor Current Ripple		30% (Peak to Peak)	
Switching Frequency		20kHz	
	Half Bridge	CUK	SEPIC /LUO
L1 (mH)	1.12	1.46	1.46
L2 (mH)	-	3.04	3.04

Non-isolated bi-directional DC-DC converters generally have advantages of simple structure, high efficiency, low cost, high reliability, etc. Several non-isolated bi-directional DC-DC converters that are reported in literature [11-14] are shown in Fig.2. They can be categorized into basic topologies such as Half-bridge converter, Cuk converter, SEPIC/Luo converter and derived topologies such as cascaded half-bridge converter and interleaved half-bridge converter. One widely used topology is Half-bridge converter which operates either in Buck or in Boost mode. Cuk and SEPIC/Luo can convert power bi-directionally by using two active switches. The cascaded half bridge and interleaved half bridge can be considered as derived topologies from the basic half bridge, and their performance can be evaluated based on the performance of half bridge converter. Therefore, Half Bridge, Cuk and SEPIC/Luo converters are compared based on the

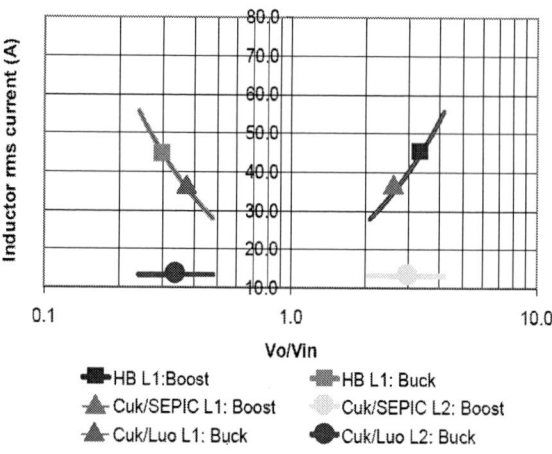

Figure 3: Comparison of Inductor RMS Current in Half Bridge, Cuk and SEPIC/Luo Converters

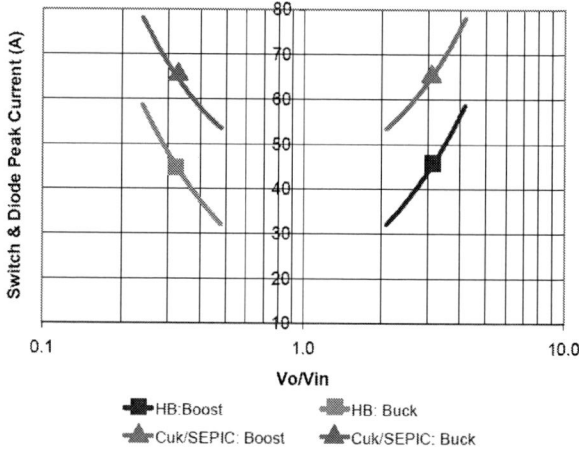

Figure 4: Comparison of Switch Current Stress in Half Bridge, Cuk and SEPIC/Luo Converters

example specification of bi-directional DC-DC converters in charging station.

The example specification and converter inductor values are listed in Table I. The power rating is 10kW. The high voltage DC link is 750V and battery side voltage ranges from 180-360V. The inductor size is calculated based on 20kHz switching frequency and 30% maximal inductor current ripple referred to the inductor DC current. It can be found that Cuk and SEPIC/Luo converter use two larger inductors and one more capacitor compared with Half Bridge converter.

For all three basic topologies shown in Fig.2, the battery is connected to C1 through a common mode choke and ground fault interrupter (GFI) to limit the leakage current. C2 is connected to DC link side. When bi-directional DC-DC converters work in charging mode, the ratio of output voltage V_o to input voltage V_{in} is lower than 1; when they work in discharging or V2G mode, the ratio of V_o to V_{in} is larger than1. The inductor RMS current and switch current stress are shown in Fig. 3 and 4 respectively. The RMS current in inductor L1 is similar for three topologies, but inductor L2 in Cuk and SEPIC/Luo converters consume additional power, although the current stress is much lower than that in L1. From Fig.4, the current stress for active switches and diodes in Cuk and SEPIC/Luo converter are larger than that in Half Bridge converter under same input/output voltage and power conditions. Therefore, Half Bridge is expected to be more efficient and it also has less number of inductor and capacitors. Half Bridge converter is a better candidate in this scenario.

IV.   VFPWM SCHEME FOR HALF BRIDGE CONVERTER EFFICIENCY IMPROVEMENT

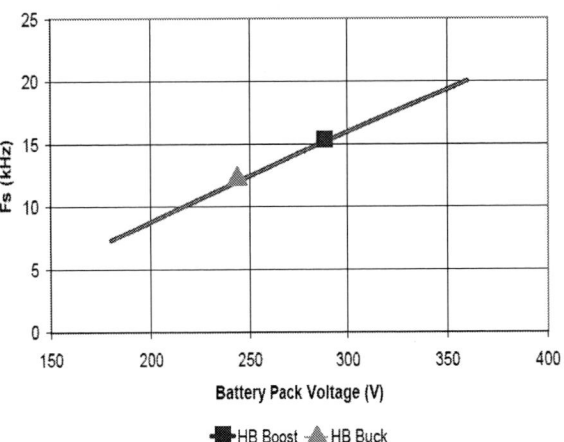

Figure 6: Switching Frequency Dependency on Battery Pack Voltage in VFPWM Half Bridge

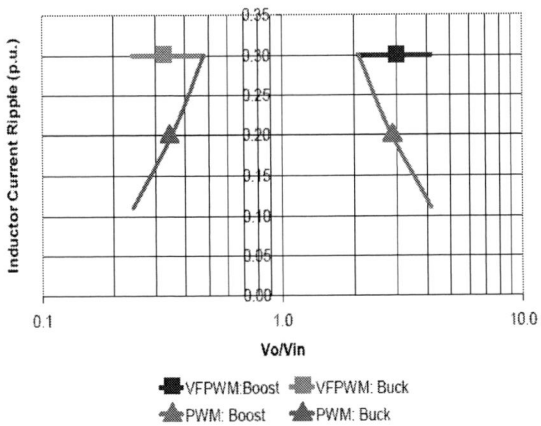

Figure 5: Comparison of Inductor Current Ripple Ratio in Conventional PWM and VFPWM

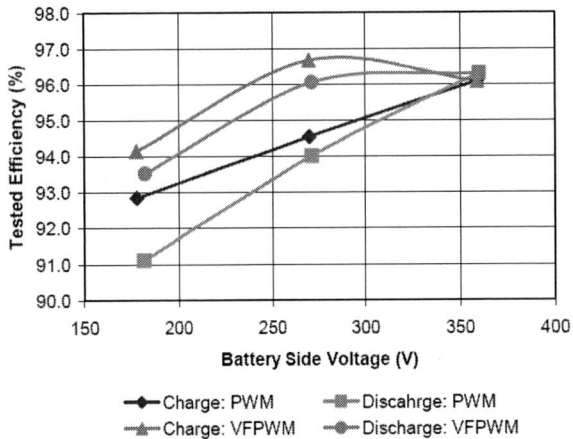

Figure 7: Experimentally Tested Converter Efficiency of PWM and VFPWM Half Bridge Converter

The battery side voltage can change in wide range from 180V to 360V (2:1). When battery pack voltage is high, the efficiency of Half Bridge is better because the current stress for inductor L1 and switches are lower. However, if battery side voltage is low and constant power (10kW) charging is assumed to be the worst scenario, current stress in converter increases and the efficiency drops quickly. It is even worse for low power conversion efficiency in lower battery voltage range because during charging process the major portion of charge is injected to battery in lower voltage range. This means that a big portion of energy is delivered to battery pack with low efficiency.

This problem can be mitigated by the proposed variable frequency pulse width modulation (VFPWM) scheme. Fig.5 shows the inductor current ripple ratio, which is referred to the DC current component of inductor L1, in the wide battery pack voltage range at constant full power (10kW). If the switching frequency is fixed at 20kHz, in low battery voltage range the current ripple ratio is reduced by roughly half and much lower than the specified maximal 30% current ripple ratio. This is not necessary because there is little benefit, and current stress of output capacitor is already very low because only the inductor ripple current goes through capacitor C1. Instead, maintaining a constant inductor current ripple ratio of 30% in full voltage range at 10kW allows lower switching

frequency in low battery voltage range, which can help reduce switching loss and improve efficiency of Half Bridge converter in wide voltage range. The switching frequency is dependent on battery pack voltage in VFPWM scheme and is shown in Fig.6. The inductor ripple current will increase due to the reduced switching frequency in lower voltage range. Since the inductor ripple current is filtered by battery side capacitor C1, the increased ripple current will have no adverse impact on PHEV battery pack.

Experiment setup is built to test the efficiency of conventional fixed frequency (20kHz) PWM and proposed VFPWM Half Bridge converter. Two 1200V/300A IGBT's APTGF300A120G are used in the half bridge converter prototype and inductor value is L1=1.12mH. The DC choke is built with KoolMu magnetic core material and E-E shape with distributed air gap. Two pairs of 00K145LE E-E cores are used and the relative core permeability is 26. DC-link voltage is 750V. The experimentally tested efficiency in wide battery side voltage range from 180V to 360V at full load 10kW is shown in Fig.7. The efficiency curves for bi-directional operation, charging or Buck mode, and discharging or Boost mode, are provided. Full load efficiency is more important because most of the energy injected into battery pack is during large current and high power charging stage. Experiment result shows that the efficiency of Half Bridge converter in both Buck/Charge mode and Boost/Discharge mode is improved by 1-2.5% by proposed VFPWM scheme in lower battery voltage range, compared to conventional PWM scheme with fixed switching frequency.

V. BI-DIRECTIONAL THREE-LEVEL CONVERTER

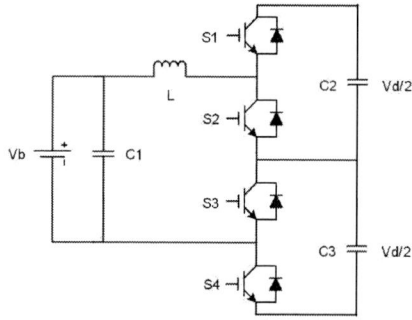

Figure 8: A Neutral Point Clamped Three Level Bi-directional DC-DC Converter

Three-level converters have been proposed for high DC-link voltage applications [15-18]. Three-level converters are characterized of low switch voltage stress and smaller energy storage devices such as inductor and capacitor. Topologies of three-level converter can be categorized into Neutral Point Clamped (NPC), Flying Capacitor and Diode Clamped converter, etc.

The topology of a neutral point clamped three-level bi-directional converter is shown in Fig.8. The operation waveforms of this three-level (TL) bi-directional DC-DC converter for Buck or charging mode, and Boost or

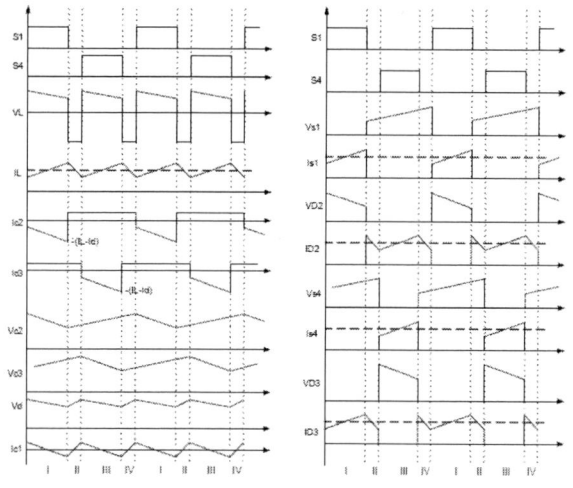

Figure 9: Buck Mode Waveforms for TL Converter

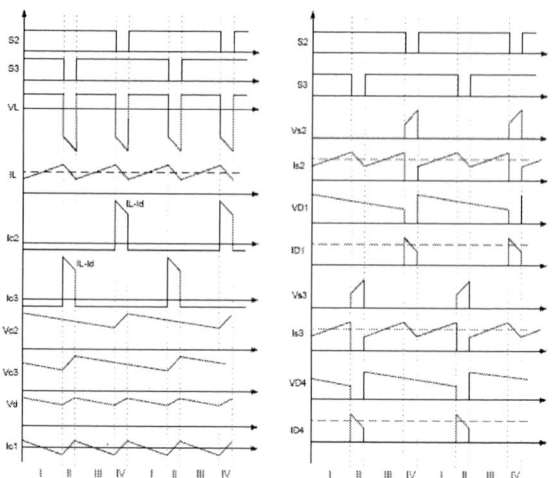

Figure 10: Boost Mode Waveforms for TL Converter

discharging mode are shown in Fig.9 and Fig.10, respectively.

The Buck mode equations are listed from (1) to (6):

$$\text{The switch duty ratio}: D_s = \frac{V_B}{V_D} = \frac{D_L}{2} \tag{1}$$

$$\text{The effective duty ratio}: D_L = \frac{2 \cdot V_B}{V_D} \tag{2}$$

$$\text{The inductor ripple current}: \frac{V_d \cdot D_s \cdot (1 - 2 \cdot D_s)}{L \cdot (2 \cdot f_s)} \tag{3}$$

$$\text{The inductor DC current}: \frac{2 \cdot I_d}{D_L} = \frac{I_d}{D_s} \tag{4}$$

$$\text{The switch RMS current}: \frac{I_d}{D_s} \cdot \sqrt{D_s} \cdot \sqrt{1 + \frac{r_L^2}{12}} \tag{5}$$

$$\text{The diode RMS current}: \frac{I_d}{D_s} \cdot \sqrt{1 - D_s} \cdot \sqrt{1 + \frac{r_L^2}{12}} \tag{6}$$

Boost mode equations are listed from (7) to (12):

$$\text{The switch duty ratio}: D_s = 1 - \frac{1}{V_D / V_B} = \frac{1 + D_L}{2} \tag{7}$$

$$\text{The effective duty ratio}: D_L = 1 - \frac{2}{V_D / V_B} \tag{8}$$

$$\text{The inductor ripple current}: \frac{V_d \cdot (1 - D_s) \cdot (2 \cdot D_s - 1)}{L \cdot (2 \cdot f_s)} \tag{9}$$

$$\text{The inductor DC current}: \frac{2 \cdot I_d}{1 - D_L} = \frac{I_d}{1 - D_s} \tag{10}$$

$$\text{The switch RMS current}: \frac{I_d}{1 - D_s} \cdot \sqrt{D_s} \cdot \sqrt{1 + \frac{r_L^2}{12}} \tag{11}$$

978-1-4244-4782-4/10 $26.00 © 2010 IEEE   1149

TABLE II. COMPARISON OF SPECIFICATION OF HB AND TL

Rated Power	10kW	
DC-link Voltage	750V	
Battery Pack Voltage	180V-360V	
Maximal Inductor Current Ripple	30% (peak to peak)	
	HB	TL
Switching Frequency	20kHz	10kHz
Inductor Size	1120uH	348uH

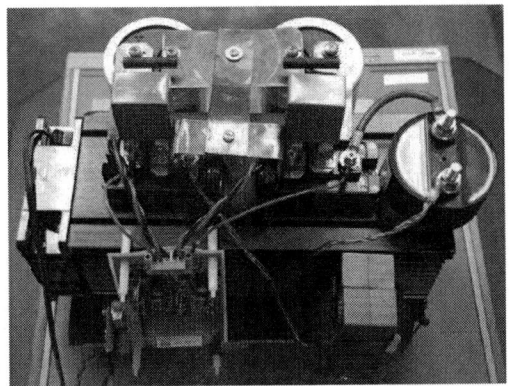

Figure 11: 10kW Prototype of Bi-directional Three-level Converter

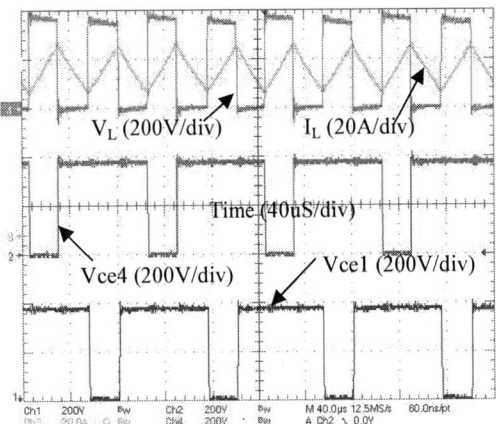

Figure 12: Test Waveforms of TL converter in Buck Mode

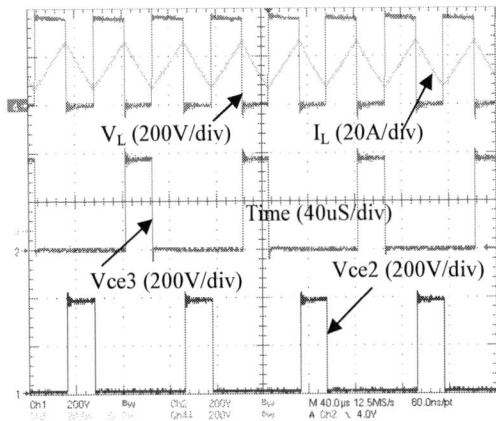

Figure 13: Test Waveforms of TL converter in Boost Mode

The diode RMS current : $\dfrac{I_d}{1-D_s} \cdot \sqrt{1-D_s} \cdot \sqrt{1+\dfrac{r_L^2}{12}}$ (12)

To meet the specifications which are listed in Table I, parameters of three-level (TL) converter are listed in Table II and compared to conventional half-bridge (HB) converter as well.

A 10kW experiment prototype, as showed in Fig.11, was built to evaluate the performance of three-level bi-directional DC-DC converter in municipal parking deck charge station application. Due to reduced switch voltage stress, 600V IGBT can be employed instead of 1200V IGBT in half bridge converter. Generally 600V IGBT has much lower on-state voltage and switching loss than 1200V IGBT. Two CM150DY-12NF IGBT half-bridge modules are used in the prototype. Also the total IGBT's cost is lower compared to 1200V IGBT's. The inductor size is only one third of that in half bridge converter. This is a great benefit for high power converters because inductors with higher inductance and high current are very bulky, expensive and inefficient.

The experiment waveforms for 180V battery side voltage and 10kW power are shown in Fig.12 and 13, which are corresponding to Buck and Boost mode respectively.

The efficiency of three-level bi-directional DC-DC converter prototype is measured with full power rating 10kW

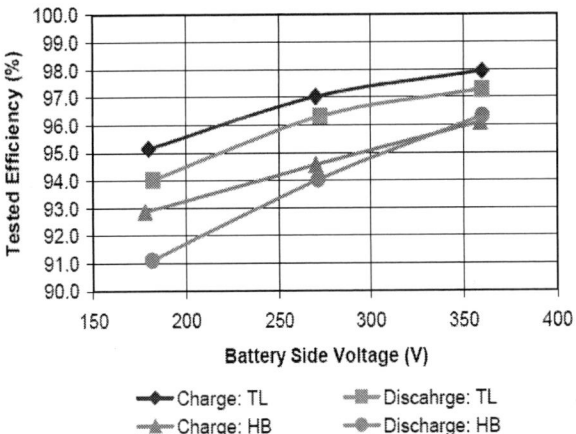

Figure 14: Efficiency Test Results for TL and HB

and wide battery side voltage range from 180V to 360V, as shown in Fig.14. The efficiency of three-level converter in charge mode varies from 98% to 95%. The discharge efficiency is about 1% lower than charging mode because IGBT's conduct for a longer time than diodes in each switching cycle. The tested efficiency of half bridge converter in the same condition is also shown in Fig.14. Test results indicate that about 2-3% efficiency improvement can be achieved by three-level DC-DC converter for Buck mode and Boost mode, respectively.

VI. CONCLUSION

Several low cost non-isolated bi-directional DC-DC converters suited for municipal packing deck charge station applications have been reviewed and compared. Half bridge converter is better than Cuk and SEPIC/Luo converter in this scenario due to smaller number of passive components, lower switch current stress and higher efficiency. However, in wide battery side voltage range, efficiency of half bridge converter drops quickly with lower battery pack voltage. This problem can be mitigated by proposed VFPWM scheme which obtains 1-2.5% improvement. Three-level bi-directional DC-DC converters have been investigated for charge station application. Experiment results show 2-3% higher efficiency than that of half bridge. In addition, much smaller inductor is required and there is no audible noise in three-level converter compared with VFPWM half bridge converter, either. Therefore, the bi-directional three-level DC-DC converter is recommended for municipal parking deck charge station application.

ACKNOWLEDGMENT

This work is also supported by Advanced Transportation Energy Center (ATEC) at North Carolina State University and we wish to acknowledge both FREEDM Systems Center and ATEC for the financial support.

REFERENCES

[1] C. Roe, J. Meisel, A.P. Meliopoulos, F. Evangelos, T. Overbye, "Power System Level Impacts of PHEVs," in 2009 42nd Hawaii International Conference on System Science, 2009.

[2] S.G. Wirasingha, N. Schofield, A. Emadi, "Plug-in hybrid electric vehicle developments in the US: Trends, barriers, and economic feasibility," in IEEE 2008 Vehicle Power and Propulsion Conference, 2008.

[3] Morgan Stanley Report, "Autos and Auto-Related, Plug in Hybrids: The Next Automotive Revolution," March 11, 2008, pp.13.

[4] Pike Research Report, "The Global Outlook for PHEVs: Business Issues, Technology Issues, Key Players, and Market Forecasts," third quarter, 2009

[5] Lisheng Shi, A. Meintz, M. Ferdowsi, "Single-phase bidirectional AC-DC converters for plug-in hybrid electric vehicle applications," in IEEE 2008 Vehicle Power and Propulsion Conference, 2008.

[6] Y. Gurkaynak, A. Khaligh, "Control and Power Management of a Grid Connected Residential Photovoltaic System with Plug-in Electric Vehicle (PHEV) Load," in IEEE 2009 Applied Power Electronics Conference, 2009, pp.2086-2091.

[7] V. Marano, G. Rizzoni, "Energy and Economic Evaluation of PHEVs and their Interaction with Renewable Energy Sources and the Power Grid," in IEEE 2008 Proceedings of the International Conference on Vehicular Electronics and Safety, 2008, pp. 22-24.

[8] S. D. Jenkins, J.R. Rossmaier, M. Ferdowsi, "Utilization and Effect of Plug-in Hybrid Electric Vehicles in the United States Power Grid," in IEEE 2008 Vehicle Power and Propulsion Conference, 2008.

[9] P. Kulshrestha, L. Wang, M.Y. Chow, S. Lukic, "Intelligent Energy Management System Simulator for PHEVs at Municipal Parking Deck in a Smart Grid Environment", in proc IEEE Power and Energy Society General Meeting Conference, 2009.

[10] M. Bojrup, P. Karlsson, M. Alakula, B. Simonsson, "A dual purpose battery charger for electric vehicles," in IEEE 1998 Power Electronics Specialists Conference, 1998, vol.1, pp.565-570.

[11] M. Ortúzar, J. Dixon, J. Moreno, "Ultracapacitor-Based Auxiliary Energy System for an Electric Vehicle: Implementation and Evaluation" in IEEE Transactions on Industrial Electronics, vol.54, issue 4, Aug.2007, pp.2147-2156.

[12] R. M. Schupbachj, C. Bald, "Comparing DC-DC Converters for Power Management in Hybrid Electric Vehicles," in IEEE 2003 International Electric Machines and Drives Conference, 2003, vol.3, pp.1369-1374.

[13] J. Czogalla, J. Li, C.R. Sullivan, "Automotive Application of Multi-Phase Coupled-Inductor DC-DC Converter," in IEEE 2003 Industry Applications Conference, 2003, vol.3, pp.1524-1529.

[14] M. Gerber, J. A. Ferreira, N. Seliger, I. W. Hofsajer, "Design and Evaluation of an Automotive Integrated System Module," in IEEE 2005 Industry Applications Conference, 2005, vol.2, pp.1144-1151.

[15] X. Ruan, B. Li, Q. Chen, S.C. Tan, C. K. Tse, "Fundamental Considerations of Three-Level DC–DC Converters: Topologies, Analyses, and Control," in IEEE Transactions on Circuits and Systems, vol.55, no.11, Dec.2008, pp.3733-3743.

[16] R.M. Cuzner, A.R. Bendre, P.J. Faill, B. Semenov, "Implementation of a Non-Isolated Three Level DC/DC Converter Suitable for High Power Systems," in IEEE 2007 Industry Applications Conference, 2007, pp.2001-2008.

[17] Y. Shi, Z. Jin, X. Cai, "Three-Level DC-DC Converter: Four Switches Vsmax=Vin/2, TL Voltage Waveform before LC Filter," in IEEE 2006 Power Electronics Specialists Conference, 2006.

[18] V. Yousefzadeh, E. Alarcon, D. Maksimovic, "Three-Level Buck Converter for Envelope Tracking in RF Power Amplifiers," in IEEE Transactions on Power Electronics, Mar. 2006, vol.21, issue 2, pp.549-552.

Control of Plug-in Hybrid Electric Vehicles for Mobile Power Generation and Grid Support Applications

Gui-Jia Su and Lixin Tang
Energy and Transportation Science Division
Oak Ridge National Laboratory
Knoxville, U.S.A.
sugj@ornl.gov

Abstract—**The use of an onboard electrical propulsion system in a hybrid electric vehicle (HEV) to provide plug-in charging, mobile power generation, and vehicle-to-grid support capabilities is discussed. Pulse width modulation (PWM) control methods are examined for operating such a plug-in HEV in mobile power generation and vehicle-to-grid support applications to assess their impacts on the power conversion efficiency. A novel unipolar/interleaved PWM switching scheme is proposed, which can significantly reduce the switching losses and substantially increase the efficiency. A dual inverter drive prototype was assembled to test the PWM control schemes and test results are included to verify the effectiveness of the proposed control scheme.**

I. INTRODUCTION

The success of gasoline–electric hybrid vehicles in the market has made it well recognized that using an electric drive to augment the internal combustion engine (ICE) can effectively reduce pollutant emissions and increase gas mileage [1][2]. In an HEV, while the ICE still plays a dominant role for propulsion, the electric motor drive enables downsizing the ICE, operating the engine in its optimal efficiency regions, capturing and reusing the kinetic energy obtained during braking, and switching off the engine when the vehicle is stopped at red lights or in traffic jams, thus boosting the fuel efficiency significantly. This improvement on fuel economy is especially effective in city driving, where vehicles are operated frequently in stop and start situations.

To further promote reductions of oil consumption and pollutant emissions by automobiles, interest has recently been increasing in research and development on plug-in HEVs (PHEVs), in which a larger battery pack is used so that it enables the vehicle to operate on the electric drive system alone, without running the ICE for a distance of several tens of kilometers (this is called charge-depletion or pure EV mode) [3]. The battery is charged by plugging into the grid; ideally done at home during night time to take the advantage of lower off-peak electricity rates. A PHEV thus operates first as a pure battery-electric vehicle and then switches to the conventional series HEV mode after the battery is discharged to a predefined level of state-of-charge. At that point, the ICE kicks in and operates as a generator to maintain the battery state-of-charge within a narrow band (charge sustaining mode). This type of HEV is specifically suitable for daily commutes within the electric drive ranges of the HEVs because no engine operations are needed over the entire commuting drive. Being capable of achieving zero emissions and substantially higher miles per gallon in the pure EV mode, PHEVs can reduce emissions and boost gas mileage more significantly than conventional HEVs [4].

Because both the size of the batteries for PHEVs and the amount of energy that can be stored in the batteries remain limited, to maximize the range for a PHEV in the EV mode, it is desired to charge the batteries whenever the vehicle is parked and as fast as they can be charged. Hence, an onboard charger with a high power rating is highly desirable. In addition, chargers with bidirectional power flow capability can enhance the functionality and value of the vehicles by providing additional capabilities. As the number of PHEVs continues to increase, these vehicles can collectively serve as a load leveling energy storage device to the grid; supplying power to the grid during peak demands and drawing power from the grid to charge the batteries during off-peak periods. This can also help facilitate the integration into the grid of renewable energy sources with an intermittent power generating nature such as wind and solar power plants. Moreover, a bidirectional charger can enable the vehicle to function as a mobile power generator. The requirements for this bidirectional charger are similar to those for other automotive parts—low cost, low volume and weight, and high efficiency and reliability. To reduce the cost of such a bidirectional charger, several methods of utilizing the onboard electrical propulsion system to fulfill the charging/power generating function are reported [5-9].

To maximize the fuel saving and emission reduction, the efficiency of the bidirectional charger needs to be enhanced. In this paper, pulse width modulation switching schemes uniquely suited for control of PHEVs that utilize onboard electrical propulsion inverters and motors for mobile power

---

This manuscript has been authored by UT-Battelle, LLC, under contract DE-AC05-00OR22725 with the U.S. Department of Energy. The United States Government retains and the publisher, by accepting the article for publication, acknowledges that the United States Government retains a nonexclusive, paid-up, irrevocable, world-wide license to publish or reproduce the published form of this manuscript, or allow others to do so, for United States Government purposes.

978-1-4244-4782-4/10 $26.00 © 2010 IEEE

generation and grid support applications are proposed and examined experimentally. In particular, it is found that a novel unipolar/interleaved PWM switching scheme can significantly reduce the switching losses and substantially increase the efficiency. A dual inverter drive prototype was assembled to test the PWM control schemes and experimental results are included to verify the effectiveness of the proposed control scheme.

II. SYSTEM DESCRIPTION AND CONTROL

A. System Description

Fig. 1 shows a schematic of an HEV traction drive system configured for battery charging, mobile power generation, and grid support applications. The system consists of two motor/generators, MG1 and MG2, with MG2 coupled to the ICE, and two inverter/converters, INV/CONV1 and INV/CONV2. The basic idea is that by connecting the neutral points of the motors to the grid or external loads, the motor/generators can be used as inductors to eliminate the need for an extra external inductor and the two three-phase inverters can be operated as a single-phase ac-to-dc or dc-to-ac converter for charging the battery, feeding power to the grid or loads. Two sets of contact switches are used for safety and proper vehicle operation considerations; one for disconnecting the high-voltage (HV) battery pack from the drive system when the vehicle is not in operation and the other for disconnecting the charging socket and the filter capacitor, C_{acf}, when the vehicle is not operated in charging or generation modes. It is noted that the configuration shown in Fig. 1 is readily applicable to the series hybrid and power-split hybrid configurations. For the parallel hybrids, vehicle accessory drives such as the air-conditioning compressor drive can be used to replace INV/CONV1 and MG1.

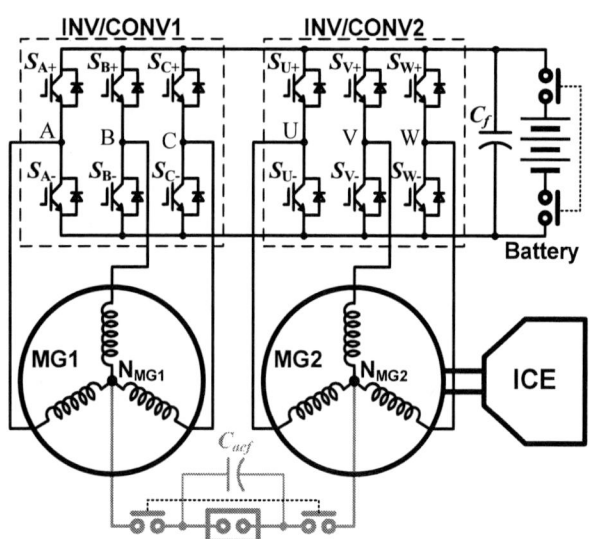

Figure 1. Schematic of utilizing the onboard traction drive system for battery charging, mobile power generation, and grid support applications.

All the switch legs in each inverter/converter collectively function as a single switch leg and the motor/generator as an inductor. Together, the two drive units form a single-phase converter/inverter to charge/discharge the battery, when operating in the grid support mode. When operating in the generation mode, the drive units form a single-phase inverter to supply the external loads. In this mode, the motor/generator (MG2) of the drive unit coupled to the engine shaft is driven by the engine to generate electrical power for supplying the dc bus and ultimately the external loads. Furthermore, power can be drawn from the HV battery for short time interval operations.

Fig. 2 shows an equivalent circuit of the electrical drive system for operating in the mobile generation mode, where electrical power is generated by MG2 that is driven by the engine. In this case, the three switch-legs in INV/CONV1 (A, B, and C) collectively function as the first single switch leg of the single-phase inverter and the MG1 as an impedance net work – its stator zero-sequence (ZS) impedance networks, ZSIN1. ZSIN1 consists of three identical branches and each branch is comprised of the stator winding phase resistance, R_{ms1}, and the phase leakage inductance, l_{m0s}. The other inverter/converter, INV/CONV2, has dual functions. It first operates as a three-phase converter to regulate the dc bus voltage, V_{dc}, by controlling the power generated by the generator, MG2. At the same time, its three phase-legs collectively form the second switch leg of the single-phase inverter to regulate the load voltage, v_{Load}. While the three-phase generator currents pass through its positive network and generate a breaking torque to the engine, the external load current splits into three equal parts and each part flows in one branch of its ZS network, ZSIN2, and as such does not produce air-gap flux or generate any torque.

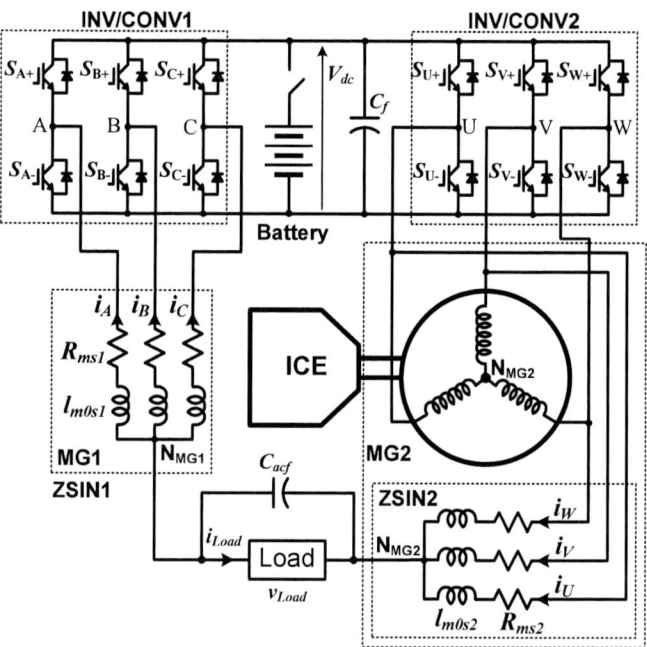

Figure 2. Equivalent circuit of the traction drive system for operating in the mobile power generation mode.

Fig. 3 shows an equivalent circuit of the electrical drive system for operating in the grid support mode. All the three switch-legs (A, B, and C or U, V, and W) in each of the two inverters/converters collectively function as a single switch leg and the motors/generators as two ZS impedance networks,

978-1-4244-4782-4/10 $26.00 © 2010 IEEE

ZSIN1 and ZSIN2. Together, the two drive units form a single-phase converter to regulate the battery voltage, V_{bat}, or the charging current, I_{bat}, for charging the battery. Alternately, the two drive units form a single-phase inverter to discharge the battery to feed the grid for peak load support. Normally, the single-phase converter is controlled in such a way to maintain a unity power factor. It can also be used as a reactive power source for grid voltage support.

B. System Control

Fig. 4 shows control block diagrams for both the mobile power generation (a) and grid support operations (b). In (a), a field orientation based control is implemented for the generator, MG2, to maintain the dc bus voltage at a commanded level. The load voltage, v_{Load}, is controlled by the other inverter with a current control inner loop. In (b), the grid current is controlled by both inverters with the commanded current determined by grid demands and the battery charging/discharging profile.

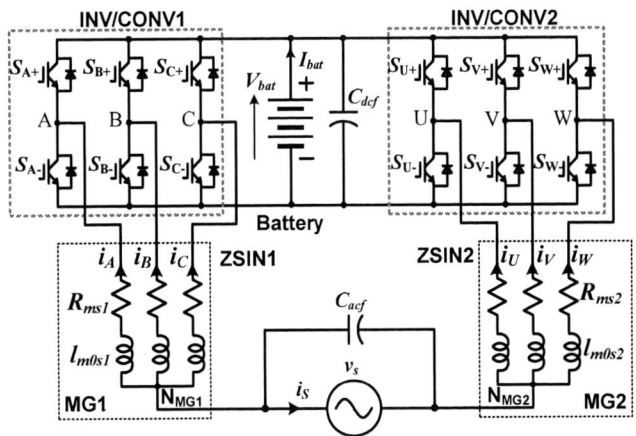

Figure 3.   Equivalent circuit of the traction drive system for operating in vehicle-to-grid applications.

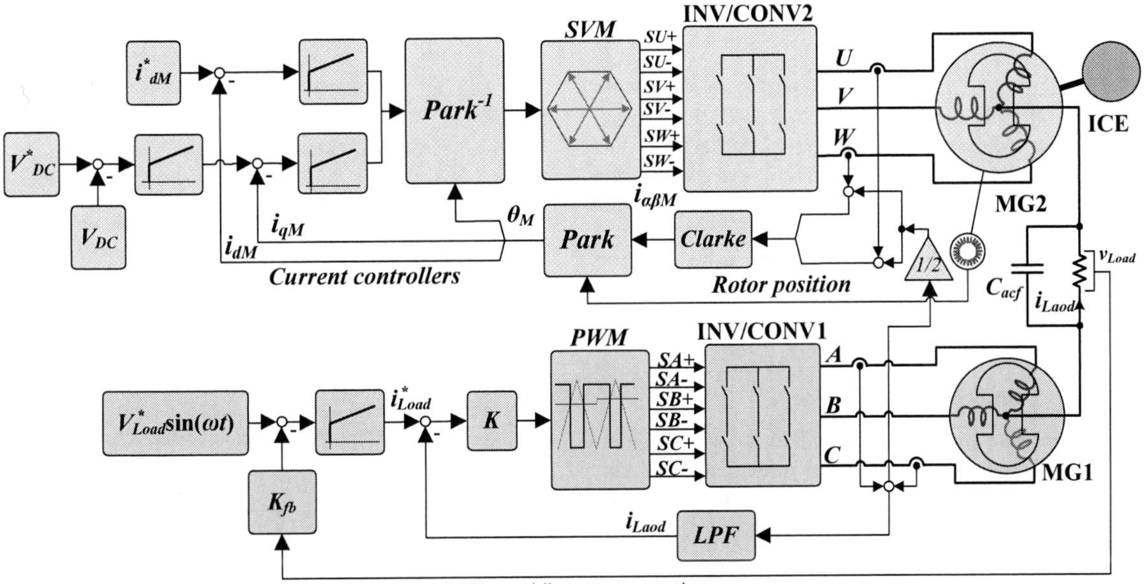

(a) For mobile power generation

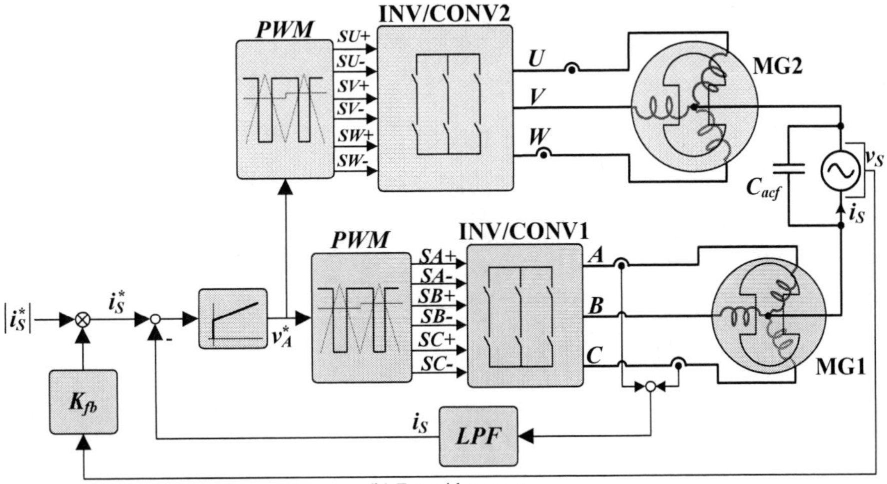

(b) For grid support

Figure 4.   Control block diagrams.

Usually, only two phase currents (i_U and i_W or i_A and i_B in Fig. 3) of each motor/generator are sensed and fed back to the drive system controller for operating in the traction drive mode. Assuming each stator phase winding of each motor carries one-third of the load current, the torque producing currents of MG2, $i_{U(Te)}$, $i_{V(Te)}$, and $i_{W(Te)}$, can be derived from the measured motor currents by

$$\begin{cases} i_{U(Te)} = i_U - (i_A + i_C)/2 \\ i_{W(Te)} = i_W - (i_A + i_C)/2 \ , \\ i_{V(Te)} = -i_{U(Te)} - i_{W(Te)} \end{cases} \quad (1)$$

where i_U and i_W are measured phrases U and W currents of MG2, and i_A and i_B are measured phrases A and C currents of MG1, respectively. Similarly, the load current can be reconstructed by

$$i_{Load} = 3(i_A + i_C)/2. \quad (2)$$

For mobile power generation, the torque producing currents of MG2, $i_{U(Te)}$, $i_{V(Te)}$, and $i_{W(Te)}$, are projected to the rotor synchronous reference frame as the d and q axis currents, i_{dM} and i_{qM}, through the Clarke and Park transformation using the rotor position information detected by a shaft encoder. Using the field orientation technique and space vector PWM method, the d and q axis currents, i_{dM} and i_{qM}, are controlled through two proportional and integral (PI) controllers with the motor inverter (INV/CONV2) to maintain the inverter dc bus voltage at the commanded level, V^*_{DC}. The other inverter (INV/CONV1) is used to control the load voltage with the help of an inner current loop to improve the dynamic response. The inner current loop is carried out through a high gain proportional controller, while both the dc bus voltage and load voltage are controlled via PI controllers.

To investigate the impact on the efficiency by switching schemes that take the advantages of the unique circuit configurations, the following PWM schemes were used for the inverter, INV1: (1) synchronized switching, in which all three legs switch at the same time, and (2) interleaved switching, in which each phase leg is controlled by its own carrier, whose frequency is one-third of that for the synchronized PWM, and the three carrier signals are phase-shifted by 120 electrical degrees. Due to limitations of the PWM generation hardware in the digital signal processor (DSP) employed in the test set-up, a software implementation of the interleaved PWM based on a single carrier was developed to eliminate the need for external PWM generation logic circuits.

In Fig. 4(b), the system operates in grid support mode using the HV battery as the energy source. Both inverters, INV/CONV1 and INV/CONV2, are used to control the output current. The following combinations of PWM schemes were used for the inverters to assess the impact on the efficiency by the switching schemes: (1) synchronized PWM for both inverters, (2) interleaved PWM for both inverters, and (3) a novel unipolar/interleaved PWM, in which INV/CONV2 is switched at the fundamental frequency while an interleaved PWM is used for INV/CONV1. It is found that the novel unipolar/interleaved PWM scheme can significantly reduce the switching losses while still maintaining a low current

distortion factor. Again, the interleaved PWM schemes were implemented using only the built-in PWM hardware designed for three-phase motor control in the DSP.

III. EXPERIMENTAL RESULTS

A prototype electrical drive inverter system was built to test the control schemes. It is comprised of a 55 kW and a 30 kW inverter for the traction motor and generator, respectively, and is configured to provide plug-in charging, mobile generation, and grid support capabilities. Fig. 5 shows a photo of the prototype. The 55 kW inverter was implemented with a 600V/600A six-pack IGBT module, and the 30 kW inverter was implemented with a 600V/300A six-pack IGBT module. The dc bus capacitor bank was constructed using four 375 μF film capacitors rated at 600 V_{DC}. These components are mounted on a 30.48 cm x 17.78 cm water cooled cold plate.

Figure 5. A photo of a traction drive prototype consisting of a 55 kW motor inverter and a 30 kW generator inverter configured for providing plug-in charging and mobile power generation capabilities.

An induction motor of 10.9 kW and a PM motor (used as a generator) rated at 8.2 kW were used in the tests. Table I lists the measured ZS resistances of the motors along with those of the Toyota Camry motors for comparison. Due to the large resistance value of the relatively low power induction motor, the combined ZS resistances of the test motors are quite high at 189.54 mΩ. For comparison, the combined resistance of the two motors in the Camry is 32.32 mΩ, less than 1/5 that of the test motors. This will have a significant impact on the efficiency for the grid support tests.

TABLE I.   MEASURED MOTOR ZERO-SEQUENCE RESISTANCE

	Test Setup	Toyota Camry
Generator	23.8 mΩ	21.58 mΩ
Motor	165.74 mΩ	10.75 mΩ
Combined	189.54 mΩ	32.32 mΩ

The prototype was tested in the mobile generation mode for an output voltage of 120V and in the grid support mode for both 120V and 240V. Fig. 6 illustrates test results showing operating waveforms at output power levels of 3.0 kW and 5.0 kW in the engine-powered generator mode. The near sinusoidal waveforms of the output voltage and current and a flat DC bus voltage indicate good performance of the controller. Since one-third of the 60 Hz load current is superimposed as a ZS component on the original generator currents, the generator phase currents are no longer sinusoidal.

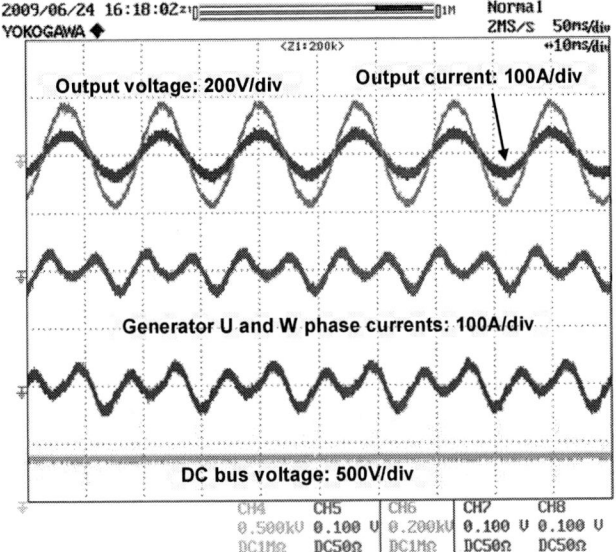

(a) Output power: 3.0 kW at 120V

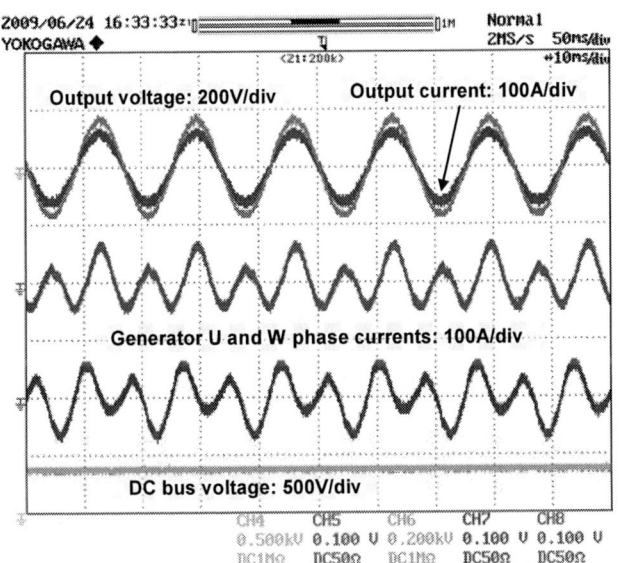

(b) Output power: 5.0 kW at 120V

Figure 6. Typical voltage and current waveforms in the mobile generator mode.

Fig. 7 plots measured total efficiency against output power. Due to the relatively low efficiency (<85%) of the small PM generator, the system efficiency does not exceed 80%. However, it is estimated a maximum efficiency of greater than 90% could be achieved with a production PM generator with efficiency of 93% since the inverter efficiency can reach 97% as will be shown later. In addition, one percentage point improvement is attained with the interleaved switching over the synchronized switching scheme.

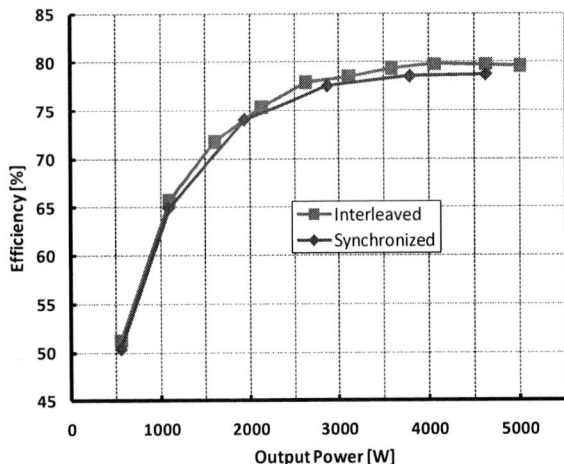

Figure 7. Measured total efficiency in the mobile generator mode.

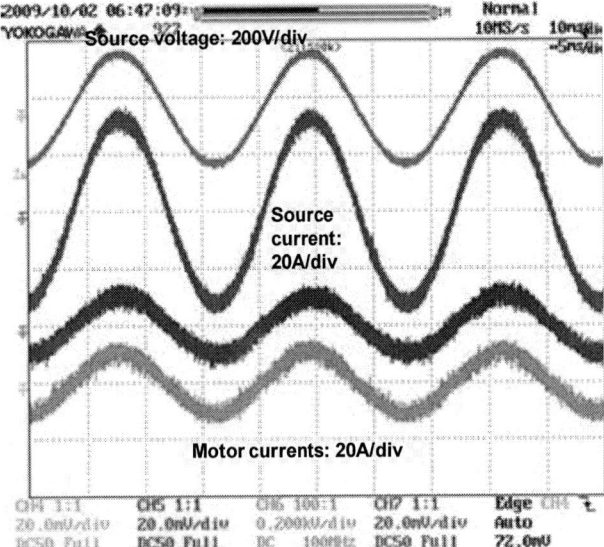

Figure 8. Typical voltage and current waveforms in the grid support mode: supplying power of 3.0 kW at 120V.

Figs. 8 and 9 show operating waveforms in the grid-support mode at the grid (source) voltage of 120V and 240V, respectively. Again, the good sinusoidal output voltage and current waveforms in both cases indicate excellent performance of the control method. In this operation mode, both motors are used as an inductor and one-third of the 60 Hz grid current is flowing in each phase winding as a ZS component, as indicated by the identical two motor phase currents in the test results.

Fig.10 plots measured total efficiency versus output power for the three PWM methods: synchronized, interleaved, and a combination of unipolar and interleaved switching schemes. Estimated efficiency using the resistance values of the Camry motors driven with the proposed novel switching technique is also given to illustrate the impact of the high resistances of the test motors on the efficiency. At the grid voltage of 240V as shown in (a), significant efficiency improvements are achieved with the interleaved over the synchronized switching, and again the interleaved/unipolar switching over

978-1-4244-4782-4/10 $26.00 © 2010 IEEE

the interleaved PWM scheme, due to the significant reduction of switching losses through the lower switching frequencies in the interleaved/unipolar switching methods. The efficiency improvements are the largest at the beginning since the switching losses are dominant and drop as the output power increases due to the quadratically increasing copper losses at high load current in the high stator resistance of the test motors. It is therefore expected and is indicated by the estimated curve that much better efficiency could be attained at high output power levels with production PM motors whose resistances are much lower than those of the test motors. The maximum measured efficiency is 97.1% at the output power of 6.2kW while the estimated maximum efficiency reaches 98.6%.

Fig. 10(b) shows the measured and estimated total efficiency at the grid voltage of 120V. The measured and estimated maximum efficiencies in this case are 93.7% with an output power of 1.9kW and 95.4% at 2.6kW, respectively.

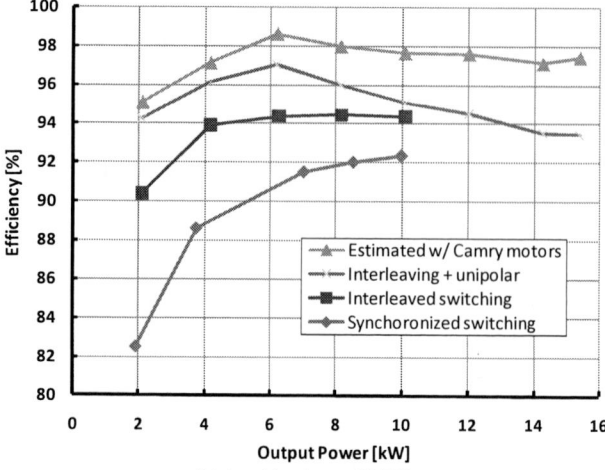

(a) At grid voltage of 240V

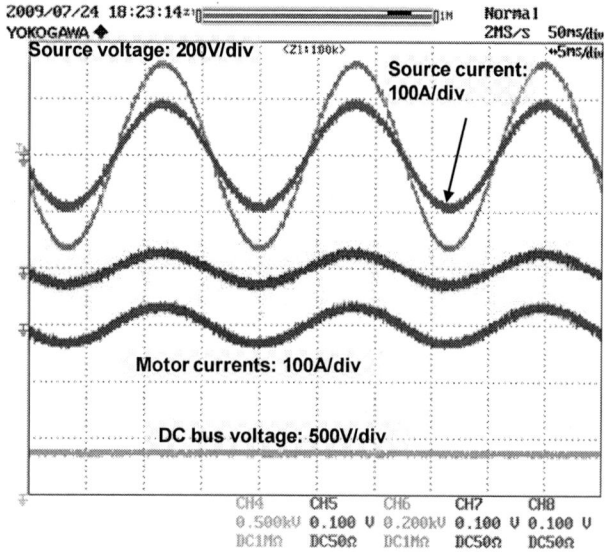

Figure 9. Typical voltage and current waveforms in the grid support mode: supplying power of 14.7 kW at the grid voltage of 240V.

(b) At grid voltage of 120V

Figure 10. Measured and estimated total efficiency in the grid support mode.

IV. CONCLUSIONS

Using the onboard electrical drive systems in HEVs to provide plug-in charging capability offers many benefits, including (1) significantly reducing the cost and volume of battery chargers for PHEVs, (2) providing rapid charging capability, and (3) enabling the use of the PHEVs as mobile generators and grid support energy storage devices. In this paper, PWM switching schemes uniquely suited for control of this type of PHEVs for mobile power generation and grid support applications are proposed and examined experimentally. In particular, it is found a novel unipolar/interleaved PWM switching scheme can significantly reduce the switching losses and substantially increase the efficiency.

ACKNOWLEDGMENT

The authors thank Mr. Cliff White for his assistance in assembling the hardware test setup.

REFERENCES

[1] U.S. Department of Energy, "HEV Sales by Model – Trend of sales by HEV models from 1999-2008," http://www.afdc.energy.gov/afdc/data/docs/hev_sales.xls.

[2] Toyota Motor Corporation, "Sustainability Report 2008," http://www.toyota.co.jp/en/csr/report/08/download/index.html.

[3] The California Cars Initiative (CalCars.org), "Plug-In Hybrids: State Of Play, History & Players," http://www.calcars.org/history.html.

[4] U.S. Dept. of Energy, "Plug-In Hybrid Electric Vehicle Benefits," http://www.afdc.energy.gov/afdc/vehicles/plugin_hybrids_benefits.html.

[5] D. Thimmesch, "An SCR inverter with an integral battery charger for electric vehicles," IEEE Trans. on Industry Applications, vol. IA-21, no. 4, pp.1023–1029, Jul./Aug. 1985.

[6] S.-K. Sul and S.J. Lee, "An integral battery charger for four-wheel drive electric vehicle", IEEE Transactions on Industry Applications, vol. 31, no. 5, pp. 1096–1099, Sept./Oct. 1995.

[7] T. Ishikawa, T. Sekimori and Y. Hotta, "Development of a Traction Inverter with Charging Function", EVS-14, Orlando, FL, Dec. 11–17, 1997.

[8] L. Tang and G.J. Su, "A Low-Cost, Digital-Controlled Charger for Plug-In Hybrid Electric Vehicles," in IEEE Energy Conversion Conference & Exposition (ECCE), Sept. 20-24, 2009, San Jose, CA, pp. 3923–3929.

[9] H. Oyobe, M. Nakamura, T. Ishikawa, S. Sasaki, Y. Minezawa, Y. Watanabe, and K. Asano, "Development of ultra low-cost, high-capacity power generation system using drive motor and inverter for hybrid vehicle," in IEEE 40th IAS Annual Meeting, 2005, vol. 3, pp. 2029–2034.

Interface Issues of Mining Haul Trucks Operating on Trolley Systems

Joy Mazumdar and Walter Koellner
Industry Solutions Division – Mobile Mining
Siemens Industry Inc
Alpharetta, GA USA
joy.mazumdar@siemens.com
walter.koellner@siemens.com

Rohit Moghe
School of Electrical and Computer Engineering
Georgia Institute of Technology
Atlanta, GA USA
rohitmoghe@gatech.edu

Abstract—**A typical mining facility uses off-road mining trucks to haul materials, such as coal, ore, and overburden from the pit to a stockpile where the material can be stored or processed. A conventional haul truck operates with diesel engine, adding considerable fuel expenses to the entire mining operation. Mining haul trucks are one of the most challenging applications of power electronics in vehicular systems. It is established that the use of inverter fed three-phase induction motors with vector control is the preferred solution to reach the required high starting torque and a good dynamic performance. In addition, the system is customized to meet the high shock and vibration limit and a special cooling system because the air has high dust level including conductive and corrosive materials. The objective of the mines is to achieve the movement of the highest possible payload per hour while minimizing operating costs over the lifetime of the machines. Operating the trucks with power from the overhead trolley lines, especially while going up on a grade significantly increases productivity, reduces fuel costs and qualifies as a green solution. This paper presents the overhead power systems interface issues of the trolley assisted haul trucks and their solutions through simulation study and field experimental results.**

I. INTRODUCTION

Mining industry is one of the most demanding areas for automotive electronic systems concerning mining haul trucks employed for massive ore handling and transportation, within hard conditions given by the harsh environment, safety issues and production requirements for economy of scale.

In addition, vehicular transportation systems are one of the most challenging fields of electronics systems design and applications, with embedded software for communication and control of sensors, actuators, microprocessors and networks [1]. Recent advances in electronics systems state new system-approaches for dealing with the new challenges [2]. Low and high-speed networks are employed for non-critical and critical applications including real-time functions [3]. Power electronics technology intended for traction drives has been driven by power switches (SCR's, GTO's, IGBT's, IGCT's),

AC adjustable drives and machine control algorithms like field-oriented [4]. Next improvements can be achieved through a system-level approach by developing smarter power electronics modules, intelligent diagnostic systems and reliable data communication and control networks with fault-tolerance [5] – [8].

In traditional haul trucks, power is transferred to the rear wheels via either a mechanical transmission or a DC electric drive system. A drawback of mechanical drive haulers is the high maintenance and repair costs associated with their sophisticated and complex mechanical transmission system. Elimination of the mechanical torque converter, transmission and differential is still one of the major advantages of electric haul trucks.

High power AC inverter fed induction motor drives employed in locomotives were found to be a solution to meet the stringent demands of this mining application [9]. Ruggedness and overloading capacity of three-phase induction motors, together with GTO/ IGBT inverters offer a compact drive with bigger power and high performance for meeting high reliability, continuous operation, high shock and vibration levels, broad range of temperature fluctuations and cooling air with high dusty particulate levels [10], [11]. Vector controlled AC-Drives offer a smooth step less full torque operation by motoring or braking at full operating speeds. A high starting torque and high top speed allow more ore can be transported in the same time giving the operator greater capacity from trucks carrying the same load. Smooth and efficient electric braking offer improved safety and reduces the wear rates of the mechanical brakes. Powerful retardation is made with fine control down to standstill, with automatic blending between electric and service brakes. So, application of state of the art electromechanical design, power electronics and packaging enable powerful haul trucks.

The trucks typically travel up and down a gradient several times a day. In order to achieve a lower cost per ton of material moved, it is desired to have a system such that the trucks can travel faster on a upgrade and consume less fuel. Although trolley systems have a high initial capital investment, feasibility studies [12] have shown promising results in terms of payback period and increased productivity. This paper presents the scenarios when multiple trucks engage on the trolley lines and its impact on the grid [13]. In addition, some field results are presented to validate the simulations. Economic benefits analysis is included to demonstrate the increased productivity.

II. REQUIREMENTS FOR THE TRUCK

Bigger haulage distances and grades state a challenge for more powerful transporting systems. The optimum loading target for shovels is three passes per truckload. Diesel-electric haul trucks with DC drives are popular and built up to 240 tons. Trucks with payloads of 280, 320, 360 and 400 tons have adopted the AC technology. A 120 ton AC rope shovel can load 320 to 360 ton trucks in 3 passes and 400 ton trucks in 4 passes.

The payload requirements need a huge power source (around 2200 kW) that has to be efficiently used. Keeping haul trucks as small as possible is also desired, so power consumption goes to ore transport only. These conditions make electronic packaging a fundamental part in haul trucks design not only because of the small room left to place an adequate power source drivers, extreme environmental conditions have to be taken into account too.

Size and capacity are not the only necessities that have to be satisfied. There are other operational and environmental requirements:

- Very high reliability: 24 months mean time between failure (MTBF)

- Continuous operation: with >95% electrical availability

- Reduced maintenance: no brushes, filters or regular lubrication.

- High shock and vibration limits: 2g in all directions

- Operation between –40 C and +55 C (-40 F and 122 F)

- Forced air cooling with unfiltered air and very high dust levels (including coal, ore, salt, conductive and corrosive materials)

The most important components of the truck drive system are shown in figure 1.

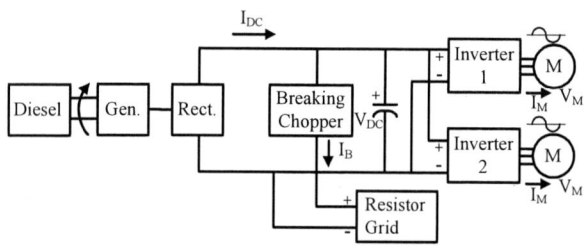

Figure 1. Block diagram of the truck system

One of the advantages of using electrical motors in haul trucks instead of diesel engines is the possibility to draw power from overhead lines called trolley lines. These trolley lines are usually installed in mines that have single road coming out of a deep pit or where fuel costs are much higher than the cost of electricity. The principal characteristic of trolley line operation is that the power available for propel motors are greatly increased. This means that a truck working on grade can move faster because of the extra power reducing the time needed to transport the mineral and thus, increasing production rates [14].

III. TROLLEY SYSTEM SIMULATION

The power system configuration of the truck trolley interface is shown in figure 2. Figure 3 shows the simplified truck model used in the simulation.

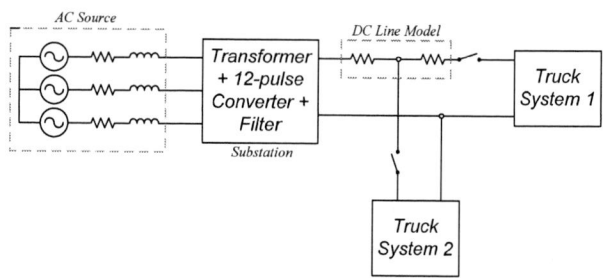

Figure 2. Block diagram of truck trolley interface

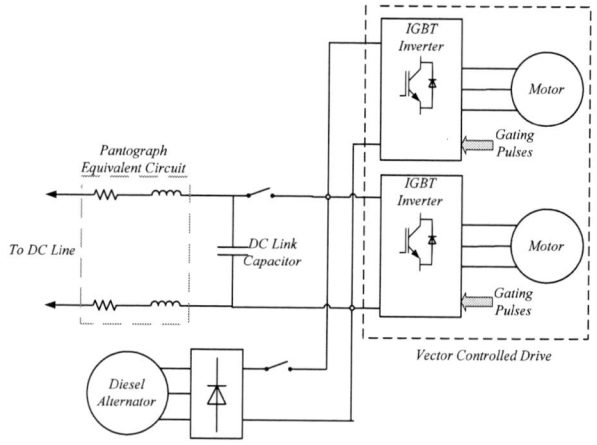

Figure 3. Simplified truck model

Although trolley systems are beneficial, there are certain design issues that needs careful investigation. Whenever a new truck engages to the line, certain current oscillations due to the interaction between inductance of the line with the dc link capacitor of the motors may lead to excessive current overshoots that may trip the protective devices and render the system non-operational.

Given a substation, it needs to be determined how many trucks can engage per segment of the line. This case was simulated in Matlab using Simulink. The parameters used in the simulation are mentioned in Table 1.

978-1-4244-4782-4/10 $26.00 © 2010 IEEE   1159

TABLE I. SIMULATION PARAMETERS

Substation

Input AC source voltage	33,000	V
Line frequency	50	Hz
3-phase short circuit level	150	MVA
X/R ratio	10	
DC bus voltage (No Load)	2590	V

Impedance between transformer and rectifier

Series resistance	0.5	mΩ
Series inductance	1.0	μH
DC Filter at Rectifier Output		
Inductance	1	mH
ESL	0.2	mΩ
Capacitance	200	μF

Transformer (three phase wye-delta connection)

Capacity	5.5	MVA
Input/output voltage	33,000 / 950	V

Primary Winding

Resistance	60.5	mΩ
Leakage inductance	4.2319	mH
Secondary Winding		
Resistance	2.16	mΩ
Leakage inductance	0.15109	mH
Magnetizing Branch		
Resistance	15.126	kΩ
Inductance	48.146	H

Transformer (three phase delta-delta connection)

Capacity	5.5	MVA
Input/output voltage	33,000 / 950	V

Primary Winding

Resistance	181.5	mΩ
Leakage inductance	12.696	mH
Secondary Winding		
Resistance	2.16	mΩ
Leakage inductance	0.15109	mH

Magnetizing Branch

Resistance	45.377	kΩ
Inductance	144.44	H

Rectifier

On-state resistance	1	mΩ
Diode forward voltage	0	V
Snubber resistance	100	kΩ
Snubber capacitance	Inf.	μF

DC Line (Trolley Line)

Resistance	0.1125	Ω/km

Truck

Pantograph Impedance

Resistance	0.9375	mΩ
Inductance	7.5	μH

DC Link Capacitor

Series resistance	0.167	mΩ
Capacitance	9000.0	

Inverter

On-state resistance	1	mΩ
Device forward voltage	0.8	V
Diode forward voltage	0.8	V
Snubber resistance	1000	Ω

Motor Load

Frequency	50	Hz
Poles	6	
Stator resistance	32.3	mΩ
Stator inductance	0.6	mH
Rotor resistance	58.82	mΩ
Rotor inductance	1	mH
Mutual Inductance	12	mH
Inertia	100	Kg.m2

A. Results with no current limiting reactor

The ac source is at 11 kV l-l which is rectified using a 12 pulse converter in the substation. The truck is initially assumed to operate using the onboard engine which charges the DC link to 1800 V. The simulation started with no trucks connected to the dc bus of the rectifier. Then, at t=8 sec the first truck was engaged to the dc line at a distance of 500m from the substation. Finally, at t=9.5 sec the second truck was engaged to the line at a distance of 250 m from the substation. The truck motors were vector controlled to run at a fixed speed of 700 rpm and the load torque on each motor was selected to be 17000 Nm. The trucks are connected to the overhead lines with the pantograph. Figures 4 to 9 shows the simulation results.

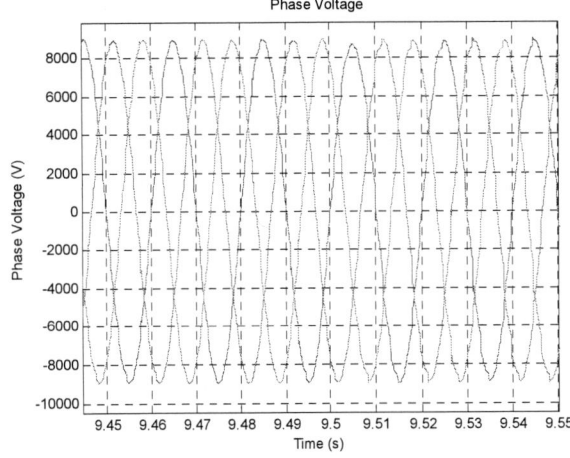

Figure 4. Phase voltage measured without the truck reactors

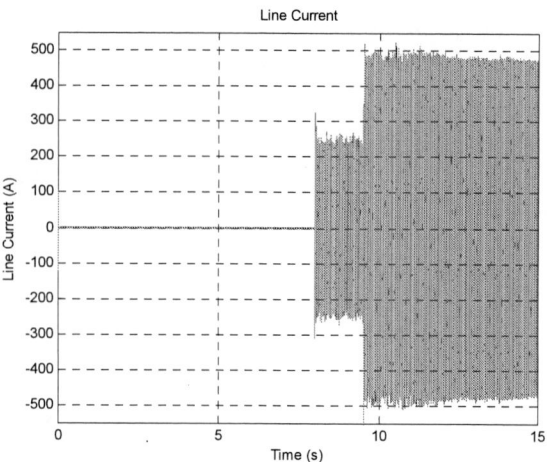

Figure 5.   Line current measured without the truck reactors

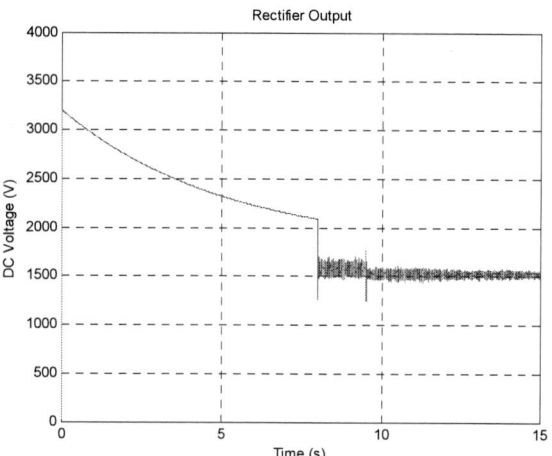

Figure 6.   Overhead DC line voltage without the truck reactors

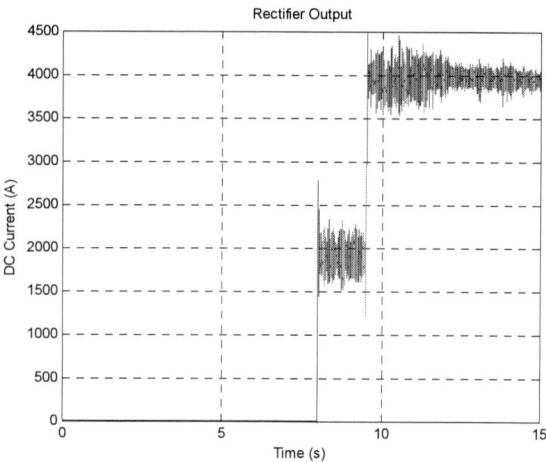

Figure 7.   Overhead DC line current with trucks engaging

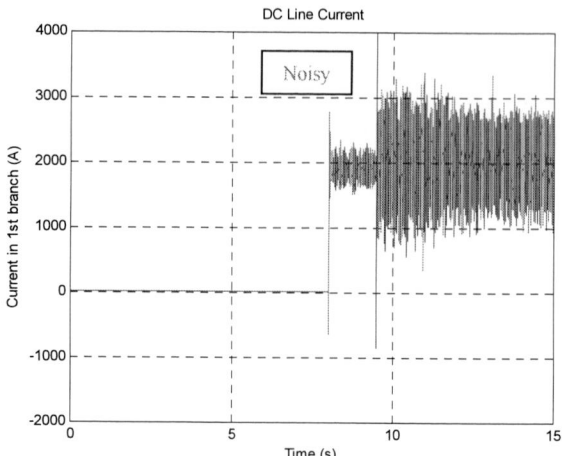

Figure 8.   Truck 1 current without reactor

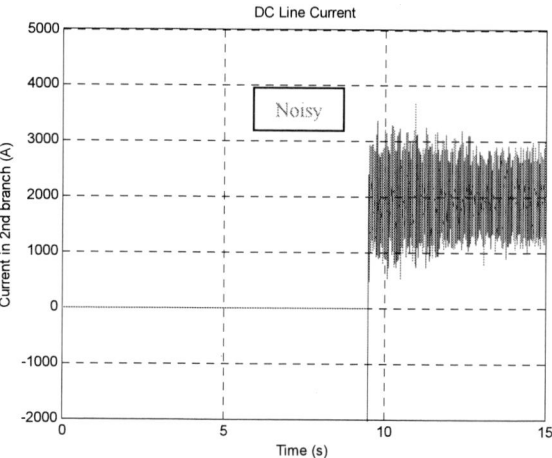

Figure 9.   Truck 2 current without reactor

The DC line currents in figures 7, 8 and 9 show considerable oscillation whenever a truck is engaged to the DC line i.e. when the truck shifts from onboard diesel alternator to the trolley line. The high current peaks triggers the protective devices and impede any further operation of the machine in practice. Without the reactor, the input side line current had a THD of 5.31% and the dominant harmonics at 4.25% (11th) and 2.86% (13th).

B.   Results with current limiting reactor

The use of a current limiting reactor is expected to considerably reduce the current peaks. In this case a current limiting reactor of 0.5 mH having a damping resistance of 0.1 Ω is used along with the pantograph to engage to the overhead lines. Figures 10 to 15 show the simulation results.

As can be seen from figures 13 to 15, the peak current oscillations have reduced. This would keep the protective devices from tripping and thus increase the reliability of the truck.

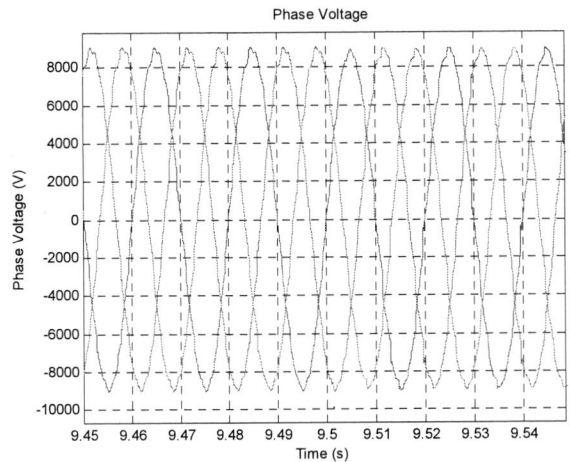

Figure 10. Phase voltage measured with the truck reactors

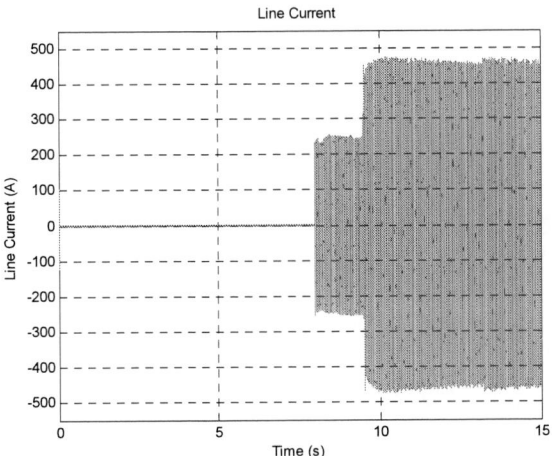

Figure 11. Line current measured with the truck reactors

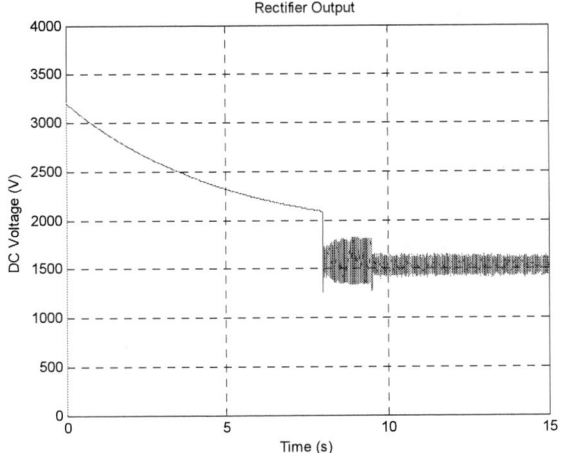

Figure 12. Overhead DC line voltage with the truck reactors

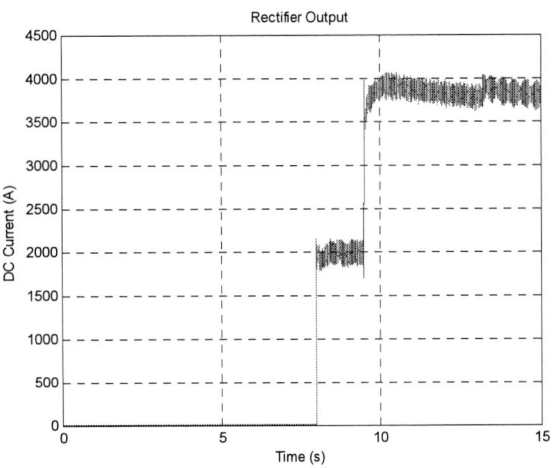

Figure 13. Overhead DC line current with trucks engaging

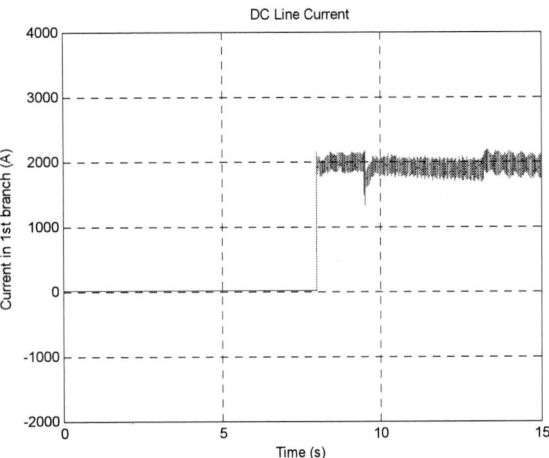

Figure 14. Truck 1 current with reactor

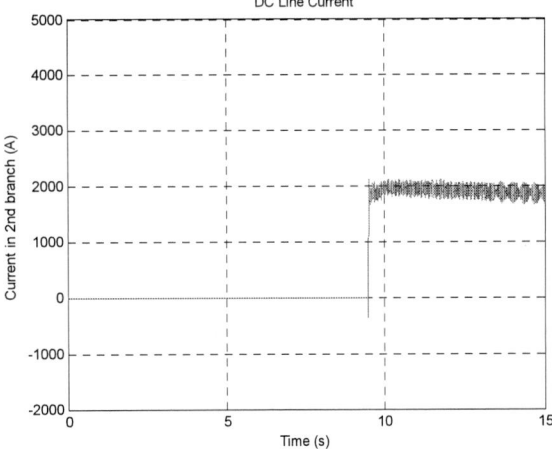

Figure 15. Truck 2 current with reactor

THD of 5.41% with the dominant harmonics at 4.41% (11th) and 2.85% (13th) is seen in the input current.

C. Advantages with the current limiting reactor

With no current limiting reactor, when the first truck engages, the peak DC current is 2.8 kA. When the second truck engages, the peak DC current is 4.9 kA.

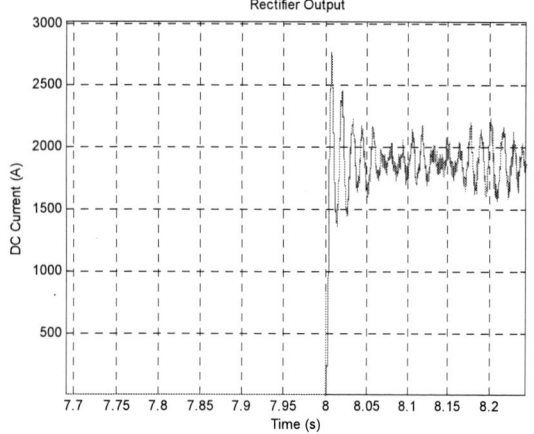

Figure 16. Peak current with Truck 1 engaging without reactor

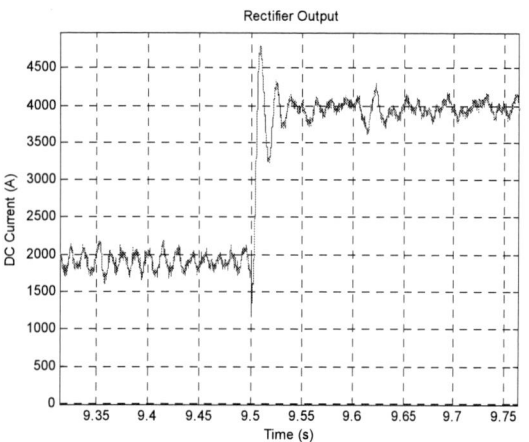

Figure 17. Peak current with Truck 2 engaging without reactor

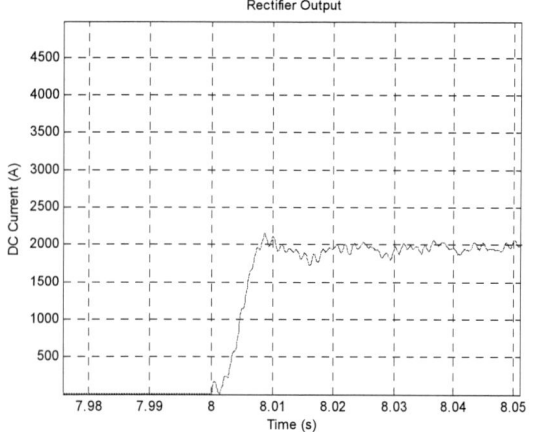

Figure 18. Peak current with Truck 1 engaging with reactor

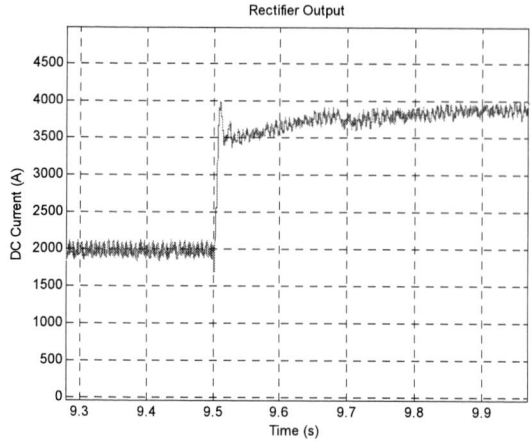

Figure 19. Peak current with Truck 2 engaging with reactor

With the reactor, when the first truck engages, the peak DC current is smooth at 2 kA. When the second truck engages, the peak DC current is only about 4 kA.

IV. FIELD EXPERIMENTS

Figures 20 and 21 show the field results with and without the reactor. The current spikes reduce considerably when using the line reactors.

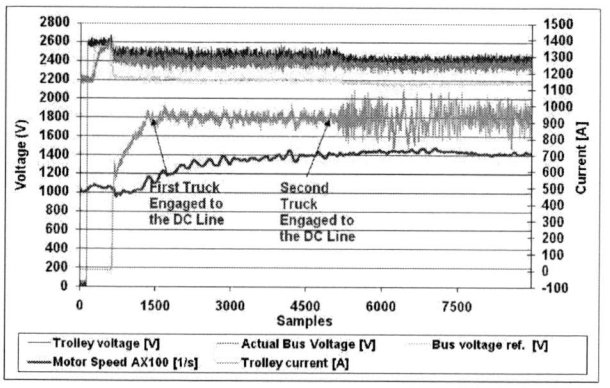

Figure 20. Experimental results without a reactor

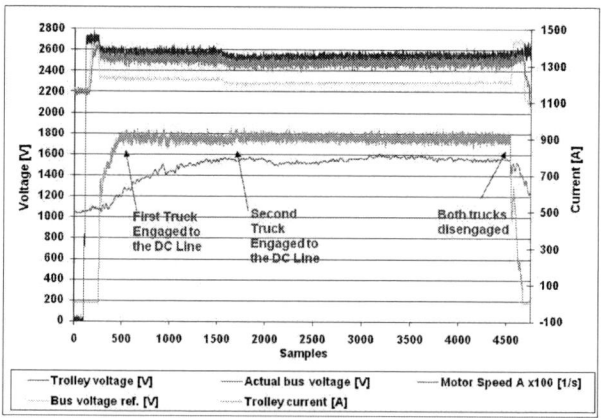

Figure 21. Experimental results with a reactor

V. BENEFITS ANALYSIS

A case study is presented with step by step calculations to show the increased velocity on a grade. A truck with a payload of 360 short tons (US unit) is chosen.

Payload: 360 tons

Gross vehicle weight (GVW) of truck with trolley hardware (lb): 1274500

Rated engine brake power (hp): 3500

Rated engine flywheel power (hp): 3346

Overall efficiency from flywheel to ground (%): 85.8

HP at ground = 2870

Road Resistance (lb/ton): 30

Vehicle Rolling Resistance (lb): 19117.5 (1.5 %)

Grade (lb): 127450 (10 %)

Tractive Effort (lb): 19117.5 + 127450 = 146567.5

Diesel speed on grade (miles/ hrs):

HP at ground/ Tractive effort = 7.34

Trolley line voltage (V): 1800

Trolley current (A): 2200 A

Trolley power at ground (hp): 4766

Trolley speed on grade (miles/hrs): 12.19

Speed increase on trolley: 66 %

The above calculations could be performed on any other payload. Significant savings are observed in the form of increased velocity on grade as well as lower fuel rate on trolley. Figure 22 shows haul trucks operating on a trolley.

Figure 22. Trucks operating on a trolley in a mine

VI. CONCLUSIONS

The haul trucks technology has shown an enormous development in the last years, producing an important improvement in the performance with increasing payload capacity. Electric haul trucks outfitted with pantographs can pull power from an overhead trolley line. The trucks run on diesel power in the pit, around the crusher and on level segments. The trolley line provides power on the grade resulting in increased truck speed, extending intervals between engine overhauls and reducing energy costs.

A mining diesel electric operating on trolley line has significant advantages in terms of increased speed on a grade and reduced fuel consumptions. However, the system needs to be designed such that at the grid interface there are no current oscillations, voltage drops or harmonic current injections. In addition, the system should be stiff enough such that greater number of trucks can be engaged. In this paper, implementation challenges of a trolley assist system in the field were addressed using simulation study and actual field experiments. A reactor based solution was provided which considerably reduced any kind of current oscillations during the engaging process of the truck. Field experiments have shown similar results.

ACKNOWLEDGMENT

The authors would like to acknowledge the technical help and data provided by the Siemens mobile mining engineering group.

REFERENCES

[1] G. Leen, D. Heffernan, "Expanding automotive electronic systems", IEEE Computer, Vol. 35, Issue 1, pp. 88 –93, January, 2002.

[2] T.M. Jahns, V. Blasko, "Recent advances in power electronics technology for industrial and traction machine drives," Proceedings of the IEEE , Vol. 89, Issue 6, pp. 963 –975, Jun 2001.

[3] R. Hanxleden, A. Botorabi, S. Kupczyk, "A Co-Design Approach for Safety-Critycal Automotive Applications", IEEE Micro, Special Issue on Embedded Fault-Tolerant Systems, Vol. 18, No.5, pp.66-79, Sep./Oct. 1998.

[4] J.D. van Wyk, F.C. Lee, D. Boroyevich, "Power Electronics Technology: Present Trends and Future Developments", Proceedings of IEEE, Vol. 89, No.6, pp.799-802, June 2001.

[5] F.C. Lee, J.Daan van Wyk, D. Boroyevich, P. Barbosa, "An integrated approach to power Electronics systems", Proceedings of Power Conversion Conference PCC-Osaka 2002, Vol. 1, pp. 7-12.

[6] H. Contreras, "GPS+ GLONASS Technology at Chuquicamata Mine, Chile", Proceedings of ION GPS 98, 11th International Technical Meeting of the Satellite Division of the Institute of Navigation, September 15-18, 1998, Nashville, Tennessee, pp.93-97.

[7] J.A. Flinn, C. Waddell, M.A. Lowery, Phelps Dodge Morenci, Inc., "Practical Aspects of GPS Implementation at the Morenci Copper Mine", Proceedings of ION GPS-99, 12th International Technical Meeting of the Satellite Division of the Institute of Navigation, September 14-17, 1999, Nashville, Tennessee, pp.915-919.

[8] J.A. Flinn, S.M. Shields, "Optimization of GPS on Track-Dozers at a Large Mining Operation", Proceedings of ION GPS-99, 12th International Technical Meeting of the Satellite Division of the Institute of Navigation, September 14-17, 1999, Nashville, Tennessee, page 927-932.

[9] S. Bernett, "Recent developments of high power converters for industry and traction applications", IEEE Trans. on Power Electronics, Vol.15, No.6, Nov 2000, pp.1102-1117.

[10] W. G. Koellner, G. M. Brown, J. Rodriguez, J.Pontt, P. Cortes and H. Miranda, "Recent advances in mining haul trucks," IEEE Transactions on Industrial Electronics, Vol. 51, Issue 2, pp. 321-329, April 2004.

[11] G.M. Brown, W.G. Koellner, "A GTO Powered AC Drive System Increases the Performance of Off-Highway Haul Trucks", in

Proceedings of the IEEE Industry Applications Society Annual Meeting (IAS 1999), Vol. 1, pp. 222 – 233, Oct. 3-7, 1999.

[12] J. Mazumdar, W. Koellner and F. Gerdts, "Trolley assist to counter high fuel costs and enhanced operational productivity," in Proceedings of the Load and Haul Asia, Nov. 17 – 19, 2008.

[13] K. J. Kutsmeda, K. G. Fehrle and P. J. Trick, "Computer modeling, simulation, and validation by field testing of a traction power system for electric trolley buses," Proceedings of IEEE/ASME Joint, Apr 1995.

[14] G.M. Brown, B.J. Ebacher and W.G. Koellner, "Increased Productivity with AC Drives for Mining Excavators and Haul Trucks", in Proceedings of the IEEE Industry Applications Society Annual Meeting (IAS 2000), Vol. 1, pp. 22 – 37, Oct. 8-12, 2000.

Regenerative AC Electronic Load with One-Cycle Control

In Wha Jeong, Mikhail Slepchenkov, Keyue Smedley, and Franco Maddaleno
Power Electronics Laboratory
Dept. of Electrical Engineering and Computer Science
University of California, Irvine
Irvine, CA 92697, USA

Abstract—**Variable linear or nonlinear loads are critical for testing AC power supplies and power equipment in various operating conditions. At present, most testing loads are based on bulky resistors, capacitors, and inductors on which the testing energy is consumed to generate excessive heat, and moreover, the impedance value is limited by finite combination of these bulky loads. Some commercial electronic loads are available, which provide flexible impedance control but the energy absorbed by the load is still dissipated as heat. The work presented in this paper is a continuous effort of UCI Power Electronics AC load project. A Regenerative AC Electronic Load (RACEL) is developed with One-Cycle Controller (OCC), which doesn't require any Digital Signal Processor (DSP) and software in the control loop. Using a reconfigurable Field Programmable Analog Array (FPAA), the OCC can be integrated into a single FPAA with easy reconfiguration and simple interface structure. The proposed RACEL can emulate any impedance load, linear or nonlinear as well steady or dynamic with the regeneration of the testing energy. Simulation and experimental results are presented to verify the proposed RACEL performance.**

I. INTRODUCTION

At present, passive dissipative loads are used for most power equipment testing. In these situations, the testing energy is wasted. At high power levels, heat management of the bulky load becomes very challenging. Moreover, the testing impedance cannot be adjusted smoothly. When a different load configuration is required, reconnection of the load bank has to be performed to achieve different load parameters. Some commercial electronic loads are available for relatively low power rating in the range of 1-5kW, which provides flexible impedance control but the energy absorbed by the load is still dissipated as heat. What in great need is a Regenerative AC Electronic Load (RACEL). RACEL is a piece of power electronic equipment that emulates any physical impedance and sends the testing energy back to the power grid. The RACEL may be used in the burn-in test of AC power supplies, including UPSs, AC motor drives, and protection facilities.

Some researchers proposed RACELs with regenerative function based on hysteresis current control and repetitive control using Digital Signal Processors (DSPs) [1-2]. Those control methods require real time calculation and some complex circuit to implement the main controllers. The UCI Power Electronics Laboratory proposed a regenerative AC electronic load in [3-4]. In this effort, One-Cycle Control (OCC) was used to realize a three-phase AC electronic load. Due to time limitation of the MS thesis, the regenerative operation was not completed. The work presented here is the continuation of the effort in [3-4].

In this paper, a single-phase RACEL is implemented with One-Cycle Controller. Based on a back-to-back configuration, the proposed RACEL can emulate linear or nonlinear load in steady or dynamic condition with energy regeneration capability.

II. REGENERATIVE RACEL CONFIGURATION

Fig. 1 shows the single-phase regenerative RACEL with the OCC controller. The Equipment Under Test (EUT) is connected to the input rectifier, which is connected to the output inverter in a back-to-back configuration. The control objectives of the OCC controller are to draw the required active and reactive power from the testing AC power source for the input rectifier and make the output current synchronous with the grid voltage for the output inverter. The current reference generator produces the current reference signal V_{iref} for the OCC controller to emulate the user-defined linear or nonlinear load.

The power stage of the single-phase RACEL based on OCC control is shown in Fig. 2. The input rectifier and the output inverter are all based on the standard H-bridges. The input rectifier emulates the specified load and the output inverter feeds back the absorbed energy to the utility grid with unit power factor. When the input of the RACEL is connected to the AC power supplies or the power equipment under test, the real impedance load can be emulated losslessly by controlling the amplitude and phase of the input current and feedback of the energy to the power grid.

978-1-4244-4782-4/10 $26.00 © 2010 IEEE

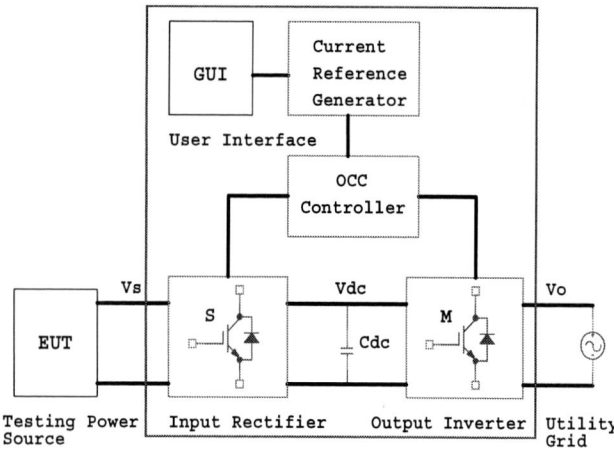

Figure 1. Regenerative RACEL implementation with OCC control.

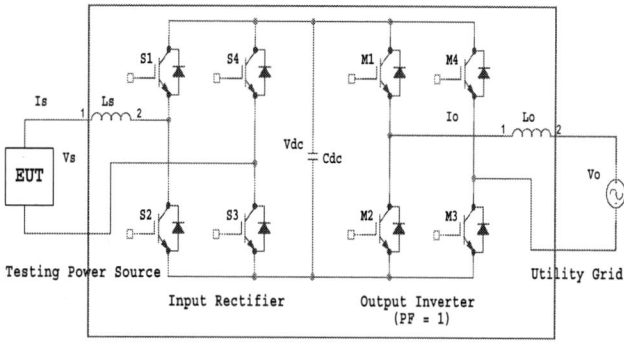

Figure 2. Single-phase regenerative RACEL power stage configuration.

In addition, the proposed RACEL can be used in dynamic load emulation. Since the OCC controller has fast and stable response as well as simple circuit, it can perfectly be used in the RACEL applications.

III. OCC CONTROL OF RACEL

The proposed RACEL has three working modes: linear load mode, rectified load mode, and user-defined load mode. In linear load mode, the RACEL can exactly emulate the load whose impedance value can be adjusted continuously. In rectified load mode, the RACEL can emulate nonlinear rectified loads for a wide range of testing applications. Finally, in user-defined load mode, if user has a complicated load model and needs a specific dynamic test, the RACEL can provide unique capability to emulate any complex load. The proposed RACEL works in regenerative operation for all the operating modes.

The OCC controller of the single-phase RACEL requires one current reference signal (V_{iref}) with the phase shift from the line voltage (V_s) of the EUT depending on the emulated load. The input rectifier accurately controls the input current to track the current reference signal and make the RACEL to function as the specified linear or nonlinear impedance load.

From [5], the active and reactive testing power processed by the input rectifier can be directly controlled by changing

the values of k_h and k_v, respectively. On the other hand, the output inverter of the RACEL can operate with the fixed amplitude of the saw-tooth carrier, V_m and the regenerated testing power can be controlled by only adjusting k_h with unit power factor.

The proposed RACEL uses a FPAA-based OCC controller. Using a statically or dynamically reconfigurable Field Programmable Analog Array (FPAA), the OCC controller has a simple integrated structure and enough flexibility to implement variable internal parameters and multiple operation functions for the regenerative RACEL. The Anadigm 5V 44pin FPAA is used to realize the single-FPAA OCC controller with the help of AnadigmDesigner2 software.

IV. CURRENT REFERENCE GENERATOR OF RACEL

Fig. 3 shows the Graphical User Interface (GUI) window of the current reference generator for the regenerative RACEL using GUIDE, the GUI development environment of MATLAB. This current reference generator particularly has unique function to emulate any complex or dynamic load with the help of simulation tools, for example, Simulink. It can open the simulation software which includes the schematic diagram of the complicated or dynamic load model, execute the simulation program, and receive the simulated data of the dynamic load to accurately generate the current reference signals.

The stand-alone current reference generator of the regenerative RACEL uses a 340 x 240 resolution touch screen and an 8-bit microcontroller, ATmega328P. With the help of the user-friendly interface, any testing condition or any impedance load can easily be selected and realized by generating the proper current reference signals. In linear load mode, the RACEL can emulate the linear impedance load by the specified R, L, and C values or the testing active and reactive power values, P and Q.

For the single-phase regenerative RACEL, the current reference signal V_{iref} of the OCC controller can be calculated by the following equations in each case.

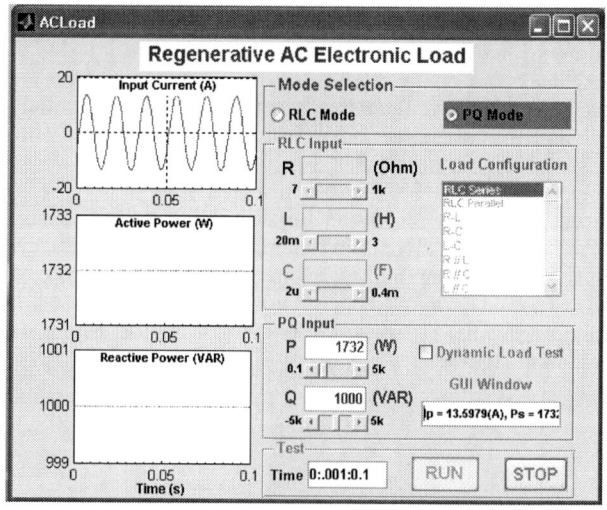

Figure 3. GUI-type current reference generator of RACEL.

978-1-4244-4782-4/10 $26.00 © 2010 IEEE   1167

Assuming that the input line voltage V_s and input current i_s from the EUT are sinusoidal, the input voltage V_s and current i_s of the input rectifier can be described as:

$$V_S = V_{SM} \sin \omega t$$
$$i_S = I_{SM} \sin(\omega t - \varphi) \tag{1}$$

where V_{SM} and I_{SM} are the voltage and current amplitudes of the tested AC power supply, φ is the phase difference between the input voltage and current or the phase angle of the impedance load, and ω is the angular frequency of the tested AC power supply.

A. Resistive Load Emulation

. In this case, the RACEL appears pure resistive characteristic. The current reference signal V_{iref} can be described as follows:

$$\varphi_{ref} = 0$$
$$I_{SMref} = \frac{2P_{Sref}}{V_{SM}} = \frac{V_{SM}}{R}$$
$$V_{iref} = R_S \cdot i_{ref} = R_S \cdot I_{SMref} \sin(\omega t) \tag{2}$$

where R is the required load resistance and R_s is the input current sensing resistance.

B. Inductive Load Emulation

The RACEL works as a RL impedance load during this test condition.

$$\varphi_{ref} = \tan^{-1}\left(\frac{\omega L}{R}\right)$$
$$I_{SMref} = \frac{2P_{Sref}}{V_{SM} \cos\varphi_{ref}} = \frac{V_{SM}}{\sqrt{R^2 + (\omega L)^2}} \tag{3}$$
$$V_{iref} = R_S \cdot i_{ref} = R_S \cdot I_{SMref} \sin(\omega t - \varphi_{ref})$$

where R and L are the resistance and inductance of the impedance load, respectively.

C. Capacitive Load Emulation

In this test condition, the RACEL shows a RC impedance characteristic.

$$\varphi_{ref} = -\tan^{-1}\left(\frac{1}{\omega CR}\right)$$
$$I_{SMref} = \frac{2P_{Sref}}{V_{SM} \cos\varphi_{ref}} = \frac{V_{SM}}{\sqrt{R^2 + \left(\frac{1}{\omega C}\right)^2}} \tag{4}$$
$$V_{iref} = R_S \cdot i_{ref} = R_S \cdot I_{SMref} \sin(\omega t - \varphi_{ref})$$

where R and C are the resistance and capacitance of the impedance load, respectively.

D. Active and Reactive Power Load Emulation

From (1), the average active power P_S and reactive power Q_S can be represented as follows:

$$P_S = \frac{V_{SM} I_{SM}}{2} \cos\varphi$$
$$Q_S = \frac{V_{SM} I_{SM}}{2} \sin\varphi \tag{5}$$

If the user wants to emulate the active power P_{Sref} and reactive power Q_{Sref} of the specified impedance load, it is clear that the amplitude I_{SMref} and phase angle φ_{ref} of the input current reference can be calculated by (5).

$$\varphi_{ref} = \tan^{-1}\left(\frac{Q_{Sref}}{P_{Sref}}\right)$$
$$I_{SMref} = \frac{2P_{Sref}}{V_{SM} \cos\varphi_{ref}} \tag{6}$$
$$V_{iref} = R_S \cdot i_{ref} = R_S \cdot I_{SMref} \sin(\omega t - \varphi_{ref})$$

E. Nonlinear Load Emulation

The RACEL operates as a nonlinear rectified load or a distorted load for the harmonic test. For three-phase applications, various unbalanced loads can also be emulated by the three-phase RACEL.

For the output inverter of the RACEL, the control goal is to make the output current i_o of the RACEL synchronous with the grid voltage V_o and contribute to the unit power factor injection of the recycled testing energy with the highest efficiency. In this condition, the control requirement is just realizing unit power factor operation and stabilizing the dc link voltage. Thus the dynamic performance of the output inverter is not strict.

V. SIMULATION RESULTS

The single-phase regenerative RACEL with the OCC controller has been modeled and tested by PSIM to verify the steady and dynamic performance of the proposed regenerative RACEL. In the simulation model, the input line voltage V_s and the output grid voltage V_o have the same amplitude, 208Vrms and phase angle. The dc link voltage V_{dc} is 400Vdc and the dc link capacitor C_{DC} is 5mF.

In the input rectifier, the input inductor Ls is 2mH and the switching frequency is 16kHz. In the output inverter, the inductor Lo is 3mH and the switching frequency is the same value, 16kHz.

Fig. 4 shows the simulation result for the RL load emulation with 60° phase angle. The input current i_s has 40A peak amplitude and 60° phase delay compared to the input line voltage V_s, exactly following the current reference signal with the dc link voltage control. In addition, the output current i_o is in 180° phase shift with the grid voltage V_o. The output inverter feeds back the testing energy to the utility grid with unit power factor.

978-1-4244-4782-4/10 $26.00 © 2010 IEEE

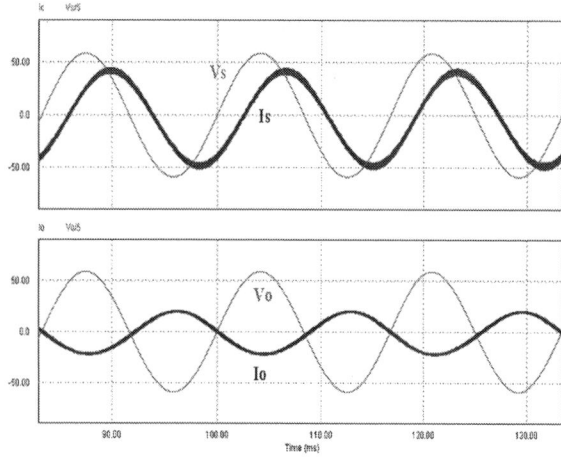

Figure 4. Input and output waveforms with RL load, 60°.

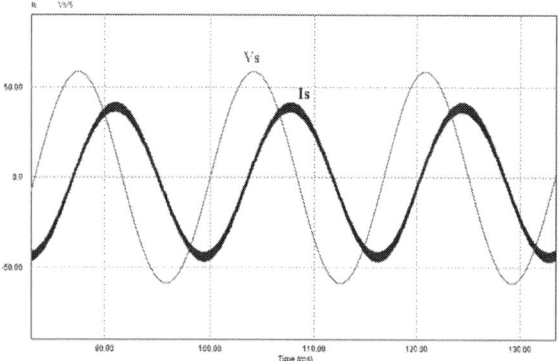

Figure 5. Input waveforms with L load (20mH), 90°.

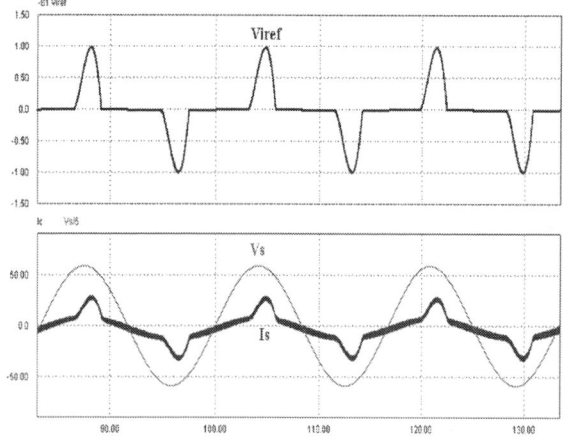

Figure 6. Current reference and input waveforms with diode rectified load.

Fig. 5 shows the input line voltage and current waveforms for the L load emulation with 90° phase angle. In this case, the RACEL works as a 20mH inductor which has a huge size and a massively expensive price for high power applications. Since there is the power dissipation in the semiconductor switches of the input rectifier, the pure inductor emulation may not be possible to realize by using the input current control of the input rectifier with 90° phase-shifted current reference signal. But since a real inductor has some resistance, the practical inductor emulation would not be purely reactive.

Fig. 6 shows the current reference signal, the input line voltage, and the input current waveforms when the RACEL emulates a nonlinear rectified load with a diode bridge rectifier, a 1.5mH dc link inductor, a 1mF dc link capacitor, and a 100Ω resistor load. The actual input current has a long tail between two sharp conduction periods but looks almost identical to the waveform of the current reference signal.

VI. EXPERIMENTAL RESULTS

On the basis of the simulation results, a 5kVA single-phase regenerative RACEL with the OCC controller is designed and built to verify the effectiveness and performance of the OCC based RACEL. The design specification of the RACEL is the same as the simulation condition.

A variable transformer is used to change the input line voltage V_s. To realize the OCC controller, a single FPAA, AN221E04 from Anadigm, is used to integrate all the OCC core circuit into a single IC with easy reconfiguration and simple interface structure.

Fig. 7 shows the current reference signal and the input voltage waveforms for the RL load emulation with 30° phase angle. In this test, the active power P_{Sref} is 1732W and the reactive power Q_{Sref} is 1000VAR. Using (6), the input current is 13.6A peak and the required amplitude of the input current reference is 0.05x13.6A = 680mV which is almost the same as 620mV in Fig. 7.

Fig. 8 shows the experimental result for a nonlinear load emulation with a diode bridge rectifier. Since the peak amplitude of the input current reference is 780mV, the RACEL prototype can provide the specified test condition of the diode rectified load with 15.6A peak input current. This test condition can be applicable to the harmonic test for typical power equipment such as power transformers.

Fig. 9 and 10 show the input line voltage and current waveforms for the RL load emulation with 20° and 45° phase angle, respectively. The input current i_s has about 7A peak and 6A peak amplitude with 20° and 45° phase delay, respectively compared to the input line voltage V_s.

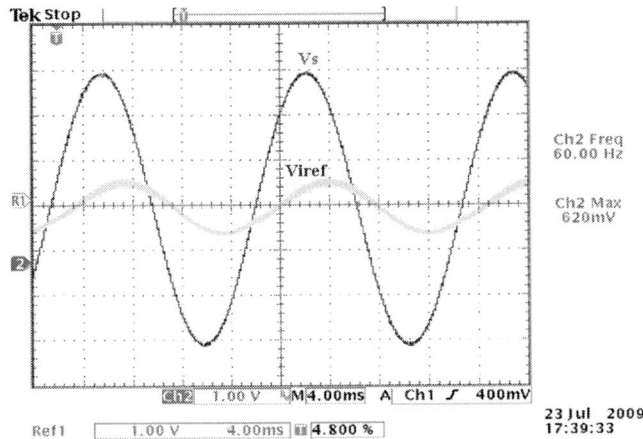

Figure 7. Input voltage and current reference waveforms with RL load, 30°.

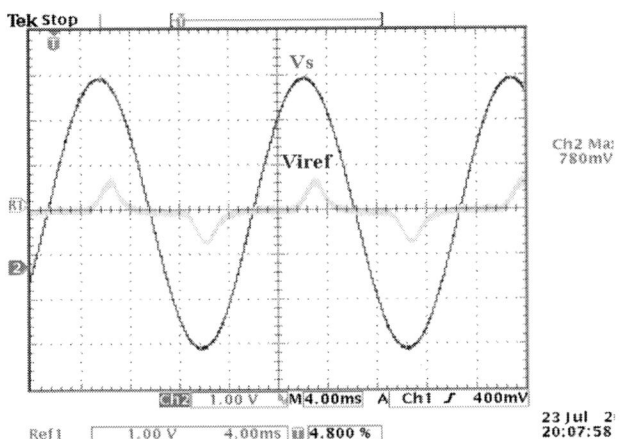

Figure 8. Input voltage and current reference waveforms with diode rectified load.

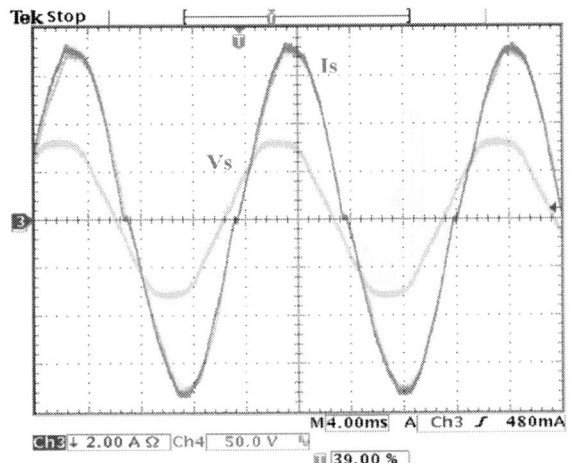

Figure 9. Input waveforms with RL load, 20°.

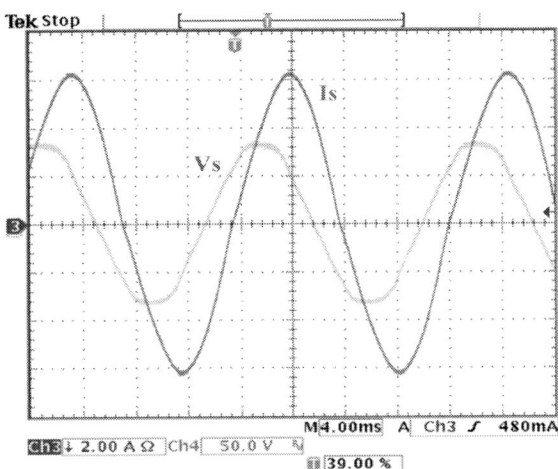

Figure 10. Input waveforms with RL load, 45°.

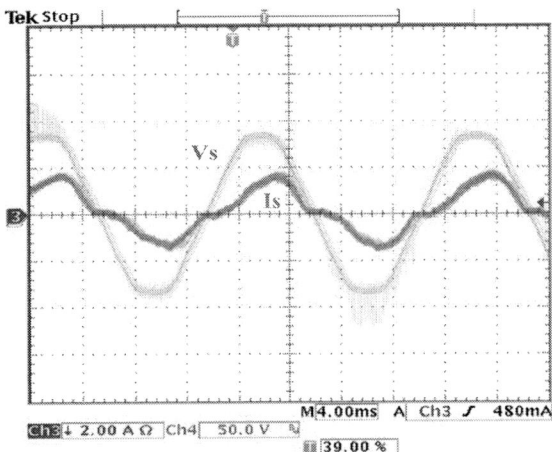

Figure 11. Input waveforms with distorted load.

Fig. 11 shows a nonlinear load emulation with harmonic distortion. This current distortion affects the AC power supplies or the testing power equipment, and may cause the testing power transformers to overheat from the harmonic loss. The proposed RACEL can provide special test conditions to emulate this kind of nonlinear loads for a wide range of testing applications.

Fig. 12 shows the experimental result for regenerative operation of the proposed RACEL. The input rectifier emulates the RL load with some phase delay and the output inverter feeds back the absorbed testing energy to the utility grid as seen in Fig. 12.

978-1-4244-4782-4/10 $26.00 © 2010 IEEE   1170

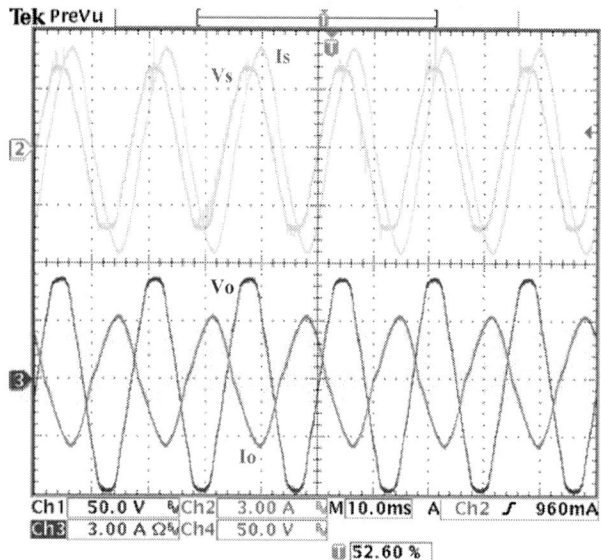

Figure 12. Input and output waveforms with regenerative operation.

VII. CONCLUSIONS

In this paper, a RACEL is implemented with a FPAA-based One-Cycle Control (OCC) to emulate any linear or nonlinear impedance load. For the proposed RACEL with energy recycling capability, the testing energy absorbed from the tested AC power equipment can be fed back to the utility grid with unit power factor. In addition, if necessary, the proposed RACEL can operate as a variable AC or DC power source by the back-to-back configuration.

The control method and controller design of the RACEL are simple and flexible with the use of the FPAA-based OCC controller instead of using separated integrator, comparator, and compensator. With the dynamic reconfigurability, the FPAA-based OCC controller can implement multiple operation functions and maintain precision operation as various AC loads. The proposed RACEL can easily be extended from single-phase to three-phase applications with the help of the three-phase OCC control. Furthermore, the regenerative RACEL can emulate and change any complicated test load with the developed GUI program and user's simulation tool. There is only small power dissipation in the semiconductor switches to realize the test load condition.

Simulation and experimental results based on the 5kVA single-phase RACEL prototype are presented to verify the regenerative RACEL performance as the linear and nonlinear loads. In the future, an OCC multilevel regenerative RACEL for high voltage and high power applications will be discussed and demonstrated with the emulation capability of various steady state and dynamic loads.

REFERENCES

[1] Jian-feng Zhao, Shi-feng Pan, and Xun Wang, "High power energy feedback AC electronic load and its application in power system dynamic physical simulation," in Proc. IEEE IAS, 2007, pp. 2303-2310.

[2] Xu She, Yunping Zou, Chengzhi Wang, Lei Lin, Jian Tang, and Jian Chen, "Research on power electronic load: topology, modeling, and control," in Proc. IEEE APEC, 2009, pp. 1661-1666.

[3] Simone Primavera, Giuseppe Rella, Franco Maddaleno, and Keyue Smedley, "Three-Phase OCC AC Load Emulator," UCI Power Electronics Laboratory Internal Report, March 2007.

[4] Simone Primavera and Giuseppe Rella, "Three-Phase Electronic Load," MS Thesis, POLITECNICO DI TORINO, 2007.

[5] Taotao Jin, Lihua Li, and Keyue Ma Smedley, "A universal vector controller for four-quadrant three-phase power converters," IEEE Trans. Circuits and Systems, vol. 54, no. 2, pp. 377-390, Feb. 2007.

[6] Guozhu Chen and Keyue M. Smedley, "Steady-state and dynamic study of one-cycle-controlled three-phase power-factor correction," IEEE Trans. Ind. Electron., vol. 52, no. 2, pp. 355-362, Apr. 2005.

[7] Yang Chen and Keyue Ma Smedley, "A cost-effective single-stage inverter with maximum power point tracking," IEEE Trans. Power Electron., vol. 19, no. 5, pp. 1289-1294, Sep. 2004.

[8] N. Femia, D. Granozio, G. Petrone, G. Spagnuolo, and M. Vitelli, "Optimized one-cycle control in photovoltaic grid connected applications," IEEE Trans. Aerospace and Electronic Systems, vol. 42, no. 3, pp. 954-971, Jul. 2006.

[9] Jose R. Rodriguez, Juan W. Dixon, Jose R. Espinoza, Jorge Pontt, and Pablo Lezana, "PWM regenerative rectifiers: state of the art," IEEE Trans. Ind. Electron., vol. 52, no. 1, pp. 5-22, Feb. 2005.

[10] Kishore Chatterjee, Ambrish Chandra, Kamal Al-Haddad, and Pierre Jean Lagace, "A PLL less VAR generator based on one-cycle control," in Proc. Inter. Conf. Harmonics and Quality of Power, 2004, pp. 512-518.

A High Efficiency Regulated Charge Pump over Wide Input and Load Range

Rong Guo[1], Liyu Yang[1], Alex Huang[1] and John Endredy[2]
[1]North Carolina State University, Raleigh, NC 27695
[2]RF Micro Devices, 7628 Thorndike Rd., Greensboro, NC 27409

Abstract- **A multi-modes regulated charge pump with improved efficiency over wide input and load range is proposed in this paper. The output voltage is regulated to 5V when input is 4.5-2.7V. By alternating the switch network, it can realize 2x, 1.5x, and 1.33x conversion modes, which improves the efficiency over wide input range. During each mode, the output voltage is regulated by a controlled current source. The regulated charge pump is designed for a very wide load range (10mA to 500mA). A hybrid control scheme is proposed and analyzed to achieve a high efficiency over desired load range. Both of the simulation and test results are provided to verify the control scheme. The results show that the total loss at light load can be reduced by 56% to 83%.**

I. INTRODUCTION

In DC-DC converters, charge pump regulators are well known for their compact sizes since they do not have large, and expensive magnetic components. Unlike LDO regulators, which can only step-down the input voltage, charge pump regulators can either boost or buck the input voltage by using different topologies. As a result, charge pump regulators are usually considered as an ideal solution in low power portable applications.

However, traditional charge pump, like voltage doubler, is not a good solution for applications where the input voltage varies a lot, such as the battery, but the output need to be regulated. It only has a high efficiency when the desired output voltage is close to two times input voltage. The power loss will be particularly high when the input voltage is close to the output voltage. Multi-modes charge pumps were proposed by many researchers to improve the efficiency at wide input range. By alternating the switch network, parallel or series flying capacitors during charging and discharging period, multiple conversion ratios can be obtained [1]. Once the optimum mode is selected for one input voltage, the output voltage can be regulated using on-resistance regulation, duty cycle modulation, etc. [2]

On the other side, when charge pump needs to operate over a wide load range, the light load efficiency is very important, especially for applications, like cell phone, which mostly work in standby mode. Normally, the devices sizes are optimized to reduce conduction loss and maintain regulation at maximum load; therefore, they are big devices with small on-resistance. However, small on-resistance will not contribute to the light load efficiency because the conduction loss is not dominant anymore. On the contrary, those big devices still need a lot of power to drive them since the gate driver loss does not scale down with the load current. In order to improve the light load efficiency, the driving loss needs to

be as low as possible. This is also true for those inductor-based DC-DC converters. Many solutions have already been proposed, by using either advanced control or improved power stage, such as pulse frequency modulation, pulse skipping modulation, device segmentation, etc. [3][4][5]

This paper proposed a high efficiency regulated multi-modes charge pump over wide input and load range. The input voltage is a single Li-ion battery with voltage between 4.5V and 2.7V, while the output voltage needs to be regulated to 5V. The charge pump can automatically adjust the conversion modes: 2x, 1.5x, and 1.33x according to different input voltage, therefore enhance the maximum efficiency the converter can achieve. The maximum load current is 500mA, but can go as low as 10mA. When load current decreases to light load condition, device segmentation control is used, and when the load further decreases, pulse skipping modulation is used in addition to device segmentation. By using the proposed hybrid control scheme, the gate driver and switching loss at light load can get greatly reduced, and thus improve light load efficiency.

The organization of this paper is as follows: The operation modes and regulation for designed multi-modes charge pump are explained in Section II. In Section III, methodologies for light load efficiency improvement are analyzed. Based on the analysis of Section II and Section III, the whole system is implemented using IBM7RF 5V process and simulated in Cadence Spectre. A PCB demo board is also built using discrete components. Simulation and test verifications are included in Section IV.

II. OPERATION PRINCIPAL OF MULTI-MODES CHARGE PUMP

The designed multi-modes charge pump is composed of 10 transistors and 3 flying capacitors, shown in Fig.1. Transistors MP1-MP9 are PMOS, and they operate as switches with fixed on-resistance. Transistor MN1 is NMOS, and works as a current source which is used to charge the flying capacitors.

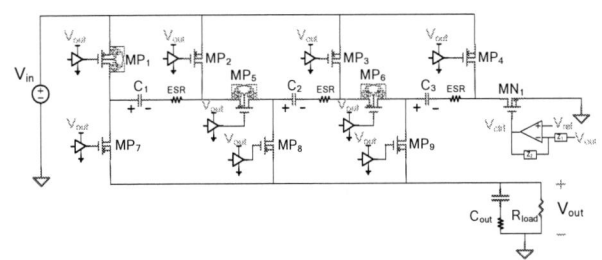

Fig. 1. Power Stage of Multi-Modes Regulated Charge Pump

Three different conversion ratios can be realized by alternating the arrangement of switches and capacitors.

1) Mode 2x: In half cycle, C3 is in series with MP3, MP6, and charged by the current source MN1. During the other half cycle, C3 is in series with MP4, MP9, and Vin, and discharge energy to the output.

2) Mode 1.5x: In half cycle, C2, C3 are in series connection with MP2, MP5, MP6, and charged by MN1. During the other half cycle, C2 is in series with MP3, MP8, and Vin; C3 is in series with MP4, MP9, and Vin; the branch circuits of C2 and C3 are in parallel to discharge energy to the output.

3) Mode 1.33x: In half cycle, C1, C2, C3 are in series connection with MP1, MP5, MP6, and charged by MN1. During the other half cycle, C1 is in series with MP2, MP7, and Vin; C2 is in series with MP3, MP8, and Vin; C3 is in series with MP4, MP9, and Vin; the branch circuits of C1, C2 and C3 are in parallel to discharge energy to the output.

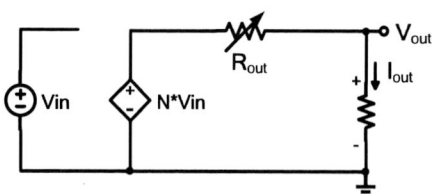

Fig.2. Simplified DC model of Charge Pump

Fig.2 shows a simplified DC model of charge pump without considering gate driver loss and leakage loss. N is the mode conversion ratio, and R_{OUT} is the output impedance of charge pump. When Vout is unregulated, R_{OUT} is fixed and usually designed to be small to reduce conduction loss. Therefore, Vout is always close to N*Vin. When output needs to be regulated to a fixed value, R_{OUT} will vary with different V_{IN}, load current I_{OUT}, and N. The voltage difference between N*Vin and Vout will be dropped on the output impedance of charge pump in order to regulate Vout, as shown in (1).

$$V_{out} = N * V_{in} - I_{out} * R_{out} \qquad (1)$$

Assume the nine unregulated switches have the same on-resistance Rsw, the regulated transistor has the impedance Rt, and all capacitors have the same ESR. The DC output impedance of charge pump can be derived. Fig.3 shows the charging and discharging paths of the multi-modes charge pump in 2x mode. In order to maintain the charging balance, the integration of the current through capacitors C3 and Cout over one cycle should be: $\langle I_{C3}\rangle_{Ts}=0$ and $\langle I_{Cout}\rangle_{Ts}=0$. Since duty cycle is fixed to 0.5, the average value of charging and

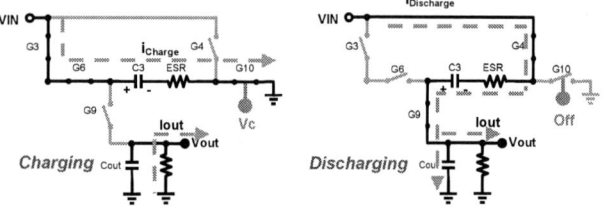

Fig.3. Multi-modes charge pump in 2x mode

discharging current for C3 and Cout should be equal, as: $I_{Charge}=I_{Discharge}$, and $I_{Discharge}=2I_{out}$. Therefore, $I_{Charge}=2Iout$. The transistor conduction loss P_{Con}, flying capacitor loss P_{Cf}, and ESR loss P_{ESR} can be derived using (2), (3), and (4). The total loss is the sum of (2), (3), (4), gate driver loss and leakage loss, as shown in (5).

$$P_{Con} = (2I_{out})^2 \times 2R_{SW} + (2I_{out})^2 \times R_t \times 0.5 = 4I_{out}^2 \times (2R_{SW} + 0.5R_t) \qquad (2)$$

$$P_C = I_{out}^2 / fC \qquad (3)$$

$$P_{ESR} = (2I_{out})^2 \times ESR + I_{out}^2 \times ESR \qquad (4)$$

$$P_{Total} = (P_{Con} + P_C + P_{ESR}) + I_q V_{IN} + P_{gate} = R_{out} I_{out} + I_q V_{IN} + P_{gate} \qquad (5)$$

Therefore, the DC output impedance Rout can be derived using loss equations for each mode, as shown in table I. Once Rsw, ESR, and capacitors are fixed, the output impedance of charge pump R_{OUT} will be determined by Rt.

Table I. Output Impedance of Multi-modes Charge Pump

Mode	Output Impedance (Rout)
2x	8Rsw+5ESR+1/fC+2Rt
1.5x	3.5Rsw+3ESR+1/(2fC)+0.5Rt
1.33x	2Rsw+2.33ESR+1/(3fC)+0.33Rt

III. LIGHT LOAD EFFICIENCY IMPROVEMENT

The power loss in a charge pump circuit includes conduction loss, gate driver loss, switching loss, and leakage loss; therefore the efficiency can be roughly expressed using (6):

$$\eta = \frac{V_{out} * I_{out}}{V_{in} * N * I_{out} + P_{gate} + P_{leak}} \qquad (6)$$

At heavy load, the gate driver loss P_{gate} and leakage loss P_{leak} are far less than Vin*N*Iout, so the maximum efficiency it can achieve is close to Vout/(N*Vin). However, when the load current decreases until the gate driver loss and leakage loss can not be neglected, the efficiency is very bad. Improving the light load efficiency is very important when the charge pump needs to work over a very wide load range. According to the gate driving loss equation shown in (7), there are several methods to reduce the gate driver loss.

$$P_{gate} \approx C_{iss} V_{GS}^2 f_s \qquad (7)$$

Pulse skipping modulation can reduce the average switching frequency, and thus reduces the gate driver loss, switching loss and leakage loss. The gate driver loss can be further reduced by reducing the capacitance associated with the transistor gates according to (7). Since the transistors sizes are optimized to reduce conduction loss and maintain regulation at maximum load condition, they are usually big devices with small on-resistance. The gate capacitances of those devices are very large. However, for the same input voltage, the required output impedances of charge pump Rout are different for high and low load current according to the DC model equation shown in (1). In light load condition, the Rout can be larger than that in heaving load condition.

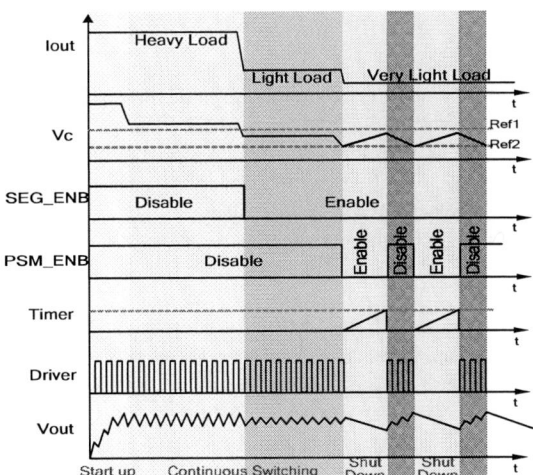

Fig. 4. Hybrid Regulation Scheme for Improving Light Load Efficiency

Therefore, small devices with less gate capacitance will be enough to maintain regulation at light load condition and it can help reduce gate driver loss. Ref.[5] proposed the concept of device segmentation in Buck converter. By segmenting each large transistor into several baby-transistors, and shut down the gate driver of part of those baby-transistors when the load current decreases, the gate driver loss can be reduced. The same concept can be used in charge pump as well and get better results because there are much more transistors in charge pump than in Buck converter.

In this design, a hybrid regulation is used for efficiency and ripple tradeoff. At heavy load, the charge pump is continuously running at 1MHz to reduce output ripple. At light load, switch sizes are reduced to 1/10 of that in the heavy load condition by shutting down 9/10 of those baby-transistors, thus reduce gate driver loss. At very light load, pulse skipping modulation (PSM) is used in addition to device segmentation. By shutdown the controller intermittently, the gate driver loss, switching loss and leakage loss can be reduced. The shutdown frequency of the controller is designed to be within 50k-100kHz, which is out of the audible band.

Fig.4 shows the designed control diagram. Output voltage is regulated using a PI compensator. Compensator output Vc is also used as the gate voltage of current source MN1, so it is proportional to the current flowing through MN1 when it is on. 'Ref1' and 'Ref2' are two references for deciding the load condition. Fixed off time generator controls the shut-down time in pulse skip mode using the capacitor-charging circuitry. Voltage detector monitors the output voltage Vout. When Vout is less than 4.9V, the voltage detector will disable all the light load control. Light-load control signal generator is responsible for generating the PSM enable (PSM_ENB) and segmentation enable (SEG_ENB) signals based on the circuit condition. Modes decision and driving sequence generator will output the driving signals for ten transistors according to the Vin, PSM_ENB, and SEG_ENB.

The primary nodes waveforms are shown in Fig.5. During start-up, Vout increases from zero and driving pulses are continuous. Vc decreases when Vout reaches 5V. Load condition is judged by sensing Vc. At heavy load condition,

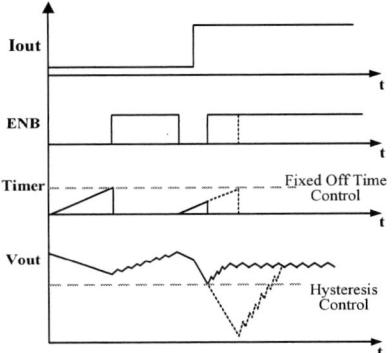

Fig. 5. Key waveforms in hybrid regulation scheme

Vc is higher than Ref1. When Vc is lower than Ref1 but higher than Ref2, charge pump operates in light load condition. Device segmentation is enabled but the driving frequency is still 1MHz. Ripple is smaller than that in heaving load. When Vc drops below Ref2, charge pump enters very light load condition. Controller will be shut down for a fixed off time and then restart. Eq. (8) gives the equation of off time. During the off time, load current is only provided by discharging the output capacitor, and the output voltage drop can be derived using (9). Therefore, smaller load current results in less voltage ripple.

$$t_{\text{off}} = C_C * V_{\text{ref}} / I_{\text{Charge}} = \text{Constant} \qquad (8)$$

$$\Delta V\text{-} = I_{\text{out}} * t_{\text{off}} / C_{\text{out}} \qquad (9)$$

When the controller is shut down, it is still able to response to a load transient quickly using the voltage detector. Since the decreasing slope of output voltage is proportional to the load current during the off time, the output voltage should drop more sharply and results in more voltage drop if a load step up happens. Without waiting for the end of the off time, the controller will wake up once the output voltage touches the threshold. The primary waveforms during load transient are shown in Fig.6.

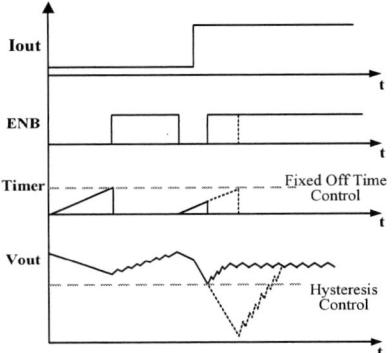

Fig. 6. Fast Light-Heavy Load Transient Response

IV. SIMULATION AND TEST VERIFICATIONS

The designed regulated multi-modes charge pump with improved light load efficiency is implemented using IBM7RF 5V process and simulated in Cadence Spectre. The on-resistance of MP1-MP9 is 90mOhm. Three flying capacitors C1-C3 are 2.2µF and the output capacitor Cout is 4.7µF. Each capacitor has a 10mOhm ESR resistance. The whole system will be fit in a 4*4mm² package.

Fig.7. shows the slow transition from very light to light to heavy load condition when Vin is 3.7V. From the simulation result: for current below 205mA, but above 70mA, only device segmentation is used; for current below 70mA, both of PSM and segmentation are used. The output voltage is regulated at 5V over all load conditions. Fig.8 shows the frequency and output ripple change with the load current at Vin=3.7V. The frequency is 1MHz when load current is above 70mA, but will be 70k-100kHz for even lower current. The maximum ripple 78mV happens at the largest current in very light load condition. However, the ripple decreases with load current no matter it is in heavy, light, or very light load condition. The ripple also decreases with more output capacitances.

A PCB demo with discrete components is also built to verify proposed methods. Fig.9 (a) and (b) are experiment results for fast load transition between different load conditions. In Fig.9(a), load steps down from 300mA to 100mA, and SEG_ENB is automatically enabled after transition while PSM_ENB is always disabled. In Fig.10(b), load steps down from 100mA to 25mA, and PSM_ENB is automatically enabled after transition while SEG_ENB is always enabled. Fig.10 shows the output voltage ripple, two enable signals in steady state at different load conditions. Fig.11 shows that the total loss comparison using different light load improvement methods. The total loss of charge pump without any light load improvement has been normalized to 1. The experiment results show that the charge pump with only device segmentation saves the loss by 47%-70%, while the one that combining both of device segmentation and PSM can reduce the total loss by 56% to 83%.

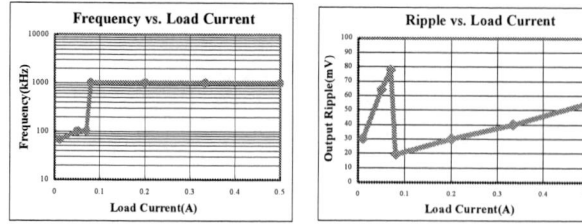

Fig.8. (Left) Frequency vs. Load Current; (Right) Ripple vs. Load Current

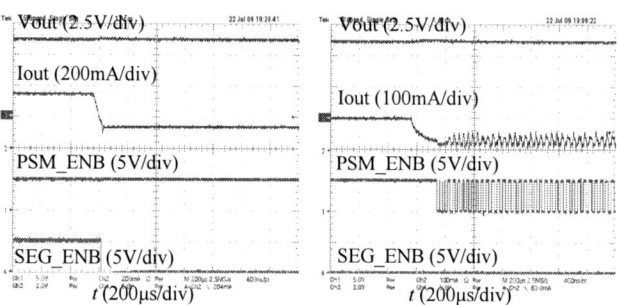

(a) Heavy->Light Load   (b) Light-> Very Light Load

Fig.9. Test Results for Load Transition from Very light->light->heavy load @Vin=3.7V

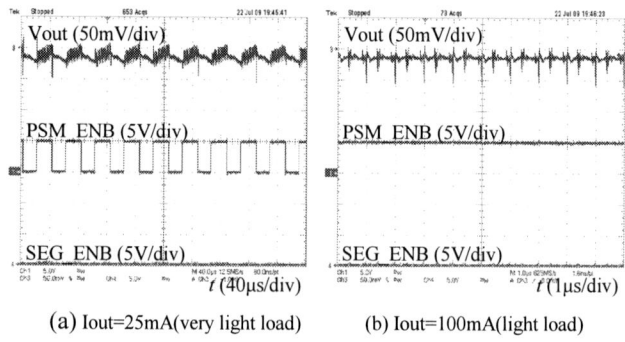

(a) Iout=25mA(very light load)   (b) Iout=100mA(light load)

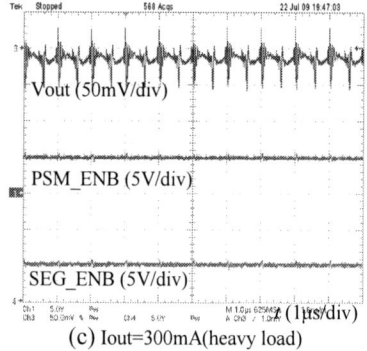

(c) Iout=300mA(heavy load)

Fig.10.Steady State at different load current when Vin=3.7V

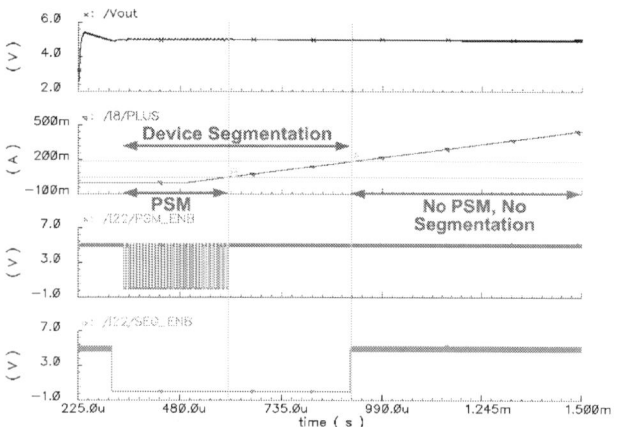

Fig.7. Load Transition from 50 mA to 500mA@Vin=3.7V

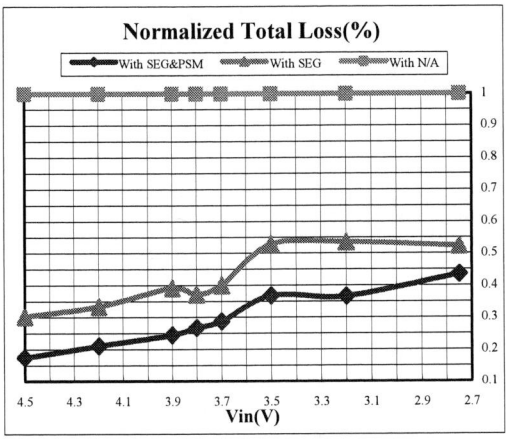

Fig.11. Total Loss Comparison @Iout=25mA

V. CONCLUSIONS

This paper proposed a high efficiency regulated multi-modes charge pump over wide input (2.7-4.5V) and load range (10mA-500mA). The multi-modes charge pump is designed to have 3 modes: 2x, 1.5x, and 1.33x to enhance efficiency over desired input range. Pulse skipping regulation and device segmentation are used to improve light load efficiency, especially for very light load conditions. The whole system is implemented using IBM 7RF 5V process and simulated in Cadence Spectre. The chip layout and fabrication are on the way. A PCB demo is built using discrete components. Both of the simulation and experiment results verified that the proposed solution can greatly reduce the total loss and improve efficiency.

ACKNOWLEDGMENTS

This project is supported by RF Micro Devices.

REFERENCES

[1] Analogic Tech , "A Simple Model for DC/DC Charge Pumps", Application Note, AN-113.

[2] Brigitte Kormann, Dipl.-Ing, "High-Efficiency, Regulated Charge Pumps for High-Current Applications", Texas Instrument.

[3] Ping Luo, Zhaoji Li, Bo Zhang, "Analysis of PSM mode in switching converter and its improved mode", Communications, Circuits and Systems, Volume 2, 27-30 May 2005 Page(s): 1385-1389.

[4] Chi-Chang Wang, Jiin-Chuan Wu, "Efficiency improvement in charge pump circuits", Solid-State Circuits, IEEE Journal of, Volume 32, Issue 6, June 1997 Page(s):852 – 860.

[5] Musunuri, S., Chapman, P.L "Improvement of Light-Load Efficiency Using Width-Switching Scheme for CMOS Transistors", Power Electronics Letters, IEEE, Volume 3, Issue 3, Sept. 2005 Page(s):105 – 110.

High Performance, High-Power Capacitor Charging: Focus on Pulse-to-Pulse Repeatability

A. Pokryvailo, C. Carp, and C. Scapellati

Spellman High Voltage Electronics Corporation
475 Wireless Boulevard, Hauppauge, NY 11788
Email: apokryva@spellmanhv.com

Abstract— **Pulse-to-pulse repeatability (PPR) is an important parameter in capacitor charging applications. This paper studies the subject both theoretically and experimentally using as a platform a novel low-cost, high performance 20-kJ/s, 10-kV, 1-kHz repetition-rate charger based on an energy-dosing converter topology. It is stated that PPR is affected by quality of the HV measurement and depends on the ratio of the energy stored in the HV transformer magnetic system, its parasitic capacitance, etc., to that stored in the load capacitor. The influence of the second factor is studied. It came clear that empiric formulae usually used in literature are too simplistic to account for the complex electromagnetic processes; they may yield overstated values. The theoretical findings are compared with both PSpice simulations and experimental results. Both PSpice simulations and analytical tools predicted pulse-to-pulse repeatability of ±0.15 %; the measured figures are ±0.4 % and ±0.5 % for short- and long-term operation, respectively, at peak charging and repetition rates.**

I. INTRODUCTION

Switch-mode power supplies (PS) almost universally superseded line-frequency PS in capacitor charging. For the charging rate above several kJ/s, soft-switching topologies prevail [1]-[7], at least in commercial products (see, e.g., papers from General Atomics [1] and Lambda [2]). Series-resonant topologies seem to dominate this niche.

It is difficult to realize a combination of high voltage, high charging rate (tens of kJ/s and higher), high pulse repetition rate (PRR), compactness, high efficiency and good pulse-to-pulse repeatability (PPR). Putting constraints of low-cost and good manufacturability makes the charger development even more challenging.

PPR is an important parameter in capacitor charging. It influences the stability of various physical processes ranging from lasing to pulsed microwave and X-ray radiation to plasma chemistry applications. PPR can be defined as

$$R = \frac{V_{C\max} - V_{C\min}}{V_{Cavg}}, \qquad (1)$$

where V_{Cmax}, V_{Cmin} and V_{Cavg} are maximum, minimum and average values of the voltage across the storage capacitor, C_s,

for a batch of pulses. Usually the charging does not involve predictive algorithms. This means that when the output voltage reaches the programmed value, the inverter is shut down. Then the remnant energy, i.e., energy stored in the HV transformer parasitics, resonant capacitors, etc., E_{rem}, may flow wholly or partially to the storage capacitor, and the output overshoots. This is one of the main factors degrading PPR. In fact, it seems to be the only factor discussed in literature. It might seem that the repeatability can be estimated easily, assuming that all this energy can be transferred to C_s; then R would be proportional to E_{rem}. More precisely it can be given by a formula derived in [8]

$$R = \sqrt{1 + \frac{E_{rem}}{E_c}} - 1 \qquad (2)$$

where E_c is the energy stored in C_s. For low charge voltages, E_{rem} is comparable with Ec. Application notes [10], [11] and [2] provide similar simplistic estimates.

Much effort was put to improving PPR, especially at high PRR. A common approach is decreasing the rate of charge by an order of magnitude or so towards the end of charge (EOC). Thus, the charge buckets carried in each period are smaller, and the energy delivery to the storage capacitor can be controlled tighter. This technique in various implementations is used in commercial products (see, e.g., [1], [2], [12]). The shortcoming of such an approach is overrating the charger power.

This paper describes the development and testing of a high-power charger satisfying the above-listed contradicting requirements within the constraints of low-cost proven technology. A focus is on theoretical and experimental investigation of PPR.

II. CONVERTER TOPOLOGY AND PRINCIPLE OF OPERATION

A charger block-diagram is shown in Figure 1. The charger comprises a 3-phase input rectifier, a converter module (CM), an HV divider and control means. Triggered by an external source, the charger charges capacitor C_s that is discharged onto a dummy load via a high-power switch DSw. CM comprises an inverter INV, an HV transformer wound on popular U100/57/25 ferrites, and a rectifier R. The CM's heart is a half-bridge quasi-resonant inverter with energy dosing capacitors (Figure 2) [13]-[15]. Work [14] provides the principle and theory of operation (its content is partially reiterated, cleaned of misprints and expanded in [8], [9]).

In normal mode, one of the divider capacitors, Cdiv, is charged to the rail voltage V_r. When the corresponding switch closes, Cdiv discharges through the primary winding, while it counterpart recharges to the rail voltage. If the current path contains an inductance, a sine waveform is generated, and ideally, all the energy stored in both capacitors Cdiv is transferred to the secondary side. If Cdiv discharges fully, but the current does not fall to zero, the freewheeling diode (FWD), which is connected in parallel to the capacitor, conducts acting as a clamp and preventing voltage reversal. Thus, the remainder of the energy stored in the circuit inductance is transferred to the output. The benefits of the energy-dosing are tight control of the energy transfer and inherent limitation of the short circuit current and voltages across the converter components. This topology is a trade-off between constant current and constant voltage charging.

The maximum frequency, at which the operation is possible at a certain load voltage V_l with zero-current crossing (ZCC), in a normalized form, is given by the equation [14]

$$f_N(V_l, V_r) = \frac{1}{\frac{2}{\pi}\left[\frac{1}{2}a\cos\left(\frac{V_l}{V_l - V_r}\right) + \frac{V_r}{2V_l}\sqrt{1 - \frac{2V_l}{V_r}}\right]}, \quad (3)$$

where V_r is the rail voltage, and both the rail and the load voltages are referenced to the same side of the transformer. The conversion frequency f is normalized to the resonant frequency f_0 of the loop formed by the leakage inductance and the resonant capacitors: $f_N = f/f_0$.

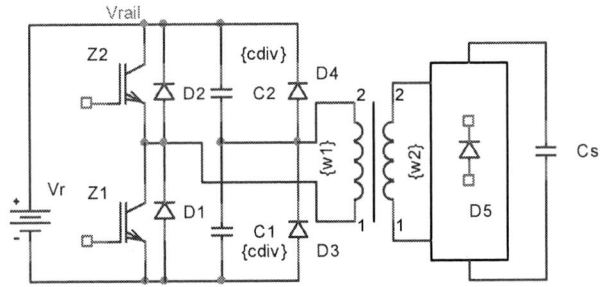

Figure 1. Figure 2. Halfbridge inverter with energy dosing capacitors. Transformer is actually comprised of two transformers, whose primaries are connected in parallel, and secondaries in series.

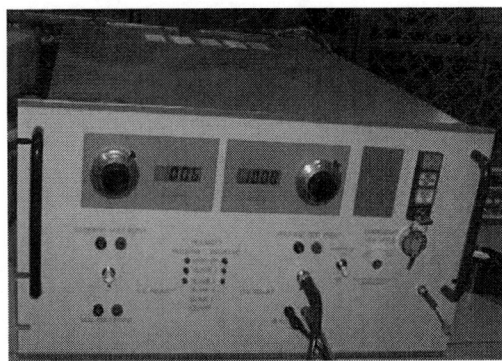

Figure 3. Charger – front view.

The FPGA-based controls adapt the switching frequency, following (3), from 12.5 kHz to 55 kHz in such a way as to ensure zero-current switching (ZCS) for all possible scenarios, keeping maximum duty cycle for high power.

The packaging was made in a 19" rack-mounted chassis, 10½" tall, 24" deep. The parasitics of the HV transformer together with the capacitors Cdiv are integrated into the resonant tank circuit, so no external chokes are needed. The circuit breaker and an HV connector are mounted on the rear panel. On the front view (Figure 3), the front panel borrowed from the ubiquitous SR6 series [16] is seen.

III. REPEATABILITY ANALYSIS

Pulse-to-pulse *variability* evolves from several factors:

- Converter remnant energy, E_{rem}, at EOC. E_{rem} may flow wholly or partially to the storage capacitor, so the output will overshoot.

- Error in generating EOC signal. This may be caused by a poor-quality feedback, noise, unstable reference voltage, etc.

- Delay, t_d, between EOC and the actual IGBT turn-off. It comprises digital delays, optocouplers delay, and the IGBT turn-off delay.

With our broad definition, E_{rem} can be stored anywhere in the system: in the leakage and magnetizing inductances and parasitic capacitance of the HV transformer and rectifier, in

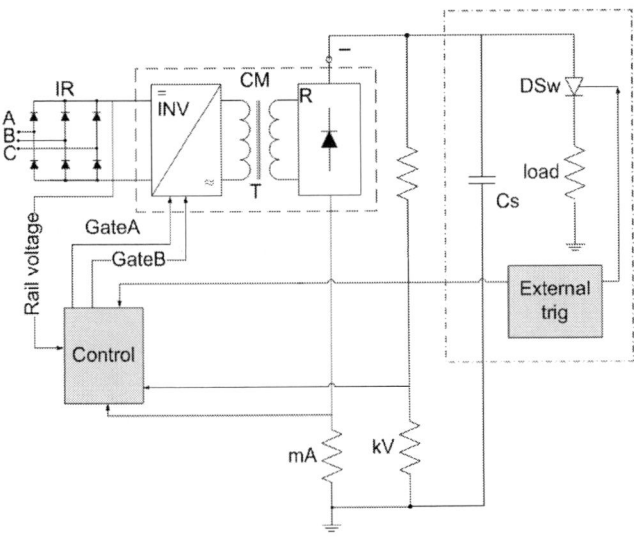

Figure 1. Charger block-diagram.

the parasitic inductances of the busbars and connections, etc. For the simplicity sake, we will limit the analysis to the case of E_{rem} stored in the leakage inductance, L_s, only.

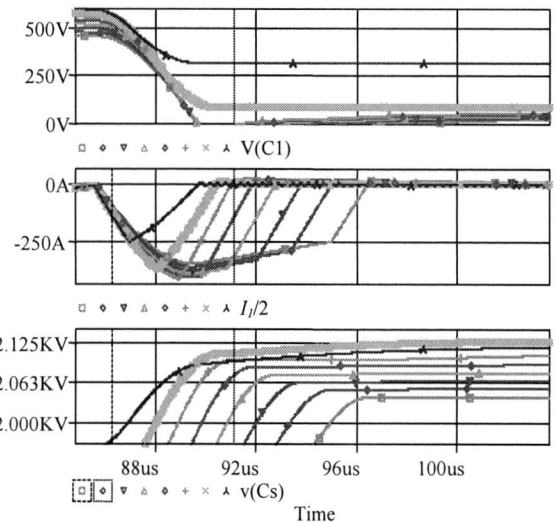

Figure 4. Programmed charge voltage 2 kV. V_r =460 V, 480 V, ..., 600 V. C_s=200 nF. Primary current I_1 is halved. Curves corresponding to largest overshoot are in thicker line.

In the circuit in question, E_{rem} flows not only into C_s but is recovered partially in the DC rail power supply and, depending on the initial conditions (IC), may be directed to Cdiv; part of it is lost in form of heat. Sample PSpice waveforms for a low charge voltage of 2 kV (the charger is rated for 10 kV) are shown in Figure 4; they are useful as an empirical guide for analytical work.

In this parametric run, the source of the variability was the DC rail voltage, V_r, swept from 460 V to 600 V in 20-V increments, which corresponds to common variations of a 400 VAC 3-phase line. We note that such variations at the AC side are usually slow; V_r, however, can oscillate at a frequency up to several kHz, the V_r ripple being visible even within one charging cycle. EOC corresponds to the chopping of the primary winding current I_1.

It is seen that the maximum overshoot takes place not at the maximum chopping current. Moreover, the same chopping current (same amount of energy stored in L_s) may result in very different overshoots depending on the EOC timing, as follows from the comparison of the first and the last curves (V_r=460 V, V_r=600 V, respectively). Calculating by 0 the overshoot ΔV above the programmed voltage of 2 kV, that would result from E_{rem} delivered wholly to C_s, we obtain

$\Delta V \approx 450$ V. In this example, $E_{rem} = \dfrac{L_s I_2^2}{2} = 0.2$ J for I_1=500 A corresponding to V_r=460 V, V_r=600 V. It is seen that the overvoltages are much lower than the above estimate even at higher currents. Thus, only part of E_{rem} reaches C_s, the rest being recuperated mainly in the DC rail source.

Linearizing the circuit piecewise and using corresponding equivalent circuits allows full analytical description of the

electromagnetic processes occurring after EOC [8]. It is limited to a worst case of EOC occurring at any time from the primary current onset to its maximum. Full rigorous analysis and a predictive control algorithm derived from it will be reported separately. For the converter parameters corresponding to our experimental setup and simulations, the repeatability R is plotted in Figure 5 versus EOC time, t_c (subscript "c" stands for Chopping). The load voltage, V_L, serves as a parameter and is given as a fraction of the rail voltage; both V_L and V_r are reflected to the same side of the HV transformer. Since a halfbridge is involved, the nominal load voltage is $V_L \approx V_r/2$ at *low* line. Considering V_r variation from 460 V (low line) to 590 V (high line), we note that the repeatability is worse at highline, whereas the nominal load voltage is V_L=10 kV$\approx V_r/2 \ast 460/590 \ast k_{tr}$=0.39$k_{tr}V_r$.

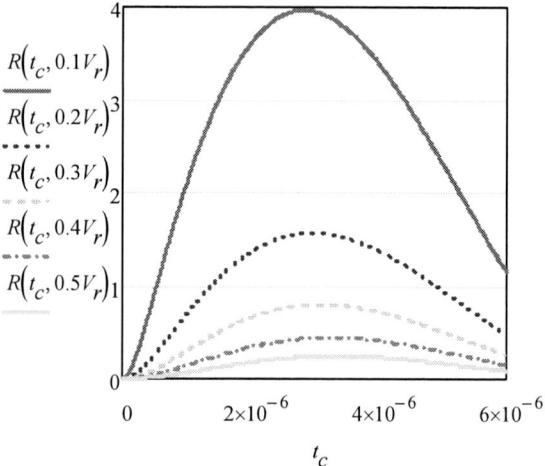

Figure 5. Repeatability, R, %, versus EOC point, t_c, with load voltage, V_L, as a parameter (in fractions of the rail voltage, both reflected to the same side of the HV transformer). Nominal load voltage (10 kV) is $V_L \approx V_r/2$ for low line (V_r=460 V at primary side). C_s =200 nF, Cdiv=2 μF. Horizontal scale is given in seconds. Compare to Figure 4.

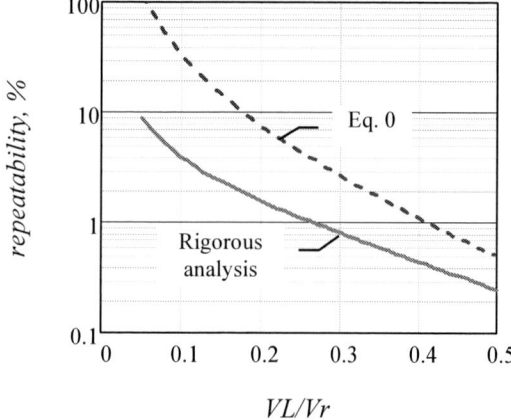

Figure 6. Repeatability as a function of charge voltage, C_s =200 nF, Cdiv=2 μF, worst case.

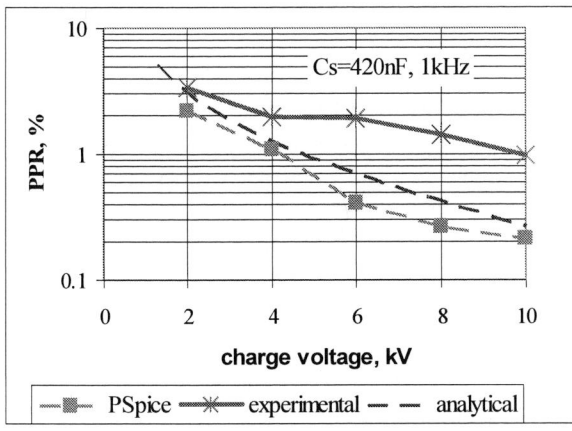

Figure 7. Repeatability as a function of charge voltage – summary of PSpice and analytical calculations and experimental results.

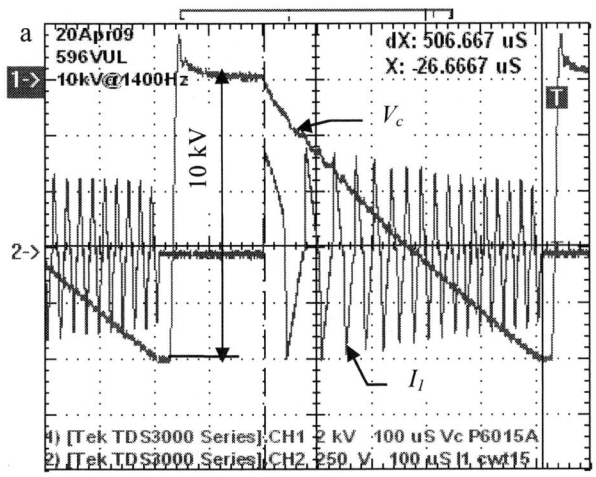

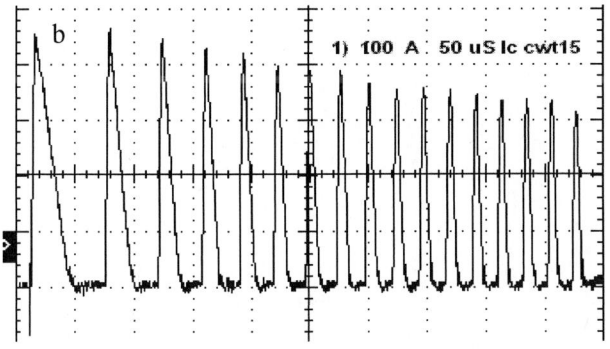

Figure 8. Typical waveforms at highline. PRR=1400 Hz in burst, charge time is 507 µs. a – load capacitor voltage and primary winding current; b – collector current.

The maxima of the curves indicate the worst case of the most unfavorable EOC timing and are plotted separately in Figure 6 together with a plot of (2). It is seen that the overshoot derived empirically is several times larger than that predicted by the rigorous analysis. Finally, Figure 7 summarizes PSpice and analytical calculations and experimental results. The latter are described in more detail in Section IV. Notably, the analytical curve lies very close to its PSpice counterpart, always above it, as it should, because the PSpice parametric sweeps performed in 20-V V_r increments do not necessarily find precisely the worst-case EOC time.

IV. EXPERIMENTAL

A. Waveforms

One of the main goals of this work was realizing as high efficiency as possible by enforcing lossless switching in all possible scenarios at all charge levels and repetition rates. The noise immunity of the control circuitry in this sense is also an important issue. A thorough experimental investigation, side by side with PSpice modeling, was performed. We found that under no circumstances ZCS was disturbed. Figure 8a shows V_c and primary winding current I1 in burst operation at a PRR of 1400 Hz for C_s=420 nF (charge rate of 29.4 kJ/s), with the collector current, I_c, of one of the transistors displayed on an expanded scale in Figure 8b. The information on horizontal and vertical scales per division here and in further plots is indicated in the waveforms' annotations. As vividly seen in Figure 8b, the conversion frequency adapts to keep high duty cycle yet maintaining ZCS. The highest conversion frequency is 55 kHz at low line, with very large margin guaranteeing ZCS without any shoot-through currents even at abnormal line sags.

At full power, the efficiency was about 0.92, and power factor, PF, was 0.94. At 5% of the rated power, these figures were around 0.7. The efficiency values are lower by 1-2 % than expected and what could be deducted from the loss measurement (see [17] for methods of the IGBT loss measurement), and intuitively from the amount of the dissipated heat. We note that the IGBTs baseplate overheat was less than 40 ^{0}C in all operational modes.

At the time of writing this paper, the charger has generated 10^{10} 10-kV shots at 1 kHz, with the test ongoing.

B. Repeatability

We will distinguish here between short-term and long-term PPR. The former is defined as that derived from N consecutive pulses. In our measurements, N=80, sampled from the 121st to the 200th pulse. Thus, the short-term PPR is not influenced by thermal drifts, aging of components, etc. It is affected by V_r variations to the extent of the high-frequency rail voltage ringing, excluding slow input changes. Long-term PPR is also influenced by V_r variation in the full specified range, for instance, from 460 VDC to 590 VDC (corresponding to $400VAC_{-14\%}^{+10\%}$). Here, the reference to long-term PPR is made in the light of such variations, other parameters being not controlled.

PPR measurements were taken using the FastFrame capability of a DPO7054 scope. Up to four signals were monitored simultaneously. The load voltage, V_c, was measured by a P6015A probe on a 100-mV scale with a 10-V offset allowing the signal at EOC to fit the screen. The scope was triggered by the EOC event. In these experiments, the discharge switch DSw was fired on in 20 µs after EOC. The first 800 shots were collected with a 500-point resolution on a

4-μs/div scale. The waveforms were saved as screen captures, and 80 frames, starting from the 121st frame, were saved in the csv format. An Excel spreadsheet was designed, in which 79 shots were processed.

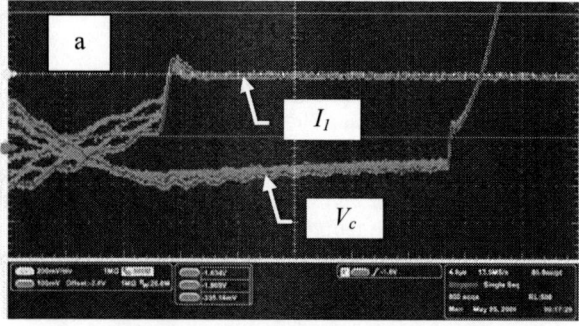

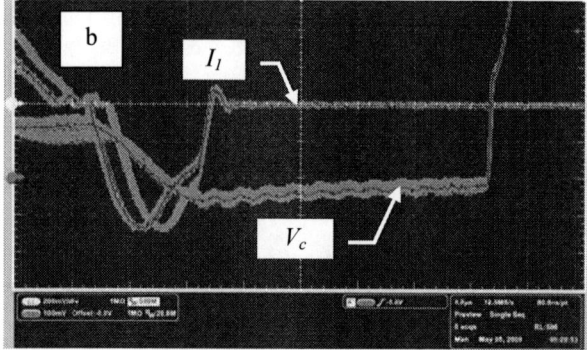

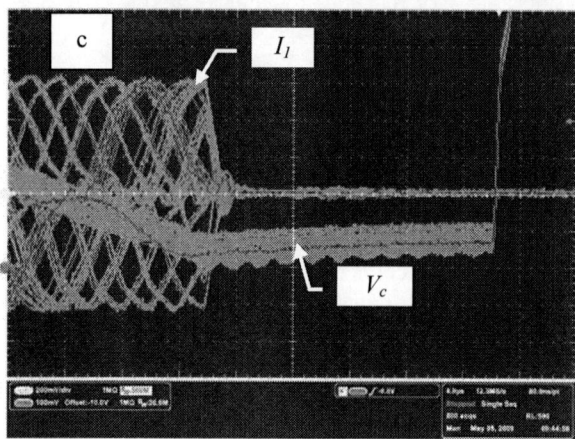

Figure 9. Overlay of 80 frames (V_c - 100V/div, I_l – 100A/div) for: a) high line, 2kV@1kHz. b) nominal line, 6kV@1kHz. c) High line, 10kV@1kHz. Horizontal 4 μs/div.

Three typical screenshots of the overlays of 80 frames are shown in Figure 9. They quite vividly show wherefrom the variability, at least partially, evolves. At EOC, the primary current is chopped at random. If there is a certain pattern (as seen at 2-kV and 6-kV settings), PPR is better. When the current is chopped at an arbitrary time point (10-kV setting, Figure 9c), PPR deteriorates. It still remains below ±0. 5% at the maximum voltage and PRR, owing to specifics of the used converter topology and high conversion frequency. For three rail voltage settings, namely 460 VDC, 520 VDC and

590 VDC, PPR was calculated by the formula that looks in an Excel convention as follows:

$$=\text{MIN(A2:C80)-MAX(A2:C80))/AVERAGE(A2:C80)*100, (4)}$$

where columns A-C each contain V_c values for 79 consecutive pulses, for 460, 520 and 590 VDC, respectively. Alternatively, we varied the line voltage continuously from the low to high level, looking for the least stable operation, i.e., for the largest V_c variation. For this method, PPR was calculated by (1) using V_{Cmax}, V_{Cmin} values from the whole measurement range; it is long-term PPR.

The short- and long-term PPR are plotted in Figure 10, Figure 11, respectively. The experimental curves shown in Figure 11 are calculated by (1), (3); they are labeled as "overall experimental cont rails", and "3 rail experimental", respectively. The variability is larger than that predicted by the theory accounting for the Factor 1 only ("analytical" curve). This discrepancy can be attributed to the measurement errors and propagation delays (Factors 2, 3). Overall, the achieved PPR can be considered as quite good for the cited charging rate and PRR.

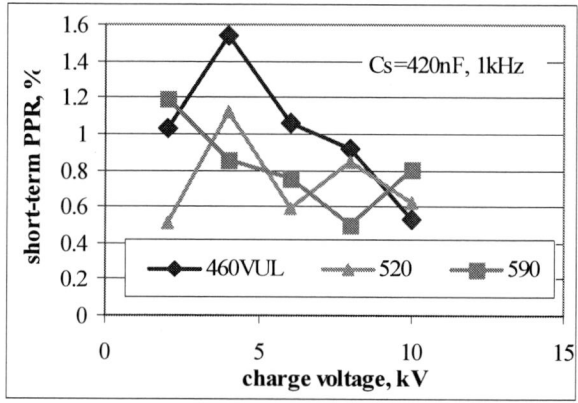

Figure 10. Short-term repeatability.

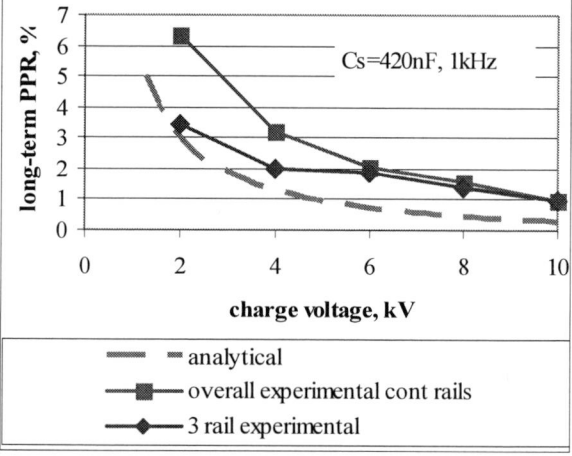

Figure 11. Long-term repeatability as a function of charge voltage – summary of analytical calculations and experimental results.

REFERENCES

[1] J. Jichetti, A. Bushnell, R. McDowell, "Precision Capacitor Charging Switching Power Supplies", Proc. 14th IEEE Pulsed Power Conference, June 2003, pp. 522 - 525.

[2] G. L. Bees and A. Tydeman, "Capacitor Charging Power Supply Design for High Pulse to Pulse Repeatability Applications", 12th Proc. IEEE Pulsed Power Conference, June 1999, pp. 397-398.

[3] J. Biebach, P. Ehrhart, A. Müller, G. Reiner, and W. Weck, "Compact Modular Power Supplies for Superconducting Inductive Storage and for Capacitor Charging", IEEE Transactions ON Magnetics, Vol. 37, No. 1, January 2001, pp. 353-356.

[4] M. M. McQuage, V.P. McDowell, F.E. Peterkin, and J. A. Pasour, "High Power Density Capacitor Charging Power Supply Development for Repetitive Pulsed Power", Proc. PM06, pp. 368-371.

[5] A. K. Jaint, C. P. Henze, C. B. Henze, K. Conroy, "Development of a 350kW, 10kV Pulse Power Converter for Capacitor Charging", Proc. 22nd Applied Power Electronics Conference (APEC), Feb. 25 -March 1, 2007, pp. 1164 – 1170.

[6] J. S. Oh†, S. D. Jang, Y. G. Son, M. H. Cho, W. Namkung, "Development AND Application of an Inverter Charging Supply to a Pulse Modulator", Proceedings of LINAC2002, Gyeongju, Korea, pp. 205-207.

[7] H. J. Ryoo, S. R. Jang, J. S. Kim, Y. B. Kim, "Design and Testing of the High Voltage Capacitor Charger for 150kJ Pulsed Power Application", Presented at 17th Int. Pulsed Power Conf., June 28-July 2, Washington DC, 2009.

[8] A. Pokryvailo, C. Carp and C. Scapellati, "Repeatability Analysis in HV Capacitor Charging Applications", Presented at 17th IEEE Int. Pulsed Power Conference, 2009.

[9] A. Pokryvailo, C. Carp and C. Scapellati, "High Power, High Efficiency, Low Cost Capacitor Charger Concept and Demonstration", Presented at 17th IEEE Int. Pulsed Power Conference, 2009.

[10] "Regulation and Repeatability", APP Note 509, A.L.E. Systems, Available: http://www.us.tdk-lambda.com/hp/product_html/high_volt.htm

[11] "High Power, High Voltage Power Supply Voltage Regulation" Available: http://www.gaep.com/voltage-regulation.html

[12] R. Ness, P. Melcher, G. Ferguson, and C. Huang, "A Decade of Solid State Pulsed Power Development at Cymer Inc.", Proc. 26th Power Modulator Symposium, 23-26 May 2004, pp. 228 – 233.

[13] B.D. Bedford and R.G. Hoft, 'Principles of Inverter Circuits", Wiley, NY, 1964.

[14] B. Kurchik, A. Pokryvailo and A. Schwarz, "HV Converter for Capacitor Charging", Pribory i Tekhnika Experimenta, No. 4, pp.121-124, 1990, Translation to English Plenum Publishing Corp.

[15] M. Wolf and A. Pokryvailo, "High Voltage Resonant Modular Capacitor Charger Systems with Energy Dosage", Proc. 15th IEEE Int. Conf. on Pulsed Power, Monterey CA, 13-17 June, 2005, pp. 1029-1032.B.D. Bedford and R.G. Hoft, 'Principles of Inverter Circuits", Wiley, NY, 1964.

[16] http://www.spellmanhv.com/Products/Rack-Supplies/SR.aspx

[17] A. Pokryvailo and C. Carp, "Accurate Measurement of On-State Losses of Power Semiconductors", 28th Int. Power Modulators Symp., Las Vegas, 27-31 May, 2008, pp. 374-377].

Generalized AC-DC Single-Phase Boost Rectifier

C. B. Jacobina, E. C. dos Santos, N. Rocha

Electrical Engineering Department (DEE)

Federal University of Campina Grande (UFCG)

58.429-900 Campina Grande - PB - Brazil

Emails: (jacobina, euzeli)@dee.ufcg.edu.br

and nady.rocha@ee.ufcg.edu.br

Abstract—**In this paper, a single-phase ac-dc converter is proposed. It is composed of N single-phase boost rectifiers in a parallel connection scheme. The proposed topology permits to reduce: the current and consequently the power ratings of the switches and the switching stress in dc-link capacitor. Also, it is proposed a PWM strategy, which allows reducing the THD of the grid current when compared to the conventional single-phase ac-dc converter. Suitable modeling and control strategy of the system, including the unbalanced case, are developed. Simulated and experimental results are presented.**

I. INTRODUCTION

There have been a considerable increase in the use of parallel converters to improve the power capability, reliability, efficiency, redundancy, and decrease the cost [1–5]. Usually the operation of converters in parallel requires a transformer for isolation. Weight, size, and cost associated with the transformer may make such a solution undesirable [6]. However, when connecting converters in parallel without isolation transformer, closed loops between different converters produce current differences in the converter stages due to i) the current unbalance of the converter stages and/or ii) the circulating currents among different converter stages.

To avoid the circulating current, the usual approaches are the use of isolation techniques [7], [8] (e.g., separate power supplies, isolation transformer), inter-phase reactors [9], [10], and special control techniques [1–5], [11–13].

In this paper, we investigate a system based in parallel single-phase ac-dc converters topology, in which single-phase boost rectifiers are in paralleled without isolation transformer. The proposed topology is composed of N parallel single-phase boost rectifiers. The proposed systems are depicted in Fig. 1(a). This work, besides introducing a new topology, presents the system modeling and control strategies, including the unbalanced case.

Comparatively to the conventional configuration [Fig. 1(b)], the proposed one permits to decrease the harmonic content in the grid without increasing the switching frequency, and reduces the switching stress in the dc-link capacitor and total losses of the converter.

II. SYSTEM MODEL

The proposed configuration comprises N converters, inductor filters (L_{na} and L_{nb} with $n = 1, 2, 3, ...N$) and a dc-link capacitor bank. Converters legs are constituted by top switches q_{na} and q_{nb} and complementary bottom switches $\bar{q}_{na}$ and

$\bar{q}_{nb}$. The conduction state of all switches is represented by homonymous binary variable, where $q = 1$ indicates a closed switch while $q = 0$ indicates an open one. Inductors filters are initially considered as having different values to demonstrate the influence of asymmetry in system model.

The following equations can be derived from the system in Fig. 1(a)

$$e_g = Z_{na}i_{na} + Z_{nb}i_{nb} + v_n \qquad (1)$$

$$i_g = \sum_{n=1}^{N} i_{na} \qquad (2)$$

$$Z_{ka}i_{ka} - Z_{wa}i_{wa} + v_{ka0} - v_{wa0} = 0 \qquad (3)$$

$$-Z_{kb}i_{kb} + Z_{wb}i_{wb} + v_{kb0} - v_{wb0} = 0 \qquad (4)$$

where $v_n = v_{na0} - v_{nb0}$, v_{na0} and v_{nb0} are the rectifier pole voltages, $Z = r + lp$, $p = d/dt$, the symbols r and l represent resistances and inductances of the inductors L_{na} and L_{nb}, $k = 1, 2, 3, ..., N - 1$ and $w = k + 1$.

Adding (3) and (4) leads to the following relationship for circulating voltages

$$v_{okw} = -Z_{ka}i_{ka} + Z_{kb}i_{kb} + Z_{wa}i_{wa} - Z_{wb}i_{wb} \qquad (5)$$

with $v_{okw} = v_{ka0} + v_{kb0} - v_{wa0} - v_{wb0}$.

From i_{na} and i_{nb}, the circulating currents of each converter are defined by

$$i_{no} = i_{na} - i_{nb}. \qquad (6)$$

When there are N parallel converters we can write $N - 1$ circulating currents between rectifiers. Thus, the currents i_{no} of each rectifiers can be write as function of circulating current between rectifiers, that is

$$i_{no} = k_1 i_{on\alpha} - k_2 i_{o\beta n} \qquad (7)$$

where $\alpha = n + 1$, $\beta = n - 1$, $k_1 = 0$ if $\alpha > N$, $k_1 = 1$ if $\alpha \leq N$, $k_2 = 0$ if $\beta < 1$ and $k_2 = 1$ if $\beta \geq 1$.

From i_{no} defined from equations (6) and (7), relations (1) and (5) are written as follows

$$e_g = (Z_{na} + Z_{nb})i_{na} - k_1 Z_{nb}i_{on\alpha} + k_2 Z_{nb}i_{o\beta n} + v_n \qquad (8)$$

$$v_{okw} = -(Z_{ka} - Z_{kb})i_{ka} + (Z_{wa} - Z_{wb})i_{wa} - (Z_{kb} + Z_{wb})i_{okw} + k_3 Z_{wb}i_{ow\gamma} + k_4 Z_{wb}i_{o\delta k} \qquad (9)$$

978-1-4244-4782-4/10 $26.00 © 2010 IEEE

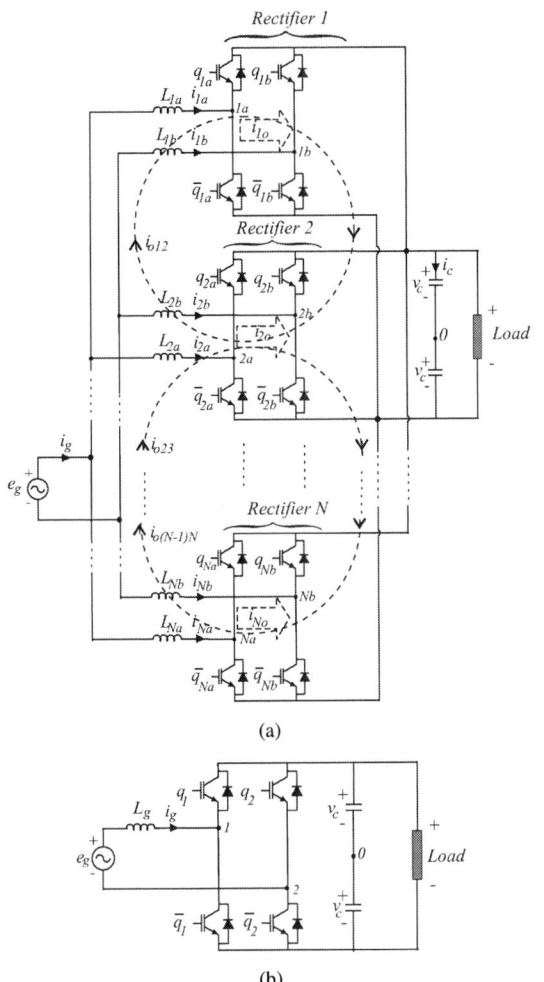

(a)

(b)

Fig. 1. Single-phase boost rectifier. (a) Proposed configuration. (b) Conventional configuration.

where $\gamma = w + 1$, $\delta = k - 1$, $k_3 = 0$ if $w = N$, $k_3 = 1$ if $w < N$, $k_4 = 0$ if $\delta < 1$ and $k_4 = 1$ if $\delta \geq 1$.

Relations (8), (9) and (2) constitute the rectifier dynamic model. Therefore, we use $v_n = v_{na0} - v_{nb0}$ to regulate the currents i_{na} and $v_{owk} = v_{ka0} + v_{kb0} - v_{wa0} - v_{wb0}$ to regulate the currents i_{okw}. The reference currents i_{na}^* are chosen equal to i_g^*/N and the reference circulating currents i_{owk}^* are chosen equal to 0.

When the system is balanced (i.e., $Z_{na} = Z_{nb} = Z_g'$), the system model (8) and (9) become

$$e_g = 2Z_g' i_{na} - k_1 Z_g' i_{on\alpha} + k_2 Z_g' i_{o\beta n} + v_n \quad (10)$$

$$v_{okw} = -2Z_g' i_{okw} + k_3 Z_g' i_{ow\gamma} + k_4 Z_g' i_{o\delta k}. \quad (11)$$

From (2), (10) and (11) we can also write the model for the grid current as follows

$$e_g = \frac{2}{N} Z_g' i_g + \frac{1}{N} \sum_{n=1}^{N} v_n. \quad (12)$$

III. PWM STRATEGY

The rectifier pole voltages v_{na0} and v_{nb0} depend on the conduction states of the power switches, that is

$$v_{na0} = (2q_{na} - 1)\frac{v_c}{2} \quad (13)$$

$$v_{nb0} = (2q_{nb} - 1)\frac{v_c}{2} \quad (14)$$

where v_c is the dc-link voltage. Considering that v_n^* and v_{okw}^* denote the reference voltages determined by the controllers (see section IV), we found that

$$v_n^* = v_{na0}^* - v_{nb0}^* \quad (15)$$

$$v_{okw}^* = v_{ka0}^* + v_{kb0}^* - v_{wa0}^* - v_{wb0}^*. \quad (16)$$

We have $2N - 1$ reference voltages, equations (15) and (16). Such equations are not sufficient to determine the $2N$ pole voltages, v_{1a0}^* to v_{Na0}^* and v_{1b0}^* to v_{Nb0}^*, then an auxiliary variable v_x^* is introduced, making $v_x^* = (v_{1a0} + v_{1b0})/2$ from (15) and (16) the reference poles voltages can be written

$$v_{na0}^* = \frac{1}{2}v_n^* + k_5\frac{1}{2}\sum_{i=1}^{n-1} v_{oij} + v_x^* \quad (17)$$

$$v_{nb0}^* = -\frac{1}{2}v_n^* + k_5\frac{1}{2}\sum_{i=1}^{n-1} v_{oij} + v_x^* \quad (18)$$

where $j = i + 1$, $k_5 = 0$ if $n = 1$ and $k_5 = 1$ if $n > 1$.

Voltage v_x^* can be obtained from their maximum, minimum values

$$v_{x\max}^* = v_c^*/2 - v_{r\max}^* \quad (19)$$

$$v_{x\min}^* = -v_c^*/2 - v_{r\min}^* \quad (20)$$

where v_c^* is the reference dc-link voltage, $v_{r\max}^* = \max V_r$, $v_{r\min}^* = \min V_r$ with

$$V_r = \{\frac{1}{2}v_n^* + k_5\frac{1}{2}\sum_{i=1}^{n-1} v_{oij}, -\frac{1}{2}v_n^* + k_5\frac{1}{2}\sum_{i=1}^{n-1} v_{oij}\}.$$

Voltages v_x^* can be chosen equal to

$$v_x^* = \mu v_{x\max}^* + (1 - \mu) v_{x\min}^* \quad (21)$$

where $0 \leq \mu \leq 1$. Note that, when the maximum ($\mu = 1$) or minimum ($\mu = 0$) value are selected, one of the converter legs operates with zero switching frequency. On the other hand, operation with the average values ($\mu = 0.5$) generates pulse voltage centered in the sampling period, improving THD of voltages.

Gating signals can be obtained by comparison of the pole voltages with a high frequency triangular carrier signal [14–17]. In this paper, gating signals were obtained comparing pole voltage v_{na0}^* and v_{nb0}^* with a single or multi high frequency triangular carrier signals, i.e., single-carrier-based PWM or multi-carrier-based PWM. In the case of multi-carrier-based PWM, the phase shift of the triangular carrier signals is $360°/N$. For example, with two rectifiers in parallel ($N = 2$) the phase shift of the two triangular carrier signals

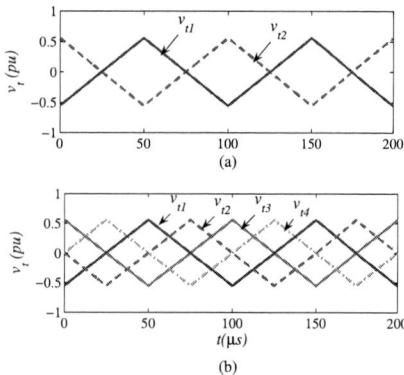

Fig. 2. Multi triangular carrier signals. (a) Two triangular carrier signals. (b) Four triangular carrier signals.

is 180° (where v_{t1} and v_{t2} are comparing with pole voltages of rectifier 1 and 2) [see Fig. 2(a)] and with four rectifiers in parallel ($N = 4$) the phase shift of the four triangular carrier signals is 90° (where v_{t1}, v_{t2}, v_{t3} and v_{t4} are comparing with pole voltages of rectifiers 1, 2, 3 and 4, respectively) [see Fig. 2(b)].

IV. CONTROL STRATEGY

Fig. 3 presents the control block diagram of the system in Fig. 1(a), highlighting the control of the rectifier. The rectifier circuit of the proposed system has the same objectives of that in Fig. 1(b), i.e., to control the dc-link voltage and to guarantee the power factor close to one. Additionally, the circulating currents i_{okw} of the proposed system rectifier need to be controlled.

In this way, the capacitor dc-link voltage v_c is adjusted to its reference value v_c^* using the controller R_C (conventional PI type controllers). Those controllers provide the amplitude of the reference currents i_{na}^*, whose amplitude are equal to I_g^*/N which means that, each rectifier supplies $1/N$ of the grid current. To control power factor, the instantaneous reference currents i_{na}^* are synchronized with the grid voltages e_g. This is obtained via block Gig, based on a PLL scheme. The control of the rectifiers currents are implemented using controllers indicated by blocks R_n. These controllers are implemented using in the two sequences synchronous controller described in [18]. The current controllers define the input reference voltages v_n^*.

The homopolar currents are measured i_{okw} and are controlled using controllers PI indicated by blocks R_{okw}, that determines the voltages v_{okw}^*. The calculation of voltage v_x^* is given from (19)-(21) as a function of μ, selected as shown in the section V.

V. HARMONIC DISTORTION

The harmonic distortion has been computed by using

$$WTHD(h) = \frac{100}{a_1}\sqrt{\sum_{h=2}^{N_h}\left(\frac{a_h}{h}\right)^2} \qquad (22)$$

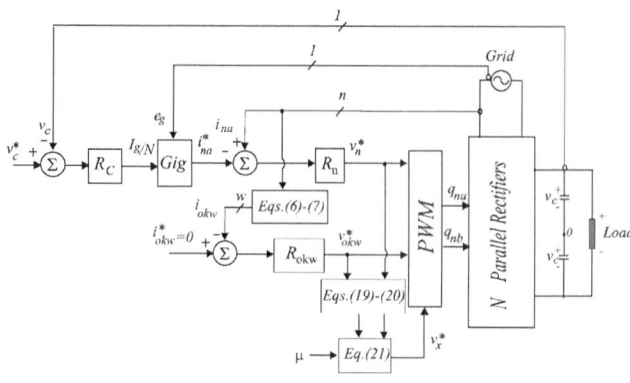

Fig. 3. System block control diagram.

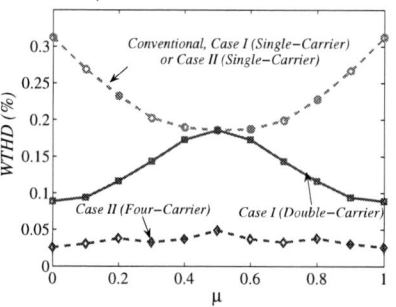

Fig. 4. $WTHD$ of the rectifiers voltages ($v_g = (v_1 + v_2)/2$ and $v_g = (v_1 + v_2 + v_3 + v_4)/4$ for proposed configuration and $v_g = v_{10} - v_{20}$ for conventional configuration) as a function of μ.

where a_1 is the amplitude of the fundamental voltage, a_h is the amplitude of hth component harmonic and N_h is the number of harmonics taken into consideration.

For comparison between conventional and proposed topologies, two cases were considered for proposed topology: case I with two rectifiers in parallel and case II with four rectifiers in parallel.

Fig. 4 shows the $WTHD$ of voltage generated by rectifier ($v_g = (v_1 + v_2)/2$ for case I, $v_g = (v_1 + v_2 + v_3 + v_4)/4$ for case II and $v_g = v_{10} - v_{20}$ for conventional configuration) as a function of μ obtained for a switching frequency of 10kHz. The voltage generated by rectifier is responsible to control i_g, which means that, this voltage is used to regulate the harmonic distortion of the utility grid, as observed in equation (12) for both cases.

In both cases, applying a single-carrier-based PWM the $WTHD$ is identical to $WTHD$ of conventional topology. But with double-carrier-based PWM (case I), and, when $\mu = 0.5$ the $WTHD$ is the same of that in the conventional topology, but for the other values of μ the $WTHD$ of the proposed system is lower. On the other hand with four-carrier-based PWM (case II) the $WTHD$ is lower for any value of μ. The WTHD of proposed topology case I (using double-carrier with $\mu = 1$) is close to 47% of that in the conventional topology (with $\mu = 0.5$) and to case II (using four-carrier $\mu = 1$) the $WTHD$ is 14% of that in the conventional topology (with $\mu = 0.5$).

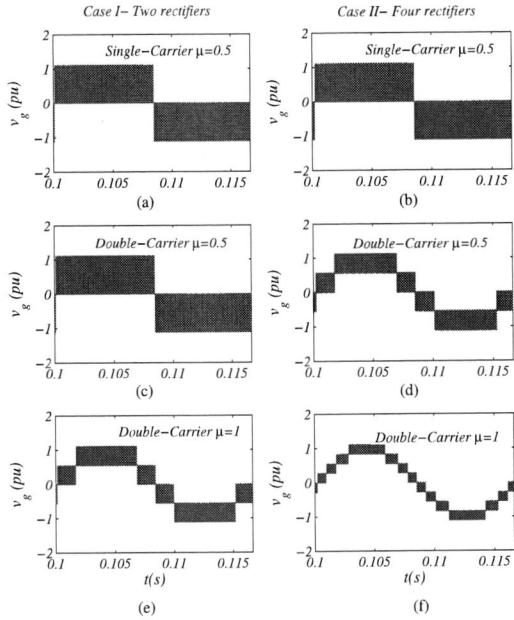

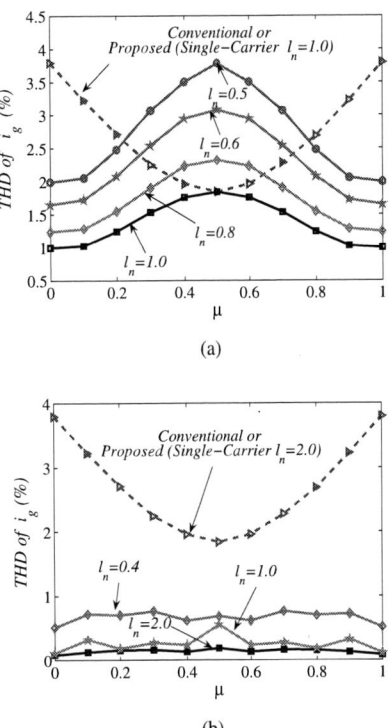

Fig. 5. *WTHD* of the rectifiers voltages ($v_g = (v_1 + v_2)/2$ and $v_g = (v_1 + v_2 + v_3 + v_4)/4$ for proposed configuration and $v_g = v_{10} - v_{20}$ for conventional configuration) as a function of μ.

Fig. 6. Inductor specification in terms of THD of i_g and μ. (a) With two rectifiers in parallel (case I). (b) With four rectifiers in parallel (case II).

The explanation of the *WTHD* behavior of Fig.4 can be obtained from Fig.5, which depicts the voltages of rectifier v_g ($v_g = (v_1 + v_2)/2$ for case I and $v_g = (v_1 + v_2 + v_3 + v_4)/4$ for case II) obtained using single-carrier ($\mu = 0.5$), double-carrier ($\mu = 0.5$) and double-carrier ($\mu = 1$) for case I and using single-carrier ($\mu = 0.5$), four-carrier ($\mu = 0.5$) and four-carrier ($\mu = 1$) for case II. Note that, the voltage v_g using double-carrier with $\mu = 0.5$ (case I) [Fig.5(c)] is equal to voltage using single-carrier (case I and case II) [Figs. 5(a) and 5(b)]. However, using double-carrier with $\mu = 1$ (case I) [Fig. 5(e)] and four-carrier with $\mu = 0.5$ and $\mu = 1$ (case II) [Figs. 5(d) and 5(f)] has increased the number of levels of voltage generated by the rectifiers; this means a reduction in the *WTHD* (as show Fig. 4). For four-carrier with $\mu = 1$ has the highest number of levels of voltage generated by rectifiers. The other hand, using the double or four carrier-based PWM the WTHD of the rectifiers currents (i_{1a}-i_{2b} for case I and i_{1a}-i_{4b} for case II) is worse than the WTHD with single-carrier-based PWM.

VI. INPUT INDUCTORS

The PWM strategy with double-carrier (case I) or with four-carrier (case II) reduces the WTHD of the resultant rectifier voltage v_g, as observed in Fig. 4. When the input inductors of the proposed topology (case I) (L_g') are equal to that of the conventional topology (L_g), the THD reduction of the grid current is directly indicated in Fig. 4. On the other hand, for case II is necessary double the inductance of the conventional topology, i.e., $L_g' = 2L_g$.

Fig. 6 depicts the THD of the grid current as a function of μ for different values of l_n [the inductances of proposed rectifier (l_g') referred to that of the conventional configuration (l_g), i.e., $l_n = l_g'/l_g$]. Fig. 6(a) illustrated the THD of the grid current of the proposed topology with two rectifiers in parallel (case I) using double-carrier PWM. Notice that for $l_n \geq 0.6$ ($l_g' \geq 0.6l_g$) the THD of i_g is smaller than that of the conventional topology. Fig. 6(b) show the THD of the grid current for proposed topology with four rectifiers in parallel (case II) using four-carrier PWM. Notice that the THD of i_g is always lower than that conventional topology (for $l_n \geq 0.4$).

However, In both cases the harmonic distortion of the rectifier currents (i_{1a}-i_{2b} for case I and i_{1a}-i_{4b} for case II) and circulating currents (i_{o12} for case I and i_{o12}-i_{o34} for case II) is higher than that of the grid current i_g. The using of common-mode inductors is good way to reduce the harmonic distortion of these currents [10].

VII. DC-LINK CAPACITOR

The power losses in the dc-link capacitor is calculated by [19]

$$P_{loss} = \sum_{h=1}^{N_h} ESR(h)I_c^2(h) \qquad (23)$$

where ESR is equivalent series resistance and $I_c(h)$ is capacitor harmonic component. The ESR decreasing with follows parameters: increasing of frequency, the ripple of current and ambient temperature. The ESR can be considered constant for frequency higher than 3kHz [10]. Its equal a 0.45 times ESR value for 100Hz. Thus of (23) becomes

$$P_{loss} = P_{loss}^{lf} + P_{loss}^{hf} \qquad (24)$$

with

$$P_{loss}^{lf} = \sum_{h=1}^{50} ESR(h)I_c^2(h) \qquad (25)$$

and

$$P_{loss}^{hf} = 0.45ESR_{(100Hz)} \sum_{h>50}^{N_h} I_c^2(h) \qquad (26)$$

P_{loss}^{lf} is power losses into low-order harmonic (caused mainly by the control and connection with single-phase) and P_{loss}^{hf} is power losses in high-order frequency (caused by the switching frequency).

The current RMS of the dc-link is defined by [19]

$$I_{c,rms} = \sqrt{\sum_{h=1}^{N_h} I_c(h)} \qquad (27)$$

soon of (26) found [20]

$$P_{loss}^{hf} = 0.45ESR_{(100Hz)}(I_{c,rms}^{hf})^2 \qquad (28)$$

where $I_{c,rms}^{hf}$ is component of high-order of the current RMS of the dc-link. As ESR is almost constant the higher frequency, then P_{loss}^{hf} depends only of the component $I_{c,rms}^{hf}$. This means the reduction of the power loss of the dc-link capacitor is determined by ripple current RMS.

Figs. 7 and 8 illustrates the harmonic spectrums of the dc-link capacitor current obtained simulation for the conventional rectifier $\mu = 0.5$ [Figs. 7(a) and 8(a)]; for the proposed rectifier case I using single-carrier with $\mu = 0.5$ [Fig. 7(b)], double-carrier with ($\mu = 0.5$) [Fig. 7(c)] and double-carrier with ($\mu = 1$) [Fig. 7(d)]; and for the proposed rectifier case II using single-carrier with $\mu = 0.5$ [Fig. 8(b)], four-carrier with ($\mu = 0.5$) [Fig. 8(c)] and four-carrier with ($\mu = 1$) [Fig. 8(d)].

The proposed rectifier using double-carrier (case I) $\mu = 1$, four-carrier (case II) with $\mu = 0.5$ and four-carrier (case II) with $\mu = 1$ provides the reduction of the high frequency harmonics when compared with conventional topology. The highest reduction of high-order frequency harmonics is obtained for the proposed rectifier (case II) using four-carrier with $\mu = 1$

Table I (obtained from Figs. 7 and 8) presents the $I_{c,rms}^{hf}$ of the proposed rectifier [$I_{c,rms}^{hf}(p)$] referred to the $I_{c,rms}^{hf}$ of the conventional rectifier [$I_{c,rms}^{hf}(c)$]. The highest reduction of $I_{c,rms}^{hf}$ is obtained for the rectifier using four-carrier with $\mu = 1$. The $I_{c,rms}^{hf}$ obtained for $\mu = 0$ is equal to that for $\mu = 1$ for both case.

Experimental result are performed on the laboratory as described in section VIII. Only the spectrum of current of the capacitor for proposed rectifier case I are show in Fig. 9. The harmonic spectrums of the dc-link capacitor current illustrates in this figure are: for single-carrier with $\mu = 0.5$ [Fig. 9(a)], double-carrier with $\mu = 0.5$ [Fig. 9(b)] and double-carrier with $\mu = 1$ [Fig. 9(c)]. There is a difference between harmonic

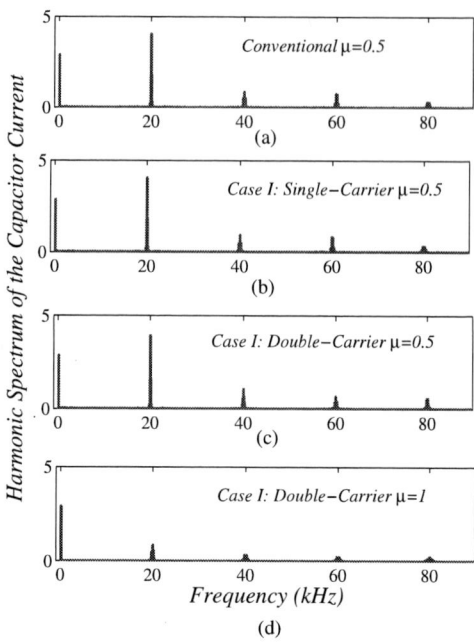

Fig. 7. Simulation result of the harmonic spectrum of the dc-link capacitor current. (a) Conventional converter ($\mu = 0.5$). (b) Proposed rectifier case I with single-carrier ($\mu = 0.5$). (c) Proposed rectifier case I with double-carrier ($\mu = 0.5$). (d) Proposed rectifier case I with double-carrier ($\mu = 1$).

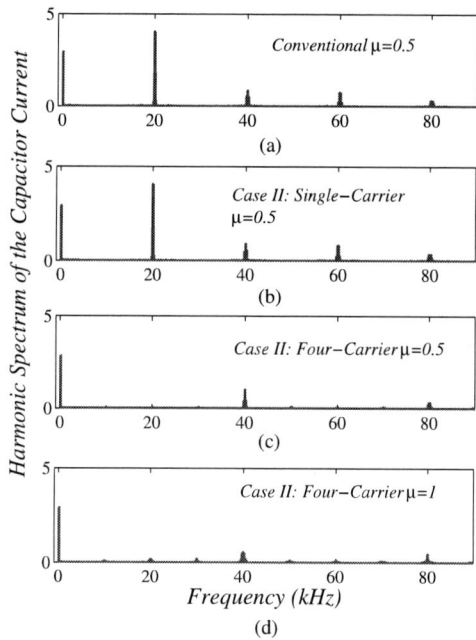

Fig. 8. Simulation result of the harmonic spectrum of the dc-link capacitor current. (a) Conventional converter ($\mu = 0.5$). (b) Proposed rectifier case II with single-carrier ($\mu = 0.5$). (c) Proposed rectifier case II with four-carrier ($\mu = 0.5$). (d) Proposed rectifier case II with four-carrier ($\mu = 1$).

TABLE I

SIMULATION RESULT: NORMALIZED $I_{c,rms}^{hf}$ OF DC-LINK CURRENTS OF THE PROPOSED CONFIGURATION. S-C: SINGLE-CARRIER, M-C DOUBLE-CARRIER (CASE I) OR FOUR-CARRIER (CASE II)

Topology	$I_{c,rms}^{hf}(p)/I_{c,rms}^{hf}(c)$		
	S-C($\mu=0.5$)	M-C($\mu=0.5$)	M-C($\mu=1$)
Case I	1.0	0.96	0.36
Case II	1.0	0.30	0.26

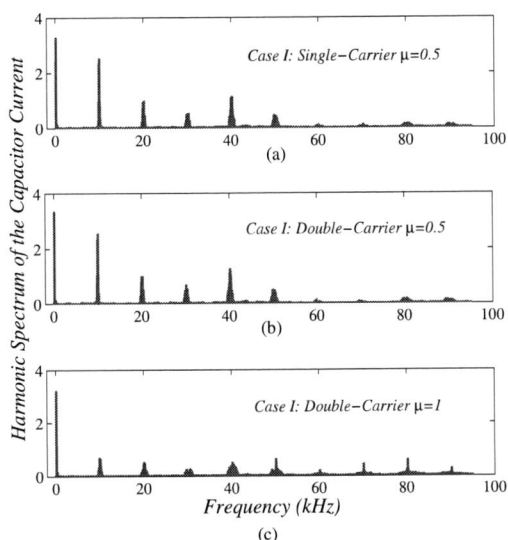

Fig. 9. Experimental result of the harmonic spectrum of the dc-link capacitor current to proposed converter case I. (a) Single-carrier ($\mu = 0.5$). (b) Double-carrier ($\mu = 0.5$). (c) Double-carrier ($\mu = 1$).

spectrums of simulation result compared with experimental result, because the experimental result were obtained using a high sawtooth frequency signals instead the high triangular frequency.

Harmonic spectrum of the conventional topology is equal to the proposed topology using single-carrier with $\mu = 0.5$. Note that the highest reduction of high-order frequency harmonics is obtained for case using double-carrier with $\mu = 1$. Table II (obtained from Figs. 9) presents the $I_{c,rms}^{hf}$ calculated from spectrum of Fig. 9. The reduction of current $I_{c,rms}^{hf}$ in the rectifier using double-carrier with $\mu = 1$ is almost 50% as compared to the proposed with single-carrier ($\mu = 0.5$). In this scenario, the power losses of high-order frequency in the capacitor are lower for the proposed topology using double-carrier with $\mu = 1$.

TABLE II

EXPERIMETAL RESULT: CURRENT $I_{c,rms}^{hf}$ OF DC-LINK CURRENTS OF THE PROPOSED CONFIGURATION FOR CASE I. S-C: SINGLE-CARRIER, D-C DOUBLE-CARRIER

Topology	$I_{c,rms}^{hf}$ (A)		
	S-C ($\mu=0.5$)	D-C ($\mu=0.5$)	D-C ($\mu=1$)
Case I	3.90	3.96	2.13

VIII. EXPERIMENTAL RESULTS

The system showed in Fig. 1(a) has been implemented in the laboratory. The set-up used in the experimental tests is based on a microcomputer equipped with appropriate plug-in boards and sensors. The system operates with a switching frequency of 10kHz and capacitance of the dc-link from $2200\mu F$. Steady state with two rectifiers in parallel with single-carrier-based PWM and double-carrier-based PWM for different values of μ have been considered in the experimental testes.

The steady state experimental result was obtained using single-carrier PWM ($\mu = 0.5$) are show in Fig. 10. The waveforms in this figure are: a) voltage and current of the grid (e_g and i_g) and dc-link voltage (v_c), b) currents of the rectifiers 1 and 2 (i_{1a} and i_{2a}), and c) current of rectifier 1 (i_{1b}) and circulating current (i_{o12}).

Notice that, the control guarantees sinusoidal grid current with power factor close to one and dc-link voltage under control and observe that there is a component of second harmonic in dc-ink voltage due the connection with of the single-phase grid [see Fig. 10(a)]. Furthermore, the proposed configuration provides current reduction in the rectifier (half of the standard topology current) [see Figs. 10(b) and 10(c)], which provides losses reduction. Also, the control guarantees the circulating current close to zero [see Fig. 10(c)].

Steady state experimental results presented in Figs 11 and 12 show the behavior of variables of the proposed system when using double-carrier PWM. In this case the results were obtained with $\mu = 0.5$ and $\mu = 1$. The waveforms in this figure are: a) voltage and current of the grid (e_g and i_g) and dc-link voltage (v_c), b) currents of the rectifiers 1 and 2 (i_{1a} and i_{2a}), and c) current of the rectifier 1 (i_{1b}) and circulating current (i_{o12}).

The results presented in Figs. 11 and 12, notice that the control guarantees sinusoidal grid current with power factor close to one, dc-link voltage under control and circulating current close to zero. Beside, has an increase in current ripple in circulating current due to use of double-carrier PWM. However, choosing $\mu = 1$ was possible to decrease ripple of the circulating current [see Fig. 12(c)].

Figs. 11(b) and 11(c) shows the currents of rectifiers using double-carrier PWM with $\mu = 0.5$, in this case has an increase of ripple the currents (i_{1a}, i_{2a} and i_{1b}) when compared with single-carrier PWM [see Figs. 10(b) and 10(c)]. However, the current of the grid is similar to that of single-carrier PWM. The increasing of the THD to current i_{1a} is 40% but for current i_g has a reduction of 7% in the THD.

When $\mu = 1$ the current of the grid (i_g) present the ripple smaller than that it for the single-carrier PWM [see Fig. 12(a)]. Notice that reduction the ripple of currents of the rectifier during semicircle positive the currents i_{1a} and i_{2a} [see Fig. 12(b)] and in semicircle negative for current i_{1b} [see Fig. 12(c)] (the behavior of current i_{2b} is similar the current i_{1b}). The current of the grid obtained for $\mu = 0$ is similar to that for $\mu = 1$.

The reduction of the THD to current i_g is 23% but for

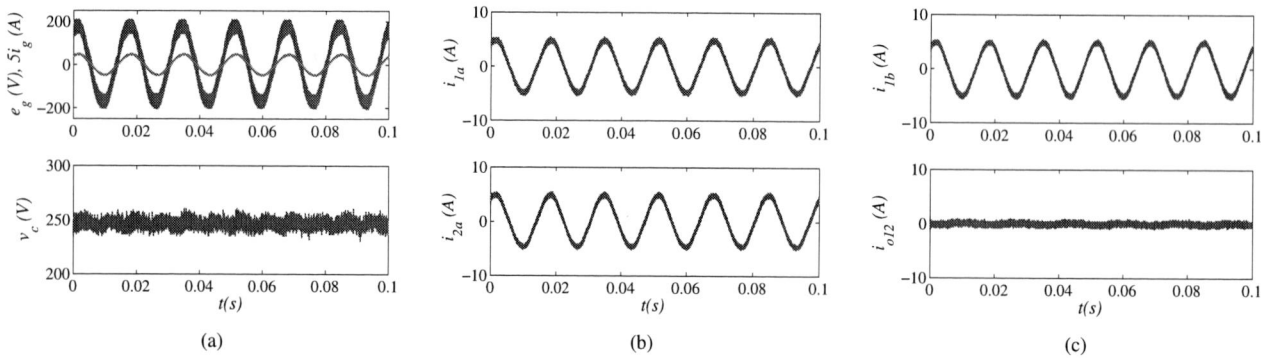

Fig. 10. Experimental results of the proposed configuration using single-carrier PWM and $\mu = 0.5$. (a) Voltage e_g and current i_g of the grid (top) and dc-link voltage v_c (bottom). (b) Currents of the rectifiers 1 i_{1a} (top) and 2 i_{2a} (bottom). (c) Current of the rectifier 1 i_{1b} (top) and circulating current i_{o12} (bottom).

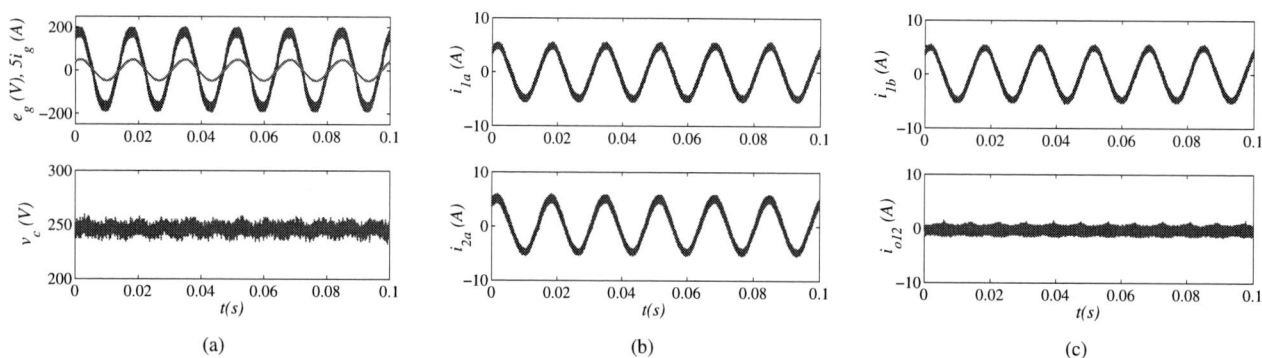

Fig. 11. Experimental results of the proposed configuration using double-carrier PWM and $\mu = 0.5$. (a) Voltage e_g and current i_g of the grid (top) and dc-link voltage v_c (bottom). (b) Currents of the rectifiers 1 i_{1a} (top) and 2 i_{2a} (bottom). (c) Current of the rectifier 1 i_{1b} (top) and circulating current i_{o12} (bottom).

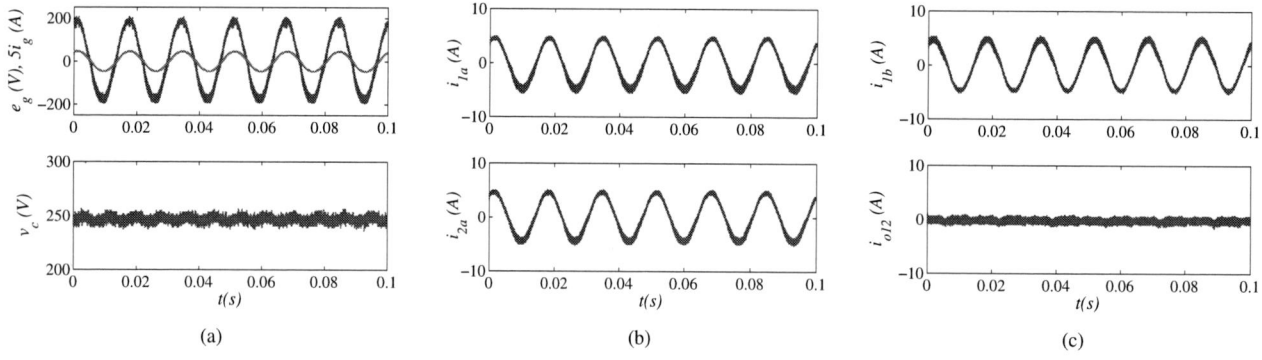

Fig. 12. Experimental results of the proposed configuration using double-carrier PWM and $\mu = 1$. (a) Voltage e_g and current i_g of the grid (top) and dc-link voltage v_c (bottom). (b) Currents of rectifiers 1 i_{1a} (top) and 2 i_{2a} (bottom) . (c) Current of rectifier 1 i_{1b} (top) and circulating current i_{o12} (bottom).

current i_{1a} has a increasing of 14% in the THD when compared with single-carrier PWM.

IX. CONCLUSION

A single-phase ac-dc converter has been proposed in this paper. It is composed of N parallel boost rectifiers without isolation transformers. The proposed topologies permits to reduce the current and power ratings of the power switches

and switching stress on the dc-link capacitor when compared to the conventional single-phase ac-dc converter. Moreover, the strategy PWM applied enabled the reduction the THD of the grid current. Experimental results were presented.

ACKNOWLEDGMENT

The authors would like to thank the financial support provided by the National Council for Scientific and Tech-

978-1-4244-4782-4/10 $26.00 © 2010 IEEE

TABLE III

EXPERIMENTAL RESULTS: THD OF CURRENTS FOR PROPOSED CONFIGURATION.

Currents	Single-carrier PWM	Double-carrier PWM	
	$\mu = 0.5$	$\mu = 0.5$	$\mu = 1$
i_g	3.52%	3.22%	2.67%
i_{r1a}	3.89%	5.18%	4.20%
i_{r2a}	3.66%	5.04%	4.09%
i_{r1b}	3.80%	5.42%	4.43%

nological Development (CNPq), by the Coordination for the Improvement of Higher Education Personnel (CAPES), and by Foundation for Research Support of the State of Paraba (FAPESQ).

REFERENCES

[1] L. MatakasJr. and W. Kaiser, "Low harmonics, decoupled histeresis type current control of a multiconverter consisting of a parallel transformerless connection of VSC converters," in *Conf. Rec. IEEE-IAS Annu. Meeting*, 1997, pp. 1633–1640.

[2] Z. Ye, D. Boroyevich, J.-Y. Choi, and F. C. Lee, "Control of circulating current in two parallel three-phase boost rectifiers," *IEEE Trans. Power Electron.*, vol. 17, no. 5, pp. 609–615, Sep. 2002.

[3] S. K. Mazumder, "A novel discrete control strategy for independent stabilization of parallel three-phase boost converters by combining space-vector modulation with variable-structure control," *IEEE Trans. Power Electr.*, vol. 18, no. 4, pp. 1070–1083, July 2003.

[4] Z. Ye, P. Jain, and P. Sen, "Circulating current minimization in high-frequency AC power distribution architecture with multiple inverter modules operated in parallel," *IEEE Trans. Ind. Electron.*, vol. 54, no. 5, pp. 2673–2687, Oct. 2007.

[5] M. Baumann and J. Kolar, "Parallel connection of two three-phase three-switch buck-type unity-power-factor rectifier systems with dc-link current balancing," *IEEE Trans. Ind. Electron.*, vol. 54, no. 6, pp. 3042–3053, Dec. 2007.

[6] J.-K. Park, J.-M. Kwon, E.-H. Kim, and B.-H. Kwon, "High-performance transformerless online UPS," *IEEE Trans. Ind. Electron.*, vol. 55, no. 8, pp. 2943–2953, Aug. 2008.

[7] J. W. Dixon and B. T. Ooi, "Series and parallel operation of hysteresis current-controlled PWM rectifiers," *IEEE Trans. Ind. Appl.*, vol. 25, no. 5, p. 644651, July/Aug. 1989.

[8] Y. Komatsuzaki, "Cross current control for parallel operating three-phase inverter," in *Proc. IEEE PESC*, 1994, p. 943950.

[9] K. Matsui, "A pulse width modulated inverter with parallel-connected transistors by using sharing reactors," in *Conf. Rec. IEEE-IAS Annu. Meeting*, 1985, p. 10151019.

[10] L. Asiminoaei, P. Aeloiza, E. Enjeti, and F. Blaabjerg, "Shunt active-power-filter topology based on parallel interleaved inverters," *IEEE Trans. Industrial Electronics.*, vol. 55, no. 3, pp. 1175–1189, March. 2008.

[11] S. Ogasawara, J. Takagaki, H. Akagi, and A. Nabae, "A novel control scheme of a parallel current-controlled PWM inverter," *IEEE Trans. Ind. Appl.*, vol. 28, no. 5, pp. 1023–1030, Mar./Apr. 1992.

[12] K. Xing, F. C. Lee, D. Borojevic, Z. Ye, and S. Mazumder, "Interleaved PWM with discontinuous space-vector modulation," *IEEE Trans. Power Electr.*, vol. 14, no. 5, pp. 906–917, Sep. 1999.

[13] X. Sun, L.-K. Wong, Y.-S. Lee, and D. Xu, "Design and analysis of an optimal controller for parallel multi-inverter systems," *IEEE Trans. Circuits and Systems II*, vol. 53, no. 1, pp. 56–61, Jan. 2006.

[14] J. Holtz, "Pulsewidth modulation for electronic power conversion," *Proceedings of the IEEE*, vol. 82, no. 8, pp. 1194–1214, Aug. 1994.

[15] A. M. Trzynadlowski, R. L. Kirlin, and S. F. Legowski, "Space vector PWM technique with minimum switching losses and a variable pulse rate," *IEEE Trans. Ind. Electron.*, vol. 44, no. 2, pp. 173–181, Apr. 1997.

[16] O. Ojo and P. M. Kshirsagar, "Concise modulation strategies for four-leg voltage source inverters," *IEEE Trans. Power Electron.*, vol. 19, no. 1, pp. 46–53, Jan. 2004.

[17] C. B. Jacobina, A. M. N. Lima, E. R. C. da Silva, R. N. C. Alves, and P. F. Seixas, "Digital scalar pulse-width modulation: a simple approach to introduce non-sinusoidal modulating waveforms," *IEEE Trans. Power Electron.*, vol. 16, no. 3, pp. 351–359, May 2001.

[18] C. B. Jacobina, M. B. de R. Correa, R. F. Pinheiro, E. R. C. da Silva, and A. M. N. Lima, "Modeling and control of unbalanced three-phase systems containing PWM converters," *IEEE Trans. Ind. Appl.*, vol. 37, no. 6, pp. 1807–1816, Nov./Dec. 2001.

[19] F. Kieferndorf, M. Forster, and T. Lipo, "Reduction of dc-bus capacitor ripple current with pam/pwm converter," *IEEE Trans. Industry Applications.*, vol. 40, no. 2, pp. 607– 614, March-April 2004.

[20] K. J.W and R. S.D, "Analytical calculation of the rms current stress on the dc link capacitor of voltage pwm converter systems," in *Proc. IEE Electr. Power Appl.*, 2006, pp. 535–543.

Parallel Connection of Two Shunt Active Power Filters with Losses Optimization

E. C. dos Santos Jr., C. B. Jacobina, A. M. Maciel

Departamento de Engenharia Elétrica

Universidade Federal de Campina Grande

Caixa Postal 10.105

58109-970 Campina Grande - PB - Brazil

E-mails: [euzeli,jacobina]@dee.ufcg.edu.br

Abstract- **This paper describes a technique to reduce the converter losses in the shunt active power filter operation. In such technique, the reactive and harmonics compensation are separated in two shunt active power filters operating at different frequencies. The reactive power of the load is compensated by low frequency filter, while the power related with harmonics are compensated by high frequency filter. This technique provides some advantages in terms of losses and costs reduction when compared with standard solution. Both filters (low frequency and high frequency filters) are coupled in a combined topology in which they are connected in parallel connection scheme. Additionally, this paper gives the filter models, pulse-width modulation technique and control strategies are investigated, as well. Simulated and experimental results verify the theoretical studies.**

I. INTRODUCTION

Electrical power quality has become an important technical issue, and it has stimulated the use of active power compensation schemes. There is a wide class of equipment to improve the power quality, e.g., transient suppressors, line voltage regulators, UPS, active filters, and hybrid filters [1–9]. Different configurations have been considered to implement active power filters, and when it is necessary to compensate load harmonic currents and load reactive power, the shunt-type active filter can be employed [10–12].

In the conventional shunt active power filter, the reactive and harmonics of the load are compensated by using the same power converter, as observed in Fig. 1(a), which may represent some drawbacks in terms of converter losses, especially when the level of currents through the filter and the switching frequency are high.

In [13] is proposed active power filters connected in cascade, each one operating at a different switching frequency, while [14] describes the control and parallel operation of

two active power filters with both inverters operating at the same frequencies.

This paper describes a technique to reduce the converter losses in the shunt active power filter operation, which is obtained connecting in parallel two filters, as in Fig. 1(b). The reactive and harmonics compensation are separated in two shunt active filters operating at different frequencies. The reactive power of the load is compensated by the low frequency filter, while the harmonics are compen-

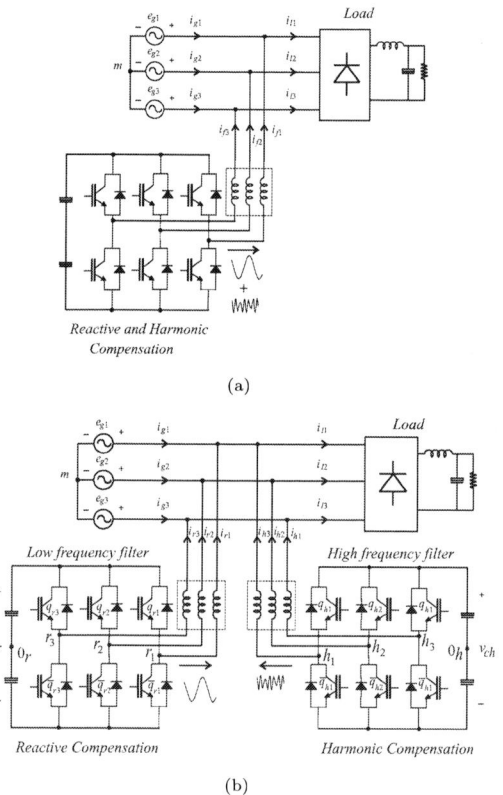

Fig. 1. Parallel Active Power Filter. (a) Conventional solution. (b) Proposed solution.

sated by high frequency filter. This technique provides some advantages in terms of losses reduction when compared with standard solution [Fig. 1(a)]. Even increasing the total number of components, the topology of Fig. 1(b) permits cost reduction, especially when the conventional high power shunt filter requires high price fast switching devices.

Furthermore, the paper presents: *i)* converter power losses comparison; *ii)* pulse-width modulation techniques; *iii)* control strategies for providing harmonic and reactive compensation; and *iv)* simulated and experimental results.

II. Shunt Filter Model

The model of the shunt active power filters (low frequency and high frequency filters) can be done approximately with a simple ideal circuit, as observed in Fig. 2. From Fig. 2, the equations of the low and high frequency filters are given respectively by

$$e_{g123} - r_r i_{r123} - l_r \frac{di_{r123}}{dt} - v_{r123} + v_{n_r m} = 0 \quad (1)$$

$$e_{g123} - r_h i_{h123} - l_h \frac{di_{h123}}{dt} - v_{h123} + v_{n_h m} = 0 \quad (2)$$

where r_r, r_h and l_r, l_h are the resistences and inductors of filters Z_r and Z_h, respectively.

III. PWM Strategy

The converters pole voltages depend on the conduction states of the power switches. For example, $v_{r_1 0_r}$ is given by

$$v_{r_1 0_r} = (2q_{r_1} - 1)\frac{v_{cr}}{2} \quad (3)$$

where v_{cr} is the dc-link voltage of low frequency filter and q_{r_1} is the conduction state of the power switch ($q_{r_1} = 1$ means closed switch and $q_{r_1} = 0$ means open switch).

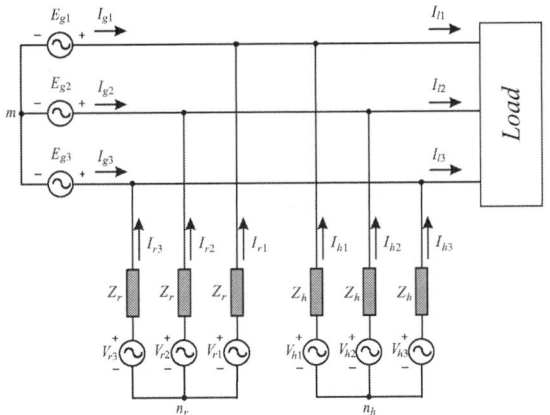

Fig. 2. Ideal equivalent circuit of the filters in a parallel connection scheme.

Considering that v^*_{r123} and v^*_{h123} denote the reference voltages requested by the controllers of reactive and harmonic filters, respectively, we have

$$v^*_{r_1 0_r} = v^*_{r1} + v^*_{\mu r} \quad (4)$$
$$v^*_{r_2 0_r} = v^*_{r2} + v^*_{\mu r} \quad (5)$$
$$v^*_{r_3 0_r} = v^*_{r3} + v^*_{\mu r} \quad (6)$$

and

$$v^*_{h_1 0_h} = v^*_{h1} + v^*_{\mu h} \quad (7)$$
$$v^*_{h_2 0_h} = v^*_{h2} + v^*_{\mu h} \quad (8)$$
$$v^*_{h_3 0_h} = v^*_{h3} + v^*_{\mu h} \quad (9)$$

Such equations are not sufficient to determine the pole voltages $v^*_{r_1 0_r}$, $v^*_{r_2 0_r}$, $v^*_{r_3 0_r}$ (reactive compensation filter) and $v^*_{h_1 0_h}$, $v^*_{h_2 0_h}$, $v^*_{h_3 0_h}$ (harmonic compensation filter). In this way, the voltages $v^*_{\mu r}$ and $v^*_{\mu h}$ can be calculated by taking into account the apportioning factors μ_r and μ_h, respectively. In the case of the high frequency active power filter, $v^*_{\mu h}$ is given by

$$v^*_{\mu h} = E\left(\mu_h - \frac{1}{2}\right) - \mu_h v^*_{h \max} + (\mu_h - 1)v^*_{h \min} \quad (10)$$

where $v^*_{hmax} = \max V_h$ and $v^*_{hmin} = \min V_h$ where $V_h = \{v^*_{h1}, v^*_{h2}, v^*_{h3}\}$. Expression (10) was derived by using the same approach used to obtain the corresponding equation for the three-phase PWM modulator [15], [16].

The apportioning factor μ_h ($0 \leq \mu_h \leq 1$) is given by

$$\mu_h = t_{oih}/t_{oh} \quad (11)$$

and indicates the distribution of the general free-wheeling period t_{oh} (period in which voltages $v_{h_1 0_h}$, $v_{h_2 0_h}$ and $v_{h_3 0_h}$ are equals) between the beginning ($t_{oih} = \mu_h t_{oh}$) and the end ($t_{ofh} = (1 - \mu_h)t_{oh}$) of the switching period.

The same equation can be used to derive $v^*_{\mu r}$, that is,

$$v^*_{\mu r} = E\left(\mu_r - \frac{1}{2}\right) - \mu_r v^*_{r \max} + (\mu_r - 1)v^*_{r \min} \quad (12)$$

where $v^*_{r \max} = \max V_r$ and $v^*_{r \min} = \min V_r$ where $V_r = \{v^*_{r1}, v^*_{r2}, v^*_{r3}\}$.

The gating signals can be generated by comparing the pole voltage - $v^*_{r_\alpha 0_r}$ and $v^*_{h_\alpha 0_r}$ ($\alpha = 1, 2, 3$) - with a high frequency triangular carrier signal.

IV. Control Strategy

Fig. 3(a) presents the control block diagram of the low frequency filter, which furnish reactive compensation. The capacitor dc-link voltage v_{cr} is adjusted to a reference value by using the controller R_c, which is a standard PI type controller. This controller provides the amplitude of the reference current I^*_{ra}, this current (I^*_{ra}) is associated

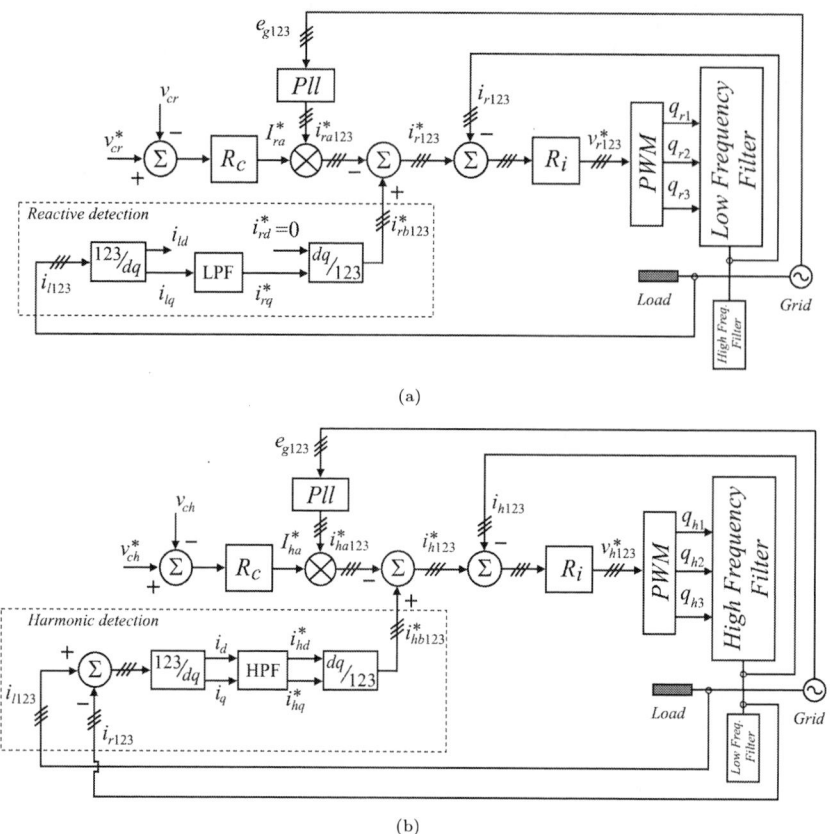

(a)

(b)

Fig. 3. Control block diagram of the shunt active power filter. (a) Control of the low frequency filter (reactive compensation). (b) Control of the high frequency filter (harmonics compensation).

with active power of the load. The instantaneous reference current i^*_{ra123} (i^*_{ra1}, i^*_{ra2} and i^*_{ra3}) must be synchronized with voltages e_{g123} (e_{g1}, e_{g2} and e_{g3}). This is obtained via block Pll, which indicates the instantaneous phase of voltages e_{g123}. In terms of power factor control, the reactive power demanded by the load is given by the low frequency filter. This is obtained as observed in Fig. 3(a), i.e., the current i^*_{l123} (i^*_{l1}, i^*_{l2} and i^*_{l3}) is converted to i^*_{ldq} (i^*_{ld} and i^*_{lq}) using the Park's transformation, the low-pass filter (LPF) is applied in the current i^*_{lq}, which gives the current of the filter (i^*_{rq}) with the same frequency of the grid. Another conversion from dq to 123 is required with $i^*_{rd} = 0$, which gives the current i^*_{rb123} furnished by the low frequency filter to compensate the reactive power of the load. The current controller is implemented by using the controller indicated by block R_i. This current controller defines the reference voltages v^*_{r123}, which has been used in the PWM strategy defined in the Section III.

Fig. 3(b) presents the control block diagram of the high frequency filter, which furnish harmonic compensation. Notice that, the difference between the strategy of both filters is in terms of the use of low-pass filter block (LPF) for

the reactive compensation [Fig. 3(a)] and high-pass filter block (HPF) for the harmonic compensation [Fig. 3(b)]. Furthermore, as the current of the high frequency filter needs to compensate the harmonics of the critical load, and the eventual harmonic caused by the low frequency filter, it is needed to measure the currents of the reactive filter.

V. Losses of the Shunt Filter

The loss estimation is obtained through regression model, which has been achieved by experimental tests, as presented in [17]. The power switch used in the experimental tests was: IGBT dual module CM50DY-24H (POWEREX) driven by SKHI-10 (SEMIKRON). That switch losses model includes: a) IGBT and diode conduction losses, b) IGBT turn-on losses, c) IGTB turn-off losses, and d) diode turn-off energy.

Fig. 4 shows the total converter losses (conduction + switching losses) of the shunt active power filter, as a function of the load power, for both systems: proposed [see Fig. 1(b)] and standard one [see Fig. 1(a)].

For the proposed system, the conditions employed in the

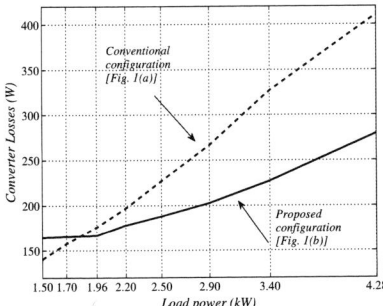

Fig. 4. Losses of the filter as a function of the load power.

tests of Fig. 4 were:

- Switching frequency of the low frequency filter: $1kHz$;
- Switching frequency of the high frequency filter: $10kHz$;
- Dc-link voltage for both filters: $410V$;
- Inductor specification of the low frequency filter: $6.5mH$;
- Inductor specification of the high frequency filter: $2.1mH$.

The conditions employed in the tests presented in Fig. 4 for the conventional system were:

- Switching frequency of the filter: $10kHz$;
- Dc-link voltage for both filters: $410V$;
- Inductor specification of the filter: $2.1mH$.

Notice that, for values of load power lower than $P_l = 1.83kW$ the conventional configuration is more attractive than the proposed one, but for values of load power higher than $P_l = 1.83kW$ the proposed solution becomes the best choice in terms of converter losses.

The reduction in the converter losses of the proposed system ($P_l > 1.83kW$) can be explained as follows: the reactive power processed by the low frequency filter becomes significantly higher than that of the harmonic power processed by the high frequency filter, i.e., the power converter with $1kHz$ processes a fraction of the load power associated with reactive power (high power) and the power converter with $10kHz$ processes a fraction of the load power associated with harmonic power (low power).

VI. SIMULATION RESULTS

Fig. 5 presents simulated results of the proposed configuration operating with two load powers: $P_l = 1.7kW$ [Fig. 5(a)] and $P_l = 2.9kW$ [Fig. 5(b)]. These variables are: load current (i_{l1}), grid current (i_{g1}), current of the low frequency filter (i_{r1}) - with switching frequency equal to $1kHz$, and current of the high frequency filter (i_{h1}) - with switching frequency equal to $10kHz$. The same conditions presented in Section VI were used in these results (Fig. 5).

Notice that, as proposed in the Control Stratefy Section

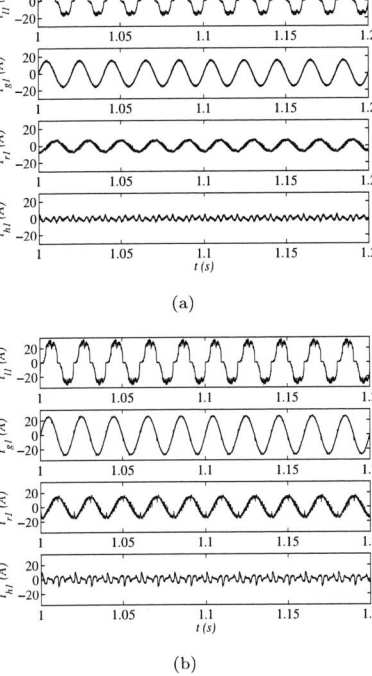

Fig. 5. Simulated results of the proposed configuration - variables of the system with: a) $P_l = 1.7kW$, b) $P_l = 2.9kW$.

the low frequency filter is used to compensate the current with the same frequency of the grid, since it is conceived to control the reactive power. On the other hand, the high frequency filter is used to compensate the current with frequencies demanded by the critical load, since it is conceived to control the power associated with harmonics.

Fig. 6 shows simulated results of the conventional configuration [see Fig. 1(b)] operating with the same load power of that in Fig. 5 (proposed configuration - with switching frequency equal to $10kHz$). In this case, both reactive and harmonic power are compensated by the same power converter.

VII. EXPERIMENTAL RESULTS

The system in Fig. 1(b) has been tested in the laboratory by using a microcomputer-based system. In the experimental tests the dc-link capacitors were selected as $C = 2200\mu F$, and the switching frequency employed was $10kHz$ to the high frequency filter and $5kHz$ to the low frequency filter. Fig. 7 shows the photo of the experimental set-up used in this work, highlighting the converters employed to reactive and harmonic compensation, i.e. low and high frequency converters, respectively.

Fig. 8 shows the currents of the system with high frequency filter operating at $10kHz$ and low frequency filter operating at $5kHz$, those variables are: grid current (i_{g1}),

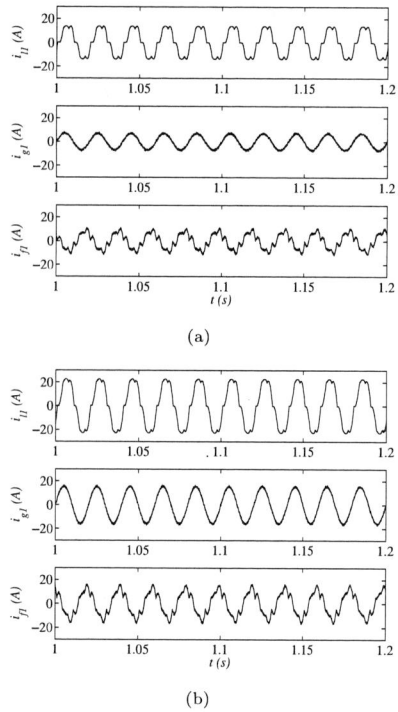

(a)

(b)

Fig. 6. Simulated results of the conventional configuration - variables of the system with: a) $P_l = 1.7kW$, b) $P_l = 2.9kW$.

load current (i_{l1}), current of the high frequency filter (i_{h1}), and current of the low frequency filter (i_{r1}).

Fig. 9 shows the variables under control: a) Dc-link voltages (v_{ch} - top, and v_{cr} - bottom), b) reference and measured currents of the high frequency filter (i_{h1}^* - top and i_{h1}- bottom), c) reference and measured currents of the low frequency filter (i_{r1}^* - top and i_{r1}- bottom).

As expected, the control of reactive and harmonics has been separated in two converters, which implies in reduction of power converter losses since part of current is pro-

Fig. 7. Photo of the experimental set up.

cessed by the converter with low frequency operation.

VIII. CONCLUSIONS

This paper proposes a technique to reduce the converter losses in the shunt active power filter. The optimization of the losses is obtained by the connection of two shunt active power filter operating in a parallel connection scheme. With this strategy, it is show that there is a wide range of load power which the losses of the proposed configuration is lower than that of the conventional system.

ACKNOWLEDGMENT

The authors would like to thank the financial support provided by the Conselho Nacional de Desenvolvimento Científico e Tecnológico (CNPq) and the Coordenação de Aperfeiçoamento de Pessoal de Nível Superior (CAPES) of Brazil.

REFERENCES

[1] B. Singh, K. Al-Haddad, and A. Chandra, "A review of active filters for power quality improvement," *IEEE Trans. Ind. Electron.*, vol. 46, pp. 960–971, Oct. 1999.

[2] H. Akagi, "Trends in active power line conditioners," *IEEE Trans. Power Electron.*, vol. 9, pp. 263–268, May 1994.

[3] Z. Pan, F. Peng, and S. Wang, "Power factor correction using a series active filter," *IEEE Trans. Power Electron.*, vol. 20, pp. 148–153, Jan. 2005.

[4] L. Asiminoaei, F. Blaabjerg, and S. Hansen, "Detection is key - harmonic detection methods for active power filter applications," *IEEE Ind. Appl. Magazine*, vol. 13, pp. 22–33, July-Aug. 2007.

[5] Y. W. Li, F. Blaabjerg, D. Vilathgamuwa, and P. C. Loh, "Design and comparison of high performance stationary-frame controllers for DVR implementa-

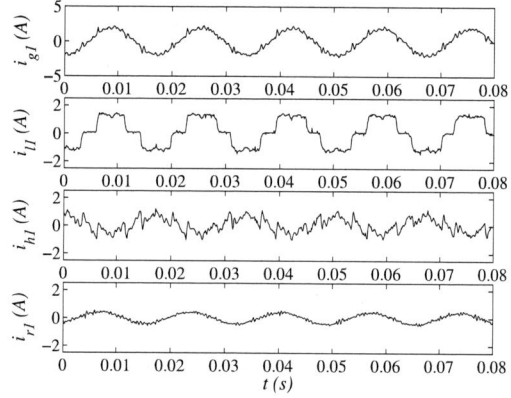

Fig. 8. Experimenatl results - currents of the system with high frequency filter operating at $10kHz$ and low frequency filter operating at $5kHz$.

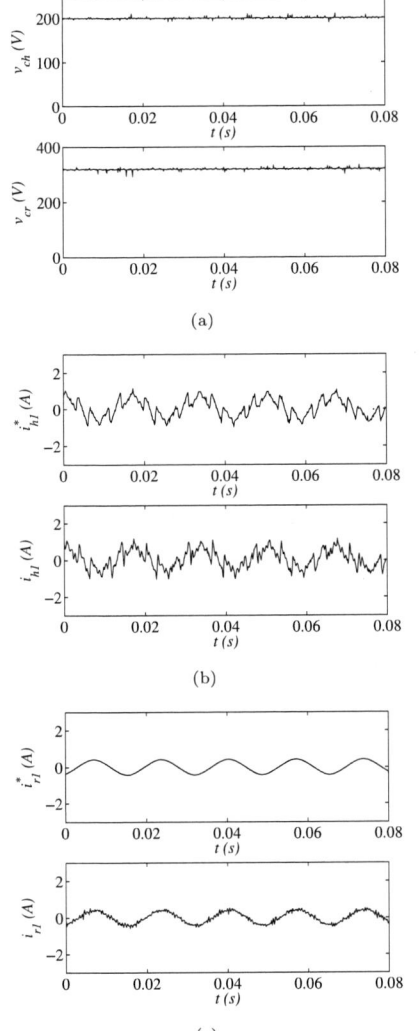

Fig. 9. Experimental results of variables under control: a) Dc-link voltages (v_{ch} - top, and v_{cr} - bottom), b) reference and measured current of the high frequency filter (i_{h1}^* and i_{h1}), c) reference and measured current of the low frequency filter (i_{r1}^* and i_{r1}).

tion," *IEEE Trans. Power Electron.*, vol. 22, pp. 602–612, Mar. 2007.

[6] Z. Pan, F. Z. Peng, and S. Wang, "Power factor correction using a series active filter," *IEEE Trans. Power Electron.*, vol. 20, pp. 148–153, Jan. 2005.

[7] B. S. amd K. Al-Haddad and A. Chandra, "A review of active filters for power quality improvement," *IEEE Trans. Ind. Electron.*, vol. 46, pp. 960–971, Oct. 1999.

[8] S. Jain and V. Agarwal, "A single-stage grid connected inverter topology for solar PV systems with maximum power point tracking," *IEEE Trans. Power Electron.*, vol. 22, pp. 1928–1940, Sept. 2007.

[9] K. Borisov, H. L. Ginn, and A. M. Trzynadlowski,

"Attenuation of electromagnetic interference in a shunt active power filter," *IEEE Trans. Power Electron.*, vol. 22, pp. 1912–1918, Sept. 2007.

[10] H. Fujita and H. Akagi, "Voltage-regulation performance of a shunt active filter intended for installation on a power distribution system," *IEEE Trans. Power Electron.*, vol. 22, pp. 1046–1053, May 2007.

[11] G. Escobar, A. Valdez, R. Torres-Olguin, and M. Martinez-Montejano, "A model-based controller for a three-phase four-wire shunt active filter with compensation of the neutral line current," *IEEE Trans. Power Electron.*, vol. 22, pp. 2261–2270, Nov. 2007.

[12] T. Tanaka, E. Hiraki, K. Ueda, K. Sato, and S. Fukuma, "A novel detection method of active and reactive currents in single-phase circuits using the correlation and cross-correlation coefficients and its applications," *IEEE Trans. Power Delivery*, vol. 22, pp. 2450–2456, Oct. 2007.

[13] L. A. Moran, L. Fernandez, J. W. Dixon, and R. Wallace, "A simple and low-cost control strategy for active power filters connected in cascade," *IEEE Trans. Ind. Elect.*, vol. 44, pp. 621–629, Oct. 1997.

[14] L. Asiminoae, C. Lascu, F. Blaabjerg, and I. Boldea, "Performance improvement of shunt active power filter with dual parallel topology," *IEEE Trans. Power Electron.*, vol. 22, pp. 247–259, Jan. 2007.

[15] V. Blasko, "Analysis of a hybrid PWM based on modified space-vector and triangle-comparison methods," *IEEE Trans. Ind. Appl.*, vol. 33, pp. 756–764, May/June 1996.

[16] C. B. Jacobina, A. M. N. Lima, E. R. C. da Silva, R. N. C. Alves, and P. F. Seixas, "Digital scalar pulse-width modulation: a simple approach to introduce non-sinusoidal modulating waveforms," *IEEE Trans. Power Electron.*, vol. 16, pp. 351–359, May 2001.

[17] M. C. Cavalcanti, E. R. C. da Silva, D. Boroyevich, W. Dong, and C. B. Jacobina, "A feasible loss model for igbt in soft-switching inverters," in *Proc. IEEE PESC*, pp. 1845–1850, 2003.

Design and Implementation of an Improved Controller for Parallel-Connected 400 Hz Frequency Converters

B. Tamyurek

Department of Electrical Engineering
Eskisehir Osmangazi University
Eskisehir, 26480 Turkey
btamyurek@ogu.edu.tr

E. Birdane, A. Ceyhan

Kaynak Electronic Machine Industry and Trade Co. Ltd.
Eskisehir, Turkey
{birdane, aceyhan}@kemsan.com.tr

Abstract—This paper presents the design and implementation of parallel connected 400 Hz frequency converters that supply power to military equipment. Achieving tight steady-state and transient voltage regulations and high efficiency in inverters with high fundamental frequency presents great challenge both on the control system design and the power stage design. This study develops and implements a controller with paralleling functions that achieves the stringent power quality requirements. It also includes developing the circuit models of the controller and the converter stage, and finding the optimum design based on the simulation results. Then, experimental work is performed on the converters built at the rated power to validate the design. Finally, the test results demonstrate that the developed control system and the converter design achieve the desired specifications demanded by the military applications.

I. INTRODUCTION

Some equipment used in aircrafts and naval ships need frequency converters to interface to the ground electric power system which is either fed from 50/60 Hz utility mains or from a variable frequency diesel generator. These converters are installed to provide an output voltage of 115 V or 440 V at a constant frequency of 400 Hz. The type of power defined in the interface standard for ship board systems is classified into three types, based on the frequency of the output voltage: type I is 60 Hz power and type II and III are 400 Hz power [1]. According to the standard, the difference between type II and type III power is that type III has more stringent steady-state and transient voltage regulation requirements than the other. Achieving tight

regulations and high efficiency in inverters with high fundamental frequency presents great challenge both on the control system design and the power stage design of the converter. However, the objective of this study is to realize such a system and produce a product that supplies high quality AC power to the military loads. Moreover, direct paralleling of converters is desired in most applications. So, this study also includes developing and implementing an effective and reliable paralleling strategy.

The single line diagram of a typical parallel connected type III frequency converter system, which is also the system to be implemented in this work, is shown in Fig. 1. The major difficulty in the design of existing 400 Hz frequency converters is to achieve a low total harmonic distortion (THD) while maintaining a tightly regulated sinusoidal output voltage [3]. Also, a good waveform quality with perfect synchronism is important if direct paralleling of converters in voltage mode is desired.

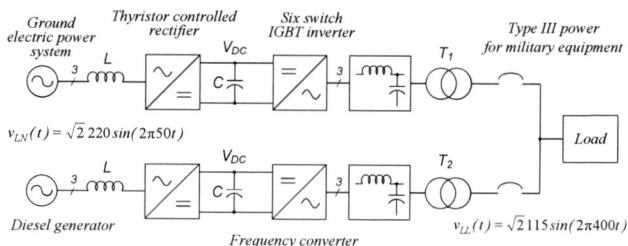

Fig. 1. The single line diagram of a ground electric power interface system including two parallel connected 400 Hz frequency converters.

978-1-4244-4782-4/10 $26.00 © 2010 IEEE

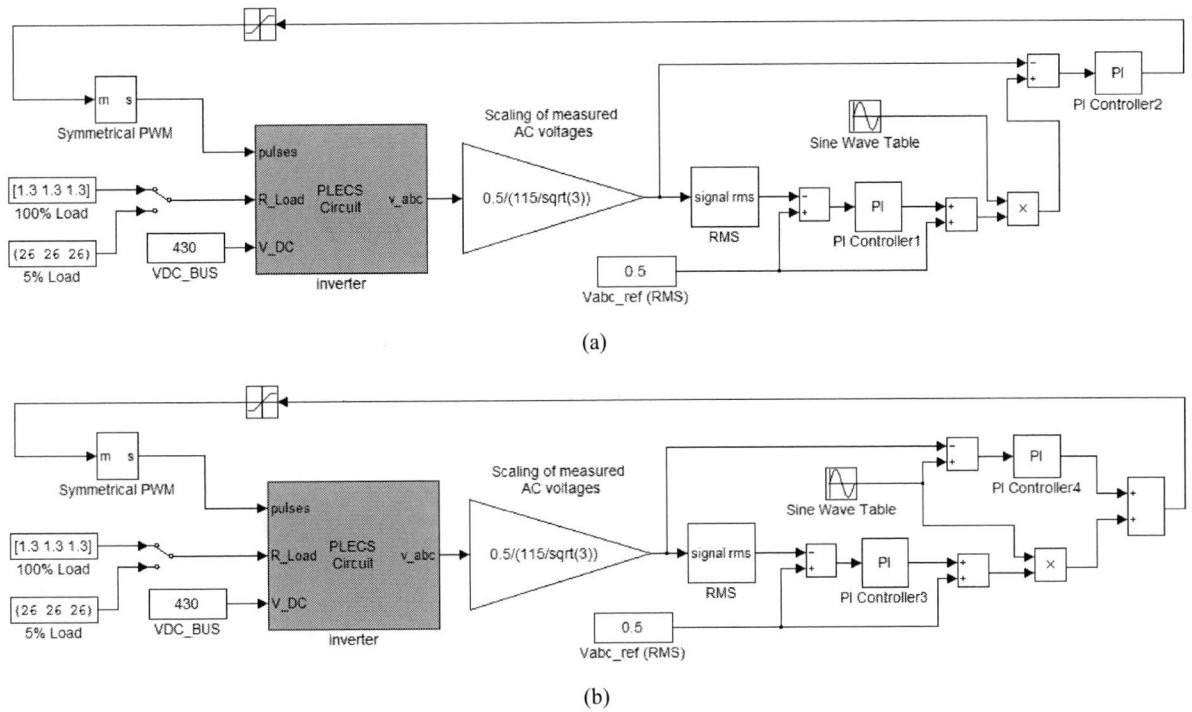

(a)

(b)

Fig. 2. The Simulink model of the control system: (a) the classical SPWM method and (b) the improved SPWM controller.

Research so far has shown that some improvements must be made on the classical control techniques, such as the sinusoidal PWM (SPWM) and the space vector PWM (SVPWM) to meet the regulation requirements of the mid frequency converters [3]-[11]. The frequency modulation ratio m_f is typically less than 25 for frequency converters. The excess switching losses in IGBTs put an upper limit to the maximum switching frequency [2]. So, the design of a controller with low m_f results in a lower controller bandwidth which causes steady state error in the output voltage and a slower dynamic response. Some research has effectively increased the switching frequency by using full-bridge topology for each phase leg [3]-[4]. Unipolar voltage PWM switching of the bridges doubles the switching frequency in the inverter legs. The drawback of this technique is the usage of more semiconductor devices [5]-[6]. We propose an improved controller, which is shown in Fig. 2, to increase the bandwidth without increasing the switching frequency and changing the topology. Additionally, the proposed controller strategy is more convenient for digital implementation made with fixed point DSPs than the classical method. The next section discusses the control system and its advantages.

II. CONTROL SYSTEM

Specifications require that the converter should be operable both as a three-phase and also as a single-phase power source. For this reason the sinusoidal PWM method is used for the control system, since it allows independent control of each phase. Fig. 2 shows the Simulink models of the classical SPWM and the proposed improved SPWM methods. The problem with the low bandwidth at 400 Hz designs can be improved by modifying the inner AC voltage loop of the classical system. The following sections will describe each method.

A. Description of Classical SPWM Control System

As shown in Fig. 2(a), in the classical two loop control method, the output of the outer voltage loop, which are the compensated RMS voltage errors for each phase, is multiplied by the reference sine wave tables, and reference AC signals are generated. Then, these signals are compared to the measured AC voltages and errors are detected. Finally, the detected errors are fed into PI regulators which produce the modulating control signals for each phase. The modulating signals are the inputs to the symmetrical PWM block as shown in Fig. 2(a), where they are compared to the triangular waves for PWM generation.

B. Description of Improved SPWM Control System

In the proposed improved control system, as shown in Fig. 2(b), the measured AC voltages are directly compared to the reference sine waves and the detected errors are fed into PI regulators, which generate the control signals that are responsible for the waveform correction of the output voltages. In other words, these signals independently affect and control the THD of the output voltages and correct the phase difference in the waveform. The RMS magnitude correction comes from the outer voltage loop. Finally, the two control signals are added together to generate the main modulating control signals for the PWM generation.

The bandwidth improvement in this control system is obtained by separately controlling the RMS magnitude of the output voltages using the outer DC feedback loop and separately controlling the THD using the inner AC feedback loop with the help of the reference sine wave tables. The only criterion is that under the nominal steady-state operating conditions, the magnitude of the sine waves in the table must be equal to the scaled magnitude of the desired output voltages. For example, the magnitude of the measured AC voltages in the simulation as shown in the Simulink model of Fig. 2 is 0.7071 after the scaling block when the actual voltage is $115\,V_{LL}$ (RMS), and so the magnitudes of the sine waves in the table must be equal to this value.

Separating the control loops also provides a great deal flexibility and simplification during the digital implementation of the control system with fixed point DSPs. When the waveforms in Figs. 3 and 4 are compared, it is explained how the digital implementation benefits from the improved control structure. Fig. 3 shows the simulated waveforms of the classical control, where (a) are the signals coming into the PI controller2 and (b) are the signals generated. The signals in (b), which indicate a voltage modulation ratio (m_a) of 0.8, are compared to triangular waves to generate PWM pulses. The magnitude of the ripple in (a) is multiplied by the gain of the PI controller, and any attempt to modify the ripple affects the steady-state behavior and the dynamic response of the system. An attempt to reduce the ripple at the output voltage results in slower dynamic response and poor voltage transient regulation. However, we are able to eliminate this problem using the proposed control system.

Fig. 4 shows the simulated waveforms of the proposed controller, where (a) shows the signals coming into the PI controller4 (see Fig. 2(b)), which are mainly the ripple part of the measured AC voltages, but also

include information about the phase difference, (b) shows the signals generated by the PI controller4, (c) is the output of the outer voltage loop indicating an m_a of 0.8, and finally (d) shows the modulating control signals. Since the modulating control signal is the addition of compensated AC error (ripple) and the compensated RMS voltage error in the improved method, the independent control of ripple without sacrificing the system dynamic response is possible. First, the steady-state error and the transient voltage regulation are optimized by adjusting the parameters of the PI controller3, which compensates the RMS voltage error. Then, parameters of PI controller4 are adjusted to obtain the best ripple and waveform quality without affecting the dynamic response significantly.

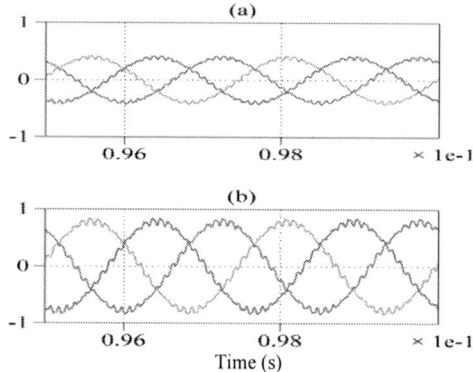

Fig. 3. The simulated waveforms of classical control system: (a) the signals coming into and (b) the signals generated the PI controller2.

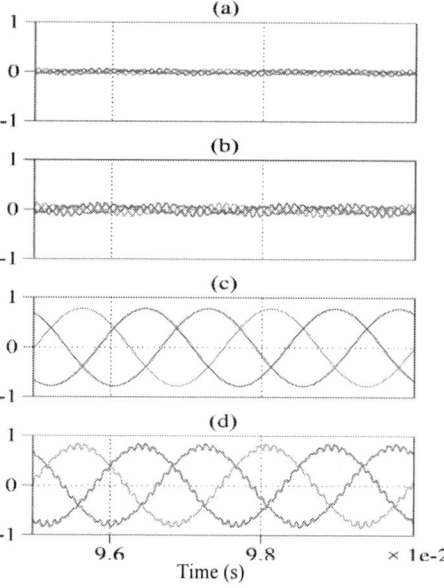

Fig. 4. The simulated waveforms of proposed control system: (a) the signals coming into and (b) the signals generated by the PI controller4; (c) output of the outer voltage loop and (d) the modulating control signals for each phase.

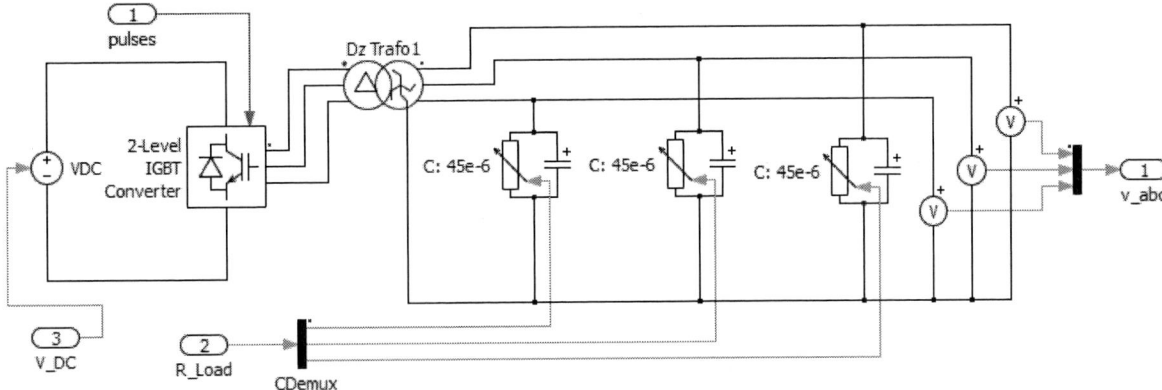

Fig. 5. The PLECS circuit model of the inverter, the transformer, and the output LC filter used in the simulations.

C. Paralleling Strategy

There are various methods available to use for direct paralleling of converters. However, we prefer the master-slave approach because it is easier to implement and provides very good load sharing. In this approach, both units behave like two voltage sources. The drawback of this approach is that there has to be a communication link between the paralleled units. One more problem, as also reported in the literature, is the requirement for the continuous supply of load in case of an error, such as failing of the master unit and/or broken communication link. Nevertheless, the needed precautions are implemented within our proposed control system. When the control system detects a fault in the master unit, the slave unit quickly becomes master and keeps supplying power to the load. Also the main controller always checks for the quality of the data transfer across the communication link. When a broken communication link is detected, the main controller maintains the system to operate independently as unparalleled units. In other words, the converters are not shut down, but they work as two master units. In this case, appropriate switching elements installed externally will physically isolate the output power of the units from each other. The following paragraphs describe the communication and paralleling strategy in detail.

The cable placed on the top of the converters as shown in Fig. 6 is a RS485 interface. The length of this cable is 50 meters. On this interface, we transmit the following signals:

The modulating control signals for each phase, which are the signals compared to the triangular waves for PWM generation. The data are transferred at every 100 μs and with a 1.5Mbps transfer rate. In this paralleling strategy, both units work as voltage sources

and the outputs of the converters are directly connected. For this reason generating sinusoidal voltages with perfect synchronism is important to achieve good power and current sharing. This is only obtained when the triangular waveforms on both units are synchronized. At the end of the every data transfer, when the last data bit has arrived, the controller of the slave unit synchronizes its triangular waveforms with the master.

Besides synchronization of triangular waveforms, the magnitudes of the output voltages must be the same to provide equal power sharing. The output voltages will not be equal when the DC bus voltages are not the same. Therefore, the next data that must be transmitted is the DC bus voltage information. The correct voltage modulation ratio for the slave unit is calculated using the DC bus voltage of the master unit. The modulation ratio of the slave unit is divided by the ratio of the DC bus voltages for the equalization of output voltages.

Finally, the last data is the unit's fault information. When any one of the units receives such information, that unit automatically becomes a master and continues to operate as long as the supplied load is within the rated limits of the converter. Moreover, when the faulted unit is reenergized or the error is cleared, the unit first monitors the communication port to see the existence of any master unit by checking the data on the bus. If the presence of a master unit has been detected through the received data, the unit will start as slave; otherwise it will start as master.

III. DESIGN OF THE CONVERTER

A. Converter Design Specifications

This project requires two 10 kVA converters operating in parallel mode. The converters will be installed on a warship, and the distance between the units will be 50 m. Each unit requires an isolation transformer at the output. The voltage at the load side is 115 V line-to-line. The type III power specifies the worst case voltage tolerance as ±5% around the nominal voltage if the event duration is less than 250 ms and specifies the user voltage tolerance as ±1% with no limitation on time. Moreover, the frequency tolerance of the output voltages should be ±0.2% and the THD should be less than 3% [1].

B. Description of Converter Power Stage

As shown in Fig. 1, the front end of the converter is a three-phase thyristor controlled rectifier which regulates a constant voltage at 430 V. This voltage yields an m_f around 0.756 to regulate 115 V_{LL} at the output and it is the optimum to get a good transient response. The converter uses a 2-level three-phase IGBT inverter for frequency conversion. The modulated DC bus voltages are applied to the primary side of the isolation transformer as shown in Fig. 5. The transformer is a delta/zig-zag connected three-phase custom wound transformer. The transformers are wound in such a way that the internal leakage inductance becomes equal to the desired output filter inductance of 150 μH. The filter capacitance value is 45 μF per phase. These values provide a corner frequency of 1937 kHz for the output LC filter. Finally, the switching frequency of the PWM operation is selected as 10 kHz to meet the efficiency expectations. The control system was implemented with a Texas Instruments fixed point DSP, which is TMS320F2808.

IV. EXPERIMENTAL RESULTS

The two 10 kVA converters as shown in Fig. 6 have been specifically built and tested to investigate the performance of the proposed control system and the paralleling method. A HIOKI 3196 power quality (PQ) meter has been used to monitor and evaluate the system. Fig. 7 shows the transient response of the converter to various sag and swell events recorded by the PQ meter when a 8 kW resistive load has switched in and out during dynamic response tests. Notice that almost all the events fall within the limits of the type III power. However, there are 2 events violating but very close to the limits. Fig. 8 shows the three-phase voltages and currents, (a) when the load is applied, and (b) when the load is removed.

Fig. 9 shows the profile of the RMS voltage (a), the THD of the voltage (b), and the RMS current (c) as measured by the PQ meter for one phase of the converter

during the dynamic response test. As shown in Fig. 9(a), the RMS voltage stays within the steady-state limits during sudden load changes, which is ±1% of 66.4 V. Similarly, in Fig. 9(b), the THD of the voltage is always below 2.4% when the converter is unloaded, and it is even better when under load. These values are lower than 3% specification.

Fig. 6. The experimental step up for two 10 kVA parallel connected frequency converters.

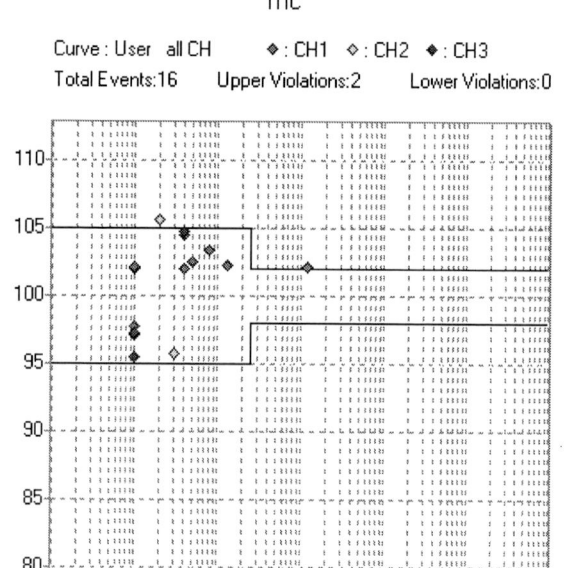

Fig. 7. The records of events occurred during dynamic response tests.

Figs. 10 and 11 show the output voltages and the currents for the same phase of the converters during the parallel operation functionality tests. Fig. 10 shows the waveforms while the master unit supplies the full load power, the slave unit is energized and shares the load current. As shown in Fig. 10, a perfect load sharing is achieved. Fig. 11 shows the waveforms during the fault tests; in (a) a fault occurs in the slave unit and it shuts off, and in (b) a fault occurs in the master unit and it shuts off. The results confirm that the load is always supplied when one of the units fails or shuts down for any reason. However, there is an unavoidable delay in the control when the master unit is shutting off as shown in Fig. 11(b). Nevertheless, this delay only lasts for one cycle, which is 2.5 *ms*. Note that the small currents drawn by the units after they shut off are due to the filter capacitors at the output.

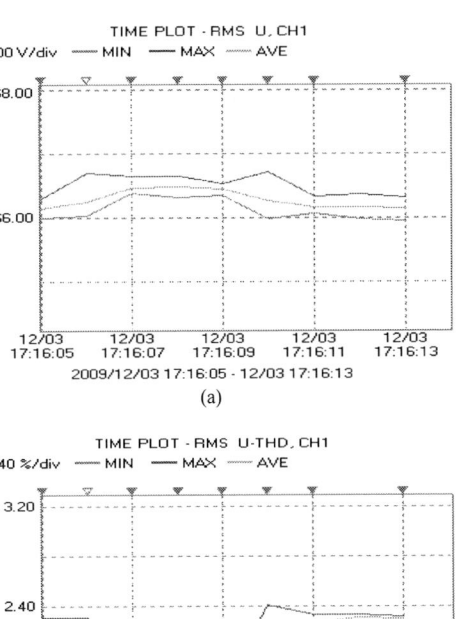

(a)

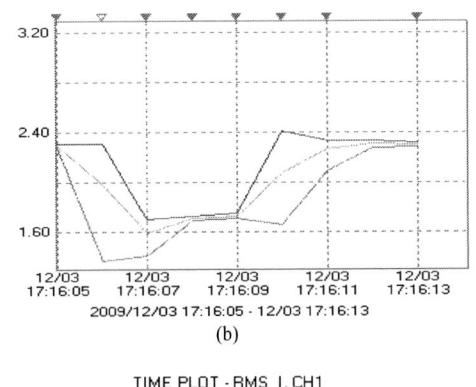

(b)

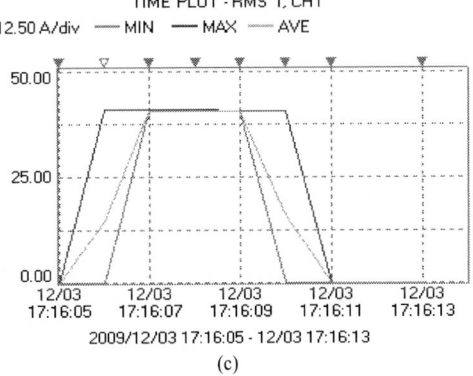

(c)

Fig. 9. The PQ meter recordings during a dynamic response test: (a) the profile of the RMS output voltage for one phase, (b) THD of the voltage for the same phase, and (c) the RMS current.

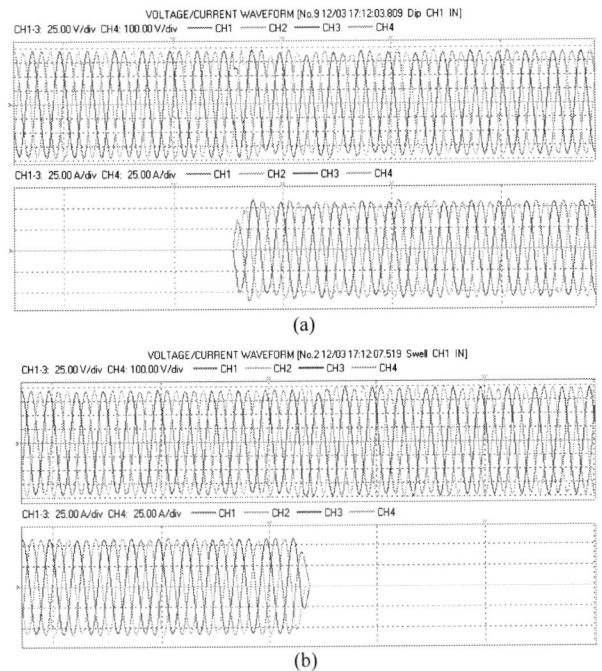

(a)

(b)

Fig. 8. The three-phase voltages and currents: (a) when the load is applied, (b) when the load is removed.

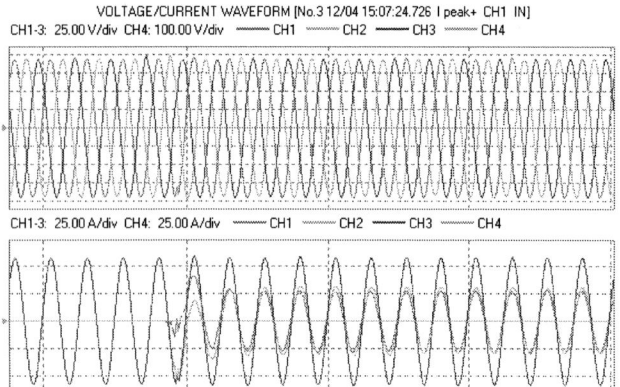

Fig. 10. The voltages (upper trace) and the currents (lower trace) when slave unit is energized while the master is supplying the full load. The CH1 is the slave current, CH2 is the master current, and CH3 is the total load current.

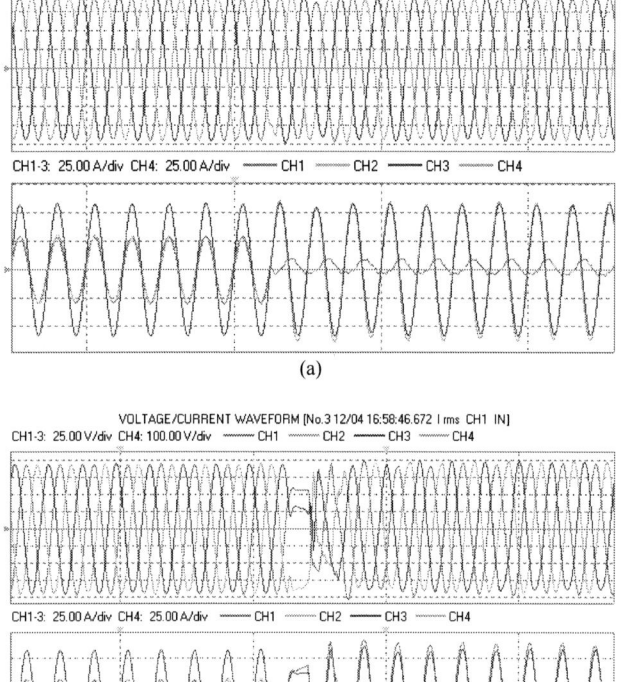

(a)

(b)

Fig. 11. Parallel mode failure tests: (a) The fault in the slave unit and (b) the fault in the master unit.

V. CONCLUSIONS

The objective of this study is to design digitally controlled parallel connected 400 Hz frequency converters for the military applications that need type III

power. The converters provide an interface between the military equipment and either the ground electric power system or a diesel generator. The high fundamental frequency of the output makes the converter and the controller design a challenging job. So, the study includes developing a control system and finding an optimum power circuit design that achieves the stringent requirements of type III power. Controller design is done by MATLAB software, using the developed Simulink model of the control system and the PLECs model of the inverter. Then, a master-slave paralleling strategy is implemented and tested for optimum power sharing and maximum reliability. It is important to operate parallel connected systems in such a way that the power supplied to the load is not interrupted when paralleling functionality has been broken. Finally, the design is verified experimentally on two parallel connected 10 kVA converters. The results demonstrate that proposed control system, the converter design, and the paralleling strategy achieve the desired specifications and that the converters can serve the need for the intended application.

REFERENCES

[1] MIL-STD-1399 (Navy) section 300A, "Interface standard for ship board systems," 13 October 1987.

[2] N. Mohan, T. M. Undeland, W. P. Robbins, *"Power electronics-converters, applications and design,"* Second edition, John Wiley & Sons Inc., 1995.

[3] Liu, Chunxi; Ma, Weiming; Sun, Chi; Hu, Wenhua; "Research on digital control design of high power middle frequency 400 Hz inverter power," *IEEE 6th International Power Electronics and Motion Control Conference*, 2009, pp. 1544 – 1549.

[4] Chunxi Liu, Chi Sun, Wenhua Hu, "Proportional-Resonant Controller of High Power 400 Hz Inverter in Stationary Frame," *International Conference on Electrical Machines and Systems*, 17-20 Oct. 2008, pp. 1772-1777.

[5] U. B. Jensen, F. Blaabjerg, and K. Pedersen, "A new control method for 400-Hz ground power units for airplanes," *IEEE Trans. Ind. Appl.*, vol.36, no. 1, pp. 180–187, Jan./Feb. 2000.

[6] U. Borup, F. Blaabjerg, and P. N. Enjeti, "Sharing of nonlinear load in parallel-connected three-phase converters," *IEEE Trans. Ind. Applications.*, vol. 37, pp. 1817–1823, Nov./Dec. 2001.

[7] G.L. Basile, S. Buso, S. Fasolo, P. Tenti, P. Tomasin, "A 400 Hz, 100 kVA, Digitally Controlled UPS for Airport Installations," *IEEE Industry Applications Conference*, vol. 4, Oct 2000, Rome, Italy, pp. 2261-2268.

[8] M. H. Taha, "Power electronics for aircraft application," *IEE Colloquium on Power Electronics for Demanding Applications* (Ref. No. 1999/059), pp. 7/1-7/4, London, UK, 1999.

[9] Seok-Eon Joung, Dong-Seok Hyun , "A Low Cost and High Reliability Control Scheme In Parallel Operation for 400 Hz Power Supply," *7th Internatonal Conference on Power Electronics*, pp. 941-944, 2007.

[10] M. Macellari, U. Grasselli, L. Schirone, "An On-board Inverter Controlled by the Waveform Switching Technique," *IEEE Aerospace Conference*, 2006.

[11] L. Mihalache, "DSP control of 400 Hz inverters for aircraft applications," *37th Industry Applications Conference*, vol.3, 2002, pp. 1564- 1571.

Study on the Impact of the Complex Impedance on the Droop Control Method for the Parallel Inverters

Wei Yao,   Mingzhi Gao,   Zheng Ren,   Min Chen,   Zhaoming Qian

Power Electronics Institute, College of Electrical Engineering

Zhejiang University, Hang Zhou 310027 China

Abstract- **In this paper, a new droop controller for parallel connected UPS inverters is presented. The proposed controller is based on the droop method in order to avoid the using of communication signals between the modules. The paper studies the control of the active and reactive power flow through the analysis of the output impedance of inverters and its impact on power sharing. As a consequence, we propose a novel control scheme which is based on virtual complex impedances. As opposed to the conventional methods, the controller proposed can achieve accurate power sharing, good dynamic performance and be more adaptive to the different line impedance situations. Design method and experimental results are presented in order to show the validity of the proposed control technique.**

I. INTRODUCTION

Power energy distribution in a system formed by several uninterruptible power supplies (UPS) operating in parallel can be performed today with different methods, either with or without communication between the power elements. Several control strategies have been adopted to achieve this goal in parallel UPS inverter systems [1]-[5]. Over these, the droop control method is very interesting because it is only based on UPS local information, does not rely on intercommunication signals between the inverters, and constitutes a truly distributed and redundant system.

The transient and steady state behavior of droop method is highly dependent on the system impedances: the inverter output impedance and the line impedance of the wires used to do UPS connection [6], [7]. When using the conventional droop method, the output impedance and line impedance are considered to be mainly inductive, that is often justified by the large inductor value or by the long distances between the units. However in many practical applications, complex system impedance can not be avoided which present non-negligible inductive and resistive impedance [8], [9]. In these cases of complex impedance, the performance of the conventional droop method is poor due to coupled active and reactive power characteristic of system [10], [11]. Moreover, the classical droop methods can only work well when linear loads are shared [12], since the amount of distorted power demanded by the nonlinear loads is not taken into account. Besides, the droop method exhibits slow dynamic response, since it requires low-pass filters with a reduced bandwidth to calculate the value of the active and reactive power [13].

In this paper the characteristics of active and reactive power sharing under purely inductive, purely resistive and complex output impedances are analyzed. It is shown that the conventional droop method can not achieve good power sharing in the case of complex impedance. In order to overcome the mentioned drawbacks and minimize the circulating current under different impedance conditions, a novel droop controller with considering the impact of complex system impedance is proposed in this paper. The controller proposed can simplifier the coupled active and reactive power relationship under complex impedance condition, and achieve good power sharing with unbalance line impedance. The principles of design scheme are presented in this paper and experimental evaluations with two single-phase UPS inverters are also reported in this paper.

II. THEORETICAL ANALYSIS OF POWER SHARING

For the analysis of power sharing behaviors between parallel inverters, Fig.1 depicts the power supply system with several parallel connected single-phase inverter modules and loads connected to the ac common bus through wires. The equivalent circuit of an inverter can be derived from Thevenin's Theory as a voltage source with output impedance in series, as shown in Fig. 2.

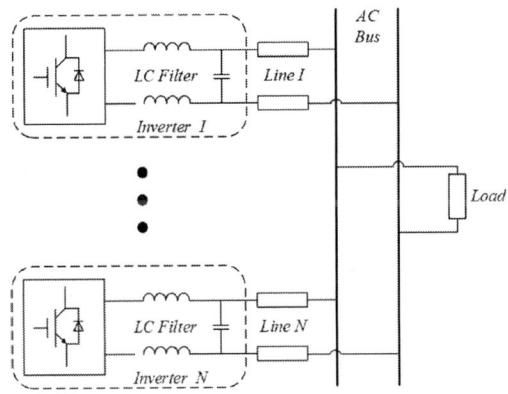

Fig. 1. Power supply system with the parallel inverters

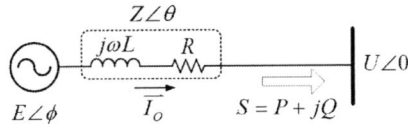

Fig. 2. Equivalent circuit of the parallel inverter

Where $E\angle\phi$ is the inverter's open circuit voltage, $\overline{i_o}$ is the inverter's output current, $U\angle 0$ is the load voltage, $R+j\omega L$ is the output impedance of inverter, Z and θ are the magnitude and phase of output impedance. By using

978-1-4244-4782-4/10 $26.00 © 2010 IEEE

the output current and the load voltage, the apparent powers injected to the ac common bus by every inverter is $\overline{S} = \overline{U} \cdot \overline{I}_O^* = P + jQ$, the active and reactive powers can be expressed as

$$P = \frac{U}{Z}\left[(E\cos\phi - U)\cos\theta + E\sin\theta\sin\phi\right] \quad (1)$$

$$Q = \frac{U}{Z}\left[(E\cos\phi - U)\sin\theta - E\cos\theta\sin\phi\right] \quad (2)$$

Assuming that the phase differences ϕ between the output voltage and the load voltage are always very small because of the synchronization achieved by the phase lock loop (PLL) techniques ($\sin\phi \approx \phi$, $\cos\phi \approx 1$), (1) and (2) can be simplified to (3), and (4).

$$P = \frac{U}{Z}\left[(E - U)\cos\theta + E\phi\sin\theta\right] \quad (3)$$

$$Q = \frac{U}{Z}\left[(E - U)\sin\theta - E\phi\cos\theta\right] \quad (4)$$

In order to facilitate the derivation of the relationship between the output voltage and the power sharing, the output powers of inverter with different impedances are illustrated using polar coordinates in Fig.3. Note that only the shaded parts are valid in a real system because the power angle ϕ is always small. All the parameters are chosen to better illustrate the relative variation of P and Q.

When considering that the output impedance is purely inductive, from Fig.3, if the amplitude of parallel inverter E increases, the active and reactive powers of inverters are both increased. The difference is that the active power also increases together with the power angle ϕ as solid line in Fig. 3. However the polar radius of reactive power circle is invariable when power angle ϕ changes as dashed line shown in Fig. 3. On the contrary, if the unified impedance is resistive the variations of active power and reactive power are exchanged as shown in Fig. 4.

Usually, the analyses presented in literatures are based on the assumption of inductive impedance condition. However this is not always true since the output impedance of the inverters is depended on the control strategy and parameters of LC filters. Mostly at the concerned frequency it is composed by both resistance and inductance. Fig. 5 illustrates the output power of parallel inverters particularly with complex output impedance. When increasing power angle ϕ from $0°$ to $360°$, the active power increases together with ϕ in clockwise direction. Otherwise, the reactive power increases when ϕ is from $-180°$ to $180°$ in clockwise direction. And when the complex impedance angle θ increase which can be seen as more inductive, the polar radius of reactive power curve diminishes as shown in Fig.5 dashed line. Based on the analysis above, it can be seen that the output power of parallel inverters with complex impedance is completely different from the other two conventional situations. The droop control methods, which are based on the assumption of inductive or resistive impedance, can not satisfy well the power sharing under

complex impedance situation. And this issue has not been included in the conventional techniques properly.

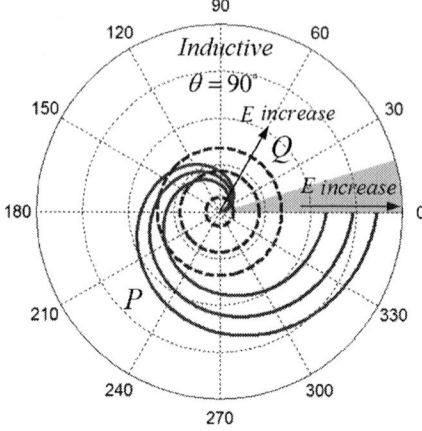

Fig. 3. Polar plot depicting the P/Q behaviors of parallel inverter with pure inductive impedance

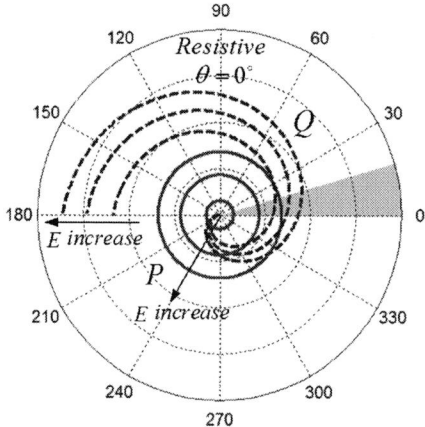

Fig.4. Polar plot depicting the P/Q behaviors of parallel inverter with pure resistive impedance

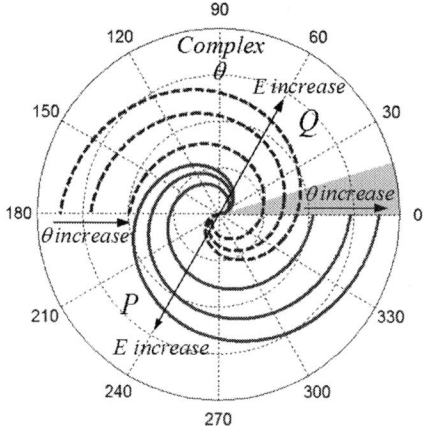

Fig. 5. Polar plot depicting the P/Q behaviors of parallel inverter with complex impedance

III. DESIGN AND IMPLEMENTATION OF THE PROPOSED DROOP CONTROLLER

In this paper we redesign the output impedance of the inverter by adding virtual complex impedance as shown in equation (5). The parameters according to real system are also listed in Table I, and Fig 6 depicts the frequency behaviors of native output impedance of inverter used in this paper and redesigned one by adding virtual complex impedance part. As a result, the redesigned output impedance has the following advantages: (1) The proposed complex impedance approach for circulating current is effective for both linear load and nonlinear load, which is even less sensitive to the complex line impedance unbalance or system parameters mismatching; (2) To avoid the noise from the derivative of virtual inductor part, a low pass filter is added, in which the cut-off frequency is defined as ω_v. Consequently the inductive only work near the fundamental circulating current, and it will not increase with the harmonic frequency to degrade the output voltage; 3) The solid line shown in Fig. 6 shows that the redesigned output impedance is more resistive at harmonic frequency, and the load harmonic current can be shared equally and automatically by the inverters without any extra control. Moreover, the implementation of complex impedance can alleviate the effects of voltage drop which is produced by only a virtual resistor; 4) Based on the precise virtual complex impedance, it can be adopted to design the impedance angle θ described in section II, and decouple the active and reactive relationship more easily.

Table I
Parameters of the controller for single inverter

Items	Value
Output LC filter	L_f=1.36mH, C_f=11μF,
Voltage PI controller	K_P=0.25, K_I=350
Parasitic resistor of L	r_f=0.3Ω
Current loop gain	G_C=4.5
Fundamental frequency	f_O=50Hz
Virtual complex	R_V=0.19Ω, L_V=535μH
Impedance parameters	ω_V= 350×2π rad/s

As we stated previously, the ouput impedances will greatly affect the power sharing behaviors among parallel inverters. And it is observed that the relationship between P/Q and E/ϕ completely changed with complex impedance comparing to the traditional situations. Moreover, the complex impedance conditions almost can not be avoided in the real parallel inverters system. In order to achieve P/Q droop control effectively and minimize the circulating current, as shown in Fig. 6 solid line, notice that the angle of new output impedance has been redesigned to $\pi/4$. As a result, the

new droop control scheme according on the power sharing relationship (as shown in equation 3 and 4) can be obtained as follows:

$$\omega = \omega_0 - m(P - Q) \tag{5}$$

$$E = E_0 - n(P + Q) \tag{6}$$

Where m and n are the droop coefficients of frequency and amplitude. Increasing droop coefficient will result in worse deviation but better sharing performance. So they should be choosed with compromise between sharing precision and the allowed deviation.

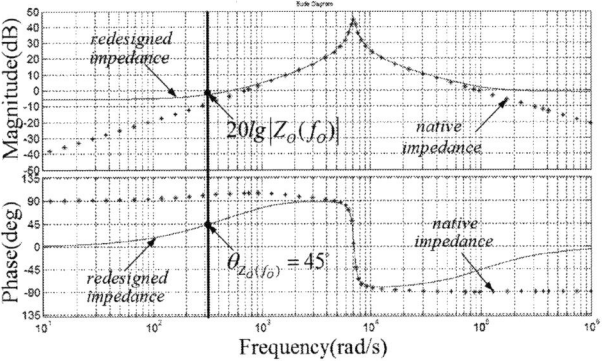

Fig. 6. Bode diagram of output impedance (dots line: native output impedance; solid line: redesigned output impedance with virtual complex impedance)

Fig. 7 depicts the realization of the proposed parallel operation controller, the droop controller are based on multiple-loop feedback scheme, which can ensure the excellent steady state and dynamic performance. Single-phase DQ transformantion are used to increase the bandwidth of power calculation in order to improve the dynamic performance. Crossed active and reactive power sharing controller are proposed based on the virtual complex impedance, which can effectively reduce the circulating current, regardless of unblanced line impedance.

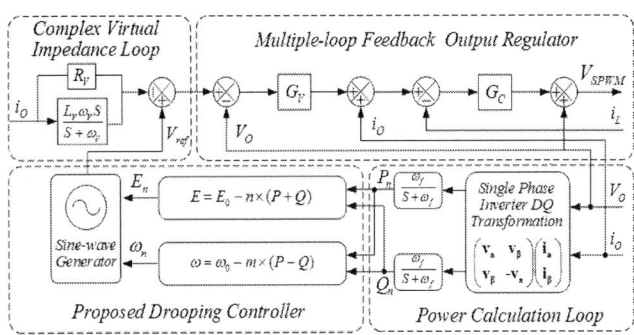

Fig. 7. Realization of the proposed power sharing control scheme

$$Z_{redesigned}(s) = Z_O(s) + Z_{Virtual}(s) = \frac{L_f s^2 + r_L s}{L_f C_f s^3 + (G_C C_f + r_L C_f)s^2 + K_P G_C s + K_I G_C} + R_v + \frac{\omega_v}{s + \omega_v}sL_v \tag{5}$$

IV. EXPERIMENTAL RESULTS

Two 2kVA UPS inverters were built and tested. Each inverter consisted of a single-phase IGBT full-bridge and a LC output filter, using the parameters listed in Table II. The line impedances in practical prototype system were intentional unbalanced which are $Z_{L1}=0.08+j0.05\Omega$, $Z_{L2}=0.01+j0.01\Omega$. The controllers of these inverters were implemented on a TMS320F2811 DSP, 32-bit fixed-point 150 MHz digital signal processor (DSP) from Texas Instruments. The controller also includes the phase lock loop (PLL) which is constituted by a second order generalized integrators (SOGI) and the frequency lock loop (FLL), in order to synchronize the inverter with the common AC bus at the beginning and can work for hot-swap.

TABLE II
Parameters of implemented system

Items	Symbol	Value
Initial frequency	ω_O	$2\pi\times50$ rad/s
Initial frequency	V_O	311 V
Frequency droop coefficient	m	0.0005
Amplitude droop coefficient	n	0.001
Bandwidth of power calculation	ω_f	$2\pi\times10$ rad/s
Line impedance I	n/a	$0.08+j0.02$
Line impedance II	n/a	$0.15+j0.11$

Fig 8(a) depict the inverter's output current showing the time response to a connection and disconnection of a linear load. Fig 8(b) depict the inverter's output currents time response for the case in which one of the inverters is connected and disconnected from the parallel system. Fig 8(c) shows a detail of the time response of the inverter's output current when one inverter is feeding a nonlinear load and a second converter is connected to the system to share this load. Note that the nonlinear load is precisely shared between the two inverters. Fig 8(d) shows the output voltage and output current waveforms in detail when the inverters are sharing a nonlinear load. It can be seen that the inverter synchronizes smoothly, while the currents stabilize after a transient time to almost perfect current sharing. These experimental results demonstrate that the proposed controller achieves good sharing and dynamic behaviors under both linear and nonlinear loads.

V. CONCLUSION

This paper presented a novel control scheme for the parallel inverters with no intercommunication signals, considering the impact of the complex impedance on the power sharing. Applying the proposed controller, the parallel UPS inverters are capable to deliver high quality power even in presence of the nonlinear load with unbalanced line impedances. The analyses clarify the relationship between the impact of the impedance and the power sharing in a parallel system. The virtual complex impedance module is also proposed and well researched in this paper. Experimental results confirmed the validity of the proposed controller.

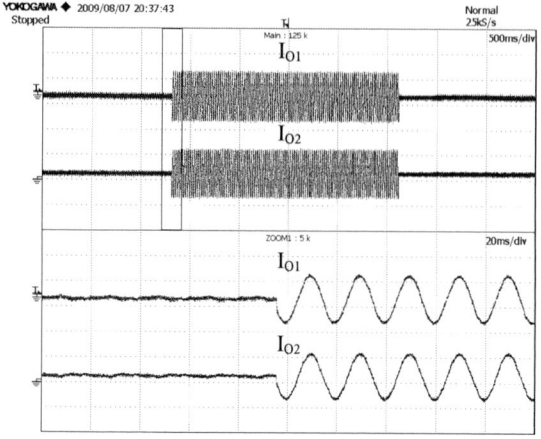

(a) The linear load connected and disconnected (X-axis: top 500 ms/div, bottom: 20 ms/div; Y-axis: 10A/div).

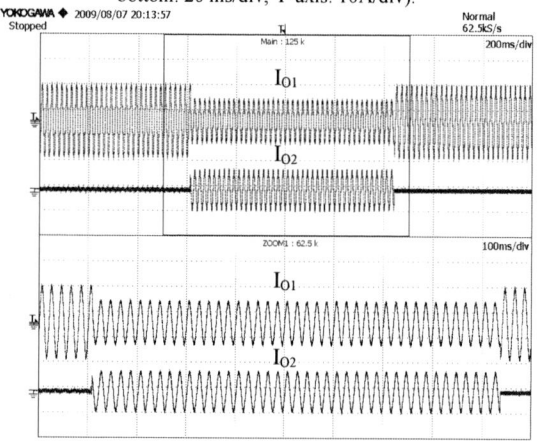

(b) One inverter added to and then removed from the parallel system (X-axis: top 100 ms/div, bottom: 100 ms/div; Y-axis: 10A/div)

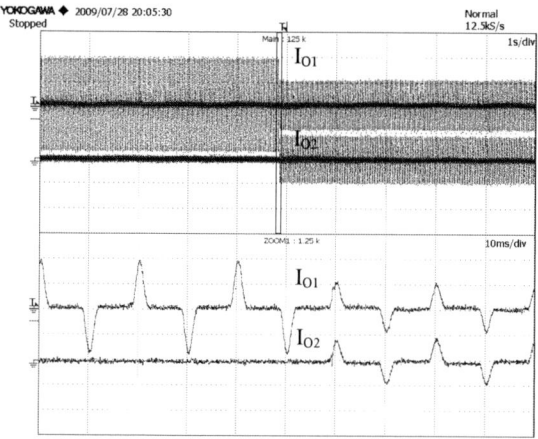

(c) One inverter add to the parallel system when supplying a nonlinear load (X-axis: top 1s/div, bottom: 10 ms/div; Y-axis: 10A/div)

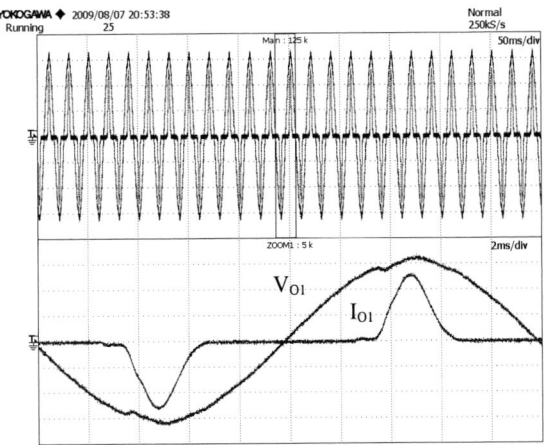

(d) Output voltage and output current when sharing a nonlinear load (X-axis: top 50ms/div, bottom: 2 ms/div; Y-axis: 10A/div, 100V/div)

Fig. 8 Experimental results of the parallel inverters using the proposed controller.

VI. ACKNOWLEDGEMENT

The author would like to thank Prof. Jose Matas and Prof. Josep M. Guerrero from Technical University of Catalonia (Spain) for the support during this project.

REFERENCE

[1]. Chandorkar, M. C.; Divan, D. M.; Adapa, R., "Control of parallel connected inverters in standalone ac supply systems", IEEE Trans. Ind. Applicat., vol. 29, pp. 136-143, Jan./Feb. 1993.

[2]. Holtz, J., Werner, K.-H., "Multi-inverter UPS system with redundant load sharing control", IEEE Trans. Ind. Electron., vol. 37, pp. 506-513, Dec. 1990.

[3]. Jiann-Fuh Chen, Ching-Lung Chu, "Combination voltage controlled and current-controlled PWM inverters for ups parallel operation", IEEE Trans. Power Electron., vol. 10, pp.547-558, Sep. 1995.

[4]. Van Der Broeck H., Boeke, U., "A simple method for parallel operation of inverters", in Proc. IEEE INTELEC'98, 1998, pp. 143-150.

[5]. Tajao Kawabata, Shigenori Higashino, "Parallel operation of voltage source inverter", IEEE Trans. Ind. Applicat., vol. 24, 1988.

[6]. Tuladhar, H. Jin, T. Unger, and K. Mauch, "Control of parallel inverters in distributed ac power systems with consideration of line impedance," IEEE Trans. Ind. Applicat., vol.36, pp.131-138, Jan./Feb. 2000.

[7]. Coelho, E.A.A., Cortizo, P.C., Garcia, P.F.D., "Small-signal stability for parallel-connected inverters in stand-alone AC supply systems", IEEE Trans. Ind. Applicat., vol 38, pp. 533-542, March./April. 2002.

[8]. Corradini L., Mattavelli P., Corradin M., Polo F., "Analysis of Parallel Operation of Uninterruptible Power Supplies Loaded through Long Wiring Cables", in Proc. IEEE APEC'09, 2009, pp.1276-1282.

[9]. S. J. Chiang, C. Y. Yen, K. T. Chang, "A Multimodule Parallelable Series-Connected PWM Voltage Regulator", IEEE Trans. Ind. Electron., vol 48, pp.506-516, June 2005.

[10]. De Brabandere, K., Bolsens, B., Van den Keybus, J., Woyte, A., Driesen, J., Belmans, R.A, "Voltage and Frequency Droop Control Method for Parallel Inverters", IEEE Trans. Power Electron., vol 22, pp. 1107-1115, July 2007.

[11]. Meng Xie, Yaohua Li, Kun Cai, Ping Wang, Xiaosong Sheng, "A novel controller for parallel operation of inverters based on decomposing of output current", in Proc. IEEE IAS'05, 2005, pp. 1671-1676.

[12]. L. Chen, X. Xiao, C. Gong, and Y. Yao, "Circulating current's characteristics analysis and the control strategy of parallel system based on double close-loop controlled VSI", in Proc. IEEE PESC'04, 2004, pp. 4791-4797.

[13]. Prodanovic, M., Green, T.C., "High-Quality Power Generation Through Distributed Control of a Power Park Microgrid", IEEE Trans. Ind. Electron. vol 53, pp. 1471-1482, Oct. 2006.

A Three-Phase Adaptive Approach to Extract Harmonic and Reactive Currents

D. Yazdani, *Member, IEEE*, A. Bakhshai, *Senior Member, IEEE*, and P. K. Jain, *Fellow, IEEE*

Abstract— **This paper introduces a new three-phase adaptive notch filtering approach for the extraction of harmonic and reactive current components for use in grid connected converters. The main function of this PLL-less method is to provide synchronized harmonic and reactive current components for control purposes. Moreover, this method is capable of detecting a selective order of harmonics. This feature is useful where elimination of certain harmonics is of concern. The theoretical analysis is presented and the performance of the method is experimentally evaluated. The methodology is applicable for a wide range of equipment like regenerative converters, distributed generation systems, as a basis for detection of the reference signals.**

Index Terms— **Grid-connected converters, power quality, adaptive filters, harmonics and reactive power compensation.**

I. INTRODUCTION

Harmonic distortion and reactive currents injected by power semi-conductor devices into the utility system have become a major concern for power system specialists. Therefore, harmonic and reactive current extraction and decomposition has become a major concern for power system specialists. Advanced signal processing methods in both time- and frequency-domains have been developed to detect and extract the compensating signal [1-18].

Notch filters are also being used to extract the harmonic content of a distorted signal [21]. However, a fast transient response and good harmonic rejection is not attainable using a single notch filter [21].

More advanced control and signal processing techniques including, fuzzy logic control [22], neural network theory [23-24], sliding mode control [25], and adaptive signal processing [26-27], have also been applied. Most of these algorithms are quite accurate and have a much better dynamic response than the FFT, but they require a large amount of calculation and none is reported to demonstrate good performance in frequency-varying environments. In frequency-varying environments, the ideal solution is the use of adaptive approaches [28]. Such adaptive approaches might suffer from low convergence speed and lack of robustness with respect to input frequency variations [26]. The adaptive approach given in [27] overcomes these drawbacks. However, it is only suitable for harmonic detection and is not able to extract the reactive current component. The reader is referred to [3, 4, 11] for a comprehensive review of the existing signal processing techniques for harmonics detection and extraction.

A new three-phase adaptive notch filtering approach for the extraction of harmonic and reactive current components is introduced in this paper. The prominent features of the proposed technique are: i) its simplicity that provides a major advantage for its implementation within embedded controllers, ii) the lack of a need for a synchronizing tool like a PLL, and iii) simultaneous extraction of harmonics and all useful information embedded in a signal such as frequency, amplitude, and phase angle. Moreover, the proposed method can successfully detect and track the variations in the frequency of the signal and extract the time-varying harmonics.

II. HARMONIC/REACTIVE-CURRENT EXTRACTION

For power quality purposes, all grid connected converters require an advanced phase-detecting scheme that might further be employed to detect current harmonics and extract the reactive current components. When voltage is pure sinusoidal, only the fundamental component of the current contributes in power transfer (real power transfer). The load current can be decomposed into the fundamental and harmonic current components, as shown in (1).

$$i(t) = I_1 \sin\phi_{i1} + \sum_{h=2}^{\infty} I_h \sin\phi_{ih} \tag{1}$$

For the given input voltage, the active, i_a, and reactive components, i_r, of the current are given by (refer to Fig. 1):

$$i_a = I_1 \cos(\phi_{i1} - \phi_v)\sin\phi_v$$

$$i_r = I_1 \sin(\phi_{i1} - \phi_v)\cos\phi_v + \sum_{h=2}^{\infty} I_h \sin\phi_{ih} \tag{2}$$

In general, when the input voltage/current are distorted:

$$v(t) = \sum_{h=1}^{\infty} V_h \sin\phi_{vh} \tag{3}$$

978-1-4244-4782-4/10 $26.00 © 2010 IEEE

$$i(t) = \sum_{h=1}^{\infty} I_h \sin \phi_{ih} \tag{4}$$

The active and reactive components are calculated by:

$$i_a = \sum_{h=1}^{\infty} I_h \cos(\phi_{ih} - \phi_{vh}) \sin \phi_{vh} \tag{5}$$

$$i_r = i - i_a = \sum_{h=1}^{\infty} I_h \sin(\phi_{ih} - \phi_{vh}) \cos \phi_{vh}$$

Note that the calculation of the sin and cos functions of the phase angle difference between the voltage and current can be simply performed by:

$$\cos(\phi_i - \phi_v) = \cos(\phi_i)\cos(\phi_v) + \sin(\phi_i)\sin(\phi_v) \tag{6}$$
$$\sin(\phi_i - \phi_v) = \sin(\phi_i)\cos(\phi_v) - \cos(\phi_i)\sin(\phi_v) \tag{7}$$

Note that ϕ_i and ϕ_v are $2\pi f t + \varphi_i$ and $2\pi f t + \varphi_v$.

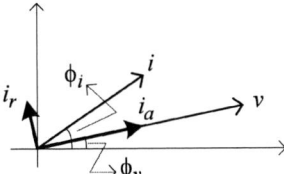

Fig. 1. Phasor diagram.

The objective of the next section is to find an algorithm that receives a three-phase voltage/current and extracts: (i) the amplitude, phase angle and frequency of the fundamental component, (ii) sin and cos functions of phase angles of the voltage/ current which are used to calculate the phase angle difference between the phase voltage and phase current by (6-7), (iii) the harmonic content of the input signal, and (iv) the amplitude, and sin and cos functions of phase angles of the individual harmonics of the voltage/ current.

III. PROPOSED THREE-PHASE ANF-BASED APPROACH

For three-phase applications, three ANFs can be employed to provide the fundamental component, its phase angle and its frequency for each $u_a(t)$, $u_b(t)$ and $u_c(t)$ [35]. Each ANF is a third-order dynamic system, hence the three-phase system in [35] has a dynamic of order nine. Since the three-phase signals have a common frequency, ω_o, there is no need to separately estimate the frequency of each phase independently. In other words, only one fundamental frequency is required. Therefore, a common frequency estimation law based on the output information of all three ANFs is proposed here [36]. Consider three identical ANFs:

$$\ddot{x}_\alpha = -\theta^2 x_\alpha + 2\zeta_\alpha \theta e_\alpha(t), \quad \alpha = a,b,c$$
$$e_\alpha(t) = u_\alpha(t) - \dot{x}_\alpha \tag{8}$$

where θ is an estimate for ω_o. To derive an equation for estimating ω_o, we note that i) ω_o is the common frequency of three-phase signals, therefore, information of all three sub-filters must be incorporated into the update law for frequency estimation, and ii) the regressor signal $x(t)$ and the error signal $e(t)$ incorporate into the θ update law in (9). The term θ is for scaling. Therefore, the update law for frequency estimation is turned to:

$$\dot{\theta} = -\gamma\theta \sum_{\alpha=a,b,c} x_\alpha e_\alpha(t) \tag{9}$$

For a three-phase sinusoidal signal $u(t)$, the dynamical system given by (8) and (9) has a unique periodic orbit located at

$$P(t) = \begin{pmatrix} P_a(t) \\ P_b(t) \\ P_c(t) \\ \bar{\theta} \end{pmatrix} \tag{10}$$

where, $P_\alpha(t)$ is given by,

$$P_\alpha(t) = \begin{pmatrix} x_\alpha \\ \dot{x}_\alpha \end{pmatrix} = \begin{pmatrix} -\dfrac{A_\alpha}{\omega_o}\cos(\omega_o t + \delta_\alpha) \\ A_\alpha \sin(\omega_o t + \delta_\alpha) \end{pmatrix} \tag{11}$$

and $\bar{\theta} = \omega_o$. For the ANF_α in the steady state, the defined outputs $\dot{x}_\alpha$ and θx_α are

$$\dot{x}_\alpha = A_\alpha \sin(\omega_o t + \delta_\alpha)$$
$$-\theta x_\alpha = A_\alpha \cos(\omega_o t + \delta_\alpha) \tag{12}$$

This means that the αth fundamental component of the input signal and its $90°$ phase shift are made available by ANF_α at its outputs, $\dot{x}_\alpha$ and θx_α.

A structural block diagram of the proposed algorithm is shown in Fig. 2. The detailed implementation block diagram can be found in [36]. Note that $u_{e\alpha} (\alpha = a,b,c)$ in Fig. 2 is the estimated fundamental component of the input signal associated to each phase, $\dot{x}_\alpha (\alpha = a,b,c)$.

To extract a selective order of harmonics, the structure of the sub-filter replaced by the multi-block ANF, as shown in Fig. 3. In this configuration the inputs $e_\alpha (\alpha = a,b,c)$ and θ are coming from the frequency estimator and the outputs $u_{e\alpha} (\alpha = a,b,c)$ and $\theta x_\alpha (\alpha = a,b,c)$ are fed back to the frequency estimator.

When the three-phase input signal contains harmonics, the first sub-filter outputs the fundamental component of the input signal and ith sub-filter outputs the ith harmonic components of the input signal. In addition,

$$A_i = \left(i^2 \theta^2 x_i^2 + \dot{x}_i^2 \right)^{1/2}$$

is the amplitude of the ith harmonic component of the input signal. In other words, the algorithm can further be furnished to estimate the amplitudes of the harmonics using arithmetic units that compute the right-hand side of A_i. Note, since $i\theta x_i$ is available, the right-hand side of A_i can simply be calculated by two multiplications, a sum and a square-root computation.

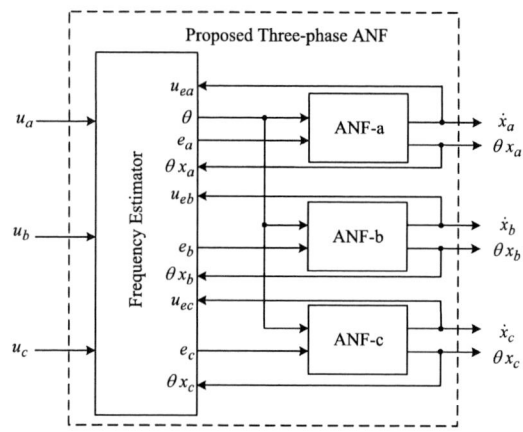

Fig. 2. The proposed three-phase system.

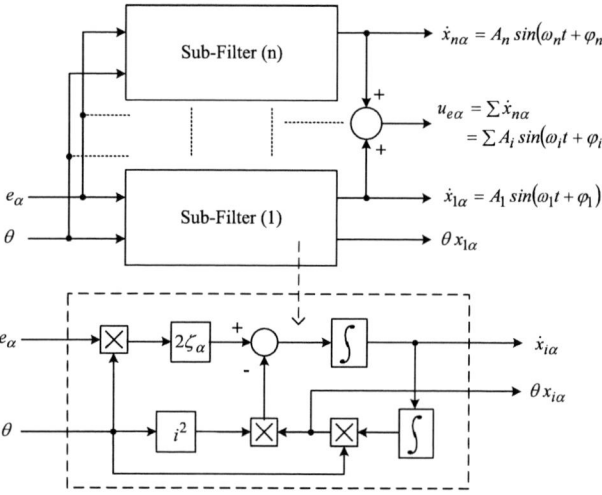

Fig. 3. Modified structure of the αth sub-filter.

This configuration guarantees a fast (one cycle) and precise extraction of the individual harmonics, which can be employed for further harmonic analysis or elimination purposes. The input signal is assumed to be periodic, which is the case in many power electronics grid-connected converter applications. No further information about the input signal is required. This structure outputs useful signal information such as the fundamental component, its 90 degrees phase-shift, its amplitude, its frequency, the sin/cos functions of its phase angle, and its harmonics.

A close observation of the proposed method reveals that the features of the proposed structure when compared to its counterparts are: (i) adaptive structure that can track the signal variations, (ii) simplicity that provides major advantage for its implementation within embedded controllers, ii) the lack of a need for a synchronizing tool like a PLL, iii) simultaneous extraction of harmonics and all useful information embedded in a signal such as frequency, amplitude, and phase angle, and iv) adjustable accuracy and speed of response.

IV. HARMONIC AND REACTIVE CURRENT EXTRACTION

A. Proposed ANF-based Harmonic/Reactive-Current Extraction Technique

In this section, the proposed structure in Fig. 2 is used to develop a simple mechanism for extracting the harmonic and reactive current components of a measured three-phase signal. Fig. 4 shows the single-phase diagram of the three-phase harmonic/reactive-current component extractor. The two identical ANF units are to extract voltage and current information such as their harmonic contents, peak fundamental components, frequencies, and sin and cos functions of phase angles. Since I_1, $\cos\phi_i$, $\sin\phi_i$, $\cos\phi_v$, and $\sin\phi_v$ are all available at the output of the two ANFs, the phase angle difference between the phase voltage and phase current is simply obtained by (6-7) and thus active and reactive current components can be simply obtained based on (2). The approach, although intended to extract harmonic and reactive current components and useful signal information, can output the total harmonic distortion and power factor with more modifications.

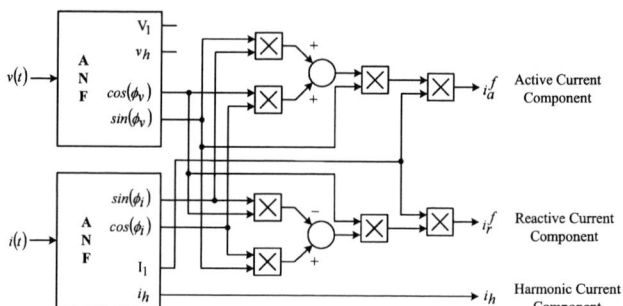

Fig. 4. Single phase presentation of three-phase harmonic/reactive current extractor.

Without loss of generality, the proposed method detailed in Figs. 2 and 3, can be used in Fig. 4 only to extract the fundamental component, $\dot{x}_1$, its amplitude and the total harmonics, e. In this case, we just need one sub-filter. Other sub-filters are added if the voltage is distorted and reactive components of the load current corresponding to each harmonic is required or if elimination of any individual harmonic is needed.

978-1-4244-4782-4/10 $26.00 © 2010 IEEE   1211

Although the proposed approach is intended to extract the harmonic and reactive current components, it also outputs useful information such as the amplitude, the phase angle and the frequency of the fundamental component. With more modifications it can also output the total harmonic distortion (THD) and the power factor.

B. Performance Evaluation

The proposed ANF-based method of Fig. 4 is employed, as shown in Fig. 5, as a harmonic/reactive- current component extractor. Note that the proposed method is not limited to this application. The extraction unit is implemented in dSPACE 1103 hardware platform.

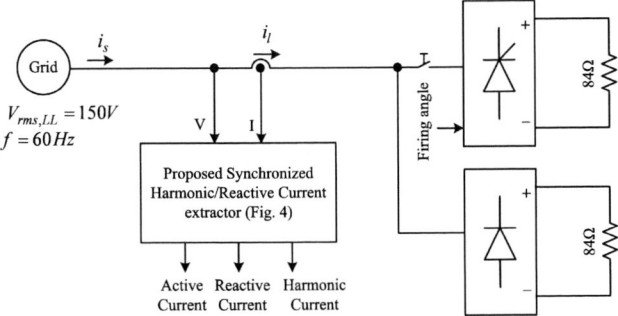

Fig. 5. Proposed technique for extracting reactive/harmonic currents of a nonlinear load.

In the structure of Fig. 5, the voltage and load current are sensed at the point of connection and fed to the detection block, which is a complementary part of the control system and a synchronization tool. This block uses the mechanism introduced in sections III and IV-A, and extracts the harmonic and reactive current components to generate the reference signal. The nonlinear load is a three-phase thyristor rectifier. The load current and source voltage are measured and fed to the extractor unit of Fig. 4. The firing angle of the thyristor rectifier is changed from $0°$ to $60°$ to provide an inductive load. Fig. 6 and 7 show that the scheme successfully extracts the active and reactive components of the load current, and the load harmonic current within one cycle of the system voltage.

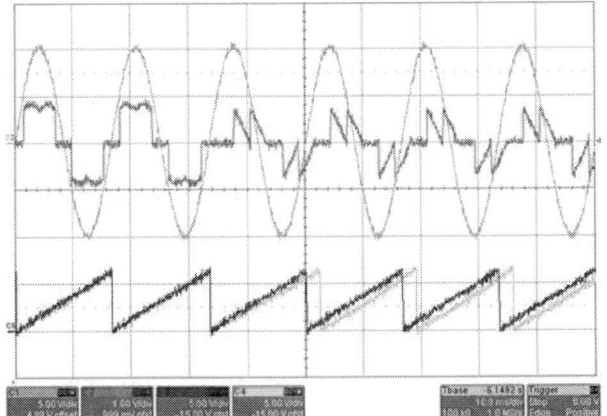

Fig. 6. Response of the proposed system to a step change in load: the source voltage (CH1), its phase angle (CH3), load current (CH 2) and its phase angle (CH4).

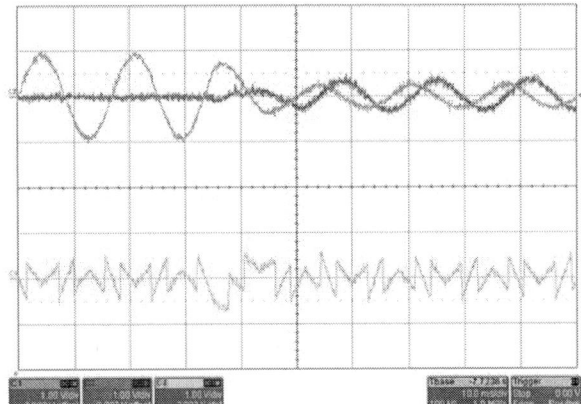

Fig. 7. Response of the proposed system to a step change in load: the active current (CH1), reactive current (CH 2) and harmonics current (CH4).

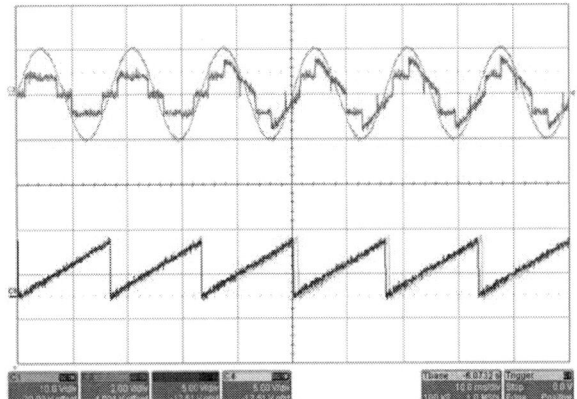

Fig. 8. Response of the proposed system to a step change in load: the source voltage (CH1), its phase angle (CH3), load current (CH2) and its phase angle (CH4).

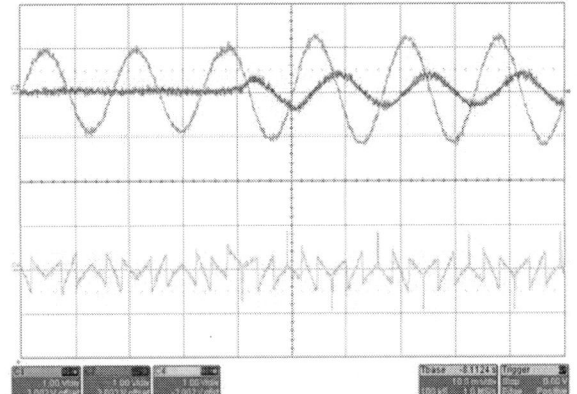

Fig. 9. Response of the proposed system to a step change in load: the active current (CH1), reactive current (CH 2) and harmonics current (CH4).

In another test, initially the load is a three-phase full-bridge rectifier. The fundamental current is in phase with the voltage (no reactive current component), Figs. 8 and 9. Then a three-phase thyristor rectifier is connected in parallel with the full-bridge diode rectifier. The firing angle of the thyristor rectifier is set to $60°$ to provide an inductive load. Fig. 9 shows that the scheme successfully extracts the active and reactive components of the load current, and the load harmonic current within one cycle of the system voltage. These extracted components are traditionally used to generate reference currents in grid connected converters. The experimental results are in agreement with the theoretical discussion.

V. CONCLUSION

An adaptive notch filtering approach for the synchronous detection and extraction of load harmonics and reactive current components is presented in this paper. A mathematical description of the filter was introduced and its outstanding features were experimentally examined.

REFERENCES

[1] L. Asiminoaei, F. Blaabjerg, and S. Hansen, "Detection is key—Harmonic detection methods for active power filter applications," *IEEE Ind. Appl. Mag.*, vol. 13, no. 4, pp. 22–33, Jul./Aug. 2007.

[2] H Akagi, "Active harmonic filters," *Proceedings of IEEE.*, vol. 93, no. 12, pp. 2128-2141, Dec. 2005.

[3] B. Singh, K. Al-Haddad, and A. Chandra, "A review of active filters for power quality improvement," *IEEE Trans. Ind. Electron.*, vol. 46, no. 5, pp. 960–971, Oct. 1999.

[4] M. El-Habrouk, M. K. Darwish, and P. Mehta, "Active power filters: a review," *IEE Proc.-Electr. Power Appl.*, vol. 147, no. 5, pp. 403-413, September 2000.

[5] R. S. Herrera, P. Salmeron, and H. Kim, "Instantaneous reactive power theory applied to active power filter compensation: different approaches, assessment, and experimental results" *IEEE Trans. Ind. Electron.*, vol. 55, no. 1, pp. 184196, Jan. 2008.

[6] M. Cirrincione, M. Pucci, and G. Vitale, "A Single-Phase DG Generation Unit With Shunt Active Power Filter Capability by Adaptive Neural Filtering," *IEEE Trans. Ind. Elec.*, vol. 55, no. 5, pp. 2093-2110, May 2008.

[7] L. Asiminoae, P. Rodriguez, F. Blaabjerg, and M. Malinowski, "Reduction of Switching Losses in Active Power Filters With a New Generalized Discontinuous-PWM Strategy," *IEEE Trans. Ind. Elec.*, vol. 55, no. 1, pp. 467-471, Jan. 2008.

[8] S. Zeliang, G. Yuhua, and L. Jisan, "Steady-State and Dynamic Study of Active Power Filter With Efficient FPGA-Based Control Algorithm," *IEEE Trans. Ind. Elec.*, vol. 55, no. 4, pp. 1527-1536, April 2008.

[9] S. Buso, L. Malesani, and P. Mattavelli, "Comparison of current control techniques for active filter applications," *IEEE Trans. Ind. Electron.*, vol.45, no. 5, pp. 722–729, Oct. 1998.

[10] X. Yuan, W. Merk, H. Stemmler, and J. Allmeling, "Stationary-frame generalized integrators for current control of active power filters with zero steady-state error for current harmonics of concern under unbalanced and distorted operating conditions," *IEEE Trans. Ind. Appl.*, vol.38, no. 2, pp. 523-532, March/April 2002.

[11] W. M.Grady, M. J.Samotyj, and A. H.Noyola, "Survey of active power line conditioning methodologies," *IEEE Trans. Power Delivery*, vol. 5, pp. 1536-1542, 1990.

[12] A. A. Girgis, W. B. Chang, and E. B. Makram, "A digital recursive measurement scheme for on-Line tracking of power system harmonics," *IEEE Trans. Power Del.*, vol. 3, pp. 1153-1160, Jul. 1991.

[13] M. EI-Habrouk and M. K. Darwish, "Design and implementation of a modified Fourier analysis harmonic current computation technique for power active filters using DSP's," *Proc. Inst. Elect. Eng.—Elect. Power Appl.*, vol. 148, no. 1, pp. 21-28, Jan. 2001.

[14] J. M. M Ortega, M. P. Steve, M. P. Payan, A. G. Exposito, and L. G. Franquelo, "Reference current computation method for active power filter: accuracy assessment in the frequency domain," *IEEE Trans. Power Electron.*, vol. 20, no. 2, pp. 446-456, March 2005.

[15] H.Akagi, Y.Kanazawa, and A.Nabae, "Instantaneous reactive power compensators comprising switching devices without energy storage components," *IEEE Trans. Ind. Appl.*, vol. 20, pp. 625-630, May/June 1984.

[16] A. Nabae and T. Tanaka, "A new definition of instantaneous active-reactive current and power based on instantaneous space vectors on polar coordinates in three-phase circuits," *IEEE Trans. Power Del.*, vol. 11, pp. 1238-1243, July 1996.

[17] Z. P. Fang and J. S. Lai, "Generalized instantaneous reactive power theory for three-phase power systems," *IEEE Trans. Instrum. Meas.*, vol. 45, pp. 293-297, Feb. 1996.

[18] V. Soares, P. Verdhelo, and G. D. Marques, "An instantaneous active and reactive current component method for active filters," *IEEE Trans.Power Electron.*, vol. 15, no. 4, pp. 660–669, Jul. 2000.

[19] Y. Sato, T. Ishizuka, K. Nezu, and T. Kataoka, "A new control strategy for voltage-type PWM rectifiers to realize zero steady-state control error in input current," *IEEE Trans. Ind. Applicat.*, vol. 34, no. 3, pp. 480–486, May/Jun. 1998.

[20] M. J. Newman, D. N. Zmood, and D. G. Holmes, "Stationary frame harmonic reference generation for active filters," *IEEE Trans. Ind. Applicat.*, vol. 38, no. 6, pp. 1591–1599, Nov./Dec. 2002.

[21] S. D. Round and D. M. E. Ingram, "An evaluation of techniques for determining active filter compensating currents in unbalanced systems," in *Proc. EPE Trondheim*, Norway, vol. 4, pp. 767-772, 1997.

[22] S. K. Jain, P. Agrawal, and H. O. Gupta, "Fuzzy logic controlled shunt active power filter for power quality improvement," *Proc. Inst. Elect. Eng.—Elect. Power Appl.*, vol. 149, no. 5, pp. 317-328, Sep. 2002.

[23] J. R. Varquez and P. Saimeron, "Active power filter control using neural network technologies," *Proc. Inst. Elect. Eng.—Elect. Power Appl.*, vol. 150, no. 2, pp. 139-145, Mar. 2003.

[24] L. H. Tey, P. L. So, and Y. C. Chu, "Improvement of power quality using adaptive shunt filter," *IEEE Trans. Power Del.*, vol. 20, no. 2, pp. 1558-1568, April 2005.

[25] S.Saetieo, R.Devaraj, and D. A.Torrey, "The design, implementation of a three-phase active power filter based on sliding mode control," *IEEE Trans. Ind. Applicat.*, vol. 31, pp. 993-1000, Sept./Oct. 1995.

[26] S. Lou and Z. Hou, "An adaptive detecting method for harmonic and reactive currents," *IEEE Trans. Ind. Electron.*, vol. 42, pp. 85-89, Feb. 1995.

[27] M. Karimi-Ghartemani, H. Mokhtari, R. Iravani, and M. Sedighy, "A signal processing system for extraction of harmonics and reactive currents of single phase systems," *IEEE Trans. Power Delivery*, vol. 19, no. 3, pp. 979-986, July 2004.

[28] H. Karimi, M. Karimi-Ghartemani, R. Iravani and A. Bakhshai, "An adaptive filter for synchronous extraction of harmonics and distortions," *IEEE Trans. Power Delivery*, vol. 18, no. 4, pp. 1350-1353, Oct. 2003.

[29] M. Mojiri and A. Bakhshai, "An adaptive notch filter for frequency estimation of a periodic signal," *IEEE Trans. Autom. ontrol*, vol. 49, no. 2, pp. 314–318, Feb. 2004.

[30] M. Mojiri, M. Karimi-Ghartemani and A. Bakhshai, "Time domain signal analysis using adaptive notch filter," *IEEE Trans. Signal Processing*, vol. 55, no. 1, pp. 85-93, Jan 2007.

[31] M. Mojiri and A. Bakhshai, "Estimation of n frequencies using adaptive notch filter," *IEEE Trans. Circuit and Systems II*: Express Briefs, vol. 54, no. 4, pp. 338-342, April 2007.

[32] F. Blaabjerg, R. Teodorescu, M. Liserre, and A. V. Timbus, "Overview of control and grid synchronization for distributed power generation systems," *IEEE Trans. Ind. Electron.*, vol. 53, no. 5, pp. 1398-1409, Oct. 2006.

[33] V. Blasko, "A novel method for selective harmonic elimination in power electronic equipment," *IEEE Trans. Power Electron.*, vol. 22, no. 1, pp. 223-228, Jan. 2007.

[34] P. Mattavelli, "A close-loop selective harmonic compensation for active filters," *IEEE Trans. Ind. Appl.*, vol. 37, no. 1, pp. 81-89, Jan./Feb. 2001.

[35] D. Yazdani, A. Bakhshai, G. Joos, and M. Mojiri, "A nonlinear adaptive synchronization technique for grid-connected distributed energy sources," *IEEE Trans. Power Electron.*, Vol. 23, no. 4, pp. 2181-2186, July 2008.

[36] D. Yazdani, M. Mojiri, A. Bakhshai, and G. Joos, "A Fast and Accurate Synchronization Technique for Extraction of Symmetrical Components," *IEEE Trans. Power Electron.*, Vol. 24, no. 3, pp. 674-684, March 2009.

Analysis, Optimized Design and Adaptive Control of a ZCS Full-bridge Converter without Voltage Over-Stress on the Switches

Xin Zhang	Henry Shu-hung Chung	Xinbo Ruan	Adrian Ioinovici
Student Member, IEEE	*Senior Member, IEEE*	*Senior Member, IEEE*	*Fellow, IEEE*
Nanjing University of Aeronautics and Astronautic No. 29, Str. YuDao, Nanjing, 210016, Jiangsu, China	City Univ. of Hong Kong Tat Chee Avenue, Kowloon Tong, Kowloon, Hong Kong eeshc@cityu.edu.hk	Huazhong University of Science and Technology Wuchang, Wuhan, 430074, Hubei, China	Holon Institute of Technology Holon, 58102, Israel adrian@hit.ac.il

Abstract –This paper presents the analysis and design of a new current-driven soft-switched full-bridge converter, in which all main switches are zero-current-switched (ZCS), and the switches in a switched-capacitor snubber are zero-voltage-switched (ZVS). For each value of the supply and load, the snubber capacitor is adaptively charged at the minimum necessary value for assuring soft-switching; thus, less resonant energy is circulated. There is no extra voltage stress on the switches. The current through the switches is limited to the input current. A dc analysis led to the derivation of the voltage conversion ratio. The resonant elements of the snubber circuit are optimally designed by trading-off the soft-switching range the duty-cycle loss over a wide supply and load variation. The calculation of the ac small-signal transfer functions allowed for the design of the controller, which was implemented by DSP. A 530V/15kV, 5kW prototype has been built and evaluated.

Index Terms- Adaptable soft-switching, dc-dc conversion, full-bridge converter, high voltage converter, zero-current switching

I. INTRODUCTION

Pulse-width modulated (PWM) full-bridge (FB) converters have been popularly used for high-power dc/dc conversion. There is a large literature on ZVS PWM-FB converters. However, the ZVS technique is more suitable for converters using majority-carrier type switching devices, such as MOSFET. For high-voltage and power applications, the suitable switch is the insulated gate bipolar transistor (IGBT) . For such switches, ZCS is preferable, for dealing with the tail current at turn-off. Although a few ZCS switching schemes have been proposed [1-9], they present different drawbacks, such as large current and/or voltage over-stresses due to resonant peaks, or use of much resonant energy even when it is not necessary .

The work described in this paper was fully supported by a grant from the Research Grants Council of the Hong Kong Special Administrative Region, China (Project No.: CityU 112406).

Recently, a new current-driven full bridge converter was proposed [10]. It contains a switched-capacitor snubber connected in parallel to the primary winding. All main switches are ZCS and the snubber switches are ZVS. The parasitic elements of the coupling transformer are used in the resonant operation. The resonant energy used to assure soft-switching is self-adaptable for each actual value of the input/load current. The purpose of this paper is to present a detailed dc analysis of this converter. The voltage conversion ratio formula is found, allowing for an analysis of the duty-cycle loss. The conditions necessary for achieving ZCS of the primary-side switches are derived analytically, and a knowledge design of the resonant elements is proposed. A trade-off between the duty-cycle loss and the soft-switching range gave the design equations. The small-signal ac transfer functions of the converter are found , serving to the design of the controller. The controller is implemented by DSP. A 530V/15kV, 5kW prototype has been built and evaluated, the good measured efficiency proving the advantages of the solution.

II. ZCS FB CONVERTER

Fig. 1 shows the circuit schematic of the proposed current-driven FB converter. It consists of a front-stage dc/ac converter formed by the switches $S_1 \sim S_4$ and an output-stage rectifier formed by $D_A \sim D_D$. The two stages are interconnected by a coupling transformer T_r with the turns ratio $n : 1$, leakage inductance L_{lk}, and parasitic capacitance C_P. The input side of the converter is supplied through an inductor L_I. A snubber formed by a resonant inductor L_r and a switched-capacitor circuit consisting of two MOSFETs, S_{a1} and S_{a2}, and a resonant capacitor C_r is connected across the primary side of the coupling transformer. It can make $S_1 \sim S_4$ switch at zero current. Fig. 2 shows the theoretical voltage and current waveforms and timing diagram of the converter. The output voltage of the converter is controlled by adjusting the angle ϕ between the switch pairs (S_1, S_2) and (S_3, S_4). S_1 (S_3) and S_2 (S_4) are operated in anti-phase.

There are twelve operating modes from t_0 to t_{12} in one switching period T_s. However, as the operation is symmetrical in every one half of the switching cycle, only the operation from t_0 to t_6 is described

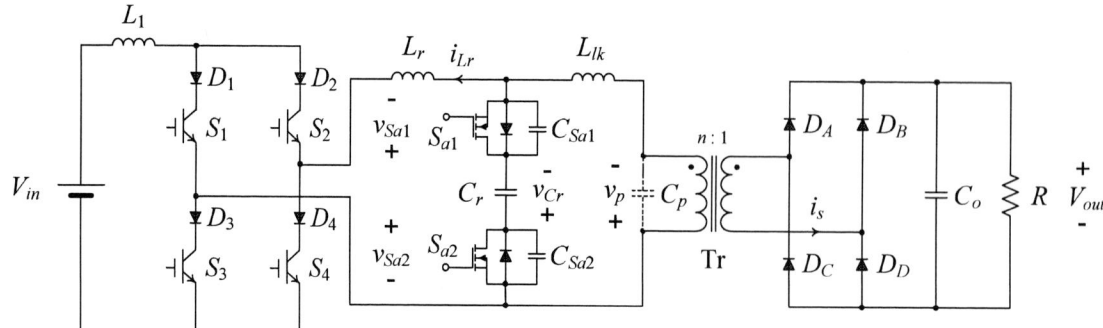

Fig. 1 Proposed ZCS current-fed FB converter.

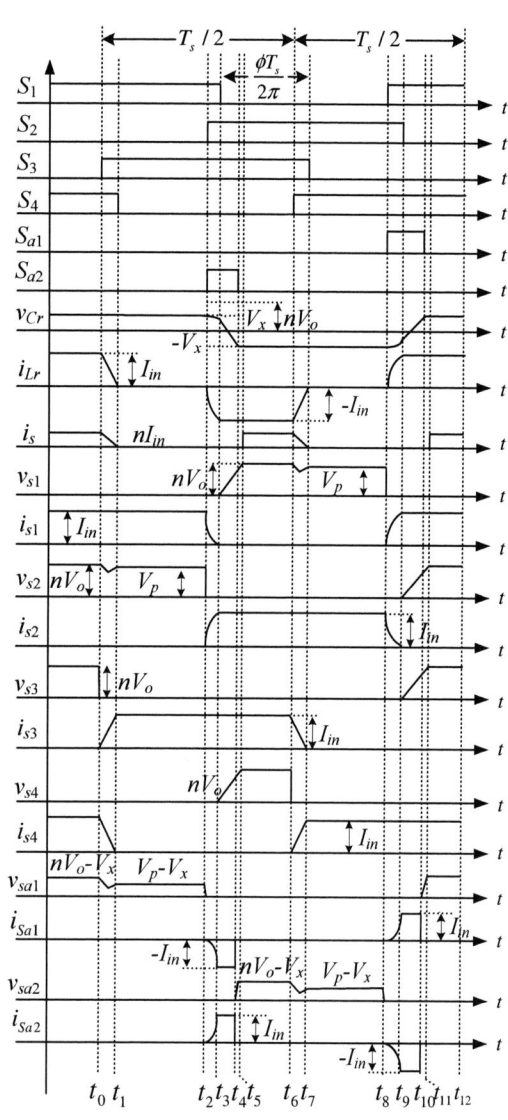

Fig. 2 Timing diagram and key waveforms of the converter.

1. Mode 0 [Before t_0]

Before t_0, the energy is transferred from the input to the output via L_1, S_1, S_4, T_r, D_B and D_C. C_r is charged at the minimum necessary voltage V_x for assuring the main switches switch at zero current, It will be shown later that the value of V_x is chosen to satisfy the following criterion

$$\frac{1}{2} C_r V_x^2 \ge \frac{1}{2} L_r I_{in}^2 \tag{1}$$

2. Mode 1 [t_0 - t_1]

At t_0, S_3 is turned on with ZCS. A resonance path is formed by L_r, L_{lk}, C_{Sa1}. The current through S_4, i_{S4}, decreases while the current through S_3, i_{S3}, increases. This mode ends at t_1 when $i_{Lr}(t_1) = i_{S4}(t_1) = 0$, $i_{S3}(t_1) = I_{in}$, and S_4 is turned-off with ZCS.

$$i_{Lr}(t) = I_{in} - \frac{n V_{out}}{L_r + L_{lk}} \left[t + \frac{L_{lk}}{\omega_T L_r} \sin \omega_T (t - t_0) \right] \tag{2}$$

where $\omega_T = \dfrac{1}{\sqrt{L_T C_{Sa1}}}$ and $L_T = \dfrac{L_r L_{lk}}{L_r + L_{lk}}$.

As the resonant component of the current in (2) is small,

$$t_{01} = t_1 - t_0 \cong I_{in} \frac{L_r + L_{lk}}{n V_{out}} \tag{3}$$

3. Mode 2 [t_1 - t_2]

L_1 undergoes charging. At t_2, the controller dictates the turn-on of S_2 (ZCS) and S_{a2} (ZVS).

4. Mode 3 [t_2 - t_3]

A resonant path formed by S_2, L_r, S_{a1}, C_r, S_{a2}, and S_1 is created.

$$v_{Cr}(t) = V_x \cos \omega_1 (t - t_2) \tag{4}$$

$$i_{S2}(t) = -i_{Lr}(t) = \frac{V_x}{Z_1} \sin \omega_1 (t - t_2) \tag{5}$$

978-1-4244-4782-4/10 $26.00 © 2010 IEEE

$$i_{S1}(t) = I_{in} - i_{S2}(t) \qquad (6)$$

where $\omega_1 = \dfrac{1}{\sqrt{L_r C_r}}$ and $Z_1 = \sqrt{\dfrac{L_r}{C_r}}$.

This mode ends at t_3 when $i_{Lr}(t_3) = -I_{in}$ and $i_{S1}(t_3) = 0$. S_1 is turned off with zero current. Therefore,

$$t_{23} = t_3 - t_2 = \frac{1}{\omega_1} \sin^{-1} \frac{I_{in} Z_1}{V_x} \qquad (7)$$

In order to make i_{Lr} reach $-I_{in}$, based on (5), it is necessary to ensure that $V_x / Z_1 > I_{in}$, giving the ZCS criterion in (1). This means that there is enough energy in the resonant capacitor C_r to charge the resonant inductor L_r. It can be noted that for the designed values of C_r and L_r, the necessary energy stored in C_r for this purpose depends on the value of the input current. At a small value of I_{in}, less energy is needed. The maximum energy is required at the high end of the range of the input current (i.e., at the low value of the range of the input voltage and at high load current). So, V_x can be calculated for each actual value of I_{in}. It will take a higher value only when needed. This gives the adaptability characteristic of the proposed solution : it allows for using at each actual value of the input voltage and load the minimum necessary of resonant energy that can assure ZCS of the primary switch. Consequently, the resonant energy and conduction loss are reduced.

5. *Mode 4 [t_3 - t_4]*

C_r is discharging and then charged by I_{in}, so that its voltage is reversed. This mode ends at t_4 when the voltage on C_r reaches $-V_x$. S_{a2} is switched off at zero voltage.

$$t_{34} = t_4 - t_3 = \frac{C_r}{I_{in}} [v_{Cr}(t_3) + V_x] \qquad (8)$$

The voltage on C_r is sensed. When it reaches the calculated value V_x, the control circuit dictates the end of this mode ,by turning S_{a2} off .

6. *Mode 5 [t_4 - t_5]*

The junction capacitance of S_{a2} is charged by I_{in} until $v_{Sa2}(t_5) = n V_{out} - V_x$.

7. *Mode 6 [t_5 - t_6]*

When the voltage across the primary side of T_r reaches nV_{out}, the energy is transferred from the input to the load via S_2, L_r, T_r, D_A, D_D.

III. STEADY-STATE ANALYSIS AND DESIGN CONSIDERATION

A. ZCS conditions

1. *For the leading switches*

In order to ensure soft-switching of the leading switches, it is necessary to ensure that the switch pairs have sufficient dead time $t_{d,lead}$ for current transfer. In Mode 1, $t_{d,lead}$ should be long enough for the current through S_3 to increase from zero to I_{in} and the current through S_4 to decrease from I_{in} to zero. Thus, based on (3), $t_{d,lead}$ should satisfy the criterion

$$t_{d,lead} > I_{in} (L_r + L_{lk}) / n V_o \qquad (9)$$

2. *For the lagging switches*

As shown in Mode 3, C_r provides the required energy for achieving zero-current switching of the lagging switches. Thus, based on (5), the dead time of the lagging switches $t_{d,lag}$ should satisfy the criterion

$$t_{d,lag} > \frac{1}{\omega_1} \sin^{-1} \frac{I_{in} Z_1}{V_x} \qquad (10)$$

According to (1), for the designed value of Z_1, V_x depends solely on the value of I_{in}, I_{in} is continuously sensed and the minimum necessary capacitor voltage for ensuring ZCS at the measured value of the input current, V_x, is calculated with (1). When the sensed value of v_{Cr} reaches the calculated value of V_x, S_{a1} (or S_{a2}) will be turned off, marking the end of Mode 4.

B. DC voltage conversion ratio

As shown in Fig. 2, ϕ is defined as follows

$$\varphi_T = \frac{\varphi T_s}{2 \pi} = t_{34} + t_{45} + t_{56} \qquad (11)$$

where ϕ_T is the time duration of the phase shift.

From the conservation of energy in a half switching cycle

$$\frac{T_s}{2} V_{in} I_{in} = \sum_{k=1}^{5} i_k V_o \qquad (12)$$

where $i_k = \int_{t_{k-1}}^{t_k} |i_s(t)| dt$

t_{01}, t_{23}, and t_{34} are obtained from (3), (7), and (8).

$$t_{12} = t_2 - t_1 = \frac{T_s}{2} - \varphi_T + t_{56} - t_{23} - t_{01}$$

$$t_{45} = t_5 - t_4 = \frac{T_s}{2} - t_{01} - t_{12} - t_{23} - t_{34} = \varphi_T - t_{01} - t_{34}$$

It should be noted that t_{56} equals t_{01}. It can be shown that

$$i_1 = \frac{1}{2} n I_{in} t_{01} = \frac{1}{2} I_{in}^2 \frac{L_r + L_{lk}}{V_o} \qquad (13)$$

$$i_2 = i_3 = i_4 = 0 \qquad (14)$$

$$i_5 = n\,I_{in}\,t_{45} = n\,I_{in}\,(\varphi_T - t_{01} - t_{34}) \qquad (15)$$

By substituting (13)-(15) into (12), the dc conversion ratio M is obtained

$$M = \frac{T_s}{n\,(2\,\varphi_T - t_{01} - 2\,t_{34})} \qquad (16)$$

C. Design of the component values

1. Value of L_r

Based on (9), in order to assure that the leading switches can complete the current transfer process within t_{01}, the maximum value of L_r, $L_{r,max}$, is found as

$$L_{r,max} = \frac{n\,V_o}{I_{in}}\,t_{01} - L_{lk} \qquad (17)$$

$L_{r,max}$ is designed at the rated load condition. Fig. 3 shows the relationships between t_{01} and $L_{r,max}$ with the parameters of the experimental prototype. The parameters are given in Table I. In Fig. 3, 0.14 μs was chosen to be a reasonable value for t_{01}.

The minimum value of L_r, L_{r_min}, is determined by considering the maximum rate of rise of the leading switch current $di/dt|_{max}$. Thus, in Mode 1,

$$\frac{n\,V_o}{L_r + L_{lk}} \le \frac{di}{dt}|_{max}$$

giving

$$L_{r,min} = n\,V_o\,(di/dt\,|_{max})^{-1} - L_{lk} \qquad (18)$$

A value of $di/dt|_{max}$ =80 A/μs is used for shaping the switching trajectory and avoiding high electromagnetic interference.

2. Value of C_r

Based on (1), and, as $V_x < nV_o$, it results

$$C_r \ge \frac{L_r\,I_{in}^2}{n^2\,V_o^2} \qquad (19)$$

The time taken for reversing the polarity of the snubber capacitor voltage in Modes 3 and 4 from V_x to $-V_x$ is equal to $t_{Cr} = t_{23} + t_{34}$. By using (7) and (8), it can be shown that

$$t_{Cr} = t_{23} + t_{34} = \frac{1}{\omega_1}\sin^{-1}\frac{I_{in}\,Z_1}{V_x} + \frac{\sqrt{V_x^2 - I_{in}^2\,Z_1^2} + V_x}{I_{in}}\,C_r \quad (20)$$

Fig. 4 shows the relationships between t_{Cr} and C_r for different values of V_x calculated for the parameters given in Table I. The value of C_r is determined by considering the required values of t_{Cr} and V_x. As shown in Fig. 2, the voltage stress on S_{a1} and S_{a2} is $nV_o - V_x$. Increasing V_x will reduce the voltage stress on S_{a1} and S_{a2}, but will increase t_{34}, and thus, t_{Cr} is chosen 1.5 μs with $V_x = 50$ V, then, C_r is 100 nF.

D. Self-Adaptable Voltage for C_r

According to (1), the required value of V_x is maximum when I_{in} is maximum

$$V_{x,max} = \sqrt{\frac{L_r}{C_r}}\,\frac{P_{max}}{V_{in,min}} \qquad (21)$$

Fig. 5 shows the value of V_x versus the output load current with the parameters given in Table I. For example, when V_{in} is 530 V, at the rated load, V_x is about 30 V. If V_{in} is changed to 424 V, V_x becomes about 48 V, because I_{in} is increased.

E. Soft-switching range of the proposed converter

Based on (11),

$$\varphi_T = \frac{T_s}{2} - t_{12} - t_{23} = t_{01} + t_{34} + t_{45} \qquad (22)$$

Thus, the soft-switching range of the converter can be expressed as

$$t_{01} + t_{34} < \varphi_T \le \frac{T_s}{2} - t_{23} \qquad (23)$$

Thus, by using (3), (7) and (8), and substituting the above boundaries of ϕ_T into (16), it is possible to calculate the maximum (M_max) and minimum (M_min) values of M. Fig. 6 shows the relationships of M_max and M_min versus the normalized output load current with the parameters given in Table I. The interception point between M_max and M_min for the minimal input voltage of 424 V gives the minimum load for which soft-switching is assured. For the specified input voltage range, both output voltage regulation and soft-switching operation begins at 7% of the rated load, and it is assured in all load range, including heavy load.

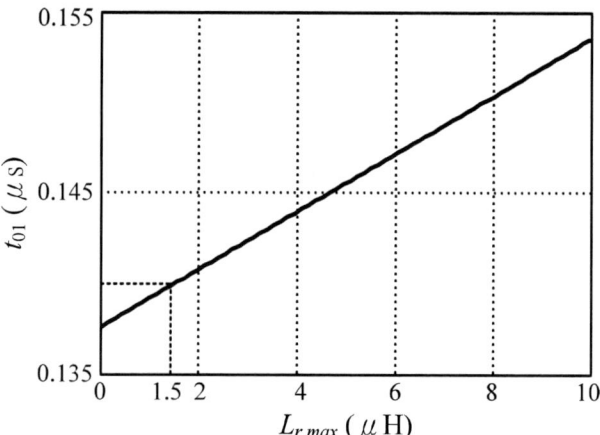

Fig. 3 t_{01} versus $L_{r,max}$.

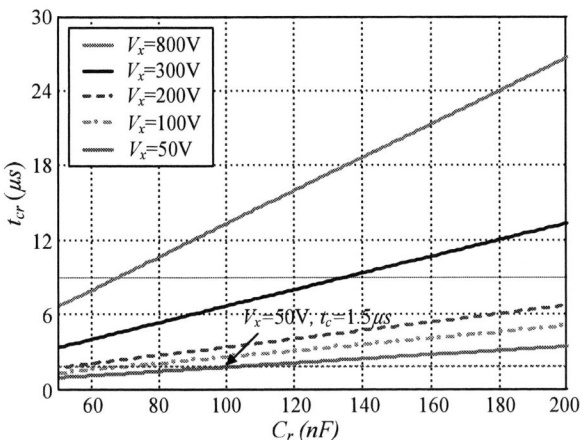

Fig. 4 Resonant time t_{Cr} versus C_r for different values of V_x.

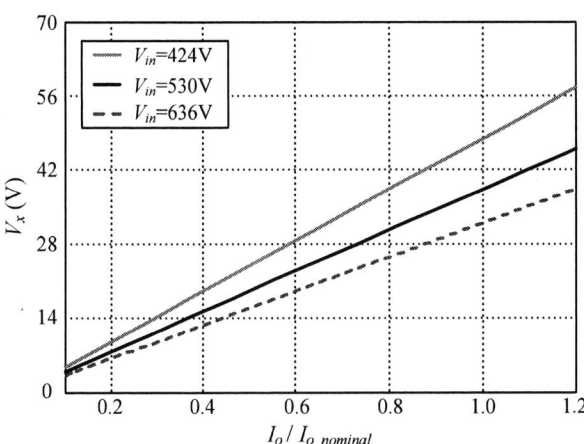

Fig. 5 V_x versus per-unit I_o.

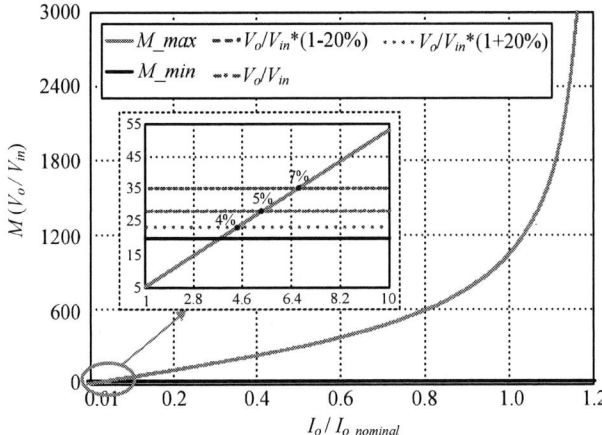

Fig.6 ZCS assured load range of the proposed converter.

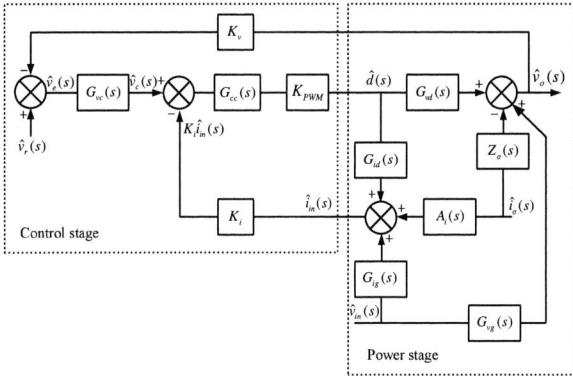

Fig.7 Closed-loop small-signal ac equivalent model of proposed converter

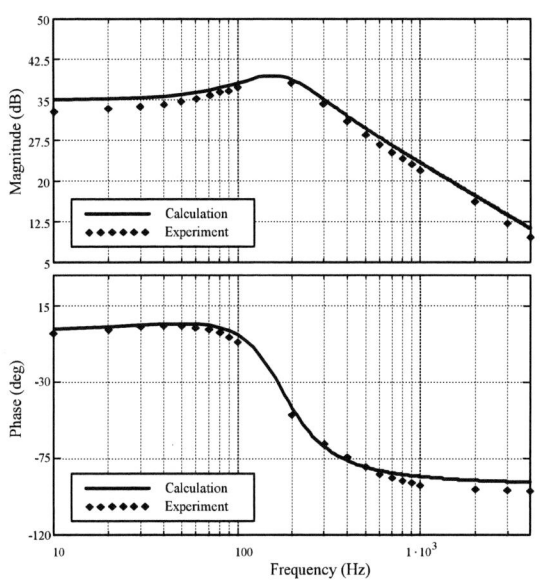

Fig.8 Bode diagram of $G_{id}(s)$.

IV. ADAPTIVE CHARACTER OF THE ZCS-ASSISTED RESONANT ENERGY

According to (1), V_x is calculated for each measured value of the input current, such that to ensure ZCS for all the specified range of input/load currents. At a low input/load current, less resonant energy is needed. The controller will determine the instant t_4 for which the actual $v_{Cr}(t)$ reached the calculated value V_x.

V. CONTROLLER DESIGN

A current-mode control of the proposed converter is used. The equivalent small-signal ac model of the closed-loop regulator is given in Fig. 7.,where $\hat{v}_e(s)$ - Small-signal error voltage, K_v- Output voltage scaling factor. K_i- Inductor current scaling factor. $G_{vc}(s)$ -Transfer function of voltage loop compensator. $G_{cc}(s)$ -Transfer function of current loop compensator.

According to Fig.7, the current loop gain, $T_i(s)$, results in

$$T_i(s) = K_i K_{PWM} G_{cc}(s) G_{id}(s) \qquad (24)$$

The parameters K_i and K_{PWM} have been designed as: K_i =0.265, K_{PWM} =0.303. and $G_{cc}(s)$ was designed as a PI controller:

$$G_{cc}(s) = K_{ip} + K_{ii}/s \qquad (25)$$

The values of K_{ip} and K_{ii} were designed as K_{ip}=30, K_{ii}=30000. Fig.8 presents the Bode diagram of $G_{id}(s)$ which provides the information for the *PI* controller design.

The voltage loop gain, T_v, results in:

$$T_v(s) = \frac{K_{PWM} K_v G_{vc}(s) G_{vd}(s) G_{cc}(s)}{1 + T_i(s)} \qquad (26)$$

where K_v is chosen as K_v=1/5000, and $G_{vc}(s)$ is designed as a PI controller:

$$G_{vc}(s) = K_{vp} + \frac{K_{vi}}{s} \qquad (27)$$

With K_{vp}=2.5, K_{vi}=8000, Fig. 9 shows the Bode diagram of the voltage-loop with and without compensator G_{vc}. The designed cross frequency of the voltage-loop is 650Hz. The phase margin is 50 degree.

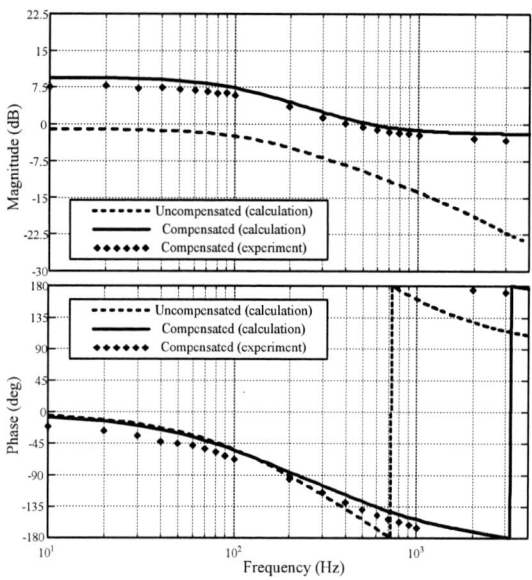

Fig.9 Bode diagram of the voltage-loop gain $T_v(s)$.

The voltage and current loop controllers $G_{vc}(s)$ and $G_{cc}(s)$ are implemented by DSP. The bilinear transformation $s \rightarrow \dfrac{2}{T} \dfrac{Z-1}{Z+1}$ was applied to render discrete the above two continuous compensators

$$G_{cc}(z) = (30.075z - 29.925)/(z - 1) \qquad (28)$$
$$v_n = v_{n-1} + 30.075(e_n - e_{n-1}) + 0.15 e_n \qquad (29)$$
$$G_{vc}(z) = (2.52z - 2.48)/(z - 1) \qquad (30)$$
$$v'_n = v'_{n-1} + 2.52(e'_n - e'_{n-1}) + 0.04 e'_n \qquad (31)$$

VI. ADAPTIVE CONTROL CIRCUIT

The control system was implemented by a digital signal processor (DSP) TMS320F28335 with a soft-start scheme. The input inductor current i_{L1} (i.e. the input current I_{in}) and output voltage V_{out} were sampled to provide the necessary

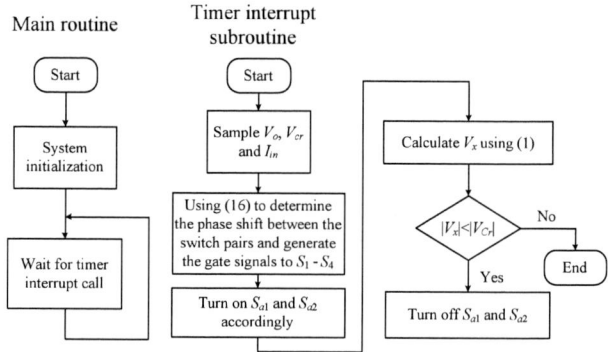

Fig. 10 Software flowchart of the control algorithm.

information for the current-mode control and calculate the minimum snubber voltage V_x for achieving zero-current switching [Eq. (1)]. The instantaneous value of the snubber capacitor voltage v_{Cr} is also sensed and compared with the calculated value of V_x. The auxiliary switches S_{a1} and S_{a2} are synchronized to switch on with S_1 and S_2, respectively. When the value of v_{Cr} equals V_x or $-V_x$, S_{a1}, respectively, S_{a2} are turned off. Fig. 10 shows the flowchart of the control mechanism.

Table I Main components of the prototype

Component	Value	Component	Value
$S_1 \sim S_4$	FF150R17ME3G	C_r	100nF
$S_{a1} \sim S_{a2}$	IPW90R340C3	L_{lk}	8.6μH
IGBT driver	2SD106AI-17	C_o	0.2μF / 30kV
$D_1 \sim D_4$	30kV / 3A	T_r	0.05 : 1
		L_r	1.5 μH

VII. PROTOTYPE AND EXPERIMENTAL RESULTS

A prototype (V_{in}=530V, V_o =15kV, P=5kW, f_s=20kHz) was realized, with C_r= 100nF, L_{lk} =8.6 μH, C_o= 0.2μF/30kV, L_r=1.5uH. Table I gives the components of the prototype.

The experimental time-domain steady-state voltage and current waveforms of the primary switches S_1, S_4, and S_{a2} at full load are given in Fig11. It proves the ZCS of the main switches and ZVS of the auxiliary switches .

Fig. 12 presents the snubber capacitor voltage $v_{Cr}(t_4)$ (in percentage of $v_{Cr,\max}$) under different line/load conditions (a reduced input voltage attracts an increased input current at a constant output power).

The experimental efficiency versus the load is given in Fig. 13. It can be seen that an efficiency of almost 94 % is obtained at full load.

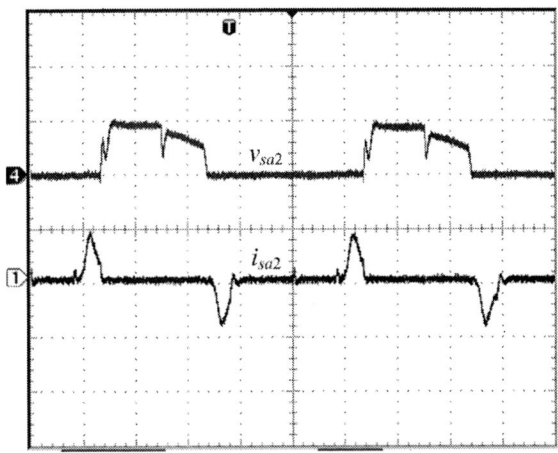

(c) Waveforms of v_{Sa2} and i_{Sa2} (v_{Sa2} :800V/div and i_{Sa2} : 10A/div).

Fig. 11 Voltage and current waveforms of the switches S_1, S_4 , and S_{a2} at full load condition (Timebase: 10μs/div).

VIII. CONCLUSIONS

The design and analysis of a current-driven full-bridge converter suitable for high voltage applications has been presented. By using a simple snubber, it can realize ZCS of the main switches in a large input/load range, and ZVS of the auxiliary switches. The resonant energy used for achieving ZCS is self-adaptable. The circulation of resonant energy is reduced at the minimum necessary for achieving ZCS of the main switches. The voltage stress on the switches never surpasses nV_o . There is no additional current stress, as the current through the primary switches never overpasses the nominal value. A trade-off design allowed for minimizing the

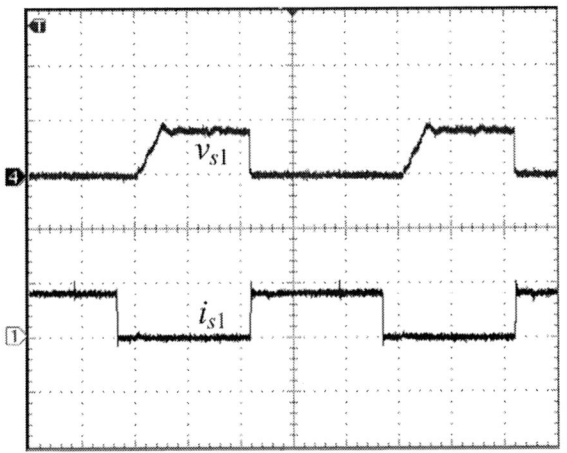

(a)   Waveforms of v_{S1} and i_{S1} (v_{S1} : 1kV/div and i_{S1} : 10A/div).

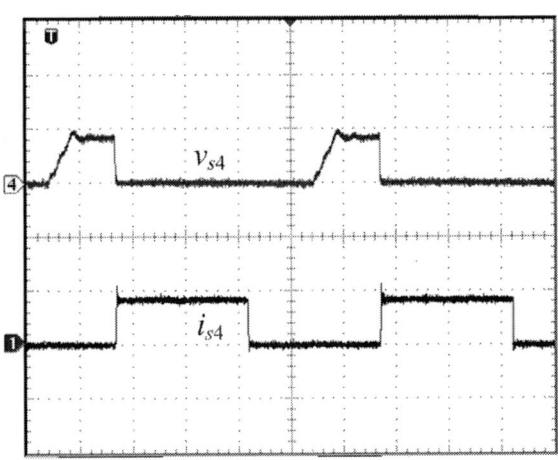

(b)   Waveforms of v_{S4} and i_{S4} (v_{S4} : 1kV/div and i_{S4} : 10A/div).

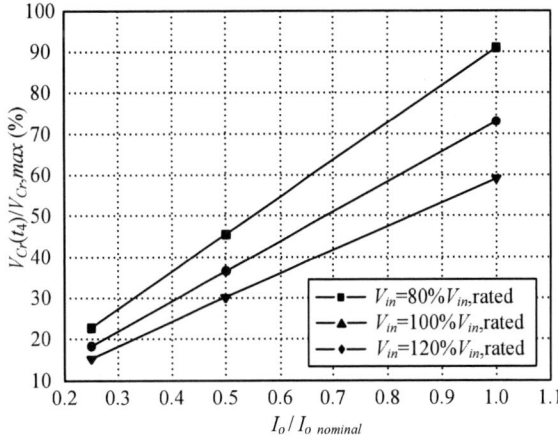

Fig. 12 Measured $v_{Cr}(t_4)$, i.e., actual V_x , in percentage from its maximum possible value under different line/load conditions.

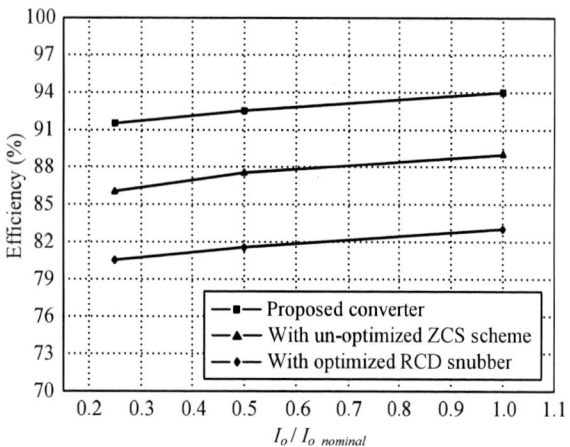

Fig. 13 Experimental efficiency versus load power.

duty-cycle loss resulted from the resonant process , and concomitantly assuring soft-switching and output voltage regulation from a low value of the load (7 % of the rated value). Soft-switching and output voltage regulation is then assured in all range of the load, including heavy load. The measured efficiency of the proposed soft-switching converter is very high :94 % at full load, being much higher than the value for a full-bridge converter with RCD snubber.

REFERENCES

[1] X. H. Wu, D. M. Xu, J. H. Kong, C. Yang, and Z. Qian, "High power high frequency zero current transition full bridge dc/dc converter," in *Proc. APEC'98 Conf.*, 1998, pp. 823–827.

[2] M. Marx and D. Schroder, "A novel zero-current-transition full-bridge dc-dc converter," in Proc. PESC'96 Conf., 1996, pp. 664–669.

[3] C. Iannello, S. Luo and I. Batarseh, "Small-signal and transient analysis of a full-bridge, zero-current-switched PWM converter using an average model," *IEEE Trans. Power Electron.*, vol.18, no. 3, pp. 793-801, May 2003.

[4] J. Chen, R. Chen, and T. Liang, "Study and implementation of a single-stage current-fed boost PFC converter with ZCS for high voltage applications," *IEEE Trans. Power Electron.*, vol. 23, no. 1, pp. 379-386, Jan. 2008.

[5] A. Leung, H. Chung, and K. Chan "A ZCS isolated full-bridge boost converter with multiple inputs," in *Proc. IEEE Power Electron. Spec. Conf.*, 2007, pp. 2542-2548

[6] S. Atoh and H. Yoshike, "PWM dc-dc converter with a resonant commutation means," in *Proc. INTELEC'91 Conf.*, 1991, pp. 308–313.

[7] D. M. Xu, X. H. Wu, J. M. Zhang, and Z. Qian, "High power high frequency half-wave-mode ZCT-PWM full bridge dc/dc converter," *in Proc. IEEE Appl. Power Electron. Conf.*, 2000, pp. 99-103.

[8] J. Zhang, X. Xie, X. Wu, G.Wu, and Z. Qian, "A novel zero-current-transition full bridge dc/dc converter," *IEEE Trans. Power Electron.*, vol. 21, no. 2, pp. 354-360, Mar. 2006.

[9] L. Qin, S. Xie and H. Zhou, "A novel family of PWM converters based on improved ZCS switch cell" in *Proc. IEEE Power Electron. Spec. Conf.*, 2007, pp. 2725-2730.

[10] X. Zhang, H. Chung, X. Ruan and A. Ioinovici, "A ZCS Full-Bridge Converter without Voltage Over-Stress on the Switches", *Proc. IEEE ECCE* 2009, pp. 1991-1998.

Analysis and design of a Novel ZVS-PWM DC-DC Converter for Bidirectional Applications with Steep Conversion Ratio

Pritam Das
Department of Electrical and
Computer Engineering
University of Western Ontario
London, Ontario, Canada
pdas2@uwo.ca

Ahmad Mousavi
Department of Electrical and
Computer Engineering
University of Western Ontario
London, Ontario, Canada
mseyeda@uwo.ca

Gerry Moschopoulos
Department of Electrical and
Computer Engineering
University of Western Ontario
London, Ontario, Canada
gmoschopoulos@uwo.ca

Abstract—A new bidirectional dc-dc PWM converter with a steep voltage conversion ratio is proposed in the paper. The proposed converter makes use of coupled inductors to step up small dc voltage to large ones or to step down large dc voltages to small ones. Its switches operate with zero-voltage switching due to a very simple auxiliary circuit that contains a single active switch and a few passive elements. In the paper, the proposed converter is presented, its operation is explained in detail and analyzed, and a systematic design procedure is derived and used to design a prototype converter. The feasibility of the converter is confirmed with results obtained from the lab prototype.

I. INTRODUCTION

Power electronic converter systems for applications such as telecom, automotive and space can have dc voltage buses that are backed up with batteries. The batteries are connected to the buses with bidirectional dc-dc converters that allow them to discharge or to be charged whenever necessary. The bidirectional dc-dc converters may be transformer isolated [1]-[3] or non-isolated [4]-[11], depending on the application. The main focus of the paper is on non-isolated bidirectional dc-dc converters.

Dc-dc converters (unidirectional or bidirectional) are typically implemented with coupled inductors [12]-[14] in applications where the low side voltage is much lower than the high side voltage so that the converter duty cycle is not too narrow, which would lead to undesirable peak component stresses. Most non-isolated bidirectional dc-dc converters are based on a boost/buck converter structure. The converter shown in Fig. 1 is an example of a typical bidirectional buck/boost converter that can be used in an application where the battery voltage is very low relative to the bus voltage. S_1 operates like a boost switch and S_2 operates as a boost diode when energy is transferred from the low-side source V_{lo} to the high-side source V_{hi} and S_1 operates like a buck diode and S_2 like a boost diode when energy is transferred from V_{hi} to V_{lo}.

Implementing non-isolated bidirectional converters with zero-voltage switching (ZVS), regardless of whether or not they are implemented with coupled inductors, is challenging as they do not use inductive energy stored in a transformer to discharge the capacitance across their switches as isolated bidirectional half-bridge and full-bridge dc-dc converters do. Previously proposed techniques to implement soft-switching in non-isolated bidirectional dc-dc converters can be categorized as follows:

- Converters that are based on a buck/boost topology and are made to operate with an inductor (L_{in}) current that flows in both directions during each switching cycle [4]. Both converter switches are on (never simultaneously) sometime during each cycle so that the energy stored in the inductor when one switch is on is used to turn on the other switch with ZVS after the switch is turned off. The main drawback of this technique is that the inductor current has a lot of ripple with a very high peak as it must flow in both directions during each switching cycle. This results in high turn-off losses that counteract the improvement in efficiency due to the ZVS turn-on.

- Another approach to implementing soft-switching in a non-isolated bidirectional dc-dc converter is to use quasi-resonant or multi-resonant techniques. Doing so, however, results in the converter having high peak voltage and/or current stresses and forces the converter to be operated with variable switching

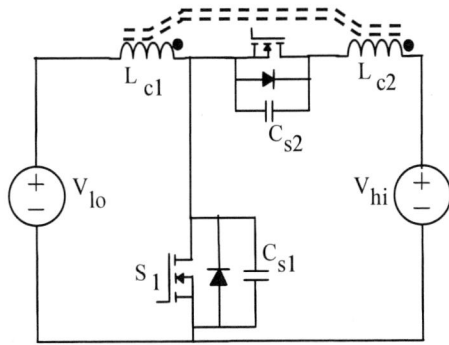

Fig. 1. Standard boost/buck converter structure for bidirectional dc-dc power conversion.

frequency control, which complicates the design of the converter - especially the design of the magnetic and filtering elements as a wide range of switching frequencies must be considered. In the case of a converter such as the one proposed in [5], the converter can be made to operate with constant switching frequency, but the switch stresses remain. A fixed frequency resonant-type bidirectional converter was proposed in [6], but this converter was very costly and sophisticated as it required two half-bridge converters in series.

- A third approach has been to use auxiliary circuits to assist the switches to operate with ZVS as is done in zero-voltage transition (ZVT) converters. This has been done with converters such as the ones proposed in [7]-[12]. Although this approach is an improvement over the other two approaches, it can be costly and complex. This is because a separate independent auxiliary circuit is needed for each main power switch so that the converter must be implemented with four active switches. Converters such as the ones shown in Fig. 2(a) and 2(b) have a single auxiliary circuit, but require two auxiliary switches to make the circuit bidirectional.

A new ZVS-PWM non-isolated bidirectional dc-dc converter with a steep conversion ratio is proposed in the paper. The proposed converter, shown in Fig. 3, is very similar to the conventional converter in Fig. 1 except that auxiliary switch S_a, capacitor C_r and inductors L_{r1} and L_{r2} have been added. These components make up a simple active clamp circuit that can be used to ensure that the main power switches, S_1 and S_2, operate with ZVS regardless of whether the converter is operating in a boost or buck mode. The proposed converter can operate with a continuous inductor current, fixed switching frequency, steep conversion ratio,

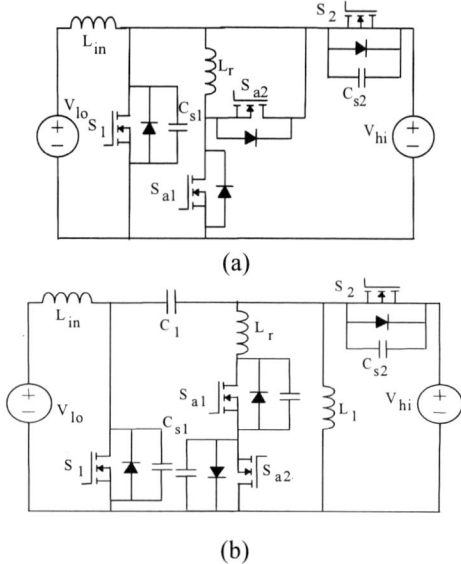

(a)

(b)

Fig. 2. (a) Converter proposed in [10]. (b) Converter proposed in [11].

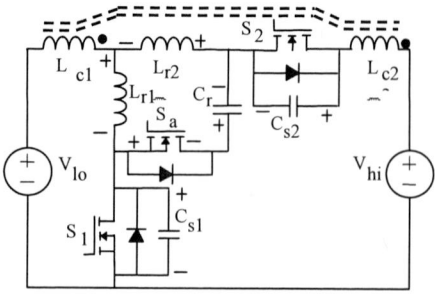

Fig. 3. Proposed Converter

and the switch stresses of a conventional coupled inductor converter regardless of the direction of power flow.

In the paper, the converter's operation is explained and analyzed. The analysis is used as the basis of a design procedure that is demonstrated with an example. Experimental results obtained from a prototype, which confirm the feasibility of the proposed converter, are also presented.

II. CONVERTER OPERATION

The modes of operation that the proposed converter goes through when it is operating as a boost or as buck converter are given below along with typical converter waveforms for the boost mode are shown in Fig. 4. Equivalent circuit diagrams for each mode of operation in boost and buck operations are shown in Figs. 5 and 6.

Referring to Fig. 3 the inductor current I_{Lr1} is positive if it enters the inductor through its positive terminal and the same goes to inductor current I_{Lr2}. The currents through switches S_1 and S_2 are considered positive if they flow into their respective switches through the drain. The current through Sa is considered to be positive if it flows from the source of the switch through its drain. Any current flowing into the positive terminal of the capacitor C_r is considered to be positive. Voltage measured from drain to source is considered positive for all the switches.

A. Operation in Boost Mode

The operation of the converter in boost mode is explained in this section. For boost mode operation, switch S_1 is the main power switch and S_2 acts as the freewheeling diode. Typical waveforms describing the boost mode of operation are shown in Fig. 4, while Fig. 5 shows the equivalent active part of the converter in each mode of operation as a boost converter and fig.6 shows active parts in buck mode.

Mode 0 (t < t₀): Before time t = t₀, the converter operates as a standard coupled inductor PWM boost converter with switch S_1 on and the currents through L_{C1} rising.

Mode 1 (t₀ < t < t₁): At t = t₀, switch S_1 is turned off and the rise in voltage across it is limited by C_{s1}. The current through L_{r1} charges up C_{s1} and begins to flow through the body diode of S_a, charging up C_r. Also during this mode, input current begins to be diverted to L_{r2} and the capacitance across S_2, C_{s2}, begins to be discharged. During this mode the auxiliary capacitor voltage is also reflected across the switch

978-1-4244-4782-4/10 $26.00 © 2010 IEEE   1223

Mode 2 ($t_1 < t < t_2$): This mode is a continuation of Mode 1 except that C_{s2} is completely discharged at $t = t_1$ and current flows through the anti-parallel diode across S_2. At the end of this mode the body diode of the auxiliary switch cease to conduct and the main switch voltage drops back to its steady state off-state voltage. The key equations describing Modes 1 and 2 are

$$V_{Cr}(t) = I_{in,lo} Z_1 \sin\omega_1 (t - t_2) \qquad (1)$$

where V_{Cr} is the voltage across the clamp capacitor C_r,

$Z_1 = \sqrt{\dfrac{L_{r1}}{C_r}}$, $\omega_1 = \sqrt{\dfrac{1}{L_{r1}C_r}}$ and $I_{in,lo}$ is the input current from

the low side source. At the end of this mode the auxiliary capacitor voltages reaches a maximum value of

$$V_{Cr,max} = I_{in,lo} Z_1 \qquad (2)$$

Mode 3 ($t_2 < t < t_3$): At $t = t_2$, current stops flowing through the auxiliary active clamp circuit and the converter operates as a standard coupled inductor PWM boost converter. The current through L_{C1} decreases during this mode as a negative voltage is impressed across L_{C1} and so does current in L_{C2}. During this mode energy trapped in the low side coupled inductor is transferred to the high side through the magnetic coupling between the inductors L_{C1} and L_{C2}.

Mode 4 ($t_3 < t < t_4$): Some time before switch S_1 is to be turned on, at $t = t_3$, switch S_a is turned on with ZCS. Capacitor C_r begins to discharge through L_{r1} and L_{r2}, as currents in the coupled inductors continue to decrease. The resonant capacitor voltage will also be reflected across the switch in this mode. The key equations describing this mode are:

$$i_{Lr1}(t) = -\frac{V_{Cr}(0)}{Z_a} \sin\omega_a (t - t_3) \qquad (3)$$

$$v_{Cr}(t) = V_{Cr}(0)\cos\omega_a (t - t_4) \qquad (4)$$

$$V_{S1}(t) = \frac{V_o}{1 - D_1} + V_{Cr,max}\cos\omega_a (t - t_4) \qquad (5)$$

where $\omega_a = \sqrt{\dfrac{1}{(L_{r1} + L_{r2})C_r}}$ and $Z_a = \sqrt{\dfrac{L_{r1} + L_{r2}}{C_r}}$

Mode 5 ($t_4 < t < t_5$): Switch S_a is turned off at $t = t_4$. The current in L_{r1} is used to discharge C_{s1}. The key equations describing this mode are

$$i_{Lr1}(t) = -\frac{V_{Cr}(0)}{Z_a}\cos\omega_S (t - t_4) \qquad (6)$$

$$v_{S1}(t) = V_{S1}(0) - \frac{V_{Cr}(0)}{Z_a} Z_{S1}\sin\omega_{S1}(t - t_4) \qquad (7)$$

where $\omega_{S1} = \sqrt{\dfrac{1}{(L_{r1} + L_{r2})C_{S1}}}$ and $Z_{S1} = \sqrt{\dfrac{L_{r1} + L_{r2}}{C_{S1}}}$

Mode 6 ($t_5 < t < t_6$): At $t = t_5$, capacitor C_{s1} has been completely discharged and the anti-parallel diode across S_1 begins to conduct. S_1 can be turned on while this diode is conducting.

Mode 7 ($t_6 < t < t_7$): Some time after S_1 has been turned on, at $t = t_6$, the current through L_{r1} will begin to reverse direction and the transfer of current from L_{r2} to S_1 will begin. This mode of operation will continue until current has been completely transferred to S_1 and the converter enters Mode 0 at $t = t_7$. The key equations describing this mode are

$$t_{C1} = \frac{I_{in,lo}(L_{r1} + L_{r2})}{(V_{in} + \dfrac{V_o}{n})} \qquad (8)$$

where t_{C2} is the time in which the current in the inductor L_{r2} is completely transferred to L_{r1}.

B. Operation in Buck Mode

The operation of the converter in the buck mode is described in this section of the paper. For buck mode operation, switch S_2 is the main power switch and S_1 acts as the freewheeling diode. It should be noted that some of the equations that describe each of the following modes are identical to corresponding modes of operation when the converter is in boost mode.

Fig. 6 shows the equivalent active part of the converter in each mode of operation as a buck converter. The waveforms that describe the buck mode of operation are almost identical to those of the boost mode of operation and thus are not shown.

Mode 0 ($t < t_0$): Before time $t = t_0$, the converter operates as a standard coupled inductor PWM buck converter with switch S_1 on and the currents through L_{C2} rising. At the same time,

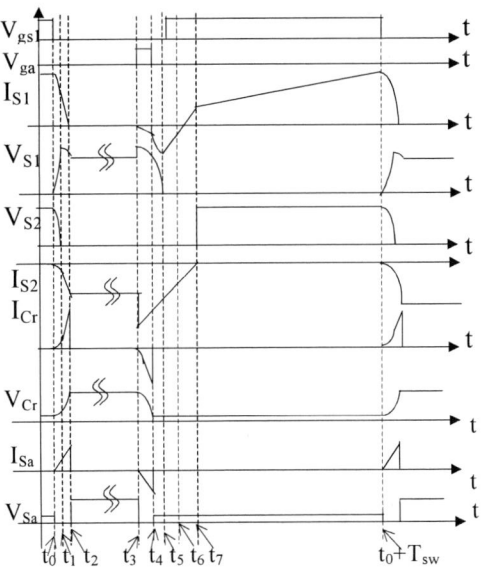

Fig. 4. Typical converter waveforms for boost mode operation.

the low side-coupled inductor L_{C1} gets energized by the current reflected from the high side coupled inductor L_{C2} so that current through L_{C1} also rises.

Mode 1 ($t_0 < t < t_1$): At $t = t_0$, switch S_2 is turned off and the rise in voltage across it is limited by C_{s2}. The current through L_{r2} charges up C_{s1} and begins to flow through C_r. Also during this mode, input current begins to be diverted to L_{r1} and the capacitance across S_1, C_{s1}, begins to be discharged.

Mode 2 ($t_1 < t < t_2$): This mode is a continuation of Mode 1 except that C_{s1} is completely discharged at $t = t_1$ and current is flowing through its body diode. At the end of this mode, C_{s2} is completely charged and current stops flowing through C_r branch. The key equations describing this mode are:

$$V_{Cr}(t) = 2I_{o,lo}Z_2\sin\omega_1(t-t_3) \qquad (9)$$

where V_{Cr} is the voltage across the clamp capacitor C_r, $Z_2 = \sqrt{\dfrac{L_{r2}}{C_r}}$, $\omega_1 = \sqrt{\dfrac{1}{L_{r2}C_r}}$ and $I_{in,lo}$ is the input current from the low side source.

Mode 3 ($t_2 < t < t_3$): At $t = t_2$, current stops flowing through the auxiliary active clamp circuit and the converter operates as a standard PWM buck converter. The current through L_{C1} decreases during this mode as a negative voltage is impressed across L_{C1} and so does current in L_{C2}. During this mode energy trapped in the high side coupled inductor is transferred to the low side through the magnetic coupling between the inductors L_{C1} and L_{C2}.

Mode 4 ($t_3 < t < t_4$): Some time before switch S_2 is to be turned on, at $t = t_3$, switch S_a is turned on with ZCS. Capacitor C_r begins to discharge through L_{r1} and L_{r2}, as currents in the coupled inductors continue to decrease. The key equations describing this mode are:

$$i_{Lr2}(t) = -\frac{V_{Cr}(0)}{Z_a}\sin\omega_a(t-t_3) \qquad (10)$$

$$v_{Cr}(t) = V_{Cr}(0)\cos\omega_a(t-t_4) \qquad (11)$$

Mode 5 ($t_4 < t < t_5$): Switch S_a is turned off at $t = t_4$. The current in L_{r2} is used to discharge C_{s2}. The key equations describing this mode are:

$$i_{Lr2}(t) = -\frac{V_{Cr}(0)}{Z_a}\cos\omega_S(t-t_4) \qquad (12)$$

$$v_{s2}(t) = V_{S2} - \frac{V_{Cr}(0)}{Z_a}Z_{S2}\sin\omega_{S2}(t-t_4) \qquad (13)$$

$$\omega_{S2} = \sqrt{\frac{1}{(L_{r1}+L_{r2})C_{S2}}} \qquad (14)$$

$$Z_{S2} = \sqrt{\frac{(L_{r1}+L_{r2})}{C_{S2}}} \qquad (15)$$

Mode 6 ($t_5 < t < t_6$): At $t = t_5$, capacitor C_{s2} has been completely discharged and the anti-parallel diode across S_2 begins to conduct. S_2 can be turned on while this diode is conducting.

Mode 7 ($t_6 < t < t_7$): Sometime after S_2 has been turned on, at $t = t_6$, the current through L_{r2} begins to reverse direction and the transfer of current from L_{r1} to S_2 will begin. This mode of operation will continue until current has been completely transferred to S_2 and the converter enters Mode 0 at $t = t_7$. The key equation describing this mode of operation is

$$t_{C2} = \frac{I_{in,lo}(L_{r1}+L_{r2})}{(V_{in}+\dfrac{V_o}{n})} \qquad (16)$$

By symmetry, $t_{C1} = t_{C2}$.

III. CONVERTER DESIGN PROCEDURE AND EXAMPLE

A procedure for the design of the proposed bidirectional ZVS converter with coupled inductor is presented in this section of the paper along with a design example. It should be noted that the auxiliary circuit must be designed so that it can create a ZVS turn-on for the device that acts as the main power switch, regardless of direction of power flow.

The coupled inductor being designed will have the following basic specifications:

- low-side voltage of $V_{lo} = 24\text{-}30\text{V}$,
- high-side voltage of $V_{hi} = 200$ V,
- maximum power $P_{o,max} = 200$ W,
- switching frequency $f_{sw} = 66$ kHz.

A. Designing the basic coupled inductor converter

In designing the basic converter the actual voltage conversion ratio in each direction of power flow and voltage and current stresses of the devices need to be considered. Due to the existence of Mode 7 in each direction of power flow, the effective duty ratio will be reduced since in this mode, commutation of current from L_{r2} to L_{r1} in boost mode and vice-versa in buck mode occurs. The actual steady state voltage conversion ratios in each mode will be given by

$$M_{real,boost} = \frac{V_{hi}}{V_{lo}} = \frac{n(D_1 - T_{cn})+1}{1-(D_1 - T_{cn})} \qquad (17)$$

and

$$M_{ideal,buck} = \frac{V_{lo}}{V_{hi}} = \frac{D_2 - T_{cn}}{1+n(1-D_2-T_{cn})} \qquad (18)$$

where $T_{cn} = t_c/T_{sw}$ is the normalized commutation time t_c, which is same as the duration of Mode 7 in each direction of power flow and T_{sw} is the switching frequency.

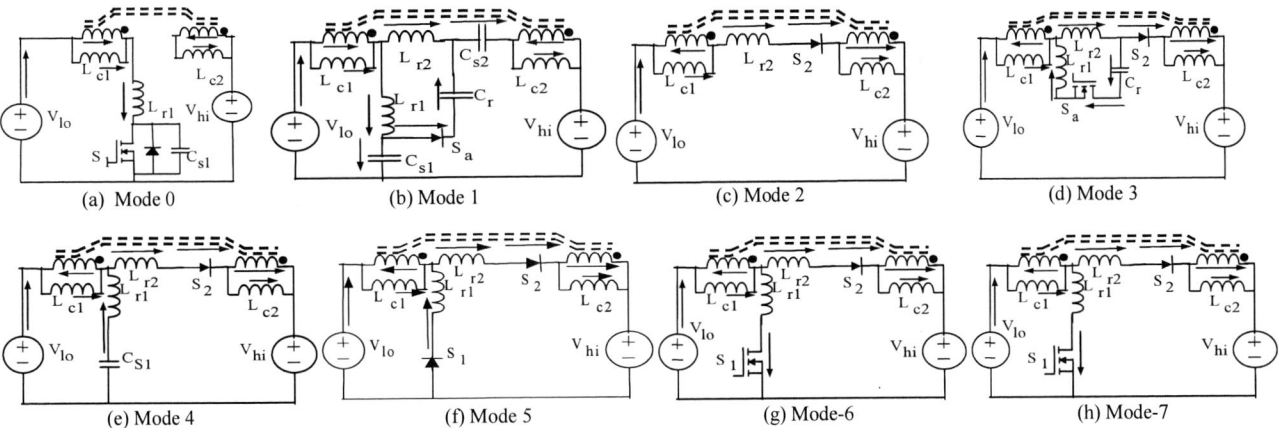

Fig. 5. Equivalent circuit diagrams for boost mode of operation.

The peak voltages across the devices without the auxiliary circuit are given by

$$V_{S1} = \frac{V_o}{1-D_1} \qquad (19)$$

$$V_{S2} = V_{hi} + nV_{in} \qquad (20)$$

The peak current in S_1 is given by

$$I_{S1,pk} = I_{in} + \frac{V_{in}}{L_{C1}} \frac{D_1 T_{sw}}{2} \qquad (21)$$

while the peak current through switch S_2 in boost mode is given by

$$I_{S2,pk} = \frac{1}{n}(\frac{V_{in}}{L_{C1}} \frac{D_1 T_{sw}}{2}) + \frac{2i_o}{1-D_1} \qquad (22)$$

Due to the symmetry of the converter, the chosen value of

D_1 in boost mode will be equal to $(1-D_2)$ in buck mode and that any current through the switch in boost mode is equivalent to the same current in through the corresponding current through body diodes in buck mode and vice-versa.

The value of n must be such that the converter can operate with the specified low side and high side voltages without the duty ratios of the switches being too small or too large. Moreover, n should be such that the voltage and peak current stresses of each switch are within some appropriate limits, regardless of the direction of power flow in the converter. If n is too small, then the voltage stress across S_1 will be too large, and this will restrict the use of low voltage devices with low values of $r_{ds,on}$ for this switch. If n is too large then so too will be the voltage stress across S_2. A trade-off must therefore be made in selecting the value of n so that the voltage stress across each switch is reasonable.

The example converter has a maximum a boost mode conversion ratio of

$$M_{realboost} = V_{hi}/V_{lo} = 200/24 = 8.5$$

and a maximum buck mode conversion ratio of:

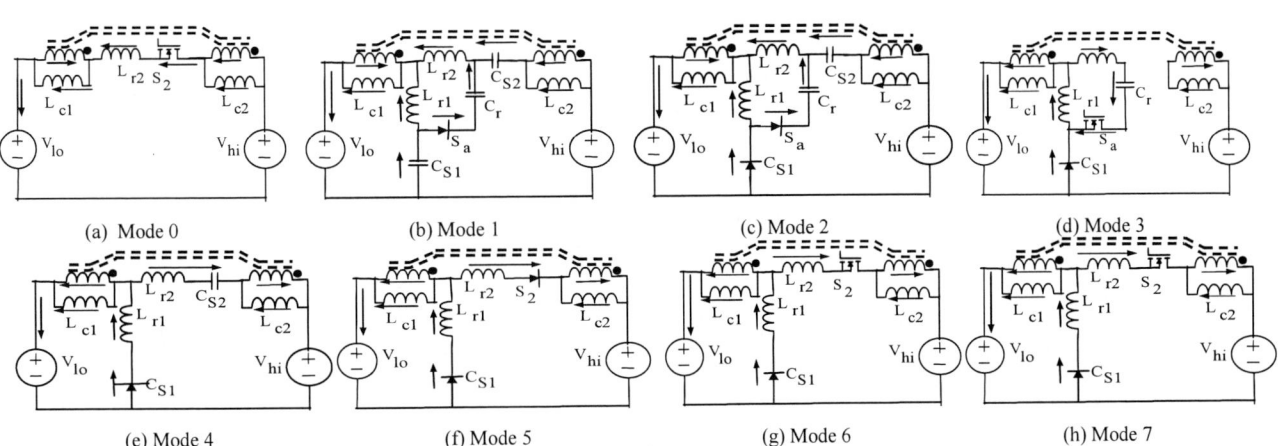

Fig. 6. Equivalent circuit diagrams for buck mode of operation.

$$M_{realbuck} = V_{lo}/V_{hi} = 30/200 = 0.15$$

Values of D_1=0.3 and n=18 are chosen for the example converter which results in the maximum peak value of the voltage across S_1 with a 30V input is 40V, while the maximum peak voltage across S_2 is 650V.

B. Designing the auxiliary circuit:

1) Determine condition of ZVS in both directions: The minimal conditions needed to ensure the ZVS turn-on of the main power switch in both boost and buck modes are derived in this section. For simplicity of design, the values of L_{r1} and L_{r2} will be chosen to be the same. It was found that the main switch voltage reduces to zero following eqn. (10), during Mode 5 of boost operation. The condition for the ZVS turn-on of the main switch in boost mode can therefore be determined from eqn. (10) and the fact that the energy stored in the resonant inductors at the end of mode-4 must be larger than the energy stored in the switch capacitor C_{S1} which leads to the following ZVS condition in boost mode:

$$\frac{1}{2}(L_{r1} + L_{r2})(I_{in,lo}\frac{Z_1}{Z_a})^2 > \frac{1}{2}C_{S1}(\frac{V_{lo}}{1-D_1})^2 \quad (23)$$

Similar relations can be derived for ZVS from Mode 4 of buck mode of operation

$$\frac{1}{2}(L_{r1} + L_{r2})(2I_{o,lo}\frac{Z_2}{Z_a})^2 > \frac{1}{2}C_{s2}(V_{hi} + nV_{lo})^2 \quad (24)$$

It should also be mentioned that the choice of inductors will be such that the values of $L_{r1}=L_{r2}$ to preserve the symmetry of the circuit and also simplicity of designing and manufacturing. Too low a value of L_{r1} and L_{r2} will restrict the range of ZVS operation in terms of load current. This is especially true for S_2 when the converter is in buck mode as the voltage across it is higher than that of S_1. Too high a value of L_{r1} and L_{r2} will increase the circulating current losses in the converter especially in boost mode of operation and increase the operation time of the auxiliary circuit, which impact the operation of the main converter.

For the converter under consideration, a characteristic graph of the value of L_{r1} necessary for the ZVS turn on of switch S_1 when the converter is operating in boost mode is plotted in Fig. 7(a) for different values of average input current and input low side voltage using eqn (23). The same is done in Fig. 7(b) for buck mode operation using eqn. (24). From these graphs, values of $L_{r1}=L_{r2}=2\mu H$ are chosen which will ensure ZVS turn on of S_2 between 55% of full load to full load for V_{lo}=24V and between 60% of full load to full load for V_{lo}=30V in buck mode. This value of L_{r1} will correspond to ZVS range of about 10% of full load to full load to full load for V_{lo}=24V in boost mode.

2) Choice of resonant capacitor: For choosing resonant capacitor C_r, the maximum voltage across switch S_1, this is given by

$$V_{S1}(t) = \frac{V_o}{1-D_1} + V_{Cr,max} \quad (25)$$

should be considered. This voltage is proportional to the load in both boost and buck modes and must be evaluated at maximum load when choosing C_r.

A value of C_r=330nF is chosen as the resonant capacitor, so that the peak voltage across S_1 given by eqn. (25), is about 50 V. This ensures that a 100V device with relatively low $r_{ds,on}$ can be used for S_1, which, in turn, helps reduce conduction losses and improve overall efficiency when the converter is in boost mode.

IV. EXPERIMENTAL RESULTS

An experimental prototype of the proposed converter was built to confirm its feasibility. The converter was built to operate with the following specifications and components:

- low-side voltage of V_{lo} = 24-30V,
- high-side voltage of V_{hi} = 200 V,
- maximum power $P_{o,max}$= 200 W,
- switching frequency f_{sw}= 66 kHz.
- $L_{r1}=L_{r2}$=1.5µH,

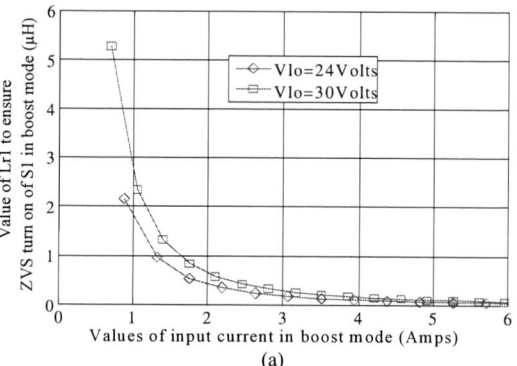

(a)

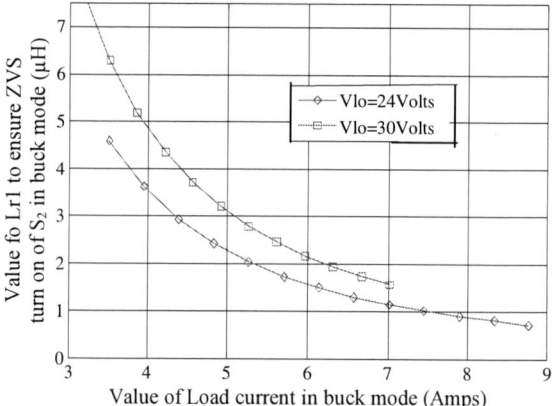

(b)

Fig. 7 Characteristic graphs of L_{r1} to ensure ZVS turn on of S_1 in boost mode and S_2 in buck mode.

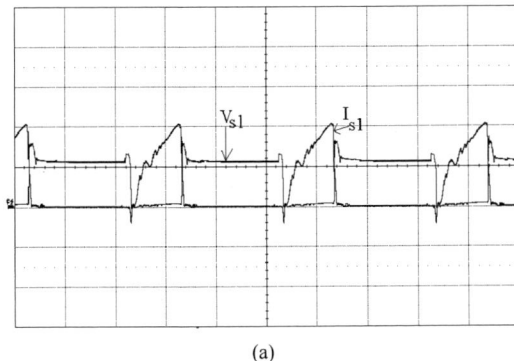

(a)

Fig. 8. Voltage and current waveforms for S_1 with converter working in boost mode with V_{in}=24V, V_o=200V, P_o=200W (V: 30 V/div., I: 15 A/div., t: 5 μs/div.).

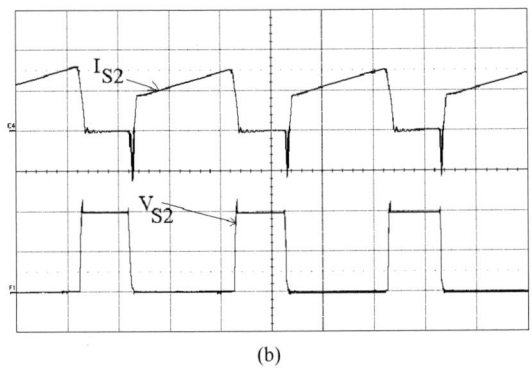

(b)

Fig. 9. Voltage and current waveforms for S_2 with converter working in buck mode with V_{in}=200V, V_o=24V, P_o=200W (V: 300 V/div., I: 2 A/div., t: 5 μs/div.)

- C_r=330 nF, 250V.

- Coupled inductors L_{c1}=5μH and L_{c2}=1.6mH.

- S_1: IRF540 device

- S_2: STW10NK80Z device

- S_a: IRL3215 device

Typical converter waveforms are shown in Figs. 8 and 9. Fig. 8 shows the voltage and current waveforms of S_1 when it is turning on and the converter is operating in boost mode. Fig. 9 shows the current and voltage of S_2 in buck mode. It can be seen that in both cases, the converter switches can operate with a ZVS turn-on. Graphs of converter efficiency for both boost and buck modes of operation are shown in Fig. 10. It can be seen that the efficiency curves of the ZVS and the hard-switched converter diverge at heavier loads. This is because the active clamp auxiliary circuit significantly improves converter efficiency regardless of the direction of power flow. The auxiliary circuit reduces the losses due to switching transitions, which still exist in the hard-switched converter and become more dominant at heavier loads.

V.   CONCLUSION

A soft-switched bidirectional dc-dc converter with coupled inductors for steep conversion ratios was proposed in

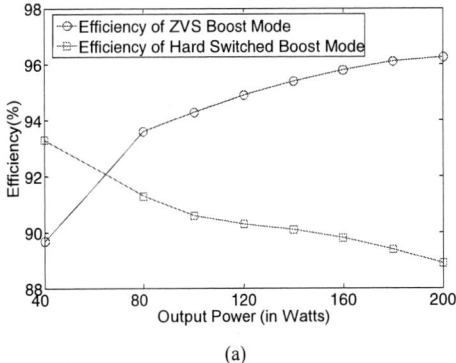

(a)

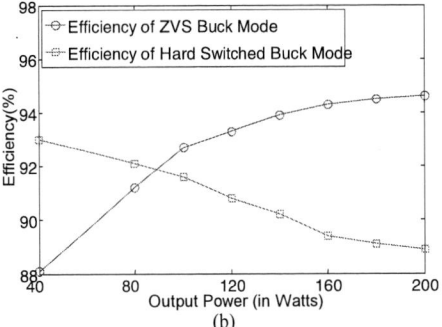

(b)

Fig.10. Experimental efficiency graphs. (a) Efficiency of ZVS and hard-switched converter in boost mode. (b) Efficiency of ZVS and hard-switched converter in buck mode.

the paper. The proposed converter can operate with zero voltage-switching, a continuous inductor current, fixed switching frequency, and the switch stresses of a conventional PWM coupled inductor converter regardless of the direction of power flow. These features are due to a very simple auxiliary circuit that is operational regardless of the direction of power flow.

In the paper, the converter's operation and design were discussed, and results obtained from an experimental prototype that confirm the feasibility of the converter were presented. It was shown how the ZVS turn-on of both switches is achieved with just a single, common auxiliary circuit that is operational regardless of whether the converter is operating in boost mode or in buck mode.

ACKNOWLEDGMENT

The authors would like to acknowledge funding from the National Sciences and Research Council of Canada (NSERC) for financial support during the course of this work.

REFERENCES

[1]   R. Li, et al, "Analysis and design of improved isolated full-bridge bidirectional DC-DC converter," *IEEE PESC Conf. Rec.*, 2004, pp. 521 – 526.

[2]   L. Zhu, "A novel soft-commutating isolated boost full-bridge ZVS-PWM DC-DC converter for bidirectional high power applications," *IEEE PESC Conf. Rec.*, 2004, pp. 2141 - 2146.

[3]   S. Inoue and H. Akagi, "A bidirectional isolated DC–DC converter as a core circuit of the next-generation medium-voltage power conversion system," *IEEE Trans. on Power Elec.*, vol. 22, no. 2, pp. 535-542, Mar. 2007.

[4] C.P. Henze, et al., "Zero voltage switching in high frequency power converters using pulse width modulation," *IEEE APEC Conf. Rec.* 1988, pp. 33-40.

[5] Z. R. Martinez and B. Ray, "Bidirectional DC/DC power conversion using constant frequency multi-resonant topology," *IEEE APEC Conf. Rec.*, 1994, pp. 991 – 997.

[6] C.-E. Kim, et al, "A new high efficiency ZVZCS bidirectional DC/DC converter for 42V power system of HEVs," *IEEE PESC Conf. Rec.*, 2006, pp. 792 – 797.

[7] K. T. Chau, et al., "Bidirectional soft-switching converter-fed DC motor drives," *IEEE PESC Conf. Rec.* 1998, pp. 416-422.

[8] H. Shiji, et al, "A zero-voltage-switching bidirectional converter for PV systems," *IEEE INTELEC Conf. Rec.* 2003, pp. 14 - 19.

[9] E. Sanchis-Kilders, et al. "Soft switching bidirectional converter for battery discharging-charging," *IEEE APEC Conf. Rec*, 2006, pp. 603-609.

[10] J. Zhang, R.-Y. Kim, and J.-S. Lai, "High-power density design of a soft-switching high-power bidirectional DC-DC converter," *IEEE PESC Conf. Rec.*, 2006.

[11] L. Schuch, et al, "Analysis and design of a new high-efficiency bidirectional integrated ZVT PWM converter for DC-bus and battery-bank interface," *IEEE Trans. on Ind. Appl.*, vol. 42, no. 2, pp. 1321-1332, Sept. 2007.

[12] E. S.-Kilders, A Ferreres, E. Maset, J. B. Ejea, V. Esteve, J. Jordan, A. Garrigos, and J. Calvente, "Soft switching bidirectional converter for battery discharging-charging," IEEE APEC Conf. Rec, 2006, pp. 603-609.

[13] D. M. Van de Sype, K. De Gusseme, B. Renders, A. R. Van den Bossche, and J. A. Melkebeek, "A single switch boost converter with a high conversion ratio," *IEEE APEC Conf. Rec.*, 2005, pp. 1581-1587.

[14] W. Lin, J. Wang, J. Huang, Y. Xu, "A novel tapped inductor bi-directional buck-boost topology", *IEEE INTELEC Conf. Rec.* 2008, paper P.7.

Three-Level Phase-Shift ZVS-PWM DC-DC Converter with High Frequency Transformer for High Performance Arc Welding Machines

Tomokazu Mishima[1], Hisayuki Sugimura[2], Khairy Fathy Sayed[3], Soon Kurl Kwon[3], and Mutsuo Nakaoka[3,4]

[1]Kure National College of Technology
Hiroshima, Japan
mishima@kure-nct.ac.jp
[2]Daihen Corporation
Kobe, Japan
[3]Kyungnam University
Masan, Republic of Korea
[4]Yamaguchi University
Yamaguchi, Japan

Abstract— **This paper is concerned with some experimental results and practical evaluations of a three-level phase shifted ZVS-PWM DC-DC converter with neutral point clamping diodes and flying capacitor for a variety of gas metal arc welding machines. This new DC-DC converter suitable for high power applications is implemented by modifying the high-frequency-linked half bridge soft switching PWM DC-DC converter with two active edge resonant cells (AERCs) in high side and low side DC rails, which is previously-developed put into practice by the authors. The operating principle of the three-level phase-shift ZVS-PWM DC-DC converter and its experimental and simulation results including power regulation characteristics vs. phase-shifted angle and power conversion efficiency characteristics in addition to power loss analysis are illustrated and evaluated comparatively from a practical point of view, along with the remarkable advantageous features as compared with previously-developed one.**

Index Terms--*DC-DC power converter, neutral point clamping diodes and flying capacitor hybrid topology, Three-level half bridge, high frequency transformer link, primary side phase-shift PWM, Soft switching commutation, Gas metal arc welding*

I. Introduction

In, recent years, the authors have demonstrated the half-bridge and full-bridge prototype of the high frequency transformer-linked soft-switching PWM DC-DC converters with simple active edge resonant cells (AERCs) in the input bus-lines of high side and low side rails for gas metal arc

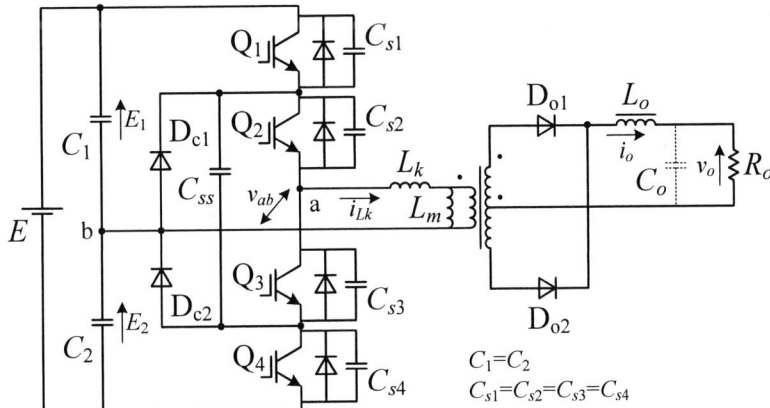

Fig. 1. Three-level phase-shift ZVS-PWM DC-DC high power converter with a high frequency transformer link.

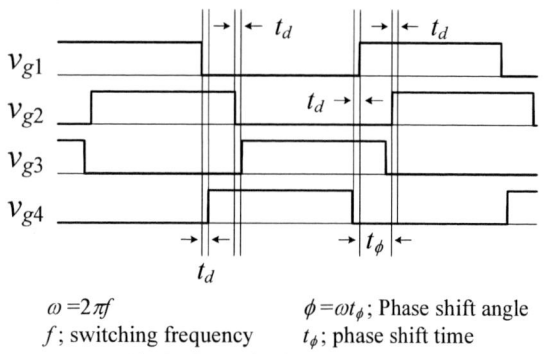

$\omega = 2\pi f$ $\phi = \omega t_\phi$; Phase shift angle

f; switching frequency t_ϕ; phase shift time

Fig. 2. Gate pulse timing sequences.

welding. These DC-DC converter topologies could basically achieve their high efficiency, high performance and high power density [1]-[4]. These prototypes of soft-switching symmetrical PWM DC-DC high power converter using the latest IGBT power modules has been partially put into practical use for low voltage large current application in industry. However, the previously-proposed DC-DC converter operates even at hard switching in the power setting ranges of low voltage low current DC outputs. In this case of light load conditions, the DC-DC converter topologies should utilize the allowable switching power semiconductor devices with higher rated voltage because the maximum peak voltage stress applied in some power devices causes through hard switching operation. Though, the power losses of the switches due to hard switching operation are relatively lower than those of snubberless circuit, but considering electromagnetic interference is needed for the DC-DC converter. In order to improve some issues such as high surge voltage and large condition losses due to the high voltage rating switches, this paper deal with three-level phase shifted ZVS-PWM DC-DC converter with neutral point clamping diodes and flying capacitor, which is suitable for wide varying voltage/current applications . In this paper, its operating principle and DC output characteristics are evaluated and discussed on the basis of experimental results.

II. PROPOSED SOFT-SWITCING DC-DC CONVERTER

A. Circuit Description

Figure 1 shows a three-level phase-shift ZVS-PWM DC-DC high power converter with a high frequency transformer. This DC-DC converter topology consists of DC power source E delivered by three phase PFC converter with three phase diode rectifier, the smoothing capacitors C_1, C_2 for dividing input DC voltage, IGBTs Q_1-Q_4, lossless snubbing capacitors C_{s1}-C_{s4} in parallel with the switches in the outer bridge arms and inner bridge arms, flying capacitor C_{ss} and neutral point voltage clamping diodes D_{c1}, D_{c2} in the primary side high frequency transformer with its turn ratio N_1, N_2, N_3 ($N_2 = N_3$), center-tapped high frequency rectifier represented by the lumped L type equivalent circuit due to leakage inductance L_k and magnetizing inductance L_m, rectifier fast recovery diode rectifier D_{o1}, D_{o2} in the secondary side of high frequency transformer, output inductor filter L_o; in some cases, the output inductor L_o and Capacitor C_o filters, and load resistor R_o.

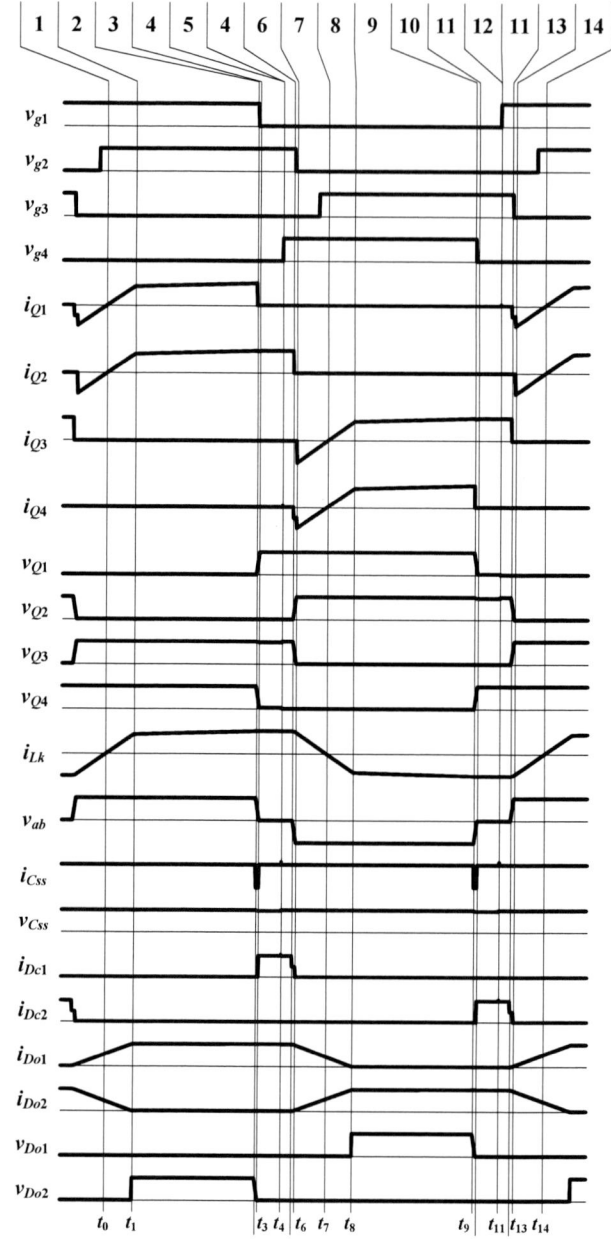

Fig. 3. Relevant operating voltage and current waveforms.

B. Gate Pulse Timing Signals

The DC output voltage of proposed DC-DC converter shown in Fig. 1 can be continuously regulated through phase shift PWM scheme. Figure 2 illustrates the gate pulse sequences of phase shift PWM. The phase shift PWM IC used to regulate the DC output power or DC voltage/current by retarding the pulse timing of control phase inner arm switches (Q_2, Q_3) with respect to reference phase outer arm switches (Q_1, Q_4). The outer switches, high side arm switch Q_1 and low side arm switch Q_4 as the reference phase, and the inner arm switches high side arm switch Q_2 and low side arm switch Q_3

as the control phase are used to determine the gate pulse timing patterns for the proposed DC-DC converter circuit.

C. Principle of Circuit Operation

The newly proposed DC-DC converter treated here can operate with 16 state modes during one switching cycle. The circuit operating voltage and current waveforms, and the 16 switching mode transitions as well as the corresponding switching mode equivalent circuits are depicted in Fig. 3 and Fig. 4, respectively. The 8 operation modes in the first half cycle are described below because the second half cycle 8 operation modes (Mode 9-16) become the similar circuit

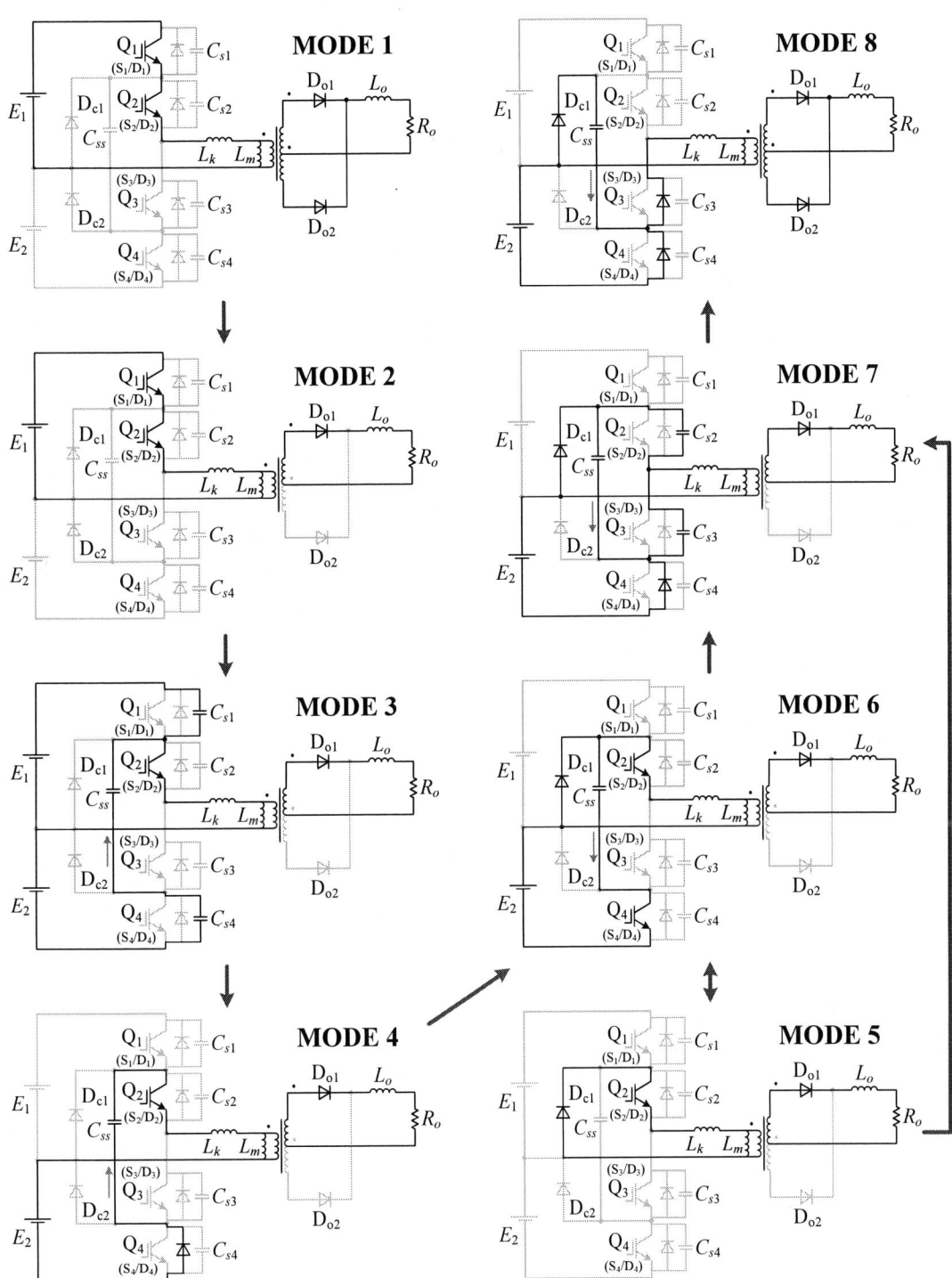

Fig. 4. Switching mode transitions and switching mode equivalent circuits during the first half cycle period.

operation as those in the 8 operation modes during the first half cycle.

[MODE 1 (t_0-t_1)] MODE 1 starts from turning on the switches S_1 of Q_1 and S_2 of Q_2 with ZVS and ZCS at the synchronous same pulse timing. The secondary side of high frequency transformer operates freewheeling mode because the rectifier diodes D_{o1} and D_{o2} are conducting under the overlapping current modes or freewheeling mode. This mode ends when D_{o2} is turned off naturally or the diode current reaches the load current.

[MODE 2(t_1-t_2)] In MODE 2, the high side switches S_1, S_2 and diode D_{o1} conduct as soon as the diode D_{o2} is turned off. MODE 2 ends when the switch S_1 is turned off.

[MODE 3(t_2-t_3)] In MODE 3, the switch S_1 is turned off with ZVS and S_2 and D_{o1} still conduct continuously. The lossless snubbing capacitor C_{s1} is charged and C_{s4} is discharged simultaneously through the flying capacitor C_{ss} in this mode. This mode ends when the charging of C_{s1} and the discharging of C_{s2} complete.

[MODE 4 (t_3-t_4)] The switch S_2 and the rectifier diode D_{o1} keep conducting and diode the D_4 of Q_4 is turned on in MODE 4. The flying capacitor C_{ss} is still discharged continuously. MODE 4 ends when the current through C_{ss} reaches zero.

[MODE 5 (t_4-t_5)] In MODE 5, the low side diode D_4 is turned off and the switch S_2, the diodes D_{c1} and D_{o1} keep conducting. In this mode, the circulating current flows through D_{c1} and S_2 because the inverter output voltage in the primary side of high frequency transformer becomes nearly zero due to the existing magnetizing inductance L_m. This mode ends when the switch S_4 of Q_4 is turned on by the external gating pulse.

[MODE 6 (t_5-t_6)] If the gate signal v_{g4} is delivered to the low side switch S_4 during MODE 5, the circuit operating mode moves to MODE 6. In this mode, although the switch S_4 begins to be turned on, and the switch S_2 and the diodes D_{c1} and D_{o1} keep conducting. This mode ends when C_{ss} is charged up and the circuit mode returns to MODE 5 in a certain case.

[MODE 7 (t_6-t_7)] In MODE 7, the switch S_2 is turned off with ZVS by the gate timing pulse rejection, the antiparallel diode D_4 is turned on and the both diodes D_{c1}, D_{o1} conduct. Lossless snubbing capacitor C_{s2} is charged and C_{s3} in parallel with Q_3 is discharged simultaneously in this mode. This mode ends when charging state of C_{s2} and discharging state of C_{s3} has just completed.

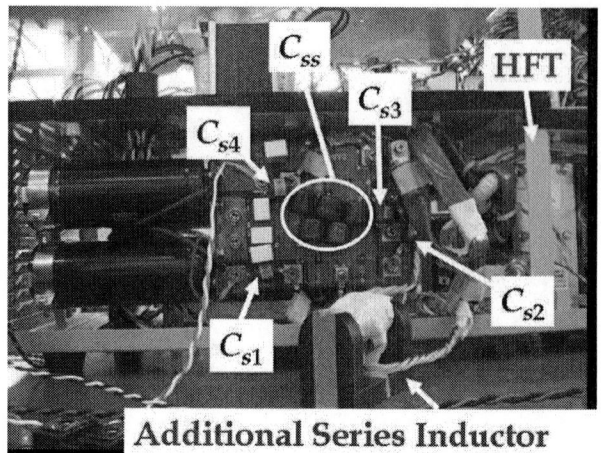

(a)   Prototype of DC-DC converter

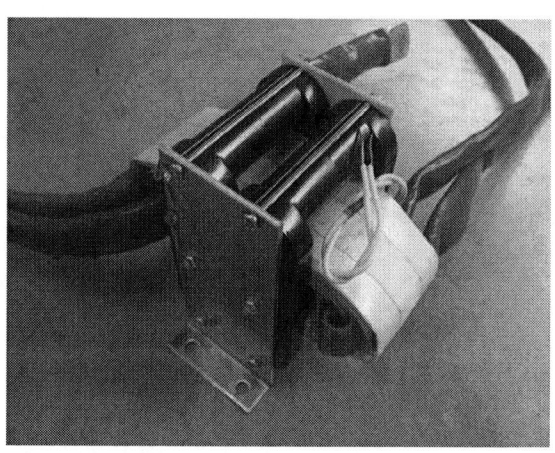

(b)   High frequency transformer

Fig. 5. Experimental setup.

[MODE 8 (t_7-t_8)] The inner low side switch D_3 of Q_3 is turned on, the diodes D_4, D_{c1}, D_{o1}, D_{o2} conduct in MODE 8. The secondary side of high frequency transformer puts into freewheeling mode. This mode ends when the currents through D_3 and D_4 reach zero. In this time, two low side switches S_3 and S_4 are turned on naturally with ZVS&ZCS for delivered the pulse v_{g3} to the gate terminals of Q_3, Q_4.

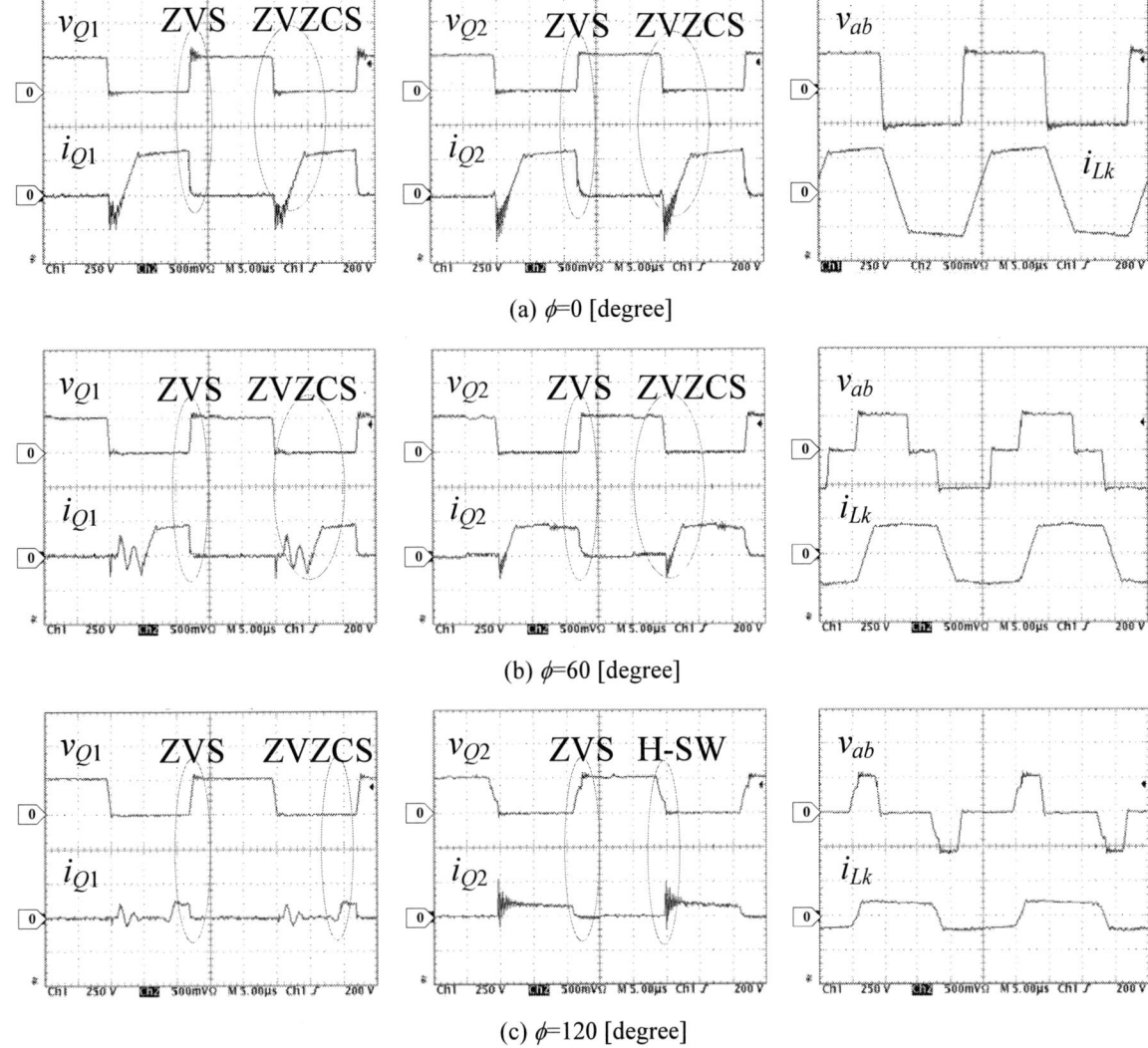

(a) ϕ=0 [degree]

(b) ϕ=60 [degree]

(c) ϕ=120 [degree]

Fig. 6. Observed voltage and current waveforms. (Voltage: 250V/div., Current: 50A/div., Time, 5μsec/div.)

[MODE 9-16] These 8 operating modes described above reveal the similar operation as those of MODE 1 to MODE 8.

This three-level soft-switching PWM DC-DC converter can operate with these 16 operating modes repeated periodically.

III. EXPERIMENTAL RESULTS AND DISCUSSIONS

Table I lists the circuit parameters and the design specifications in experiment for the proposed three-level DC-DC high power converter with a high frequency transformer and center tapped rectifier. In this case, the phase shift PWM control IC conveniently uses the controller IC; UCC3895 (Texas Instruments). The circuit load in experiment makes use of the electronic power load Takasago RL-6000L (6kW unit, 4 parallel).

Figure 5 shows the photo of experimental setup. IGBT modules are mounted on heat sink behind of PCB.

A. Measured Operating Waveforms

In Fig. 6, the measured voltage and current waveforms under the conditions of phase shift angle ϕ=0°, ϕ=60° and ϕ=120°are respectively depicted for design specifications in Table I. As can be seen from Fig. 5(a)(b), all the active switches in the outer and inner bridge arms can achieve soft switching transition commutation. In addition to this, the peak voltage across Q_1 and Q_2 are clamped to the half of the input DC voltage 270V. In Fig. 5(c), the inner high side and low side switches Q_2 and Q_3 operate hard switching turn on. However, the peak voltage across Q_2 and Q_3 are clamped to the half of the input DC voltage, 270V. In this case, the outer high side and low side switches Q_1, Q_4 can perform soft switching transitions. Therefore, this three-level DC-DC converter can substantially reduce the semiconductor switch conduction losses by using the lower voltage rated IGBT modules due to the voltage clamping effectiveness.

TABLE I
DESIGN SPECIFICATIONS AND CIRCUIT PARAMETERS

Item	Symbol	Value
Input Voltage	v_{AC}	$3\phi\,400\text{Vrms}$
Switching frequency	f	40kHz
Ripple frequency	f_r	80kHz
Dead time of Q_1&Q_4 and Q_2&Q_3	t_d	1.4μsec
Input side capacitor (Electrolytic DC capacitors)	C_1, C_2	2.2mF
Lossless snubbing capacitors	C_{s1}-C_{s4}	22.0nF
Flying capacitor	C_{ss}	500nF
Leakage inductance of HF transformer	L_k	2.0μH
Magnetizing inductance of HF transformer	L_m	200μH
Additional Series resonant inductor	L_r	6.0μH
Turn ratio of HF transformer	$N_1 : N_2 : N_3$	5:1:1
Output filter inductance	L_o	60μH
Output resistance	R_o	100mΩ
IGBT Q_1, Q_4	CM150DY-12NF V_{ces}=1200V, I_c=150A(25℃)	
IGBT Q_2, Q_3	SKM150GB128D V_{ces}=600V, I_c=150A(25℃)	
Diode D_{c1}, D_{c2}	FRG25CA120 V_{ces}=1200V, I_c=25A(78℃)	
Diode D_{o1}, D_{o2}	DSE12x101-06A V_{ces}=1200V, I_c=96A*5(70℃) (5 in parallel)	

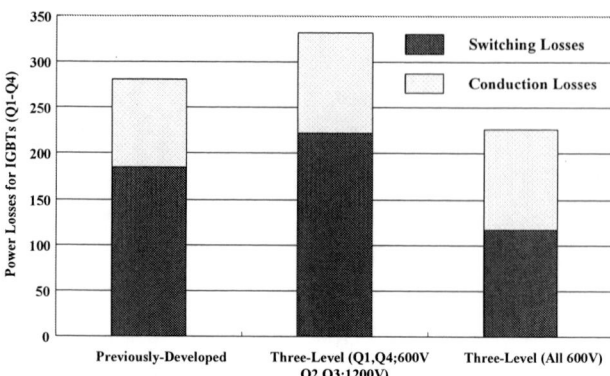

Fig. 8. Power loss analysis by using IGBT rated 600V (Q_1, Q_4) and 1200V (Q_2, Q_3).

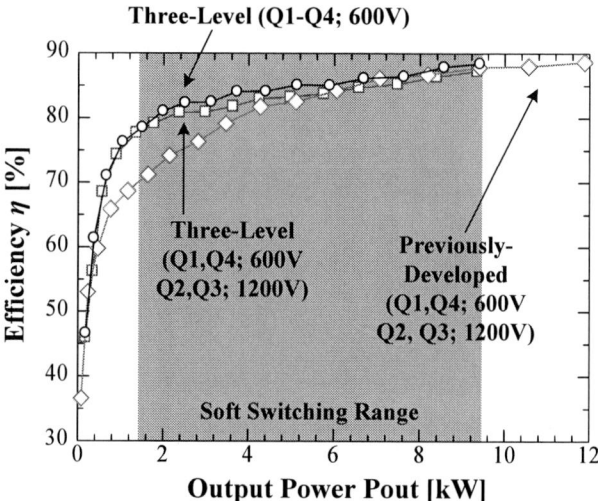

Fig. 9. Comparisons of power conversion efficiency (actual efficiency) between previously-and newly-developed converters.

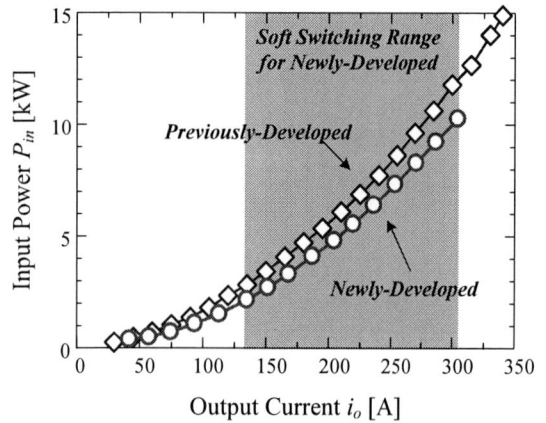

Fig. 7. Output power regulation vs. output current characteristics in open loop.

B. Power Regulation Characteristics

Figure 7 plots the input DC power regulation characteristics in open-loop control scheme compared with that of the previously-developed symmetrical ZVS-PWM DC-DC converter with active edge resonant cells (AERCs) and that of the proposed three-level phase-shift ZVS-PWM DC-DC converter. As can be seen, the newly-proposed DC-DC high power converter can respectively regulate input power by phase-shift PWM control strategy from 10kW to 0.2kW, 10kW to 2kW under soft switching operation. On the other hand, the rated input DC power is 15kW in the previously developed DC-DC converter because proposed DC-DC converter is inserted series resonant inductor L_r in the primary side of high frequency transformer for soft switching operation for the newly-proposed DC-DC converter the output

power more than 10kW can be outputted by adjusting the input DC voltage.

C. Power Loss Calcurations

Figure 8 depicts the quantitative results of power loss analysis for IGBTs (Q_1-Q_4) by using 600V (Q_1, Q_4) and 1200V (Q_2, Q_3) voltage rated IGBTs in the three-level DC-DC converter. The power losses for three-level (Q_1, Q_4; 600V, Q_2, Q_3; 1200V) is larger than previously-developed one, however, the proposed converter replaced IGBTs (Q_1-Q_4; 600V) is lower than previously developed one.

D. Power Conversion Efficiency

Figure 9 illustrates actual power conversion efficiency characteristics compared with those of previously-developed DC-DC converter and selected power semiconductor devices in high side and low side arm inner switches Q_2 and Q_3 from

1200 rated voltage in the replacement of 600V under the open loop control scheme. Observing this figure, the measured actual power conversion efficiency of the proposed DC-DC converter (see Fig. 1) can achieve 87% in the case of $\phi=0°$ and over 80% under soft switching operation. In addition, the power conversion efficiency of three-level DC-DC converter could achieve higher than that of the previously-developed DC-DC converter in the low current output setting ranges. Furthermore, the proposed DC-DC converter using two types of IGBTs; 1200V, 600V rated voltage devices for Q2 and Q3 can achieve the actual efficiency 89% or more for the rated power settings. The actual efficiency of the proposed DC-DC converter 600V IGBT power devices in Q2 and Q3 can be effectively higher than the previously developed DC-DC converter.

IV. CONCLUSIONS

This paper presented a three-level phase shifted ZVS-PWM DC-DC converter with neutral point clamping diodes and flying capacitor hybrid configuration for a variety of gas metal arc welding, which includes lossless snubbing capacitor in parallel with each reverse conducting power switches. Its operating principle and output steady-state characteristics were discussed and evaluated in simulation results and feasibility study. From these experimental results, the DC output power/voltage regulation characteristics of the proposed three-level soft switching DC-DC converter with a high frequency transformer and center-tapped rectifier were illustrated under a condition of soft switching commutation. Furthermore, the maximum power conversion efficiency 88% could be performed by the experimental setup developed previously. The power losses for power devices were significant for the previously-proposed DC-DC converter. Furthermore, the actual power conversion efficiency for proposed DC-DC converter using the 600V-IGBTs for inner high side arm and low side arm switches Q_2 and Q_3 could achieve 90% higher or more than all the power setting ranges compared with the previously developed DC-DC converter. As for the other results, the voltage surges and stresses applied to IGBTs were effectively lowered even under hard switching operating ranges. It is noted that the newly-proposed three-level soft switching DC-DC converter could be practically cost effective and the conduction loss reduction and switching losses by using lower rated voltage power switching devices and power modules.

In the future, the following items for practical usage in industry should be considered; (i) detail power loss analysis for each power devices and DC filter component, (ii) SiC-SBD/GaN-SBD could be introduced as the neutral clamping diodes, (iii) the minimizations of phase shift related circulating current in the three-level PWM DC-DC converter should be evaluated by introducing (a) passive lossless voltage clamped topology, (b) tapped inductor DC filter with freewheeling diode, and (c) power recovery circuit due to tapped inductor in primary side, (iv) efficiency practical discussions on current doubler rectifier as well as voltage doubler rectifier, (v)

improvement based on the output DC filter inductor such as amorphous core and powder metal core.

REFERENCES

[1] T.Etoh, T.Doi, K.Morimoto, T.Ahmed, H.Sugimura, B.Saha, M. Nakaoka, "Practical Development of Half-Bridge High Frequency PWM Inverter-Linked DC-DC Converter with Diode-Clamped Active Edge Resonant Cells and Extended Topologies," Proc. of IEEE International Telecommunications Energy Conference (INTELEC '08), pp.585-592, September, 2008

[2] T.Etoh, T. Doi, K. Morimoto, T.Mishima and M.Nakaoka, "A Novel Input DC Rails Side Diode Clamped-Active Edge Resonant Cells PWM DC-DC Converters with High Frequency Planar Transformer Link," Proc. of Japan Industry Applications Society Conference 2008, Kochi, pp.I-335-340

[3] T.Doi, T.Etoh, K.Morimoto, H.Manabe, K. Yamaguchi, H.Sugimura and M.Nakaoka, "DC Rails Side Diode Clamped-Active Edge Resonant PWM DC-DC Converters with HF Link," Proc. of 39th IEEE Power Electronics Specialists Conference (PESC), pp.4149-4145, Island of Rhodes, Greece, June, 2008

[4] T.Doi, K.Morimoto, K.Yamaguchi, and T.Etoh, "New DC Rail Side Diode-Clamping AERC-Assisted Soft- Switching PWM DC-DC Converter with High-Frequency Link for High Performance Arc Welding," Proc. of 4th IET International Conference on Power Electronics, Machines and Drives (PEMD 2008), pp.737-741, York, UK, 2-4 April 2008

[5] H.Sugimura, K.Fathy, S.P.Mun, T.Doi, T,Mishima, M. Nakaoka, "Three-Level Phase Shifted Soft Transition PWM DC-DC Power Converter with High Frequency Link for Arc Welders and Its Extended Version," Proc. of International Power Electronics and Motion Control Conference (IPEMC '09), pp.1288-1294, May 2009.

[6] Z.L.Lou and Z.S.Wang, "An Improved Three-Level Soft-Switching DC/DC Converter," Proc. of The 5th International Power Electronics & Motion Control Conference (IPEMC), pp.373-377, Shanghai, China, August, 2006.

[7] P.M.Barbosa, F.Canales, J.M.Burdío and F.C.Lee, "A Three-Level Converter and Its Application to Power Factor Correction," IEEE-Trans. on Power Electronics, vol.20, no.6, pp.1319-1327, November 2005.

[8] F.Canales, P.Barbosa and F.C.Lee, "A Zero-Voltage and Zero-Current Switching Three-Level DC/DC Converter," IEEE-Trans. on Power Electronics, Vol.17, No.6, pp.898-904, November 2002.

[9] X.Ruan, B.Li, Q.Chen, "Three-Level Converters—A New Approach for High Voltage and High Power DC-to-DC Conversion," Proc. of Power Electronics Specialists Conference (PESC '02), Vol.2, pp.663-668, June 2002.

[10] S.J.Jeon, F.Canales, P.M.Barbosa, F.C.Lee, "A Primary-Side-Assisted Zero-Voltage and Zero-Current Switching Three-Level DC-DC Converter with Phase-Shift Control," Proc. of The Applied Power Electronics Conference and Exposition (APEC '02), Vol.2, pp.641-647, 2002.

[11] X.Ruan, L.Zhou, Y.Yan, "Soft-Switching PWM Three-Level Converters," IEEE-Trans. on Power Electronics, Vol.16, No.5, pp.612-622, September 2001.

[12] N.Fröhleke, M.Schnidermann, "Enhanced Analysis and Design Issues of a 3-Level DC/DC Converter with Zero Voltage and Zero Current Switching," Proc. of 9th European Conference on Power Electronics and Applications, pp.1-10, Leoben, Austria, 2001.

[13] F.Canales, P.M.Barbosa, F.C.Lee, "A Zero Voltage and Zero Current Switching Three Level DC/DC Converter," Proc. of Applied Power Electronics Conference and Exposition (APEC '00), Vol.1, pp.314-320, 2000.

[14] J.R.Pinheiro, I.Barbi, "Three-Level Zero-Voltage-Switching PWM DC-DC Converters – A Comparison," Pro. of Power Electronics Specialists Conference (PESC '95), Vol.2, pp.914-919, 1995

[15] J.R.Pinheiro, I.Barbi, "The Three-Level ZVS-PWM DC-to-DC Converter," IEEE-Trans. on Power Electronics, Vol.8, No.4, October 1993.

[16] J.Renes Pinheiro, Ivo Barbi, "Wide Load Range Three-Level ZVS-PWM DC-to-DC Converter," Proc. of Power Electronics Specialists Conference (PESC '93), pp.171-177, 1993.

[17] A.Jangwanitlert, J.Songbonkaew, W.Thammasirioj, J.C.Balda, "Analysis of Three-Level ZVS PWM Inverter for Induction Heating Applications," Proceedings of 5th International Power Electronics and motion Control Conference (IPEMC) 2006, Vol.2, pp.1-5

Fully Soft-switched Bidirectional Resonant DC-DC Converter with A New CLLC Tank

Wei Chen[1,2], Siran Wang[1], Xiaoyuan Hong[1], Zhengyu Lu[1] (Senior member, IEEE)   Shaoshi Ye[2]

[1]College of Electrical Engineering
Zhejiang University
Hangzhou, Zhejiang, China

[2]Delta Electronics (Shanghai) Co., LTD.
No. 238 Minxia Road, Pudong
Shanghai, 201209, China

Abstract-A bidirectional dc-dc converter with a new CLLC type resonant tank, which features in ZVS for the input inverting port switches and ZCS for the output rectifier port switches, regardless of the direction of the power flow, is proposed. If the MOSFETs are implemented as all of the main switches, the proposed converter has a minimized switching loss. The detail operation principles, as well as the design considerations, are presented. A prototype, which interfacing the 400V/48V DC buses for the UPS system with a power rating of 500VA, whose highest conversion efficiencies for the bidirectional mode both exceed 96%, was developed to verify the validity and applicability of this proposed converter.

Keywords-Bidirectional dc-dc converter, ZVS, ZCS, Topology combination

I. INTRODUCTION

Nowadays, plenty of soft switching Bidirectional Dc-dc Converters (BDC) with focus on eliminating the switching loss, reducing the electromagnetic interference (EMI), and achieving an attainable high frequency operational ability and thereby power density without sacrificing the efficiency, have been praised and extensively reported in the open literature [1]-[7]. As implemented with the power MOSFETs, the zero-voltage-switching (ZVS) technique is desirable when the MOSFETs are employed as the input port inverting device, since the turn-on loss due to the energy discharging of the output capacitance is large enough. However, if the MOSFETs used as the output port rectifier with SR, the zero-current-switching (ZCS) technique is very suitable to mitigate the reverse-recovery problem exhibiting in the intrinsic diode of the MOSFET. Consequently, a perfect soft switching BDC should have the features as ZVS for the input port inverting switches and ZCS for the output port rectifier switches, in either power flow direction, to substantially eliminate the switching loss as low as possible. [1] introduces a BDC operating in the phase-shifted manner, which can realize ZVS for the voltage-fed side switches. However, the switches on the current-fed side remain operating under hard switching conditions. In [2]-[4], several BDCs are presented, where the ZVS attribute is enabled for the switches operate in the input inverting stage, according to the power flow direction. However, on the corresponding output port, the rectifier switches snap off and high voltage surge is observed, where the reverse-recovery loss remains unsolved and the RCDi snubber is required [2]. To alleviate the reverse-recovery issue, several ZCS BDCs are presented [5]-[7]. In these ZCS BDCs, both of the input and output port

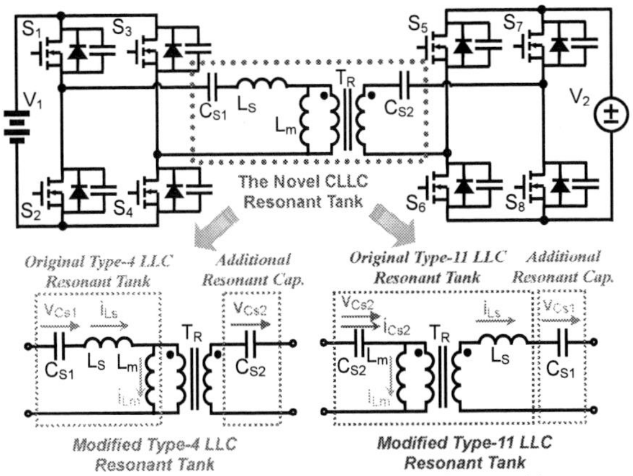

Figure. 1 The proposed BDC and its topology decomposition

switches realize ZCS regardless of the direction of the power flow, which have a significant reduction of the reverse-recovery power loss for rectifier stage. Unfortunately, the turn-on loss for the inverting switches of the input port is considerable and unsolved.

This paper presents a novel bidirectional resonant dc-dc converter, which possesses the aforementioned soft switching advantages as ZVS for the inverting switches and ZCS for the rectifier switches for both of the power flow directions. Thus, since the turn-on loss and reverse-recovery loss for the applied MOSFET are both eliminated, the proposed converter is totally snubberless. Following the introduction, the converter operations and design considerations are described respectively. The experimental results are presented from a 500VA prototype to confirm validity and applicability of the proposed converter.

II. TOPOLOGY DERIVATION AND OPERATION PRINCIPLES

The circuit configuration of the proposed converter is presented in Fig. 1, which can be recognized as a combinatorial resonant tank merging the Type-4 and Type-11 LLC resonant converter presented in [8], [9] together. With the ZVS+ZCS feature for the two resonant sub-tanks, the proposed CLLC resonant tank achieves the desired soft switching feature of ZVS+ZCS.

Both of the directions of the power flow are modulated under the variable Frequency Modulation (FM) above resonance. The driving signals for the inverting stage switches and the main principle waveforms are shown in Fig. 2. In either mode, there are 8 operation stages during a switching period. The equivalent circuits in the former 4 operation

Project supported by National Natural Science Foundation of China (50677063)

978-1-4244-4782-4/10 $26.00 © 2010 IEEE

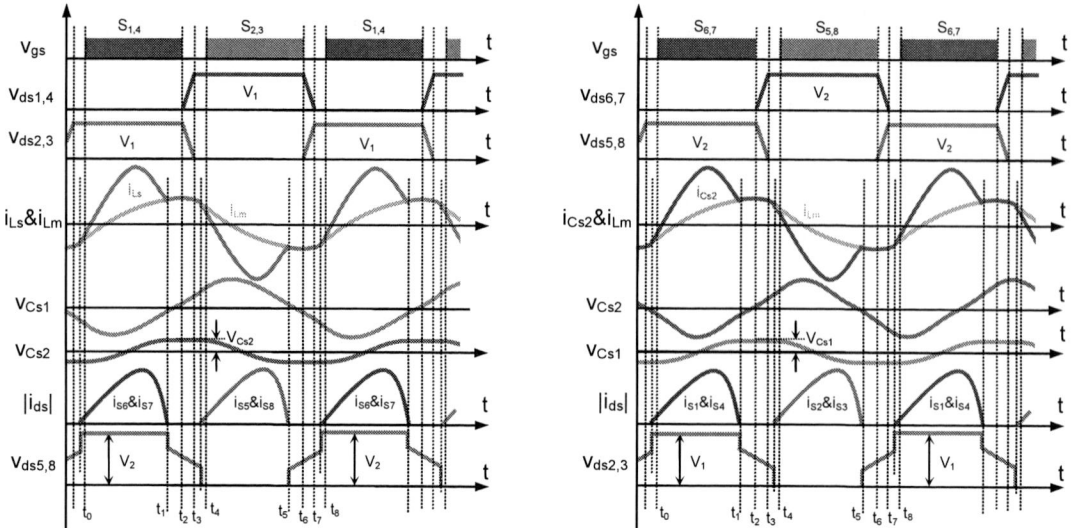

(a) Operation principles of converter with the *Modified Type-4 LLC* tank (b) Operation principles of converter with the *Modified Type-11 LLC* tank

Figure. 2 Operation principles of the proposed CLLC resonant tank

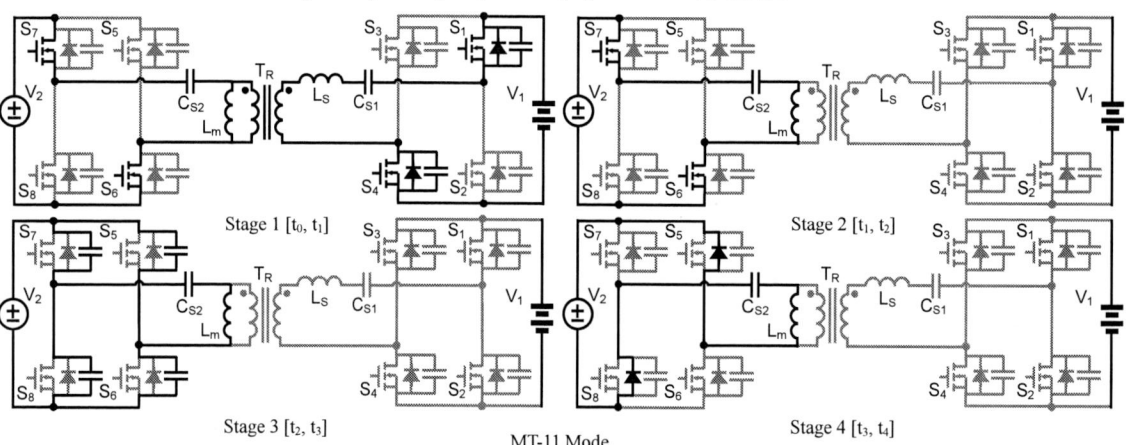

MT-11 Mode

Figure. 3 Equivalent circuits for each stage under the both directions of the power flow

stages in a half switching cycle for MT-11 mode are given by Fig. 3. Other 4 stages in each mode are symmetrical to those of the previous half cycle. The detailed operating processes of the MT-11 Mode are presented here in the following text:

Stage 1 [t₀, t₁]: At the time of t_0, S_6 and S_7 are conducting. The resonant current i_{Cs2} will increase in a sine-wave shape. On the secondary side, the rectified current i_{S1} and i_{S4}, are proportional to the difference between i_{Cs2} and i_{Lm}.

Stage 2 [t₁, t₂]: i_{Cs2} equals to the increasing magnetic current i_{Lm} at time t_1. S_1&S_4 turn off under ZCS conditions. The voltage on the resonant capacitor C_{S1} also resonates to its peak with the absolute value of V_{Cs1} and will keep unchanged, unless excites again in the next resonance.

Stage 3 [t₂, t₃]: At time t_2, S_6 and S_7 are turned off. i_{Cs2} begins to charge the parasitic capacitors of S_6&S_7 and discharge those of S_5&S_8 symmetrically. This stage ends up with $v_{ds6,7}$ reaching input voltage and $v_{ds5,8}$ decreasing to zero.

Stage 4 [t₃, t₄]: After the fully discharge of S_5 and S_8, the resonant current of i_{Cs2} immediately flows through the body diodes of S_5 and S_8 and feeds back to the input source. At time

t_4, S_5 and S_8 turn on under ZVS conditions, the first-half switching period ends and the converter enters the next half.

From the operation principles, it can be comprehended that during the MT-11 Mode, the additional resonant capacitor C_{S1} appears and disappears with the load at the same time. As a consequence, the insertion of the C_{S1} to the original Type-11 resonant tank will only slightly modify the resonant frequency when all of the four resonant elements are involved, but not influence the fundamental of the original resonant tank.

III. DESIGN CONSIDERATIONS

A. Behavioral Difference for the MT-4 and MT-11 Tanks

The characteristics of the MT-4 and MT-11 modes can be investigated rapidly by the frequency-domain methodology of Fundamental Mode Approximation (FMA) [10]. The

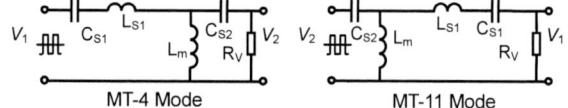

Figure. 4 Equivalent circuits for Forward and Reverse Modes

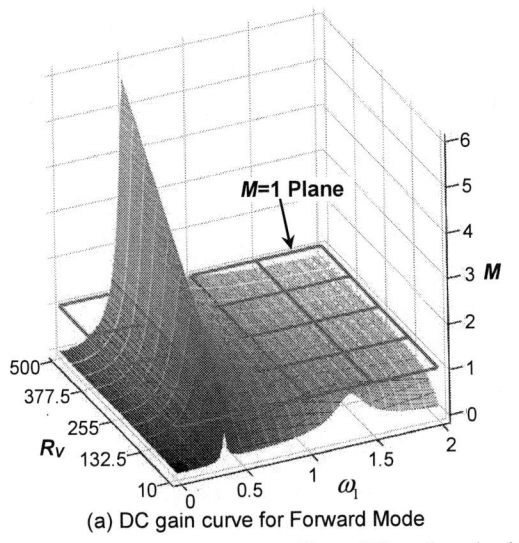

(a) DC gain curve for Forward Mode

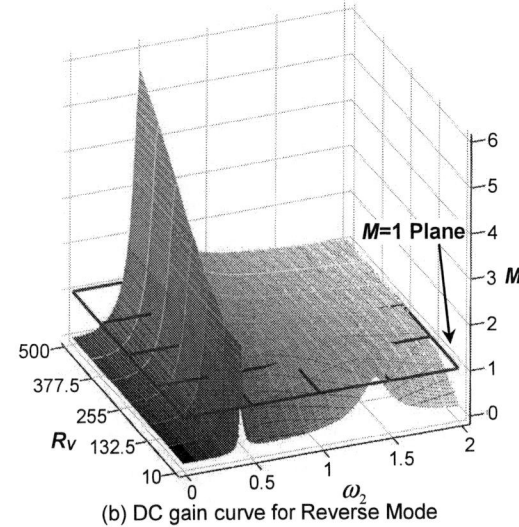

(b) DC gain curve for Reverse Mode

Figure. 5 Three-dimensional DC gain contour over switching frequency

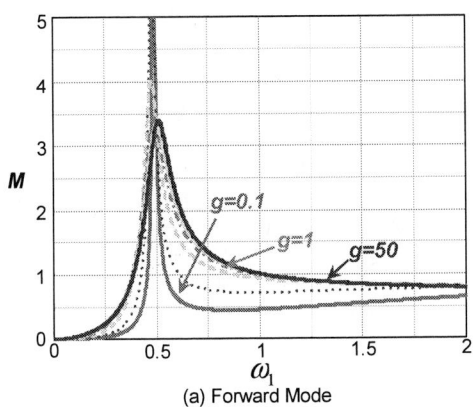

(a) Forward Mode

(b) Reverse Mode

Figure. 6 Parameter sweep against g under two modes

equivalent circuits for the MT-4 and MT-11 resonant tanks are shown by Fig. 4, respectively, where R_V is the equivalent load resistance. The equations of the DC gain for both of the MT-4 and MT-11 modes can be expressed as follows:

$$M_{MT4} = \left| \frac{(R_V + Z_{Cs2})//Z_{Lm}}{Z_{Cs1} + Z_{Ls1} + (R_V + Z_{Cs2})//Z_{Lm}} \cdot \frac{R_V}{R_V + Z_{Cs2}} \right| \quad (1)$$

$$M_{MT11} = \left| \frac{(R_V + Z_{Ls1} + Z_{Cs1})//Z_{Lm}}{(R_V + Z_{Ls1} + Z_{Cs1})//Z_{Lm} + Z_{Cs2}} \cdot \frac{R_V}{R_V + Z_{Ls1} + Z_{Cs1}} \right| \quad (2)$$

Define $\quad Q_1 = \dfrac{\sqrt{\dfrac{L_S}{C_{S1}}}}{R_V}$, $\quad Q_2 = \dfrac{\sqrt{\dfrac{L_S}{C_{S2}}}}{R_V}$, $\quad \omega_{T\_4} = \dfrac{1}{\sqrt{L_S \cdot C_{S1}}}$,

$\omega_1 = \dfrac{\omega_S}{\omega_{T\_4}}$, $\quad \omega_{T\_11} = \dfrac{1}{\sqrt{(L_S // L_m) \cdot C_{S2}}}$, $\quad \omega_2 = \dfrac{\omega_S}{\omega_{T\_11}} \cdot \sqrt{\dfrac{h}{1+h}}$,

$h = \dfrac{L_m}{L_S}$, $\quad g = \dfrac{C_{S2}}{C_{S1}}$, where ω_S is the switching frequency, then (1) and (2) can be simplified as:

$$M_{MT4} = \left| \frac{1}{a_1 - j \cdot b_1} \right| = \frac{1}{\sqrt{a_1{}^2 + b_1{}^2}} \quad (3)$$

where: $\quad a_1 = \dfrac{1}{h} + 1 - \dfrac{1}{h \cdot \omega_1^2}$,

$$b_1 = Q_1 \cdot \left(\frac{1}{\omega_1} - \omega_1 \right) + \frac{Q_1 \cdot (1+h)}{g \cdot h \cdot \omega_1} - \frac{Q_1}{g \cdot h \cdot \omega_1^3} .$$

and: $\quad M_{MT11} = \left| \dfrac{1}{a_2 - j \cdot b_2} \right| = \dfrac{1}{\sqrt{a_2{}^2 + b_2{}^2}} \quad (4)$

where: $\quad a_2 = 1 - \dfrac{1}{h \cdot \omega_2^2}$,

$$b_2 = Q_2 \cdot \left(\frac{1}{\omega_2} - \omega_2 \right) + \frac{Q_2 \cdot (1 + h \cdot g)}{h \cdot \omega_2} - \frac{g \cdot Q_2}{h \cdot \omega_2^2} .$$

According to (3) and (4), the normalized DC gain over switching frequency can be plotted in Fig. 5.

From Fig. 5, it can be observed the frequency of resonance for both two modes have not been the ω_{T_4} or ω_{T_11} any more, since the extra capacitor for each mode actually participates in the resonance and substantially changes the resonance frequency. Nevertheless, the new resonance frequency is no mean exact to be the frequency

978-1-4244-4782-4/10 $26.00 © 2010 IEEE

of $\dfrac{1}{\sqrt{L_S \cdot (C_{S1} // C_{S2})}}$. This is because the resonant inductor L_m still needs some portion of energy to excite during the stages that transfer energy to load. The precise symbolic solutions for the resonance frequency of the two modes can be solved by let the image part of (1) and (2) to be zero. However, for the complexity, the expressions are omitted here.

B. Design Procedure for the CLLC Resonant Tank

Due to the difficulties when it comes to solve all of the 4 elements at the same time, it becomes more convenient to design a basic Type-4 or Type-11 resonant tank first, then cope with the last remaining resonant capacitor lastly. It is more suitable to start with the design of a Type-4 LLC resonant tank firstly than with the Type-11 resonant tank. The design procedure can be outlined by the following 7 steps:

1). Choose a proper value for resonance frequency ω_{T_4} (or f_{T-4}).

$$f_{T-4} = \frac{1}{2\pi\sqrt{C_{S1} \cdot L_S}} \qquad (5)$$

2). Parameter of h should be specified as large as possible, e.g., $h=3$, to accommodate a wide input range adaptability.

3). The inductance of L_m can be determined by the optimal design methodology provided by [11]:

$$L_m = \frac{t_{dead}}{8 \cdot f_S \cdot C_{OSS}} \qquad (6)$$

where t_{dead} is the dead time between the two switches in the same bridge leg, f_S is the switching frequency and C_{OSS} is the equivalent output capacitance of the inverting switches.

According to (5) and (6), L_S, L_m and C_{S1} can be determined. If the attainable maximum DC gain under this set of parameters can not fulfill the requirement of the input range, go back to Step 2 to reset a new h.

5). Sweep parameter of g under the heaviest load of Q_1. Fig. 6 shows a set of typical DC gain curves for the both two modes. The appropriately choosing of parameter g highly depends on how similar the curves of DC gain of the two modes are. A fine g should make the two DC gain curves as the same as possible.

6). Having the parameter g been known, the second resonant capacitor C_{S2} can be calculated by $g = \dfrac{C_{S2}}{C_{S1}}$, which has been mentioned above. Check whether the DC gain can meet the variation of load and line change. If not, go back to Step 6 to re-choose a more proper g.

7). Determine the DC gain of MT-4 mode at the resonance frequency, assuming M_{MT4_R}, then the turn ratio n (MT-4 mode) of the transformer can be obtained directly by:

$$n = M_{MT4\_R} \cdot \frac{V_{in}}{V_O} \qquad (7)$$

With n was presented, the design procedure for the proposed CLLC resonant tank completes.

C. Topology Extension

Two Type-4 LLC resonant tanks can also be combined and create a new resonant tank with the ZVS+ZCS feature. However, the magnetic integration technique should be adopted, shown as Fig. 7. The leakage inductance between the primary and secondary windings functions as the resonant inductor for the corresponding power flow directions.

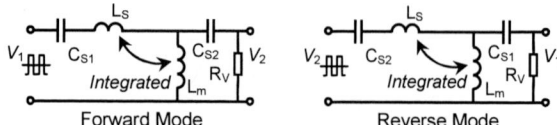

Figure. 7 The combinatorial ZVS+ZCS bidirectional resonant tank with two integrated Type-4 LLC resonant tanks

IV. EXPERIMENTAL RESULTS

A 500VA prototype verifies its operation principle of the proposed converter. The specification and main parameters are specified as follows:

Forward Mode:

V_{in}: 200-400 Vdc; V_O: 48 Vdc; I_O: 0-10 A.

Reverse Mode:

V_{in}: 24-48 Vdc; V_O: 400 Vdc; I_O: 0-1.2 A.

The components of the power stage shown in Fig. 1 are listed as follows:

High Voltage Fed Side Switches, $S_1 \sim S_4$: 2SK3934; *Low Voltage Fed Side Switches $S_5 \sim S_8$*: IPP070N08N3; T_R: n=35:5; L_m: 228 uH; L_S: 76 uH; C_{S1}: 33 nF/630V; C_{S2}: 3·470nF ($g \approx 0.9$)

Operation waveforms for the Forward Mode of full load under 200 V and 400 V input are shown in Fig. 8. Fig. 8 (a)-(b) show the ZVS feature of S_2. Fig. 8 (c)-(d) show the waveforms obtained from the MT-4 resonant tank. Fig. 8 (e)-(f) show the ZCS feature for the rectifier switches.

The operation waveforms for the Reverse Mode are also given by Fig. 8 (g)-(l). A simple way to appreciate these waveforms for the MT-11 resonant tank is just considering the power flows through the MT-4 resonant tank backward.

Fig. 9 shows the conversion efficiency for both Forward Mode and Reverse Mode under all conditions. The highest efficiencies for the two modes are both exceeding 96% under high line input. Fig. 10 shows the switching frequency versus load under different voltage input.

V. CONCLUSIONS

The new bidirectional dc-dc converter proposed in this paper, which has a combinatorial CLLC resonant tank, possesses the optimal soft-switching features as ZVS for the inverting stage switches and ZCS for the rectifier switches, regardless of the direction of the power flows. Thus the switching loss of the proposed converter has been minimized if the MOSFETs are implemented as the main switches. The experimental results verify the theoretical analysis and the merits of the proposed converter. Therefore, the proposed converter provides designers with an alternative choice for wide input range, high efficiency and high power density bidirectional dc-dc conversion applications in industry.

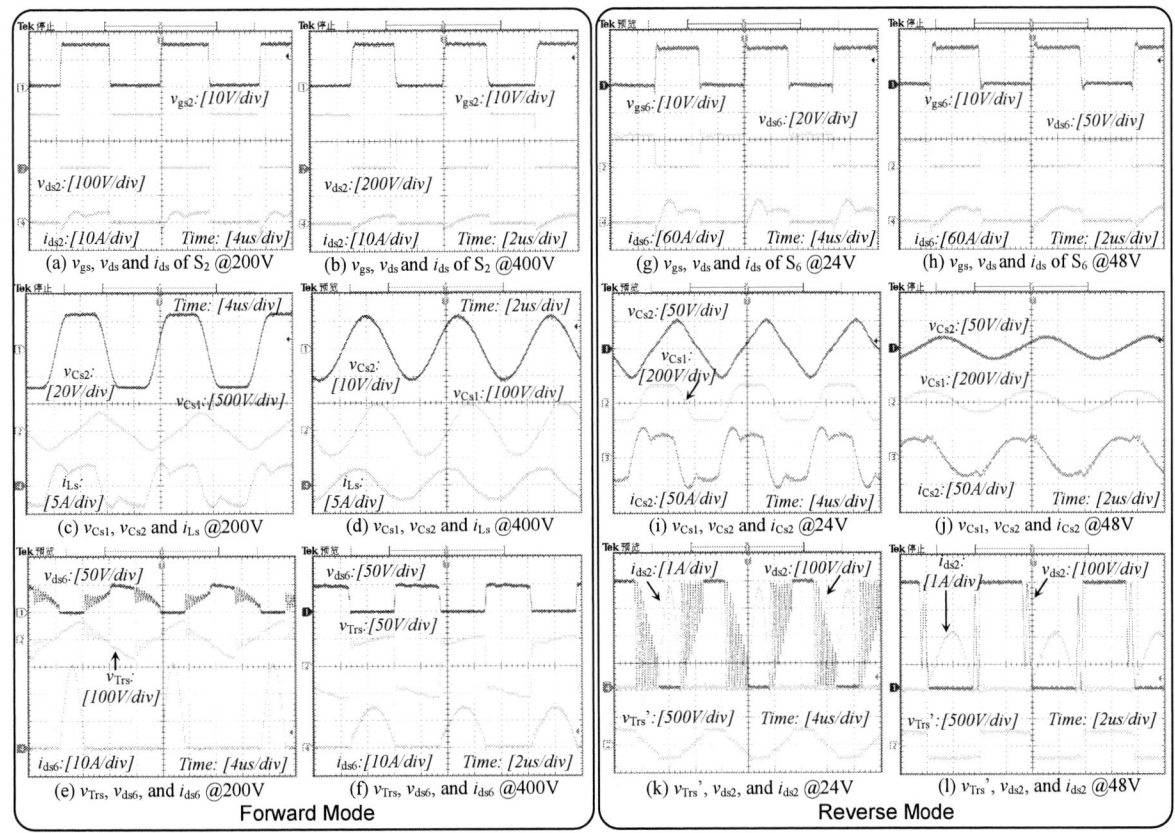

Forward Mode   Reverse Mode

Figure. 8 Experimental Waveforms

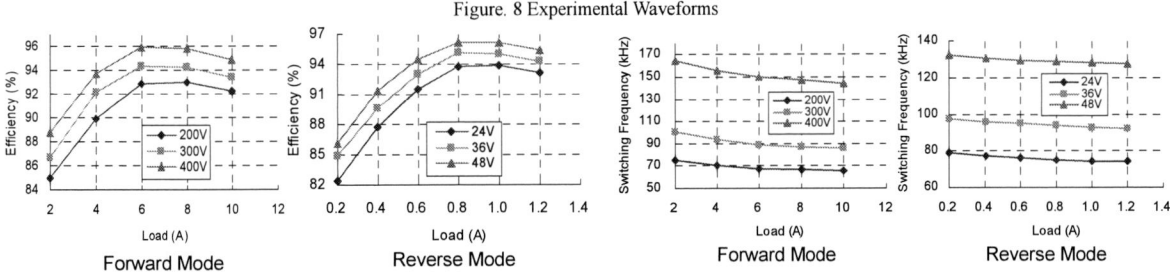

Forward Mode   Reverse Mode

Figure. 9 Measured efficiency at different loads

Forward Mode   Reverse Mode

Figure. 10 Switching frequency versus load

REFERENCE

[1] L. Zhu, "A Novel Soft-Commutating Isolated Boost Full-Bridge ZVS-PWM DC–DC Converter for Bidirectional High Power Applications," IEEE Trans.on, Power Electron., vol. 21, issue 2, pp. 422-429, March 2006.

[2] E. Hiraki, K. Yamamoto and T. Mishima, "An Isolated Bidirectional DC-DC Soft Switching Converter for Super Capacitor Based Energy Storage Systems," in Proc. IEEE PESC 2007, pp. 390-395, 17-21 June 2007.

[3] Ma G., Qu W., Yu G., et al, "A Zero-Voltage-Switching Bidirectional DC–DC Converter With State Analysis and Soft-Switching-Oriented Design Consideration," Industrial Electronics, IEEE Transactions on, vol. 56, Issue 6, pp. 2174-2184, June 2009.

[4] H. F. Xiao and S. J. Xie, "A ZVS Bidirectional DC–DC Converter With Phase-Shift Plus PWM Control Scheme," IEEE Trans. Power Electron., vol. 23, issue 2, pp. 813-823, Mar. 2008.

[5] Y. S. Lee and G. T. Cheng, "Quasi-Resonant Zero-Current-Switching Bidirectional Converter for Battery Equalization Applications," IEEE Trans. Power Electron., vol. 21, issue 5, pp. 1213-1224, Sept. 2006.

[6] H. S. -H. Chung, W. L. Cheung and K. S. Tang, "A ZCS bidirectional flyback DC/DC converter," IEEE Trans. Power Electron., vol. 19, issue 6, pp. 1426-1434, Nov. 2004.

[7] Y. S. Lee, Y. S. Lin, S. H. Hsiao and Y. P. Ko, "Multiple Output Zero-Current Switching Bi-directional Converter," in IEEE Proc. IECON 2007, pp. 2058-2063, Nov. 2007.

[8] Severns R.P., "Topologies for three-element resonant converters,." Power Electronics, IEEE Transactions on, vol. 7, Issue 1, pp. 89-98, Jan. 1992.

[9] Wei Chen, Zhengyu Lu, "Investigation on topology for Type-4 LLC resonant Dc-Dc converter," in IEEE PESC 2008, pp. 1421-1425, 15-19 June, 2008.

[10] Foster M.P., Sewell H.I., Bingham C.M., et al, "Methodologies for the design of LCC voltage-output resonant converters," Electric Power Applications, IEE Proceedings-, vol. 153, Issue 4, pp. 559-567, July 2006.

[11] Bing Lu, Wenduo Liu, Yan Liang, et al, "Optimal design methodology for LLC resonant converter," in IEEE APEC 2006, pp. 6-12, 19-23 March 2006.

Optimal Phase Changing Frequency Determination for Multiphase Voltage Regulator Modules

Anand Ramamurthy, Subhashish Bhattacharya
Department of Electrical & Computer Engineering.
North Carolina State University
Raleigh, NC, USA

Chris Thompson, Jon Day
Intersil Corporation
Milpitas, CA, USA

Abstract— **Modern processers have the capability of indicating the power state of the processor to the Voltage Regulator (VR) PWM controller so that it can change its operating state to maximize efficiency at light loads and to flatten out its efficiency curve for idle power reduction. The CPU worst case assert and de-assert frequency can be very high for the PWM controller and for the VR to follow. Thus for the VR to take advantage of the low power state signal from the CPU, the signal has to be passed through an analog/digital low-pass filter. The optimum frequency for this filter design is determined in this paper. This filtered frequency with which the VR drops its phases optimizes the overall efficiency of the system. The experimental results are given for a four-phase VRM. It is also shown in this paper that the transient efficiency is as vital as the steady state efficiency considering the load profile of modern CPUs.**

I. INTRODUCTION

Multiphase systems have advantages of improved thermal distribution between phases and with increased number of phases the output voltage ripple reduces. The efficiency improvement in multiphase converter by changing dynamically the number of phases is already established [1]. The efficiency of the interleaved buck converter is typically low at light load conditions because of the switching losses. Improvements in the converter efficiency can be achieved by dynamically changing the number of phases during light loads. The current sharing control between the phases is important to maintain the appropriate amount of power handled by each phase in order to maintain the expected design performance and thermal distribution between phases [2]. Although [2] discusses the effect on efficiency due to active current sharing between phases, it focuses only during steady state conditions.

The efficiency of DC/DC converter is defined as total output power divided by total input power under the specified working conditions. The widely used efficiency curve considers the load current on one axis and the efficiency in the other. However this curve does not consider the load changing conditions. The load profile of modern processors suggests that the efficiency for a dynamically changing load must also be considered.

Intel's VR 11.1 design guide outlines the worst case step changes that the VR must handle and the repetition rate of the load changes. If studied closely the impact on efficiency for different repetitive rate of the load changes is also one of the major criteria. In this paper, the effect of current sharing during transient loads and steady state loads and its impact on efficiency is discussed. The optimum phase changing frequency is determined by studying in detail the effects of the transients on current sharing and its impact on efficiency.

The current sharing loop bandwidth effect on the dynamic performance of transients is discussed in [3], but the effect of control bandwidth and its impact on transient efficiency is not been investigated. A predictive current sharing method is proposed in [4] and in most multiphase VR controllers commercially available use this current sharing scheme. The performance of the current sharing schemes during phase changing and the optimal frequency with which the phase changing can be done is studied in this paper.

II. POWER STATE INDICATOR

The scheme of connecting and disconnecting the phases are a function of the load. The Intel VR11.1 platform specifies that the CPU also indicates to the VR controller when it is at a low power state (this saves a current sensing scheme for the load). The PSI (power state indicator) is an input logic signal sent by the CPU which indicates that the CPU is in a low power state. The load at which the PSI asserts (see Fig.1) is set to 20A and below. It is shown in Fig.2 that with a load below 20A the losses of a four phase system exceed that of a two phase system. Thus dropping two phases below 20A helps the overall efficiency. The PSI signal sent by the processor is used by the VR PWM controller to change its operating state to maximize energy efficiency at light loads or flatten out its efficiency curve for idle power reduction. Per Intel Xeon processor CPU specification, PSI minimum dwell time is 7.5 ns and the CPU's worst case, maximum assert/ de-assert frequency is 133 MHz [9]. The CPU is capable of this very short minimum assertion time, which is too fast to be useful to the

---

Work supported by Intersil Corporation.

978-1-4244-4782-4/10 $26.00 © 2010 IEEE

VR, so the VR's PSI input may need to have installed a low-pass filter circuit.

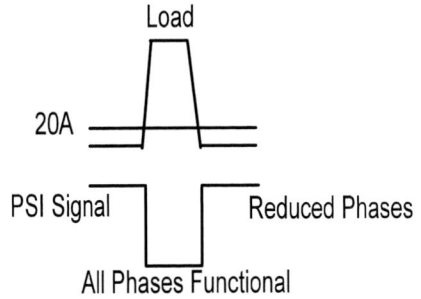

Figure 1: Power State Indicator Functionality

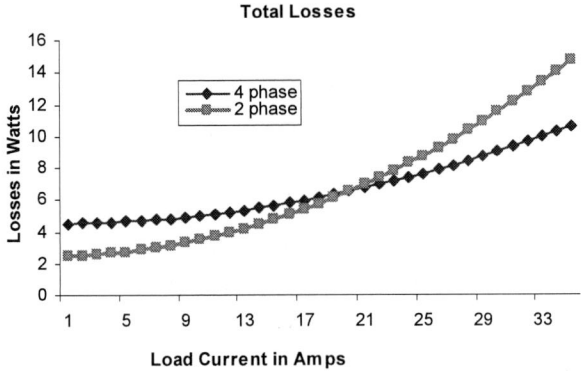

Figure 2 Losses versus number of phases

In Fig.3 the LC filter stages of a four phase VR is shown. The CPU sends out the PSI signal which goes through the low pass filter to shed some phase during low power operating state of the CPU.

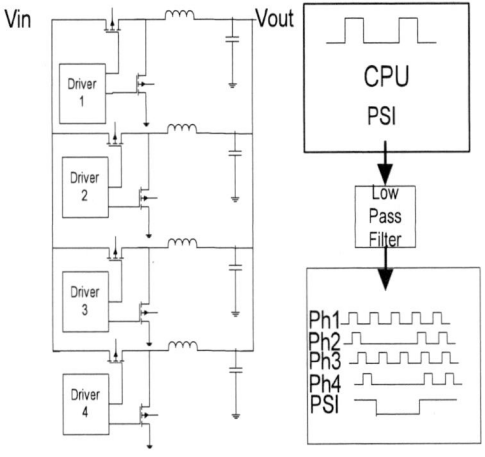

Figure 3 Four Phase VR with PSI & low pass filter

III. TRANSIENT EFFICIENCY

A. Formulation

Efficiency is always determined with a steady state load but in the case of a CPU, its load profile may contain load transients of several varying frequencies. Thus the total efficiency is thus defined as the total output energy divided by the total input energy of a converter over a given period of time or number of cycles. To simplify the analysis, assume the load changes from zero to full load with a fixed duty D and fixed period T as shown in Fig. 4.

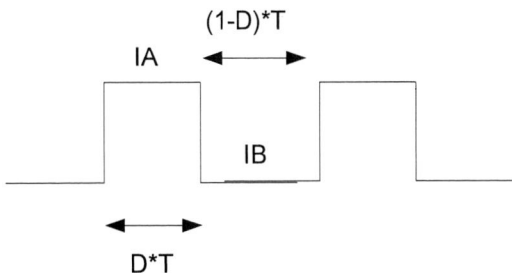

Figure 4 Load Changing Scheme

The total efficiency is given by eq. (1)

$$\eta = \frac{\int_0^T P_o(t).dt}{\int_0^T P_o(t).dt + \int_0^T P_{losses}(t).dt} \tag{1}$$

Where P_o refers to the output power and P_{losses} refers to the various conduction and driving losses in the entire period of the duty cycle (including different load states). To have a fair comparison output energy consumed (the average load) is kept constant. Thus efficiency for a four phase VR system dropping two phases during light load is given in eq. (2).

$$\eta_{4to2} = \frac{\int_0^T P_o(t).dt}{\int_0^T P_o(t).dt + \int_0^{T \cdot D} P_{losses@IA}(t).dt + \int_0^{T \cdot (1-D)} P_{losses@IB}(t).dt} \tag{2}$$

Where $P_{losses@IA}$ refers to the power loss (including conduction and driving losses) at the load current IA and $P_{losses@IB}$ refers to the power loss at load current IB. Calculating this transient efficiency will be a tedious task since the duty cycle and the frequency of the load transient are the function of the use of the CPU resources. In other words the CPU's load profile is an unknown parameter but efficiency analysis is done for worst case scenario, both for duty cycle and for load frequency. The analytical study of combining the change of both the parameters might be too complex for this paper but certainly that would be the ideal case study using a real time load profile. The load profile is usually software generated and it will be possible to create a load profile of choice and also a realistic one.

B. Duty cycle variation

The efficiency analysis with duty cycle variation is done using simulation in Saber simulation software. Even though the simulation uses ideal control components most of the loss elements were modeled and included in the simulation and the transient waveforms were used to calculate the total energy dissipated over a fixed period of time.

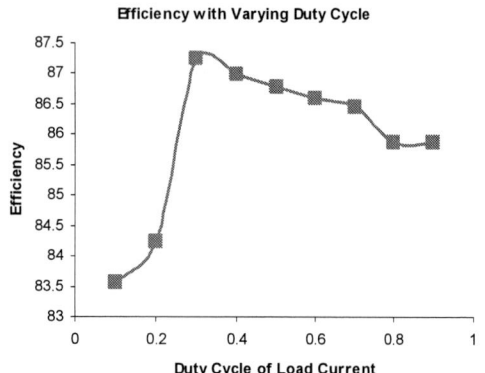

Figure 5 Efficiency versus load current duty cycle

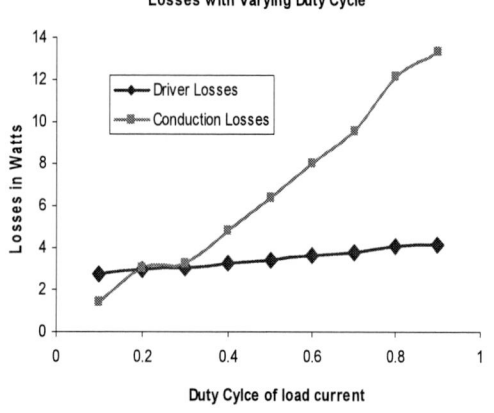

Figure 6 Efficiency versus losses breakdown

The simulation is done with a four phase VR module with input voltage of 12V and output voltage of 1.1V. The VR was run at 500 kHz switching frequency and the phases are dropped from four to two when the load current is below 20A. The phase changing frequency is kept constant at 1 kHz and the duty cycle was varied. The variation of duty cycle leads to the change in the average load current and effectively the efficiency analysis is close to the steady state efficiency analysis. Fig. 5 shows the efficiency variation with duty cycle, and it is seen that efficiency peaks at 87.2% with an average load current of 40A at a duty cycle of 30%. The load transient step is from 10A to 110A for all duty cycle. In Fig. 6 the break down of the losses shows that the conduction losses have a steeper slope than the driving losses.

C. Experimental results with phase changing frequency variation

The phase changing frequency variation is usually synchronized with the load frequency. The faster the load step repetitive rate the faster the phases will drop. The experiment for determining the efficiency variation with the phase dropping frequency is done on a four phase coupled inductor based voltage regulator. For a range of high frequency load Intel's VR11 LGA775 VTT tool for transient testing with LGA1366 interposer is used. The load is applied to a four phase coupled inductor based Intersil's VR11.1 evaluation board as shown in Fig. 7. Intersil's ISL6334 & ISL662 were used as PWM controller and gate driver. The evaluation board was set to a VID of 1.1V and a load line of 0.3mohm. The Infineon BSC059 and BSC016 were used as the top and the bottom switch for a synchronous buck topology.

Figure 7 Four Phase Coupled Inductor based VR 11.1 Evaluation Module from Intersil.

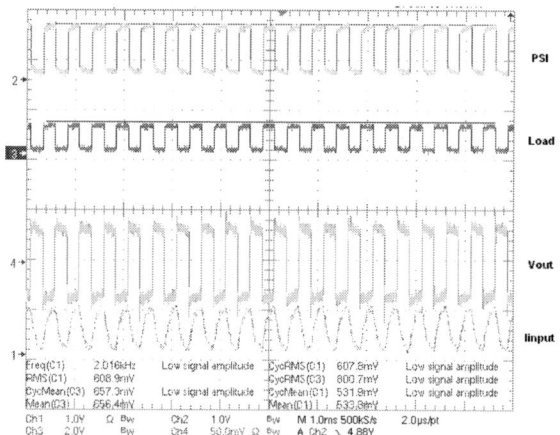

Figure 8 PSI Waveforms at 5 kHz frequency

The efficiency is measured with load frequency swept from .5 kHz to 50 kHz which synchronizes with the PSI signal. The duty cycle of the load is always adjusted to 50% to have a fair comparison. The voltage and the current waveforms of the resulting setup for a PSI/load frequency of 5 kHz are shown in Fig. 8. The waveforms corresponding to 10 kHz PSI frequency is shown in Fig. 9.

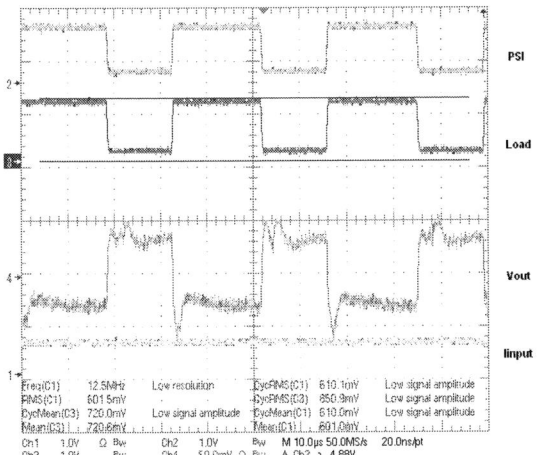

Figure 9 PSI Waveforms at 10 kHz frequency

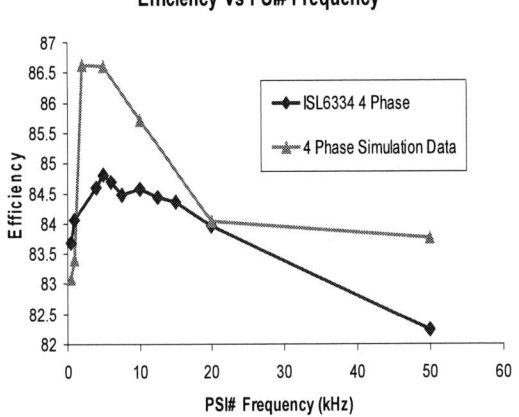

Figure 10 PSI Frequency versus Efficiency

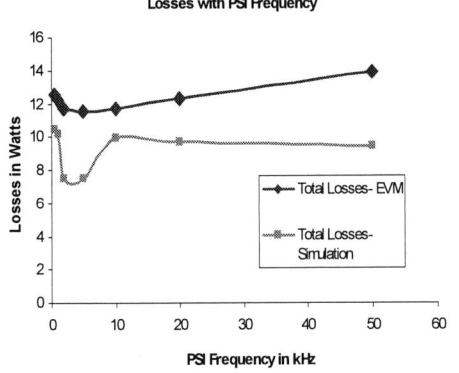

Figure 11 Losses versus PSI Frequency

The experimental and simulation results given in Fig.10 show that efficiency peaks at 5 kHz and reduces at around 10 kHz. Efficiency curve further goes down with increase in load frequency. The total losses for simulation and experimental results are shown in Fig. 11. The 5 kHz peak efficiency point would be a choice for the PSI low pass filter corner frequency.

IV. DISCUSSION OF RESULTS

As shown in the previous section the efficiency of the converter may be affected by the phase dropping scheme over a certain frequency. Even though the increase in losses at very high frequency may be impossible to avoid but it can be overcome by following several design guidelines. The multiphase buck converter control loop is based on adaptive voltage positioning (AVP) [4-8]. Compensation for AVP can be specifically tuned for high frequency load transients. High frequency load transients can result in overshoots beyond the necessary specifications and it can even lead to a reduced load line. The frequency of PSI signal follows the load transient frequency and hence at very high load frequencies the PSI functionality is disabled to operate at the maximum efficiency.

The AVP control involves two loops, the voltage control loop and current control loop. A type 3 compensator is essential for high frequency load transients; where the first zero is placed before the double poles. This zero helps in reducing load release ring back during high frequency transients. The second zero is placed after loop gain bandwidth and current control loop bandwidth which increases the phase margin of the system. The integrator increases gain and helps in bandwidth improvement. The high frequency pole improves jitter and helps to increase high frequency droop. By designing a fast control loop the inductor current catch up to the load current much quickly.

One reason for the tapering of the efficiency curve with high frequency load transients is uneven current sharing. Even though ISL6334 has a current share control inbuilt in the controller, due to the lack of fast recovery from the transient, it still suffers from conduction losses due to imbalance of channel currents. The phenomenon can further be substantiated by the fact that the 5 kHz load transient is just before the resonant frequency of the LC filter which is 10 kHz in this case. The resonant pole and the compensation poles and zeros play a role in modifying the output impedance of the converter which affects the transient response to different load frequencies.

V. CONCLUSION

The Intel's VR11.1 specification requires its PWM vendor to take advantage of the PSI signal that is given out by the Intel processor, but its frequency could be as high as 100MHz which cannot be used by the VR controller to its advantage. This paper discusses the determination of the corner frequency of the low pass filter that needs to be installed in between the processor and the VR PWM controller. Higher frequency load transients and hence higher frequency of shifting between phases leads to increased conduction losses and affects the efficiency unfavorably. This is experimentally shown in this paper.

ACKNOWLEDGMENT

I thank Chris Thompson, Tony Robles from Intersil for their support and guidance. Jon Day for giving me an opportunity to work with Intersil. Dr. Alex Huang for teaching me about VRM control aspects.

REFERENCES

[1] Zumel, P. Fernnndez, C. de Castro, A. Garcia, O.; "Efficiency improvement in multiphase converter by changing dynamically the number of phases" Power Electronics Specialists Conference, 2006. PESC '06 page(s): 1-6

[2] Abu Qahouq, J.A. Huang, L. Huard, D, "Efficiency-Based Auto-Tuning of Current Sensing and Sharing Loops in Multiphase Converters", Power Electronics, IEEE Transactions on March 2008 page(s): 1009-1013.

[3] Y. Panov and M. M. Jovanovic "Stability and dynamic performance of current sharing control for paralleled voltage regulator modules," *IEEE Trans. Power Electron.*, vol. 17, pp. 172, Mar. 2002.

[4] J. Luo , N. Pongratananukul , J. A. Abu-Qahouq and I. Batarseh "Time-varying current observer with parameter estimation for multiphase low-voltage high-current voltage regulator modules," *Proc. APEC* vol. 1, Feb. 2003, p. 444.

[5] Kaiwei Yao; Yuancheng Ren; Julu Sun; Kisun Lee; Ming Xu; Jinghai Zhou; Lee, F.C.; "Adaptive voltage position design for voltage regulators", APEC '04. Nineteenth Annual IEEE ,Volume: 1 ,22-26 Feb. 2004,Pages:272 – 278

[6] Yao, K.; Lee, K.; Xu, M.; Lee, F.C, "Optimal design of the active droop control method for the transient response", APEC '03. Eighteenth Annual IEEE ,Volume: 2 ,9-13 Feb. 2003,Pages:718 - 723 vol.2.

[7] Kaiwei Yao; Yu Meng; Lee, F.C, "Control bandwidth and transient response of buck converter", Power Electronics, IEEE Transactions on ,Volume: 17 ,Issue: 4 , July 2002,Pages:485 – 492.

[8] Pit-Leong, Fred C.Lee, Peng Xu and Kaiwei Yao, "Critical inductance in voltage regulation modules", Power Electronics, IEEE Transactions on ,Volume: 17 ,Issue: 4 , July 2002,Pages:485 – 492.

[9] VRM / EVRD 11.1 Design Guide (www.intel.com)

A New Digital Adaptive Voltage Positioning Technique with Dynamically Varying Voltage and Current References

S. Pan, *Member, IEEE*   P. K. Jain, *Fellow, IEEE*

Department of Electrical and Computer Engineering

Queen's University

Kingston, Ontario, Canada

Email: Shangzhi.Pan@queensu.ca

Abstract -A digital adaptive voltage positioning (Digital AVP) technique with fast transient response for voltage regulators (VRs) is proposed in this paper. In this proposed digital control architecture, two digital-to-analog converters (DACs) are used instead of analog-to-digital converters (ADCs), thus significantly reducing system complexity. Both the voltage and current references are changed dynamically at DAC clock frequency resulting in fast transient response. Different from other control methods (the output voltage tracks the voltage reference), the generated voltage reference is always trying to track the output voltage. A straightforward control law is used, which does not require compensator. Moreover, the proposed control technique automatically avoids the limit cycle by automatic dithering method. Further, it allows the use of non-linear control methods which significantly improve transient response speed. Finally, a two-phase 12V-to-1V, 40A, 250 kHz synchronous buck converter with the proposed digital controller was designed to verify the theoretical analysis by simulation and experimental results.

I. INTRODUCTION

Digital control of switching power converters is receiving increased attention in these years, due to its salient advantages including flexibility, fewer external components and reduced overall cost as compared to conventional analog techniques [4-6, 9-11]. When digital control is applied in a processor-dedicated voltage regulator (VR), the most significant technical limitation is the delay associated with the sampling process and discrete-time computation negatively impacting the dynamic performance. High dynamic performance is necessary for voltage regulators to deliver a very large current to the processors at extremely high slew rate while maintaining a tightly regulated low supply voltage. High dynamic characteristics of the processors impose large-step load oscillations (up to 100A load transient step at the slew rate up to 2A/ns, from 100 Hz to 1 MHz oscillation frequency) to the voltage regulators, making it more difficult to maintain accurate voltage regulation. As results, the voltage regulator will need to use a large number of output bulk capacitors to sustain the output voltage, which increases size and cost.

Therefore, in most of existing digital controller for VRs [5-9], a high speed analog-to-digital converter (ADC) is required to improve VR transient response speed, thus to reduce the number of bulk capacitors. Moreover, to prevent limit cycle [10] and achieve tight voltage regulation, high resolution ADC and digital pulse width modulator (DPWM) are required, resulting in large silicon area and power consumption.

In [4], a digital controller with two digital-to-analog converters (DACs) was proposed, which was based on variable frequency peak current mode control. Two low-resolution DACs (7-bit) were used to replace the high-resolution ADC and DPWM (10-bit), thus significantly reducing system complexity and die area. The basic idea behind this control method was trying to generate a voltage saw-tooth waveform in each period or phase period, compared to the feedback error voltage (fixed voltage reference) to generate control signals, which was like its analog counterpart. Main limitations of this digital type of controllers are summarized as: 1) low dynamic performance: the peak current reference is updated every switching cycle, which cannot take advantage of high speed DAC clock; 2) high-speed clock required; 3) variable frequency operation.

This paper presents a new digital adaptive voltage positioning (digital AVP) technique, which overcomes all the above mentioned limitations with reduced system complexity and die area. In this proposed digital control architecture, two low-speed (8MHz) and low-resolution (7-bit) digital-to-analog converters (DAC) was used instead of two high-speed (up to 50MHz [9]) high-resolution (at least 10-bit) analog-to-digital converters. Both the voltage and current references are changed dynamically at DAC clock frequency resulting in fast transient response. Different from other control methods (the output voltage tracks the voltage reference), the generated voltage reference is always trying to track the output voltage. A fixed-frequency peak current mode control is used, and the peak current reference is updated every DAC clock. A straightforward control law, which does not require compensator, is implemented by only using the addition computation and the simple control logic (Only 106 combinational logic units and 57 registers are used in Altera's FPGA). This greatly reduces the computational delay. Moreover, it allows the use of non-linear control methods which significantly improve transient response speed. Dynamic behavior of the voltage regulator with this digital AVP controller under large load transient is analyzed. Then, the transient response is optimized based on derived equations. Its excellent transient performance under large load transient oscillations (from 100 Hz to 1 MHz) was verified by both simulation and experiments, making it a valuable candidate as a high-volume, low-cost controller for processor voltage regulators.

978-1-4244-4782-4/10 $26.00 © 2010 IEEE

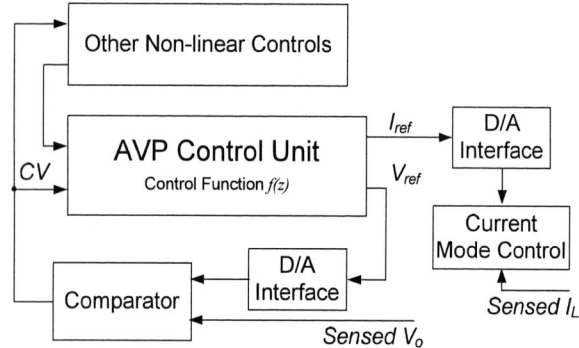

Fig. 1 Block diagram of the proposed control idea

II. PROPOSED DIGITAL CONTROL ARCHITECTURE

The AVP feedback control will not always keep the output voltage fixed at the same level regardless of the load current, but rather will position the output voltage at different levels according to the load, a slightly lower level at heavy load and a slightly higher at light load, thus the whole voltage tolerance range can be used for the voltage overshoot or undershoot limit during the transient, resulting in fewer output capacitors [1-5].

The nature of AVP is to design the output impedance (R_o) of the VR to be a constant value. In the digital form, the output impedance can be expressed as:

$$R_o = -\frac{\Delta V_o}{\Delta I_o} \tag{1}$$

The basic idea of the proposed digital control is to separately adjust the voltage reference V_{ref} and the current reference I_{ref} according to AVP control. The voltage reference V_{ref} should decrease while the current reference I_{ref} should increase, that is, $V_{ref} = V_{omax} - I_{ref} R_o$. By closing the current/voltage control loop, the output voltage can be regulated by the relationship $V_o = V_{ref} = V_{omax} - I_{ref} R_o = V_{omax} - I_o R_o$, where R_o is the desired output impedance and V_{omax} is the maximum output voltage.

Fig. 1 shows a block diagram of the AVP control unit, which generates voltage and current references according to the desired output impedance. Either peak current mode control or average current mode control can be applied. A control function $f(z)$ is generalized to represent the relationship function between the voltage reference variation step ΔV_{ref} and the current reference variation step ΔI_{ref}, which is expressed as

$$\Delta I_{ref} = f(z) \cdot \Delta V_{ref} \tag{2}$$

$$f(z \to 1) = -1/R_o \tag{3}$$

Obviously, the simplest control function is a constant, that's, $-1/R_o$.

A. Controller Configuration and Steady Operation

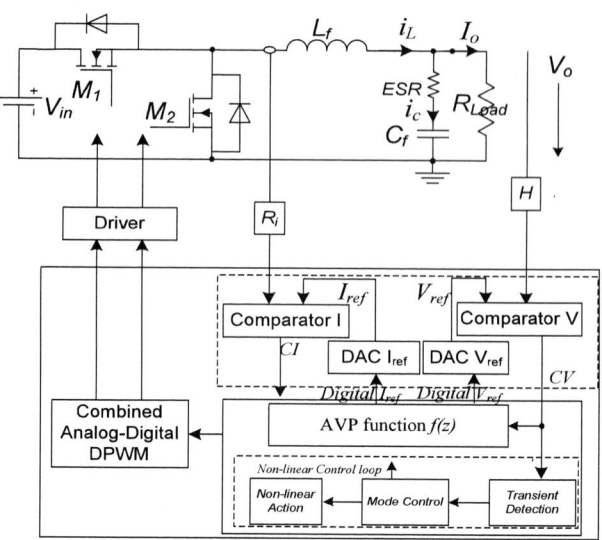

Fig. 2 A single phase buck converter with the proposed digital controller

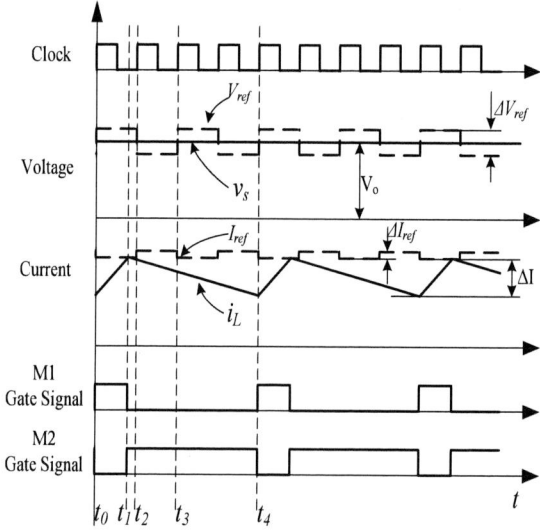

Fig. 3 Typical steady-state operation waveforms

Fig. 2 shows the block diagram of the proposed digital controller with the AVP control unit, along with a single phase synchronous buck converter. The digital controller consists of a digital control algorithm block, a combined analog-digital pulse-width-modulation (ADPWM) block, two digital-to-analog converters (*DAC I_{ref}* and *DAC V_{ref}*), and two comparators (*Comparator I* and *Comparator V*). The control algorithm is implemented in the digital control algorithm block to obtain the desired reference current data and the desired reference voltage data. The DAC (*DAC I_{ref}*) receives the reference current data from the digital control algorithm block to generate the peak current reference I_{ref}. The other DAC (*DAC V_{ref}*) receives the reference voltage data from the digital control algorithm block to generate the voltage reference V_{ref}. Comparator *I* compares the generated peak current reference I_{ref} with the inductor current i_L which is sensed by the R_i block (which is the DCR of inductor L_f, scaled by an amplifier). The comparison result CI is sent to the digital control algorithm block and the ADPWM block, to generate the gate driving signals to turn off Switch M_1 and turn on Switch M_2. The ADPWM only needs to schedule timing of the turn-on of Switch M1 and the turn-off of Switch M2 (e.g., a 2-bit counter is enough for a 4-phase controller). The resolution of PWM is determined by the resolution of the digital current reference I_{ref}. Comparator V compares the generated voltage reference V_{ref} with the output voltage V_o that is sensed by the H block. The comparison result CV is sent to the digital control algorithm block which adjusts the current reference data and the voltage reference data according to the control law.

According to (3), the straightforward adjustment control law ($f(z) = -1/R_o$) to describe the behavior of the AVP control function block is summarized as follows:

1) At each DAC clock, if logic signal CV is high (CV =1), that is, the output voltage v_o exceeds the voltage reference V_{ref} set by the voltage DAC, the voltage reference V_{ref} will increase by a small amount ΔV_{ref} and the peak current reference I_{ref} will decrease by a small amount ΔI_{ref}.

2) At each DAC clock, if logic signal CV is low (CV=0), that is, the output voltage v_o is smaller than the voltage reference V_{ref}, the voltage reference V_{ref} will decrease by a small amount ΔV_{ref} and the peak current reference I_{ref} will increase by a small amount ΔI_{ref}.

Obviously, this control law does not require any compensator, which can be implemented using only the addition computation and the simple control logic.

Fig. 3 shows the steady-state operation waveforms of the buck converter with the proposed digital controller, under peak current mode control. Under the control law, both the voltage reference V_{ref} and the current reference I_{ref} vary by steps (ΔV_{ref} and ΔI_{ref}) according to the logic comparison result CV in the previous clock period. At t_0 and t_4, switch M_1 is turned on and switch M_2 is turned off as scheduled by the ADPWM module. At t_1, switch M_1 is turned off, switch M_2 is turned on, since the sensed inductor current i_L reaches the peak current reference I_{ref}.

It is shown that the digital controller updates the voltage reference and the current reference at every DAC clock cycle. Normally, the DAC clock is much faster than the switching clock. The transient response at the falling edge of the control pulse signal (load transient down) can be significantly improved, which is the dominant limitation in low-conversion-ratio VR applications. Hence, this digital controller exhibits big advantages in load transient response performance over other digital controllers presented in [4-6], which only update the duty cycle at every switching clock cycle.

B. Steady State Analysis

For a buck converter in continuous current mode (CCM), the average inductor current $< I_L >$ is estimated as

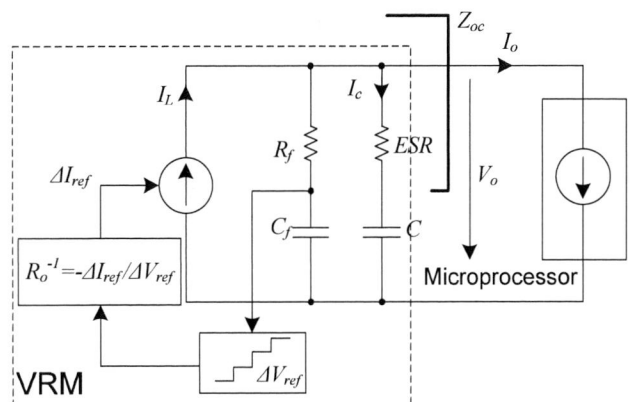

Fig. 4 Current source equivalent circuit of the buck converter with the proposed digital controller

$$I_o = (< I_L >) = I_{pk} - \frac{\Delta I}{2} = I_{pk} - \frac{V_o \cdot (1-D) \cdot T_{sw}}{2 L_f} \quad (4)$$

where D is the duty ratio, T_{sw} is the switching period, and ΔI is the current ripple amplitude.

The circuit delay from command to action results in a current error between the peak current reference I_{ref} and the peak inductor current I_{pk}, calculated as

$$I_{err} = I_{ref} - I_{pk} = -t_d \frac{V_o \cdot (1-D)}{D \cdot L_f} \quad (5)$$

where this circuit delay t_d includes the high-side MOSFET M_1 turn-off delay, current detection delay, and controller computing delay.

For a given design, which is assumed to have constant input voltage and output voltage, the duty cycle D is constant; therefore, combining (4) and (5) and differentiating, the variations on the output current I_o and the current reference I_{ref} are given as

$$\Delta I_o = (\Delta < I_L >) = \Delta I_{pk} = (\Delta < I_{ref} >) = \Delta I_{ref} \quad (6)$$

With the outer voltage loop closed, the voltage reference follows the output voltage. Hence, variations on the output voltage and the voltage reference can be calculated as

$$\Delta V_o = (\Delta < V_o >) = (\Delta < V_{ref} >) = \Delta V_{ref} \quad (7)$$

Therefore, the output impedance R_o within the control bandwidth is approximated as

$$R_o = \frac{\Delta V_o}{\Delta I_o} = \frac{\Delta V_{ref}}{\Delta I_{ref}} \quad (8)$$

Fig. 4 shows the current source equivalent circuit of the buck converter with the digital controller. An R_f-C_f low-pass filter is used for this sensing network with a pole at $\omega_{ESR} = 1/(ESR \cdot C)$, which ideally cancels the capacitor ESR zero and chooses $C >> C_f$. This low-pass filter can be easily integrated into the digital controller and be implemented in digital words. When we include the effect of the output capacitors, the closed-loop impedance Z_{oc}, obtained from Fig. 4, is found to be

$$Z_{oc} = \cfrac{1}{\cfrac{s \cdot C}{1+s \cdot ESR \cdot C} + \cfrac{1+s \cdot C_f \cdot R_o}{(1+s \cdot R_f \cdot C_f) \cdot R_o}} \qquad (9)$$

$$= R_o \frac{1+s \cdot ESR \cdot C}{1+s \cdot R_o \cdot (C+C_f)} \approx R_o \frac{1+s \cdot ESR \cdot C}{1+s \cdot R_o \cdot C}$$

When the low-pass filter is implemented in the digital controller, the AVP control function in (3) can be expressed as

$$f(z) = \frac{-1}{R_o} \cdot \frac{T_s \cdot (z+1)}{(T_s+2 \cdot ESR \cdot C) \cdot z^2 + (T_s-2 \cdot ESR \cdot C) \cdot z} \qquad (10)$$

Where T_s is the sampling frequency and Tustin Approximation is applied to convert s-domain transfer function into z-domain transfer function. In the steady state ($z \rightarrow 1$), this AVP control function also has the same expression as (3). But in this case more complicated computation is required, so an external filter is implemented in this paper.

C. Effect of Quantization and Dithering

The voltage regulation tolerance is determined by the least significant bit (LSB) of the voltage DAC (which has n bits), which is expressed as

$$\Delta V_{refLSB} = \frac{\Delta V_{tol}}{2^n} \qquad (11)$$

$$\Delta V_{ref} = N \cdot \Delta V_{refLSB} \qquad (12)$$

where ΔV_{tol} is the maximum allowed voltage tolerance.

The minimum number n of bits required in the voltage DAC is given by

$$n = \text{int}\left[\log_2\left(\frac{\Delta V_{tol}}{\Delta V_{reg}} \right) \right] \qquad (13)$$

where the function $int[~]$ takes the upward-rounded integer value of its argument and ΔV_{reg} is the voltage regulation tolerance requirement. For example, if the allowed output voltage tolerance is 120mV and the required voltage regulation resolution is 1mV, a 7-bit DAC should be chosen by (13).

To meet the voltage regulation tolerance requirement, the minimum adjustment step ΔV_{ref} of the voltage reference should be less than ΔV_{reg}. Fig. 5 shows an enlarged view of the shape of the voltage reference (from simulation results); note that the voltage reference also will have the same shape as the voltage ripple and the difference between their average values is always less than ΔV_{ref}.

Because of the quantization of the current reference DAC, the average inductor current $<I_L>$ only has a finite number of discrete levels defined by the least significant bit (LSB) of the current DAC (which has m bits), which is expressed as

$$\Delta I_{refLSB} = \frac{I_{ref\,max}}{2^m} \qquad (14)$$

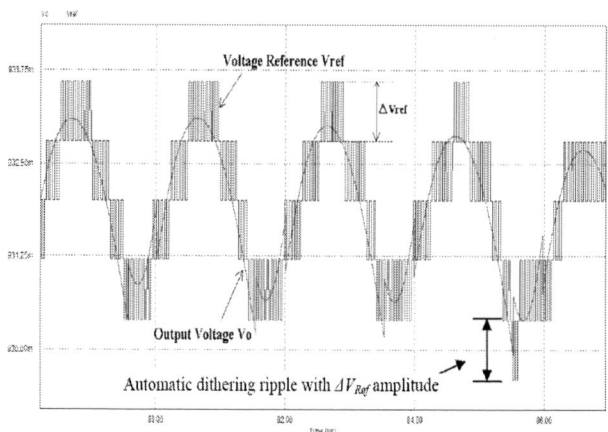

Fig. 5 Enlarged view of the output voltage and voltage reference (simulation results) (Voltage scale: 1.25mV/div, time scale: 2µs/div)

$$\Delta I_{ref} = M \cdot \Delta I_{refLSB} \qquad (15).$$

Therefore, a current difference exists between I_o and $<I_L>$. The maximum current difference is ΔI_{ref}, which will be absorbed by the output capacitors. If the inductor current ripple is well averaged and not considered here in this analysis, the DC current I_c through the filtering capacitor C is given by

$$I_c = <I_L> - I_o \leq \Delta I_{ref} \qquad (16)$$

where we know that

$$-|\Delta I_{ref}| \leq <I_L> - I_o \leq |\Delta I_{ref}| \qquad (17)$$

Thus, the voltage variation on the filtering capacitor is calculated by

$$\Delta V_o = ESR \cdot I_c + \frac{1}{C}\int_0^T I_c \cdot dt \qquad (18)$$

In the steady state, over several DAC sampling periods (assume a total of i periods), when the accumulated voltage variation ΔV_o reaches ΔV_{ref}, the (average) voltage reference V_{ref} is adjusted by one step ΔV_{ref}, and I_{ref} also is adjusted by one step ΔI_{ref}. After that, the output voltage changes in the opposite direction until ΔV_o reaches ΔV_{ref} again (assume a total of j periods). The effective inductor current step ΔI_{Leff} over the $(i+j)$ periods is given as:

$$\Delta I_{Leff} = \frac{i \cdot \Delta I_{ref}}{i+j} \qquad (19)$$

where i and j depend on various factors such as ΔV_{ref}, ΔI_{ref}, and I_c.

The low-frequency automatic oscillation described above is similar to the dithering method introduced in [10], which increases the effective duty cycle resolution to avoid limiting cycle without increasing the hardware cost. Therefore, low-resolution DACs can be used. Moreover, the automatic dithering method here is passive, resulting as a side effect of the control law itself, so it does not require any extra computation. Another advantage over other dithering method [10-11] is that the low frequency oscillation caused by this

978-1-4244-4782-4/10 $26.00 © 2010 IEEE

automatic dithering has a well-controlled amplitude (ΔV_{ref}), as shown in Fig. 5, which can be made negligibly small. However, the oscillation frequency is uncertain, depending on various factors such as ΔV_{ref}, ΔI_{ref}, and I_c.

D. Extended to Multiphase Structure

The proposed digital controller can be easily extended to an interleaved multiphase buck converter. The timing of the control signals is easily arranged using the digital method and an active current sharing ability is inherently achieved at no extra cost due to the peak current mode control.

Fig. 6 shows the circuit configuration of an N-phase buck converter with the proposed controller, where a total of $(N+1)$ comparators are used. One comparator is needed for the current loop of each phase and one comparator is needed for the voltage loop. The inductor current of each phase is sensed separately and compared to the peak current reference I_{ref} produced by the current DAC. The peak current reference I_{ref} is set as the phase peak current reference. The logic signal of the comparator of each phase is used to turn off its own high-side MOSFET and turn on its own low-side MOSFET, whereas the high-side MOSFET and the low-side MOSFET are turned on and off respectively scheduled by the ADPWM. The ADPWM is implemented using the counter-comparator method, combined with asynchronous comparison signal CI. For an N-phase buck converter, the counter bit n is given as

$$n = \mathrm{int}[\log_2 N]$$

where $int[]$ takes the upward-rounded integer value of its argument.

The operating clock of the target FPGA (or ASIC) will not increase significantly with increasing phase number; the only requirement is that it is high enough to generate turn-on signals for an n-phase converter , such as a 16MHz clock is enough for a 4-phase 1MHz/phase application.

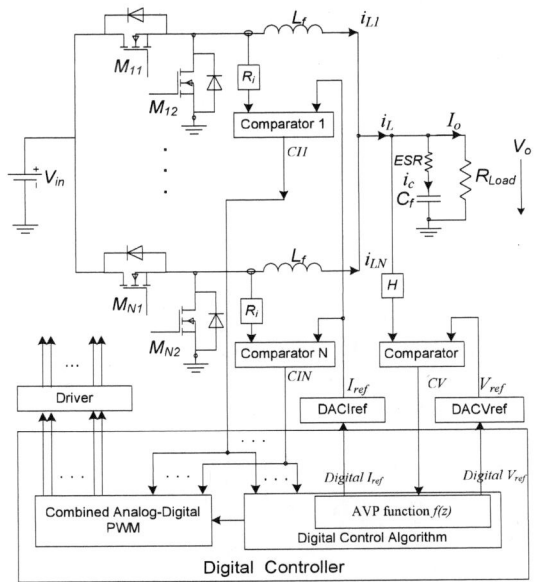

Fig. 6 Circuit configuration of an N-phase buck converter with the proposed controller

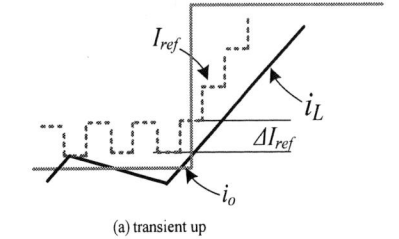

(a) transient up

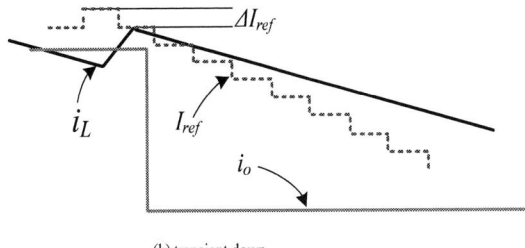

(b) transient down

Fig. 7 Current reference adjustment during load transient

III. NON-LINEAR CONTROL STRATEGY

When load transient-up occurs, and the output voltage v_o starts to drop. v_o is smaller than the voltage reference V_{ref}, and by the control law, V_{ref} decreases by ΔV_{ref} step by step and the peak current reference I_{ref} increases by ΔI_{ref} step by step, as shown in Fig. 7 (a). When load transient-down occurs, and the output voltage v_o begins to rise. The output voltage v_o is larger than the voltage reference V_{ref}, so V_{ref} increases and I_{ref} decreases step by step, as shown in Fig. 7 (b).

Hence, to judge whether the power system is in the transient state or not, a counter TU is used to count the consecutive increase of the current reference, and another counter TD is used to count the consecutive decrease of the current reference. The counter TU is reset to zero if the current reference decreases and the counter TD is reset to zero if the current reference increases. If the counter TU reaches a limitation value (LMT_TU, predefined in the digital controller), it will judge that there is a load transient-up. If the counter TD reaches the limitation value (LMT_TD, predefined in the digital controller), it will judge that there is a load transient-down.

Fig. 8 shows control strategy of this non-linear control, which includes a mode control unit to perform operation mode judgment. In the steady state, peak current mode control (linear control) is applied, named normal operation mode. A small adjustment step of the voltage/current reference is used. But, once the control unit asserts load transient-up, the peak current mode control will be disabled and the controller will enter transient-up mode and all high-side switches will be turned on to obtain maximum inductor current ascending slew rate. Once the control unit asserts load transient-down, the peak current mode control will be disabled and the controller will go into transient-down mode and all high-side switches will be turned off to obtain maximum inductor current descending slew rate. Therefore, the transient-assertion-to-delay is minimized in this manner, which offers very fast load transient speed, especially in multi-phase VRs.

978-1-4244-4782-4/10 $26.00 © 2010 IEEE

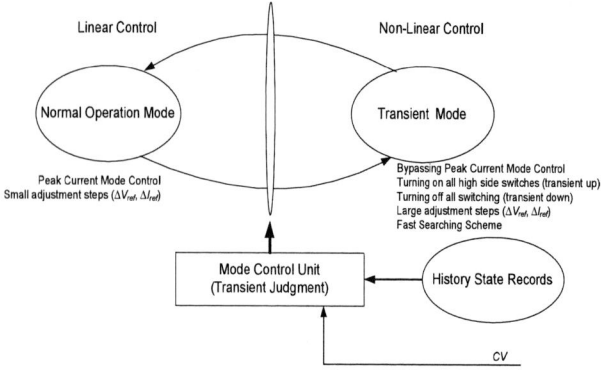

Fig. 8 Control Strategy of non-linear control

the relationship between the voltage reference and the output voltage in Fig. 5, appropriate minimum limitation value constraints can be determined to be

$$LMT\_TU \geq \min\left\{ \mathrm{int}\left[\frac{f_{clk}}{2f_{sw}}\right] \;,\; \mathrm{int}\left[\frac{\Delta V_{o,r}}{\Delta V_{ref}}\right]\right\}+1 \tag{20}$$

$$LMT\_TD \geq \min\left\{ \mathrm{int}\left[\frac{f_{clk}}{2f_{sw}}\right] \;,\; \mathrm{int}\left[\frac{\Delta V_{o,r}}{\Delta V_{ref}}\right]\right\}+1 \tag{21}$$

where D is the duty cycle, f_{sw} is the switching frequency, f_{clk} is the operating clock frequency of the target DACs, $\Delta V_{o,r}$ is the peak-to-peak output ripple voltage, $int[x]$ is the function that takes the smallest integer larger than x, and $min[x,y]$ is the function that takes the smaller value of x and y.

The selection of the magnitude of the reference adjustment step is important to achieve good dynamic performances, such as small voltage spikes and a reduced voltage ring-back oscillation. During the load transient-up stage, the voltage reference should be larger than the output voltage before the output voltage reaches its minimum, and then it will stay within transient mode and force the inductor current to continue increasing. So, the constraint on the reference adjustment amount can be derived as (22). M for the transient-up state stage can be chosen to be the largest integer number that meets constraint.

Similarly, during the load transient-down stage, the voltage reference will be smaller than the output voltage before the output voltage reaches its maximum, and then it will stay in transient mode and as the inductor current continues to decrease. So the constraint on the reference adjustment amount can be expressed as (23). M for the transient-down state stage can be chosen to be the largest integer number that meets this constraint.

In a transient state, the adjustment step of the voltage/current reference will also increase to a larger amount (predefined) than in the steady state. Therefore, the voltage reference and the current reference will reach the new balance value quickly at the same operating clock. During the transient mode, once the reference voltage reaches the output voltage, the transient mode will end. Binary search algorithm is used to reach the new steady state smoothly and quickly, without a large voltage ring-back. The transition time during load transient will be roughly adaptive for different load transient steps. It can be verified that smaller load transient will leave the transient mode earlier in this manner.

Choosing the limitation values (LMT_TU and LMT_TD) is very important to achieve good performance. Limitation values that are too large slow down the transient response and limitation values that are too small affect steady-state operation and impair steady-state performance. By observing

$$M \leq \begin{cases} \dfrac{\dfrac{\Delta I_{o\max} \cdot t_D}{C} + ESR \cdot \Delta I_{o\max} - \Delta V_{ref} \cdot LMT\_TU}{\Delta V_{ref} \cdot \mathrm{int}\left[(t_{dc}+T_m)f_{clk}\right]} \\[4ex] \dfrac{\dfrac{\Delta I_{o\max} \cdot t_D}{C} + \dfrac{ESR^2 \cdot C^2 \cdot (V_{in}-V_o)^2 + \Delta I_{o\max}^2 \cdot L_f^2}{2L_f \cdot C \cdot (V_{in}-V_o)} - \Delta V_{ref} \cdot LMT\_TU}{\Delta V_{ref} \cdot \mathrm{int}\left[(t_{dc}+T_m)f_{clk}\right]} \end{cases} \qquad \begin{aligned} & L_f < L_{crit} \; \frac{ESR \cdot C \cdot (V_{in}-V_o)}{\Delta I_o} \\[3ex] & L_f \geq L_{crit} = \frac{ESR \cdot C \cdot (V_{in}-V_o)}{\Delta I_o} \end{aligned}$$

$$T_m = \frac{\Delta I_{o\max} \cdot L_f}{V_{in}-V_o} - ESR \cdot C \tag{22}$$

Where t_{dc} represents the circuit delay, including the turn-on/off delay of the switches and the gate driver delay; t_D represents maximum action delay including t_{dc} and transient-state judgment time; ΔI_{omax} is maximum load transient step.

$$M \leq \begin{cases} \dfrac{\dfrac{\Delta I_{o\max} \cdot t_D}{C} + ESR \cdot \Delta I_{o\max} - \Delta V_{ref} \cdot LMT\_TD}{\Delta V_{ref} \cdot \mathrm{int}\left[(t_{dc}+T_m)f_{clk}\right]} \\[4ex] \dfrac{\dfrac{\Delta I_{o\max} \cdot t_D}{C} + \dfrac{ESR^2 \cdot C^2 \cdot V_o^2 + \Delta I_{o\max}^2 \cdot L_f^2}{2L_f \cdot C \cdot V_o} - \Delta V_{ref} \cdot LMT\_TD}{\Delta V_{ref} \cdot \mathrm{int}\left[(t_{dc}+T_m)f_{clk}\right]} \end{cases} \qquad \begin{aligned} & L_f < L_{crit} = \frac{ESR \cdot C \cdot V_o}{\Delta I_o} \\[3ex] & L_f \geq L_{crit} = \frac{ESR \cdot C \cdot V_o}{\Delta I_o} \end{aligned}$$

$$T_m = \frac{\Delta I_{o\max} \cdot L_f}{V_o} - ESR \cdot C \tag{23}$$

IV. SIMULATION VERIFICATIONS AND EXPERIMENTAL RESULTS

A two-phase interleaved synchronous buck converter for 12V-1V/40A VRs was designed and simulations were conducted in the co-simulation environment consisting of PSIM, ModelSim and Simulink. The phase switching frequency is 250 KHz, the phase inductor is 400nH and the output capacitors are two 330μF OSCON capacitors in parallel (ESR: 5mΩ/2, ESL: 1.6nH/2). The desired output impedance is set at 2mΩ. The minimum voltage reference step ΔV_{ref} is 0.84mV and the minimum current reference step ΔI_{ref} is 0.21A. During load transient-up stage, the voltage reference step $N \cdot \Delta V_{ref}$ is 16×0.84mV=13.44mV and the current reference step $M \cdot \Delta I_{ref}$ is 16×0.21A=3.36A. During load transient-down stage, the voltage reference step $N \cdot \Delta V_{ref}$ is 2×0.84mV=1.68mV and the current reference step $M \cdot \Delta I_{ref}$ is 2×0.21A=0.42A.

Fig. 9 shows the extended transient response with the non-linear control. The load transient step is 27A at the slew rate of 2A/ns. Only a very small output voltage ring-back exists after a large load transient; nearly perfect AVP control is achieved with the proposed digital controller. The transient-assertion-to-action delay is minimized because of the non-linear control. The number of bulk capacitors is reduced from 3 to 2, compared with the digital controller without the non-linear control. Actually, the number of bulk capacitors can be further reduced to 1; however, a large ring-back after load transient-up will occur, which is not accepted under high dynamic load oscillation.

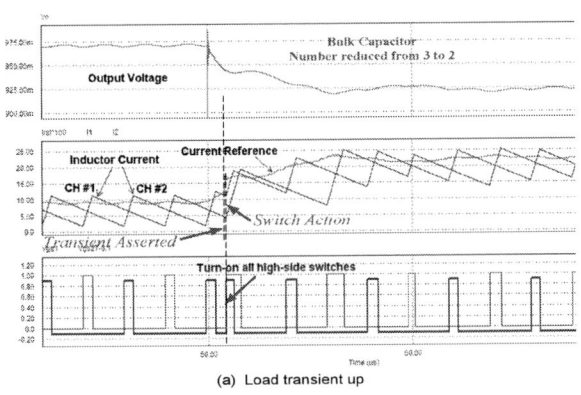

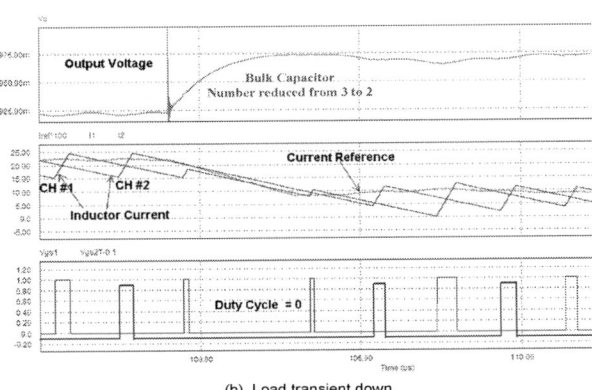

Fig. 9 The transient response with non-linear control (voltage scale: 25mV/div, current scale: 5A/div, time scale: (a) 10μs (b) 5μs)

Table 1 MINIMUM OUTPUT BULK CAPACITOR

	Phase Switching Frequency	Phase Inductance	Minimum number of bulk Capacitors	Improvement
Proposed digital controller	250KHz	400nH	2 x 330uF	50%
Active-droop analog controller in [3]	250KHz	400nH	3 x 470uF (required for matching ESR with R_o)	0%
Digital Controller in [4]	250KHz	400nH	4 x 330uF	0%

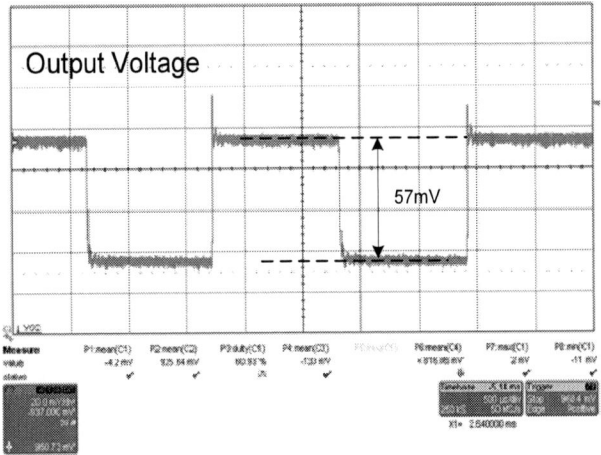

Fig. 10 The experimental transient response with non-linear control while load current transient step is 28A and load oscillation frequency is 500Hz (voltage scale: 20mV/div, time scale: 500μs/div)

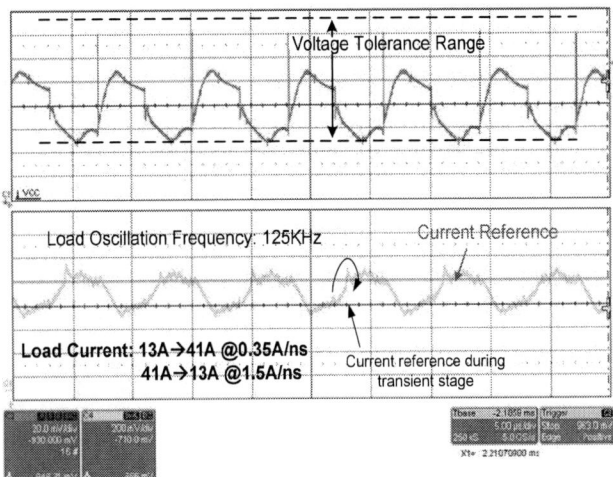

Fig. 11 the tested output voltage and current reference waveforms with 125 kHz load oscillation (voltage scale: 20mV/div, current scale: 20A/div, time scale: 5μs/div)

Table 1 shows the minimum output bulk capacitor requirement comparison. Compared to the controllers proposed in [3] and [4], at least two bulk capacitor (330uF) has been saved. For the active-droop analog controller present in [3], the ESR of bulk capacitor is required to match with R_o.

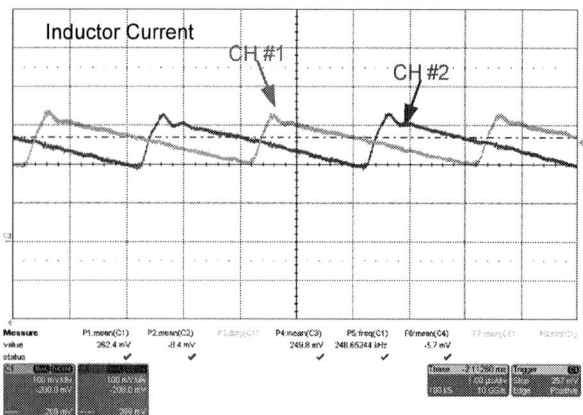

Fig.12 Tested phase inductor current at full load (40A)
(current scale: 10A/div, time scale: 1μs/div)

The digital controller was tested on a prototype of a two-phase interleaved synchronous buck converter with an FPGA board (Stratix II EP2S60 from Altera). The control algorithm was developed in VHDL (Only 106 combinational logic units and 57 registers are used). Two digital-to-analog converters (DAC2904) were used to generate the peak current reference and voltage reference. The current DAC operates at 7 bits, and the voltage DAC operates at 10 bits with an effective 7-bit voltage variation range, both at 8MHz. The circuit and controller parameters are same as those in simulations.

A load transient circuit was used to generate a large-step transient load current at different oscillation frequencies (from 100Hz to 1MHz) to verify the transient performance of the proposed digital controller. The load transient step was 28A (13A~41A) at the slew rate of 1.5A/ns for load transient-down and at the slew rate of 350A/μs for load transient-up. Fig. 10 showed the output voltage during load transient-up/down at 500Hz load oscillation, showing that the accurate AVP operation was realized as expected: 2.04mΩ output impedance. And an example of transient response at 125 kHz load oscillation was shown in Fig.11. The output voltage was always maintained within the voltage tolerance range (0.92~1.026V @ 41~13A). Current reference step variation from the steady-state stage to the transient stage could be observed apparently, shown in Fig.11.

Fig. 12 showed the tested phase inductor current at full load (40A). It was verified again that phase inductor currents were well balanced between two phases because of the inherent active current sharing ability.

V. CONCLUSIONS

This paper presents a digital AVP control architecture for powering the processors, which need a high-dynamic large current at a tightly-regulated low voltage. The proposed digital controller, which has a reduced-complexity structure, realizes the adaptive voltage positioning control by generating dynamic voltage and current references.

Main merits of this digital controller include: 1) Voltage and current references are updated at every DAC clock, thus significantly reducing the controller delay and offering fast transient response speed; 2) A straightforward control law is applied for the AVP control unit; no compensator, and only simple addition computation and control logic required; 3) Dynamic reference step adjustment method accelerates transient response speed without degrading steady-state performance, while reducing the high speed requirement on reference updating clock.; 4) Non-linear control, including operation state recognition, multi-mode operation, decision-making and multi turn-on/turn-off control schemes, is used to improve the transient performance, which minimizes the transient-assertion-to-action delay and maximizes the inductor current slew rate; 5) The use of low-complexity DACs and the straightforward control law makes the proposed digital controller a good candidate for high-volume low-cost low-dissipated- power integrations.

Analysis and design of the proposed digital controller has been given and verified experimentally. The proposed controller is equally applicable for the voltage regulator designs with zero load-line impedance.

REFERENCE

[1]. Gabriel Eirea, and Seth R. Sanders, "Adaptive Output Current Feedforward Control in VR Applications", IEEE Transactions on Power Electronics, Vol. 23, Pp.: 1880-1887. July 2008.

[2]. Martin Lee, Dan Chen, and Kevin Huang, etc., "Modeling and Design for a Novel Adaptive Voltage Positioning (AVP) Scheme for Multiphase VRMs", IEEE Transactions On Power Electronics, Vol. 23, No. 4, Pp. 1733-1742, July 2008.

[3]. K. Yao, K. Lee, M. Xu and F. C. Lee, "Optimal Design of the Active Droop Control Method for the Transient Response", IEEE Applied Power Electronics Conference and Exposition, 2003, Vol.2, pp.718-723.

[4]. S. Saggini, M. Ghioni, and A. Geraci,"An Innovative Digital Control Architecture for Low-Voltage, High-Current DC–DC Converters with Tight Voltage Regulation", IEEE Transactions On Power Electronics, Vol.19, No.1, pp.210-218, January 2004.

[5]. L. Ma, M. Xu, A. Q. Huang, "Digital Implementation of Adaptive Voltage Position (AVP) for Voltage Regulator", semiconductor power electronics center (SPEC 2005), pp. 226-229.

[6]. A. V. Peterchev, J. Xiao, S. R. Sanders, "Architecture and IC implementation of a digital VR controller," IEEE trans. on power electronics, Vol. 18, No.1 Jan. 2003, pp.356-364.

[7]. J. Quintero, A. Barrado, M. Sanz, C. Raga, A. Laizaro, "Bandwidth and Dynamic Response Decoupling in a Multi-phase VRM by applying Linear-Non-Linear", 2007 IEEE International Symposium on Industrial Electronics, pp. 3373-3378, June 2007.

[8]. G.M. Di Blasi, V. Boscaino, P. Livreri F. Marino, and M. Minieri, "A novel linear-non-linear digital controlfor DC/DC converter with fast transient response", IEEE Applied Power Electronics Conference and Exposition, 2006, pp. 705-711, March 2006.

[9]. Abu-Qahouq, J.A., Pongratananukul, N., Batarseh, I, and Kasparis, T., "Multiphase voltage-mode hysteretic controlled VRM with DSP controland novel current sharing", IEEE Applied Power Electronics Conference and Exposition, 2002, Vol.2, pp663-669.

[10]. A. V. Peterchev, S. R. Sanders, "Quantization resolution and limit cycle in digitally controlled PWM converters," IEEE Transactions on Power Electronics, Vol. 18, No. 1,pp. 301-308, January 2003.

[11]. Z. Lukic, N. Rahman, A. Prodic, "Multibit S-D PWM Digital Controller IC for DC-DC Converters Operatingat Switching Frequencies Beyond 10 MHz", IEEE Transactions on Power Electronics, Vol. 22, No. 5,pp. 1693-1707, Sept. 2007.

A Three-Level Buck Converter and Digital Controller for Improving Load Transient Response

Zhenyu Zhao

Toronto Design Centre, Exar Corporation
1505-180 Dundas Street West
Toronto, ON, CANADA M5G 1Z8
E-mail: Zhenyu.zhao@exar.com

Aleksandar Prodić

Laboratory for Power Management and Integrated SMPS,
ECE Department
University of Toronto, 10 King's College Road
Toronto, ON, CANADA M5S 3G4
E-mail: prodic@ power.ele.utoronto.ca

Abstract— **A method for improving heavy-to-light load transient response in low-power switch-mode power supplies is presented in this paper. It relies on the utilization of a 3-level 5-switch buck converter and a digital controller with an extended duty ratio control value. The method is experimentally verified with a 6 V to 1.3 V, 400 kHz, 8 W digitally controlled system showing 3 times faster response than that of a conventional buck.**

I. INTRODUCTION

Numerous methods for improving load transient response of dc-dc converters and, thus, reducing their overall size have been proposed [1]-[8]. In general, the methods can be divided in control based and topology based. The control methods usually utilize advanced nonlinear techniques [1]-[4], with ultimate goal to achieve time-optimal response, limited by the converter physical parameters only. The response speed in these control methods is limited by the values of the power stage's inductor and capacitor, as well as with the values of input and output voltages. In modern dc-dc converters with a significant step-down ratio, these limitations make heavy-to-light load transient response significantly slower than the response to the opposite transients causing significant voltage overshoot and/or requiring larger output capacitors. To minimize this problem, in most topology-based methods, filter inductance is altered or bridged during transient to result in a higher current slew rate than in steady state [5]-[7].

In this paper we propose a different approach that does not alter any output filter parameter but extend the input voltage of the converter to a negative value, to improve the slew-rate during heavy-to-light load transients.

II. PRINCIPLE OF OPERATION

Theoretically, for a given converter, a linear compensator can be designed to achieve arbitrarily high control bandwidth. However, in reality, the achievable bandwidth is limited by

This work is supported by Natural Science and Engineering Research Council of Canada (NSERC)

the slew rate of the inductor current.

In a conventional buck converter with a relatively large step down ratio, i.e. V_{in} significantly larger than $2V_{out}$, inductor current slew rates during step-up and step-down load changes have considerably different values. During a heavy-to-light load transient the current slew rate is limited by $(V_{in}-V_{out})/L$, while during the opposite load change the rate is $-V_{out}/L$, where

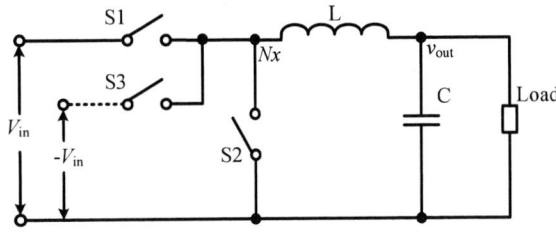

Fig. 1. Conceptual block diagram of a 3-level buck converter

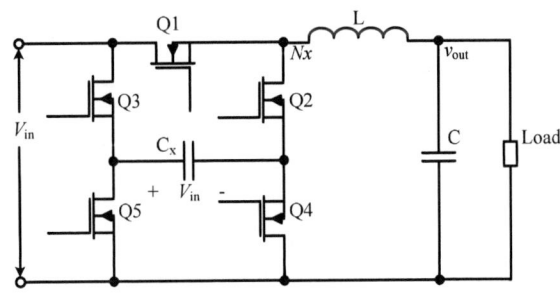

Fig. 2. Practical implementation of the 3-level buck presented in Fig.1.

L is the inductance value.

Because of the much smaller rate, the converter needs more time to suppress heavy-to-light load transients. In modern dc-dc converters supplying electronic devices this problem becomes evident, as the trend of reducing the output voltage, i.e. increasing step down ratio, of converters to a sub 1 V range [9] continues. Consequently, during step down load changes, the converter experience longer transients and larger overshoots. This phenomenon is investigated in [8] and the critical inductance concept for selecting small enough inductance value, to achieve symmetric load transients is proposed. In some applications magnetic structures that can dynamically change their inductance are not readily available limiting applicability of this method.

As an alternative, we propose the use of a modified 3-level buck converter of Fig. 1 that increases the slew rate during heavy-to-light load transients. An implementation of the conventional 3-level 4-swith buck converter [10] usually used in high power applications was proposed for envelope tracking in low-power high-bandwidth amplifiers [10]. There the converter producing inputs of 0, $V_{in}/2$, and V_{in} is used to minimize the output voltage ripple and increase efficiency.

In this case, we show how a 5-switch implementation of a 3-level buck of Fig.2, can be used to improve heavy-to-light load transient response. The input voltage of the converter in this case has three levels, $+V_{in}$, 0, and $-V_{in}$. The idea is that during heavy-to-light transients connect the inductor input node, N_X, to the negative voltage rail, $-V_{in}$, instead of 0. In this way the absolute value of slew rate for heavy-to-light load transients is increased to $(V_{out} + V_{in})/L$. It should be noted that not all of the transistors added to this topology need to have high current ratings. In steady state Q_1 and $(Q_2 + Q_4)$ operate as complementary switches and the converter behaves as a conventional buck. During heavy-to-light load transients only transistors Q_3 and Q_5 are active requiring much smaller average current compared to that of the other switches. The capacitor C_x, used to generate a negative V_{in} at N_X also serves as an input filtering capacitor during steady state operation.

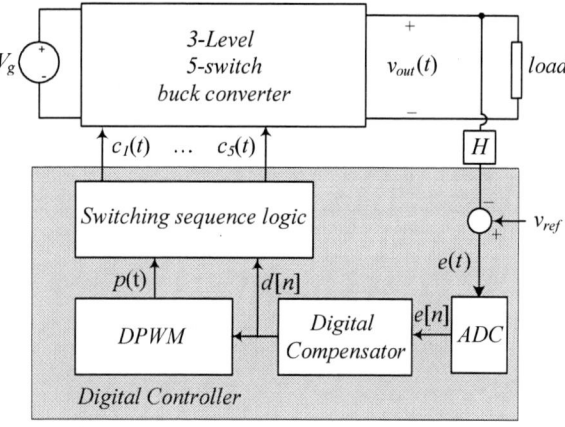

Fig. 3. Block diagram of a digitally controlled 3-level buck converter

III. DIGITAL CONTROLLER

The operation of the digital controller governing operation of the proposed converter can be illustrated with the block and

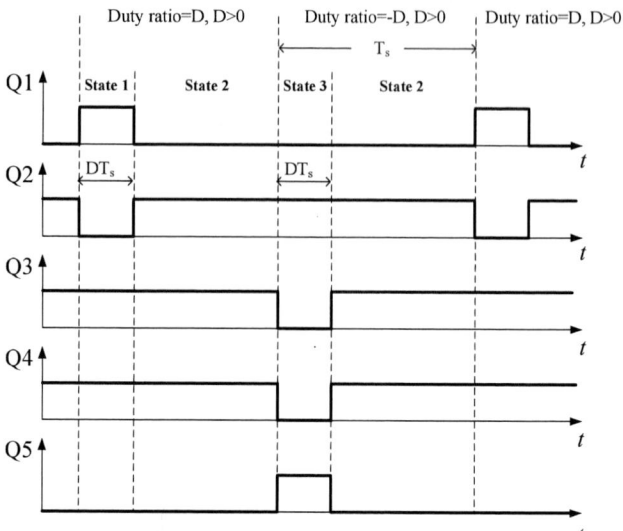

Fig. 4. Timing diagram describing switching sequences of the proposed system.

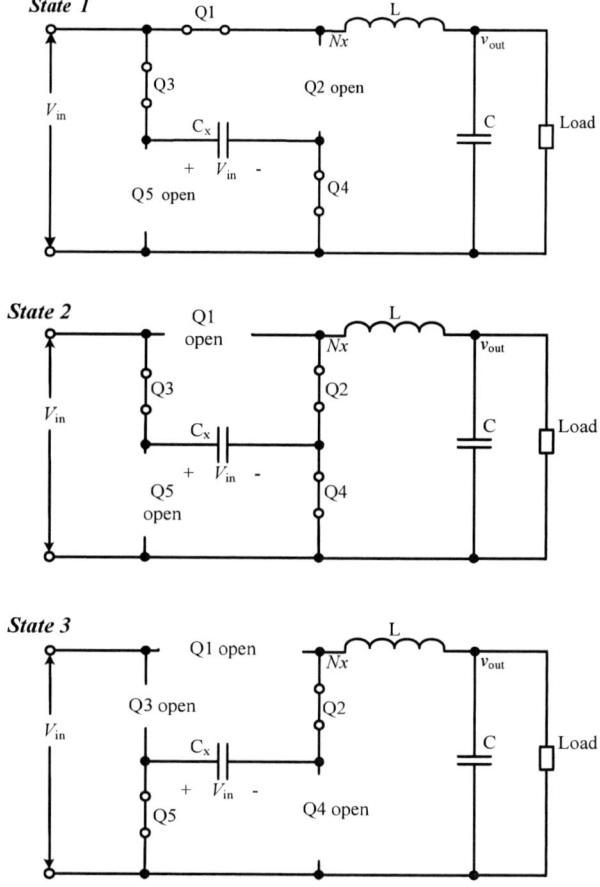

Fig. 5. Modes of operation of the proposed 3-level buck converter

timing diagrams of Fig.3 and Fig.4, respectively.

The digital controller is quite simple and requires minor modification compared to that of a conventional voltage mode pulse-width modulated structures [11]. The main difference is that in this case the duty ratio control variable $d[n]$, produced by the PID compensator, is not limited between 0 and 1 but has an extended range of $-1 < d[n] < 1$. When the output voltage increases, due to a heavy-to-light load transient or some other disturbance, instead being limited to 0, $d[n]$ can become negative. This negative value is automatically produced by the PID compensator.

Unlike in a conventional buck converter, where the negative value $d[n]$ is discarded, due to the physical limitation of the topology, in this topology it is used to initiate change of the converter configuration. The value of $d[n]$ is always monitored by the block named *Switching Sequence Logic* that selects the active transistors sequence based on the sign of duty ratio command. The operation of the logic is described by the timing diagram of Fig.4 and corresponding circuit configuration diagrams of Fig. 5.

As we can see, when duty ratio control variable is positive, converter operates as a conventional buck switching between states 1 and 2 and disabling Q_5. In this case, C_x behaves as an input filtering capacitor and is charged to the voltage V_{in}.

When a negative duty ratio command is created, the converter configuration of the state 2 is replaced by that of state 3 where the orientation of the capacitor C_x is reversed by disconnecting Q_3, Q_4 and shorting Q_5 to impose a $-V_{in}$ voltage at N_X node. In this way, an undisturbed feedback regulation is achieved. In this mode, sufficient non-overlapping times are necessary before the turn-on of Q_5 to prevent current shoot-through.

Compared to the response in the conventional case where duty ratio is limited at zero, the proposed method does not introduce nonlinearity due to the saturation action of the duty ratio control. Therefore, during the step-up and step-down transient response the same range of duty ratio values can be produced.

IV. DESIGN ISSUES AND TRADEOFFS

The extra components of the proposed topology of Fig.2 introduce design tradeoffs between the converter efficiency and speed of its response, i.e. size of the filtering components.

The extra switch Q_4 in the power path introduces additional conduction losses and the implementation requires a large number of components. However, it is worth noting that by providing faster load transient response the bulky output capacitor can be drastically reduced and, potentially, fully compensate for the extra components introduced.

Another design issue is the added component stress on switch Q_1. It can be alleviated if another capacitor is introduced in series with C_x as part of a voltage divider where Q_5 is instead connected to the midpoint of the two capacitors. In this way, the voltage across C_x can be reduced and as well as the stress on Q_1.

Finally, for Q_1 and Q_4 to remain off in state 3 as shown in Fig. 5, special gate drivers capable of providing negative gate drive signals are required. In this work, to demonstrate the effective of the method, off-shelf gate drivers are used. In practice, they can be integrated on the same chip with the digital controller. This might imply more design work and extra cost. However, as silicon cost continues to decrease and transient response requirement more stringent, this cost might as well be compensated by the saving in the output capacitors.

V. EXPERIMENTAL RESULTS

An experimental setup realizing the system of Figs. 2 and 3 was built and tested. The power stage is a three-level 6 V to 1.3 V, 400 kHz, 8 W. The converter parameters are: inductor L=3.3 µH, output capacitor C=140 µF, switching frequency f_{sw}=400 KHz. To provide the negative input voltage a relatively small capacitor C_x = 30 µF was used. The controller is implemented with an Altera DE2 FPGA board and an off-shelf ADC.

To demonstrate advantages of the proposed converter its response is compared with that of a conventional buck, for the step-down conversion ratio from 6 V to 1.3 V. In both cases the same compensator was used and the control bandwidth is set to be 1/10 of f_{sw} i.e. 40 kHz. First, a conventional operation of a buck converter is realized by limiting the duty ratio command between (0, 1). It is shown in Fig 6 that symmetric load changes between 0A and 3A causes asymmetric load transients. The zoomed in view of the step-down transient shown in Fig. 8 has a 3 times longer settling time and 50mV greater voltage deviation than those of step-up transient. This is due to the duty cycle saturation during load step-down, which can be seen in the N_X node voltage waveform in Fig. 8.

Next, the duty ratio limit is relaxed to $(-1, 1)$ and the flying capacitor is in action. With the exact same test condition the step-down load transient shown in Fig. 9 is

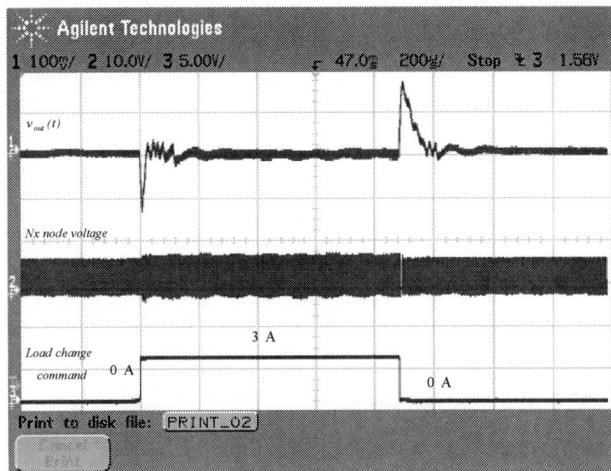

Fig. 6. Asymmetric transient responses of a 6V-1.3V buck converter to a 3 A load change. Ch.1: Output voltage (100 mV/div); Ch.2: Node Nx voltage (10V/div); Ch.3: Load transient control signal. Time scale is 200us/div.

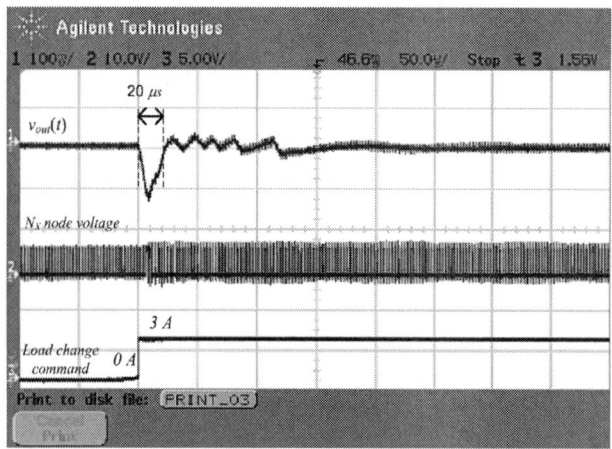

Fig. 7. Enlarged view of a 0A to 3A step-up transient with a settling time of 20us and a voltage deviation of 120 mV. . Ch.1: Output voltage (100 mV/div); Ch.2: Node Nx voltage (10V/div); Ch.3: Load transient control signal. Time scale is 50µs/div.

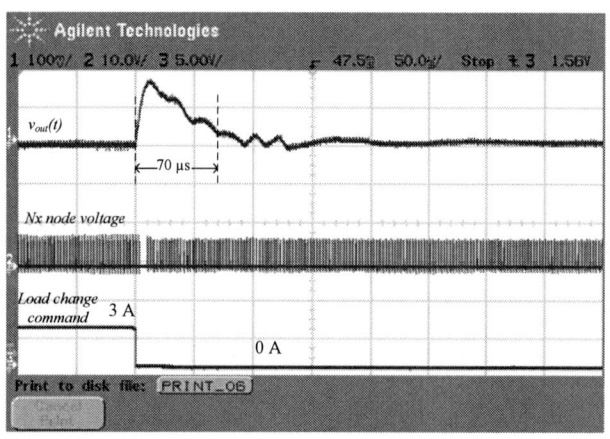

Fig. 8. Enlarged view of the 3A to 0A step-down transient with a settling time of 70us and a voltage deviation of 170 mV. . Ch.1: Output voltage (100 mV/div); Ch.2: Node Nx voltage (10V/div); Ch.3: Load transient control signal. Time scale is 50 µs/div.

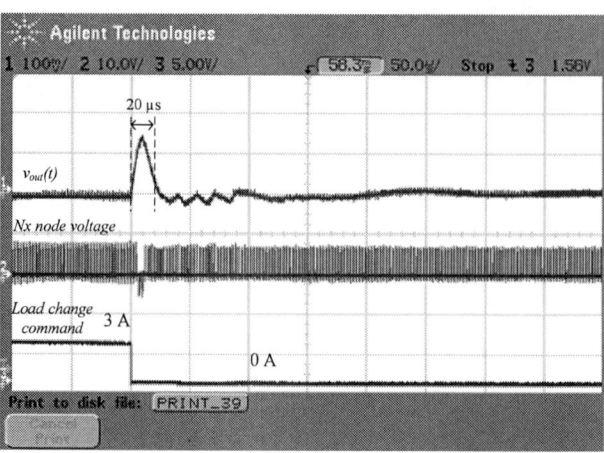

Fig. 9. Improved 3A to 0A step-down transient with a settling time of 20us and a voltage deviation of 130 mV. . Ch.1: Output voltage (100 mV/div); Ch.2: Node Nx voltage (10V/div); Ch.3: Load transient control signal. Time scale is 50µs/div.

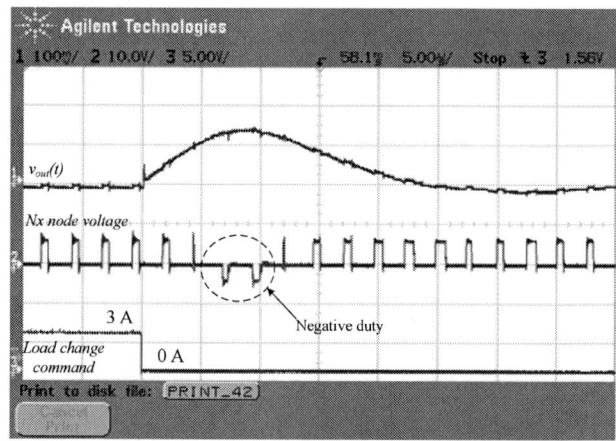

Fig. 10. Detailed view of the improved step-down transient with the negative Nx node voltage shown. . Ch.1: Output voltage (100 mV/div); Ch.2: Node Nx voltage (10V/div); Ch.3: Load transient control signal. Time scale is 50 µs/div.

observed. Compared to that in Fig 8, the proposed converter achieves a 3 times shorter settling time and a 40 mV smaller voltage overshoot. Compared to the step-up transient in Fig.7 the improved step-down transient has approximately the same response, in other words, symmetrical response. An enlarged view of the improved response is shown in Fig. 10 and a negative voltage at the N_X node is observed when duty command becomes negative during the transient. It is can be seen that the presented converter topology is able to enlarge the operating duty ratio range and drastically improve converter performance during step-down load transients.

VI. CONCLUSIONS

A method for improving load transient response through active change of the inductor current slew-rate is presented. The slew rate is improved by using a modified 3-level 5-switch buck converter and digital controller changing the input voltage of power stage to negative value during heavy-to-light load transients. The effectiveness of the method is experimentally verified and 3 times faster response compared to the conventional buck is obtained.

ACKNOWLEDGMENT

The authors would like to thank Mr. Zdravko Lukic for his kind help on the design of experimental setup.

REFERENCES

[1] V. Yousefzadeh, A. Babazadeh, B. Ramachandran, L Pao, D. Maksimović and E. Alarcon, "Proximate Time-Optimal Digital Control for DC-DC Converters," in Proc. IEEE Power Electronics Specialists Conference, 2007, pp. 124 - 130.

[2] Zhenyu Zhao and A. Prodić, "Continuous-Time Digital Signal Processing Based Controller for High-Frequency DC-DC Converters," IEEE Trans. on Power Electronics , Vol. 23, Issue 2, pp. 564 – 573, March 2008.

[3] E. Meyer, Zhiliang Zhang and Yan-Fei Liu, "An Optimal Control Method for Buck ConvertersUsing a Practical Capacitor ChargeBalance Technique Power Electronics," IEEE Transactions on Power Electronics, Volume 23, Issue 4, pp.1802 - 1812, July 2008.

[4] Haitoa Hu, V. Yousefzadeh, and D. Maksimović."Nonlinear Control for Improved Dynamic Response of Digitally Controlled DC-DC Converters," in Proc. IEEE Power Electronics Specialists Conference, 2006, pp.1 - 7.

[5] D. D. -C. Lu, J. C. P. Liu, F. N. K. Poon, and B. M. H. Pong, "A single phase voltage regulator module (VRM) with stepping inductance for fast transient response," IEEE Trans. Power Electron., vol. 22, pp. 417-424, Mar. 2007.

[6] G. E. Pitel and P. T. Krein, "Transient reduction of Dc-Dc converters via augmentation and geometric control," in Proc. IEEE Power Electronics Specialists Conference, 2007, pp. 1652–1657.

[7] P. Xu, J. Wei and F. C. Lee, "Multiphase coupled-buck converter—a novel high efficient 12 V voltage regulator module," IEEE Trans. Power Electron., vol. 17, pp. 74–82, Jan. 2003.

[8] P. Wong, F. C. Lee, P. Xu, and K. Yao, "Critical Inductance in Voltage Regulator Modules," IEEE Trans. Power Electron., vol. 17, pp. 485–492, Mar. 2002.

[9] Intel VR Technology Roadmap, Intel Technology Symposium. 2001.

[10] V. Yousefzadeh, E. Alarcon, and D. Maksimović, "Three-level buck converter for envelope tracking applications," IEEE Trans. Power Electron., vol. 21, pp. 549-552, Mar. 2006.

[11] Z. Zhao, H. Li, A. Feizmohammadi, and A. Prodić, "Limit-cycle based auto-tuning system for digitally controlled low-power SMPS," in Proc. IEEE Applied Power Electronics Conference, 2006, pp. 1143–1147.

Trends in MW-Rated VSI Technology and Reliability for Adjustable Speed Drives

Hiromi Hosoda, Mostafa Al Mamun and Teruo Yoshino
Drive Systems Department, Power Electronics System Division
Toshiba Mitsubishi-Electric Industrial Systems Corporation (TMEIC)
1, Toshiba-cho, Fuchu-shi, Tokyo 183-8511, Japan
E-mail: {HOSODA.hiromi, MOSTAFA.mamun, YOSHINO.teruo}@tmeic.co.jp

Abstract— **In recent years, the drive equipment capacity has been developed to more than megawatt based on demands in various industries. High reliability is indispensable for such large scale equipment to make the total system reliability high. The paper reports trends of the voltage source inverter now applied to very large scale drive equipment. The paper also reports trends of the high capacity self-turn-off semiconductor devices and the inverter circuit configuration, which realize both large and reliable drive equipment. In the paper, reliability factors in manufacturing process of drive equipment are discussed and an example of field experiences is introduced.**

I. INTRODUCTION

In various industries, for higher production capacity, better quality and higher efficiency are required and they are realized by increasing the system scale. The typical examples are the iron steel mills and the oil and gas plants. Along with the scaling up of the systems, the larger capacity is required for the ac motor drive equipment for contributing to energy saving and high efficiency [1]. However, the effect of the reliability of large scale equipment becomes relatively larger comparing to small capacity equipment. Thus, the equipment reliability is one of the important issues for reliable operation of the large scale systems.

II. LARGE SCALE PLANTS AND THEIR RELIABILITY IMPROVEMENT

A. Architecture of A Steel Plant

Fig. 1 shows the basic configuration of drive equipment of a recent steel plant of hot strip mill. The configuration is a cascade system from viewpoint of the reliability theory. Since the rolling mills are arranged in a line, a single contingency stops all of the lines. The reliability depends on the number of equipment arranged in the line. The redundant option is considered as one of measures to improve the reliability. However, the redundancy may not be an economical solution for the large systems. The alternative solution is to keep the

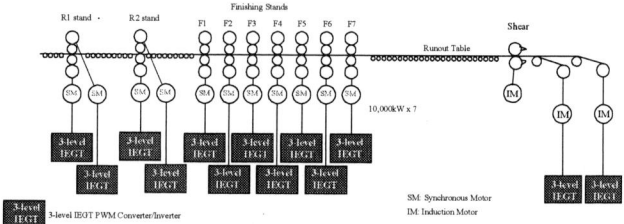

Figure 1. Configuration of a typical hot strip mill

restoration time to be the minimum. The short restoration time improves the system availability sufficiently high [2].

The points of restoration are to find out the reason for the failure to specify the faulted part and to restore the faulted part as soon as possible. For the quick restoration, the trace back function helps the fault analysis by showing the equipment operation information before and after the fault.

Fig. 2 shows a large scale drive equipment. Every phase legs of the inverter is set as a unit in the drive equipment and it has convenient access for the replacement. Moreover, the replacement is possible by the unit base and it makes the restoration work quick and easy.

Figure 2. Typical main drive equipment

B. A Large Scale LNG Plant

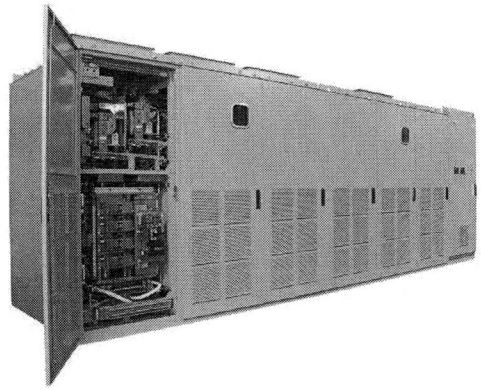

Figure 3. Development of a very large VSI system

The demand of LNG is increasing due to the comparatively less emission of CO_2 among the fossil fuels. For this reason, the large scale LNG facilities are built and planned over the world. Although the compressors in the LNG facility have been operated by gas turbines in general, the electric drive systems are now being applied since they offer higher availability of the system originated from the short maintenance period. As one of the largest drive systems, a 75MW drive system is used in a large scale LNG plant these days.

Fig. 3 shows the recently developed very large Voltage Source Inverter (VSI) system. The VSI is configured with 3 pairs of single phase 3-level GCT[1] inverters which operate as a 3-phase 5-level inverter. The equipment has been developed for 30MVA and 7.2kV output. With this 30MVA GCT inverter, it is possible to obtain 120MVA by connecting four banks together.

III. RECENT DEVELOPMENT OF DRIVE EQUIPMENT

In this chapter, the key factors for high reliability built by Japanese practices are introduced at every stage of the development and manufacturing. The key factors for the high reliability are briefly shown in Fig. 4.

The first factor is the circuit design process. In the process, these days owing to the good simulation tools, the voltage and the current stresses of the parts in the drive equipment can be calculated in detail with sufficient accuracy. Then, the appropriate parts can be selected with appropriate design margins.

The second is the part selection process. The basic idea for high reliability is the minimization of the parts in the equipment. Then, the large semiconductor devices are developed and applied for the large capacity drive. The key parts associated with the device are also developed based on the cooperation between the drive equipment manufacturer and its part vendors.

---

[1]GCT: Gate Commutated Turn-off
(Developed by Mitsubishi-Electric Corporation)

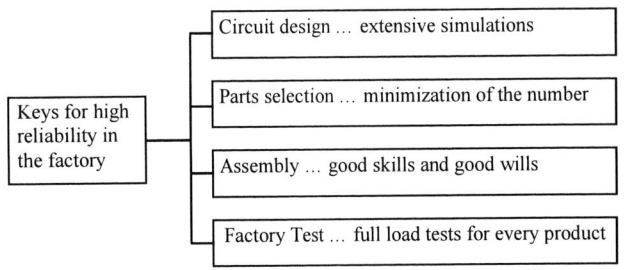

Figure 4. Key factors for high reliability

The third is the reliability built in at the assembly process. As well known, in Japan, the members at the assembly line also participate to realize high quality and high reliability. It should be reminded that the number of connections in the equipment. Then, the assembly quality is one of the key factors.

The fourth is the factory tests. The tests are the final gate for the high reliability. The test under full load condition, which the products will experience in the actual field, is adapted as one of good measures.

A. Reliability Improvement by Enlargement of Device

The reliability of equipment depends on the reliability of every used part, number of the parts and the application condition. If a large scale equipment is built with the parts which is the same to those for the small scale equipment, the numbers of parts are increased because the parts should be connected in series or in parallel to realize the large capacity. In such case, the total reliability will be decreased. In order to improve reliability, one of solutions is to apply large capacity device and to reduce the number of parts as low as possible. From viewpoint of the part operation condition in the equipment, appropriate de-rating improves the reliability.

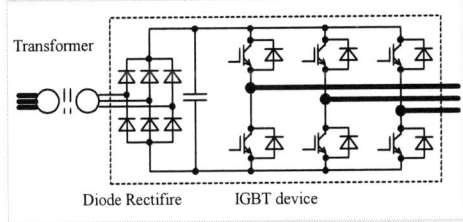

a) General voltage source inverter circuit configuration

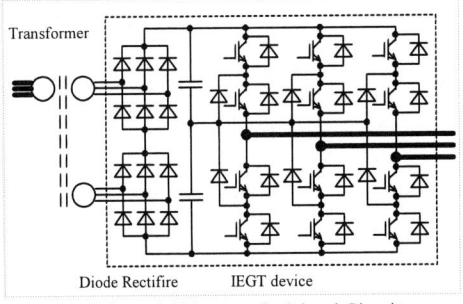

b) Schematic Diagram of a 3-level Circuit

Figure 5. Enlargement of device in the main circuit to improve reliability

Fig. 5a shows the general configuration of voltage source inverter. In this configuration, there are 6 IGBTs[2] and 12 diodes for both inverter and converter portion. This type of configuration is generally used for the drive equipment up to around 1000kVA.

Fig. 5b shows the 3-level configuration and this offers higher output voltage and larger capacity compared with the general configuration in Fig. 5a. The 3-level inverter is configured with 12 IEGTs[3] or GTCs and 30 diodes. With this configuration and with large-capacity devices, it is possible to build drive equipment around 8MVA to 12MVA with small number of devices and parts. Then, this type of drive system is expected to show high reliability.

B. Design Improvement

As an example of the reliability consideration in the design stage, the capacitor selection is discussed in this section. In the drive equipment, the following measures are considered in the design stage to reduce the maintenance works on the electrolytic capacitors.

- The capacitor installed in the power supply input section of the control board can be substitute by high power ceramic capacitor and it is possible to make it maintenance free.

- As there is no substitute device of the electrolytic capacitor for power supply unit, it is mounted in such a way so that it can be exchanged easily.

- The ceramic capacitor can be applied because of the compact configuration of the IGBT or IEGT gate drive unit and it is possible to make it maintenance free.

Figure 6. GCT power device

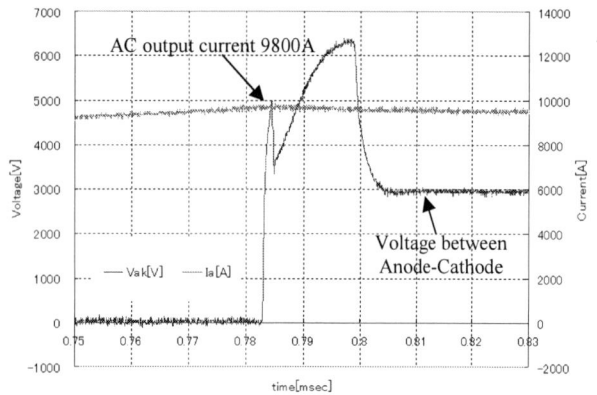

Figure 7. Limit of cut-off current test waveform of GCT at ambient temperature

Large gate current is necessary for the GCT thyristor. To ensure the current, gate drive circuits are placed close to the GCT thyristor. By optimizing the drive circuit arrangement, a new gate drive unit (GDU) has been developed successfully with the ceramic capacitors [3].

Further to ensure the reliability, design evaluation test has been conducted by combining the GCT thyristor with the developed GDU. The maximum cut-off current of 9800A has been proven as shown in Fig. 7. Although the nominal cut-off current rating is 6000A, the evaluation test showed that the GDU helped to extend the cut-off current more than 1.5 times. In other words, by applying the developed GDU, the GCT thyristor has much more cut-off current margin and expected to improve the operation reliability.

C. Devices Verification and Quality Check

The reliability of each semiconductor device is one of the important factors for equipment reliability. Then it is preferable to check each device one by one during the manufacturing process. The fundamental tests like operation tests are usually performed. However, additional check processes sometimes are effective to ensure the device reliability.

Fig.8 shows the typical check flowchart in the manufacturing and the test processes. Careful investigation is performed at every step to make sure that all the devices have sufficient characteristics and show good performance.

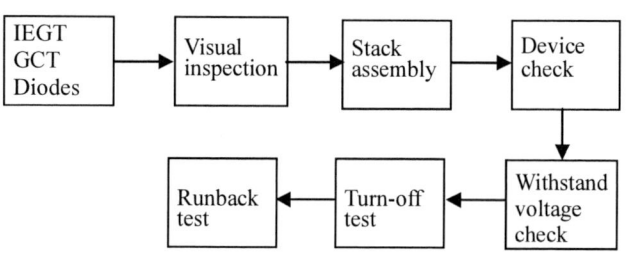

Figure 8. Devices verification flowchart

D. Reliability Improvement by Reformed Verification

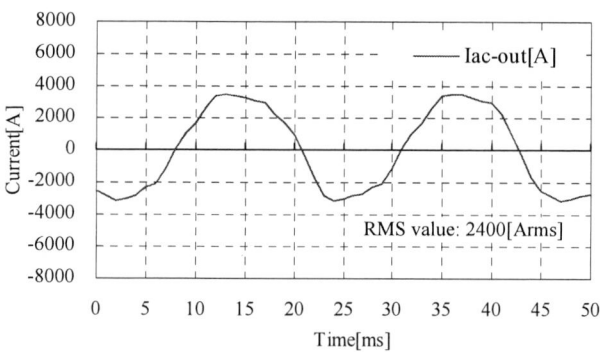

Figure 9. An example of the rated current test waveform (Runback Test)

---

[2]IGBT: Insulated Gate Bipolar Transistor
[3]IEGT: Injection Enhanced Gate Transistor
(Developed by TOSHIBA Corporation)

The test with low voltage and rated current and the test with the rated current and small current are separately performed for the thyristor equipment as a common practice. However, this practice is not sufficient for the inverters made of self turn-off devices, IGBT, GCT or IEGT. The current and voltage stresses of these devices depend on the output voltage, current, frequency and power factor. It is appropriate to test them according to the actual operation condition. A special test system, Runback Test system, has been developed for this purpose. The Runback Test is done for all the IEGT and GCT stacks before shipment.

The inverter and converter are controlled as if they are in the actual field operation. The output of the inverter, frequency, voltage, current and power factor can be controlled. So, all kinds of operation mode including low frequency large current test can be performed at the factory.

In the developed test circuit, most of the test power is circulated and thus even a very large capacity equipment can be tested at the factory. Fig. 9 shows an example of the rated current test waveform of the Runback Test system. It shows one of the output currents of the inverter running at 45Hz and 100% load current.

E. Reliability Built in at the Assembly Line

The reliability discussions are sometimes made based on failure rate calculation from the part numbers and failure rates. In the calculation, the redundant configuration and the time to restoration are also considered. However, one of important factors may be missing. It is the assembly line quality. In other words, it means the good skills and good wills of the assembly line.

Careful observation of equipment indicates the fact that the numbers of connections are larger than the number of the parts. For example, a resistor has two terminals and they are connected by screws or solder to the wires or the bus bars. For main current conduction, the cables or bus bars are necessary. Then bolts and nuts are also necessary to connect them each other or to the terminal blocks. Furthermore, looking onto the electronic cards, the same situation is also found. The number of connection from a part to the circuit board is larger than the number of the parts.

Only by these simple examples, it can be easily understood that the number of connections in equipment is lager than the number of the parts. Although these connections are invisible in the electric circuit diagram, however, in the actual world, a bad connection results in the visible trouble.

In Japan, the members of the assembly line participate in quality control. The most famous actions taken by the assembly line is known as "Kaizen". They develop the quality process by themselves and sometimes give feedbacks to the design sections with good wills. They also keep their good skills by periodical educations and skill license tests. These efforts realize very good reliability of the connections. Their quality control activity is based on Japanese industrial culture beyond the international quality standards.

Most of the development engineers stay with the manufacturer for their life in Japan. This makes another difference. Namely, the feedbacks from the assembly line can be firmly reflected to the design to improve the performance and the reliability. This plan-do-see cycle improves the product reliability. The feedbacks from the field also improve the products by the original development engineer.

The actions above are also repeated at the assembly line of the part vendors in Japan. Taking the high voltage capacitor as an example, it is not just a single block. Inside of the can, several elements are connected with leads by soldering. Then, even for the part, the assemble quality is very important. In Japan, the equipment manufacturer and the part vender work together to make the part better with information exchange including the assembly line quality control process.

IV. FIELD EXPERIENCES OF VSI MOTOR DRIVES

The VSI technology plays main roles these days in various fields and has the history more than 30 years. Compared with the early stage of VSI, the inverter reliability is improved by the development of semiconductor device.

At the early stage, GTO was the major player of VSI and then IGBT was developed. However, they are used for rather small capacity and the new generation of high-capacity semiconductor devices is developed. They are GCT and IEGT. They offer circuit simplification resulting in smaller number of parts in the circuit compared with the old generation.

Fig. 10 shows the cumulative numbers of the large drive equipment based on the VSI technology manufactured in a factory from the year 1990. From the figure, it is clearly indicated that production of large VSI-based drive equipment increases year by year. Up to now, more than 1000 banks of VSI based large drive equipment rated at more than 8MVA have been manufactured at the factory taken as an example. After 2000, the VSI production occupies the majority of the factory. The trend is considered to be endorsed by the maturity of the VSI technology and the establishment of reliability of the large capacity drive equipment.

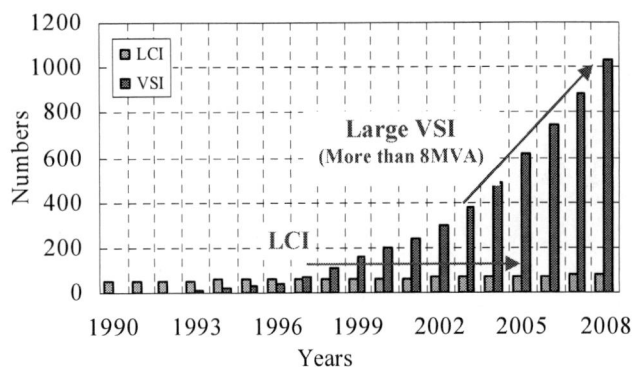

Figure 10. Cumulative number of supplied drive systems by years

Regarding the VSI equipment, more than 800 are operating in the market at present and the Mean Time Between Failure (MTBF) is expected more than 30 years based on the statistical calculation.

V. REDUNDANT OPTION

A. (n+1) Device Series Connection Redundancy

The reliability of the drive system can be improved using the redundant option. In case of the series connected thyristors, the option of (n+1) redundant series connection can be applied owing to its circuit characteristics. The VSI has different circuit characteristics and then special cares are necessary for the series connection. Then, for some cases, the (n+1) redundancy may not be a practical solution.

B. Bank Redundancy

Fig. 11 shows a large VSI system consists of 4 parallel connection of the inverter. All those four inverters operate in parallel and if a fault occurs in an inverter, the faulted inverter can be disconnected and the system continues operation by the following process.

1) Fault detection → 2) Inverter operation stop → 3) Disconnection of the faulted inverter → 4) Operation with remaining three inverters.

C. Control Circuit Redundancy

Two sets of automatic control panels, C1 and C2 are also shown in Fig. 11. The two panels configure the stand-by redundant control system. The inverter is controlled by the controller selected to be in service. The inverter cubicle is equipped with the signal selection circuit. The selection circuit selects one of the gate signals from the two automatic control panels. The selection circuit usually selects the gate signal from the control panel in service. When the control panel in service fails, a change over signal is issued and the selection circuit will select the other signal from the stand-by control panel. Then, the inverters can continue the operation with the stand-by controller.

Meanwhile, the faulted control panel will be turned off for maintenance and restored. The restoration can be done while the system is operating, then the restoration time can be considered practically zero. Thus, the redundant control improves the total system reliability.

VI. CONCLUSIONS

The recent plant systems are getting larger and larger to improve their production efficiency. However, in order to keep the operation availability sufficiently high, the system reliability has significant importance. For such large plants, the capacity of the drive equipment is also required to be increased to megawatt or tens of megawatt range with high reliability.

In this paper, the recent technical trend in the field of the large drive equipment based on VSI built with the large capacity semiconductor devices. The large capacity semiconductor and its application technology help to realize simple inverters with small number of parts. This type of drive equipment is expected to be reliable and the field experience showed the expectation.

Other key factors for the reliability are also introduced from viewpoint of the manufacturing process to build the reliability in equipment.

REFERENCES

[1] M. Tsukakoshi, M. A. Mamun, K. Hashimura and H. Hosoda, "Performance evaluation of a large capacity VSD system for oil and gas industry," 2009 IEEE Energy Conversion Congress and Exposition (ECCE 2009), San Jose, CA,USA, pp. 3485-3492, September 2009.

[2] H. Hosoda, S. Wada, S. Kodama and L. Junfeng, "Recent hot strip mill in China," The Sixth International Conference on Power Electronics and Drive Systems, Kuala Lumpur, Malaysia, Session 4A P0286, pp. 1-6, November 2005.

[3] M. Tsukakoshi, M. A. Mamun, K. Hashimura and H. Hosoda, "Study of large VSI drive system for Oil & Gas industry," 38th Turbomachinery Symposium, George R. Brown Convention Center, Houston, TX,USA, pp. 213-217, September 2009.

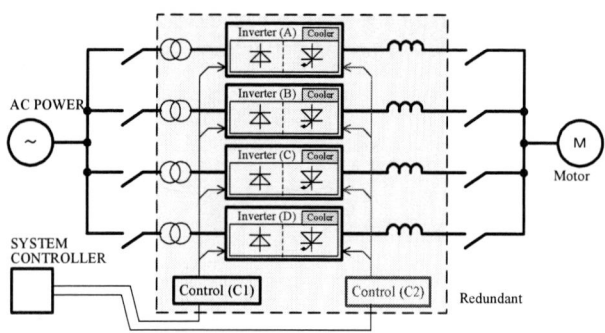

Figure 11. Example of a large VSI system

Development of a Compact 750KVA Three-phase NPC Three-level Universal Inverter Module with Specifically Designed Busbar

Jun Wang, Binjian Yang, Jing Zhao, Yan Deng,
Xiangning He
College of Electrical Engineering
Zhejiang University
Hang Zhou, China
cooldynamic@gmail.com

Xu Zhixin
School of Science
Zhejiang University of Science & Technology
Hang Zhou, China

Abstract—**This paper presents the bus bar design of a 750kVA three-phase Neutral Point Clamped three-level universal inverter module (PEBB) of high power density. ANSOFT Q3D is used in comparative evaluation for three types of multi-layer bus bar structures and a novel low inductive laminated configuration with divided connections is proposed for this inverter modules. Combined with specially arranged snubber capacitors distribution, reduced voltage overshoots across IGBTs at turned-off and lower loss within the electrolytic capacitors are achieved without any RC snubbers, along with optimized high frequency current equalization. The stray inductance of the bus bar is simulated as a contrast to the test results. Feasibility and improved performance are confirmed by experiments.**

I. INTRODUCTION

High power NPC three-level topology has been widely used in medium-voltage AC drives nowadays. Developing such converters requires specialized considerations and experience. They often have to be redesigned to meet the custom's various demands, which are costly and time-consuming [1]. Thus we have developed a compact three-phase NPC three-level inverter module in the concept of Power Electronics Building Block (PEBB) for flexible applications like wind turbines, ships and tractions. The specifically designed busbar is used to minimize stray inductance and to prevent the electrolytic capacitors from overheat. In this paper a prototype rated 750KVA has been successfully developed.

II. MAIN CIRCUIT CONFIGURATION AND SYSTEM DESCRIPTION

As shown in Fig.1, the main circuit of the inverter module consists of DC capacitors, three-phase NPC three-level devices and braking devices. Eight terminals stretch out as power interfaces, of which P, O, N are dc inputs and U, V, W

China's National Natural Science Foundation (50737002)
Science and Technology Department of Zhejiang Province (2006C11005)

are outputs. Dynamic braking resistors should be connected with P-X and N-Y in case of the overvoltage of DC buses. Fig.2 shows that the control system is integrated in the module where signals sampled from four voltage sensors and three current sensors are used for feedback control and system protection. The specification is as follows:

- Rated Input voltage: 2000 V dc
- Rated output voltage: 1140 V rms
- Rated output current: 380 A rms
- Output frequency: 3~60 Hz
- Switching frequency: 1 kHz
- IGBT rating: 1700V, 800A
- FRD rating: 1700V, 600A
- Dimensions: 1000×700×500 mm (L×W×H)
- Total weight: 180 kg
- Water Cooling

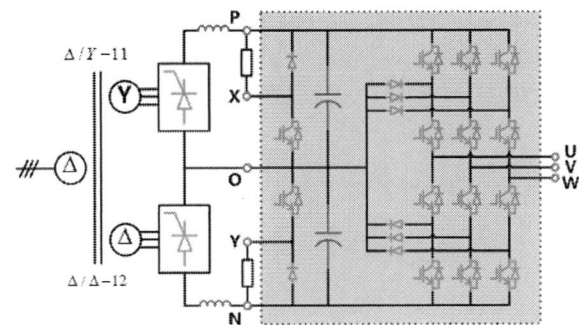

Figure 1. Main circuit configuration of developed inverter module of NPC three-level topology.

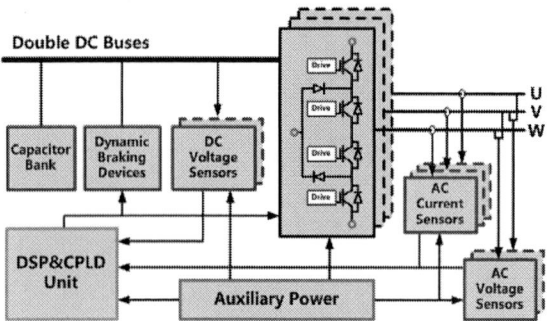

Figure 2.   Control block diagram

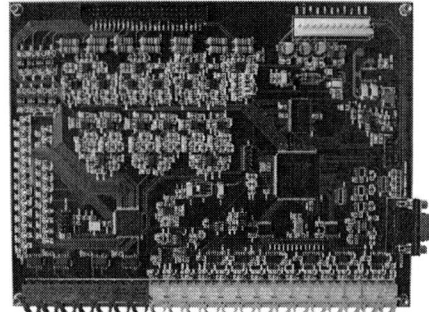

Figure 3.   DSP&CPLD board

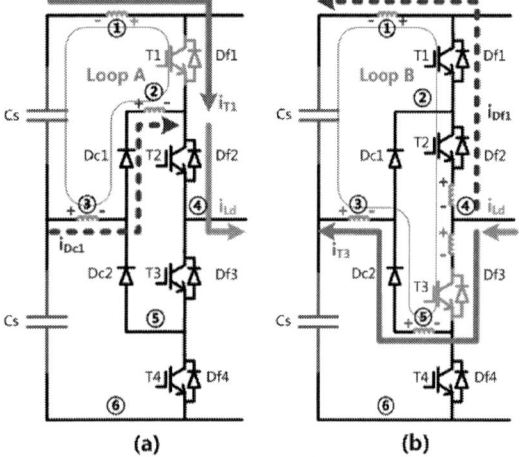

Figure 4.   Basic commutation loop of NPC three-level topology

III.   STRUCTURE AND LAMINATED BUSBAR DESIGN

A.   Effect of Stray Inductance of Commutation Loop

In the NPC three-level topology, there are eight forced commutation modes and six natural commutation modes. Because of the symmetry of the two dc voltages, four forced commutation modes are representative when the output voltages are positive. The study could be focused on the two of the four when IGBTs are being turned off. One is shown in Fig.2a while T1 is being turned off and T3 on with the output current from the inverter to the load. The other is shown in Fig.2b while T3 is being turned off and T1 on with the output

current from the load to the inverter. During commutation the red current i_{T1} or i_{T3} is reducing and blue current i_{Dc1} or i_{Df1} increasing very fast to induce voltages on the stray inductance of Loop A or Loop B, which lead to an overshoot voltage on the IGBT that is being off. The overshoot voltage may cause destruction of IGBTs, more switching loss and severe common mode EMI [2], so it is crucial to minimize the stray inductance of commutation loops.

B.   Study on Multi-layer Laminated Busbar and Contrast of Different Lamination Schemes

Using laminated busbar has been proved a good way to minimize the stray inductance. As shown in Fig.5, according to proximity effect, when w>>h the high frequency current distribution tends to concentrate on the adjacent surfaces of the signal trace and the ground plane. The opposing magnetic flux lines set up by oppositely directed bus currents lead to desirable subtractive radiated emissions [3], which means the stray inductance of the commutation loop that consists of bars is much lower [4].

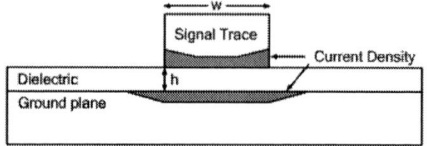

Figure 5.   High frequency current concentration on adjcent conductors

When it comes to NPC three-level topology, more than two bars are connected among power devices and capacitors in a commutation loop and in that case the number of lamination layers increases to three or more. The study here on the high frequency current trace and stray inductance of multi-layer laminated busbar is based on the simulation of three-layer lamination structure in ANSOFT Q3D. In Fig.6, 8, 10, the current flows are directed by arrows and the proximal real line and dashed compose a couple of image current traces. Power devices are marked as double lines. The shape and terminals of busbar in all simulations are simplified to identical extent and power devices are supposed to be in the same position to guarantee that the lamination scheme is the only factor which influence the inductance value. The dimensions of copper busbar are as follows (mm): length 200, width 100, thickness of bar 1.5, thickness of dielectric 1.

As shown in Fig.6, in the first type of scheme, the current flows into the outer layer and out from the other outer one. The image current is induced as a closed loop on the surface of the left part of inner layer where low frequency current is not supposed to be. That is validated by Fig. 7a in which the red circles indicate the current distributions of both side of Fig.6 that are magnified in Fig. 7b and Fig. 7c.

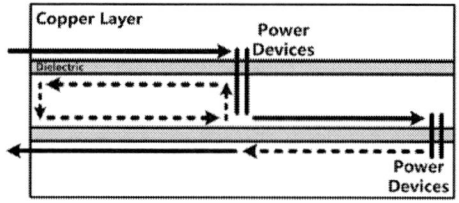

Figure 6.   Laminating scheme and current traces of Type I

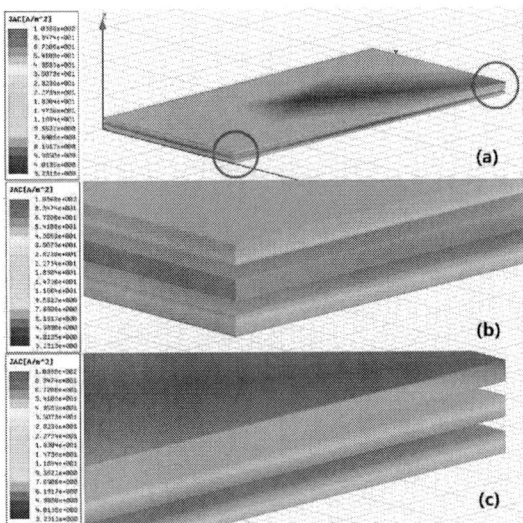

Figure 7. Surface current distribution of Type I

As shown in Fig.8, in the second type of scheme, the current flows into the inner layer and out from an outer one. The image current is induced on the surface of the right part of inner layer, which is validated by Fig. 9a where the red circles indicate the current distributions of both side of Fig.8 that are magnified in Fig. 9b and Fig. 9c.

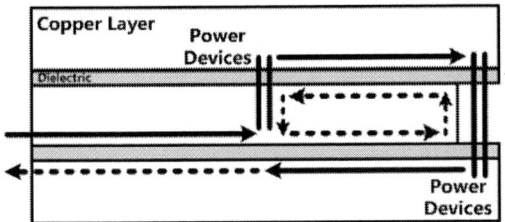

Figure 8. Laminating scheme and current traces of Type II

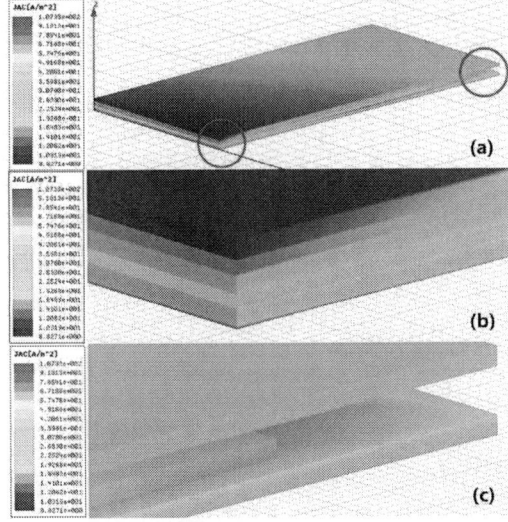

Figure 9. Surface current distribution of Type II

These two simulations illuminate that when three layers are laminated, image current could be induced on the surface of the inner layer where no current should be if its frequency

is not high enough. The opposing magnetic flux lines to the inner surfaces of outer layers, set up by the induced current in inner layer, respectively attenuate the radiated emissions, which lead to a minimized inductance of the three-layer busbar. This phenomenon still exists if more layers are laminated.

As shown in Fig.10, in the third type of scheme, two plates of copper bar are in the same plane. Therefore, three-layer lamination is simplified as a two-layer one and the current distribution is almost the same.

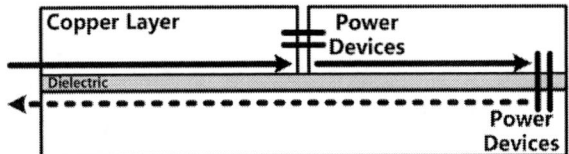

Figure 10. Laminating scheme and current traces of Type III

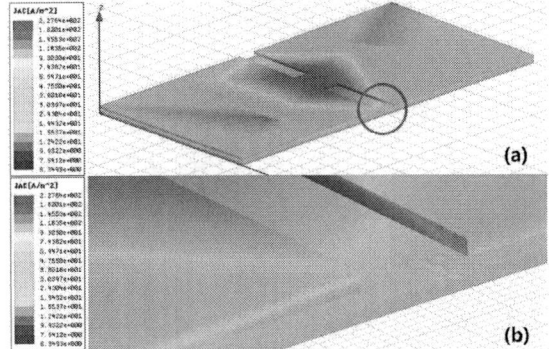

Figure 11. Surface current distribution of Type III

If the busbar is not laminated, the current distributes as what Fig.12 shows. The magnetic flux lines cannot be counteracted although the current tends to concentrate adjacently as well.

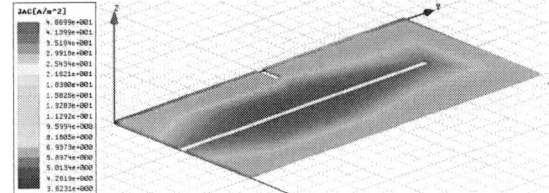

Figure 12. Surface current distribution of non-laminated scheme

TABLE I.   SIMULATION RESULT OF FOUR TYPES OF BUSBAR

Simulation Result	Lamination Schemes			
	I	II	III	Non-laminated
Stray Inductance/nH	3.36	3.31	4.30	84.54

It is evidently shown from the simulation results in Tab.I that the stray inductance is significantly reduced when the busbar is laminated. Moreover, lamination scheme of Type II is slightly better than Type I and Type III and supposed to be the first choice of three-layer laminations. Positions of capacitors and power devices should also be taken into consideration when the lamination scheme is being made.

C. Configuration of 3-D Layout and Snubber Capacitors Arrangement

Fig.13 shows the 3-D design of the inverter module. The capacitor bank put in a steel container is positioned at the bottom because it is the heaviest part of the module. The drawer-like design makes it easier to pull out the capacitor bank when maintenance. Seven IGBT modules, eight fast recover diodes, three balance resistors and four voltage sensors are placed on the water-cooling heat sink as shown in Fig.14. There are also several temperature switches on the heat sink to provide protect signals to the control board. The water inlet and outlet are at the opposite side of the electrical terminals to separate water pipes and electrical connections. The channel routes inside the heat sink on which all the power devices are passed are marked with pink arrows.

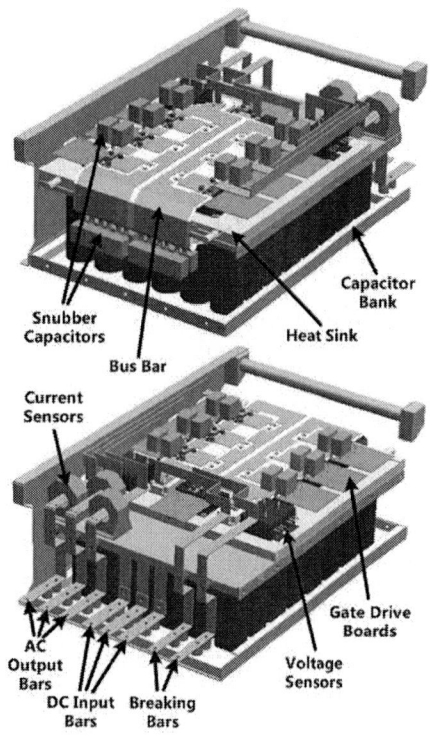

Figure 13. 3-D layout of inverter module

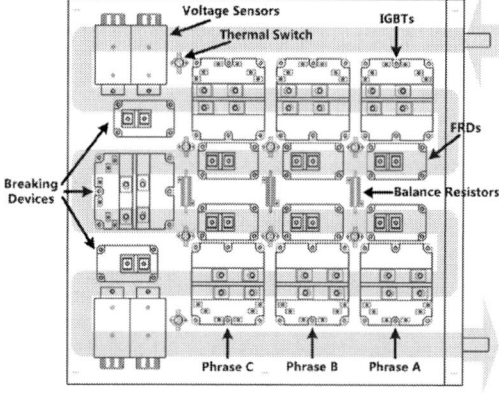

Figure 14. Placement of power devices

The stray inductance of the main circuit is shown in Fig.15. A lot of connection inductance L_{Sd} is produced when electrolytic capacitors are series-connected and paralleled. The large capacitor dimension makes distinctively different distances between electrolytic capacitors and power devices and thus the capacitors closest to the power devices undertake more high frequency current that may lead to an overheat destruction [5].

Accordingly, connecting inductance L_{Sd} and commutation loop inductance L_{SA} and L_{SB} ought to be diminished by using laminated busbar. The low inductive busbar in this system is divided into two parts, one for power devices and the other for electrolytic capacitors. The two parts are connected by tooth-like terminals on which twelve snubber capacitors C_m are mounted to suppress overshoot voltage produced by L_{Sd}. They can also undertake some high frequency current for the electrolytic capacitors. Snubber capacitors C_s are put very close to power devices and equivalently the lengths of all the commutation loops are dwindled.

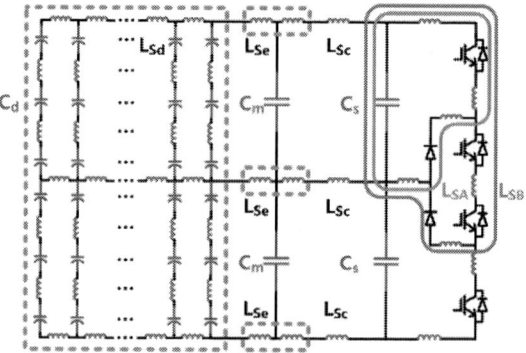

Figure 15. Stray inductance distribution of the main circuit

D. Low Stray Inductance Busbar Design

The neutral point copper layer participates in all of the commutation loops that could be inferred from the basic loop A and B. Therefore, it is the best way to laminate the neutral point copper layer as a mutual platform which could not only brace the other layers but also provide a path for induced image current. In respect that the positive (red) and negative (blue) bus are no included in a same commutation loop, as shown in Fig.16, it is not necessary to laminate them together. They are placed on the neutral point layer in the same plane. Type II of lamination scheme is chosen here because it could provide the lowest inductance and is easy to assemble. The copper bars with numbers on Fig.16 correspond with the ones in Fig.4; the marks of current and the image of it are the same as marks in Fig.6. Fig.17 is the 3-D configuration of laminated busbar for power devices.

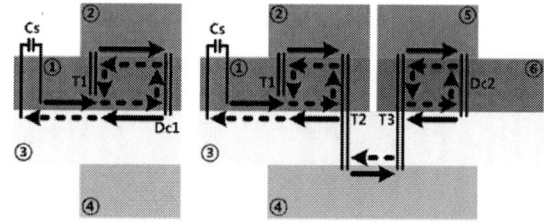

Figure 16. Cross section of laminated bus bar for devices and high frequency current traces

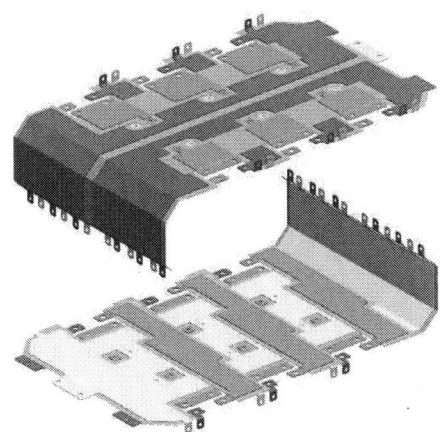

Figure 17. 3-D configuration of laminated bus bar for devices

As shown in Fig.18 Type III of lamination scheme is selected for electrolytic capacitors due to the mass connections of them. In the design the positive (red) and negative (blue) bus are separately made provide the current path for the other connections in two commutation loops. Fig.19 is the 3-D configuration of laminated busbar for electrolytic capacitors.

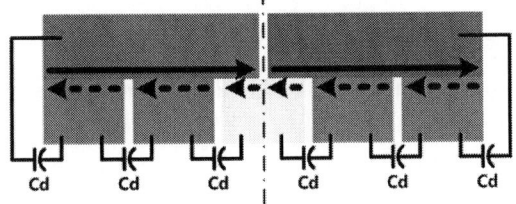

Figure 18. Cross section of the laminated bus bar for capacitor bank and high frequency current traces

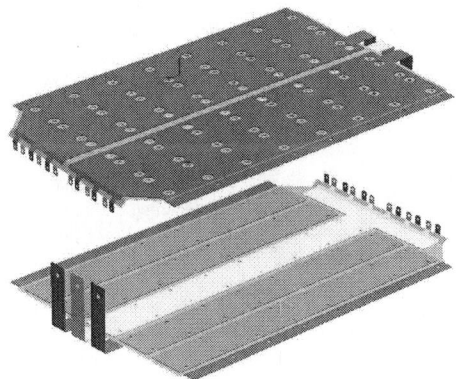

Figure 19. 3-D configuration of the bus bar for capacitor bank

IV. SIMULATION AND EXPERIMENTAL RESULTS

Fig.20 shows the surface current concentration on the neutral point copper layer. It can be seen that the current doesn't distribute uniformly because the proximity effect makes it corresponding with that of the positive and negative bus. The stray inductances are extracted from simulations in ANSOFT Q3D: L_{tA}=38.7nH, L_{tB}=117.2nH, and the loop inductances are:

$$L_{SAs} = L_{tA} + L_{sCE} + L_{sAK} = 73.3nH \qquad (1)$$

$$L_{SBs} = L_{tB} + 3L_{sCE} + L_{sAK} = 192.2nH \qquad (2)$$

L_{sCE} is the inductance of IGBT and L_{sAK} is the inductance of diode.

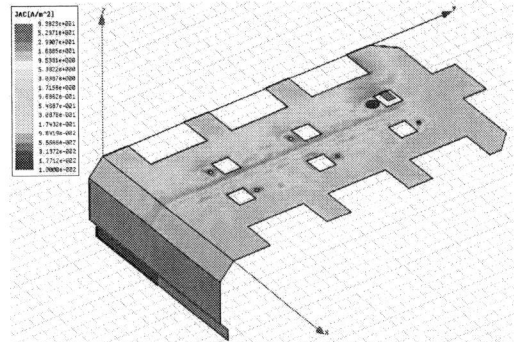

Figure 20. Current distribution on the surface of neutral point busbar

Photos of the prototype are shown in Fig.21. In the experiment the power into the prototype is supplied by double dc voltage sources that are produced by a 12-pulses rectifier, on the bottom of which the prototype is installed.

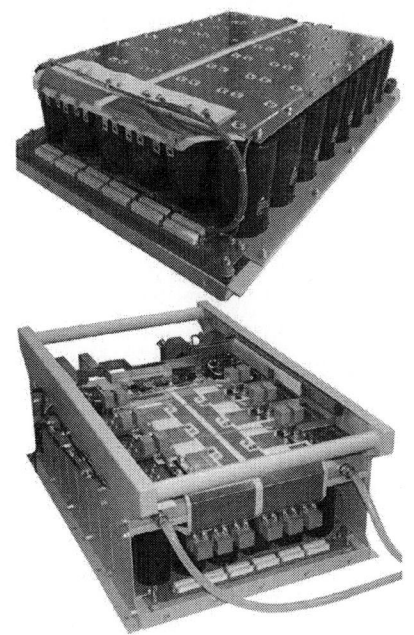

Figure 21. Photos of capacitor bank and inverter module

The maximum of rated current on T1 is about I_{T1off}=500A and that of T3 is about I_{T3off}=250A that is calculated by the rated power factor φ=0.86 according to the commutation fundamental in Fig.4b. The waveforms of T1 and T3 at turned-off are shown in Fig.20 and 21 in which U_{T1ov}=150V and U_{T3ov}=200V can be read. The fall time at turned-off is about t_{off}=260ns and the loop inductances are calculated as:

$$L_{SAe} = U_{T1ov}\, t_{off} / I_{T1off} = 78nH \qquad (3)$$

$$L_{SBe} = U_{T3ov}\, t_{off} / I_{T3off} = 208nH \qquad (4)$$

TABLE II.   CONTRAST OF THE LOOP INDUCTANCE OF LOOP A AND B

Loop	Loop Inductance/nH	
	Simulation	Experiment
Loop A	73.8	78
Loop B	192.2	208

In the contrast in Tab.II, the experimental result is very close to that of simulation. The design target is met that the overshoot voltages are below 200V in rated condition and none of RC snubbers are required.

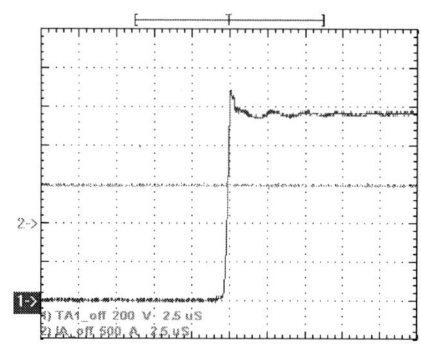

Figure 22. Overshoot voltage on T1 at turned-off (500A)

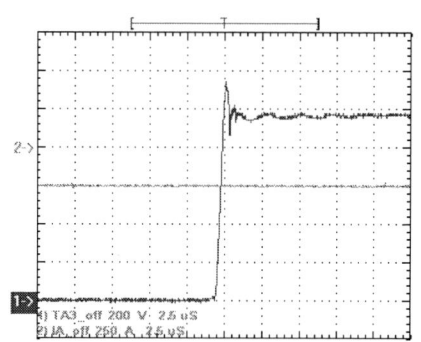

Figure 23. Overshoot voltage on T3 at turned-off (250A)

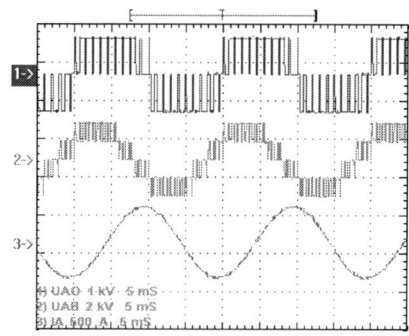

Figure 24. Output voltage U_{AO}, U_{AB} and I_A

Fig.24 shows the output waveforms of the inverter module in three-phrase operation under the rated power of 1140V rms and 350A rms. In the experiment, the temperature of electrolytic capacitors settled down at 38℃ and power devices 42℃ after 30 minutes. The coolant temperature rise was no more than 6℃. These statistics illustrated a successful design of prototype.

V.   CONCLUSIONS

A compact three-phrase NPC three-level inverter module of 750KVA has been developed and several conclusions are drawn:

- The specifically designed laminated busbar extenuates the overshoot voltage on every power device without any RC snubbers and lowers the temperature of electrolytic capacitors under rated condition, and meantime a compact configuration and excellent EMC are obtained.

- There are several schemes for multi-layer lamination. The placement of components should be taken into account when the scheme is being selected.

- The snubber capacitors could suppress some the overshoot voltage on power devices and undertake some of the high frequency current for electrolytic capacitors to lower the temperature of them.

REFERENCES

[1] F. Z. Peng, Jin Wang, Fan Zhang, Mingqian Zhao, "Development of a 1.5 MVA universal converter module for traction drive and utility applications," IEEE 2005 Power Electronics Specialists Conference, 2005, pp.2290-2295.

[2] M. C. Caponet, F. Profumo, R. W. De Doncker, A. Tenconi, "Low stray inductance bus bar design and construction for good EMC performance in power electronic circui," IEEE Trans on Power Electronics, 2002, 17(2), pp.225-231.

[3] D. Moongilan, "Skin-effect modeling of image plane techniques for radiatedemissions from PCB traces," IEEE 1997 International Symposium on Electromagnetic Compatibility, 1997, pp.308-313.

[4] G. L. Skibinski, D. M. Divan, "Design methodology & modeling of low inductance planar bus structures," Fifth European Conference on Power Electronics and Applications, 1993, 3, pp.98-105.

[5] E. Clavel, J. Roudet, Y. Marechal, "Design of a commutation cell of a high power IGBT inverter-the contribution of the simulation," IEEE 1997 Industry Applications Conference, 1997, 2, pp.1014-1021.

Common Mode Voltage in DC-Fed Motor Drive System and its Impact on the EMI Filter

Fang Luo[1,4], Shuo Wang[3], Fred Wang[2], Dushan Boroyevich, Nicolas Gazel, Yong Kang

1 Center for Power Electronics Systems Virginia Polytechnic Institute and State University Blacksburg, VA, USA, 24060 Fangluo@vt.edu	2 Department of Electrical Engineering and Computer Science The University of Tennessee and Oak Ridge National Lab, Knoxville, TN 37966, USA fred.wang@utk.edu	3 Electrical Power Systems, GE Aviation Systems, Vandalia, OH 45377 Shwang6@vt.edu	4 College of Electrical and Electric Eng. Huazhong Univ of Sci & Tech Wuhan China 430074

Abstract—**Common mode (CM) choke saturation is a practical problem in common-mode filter applications. It's generally believed that the leakage inductance of common-mode chokes makes the core saturated. This paper analyzes two new mechanisms for common-mode choke saturation due to common-mode voltage, and these mechanisms are verified in experiment. Common-mode choke saturation is particularly important for motor drive systems, which have a high common-mode voltage and comparably higher stray grounding capacitance. A model is established to describe the relationship between the common-mode voltage and the volume of the common-mode magnetic components. According to the analysis, LISNs (line impedance stabilization networks) play an important role in the design of CM magnetic components.**

I. INTRODUCTION

Common-mode filters play an important role for noise reduction in power electronic systems, but the common-mode magnetic component can take up to 25% of the system volume. The saturation of the CM inductor core is a complex and important issue in CM filter design and the volume of CM inductors will highly impact power density of converters [12-15]. In previous literature [1, 2, 3], it has been proven that saturation can be caused by the leakage inductance. But in practice, the common-mode choke can still be saturated even if it's designed with a leakage inductance constraint. Although overdesigning the magnetic component is a safe way to prevent the core from saturation, the resulting bulky size is not desirable for high-power-density converter design. References [4, 5, and 6] determined that for three-phase AC-DC-AC motor drive systems, common-mode voltage can be a factor determining flux density inside common-mode magnetic cores. However, it is still unknown how this common-mode voltage impacts the common mode choke. This paper describes two new mechanisms of CM choke saturation caused by CM voltage and low grounding impedance, which is dominated by the LISNs in the test. Based on this knowledge, the CM choke volume can be related to CM voltage, so a new CM choke core selection method is proposed. The proposed method provides a criterion for high-power-density EMI (electromagnetic interference) filter design.

This project is supported by SAFRAN group
The first author is also supported by Chinese Scholar Council

II. II. ANALYSIS OF COMMON-MODE EMI PATH

An EMI test setup for a motor drive system with an LC common mode filter is shown in Figure. 1. C_y is the common mode capacitor (100 nF in this application); $C_{s-heatsink}$ is the stray capacitance between the power stage and ground (including cable CM stray capacitance), which is usually in the tens of pF; $C_{s-chassis}$ is the stray capacitance between the motor windings and chassis, which is several nF in this case. For safety concerns, the motor chassis is grounded in practical applications. Point O is the equivalent AC mid-point of a three-phase load. The voltage of point O varies due to different switch combinations during the modulation, as shown in Figure. 1. Here the LISN is defined by military standard MIL-STD-461E, and the EMI measurement frequency ranges from 10 kHz to 10MHz. The schematic of the LISN is shown in Figure. 2. Here we assume the motor drive is using center alignment for space vector modulation.

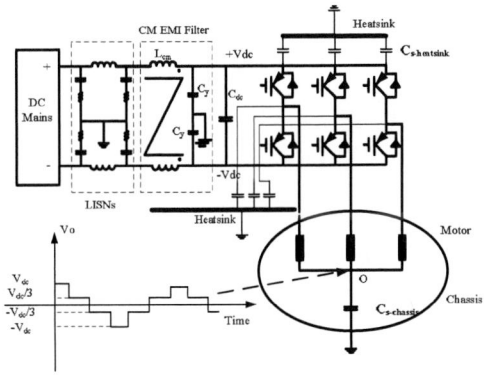

Figure 1. Motor drive EMI test setup

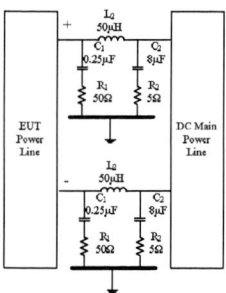

Figure 2. LISN schematic from MIL-461 standard

Figure 3 shows the equivalent circuit for the common-mode noise path showed in Figure 1. In Figure 3, $C_s = C_{s\text{-chassis}} + C_{s\text{-heatsink}}$, and C_{LG} is the line-to-ground stray capacitance inside the DC main, which is usually very small. To conduct CM noise, two LISNs are in paralleled. Figure 3 has - two practical paths for common-mode current; P1 (red) and P2 (blue). The calculated path impedances are shown in Figure 4. The figure shows that the impedance of P2 is lower than P1 at frequencies below 150kHz. The fundamental frequency of common-mode voltage V_o is equal to the switching frequency [7] at around 10 kHz, which is much lower than 150kHz. Thus the CM noise path can be further simplified into the schematic in Figure 65. Furthermore, components V_o, C_s and C_y can be replaced by a Thevenin's equivalent circuit, as shown in

Figure 6. In real cases, because $C_2 >> 2*C_y >> C_s$, $C_{eq} \approx C_y$.

Similar to C_{eq}, when $L_{cm} >> L_2$, $L_{eq} \approx L_{cm}$. At frequencies beyond 150kHz, R_2 becomes R_1, and C_{eq} is equal to C_y in series with C_1.

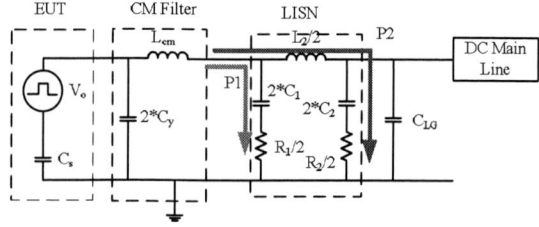

Figure 3. Equivalent CM path for motor drive system

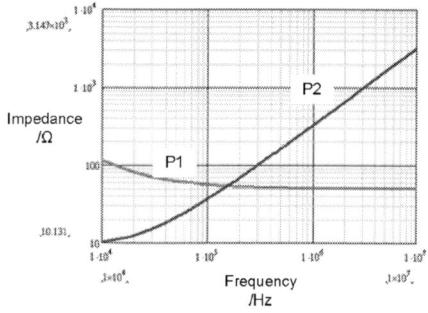

Figure 4. Impedance of Path 1 and Path 2

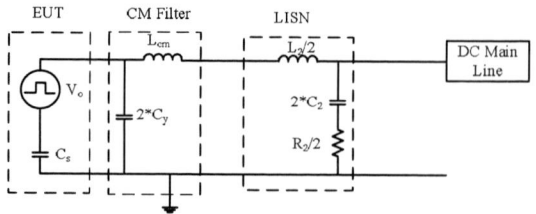

Figure 5. Simplified CM path below 150 kHz

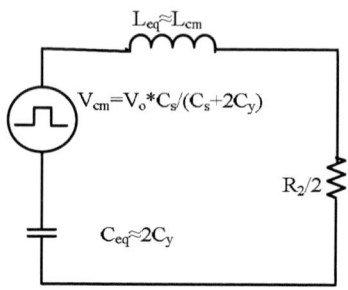

Figure 6. Equivalent loop of CM path

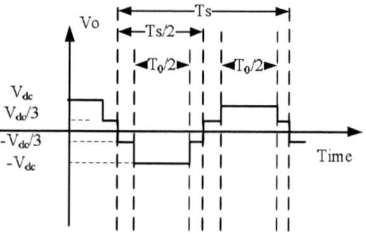

Figure 7. Common-mode voltage time-domain waveform

III. SATURATION OF CM INDUCTOR DUE TO THE RESONANCE IN CM EMI PATH

Based on the analysis of the CM noise path, the common-mode voltage is applied to the CM inductor through the CM path. The CM noise propagation path in Figure 6 is a typical RLC resonate circuit. The real common-mode voltage V_o waveform is shown in Figure 7, and the amplitude of equivalent common-mode voltage V_{cm} in Fig. 6 can be represented by (1), where the volt-second of the CM voltage has a direct impact on the magnetic flux density inside the CM inductor. Here T_s is the time of the switching cycle and T_0 is the zero vector time.

The volt-second on the CM inductor determines the CM

$$|V_{cm}| = \frac{\left(T_0 \, |V_{dc}| + (T_s - T_0) \, | \frac{1}{3} V_{dc} | \right)}{T_s} \cdot \frac{C_s}{C_s + 2C_y} \qquad (1)$$

flux density inside the CM magnetic core. The effects of the volt-second on the inductor can be represented by a sine wave, which has the same average value (as shown in (1)) within a half-cycle. The CM voltage on the CM inductor is given in

(2). Here the CM voltage is represented by a sine wave. The CM volt-second across the CM inductor is represented by (3).

$$|V_L| = \frac{\pi \, |V_{cm}| \, \omega L_{cm}}{2\sqrt{R_{eq}^2 + \omega^2 L_{cm}^2 (1 - \frac{\omega_0^2}{\omega^2})^2}} \qquad (2)$$

$$|VS_L| = \int_0^{Ts/2} |V_L| \sin(\omega t) dt = \frac{\pi \, |V_{cm}| \, L_{cm}}{\sqrt{R_{eq}^2 + \omega^2 L_{cm}^2 (1 - \frac{\omega_0^2}{\omega^2})^2}} \qquad (3)$$

$$\omega_0 = \frac{1}{\sqrt{L_{eq} C_{eq}}}$$

Here $\omega_0 = \frac{1}{\sqrt{L_{eq}C_{eq}}}$, and ω is the angular frequency of common-mode voltage source V_{cm}.

According to (2) and (3), we can calculate the CM volt-second over the CM inductor, as shown in Fig. 8. Here and in the following simulation, it's assumed that $L_{cm}=3$ mH and $C_s=5$ nF. The curve in Fig. 8 shows that the CM volt-second is higher as the CM voltage frequency gets closer to the resonant frequency of the CM path loop; this is especially true for a high-Q network. If the leakage of the CM choke is designed to be within a given range, the CM flux inside the core dominates in core saturation. The CM inductor design should meet the condition in (4). The common-mode inductance value is determined by (5).

$$B_{max} \geq \frac{V_L * \Delta T}{N * Ac} \qquad (4)$$

$$L_{cm} = \frac{\mu_0 \mu_r Ac N^2}{le} \qquad (5)$$

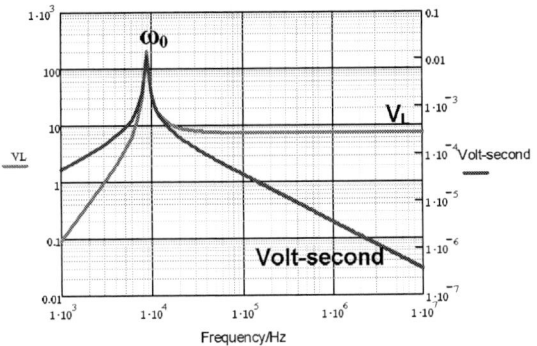

Figure 8. CM inductor voltage and volt-second vs frequency

Here N is the winding number of turns, Ac is the cross-section area of magnetic core, le is the length of the magnetic path, and B_{max} is the saturation flux density of the magnetic material.

IV. SATURATION OF CM INDUCTOR DUE TO DIFFERENT MODULATION INDEXES

According to the analysis in Section II, the amplitude of CM voltage is another issue that impacts the CM magnetic component saturation. The amplitude of the CM voltage is related to the length of zero vectors, which is linked to the motor drive modulation index and operating point. The motor drive operating point is always changing according to the variation of input voltage/output current. This is directly related to different modulation indexes. Different modulation indexes lead to different zero vector lengths, which result in different CM voltage amplitudes. For example, when the motor drive is starting up, the modulation index is low to limit the starting current, so the CM voltage will be much higher; when the motor drive is running at the rated speed, the modulation index is higher, so the CM voltage is relatively lower. The variation of the CM voltage at different modulation indexes is shown in Figure 9. According to (1), lower modulation index corresponds to longer T0, while a higher modulation index corresponds to shorter T0. The modulation index changes the CM noise waveform and CM noise volt-second. For an extreme case, the CM volt-second when the modulation index is zero is twice as large as the volt-second when the modulation index is 1.

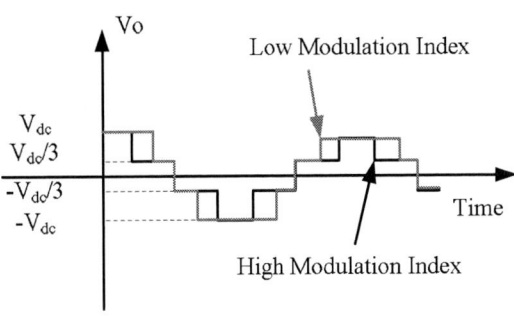

Figure 9. Different CM voltages at different modulation index

Understanding this can help determine how to design EMI filters for different conditions. Some EMI filters work well at the rated operating point, but do not work well for light load/low speed operation due to the higher CM voltage.

V. DESIGN GUIDELINES FOR CM INDUCTORS BASED ON CM NOISE CONSTRAINT

With the information of the CM voltage and CM noise path, the minimum volume of the CM magnetic component needed to avoid saturation can be estimated. From (3), (4), and (5), the relationship of total magnetic volume to loop impedance (resonant frequency) and CM voltage frequency is derived and represented by (6). A visual representation is shown in Figure 10. Equation (6) can be also used as a criterion in common-mode choke core selection.

According to Figure 10, when designing a CM choke for a high-power-density application, the design should avoid the resonant peaks to avoid high CM volt-second and high magnetic volume.

Knowing the relationship between the modulation index and the CM volt-second, EMI filters should be designed to fit different operating conditions. When designing CM inductors, the lowest modulation index should be taken as the worst case for the CM volt-second constraint.

$$Ac * le \geq \frac{\pi \mu_0 \mu_r L_{cm} |V_{cm}|^2}{B_{max}^2 (R_{eq}^2 + \omega^2 L_{cm}^2 (1 - \frac{\omega_0^2}{\omega^2})^2)} = \left(\frac{|VS_L|}{B_{max}}\right)^2 \frac{\mu_0 \mu_r}{L_{cm}} \quad (6)$$

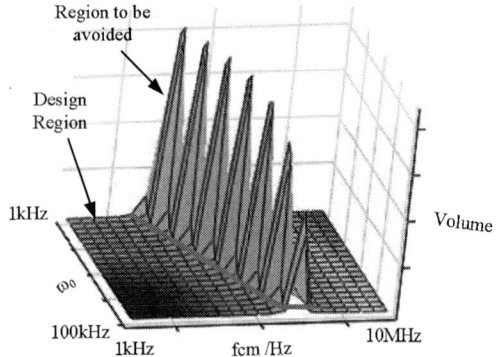

Figure 10. Relationship among total magnetic volume and resonate frequency and CM voltage frequency

VI. EXPERIMENTAL RESULT

An experiment is established following Figure 1, with a 2 kW motor drive, a 2 kW fan load, and two LISNs defined by MIL-STD-461. The drive is powered by a 300VDC source. The common-mode noise is extracted from the total noise using a noise separator [8, 9]. An LC CM EMI filter is designed based on the CM bare noise, and the C_y capacitor is 200 nF, as shown in Fig. 2-1. The CM choke core is a Finemet nano-crystalline core [8]. Three chokes are built, and are shown in Figure 11. Choke 1 and Choke 2 use different cores with the same inductance value (3mH). The core size of Choke 2 is approximately half that of Choke 1. Choke 2 and Choke 3 use the same small cores. Choke 3 has twice the number of winding turns as Choke 2. Thus the inductance of Choke 3 (12mH) is four times greater than that of Choke 2. The parameters of the cores are shown in Table 1.

Table 1. Finemet core parameters used in the experiments

	Core Part Number	Ac (mm²)	Le (mm)	Volume (mm³)	Number of Turns
Choke 3	F1AH0654	11.3	47.1	2596	28*2
Choke 2	F1AH0654	11.3	47.1	2596	12*2
Choke 1	F1AH0695	22.2	56.5	4876	11*2

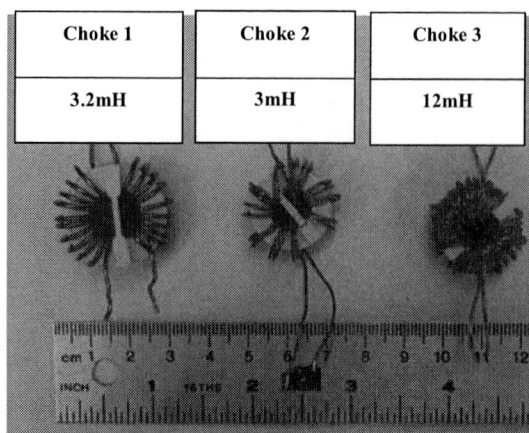

Figure 11. CM Inductor using different core / different inductance

Figure 12 shows the time-domain waveforms of the CM filter during the motor drive start up, as tested with Choke 1. Since the motor drive is using vector modulation, at start up at low speed, the zero vectors are longer. This leads to a higher CM volt-second compared with the rated operating point. Thus the CM choke is saturated at low speeds when the motor drive starts up, even if the DM current is very low. This is the phenomenon analyzed in section IV. When the motor drive is running at the rated speed, the zero vector and the CM volt-second are reduced; the CM choke is then de-saturated and resumes working. Figure 13 shows the CM time domain waveform tested with Choke 2 (3mH, small core) in two cases: with LISNs and without LISNs. The case with LISNs has been analyzed in sections II and III, and has been shown to contain a low impedance path to ground, so that it may saturate the CM inductor when f_{cm} is close to the resonant frequency. With LISNs, the CM voltage on the CM inductor is higher, so the CM inductor is saturated. When LISNs are removed, there's no low impedance path between ground and the line side; furthermore, the line-to-ground capacitance of the DC main side is low (C_{LG}). This results in a very high loop resonant frequency that is far away from f_{cm}; as a result the CM voltage on the CM inductor is suppressed. In the experiment, the CM flux is measured with a CM transformer [9]. The CM voltage in Figure 12 and Figure 13 are measured on the C_y capacitors. When the core is saturated, the switching combination and vector length are still the same, but the saturated inductor will have resonance with the C_y capacitor, which makes the CM voltage higher. Here the CM voltage on the C_y capacitors is the same as the voltage across the inductor since the loop resistance is small. The increasing of the CM voltage is further analyzed in the following paragraph.

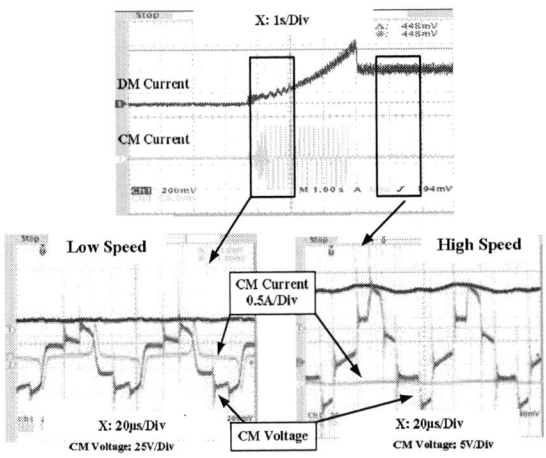

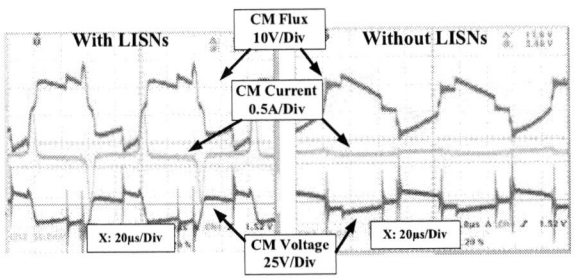

Figure 12. Time domain waveforms of CM filter performance during motor drive start up

Figure 13. Time domain waveforms with and without LISNs

are positively added, which makes the CM choke voltage increase.

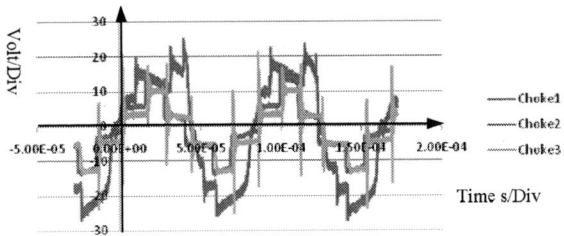

Figure 14. CM inductor voltage comparison of different chokes using CM transformer

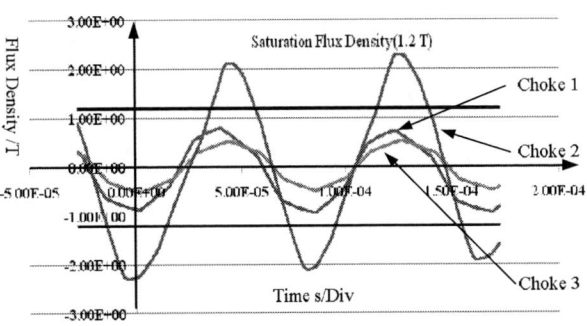

Figure 15. Time domain waveform of flux density inside core

Figure 14 shows the measured CM inductor time-domain voltage waveform when using the CM transformer. The flux density inside the core can be calculated based on the CM inductor volt-second measurement. According to Figure 14, the CM voltage can be approximated like Fig. 7, where T_s=83μs and T0=22.3μs. The equivalent CM voltage amplitude can be represented by (1). The CM volt-second across the CM inductor can be calculated using (1), (2), and (3). The flux density can be extracted by numerical integration of the waveforms in Figure 145.

Figure 15. shows the flux density time domain waveforms based on the measurement in Figure 14 According to the waveforms, Choke 2 is saturated. The saturation flux density limits are also shown in the figure. Theoretically Choke 2 has almost the same volt-second as Choke 1, but Choke 2 has fewer turns and a much smaller cross-section area, which makes the B_{max} in Choke 2 higher.

In Figure 145, the voltage across the CM inductor is higher when the CM choke is saturated. This phenomenon can be explained using the equivalent circuit in Figure 146. The current going through the CM path is equal to the current going through the CM choke, which is also charging C_{eq} in a CM voltage half-cycle; the voltage across the CM inductor is equal to the sum of Vc_{eq}, V_{cm} and V_{Req}. When the CM choke is saturated in one of the CM voltage half-cycles, the voltage of C_{eq} increases as a result of the increasing in the inductor current. In the following half-cycle, the voltages Vc_{eq} and V_{cm}

Table 2 shows the comparisons of calculated result using (6) and the product of Ac * le in the real core. The product of Ac*le is the product of the cross-section and the magnetic length from the datasheet; and Cal_result (3.2 mH) is the calculated Ac*le product using (6) when L_{cm} is 3.2 mH. This result shows the required minimum Ac*le when L_{cm} is 3.2mH. Similarly, Cal_result (12 mH) is the calculated Ac*le product using (6) when L_{cm} is 12 mH. The equivalent permeability of these two cores (F1AH0654 and F1AH0695) has slight differences according to the real measurement in the datasheet, which results in two different calculated Ac*le products. According to these results, when L_{cm}=3.2 mH, the minimum required Ac*le is $1.59*10^{-6}m^3$, which is larger than the Ac*le of the smaller-core F1AH0654, and the big-core F1AH0695 can avoid saturation. When L_{cm} is increased to 12mH, the minimum required Ac*le is around $4.25*10^{-7}$ m³, which is smaller than the product of Ac*le in the F1AH0654 core, so we can choose the smaller-core F1AH0654 instead of the big-core F1AH0695 for the CM inductor.

Table 2. Comparison of Ac*le product

		Ac*le(m³)	Cal_result (3.2mH) (m³)	Cal_result (12mH) (m³)
Choke 2 and Choke 3	F1AH0654	$5.32*10^{-7}$	$1.59*10^{-6}$	$4.25*10^{-7}$
Choke 1	F1AH0695	$1.254*10^{-6}$	$1.46*10^{-6}$	$3.89*10^{-7}$

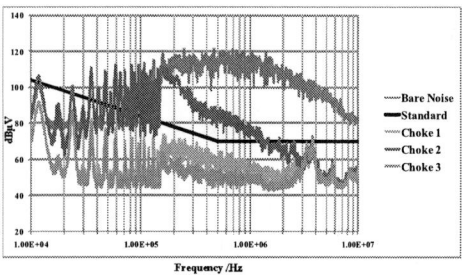

Figure 16. In circuit noise measurement

Figure 16 shows the CM noise measurement with different CM chokes. Although Choke 1 and Choke 2 have approximately the same small-signal impedance, their in-circuit performances are totally different, because Choke 2 is saturated with a smaller core. The in-circuit test also shows that even with a smaller core, Choke 3 can still perform well with the increased L_{cm} by avoiding the resonant frequency in the CM path loop close to f_{cm}. Applying (6), with a given converter, if the CM voltage V_o is known, we can determine the minimum volume for the CM magnetic core.

VII. CONCLUSIONS

In this chapter, a new mechanism for CM inductor saturation and their impacts on the CM magnetic component volume are analyzed. The CM noise path resonance plays an important role here. It can cause the CM noise voltage to resonate in the loop, and makes the CM volt-second over the CM inductor high enough to saturate the CM inductor. The CM voltage source is related to the modulation index and DC bus voltage such that the lower the modulation index is, the higher the CM voltage volt-second will be. The LISNs in the CM loop create a low grounding impedance path that brings the CM noise path to the resonant frequency. This is closer to the CM voltage frequency f_{cm}, which makes the CM inductor suffer from higher a CM volt-second and results in a higher CM inductor volume. The experiments carried out verify the analysis. A new criterion is proposed for high power density common-mode chokes design based on above knowledge.

VIII. ACKNOWLEDGEMENT

The authors want to thank SAFRAN group for founding this project. Author Fang Luo is also supported by Chinese Scholarship Council.

REFERENCES

[1] H. Chen, Z. Qian, S. Yang, C. Wolf, "Finite-Element Modeling of Saturation Effect Excited by Differential-Mode Current in a Common-Mode Choke", IEEE Transactions on Power Electronics, vol. 24, no. 3, pp. 873-877, Mar. 2009.

[2] R. Anne, B. Hans, D. Zhao, B. Ferreira, L. Frank, "A new behavioral model for performance evaluation of common mode chokes", in 2007 International Zurich Symposium on Electromagnetic Compatibility, 2007, pp. 501 - 504.

[3] H. Hemphill, B. Wallertz, "Critical core parameters in the design of common mode suppression chokes", in 1995 International Zurich Symposium on Electromagnetic Compatibility, 1995, pp. 334 - 336.

[4] Y. Sun, A.Esmaeli, L. Sun, E.Kang, " Investigation and Suppression of Conducted EMI and Shaft Voltage in Induction Motor Drive System", in 2006 World Congress on Intelligent Control and Automation, 2006, pp. 8256 - 8259.

[5] A. Esmaeli, B. Jiang, L. Sun, "Modeling and suppression of PWM inverter's adverse effects", in 2006 International Symposium on Systems and Control in Aerospace and Astronautics, 2006, pp. 1450-1454.

[6] A. Esmaeli, K. Zhao, L. Sun, Q. Wu, "Investigation and Suppression of the Adverse Effects of PWM Inverter through Passive Filter Technique", in IEEE 2006 Conference on Industrial Electronics and Applications, 2006, pp. 1-4.

[7] G.L. Skibinski, R. J. Kerkman, D. Schlegel, "EMI emissions of modern PWM AC drives", IEEE Industry Applications Magazine, Vol. 5, No. 6, pp. 47-80, Nov. 1999

[8] http://metglas.com/downloads/finemet/FMCMC190 804.pdf

[9] C.Khun, W. Khan-ngern, M. Kando, "Passive EMI Filter Performance Improvements with Common Mode Voltage Cancellation Technique for PWM Inverter", in IEEE 2007 International Conference on Power Electronics and Drive Systems, 2007, pp. 257-261.

[10] S. Wang, Y. Maillet, F. Wang , R. Lai, R. Burgos, F. Luo, " Investigating the Grounding of EMI Filters in Power Electronics Systems", in IEEE 2008 Power Electronics Specialists Conference. 2008, pp. 1625 - 1631.

[11] Shuo Wang, " Characterization and Cancellation of High-Frequency Parasitics for EMI Filters and Noise Separators in Power Electronics Applications", PhD dissertation, Dept. of Elect. Eng., Virginia Polytechnic Institute and State University, Blacksburg, VA, 2005.

[12] D. Huang, D. Fu, Fred C. Lee, "High Switching Frequency, High Efficiency CLL Resonant Converter with Synchronous Rectifier", in Proc. IEEE ECCE, 2009, pp. 804-809.

[13] D. Fu, Y. Liu, F.C Lee, M. Xu, "An Improved Novel Driving Scheme of Synchronous Rectifiers for LLC Resonant Converters", in Proc. IEEE APEC, 2008, pp.510-516.

[14] D. Fu, B. Lu, F.C. Lee, "1MHz High Efficiency LLC Resonant Converters with Synchronous Rectifier," in Proc. IEEE PESC, 2007, pp. 2404-2410.

[15] Y. Qiu, B. Lu, B. Yang, D. Fu, F.C. Lee, "A high-frequency high-efficiency three-level LCC converter for high-voltage charging applications", in Proc. IEEE PESC, 2004, pp. 4100-4106.

Black-Box Modeling of Three Phase Voltage Source Inverters Based on Transient Response Analysis

V. Valdivia, A. Lázaro, A. Barrado, P. Zumel, C. Fernández, M. Sanz

Carlos III University of Madrid
Electronics Technology Department
Power Electronics Systems Group
Leganés, Madrid, Spain
virgilio.valdivia@uc3m.es

Abstract— **Nowadays, black-box behavioral models of power converters are becoming interesting for system-level analysis. These models can be used to evaluate the response of power electronics systems which are composed of commercial converters, since they can be fully parameterized by analyzing the actual converter response. First black-box models of power converters have been recently proposed, but all of them are oriented to DC-DC converters. However, three-phase voltage source inverters are usually applied in current power electronics-based systems, such as aircraft power systems, and a black-box modeling method of this kind of converters has not yet been proposed. In this paper a large-signal black-box modeling method of three-phase voltage source inverters is proposed. The identification of the model is based on the analysis of the converter transient response, which is obtained by means of a set of simple experiments and easily usable fitting algorithms. An experimental validation of the proposed method has been carried out on a 5 kW actual inverter applied on an aircraft power system test bench.**

I. INTRODUCTION

Nowadays, power electronics systems composed of multiple converters are growing; e.g. the system of the "more electric aircraft". Modeling and simulation are essential tools for the system-level analysis [1]-[6]. However, power electronics systems are usually composed of commercial converters and, owing to the confidentiality of manufacturers, the system designer does not have access to all the necessary data to parameterize conventional models (such as average models and switching models) because they are based on the internal structure of the converter (topology, control, etc). Hence behavioral models which are based on the external behavior of the converter are required, as well as identification procedures of them (a black-box approach). These models only compute the signals of interest regarding system level analysis and can be fully parameterized by analyzing the actual converter response. Therefore, they can be used to model the behavior of commercial converters looking for system-level analysis.

First black-box models of power converters and their identification procedures have been recently proposed, but all

of them are oriented to DC-DC converters [7]-[12]. However, the voltage-controlled three phase voltage source inverter VSI (Fig. 1) is usual in current power electronics systems, such as new aircraft power distribution systems [1], [2].

In this paper, a large-signal black-box behavioral modeling technique of three-phase VSI is proposed. On the one hand, the proposed model is simple; therefore a suitable simulation time can be achieved. On the other hand, the identification procedure is easy, since the parameters of the model can be identified by means of simple experiments, which can be carried out using low cost equipment and well known and easily usable fitting algorithms. Along the paper, *uppercase* signal names mean steady-state values and *lowercase* signal names mean varying values.

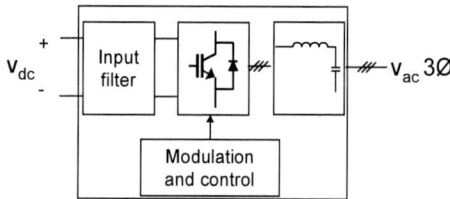

Figure 1. Voltage-Controlled Three Phase Voltage Source Inverter

II. BLACK-BOX MODEL

A. Model structure

The proposed model (Fig. 2) computes both the large-signal dynamic and the static behavior of the VSI. It is an extension of the g-parameters model proposed in [6], [9] for DC-DC converters. Although a first approach to this structure has been already proposed [4], an identification method has not been yet reported. This model is defined in the synchronous reference frame *d-q*, because three-phase AC magnitudes become DC signals at steady-state [13]. Hence, both the modeling and the identification tasks become easier.

The model is composed of the following elements:

a) *Input admittance Y_i:* Dynamic system modeling the input current dependence on the input voltage.

978-1-4244-4782-4/10 $26.00 © 2010 IEEE

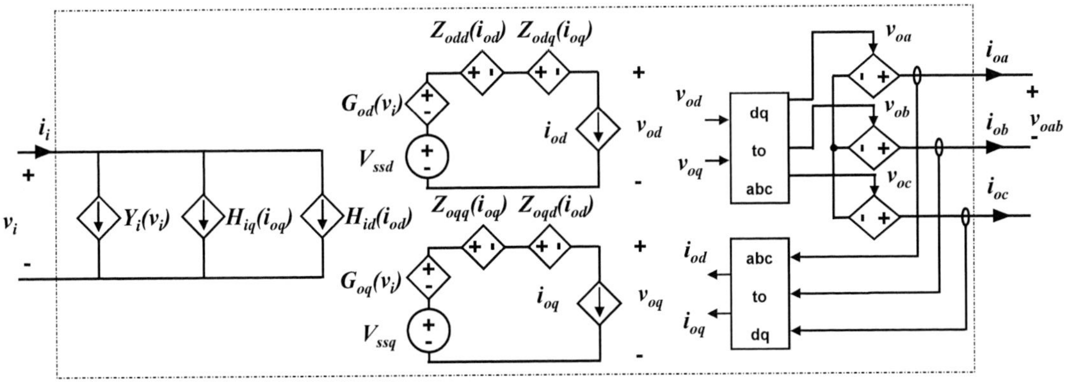

Figure 2. Behavioral Black-box model structure of the three-phase voltage source inverter

b) Back current gain H_{id}, H_{iq}: Dynamic systems modeling the input current dependence on the output current.

c) DC voltage sources V_{ssd} and V_{ssq}: DC sources modeling the output voltage at steady state in *d-q* coordinates.

d) Audio-susceptibility G_{od}, G_{oq}: Dynamic systems modeling the output voltage dependence on the input voltage.

e) Output impedance Z_{odd}, Z_{odq}, Z_{oqd}, Z_{oqq}: Dynamic systems modeling the output voltage dependence on the output current.

f) Reference frame transformation blocks dq - abc: They interface the model with the three-phase AC bus.

This proposal is focused on balanced conditions and can be extended to model also both negative and zero sequence components.

B. Dynamic systems modeling

The dynamic systems composing the model can exhibit either linear or nonlinear behavior. If they can be considered linear into the operating range of the converter, they can be modeled using one single-input single-output (SISO) linear-time-invariant (LTI) model (Fig. 3a), defined by a single transfer function. If nonlinear effects are found to be significant, Wiener and Polytopic structures are proposed.

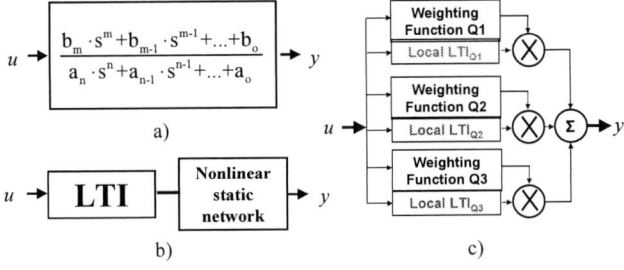

Figure 3. Dynamic systems which compose the black-box model. u ≡ input, y ≡ output. a) LTI model b) Wiener model c) Polytopic model

Wiener structure (Fig. 3b) is composed of a nonlinear static model cascaded with a SISO LTI model [14], [15]. In practice, it is useful to model the nonlinear large-signal static behavior of the converter as a function of the operating point. The "constant power load" behavior of regulated inverters (the input current decreases when the input voltage increases if the

load is kept constant) as well as the efficiency can be taken into account by modeling the back current gain using it.

Polytopic structure (Fig. 3c) is composed of a set of local "small signal" SISO LTI models [8]. Each local model is parameterized and weighted as a function of the operating point, and the response of all of them is combined to compute the model response. Therefore, Polytopic structures are useful to model dynamic dependence on the operating point. For instance, if the inverter is fed by a wide input voltage range and significant audio-susceptibility dependences on the input voltage operating point are found, this structure is useful to have them taken into account. The example shown in Fig. 3c corresponds to the use of three local linear models and weighting functions driven by the input signal.

The applicability of this structures on DC-DC converters behavioral modeling can be found in [8]-[12].

III. IDENTIFICATION EXPERIMENTAL TESTS

The model shown in Fig. 2 is an un-terminated model, thus the dynamic systems composing it only model the "internal" converter dynamics. In order to parameterize these dynamic systems, the converter has to be properly tested. The selected experiments have to be "informative enough", meaning that the obtained data allow an accurate fitting of the dynamic model in the frequency bandwidth of interest. Regarding measurements on AC systems, an impedance measurement test for three-phase AC systems has been proposed in [16], based on the injection of an unbalanced line-to-line current. In this paper step tests are proposed, owing to the following reasons:

- Step tests can be easily carried out in power electronics laboratories. The proposed experiments can be carried out with low cost equipment: oscilloscopes, switches, DC voltage sources and passive loads.

- The measured transient waveforms can be used to accurately fit the parameters of the dynamic systems that compose the black-box model, as shown in [11], [12] on DC-DC converters modeling. The step slew rate limits the identified frequency bandwidth.

Regarding the tests, it is important to note that if the dynamic system can be supposed linear, the LTI model can be identified from a single step response. However, if significant

978-1-4244-4782-4/10 $26.00 © 2010 IEEE   1280

nonlinearities are found, the nonlinear model should be identified as follows:

a) If a Wiener structure is used, first the static nonlinear model is identified by analyzing the steady-state response of the converter at different operating points. After that the effect of the static network is subtracted from the measured response and finally the LTI model is identified.

b) If a Polytopic structure is chosen, small magnitude steps are performed on different operating points in order to identify each local LTI model.

Next, the proposed step tests are described. Further on, the identification algorithms, which are applied on the measured transient responses, are described. Two kinds of step test are proposed: output current step test and input voltage step test.

A. Output current step

This test is applied to identify both the output impedance and the back current gain. The proposed tests are shown in Fig. 4 and Fig. 5. Both balanced resistive load and balanced capacitive load are needed.

1) *Resistive load connection or disconnection:* The resistive load step (Fig. 4) is applied to identify Z_{odd}, Z_{oqd} and H_{id}. Both the output voltage and the output current are transformed into the d-q synchronous reference frame, by setting the d-q axis in a way that $V_{oq}{\approx}0$ at steady state. Since a resistive load is used $I_{oq}{\approx}0$; hence the output current step is reflected as an i_{od} step and the variation of i_{oq} is much lower. Therefore, the transient evolutions of v_{od}, v_{oq} and i_i are mainly due to the i_{od} step (see Fig. 6) and Z_{odd}, Z_{oqd} and H_{id} can be identified as expressed in (1).

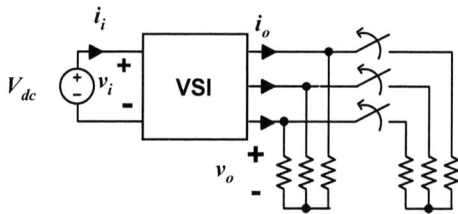

Figure 4. Resistive load disconnection (a connection is also valid). Identification of Z_{odd}, Z_{oqd} and H_{id}

$$Z_{odd} = -\frac{v_{od}}{i_{od}}\bigg|_{v_i,i_{oq}\approx0} \quad Z_{oqd} = -\frac{v_{oq}}{i_{od}}\bigg|_{v_i,i_{oq}\approx0} \quad H_{id} = \frac{i_i}{i_{od}}\bigg|_{v_i,i_{oq}\approx0} \quad (1)$$

2) *Capacitive load disconnection:* The capacitive load disconnection (Fig. 5) is applied to identify Z_{oqq}, Z_{odq} and H_{iq}. Both the output voltage and the output current are transformed into the *d-q* synchronous reference frame as explained above ($V_{oq}{\approx}0$). In this case, the output current step is reflected as an i_{oq} step, meaning that the variation of i_{od} is much lower. Thus, the transient evolutions of v_{od}, v_{oq} and i_i are now mainly due to the i_{oq} step (see Fig. 7) and Z_{oqq}, Z_{odq}, H_{iq} can be identified as expressed in (2).

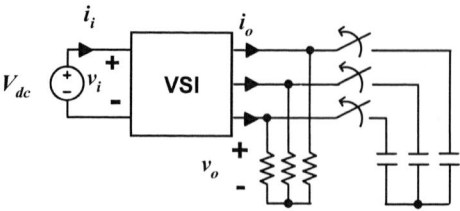

Figure 5. Capacitive load disconnection. Identification of Z_{oqq}, Z_{odq} and H_{iq}

$$Z_{oqq} = -\frac{v_{oq}}{i_{oq}}\bigg|_{v_i,i_{od}\approx0} \quad Z_{odq} = -\frac{v_{od}}{i_{oq}}\bigg|_{v_i,i_{od}\approx0} \quad H_{iq} = \frac{i_i}{i_{oq}}\bigg|_{v_i,i_{od}\approx0} \quad (2)$$

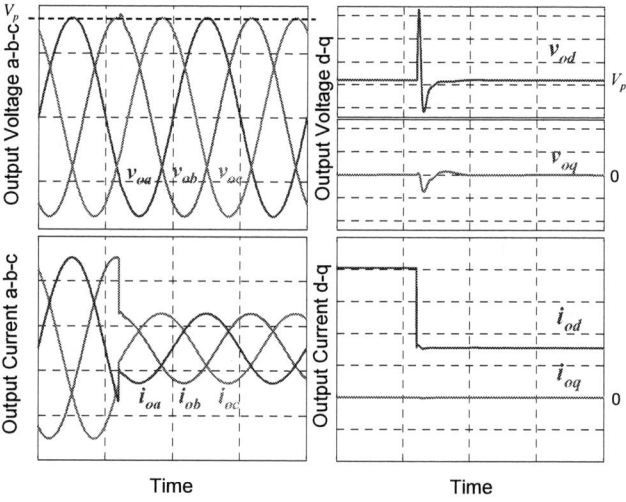

Figure 6. Resistive load disconnection

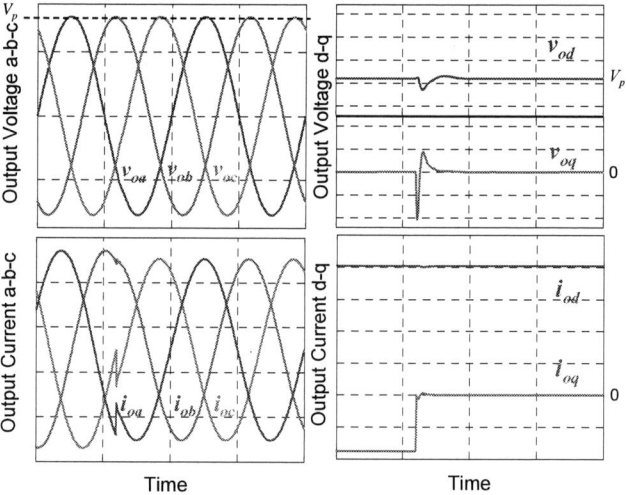

Figure 7. Capacitive load disconnection

Input voltage perturbations have to be minimized during the test in order to assure that the audio-susceptibility and the input admittance influence on the measured transient waveforms can be neglected. Thus, a DC voltage source with low enough output impedance is recommended. Anyway, if this dependence is significant, the output impedance of the voltage source can be characterized (by following the procedures proposed in [7]-[12]) and its influence can be subtracted from the identified subsystems, in a similar way to that illustrated in [7] on DC-DC converters.

B. Input voltage step

This test allows the identification of G_{od}, G_{oq} and Y_i. Two circuits are proposed in this paper. The test shown in Fig. 8a is based on the interconnection of an array of diodes in parallel with a switch. When the switch is off, there is a voltage drop in the diodes V_d. When the switch is turned on, a voltage step with magnitude $\approx V_d$ appears at the input of the converter. The test shown in Fig. 8b is based on the use of two voltage sources ($V_{dc1} > V_{dc2}$). When the switch is turned on, a voltage step with magnitude $\approx V_{dc1} - V_{dc2}$ is performed.

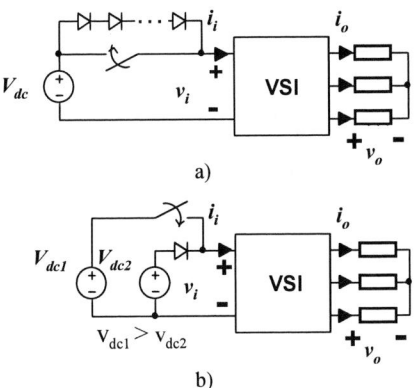

Figure 8. Input voltage step. Identification of G_{od}, G_{oq} and Y_i a) Array of diodes based test. b) Two DC voltage sources based test

$$G_{od} = \frac{v_{od}}{v_i}\bigg|_{i_{od}, i_{oq} \approx 0} \quad G_{oq} = \frac{v_{oq}}{v_i}\bigg|_{i_{od}, i_{oq} \approx 0} \quad Y_i = \frac{i_i}{v_i}\bigg|_{i_{od}, i_{oq} \approx 0} \quad (3)$$

Output current perturbations have to be minimized during the tests in order to assure that the output impedance and the back-current gain influence on the measured transient waveforms can be neglected. Therefore, if the audio-susceptibility does not significantly depend on the output current, the best choice is to carry out the test in no-load conditions. Anyway, if this dependence is considerable, the test is performed in load conditions and significant output current perturbations are found, the load can be characterized and its effect can be subtracted from the identified subsystems as appointed before. Notice that the output voltage is transformed into the d-q synchronous reference frame by setting the d-q axis in a way that $V_{oq} \approx 0$ at steady state, as previously explained.

C. Static tests

The efficiency can be characterized by measuring the static response of the converter into its working range, at steady-state conditions. The efficiency is included within the model using a Wiener structure to model the back-current gain.

IV. IDENTIFICATION ALGORITHM

After measuring the transient response of the converter, the next task is to identify the corresponding dynamic systems. As mentioned in Section II, LTI SISO models are used to model the dynamic systems in both the linear and the nonlinear case. These models can be fully defined by means of a transfer function. Parametric methods are applied to identify models which are defined by a set of parameters. They are based on the least-squares method and the extensions of it (for instance, the prediction error method), and they can be applied to fit transfer function models from transient response and frequency response data [14]-[15].

The present paper focuses on parametric methods applied to identify transfer function models from transient response data. There are several structures of transfer function models, and they differ on how they model the dynamic relation between input, output and disturbance [14]-[15]. Two common structures are shown in Fig. 9.

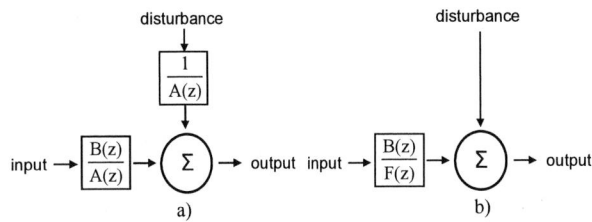

Figure 9. Common transfer function models. a) Autoregreesive with exogenous input model. b) Output Error model

Selecting a proper model structure is the first task in the identification procedure, as well as the number of parameters of the model to be identified. A trade-off between simplicity and performance (i.e. fitting accuracy) may be considered. Moreover, the order of the model to be identified has to be selected before running the fitting algorithm (i.e. the number of parameters which define the LTI model).

The Output Error OE model is suggested in this paper. This model is characterized for a disturbance model independent of the input to output transfer function. Although white noise characteristics are assumed for the disturbance, just two polynomial orders ($B(z)$ and $F(z)$) have to be selected. Moreover the fitting accuracy is good in practice, as will be shown along the following sections. Identification algorithms of OE models have been widely discussed in the literature [14]-[15] and there are commercial tools that allow the identification of them straightforwardly (e.g. MATLAB System Identification Toolbox [17]). An OE model can be identified using MATLAB by means of the "oe" function, which performs the fitting using a prediction-error method. Nevertheless, if the OE model does not provide good identification results, other structures should be tried.

Notice from Fig. 9 that discrete models are identified. They can be transformed into continuous time domain for a simpler implementation in simulators. For instance, the conversion can be performed using the "d2c" function of

MATLAB. The sampling time of the transient data should be high enough to avoid inaccuracy problems when transforming the models.

The identification method proposed in [11] can be also applied to perform the identification from the step response.

V. SIMULATION BASED VALIDATION

The scope of this section is to evaluate the performance of the proposed black-box modeling method by means of simulation. In order to carry out this analysis, a three phase VSI has been modeled (switching model) and simulated using PSIM [18], and the procedure exposed in the flowchart shown in Fig. 10 has been performed. On the one hand, the performance of the identification procedure has been evaluated by identifying LTI models from the step response of the inverter and by comparing the frequency response of both the identified LTI models and the small-signal frequency response of the modeled inverter, which has been measured using ac-sweep capabilities of PSIM. On the other hand, the performance of the black-box model has been evaluated by applying large-signal perturbations on it and the switching model and by comparing the response of them.

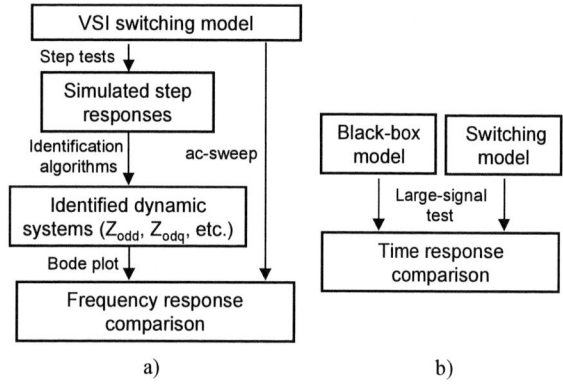

Figure 10. Black-box modeling procedure evaluation. a) Identification procedure. b) Large-signal behavior of the model

The modeled inverter is shown in Fig. 11. Its main characteristics are: $V_i =1\ kV\pm10\%$, $V_o =220/380\ V_{ac}$, $f_o =50$ Hz, $P_{max}=30\ kW$, $f_{sw}=6\ kHz$, multi-loop control structure implemented in the stationary reference frame $\alpha\beta$ and based on an outer Proportional-Resonant compensator PR and an inner Proportional-Integral compensator PI [19]. An output voltage feed-forward loop improves the dynamic performance.

Fig. 12 shows the resulting fitting of v_{od}, v_{oq} when a resistive load step Fig. 4) from full-load to no-load is applied (i_{od} step while $i_{oq}\approx0$). The resulting LTI models correspond to Z_{odd} and Z_{oqd}. As can be seen, highly accurate fits are achieved. The frequency response of the identified models is compared with the small-signal frequency response of the switching model in Fig. 13 (small-signal perturbations have been introduced on i_{od} while i_{oq} has been kept constant). It can be seen that the frequency responses of the fitted LTI models and the switching model match properly up to approximately half the switching frequency f_{sw}. Thus, it has been concluded that the proposed identification procedure exhibits a good performance.

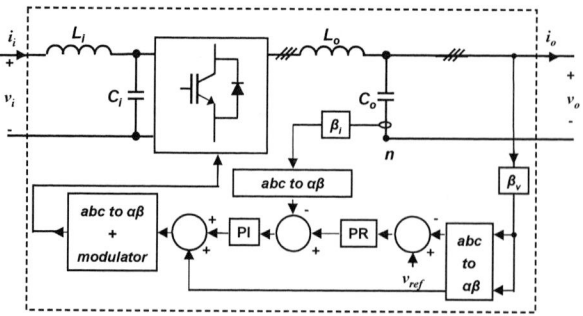

Figure 11. Modeled three phase VSI for simulation based validation

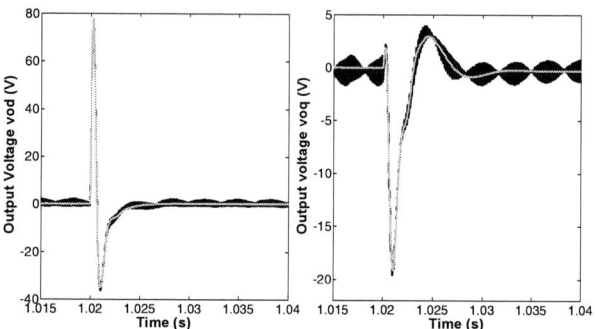

Figure 12. Resistive load step transient response fitting. Black-lines: Swithcing model response. Grey-lines: Identified model response. The applied input signal i_{od} corresponds to the switching model in both cases

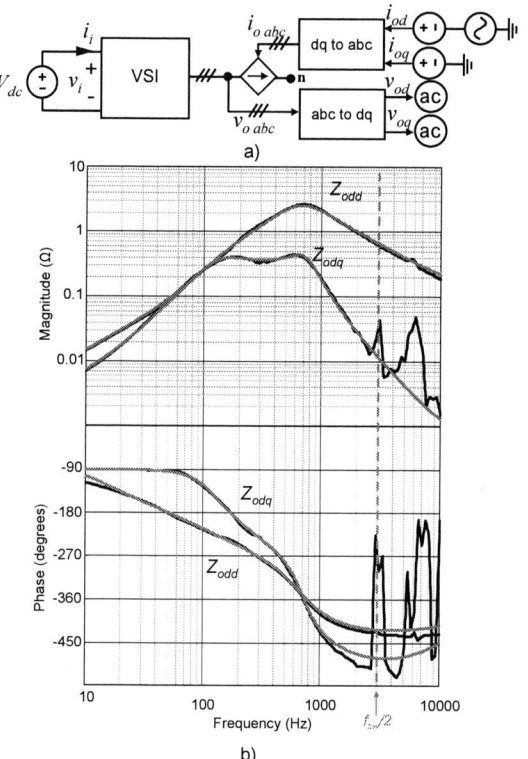

Figure 13. Small signal validation. a) ac-sweep test carried out using PSIM. b) Comparison of the switching model frequency response (black line) with the identified model frequency response (grey line). Z_{odd} and Z_{odq}

After applying the step tests, no significant dynamic systems dependence on the operating point has been found. This is because the input voltage operating range is narrow and no significant dependences on the output current were found, since a linear control scheme is used and the input filter does not interact with the inverter dynamics in this example. Therefore, every dynamic system has been modeled using LTI models except H_{id}, which has been modeled using a Wiener structure, in order to consider both the "constant power load" behavior (explained before) and the efficiency of the converter η. This efficiency was modeled as a function of i_{od} by means of a 1-D look-up table. The derived model is shown in Fig. 14.

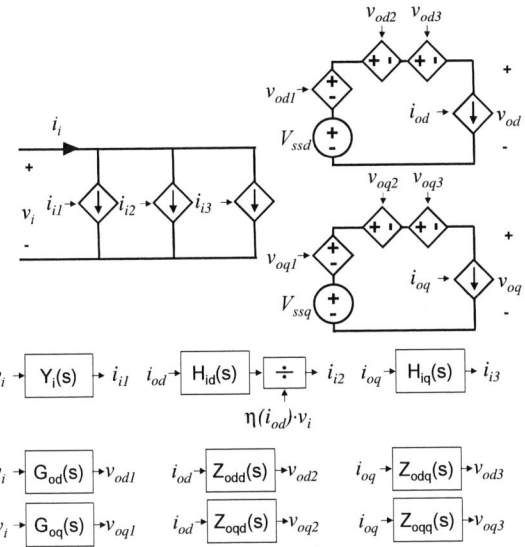

Figure 14. Black-Box model of the simulated inverter (dq-abc and abc-dq transformation blocks are not shown)

The test shown in Fig. 15 has been performed to compare the large-signal behavior of both the switching model and the identified black-box model. Initially a nonlinear load of 3.6 kW is fed by the inverter. A resistive load of 9.6 kW is suddenly connected at 200 ms. A resistor is also placed in series at the input to perturb the input voltage during the entire test. Thus, both the input voltage and the output current are simultaneously disturbed and these disturbances are different from those used for identification.

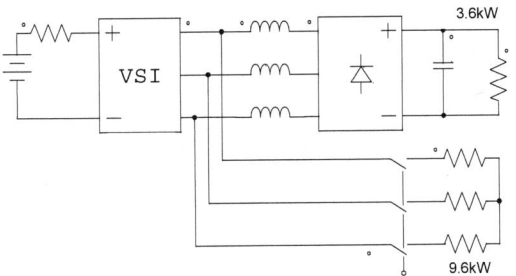

Figure 15. Large signal validation. Simulated test

Both responses are very close (see Fig. 16 and Fig. 17), so it is concluded that the model is suitable to reproduce the large signal behavior of three phase voltage source inverters.

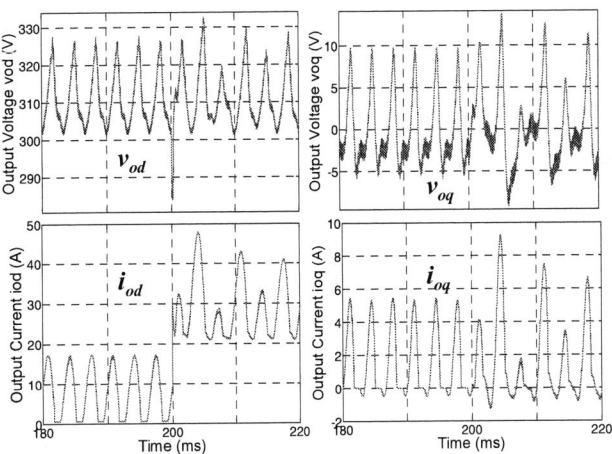

Figure 16. Large signal simulation results. Output voltage and output current in d-q coordinates. Blue line: Switching model. Red line: Black-box model

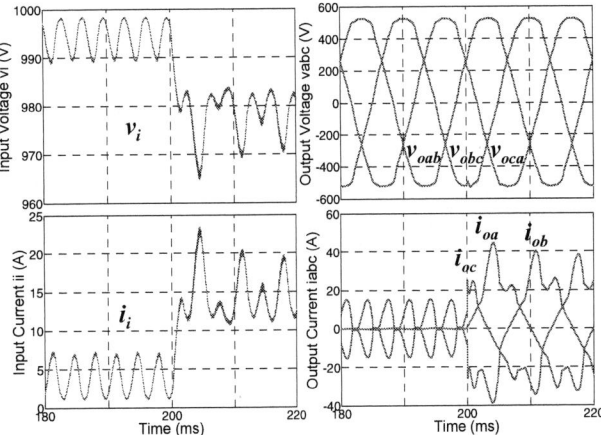

Figure 17. Large signal simulation results. Input-output voltage and current. Blue line: Switching model. Red line: Black-box model

VI. EXPERIMENTAL VALIDATION

An actual 5 kW regulated inverter has been tested and modeled using the proposed procedure. This inverter is applied in a test bench of an aircraft power system. It is fed from a 270V_{dc} bus and generates a 115/200 V_{ac}, 400 Hz three-phase bus. The measurements have been taken using two TEKTRONIX oscilloscopes MSO4104 and TDS5104, hall-effect current probes TCPA300 and TCP202, common mode voltage probes P6319 and differential mode voltage probes P5205. The sampling frequency has been set to 1 MHz.

A. Modeling and identification

First, the dynamic systems dependence on the operating point has to be evaluated, in order to select either a linear or a nonlinear model for each of them. The allowed input voltage range is narrow in this application, according to MIL-STD-704C regulation [20]. Hence, for the sake of simplicity, dynamic systems dependence on the input voltage operating point has not been considered. In order to evaluate the dynamic systems dependence on the operating point, several load steps have been carried out.

Although a certain dynamic systems dependence on the output current has been measured, it was found to be low and was neglected. Thus, every dynamic system has been modeled using LTI models except H_{id}, which has been modeled using a Wiener structure to consider the efficiency and the "constant power load" behavior explained before. Thus, the structure of the model is the same as the structure derived in the previous section (Fig. 14). However G_{oq} has been neglected, since the perturbation of v_{oq} is much lower than the perturbation of v_{od} when v_i steps are performed (as will be shown further on) and therefore G_{oq} is negligible with respect to G_{od}.

B. Validation

The black-box model has been validated by reproducing the experimental tests in PSIM and by comparing the responses of the model and the actual inverter (Fig. 18, Fig. 19 and Fig. 20). The loads were modeled using passive elements and the output impedance of the input voltage source was identified by measuring its load step transient response and by applying parametric identification [12]. The time span of the a-b-c frame signals plots is 10 ms and the time span of both the d-q frame signals plots and input signals plots is 60 ms.

Fig. 18 shows the comparison of the actual converter response with the model response when a resistive load step from 1 kW to 3 kW is applied. Both of them match properly. Note that not only the low frequency transient oscillation but also the high frequency oscillation exhibited at the beginning of the v_{od} transient response have been properly fitted. In order to achieve this fine fit, the identification based on the OE model was complemented with the method proposed in [11].

Fig. 19 shows a comparison of the actual converter response with the model response when a 16.5 μF capacitive load is disconnected while the inverter supplies 5 kW (this capacitor implies a power factor cosφ = 0.95 and corresponds to the maximum capacitive load specified for this inverter). The current exhibits high ripple content before the capacitive load is disconnected, which is related to the high frequency current flowing through the capacitive load. This ripple has to be filtered when this response is used to identify the model. A certain time delay was found between the switch-off time of the test bench switches used to carry out both load steps (see a-b-c frame currents). A higher delay was found in case of disconnecting the capacitive load. It was taken into account in the simulations and it causes the high frequency oscillation found at the beginning of the i_{od} and i_{oq} transient waveform.

Fig. 20a corresponds to an input voltage step carried out on no load conditions, showing that the proposed test (Fig. 8a, implementation shown in Fig. 20b) allows achieving voltage steps with high enough slew rate. The simulated and the measured waveforms match properly.

It is worth noticing that negative-sequence component has been found in the measurements, especially in those corresponding to the capacitive load disconnection, because the loads are not absolutely balanced, as well as the wires length and paths are not equal in all phases. A small zero sequence was also found at the output voltage. These effects were found to be low and therefore neglected, but they could be considered by extending the model shown in Fig. 2.

Since the measured and the simulated responses match in all cases, the proposed modeling procedure has been validated.

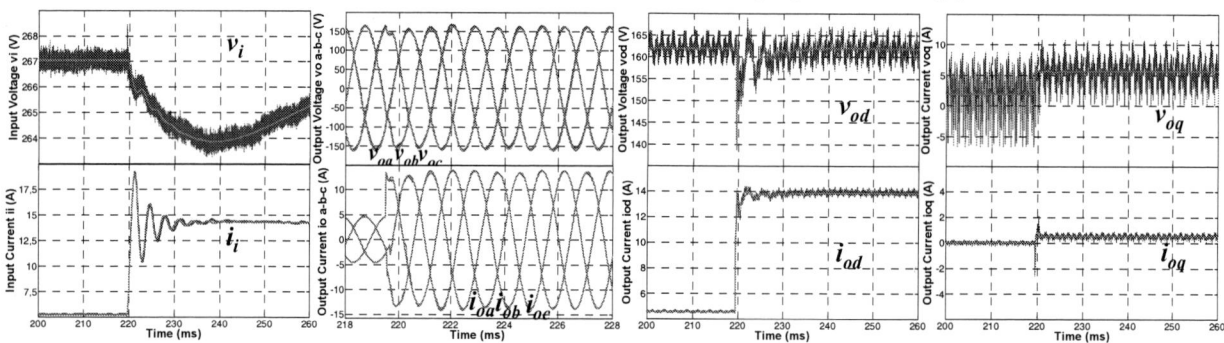

Figure 18. Resistive load connection response. Actual measurements (blue lines) vs black-box model simulated response (red lines)

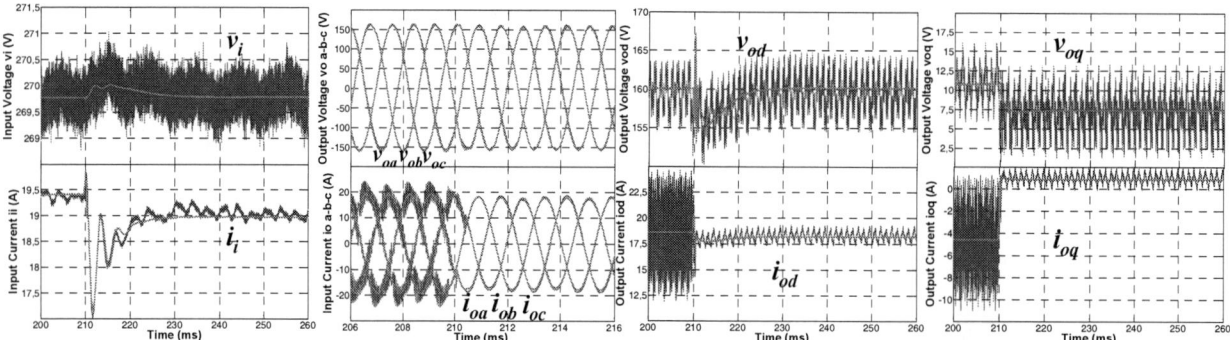

Figure 19. Capacitive load disconnection response. Actual measurements (blue lines) vs black-box model simulated response (red lines)

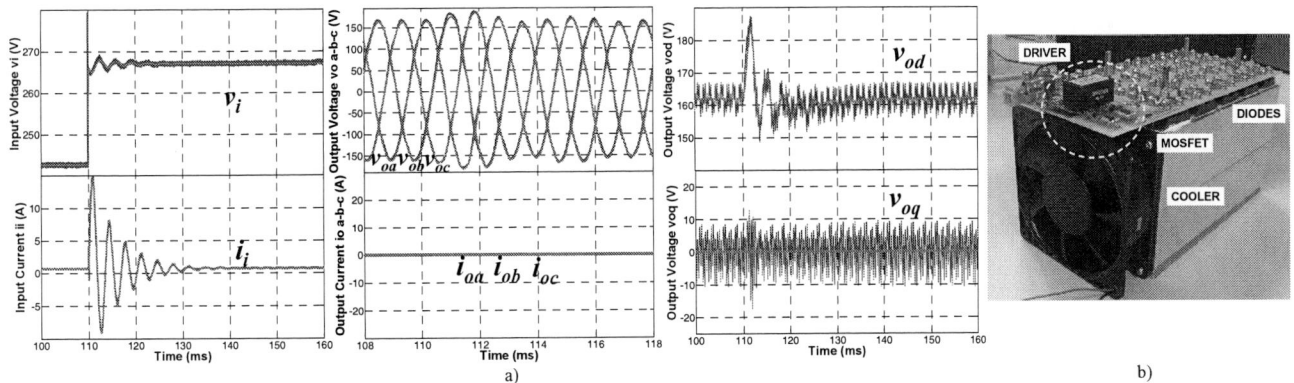

Figure 20. Input voltage step response. a) Actual measurements (blue lines) vs black-box model simulated response (red lines). b) Actual implementation of the array of diodes based test to create voltage steps at the input of the inverter

VII. CONCLUSIONS

In this paper a black-box modeling procedure of three-phase voltage source inverters, based on transient response analysis, has been proposed. The proposed model can be used to evaluate the large-signal response of power electronics systems composed of commercial converters. The identification of the model is carried out by means of simple experiments and easily usable fitting algorithms. First, a simulated-based validation has been carried out to evaluate the performance of the proposal (both large signal and small signal behavior have been evaluated). Finally, an actual 5 kW inverter applied on an aircraft power system has been modeled using the proposed procedure. All the validations performed show that the proposed black-box modeling method works properly.

ACKNOWLEDGMENT

This work has been supported by the Spanish Ministry of Education and Science, through the research projects "Diseño y Modelado de Sistemas Aeroespaciales, Nivel Subsistema" (Code: DPI2006-14866) and "Sistemas de Alimentación Para Aplicaciones Embarcadas y Portátiles Basadas en Fuentes y Dispositivos de Almacenamiento de Energía Emergentes" (Code: DPI2009-12501) and by private contract with EADS-CASA through the research project "HVDC Load Distribution System" (Code: 04-AEC0527-000050/2005) financed by The European Regional Development Fund via the Aerospace Sector Plan of the Community of Madrid.

REFERENCES

[1] A. Emadi, M. Ehsani, "Aircraft power systems: technology, state of the art and future trends", *IEEE Aero. Electron. Systems Maganize, vol. 15, no. 1, pp. 28-32, Jan. 2000.*

[2] R. Jayabalan, B. Fahimi, A. Koenig, S. Pekarek, "Applications of power electronics-based systems in vehicular technology: state of the art and future trends", *in Proc. of IEEE Power Electron. Specialists Conf. (PESC), pp. 1887-1894, 20-25 Jun. 2004.*

[3] J.A. Rosero, J.A. Ortega, E. Aldabas, and L. Romeral, "Moving towards a more electric aircraft", *IEEE Aero. Electron. Systems Maganize, vol. 22, no. 3, pp. 3-9, Mar. 2007.*

[4] D. Boroyevich, R. Burgos, R., L. Arnedo, Fei Wang; "Synthesis and Integration of Future Electronic Power Distribution Systems", *in Proc. of Power Conv. Conf. - Nagoya, (PCC), pp: 1-8, 2-5 Apr. 2007.*

[5] R. Burgos, S. Rosado, F. Wang, D. Boroyevich, Z. Lewis, K. Karimi; "Modeling considerations and stability analysis of aerospace power systems with hybrid ac/dc distribution", *SAE Pow. Sys. Conf., Nov. 2006.*

[6] B.H. Cho, and F.C. Y. Lee, "Modeling and analysis of spacecraft power systems," *IEEE Trans. Power Electron., vol. 3, no. 1, pp. 44–54, May. 1988.*

[7] L. Arnedo, D. Boroyevich, R. Burgos, F. Wang, "Un-terminated frequency response measurements and model order reduction for black-box terminal characterization models", *in Proc. of IEEE Appl. Power Electron. Conf. (APEC), pp. 1054 - 1060, 24-28 Feb. 2008.*

[8] L. Arnedo, D. Boroyevich, R. Burgos, F. Wang, "Polytopic Black-Box modeling of DC-DC converters", *in Proc. of IEEE Power Electron. Specialists Conf. (PESC), pp. 1015 - 1021, 15-19 Jun. 2008.*

[9] L. Arnedo, R. Burgos, D. Boroyevich, F. Wang, "System-Level Black-Box Dc to Dc Converter models", *in Proc. of IEEE Appl. Power Electron. Conf. (APEC), Page(s):1476 – 1481, 15-19 Feb. 2009.*

[10] J. Oliver, R. Prieto, J. Cobos, P. Alou, and O. Garcia, "Hybrid Wiener-Hammerstein Structure for Grey-Box Modeling of DC DC Converters," *in Proc. of IEEE Appl. Power Electron. Conf. (APEC), Page(s):280-285, 15-19 Feb. 2009.*

[11] V. Valdivia, A. Barrado, A. Lazaro, P. Zumel, C. Raga, C. Fernandez, "Simple Modeling and Identification Procedures for "Black-box" Behavioral Modeling of Power Converters Based on Transient Response Analysis", *IEEE Trans. Power Electron., vol. 24, no. 12, Dec. 2009.*

[12] V. Valdivia, A. Barrado, A. Lazaro, C. Fernandez, P. Zumel, "Black-Box Modeling of DC-DC Converters Based on Transient Response Analysis and Parametric Identification Methods", *in Proc. of IEEE Appl. Power Electron. Conf. (APEC), 21-25 Feb. 2010.*

[13] S. Buso and P. Mattavelli, "Digital Control in Power Electronics" (Synthesis Lectures on Power Electronics), San Rafael, CA: *Morgan & Claypool Publisher, 2006.*

[14] L. Ljung, "System Identification: theory for the user, 2nd ed.", *Prentice Hall, 1999.*

[15] O. Nelles, "Nonlinear System Identification", *Springer 2001.*

[16] Jing Huang; K.A. Corzine, M. Belkhayat, "Small-Signal Impedance Measurement of Power-Electronics-Based AC Power Systems Using Line-to-Line Current Injection.; *IEEE Trans. Power Electron., vol. 24, no. 2, pp. 445 – 455, Feb. 2009.*

[17] MATLAB R2008b *http://www.mathworks.com/*

[18] PSIM 8.0 *http://www.powersimtech.com/*

[19] P. C. Loh, M. J. Newman, D. N. Zmood, and D. G. Holmes, "A comparative analysis of multi-loop voltage regulation strategies for single and three-phase UPS systems," *IEEE Trans. Power Electron., vol. 18, no. 5, pp. 1176-1185, Sep. 2003.*

[20] MIL-STD-704C, Military Standard, Aircraft Electrical Power Characteristics

Digital Autotuning of dc-dc Converters Based on Model Reference Impulse Response

A. Costabeber[*], P. Mattavelli[**], S. Saggini[***], A. Bianco[****]

[*]Dept. of Information Engineering – University of Padova, Italy. E-mail: alessandro.costabeber@dei.unipd.it
[**] DTG– University of Padova, Vicenza, Italy. E-mail: paolo.mattavelli@unipd.it
[***] DIEGM, University of Udine, Udine, Italy. E-mail: stefano.saggini@uniud.it
[****]DORA S.p.a., Italy

Abstract—**This paper proposes an autotuning technique for digitally controlled voltage-mode dc-dc converters based on the model reference approach. The proposed solution uses the difference between the measured system impulse response and the reference one, determined by the desired dynamic performances, and minimizes the error function acting on the regulator parameters. Two different approaches have been investigated for the evaluation of the impulse response: a deterministic one, based on duty cycle impulse, and a statistical one, based on white noise injection. Compared to existing approaches, this solution has the advantage of simplicity, small-signal processing and on-line tuning capabilities. Experimental investigation has been performed on a 20 A, 1.5V synchronous buck converter, and both simulation and experimental results confirm the effectiveness of the proposed solution.**

Keywords – **dc-dc converters, autotuning, model reference, impulse response, white noise, on-line tuning**

I. INTRODUCTION

Interest in digital control solutions for Switch Mode Power Supplies (SMPS) is increasing due to availability of digital controller ICs developed for high-frequency switching converters [1-3] and due to some attractive properties not available in analog control as flexibility, immunity to external influences and availability of automated design tools. Moreover, the digital controller is able to include some self-tuning capabilities and the automatic adjustments of controller parameters based on the desired dynamic performance.

Autotuning techniques [4], developed for electrical drives or high power electronic applications, usually show a degree of complexity which is not compatible with the limits of power consumption and computational efforts in digitally controlled SMPS. This has driven the research toward the study and the development of simpler autotuning algorithms for SMPS. Recently several autotuning solutions [5-10] for high-frequency dc-dc converters have been presented. In [5-6] autotuning is based on the relay feedback technique. In [5] the tuning is performed measuring amplitude and phase of the output voltage oscillations during the start-up, while [6] is based on limit-cycle oscillations. In [7] a solution based on the model-reference approach [4] is presented, by injecting a

perturbation at the desired crossover frequency into the feedback loop. In [8] a continuous monitoring of the crossover frequency and phase margin is performed and a multi-input-multi-output control loop continuously tunes the regulator parameters. In [9] a non-parametric method is proposed that requires process identification based on Pseudo-Random Binary Sequence (PRBS) injection. In [10] a tuning technique based on adaptive prediction error filters is addressed, permitting on-line system identification and controller tuning.

This paper proposes an autotuning technique, based on the model reference approach [4], that has a low degree of complexity and does not present any specific limitation on the controller structure and converter topology. The proposed algorithm is based on the regulation of loop gain and phase margin at the desired control bandwidth, by injecting a duty-cycle impulse, measuring the impulse response, and comparing it with the model reference impulse response. The energy of the difference between the two signals is minimized through a numerical search algorithm. The same approach is then applied for a white-noise injection on the duty-cycle, leading to the possibility of an on-line tuning of the controller, tracking the parametric variations during the normal operation. This second solution guarantees smaller output deviation, but longer convergence time.

The paper is organized as follows: in Section II the principle of proposed tuning technique is described, including the numerical search algorithm, the statistical approach and some issues related with measurement noise, while Section III and IV report the simulation and experimental verifications, respectively.

II. PROPOSED TUNING TECHNIQUE

The tuning proposed in this paper is based on the model reference approach [4] where closed-loop system properties are derived from the measurement of the impulse response. The technique is here discussed for the voltage control of a dc-dc buck converter and it can be easily extended to other control configurations.

Following Fig.1, a duty-cycle impulse perturbation d_P is added to the output of the voltage-mode regulator $C(z)$. The tuning algorithm is based on the comparison between the real

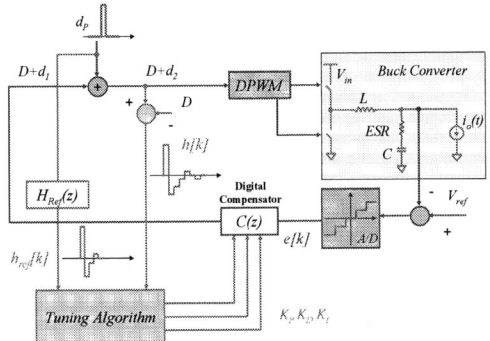

Fig. 1 – Proposed tuning scheme for impulsive perturbation d_P

system impulse response measured in d_2 and the reference one. This case is here denoted as *deterministic tuning*. In Fig. 2 a similar approach is reported where the impulse response is measured via a cross-correlation algorithm, injecting a white noise signal in d_P instead of an impulsive excitation and evaluating the impulse response using a correlation algorithm. This case is instead denoted as *statistical tuning*. In both cases, the transfer function that represents the impulse response from d_p to d_2 is:

$$H(z) = \frac{d_2(z)}{d_P(z)} = \frac{1}{1+C(z)G(z)} = \frac{1}{1+T_v(z)} \qquad (1)$$

where $G(z)$ is the sampled-data transfer function between the duty-cycle $d_2(z)$ and the output voltage $v_o(z)$ and $T_v(z)$ is the voltage loop gain. Comparing the measured impulse response $h[k]$

$$h[k] = Z^{-1}(H(z)) \qquad (2)$$

with the reference one $h_{REF}[k]$

$$h_{REF}[k] = Z^{-1}(H_{REF}(z)) = Z^{-1}\left(\frac{1}{1+T^*(z)}\right) \qquad (3)$$

an error ε is computed. The error function ε is defined as the energy of the difference between the reference impulse

response $h_{REF}[k]$ and the measured one $h[k]$:

$$\varepsilon = \sum_{k=1}^{N}(h([k]-h_{REF}[k])^2 \qquad (4)$$

where N is the number of sampling periods that includes the most relevant part of the impulse response.

The algorithm is aimed to tune the controller parameters to minimize the error function ε, which is a function of the controller parameters. Tuning targets are given in terms of desired loop gain $T^*_v(z)$ with specified crossover frequency and phase margin. From the desired loop gain $T^*_v(z)$, $H_{REF}(z)$ and $h_{REF}[k]$ are derived. Although the proposed approach is not restricted to a specific controller structure, a PID digital regulator is considered, i.e.:

$$C(z) = PID(z) = K_P + \frac{K_I}{1-z^{-1}} + K_D(1-z^{-1}) \qquad (5)$$

where K_P, K_D, K_I are proportional, integral and derivative gains, respectively.

Fig. 3 summarizes the main features of the proposed solution, considering the *deterministic case*. The duty cycle impulsive perturbation is injected into the regulator $C(z)$ output, and the closed loop transfer function impulse response $h[k]$ is measured after the injection node. The reference impulse response $h_{REF}[k]$ is stored in a Look-Up-Table (LUT) in the digital controller (DSP or FPGA). The measured response $h[k]$ and the reference one $h_{REF}[k]$ are then used to compute the error (4). The core of the tuning algorithm is the error function minimization block, where a minimum search algorithm is implemented to find the regulator parameters that minimize the distance between $h[k]$ and $h_{REF}[k]$.

For the sake of explanation, an additional simplification on the controller structure is assumed: in the next paragraphs only a two parameters PD (Proportional-Derivative) tuning is addressed, assuming a fixed integral part. Thus, a more intuitive two variables representation is used for the error function analysis and for the minimum search method. Nevertheless, the same approach holds for the more general three parameters tuning, as reported in the experimental results.

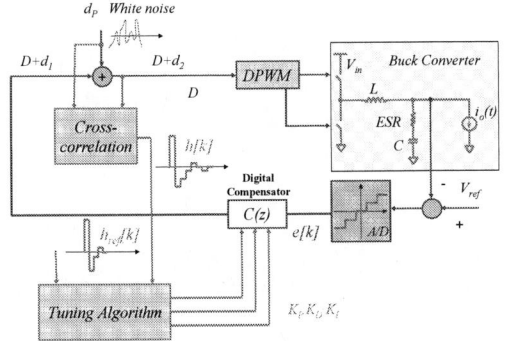

Fig. 2– Proposed tuning scheme for white noise perturbation d_P

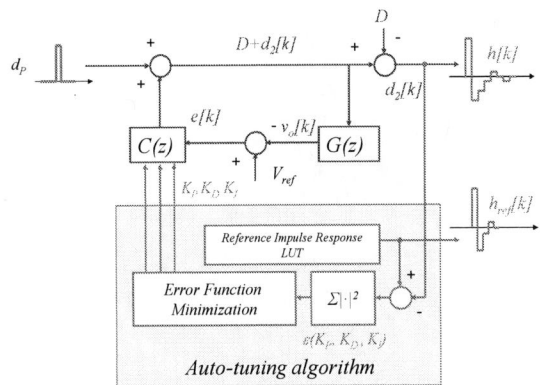

Fig. 3 – Proposed method architecture

978-1-4244-4782-4/10 $26.00 © 2010 IEEE   1288

Having fixed the integral gain, the error function in (4) becomes a function of two variables. In Fig.4 the two parameters error function (4) is reported for the power stage parameters and reference transfer function used in section III for simulations, with K_P and K_D normalized by their values in the minimum point K_{PNOM} and K_{DNOM}. The numerical search algorithm is aimed to find the minimum, and thus the set of parameters (K_{PNOM}, K_{DNOM}) for which equation (6) holds:

$$\frac{\partial \varepsilon}{\partial K_P}\bigg|_{K_{PNOM}} = 0, \quad \frac{\partial \varepsilon}{\partial K_D}\bigg|_{K_{DNOM}} = 0 \qquad (6)$$

Ideally, the minimum point could be zero, meaning that the measured impulse response exactly matches the reference one. In practice the exact loop gain of the system is assumed to be unknown, and thus the minimum point is where the distance between the reference response and the measured one is minimum.

A. Minimum search method

The auto-tuning solution is based on the numerical minimization of the error function ε. In literature there are several algorithms to numerically find the minimum of an n-variables function. In this paper a simple algorithm has been selected and described hereafter. It is worth to point out that the tuning principle is independent on the specific minimization technique, and a different search method can be adopted.

The basic principle of the proposed minimum search algorithm [12-13] is to test the four possible directions (Fig. 5) of movement from the starting point (K_{P0}, K_{D0}), measuring for each of them the error function value. Δ is defined as the step in the increase or decrease of the parameters, and it is assumed to be the same for K_P and K_D. The coordinates that guarantee the maximum decrease in the error function are selected as the new point from which the search is restarted. If none of the directions show a decrease in the function, the search step Δ is reduced. This algorithm generally presents slow convergence speed, but, unlike faster search solutions, its convergence is guaranteed under suitable properties of the error function [12]. The numerical search algorithm is now described after the introduction of some definitions.

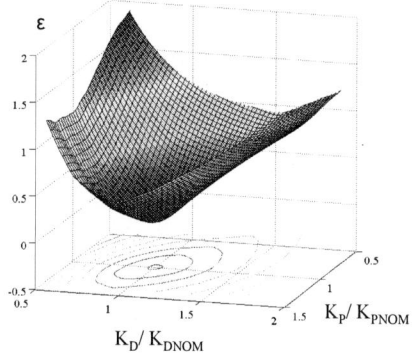

Fig. 4 – Two-parameters K_P, K_D error function

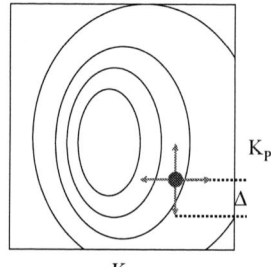

Fig. 5 – Minimum search algorithm

$p_0 = (K_{P0}, K_{D0})$ is the initial point, Δ_0 is the initial search step, $S = \{s_{+K_P}, s_{-K_P}, s_{+K_D}, s_{-K_D}\}$ is the set of search directions in the K_P, K_D plane, where s_i are unit vectors.

For each iteration n=1,2,.. one of the next two actions is performed. s_n is the search direction for the n-th iteration, p_n are the starting coordinates for the n-th iteration and Δ_n the n-th iteration search step.

- If $s_n \in S$ such that $\varepsilon(p_n + \Delta_n s_n) < \varepsilon(p_n)$

$$p_{n+1} = p_n + \Delta_n s_n \qquad (7)$$
$$\Delta_{n+1} = \Delta_n \qquad (8)$$

- Else if for each $s_n \in S$ $\varepsilon(p_n + \Delta_n s_n) > \varepsilon(p_n)$

$$p_{n+1} = p_n \qquad (9)$$
$$\Delta_{n+1} = \frac{\Delta_n}{R} \quad R > 1 \qquad (10)$$

Thus, two kind of iterations are possible: *1) successful iterations*, where a direction of error decrease is found, the step is unchanged and the parameters K_P, K_D are updated to the decrease direction; *2) unsuccessful iterations*, where all the directions lead to an error function increase and the parameters are unchanged while the search step is reduced of a factor R.

The convergence of this algorithm is proved for each p_0, Δ_0 and R, assuming some properties of the error function ε [13]. The convergence result is resumed in (11).

$$\lim_{n \to \infty} \|\nabla \varepsilon(p_n)\| \leq \lim_{n \to \infty} \Delta_n = 0 \qquad (11)$$

where ∇ is the gradient of the error function ε. The key of this convergence relies in the unsuccessful iterations, where the reduction of the search step Δ_n guarantees

$$\lim_{n \to \infty} \Delta_n = 0 \qquad (12)$$

In the proposed tuning algorithm, a slightly modified version of this basic search algorithm is used, where the step reduction is limited to Δ_{min}, and, at each iteration, not only the four directions are tested but also the starting point is remeasured and compared with the other directions. These variations address to two important issues. The first is that a noisy measurement of h[k] could lead the algorithm to converge to a non-minimum point. This is because some instantaneous local minima may appear due to sampling noise. Leaving the algorithm running with fix small step and remeasuring at each iteration the initial point, the convergence in the exact minimum point is guaranteed. The second is to add the on-line tuning feature. With a non-zero final step, and remeasuring the iteration initial point, the algorithm sensitivity to the error function modifications caused for example by power stage parameters variations, is maintained and the PID regulator parameters are continuously updated.

B. Statistical tuning algorithm

The second solution is based on the same principle of function ε minimization, but injecting in the control loop only a small amplitude uncorrelated white noise added to the duty cycle. A cross-correlation algorithm is then performed to statistically measure the impulsive response. The main advantage of this solution is that the complexity is similar to the deterministic solution one, but the output voltage deviation is smaller, being the additional noise effect similar to a limit cycle oscillation. The major drawback of this proposal is the tuning time which is much longer.

The statistical measurement of the impulse response is performed through the cross-correlation approach in [11]. In Fig. 6 an ideal white noise disturbance $d_P(k)$ is injected between the regulator $C(z)$ and the process $G(z)$, where the output voltage $v_o(k)$ is affected also by current sampled value $i_{out}(k)$. Considering the system output $d_2(k)$, and introducing the buck converter open loop output impedance $Z_{out\_ol}(z)$, the corresponding transfer functions are:

$$H(z) = \frac{d_2(z)}{d_P(z)} = \frac{1}{1+C(z)G(z)} \qquad (13)$$

$$Q(z) = \frac{d_2(z)}{i_{out}(z)} = \frac{Z_{out\_ol}(z)C(z)}{1+C(z)G(z)} \qquad (14)$$

In the time-domain, the signal $d_2(k)$ is the sum between the convolution of $d_p(k)$ with the impulsive response h(k) of H(z) and the convolution of $d_p(k)$ with the impulsive response q(k) of Q(z). The convolution is identified with the symbol *.

$$d_2(k) = (d_P * h)(k) + (i_{out} * q)(k) =$$
$$= \sum_{n=-\infty}^{+\infty} d_P(n)h(k-n) + \sum_{n=-\infty}^{+\infty} i_{out}(n)q(k-n) \qquad (15)$$

The input-output cross-correlation for k greater or equal than zero is:

$$\phi_{d_P d_2}(k) = (d_P \oplus d_2)(k) =$$
$$= \lim_{n \to +\infty} \frac{1}{N} \sum_{n=0}^{N} d_P(n-k)d_2(n) \qquad (16)$$

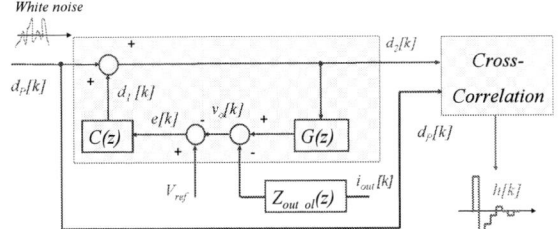

Fig. 6 – Impulse response measurement through cross-correlation

where $\oplus$ is the symbol for the correlation. Substituting (15) into (16), the cross-correlation is:

$$\phi_{d_P d_2}(k) = (d_P \oplus (d_P * h + i_{out} * q))(k) =$$
$$= (\phi_{d_P d_P} * h)(k) + (\phi_{d_P i_{out}} * q)(k) \qquad (17)$$

The second convolution in (17) is constant if $d_P(k)$ and i_{out} are independent and, in particular, it is zero if at least one of the two signals has zero mean. In (17) $d_P(k)$ is assumed to be a white noise signal, with zero mean. In the hypothesis of an ideal white noise with power spectral density P, the auto-correlation of $d_P(k)$ is an impulse:

$$\phi_{d_P d_P}(k) = \frac{1}{P}\delta(k) \qquad (18)$$

Then (17) becomes:

$$\phi_{d_P d_2}(k) = (\frac{1}{P}\delta * h)(k) = \frac{1}{P}h(k) \qquad (19)$$

where *h(k)* is exactly the impulse response. It is worth to note that in (17) i_{out} can be considered as a disturbance in the impulse response measurement, but the correlation guarantees immunity to any uncorrelated noise. The drawback of the correlation is that the measured impulse response is exact only if the observation time N tends to infinite. A finite observation window N leads to measurements errors, which are limited by using a long observation period.

C. Measurement Noise Effects

Both in deterministic and statistical case the previous analysis assumes a noiseless measurement of the impulse response. In the real system a noisy measurement is due to different causes. For any case, the possible presence of a limit cycle oscillation is translated in an error in the response measurement. In the statistical case, another noise source is the limited observation window in cross-correlation algorithm. Sampling noise coming from the prototype adds to these terms. Neglecting the exact form of the noise sources, it is important to observe that the measured response $h_{meas}[k]$ results in an aleatory variable given by:

$$h_{meas}[k] = h[k] + H_N[k] \qquad (20)$$

where *h[k]* is the impulse response without noise and *H_N[k]* a zero mean stochastic function accounting for the measurement noise. Substituting (20) in the error function (4), results in a noisy error function. The search algorithm is modified as described in section II.A, to account for the noise. In particular, a limit in the search step decrement is imposed to ensure a continuous cycling of the search that is required to average the noise effect, at the cost of a precision loss in the minimum detection. The result is a small steady-state oscillatiton of the regulator parameters near the minimum point.

III. SIMULATION RESULTS

The proposed solution has been simulated in Simulink/Matlab® for the voltage mode control of a dc-dc buck converter with power stage parameters: V_{in}=5 V, V_o=1.3 V, L=0.8 uH, C=2000 uF, f_{sw}=500 kHz. Due to the absence of ESR and assuming one step control delay in the control implementation, a PID regulator is required. This regulator is designed to have a control bandwidth equal to $f_{sw}/15$=33.3 kHz with a phase margin of 50°. The integral part $K_{I\,NOM}$ is supposed to be fixed for simplicity.

Under these design constraints, nominal PD parameters K_{PNOM} and K_{DNOM} have been designed, assuming the knowledge of all the converter parameters, and the reference impulse response (3) has been obtained. To validate the proposed tuning algorithm, the PD controller gains are shown to converge to their final nominal values K_{PNOM} and K_{DNOM} starting from an arbitrary initial condition and applying the *deterministic* tuning. Regulator gains are initialized to K_P=0.01K_{PNOM} and K_D=0.05K_{DNOM} and in spite of their mismatch, the proposed algorithm succeeds in leading them to their normalized nominal values in about 15 ms, as shown in Fig. 7. In Fig. 8 the pre-tuning and post-tuning loop gains are reported, showing a good match with the tuning targets: before the tuning the crossover frequency is f_C=1.42kHz and the phase margin m_Φ=80°, while after the PD tuning f_C=35.5kHz and m_Φ=52°, with a 6% error in the target frequency and a 4% error in the target margin.

The same simulations have been performed for the

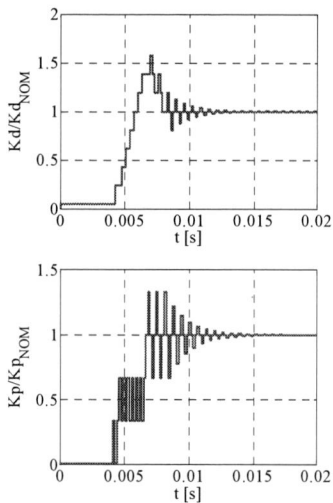

Fig. 7– Normalized derivative gain K_d/K_{dNOM} tuning, normalized proportional gain K_p/K_{pNOM} tuning

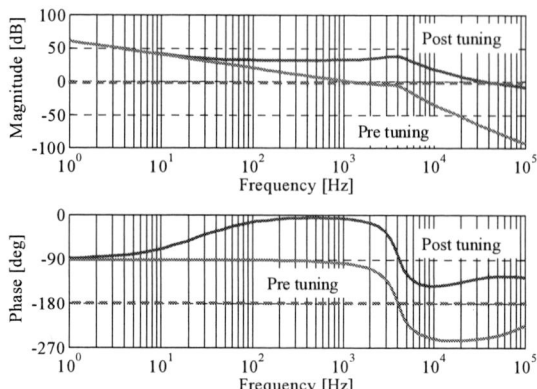

Fig. 8– Loop gain Bode plots before and after the PD tuning

complete three parameters PID tuning, starting from a low integral part K_I=0.001K_{INOM} and confirming the convergence to the precalculated parameters. The same results were obtained with the *statistical* tuning.

In Fig. 9 the effect of a uniformly distributed noise ±1% superposed to the measured duty cycle response is presented, showing that the original algorithm fails to converge for both the parameters (lines a) and b) are the convergence values with the original algorithm), while the modified version converges successfully, accepting an oscillation around the minimum point, caused by the non-zero final step size. Times t_1 and t_2 are the instants at where the search step saturates at its minimum value.

IV. EXPERIMENTAL RESULTS

The proposed algorithm has been tested on a voltage-mode control of a synchronous buck. The parameters of the experimental prototype are: V_{in}=5 V, V_o=1.5 V, L=0.8 μH, C= 660 uF, f_{sw}=200 kHz, I_{oNOM}=20 A. For the sake of rapid prototyping, the control has been implemented in a digital signal processor (TMS320F2808), which is suitable for high-frequency dc-dc converters having an embedded Digital Pulse Width Modulator (DPWM) with resolution of 150 ps. The control loop targets of bandwidth and phase margin are respectively f_C=17 kHz and m_Φ=60°. Starting from these

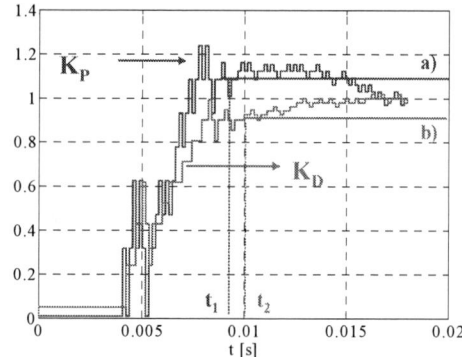

Fig. 9– Parameters K_P, K_D normalized to their final values during deterministic tuning with noisy measurement

specifications an approximate reference loop gain transfer function $T_v^*(z)$ has been derived:

$$T_v^*(z) = G \frac{z - A}{(z - 1)^2} \qquad (21)$$

where the gain G and the zero A are derived from bandwidth and phase margin constraints. Once $T_v^*(z)$ is known, the reference impulse response follows from equation (3).

Fig.10 shows the parameters K_P, K_D during the PD tuning with fixed integral part, normalized to their final values, the error function, normalized to the impulse perturbation amplitude. Fig.11 reports the output voltage and inductor current during the *deterministic* tuning. The parameters K_P and K_D are both initialized to zero. The voltage deviation due to the duty cycle impulse is $\Delta v_o = 60$ mV, and the distance between two consecutive duty impulses is the tuning period $T_{TUNING} = 250$ μs. From Fig.10, after 60 tuning periods, corresponding to 15ms, the tuning reaches the step saturation, where the error for K_P is 20% and for K_D is 10%. This is caused by the noisy measurement, and leaving the minimum search algorithm cycling with fixed step this error is reduced, leading to the post tuning loop gain in Fig.12, measured for $V_{in} = 5$ V, that is very closed to the target of $f_C = 17$ kHz and $m_\Phi = 60°$. The final measured parameters errors are 4% for K_P and 6% for K_D.

In Fig. 13-14-15-16 the experimental results with the *statistical* method are reported. Fig.13 addresses the most important limitation of the statistical method, which is the long tuning time. In Fig. 13 K_P and K_D oscillate around their nominal value and the error approaches zero in about 60 periods. Being the duration of each period $T_{TUNING} = 25$ ms the total tuning time to compute a good impulse response through cross-correlation is 0.5 s, making this approach unsuitable for a fast start-up tuning. In Fig.14 the small output voltage deviation ±10 mV during the tuning can be appreciated, showing the feasibility of an on-line tuning. Fig.15 reports the tuning result for $V_{in} = 5$ V, with the same targets of the deterministic case, while in Fig. 16 the input voltage was changed from 5 to 8 V. Despite its intrinsic slowness, this solution can be active continuously to monitor the error minimization and thus tracking the process variations.

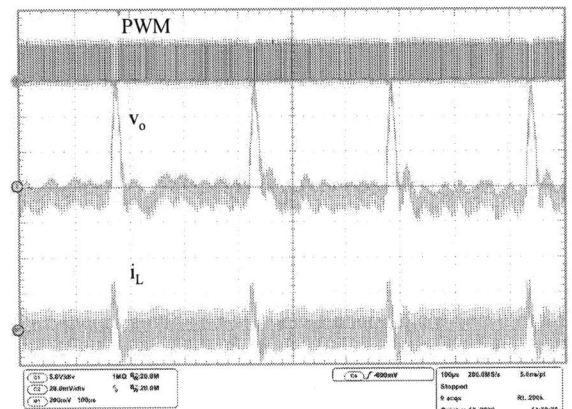

Fig. 11– Output voltage during *deterministic* tuning, $V_{in} = 5V$ (v_o 20mV/div, i_L 0.5A/div, time 100μs/div)

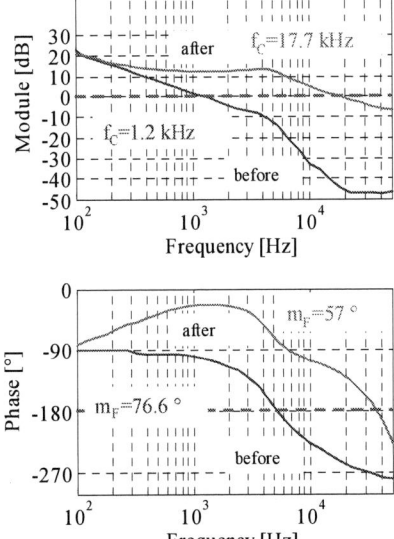

Fig. 12– Bode diagram before and after the PD tuning for the *deterministic* case with $V_{in} = 5V$

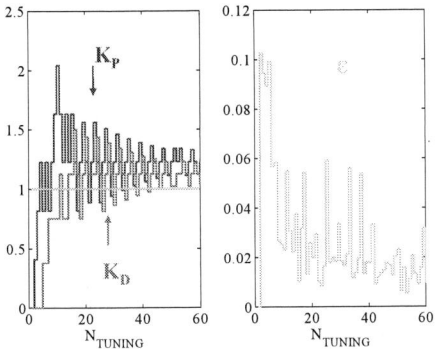

Fig. 10– Parameters K_P, K_D normalized to their final values and tuning error normalized to the duty cycle impulse amplitude during *deterministic* tuning. N_{TUNING} is the number of tuning periods T_{TUNING}

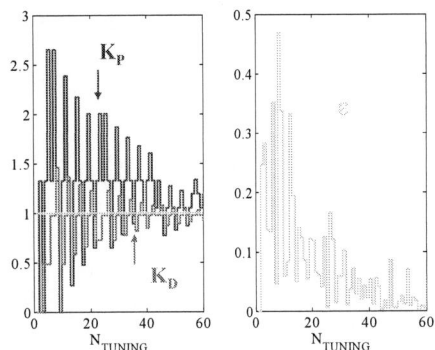

Fig. 13– Parameters K_P, K_D normalized to their final value and tuning error normalized to the duty cycle impulse amplitude during *statistical* tuning N_{TUNING} is the number of tuning periods T_{TUNING}

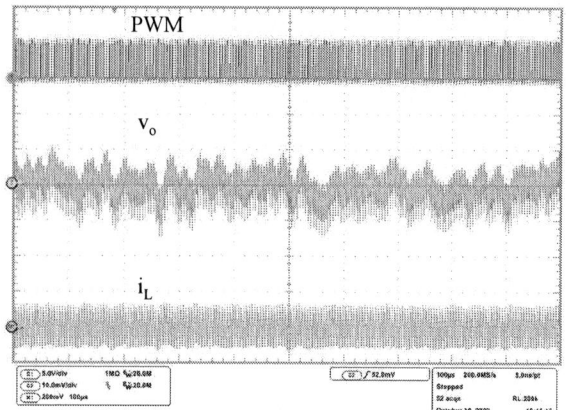

Fig. 14– Output voltage during statistical tuning, V_{in}=5V(v_o 10mV/div, i_L 0.5A/div, time 100µs/div)

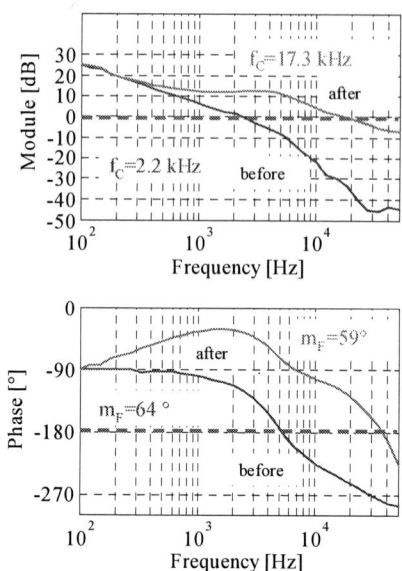

Fig. 16– Bode diagram before and after the PD tuning for the *statistical* case with V_{in}=8V

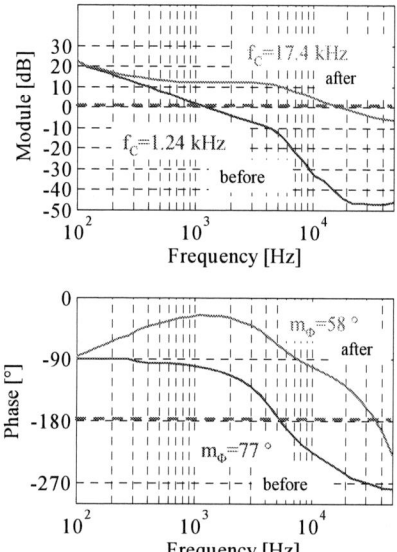

Fig. 15– Bode diagram before and after the PD tuning for the *statistical* case with V_{in}=5V

In Fig. 17 a complete three parameters PID tuning is presented, computed through the statistical method and with the same target bandwidth and phase margin. Similar results were obtained with the *deterministic* tuning of 3-parameters PID.

V. CONCLUSIONS

In this paper a closed loop autotuning technique is proposed. The tuning is performed injecting a disturbance in the duty cycle suitable to measure in either a deterministic or a statistical way the impulse response of the system. The impulse response is compared with the reference one, obtained from the desired loop frequency response and the energy of the error signal is minimized through a numerical search method to meet the tuning targets. The experimental results for the two proposed solutions show good agreement with the theoretical expectations, confirming that the deterministic

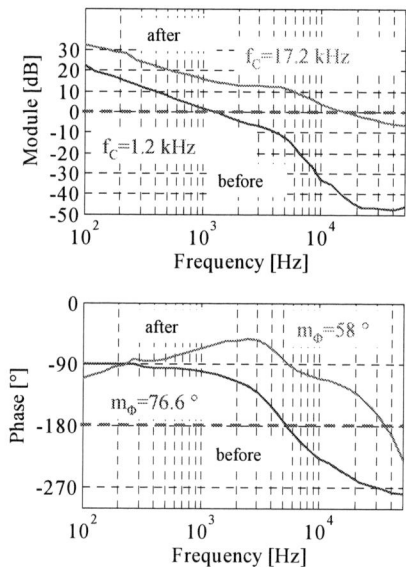

Fig. 17– Bode diagram before and after the PID tuning for the *statistical* case with V_{in}=5V

solution guarantees faster tuning. Instead, the statistical solution presents smaller output voltage deviation and on-line tuning capabilities, but much longer tuning time.

REFERENCES

[1] B.J. Patella, A. Prodic, A. Zirger, D. Maksimović, "High-frequency Digital Controller IC for dc/dc Converters", IEEE Applied Power Electronics (APEC), Dallas, March 2002, pp. 374-380.

[2] J. Xiao, A.V. Peterchev, S.R. Sanders, "Architecture and IC implementation of a digital VRM controller", IEEE Trans. on Power Electronics, Vol. 18, No. 1, January 2003, pp. 356-364.

[3] S. Saggini, M. Ghioni, A. Geraci, "An innovative digital control architecture for low-voltage, high-current dc-dc converters with tight voltage regulation", IEEE Trans. on Power Electronics, vol. 19, January 2004, pp. 210-218.

[4] K. J. Åström, T.Hagglund, "Automatic tuning of PID controllers", Instrument Society of America, (ISBN 1-55617-081-5), 1988.

[5] W. Stefanutti, P. Mattavelli, S. Saggini, and M. Ghioni, "Autotuning of digitally controlled buck converters based on relay feedback," IEEE Trans.Power Electron., vol. 22, no. 1, pp. 199–207, Jan. 2007.

[6] L. Corradini, P. Mattavelli, and D. Maksimovic, "Robust relay-feedback based autotuning for DC-DC converters," in Proc. IEEE Power Electron.Spec. Conf., Jun. 2007, pp. 2196–2202.

[7] L. Corradini, P. Mattavelli, W. Stefanutti, and S. Saggini, "Simplified model reference-based autotuning for digitally controlled SMPS," IEEE Trans. Power Electronics, vol. 23, no. 4, pp. 1956–1963, Jul. 2008.

[8] J. Morroni, R. Zane, D. Maksimovic, "Design and Implementation of an Adaptive Tuning System Based on Desired Phase Margin for Digitally Controlled DC–DC Converters", IEEE Trans. Power Electronics, vol. 24, no. 2, pp. 559-564. Feb. 2009.

[9] B. Miao, R. Zane, D. Maksimovic, "System identification of power converters with digital control through cross-correlation methods," IEEE Trans. Power Electronics, vol.20, no. 5, pp. 1093-1099, Sept. 2005.

[10] A. Kelly and K. Rinne, "A self-compensating adaptive digital regulator for switching converters based on linear prediction," in Proc., IEEE APEC 2006, pp. 712-718.

[11] R.J. Polge, E.M. Mitchell, "Impulse Response Determination by Cross Correlation" IEEE Trans. Aerospace and Eletronic Systems, Vol. AES-6, pp. 91-97.

[12] V. Torczon, "On the convergence of pattern search algorithms", SIAM J. Optim., 7 (1997), pp. 1–25

[13] T.G. Kolda, R.M. Lewis, V. Torczon "Optimization by Direct Search: New Perspectives on Some Classical and Modern Methods", SIAM Review, Vol. 45, No. 3. (2004), pp. 385-482.

High-fidelity and High-speed Modeling and Simulation for Power Conversion Systems

Chunchun (Emily) Xu , Luis Garces , Paul Szczesny

GE Global Research Center

1 Research CircleNiskayuna, NY, 12309, US

Abstract—High fidelity and high-speed modeling and simulation are essential analytical tools for the design of sophisticated large-scale power conversion systems, like Oil and Gas compressor drives, Wind and Solar converters, and MRI gradient drivers. This paper presents various approaches to the dynamic modeling of power electronics components, new topologies and their control applicable to their system design and analysis. Discussions on the practical technical issues in the system modeling and simulation will also be presented, with a few demonstrative examples of the power converters in as used in power generation and in compressor drives for Oil and Gas.

I. INTRODUCTION

The worldwide strong demand for energy raises more and more challenges to the power conversion systems design. Especially their very high cost, the required efficiency and the extremely high reliability requirements are driving the explorations of new high fidelity, high-speed yet cost-efficient approaches to improve the design, test and validation of power converter system of larger sizes and voltages.

To server the different needs of a large variety of power conversion applications in aircrafts, naval vessels, renewable and conventional power generation, transportation, and medical systems, etc [1-2], full set of tools and libraries has been developed. They have been applied for the development and testing of the controllers, new circuit topologies verification, faults detection, to run Design of Experiments (DOE), and not the least, to train the final users [3-7].

The "ideal" solution is one that permits to embed a faithful simulation of the controlled plant within the client's hardware controller. In the optimal case, this should allow the operation of the control and of the simulated system as close to real time as possible. Presently, it is possible to reach this goal only for a very small set of controllers. Most of the time, the system has to be simulated off-line using special purpose simulation packages, or in a combination of the real hardware with parts of the system simulated in another computer by software and not in real-time. In what follows, a short description of the many difficulties and limits that have been encountered, as well as some of the successful examples will be presented.

II. COMPONENT MODELING AND SIMULATIONS

A. Main components in a power conversion system and their models

Power devices (Diodes, MOSFET, IGBT, IGCT, SCR, etc.), sources (Grid, Battery, Fuel Cell, etc.), loads (electric machines, MRI coil, battery, etc.), filters and thermal management systems are some of the main components of a typical power conversion system. Because these components have a wide range of time constants, and since the simulation speed is decided by the smallest simulation step required, one of the first steps in the planning of a system model is to determine the level of details that will be required.

To avoid extremely long computations, many times the converters will be represented using "average models", thus avoiding the very short computation intervals required by the complete switching models. However, these cannot be avoided if the simulation is to be used for the study of losses in the devices, their operation under faults or other fast contingencies. Saber, Matlab, PSpice and LTspice are a few of many primary modeling and simulation tools widely used for power conversion system designs and validations.

Saber has extensive libraries for both electrical, thermal and mechanical components, and it is suitable for very complex high power application designs. It has a dual set of clocks, one dedicated to the analog and other to the digital part of the simulation. This characteristic allows a very high resolution in the timing of interrupts and switching events without slowing the simulation of the analog part of the model.

Matlab/Simulink is a very widely used engineering and scientific tool. It has toolboxes and third-party plug-ins dedicated to modeling power devices, electrical machines and many other electro-mechanic components. These provide the users with the flexibility to mix and match the models with very consistent numerical mechanisms. Furthermore, it provides tools to convert models directly into software that can be executed in many DSP's and processors.

Psipce and LTspice are mixed signal simulators supporting analog circuit, digital logic and device modeling, and they are tools that can be used for quick and easy evaluations of low and medium power applications. All the simulation packages offer now the descriptive modeling as an alternative to ready made primitives or element-based methods. Scripts and high-level languages like C, can be used to build the models. Moreover, most of the simulation packages now provide the Interface to databases and DLL's (Dynamic-Link Library), which could be written in C/C++ or Fortran. Some simulation packages also provide co-simulation options with each other, such as the SaberSimulinkCosim in Saber.

B. Average System Model

Up to the system level, average model is a good option for a quick and simplified simulation, focused on the overall system behavior or performance. Figure 1 and Figure 2 show a three-phase PWM two-level converter with the equivalent representation of one of its legs. This model can be fed directly with the commands for the PWM modulation index generating at its output the average voltage. It has in its structure the calculation for the average current flowing into the DC link capacitor which allows its use in a back to back configuration or when fed from a battery. Obviously, the same model can also be used for three and more level converter legs.

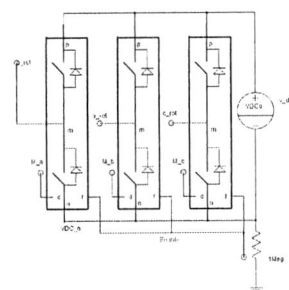

Figure 1. Two-level Converter Circuit Diagram

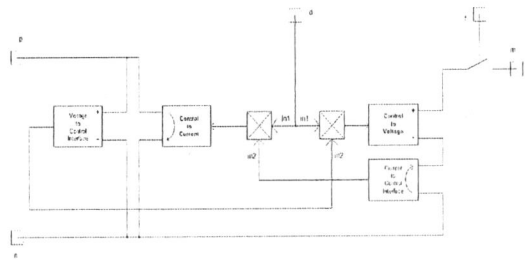

Figure 2. One Phase Leg Equivalent Diagram

For an even simpler and faster representation, Figure 3 shows a simplified average model circuit diagram of 3-phase PWM converter connected to the grid, where V_1, V_2, V_3 are grid voltages, V_a, V_b, V_c the bridge voltages, and V_{cm} is the system common mode voltage.

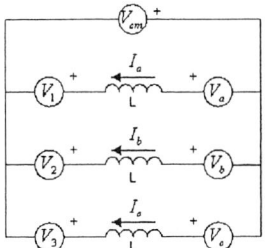

Figure 3. Line Converter Circuit Diagram

The digital implementation of the line currents can then be derived from equation (1), which can be easily modeled using MAST in Saber or an S-function in Matlab.

$$
\begin{aligned}
I_{a(n)} &= I_{a(n-1)} + \frac{(V_a + V_{cm} - V_1)}{L} * \Delta t \\
I_{b(n)} &= I_{b(n-1)} + \frac{(V_b + V_{cm} - V_2)}{L} * \Delta t \\
I_{c(n)} &= I_{c(n-1)} + \frac{(V_c + V_{cm} - V_3)}{L} * \Delta t
\end{aligned}
\tag{1}
$$

C. System Modeling Example Case

When it is necessary to get more details of the design, waveform level simulations are required for the modeling of sophisticated systems. One example is the PV array emulator design. This is a circuit capable to behave electrically similar to PV arrays but without depending on weather conditions.

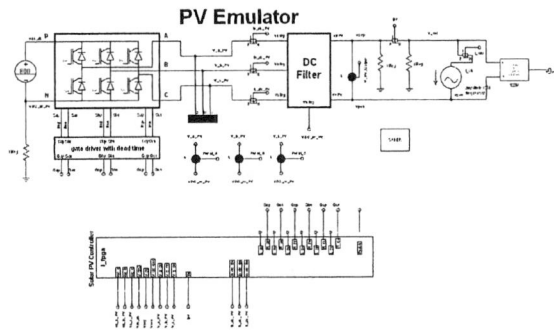

Figure 4. PV Emulator Simulation

They are being developed due to the increased in the high power solar inverter design activities which require appropriate test equipments for their development and manufacturing. In order to fulfill different test modes, like converter conversion efficiency test, static and dynamic MPPT efficiency test, Grid features validation, etc., these emulators have to provide current-voltage characteristics (I-V curve) representative of a variety of PV array technologies, while meeting the power output and dynamic response requirements.

Figure 4 is a detailed PV array emulator model, which includes the complete power bridge and its control. In the simulation, it was assumed that multiple identical PV arrays can be connected in series or in parallel in order to get the desired voltage and power level. The model faithfully represents the real laboratory hardware configuration, while the controller covers the actual firmware used for the final design. These allowed the capture of any changes and refinements during the simulation testing and design. Figure 5 shows the results obtained from the waveform level simulation of the PV array emulator compared to the ideal PV array numeric model simulated using Matlab. The results indicate that the PV Array emulator design closely follows the electrical behavior of the real PV array.

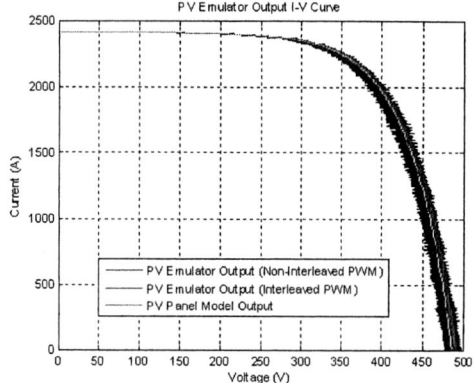

Figure 5. PV Array Simulation Results

D. Model-Based Control Design

Recent years, with the continuous improvement of auto-code generation and optimization tools, model based control design have gained more and more momentum in the power conversion control design area. Figure 6 is one example of the model-based control design flow.

The signal processing, regulator, PID compensator, modulator,etc. can be highly modularized and built in Matlab/Simulink, then real-time workshop generates the C/C++ code automatically. Depending on the target controller, various compilers can be selected to build the controller executables. This method is only one of many for model-based design, which is still a relative new area with a lot of potential growth in the near future.

Reusability and modularization are the keys of model-based control designs, so each control part has to be carefully partitioned into a few subsystems of single or multiple layers. Each subsystem is designed and tested individually, and the interface (inputs and outputs) between the control parts can be standardized.

In a very similar way, there are more and more tools available on the market that provide automatic code generation for FPGA's. Lab View Real-Time is one of these tools, it allows a fast control designing and prototyping without the need of knowing all the details of the FPGA hardware layout and programming. The capabilities of running processes in parallel allows high performance FPGA's to be configured to run much faster than conventional a PC, and their strong I/O

interface capabilities minimizes the data transmission time between the FPGA and its peripherals, opening up a new space for real-time hardware-in-the-loop simulation.

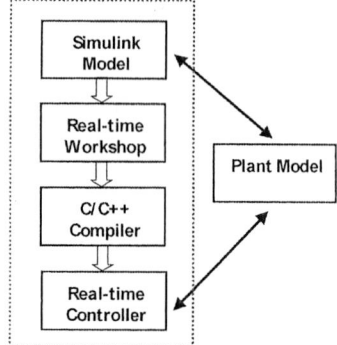

Figure 6. Model-based Control Design Flow

III. (HIL)HARDWARE-IN-THE-LOOP SIMULATION

A. HIL Simulation Platform

Leveraging the fast growing computer and communication technology, hardware-in-the-loop simulations provide a quick, safe and high fidelity way to test the real-time control system. In the hardware-in-the-loop simulation, the controller hardware and software are the same that will be used in the final product, while the plant (power converter, machines, grid, loads, disturbances, etc.) can be simulated in different ways and at desired detail levels.

Figure 7 is the overview of a typical HIL simulation platform compared to the scale down lab setup. In the HIL simulation platform, the controller hardware is connected to the workstation via an Ethernet connection and to the SABER simulation through a template written in the MAST modeling language containing a foreign DLL with the communication software.

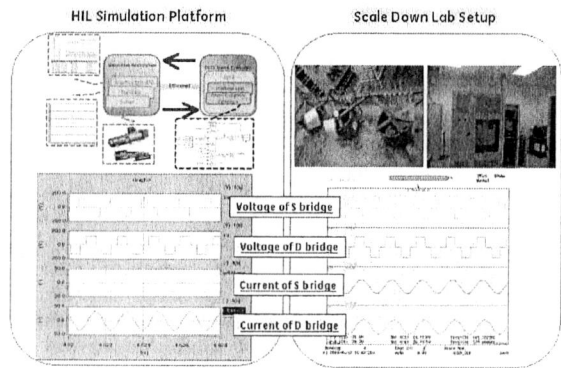

Figure 7. HIL Simulation Platform for System Validation

Figure 8 shows the detail of the HIL simulation platform. The controller is set to operate at a sampling time given by an interrupt generated by the simulator and send through the communication channel. The length of the interrupt is set by the SABER simulation and represents in model time, a fixed time interval. The controller sends to the model, the time to fire for each of the switch device of the power converter, receiving in return the system variables required by the control (currents, voltages, speed, etc.) Although the system does not run in real time, the use of the actual hardware and software for the control gives another level of confidence of the validity of the simulation. Moreover, the controller allows the interaction with the user through its HMI (Human Machine Interface) as when operating with the actual system, and can, at its turn, be connected to supervisory controllers to verify its operability under all case of contingencies.

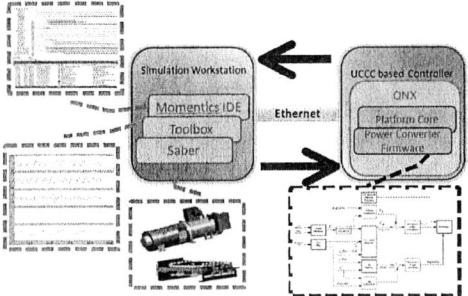

Figure 8. HIL Simulaiton Platform

There are a few metrics of HIL simulation. First of all, it is relatively easy to incorporate very complex systems into the plant models. In other words, before building the potentially very expensive hardware of the power conversion system, the control plant can be modeled and tested in full details with the actual controller in the HIL platform, thus reducing the time to market of new products. Secondly, for large-scale systems, especially like the high voltage and high power applications, it is very costly to conduct validations and tests on the real hardware. The HIL simulation provides lower cost risk mitigation methods for these activities. It also provides a common and consistent platform across the product development, validation and testing, field engineering, and training.

B. HIL simulation Example Case

High power systems for oil & gas, wind, solar, and other power generating applications represent challenges to high-fidelity HIL simulation [1]. Their large sizes and the time constants of the power sources and loads could lead to very long simulation periods. However, the fast switching of the PWM generation still has to be included in the models. They could interact among themselves as in the case of inverters in parallel or in series, and also can introduce undesired low frequency beating frequencies or excite resonances with the rest of the systems that can only be identified after very long simulation runs.

One of such cases was found in the development of the control for a 32 MW medium voltage inverter-driven synchronous machine connected to a turbine-driven gas compressor, as shown in Figure 9. The system comprises four three-level NPC AC/DC/AC voltage source converters connected in parallel at the machine side. Each of the converters was connected to the secondary of a transformer at the grid side.

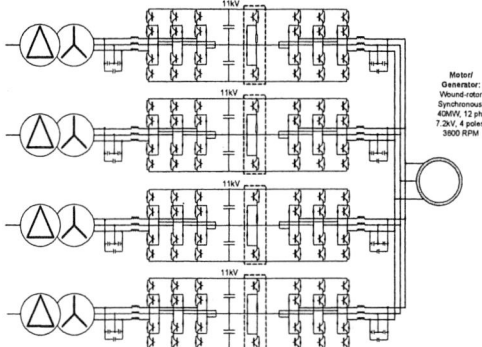

Figure 9. 32 MW Medium Voltage Compressor Drive

The real-time control software was structured in the master-slave mode, which resides in one physical Intel 86 family controller. In the HIL the converters, the machine with its exciter and the transformers were simulated in a very detailed SABER simulation. The controller sent the firing times for all the power devices and received the electrical and mechanic feedbacks every $250\mu s$ simulated computation time.

Extreme care was taken in the simulation of all possible identified critical and fault cases, which also included for some of the runs, a Matlab/Simulink based mechanic-thermal model of the coupled turbine. These runs allowed the identification of critical power elements such as the crowbar size required to provide protection to the IGCT's in case of a short at the machine terminals. It also allowed the optimization of the PWM modulation strategy insuring that the torque harmonics produced by the switching did not excite undesired resonances of the mechanical shaft of the turbine. These analyses required a close interaction of teams of very diverse fields whose expertise and experience contribute to the validation of the results.

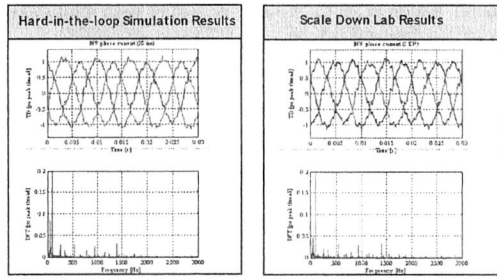

Figure 10. Current Waveform Comparisons

Together with the HIL simulation, a scaled down prototype of the system was build and simulated, allowing a second and unfortunately, still required opportunity to validate the simulation results. Figure 10 shows one of many comparisons of current waveforms with their frequency spectrums obtained through the HIL simulation, with the actual scope pictures obtained with the scale down model. These comparisons are part of a model validation strategy based on the Six_Sigma methodology that, although time consuming, allows the reduction of the time and cost that would be necessary if these tests were to be done using the full size prototype.

IV. CONCLUSION

The wide range of the time constants of different components in the power conversion system imposes a major challenge to the simulation speed and its fidelity. A successful dynamic simulation of power electronics requires a proper choice of the simulator package and platforms. A variety of dynamic simulation approaches of power electronic systems are presented, and demonstrative examples and results are shown to demonstrate the need of a systematic approach to this problem.

REFERENCES

[1] R. Baccani, R. Zhang, T. Toma, A. Iuretig, and M. Perna, "Electric systems for high power compressor trains in Oil and Gas applications- System design, validation approach and performance", Proceedings of the thirty-six Turbomachinery Symposium, 2007.

[2] J. Sabate, L. Garcés, Q. Li, and W. Wirth, "High-bandwidth high power gradient driver for magnetic resonance imaging with digital control", IEEE Applied Power Conference and Exposition, 2005.

[3] L. Garcés, C. Xu, X. Huang, and P. Szczesny, "System modeling for power electronics" International Simulation Multiconference, June 16-19, 2008.

[4] D. Maksimovic, A. M. Stankovic, V. J. Thottuvelil, and G. C. Verghese, "Modeling and simulation of power electronic converters," Proceedings of the IEEE, Vol. 89, No. 6, pp. 898-912, June 2001.

[5] N. Mohan, W. P. Robbins, T. M. Underland, and O. Mo, "Simulation of power electronics and motion control systems – An overview," Processing of the IEEE, Vol. 82, pp. 1287-1302, Aug. 1994.

[6] S. R. Sanders, J. M. Noworolski, X. Z. Liu, and G. C. Verghese, "Generalized averaging method for power conversion circuits," IEEE Transactions on Power Electronics, Vol. 6, pp. 251-259, Apr. 1991.

[7] L. Garcia de Vicuna, A. Poveda, LF. Guinjoan, and J. Majo, "Computer-aided discrete-time large-signal anaysis of switching regulators," IEEE Transactions on Power Electronics, Vol. 7, pp. 75-82, Jan. 1992

[8] Iov, F.; Blaabjerg, F.; Zhe Chen; Hansen, A.D.; Sorensen, P.; "A new simulation platform to model, optimize and design wind turbine" IECON 02

[9] Zhenhua Jiang; Dougal, R.A.; Leonard, R.; Figueroa, H.; Monti, A.; "Hardware-in-the-Loop Testing of Digital", APEC '06. Twenty-First Annual IEEE

[10] R. M. Nelms, B. W. Evans, and L. L. Grigsby, "Simulation of ACspacecraft power systems", IEEE Transactions on Industrial Electronics, Vol.36, No.3, pp.398-402, Aug. 1989.

Electrical Power Distribution System (HV270DC), for Application in More Electric Aircraft

D. Izquierdo[1], R. Azcona[1], F. J López del Cerro[2], C. Fernández[1], B. Delicado[2]

[1] Military Air Systems - Defence and Security Division (EADS)
[2] Airbus Military (EADS)
Paseo John Lennon, s/n.
28906 Getafe (Madrid) ESPAÑA
Tel.: +34-91-624 7130; FAX: +34-91-624 2411
E-mail: Daniel.Izquierdo@eads.com

Abstract - In the new designs of military aircraft and unmanned aircraft there is a clear trend towards increasing demand of electrical power. This fact is mainly due to the replacement of mechanical, pneumatic and hydraulic equipments by partially or completely electrical systems. Generally, use of electrical power onboard is continuously increasing within the areas of communications, surveillance and general systems, such as: radar, cooling, landing gear or actuators systems. To cope with this growing demand for electric power, new levels of voltage (270 V_{DC}), architectures and power electronics devices are being applied to the onboard electrical power distribution systems.

The purpose of this paper is to present and describe the technological project HV270DC. In this project, one Electrical Power Distribution System (EPDS), applicable to the more electric aircrafts, has been developed. This system has been integrated by EADS in order to study the benefits and possible problems or risks that affect this kind of power distribution systems, in comparison with conventional distribution systems.

I. INTRODUCTION

There is a strong trend in new designs of military aircraft and unmanned aircrafts towards the More Electric Aircraft (MEA) concept. On the one hand this is a consequence of substitutions of conventional equipments, which depend on pneumatic, mechanic and hydraulic power, by equipments that depend on electrical power. This factor has provoked increase of equipments, which require electric power. Besides, these changes provide a better system performance due to increase of reliability, less maintenance, efficiency on energy conversion and therefore also higher efficiency of aircraft in general. On the other hand, there is an increase of the number of equipments that depend on electric power and form part of the different aircraft systems (avionics, communications, surveillance...), such as: radar, infrared and electro-optical cameras, radios, etc [1] [2] [3] [4] [5] [6].

These beforehand mentioned factors have caused a considerable increase of the demand for electrical power. To cope with this increase of electrical power in the new distribution architectures, in the area of military aircraft, the traditional voltage levels of 28 V_{DC} and 115 V_{AC} [7] have evolved to 270 V_{DC} [8] [9].

One of the main benefits of these high voltage systems is that current levels are 10 times lower than the conventional low voltage levels; therefore it is possible to reduce the section of the wires and by that, also the weight of the wiring. On the other hand, the major electrical loads are powered directly from direct current instead of alternating three-phase current, which means decrease in number of the wires that are used to connect the different loads. This considerable growth of the number of electric loads in these new electric distribution architectures has contributed to an increase of the quantity of the electric and electronics components, which could conduce to instability of whole system due to the interactions between the different equipments that compose the system [10]. Also, raising the level of voltage provokes to appear new problems regarding the function of some devices, such as conventional protections [11]; and other inconveniences originated by physical effects in the wires with the new levels of voltage: corona effect, arc fault and others [12]. Therefore, before being applied in the future unmanned aircraft programs developed by EADS, see Fig. 1., it is necessary to study the behavior of the high voltage electrical power distribution systems, as well as the components that form a part of these systems.

Fig. 1: Future unmanned aircraft program developed by EADS

This paper describes a Electrical Power Distribution System (EPDS) architecture based on 270 V_{DC}, and equipments that are part of it. The architecture has been developed and integrated by EADS in collaboration with various national companies from different sectors. The proposed EPDS has a smart electrical load management control, which is responsible for managing the connection and disconnection of the loads depending on the operational mode and available power sources. Besides, in the laboratory and virtual test bench have been identified and checked possible problems and risks of this kind of architectures.

II. TECHNICAL APPROACH

In the concept of MEA, the new EPDS designs are the key element. These new power distribution systems are based on two voltage levels, depending on whether the application is Alternating Current (AC) or Direct Current (DC):

- In case of AC systems in commercial aircrafts, with voltage level of 230 V_{AC}, currently used in civil platforms such as A380 and future designs, such as A350 and B787 [13].

- In case of DC with voltage level of 270 V_{DC}, normally used in military aircrafts in platforms like F-22 or F-35 [8] [9] [10] [14].

The origin of the architectures with 270 V_{DC} voltage level is a consequence of the rectification of conventional 115 V_{AC}, available in the onboard power sources. There are two main reasons for this new voltage level. Firstly, it is possible to obtain a 270 V_{DC} generator from a conventional generator, by means of integration of natural rectifier built from diodes. Secondly, some of the equipments, that are powered from 115 V_{AC}, have integrated blocks of rectification, that convert the 115 V_{AC} levels to 270 V_{DC}, as in case of some radars and electrical actuator controllers, which are using an internal high voltage bus. In the aforementioned EPDS, it is possible to reduce weight of some electrical loads by eliminating aforementioned conversion block.

In these architectures, due to the new voltage levels, the conventional wiring protection and load control systems, such as fuses, circuit breakers and relays are causing certain limitations in their functionality. Therefore, new components that permit substitution of the mechanical devices have to be introduced. The Solid State Power Controllers (SSPC) permit replacing of the relay and circuit breakers, elements of mechanic nature, for a single power electronic device as SSPC [11]. These SSPC allow connecting loads to a main bus and provide the function of protecting the electric installations from overloads and short circuits, exactly as it is done by relays and Circuit Breakers (CB).

In the field of military applications, to keep up with the development of the new EPDS, several international companies of the aeronautic and military sectors have developed or are currently developing projects that study the problems associated with these kinds of electrical power distribution systems. Among the most important technological projects that have centered round this problem, is the MADMEL project, developed in the United States by Military aircraft division of the Northrop/Grumman at the end of the nineties [8]. There also have been some less relevant initiatives in Europe, as the TIMES program (Totally Integrated More Electric Systems) [15] or the DEPMA project (Distributed Electrical Power Management Architecture) [16].

In 2005, EADS started the HV270DC project, which is based on the development of one electric power distribution architecture of 270 V_{DC} and its test bench, for its application in future aircraft designs. In this architecture, the primary and secondary distribution systems are designed with a voltage level of 270 V_{DC}, and two additional power converters have been introduced into the final design. These power converters provide conventional voltage levels (28 V_{DC} y 115 V_{AC}), coming from the primary distribution busbars, with a purpose to supply electric power to the equipments that proceed from conventional EPDS and which are powered with the aforementioned voltage levels.

The main objective of the project is to integrate the different equipments and components that contain EPDS of one MEA and test the associate functionality. The main computer constantly controls EPDS, which also is in charge of monitoring the state of the whole EPDS. In this manner, computer configures the system for several operative conditions, depending on the different power sources and operative sceneries. This project is designed with an objective to identify future problems that could arise from the use of this type of EPDS, such as: the problems caused by the new components, like the SSPC and their application on critical loads control, the physical effects generated in the wiring, as a consequence of the new high voltage levels and the interferences that could be introduced in the conventional electric distribution lines, in case both of them co-exist on the same platform.

III. EPDS ARCHITECTURE DESCRIPTION

The EPDS architecture is divided in five main blocks, depending on functionality of each part, see Fig. 2:

- Power Generation area, provides the electrical power to the rest of the system and loads.

- Primary distribution block distributes the electric power to all the points and also protects the connected blocks from overcurrent.

- Conversion area, in charge of converting the 270 V_{DC} level to conventional distribution levels, such as 28 V_{DC} and 115 V_{AC}.

- Secondary distribution area, which provides electric power to the loads. In this EPDS, the secondary distribution is based on power electronic modules, instead of conventional devices, such as relays and circuit breakers.

- Block of management and monitoring, controls the correct function of all the equipment and devices, depending on the operative sceneries.

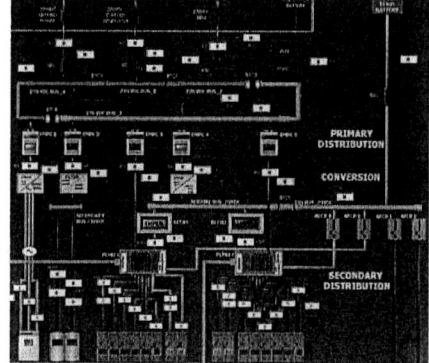

Fig. 2: Proposed EPDS Architecture based on SSPC

The following paragraphs include a brief description of some of the characteristics and equipments that compose each of the areas aforementioned.

A. Generation Block.

In the generation area, the following equipments have been integrated:

- The Lithium Ion DC battery supplies the initial electric power for the internal control systems which are integrated inside the equipments that compose the architecture. The battery also energizes the contactors and the Remote Controlled Circuit Breakers (RCCB).

- The generation source of 270 V_{DC}, which has three outputs and permits to emulate several generation sources of the aircraft: Ground Power Unit (GPU), Auxiliary Power Unit (APU) and Main Generator. This power source has been adjusted according to military standard, MIL-STD-704 [17].

B. Primary distribution Block.

The primary distribution block consists of primary distribution busbars, contactors and protections that connect this block with the conversion and secondary distribution areas. All these devices work with 270 V_{DC}. The circular configuration of the primary distribution busbars, see Fig. 2, comparing with the conventional architectures [18], permit a higher number of reconfiguration options. The protections installed in the primary distribution level are the Electro-Magnetic Power Controllers (EMPC). These devices measure the charge and interrupt the current when charge exceeds the allowed threshold, avoiding the damage to the cables or downstream equipments.

C. Conversion area.

The conversion area is composed of various equipments, which convert the new levels of primary distribution to the levels of conventional distribution of 28 V_{DC} and 115 V_{AC}. These levels have been kept for providing electrical power to the installed equipments in the conventional aircrafts. This is a consequence of the tendency in the market to produce new equipments designed for this SPDE.

Therefore, it is possible to apply the architecture to different platforms, including this kind of power converters. In this block, two power converters have been integrated:

- The DC-DC converter 270 V_{DC}/28 V_{DC}, provides electrical power to the bus bar of the 28 V_{DC} and charges the battery, depending on the electric power that the primary area provides. The equipment has been designed by the *Greenpower* Company, with nominal power of 3,5 Kw, in conformity to the EADS technical specifications.

- The inverter 270V_{DC}/115V_{AC}, generates a signal in CA starting from the level of 270V_{DC}, see Fig. 4. This equipment has been designed by *Ingeniería Viesca* Company, with a nominal power of 5 Kw. This inverter has been designed in conformity to military standard, MIL-STD-704 [17]

Fig. 3: DC Converter developed by Greenpower ©

Both equipments include discrete signals that inform about correct function and differentiate between several failures categories.

Fig. 4: Inverter developed by Ingeniería Viesca ©

D. Secondary distribution block.

In this block have been integrated the equipments and devices that connect the loads to the EPDS.

- The Power Load Management Unit (PLMU), see Fig. 5. This equipment is based on SSPC modules [11], which permit remote electrical load control and cable protection. Indra Company has developed PLMU in different phases in conformity to the EADS technical specifications.

- Remote Control Circuit Breaker protections (RCCB) are similar to conventional circuit breakers, but they also have the capability to control and to monitor protection status remotely.

It is important to point out, that in this area 270 V_{DC} coexists with the conventional voltage levels of 28 V_{DC} and 115 V_{AC}.

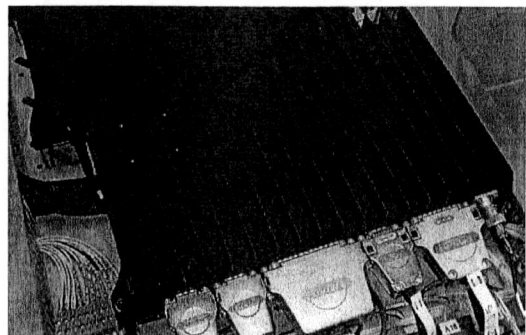

Fig. 5: PLMU developed by Indra ©

E. EPDS control and monitoring block.

Last block manages the EPDS depending on the operative conditions of connected loads, available power sources and operative sceneries. It is composed of the following equipments:

- The Load Management Computer (LMC) runs the software that controls the EPDS. This computer takes decisions according to the information that is received form the data bus coming form the all equipments that are connected to the MIL-STD-1553B data bus. This equipment has been developed by Indra, according to EADS specifications.

- The Remote Interface Unit (RIU), see Fig. 6, is responsible for monitoring all the available information from the analogue and discrete EPDS signals, with objective to maintain all this data accessible in the data bus. It also controls the contactors, RCCB and EMPC of the system according to the indications that LMC receives trough the data bus.

Within the particular architecture diverse communication data bus options can be used, such as: RS-232, RS-422 or RS-485; however none of them is in conformity with military standards. The 1553B data bus satisfies military standards and dual communications. From other side, most of the equipments developed during the project come from past projects, which already include this type of data bus, frequently used in the on-board systems [19].

Fig. 6: Remote Interface Unit

The LMC communicates through MIL-STD-1553B data bus with various equipments of the EPDS, such as: PLMU, RIU and other equipments, that belong to other aircraft systems that could have interface with the electrical system: Cockpit Display (CD), Aircraft Computer (A/C) or Maintenance Data Panel (MDP), see Fig. 7.

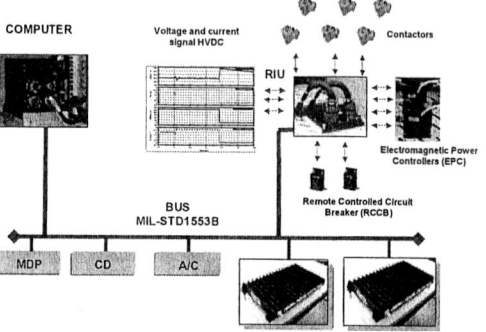

Fig. 7: Equipment connected to 1553B data bus

LMC has bidirectional communication with all of the real or simulated equipments that are connected to the data bus. This way each equipment send the information to LMC and the computer can process the data, and send the commands of control to each of the equipments connected to the 1553B data bus. Therefore, it is possible that LMC warns the other aircraft systems about the EPDS failures, or disconnects automatically the loads during emergency conditions by means of SSPC modules.

IV. DEVELOPMENT OF THE HV270DC PROJECT

The HV270DC project has been designed according to a functional specification for its application in the more electric aircrafts [20]. This document is a reference for the rest of the project's documents. The specification document describes the functional and operational requirements that must be taken into account during design phase, for integration into an aircrafts platform.

Taking into account the document that describes systems function, was developed a specific test bench that includes all the elements that shape the architecture and the different interfaces, among them all the equipments that form a part of the EPDS. The equipments that have been integrated in the test bench have been specified in conformity to the military standards [17].

Also, software functional requirements document of the HV270DC Systems control has been defined. This software resides in the LMC and has been developed by the Department of Software EADS. In order to facilitate the integration tasks, there has been developed a simulated environment that reproduces the behavior of real communication bus. This permits to perform the first tests of the systems software integration in a virtual bench, prior to the systems integration in the real HV270DC demonstration bench. The identified software problems are reported and stored in database; therefore it is possible to trace a problem from its origin until its final solution. In order to verify EPDS functionality and solve integration problems, an electrical virtual test bench has been implemented in PSIM®, see Fig. 8. Modelling and simulation have been the key tools in the EPDS development [21] [22].

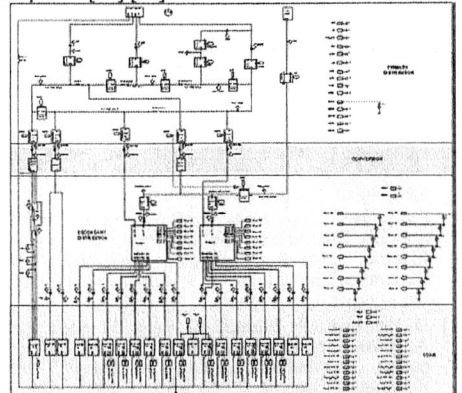

Fig. 8: *Electrical virtual test bench*

Once the architecture was integrated, test requirements were developed to validate the correct function of the EPDS and perform the necessary adjustments to the equipments and smart control software.

The demonstrator test bench includes a monitoring system that permits acquisition and monitoring of all the EPDS parameters: voltage, currents, condition of the device, etc. for processing and analysis. All the information is visible on the screen in the panels generated in the "Labview ©", that permit to check the system status real time, see Fig 2.

V. IDENTIFIED PROBLEMS

During the test phase in the real test bench, there have been identified some problems, which are described below.

The data bus MIL-STD-1553B imposes a limitation of the minimum times of the EPDS control, which considerably slow down the response of the system, specially when the commands proceeding from the LMC.

There have been detected limitations in the SSPC during the connection and disconnection of specific type of charges. Problems were detected during the connection of highly capacitive loads, and during disconnection of the inductive loads. Other limitation of the devices is in the range of the currents they are able to manage; therefore for high levels of current it is necessary to use conventional devices. Another problem that could appear is the internal fusion of the SSPC as a consequence of elevated temperatures that are provoked by uncontrolled high current circulation through the component.

The physical effects, which can appear in the components and wiring of this kind of architecture, are even more critical if we compare them with the conventional EPDS, increasing the probability of the corona effect and arc fault in the wiring due to the high voltage level.

VI. EXPERIMENTAL RESULTS

The following paragraph includes some of the experimental results obtained during the test phase in the real test bench.

Tests with different loads of 270 V_{DC} have been done, with objective to detect possible problems of whole systems stability.

During the consecutive closures of the contactors of 270 V_{DC}, the ripple of the voltage in the bus bars of principal distribution was measured. This ripple can cause instability of the system and provoke problems in the downstream equipments operation.

By help of the loads test bench, it was possible to verify the correct function of the I^2t curve in the SSPC for different levels of current and components of different manufacturers. In Fig. 9, it can be seen how for different current levels in the SSPC reacts the programmed protection of SSPC, in correspondence with the curve defined in the manufacturer's datasheet.

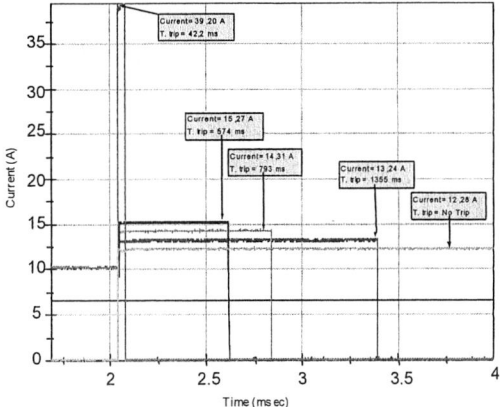

Fig.9: *SSPC I^2t- curve validation*

During the normal operation of the DC/DC Converter, it was detected that the device transmits downstream the ripple that is present in the voltage of the primary distribution bars. This ripple provokes non-compliance regarding standard harmonic distortion measurements [17].

Test results show that the system is out of the distortion spectrum normative in the 270V_{DC} primary and secondary distribution bars when DC converter is activated, see Fig. 10. Besides, it has been checked that when DC converter is disconnected, harmonic distortion is reduced, see Fig. 11.

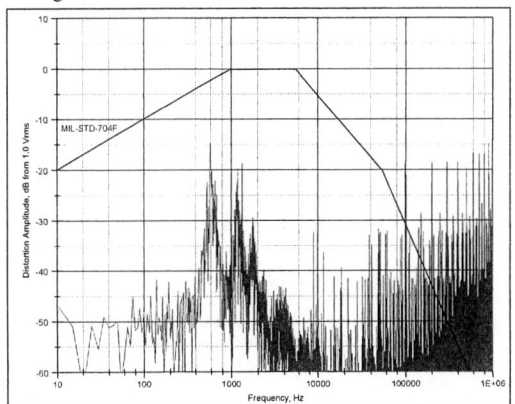

Fig. 10: *Measured distortion amplitude without DC Converter*

978-1-4244-4782-4/10 $26.00 © 2010 IEEE

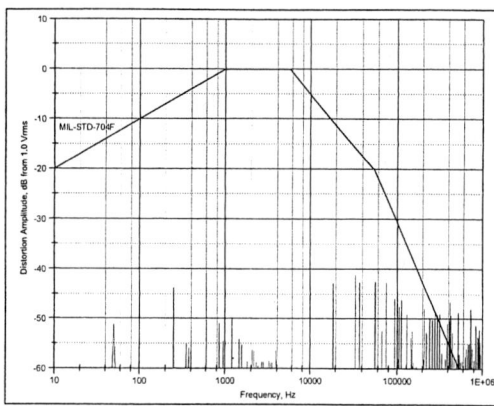

Fig.11: Measured distortion amplitude with DC Converter

VI. CONCLUSIONS

This paper describes an Electrical Power Distribution System (EPDS) of 270V$_{DC}$ for its use in more electric aircrafts. EADS has designed the architecture with collaboration of several national and international companies. This electric power distribution architecture is controlled by a totally autonomous central processor, thanks to software that controls EPDS during function of the different operative sceneries.

In this architecture, the different areas that constitute the EPDS are taken into account, such as: power sources, primary distribution, conversion area and secondary distribution. Secondary distribution block is based on the power distribution units, which are composed of Solid State Power Controllers (SSPC) modules. The EPDS has been installed in a test bench where different tests were done to validate the function of the system and identify problems in this type of architectures.

ACKNOWLEDGMENT

This work has been supported by The European Regional Development Fund (ERDF) via the Aerospace Sector Plan of the Community of Madrid, through the research project "HVDC Load Distribution System" (Code: 04-AEC0527-000050/2005).

REFERENCES

[1] Spitzer, C.R. "The All-Electric Aircraft: A Systems View and Proposed NASA Research Programs" NASA Langley Research Center. Aerospace and Electronic Systems, IEEE Transactions. Volume AES-20, Issue 3. May 1984. Page(s): 261-266

[2] Lester F. Faleiro," Trends towards a more electrical aircraft", Proc. ICAS 2006.

[3] M. Howse, "All Electric Aircraft", Power Engineer, vol. 17, pp. 35-37, 2003.

[4] David Blanding, "Subsystem design and integration for the More Electric Aircraft". Proc. ICAS 2006.

[5] Rosero, J.A.; Ortega, J.A.; Aldabas, E.; Romeral, L. "Moving towards a more electric aircraft" Aerospace and Electronic Systems Magazine, IEEE Volume 22, Issue 3, March 2007.

[6] M. David Kankam "A survey of Power Electronics applications in Aerospace Technologies". 36th Intersociety Energy Conversion Engineering Conference

cosponsored by the ASME, IEEE, AIChE, ANS, SAE, and AIAA Savannah, Georgia, July 29–August 2, 2001.

[7] E. H. J. Pallet; "Aircraft Electrical System". Third Edition, Pearson, Prentice Hall. 1991.

[8] Miguel A. Maldonado et al. "Power Management and Distribution System for a more-Electric Aircraft (MADMEL)- Program Status". Intersociety Energy Conversion Engineering Conf., August 1997, pp.274-279.

[9] Emadi, K.; Ehsani, M.;"Aircraft power systems: technology, state of the art, and future trends". Aerospace and Electronic Systems Magazine, IEEE Volume 15, Issue 1, Jan. 2000.

10] Liqiu Han, Jiabin Wang and David Howe " Small-signal Stability Studies of 270V DC Power System for More Electric Aircraft Employing Switched Reluctance Generator Technology" 25th International Congress of the Aeronautical Sciences ICAS 2006.

[11] D. Izquierdo, A. Barrado, C. Raga, M. Sanz, P. Zumel, A. Lázaro. "Protection devices for aircraft electrical power distribution systems: a Survey" Industrial Electronics Society, 2008. IECON 2008. 35th Annual Conference of IEEE. Orlando November 11-13, 2008.

[12] I. Cotton, M. Husband; "Higher Voltage Aircraft Power Systems". IEEE Aerospace and Electronics Systems Magazine, February 2008.

[13] AMC-Plane Talk. Aeronautical Radio INC. Maryland (United States). Industry Highlights "Electrical Protection devices". February 2005. Vol XVI, No. 2.

[14] Allen J. Lockyer et al. "Power systems and requirements for integration of smart structures into aircraft" SAGE Publications, Journal of Intelligent Material Systems and Structures, Vol. 15, No. 4, 305-315, 2004.

[15] Cutts, S.J.; "Totally Integrated More Electric Systems (TIMES) Refer to: A collaborative approach to the More Electric Aircraft"; Power Electronics, Machines and Drives, 2002. International Conference on 4-7 June 2002 Page(s):223 – 228.

[16] Bailey, M.; Hale, N.; Ucerpi, G.; Hunt, J.-A.; Mollov, S.; Forsyth, A.; "Distributed Electrical Power Management Architecture (DEPMA) Electrical Machines and Systems for the More Electric Aircraft (Ref. No. 1999/180), IEE Colloquium on 9 Nov. 1999 Page(s):7/1 - 7/4.

[17] "MIL-STD-704, Aircraft Electric Power Characteristics", Military Standard.

[18] S. A. Long and D. R. Trainer, "Ultra-compact Intelligent Electrical Networks,". SEAS DTC Technical Conference, Edinburgh, 2006.

[19] F. Dafeng, X. Yan, Y. Shanshui, Y Yangguang, "Electrical Power Distribution System Based on 1553B Bus for Advanced Aircraft"; Electrical Machines and Systems. ICEMS. Proceedings of the Eighth International Conference on; 27-29 Sept. 2005

[20] L. Pintos, D. Izquierdo, M. sierra; NT-T-MS2-09002; "Descripción Funcional del Sistema de Distribución de Potencia Eléctrica de Alto Voltaje HV270DC y Gestión Inteligente de cargas"; Feb 2007.

[21] A. Barrado, D. Izquierdo, C. Raga, M. Sanz, A. Lázaro, "SSPC Model with Variable Reset Time, Environmental Temperature Compensation and Thermal Memory Effect". Applied Power Electronics Conference and Exposition 2008. APEC '08. 23rd Annual, 24-28 February 2008. Page(s): 1716-1721.

[22] A. Barrado, D. Izquierdo, C. Raga, M. Sanz, A. Lázaro. "Behavioral Modeling of Solid State Power Controllers (SSPC) for Distributed Power Systems". Applied Power Electronics Conference and Exposition 2009. APEC '09. 24th Annual, 16-19 February 2009.

Supercapacitor-Based Energy Management for Future Aircraft Systems

R. Todd, D. Wu, J.A. dos Santos Girio, M. Poucand, A.J. Forsyth

Rolls-Royce UTC
School of Electrical & Electronic Engineering
The University of Manchester, UK
rebecca.todd@manchester.ac.uk

Abstract— A power-based control method is proposed and analysed for a supercapacitor energy storage device. The performance of the energy storage device is examined by simulation and experimentally when operating on a high voltage DC bus with a multi-phase, fault-tolerant generator and a high power, pulsed load such as an actuation or avionic system. In the practical system the energy storage device is emulated using a bi-directional electronic load and a real-time simulation platform. The energy storage device is shown to minimise the DC bus transients and virtually eliminate the torque pulsation on the generator shaft. The system design and performance trade-offs are analysed. The experimental work uses a 70kW generator and 30kW programmable load emulation devices.

I. SYMBOLS

C_{bus}	Combined filter capacitance on the DC bus
D	Duty-ratio
I_{bus}	Total bus current
I_{ESD}	Energy storage device output current
I_{gen}	Total generator current
I_{Load}	Load current of the active load system
I_{sc}	Current in the supercapacitor bank
$\tilde{I}_{sc}$	I_{sc} error ($I_{sc\text{-}L}\text{-}I_{sc\text{-}rc}\text{-}I_{sc}$)
$I_{sc\text{-}L}$	Supercapacitor load current demand
$I_{sc\text{-}rc}$	Supercapacitor recharge current demand
k_p	I_{sc} controller proportional gain
k_{rc}	Recharge control function
L	DC/DC converter inductor
$R_{ESR/EPR}$	Series resistance of supercapacitor / leakage losses
T_e	Torque
T_I	I_{sc} controller integral gain
$V_{ESD}{}^*$	Control variable
V_{sc}	Overall terminal voltage of the supercapacitor bank
$V_{sc}{}^*$	Reference value for V_{sc}
$\tilde{V}_{sc}$	V_{sc} error ($V_{sc}{}^*\text{-}V_{sc}$)
V_{sc0}	Initial value of V_{sc}

II. INTRODUCTION

The increasing levels of electrically powered equipment on aircraft, especially highly dynamic loads such as electric actuators and advanced avionic systems, are placing

---

The authors would like to thank Rolls-Royce plc for promoting and funding the Intelligent Electrical Power Network Evaluation Facility (IEPNEF) which is housed at The University of Manchester in the School of Electrical and Electronic Engineering. IEPNEF has been developed to evaluate ultra-compact and intelligent electrical networks for aerospace, marine and energy applications. It is part of the Rolls-Royce University Technology Centre (UTC) for Electrical Systems in Extreme Environments within the Power Conversion Group.

increasingly severe demands on the engine mounted generators. The problems are particularly acute on small platforms where the non-propulsive power taken from the engine is proportionately very high, and consequently there is a risk that rapid electrical load transients could result in an engine surge, which may compromise the mission objectives or cause catastrophic failure of the platform.

To mitigate these effects, this paper examines the use of a supercapacitor-based energy storage device (ESD) connected to the DC distribution bus of an experimental aircraft electrical system. Supercapacitors are used as the energy source owing to their power density and cyclability. The basic configuration of the system is illustrated in Fig. 1, which shows a five-phase, fault-tolerant engine-embedded generator [1] supplying a high voltage DC bus; the main elements on the bus are the ESD, a generic dynamic load representing an actuator or a high power avionic system, and a background load.

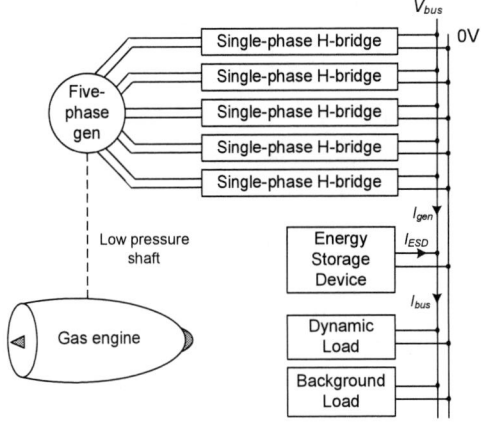

Figure 1. Basic system configuration

The variable amplitude/frequency outputs of the separate five generator phases are connected to the DC distribution bus through single-phase, H-bridge active rectifiers. The generator used in this work is rated at 70kW and the DC bus voltage is regulated to be nominally 540V by the generator control unit. Due to reliability concerns, non-electrolytic capacitors are used for the DC output of the generator system resulting in a relatively small output capacitor value, making the dynamic regulation of the DC bus voltage more difficult.

The ESD in this work is designed to meet instantaneous load transients on the DC bus thereby limiting the rate-of-change of torque applied by the generator on the gas turbine. The control system for the ESD must satisfy two opposing requirements: track the load changes sinking/sourcing a compensating current to/from the DC bus, and secondly manage the supercapacitor state-of-charge. A power-based control method is proposed to fulfil the two requirements and its performance is assessed using typical avionic load profiles.

In the following sections the ESD is described along with the supercapacitor model and an averaged-value model of the DC/DC converter. The control structure is then explained followed by simulation and experimental results. In the experimental work the ESD is emulated using a bi-directional electronic load and a real-time simulator.

III. ENERGY STORAGE DEVICE MODELLING

The ESD, Fig. 2, comprises the supercapacitor bank and the DC/DC converter.

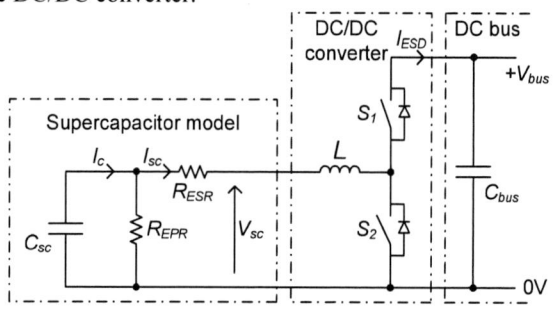

Figure 2. Energy storage device (ESD)

A. Supercapacitors

An equivalent electric circuit model [2] is used for the supercapacitor bank, shown in Fig. 2, as it provides a good compromise between complexity, accuracy and simulation time for system level studies [3].

The variable V_{sc} is the overall terminal voltage on the supercapacitor bank, I_c, is the current flowing in the ideal supercapacitor module, and I_{sc} is the terminal current. R_{ESR} and R_{EPR} represent the series resistance and leakage losses. The overall terminal voltage of the supercapacitor bank, V_{sc}, is given by (1).

$$V_{sc} = V_{sc0} - \frac{1}{C_{sc}} \int I_c dt - R_{ESR} I_{sc} \qquad (1)$$

where V_{sc0} is the initial voltage on the supercapacitor module.

B. DC/DC Converter

A simple, unisolated bi-directional DC/DC converter, shown in Fig. 2, is used to interface the variable supercapacitor terminal voltage to the fixed voltage DC bus.

An averaged-value model of the power electronics is used to enable long duration transients to be examined easily.

The switches S1 and S2 operate in a complementary manner with variable duty-ratio; the duty-ratio of S2 is D. Assuming ideal switches and lossless operation, the averaged-value differential equations for the converter are:

$$\dot{I}_{sc} = \frac{V_{sc}}{L} - \frac{V_{bus}}{L}(1-D) \qquad (2)$$

$$\dot{V}_{bus} = \frac{I_{ESD} - I_{bus} + I_{gen}}{C_{bus}} \qquad (3)$$

where I_{bus} is the total load current, I_{gen} is the total generator current, I_{ESD} is the ESD output current, $I_{sc}(1-D)$, and C_{bus} is the combined filter capacitance on the DC bus.

IV. SYSTEM CONTROLLER

The cascaded ESD control structure, the multiple inputs used to form the supercapacitor current reference and the overall system design are described in this section.

A. Supercapacitor Current Control

The closed-loop controller used to regulate the supercapacitor current, I_{sc}, is shown in Fig. 3.

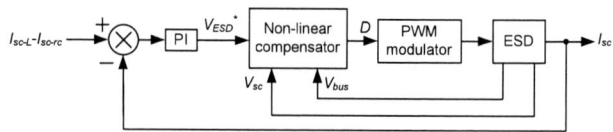

Figure 3. Converter current controller

The output signal from the PI controller, V_{ESD}^*, is passed through a non-linear compensator to form the transistor duty ratio, D. The compensator is designed such that there is a linear, first-order relationship between the control signal, V_{ESD}^*, and I_{sc}. By considering the averaged-value differential equation for the converter inductor current (2) which is equal to I_{sc} the non-linear compensator is determined to be given by:

$$D = 1 - \frac{V_{sc} - V_{ESD}^*}{V_{bus}} \qquad (4)$$

The PI controller has a transfer function of the form:

$$PI(s) = k_p \frac{(1+T_I s)}{T_I s} \qquad (5)$$

where T_I would typically be chosen to place the zero an order of magnitude below the switching frequency.

The maximum and minimum values of D are limited in the PWM modulator to have values of 0.95 and 0.05 respectively.

B. Supercapacitor Current Reference

The overall reference signal for the supercapacitor current controller has two components, the first I_{sc-L} is determined by the requirement to track the instantaneous load changes on the DC bus, whereas the second, I_{sc-rc}, arises from the requirement to manage the supercapacitor state-of-charge. Supercapacitor state-of-charge schemes have been proposed in the literature [4-6] for other applications, though these controllers directly define the supercapacitor state-of-charge, whereas the control presented in this section operates indirectly, by means of a droop characteristic; the reference value for the supercapacitor voltage is fixed. The calculation of the two current reference components is shown in Fig. 4.

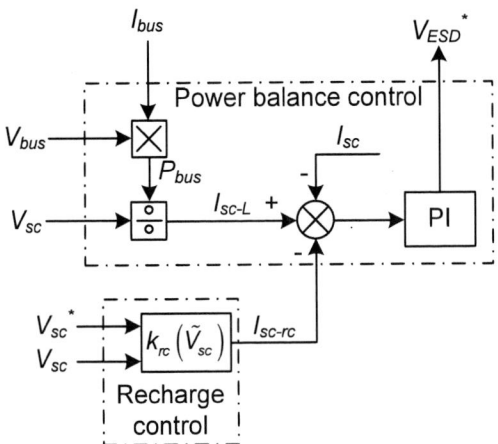

Figure 4.   Power-balance and recharge control for ESD

To ensure that the ESD responds rapidly to load changes, I_{sc-L}, is determined using an instantaneous power calculation which assumes a lossless DC/DC converter [7]; the instantaneous bus power, $V_{bus}I_{bus}$, is divided by the instantaneous supercapacitor voltage, V_{sc}, to give I_{sc-L}.

The signal I_{sc-rc} controls the rate at which the supercapacitor is recharged and is based on the error, $\tilde{V}_{sc}$, between V_{sc} and the set-point voltage, $V_{sc}{}^{*}$. Due to the size of the supercapacitors the I_{sc-rc} signal tends to change much more slowly than the I_{sc-L} signal.

The recharge control function, k_{rc}, must be chosen to satisfy the conflicting requirements of maintaining the supercapacitor voltage within an acceptable working range whilst limiting the maximum rate-of-change of load that is experienced by the generator.

Due to the operation of the PI controller, under steady-state conditions the two components of the reference for the supercapacitor current, I_{sc-rc} and I_{sc-L}, will be equal, that is

$$k_{rc}\left(\tilde{V}_{sc}\right) = \frac{V_{bus}I_{bus}}{V_{sc}} \qquad (6)$$

As a result the steady-state supercapacitor voltage will tend to drop with increased levels of DC bus power, thereby increasing the capability of the supercapacitor bank to absorb energy transiently when there is a sudden reduction of power on the DC bus.

The k_{rc} function could be a simple linear function as has been described in [8], however the use of a non-linear function was found in this work to offer greater flexibility over limiting the generator load transients whilst maintaining the supercapacitor voltage within acceptable limits. A simple non-linear function is described in the following section.

C.   System Design

To demonstrate the system operation the ESD was designed to compensate the load power drawn by typical radar equipment used for remote sensing and mapping. The load profile is shown in Fig. 5 and comprises a relatively low quiescent power, 4kW, with a superimposed transient profile comprising an additional 1kW load for 25s in mode 1, then a 17kW load for 20s in mode 2 and 3kW for 50s in mode 3.

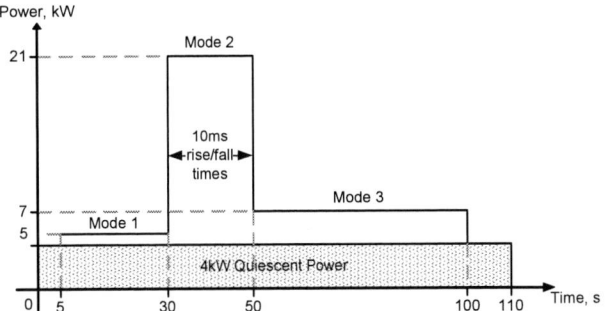

Figure 5.   Hypothetical model of a multi-functional radar

A 55F, 145V supercapacitor bank was selected, providing a usable energy storage capacity of approximately 0.4MJ. R_{ESR} and R_{EPR} were 7.1mΩ and 10kΩ respectively [9]. The bi-directional DC/DC converter was chosen to have a switching frequency of 30kHz and an inductor value of 100μH. The current control loop bandwidth was 8kHz. The total capacitance on the 540V DC distribution bus was 800μF.

To maintain the supercapacitor voltage within its working range whilst limiting the maximum rate of change of load applied to the generator shaft, a non-linear function was chosen for the recharge control loop in which

$$k_{rc} = 0.64\tilde{V}_{sc}{}^{1.5} \qquad (7)$$

The function, (7), was selected to give a drop in the steady-state supercapacitor voltage of 75V corresponding to a load power of 25kW as shown in Fig. 6. If the gain or exponential terms in (7) are reduced then the ESD recharge rate is insufficient for the load profile, Fig. 5. Higher gain or exponential terms in (7) results in larger deviations in DC bus voltage during load transients.

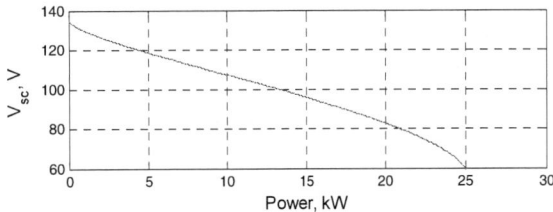

Figure 6.   Supercapacitor voltage droop characteristic

The reference supercapacitor voltage, $V_{sc}{}^{*}$, was set to 135V.

V. EXPERIMENTAL TEST SYSTEM

The laboratory system is shown in Fig. 7 and comprises a 70kW, five-phase generator and two programmable, 30kW, bi-directional electronic load units [10]. The generator speed was 1006rpm throughout the tests.

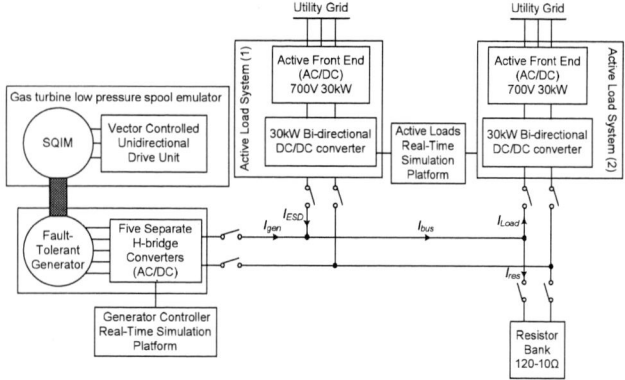

Figure 7. Test set-up

The structure of the programmable active load systems (ALSs) is shown in Fig. 7; a high-bandwidth bi-directional converter operating under the control of a real-time simulation system draws/delivers power from/to the high voltage DC bus and the power is re-generated to/sourced from the utility using a PWM converter.

Active load system ALS(1) was used to emulate the ESD whilst active load ALS(2) was programmed to follow the power steps of the load profile, Fig. 5.

In these proof of principle experiments the measurement of I_{bus} did not include the current in the resistor bank in Fig. 7 and is referred to as I_{Load}.

VI. EXPERIMENTAL VALIDATION

A Simulink-based average-value simulation of the entire system has been used to assist in the controller design. Results from the simulation are included in this section along with experimental results. In the experimental system an averaged-value model of the ESD was used to enable long-duration power profiles to be examined without compromising the 100kHz sample rate of the real-time platform.

A. Load Profile with ESD Inactive

Fig. 8 shows the response of the generator to the load profile, Fig. 5, when the ESD is offline. Measured test data and simulation results are overlaid to demonstrate the close match between the data sets. Large deviations are apparent in the DC bus voltage, V_{bus}, when the load power changes, the voltage reaches a minimum of 189V at t=30s when the power increases from 5kW to 21kW and a maximum of 680V at t=50s when power drops to 7kW from 21kW. These large bus voltage deviations of over 100V are due to the small size of the bus capacitors [11]. At t=50s the high voltage transient on the DC bus caused the active load unit to trip out and is the reason for the discrepancy between the simulation model and measured data beyond t=50s. The measured voltage is

seen decaying towards zero, however the generator remained online.

As the generator is the only power source available on the bus, the generator torque, T_e, bottom plot in Fig. 8, varies directly with the load profile, Fig. 5. This torque variation would be transmitted directly to the gas-turbine in an aircraft system. The severe rates-of-change of torque could compromise the engine operation or trigger mechanical resonances. The measured torque is slightly larger than the simulation value due to losses in the practical system that were not included in the simulation model.

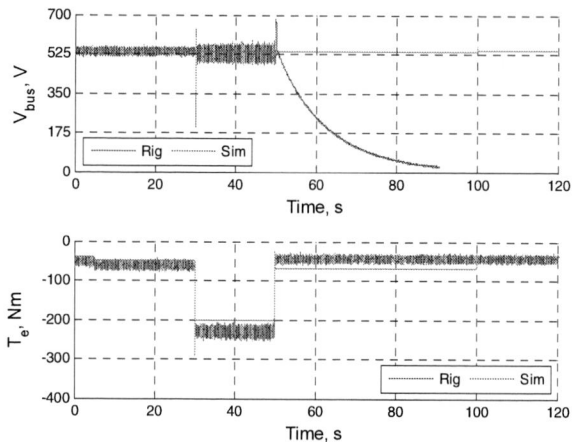

Figure 8. Generator response due to load profile with ESD inactive

B. Load profile with ESD Active

Fig. 9 shows the system response to the load profile, Fig. 5, when the ESD is active and the recharge function detailed in section IV.C, (7), is used. An extremely good correlation is evident between the measured test data and the simulation data.

The first and second plots in Fig. 9 show bus voltage, V_{bus}, and generator torque, T_e. The large deviations apparent in V_{bus} in Fig. 8 have been virtually eliminated in Fig. 9 by the use of the ESD. T_e, the second plot in Fig. 9, also shows a much more limited rate-of-change in response to the load transients, when compared to Fig. 8. The more gradual torque changes on the generator shaft are the result of the variable k_{rc} function which enables a slow ESD response to small power steps at low power, t=5s and 100s, but clearly enables a fast ESD response to large steps in power, t=30s and 50s. The load current, I_{Load}, and the ESD output current, I_{ESD}, are shown in the third plot in Fig. 9. I_{ESD} responds instantaneously to the load changes in an equal but opposite manner, then tends to fall to zero in about 20 or 30s after the load transient, thereby smoothing the load changes on the generator.

The fourth and fifth plots, in Fig. 9, show the supercapacitor voltage and current, V_{sc} and I_{sc}. V_{sc} is seen to reach its lowest value of just below 100V at the end of the high power pulse. Furthermore the voltage level at which V_{sc} tends to settle after a load transient is seen to depend on the load power. During strip mode 2, 50<t<100s, the load power from ALS(2) is 18.6kW (in parallel with a 2.4kW resistive load) and as predicted by Fig. 6, V_{sc} tends to settle at 86.8V. The reduction in load power at t=50s and 100s results in the ESD absorbing power and so recharging the supercapacitors at differing rates due to the variable k_{rc} function.

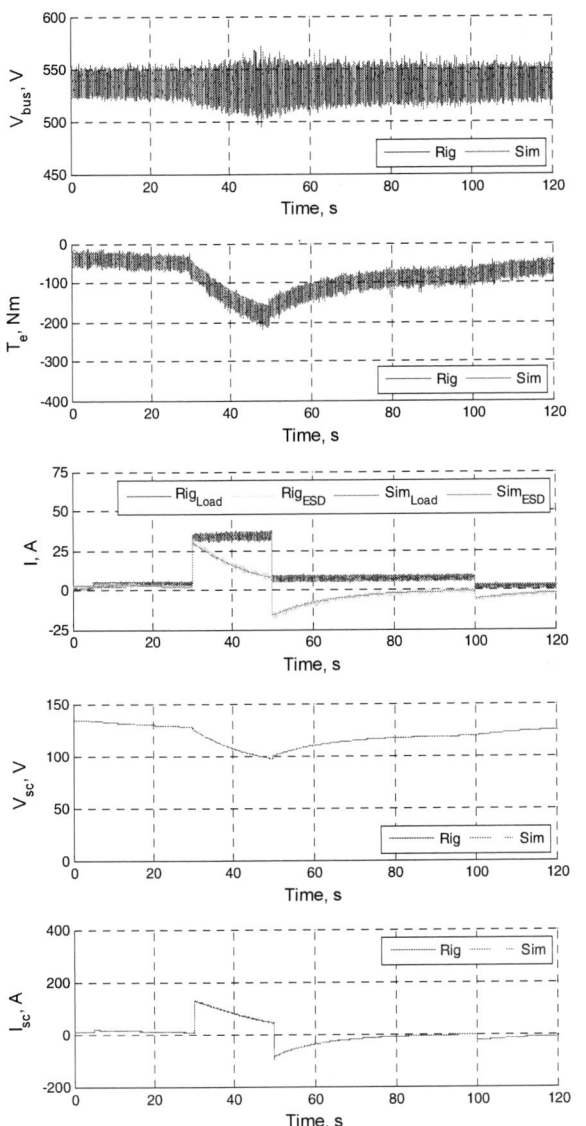

Figure 9. Overall system response due to load profile with ESD active and variable gain recharge

C. High Frequency Load ESD Inactive

To examine the performance of the system to higher frequency load changes, the response is considered to a load that switches between 23kW and 32kW, using ALS(2), with a 0.5 duty-ratio at 0.5Hz.

Fig. 10 shows the system response of the simulation and test-rig with only the generator active; no ESD. Severe deviations are evident in V_{bus} (upper waveform Fig. 10) with a minimum of 440V when the load increases, t=1s, 3s, and a maximum of 612V for the step down in load, t=0s, 2s; the

recovery time of V_{bus} is less than 0.1s for either transient. The varying load profile is reflected in T_e (lower waveform in Fig. 10) and presents an 85Nm torque step which may affect the gas turbine. The comparison between simulation and test-rig data shows an excellent correlation for V_{bus} and a slight difference, due to the omission of losses from the simulation model, in T_e.

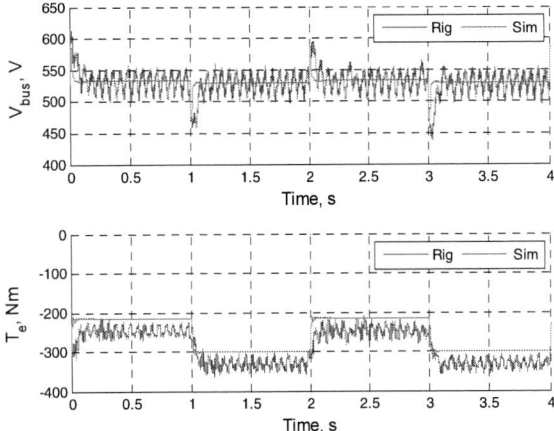

Figure 10. Generator response due to high frequency load with ESD inactive

D. High Frequency Load ESD Active

Fig. 11 shows the system behaviour in response to the high-frequency load, as used in Fig. 10, with the ESD active. The deviations in V_{bus} (top waveform Fig. 11) are almost eliminated at the load steps when compared to Fig 10. T_e (second waveform in Fig. 11) exhibits a very mild 6Nm variation during the 2s pulse cycle compared with the abrupt change in Fig. 10. Again, the error between the simulated and measured torque is due to mechanical losses being neglected in the simulation model.

The third set of waveforms in Fig. 11 show I_{ESD} closely following the variation in I_{Load}, and there are corresponding variations in the supercapacitor voltage, V_{sc}, fourth plot and supercapacitor current, I_{sc}, fifth plot. The step changes in V_{sc} are due to the equivalent series resistance of the supercapacitors, R_{ESR}. It is evident from the waveforms that the system has not quite reached steady-state and the supercapacitor voltage is drifting down to a steady-state value of 117V as dictated by the droop profile in Fig. 6.

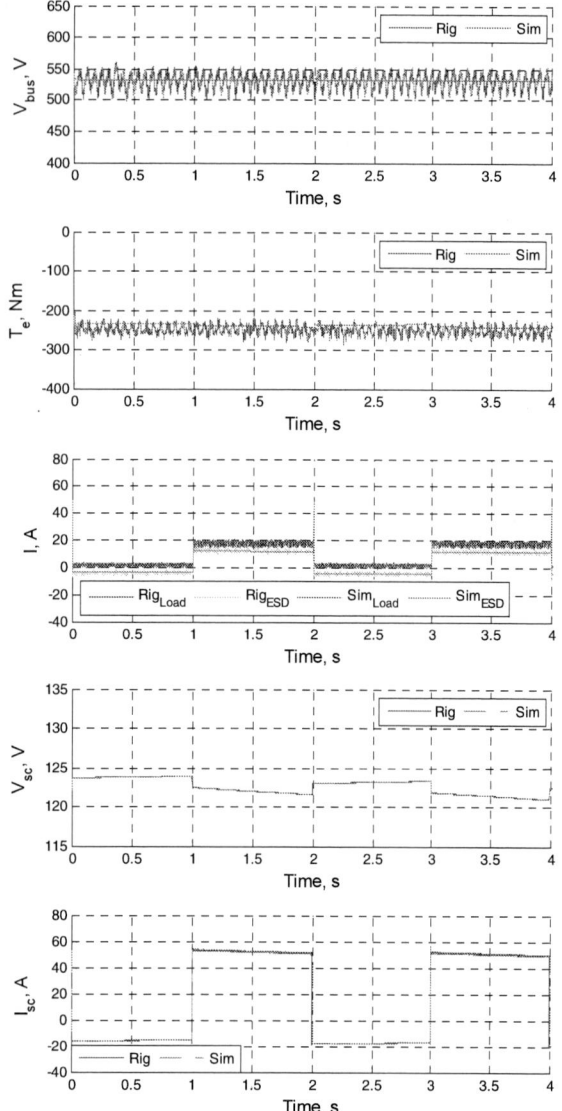

Figure 11. Overall system response due to high frequency load with ESD active and variable gain recharge

E. Load profile ESD Failure

The hardware-in-the-loop emulation of the ESD described in this paper enables the system behaviour to be examined over a wide range of conditions and parameters including abnormal operation and failure. This allows the limitations of the device to be established without risking damage to the supercapacitors or transistors.

Fig. 12 shows an example case where the recharge control of the ESD is just a proportional gain of 0.5, which, as is shown in Fig. 12, is insufficient to maintain the supercapacitor voltage within the working range. As a result of the low value of gain in the recharge controller, the waveforms in Fig. 12 show the ESD current following load current more closely than in Fig. 9 (third plot), the variation in generator torque is more limited than in Fig. 9 (second

plot), however the increased energy taken from the supercapacitor results in V_{sc} falling to unacceptably low levels (fourth plot), whilst the supercapacitor current becomes excessively large (fifth plot). At t=67s the system fails since the supercapacitor voltage is too low to step up to V_{bus} and the ESD trips out. The generator must then provide the entire load power.

The correlation between the test-rig and simulation data for all variables shown in Fig. 12 is again very good, further confirming the accuracy of the technique.

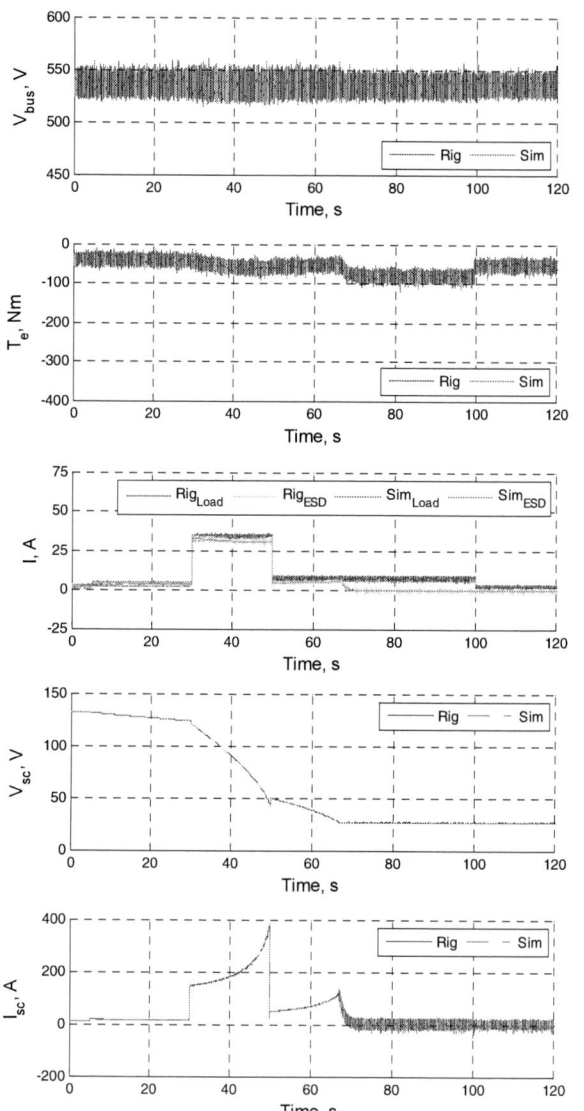

Figure 12. Overall system response due to load profile with ESD active and $k_{rc} = 0.5$

VII. CONCLUSIONS

A power-based method has been shown to provide effective control of a supercapacitor-based energy storage device operating on a high voltage DC bus. The energy storage device responds instantaneously to load changes, protecting the engine-driven generator from sudden transients, and virtually eliminating voltage transients from the DC bus. A slower-acting control loop manages the supercapacitor state-of-charge, ensuring that the voltage remains within the working range, but dropping at higher steady-state load powers to enable the energy storage device to absorb energy from the DC bus in the event of a sudden load reduction. A typical radar profile and a higher frequency switching load were used to demonstrate the system performance.

A hardware-in-the-loop technique was used successfully to study the system level performance of the energy storage device whereby a bi-directional load unit operating under the control of a real-time simulation platform emulated the energy storage device, allowing the effects of parameter changes to be readily examined.

VIII. REFERENCES

[1] A. J. Mitcham and J. J. A. Cullen, "Permanent Magnet Generator Options for the More Electric Aircraft," in *International Conference on (Conf. Publ. No. 487) Power Electronics, Machines and Drives (PEMD)*, 2002, pp. 241-245.

[2] R. L. Spyker and R. M. Nelms, "Classical Equivalent Circuit Parameters for a Double-Layer Capacitor," *IEEE Transactions on Aerospace and Electronic Systems*, vol. 36, pp. 829-836, 2000.

[3] M. Uzunoglu and M. S. Alam, "Dynamic Modeling, Design, and Simulation of a Combined PEM Fuel Cell and Ultracapacitor System for Stand-Alone Residential Applications," *IEEE Transactions on Energy Conversion*, vol. 21, pp. 767-775, 2006.

[4] E. Schaltz, A. Khaligh, and P. O. Rasmussen, "Influence of Battery/Ultracapacitor Energy-Storage Sizing on Battery Lifetime in a Fuel Cell Hybrid Electric Vehicle," *IEEE Transactions on Vehicular Technology*, vol. 58, pp. 3882-3891, 2009.

[5] P. Thounthong, P. Sethakul, S. Rael, and B. Davat, "Performance Investigation of Fuel Cell/Battery and Fuel Cell/Supercapacitor Hybrid Sources for Electric Vehicle Applications," in *4th IET Conference on Power Electronics, Machines and Drives (PEMD)*, 2008, pp. 455-459.

[6] J. Dixon, M. Ortúzar, and E. Wiechmann, "Regenerative Braking for an Electric Vehicle using Ultracapacitors and a Buck-Boost Converter," in *Proc. 17th Int. Electr. Vehicle Symp., Montreal, QC, Canada, CD-ROM*, Oct. 13-18, 2000.

[7] B. J. Arnet and L. P. Haines, "High Power DC-to-DC Converter for Supercapacitors," in *IEEE International Electric Machines and Drives Conference (IEMDC 2001)*, 2001, pp. 985-990.

[8] P. Thounthong, V. Chunkag, P. Sethakul, B. Davat, and M. Hinaje, "Comparative Study of Fuel-Cell Vehicle Hybridization with Battery or Supercapacitor Storage Device," *IEEE Transactions on Vehicular Technology*, vol. 58, pp. 3892-3904, 2009.

[9] Maxwell Technologies, "Maxwell Technologies BMOD0083-48.6V Ultracapacitors." vol. 2009: http://www.maxwell.com/ultracapacitors/products/modules/bmod0083-48-6v.asp, Nov. 2009.

[10] F. J. Chivite-Zabalza, A. J. Forsyth, D. R. Trainer, J. Calvignac, S. Long, and R. Todd, "Control and Development of an Electrical Systems Evaluation Platform for Uninhabited Autonomous Vehicles," in *34th Annual Conference of IEEE Industrial Electronics (IECON)*, 2008, pp. 1479-1486.

[11] R. Todd, A. A. Abd Hafez, A. J. Forsyth, and S. A. Long, "Single-Phase Controller Design for a Fault-Tolerant Permanent Magnet Generator," in *IEEE Vehicle Power and Propulsion Conference (VPPC)*, 2008, pp. 1-6.

978-1-4244-4782-4/10 $26.00 © 2010 IEEE

Buck Boost Regulator (B²R)
for Spacecraft Solar Array Power Conversion

Olivier Mourra, Arturo Fernandez, Ferdinando Tonicello
European Space Agency
ESTEC – Keplerlaan 1 – P.O
2200 AG Noordwijk ZH – The Netherlands
Olivier.mourra@esa.int, Arturo.fernandez@esa.int, Ferdinando.tonicello@esa.int

Abstract—**Solar Array Regulators (SAR) in a satellite Power Conditioning Unit (PCU) are implemented to transfer the power during sunlight from the solar arrays to the power bus. The objective of this paper is to present how Buck Boost Regulators (B²R) could be used as Step-Up, Direct Energy Transfer and Step-Down Solar Array Regulators. Control circuitries of a new kind are introduced for regulated and unregulated buses, illustrated by practical results demonstrating that such regulators can be used in satellite power systems.**

I. INTRODUCTION

Today two different Solar Array Regulator (SAR) techniques are commonly applied on spacecraft platforms. The first family is called Direct Energy Transfer (DET) [1][2][3]. It applies to topologies where the Solar Array Power (PSA) is transferred to the power bus by directly connecting the Solar Array (SA) to the power bus, when power is requested. When no power is requested, the solar array is either shunted or disconnected from the power bus (open circuit), and therefore no solar array power is transferred to the bus. The Sequential Switching Shunt Regulator (S3R) [4][5][6] is probably the most popular DET solar array regulator thanks to its simplicity, good power density and its excellent power in versus power out efficiency when full solar array power is required for the power bus (only diodes and harnesses/connectors conduction losses).

Unfortunately DET power systems require specific adaptation of the regulator electronics according to the solar array design or vice versa. Moreover when a section is attached to the bus the solar array power delivered to the bus depends on both solar array characteristics and power bus voltage level. This is particularly a negative point for space missions where the solar array characteristics vary widely (due to temperature - as seen in Fig. 2 - , ageing, radiation, solar aspect angle, sun intensity) or when a the battery is directly connected to the power bus which presents some voltage variation.

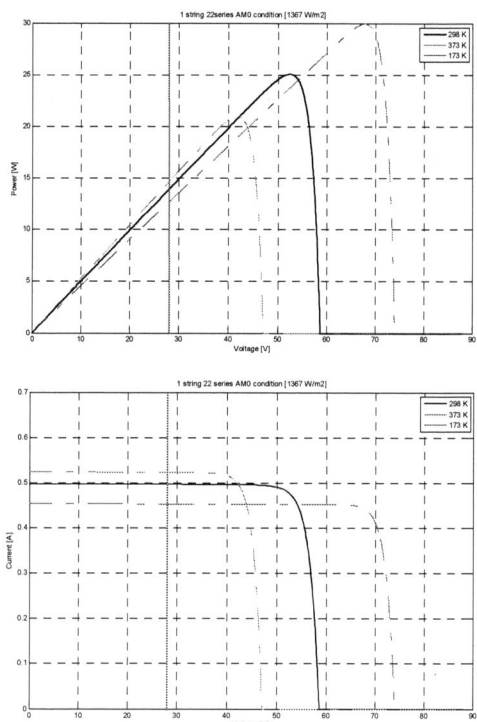

Figure 2. Example of {PSA, VSA}{ISA, VSA} Solar Array characteristics
- 1 string of 22 cells in series at 28 degC and +/-100degC (BOL)

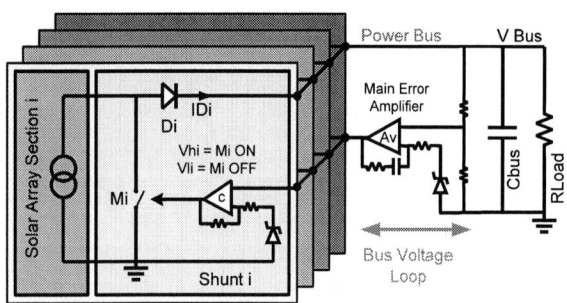

Figure 1. S3R with 4 solar array sections

978-1-4244-4782-4/10 $26.00 © 2010 IEEE

As a consequence, for these missions (e.g. deep space missions or Low Earth Orbit missions), DET techniques might be conveniently replaced by Pulse Width Modulation (PWM) switching SAR with Maximum Power Point Tracking (MPPT), capable of extracting the solar array maximum power. To the author's knowledge so far only step-down conversion where the solar array Maximum Power Point (MPP) voltage is higher than the bus voltage (e.g. Globalstar 1 [7], Rosetta, Mars Express, Venus Express [8][9][10], Goce and future missions as Swarm [11], Lisa Pathfinder [12], GAIA), or step-up conversion where the solar array MPP voltage is lower than the bus voltage (e.g. Bepi-Colombo [13][14], Globalstar 2) have been considered for European spacecrafts. The main advantage of both solutions is the fact that the same solar array regulator design can be used for different missions. On the other hand the main drawback of MPPT-SAR is that the switching power converter introduces significant losses (a few percents), which can completely offset the advantage provided by the operation at the maximum power point. Furthermore the efficiency of the SAR drops when the difference between the solar array voltage and the bus voltage increases: the further away the MPP voltage (VMPP) is from the bus voltage, the lower the SAR efficiency is (this is true both for step-up and step-down converters).

In fact the best possible Power Conditioning Unit (PCU) would be one using converters capable to track the maximum power from the solar array in any condition, with an efficiency equal or very close to the efficiency of a power conditioning unit based on DET concept and flexible such that it could be used for different types of missions without any design modifications, reducing thus its cost. The sizing of the system would be defined at the mission conditions with the minimum power margin (e.g. when the available SA power minus the output power demand is minimum) with the SAR operating in DET or close to DET such that the best efficiency is achieved.

II. STEP-UP DET AND STEP-DOWN CONVERSION WITH BUCK BOOST REGULATOR (B²R)

The idea of using a PWM SAR based on step-up and step-down conversion with an MPPT is born from the problematic previously enounced. One has to realize that only a few topologies [15] [16] [17] using two power switching cells have the possibility to operate in step-up, step-down and also in DET conversion mode. Based on this remark the five topologies introduced shown in Fig. 3 have been selected. Their known performances (efficiency), simplicity and frequency responses in continuous mode with a traditional switching frequency (100 KHz) render them suitable for a use in spacecraft power system with either a constant regulated bus voltage or a variable bus voltage with the battery directly connected to the bus. By using such converters the solar array Maximum Power Point Voltage (VMPP) is in all conditions closer to the bus voltage than if simple buck or boost converters are used. Consequently the converter operates always in DET or close to it, improving greatly the efficiency (minimization of the losses).

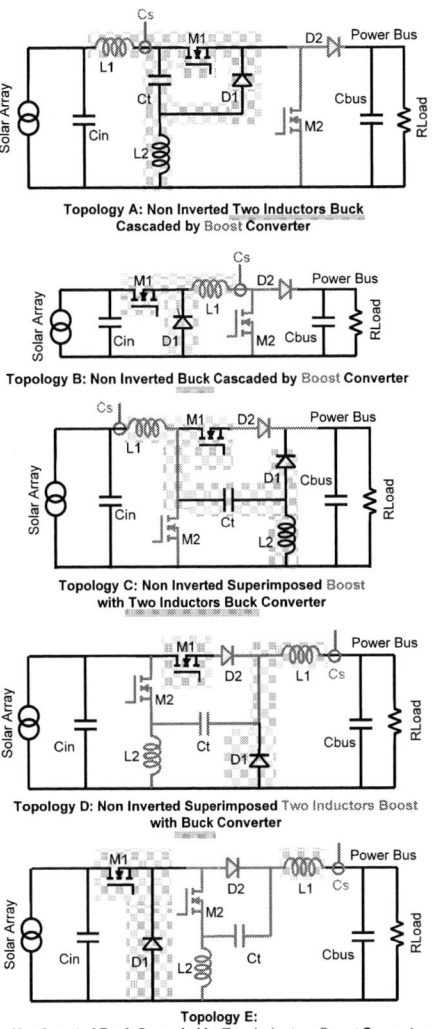

Figure 3. B²R topologies

III. B²R CONTROL LOOPS DESCRIPTION

At the time of the Rosetta development, several control schemes were studied in the European Space Agency Power Lab for the PCDU Step-Down solar array power converter in MPPT mode. The final decision was to implement two inductors buck as SAR with a current loop controlling the output current of the SAR and an input voltage loop embedding the output current loop. The input voltage loop has the main advantage of being stable on the voltage (VSA>VMPP) and current region (VSA<VMPP) of the solar array characteristics. In case of a power deficit the MPPT becomes active and provides to the input voltage loop a reference signal to let the SAR extract the SA maximum power by oscillating around the MPP. More details are provided concerning the design of this input voltage loop in [8][18][19]. The control of the input voltage provides also some interesting capabilities: indeed for the Rosetta, Mars Express and Venus Express PCUs it is possible by command to set a minimum SA Voltage (VSA) limit with respect to the

relevant solar array {I,V} curve, while the MPPT feature can be enabled or disabled. This feature has been successfully used during the Mars Express mission, where it has been possible to recover the spacecraft power despite a wrong inter-connection between the SA and the PCU [10].

Based on this experience it was decided in an early phase of this project to try to implement a similar control scheme as the one used in Rosetta satellite, capable of controlling the input voltage of the step-up DET and step-down topologies of Fig.3 when operating in MPPT mode. Several simulations with average models of the converters [20][21] were performed to find a control scheme respecting the stability requirements applicable to space power equipments (60° Phase and 10dB gain margins) [22][23]. Fig.5 and Fig.6 present two final control schemes for the 5 topologies of Fig. 3 either for battery bus or for a regulated bus spacecraft.

For both architectures two comparators are implemented to generate the buck and boost PWM duty cycle signals. A single current amplifier is connected to the two PWM comparators. Such implementation is an asset to use several B²R in parallel (current sharing) and drive more power and/or for redundancy purposes. As it can be seen on Fig.4 the buck sawtooth is always lower than the boost sawtooth such that the buck and boost cells never switch at the same time (in boost mode the buck switch is always ON, and in buck mode the boost switch is always OFF).

The current loop controls the current in the inductor where the placed on the positive power path. On Fig.3 the current sensing is indicated by the symbol CS. The inductor current loop is then embedded in an input voltage loop, which allows the system to control the input voltage of the SAR in MPPT mode.

It is noted that the topologies A, B, and C contain a boost. It is well know that boost input to output transfer function presents typically a Right Half Plane Zero (RHPZ). The presence of the RHPZ at mid frequencies causes difficulties to achieve high output loop bandwidths. The A, B, and C topologies were therefore primarily studied for unregulated buses where a low battery voltage and current management loop bandwidth (necessary to avoid oscillation with the RHPZ) is not an issue since the battery (low impedance) is directly connected to the power bus.

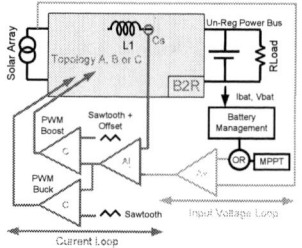
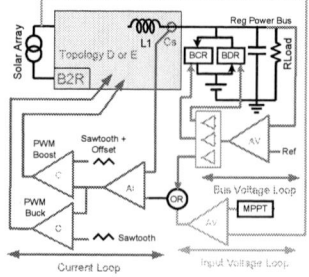

Figure 5. Control scheme for Battery Bus Architecture

Figure 6. Control scheme for Regulated Bus Architecture

As it can be seen on Fig. 5 the input voltage loop is used either by the battery management (when the available SA power is higher than the battery power bus needs) or by the solar array MPPT (operative in case of SA power deficit).

Topology E or D, containing a two inductors boost, might be more conveniently used for a regulated bus architecture. In excess of SA power an output voltage loop controls the regulated power bus with good performances (no RHPZ thanks to the two inductors boost). In case of SA power deficit, the MPPT controls the input voltage of the converters and the power bus is regulated thanks to the bus voltage loop controlling a battery discharge regulator (BDR).

IV.   BREADBOARDS RESULTS

Topologies B and D have been bread-boarded and tested with the control schemes presented on Fig. 5 and Fig.6 (topology B 250W power stage: Cin=49uF, L1=50uH, Cbus=22uF//battery; topology D 450W power stage: Cin=49uF, Li=60uH, Cbus=500uF, Ct=22uF//78uF+1Ω).

For each converter the stability has been studied by measuring the frequency responses of all loops. In particular the Phase Margin (PM), Gain Margin (GM), and Bandwidth (BW) of the input voltage loop were measured in buck mode, boost mode, on the voltage side, current side of the SA characteristics and at the SA maximum Power Point. Fig. 7 and Fig. 8 illustrate the six different operating points at which the open frequency response of the input voltage loop was measured. All measurements confirmed the stability and performances of the input voltage loop in the current and voltage regions of the solar array characteristics, in buck and boost modes for the two breadboards. The results are presented from Fig. 9 to Fig. 20.

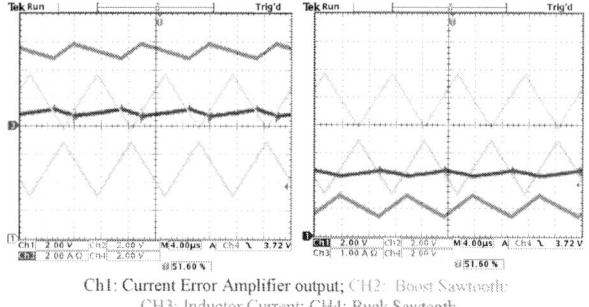

Ch1: Current Error Amplifier output; CH2: Boost Sawtooth;
CH3: Inductor Current; CH4: Buck Sawtooth

Figure 4.   Current Loop waveforms in Buck and Boost mode

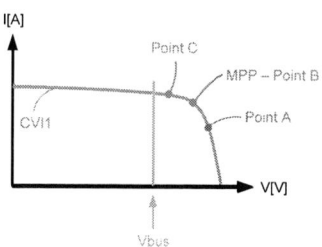
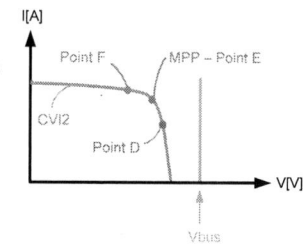

Figure 7. Static Operating points of the SAR in Buck mode when measuring the frequency loop responses

Figure 8. Static Operating points of SAR in Boost mode when measuring the frequency loop responses

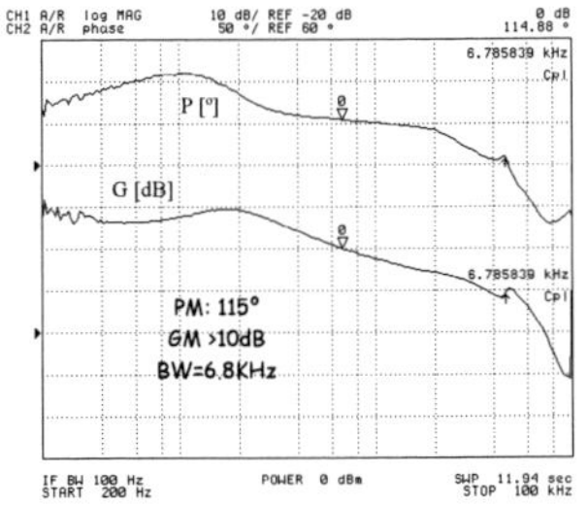

Figure 9. OVL(s) – Buck Mode – Point A – Topology B

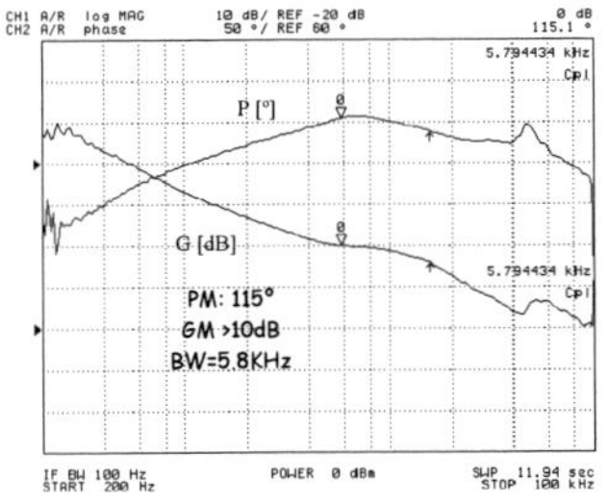

Figure 10. OVL(s) – Buck Mode – Point B – Topology B

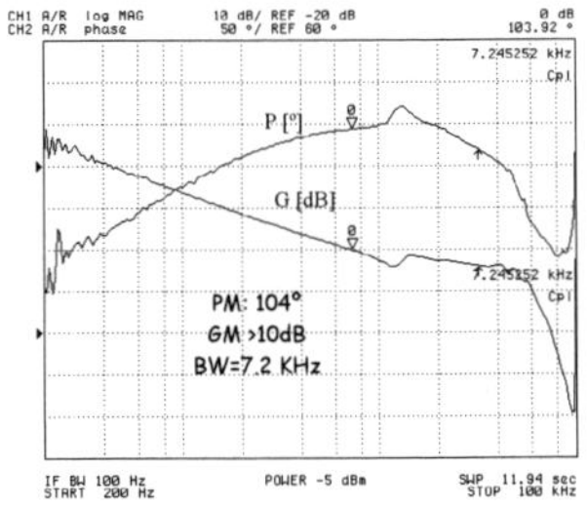

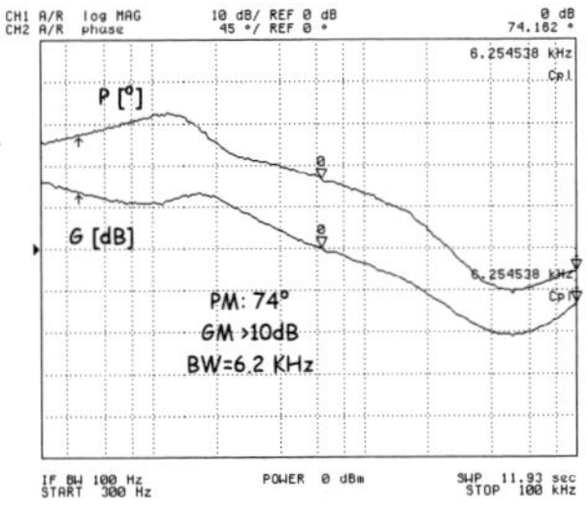

Figure 15. OVL(s) – Boost Mode – Point D – Topology D

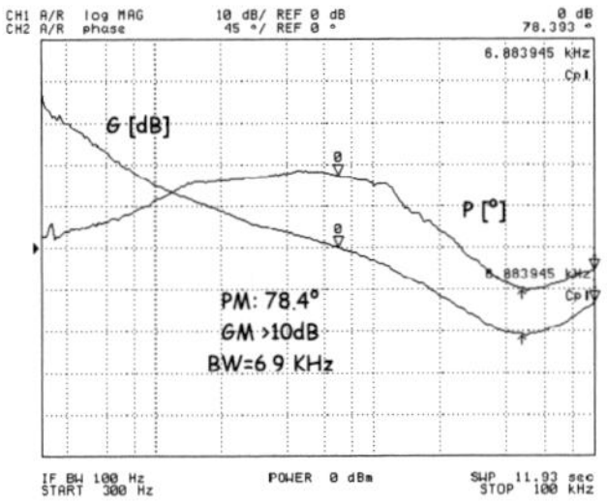

Figure 16. OVL(s) – Boost Mode Point E – Topology D

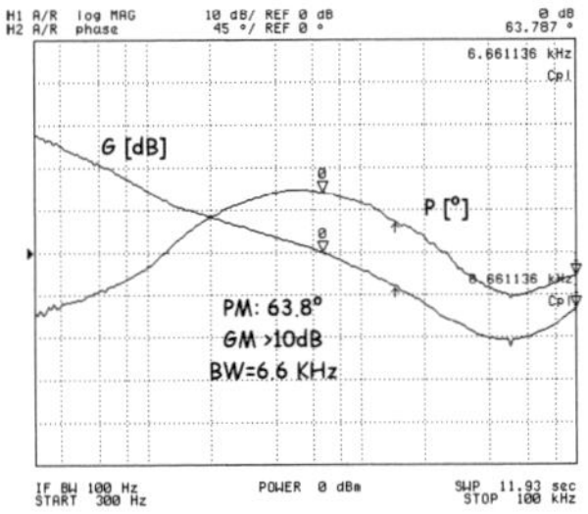

Figure 11. OVL(s) – Buck Mode – Point C – Topology B

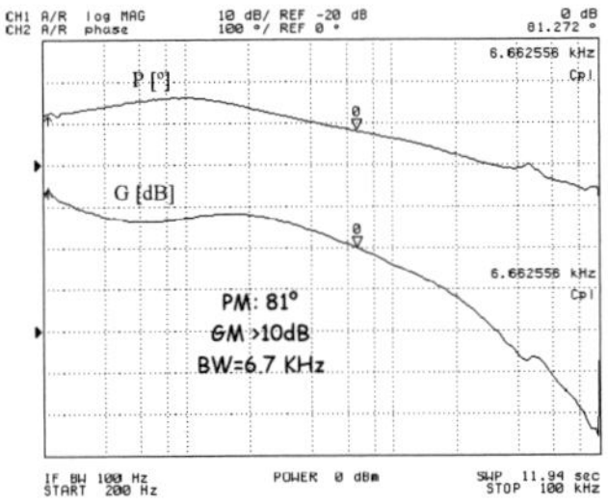

Figure 12. OVL(s) – Boost Mode – Point D – Topology B

Figure 17. OVL(s) – Boost Mode – Point F – Topology D

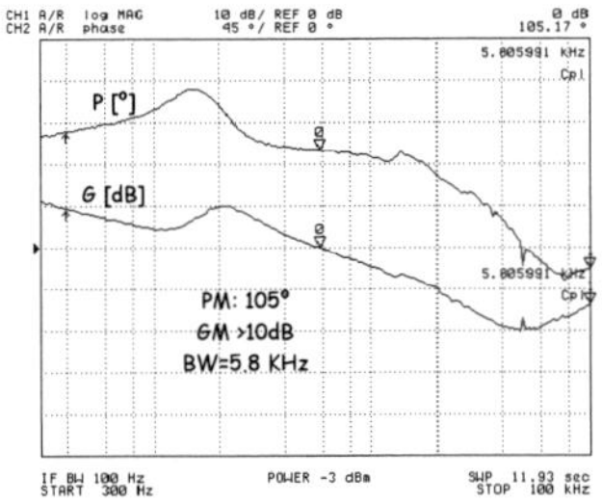

Figure 18. OVL(s) – Buck Mode – Point A – Topology D

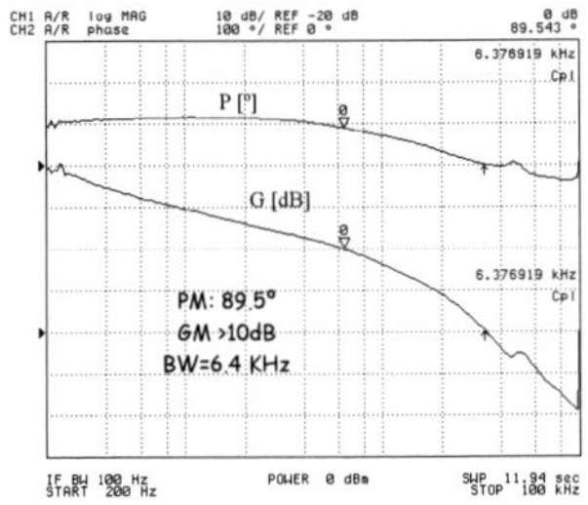

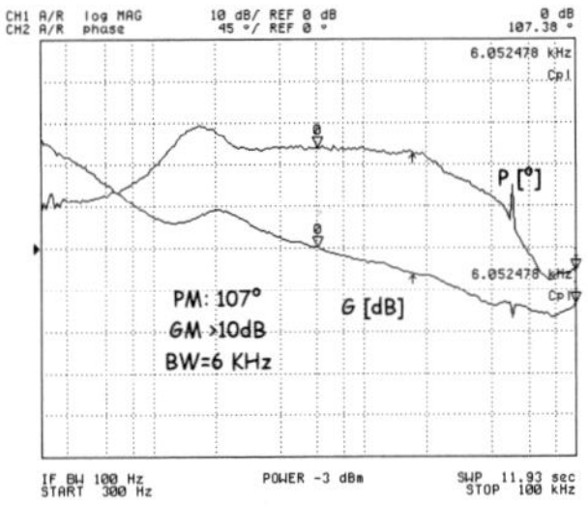

Figure 13. OVL(s) – Boost Mode – Point E – Topology B

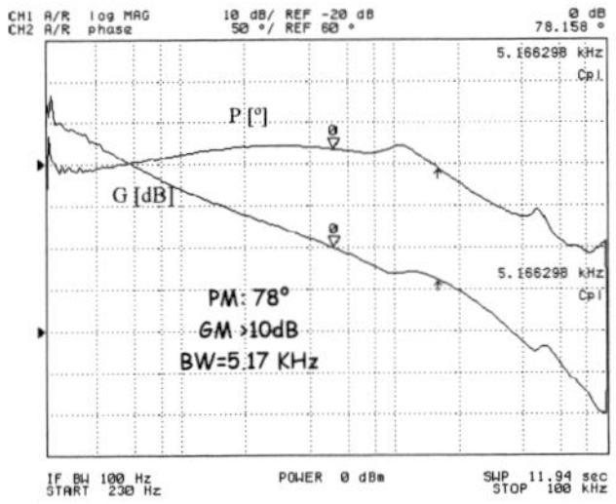

Figure 14. OVL(s) – Boost Mode – Point F – Topology B

Figure 19. OVL(s) – Buck Mode – Point B – Topology D

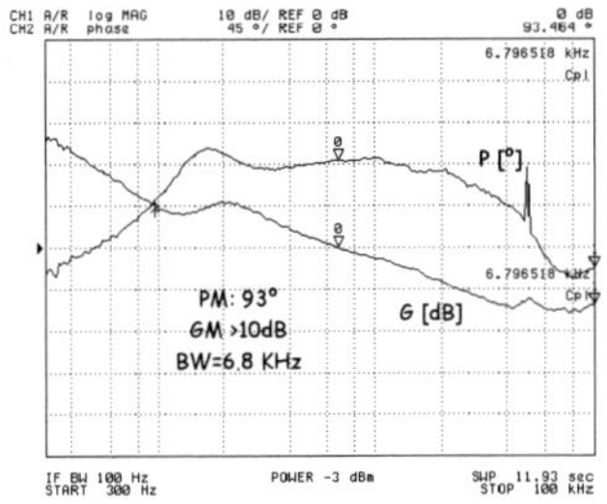

Figure 20. OVL(s) – Buck Mode Point C – Topology D

The input voltage loop performances (bandwidth of several KHz) make this loop compatible for MPPT operation with a traditional tracking frequency of several hundreds of Hz.

Since the particularity of the first B²R breadboard based on the topology B was to use the input voltage loop also for the battery management, the stability of the battery management was measured. The battery management consists in two loops: a current loop in order to charge the battery at a maximum preset current value and a voltage loop reducing the current charge when the battery voltage reaches the preset end of charge voltage. On Fig. 21 and Fig. 22 the Open Battery Current Loop (OBCL) frequency responses are presented when the SAR is in buck and in boost mode. It can be clearly seen on Fig. 22 the effect of the RHPZ in boost mode where the phase goes down and the gain goes up above 1KHz. Due to this effect the bandwidth of the battery current loop has been voluntary reduced to 430Hz to avoid any possible oscillation and meet the stability requirements [22].

Following the frequency responses measurements the B²R breadboards were tested in time domain: as can be seen on Fig. 23, Fig.24 and Fig.25 the B²R is capable to operate in step-up, DET and step-down modes. These results were identical for both breadboards (B and D topologies). VSA represents the solar array voltage, VBUS is the power bus voltage (B²R output voltage) and PSA is the solar array power extracted by the B²R.

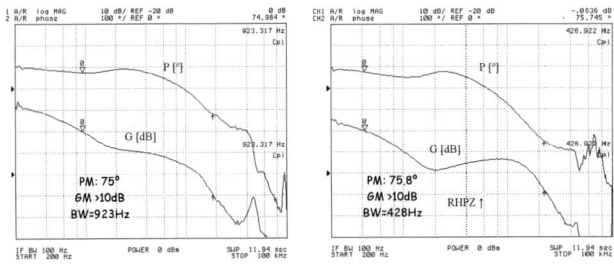

Figure 21. OBCL(s) in buck mode – Point A – Topology B

Figure 22. OBCL(s) in boost mode – Point D – Topology B

For the three figures, the B²R operates within the voltage and current regions of the SA: it oscillates around the maximum power point. On Fig. 25 and Fig. 26 VMPP is equal to the bus voltage. The internal waveforms at the PWM comparators level are plotted on Fig. 26, when the B²R has to operate alternately in buck DET and boost modes. One can see the change of operating domain of the current amplifier output between these three modes.

Finally the efficiencies of the two B²R breadboards have been measured for:

· a 100 KHz switching frequency
· an output power from 20W to 250W,
· a minimum and maximum solar array voltages corresponding to the MPP voltages of a solar panel composed of sections with 15 cells in series at +100 °C and -100 °C (26V - 45V)
· a minimum and maximum output voltage corresponding to the classical values of a battery bus variation in nominal operation (34V - 26V)

On Fig. 27 and Fig. 28 the breadboards efficiency results including the driver losses are reported:

· with the B²R in DET mode for the two extreme bus voltages
· with the B²R in buck mode with an input SA voltage of 45V and a bus voltage of 26V
· with the B²R in boost mode with an input SA voltage of 26V and a bus voltage of 34V

The efficiency is greater than 96% including the worst case for the buck and boost modes with the highest difference between the MPP SA voltage and the battery bus voltage. Please note that no protection was present on the B²R breadboards, which would increase the losses and reduce the efficiency in a real application.

V. B²R APPLICATIONS

One aspect that has not been treated until now is the use of B²R in a Single Point Failure Free (SPFF) spacecraft power system. A Single point failure free design is characterized by the fact that no degradation is foreseen after a single failure.

Fig. 29 illustrates the case of a battery bus architecture using as SAR the topology B of Fig.3 in a satellite having two SA wings with different characteristics. On this figure four SARs are used: two SARs are in hot redundancy for each wing such that if one fails, it is disconnected thanks to the protection P1. One can see that an extra diode (D3) has been placed in series with the boost diode D2 in order to protect the bus in case of a short circuit failure of the diode D2.

Dedicated MPPT and Input Voltage Loop (IVLA and IVLB) regulation are implemented for each wing. These loops are triple majority voted with majority voters (MV) in order to guarantee the input voltage control even after a failure.

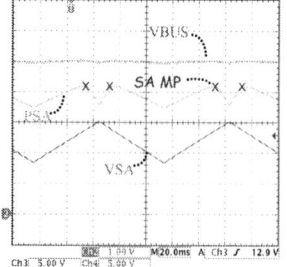

Figure 23. Input Voltave Modulation, with the SA MP Voltage below the Bus voltage

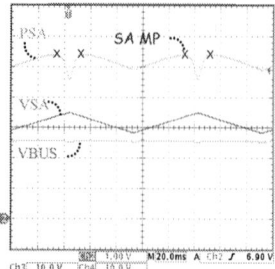

Figure 24. Input Voltave Modulation, with the SA MP Voltage above the Bus voltage

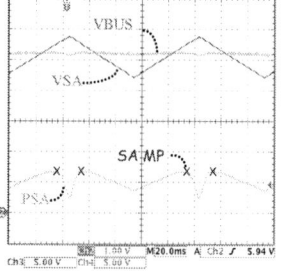

Figure 25. Input Voltave Modulation, with the SA MP Voltage equal to the Bus voltage

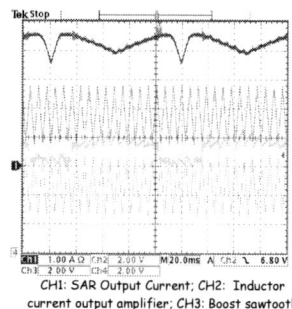

CH1: SAR Output Current; CH2: Inductor current output amplifier; CH3: Boost sawtooth CH4: Buck sawtooth

Figure 26. Input Voltave Modulation, with the SA MP Voltage equal to the Bus voltage

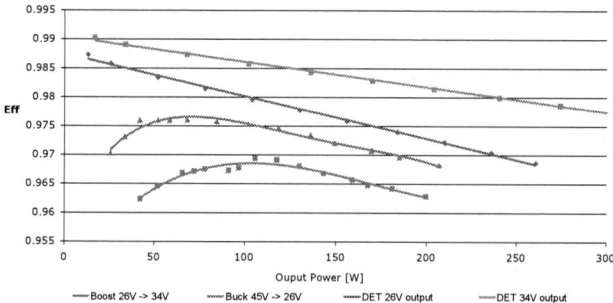

Figure 27. Efficiency measurements of the B²R breadboard with the topology B

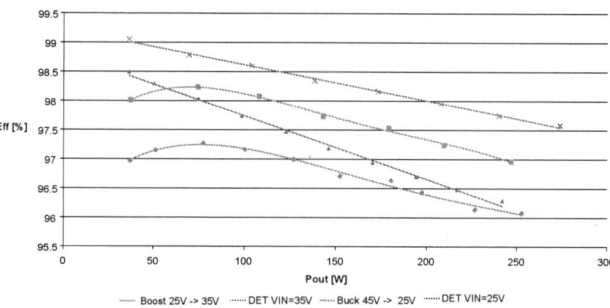

Figure 28. Efficiency measurements of the B²R breadboard with the topology D

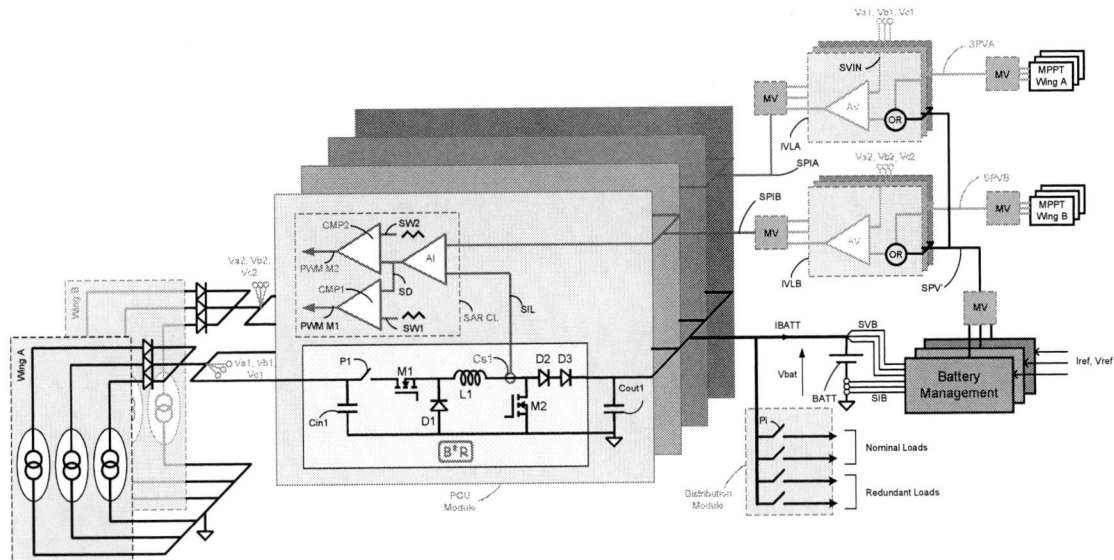

Figure 29. B²R in a Single Point Failure Free Battery Bus Power Architecture

The battery management (also triple majority voted for reliability reason) controls the Input Voltage Loop (IVLA and IVLB) when the spacecraft is in excess of SA power. Note that in this condition the operating point of the SARs is in the voltage region of the solar array characteristics. In case of a SA power deficit, the MPPT associated to the wing starts to operate and oscillates around the MPP by controlling the input voltage of the SAR.

It may be clearly seen that the system has the capability of having one group of SAR connected to the first wing in MPPT mode and the other group of SAR connected to second wing in Battery management mode when the two spacecraft solar array wings have different MPP.

Fig. 30 illustrates the case of a regulated bus architecture by using topology D of Fig. 3 as SAR. On this figure the satellite has also two SA wings with different characteristics. Two SARs are in parallel for each wing. Should one SAR fail, it would be disconnected thanks to the protection switches P1 and PGND.

In case the SA power excess, the SARs regulate the main power bus thanks to the triple majority voted Output Voltage Loop (OVL) controlling the Solar Array Regulator Current Loop (SARCL) with the MEA signal. Note that in this condition the operating point of the SARs is in the voltage region of the solar array characteristics.

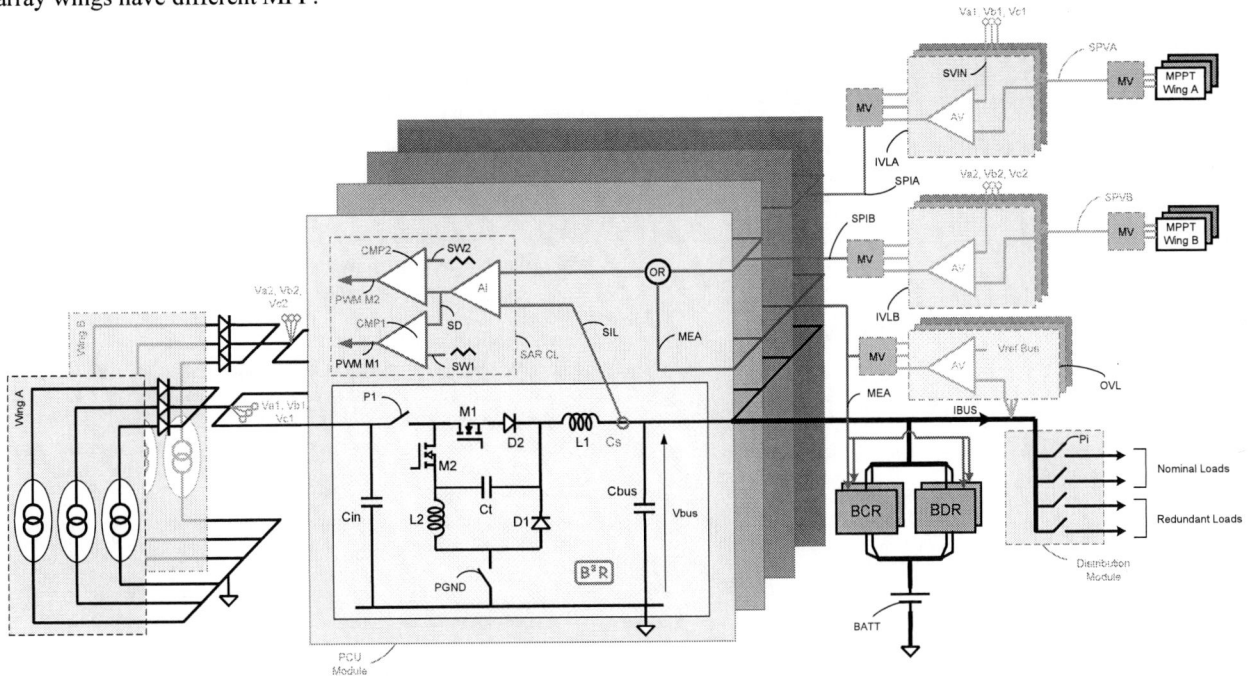

Figure 30. B²R in a Single Point Failure Free Regulated Bus Power Architecture

978-1-4244-4782-4/10 $26.00 © 2010 IEEE   1318

When SA power is in excess, the SARs regulate the main power bus thanks to the triple majority voted OVL controlling the SARCL with the MEA signal. Note that in this condition the operating point of the SARs is in the voltage region of the solar array characteristics.

In case of a power deficit, the MPPT associated to the solar array wing starts to operate and controls the input voltage of the SAR. If both wings are in MPPT mode, the bus regulation is guaranteed by the OVL controlling the Battery Discharge Regulator (BDR) with the MEA signal.

Finally another interesting property of using B²R is that for both architectures it is possible to characterize the solar arrays during Assembly Integration and Test (AIT) and during flight. As it can be seen on Fig. 31 and Fig. 32, by directly accessing the control circuitry of the step-up, DET and step down SAR breadboards, the solar arrays have been characterized from 0V up to their Open Circuit Voltage (VOC). Such a feature might be an interesting option in order to check the health of the spacecraft solar arrays and to learn more about their characteristic evolutions through a mission.

VI. CONCLUSION

Buck-Boost Regulators (B²R) with a simple control of a new kind have been studied and tested. Practical measurements have confirmed the good efficiency and performances of the proposed control scheme rendering B²R suitable for solar array maximum power point tracking. Two single point failure free satellite power architectures have been introduced to the reader to illustrate how B²R can be used in a space application.

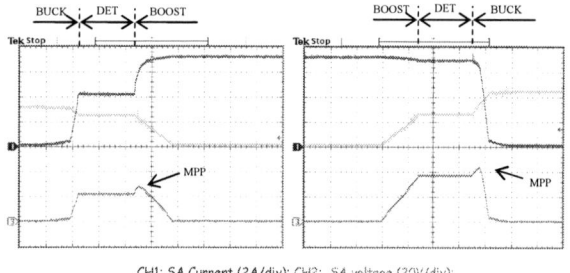

CH1: SA Current (2A/div); CH2: SA voltage (20V/div);
CH3: SAR output current (2A/div)

Figure 31. Charaterisation of the SA with the MPP Voltage below the Bus voltage

Figure 32. Charaterisation of the SA with the MPP Voltage above the Bus voltage

REFERENCES

[1] D. Loche, "The Pleiades Electrical Power Subsystem", Proceedings of the Seventh European Space Power Conference, Stresa, 2005.

[2] Perol Philippe, Olsson Dan, Haines James Edward, "Device for Generating Electrical Energy for a Power Supply Bus" US Patent 6181115 B1, 2001, Jan 30.

[3] Tonicello Ferdinando, Hans Jensen "Single Point Failure Free Sequential Serial and Shunt Regulator (S4R) with respect to Battery Overcharge", Proceedings of the European Space Power Conference held in Konstanz, 2008.

[4] Weinberg Alan, O Sullivan Dermot, "The sequential Serial Shunt Regulator -S³R", Proceedings of the Third ESTEC Spacecraft Power Conditioning Seminar, Noordwijk, The Netherlands, 21-23 September 1977.

[5] Weinberg Alan, O Sullivan Dermot, "The sequential Serial Shunt Regulator -S³R" US Patent 4186336, January 29, 1980.

[6] Hans Jensen, "Shunt Regulator Module", Proceedings of the European Space Power Conference held in Konstanz, 2008.

[7] W. Denzinger "Electrical Power System of Globalstar", Proceedings of the European Space Power Conference held in Poitiers, 1995.

[8] Hans Jensen, Johnny Laursen "Power Conditioning Unit for Rosetta-MarsXpress", Proceedings of the European Space Power Conference held in Porto, 2002.

[9] Tonicello Ferdinando, Hans Jensen, Johnny Laursen "Power Control Unit for Rosetta- MarsXpress – Lessons Learnt", Proceedings of the European Space Power Conference held in Porto, 2002.

[10] D. Loche "Mars Express and Venus Express Power Subsystem in Flight Behaviour", Proceedings of the European Space Power Conference held in Konstanz, 2008.

[11] N. Breier, B. Kiewe & O. Mourra "The Power Control and Distribution Unit for the SWARM Satellites", Proceedings of the European Space Power Conference held in Konstanz, 2008.

[12] A. De Luca, R. Gray & J. Otero "LISA Pathfinder Power System", Proceedings of the European Space Power Conference held in Konstanz, 2008.

[13] M.M. Hansen "Power Conditioning Unit for BepiColombo Transfer Module", Proceedings of the European Space Power Conference held in Konstanz, 2008.

[14] E. Lapeña, P. Rueda, O. García, J. Rancaño et al. "Boost-Based MPPT for the MTM PCDU of the BepiColombo Mission", Proceedings of the European Space Power Conference held in Konstanz, 2008.

[15] Severns R.P., Bloom G.E., "Modern DC to DC Switch mode Power Converter Circuits", Van Nostrand Reinhold, 1985.

[16] Jingquan Chen, Dragan Maksimovic and Robert Erickson, "Buck-Boost PWM Converters Having Two Independently Controlled Switches", Proceedings of the Power Electronics Specialists Conference, 2001.

[17] White J.L., Muldoon W.J., "Two Inductors Boost and Buck Converter", Proceedings of the Power Electronics Specialists Conference, 1987.

[18] Tonicello Ferdinando, "The Control Problem of Maximum Point Power Tracking in Power Systems", Proceedings of the European Space Power Conference held in Stresa, 2005.

[19] Tonicello Ferdinando, S. Vazquez del Real "Maximum Power Point Tracker Approach for Regulated Bus", Proceedings of the European Space Power Conference held in Tarragona, 1998.

[20] Vorperian V. & al, "Simplified analysis of PWM converters using model of the PWM-switch" IEEE transaction on Aerospace and Electronic Systems, 26, 3 pp490-505, 1990

[21] Vorperian V. & al, "Equivalent circuits model for resonant and PWM switches" IEEE transaction on Power Electronics, 4, 2, pp205-214, 1989.

[22] Space Engineering Electrical and Electronic Standard - ECSS EST 20C – European Coordination on Space Standardization, 31 July 2008.

[23] Electrical Power Systems for Unmanned Spacecraft – AIAA S-122-2007 – American Institute of Aeronautics and Astronautics (AIAA), 5 January 2007.

Quadratic Power Conversion for Industrial Applications

Gerry Moschopoulos
University of Western Ontario
Department of Electrical Engineering
London, ON, Canada, N6A 5B9
gmoschopoulos@eng.uwo.ca

Abstract—**Quadratic converters have the voltage conversion ratios of two combined cascaded converters, but with fewer switches. They are not widely used, however, because they are generally less efficient than other types of converters, but there are applications where they can be useful. In the paper, the characteristics of quadratic converters are explained and several examples of where they can be used in industrial applications are given.**

I. INTRODUCTION

Quadratic converters can be synthesized by cascading two converters in series then eliminating any redundant switches, passive elements, and controllers [1]. They can have the output-to-input voltage conversion ratio M of cascaded converters, but at a lower cost, as they have fewer components. This allows them to operate with wider ranges of M than those of conventional single-switch converters, and thus avoid problems due to extreme current stresses, poor component utilization, and very low efficiency. For applications where M is restricted, quadratic converters can operate with narrower variations in D than conventional converters, which can allow for better optimization of their design and better performance.

Examples of dc-dc quadratic boost and buck converters are shown in Figs. 1(a) and (b) respectively [1]. The quadratic dc-dc boost converter has been synthesized by the cascading of two dc-dc boost converters. If the converter is assumed to operate with both inductor currents being continuous, then it has a conversion ratio of $M = 1/(1-D)^2$, where D is the duty cycle of the switch. When the switch is turned on, a positive voltage is impressed across both inductors and their currents rise while they flow through the switch. When the switch is turned off, the currents in L_{in} and L_1 flow through D_1 and D_3 respectively.

Similarly, the quadratic dc-dc buck converter shown in Fig. 1(b) has been synthesized by the cascading of two dc-dc buck converters. If the converter is assumed to operate with both inductor currents being continuous, then it has a conversion ratio of $M = D^2$. When the switch is turned on, diode D_1 becomes reversed biased and D_2 becomes forward biased. The forward biasing of D_2 allows current to flow out of C_b so that the current through L_o rises. The current flowing through inductor L_b also rises as the difference between the input voltage and the voltage across C_b is impressed across it. When the switch is turned off, the current in L_b freewheels through D_1 and the current in L_o freewheels through D_3.

Quadratic converters, however, are less efficient than regular converters such as the boost and the buck. This is because they are essentially two cascaded converters and, thus, process power twice instead of once. The lower efficiency has limited their use so that their properties and characteristics are not well-known. Whatever information exists is scattered about in publications and is difficult to find. The main objectives of the paper are

- to familiarize readers with quadratic power converters, consolidate the information that exists about them and put it in a useful form,

- to show how quadratic power conversion can be used advantageously in certain industrial applications.

II. ALTERNATIVES TO QUADRATIC CONVERTERS

Quadratic converters are not the only type of converters with wide ranges of output to input voltage conversion ratio M. The two most popular alternatives are current-fed boost full-bridge converters and coupled inductor converters. These are reviewed here.

A. Current-Fed Full-Bridge Boost Converters

Current-fed full-bridge boost converters like the one shown in Fig. 2(a) are very attractive in applications where an output dc voltage that is considerably larger than the input voltage is needed. Such applications include medical power supplies and power supplies for electrostatic applications where extremely high output voltages are required and fuel cell and photovoltaic applications where the input voltage is

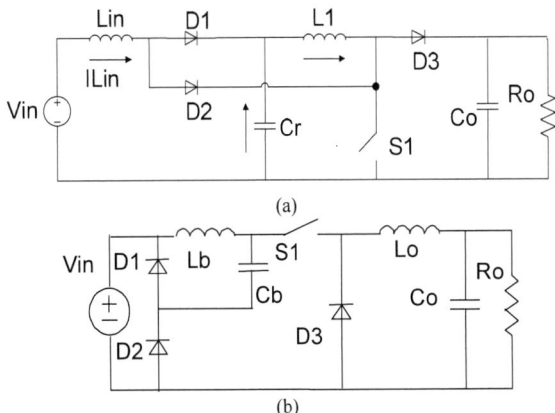

Fig. 1. (a) Single-switch dc-dc quadratic boost converter. (b) Single-switch dc-dc quadratic buck converter.

very low.

Current-fed full-bridge boost converters can have extremely high values of M because they are boost converters with step-up transformers. They operate in "boost" mode whenever all the switches are on and operate in "energy-transfer" mode whenever a pair of diagonally opposed pair of switches is on (Q_1 and Q_4 or Q_2 and Q_3). They do not, however, have an energy storage capacitor connected to their transformer primary-side dc bus. The leakage inductance of the transformer interacts with the output capacitances of the converter switches whenever they turn off. High voltage overshoots and ringing can appear across the switches since there is no bulk capacitor at the dc bus to suppress them. This means that they cannot be implemented unless some sort of snubber, either passive or active, is added across the dc bus.

To avoid compromising efficiency, the current-fed full-bridge boost converter is implemented using an active snubber, like the one shown for the zero-current-switched (ZCS) converter in Fig. 2(b) [2]. Doing so, however, adds to the cost of the converter, which is already expensive when compared to the single-switch quadratic boost converter shown in Fig. 1(a).

B. Coupled Inductor Converters

Converters with coupled inductors have inductors in their circuit that are coupled together so that certain attractive performance features can be obtained. Among these features are extremely low or extremely high output-to-input voltage conversion ratios of M.

One popular application where extremely low values of M are needed is for voltage regulator modules (VRMs). A very common requirement for VRMs is that they convert voltages such as 12 Vdc to very low voltages such as 1.5 Vdc. Consider the coupled inductor VRM buck converter shown in Fig. 3(a) [3]. The output filter inductor of the buck converter is replaced by a coupled inductor, the freewheeling diode is replaced by a low-voltage MOSFET to reduce conduction losses, and a passive clamp consisting of diodes D_{s1}, D_{s2}, and C_s is added to absorb leakage inductance spikes.

The coupled inductor allows for different inductances between the charging and discharging periods. The inductance of winding n_1 and n_2 is effective during the charging period, while the inductance of only winding n_1 is effective during the discharging period. This introduces an extra degree of design freedom. Increasing the turns ratio n of the coupled inductors, where n is defined as $n = (n_1 + n_2)/n_1$ can help extend the duty cycle, reduce the bottom switch voltage stress, and reduce the top-switch switching current. The smaller switching current means less switching loss in the top switch. Lower voltage stress means less body-diode reverse-recovery losses in the bottom switch.

Most important of all, the converter has a very low conversion ratio M and, thus, does not have the problem of small duty cycle that conventional buck converters have. Since coupled inductor converters do not have any outstanding flaws that would limit their use, they are the preferred choice over quadratic buck converters in VRMs, and quadratic converters are never used in this application.

Coupled inductors can also be implemented in boost converters to increase their conversion ratio M. An example of a high step-up boost converter with a coupled inductor is shown in Fig. 3(b) [4]. The primary winding of the coupled inductor is used as the filter inductor. The second winding operates as a voltage source in series to the power branch to extend the voltage gain and reduce the switch voltage stress. Extreme duty cycles are avoided by the proper design of the turns ratio of the coupled inductor. The leakage energy of the coupled inductor can be absorbed and the turn-off voltage ringing on the MOSFET can be suppressed by the passive lossless clamp circuit, which is composed of a clamp diode

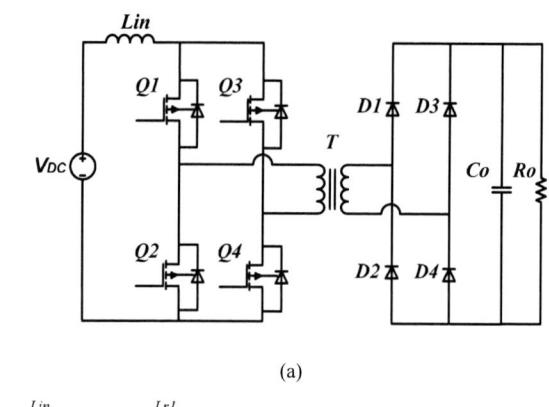

(a)

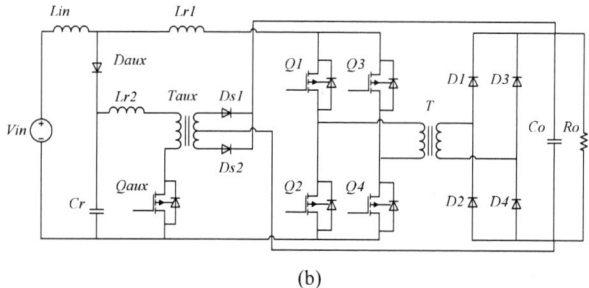

(b)

Fig. 2. (a) Current-fed PWM full-bridge boost converter. (b) ZCS full-bridge boost converter [2]

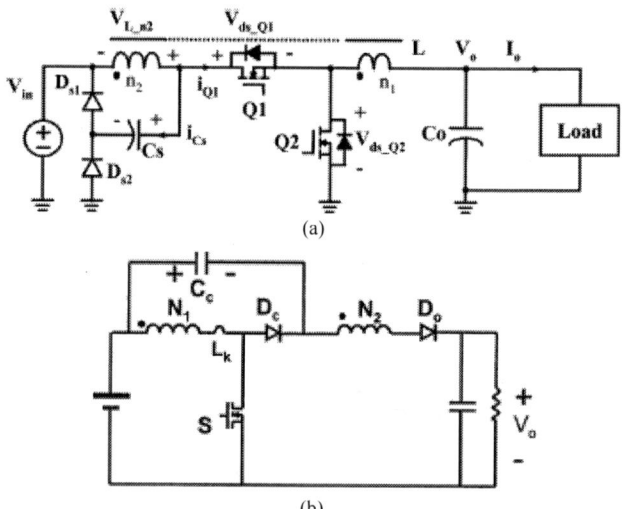

Fig. 3 (a) Coupled inductor buck converter for VRM [3]. (b) Coupled inductor boost converter [4].

978-1-4244-4782-4/10 $26.00 © 2010 IEEE

D_c and a small clamp capacitor C_c.

Coupled inductor boost converters such as the one shown in Fig. 3(b) have significantly lower peak switch voltage stresses and higher switch current stresses than a conventional boost converter. The ripple of the current that is drawn from the input voltage source is also considerably higher than that in a conventional boost converter.

III. ALTERNATIVE ENERGY POWER SYSTEMS

Converters for fuel cell or solar power systems typically need some sort of boost converter to convert a low level dc voltage obtained from the fuel or solar cells into a high level dc voltage that is fed to a dc-ac inverter that is connected to the grid. In applications where transformer isolation is not a mandatory requirement, quadratic dc-dc boost converters such as the one shown in Fig. 1 are a very attractive option. This is because of their $M = 1/(1-D)^2$ voltage conversion ratio, and because they have lower current stresses and less input ripple than couple inductor converters. The latter is particularly important when the input voltage is produced by fuel cells [5].

As was mentioned in the Introduction of this paper, above, the fact that the quadratic boost converter must process energy twice results in a reduction of converter efficiency. In order to maintain the benefits of quadratic boost converters, but with more efficiency, modifications to the basic quadratic boost structure have been proposed. Two such modifications are briefly reviewed here.

A. Three-Level Quadratic Boost Converter

The converter shown in Fig. 4 is a three-level quadratic converter that was proposed in [6] for use in fuel cell applications. The converter operates with the following modes of operation, as shown in Fig. 5.

Mode 1 ($t_0 - t_1$): Mode 0 begins when S_2 is turned on. Current i_{L2} remains the same (= I_{L2min}) because the voltage across inductor L_2 is zero. The current through inductor L_1 decreases linearly while L_1 and input voltage V_i provide energy to the intermediate capacitor, C_{oint}.

Mode 2 ($t_1 - t_2$): When S_1 is turned on, i_{L1} and i_{L2} increase. Inductors L_1 and L_2 receive energy from the input source and C_{oint} respectively. D_1 is reverse biased during this mode.

Mode 3 ($t_2 - t_3$): This mode is exactly the same as Mode 1.

Mode 4 ($t_3 - t_4$): When S_2 is turned off, current i_{L2} flows through D_2 and the load, while i_{L1} decreases linearly. This is the only stage when energy is transfer to the load and it ends when S_2 is turned on and another switching cycle begins.

It can be seen in Fig. 5 that each converter switch is never exposed to more than half of the output voltage. It was shown in [6] that this converter is at least 2% more efficient than the cascaded boost and conventional quadratic boost converters. The main reason for the increase in efficiency is that the switches are exposed to lower voltages, which allows for two low-voltage MOSFETs with low R_{DSon} values to be used.

B. Quadratic Boost Converter with Direct Boost Stage

Another approach that can be used to improve efficiency is to ensure that at least some of the energy that passes through a quadratic boost converter is processed only once. This is the approach taken in the quadratic boost converter that was proposed in [7] and shown in Fig. 6.

Components D_1, L_1, T, D_3, and C_1 form a conventional boost converter. Components L_2, C_2, T, D_3, and C_1 form a second boost converter with an intermediate energy storage capacitor. Diode D_2 is used to discharge C_2 through L_1. Energy is placed in L_1 and L_2 and capacitor C_2 also charges up when switch T is on, and energy is transferred to the output switch T is off.

In [7], experimental results comparing the performance of the conventional boost converter and the modified quadratic boost converter were made. Both converters were implemented with an input voltage of $V_g = 12$ V and a resistive load of R = 500 Ω and were operated with 20 kHz switching frequency. Graphs of output-to-input voltage conversion ratio vs. duty ratio and efficiency vs. conversion ratio are shown in Figs. 7(a) and 7(b) respectively. The modified quadratic boost converter is referred to as the "two-boost" converter in Fig. 7.

It can be seen from the graphs in Fig. 7 that the efficiency of the modified two-boost quadratic converter closely matches that of the conventional boost converter for most conversion ratios.

III. SINGLE-STAGE AC-DC CONVERSION

Ac-dc power conversion with transformer isolation is typically done with two converter stages. The first stage is an ac-dc converter that converts input ac voltage onto an intermediate dc bus voltage. This front-end converter is usually a boost converter that also performs input power factor correction (PFC), to shape the input current so that it is sinusoidal and in phase with the input voltage. The dc bus voltage is fed to a second converter stage that converts it into the required, isolated, output dc voltage. This second stage is typically a flyback or a forward converter for low power applications and a dc-dc full-bridge converter for higher power applications.

In essence, conventional two-stage ac-dc converters are cascaded converters. As with other cascaded converters, researchers have tried to reduce their cost and complexity by using approaches such as the ones described in this paper. Just as current-fed isolated boost converters, coupled inductor converters, and quadratic converters are used to avoid the use of two independent switch-mode converters, so-called ac-dc

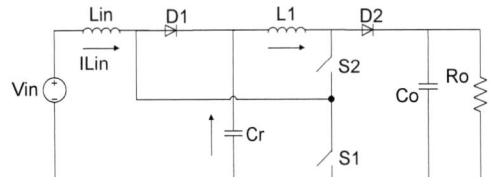

Fig. 4 Three-level quadratic boost converter [6].

978-1-4244-4782-4/10 $26.00 © 2010 IEEE

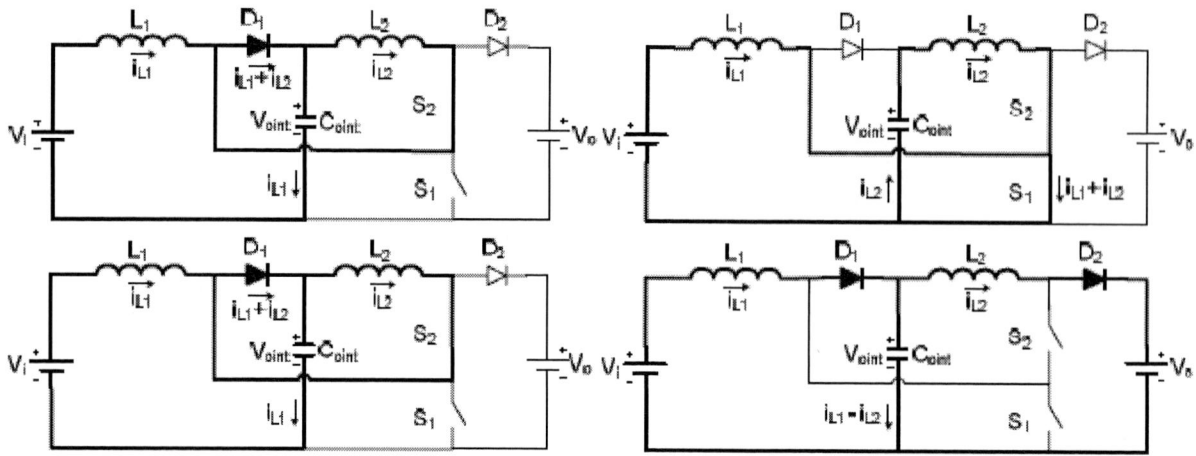

Fig. 5. Modes of operation of a three-level quadratic boost converter.

single-stage converters [8]-[9] can be used to avoid the use of two separate and independent converter stages. Single-stage converters combine an ac-dc converter stage with a transformer isolated dc-dc converter stage in a single converter to save on the cost of two separate converters.

There are generally two types of ac-dc single stages converters - those based on isolated boost converters and those based on cascaded converters. Each type is described below.

A. Ac-Dc Single-Stage Isolated Boost Converters

The most popular type of ac-dc converter for input PFC is the boost converter, as it is simple and can be operated to produce an input current that is either continuous or discontinuous. Like single-switch dc-dc boost converters, isolated boost converters such as the current-fed full-bridge boost converter shown in Fig. 2 can be implemented with an ac input voltage and a diode bridge rectifier instead of a dc input voltage. They can perform PFC by shaping the input current in almost the same way as a single-switch boost converter can. They are attractive as ac-dc PFC converters because they can simultaneously perform ac-dc power conversion and PFC to produce a transformer isolated voltage using only a single converter stage instead of two.

Isolated boost converters such as converters that are based on current-fed boost full-bridge topologies, however, do not have an energy storage capacitor connected to their transformer primary-side dc bus. The lack of such a capacitor creates two fundamental drawbacks that have limited their use as ac-dc PFC converters. The first is that a low frequency 120 Hz ripple appears at the output that requires considerable

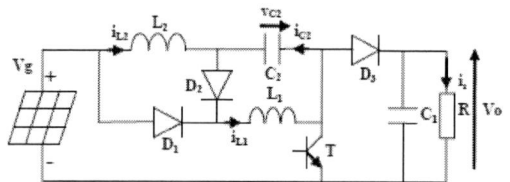

Fig. 6. Quadratic boost converter with direct power stage [7].

output filtering to minimize, and the second is that there is no hold-up time capability whenever there is a temporary loss of ac input. Isolated boost converters tend to be used as battery-chargers, which is an application where low frequency output ripple and hold-up time are not issues.

B. Ac-Dc Single-Stage Converters Based on Cascaded Converters

Another approach to ac-dc single-stage power conversion is to use converters that are based on cascaded converters. For example, consider the two-stage ac-dc converter shown in Fig. 8(a). This converter consists of an ac-dc boost stage and a dc-dc forward stage. These two stages can be combined to form the single-stage converter shown in Fig. 8(b). In this converter, both ac-dc PFC and dc forward conversion are performed by the turning on and off of the converter switch.

The cascade connection of two separate converters in Fig. 8(b) makes the resulting converter like a quadratic converter except that the first converter is an ac-dc converter instead of a dc-dc converter. Although single-stage and quadratic converters are similar, the lack of knowledge about quadratic converters has led researchers to "reinvent the wheel", proposing single-stage converter that are just versions of dc-dc quadratic topologies that were previously proposed. This can be seen in Fig. 9, for example, by comparing the two quadratic topologies that were presented in [1], which was published around 1990, to the single-stage topologies, which were presented in [10] and [11], both almost 10 years later.

Since ac-dc converters are very similar to quadratic converters, therefore, they share many of the strengths and weaknesses of these converters. The main strengths are (i) they can achieve converter performance similar to that of two separate converters connected in series at lower cost, (ii) they can be implemented with energy-storage capacitors at their transformer primary-side dc bus so that they do not have the drawbacks that isolated boost single-stage converters have. Their main disadvantage is that they are less efficient than conventional two-stage converters as they are implemented with fewer switches so that there are fewer degrees of freedom in their design and control. Several other properties of single-stage converter are briefly reviewed here.

978-1-4244-4782-4/10 $26.00 © 2010 IEEE 1323

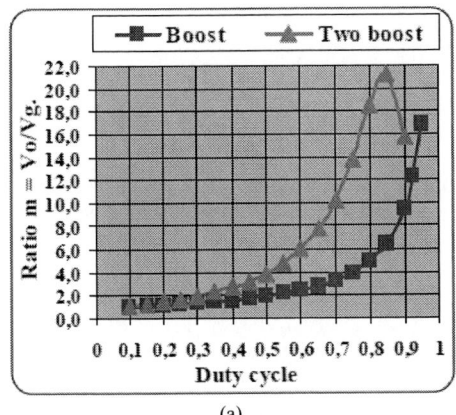

(a)

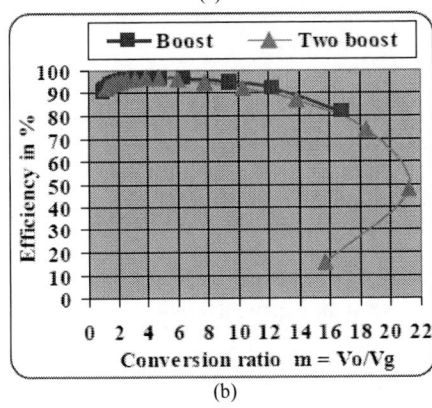

(b)

Fig. 7. (a) Variations of conversion ratio m =Vo /Vg. (b) Variations of efficiency η = Po / Pg.

1) The use of auxiliary windings in single-stage converters to improve performance: It is standard practice in single-stage boost-type converters to use one or more windings taken from the main power transformer to reduce the dc bus voltage. An example of how this can be done is shown in Fig. 10 [12] where the winding is placed between the input inductor L_{in} and the switch.

When the switch is on, voltage is impressed across the main transformer winding and thus appears across the auxiliary winding. This voltage counters the input voltage so that the full input voltage is not impressed across the input

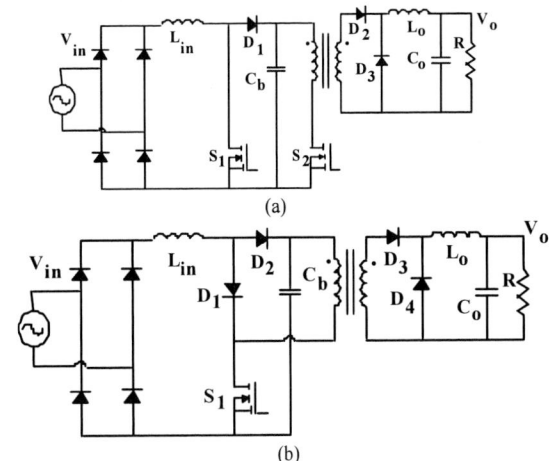

Fig. 8. (a) Two-stage ac-dc boost/forward converter (b) Single-stage converter ac-dc boost/forward converter. (Note: Reset mechanism not shown).

inductor, which results in less energy being put into the input inductor than would otherwise be and thus less energy is transferred into the dc bus capacitor C_{bus} when the switch is turned off. The presence of the auxiliary winding in the converter therefore affects the steady-state energy equilibrium in the dc bus capacitor so that it causes a reduction in the dc bus voltage.

2) Comparison of single-stage boost and non-boost converters: One of the features of a single-stage non-boost-type converter (Figs. 9(b) and (d) is that the primary-side dc bus voltage can be less than the input voltage, unlike a boost-type converter in which this voltage must be greater than the input voltage. This is because the input section operates like a step-down converter, making the dc bus voltage lower than the peak input ac voltage. The lower dc bus voltage allows the switch to be a lower voltage rated device that can operate with lower turn-on losses than the switch in a boost-type converter, which is advantageous, but the current flowing through the switch, however, is proportional greater.

Consider an operating condition in a single-stage boost-type flyback converter under which the dc bus voltage is a certain voltage and the switch current is a certain current. If a

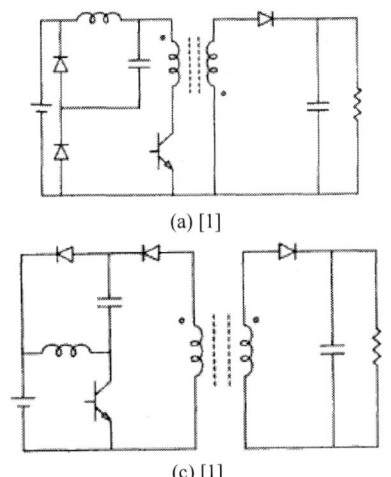

(a) [1]

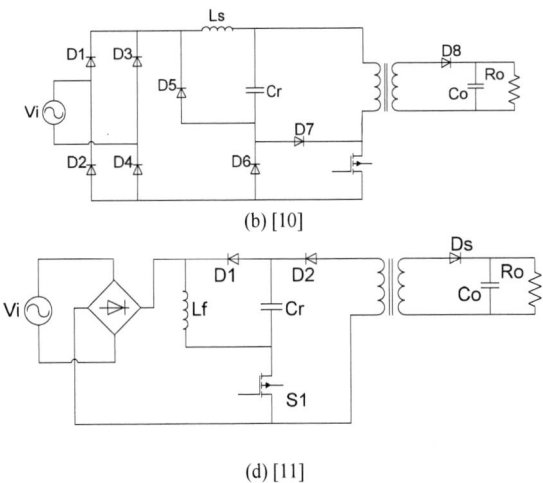

(b) [10]

(c) [1]

(d) [11]

Fig. 9. Quadratic dc-dc converters and single-stage converters.

978-1-4244-4782-4/10 $26.00 © 2010 IEEE

single-stage non-boost type converter operating under the same condition has a dc bus voltage that is half that of the boost-type converter, the switch current will be approximately twice that in the boost-type converter. The higher current will result in higher conduction losses and higher turn-off losses, and, as current is increased, these losses will become more predominant than the turn-on losses. As a result, non-boost type converters are generally better suited in applications where is load is relatively light, while boost-type converter are generally better suited in heavier load applications.

3)Non-isolated single-stage converters: For applications where a low, non-isolated, output dc voltage needs to be obtained from an ac input, single-switch buck–boost, SEPIC, and Cuk converters can be used. If a very low output voltage is needed, however, then these converters must operate with extremely low duty ratios to achieve very low values of M. Doing so, however, means that they will have problems due to extreme current stresses, poor component utilization, and very low efficiency.

In order to achieve a very low value of M, the converter shown in Fig. 11 was proposed in [13]. It is a quadratic converter formed by cascading a front-end buck–boost converter with an output buck converter. The buck–boost converter is selected due to its capability of providing a step-down voltage conversion and a high power factor when it is operating in the discontinuous conduction mode (DCM). The buck converter is selected due to its step-down capability. The converter has the following modes of operation, as shown in Fig. 12:

Mode 1 [t_0- t_1]: Switch S_1 is turned on at t = t_0. Diode D_y becomes forward biased, and currents i_{L1} and i_{L2} begin to increase linearly.

Mode 2 [t_1-t_2]: When the switch is turned off at t = t_1, diode D_y becomes reverse biased. Current i_{L1} linearly decreases through diode D_x, while current i_{L2} linearly decreases through the freewheeling diode, D_F. This mode ends when i_{L1} reaches zero; diode D_L prevents current i_{L1} from becoming negative.

Mode 3 [t_2-t_3]: In this mode, current i_{L2} continues to decrease through diode D_F until it becomes zero. The converter stays in this mode until the switch is turned on again.

The operation of the buck output cell in Fig. 11, in either the continuous conduction mode (CCM) or DCM, has no effect on the quality of the input current. Operating the buck cell in the DCM, however, gives several advantages

including: a zero-current switch turn-on, reduced reverse recovery current in the diodes, and low-voltage stress on both the energy storage capacitor C and the active switch S_1.

IV. THREE-PHASE AC-DC SINGLE-SWITCH QUADRATIC BUCK CONVERTERS

Three-phase ac-dc rectifiers typically use six switches that are turned on and off in a manner that ensures that ac-dc conversion occurs with sinusoidal input line currents that are in phase with the input voltages. These rectifiers can operate with an excellent input power factor, but are expensive and complicated. An alternative for lower power three-phase applications is using single-switch converters such as three-phase buck converter shown in Fig. 13(a) [14]. This converter operates like a dc-dc buck converter, but if the input section is designed so that the voltage across the input capacitors are discontinuous as shown in Fig. 13(b), then the input inductors currents will be sinusoidal.

This, however, is not easy to achieve because whether the input capacitor voltages are continuous or discontinuous is dependent on the converter's operating conditions and on D. Wide variations in D make it difficult to ensure discontinuous input capacitor voltages and thus good power factor operation. The variation in D can be restricted if the quadratic converter proposed in [15] is used (Fig. 14). This converter has a value of M that is dependent on D^2 instead of D.

The preferred approach to soft-switching for a single-switch buck converter is to use some sort of zero-current-switching (ZCS) quasi or multi-resonant technique, as proposed in [16]. This is because it requires that only a few passive components be added to the converter as opposed to zero-current-transition (ZCT) PWM techniques that require an active auxiliary circuit. The use of quasi and multi-resonant techniques is in line with the general philosophy of using single-switch converters for three-phase PFC, which is to maintain low cost.

The main drawback with these techniques, however, is that variable frequency control must be used to control the converter, which complicates the design optimization of magnetic components and EMI filtering and requires bulky magnetics for the lowest switching frequency. The drawbacks that result from variable switching control can be avoided by limiting the line and load range over which the converter operates so that the variation in switching frequency is reduced.

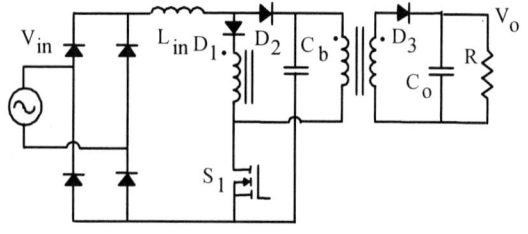

Fig. 10. An ac-dc single-stage, voltage-fed converter with an auxiliary transformer winding [12].

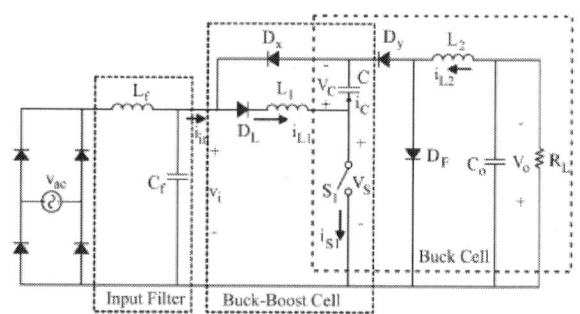

Fig. 11. Non-isolated single-stage buck-boost/buck quadratic converter [13].

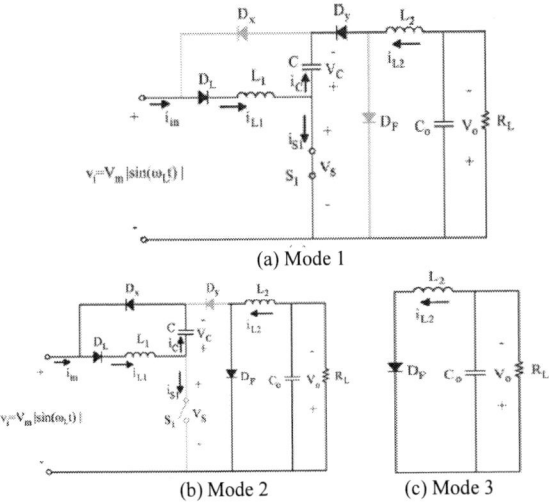

(a) Mode 1

(b) Mode 2   (c) Mode 3

Fig. 12. Modes of operation of the converter in Fig. 11.

In order to overcome the limitations of the conventional single-switch buck-type converter, the three-phase, single-switch soft-switched rectifier shown in Fig. 15 was proposed in [17]. This converter operates with a variable frequency quasi-resonant ZCS technique for soft-switching. The key feature of the proposed converter is that its output to input voltage conversion ratio has a quadratic dependence on the switching frequency so that the variation in switching frequency is significantly reduced. As a result, the following improvements are made

- The design of magnetic components and EMI filtering is simplified as the switching frequency range that must be considered is reduced.

- The magnetic components and thus the converter itself can be made to be less bulky since the converter does not have to operate with switching frequencies that are as low as those of the conventional buck converter.

- The converter can perform PFC over a wider range of line and load conditions. Limitations to the operating range due to the wide variation in switching frequency of the conventional topology are reduced.

An experimental prototype of the proposed converter was built to confirm its feasibility. The converter was designed to operate with an input line voltage of 200-240 V_{LLrms}, an output voltage of 100V, a maximum output power of 2 kW, and a resonant frequency of 50 kHz. The converter was implemented with main circuit components $L_a=L_b=L_c=1.2$ mH, $C_a=C_b=C_c=215$ nF, $L_m=350\,\mu$H, $C_m=82\,\mu$F, $L_o=100\,\mu$H, $C_o=164\,\mu$F. The resonant components were $L_r=11.78\mu$H and $C_r=0.86\,\mu$F. Fig. 16 shows typical experimental waveforms.

It can be seen from Fig. 16(a) that a nearly sinusoidal input current waveform can be obtained. It can be seen from Fig. 16(b) the converter switch can operate with ZCS as the resonant components delay the rise in switch current and make it fall to zero. It should be noted that the voltage across the switch is triangular, which is characteristic of single-

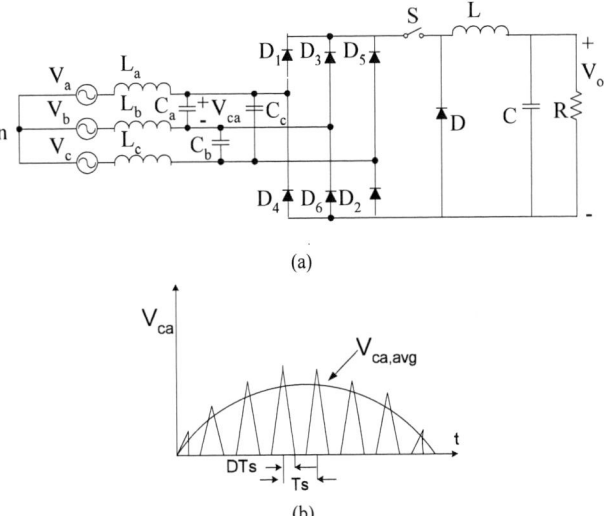

(a)

(b)

Fig. 13. (a) Three-phase single-switch buck rectifier [14]. (b) Input capacitor voltage (half cycle).

switch buck converters as the voltage across each input ac side capacitor drops to zero due to charging and discharging. It can be seen from Fig 16(c) that that the output diode current has no reverse recovery current due to ZCS operation.

V. CONCLUSION

The characteristics and uses of quadratic power converters in industrial applications were reviewed in this paper. Three main applications were reviewed: alternative energy power systems, ac-dc single-stage power factor correction, and three-phase ac-dc buck power conversion. The main points of the paper can be summarized as follows:

- Modified versions of the basic quadratic boost converter are suitable for alternative energy power systems where transformer isolation is not required. They are more efficient than the basic quadratic boost converter, cheaper than isolated current-fed boost converters, and have less input current ripple than coupled inductor boost converters.

- Ac-dc single-stage converters are variations of quadratic converters and thus share many properties. Topologies that have been presented as new single-stage converters are just modified versions of quadratic converters, with the only difference being that the input section is derived from an ac-dc converter instead of from a dc-dc converter.

- Quadratic power conversion can be used in single-

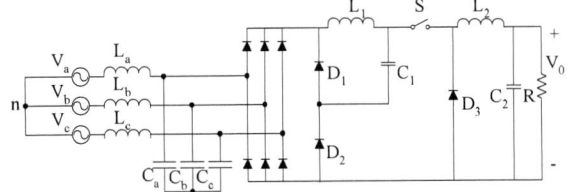

Fig. 14. Three-phase quadratic buck rectifier [15].

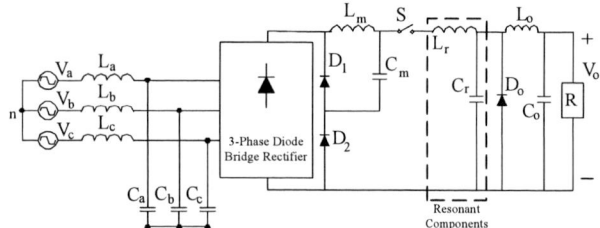

Fig. 15. Three-phase single-switch ZCS quasi-resonant rectifier [17].

switch, three-phase, ac-dc converters to improve performance. This is especially true when these converters are implemented with some sort of quasi-resonant or multi-resonant technique to reduce switching losses. The use of quadratic power conversion can limit the range of switching frequency variation, and simplify the design of the converter and its magnetics.

ACKNOWLEDGMENT

The author would like to acknowledge technical assistance from Sondeep Bassan, Pritam Das, and Ahmad Mousavi, and financial assistance from the National Sciences and Research Council of Canada (NSERC) for this work.

REFERENCES

[1] D. Maksimovic, and S. Cuk, "Switching converters with wide DC conversion range," *IEEE Trans. on Power Elec.*, vol. 6, no. 1, pp 151-157, Jan. 1991

[2] A. Mousavi, P. Das, and G. Moschopoulos, " Analysis and design of a new ZCS-PWM full-bridge fuel cell converter, in IEEE ECCE Conf., 2009, pp. 2037-2042.

[3] K. Yao, M. Ye, M. Xu, and F. C. Lee, "Tapped-inductor buck converter for high-step-down DC-DC conversion," in *IEEE Trans. on Power Elec.*, vol. 20, no. 4, pp. 775-780, Jul. 2005.

[4] Q. Zhao and F. C. Lee, "High-efficiency, high step-up DC-DC converters," in *IEEE Trans. on Power Elec.*, vol. 18, no.1, pp. 65-73, Jan. 2003.

[5] R. S. Gemmen, "Analysis for the effect of inverter ripple current on fuel cell operation condition," *ASME Intcr. Mech. Eng. Cong. and Expo.* 2001.

[6] Y, de Novaes, A. Rufer, and I. Barbi, "A new quadratic, three-level DC/DC converter suitable for fuel cell applications," PCC Conf. 2007, pp. 601-607.

[7] J.-P. Gaubert and G. Chanedeau, "Evaluation of DC-to-DC converters topologies with quadratic conversion ratios for photovoltaic power systems," *Euro. Conf. on Power Elec and Appl. (EPE)*, 2009

[8] R. Redl, L. Balogh, and N. O. Sokal, "A new family of single-stage isolated power-factor correctors with fast regulation of the output voltage", *IEEE PESC Conf. Rec.*, 1994, pp. 1137 - 1144.

[9] L. Huber, J. Zhang, M. M. Jovanovic, and F. C. Lee, "Generalized topologies of single-stage input-current-shaping circuits," *IEEE Trans. Power Elec.*, vol. 16, no. 4, pp. 508-513, Jul. 2001.

[10] H.-W. Kim, G.-W. Moon, K.-Y. Cho, and M.-J. Youn, "Single-stage high power factor converter with single-switch for universal input," in *IEEE IECON Conf.*, 2001, 931 – 936.

[11] T. F. Wu and Y-K. Chen, "Analysis and design of an isolated single-stage converter achieving power-factor correction and fast regulation," in *IEEE Trans. on Ind. Elec.* vol. 46, no. 4, pp. 759-767, Aug. 1999.

[12] F.-S. Tsai, P. Markowski, and E. Whitcomb, "Off-line flyback converter with input harmonic current correction," *IEEE INTELEC Conf.*, 1996, pp. 120-124.

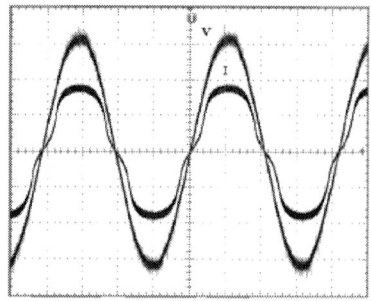

(a)

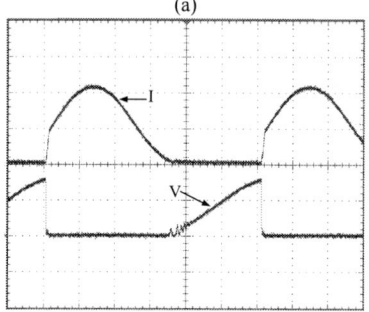

(a)

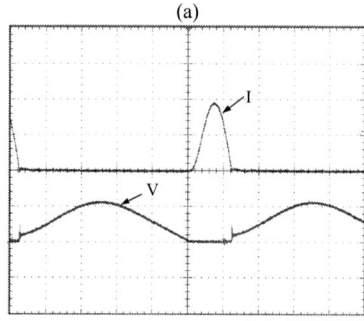

Fig. 16 (a) Input phase current and phase voltage (V: 50 V/div, I: 5 A/div, t: 4 ms/div) (b) Switch current and voltage waveforms (I: 20 A/div, V: 500 V/div, t: 4 μ s/div). (c) Output diode current and voltage waveforms (I: 10 A/div, V: 200 V/div, t: 4 μ s/div).

[13] E. H. Ismail, A. J. Sabzali, and M. A. Al-Saffar, " Buck–boost-type unity power factor rectifier with extended voltage conversion ratio," in *IEEE Trans. on Ind. Elec.*, Mar. 2005, vol. 55, no. 3, pp. 1123-1132.

[14] E. Ismail and R. W. Erickson, "Single-switch 3φ PWM low harmonic rectifiers", *IEEE Trans. on Power Elec.*, vol. 11, no. 2, pp. 338-346, Mar. 1996.

[15] J. Shah and G. Moschopoulos, "A novel three-phase single-switch buck-type rectifier" in *IEEE APEC Conf.*, 2005, pp. 515 – 521.

[16] Y. Jang and R. W. Erickson, "New single-switch three-phase high power factor rectifiers using multi-resonant zero current switching," in *IEEE Trans. on Power Elec.*, vol. 13, no.1, pp 194-201, Jan. 1998.

[17] S. Bassan and G. Moschopoulos, "A three-phase single-switch high power factor buck-type converter operating with soft-switching," *IEEE PESC Conf.*. 2007, pp. 3053-3059.

Multiple-Output Resonant Inverter Topology for Multi-Inductor Loads

O. Lucía, J.M. Burdío, I. Millán, and J. Acero

Department of Electronic Engineering and Communications
University of Zaragoza
50018 Zaragoza. SPAIN.
E-mail: {olucia, burdio, millan, jacero}@unizar.es

Abstract—**Multiple-inductor systems are widely present in the current technology. These systems require controlling either the supplied voltage or power to several loads with different requirements simultaneously. As a consequence, the cost and size of the power stage may increase beyond the admissible limits for certain applications.**

Considering multiple-inductor loads, a novel multiple-output series resonant inverter (MOSRI) topology is proposed to obtain a cost-effective and high-power-density solution. The converter is based on a common inverter block and a resonant-load block. The performed analysis includes the description of the operation modes and the control strategy analysis. Domestic induction heating has been considered for application due to its special cost and size requirements, and the extensive inductor use. The proposed converter has been designed and validated experimentally through a prototype, which includes the power converter and the FPGA-based control architecture.

I. INTRODUCTION

Multiple inductor systems are widely present in the current technology [1]. These systems require supplying either a fixed voltage or power for each load. Domestic induction heating appliances are an example where different output power is required for each load. Besides, it requires a cost-effective and high-power density implementation to compete in nowadays induction heating market.

Recent development trends suggest a performance improvement through adjustable cooking surfaces. These architectures make an intensive use of multi-inductor configurations to provide reconfigurable cooking surfaces. Different approaches to achieve this functionality are possible. Fig. 1 summarizes the main possibilities. Concentric inductors can be used to adapt the induction system to the pan size and improve the magnetic coupling (a). The same effect can be achieved by means of series/parallel-connected non-concentric inductors, which can adapt to more complex pan shapes (b). Finally, the whole surface can be composed of inductors providing a total-active surface solution (c).

Traditional power converters are based on a dc-link resonant inverter, intended to supply a single load [2], [3]. The development of new multi-windings appliances requires the design of new power converters able to supply several loads at the same time. The aim of this paper is to propose a novel multiple-output resonant inverter topology suitable for domestic induction heating application with a high number of loads. The main features of the converter have to include a reduced number of devices and high power density, proper power controllability for each load, and high efficiency.

This paper is organized as follows. Section II describes the current multi-load inverter alternatives and Section III presents the novel multiple output series resonant inverter proposed, including its operation mode. The basic control strategy is presented in Section IV and a comparative evaluation is carried out in Section V. Finally the experimental results based on a 4-load prototype are shown in Section VI and the main conclusions of this paper are outlined in Section VII.

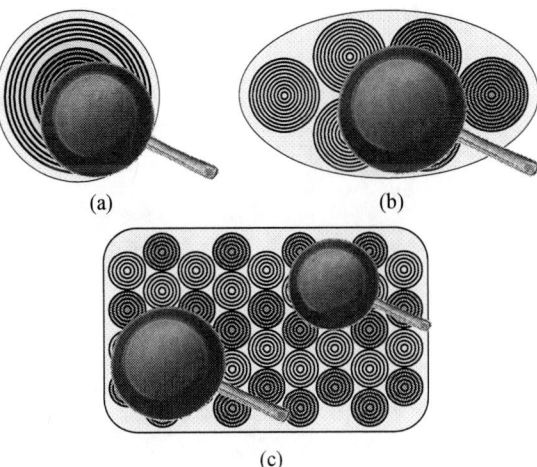

(a)   (b)

(c)

Fig. 1. Multi-windings arrangements for domestic induction heating: (a) concentric-windings hob, (b) multi-inductor hob, and (c) total-active surface concept.

This work was supported in part by the Spanish Ministry of Science and Innovation (MICINN) under Project TEC2007-64188 and FPU Grant AP2007-03276, in part by Diputación General de Aragón (DGA) under Project PI008/08, and in part by the Bosch and Siemens Home Appliances Group.

The authors are with the Department of Electronic Engineering and Communications, University of Zaragoza, and the Aragon Institute for Engineering Research (I3A).

II. POWER CONVERTERS APPLIED TO MULTI-INDUCTOR DOMESTIC INDUCTION SYSTEMS

Different approaches are possible when designing multi-load power converters for domestic induction systems. The proposed solutions can be classified according to their functional operation. Thus, three different approaches are found: paralleled power converters, load-multiplexation technique and multiple-output inverters. These solutions are represented in Fig. 2.

The paralleled power converter solution (Fig. 2 (a)) is the most intuitive and flexible approach. However, it implies a high cost due to redundant elements and low power density. Multiplexation technique has been proposed to achieve a cost-effective solution. Load-configuration through electro-mechanical switches for a limited amount of loads is proposed in [4], and frequency-load-multiplexation is proposed in [5]. The main drawbacks of these solutions are the limited and complex power control, limited amount of loads, the wide frequency operation range required, the acoustic noise, and the low speed of electro-mechanical switches. Finally, multiple output inverters are also proposed as the most efficient method to supply power to several loads. Their benefits include higher utilization ratio of electronics, higher power density and reduction of components. Different proposals for a low number of loads include a dual full-bridge topology [6], dual half-bridge topology [7] and a variable-dc-link half-bridge topology [4].

Considering the alternatives, the multiple-output inverter is the most appropriate solution to design a new induction heating architecture with a high number of loads. For this reason the topology proposed in this work will focus on this alternative. Next Section describes the proposed multiple-output series resonant inverter (MOSRI) topology.

III. PROPOSED RESONANT MULTIPLE-OUTPUT INVERTER

A. Power Converter Topology

The proposed multiple-output series resonant inverter topology for n loads is a DC-link inverter based on a common inverter block and a resonant-load block (Fig. 3). The basic inverter topology chosen is the half bridge, due to its appropriate balance between number of switches, electrical stress and performance [8]. The resonant-load block has been implemented through semiconductor switches S_i and the resonant tanks associated to each load. Then the total number of switches N_s for a certain number of loads n is:

$$N_s = 2 + n \qquad (1)$$

The power switches are composed of a transistor (T_i) with anti-parallel diode (D_i) to simplify the drive circuit. The association of these switches with the resonant capacitor makes possible to ensure the proper inductor disconnection. Since the average current through a capacitor in steady state is null, if the current in one direction through the transistor is blocked then the load is effectively disconnected. This allows a simplified implementation of the resonant-load block devices in order to achieve a cost-effective solution.

The induction loads are modeled as the series connection of an equivalent resistor $R_{eq,i}$ and an equivalent inductor $L_{eq,i}$ [2], [9]. They are usually classified according to their quality factor Q_i defined as usual in (2). Series resonance has been selected to avoid the need of additional inductors. Each load is provided with a series-connected resonant capacitor $C_{r,i}$ to ensure independent power control, and it determines the resonant frequency for each load (3).

$$Q_i = \frac{X_{eq,i}}{R_{eq,i}} = \frac{\omega L_{eq,i}}{R_{eq,i}} \qquad (2)$$

$$f_{o,i} = \frac{\omega_{o,i}}{2\pi} = \frac{1}{2\pi\sqrt{L_{eq,i}C_{r,i}}} \qquad (3)$$

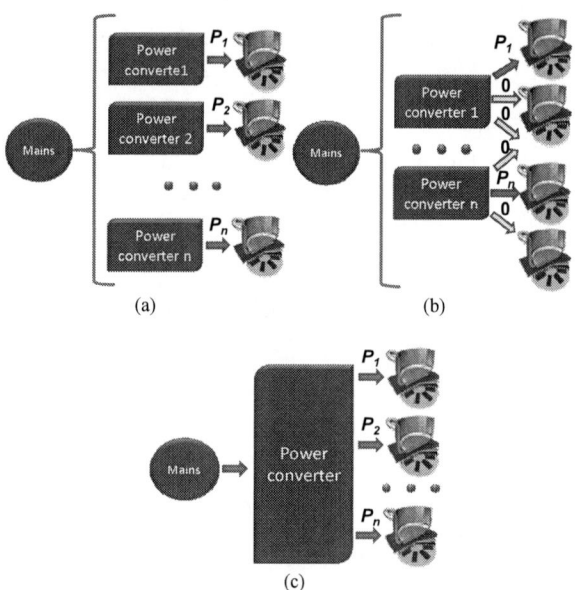

Fig. 2. Power converters for multi-load systems classification: (a) paralleled power converters, (b) load-multiplexation technique, and (c) multiple-output converter.

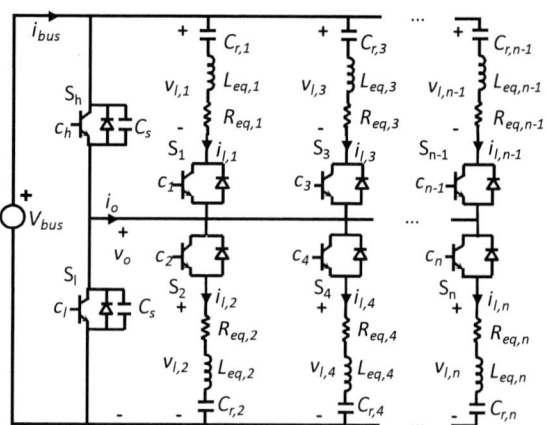

Fig. 3. Proposed multiple-output series resonant inverter topology (MOSRI).

In order to improve the bus current consumption, loads are connected both to the positive and the negative rail bus. This allows reducing the magnitude of bus current although it increases the fundamental frequency. The sum of all load currents also determines the inverter output current as given in (4). As a consequence, the zero voltage switching (ZVS) condition for the half bridge depends on the whole resonant-load block operation.

$$i_o = \sum_n \left[(-1)^n i_{l,n} \right] \qquad (4)$$

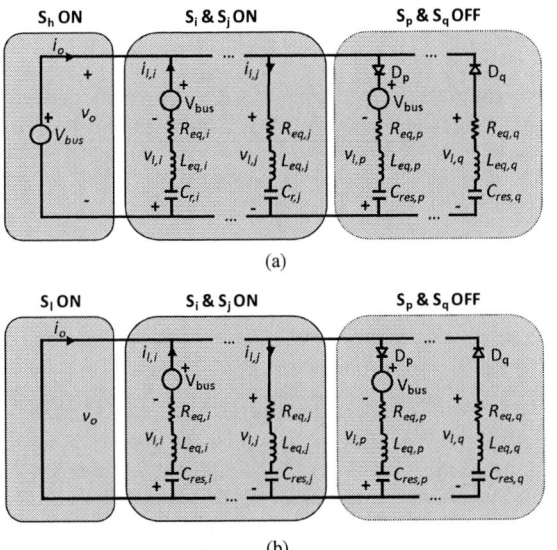

(a)

(b)

Fig. 4. Equivalent circuits for the multiple output series resonant inverter: (a) $v_o = V_{bus}$ y (b) $v_o = 0$.

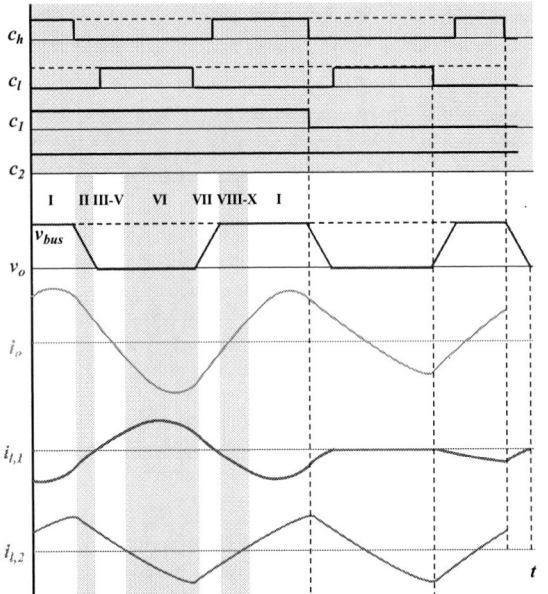

Fig. 5. Control signals and main waveforms for the multiple output series resonant inverter.

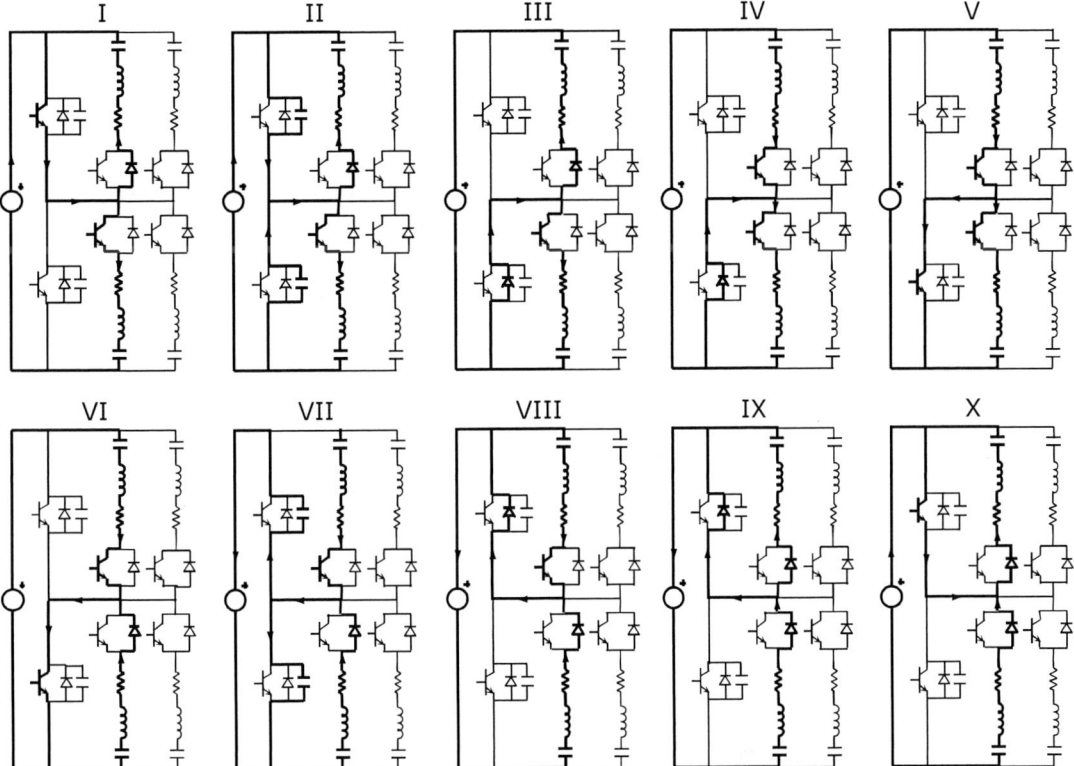

Fig. 6. Multiple-Output Series Resonant Inverter configurations with one high-side load and one low-side load active.

IV. POWER CONVERTER OPERATION MODE

The considered operation mode is based on the voltage-source operation of the half-bridge inverter. Thus, each load operates in parallel with a voltage source and the operation analysis is simplified. Fig. 4 shows the equivalent circuits for the whole converter depending on the switches state. When S_h is switched on then $v_o = V_{bus}$ and as a consequence the low-side loads are fed with V_{bus} voltage while high-side loads are in short circuit. The reciprocal behavior happens when S_l is switched on. In both cases the loads whose switches S_i are switched off remains disconnected once the remaining charge of $C_{r,i}$ has circulated through S_i diode.

Fig. 5 and Fig. 6 show the converter main waveforms and configurations when one high-side load and one low-side load are active. The sequence starts with circuit state I in which the current flows through T_h to the loads through D_1 and T_2 respectively. Then, T_h is switched off and the snubber capacitors start discharging (state II). After that D_1 starts conducting (state III). During this period T_1 is activated and current through the devices and half-bridge crosses zero (states IV-VI). In state V T_1 starts conduction with ZVS, and the reciprocal sequence continues until state X.

Besides, Fig. 5 shows the inductor $L_{eq,1}$ disconnection process. The transistor is switched off and then the soft inductor disconnection is achieved when diode current reaches zero. Finally, if there is any remaining charge in the resonant capacitor it is discharged in the next half-bridge switching cycle. After that, the load is completely disconnected.

V. MODULATION STRATEGY

The proposed control strategy for the MOSRI topology is based on the Variable Frequency Duty Cycle (VFDC) control [10], [11] and Pulse Density Modulation (PDM) [12], [13]. VFDC is applied to the half-bridge block in order to control the output power through the period T_{hb} and the duty cycle D_{hb}. Considering the first harmonic approximation, the output power for each load operating in continuous mode can be calculated as explained in [10].

$$P_{o,i} \cong \frac{\hat{V}_{o1}^2}{2 R_{eq,i}\left(1 + Q_i^2\left(\omega_{n,i} - \frac{1}{\omega_{n,i}}\right)^2\right)} \tag{5}$$

where $\hat{V}_{o1}$ is the amplitude of the output voltage first harmonic and the normalized angular frequency $\omega_{n,i}$ is a function of the angular switching frequency $\omega_{s,i}$:

$$\omega_{n,i} = \frac{\omega_{s,i}}{\omega_{o,i}} \tag{6}$$

Then, the whole converter output power can be calculated as:

$$P_o = \sum_{i=1}^{n} P_{o,i} \tag{7}$$

In addition to the half-bridge VFDC modulation, PDM is applied to S_i to control the power supplied to each load accurately. Then, output power per load operating in PDM mode ($P_{o,i,pdm}$) can be approximated as a function of the equivalent output power per load $P_{o,i}$ with continuous excitation. Provided that the considered load is in steady state, the output power can be calculated as

$$P_{o,i,pdm} = D_i P_{o,i} \tag{8}$$

This control strategy enables to control power individually from the nominal power, at resonant frequency, to low power without increasing switching frequency by applying PDM strategy. The main waveforms and control parameters are shown in Fig. 7.

VI. COMPARATIVE EVALUATION

To evaluate the viability of the proposed converter, a comparative evaluation with the paralleled-half-bridge converter has been carried out. MOSRI topology presents the same output power per load, but lower output power for the whole converter. This provides a more efficient solution with better semiconductor utilization ratio for the specific domestic induction heating requirements. Besides, the number of switching devices is reduced, which also includes their auxiliary circuits, snubber capacitors and drivers, obtaining a smaller and cost-effective converter with higher power density.

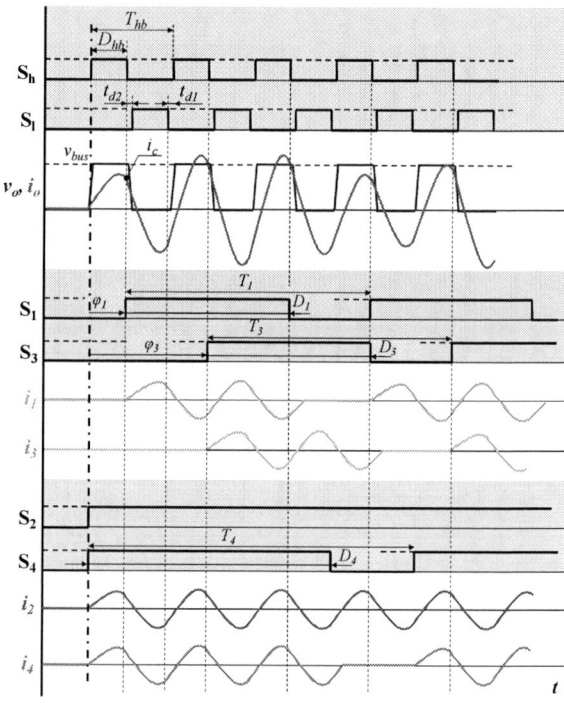

Fig. 7. Control parameters for the multiple output series resonant inverter.

978-1-4244-4782-4/10 $26.00 © 2010 IEEE

TABLE I
COMPARISON BETWEEN CLASSICAL HALF-BRIDGE
AND PROPOSED MOSRI TOPOLOGY

Parameter	Half bridge		MOSRI
Power per load	P_o		P_o
Power per converter	nP_o		$< nP_o$
Swithes and drive circuit	$2n$		$2 + n$
Resonant capacitors	n		n
Snubber capacitors	$2n$		2
Power density	Poor		Good
Switching losses	$n \cdot f(i_o, C_s, T_{hb})$	$>$	$f(i_o, C_s, T_{hb})$
Conduction losses	$\propto n \cdot k \cdot i_{o,rms}$	$<$	$k' \cdot i_{o,rms} + \sum_n k'' \cdot i_{li,rms}$
Controlability	Poor		Good

The converter efficiency depends on the balance of switching losses, decreased due to PDM modulation, and conduction losses, increased due to the additional switch per load. As a result, similar efficiency is expected with a proper design. Finally the converter controlability is improved due to the simplified power control strategy. It also avoids intermodulation acoustic noise present in classical multi-half-bridge configurations. Table I summarizes the comparative evaluation between the paralleled-half-bridge configuration and the proposed MOSRI topology.

VII. EXPERIMENTAL TEST-BENCH

A converter prototype has been designed and implemented to validate the simulation results. A 4-load inverter has been considered, composed of 6 power switches, to analyze the converter operation. The converter has been implemented following a modular approximation which allows its fast reconfiguration in order to evaluate different topologies. It has been selected the HGTG20N60 IGBT with antiparallel diode to obtain a robust solution.

For this work it has been considered a 150 μH and 10 Ω induction load with a 44 nF resonant capacitor. The bus voltage is 230 V rms and the switching frequency range is between 20 kHz and 150 kHz. A 3.3-nF-snubber capacitor has been also used to reduce the switching losses.

The converter gating signals are generated using an FPGA-based control architecture [14]. It is based on an embedded microprocessor (μP) and specific-purpose digital hardware. It allows easy implementation of high-level tasks through μP firmware and performing specific modulations through ad-hoc digital pulse width modulators.

Fig. 8 (a) shows the experimental test-bench including the power converter, the induction load and the used instrumentation. In addition, Fig. 8 (b) shows the control signals and current waveforms for the converter operating with one high-side load and one low-side load. It is shown the proper operation in both continuous excitation and PDM control for both induction loads.

VIII. CONCLUSIONS

In this paper, a novel multiple-output series resonant inverter (MOSRI) topology has been presented. It provides a versatile and cost-effective solution for multi-inductor systems. The proposed topology has general application for multi-load systems, although it has been particularized for the new domestic induction heating architectures.

The proposed topology is derived from the half-bridge topology and its control strategy is based on both the Variable Frequency Duty Cycle control and the Pulse Density Modulation. These strategies allow obtaining accurate power control for each load with a reduced number of switches.

A 4-load prototype has been designed and implemented to validate the simulation results. Experimental measurements confirm the expected results and, as a consequence, the MOSRI topology is proposed as a cost-effective solution for multi-load systems with high power density.

(a)

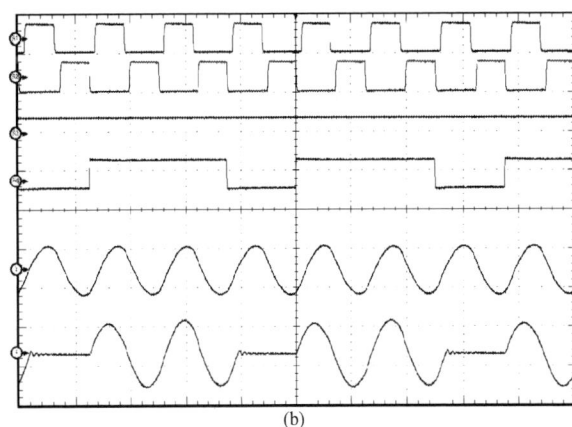

(b)

Fig. 8. Experimental test-bench: (a) MOSRI prototype and (b) experimental waveforms for 2 loads. From top to bottom: half-bridge and auxiliary switches gate signals (30 V/div) and inductor current waveforms (10 A/div) with 10 μs/div.

References

[1] P. J. Grbovic, "Master/slave control of input-series- and output-parallel-connected converters: concept for low-cost high-voltage auxiliary power supplies," *IEEE Transactions on Power Electronics,* vol. 24, no. 2, pp. 316-328, February 2009.

[2] F. P. Dawson and P. Jain, "A comparison of load commutated inverter systems for induction heating and melting applications," *IEEE Transactions on Power Electronics,* vol. 6, no. 3, pp. 430-441, July 1991.

[3] H. W. Koertzen, J. D. v. Wyk, and J. A. Ferreira, "Design of the half-bridge series resonant converters for induction cooking," in *IEEE Power Electronics Specialist Conference Records,* 1995, pp. 729-735.

[4] F. Forest, S. Faucher, J.-Y. Gaspard, D. Montloup, J.-J. Huselstein, and C. Joubert, "Frequency-synchronized resonant converters for the Supply of multiwindings coils in induction cooking appliances," *IEEE Transactions on Industrial Electronics,* vol. 54, no. 1, pp. 441-452, February 2007.

[5] F. Forest, E. Labouré, F. Costa, and J.-Y. Gaspard, "Principle of a multi-load/single converter system for low power induction heating," *IEEE Transactions on Industrial Electronics,* vol. 15, no. 2, pp. 223-230, March 2000.

[6] J. M. Burdio, F. Monterde, J. R. Garcia, L. A. Barragan, and A. Martinez, "A two-output series-resonant inverter for induction-heating cooking appliances," *IEEE Transactions on Power Electronics,* vol. 20, no. 4, pp. 815-822, July 2005.

[7] Y.-C. Jung, "Dual half bridge series resonant inverter for induction heating appliance with two loads," *Electronics Letters,* vol. 35, no. 16, pp. 1345-1346, May 1999.

[8] S. Llorente, F. Monterde, J. M. Burdio, and J. Acero, "A comparative study of resonant inverter topologies used in induction cooking," in *Applied Power Electronics Conference and Exposition,* 2002, pp. 1168-1174.

[9] J. Acero, R. Alonso, J. M. Burdío, L. A. Barragán, and D. Puyal, "Analytical equivalent impedance for a planar circular induction heating system," *IEEE Transactions on Magnetics,* vol. 42, no. 1, pp. 84-86, January 2006.

[10] J. M. Burdio, L. A. Barragan, F. Monterde, D. Navarro, and J. Acero, "Asymmetrical voltage-cancelation control for full-bridge series resonant inverters," *IEEE Transactions on Power Electronics,* vol. 19, no. 2, pp. 461-469, March 2004.

[11] O. Lucía, J. M. Burdío, I. Millán, J. Acero, and S. Llorente, "Efficiency optimization of half-bridge series resonant inverter with asymmetrical duty cycle control for domestic induction heating," in *European Conference on Power Electronics and Applications EPE09,* 2009.

[12] K. Young-Sup, Y. Sang-Bong, and H. Dong-Seok, "Half-bridge series resonant inverter for induction heating applications with load-adaptive PFM control strategy," in *Applied Power Electronics Conference and Exposition, 1999. APEC '99. Fourteenth Annual,* 1999, pp. 575-581 vol.1.

[13] O. Lucía, J. M. Burdío, I. Millán, J. Acero, and D. Puyal, "Load-adaptive control algorithm of half-bridge series resonant inverter for domestic induction heating," *IEEE Transactions on Industrial Electronics,* vol. 56, no. 8, pp. 3106-3116, August 2009.

[14] O. Lucía, O. Jiménez, L. A. Barragán, I. Urriza, D. Navarro, and J. M. Burdío, "System-on-programmable-chip-based versatile modulation architecture applied to domestic induction heating," in *35th Annual Conference of the IEEE Industrial Electronics Society IECON09,* 2009.

Variable Frequency Pulse Density Modulation[1] for Efficient High Frequency Operation of Series Resonant Converters Operating as Voltage Regulators

Darryl J. Tschirhart, *Grad. Student Member, IEEE,* and Praveen K. Jain, *Fellow, IEEE*

Centre for Power Electronics Research (ePOWER)
Dept. of Electrical & Computer Engineering, Queen's University
Kingston, Ontario, Canada
darryl.tschirhart@ieee.org, praveen.jain@queensu.ca

Abstract— **Voltage regulators (VRs) powering microprocessors face stringent requirements on size; response; and efficiency. Presently, buck converters are used exclusively as VRs for their simple, well-understood behaviour despite their many shortfalls; while more suitable resonant converters are overlooked due to size and performance issues associated with their control methods. In this paper, variable frequency pulse density modulation (VF-PDM) control of the series resonant converter (SRC) is presented to enable resonant converter benefits to be realized in VR applications. It achieves efficient multi-megahertz switching frequency, ultra-fast transient response with minimal filter capacitance, and is inherently stable. It will be shown that the SRC under VF-PDM can outperform a two phase buck converter in every aspect including size, speed, and efficiency.**

I. INTRODUCTION

The simple, well-understood operation of the buck converter has made it the topology of choice for voltage regulators (VRs) in notebooks, desktops, and servers. To meet the increased power demand, multiple phases are operated in parallel to maintain acceptable loss, and improve the transient performance. The source of the problems associated with the buck converter is its inductive output filter which limits the achievable current slew rate, and hence the ability to respond to load steps. Increasing the switching frequency to reduce the size of the inductor increases switching loss, and therefore has an upper limit. Attempts to increase the response of buck converters through topology modifications have been made to mitigate the inherent flaws of the topology [1],[2]. However, they come at the expense of increased cost, complexity, and size. To meet the demands of modern VRs, a different topology must be adopted; along with a suitable control scheme that does not introduce tradeoffs.

Resonant converters have many advantages over PWM topologies due to their ability to achieve near lossless switching. This enables miniaturization through reduced heat sink requirements, and smaller reactive components through high frequency operation. Of the two types of resonant converters, current-type, like the series resonant converter (SRC) are advantageous in applications with highly dynamic loads because of their capacitive output filter. However, one of the main stumbling blocks in adoption of the SRC is its control. Regulation is lost at light load with variable frequency control [3]. Constant frequency control solves the regulation issue, but has its own problems. If an asymmetric drive train is employed, there are difficulties with gate pulse generation at reduced load; which limits the practical operating frequency [4],[5]. Pulse density modulation (PDM) is another constant frequency technique where a SRC operates for a portion of the PDM period, and is off for the rest of the time. The greatest benefit of this technique is its nearly flat efficiency curve across the entire load range. However, due to the low PDM frequency, it does suffer some serious setbacks such as poor transient performance, and large filter size with high voltage ripple despite high switching frequency [6]. Digital implementations for induction heating or CCFL inverters typically use a lookup table to generate the gating signals [7]-[9]. To reduce power ripple, the off-time is dispersed throughout the PDM period, instead of grouping it all together at the end [8],[9]. However, the output power is limited to discrete values, and the response is still limited by the PDM period; making present PDM implementations impractical in tightly regulated, dynamic systems.

In this work, variable frequency PDM (VF-PDM) is introduced to enable series resonant converters to operate as voltage regulators by achieving the efficiency benefits of standard PDM, while overcoming the demerits of poor transient response and large filter size.

II. CIRCUIT DESCRIPTION

The series resonant converter operating under VF-PDM control is shown in Fig. 1. A hysteretic comparator asserts a high command signal when the output voltage falls below a

---

[1] Canadian and U.S. patents pending
Funding provided by Ontario Centres of Excellence, Ontario Research Funds, and Natural Sciences and Engineering Research Council of Canada

desired value, and sets the signal low when the voltage rises above a specified value. The controller; implemented with a field programmable gate array (FPGA) in this work, ensures a symmetric drive signal is sent to the power train while the command signal is high. If the command signal goes low in the middle of a cycle, the controller ensures proper tank behaviour by completing the cycle at the desired switching frequency.

Representative waveforms are given in Fig. 2; where the converter is on for two different times lasting different durations in the PDM period. The on intervals start and end with zero current transitions, while maintaining zero voltage switching in the middle. Usually ZCS is suboptimal because it still leads to frequency-dependent output capacitance loss. However, with this control method, the frequency at which ZCS occurs is much lower than the switching frequency, so the loss is almost negligible. By allowing the converter to dictate when energy is required, reduction of the low frequency ripple is achieved without sacrificing response.

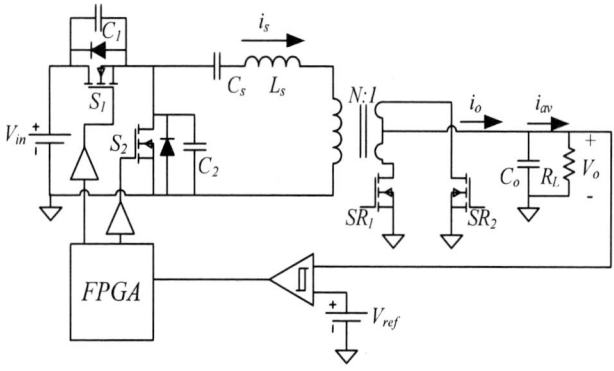

Fig. 1: Schematic of SRC under VF-PDM control

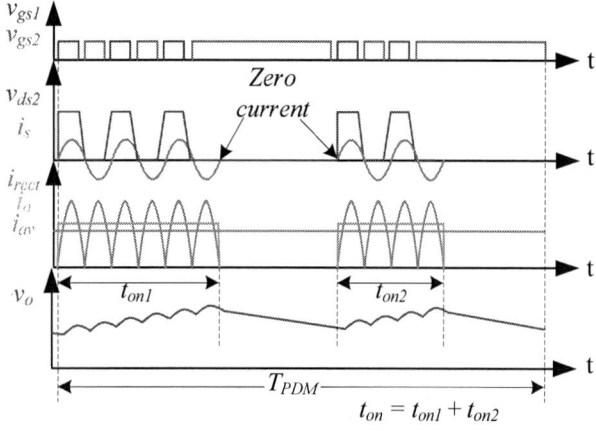

Fig. 2: Sample waveforms of the SRC under VF-PDM control

III. ANALYSIS

To analyze the SRC under this control method, the output currents have to be defined according to Fig. 1. The pulse density duty cycle is defined by (1) as the ratio of the total on-time to the total on- and off-time. This is equal to the ratio of load current i_{av} to the per-cycle average of the rectified resonant tank output current i_o.

$$D_{PDM} = \frac{\sum_k t_{on,k}}{\sum_k t_{on,k} + \sum_k t_{off,k}} = \frac{i_{av}}{i_o} \qquad (1)$$

The equivalent ac resistance is a function of the pulse-density duty cycle, and defined by (2), where $R_{ac0} = \dfrac{8N^2 R_L}{\pi^2}$, and N is the transformer turns ratio. The transfer function is given by (3), with the variables in the equation defined as $Q = \dfrac{\omega_r L}{R_{ac0}}$; $\omega_r = \dfrac{1}{\sqrt{L_s C_s}}$; and

$\omega = \dfrac{\omega_0}{\omega_r}$; where ω_0 is the radian switching frequency. The definitions of the resonant tank parameters Q and ω are identical to those used in standard converter analyses. Thus, setting D_{PDM} to unity will result in identical voltage transfer characteristics.

$$R_{ac} = \frac{8N^2 D_{PDM} R_L}{\pi^2} \qquad (2)$$

$$R_{ac} = D_{PDM} R_{ac0}$$

$$\frac{V_o}{V_{in}} = \frac{D_{PDM}}{2N\left[D_{PDM} + jQ\left(\omega - \dfrac{1}{\omega}\right)\right]} \qquad (3)$$

The voltage stress of the resonant components normalized to V_{in} are given by (4) and (5). Both are inversely proportional to D_{PDM} and are equal to the values under traditional control methods at unity pulse density duty cycle.

$$\frac{V_{Cs}}{V_{in}} = \frac{2Q}{D_{PDM}\pi\left(Q(\omega^2 - 1) + j\omega\right)} \qquad (4)$$

$$\frac{V_{Ls}}{V_{in}} = \frac{j2Q\omega}{D_{PDM}\pi\left(1 + jQ\left(\omega - \dfrac{1}{\omega}\right)\right)} \qquad (5)$$

IV. DESIGN CONSIDERATIONS

To highlight the merits of VF-PDM control, a converter will be designed following the specifications of Table I. The output voltage specifications are chosen to correspond to a 1V, 40A VR with 2mΩ load line.

978-1-4244-4782-4/10 $26.00 © 2010 IEEE

TABLE I: 12V RESONANT VOLTAGE REGULATOR SPECIFICATIONS

Parameter	Value
Input Voltage (V_{in})	12V +/-10%
Output Voltage (V_o)	0.96V +/-40mV
Output Current (i_{av})	40A
Maximum Load Step ($\Delta i_{av,max}$)	27A (40A↔13A)
Switching Frequency (f_0)	5MHz

A. Transfer Characteristics

The voltage transfer characteristics are shown in Fig. 3 for different circuit parameters. It is shown that the gain of the circuit reduces with increased quality factor and relative operating frequency. It is also observed that the influence of Q is reduced in circuits operated close to the resonant frequency. From these curves, it is desirable to have a moderate value of ω and a fairly high Q to increase the range of duty cycle required for regulation.

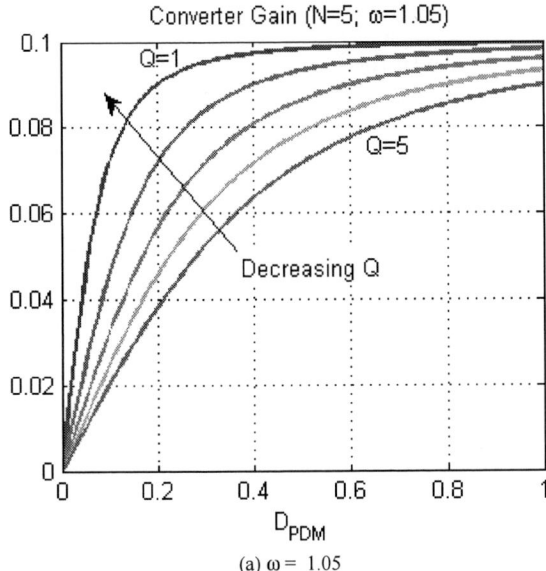

(a) ω = 1.05

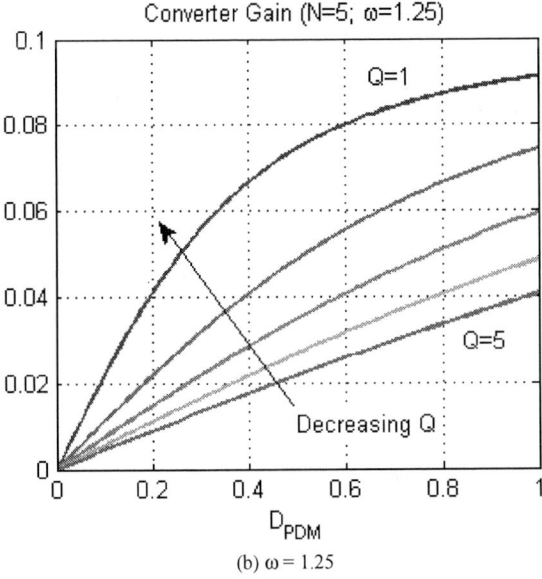

(b) ω = 1.25

Fig. 3: Voltage transfer curves under VF-PDM

B. Component Stress

The component voltage stresses of the resonant tank at full-load are shown in Fig. 4 and Fig. 5. In general, the stress increases with Q, and the peak stress reduces as the operating frequency increases away from the resonant frequency. Therefore, low Q and high ω are desirable for minimizing component stress.

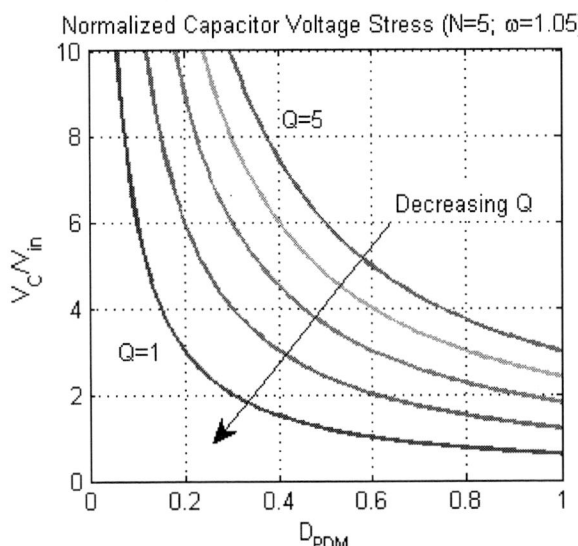

(a) ω = 1.05

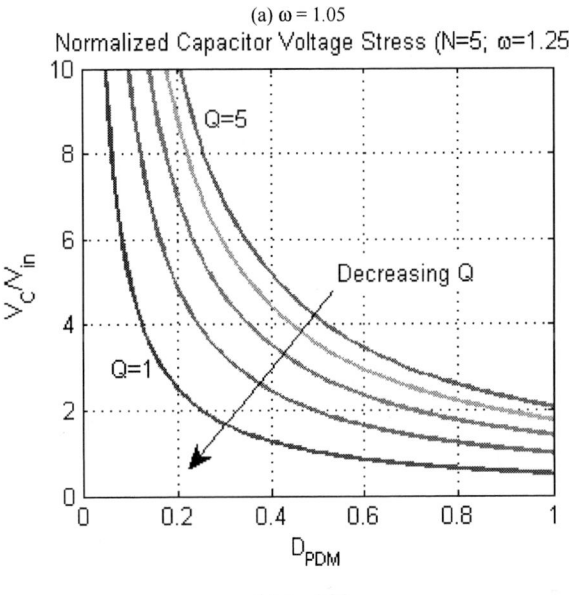

(b) ω = 1.25

Fig. 4: Resonant capacitor voltage stress

978-1-4244-4782-4/10 $26.00 © 2010 IEEE

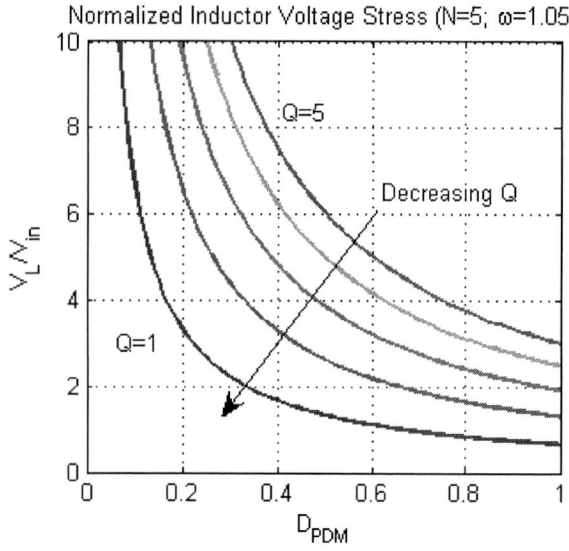

(a) $\omega = 1.05$

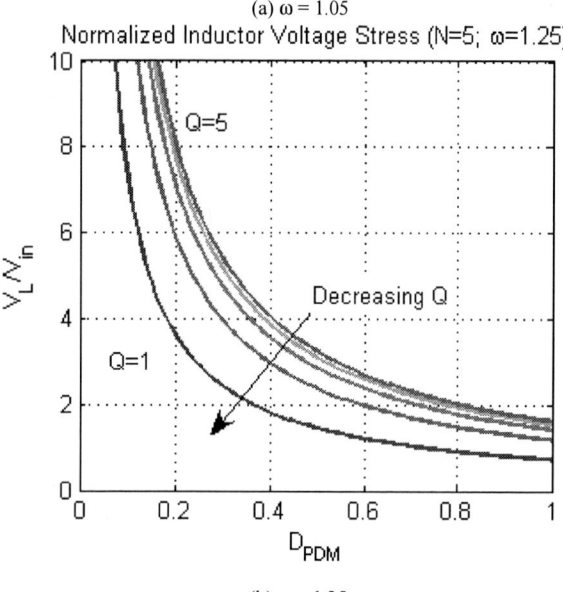

(b) $\omega = 1.25$

Fig. 5: Resonant inductor voltage stress

C. Tank Design

From the above discussions, design of the resonant tank is a trade-off between component stress and transfer capability. However, the voltage stress curves are somewhat misleading; as it appears that the stress tends to infinite at low duty cycle, which would result in extremely oversized components. For a given set of tank parameters, the PDM duty cycle should be selected close to unity under the worst-case operating conditions. That way, the full-load stress is approximately the same as traditionally controlled resonant converters, and when the duty cycle reduces with load, the peak stress remains roughly constant. Thus, component ratings will not increase beyond those for other control methods. Resonant parameters of $\omega = 1.15$ and $Q = 1.8$ are selected for the design. At 5MHz switching frequency, these parameters translate to component values: $C_s = 42nF$ and $L_s = 31nH$.

D. Filter Capacitance

The ideal filter design depends on the worst-case load transient and allowable voltage deviation, as given by (6). In the equation, the required capacitance is determined by a worst-case scenario of the off signal being asserted the instant after a switching cycle has begun. T_s is the switching period; ΔV_o is the allowable voltage deviation; and $\Delta i_{Co} = \frac{i_{av,FL}}{D_{PDM}} - \Delta i_{av,max}$ is the current through the filter capacitor under the maximum load step.

$$C_o = \frac{\Delta i_{Co} T_s}{\Delta V_o} \qquad (6)$$

However, the combination of high operating frequency and high current pushes the limits of present-day capacitor technology. As such, the effect of the equivalent series inductance (ESL) is more pronounced. In general, ESL is a function of the geometry of the capacitor; meaning larger packages will have greater ESL, as will larger capacitor values with the same package designation. With the given specifications, an ideal filter requirement of 180µF is calculated. A number of 22µF capacitors (C_{std}) in 0805 packaging with 1.1nH ESL (l_{std}) are to be used in parallel to form the foundation of the filter. However, at 10MHz, the ripple frequency of the converter, the supplied capacitors behave as inductors. To overcome this, it is necessary to add low-ESL capacitors in parallel with the standard devices to create a 'capacitor cell' with a self-resonant frequency (SRF) that is greater than the ripple frequency. A number of capacitor cells can then be used to form the output filter. The two options for low-ESL capacitors are reverse geometry or multi-terminal caps; with the latter being chosen for their superior reduction of ESL. Using (7), and low-ESL capacitor parameters $C_{lowESL} = 2.2µF$, $l_{lowESL} = 45pH$, the curve of Fig. 6 is produced. Fig. 6 is used to determine the minimum number of low-ESL capacitors required in each cell by finding the point on the curve with a y-coordinate equal to the minimum acceptable SRF value. The x-coordinate of this point or the next highest integer value in the event the point lies between two integers, is the minimum number of low-ESL capacitors required per cell. The number of cells required to at least meet the required filter capacitance value is found with Fig. 7 simply by using the result from Fig. 6 along the x-axis and reading the corresponding y-value. In this design six low-ESL capacitors are required per capacitor cell, and six cells are required for the filter.

$$SRF_{cell} = \frac{1}{2\pi \sqrt{\left(C_{std} + nC_{low-ESL}\right)\left(l_{std} \,//\, \frac{l_{low-ESL}}{n}\right)}} \qquad (7)$$

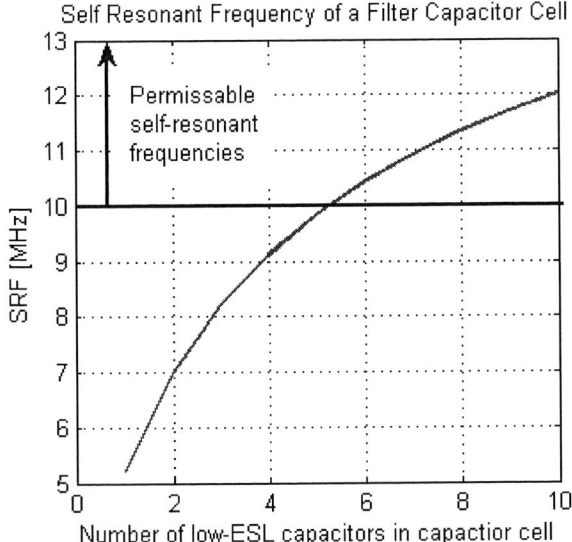

Fig. 6: Self resonant frequency of a filter capacitor cell to find the required number of low-ESL capacitors

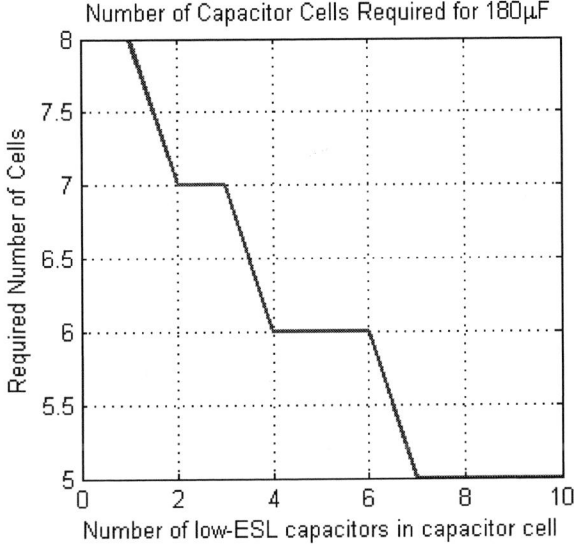

Fig. 7: Required number of capacitor cells to meet filter requirements

V. RESULTS

Simulation results of a series resonant converter fulfilling the specifications of Table I are shown in Fig. 8. A total of 211.2µF of filter capacitance is used to achieve these results; which is only that high due to limitations of current capacitor technology. As such, six low ESL capacitors have to be used in parallel with six 22µF capacitors to push the self resonant frequency of the filter above 10MHz. Improvements in capacitor technology will enable further miniaturization through filter size reduction with this control technique. That said, in the ideal case, a two-phase buck converter (L=300nH) would require 700µF of capacitance to obtain the same results. That's 3.3 times more capacitance than a practical VF-PDM

SRC! In Fig. 9, the semiconductor efficiency results are shown for the simulated converter across the load range; assuming full recovery of the gate energy of the synchronous rectifiers. Soft-switching and pulsed operation maintains efficiency greater than 91% across the load range. Further, the efficiency increases with load reduction, which is the contrary to typical converter behaviour. This can be explained by the graph in Fig. 10 where the duty cycle is shown across the operating range of the converter. At light load, the conduction loss is reduced through natural current reduction and increased off-time of the converter; as well as through the maintenance of soft-switching. While transformer losses, and conduction loss of the traces and resonant and filter capacitors are not accounted for, symmetric constant frequency operation of the converter will minimize these as well.

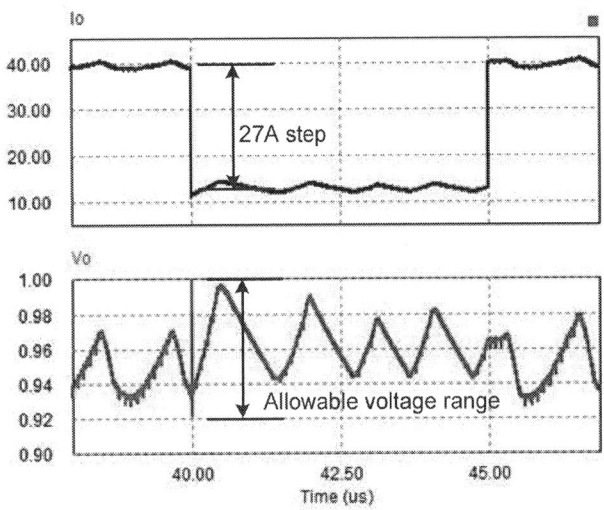
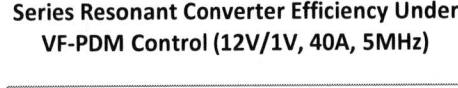

Fig. 8: Simulated transient results of SRC under VF-PDM control

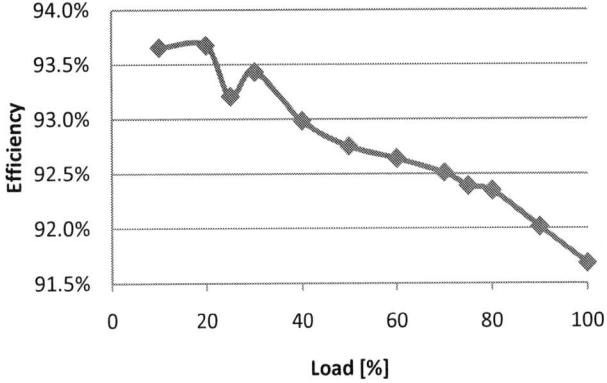

Fig. 9: Semiconductor efficiency results of SRC under VF-PDM control

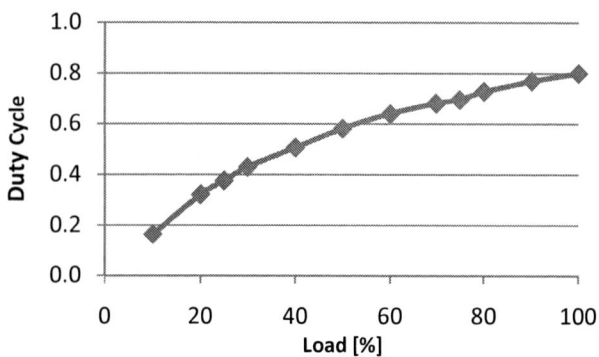

Fig. 10: PDM Duty cycle across the load range at 12V input

An Altera UP3 education board (with EP1C6Q240C8 FPGA) has been programmed to implement VF-PDM control. As illustrated in the experimental results of Fig. 11, the PWM signal (third trace) is active when the command signal (bottom trace) is high; and inactive otherwise. To highlight the speed of the controller, the results of a 1.5MHz command signal is shown in Fig. 12.

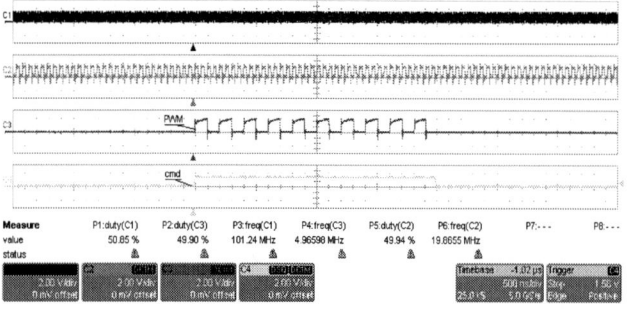

Fig. 11: Open loop results of FPGA programmed to implement VF-PDM

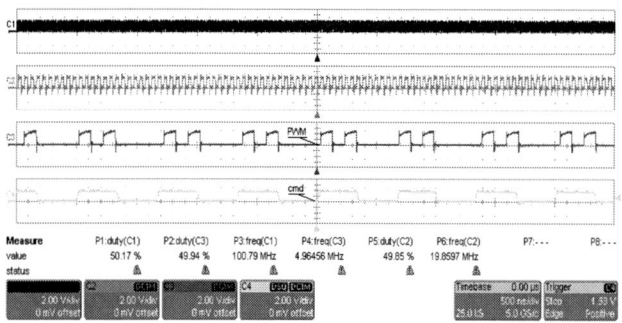

Fig. 12: Controller output with a 1.5MHz command signal

VI. CONCLUSION

In this paper a new form of pulse density modulation was presented to overcome all the negative issues associated with existing resonant converter control methods. By exciting the tank with a constant frequency, symmetric voltage, high efficiency and load regulation is maintained across the load range; component optimization at multi-megahertz switching frequencies is possible; and load transients are almost undetectable. All these merits make this control method suitable for voltage regulator applications; particularly in mobile systems. It was shown with a reference design that a series resonant converter under VF-PDM uses less than one third of the capacitance required by the industry-standard two-phase buck converter.

REFERENCES

[1] D.D-C Lu, J.C.P. Liu, F.N.K. Poon, and B.M.H. Pong, "A Single phase voltage regulator module (VRM) with stepping inductance for fast transient response", *IEEE Trans. Power Elec.*, vol. 22, pp. 417-424, March 2007.

[2] X. Wang, I. Batarseh, S.A. Chickamennahalli, and E. Standford, "VR transient improvement at high slew rate load-active voltage transient voltage compensator", *IEEE Trans. Power Elec.*, vol. 22, pp. 1472-1479, July 2007.

[3] R.L. Steigerwald, "A Comparison of Half-Bridge Resonant Converter Topologies," *IEEE Trans. Power Elec.*, vol. 3, pp. 174-182, April 1988.

[4] P.K. Jain, A. St-Martin, and G. Edwards, "Asymmetrical pulse-width-modulated resonant DC/DC converter topologies," *IEEE Trans. Power Elec.*, vol. 11, pp. 413-422, May 1996.

[5] D.J. Tschirhart, and P.K. Jain, "A CLL resonant asymmetrical pulsewidth-modulated converter with improved efficiency," *IEEE Trans. Industrial Elec.*, vol. 55, pp. 114-122, January 2008.

[6] S. Dalapati, S. Ray, S. Chaudhuri, and C. Chakraborty, "Control of a series resonant converter by pulse density modulation," in *Proc. IEEE India Annual Conf. (INDICON)*, 2004, pp. 601-604

[7] H. Fujita, H. Akagi, K. Sano, K. Mita, and R.H. Leonard, "Pulse density modulation based power control of a 4 kW 400 KHz voltage-source invertor for induction heating applications," in *Proc. Power Conversion Conf.*, 1993, pp. 111-116.

[8] H. Koizumi, K. Kurokawa, and S. Mori, "Analysis of class d inverter with irregular driving patterns," *IEEE Trans. Circuits and Sys.*, vol. 53, pp. 677-687, March 2006.

[9] Y.-H. Liu, S.-C. Wang, Y.-F. Luo, "Digital dimming control of CCFL drive system using pulse density modulation technique," in *Proc. IEEE Region 10 Conf.*, 2007, pp. 1-4.

Flexible-Controlled High Power-Density Automotive HID Electronic Ballast Using Full-Digital Control Mode

Xinyi Yang, Biwen Xu, Chongguang Ma, Min Chen, Zhaoming Qian

College of Electrical Engineering

Zhejiang University

Hangzhou,China

Abstract: A flexible-controlled high power density electronic ballast for automotive high intensity discharge (HID) lamp is proposed. This paper also introduces a decreased power control method for the startup phase of the headlamp. Conventional digital controlled automotive HID electronic ballast needs extra analog IC and realizes the power control by current control means. The proposed one uses one high speed microprocessor to realize full digital control. As a result, it not only has the advantages of low-cost and small-size, but also easily realizes voltage control, current control and power control. Experimental results demonstrate that both the design of hardware stage and software stage are successful.

I. INTRODUCTION

High intensity discharge (HID) lamp has been becoming more and more popular. It has been accepted as a good lighting source for automotive or high-intensity illumination due to its superior performance (energy saving, long life time, high luminous efficacy and good color rendering) over conventional halogen system [1] [2]. However, automotive HID lamp has very complicated characteristics, it has different control phases from ignite to steady state, so the design of a low cost, high efficiency, compact size and high reliability automotive HID ballast is not easy to realize. Due to the automotive HID lamp's complicated characteristics, control circuit built with the analog platform is of course cumbersome resulting in high cost, unstable and inflexible.

Fortunately, digital control, such as DSP and microprocessor, has brought great convenience for control stage design. For low cost HID electronic ballast development, the most powerful digital IC doesn't mean it is the best one. DSP has the highest capacity of data processing which can help to improve the performance of HID electronic ballast, however, due to its high price, it is not appropriate for developing a low cost HID ballast. Microprocessor is a good choice and has been widely used in automotive HID electronic ballast recently. Microprocessor has many advantages over the analog controller, such as cheap, easy to use, robust, and owning a certain extent flexibility. However, microprocessor's clock rate is not enough fast and the data width is not helpful enough in the application of power electronics system. Usually, HID ballast based microprocessor has to rely on additional PWM control chips to accomplish the PWM switching function and an interface a digital-to-analog(DAC) is needed, such kind of control stage can be called a mixture control stage which is the combination of digital control and analog control. If there is an appropriate microprocessor, the control system still has space to be optimized. For instance, taking off the additional PWM control chip, the DAC and external oscillator, the cost doesn't increase too much. The digital control IC named DSPIC 33F series made by Microchip Company is just the one suitable for HID electronic ballast applications. The MCU contains all the following features: 1: Internal oscillator up to 40MIPS; 2: 16-Bit Wide Data Path; 3: 24-Bit Wide

978-1-4244-4782-4/10 $26.00 © 2010 IEEE

Instructions; 4: DSP engine; 5:High-Speed PWM Module; 6:High-Speed 10-Bit ADC up to 2MIPS; 7.High-speed comparator and an associated 10-bit DAC that provides a programmable reference voltage to the inverting input of the comparator. All the features of the microprocessor list above make it very possible to develop a full digital controlled automotive HID electronic ballast with simplest control topology, high control flexibility, low cost and high reliability. The object of the paper is to develop such kind of ballast and present the successful design of the full digital controlled automotive HID electronic ballast.

II. HARDWARE STAGE

Compared to conventional mixed digital controlled automotive HID electronic ballast shown as Fig.1, the full digital controlled automotive HID electronic ballast shown as Fig.2 omit the additional PWM control chip, the DAC and external oscillator, it can easily realize complicated control phase in mathematical calculation way. An automotive HID electronic ballast for operating the headlamp is adopted and is shown in Fig.3, the ballast consists of an input reversed polarity protection circuit, an EMI filter, a DC/DC flyback converter[3], a DC/AC inverter, a full digital control microprocessor, MOSFET driver circuit and an igniter. Because the DC/DC converter responses slowly, the external current path is also needed [4].

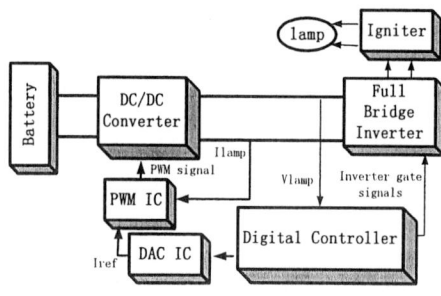

Fig.1.Conventional digital controlled ballast

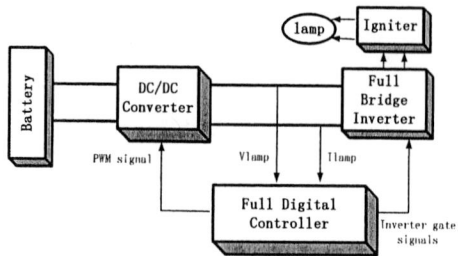

Fig.2.Proposed full digital controlled ballast

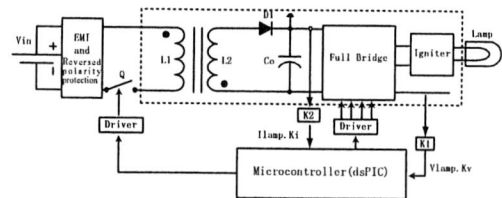

Fig.3. Schematic diagram of proposed ballast

Active low pass filter: The measurements of lamp current and voltage are the key control factors, the accuracy of the feedback is so important that the right result should be guaranteed . The high frequency switch will bring high frequency noise to the current and voltage feedback signals, if the noise is fed into AD pins, the control reference signals will be a terrible mistake, so a effective solution to avoid these noises is needed. Here, a Butterworth low pass filter is adopted as shown in Fig.4.

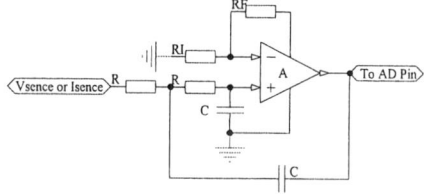

Fig.4. Active low pass filter for sense

The gain is defined by the following formula, W0=1/(R×C).

$$A_v(j\omega) = \frac{V_o(j\omega)}{V_i(j\omega)} = \frac{3-\sqrt{2}}{(\frac{j\omega}{\omega_0})^2 + \sqrt{2}\frac{j\omega}{\omega_0} + 1}$$

Ignited Circuit: The proposed automotive HID lamp ballast adopt an igniter shown in Fig.6 driven by a dual-frequency inverter [5]. Compared to conventional igniter with a voltage doubler shown in Fig.5, it has two main advantages: one is a large

978-1-4244-4782-4/10 $26.00 © 2010 IEEE   1341

igniter capacitor C1 which can be replaced by a much more smaller one, the other one is the maximum igniter voltage can be reached that improves the ignite success rate, for the peak igniter voltage is the sum of DC bus voltage and the igniter voltage.

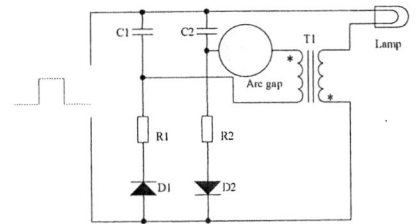

Fig.5.Voltage doubler igniter

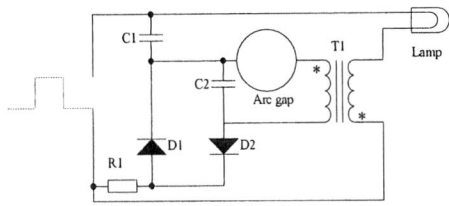

Fig.6.Igniter driven by dual-frequency

Before successful ignition, the ignition frequency is determined by formula:

$$f_{ignition} = \frac{1}{2 \times T_{c1\_charge}}$$

Where: Tc1_charge=5×R1×C1, once the lamp has been successful ignited, the frequency of full bridge is totally determined by the lamp state.

III. CONTROL SYSTRM

Complicated characteristic of automotive head lamp requires complicated startup to steady state control [6][7][8][9][10],the control ideas can be shown in Fig.7.

DC/DC converter is the main part for the operation of HID lamp, in order to minimize the size of the inductor, the switch frequency for DC/DC convert has been chosen at 200KHz.The DC/DC converter will be controlled at different control ways by the state of HID lamp voltage during operation.

For the successful ignite of the HID lamp, the DC/DC converter need to boost the output voltage to reach the open circuit output voltage (380V~400V).

At this phase, the ballast needs to be operated at constant voltage control mode. Right after successful ignite, sufficient current to transit the glow discharge state to arc discharge state is needed. So the constant voltage control mode should be replaced by current control mode. There is a problem that once the lamp ignited, the DC/DC converter could not response quickly enough, the capacitor in external current path will take the responsibility of DC/DC converter during a very short time (usually 300us, called take over phase). As soon as the converter start to response, dc current operation last a certain time (called warm up phase) is very important for it is a effective way to avoid lamp extinguish during startup.

After current control, 75W to 35W power is needed to meet the requirement of startup, when the lamp voltage enter into its rated voltage range (68~102), the power supplied to lamp should be maintained at 35W. During these phase, the DC/DC convert should be regulated at power control mode consists of decreased power control and constant power control. To prevent the acoustic resonance of HID lamp, the lamp has to be driven by an ac current to make sure the consumption of each electrode is equal. This task is accomplished by the full bridge inverter operated at the frequency about 200Hz [11]. The control flow chart of the whole control system is shown in Fig.8.

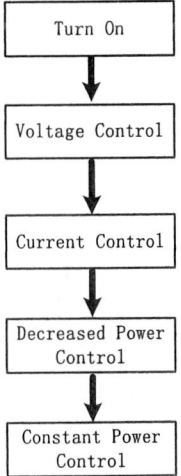

Fig.7.Shematic control diagram

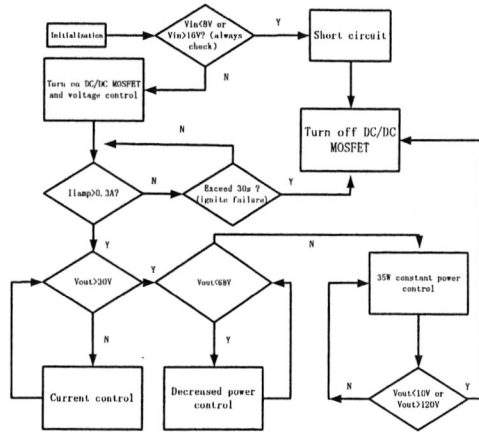

Fig.8. Control flow chart of whole system

IV. SOFTWARE STAGE

In the conventional digital control electronic ballast, realizing such kind of complicated control phase is not easy. Both the current control and power control are controlled by means of setting the current reference of the extern PWM IC. The reference current is the output of DAC IC, which will result in a certain delay in control loop, and the reference current usually is the operation result of table inquiring, it is very inconvenient and limited to the data space. The proposed ballast will take full use of the MCU's high performance, all the control phases are controlled by the high speed CPU core of the MCU using mathematics algorithm, resulting in simplest control circuit with least component, no table inquiring and no data space limitation, all the control modes will be implemented in the most direct way. Full digital control ideas for each control mode are detailed as follows: the DC/DC PWM duty cycle is generated by mathematics algorithm, and sent to MOSFET from MCU directly. Gv, Gi, Gp blocks is the digital error compensator, Ko is the correction term. Kv, Ki are the proportionality constants for lamp voltage and current sense respectively

Constant Voltage Control: The control schematic diagram is shown as Fig.9. Vref is the constant voltage control reference, Vset is the set ignition voltage for igniter. Vmeasured is the measured results by MCU. The MCU uses its high performance CPU to call the voltage control code and calculate the

respectively PWM duty cycle based on the Vref and Vmeasured.

Vref=Vset×Kv

Vmeasured=Vlamp×Kv

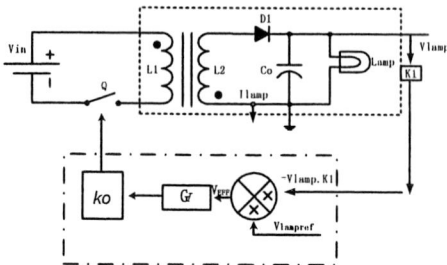

Fig.9. Schematic diagram of voltage control

Constant Current Control: The control schematic diagram is shown as Fig.10.Iset is the set current value. Imearsured is the measured results by MCU. The MCU uses its high performance CPU to call the current control code and calculate the respectively PWM duty cycle based on the Iref and Imeasured.

Iref=Iset×Ki

Imeasured=Ilamp×Ki

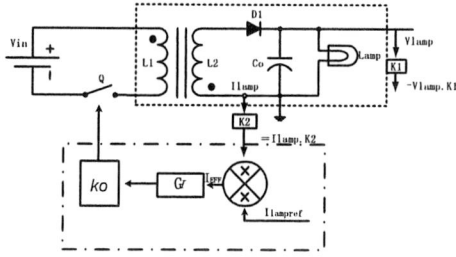

Fig.10. Schematic diagram of current control

Decreased Power Control and Constant Power Control: The control schematic diagram is shown as Fig.11, Pset is set power control reference, Pmeasured is the measured results by MCU. Pset is totally determined by the lamp voltage. The MCU uses its high performance CPU to call the power control code and calculate the respectively PWM duty cycle based on the Pref and Pmeasured. Here, a unique lamp voltage determines a unique Pset without table inquiring way during startup phase, the control valve is set for decreased power control is shown as Fig.12.This method brings high accuracy power control for lamp's decreased power control and

978-1-4244-4782-4/10 $26.00 © 2010 IEEE   1343

constant power control, and effectively optimized the startup procedure.

Pref= Pset×Kv ×Ki

Pmeasured= Plamp×Kv ×Ki

Plamp= Vlamp×Iamp

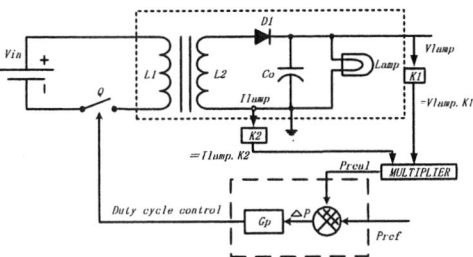

Fig.11. Schematic diagram of power control

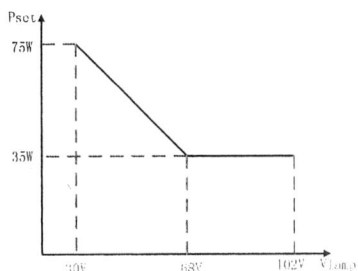

Fig.12. Schematic diagram of Vlamp and Pset

V.   EXPERIMENTAL RESULTS

For the verification, a 35W automotive HID electronic ballast was implemented. Fig.13 (a) and (b) show the prototype of the ballast based on full digital control and magnetics planer core. Fig.14 shows the igniter pulse voltage (28KV) which can meet the requirement of both cold start and hot start. Fig.15 shows the drive signals and Vds (peak Vds is around 30V, where Vin is 13.5V) of DC/DC MOSFET.Fig.16 shows the decreased power control procedure, the power controlled from high state to constant state falls down smoothly which result in no flicker during startup. Fig.17 shows the lamp current and voltage during 35W constant power control.

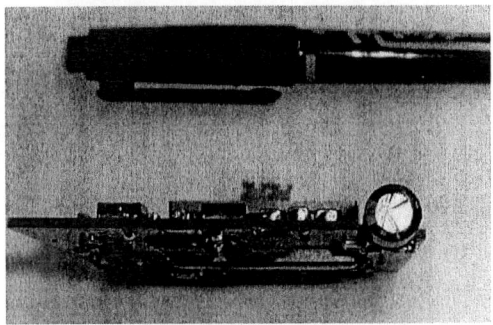

Fig.13. (a) (b) Photograph of the full digital controlled automotive HID electronic ballast

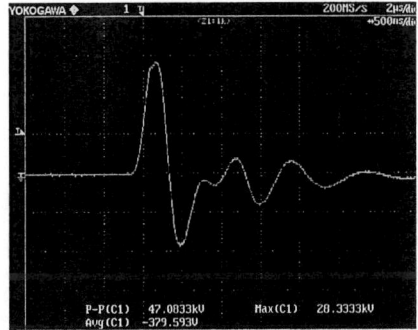

Fig.14 .Voltage pulse of igniter

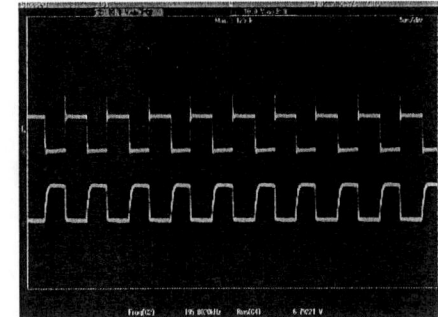

Fig.15.Vgs(down,10V/DIV)and Vds(up,20V/DIV)

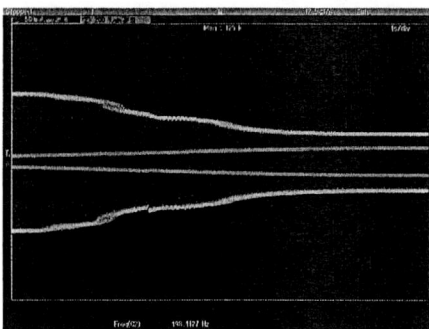

Fig.16.Vlamp(inside,200V/DIV)and Ilamp

outside,0.5A/DIV)during power control

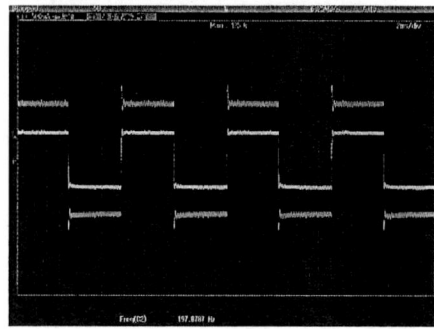

Fig.17.Vlamp(outside,50V/DIV) and

Ilamp(indide,0.5A/DIV)during stable state

VI. CONCLUSION

A full digital controlled automotive HID headlamp electronic ballast and a control method for both decreased and constant power control have been proposed. The prototype has compact size, low cost, high control flexibility and reliability. Due to appropriate design of magnetics planer transfer, large leakage inductances and high voltage spikes on power switch have been avoided. The experiment results successfully validate the development of the prototype and the control idea.

REFERENCE

[1] Wang, D.H.; Cheng, K.W.E.; Divakar, B.P.; Dong, P.; Chan, W.W.; Xue, X.D.; Ding, K.; Che, Y.B.; Xu, C.D.; "Power Electronics Systems and Applications",2006.ICPESA '06. 2nd International Conference on 12-14 Nov. 2006 Page(s):229 - 233.

[2] Kyu-Chan Lee; Bo Hyung Cho; "Design and analysis of automotive high intensity discharge lamp ballast using micro controller unit", Power Electronics, IEEE Transactions on Volume 18, Issue 6, Nov. 2003 Page(s):1356 – 1364.

[3] H.-J.Hans-Juergen Faehnrich and E.Erhard Rasch, "Electronic ballasts for metal halide lamps," J. Illum. Eng. Soc., pp. 131–140, Summer 1988.

[4] [H. Konishi, T. Nakamura, T. Kambara, M. Kotani, and T. Tanaka, "Ballast for a discharge lamp",US Patent 6693393, Feb. 17, 2004.

[5] Hu, Yuequan; Jovanovic, Milan M. "High-Intensity-Discharge Lamp Ballast With Igniter Driven by Dual-Frequency Inverter", Applied Power Electronics Conference, APEC 2007 - Twenty Second Annual IEEE Feb. 25 2007-March 1 2007 Page(s):268 – 273.

[6] D. Wolfgang, and W. F. Karl, "Ballast for starting and operating high pressure discharge lamps",ES2077953T, Dec. 01, 1995.

[7] Microchip Technology inc., "dsPIC33FJ06GS 101/X02 and dsPIC33FJ 16GS X02/X04 Data Sheet", 2009 Microchip Technology inc.

[8] J. Osterried, A. Huber, A. Veser, and F. Hansmann,"Ignitional run circuit that immediately applies only a DC voltage after lamp ignition but before the main AC potential is applied", United States Patent 5770924, June 23, 1998.

[9] J. Hollander, "Method and apparatus for igniting a gas discharge lamp," United States Patent 6172468, Jan. 09, 2001.

[10] K. Toyama, K. Kato, and K. Aida,"Discharge lamp lighting device",United States Patent 6291945, Sept. 18, 2001.

[11] R.Ron Fiorello, "Lamp Ignitor Circuit", Tech. Rep., 1997.

A "Class-A2" Ultra-Low-Loss Magnetic Ballast for T5 Fluorescent Lamps

S.Y.R. Hui, *Fellow IEEE*, D.Y. Lin[*], W.M. Ng and W. Yan, *Member IEEE*

Centre for Power Electronics
City University of Hong Kong
Hong Kong, China
Email: eeronhui@cityu.edu.hk

*School of Physics & Information Engineering
Jianghan University
Wuhan, Hubei, China

Abstract—**This paper shows that magnetic ballasts for high-voltage T5 lamps are not only feasible but their luminous and ballast-loss performance can be better than those of electronic ballasts. A computer-aided design and practical implementation of an ultra-low-loss magnetic ballast system for T5 28W lamps is presented. The high-voltage & low-current features of T5 lamps imply that the copper and core losses of the magnetic ballast can be greatly reduced. Its total system power is 30W (with a ballast loss of only 2.42W) is less than the 32W upper limit specified for Class A2 of energy-efficient electronic ballast for T5 28W lamps. High luminous efficacy of 74.8 lumen/W can be achieved. This important breakthrough has the potential of reversing the existing trend of using electronic ballasts as the energy-saving technology in lighting industry. With a better luminous efficacy, lower product and maintenance costs, much longer lifetime and the use of recyclable metallic materials over its electronic counterparts, this patent-pending proposal provides a truly sustainable lighting solution to the lighting industry.**

I. INTRODUCTION

T5 fluorescent lamps have been increasingly used in both office and industrial applications. Compared with T8 lamps, T5 lamps have smaller diameters and higher luminance. Moreover, some T5 lamps have an improved phosphor coating that prevents mercury from being absorbed into the phosphor and the bulb glass. T5 lamps can, therefore, have less impact on the global environment than T8 lamps [1]. Unlike LED tubes which still have the binning problems (i.e. the color temperature of LED tubes varies from one manufacturer to another and a lack of common color standard), T5 lamps have common color temperature standard and so users do not have to worry about future color mismatch problem in lamp replacement. However, T5 lamps with small tube diameters require a higher lamp voltage to sustain the lamp current. Therefore, for power higher than 28W, it has been traditionally accepted in lighting industry that only electronic ballasts, which can generate a high voltage using resonant circuit, can be used to operate the T5 lamps. Electronic ballasts, limited by the short lifetime of the electrolytic capacitors, will eventually become electronic waste and thus resulting in long-term pollution of soil and water.

Environmental study [2,3] on the amount of PBDE, which is the flame-retardant chemical used in printed-circuit-boards in South China has confirmed that both fresh-water and sea-water fish in this region has high level of PBDE several times higher than fish in other countries.

The sustainable lighting technology should satisfy the criteria of (1) energy saving, (2) long lifetime, (3) recyclability. Recyclability inherently implies reduction of electronic waste and further mining that unavoidably cause damage to the environment. Comparison of magnetic ballasts and electronic ballasts are reported in [4,5,6]. It should be noted that dimmable magnetic ballasts are possible [5-8]. This paper presents a novel T5 electromagnetic ballast that meets the three criteria for sustainable lighting technology. Despite the fact that high-frequency operation can produce about 10% extra light from florescent lamps, experimental results have confirmed that the ultra-low-loss T5 magnetic ballast system has higher system efficiency and efficacy than their electronic counterparts for T5 lamp. Detailed experimental and simulation results are included to confirm the theory.

II. ULTRA LOW LOSS FEATURES OF MAGNETIC BALLASTS FOR T5 LAMPS

T5 lamps were originally designed to be driven by electronic ballasts which can use the resonant tanks to generate high ignition voltage. For T5 28W and 35W, the on-state lamp voltages at high-frequency operation are 167V and 209V respectively. These high voltage levels are close to the mains voltage of 220V-240V. Traditionally magnetic ballasts were thought to be not suitable for driving high voltage lamps such as T5 lamps. The technical challenges for developing magnetic ballasts that can outperform electronic ballasts for T5 lamps are:

(i)   Sufficient ignition voltage.

(ii)   End of life detection for aged or faulty lamps.

(iii) Provision of high lamp voltage to sustain the lamp arc after lamp ignition

(iv)   Less ballast loss than the electronic counterparts.

Center for Power Electronics, City University of Hong Kong.

The requirements (i) and (ii) can be met by using electronic starters [9-13]. Electronic starters developed in late 1990's for T8 lamps can be applied to T5 lamps in general. End-of-life detection has been a common feature among some electronic starters [9,11]. Since T5 lamps have on-state voltage close to the mains voltage, series inductive-capacitive (LC) ballasts (Fig.1) previously suggested for high-voltage lamps can be used [9,10]. As the voltage vector of the capacitor is opposite to that of an inductor, the voltage drop across the inductor can be partially or totally cancelled by the voltage vector of the capacitor. Thus, requirement (iii) can be met with a LC ballast.

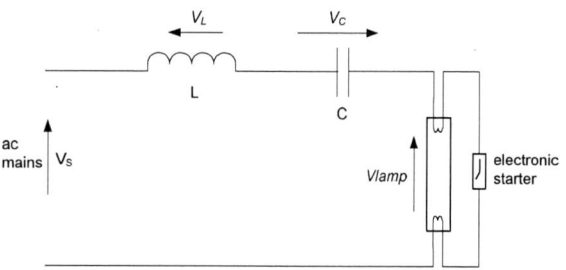

Fig.1 Circuit diagram of the magnetic ballast (LC ballast) for T5 28W lamps

Since T8 36W lamps are being replaced by T5 28W lamps, it is meaningful to use them for comparison. Table 1 contains a comparison of typical manufacturers' data for T5 and T8 lamps. It can be seen that high-voltage T5 lamps have high on-state voltage and low on-state current when compared with T8 lamps of similar power. For magnetic ballasts, the power losses include the conduction loss and core loss. Since conduction loss is proportionally to the square of the current, the low-current feature of T5 lamps enables huge reduction of the conduction loss.

Table 1   Initial assessment of loss components

Lamp Type	T8 36W	T5 28W
Rated voltage (Vrms)	103	167
Rated current (Arms)	0.44	0.175
Conduction loss (i^2R)	100%	16%
Core loss ($\propto I$ or ϕ)	100%	40%

In Table 1, the conduction loss of a T8 magnetic ballast is used as a reference (100%). Assuming that the winding resistance of the magnetic ballasts for T5 and T8 lamps are identical, the conduction loss of the T5 magnetic ballast is only 16% that of T8 magnetic ballast. This is an 84% reduction in conduction loss. The core loss is proportional to the magnetic flux, which in turn is proportional to the current in the magnetic ballast. In this regard, the core loss of a T5 28W magnetic ballast is only 40% that of a T8 36W ballast, resulting in 60% reduction in core loss. Based on this theoretical assessment, significant reduction in both

conduction and core losses can be achieved in magnetic ballasts for T5 lamps. Therefore, it is worthwhile to practically evaluate the energy saving potential of magnetic ballast for T5 lamps, particularly knowing that magnetic ballasts can last for tens of years and can be recycled without creating toxic and non-biodegradable electronic waste. The information in Table 1 provides the ground for developing magnetic ballasts that are more efficient than electronic ones in order to meet the requirement (iv).

III.   CAD DESIGN OF LC BALLAST FOR T5 28W LAMPS

A.   Physical discharge lamp model

Fig.1 shows the circuit diagram of the ballast system. It consists of an inductor (choke), a capacitor and an electronic starter with end-of-life detection function. The steady-state circuit equation for the circuit in Fig.1 can be expressed as:

$$V_s = I_s\left(r + j\omega L - 1/j\omega C\right) + V_{lamp} \qquad (1)$$

where V_s is the ac input voltage, I_s is the input current, r is the winding resistance of the inductor L, C is the capacitance and V_{lamp} is the total voltage across the lamp. For T5 high-efficient lamps, the rated current is specified at 175mA. From (1), the lamp current equation is:

$$I_s = \frac{\left(V_s - V_{lamp}\right)}{\left(r + j\omega L - 1/j\omega C\right)} \qquad (2)$$

Since V_{lamp} is a highly non-linear function when the lamp is operated at mains frequency, it is appropriate to use an accurate lamp model.

The design of the magnetic ballast for high-voltage lamps is based on the inductive-capacitive ballast topology previously proposed for high-voltage fluorescent lamps [9,10]. To precisely determine the LC ballast parameters for a T5 28W lamp, the physical discharge lamp model reported in [14,15] is adopted for this study. This lamp model has the unique feature of including the electron density as a variable. This allows the model to predict the "turn-on voltage spike" of the discharge lamps. For high-voltage lamps with lamp voltage close to the mains voltage, such as T5 28W lamps, the presence of such voltage spike is an indication that sufficient voltage is available to sustain the lamp arc in each half cycle of the ac mains.

The lamp model equations are listed as follows:

$$\frac{dT_e}{dt} = a_1\left(i^2R - P_{rad} - P_{con}\right)/n_e \qquad (3)$$

$$P_{rad} = a_2 n_e \exp\left(-ea_3/kT_e\right) \qquad (4)$$

$$P_{con} = a_4\left(T_e - T_g\right)*n_e \qquad (5)$$

$$\frac{dn_e}{dt} = n_e\left(v_i - v_{diff}\right) \qquad (6)$$

$$R = \frac{l}{n_e e \mu_e s} \tag{7}$$

$$v_i = a_5 \exp(-ea_6/2kT) + a_7 E \tag{8}$$

$$V(t) = a_8 L \frac{di}{dt} + ri + Ri + V_{ele} \tag{9}$$

$$v_{diff} = (2.4/R_a)^2 D_a \tag{10}$$

$$\mu_e = \frac{6701960}{T_e} \tag{11}$$

$$E = i/(n_e e \mu_e s) \tag{12}$$

$$D_a = \mu_i k(T_e + T_g)/e \tag{13}$$

$$\mu_i = a_9 * 0.14 \tag{14}$$

$$T_g = a_{10} i^2 R + T_0 \tag{15}$$

$$\frac{dV_c}{dt} = \frac{i}{C} \tag{16}$$

where T_e is the electron temperature, i is the lamp current, R is the lamp resistance, P_{rad} is the radiation loss in the lamp, P_{con} is the thermal conduction loss in the lamp, n_e is the electron density, k is the Boltzmann constant, e is the charge on an electron, v_i is the ionization frequency of every electron, $V(t)$ is the power supply voltage, L is the ballast inductance, r is the ballast resistance, V_{ele} is the electrode voltage drop in the lamp, v_{diff} is the diffusion frequency of every electron, μ_e is the electron mobility, μ_i is the ion mobility, T_g is the gas temperature, T_0 is the lamp wall temperature, C is the ballast capacitance and V_c is the voltage across the capacitor in the LC ballast.

The parameters for the T5 28W lamp can be obtained by feeding the measured lamp voltage and current data into the model equations in an iterative manner. With the help of evolutionary algorithm such as Genetic Algorithm (GA), the error function can be minimized so that the parameters a_1 to a_{10} of each time of lamp can be obtained. Details of this physical lamp modeling technique can be found in [14,15].

Based on this method, the parameters for T5-28W lamps are shown in table 2:

Table2: the parameters for T5-28W lamps

a1	a2	a3	a4	a5
3.106e22	6.601e-12	8.23	7.4e-19	3.173e10
a6	a7	a8	a9	a10
23.5	1.64	1.5e15	1.03	114.5

B. Computer-aided Studies

To operate the T5 lamp with electro-magnetic ballast, there are two possible circuit topologies: conventional inductor ballast circuit shown in Fig. 2, and an inductive-capacitive (LC) ballast shown in Fig. 3.

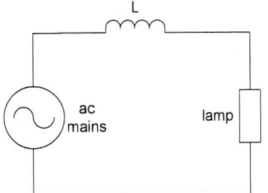

Fig.2 An inductor based magnetic ballast circuit

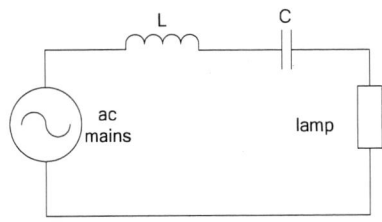

Fig.3 An inductive-capacitive ballast

(i)Simulation of the inductive ballast driven system

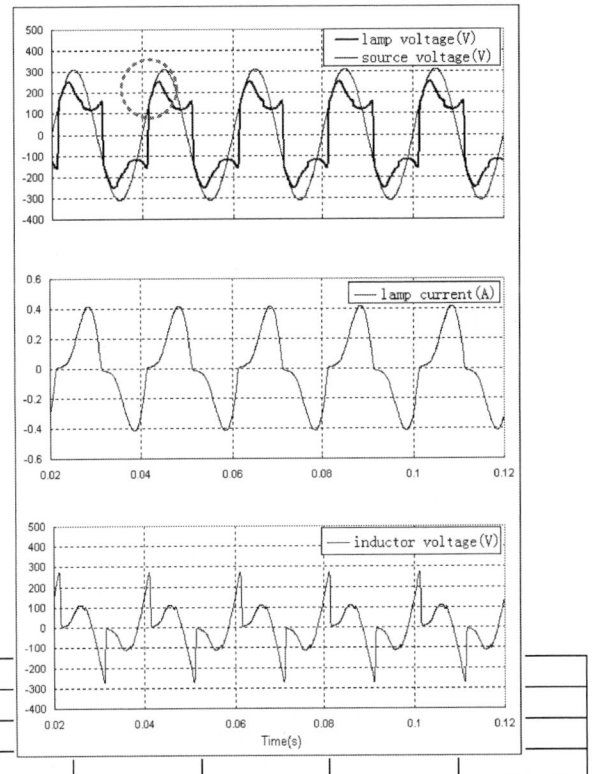

Fig. 4 Simulation results of the Inductor Ballast for a T5 28W lamp: source voltage (top thin curve) and lamp voltage (top thick curve); lamp current (middle); inductor voltage (bottom).

Based on the physical discharge lamp model, simulation of a T5 28W lamp using the inductor ballast has been conducted. V_s=230V, L = 1.36H and r (the resistor of the inductor) = 49.1Ω for the circuit shown in Fig.2. The simulation waveforms for lamp voltage, lamp current and the source voltage are shown in Fig. 4. It is important to note in the dotted circle that the sharp ignition voltage spike does not exist in the simulated voltage waveform of the T5 lamp. The absence of this sharp voltage spike usually means that the lamp arc cannot be sustained. This has been proven in our experiment that this magnetic ballast cannot drive the T5 28W lamp satisfactorily. The lamp arc distinguishes soon after initial ignition.

The inductor ballast is not suitable for high voltage lamps with on-state voltage close to ac mains voltage. This circuit topology is not suitable for T5 28W lamps which has typical on-state lamp voltage of 167V (high-frequency) and 180V (mains frequency) that is close to the mains voltage of 230V. It is important to note that the lamp voltage is limited by the ac mains voltage. In practice, it means that the lamp may not be ignited because the voltage available to the lamp arc may not be sufficient.

(ii) Simulation of the LC ballast driven system

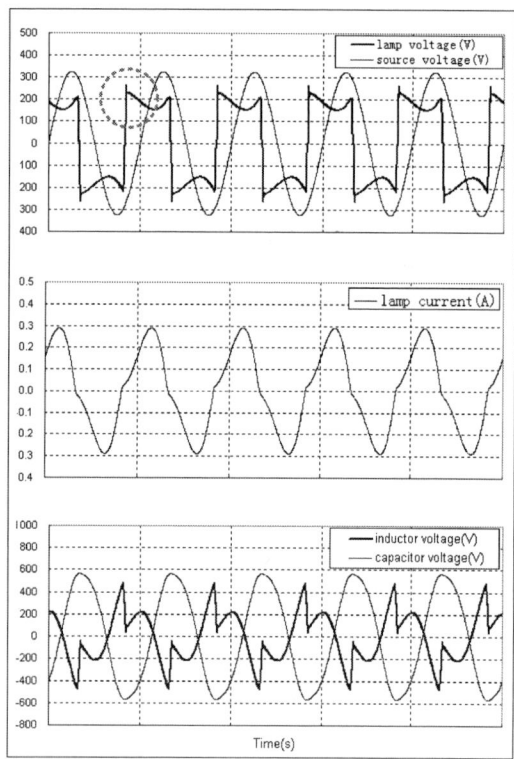

Fig. 5 Simulation results of the LC Ballast for a T5 28W lamp: source voltage (top thin curve) and lamp voltage (top thick curve); lamp current (middle curve); capacitor voltage (bottom thin curve) and inductor voltage (bottom thick curve)

The LC ballast driven system in Fig.3 has been simulated with: V_s=230V, L = 3.2H, C=1.45uF and r (the resistor of the inductor) = 49.1Ω. The simulation results are included in Fig.5. It is important to note that the high ignition voltage

spike can be observed in the dotted circle of the lamp voltage waveform in each half cycle. The circuit takes the advantage of the opposite phase of the inductor voltage and the capacitor voltage. The inductor voltage and capacitor voltage are found to be out of phase. The use of the capacitor allows the voltage drop across the inductor to be reduced so that more voltage will be available for the lamp arc. This important feature can be confirmed by both simulation and practical measurements.

IV. EXPERIMENTAL VERIFICATION

The inductor-capacitor magnetic ballast has been applied to a 28W T5 lamp. The experimental results for full power operation at input voltage of 230V are shown in Fig.6 and Fig.7. It is important to note from Fig.7 that the lamp ignition voltage spike occurs at every half cycle as predicted in the simulated lamp voltage in Fig.5. Stable lamp operation has been observed at V_s=230V.

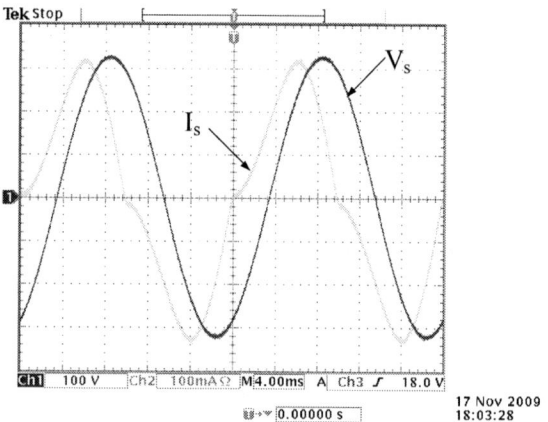

Fig.6 Measured input voltage Vs & input current I_s under full power operation at V_s=230V.

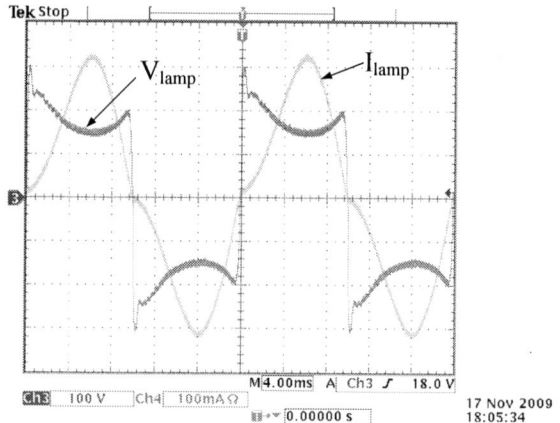

Fig.7 Measured lamp voltage V_{lamp} & lamp current I_{lamp} under full power operation at V_s=230V.

Experiments are then repeated to operate the LC ballast down to Vs=180V in order to evaluate its performance under dimming operation. The corresponding measured waveforms are shown in Fig.8 and Fig.9. The lamp voltage at 180V operation remains almost the same as that 230V operation, but the lamp current is reduced at a Vs of 180V. However, stable lamp operation has also been observed under this dimming condition.

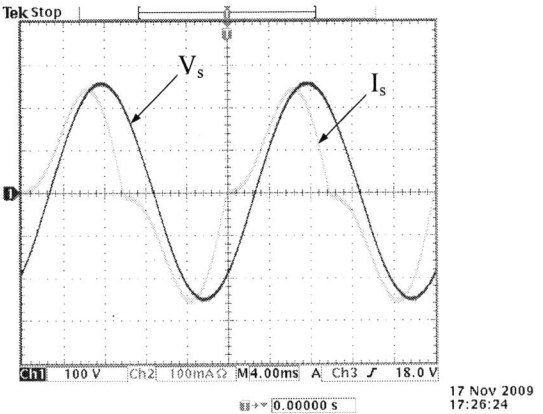

Fig.8 Measured input voltage V_s & input current I_s under reduced power operation (20W) at V_s=180V.

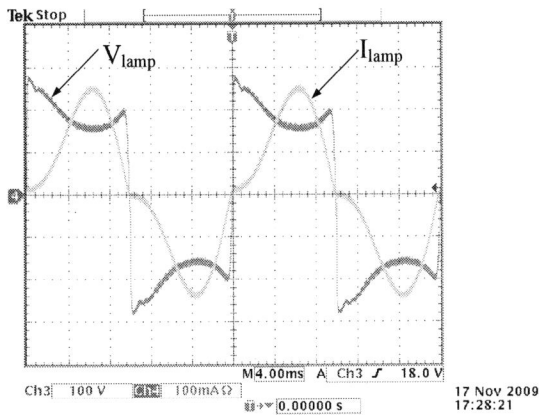

Fig.9 Measured lamp voltage V_{lamp} & lamp current I_{lamp} under reduced power operation (20W) at V_s=180V.

The electrical and luminous performance of the ultra-low-loss LC ballast has also been compared with electronic ballasts commercially available in the market. Using the same Philips TL5 28W/865 lamp, the LC ballast and two electronic ballasts are tested and compared. All tests are carried out in a photocolorimeter with an integrating sphere. Table 3 includes the measurements obtianied when the three systems are tested at V_s=230V using the same lamp. While the input power values of the three systems are almost the same, it is noted

that the ballast loss of the LC ballast (2.42W) is much smaller than those of the two electronic ballasts (4.65W and 3.28W). This practical result confirms the inherent low loss feature of using magnetic ballast for high-voltage and low-current lamps because the "low-current" feature of the lamp enables low conduction and core losses in the magnetic choke. Consequently, the energy efficiency (92.2%) and luminous efficacy (74.76 Lumen/Watt) of the LC ballast are higher than those of the other two electronic counterparts, despite that fact that high-frequency operation of the lamp would yield a higher luminous output.

V. CONCLUSION

Explanations, simulation and practical verification are provided in this paper to confirm that LC ballast can match or even outperform electronic ballast in terms of energy efficiency and luminous efficacy for high-voltage and low-current lamps such as T5 28W fluorescent lamps. Based on a physical discharge lamp model, an ultra-low-loss LC ballast has been designed for a T5 28W fluorescent lamps. The ultra-low-loss LC ballast developed here takes advantage of the inherent low-current feature of the T5 28W lamp, which enables a significant reduction in both conduction and core losses in the magnetic choke. In this experiment, a ballast loss of 2.42W, a high energy efficiency of 92.2% and a high luminous efficacy of 74.76 Lumen/Watt have been achieved in a standard Philips T5 28W fluorescent lamp. Based on CELMA guide [17], electronic ballasts for T5 28W lamps with total power consumption not exceeding 32W (i.e. ballast loss <4W) can be classified as Class A2, which is the highest class for non-dimmable electronic ballasts. (Class A1 is for dimmable electronic ballasts). Although the LC ballast is not an electronic ballast, its total power consumption of 31.01W (with a ballast loss of only 2.42W) has met the Class A2 standard. Since inductor and non-electrolytic capacitor are used to form the LC ballast (i.e no electronic switches, auxiliary power supply, control integrated circuits are needed), the proposal has the advantages of long lifetime, circuit simplicity, low maintenance cost and high reliability. With the increasing electronic waste problem caused by the wide-spread applications of electronic ballasts, it may be time for the lighting industry and international regulatory bodies to turn to a more sustainable and environmentally-friendly lighting solution.

Table 3
Comparison of electrical and luminous performance based on the use of the same Philips TL5 28W/865 lamp

Model	Input Power (W)	Lamp Power (W)	Ballast Loss (W)	Luminous Flux (Lumen)	Energy Efficiency (%)	System Luminous Efficacy (Lumen/Watt)
Ultra-low-loss LC ballast	31.01	28.59	2.42	2318.3	92.20	74.76
Philips EB-S128 TL5 230	30.95	26.30	4.65	2188.1	84.98	70.70
Osram QT-FH 1X14-35 230240 CW	30.90	27.62	3.28	2263.8	89.39	73.26

ACKNOWLEDGMENT

The authors would like to thank the Centre for Power Electronics, City University of Hong Kong for the support provided for this project.

REFERENCES

[1] "T5 Fluorescent lamp system", National Lighting Information Program, Vol. 6, Issue 1, Rensselaer Polytechnic Institute, Troy, NY 12180 USA, website http://www.lrc.rpi.edu/programs/nlpip/publications.asp

[2] Qian Luo, Zong Wei Cai, Ming Hung Wong, "Polybrominated diphenyl ethers in fish and sediment from river polluted by electronic waste", Sci Total Environ. 2007 Jun 7;: 17560632

[3] Coby S C Wong, Nurdan S Duzgoren-Aydin, Adnan Aydin, Ming Hung Wong, "Evidence of excessive releases of metals from primitive e-waste processing in Guiyu, China", .Environ Pollut. 2007 Jan 18;: 17240013

[4] E. Persson and D. Kuusito, "A performance comparison of electronic vs. magnetic ballast for power gas-discharge UV lamps," Rad Tech'98, Chicago, 1998, pp. 1-9.

[5] Chung, H. S.-H.; Ho, N.-M.; Yan, W.; Tam, P. W.; Hui, S. Y., "Comparison of Dimmable Electromagnetic and Electronic Ballast Systems—An Assessment on Energy Efficiency and Lifetime", IEEE Transactions on Industrial Electronics, Vol. 54, No. 6, 2007, pp. 3145-3154.

[6] Hie-Sik Kim; Taek-Ju Nam; "Saving power by discharge lamps with high efficient magnetic ballast", Proceedings. 1998 International Conference on Power Electronic Drives and Energy Systems for Industrial Growth, 1998. Volume 2, 1-3 Dec. 1998 Page(s):700 - 704 Vol. 2

[7] B. Szabados, "Apparatus for dimming a fluorescent lamp with a magnetic ballast," US patent 6,538,395, Mar. 25, 2003.

[8] W. Yan, S.Y.R Hui. and H. Chung, "Energy Saving of Large-Scale High-Intensity-Discharge Lamp Lighting Networks using a Central Reactive Power Control System", IEEE Transactions on Industrial Electronics, Volume 56, Issue 8, Aug. 2009 Page(s):3069 – 3078

[9] Dan E Rothenbuhler;. Samuel A. Johnson; Glenn A. Noble; Jon P. Seubert; "Preheating and starting circuit and method for a fluorescent lamp" US patent 5,736,817, April 7, 1998.

[10] Dan E Rothenbuhler and. Samuel A. Johnson, "Resonant voltage multiplication, current-regulating and ignition circuit for a fluorescent lamps", US patent 5,708,330, Jan. 13, 1998

[11] In-Seon Yeo, Dong-Ho Lee, and Sang-Bin Song, "A Simple Electronic Starter Capable of End-Of-Life Protection for Fluorescent Lamps",

[12] Chuan-Sheng Liu, Liang-Rui Chen, Neng-Yi Chu, Jieh-La Jaw, "Design of an Adaptive Electronic Starter for Fluorescent Lamps", IEEE ISIE 2005, June 20-23, 2005, Dubrovnik, Croatia

[13] Liang-Rui Chen, Neng-Yi Chu and Chuan-Sheng Liu, "Adaptive fluorescent lamp start control strategy for magnetic ballast system", Journal of the Chinese Institute of Engineers, Vol. 31, No. 6, pp1-6.

[14] W. Yan and S.Y.R. Hui;, ""A semi-theoretical fluorescent lamp model for time-domain transient and steady-state simulations", IEEE Transactions on Power Electronics, Volume 22, Issue 6, Nov. 2007 Page(s):2106 – 2115

[15] D. Lin, W. Yan, G. Zissis, and S. Y. R. Hui, "A simple physical low pressure discharge lamp model," in Energy Conversion Congress and Exposition, 2009. ECCE. IEEE, 2009, pp. 2051-2058.

[16] S.Y.R. Hui, D.Y. Lin and W.M. Ng, "A Passive LC Ballast and Method of Manufacturing a Passive LC Ballast", Patent Application No. PCT/IB2009/007289, 03 Nov 2009

[17] Federation of National Manufacturers Associations for Luminaires and Electrotechnical Components for Luminaires in the European Union (CELMA) Guide for the Application of Directive 2000/55/EC on Energy Efficiency for Ballasts for Fluorescent Lighting, Annex III, Issue 3.1, July 2007, page 14

Simple Triac Dimmable Compact Fluorescent Lamp Ballast and Light Emitting Diode Driver

Andre Tjokrorahardjo
Lighting Systems and Applications
International Rectifier
El Segundo, CA, USA
Email: atjokro1@irf.com

Abstract— Compact Fluorescent Lamp (CFL) and Light Emitting Diode (LED) are energy efficient light sources. However, they have a disadvantage that they cannot be directly dimmed with a triac dimmer. This article briefly explains the incompatibility issues, and suggests a CFL ballast circuit to overcome this disadvantage. The ballast is based on resonant mode topology, with additional circuitry for triac interface and dimming control. Furthermore, this article also presents, with few modifications to the CFL ballast circuit, an LED driver circuit that is dimmable with triac dimmer. Experimental results show that both proposed circuits are able to be dimmed down close to 10% of the light output.

I. INTRODUCTION

A Compact Fluorescent Lamp (CFL) has 5 times better efficiency compared to an incandescent light bulb and is often used as a direct replacement for energy-saving purpose. Light Emitting Diode (LED) is another type of energy efficient light sources that starts to gain broad acceptance in the market. Technological advancement in recent years makes it possible for a high-brightness LED to be used as a replacement for an incandescent bulb.

A dimming system is beneficial for every lighting system, not only for energy-saving purpose, but also for electricity demand reduction, visual comfort for the room occupants, and better productivity at work place according to some researches [1]. Virtually all domestic and architectural dimming systems are based on triac (phase-cut) dimmers. These dimmers work very well with a resistive load such as an incandescent light bulb and use existing power wiring to communicate control signals so no additional wiring needed between the control device and the ballast [2].

CFL and LED with their traditional ballasts, however, cannot be directly used with this type of dimmer. This article briefly explains the operation of the dimmer circuit and the incompatibility issues, and proposes a CFL ballast circuit to overcome this disadvantage. The ballast is based on resonant mode topology, with additional circuitry for triac interface and

dimming control. Furthermore, this article also presents, with few modifications to the CFL ballast circuit, an off-line LED driver circuit that is dimmable with triac dimmer. Experimental results show that both proposed circuits are able to be dimmed down close to 10% of the light output.

II. TRIAC DIMMER CIRCUIT AND COMPATIBILITY ISSUES WITH CFL ELECTRONIC BALLAST

Fig. 1 shows the schematic of a triac dimmer. There are two modes of operation that can be classified according to conduction of the triac device. The first mode of operation is cut-off mode where the triac is not conducting. Current flows from the line input to the load through resistor R1, variable resistor R2, and capacitor C1. The value of variable resistor R2 can be controlled by the control-knob of the dimmer, and this value determines the charging time of C1. When voltage across C1 reaches the breakover voltage of the diac, the triac starts to conduct. The charging time of C1 is the mean to control the firing angle of the triac and therefore the power delivered to the load. After the triac has fired, the dimmer is in conduction mode. Current flows from the line input to the load through the triac device. Inductor L1 and capacitor C2 are additional components used to filter noise generated by the dimmer.

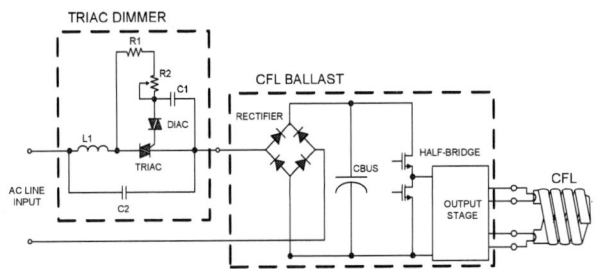

Figure 1. Schematic of a triac dimmer connected to a traditional CFL ballast

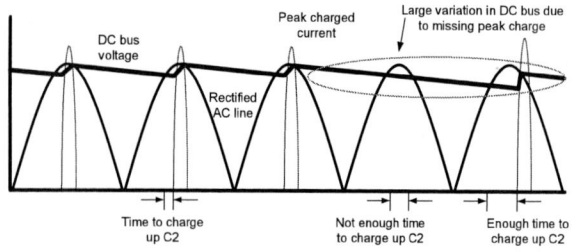

Figure 2. Key waveforms of traditional CFL ballast with triac dimmer

Problems arise when a traditional CFL ballast is connected to the dimmer. When there is no power factor correction, a CFL ballast consists of a rectification stage followed by a large bus capacitor and a half-bridge inverter. In this system, the only time period to charge up C1 is when the line input voltage is higher than the DC bus voltage (Fig. 2). This is not a problem when the dimmer is set at high dimming level as R2 is small enough to allow C1 to be charged up in small period of time. When R2 is increased for dimming, however, capacitor C1 takes longer to charge to the breakover voltage of the diac, and the triac may not be turned on within the line input half cycle. As a result, there is no peak charge current and capacitor CBUS continues to be discharged by the load. In the next line input half-cycle, the time period where the line input voltage is greater than the DC bus is longer, and C1 has enough time to be charged up to the breakover voltage of the diac. Due to missing peak charge, there is a large variation in the DC bus voltage which causes instability in lamp power.

Another compatibility issue is related to the maintenance

of the triac's conduction. After the triac is fired, the current through it has to be maintained above the holding current to sustain conduction of the triac. This is not an issue with a resistive load such as an incandescent bulb since current is drawn continuously and the triac will continue to conduct until the end of the line half-cycle. With a CFL ballast as a load, when the capacitor CBUS is charged to the same voltage as the line input voltage, current will immediately drop below the holding current and the triac will switch off. Capacitor C1 begins charging again through R1 and R2, and if there is enough time remaining in the line half cycle, the triac will fire again. This can occur several times during each line half-cycle, resulting in severe lamp flicker and loss control of lamp power.

III. PROPOSED CFL BALLAST

Fig. 3 shows the proposed CFL ballast which is able to operate with minimal flicker over a considerable portion of adjustment range of the dimmer. The ballast consists of typical CFL ballast components: EMI filter (L1 and C1) to block ballast-generated noise, a full-bridge rectifier (D1-4) and bus capacitor (CBUS) to convert AC line input into a DC Bus voltage, control IC (IC1) and half-bridge (MHS and MLS) to produce high-frequency square-wave voltage, and resonant output stage (LRES and CRES) for prehating, igniting and dimming the lamp. The triac interface circuitry is an additional circuitry to maintain triac conduction above the holding current after it is fired. The triac angle detection and the current sense resistor are parts of dimming circuitry used to dim the lamp based on the triac firing angle. The ballast utilizes the new 8-pin dimming ballast control IC IRS2530D from International Rectifier.

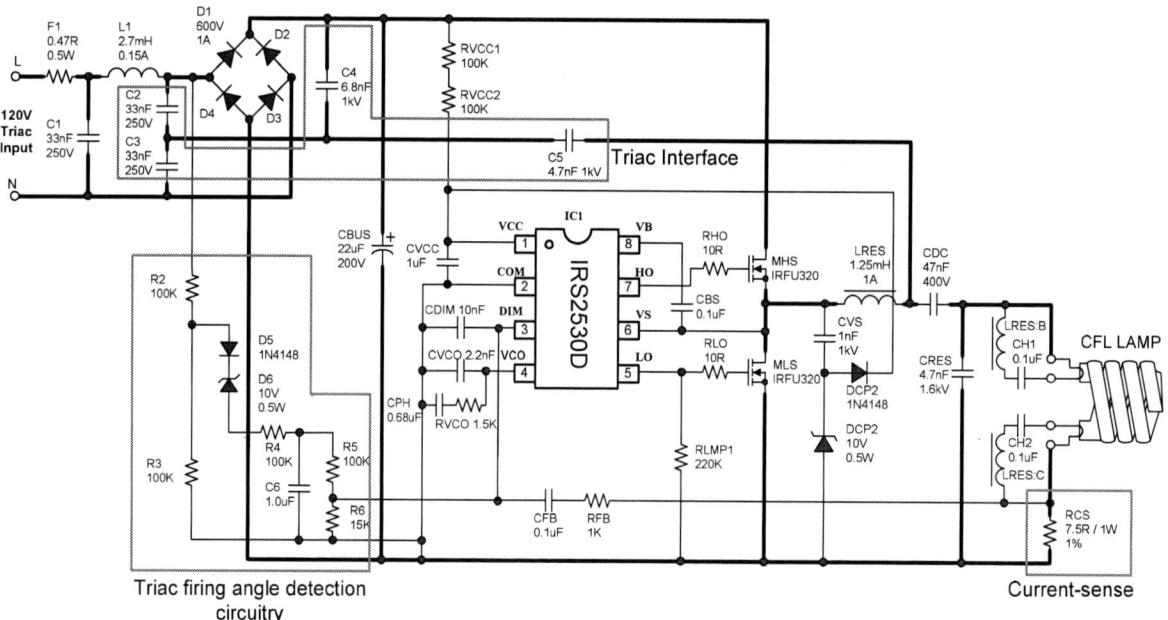

Figure 3. Schematic of proposed triac dimmable CFL ballast

A. Preheat, Ignition and Dimming

The operation of a CFL needs several requirements. First, the filaments of the lamp need to be preheated before the lamp ignites to the correct emission temperature within a defined time. Preheating the filaments before ignition extends the lifetime of the lamp and increases the number of starts it can achieved. The second requirement is a high-voltage for ignition to establish arc across the lamp. The next requirement is the lamp should be driven at the manufacturer's recommended lamp power and voltage during running [4]. All of these requirements are accomplished with the resonant output stage and controlled operating frequency by the control IC.

When the supply voltage of the IC VCC, goes above the Under Voltage Lock-Out (UVLO) threshold, the half-bridge starts to oscillate at the maximum frequency. An internal current source charges up the capacitor at VCO pin (pin #4) to increase the voltage linearly and decrease the oscillating frequency of the half-bridge. The output stage is a resonant tank with high Q-value before the lamp ignites. As frequency decreases toward the resonant frequency, the voltage across the lamp increases, and at the same time the filaments of the lamp are preheated using secondary windings of the resonant inductor. The frequency keeps decreasing until the voltage across the lamp reaches a high-enough voltage to ignite the lamp. After the lamp is ignited, lamp current begins to flow and closed-loop feedback circuit controls the lamp power to the desired dimming level.

B. Triac Interface

The front end of the proposed ballast has been designed so that when the triac in the dimmer has fired, it will remain on continuously until almost the end of the line half-cycle. This is achieved by ensuring that the ballast draws current exceeding the holding current of the triac. The triac interface circuitry is a capacitor network consisting of C2, C3, C4 and C5. When the ballast is operating, the point between LRES and CDC swings low when MLS is on. This charges C2 and C5 during the positive half-cycle of the line voltage. When MLS switches off and MHS switches on, the voltage between LRES and CDC swings high causing C5 to be discharged through the capacitor C3. During the negative half-cycle of the line voltage, the opposite happens between C2 and C3. The result is a continuous series of current pulses drawn from the input during the period when the triac has been fired until close to the end of the mains half-cycle.

The inductor L1 ensures that a continuous current is drawn from the input and that the triac does not switch off between pulses. In order to do this the inductor must store energy when current is being drawn to charge C5, and release this energy during the period when C5 is discharging.

C. Triac Firing Angle Detection

The triac angle detection circuitry detects the firing angle of the triac and generates a DC reference voltage for the dimming feedback loop. This circuitry consists of resistors R2, R3, R4 and R5 together with diode D5, a zener diode D6 and a capacitor C6.

The voltage waveform at the junction of D1 and D3, ignoring high frequency, is equivalent to the output voltage of the dimmer. With respect to the ground this will be a phase-cut approximate sine wave with a DC offset such that the negative peak is at ground. This is reduced by the voltage divider network of R2 and R3, and then fed into the anode of D5. Only the signal representing the positive half-cycle of the line input is left at the anode of D6, which is then converted to a DC voltage via the filter of R4 and C6. Because the minimum dimming level occurs at a point where the dimmer is still capable of providing enough output for the ballast to operate, this voltage will never actually be zero. The DC voltage is further reduced to the appropriate level with the voltage divider network of R5 and R6, and is used as the dimming reference.

D. New Dimming Control Method

A closed-loop system is needed to regulate lamp current due to the non-linear electrical characteristics of the fluorescent lamp. The new dimming control method in the proposed CFL ballast allows lamp current to be constantly controlled over line, lamp and temperature variations at various dimming levels.

The AC lamp current is sensed by the resistor RCS, and the resulting AC voltage is coupled with the DC dimming reference voltage from triac angle detection circuitry through feedback resistor (RFB) and feedback capacitor (CFB). The DC + AC voltage is then fed to the DIM pin (pin #3) of the IRS2530D and is regulated by the control loop such that the valley of the AC voltage always stays at COM. When the DC reference voltage is decreased for dimming, the valleys of the AC voltage are pushed below COM. The dimming control circuit detects this and increases the frequency to decrease the AC lamp current until the AC valleys at the DIM pin are at COM again. When the DC reference is increased, the valleys of the AC voltage increase above COM. The control loop decreases the frequency to increase the AC lamp current until the AC valleys at the DIM pin are at COM again. In this way, the dimming control circuit keeps the AC lamp current peak-to-peak amplitude regulated to the desired value at all DC dim level settings.

This dimming control method also regulates the lamp current against variation in the DC bus voltage. The control IC will adjusts the frequency of the half-bridge every switching

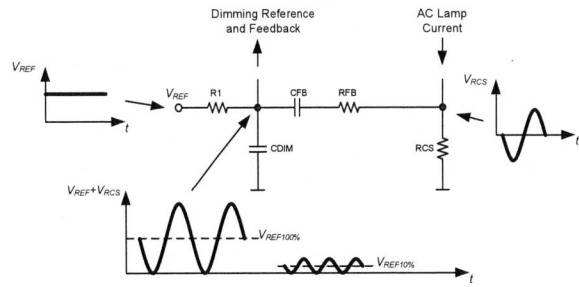

Figure 4. New AC + DC dimming control method

cycle to keep lamp current constant for a given dimming level. When the DC bus voltage drops due to the triac missing peak charge, the lamp current decreases and the AC valley at the DIM pin is increased above COM. The control loop decreases the frequency to increase the lamp current until the valley back at COM again. The opposite happens when the DC bus increases after the peak charge current charges the bus capacitor. The result is a stable lamp current despite DC bus voltage variation at all dimming level.

IV. EXPERIMENTAL RESULT

A CFL ballast circuit based on the above schematic has been built and tested in the lab. Because of the simplicity of the triac interface and dimming control loop, the hardware uses considerably lower number of components compared to other solutions available in the market.

Fig. 5 shows a picture of the proposed CFL ballast. The experimental result waveforms are shown in Fig. 6A, B and C. Fig. 6A shows the ballast goes through preheat, ignition, and running mode. After the ballast is switched on, voltage at VCO pin increases linearly, the frequency decreases toward resonant and the voltage across the lamp increases until the lamp ignites. After the lamp ignites, current starts to flow through the lamp. Fig. 6B shows the voltage across the lamp, the current through the lamp and the power delivered to the lamp during running condition for maximum dimming level. Fig. 6C shows these waveforms for the minimum dimming level.

The light output of the lamp is directly related to the current that goes through the lamp. The lamp current at maximum dimming level is 150mArms, and at the minimum dimming level is 20mArms. This gives a dimming level of about 13%. The light output cannot go down below this dimming level since the bus voltage becomes too low for the ballast to operate. The ballast also possesses a hysteresis behavior where the ballast is turned on at certain angle about midway between the turn-off point and the maximum point. This is necessary to ensure that the bus voltage is high enough to ignite the lamp and allows the ballast to operate sufficiently.

Figure 5. Photo of the proposed CFL ballast

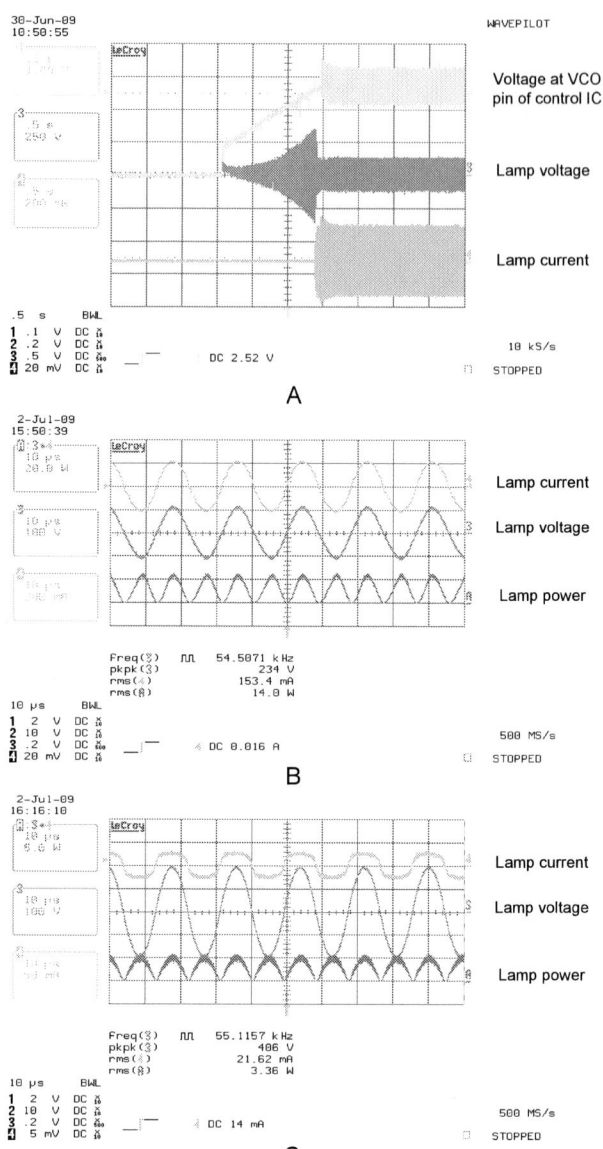

Figure 6. CFL ballast measured waveforms

V. OFFLINE TRIAC DIMMABLE LED DRIVER

LED works on a completely different principle. Unlike fluorescent lamp which is driven with an alternating current, LED requires constant current control and do not need to be preheated or ignited. The proposed triac dimmable CFL ballast can be used to drive and dim LED with only slight modifications. The LED driver has the same EMI filter, rectifier, bus capacitor, control IC, half-bridge, triac interface and triac angle detection circuitry. Since it is no longer necessary to preheat and ignite the load, the resonant output stage has been modified to become series L-C-LED resonant circuit (Fig. 7). The resulting square-wave AC voltage at the output is then converted to positive full-wave rectified voltage using an additional full-bridge rectifier.

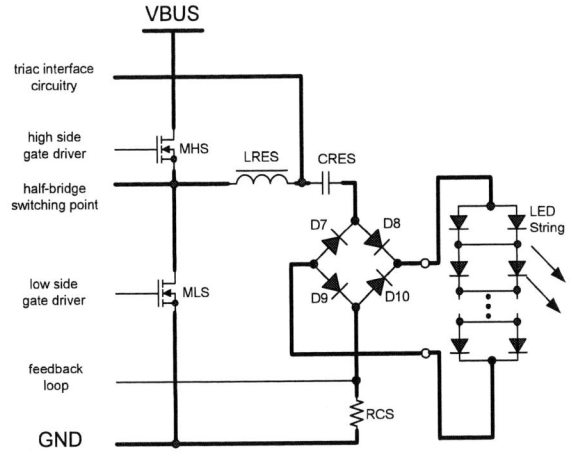

Figure 7. Modified output stage of the LED driver

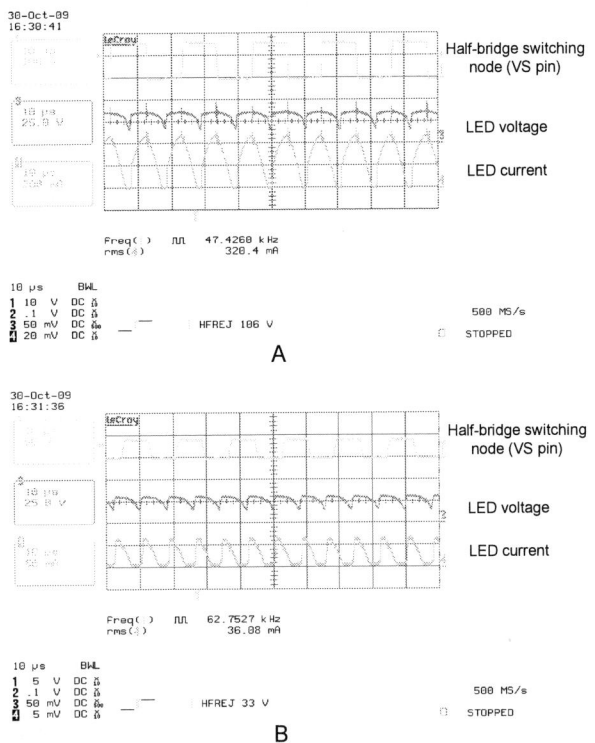

Figure 8. LED driver measured waveforms

The AC current sensing is still performed by the current-sensing resistor (RCS). This gives a direct AC measurement of the full-wave rectified LED current amplitude. This AC signal is then coupled with the DC voltage from the triac angle detection onto the DIM pin of the control IC. The dimming control loop keeps the amplitude of the LED current regulated by continuously adjusting the frequency of the half-bridge switching circuit such that the nominal rms LED current is maintained within the manufacturer's specifications. When the DC reference voltage is decreased for dimming, the dimming loop increases the frequency to decrease the gain of the resonant tank circuit and thus decrease the LED current. This control scheme keeps the LED current constant over line, load and temperature variations for any given dimming reference input, and will work for any number of LEDs in series.

A prototype of the off-line triac dimmable LED driver has been built and tested in the lab. The experimental result waveforms are shown in Fig. 8A and B. Fig. 8A shows the half-bridge switching node, voltage across LED and LED current at maximum dimming level. The LED current is full-wave rectified and operates at twice the frequency of the half-bridge switching node. Fig. 8B shows these waveforms at the minimum dimming level. The driver is able to be dimmed down to 11% of the maximum light output. Similar to the CFL ballast, the driver cannot be dimmed down below this level and possesses a hysteresis behavior.

VI. CONCLUSION

The incompatibility issues between a triac dimmer and a traditional CFL ballast come from the missing peak charge and the inability of the ballast to maintain conduction above the holding current. This article presents a CFL ballast and an LED driver that are compatible with a triac dimmer. The proposed circuits use triac interface circuitry to maintain conduction above holding current, and triac angle detection circuitry to detects the firing angle of the triac and dim the load accordingly. Because of simplicity of dimming control of the control IC, the circuit uses considerable lower number of components compared to other solutions. Both circuits have been built and tested in lab. The circuits are able to be dim

down close to 10% of the maximum light output with minimal flicker. Possible future improvements include lower dimming level and reduction in the hysteresis of the system.

ACKNOWLEDGMENT

The author would like to acknowledge Tom Ribarich and Peter Green at International Rectifier, Lighting Systems and Applications Group for their support and guidance.

REFERENCES

[1] R. Leslie, R. Raghavan, O. Howlett, and C. Eaton, "The potential of simplified concepts for daylight harvesting," Lighting Research and Technology 37 (1), 2005, pp. 21-40.

[2] C. DiLouie, Lighting Controls Handbook, 1st ed. The Fairmont Press, Inc.: Lilburn, 2007, pp.102–104.

[3] J. Smith, J. Speakes and M. H. Rashid, "An overview of the modern light dimmer: design, operation, and application," Proceedings of the 37th Annual North American Power Symposium, pp. 299-303, Octover 2005.

[4] D. Rand, B. Lehman and A. Shteynberg, "Issues, models and solutions for triac modulated phase dimming of LED lamps, " IEEE Power Electronics Specialists Conference, pp. 1398-1404, June 2007.

[5] P.W. Tam, S.Y.R. Hui and S.H. Chung, "An analysis and practical implementation of a dimmable compact fluorescent lamp ballast circuit withouth integrated circuit control," 37th IEEE Power Electronics Specialists Conference, pp. 1-8, June 2006.

[6] T. Ribarich and J. Ribarich, "A new procedure for high-frequency electronic ballast design," IEEE Industry Applications Society Annual Meeting, pp. 271–350, New Orleans, Louisiana, October 1997.

[7] A. Tjokrorahardjo, "AN-1153: Low cost triac dimmable CFL ballast using IRS2530D," http://www.irf.com/technical-info/appnotes/an-1153.pdf.

[8] IRS2530D Dimming Ballast Control IC datasheet, www.irf.com.

High Efficiency Soft-switched Step-up DC-DC Converter with Hybrid Mode LLC+C Resonant Tank

Wei Chen[1,2], Xiaoyuan Hong[1], Siran Wang[1], Zhengyu Lu[1] (Senior member, IEEE)   Shaoshi Ye[2]

[1]College of Electrical Engineering
Zhejiang University
Hangzhou, Zhejiang, China

[2]Delta Electronics (Shanghai) Co., LTD.
No. 238 Minxia Road, Pudong
Shanghai, 201209, China

Abstract-To improve the efficiency performance for the dc-dc converter used in extra wide input range applications, a novel resonant converter with a combinatorial LLC+C tank is proposed in this paper. The proposed resonant tank has a hybrid operational mode, which can operates as LLC resonant tank for high line input or as LLCC resonant tank for the low line input, realized by the topology shift through the control of a cost-effective shifting switch. By being carefully designed and with the ZVS+ZCS soft switching feature, the proposed converter can accommodate the extra wide input range applications without sacrificing the overall efficiency. The details for the design procedure are presented, small signal compatibility is discussed, and the control strategy for topology shifting is also stated. A prototype with 100V-400V input voltage range, 24V/10A output power rating was built to verify the validity and applicability of this proposed converter. A comparison with a single topology LLC converter with the same specifications was also made to show the superiority of the proposed converter when applied for the extra wide input range applications.

Key words: Dc-dc converter, Voltage gain, Soft switching, Topology combination

I. INTRODUCTION

To date, modern power electronics systems continue calling for the state-of-the-art advanced high power conversion efficiency wide input range (usually more than twice as the top to its bottom, e.g., 100Vdc-400Vdc) dc-dc converters to enable high power density, high system reliability, and being systematic cost-effective [1]-[11]. In a typical Distributed Power System (DPS) or a telecom/server power supply, which has a certain hold-up time requirement up to 20ms after the line dropout, the wide input range dc-dc converters reduce the total amount of hold-up time capacitors and achieve size-minimization [3]-[6]. In the Uninterrupted Power Supply (UPS) systems, renewable energy systems, or hybrid Electric Vehicles (EVs), where the low-voltage energy-storage devices, such as the battery sources, the fuel cell stacks, the photovoltaic arrays or the super capacitors, are extensively installed, the wide input range dc-dc converters are expected as the key component to actively provide clean and stable power to load, extend energy-storage device lifecycle, and efficiently improve the flexibility of system reconfiguration ability [2]. However, for the basic dc-dc converters, enlarging the operation range usually is at the expense of the sacrifice of normal operation efficiency. For PWM converters, the extreme narrow duty cycle will cause

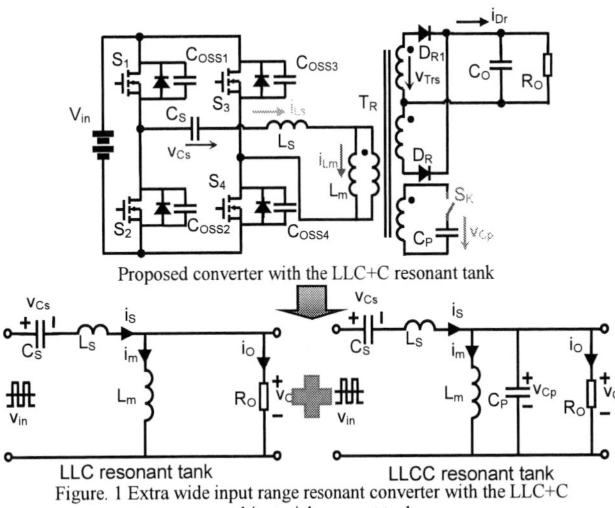

Project supported by National Natural Science Foundation of China (50677063)

Proposed converter with the LLC+C resonant tank

LLC resonant tank   LLCC resonant tank
Figure. 1 Extra wide input range resonant converter with the LLC+C combinatorial resonant tank

excessive switching loss. For resonant converters, the high quality factor Q for a high voltage gain will unavoidably increase the reactive power and circulating loss for the converter.

So far, there are three techniques proposed to extend the input range for the dc-dc converters. The first and most straightforward solution is to employ the cascade structure, viz., two-stage converter, for the extreme voltage gain applications [3]-[4]. However, the overall efficiency suffers since an extra conversion stage should be taken into account and, circuit configuration is comparatively complex and consequently costly. In [5]-[8], the concept of range winding by changing the transformer turns-ratio through the control of a range switch has been proposed and implemented without sacrificing normal operation efficiency. But mostly in these converters, the transformer is usually difficult to be fabricated in the optimum way, since the extra range winding will inevitably occupies the window area of the magnetic component, which requires an upgrade in size for the transformer, making it bulky and eventually losing the benefits of high frequency. The approach of multi-phase converters interleaving was proposed in [9]-[11] to effectively extend the voltage gain, where another feasible benefit is the primary conduction loss can be halved. However, the overall circuit configuration remains complex due to nearly two sets of components are required, which significantly degrades the attainable power density and increases the manufacturing cost.

In order to achieve the high voltage gain for the wide input voltage range applications, a novel LLC+C resonant tank with hybrid operational modes is proposed in this paper, shown as Fig. 1. In this proposed resonant tank, a conventional LLC resonant tank is implemented with an additional controllable range capacitor devoted to reconfigure the LLC resonant tank into a LLCC resonant tank, which operates basically on a high voltage gain LCC type resonant tank [12]. Thus, the proposed converter has two modes to accommodate both of the high line and low line input conditions, in the manner of LLC for the high line and LLCC for the low line. An optimized overall efficiency can also be achieved since the soft-switching feature of ZVS for the primary switches and ZCS for the rectifiers is possessed in either mode. The operation principles as well as detail design considerations are presented. A 500W prototype with 100V-400V input operation ability validates its advantage which surpassing the conventional LLC converter with the same input range.

II. Operation Principles

The power stage of the proposed converter can be either with a full-bridge conception or a half-bridge conception. The full bridge circuit configuration was implemented in Fig. 1. S_1~S_4 are main switches, where C_{OSS1}~C_{OSS4} represent parasitic capacitors of S_1-S_4 correspondingly. The LLC+C resonant tank, which developed from a conventional LLC resonant tank, adds an additional resonant capacitor of C_P with the tertiary winding on the secondary side according to the status of the topology shifting switch S_K in green color. When S_K keeps OFF state, the LLC+C resonant tank returns to the conventional LLC resonant tank to obtain an optimized efficiency with a properly designed low Q. If S_K switches ON,

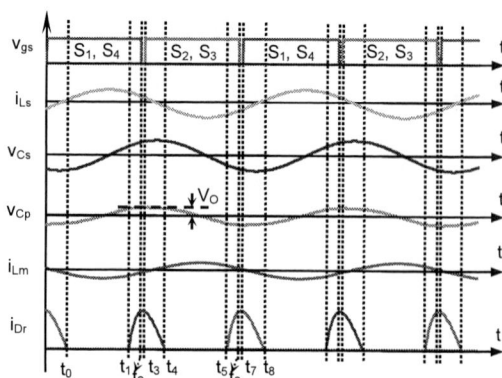

Figure. 2 Operation principles for the LLCC resonant mode

a high voltage gain LLCC resonant tank enables, then the utility of low voltage power is achieved. The status of the shifting switch S_K is controlled according to the present input voltage condition. Finally, for the rectifier stage, D_{R1} and D_{R2} constructs the center-tapped structure with the coaxially wound secondary.

The proposed converter operates under variable frequency modulation for both conversion modes, i.e., $S_1\&S_4$ and $S_2\&S_3$ run at 50% duty cycle, 180° out of phase. Due to the extensive research for the conventional LLC converter, it will not be addressed in detail. However, the operational principle for the LLCC mode will be described here, where the driving signals for main switches and principle waveforms for the LLCC mode are shown in Fig. 2. The LLCC mode has 8 operation stages during a switching period. The waveforms and equivalent circuits in the former 4 operation stages in a half switching cycle are given by Fig. 3. Other 4 stages are just similar to the previous half cycle. To simplify the process

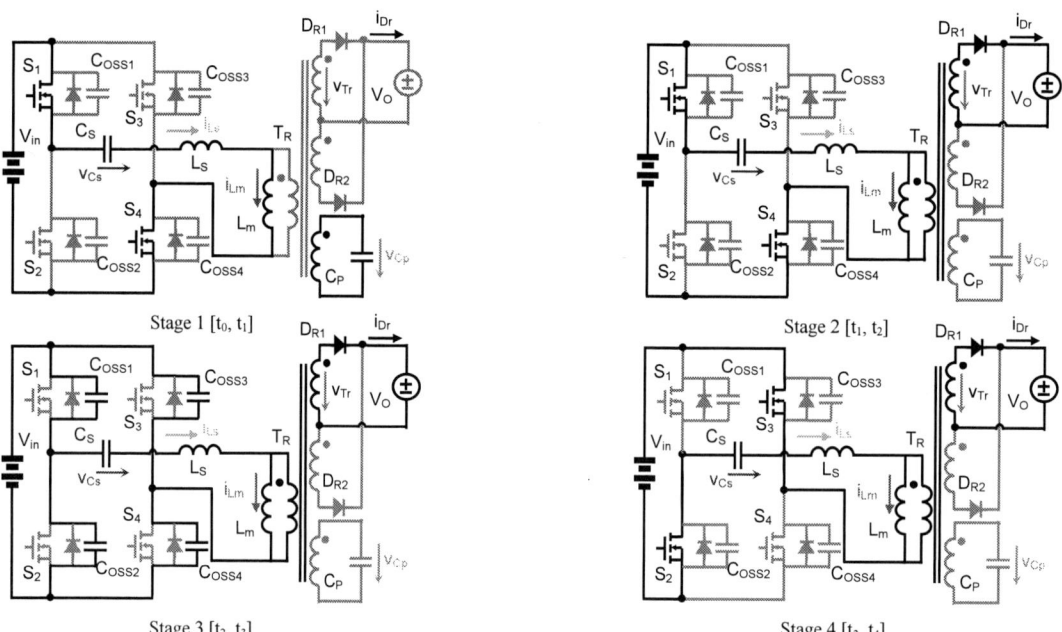

Figure. 3 Equivalent circuits for each stage for the LLCC resonant mode

of the analysis, it is assumed that the converter is under steady operation state. The output capacitor C_O is large enough to be considered as a voltage source V_O. The detailed processes for each operation mode are described as follows:

Stage 1 [t_0, t_1]: At the time of t_0, S_1 and S_4 are conducting. On the secondary side, the rectifier conducting current i_{Dr2} resonate to zero with a comparative low di/dt, D_{R2} realizes ZCS off. Since i_{Dr2} is proportional to the difference value between the primary resonant current i_{Ls}, magnetizing current i_{Lm} and parallel capacitor resonant current i_{Cp}, if i_{Dr2} deceases to zero, the magnetizing inductor L_m and the parallel resonant capacitor C_P will free from the clamping effect of the output and participate the resonance with the other two resonant elements. Thus, the resonant tank consists all of the 4 resonant elements, where the corresponding resonant voltages and currents inside the resonant tank are in the sine shape waveforms.

Stage 2 [t_1, t_2]: It is some time of resonance before the voltage of the parallel capacitor v_{Cp} increases to the output voltage V_O again. As a result, the rectifier D_{R1} on the secondary side conducts, and power is delivered to the load again. As like as i_{Dr2}, the rectifier current i_{Dr1} is also proportional to the difference of the primary resonant currents. The parallel capacitor C_P and the magnetizing inductor L_m are clamped by the output again. The voltage applied on C_P is exactly the output voltage, but the magnetizing current will increase linearly. At this time, the resonant tank only consists of C_S and L_S.

Stage 3 [t_2, t_3]: At time t_2, main switches S_1 and S_4 turn off simultaneously. The primary resonant current i_{Ls} in the resonant tank begins to charge C_{OSS1}&C_{OSS4} and discharge C_{OSS2}&C_{OSS3}, correspondingly. To simplify the analysis, it is assumed that the values of C_{OSS1}-C_{OSS4} are all small enough so that this period could be very short. When the voltage of S_1 and S_4 reach the input voltage, at the same time, the voltage applied on S_2 and S_3 decrease to zero. Thus, the body diodes of S_2 and S_3 conduct naturally.

Stage 4 [t_3, t_4]: The gate signal for S_2 and S_3 come at time t_3. Since the voltages of the two switches are zero, they achieve ZVS on. Although the polarity of the voltage applied on the resonant tank has changed, however, due to the difference value for i_{Ls}, i_{Lm} and i_{Cp} are not zero, power continues flowing to load. This stage lasts until time t_4, then the 4 stages in a half switching period end, the converter enters into the next half switching period.

From the above discussion, it can be comprehended the LLCC resonant tank basically operates under the conventional LCC resonant manner. Thus, a similar operational characteristic with the LCC resonant tank for the LLCC resonant tank should be expected, which implies the methodologies to analyze the LCC resonant tank are also valid for the LLCC resonant tank to facilitate the design.

III. DESIGN CONSIDERATIONS

A. Design Procedure for the LLC+C Resonant Tank

The fundamental goal for the proposed LLC+C resonant converter is expected as an overall efficiency improvement regardless of the change of the input line voltage. Thus, the design for the LLC+C resonant tank consequently exhibits as the most important part to aim for the optimized efficiency. Basically, the efficiency improvement for the resonant converters is always associated with the reduction of the reactive power circulating in the resonant tank, which implies a low Q always deserves being paid great effort on. Due to the complexity of the approach to directly solve all of the 4 resonant elements at the same time, a much convenient method is firstly solving the LLC resonant tank, then coping with the parallel resonant capacitor C_P. The equivalent circuits for the LLC and LLCC resonant tanks have been shown in Fig. 1. Their voltage gain expressions deriving from the fundamental mode approximation approach are given by (1) and (2):

$$M_{LLC} = \frac{1}{\sqrt{(1 + \frac{1}{h} - \frac{1}{\omega^2 \cdot h})^2 + Q_L^2 \cdot (\frac{1}{\omega} - \omega)^2}} \tag{1}$$

$$M_{LLCC} = \frac{1}{\sqrt{(1 + g - g \cdot \omega^2 + \frac{1}{h} - \frac{1}{\omega^2 \cdot h})^2 + Q_L^2 \cdot (\frac{1}{\omega} - \omega)^2}} \tag{2}$$

where $Q_L = \frac{\pi^2}{8 R_O} \sqrt{\frac{L_S}{C_S}}$, $h = \frac{L_m}{L_S}$, $g = \frac{C_P}{C_S}$, $\omega = \frac{\omega_S}{\omega_O}$, R_O is the normalized load resistance, the resonant frequency is defined as $\omega_O = \frac{1}{\sqrt{L_S \cdot C_S}}$, ω_S is the switching frequency.

Thus a design procedure for the LLC+C resonant tank can be outlined by the following 4 steps:

1). Design the LLC resonant tank with an optimized set of resonant elements based on [13], to accommodate an input voltage range as 200V-400V, at all load conditions.

2). The design for the parallel resonant capacitor C_P should satisfy 3 requirements: high DC gain, optimized efficiency for all conditions, and, insensitivity for variation of the switching frequency with respect to the line and load changes. Since the LLCC resonant tank behaviors more like the LCC tank, therefore the magnetizing inductor L_m can be removed during the design for C_P to simplify the analysis. The voltage gain will be employed for the design procedure for the LCC resonant tank can be expressed as follows:

$$M_{LCC} = \frac{1}{\sqrt{(1 + g - g \cdot \omega^2)^2 + Q_L^2 \cdot (\frac{1}{\omega} - \omega)^2}} \tag{3}$$

Fig. 4 plots the normalized voltage gains for all of the LLC, LLCC, and LCC resonant tanks, in versus with the normalized switching frequency. It can be appreciated that the LLCC and LCC resonant tanks behavior very much like with each other when operating in the area around the resonant frequency. Therefore, the removal of L_m will not lose the necessary accuracy for the calculation of C_P.

3). Assuming a candidate g has been chosen, Then, if using the ratio between the peak value for the present resonant current I_{LLC-p}, and the peak value for the resonant current at

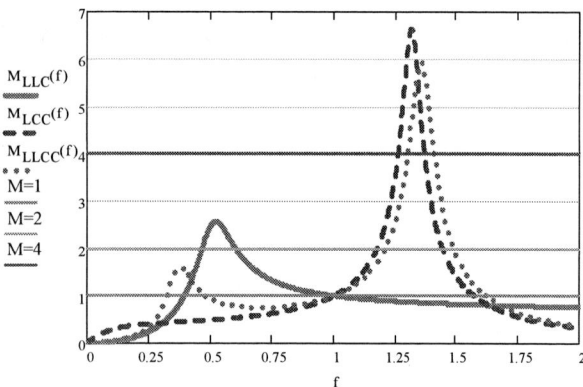

Figure. 4 The typical normalized voltage gain curves versus switching frequency for LLC, LLCC, and LCC resonant tanks

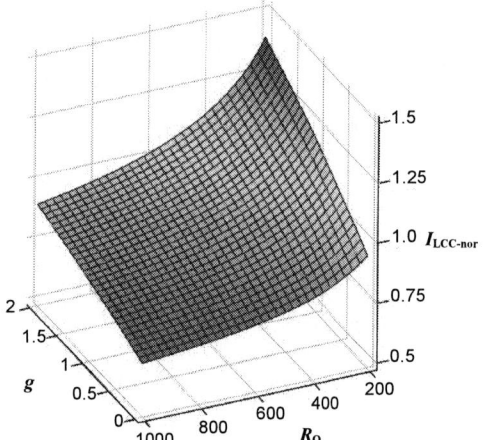

Figure. 5 Three- dimensional contour of $I_{LLC-nor}$ as the function of g and R_O

the resonant frequency under full load I_R, the quality factor Q can also be evaluated by the normalized resonant current of $I_{LCC-nor}$, which can be expressed as follows:

$$I_{LCC-nor} = I_{LCC-p}/I_R \qquad (4)$$

According to [12], (4) can be simplified as:

$$I_{LCC-nor} = \frac{1}{8\sqrt{R_O \cdot f_O \cdot L_S \cdot (M-1)}} \cdot (a+b) + 0.5 \qquad (5)$$

where

$$a = \frac{(sgn(4gM^2 - 4gM + 4M^2 - 4M + 1) + 1) \cdot (4gM^2 + 4M^2 - 1)}{\sqrt{g+1}}$$

$$b = \frac{(1 - sgn(4gM^2 - 4gM + 4M^2 - 4M + 1)) \cdot \left|4gM^2 + 4M^2 - 4M + 1\right|}{\sqrt{g}}.$$

From (5), the curves for $I_{LCC-nor}$ in a full vision, over with parameter g and the load R_O under a given voltage gain, are plotted in Fig. 5. When an input voltage gain is specified, as 100V-200V, the preliminary chosen g can be examined by checking the value of $I_{LCC-nor}$. The parameter g should be evaluated and rechosen through Fig. 5, until $I_{LCC-nor}$ seems relative low and optimized for all load conditions, with the desired voltage gain.

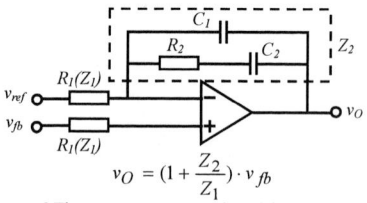

$$v_O = (1 + \frac{Z_2}{Z_1}) \cdot v_{fb}$$

Figure. 6 The compensator employed for the prototype

4). If 3) is fulfilled, the parameter g should substitute into (2) to double check whether the voltage gain is suitable for the assigned input voltage range.

With the parameter of g is determined, C_P can be obtained by $C_P = g \cdot C_S$. Then, all of the resonant elements have been appropriately calculated and designed for the high efficiency LLC+C resonant tank with an alterable voltage gain for the extra input range applications.

B. Small Signal Compatible Considerations for LLC and LLCC Operational Modes

Generally, the two operational modes are both under frequency modulation control and with the same gain-frequency logic if operated above resonance. Thus, a typical PI compensator with the feedback signal connecting to the non-inverting terminal of the Op-Amp can be employed to accommodate both the two modes. The voltage gain for the compensator is given by Fig. 6. However, the design for the compensator to achieve the small signal compatibility is still depends on the small signal response characteristics of the LLC and LLCC resonant tanks. An analysis based on the Extended Describing Function (EDF) [14] has been performed and the small signal circuit models for the two resonant tanks are derived, in which the LLCC resonant tank is replaced by the LCC resonant tank to achieve simplicity. The detail circuit models are shown in Fig. 7, and the theoretical and practical measured results, which explain the small signal compatibility feature, will be given in the section of experimental result analysis later in this paper.

C. Control Strategy for Topology Shifting

As was stated previously, the proposed LLC+C resonant tank can automatically transform its interior circuit configuration by employing a topology shifting switch S_K, to suit to the present input voltage status. However, the abrupt change of the shifting switch may cause out-of-spec voltage/current fluctuations and will damage the power semiconductor devices, since the converter is close-looped and will find the new steady state operation point in a very hard way. To counteract this issue, a topology shifting control strategy is proposed. This control strategy can be referred as a soft-start control, similar with the soft-start function for the control ICs for the resonant type converters [15]. Before the action of S_K being taken, a digital MCU monitors the status of the input voltage, judging whether the shifting threshold is triggered. If the present converter operational mode is no longer suitable for the new status of the input voltage, firstly, the MCU instructs the switching frequency increases to high enough to force the converter enters into open loop operation,

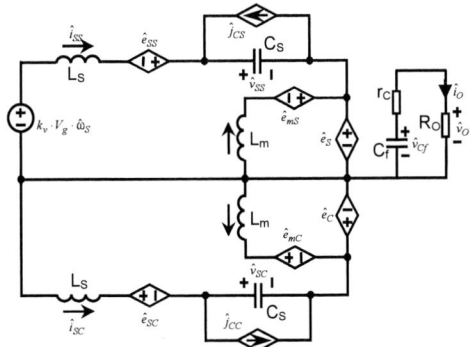
(a) For LLC resonant tank

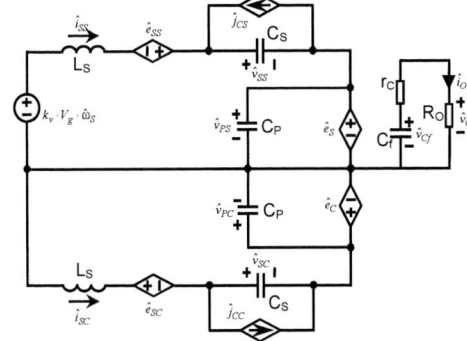
(b) For LCC resonant tank

Figure. 7 Small signal circuit models for the LLC and LCC resonant tanks

regardless what previous mode was that the converter operates under with. Secondly, S_K switches ON or OFF according to the desired topology configuration, but the switching frequency is still so high that the converter is still in the open loop status. Thirdly, MCU instructs the switching frequency sweeping back to the low area gradually to realize a soft-start process, and the converter will finally operate under normal state after some transient response time. The experimental results examine the validity for the proposed control strategy for topology shifting, which will be illustrated in the next section.

IV. EXPERIMENTAL RESULTS

A prototype, which can be used in the 100V-400V extra wide input range applications with a rated output power of 240W, verifies the operation principles of the proposed converter and the characteristic of ZVS and ZCS. The transformer is fabricated by the technique of magnetic integration, where the leakage inductance is utilized as the resonant inductor L_S. The specification and main parameters are specified as follows:

V_{in}: 100-400 Vdc, V_O: 24 Vdc, I_O: 0-10 A.

The components of the power stage are listed as follows:

$S_1 \sim S_4$: 2SK3934, D_{R1}&D_{R2}: S60SC6M, T_R: n=34:2:2:17 (RM14), L_S: 150 uH, L_m: 450 uH; C_S: 3·2.2 nF/1200V=6.6 nF, C_P: 2·19 nF/1400V=38 nF (g=1.44).

The input voltage boundary for topology shifting is set as 200V, with a hysteresis of 20V to avoid a fault trigger. The operational topology for 200-400V input range is the LLC resonant tank to achieve an attainable high efficiency, and the LLCC resonant tank is used for the 100-200V input range to obtain a high voltage gain with the same transformer turns-ratio without significant efficiency deterioration.

Operation waveforms for the LLC and LLCC resonant tanks under corresponding 300 V and 150 V input conditions at full load are shown in Fig. 8, respectively. Fig. 8 (a) and (d) show the ZVS feature for the primary switch of S_1 under the two operational modes. Fig. 8 (b) and (e) are the waveforms obtained from the resonant tanks for LLC and LLCC. Fig. 8 (c) and (f) show the ZCS feature for the secondary rectifiers, also, applicable for the both two operational modes.

Fig. 8 (g) and (h) are the transient waveforms obtained for the topology shifting. The traces in the waveforms are input voltage V_{in}, shifting signal v_{shift}, and output voltage V_O, from top to bottom, correspondingly. Fig. 8 (g) is the transient

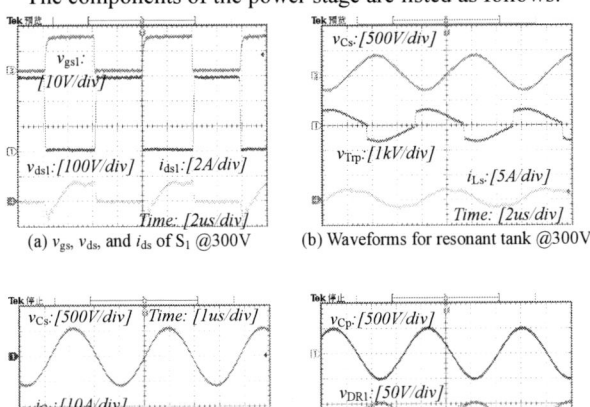

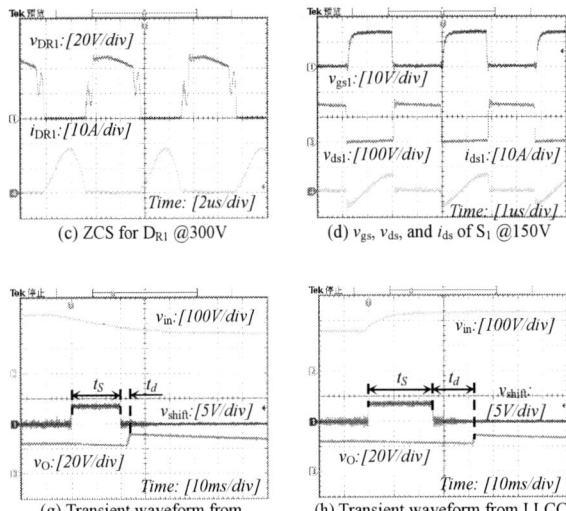

(a) v_{gs}, v_{ds}, and i_{ds} of S_1 @300V

(b) Waveforms for resonant tank @300V

(c) ZCS for D_{R1} @300V

(d) v_{gs}, v_{ds}, and i_{ds} of S_1 @150V

(e) Waveforms for resonant tank @150V

(f) ZCS for D_{R1} @150V

(g) Transient waveform from LLC to LLCC

(h) Transient waveform from LLCC to LLC

Figure. 8 Experimental waveforms

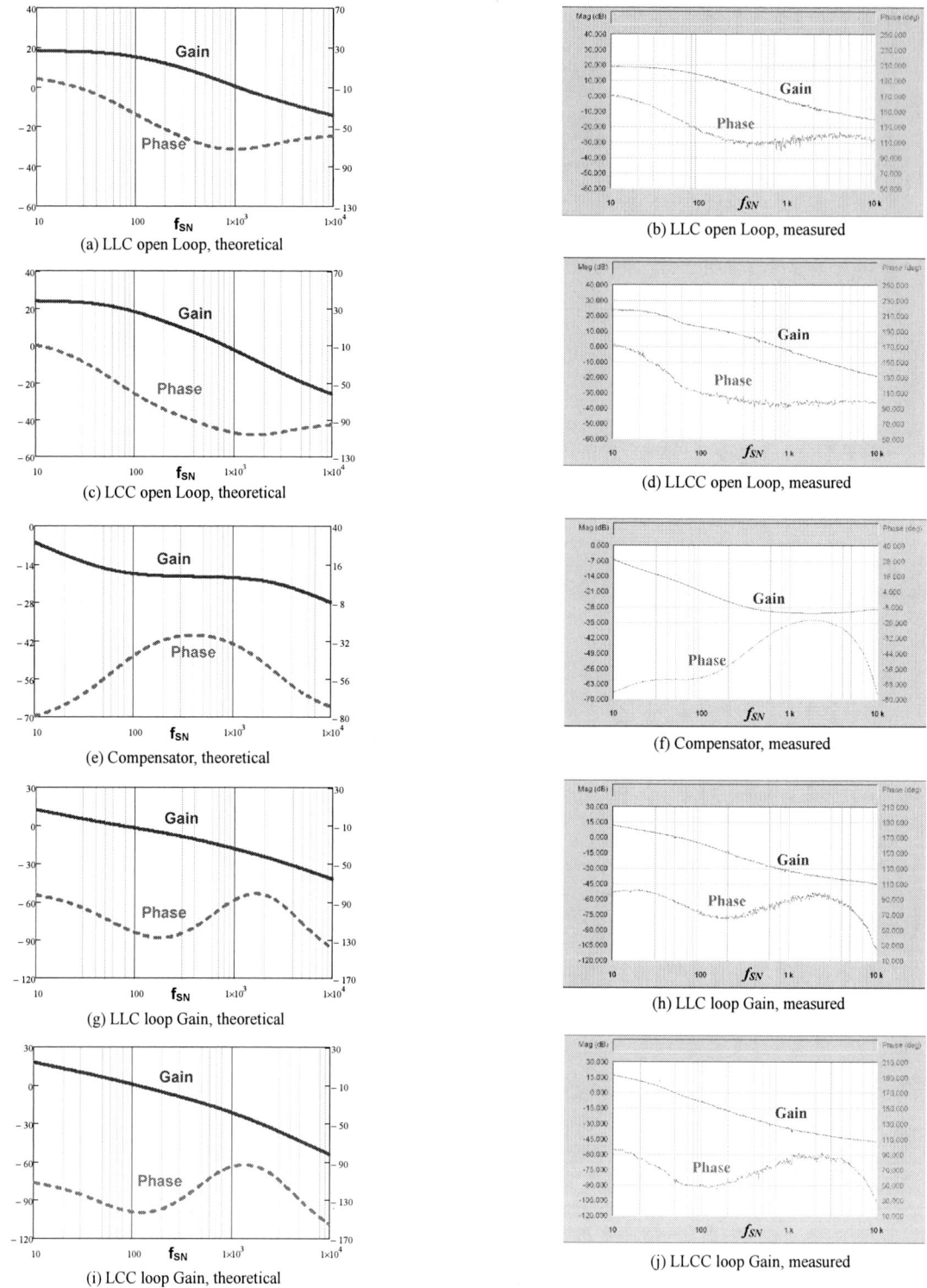

Figure. 9 Theoretical and measured bode plots

waveform for shift from LLC to LLCC. It can be observed the shifting threshold is approximately at 220V. The switching frequency increases to a relative high area about 300 kHz to force the converter open-looped. During the delay time t_S, shifting switch S_K closes and C_P participates the resonance with the other 3 resonant elements. At the end of the delay time period t_S, the switching frequency gradually falls to the low frequency region and the soft-start process occurs. The converter can finally close-looped and operated at a new steady state under the topology of LLCC for low line input. A similar manner can be observed in Fig. 8 (h), but without the same time period of t_S and t_d. The difference for t_S and t_d is mainly concerned about the difference for the circuit characteristics for the two resonant tanks, and as a result the t_S and t_d should be set and optimized individually. The output voltage fluctuation is about 10V(42%). Provided a shorter t_S and a longer t_d, this voltage fluctuation can be further minimized.

Fig. 9 shows the bode plots for the theoretical analysis and the practical measured ones, with the same compensator shown as Fig. 6. It can be seen these measured bode plots are very close to the theoretical analysis except for the high frequencies, where the parasitic effects associated with the actual devices become much more dominant at this region. Another aspect which can be comprehended is that, although the LCC resonant tank is employed for analysis instead of the actually LLCC resonant tank, however, the measured results for LLCC are close to those of LCC. Thus, it proves the validity to use LCC tank for the design instead of the LLCC tank to simplify the process.

Fig. 10 shows the conversion efficiency of all conditions. The highest efficiency is 97.1% under 400 V input, and the lowest one is 90.1% under 100 V input. A comparison with a single topology LLC converter with the same 100-400V input voltage range and the same power rating has been made. It can be clearly seen the proposed converter surpasses the single topology LLC converter at every efficiency point. It should be noted if the power rating increases higher, e.g., 500W, the single topology LLC converter can not possess the desired voltage gain any more and is no longer suitable for the 100V input application, no matter how high the Q is tuned. Contrastively, the proposed converter will still functionalize properly under such power rating condition, by a new set of carefully designed resonant parameters.

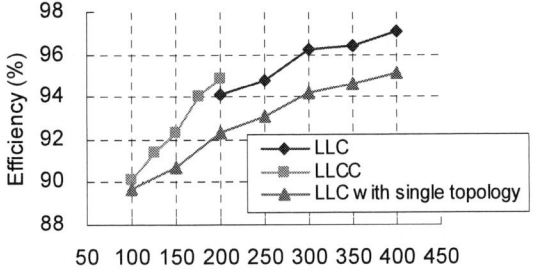

Figure. 10 Efficiency curves for the prototype and comparison with the single topology LLC converter

V. CONCLUSIONS

The new resonant converter proposed in this paper, which features in a combinatorial LLC+C resonant tank with the hybrid operational modes of LLC and LLCC resonant tanks, possesses the ability of high voltage gain by topology shift to accommodate the 100V-400V extra wide input range applications without sacrificing the efficiency neither at the high line nor the low line input condition. The soft switching feature as ZVS for the primary switches and ZCS for the output rectifiers can be achieved and maintained unconditionally. Besides, the proposed LLC+C resonant tank is simple and easy to be implemented, which realizes cost-effective for practical use. The experimental results and comparison with a single topology LLC converter verify the effectiveness and advantages of the proposed converter. Therefore, this proposed topology is very attractive for the extra wide range and high efficiency applications in industry.

REFERENCE

[1] Lee F.C., Shuo Wang, Pengju Kong, et al, "Power architecture design with improved system efficiency, EMI and power density," in *IEEE PESC 2008*, 15-19 June 2008, pp: 4131-4137.

[2] Marchesoni M., Vacca, C., "New DC–DC Converter for Energy Storage System Interfacing in Fuel Cell Hybrid Electric Vehicles," *Power Electronics, IEEE Transactions on*, vol. 22, Issue 1, Jan. 2007, pp: 301-308.

[3] Yan Xing, Lipei Huang, Xuansan Cai, et al, "A combined front end DC/DC converter," in *IEEE APEC 2003*, vol. 2, 9-13 Feb. 2003, pp: 1095-1099.

[4] Jang Y., Jovanovic M.M., Dillman D.L., "Hold-up time extension Circuit with integrated magnetics," *Power Electronics, IEEE Transactions on*. 2006, 21(2), pp: 394–400.

[5] Wang X.C., Tian F., Batarseh I, "High Efficiency Parallel Post Regulator for Wide Range Input DC–DC Converter," *Power Electronics, IEEE Transactions on*. 2008, 23(2), pp: 852-858.

[6] Yang B., Xu P., Lee F.C., "Range winding for wide input range front end DC/DC converter," in *IEEE APEC 2001*, vol. 1, pp: 476-479, Anaheim, California, USA.

[7] Ayyanar R., Mohan N., "A novel full-bridge DC-DC converter for battery charging using secondary-side control combines soft switching over the full load range and low magnetics requirement," *Industry Applications, IEEE Transactions on*, vol. 37, Issue 2, March-April 2001, pp: 559-565.

[8] Zubieta L., Panza G., "A wide input voltage and high efficiency DC-DC converter for fuel cell applications," in *IEEE APEC 2005*, vol. 1, 6-10 March 2005, pp: 85-89.

[9] Bing Lu, Ming Xu, Chunayin Wang, et al, "A High Frequency ZVS Isolated Dual Boost Converter with Holdup Time Extension Capability," in *IEEE PESC 2006*, 18-22 June 2006, pp: 1-6.

[10] Xinke Wu, Wei Lu, Junming Zhang, et al, "Extra Wide Input Voltage Range and High Efficiency DC-DC Converter Using Hybrid Modulation," in *IEEE IAS 2006*. Vol. 2, 8-12 Oct. 2006, pp: 588-594.

[11] Feng Han, Xu Dehong, Matsui M., "A novel ZVT circuit for interleaving two-transistor forward converter," in *IEEE APEC 2000*, vol. 2, 6-10 Feb. 2000, pp: 754-759.

[12] Gilbert A.J., Bingham C.M., Stone D.A., et al, "Normalized Analysis and Design of LCC Resonant Converters," *Power Electronics, IEEE Transactions on*, vol. 22, Issue 6, Nov. 2007, pp: 2386-2402.

[13] Bing Lu, Wenduo Liu, Yan Liang, et al, "Optimal design methodology for LLC resonant converter," in *IEEE APEC 2006*, 19-23 March 2006, pp: 6-12.

[14] Yang E.X., Lee F.C., Jovanovic M.M., "Small-signal modeling of series and parallel resonant converters," in *IEEE APEC 2002*, 23-27 Feb. 1992, pp: 785-792.

[15] L6599 datasheet, www.st.com.

A Family of Zero Current Switching Switched-Capacitor DC-DC Converters

Dong Cao
Department of Electrical & Computer Engineering
Michigan State University
East Lansing, MI 48824, USA
caodong@msu.edu

Fang Zheng Peng
Department of Electrical & Computer Engineering
Michigan State University
East Lansing, MI 48824, USA
fzpeng@egr.msu.edu

Abstract— **This paper presents a new zero current switching (ZCS) technique for a family of switched-capacitor dc-dc converters. Compared to the traditional ZCS switched-capacitor dc-dc converters by inserting a magnetic core in the circuit, these new ZCS switched-capacitor dc-dc converters employ the stray inductance present in the circuit as the resonant inductor and provide soft switching for the devices. These ZCS switched-capacitor dc-dc converters do not utilize any additional components to minimize switching loss and reduce the current and voltage spike, thus leading to high efficiency and reliable benefit over traditional switched-capacitor dc-dc converters. Moreover, the bulky capacitor bank existing in traditional switched-capacitor circuits for high power high current application to achieve high efficiency was reduced significantly. Small size, low capacitance, low ESR, high current rating and high temperature rating ceramic capacitors can be employed. Therefore, by using proposed ZCS technique, small size, high power density, high efficiency, high temperature rating, and high current rating switched-capacitor dc-dc converter could be built. Simulation and experimental results are given to demonstrate the validity and features of the soft switching switched-capacitor dc-dc converters.**

I. INTRODUCTION

With the technological development of silicon carbide (SiC) and ceramic materials, very high temperature (250 °C) SiC switching devices and ceramic capacitors will be available. But the magnetic cores become dysfunctional at very high temperature. Therefore, magnetic-less switched-capacitor dc-dc converters operating at very high temperatures become possible and attractive with the adoption of natural air cooling by eliminating bulky heat sinks and magnetic cores. There are three main basic structures of traditional switched-capacitor dc-dc converters that is most popular, Marx generator type switched-capacitor dc-dc voltage multiplier shown in Fig. 1 [1-3], charge pump type multi-level modular switched-capacitor dc-dc converter shown in Fig. 2 [4, 5], and generalized multi-level type switched-capacitor dc-dc converter shown in Fig. 3 [6-9]. The derivation circuits of generalized multi-level type switched-capacitor dc-dc converter are also popular in automotive applications [10-14].

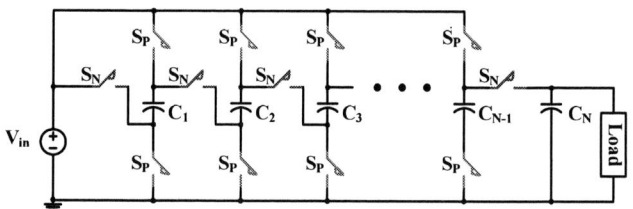

Fig. 1.   Switched-capacitor dc-dc voltage multiplier.

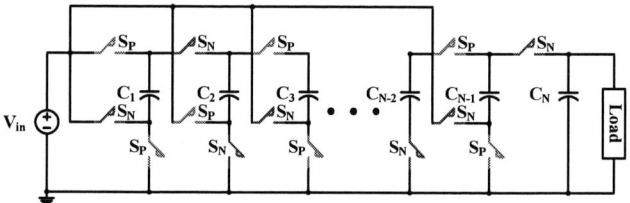

Fig. 2.   Multilevel modular switched-capacitor dc-dc converter.

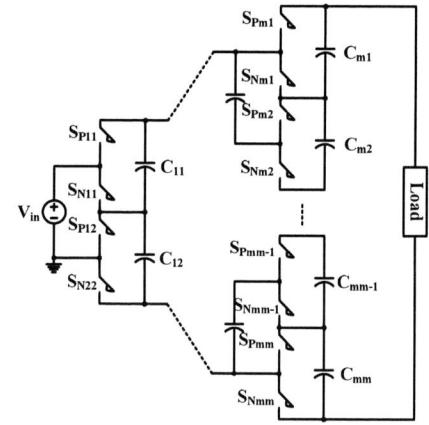

Fig. 3.   Generalized multi-level switched-capacitor dc-dc converter.

However, there are many common drawbacks in traditional switched-capacitor dc-dc converters for high power and high current automotive application. A huge capacitor bank with high capacitance has to be utilized in order to

978-1-4244-4782-4/10 $26.00 © 2010 IEEE

reduce the voltage ripple of the capacitors and achieve high efficiency, which will undoubtedly increase the size of the converter [7, 10, 13, 15]. Besides, with the increase of current rating, huge turn off current leads to unneglectable switching loss of the device. High voltage overshoot is also caused by large turn off current. Moreover, EMI problems caused by the di/dt and dv/dt generated during the switching transient are undesirable in automotive applications.

In order to reduce the switching loss, voltage spikes, and EMI, several ZCS methods have been proposed by inserting an inductor in series with the capacitor [16-21]. But these methods require a relatively big resonant inductor (larger than 1 μH) which is not viable to be achieved by the circuit stray inductance as claimed. By inserting a magnetic core in the switched-capacitor circuit to achieve ZCS is a contradiction by itself. The switched-capacitor circuit with a magnetic core is not a switched-capacitor circuit anymore. Many good features of switched-capacitor will lose including good integration capability, small size and high temperature operation. Also, to utilize the parasitic inductance in the circuit, the switching frequency has to be pushed to tens of megahertz by using these methods, which is not technically feasible [22-24].

This paper presents a new ZCS strategy for a family of switched-capacitor dc-dc converters that is able to overcome the aforementioned drawbacks of traditional switched-capacitor dc-dc converters without inserting big magnetic cores or increasing switching frequency to megahertz level. This new strategy can achieve the ZCS for all the switches without losing any the good features of switched-capacitor circuit compared to the traditional ZCS strategy. This ZCS strategy employs the distributed parasitic inductance in the circuit resonating with the capacitors to provide zero current transition to the switching devices. So, the switching loss is minimized, voltage spike is reduced and EMI is limited. In case the stray inductance is not sufficiently large or equalized as expected, small air core inductors (less than 100n) or layout PCB wires can be used to meet the requirement of the stray inductance. By considering the influence of stray inductance, the power loss is only related to the conduction loss of capacitors and switching devices without considering gate drive and control loss, which is different from the traditional switched-capacitor circuits. Therefore, bulky capacitor banks with high capacitance to reduce voltage ripple and achieve high efficiency are no longer needed, while small size MLCC capacitors with low ESR and low capacitance can be adopted to increase power density and reduce the converter size. No extra components are added in the presented soft switching strategy, thus making it reliable and low in cost. A 160 W ZCS switched-capacitor dc-dc multiplier is analyzed and simulated to confirm the proposed ZCS strategy. A 12 V input 480 W ZCS voltage doubler prototype is also designed and built to confirm the operation, simulation and experiment results are provided.

II. PROPOSED ZCS SWITCHED-CAPACITOR DC-DC CONVERTER FAMILY

Fig. 4, Fig. 5, and Fig. 6 show the proposed ZCS switched-capacitor family utilizing the distributed parasitic inductance existing in the circuit. Fig. 4 shows the N-level ZCS switched-

capacitor dc-dc voltage multiplier. Fig. 5 shows the N-level ZCS multi-level modular switched-capacitor dc-dc converter (ZCS-MMSCC) which has already been discussed in detail in [25]. Fig. 6 shows the N-level ZCS generalized multi-level switched capacitor dc-dc converter. The parasitic inductance mainly includes the stray inductance due to the circuit layout, MOSFETs package parasitic inductance and the capacitor ESL. The stray inductance caused by circuit layout is usually the dominant part. By designing the circuit layout properly, the inductance needed for resonant can be achieved. In case the circuit layout is not well designed or equalized as expected, an air core inductor could be inserted to achieve the prerequisite distributed stray inductance. Because resonant inductors are distributed in the circuit, only small resonant inductance is needed for resonant. Therefore, ZCS of all the switching devices can be achieved by the resonant of the capacitor and the distributed stray inductance in the circuit.

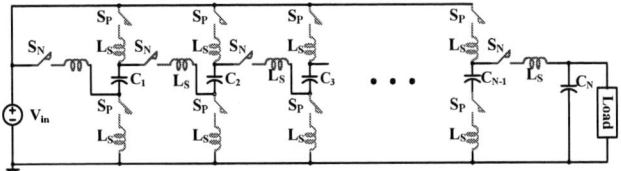

Fig. 4.   ZCS switched-capacitor dc-dc voltage multiplier.

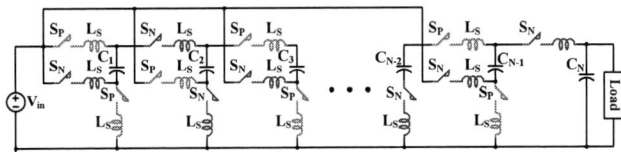

Fig. 5.   ZCS multilevel modular switched-capacitor dc-dc converter.

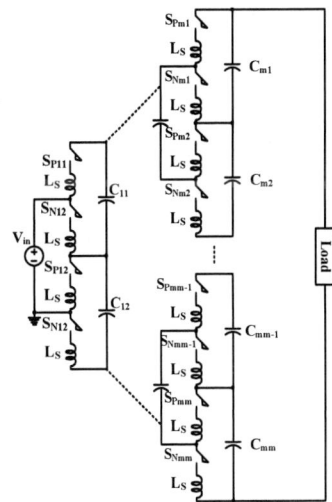

Fig. 6.   ZCS generalized multi-level switched-capacitor dc-dc converter.

III. OPERATION PRINCIPLE

A. ZCS Switched-Capacitor DC-DC Voltage Multiplier

Fig. 7 shows the four-level ZCS switched-capacitor dc-dc voltage multiplier as an example. Similar to the four-level ZCS-MMSCC, this circuit also works as a four times step-up dc-dc converter. V_{in} represents the ideal input voltage source.

$L_{S1} - L_{S7}$ represents the equivalent stray inductance present in the circuit. L_{S2}, L_{S4} and L_{S6} are the equivalent stray inductance when the capacitor is charged in parallel. L_{S1}, L_{S3}, L_{S5} and L_{S7} are the equivalent stray inductance between each capacitor when the capacitor discharges in series. S_P and S_N are the same switching devices controlled complementary at 50% duty cycle. C_1 to C_3 are the capacitors with the average voltage of input voltage, while C_4 has the average voltage of four times of input voltage. L_S does not have to be in the position drawn in Fig. 7, it could be distributed anywhere in series with switches or capacitors in the circuit. Similar to the ZCS-MMSCC, the total equivalent stray inductance is the sum of the connection wire parasitic inductance, capacitor parasitic inductance and the MOSFETs package parasitic inductance. Usually the connection wire parasitic inductance is the major part of the stray inductance. For the analysis convenience, only one equivalent L_S is used to represent the total stray inductance present in each parallel resonant loop. And one equivalent L_S is used between two capacitors in the series resonant loop.

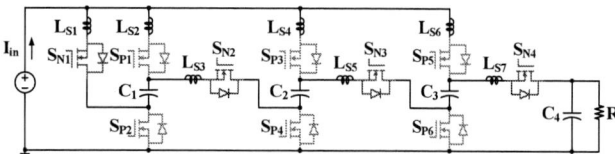

Fig. 7.   Four-level ZCS switched-capcitor voltage multiplier main circuit.

Fig. 8 shows the idealized waveform of proposed ZCS switched-capacitor dc-dc voltage multiplier under steady-state conditions. $V_{GS}\_S_P$ and $V_{GS}\_S_N$ are the gate signal of switch S_P and S_N with 50% duty cycle. $V_{DS}\_S_P$ and $V_{DS}\_S_N$ are the drain source voltage of the switch S_P and S_N. Actually, the drain source voltage of different S_P and S_N are different in value but similar in shape due to the characteristics of voltage multiplier circuit. I_S_P and I_S_N are the drain source current of the switch S_P and S_N. All the switches of S_P or S_N have the same current waveforms. Assume input voltage is an ideal voltage source. By considering the stray inductance present in the circuit, when the switches are turned on, the current through the stray inductance, capacitors and the switch will begin to resonate from zero. By adjusting switching frequency to the resonant frequency, the current through switches S_P, S_N and stray inductance will decrease to zero when the switches are turned off. The half switching period is a half sinusoidal waveform. Therefore, the ZCS of all the switches is achieved in both turn on and turn off. The capacitor C_1 C_2 and C_3 are charged in parallel in the half-period with the sinusoidal current waveform when S_P is on. And they are discharged in series with the input voltage source in another half-period also in the sinusoidal shape when S_N is on. $I\_C_{1,2,3}$ are the current through capacitor C_1, C_2, and C_3, which is the sum of the current through the switch with the sinusoidal shape. When the capacitor is charged, the current is the positive part; when the capacitor discharges, the current is negative. The capacitor C_4 is charged when other capacitors are in series with input voltage when S_N is on. And it is discharged to the load when S_P is on. I_C_4 is the current through capacitor C_4, which is a sinusoid waveform when S_N is on When S_P is on, the current through C_4 is the negative dc load current with small ripple.

$V\_C_{1,2,3}$ is the voltage across the capacitor C_1, C_2 and C_3 with the input voltage dc offset and a sinusoidal ripple. The voltage ripple is determined by the capacitor current and capacitance. V_C_4 is the voltage across the capacitor C_4, with the dc offset four times input voltage and a voltage ripple also determined by the capacitor current and capacitance. The operation of this circuit can be described in two states shown in Fig. 9 and Fig. 10 with different switches turned on.

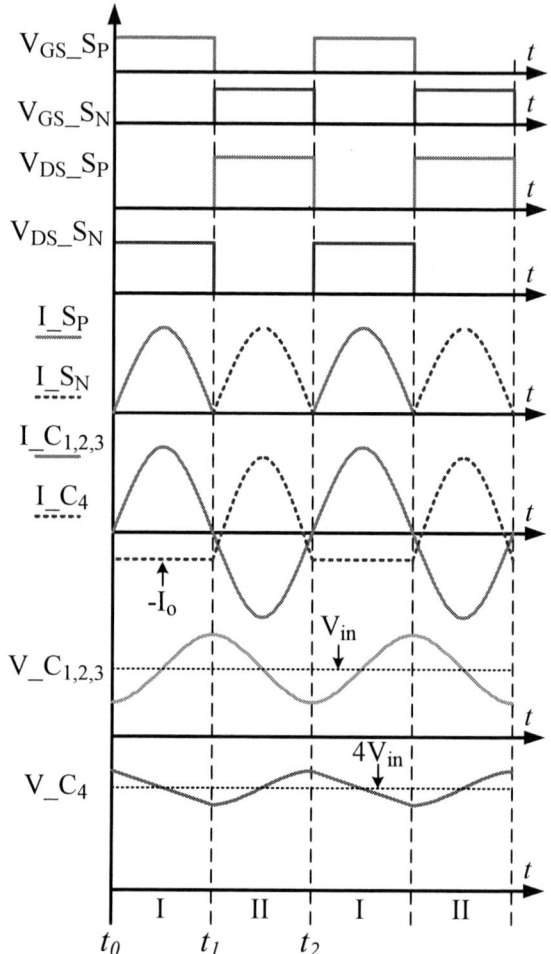

Fig. 8.   Ideal waveforms of four-level ZCS voltage multiplier.

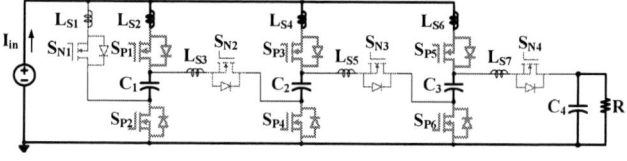

Fig. 9.   Operation modes of state I when S_P is on.

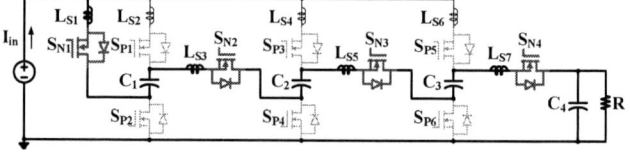

Fig. 10.   Operation modes of state II when S_N is on.

1) State I [t₀,t₁]

1) State I [t_0, t_1]

Fig. 9 shows the state when S_P is turned on at $t = t_0$ while S_N is off. During this state C_1, C_2 and C_3 are charged by V_{in}. Fig. 11 show the three simplified equivalent circuits of state I. Fig. 11(a) shows the situation when V_{in}, L_{S2}, S_{P1}, C_1 and S_{P2} form a resonant loop. Fig. 11(b) shows the situation when V_{in}, L_{S4}, S_{P3}, C_2 and S_{P4} form a resonant loop. Fig. 11(c) shows the situation when V_{in}, L_{S6}, S_{P5}, C_3 and S_{P6} form a resonant loop. Because of the presence of the stray inductance L_S, before the switch is turned on, the current through L_S already decreases to zero. The current through S_P will increase from zero when the switch is turned on, so S_P is turned on at zero current. For the case shown in Fig. 11(a), after L_{S2} and C_1 resonate for half cycle, the current through S_{P1} and S_{P2} falls to zero. Therefore S_{P1} and S_{P2} turn off at zero current. Similarly, for the case shown in Fig. 11(b) and Fig. 11(c) the current through S_{P3}, S_{P4}, S_{P5} and S_{P6} will realize zero current turn on and turn off too. So, ZCS is achieved on all the switches. The required stray inductance value is the same for the three circuit and is easy to be achieved by the circuit layout due to the symmetry of the circuit, assuming the capacitance are the same.

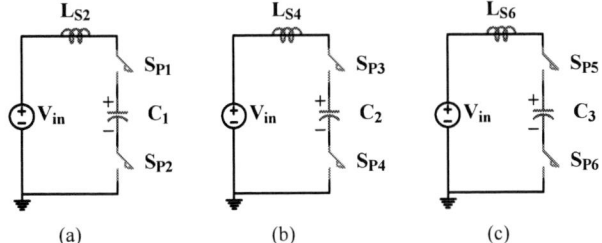

Fig. 11. Simplified equivalent circuits of state I when S_P is on.

Without loss of generality, the following assumptions have been made for the analysis: all the switches are ideal, i.e. no conduction resistance is considered; input voltage source is ideal, i.e. constant and no internal impedance; the capacitor ESR is zero. Assuming $C_1 = C_2 = C_3 = C_4$, so in order to have the same resonant frequency, $L_{S2} = L_{S4} = L_{S6}$, Assume $L_{S1} = L_{S3} = L_{S5} = L_{S7}$. The state equations of Fig. 11(a) are:

$$V_{in} = L_{S2} \frac{di_{S2}}{dt} + v_{C_1} \qquad (1)$$

$$i_{L_{S2}} = C_1 \frac{dv_{C_1}}{dt} \qquad (2)$$

The solutions are:

$$i_{S2(t)} = \frac{\pi P_o}{4V_{in}} \sin \omega_r t \qquad (3)$$

$$v_{C_1}(t) = V_{in} - \frac{\pi P_o}{4V_{in} C_1 \omega_r} \cos \omega_r t \qquad (4)$$

Where V_{in} is the value input voltage, L_{S2} is the value of stray inductance, ω_r is the resonant frequency equals to $1/\sqrt{L_{S2}C_1}$, and P_o is the output power. After half cycle, the capacitor voltage is charged to:

$$v_{C_1}(t_1) = V_{in} + \frac{\pi P_o}{4V_{in} C_1 \omega_r} \cdot \qquad (5)$$

The capacitor voltage ripple is:

$$\Delta v_{C_1} = \frac{\pi P_o}{4V_{in} C_1 \omega_r} \qquad (6)$$

The state equations of Fig. 11(b) and Fig. 11(c) are similar to the state equations of Fig. 11(a). The voltage across C_2 and C_3 are the same as the voltage across C_1. The voltage ripples of C_2 and C_3 are also the same as Δv_{C1}.

2) State II [t_1, t_2]

Fig. 10 shows the state when S_N is turned on at $t = t_1$ while S_P is off. During this state, C_4 is charged by the V_{in} C_1 C_2 and C_3 in series. Assume the load is zero. Fig. 12 shows the simplified equivalent circuits of state II. It shows the situation when V_{in}, L_{S1}, S_{N1}, C_1, L_{S3}, S_{N2}, C_2, L_{S5}, S_{N3}, C_3, L_{S7}, S_{N4}, and C_4 form a resonant loop. Similar as the last state, because of the presence of the L_S, the current through S_N will also increase its value from zero in a resonant manner. The zero current turn-on of S_N can be achieved. After stray inductance L_{S1}, L_{S3}, L_{S5} and L_{S7} in series resonate with capacitor C_1, C_2, C_3 and C_4 in series for half cycle at $t = t_2$, S_{N1}, S_{N2}, S_{N3} and S_{N4} will have a zero current turn-off when the current through them decreases to zero. Hence, zero current switching can be achieved on these switches.

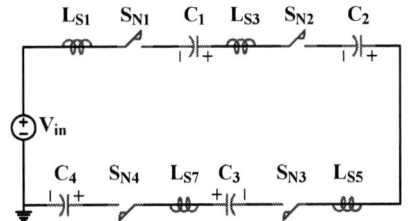

Fig. 12. Simplified equivalent circuits of state II S_N is on.

The state equations of Fig. 12 are:

$$V_{in} + v_{C_1} + v_{C_2} + v_{C_3} = 4L_{S1} \frac{di_{L_{S1}}}{dt} + v_{C_4} \qquad (7)$$

$$i_{L_{S1}} = C_4 \frac{dv_{C_4}}{dt} \qquad (8)$$

The solutions are:

$$v_{C_4}(t) = 4V_{in} - \frac{\pi P_o}{4V_{in} C_4 \omega_r} \cos \omega_r t \qquad (9)$$

$$i_{L_{S1}}(t) = \frac{\pi P_o}{4V_{in}} \sin \omega_r t \qquad (10)$$

After half cycle, the capacitor voltage is charged to

$$v_{C_4}(t_2) = 4V_{in} + \frac{\pi P_o}{4V_{in} C_4 \omega_r} \qquad (11)$$

The capacitor voltage ripple is:

$$\Delta v_{C4} = \frac{\pi P_o}{4 V_{in} C_4 \omega_r} \qquad (12)$$

If zero load is considered, the required stray inductance in the series resonant loop is the same as the parallel resonant loop shown in Fig. 11 which means $L_{S1} = L_{S3} = L_{S5} = L_{S7} = L_{S2} = L_{S4} = L_{S6}$. After considering the load effect with a negative dc offset in the capacitor C_4 current. The required stray inductance of series resonant loop will be a little bit smaller than the parallel resonant loop. Because the resonant frequency of a pure LC network is different with the LCR network.

B. ZCS Voltage Doubler

Fig. 13(a) shows the two-level ZCS generalized multi-level switched-capacitor dc-dc converter as an example. This circuit is called ZCS voltage doubler in short because the output voltage is twice as much as the input. V_{in} represents the ideal input voltage source. L_{S1}, L_{S2} L_{S3} and L_{S4} can be considered as the distributed stray inductance in series with the switches in the circuit layout. S_P and S_N are the same switching devices controlled complementary at 50% duty cycle. C_1 and C_2 are the capacitors with the average voltage the same as the input voltage. Because of the symmetry of the circuit, the distributed stray inductance L_{S1}, L_{S2} L_{S3} and L_{S4} can be replaced with an equivalent stray inductance L_S in the input side shown in Fig. 13(b) for the analysis convenience. Only one equivalent stray inductance L_S is needed to represent the total stray inductance in two resonant loop.

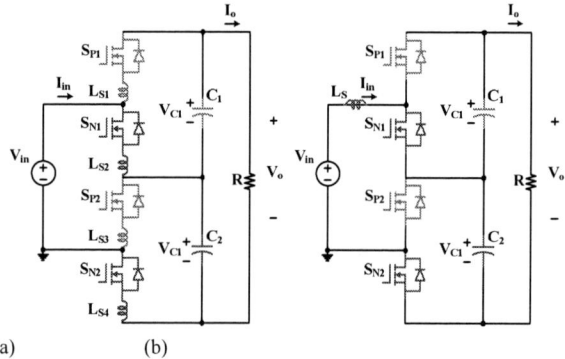

(a) (b)

Fig. 13. ZCS voltage doubler and its topology simplification.

Fig. 14 shows the idealized waveform of proposed ZCS voltage doubler under steady-state conditions. $V_{GS}\_S_P$ and $V_{GS}\_S_N$ are the gate drive signal of switch S_P and S_N complementary with 50% duty cycle. $V_{DS}\_S_{P1}$ to $V_{DS}\_S_{N2}$ are the drain source voltage of switch S_P and S_N. I_S_P and I_S_N are the drain source current of the switch S_P and S_N. All the switches of S_P or S_N have the same current waveforms. Assume input voltage is an ideal voltage source. By considering the equivalent stray inductance in the input side, when the switches are turned on, the current through the stray inductance, capacitors and the switch will begin to resonate from zero. By adjusting switching frequency to the resonant frequency, the current through switches S_P, S_N and stray inductance will decrease to zero when the switches are turned off, which is the half period of the sinusoidal waveform. Therefore, the ZCS of all the switches is achieved in both turn

on and turn off. I_C_1 and I_C_2 are the current through capacitor C_1 and C_2. The capacitor C_1 and C_2 are charged in each half-period with the sinusoidal current waveform when S_P or S_N is on. And they are always discharged in series with load current. So, the current through the capacitor C_1, and C_2, is the current through the switch subtracts the output current. V_C_1 and V_C_2 are the voltage across the capacitor C_1 and C_2. Vo is the output voltage with the voltage ripple half of the voltage ripple of the capacitor. Because of the 180 degree phase shift of the capacitor voltage ripple, the capacitor voltage ripple will cancel out with each other. The operation of the circuit can be described in two states as shown in Fig. 15(a) and Fig. 15(b) with different switches turned on.

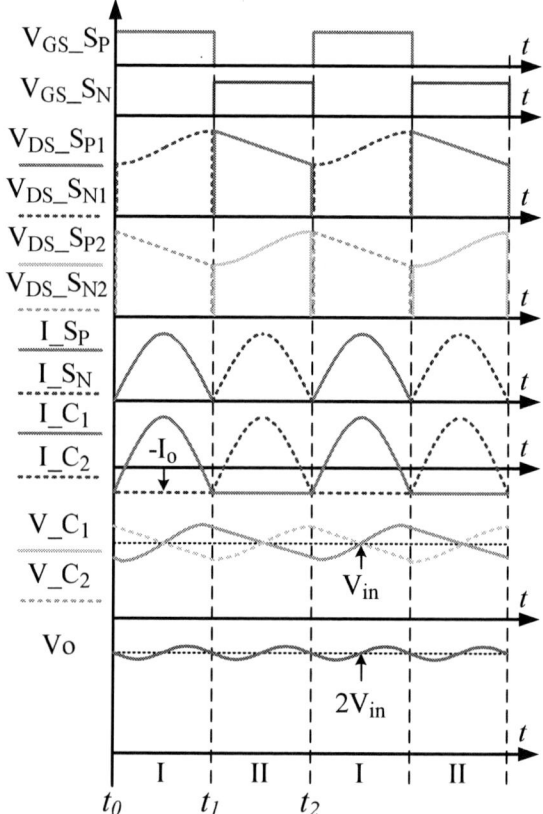

Fig. 14. Ideal waveforms of ZCS voltage doubler.

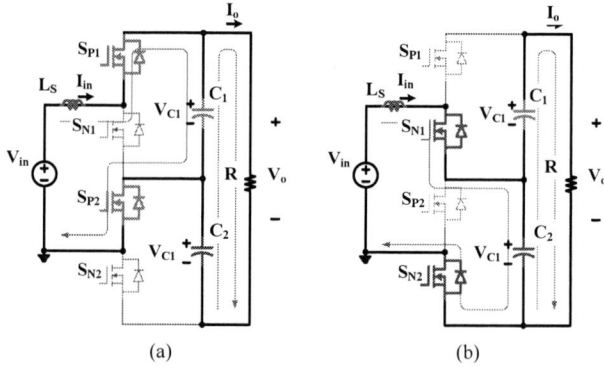

(a) (b)

Fig. 15. (a)Operation modes of state I when S_P is on. (b) Operation modes of state II when S_N is on.

IV. Design Guidelines

The 480 W ZCS voltage doubler design procedure is shown here as an example. The specification of the prototype converter is V_{in}=12 V, V_o=24 V, P_o=480 W, f_S=44.25 kHz.

A. Capacitance

Capacitance value should be chosen according to the voltage ripple of the capacitor using the following equation.

$$C = \frac{\pi P_o}{2 V_{in} \Delta v_C \omega_r} \qquad (13)$$

Because of the special structure of the circuit, the voltage ripple across all the capacitors are the same. And the value of the voltage ripple is determined by the output power, input voltage, capacitance and the resonant frequency. The capacitor voltage ripple should be chosen smaller than the input voltage to prevent the voltage across the capacitor from resonating to the negative region and lose the zero current switching. The output voltage ripple is about five times smaller than the capacitor voltage ripple because of the 180 degrees phase shift operation of the switches. And the capacitor voltage ripple will cancel out and the output voltage is two times bigger than one capacitor voltage. The output ripple is chosen around 8% of output voltage. So the capacitor voltage ripple is 40% of the capacitor voltage. According to the (13), the capacitance is chosen 47 µF in the simulation.

B. Stray inductance

The circuit layout should be designed carefully, in order to make the stray inductance of each equivalent circuit equal. Assume all the stray inductance in the circuit is equalized. And the required stray inductance can be satisfied by using proper input inductance. Required stray inductance for ZCS can be determined by the resonant LCR network.

C. Switching frequency

Switching frequency should be set the same as resonant frequency in order to promise the zero current switching. Usually, the stray inductance of the circuit is already determined when the circuit layout is finished. The capacitance is also determined by the required voltage ripple. Assume the air core inductor is used for the input inductance, and the switching frequency can be determined by the input inductance value. If the input inductance is determined, the switching frequency can be twisted by monitoring the power device current when the switch is turned off. The switching frequency should be set exactly at the point when the switch current resonates to zero when the switch is turned off.

V. Simulation and Experiment Results

A. ZCS Voltage Multiplier

Fig. 16 shows saber® simulation waveforms of a 160 W four-level ZCS voltage multiplier, where Vp and Vn are the switch gate-source control voltage, Vds_Sp and Vds_Sn are the switch drain-source voltage, I_Sp and I_Sn are the switch drain-source current. I_C1,2,3 is the current through capacitor C_1, C_2 and C_3. I_C4 is the current through capacitor C_4. V_C1,2,3 is the voltage across the capacitor C_1, C_2 and C_3. Vin is the input voltage. Vo is the output voltage or the

voltage across the capacitor C_4. The input voltage is 5 V. Switching frequency is about 70 kHz. Capacitance is 47 µF and the value of the stray inductance $L_{S2} = L_{S4} = L_{S6} = 108$ nH, $L_{S1} = L_{S3} = L_{S5} = L_{S7} = 98$ nH. The series resonant stray inductance is smaller caused by the load effect. The simulation results is consistent with the theoretical analysis, which verifies the above analysis.

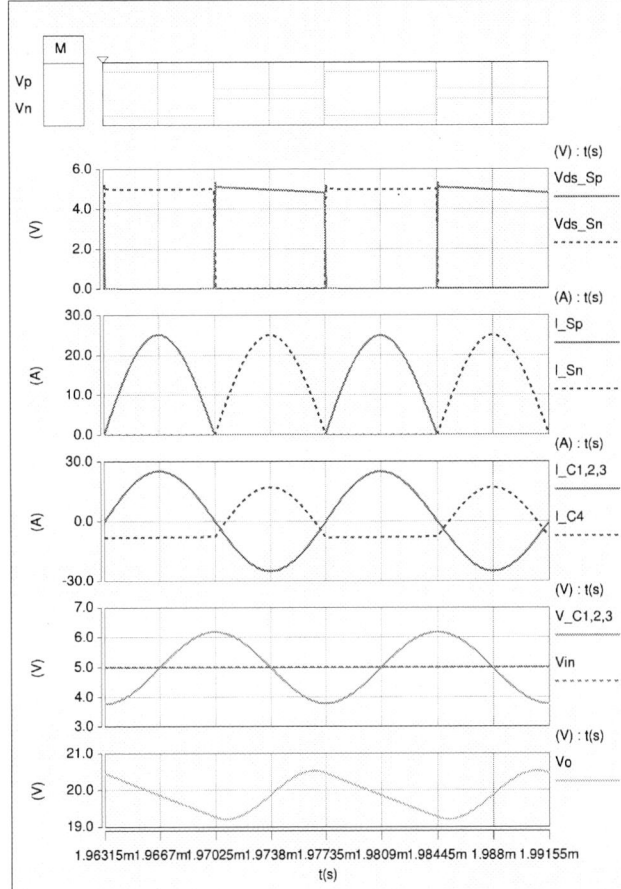

Fig. 16.   Simulation results of 160 W ZCS voltage multiplier.

B. ZCS Voltage Doubler

Fig. 17 and Fig. 18 shows saber® simulation waveforms of a 480 W ZCS voltage doubler, where Sp and Sn are the switch gate-source control voltage, Vds_Sp1, Vds_Sn1, Vds_Sp2, Vds_Sn2 are the switch drain-source voltage of switch S_{P1}, S_{N1}, S_{P2} and S_{N2}. I_Sp1 and I_Sn1 are the switch drain-source current of switches S_{P1} and S_{N1}. which is omitted here. The current through S_{P2} is the same as S_{P1}, the current through S_{N2} is the same as S_{N1}. Iin is the current through the input stray inductance L_S. I_C1 and I_C2 are the current through capacitor C_1 and C_2. V_C1 and V_C2 are the voltage across the capacitor C_1 and C_2. Vin is the input voltage. Vo is the output voltage. The input voltage is 12 V. Switching frequency is about 44 kHz. Capacitance is 47 µF and stray inductance $L_S = 165$ nH. The simulation results is consistent with the theoretical analysis, which verifies the analysis.

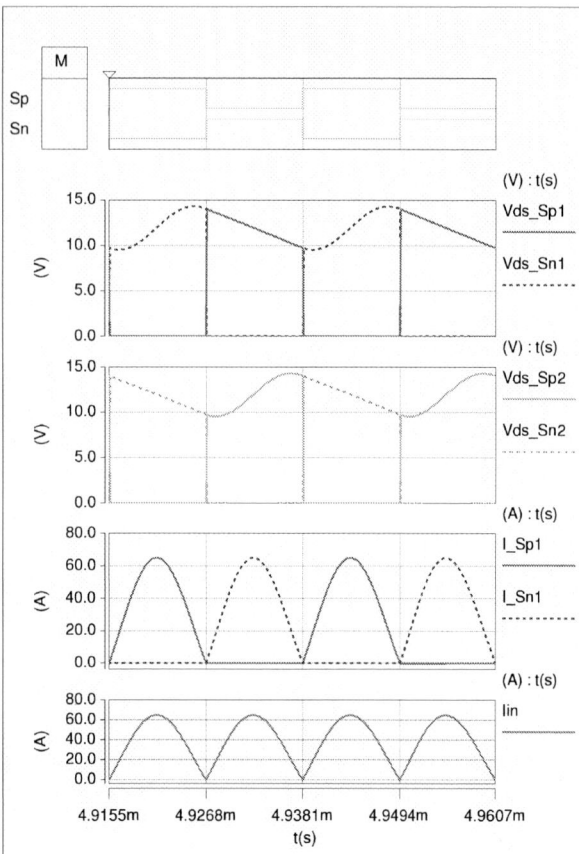

Fig. 17.   Simulation results of 480 W ZCS voltage doubler.

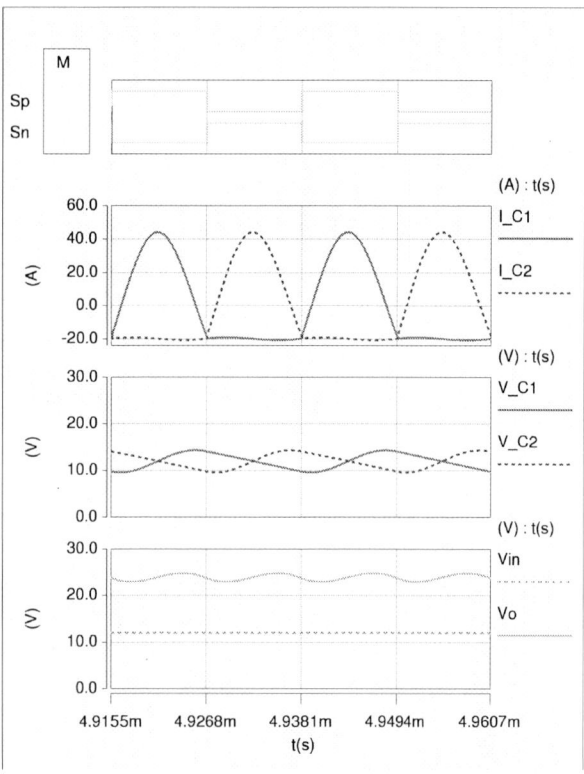

Fig. 18.   Simulation results of 480 W ZCS voltage doubler.

In the 480 W ZCS voltage doubler prototype, the switching devices are two 30 V 180 A MOSFETs IPB009N03L from infineon connected in parallel. Resonant capacitor are ten 100 V 4.7 µF MLCC capacitors C5750X7R2A475K from TDK connected in parallel. Input capacitor are twelve 16 V 470 µF conductive polymer aluminum solid electrolytic capacitors PLG1C471MDO1 from nichicon connected in parallel. Gate drive voltage is 7 V. The switching frequency is about 40 kHz. The input stray inductance is around 200 nH.

Fig. 19 and Fig. 20 show the experiment waveforms of the 480 W ZCS voltage doubler prototype with the parts mentioned above. Fig. 19 shows the complementary gate drive signals of S_P and S_N with duty cycle about 49% due to some dead time. $V_{GS\_S_P}$ and $V_{GS\_S_N}$ are the complementary gate-source control voltage.

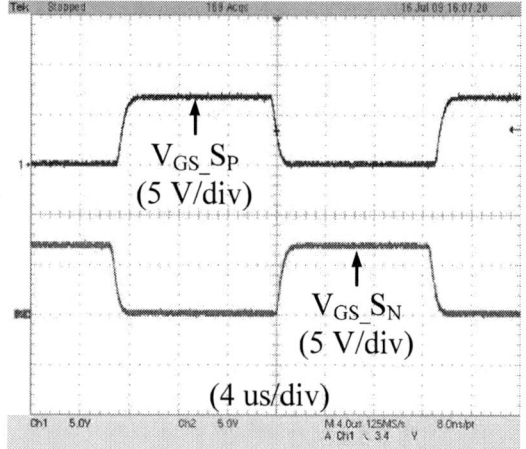

Fig. 19.   Gate drive signal of ZCS voltage doubler.

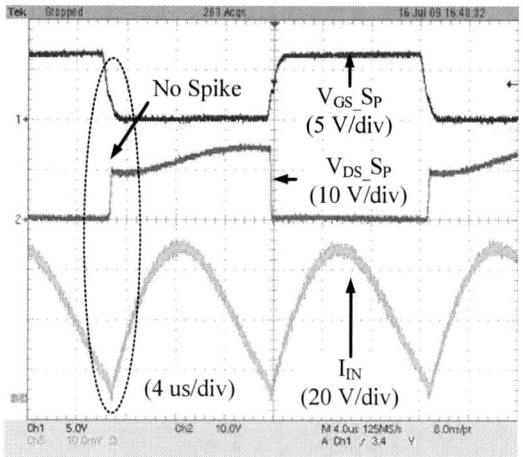

Fig. 20.   Switch voltage and input current waveform of ZCS voltage doubler.

VI. CONCLUSION

In this paper, a family of zero current switching switched-capacitor dc-dc converters is proposed. By eliminating the bulky, lossy inductive component with a magnetic core, the converter is able to operate at very high temperatures with high efficiency. By using the proposed soft-switching strategy, switching loss has been minimized and EMI has been restricted; the size of capacitors has been reduced, thus making the converter small and light. Owing to the utilization of distributed stray inductance or distributed air core inserting technique, the big inductor is avoided to achieve ZCS. Hence, the proposed ZCS switched-capacitor converter family shows great potential in high temperature, high power future automotive applications.

REFERENCES

[1] F. Ueno, T. Inoue, I. Oota and I. A. H. I. Harada, "Power supply for electroluminescence aiming integrated circuit," in *Circuits and Systems, 1992. ISCAS '92. Proceedings., 1992 IEEE International Symposium on*, 1992, pp. 1903-1906 vol.4.

[2] M. S. Makowski and D. Maksimovic, "Performance limits of switched-capacitor DC-DC converters," 1995, pp. 1215-1221 vol.2.

[3] M. On-Cheong, W. Yue-Chung and A. Ioinovici, "Step-up DC power supply based on a switched-capacitor circuit," *Industrial Electronics, IEEE Transactions on*, vol. 42, pp. 90-97, 1995.

[4] J. F. Dickson, "On-chip high-voltage generation in MNOS integrated circuits using an improved voltage multiplier technique," *Solid-State Circuits, IEEE Journal of*, vol. 11, pp. 374-378, 1976.

[5] F. H. Khan and L. M. Tolbert, "A Multilevel Modular Capacitor Clamped DC-DC Converter," in *Industry Applications Conference, 2006. 41st IAS Annual Meeting. Conference Record of the 2006 IEEE*, 2006, pp. 966-973.

[6] F. Z. Peng, "A generalized multilevel inverter topology with self voltage balancing," *Industry Applications, IEEE Transactions on*, vol. 37, pp. 611-618, 2001.

[7] F. Z. Peng, F. Zhang and Z. Qian, "A magnetic-less DC-DC converter for dual-voltage automotive systems," *Industry Applications, IEEE Transactions on*, vol. 39, pp. 511-518, 2003.

[8] M. Shen, F. Z. Peng and L. M. Tolbert, "Multilevel DC-DC Power Conversion System With Multiple DC Sources," *Power Electronics, IEEE Transactions on*, vol. 23, pp. 420-426, 2008.

[9] T. Umeno, K. Takahashi, I. Oota, F. Ueno and T. Inoue, "New switched-capacitor DC-DC converter with low input current ripple and its hybridization," 1990, pp. 1091-1094 vol.2.

[10] Z. Pan, F. Zhang and F. Z. Peng, "Power losses and efficiency analysis of multilevel dc-dc converters," in *Applied Power Electronics Conference and Exposition, 2005. APEC 2005. Twentieth Annual IEEE*, 2005, pp. 1393-1398 Vol. 3.

[11] F. Zhang, L. Du, F. Z. Peng and Z. Qian, "A New Design Method for High-Power High-Efficiency Switched-Capacitor DC-DC Converters," *Power Electronics, IEEE Transactions on*, vol. 23, pp. 832-840, 2008.

[12] W. Qian, F. Z. Peng, M. Shen and L. M. Tolbert, "3X DC-DC Multiplier/Divider for HEV Systems," in *Applied Power Electronics Conference and Exposition, 2009. APEC 2009. Twenty-Fouth Annual IEEE*, 2009.

[13] F. Z. Peng, F. Zhang and Z. Qian, "A novel compact DC-DC converter for 42 V systems," 2003, pp. 33-38 vol.1.

[14] X. Ming, J. Sun and F. C. Lee, "Voltage divider and its application in the two-stage power architecture," in *Applied Power Electronics Conference and Exposition, 2006. APEC '06. Twenty-First Annual IEEE*, 2006, p. 7 pp.

[15] F. Zhang, L. Du, F. Z. Peng and Z. Qian, "A new design method for high efficiency DC-DC converters with flying capacitor technology," in *Applied Power Electronics Conference and Exposition, 2006. APEC '06. Twenty-First Annual IEEE*, 2006, p. 5 pp.

[16] K. W. E. Cheng, "Zero-current-switching switched-capacitor converters," *Electric Power Applications, IEE Proceedings -*, vol. 148, pp. 403-409, 2001.

[17] K. W. E. Cheng, "New generation of switched capacitor converters," in *Power Electronics Specialists Conference, 1998. PESC 98 Record. 29th Annual IEEE*, 1998, pp. 1529-1535 vol.2.

[18] Y. P. B. Yeung, K. W. E. Cheng, D. Sutanto and S. L. Ho, "Zero-current switching switched-capacitor quasiresonant step-down converter," *Electric Power Applications, IEE Proceedings -*, vol. 149, pp. 111-121, 2002.

[19] Y. P. B. Yeung, K. W. Cheng, S. L. Ho and D. Sutanto, "Generalised analysis of switched-capacitor step-down quasi-resonant converter," *Electronics Letters*, vol. 38, pp. 263-264, 2002.

[20] Y. P. B. Yeung, K. W. Cheng, S. L. Ho, K. K. Law and D. Sutanto, "Unified analysis of switched-capacitor resonant converters," *Industrial Electronics, IEEE Transactions on*, vol. 51, pp. 864-873, 2004.

[21] K. K. Law, K. W. E. Cheng and Y. P. B. Yeung, "Design and analysis of switched-capacitor-based step-up resonant converters," *Circuits and Systems I: Regular Papers, IEEE Transactions on [see also Circuits and Systems I: Fundamental Theory and Applications, IEEE Transactions on]*, vol. 52, pp. 943-948, 2005.

[22] A. Ioinovici, "Switched-capacitor power electronics circuits," *Circuits and Systems Magazine, IEEE*, vol. 1, pp. 37-42, 2001.

[23] A. Ioinovici, C. K. Tse and H. S. H. Chung, "Comments on "Design and analysis of switched-capacitor-based step-up resonant Converters"," *Circuits and Systems I: Regular Papers, IEEE Transactions on [see also Circuits and Systems I: Fundamental Theory and Applications, IEEE Transactions on]*, vol. 53, p. 1403, 2006.

[24] A. Ioinovici, S. H. C. Henry, S. M. Marek and K. T. Chi, "Comments on Unified Analysis of Switched-Capacitor Resonant Converters," *Industrial Electronics, IEEE Transactions on*, vol. 54, pp. 684-685, 2007.

[25] D. Cao and F. Z. Peng, "Zero-Current-Switching Multilevel Modular Switched-Capacitor DC-DC Converter," in *IEEE Energy Conversion Congress and Exposition, 2009. ECCE 2009. San Jose*, 2009.

978-1-4244-4782-4/10 $26.00 © 2010 IEEE

Analysis and Design of the Half Bridge Magnetizing Inductor Resonant(L_mC) DC/DC Converter

B.-C. Hyeon and B.-H. Cho

Department of Electrical Engineering, Seoul National University
San 56-1, Sillim-dong, Kwanak-gu, Seoul, 151-742, Korea
E-mail: novasa@naver.com

Abstract—**Half bridge magnetizing inductor (L_mC) resonant dc/dc converter reduces the high current ripple of the output capacitor in LLC resonant converter maintaining narrow switching frequency and soft switching. This paper presents analysis and design of the half bridge L_mC resonant converter. Using the extended describing function (EDF), the steady state analysis is carried out. Based on the results, the design considerations are suggested. The hardware prototype with 24-V/10-A is built and tested to verify the analysis results.**

I. INTRODUCTION

In recent years, LLC resonant converter is widely used for an isolated dc/dc converter due to its attractive features such as: 1) zero voltage switching (ZVS) of MOSFETs and zero current switching (ZCS) of the rectifier diodes; 2) narrow switching frequency range according to the load current; 3) simple circuitry and low cost; 4) utilization of the parasitic elements for the soft switching [1]-[5].

However, since LLC resonant tank has current source characteristic and it is connected to a capacitive output filter, the current ripple in the output capacitor is large. Hence, many capacitors are needed to satisfy the maximum current rating of the electrolytic capacitor and to reduce the voltage ripple especially in a high output current application. Therefore, multi phase LLC resonant converter is suggested in [1, 2]. Although N-phase converter reduces the current ripple of the output capacitor to 1/N, active and passive elements of the converter are increased with N times. Furthermore, an additional control is required for the current sharing.

To reduce the current ripple without additional elements, the half bridge magnetizing inductor (L_mC) resonant dc/dc converter was proposed as a topological approach [6]. The circuit diagram of the L_mC resonant converter is shown in Fig. 1. While it is similar to the LLC half bridge resonant converter in Fig. 2, there are two differences. The series resonant inductor (L_r) is eliminated in primary side of the transformer (T_1) and the main resonant tank which transfers the power to the load is changed. The output filter inductor (L_o) is used to reduce the current ripple of the output capacitor (C_o) and it is decreased to 1/6 maintaining the advantages of LLC resonant converter such as soft switching, narrow switching frequency

range. Thus, the converter is suitable for the high current application with low cost and high power density.

In this paper, the steady state characteristics of the half bridge L_mC resonant converter such as voltage gain, peak value of the resonant elements are analyzed. Since a number of techniques have been developed to explore the resonant converter [7]-[10], the extended describing function (EDF) method with high order terms is used for the analysis. Based on the result, the design considerations are proposed for the efficient power stage. To show the validity of the analysis results, the simulation and experiment with 24-V/10-A hardware prototype are carried out for SMPS of the laser printer.

II. STEADY STATE ANALYSIS USING EDF METHOD

Typical waveforms of the L_mC resonant converter are shown in Fig. 3. The half bridge switch network generates v_a voltage to the input of the resonant tank. The duty ratio of S_H is fixed to 50% and the short dead time is neglected in the analysis. ZVS on is achieved by resonant current during the commutation which is occurred after switches are turned off.

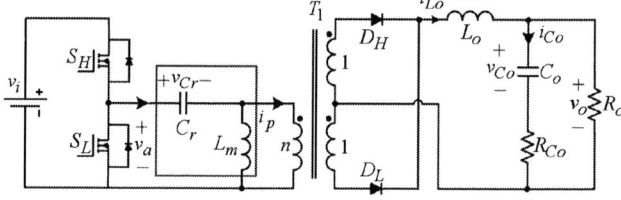

Figure 1.   The magnetizing inductor (LmC) resonant dc/dc converter.

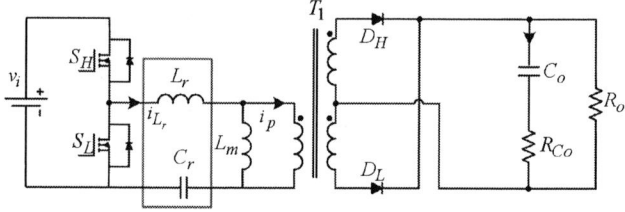

Figure 2.   The LLC half bridge resonant dc/dc converter.

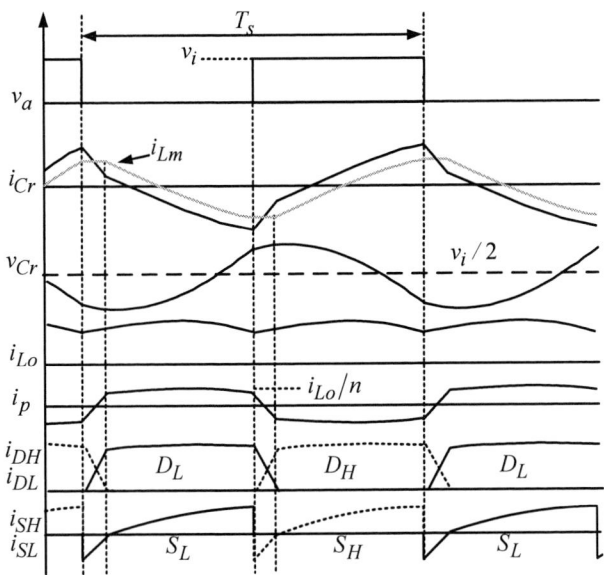

Figure 3. The typical waveforms of L_mC resonant converter.

Since L_m is a main resonant inductor, its current shape is sinusoidal. Assume that the current of L_o is continuous and almost constant in rated load condition for the simple analysis. Thus, the current of the primary side of T_1 (i_p) is square wave due to the center-tapped rectifier. The voltage across L_m is rectified and filtered for the output voltage (v_o). The detail modes of operations are explained in [6].

To find steady state characteristics of the converter, EDF is applied because it is useful regardless of the resonant scheme (series, parallel or series-parallel etc.) and resonant region (above, below etc.). Since the switching frequency is close to the resonant frequency of (1), the states $\{i_{Lm}, v_{Cr}, v_{Cd}\}$ are quite similar to the sinusoidal. Therefore, they can be approximated to the combination of sin and cosine terms like (2).

$$\omega_o \left(resonant\ frequency\right) = \frac{1}{\sqrt{L_m C_r}} \qquad (1)$$

$$v_{Cr}(t) \cong v_{Crs}(t)\sin \omega_s t + v_{Crc}(t)\cos \omega_s t$$
$$v_{Cd}(t) \cong v_{Cds}(t)\sin \omega_s t + v_{Cdc}(t)\cos \omega_s t \qquad (2)$$
$$i_{Lm}(t) \cong i_{Lms}(t)\sin \omega_s t + i_{Lmc}(t)\cos \omega_s t$$

The nonlinear waveforms such as v_a, i_p are described to the combination of sin and cosine term with functional coefficient $\{f_1, f_2, f_3, f_4\}$.

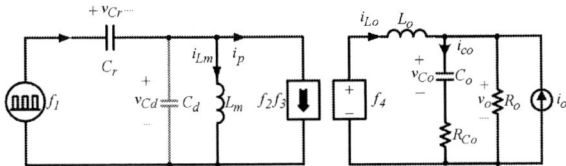

Figure 4. The equivalent circuit model for EDF

Since v_o is generated by the voltage across L_m, the dummy state, v_{Cd}, is inserted to express it as a state. To minimize the effect of dummy state, small C_d is inserted in final result. The steady state operation is modeled to the equivalent circuit for EDF and it is shown in Fig. 4. The corresponding state equations are given by:

$$C_r \frac{dv_{Cr}}{dt} - C_d \frac{dv_{Cd}}{dt} = i_{Lm} + \frac{sgn(v_{Cd})i_{Lo}}{n}$$

$$\left(1 + \frac{R_{Co}}{R_o}\right)C_o \frac{dv_{Co}}{dt} + \frac{v_o}{R_o} = i_{Lo} + i_o , \ L_m \frac{di_{Lm}}{dt} = v_{Cd} \qquad (3)$$

$$L_o \frac{di_{Lo}}{dt} = \frac{|v_{Cd}|}{n} - v_o , \ v_o = (i_{Lo} + i_o)r_c + \left(1 - \frac{r_c}{R_o}\right)v_{Co}$$

$$v_a = v_{Cr} + v_{Cd} \qquad where, \ r_c = (R_{Co} // R_o)$$

Since the effective values of L_o and C_o are large enough, the cut off frequencies are much lower than switching frequency. The nonlinear voltage and current in (3) are substituted to describing functions with the fundamental components of the switching frequency or dc term like (4).

$$v_a(t) \approx f_1(v_i)\sin \omega_s t$$
$$sgn(v_{Cd})i_{Lo} \approx f_2(v_{Cds}, v_{Cdc}, i_{Lo})\sin \omega_s t$$
$$+ f_3(v_{Cds}, v_{Cdc}, i_{Lo})\cos \omega_s t \qquad (4)$$
$$|v_{Cd}| \approx f_4(v_{Cds}, v_{Cdc})$$

To describe the nonlinear waveforms in Fig. 3 to the sinusoidal function, Fourier series expansion is applied. The coefficients of (4) are obtained as (5).

$$f_1 = \frac{2v_i}{\pi}, \ f_2 = \frac{4i_{Lo}v_{Cds}}{n\pi A_p}, \ f_3 = \frac{4i_{Lo}v_{Cdc}}{n\pi A_p}, f_4 = \frac{2A_p}{\pi}$$
$$where, \ A_p = \sqrt{v_{Cds}^2 + v_{Cdc}^2} \qquad (5)$$

By substituting (2), (4), (5) into (3) and by equating the coefficients of dc, sine, cosine terms respectively, the equations of (6) are obtained.

$$f_1 = v_{Crs} + v_{Cds}, \qquad 0 = v_{Crc} + v_{Cdc}$$

$$C_r\left(\frac{dv_{Crs}}{dt} - \omega_s v_{Crc}\right) - C_d\left(\frac{dv_{Cds}}{dt} - \omega_s v_{Cdc}\right) = i_{Lms} + f_2$$

$$C_r\left(\frac{dv_{Crc}}{dt} + \omega_s v_{Crs}\right) - C_d\left(\frac{dv_{Cdc}}{dt} + \omega_s v_{Cds}\right) = i_{Lmc} + f_3 \qquad (6)$$

$$L_m\left(\frac{di_{Lms}}{dt} - \omega_s i_{Lmc}\right) = v_{Cds}, \ L_m\left(\frac{di_{Lmc}}{dt} + \omega_s i_{Lms}\right) = v_{Cdc}$$

$$L_o \frac{di_{Lo}}{dt} = f_4 - v_o, \ v_o = (i_{Lo} + i_o)r_c + \left(1 - \frac{r_c}{R_o}\right)v_{Co}$$

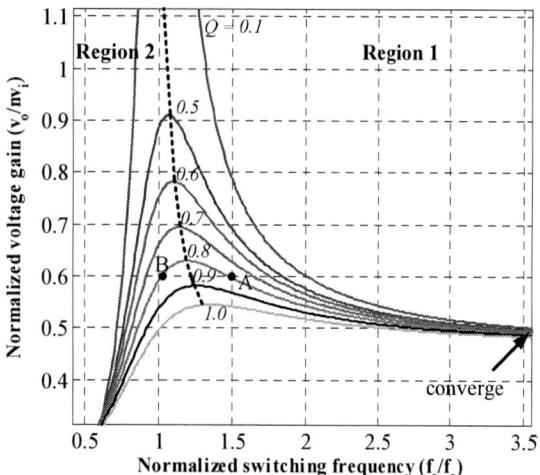

Figure 5. The voltage gain curves according to f_s/f_o.

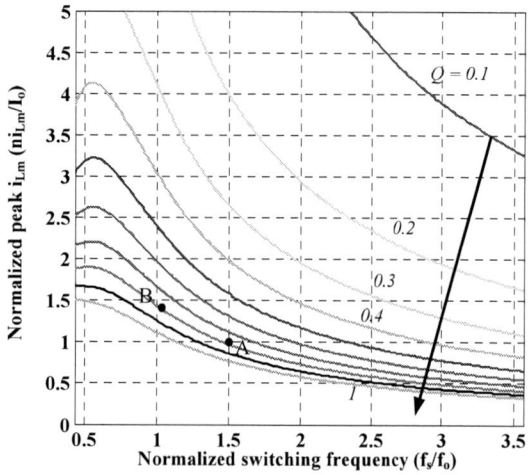

Figure 6. The normalized peak magnetizing current.

Since the coefficients (or envelope) of the states in (2) have slow time varying characteristic, they are treated as a constant for the switching period and the steady state solutions of (6) are given by:

$$V_{Cdc} = \frac{2 R_e \beta V_i}{\pi \left(1 + (\alpha - \beta)^2 R_e^2\right)}, V_{Cds} = \frac{2(\alpha - \beta) R_e^2 \beta V_i}{\pi \left(1 + (\alpha - \beta)^2 R_e^2\right)},$$

$$I_{Lmc} = -\frac{V_{Cds}}{\beta}, I_{Lms} = -\frac{V_{Cdc}}{\beta}, V_{Crc} = -V_{Cdc}, V_{Crs} = \frac{2V_i}{\pi} - V_{Cds} \quad (7)$$

$$V_o = \frac{2 A_p}{n\pi}, \quad \text{where } R_e = \frac{n^2 \pi^2}{8} R_o, \ \alpha = C_r \omega_s, \ \beta = \frac{1}{L_m \omega_s}$$

The dc voltage gain curves according to the load resistance and normalized switching frequency are shown in Fig. 5. As f_s increases, two interesting features are observed. First, the voltage gains converge to 0.5 due to the half bridge inverter with 50% duty ratio. For the much higher frequency region,

PFM method is not applicable because the gain slope according to f_s becomes zero. Second, the voltage gain variations according to Q factor become small. This means that the switching frequency range respect to the load current is narrow. The Q factor is defined in (8).

$$Q = \frac{1}{R_e} \sqrt{\frac{L_m}{C_r}} \ @\ n = 1 \quad (8)$$

The normalized peak current of i_{Lm} is shown in Fig. 6. From the curves, two interesting facts are also observed. First, there is no peak point near the resonant frequency. In Fig. 5, the voltage gain is decreased as f_s decreases in region 2. However, the normalized peak current of i_{Lm} is increased. The second interesting fact is that the normalized peak current is decreased as Q factor is decreased. This means that the relative increment of peak current becomes small along with the load current is increase. Thus, the converter is suitable for the high current application and it is advantageous for the transformer design.

III. DESIGN CONSIDERATIONS

Based on the analysis result, the design considerations for higher efficiency are as follows.

First, the operating of L_mC resonant converter in Region 1 is recommended. Even though Region 2 offers ZCS off of MOSFET and smaller f_s range for the load current than Region 1, the peak current is much higher. At Point A and B in Fig. 5, two operating points provide same voltage gain when $Q = 0.9$. While, the peak value of Point B is higher than Point A in Fig. 6. This means that the circulating current is large. Thus, the conversion efficiency is deteriorated in this region.

Fig. 5 shows that the slope of the gain curve becomes zero at high frequency region. This means that the input characteristic of the resonant tank becomes resistive and the phase of resonant current is not lagged. Thus, ZVS switching is failed in this area. To achieve ZVS at full load, it is recommended that f_s/f_o is less than 3. For the regulation of the output voltage in light or no load, the burst mode operation is used like LLC resonant converter.

In general, the winding loss of L_o affects to the conversion efficiency especially for low voltage high current application. It is verified in [6] that the small inductance value 3.5uH is enough to reduce the current ripple effectively (1/6). Therefore, the small bar type core is proper for the high efficiency and high power density. The minimum operating point is given and (9) should be satisfied to reduce the effect of L_o during the resonant operation.

$$f_{min} = \frac{k}{2\pi \sqrt{\dfrac{n^2 L_o L_m}{n^2 L_o + L_m} C_r}} \quad (9)$$

The value of k is recommended to be 1.1~1.3 by the empirical data through the simulation.

978-1-4244-4782-4/10 $26.00 © 2010 IEEE

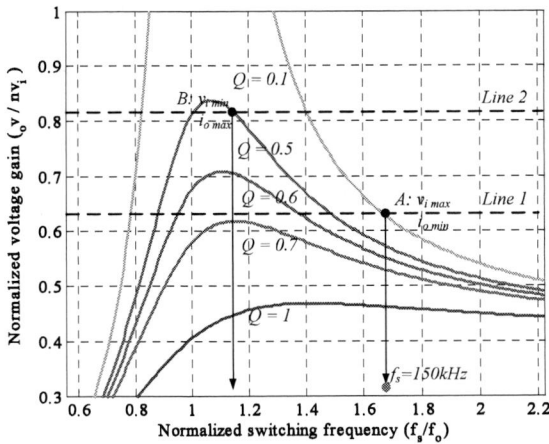

Figure 7. The detail voltage gain curves according to Q @ n=1.

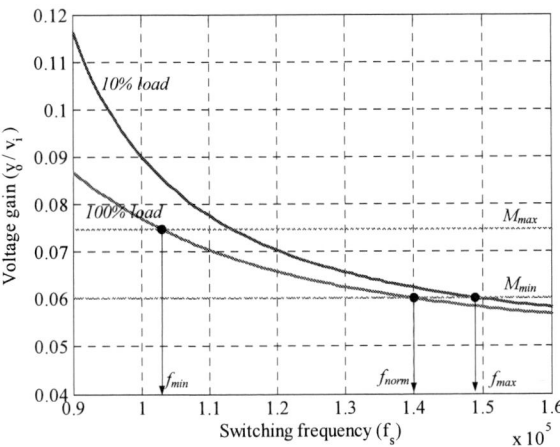

Figure 8. The operating region of the designed converter.

IV. VERIFICATION

To verify the steady state analysis results, the converter is designed for $320V_{dc} \sim 400V_{dc}$ input voltage, 24V output voltage, 10A output current. The value of L_o is selected to 3.2uH with $l=35mm$, $\phi=6mm$ bar type core. To satisfy the required voltage gain, the turn ratio (n) is selected first. The detail gain curves according to Q factor are shown in Fig. 7.

The required maximum gain (M_{max}) is 24/320=0.075 and the minimum gain (M_{min}) is 24/400=0.06. Thus, M_{max}/M_{min}=1.25. To make the maximum switching is smaller than 150kHz with f_o = 90kHz, the maximum normalized switching frequency is 1.67 when the maximum input and minimum load. Since M_{min} is 0.06, the turn ratio is selected to 10:1. For the minimum input voltage, *Line 2* gain is required because M_{max} = 1.25*M_{min}*1.1(10% margin). Therefore, the available Q is selected to 0.5 for the given input voltage range. The resonant frequency was selected as 90kHz and set k=1.2, f_{smin}=100kHz. Then the resonant tank is calculated to L_m=300uH, C_r=10.4nF by (1), (8) and (9). The operating frequencies according to the load are shown in Fig. 8. The other circuit parameters and devices are listed in Table 1. The

TABLE I. PARAMETERS OF THE HARDWARE PROTOTYPE

Device	Parameter	Device	Parameter
L_r	300uH	R_{Co}	90mΩ
C_r	16.5nF	L_o	3.2uH
C_o	35V/1500uF	n:1	10:1
S_H, S_L	5R199P	D_H, D_L	KCH30U10
T_I	EER4950s	IC	L6599, ST

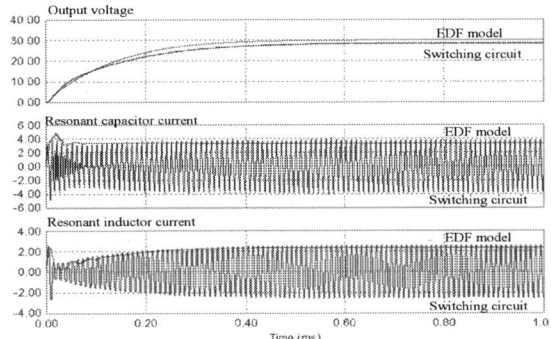

Figure 9. Simulation result of EDF model and switching circuit model.

circuit model using EDF solutions of (7) is built and it is compared to the switching circuit model simulation in Fig. 9 when v_i=400V, R_o=2.4Ω, f_s=125kHz.

In Fig. 9, the output voltage, peak value of resonant current and voltage by EDF and switching circuit model show good agreement during the initial transient and steady state. In simulation, to reduce the steady state error, 3th harmonic component is included in calculation of (6). The high order harmonics are obtained using the relation of (10)

$$state: x\big|_{nth} = \sum_{k=1,3,5...n} x_s\big|_k \cdot \sin k\omega_s + x_c\big|_k \cdot \cos k\omega_s$$
$$fourier\ series: f_a\big|_{nth} = \sum_{k=1,3,5...n} \frac{1}{k}\sin k\omega_s \quad (10)$$

Applying (10) to (2) and (4), the steady state solutions by EDF considering high order terms can be calculated.

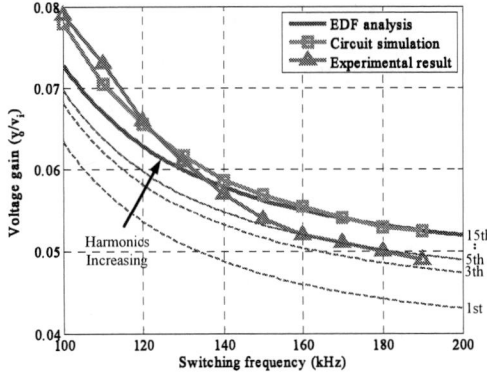

Figure 10. Comparison of the voltage gain when R_o=2.4Ω.

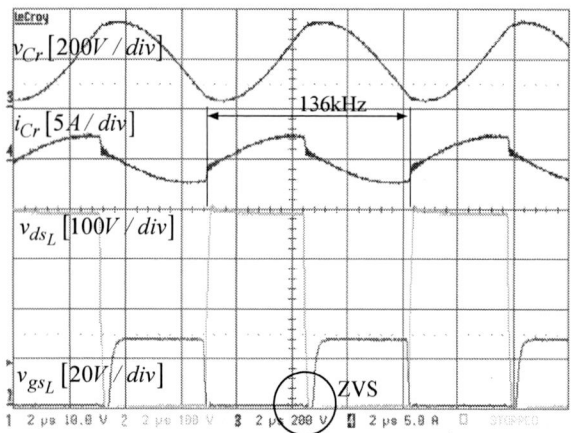

Figure 11. The operating waveform when v_i=400V, R_o=2.4 Ω, f_s=136kHz.

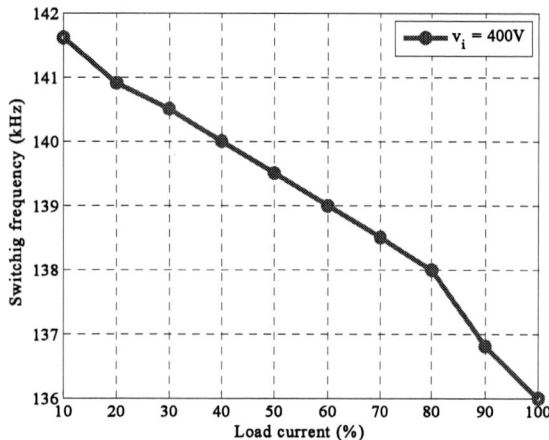

Figure 12. The switching frequency range according to load.

The voltage gain according to switching frequency is shown in Fig. 10 with a fixed load resistance. The obtained gain with fundamental component shows similar curve to the experimental result while there is offset error. It is observed that the error is reduced as the high order harmonics are included. There is a trade off between the accuracy and calculation time and it is recommended considering 3th order term in this application. Fig. 11 shows the operating waveforms of the designed L_mC resonant converter. ZVS on is observed and the converter is operated at 136kHz and it is equal to f_{nom} in Fig. 8. The operation range according to load is shown in Fig. 12 and it is similar to the range from f_{nom} to f_{max} in Fig. 8.

The maximum efficiency of the converter is measure to 94% when the maximum input voltage and the maximum load current.

V. CONCLUSION

This paper presents the steady state analysis and design considerations of the half bridge L_mC resonant converter. The large signal characteristics such as voltage gain, peak stress, frequency ranges are founded from the steady state solutions

using EDF method. The accuracy increasing of the analysis result according to the high order terms is shown. Based on the result, the considerations for practical circuit design under the given specification are suggested. The simulation and experiment using 24V/10A hardware prototype for the laser printer application are carried out to show the validity of the analysis results. Even though the obtained EDF model shows small amount of error, it is applicable to the initial design procedure of the converter.

ACKNOWLEDGMENT

This study is partly supported by Digital Printing Division in SAMSUNG DMC.

REFERENCES

[1] T. Jin and K. Smedley, "Multiphase LLC Series Resonant Converter for Microprocessor Voltage Regulation", in *Industry Applications Conference, 2006.41st IAS Annual Meeting.Conference Record of the 2006 IEEE, 2006, pp.2136-2143.*

[2] Kang-Hyun Yi and Gun-Woo Moon, "A Novel Two Phase Interleaved LLC Series Resonant Converter using a Phase of the Resonant Capacitor", *JOURNAL OF POWER ELECTRONICS*, vol.8, pp.275-279, 2008.

[3] Bo Yang, F. C. Lee, A. J. Zhang, and G. Huang, "LLC resonant converter for front end DC/DC conversion," in *Applied Power Electronics Conference and Exposition, 2002. APEC 2002. Seventeenth Annual IEEE*, 2002, pp. 1108-1112 vol.2.

[4] Hang-Seok Choi, "Design Consideration of Half-Bridge LLC Resonant Converter," *JOURNAL OF POWER ELECTRONICS*, vol. 7, pp. 13-20, 2007.

[5] Jee-Hoon Jung, Jong-Moon Choi and Joong-Gi Kwon, "Design Methodology for Transformers Including Integrated and Center-tapped Structures for LLC Resonant Converters," *JOURNAL OF POWER ELECTRONICS*, vol. 9, pp. 215-223, 2009.

[6] B. C. Hyeon, J. T. Kim and B. H. Cho, "A Half Bridge LC Resonant Converter with Reduced Current Ripple of the Output Capacitor", *Telecommunications Energy Conference, INTELEC09, 31st Annual IEEE, 2009.*

[7] M. E. Elbuluk, G. C. Verghese, and J. G. Kassakian, "Sampled-data modeling and digital control of resonant converters," *Power Electronics, IEEE Transactions on*, vol. 3, pp. 344-354, 1988.

[8] E. X. Yang, B. Choi, F. C. Lee, and B. H. Cho, "Dynamic analysis and control design of LCC resonant converter," in *Power Electronics Specialists Conference, 1992. PESC '92 Record., 23rd Annual IEEE*, 1992, pp. 362-369 vol.1.

[9] M. Castilla, L. G. de Vicuna, M. Lopez, and V. Barcons, "An averaged large-signal modeling method for resonant converters," in *Industrial Electronics, Control and Instrumentation, 1997. IECON 97. 23rd International Conference on*, 1997, pp. 447-452 vol.2.

[10] J. A. Oliver, C. Fernandez, R. Prieto, and J. A. Cobos, "Circuit Oriented Model of Rectifiers for Large Signal Envelope Simulation," in *Power Electronics Specialists Conference, 2005. PESC '05. IEEE 36th*, 2005, pp. 2771-2776.

A High Power Density Single Phase PWM Rectifier with Active Ripple Energy Storage

Ruxi Wang*, Fred Wang**, Dushan Boroyevich*, Puqi Ning*

*Center for Power Electronics Systems

Virginia Polytechnic Institute and State University

Blacksburg, VA, 24061, USA

ruxi@vt.edu

** The University of Tennessee and Oak Ridge National Lab

Knoxville, TN, 37966, USA

fred.wang@utk.ed

Abstract:–It is well known that there exist second-order harmonic current and corresponding ripple voltage on dc bus for single phase PWM rectifiers. The low frequency harmonic current is normally filtered using a bulk capacitor in the bus which results in low power density. This paper proposed an active ripple energy storage method that can effectively reduce the energy storage capacitance. The feed-forward control method and design considerations are provided. Simulation and 15kW experimental results are provided for verification purposes.

Index Terms:--High power density converter, Single phase rectifier, Ripple energy, Capacitive energy storage, Active energy storage

I. INTRODUCTION

Fault-tolerant multi-phase converter systems have been extensively researched for aircraft application because of their inherent fault tolerance capability [1]. Accordingly, high power density single phase converter modules are desirable for such systems. One of the important characteristics of the single-phase system is the low-frequency ripple on the dc link when the ac input voltage and current are sinusoidal. To limit this low-frequency ripple, a bulk electrolytic dc-link capacitor is usually required, which results in large converter volume, low power density and poor life-time due to the electrolytic capacitors needed. To improve the power density of a single-phase converter, it is essential to reduce the dc-link capacitor required for filtering the low-frequency ripple energy [2]. Some active methods for ripple reduction are summarized and classified in a previously published work [3]. In addition, the previous work verified the feasibility of increasing the system power density by using active ripple energy storage method. In this paper, a high power density single phase PWM rectifier is proposed and a feed-forward control method is provided. This feed-forward method can help the auxiliary active energy storage circuit working as a parallel active power filter for filtering out the low frequency ripple current from the H-bridge rectifier. The detailed design considerations are provided. Finally, a 15kW hardware prototype is developed and the experimental results are provided for verification purposes.

II. PROPOSED SINGLE PHASE PWM RECTIFIER

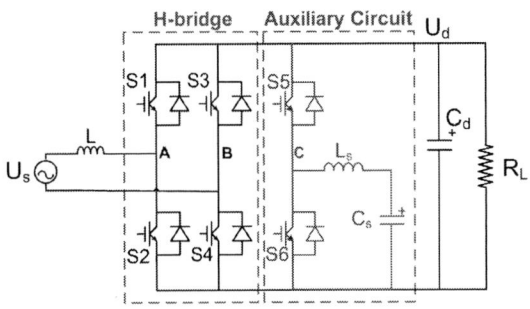

Fig. 1. Proposed single phase PWM rectifier.

The proposed topology of the ripple energy storage method is depicted in Fig.1. A bidirectional buck-boost converter is connected as auxiliary circuit at the output of a typical single-phase bidirectional PWM rectifier. Since second-order harmonic current is generated from the single phase H-bridge rectifier, the auxiliary circuit is used as a parallel current filter. An auxiliary capacitor, with capacitance C_s, is used as an energy storage element; while the inductor L_s is used as an energy transfer component. A dc-link capacitor, with capacitance C_d, is still needed at the output of the PWM rectifier to filter the switching ripple energy and the residual second-order harmonic ripple energy not fully absorbed by the auxiliary capacitor C_s. S5 is controlled as a buck switch for charging and S6 is controlled as a boost switch for discharging. The current of switch S5 is discontinuous, so this auxiliary circuit can only be used as low frequency current filter which is typical for single phase. Meanwhile, there is no voltage higher than the dc bus existing in this system and the auxiliary circuit can be integrated together with the main circuit easily as one additional phase leg.

The single phase rectifier parameters for the sample system are summarized in TABLE.I.

978-1-4244-4782-4/10 $26.00 © 2010 IEEE   1378

TABLE I. Parameters of the single phase rectifier

Parameter	Value
Voltage Source (Peak)	214 V
AC supply Frequency	233 Hz
Input Inductor	1.24 mH
DC Link Voltage	540±2%V
Output Power	15 kW

III. CONTROL ANALYSIS

As mentioned above, there exists second-order ripple power in the single phase system. The ripple power after the H-bridge can be expressed as:

$$P_r = P_{r\_peak} \sin(2\omega t) \qquad (1)$$

where ω is the supply frequency.

Assume all the ripple energy is stored in the auxiliary capacitor.

$$\frac{dU^2_{cs}}{dt} = \frac{2 \cdot P_{r\_peak}}{C_s} \sin 2\omega t \qquad (2)$$

With this, the low frequency ripple current and ripple voltage in the capacitor are shown in (3), (4) and Fig.2.

$$U_{cs} = \sqrt{Const - \frac{P_{r\_peak}}{C_s \omega} \cos 2\omega t} \qquad (3)$$

$$i_{cs} = \frac{P_{r\_peak} \sin 2\omega t}{\sqrt{Const - \frac{P_{r\_peak}}{C_s \omega} \cos 2\omega t}} \qquad (4)$$

where: $Const = k \times \frac{P_{r\_peak}}{C_s \omega}$ $(k \geq 1)$. If we consider total

charge and discharge, $Const = \frac{P_{r\_peak}}{C_s \omega}$ is desired.

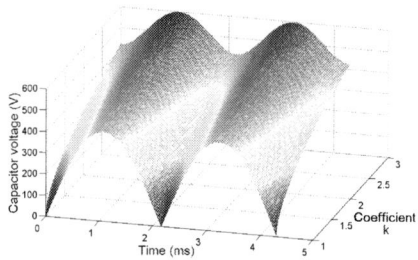

(a). Low-frequency capacitive voltage

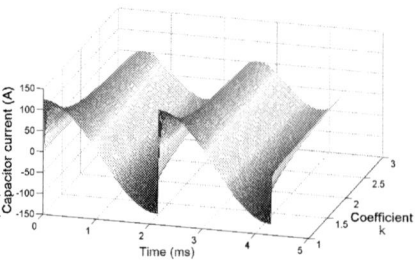

(b). Low-frequency capacitive current

Fig.2. Low-frequency capacitive voltage and capacitive current

The energy storage capacitor C_s is selected as 140μF to meet the minimum requirement [3] and the energy transfer inductor L_s is designed as 40 μH according to (15) and (16) in the design consideration section. Due to that, the switching frequency 20 kHz is much higher than the L_s and C_s resonant frequency. At each switching period, the dc-link voltage and auxiliary capacitor voltage can be considered as quasi-static. This means the inductor charging slope and discharging slope can each be considered as a fixed value within each switching period.

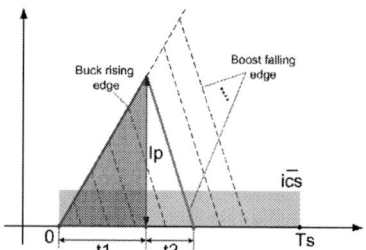

(a) Charging phase

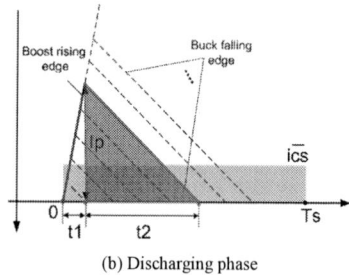

(b) Discharging phase

Fig.3. Charging and discharging phases duty-cycle generation strategy

The inductor current in Fig.3 (a) shows the charging phase for one switching period, and Fig.3 (b) shows the discharging phase for one switching period. The aim is to control the average inductor current (the inductor and capacitor currents are same) to match the reference which is derived in (4). Defining the boost and buck slopes as (5) and (6):

$$Boost\_slope = \frac{U_{cs}}{L} \qquad (5)$$

$$Buck\_slope = \frac{U_d - U_{cs}}{L} \qquad (6)$$

For the charging phase, the time interval relationship between t1 and t2 can be expressed as:

978-1-4244-4782-4/10 $26.00 © 2010 IEEE   1379

$$t2 = \frac{Buck\_slope}{Boost\_slope} \times t1 \qquad (7)$$

According to the control objective, the average current within one switching cycle should be equal to the current reference, that is:

$$\frac{1}{2}(t1 + \frac{Buck\_slope \cdot t1}{Boost\_slope}) \cdot Buck\_slope \cdot t1 = \bar{i}_{cs} \cdot T_s \qquad (8)$$

Then, the duty cycle for the charging and discharging phases can be derived as (9) and (10). By using these equations, the second-order ripple energy can be accurately filtered out from the H-bridge.

$$D1 = \sqrt{\frac{2 \cdot \bar{i}_{cs} \cdot f_s}{(1 + \frac{Buck\_slope}{Boost\_slope}) \cdot Buck\_slope}} \qquad (9)$$

$$D1 = \sqrt{\frac{2 \cdot \bar{i}_{cs} \cdot f_s}{(1 + \frac{Boost\_slope}{Buck\_slope}) \cdot Boost\_slope}} \qquad (10)$$

Equations (9) and (10) determine the duty cycle control laws for the charging and discharging operating modes. For a practical implementation, it is not easy to determine the auxiliary capacitor current reference in (4). A more straightforward, but similar current filter method, is shown in Fig.4. The compensation current is used to regulate the low frequency ripple current. In Fig.3, the triangular shaded area is the current waveform of the compensation current. Using the previous method, the average compensation current within one switching period should be equal to the low frequency ripple current, then the duty cycle for the charging and discharging phases are derived as (11) and (12).

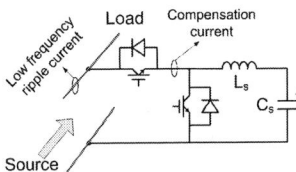

Fig.4. Auxiliary circuit working as a parallel active ripple current filters

$$D1 = \sqrt{\frac{2 \cdot \bar{i}_{comp} \cdot f_s}{Buck\_slope}} \qquad (11)$$

$$D1 = \sqrt{\frac{2 \cdot \bar{i}_{comp} \cdot f_s \cdot Buck\_slope}{Boost\_slope^2}} \qquad (12)$$

The control schematic of the system is shown in Fig.5. The rectifier duty cycle and the measured ac-side current are used to generate the ripple current reference for the auxiliary circuit. The dc link voltage and auxiliary capacitor voltage are sensed to generate the duty cycle for both charging and discharging phases. Within the duty cycle generation block, if the compensation current is positive, the auxiliary circuit is controlled in buck mode to assimilate the ripple power from the dc link charging the auxiliary energy storage capacitor. Similarly, when the compensation current is negative, the auxiliary circuit is controlled in boost mode to release the ripple energy stored

back into the dc link.

The auxiliary capacitor mean voltage control loop is required to prevent the C_s from over charging or under charging. The PLL block is designed as shown in [4].

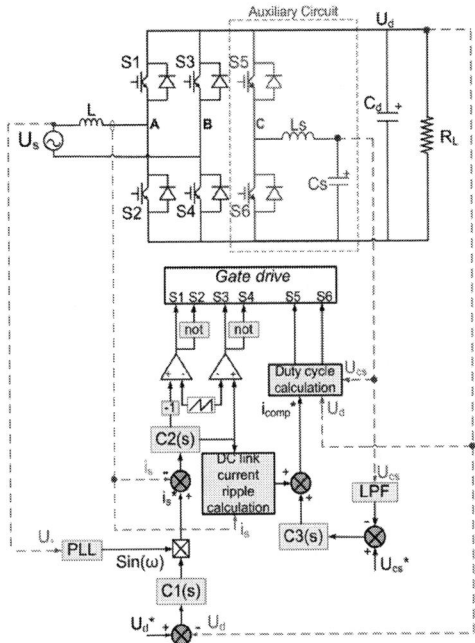

Fig.5. Control schematic figure for the system

IV. DESIGN CONSIDERATIONS

For single phase H-bridge rectifier, the modulation method is specified to achieve both the minimum loss and balanced temperature distribution. For the auxiliary circuit, the auxiliary capacitor is selected according to the ripple energy requirements and the auxiliary inductor is designed as below.

A. Modulation method

Fig.6 shows the single phase discontinuous PWM modulation method [5]. One phase leg will not switch within half of the supply frequency. It can lead to the minimum switching loss.

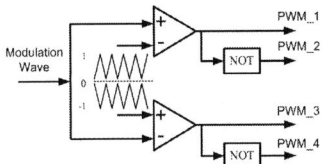

Fig.6. Single phase discontinuous PWM modulation method

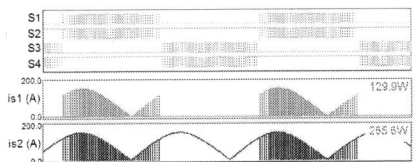

Fig.7. Discontinuous PWM method simulation results

Fig.7 shows the simulation results of modulation method in [5]. During the discontinuous current period, switch 2 is always conducted. Then, the loss distribution is not equal for the top and bottom switches.

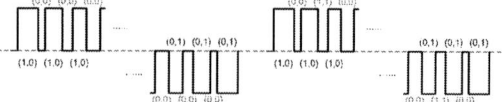

(a). Asymmetrical zero vector   (b). Symmetrical zero vector

Fig.8. Terminal voltages and voltage vectors

The reason for the unequal loss distribution is shown in Fig.8 (a). The only zero vector (0,0) is inserted and causes this asymmetrical problem. To prevent this, symmetrical zero vectors are required as shown in Fig.8 (b). Zero vectors (0,0) and (1,1) are inserted alternatively. This will leads to the balanced thermal performance.

B. Auxiliary inductance selection

There are two criteria for selecting the auxiliary inductance: Peak current boundary and the Discontinuous Current Mode (DCM) boundary.

The maximum current in the auxiliary circuit must be smaller than the peak current requirement of the selected power semiconductor:

$$Buck\_slope \times D1 \times T_s \le I_{peak} \qquad (13)$$

$$Boost\_slope \times D1 \times T_s \le I_{peak} \qquad (14)$$

Then, the auxiliary inductance selection based on the peak current requirement is calculated as:

$$L_s \ge \frac{2 \cdot \bar{i}_{cs} \cdot T_s}{I_{peak}^2} \cdot \frac{U_d U_{cs} - U_{cs}^2}{U_d} \qquad (15)$$

Meanwhile, as shown in Fig.3, in order to maintain DCM operation, subinterval t1 plus subinterval t2 should smaller than switching period Ts.

Then, the auxiliary inductance selection based on the DCM requirement is calculated as:

$$L_s \le \frac{T_s}{2 \cdot \bar{i}_{cs}} \cdot \frac{U_d U_{cs} - U_{cs}^2}{U_d} \qquad (16)$$

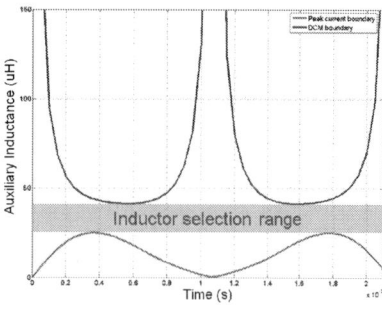

Fig.9 Auxiliary inductance selection

Fig.9 shows the auxiliary inductance selection range, 40μH inductance is selected as our auxiliary inductor.

V. SIMULATION AND EXPERIMENT RESULTS

Fig.10. shows the simulation result. After 0.6 s, the auxiliary circuit is inactive. A large dc-link voltage ripple can be observed due to the small dc-link capacitor C_d. Before 0.6 s, the auxiliary circuit is active. Most of the ripple energy is stored in the auxiliary energy storage capacitor. As such, the dc-link voltage ripple is within the required 2% limit even with a 140 μF auxiliary energy storage capacitance Cs compared with 1.6 mF needed to meet the same requirement using the conventional method. The voltage range of the auxiliary capacitor is between 320 V and 520 V.

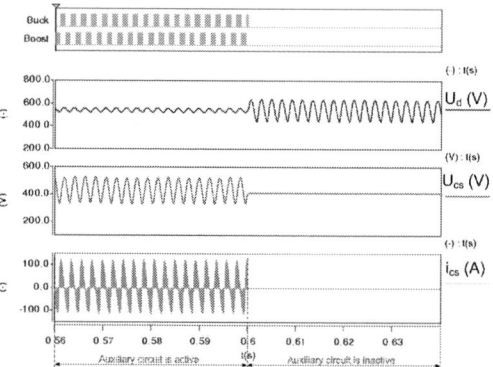

(a) Simulation waveforms with the effect of the auxiliary circuit

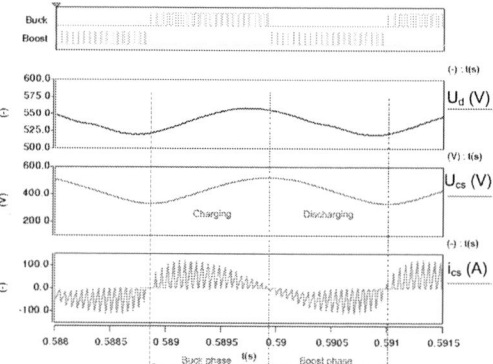

(b) Simulation waveforms when the auxiliary circuit is active

Fig.10. Simulation waveforms.

A 15kW prototype as shown in Fig.11 is developed to verify the proposed active ripple energy storage method. A three-phase Intelligent Power Module (IPM) from Powerex is used as the active components both for the H-bridge and the auxiliary circuit. A 40μH amorphous core copper foil inductor is built as the auxiliary inductor. The inductor core loss calculation is based on method in [6]. A pin heatsink and its fans from UltraCool are selected as the cooling system. The thermal resistance of the pin heatsink is 0.08°C/W. Two 100μF film capacitors from Electronic Concepts are selected as the auxiliary energy storage capacitor and another two 100μF film capacitors are selected as the dc link capacitor to filter the high-frequency ripple current. Four layers PCB bus board is designed. The protection and sensor board is used to provide dc-dc power supply, voltage and current sensing and over-current and over-voltage protection.

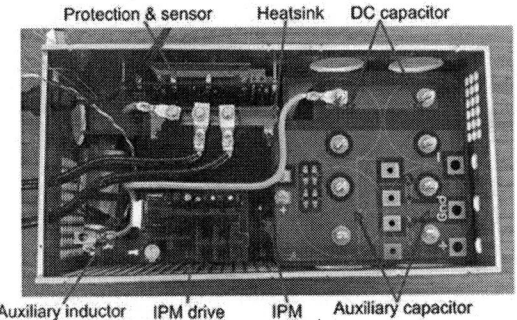

Fig.11. Photo of the 15kW experimental hardware.

Fig.12. shows the test results of 15kW output. The dc bus voltage is controlled as 540V by H-bridge rectifier. As shown in Fig.12 (a), without the active method, the ripple energy is stored on the 200μF dc bus capacitors, which leads to the 148V peak to peak voltage ripple.

With the active method as shown in Fig.12 (b), most of the ripple energy is stored in the auxiliary capacitors, and the dc bus peak to peak voltage ripple decreases from 148V to 28V. The mean value of the auxiliary capacitor voltage is controlled by another voltage control loop which is 350V.

Fig.12 (c) also shows the waveforms when the auxiliary circuit is active. The mean value of the auxiliary capacitor voltage is controlled as 300V.

Fig.12 (d) shows the auxiliary capacitor current and its reference indicated by the red dashed line.

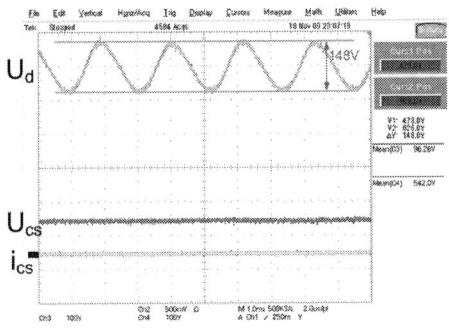

(a). Auxiliary circuit is inactive

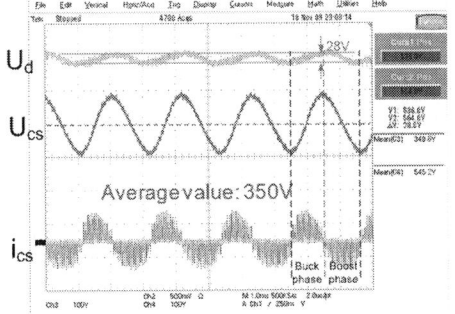

(b). Auxiliary circuit is active

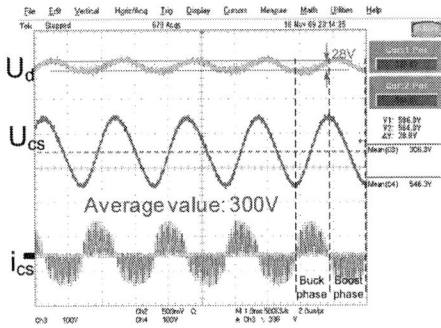

(c). Auxiliary circuit is active

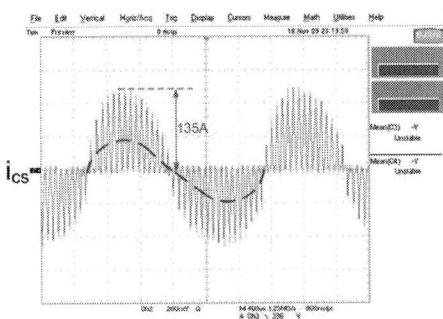

(d). Auxiliary capacitor current and reference

Fig.12. Experiment results

VI. CONCLUTIONS

In this paper, an active ripple energy storage method is proposed to increase the single phase PWM rectifier's power density. Based on analysis, simulation and experiment, the following conclusions can be drawn:

Firstly, the proposed auxiliary circuits will bring no voltage higher than the dc bus in the system and it can be easily integrated together with the H-bridge rectifier as an additional phase leg. Different from the traditional parallel active power filter, the auxiliary circuit compensation current is in discontinuous current mode so that it can only filter out the low frequency ripple current that dominates in single phase rectifier system.

Secondly, the proposed feed-forward control method can generate the compensation current reference as fast as one switching period and can effectively filter out the low frequency ripple current from the H-bridge rectifier.

Thirdly, although the total capacitance will decrease dramatically compared with traditional method, the total ripple current in capacitors will increase by using the active method.

Finally, simulation and 15kW hardware prototype experimental results were provided for verification purposes.

ACKNOWLEDGMENT

The authors would like to thank Rolls-Royce Corporation and Dr. Kaushik Rajashekara for providing support for the project.

REFERENCES

[1] G. Jack, B. C. Mecrow, and J. A. Haylock, "A comparative study of permanent magnet and switched reluctance motor for high-performance fault-tolerance applications", IEEE Trans. on Ind. Appl., vol. 32, 1996

[2] J. W. Kolar, U. Drofenik, J. Biela, M . L. Heldwein, H. Ertl, T. Friedli, and S. D. Round, "PWM Converter Power Density Barriers". Power Conversion Conference Nagoya, 2007. PCC '07

[3] Ruxi Wang, Fred Wang, Puqi Ning, Rixin Lai, Rolando Burgos, Dushan Boroyevich. "Study of Energy Storage Capacitor Reduction for Single Phase PWM Rectifier" . IEEE Applied Power Electronics Conference. 2008

[4] Timothy Thacker, Ruxi Wang, Dong Dong, Rolando Burgos, Fred Wang, Dushan Boroyevich. "Phase-Locked Loops using State Variable Feedback for Single-Phase Converter Systems". IEEE Applied Power Electronics Conference. 2008

[5] D.Grahame Holmes, Thomas A. Lipo "Pulse Width Modulation for Power Converters: Principles and Practice". A John Wiley & Sons, INC. Publication. 2003

[6] Venkatachalam, K. Sullivan, C.R. Abdallah, T. Tacca, H. "Accurate prediction of ferrite core loss with nonsinusoidal waveforms using only Steinmetz parameters" . Computers in Power Electronics, 2002. Proceedings. 2002 IEEE Workshop on.

Design Considerations for High Efficiency Buck PFC with Half-Bridge Regulation Stage

Bernard Keogh, George Young, Hagen Wegner, Colin Gillmor
Power Supply Control,
Texas Instruments (Cork) Ltd.,
Cork, Ireland

Abstract— AC-DC converters for applications such as notebook adapters require operation over universal input mains range, and are challenged to achieve high efficiency over the entire range. Given the enclosed construction, and the demand for smaller, thinner adapters, they must be designed for high, yet relatively flat efficiency over the full mains range, to meet stringent thermal constraints. A novel two-stage architecture has been developed for high density, high efficiency notebook adapters. A buck regulator is used for the PFC front-end stage. The output isolation & regulation stage is implemented using a Half-Bridge topology operated in an Integral-Cycle-Controlled mode. Smart burst modes are deployed in both stages, delivering excellent efficiency also across the full load range – 93% efficiency at 100% loading, and >80% efficiency at 5% loading. The Buck PFC delivers Power Factor performance that greatly exceeds EN61000-3-2 & meets Energy Star requirements, with Power Factor >0.93 at 115Vac (low line). This paper details the design approach, practical issues and their resolutions, and benefits of the approach. Performance data is presented.

I. INTRODUCTION

Notebook adapters are a challenging design area – they are severely cost-challenged, and effectively are a commodity product. They must operate worldwide over universal mains range, without range switching. Market trends are driving smaller, lighter and most importantly low profile slim-line packaging of adapters. Requirements for smaller size and lower profile means much less surface area for dissipation of the heat generated internally, and consumers and industry are becoming more sensitive to the thermal performance of consumer products – there have been many widely publicised cases of accidental burns from over-heating laptops, batteries, adapters, etc. These requirements can only be met by developing ac-dc solutions that deliver very high levels of efficiency, to ensure lower temperature rise of adapter packages that are shrinking in surface area. Universal mains operation demands that the ac-dc solution must also deliver high efficiency performance from 90V to 264V, in order to fulfil maximum case surface temperature rise constraints (typically 45C to 50C maximum temperature rise is allowed for plastic cases). To further compound design difficulties, high efficiency performance right across the load range is also

being mandated through the efforts of organisations and initiatives such as Energy Star, European Code of Conduct, 80-Plus, Climate Savers, etc [1,2,3,4]. Industry has recognised that most electronic equipment spends most time operating at perhaps 10-20% of full rated load, so there is a greater focus on efficiency and power consumption at the very low load ranges.

To summarise, the notebook adapter must be packaged in increasingly small case sizes; it must operate with extremely high efficiency at full load, to meet thermal limitations; it must be able to meet these limitations universally from minimum to maximum line voltage, worldwide; it must also operate at high efficiency at light loading to allow the notebook + adapter system to meet Energy Star limits for power consumption when in off, standby and idle modes; there are increasingly tough demands to drive the adapter standby power to effectively zero power consumed at zero load. The sum of these design challenges has led to the development of the two-stage architecture that is outlined in detail in this paper.

II. HIGH EFFICIENCY DESIGN CHALLENGES

The myriad challenges outlined above challenge conventional design approaches. Many approaches and topologies have been used to implement Power Factor Correction (PFC) [5,6,7]. Many typically used two-stage adapter topologies, including the Boost PFC Front-end with Flyback regulation/isolation stage, Quasi-Resonant (QR) variations of this, Boost PFC Front-end with LLC Resonant Half-Bridge (HB), QR Boost with LLC, all have their own inherent or unique advantages, but also have their inherent limitations and disadvantages. In particular, Boost Front-end stages have suffered poor low line efficiency, where the efficiency performance at low line 90V is typically 1-3% lower than at 230V [8]. Tracking Boost PFC stages address this limitation to some extent, but the wide PFC bus variation does not lend itself to some second stage topologies such as LLC, and wide input range for the second stage compromises potential efficiency performance of that stage. Flyback regulation stages can easily tolerate wide input voltage range, however the Flyback is challenged to deliver high efficiency, with high peak and rms currents, significant transformer air-gap fringing flux effects, and synchronous rectifier (SR) drive complexities. In summary, the two-stage adapter topologies that are currently widely used are challenged to deliver high efficiency over the wide line/load envelope of operation, whilst also maintaining very low standby power.

Thus, a novel two-stage architecture was developed to suit high efficiency notebook adapter applications This uses a Buck PFC Front-end, delivering a low voltage bus (~84V) to a Half-Bridge (HB) isolation/regulation stage, as shown in figure 1 [9]. The HB stage is operated in an Integral-Cycle-Controlled (ICC) fashion to maintain high efficiency over the entire load range from no load to full load. ICC operation is detailed further below. The unique synergy that is realised by the combination of these stages results in superior overall efficiency performance across both the universal mains range and the entire load range.

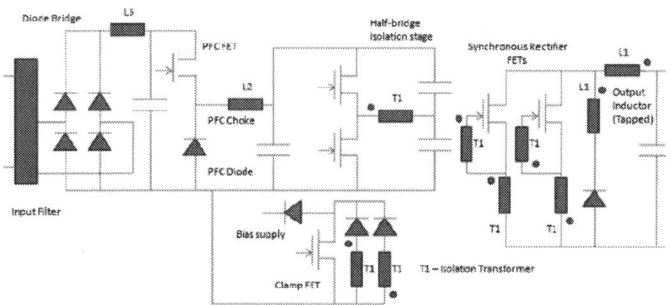

Figure 1.   Buck PFC + ICC Half-Bridge Adapter Implementation

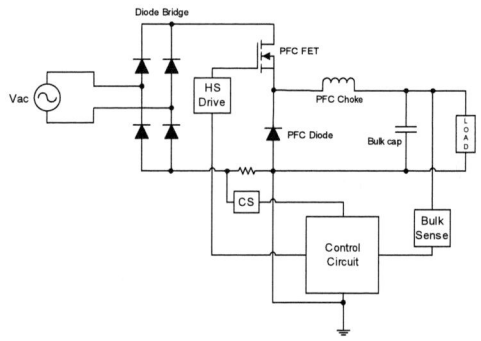

Figure 2.   Buck PFC Basic Power Circuit

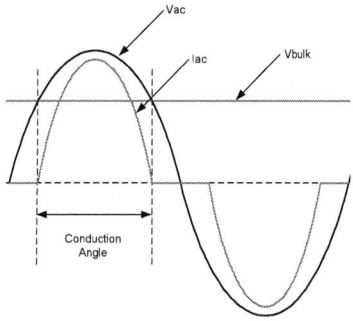

Figure 3.   Buck PFC illustrative line current at 100Vac, with 80Vdc output

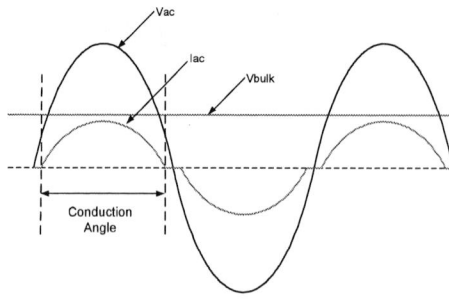

Figure 4.   Buck PFC illustrative line current at 230Vac, with 80Vdc output

III.   BUCK PFC FRONT-END

A.   Basic Operation

The basic circuit of the Buck PFC power stage is shown in figure 2 above. This is basically a conventional buck (step-down) converter connected to an AC voltage source and bridge rectifier. The output bus voltage is set at a level somewhat less than the peak AC voltage at lowest line. When the instantaneous AC input voltage is greater than the output bus voltage, the PFC stage is forward biased and current can be drawn from the AC input. When the AC input voltage falls below the bus voltage level, the diode bridge rectifiers become reverse biased and no power can be drawn from the AC line. Hence there will be an inherent "cross-over" distortion in the AC line current when the Buck PFC stage is reverse biased. But in many applications, this distortion can be made acceptable, delivering adequate line current Total Harmonic Distortion (THD) and Power Factor (PF) performance. Some putative AC line voltage and current waveforms are shown below at 100Vac (figure 3) and at 230Vac (figure 4) for a typical bus voltage level of 80Vdc. It can be seen that at higher input voltage, the cross-over distortion becomes much less significant, with correspondingly lower THD and higher PF, as one would expect.

In a similar fashion to more conventional Boost PFC stages, the Buck PFC is typically controlled using an outer voltage control loop to regulate the bus voltage, with an inner current control loop to control the average current shape. The outer loop adjusts the demand applied to the inner loop to maintain regulated average bus voltage in response to line and load changes. This loop is typically designed for slow response, low bandwidth, to maintain good PF of the input current. The inner current loop controls the PWM duty cycle of the PFC mosfet over the AC half-cycle, in response to the demand from the voltage loop, in order to control the average AC line current to follow a pseudo-sinusoidal shape, to deliver the required PF performance.

B.   Buck PFC Advantages

The high voltage AC input is immediately bucked to a lower voltage level by the PFC Front-end:

- Eases functional safety spacing in the subsequent regulation/isolation stage
- Lower downstream voltages can improve robustness and reliability

- Low output voltage results in lower Common-Mode (CM) noise
- Lower voltage on downstream stage, allows more efficient design, lower voltage & lower cost MOSFETS with better figure of merit.

Lower input-to-output voltage differential on PFC choke:

- Allows use of a lower inductance value when compared to a Boost
- Input-output differential voltage across the choke varies in the opposite direction compared to the Boost – lowest voltage differential at low line where efficiency is most challenged
- The buck has to work hardest at high line where the currents are less, whereas the boost has to work hardest when it boosts the most – at low line where the current is highest

Inherent "free" inrush limitation at start-up, no dedicated inrush mechanism is required.

Compatible with low cost:

- Low output bus voltage allows downstream stage to make use of lower cost, low voltage switches compared to conventional Boost PFC bus voltage of ~400V

C. Buck PFC Disadvantages

As with every circuit topology, there are always downsides and limitations. The Buck PFC will not be an appropriate solution for all applications. But in many cases, the limitations can be sufficiently managed to offer a solution with distinct performance advantages over Boost PFC based designs.

- Inherent AC line current "cross-over" distortion – limits achievable THD and PF performance.
- Requires a high-side element – either drive for the Buck PFC Switch, or high side bus voltage sense
- No direct forward path from AC input to bulk storage capacitor – complicates lightning surge management, requires use of dedicated surge suppression techniques
- Lower output voltage results in less efficient bulk capacitor energy storage, so bulk capacitors need to be larger and/or hold-up time is lower
- Higher percentage bus voltage ripple compared to Boost PFC, requires voltage loop with lower bandwidth, slower transient response
- Limitation of available control IC's, expertise and design/analysis literature to aid the designer

D. Buck Power Factor Analysis

Extensive analysis of the Buck PFC has been published utilising various methods of control and various current wave-shapes [10,8,11,12,13]. Many of these analyses have primarily explored the possible THD and line current harmonics performance achievable, and maximum power levels that can be delivered within the limits of EN61000-3-2 [14]. However, EN61000-3-2 does not impose any requirement for a minimum PF. It can be shown that a current wave-shape containing 90% of the maximum allowed harmonic current for each and every harmonic covered by the standard would have THD of 85%, resulting in a PF as low as 0.76 (assuming displacement factor of unity). More recent initiatives such as Energy Star EPS 2.0 have started mandating a minimum PF value of 0.9, which actually imposes much tougher requirements on THD. This puts a different perspective and constraint on the Buck PFC performance – however, 0.9 PF is readily achievable with the Buck PFC, even at low line

Clearly, as already discussed, the basic Buck PFC Front-end cannot deliver unity power factor, i.e. deliver a perfectly sinusoidal line current waveform. The inherent dead-time in the current waveform at the AC line cross-over will always limit power factor. So choice of the output bus voltage is probably the most important design parameter of the Buck PFC, and is the starting point for any design. Fundamentally, the bus voltage level must be chosen to be lower than the minimum rated AC line peak voltage, similar to the requirement for Boost bus voltage to be greater than the maximum rated AC line peak voltage. However, the bus voltage must be chosen sufficiently lower than the minimum AC line peak to allow a reasonable conduction angle.

Figure 5 below illustrates the trade-off between bus voltage level and conduction angle. Conduction angle can be calculated as follows, as the number of degrees out of each 180° half cycle that the Buck PFC is forward biased, during which line current can be drawn.

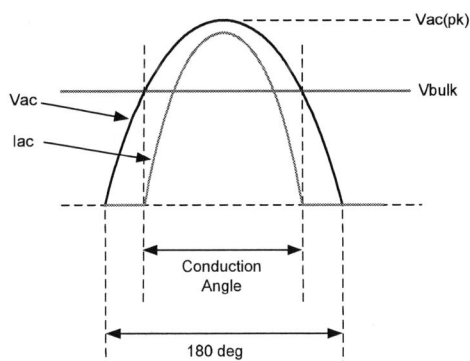

Figure 5. Buck PFC line current conduction angle

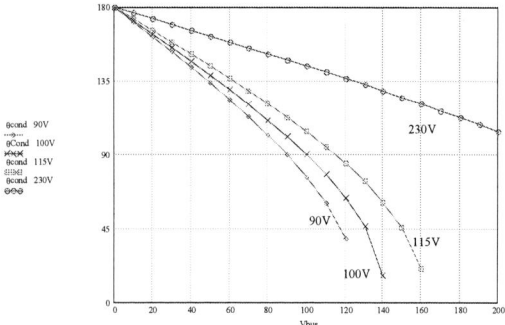

Figure 6. Buck PFC conduction angle vs. Vbus & AC line

Figure 6 illustrates the conduction angle in degrees versus bus voltage level for a variety of AC line voltages. Clearly, for operation down to 90Vrms, a maximum bus voltage level of 90Vdc is appropriate to ensure a conduction of at least 50%, or 90° out of 180° half-cycle. In practice, bus voltage should be chosen somewhat lower than this limiting value.

In the basic topology overview, some putative current waveforms were drawn. Three such typical waveforms have been simulated and analysed for harmonic content, distortion factor and maximum possible power factor. The simulation assumed a nominal bus voltage of 80Vdc for the buck output, assumed infinite buck output capacitor (i.e. zero ripple on the buck output). Figure 7 below illustrates the different waveforms, referred to as A (modified sine, where current is proportional to input-output voltage differential during the conduction angle), B (truncated sine, where current is proportional to input voltage during the conduction angle) and C (clamped current "trapezoid", where current is constant during the conduction angle). The resulting potential harmonic content and power factor (neglecting the effects of phase displacement) are summarised in Table I. The analysis was confined to low line 100Vac, since this is the operating point where PF is most challenged. At 230Vac, cross-over distortion is much less, so that good PF is readily achieved.

TABLE I.   SUMMARY OF WAVEFORM THD & PF

	Waveform A	Waveform B	Waveform C
Irms(tot)	1.0886	1.0362	1.0527
I1rms	0.9953	0.9866	1.0008
Real Power (W)	99.53	98.66	100.1
THD	44.3%	31.9%	32.62%
Distortion Factor	0.914	0.953	0.9507
Max Possible PF (no phase displacement)	0.914	0.953	0.9507

From the above analysis, it can be seen that there is a trade-off between the PF and individual harmonic content, depending on the current control technique used, and the shape of the AC line current. Waveform A (modified sine) has very low harmonic content at higher orders, but is dominated by a large third harmonic. This waveform passes the requirements of EN/JIS-61000-3-2 very easily, since those standards permit a high degree of third harmonic. However, the third harmonic results in quite high THD of 44%, limiting potential PF performance at 100Vac to 0.91 maximum (neglecting phase shift and other sources of distortion). In effect, the slow rate of rise and fall of current within the conduction angle is good for limiting high order harmonics, but results in poor utilisation of the already limited conduction angle of the Buck PFC, with limited PF as seen.

Waveform B (truncated sine) has a faster rate of rise and fall of current within the conduction angle, so makes better use of the available window, and consequently has much lower third harmonic content than waveform A. Within the conduction angle, the current shape is approximately sinusoidal, in proportion to instantaneous AC line voltage. This results in far better 32% THD, corresponding to a potential maximum PF performance of 0.952. However, this waveform has a greater content of higher order harmonics due to the higher di/dt of the current at the "shoulders" of the waveform. This simulation assumed a di/dt of 4A/ms, which is quite steep, and shows some harmonics that are close to the JIS61000-3-2 limits, particularly the 15th. Thus, careful design is required to ensure that all harmonics are kept within the harmonics limits, while still maximising PF. In practice, besides the action of the current control loop, the di/dt of the current is also limited by circuit practicalities such as filter impedance, PFC choke inductance, control circuit Dmax, etc, so that slower di/dt values will usually occur.

Waveform C (clamped current trapezoid) is a variation of the truncated sine waveform B, where the sinusoidal variation is instead clamped, resulting in a flat-topped trapezoidal waveform. This type of waveform has been previously analysed in [10]. This also has a fast but limited rate of rise and fall of current within the conduction angle. This waveform has a very similar 32% THD and 0.952 PF, but the clamped current nature can offer an efficiency advantage since both average and rms current flowing in the bridge rectifiers, PFC MOSFET and choke will be lower. However, it should be noted that while the THD and PF are similar to waveform B, the harmonic distribution is different – in this case there is a greater higher frequency content, where the EN/JIS61000-3-2 limits are much lower. Consequently, this waveform actually fails harmonics limits at some frequencies (9th, 15th and 19th), and comes very close at several others (11th, 25th, 29th).

978-1-4244-4782-4/10 $26.00 © 2010 IEEE

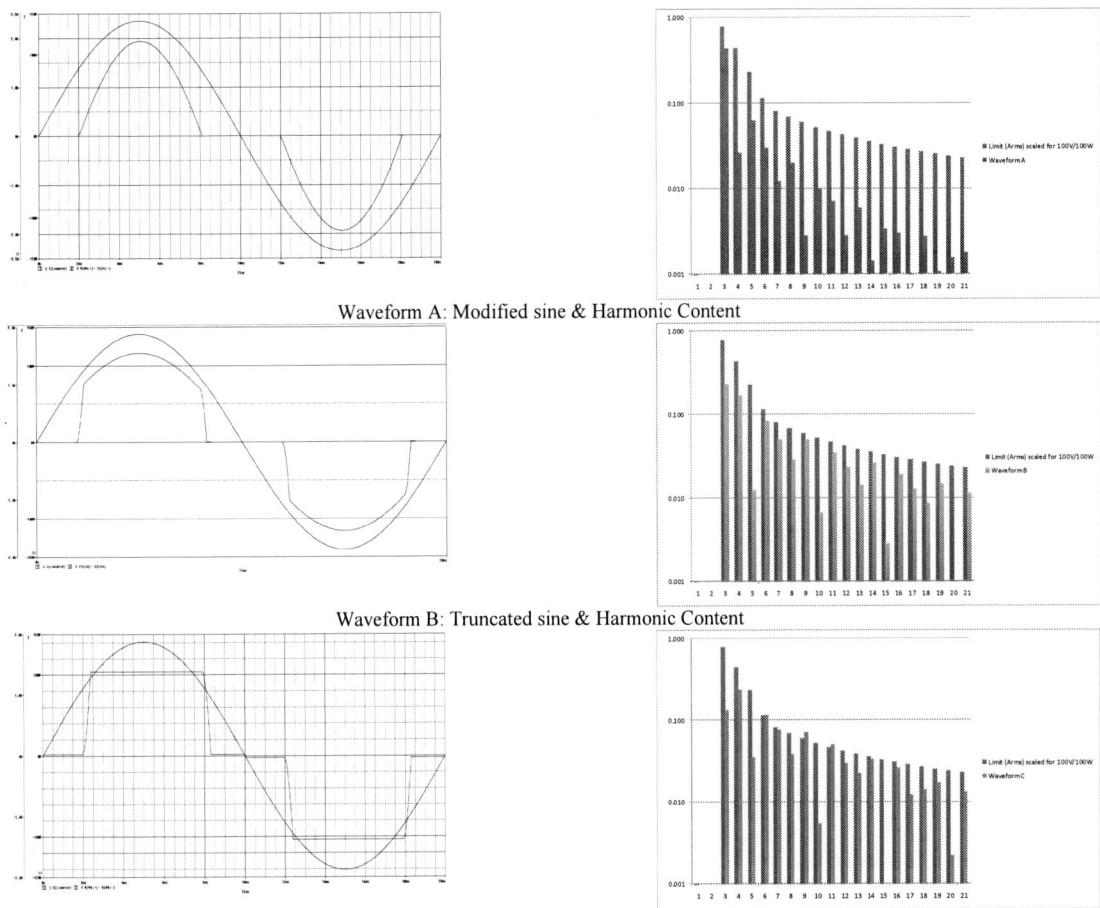

Waveform A: Modified sine & Harmonic Content

Waveform B: Truncated sine & Harmonic Content

Waveform C: Clamped-Current Trapezoid & Harmonic Content

Figure 7. Buck PFC current waveforms and their hamronic analyses

Some interesting conclusions can be drawn from these results. THD and PF are heavily influenced by third harmonic, as evidenced by waveform A analysis. Waveforms with low third harmonic, but quite high levels of higher order harmonics can result in better PF performance, as evidenced by waveform B. Somewhat similar waveforms, with similar THD and PF, can result in quite different harmonic content – waveform B passes EN61000-3-2, while waveform C fails, yet both have 0.952 PF.

Practical Issues That Degrade Buck PFC PF

Besides the current wave-shape, there are many other practical factors that affect PF. These can cause either phase displacement of the fundamental component of the current, or increase distortion and harmonic content of the current. Many of these are problematic for all PFC topologies. Where noted, some are specific to the Buck PFC.

- EMC Filter film capacitor displacement current – problematic for all topologies.
- Voltage loop bandwidth and resultant control signal ripple at 100/120Hz – more of a problem for the Buck PFC due to higher % bus ripple
- Bulk capacitor ripple phase shift effect – particular to the Buck, where the bus ripple causes the

conduction to be phase shifted, adding to filter capacitor displacement effects
- Circuit practicalities that limit current di/dt at the start/end of the conduction angle:
- Source impedance
- PFC choke inductance value
- Control circuit maximum duty cycle limit.

IV. HALF-BRIDGE ISOLATION STAGE

A. Basic Operation

The isolation stage as shown is a half-bridge primary, centre-tapped (CT) secondary, with self-driven synchronous rectifiers. Typically adapters of this type feed a downstream buck converter (within the notebook computer), and a battery is present in the computer. Holdup requirements are thus limited, and it is possible to operate the isolation stage with reduced headroom, corresponding to a high duty cycle. With this type of design, the "component load factor" [15] is close to optimal.

A control approach that is well suited in this environment is the Integral-Cycle-Control (ICC) approach. This ICC approach allows a very simple and consistent control technique to be deployed between no-load and full-load operation. In operation, a half-bridge controller needs to be

978-1-4244-4782-4/10 $26.00 © 2010 IEEE

operated in a balanced condition with approximately equal voltage across the capacitors. A balanced pulse-set with zero net magnetising current is thus very desirable. A simple case of such a pulse set is where the lower FET is driven initially for ¼ period, then the upper FET for ½ period and the lower FET then for ¼ period. This ensures near-zero net magnetising current. Most experimental work has been undertaken at switching frequencies around 125KHz, and thus an 8us period for the pulse-set is implied.

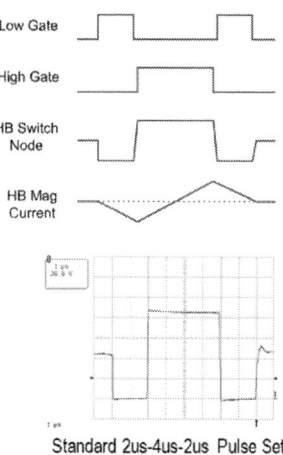

Figure 8. ICC HB typical pulse set

The magnetising current is as shown, with the current reversing during the longer pulse to give a net-zero value. An important aspect is that the transformer waveforms are well-suited to direct drive of synchronous rectifiers. Headroom of operation is usually such that freewheeling requirements are addressed using a schottky diode. The effectiveness of the schottky diode is enhanced – and output ripple currents and voltages reduced – if the output inductor is tapped as shown in figure 1. This reflects the fact that the converter is operating normally with low-headroom, and thus the voltage applied to the inductor during any off period will be materially higher than in on periods.

When operating with synchronous rectifiers in this fashion, it is important that the waveform be clamped in the off- time. If this is not the case, a "Royer Oscillator" condition may occur where the synchronous rectifiers can drive an oscillatory condition. The simplest way of effecting clamping is usually by loading a winding or windings which can also be used for bias supply.

B. Hysteretic Mode Control

The basis of control is very simple and consistent across load. If the output voltage is below a reference level, a pulse-set is provided to the FETs in the half-bridge stage, and if the output voltage is above the level, then the output is blanked, i.e. switching cycles are dropped. This hysteretic approach allows for rapid transient response, and is particularly effective in protecting against transient over-voltages as may occur during load-dump conditions.

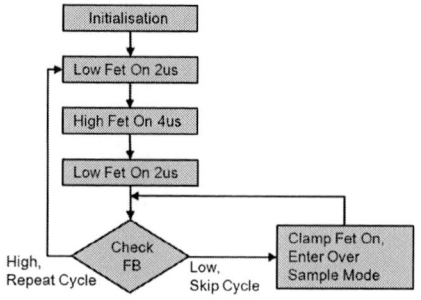

Figure 9. ICC HB Control "Flowchart"

The basis of operation of the control approach is as shown in figure 9. When a low output condition is detected, the 8us pulse-set is provided, and if the output is detected as high, pulse-sets are dropped, and repeated sampling at a rate of typically 2us is undertaken until a pulse-set is commanded.

- **Typical ICC Half-Bridge Waveforms**

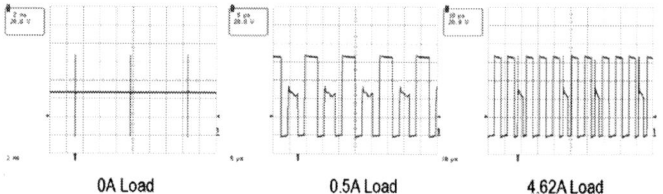

Figure 10. ICC HB pulse train at various loads

The ICC algorithm, with appropriate design of drive timings, achieves zero-voltage switching on a majority of the transitions. The approach is however particularly well suited to usage with a buck PFC stage where operation is from approximately 80V, and partial loss of ZVS operation on some transitions does not result in excessive efficiency penalty.

C. Performance Advantages

There are many advantages that result from ICC operation of the half-bridge stage. In essence, it operates effectively as a bus-converter, using burst mode "mark-space" ratio to effect output regulation. Thus, since it operates in burst mode at all load levels, drive, switching and core losses are modulated with load – rather than a fixed standing loss that becomes an increasing penalty with decreasing load, with ICC the standing loss also decreases with decreasing load. This helps to maximise efficiency at all load levels. Moreover, since the stage operates like a bus converter, it also operates with high efficiency when switching. Each switching cycle operates with high duty cycle square-wave pulses, with a small switching dead-time to provide for Zero-Voltage-Switching (ZVS) using the energy stored in the transformer leakage inductance. Since the bus voltage provided by the Buck PFC stage is much lower than the more conventional Boost stage, there is a much less severe efficiency penalty for partial loss of ZVS. This makes the design far more robust to dead-time and leakage inductance variability. The high duty cycle square-waves result in good utilisation of transformer and rectifier devices. This high duty cycle characteristic also readily allows for very simple and low-cost self-driven Synchronous Rectifiers (SR's). The SR

978-1-4244-4782-4/10 $26.00 © 2010 IEEE   1389

drive comes virtually for free using a transformer drive winding and a gate resistor – there is no requirement for complex drive or current sense circuitry.

V. DESIGN EXAMPLE

TABLE II.   90W HIGH DENSITY ADAPTER SPECIFICATION

90WHD Low Profile Adaptor	
Input Voltage	90Vac to 264Vac, 47 to 63Hz
Output Power	90 W
Adapter Dimensions (LxWxH)	90mm x 68mm x 16mm
EMI-conducted and radiated	EN55022 Class B
Output Voltage	19.5V ±1V
Output Current	0 to 4.62A Continuous
Efficiency (at 100% load, W.C. line)	92.50% min (Not including cable losses)
Energy Star	Eff >90% average over 25%, 50%, 75% and 100% Load No Load power < 0.3W; Eff at 0.5W load > 60%
PF at 115Vac 60Hz (90W)	0.94 typ
PF at 230Vac 50Hz (90W)	0.96 typ
Output Ripple/Noise	250mV typ, 400mV max
Turn on AC input to Vo	1 sec typ at 90V
Vo rise time	< 100msec (10% to 90%)
Vo fall time	< 100msec

A. PFC Stage Design Choices

The design process is outlined briefly as follows using the above specification. For 19.5V output, a transformer turns ration of 4:1 was assumed for the HB stage. Thus a suitable Buck PFC bus voltage was chosen based on a required minimum of 78V. Allowing for +/-5% bus ripple, and suitable headroom, the bus voltage was chosen as 84V nominal. Given the low profile requirement of 16mm packaged height, 12.5mm bus capacitors were chosen, using 2x470uF 100V parts in 12.5mm x 40mm package. This results in 8Vpp bus voltage ripple at low line. From eqn x, the required PFC choke inductance was calculated to be approximately 90uH to set the boundary of CCM/DCM operation at mid line of 160Vac. To meet the required efficiency, the PFC MOSFET was chosen as a 600V 199m-ohm super-junction device. This allows for a good compromise between low conduction loss at low line, and low switching loss at high line due to the low gate charge and fast switching capability. The PFC diode was chosen as a cost effective hyper-fast 6A-600V type – again providing a good trade-off between reasonably fast reverse-recovery at low line and low forward voltage drop at high line. A dedicated Buck PFC controller IC UCC29910 [16] was utilised to implement the control of the PFC stage.

B. HB Stage Design Choices

In order to minimise transformer turns count to maintain good efficiency, a core with Ae of approximately 100mm2 was chosen. To keep core loss to an acceptable level, the peak flux swing needs to be kept as close as possible to 100mT. For the chosen 84V bus level, and 125kHz switching frequency, a primary turns count of 8 yields 5.25V/turn on the transformer, leading to approximately 105mT peak flux. If a greater number of turns is used, then flux and core loss will be lower, but at the expense of copper loss. Similarly, a lower turns count would lower copper loss, but the increased flux level would drastically increase core loss. Thus 8T was

found to be a good compromise. To deliver 19.5V output, a 4T secondary was chosen (21V reflected voltage).

With 84V primary bus level, 100V MOSFETS were chosen, using compact efficient SOIC8 packaged parts. Since the transformer secondary is centre-tapped, 80V MOSFETS were chosen for the SR devices, to allow margin for voltage spiking due the leakage inductance between each half of the centre-tap windings. A dedicated ICC controller IC UCC29900 [17] was used for the HB stage control.

VI. PERFORMANCE DATA

A. Buck PFC Stage

The Buck PFC has been shown to offer superior efficiency performance when operated over universal mains range [8]. In addition, the Buck PFC can also maintain high efficiency over a very large portion of the load range. Figure 11 illustrates the efficiency performance vs line and load of the 95W Buck PFC front-end stage of this adapter design

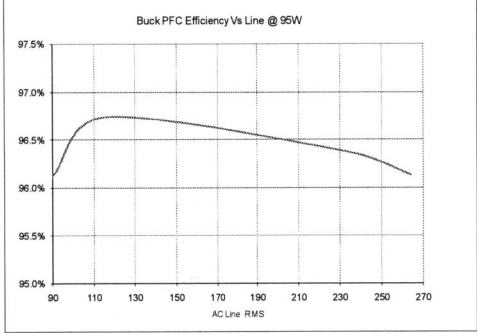

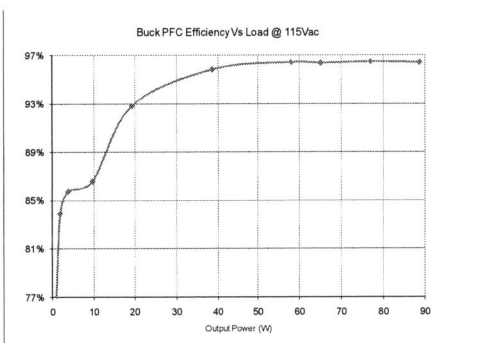

Figure 11. Buck PFC stage efficiency performance

B. ICC Half-Bridge Stage

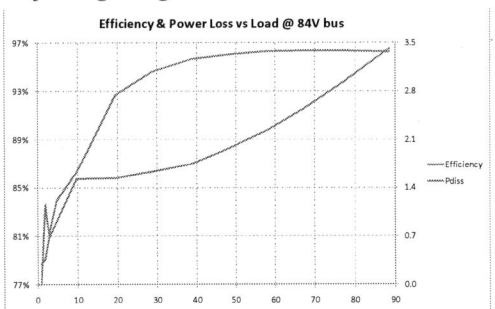

Figure 12. ICC HB stage efficiency performance

C. Overall Performance

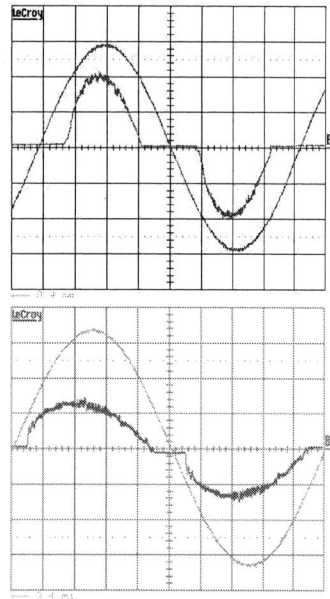

Figure 13. Buck PFC full load input current at 100V (1A/div) & 230V (0.5A/div)

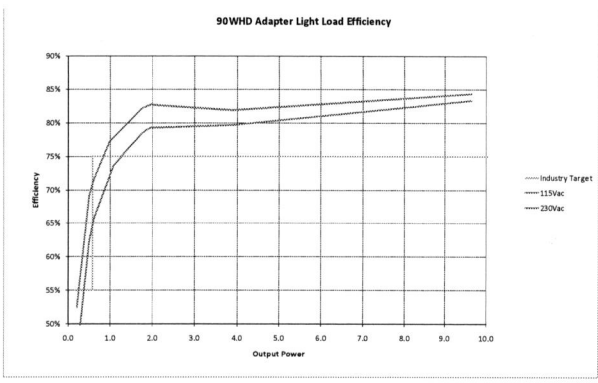

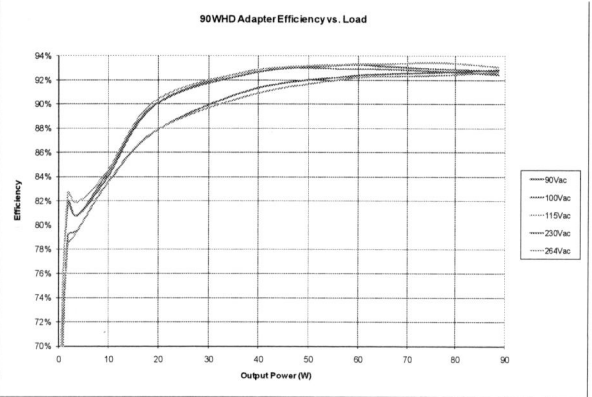

Figure 14. Overall Two-Stage Efficiency Performance

VII. CONCLUSIONS

Usage of the two-stage buck PFC + ICC HB topology has demonstrated excellent full load efficiency of 93% at nominal line, with high efficiency maintained over universal line.

Minimum full load efficiencies at the line extremes of 90V and 264V are 92.5% and 92.7% respectively. This enables design of very small format, low profile notebook adapters with power densities >16W/in^3. The topology enables high efficiency over the full load range – the addition of smart burst mode control further enhances performance close to zero load, demonstrating > 80% efficiency at 5% loading.

ACKNOWLEDGMENTS

The authors would like to thank the entire PSC team at Texas Instruments (Cork), for their valuable inputs and support throughout this work.

REFERENCES

[1] (2008) Environmental Protection Agency (EPA). [Online]. http://www.energystar.gov/ia/partners/prod_development/revisions/downloads/computer/Version5.0_Computer_Spec.pdf

[2] European Commission. [Online]. http://sunbird.jrc.it/energyefficiency/pdf/Workshop_Nov.2004/PS%20meeting/Code%20of%20Conduct%20for%20PS%20Version%202%2024%20November%202004.pdf

[3] http://www.80Plus.org/.

[4] Technical Specs. [Online]. http://www.climatesaverscomputing.org/about/tech-specs/

[5] A. Fernandez, J, Sebastian, M. Hernando, P. Villegas, and J. Garcia, "Helpful Hints to Select a Power-Factor-Correction Solutionfor Low- and Medium-Power Single-Phase Power Supplies," *IEEE Transactions on Industrial Electronics*, pp. 46-55, 2005.

[6] H, Wei and I. Batarseh, "Comparison of Basic Converter Topologies For Power Factor Correction," in *IEEE Southeastcon*, 1998.

[7] C, Qiao and K. Smedley, "A Topology survey of Single-Stage Power Factor Corrector With a Boost Type Input-Current-Shaper," in *IEEE Applied Power Electronics Conference*, 2000.

[8] L. Huber, L. Gang, and M. Jovanovich, "Design-Oriented Analysis And Performance Evaluation of Buck PFC Front-End," in *IEEE Applied Power Electronics Conference*, 2009.

[9] G. Young, G. Tomlins, and A. Keogh, "An ACDC Converter," World Intellectual Property Organisation, International Publication WO 2006/046220 A1, May 4, 2006.

[10] R. Redl, A.S. Kislovski, and B.P. Erisman, "Input-Current-Clamping: An Inexpensive Novel Control Technique to Achieve Compliance With Harmonic Regulations," in *IEEE Applied Power Electronics Conference*, 1996, pp. 145-151.

[11] R. Redl and L. Balogh, "RMS, DC, Peak, And Harmonic Currents in High-Frequency Power-Factor-Correctors with Capacitive Energy Storage," in *IEEE Applied Power Electronics Conference*, 1992.

[12] G. Spiazzi, "Analysis of Buck Converters Used As Power Factor Preregulators," in *IEEE Power Electronics Specialists Conference*, 1997.

[13] L. Xiangrong, X. Dianguo, and Z. Xiangjun, "Low Cost Electronic Ballast with Buck Converter as PFC Stage," in *IEEE Power Electronics & Motion Conference IPEMC*, 2006.

[14] Electromagnetic Compatability (EMC) - Part 3-2: Limits. Limits for harmonic current emissions (equipment input current ≤ 20A per phase), 2005.

[15] B. Carsten, "Converter Component Load Factors; A Performance Limitation of Many Topologies ," in *PCIM*, 1988.

[16] "UCC29910: Buck PFC PWM Controller Datasheet," Texas Instruments, 2009.

[17] "UCC29900 Integral Cycle Controller Datasheet," Texas Instruments, 2009.

A Novel Variable Frequency soft switching Method for Flyback Converter with Synchronous Rectifier

Xiucheng Huang, Weijing Du, Wei Yuan, Junming Zhang, Zhaoming Qian

College of Electrical Engineering, Zhejiang University
38 Zheda Road, Hangzhou, China 310027
E-mail: zhangjm@zju.edu.cn

Abstract-This paper proposes a variable frequency zero-voltage-switching (ZVS) control method for Flyback converter with synchronous rectifier. A lossless snubber capacitor is paralleled to the drain and source of primary switch to absorb the leakage inductance energy. The conventional passive clamp circuit with RCD is omitted, and no auxiliary switch or winding is needed. Synchronous rectifier is turned on additionally for a short time before primary switch is on. The negative secondary current transferred to primary side is used to discharge the auxiliary capacitor and achieve ZVS operation. Experimental results from a 16V/4A prototype circuit are provided to demonstrate the converter performance. Both Soft-switching and high efficiency are achieved in the entire input voltage range at both full load and light load conditions.

I. INTRODUCTION

Due to the simplicity of the power stage, Flyback converters are widely adopted in the switching mode power supply design to provide regulated output voltages for low-power applications. The main power loss in Flyback converter includes transformer loss, primary switch (SW) and secondary rectifier (SR) conduction loss, SW switching loss and RCD clamp circuit loss. The conventional RCD clamp circuit absorbs the leakage energy and dissipates it in the snubber resistor which deteriorates the efficiency. To reduce clamp circuit loss, the lossless snubber for single ended converter was proposed to recycle the leakage energy, but the difficulties in parameters optimization and the large circulating energy limit the efficiency improvement [1]. The active clamp technique [2]-[8] has been proposed to absorb the energy stored in the leakage inductance lossless and provides a way of achieving ZVS for the SW so the switching loss is eliminated. Active clamp Flyback converter has good performance at full load condition but poor behavior at light load condition due to its complementary gate signals of main switch and auxiliary switch with constant frequency control method. Many ZVS approaches for Flyback converter are presented and studied [9]-[11], but they usually has poor light load efficiency. ZVS DCM Flyback converter with synchronous rectifier (SR) is discussed in [12]. However, the control method has disadvantages: the turn-on delay of the SR causes more conduction loss, and the system frequency will increase dramatically on light load condition to achieve ZVS operation, which affects other losses and lowers the average

efficiency.

This paper proposes a variable frequency ZVS control method for Flyback converter with synchronous rectifier to achieve soft switching and high efficiency in whole load range. RCD clamp circuit is omitted and a proper capacitor paralleled to SW is employed to absorb the leakage inductance energy and provides a way to achieve ZVS operation for SW. In the proposed control method, the SR is turned on additionally for a short time before the primary switch is on, which can be adopted in variable-frequency control to greatly improve efficiency at light load condition. Experimental results from a 64W prototype demonstrate the performance characteristics of the proposed converter.

II. OPERATION PRINCIPLES

Fig.1 shows the circuit configuration of the proposed ZVS Flyback converter with synchronous rectifier. The RCD clamp circuit is removed, and a snubber capacitor Ca (usually much larger than the output capacitor of SW) is paralleled between the drain and source of SW. SW is the primary switch, and SR is the output synchronous rectifier. L_m is the transformer magnetizing inductance and L_k is the transformer leakage inductance. The transformer turns ratio is N. The ZVS operation of SW can be realized by the resonance between L_m and C_a.

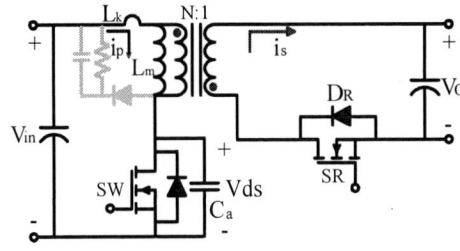

Fig.1 Proposed ZVS flyback converter with SR

Fig.2 shows the steady state waveforms of the proposed converter. Based on the waveforms, there are eight operating stages during one switching period. Fig.3 shows the equivalent circuits for each stage of the ZVS Flyback converter during one switching period. Each operation mode is simply described below.

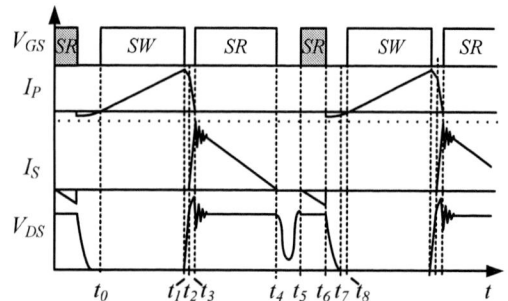

Fig.2 Staedy state waveforms of proposed converter

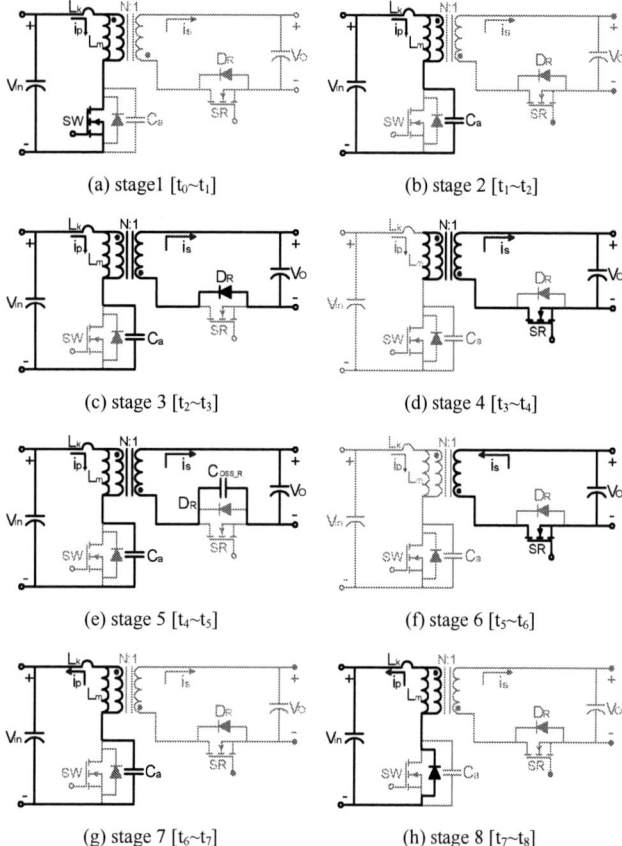

(a) stage1 [$t_0 \sim t_1$] (b) stage 2 [$t_1 \sim t_2$]

(c) stage 3 [$t_2 \sim t_3$] (d) stage 4 [$t_3 \sim t_4$]

(e) stage 5 [$t_4 \sim t_5$] (f) stage 6 [$t_5 \sim t_6$]

(g) stage 7 [$t_6 \sim t_7$] (h) stage 8 [$t_7 \sim t_8$]

Fig.3 Equivalent circuits in steady state operation

a) Mode 1 [t_0-t_1]

In this mode, primary side switch SW is turned on and the primary side current I_p increases linearly. The energy is stored in the magnetizing inductor, which is the same as the conventional flyback converter.

b) Mode 2 [t_1-t_2]

At t_1, when SW turns off, C_a is charged up by the magnetizing current. Due to relative large magnetizing inductance, the voltage V_{ds} is increased linearly.

c) Mode3 [t_2-t_3]

At t_2, the voltage V_{ds} reaches $V_{in}+(1+m)NV_o$, where m is

the ratio of the leakage inductance and the magnetizing inductance, the anti-paralleled diode of synchronous rectifier turns on. The remaining leakage inductor energy charges the auxiliary capacitor C_a, but the voltage spike on SW is much smaller due to the larger value. During this mode, the difference between the magnetizing current and primary current is delivered to secondary side. This mode can be treated as a primary to secondary commutation period. As soon as the current in the leakage inductance reaches zero, this mode is finished.

d) Mode 4 [t_3-t_4]

At t_3, a resonance formed by leakage inductance and auxiliary capacitor occurs, but the magnitude is much smaller. SR is turned on during this state, and the turn on delay time is depended on the SR drive circuit. The magnetizing energy is delivered to the secondary side as conventional flyback converter and the magnetizing current decreases linearly.

e) Mode 5 [t_4-t_5]

At t_4, magnetizing current decreased to zero, and SR turns off. A parasitic resonance occurs between L_m and C_a as conventional flyback at DCM condition except the resonant period being longer.

f) Mode 6 [t_5-t_6]

At t_5, SR is turned on again. The current through secondary winding I_{sec} increases reversely and the magnitude is proportional with SR auxiliary turn on time T_{SR_aux}. These negative current is used to achieve ZVS of primary switch SW. The voltage across the magnetizing inductance is clamped to $-NV_o$, and drain to source voltage of SW is clamped to $V_{in}+NV_o$.

g) Mode 7 [t_6-t_7]

At t_6, SR turns off. The negative secondary current I_{sec} transfers to primary side and discharges the auxiliary capacitor C_a. To achieve ZVS in the entire input-voltage range, the energy stored in magnetizing inductance by the negative secondary current must be large enough to discharge C_a from $V_{in}+NV_o$ down to zero.

h) Mode 8 [t_7-t_8]

At t_7, the auxiliary capacitor C_a voltage decreased to zero and the anti-parallel diode of primary switch SW turns on. SW should be turned on before the primary current I_p changes the polarity.

III. DESIGN CONSIDERATIONS

The proposed circuit can be applied to any control scheme to achieve ZVS operation such as constant frequency or variable frequency. From steady state operation described above, there is an extra reverse power deliver period when the SR is on before SW turns on. But usually the energy delivered to the primary side is quite little, which will not affect the output. Therefore, the design considerations for the power stage, such as transformer, primary switch and secondary rectifier is almost the same as conventional Flyback converter, which will not be elaborated here.

A. Select auxiliary capacitor C_a

The capacitor C_a is an important parameters to prevent SW from overvoltage and achieve ZVS operation. Also, it determines the circulating energy. Too small value may cause damage to SW due to the high voltage spikes when SW is off. Meanwhile, a large capacitance needs larger negative current to fully discharge it during the dead time and finally lead to larger secondary conduction loss. In practical design, the voltage spike on SW is limited to 90% of its breakdown voltage. Therefore C_a can be designed as:

$$C_a \geq \frac{L_k \cdot I_{p\_peak}^2}{0.9(V_{(BR)DSS\_SW} - V_{in})^2} \quad (1)$$

Where $V_{(BR)DSS\_SW}$ is the Drain-source breakdown voltage of SW. A minimal value is preferred to minimize the circulating energy.

B. ZVS operation

For the purpose of ZVS of SW, there must be sufficient reverse energy stored in the magnetizing inductance to completely discharge the snubber capacitor Ca. This requirement is valid at t_6 (when SR is turned off):

$$\frac{1}{2}L_m \cdot I_{p\_neg}^2 \geq \frac{1}{2}C_a \cdot (Vin + NVo)^2 \quad (2)$$

Where I_{p_neg} is negative peak value of magnetizing current. Obviously, even insufficient energy stored in magnetizing inductance, the switching loss can still be minimized due to valley switching.

In this circuit, primary switching loss is traded off for the secondary conduction loss and drive loss caused by the additional turn on stage of SR. The negative secondary current increases linearly during auxiliary SR on time. To meet the condition set in (2), the turn on time of SR during stage 6 is given by:

$$T_{SR\_aux} = \frac{L_m \sqrt{[V_{in\_max}^2 - (N \cdot Vo)^2]\dfrac{C_a}{L_m}}}{N \cdot Vo} \quad (3)$$

To achieve ZVS operation, the SW must be turned on during the t_7 to t_8 interval as shown in Fig 2. Otherwise, the primary resonant will charge up C_a, and ZVS is lost (or at least partially lost). Therefore, the delay time between the turn off of SR and the turn on of SW is critical to ZVS operation. The optimum value of this delay is one quarter of the resonant period formed by L_m and C_a.

$$T_{delay} = \frac{\pi}{2} \cdot \sqrt{L_m \cdot C_a} \quad (4)$$

C. Auxiliary turn-on moment of SR

The auxiliary turn-on moment of SR decides the secondary current initial value and voltage spikes on SW. A typical waveforms of Flyback converter at DCM operation is shown in Fig.4, where secondary resonant current is amplified to elaborate the issues. If auxiliary turn-on moment of SR is selected between A and C interval, a negative secondary current is built more easily as the initial value is negative. This improves the soft switching characteristic greatly. Meanwhile, turning on SR at A or E point minimizes the voltage spike on SW as given in (5). A resonance occurs between L_k and C_a when SR turns on at t_5, and the voltage of C_a can be expressed as:

$$V_{ds}(t) = V_{in} + NV_o + (V_{in} + NV_o - V_{ds}(t_5))\cos\omega(t - t_5) \quad (5)$$

where $\omega = \sqrt{\dfrac{1}{L_k \cdot C_a}} \quad (6)$

If turning on SR at C point, the peak voltage of C_a could reach to $V_{in}+3NV_o$ which may cause damage to SW at high line. Taking both current building and the voltage spike into consideration, the period during A and B or E and F is preferred for circuit implementation, which depend on input voltage and load condition

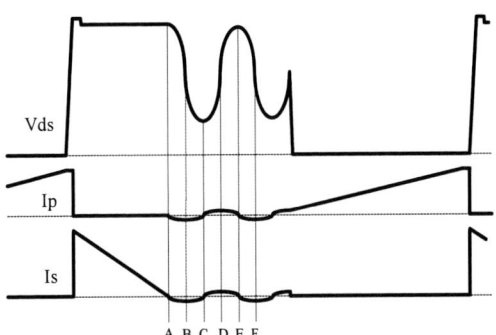

Fig .4　Conventional DCM Flyback converter waveform

IV EXPERIMENTAL RESULTS

To evaluate the performance of the proposed control method, a 64W prototype circuit with AC input voltage of $90-265V_{rms}$, output of 16V/4A are implemented and tested. The key prototype parameters are listed in Table.1. The auxiliary turn on time T_{SR_aux} for SR and delay time for SW are calculated by (3) and (4) respectively. The transformer core is PQ26/20 with PC40 equivalent material. Off time control is adopted in the prototype to achieve high average efficiency and light load efficiency.

Fig. 5 shows the gate drive signals, drain to source voltage of SW and the primary side current I_p at different loads and input conditions. It is clear that the soft switching for SW is achieved at any load conditions. The voltage spike is limited to an acceptable value with the snubber capacitor.

Table. 1. Key parameters of the experimental prototype

Parameters	Symbol	Value
AC input voltage	V_{in}	AC 90-265V (RMS)
Output	V_o, I_o	16V/4A
Primary side switch	SW	SPA11N60C3
Synchronous rectifier	SR	FDB3652
Bridge rectifier		KBJ406
Transformer turns ratio	$n=N_p:N_{s1}:N_{s2}$	24:4:3
Magnetizing inductance	L_m	280uH
Leakage inductance	L_k	1.5 uH
Auxiliary capacitor	Ca	2.2nF
Controller		NCP1351B
Auxiliary switch on time	T_{SR_aux}	1.3us
Dead time	Td	1us

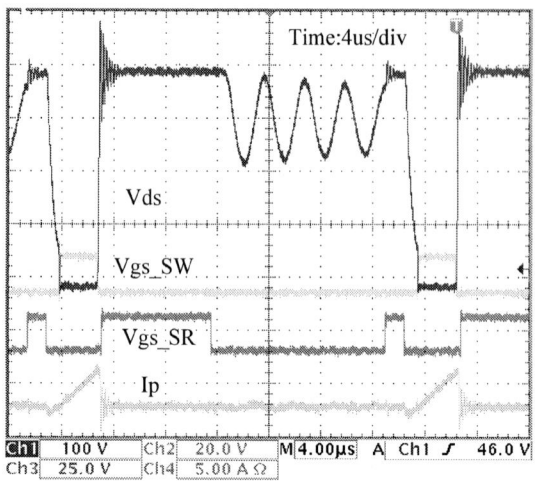

(c) Vin=240VAC & full load (Io=4A)

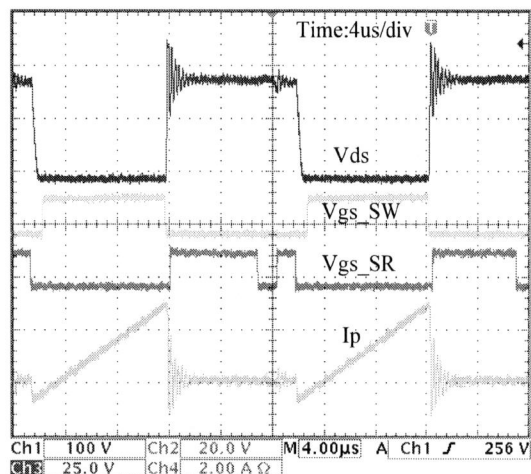

(a) Vin=90VAC & full load (Io=4A)

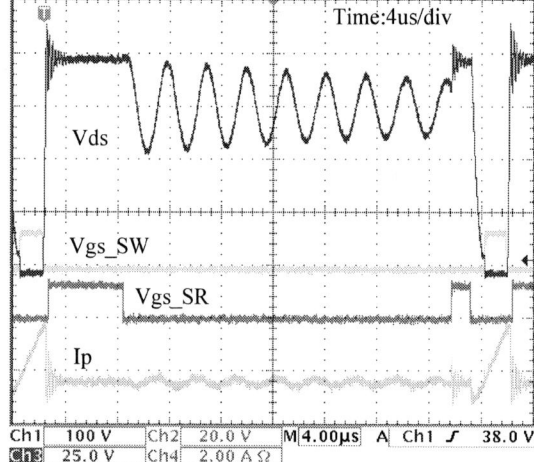

(d) Vin=240VAC & light load (Io=1A)

Fig.5. Key waveforms of the proposed converter

The measured efficiency of the prototype is shown in Fig. 6. The average efficiency is measured and calculated at 25%/50%/75%/100% load conditions. Also, the efficiency of conventional RCD clamp circuit (R=100K, C=2.2nF) at the same condition is presented for comparison. The proposed converter efficiency at high line and light load conditions is much higher than conventional one.

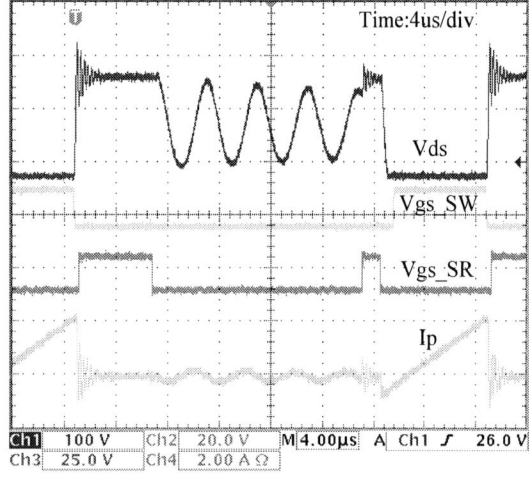

(b) Vin=90VAC & light load (Io=1A)

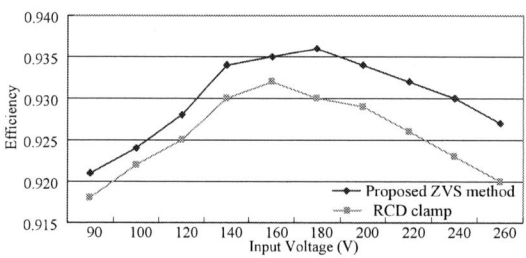

a. Average efficiency vs. input voltage

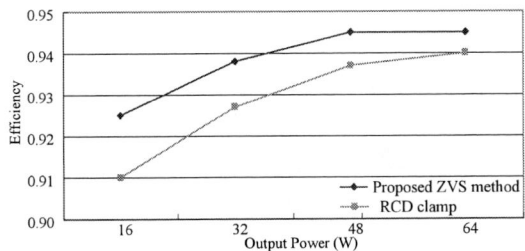

b. Efficiency vs. output power at Vin=220Vrms

Fig. 6. Efficiency comparison between Flyback with proposed control
method and conventional RCD clamp

V CONCLUSIONS

A novel variable frequency ZVS control method for
Flyback converter with synchronous rectifier has been
presented. The conventional passive clamp circuit with RCD
is omitted. A lossless snubber capacitor paralleled between
the drain and source of SW is employed to achieve ZVS
operation and absorb the leakage inductance energy. The
control method can be easily implemented by adjusting the
driver logic. Experimental results on a 64W prototype are
recorded to verify the theoretical analysis. Based on the
experimental results, ZVS operation and high efficiency are
achieved at any load and input conditions with the proposed
control method.

ACKNOWLEDGEMENT

This work is supported by China National Science Fund,
No. 50907061.

REFERENCES

[1] Tamotsu Ninomiya, et al, "Analysis and Optimization of a
Nondissipative LC Turn-Off Snubber", *IEEE Trans. Power Electronics*,
vol. 3, No.2, April 1988.

[2] E.H. Wittenbreder, "Zero Voltage Switching Pulse with Modulated
Power Converters", U.S. Patent 5 402 329, March 1995.

[3] Robert Watson, Fred C. Lee, Guichao C. Hua, "Utilization of an
Active-Clamp Circuit to Achieve Soft Switching in Flyback
Converters", *IEEE Trans. Power Electronics*, vol. 11, pp. 162-169,
January 1996.

[4] David A. Cross, "Clamped Continuous Flyback Power Converter", U.S.
Patent 5 570 278, Oct. 29, 1996.

[5] Choi, C.T., Li, C.K., and Kok, S.K., "Control of an Active Clamp
Discontinuous Conduction Mode Flyback Converter", in *Proc. 1999
IEEE Power Electronics and Drive Systems Conf.*, vol. 2, pp.
1120–1123.

[6] Gwan-Bon Koo, Myung-Joong Youn, "A New Zero Voltage Switching
Active Clamp Flyback Converter", in *Proc. 2004 IEEE PESC Conf.*, pp.
508-510.

[7] P. Alou, et al, "A Low Power Topology Derived from Flyback with
Active Clamp Based on a very Simple Transformer", in *Proc. 2006
IEEE APEC Conf.*, pp. 627-632

[8] Yu-Kang Lo, and Jing-Yuan Lin, "Active-Clamping ZVS Flyback
Converter Employing Two Transformers," *IEEE Trans. Power
Electronics*, vol. 22, No. 6, November 2007.

[9] N.Nagagata and H.Hiromura, "Switching Power Supply", U.S. Patent 4
958 268, Sep. 18, 1990.

[10] Sung Yun Lee, Suwon, "Synchronous rectifier flyback circuit for zero
voltage switching", U.S. Patent 6 198 638, Mar. 6, 2001.

[11] J.Claassens, I.W.Hofsajer, K.de Jager, "Passive Component Integration
in A ZVS Flyback Converter", in *Proc. 2006 IEEE APEC Conf.*, pp.
539-544

[12] Michael T. Zhang, Milan M.Jovanovic and Fred C.Y.Lee, "Design
Consideration and Performance Evaluations of Synchronous
Rectification in Flyback Converters", *IEEE Trans. Power Electronics*,
vol. 13, pp. 538-546, May 1998.

Optimal Design of a Compact 99.3% Efficient Single-Phase PFC Rectifier

J. Biela and J.W. Kolar
Power Electronic Systems Laboratory
ETH Zurich, Physikstrasse 3, CH-8092 Zurich, Switzerland
Email: biela@lem.ee.ethz.ch;

G. Deboy
Infineon Technologies Austria AG
P.O. Box 173, A-9500 Villach, Austria
Email: gerald.deboy@infineon.com

Abstract—Due to rising energy costs the efficiency of power electronics converter systems is of higher and higher importance, especially for applications with continuous operation as e.g. server power supplies. For designing such supplies numerous parameters like the switching frequency and the characteristic values and the geometry of magnetic components must be determined. In this paper, an optimisation procedure, which automatically determines the parameter values of a single-phase bridgeless PFC rectifier for maximum efficiency is presented. There, continuous and discontinuous operation mode, as well as a concept for magnetic integration of the CM and DM filter inductors is included. For verifying the considerations, a prototype with 99.3% efficiency and a power density of 1.35kW/dm^3 is presented.

I. INTRODUCTION

In the last decades, increasing the power density ρ has been one of the most important design criteria in power electronics [1]–[4] besides cost reduction. There, an increase of the efficiency η was only indirectly required, as a lower system volume results in a smaller surface for power loss dissipation. However, due to environmental concerns, efficiency became more and more important, so that now at least two design targets, i.e. high power density and high efficiency, have to be met at the same time. This results in a multi-objective optimisation problem, where the design parameters, as e.g. the component values or the geometry of magnetics, are selected under consideration of multiple performance indices like costs, volume, efficiency, etc.

In [5] single-phase PFC rectifier systems have been optimised for efficiency and alternatively for power density. Furthermore, the trade-off between the two competing design criteria has been investigated. Based on analytical expressions describing the relation between maximal achievable efficiency and power density, the limit (Pareto-Front) in the efficiency-

Fig. 1: Laboratory prototype of the proposed ultra-efficient 3.2kW dual-boost PFC rectifier composed of two interleaved 1.6kW units with f_P=16kHz; overall dimensions: 190mm×188mm×65mm; output power density: 1.35 kW/dm^3, efficiency: 99.3%.

power density performance space (η-ρ-plane) has been derived and the sensitivity of the Pareto-Front on the applied technologies has been investigated. Furthermore, a prototype with an efficiency of 99.09% and a power density of $\approx$1.1kW/dm^3 at f_P=33kHz switching frequency has been presented.

However, the structure of the optimisation procedure has been only briefly explained. In this paper, the analytical models of the converter and of the magnetic components as well as the efficiency optimisation procedure are described in detail (section III). There, in contrary to [5] discontinuous (DCM) and continuous (CCM) conduction mode in sections of the mains period is covered. Furthermore, a magnetic integration of the common mode (CM) EMI filter inductor and the boost inductor into a single magnetic component is proposed (section II) which increases the system power density as well as the efficiency compared to [5]. Moreover, the system design and the cooling concept are improved.

Using the optimisation procedure, the parameter values and the combination of DCM and CCM resulting in maximum efficiency are determined in section IV Finally, in section V a prototype system and details of the system design are presented.

TABLE I: Specifications of the proposed ultra-efficient, bridgeless single-phase PFC rectifier with two interleaved systems.

Output Power P_O	2×1.6kW
Line Voltage U_N	230±10%
Output Voltage U_O	365V – 400V
Ambient Temperature	45°C
Power Density	1.35kW/dm^3
Efficiency	99.3%

II. HIGH-EFFICIENCY SINGLE-PHASE PFC RECTIFIER TOPOLOGY

The basis of efficiency optimisation is the choice of the most suitable circuit topology that enables minimum semiconductor losses at low semiconductor chip area requirements, i.e. low realisation costs. The most used single-phase PFC boost rectifier concepts are the conventional boost, the bridgeless/dual boost and the dual-boost AC-switch PFC rectifier [5], [6]. There, the freewheeling boost diodes are frequently realised with SiC Schottky diodes in order to minimise the transistor turn-on losses. Accordingly, relatively complex soft-switching topologies as e.g. presented in [7], [8] do not provide an advantage concerning efficiency.

With the standard bridgeless PFC rectifier a switching frequency CM voltage occurs at the system output. According to [9] this can be avoided by adding clamping diodes from the negative rail n to the two mains connections. For reducing the volume of this approach a configuration with two inversely coupled DM/CM inductors with equal turn numbers and equal inductances of all windings has been presented in [10]. There, always one winding of each of the two series connected inductors is short circuited via the constantly turned on switch and the conducting clamping diode. Most of the return current flows via the clamping diode, so that the turned on switch and also the short circuited windings of the two inductors are not utilised during this interval. However, both cores are fully utilised, what is the benefit of this concept in contrast to two separate inductors [10].

An alternative solution, which has been proposed in [5], is to connect the midpoint or one of the two rails of the output via two capacitors to the two mains connections as shown in Fig. 2. This solution has a similar performance with respect to EMI as the solution with the two additional clamping diodes, but the chip area required for the clamping diodes and the related conduction losses are saved as most of the return current flows via the constantly turned on MOSFET. There it is important to note that with this concept one inductor is operating as CM inductor and one as differential mode (DM) boost inductor and both usually do not have the same number of turns and/or inductance values. This is in contrast to the diode clamped bridgeless converter with optimised inductors.

For reducing conduction losses, diodes commutated with

mains frequency should be replaced by MOSFETs operating as synchronous rectifiers. A replacement of the freewheeling diodes is not possible as the nonlinear parasitic output capacitance of superjunction MOSFETs would significantly increase the switching losses. Considering the four above mentioned topologies, different amounts of silicon area are required for achieving the same conduction losses [5]. In the bridgeless PFC boost rectifier with capacitive coupling the current flow in the turn-on interval is only via two MOSFETs and via a fast diode and a power MOSFET in the turn-off/boost interval, if S_2 is continuously turned on for $u_N > 0$ (and/or S_1 for $u_N < 0$).

In the bridgeless PFC boost rectifier with clamping diodes, the clamping diodes must be replaced by synchronous rectifiers, in order to achieve the same conduction losses as for the bridgeless system with capacitive coupling. Thus, the total required chip area is $4\,A_{Chip}$ (A_{Chip} is the chip area of a single boost transistor of the bridgeless PFC) for the MOSFETs plus 2 fast commutated diodes for the diode clamped and $2\,A_{Chip}$ plus 2 fast commutated diodes for the capacitively coupled bridgeless PFC (cf. Fig. 3). There, it is assumed that the total line-frequency return current flows via the respective clamping diode and is not shared between the clamping diode and the inactive switch/inductor [6] as approximately given for most designs [10]. In case the current is dividing, the required chip area could be reduced.

If for a conventional PFC boost rectifier equally low conduction losses as for the capacitively coupled bridgeless PFC rectifier should be achieved, four power MOSFETs for synchronous rectification are required as replacement of the four mains-commutated input rectifier diodes. Consequently, a MOSFET chip area of $9\,A_{Chip}$, as well as one SiC freewheeling diode are required for the conventional PFC system (cf. Fig. 3 [5]). In the AC-switch PFC converter two of the four rectifier diodes could be replaced by MOSFETs. The other two diodes are commutated with switching frequency and have to be realised with SiC Schottky diodes. Consequently, for the

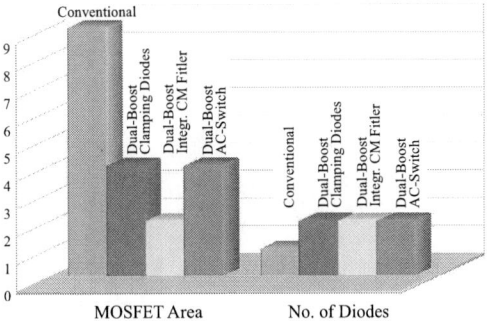

Fig. 3: *Left:* Total MOSFET chip area required for equal conduction losses of the conventional boost, the bridgeless/dual boost and the dual boost AC-switch PFC rectifier [5]. *Right:* Number of required fast recovery (SiC) freewheeling diodes. In the topologies all line-commutated diodes are replaced by MOSFETs for reducing the conduction losses by synchronous rectification.

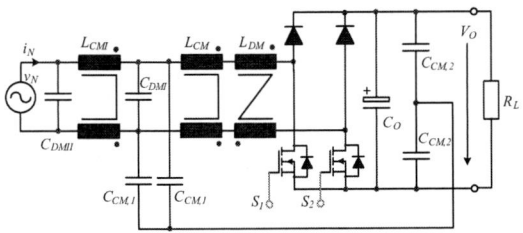

Fig. 2: Circuit diagram of the dual boost (bridgeless) rectifier with capacitive coupling of the output voltage to the mains and/or earth and EMI-filtering stage.

AC-switch PFC rectifier a MOSFET chip area of 4 A_{Chip} and 2 fast SiC diodes are required.

In summary, with regard to high efficiency at given total semiconductor chip area, the capacitively coupled bridgeless PFC boost rectifier with integrated CM filter as shown in Fig. 2 is clearly preferable.

A. Integrated Magnetics

In the original design of the capacitively coupled bridgeless PFC (cf. Fig. 2) two independent inductors with two windings are utilised. One of the inductors operates as CM filter inductor and the other one realises the DM boost inductor. In normal operation switch S_2 is constantly turned on if the mains voltage u_N is positive, and S_1 in combination with D_1 and L_{DM}/L_{CM} performs the boost function. Due to capacitors $C_{CM,1}$ and $C_{CM,2}$ always one branch of the series connected inductors L_{DM}/L_{CM} is short circuited for higher frequencies via the constantly turned on MOSFET. Consequently, the

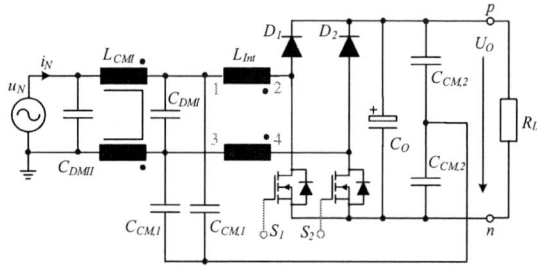

Fig. 4: Bridgeless PFC with integrated magnetics (w/o EMI filter).

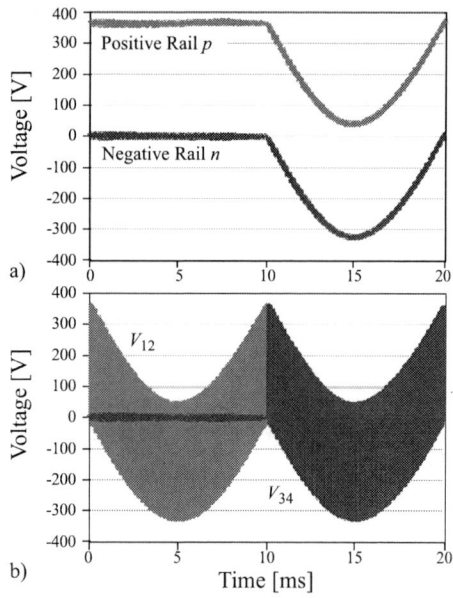

Fig. 5: a) Voltage of the positive p and negative n rail (cf. Fig. 4) to earth. b) Voltage across the two windings of the integrated magnetic device.

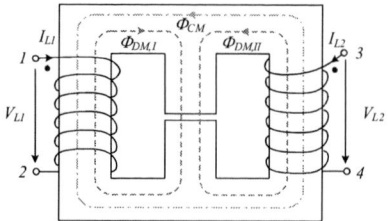

Fig. 6: Integrated magnetic device with CM and DM flux directions for positive mains voltage.

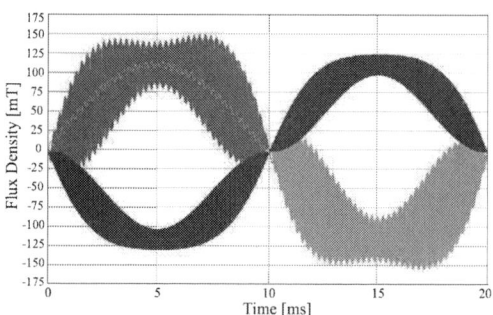

Fig. 7: Simulated flux density waveforms for the integrated magnetic device based on the design parameters given in section V. There the cross-sectional area of the middle leg is twice as large as for the outer legs ($2 \times$ E70 core).

negative rail n of the output voltage mainly varies at low frequencies (cf. Fig. 5). The high-frequency component of the CM voltage drops across the CM inductor, which also limits the high-frequency CM current flowing in the loop formed by L_{DM}/L_{CM}, $C_{CM,1}/C_{CM,2}$ and the MOSFETs. This also could be seen in the CM equivalent circuit given in [5]. The LF part of the CM voltage drops across capacitors $C_{CM,1}/C_{CM,2}$ and result in a LF variation of the CM output voltage as could be seen in Fig. 5a), where the voltage of the positive p and the negative rail n of the output voltage against earth is shown.

For integrating both inductors, the two windings are put on one core and paths for the CM and the DM flux are provided as shown in Fig. 6. The DM flux flows always through one of the two windings and the middle leg, where an air gap is used to determine the DM/boost inductance. The CM flux flows through both windings and the two outer legs of the core. By adjusting the number of turns and/or an air gap also in the outer legs the CM inductance value could be adjusted. However, usually a CM inductance as large as possible is required, so that no air gap is necessary in the outer legs, except for limiting the CM flux.

In Fig. 7 the simulated flux time behaviour for one mains period is shown for the design described in section V. It can be clearly seen, that the DM flux is proportional to the input current and that a HF flux is always present in only one of the two outer legs whereas and in the other leg only a LF flux with a small ripple occurs.

III. MODELS AND OPTIMISATION PROCEDURE

In the following analytical models for the capacitively coupled and magnetically integrated bridgeless PFC rectifier are derived, which are used to optimise the system parameters for minimal losses. The degrees of freedom in the optimisation are the switching frequency f_P, the geometry of the boost inductance, and the power semiconductor chip area A_{Chip}, i.e. the number of power MOSFETs and SiC diodes connected in parallel for realisation of a power transistor S_1, S_2 and/or a freewheeling diode D_1, D_2. Furthermore, the value of the boost inductance is varied, what influences also the operation mode. In the algorithm CCM, and DCM as well as a combination of both modes in sections of the mains period is considered.

The optimisation is carried out for nominal power and nominal input and output voltages, but basically any power level and/or input/output voltage could be considered in the algorithm. Even a combination of the efficiencies at different operating points, e.g. at 100% and 50% load, could be used as quality criteria.

In the optimisation the volume of the boost inductor L_{DM} is limited to a fixed value, since the inductor volume would grow in the course of the optimisation above all limits, because the losses monotonically decrease with increasing inductor volume [5].

In Fig. 8 a flowchart of the developed procedure for optimising the design variables (f_P, chip area of MOSFETs and diodes, number of turns and geometry of boost inductor) is shown.

The starting point of the procedure is the specification of the converter system, including input/output voltages and the output power but also component limits as e.g. the maximal allowed flux density or the maximal junction temperature of the MOSFETs, which are the constraints during the optimisation. Also the starting values of the design variables are set. With these values the currents/voltages of all the components and the losses in the semiconductor elements are calculated. There, for each switching action the operation mode is determined, so that also combined DCM/CCM operation can be considered.

Furthermore, the flux variation in each pulse interval is calculated. This is used in an inner optimisation loop, which determines the number of turns and the geometry of the core and the winding resulting in minimum boost inductor losses. Since the aim of the optimisation is an ultra-high efficiency, no thermal models for the core and the windings is required, which would limit the design in case the volume of the inductor would be minimised. In the ultra-high efficiency system all components are operating well below their thermal limits, what is also advantageous with respect to life time and reliability.

The global optimisation algorithm adds the losses of the boost inductor, the CM filter inductor, the semiconductors and varies then the free parameters, so that the overall system losses are minimised. In the system losses also the losses in the control, the output capacitor and the EMI filter, which are assumed to be constant, are included. Furthermore, the gate drive losses and the auxiliary supply losses which slightly depend on the MOSFET's chip area are added.

In the following briefly the equations for the currents and voltages, the losses of the semiconductors and the magnetic components as well as the auxiliary power are summarised.

A. Semiconductors

Besides the fundamental component of the input current with amplitude $\widehat{I}_N$, also the ripple current

$$i_{N,r} = \frac{1}{2} \frac{U_O T_P}{L_{DM}} \frac{1}{M} \sin(\omega_N t) \sin\left(1 - \frac{1}{M}\sin(\omega_N t)\right) \quad (1)$$

must be considered, where the modulation index M is defined as $M = U_O/\left(\sqrt{2}U_N\right)$. With the input current fundamental and the ripple current as well as the duty cycle

$$d = 1 - \frac{1}{M}\sin(\omega_N t) \quad (2)$$

the RMS current in a MOSFETs and the average current in the rectifier diodes are calculated for CCM.

Also, in the DCM/mixed DCM and CCM operation the average current in each switching cycle must be equal to the instantaneous value of the input current reference value. This results in

$$d_1 = \frac{\sqrt{2\widehat{I}_N L_{DM}\left(U_O - \widehat{U}_N\sin(\omega_N t)\right)}}{\sqrt{\widehat{U}_N T_P U_O}} \quad (3)$$

$$d_2 = d_1 \frac{\widehat{U}_N\sin(\omega_N t)}{U_O - \widehat{U}_N\sin(\omega_N t)} \text{ and } d_3 = 1 - d_1 - d_2$$

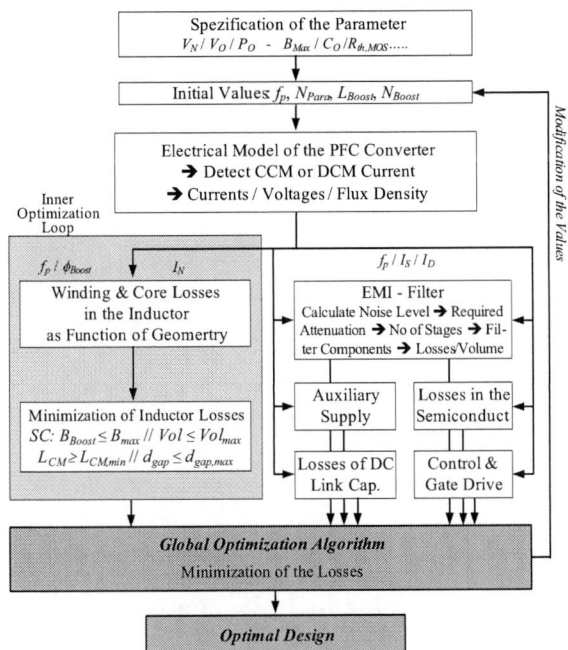

Fig. 8: Flow chart of the optimisation procedure for maximising the efficiency.

for the relative on-time d_1 of the switching MOSFET, the relative freewheeling time d_2 of the diode and the zero current time d_3, where no current flows from the input if the parasitic capacitances of the MOSFETs/diodes are neglected. By setting d_3 to zero, the currents for DCM/mixed DCM and CCM operation with variable switching frequency of the PFC converter could be calculated.

Based on the currents the conduction losses of the semiconductors are determined. Equation

$$R_{DSon} = R_{DSon,25} + \frac{R_{DSon,125} - R_{DSon,25}}{125^\circ C - 25^\circ C}(T_j - T_a) \quad (4)$$

is used for calculating the conduction losses in dependency of the junction temperature T_j (T_a is the ambient temperature). By solving

$$T_j - T_a = \frac{R_{th,MOS}}{N_{P,MOS}}\left(P_{on} + P_{off} + R_{DSon}I_{MOS,RMS}^2\right) \quad (5)$$

for T_j, the losses can directly be calculated, without any iteration. There, $N_{P,MOS}$ is the number of parallel connected MOSFETs and P_{on}/P_{off} are the switching losses, which are calculated below. A similar equation is used for determining the junction temperature of the rectifier diodes, which influences the forward voltage drop.

The power transistors are controlled such, that each transistor is switching during half a mains period where the other transistor is turned-on continuously for minimising the conduction losses.

In a next step, the current values at the turn-on and the turn-off of the switching MOSFET are determined based on the above calculated input current. The switched currents are required for calculating the switching losses based on measured loss curves.

Due to the large parasitic output capacitance of the parallel connected MOSFETs ZVS conditions are given during turn-off, so that the turn-off losses are very small and neglected in the considerations. This is also true for the switching losses of the SiC Schottky diodes employed as freewheeling diodes. Thus, the switching losses are mainly occurring during the turn-on of the MOSFETs.

B. Magnetic Components

Besides the semiconductors, the boost inductor is one of the major loss contributors. In the considered system the inductor is realised with foil windings and the geometric design of the inductors is determined by the four variables a, b, c and d as explained in [2]. With these variables for example the cross sectional area of the magnetic core or the core window area could be expressed and the losses in the core or the winding can be determined as function of these variables. This relation between the geometry and the losses then enables an optimisation of the number of turns and the geometry for minimal losses.

For calculating the winding losses first the harmonics of the winding current are calculated by Fourier analysis. The time behaviour of the inductor current is determined with the converter model and as worst case approximation it is assumed,

that the complete return current flows via the inductor winding and not the coupling capacitors $C_{CM,\nu}$. With the amplitude and the frequency of the harmonics, the skin and proximity effect losses at each frequency are calculated with an 1-D approximation as e.g. presented in [11]–[13] and then – based on the orthogonality of the losses [14] – the losses at the single harmonics $\hat{I}_{L(i)}$ are added [5].

For calculating the proximity effect losses in the inductor, it is assumed that there is an air gap only in the middle leg, i.e. the H-field ramps from 0 to $+H_{max}$. Furthermore, the losses in the winding due to the fringing field of the gap are neglected in order to simplify the calculations, what does not result in a too large error in the considered case due to a relatively large distance between foil winding and air gap as FEM simulations have proven. In all the calculations the losses are expressed as function of the variables a, b, c and d, so that it is possible to optimise the geometry of the core and the winding for minimal losses.

For the core losses, the flux density time behaviour in the core must be determined (cf. Fig. 7). The DM flux density follows a 50Hz major loop and minor magnetisation loops with switching frequency in case of CCM and only minor loops in case of DCM. In both cases the core losses can be calculated based on the method proposed in [15], where the Steinmetz coefficients [16] are utilised for characterising the core material and where the rate of change of the flux density (dB/dt) is the basis for the loss calculation [5].

With an appropriate choice of the CM inductance and coupling capacitance value, the CM flux has a relatively high mains frequency fundamental and only small high frequency harmonics (cf. Fig. 7). Both result in relatively small losses compared to the DM flux and are therefore neglected here. But the peak value of the CM flux must be considered together with the peak DM flux in the optimisation in order to avoid saturation. The peak flux density $B_{m,max}$ in the middle leg is given by

$$B_{m,max} = \frac{L_{DM}I_{DM,max}}{NA_{core,m}} \quad (6)$$

and the $B_{o,max}$ of the outer leg by

$$B_{o,max} = \frac{L_{DM}I_{DM,max}}{2NA_{core,o}} + \frac{L_{CM}I_{CM,max}}{NA_{core,o}}. \quad (7)$$

Current $I_{CM,max}$ is the CM mode current flowing in L_{CM}. The second constraint in the optimisation is the volume of the magnetic component, which must be limited as explained above [5].

C. Output Capacitor, EMI-Filter & Auxiliary Supply

For achieving a very high efficiency also minor loss contributions must be considered and minimised. The losses in the output capacitor can be calculated with the RMS current in the output capacitor and the ESR usually given in the data sheet.

The losses in the auxiliary supply are calculated with the efficiency curve of the auxiliary supply and the auxiliary power consumption of the controller, gate drives, measurement circuits, etc. In order to minimise these losses an ultra-low power controller and low power sensors are employed. Furthermore, the design of the auxiliary supply is optimised for minimal losses. All these losses are independent of the converter design except for the gate drive power, which depends on the chip area of the MOSFETs, i.e. the total gate charge.

The DM EMI filter is designed with the simplified approach presented in [17], where a symmetrical arrangement of the filter inductors is chosen in order to attenuate also mixed mode noise. Furthermore, all filter stages have the same inductance and capacitance value, since this results in the minimum filter volume. For calculating the required attenuation, it is assumed, that the total noise current $I_{Noise,RMS}$ (which consists of several harmonics at multiples of the switching frequency with according sidebands) would appear as a single peak only at the switching frequency. This peak would then cause a noise voltage U_{LISN} at the test receiver. The noise current and U_{LISN} are given by

$$I_{Noise,RMS} = I_{DM,RMS} - I_{N,RMS} \qquad (8)$$

$$U_{LISN} = 50\Omega \cdot I_{Noise,RMS}. \qquad (9)$$

With U_{LISN} the required attenuation could be calculated by comparing the noise voltage with the limit value. In case the switching frequency is below 150kHz, the required attenuation is defined by the first harmonic above 150kHz. There, the decay of the amplitude of the harmonics with increasing frequency must be considered.

With the required attenuation, the component values for the DM inductor and capacitor are determined with equations, which have been derived by minimising the filter volume. There, the volume is given as empirical function of the component value and the current/voltage. The volume is calculated for different numbers of filter stages and the solution with the lowest volume is chosen.

The required attenuation for CM depends significantly on the mechanical design, which determines the parasitic capacitances to earth. In order to limit the modelling effort and the complexity of the model, empirical values for the volume and the losses obtained with the prototype presented in [5] are used.

IV. OPTIMISATION RESULTS

Based on the procedure described in the previous section, a dual-boost bridgeless PFC rectifier with integrated magnetics (cf. Fig. 4) has been optimised for minimal losses at nominal output power and the specifications given in Table I. A result of the calculations, where the global optimisation algorithm has been replaced by a for-next loop for varying the operating frequency and the boost inductance value, is shown in Fig. 9. There, the optimised efficiency is shown in dependency of the switching frequency and the inductance of the boost inductor, so that besides the optimal operating point also the sensitivity of the operating point to frequency/inductance value variations is shown.

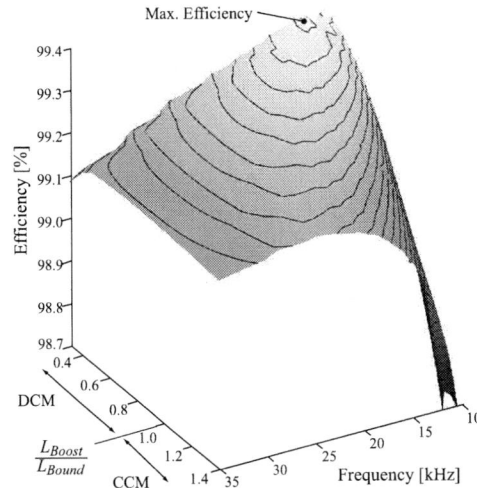

Fig. 9: Optimal efficiency and power density of the dual boost PFC in dependency of the switching frequency f_P and the inductance of the boost inductor. A value of 1 on the inductance axis is for an inductance value L_{Bound} resulting in an operation at the border of CCM. Decreasing the inductance value results in CCM and DCM operation in sections of the mains period.

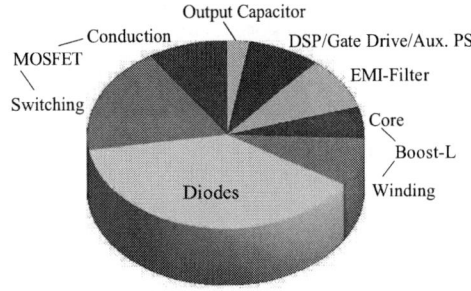

Fig. 10: Loss distribution of the optimised converter system with a switching frequency of 16kHz and a boost inductor of 400μH.

It is important to note, that the volume of the inductor is limited to maximal 0.3dm³ for all considered operating points. This limit comes from practical considerations and available core shapes and sizes.

An optimal efficiency of 99.33% is achieved for an operating frequency of approximately 9.3kHz ($L_{DM} = 470\mu H$). There, the theoretical power density is roughly 2.8kW/dm³. With decreasing frequency the efficiency drops relatively rapidly due the limit of the inductor volume. This limit results in increasing losses of the inductor, since a growing inductance value must be realised in a limited volume. Without this limitation, the optimal efficiency would be theoretically at $f_P = 0$ for an infinitely large inductor, what is not practical.

For higher switching frequencies, the losses in the semiconductors increase. First, the switching losses increase with increasing frequency and second, the optimal chip area resulting in minimal semiconductor losses decreases, so that also the conduction losses are increasing with switching frequency.

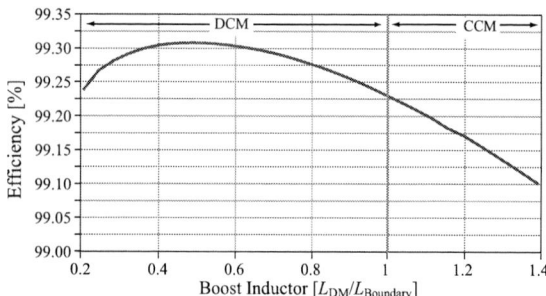

Fig. 11: System efficiency as function of the inductance value of the boost inductor L_{DM} related to the inductance $L_{Boundary}$ for which the limit of pure CCM operation is reached. The switching frequency is 16kHz.

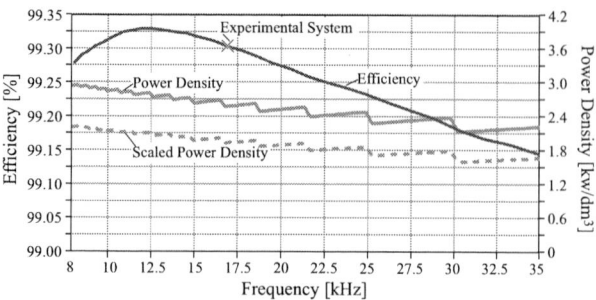

Fig. 12: System efficiency as function of the switching frequency for a fixed inductance $L_{DM} = 400\mu H$ of the boost inductor. Furthermore, the power density obtained for considering the net component volume and the scaled power density, where also the volumes required for mounting of components, not matching component shapes, etc. is considered, are shown.

By shifting the operating frequency to 16kHz, i.e. out of the audible range, the efficiency just slightly decreases to 99.31%. Limiting the operation to the CCM results in a maximal efficiency of 99.24% at frequency slightly above 20kHz ($L_{Boost} \approx 800\mu H$, 2.8kW/dm³ theoretical power density based on net volumes).

The distribution of the losses for 16kHz switching frequency and $L_{DM}=400\mu H$ is shown in Fig. 10. There, it could be seen that for the optimal system, the semiconductors cause the largest share of the system losses and that the forward voltage drop of the output diode has a significant influence on the efficiency. For the MOSFET losses it is important to note, that in the considered case the switching and the conduction losses are not equal for the optimal chip area, since additional effects as for example the parasitic capacitance of the freewheeling diodes are considered in the optimisation. Furthermore, it could be seen that the passive components and the auxiliary supply/control have a relatively low influence on the achievable efficiency.

In Fig. 11 the system efficiency as function of the boost inductance L_{DM} is shown for a switching frequency of 16kHz. A value of 1 on the x-axis means that the converter is operating at the boundary of CCM. This figure clearly shows that a mixed mode operation, i.e. a combination of CCM and DCM in sections of the mains period results in the highest efficiency at nominal load.

In Fig. 12 the system efficiency at nominal power is given as function of the switching frequency for a fixed inductance of $400\mu H$ of the boost inductor. Furthermore, the power density is shown as solid line and the scaled power density as dashed line. In the scaled power density also the additional volume resulting from mounting of components, not matching component shapes (e.g. cylindrical capacitors, rectangular magnetic cores) are considered besides the net component volumes. The steps in the power density result from the volume of the EMI filter, which increases with increasing switching frequency since lower order harmonics of higher amplitude move into the frequency band of 150kHz to 30MHz where the noise limits has to be met.

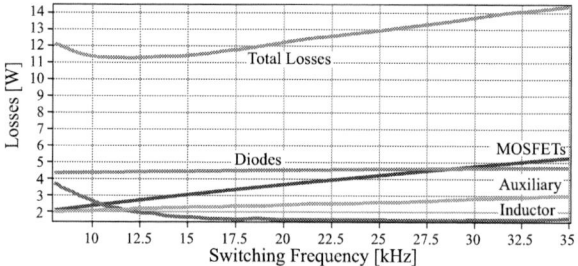

Fig. 13: Losses of the power semiconductors, the boost inductor L_{DM} and the auxiliary components in dependency on the switching frequency f_P. The boost inductor volume is set to a constant value and the inductance is adapted inversely proportional to f_P. Therefore, for increasing f_P a lower inductance value has to be realised in the same volume resulting in lower boost inductor losses.

V. EXPERIMENTAL RESULTS

In order to verify the results of the optimisation, an ultra-efficient single-phase PFC boost rectifier system was realised. The system specifications were set to the values given in Table I. Two parallel subsystems are arranged, each with 1.6kW rated output power. To lower the level of EMI emissions, both systems include a triangular variation of the switching frequency between 17kHz and 20kHz with a period of 45ms, corresponding to the averaging time constant of the EMI measurement according to CISPR 22. This assures that the EMI measurement acquires the spectrum broadened by frequency modulation and reduced in amplitude.

The output capacitance is realised with parallel connected electrolytic capacitors with low ESR and low leakage currents, which also decrease during operating due to self healing effects. The output capacitance is $816\mu F$ resulting in a ripple voltage amplitude of 17V at nominal power.

Because of the low semiconductor losses, no heat sink is required and/or the cooling can take place directly via natural convection. Nevertheless, small heat sinks have been added in order to reduce the junction temperature for a lower $R_{DS,on}$. In connection with a current measurement with low intrinsic power consumption (75mW each) and the reduction of the

TABLE II: Components of the ultra-efficient bridgeless capacitively coupled and magnetically integrated PFC rectifier.

	Components
MOSFETs	Per Leg 0.011Ω
Diodes	600V SiC Schottky
Integrated Inductor	2 × E70 Core / 2 × 18turns
L_{DM}	2 × 200μH
L_{CM}	2.4mH
$C_{CM1}=C_{CM2}$	220nF (Foil)
Output Cap.	12×450V/68μF
Capacitor	Vishay BC2222198
C_{DMI}	2μF
L_{CMI}	1.2mH
C_{CMI}	22nF
C_{DMII}	2μF

calculating capacity of the DSP TI TMS 320 LF 2808 used for control from 100MIPS to 50MIPS, the overall auxiliary power consumption for the 3.2kW system can be limited to less than 2W. In the next step an ultra-low power micro controller (TI MSP430) will be used to reduce the control power further. The power components employed for realising the system (Fig. 1) are listed for one 1.6kW subsystem in Table II.

Unfortunately, the new generation of SiC Schottky diodes, which were assumed in the optimisation, have not been delivered in time, so that in the prototype system replacement diodes had to be used for the measurements. With these diodes, the system efficiency was determined to be 99.17% at the nominal operating point by means of a highly precise calorimeter measurement setup. This agrees very well with the calculation results of 99.2% if the data of the replacement diodes is inserted also in the calculations. The inaccuracy of the efficiency measurement is ±0.03% in worst case. The calorimeter was calibrated with comparative DC measurements and ultra-precise shunts and multimeters.

Overall, the 3.2kW system shows dimensions of 190mm×188mm×65mm and thus a power density of 1.35kW/dm^3.

VI. CONCLUSION

In the paper analytical models for calculating the losses of a bridgeless PFC rectifier with integrated magnetics are presented. With the models an optimisation of the converter for minimal losses is performed. The resulting optimal loss distribution is discussed and the sensitivity of the efficiency on the switching frequency and the operation mode – DCM or CCM – is investigated. Highest efficiency is achieved for mixed operation mode, i.e. a combination of CCM and DCM in sections of the mains period.

A maximal efficiency of more than 99.33% could be achieved for a switching frequency of 9.3kHz and of 99.31% for mixed DCM and CCM operation at 16kHz. For pure CCM mode operation at nominal power the maximal efficiency is 99.24% at a switching frequency slightly above 20kHz. For validating the results a prototype system operating at 16kHz has been built, which could achieve an efficiency of 99.3% at a power density of 1.35kW/dm^3. The measured losses agree very

well with the theoretical predictions and the systems meets the required EMI standards.

ACKNOWLEDGMENT

The Authors would like to acknowledge the support of Dr. Johann Miniboeck and Mr. Dominik Hassler in connection with the practical realisation and experimental analysis of the system prototype.

REFERENCES

[1] H. Ohashi, "Research activities of the power electronics research center with special focus on wide band gap materials," *Proceedings of the 4th Conference on Integrated Power Systems, Napels, Italy*, pp. 153–156, 2006.

[2] J. Kolar, U. Drofenik, J. Biela, M. Heldwein, H. Ertl, T. Friedli, and S. Round, "PWM converter power density barriers," in *Proceedings of the Power Conversion Conference - Nagoya PCC*, 2-5 April 2007, pp. pp.9–29.

[3] J. Biela, U. Badstuebner, and J. W. Kolar, "Design of a 5-kW, 1-U, 10-kW/dm^3 resonant DC-DC converter for telecom applications," *IEEE Transactions on Power Electronics*, vol. 24, no. 7, pp. 1701–1710, 2009.

[4] U. Badstuebner, J. Biela, B. Faessler, D. Hoesli, and J. W. Kolar, "An optimized 5 kW, 147 W/in^3 telecom phase-shift DC-DC converter with magnetically integrated current doubler," in *Proceedings of the 24th IEEE Applied Power Electronics Conference and Exposition (APEC)*, 2009, pp. 21–27.

[5] J. Kolar, J. Biela, and J. Miniboeck, "Exploring the pareto front of multi-objective single-phase pfc rectifier design optimization - 99.2% efficiency vs. 7kW/dm^3 power density," in *Proceedings of the 6th IEEE International Power Electronics and Motion Control Conference (IPEMC)*, 2009.

[6] L. Huber, Y. Jang, and M. M. Jovanovic, "Performance evaluation of bridgeless pfc boost rectifiers," *IEEE Transactions on Power Electronics*, vol. 23, no. 3, pp. 1381–1390, 2008.

[7] M. M. Jovanovic and Y. Jang, "State-of-the-art, single-phase, active power-factor-correction techniques for high-power applications - an overview," *IEEE Transactions on Industrial Electronics*, vol. 52, no. 3, pp. 701–708, 2005.

[8] B. Feng and D. Xu, "1-kw PFC converter with compound active-clamping," *IEEE Transactions on Power Electronics*, vol. 20, no. 2, pp. 324–331, 2005.

[9] A. F. de Souza and I. Barbi, "High power factor rectifier with reduced conduction and commutation losses," in *Proceedings of the 21st International Telecommunications Energy Conference (INTELEC)*, 1999, Paper 8-1.

[10] Y. Jang and M. M. Jovanovic, "A bridgeless pfc boost rectifier with optimized magnetic utilization," *IEEE Transactions on Power Electronics*, vol. 24, no. 1, pp. 85–93, 2009.

[11] P. Dowell, "Effects of eddy current in transformer windings," in *Proceedings of the IEE*, vol. 113, no. 8, 1966, pp. 1387–1394.

[12] W. G. Hurley, E. Gath, and J. G. Breslin, "Optimizing the ac resistance of multilayer transformer windings with arbitrary current waveforms," *IEEE Transactions on Power Electronics*, vol. 15, no. 2, pp. 369–376, 2000.

[13] A. Van den Bossche and V. Valchev, *Inductors and Transformers for Power Electronics*. CRC Taylor & Francis Group, London, New York, 2005.

[14] J. A. Ferreira, "Improved analytical modeling of conductive losses in magnetic components," *IEEE Transactions on Power Electronics*, vol. 9, no. 1, pp. 127–131, 1994.

[15] K. Venkatachalam, C. R. Sullivan, T. Abdallah, and H. Tacca, "Accurate prediction of ferrite core loss with nonsinusoidal waveforms using only steinmetz parameters," in *Proceedings of the IEEE Workshop on Computers in Power Electronics (COMPEL)*, 2002, pp. 36–41.

[16] C. Steinmetz, "On the law of hysteresis," in *Proc. IEEE*, vol. 72, 1984, pp. 196–221.

[17] K. Raggl, T. Nussbaumer, and J. W. Kolar, "Guideline for a simplified differential mode emi filter design," *IEEE Transactions on Industrial Electronics*, no. 99, 2009.

DCM Boost PFC Converter with High Input PF

Kai Yao, Xinbo Ruan, Xiaojing Mao
Aero-Power Sci-Tech Center
Nanjing University of Aeronautics and Astronautics
Nanjing, Jiangsu   210016   China

Zhihng Ye
Lite-on Technology Corp.
Nanjing, Jiangsu   210019   China

Abstract- **Discontinuous current mode (DCM) boost power factor correction (PFC) converter features zero-current turn-on for the switch, no reverse recovery in diode and constant frequency operation, however, the input power factor (PF) is relatively low when the duty cycle is constant in a half line cycle. This paper derives the expressions of the input current and PF of DCM boost PFC converter, and based on which, a variable duty cycle control is proposed so as to improve the PF to nearly unity in the whole input voltage range. A method of fitting the duty cycle is further proposed for simplifying the circuit implementation. Besides a higher PF, the proposed variable duty cycle control achieves a lower output voltage ripple and a higher efficiency over the constant duty cycle control. The experimental results from a 120W universal input prototype are given to verify the effectiveness of the proposed method.**

I. INTRODUCTION

Power factor correction (PFC) converters have been widely used in ac-dc power conversions to achieve high power factor (PF) and low harmonic distortion. The methods of achieving PFC can be classified into active and passive types. Compared with passive PFC converter, active PFC converter can achieve a high PF and a small size [1]. There are different topologies for implementing active PFC techniques [1]-[20], among which, boost converter is the representing topology because it exhibits many advantages such as small input current ripple due to the series connection of the inductor at the input side, high PF over the whole input voltage range, small size in output capacitor due to its high voltage, and simple circuit. Depending on the inductor current to be continuous or not, the boost PFC converter may be designed to operate in three modes: continuous current mode (CCM), critical current mode (CRM) and discontinuous current mode (DCM).

When boost PFC converter operates in CCM, the inductor current ripple is very small, leading to low root-mean-square (RMS) currents on the inductor and switch, and low EMI [21]. However, the switch always operates at hard switching, and the diode suffers reverse recovery. CCM boost PFC converter is mainly used in high and medium power applications. CRM boost PFC converter has such advantages as zero-current turning on of switch, no reverse recovery of diode and high PF. However, the switching frequency is variable, resulting in difficulty in the design of the inductor and EMI filter [22]. CRM boost PFC converter is mainly used in medium and low power applications.

Same as CRM approach, DCM boost converter features zero-current turning on for the switch and no reverse recovery

of diode, but it operates in constant switching frequency, which is benefit for designing the inductor and EMI filter. The main drawback is that the PF is not so high, especially at high input voltage [23].

In [24], a technique was proposed to eliminate the input current harmonics and achieve high PF by regulating the power switch on-time. The peak current control is adopted, and the current reference is not the sinusoidal waveform proportional to rectified input voltage any longer, but a 3rd harmonic is injected into it, thus the PF can be improved to nearly unity. The content of the injected 3rd harmonic is dependent on the input voltage, and it should be adjusted online. So, the realization circuit is a little complicated.

The objective of this paper is to propose a method of achieving a near unity PF for the DCM boost PFC converter over the whole input voltage range of 90-264 VAC. Section II analyzes the input current and PF of DCM boost PFC converter in detail. In Section III, the concept of the variable duty cycle control is presented so that the PF is improved to unity over the whole input voltage range, and a method of fitting the duty cycle is further proposed for simplifying the circuit implementation. In section IV, the comparison between the proposed variable duty cycle control and the traditional constant duty cycle control is made in terms of the input current harmonics, the output voltage ripple and the design of inductor. The analytical results show that the variable duty cycle control can reduce the input current harmonics and output voltage ripple, and improve the efficiency. A 120 W prototype has been built and tested, and the experimental results are presented in Section V.

II. DCM BOOST PFC CONVERTER WITH CONSTANT DUTY CYCLE CONTROL

Figure 1 shows the main circuit of a boost PFC converter. For simplicity, the following assumptions are made: 1) all the devices and components are ideal; 2) the ripple of the output voltage is much small to be neglected; 3) the switching frequency is much higher than the line frequency.

The input voltage is defined as

$$v_{in}(t) = V_m \sin \omega t \tag{1}$$

where V_m is the amplitude of the input voltage, ω is the angular frequency of the input voltage.

Then the rectified voltage is

$$v_g = V_m \left| \sin \omega t \right|. \tag{2}$$

The work was supported by Lite-on Technology Corp.

978-1-4244-4782-4/10 $26.00 © 2010 IEEE

Figure 2 shows the inductor current waveform in a switching cycle when the converter operates at DCM. In a switching cycle, the inductor peak current i_{Lb_pk} is

$$i_{Lb\_pk}(t) = \frac{v_g}{L_b} D_y T_s = \frac{V_m |\sin \omega t|}{L_b} D_y T_s \qquad (3)$$

where D_y is the duty cycle and T_s is the switching cycle.

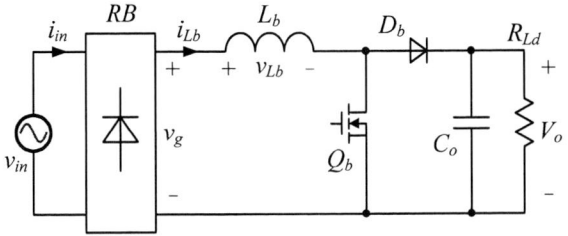

Figure 1.   Main circuit of boost PFC converter.

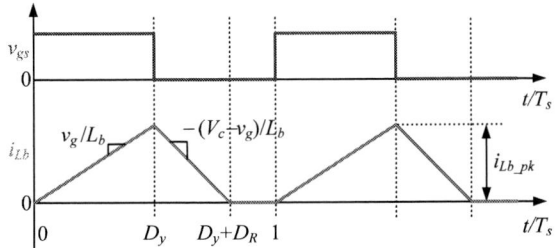

Figure 2.   Boost inductor current waveform in a switching cycle.

In each switching cycle, the inductor has the volt-second balance, i.e.,

$$v_g D_y T_s = \left(V_o - v_g\right) D_R T_s \qquad (4)$$

where V_o is the output voltage, D_R is the duty cycle corresponding to the reset time of the inductor current.

Eq. (4) can be rewritten as

$$D_R = \frac{v_g}{V_o - v_g} D_y = \frac{V_m |\sin \omega t|}{V_o - V_m |\sin \omega t|} D_y . \qquad (5)$$

From (3) and (5), the average inductor current in a switching cycle can be derived as

$$\begin{aligned} i_{Lb\_av}(t) &= \frac{1}{2} i_{Lb\_pk}(t) \cdot \left(D_y + D_R\right) \\ &= \frac{V_m D_y^2}{2 L_b f_s} \cdot \frac{|\sin \omega t|}{1 - \frac{V_m}{V_o} |\sin \omega t|} \end{aligned} \qquad (6)$$

where f_s is the switching frequency.

So the input current is

$$i_{in}(t) = \frac{V_m D_y^2}{2 L_b f_s} \cdot \frac{\sin \omega t}{1 - \frac{V_m}{V_o} |\sin \omega t|} . \qquad (7)$$

When D_y is constant, according to (3) and (6), Figure 3 shows the instantaneous waveform, the peak value envelope and the average value of the inductor current. It can be seen that the shape of the peak inductor current is sinusoidal, however the shape of the average inductor current is not sinusoidal and there is distortion in it.

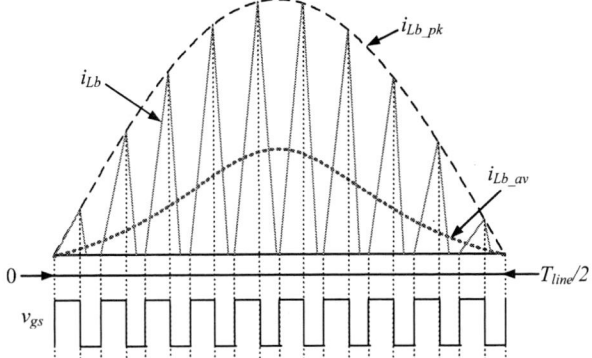

Figure 3.   Inductor current waveform in a half line cycle.

For the analysis simplicity, the average inductor current is normalized with the base of $\dfrac{1}{2} \dfrac{V_m \cdot D_y^2}{L_b \cdot f_s} \cdot \dfrac{1}{1 - \dfrac{V_m}{V_o}}$, so (7) is rewritten as

$$i_{in}^*(t) = \left(1 - \frac{V_m}{V_o}\right) \frac{\sin \omega t}{1 - \frac{V_m}{V_o} |\sin \omega t|} . \qquad (8)$$

According to (8), the normalized average inductor current is plotted in Figure 4, from which, it can be seen that the shape of the average inductor current is only dependent on V_m/V_o, and the smaller V_m/V_o is, the closer to sinusoidal the current shape is. This can be explained as follows. As the peak value of the inductor current is in the sinusoidal shape, and the duty cycle is constant in a line cycle, the average value of the inductor current in the rising period is sinusoidal. However, the falling time of the inductor current is depended on the value of ($V_o - v_g$), and it varies with v_g, so the average value of the inductor current in the falling period is not sinusoidal. Thus, the average value of the inductor current in a switching cycle is not sinusoidal. The smaller V_m/V_o is, the shorter the falling time of the inductor current is, thus the closer to sinusoidal shape the average inductor current in a switching cycle is.

From (1) and (7), the average input power in a half line cycle can be calculated as

$$\begin{aligned} P_{in} &= \frac{1}{T_{line}/2} \cdot \int_0^{T_{line}/2} v_{in}(t) i_{in}(t) dt \\ &= \frac{V_m^2 D_y^2}{2 L_b f_s} \cdot \frac{1}{\pi} \cdot \int_0^\pi \frac{\sin^2 \omega t}{1 - \frac{V_m}{V_o} |\sin \omega t|} d\omega t \end{aligned} \qquad (9)$$

Assuming the efficiency of the converter is 100%, that is to say $P_{in} = P_o$, then the duty cycle is

$$D_y = \frac{1}{V_m} \cdot \sqrt{\frac{2\pi L_b f_s P_o}{\int_0^\pi \frac{\sin^2 \omega t}{1 - \frac{V_m}{V_o}|\sin \omega t|} d\omega t}} \,. \tag{10}$$

From (7) and (9), the PF can be derived as

$$PF = \frac{P_{in}}{\frac{1}{\sqrt{2}} V_m I_{in\_rms}} = \frac{P_{in}}{\frac{1}{\sqrt{2}} V_m \sqrt{\frac{1}{\pi} \int_0^\pi (i_{in}(t))^2 d\omega t}}$$

$$= \frac{\sqrt{\frac{2}{\pi}} \cdot \int_0^\pi \frac{\sin^2 \omega t}{1 - \frac{V_m}{V_o}|\sin \omega t|} d\omega t}{\sqrt{\int_0^\pi \left(\frac{\sin \omega t}{1 - \frac{V_m}{V_o}|\sin \omega t|} \right)^2 d\omega t}} \tag{11}$$

where I_{in_rms} is the RMS value of the input current.

According to (11), the input PF can be plotted as shown in Figure 5. It can be seen that the larger V_m/V_o is, the lower the PF is. If V_m/V_o is greater than 0.9, then PF is less than 0.9. When the output voltage is 400 V, the PF is only 0.865 at the input voltage of 264 VAC, which is not satisfied with the regulation standard such as IEC61000-3-2.

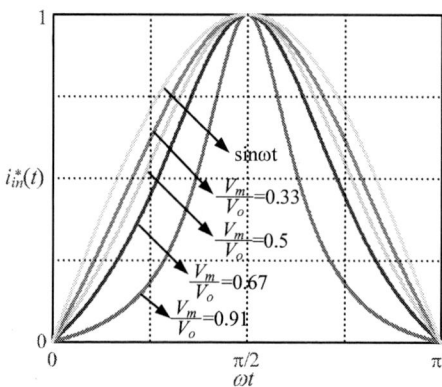

Figure 4.   Normalized input current waveform in a half line cycle.

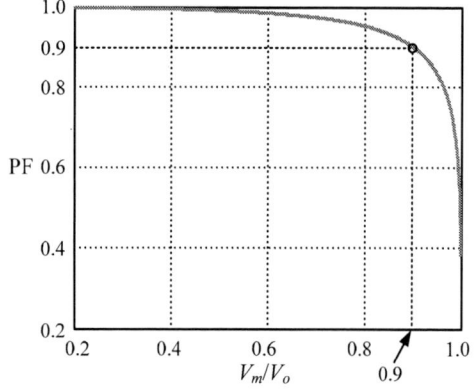

Figure 5.   Relationship between the input PF and V_m/V_o.

III. VARIABLE DUTY CYCLE CONTROL TO IMPROVE INPUT POWER FACTOR

A.   Variable Duty Cycle for PF=1

By observing (7), we can let the duty cycle varies as

$$D_y = D_0 \cdot \sqrt{1 - \frac{V_m}{V_o}|\sin \omega t|} \tag{12}$$

where D_o is a coefficient, which will be explained later.

Substitution of (12) into (7) leads to

$$i_{in}(t) = \frac{V_m D_0^2 \sin \omega t}{2 L_b f_s} \,. \tag{13}$$

From (13) we can see that if the duty cycle varies as (12), the shape of the input current is sinusoidal and the PF is unity.

From (1) and (13), the average input power is derived as

$$P_{in} = \frac{1}{2} \cdot V_m \cdot \frac{V_m D_0^2}{2 L_b f_s} = \frac{V_m^2 D_0^2}{4 L_b f_s} = P_o \,. \tag{14}$$

From (14), we have

$$D_0 = \frac{2 \cdot \sqrt{P_o L_b f_s}}{V_m} \,. \tag{15}$$

Substitution of (15) into (12) leads to

$$D_y = \frac{2 \cdot \sqrt{L_b f_s P_o}}{V_m} \cdot \sqrt{1 - \frac{V_m}{V_o}|\sin \omega t|}$$

$$= \frac{2 \cdot \sqrt{L_b f_s P_o}}{V_m} \cdot \sqrt{1 - \frac{v_g}{V_o}} \,. \tag{16}$$

From (16) we can see that if the duty cycle is a function of the rectified input voltage v_g, the PF of the DCM boost PFC converter is unity.

B.   The Fitting Duty Cycle

The duty cycle expressed in (16) is complicated to be implemented because a multiplier, a divider and a square root extractor are needed, it is necessary to seek a function that fits (16), which can be more easily implemented.

Defining $a = V_m/V_o$, $y = |\sin \omega t|$, (12) can be simplified as

$$D_y = D_0 \cdot \sqrt{1 - a \cdot y} \,. \tag{17}$$

Based on Taylor's series

$$f(x) = f(x_0) + f'(x_0) \cdot (x - x_0)$$
$$+ \frac{1}{2!} \cdot f''(x_0) \cdot (x - x_0)^2$$
$$+ \cdots + \frac{1}{n!} \cdot f^{(n)}(x_0) \cdot (x - x_0)^n + \cdots \tag{18}$$

Eq. (17) can be expressed as

$$D_y = D_0 \cdot \left[\begin{array}{c} \sqrt{1-a \cdot y_0} - \dfrac{a}{2} \cdot (1-a \cdot y_0)^{-\frac{1}{2}} \cdot (y-y_0) \\ -\dfrac{1}{2!} \cdot \dfrac{a^2}{4} \cdot (1-a \cdot y_0)^{-\frac{3}{2}} \cdot (y-y_0)^2 + \cdots \end{array} \right]. \quad (19)$$

Reserving only the first derivative item, we can approximate (19) as

$$D_{y\_fit} = D_0 \cdot \left[\sqrt{1-a \cdot y_0} - \frac{a}{2} \cdot (1-a \cdot y_0)^{-\frac{1}{2}} \cdot (y-y_0) \right]$$
$$= D_1 \cdot \left(1 - \frac{a}{2-a \cdot y_0} \cdot y \right) \quad (20)$$

where $D_1 = \dfrac{D_0 \cdot (2-a \cdot y_0)}{2\sqrt{1-a \cdot y_0}}$.

Substituting (20) into (7) and (9), respectively, we have

$$i_{in}(t) = \frac{V_m D_1^2}{2L_b f_s} \cdot \frac{\sin \omega t}{1-a|\sin \omega t|} \cdot \left(1 - \frac{a}{2-a \cdot y_0}|\sin \omega t| \right)^2 \quad (21)$$

$$P_{in} = P_o = \frac{V_m^2 D_1^2}{2\pi L_b f_s} \cdot \int_0^\pi \frac{(\sin^2 \omega t) \cdot \left(1 - \dfrac{a}{2-a \cdot y_0}|\sin \omega t| \right)^2}{1-a|\sin \omega t|} d\omega t. \quad (22)$$

From (21) and (22), the PF is calculated as

$$PF = \frac{P_{in}}{V_{in\_rms} \cdot I_{in\_rms}} = \frac{P_{in}}{\frac{1}{\sqrt{2}} \cdot V_m \cdot \sqrt{\frac{1}{T_{line}/2} \cdot \int_0^{T_{line}/2} (i_{in}(t))^2 \cdot dt}}$$

$$= \frac{\sqrt{\dfrac{2}{\pi}} \cdot \int_0^\pi \dfrac{\sin^2 \omega t \cdot \left(1 - \dfrac{a}{2-a \cdot y_0}|\sin \omega t| \right)^2}{1-a \cdot |\sin \omega t|} d\omega t}{\sqrt{\displaystyle\int_0^\pi \dfrac{\sin^2 \omega t \cdot \left(1 - \dfrac{a}{2-a \cdot y_0}|\sin \omega t| \right)^4}{(1-a|\sin \omega t|)^2} d\omega t}}. \quad (23)$$

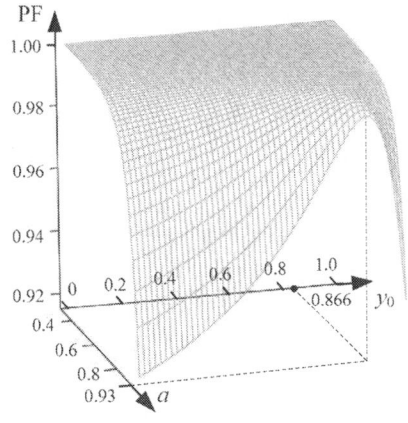

Figure 6. The surface of PF as the function of a and y_0.

The PF is depicted in Figure 6 according to (23). It can seen that when a is small, the PF is almost constant and is very close to unity as y_0 varies from 0 and 1. When a increases, the influence of y_0 on PF increases. So we choose the value of y_0, which enables the maximum PF at the highest input voltage, to ensure the PF to be close to 1 over the whole input voltage range. For the input voltage range of 90-264 VAC, the output voltage is set to 400 V. Substituting $a = 264\sqrt{2}/400$ into (23), and then differentiating (23) with y_0, and setting it to zero, we have $y_0 = 0.866$.

Substituting $y_0 = 0.85$ into (20), the duty cycle is expressed as

$$D_{y\_fit} = D_1 \cdot \left(1 - \frac{a}{2-y_0 \cdot a} \cdot y \right) = D_1 \cdot \left(1 - \frac{V_m \cdot |\sin \omega t|}{2 \cdot V_o - 0.85 \cdot V_m} \right)$$
$$= D_1 \cdot \frac{2 \cdot V_o - 0.866 \cdot V_m - V_m \cdot |\sin \omega t|}{2 \cdot V_o - 0.866 \cdot V_m}. \quad (24)$$

C. Implementation of the Control Circuit

The control circuit can be implemented as shown in Figure 7. The reasonable resistance selection of the feedforward circuit can enable the voltage of D point as follows: $v_D = v_{EA} \cdot \dfrac{2V_o - 0.866V_m - V_m|\sin \omega t|}{2V_o - 0.866V_m}$. v_D is sent to the PWM comparator and compared with the sawtooth carrier, and output of the PWM comparator determines the duty cycle, which varies as expressed in (24).

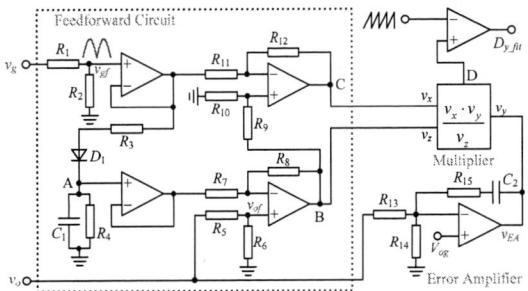

Figure 7. The control circuit of the variable duty cycle control.

IV. ADVANTAGES OF THE VARIABLE DUTY CYCLE CONTROL

A. The Improvement of PF

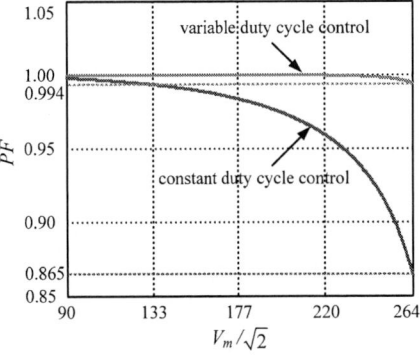

Figure 8. Comparison of the PF between constant and variable duty cycle control.

978-1-4244-4782-4/10 $26.00 © 2010 IEEE

According to (11) and (23), the input PF with constant duty cycle control and variable duty cycle control are depicted in Figure 8. We can see that the PF is greatly improved with the variable duty cycle control. When the input voltage is 264 VAC, the PF is increased from 0.865 to 0.994.

B. The Reduction of Output Voltage Ripple

When the DCM boost PFC converter is operated with the constant duty cycle control, from (1), (7) and (10), the normalized instantaneous input power is derived as

$$p_{in\_1}^{*}(t) = \frac{v_{in}(t) \cdot i_{in}(t)}{P_o} = \frac{\dfrac{\sin^2 \omega t}{1 - a|\sin \omega t|}}{\dfrac{1}{\pi} \cdot \displaystyle\int_0^\pi \dfrac{\sin^2 \omega t}{1 - a|\sin \omega t|} d\omega t}. \qquad (25)$$

Similarly, when the DCM boost PFC converter is operated with the variable duty cycle control, from (1), (21) and (22), the normalized instantaneous input power is derived as

$$\begin{aligned}p_{in\_2}^{*}(t) &= \frac{v_{in}(t) \cdot i_{in}(t)}{P_o} \\ &= \frac{\dfrac{(\sin \omega t)^2}{1 - a|\sin \omega t|} \cdot \left(1 - \dfrac{a}{2 - 0.866 \cdot a}|\sin \omega t|\right)^2}{\dfrac{1}{\pi} \cdot \displaystyle\int_0^\pi \dfrac{(\sin \omega t)^2 \cdot \left(1 - \dfrac{a}{2 - 0.866 \cdot a}\sin \omega t\right)^2}{1 - a \cdot \sin \omega t} d\omega t}.\end{aligned} \qquad (26)$$

According to (25) and (26), the curve of the normalized instantaneous input powers with the two control methods in a half line cycle are depicted in Figure 10.

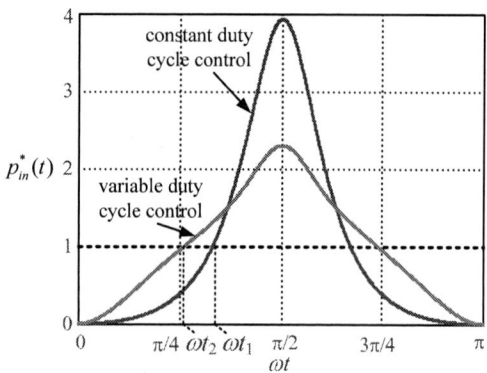

Figure 9. The normalized instantaneous input power in a half line cycle with the two control methods.

When $p_{in}^{*}(t) > 1$, the storage capacitor C_o is charged, and when $p_{in}^{*}(t) < 1$, C_o is discharged. The energy discharging C_o (which equals the charged energy) in a half line cycle with the two duty cycle control methods are

$$\Delta E_1^{*} = 2\int_0^{t_1}\left[1 - p_{in\_1}^{*}(t)\right] \cdot dt \qquad (27)$$

$$\Delta E_2^{*} = 2\int_0^{t_2}\left[1 - p_{in\_2}^{*}(t)\right] \cdot dt \qquad (28)$$

where t_1 and t_2 are the time instants when $p_{in}^{*}(t)$ crosses 1 with the constant duty cycle control and the variable duty cycle control, respectively.

ΔE_1^{*} and ΔE_2^{*} can be also expressed as

$$\Delta E_1^{*} \approx \frac{\dfrac{1}{2}C_o\left(V_o + \dfrac{\Delta V_{o1}}{2}\right)^2 - \dfrac{1}{2}C_o\left(V_o - \dfrac{\Delta V_{o1}}{2}\right)^2}{P_o \cdot \dfrac{T_{line}}{2}} = \frac{2C_oV_o \cdot \Delta V_{o1}}{P_oT_{line}} \qquad (29)$$

$$\Delta E_2^{*} \approx \frac{\dfrac{1}{2}C_o\left(V_o + \dfrac{\Delta V_{o2}}{2}\right)^2 - \dfrac{1}{2}C_o\left(V_o - \dfrac{\Delta V_{o2}}{2}\right)^2}{P_o \cdot \dfrac{T_{line}}{2}} = \frac{2C_oV_o \cdot \Delta V_{o2}}{P_oT_{line}} \qquad (30)$$

where ΔV_{o1} and ΔV_{o2} are the output voltage ripple with the constant duty cycle control and the variable duty cycle control, respectively.

From (27) to (30), we have

$$\Delta V_{o1} = \frac{P_oT_{line} \cdot \displaystyle\int_0^{t_1}\left[1 - p_{in\_1}^{*}(t)\right]dt}{C_oV_o} \qquad (31)$$

$$\Delta V_{o2} = \frac{P_oT_{line} \cdot \displaystyle\int_0^{t_2}\left[1 - p_{in\_2}^{*}(t)\right]dt}{C_oV_o} \qquad (32)$$

According to the specifications of the converter, which will be given in Section V, ΔV_{o1} and ΔV_{o2} can be figured out, as shown in Figure 10. It can be seen that with the variable duty cycle control, the output voltage ripple is reduced compared to the constant duty cycle control.

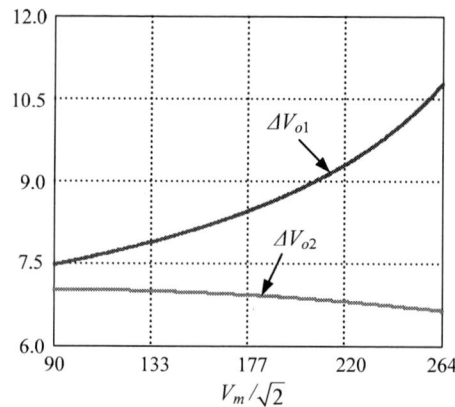

Figure 10. The output voltage ripple with two control methods.

C. The Reduction of the Inductor Current Ripple

Referring to Figure 2, in order to ensure the PFC converter operates in DCM, the following condition should be met.

$$D_y + D_R \leq 1 \qquad (33)$$

Substitution of (5) into (33), yields

$$D_y \cdot \left(1 + \frac{V_m |\sin \omega t|}{V_o - V_m |\sin \omega t|}\right) \le 1 . \tag{34}$$

It can be seen from (34) that the boost inductor current is apt to be continuous at the peak point of the input voltage, so the condition to ensure DCM is rewritten as

$$D_y \cdot \left(1 + \frac{V_m}{V_o - V_m}\right) \le 1 . \tag{35}$$

Substituting (10) into (35), the critical boost inductor with constant duty cycle control is obtained as

$$L_{b1} \le (1-a)^2 \cdot \frac{V_m^2}{2 P_o f_s} \cdot \frac{1}{\pi} \cdot \int_0^\pi \frac{\sin^2 \omega t}{1 - a |\sin \omega t|} d\omega t \tag{36}$$

Eq. (22) can be rewritten as

$$D_1 = \frac{1}{V_m} \cdot \sqrt{\frac{2 \cdot \pi \cdot L_b \cdot f_s \cdot P_o}{\displaystyle\int_0^\pi \frac{(\sin^2 \omega t) \cdot \left(1 - \dfrac{a}{2 - 0.866 \cdot a} \cdot |\sin \omega t|\right)^2}{1 - a \cdot |\sin \omega t|} \cdot d(\omega t)}} \tag{37}$$

Substituting (37) into (24) leads to

$$D_{y\_fit} = \frac{\dfrac{1}{V_m} \sqrt{2\pi L_b f_s P_o} \left(1 - \dfrac{V_m |\sin \omega t|}{2 V_o - 0.866 V_m}\right)}{\sqrt{\displaystyle\int_0^\pi \dfrac{(\sin^2 \omega t)\left(1 - \dfrac{a}{2 - 0.866 a}|\sin \omega t|\right)^2}{1 - a |\sin \omega t|} d\omega t}} \tag{38}$$

and then substituting (38) into (35), we obtain the critical boost inductor with variable duty cycle control, i.e.,

$$L_{b2} \le \frac{1}{P_o f_s} \left(\frac{V_m}{2} \cdot \frac{2 V_o - 0.866 V_m}{2 V_o - 1.866 V_m} \cdot \frac{V_o - V_m}{V_o}\right)^2$$
$$\cdot \frac{2}{\pi} \cdot \int_0^\pi \frac{(\sin \omega t)^2 \left(1 - \dfrac{a}{2 - 0.866 a} \sin \omega t\right)^2}{1 - a \cdot \sin \omega t} d\omega t \tag{39}$$

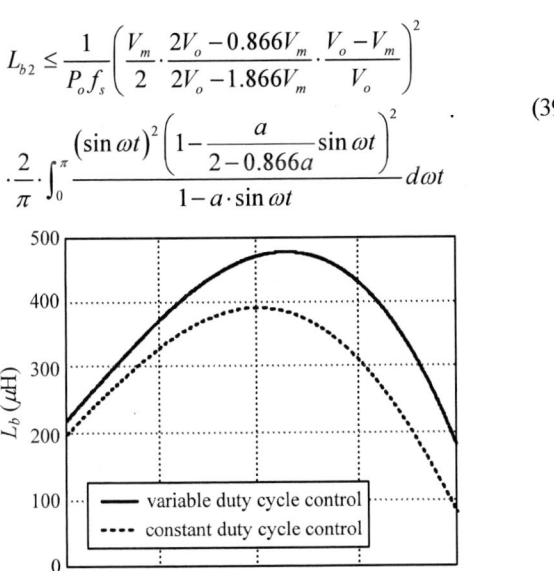

Figure 11. The critical inductors over the input voltage range.

According to the specifications of the converter, which will be given in Section V, the critical boost inductor over the input

voltage range with the two control methods are depicted in Figure 11, from which we choose $L_{b1} = 80\ \mu$H and $L_{b2} = 180\ \mu$H.

The rms value of the inductor current in a switching cycle is

$$i_{Lb\_rms}(t) = \sqrt{\frac{1}{T_s} \cdot \int_0^{T_s} i_{Lb}^2(t) \cdot dt}$$
$$= \sqrt{\frac{1}{T_s} \cdot \left[\begin{array}{l} \displaystyle\int_0^{D_y T_s} \left(\frac{v_g}{L_b} \cdot t\right)^2 \cdot dt \\[6pt] + \displaystyle\int_{D_y T_s}^{(D_y + D_R) T_s} \left[i_{Lb\_pk}(t) - \frac{V_o - V_g}{L_b} \cdot (t - D_y T_s)\right]^2 \cdot dt \end{array}\right]} \tag{40}$$
$$= \frac{V_m \cdot |\sin \omega t| \cdot T_s}{L_b} \sqrt{\frac{V_o \cdot D_y^3}{3 \cdot (V_o - V_m |\sin \omega t|)}}$$

So the rms value of the inductor current in a line cycle is

$$I_{Lb\_rms} = \sqrt{\frac{2}{T_{line}} \cdot \int_0^{\frac{T_{line}}{2}} (i_{Lb\_rms}(t))^2 \cdot dt}$$
$$= \frac{V_m \cdot T_s}{L_b} \cdot \sqrt{\frac{2}{T_{line}} \cdot \int_0^{\frac{T_{line}}{2}} \frac{V_o \cdot D_y^3 \cdot (\sin \omega t)^2}{3 \cdot (V_o - V_m |\sin \omega t|)} \cdot dt} \tag{41}$$

Substituting (10) and L_{b1}=80 μH into (41), we can obtain the rms value of the inductor current I_{Lb1_rms} with constant duty cycle control; Similarly, substituting (38) and L_{b1}=180 μH into (41), we can obtain the rms value of the inductor current I_{Lb2_rms} with variable duty cycle control. I_{Lb1_rms} and I_{Lb2_rms} is plotted in Figure 12. As seen, I_{Lb2_rms} is lower than I_{Lb1_rms}, especially at low line. Thus, the conduction loss of the switch can be reduced, leading to a higher efficiency.

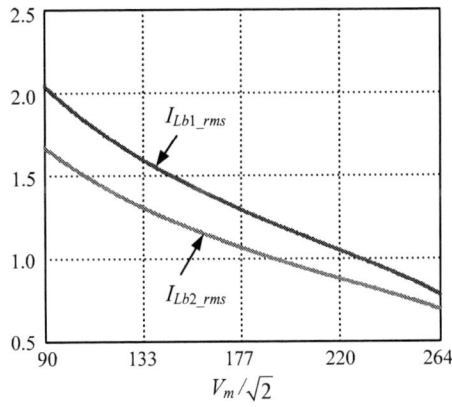

Figure 12. The rms value of the inductor current with two control methods.

V. EXPERIMENTAL VERIFICATION

In order to verify the validity of the proposed variable duty cycle control, a prototype has been built and tested in the lab. The specifications of the prototype are as follows. Input voltage: $v_{in} = 90 \sim 264$ VAC / 50 Hz; output voltage: $V_o = 400$ VDC; output power: $P_o = 120$ W; switching frequency: $f_s = 100$ kHz. output capacitor: $C_o = 136$ μF.

Figures 13 to 14 show the experimental waveforms of the rectified input voltage, rectified input current, output voltage and inductor current with two control methods at 90 VAC, 220 VAC and 264 VAC input respectively. Figure 15 and 16 show the measured PF and output voltage ripple curve with different input voltage. It can be seen from Figures 13 to 16, with the variable duty cycle control, the PF is improved to near unity over the input voltage range of 90-264 VAC, and the output voltage ripple is reduced. The improvement of PF and reduction of output voltage ripple are much more obvious at high line. Figure 17 shows the efficiency comparison of two control methods, from which it can be seen that the efficiency is increased with variable duty cycle control, especially at low line, where the conduction loss is the dominative loss.

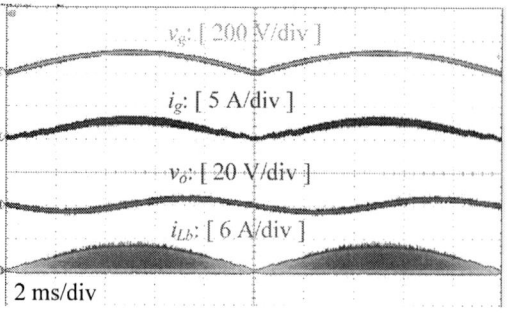

(a)   90 VAC

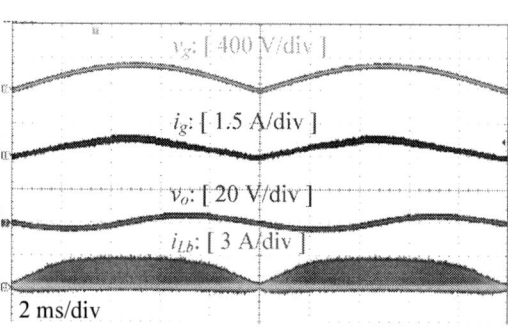

(b)   220 VAC

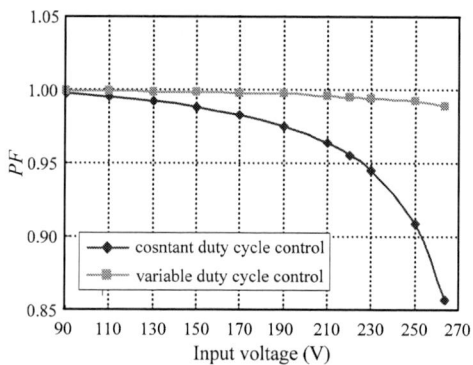

(c)   264 VAC

Figure 14.   Experimental waveforms of input voltage, input current, output voltage and inductor current with variable duty cycle control

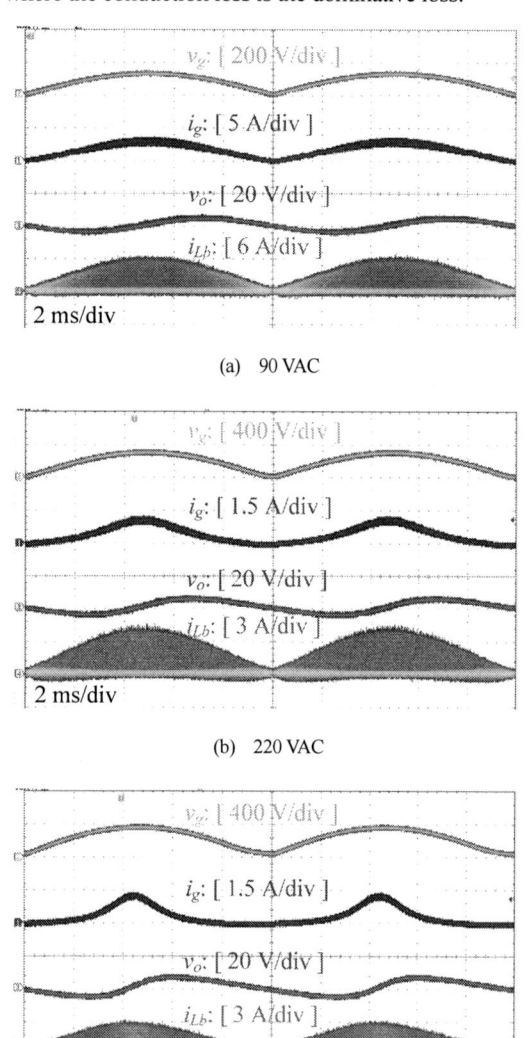

(a)   90 VAC

(b)   220 VAC

(c)   264 VAC

Figure 13.   Experimental waveforms of input voltage, input current, output voltage and inductor current with constant duty cycle control

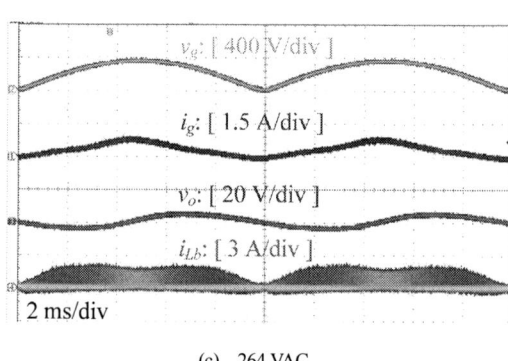

Figure 15.   Measured PF over the input voltage range.

978-1-4244-4782-4/10 $26.00 © 2010 IEEE   1411

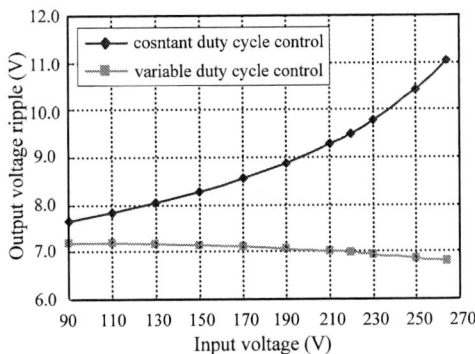

Figure 16.   Measured output voltage ripple over the input voltage range.

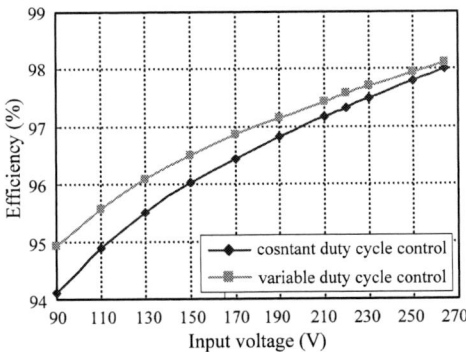

Figure 17.   Measured efficiency over the input voltage range.

VI.   CONCLUSIONS

DCM boost PFC converter has the merits of zero-current turn on for the power switch, no reverse recovery of the diode and constant frequency operation. However, when the duty cycle is kept constant in a line cycle, the average input current is not proportional to the input voltage, thus the input power factor is not so high, especially at high input voltage. In order to improve the power factor to nearly unity over the whole input voltage range, this paper proposes a variable duty cycle control scheme, and furthermore, a method of fitting the duty cycle is proposed for simplifying the circuit implementation. As the PF is improved, especially the dominative 3rd harmonic with a phase of 180° is reduced significantly, the pulsation of the instantaneous input power is reduced, leading to the reduction of the output voltage ripple. Meanwhile, when the variable duty cycle control is adopted, the critical inductance can be increased, thus the inductor current ripple is reduced, leading to a higher efficiency. The experimental results from a 120 W universal input prototype are presented to verify the effectiveness of the proposed method.

REFERENCES

[1]   O. Garcia, J. A. Cobos, R. Prieto, P. Alou, and J. Uceda, "Single phase power factor correction: a survey," *IEEE Trans. Power Electron.*, vol. 18, no. 3, pp. 749-755, May 2003.

[2]   K. Raggl, T. Nussbaumer, G. Doerig, J. Biela, and J. W. Kolar, "Comprehensive Design and Optimization of a High-Power-Density Single-Phase Boost PFC," *IEEE Trans. Ind. Electron.*, vol. 56, no. 7, pp. 2574-2587, July 2009.

[3]   D. D.-C. Lu, H. H.-C. Iu, and V. Pjevalica, "Single-Stage AC/DC Boost–Forward Converter With High Power Factor and Regulated Bus and Output Voltages," *IEEE Trans. Ind. Electron.*, vol. 56, no. 6, pp. 2128-2132, June 2009.

[4]   J. A. Villarejo, J. Sebastian, F. Soto, and Esther de Jodar, "Optimizing the Design of Single-Stage Power-Factor Correctors," *IEEE Trans. Ind. Electron.*, vol. 54, no. 3, pp. 1472-1482, June 2007.

[5]   M. M. Jovanovic and Y. Jang, "State-of-the-art, single-phase, active power-factor-correction techniques for high-power applications - an overview," *IEEE Trans. Ind. Electron.*, vol. 52, no. 3, pp. 701-708, June 2005.

[6]   L. Huber, B. T. Irving, and M. M. Jovanovic, "Review and Stability Analysis of PLL-Based Interleaving Control of DCM/CCM Boundary Boost PFC Converters," *IEEE Trans. Power Electron.*, vol. 24, no. 8, pp. 1992-1999, Aug 2009.

[7]   D. L. O'Sullivan, M. G. Egan, and M. J. Willers, "A Family of Single-Stage Resonant AC/DC Converters With PFC," *IEEE Trans. Power Electron.*, vol. 24, no. 2, pp. 398-408, Feb 2009.

[8]   Min Chen, A. Mathew, and Jian Sun, "Nonlinear Current Control of Single-Phase PFC Converters," *IEEE Trans. Power Electron.*, vol. 22, no. 6, pp. 2187-2194, Nov. 2007.

[9]   Zhonghui Bing, Min Chen, S. K. T. Miller, Y. Nishida, and Jian Sun, "Recent Developments in Single-Phase Power Factor Correction," in *Proc. IEEE PCC* 2007, pp. 1520-1526.

[10]   L. Solero, V. Serrao, M. Montuoro, and A. Romanelli, "Low THD variable load buck PFC converter," in *Proc. IEEE PESC* 2008, pp. 906-912.

[11]   J. Sebastian, D. G. Lamar, M. M. Hernando, A. Rodriguez, and A. Fernandez, "Steady-State Analysis of Power Factor Correctors with a Fast Output-Voltage Feedback Loop," in *Proc. IEEE APEC* 2009, pp. 970-976.

[12]   C. Qiao and K. M. Smedley, "A topology survey of single-stage power factor corrector with boost type input-current-shaper," *IEEE Trans. Power Electron.*, vol. 16, no. 3, pp. 360-368, May 2001.

[13]   J. Zhang, M. M. Jovanovic, and F. C. Lee, "Comparison between CCM single-stage and two-stage boost PFC converters," in *Proc. IEEE APEC 1999*, pp. 335-341.

[14]   O. Garcia, J. A. Cobos, P. Alou, R. Prieto, J. Uceda, and S. Ollero, "A new family of single-stage ac/dc power factor correction converters with fast output voltage regulation," in *Proc. IEEE PESC 1997*, pp. 536-542.

[15]   K. Matsui, I. Yamamoto, T. Kishi, M. Hasegawa, H. Mori, and F. Ueda, "A comparison of various buck-boost converters and their application to PFC," in *Proc. IEEE IECON 2002*, pp. 30-36.

[16]   C. K. Tse, and M. H. L. Chow, "A theoretical examination of the circuit requirements of power factor correction," in *Proc. IEEE PESC 1998*, pp. 1415-1421.

[17]   H. Wei, and I. Batarseh, "Comparison of basic converter topologies for power factor correction," in *Proc. IEEE Southeastcon 1998*, pp. 348-353.

[18]   A. Lázaro, A. Barrado, M. Sanz, V. Salas, and E. Olías, "New power factor correction AC-DC converter with reduced storage capacitor voltage," *IEEE Trans. Power Electron.*, vol. 54, no. 1, pp. 384-397, Feb. 2007.

[19]   M. Madigan, R. Erickson, and E. Ismail, "Integrated high quality rectifier regulators," in *Proc. IEEE PESC 1992*, pp. 1043-1051.

[20]   Q. Zhao, M. Xu, F. C. Lee, and J. Qian, "Single-switch parallel power factor correction AC-DC converters with inherent load current feedback," *IEEE Trans. Power Electron.*, vol. 19, no. 4, pp. 928-936, July 2004.

[21]   L. H. Dixon, "High power factor pre-regulators for off-line power supplies," Unitrode Switching Regulator Power Supply Design Seminar Manual, 1990, Paper 12, SEM-700.

[22]   J. S. Lai, and D. Chen, "Design consideration for power factor correction boost converter operating at the boundary of continuous conduction mode and discontinuous conduction mode," in *Proc. IEEE APEC 1993*, pp. 267-273.

[23]   K. H. Liu, and Y. L. Lin, "Current waveform distortion in power factor correction circuits employing discontinuous-mode boost converters," in *Proc. IEEE PESC 1989*, pp. 825-829.

[24]   D. S. Schramm, and M. O. Buss, "Mathematical analysis of a new harmonic cancellation technique of the input line current in DICM Boost converters," in *Proc. IEEE PESC 1998*, pp. 1337-1343.

Interleaved Forward Converter with Ripple-free Circuit for Humane Killer Poultry Applications

S.-Y. Tseng   T.-Y. Chiang   K.-C. Wang   S.-A. Chuang

GreenPower Evolution Applied Research Lab.
(G-PEARL)
Department of Electrical Engineering
Chang Gung University
Kwei-Shan, Tao-Yuan, Taiwan, R.O.C
E-mail: sytseng@mail.cgu.edu.tw
Tel: 886-3-2118800
Fax:886-3-2118026

Abstract—**The paper proposes a novel application for poultry of humane killing. The proposed poultry killer consists of two forward converter and two half-bridge inverters. In each forward converter, a *ripple*-free circuit is introduced into the one to achieve output current with ripple-free and to generate a high pulse voltage which is used to induce breakdown of poultry skin, feather or scale in feet. With the approach, forward converter with ripple-free circuit can reduce volume and size of output filter and rapid generate the breakdown of poultry skin to speed up poultry killing. Moreover, half-bridge inverter is used to chop dc voltage into pulse voltage at low frequency for sustaining the desired killing current during poultry killing time. Finally, a prototype under maximum skin breakdown voltage of 10kV, killing voltage of 150 V, killing current of 1A and killing time of 6s has been implemented to verify the feasibility of the proposed poultry killer.**

Keyword : humane killing, poultry killer, forward converter, ripple-free circuit and half-bridge inverter

I. INTRODUCTION

In recent years, the issue related to welfare of animals has attracted much attention. In particular, livestock and poultry must be rendered unconscious and insensible to pain before they are killed when they are infected diseases. In many countries, carbon dioxide (CO_2) and manual electrical killing are the two methods which are always used to kill poultry, such as chicken or turkey. Since CO_2 method is subjected to many limitations and requires higher cost, the manual electrical killing method has been used more popular.

Poultry killing is to cause unconsciousness by generating a ventricular fibrillation [1]. Degrees of a ventricular fibrillation are dependent upon the amount of current passing the nerves. The minimum current and voltage required for killing a chicken are respectively about 100mA and 150V, and they must sustain at least 3 s. Conventionally, to generate the specified electrical waveforms, a high step-up converter is used to boost input voltage to several kVs for supplying voltage source of

full-bridge inverter and help skin breakdown, as shown in Fig. 1(a). With this approach, the poultry humane killer needs a high step-up converter and full-bridge inverter requires switches with high voltage rating [2]-[5]. As a result, it is difficult to design for high isolation between each component, leading to a higher cost, lower conversion efficiency and larger volume and weight. In order to resolve this problem, the humane killer adopts a step-up converter and full-bridge inverter to sustain voltage of killing poultry and uses an ignitor to generate high narrow pulsed voltage for helping skin breakdown of poultry, as shown in Fig. 1(b). Its circuit structure is shown in Fig. 2. As mentioned above, the poultry humane killer needs a flyback converter associated with spark gap to form ignitor for generating high narrow pulsed voltage.

Since the poultry humane killer belongs to low power level applications, a DC/DC converter, such as forward or flyback converter, is usually adopted as a step-up converter. To reduce output current ripple-free technique is used to achieve ripple free of output current [6]-[10]. When the ripple-free circuit is used to reduce output current ripple, it also needs spark gap to generate high narrow pulsed voltage as a ignitor. Its block diagram is shown in Fig. 3. In order to implement this algorithm, two forward converters with ripple-free circuit combined with two half-bridge inverters are proposed to achieve poultry killing, as shown in Fig. 4. With this circuit structure, sizes of output inductor and capacitor filters can be reduced and component counts of the proposed poultry killing system can be decreased. Therefore, the proposed killing system can reduce weight, size, volume and cost, and increase conversion efficiency, significantly.

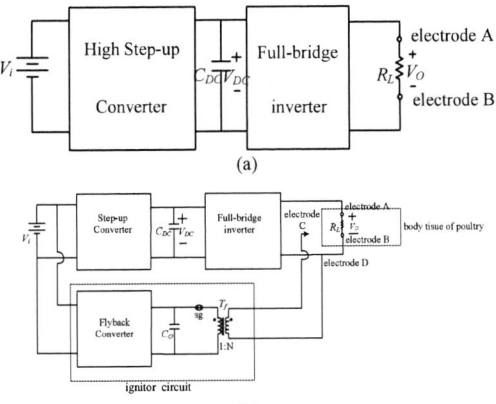

(a)

(b)

Fig.1. Block diagram of two type conventional humane killers for poultry applications (a) with high step-up converter and (b) with ignitor circuit.

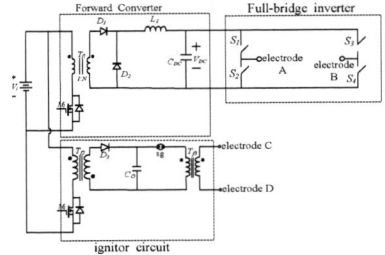

Fig.2. schematic diagram of the conventional poultry humane killer with ignitor circuit.

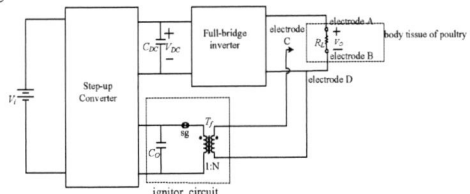

Fig. 3. Block diagram of the proposed poultry humane killer with ignitor circuit.

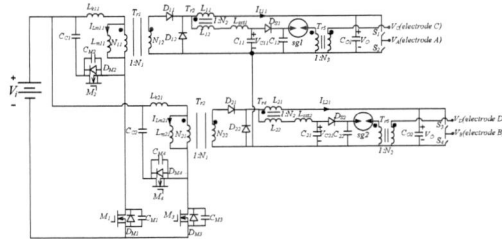

Fig. 4. Schematic diagram of the proposed poultry killing system with two forward converters.

II. MECHANISM OF POULTRY KILLING

Killing effectiveness of poultry is highly dependent on the parameters of electrical killing. For determining the desired electrical parameters, mechanism of poultry killing is briefly described. Poultry killing must cause a deep ventricular fibrillation in poultry to achieve an effective killing poultry. Since degrees of an ventricular fibrillation are dependent upon the amount of current passing the brain, an enough voltage source applied to poultry skin must overcome its skin impedance to attain an enough killing current for achieving an effective killing poultry. In the research of electrical properties for poultry skin, an equivalent circuit composed of resistor R_s series with the

parallel combination of resistor R_{sc} and capacitor C_{sc} have been proposed by many investigators, as shown in Fig. 5. The resistor R_{sc} and capacitor C_{sc} represent the total impedances of poultry skin, feather and scale in the beet, while R_s is the tissue of poultry body. Moreover, impedance R_{sc} is much greater than R_s. Both skin capacitor C_{sc} and resistance R_{sc} have been shown to be proportional to contact area [17]. The skin impedance versus voltage is shown in Fig. 6, illustrating that the skin impedance is inversely proportional to the applied voltage [12].

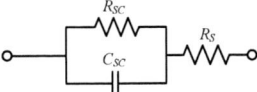

Fig. 5. An equivalent impedance of skin

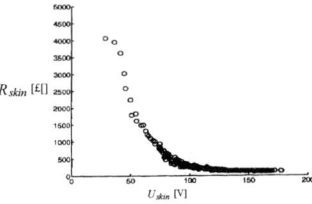

Fig. 6. Plot of the skin impedance of poultry versus the applied voltage

In general, there are three main parts to form neurons: soma, axon and dendrite. A stimulation signal is sensed by sensory receptors which exist in dermis or subcutaneous layer. When sensory receptors receive into nerve impulse, its propagation direction is, in turn, through one sensory neuron to the other sensory neurons, in which they are connected in series by synapses, until the nerve impulse propagation is transmitted to a receiver of brain. Moreover, brain can generate a excited nerve impulse through motor neurons, autonomic neurons system and sympathetic neurons system to heart of poultry for inducing a ventricular fibrillation in the poultry, as illustrated in Fig. 7. Finally, poultry will die.

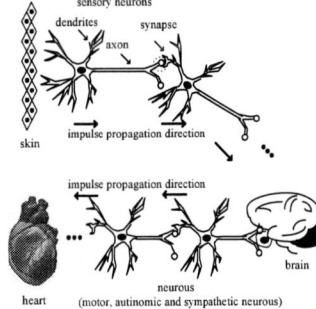

Fig. 7. Block diagram of procedure for impulse propagation between neurons and heart.

To explain the relationship between electrical killing for poultry and an excitation of a ventricular fibrillation in poultry, an equivalent circuit for describing the impulse signal propagation between neurons and heart of poultry is shown in Fig. 8. In Fig. 8, sensory and central neurons can be regarded as a current controlled voltage source to respectively control current I_B passing central neurons in brain and switch Q_1 which represents synapses in the sympathetic neurons since synapses play a role of

978-1-4244-4782-4/10 $26.00 © 2010 IEEE   1414

transducer. When a voltage E_i is applied to poultry, its skin impedance Z_o enters a breakdown state, and then a current I_1 will pass the body of the poultry. In the killing duration, if current I_1 is large enough, it will induce a high potential V_1 to generate a large current I_B passing the central neurons in brain of poultry through propagation impedance R_{t1}. When a large enough I_B passes central neurons in brain of poultry, it will induce a high voltage V_2 to drive switch Q_1 through propagation impedance R_{t2} and to turn on one. In this condition, the stimulation signal SVF is applied to heart of poultry and induce a ventricular fibrillation in poultry. Finally, poultry will be death. As described previously, it can be observed that electrical killing can induce a ventricular fibrillation in poultry to kill the poultry.

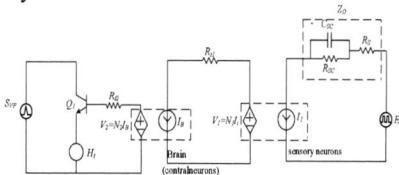

Ht: Heart of poultry
E_i: killing voltage
Z_o: equivalent impedance of poultry
I_1: current passing poultry body
Q_1: synapse in sympathetic neurons
I_B: current passing poultry central neurons in brain
V_1: potential of postsynaptic membrane in sensory neurons
V_2: potential of postsynaptic membrane in central neurons
R_{t1}: propagation impedance of sensory neurons
R_{t2}: propagation impedance of motor, autonomic and sympathetic neurons
S_{VF}: stimulation signal for inducing ventricular fibrillation

Fig. 8. An equivalent circuit for describing the impulse signal propagation between neurons and heart.

III. OPERATIONAL PRINCIPLE OF THE POULTRY KILLING SYSTEM

In the proposed poultry killing system, two forward converters with ripple-free circuit combined with half-bridge inverter are adopted to generate voltage of poultry killing. In order to generate a breakdown voltage of poultry skin, feather or scale in the feet, the ripple-free circuit be associated with spark gap to produce the breakdown voltage. Since output voltage V_A (or V_B) of each forward converter is approximated to output voltage V_{C11} (or V_{C21}) of each ripple-free circuit. As mentioned above, output voltage V_A of forward is regulated with two values. One value is the breakdown voltage V_{sg} of spark gap before skin impedance of poultry enters the breakdown state. The other value is the killing voltage V_K after skin impedance of poultry enters the breakdown state. Since two forward converters combined with two half-bridge inverters can respectively generate four set voltages V_A, V_B, V_C and V_D to apply to poultry body, its illustration of poultry killing is shown in Fig. 9. The voltage V_A and V_B are respectively generated by two inverters, while voltage V_C and V_D are separately produced by ripple-free circuit associated with spark gaps which are regarded as switches S_1 and S_2. According to the illustration of the proposed poultry killing system, its operational intervals over one poultry killing period can be divided into 3 intervals, as shown in Fig. 10. Fig. 11. depicts conceptual waveforms of the proposed system. In the following, each operational interval over one poultry killing period is briefly described.

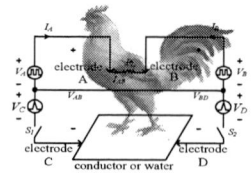

Fig. 9. An illustration of poultry killing during killing time interval.

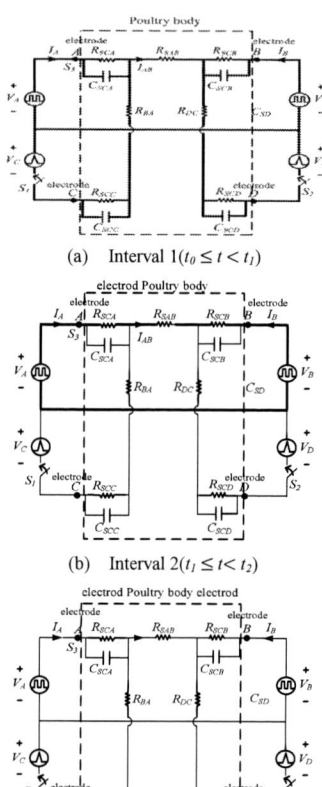

(a) Interval 1 ($t_0 \leq t < t_1$)

(b) Interval 2 ($t_1 \leq t < t_2$)

(c) Interval 3 ($t_2 \leq t < t_3$)

Fig. 10. Equivalent illustration of the proposed poultry killing system over one poultry killing period.

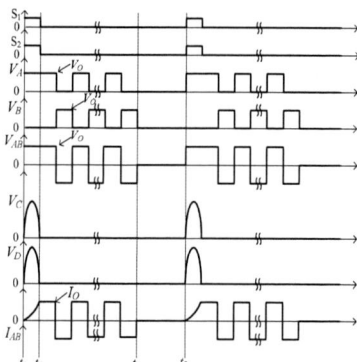

Fig.11 Conceptual waveforms of poultry killing system over one killing poultry period.

978-1-4244-4782-4/10 $26.00 © 2010 IEEE   1415

Interval 1 [Fig. 10(a);$t_0 \leqq t < t_1$]: Before t_0, switches S_1 and S_2 are in the off state. Moreover, voltage V_A, V_B, V_C and V_D are equal to 0. In this interval, current I_{AB} is equal to 0. The poultry killing system is in the rest interval. When $t = t_0$, switches S_1 and S_2 are turned on. The high output voltages V_C and V_D are from 0 to the maximum value, and then from the maximum value to 0. At the same time, voltage V_A is equal to V_o, which is generated by forward converter, while V_B is equal to 0. Thus, voltage V_{AB} is equal to V_o. Within this time interval, since poultry skin impedance R_{sc} and C_{sc} are from no breakdown to breakdown states, values of R_{sc} and C_{sc} are from a maximum value to an approximated 0. Therefore, current I_{AB} is from 0 to a desired value I_o.

Interval 2 [Fig. 10(b);$t_1 \leqq t < t_2$]: When $t = t_1$, poultry skin enters the breakdown state. At this moment, switches S_1 and S_2 are turned off. During this time interval, voltage V_A and V_B are in the alternative changes which are between 0 and V_o. Moreover, current I_{AB} flows through resistor R_{SAB} since R_{SCA}, R_{SCB}, C_{SCA}, and C_{SCB} are approximated to 0, and its value is kept on the desired value I_o. In this interval, it will sustain several seconds untill poultry is killed.

Interval 3 [Fig. 10(c);$t_2 \leqq t < t_3$]: When $t = t_2$, poultry is killed. At the same time, voltage V_A and V_B are turned off and their values are equal to 0. In this interval, current I_{AB} is also equal to 0. When switch S_1 and S_2 are turned on at end of interval 3, a new poultry killing period will start.

IV. DESIGN OF THE PROPOSED CONVERTER

The proposed poultry killing system consists of two forward converters with active clamp and ripple-free circuit. Since each forward converter has the same design procedure, design of single forward converter is only presented in this paper. Schematic diagram of single forward converter is shown in Fig. 12. Its conceptual waveforms of the one are illustrated in Fig. 13. In order to design the proposed converter systematically, determination of duty ratio D, transformers T_{r1} and T_{r2}, and active clamp capacitor C_c are presented in the following.

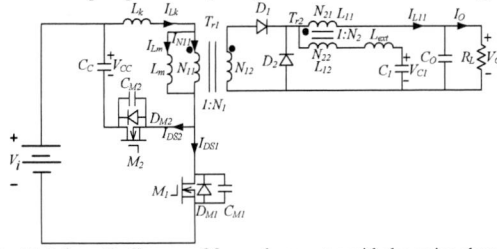

Fig. 12. Schematic diagram of forward converter with the active clamp and ripple-free circuits.

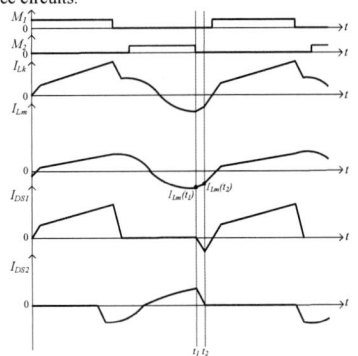

Fig.13 Conceptual waveforms of forward converter with the active clamp circuit.

A. Duty Ratio D

To determine duty ratio D, it needs to first obtain input to output voltage transfer ratio M. Since the active clamp and ripple-free circuits only help switches M_1 and M_2 to achieve soft-switching features and output inductor current I_{L11} to reach ripple-free feature respectively, they do not affect transfer ratio M of the proposed forward converter. That is, transfer ratio M will be the same as the conventional one. Moreover, since output voltage V_o of the proposed one has two voltage regulation ranges, they are separately V_o and V_{sg}, and voltage V_{sg} is greater than V_o. Output voltage is regulated to V_{sg} before poultry skin enters the breakdown state. In this state, impedance of poultry skin is very large. Therefore, its required power is very low. The proposed forward converter is only operated in discontinuous conduction mode (DCM) to generate the voltage V_{sg}. As mentioned above, output voltage is only design to regulate the value of V_o when the proposed one is operated in continuous conduction mode (CCM). According to volt-second balance principle of inductor L_{11}, the following equation can be obtained by

$$(N_1 V_i - V_o)DTs + (-V_o)(1-D)T_s = 0, \tag{1}$$

where N_1 is the turns ratio of transformer T_{r1}. From (1), it can be found that M can be indicated as

$$M = \frac{V_O}{V_i} = N_1 D \cdot \tag{2}$$

According to (2), once transfer ratio M and turns ratio N are specified, duty ratio D can be determined.

B. Transformer T_{r1}

Once the duty ratio D is selected, turns ratio N_1 of transformer T_{r1} can be determined from (2), which yields

$$N_1 = \frac{V_O}{DV_i} \cdot \tag{3}$$

In the forward converter, since magnetizing current I_{Lm} is less than $0.1 I N_{11} (=N_{11} L_{11})$, flux of magnetizing inductor is much less than the working flux which supplies energy to load through output inductor L_{11}. Current ΔI_{Lm} can be expressed by

$$\Delta I_{Lm} = \frac{V_i DT_S}{L_{mf}} < \frac{(N_1 V_i - V_O)DT_S}{10 L_{11}} \cdot \tag{4}$$

According to (4), $V_i DT_s$ must be limited and it must be satisfied in the following inequality:

$$VDT_S < \frac{1}{10} \frac{L_{mf}(N_1 V_i - V_O)DT_S}{L_{11}} \cdot \tag{5}$$

Therefore, by applying the Faraday's law, the number N_{11} of turns at the primary winding can be determined as

$$N_{11} = \frac{V_i DT_S}{A_C \Delta B}, \tag{6}$$

Where Ac is the effectively cross-section area of the transformer core and ΔB is the working flux density of magnetizing inductor Lmf. According to (3) and (6), N_{12} can be therefore determined.

C. Transformer T_{r2}

Since the proposed converter is operated in CCM, output inductor L_{11} must be greater than the boundary inductor L_{11B}. Furthermore, average current $I_{L11(av)}$ is equal to output current Io. When the converter is operated in the boundary between CCM and DCM, the boundary inductor L11B can be expressed by

$$L_{11B} = \frac{(N_i V_i - V_O)(1-D)T_S}{2I_O}. \tag{7}$$

To meet the operational condition of CCM, output inductor L_{11} must be greater than the boundary inductor L_{11B}, as listed in (7).

In order to achieve ripple-free features in output inductor current I_{L11}, the external inductor L_{ext} must be properly designed. According to design principle of ripple-free technique, external inductor $Lext$ can be determined as

$$L_{ext} = L_{11}(\frac{N_{21}}{N_{22}} - 1), \tag{8}$$

Where N_{21} and N_{22} is the number of turns of primary and secondary windings of transformer T_{r2}. According to (8), turns ratio N_2 must be greater than 1. Since inductance $Lext$ is proportional to turns ratio N_2, N_2 must be designed at between 1 and 1.1. When turns ratio N_2 is selected, external inductor L_{ext} is also determined.

D. Active Clamp capacitor Cc

The active clamp capacitor C_c is used to achieve soft-switching features. Since the energy stored in magnetizing inductor L_m is much greater than that stored in leakage inductor L_k. The energy stored in leakage inductor L_k can be neglected. To achieve a ZVS feature, the energy stored in inductor Lm must satisfy the following inequality:

$$\frac{1}{2}L_m(I_{Lm(t1)} - I_{Lm(t2)})^2 \geq \frac{1}{2}(C_{M1} + C_{M2})^2 V_{DS(\max)}^2 \tag{9}$$

where $I_{Lm(t1)}$ is the magnetizing inductor current at time t1, $I_{Lm(t2)}$ is that at time t_2, C_{M1} and C_{M2} are respectively the junction capacitors of switches M_1 and M_2, and $V_{DS(max)}$ is the voltage across switch M1. The voltage $V_{DS(max)}$ is equal to $(V_i + V_{cc})$ where V_{cc} is the voltage across capacitor C_c. Once C_{M1}, C_{M2}, $I_{Lm(t1)}$ and $I_{Lm(t2)}$ are specified, magnetizing inductor Lm can be determined as

$$L_m \geq \frac{(C_{M1} + C_{M2})V_{DS(\max)}^2}{(I_{Lm(t1)} - I_{Lm(t2)})^2}. \tag{10}$$

To achieve ZVS feature using the active clamp circuit, one half of the resonant period formed by Lm and C_c should be equal to or greater than the maximum off time of switch M_1. Thus, capacitor C_c must satisfy the following inequality:

$$\pi\sqrt{L_m C_C} \geq t_{off} = (1-D)T_s \tag{11}$$

From (11), when L_m is specified, the capacitance range of the clamp capacitor C_c can be determined as

$$C_C \geq \frac{(1-D)^2 T_S^2}{\pi^2 L_m} \tag{12}$$

V. EXPERIMENTAL RESULTS

To verify the performances of the proposed poultry killing system, as shown in Fig. 4, a prototype with the following specifications was implemented.

A. Each forward converter with ripple-free circuit

- Input voltage V_i : AC 110 V,
- Switching frequency f_{s1} : 50 kHz,
- Output voltage V_O : 180~300 V, and
- Maximum output current $I_{o(max)}$: 1A.
- Each half-bridge inverter
- Input voltage V_O : 180V~300 V,

- Switching frequency f_{S2} : 400 Hz
- Output voltage V_A (or V_B) : $\pm180 \sim 300$ V,
- Maximum output current $I_{A(max)}$ (or $I_{B(max)}$): ±1 A, and
- Maximum output power $P_{O(max)}$: 300 W

B. Ignitor (ripple-free circuit)
- Input viltage V_{C11}(or V_{C21}) : DC 180 ~ 250 V, and
- Output viltage V_C (or V_D) : 3 ~12 kV.

According to the specifications, components of the active clamp forward converter with ripple-free circuit are determined as follows:
- Turns ratio of T_{r1}(or T_{r2}) : 4,
- Turns ratio of T_{r3}(or T_{r4}) : 1.05,
- Turns ratio of T_{r5}(or T_{r6}) : 40,
- Magnetizing inductor L_{m11} (or L_{m21}) : 5.3 mH,
- Leakage inductor L_{K11} (or L_{K21}) : 8.55 μH,
- Transformer T_{r3}, T_{r4} core : EE-33,
- Magnetizing inductor L_{11} (or L_{21}) : 5.8 mH,
- Transformer T_{r3}, T_{r4} core : EE-30,
- Switches M_1~M_4 : IRF840,
- Capacitors C_{c1}, C_{c2} : 30 nF,
- Inductors L_{ext1}, L_{ext2} : 0.29 mH,
- Diodes D_{11}~D_{22} : UF304,
- Capacitors C_{11}, C_{21} : 0.1 μF/400 V,
- Transformer T_{r5}, T_{r6} core : EE-28,
- Capacitor C_{o1}, C_{o2} : 470 uF/400 V, and
- Spark Gap $sg1$, $sg2$: LSE 301LX

To generate ac voltage waveforms, the switches of two half-bridge inverters are also determined as S1~S4: IRF840.

In order to prove the feasibility of the proposed converter, experimental results of one forward converter are shown in this paper. Measured waveforms of V_{DS} and I_{DS} of switches M_1 and M_2 are shown in Fig. 14. From Fig. 14, it can be seen that the active clamp forward converter can achieve ZVS features. Fig 15, depicts measured waveforms of output current I_{L11} and output voltage V_o. Fig. 15(a) shows those waveforms without ripple-free circuit, while Fig. 15(b) illustrates those waveforms with one. From Fig. 15, it can be found that ripple of output current I_{L11} are reduced and limited within 5% of output current ratings when the proposed converter adopts ripple-free circuit. Fig. 16 shows measured waveforms of voltage V_c is generated, it will lead to breakdown of poultry skin and induce a desired killing current I_{AB} for killing chicken. Measured killing voltage V_{AB} and current I_{AB} waveforms of the proposed killing system under the full load of ±1 A, output frequency of 400 Hz and duty ratio of 50% is shown in Fig. 17, illustrating that an alternative voltage V_{AB} and current I_{AB} are adopted to kill chicken. From practical experimental results for killing a chicken, it can be observed that killing a chicken under voltage of 180 V, current of 600 mA, and killing time of 5 s can effectively achieve chicken killing.

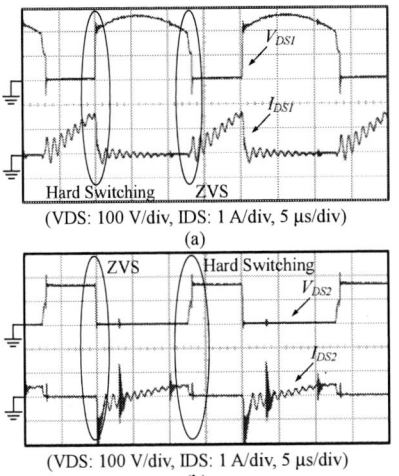

(VDS: 100 V/div, IDS: 1 A/div, 5 μs/div)

(a)

(VDS: 100 V/div, IDS: 1 A/div, 5 μs/div)

(b)

Fig. 14. Measured voltage V_{DS} and current I_{DS} waveforms of (a) switch M_1, and (b) switch M_2 in forward converter with the active clamp circuit.

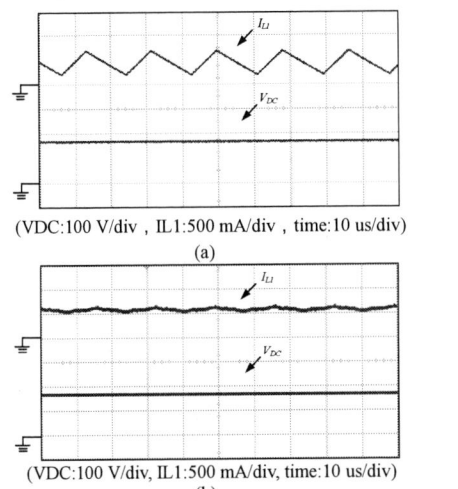

(VDC:100 V/div , IL1:500 mA/div , time:10 us/div)

(a)

(VDC:100 V/div, IL1:500 mA/div, time:10 us/div)

(b)

Fig. 15. Measured waveforms of output currents IL1 and output voltage VDC (a) without ripple-free circuit, and (b) with ripple-free circuit.

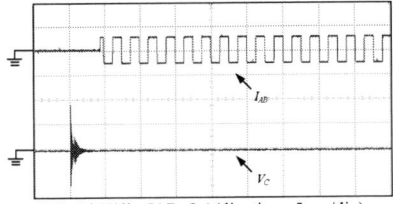

(VC: 2 kV/div, IAB: 2 A/div, time: 5 ms/div)

Fig. 16. Measured waveforms of voltage VC and current IAB during killing chicken.

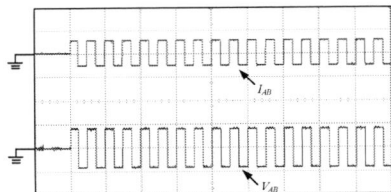

(VO : 200 V/div , IAB : 2 A/div , time : 5 ms/div)

Fig. 17. Measured ac-link voltage VAB and current IAB waveforms of the proposed killing system under a load of 1 A, output frequency of 400 Hz and duty of 50%.

VI. CONCLUSION

In this paper, killing mechanism of poultry with electrical stunning has been briefly reviewed. Operational principle, steady-state analysis and design of the proposed poultry killing system have been implemented to generate killing electrical parameters, in which its current is 600 mA and its voltage is 180 V. Moreover, features of forward converter with the active clamp and ripple-free circuit have been also accomplished to achieve ZVS features and ripple free features of output current I_{L11} and I_{L21}. From experimental results, it can also found that the proposed one with killing voltage of 180 V and killing current of 600 mA has been adopted to achieve killing chicken, which meets the regulation of animal welfare.

REFERENCES

[1] H. A. Channon, A. M. Payne and R. D. Warner, "Comparison of CO2 Stunninng with Manual Electrical Stunning (50Hz) of Pig on Carcass and Meat Quality," Trans. on Meat Science, 2002, pp.63–68.

[2] W Li, Y Zhao, J Wu and X He, "Interleaved High Step-Up Converter With Winding-Cross-Coupled Inductors and Voltage Multiplier Cells," Transactions on Power Electronics , 2009, pp.1 – 1.

[3] Wai Rong-Jong and Wang Wen-Hung, "Design of Grid-Connected Photovoltaic Generation System with High Step-Up Converter and Sliding-Mode Inverter Control," Transactions on ISIC, 2007, pp.1179 – 1184.

[4] S. Y. Tseng, S. H. Tseng, and J. Z. Shiang, "High Step-up Converter Associated with Soft-Switching Circuit with Partial Energy Processing for Livestock Stunning Applications," Transactions on IPEMC, Volume 2, 2006, pp.1 – 5.

[5] C Pan and C Lai, "A High Efficiency High Step-Up Converter with Low Switch Voltage Stress for Fuel Cell System Applications," Transactions on Industrial Electronics, 2009, pp.1 – 1.

[6] Do Hyun-Lark and Kwon Bong-Hwan, "Single-stage line-coupled half-bridge ballast with unity power factor and ripple-free input current using a coupled inductor," Transactions on Industrial Electronics, Volume 50, 2003, pp.1259 – 1266.

[7] D McDonnall, G. A. Clark and R. A. Normann, "Interleaved, multisite electrical stimulation of cat sciatic nerve produces fatigue-resistant, ripple-free motor responses," Transactions on Neural Systems and Rehabilitation Engineering, Volume 12, 2004, pp.208 – 215.

[8] Lee Jong-Jae and Kwon Bong-Hwan, "Active-Clamped Ripple-Free DC/DC Converter Using an Input–Output Coupled Inductor," Transactions on Industrial Electronics, Volume 55, 2008, pp.1842 – 1854.

[9] F Causa and L Burrow, "Ripple-Free High-Power Super-Luminescent Diode Arrays," Transactions on Quantum Electronics, Volume 43, 2007, pp.1055 – 1059.

[10] J. F. Silva, A. Galhardo and J. Palma, "High-efficiency ripple-free power converter for nuclear magnetic resonance," Transactions on PESC, 2000, pp.384 – 389.

[11] B.-Y. Shmuel, "Electrical Evaluation of the Taser M-26 Stun Weapon Final Report," The Israel National Police, R&D Division, 2006.

[12] Henneman, C. Olson, "Relations between structure and function in the design of skeletal muscle," Trans on Neurophysiology, 1965, pp. 581-589.

The Optimal Control Strategy for Rectifier Side of Low Switching Frequency Back-to-Back Converter

Kai Tan[1,2], Qiongxuan Ge[1], Zhenggang Yin[1,2], Congwei Liu[1], Yaohua Li[1]

1. Institute of Electrical Engineering of Chinese Academy of Sciences
2. Graduate University of Chinese Academy of Sciences
Beijing, China
E-mail: tank@mail.iee.ac.cn

Abstract—**The industry application of High-power medium-voltage drives converter is an appealing spot in Power electronics technology in recent years. In this paper, an optimal control strategy for rectifier side of three-level NPC back-to-back converter under low switching frequency conditions have been presented. First, the power feed forward is applied for enhancing the dynamic characteristics in the circumstance of bidirectional energy flow. And the demand of three-level DC neutral point control is solved by using a modified asymmetrical regular sampling SVPWM whenever the energy is being used or feed backed. Simulation and experimental results based on a rectifier side of a 75kVA three-level NPC back-to-back experimental prototype converter with a 400Hz device switching frequency constraint show satisfactory performance.**

I. INTRODUCTION

The pressure of energy-saving and ejection-decreasing is a world-wide hot issue. The power electronics technology is already widely used in the low voltage and capacity AC converters for energy-saving. In recent years, the rapid development of the capacity and the switching frequency of power semiconductor devices and continuous advance of power electronics technology lead the new industry applications in high-power area. The high-power medium-voltage drives have found widespread applications in industry. They are used for steel main rolling mills in the metal industry [1], pipeline pumping stations in the petrochemical industry [2], pumps in water pumping station [3], fans in the cement industry [4], traction applications in the transportation industry [5], etc.

This paper proposes an optimal control strategy for rectifier side of three-level NPC back-to-back converter based on the research and development of a 15MVA IGCT converter system for high-speed maglev propulsion control system. The technical requirements and challenges for medium-voltage drive are different in many aspects from those for the low-voltage ac drives. These requirement and challenge of rectifier side can be generally divided as follows: the requirements related to the power quality of the line-side circumstance, the challenges associated with the dynamic and stable performance of DC bus, the switching frequency constraints of the high power switching devices.

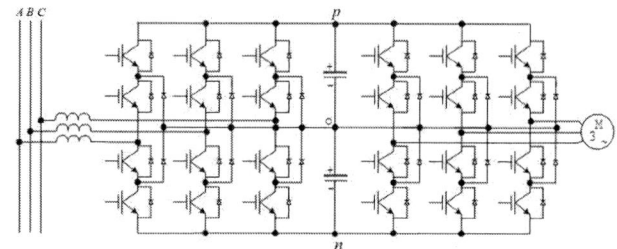

Figure 1. Three-level Back-to-Back Converter Topology

Based on an overall consideration of various requirements, we choose the three-level back-to-back neutral-point clamped (NPC) converter configuration [6]. Fig. 1 shows circuit diagram of a three-level back-to-back NPC converter

Actually, the line-side rectifier and the motor-side inverter have the same three-level NPC configurations form a back-to-back circuit for achieving bidirectional energy flow function.

II. RECTIFIER CONTROL STRATEGY

The aims of the rectifier side control include lower input current harmonic, power factor correction, bidirectional energy flow and stable DC bus voltage and balance for inverter in a 400 Hz switching frequency circumstance. In our strategy, classical d-q coordinates control is utilized as a fundamental frame to satisfy the power factor correction [7].

A. Rectifier Side d-q Coordinates Mathematical Model

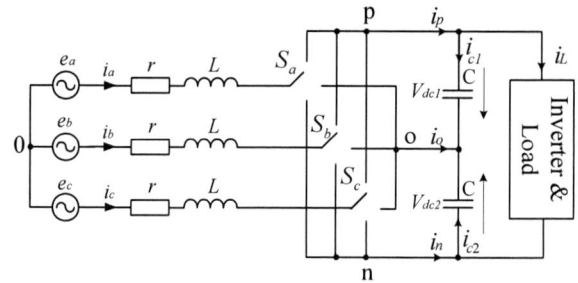

Figure 2. Three-level PWM Rectifier Model

The three-level PWM rectifier model is shown in Fig. 2. Assuming a balanced three-phase system without the neutral connection and neglecting the resistance of the power switches, the three-phase rectifier can be modeled as

$$L\frac{di_k}{dt} = e_k - ri_k - (S_{kp} - \frac{1}{3}\sum_{k=a}^{c} S_{kp})V_{dc1} - (S_{kn} - \frac{1}{3}\sum_{k=a}^{c} S_{kn})V_{dc2}$$

$$C\frac{dV_{dc1}}{dt} = \frac{1}{3}\sum_{k=a}^{c} S_{kp}i_k - i_o$$

$$C\frac{dV_{dc2}}{dt} = \frac{1}{3}\sum_{k=a}^{c} S_{kn}i_k + i_o \qquad (1)$$

$$\sum_{k=a}^{c} e_k = \sum_{k=a}^{c} i_k = 0$$

$$S_k = \begin{cases} 1, as.S_{kp}=1, S_{ko}=0, S_{kn}=0 \\ 0, as.S_{kp}=0, S_{ko}=1, S_{kn}=0 \\ -1, as.S_{kp}=0, S_{ko}=0, S_{kn}=1 \end{cases}$$

where k means a to c phase, S_{kp}, S_{kn} means switching functions, other parameters illustrated in Fig. 2.

A two-phase coordinate system (d − q) is defined in Fig. 3 along with a three-phase system (a, b, c). The transformation of variables between these two coordinate systems obeys the.

A two-phase reference-frame d − q model (2) of the three-level rectifier is obtained by applying Clarke transform and Park transform to (1):

$$L\frac{di_d}{dt} = -ri_d + \omega Li_q - S_{dp}V_{dc1} - S_{dn}V_{dc2} + V_d$$

$$L\frac{di_q}{dt} = -ri_q - \omega Li_d - S_{qp}V_{dc1} - S_{qn}V_{dc2}$$

$$C\frac{dV_{dc1}}{dt} = S_{dp}i_d + S_{qp}i_q - i_L \qquad (2)$$

$$C\frac{dV_{dc2}}{dt} = S_{dn}i_d + S_{qn}i_q + i_L$$

The requirement of power factor in our rectifier equals 1 which means $i_q = 0$. So the steady-state equation of two-phase d − q frame is [8]:

$$V_{d\_ref} = S_{dp}V_{dc1} + S_{dn}V_{dc2} = V_d - ri_d$$

$$V_{q\_ref} = S_{qp}V_{dc1} + S_{qn}V_{dc2} = -\omega Li_d \qquad (3)$$

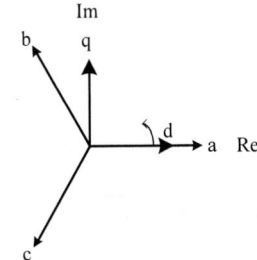

Figure 3. Orientation of three-phase and d–q frame in a complex plane

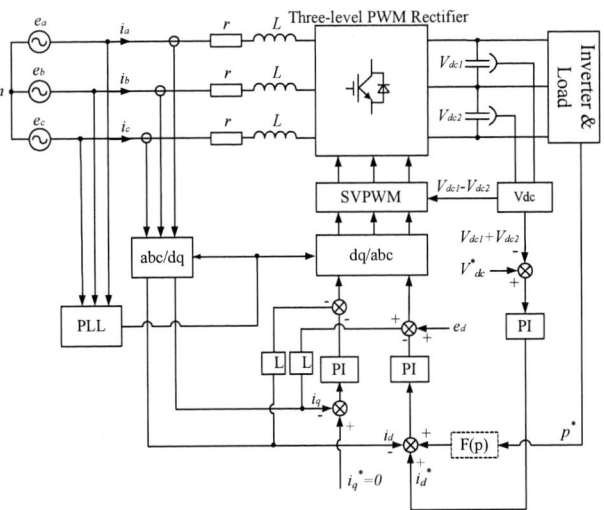

Figure 4.   Control Strategy Based on d-q Coordinates Frame

B.   Control Strategy Based on d-q Coordinates Frame

The optimal control strategy reported in this paper shown as Fig. 4 bases on the classical d-q rotating coordinates control due to the requirements of power factor correction, bidirectional energy flow. The neutral point voltage balance control is also added which is demanded by the three-level topology. Based on voltage and current loops, Classical d-q rotating coordinates control could well fulfill the requirements of power factor correction, bidirectional energy flow due to its d-coordinate active power control and q-coordinate reactive power control.

All waveforms recorded on the 380V, 50Hz, 75Kva, 400Hz, DC bus 600V prototype back-to-back converter if there is no special note. Fig. 5 (a) and Fig. 5 (b) separately show the performance of power factor correction at stable and dynamic moment. The dynamic experiment is implemented by brake a rotating induction motor in inverter side in 2s. It is worthwhile to note that the power feed forward is not be used to verify the basic d − q power factor correction capability. The dc bus rises to 624 when the power from inverter side flows back. And the dynamic adjust time is 22ms.As the motor is still with load while braking, the power feed back phenomena is just last for 2-3 periods.

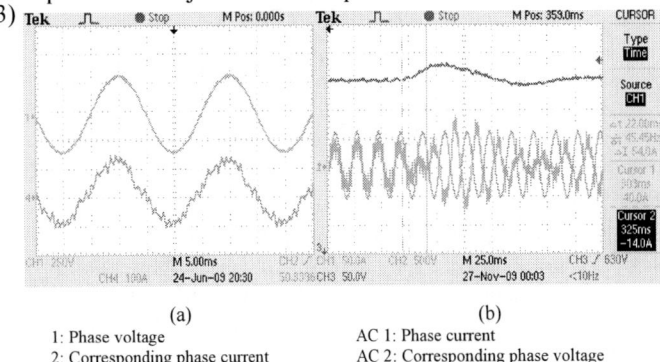

(a)

1: Phase voltage
2: Corresponding phase current

(b)

AC 1: Phase current
AC 2: Corresponding phase voltage
DC 3: DC bus voltage

Figure 5.   Stable & Dynamic Power Factor Correction and Bidirectional Flow

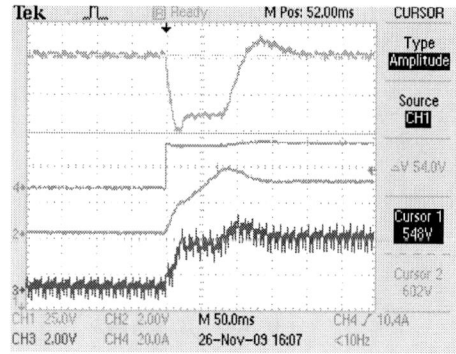

Top: DC bus voltage
2nd: Sudden changed power signal
3rd : id*+0 (no power feedforward)
Bottom: id

Figure 6. Dynamic Performance without Power Feedforward

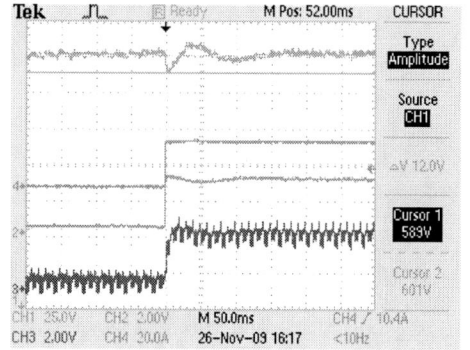

Top: DC bus voltage
2nd: Sudden changed power signal
3rd : id*+power signal (power feedforward implement)
Bottom: id

Figure 7. Dynamic Performance with Power Feedforward

The dynamic performance of basic d – q coordinates frame is not satisfied, even causes out of control in high-power area. So, the power feedforward strategy is proposed for rectifier side control to enhance the dynamic performance.

Fig. 6 and Fig. 7 show the results of comparative experiment which without or with the power feedforward implement. Same 20kW load is sudden added. The max DC bus fall is 54V and 12V. The recover time is 170ms and 82ms. The performance of current loop id track is better.

III. MODIFIED SVPWM

As the grid frequency is a relative constant than the situation of inverter side, we could regulate the switching frequency to form synchronous modulation. However, even the synchronous modulation is used, the input current harmonic is not satisfied due to the switching frequency constraint of high power devices.

Indeed, the SHEPWM is more convenient to eliminate particular harmonic than SVPWM [9]. However, in three-level topology, the neutral point balance control of DC side is very important. The SVPWM has a strong capability for neutral balance control by using its alternative P-N small vector which is difficult for SHEPWM method by using its determined notch angles. SHEPWM method also needs more

memory for its notch angles lookup table. As mentioned above, a modified SVPWM method with asymmetrical regular sampling and even-order harmonic elimination is proposed to optimize the input current harmonics.

A. Asymmetrical regular sampling SVPWM

The line to line voltage waveform produced by the conventional SVPWM contains even-order harmonics. The even-order harmonics may not have a significant impact on the motor in the inverter side. However, when it is used in the rectifier side, the line current THD should comply with harmonic standards such as IEEE 519-1992 [10]. Because these standards have more stringent stipulation on even-order harmonics than on odd-order ones, the strategy proposes a modified P-N alternative SVPWM scheme with even-order harmonic elimination by producing the line to line voltage waveform half-wave symmetrical.

Asymmetrical regular sampling means have two sampling points in each switching interval. This sampling method could double the PWM control frequency which is especially useful under low switching frequency condition. It could also provide a harmonic improvement [11].

Fig. 8 and Fig.9 show the stable waveform and THD analysis separately using conventional P-N SVPWM and using modified asymmetrical regular sampling P-N SVPWM.

Two P-N SVPWM algorithms eliminate the even-order harmonics. The remaining harmonics display at 6k+1, 6k-1.

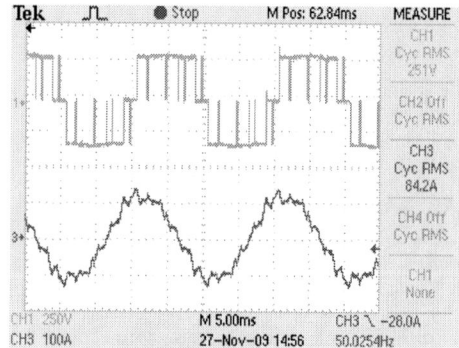

Top: rectifier phase voltage
Bottom: input current

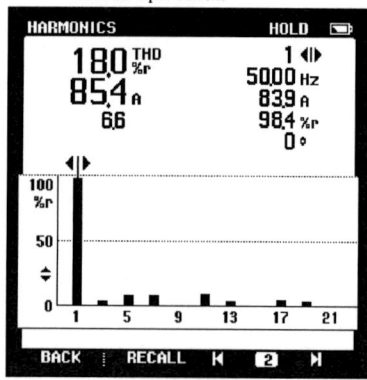

Input current THD

Figure 8. Phase Voltage and Input Current by Conventional SVPWM

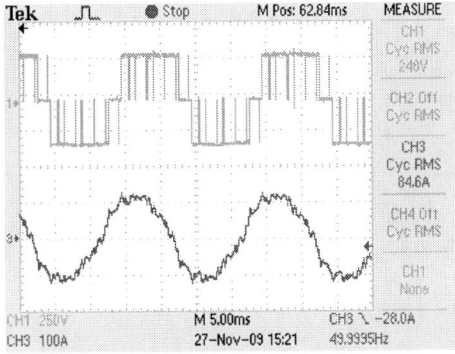

Top: rectifier phase voltage
Bottom: input current

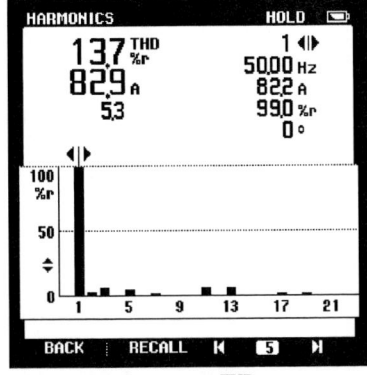

Input current THD

Figure 9. Phase Voltage and Input Current by Asymmetrical SVPWM

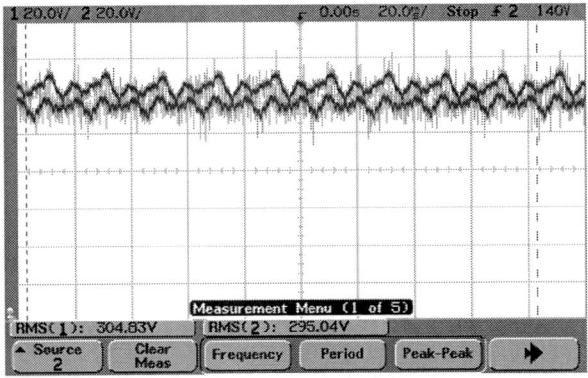

Asymmetrical regular sampling P-N SVPWM natural balancing

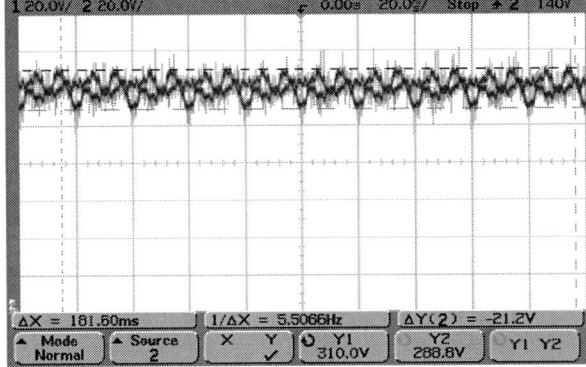

Asymmetrical regular sampling P-N SVPWM controlled balancing

Figure 11. Stable Performance of DC bus by Asymmetrical SVPWM

The experiments results indicate that the modified asymmetrical regular sampling P-N SVPWM could provide the input current improvement while eliminate the even-order harmonics. However, the improvement also increases the device switching frequency as negative effect. The additional frequency equals the fundamental frequency of the power grid in the rectifier side. It means that rises from 350Hz to 400Hz in our case.

This increase is caused by the additional sampling point in the middle of each switching interval. Fig. 10 explains the causes clearly.

The doubled control frequency of asymmetrical regular sampling SVPWM also is a significant feature which worthy to consider, especially in the high-power area under the low switching frequency constraint. The merit and the negative effect should be balanced in different conditions.

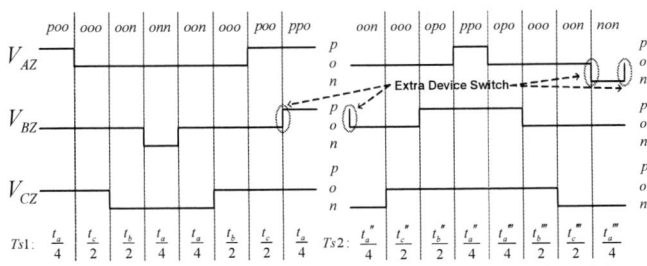

Figure 10. The Extra Device Switch by Modified Asymmetrical SVPWM

B. Neutral Point Balancing

As mentioned above, the neutral point of DC bus control is a fundamental requirement of inverter side and an important task of rectifier side in three-level NPC back-to-back converter. And this task is especially important in high-power area and harder as the low switching frequency constraint.

For stable performance, the modified asymmetrical regular sampling P-N SVPWM has a natural balancing feature due to its P-N alternative selected. The stable DC bus waveform Fig. 11 shows the steady performance of natural balancing and controlled balancing.

As the high-power requirement and the low switching frequency constraint, we propose different section scheme judged by Vdc1-Vdc2 for enhancing the dynamic performance. The balance strategy is 1.natural balancing without control; 2.proportional linear balance control; 3.bang-bang balance control; 4.safety protection which divided by 3 threshold. The prototype converter experiment indicate this rectifier side neutral point balancing strategy could control the abs(Vdc1-Vdc2) <5% DC bus even better which decided by the inverter side unbalanced load conditions and the magnitude of rectifier side input current.

Fig. 12 shows the dynamic performance of neutral point balancing algorithm. The experiment condition is start without neutral point balancing, and then sudden takes unbalanced load in, at last start the neutral point balancing algorithm by

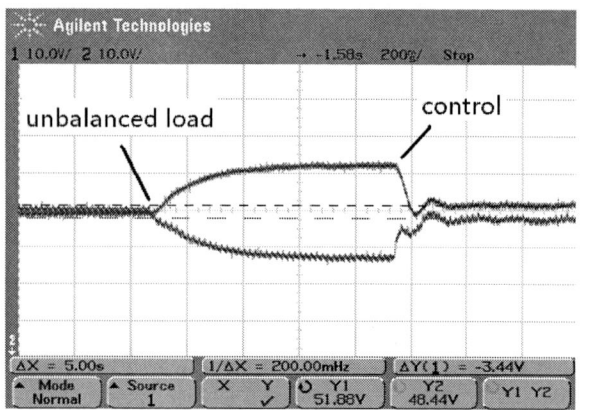

Figure 12. Dynamic Performance of DC bus by Asymmetrical SVPWM

manual command to observe the dynamic balancing capacity. The DC bus is 200V in this experiment.

The experiment indicate that the algorithm at least have the capability to balance the voltage from time to 12% (25V/200V) to 1.72% (3.44V/200V) in 120ms.

IV. CONCLUSION AND FUTURE WORK

In this paper, an optimal control strategy for rectifier side of three-level NPC back-to-back converter based on the research and development of a 15MVA IGCT converter system for high-speed maglev propulsion control system is proposed for apply to low switching frequency condition. Power feedforward, neutral point sectional balancing control, modified asymmetrical regular sampling P-N SVPWM are presented above and display well performance both in dynamic and stable moment from the experiment results of 75kVA prototype as mentioned above.

The minimum pulse width and dead time of high power device is a critical parameter that bring in a lot of problems both dynamic and stable moment. We are going to have more experiments in this part then.

REFERENCES

[1] H. Okayama, M. Koyama, et al., Large Capacity High Performance 3-level GTO Inverter System for Steel Main Rolling Mill Drives, IEEE Industry Application Society (IAS) Conference, pp. 174–179, 1996.

[2] W. C. Rossmann and R. G. Ellis, Retrofit of 22 Pipeline Pumping Stations with 3000-hp Motors and Variable-Frequency Drives, IEEE Transactions on Industry Applications, Vol. 34, Issue: 1, pp. 178–186, 1998.

[3] B. P. Schmitt and R. Sommer, Retrofit of Fixed Speed Induction Motors with Medium Voltage Drive Converters Using NPC Three-Level Inverter High-Voltage IGBT Based Topology, IEEE International Symposium on Industrial Electronics, pp. 746–751, 2001.

[4] R. Menz and F. Opprecht, Replacement of a Wound Rotor Motor with an Adjustable Speed Drive for a 1400 kW Kiln Exhaust Gas Fan, The 44th IEEE IAS Cement Industry Technical Conference, pp. 85–93, 2002.

[5] S. Bernert, Recent Development of High Power Converters for Industry and Traction Applications, IEEE Transactions on Power Electronics, Vol. 15, No. 6, pp. 1102–1117, 2000.

[6] A. Nabae, I. Takahashi and H. Akagi, A New Neutral Point Clamped PWM Inverter, IEEE Transactions on Industry Applications, Vol. 17, No. 5, Sep./Oct. 1981, pp. 518-523.

[7] Blasko, V.; Kaura, V., A new mathematical model and control of a three-phase AC-DC voltage source converter, IEEE Transactions on Power Electronics, Vol. 12, Issue. e, pp. 116-123, Jan. 1997.

[8] Lixiang Wei, Research on Dual-PWM Three-Level Flux Orientation Induction Motor Control System, Beijing:Tsinghua University, 2000, pp. 14-21.

[9] Bimal K. Bose, Modern Power Electronics and AC Drives, Prentice Hall PTR, 2001, pp. 218-223.

[10] Bin Wu. High-power converters and AC drives. IEEE PRESS, 2006, pp. 112-115.

[11] D. Grahame Holmes and Thomas A. Lipo. Pulse Width Modulation For Power Converters. IEEE PRESS, 2003, pp. 139-146.

Transformer Structure and Its Effects on Common Mode EMI Noise in Isolated Power Converters

Pengju Kong and Fred C. Lee
Center for Power Electronics Systems
Virginia Polytechnic Institute and State University
Blacksburg, VA, 24060
pjkong@vt.edu

Abstract—This paper investigates the effects of transformer structure and parasitic on common mode (CM) EMI noise of isolated power converters. Study on typical transformer winding structures indicates that the distribution of inter-winding capacitances and the distribution of voltage potentials on windings are the two critical factors that determine the CM noise levels of the converter in switching frequency ranges. CM noise reduction methods by improving the transformer structure and by compensation are proposed. At high frequencies, leakage inductances of the transformer resonate with the inter-winding capacitances or the junction capacitances of the secondary side switches. It may result in severe high frequency CM noise peaks. Methods of controlling such peaks are also discussed.

I. INTRODUCTION

One can observe in today's power converter products that significant efforts, such as additional large CM chokes and shielded EMI filters, are made in order to filter the CM noise [1]. Reducing the original CM noise emission of power converters can help to simplify the filter design and reduce the total filter size. In most applications, CM noise of a power converter is dominated by displacement currents which are the results of voltage pulsating (dv/dt) on the converter parasitic capacitances. Several noise reduction techniques have been proposed to reduce the CM noise of power converters by limiting the voltage pulsating of the converter [2], reducing the parasitic capacitances of the converter [3], or introducing out-of phase displacement currents to cancel the original noise currents [3-7].

Identifying the characteristics of voltage potentials and parasitic capacitances of the converter is the basis for applying the most suitable noise reduction technique. In an isolated power converter, the inter-winding capacitances of power transformer are critical parasitic capacitances of the converter in the CM noise perspective. Such capacitances are distributed along the windings with different voltage pulsating (dv/dt). Different transformer structures or winding/termination arrangements may have different inter-winding capacitance distribution and voltage pulsating distribution. This may lead to significant different CM noise characteristics [7]. In this paper, effects of transformer structure and winding arrangement on the characteristics of inter-winding capacitances and voltage distribution are analyzed. Winding-to-winding capacitances and winding-to-core capacitances are discussed respectively.

Shielding is a widely adopted CM noise reduction technique in isolated power converters [3]. It reduces the CM noise by reducing the effective parasitic capacitances associated with the most severe voltage pulsating nodes. However, inserting shielding layers in a power transformer introduces additional power loss. It is also not practical in fully interleaved transformer winding structure since too many shielding layers are needed.

CM noise of isolated power converters can also be reduced by introducing out-of-phase displacement currents [3, 5, 7]. Such currents are introduced by the out-of-phase voltage pulsating on parasitic capacitances. In order to achieving fully cancellation, either the magnitude of the voltage pulsating or the amount of parasitic capacitances should be controlled. Reference [3] proposed to control the parasitic capacitances with optimum shielding to minimize the CM noise. In this paper, instead of controlling the parasitic, the voltage pulsating distribution in the transformer winding is controlled to achieve CM noise reduction in order to avoid the need of shielding. To further reduce the CM noise, compensation capacitor is introduced to cancel the effect of winding-to-core capacitances.

The leakage inductance of transformer is another critical parasitic in the CM noise perspective. It resonates with the inter-winding capacitances. At the resonant frequency the CM noise propagation path has low impedance and results in relatively high CM noise peak. The leakage inductance also resonates with the junction capacitances of the secondary side diode (or synchronous rectifier). It introduces a high frequency noise source on the secondary side and results in a CM noise peak. Methods of reducing such noise peaks are discussed.

II. INTER-WINDING CAPACITANCES AND ITS EFFECTS ON CM NOISE

A. CM noise sources in isolated power converters

The switching of MOSFETs/diodes is the major source of voltage pulsating. Voltages across these devices can be treated as the CM noise voltage sources. Topology of the flyback converter is shown in Fig. 1. S_1 on primary side introduces severe dv/dt at node A. The secondary side diode also has dv/dt too. They can be modeled as voltage sources [3, 6], as shown in Fig. 2. The input capacitor and output capacitor can be treated as short-circuit in the conducted EMI frequency range. It can be observed from Fig. 2 that v_A and v_D determine the transformer winding terminal voltages, and their voltage ratio should follow the turns ratio N when perfect coupling of transformer windings is assumed. In most applications, output of the converter is grounded for safety reasons. CM noise current generated by the noise sources propagates through inter-winding capacitance of the transformer and goes into Line Impedance Stabilization Network (LISN) through the grounding wire of secondary side output, as shown in dotted line in Fig. 2.

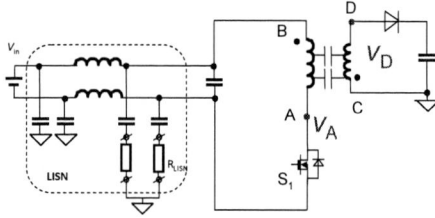

Figure 1.   Topology of the flyback converter

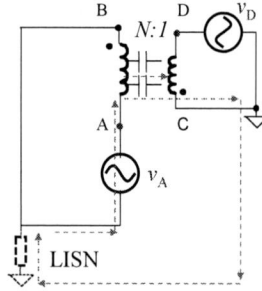

Figure 2.   CM noise sources in the flyback converter

B. Characteristics of Inter-winding capacitances

Fig. 3 shows a transformer structure implemented with EE core. Fig. 4 shows the half window of transformer in Fig. 3 with the inter-winding capacitances. There are four layers of windings in the structure. The inner layer P_s is the secondary winding. Layers P_{p1}, P_{p2} and P_{p3} are the primary winding layer. There are distributed parasitic capacitances between every two layers. Those capacitances between two primary winding layers do not contribute to CM noise because the displacement currents between them are confined within the primary side of the converter. The distributed parasitic capacitances between L_{p3} and L_s provide the major paths for the CM noise current from primary to secondary side of the converter. There are also distributed parasitic capacitances between P_{p1} and core, as well as between P_{p3} and core. Since the impedance of the core is much smaller than the winding-to-core capacitances, these two capacitances are in series and provide CM noise current paths between P_{p1} and P_s. In fig. 4, C_{p-s} denotes the sum of the capacitances between P_{p3} and P_s. C_{P-core} denotes the sum of the capacitances between P_{p1} and core. C_{S-core} denotes the sum of the capacitances between P_s and core.

It is necessary to quantify these two capacitances in order to study their effects on CM noise. Parasitic capacitances are reverse proportional to the distance between two layers or layer and core. Primary and secondary winding layers are close in design to avoid excessive leakage. The core, on the other hand, is away from the windings due to the existence of bobbin and insulation. So the winding-to-winding capacitances are much larger than the winding-to-core capacitances. A transformer with the structure shown in Fig. 3 is built for the flyback converter prototype. The total capacitance between primary and secondary winding is measured with an impedance analyzer. It is 34pF. It is the sum of C_{P-S} and the serial capacitance of C_{P-core} and C_{S-core}. The core is then removed and the capacitances between primary and secondary are measured a second time. The result reflects only C_{P-S} and it is 32.5pF. The capacitance of the series of C_{P-core} and C_{S-core} can be then calculated as 1.5pF, less than 5% of the winding-to-winding capacitances.

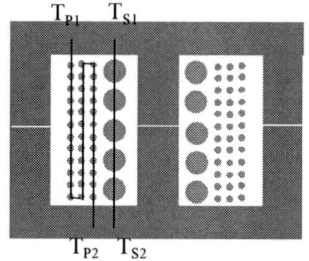

Figure 3.   Power transformer for flyback converter

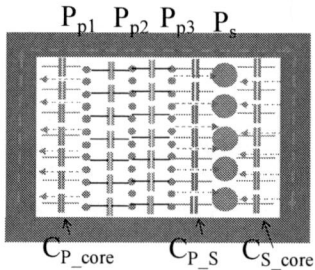

Figure 4.   Inter-winding capacitances in the power transformer

In some transformer structures, winding-to-core capacitances do not contribute to CM noise. Since the inter-winding capacitances associated with the core are the serial capacitances of the primary winding-to-core capacitances and the secondary winding-to-core capacitances, eliminating either capacitance will eliminate the effects of the core. Fig. 5 shows a transformer structure with interleaved windings. The secondary winding is between two primary winding layers. Its capacitance to the core can be ignored. In this case the core does not participate as an electric coupling path between primary and secondary windings.

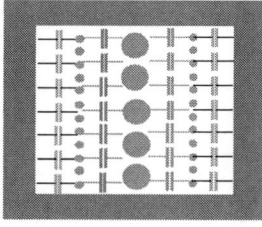

Figure 5.   Transformer structure with interleaved windings

978-1-4244-4782-4/10 $26.00 © 2010 IEEE

C. Distribution of the voltages along windings

The total CM noise current going through the inter-winding capacitances is determined by the distribution of voltage pulsating along the windings [3, 7]. It is determined by both the voltage pulsating on the transformer terminals and the winding structure of the transformer.

The voltage pulsating on the transformer terminals is determined by the converter characteristics and how the transformer is connected to the converter. In Fig. 1 and Fig. 3, when we connect T_{P1} to A, T_{P2} to B, T_{S2} to C and T_{S1} to D, the terminal voltages of the transformer can be simulated or measured. According to the CM noise source model in Fig. 2, they are v_A, 0, 0 and v_D respectively.

Voltage distribution along windings can be calculated by assuming the voltages are evenly distributed along windings. Fig. 6 shows the voltage distribution of the 4 layers shown in Fig. 4. N_P and N_S are turns number of primary and secondary windings. N_{P2} and N_{P3} are turns number of the winding layer P_{P2} and P_{P3}.

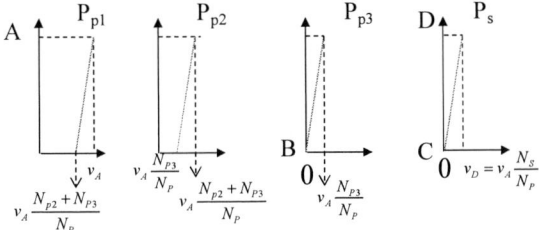

Figure 6. Voltage distribution of winding layers

Voltage distribution is determined by two aspects in a transformer structure. One is the number of layers in series. The other is the number of turns in each layer.

D. Distribution of the voltages along windings

CM noise currents introduced by the voltage pulsating can be calculated as in

$$i_{CM} = \sum C \frac{dv}{dt}. \tag{1}$$

In the flyback converter case, the total CM noise current consists of two components. One component is the currents between P_{P3} and P_S. It can be calculated as in

$$i_{winding-winding} = \int \frac{C_{P-S}}{H} \frac{d}{dt} (\frac{x}{H} v_A \frac{N_{P3}}{N_P} - \frac{x}{H} v_A \frac{N_S}{N_P}) dx, \tag{2}$$

$$= C_{P-S} \frac{d}{dt} \frac{v_A}{2 N_P} (N_{P3} - N_S)$$

where H is the height of the winding. The other component is the currents between P_{P1} and P_S through core. It can be calculated as in

$$i_{winding-core} \tag{3}$$

$$= \int \frac{1}{H} \frac{C_{P-core} C_{S-core}}{C_{P1-core} + C_{S-core}} \frac{d}{dt} [(v_A \frac{N_{P2} + N_{P3}}{N_P} + \frac{x}{H} v_A \frac{N_{P1}}{N_P}) - \frac{x}{H} v_A \frac{N_S}{N_P}] dx$$

$$= \frac{C_{P-core} C_{S-core}}{C_{P-core} + C_{S-core}} \frac{d}{dt} \frac{v_A}{N_P} [N_{P2} + N_{P3} + \frac{1}{2}(N_{P1} - N_S)]$$

where N_{p1} is the turns number of winding layer P_{P1}. The total CM noise of the flyback converter is

$$i_{CM\_total} = i_{winding-winding} + i_{winding-core}. \tag{4}$$

III. CM NOISE REDUCTION TECHNIQUES

A. Effects of termination

CM noise of the converter can be significantly different with different transformer termination [7]. In the flyback converter, when the outer layer P_{p1} is connected to node A in the converter, its voltages along windings is much higher than P_{p3} in which node B is connected to voltage 0. Since the total capacitance between P_{p1} and P_s is very small, this high voltage does not cause significant CM noise.

If layer P_{p3} is connected to node A, the result will be different. The total inter-winding capacitance (C_{p-s}) between P_{p3} and P_s is large. The high voltage pulsating on layer P_{p3} will introduce relatively high CM noise current through C_{p-s}.

In the case of multiple layers for primary windings, it is always preferred to connect the layer adjacent to the secondary side winding with low voltage pulsating node in the converter.

B. Adjusting the winding structure

Since C_{p-s} is much larger than the series of $C_{p1-core}$ and C_{s-core}, in Eq. (4) the first term which represents the displacement current between P_{p3} and P_s dominates the total CM noise current. One can observe from Eq. (2) that if the turns number of layer P_{p3} equals to that of the secondary winding, the net current becomes 0. There is no net current between P_{p3} and P_s.

When the turns number of layer P_{p3} (N_{p3}) is smaller than that of the secondary winding (N_s), the value of $i_{winding-winding}$ becomes negative. It can partially cancel $i_{winding-core}$. The best N_{p3} in the CM noise reduction sense can be solved with the criterion that the absolute value of Eq. (4) should be minimized. For example, assuming N_{p1}, N_{p2} and N_{p3} are all equal to 1/3 of N_p, the required N_{p3} for complete CM noise cancellation ($i_{CM\_total} = 0$) can be calculated in

$$N_{p3} = N_S - N_P \frac{C_{p-core} C_{s-core}}{(C_{p-core} + C_{s-core}) C_{p\_s}} (\frac{5}{3} - \frac{N_s}{N_p}). \tag{5}$$

In practice it is possible that the solved N_{p3} is minus. One possibility is to introduce reverse coupled windings in the primary side. However, this method will increase the conduction loss especially when the turns number is large.

A better solution is to let N_{p3} equals to 1 and introduces additional cancellation capacitor to cancel the remained CM noise current, as shown in Fig. 7. It introduces a CM noise current calculated as in

$$i_{cm\_com} = C_{add} \frac{d}{dt} v_D. \tag{6}$$

The required capacitance can be calculated as in

$$C_{add} = \frac{C_{p1} C_s}{C_{p1} + C_s} \cdot (\frac{5}{6} \frac{N_p}{N_s} - \frac{1}{2}). \tag{7}$$

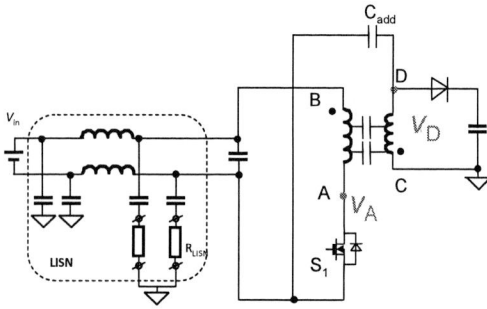

Figure 7. Flyback converter with compensation capacitor

C. Experiment validation

A flyback converter prototype is built to verify the proposed CM noise reduction techniques. It operates at discontinuous current mode. In the experiment its input is 110Vac and output is 12.5Vdc with the switching frequency around 130kHz. The output power is 30W. The transformer uses 00k3007 EE core from Magnetics with the permeability 90. The original winding structure is shown in Fig. 3 with 118 turns on the primary side and 5 turns on the secondary side. In order to reduce the CM noise, a new transformer is built with the winding structure shown in Fig. 8. The turns number of P_{P3} is the same as that of P_s. The P_{P3} layer is implemented with copper foil with proper height in order to cover the whole window height. With this structure, the CM noise between P_{p3} and P_s is eliminated. However, there are still CM noise currents from primary to secondary side windings via the core.

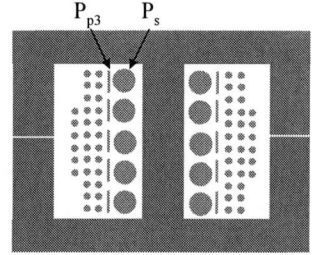

Figure 8. New transformer structure, $N_{p3} = N_s$

Fig. 9 shows the CM noise reduction by using the new transformer structure. CM noise is significantly reduced.

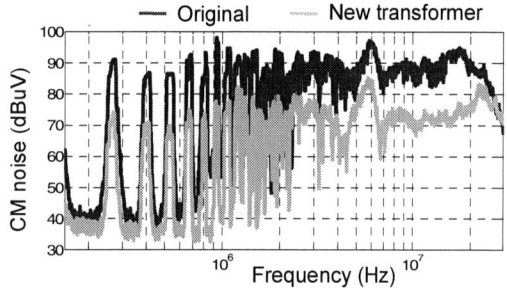

Figure 9. CM noise reduction with new transformer structure

In order to further reduce the CM noise, a transformer with only one turn in the P_{P3} layer is built. A cancellation capacitor

of 22pF is added to achieve best CM noise reduction. Fig. 10 shows the CM noise after compensation. CM noise is further reduced.

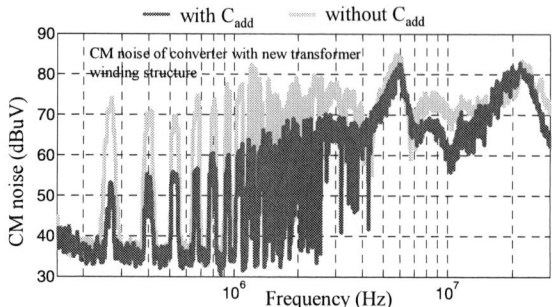

Figure 10. CM noise reduction with compensation

IV. EFFECTS OF LEAKAGE INDUCTANCE

Although the low frequency CM noise of flyback converter is significantly reduced, the high frequency noise reduction is not as satisfactory. The simplified converter and transformer model in Fig. 2 and Fig. 4 cannot predict the noise peaks at high frequencies.

High frequency issues exist in other isolated power converters with noise reduction techniques too. Reference [7] investigated the CM noise in two-switch forward converter and proposed noise reduction techniques. Fig. 11 shows the effects of CM noise reduction. There are also noise peaks at high frequencies.

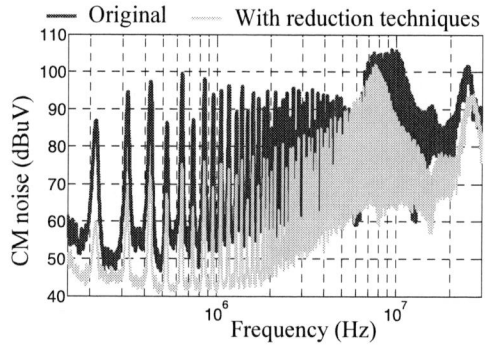

Figure 11. CM noise of two-switch forward converter

These high frequency noise peaks are the results of resonant in the converter. Fig. 12 shows the two-switch forward converter topology with important parasitic inductances and capacitances. L_{k1} and L_{k2} are lumped equivalent leakage inductances of primary and secondary windings. C_{S1} and C_{S2} are junction capacitances of S_1 and S_2. C_{D1}, C_{D2}, C_{D3} and C_{D4} are junction capacitances of D_1, D_2, D_3 and D_4. C_t denotes the inter-winding capacitances.

There are two types of resonant associated with the leakage inductance. One is caused by the inter-winding capacitances. The other is caused by the junction capacitances of the secondary device D_3 and D_4.

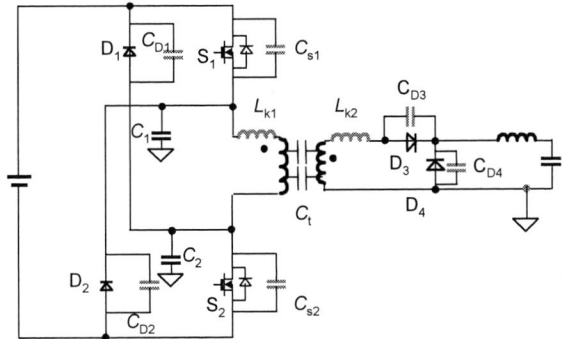

Figure 12. Two-switch forward converter with critical parasitics

A. Resonant with inter-winding capacitances

Fig. 13 shows the CM noise model of two-switch forward converter. The input line and grounding wire inductance is included in the model as L_g. According to Thevenin's theorem, effect of each noise source on CM noise can be represented in an equivalent circuit shown in Fig. 14.

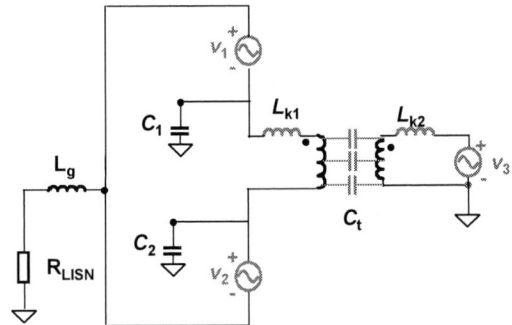

Figure 13. CM noise model of Two-switch forward converter

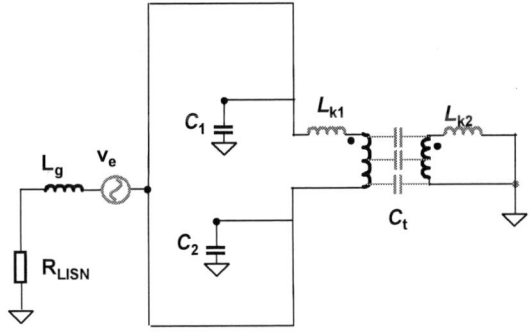

Figure 14. Theorem equivalent circuit

v_e in the equivalent circuit is the equivalent voltage source of the model circuit in Fig. 13. L_g and the impedance of the transformer can be treated as the equivalent source impedances of the circuit. R_{LISN} is the load since its voltage represents the CM noise. In the equivalent circuit, L_g, L_{k1} and L_{k2} resonate with the inter-winding capacitances. At the series resonant frequency the noise source impedance is approximately 0. Noise source v_e applies to R_{LISN} directly and causes a high noise peak.

Fig. 15 shows the transformer structure of the two-switch forward converter prototype. There are four primary winding layers and one secondary winding layer. Every winding layer has its leakage inductance. Inter-winding capacitances are distributed between layer Q_{p2} and Q_s, as well as Q_{p3} and Q_s.

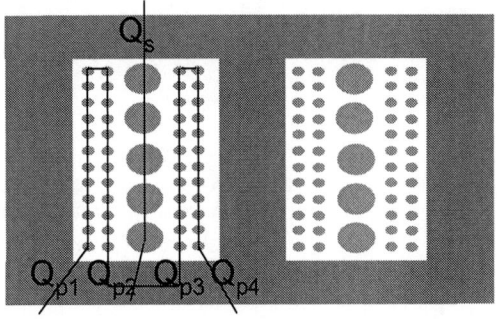

Figure 15. Transformer structure in two-switch forward converter

According to the equivalent circuit in Fig. 14, both primary winding and secondary winding of the transformer are short-circuit. The equivalent impedance network of the source impedance is shown in Fig. 16, where L_{p1}, L_{p2}, L_{p3}, L_{p4} are leakage inductances of primary winding layers. They are in total 15uH in measurement. L_s is the leakage inductance of secondary winding. Its impedance can be ignored. C_{e1}, C_{e2} and C_{e3} are lumped equivalent capacitances between layers. They are in total 118pF.

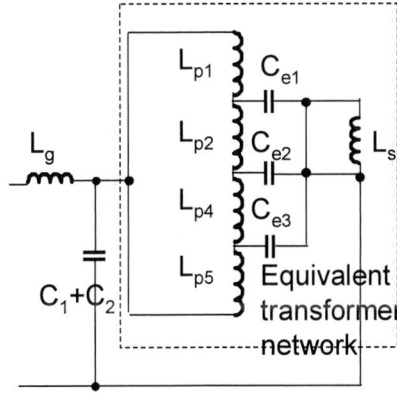

Figure 16. Noise source impedance network

The total noise source impedance versus frequency is shown in Fig. 17. At low frequency the inter-winding capacitances dominates the impedance. At around 7 MHz, there is a valley where the impedance is less than 1 Ohm. Correspondingly CM noise of the converter has a peak at that particular frequency. On the contrary, the impedance peak at around 13 MHz results in a valley in CM noise.

In the two-switch forward converter prototype, L_g is around 0.5uH, much smaller than the leakage inductances. C_1 and C_2 are also much smaller than the inter-winding capacitances. The peaks and valleys of the noise source impedance is mainly determined by the resonant between transformer leakage inductances and inter-winding capacitances.

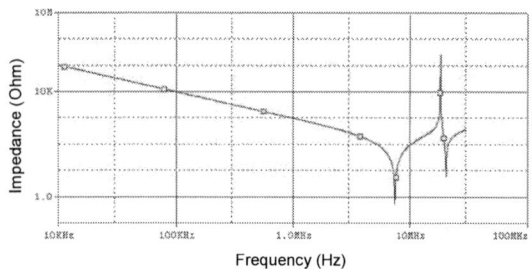

Figure 17. Noise source impedance

B. Resonant with junction capacitances of secondary side devices

During the turn-off transient of S_1 and S_2, secondary side diode D_3 is forced to turn off. The leakage inductances resonate with the junction capacitances of all the devices. It should be noticed that in the calculation of resonant frequency, the primary inductances and capacitances should be reflected to secondary side. As a result, the reflected total leakage inductance is much smaller than its net inductance. This results in relatively high resonant frequency.

The effect of reverse recovery on this resonant will be an interesting topic for further study. Large reverse recovery current results in large initial current in the leakage inductance at the beginning of resonant. It may result in higher magnitude of the resonant peak and causes high CM noise.

Fig. 18 shows the junction voltage on D_3. The resonant frequency is around 23 MHz. This frequency matches the second noise peak in Fig. 11. A general practice for reduction the resonant peak is to add snubbers for the diodes. With snubbers, the resonant peak in time domain and CM noise peak in frequency domain at 23 MHz are both reduced.

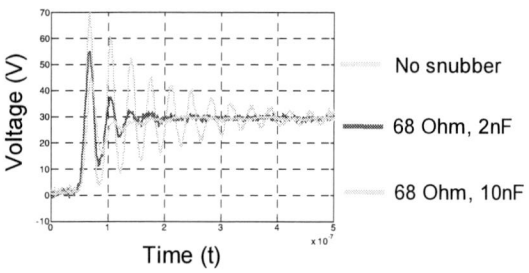

Figure 18. Juction voltage on D_3

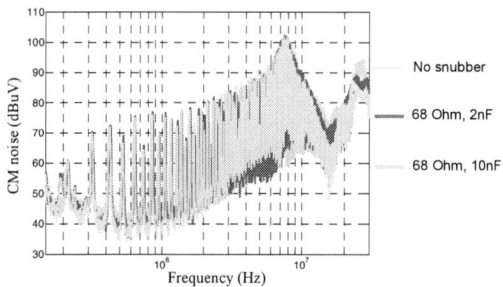

Figure 19. CM noise peak reduction by adding snubbers

C. Controlling the high frequency CM noise peaks

The above-mentioned resonant due to the leakage inductances exists in almost every isolated power converters. By reducing the leakage inductances, it is possible to push the resonant frequencies above 30 MHz so that the conducted CM noise of the converter can be improved. Their effect on radiated EMI is another interesting topic for further study.

V. CONCLUSION

In this paper the effects of power transformer on CM EMI noise of isolated power converters are analyzed. Characteristics of inter-winding capacitances are first summarized. Effects of transformer winding structure are discussed and CM noise reduction by controlling the winding structure is proposed. Effects of transformer leakage inductances on high frequency CM noise are also classified and analyzed. Experimental results validate the analysis and the proposed noise reduction methods.

REFERENCES

[1] R. Lai, Fred Wang, R. Burgos, Y. Pei, D. Boroyevich, B. Wang, T.A. Lipo, V. D. Immanuel and K. J. Karimi, "A systematic topology evaluation methodology for high-density three-Phase PWM AC-AC converters", IEEE Transactions on Power Electronics, vol. 23, Iss. 6, pp. 2665-2680 Nov. 2008.

[2] P. Kong , S. Wang, F. C. Lee. "Modeling and applications of symmetry and balance techniques in reducing common mode EMI of PFC converters". in Proc. 2008 CPES Power Electronics Seminar, pp. 128-135.

[3] S. Lin, M. Zhou, W. Chen and J. Ying, "Novel methods to reduce common-mode noise based on noise balance," in Proc. 2006 IEEE Power Electronics Specialist Conf., pp. 2728 – 2733.

[4] M. Shoyama, L. Ge and T. Ninomiya, "Balanced switching converter to reduce common mode conducted noise," IEEE Trans. Industrial Electronics, 50(6), pp. 1095-1099, Dec. 2003.

[5] M. Borage, S. Tiwari, and S. Kotaiah, "Common-mode noise source and its passive cancellation in full-bridge resonant converter," in proc. 2003 IEEE 8th International Conference on Electromagnetic Interference and Compatibility, pp. 9 – 14.

[6] S. Wang, P. Kong, F. C. Lee. "Common mode noise reduction for boost converters using general balance technique". IEEE Trans. on Power Electronics, 2007, 22(4), pp1410-1416.

[7] P. Kong, S. Wang, F. C. Lee and Z. Wang, "Reducing common mode EMI noise in two-switch forward converter", in Proc. 2009 IEEE Energy Conversion Congress & Expo., pp. 3622 – 3629.

978-1-4244-4782-4/10 $26.00 © 2010 IEEE

A Single-Stage Single-Phase Bi-Directional Grid Interface Circuit with Digital Lookup Table Based Control

Evan Reutzel and Seth Sanders
Department of Electrical Engineering and Computer Science
University of California, Berkeley
Berkeley, CA
emr143@eecs.berkeley.edu, sanders@eecs.berkeley.edu

Abstract—**Emerging load-side commercial and residential building technologies are driving the need for high performance, low cost line-interface systems. This paper reports on the design and test of a single-stage single-phase bi-directional line interface circuit and controller that provides high quality AC line waveforms along with load-line regulation of the DC bus. The circuit is based on a series-parallel connected interleaved flyback topology, with a single integrated coupled magnetic device. A digital controller relies on a lookup table based approach, which utilizes either on-line training or pre-training of the controller so that the internal current waveform loop is effectively run open-loop within a given half-line cycle.**

I. INTRODUCTION

Emerging technological and societal trends such as renewed interest in distributed photovoltaic generation, electric vehicles, consumer electronics with high full-spectrum efficiency (including low power and standby modes), and other "smart grid" applications are driving the need for a new generation of versatile, high performance, low-cost line-interface converters. In order to meet these needs, power interface system architectures for everything from consumer appliances to electric vehicle battery chargers are moving toward standard two-stage architectures [1] [2] [3] [4].

In this type of architecture, the first stage is generally an AC-DC conversion step, which is necessary to achieve an appropriate interface with the grid such as power factor correction (PFC), as required by standards such as IEC/EN 61000-3-2. However, in the age of the "smart grid" this block is likely to encompass more complex functions such as bi-directional power flow and reactive and harmonic power compensation. This stage is usually itself a two-stage process in order to provide a power-ripple-free DC bus. The second stage in the architecture typically carries out a DC side load regulation function. Here the goal might be as simple as providing galvanic isolation, and a stable noise-free DC voltage, or more complex functions such as load-line regulation [5] [6] or the generation of high-frequency AC [1] [3].

In the case of a rectifier, the front-end AC/DC stage is often accomplished with a boost converter for PFC followed by an isolated DC-DC converter to block the 100Hz or 120Hz ripple. An alternative arrangement that has been proposed in [7] is a single AC/DC stage that incorporates a third power port in order to reduce the size of the DC-link capacitance. This arrangement can be viewed as a variation of the two-stage approach where the second stage lies tangential to the main power path rather than in the direct path. However, in order to achieve both PFC and a regulated output, another power stage must be introduced somewhere in the frontend architecture.

Another possible arrangement for the front-end involves a single-stage converter that utilizes the 120 Hz power ripple for a useful purpose rather than attempting to eliminate it. This is the approach taken in the Enphase Microinverter [8] where the power ripple is used to perform the maximum power point tracking (MPPT) function for the attached solar panel. While this is an exciting approach, since it eliminates the need for a second stage load-side regulation step, it is not easily transferrable to other applications. Another potential application that may tolerate some 120 Hz power ripple is in a battery interface architecture where the battery handles some of the ripple power.

The focus of this work is on a single-stage converter and control methodology to meet the needs of the first stage AC-DC conversion block in the two-stage architecture. Rather than attempt to eliminate the power ripple in the frontend stage, it is presumed that this power ripple will be eliminated in a second point-of-load stage of the architecture. A 1kW interleaved flyback converter topology with coupled magnetics is designed and tested to meet this goal. The circuit is controlled using a feed-forward lookup table based method. Simulation and experimental results are presented below.

II. CONVERTER TOPOLOGY

A simplified schematic of the proposed topology is shown in Fig. 1. The topology is based on the two-switch flyback

---

This research was supported by National Semiconductor Corporation.

978-1-4244-4782-4/10 $26.00 © 2010 IEEE

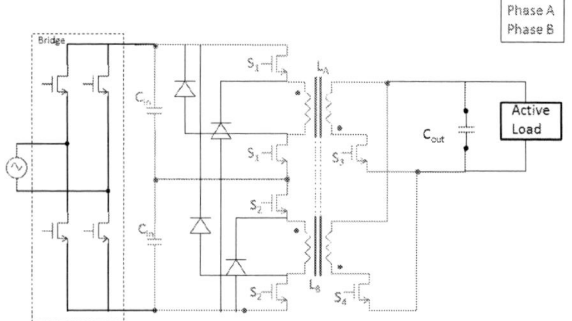

Figure 1. Series/parallel interleaved flyback topology. The two flyback transformers are magnetically coupled.

circuit [9]. Two of these cells are connected in series on the primary and in parallel on the secondary. Much work has been published on active balancing of interleaved power converters [10] [11] [12]. In this circuit, power balance is achieved passively between the two halves, as easily explained here for operation in continuous conduction mode (CCM). In CCM, the voltage and current conversion ratios across a cell are set solely by the duty cycle and the turns ratio. Since the two phases are controlled with identical duty cycles (perhaps phase-shifted) and have matched turns ratios, the parallel secondary connection forces primary side voltage balance, and similarly, the series primary connection forces secondary side current balance. This argument can easily be extended to discontinuous conduction modes.

The two-switch version of the flyback is preferred in this case because the diode clamps (shown in the figure) can be used to protect the primary switches and recover the leakage energy in a lossless manner. Connecting the clamp circuit across the two series-connected phases allows this circuit to achieve DC duty cycles as high as two-thirds, referred to the primary (AC) side without compromising the transformer reset time.

This clamping circuit imposes a fundamental limit on the converter that must be accounted for during the design process. For the flyback topology the output voltage (V_o) to input voltage (V_i) conversion ratio is shown in (1), where D represents the commanded duty cycle. A maximum duty cycle of two-thirds implies that the maximum conversion ratio is two for a unity turns ratio transformer. In the context of AC line rectification, where the AC-side input is varying while the DC output remains relatively constant, this implies that no energy can be transferred when the input is below one half of the (reflected) output voltage. This limit manifests itself as a limited conduction angle during each line cycle, which is important when considering the harmonic content of the current drawn from the grid. Care should be taken to restrict the lower harmonics to below the limits set by the appropriate standard for the given application. Higher harmonics are less of a concern as they can be controlled with appropriate input filtering.

$$V_o/V_i = D/(1-D) \qquad (1)$$

The two circuit cells can be operated with a 180° phase shift between their switching waveforms in order to achieve

partial ripple cancellation. The extent of the cancellation is duty cycle dependent with the most benefit occurring at a duty cycle of one-half. Further, the use of a coupled magnetic structure implemented by winding both transformers on a common core (see Fig. 2), with a shared gap, leads to further ripple cancellation and potential cost savings in the core. Further current ripple cancellation within the transformer occurs because the MMFs appearing across each transformer outer leg are forced to be equal modulo finite core permeability, which ensures magnetizing ripple current balance between the two stages. Additionally, ripple flux cancellation is achieved in the middle leg, enabling further reduced core losses in this portion of the core.

As explained above, the interleaved flyback topology is most efficient when operated at a duty cycle of one-half. However, to achieve PFC the circuit must be operated over a wide range of conversion ratios. The maximum ratio was set by the use of the clamp circuit at two ($D = 2/3$). The minimum ratio occurs at the maximum input voltage (the peak of the line). In order to achieve high efficiency the commanded duty cycles should nominally be balanced about $D =$ one-half. It is assumed that the grid voltage may vary by as much as 20%. Since the minimum input voltage implies both the highest currents (lowest efficiency) and the smallest conduction angle, this operating point is used to design the primary-referred output voltage. The actual output voltage can then be matched to the desired load using the turns ratio of the transformer. Fig. 3 illustrates the desired design point using the nominal single-phase voltages of the United States. Note that in the figure only half the line voltage is used since this is the input voltage seen by each (half) flyback phase of the circuit.

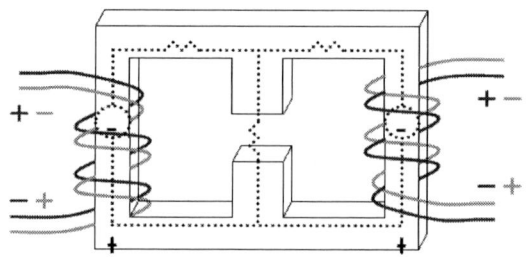

Figure 2. Winding structure of a two phase coupled flyback transformer.

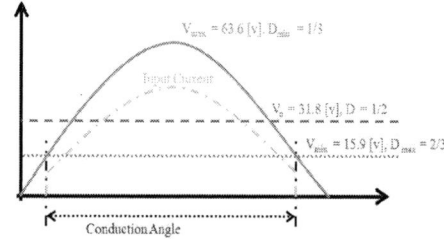

Figure 3. Design of the output voltage referred to the primary. Top curve represents half of the line voltage which is the full voltage seen by each flyback converter. Second curve represents input current waveform. Third curve represents desired output voltage (referred to primary). Fourth curve represents minimum input voltage at which flyback converter will operate due to maximum duty cycle limit.

The topology is capable of bi-directional power flow as well. In order to achieve this, the secondary switches must be implemented with synchronous FETs and the input bridge must be implemented with 60 Hz synchronous switches, as well.

Another consideration is the clamping circuit for the DC-side transistors during inverter operation. Though not shown in Fig. 1, a clamping circuit must be implemented.

The coupled magnetic configuration introduces a more complex set of operating modes if the secondary switches are implemented as diodes or synchronous operated to emulate diodes. In this case, there are two distinct discontinuous conduction modes along with the continuous conduction mode. This multi-mode operating characteristic along with the need to achieve high quality AC waveforms demands that special attention be given to the control scheme. This is the subject of the next section.

III. CONTROL DESIGN

The trend in control research and practice for PFC and other line-connected applications has been toward fully digital implementations [13] [14] [15] [16]. Much of the effort has been on minimizing sensing requirements, data conversion steps, and the number of calculations that must be performed in the current control loop. In this work a digital technique is proposed and developed that relies on lookup tables to minimize current sensing requirements, data conversion steps, and the amount of calculation that needs to be done in real-time. This is similar to the predictive techniques discussed in [17] [18]. These authors attempt to limit the calculations so that they occur only once per half-line cycle. In the technique developed here, the on-line calculations are eliminated all together. Instead, the necessary duty cycles are pre-programmed using a dedicated on-line or off-line training algorithm.

This control technique works as follows. A lookup table holding the duty cycle trajectory of a single half AC line cycle is used to achieve the current waveform control function. This lookup table could, in principle, have the full resolution of the switching frequency of the circuit. In practice, to save space it is only necessary to have as much resolution as needed to effect control. Typically, this is about 20-30 times the fundamental or ~1-2 kHz of resolution. An interpolating algorithm is used to determine the individual duty cycle commands at each switching instance.

The lookup table is indexed off of the line angle and is populated using a separate training algorithm to be discussed shortly. In addition to the PFC function, it is desired that the output voltage be controlled as well. Since this circuit topology involves a single power conversion stage, removing the 120Hz ripple while achieving PFC is impossible. Instead, a point-of-load regulation stage is responsible for insulating the load from the remainder of the double-line-frequency power ripple.

It is useful to be able to control the DC component of the output. It may be that a specific stiff dc bus voltage is desired, or it may be that more complex behavior is desirable such as load line regulation. This is easily achieved with a relatively slow outer control loop. This loop can operate as slow as 120Hz, in which case, synchronous sampling is effective to alias out the 120Hz ripple. If faster performance is desired, the output ripple can be estimated and removed from the output sampling through digital filtering.

In order to achieve the outer control loop functions while maintaining PFC, the lookup table is expanded into two dimensions where the new axis is indexed off of the output power. At each power level, a full half-line duty cycle trajectory is stored that achieves the PFC function on the input while providing the appropriate output voltage and current. A third dimension may be added as well that is indexed off of the line voltage magnitude. This is particularly useful if the circuit is intended for universal line input. The difference between stiff voltage regulation and load-line techniques is simply utilizing a different set of stored trajectories. Movement between trajectories can be achieved with basic PID control techniques.

The training algorithm involves simple feedback of the input current in order to update the trajectory for the next half-line cycle. In this manner the duty cycle path quickly converges to a fixed periodic trajectory. A trajectory must be stored for each load level; however, interpolation can once again be used to minimize the amount of data that must be stored. Once the correct trajectories are stored, the inherent open-loop stability of the circuit ensures that the control will remain stable without sampling of the input current provided the controller is synchronized to the line. Due to robustness against moderate parameter variation, this training operation can occur purely in off-line simulation with the resulting lookup table applied to the actual circuit. Both control algorithms – operational and training – are shown schematically in Fig. 4.

IV. SIMULATION AND EXPERIMENTAL RESULTS

A 1kW converter was designed and tested. The estimated loss budget for the primary components of the design is summarized in Table I. The circuit is designed to operate at a

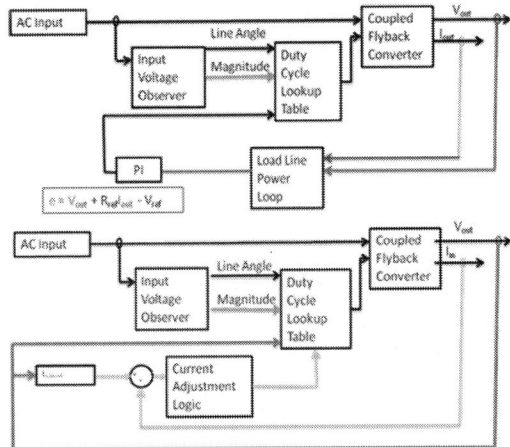

Figure 4. (top) Primary control loop. This control loop uses voltage and/or current feedback to regulate the dc output characteristic. Stiff voltage control or load-line control are possible. (bottom) This control loop implements the training algorithm. The training can be achieved off-line using simulation or one time per nominal design in production.

TABLE I.   ESTIMATED LOSS BUDGET

Component	1 kW Input Power	
	Loss (W)	Loss (%)
Bridge	8.26	0.83
Primary Switches	16.53	1.65
Secondary Switches	12.53	1.25
Magnetics	18.18	1.82
TOTAL	55.50	5.52

125 kHz switching frequency. The bridge and AC side switches are implemented using two parallel CoolMOS IPP60R099 transistors at each switch. The DC side switches are implemented using two parallel Fairchild FDB28N30 MOSFETs.   The coupled flyback transformers were constructed using the Magnetics Inc. core OP49928EC. The windings are constructed using Litz wire with a 14:9 AC to DC side turns ratio.   Simulation of the circuit was accomplished using the PSIM power circuit simulator while the control was co-simulated in Matlab using Simulink. Fig. 5 shows the steady state response after training is completed for an average output voltage of 20 V and an output current of 20

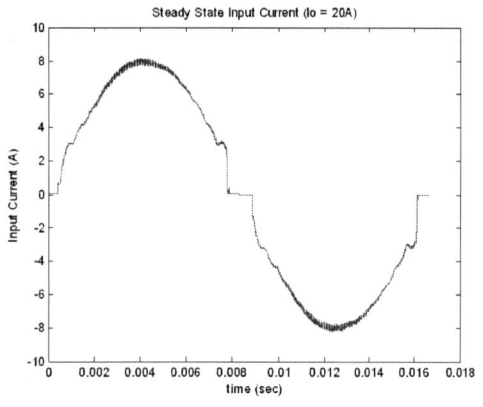

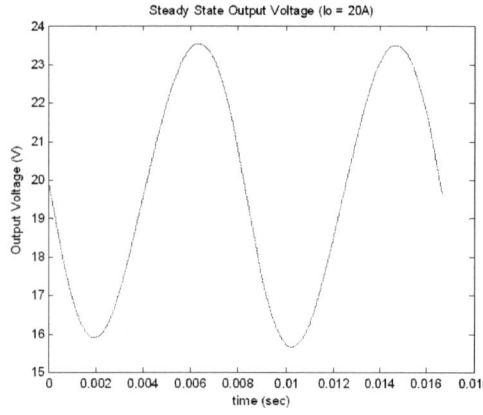

Figure 5. Simulated steady state response for an output current of 20A and an input voltage of 110Vrms. (top) Input Current. (bottom) Output Voltage.

A. In all of the simulation and experimental results reported here the load was assumed to be a constant current type. Fig. 6 and 7 show the transient response (input current and output voltage) of the circuit during a step change in the output load. In this case the output current was suddenly switched from 10 A to 20 A.

A prototype of the circuit has been built and is currently being tested. Fig. 8 shows the steady state operation for an input of 110V$_{rms}$ and an output current of 15 A. The input voltage, input current and output voltage are all shown. Fig. 9 shows the harmonic content of the input current for the case when the output current is 15 A. Fig. 10 shows the duty cycle trajectories for several operating points.

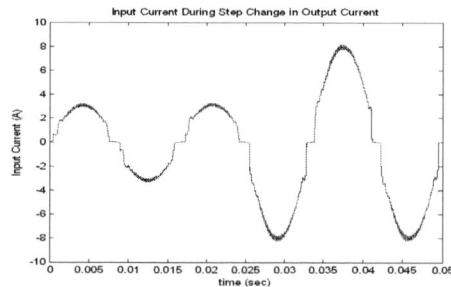

Figure 6. Input current response to a step change in the output current (10A to 20A). The (average) output voltage is constant at 20V.

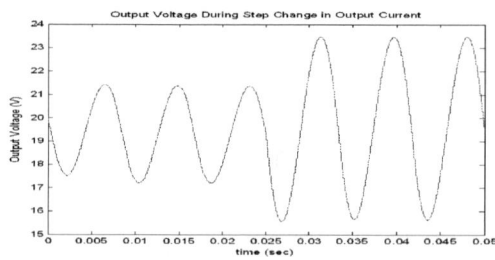

Figure 7. Output voltage response to a step change in the output current (10 A to 20 A).

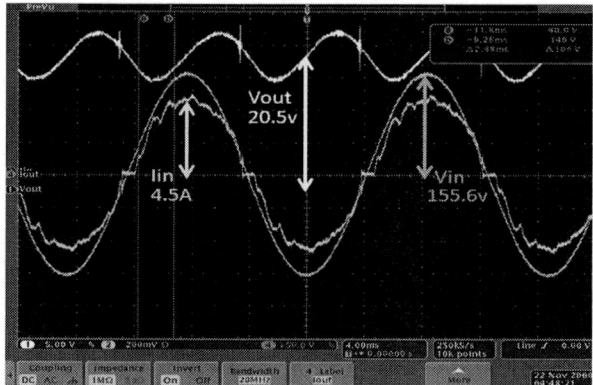

Figure 8. Experimental result showing steady-state operation at for an input voltage 110 Vrms and an output current of 15 A. The top curve shows the output voltage (5V/div). The middle curve shows the input voltage (50V/div). The lowest curve represents the input current (2A/div).

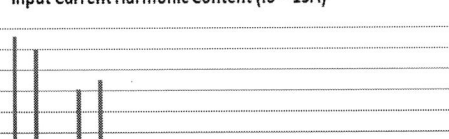

Input Current Harmonic Content (Io = 15A)

Figure 9. Harmonic content as a percentage of the fundamental of the input current during steady state operation of the circuit with a 15 A load. The input voltage is 110 Vrms.

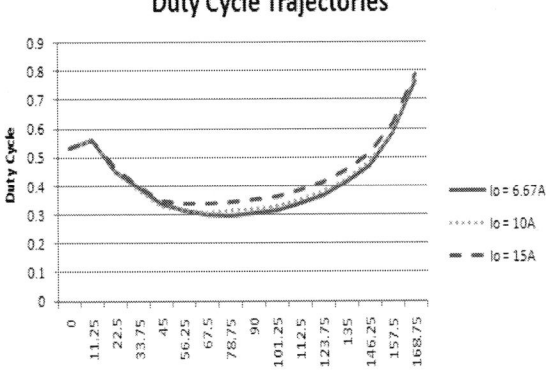

Figure 10. Several steady-state duty cycle trajectories as a function of line angle. Each trajectory has been trained to produce a 20V output at the given output current level. From top to bottom they are 15A, 10A, 6.67A.

V. CONCLUSIONS AND FURTHER WORK

This work demonstrates a circuit topology and its control scheme for single-phase PFC rectification and inversion using the flyback converter as a building block. A control technique based on lookup table "open-loop" PFC with closed loop power control is demonstrated. The lookup table is trained in advance either using simulation tools or another control loop implemented in hardware.

Work is continuing on the experimental platform. In the future we will demonstrate the closed loop control as well as experimentally verify the efficiency of the circuit. Bi-directional power flow will also be demonstrated.

REFERENCES

[1] Yong Li and Toshio Takahashi, "A Digitally Controlled 4-kW Single-Phase Bridgeless PFC Circuit for Air Conditioner Motor Drive Applications," in *Power Electronics and Motion Control Conference 2006*, vol. 1, 2006, pp. 1-5.

[2] Ali Emadi and Sheldon S., Khaligh, Alireza Williamson, "Power Electronics Intensive Solutions for Advanced Electric, Hybrid Electric, and Fuel Cell Vehicular Power Systems," *Power Electronics, IEEE Transactions on*, vol. 21, no. 3, pp. 567-577, May 2006.

[3] Yuki Kawaguchi, Toshihiko Tanaka, Mutsuo Nakaoka, Atsushi Fujita, and Hideki Omori, "Feasible Evaluation of a Full-Bridge Inverter for Induction Heating Cooking Appliances with Discontinuous Current Mode PFC Control," in *Power Electronics Specialists Conference*, 2008, pp. 2948-2953.

[4] Francisco J. Azcondo, Christian Branas, Rosario Casnueva, and Salvador Bracho, "Power-Mode-Controlled Power-Factor Corrector for Electronic Ballast," *Industrial Electronics, IEEE Transactions on*, vol. 52, no. 1, pp. 56-65, February 2005.

[5] A. V. Peterchev and S. R. Sanders, "Load-Line Regulation With Estimated Load-Current Feedforward: Application to Microprocessor Voltage Regulators," *Power Electronics, IEEE Transactions on*, vol. 21, no. 6, pp. 1704-1717, November 2006.

[6] G Eirea and S. R. Sanders, "High Precision Load Current Sensing Using On-Line Calibration of Trace Resistance," *Power Electronics, IEEE Transactions on*, vol. 23, no. 2, pp. 907-914, March 2008.

[7] P. T. Krein and R. S. Balog, "Cost-Effective Hundred-Year Life for Single-Phase Inverters and Rectifiers in Solar and LED Lighting Applications Based on Minimum Capacitance Requirements and a Ripple Power Port," in *Applied Power Electronics Conference and Exposition 2009*, 2009, pp. 620-625.

[8] Fornage and Martin, "Method and apparatus for maximum power point tracking in power conversion based on dual feedback loops and power ripples ," 20090080226, March 26, 2009.

[9] Robert W. Erickson and Dragan Maksimovic, *Fundamentals of Power Electronics*, 2nd ed. New York, NY, USA: Springer Science and Business Media, 2001.

[10] K. Siri and M. A. Willhoff, "Uniform Current/Voltage-Sharing for Interconnected DC-DC Converters," in *Aerospace Conference, 2007 IEEE*, 2007, pp. 1-17.

[11] K. Siri, K. A. Conner, and C. H. Truong, "Uniform Voltage Distribution Control for Paralleled-Input, Series-Output Connected Converters," in *Aerospace Conference, 2005 IEEE*, 2005, pp. 1-11.

[12] Raja Ayyanar, Ramesh Giri, and Ned Mohan, "Active Input-Voltage and Load-Current Sharing in Input-Series and Output-Parallel Connected Modular DC-DC Converters Using Dynamic Input-Voltage Reference Scheme," *Power Electronics, IEEE Transactions on*, vol. 19, no. 6, pp. 1462-1473, November 2004.

[13] Wanfeng Zhang, Guang Feng, Yan-Fei Liu, and Bin Wu, "New Digital Control Method for Power Factor Correction," *Industrial Electronics, IEEE Transactions on*, vol. 53, no. 3, pp. 987-990, June 2006.

[14] Wanfeng Zhang, Guang Feng, and Yan-Fei Liu, "A Direct Duty Cycle Calculation Algorithm for Digital Power Factor Correction (PFC) Implementation," in *Power Electronics Specialists Conference*, vol. 3, 2004, pp. 2326-2332.

[15] Wanfeng Zhang, Yan-Fei Liu, and Bin Wu, "A New Duty Cycle Control Strategy for Power Factor Correction and FPGA Implementation," *Power Electronics, IEEE Transactions on*, vol. 21, no. 6, pp. 1745-1753, November 2006.

[16] Xin Chen, Chun Ying Gong, and HuiZheng Wang, "The Full Digital Control based Switched Mode Power Supply," in *Computers in Power Electronics, IEEE Workshops on*, 2006, pp. 250-254.

[17] C. Attaianese, V. Nardi, F. Parillo, and G. Tomasso, "Predictive Control of Parallel Boost Converters," in *Industrial Electronics, 34th Annual Conference of IEEE*, 2008, pp. 892-898.

[18] Wanfeng Zhang, Guang Feng, Yan-Fei Liu, and Bin Wu, "DSP Implementation of Predictive Control Strategy for Power Factor Correction (PFC)," in *Applied Power Electronics Conference and Exposition*, vol. 1, 2004, pp. 67-73.

A High Performance Dual Output DC-DC Converter Combined the Phase Shift Full Bridge and LLC Resonant Half Bridge with the Shared Lagging Leg

Yu Chen, Xuejun Pei, Li Peng, Yong Kang
College of Electrical and Electronic Engineering
Huazhong University of Science and Technology
Wuhan, China
Email: ayu03@163.com

Abstract—a new dual output dc-dc converter is proposed in this paper. The first output works as a conventional full bridge converter; by connecting a LLC resonant tank to the lagging leg, a half bridge resonant converter is structured, and the second output is obtained. The first output is regulated by the phase shift and the second output by the switching frequency. Both of the outputs are fully regulated and independent. The narrow frequency variation of the LLC part makes the magnetic components easy to optimize. The ZVS of the leading leg is easy to achieve by the phase shift operation, the ZVS range of the lagging leg is extended down to empty load by the shared structure. Moreover, compared to the conventional full bridge phase shift converter, the shared lagging leg offers a loose dead-time limitation. Therefore the proposed converter is low cost, high reliability, high efficiency and high performance. A 300V input, 36V/10A, 24V/10A outputs prototype is build to verify the design.

I. INTRODUCTION

Dual output converters are widely used in many applications. For an ideal dual output converter, following requirements should be satisfied: (1) accurate voltage regulation; (2) uncoupling between the outputs; (3) high power density and efficiency; (4) simply control method and (5) simple topology for compact size and low cost.

As a special case of multiple-output converter, dual output converter can be constructed using many methods, such as post regulation [1, 2], PWM-PD (phase delay) [3, 4] and PWM-PS (phase shift) modulation [5, 6]. PWM-FM (frequency modulation) or PS-FM is another way for dual output application. A single-switch two-output converter using PWM-FM method is presented in [7]. A full bridge dual output converter using PS-FM method is presented in [8]. However, none of them meets all the five requirements above.

In order to satisfy all the five requirements above, a new dual output converter based on PS-FM is presented in this

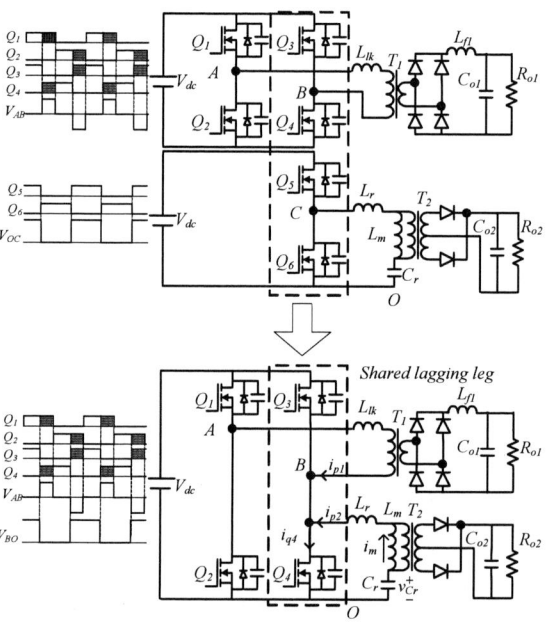

Fig. 1 Derivation of the proposed topology

paper. As shown in Fig. 1, a conventional phase shift full bridge (PSFB) converter is merged with a half bridge LLC (HBLLC) resonant converter by sharing the lagging leg. Since the PS and FM operation have the same 180° complementary gating pulse, the standard PS modulator is still available. Both the PSFB part and HBLLC part have full regulation ability. And, as will be discussed in this paper, the cross-regulation problem between the outputs is overcome naturally. The narrow frequency variation of the LLC part makes the magnetic components easy to optimize. Moreover, by sharing the lagging leg, the ZVS range of the lagging leg is extended to empty load without any auxiliary circuit, the proposed converter can work in high switching frequency and high efficiency.

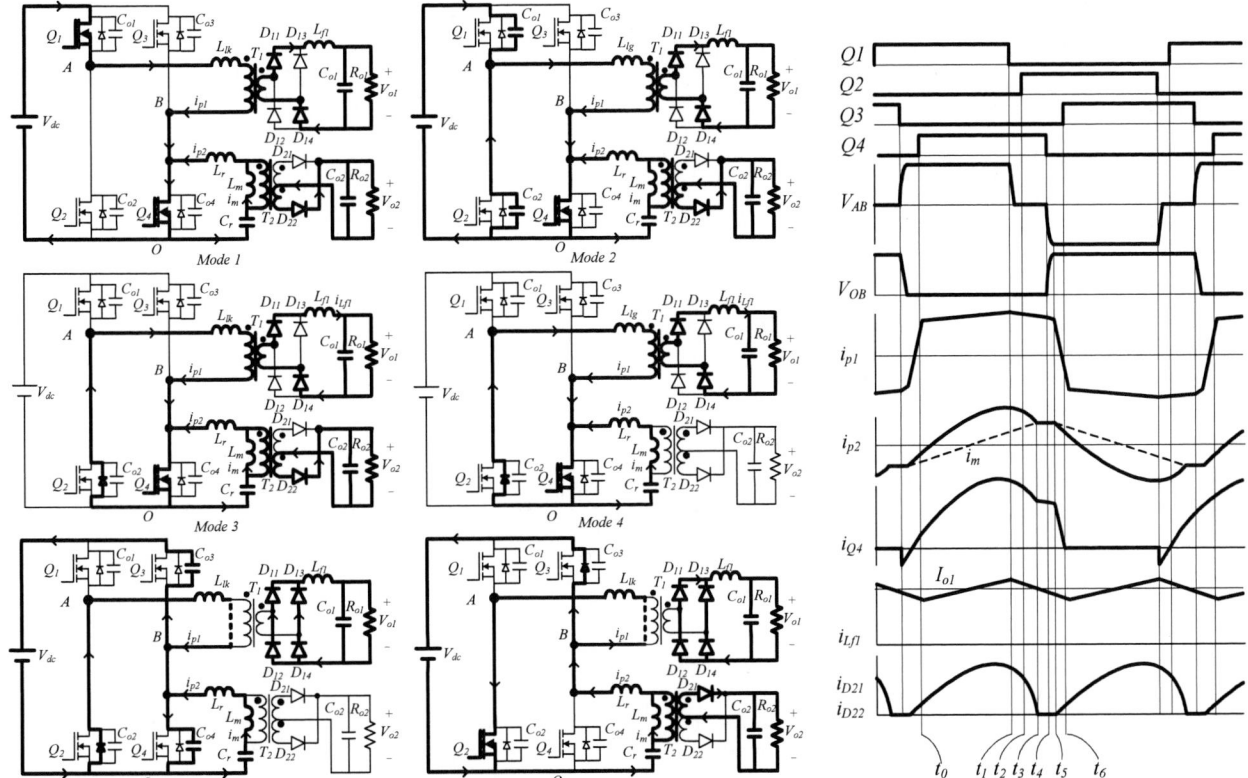

Fig. 2 The operational principle

Fig. 3 The key waveforms

II. OPERATIONAL PRINCIPLE OF THE PROPOSED CONVERTER

Fig. 2 shows the normal operational modes of the proposed converter. Before the analysis, some assumptions are made first: (a) all switches, diodes and magnetic components are ideal; (b) C_{o1}-C_{o4} are the equal ZVS capacitances, including the parasitic capacitor of the switches and the additional capacitors; (c) the two output voltages are constant. There are six modes in half of one switching period. The analysis begins at t_0, when Q_1 and Q_4 are turned on.

Mode 1(t_0-t_1): Q_1 and Q_4 are turned on. In the primary side of the PSFB part, current $i_{p1}(t)$ flows through Q_1 and Q_4, in its secondary side, D_{11} and D_{14} conduct, the PSFB part is in the powering mode:

$$V_{dc} - L_{lk}\frac{di_{p1}(t)}{dt} = K_1 L_{f1}\frac{di_{Lf1}(t)}{dt} + K_1 V_{o1} \qquad (1)$$

where $L_{lk} = L_{lkp1} + K_1^2 L_{lks1}$ is the equal leakage inductor of the PSFB part.

In the HBLLC part, the primary current $i_{p2}(t)$ flows through Q_4, in its secondary side, D_{22} conducts, the HBLLC part is also in the powering mode:

$$v_{Cr}(t) - K_2 V_{o2} = L_r \frac{di_{p2}(t)}{dt}$$
$$i_{p2} = -C_r \frac{dv_{Cr}(t)}{dt}, \qquad K_2 V_{o2} = L_m \frac{di_m(t)}{dt} \qquad (2)$$

The resonant period of the HBLLC part is

$$T_{r1} = 2\pi\sqrt{C_r L_r} . \qquad (3)$$

where $L_r = L_{r'} + L_{lkp2} + K_2^2 L_{lks2}$ is the equal resonant inductor, including the additional inductor $L_{r'}$, the primary leakage inductor L_{lkp2} and the secondary leakage inductor L_{lks2}.

Mode 2(t_1-t_2): Q_1 is turned off at t_1. In the PSFB part, $i_{p1}(t)$ begins to charge and discharge the ZVS capacitors C_{o1} and C_{o2}. Since the diodes D_{11} and D_{14} still conduct, L_{lk} and $K_1^2 L_{f1}$ are in series. $i_{p1}(t)$ is large enough to be treated as a constant current source I_{o1}/K_1. The state equation of the leading leg thus can be expressed as:

$$\frac{I_{o1}}{K_1} = 2C_{lead}\frac{dv_{DS1}(t)}{dt} \qquad (4)$$

where $v_{DS1}(t)$ is the drain-source voltage of Q_1, and $C_{o1}=C_{o2}=C_{lead}$ is the ZVS capacitors of the leading leg. The ZVS condition of the leading leg is the same as the conventional PSFB converter, it is easy to achieve:

$$I_{o1\_zvs} = 2C_{lead}K_1 V_{dc}/T_{lead} \qquad (5)$$

where T_{lead} is the dead-time of the leading leg and I_{o1_zvs} is the load current with which the leading leg achieves ZVS critically.

In the HBLLC part, since Q_4 is still turned on, it keeps the state as in mode 1.

978-1-4244-4782-4/10 $26.00 © 2010 IEEE

Mode 3(t_2-t_3): After the drain-source voltage of Q_2 decreases to zero, Q_2 can be turned on in ZVS at t_2. In the PSFB part, D_{11} and D_{14} still conduct, the freewheeling mode begins, and the state equation of the PSFB part can be expressed as:

$$-L_{lk}\frac{di_{p1}(t)}{dt} = K_1 L_{f1}\frac{di_{Lf1}(t)}{dt} + K_1 V_{o1} \qquad (6)$$

And, since Q_4 is still turned on, the state in the HBLLC part doesn't change.

Mode 4(t_3-t_4): In this mode, Q_4 is still turned on; the freewheeling mode is persisted in the PSFB part. In the HBLLC part, the state changes because the primary current $i_{p2}(t)$ equal to the magnetizing current $i_m(t)$ at t_3 and D_{22} blocks naturally, the state equations thus become

$$v_{Cr}(t) = (L_r + L_m)\frac{di_{p2}(t)}{dt}$$
$$i_{p2}(t) = -C_r\frac{dv_{Cr}(t)}{dt}, \qquad i_{p2}(t) = i_m(t) \qquad (7)$$

In this mode, the resonant period of the HBLLC part becomes

$$T_{r2} = 2\pi\sqrt{C_r(L_r + L_m)}. \qquad (8)$$

It is emphasized that the state change of the HBLLC part may happen in anytime during mode 1 to mode 3, depended on the load conditions.

Mode 5(t_4-t_5): at t_4, Q_4 is turned off. Since Q_4 is the common component shared by the PSFB and HBLLC parts, the ZVS process in the lagging leg shows the new features. The ZVS capacitors are charged and discharged by $i_{p1}(t)$ and $i_{p2}(t)$ at the same time:

$$2C_{lag}\frac{dv_{DS4}(t)}{dt} = i_{p1}(t) + i_{p2}(t) \qquad (9)$$

where $v_{DS4}(t)$ is the drain-source voltage of Q_4, and $C_{o3}=C_{o4}=C_{lag}$ is the ZVS capacitors of the lagging leg.

In the PSFB part, the $i_{p1}(t)$ is deficient to feed the load, all the diodes D_{11}-D_{14} conduct to commutate the currents. The state equation of the PSFB part is:

$$v_{DS4}(t) + L_{lk}\frac{di_{p1}(t)}{dt} = 0 \qquad (10)$$

In the HBLLC part, the relations become:

$$v_{Cr}(t) - v_{DS4}(t) = (L_r + L_m)\frac{di_{p2}(t)}{dt}$$
$$i_{p2}(t) = -C_r\frac{dv_{Cr}(t)}{dt}, \qquad i_{p2}(t) = i_m(t) \qquad (11)$$

where $C_{o1}=C_{o2}=C_{lag}$ is the equal ZVS capacitor of the lagging leg and $v_{DS4}(t)$ is the drain-source voltage of Q_4.

The equations (9)-(11) make up of the fourth order system, its equivalent circuit is shown in Fig. 4 (a). It is difficult to solve. However, noticing that $C_r \gg 2C_{lag}$ and $L_r + L_m \gg L_{lk}$, the HBLLC part can be treated as a constant current source $I_{p2}(t_4)$ during this short period, the equivalent circuit thus can be simplified as Fig. 4(b). The solutions are:

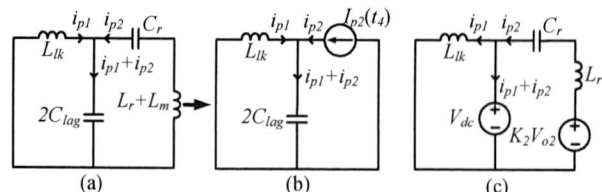

Fig. 4 The equivalent circuit during mode 5 and mode 6

$$i_{p1}(t) = [I_{p1}(t_4) + I_{p2}(t_4)]\cos\omega_{r3}t - I_{p2}(t_4)$$
$$v_{DS4}(t) = [I_{p1}(t_4) + I_{p2}(t_4)]\sqrt{L_{lk}/2C_{lag}}\sin\omega_{r3}t \qquad (12)$$

where $\omega_{r3} = \sqrt{2C_{lag}L_{lk}}$ is the resonant frequency during this period, $I_{p1}(t_4)$ and $I_{p2}(t_4)$ is the value of $i_{p1}(t)$ and $i_{p2}(t)$ at the moment of t_4.

Mode 6(t_5-t_6): at t_5, $v_{DS4}(t)$ has reached V_{dc}. In the PSFB part, since the commutation of the currents flowing through D_{11}-D_{14} hasn't finished yet, V_{dc} impresses on L_{lk} directly and $i_{p1}(t)$ decreased linearly and rapidly. The state equation becomes

$$V_{dc} + L_{lk}\frac{di_{p1}(t)}{dt} = 0. \qquad (13)$$

In the HBLLC part, D_{21} conducts, L_r begins to resonate with C_r. The state equations thus can be expressed as:

$$V_{dc} - v_{Cr}(t) - K_2 V_{o2} = -L_r\frac{di_{p2}(t)}{dt}$$
$$i_{p2} = -C_r\frac{dv_{Cr}(t)}{dt}, \qquad K_2 V_{o2} = -L_m\frac{di_m(t)}{dt} \qquad (14)$$

The resonant period changes from T_{r2} to T_{r1}. The equivalent circuit in this mode is shown in Fig.4 (c). During this mode, $i_{p1}(t)$ decreases to zero and goes to negative. However, the actual current $i_{p1}(t)+i_{p2}(t)$ doesn't change its polarity because $i_{p2}(t)$ is still positive. $v_{DS4}(t)$ is hold in V_{dc} until t_6, when $i_{p1}(t)=-i_{p2}(t)$ and the actual current $i_{p1}(t)+i_{p2}(t)$ equals to zero. Q_3 can be turned on in ZVS in anytime before t_6. Later, D_{11}, D_{14} block, D_{12}, D_{13} conduct and $-i_{p1}(t)=i_{Lf1}(t)/K_1$. Another symmetrical half switching period begin.

According to the analysis, several consequences are summarized here: **(a)** during mode 1 to mode 4, the common switch Q_4 doesn't change its state, the PSFB part and the HBLLC part are independent of each other; **(b)** when Q_4 is turned off in mode 5, both the two parts contribute to the ZVS condition of the lagging leg, compared to the conventional PSFB converter, the ZVS of lagging leg is achieved easily; **(c)** the ZVS state is persisted during mode 6 even when $i_{p1}(t)$ changes its polarity. Compared to the conventional PSFB converter, the proposed converter has a much looser dead-time limitation.

III. ZVS RANGE EXTENDING OF THE LAGGING LAG

The ZVS condition of the conventional PSFB converter is rewritten here:

$$L_{lk}(I_{o1}/K_1)^2 \geq 2C_{lag}V_{dc}^2 \qquad (15)$$

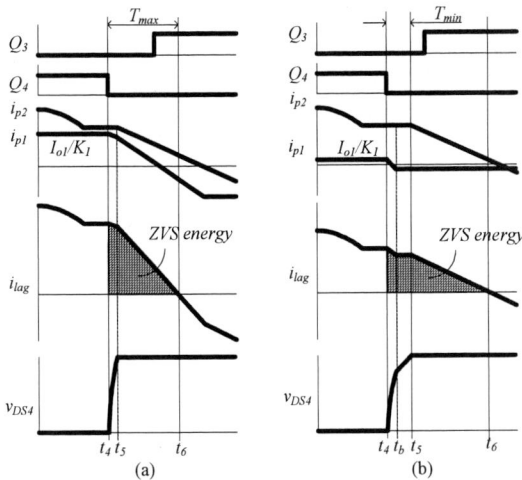

Fig. 5 The ZVS waveforms of the lagging leg

In the conventional PSFB converter, L_{lk} is usually small so the ZVS energy is deficient. The problem is solved by sharing the lagging leg with the HBLLC part in the proposed converter. The ZVS of the lagging leg is achieved in different ways, depending on the load condition of the PSFB part.

A. The PSFB part is in heavy load to medium load

When the PSFB part is in heavy to medium load, the commutation of the currents flowing through D_{11}-D_{14} won't be completed during the whole ZVS interval. Equation (12) is always true. The ideal ZVS process can be expressed as Fig. 5(a). Once Q_4 is turned off at t_4, $v_{DS4}(t)$ increases to V_{dc} sinusoidally and rapidly, as described in (12). Ignore the output current ripples, $I_{p1}(t_4) = I_{o1}/K_1$. The ZVS condition can be expressed as:

$$L_{lk}\left[I_{o1}/K_1 + I_{p2}(t_4)\right]^2 \geq 2C_{lag}V_{dc}^2 \quad (16)$$

It's seen that $I_{p2}(t_4)$ contributes to the ZVS condition a lot, the ZVS energy is always enough even when L_{lk} and I_{o1} are small.

B. The PSFB part is in light load to empty load

When the PSFB part is in light load to empty load, the commutation of the current flowing through D_{11}-D_{14} will have completed before $v_{DS4}(t)$ is fully charged. In this situation, the output inductor L_f is series with the primary side again, and the primary current of the PSFB part is hold at $-I_{o1}/K_1$. Since $v_{DS4}(t)$ hasn't reached V_{dc} yet, it will be charged by the constant current source $I_{p2}(t_4) - I_{o1}/K_1$ continuously, as shown in Fig. 5(b). Especially, if the PSFB part is empty load, the commutation period is zero and the ZVS capacitors are always charged by the constant current source $I_{p2}(t_4)$. It means all the ZVS energy is provided by the HBLLC part. In this extreme situation, the ZVS condition can be expressed as:

$$\frac{I_{p2}(t_4)T_{lag}}{2C_{lag}} \geq V_{dc} \quad (17)$$

where T_{lag} is the dead time of the lagging lag.

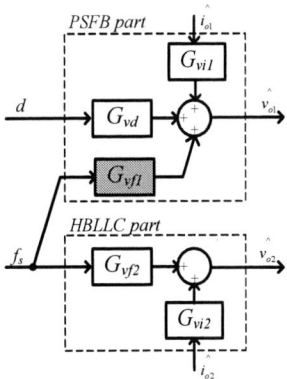

Fig. 6 The Small signal model of the proposed converter

C. The full ZVS conditions

$I_{p2}(t_4)$ plays an important role in the ZVS process of the lagging leg. Since the circulate current always exists in the resonant converter, $I_{p2}(t_4)$ is usually enough. To achieve the full range ZVS in the lagging leg, considering the extreme situation that $I_{o1}=0$ in both (16) and (17), the full ZVS condition thus can be obtained:

$$\begin{cases} I_{p2}(t_4) \geq \sqrt{2C_{lag}V_{dc}^2/L_{lk}} \\ T_{lag} \geq 2V_{dc}C_{lag}/I_{p2}(t_4) \end{cases} \quad (18)$$

In fact, a transient process exists when I_{o1} changes from full load to empty load. However, if (18) is satisfied, the ZVS will be always achieved because the extreme situation has been considered.

IV. OUTPUT VOLTAGES INTERACTION

Since the phase shift operation doesn't change the lagging leg's gating pulse width, HBLLC part won' be affected. The HBLLC part is independent of the PSFB part completely. However, frequency variation will affect the PSFB part. Based on [9], the small signal model of the proposed converter with the frequency variation consideration can be expressed as Fig. 6. It is clear that the interaction is brought in by G_{vf1}, the frequency-to-output transfer function. The coefficient of G_{vf1} can be expressed as:

$$k_{vf1} = \frac{4L_{lk}I_o}{V_{dc}K_1^2} \quad (19)$$

It is possible to design the controller to compensate the effect of G_{vf1} completely, but it will result in a complex control scheme. In fact, the effect of frequency variation can be minimized by the appropriate power stage design.

To minimize the interaction, the simplest way is to minimize the coefficient k_{vf1}. For example, we can minimize the leakage inductor L_{lk} since it is no longer the key factor of the ZVS condition.

Moreover, we can minimize the frequency variation of the HBLLC part; this can be realized by designing the resonant parameters of the HBLLC part appropriately. Fig. 7(a) shows the normal design principle of the conventional HBLLC converter. To keep the rectifier of the converter

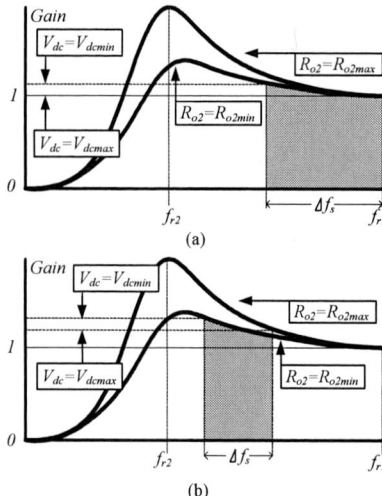

Fig. 7 The operational region design of the HBLLC part

always works in ZCS mode, the switching frequency should be below the resonant frequency f_{r1}. To realize this, traditionally, when V_{dc} is maximal and the output is empty load (i.e., $V_{dc}=V_{dcmax}$ and $R_{o2}=R_{o2max}$), the switching frequency f_s is configured at f_{r1}, where the dc gain is unit. With this implementation, when V_{dc} is low and output is heavy load (i.e., $V_{dc}=V_{dcmin}$ and $R_{o2}=R_{o2min}$), the switching frequency could decrease to increase the dc gain, so as to keep the output regulated. However, this design results in a related wide frequency variation in the entire load range, as the shaded region of Fig. 7(a). Noticing the dc characteristics curves located far away from f_{r1} has a much steeper slope than near f_{r1}, it is better to reset the frequency range as Fig. 7(b) so as to minimize the frequency variation range. With this implementation, when V_{dc} and R_{o2} change, the switching frequency only needs to change slight because the slope of the curve is much steeper than Fig. 7(a).

V. PROTOTYPE DESIGN

A. The prototype design

A prototype with 280-320V input, 36V/10A of the PSFB part and 24V/10A of the HBLLC part is build. The main parameters are listed in Table I. The ZVS parameters are also given. In the leading leg, the ZVS range can be calculated according to (5), the ZVS will be still achieved when $I_{o1}=1.2A$. To determine the ZVS parameters of the lagging leg, the parameter $I_{p2}(t_4)$ in (18) should be known first. It is possible to compute $I_{p2}(t_4)$ based on the circuit parameters, but it leads to a complex calculation. We propose a simple approach here: we let the HBLLC part to work solely since the HBLLC part is independent, the minimal $I_{p2}(t_4)$ thus can be measured as 1.4 A. The ZVS capacitors of the lagging leg C_{lag} are chosen as 300pF first, and the first condition of (18) can be verified as:

$$I_{p2}(t_4)=1.4A > \sqrt{2C_{lag}V_{dc}^2/L_{lk}}=0.94A$$

It is seen that the ZVS condition is very easy to satisfy. Then the dead-time T_{lag} limitation can be calculated according to the second condition of (18):

TABLE I. PROTOTYPE PARAMETERS

PSFB part		RHB part		Other parameters	
L_{lkp1}	38uH	L_m	316uH	MOSFET	IRFP450
L_{lks1}	0.8uH	L_{lkp2}	18uH	D_{11}-D_{14}	MBR2060
K_1	6.25	L_{lks2}	0.7uH	D_{21}-D_{22}	MBR2060
L_f	60uH	L_r	30uH	C_{o1}-C_{o2}	180pF
C_{o1}	470uF	K_2	8.6	C_{o3}-C_{o4}	300pF
		C_r	22nF	T_{lead}	600ns
		C_{o2}	4mF	T_{lag}	350ns

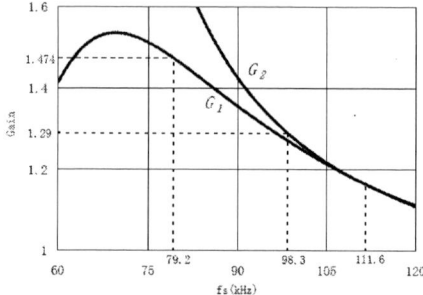

Fig. 8 The DC gain characteristic of the HBLLC part

$$T_{lag} \geq 2V_{dc}C_{lag}/I_{p2}(t_4)=137ns$$

Considering the tolerant error of measurement and the security of leg, $T_{lag}=350ns$ is chosen finally.

The DC gain characteristics of the HBLLC part are shown in Fig. 8. G_1 is the DC gain curve under the full load condition, and G_2 is the DC gain curve under the empty load condition. With the turning ratio $K_2=8.6$, the DC gain should be 1.29 when $V_{dc}=320V$, and 1.47 when $V_{dc}=280V$. The maximal and minimal switching frequencies are 98.3kHz and 79.2kHz. Specially, in the normal input voltage condition $V_{dc}=300V$, a DC gain 1.376 is required, the frequency will change from 88.1kHz to 92.6kHz when the output of the HBLLC part change from full load to empty load. Only a frequency variation of 4.5kHz is required, the effect of the frequency variation on the PSFB part is minimized.

The conventional phase shift controller UC3875 is used to control the proposed converter. Two independent control loops are design. The output voltage of the PSFB part is fed back to control the phase shift and the output voltage of the HBLLC part is fed back to control the frequency. By replacing the oscillated resister with a transistor, a voltage controlled oscillator is obtained and the FM is realized.

VI. EXPERIMENTAL RESULTS

A. Experimantal results

The steady waveforms of the primary side are shown in Fig. 9. Fig. (a) and (b) correspond to the situations when both the PSFB and HBLLC parts are full loads ($I_{o1}=10A$, $I_{o2}=10A$) and light loads ($I_{o1}=0.5A$, $I_{o2}=0.5A$). V_{ds4} and V_{ds2} are the drain-source voltages of S_4 and S_2. i_{p1} and i_{p2} are the primary currents of the PSFB and RHB parts. It's seen clearly that the two parts are independent of each other and work as the conventional PSFB and HBLLC converters respectively, as discussed in section II.

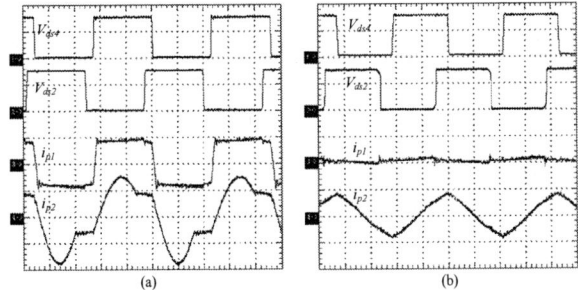

Fig. 9 The steady state waveforms
(V_{ds4}, V_{ds2}:200V/div, i_{p1}, i_{p2}:2A/div, t:2.5us/div)

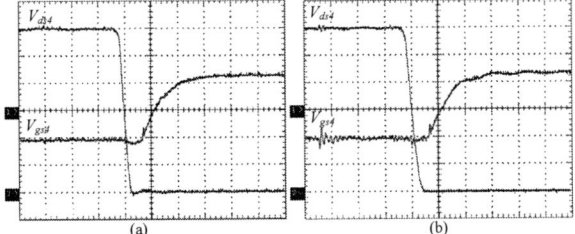

Fig. 10 The ZVS performance
(V_{ds4}: 50V/div V_{gs4}:10V/div, t:250ns/div)

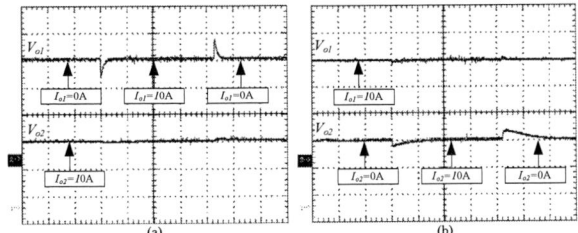

Fig. 11 The transient performance
(V_{o1}, V_{o2}:10V/div; t:25ms/div)

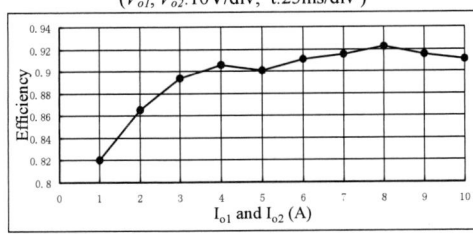

Fig. 12 The efficiency

The ZVS performances of the lagging leg are shown in Fig. 10. V_{gs4} and V_{ds4} are the gating pulse voltage and the drain-source voltage of S_4. The HBLLC part keeps working in empty load. In Fig. 8(a), the PSFB part is full load (I_{o1}=10A). It is seen that the ZVS of S_4 is achieved sinusoidally and rapidly, because the secondary side of the PSFB part is short and the ZVS energy comes from both the PSFB and HBLLC part. Fig. 8(b) shows the situation when the PSFB part is empty load, V_{ds4} decreases linearly and slowly, because the secondary side of the PSFB part is open and the ZVS energy only comes from the HBLLC part.

Fig. 11 shows the output voltage waveforms. Fig. 9(a) shows that load switching in PSFB part (0A-10A-0A) has no effect on the HBLLC part, because the phase shift operation is independent of the frequency variation. In Fig. 9(b), the load switching in HBLLC part (0A-10A-0A) has a very slight effect on the PSFB part, because the frequency variation affects the effective duty cycle of the PSFB part slightly. However, the effect has been minimized by the narrow frequency variation feature of the LLC resonant tank. Moreover, it is seen that one output is fully regulated even when the other is in the extreme load situation. Both the two outputs in the proposed converter are fully regulated.

Fig. 12 shows the efficiency performance. The efficiency is measured when the PSFB and HBLLC parts have the same output currents. The maximal measured efficiency appears when both I_{o1} and I_{o2} are 8A, the total efficiency is 92.2%. It's seen the efficiency in the entire load is satisfactory; it is mainly benefited by the full ZVS characteristic and the compact shared leg structure.

VII. CONCLUSIONS AND FUTURE WORK

The proposed dual output converter has a fully regulation ability; the shared leg structure results to a simply topology; both the advantages of PSFB and HBLLC are included in the proposed converter; the ZVS range of lagging leg is extended to empty load without any auxiliary circuit; the two outputs have little interaction and can be regulated independently; the narrow frequency variation characteristic of the LLC tank makes the magnetic components easy to design. Therefore the proposed dual output converter is simply topology, compact size, low cost, high power density, high efficiency and high performance.

REFERENCES

[1] S. Havanur, "Combining synchronous rectification and post regulation for multiple isolated outputs," in *Proc. IEEE Appl. Power Electron. Conf. (APEC)*, 2004, pp. 872-877.

[2] C.C. Wen, C.L. Chen, w. Chen, and J. Jiang, "Magamp post regulation for flyback converter," in *IEEE Power Electron. Spec. Conf. (PESC)*. vol. 1, 2001, pp. 333-338.

[3] A. Barrado, E. Olias, A. Lazaro, and J. Pleite, "PWM-PD multiple output DC/DC converters without transformer," in *Proc. IEEE Appl. Power Electron. Conf. (APEC)*, 2000, pp. 748-753.

[4] A. Barrado, E. Olias, A. Lazaro, J. Pleite, and R. Vazquez, "PWM-PD multiple output DC/DC converters: operation and control-loop modeling," *IEEE Trans. Power Electron.*, vol. 19, no.1, pp. 140-149, Jan. 2004.

[5] Y.J. Zhang and D.H. Xu, "design and implementation of an accurately regulated multiple output ZVS DC-DC converter," *IEEE Trans. Power Electron.*, vol. 22, no.5, pp. 1731-1742, Sept. 2007.

[6] Y.J. Zhang, Y.J. Zhang, D.H. Xu, F.C. Gao, Y. Han, and Z. Du, "design of an accurately regulated multiple output ZVS DC/DC converter," in *IEEE Power Electron. Spec. Conf. (PESC)*, 2006, pp. 1-6.

[7] H.E. Tacca, "Single-switch two-output flyback-forward converter operation," *IEEE Trans. Power Electron.*, vol. 13, no.5, pp. 903-911, Sept. 1998.

[8] H.H. Seong, D.J. Kim, and G.H. Cho, "A new ZVS DC/DC converter with fully regulated dual outputs," in *IEEE Power Electron. Spec. Conf. (PESC)*, 1993, pp. 351 - 356.

[9] V. Vlatkovic, J.A. Sabate, R.B. Ridley, F.C. Lee, and B.H. Cho, "Small-signal analysis of the phase-shifted PWM converter," *IEEE Trans. Power Electron.*, vol. 7, no.1, pp. 128-135, Jan. 1992.

Dual Output DC-DC Converter with Shared ZCS Lagging Leg

Yu Chen, Li Peng, Xuejun Pei, Yong Kang
College of Electrical and Electronic Engineering
Huazhong University of Science and Technology
Wuhan, China
Email: ayu03@163.com

Abstract—a dual output dc-dc converter with shared ZCS lagging leg is presented in this paper. In this proposed converter, the two outputs are formed by their independent leading legs and the common lagging leg. Each output is regulated by the phase shift between their leading legs and the shared lagging leg. The leading legs work in zero-voltage-switching (ZVS) mode, and the ZVS condition is easy to satisfy; the shared lagging leg works in ZCS mode, and the ZCS condition is independent of the loads. Therefore the proposed topology is high switching frequency, high efficiency, simply structure and easy to control. The operational principle of the proposed converter is analyzed, and the new features are discussed. A prototype with 200V input, 24V/5A and 15V/15A outputs is build to verify the analysis.

I. INTRODUCTION

Multiple output converters are widely used in many applications. There are many methods to structure a multiple output converter. Pulse width modulation-pulse frequency modulation (PWM-PFM) and phase shift modulation-pulse frequency modulation (PS-PFM) have simply topologies [1, 2], but only two outputs is obtained; post regulation (PR) is presented in [3, 4], but the auxiliary components such as linear regulator, mag-amp and synchronous switch increase the complexity and decrease the efficiency; a multiple output converter with pulse width modulation-phase shift (PWM-PS) is presented in [5, 6], but the outputs are coupled and the modulation is complex. Pulse width modulation-pulse delay (PWM-PD) [7, 8] has a special topology based on flyback or forward converter, but the output capability is low.

A dual output converter with shared ZVS leg topology is presented in [9]. As shown in Fig. 1(a), two full bridge converters are combined by sharing one of the bridge legs. Each output is regulated by the phase shift between their independent leading legs and the common lagging leg. This implementation not only decreases the amount of switches, simplifies the modulation but also remains the good features of the full bridge topology. Moreover, by configuring the shared leg as the lagging leg, the energy stored in all the leakage inductors can be utilized to achieve ZVS. Compared

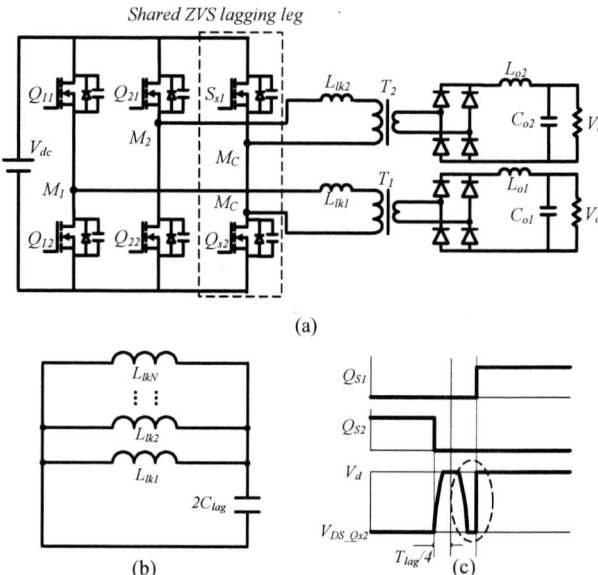

Fig. 1 The shared ZVS leg topology (a) the power stage (b) the equivalent circuit during the ZVS period of the lagging leg and (c) the waveforms during the ZVS period of the lagging leg

to the conventional phase shift full bridge (PSFB) converter, the ZVS range of the lagging leg is extended. However, two inherent deficiencies limit its output extending. First, if the output branches increase to N, when the lagging leg switch Q_{s2} is turned off, the equivalent circuit becomes Fig. 1(b). The resonant period can be expressed as:

$$T_{lag} = 2\pi\sqrt{2(L_{lk1} \| L_{lk2} \| \ldots \| L_{lkN})C_{lag}} \qquad (1)$$

Though the ZVS energy is enough, Q_{s1} should be turned on within a short dead-time interval of $T_{lag}/4$, or the ZVS will lose again, as shown in Fig. 1(c). However, when the output branches increase, more leakage inductors are in parallel. According to (1), T_{lag} will decrease and a shorter dead-time is required. The excess short dead-time endangers the security of the bridge leg. This is the main deficiency that limits the

978-1-4244-4782-4/10 $26.00 © 2010 IEEE

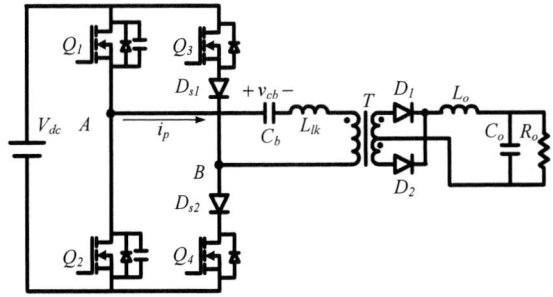

Fig. 2 The original single output ZVZCS converter

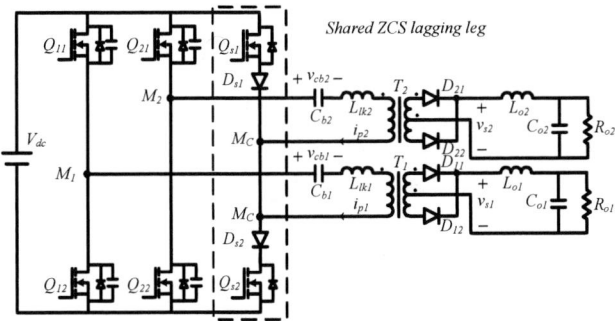

Fig. 3 The proposed dual output ZVZCS converter

output extending. Second, through the MOSFETs especially suit for the ZVS operation, their capabilities are usually low and don't adapt to the high power capability requirement of the shared lagging leg.

To keep all the advantages of the shared ZVS leg structure and overcome its inherent deficiencies, the conception of shared ZCS lagging leg is proposed in this paper. In section II, the derivation of the proposed topology is presented. In section III, the principle of the proposed dual output converter is analyzed. In section IV, the new features of the proposed converter are discussed. In section V, a prototype with 200V input, 24V/5A and 15V/15A outputs is built to verify the analysis. In section VI, the main conclusions are given. Based on the conclusions, a new type of multiple-output converter with shared ZCS lagging leg is presented as the future work.

II. DERIVATION OF THE PROPOSED TOPOLOGY

A. Review of the single output ZVZCS converter

To overcome the inherence deficiencies of the shared ZVS topology, the key is to remove the effect of the variable resonant period T_{lag} but still remains the soft switching characteristic. For this reason, another type of soft switching technology, the ZCS technology, is taken into consideration.

Fig. 2 shows the single output phase shift full bridge ZVZCS converter with the primary series-block-diodes structure[10], which realizes ZVS for the leading leg and ZCS for the lagging leg. In this converter, the blocking capacitor C_b is added in series with the primary winding of the transformer to make the primary current decay to zero during zero state (i.e., the period when both the upper or

lower switches are turning on) to ensure ZCS for the lagging leg. In order to prevent the primary current from reversing during zero state, two diodes D_{s1} and D_{s2} in series with the lagging leg are added. This topology is simply, the leading leg can achieve ZVS in a wide load range because of the contribution of the energy stored in the output inductor L_o; the lagging leg can achieve ZCS, and the ZCS operation is independent of the load condition.

B. Derivation of the proposed dual output converter

Based on the mentioned single output ZVZCS converter, the proposed dual output converter is deduced, as shown in Fig. 3. It can be treated as the extending version of the original one: by adding an independent leading leg and the output branch, the new output is obtained. The proposed converter remains the advantages of its original version, and overcome the drawbacks of the shared ZVS leg topology:

- The ZVS of the lagging leg is replaced by ZCS, the effect of T_{lag} is removed. And, as will be discussed, the ZCS operation is independent of the load conditions.

- Besides the shared lagging switches, the two series-block diodes D_{s1} and D_{s2} are also located in the shared leg, they are also shared by all the output branches, thus minimize the amount of required components.

- The leading legs process their own output power and work in ZVS mode; the ZVS is easy to achieve because the contribution of the energy stored in their output inductors, both the low power capability requirement and the soft switching operation of the leading legs especially suits for the MOSFETs applications.

- The shared lagging leg processes all the output power of the branches, the switches in the lagging leg thus require the high power capabilities, and, the lagging leg works in ZCS mode, both the capability requirement and the soft switching mode especially suit for the IGBTs application.

III. OPERATIONAL PRINCIPLE

There are ten typical modes in half of the operational period, as shown in Fig. 4. The key waveforms are shown in Fig. 5. The following assumptions are made in the analysis: (a) all the power devices and diodes are ideal; (b) all capacitors and inductors are ideal; (c) output filter inductors L_{o1} and L_{o2} are large enough to be treated as the constant current sources during a switching period.

Mode 1(t_0-t_1): Q_{11}, Q_{21} and Q_{s2} are turned on. In the secondary sides, D_{11} and D_{21} conduct. The large output inductors L_{o1} and L_{o2} are series with their primary sides so the primary currents $i_{p1}(t)$ and $i_{p2}(t)$ can be treated as the constant current sources I_{o1}/K_1 and I_{o2}/K_2. Both the 1st and 2nd branches are in the powering mode. The voltages across C_{b1} and C_{b2} can be expressed as:

$$v_{Cb1}(t) = V_{Cb1}(t_0) + I_{p1}(t-t_0)/C_{b1} \qquad (2\text{-}a)$$

$$v_{Cb2}(t) = V_{Cb2}(t_0) + I_{p2}(t-t_0)/C_{b2} \qquad (2\text{-}b)$$

where $I_{p1}=I_{o1}/K_1$ and $I_{p2}=I_{o2}/K_2$.

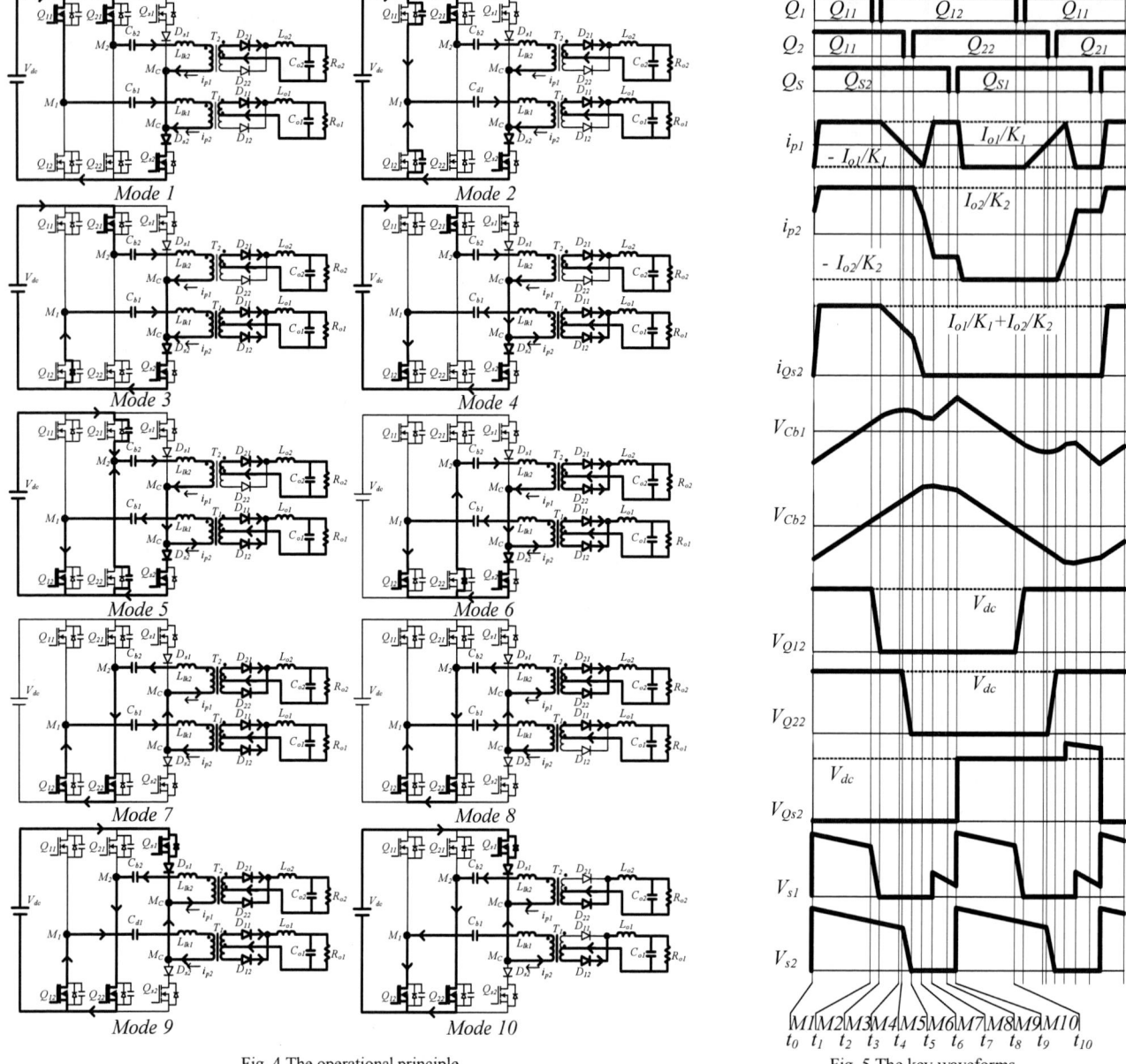

Fig. 4 The operational principle

Fig. 5 The key waveforms

Mode 2(t_1-t_2): Q_{11} is turned off but Q_{21} and Q_{s2} are still turned on. In the 1st branch, $i_{p1}(t)$ begins to charge and discharge the output capacitors of Q_{11} and Q_{12}. Since D_{11} still conducts, L_{o1} is still series with its primary side. And, since the output capacitors are small, $i_{p1}(t)$ is considered keeping constant. The output capacitors are charged and discharged linearly:

$$v_{DS\_Q11}(t) = \frac{I_{p1}(t-t_1)}{2C_{lead1}} \quad (3)$$

where $v_{DS\_Q11}(t)$ is the drain-source voltage of Q_{11} and C_{lead1} is the equal ZVS capacitor of the 1st leading leg.

Mode 3(t_2-t_3): The ZVS condition (3) is easy to achieve. Once $v_{DS\_Q11}(t)$ reaches V_{dc}, $i_{p1}(t)$ begins to flow through the

anti-parallel diode of Q_{12}. Q_{12} can be turned on in ZVS now. Since $i_{p1}(t)$ is deficient to feed the load, D_{11} and D_{12} conduct to commutate the currents. The block capacitor C_{b1} connects to L_{lk1} directly, $i_{p1}(t)$ decreases rapidly and $v_{Cb1}(t)$ increase slightly. Considering that $i_{p1}(t)$ decreases with the original slope $V_{Cb1}(t_2)/L_{lk}$ during mode 3, $i_{p1}(t)$ can be expressed as:

$$i_{p1}(t) = I_{P1} - \frac{V_{Cb1}(t_2)}{L_{lk1}}(t-t_2) \quad (4)$$

Mode 4(t_3-t_4): According to (4), $i_{p1}(t)$ decreases and reaches zero at t_3. After t_3, since the 2nd branch is still in the powering mode, D_{s2} is conducting. C_{b1} is keeping connecting to L_{lk1} directly and (4) is still true. $i_{p1}(t)$ will change its polarity and goes to negative. Noticing that following relation is always true:

978-1-4244-4782-4/10 $26.00 © 2010 IEEE   1443

$$i_{p1}(t) + i_{p2}(t) = i_{Qs2}(t) \qquad (5)$$

$i_{Qs2}(t)$, the current flowing through Q_{s2}, is decreasing now. Physically, it is because part of the current coming from the 2nd branch is sorbed by the 1st branch, and only the rest part is flowing through Q_{s2}.

Mode 5(t_4-t_5): Q_{21} is turned off. Similar to mode 2, in the 1st branch, $i_{p2}(t)$ begins to charge and discharge the output capacitors of Q_{21} and Q_{22} linearly:

$$v_{DS\_Q21}(t) = \frac{I_{p2}(t-t_4)}{2C_{lead2}} \qquad (6)$$

where $v_{DS\_Q21}(t)$ is the drain-source voltage of Q_{21} and C_{lead2} is the equal ZVS capacitor of the 2nd leading leg.

Mode 6(t_5-t_6): Once the output capacitor of Q_{21} is charged to V_{dc}, $i_{p2}(t)$ begins to flow through the anti-parallel diode of Q_{22}. Q_{22} can be turned on in ZVS now. And $i_{p2}(t)$ is deficient to feed its load, D_{21} and D_{22} conduct to commutate the currents. The block capacitor C_{b2} applied to L_{lk2} directly, $v_{Cb2}(t)$ increases slightly and $i_{p2}(t)$ decreases rapidly. Similar to (4), $i_{p2}(t)$ can be expressed as:

$$i_{p2}(t) = I_{p2} - \frac{V_{Cb2}(t_5)}{L_{lk2}}(t-t_5) \qquad (7)$$

Mode 7(t_6-t_7): Since $i_{p1}(t)$ has changed its polarity and $i_{p2}(t)$ are decreasing, according to (5), once $i_{p2}(t)=-i_{p1}(t)$ at t_6, i_{Qs2} will decrease to zero. The primary-series diode D_{s2} blocks the reverse path of $i_{Qs2}(t)$, $i_{Qs2}(t)$ keeps at zero. The 1st and the 2nd branches begin to commutate the currents because they are connected together. The current relation can be expressed approximately:

$$i_{p1}(t) = I_{p1}(t_6) + \frac{V_{Cb2}(t_6)-V_{Cb1}(t_6)}{L_{lk1}+L_{lk2}}(t-t_6) \qquad (8\text{-}a)$$

$$i_{p1}(t) = -i_{p2}(t) \qquad (8\text{-}b)$$

According to (8), whether $i_{p1}(t)$ and $i_{p2}(t)$ will increase or decrease depends on the values of $V_{Cb2}(t_6)$ and $V_{Cb1}(t_6)$. In this analysis, the assumption $V_{Cb2}(t_6) > V_{Cb1}(t_6)$ is made so $i_{p1}(t)$ increases and $i_{p2}(t)$ decreases.

Mode 8(t_7-t_8): If $i_{p1}(t)$ increases to I_{o1}/K_1 again, it is large enough to feed its load. The commutation of the currents flowing through D_{11} and D_{12} is completed. D_{11} conducts and D_{12} is block. In this situation, the current relation becomes:

$$i_{p1}(t) = -i_{p2}(t) = I_{o1}/K_1 = I_{p1} \qquad (9)$$

In the secondary side of the 1st branch, an additional pulse appears. This is the new feature exists in the proposed converter; we call it the re-conduction in this paper. In the primary side of the 2nd branch, $i_{p2}(t)=-I_{p1}$, it's still deficient to feed the its load, D_{21} and D_{22} keep conducting.

Mode 9(t_8-t_9): Though the circle currents exist between the two branches, $i_{Qs2}(t)$ is still zero. Q_{s2} can be turned off at t_8 in ZCS. And, with a short delay, the upper switch Q_{s1} can be turned on.

Mode 10(t_9-t_{10}): Q_{s1} is turned on at t_9. Since L_{lk1} and L_{lk2} limit $di_{p1}(t)/dt$ and $di_{p2}(t)/dt$, Q_{s1} is turned on in ZCS. All the diodes in the secondary sides conduct again as $i_{p1}(t)$ and $i_{p2}(t)$ are deficient to feed their load. Since this mode is very short, C_{b1} and C_{b2} can be treaded as the constant voltage sources, $i_{p1}(t)$ and $i_{p2}(t)$ thus rise linearly in the negative direction:

$$i_{p1}(t) = I_{p1} - \frac{V_{dc}+V_{cb1}(t_9)}{L_{lk1}}(t-t_9) \qquad (10\text{-}a)$$

$$i_{p2}(t) = -I_{p1} - \frac{V_{dc}+V_{cb2}(t_9)}{L_{lk2}}(t-t_9) \qquad (10\text{-}b)$$

At t_{10}, $i_{p1}(t)$ and $i_{p2}(t)$ are large enough to feed their load again, $i_{p1}(t_{10})=-I_{p1}$ and $i_{p2}(t_{10})=-I_{p2}$, another symmetrical half period begins.

IV. DISCUSSION OF THE PROPOSED TOPOLOGY

A. The peak voltage of the block capacitors

Be different from the single output ZVZCS converter, the voltages across the blocking capacitors C_{b1} and C_{b2} won't

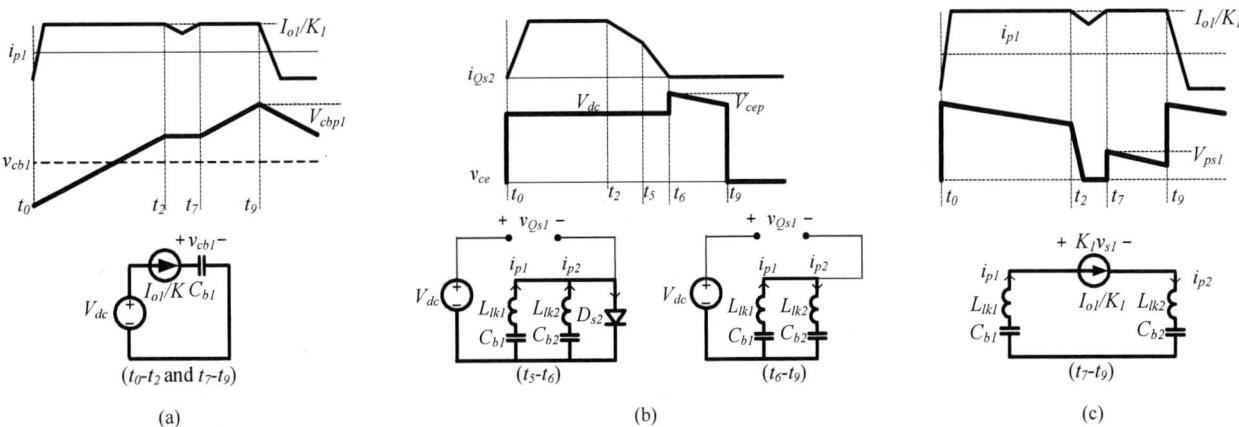

Fig. 6 The ideal waveforms of the proposed converter (a) variation of the blocking capacitors (b) the voltage stress of the IGBTs and (c) the additional pulse width in the secondary sides

keep in constant during the ZCS period, because the circle current exists between the two branches. Take the 1st branch as the example. As shown in Fig. 6(a), during the period $T_{02}=t_2-t_0$, C_{b1} is charged linearly by I_{o1}/K_1. During $T_{27}=t_7-t_2$, $i_{p1}(t)$ is variable, as described in mode 3 and mode 7. To simplify the analysis, the slight voltage variation across C_{b1} during T_{27} is omitted. Furthermore, if the re-conduction happens, $i_{p1}(t)$ will reach I_{o1}/K_1 again and C_{b1} is charged continuously during the period $T_{79}=t_9-t_7$. The peak voltage of $v_{cb1}(t)$ during one operation period $T_s/2=t_9-t_0$ can be can be expressed as:

$$V_{cbp1} = \frac{I_{p1}T_s}{4C_{b1}}(D_1 + D_{re1}) \qquad (11)$$

where $D_1=2T_{02}/T_s$ is the duty cycle and $D_{re1}=2T_{79}/T_s$ is the re-conduction duty cycle of the 1st branch.

B. The voltage stress of the IGBTs

Fig. 6(b) focuses on the voltage stress of the IGBTs. As discussed in section III, before mode 7, $i_{Qs2}(t)$, the current flowing through Q_{s2}, hasn't decreased to zero yet, the voltage stress of the upper IGBT Q_{s1} is V_{dc}. after t_6, $i_{Qs2}(t)$ decrease to zero and D_{s2} blocks its reverse path, the current begins to commutate between the two branches. After t_6, $i_{p1}(t)=-i_{p2}(t)$, and according the assumption $V_{Cb2}(t_6)>V_{Cb1}(t_6)$ in mode 7, $i_{p1}(t)$ increases, as described in (8). The voltage stress across Q_{s1} can be expressed as:

$$v_{ce}(t) = V_{dc} + v_{cb2}(t) - \frac{L_{lk2}[v_{cb2}(t)-v_{cb1}(t)]}{L_{lk1}+L_{lk2}} \qquad (12)$$

C. The additional pulse in the secondary sides

Once $i_{p1}(t)$ increases to I_o/K_1 again at t_7, the re-conduction happens. According to mode 8, the key waveforms and the equivalent circuit can be expressed as Fig. 6(c). According to the equivalent circuit, the additional pulse width in the secondary side of the 1st branch can be expressed as:

$$v_{s1}(t) = \frac{v_{cb2}(t)-v_{cb1}(t)-(L_{lk1}+L_{lk2})\dfrac{di_{p1}(t)}{dt}}{K_1} \qquad (13\text{-a})$$

Noticing that $i_{p1}(t)=I_o/K_1$ is near constant, we have:

$$(L_{lk1}+L_{lk2})\frac{di_{p1}}{dt} \approx 0 \qquad (13\text{-b})$$

And (13-a) can be simplified as:

$$v_{s1}(t) = \frac{v_{cb2}(t)-v_{cb1}(t)}{K_1} \qquad (13\text{-c})$$

D. The ZCS margin

As shown in Fig. 7, the period $T_{69}=t_9-t_6$ is the ZCS margin of the proposed converter, because during this period $i_{Qs2}(t)$ has decreased to zero and Q_{s2} can be turned off in ZCS. To survey the minimal ZCS margin, consider the extreme situation first. As shown in Fig. 7(a), $i_{p1}(t)$ and $i_{p2}(t)$ decrease to zero at t_6 at the same time. $i_{Qs2}(t)$ also decreases to zero at t_6. In this situation, we have:

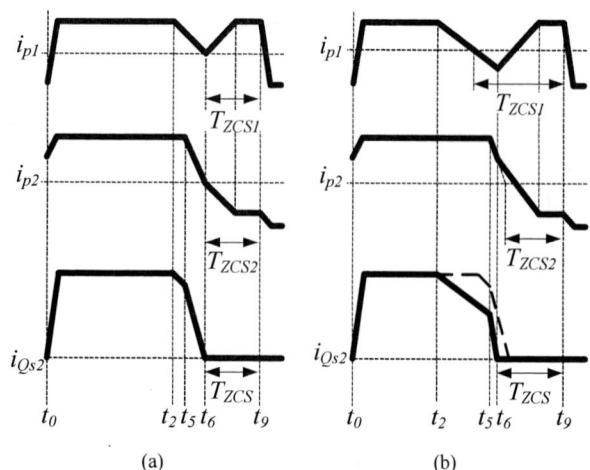

Fig. 7 The ZCS condition of the proposed converter

$$T_{ZCS1} = T_{ZCS} = T_{ZCS2} \qquad (14\text{-a})$$

T_{ZCS1} and T_{ZCS2} are the ZCS margins when the outputs are working solely:

$$T_{ZCSi} = (1-D_i-D_{reseti})T_s/2 \qquad (i=1,2) \qquad (14\text{-b})$$

$$D_{reseti} = \frac{8L_{lki}C_{bi}}{T_s^2 D_i} \qquad (i=1,2) \qquad (14\text{-c})$$

Generally, $i_{p1}(t)$ and $i_{p2}(t)$ won't decrease to zero at the same time. As shown in Fig. 7(b), if $i_{p1}(t)$ goes to zero earlier than $i_{p2}(t)$, it will change its polarity. As a result, part of the current flowing Q_{s2} is sorbed by $i_{p1}(t)$, $i_{Qs2}(t)$ thus decreases to zero much quickly:

$$T_{ZCS2} < T_{ZCS} < T_{ZCS1} \qquad (14\text{-d})$$

V. THE CRITICAL CONDITIONS OF THE PROPOSED CONVERTER

When the load conditions change, the states of the converter are variable. The exact solutions of (11)-(14) are difficult to obtain. However, the critical maximal or minimal values can be deduced from the analysis. If these critical limitations can be satisfied, the proposed converter can work steadily in any load conditions.

As described in (11), the re-conduction duty cycle D_{re1} is difficult to determine because it not only depends on the circuit parameters but also the load conditions of the two outputs. Consider the most serious situation: $T_{27}\approx0$, $D_1+D_{re1}\approx1$, the voltage across C_{b1} thus has the maximal value:

$$V_{cbp1-\max} = \frac{I_{p1}T_s}{4C_{b1}} \qquad (15\text{-a})$$

In (12), noticing that during the period $T_{69}=t_9-t_6$, $v_{cb2}(t)$ is decreasing because $i_{p2}(t)$ is in the direction of discharging, the relation $v_{cb2}(t)<V_{cbp2}$ is always true. Furthermore, by ignoring the third term in (12), the maximal possible voltage stress across the IGBTs can be modified as:

$$V_{ce-\max} = V_{dc} + V_{cbp2-\max} \qquad (15\text{-c})$$

TABLE I. THE CRITICAL CONDITIONS

Limitations	Symbols	Expressions (i=1 and 2)
The maximal voltage across the blocking capaciotrs	$V_{cbp1\text{-}max}$ $V_{cbp2\text{-}max}$	$V_{cbpi\text{-}max} = \dfrac{I_{pi}T_s}{4C_{bi}}$
The maximal voltage stress of the IGBTs	$V_{ce\text{-}max}$	$V_{ce\text{-}max} = V_{dc}$ $+ \max(V_{cbp1\text{-}max}, V_{cbp2\text{-}max})$
The maximal additional pulse in the secondary side	$V_{s1\text{-}max}$ $V_{s2\text{-}max}$	$V_{si\text{-}max} =$ $\max(V_{pcb1\text{-}max}, V_{pcb2\text{-}max})/K_1$
The minimal ZCS margin of the proposed converter	$D_{ZCS\text{-}min}$	$D_{ZCS\text{-}min} \geq \min(D_{ZCS1}, D_{ZCS2}) > 0$

TABLE II. PROTOTYPE PARAMETERS

The 1st output (24V/5A)	$K_1(n_p{:}n_s)$	4.6(32:7)
	L_{lk1}	23.4uH
	C_{b1}	69nF
	L_{f1}	140uH
	C_{f1}	470uF
	C_{lead_1}	85pF
	MOSFETs	IRFP450
	D_{11}, D_{12}	MBR20150CT
The 2nd output (15V/15A)	$K_2(n_p{:}n_s)$	7(35:5)
	L_{lk2}	48.5uH
	C_{b2}	47nF
	L_{f1}	130uH
	C_{f1}	470uF
	C_{lead_2}	85pF
	MOSFETs	IRFP450
	D_{21}, D_{22}	MBR2060CT
The shared ZCS leg	IGBTs	IKW20N60T
	D_{s1}, D_{s2}	MUR3060PT

In (13-c), considering the extreme situation that $v_{cb2}(t)= V_{cbp2\text{-}max}$ and $v_{cb1}(t)=0$, the maximal additional pulse in the secondary side of the 1st branch can be obtained:

$$V_{s1\text{-}max} = \frac{V_{pcb2\text{-}max}}{K_1} \qquad (15\text{-}d)$$

According to (14-a) and (14-d), it's seen that T_{ZCS} always locates between T_{ZCS1} and T_{ZCS2}. If the ZCS margins exist in both the two branches ($T_{ZCS1}>0$ and $T_{ZCS2}>0$), the ZCS margin of the proposed converter ($T_{ZCS}>0$) also exists. Since $T_{ZCS1}>T_{ZCS2}$ in the discussed situation, the ZCS margin of the proposed converter always exists if:

$$D_{ZCS2} > 0 \qquad (15\text{-}e)$$

where $D_{ZCS2}=2T_{ZCS2}/T_s$ is the ZCS margin duty cycle.

Moreover, though the analysis presented above focus on one output branch, it suits for the other branch because the two branches in the proposed converter have the same structure. The general critical limitations of the proposed converter thus can be summarized, as list in Table I.

VI. EXPERIMENTAL RESULTS

A laboratorial prototype is built. The input is 200V, the 1st output is 24V/5A and the 2nd output is 15V/15A, the switching frequency is 100kHz.

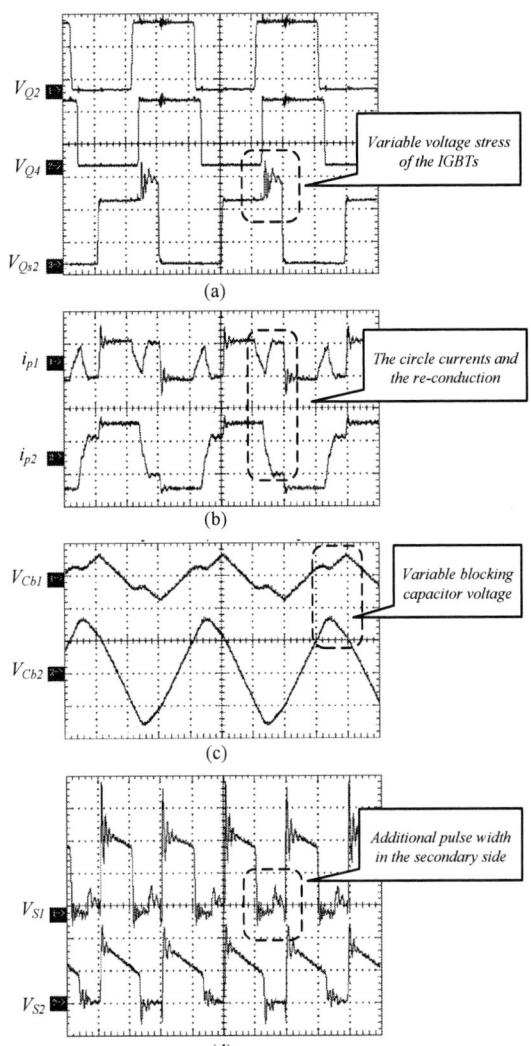

Fig. 8 Key waveforms when both the outputs are full loads
(V_{Q2}, V_{Q4}, V_{S2}:100V/div; i_{lead}, i_{lag}, i_{mid}: 2A/div; V_{Cb1}, V_{Cb2}:50V/div; V_{s1}, V_{s2}:20V/div; t:2.5us/div)

The power state design follows the guidelines presented in [10] first, then, the parameters in Table I should be checked so as they can meet the extreme situations. The parameters are shown in Table II.

In Fig. 8, both the outputs are full loads (I_{o1}=5A and I_{o2}=15A). As shown in Fig. 8(a), V_{Q2}, V_{Q4} and V_{Qs2} are the voltages across of the MOSFETs and IGBT. During the ZCS period, V_{Qs2} is variable; in Fig. 8(b), it's seen that i_{p1} reaches I_{o1}/K_1 again and the re-conduction exists in the 1st branch; in Fig. 8(c), it is seen distinctly that the voltages across C_{b1} and C_{b2} are variable and in Fig. 8(d), since the 1st branch re-conducts, the additional pulses appear in secondary side of the 1st branch, as shown in the waveform of V_{s1}. Also, all the waveforms agree well with the analysis in section III.

In Fig. 9, the 1st outputs is full load (I_{o1}=5A) and the 2nd output is medium load (I_{o2}=7.5A). As shown in Fig. 9(c), the peak voltages across C_{b1} and C_{b2} voltages happen to have the

TABLE III. THE CALCULATED AND MEASURED VALUES

Symbols	Calculation	Measurement	
		Fig. 8	Fig. 9
$V_{cbp1-max}$	40V	30V	24V
$V_{cbp2-max}$	114V	83V	26V
V_{ce-max}	314V	290V	220V
V_{s1-max}	24.9V	16V	0V
V_{s2-max}	6.3V	0V	0V

such as the commutation of the currents in the primary side during the ZCS period and the re-conduction of the rectifier in the secondary side, are discussed. The analysis is verified by the experiments results.

Based on the analysis, the outputs can be extended to N, thus a new type of multiple-output converter with the shared ZCS lagging leg can be deduced. However, some issues must be considered when the output increasing: (a) the effect of the commutated current existed in the output branches; (b) voltage and current stress of the shared leg components; (c) the effect of re-conduction on the voltage regulation; and (d) the stability of the system when the number of output increasing. All of these will be done as the future work.

REFERENCES

[1] H.H. Seong, D.J. Kim, and G.H. Cho, "A new ZVS DC/DC converter with fully regulated dual outputs," in *IEEE Power Electron. Spec. Conf. (PESC)*, 1993, pp. 351 - 356.

[2] H.E. Tacca, "Single-switch two-output flyback-forward converter operation," *IEEE Trans. Power Electron.*, vol. 13, no.5, pp. 903-911, Sept. 1998.

[3] L.H. Hang, Y.L. Gu, Z.Y. Lu, Z.M. Qian, and D.H. Xu, "Magamp post regulation for LLC series resonant converter with multi-output," in *Proc. IEEE Ind. Electron. Soc. Annu. Conf. (IECON)*, 2005, pp. 628-631.

[4] S. Havanur, "Combining Synchronous Rectification and Post Regulation for Multiple Isolated Outputs," in *Proc. IEEE Appl. Power Electron. Conf. (APEC)*, 2004, pp. 872-877.

[5] Y.J. Zhang, D.H. Xu, F.C. Gao, Y. Han, and Z. Du, "Accurately regulated multiple output ZVS dc-dc converter," in *Proc. IEEE Appl. Power Electron. Conf. (APEC)*, 2006, pp. 517-522.

[6] Y.J. Zhang and D.H. Xu, "Design and implementation of an accurately regulated multiple output ZVS DC-DC converter," *IEEE Trans. Power Electron.*, vol. 22, no.5, pp. 1731-1742, Sept. 2007.

[7] A. Barrado, E. Olias, A. Lazaro, and J. Pleite, "PWM-PD multiple output DC/DC converters without transformer," in *Proc. IEEE Appl. Power Electron. Conf. (APEC)*, 2000, pp. 748-753.

[8] A. Barrado, E. Olias, A. Lazaro, J. Pleite, and R. Vazquez, "PWM-PD multiple output DC/DC converters: operation and control-loop modeling," *IEEE Trans. Power Electron.*, vol. 19, no.1, pp. 140-149, Jan. 2004.

[9] J. Sabate, Y. Liu, and M. Wiza, "Power supply with independently regulated multiple outputs," in *Proc. Power Electron. and Appl. Conf.*, 2007, pp. 1-8.

[10] X.B. Ruan and Y.G. Yan;, "A novel zero-voltage and zero-current-switching PWM full-bridge converter using two diodes in series with the lagging leg," *Trans. Power Electron.*, vol. 48, no.4, pp. 777-785, Aug. 2001.

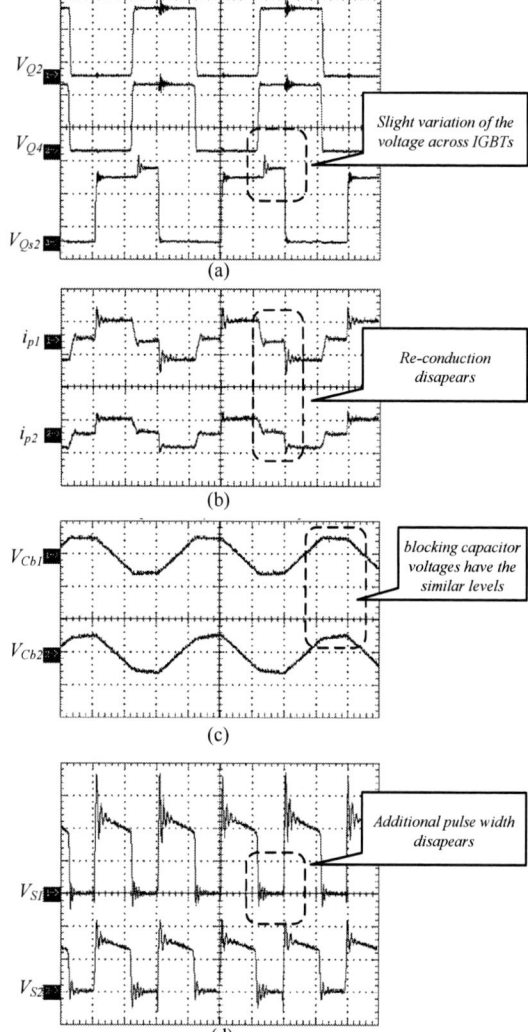

Fig. 9 Key waveforms when the 2nd output is half load
(V_{Q2}, V_{Q4}, V_{S2}:100V/div; i_{lead}, i_{lag}, i_{mid}: 2A/div; V_{Cb1}, V_{Cb2}:50V/div;
V_{s1}, V_{s2}:20V/div;   t:2.5us/div)

similar levels, according to (8), i_{p1} and i_{p2} in Fig. 9(b) thus change slowly and won't reach I_{o1}/K_1 or I_{o2}/K_2 during the whole ZCS period, the re-conduction thus disappear. And, as can be seen in Fig. 9(a), V_{Qs2} also changes slightly; in Fig. 9(d), the additional pulse width disappears.

Table III shows the calculated maximal or minimal critical values mentioned in Table I and the measured values in the two load conditions. It's seen that the measured values are always smaller than the calculated ones. Since the critical stresses have been taken into consideration when choosing the components, the converter can work steadily in any load situations.

VII. CONCLUSIONS

The conception of shared ZCS lagging leg structure is presented in this paper. The dual output converter with the shared ZCS lagging leg is analyzed. The ZCS operation is remained in the proposed topology. Some new characteristics,

A Novel ZVS Full-Bridge Converter with Auxiliary Circuit

Zhong Chen, Biao Ji, Feng Ji, and Lei Shi

Aero-Power Sci-tech Center
Nanjing University of Aeronautics & Astronautics
Nanjing, 210016, China
Email: chenz@nuaa.edu.cn

Abstract—**A novel full-bridge (FB) pulse-width-modulated (PWM) converter features zero-voltage-switching (ZVS) of all primary switches in the entire line and load range is described. In contrast to conventional full bridge converter, ZVS is achieved by utilizing energy stored in inductive components of the auxiliary circuit. The auxiliary inductors do not appear as series inductances in the power transfer path, so they do not cause severe voltage ringing across the output rectifier or duty cycle loss. The operation principle of the circuit is analyzed and design considerations of the converter are discussed. Finally, the experimental results from a 1-kW (54-V/20-A) prototype are presented to confirm the operation, validity, and features of the proposed converter.**

I. INTRODUCTION

The full-bridge (FB) zero-voltage-switching (ZVS) pulse-width-modulated (FB ZVS-PWM) converter, [1]-[11], is applied in medium to high power conversion due to high power density, ZVS operation, high efficiency, low electromagnetic interference (EMI) and moderate device stresses. In a conventional phase shifted full bridge (PSFB) converter, ZVS is achieved by inserting an inductor in series with the transformer [1]-[4]. However, full ZVS operation can only be achieved with a limited load and input-voltage range. The loss of ZVS will result in increased switching loss and electromagnetic interference. Intentionally by increasing the leakage inductance or by adding a large external inductance in series with the primary transformer can help extend the ZVS range of the lagging leg switches. But the larger series inductance has a detrimental effect on the performance of the converter since it results in increased loss of duty cycle, as well as severe voltage ringing across secondary-side rectifier diodes due to the resonance between the inductance and the junction capacitance of the rectifier. To suppress the ringing, a snubber circuit is required. If a conventional RC or RCD snubber is used, the conversion efficiency of the circuit may be significantly degraded [5]. And the large external inductance will cause large circulating current at full load which adversely affects the conversion

efficiency. As a nondissipative method, the secondary-side voltage oscillation in the FB ZVS converter is virtually eliminated by employing an active switch in the secondary side, but an added active switch increases the system complexity and causes additional switching loss [6]. For implementations with an external primary inductor, the ringing can also be effectively controlled by employing primary-side diodes. While the approaches in [7] offer practical solutions to the secondary-side ringing problem, they do not offer any improvement of the secondary-side duty-cycle loss. Using a saturable external inductor instead of a linear inductor, ZVS range can be increased without significantly losing the duty ratio [8]. However, a large-size core is required to eliminate the thermal problem. And at very light load current, ZVS operation can still be lost [8]-[9]. The energy stored in the magnetizing inductance can also be used to extend the ZVS range. In conventional PSFB converter, this results in significant increase in the rms switch current and the conduction loss [10]-[11].

In this paper, a novel full-bridge converter that achieves ZVS right down to no load without serious conduction loss penalty is proposed. This constant-frequency, FB ZVS converter employs an asymmetrical auxiliary circuit consisting of a few passive components. The proposed converter and its operating principle are described in Section II. Section III presents the optimal design considerations for the proposed converter. The experimental results are presented in Section IV to verify the validity of the proposed converter.

II. OPERATION PRINCIPLE

Fig. 1 shows the proposed ZVS full-bridge converter topology. The primary side of the converter consists of two bridges Q_1-Q_3 and Q_2-Q_4 connected through two capacitors C_{a1} and C_{a2} to the connection of the power transformer T_r and the auxiliary inductors L_{a1} and L_{a2}. The two primary side capacitors are used to prevent the saturation of the power transformer and the auxiliary inductor cores by blocking the flow of any dc current through T_r, L_{a1} and L_{a2}. The switching

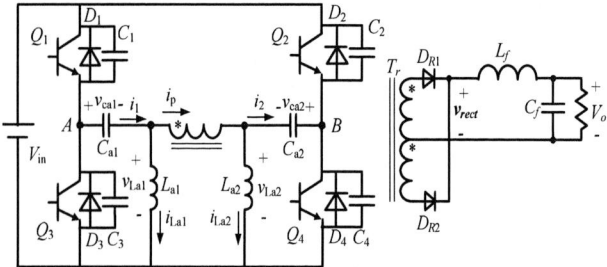

Fig. 1. Proposed full bridge ZVS converter.

transition of the switches in the Q_2-Q_4 leg of the bridge is delayed, phase-shifted, with respect to the switching transition of corresponding switches in the Q_1-Q_3 leg. Generally, these two auxiliary capacitors are selected large enough so that their voltages are approximately constant during a switching cycle. Because the average voltages of the auxiliary inductors and the power transformer during a switching cycle are zero and the pair of switches in each bridge leg operate with 50% duty cycle, the magnitude of voltage sources v_{ca1} and v_{ca2} are equal to $1/2V_{in}$, i.e., $v_{ca1}= v_{ca2}=1/2V_{in}$. In addition, the auxiliary inductor stores energy which only be used to achieve ZVS, its size can be small.

The output side of the converter is implemented with a full-wave rectifier with a tapped secondary. Also, any other implementation of the secondary side rectification stage is possible. In terms of power transfer from the input to load, the power circuit operates in exactly the same way as does a conventional phase-shifted full-bridge converter, and the auxiliary circuit hardly interferes with its power transfer. However, the auxiliary circuit removes the switching losses from all the switches.

The key waveform of the ZVS PWM full-bridge converter is shown in Fig. 2. To perform the steady state analysis, the following assumptions are made.

1) All components and devices have ideal properties and characteristics.
2) $C_1= C_3= C_{lead}$, $C_2= C_4= C_{lag}$.
3) The output filter capacitor is large enough to be treated as a constant voltage source with a magnitude equal to V_o.
4) The turns ratios of the power transformer is the primary winding: the secondary winding=N_p: N_s. (where N_p: N_s=K)
5) The switching frequency f_s is fixed and the inverse of the switching period T_s.

Fig.3. shows the topological stages of the converter during a half period. The second half period is similar to the first half period.

1) Stage 1 [t_0, t_1] [Refer to Fig. 3(a)]

In the last interval of the previous cycle, diagonal switches Q_1 and Q_4 are conducting, primary voltage is positive so that load current I_o flows through D_{R1} and the upper secondary of transformer T_r. After switch Q_1 is turned off at t=t_0, current i_1 starts charging output capacitance C_1 of switch Q_1 and discharging output capacitance C_3 of switch Q_3, where i_1 is the sum of the current i_{La1} and the primary current i_p=I_o/K. During this interval, the current of the inductor L_{a1} keeps at I_{La1} which can be calculated as (1). The voltage of Q_1 rises slowly owing to C_1. After the capacitor C_3 is fully

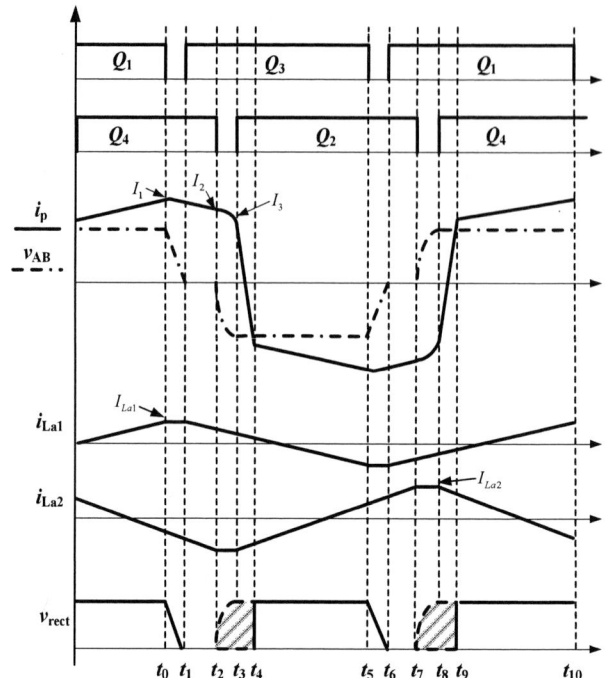

Fig. 2. Key waveforms of proposed converter power stage.

discharged, current i_1 continues to flow through the anti-parallel diode of switch Q_3. The voltage across switch Q_3 is given in (2) during this interval.

$$I_{La1} = \frac{V_{in}}{8L_{a1} \cdot f_s} \tag{1}$$

$$v_{c3}(t) = V_{in} - \frac{I_o / K + I_{La1}}{2C_{lead}}(t - t_0) \tag{2}$$

2) Stage 2 [t_1, t_2] [Refer to Fig. 3(b)]

Q_3 can be turned on at zero voltage when D_3 conducts. In this topological stage, the potential of point A becomes zero. Voltage across the power transformer also becomes zero since the transformer is shorted by the simultaneous conduction of the body diode of Q_3 and switch Q_4. The voltage applied across L_{a2} is $-V_{in}/2$. Due to this voltage, the current i_{La2} still decreases until t_2. At t_2, i_{La2} reaches its minimum value $-I_{La2}$. And I_{La2} can be expressed as (3).

$$I_{La2} = \frac{V_{in}}{8L_{a2} \cdot f_s} \tag{3}$$

And with real components, the primary current will decrease because of the forward voltage of power switch $V_{forward}$ and parasitical impedance. The primary current can be expressed as

$$i_p(t) = \left[i_p(t_1) + \frac{V_{forward}}{r} \right] e^{-\frac{r}{L_k}(t-t_1)} - \frac{V_{forward}}{r} \tag{4}$$

where r is the equivalent series resistor for the circuit, L_k is the leakage inductance of power transformer . This interval is ended when the switch Q_4 is turned off.

3) Stage 3 [t_2, t_3] [Refer to Fig. 3(c)]

Q_4 is turned off in ZVS at t_2. Current i_2 starts charging output capacitance C_4 of switch Q_4 and discharging output

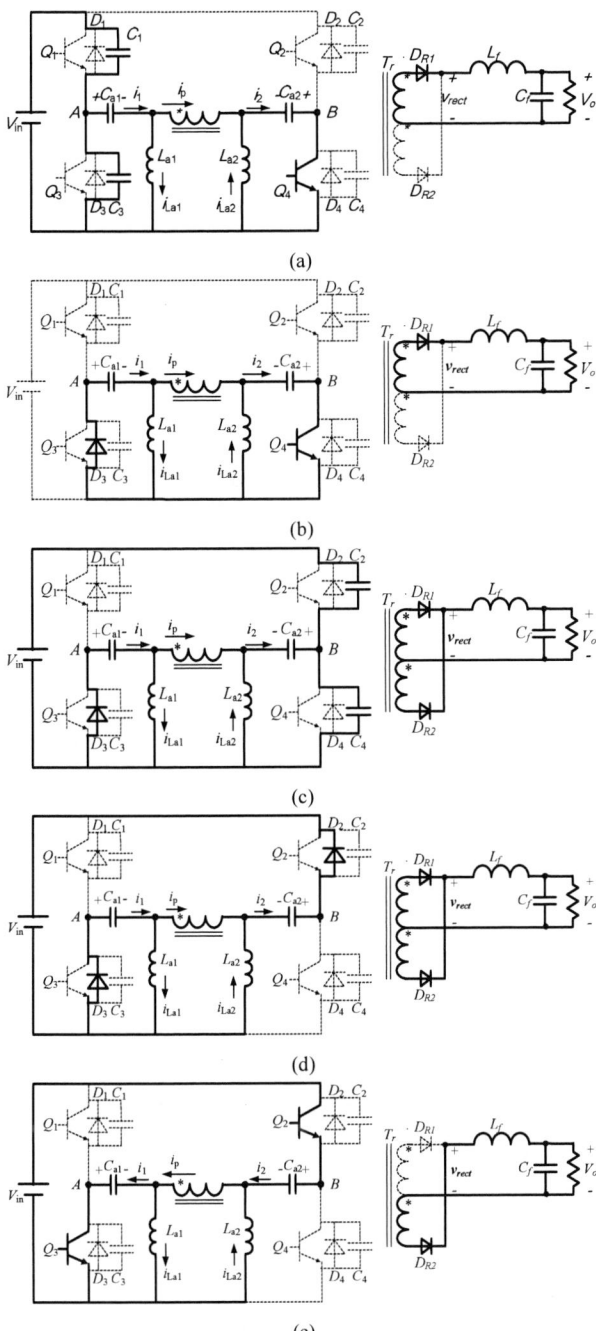

Fig. 3. Topological stages of proposed converter power stage.

$$i_p(t) = (\frac{I_o}{K} + I_{La2})\cos \omega_r (t - t_2) - I_{La2} \qquad (5)$$

$$v_{c4}(t) = (I_{La2} + \frac{I_o}{K})Z_r \sin \omega_r (t - t_2) \qquad (6)$$

where $Z_r = \sqrt{L_k/2C_{lag}}$, $\omega_r = 1/\sqrt{2L_kC_{lag}}$

4) Stage 4 [t₃, t₄] [Refer to Fig.3(d)]

D_2 conducts naturally when v_{c2} decays to zero, and Q_2 can be turned on at zero voltage. At the same time, load current I_o commutates from the upper secondary and rectifier D_{R1} into the lower secondary and corresponding rectifier D_{R2}. The rate of change of the primary current is given by

$$\frac{di_p}{dt} = -\frac{V_{in}}{L_k} \qquad (7)$$

Since the leakage inductance of the power transformer is smaller, the duty-cycle loss in the converter is decreased.

5) Stage 5[t₄, t₅] [Refer to Fig.3(e)]

The commutation of the load current from the upper to the lower secondary is completed at t=t₄, the primary current commutation from the positive to negative direction is also finished so that the primary is i_p=−I_o/K. The input energy is transferred into the secondary side by power switches Q_2 and Q_3 and T_r. This stage ends at t_5 when switch Q_3 turns off. Then the second half of the switching period begins. In the second half of the switching period, the operation of the circuit is exactly the same as the operation in the first half of the switching period.

III. DESIGN CONSIDERATIONS

Auxiliary inductors are required in the proposed converter to achieve ZVS. However, the auxiliary inductor current will result in additional conduction losses. This can be achieved by using an asymmetrical auxiliary arrangement.

A. ZVS Range for the Leading leg and the selection of L_{a1}

In order to achieve ZVS turned on, the output capacitance of the switch shall be completely discharged within the dead time period $t_{d(lead)}$ under all operating conditions. Neglecting the capacitances of the transformer's windings, the value of the current required to achieve ZVS of leading-leg switches I_{charge} is calculated as

$$I_{charge} \cdot t_{d(lead)} = \frac{I_{La1}}{2} \cdot t_{d(lead)} \geq C_{lead} \cdot V_{in} \qquad (8)$$

Expressions (1) and (8) can be used to estimate the required value of the auxiliary inductance L_{a1}. From (1) and (8), we can get the following equation

$$L_{a1} \leq \frac{t_{d(lead)}}{16 \cdot C_{lead} \cdot f_s} \qquad (9)$$

Fig. 4 shows an example of the selection curves of the auxiliary inductor L_{a1} as functions of $t_{d(lead)}$ at different output capacitances. It is seen that with a fixed dead time, smaller L_{a1} will achieve ZVS much more easily. When the output capacitance is decided, the dead time can be selected shorter with smaller L_{a1}. The dead time for the leading leg should be selected together with auxiliary inductor.

capacitance C_2 of switch Q_2. Since the potential of point B increases from zero toward V_{in}, while the potential of point A is constant at zero. The secondary windings are also shorted so that rectifiers D_{R1} and D_{R2} can conduct the load current simultaneously. v_{AB} is fully applied to the L_k. Because the auxiliary inductor L_{a2} is much larger than the leakage inductance L_k, it can be treated as a constant current source during the transition time. This stage finishes when v_{c4} rises to V_{in} and v_{c2} falls to zero at t_3. The primary current and the voltage across Q_4 are given by (5) and (6).

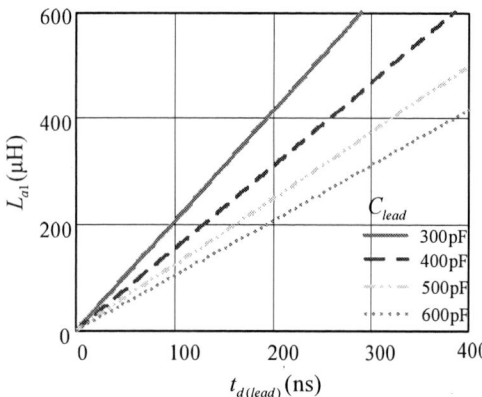

Fig. 4. Example design curves for selecting the auxiliary inductors vs. dead time.

B. ZVS Range for the lagging leg and the selection of L_{a2}

Before analyzing the ZVS condition for the lagging leg, we suppose the dead time $t_{d(lag)} = T_c/4$, where $T_c = 2\pi\sqrt{2 \cdot L_k \cdot C_{lag}}$. The output current ripple is neglected, so the primary current can be expressed as I_o/K.

Unlike the transition of leading leg, the transition of lagging leg switches can be divided into two cases according to the load current I_o as shown in Fig. 5.

Under most load current conditions, the transition may only go through stage I, as shown by Fig. 5(a). The primary current decreases as the voltage of C_4 increases. When the voltage of C_4 reaches V_{in}, D_2 conducts, so Q_2 can be turned on with ZVS. When the transition starts at $t=t_2$, the voltage of C_4 can be calculated as (6).

Since the leakage inductance is much smaller compared with conventional full bridge converter, the resonant time is relatively shorter. So the load current may complete the commutation from the upper secondary to the lower secondary before the voltage of C_4 reaches V_{in}. Then the converter will enter stage II, as shown by Fig. 5(b). The current through L_{a2} will go on discharging C_2 and charging C_4. Simultaneously this current will compensate the output current I_o.

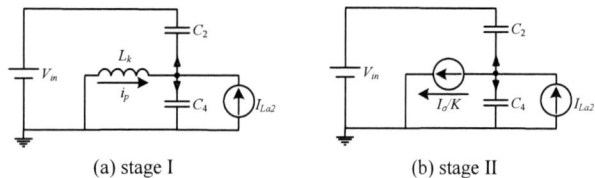

(a) stage I   (b) stage II

Fig. 5. Transition equivalent circuit of lagging leg.

When the transition enters stage II, the voltage across Q_4 can be expressed as (10).

$$v_{c4}(t) = \left(\frac{I_o}{K} + I_{La2}\right)Z_r \sin\omega_r(t_2' - t_2)$$

$$+ \frac{\left(-\frac{I_o}{K} + I_{La2}\right)\cdot(t - t_2')}{2 \cdot C_{lag}} \quad (10)$$

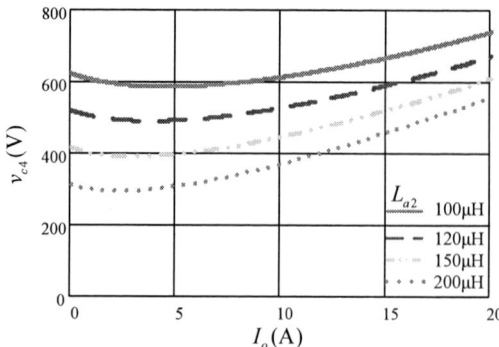

Fig. 6. Voltage across the parasitic capacitor of the lagging leg versus load current at different auxiliary inductances. (V_{in}=400V, L_k=6μH)

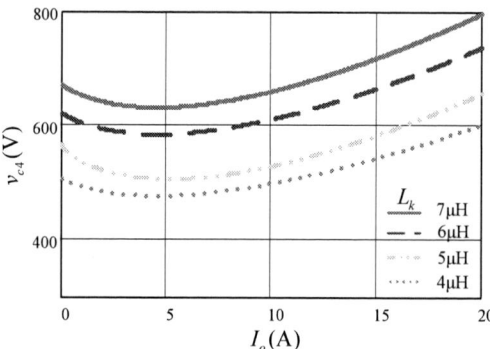

Fig. 7. Voltage across the parasitic capacitor of the lagging leg versus load current at different leakage inductances.(V_{in}=400V, L_{a2}=100μH)

where $t_2' = t_2 + \left[\arccos\left(\dfrac{-\dfrac{I_o}{K} + I_{La2}}{\dfrac{I_o}{K} + I_{La2}}\right)\right]\Bigg/\omega_r$. t_2' is the moment when the load current completes commutation.

According to (6) and (10), Fig. 6 and Fig. 7 can be plotted. Fig. 6 shows the voltage across Q_4 versus load current I_o with definite leakage inductance under different auxiliary inductances. Fig. 7 shows the voltage across Q_4 in a function of load current I_o with determined auxiliary inductance under different leakage inductances. From Fig. 6 and Fig. 7, we can know ZVS can only be achieved when the voltage across Q_4 is higher than V_{in}. As the leakage inductor becomes smaller, the value of auxiliary inductance L_{a2} should be made smaller. Smaller L_{a2} will in turn result in larger I_{La2}. And this will increase conduction losses in the switches.

If the leakage inductance is minimized, the load current will commutate fast, as shown in Fig. 5(b). And the current through auxiliary inductance should be made larger than the reflected load current. It results in increased conduction losses. So the leakage inductance should be optimally selected to simultaneously achieve the entire range ZVS operation and improve overall efficiency over the entire conversion range.

C. The selection of C_{a1} and C_{a2}

The two capacitors are employed to establish dc voltages and block the flow of any dc current through the magnetic components. The permitted ripple voltage on these two capacitors is about 1% of the maximum voltage.

D. Turns ratio of the transformer

Finally, to achieve maximum efficiency improvement, the turns ratio of the transformer must be maximized. In fact, since the duty-cycle loss in the converter is negligible due to the smaller leakage inductance of the transformer, the converter can be designed with a larger turns ratio compared to the conventional PSFB converter. Moreover, the secondary–side ringing between the leakage inductance of the transformer and the junction capacitance of the rectifier is significantly reduced because of the smaller leakage inductance. Any residual parasitic ringing can be damped by a simple RCD-snubber circuit.

IV.   EXPERIMENTAL RESULTS

In order to verify the operation principle of converter, a prototype converter was built in the lab with the following parameters.

- Input voltage V_{in} : 300-400VDC;
- Output voltage V_o : 54VDC;
- Maximum output current I_o : 20A;
- Switching frequency f_s : 100kHz;
- Main switches $Q_1 \sim Q_4$: IRFP460 (International Rectifier), and its $R_{DS(on)}$=0.3Ω;
- Rectifier diodes (full-wave rectifier) : DSEI30-06A (IXYS);
- Turns ratio K : 14:3:3;
- Auxiliary inductor L_{a1} : 250μH (core: PQ2020);
- Auxiliary inductor L_{a2} : 100μH (core: PQ2020);
- Auxiliary capacitors C_{a1}&C_{a2}:   2.2μF/250V (polypropylene).

The phase-shift control circuit was implemented using a UCC3895 controller.

Fig. 8 shows the gate signals of PWM switches Q_3 and Q_2 along with their drain-to-source voltage waveforms at 5% load. As seen, the drain voltage reaches zero before the gate reaches its threshold demonstrating zero-voltage turn-on.

Fig. 9 shows the currents flowing through the two auxiliary inductors with different peak value. The changing current can help realize ZVS for both legs from no load to full load.

Fig. 10 shows the key waveforms of the proposed FB ZVS converter. As can be seen from the corresponding waveform in Fig. 10, the proposed converter has a very small duty cycle loss as well as a very much decreased parasitic ringing because of the smaller leakage of the power transformer. And we can see from the active state to the passive state the primary current has a downward step caused by the junction capacitance of output rectifier diodes discharging.

Fig. 11 shows the measured efficiencies as functions of output current at V_{in}=400V. Generally, the efficiency improvement is more pronounced at light loads where the conventional FB ZVS converter operates with hard switching. Meanwhile, the ZVS operation is helpful to reduce EMI problem at very light load. By increasing the turns ratio of the transformer, both the conduction loss of the primary switches and the voltage stress on the components at the secondary side is decreased.

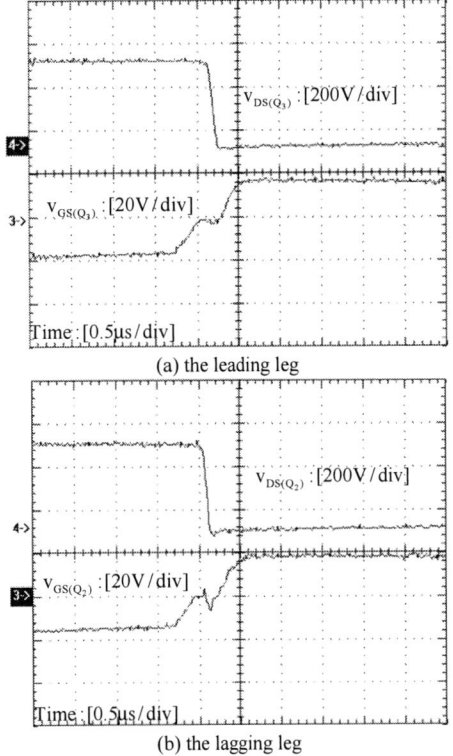

(a) the leading leg

(b) the lagging leg

Fig. 8. Drive voltage and drain to source voltage of the leading lag and lagging leg at 5% load current.

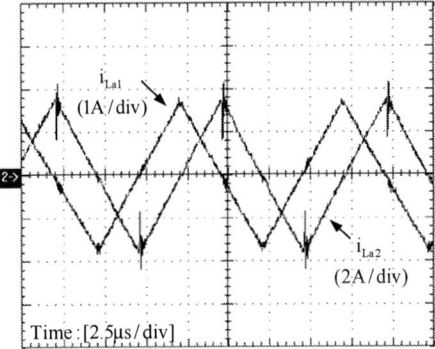

Fig. 9. Current through auxiliary inductors at 5% load current.

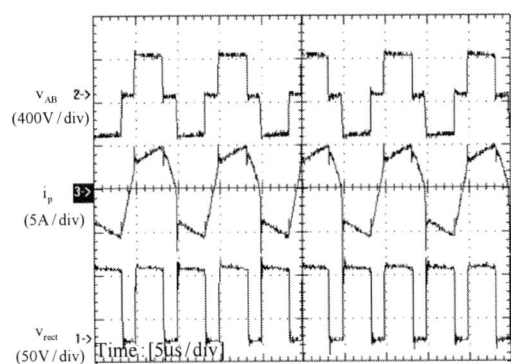

Fig. 10. Experimental waveforms of v_{AB} and secondary voltage v_{rect} at full load.

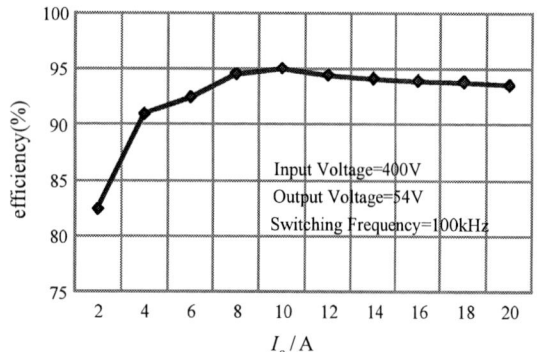

Fig. 11. Measured efficiency of proposed converter as a function of output current.

V. CONCLUSION

In this paper, a novel FB ZVS converter has been proposed, which employs auxiliary network to achieve ZVS in a wide range of load current and input voltage. The auxiliary network has a simple structure which includes two auxiliary inductors and two auxiliary capacitors. The circulating energy and conduction losses have been substantially reduced. Since the two auxiliary inductors do not appear as series inductances, they will not cause the duty cycle loss or severe voltage ringing across the output rectifiers. In addition, it offers improved electromagnetic compatibility and higher efficiency. The operation principle and characteristic of the converter are analyzed in detail. Experimental results from a 1kW/100-kHz prototype confirm the advantages of the proposed configuration.

REFERENCES

[1] O. D. Petterson and D. M. Divan, "Pseudo-resonant full bridge dc/dc converter," in *Proc. IEEE Power Electron. Sepc. Conf.* 1987, pp. 424-430.

[2] L. H. Weene and C. A. Wright, "A 1-kW, 500-kHz front-end converter for a distributed power supply system," in *Proc. IEEE Appl. Power Electron. Conf. (APEC'89),* 1989, pp. 423-432.

[3] W. Chen, F. C. Lee, M. M. Jovanovic, and J. A. Sabate, "A comparative study of a class of full bridge zero-voltage-switched PWM converters," in *Proc. IEEE Appl. Power Electron. Conf. (APEC'95),* 1995, pp. 893-899.

[4] B. P. McGrath, D. G. Holmes, P. J. McGoldrick, and A. D. McIve, "Design of a soft-switched 6-kW battery charger for traction applications," *IEEE Trans. Power Electron.,* vol. 22, no. 4, pp. 1136-1144, Jul. 2007.

[5] S. Y. Lin and C. L. Chen, "Analysis and design for RCD clamped snubber used in output rectifier of phase shifted full-bridge ZVS converters," *IEEE Trans. Ind. Electron.,* vol. 45, no. 2, pp. 358-359, Apr. 1998.

[6] J. A. Sabate, V. Vlatkovic, R. B. Ridley, and F. C. Lee, "High-voltage, hign-power, ZVS, full-bridge PWM converter employing an active snubber," in *Proc. IEEE Appl. Power Electron. Conf. (APEC'91),* 1991, pp. 158-163.

[7] R. Redl, N. O. Sokal, and L. Balogh, "A novel soft-switching full-bridge DC/DC converter: analysis, design considerations, at 1.5kW, 100kHz," *IEEE Trans. Power Electron.,* vol. 6, no. 3, pp. 408-418, Jul. 1991.

[8] G. Hua, F. C. Lee, and M. M. Jovanovic, "An improved full-bridge zero-voltage-switched PWM converter using a saturable inductor," *IEEE Trans. Power Electron.,* vol. 8, no. 4, pp. 530-534, Oct. 1993.

[9] R. Waston and F. Lee, "Analysis, design, and experimental results of a 1-kW ZVS-FB-PWM converter employing magamp secondary-side control," *IEEE Trans. Ind. Electron.,* vol. 45, no. 5, pp. 806-813, Oct. 1998.

[10] R. Ayyanar and N. Mohan, "Novel soft-switching DC-DC converter with full ZVS-range and reduced filter requirement—Part I: regulated-output applications," *IEEE Trans. Power Electron.,* vol. 16, no. 2, pp. 184-192, Mar. 2001.

[11] K. Park, C. Kim, G. Moon, and M. Youn, "Voltage oscillation reduction technique for phase-shift full-bridge converter," *IEEE Trans. Ind. Electron.,* vol. 54, no. 5, pp. 2779-2790, Oct. 2007.

978-1-4244-4782-4/10 $26.00 © 2010 IEEE

An Active Clamp ZVT Converter with Input-parallel and Output-series Configuration

Yi Zhao, Wuhua Li, Weichen Li, Xiangning He
College of Electrical Engineering, Zhejiang University
Hangzhou, 310027, P.R. China
Email: woohualee@zju.edu.cn

Abstract—**A novel converter is proposed in this paper to satisfy the high power, high step-up and isolated conversion requirements. In the proposed converter, the input-parallel configuration is adopted to share the large input current and reduce the conduction losses, while the output-series configuration is adopted to get a high voltage gain. Consequently, a transformer with a low turns ratio can be applied, which makes the transformer optimized easier. Moreover, the active clamp circuits are employed here to reduce the voltage stress of the switches and recycle the energy stored in the leakage inductance, and ZVT is achieved during the whole switching transition for all the active switches, so the switching losses can be reduced greatly. Furthermore, the diode reverse-recovery problem is partly solved due to the leakage inductance. Finally, a 40V-input 400V-output 1kW prototype is built to demonstrate the theoretical analysis, and the maximum efficiency of 94.9% and 93.3% at full load are achieved.**

I. INTRODUCTION

Nowadays, high power, high step-up and high efficiency DC-DC converters with isolation are widely required in renewable generation system and uninterruptable power supply (UPS). In respect that the sources like photovoltaic arrays, fuel cells and acid batteries used in the situation mentioned above are usually below 60V while a 380V DC bus voltage is needed to get a 220V AC output [1~3], and additionally, galvanic isolation is usually required to meet the safety standards in these applications. Therefore, high-frequency transformers are commonly employed in the converter [4, 5].

To obtain good performances in these applications, the converter should handle the high input current and the high output voltage. Clearly, classic hard switching flyback or forward converters are not good candidates for these applications. The drawbacks of them are the following: the input current ripple is large, the voltage stresses of the output diode are high, and it's difficult to design a good performance transformer with a large turns ratio [6].

This work is sponsored by the National Nature Science Foundation of China (50907058), the Power Electronics S&E Development Program of Delta Environmental & Education Foundation (DREM2009001) and the China Postdoctoral Science Foundation (20080440197).

In order to solve the above-mentioned problems, a novel isolated ZVT Boost converter is proposed in this paper. The input-parallel and output-series configuration is adopted to handle the high input current and high output voltage effectively [7, 8]. On the primary side of the converter, interleaved control technique is employed to share the input current, which can partly cancel the input current ripple and decrease the conduction losses. Meanwhile, the active clamp circuits are induced to depress the voltage stress of the main switches and recycle the energy stored in the leakage inductance. Both the main switches and the clamp switches are ZVT performances during the whole switching cycle, therefore, the switching losses are reduced significantly [7~9]. On the secondary side, the windings are connected in series to get a high voltage. When one coupled inductor works in the flyback mode, the other coupled inductor works in the forward mode due to the interleaved control. Thus the coupled inductors transfer the energy when the corresponding switch is in the on-state and in the off-state. So the coupled inductors are fully utilized, which reduces the volume of the magnetic materials. Besides, the secondary diodes reverse-recovery problem is partly solved because of the leakage inductance, which is in series with the diodes. The current falling rate of the secondary diodes can be controlled by adjusting the inherent leakage inductance of the coupled inductors.

II. PROPOSED CONVERTER AND PRINCIPLE OPERATION

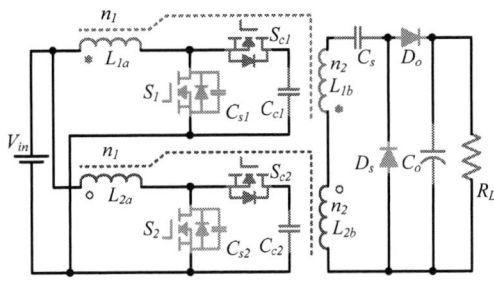

Figure 1. Proposed converter

The proposed converter is illustrated in Fig. 1. There are two coupled inductors, and each of them has two windings.

L_{1a} is coupled to L_{1b}, and L_{2a} is coupled to L_{2b}. The coupling reference is marked by "*" and "°". The turns ratio of the coupled inductors is expressed by $N=n_2/n_1$.

The coupled inductor is modeled as an ideal transformer with the original turns ratio, which is in parallel with a magnetizing inductance and then in series with a leakage inductance [10]. The equivalent circuit of the proposed converter is illustrated in Fig. 2. C_{s1} and C_{s2} are the parallel capacitors for the active switches to implement zero-voltage-switching (ZVS) soft switching performance, which includes the parasitic capacitances of the switches.

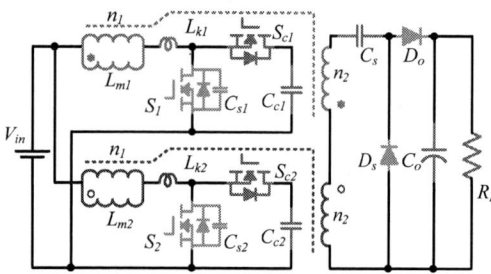

Figure 2. equivalent of the proposed converter

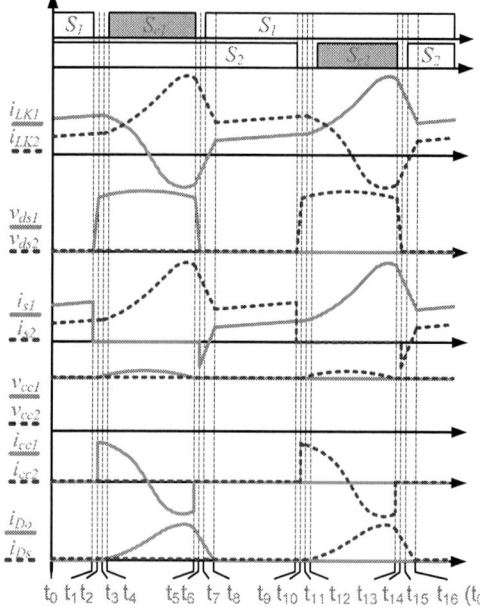

Figure 3. Key waveforms of the proposed converter

The key waveforms of the proposed converter are shown in Fig. 3. There are sixteen main subintervals in one switching period, and only eight of them are analyzed in detail to describe the operation process of one main switch and its corresponding clamp switch because the primary side circuit is symmetrical. The equivalent circuits for each subinterval are shown in Fig. 4.

Subinterval 1 [t_0~t_1]: Before t_1, the main switches S_1 and S_2 are in the on-state, and the clamp switches S_{c1} and S_{c2} are in the off-state. The secondary side diodes D_o and D_s are reverse-

biased. The input voltage causes the currents of the magnetizing inductances L_{m1}, L_{m2} and the leakage inductances L_{k1}, L_{k2} to increase linearly.

Subinterval 2 [t_1~t_2]: At t_1, the main switch S_1 is turned off, and then the capacitor C_{s1} is charged by the magnetizing current. The voltage of C_{s1} increases almost with a constant slope. Due to the capacitor C_{s1}, the main switch S_1 turns off with ZVS.

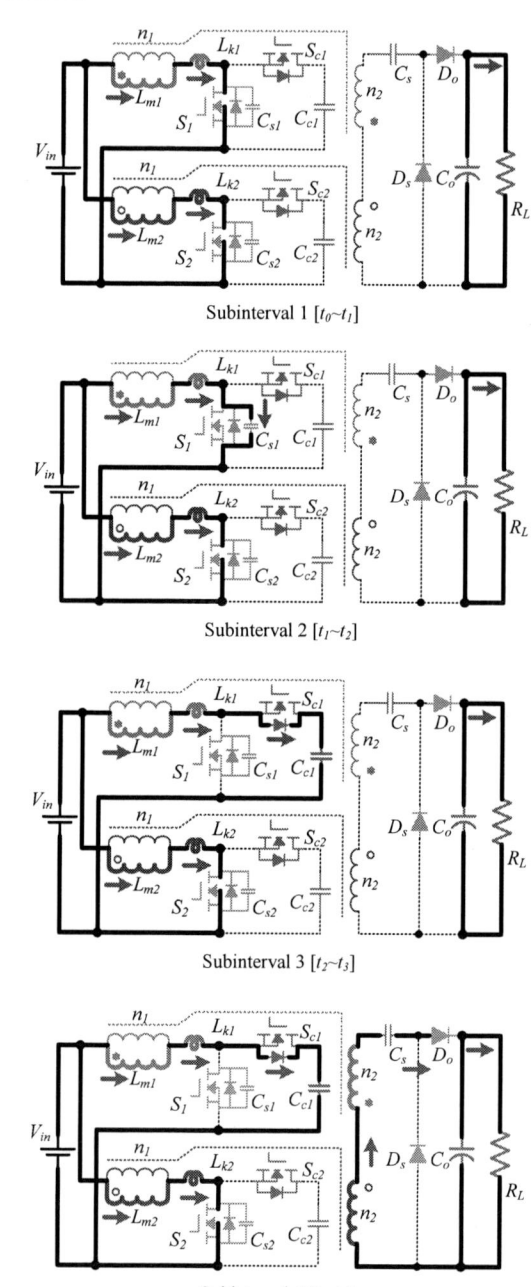

Subinterval 1 [t_0~t_1]

Subinterval 2 [t_1~t_2]

Subinterval 3 [t_2~t_3]

Subinterval 4 [t_3~t_4]

978-1-4244-4782-4/10 $26.00 © 2010 IEEE

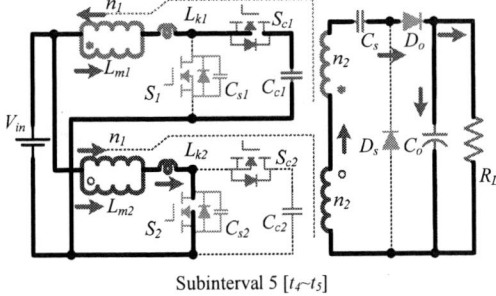

Subinterval 5 [t_4~t_5]

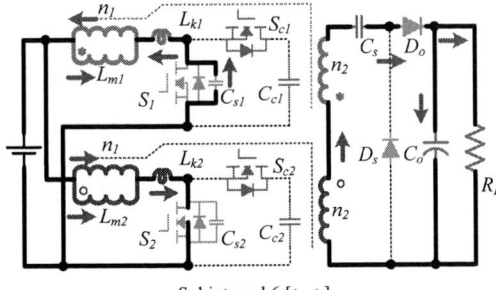

Subinterval 6 [t_5~t_6]

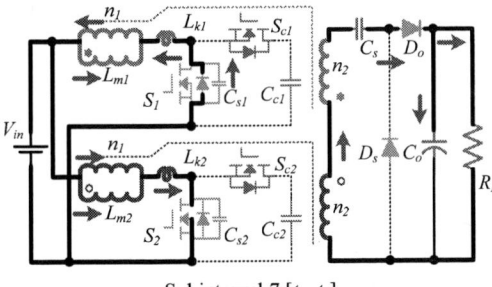

Subinterval 7 [t_6~t_7]

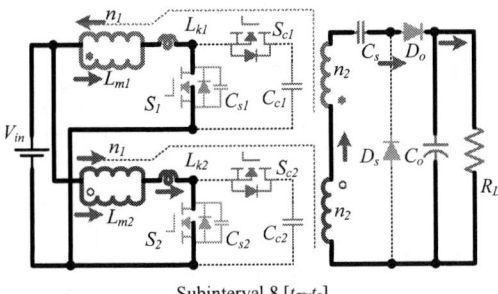

Subinterval 8 [t_7~t_8]

Figure 4. Operation processes of the proposed converter

Subinterval 3 [t_2~t_3]: At t_2, the voltage of the main switch S_1 reaches the voltage of the clamp capacitor C_{c1}, and the anti-parallel diode of the clamp switch S_{C1} begins to conduct. Then the anti-parallel diode clamps the voltage of the main switch S_1 to that of C_{c1}. Since C_{c1} is much greater than C_{s1}, most of the magnetizing current flows through C_{c1}.

Subinterval 4 [t_3~t_4]: At t_3, the output diode D_o begins to conduct since the voltage of D_o is falls to zero, and then the

energy is transferred to the load. During this period, the energy stored in the magnetizing inductance L_{m1} is transferred to the load, while the other coupled inductor transfers the energy to the secondary side as a transformer. The series capacitor C_s is discharged. The leakage inductance L_{k1} and the clamp capacitor C_{c1} resonate. The magnetizing inductance L_{m2} is still charged by the input voltage, and the current flowing through S_2 and L_{k2} contains the magnetizing inductance current and the reflected secondary winding current.

Subinterval 5 [t_4~t_5]: The turn-on signal of S_{c1} is given at t_4. The clamp switch is turned on with ZVS because its anti-parallel diode is in the on-state. The equivalent circuit of this subinterval is similar to the previous one.

Subinterval 6 [t_5~t_6]: The clamp switch S_{c1} is turned off at t_5. Then the clamp capacitor C_{c1} is disconnected from the resonant circuit, and L_{k1} begins to resonate with C_{s1}. The energy stored in C_{s1} is transferred to L_k. While the voltage of the parallel capacitor C_{s1} decreases, the voltage of S_{c1} increases from zero with the same rate. Thus the clamp switch S_{c1} is turned off with ZVS.

Subinterval 7 [t_6~t_7]: The voltage of C_{s1} falls to zero at t_6, and the anti-parallel diode of S_1 begins to conduct. C_{s1} and L_{k1} stop resonating. The output voltage causes the current through L_{k1} increases linearly, which controls the current decreasing rate of the output diode D_o.

Subinterval 8 [t_7~t_8]: The turn-on signal of the main switch S_1 is given when its anti-parallel diode is in the on-state, and S_1 is turned on with ZVS. Other parts of the converter works in the similar way as in subinterval 7 until the current through the diode D_o falls to zero at t_8, and the current through L_{k1} becomes equal to that of the magnetizing inductance Lm1 at the same time. After t_8, the output diode is reversed-biased, and L_{m1} and L_{k1} are charged by the input voltage again.

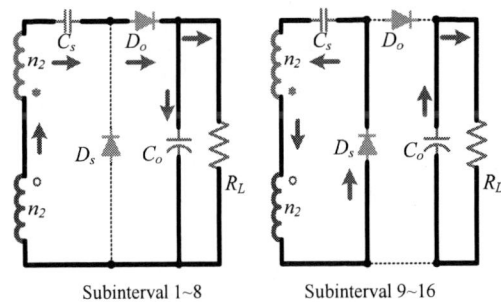

Subinterval 1~8   Subinterval 9~16

Figure 5.   Current paths on the secondary side

The operation processes are the similar in the remaining subintervals. On the primary side of the converter, the switches S_2 and S_{c2} operate coordinately, and the commutation processes between S_2 and S_{c2} can be obtained from that between S_1 and S_{c1} due to the symmetry of the circuit. On the secondary side of the converter, the current flows through D_s instead of D_o, and the series capacitor C_s is charged in these subintervals. The current paths on the secondary side in a switching period are illustrated in Fig. 5.

III. PERFORMANCE ANALYSIS

In the analysis, the parameters of the symmetrical circuits are considered the same to simplify the calculation. The voltage of the clamp capacitors and the current of the magnetizing inductances are assumed to be constant. The effects of leakage inductances and the dead time between the main switch and the clamp switch are neglected.

$$L_{m1} = L_{m2} = L_m \qquad (1)$$

$$C_{c1} = C_{c2} = C_c \qquad (2)$$

The detailed performances of the converter are analyzed as follow:

A. Voltage Gain

The principle of inductor volt-second balance is applied to obtain the voltage gain of this converter. The magnetizing inductance has two main states in one switching period. In one state, the main switch is in the on-state and the clamp switch is in the off-state, and the magnetizing inductance is charged by the input voltage.

$$V_{Lm\_charge} = V_{in} \qquad (3)$$

In the other state, the main switch is in the off-state and the clamp switch is in the on-state. Due to the symmetry of the circuit, the voltage of the series capacitor C_s is exactly half of the output voltage, and the discharge voltage across the magnetizing inductance is given by:

$$V_{Lm\_discharge} = \frac{V_{out}}{2 \cdot N} - V_{in} \qquad (4)$$

According to (3) and (4), the voltage gain of this converter can be expressed as following:

$$M = \frac{V_{out}}{V_{in}} = \frac{2 \cdot N}{1 - D} \qquad (5)$$

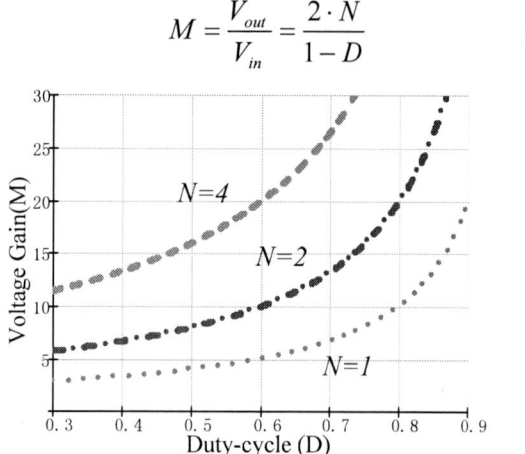

Figure 6.   Voltage conversion ratio of the converter

Due to (5), the voltage gain of the proposed converter is much higher than that of the classic flyback and forward converters. So the same voltage gain can be derived by using lower turns ratio coupled inductors. The relationship among the conversion ratio, the duty-cycle and the turns ratio are sketched in Fig. 6.

B. Voltage Stress of Switches

The voltage stress of the main switch is equal to the voltage across the clamp capacitor. The voltage stress is given by:

$$V_{stress\_main} = \frac{V_{in}}{1 - D} = \frac{V_{out}}{2 \cdot N} \qquad (6)$$

The voltage stress of the clamp switch is equal to that of the main switch. The voltage stress of the active switches is relative to the turns ratio of the coupled inductors, therefore a proper turns ratio can be designed to depress the voltage stress of the active switches, and a low on-resistance MOSFET can be adopted to increase the efficiency.

C. Effects of the Leakage inductance

The energy in the leakage inductance is recycled, and the leakage inductance causes both the main switch and the clamp switch to work with ZVS, accordingly the losses are reduced greatly. Meanwhile, the leakage inductance limits the variation rate of the current, and it would partly solve the EMI problem. Especially to the secondary-side diode, the series leakage inductance can alleviate the reverse-recovery problem significantly. The current falling rate of both D_o and D_s are controlled by the leakage inductance.

$$\frac{di_D(t)}{dt} = \frac{V_{out}}{4 \cdot N^2 \cdot L_k} \qquad (7)$$

IV. EXPERIMENT RESULT

A 1kw prototype of proposed converter is built to verify the analysis.

The parameters of the converter are shown as follow: P_{out}: 1000W; V_{out}: 400V; V_{in}: 40V; f_s: 50 kHz; n_2/n_1: 25/44; L_m: 85μH; L_k: 1.2μH; C_c: 2.2μF; C_s: 9.4μF; S_1 and S_2: two pieces of IRFP250N in parallel; S_{c1} and S_{c2}: IRFP250N; D_s and D_o: MUR1560.

The experimental result of the gate signals and the voltage stress of both the main switches and the clamp switches are shown in Fig. 7. A proper duty-cycle is obtained from the reasonable design of the coupled inductors.

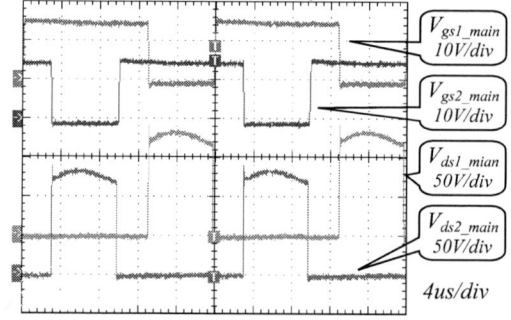

(a) Gate signals and voltage stress of the main switches

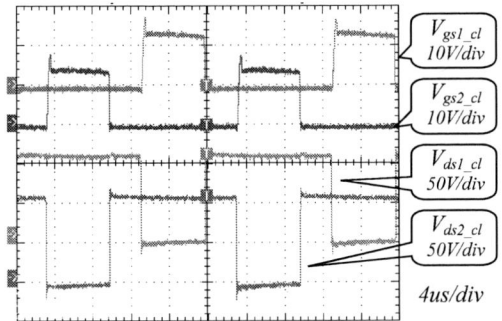

(b) Gate signals and voltage stresses of the clamp switches

Figure 7. Gate signals and voltage stresses of the active switches

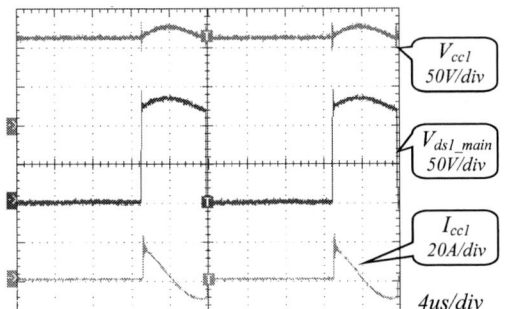

Figure 8. Performance of the clamp circuit

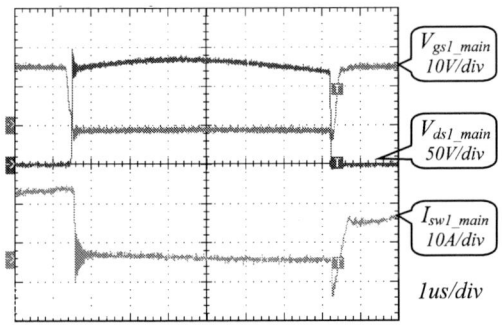

(a) ZVT performance for the main switch

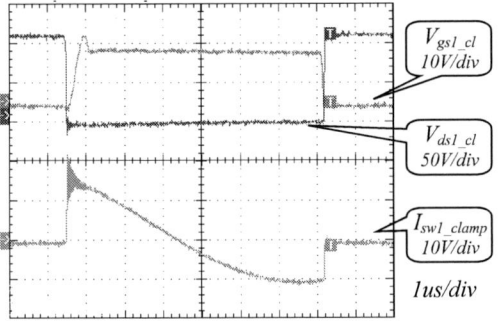

(b) ZVT performance for the clamp switch

Figure 9. ZVT performance for the active switches

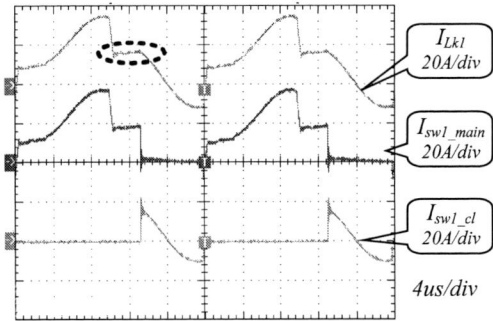

Figure 10. Currents on the primary side

Fig. 8 shows the performance of the clamp circuit. Clearly, when the main switch is turned off, the current begins to flow through the clamp circuit, and voltage of the main switch is clamped to that of the clamp capacitor, hence the voltage spikes caused by the leakage inductance are small enough to be ignored.

Fig. 9 illustrates the switching transition of the main switch and the clamp switch. All the active devices in this converter are switched with ZVS, which reduces the switching loss greatly.

Fig. 10 shows the current through the leakage inductance, the current through the main switch and the current through the clamp circuit. The current marked in this figure represents the magnetizing inductance current.

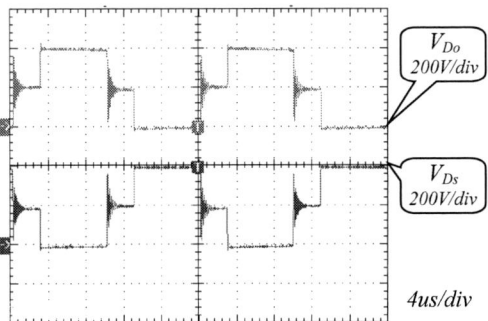

Figure 11. Voltage stress of the secondary side diodes

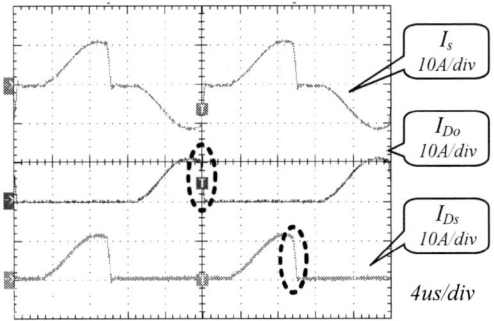

(a) The currents on the secondary side

978-1-4244-4782-4/10 $26.00 © 2010 IEEE

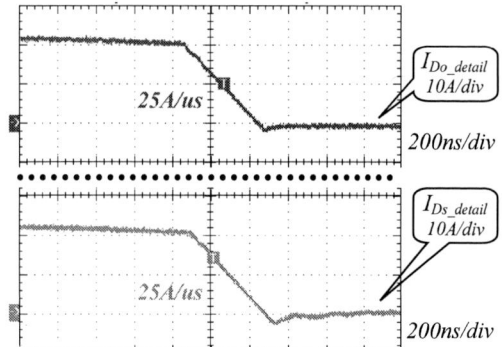

(b) Details of the current through the diodes

Figure 12. The current flow on the secondary side

The voltage stresses of the secondary-side diodes are shown in Fig. 11. The voltage stresses of both the diodes are equal to the output voltage.

Fig. 12 (a) illustrates the current through the secondary winding of the coupled inductor, the current of the switched diode and current through the output diode. The details of the current decreasing transition marked in Fig. 12 (a) are shown in Fig. 12 (b). The falling rates of the current through the diodes are limited by the leakage inductance, and the falling rate of the currents is in accord with the theoretical analysis well.

The measured efficiency of the proposed converter at different loads is sketched in Fig. 13. The maximum efficiency of 94.9% and 93.3% efficiency at full load are obtained in the experiment.

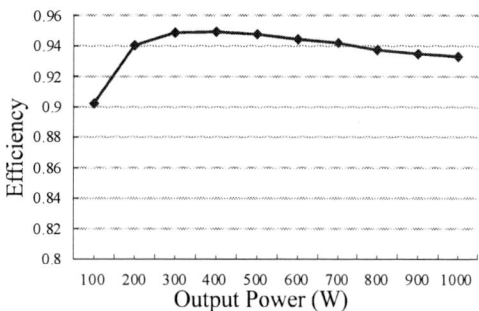

Figure 13. Measured efficiency of the proposed converter

V. CONCLUSIONS

This paper introduces a novel isolated converter with input-parallel and output-series configuration for high step-up applications. The input-parallel configuration helps to share the large input current and reduce the conduction loss and the output-series configuration helps to get a high voltage gain, meanwhile a lower turns ratio is obtained to simplify the optimized design of the transformer. The active clamp circuit suppresses the voltage spike of the main switches and recycles the energy stored in the leakage inductance, and both the main switches and the clamp switches work with ZVS. The diode reverse-recovery problem is alleviated by the series leakage inductance. A 1kW, 40V-400V high efficiency prototype converter verifies the analysis, and illustrates that the proposed converter is a competitive candidate for high power and high step-up applications with isolation requirement.

REFERENCES

[1] S. B. Kjaer, J. K. Pedersen and F. Blaabjerg, "Power Inverter Topologies for Photovoltaic Modules-A Review", In Proc. IEEE 2002 IAS Conf., 2002, vol.2, pp.272-288.

[2] O. Krykunov, "Analysis of the Extended Forward Converter for Fuel Cell Applications", In IEEE 2007 ISIE Conf., 2007, pp.661-666.

[3] J. K. Park, W. Y. Choi and B. H. Kwon, "A Step-Up DC–DC Converter With a Resonant Voltage Doubler" IEEE Transactions on Industrial Electronics, Volume.54, Issue.6, pp.3267-3275, Dec. 2007.

[4] R. P. Torrico-Bascope, D. S. Oliveira, Jr., C. G. C. Branco and F. L. M. Antunes, "A UPS With 110-V/220-V Input Voltage and High-Frequency Transformer Isolation", IEEE Transactions on Industrial Electronics, Volume.55, Issue.8, pp.2984-2996, Aug. 2008.

[5] Wuhua Li and Xiangning He, "A Family of Isolated Interleaved Boost and Buck Converters with Winding-Cross-Coupled Inductors", IEEE Transactions on Power Electronics, Volume.23, Issue.6, pp.3164-3173, Nov. 2008.

[6] H. D. Thai, J. Barbaroux, H. Chazal, Y. Lembeye, J. C. Crebier and G. Gruffat, "Implementation and Analysis of Large Winding Ratio Transformers", in IEEE 2009 Applied Power Electronics Conference and Exposition, 2009, pp. 1039-1045.

[7] Dong Wang, Yan Deng and Xiangning He, "Isolated ZVT Boost converter with Switched Capacitors and Coupled Inductors" In IEEE 2008 IECON, 2008, pp. 808-814.

[8] J. M. Kwon and B. H. Kwon, "High Step-Up Active-Clamp Converter with Input-Current Doubler and Output-Voltage Doubler for Fuel Cell Power Systems" IEEE Transactions on Power Electronics, Volume.24, Issue.1, pp.108-115, Jan. 2009.

[9] S. Y. Tseng, J. Z. Shiang, W. S. Jwo and C. M. Yang, "Active Clamp Interleaved Boost Converter with Coupled Inductor for High Step-up Ratio Application", In IEEE 2007 PEDS Conf., 2007, pp.1394-1400.

[10] Qun Zhao and F. C. Lee, "High-efficiency, High Step-up DC-DC Converters", IEEE Transactions on Power Electronics, Volume.18, Issue.1, Part.1, pp.65-73, Jan. 2003.

A Parallel Front-End LCL Resonant Push-Pull Converter with a Coupled Inductor for Automotive Applications

Yuan Yisheng
Electrical Engineering College
East China Jiaotong University
Nanchang, China
Cloudstone_yuan@yahoo.com.cn

Chen Min
Electrical Engineering College
Zhejiang University
Hangzhou, China
Calim@cee.zju.edu.cn

Qian Zhaoming
Electrical Engineering College
Zhejiang University
Hangzhou, China
qian@zju.edu.cn

Abstract—**A parallel technique with a coupled inductor is proposed for an LCL resonant push-pull converter. The coupled inductor is inserted in the front of the common resonant capacitor of two parallel push-pull converters, sharing the current of the two converters. The equivalent circuit and state equation of the proposed converter is described, and then the effects of the mismatch of the equivalent resistor and leakage inductor on sharing current is derived. The controller of the cascaded full-bridge inverter is presented. Experimental results of a 2kW inverter showed that the current difference of the two parallel units is less than 2%.**

I. INTRODUCTION

Automotive inverters supplied by a 12VDC battery source feature a low input voltage and high input current. Thus by comparison with its back-end dc-ac inverter, its front-end dc-dc converter always account for major losses in entire converter system. Because of its low cost, push-pull converters always be prior topology applied as front-end dc-dc converters. To further decrease switching losses in hard push-pull converters, some soft-switching push-pull converters have been investigated. An active-clamped push-pull converter[1] can achieve zero-voltage turn-on of main switch and auxiliary switch and clamp the voltage spike of the main switch, but its main switch and auxiliary switch still hard switch off. An LCL resonance push-pull converter[2] with a fixed duty cycle can achieve zero-voltage turn-on of the switches, and even zero-current turn-off if the resonant period is proper designed.

For higher power applications, paralleling dc-dc converter is a good choice than paralleling too many power MOSFETs. There are active and passive paralleling techniques. Droop current-sharing method[3] is the most widely adopted active paralleling technique, but it increases the controller cost and more complicated. Passive current-sharing method[4] needs a large coupled-inductor and increases size, but it is simpler and more credible.

This paper proposed a parallel technique with a coupled inductor for the LCL resonant push-pull converter. The equivalent circuit is described. The effect of the coupled inductor on current sharing is investigated. The control method of the back-end inverter is explained. This converter has been incorporated into a 2kW industrial prototype. The experimental results verified the converter operation and performance.

II. FRONT-END PUSH-PULL CONVERTER CONFIGURE AND OPERATION PRINCIPLE

The proposed converter showed in Fig.1 consists of two parallel LCL resonant push-pull converters. The converter operates in a maximum duty cycle of 0.5 for achieving a zero-current turn-on for power switches. It steps up the 12VDC voltage to a high bus voltage u_{bus}. The back-end full-bridge dc-ac inverter then modulates u_{bus} to ouput ac voltage u_{out}.

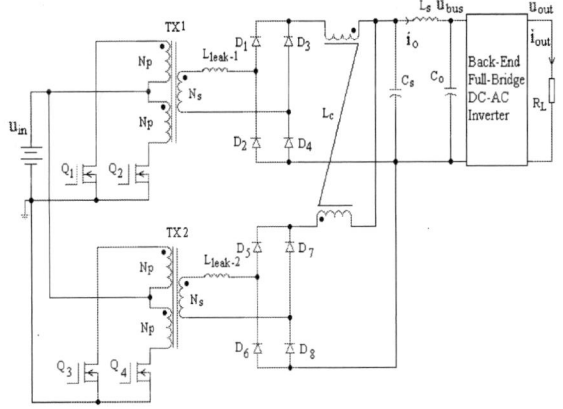

Figure 1. The proposed inverter circuit

In Fig.1, leakage inductor L_{leak-1} of transformer TX1 and C_s constitute the resonant circuit for the upper converter, and leakage inductor L_{leak-2} of TX2 and C_s is another resonant circuit for the lower converter. The two converters are connected in parallel through the coupled inductor L_c. Filter inductor L_s is far more in the inductance than L_{leak-1} and L_{leak-2}, which isolated the left resonant circuit and right bulk capacitor C_o.

The power switches are driven by fixed gate signals less than 0.5. During the power switches conducting time, input energy transfer to the secondary side of the transformer by resonance mode. Because the inductance of inductor L_s is far more than that of the leakage inductors and it has a small ripple current, so the inductor L_s can be taken placed by a current source in the equivalent circuit.

The equivalent circuit of the parallel converter is showed in Fig.2. Current source Io takes place of the inductor Ls in the figure. Voltage source us represents the secondary side of the transformer. Equivalent R includes the on-resistors of the MOSFETs and diodes, primary side winding and secondary side winding resistors of the transformer.

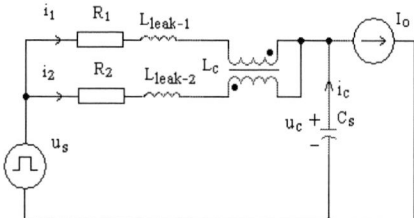

Figure 2.   Equivalent Circuit

Assuming the magnetizing inductance of the coupled inductor L_c is Lm, and $u_{c(0-)}$ is the initial voltage of resonant capacitor C_s. The state-equation of the equivalent circuit is written as followed:

$$\begin{cases} u_s = R_1 i_1(t) + \left(L_{leak-1} + L_c\right)\dfrac{di_1(t)}{dt} - L_m \dfrac{di_2(t)}{dt} + u_c(t) \\[2mm] u_s = R_2 i_2(t) + \left(L_{leak-2} + L_c\right)\dfrac{di_2(t)}{dt} - L_m \dfrac{di_1(t)}{dt} + u_c(t) \\[2mm] u_c(t) = u_{c(0-)} - \dfrac{1}{C_s}\int i_c(t)dt \\[2mm] I_o = i_1(t) + i_2(t) + i_c(t) \end{cases} \quad (1)$$

Assuming equivalent resistors R_1 and R_2, equivalent inductors L_{leak-1} and L_{leak-2} is different and their difference is showed as the following equation.

$$\begin{cases} R_1 = R - \Delta R; R_2 = R + \Delta R; L_{leak-1} = L_{leak-2} = L \\[1mm] L_{leak-1} = L - \Delta L; L_{leak-2} = L + \Delta L; R_1 = R_2 = R \end{cases} \quad (2)$$

Then the according current sharing equations are deduced respectively in Equation (3).

$$\begin{cases} \dfrac{i_1(s) - i_2(s)}{i_1(s) + i_2(s)} = \dfrac{\Delta R}{R + s(L + L_c + L_m)} \\[3mm] \dfrac{i_1(s) - i_2(s)}{i_1(s) + i_2(s)} = \dfrac{s \cdot \Delta L_{leak}}{R + s \cdot (L + L_c + L_m)} \end{cases} \quad (3)$$

Averaging Equation (3) during the resonant period T, The total current-sharing equation can be deduced in Equation (4).

$$\frac{I_1 - I_2}{I_1 + I_2} = \frac{\Delta R}{R}[1 - \exp(-\frac{R}{L + L_c + L_m} \cdot T)] + \frac{\Delta L_{leak}}{L + L_c + L_m}\exp(-\frac{R}{L + L_c + L_m} \cdot T) \quad (4)$$

It is obvious that the current difference between I_1 and I_2 is reduced as L_c increases.

III.   DESIGNMENT OF THE FRONT-END PUSH-PULL CONVERTER

For a single LCL resonant push-pull converter without coupled inductor L_c, the secondary side current $i_s(t)$ is given by Equation (5).

$$i_s(t) = 2(u_s - u_{c(0-)}) \cdot \frac{e^{-\xi t} \cdot \sin(\sqrt{\omega_n^2 - \zeta^2} \cdot t)}{\sqrt{Z_c^2 - R^2}} + I_o - I_o \cdot e^{-\xi t} \cdot \cos(\sqrt{\omega_n^2 - \zeta^2} \cdot t), \quad (5)$$

Where

$$\xi = R / 2L_{leak}) , \quad (6)$$

$$\omega_n = 1/\sqrt{L_{leak} \cdot C_s}, \quad (7)$$

$$Z_c = 2\sqrt{L_{leak}/C_s}. \quad (8)$$

By Equation (5), the resonant current $i_s(t)$ is a damping sinusoidal waveform with a positive bias.

A.   Three kinds of Resonant Mode

The resonant current $i_s(t)$ in Equation(5) has three kinds of waveforms: over damped waveform, under damped waveform and critically waveforms as showed in Fig.4. It is obviously that critically damped current is the best, which make the power switches minimum switching-losses with zero-current turn-on transition and nearly zero-current turn-off transition.

It is the best selection to make the current $i_s(t)$ operated in the critically waveform mode to reduce switching-off losses. But the Equation.5 is too complex that it is difficult to solve $i_s(t)$. So these resonant parameters have to be adjusted in practical.

B. Coupled Inductor Designment

By Equation (4), the effect factors on current-sharing degree include L_c, R, L_{leak}. Inductance L_{leak} usually estimated by engineering practical. Resistance R has many involved factors, include windings resistors, diodes on-resistors and MOSFETs on-resistors. The difference of diodes and MOSFETs on-resistors can be found in their datasheets. After R, L_{leak} and the current-sharing degree are validated, finally the coupled inductor can deduced.

IV. CONTROL METHOD OF BACK-END DC-AC INVERTER

The duty cycle of power MOSFETs in proposed front-end converter is fixed to nearly 0.5 in order to achieve LCL resonance and minimum switching losses. So bus voltage u_{bus} varied linearly with u_{in}. This varied u_{bus} make the controller of back-end dc-ac inverter complex. A controller with u_{bus} disturbance suppression is adopted as showed in Fig.5.

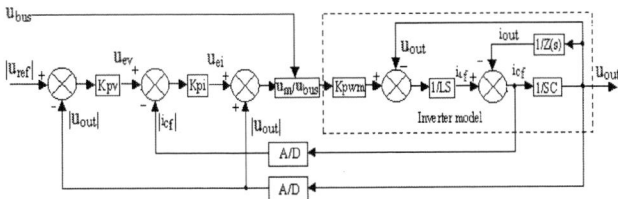

Figure 3. Control blocks of the back-end DC-AC inveter

In this controller, the most outer loop is a voltage loop that feedbacks and controls instantaneous output voltage u_{out}. In this loop, only a proportion factor K_{pv} but no integration factor is adopted to avoid oscillation.

The inner loop is a current loop. This loop control the filter capacitor current i_c but not the inductor i_L, because capacitor current i_c can more direct reflect the vary of the output voltage than inductor i_L. In this current loop, the proportion factor K_{pi} is far more than proportion factor K_{pv}, this proves the current loop has a more quickly response speed than the voltage loop.

The most inner loop is a disturbance suppression loop. In this loop, two disturbance voltage, include the output voltage u_{out} and the bus voltage u_{bus}, are feed-forward and cancelled in the final closed-loop. This simplifies the design of the controller parameters.

The closed-loop S-domain function is deduced and showed in Equation (6).

$$G_c(s) = \frac{K_{pv}K_{pi}}{LCs^2 + (\frac{L_f}{Z(s)} + K_{pi}C_f)s + K_{pv}K_{pi}} \quad (9)$$

Varied u_{bus} disappear in Equation (9). This makes the controller more simply and reliability.

V. EXPERIMENTAL RESULTS

An industrial prototype of the proposed converter is designed. The circuit parameters are listed in Table I.

TABLE I. MAIN CIRCUIT PARAMETERS

U_{in}(V)	N_p:N_s	U_{bus}(V)	U_o(V)	C_s(nF)
10.5~14.5	2:62	325~465	230	282
L_{leak-1}(uH)	L_{leak-2}(uH)	L_c(uH)	L_m(uH)	L_s(uH)
46	48	80	72	800

Each power switch in Fig.1 consists of five MOSFETs of IRF3205Z in parallel. Its typical on-resistor is 4.9mΩ and its maximum on-resistor is 6.5mΩ.

Fig.4, Fig.5 and Fig.6 show the experimental waveforms of input currents i_{in-1} and i_{in-2} of two parallel push-pull converters, inverter output current i_{out} and output voltage u_{out} under different input voltage u_{in} with the same 2kW load. From these figures, it is obvious that the current-sharing degree under different input voltage is very good. The detail test data are listed in Table II.

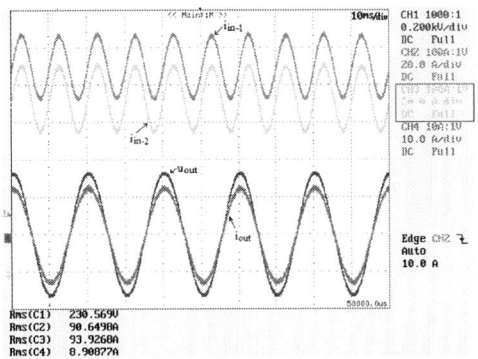

Figure 4. Test waveforms under 12V input voltage and 2kW load

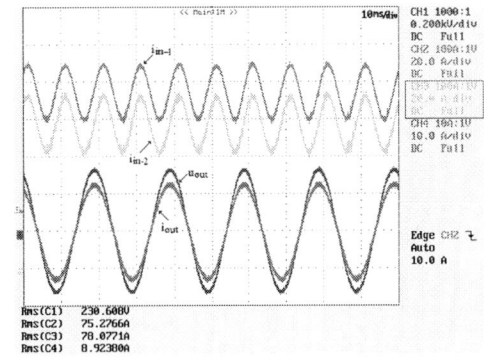

Figure 5. Test waveforms under 14.5V input voltage and 2kW load

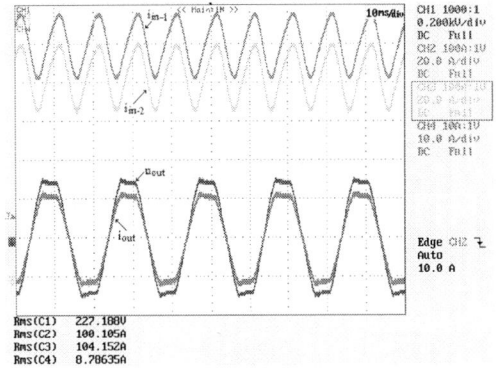

Figure 6. Test waveforms under 10.5V input voltage and 2kW load

TABLE II.   TEST DATA UNDER 2kW LOAD OF THE INVERTER

U_{in}(V)	I_{in-1}(A)	I_{in-2}(A)	Current-difference degree	Efficiency
12	90.65	93.93	1.78%	92.6%
14.5	75.28	78.08	1.83%	92.5%
10.5	100.11	104.15	1.98%	93%

By Table II, the current-difference degree of two parallel push-pull converters is less than 2%, and the inverter has a good efficiency more than 92%.

Fig.6 shows the waveforms of resonant current i_1 and i_2 of secondary side of TX1 and TX2 and their difference ($i_1 - i_2$). It is clear that the MOSFET switch on with zero-current transition and switch off with low turn-off current. So this prototype has a good efficiency by its LCL resonant soft-switching technique.

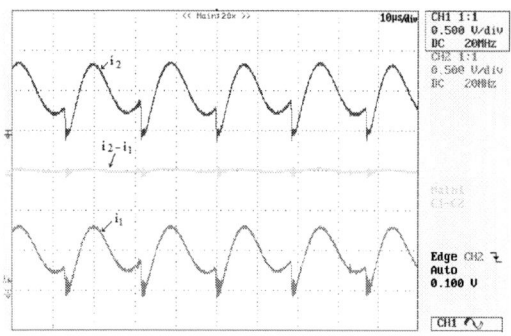

Figure 7.   Secondary side current is (6.67A/V)

VI.   CONCLUSION

A parallel soft-switching push-pull dc-dc converter applied as the front-end converter in automotive applications is proposed and development. By inserting a coupled inductor in the common resonant branch of the two parallel converters, the converter can get a good current-sharing result. Experiment waveforms verified the converter operation and performance. This converter can also be applied in other two-stage inverters or converters.

ACKNOWLEDGMENT

This work is supported by National Natural Science Foundation of China (50577025) and Science and Technology Foundation of Education Department of Jiangxi Province of China (GJJ08240). The researchers also gratefully acknowledge the instruction of Professor Liuchen Chang in the University of New Brunswick in Canada.

REFERENCES

[1]   J. C. Hung, T. F. Wu, J. Z. Tsai, C. T. Tsai and Y. M. Chen, "An Active-Clamped Push-Pull Converter for Battery Sourcing Applications", IEEE APEC 2005, pp.1186-1192.

[2]   Michael J. Ryan, William E. Brumsickle, Deepak M. Divan et al, " A New ZVS LCL-Resonant Push-Pull DC-DC Converter Topology," IEEE trans. On Industry Applications, vol.34, no.5, pp.1164-1174, Sep. 1998.

[3]   Brian T. Irving and Milan M. Jovanovic, "Analysis, Design, and Performance Evaluation of Droop Current-Sharing Method," IEEE APEC 2000, pp.235-241.

[4]   Shen Yanqun, Yao Gang, and He Xiangning, "A Parallel-parallel Combined Two Transistor Forward Converter with Coupled Inductor," Power Electrical Technology, vol.39, no.2, Xian. China, Apri. 2005, pp.53-55.

A Novel Full Bridge Dual Output DC-DC Converter with Complementary Pulse Widths and Frequency Modulation

Yu Chen, Xuejun Pei, Li Peng, Yong Kang
College of Electrical and Electronic Engineering
Huazhong University of Science and Technology
Wuhan, China
Email: ayu03@163.com

Abstract—**A novel dual output DC-DC converter is proposed in this paper. In the proposed converter, one operational period is divided into two complementary parts, and two output branches with complementary pulse widths are structured. One output is fully regulated by the pulse width modulation; the other output is affected by the complementary pulse width but compensated by the switching frequency. Using the variable frequency phase shift modulation, only one full bridge topology is needed but two fully regulated outputs are obtained. And, all switches can achieve ZVS without auxiliary circuit by the appropriate design. Therefore the proposed topology has simply structure, fully regulated ability, high switching frequency and high efficiency. This paper focuses on the steady state analysis and design of the proposed converter. The overview of the operational principle is introduced first; then, the equivalent circuits are obtained and the new characteristics are deduced; based on the discussion, the design procedure is given and a prototype with 300V input, 24V/10A and 48V/5A outputs is build to verify the analysis.**

I. INTRODUCTION

As a special case of multiple output converter, dual output converter can be structured by many multiple output technologies such as PR (post regulation) technology [1, 2], PWM-PD (pulse width modulation-pulse delay) technology [3] and PWM-PS (phase shift) modulation [4, 5]. However, if only two outputs are needed, the additional components and special modulators will increase the complexity and cost of the system distinctly. Moreover, a common cross regulation problem is found: when the main output runs into the empty load, the additional outputs lose the voltage regulation ability. PWM-FM (frequency modulation) suits for the dual output application especially. By changing the operational frequency, an extra degree of freedom is obtained without any auxiliary switches. A single switch two output converter with PWM-FM is presented in [6]. A full bridge dual output converter with PS-FM is presented in [7], However, in both of the two converters, the additional output can only be partly controlled

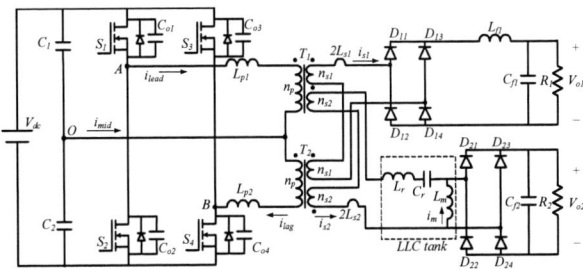

Fig. 1 The proposed dual output converter

by the frequency and affected by the main output, the cross regulation problem still exists.

To overcome the cross regulation problem, a new dual output converter with complementary pulse widths modulation is proposed. As shown in Fig. 1, two identical transformers are used ($n_{p1}=n_{p2}=n_p$, $n_{s11}=n_{s21}=n_{s1}$, $n_{s12}=n_{s22}=n_{s2}$, $K_1=n_p/n_{s1}$ and $K_2=n_p/n_{s2}$). The primary windings n_{p1} and n_{p2} are connected in series; the 1st branch is formed by connecting n_{s11} and n_{s21} in homophase; the 2nd branch is formed by connecting n_{s12} and n_{s22} in reversed phase. With this special-connected method, the voltage across one branch is enhanced while the voltage across the other branch is counteracted.

The variable frequency phase shift modulation is utilized. V_{o1} is fully regulated by the pulse width DT_s when the diagonal switches are turned on; the 2nd branch gets the complementary pulse width $(1-D)T_s$ when the upper or lower switches are turned on. A secondary-located LLC resonant tank[8, 9] is inserted into the 2nd branch to make its output frequency-sensitive. Though the 2nd branch is affected by the duty cycle, it can be compensated by the FM. Furthermore, when the duty cycle of the 1st branch changes from 0 to D_{max}, the 2nd branch always has the available pulse width from $(1-D)T_s$ to unit, the FM thus is always enabled and both outputs are fully regulated anytime.

II. OVERVIEW OF THE OPERATIONAL PRINCIPLE

A. The general characteristics

To further discuss the characteristics of the proposed converter, a general circuit is analyzed first. As shown in Fig. 2, all the components in the primary and secondary sides are summarized as Z_p, Z_{s1} and Z_{s2}. When the two input voltages have the same directions, for example, both the input voltages are positive, the general circuits of the two branches can be shown as Fig. 2(a) and (b), and the relations can be expressed as:

The 1st branch:
$$V_{dc} = 2Z_p \frac{i_{s1}}{K_1} + K_1 Z_{s1} i_{s1} \qquad (1)$$

The 2nd branch:
$$0 = 2Z_p \frac{i_{s2}}{K_2} + K_2 Z_{s2} i_{s2} \qquad (2)$$

According to (1) and (2), the equivalent circuits can be obtained, as shown in Fig. 2(c) and Fig. 2(d), all the parameters are reflected to the secondary side.

When the two voltages have the revise directions, for example, the upper input voltage is negative and the lower one is positive, the equal circuits of the two branches become Fig. 3(a) and (b), and the relations can be expressed as:

The 1st branch:
$$0 = 2Z_p \frac{i_{s1}}{K_1} + K_1 Z_{s1} i_{s1} \qquad (3)$$

The 2nd branch:
$$V_{dc} = 2Z_p \frac{i_{s2}}{K_2} + K_2 Z_{s2} i_{s2} \qquad (4)$$

The equivalent circuits can be obtained, as shown in Fig. 3(c) and Fig. 3(d). Another two symmetrical modes can be analyzed similarly. According to the analyzed, several conclusions can be obtained:

- When the two input voltages have the same directions, the 1st branch is in the powering mode, the voltage sources transfer the energy to the 1st branch; while the 2nd branch is in the freewheeling mode, the voltage sources are removed from the 2nd branch.

- When the two input voltage have the reverse directions, the 2nd branch is in the powering mode. The voltage sources are series with the 2nd branch; while the 1st branch is in the freewheeling mode, the voltage sources are removed from the 1st branch.

- The components in the primary sides of the two transformers are always series. Furthermore, they are the common components that shared by the 1st and 2nd branches.

- Through the two branches share the same primary side, they work independently. The work state in one branch won't affect the other. This feature facilitates the steady analysis a lot.

B. The oprational principle

Based on the general conclusions, the operational principle of the proposed converter can be deduced. Since we focus on the steady state analysis in this paper, the transient ZVS

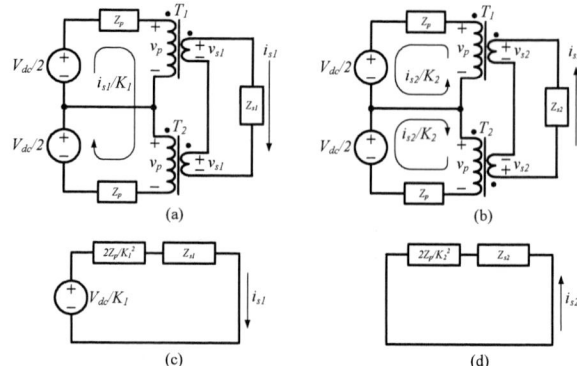

Fig. 2 The equivalent circuit when both the inputs are positive

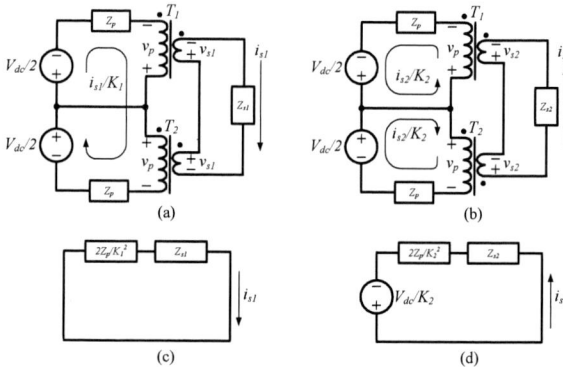

Fig. 3 The equivalent circuit when the upper input is negative while the lower one is positive

switching processes are omitted. There are four steady modes in the proposed converter, as shown in Fig. 4, and they are described concisely as follow:

Mode 1: the analysis begins when S_1 an S_4 are turned on. Both the input sources are positive. This situation is similar to Fig. 2, the 1st branch is in the powering mode and the 2nd branch is in the freewheeling mode. In the secondary side of the 1st branch, the diodes D_{11} and D_{14} are conducting. The large output inductor is series with the primary side, the secondary current i_{s1} increases linearly and slowly. In the secondary side of the 2nd branch, the diodes D_{21}-D_{24} are all blocked and the large resonant inductor L_m is series with the primary side. Its secondary current $i_{s2}(t)$ decreases slowly.

Mode 2: S_1 is turned off and S_2 is turned on. The upper voltage source becomes negative and the lower voltage source is still positive. This situation is similar to Fig. 3, the 1st branch goes into the freewheeling mode. However, the diodes D_{11} and D_{14} are still conducting, i_{s1} decreases slowly. The 2nd branch goes into the powering mode, D_{22} and D_{23} conduct. L_m is removed from the primary side, and i_{s2} changes sinusoidally and rapidly.

Mode 3: S_2 and S_4 are still turned on. The 1st branch keeps in the freewheeling mode. The 2nd branch is still in the powering mode, however, since i_{s2} equals to i_m (the current flowing through L_m) and the diodes D_{22} and D_{23} block naturally, L_m is now series with its primary side again and i_{s2} changes slowly.

978-1-4244-4782-4/10 $26.00 © 2010 IEEE   1465

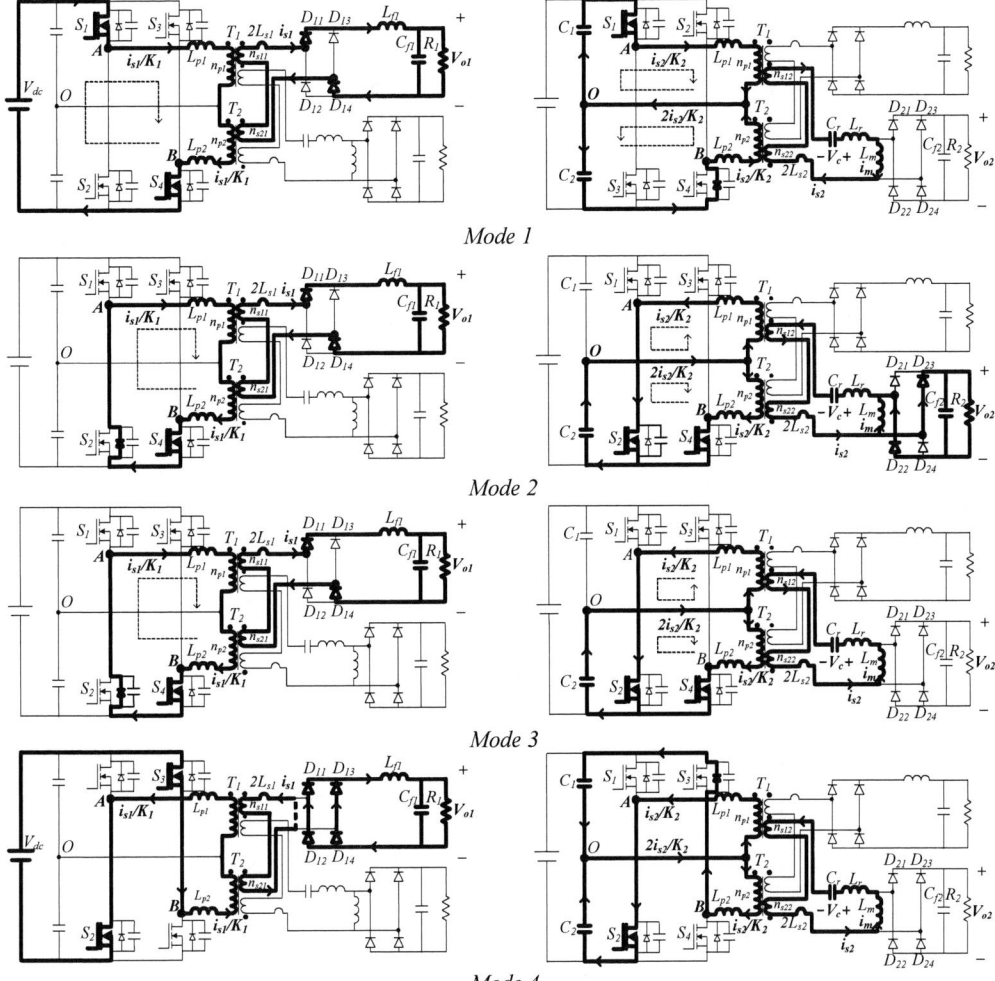

Fig. 4 The operational principle of the steady states

Mode 4: S_2 and S_3 are turned on. The 1st branch is in the powering mode again. However, since i_{s1} is deficient to supply the load current, all the diodes D_{11}-D_{14} conduct to commutate the currents. As a result, the negative pulse width impresses on the leakage inductor directly and i_{s1} decreases linearly and rapidly. The 2nd branch is in the freewheeling mode now, the voltage sources are removed from the 2nd branch. Since L_m is still in series, i_{s2} changes slowly.

When the current i_{s1} is large enough to supply the load, the commutation of currents flowing through D_{11}-D_{14} is completed, D_{12} and D_{13} conduct. The secondary side of the 1st branch is open again, another symmetrical half period begins.

III. ANALYSIS OF THE 1ST BRANCH

A. The equivalent cirtuits

According to the description, the equivalent circuits of the 1st output branches can be drawn, as shown in Fig. 5. $L_{lk}=2L_p/K_1^2+2L_{s1}$ is the equal leakage inductor reflected to the secondary side and $V_{e1}=V_{dc}/K_1$ is the equal input voltage reflected to the secondary side. The leakage inductors of the

two transformers are treated as identical, i.e., $L_{p1}=L_{p2}=L_p$, $L_{s11}=L_{s21}=L_{s1}$ and $L_{s12}=L_{s22}=L_{s2}$.

As shown in Fig. 5(a), when the diagonal switches S_1 and S_4 are turned on during mode 1, the 1st branch gets the available pulse width with DT_s. According to the analysis in section II, the variation of i_{s1} can be expressed as:

$$i_{s1}(t)=I_{s1}(t_0)+\frac{V_e-V_o}{L_{lk}+L_{f1}}(t-t_0) \tag{5}$$

When S_1 is turned off and S_2 is turned on, the 1st branch goes into the freewheeling mode:

$$i_{s1}(t)=I_{s1}(t_1)-\frac{V_o}{L_{lk}+L_{f1}}(t-t_1) \tag{6}$$

When S_4 is turned off and S_3 is turned on, the 1st branch gets the negative pulse width and it goes into the powering mode again. However, since the commutation of the currents flowing through the diodes D_{11}-D_{14} hasn't completed yet, there is no available pulse width for the secondary of the 1st branch:

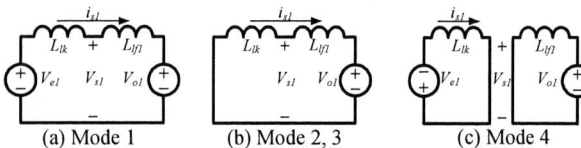

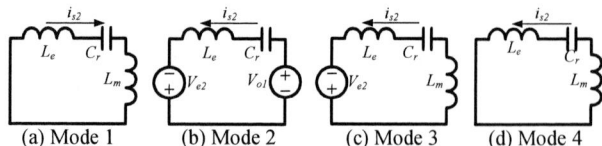

(a) Mode 1   (b) Mode 2, 3   (c) Mode 4

Fig. 5 The equivalent circuits of the 1st branch

(a) Mode 1   (b) Mode 2   (c) Mode 3   (d) Mode 4

Fig. 7 The equivalent circuits of the 2nd branch

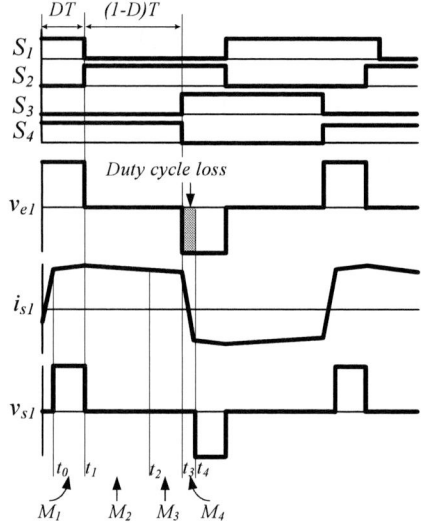

Fig. 6 The key waveforms of the 1st branch

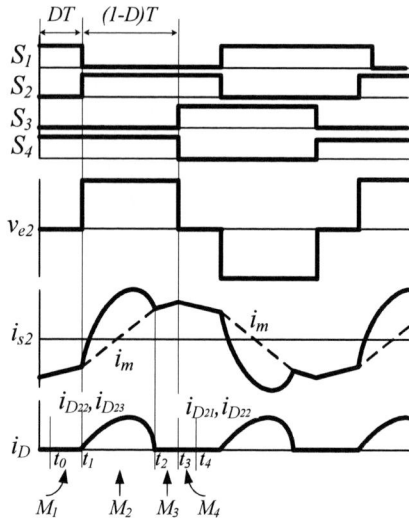

Fig. 8 The key waveforms of the 2nd branch

$$i_{s1}(t) = I_{s1}(t_3) - \frac{V_{e1}}{L_{lk}}(t-t_3) \tag{7}$$

The key waveforms of the 1st branches are shown in Fig. 6. The performance of the 1st branch is completely the same as a conventional phase shift full bridge (PSFB) converter.

B. The maximal duty cycle D_{max}

When the 1st branch is full load, the duty cycle for the 1st branch becomes D_{max} and only $(1 - D_{max})$ left for the 2nd branch. To keep the 2nd output regulated, the switching frequency f_s should be changed. If D_{max} is large, the voltage stress in the rectifier of the 1st branch is low, but the available complementary pulse for the 2nd branch is narrow. To ensure the 2nd output fully regulated, a wide frequency variation range is required. And, if D_{max} is small, the frequency variation is narrow, but the voltage stress in the rectifier of the 1st branch is high. D_{max} thus should be chosen compromisingly.

C. The duty cycle loss D_{loss}

By ignoring the current ripples of i_{s1}, the duty cycle loss can be expressed as:

$$D_{loss} = \frac{4I_{o1}L_{lk}f_s}{V_{e1}} \tag{8}$$

It's seen that when the frequency increases, D_{loss} also increases and the effective duty cycle for the 1st branch decreases. Since the switching frequency is variable in the proposed converter, the serious situation will happen when the 2nd output is empty load and the frequency increases to the

maximal value. In this situation, the 1st branch needs the maximal duty cycle to keep its output regulated.

D. The voltage gain

Compared to the filter inductor L_{f1}, the equal leakage inductor L_{lk} is small enough to be ignored. Based on this approximation, $V_{s1}=V_{e1}$ during mode 1 and $V_{s1}=0$ during mode 2 to mode 4. V_{o1} thus can be expressed as:

$$V_{o1} = V_{e1}(D - D_{loss}) \tag{9}$$

Substituting (8) in (9), the voltage gain of the 1st branch can be obtained:

$$V_{o1} = \frac{V_{dc}D}{K_1 + 4K_1 L_{lk}f_s / R_{o1}} \tag{10}$$

It is seen that V_{o1} is not only depended on the duty cycle D, but also the load condition R_{o1} and switching frequency f_s.

IV. ANALYSIS OF THE 2ND BRANCH

A. The equivalent cirtuits

According to the analysis in section II, the equivalent circuits of the 2nd branch are drawn in Fig. 7. $L_e=2L_p/K_2^2+2L_{s2}+L_r$ is the equal secondary series resonant inductor of the LLC tank and $V_{e2}=V_{dc}/K_2$ is the equal input voltage that reflected to the secondary side of the 2nd branch.

As shown in Fig. 7(a), the equivalent inductor L_e is series with L_m in mode 1, the resonant period thus can be obtained as:

$$T_{r1} = 2\pi\sqrt{C_r(L_e + L_m)} \tag{11}$$

978-1-4244-4782-4/10 $26.00 © 2010 IEEE   1467

T_{r1} is much larger than the operational period, i_{s2} can be treated as increasing linearly. As shown in Fig. 7(b), in mode 2, the large inductor L_m is removed from the circuit, the resonant period thus becomes:

$$T_{r2} = 2\pi\sqrt{C_r L_e} \qquad (12)$$

Once i_{s2} equals to i_m, the large inductor L_m is in series with L_e again, as shown in Fig. 7(c). Notice that the 2nd branch is still in the powering mode and gets the available pulse V_{e2}, i_{s2} is increasing. However, since the resonant period has returned to T_{r1}, i_{s2} increases linearly and slowly.

When the diagonal switches are turned on in mode 4, there is no available pulse for the 2nd branch, as shown in Fig. 7(d), since L_m is still in series with L_e, i_{s2} decreases linearly and slowly. The key waveforms are shown in Fig. 8. In fact, the 2nd branch works as the conventional LLC resonant converter with duty cycle and frequency control.

B. The voltage gain

The voltage gain of the resonant converter, especially the multiple resonant converters, is difficult to obtain exactly, because not only the fundamental component but also the high order harmonic components transfer the energy. To simplify the analysis, the voltage gain is deduced using the first harmonic approximation (FHA) method [10]. Furthermore, since the available duty cycle for the 2nd branch is $1-D$ and it is variable, the first harmonic component of the input voltage should be modified as:

$$V_{f1} = \frac{4V_{e2}}{\pi}\sin[\frac{\pi}{2}(1-D)] \qquad (13)$$

where V_{f1} is the first harmonic voltage component applied to the 2nd branch.

The voltage gain of the 2nd output thus can be obtained as:

$$V_{o2} = \frac{V_{dc}\sin[\frac{\pi}{2}(1-D)]}{K_2\sqrt{\left[1+\frac{1}{h}-\frac{f_{r2}^2}{f_s^2 h}\right]^2 + Q^2\left[\frac{f_{r2}}{f_s} - \frac{f_s}{f_{r2}}\right]^2}} \qquad (14)$$

where $h = L_m/L_e$, $f_{r2} = 1/T_{r2}$ and $Q = \frac{\pi^2\sqrt{(L_r + 2L_{s2} + 2L_p/K_2^2)/C_r}}{8K_2^2 R_{o2}}$.

It is emphasized that the high order harmonic components also play an important role in the LLC resonant converter. When $1-D$ decreases, the weight of the high order harmonic components increases distinctly. Since only the first harmonic component is taken into consideration, the result will deviate from the actual value. However, as the original design guidance, equation (14) is enough.

V. ANALYSIS OF THE PRIMARY SIDE

The components in the primary side are shared by the two branches, some new features thus appear. The most differences are the primary currents flowing through the leading and lagging leg. According to the Kirchhoff's current law, the primary and secondary current relations can be obtained as follow:

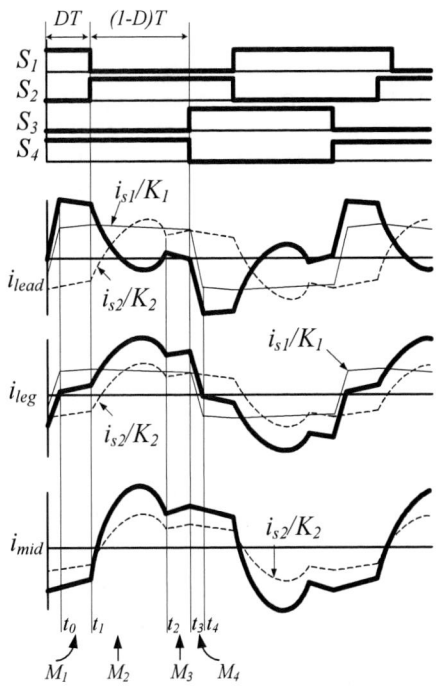

Fig. 9 The key waveforms of the primary side

TABLE I THE PROTOTYPE PARAMETERS

The full bridge part		SCTT cell	
C_1, C_2	35uF	K_1 (n_p:n_{s1})	3.875(31:8)
S_1 - S_4	IRFP450	K_2 (n_p:n_{s2})	6.2(31:5)
C_{o1}, C_{o2}	300pF	$2L_p$	33uH
C_{o3}, C_{o4}	180pF	$2L_{s1}$ /$2L_{s2}$	1.5uH/1.1uH
The 1st branch (24V/10A)		The 2nd branch (48V/5A)	
D_{11} - D_{14}	MBR20100CT	D_{21} - D_{24}	MBR2060CT
L_{f1}	90uH	C_{f2}	4mF
C_{f1}	470uF	C_r	1uF
		L_m	30uH

$$
\begin{aligned}
i_{lead} &= i_{s1}/K_1 - i_{s2}/K_2 \\
i_{lag} &= i_{s1}/K_1 + i_{s2}/K_2 \\
i_{mid} &= 2i_{s2}/K_2
\end{aligned}
\qquad (15)
$$

The reference directions of the currents have been denoted in Fig. 1. From (15), it' seen that the primary current i_{lead} and i_{lag} are combined by the reflected currents of i_{s1} and i_{s2}, and i_{mid} has the same shape as i_{s2}. According to Fig. 6 and Fig. 8, the primary currents can be drawn, as shown in Fig. 9.

VI. PROTOTYPE DESIGN

The specifications of the prototype are given first: the input voltage: 280V-300V; the 1st branch: 24V/10A; the 2nd branch: 48V/5A; the frequency: 80kHz-130kHz. The prototype parameters are listed in Table I, and the main design considerations are listed as follow:

A. Turning ratios of the 1st and 2nd branches

In the 1st branch, the maximal duty cycle D_{max} for the 1st branch is chosen as 0.5 first. When the 1st branch is full load, there is still a duty cycle of $(1-D_{max})=0.5$ available for the 2nd branch. The turning ratio of the 1st output branch can be estimated as:

$$K_1 = \frac{V_{dc\,min}(D_{max} - D_{loss})}{V_{o1} + V_D} = 3.92 \quad (16)$$

where $D_{loss}=0.15$ is the estimative maximal duty cycle loss, and $V_D=1V$ is the estimative total voltage drop.

In the 2nd branch, the turning ratio choosing follows the guideline of the conventional LLC converter: the 2nd branch should work at the unit gain when the input voltage is maximal and the 2nd branch gets the unit duty cycle. With this consideration, the 2nd branch always works below the series resonant frequency f_{r2} so as the rectifier of the 2nd branch always works in the ZCS mode. The turning ratio of the 2nd branch thus can be determined as:

$$K_2 = V_{dc\,max}/V_{o2} = 6.25 \quad (17)$$

Compromising (16) and (17), the winding turns are selected as: $n_p : n_{s1} : n_{s2} = 31:8:5$, thus $K_1=3.875$ and $K_2=6.2$. Verifying the calculated parameters using (10), a maximal duty cycle 0.6 is required when the input voltage is minimal, the switching frequency is maximal and the load is heavy ($V_{dc}=280V$, $f_s=130kHz$ and $I_{o1}=10A$). It is a little bigger than the expectation. However, considering the tolerant errors of the parameter measurement, the result is acceptable.

B. The 1st branch designing

The parameters design of the 1st branch is rather loose. The main considerations are the output ripple level and the transient responding. $L_{f1}=90uH$ and $C_{o1}=470uF$ are chosen finally in our design. However, since the maximal duty cycle of the 1st branch is configured as 0.5, the voltage stress of the diodes D_{11}–D_{14} should be considered carefully:

$$V_{stress} = \frac{V_{dc\,max}}{K_1} = 77.4V \quad (18)$$

Moreover, the secondary voltage ripples occurred by the parasitic capacitances of the diodes and leakage inductors of the transformers should be also taken into consideration. For these reasons, MBR20100CT is used to ensure the security of the rectifier.

C. The 2nd branch designing

To simplify the topology of the 2nd branch, the leakage inductors $2L_p$ and $2L_{s2}$ is utilized as the resonant inductor, so the additional inductor L_r is not needed. Setting the resonant frequency f_{r2} at 120kHz, according to (8), the resonant capacitor C_r is chosen as:

$$C_r = \frac{1}{(2\pi f_{r2})^2 (2L_p / K_2^2 + 2L_{s2})} = 0.9uF \quad (19)$$

The other resonant component L_m is determined by choosing the parameter h in (14). The main consideration is that when both the input voltage and available duty cycle reaches their minimal values and the 2nd output is full

load($V_{dc}=280V$, $D=0.5$ and $I_{o2}=5A$), V_{o2} can still be regulated at 48V. $h=9$ is chosen finally. According to (14), the switching frequency has to decrease to about 60kHz in this extreme situation. However, as mentioned in section V, the actual switching frequency will higher than 60kHz because the high order harmonic also contributes to the output voltage gain. $C_r=1uF$ and $L_m=30uH$ are chosen finally in our design.

D. The contorl circuit

The control circuit of the proposed converter can be realized by using the analog devices easily. The conventional control IC UC3875 is used. Two independent loops are design, the 1st output voltage is fed back to control the phase shift and the 2nd output is fed back to control the frequency. By replacing the oscillated resistor of the UC3875 with a transistor, the FM can be achieved easily.

VII. EXPERIMENTAL RESULTS

Fig. 10 shows steady waveforms when both the two branches are in the heavy loads situation (24V/10A and 48V/5A). The input voltage is 300V.

Fig. 10(a) shows the waveforms of the drain-source voltages of S_2 and S_4. When the 1st branch is full load, the duty cycle is $D=0.4$, so as to keep the 1st output still regulated; the 2nd branch thus gets a complementary pulse width $(1-D)=0.6$. To keep the 2nd output regulated, the switching frequency has to decrease to about 78kHz.

The secondary waveforms of the 1st branch are shown in Fig. 10(b). The duty cycle loss can be observed from the secondary voltage V_{s1}. And, the amplitude of V_{s1} is high (80V) since only a maximal duty cycle of 0.5 is available for the 1st branch. The secondary current i_{s1} is also shown in Fig. 8(b). In the powering mode, it is increasing slowly; in the freewheeling mode, it is decreasing slowly; and in the commutation period, it decreases and changes it polarity linearly and rapidly.

The secondary waveforms of the 2nd branch are shown in Fig. 10(c), since the equal voltage source V_{e2} can't be measured, the secondary voltage of the 2nd branch V_{s2} is shown. From the waveform of V_{s2}, we can seen that the 2nd branch gets the complementary pulse width with $(1-D)$ when the upper or lower switches are turned on. The secondary current i_{s2} is also shown in Fig. 8(c). It's seen that i_{s2} includes two resonant frequency, and the shape of i_{s2} is similar to the conventional LLC resonant converter.

The primary currents are shown in Fig. 10(d). It's seen that i_{lead} and i_{lag} include the DC and AC components. The waveform of i_{mid} is also shown in Fig. 8(d). It's seen that i_{mid} has the same shape as i_{s2}, as described in (15). All the waveforms agree well with the analysis.

Table II shows the load and cross regulation performance. It covers all the possible load conditions of the two outputs. It's seen that both the outputs are fully regulated anytime. It's benefited by the proposed complementary pulse width modulation. The efficiency is 92% when both the branches are full loads. It is mainly benefited by the compact topology and the full range ZVS characteristics.

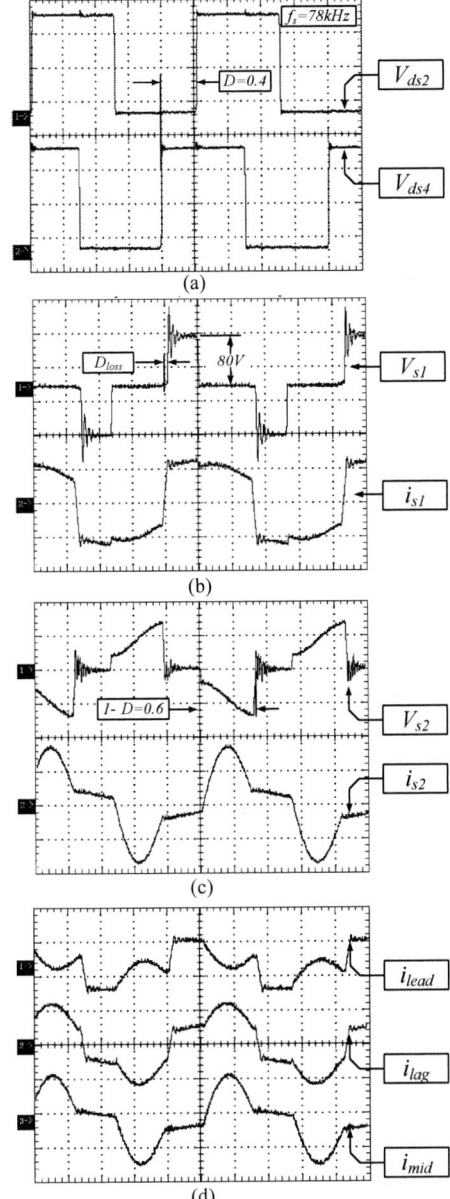

Fig. 10 The steady state waveforms
(V_{ds2}, V_{ds4}:100V/div; V_{s1}, V_{s2}:50V/div; i_{lead}, i_{lag}, i_{mid}: 5A/div;
i_{s1}, i_{s2}:10A/div;   t:2.5us/div)

TABLE II THE LOAD AND CROSS REGULATION

Load conditions (%)		V_{o1}(V)	V_{o2}(V)
The 1st branch	The 2nd branch		
0	0	24.05	48.2
0	50	24.07	48.0
0	100	24.06	47.9
50	0	24.03	48.1
50	50	24.02	48.0
50	100	24.02	48.0
100	0	24.01	48.2
100	50	24.00	48.0
100	100	24.01	47.9

CONCLUSIONS AND FUTURE WORK

A dual output converter based on the complementary pulse widths is proposed in this paper. One of the outputs is regulated by the pulse width completely; the other output is affected by the complementary pulse width, and compensated by the frequency modulation. Both the two outputs are fully regulated. Only one full bridge topology is needed, thus the proposed converter is simple and compact; the proposed converter can work in high efficiency and high efficiency.

The steady performance of the proposed converter is analysis. In the steady modes, the two outputs work independent of each other. The characteristic of each output is analysis and their voltage gains are deduced.

The design guidelines are given and a prototype is build to verify the analysis. The experimental results agree well with the analysis.

Furthermore, the full range ZVS can be achieved in all the switches without auxiliary circuit by the appropriate design. And, if the fully regulation ability of the 2nd output is not required, the LLC resonant tank can be replaced by any other resonant tank, and a new family of duty output converter based on the complementary pulse widths modulation and frequency modulation can be deduced. The ZVS analysis, topology variation and the general characteristic summaries will be analyzed in the future work.

REFERENCES

[1] L.H. Hang, Y.L. Gu, Z.Y. Lu, Z.M. Qian, and D.H. Xu, "Magamp post regulation for LLC series resonant converter with multi-output," in *Proc. IEEE Ind. Electron. Soc. Annu. Conf. (IECON)*, 2005, pp. 628-631.

[2] S. Havanur, "Combining synchronous rectification and post regulation for multiple isolated outputs," in *Proc. IEEE Appl. Power Electron. Conf. (APEC)*, 2004, pp. 872-877.

[3] A. Barrado, E. Olias, A. Lazaro, J. Pleite, and R. Vazquez, "PWM-PD multiple output DC/DC converters: operation and control-loop modeling," *IEEE Trans. Power Electron.*, vol. 19, no.1, pp. 140-149, Jan. 2004.

[4] Y.J. Zhang and D.H. Xu, "Design and implementation of an accurately regulated multiple output ZVS DC-DC converter," *IEEE Trans. Power Electron.*, vol. 22, no.5, pp. 1731-1742, Sept. 2007.

[5] Y.J. Zhang, D.H. Xu, F.C. Gao, Y. Han, and Z. Du, "Design of an accurately regulated multiple output ZVS DC/DC converter," in *IEEE Power Electron. Spec. Conf. (PESC)*, 2006, pp. 1-6.

[6] H.E. Tacca, "Single-switch two-output flyback-forward converter operation," *IEEE Trans. Power Electron.*, vol. 13, no.5, pp. 903-911, Sept. 1998.

[7] H.H. Seong, D.J. Kim, and G.H. Cho, "A new ZVS DC/DC converter with fully regulated dual outputs," in *IEEE Power Electron. Spec. Conf. (PESC)*, 1993, pp. 351 - 356.

[8] W. Chen, Z.Y. Lu, X.F. Zhang, and S.S. Ye, "A novel ZVS step-up push-pull type isolated LLC series resonant DC-DC converter for UPS systems and its topology variations," in *Proc. IEEE Appl. Power Electron. Conf. (APEC)*, 2008, pp. 1073-1078.

[9] E.S. Kim, I.S. Cha, and M.H. Kye, "A novel topology of secondary llc series resonant converter," in *Proc. IEEE Appl. Power Electron. Conf. (APEC)*, 2007, pp. 1625-1629.

[10] J. H. Jung and J. G. Kwon, "Theoretical analysis and optimal design of LLC resonant converter," in *Power Electron. and Appl. Conf.*, 2007, pp. 1-10.

Analysis and Design Considerations of An Improved ZVS Full-Bridge DC-DC Converter

Zhong Chen, Biao Ji, Feng Ji, and Lei Shi

Aero-Power Sci-tech Center
Nanjing University of Aeronautics & Astronautics
Nanjing, 210016, China
Email: chenz@nuaa.edu.cn

Abstract—An improved full-bridge (FB) pulse-width-modulated (PWM) converter is proposed featuring zero-voltage-switching (ZVS) of all primary switches in the entire line and load range. Compared to the conventional full bridge dc/dc converter, the proposed converter achieves ZVS with substantially reduced duty-cycle loss and circulating current. The operation principle of the circuit is described and design considerations of the converter are discussed. Finally, the operation and performance of the proposed converter is verified on a 1-kW (54-V/20-A) experimental PWM FB converter prototype operating at 100 kHz from a 300-400 V dc input.

I. INTRODUCTION

The full-bridge (FB) zero-voltage-switching (ZVS) pulse-width-modulated (FB ZVS-PWM) converter, [1]-[15], is extensively applied in medium to high power conversion due to its inherent electrical robustness and excellent electrical characteristics. In a conventional phase shifted full bridge (PSFB) converter, ZVS operation can only be achieved with a limited load and input-voltage range [1]-[4]. By increasing the leakage inductance or by inserting a large external inductance in series with the primary transformer, the ZVS range of the lagging leg switches can be extended. But the increased series inductance has a detrimental effect on the performance of the converter since it causes increased duty-cycle loss, as well as severe voltage ringing across secondary-side rectifier diodes. The secondary-side ringing can be suppressed with a passive RCD snubber [5]. By adding two clamp diodes and a series inductance, the ringing can also be effectively controlled, but it does not deal with the secondary-side duty-cycle-loss issue [6]. Using a saturable external inductance instead of a linear inductance, ZVS range can be increased without significantly losing the duty ratio [7]. However, a large-size core is required to eliminate the thermal problem. The energy stored in the magnetizing inductance can also be used to extend the ZVS range, but this causes significant increase in the conduction loss [8]-[9]. In the approach proposed in [10], full load range ZVS of switches are achieved by using a simple LC auxiliary circuit with a minimal duty-cycle loss and secondary-side parasitic

ringing. The energy stored in the auxiliary circuit is independent of load current, so the efficiency decreases due to the fixed conduction losses caused by the auxiliary circuit.

A number of FB ZVS converters featuring ZVS over the entire conversion range are proposed in [11]-[15]. In this paper, a novel full-bridge converter with an additional auxiliary circuit based on the concept described in [12] is proposed. The energy stored in the auxiliary circuit progressively increases as the load current decreases, so the proposed converter exhibits increased conversion efficiency. The operating principle of the proposed converter is described and the optimal design considerations are discussed. The experimental results will be given in the last section.

II. OPERATION PRINCIPLE

Fig. 1 shows the proposed ZVS full-bridge converter topology that provides ZVS for bridge switches over a wide range of load current. The primary side of the converter consists of two bridges Q_1-Q_3 and Q_2-Q_4 connected through two capacitors C_{a1} and C_{a2} to the connection of the power transformer T_r, the auxiliary transformer T_{ra} and auxiliary inductor L_a. The auxiliary transformer has a primary-to-secondary turns ratio 1:1. The phase-shift PWM technique is used to regulate the output DC voltage where Q_1 and Q_3 from the leading leg and Q_2 and Q_4 from the lagging leg. V_{in} and V_o are dc input voltage and output voltage respectively. D_{R1} and D_{R2} are output rectifier diodes.

And the parasitic capacitors of D_{R1} and D_{R2} are neglected to simplify the analysis. The magnetic current of the transformer is neglected due to its relative small value. Generally, two auxiliary capacitors are selected large enough so that their voltages are approximately constant during a switching cycle. Because the average voltages of the auxiliary inductive components and the power transformer during a switching cycle are zero and the pair of switches in each bridge leg operate with 50% duty cycle, the magnitude of voltage sources v_{ca1} and v_{ca2} are equal to $1/2V_{in}$, i.e., $v_{ca1}=v_{ca2}=1/2V_{in}$. The output side of the converter is implemented with a full-wave rectifier with a tapped secondary. Any other implementation of the secondary rectification stage such as

978-1-4244-4782-4/10 $26.00 © 2010 IEEE

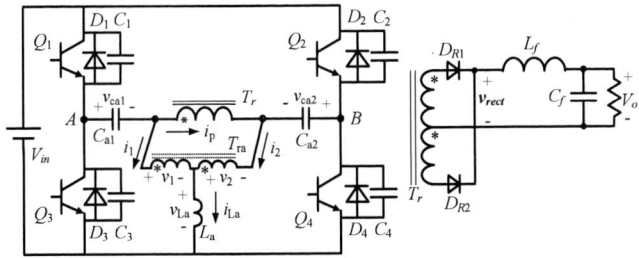

Fig. 1. Proposed full bridge ZVS converter.

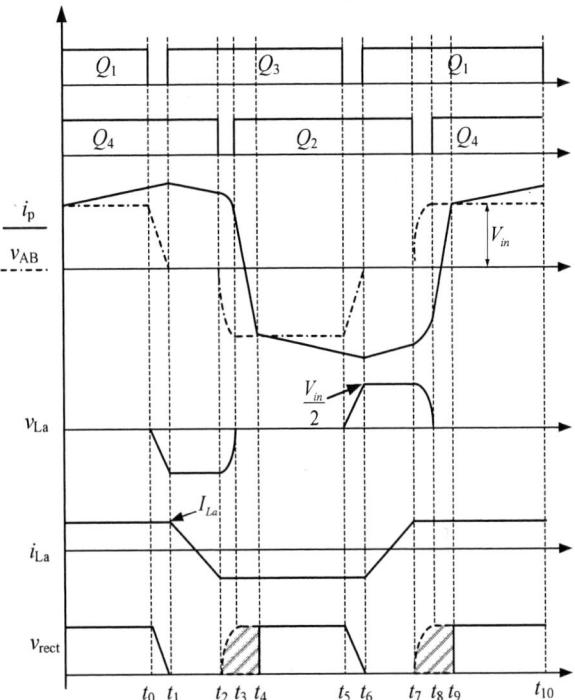

Fig. 2. Key waveforms of proposed converter power stage.

full-bridge rectifier and current-doubler-rectifier is also possible.

To further simplify the analysis, it is assumed that resistance of each conducting semiconductor switch is zero, whereas the resistance of each non-conducting switch is infinite. However, the output capacitance of each primary switch is not neglected since it is important for understanding the operation of proposed circuit and we assume $C_1 = C_3 = C_{lead}$, $C_2 = C_4 = C_{lag}$.

Fig. 2 shows key waveforms of proposed converter, whereas Fig. 3 shows topological stages during a half period. As shown in Fig. 3(a), when diagonal switches Q_1 and Q_4 are conducting and switches Q_2 and Q_3 are closed, primary voltage is positive so that load current I_o flows through D_{R1} and the upper secondary of transformer T_r. The reflected primary current i_p equals to I_o/K, where N_p: $N_s = K$ is the turns ratio of the transformer, N_p is the number of primary-winding turns, and N_s is the number of secondary-winding turns. During this topological stage, almost all the input voltage is induced across primary winding N_p of transformer T_r because the impedance of leakage inductor L_k is very small compared to the reflected output impedance across primary winding of transformer T_r. As the auxiliary transformer has 1:1 turns ratio, $v_1 = v_2 = 1/2 V_{in}$. Since $v_{ca1} = 1/2 V_{in}$, in the loop Q_1-C_{a1}-T_{ra}-L_a, we get $v_{La} = 0$. So the current through auxiliary inductor L_a keeps constant.

After switch Q_1 is turned off at t=t_0, primary current i_p plus i_1 starts charging output capacitance C_1 of switch Q_1 and discharging output capacitance C_3 of switch Q_3, where i_1 is the current flowing through the left winding of auxiliary transformer T_{ra} and equals to $I_{La}/2$, as shown in Fig. 3(b). During this interval, the current of the inductor L_a keeps at I_{La} because the transition time is short. As a result, voltage v_{C1} across switch Q_1 starts increasing toward V_{in}, whereas voltage v_{C3} across switch Q_3 decreasing toward zero. At the same time, voltage across auxiliary inductor L_a starts decreasing from zero to $-V_{in}/2$. The voltage of Q_1 rises slowly owing to output capacitance C_1 of switch Q_1. Since the energy for charging C_1 and discharging C_3 is supplied from filter inductor L_f and auxiliary inductor L_a, which generally have large inductances, the energy is large enough to completely discharge C_3 even at low currents. And with impact of L_a, the energy stored in L_a is able to discharge C_3 even at no load. After the capacitor C_3 is fully discharged at t=t_1, current continues to flow through the anti-parallel diode of switch Q_3. Q_3 can be turned on at zero voltage when D_3 conducts.

During the topological stage in Fig. 3(c), the potential of point A becomes zero, voltage across the power transformer

also becomes zero since the transformer is shorted by the simultaneous conduction of the body diode of Q_3 and switch Q_4. The voltage applied across L_a is $-V_{in}/2$. Due to this voltage, the current i_{La} decreases until t_2. At t_2, i_{La} reaches its minimum value $-I_{La}$, as illustrated in Fig. 2. And with real components, the primary current i_p may decrease because of the forward voltage of power switch $V_{forward}$ and voltage drop of body diode V_D.

After switch Q_4 is turned off at t=t_2, primary current i_p plus i_2 starts charging output capacitance C_4 of switch Q_4 and discharging output capacitance C_2 of switch Q_2, where i_2 is the current flowing through the right winding of auxiliary transformer and equals to $I_{La}/2$, as shown in Fig. 3(d). Since the potential of point B increases from zero toward V_{in}, while the potential of point A is constant at zero. The rectifiers D_{R1} and D_{R2} conduct the load current simultaneously, so the secondary windings are shorted and v_{AB} is fully applied to the leakage inductance of power transformer L_k. Because the auxiliary inductance L_a is much larger than the leakage inductance L_k, it can be treated as a constant current source during the transition time. By properly selecting auxiliary inductance L_a, the energy stored in auxiliary inductor L_a will be enough to charge output capacitance C_4 of switch Q_4 and discharge output capacitance C_2 of switch Q_2 even at no load. This stage finishes when v_{c4} rises to V_{in} and v_{c2} falls to zero at t_3.

D_2 conducts naturally when v_{c2} decays to zero, and Q_2 can be turned on at zero voltage at t=t_3. The primary current i_p starts decreasing because a negative voltage appears across leakage inductance L_k of transformer T_r, as shown in Fig. 3(e). At the same time, load current I_o commutates from the upper secondary and rectifier D_{R1} into the lower secondary and

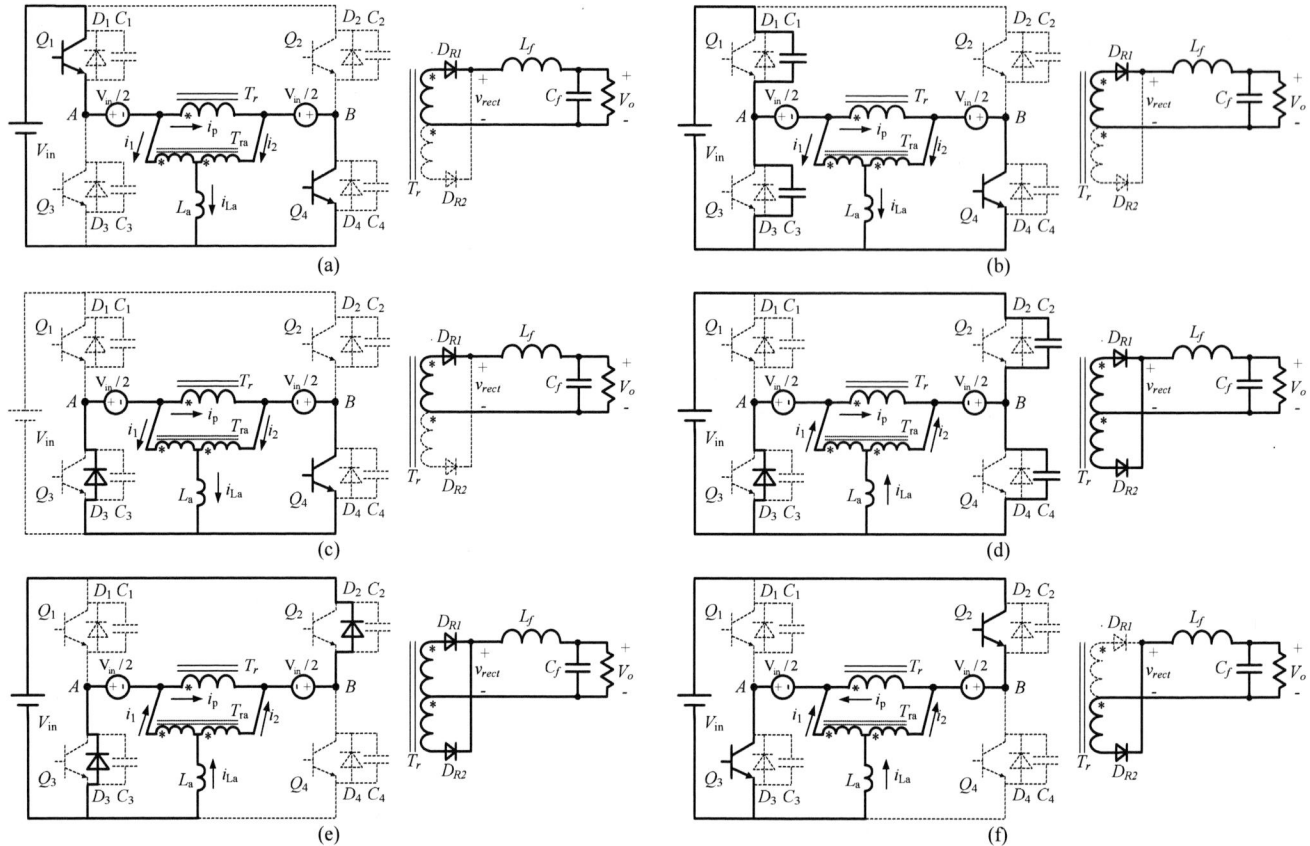

Fig. 3. Equivalent circuits of each operation stage.

corresponding rectifier D_{R2}. The rate of change of the primary current is given by (1). Since the leakage inductance of the power transformer is small, the duty-cycle loss in the converter is dramatically decreased.

$$\frac{di_p}{dt} = -\frac{V_{in}}{L_k} \qquad (1)$$

When commutation of the load current from the upper to the lower secondary is completed at $t=t_4$, the primary current commutation from the positive to negative direction is also finished so that the primary current is $i_p = -I_o/K$. The input energy is transferred into the secondary side by power switches Q_2 and Q_3 and T_r, as shown in Fig. 3(f). Voltage across L_a collapse to zero, as illustrated in Fig. 2. This stage ends at t_5 when switch Q_3 turns off. Then the second half of the switching period begins. In the second half of the switching period, the operation of the circuit is exactly the same as the operation in the first half of the switching period.

III. DESIGN CONSIDERATIONS

The charging and discharging of the capacitances of switches Q_1 and Q_3 is done by the sum of energy stored in the output filter inductor and the energy stored in the auxiliary inductor L_a. On the other hand, the charging and discharging of capacitances of switches Q_2 and Q_4 is done by the sum of the energy stored in the leakage inductance of transformer and the energy stored in the auxiliary inductor L_a. Therefore, switches Q_1 and Q_3 can achieve ZVS in a wide range of input

voltage and load current even without assistance from the energy stored in auxiliary inductor L_a since plenty of energy is available from filter inductor L_f. However, ZVS of the switches Q_2 and Q_4 is dependent on the energy stored in the auxiliary inductance L_a ultimately.

Auxiliary current i_{La} changes linearly from maximum value I_{La} to minimum value $-I_{La}$, i.e., i_{La} changes for $2I_{La}$, due to a negative voltage of $V_{in}/2$ across auxiliary inductor L_a, as shown in Fig. 2. According to Fig. 2, the time interval t_1-t_2 is approximately equal to $(1-D)T_s/2$, where D is duty cycle and T_s is a switching period, I_{La} can be calculated from

$$\frac{V_{in}}{2} = L_a \frac{2I_{La}}{(1-D)T_s/2} \qquad (2)$$

From (2), we can get

$$I_{La} = \frac{(1-D)V_{in}}{8L_a f_s} \qquad (3)$$

where $f_s=1/T_s$ is the switching frequency. As the leakage inductance is selected smaller compared to the conventional full bridge dc/dc converter, the duty cycle of the proposed converter hardly changes when it enters CCM (continuous conduction mode). The duty cycle D can be approximately expressed as

$$D \approx D_{eff} = \frac{KV_o}{V_{in}} \qquad (4)$$

where D_{eff} is the effective duty cycle. When it works in CCM,

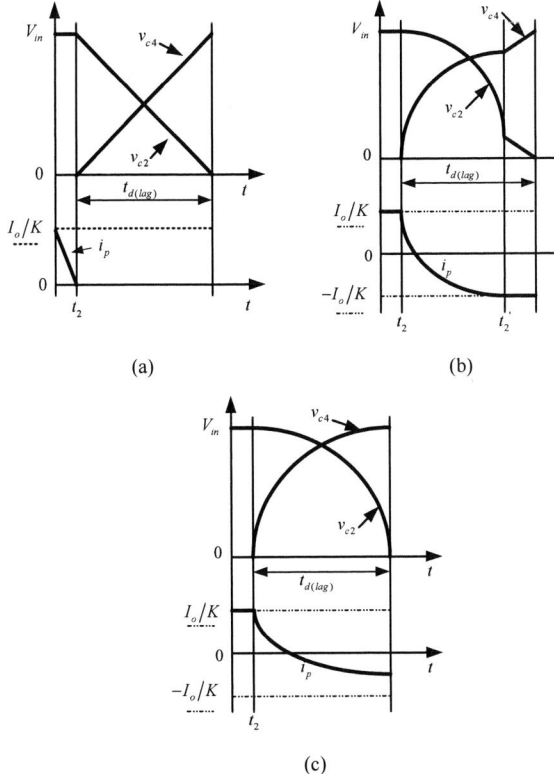

(a)   (b)

(c)

Fig. 4. the transition of the lagging leg at different load: (a) DCM; (b) CCM ($I_o/K<I_{La}/2$); (c) CCM ($I_o/K>I_{La}/2$).

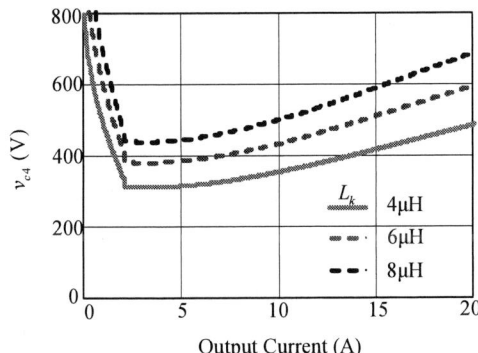

Fig. 5. Voltage across the parasitic capacitor of the lagging leg versus load current at different leakage inductances.(V_{in}=400V, L_a=30μH)

we assume the magnitude of auxiliary inductor current is constant.

When the proposed converter enters DCM (discontinuous conduction mode), the duty cycle changes obviously. The duty cycle can be expressed as

$$D = 2 \cdot \sqrt{\frac{V_o \cdot f_s \cdot I_o \cdot K^2 \cdot L_f}{V_{in} \cdot (V_{in} - KV_o)}} \tag{5}$$

So the current of auxiliary inductor will impact the lagging leg transition distinctly especially in DCM mode. The transition of lagging leg switches can be divided into

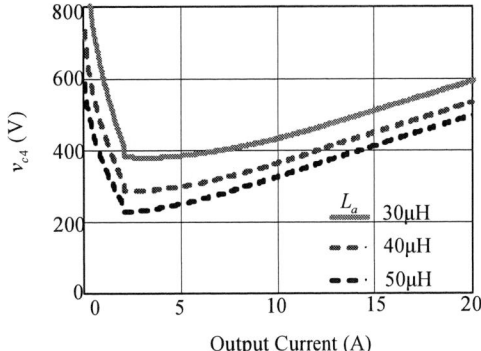

Fig. 6. Voltage across the parasitic capacitor of the lagging leg versus load current at different auxiliary inductances. (V_{in}=400V, L_k=6μH)

three cases according to the load current I_o as shown in Fig. 4[13]. The selection of the auxiliary components should guarantee ZVS operation over the entire conversion range. To simplify the analysis, we assume the dead time for the lagging leg t_{lag}=T_c/4, where $T_c = 2\pi\sqrt{2L_kC_{lag}}$.

The voltage across C_4 can be expressed in a function of load current I_o as

$$v_{c4}(t) = \begin{cases} \dfrac{I_{La} \cdot t_{d(lag)}}{4 \cdot C_{lag}} & if\ 0 \le I_o \le \dfrac{\Delta I_o}{2} \\[4mm] \left(\dfrac{I_o}{K} + \dfrac{I_{La}}{2}\right)Z_r\sin\omega_r(t_2' - t_2) \\ \quad + \dfrac{\left(-\dfrac{I_o}{K} + \dfrac{I_{La}}{2}\right) \cdot (t - t_2')}{2 \cdot C_{lag}} & if\ \dfrac{\Delta I_o}{2} < I_o \le \dfrac{I_{La}}{2} + \dfrac{\Delta I_o}{2} \\[4mm] \left(\dfrac{I_o}{K} + \dfrac{I_{La}}{2}\right)Z_r\sin\omega_r(t - t_2) & if\ \dfrac{I_{La}}{2} + \dfrac{\Delta I_o}{2} < I_o \le I_{o\max} \end{cases} \tag{6}$$

where $Z_r = \sqrt{L_k/2C_{lag}}$, $\omega_r = 1/\sqrt{2L_kC_{lag}}$. ΔI_o is output current ripple. t_2' is the moment when commutation is completed.

The leakage inductance should be optimally selected to simultaneously achieve the entire range ZVS operation and improve overall efficiency over the entire conversion range instead of choosing the minimized inductance value. The optimum design of the proposed converter is only possible with a compromise between opposing requirements. Different weights of leakage inductance and auxiliary inductance are tried to determine the optimal design of the converter together with loss analysis. As shown in Fig. 5, voltage across C_4 is plotted in a function of load current I_o with different leakage inductances. Also, in Fig. 6, voltage across C_4 is plotted in a function of load current I_o with different auxiliary inductances. Finally, we choose L_a=30μH, L_k=6μH.

Dc blocking capacitors are placed in series with the magnetic components not only blocking the flow of any dc current but also helping establish complementary voltage to create the adaptive energy for the ZVS operation. Finally, we selected C_{a1}=C_{a2}=2.2μF.

IV. EXPERIMENTAL RESULTS

The performance of the proposed converter was verified

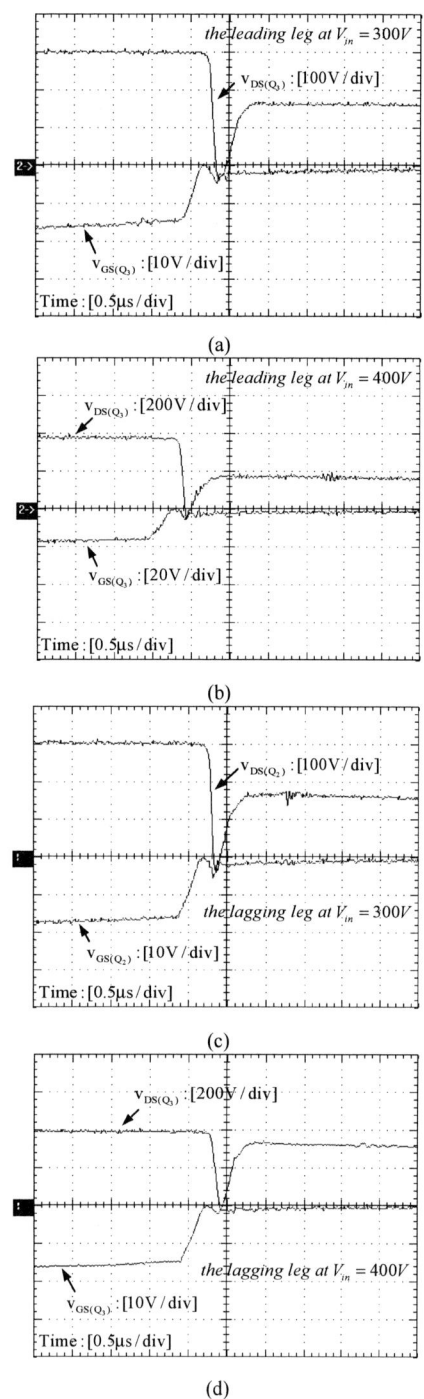

(a)

(b)

(c)

(d)

Fig. 7. Drive voltage and drain to source voltage of the leading and the lagging leg at 10% load at different input voltage.

on a 1-kW, 100-kHz prototype circuit designed to provide 54-V dc output voltage from a 300-400V dc input. All the four switches used were IRFP460. The secondary-side diodes were MUR3040PT. The phase-shift control circuit was implemented using a UCC3895 controller. The closed loop compensator was designed based on the same principles as the conventional full bridge dc/dc converter.

Fig. 7 shows the gate signals of PWM switches Q_3 and Q_2

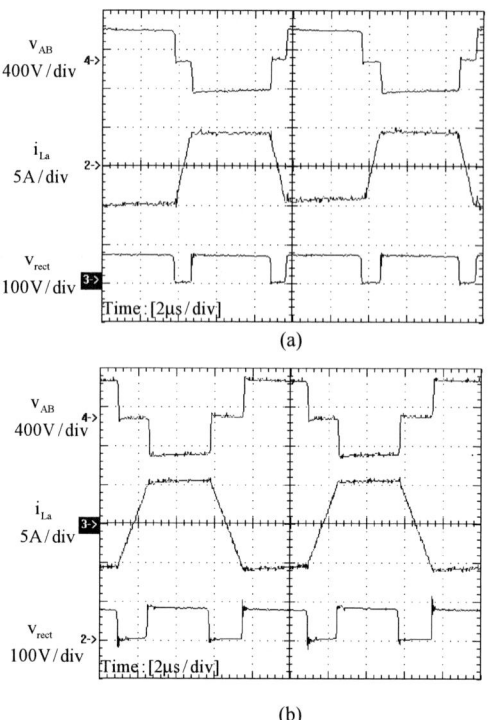

(a)

(b)

Fig. 8 Measured key waveforms at 10% load at different input voltages:(a) V_{in}=300V and (b) V_{in}=400V. From top to bottom: primary voltage v_{AB}; current of auxiliary inductor i_{La}; secondary voltage v_{rect}.

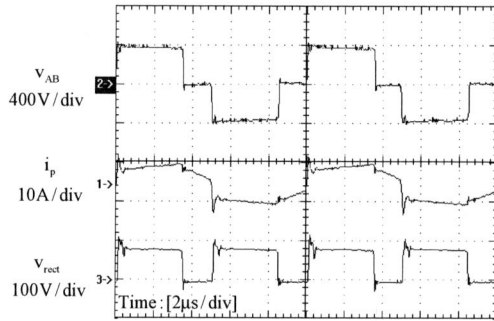

Fig. 9 Experimental waveforms of the proposed converter at full load.

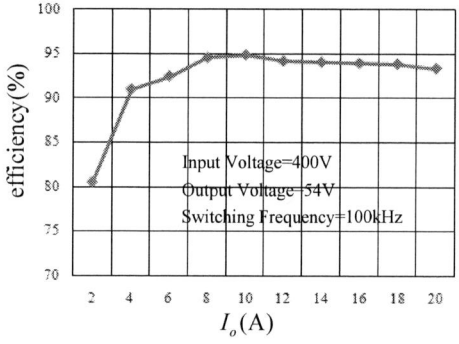

Fig. 10. The overall efficiency.

along with their drain-to-source voltage waveforms at 10% load. As seen, the drain voltage reaches zero before the gate reaches its threshold demonstrating zero-voltage turn-on.

978-1-4244-4782-4/10 $26.00 © 2010 IEEE   1475

Fig. 8 shows measured key waveforms at 10% load. Fig. 8(a) shows the waveforms for V_{in}=300V at 10% load current. Fig. 8(b) shows the waveforms for V_{in}=400V at 10% load current. As the input voltage increases, the duty cycle of v_{La} increases, thereby increasing I_{La}. The magnitude of i_{La} progressively increases with the input voltage. The auxiliary circuit provides maximum energy for ZVS at no-load, whereas it provides minimum energy at full load. Fig. 9 shows the key waveforms of the proposed converter at full load. As can be seen from the waveforms in Fig. 9, the proposed converter has a very small duty cycle loss as well as a small parasitic ringing. Fig. 10 shows the measured efficiencies as a function of output current. Generally, the efficiency is more pronounced at light loads where the conventional FB ZVS converter operates with hard switching.

V. CONCLUSION

In this paper, an improved soft-switched FB converter that achieves ZVS over the entire range of load current has been proposed. The auxiliary network employed in the FB ZVS converter is used to store energy for the ZVS operation which is not only dependent on the input voltage but also dependent on the load current. The circulating energy and conduction losses have been substantially reduced. Since the two auxiliary inductive components do not appear as series inductances, they will not cause the duty cycle loss or severe voltage ringing across the output rectifiers. The operation principle and characteristic of the converter are analyzed in detail. Experimental results from a 1-kW/100-kHz prototype confirm the advantages of the proposed configuration.

REFERENCES

[1] O. D. Petterson and D. M. Divan, "Pseudo-resonant full bridge dc/dc converter," in *Proc. IEEE Power Electron. Sepc. Conf.* 1987, pp. 424-430.

[2] L. H. Weene and C. A. Wright, "A 1-kW, 500-kHz front-end converter for a distributed power supply system," in *Proc. IEEE Appl. Power Electron. Conf. (APEC'89)*, 1989, pp. 423-432.

[3] W. Chen, F. C. Lee, M. M. Jovanović, and J. A. Sabaté, "A comparative study of a class of full bridge zero-voltage-switched PWM converters," in *Proc. IEEE Appl. Power Electron. Conf. (APEC'95)*, 1995, pp. 893-899.

[4] B. P. McGrath, D. G. Holmes, P. J. McGoldrick, and A. D. McIve, "Design of a soft-switched 6-kW battery charger for traction applications," *IEEE Trans. Power Electron.*, vol. 22, no. 4, pp. 1136-1144, Jul. 2007.

[5] S. Y. Lin and C. L. Chen, "Analysis and design for RCD clamped snubber used in output rectifier of phase shifted full-bridge ZVS converters," *IEEE Trans. Ind. Electron.*, vol. 45, no. 2, pp. 358-359, Apr. 1998.

[6] R. Redl, L. Balogh, and D. W. Edwards, "Optimal zvs full-bridge DC/DC converter with pwm phase-shift control: Analysis, design considerations, and experimental results," in *Proc. IEEE Appl. Power Electron. Conf. (APEC'94)*, 1994, pp. 159-165.

[7] G. Hua, F. C. Lee, and M. M. Jovanović, "An improved full-bridge zero-voltage-switched PWM converter using a saturable inductor," *IEEE Trans. Power Electron.*, vol. 8, no. 4, pp. 530-534, Oct. 1993.

[8] R. Ayyanar and N. Mohan, "Novel soft-switching DC-DC converter with full ZVS-range and reduced filter requirement—Part I: regulated-output applications," *IEEE Trans. Power Electron.*, vol. 16, no. 2, pp. 184-192, Mar. 2001.

[9] R. Jain, N. Mohan, R. Ayyanar, and R. Button, "A comprehensive analysis of hybrid phase-modulated converter with current-doubler rectifier and comparison with its center-tapped counterpart," *IEEE Trans. Ind. Electron.*, vol. 53, no. 6, pp. 1870-1880, Oct. 2006.

[10] P. K. Jain, W. Kang, and H. Soin, "Analysis and design considerations of a load and line independent zero voltage switching full bridge dc-dc converter topology," *IEEE Trans. Power Electron.*, vol. 17, no. 5, pp. 649-657, Sep. 2002.

[11] Y. Jang, M. M. Jovanović and Y. Chang, "A new ZVS-PWM full-bridge converter," *IEEE Trans. Power Electron.*, vol. 18, no. 5, pp. 1122-1129, Sep. 2003.

[12] Y. Jang and M. M. Jovanović, "A new family of full-bridge ZVS converters," *IEEE Trans. Power Electron.*, vol. 19, no. 3, pp. 701-708, May 2004.

[13] X. Wu, J. Zhang, X. Xie, and Z. Qian, "analysis and optimal design considerations for an improved full bridge ZVS dc-dc converter with high efficiency," *IEEE Trans. Power Electron.*, vol. 21, no. 5, pp. 1225-1234, Sep. 2006.

[14] Y. Jang and M. M. Jovanović, "A new PWM ZVS full-bridge converter," *IEEE Trans. Power Electron.*, vol. 22, no. 3, pp. 987-993, May 2007.

[15] M. Borage, S. Tiwari, S. Bhardwaj, and S. Kotaiah, " A full-bridge dc-dc converter with zero-voltage-switching over the entire conversion range," *IEEE Trans. Power Electron.*, vol. 23, no. 4, pp. 1743-1750, Jul. 2008

A New Resonant Gate Driver for Switching Loss Reduction of High Side Switch in Buck Converter

Xin Zhou, Student Member, IEEE, Zhigang Liang, Student Member, IEEE and Alex Huang, Fellow, IEEE

Department of Electrical and Computer Engineering
North Carolina State University
Raleigh, NC 27606

Abstract— **In this paper, a new resonant gate driver circuits is proposed to reduce the switching loss of high side switch in buck converter. Hard switching causes major parts of the power loss in high side switch and limits high switching frequency application of DC-DC converter. The proposed resonant gate driver behaves more like the ideal current source driver which can fast turn-on/turn-off power switch to reduce switching loss. In addition, with proposed resonant gate driver, impact of parasitic gate resistance on switching speed of power switch can be greatly reduced. Test results show that, for buck converter with 12V input voltage, 1.3V output voltage, 10A load current and 5.5Ω gate resistance, comparing to conventional driver, with the proposed resonant gate driver for high side switch, total efficiency of buck converter can be improved by more than 3.5%.**

I. INTRODUCTION

With rapid growth in mobile application, power supplies are required to have small volume and fast transient response. In order to satisfy the requirements, in recent years, switching frequency of power converter has been increased to the range of hundreds of kHz to several MHz.[1-3][9] With increasing switching frequency, the frequency dependent switching loss and gate driving loss become more and more important in high efficiency converter design. In the conventional voltage source gate driver (VSD), the complementary PMOS and NMOS pair is used to drive the power MOSFET, as shown in Fig.1(a). During switching, the driver supply voltage source charges/discharges power switches Ma through R-C network. In the resonant gate drivers (RGD), inductors are used to form the L-C resonant network to charge/discharge the power switch, thus, part of the gate driving energy can be recovered [1-4]. The resonant gate drivers can be divided into two categories. One type is the gate drivers with zero initial inductor current [4][5]. During switching, resonance between inductor and MOS gate capacitance generates current to charge/discharge the gate capacitance. The sinusoidal resonant inductor current increases from zero. This type of gate driver is more focusing on recovering of gate driving loss. Due to comparably slow turn-on/turn-off time, the switching loss may not be improved.

Another type of RGD is the one with non-zero initial inductor current [1-3]. The initial current in resonant inductor is built up before the switching transient. During switching, the pre-charged inductor current charges/discharges gate capacitance of power switch, which can reduce switching time and switching loss [1][2]. This type of driver is also called current source driver (CSD).

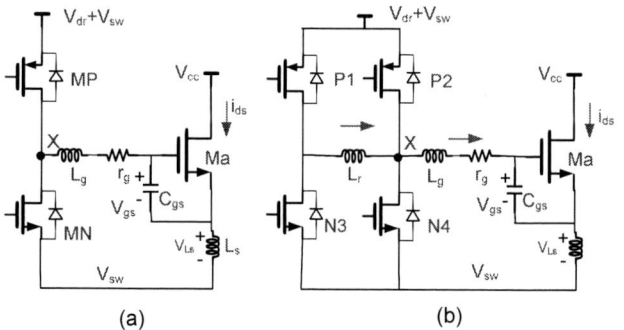

(a)   (b)

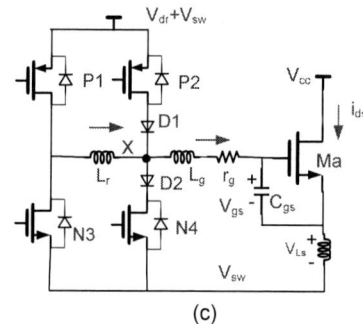

(c)

Figure 1. (a) conventional voltage source gate driver; (b) 4 switches current source gate driver; (c) proposed resonant gate driver with blocking diodes.

Parasitic gate resistance r_g and source inductance L_s on the gate driving path have great impact on the switching process of power MOSFET, as shown in Fig.1. During switching, gate current flows through r_g with voltage drop V_{rg} which reduces the real V_{gs} of Ma, especially for the cases with large r_g or

978-1-4244-4782-4/10 $26.00 © 2010 IEEE

large gate driving current in CSD. For L_s, during the turn-on process, i_{ds} of Ma flows through L_s and causes voltage drop

$$V_{Ls} = L_s \cdot di_{ds} / dt \qquad (1)$$

with polarity the same as V_{gs}, which reduces the actual V_{gs} on Ma. During the turn-off process, i_{ds} decreases and

$$di_{ds} / dt < 0 \qquad (2)$$

Thus, the polarity of induced V_{Ls} reverses and also reduces the actual V_{gs} on Ma. With the impact of r_g and L_s, the real V_{gs} of Ma is only $V_{gs}=V_{dr}-V_{Ls}-V_{rg}$. Smaller V_{gs} increases switching time and causes more switching loss.

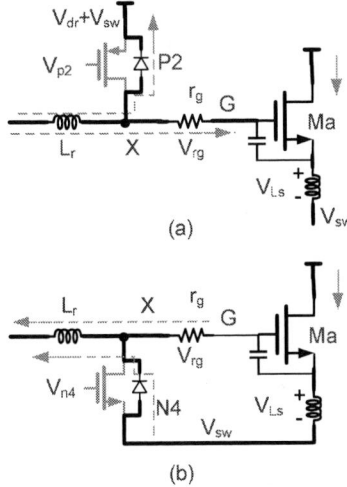

(a)

(b)

Figure 2. Impact of body diode clamping effect (BDC) on the 4 switches current source gate driver (CSD). (a) Impact of body diode of P2 on the turn-on process of Ma; (b) Impact of body diode of N4 on the turn-off process of Ma.

To reduce the impact of r_g and L_s, current source driving is a good solution. With ideal current source, voltage at node X can be increased to be higher than the driver supply voltage swing V_{dr} as in Fig.1. Thus, the voltage drop across r_g and L_s can be compensated by the voltage increasing at node X [1], as in Fig.1(b). For the power MOSFET with nearly zero r_g and very small L_s, Fig.1(b) shows very good results to reduce switching loss [1], however, for typical discrete power MOSFET, r_g is usually several ohms; for power MOSFET integrated on-chip with larger on-resistance, r_g may become even higher. So, impact from r_g during switching transient should not be neglected. In CSD, for the turn-on process with larger r_g, voltage across r_g may be too high and $V_x=V_{sw}+V_{Ls}+V_{gs}+V_{rg}$ may be larger than the critical voltage $V_{Hcl}=V_{dr}+V_{sw}+V_{fd}$. Here, V_{fd} is the forward voltage drop of the MOSFET body diode, as in Fig.2. Thus, after V_x reaches V_{Hcl}, driving current from L_r flows through body diode of P2 to driver supply voltage source instead of to C_{gs} of Ma, then the CSD behaves like the conventional voltage source driver. For the turn-off process, as V_x decreases to less than critical voltage $V_{Lcl}=V_{sw}-V_{fd}$, the body diode of N4 conducts and the

CSD behaves like the conventional voltage source driver. Thus, voltage at node X is clamped by the body diodes of P2 and N4, which can be called body diode clamping effect (BDC). With larger r_g, switching time reduction of Ma by using CSD is limited by the body diode clamp effect. To solve this problem and realize the more ideal current source gate driver, a new resonant gate driver is proposed in Fig.1(c).

II. THE PROPOSED NEW RESONANT GATE DRIVER TO REDUCE HIGH SIDE MOS SWITCHING LOSS

The proposed resonant gate driver (RGD) is used to drive the high side switch of buck converter, in which switching loss is the dominant power loss. In Fig.1(c), two schottkey diodes, D1 and D2 are added to allow the voltage at node X have more dynamic range without clamped by the body diodes of P2 and N4. Control logic signals and key waveforms are shown in Fig.3. The operation process of the new RGD can be divided into 8 phases.

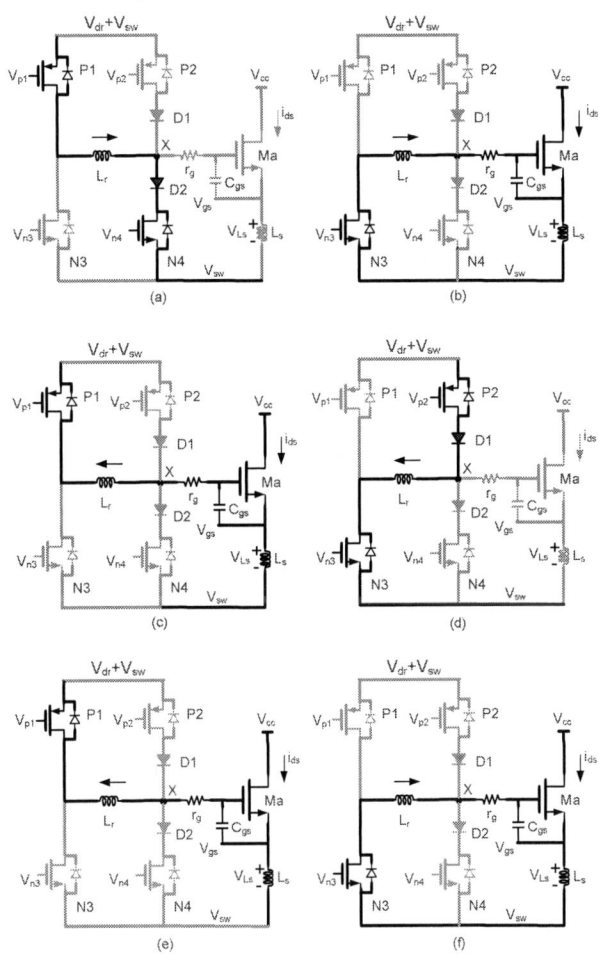

Figure 3. Operating phases of the proposed resonant gate driver

Phase 1: $t_0 \sim t_1$ in Fig.3(a). This is the turn-on pre-charging phase of Ma. Before this phase, P1, P2, N3 are off, N4 is on, V_x is connected to V_{sw} through D2 and N4. The high side switch is off. In phase 1, P1 and N4 are turned-on, current

978-1-4244-4782-4/10 $26.00 © 2010 IEEE   1478

flowing through P1, L_r and N4 pre-charges L_r to the peak current I_{pre1}

$$I_{pre1} = (V_{dr} - V_{fd2}) \cdot t_{pre1} / L_r \qquad (3)$$

Here, V_{fd2} is the forward voltage drop of D2, $t_{pre1}=t_1-t_0$ is the pre-charge time. Since N4 is kept on, high side switch Ma in this phase is kept off.

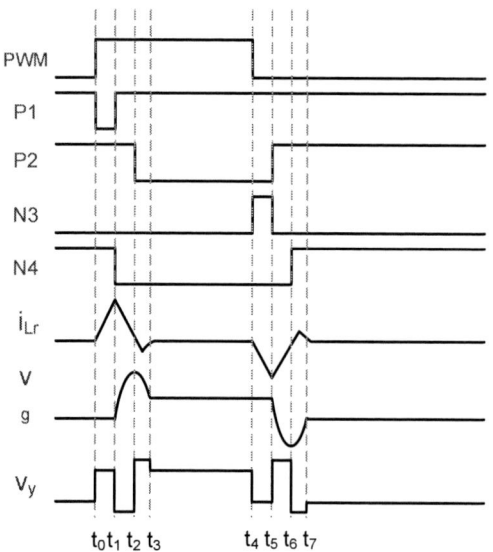

Figure 4.   Key operating waveforms of the proposed resonant gate driver

Phase 2: t_1~t_2 in Fig.3(b). This is the turn-on phase of Ma. P1 and N4 are turned-off, then, all the driver switches are kept off. Body diode of N3 conducts to form the freewheeling path for L_r inductor current i_{Lr} and $V_y=V_{sw}-V_{fd}$. L_r resonates with C_{iss} with initial current I_{pre1} and Ma gate voltage, V_g, increases. $C_{iss}=C_{gs}+C_{gd}$ is the input gate capacitance of Ma. With large r_g, as discussed before, V_x can be increased to higher than $V_{dr}+V_{fd}+V_{sw}$. With blocking diode D1, body diode of P2 can't conduct and the RGD behaves more like the ideal current source. Voltage increasing in V_x can compensate for the voltage drop on r_g and L_s, thus, fast turn-on of Ma can be realized. With V_x increasing and V_y decreasing to $V_{sw}-V_{fd}$, reverse voltage V_x-V_y is applied across L_r and i_{Lr} decreases. When i_{Lr} decreasing to zero, V_x increases to its peak value, which is determined by L_r, C_{iss}, r_g and I_{pre}. For ideal case, neglecting power loss caused by r_g and other parasitics on the charging path, peak voltage of V_x (also peak V_g) V_{gpkp} can be

$$V_{gpkp} = I_{pre1} \cdot \sqrt{L_r / C_{iss}} \qquad (4)$$

V_{gpkp} should be smaller than the voltage rating of V_{gs} of Ma.

Phase 3: t_2~t_3 in Fig.3(c).P2 ZVS turns on at the beginning of this phase. Ma is kept on. Since V_x is kept at high voltage and D1 is reverse bias, the turn-on moment of P2 doesn't need to be very accurately controlled. P1, N3 and N4 are kept off in phase 3. i_{Lr} flows from node X to Y, and discharges the gate of

Ma. Body diode of P1 conducts and forms the freewheeling path for i_{Lr} to return to driver supply, which is a gate energy recovery process. After V_x drops to $V_{dr}+V_{sw}-V_{fd}$, D1 conducts and V_g is clamped by D1 and P2.

Phase 4: t_3~t_4. P2 is kept on, P1, N3, N4 are kept off. This is the on-time for Ma, with $V_g=V_{dr}+V_{sw}-V_{fd}$.

Phase 5: t_4~t_5, in Fig.3(d). This is the turn-off pre-charging phase of Ma. P2 and N3 are turned-on, P1 and N4 are kept off. Current flowing through P2, L_r and N3 pre-charges L_r to be

$$I_{pre2} = (V_{dr} - V_{fd1}) \cdot t_{pre2} / L_r \qquad (5)$$

V_{fd1} is the forward voltage drop of D1, $t_{pre2}=t_5-t_4$ is the pre-charge time. Since P2 is kept on, the high side switch Ma in this phase is kept on.

Phase 6: t_5~t_6, in Fig.3(e). This is the turn-off phase of Ma. At the beginning of this phase, P2 and N3 are turned off, while P1 and N4 are kept off. Body diode of P1 conducts to form the freewheeling path for i_{Lr}. L_r resonates with C_{iss} with reverse initial current I_{pre2} and V_g decreases. With D2, V_x can be reduced to lower than $V_{sw}-V_{fd}$ without being clamped by the body diode of N4. Voltage decreasing in V_x can compensate for the voltage drop on r_g and L_s, thus, fast turn-off of Ma can be realized. With $V_y>V_x$, magnitude of L_r current keeps reducing in this phase. After i_{Lr} reaches zero, V_g drops to its negative peak value, which can be expressed ideally as

$$V_{gpkn} = -I_{pre2} \cdot \sqrt{L_r / C_{iss}} \qquad (6)$$

V_{gpkn} should be within the gate to source voltage rating of Ma.

Phase 7: t_6~t_7, in Fig.3(f). N4 ZVS turns on at the beginning of this phase. Ma is kept off. P1, P2, N3 are kept off. i_{Lr} flows from Y to X, which charges negative V_g through the body diode of N3. After V_x increases to be higher than $V_{sw}+V_{fd}$, D2 conducts and V_g is clamped by D2 and P2.

Phase 8: t_7~t_8. N4 is kept on, P1, P2, N3 are kept off. This is the off-time for Ma, with $V_g=V_{sw}+V_{fd}$.

III.   EXPERIMENT AND TEST RESULTS

To compare the proposed RGD with VSD and CSD, test PCB board is built up as shown in Fig.5. Three types of drivers are used to drive the high side switch of buck converter respectively. Bootstrap structure is used to provide the supply voltage for the gate driver of high side switch. To focus on the impact of different drivers to the high side switching loss, and do fair comparison between different drivers, VSD is used as the gate driver of low side power switch of buck converter in all the tests. High side power switch is SiR472DP (voltage rating V_{dsm}: 30V, current rating, I_{dsm}: 20A, on resistance R_{dson} 12mΩ) with internal r_g equals to 1.5Ω. To test the impact of different value of r_g on the switching loss, external resistor is added in series with the gate of Ma instead of using power MOSFET with different value of r_g. Thus, fair comparison can be made without impact from parameters change of different types of power MOSFET. Low side switch of buck converter is Si7446ADP (V_{dsm}: 30V, I_{dsm}: 30A, R_{dson}: 12mΩ). For the driver switches in three different gate drivers, PMOS (V_{dsm}: 20V, I_{dsm}: 1.8A, R_{dson}: 0.2Ω) and NMOS (V_{dsm}: 20V, I_{dsm}: 2.4A, R_{dson}: 0.125Ω) in Si3585DV are used for the driver PMOS (MP, P1, P2 in Fig.1) and NMOS (MN, N3, N4 in Fig.1) respectively. Power stage inductor is 330nH, switching frequency is 1MHz, L_r is 100nH. Converter dead-times in three different driver tests are kept the same.

Figure 5. Photo of power stage and high side switch gate driver on test PCB board

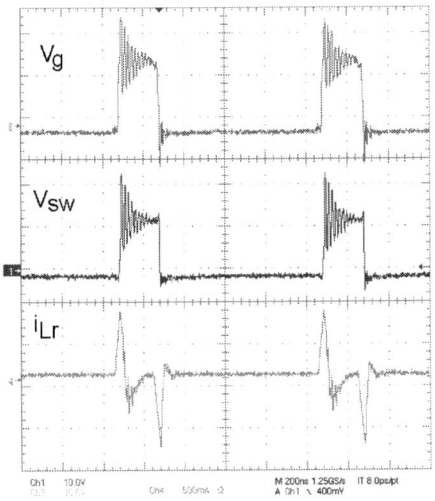

Figure 6. test waveforms of high side switch gate voltage V_g, switching node voltage V_{sw} and resonant inductor current i_{Lr}.

To compare CSD and proposed RGD, all the driver switches, L_r and turn-on/turn-off pre-charged current in L_r are the same. Fig.6 shows the test waveforms of the proposed RGD for V_g of Ma, V_{sw} and i_{Lr}, with 12V converter input voltage, 2.5V output voltage and 10A load current. Turn-on/turn-off pre-charge time is 25ns and pre-charged L_r current is 0.8A.

High side gate driving losses of three different drivers are 0.065W (VSD), 0.125W (CSD), 0.135W (proposed RGD) respectively. Two resonant drivers consume more power.

However, in the total converter efficiency as shown in Fig.7, CSD and proposed RGD have higher efficiency than VSD, which means that the switching loss reduction due to current source driving has dominant impact on power efficiency than the impact of gate driving loss increase.

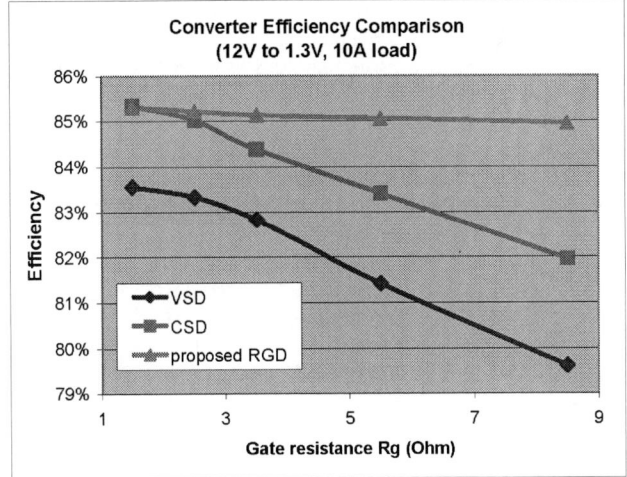

Figure 7. Comparison of the total power efficiency of buck converter with different high side switch gate resistance for three types of high side gate drivers.

To compare the impact of r_g on three different high side drivers, r_g is set to be 1.5Ω (internal r_g of SiR472DP), 2.5Ω, 3.5Ω, 5.5Ω, and 8.5Ω (with added external resistor) in different tests. Test results of the converter efficiency are shown in Fig.7. Since the same power stage is used for all the tests, all the other test parameters setup and test conditions are kept the same, the efficiency different can be considered to be the results of different high side gate drivers used. When r_g is as low as 1.5Ω, the efficiency of buck converter with CSD is almost the same as that of the converter with the proposed RGD. With smaller r_g, less V_{rg} is induced during switching and there is less impact of body diode clamping effect. Also, gate driving loss of CSD is smaller than the proposed RGD. As r_g becomes less than 1.5Ω, or close to 0Ω, converter with CSD driver may have better efficiency than the converter with proposed RGD. However, with larger r_g, converter efficiency of both VSD and CSD obviously drops due to the impact of r_g, while efficiency of the proposed RGD only slightly decreases. For converter with 12V input voltage, 1.3V output voltage, 10A load current, as in Fig.7, with r_g=3.5Ω, 5.5Ω and 8.5Ω, efficiency of converter with the proposed RGD is 0.75%, 1.65%, 3% higher than the converter with CSD, and 2.3%, 3.64%, 5.34% higher than the converter with VSD, respectively. For converter with 12V input voltage, 1V output voltage, 15A load current, with r_g=3.5Ω, 5.5Ω and 8.5Ω, efficiency of converter with the proposed RGD is 1.46%, 2.61%, 4.48% higher than the converter with CSD, and 3.28%, 4.77%, 7.27% higher than the converter with VSD, respectively. So, with proposed RGD, high side switching loss of buck converter caused by larger r_g can be obviously reduced.

IV. CONCLUSIONS

In this paper, a new resonant gate driver (RGD) is proposed to reduce the switching loss of the high side switch of buck converter. With the proposed RGD, gate driver behaves more like ideal current source, which can reduce the impact of parasitic resistance and inductance on the charging/discharging path. Thus, power MOS switching speed can be improved and switching loss can be reduced. From experiment results, with 5.5Ω gate resistance, for buck converter with 12V input voltage, 1.3V output voltage, 10A load current, converter with the proposed RGD has power efficiency 1.65% higher than the converter with the current source driver (CSD) and 3.64% higher than the converter with the conventional gate driver (VSD). For buck converter with 12V input voltage, 1V output voltage and 15A load current, the proposed RGD has converter efficiency 2.61% higher than CSD and 4.77% higher than VSD. The efficiency improvement is the results of high side switching loss reduction realized by the proposed resonant gate driver.

REFERENCES

[1] Zhiliang, Z., Eberle, W., Ping, L., Yan-Fei, L., and Sen, P.C. "A 1-MHz high efficiency 12V buck voltage regulator with a new current-source gate driver". *IEEE Transactions on Power Electronics*. vol. 23, pp.2817-2827, Nov. 2008

[2] Zhihua, Y., Sheng, Y., Yan-fei, L., "A new dual-channel resonant gate drive circuit for low gate drive loss and low switching loss". *IEEE Transactions on Power Electronics*. vol. 23, pp.1574-1583, May. 2008

[3] Zhihua, Y., Sheng, Y., Yan-fei, L., "A new resonant gate drive circuit for synchronous buck converter". *IEEE Transactions on Power Electronics*. vol. 22, pp.1311-1320, Jul. 2007

[4] Yuhui, C., Lee, F.C., Amoroso, L. and Ho-Pu, W., "A resonant MOSFET gate driver with efficient energy recovery". *IEEE Transactions on Power Electronics*. vol. 19, pp.470-477, Mar. 2004

[5] Pan, S. and Jain, P.K., "A new resonant gate driver with two half bridge structures for both top switch and bottom switch". In *Proc. IEEE Power Electronics Specialists Conference*, pp. 742-747, 2007

[6] Xiao, Y., Shah, H., Chow, T.P., and Gutmann, R. J. Analytical modeling and experimental evaluation of interconnect parasitic inductance on MOSFET switching characteristics. *Proc. IEEE Applied Power Electronics Conference and Exposition*, pp.516-521. 2004

[7] Yang, B., Zhang, J., "Effect and utilization of common source inductance in synchronous rectification". *Proc. IEEE Applied Power Electronics Conference and Exposition*, pp.1407-1411, 2005

[8] Meade, T., O'Sullivan, D., Foley, R., Achimescu, C., Egan, M. and McCloskey, P. "Parasitic inductance effect on switching losses for a high frequency DC-DC converter". *IEEE Applied Power Electronics Conference and Exposition*, pp. 3-9. 2008

[9] Ren. Y., Yao, K., Xu, M. and Lee, F. C., "Analysis of the power delivery path from the 12-V VR to the microprocessor". *IEEE Transactions on Power Electronics*, vol.19, pp. 1507-1514, Nov, 2004.

Switching Loss Analysis Considering Parasitic Loop Inductance with Current Source Drivers for Buck Converters

Zhiliang Zhang (*Member IEEE*)
Aero-Power Sci-tech Center
College of Automation Engineering
Nanjing University of Aeronautics & Astronautics
Nanjing, Jiangsu, P.R.China
Email: zlzhang@nuaa.edu.cn

Jizhen Fu (*Student Member, IEEE*), Yan-Fei Liu (*Senior, Member IEEE*) and P.C. Sen (*Life Fellow IEEE*)
Department of Electrical and Computer Engineering
Queen's University, Kingston, Ontario, Canada, K7L 3N6
jizhen.fu@queensu.ca, yanfei.liu@queensu.ca and senp@post.queensu.ca

Abstract— **In this paper, the switching loop inductance was investigated on the Current Source Drivers (CSDs). The analytical model was developed to predict the switching losses. It is noted that although the CSDs can reduce the switching transition time and switching loss greatly, the switching loop inductance still has the current holding effect on the CSDs. This results in high turn off loss for the control MOSFET in a buck converter. Thus, an improved layout was proposed to achieve minimum switching loop inductance. The experimental results verified the significant switching loss reduction owing to the proposed layout of a buck converter with 12V input, 1.3V output and 1MHz.**

I. INTRODUCTION

Voltage Regulators (VRs) with MHz switching frequencies can significantly reduce the size of the output inductances and capacitances, and improve the dynamic response during the transient events. However, the major concern of the high frequency application is excessive frequency-dependent losses including the switching loss, the gate drive loss and the body diode loss etc.

Resonant gate driver technique was proposed to recover large MOSFET drive loss at high frequency (>1MHz), especially for synchronous rectifier (SR) [1]-[4]. Actually, in a buck VR, the switching losses, especially turn off losses, are the dominant loss among the total loss breakdown due to the parasitic inductances. The effect of the common source inductance was investigated thoroughly to predict the switching loss accurately [5]. In order to reduce the switching loss, Current Source Drivers (CSDs) were proposed in [6]-[7] to reduce the switching transition time and switching loss by a constant drive current. Hybrid gate driver scheme proposed in [8] uses the CSD to reduce the high turn off loss of the control MOSFET, while drives the SR with a conventional voltage driver for the purpose of simplicity. In order to achieve optimal design, the loss model on the CSD was proposed in [9] and the current diversion problem was investigated in [10].

However, the effect of the switching loop inductances on the CSD has not been investigated carefully and analytically.

In this paper, the effect of the switching loop inductance is investigated on the CSDs. Through the mathematical modeling and simulation, it is concluded that the switching loop inductance still has the currents hold effect and thus, increases the switching loss significantly. Therefore, in order to improve the performance of the CSD, the switching loop inductance should also be minimized. Thus, an improved layout was proposed for the buck converter with the CSD.

II. IMPACT OF LOOP PARASITIC INDUCTANCE ON THE CSDs

In order to investigate the impact of the switching loop parasitics on the CSD, the basic clamp circuit as shown in Fig. 1 is used including a MOSFET in series with a diode D_1, dc input voltage V_D and an inductive load.

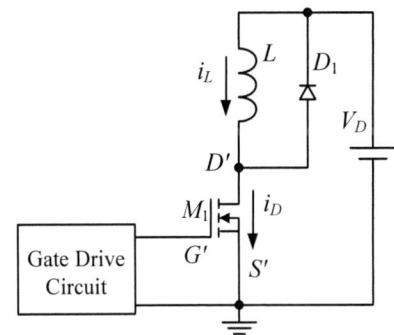

Fig. 1   Circuit with a clamped inductive load

The simplified equivalent circuit for the switching transition is shown in Fig. 2, where MOFET M_1 is represented with a typical capacitance model, the clamped inductive load is replaced by a constant current source I_L and the CSD is simplified as a current source (I_G). L_D is the switching loop inductance including the packaging inductance and any unclamped portion of the load inductance. L_S is the common source inductance.

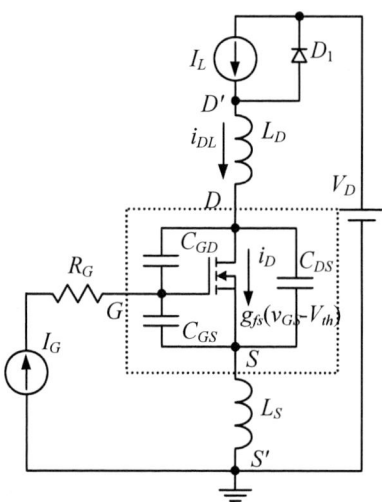

Fig. 2   Equivalent circuit of MOSFET switching transition

In order to simply the transient analysis, the following assumptions are made:

(1) $i_D = g_{fs}(v_{GS}-V_{th})$ and MOSFET is ACTIVE, provided $v_{GS} > V_{th}$ and $v_{DS} > i_D R_{DS(on)}$;

(2) For $v_{GS} < V_{th}$, $i_D = 0$, and MOSET is OFF;

(3) When $g_{fs}(v_{GS}-V_{th}) > v_{DS}/R_{DS(on)}$, the MOSFET is fully ON.

During the switching transition period, the MOSFET enters its active state and the linear transfer characteristics is assumed as given in (1) [11], where $i_D(t)$ is the instantaneous switching current and $v_{GS}(t)$ is the instantaneous gate-to-source voltage of the MOSFET:

$$i_D(t) = g_{fs}(v_{GS}(t)-V_{th}) \tag{1}$$

According the equivalent circuit in Fig. 2, the circuit equations take the form

$$I_G = C_{GD}\frac{dv_{GD}}{dt} + C_{GS}\frac{dv_{GS}}{dt} \tag{2}$$

And

$$v_{GD} = v_{GS} - v_{DS} \tag{3}$$

So

$$I_G = (C_{GS}+C_{GD})\frac{dv_{GS}}{dt} - C_{GD}\frac{dv_{DS}}{dt} \tag{4}$$

From(4), dv_{DS}/dt is solved as

$$\frac{dv_{DS}}{dt} = \frac{C_{GS}+C_{GD}}{C_{GD}} \cdot \frac{dv_{GS}}{dt} - \frac{I_G}{C_{GD}} \tag{5}$$

So $d^2 v_{DS}/dt^2$ and $d^3 v_{DS}/dt^3$ are respectively

$$\frac{d^2 v_{DS}}{dt^2} = \frac{C_{GS}+C_{GD}}{C_{GD}} \cdot \frac{d^2 v_{GS}}{dt^2} \tag{6}$$

$$\frac{d^3 v_{DS}}{dt^3} = \frac{C_{GS}+C_{GD}}{C_{GD}} \cdot \frac{d^3 v_{GS}}{dt^3} \tag{7}$$

During the switching interval, the change of the switching loop current i_{DL} induces a voltage across the parasitic inductance. The drain-to-source voltage v_{DS} is given as

$$v_{DS} = V_D - L_D\frac{di_{DL}}{dt} - L_s\frac{d(i_{DL}+I_G)}{dt}$$

$$= V_D - (L_D+L_s)\frac{di_{DL}}{dt} \tag{8}$$

And

$$i_{DL} = C_{GS}\frac{dv_{GS}}{dt} + C_{DS}\frac{dv_{DS}}{dt} + g_{fs}(v_{GS}-V_{th}) - I_G \tag{9}$$

Substituting (9) to (8) yields

$$v_{DS} = V_D - (L_D+L_s)(C_{GS}\frac{d^2 v_{GS}}{dt^2} +$$

$$C_{DS}\frac{d^2 v_{DS}}{dt^2} + g_{fs}\frac{dv_{GS}}{dt}) \tag{10}$$

Substituting (6) to (10) yields

$$v_{DS} = V_D - (L_D+L_s)$$

$$(\frac{C_{GS}C_{GD}+C_{DS}C_{GD}+C_{DS}C_{GS}}{C_{GD}} \cdot \frac{d^2 v_{GS}}{dt^2} + g_{fs}\frac{dv_{GS}}{dt}) \tag{11}$$

Differentiating (11) yields

$$\frac{dv_{DS}}{dt} = -(L_D+L_s)$$

$$(\frac{C_{GS}C_{GD}+C_{DS}C_{GD}+C_{DS}C_{GS}}{C_{GD}} \cdot \frac{d^3 v_{GS}}{dt^3} + g_{fs}\frac{d^2 v_{GS}}{dt^2}) \tag{12}$$

Substituting (5) into (12), (13) is derived

$$A\frac{d^3 v_{GS}(t)}{dt^3} + B\frac{d^2 v_{GS}(t)}{dt^2} + C\frac{dv_{GS}(t)}{dt} = I_G \tag{13}$$

where parameters A, B and C are represented in terms of the device parameters (C_{GS}, C_{GD}, C_{DS}, g_{fs} and R_G) and the equivalent circuit parameters (L_D and L_S) as $A = (L_D+L_s)(C_{GS}C_{GD}+C_{DS}C_{GD}+C_{DS}C_{GS})$, $B = g_{fs}(L_D+L_s)C_{GD}$ and $C = C_{GS}+C_{GD}$.

For turn-on transition, the initial condition for (13) is $v_{GS}(0)=V_{th}$. Then (13) solves to give either sinusoidal or exponential solutions, depending on the relative magnitudes of B^2 and AC.

When $B^2 - 4AC < 0$, sinusoidal solution occurs and $v_{GS}(t)$ takes the form:

$$v_{GS}(t) = (\frac{B^2}{2C^2 \cdot \sqrt{4AC-B^2}} - \frac{\sqrt{4AC-B^2}}{2C^2}) \cdot$$

$$I_G \cdot \exp(-\frac{t}{T_1}) \cdot Sin(\omega_1 t) + \frac{B}{C^2} \cdot I_G \cdot \exp(-\frac{t}{T_2}) \cdot Cos(\omega_1 t) \tag{14}$$

$$+ \frac{I_G \cdot t}{C} - \frac{B}{C^2} \cdot I_G + V_{th}$$

978-1-4244-4782-4/10 $26.00 © 2010 IEEE

where $T_1 = \dfrac{2A}{B}$, $\omega_1 = \dfrac{\sqrt{4AC - B^2}}{2A}$.

When $B^2 - 4AC > 0$, exponential solution occurs. Then $v_{GS}(t)$ takes the form

$$v_{GS}(t) = -\frac{(\sqrt{B^2 - 4AC} + B) \cdot I_G \cdot A \cdot \exp(-\dfrac{t}{T_2})}{(\sqrt{B^2 - 4AC} - B) \cdot C \cdot \sqrt{B^2 - 4AC}}$$

$$+\frac{(\sqrt{B^2 - 4AC} - B) \cdot I_G \cdot A \cdot \exp(-\dfrac{t}{T_3})}{(\sqrt{B^2 - 4AC} + B) \cdot C \cdot \sqrt{B^2 - 4AC}} \qquad (15)$$

$$+\frac{I_G \cdot t}{C} - \frac{B \cdot I_G}{C^2} + V_{th},$$

where $T_2 = \dfrac{2A}{B - \sqrt{B^2 - 4AC}}$ and $T_3 = \dfrac{2A}{B + \sqrt{B^2 - 4AC}}$.

Then, by substituting $v_{GS}(t)$ to (1) and (11), $i_D(t)$ and $v_{DS}(t)$ of the MOSFET can be calculated respectively. Therefore, the turn on loss is

$$P_{turn\_on} = \int_0^{t_{sw(on)\_Q1}} v_{DS}(t) \cdot i_D(t) dt \cdot f_s \qquad (16)$$

The turn-off transition is similar to the turn-on transition except for the initial condition becomes $v_{GS}(0) = V_{th} + \dfrac{I_L}{g_{fs}}$.

The turn off loss is

$$P_{turn\_off} = \int_0^{t_{sw(off)\_Q1}} v_{DS}(t) \cdot i_D(t) dt \cdot f_s \qquad (17)$$

From (16) and (17), the switching loss is

$$P_{switching} = P_{turn\_off} + P_{turn\_on} \qquad (18)$$

Fig. 3 shows the simulated the waveforms of the turn off transition with different switching loop inductances. It is observed that the drain-to-source current i_D with 4nH has a slower decay rate with longer turn off time as it is held by the loop inductances. As a result, the switching loss (turn off loss) is increased with higher loop inductance. The peak value of p_{loss} is increased from 490W to 690W (an increase of 40%) with longer turn off time.

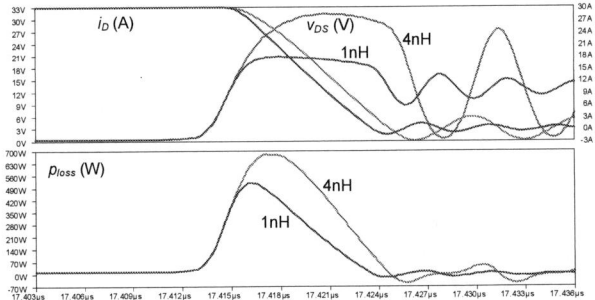

Fig. 3   Turn off transition comparison with L_D=4nH and L_D=1nH

III.   IMPROVED LAYOUT TO MINIMIZE THE PARASITIC EFFECTS

From Section II, the switching loop inductances increase the switching transition time and hold the drain current of the MOSFET. This leads to higher switching loss. Particularly, this problem becomes more serious in a high frequency buck converter since the switching loss (especially turn off loss) is the dominant loss.

Fig. 4 shows the synchronous buck converter with the loop parasitic inductance L_{d1}, L_{s1}, L_{d2} and L_{s2}. The basic idea is to reduce the switching loop inductance, and thus high turn off losses. As shown in Fig. 5, the input decoupling capacitance C_{in} is rearranged compared to Fig. 4. In this way, the parasitic inductances L_{d1} and L_{GND} can be significantly reduced. Based on this concept, two different layouts of the buck converter were implemented in the experimental test.

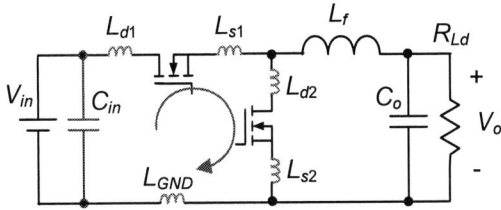

Fig. 4   Buck converter with the loop parasitics

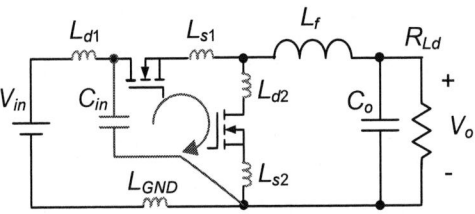

Fig. 5   Buck converter with rearranged input decoupling capacitance to reduce the loop parasitics

IV.   EXPERIMENTAL RESULTS AND DISCUSSION

In order to verify the efficiency improvement of the proposed layout arrangement, a 1MHz synchronous buck converter was built with the CSD as shown in Fig. 6. In order to reduce the dominant loss (the switching loss) in the buck converter in a cost-effective manner, a new hybrid gate drive scheme as shown in Fig. 6 is proposed for a buck converter. For the control MOSFET Q_1, the high side CSD proposed in [12] is used to achieve the switching loss reduction. For the SR Q_2, the conventional voltage source driver is used for low cost and simplicity, which is the bipolar totem-pole drive structure. PWM_SR is the signal fed into the bipolar totem-pole pair.

The specifications are as follows: input voltage V_{in}=12 V; output voltage V_o=1.3V; output current I_o=30A; switching frequency f_s=1 MHz; gate driver voltage V_c=5V. The PCB uses six-layer with 4 oz copper. The components used in the circuit are: Q_1: Si7860DP; Q_2: IRF6691; output filter inductance: L_f=300nH; current-source inductor: L_r=18nH (SMT 1812SMS-18N, Coilcraft); drive switches S_1-S_4: FDN335.

Photos of the prototype are illustrated in Fig. 7. The driver was built using discrete components and an Altera Max II EPM240 CPLD was used to generate the driver gate signals as illustrated in Fig. 7 (a). Surface mount (SMT) air core was used for the inductor as illustrated in Fig. 7 (b).

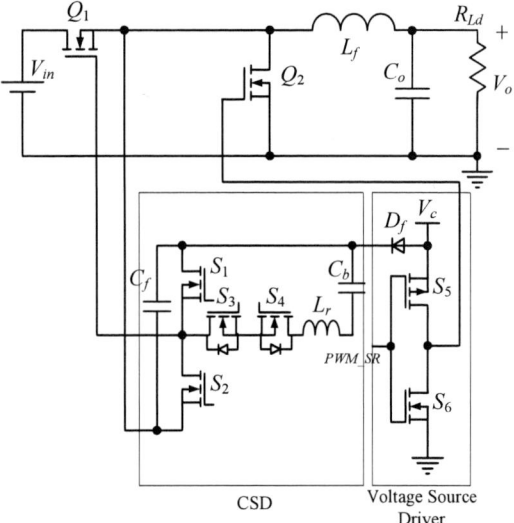

Fig. 6   Buck VR with hybrid driver scheme

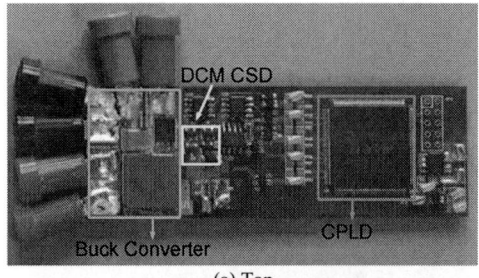

(a) Top

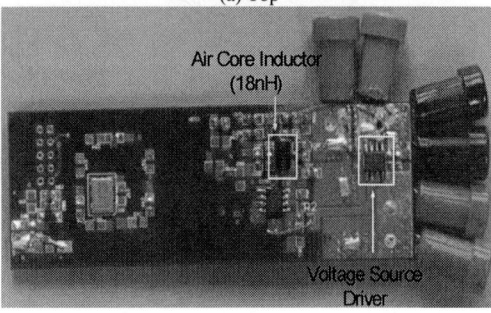

(b) Bottom

Fig. 7   Photos of the synchronous buck prototype with the hybrid gate driver

Fig. 8 shows the original power stage layout of the buck converter. The switching loop is highlighted in blue and the loop inductance is 10.5nH@1MHz measured with Agilent 4395A Analyzer. The power stage layout was rearranged to have much smaller switching loop as shown in Fig. 9. The measured loop inductance is only 3.7nH@1MHz, a reduction of 65%. The major difference between layout #1 an #2 is that input decoupling capacitances reduce the ground trace inductance and provide the transient energy, therefore, the

negative impact of the switching loop inductance is reduced greatly. This is very important for the CSD to reduce the high switching losses.

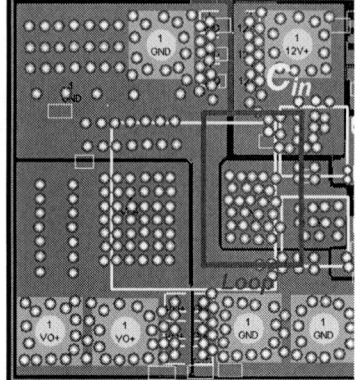

Fig. 8   Buck stage: #1

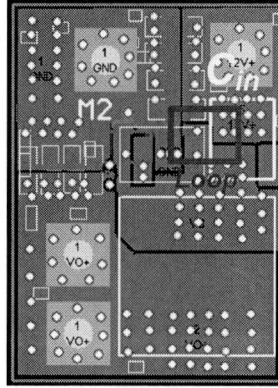

Fig. 9   Buck stage: #2

Fig. 10 and Fig. 11 illustrate the drain-to-source voltages of the SR MOSFET for layout #1 and #2 respectively. It is noted that compared to the layout #1, the layout #2 with reduced loop inductance alleviates the oscillation of the drain-to-source voltage greatly, which results from the parasitic inductance and reverse recovery of the SR body diode.

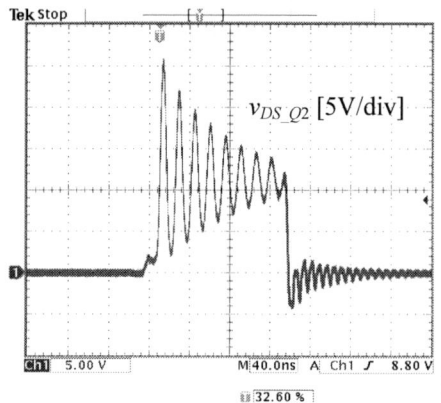

Fig. 10   Drain-to-source voltage at I_o=30A: Buck #1

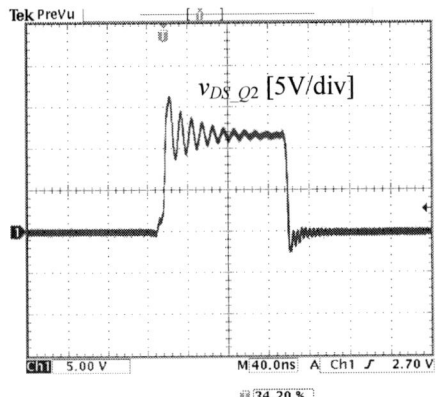

Fig. 11   Drain-to-source voltage at I_o=30A: Buck #2

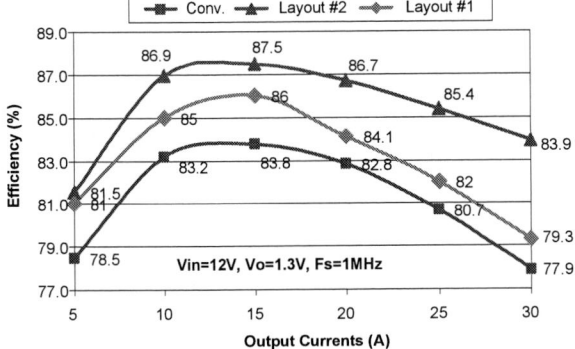

Fig. 12   Efficiency comparison: top: Buck #2; mid: Buck #1; bottom: conventional voltage driver (Conv.)

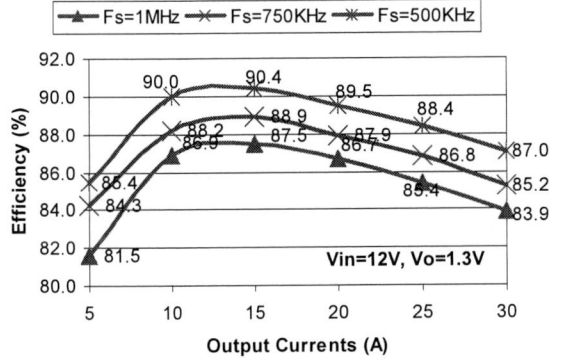

Fig. 13   Efficiency with different currents and switching frequencies at V_o=1.3V

Fig. 12 shows the measured efficiency comparison for two different layouts at 1.3V output. It is observed that at 20A, the efficiency is improved from 84.1% to 86.7% (an improvement of 2.6%) and at 30A, the efficiency is improved from 79.3% to 83.9% (an improvement of 4.6%). Higher efficiency improvement is achieved when the load current increases. This is because the switching loop inductance has stronger current holding effect with the higher load current. It is also noted the CSD with the layout #2 achieves higher efficiency than the conventional voltage gate driver. In the test, the predictive gate drive UCC 27222 from Texas Instruments was used as the conventional voltage driver.

Fig. 13 shows the measured efficiency for the CSD at different load currents and V_o=1.3V when the switching frequency changes. It is observed that at the load current of 30A, when the switching frequency changes from 1MHz to 500KHz, the efficiency is improved from 83.9% to 87%.

V.   CONCLUSION

In this paper, the switching loop inductance was investigated on the CSDs and the analytical model was developed to predict the switching losses accurately. The CSDs can reduce the effect of the common source inductance to expedite the switching speed and switching loss. However, the switching loop inductance still has the current holding effect on the CSDs. This will weaken the effectiveness of the CSDs in the sense of the switching loss reduction. Therefore, the switching loop inductance should also be minimized.

Based on this conclusion, an improved layout was proposed here to achieve minimum switching loop inductances compared to the original buck layout. The experimental results verified the efficiency improvement.

References

[1]   D. Maksimovic, "A MOS gate drive with resonant transitions," in *Proc. IEEE Power Electronics Specialists Conference (PESC)*, 1991, pp. 527-532.

[2]   L. Huber, K. Hsu, M. M. Jovanovic, D. J. Solley, G. Gurov and R. M. Porter, "1.8-MHz, 48-V resonant VRM: analysis, design, and performance evaluation," *IEEE Trans. Power Electron.*, Vol. 21, No. 1, pp.79-88, Jan. 2006.

[3]   J. R. Warren, K. A. Rosowski and D. J. Perreault, "Transistor selection and design of a VHF DC-DC power converter," *IEEE Trans. Power Electron.*, Vol. 23, No. 1, pp.27 – 37, Jan. 2008.

[4]   G. Spiazzi, P. Mattavelli and L. Rossetto, "Effects of parasitic components in high-frequency resonant drivers for synchronous rectification MOSFETs," *IEEE Trans. Power Electron.*, Vol. 23, No. 4, pp. 2082-2092, Jul. 2008.

[5]   Y. Ren, M. Xu, J. Zhou and F. C. Lee, "Analytical loss model of power MOSFET," *IEEE Trans. Power Electron.*, vol. 21, no. 2, pp. 310 - 319, Mar. 2004.

[6]   W. Eberle, Z. Zhang, Y. F. Liu and P. C. Sen, "A current source gate driver achieving switching loss savings and gate energy recovery at 1-MHz," *IEEE Trans. Power Electron.*, Vol. 23, No. 2, pp. 678-691, Mar. 2008.

[7]   Z. Yang, S. Ye and Y. F. Liu, "A new resonant gate drive circuit for synchronous buck converter," *IEEE Trans. Power Electron.*, Vol. 22, No. 4, pp.1311-1320, July 2007.

[8]   Z. Zhang, W. Eberle, Z. Yang, Y. F. Liu and P.C. Sen, "A new hybrid gate drive scheme for buck voltage regulators," in *Proc. IEEE Power Electronics Society Conference (PESC)*, June, 2008, pp.2498-2503.

[9]   Z. Zhang, W. Eberle, Z. Yang, Y. F. Liu and P.C. Sen, "Optimal design of resonant gate driver for buck converter based on a new analytical loss model," *IEEE Trans. Power Electron.*, Vol. 23, No. 2, pp.653-666, Mar. 2008.

[10]   J. Fu, Z. Zhang, Y. F. Liu and P. C. Sen, "A high efficiency current source driver with negative gate voltage for buck voltage regulators," IEEE Energy Conversion Congress and Exposition (ECCE 2009), Sep., San Jose, U. S..

[11]   Duncan A. Grant and John Gowar, "Power MOSFET theory and applications," A Wiley- Interscience Publication, Feb. 1989, pp. 97-143.

[12]   Z. Zhang, J. Fu, Y. F. Liu and P. C. Sen, "A new discontinuous current-source driver for high frequency power MOSFETs," *IEEE Energy Conversion Congress and Exposition (ECCE 2009)*, Sep., pp. 1655-1662.

Improved Asymmetric Space Vector Modulation for Voltage Source Converters with Low Carrier Ratio

Di Zhang[1], Fred Wang[2], Said El-Barbari[3], Juan Sabate[4], Dushan Boroyevich[1]

| 1 Center for Power Electronics System Virginia Polytechnic Institute and State University Blacksburg, VA 24060 USA | 2 University of Tennessee and Oak Ridge National Lab, Knoxville, TN 37996, USA | 2 GE Global Research Center Munich, Germany 3 GE Global Research Center Niskayuna, NY 12309, USA |

Abstract- **This paper presents an improved asymmetric space vector modulation (ASVM) for two level voltage source converters (VSCs) when the switching frequency is only 9 times of line frequency. By adding two pulses in each line cycle when the fundamental voltage crosses zero, the total harmonic distortion (THD) of output current can be reduced significantly with very limited penalty. The applications of improved ASVM in a single VSC or two interleaved VSCs systems are shown separately. With optimization, the ac current THD can be reduced to as low as 50% for single VSC and even lower to 20% for interleaved VSCs systems. Such THD reduction has close relationship with modulation index and interleaving angle. In addition, improved ASVM can also reduce the amplitude of circulating current which mainly determined the size of inter-phase inductors. Finally, the weights of total inductors needed to meet the same THD requirement are compared to demonstrate the benefits of improved ASVM when different PWM schemes are used. The analysis results are verified by experiments on a demo system.**

I. INTRODUCTION

Asymmetric space vector modulation (ASVM) is a popular pulse-width modulation (PWM) scheme used for voltage source converters (VSCs) operated with low carrier ratio (Rc), which is defined as the ratio between the switching frequency and operating ac frequency. When Rc is very low, ASVM, compared with symmetric SVM, has many benefits such as keeping the PWM waveforms symmetric in a line cycle, avoiding even order harmonic currents. [1]. Limited by the switching frequency of current power semiconductor devices, Rc, in high power applications, cannot be high. Especially, for applications involving high speed generator or motor, Rc will be extremely low such as around 9. In many cases, even if the devices can operate at higher switching frequency, it is also desirable to use lower switching frequency, in order to reduce loss and increase power capability. Low carrier ratio results in lower order voltage harmonics. To limit current harmonics, large ac line inductors are generally necessary, which can have cost and size penalty. Therefore, it is very desirable to improve the ASVM algorithm for better harmonic performance, so smaller or lighter passive components can be used to meet the total harmonic distortion (THD) requirement for ac currents.

Some papers have studied the performance of PWM with low carrier ratio. [2] and [3] compared four kinds of space vector modulations, including traditional asymmetric space vector modulation. [4] also investigated synchronized space vector modulation for active front-end rectifiers in high-power current-source drive. However, these studies mainly focused on the impact of space vector sequence on the performance of space vector modulation and the harmonic current performance of these space vector modulations are very similar. [5]-[7] proposed methods to optimize the harmonic performance of converters under space vector modulation control. But these methods cannot be used directly to the system with very low carrier ratio. In addition, how to maximize the current THD reduction in a paralleled VSCs system with interleaving and ASVM is still unknown.

This paper presents a method to improve the harmonic current performance of traditional ASVM with very limited penalty. Section II explains the basic idea and principle of the proposed improved ASVM. Based on that, the method to apply such improved ASVM in a single VSC is shown in Section III. With synchronized PWM, the THD of output current can be reduced by 50% compared with traditional ASVM when modulation index is high. In Section IV, the application of such improved ASVM is extended to a paralleled VSCs system with interleaving. With asymmetric interleaving [8], the THD of output current can be further reduced by more than 50% and additional benefit, reduced circulating current, is also presented. Finally, the potential benefit of such improved ASVM in increasing system power density is demonstrated by inductor physical design comparison in Section V. All analyses are verified by experimental results in Section VI.

II. PRINCIPLE OF IMPROVED ASVM

Without loss of generality, the work in this paper is based on the example system shown in Fig. 1. In Fig. 1, a simple two level VSC is used for a rectifier application. The power flows from the generator (modeled as a voltage source connected in series with a boost inductor) to the dc load (shown as a dc resistor). In the system under study, the ac line-to-neutral rms voltage is 230 V, the apparent power is 300 kVA, the dc bus voltage is 700V, the fundamental frequency varies between 500 Hz and 2 kHz, and ASVM is used.

Following the conventional wisdom, an odd triplen Rc is preferred [9-10]. In this paper, Rc is assumed to be 9. As a result, 18 vectors can be generated in one line cycle as shown in Fig.2. Since lower switching frequency usually means higher system harmonic currents, the lowest switching frequency 4.5 kHz is selected for study. The corresponding fundamental frequency is 500 kHz.

978-1-4244-4782-4/10 $26.00 © 2010 IEEE

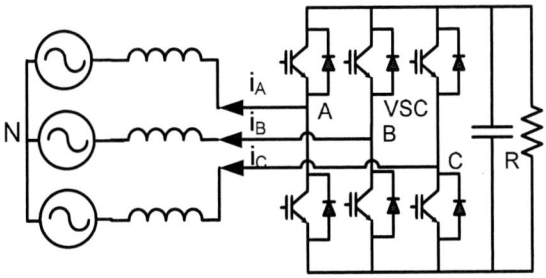

Fig. 1. Single VSC system under study.

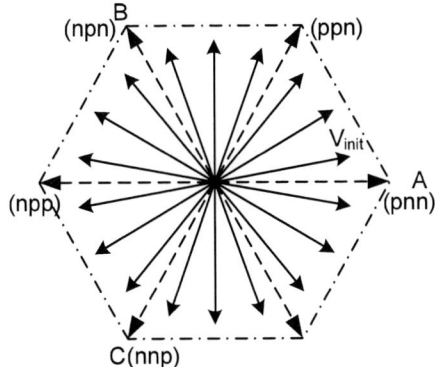

Fig. 2. Output vectors of ASVM (Rc=9)

The PWM waveform for phase A is shown in Fig. 3 when the first vector (V_{init}) is located at 10° and modulation index (M) is 0.9. In this paper, M is defined as the ratio between the line-to-line peak voltage to dc bus voltage, which is between 0 and 1.

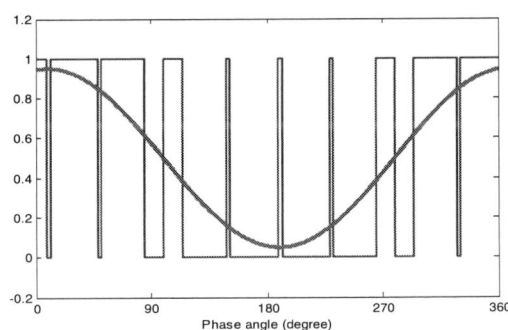

Fig. 3. PWM wave for phase A (V_{init} at 10°and M=0.9)

As presented in [10-11], the integration of the error between PWM waves and fundamental component indicates the energy of harmonic component. The integration results of such error in each half switching period (period for each vector in Fig. 2) is shown and compared in Fig. 4. From Fig. 4, it can be seen that the error is highest for the vectors located at 90°or 270°, where the fundamental component crosses zero. As a result, if additional switching is added to the system to increase the

equivalent switching frequency of harmonic voltage to reduce harmonics, such additional switching should be added there (vectors 5 and 14 in Fig. 4). When the fundamental voltage crosses zero, the fundamental current will also cross zero for unity power factor control, which is usually a desirable control strategy for grid-connected rectifier to maximize the power rating of a converter. Consequently, the penalty of additional switching loss is very limited.

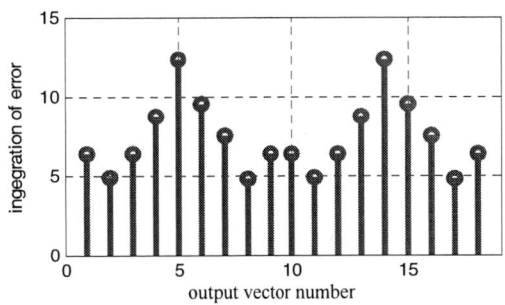

Fig. 4. Energy of harmonic voltage for each vector

For traditional ASVM, to generate each vector, two non-zero vectors (V1 and V2) and two zero vectors (V0 and V7) in each sector will be used as shown in Fig. 5 for the vector 90°. To increase the equivalent frequency of harmonic voltage, in such period part of V1 and V2 is swapped as shown in Fig. 6. In the time domain, it means one additional pulse is added to the system in phase A when fundamental voltage of phase A crosses zero. The method is the same for phase B and phase C. As a result, 11 pulses will be generated in one line cycle instead of 9 pulses used in traditional ASVM.

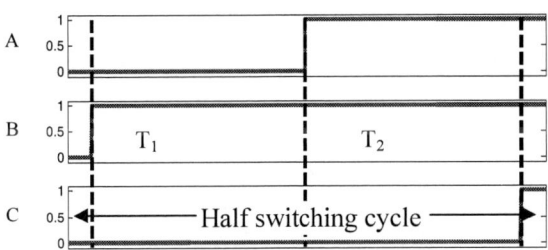

Fig. 5. Three phase PWM waves when output vector is 90°

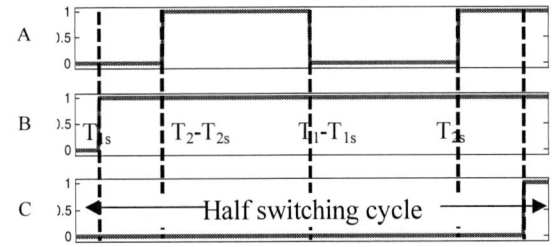

Fig. 6. Basic idea of improved ASVM

978-1-4244-4782-4/10 $26.00 © 2010 IEEE

By selecting k_1 and k_2, the duration and position of the additional pulses can be controlled. And, to keep the PWM wave symmetric in a line cycle, the sum of two k_1's used in vector 90 ° and 270° should be 1. So should the sum of k_2's.

III. IMPROVED ASVM IN A SINGLE VSC

For the improved ASVM, k_1 and k_2 can be optimized to minimize the THD of output current in a single VSC system. However, the analytical analysis for calculating the optimized k_1 and k_2 is still under study. In this paper, the optimized k_1 and k_2 are obtained by exhaustive searching method. However, the optimized k_1 is always around 0.2 and k_2 is always around 0.8 for vector 90 °. As a result, k_1 and k_2 are kept to be 0.2 and 0.8 for vector 90 °in all simulation and experiment.

The optimized k_1 and k_2 and the corresponding minimized THD of output current has close relationship with modulation index and the position of vectors as shown in Fig. 2. Modulation index is determined by the system operating condition and it is assumed to be 0.9 for the analysis in this section, corresponding to rectifier applications or inverter applications under rated conditions. The impacts of the initial vector V_{init}'s position on the THD of output current for traditional ASVM and improved ASVM with optimized k_1 and k_2 are shown and compared in Fig. 7. In Fig.7, the THD in all cases are normalized to the value when traditional ASVM is used and V_{init} is at 1°. The cases when V_{init} is at 0° and 20° are not analyzed, since in these cases the output vectors can be on the boundary of adjacent sectors which should be avoided.

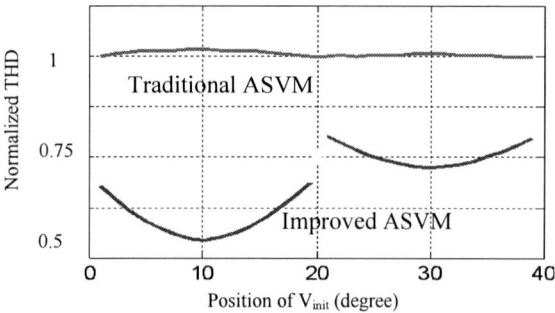

Fig. 7. Relationship between normalized THD of i_A and V_{init} for traditional and improved ASVM

From Fig. 7, the position of the first vector nearly cannot change the THD of output current for traditional ASVM. For improved ASVM, the THD of output current can be affected significantly by the position of the first vector. For the best case, when V_{init} is located at 10°, the THD of output current can be reduced by as much as about 50%. But for the worst cases, when V_{init} is located close to 20° and 40°, the THD reduction is very limited.

However, based on the study in [10] and [11], synchronized PWM should be used for the system with very low carrier ratio to avoid beat phenomenon. And synchronized ASVM can remain the positions of output vectors [12]. In other words, the position of V_{init}, practically, can be controlled to be at 10° to

maximize the benefit of the improved ASVM. The simulation results with traditional and improved ASVM are shown in Fig. 8 under the operation condition described in last section.

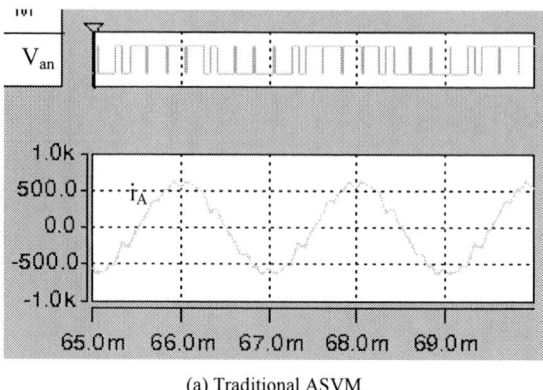

(a) Traditional ASVM

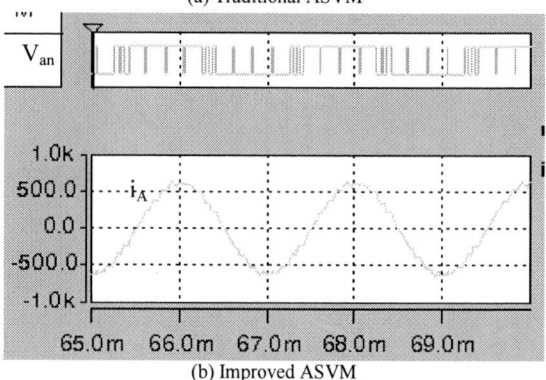

(b) Improved ASVM

Fig. 8. Simulation result of Van and i_A

As mentioned in last section, compared with traditional ASVM, the devices are switching two more times in each line cycle with improved ASVM. However, even if Rc is increased to 11, traditional ASVM can only reduce the THD of output current to 82%, not mentioning the penalty of higher switching loss and the unbalance between different phases due to non triplen Rc [10]. On the contrary, improved ASVM can reduce the current THD significantly with very limited extra loss especially for unity power factor operation.

The benefit of improved ASVM can be explained clearly in frequency domain. The spectra of output harmonic voltage for these two ASVM schemes are shown in Fig. 9 and Fig.10 where half of dc bus voltage is used as unity.

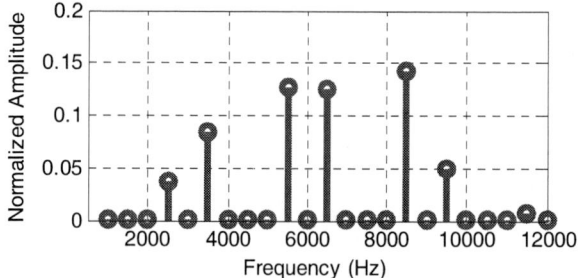

Fig. 9. Spectra of V_{AN} for traditional ASVM

978-1-4244-4782-4/10 $26.00 © 2010 IEEE

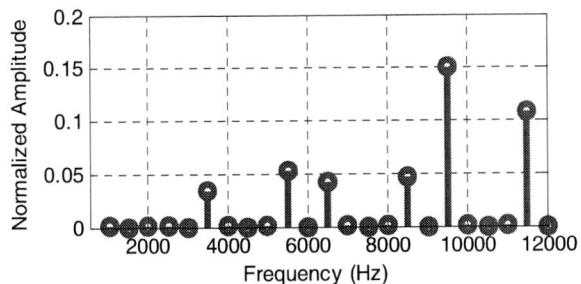

Fig. 10. Spectra of V_{AN} for improved ASVM

By comparing the voltage spectra in Fig. 9 and Fig. 10, the energy of harmonic components is pushed from around switching frequency to around twice switching frequency by the additional two pulses in improved ASVM. As a result, the equivalent impedance of boost inductor is increased and the THD of output current is reduced.

IV. IMPROVED ASVM IN INTERLEAVED VSCs SYSTEM

For paralleled VSCs system, the topology used is shown in Fig. 11 for high power density design [13]. The inter-phase inductor in simulation is realized by two coupled 100 μH inductor. Since the size of the inter-phase inductor is mainly determined by the amplitude of circulating current, it would be very desirable if the amplitude of circulating current can be reduced together with the THD of output current. Improved ASVM can also reduce the circulating current which can be seen in the following two Sections V and VI.

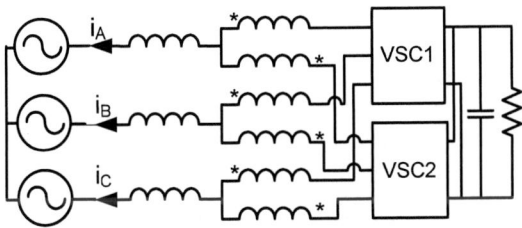

Fig. 11. Paralleled VSCs system under study.

Interleaving can further reduce the THD of output current by changing part of the output harmonic currents into circulating currents between two paralleled VSCs. and the interleaving angle (κ) should be designed to eliminate the dominant harmonic currents in output currents. [8]

From Fig. 9, the dominant harmonic currents are around switching frequency, so interleaving angle π should be used for traditional ASVM when M is high, e.g. 0.9. Based on Fig. 10, interleaving angle π/2 should be used, since for improved ASVM the dominant harmonic currents are pushed to around twice switching frequency. Considering system symmetry, V_{init} for VSC1 and VSC2 in Fig. 11 are placed at 5° and 15°.

As a summary of the analysis, the normalized THD of output current in different cases are compared in Fig. 12. In Fig. 12,

the THD of output current, when traditional ASVM is used and M is 0.9, is used as unity. From Fig.12, improved ASVM is only good when modulation index is high. When modulation index is close to 1, improved ASVM can reduce the THD almost by half. However, when modulation index is lower than 0.65, it may not be as good as traditional ASVM with 11 pulses per line cycle. For interleaved VSCs system with traditional ASVM, interleaving angle π/2 should be used for low modulation index case and interleaving angle π should be used for high modulation index case. This is similar to the case with high carrier ratio case [13]. But improved ASVM with π/2 interleaving is always the best PWM scheme.

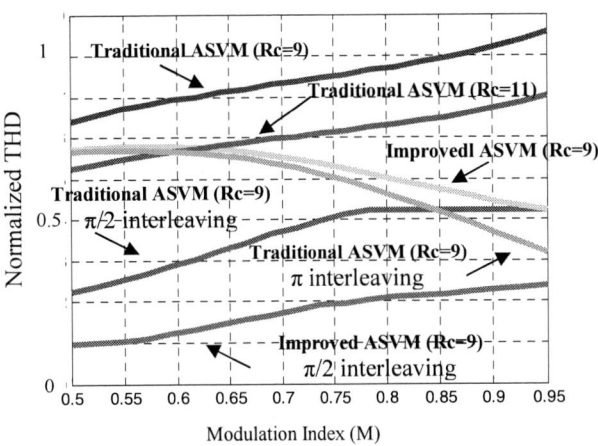

Fig. 12. THD reduction comparison

V. INDUCTOR WEIGHT REDUCTION WITH IMPROVED ASVM

For the example system shown in section IV, the simulations for paralleled VSCs topology are performed for traditional and improved ASVM. The simulation results are shown in Fig.13 and Fig. 14. From the simulation results, the THD of output current, with optimized k_1 and k_2, can be reduced from 4.3% to 2.1% with interleaving and improved ASVM. In addition, the amplitude of circulating current is also reduced from 194A to 80A.

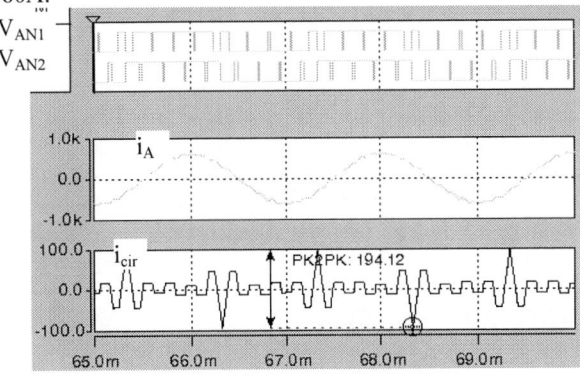

Fig. 13. Simulation results for traditional ASVM (κ=π)

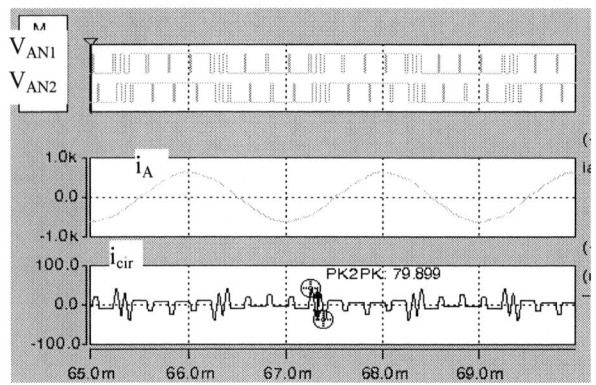

Fig. 14. Simulation results for improved ASVM (κ=π/2)

Based on the simulation results in last two sections, the inductor physical design is done and summarized in Table 1 for each case to reduce the THD of output current to the same level 2.1%. In the design, Metglas amorphous Alloy 2605SA1 is used for core design, the temperature rise is limited within 100 °C, EE core is used for three inductor design and CC core is used for inter-phase inductor design. From Table I, improved ASVM with interleaving can reduce the weight of required inductor significantly.

Table I. Comparison of Inductor Weight

Case	Single VSC with traditional ASVM	Single VSC with Improved ASVM	Interleaved Two VSCs with traditional ASVM (κ=π)	Interleaved Two VSCs with improved ASVM(κ=π/2)
Additional inductors needed	one 300uH 3φ inductor	one 100uH 3 phase inductor	one 80 uH 3 phase and three 400 uH inter phase inductors	three 400uH inter phase inductors
Weight of total inductors	140kg (100%)	47kg (34%)	70kg (50%)	16kg (11.5%)

VI. EXPERIMENTAL RESULTS

The experiment is designed to verify three key points in the analysis above. First, the improved ASVM can reduce the THD of output current dramatically. Second, interleaving with designed interleaving angle can further reduce the THD of output current. Third, the amplitude of circulating current in interleaved VSCs system can also be significantly reduced by the improved ASVM.

To simplify the experiment, the inverter configuration in Fig. 15 is used, whose voltage harmonic spectra are the same as in the case of the rectifier configuration in Fig. 11. The ac side now has a three phase resistor load in addition to inductors.

Since in the analysis, the harmonic current is assumed to be only determined by inductance and harmonic voltage, the resistor load may change the amplitude of harmonic currents especially when switching frequency is low and the power factor is high. However, the same setup is used for all PWM

schemes, so the relative THD reduction of output current is still very close to the analysis results.

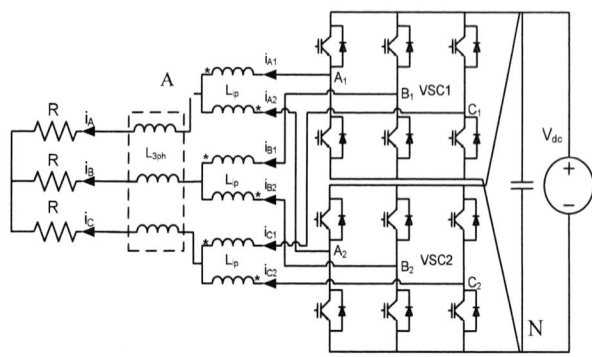

Figure 15. Schematic of the experimental setup

The experimental setup comprises of two 6-pack IGBT intelligent power modules (IPMs) from Fuji (6MBP20RH060) for two VSCs power stage, one common DSP-FPGA digital controller, three inter-phase inductors, one three phase inductor and three 6Ω resistors as load. The ac line inductor is 400μH and the inter-phase inductor used to limit circulating current is 3mH. The dc voltage is 100V, and the fundamental and switching frequency are chosen as 500Hz and 4.5 kHz respectively. Synchronized PWM and close loop control are used for all cases. The reference RMS value of fundamental current in the resistor load is 6A with corresponding modulation index of 0.9. To verify the analysis results in single VSC system, the two VSCs are controlled as one VSC without interleaving.

The experimental waveforms for single VSC case with traditional ASVM and improved ASVM are shown in Fig. 16 and Fig. 17. The current THD value and RMS value of fundamental components are summarized in Table II.

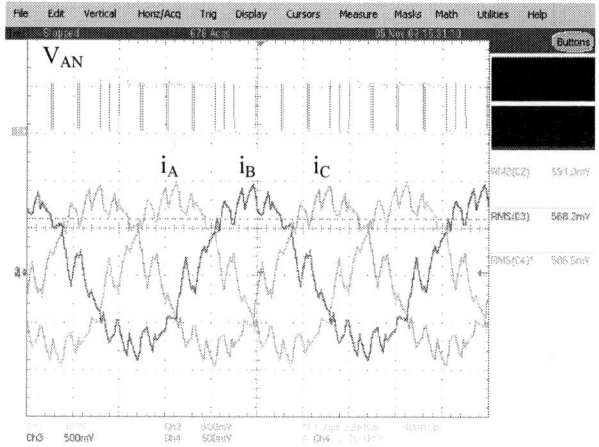

Fig. 16. V$_{AN}$, i$_A$, i$_B$, i$_C$ for traditional ASVM

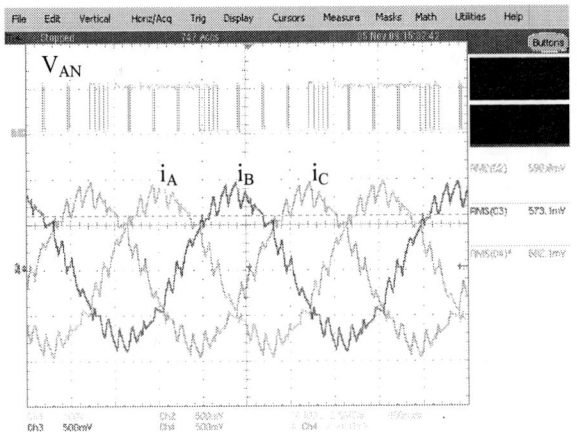

Fig. 17. V_{AN}, i_A, i_B, i_C for improved ASVM

Table II. Experimental results for single VSC

		Experimental Results	Simulation Results
Traditional ASVM	RMS of i_A	5.82A	5.94A
	THD of i_A	17.8%	19.5%
Improved ASVM	RMS of i_A	5.93A	5.94A
	THD of i_A	10.7%	11.4%

From Fig. 16, Fig. 17 and Table II, it can be seen that the RMS value of output current is controlled very close to the reference value. And by adding two pulses when voltage fundamental component crosses zero, the THD of output current can be reduced by 40%. Since the power factor in this system is very close to 1, the current amplitude is almost zero during the added pulses, introducing very limited additional power loss. The difference of i_A RMS value between experimental and simulation results is mainly caused by sampling issues. When current waveform is distorted, it is difficult to sample the real fundamental component. Such difference can be significantly reduced by improved ASVM, since ripple current is reduced. The difference of THD of i_A between experimental and simulation results is mainly caused by the system parameters difference between simulation and experiment.

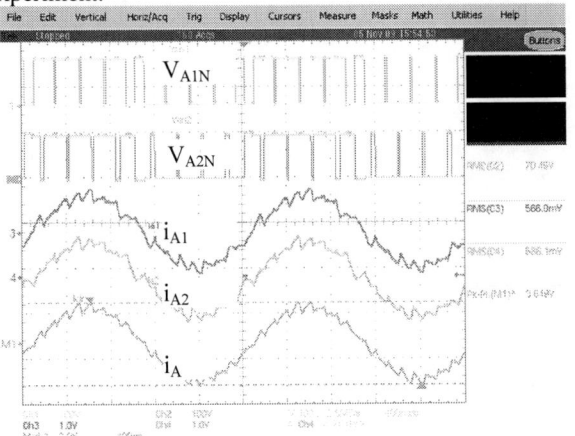

Fig. 18. V_{A1N}, V_{A2N}, i_{A1}, i_{A2}, i_A for traditional ASVM with interleaving

The experimental waveforms for interleaved VSCs case with traditional ASVM and improved ASVM are shown in Fig. 18 and Fig. 19. The current THD value and RMS value of fundamental components are summarized in Table III.

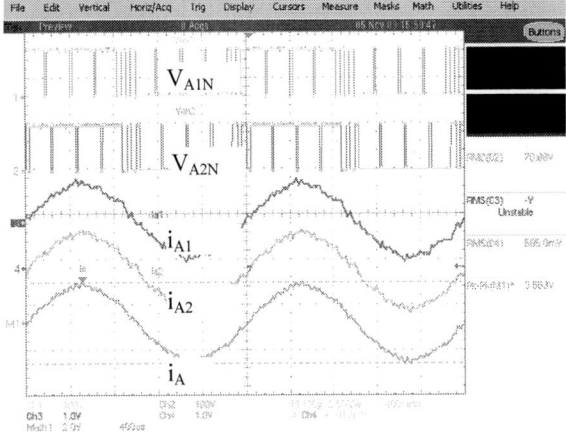

Fig. 19. V_{A1N}, V_{A2N}, i_{A1}, i_{A2}, i_A for improved ASVM with interleaving

Table III. Experimental results for interleaved VSCs system

		Experimental Results	Simulation Results
Traditional ASVM ($k=\pi$)	RMS of i_A	5.82A	5.94A
	THD of i_A	9.5%	9.5%
Improved ASVM ($k=\pi/2$)	RMS of i_A	5.94A	5.94A
	THD of i_A	5.0%	5.5%

From Fig. 18, Fig. 19 and Table III, it can be seen that interleaving can also reduce the THD of output current to 54%. And it can be further reduced to 28% with improved ASVM. Again, the additional switching is added when the output current crosses zero, introducing very small additional power loss.

To verify the impact of improved ASVM on the circulating current, the experiment for interleaved VSCs cases are repeated. The experimental waveforms including circulating currents for interleaved VSCs case with traditional ASVM and improved ASVM are shown in Fig. 20 and Fig. 21.

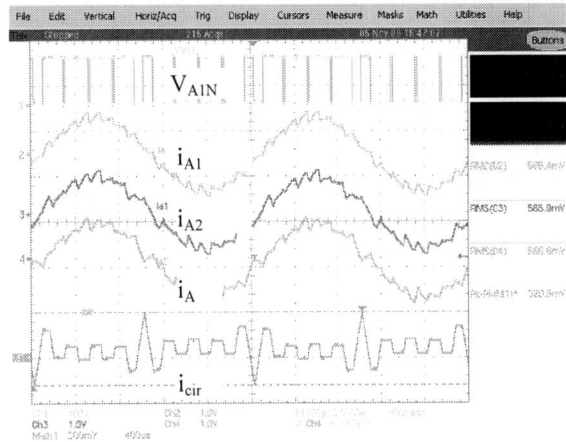

Fig. 20. V_{A1N}, i_A, i_B, i_C, i_{cir} for improved ASVM

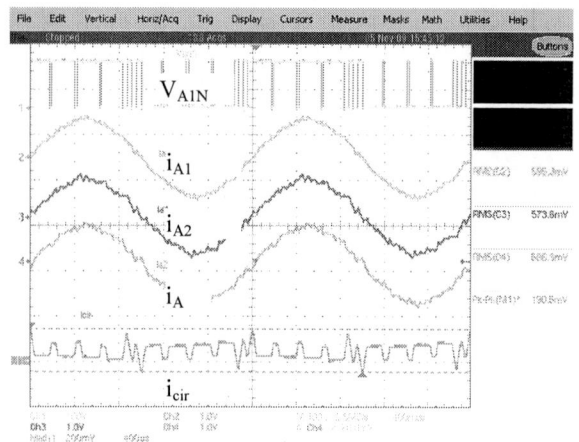

Fig. 21. V_{A1N}, i_A, i_B, i_C, i_{cir} for improved ASVM

From Fig. 20 and Fig. 21, it can be seen that the amplitude of circulating current is reduced from 1.6A to 0.9A with improved ASVM. The analysis and simulation results are verified.

To further verify the impact of modulation index on the THD reduction benefit of improved ASVM. The experiments are repeated with reduced modulation index (M=0.75). The experimental data are listed in Table IV.

Table IV. Summary of Experimental results (M=0.75 or M=0.9)

		M=0.9	M=0.75
Traditional ASVM	RMS of i_A	5.82A	5.0A
	THD of i_A	17.8% (100%)	20% (100%)
Improved ASVM	RMS of i_A	5.93A	4.95A
	THD of i_A	10.7% (60%)	14.9% (75%)
Traditional ASVM (k=π)	RMS of i_A	5.82A	4.91A
	THD of i_A	9.5% (53%)	13.9% (70%)
Improved ASVM (k=π/2)	RMS of i_A	5.94A	5.07A
	THD of i_A	5.0% (29%)	5.6% (28%)

From Table IV, higher modulation index is good for THD reduction in single VSC system with improved ASVM and interleaved VSCs with traditional ASVM. But improved ASVM with interleaving is always the best PWM scheme in THD reduction. These match the analysis results in Fig. 12.

VII. CONCLUSION

This paper presented an improved asymmetric space vector modulation (ASVM) for two level voltage source converters operated with low carrier ratio. By adding two pulses for each phase in one fundamental cycle, the THD of output current can be reduced as much as 50%. The THD can be further reduced by more than 50% if improved ASVM is used together with interleaving in a paralleled VSC system. The THD reduction depends highly on the modulation index, but improved ASVM with interleaving can always minimize the THD of output current. In addition, improved interleaving can also reduce the amplitude of circulating current between the paralleled VSCs which can help reduce the size and weight of inter-phase inductor. Since the additional pulses are added when fundamental voltage crosses zero, the resulted extra power loss

is very limited especially if system power factor is high. The benefit of improved ASVM is also demonstrated by inductor weight comparison in different PWM schemes and all of the analyses are verified by experimental results.

ACKNOWLEDGEMENT

This work was supported by a GE fellowship.

REFERENCES

[1]   D. G. Holmes and T. A. Lipo, "*Pulse Width Modulation for Power Converters-Principle and Practice,*" Wiley-IEEE Press, 2003.

[2]   G. Narayanan and V.T. Ranganathan," Synchronised PWM strategies based on space vector approach. I. Principles of waveform generation," *IEE Trans. Elec. Power Appl.*, vol. 146, no. 3, pp. 276 – 281, 1999.

[3]   G. Narayanan and V.T. Ranganathan," Synchronised PWM strategies based on space vector approach. II. Performance assessment and application to V/f drives," *IEE Trans. Elec. Power Appl.*, vol. 146, no. 3, pp. 267 – 275, 1999.

[4]   Y. W. Li, B. Wu, D. Xu and N.R. Zargari, "Space Vector Sequence Investigation and Synchronization Methods for Active Front-End Rectifiers in High-Power Current-Source Drives", *IEEE Trans. Ind. Elec.*, vol. 55, no. 3, pp. 1022-1034, Mar. 2008.

[5]   G. Narayanan, V.T. Ranganathan, D. Zhao, H. K. Krishnamurthy and R. Ayyanar, "Space Vector Based Hybrid PWM Techniques for Reduced Current Ripple", *IEEE Trans. Ind. Elec*, vol. 55, no. 4, pp. 1614 – 1627, Apr. 2008.

[6]   A. Mehrizi-Sani and S. Filizadeh, S. "An Optimized Space Vector Modulation Sequence for Improved Harmonic Performance", *IEEE Trans. Ind. Elec*, vol. 56, no. 8, pp. 2894-2903, Aug. 2009.

[7]   D. Zhao and R. Ayyanar, " Space Vector PWM with DC Link Voltage Control and Using Sequences with Active State Division", *Industrial Electronics, 2006 IEEE International Symposium on* ,vol 2, pp.1223-1228, Jul. 2006

[8]   D. Zhang, F. Wang, R. Burgos, R. Lai, T. Thacker and D. Boroyevich, "Interleaving impact on harmonic current in DC and AC passive components of paralleled three-phase voltage-source converters," *in Proc. IEEE Appl. Power Electron. Conf. (APEC)*, 2007, pp. 219 – 225.

[9] D. G. Holmes and B. P. McGrath, "Opportunities for harmonic cancellation with carrier-based PWM for a two-level and multilevel cascaded inverters," *IEEE Trans. Ind. Appl.*, vol. 37, no. 2, pp. 574–582, Mar./Apr. 2001.

[10]   F. Wang, "Reduce Beat and Harmonics in Grid-Connected Three-Level Voltage-Source Converters With Low Switching Frequencies", *IEEE Trans. Ind. Appl.*, pp. 1349-1359, 2007

[11]   Y. Iwaji, T. Sukegawa, T. T. Okuyama, T. Ikimi, M. Shigyo, and M. Tobise, "A new PWM method to reduce beat phenomenon in largecapacity inverters with low switching frequencies," *IEEE Trans. Ind. Appl.*, vol. 35, no. 3, pp. 606–612, May/Jun. 1999.

[12]   F. Blaabjerg, V. Oleschuk, and F. Lungeanu, "Synchronization of output voltage waveforms in three-phase inverters for induction motor drives," in *Proc. Power Conversion Conf.*, Osaka, Japan, 2002, vol. 2, pp. 528–533.

[13]   D. Zhang, F. Wang, R. Burgos, R. Lai and D. Boroyevich, "Interleaving Impact on AC Passive Components of Paralleled Three-Phase Voltage-Source Converters," *in Proc. IEEE Ind. Appl. Spec. (IAS)*, 2008, pp. 1 – 7.

A Hybrid Switching Scheme for LLC Series-Resonant Half-Bridge DC-DC Converter in a Wide Load Range

Woo-Young Choi[1], Bong-Hwan Kwon[2], and Jih-Sheng (Jason) Lai[1]

[1]Virginia Polytechnic Institute and State University
Future Energy Electronics Center
106 Plantation Road, Blacksburg, VA, 24061, USA
[2]Pohang University of Science and Technology, South Korea
E-mail: wychoi@vt.edu[1]

Abstract — This paper presents a hybrid switching scheme for LLC series-resonant half-bridge (SRHB) dc-dc converter. The concept of the proposed hybrid switching scheme is changing the switching mode to improve power efficiency in a wide load range. The SRHB converter operates in three different switching modes depending on the output load condition. Digital control of the system facilitates the implementation of the proposed hybrid switching scheme. High efficiency is achieved with a constant output voltage over wide load range. Experimental results on a 5 kW prototype are presented to evaluate the feasibility of the proposed switching scheme for the SRHB converter.

I. INTRODUCTION

High efficiency dc-dc converter for wide load ranges is necessary for the applications which are battery-powered and have energy consumption constrains [1], [2]. For a high efficiency dc-dc converter, the LLC series-resonant half-bridge (SRHB) converter is gaining its popularity [3]-[6]. Output voltage regulation is achieved by switching-frequency modulation [7]. However, switching frequency increases as the output load decreases. Power efficiency decreases because switching losses dominate at light load condition. To reap the benefits of the resonant converters while operating at a constant frequency, the asymmetrical pulse-width modulation (APWM) is applied to the SRHB converter [8], [9]. The APWM SRHB converter features high efficiency by zero-voltage switching (ZVS) operation. However, the disadvantage of the APWM SRHB converter is low efficiency at light loads. The switching losses constitute a major portion of the total power losses when lightly loaded. The key to achieve high efficiency under light load condition is reducing the load-independent power losses. Variable-frequency APWM method reduces switching losses by lowering the switching frequency [10]. However, it results in a higher peak current flowing through the power switches, which increases their respective current ratings and die size, eventually increases cost.

To deal with these problems, we propose a new switching scheme for the LLC SRHB converter. Hybrid switching mode is proposed for the LLC SRHB converter operating in a wide load range. The concept of the proposed switching mode is changing the switching mode according to the output load condition. The LLC SRHB converter operates in different switching modes depending on the output load condition. The proposed hybrid switching mode includes three switching modes: switching frequency modulation is used at heavy load condition, pulse frequency modulation [11] at medium-to-light load condition, and pulse duty-cycle modulation at very light load condition. Digital control of the system facilitates the

implementation of the proposed switching scheme. High efficiency is achieved in a wide output load range. The overall system configuration is briefly described. The proposed hybrid switching scheme is presented. Experimental results on a 5 kW prototype are presented to evaluate the feasibility of the proposed switching scheme for the LLC SRHB converter.

II. SYSTEM CONFIGURATION

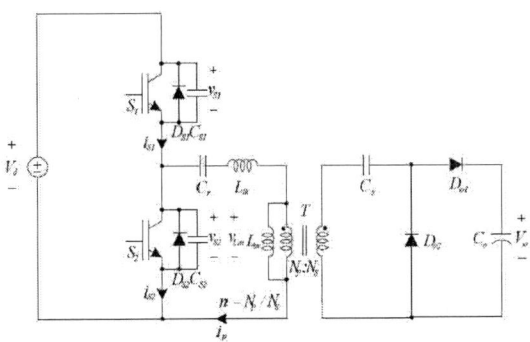

Fig. 1. Circuit diagram of the proposed LLC SRHB converter.

Fig. 1 shows the LLC-SRHB dc-dc converter with a voltage-doubler rectifier stage. V_d and V_o are the input and output voltages, respectively. The power switches S_1 and S_2 include their body diodes D_{S1} and D_{S2}, respectively. C_{S1} and C_{S2} are the output capacitors of S_1 and S_2, respectively. The transformer T is an ideal transformer which has the magnetizing inductor L_m and leakage inductor L_{lk} with the turns ratio n ($= N_p/N_s$). The capacitor C_s and the output diodes D_{o1} and D_{o2} form the voltage doubler rectifier stage. In the primary-side, the series-resonant circuit consists of the leakage inductor L_{lk} and the equivalent capacitor C_{eq} where

$$C_{eq} = \frac{n^2 C_s C_r}{C_r + n^2 C_s}.$$ (1)

The resonant frequency f_r is determined as

$$f_r = \frac{1}{2\pi\sqrt{L_{lk}C_{eq}}}.$$ (2)

In the series-resonant circuit, the switching frequency f_s is faster than the resonant frequency f_r; the primary current i_P is lagging with respect to the voltage v_{Lm} across L_m. Turn-on switching losses of the power switches are reduced under zero-voltage switching condition as the conventional LLC-SRHB dc-dc converter [3]-[6]. When S_2 is turned off, the primary current i_p starts to charge C_{S2} and discharge C_{S1}. The voltage v_{S1} across S_1 decreases and the voltage v_{S2} across S_2 increases owing to the negative primary current i_p. After i_p discharges C_{S1} completely, the body diode D_{S1} is turned on. i_p flows through D_{S1}. Then, zero-voltage turn-on of S_1 is achieved because i_p has already flown through the body diode D_{S1}. S_1 is turned off by the similar operation.

III. HYBRID SWITCHING SCHEME

The LLC SRHB converter in Fig. 1 operates in three different switching modes depending on the output load condition. Fig. 2 shows the flowchart of the hybrid switching control scheme.

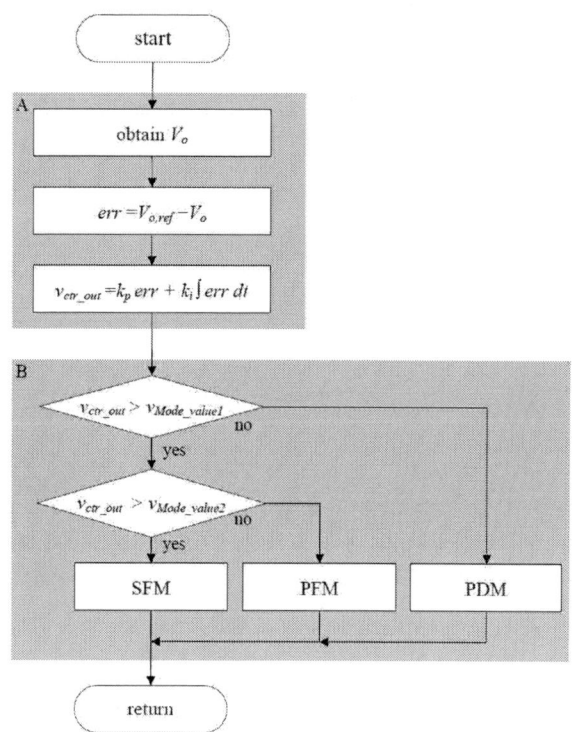

Fig. 2. Flowchart of the proposed hybrid switching scheme.

In stage A of the flowchart, the output voltage V_o is first measured. Then, the error voltage e is calculated as

$$e = V_o^* - V_o . \qquad (3)$$

where V_o^* and V_o are the reference and measured output voltages. A proportional-integral (PI) voltage controller generates the output value v_{ctr_out} as

$$v_{ctr\_out} = k_p e + k_i \int e\, dt \qquad (4)$$

where k_p and k_i are the proportional and integral gains. The output value of the PI voltage controller is proportional to the amount of output load. It is compared with output load reference values in the next stage.

In stage B of the flowchart, the switching mode of the SRHB converter is determined by comparing v_{ctr_out} with the reference values. v_{Mode_value1} and v_{Mode_value2} are the output load reference values, and $v_{Mode\_value1} > v_{Mode\_value2}$. Fig. 3 shows the gating signals of the switches for each switching mode. ξ_k is the time at which the kth cycle starts, T_k is the duration of the kth switching period, Φ_k is the on-time of each switch, and $d_k = \Phi_k / T_k$ is the duty cycle. It is assumed that each switch is turned on subsequently with a fixed dead-time.

1) Switching Frequency Modulation (SFM) Mode: When v_{ctr_out} is higher than v_{Mode_value1}, switching frequency modulation is used. Fig. 3(a) shows the gating signals of the switches in switching frequency modulation mode. The LLC SRHB converter operates at heavy load condition. The on-time of each switch is identical as $\Phi_k = T_k / 2$. The switching period T_k is modulated by adding a modulated switching period $\Delta T_{k,SFM}$ to the minimum switching period $T_{k,min}$ as

$$T_k = T_{k,\min} + \Delta T_{k,SFM} . \qquad (5)$$

The modulated switching period $\Delta T_{k,SFM}$ is represented as

$$\Delta T_{k,SFM} = (T_{k,\max} - T_{k,\min})\alpha \qquad (6)$$

where $T_{k,max}$ is the maximum switching period. The modulation index α is expressed as

$$\alpha = \frac{v_{ctr\_out} - v_{Mode\_value1}}{v_{ctr\_out}^* - v_{Mode\_value1}}, \qquad 0 \le \alpha \le 1 \qquad (7)$$

where $v_{ctr\_out}^*$ is the maximum output value of the PI voltage controller. As the output load decreases, v_{ctr_out} approaches v_{Mode_value1}, and α is close to zero. The switching period scales down with the decreasing output load. The switching frequency f_s can be faster than the resonant frequency f_r. ZVS operation is ensured at heavy load condition.

2) Pulse Frequency Modulation (PFM) Mode: When v_{ctr_out} is between v_{Mode_value1} and v_{Mode_value2}, fixed-on-time pulse frequency modulation [15] is used. Fig. 3(b) shows the gating signals of the switches in fixed-on-time pulse frequency modulation mode. The LLC SRHB converter operates at medium-to-light load condition. The on-time Φ_k of each switch is fixed to $T_{k,min} / 2$. The switching period T_k is modulated by adding a modulated switching period $\Delta T_{k,PFM}$ to the minimum switching period $T_{k,min}$ as

$$T_k = T_{k,\min} + \Delta T_{k,PFM} . \qquad (8)$$

The modulated switching period $\Delta T_{k,PFM}$ is represented as

$$\Delta T_{k,PFM} = (T_{k,\max} - T_{k,\min})\beta . \qquad (9)$$

Here, the modulation index β is expressed as

$$\beta = \frac{v_{Mode\_value1} - v_{ctr\_out}}{v_{Mode\_value1} - v_{Mode\_value2}}, \qquad 0 \le \beta \le 1. \qquad (10)$$

As the output load decreases, v_{ctr_out} approaches v_{Mode_value2}, and β is close to unity. The switching period scales up with the

978-1-4244-4782-4/10 $26.00 © 2010 IEEE

decreasing output load. The switching frequency and its associated switching losses can be reduced at light load condition.

3) Pulse Duty-Cycle Modulation (PDM) Mode: When v_{ctr_out} is lower than v_{Mode_value2}, pulse duty-cycle modulation is used. Fig. 3(c) shows the gating signals of the switches in pulse duty-cycle modulation mode. The LLC SRHB converter operates at very light load condition. The switching period T_k is fixed to $T_{k,max}$. The on-time Φ_k of each switch is modulated as

$$\Phi_k = T_{k,\min}\gamma \qquad (11)$$

where the modulation index γ is expressed as

$$\gamma = \frac{v_{ctr\_out}}{v_{Mode\_value2}}, \qquad 0 \le \gamma \le 1. \qquad (12)$$

The on-time of each switch is adjusted with the output load at a constant frequency. Power efficiency is improved with low output voltage ripple and improved dynamic response at extremely light load condition.

IV. EXPERIMENTAL RESULTS

A 5 kW experimental prototype is built. The input and output voltages are $V_d = 200$ V and $V_o = 100$V, respectively. The proposed hybrid switching modulation scheme is implemented fully in software using a single-chip microcontroller, Microchip dsPIC30F6015.

Fig. 4 shows the experiment waveforms of the LLC-SRHB converter for different output load conditions. Fig. 4(a) shows the experimental waveforms of the LLC-SRHB converter at heavy load condition. The maximum switching period T_{max} is 45 μs. The converter operates in SFM when the output power

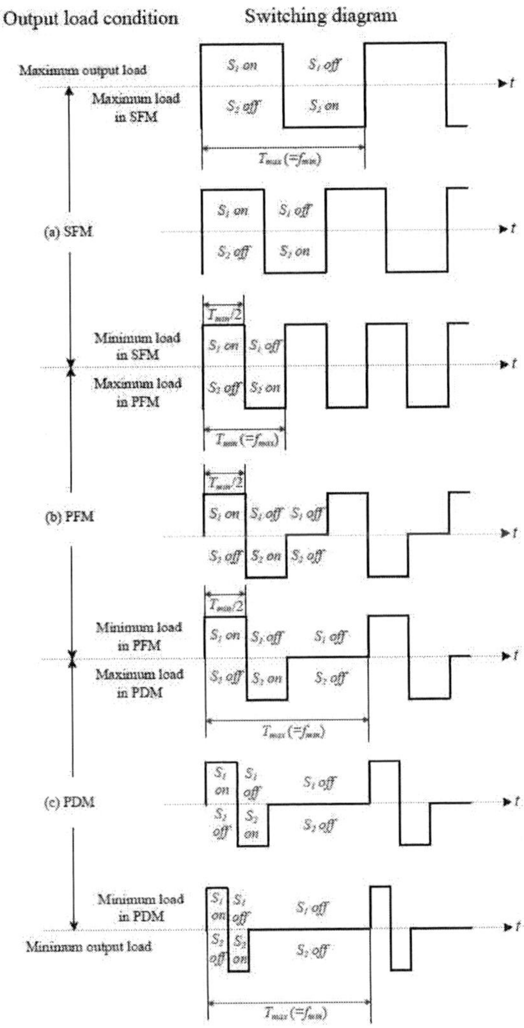

Fig. 3. Switching diagram with output load condition.

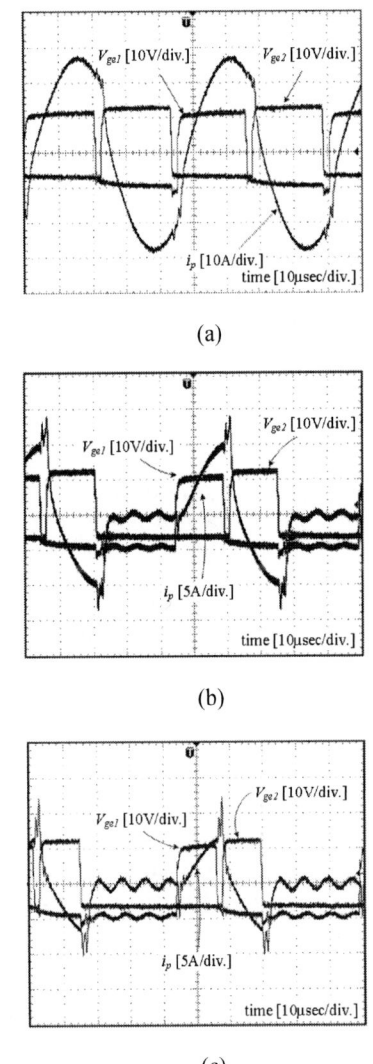

(a)

(b)

(c)

Fig. 4. Experimental results.

is 5 kW. Fig. 4(b) shows the experimental waveforms of the LLC-SRHB converter at medium load condition. The converter operates in PFM when the output power is 1 kW. Fig. 4(c) shows the experimental waveforms of the proposed LLC-SRHB converter at light load condition. The converter operates in PDM when the output power is 500 W.

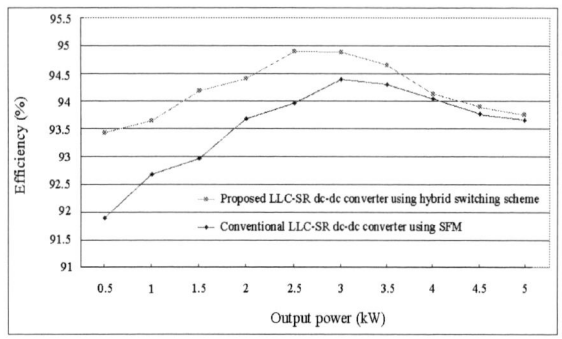

Fig. 5. Measured efficiency.

Fig. 5 shows the measured efficiencies. Compared to the conventional LLC-SRHB converter using the SFM, the proposed LLC-SRHB converter using the proposed hybrid switching scheme achieves a high-efficiency at medium-to-light load condition. The proposed scheme can be a very promising solution for improving light load efficiency for the applications which are battery-powered with energy consumption constrains

V. CONCLUSION

In this paper, a new hybrid switching scheme for the LLC SRHB dc-dc converter has been proposed for wide load ranges. The concept of the proposed hybrid switching scheme is changing the switching modes depending on the output load condition. The LLC-SRHB converter operates in three different switching modes, achieving a high-efficiency over wide load ranges. Experimental results on a 5 kW prototype are presented to evaluate the feasibility of the proposed hybrid switching scheme for the LLC-SRHB converter.

VI. REFERENCES

[1] J. P. Lee, B. D. Min, T. J. Kim, D. W. Yoo, and B. K. Lee, "A novel topology for photovoltaic series connected dc/dc converter with high efficiency under wide load range," in *IEEE 2007 Power Electronics Specialist Conference*, 2007, pp. 152-155.

[2] W. S. Yu and J. S. Lai, "Ultra high-efficiency bidirectional dc-dc converter with multi-frequency pulse-width modulation," in *IEEE 2008 Applied Power Electronics Conference and Exposition*, 2008, pp. 1079-1084.

[3] Y. T. Jang, M. M. Jovanovic, and D. L. Dillman, "Light-load efficiency optimization method," in *IEEE 2009 Applied Power Electronics Conference and Exposition*, 2009, pp. 1138-1144.

[4] D. B. Fu, Y. Liu, F. C. Lee, and M. Xu, "A novel driving scheme for synchronous rectifiers in LLC resonant converter," *IEEE Transactions on Power Electronics*, vol. 24, no. 5, pp. 1321-1329, May 2009.

[5] K. H. Yi and G. W. Moon, "Novel two-phase interleaved LLC series-resonant converter using a phase of the resonant capacitor," *IEEE Transactions on Industrial Electronics*, vol. 56, no. 5, pp. 1815-1819, May 2009.

[6] W. S. Choi and S. M. Young, "Improving system reliability using FRFET in LLC resonant converters," in *IEEE 2008 Power Electronics Specialist Conference*, 2008, pp. 2346-2351.

[7] M. Z. Youssef and P. K. Jain, "A front-end self-sustained LLC resonant converter," in *IEEE 2004 Power Electronics Specialist Conference*, 2004, pp. 2651-2656.

[8] P. K. Jain, A. St-Martin, and G. Edwards, "Asymmetrical pulse-width-modulated dc/dc converter topologies," *IEEE Transactions on Power Electronics*, vol. 11, no. 3, pp. 413-422, May 1996.

[9] D. J. Tschirhart and P. K. Jain, "A CLL resonant asymmetrical pulsewidth-modulated converter with improved efficiency," *IEEE Transactions on Industrial Electronics*, vol. 55, no. 1, pp. 114-122, Jan. 2008.

[10] M. S. Agamy and P. K. Jain, "A three-level resonant single-stage power factor correction converter: analysis, design, and implementation," *IEEE Transactions on Industrial Electronics*, vol. 56, no. 6, pp. 2095-2107, Jun. 2009.

[11] O. Abdel-Rahman, J. A. Abu-Qahouq, L. Huang, and I. Batarseh, "Analysis and design of voltage regulator with adaptive FET modulation scheme and improved efficiency," *IEEE Transactions on Power Electronics*, vol. 23, no. 2, pp. 896-906, Mar. 2008.

Industrial Servo Applications of Linear Induction Motors Based on Dynamic Maximum Force Control

Haidong Yu
Phoenix International – A John Deere Company
Fargo, ND, USA
yuhaidong@ieee.org

Babak Fahimi
University of Texas at Arlington
Arlington, TX, USA
fahimi@uta.edu

Abstract—**Linear induction motor (LIM) drives have been investigated vastly for the past several decades. Attractive features, such as simple structure and easy maintenance, have made this family of machines utilized broadly in military, transportation, and aerospace, to name a few. Especially, LIM is suitable for industrial servo applications. Up to nowadays, vector control is a dominant control strategy for LIM based on the assumption that this family of machines portrays inherent similarities with their rotary counterparts. However, due to existence of end effects and magnetic asymmetry effects, conventional vector control can not provide its expected functionality for LIM. Therefore, a novel control strategy, dynamic maximum force control, has been invented and introduced in [1]. From simulation study, it has shown that at any instant maximum force production is guaranteed. In addition, the regulation of linear speed during transient and steady state is satisfactory. Meanwhile, steady state ohmic loss is minimized. In this paper, position of linear induction motors is controlled based on dynamic maximum force control strategy. Simulation results are presented for verification of control performance.**

I. INTRODUCTION

Traditionally, vector control is a dominant control strategy for linear induction motor (LIM) drives based on the assumption that this family of machines portrays inherent similarities with their rotary counterparts. In a conventional rotor indirect field oriented control, both d-axis current and q-axis current are regulated such that the excitation frequency ω_e can be calculated analytically based on the following equation:

$$\omega_e = \frac{P}{2} \cdot \omega_r + \frac{i_{qs}^*}{\tau \cdot i_{ds}^*} \tag{1}$$

The resulted phase current amplitude and electromagnetic torque are expressed as follows:

$$I_s = \sqrt{i_{qs}^2 + i_{ds}^2} \tag{2}$$

$$T_e = k_t \cdot i_{qs} \cdot i_{ds} \tag{3}$$

However, due to existence of end effects and magnetic asymmetry effects, conventional field oriented control can not provide its expected functionality for LIM. Therefore, equations (1) and (3) have been modified to accommodate the end effects [2]. Nevertheless, the above modifications did not solve the problems of magnetic asymmetry effects. In addition, the complexity of the control has been increased significantly. It can be observed from (1) to (3) that both current amplitude and excitation frequency have to be regulated to produce desired electromagnetic torque. In addition, excitation frequency calculation is coupled with speed control due to second term on the right hand side of equation (1). In order to pursue superior transient performance and steady state efficiency, [1] has proposed dynamic maximum force control. Simulation study has shown that at any instant maximum force production is guaranteed. This provides a solid basis for the speed and position controls of LIM. Detailed structure of the control will be discussed in the next section.

II. DYNAMIC MAXIMUM FORCE CONTROL

Figure 1 illustrates a generalized torque speed curve of induction motors. The operation modes of LIM have similar behaviors, which include two regions: A) Constant thrust region. B) Constant power region. During this operation mode, field weakening should be applied. Figure 2 represents the thrust variation with excitation frequency when linear speed is 0.4 m/sec. One can notice that at any linear speed there exists a unique pair of excitation frequencies that can produce maximum thrust. These frequencies are characterized as optimum excitation frequencies. Based on this observation, [3] has utilized field reconstruction (FR) method to offline calculate the optimum excitation frequencies. Therefore, at any linear speed maximum force production or acceleration is guaranteed.

It has been observed in [4] that both thrust and normal force are proportional to square of phase current amplitude. Therefore, below base speed thrust can be expressed as:

$$F_t = c_f \cdot i_s^2 \qquad (4)$$

where c_f is force constant. In addition, the mechanical equation of the LIM system is governed by:

$$F_t - F_l = M \cdot \dot{v} + B \cdot v \qquad (5)$$

where F_l is the load force, M is the total mass of the LIM movable system, B is the viscous friction coefficient, $\dot{v}$ is the acceleration speed, and v is linear speed. Equation (4) can be further transformed as:

$$F_t = c_f \cdot \varepsilon \qquad (6)$$

As a result, one can regulate linear speed by controlling ε (square of phase current amplitude) using a single PI controller. Furthermore, position of LIM secondary can be controlled through the speed control loop.

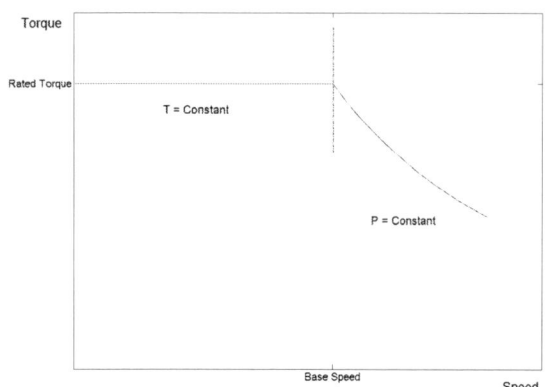

Figure 1. Generalized torque speed curve of induction motors

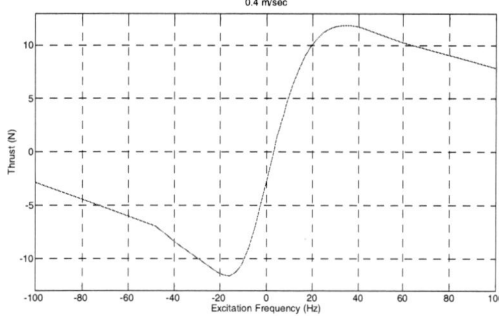

Figure 2. Thrust variation with respect to excitation frequency when linear speed is 0.4 m/sec

The complete functionality of dynamic maximum force control is illustrated in figure 3. As can be seen, both secondary position and linear speed are regulated by PI controllers. Due to speed limit and current limit (4 A of amplitude) of LIM, a saturation block is added to the output of

each PI controller. Instantaneous value of phase current amplitude can be calculated based on $\sqrt{|\varepsilon|}$. Using interpolation, relationship between linear speed and optimum excitation frequency is characterized into two lookup tables in the block diagram. It needs to be mentioned that there are several ways to obtain the optimum excitation frequency of LIM. Therefore, the calculation of the optimum excitation frequency is not in the scope of this paper. The block 'frequency selector' selects appropriate excitation frequency based on operation mode using 'Sign' function. The functionalities of 'Sign' function and 'frequency selector' are expressed as follows:

$$Sgn(\varepsilon) = \begin{cases} 1 & \text{if } \varepsilon > 0 \\ 0 & \text{if } \varepsilon = 0 \\ -1 & \text{if } \varepsilon < 0 \end{cases} \qquad (7)$$

$$f\_out_k = \begin{cases} f\_in(motoring) & \text{if } Sgn(\varepsilon) = 1 \\ f\_out_{k-1} & \text{if } Sgn(\varepsilon) = 0 \\ f\_in(generating) & \text{if } Sgn(\varepsilon) = -1 \end{cases} \qquad (8)$$

where k represents the current instant index, and $k-1$ represents the previous instant index. Finally, three phase balanced currents are regulated using hysteresis control.

III.   SIMULATION STUDIES

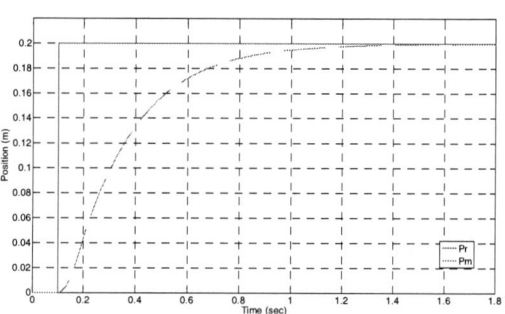

Figure 4. Position response

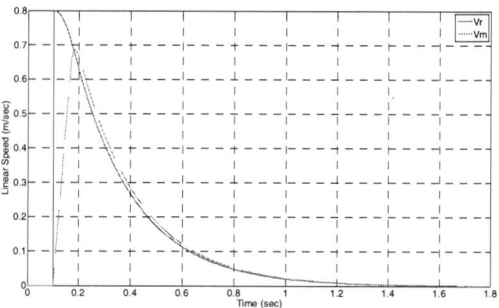

Figure 5. Linear speed response

978-1-4244-4782-4/10 $26.00 © 2010 IEEE   1499

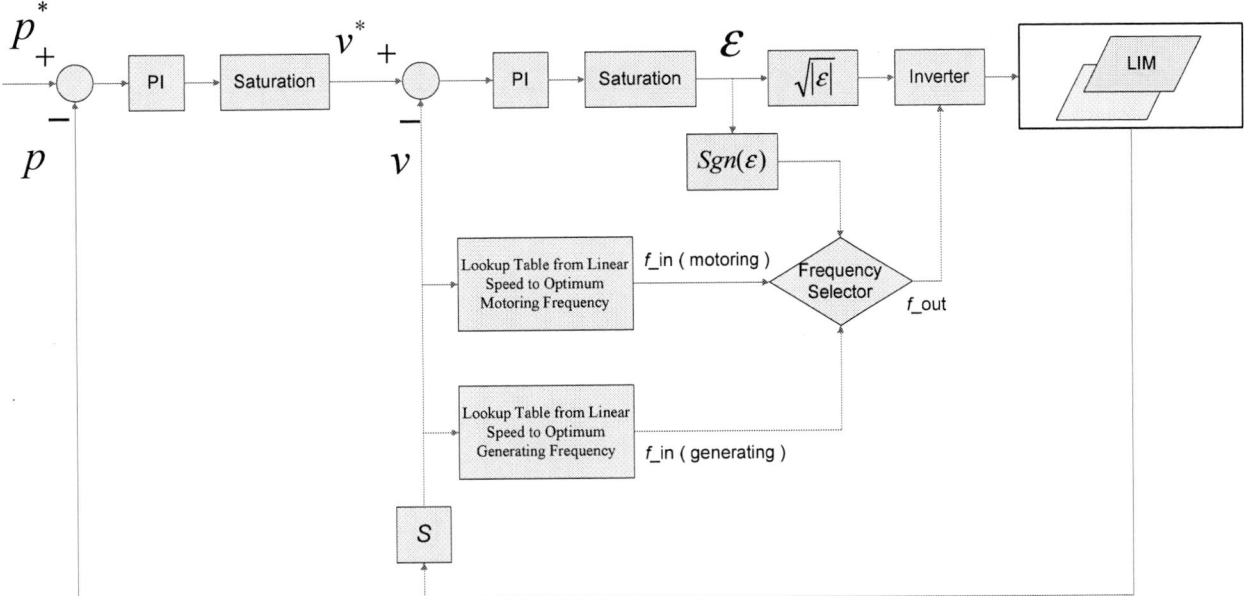

Figure 3. Complete functionality of position control of LIM based on dynamic maximum force control

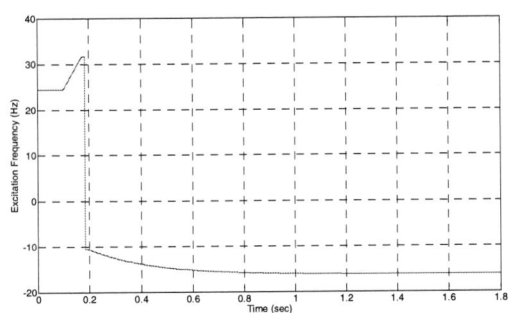

Figure 6. Excitation frequency

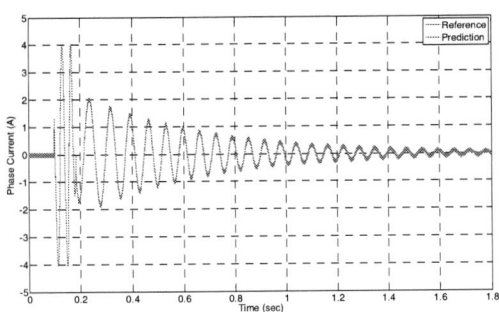

Figure 7. Phase current

In order to testify the performance of the dynamic maximum force control, a series of simulation studies have been conducted, where $M = 17kg$ and $B = 0.1N/(m/\sec)$. To balance fast transient performance and zero overshoot, the bandwidth of inner speed control loop is about 100 Hz with damping factor 1.

Figure 4 represents the position response. One can notice the steady state error of position is zero, and overshoot percentage is zero. Figure 5 depicts the linear speed response.

It is notable that the speed controller has satisfactory transient performance. Figure 6 illustrates the reference of excitation frequency. As can be seen, depending on the linear speed, excitation frequency varies correspondingly. Figure 7 shows the reference of phase current and prediction using current hysteresis control. One can observe the current control functions properly such that there is no visual difference between reference and prediction.

In a lot of industrial servo applications such as conveyer system, the mechanical system is time-varying. As a result, the dominant pole of the mechanical plant may change significantly. In order to investigate the stability of the proposed control method, at instant 3 sec, the mass of the moveable part was ramped up such that at instant of 4 sec the total mass was doubled. Figure 8-11 represent the corresponding response of the motor drive system. It can be observed that when the parameter of mechanical is dynamically changing there is no visual difference between reference and response. This verifies the robustness of the entire control structure. Other results illustrate similar satisfactory performance as in normal condition.

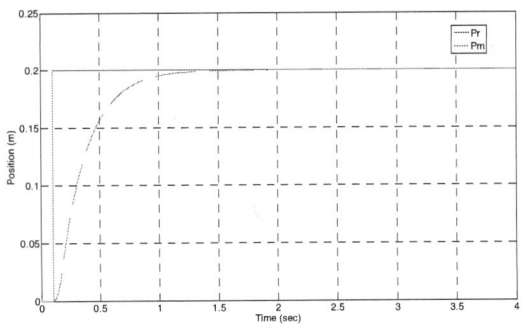

Figure 8. Position response in the time-varying mechanical system

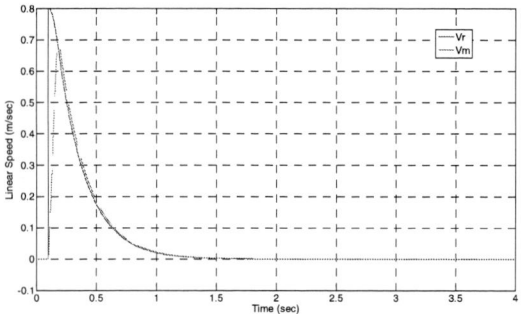

Figure 9. Linear speed response in the time-varying mechanical system

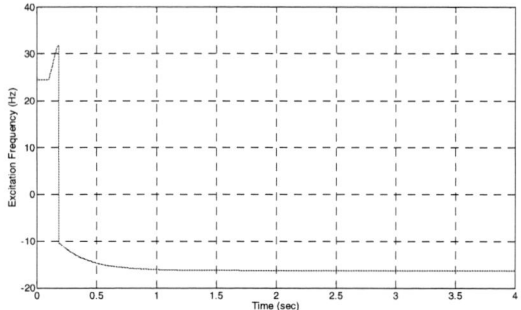

Figure 10. Excitation frequency in the time-varying mechanical system

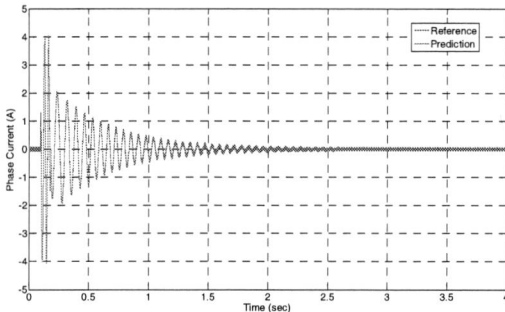

Figure 11. Phase current in the time-varying mechanical system

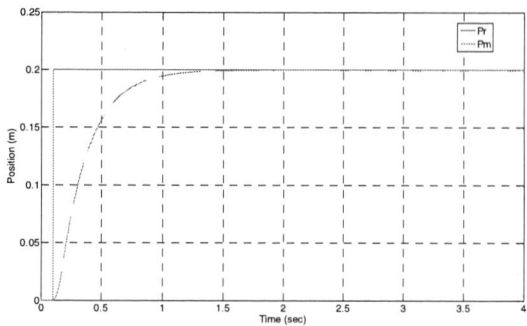

Figure 12. Position response with injection of disturbance

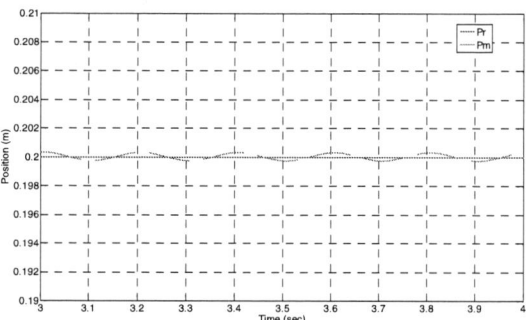

Figure 13. Zoomed the position response

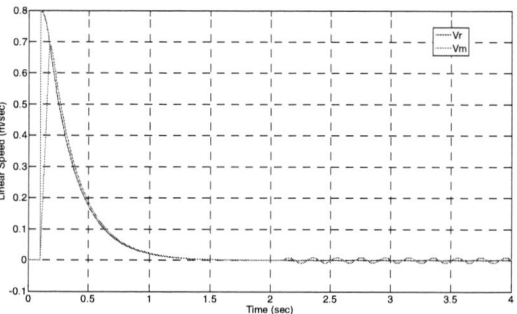

Figure 14. Linear speed response with injection of disturbance

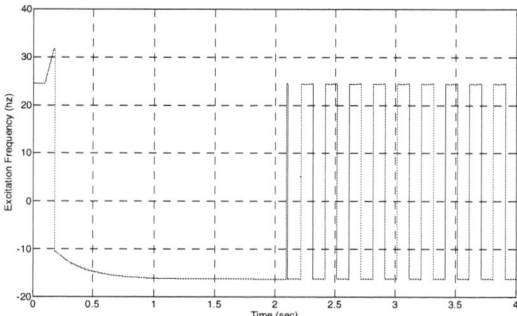

Figure 15. Excitation frequency with injection of disturbance

Finally, the disturbance rejection test has been done to verify the control stiffness of the LIM control. Starting at instant of 2.15 sec, a sinusoidal load force with amplitude of 10 N and frequency of 5 Hz was injected into the LIM system. As can be seen from figure 12, the position response reflects periodic ripples. Figure 13 zoomed in this period, which indicates the effect from disturbance is negligible. This validates the stiffness of the dynamic maximum force control. Figure 14-17 reflects the periodic phenomenon caused by the disturbance.

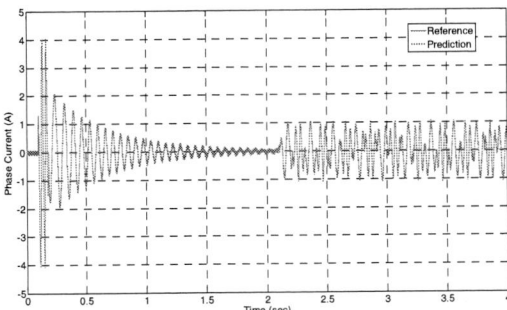

Figure 16. Phase current response with injection of disturbance

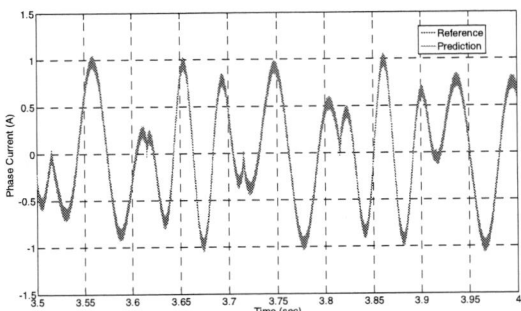

Figure 17. Zoomed phase current

IV. CONCLUSIONS

A multi-layer LIM controller based on dynamic maximum force control has been proposed for the industrial servo applications. The control structure and principle have been explored systematically. Simulation results verified the superior transient and steady state performance. In addition, control reliability and stiffness have been confirmed by satisfactory simulation results. Future work will be focused on experimental verification.

REFERENCES

[1] H. Yu, and B. Fahimi, "A Novel Control Strategy of Linear Induction Motor Drives Based on Dynamic Maximum Force Production," VPPC 2009, Dearborn, MI, 2009.

[2] J. Duncan, "Linear Induction Motor-Equivalent Circuit Model," Proc. Inst. Elec. Eng., pt. B, vol. 130, no. 1, 1983.

[3] Haidong Yu, "High Grade Control of Linear Induction Motor Drives," Ph.D. dissertation, University of Texas at Arlington, 2007.

[4] H. Yu, R. Jayabalan, M. Krishnamurthy, and B. Fahimi, "Analysis of High Speed Characteristics for Linear Induction Machines," VPPC 2006, Windsor, England, UK, 2006.

978-1-4244-4782-4/10 $26.00 © 2010 IEEE

A Soft-Switching Interleaved Three-Level Inverter

Yuan Yisheng
Electrical Engineering College
East China Jiaotong University
Nanchang, China
Cloudstone_yuan@yahoo.com.cn

Chen Min
Electrical Engineering College
Zhejiang Uninversity
Hangzhou, China
Calim@cee.zju.edu.cn

Qian Zhaoming
Electrical Engineering College
Zhejiang University
Hangzhou, China
qian@zju.edu.cn

Abstract— **In this paper, a soft-switching interleaved circuit applied in three-level inverters is proposed. An interleaved three-level inverter is realized by employing two interleaved power switches in parallel at the most upper and the lowest location in traditional three-level inverters. By further inserting a small current-limiting inductor in each interleaved switch branch, each interleaved power switch can realized zero-current switching on and a reduced reverse recovery current of the freewheeling diode. The Each operation stage and the soft-switching theory are elaborated. A 2kVA experimental prototype is produced and the test waveforms verified the circuit available.**

I. INTRODUCTION

Three-level inverters[1] have the advantages of low power switch losses, reduced output ripple current and low THD of the output voltage. Interleaved techniques[2] can reduce the losses and sizes of filter inductors and have been applied in voltage regulated modules(VRM)[3] and some other power converters[4]. But interleaved techniques cannot reduce the switching losses of the power switch if the switching frequency is fixed, so its improvement on circuit efficiency is limited. Recent years, some soft-switching interleaved converters have been investigated. Interleaved zero current transition techniques applied in buck converter[5] and three-level buck converter[6] can realized zero current turn-on of the main switches by inserting small commutation inductors. Interleaved boost converters[7][8] with a coupled inductor can achieve zero-voltage turn-on and zero-current turn-off of the main switches by auxiliary switch branches. Isolated interleaved soft-switching converters, such as interleaved active-clamped forward converters[9] have also been investigated. However, little research has been done on interleaved three-level inverters, particularly on soft-switching interleaved three-level inverters.

This paper applies the interleaved PWM technique to a proposed three-level inverter with six power switches and analyses its PWM mode. By inserting a small current-limiting inductor in each interleaved switch branch, all interleaved power switch can get zero-current turn-on, and reduce the reverse recovery current of the freewheeling diode. The operation theory is elaborated. Based on the same theory, another soft-switching NPC inverter is deduced. The

experimental results of a 2kVA prototype are given to demonstrate the viability and features of the proposed inverter.

II. PROPOSED SOFT-SWITHING INTERLEAVED THREE-LEVEL INVERTER

Fig.1 shows the power circuit of the proposed soft-switching interleaved three-level inverter. The power switches Q_1 and Q_2 construct a pair of interleaved switches. The power switched Q_5 and Q_6 make another pair of interleaved switches. Q_3 and Q_4 operate in double switching frequency. In addition, series current-limiting inductor $L_1 \sim L_4$ are assumed equal to L in inductance, which is far less than that of the filter inductor L_f. They are auxiliary components to realize zero-current turn-on of switches Q_1, Q_2, Q_5 and Q_6.

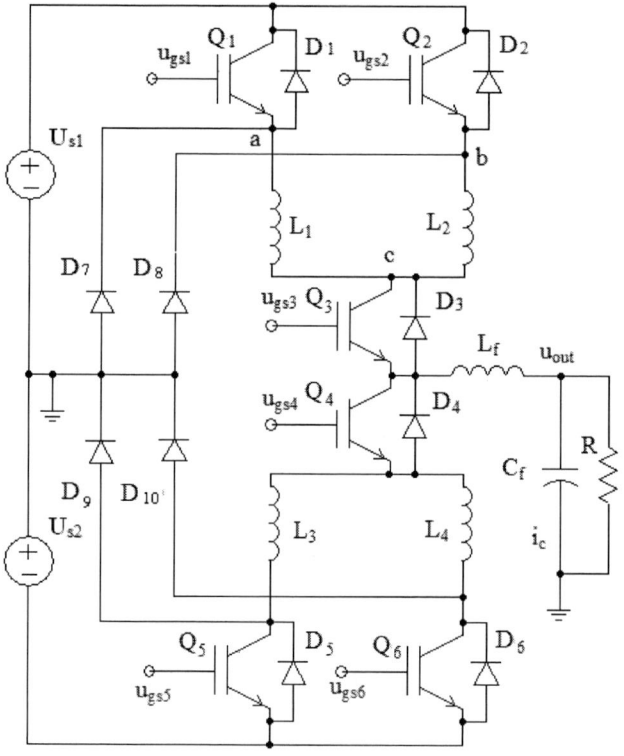

Figure 1. Inverter power circuit

978-1-4244-4782-4/10 $26.00 © 2010 IEEE

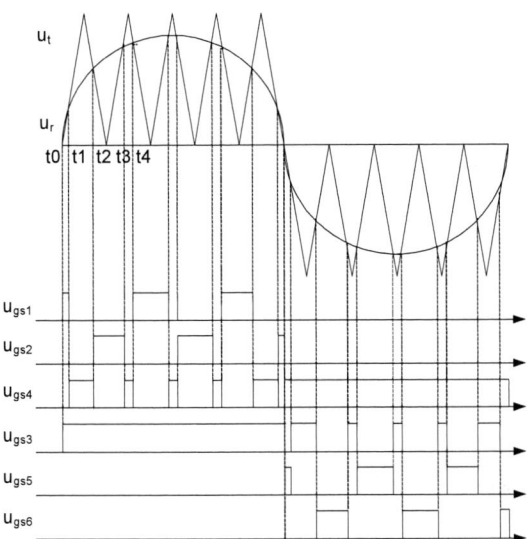

Figure 2. SPWM and gate signals

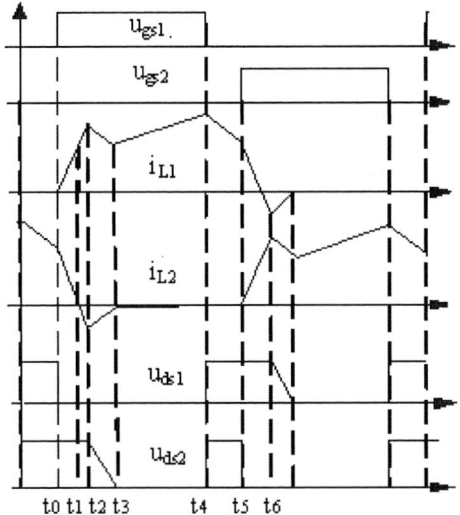

Figure 3. Operation waveforms

$$i_{L1} = \frac{U_{S1}}{2L} \cdot t \qquad (1)$$

$$i_{L2} = i_{L2(t0)} - \frac{U_{S1}}{2L} \cdot t \qquad (2)$$

So power switch Q_1 achieves zero-current turn-on in this stage.

This stage finishes while i_{L2} falls down to zero.

Stage 2: Interval [t1<t<t2]. During this stage, i_{L2} becomes negative and then reaches the peak current because of the reverse current of diode D_8. But because of the current-limiting inductor L_2, the reverse peak current of diode D_8 is reduced.

Stage 3: Interval[t2<t<t3]. During this stage, the reverse current of diode D_8 recovers from the peak value to zero. The reverse voltage of diode D_8 begins rising, so u_{ds2} begins falling. When this stage finishes, diode D_8 turns off truly and u_{ds2} falls down to nearly zero because inductor L_1 is far less than L_f.

Stage 4: Interval[t3<t<t4]. After the commutation phase finished, the switch Q_1 begins conducts normally and the current i_{L1} begin rising linearly as the following equation:

$$i_{L1} = \frac{U_{S1} - u_{out}}{L_f + L} \cdot t + i_{L1(t3)} \qquad (3)$$

Stage 5: Interval[t4<t<t5]. In this stage, switch Q_1 is turned off and diode D_7 is forced to conducts. The current i_{L1} flows through D_7, L_1, Q_3 and L_f into load. The current i_{L1} decreases linearly in this stage as the following equation:

$$i_{L1} = i_{L1(t4)} - \frac{u_{out}}{L_f + L} \cdot t \qquad (4)$$

A. PWM Mode

Fig.2 shows the relationship of the six gate signals of the six power switches, where waveform u_r represents modulation signal of the output voltage, and u_t represents the carrier signal.

In the positive half cycle region of modulation signal u_r: when u_r value is more than u_t value, the gate signal u_{gs1} of switch Q_1 and u_{gs2} of switch Q_2 turn on alternately and their duty cycle are less than 50%; when u_r value is less than u_t value, the gate signal u_{gs4} of switch Q_4 turn on; In this half region, the gate signal u_{gs3} of switch Q_3 always turns on, but the gate signal u_{gs5} of switch Q_5 and u_{gs6} of switch Q_6 always turn off.

In the negative half cycle region of modulation signal u_r: when u_r value is more than u_t value, the gate signal u_{gs5} of switch Q_5 and u_{gs6} of switch Q_6 turn on alternately and their duty cycle are less than 50%; when u_r value is less than u_t value, the gate signal u_{gs3} of switch Q_3 turn on; In this half region, the gate signal u_{gs4} of switch Q_4 always turns on, but the gate signal u_{gs1} of switch Q_1 and u_{gs2} of switch Q_2 always turn off.

B. Operation Theory

Assuming output voltage u_o is in a positive half cycle and current i_{Lf} of the filter inductor is positive, the operation waveforms are showed in Fig.3. To achieve zero-current turn-on of switches Q_1 and Q_2, there are six switching stages as showed in Fig.4.

Stage 1: In terval[t0<t<t1]. Before this stage, i_{Lf} flows through D_8, L_2, Q_3 and L_f into C_f and R. This stage begins when Q_1 is turned on conducts. Voltage u_{ds1} falls down to nearly zero. But because of two current-limiting inductors L_1 and L_2, current i_{L2} and current i_{L1} change as the following equations:

978-1-4244-4782-4/10 $26.00 © 2010 IEEE   1504

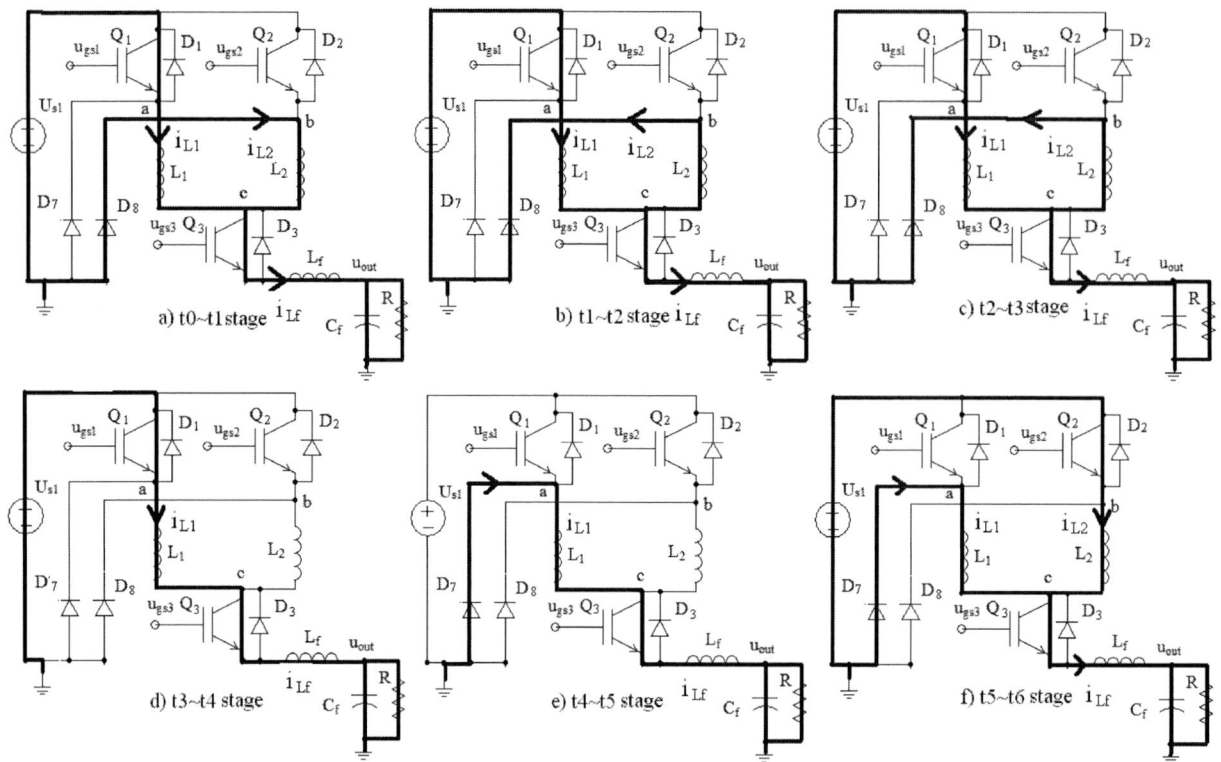

Figure 4. Operating stages

Stage 6: Interval[t5<t<t6]. This stage begins when Q_2 is turned on. The voltage u_{ds1} falls down quickly. But because of two current-limiting inductors L_1 and L_2, the current i_{L2} and current i_{L1} change as the following equations:

$$i_{L2} = \frac{U_{S1} \cdot t}{2L} \qquad (5)$$

$$i_{L1} = i_{L1(t5)} - \frac{U_{S1} \cdot t}{2L} \qquad (6)$$

By Equation 5, the switch Q_2 achieves zero-current turn-on.

When the output voltage is in the negative half cycle, the switches Q_5 and Q_6 can also realize zero-current turn-on by the same theory.

III. ANOTHER INTERLEAVED SOFT-SWITCHING NEUTRAL-POINT-CLAMPED INVERTER

Bases on the same theory, another soft-switching interleaved Neutral-Point-Clamped (NPC) inverter can be deduces as Fig.5. The switches Q_1 and Q_2, Q_3 and Q_4 make two pairs of PWM interleaved power switches. Two small current-limiting inductors L_1 and L_2 are far less than inductor L_f. In this circuit, the switches Q_1, Q_2, Q_3 and Q_4 can achieve zero-current turn-on.

The gate signals of the power switches in Fig.5 and Fig.1 have a match relationship as the following Table I .

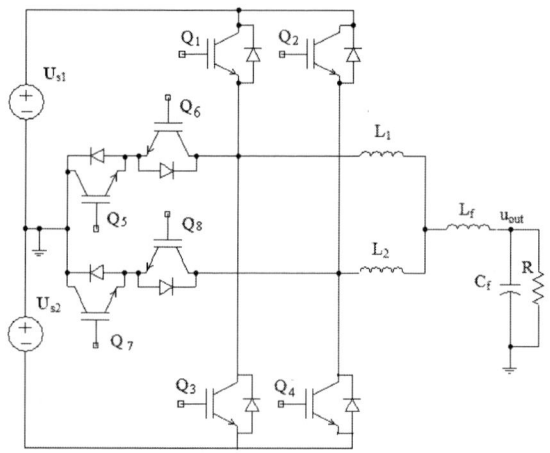

Figure 5. Another interleaved soft-switching NPC inverter

TABLE I. MATCH RELATIONSHIP OF THE GATE SIGNALS OF THE TWO PROPOSED INVERTER

Figure	Power Switches					
Fig.1	Q_1	Q_2	Q_3	Q_4	Q_5	Q_6
Fig.5	Q_1	Q_2	Q_5,Q_7	Q_6,Q_8	Q_3	Q_4

IV. EXPERIMENT AND RESULTS

An experimental prototype with 50Hz output voltage is designed and tested. Its parameters are listed in the Table II.

The type of the power switch is IRFG4PC40UD.

The choice of the current-limiting inductors L1~L4 is a key for this soft-switching inverter. If the inductance of the current-limiting inductors is too little, the commutation time will too short and loss the current-limiting effect. If the inductance of the current-limiting inductors is too big, although zero-current turn-on of the switches can be realized, but the commutation time will take too long time and lead to a larger duty cycle loss. So the inductance of the current-limiting inductor must be a proper value.

By measurement, the falling time of the voltage u_{ds} of IRG4PC40UD equals to 50nS. To make the current i_{ds} rising time reach 250nS (equal to five times time of the voltage falling time) under peak load current, the inductance of the current-limiting inductor is designed to equal to 5uH.

Fig.6 shows the waveforms of the gate signals of the main switches Q_1, Q_2 and Q_4, and inductor current i_{Lf}. The gate signals u_{gs1} and u_{gs2} are interleaved PWM signals, and the ripple frequency of inductor current i_{Lf} is a double switching frequency.

Fig.7, Fig.8 and Fig.9 showed the test waveforms of the prototype when its output voltage u_{out} and the current i_{Lf} are in the positive half cycle. Fig.7 shows the switching voltage and current waveforms of switches Q_1 and Q_2. Fig.8 shows the magnified turn-on waveforms of switch Q_1 based on Fig.7. Fig.9 shows the magnified turn-on waveforms of switch Q_2 based on Fig.7. It is obviously that the voltage u_{ds1} and u_{ds2} of the switch Q_1 and Q_2 fall to zero respectively before their currents i_{ds1} and i_{ds2} rise up. Both of interleaved power switches Q_1 and Q_2 achieve the zero-current turn-on.

TABLE II.   DESIGN PARAMETERS OF THE PROTOTYPE

U_{s1}, U_{s1} (V)	U_{out} (V)	f_s (kHz)	L_1~L_4 (uH)	L_f (uH)	C_f (uF)	P_o (W)
360	220	20	5	1500	6.8	1400

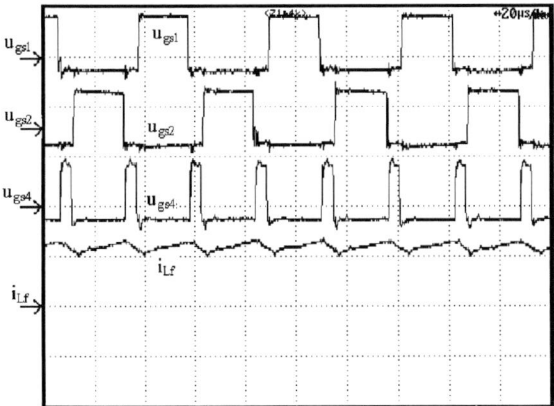

Figure 6.   Interleaved u_{gs} and i_{Lf} waveforms

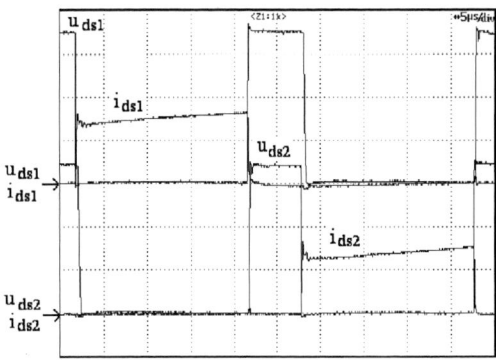

Figure 7.   Switching waveforms of Q_1 and Q_2

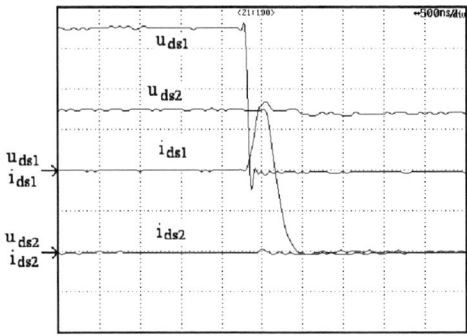

Figure 8.   Turn-on waveforms of Q_1

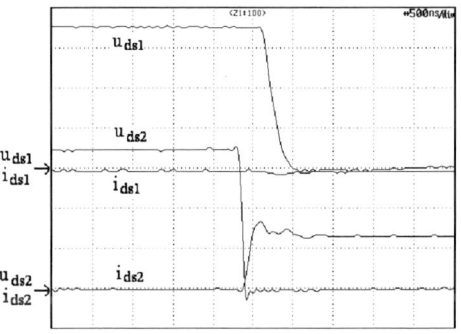

Figure 9.   Turn-on waveforms of Q_2

V. CONCLUSION

A soft-switching interleaved three-level inverter has been presented and developed, which can be used for high power application. It can achieve zero-current turn-on for the four interleaved power switches, and the reverse recovery current of the diode is reduced. Experimental results verified the circuit.

ACKNOWLEDGMENT

This work is supported by National Natural Science Foundation of China (50577025) and Science and Technology Foundation of Education Department of Jiangxi Province of China (GJJ08240). The researchers also gratefully acknowledge the instruction of Professor Liuchen Chang in the University of New Brunswick in Canada.

REFERENCES

[1] Nabae, I.Takahashi and H. Akgai, "A new Neutral-Point-Clamped PWM Inverter," IEEE trans. On Industrial Application, vol.17,no.5, pp.518-523,Sep. 1981.

[2] M.T. Zhang, M.M. Jovanovic and F.C.Y. Lee, "Analysis and Evaluation of Interleaving Techniques in Forward Converter," IEEE trans. On Power Electricals, vol. 13, no.4, pp.690-698, Jul.1998.

[3] Yungtaek Jang, Milan M. Jovanovic and Yuri Panov, "Multi-Phase Buck Converters with Extended Duty Cycle," IEEE APEC 2006, pp.38-44.

[4] Aluisio A. Bento, Edison R.da Silva and Euzeli C. Dos Santos Jr, "Reducing the Inductor Size and Current Stress by Interleaved Bidirectional Boost Rectifiers Used for Power Factor Correction," IEEE APEC 2006, pp. 273-279.

[5] Milan Ilic and Dragan Maksimovic, "Interleaved Zero Current Transition Buck Converter," IEEE APEC2005, pp.1265-1271.

[6] Milan Ilic, Bryce Hersterman and Dragan Maksimovic, "Interleaved Zero Current Transition Three-Level Buck Converter," IEEE APEC 2006, pp.72-78.

[7] Gang Yao, Haiyang He, Jiangjiang Shi, Yan Den and Xiangning He, "A ZCS PWM Switch Circuit for the Interleaved Boost Converters," IEEE PESC 2006, pp.2478-2481.

[8] Gang Yao, Yanqun Shen, Wuhua Li and Xiangning He, "A new Soft Switching Snubber for the Interleaved Boost Converters," IEEE PESC2004, pp.3765-3769.

[9] Yu-kang Lo, Tsu-Shou Kao and Jing-Yuan Lin, "Analysis and Design of an Interleaved Active-Clamping Forward Converter," IEEE trans. On Industrial Electronics, vol. 54, no.4, pp.2323-2332, Aug. 2007.

Reducing Common-Mode Voltage in Three-Phase Sine-Triangle PWM with Interleaved Carriers

Jonathan W. Kimball, *Senior Member*
Department of Electrical and Computer Engineering
Missouri University of Science and Technology
Rolla, MO, USA
kimballjw@mst.edu

Maciej Zawodniok, *Member*
Department of Electrical and Computer Engineering
Missouri University of Science and Technology
Rolla, MO, USA
mjzx9c@mst.edu

Abstract— Interleaving PWM waveforms is a proven method to reduce ripple in dc-dc converters. The present work explores interleaving for three-phase motor drives. Fourier analysis shows that interleaving the carriers in conventional uniform PWM significantly reduces the common-mode voltage. New DSP hardware supports interleaving directly with changes to just two registers at setup time, so no additional computation time is needed during operation. The common-mode voltage reduction ranges from 36% at full modulation to 67% when idling with zero modulation. Third harmonic injection slightly reduces the advantage (to 26% at full modulation). However, the maximum RMS common-mode voltage is still less than 20% of the bus voltage under all conditions. Low-voltage experimental results support the findings.

I. INTRODUCTION

Modern digital signal processors (DSPs), such as a Texas Instruments TMS320F28335, provide new flexibility in pulsewidth modulation (PWM) signal generation. Specifically, different phases of the PWM generator may have its own time base, delayed relative to the others or even at a different frequency. This feature was added to support power topologies such as one of the phase-shift zero-voltage-switching topologies discussed in [1], multi-phase dc-dc converters, or even multiple independent dc-dc converters with different switching frequencies. If motor control is implemented in an FPGA or other programmable device, similar capabilities are possible. The present work will show that interleaving the carriers also provides an advantage in a conventional three-phase motor drive.

Many motor drives use uniformly sampled PWM. All three phases use the same carrier, a triangle wave with frequency f_c. The modulating functions for the three phases, $m_1(t)$, $m_2(t)$, and $m_3(t)$, are sampled at either f_c (symmetric PWM) or $2f_c$ (asymmetric PWM) to determine the duty ratios of each phase. Alternative formulations are based on space vector modulation (SVM) [2]. In SVM, the desired voltage vector is sampled at either f_c or $2f_c$, and then available voltage vectors are chosen to construct the desired voltage vector. There are extra degrees of freedom in the basic formulation of SVM that are used to achieve a variety of goals, including reduced common-mode voltage, in alternative SVM

formulations. According to [3], many of these alternatives are impractical, and all of them include some extra computational effort.

Several schemes have been proposed where the PWM frequency is randomized [4, 5] in order to minimize sidebands and harmonics. Such a technique was shown to improve frequency spectrum characteristics of a motor drive at a cost of higher processing overhead. First, the PWM module has to be reconfigured on a cycle-by-cycle basis. Second, the drive control scheme needs to be modified to accommodate the randomness in PWM frequency [4, 5]. Moreover, the theoretical analysis lacks a hard guarantee of signal quality (that is, presence of harmonics in frequency spectrum). In contrast, the proposed scheme has low overhead since the phase shift is set only once using the new ePWM hardware, a mathematical frequency spectrum analysis is presented to guarantee high signal quality, and the motor control algorithm does not need to be modified.

The present work focuses on uniformly sampled sine-triangle PWM, rather than SVM, with interleaved carriers to achieve reduced common-mode voltage. If the carrier waves are interleaved instead of synchronized, the common-mode voltage may be reduced by as much as 67%. This improvement is achieved with a single three-phase inverter, whereas other interleaving techniques [6, 7] require multiple three-phase inverters in parallel. Since the digital hardware is already available, this reduction comes with no extra computational penalty, whereas the SVM variants and random PWM both require significant calculations. The following sections show Fourier analysis and simulations. Two cases are considered, namely, asymmetric uniform PWM with sinusoidal modulation and symmetric uniform PWM with third harmonic injection. Experimental results for a low-voltage inverter agree well with the conclusions.

II. FOURIER ANALYSIS OF ASYMMETRIC UNIFORM PWM

The common approach to PWM harmonic analysis is two-dimensional Fourier analysis [8-13]. In conventional Fourier analysis, a periodic function $F(t)$ is decomposed into frequency components by integrating the product of $F(t)$ and complex exponentials. The difficulty with PWM is

---

This work was supported by ITW Military GSE and the National Science Foundation I/UCRC on Intelligent Maintenance Systems.

formulating $F(t)$. In two-dimensional Fourier analysis, the two dimensions map time t into the period of the carrier $(x = 2\pi f_c t = \omega_c t)$ and the period of the modulating function $(y = 2\pi f_o t = \omega_o t)$. The conventional integral is replaced by a double integral over the two dimensions to find coefficients for a Fourier representation of the general form

$$F(t) = \frac{A_{00}}{2} + \sum_{m=0}^{\infty} \sum_{n=-\infty}^{\infty} \binom{A_{mn} \cos(m\omega_c t + n\omega_o t)}{+B_{mn} \sin(m\omega_c t + n\omega_o t)} \quad (1)$$

where A_{mn}, B_{mn} are amplitudes of harmonics m (of the switching frequency) and n (of the modulating frequency). The sine component can be canceled ($B_{mn}=0$ for all values of m, n) by choosing proper angle references. Also, some special conditions may apply when either m or n equal zero. The advantage of the two-dimensional approach is that $F(t)$ is reduced to a pair of limits of integration, rather than a complicated time-varying square wave.

The present objective is to determine the common-mode voltage $V_0(t)$ when the motor drive three-phase output voltages are $V_1(t)$, $V_2(t)$, and $V_3(t)$. From Kirchhoff's voltage law,

$$V_0 = \frac{1}{3}(V_1 + V_2 + V_3) \quad (2)$$

Following the notation of [10], each phase voltage is either 0 or V_{dc} at any given time, where V_{dc} is the magnitude of the source. The common-mode voltage V_0 may take four values: 0, $V_{dc}/3$, $2V_{dc}/3$, or V_{dc}. Previous methods to reduce common-mode voltage [3] eliminated the options of 0 and V_{dc}. With interleaved carriers, all four voltages are possible, but a simulation, such as Fig. 1, shows that V_0 is usually either $V_{dc}/3$ or $2V_{dc}/3$. From the simulated result, one would expect that the rms value of V_0 would be much smaller than in a conventional PWM process. The derivation to follow will verify this expectation and quantify the advantage for a given operating point.

The results of [10] for asymmetric uniform PWM with single sine wave modulation can be adapted easily with a coordinate transformation, where x and y are offset by some angle (e.g., $x = \omega_c t + \phi$). Table 1 shows the modulating function (a single-frequency sinusoid) and carrier wave offsets for all three phases. The Fourier representations of V_1-V_3 are

$$V_1(t) = \frac{V_{dc}}{2} + \sum_{m=1}^{\infty} \sum_{n=-\infty}^{\infty} \left(A_{mn} \cos(m\omega_c t + n\omega_o t)\right)$$
$$+ \sum_{n=1}^{\infty} \left(A_{0n} \cos(n\omega_o t)\right) \quad (3)$$

$$V_2(t) = \frac{V_{dc}}{2} + \sum_{m=0}^{\infty} \sum_{n=-\infty}^{\infty} \left(A_{mn} \cos\begin{pmatrix} m\omega_c t + n\omega_o t \\ + \frac{2\pi}{3}(m+n) \end{pmatrix}\right)$$
$$+ \sum_{n=1}^{\infty} \left(A_{0n} \cos\left(n\omega_o t + \frac{2n\pi}{3}\right)\right) \quad (4)$$

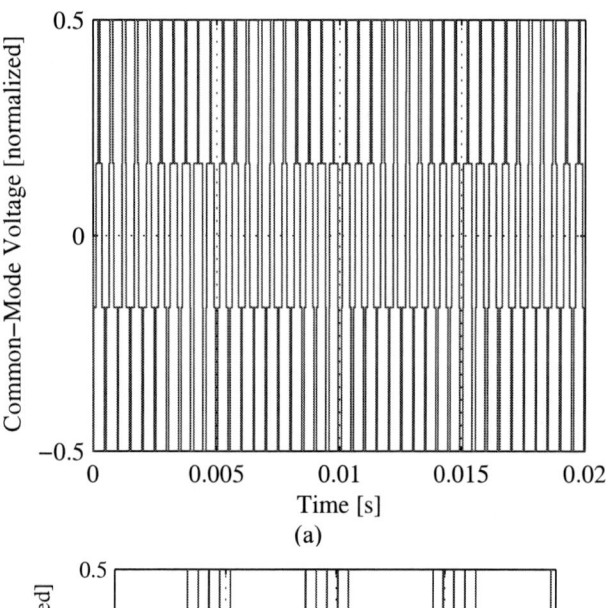

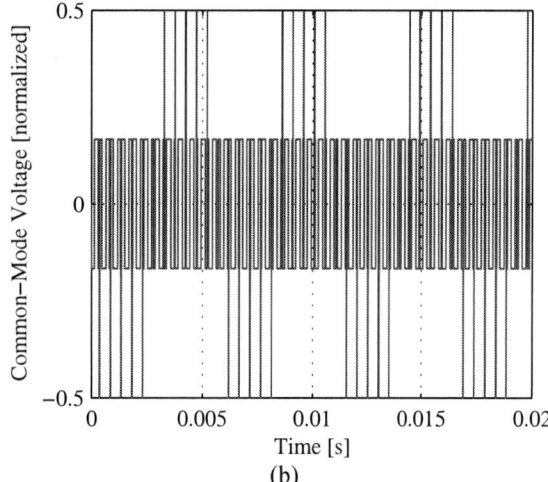

Fig. 1. Simulated common-mode voltage (normalized) for a three-phase inverter, 60 Hz output, 2 kHz switching, modulation depth 0.8: (a) conventional carrier, (b) interleaved carriers.

TABLE 1. MODULATING FUNCTIONS AND CARRIERS FOR THE THREE PHASES.

Phase	Carrier Offset (radians on ω_c scale)	Modulating Function
1	0	$M \cos(\omega_o t)$
2	$+2\pi/3$	$M \cos(\omega_o t + 2\pi/3)$
3	$-2\pi/3$	$M \cos(\omega_o t - 2\pi/3)$

$$V_3(t) = \frac{V_{dc}}{2} + \sum_{m=0}^{\infty} \sum_{n=-\infty}^{\infty} \left(A_{mn} \cos\left(\begin{matrix} m\omega_c t + n\omega_o t \\ -\frac{2\pi}{3}(m+n) \end{matrix} \right) \right)$$
$$+ \sum_{n=1}^{\infty} \left(A_{0n} \cos\left(n\omega_o t - \frac{2n\pi}{3} \right) \right) \tag{5}$$

$$A_{mn} = \frac{2V_{dc}}{q\pi} J_n\left(\frac{\pi}{2} qM \right) \sin\left((m+n)\frac{\pi}{2} \right)$$
$$q = m + n\frac{\omega_0}{\omega_c} \tag{6}$$

All three phases have the same magnitude, but vary in phase. If the carriers are not interleaved, then Eqs. (4)-(5) have phase offsets of only $\frac{2n\pi}{3}$ inside the double summation. The common-mode voltage may be found by substituting (3)-(6) into (2) and simplifying.

$$V_0(t) = \frac{V_{dc}}{2} + \frac{1}{3} \left(\begin{matrix} \sum_{m=1}^{\infty} \sum_{n=-\infty}^{\infty} \left(\begin{matrix} A_{mn}\left(1+2\cos\left(\frac{2\pi}{3}(m+n)\right)\right) \times \\ \cos(m\omega_c t + n\omega_o t) \end{matrix} \right) \\ + \sum_{n=1}^{\infty} \left(\begin{matrix} A_{0n}\left(1+2\cos\left(\frac{2n\pi}{3}\right)\right) \times \\ \cos\left(n\omega_o t + \frac{2n\pi}{3}\right) \end{matrix} \right) \end{matrix} \right) \tag{7}$$

The dc term is an artifact of the way voltages are defined. The second summation is for the case where $m = 0$. From the definition of A_{mn} in (6), $(m+n)$ must be odd for the result to be non-zero. Because of the phase delays, $(m+n)$ must also be a multiple of 3 for the terms in the double summation to be non-zero. By way of contrast, in conventional PWM, n must be a multiple of 3 for the result to be non-zero, regardless of the value of m.

The rms value of the common-mode voltage may be found from the square root of the sum of the squares of the coefficients to the cosine terms. The dc term is discarded, as it is simply an artifact of the voltage definitions. In the following equations, V_{0rms} is the rms value of the common-mode voltage with interleaving and $V_{0nonrms}$ is the rms value of the common-mode voltage without interleaving.

$$V_{0rms} = \sqrt{ \left(\begin{matrix} \sum_{m=1}^{\infty} \sum_{n=-\infty}^{\infty} \left(\frac{A_{mn}}{3}\left(1+2\cos\left(\frac{2\pi}{3}(m+n)\right)\right) \right)^2 \\ + \sum_{n=1}^{\infty} \left(\frac{A_{0n}}{3}\left(1+2\cos\left(\frac{2n\pi}{3}\right)\right) \right)^2 \end{matrix} \right) } \tag{8}$$

$$V_{0nonrms} = \sqrt{ \left(\begin{matrix} \sum_{m=1}^{\infty} \sum_{n=-\infty}^{\infty} \left(\frac{A_{mn}}{3}\left(1+2\cos\left(\frac{2n\pi}{3}\right)\right) \right)^2 \\ + \sum_{n=1}^{\infty} \left(\frac{A_{0n}}{3}\left(1+2\cos\left(\frac{2n\pi}{3}\right)\right) \right)^2 \end{matrix} \right) } \tag{9}$$

Fig. 2 shows the rms common-mode voltage for both conventional PWM and interleaved PWM as modulation depth M varies, with the following parameters: f_o = 60 Hz, f_c = 2 kHz, V_{dc} = 1 (to normalize). With interleaving, the common-mode voltage is relatively constant regardless of modulation depth. When $M = 0$, the common-mode voltages are 0.2297 when interleaved and 0.7016 with conventional modulation, a reduction of 67.26%. When $M = 1$, the common-mode voltages are 0.2299 when interleaved and 0.3596 with conventional modulation, a reduction of 36.07%.

III. FOURIER ANALYSIS OF SYMMETRIC UNIFORM PWM WITH THIRD HARMONIC INJECTION

The analysis of symmetric uniform PWM with third harmonic injection is significantly more complex than the above analysis of asymmetric uniform PWM with a single sine wave modulating function. Two problems quickly emerge. The modulating function is more complex, leading to four times as many terms in the summation. Also, more harmonics are non-zero due to the sampling algorithm. Closed-form results, while possible, are more difficult to achieve. Fortunately, numerical integration is possible and provides useful results.

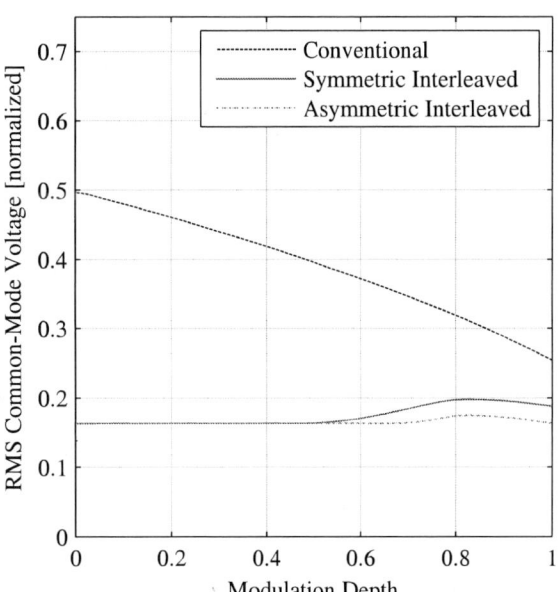

Fig. 2. RMS common-mode voltage as modulation depth varies for both conventional (non-interleaved) PWM and interleaved PWM. Conventional and symmetric interleaved curves include third harmonic injection, whereas the asymmetric interleaved curve does not.

The conventional double Fourier integral that is used to analyze PWM systems is evaluated first over the switching period, and then over the modulating period. The inner integral, evaluated over the switching period, can be performed easily. Unfortunately, terms of the generic form $e^{j\lambda \cos y}$ emerge and must be integrated over y. Closed-form solutions may be obtained from a Bessel function expansion of these terms. In the present work, numerical integration is used instead.

For symmetric uniform PWM with modulating function of the form

$$m(\theta) = \frac{1}{2}\left(1 + M \cos\theta + A \cos 3\theta\right) \qquad (10)$$

the coefficients to the Fourier summations (3)-(5) may be found by numerically evaluating

$$A_{mn} = \mathrm{Re}\left\{ \frac{V_{dc}}{2jq\pi^2} \int_{-\pi}^{\pi} e^{jny}\left(e^{jqx_f} - e^{jqx_r}\right) dy \right\}$$

$$q = m + n\frac{\omega_o}{\omega_c}$$

$$x_f = \frac{\pi}{2}\left(1 + M \cos y + A \cos 3y\right) \qquad (11)$$

$$x_r = -\frac{\pi}{2}\left(1 + M \cos y + A \cos 3y\right)$$

The result inside the braces should be real. However, due to numerical inaccuracies, there may be an imaginary part that must be discarded. Most applications of this sort use $\frac{1}{6}$ third harmonic injection, that is, $A = -\frac{M}{6}$, to achieve the maximum possible undistorted output.

Once the integration in (11) has been performed numerically, the common-mode voltages given in (8)-(9) for interleaved/non-interleaved PWM may be computed. Fig. 2 also includes a curve for symmetric modulation with interleaving. For the non-interleaved case, the common-mode voltage is affected by neither the sampling method nor the third-harmonic injection. For the interleaved case, there is a slight increase in common-mode voltage at high modulation depths (greater than about 0.5). Still, interleaving the PWM carriers substantially reduces common-mode voltage, by 67.26% at zero modulation depth (as before) and by 26.19% at full modulation.

By comparison, the methods discussed in [3] all have a common-mode voltage magnitude of $\frac{1}{6}$ (normalized), or 0.1667. The common-mode voltage magnitude of interleaved PWM with uniform sampling and third-harmonic injection ranges from 0.2297 to 0.2794. The methods of [3] are computationally intense and, in some cases, impractical. Both approaches achieve the same output voltage range. Given the ease of implementing interleaved uniform PWM, and its substantial improvement over conventional uniform PWM, few applications would require the additional complexity of the methods in [3] to further reduce common-mode voltage.

IV. SIMULATION RESULTS

The above method is easily implementable on a Texas Instruments TMS320F28335. Digital signal controllers (DSCs, also called digital signal processors or DSPs) in the 2833x family include ePWM peripherals for enhanced pulsewidth modulation. Each channel has its own time base from which two gate waveforms are generated, for the upper and lower IGBT of a leg. Three channels are used for a conventional three-phase inverter, and the channels may be synchronized. One register per channel determines the point in the carrier waveform to which the time base is reset when the synchronization pulse is received. That is, one register per channel determines whether the channels are synchronized or interleaved. This register is separate from the duty ratio and switching period registers, but must be coordinated with the switching period register in order to achieve the proper level of interleaving. Typically, one channel is set as the master and the other two synchronize/interleave from that base.

A simulation was constructed to model the interleaving method. The inverter circuit was modeled with PLECS® version 2.2.1. A deadtime of 2.5 µs and conduction characteristics of an FS100R12KT4G module from Infineon were included. The controller was modeled in Simulink®[2]. For all simulations, the switching frequency was set to 2 kHz and the modulating frequency was set to 60 Hz. Fig. 3 shows the same results as Fig. 2, with simulated results indicated. The simulation follows the same trend, and agrees almost exactly with the predicted performance for conventional modulation. For the interleaved variants, the simulated common-mode voltage is slightly higher than the prediction at high modulation depths. The most likely explanation is that the deadtime interferes with the cancellation effect of the interleaving. Deadtime compensation could be used to achieve the predicted performance. Still, the worst-case

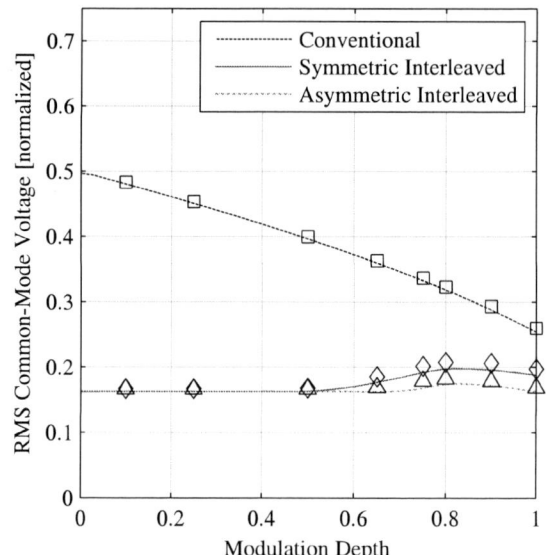

Fig. 3. As in Fig. 2, with simulated results indicated by boxes (conventional), diamonds (symmetric interleaved with third harmonic), and triangles (asymmetric interleaved without third harmonic).

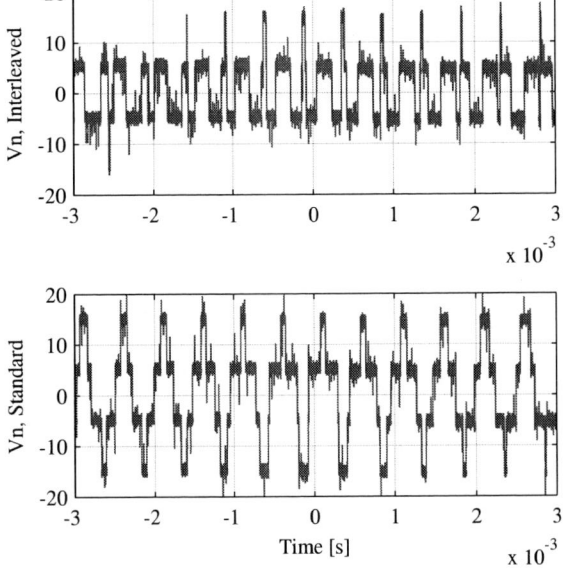

Fig. 4. Example neutral voltage waveforms for interleaved (top) and conventional (bottom) PWM. Modulation frequency was 46.665 Hz with a depth of 0.84.

improvement (at a modulation depth of 1.0 with third-harmonic injection) is 24.14%.

V. EXPERIMENTAL VALIDATION

An inverter was constructed to validate the simulations and calculations. The controller was a TMS320F28335. Switching frequency was fixed at 2 kHz. Linear volts-per-hertz modulation with boost was used, with third-harmonic injection and symmetric uniform sampling. The power stage was based on an Infineon FS100R12KT4G. For these validation experiments, the bus voltage was fixed at 30.3 V and the load motor was a Baldor VM3554T (1½ hp, 4-pole) motor in low-voltage connection. To find the common-mode voltage, all three line-to-negative-bus voltages were probed and averaged. The dc offset was then subtracted to find the ac common-mode voltage applied to the motor.

Fig. 4 shows typical waveforms for the interleaved and standard (non-interleaved) cases. Here the modulation frequency was 46.665 Hz and the modulation depth was 0.84. As in the simulation (Fig. 1), the neutral voltage in the interleaved case is usually at ±1/6 of the bus voltage, with occasional pulses to ±1/2 of the bus voltage. For the non-interleaved case, the neutral voltage is at ±1/2 of the bus voltage for a significant fraction of the time.

Fig. 5 compares the computed results to the experimental results. At high modulation depths, the experimental common-mode voltage is actually slightly less than expected. One possible explanation is that the modulation frequency was lower in the experiment than in the computation. Another possible factor is the lack of stiffness in the bus voltage. Interleaving reduced common-mode voltage by 63.7%, 56.8%, and 39.2% at modulation frequencies of 10 Hz, 30 Hz,

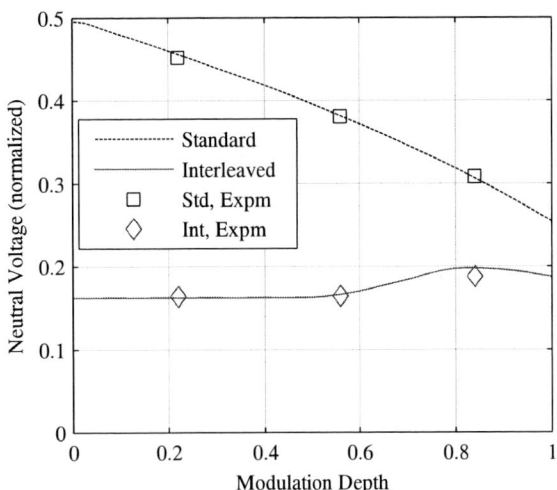

Fig. 5. Comparison of experimental neutral voltage to computed neutral voltage. Symmetric uniformly-sampled PWM with third-harmonic injection was used.

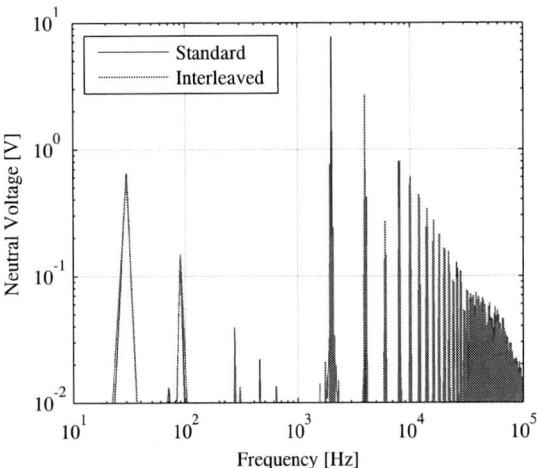

Fig. 6. FFT of neutral voltage for standard (non-interleaved) and interleaved PWM methods. Modulation frequency of 30 Hz, modulation depth of 0.56.

and 46.665 Hz, respectively. The largest common-mode voltage with interleaving was only 5.68 V, or 18.8% of the bus voltage.

Fig. 6 shows fast Fourier transform (FFT) results for the 30 Hz case. In both standard and interleaved PWM, there is some common-mode content at the fundamental and third harmonic of the modulation frequency, possibly due to power supply variations. Subharmonics (between 90 Hz and 1 kHz) are more pronounced in standard PWM. Interleaving nearly eliminates the fundamental of the switching frequency, although the sidebands (switching frequency plus multiples of the modulation frequency) and second harmonic are more pronounced.

VI. CONCLUSIONS

A simple method for achieving significantly reduced common-mode voltage in a three-phase inverter was analyzed, demonstrated through simulations, and validated with an experiment. The method relies on interleaving the carrier waves of the three phases. The advantage of the new method ranges from a low of 26% reduction at full modulation with symmetric uniform PWM and third harmonic injection, to a maximum of 67% reduction at zero modulation depth. No additional computations are needed beyond conventional sine-triangle type uniform PWM. Existing digital hardware in readily available DSCs/DSPs easily implements the interleaving. The improvement varies based on the modulating function and sampling method, but remains substantial regardless. If motor control code is already written for such a DSC with sine-triangle PWM, then the change amounts to two lines of code that change two registers during initialization. Experimental results agree well with the predicted performance and validate the approach.

VII. ACKNOWLEDGEMENTS

This work was supported by ITW Military GSE and the National Science Foundation I/UCRC on Intelligent Maintenance Systems.

VIII. REFERENCES

[1] R. W. A. A. De Doncker, D. M. Divan, and M. H. Kheraluwala, "A three-phase soft-switched high-power density dc/dc converter for high-power applications," *IEEE Transactions on Industry Applications*, vol. 27, pp. 63-73, Jan./Feb. 1991.

[2] G. Pfaff, A. Weschta, and A. F. Wick, "Design and experimental results of a brushless AC servo drive," *IEEE Transactions on Industry Applications*, vol. IA-20, pp. 814-821, 1984.

[3] A. M. Hava and E. Un, "Performance analysis of reduced common-mode voltage PWM methods and comparison with standard PWM methods for three-phase voltage-source inverters," *IEEE Transactions on Power Electronics*, vol. 24, pp. 241-252, January 2009.

[4] C. B. Jacobina, A. M. N. Lima, E. R. C. da Silva, and A. M. Trzynadlowski, "Current control for induction motor drives using random PWM," *IEEE Transactions on Industrial Electronics*, vol. 45, pp. 704-712, October 1998.

[5] Y.-S. Lai and Y.-T. Chang, "Design and implementation of vector-controlled induction motor drives using random switching technique with constant sampling frequency," *IEEE Transactions on Power Electronics*, vol. 16, pp. 400-409, May 2001.

[6] T. Beechner and J. Sun, "Harmonic cancellation under interleaved PWM with harmonic injection," in *Proc. IEEE Power Electronics Specialists Conference*, 2008, pp. 1515-1521.

[7] C. Casablanca and J. Sun, "Interleaving and harmonic cancellation effects in modular three-phase voltage-sourced converters," in *Proc. IEEE Computers in Power Electronics Workshop*, 2006, pp. 275-281.

[8] H. Deng, L. Helle, Y. Bo, and K. B. Larsen, "A general solution for theoretical harmonic components of carrier based PWM schemes," in *Proc. IEEE Applied Power Electronics Conference*, 2009, pp. 1698-1703.

[9] R. A. Guinee and C. Lyden, "A novel Fourier series time function for modeling and simulation of PWM," *IEEE Transactions on Circuits and Systems--I: Regular Papers*, vol. 52, pp. 2427-2435, November 2005.

[10] D. G. Holmes, "A general analytical method for determining the theoretical harmonic components of carrier based PWM strategies," in *Proc. IEEE Industry Applications Conference*, 1998, vol. 2, pp. 1207-1214.

[11] S. Jayawant and J. Sun, "Double-integral Fourier analysis of interleaved pulse width modulation," in *Proc. Computers in Power Electronics Workshop*, 2006, pp. 34-39.

[12] A. W. Leedy and R. M. Nelms, "Harmonic analysis of a space vector PWM inverter using the method of multiple pulses," in *Proc. IEEE International Symposium on Industrial Electronics*, 2006, vol. 2, pp. 1182-1187.

[13] J. Shen, J. A. Taufiq, and A. D. Mansell, "Analytical solution to harmonic characteristics of traction PWM converters," *IEE Proceedings - Electric Power Applications*, vol. 144, pp. 158-168, March 1997.

Dynamic DC-Bus Voltage Control Strategies for a Three-Phase High Power Shunt Active Power Filter

Zhiqiang Wang, Chuan Xie, Jing Zhang and Guozhu Chen
College of Electrical Engineering, Zhejiang University
Hangzhou, Zhejiang 310027, China
Email: gzchen@zju.edu.cn

Abstract—**Shunt Active Power Filters (SAPF) have been proved to be an effective approach to eliminate harmonic current in power grids. The DC voltage control during dynamic processes is critical for the safe and performance of the SAPF. In this paper, soft startup and impact suppression techniques are proposed to guarantee no voltage overshoots or sags in DC bus during startup and other dynamic processes. Subsequently, a novel controller for DC bus voltage regulation is put forward according to the small signal model based on the instantaneous power balance in AC and DC sides of three-phase SAPF. With good compensation effects, the proposed voltage loop controller has improved dynamic performances greatly compared with traditional PI controllers as loads change. Finally, simulation and experimental results from a 260kVA prototype verify the feasibility of the proposed techniques and control strategies.**

I. INTRODUCTION

With the rapid development of modern electronics industry and the continuous proliferation of nonlinear loads, there is an increasing harmonic level in power grids due to voltage and current waveform distortion in these years. To solve this problem, shunt active power filters are now widely researched and used, which are suitable to compensate harmonic and reactive current in power system [1] [2].

In practical application, the DC voltage control is important for voltage source SAPF whether it operates at startup or steady-state process. On one hand, there will be unexpected great DC capacitor voltage overshoots or sags and impulse current as a result of small filter inductance in high power SAPF during the startup transient process. On the other hand, when under normal operation, it will cause DC link voltage to raise, fall and fluctuate in a wide range due to the active power exchange between the AC and DC side. Consequently, soft startup techniques and regulation control strategies are necessary to control the DC link voltage.

The most commonly used soft startup techniques for voltage loop are increasing the reference of DC voltage gradually, charging with constant active current, modifying PI parameters and adopting fuzzy controllers [3]. However, all of these methods are fairly complex and usually need the operation of both voltage and current loop controllers.

Traditionally, the DC link capacitor voltage was regulated by an uncontrolled rectifier, which would not only induce DC

voltage to fluctuate along with power supply but also increase the overall cost [4]. Nowadays, DC capacitor voltage feedback with the PI controller is widely used in SAPF due to its good steady-state and dynamic performances. However, it cannot mitigate the voltage harmonics in voltage loop effectively which could interrupt the operation of current loop and deteriorate the compensation precision. In addition, paper [5] proposed a fuzzy control strategy which has gained better performances in steady-state and is easier to be designed compared with ordinary PI controllers, while its dynamic control effect is not satisfying. Some other control strategies such as adaptive filtering technique [6], nonlinear PID controllers [7] are complicated from design to practical application.

In order to solve these problems, soft startup techniques and a novel controller for DC link voltage regulation are proposed according to the small signal model based on the instantaneous power equilibrium of three-phase three-wire SAPF. Simulation and experimental results from a 260kVA prototype demonstrate that these techniques have guaranteed the DC voltage rising to the expected value smoothly and the new controller has obtained excellent steady-state and dynamic performances.

II. SYSTEM CONFIGURATION

The main circuit of three phases SAPF is shown in Fig.1. Where, u_g is the grid phase voltage, L_g is the grid side inductance, L_i is the load side inductance, C_{dc} the DC capacitor, R_L is the load resistor, L_1, L_2, C and R_d represents the inverter side inductor, grid side inductor, filter capacitors and damping resistor of the LCL filter respectively. i_L, i_2, u_s, u_{dc} stands for the load current, compensation current, system voltage and DC voltage respectively.

III. SOFT STARTUP TECHNIQUES

Before the normal operation of SAPF, soft startup in DC link is necessary to avoid high impulse current in DC capacitors which may damage devices. The soft startup is mainly composed by precharging and chopping boost process.

A. Precharging Process

Precharge in the DC link can be realized by the anti-parallel diodes of the main switches with the grid voltage. To

---

The authors would like to thank the sponsorship of NCET Program of Ministry of Education, China (#060512)

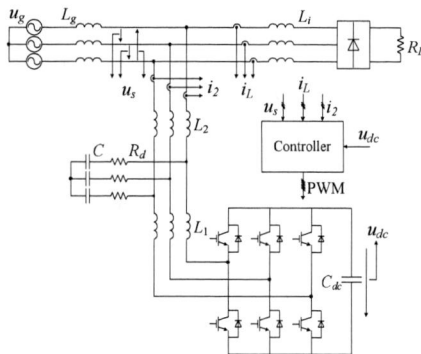

Fig. 1. Configurations of the shunt active power filter with LCL filter.

limit the surge current, a proper resistor R_s in series with the grid side inductance L_2 is required. Since the current flowing across L_1 is very small, the value of DC voltage is determined by the voltage across filter capacitor in the second-order circuit composed by R_s, L_2, R_d and C. The DC voltage can be given approximately as

$$U_{dc1} = \sqrt{2}\frac{X_c}{\sqrt{X_c^2 + R_s^2}}U_g . \tag{1}$$

Then, resistor R_s is short-circuited by the contactor in the main circuit. The DC voltage step into a new value which can be expressed as

$$U_{dc2} = \sqrt{2}U_g . \tag{2}$$

In order to inhibit the over change of DC voltage, usually the two values should be

$$U_{dc1} \geq 0.9U_{dc2} . \tag{3}$$

The maximum value of R_s is determined by (3) and the minimum value of R_s depends on the maximum acceptable surge current during the process.

B. Chopping Boost Process

In the process, the main circuit serves as a three-phase Boost-PWM rectifier to increase DC voltage to the expected value smoothly, while the control can be simplified without taking unity power factor into account. The operation principle of the circuit neglecting the filter capacitor branch with relatively high impedance is depicted in Fig. 2 and Fig. 3.

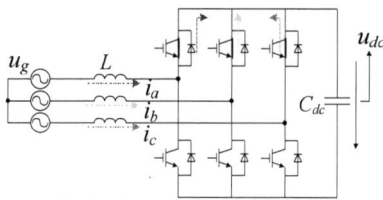

Fig. 2. Chopping boost process (a).

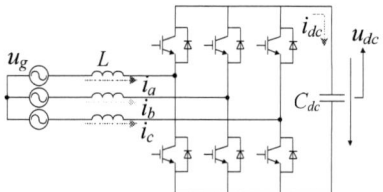

Fig. 3. Chopping boost process (b).

The circuit can be controlled in the way as follows. Firstly, turn off the three lower bridge legs while chop the three upper ones with the same duty cycle increasing periodically from 0 to a constant. Then, keep chopping with the constant duty until the DC voltage rises to the desired value which can be detected by digital controllers real-timely. The constant should be a trade-off between overshoot current flowing across DC capacitors and charging time.

IV. IMPACT SUPPRESSION TECHNIQUES

After the soft startup of DC link voltage, the system will enter into normal operation. However, impact occurs the very moment the main circuit output compensation current. The model neglecting filter capacitor can be expressed as

$$\begin{cases} \dfrac{di_a}{dt} = -\dfrac{R}{L}i_a + \dfrac{1}{L}\left[u_{sa}^{\cdot} - (S_a^* - K_a)u_{dc}\right] \\[2mm] \dfrac{di_b}{dt} = -\dfrac{R}{L}i_b + \dfrac{1}{L}\left[u_{sb} - (S_b^* - K_b)u_{dc}\right] \\[2mm] \dfrac{di_c}{dt} = -\dfrac{R}{L}i_c + \dfrac{1}{L}\left[u_{sc} - (S_c^* - K_c)u_{dc}\right] \\[2mm] C\dfrac{du_{dc}}{dt} = S_a^*i_a + S_b^*i_b + S_c^*i_c \end{cases} \tag{4}$$

where, R is the equivalent resistance of the total loss including LCL filter, converter, etc; $L = L_1 + L_2$.

In high power SAPF application, the total inductance L in LCL filters is very small to acquire high bandwidth. From (4), it can be seen that the larger the inductance, the lower di/dt and tracking ability in the system under the condition that DC link and grid voltage is constant. On the contrary, the small inductance is conducive to fast dynamic response while it will result in not only high impulse current at AC side but also voltage sag or overshoot at DC link during the transient process. In addition, the last equation indicates that the DC voltage impact is directly related with the three phase output compensation current in AC side. Consequently, some measures must be taken to suppress great DC voltage fluctuation during the starting transient process.

A. Initializing PI controller output

The current loop PI controller work under the synchronous reference frame. Usually, the initial output of the PI controller in d-q axis at the starting moment is a random value. The value in d-axis will create an active current accordingly which may lead the DC voltage fluctuate greatly. Even if the value is zero, the total loss on the SAPF will cause DC voltage reduce sharply. What's worse, the bandwidth of the current loop is usually fairly small due to the low switching frequency in high power SAPF, which is not beneficial for impact suppression during the transient process.

Setting initial value of the current loop controller aims at accelerating the transient process when main circuits start to output compensation current. Therefore, this constant value should be the output of the controller in steady-state which can be determined approximately by simulation and experimental results. In terms of the instantaneous reactive power theory, it serves as an effective way to balance the energy in DC capacitors and maintain the DC voltage at the starting moment.

B. Introducing voltage feed-forward control

Although initializing PI output plays a role in suppressing DC voltage impulse to some extent, its effects rely heavily on the initial value which is difficult to set accurately in practical application. Moreover, when the grid voltage reduces and increases greatly, PI output will also fluctuate significantly along with the grid voltage. Therefore, its suppression effects will be deteriorated seriously.

In order to solve this problem, voltage feed-forward control is introduced in the system, as shown in Fig. 4. Since it can sense and follow the rapid change of the grid voltage, the output of PI controller is nearly constant which is easy to be set in control algorithm. Another important advantage of voltage feed-forward control is that it could enhance the dynamic response speed without enlarging the bandwidth of the current loop. Actually, the role of feed forward control is similar with that of the initial value in that they both aim at suppressing DC voltage impact by adding proper active components in d-axis.

The $G_n(s)$ can be selected properly to eliminate the negative effects of grid voltage completely in theory. The influence of grid voltage to compensation current can be expressed as

$$\frac{i_2(s)}{u_g(s)} = \frac{G_n(s)G_p(s)-1}{(L_2+L_g)s+[L_1s+PI(s)]G_p(s)} \tag{5}$$

where

$$G_p(s) = \frac{1+CR_ds}{L_1Cs^2+CR_ds+1}. \tag{6}$$

The effects of gird voltage can be eliminated absolutely if

$$G_n(s) = 1/G_p(s) = \frac{L_1Cs^2+CR_ds+1}{CR_ds+1}. \tag{7}$$

However, the physical implementation of (7) is impossible due to the higher order in numerator. Considering the physical implementation and approximately complete compensation in main frequencies, (7) can be modified as

$$G_n(s) = \frac{L_1Cs^2+CR_ds+1}{(CR_ds+1)(T_ns+1)} = \frac{L_1Cs^2+CR_ds+1}{CR_dT_ns^2+(CR_d+T_n)s+1}, L_1C \ll CR_dT_n. \tag{8}$$

Practically, the time constant can be

$$T_n = L_1/10R_d. \tag{9}$$

Then, the ultimate transfer function of $G_n(s)$ is

$$G_n(s) = \frac{L_1Cs^2+CR_ds+1}{L_1Cs^2/10+(CR_d+L_1/10R_d)s+1}. \tag{10}$$

C. Delaying and modifying repetitive controller

In the system, a repetitive controller is plugged into conventional current loop with a PI controller to attain excellent steady-state precision and dynamic responses. The double loop composite controller composed by a PI inner loop and repetitive control outer loop is depicted as Fig. 5.

The two methods above could suppress DC voltage impact effectively, while current impulse in grid side inductance may still exist due to inaccurate initial value and feed-forward voltage. According to the principle of the repetitive control theory [8][9], the unexpected errors caused by the impulse current will be accumulated by the internal model of repetitive controllers and have a negative effect on DC link voltage as well as following output current waveforms. The transfer function of internal model is

$$G_{im}(z) = \frac{e_0(z)}{e_i(z)} = \frac{1}{1-Q(z)z^{-N}} \tag{11}$$

where, $Q(z)$ is attenuation filter. Usually it is a constant smaller than unit e.g. 0.9 in this system.

Difference equation can be deduced from (11) as

$$e_0(k) = e_i(k)+0.9e_0(k-N). \tag{12}$$

Equation (12) shows that the internal model accumulates the input system error with step of fundamental period one by one until the error is smaller than 0.9 times of output signal. The negative effects can be eliminated by delaying the operation of repetitive controller for a short period of time e.g. 10ms to avoid the surge current or by adopting a small $Q(z)$ e.g. 0.5 increasing periodically to 0.9.

V. SMALL SIGNAL MODELING OF DC-BUS VOLTAGE CONTROL

Considering the switching frequency and its multiple frequency of the system can be eliminated if the LCL filter is designed properly, the small signal model can be deduced by neglecting those switching ripples. Since capacitor branch mainly flows across high frequency ripples, it can be neglected during modeling process. According to Fig.1, the instantaneous power equilibrium equation between AC and DC side of SAPF can be given as [10]

$$u_a(t)i_a(t)+u_b(t)i_b(t)+u_c(t)i_c(t)-R\left[i_a^2(t)+i_b^2(t)+i_c^2(t)\right]$$
$$-\frac{1}{2}(L_1+L_2)\frac{d}{dt}\left[i_a^2(t)+i_b^2(t)+i_c^2(t)\right] = U_{dc}(t)I_{cc}(t) \tag{13}$$

where, R is the equivalent resistance of the total loss including LCL filter, converter, etc. I_{cc} is the current flow across DC capacitors.

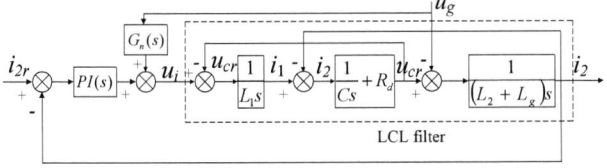

Fig. 4. The PI inner current loop with grid voltage feed-forward control.

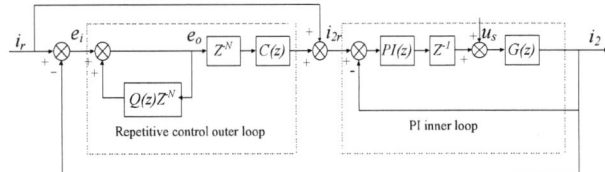

Fig. 5. The composite current loop controller.

According to the instantaneous power theory, there is

$$\begin{cases} u_a(t)i_a(t)+u_b(t)i_b(t)+u_c(t)i_c(t)=3U_sI_p(t) \\ R\left[i_a^{\,2}(t)+i_b^{\,2}(t)+i_c^{\,2}(t)\right]=3R\left[I_p^{\,2}(t)+I_c^{\,2}(t)\right] \\ I_{cc}(t)=C_{dc}\dfrac{dU_{dc}(t)}{dt} \\ \dfrac{(L_1+L_2)}{2}\dfrac{d\left[i_a^{\,2}(t)+i_b^{\,2}(t)+i_c^{\,2}(t)\right]}{dt}=\dfrac{3(L_1+L_2)}{2}\dfrac{d\left[I_p^{\,2}(t)+I_c^{\,2}(t)\right]}{dt} \end{cases} \quad (14)$$

where, $I_p(t)$ is the rms value of active current on AC side. $I_c(t)$ is the rms value of compensated current outputted by the SAPF, including harmonic and reactive current.

Substituting (14) into (13), the following equation can be obtained.

$$U_sI_p(t)-R\left[I_p^{\,2}(t)+I_c^{\,2}(t)\right]-\frac{1}{2}(L_1+L_2)\frac{d}{dt}\left[I_p^{\,2}(t)+I_c^{\,2}(t)\right]=\frac{1}{3}C_{dc}U_{dc}(t)\frac{dU_{dc}(t)}{dt}. \quad (15)$$

Applying small disturbance to I_p, I_c and U_{dc} as shown in (16) under the condition that other parameters are all constants.

$$\begin{cases} I_p(t)=I_p+\Delta I_p \\ I_c(t)=I_c+\Delta I_c \\ U_{dc}(t)=U_{dc}+\Delta U_{dc} \end{cases} \quad (16)$$

Substituting (16) into (15), the small perturbation equation neglecting second order terms can be given as

$$3U_s\Delta I_p-6R(I_p\Delta I_p+I_c\Delta I_c)-3(L_1+L_2)(I_p\frac{d}{dt}\Delta I_p+I_c\frac{d}{dt}\Delta I_c)=C_{dc}U_{dc}\frac{d}{dt}\Delta U_{dc}. \quad (17)$$

Taking the Laplace transform of (17), the disturbance linearization equation around steady-state point (I_p, I_c, U_{dc}) is

$$\Delta I_p(s)\left[3U_s-6RI_p-3(L_1+L_2)I_ps\right]-\Delta I_c(s)\left[6RI_c+3(L_1+L_2)I_cs\right]=sC_{dc}U_{dc}\Delta U_{dc}(s). \quad (18)$$

The open loop block diagram of the small signal model is built based on (18), as shown in Fig. 6.

VI. DESIGN OF THE NOVEL DC-BUS VOLTAGE CONTROLLER

The DC voltage close loop block diagram with a controller $F(s)$ can be depicted in Fig. 7.

The open loop transfer function is

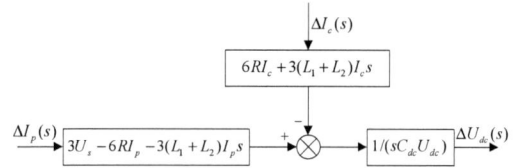

Fig. 6. Open loop block diagram of SAPF DC link voltage control

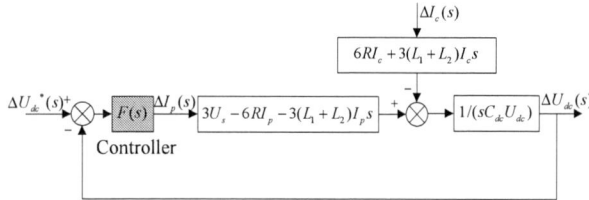

Fig. 7. Closed loop block diagram of SAPF DC link voltage control.

$$M(s)=\frac{3U_s-6RI_p-3LI_ps}{C_{dc}U_{dc}s}F(s). \quad (19)$$

The transfer function of conventional PI controller is

$$F_{PI}(s)=K_p+\frac{K_i}{s}. \quad (20)$$

The bode diagram of $M(s)$ without controller and with the PI controller are shown in Fig. 8 respectively.

From Fig. 8, it can be seen that they have similar frequency characteristics only with small difference in gain. Although the DC voltage control system has a large phase margin, attenuation at high frequency is not fast enough to suppress noises. Thus, expected controller should acquire fast attenuation at high frequency as well as enough gain at low frequency.

The external voltage loop controller is responsible for the regulation of DC voltage to an expected constant by adding or reducing active current in the reference of inner current loop. The DC bus contains $6n$ voltage harmonics due to output compensation current ($6n\pm1$ harmonic current) and these harmonics will also exist in the output of conventional PI controller which could interrupt the operation of current loop. Compensation precision will be deteriorated, especially when harmonic voltage ripples are fairly high as a result of small value of the DC capacitor.

It seems a wise choice to eliminate these undesired harmonics by substituting a low pass filter for the ordinary PI controller. With the same bandwidth as PI controller, the low pass filter could obtain better compensation performances. In other words, the bandwidth of voltage loop can be enlarged with the same compensation precision as PI controller. In addition, the tracking ability for DC voltage is also satisfying due to the internal integral element ($1/sC_{dc}U_{dc}$). A first-order or second-order low pass filter is easy to be designed and stabilized, the transfer function of which are given as

$$F_1(s)=\frac{K_1}{1+s/\omega_{c1}}. \quad (21)$$

$$F_2(s)=\frac{K_2\omega_{c2}^{\,2}}{s^2+2\xi\omega_{c2}s+\omega_{c2}^{\,2}}. \quad (22)$$

The cut-off frequency is determined by the bandwidth. The gain K should be a trade-off between tracking ability at low frequency and stability margin. The damping ratio is mainly determined by expected phase margin and attenuation ratio for high frequency ripple. In the system, the cut-off frequency and the gain are set at 85 Hz, 3 for first-order and 66 Hz, 3 for second-order low pass filter. The attenuation ratio is 2. The bode diagram of $M(s)$ with first-order and second-order low pass filter are shown in Fig. 9.

From Fig. 9, it can be seen that at high frequency they attenuate with -20dB/div and -40dB/div respectively which is conducive to suppress harmonics and noises. Their phase margin is 45°and 30°which is sufficient for the system. In addition, compared with the first-order filter, the second-order filter has a higher attenuation ratio which is expected to obtain

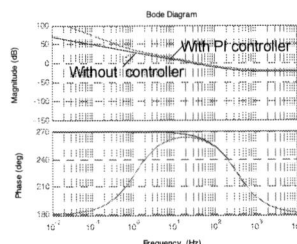

Fig. 8. The bode diagram of M(s) without controller and with PI controller.

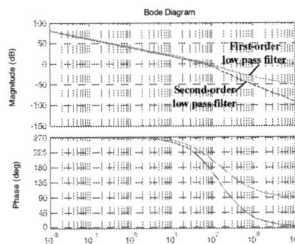

Fig. 9. The bode diagram of $M(s)$ with the first-order and second-order filter.

a better compensation effect, while a small bandwidth usually results in a lower dynamic response speed. Although their phase margin is relatively smaller than that of PI controller, taking one with another, they have better performances in DC voltage control. The comparison of their main characteristics is shown in Table 1.

VII. SIMULATION AND EXPERIMENTAL RESULTS

A. Experimental circuit and parameters

To verify the design, simulation and experiments are conducted on the three phase SAPF, as is depicted as Fig. 1. Parameters of main circuit are: u_{dc}=700V, C_{dc}=5mF, L_1=0.056mH, L_2=0.020 mH, C=120uF, R_d=0.1 Ohm, u_g =380, f_g=50 Hz, rated capacity S_c=260 kVA. The control scheme is implemented with a TMS320F2812DSP.

B. Simulation Results

Simulation has been done with MATLAB 7.1 based on the parameters and design results above.

Fig. 10 show the dynamic waveforms of DC link voltage when the load is reduced (100%-50%) and increased (100%-50%) in normal operation.

From Fig. 10, it can be seen that the new controllers have obtained better dynamic responses than PI controller. The new controller can eliminate DC link voltage errors caused by the disturbance from load with very good dynamic responses. The maximum fluctuation of DC voltage is 80V with PI controller, 50V with the first-order filter, and 60V with the second-order filter. This simulation results match with bode diagrams in Fig. 8 and Fig. 9 perfectly.

Fig. 11 show simulation waveforms with the PI controller, the first-order and second-order low pass filter respectively where the upper one is output of the controller, and the lower one is the grid current after compensation.

According to the simulation results in these figures, the output harmonics are suppressed effectively by new controllers, which play an important role in meliorating the

TABLE I. COMPARISON OF MAIN CHARACTERISTICS WITH DIFFERENT CONTROLLERS

Index / Controller	Phase margin	Cut-off frequency	Attenuation ratio
Without controller	85°	30Hz	0db/dec
PI controller	80°	40Hz	0db/dec
First-order filter	45°	85 Hz	-20dB/dec
Second-order filter	30°	66 Hz	-40dB/dec

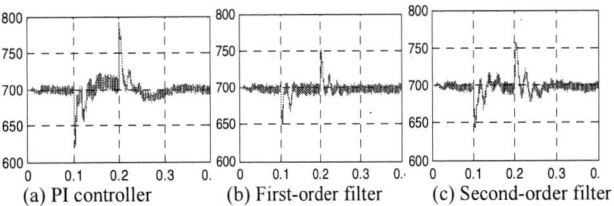

(a) PI controller   (b) First-order filter   (c) Second-order filter

Fig. 10. Comparison of DC voltage dynamic waveforms with different controllers.

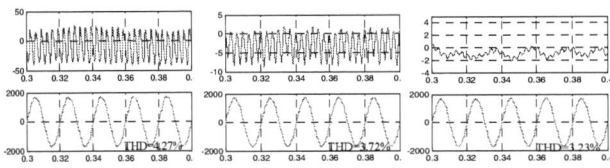

(a) PI controller   (b) First-order filter   (c) Second-order filter

Fig. 11. Comparison of controller output and grid current waveforms with different controllers.

grid current waveforms. The THD of the grid current is 4.27 % with the PI controller, 3.72% with the first-order filter and 3.23% with the second-order filter.

C. Experimental Results

Fig. 12 and Fig. 13 show the DC link voltage waveform without suppression technique and with proposed techniques during the startup process.

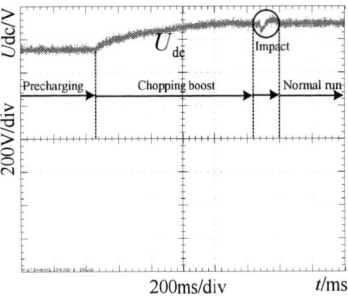

Fig. 12. The DC bus voltage without impact suppression.

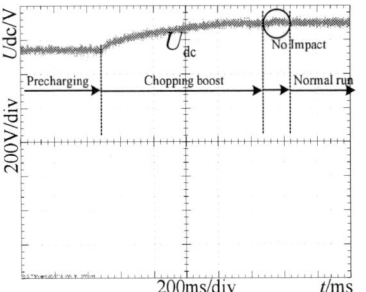

Fig. 13. The DC bus voltage with impact suppression.

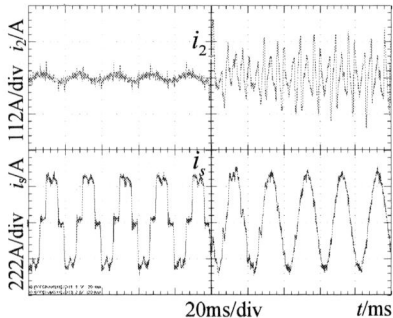

Fig. 14. The compensation and grid current during startup.

The voltage impact is 50V in Fig. 12. In sharp contrast, in Fig. 13 the voltage rises to the expected value smoothly and the voltage sag is very small the moment the main circuit output compensation current. Fig. 14 shows the compensation current and grid current during the startup process. It can be seen that the surge current is also very small and it will not increase along with the output compensation current. These waveforms indicate that soft start is realized in the system with neither voltage overshoot nor current surge.

Fig. 15, Fig. 16 and Fig. 17 show the dynamic DC voltage (AC-coupling, 700VDC) waveforms with different controllers. The voltage sag is reduced by new controllers as loads increase. Fig. 18, Fig. 19 and Fig. 20 show the grid current waveforms and spectrograms with different controllers. The THD of the grid current is 4.86% with the PI controller, 4.12% with the first-order filter and 3.73% with the second-order filter. The THD of the grid current is also improved by new controllers.

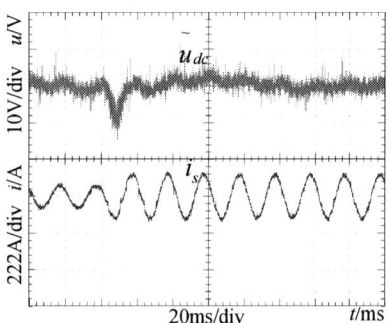

Fig. 15. Dynamic DC voltage during startup with PI controller.

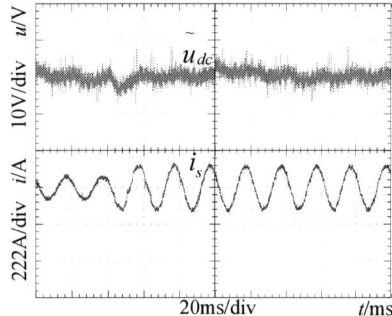

Fig. 16. Dynamic DC voltage during startup with first-order filter.

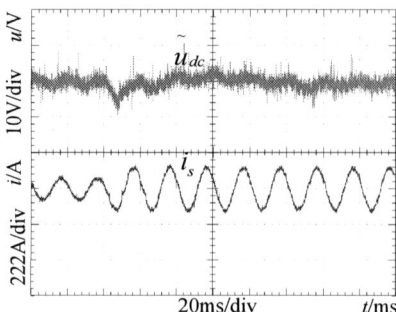

Fig. 17. Dynamic DC voltage during startup with second-order filter.

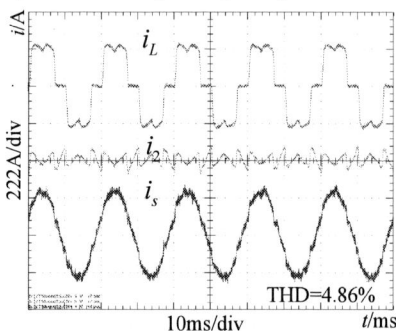

(a) current waveforms

THD=4.86%

(b) grid current spectrogram

Fig. 18. The current waveforms and spectrograms with PI controller.

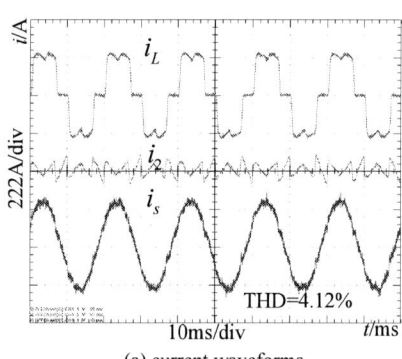

(a) current waveforms

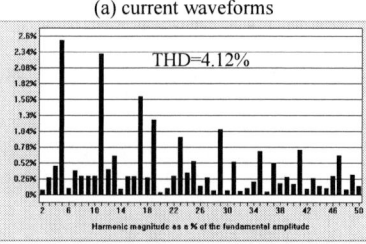

THD=4.12%

(b) grid current spectrogram

Fig. 19. The current waveforms and spectrograms with first-order filter.

978-1-4244-4782-4/10 $26.00 © 2010 IEEE

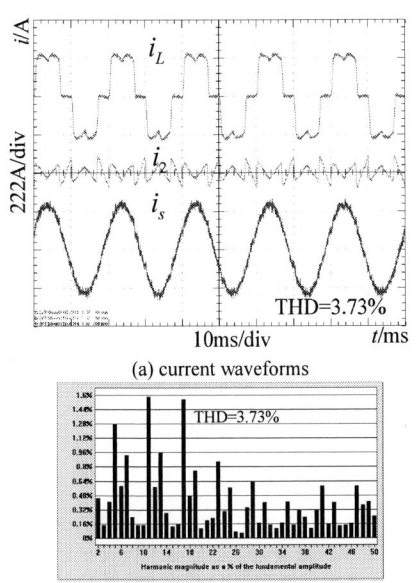

(a) current waveforms

(b) grid current spectrogram

Fig. 20. The current waveforms and spectrograms with second-order filter.

VIII. CONCLUSION

This paper proposed soft startup techniques to suppress voltage overshoots or sags in DC bus during startup dynamic processes and a novel controller for DC bus voltage regulation according to the small signal model based on the instantaneous power equilibrium. Simulation and experimental results demonstrate that these techniques have guaranteed the DC voltage rising to the expected value smoothly and the new controller has obtained better dynamic performances compared with traditional PI controllers for DC bus voltage control.

REFERENCES

[1] Akagi.H, "New trends in active filters for power conditioning, " *IEEE Transactions on Industry Application*, vol. 32, no. 6, pp. 1312-1322, 1996.

[2] Chen Guozhu, Lü Zhengyu, Qian Zhaoming, "The general principle of active filter and its application," *Proceedings of the CSEE*, vol. 20, no. 9, pp. 17-21, 2000.

[3] He Na,Wu Jian, Xu Dianguo, "Fuzzy soft-startup controller of active power filter," *Transactions of China Electrotechnical Society*, vol. 20, no. 9, pp. 115-120, 2000.

[4] Akagi. H, Kanazawa.Y, Nabae.A, "Instantaneous reactive power compensators comprising switching devices without energy storage components, " *IEEE Transactions on Industry Application*, vol. 20, no. 3, pp. 625-630, 1984.

[5] Juan Dixon, Jose Contardo, Luis Moran, "DC link fuzzy control for an active power filter, sensing the line current only," in *IEEE 1997 Power Electronics Specialists Conference*, 1997, pp. 1109-1114.

[6] Abedini.A, Nasiri.A, "An improved adaptive filter for voltage and current reference extraction," in *IEEE 2006 Power Electronics and Motion Control Conference*, 2006, pp. 1-5.

[7] Wenjie Guo, Fei Lin, Trillion Zheng, "Nonlinear PI control for three-phase PWM AC-DC converter," in *IEEE 2006 Annual Conference of the IEEE Industrial Electronics Society*, 2006, pp. 1093-1097.

[8] Aurelio García Cerrada, Omar Pinzón Ardila, Vicente Feliu Batll, Pedro Roncero Sánchez, and Pablo García-González, "Application of a repetitive controller for a three-phase active power filter," *IEEE Transactions on Power Electronics*, vol. 22, no. 1, pp. 237-246, 2007.

[9] R. Griñó, R. Cardoner, R. Costa-Castelló, and E. Fossa, "Digital repetitive control of a three-phase four-wire shunt active filter," *IEEE Transactions on Industrial Electronics*, vol. 54, no. 3, pp. 1495-1503, 2007.

[10] Luo Shiguo, "Optimal design of DC voltage close loop control for an active power filter," in *IEEE 1995 Power Electronics and Drive Systems*, 1995, pp. 565-570.

A Simplified Three Phase Three-level Zero-Current-Transition Active Neutral-Point-Clamped Converter with Three Auxiliary Switches

Jin LI[1,2], *Student Member, IEEE*   Jinjun LIU[1], *Member, IEEE*   Dushan Boroyevich[2], *Fellow, IEEE*

[1]Power Electronics and Renewable Energy Research Center

Xi'an Jiaotong University

Xi'an , Shaanxi , P.R. China, 710049

[2]Center for Power Electronics Systems

Virginia Polytechnic Institute and State University

Blacksburg, VA,USA,24060

Email: libro.jin@mail.xjtu.edu.cn

Abstract: **A novel three-level zero-current-transition active neutral-point-clamped inverter is presented in this paper. The proposed soft-switching topology employs only one auxiliary switch, one LC resonant tank to assist the main switching devices in one phase leg. The zero-current turning-off of all the main and auxiliary switches are achieved with voltage stress of all switching devices are clamped to half of the DC-link voltage. Meanwhile, The turning-on losses of switches and reverse recovery losses of diodes are reduced. The control scheme is simple and no additional sensor is needed and no need modification to the normal PWM strategy. The operation principles are analyzed and simulation result verify the analysis.**

Key words: **Three-level, Active Neutral-point-clamped converter, Zero-current-transition, Soft-switching;**

I. INTRODUCTION

Three-level neutral-point-clamped(NPC) voltage source converter(VSC) is widely used in many medium voltage and high power motor drives applications. What is more, in low voltage applications the performance of three-level NPC VSC is very attractive and competitive. Recently an active neutral-point-clamped(ANPC) VSC is proposed to increase the output power[1]. ANPC VSC uses active switches instead of diodes to clamp the neutral point. In most VSCs, which are equipped with either IGBTs or IGCTs, the switching losses are the bottleneck of increasing switching frequency, the bandwidth and power density. Particularly in high power applications, the switching frequency is limited by the switching losses to below 1kHz.

Resonant techniques are popular in AC-DC, DC-DC, DC-AC power conversions for high frequency, high power density and high efficiency applications.[2~17] The utilization of resonant soft-switching techniques to reduce the losses of NPC VSC has also received lots of attention, different resonant techniques are developed in the last ten years, like auxiliary resonant commutated pole(ARCP) technique, resonant dc link(RDCL) technique, and zero-current-transition(ZCT) technique, etc.. Among those techniques the ZCT technique, which is proposed and developed in [2-5], has many attractive features:1) achieve the zero-current turning-off of all the main switches; 2) besides the great reduction of turning-off losses, turning-on losses are also reduced, meanwhile, the auxiliary switches have no switching losses; 3) the voltage stress of all switches are kept to DC-link voltage; 4) the current rating of auxiliary switches

This work was supported by State Key Development Program Basic Research of China under award number 2009CB219705.

are much smaller than the main switches; 5) no modification to the normal PWM strategy. In the latest literature a simplified ZCT VSC topology with 3 auxiliary switches is proposed for two-level VSC[5] and extended to three-level NPC VSC with two auxiliary switches and two LC resonant tank in one phase leg[2].

In this paper, the ZCT concept is extended to ANPC VSC, a novel three phase three-level ZCT soft-switching converter based on ANPC VSC with only three auxiliary switches(ZCT-3S ANPC VSC) is presented and shown in Fig.1. As shown in Fig.2, the proposed topology employs one auxiliary switch and one LC resonant tank in one phase leg, such a phase leg can be implemented using three half-bridge modules for the main devices and one chop module for the auxiliary devices, so it is easy to modularly design.

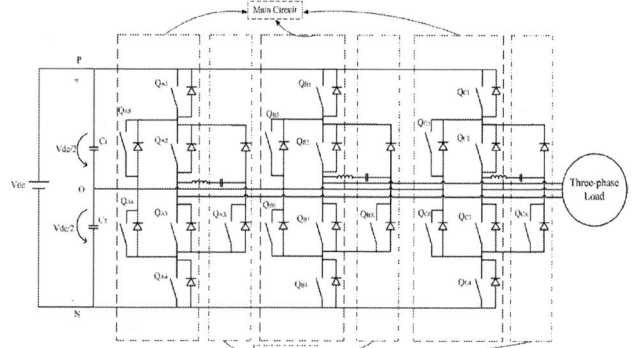

Fig.1 Proposed three phase three-level ZCT ANPC VSC

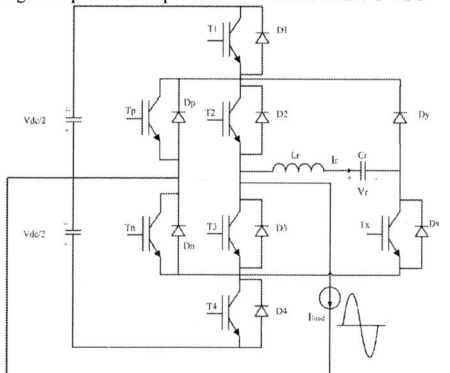

Fig.2 One phase leg circuit of the proposed topology

II. Operation Principle of Proposed ZCT ANPC VSC

A. Operation principle of ANPC VSC

A three-level ANPC VSC can yield voltage levels of Vdc/2 and –Vdc/2 in the same way like the NPC VSC. However in ANPC there are two paths to connect the output node with the neutral point no matter which direction the load current is, one is the lower path through Tn(Dn) and D3(T3), the other one is the upper path through Tp(Dp) and D2(T2)[1]. Take the level transition between Vdc/2 and 0 for example to explain the difference between the upper and lower path. Shown in Fig.3, the choice of the lower path involves a commutation between T2 and D3 when phase current is flowing out of the bridge, and a commutation between T3 and D2 when phase current is flowing into the bridge. Note that during the transition T1 and Tn are always given turning-on signals and Tp and T4 are kept off. Fig.4 shows the case of choosing the upper path. One can find that when the lower path is chosen for the Vdc/2 and 0 level transitions no matter which current direction is, the commutations only occur between the inner devices T2,T3,D2,D3, which form a two-level half-bridge cell. So the existing ZCT concept for two-level VSC can be applied to the three-level ANPC VSC.

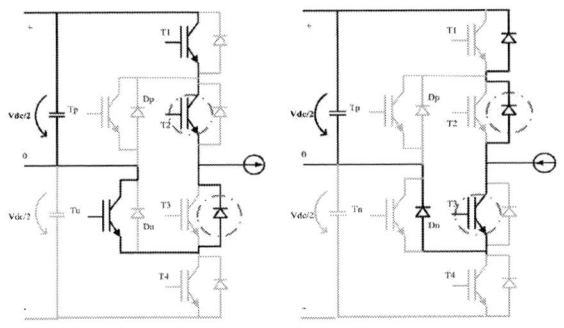

(a)Commutation between T2 and D3   (b)Commutation between T3 and D2

Fig.3 lower paths for the level transition between Vdc/2 and 0

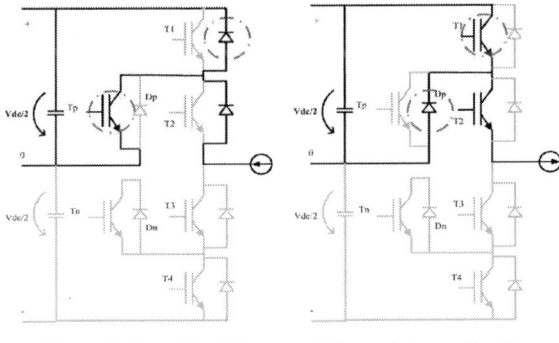

(a)Commutation between Tp and D1   (b)Commutation between T1 and Tp

Fig.4 upper paths for the level transition between Vdc/2 and 0

The complementary situation for the level transition of –Vdc/2 and 0 is that the upper path is chosen no matter which current direction is, and the T1 and Tn are kept off and T4 and Tp are kept on , so the commutations also only occur in the inner devices T2,D2,T3,D3.

B. Control timing of ZCT-3S three-level ANPC VSC

With the above described operation principle of ANPC VSC, the outer switch devices and the clamping switch devices only have one switching time in one line cycle, the switching losses stressed switch devices are only the inner switch devices.

The control timing of the auxiliary switch depends on the load current direction and the resonant cycle of LC tank Tr. Fig.5 shows the control timing of the auxiliary switch Tx for the voltage level transition between Vdc/2 and 0. When the load current is flowing out of the bridge, the commutations are between T2 and D3, the auxiliary switch Tx is turned on in advance of 3/4Tr before the turning-on and turning-off of T2 and conducts for 3/4Tr. When the load current is flowing into the bridge, the commutations are between T3 and D2, Tx is turned on in advance of 1/4Tr before the turning-on and turning-off of T3 and conducts for 3/4Tr.

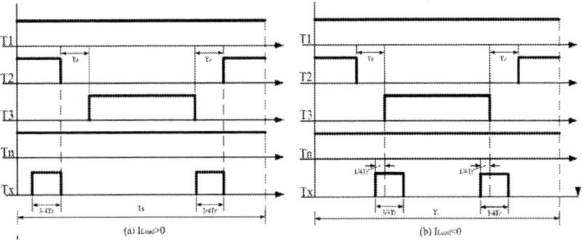

(a) ILoad>0   (b) ILoad<0

Fig.5 Control timing of switches

C. Operation Principle of ZCT-3S three-level ANPC VSC

Considering the load of VSC are usually inductive and the switching cycle is relative much smaller than the line cycle, the load current can be assumed as constant in one switching cycle in the analysis.

1. *ILoad>0 case*

Fig.7 shows the operation waveform of ZCT-3S ANPC VSC in one switching cycle for I_{Load} >0(load current flowing out the bridge). There are eight corresponding topological stages in one switching cycle, which are illustrated in Fig.6.

a) initial state: The initial state is that the output voltage level is 0 and I_{Load} flows through Tn and D3.

b) [t0~t2]:At t0 instant Tx is turned on and triggers a resonant loop including Lr, Cr, Tx and D3, then the resonant inductor current Ir begin to rise due to the existing voltage of resonant capacitor Cx. After 3/4*Tr*, at t2 instant Ir reaches its negative peak value I_{Load}, at this point the current of D3 and Tn is reduced to zero with the assumption that there is no energy loss in the resonant process. So at this instant T2 can be turned on and D3 is turned-off without reverse recovery loss meanwhile the current rising rate of T2 is limited at a resonant rate hence the turning-on losses is reduced. In actual situation Ir can not reach I_{Load} because of the

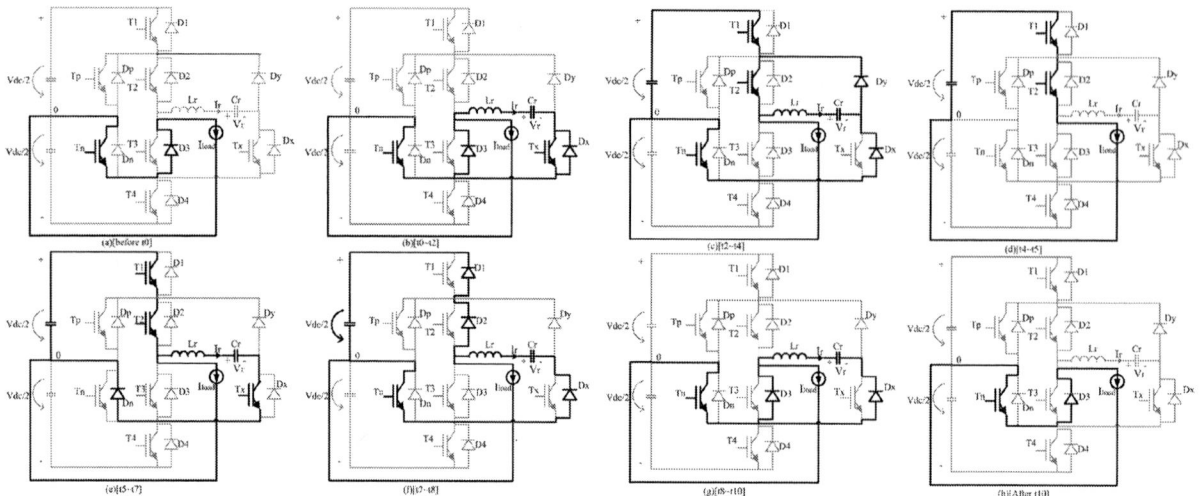

Fig.6 Topological stages of three-level ZCT-3S ANPC VSC for $I_{Load} > 0$ in one switching cycle

parasitic resistor in the resonant loop, in this case the current of D3 is not reduced to zero at t2 instant but the reverse recovery losses of D3 are greatly reduced considering the reduced turning-off current and the lower current decreasing rate. Besides the main switch losses reduction the auxiliary switch Tx is turned off at t2 instant when its freewheeling diode Dx is conducting, so there is no turning-off losses at the auxiliary switch.

c) [t2~t4]: Shown in Fig.6(c), after Tx is turned off and T2 is turned on, another resonant loop is formed which includes the upper DC capacitor, T1,T2,Lr,Cr,Dx and Tn, the resonant current Ir begin to decrease and drop to zero at t3 instant, Dx turns off naturally without reverse recovery losses and all the load current flow through T1 and T2. At this instant the voltage of resonant capacitor Vr is higher than upper DC capacitor voltage Vdc/2, so Dy is forward biased. There is another new resonant loop including Dy, T1, Lr and Cr. After half of resonant cycle at t4 instant Dy turns off naturally without reverse recovery losses.

d) [t4~t5]: The ZCT process from level 0 to level Vdc/2 ends and all the load current ILoad flows through T1 and T2, the resonant circuit does not work.

e) [t5~t7]: Shown in Fig.7 the ZCT process from level Vdc/2 to level 0 for the Iload>0 situation starts when Tx is turned on at t5 instant. Shown in Fig.6(e) there forms a resonant loop including the upper DC capacitor,T1,T2,Lr,Cr and Dn, the resonant current Ir rises and at t6 instant Ir changes its direction and flows through the freewheeling diode of Tx, at t7 instant Ir reaches the load current and the current flowing through T1 and T2 is zero.

f) [t7~t8]: Shown in Fig6(f) during t7 and t8 Ir exceeds ILoad., a part of Ir flows through D1 and D2, so T2 and

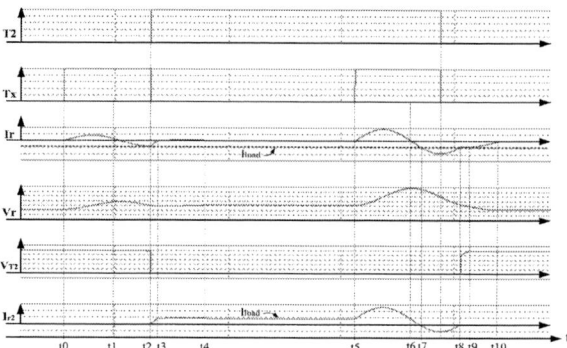

Fig.7 Operation waveforms for $I_{Load} > 0$

Tx can be turned off at zero-current and near zero-voltage condition. At t8, Ir drops to ILoad, D1 and D2 turn off naturally without reverse recovery loss.

g) [t8~t10]: Because T2 has been turned off and D3 is reverse-biased, ILoad can only flows through the resonant tank and charges the capacitor Cr linearly. At t9 instant, Vr exceeds Vdc/2 so D3 turns on, another resonant loop is formed including Dx,D3,Lr,Cr. The current of D3 begin to increase and the resonant inductor current Ir starts to decrease. At t10 instant, Ir drops to zero and Dx turns off naturally without reverse recovery losses.

h) [after t10]: The ZCT process ends, all the load current flow through Tn and D3. The auxiliary circuit does not work, the normal PWM resume.

2. ILoad<0 case

Fig.9 shows the operation waveforms of ZCT-3S ANPC VSC in one switching cycle for ILoad <0 case(flowing into the bridge). The eight corresponding topological stages in one switching cycle are illustrated in Fig.8.

a) [before t0]:Shown in Fig8(a) the initial state of ZCT process from level Vdc/2 to 0 is that the load current

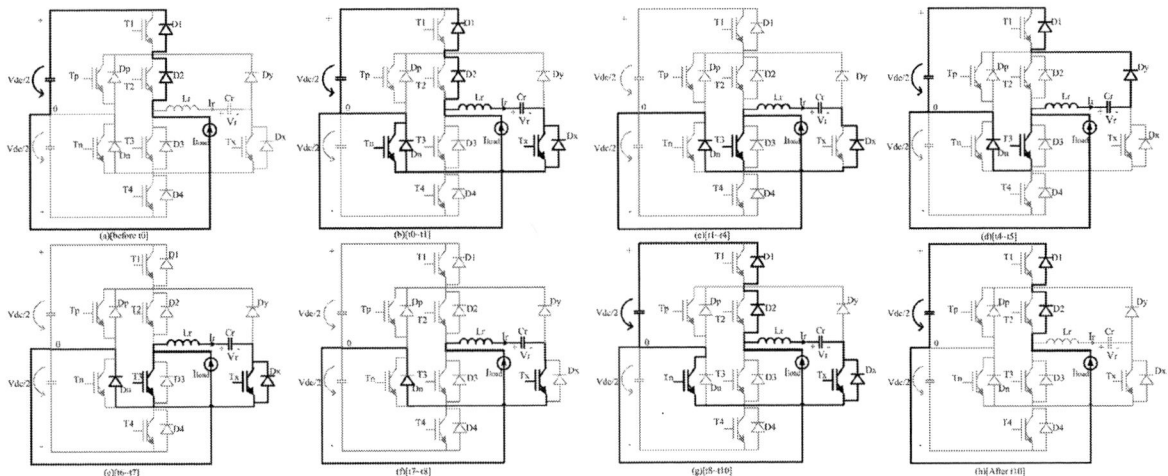

Fig.8 Topological stages of three-level ZCT-3S ANPC VSC for $I_{Load} < 0$ in one switching cycle

flows through D1 and D2.

b) [t0~t1]:At t0 instant Tx is turned on and a resonant loop is formed and shown in Fig8(b), this loop includes the upper DC capacitor, D1,D2,Lr,Cr,Tx and Dn, the resonant inductor current I_r begin to rise and reaches its positive peak value at t1 instant after 1/4Tr. If there is no losses in this resonant process this peak value can be the load current I_{Load} and the reverse recovery losses of D2 can be nearly zero with turning-on of T3 at this instant, the turning-on losses of T3 are reduced as well due to the slower current rising rate.

c) [t1~t3]:After T3 is turned on and D2 turns off, a new resonant loop is formed including Lr, Cr, Tx and T3, the load current begin to divert from the resonant tank branch to the T3, Dn branch. At t2 instant I_r drops to zero and then change direction and flow through Dx, during t2 and t3 Tx can be turned off at zero-current and nearly zero-voltage condition.

d) [t3~t4]:At t3 instant I_r drops to zero again and Dx turns off naturally without reverse recovery losses. Because the voltage of resonant capacitor is higher than the upper DC voltage, Dy and D1 are forward-biased, as a result another resonant loop including Dy,D1,the upper capacitor, Dn,T3 is formed, after half of resonant cycle at t4 instant Dy and D1 turn off naturally the resonant process ends.

e) [t4~t5]:The bridge output 0 level，all the load current flow through T3 and Dn, the normal PWM resume and the resonant circuit does not work.

f) [t5~t7]:The zero-current commutation from T3 to D2 starts at t5 instant with turning-on of Tx, I_r begin to rise and reaches its positive peak value at t6 instant. This peak value is higher than the I_{Load} so that the free-wheeling diode of T3 is conducting current which provides T3 an opportunity to turn-off at zero-current and near zero-voltage condition

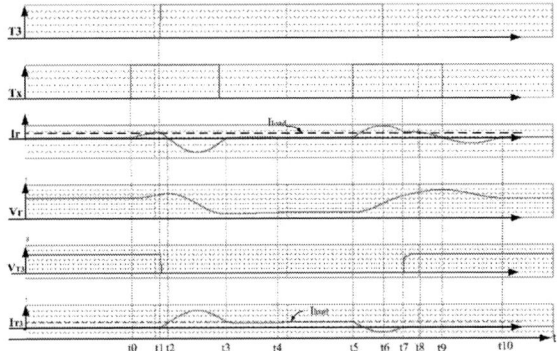

Fig.9 Operation waveforms for $I_{Load} < 0$.

g) [t7~t8]:After T3 is turned off, during t7 and t8 D1 and D2 are reverse-biased, the load current can only flow through the resonant tank branch and charge the resonant capacitor linearly.

h) [t8~t10]:At t8 instant the voltage of resonant capacitor exceeds the upper capacitor DC voltage, so the D1 and D2 are forward-biased and introduce another resonant loop including the upper capacitor ,D1,D2,Lr,Cr,Tx and Dn. The resonant inductor current I_r decreases and changes its direction at t9 instant, after that time Dx and Tn take the current instead of Tx and Dn, so Tx can be turned off without turning-off losses. At t10 instant the I_r drops to zero and Dn turns off naturally without reverse recovery losses.

i) [after t10]:The load current totally flow through D1 and D2, the resonant stop to work and the normal PWM resumes.

III. SIMULATION VERIFICATION

In order to verify the proposed three-level ZCT-3S ANPC VSC a time-domain simulation based on the software of PSIM is conducted. The main parameters in the simulation are listed as follows:1) input DC-link voltage is 800V; 2) the resonant inductor is 35.6uH, the resonant capacitor is 1.42uF; 3)switching frequency 5kHz; 4) peak phase current is 50A; 5) Modulation index is 0.7.

Fig.10 and Fig.11 show the simulation waveforms of the zero-current commutation process when the output power factor of the converter is 1. Fig.10 are waveforms in one line cycle, and Fig.11 are the waveforms in one switching cycle. One can find that during the turning-on of T2, the current of D3 reduce to zero before it bears voltage, and the current rise rate of T2 is slowed down and limited to the resonant rate. At the turning-off of T2, the current of T2 is forced to zero when the voltage stress apply to it. The voltage stress of auxiliary switch is clamped to half of DC link voltage.

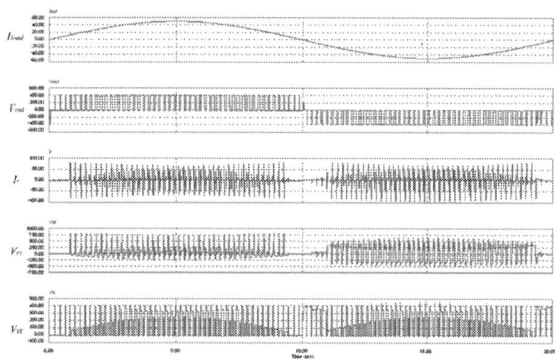

Fig.10 Simulation waveforms in one line cycle for the power factor of 1

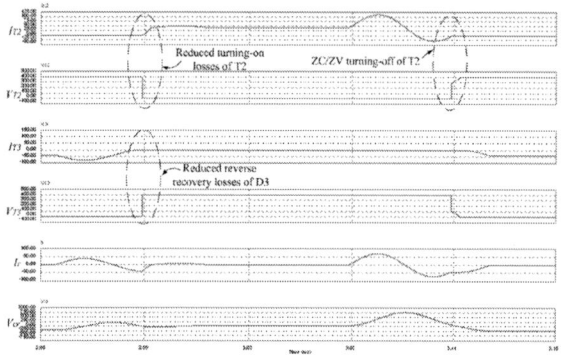

Fig.11 Simulation waveform in one switching cycle for the power factor of 1

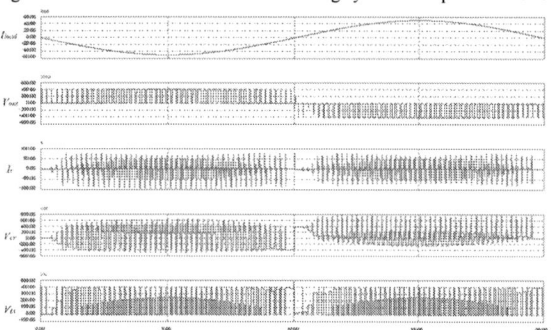

Fig.12 Simulation waveforms in one line cycle for the power factor of -1

Fig.12 and Fig.13 show the simulation waveforms of the zero-current commutation process when the power factor of the converter is -1. Fig.12 are waveforms in one line cycle, and Fig.13 are the waveforms in one switching cycle. One can also find that the zero-current transitions of T3 and D2 are realized at all the commutations.

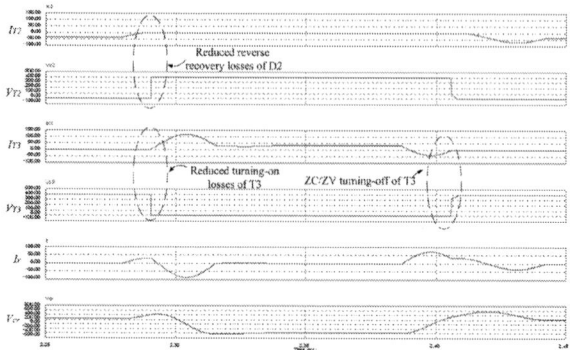

Fig.13 Simulation waveform in one switching cycle for the power factor of -1

IV. CONCLUSION

This paper proposed a novel three-level ZCT ANPC VSC topology with one auxiliary switch and one resonant tank for each phase leg. The operation principle of the proposed topology is analyzed and the simulation results verify the analysis. The proposed ZCT-3S ANPC VSC has the following features:

1) The voltage stress of all the main switch devices and auxiliary switch devices are clamped to half of the input DC voltage. The current rating of auxiliary switches are much smaller than the main devices. These switches can be easy to be implemented using modularly design.

2) The main switches and auxiliary switches are turned off at zero-current condition, the turning-on losses of main switches is reduced due to the near zero-current turning-off of diodes.

3) The control of auxiliary switches based on two set of control timing, each set of the timing is fixed for one load current direction. No need for additional current sensor, because the current information is available in the general control scheme. No need modification to the normal PWM strategy.

REFERENCES

[1] T. Bruckner, S. Bernet, and H. Guldner, "The active NPC converter and its loss-balancing control," *Ieee Transactions on Industrial Electronics,* vol. 52, pp. 855-868, Jun 2005.
[2] Y. Li and F. C. Lee, "A generalized zero-current-transition concept to simplify multilevel ZCT converters," *Ieee Transactions on Industry Applications,* vol. 42, pp. 1310-1320, Sep-Oct 2006.
[3] M. Hengchun, F. C. Lee, Z. Xunwei, and D. Boroyevich, "Improved zero-current transition converters for high power applications," in *Industry Applications Conference, 1996. Thirty-First IAS Annual Meeting, IAS '96., Conference Record of the 1996 IEEE,* 1996, pp. 1145-1152 vol.2.
[4] Y. Li, F. C. Lee, and D. Boroyevich, "A three-phase soft-transition inverter with a novel control strategy for zero-current and near

zero-voltage switching," *Ieee Transactions on Power Electronics,* vol. 16, pp. 710-723, Sep 2001.

[5] Y. P. Li, F. C. Lee, and D. Boroyevich, "A simplified three-phase zero-current-transition inverter with three auxiliary switches," *Ieee Transactions on Power Electronics,* vol. 18, pp. 802-813, May 2003.

[6] D. Fu, F. C. Lee, Y. Qiu, F. Wang, "A Novel High-Power-Density Three-Level LCC Resonant Converter with Constant-Power-Factor-Control for Charging Applications", IEEE Transaction on Power Electronics, 2008, pp. 2411-2420.

7] D. Fu, F. C. Lee, Y. Qiu, F. Wang, "A Novel High-Power-Density Three-Level LCC Resonant Converter with Constant-Power-Factor-Control for Charging Applications", IEEE Transaction on Power Electronics, 2008, pp. 2411-2420.

[8] D. Fu, Y. Liu, F.C Lee, M. Xu, "A Novel Driving Scheme for Synchronous Rectifiers in LLC Resonant Converters", IEEE Transaction on Power Electronics, 2009, pp. 1321-1329.

[9] D. Huang, D. Fu, Fred C. Lee, "High Switching Frequency, High Efficiency CLL Resonant Converter with Synchronous Rectifier", in Proc. IEEE ECCE, 2009, pp. 804-809.

[10] D. Fu, F.C. Lee, Y. Liu, M. Xu, "Novel Multi-Element Resonant Converters for Front-end DC/DC Converters" in Proc. IEEE PESC, 2008, pp. 250-256.

[11] D. Fu, P. Kong, S. Wang, F.C. Lee, M. Xu, "Analysis and Suppression of Conducted EMI Emissions for Front-end LLC Resonant DC/DC Converters", in Proc. IEEE PESC, 2008, pp. 1144-1150.

[12] D. Fu, Y. Liu, F.C Lee, M. Xu, "An Improved Novel Driving Scheme of Synchronous Rectifiers for LLC Resonant Converters", in Proc. IEEE APEC, 2008, pp. 510-516.

[13] F.C. Lee, S. Wang, P. Kong, C. Wang, D. Fu; "Power architecture design with improved system efficiency, EMI and power density", in Proc. IEEE PESC, 2008, pp. 4131-4137.

[14] D. Fu, B. Lu, F.C. Lee, "1MHz High Efficiency LLC Resonant Converters with Synchronous Rectifier," in Proc. IEEE PESC, 2007, pp. 2404-2410.

[15] D. Fu, Y. Qiu, Y. Sun, F.C. Lee, "A 700kHz High-Efficiency High-Power-Density Three-Level Parallel Resonant DC-DC Converter for High-Voltage Charging Applications," in Proc. IEEE APEC, 2007, pp. 962-968.

[16] Y. Qiu, B. Lu, B. Yang, D. Fu, F.C. Lee, "A high-frequency high-efficiency three-level LCC converter for high-voltage charging applications", in Proc. IEEE PESC, 2004, pp. 4100-4106.

Comparison and Implementation of a 3-Level NPC Voltage Link Back-to-Back Converter with SiC and Si Diodes

Mario Schweizer, Thomas Friedli, and Johann W. Kolar
Power Electronic Systems Laboratory
Swiss Federal Institute of Technology (ETH Zurich),
Physikstrasse 3,
8092 Zurich, Switzerland
Email: schweizer@lem.ee.ethz.ch

Abstract—This paper presents a high efficiency 10 kVA high-frequency input and output Si IGBT and SiC Schottky diode 3-level neutral point clamped voltage dc-link back-to-back converter (3LNPC-VLBBC). A switching frequency of 48 kHz makes the converter suitable for driving high-speed and low-inductive machines. A detailed loss analysis reveals that only four of the six diodes in a 3-level bridge-leg have to be replaced by SiC diodes to enable high efficiency operation if an appropriate modulation scheme is used. A comparison with an All-Si 3-level converter shows a reduction of the semiconductor losses by 10% at the nominal operating point. In addition, a semiconductor chip area based comparison is presented, showing the chip area partitioning of the individual semiconductor types and the corresponding costs for different implementations. The payback time for the additional costs resulting from replacing the Si diodes in the 3-level converter by SiC diodes due to energy savings is estimated. Finally, experimental results of the prototype are provided.

I. Introduction

In many current power electronic applications 3-phase voltages with high frequency and quality are required. In aircraft applications, the on-board mains grid voltage has a frequency of 400 Hz to 800 Hz, and the EMI requirements are very stringent. In high-speed drives rotational speeds of up to 60'000 rpm are common, which demand typically electrical inverter output frequencies in the range of 1 kHz. These machines are mostly low-inductive and consequently the current and torque ripple would be unacceptably high if standard (2-level) inverters with e.g. 8 kHz switching frequency were used [1]. In order to generate these fundamental frequencies and to minimize the current ripple, a high switching frequency above 25 kHz is necessary.

Unfortunately, the semiconductor losses increase with increasing switching frequencies, which reduces the overall drive efficiency and asks for a bulkier cooling system. Teichmann showed in [2] that a 3-level NPC converter, built with 600 V TrenchGate IGBTs, has lower losses than a 2-level converter built with 1200 V TrenchGate IGBTs if the switching frequency is high enough ($f_s > 10\,\text{kHz}$). This is due to the reduced switching and conduction losses of the low-voltage de-

vices which overcompensates the increased conduction losses caused by the higher number of series connected devices in the current path. Additionally, the 3-level converter generates a better voltage and current spectrum compared to the 2-level converter. This has the ability to reduce the additional PWM losses in electrical motors [3].

The silicon carbide (SiC) semiconductor technology allows utilizing SiC Schottky barrier diodes with virtually no reverse recovery effect. The implementation of SiC diodes has a positive effect on the IGBT turn-on losses, since a large fraction of typically 35% is actually caused by the reverse recovery charge of the commutating diode [4].

A detailed loss analysis in section II shows that replacing the appropriate Si diodes with SiC Schottky diodes enables an increase of the 3-level NPC converter efficiency. Since SiC Schottky diodes are expensive compared to the conventional Si diodes their use is not always justified. In section IV the required semiconductor area is investigated and the payback time for the additional costs is calculated. In section V the 10 kVA prototype for a 3LNPC-VLBBC is presented. The power circuit is built with six custom 3-level bridge-leg modules. This prototype allows for experimentally investigating the impact of utilizing SiC diodes in a 3-level converter topology (rectifier and inverter) by replacing the standard all-Si 3-level phase leg modules with pin compatible custom modules with SiC diodes.

II. Comparison of 3-level Converter Losses with Si and SiC Schottky Diodes

The calculation of the device losses in a 3-level inverter stage requires some more effort than for the well-known 2-level inverter. In previous publications the losses have been calculated approximatively for sinusoidal, carrier based PWM [5] or directly determined from time-domain simulation. In order to identify accurately the influence of different module configurations with Si and SiC diodes an appropriate method for the loss calculation with space vector modulation has been utilized. This method inherently allows for the consideration

---

978-1-4244-4782-4/10 $26.00 © 2010 IEEE

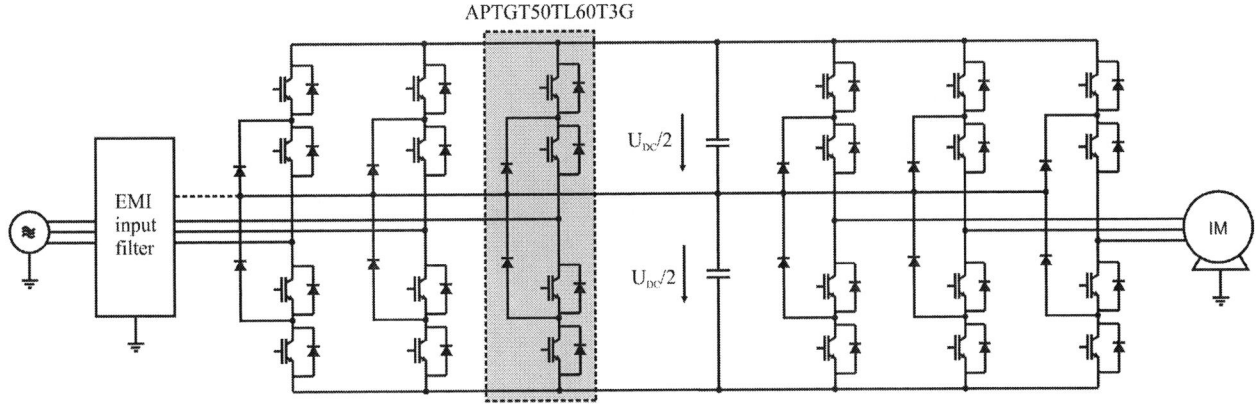

APTGT50TL60T3G

Fig. 1. Topology of the 3LNPC-VLBBC.

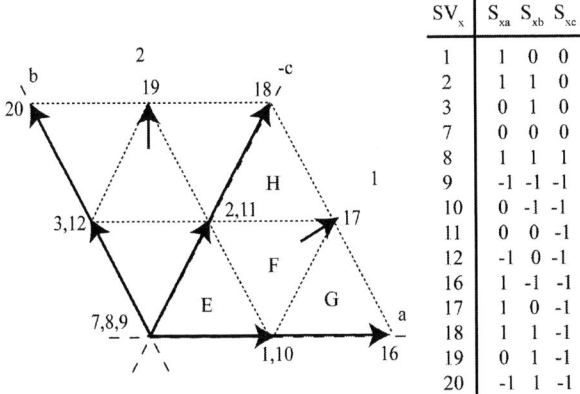

Fig. 2. Voltage vectors in the 3-level NPC topology.

SV_x	S_{xa}	S_{xb}	S_{xc}
1	1	0	0
2	1	1	0
3	0	1	0
7	0	0	0
8	1	1	1
9	-1	-1	-1
10	0	-1	-1
11	0	0	-1
12	-1	0	-1
16	1	-1	-1
17	1	0	-1
18	1	1	-1
19	0	1	-1
20	-1	1	-1

of phase clamping strategies as it incorporates the modulation strategy in the calculation.

Several modulation strategies exist for the 3-level NPC topology. For the converter presented in this paper a space vector modulation scheme, as described in [6], is implemented. The output voltage vector is always formed with the three nearest discrete voltage space vectors. Since the 3-level topology offers redundant space vectors (three equivalent zero vectors and always two equivalent vectors on the inner hexagon, cf. Fig. 2), it is possible to implement an optimal clamping strategy in order to reduce the switching losses. This is done by selecting an appropriate redundant vector and an appropriate vector sequence. The selection of the redundant vector plays also an important role in the balancing of the dc-link capacitor voltages which is discussed in chapter III.

A. Direct loss calculation with space vector modulation

In order to determine the losses, the space vector modulation scheme is repeated analytically for the rectifier and the inverter stage. Each voltage space vector defines if an output phase (a,b,c) is connected to the positive (1), neutral point (0) or the negative dc-link rail (-1). Together with the sign of the actual phase current and the vector sequence it is well-defined in which elements of the 3-level bridge leg the losses occur.

The conduction losses are approximated linearly for each device (IGBT, Si- or SiC-diode, c.f. Fig 3) depending on the device characteristics. It is possible to define a matrix giving the conduction losses of all devices in the rectifier (or inverter) stage depending on the current vector $\vec{I} = [i_a \ i_b \ i_c]$ and the actually applied discrete switching state space vector $\overrightarrow{SV} = [S_a \ S_b \ S_c]$.

$$P_{cond}(\overrightarrow{SV}, \vec{I}) = \begin{pmatrix} P_{c,T1a} & P_{c,D1a} & \cdots & P_{c,D5a} & P_{c,D6a} \\ P_{c,T1b} & P_{c,D1b} & \cdots & P_{c,D5b} & P_{c,D6b} \\ P_{c,T1c} & P_{c,D1c} & \cdots & P_{c,D5c} & P_{c,D6c} \end{pmatrix}$$
(1)

Each element of the matrix is defined as a piecewise function depending on the switching state and the output current sign. The first two elements can be written as:

$$P_{c,T1a} = \begin{cases} V_{f,T} \cdot i_a + r_{on,T} \cdot i_a^2 & i_a \geq 0 \ \& \ S_a = 1 \\ 0 & \text{otherwise} \end{cases}$$
(2)

$$P_{c,D1a} = \begin{cases} V_{f,D} \cdot (-i_a) + r_{on,D} \cdot (-i_a)^2 & i_a < 0 \ \& \ S_a = 1 \\ 0 & \text{otherwise} \end{cases}$$
(3)

With the modulation strategy, which defines the relative on-times of the discrete voltage space vectors, the averaged conduction losses over one switching period can be calculated. As an example, in sector 1E of the space vector diagram (cf. Fig. 2) the output voltage can be formed with $\overrightarrow{SV_1}$, $\overrightarrow{SV_2}$ and $\overrightarrow{SV_8}$ with the corresponding on-times d_1, d_2 and d_8.

$$P_{cond,avg}(\vec{U}, \vec{I}) = d_1 \cdot P_{cond}(\overrightarrow{SV_1}, \vec{I}) + d_2 \cdot P_{cond}(\overrightarrow{SV_2}, \vec{I}) \\ + d_8 \cdot P_{cond}(\overrightarrow{SV_8}, \vec{I})$$
(4)

It should be noted that these relative on-times are functions of the output voltage vector $\vec{U}$ and are calculated differently for each subsector E, F, G and H in Fig 2. They need to be described also as piecewise functions.

For the calculation of the switching losses, a similar approach is possible. The switching loss energy, occuring in each

978-1-4244-4782-4/10 $26.00 © 2010 IEEE

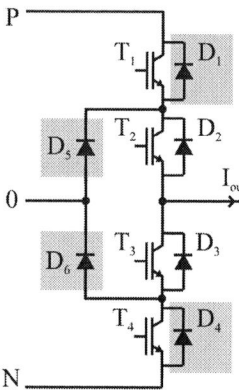

Fig. 3. 3-level bridge leg showing the placement of the SiC Schottky diodes (colored gray).

TABLE I
SWITCHING LOSS ENERGIES

Switching Transition	Generated switching loss energies
$I_{out} \geq 0$	
$1 \rightarrow -1$	$E_{T1off}, E_{T2off}, E_{D3on}, E_{D4on}$
$1 \rightarrow 0$	E_{T1off}, E_{D5on}
$-1 \rightarrow 1$	$E_{T1on}, E_{T2on}, E_{D3off}, E_{D4off}$
$-1 \rightarrow 0$	$E_{T2on}, E_{D4off}, E_{D5on}$
$0 \rightarrow 1$	E_{T1on}, E_{D5off}
$0 \rightarrow -1$	$E_{T2off}, E_{D3on}, E_{D4on}$
$I_{out} < 0$	
$1 \rightarrow -1$	$E_{D1off}, E_{D2off}, E_{T3on}, E_{T4on}$
$1 \rightarrow 0$	E_{D1off}, E_{T3on}
$-1 \rightarrow 1$	$E_{D1on}, E_{D2on}, E_{T3off}, E_{T4off}$
$-1 \rightarrow 0$	E_{T4off}, E_{D6on}
$0 \rightarrow 1$	E_{D1on}, E_{T3off}
$0 \rightarrow -1$	E_{T4on}, E_{D6off}

device in the 3-level bridge leg, is defined by the phase current and the voltage space vector transition. As an approximation, the switching energy is scaled linearly with the device current according to (5). This is not very accurate but is suitable for the comparison of two module configurations.

$$E_s(i) = E_N \cdot \frac{U_{DC}/2}{U_N} \cdot \frac{i}{I_N} \qquad (5)$$

At first sight it is not clear in which devices of a 3-level bridge leg switching losses occur. With measurements on a test setup it was determined for each switch transition which devices actually exhibit losses. Table I summarizes the switching energies occurring in one inverter leg depending on the switching transition. It can be noticed that only for direct transitions between positive (1) and negative (-1) dc-link voltage rail (or from -1 to 1) turn-off losses in the inner diodes D_2 and D_3 occur.

Now it is possible to define a switching energy matrix giving the actual losses of all inverter devices depending on the output current space vector and the transition from the old voltage

space vector $\overrightarrow{SV_o}$ to the new vector $\overrightarrow{SV_n}$:

$$E_{sw}(\overrightarrow{SV_o}, \overrightarrow{SV_n}, \overrightarrow{I}) = \\ \begin{pmatrix} E_{s,T1a} & E_{s,D1a} & \cdots & E_{s,D5a} & E_{s,D6a} \\ E_{s,T1b} & E_{s,D1b} & \cdots & E_{s,D5b} & E_{s,D6b} \\ E_{s,T1c} & E_{s,D1c} & \cdots & E_{s,D5c} & E_{s,D6c} \end{pmatrix} \quad (6)$$

Again each element of this matrix is a piecewise function returning the switching loss energies depending on the switching transition and the output current signs. The first two elements of the matrix are:

$$E_{s,T1a} = \begin{cases} E_{T1off}(i_a) & S_{a,o} = 1 \ \& \ S_{a,n} = -1 \ \& \ i_a \geq 0 \\ E_{T1off}(i_a) & S_{a,o} = 1 \ \& \ S_{a,n} = 0 \ \& \ i_a \geq 0 \\ E_{T1on}(i_a) & S_{a,o} = -1 \ \& \ S_{a,n} = 1 \ \& \ i_a \geq 0 \\ E_{T1on}(i_a) & S_{a,o} = 0 \ \& \ S_{a,n} = 1 \ \& \ i_a \geq 0 \\ 0 & \text{otherwise} \end{cases}$$
$$(7)$$

$$E_{s,D1a} = \begin{cases} E_{D1off}(\text{-}i_a) & S_{a,o} = 1 \ \& \ S_{a,n} = -1 \ \& \ i_a < 0 \\ E_{D1off}(\text{-}i_a) & S_{a,o} = 1 \ \& \ S_{a,n} = 0 \ \& \ i_a < 0 \\ 0 & \text{otherwise} \end{cases}$$
$$(8)$$

The turn-on energy of the diodes is very small and has been neglected. The space vector sequence defined by the modulation scheme allows for calculating the average switching loss over one switching period. For a symmetric space vector sequence 1-2-8-2-1 in sector 1E the average switching losses for one switching period are:

$$P_{sw,avg}(\overrightarrow{U}, \overrightarrow{I}) = \frac{1}{T_s} \cdot (E_{sw}(\overrightarrow{SV_1}, \overrightarrow{SV_2}, \overrightarrow{I}) + \\ E_{sw}(\overrightarrow{SV_2}, \overrightarrow{SV_8}, \overrightarrow{I}) + E_{sw}(\overrightarrow{SV_8}, \overrightarrow{SV_2}, \overrightarrow{I}) + \\ E_{sw}(\overrightarrow{SV_2}, \overrightarrow{SV_1}, \overrightarrow{I})) $$
$$(9)$$

Finally the average conduction and switching loss matrices contain the losses for each device in the 3-level inverter depending on the output voltage, the output current and the modulation strategy. Now the modulation in terms of the vector sequence and the relative on-times depending on the output voltage vector $\overrightarrow{U}$ has to be defined. It is sufficient to do that for the first electrical 120° because of the inherent 3-phase symmetry. With some effort this is possible with a modern mathematics software like Mathematica.

The resulting loss curves of one inverter stage bridge-leg for a switching frequency of 48 kHz are depicted in Fig. 4. An optimal clamping strategy is used so that the output phase with the highest instantaneous current value is not switched during an electric output angle of 60°. However for voltage link balancing the rectifier stage cannot do an optimal clamping scheme so that its losses will be increased slightly.

As a result of the above calculation the conduction and the switching losses averaged over a switching period are available. The mean losses over a fundamental period in each device can be obtained by integrating the corresponding

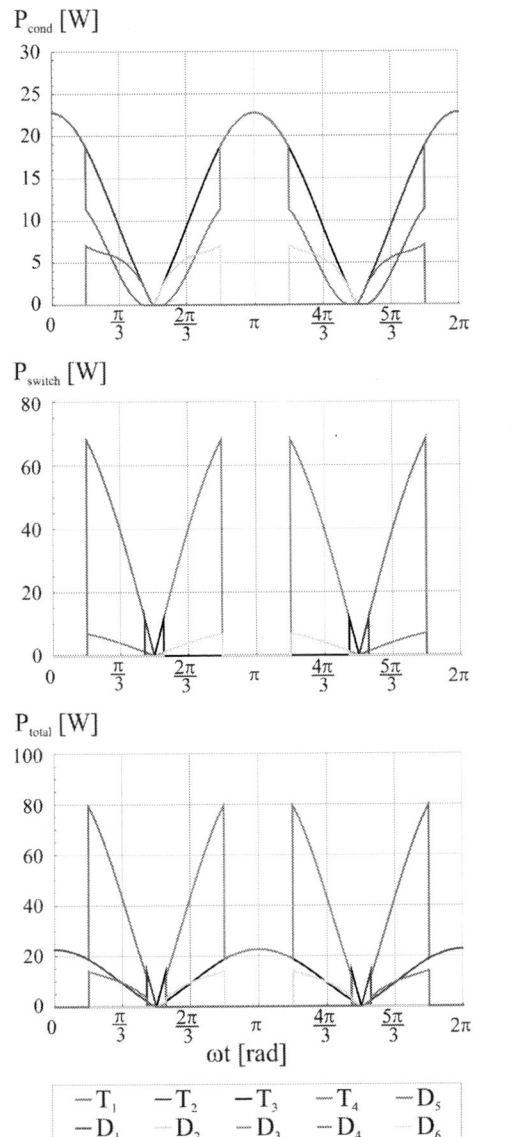

Fig. 4. 3-level inverter stage a) conduction, b) switching and c) total losses over electric output angle ($\hat{U}_2 = 325$ V, $\hat{I}_2 = 20.5$ A, $\varphi_2 = 0\,°$).

expressions over the full electric output angle of $360°$.

$$\overline{P_{cond}} = \frac{1}{2\pi} \cdot \int_0^{2\pi} P_{cond,avg} \cdot d\alpha \qquad (10)$$

$$\overline{P_{sw}} = \frac{1}{2\pi} \cdot \int_0^{2\pi} P_{sw,avg} \cdot d\alpha \qquad (11)$$

The mean losses of the devices in the rectifier and in the inverter stage are depicted in Fig. 5a and Fig. 5c. It can be observed that the inner diodes D_2 and D_3 exhibit no switching losses. This is due the implemented modulation strategy which clamps the output phase with the highest current symmetrically around $-30°\ldots + 30°$ (for a phase displacement of $0°$ between voltage and current space vector which is true for the rectifier stage and the inverter stage for low-inductive loads). As an example the vector sequences for

the subsector E are given here.

(100) - (110) - (111) - (110) - (100) for $0 \leq \alpha < 30°$
(00-1) - (0-1-1) - (-1-1-1) - (0-1-1) - (00-1) for $30° \leq \alpha < 60°$

If one has a closer look at the single switch commutations it can be recognized that no direct commutations from the positive to the negative dc-link rail or vice versa occur (1 to -1 or -1 to 1). There exist similar sequences for all subsectors E, F, G and H and for all other main sectors ($\alpha > 60°$) which exhibit only commutations from or to the neutral (0) rail [6].

As mentioned above, if for all commutations a step over the neutral dc-link rail occurs, the diodes D_2 and D_3 do not have any switching losses. Depending on the operation mode (inverter or rectifier / motor or generator operation) the semiconductors are loaded differently. For rectifier operation the diodes D_1-D_4 (cf. Fig. 5) are mainly conducting. With the increased forward voltage drop of SiC diodes compared to Si diodes, the rectifier efficiency therefore would be reduced if all Si diodes would be replaced by SiC devices. With this facts in mind, for increasing the converter efficiency a custom SiC module for the 3-level topology replacing the diodes D_1, D_4, D_5 and D_6 with SiC counterparts has been tested and employed in the 3LNPC-VLBBC. The inner diodes D_2 and D_3 remain in conventional Si technology. SiC Schottky diodes with equivalent current rating (30 A / 600 V, 3x10A SiC dies in parallel per diode) have been used for the custom module.

B. Converter efficiency with Si- and SiC module configuration

The results of the comparison between the custom SiC module with a standard Si module are depicted in Fig. 5 representatively for one rectifier and inverter bridge-leg. The calculation is based on semiconductor datasheet values for the conduction losses and on switching loss energies determined with a test setup for both the standard and the custom SiC module. The losses are averaged over one electrical period for nominal operation. It can be noticed that the diodes D_1, D_4, D_5 and D_6 exhibit nearly zero switching losses in the SiC configuration. The IGBT switching losses are also reduced. This is due to the smaller turn-on energy if the commutating diode has no reverse recovery effect as is the case with SiC devices. On the contrary, the conduction losses of the SiC diodes are slightly increased comparing to the Si version. In the inverter stage mainly the IGBTs exhibit conduction losses so that the increased conduction losses of the SiC diodes only occur in the clamping diodes D_5 and D_6.

If we compare the total converter efficiency curves (cf. Fig. 6, pure semiconductor efficiency) a considerable increase in efficiency with the custom SiC module can be achieved. The intersection point at which the common Si module exhibits smaller losses (because of the lower conduction losses) than its SiC counterpart is below 5 kHz. At the nominal switching frequency of 48 kHz and at nominal output power of 10 kVA the losses could be reduced by 10%. The converter built with the SiC module reaches a pure semiconductor efficiency of 97.0% compared to 96.6% of the standard Si version.

■ Conduction losses   ■ Switching losses (48 kHz)

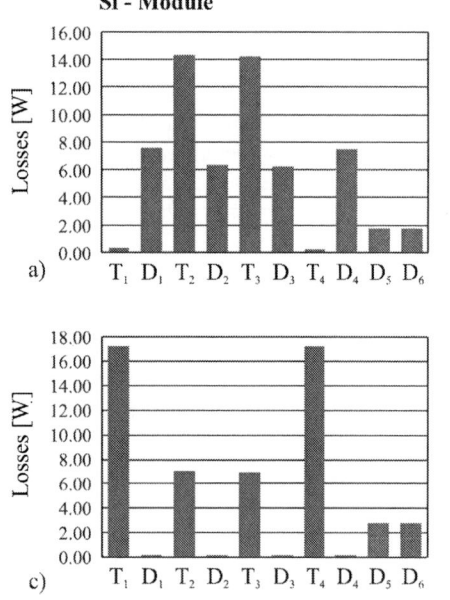

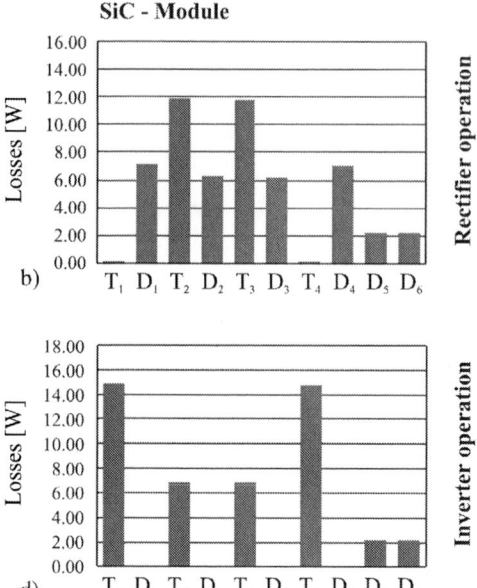

Fig. 5.   Comparison of the total losses in a bridge-leg for the rectifier and inverter stage (rectifier operating point: $\hat{U}_1 = 325$ V, $\hat{I}_1 = 21.3$ A, $\varphi_1 = 0°$ inverter operating point: $\hat{U}_2 = 325$ V, $\hat{I}_2 = 20.5$ A, $\varphi_2 = 0°$).

Compared to an equivalent 2-level converter with 1200 V IGBTs of the same generation, the reduction of the losses is 42%.

Interesting is also the very flat dependency of the efficiency on the switching frequency. This opens an additional degree of freedom for optimizing the converter. A very compact system with small filtering components because of the higher switching frequency and a low volume cooling system due to the small losses could be realized.

It should be noted that in the real implementation additional loss sources as forced-air cooling fans (15W), digital control and gate drive power (15W) and also losses in the boost inductors (30W) are present. These will sum up to additional 60W and reduce the efficiency of the SiC 3LNPC-VLBBC to approximately 96.4%.

III. CLAMPING AND DC-LINK BALANCING

In the previous section the loss calculation has been carried out neglecting the dc-link balancing. If the described clamping scheme is applied the dc-link capacitors are loaded differently depending in which sector and subsector of the space vector diagram the reference voltage vector and the current vector are located [7]. If the displacement angle between voltage and current vector is small (as is the case for the rectifier stage or the inverter stage with low inductive loads) a symmetric clamping around ±30° results in minimum switching losses.

If the dc-link capacitors are big enough as is the case with electrolytic capacitors this is no problem for the converter because the capacitor loading balances out over the fundamental period. However, since electrolytic capacitors are prone to early aging and have been the source for reliability issues in

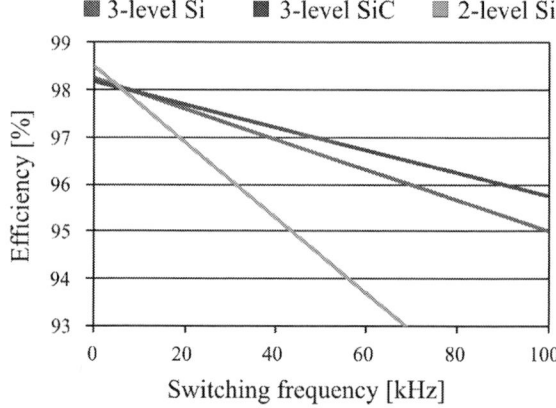

Fig. 6.   Comparison of total converter efficiency.

the past, they have been omitted in the design of the 3LNPC-VLBBC prototype. Instead a reduced dc-link capacitance has been realized with foil capacitors. If the converter is designed with a low dc-link capacitance, the balancing of the capacitors is more difficult and the modulation should be coupled between rectifier and inverter stage.

The dc-link balancing can be achieved with the rectifier stage changing the vector sequence depending on the capacitor unbalance. In sector E the vector sequence (100) - (110) - (111) - (110) - (100) solely loads the high-side capacitor while the alternative sequence (00-1) - (0-1-1) - (-1-1-1) - (0-1-1) - (00-1) solely loads the low-side capacitor. As for the nominal operating voltage of 230 V_{rms} at the rectifier side subsector E is never crossed, the reference voltage vector is located always

978-1-4244-4782-4/10 $26.00 © 2010 IEEE   1531

in the subsectors F, G and H. Using an appropriate sequence in these sectors also allows for balancing the dc-link. However, in these subsectors always both capacitors are loaded, but one slightly more than the other depending on the used sequence. The clamping of the phase with the highest current amplitude is not anymore possible, and thus the switching losses of the rectifier stage are increased.

It has to be noted that the mean capacitor loading over one switching period is solely due to active power, reactive power at the output power will not additionally unbalance the capacitors. It can be seen that the worst combination for the dc-link balancing solely with the rectifier stage is a small inverter stage voltage vector located in the inner hexagon (subsector E) and optimal clamping. This leads to an alternating loading of the upper and lower capacitor every electric 60° of the output period. As a solution the inverter stage should adopt in subsector E a changing to an alternative sequence (0-1-1)-(00-1)-(000)-(00-1)-(0-1-1). In this case optimal clamping of the phase with the highest current amplitude can still be maintained.

If dc-link balancing is done with different sequences, special care should be given on the sequence change, because this can happen frequently depending on the maximum allowed link unbalance. An intermediate sequence has to be introduced to prevent an increase of the net switching frequency due to the sequence change. Also a switching directly from positive to negative dc-link rail or in the opposite direction should be avoided during the sequence change so that the benefits of the special SiC module configuration are utilized.

IV. SEMICONDUCTOR AREA USAGE AND COST COMPARISON

Power semiconductors in SiC technology are more expensive than devices in conventional Si technology. Therfore, it is important to know if the excess expenditure will pay back with the saved energy in a reasonable time period. First, one has to know the specific partition of Si to SiC semiconductor area. In Fig. 7 the semiconductor partitioning of the conventional Si module and the custom SiC module is depicted. One can see a small increase of 18.8% in total semiconductor area for the custom module.

Considering the costs for the two modules, the custom SiC module is roughly 6.5 times more expensive than the conventional module. The converter costs accordingly will increase by roughly 300$. With a loss reduction of 10% = 32 W the energy payback time with an energy price of 0.1$/kWh [8] will be 93'700 hours, what is equivalent to 10.5 years of uninterrupted operation at nominal power. At a switching frequency of 120 kHz the payback time reduces to 3.2 years of uninterrupted operation. This is quite a long operating time and restricts the application of the SiC 3-level converter to niche areas where exceptional high switching frequencies and efficiencies is required. This is the case e.g. in aircraft applications where weight and efficiency are of major interest.

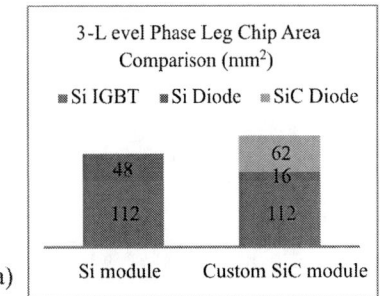

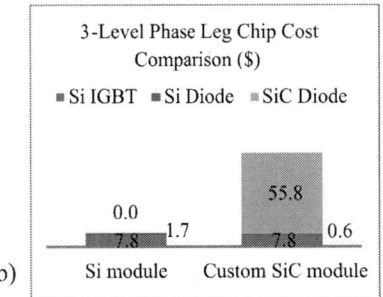

Fig. 7. Semiconductor a) area and b) cost comparison.

P_N	10 kVA
U_N	230 V_{rms}
f_N	50 Hz / 800 Hz
f_s	48 kHz
U_{DC}	2 x 66 μF in series
L_{Boost}	3 x 300 μH
η_N	96.5%
EMI filter	CISPR Class A

TABLE II
KEY PERFORMANCE DATA OF THE 3-LEVEL CONVERTER

V. 10 KVA/ 48 KHZ 3-LEVEL SIC CONVERTER PROTOTYPE

The 3LNPC-VLBBC is designed for a modular assembly (cf. Fig. 8). A gate-drive board populated with 26 isolated gate drive circuits and 5 current sensors for the input and output current is directly mounted on the 3-level modules. In order to investigate the impact of the proposed SiC configuration the custom 3-level bridge-leg module is pin compatible to an existing module from Microsemi (APTGT50TL60T3G).

A digital signal processing board with a TI DSP and a Lattice FPGA as well as filtering and amplification circuitry is mounted on top of the gate-drive board.

Three boost inductors with an inductance value of 300 μH have been employed. They limit the unfiltered current ripple at the rectifier stage to 4 A_{pp}. A minimum dc-link capacitance of 33 μF has been calculated, limiting the voltage overshoot in case of a sudden load drop from nominal power to zero to an acceptable value of 10%. Foil capacitors have been used for a reliable solution without electrolytic capacitors.

The topmost board is a compact EMI filter board which allows for an easy exchange so that both 50/60 Hz grid and 400 Hz aircraft on board grid frequency can be tested. A forced air cooling system and an optimized heatsink have

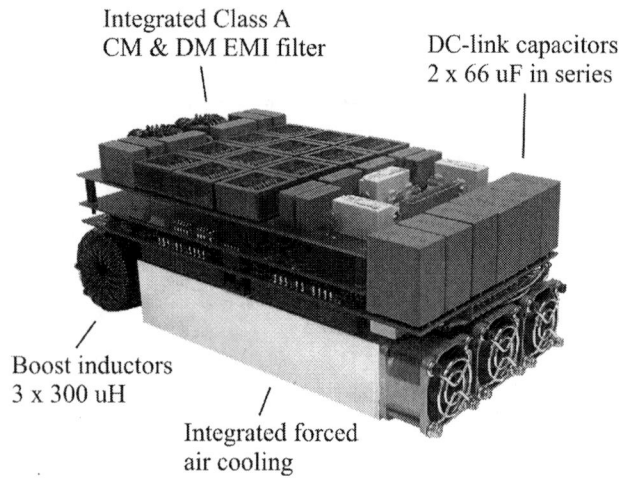

Fig. 8. Compact 10 kVA, 3LNPC-VLBBC prototype with 48 kHz switching frequency.

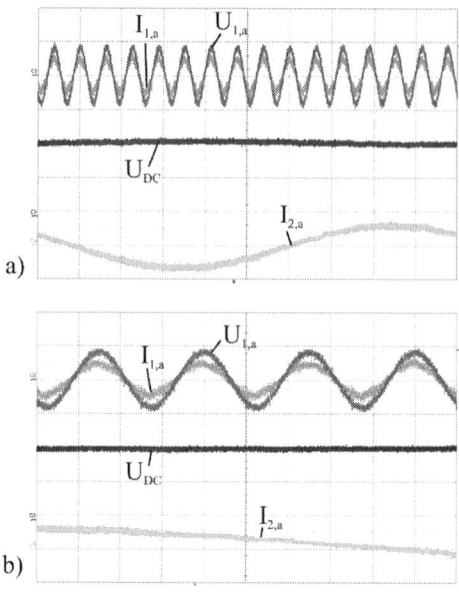

Fig. 9. Initial 3-level converter measurements operating at $f_1 = 800$ Hz, $U_1 = 115$ V$_{\rm rms}$, $U_{DC} = 400$ V, $f_2 = 50$ Hz, $P = 1.5$ kW ($U_1 = 200$ V/div, $I_1 = 10$ A/div, $I_2 = 10$ A/div, $U_{dc} = 200$ V/div, timescale a) 2 ms/div b) 500 μs/div).

been dimensioned for a maximum junction temperature of $T_j = 125\,^\circ$C. The power density of the total converter including cooling system and EMI filter is $2.9\,{\rm kW/dm^3}$. The key properties of the converter are summarized in Tab. II. The target system efficiency is 96.5% including the losses in the EMI input filter, the losses caused by the fans and the digital signal processing board.

Initial measurements at an input voltage of $115\,{\rm V_{rms}}$ and a dc-link voltage of 400 V show the basic functionality. To demonstrate the system performance measurements at 800 Hz fundamental input frequency and 50 Hz fundamental output frequency are provided in Fig. 9. A sinusoidal input current with unity power factor has been achieved. The dc-link voltage is regulated to a stable value.

VI. CONCLUSION

In this paper a 3-level topology with the anti-parallel diodes partially replaced SiC Schottky diodes is presented. This approach allows for reducing the converter losses for switching frequencies above 5 kHz and thus makes the converter suitable for driving high-speed and low-inductive machines. A loss analysis revealed that only four of the six diodes in a 3-level bridge-leg have to be replaced by SiC diodes to enable high efficiency operation. The losses have been reduced by 10% at nominal operating point.

It has been shown that with an appropriate dc-link balancing method the converter is able to operate also with a reduced link capacitance, avoiding the use of electrolytic capacitors which are sensitive to aging. A simple cost comparison showed that the additional costs for the SiC diodes are considerable.

Initial measurements at the prototype have been performed, showing the operation with unity power factor at a high fundamental input frequency of 800 Hz. The currents are sinusoidal and the dc-link voltage is stable.

REFERENCES

[1] K. Jalili, D. Krug, S. Bernet, M. Malinowski, and B. J. C. Filho, "Design and characteristics of a rotor flux controlled high speed induction motor drive applying two-level and three-level npc voltage source converters," in *Proc. of 36th IEEE Power Electronics Specialists Conference PESC '05*, pp. 1820–1826, 16 June 2005.

[2] R. Teichmann and S. Bernet, "A comparison of three-level converters versus two-level converters for low-voltage drives, traction, and utility applications," *IEEE Trans. Ind. Appl.*, vol. 41, pp. 855–865, May–June 2005.

[3] A. Boglietti, A. Cavagnino, A. M. Knight, and Y. Zhan, "Factors affecting losses in induction motors with non-sinusoidal supply," in *Proc. of 42nd IEEE IAS Annual Meeting*, pp. 1193–1199, 23–27 Sept. 2007.

[4] A. Elasser, M. H. Kheraluwala, M. Ghezzo, R. L. Steigerwald, N. A. Evers, J. Kretchmer, and T. P. Chow, "A comparative evaluation of new silicon carbide diodes and state-of-the-art silicon diodes for power electronic applications," *IEEE Trans. Ind. Appl.*, vol. 39, pp. 915–921, July–Aug. 2003.

[5] D. Krug, S. Bernet, and S. Dieckerhoff, "Comparison of state-of-the-art voltage source converter topologies for medium voltage applications," in *Proc. of 38th IEEE IAS Annual Meeting*, vol. 1, pp. 168–175, 12–16 Oct. 2003.

[6] B. Kaku, I. Miyashita, and S. Sone, "Switching loss minimised space vector pwm method for igbt three-level inverter," *IEE Proceedings - Electric Power Applications*, vol. 144, pp. 182–190, May 1997.

[7] N. Celanovic and D. Boroyevich, "A comprehensive study of neutral-point voltage balancing problem in three-level neutral-point-clamped voltage source pwm inverters," *IEEE Trans. Power Electron.*, vol. 15, pp. 242–249, March 2000.

[8] "Average Retail Price of Electricity to Ultimate Customers." http://www.eia.doe.gov/, Nov. 13, 2009.

978-1-4244-4782-4/10 $26.00 © 2010 IEEE

A Novel PWM Control Method to Eliminate the Effect of Dead Time on the Output Waveform for Hybrid Clamped Multilevel Inverters

Jing Zhao, Xiangning He, Yunlong Han, Yan Chen, Rongxiang Zhao
College of Electrical Engineering, Zhejiang University
Hangzhou, 310027, China
Email: zhdzhao@zju.edu.cn

Abstract—A novel PWM control method called AHPWM (Alternative Hybrid PWM) is proposed in this paper. By applying the AHPWM method to the hybrid clamped multilevel inverter, the unexpected levels in the output voltage waveform with the traditional PDPWM method can be eliminated. The analyses on the switching modes conversions, the reason causing the unexpected output levels, the effect of the PWM carrier waveforms on the switching modes conversions and the principle of the AHPWM method are given. Simulations and experiments show the unexpected output levels caused by the dead time can be effectively eliminated by applying the AHPWM method to the hybrid clamped multilevel inverter within broad modulation index range. The analysis on the output voltage spectrum shows the energy of lower harmonics can be remarkably reduced, which can simplify the design of the output filter and reduce its size.

I. INTRODUCTION

Multilevel converters have been a research hotspot in high-voltage and high-power applications in recent years because of low output THD, low voltage stress and low system EMI [1]. Generally speaking, diode-clamped [3, 4], flying-capacitor [5, 6] and cascaded multilevel [7] converters are the three kinds of basic multilevel topologies [1, 2].

Inserting the dead time between the complementary switching devices is necessary but causes so-called "dead-time effect", which means output waveform distortion, the increase of the output harmonics content and degradation of static or dynamic performance. Many literatures on how to eliminate or restrain the dead-time effect focus on the traditional two-level converters [10-12], and some papers discuss the dead-time effect of multilevel converters, especially three-level converters [13, 14]. Only a few papers involve the dead time for five-level or more levels inverters.

A hybrid clamped multilevel inverter topology with self-voltage balancing is presented in [8]. Figure 1 shows one leg of the five-level topology. The main switching devices Sa1, Sa2, Sa3 and Sa4 are complementary with Sa1', Sa4', Sa3' and Sa2' respectively, and Sa1 is complementary with Sac1.

This work is supported by the National Nature Science Foundation of China (50737002) and Zhejiang Province Science and Technology Funding of China (2006C11005)

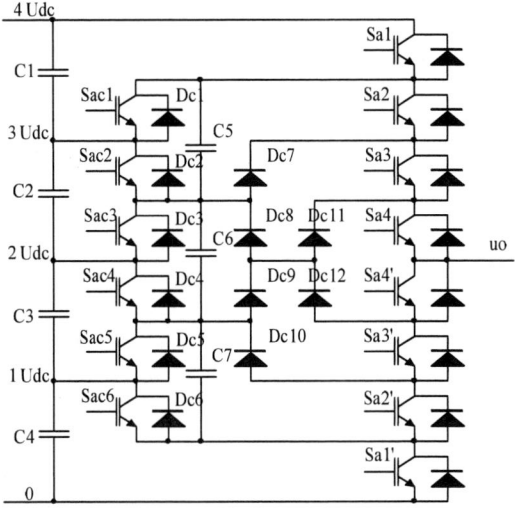

Figure 1. One leg of the hybrid clamped five-level inverter topology

TABLE I. TWO KINDS OF SWITCHING MODE COMBINATIONS

Output voltage	Number of on-state devices among the upper four devices	First kind of switching mode combination				Second kind of switching mode combination			
		Sa1	*Sa2*	*Sa3*	*Sa4*	*Sa1*	*Sa2*	*Sa3*	*Sa4*
2Ud	4	1	1	1	1	1	1	1	1
1Ud	3	0	1	1	1	1	0	1	1
0	2	0	0	1	1	1	0	0	1
-1Ud	1	0	0	0	1	1	0	0	0
-2Ud	0	0	0	0	0	0	0	0	0

"1" represents on-state of one device; "0" represents off-state ; Ud is equal to the voltage of one capacitor.

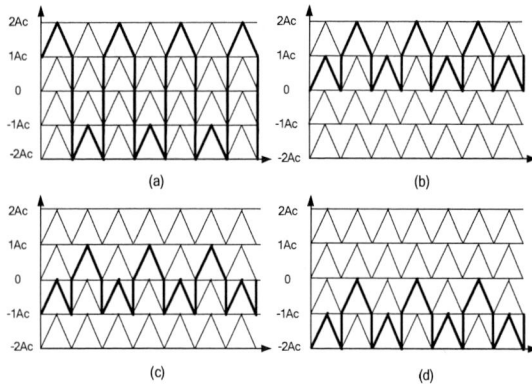

Figure 2. Carrier waveforms of the upper four main switching devices for a hybrid clamped five-level inverter with the "M" PDPWM method

(a) Carrier of Sa1   (b) Carrier of Sa2

(c) Carrier of Sa3   (d) Carrier of Sa4

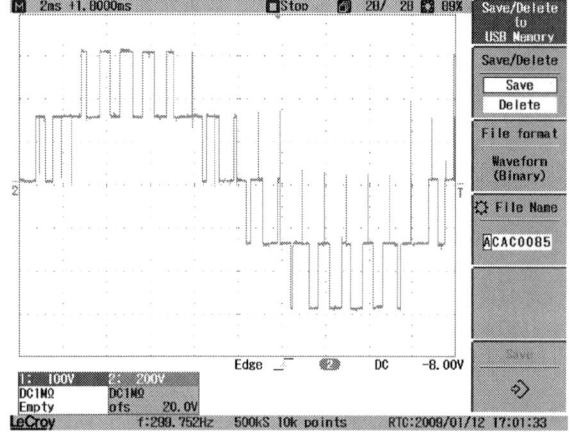

Figure 3. The unexpected levels in the experimental output voltage for a hybrid clamped five-level inverter with the PDPWM method

Among the clamping switching devices Sac1~Sac6, the adjacent switching devices are complementary. Self-voltage balancing in capacitors is realized by switching from one kind of device switching mode combination to another by turns. Two kinds of device switching mode combinations are listed in TABLE 1. Figure 2 shows the carrier waveforms of the traditional PDPWM method for a hybrid clamped five-level inverter. Figure 3 shows the experimental output voltage waveforms for a hybrid clamped five-level inverter with the PDPWM method. It can be seen that many voltage spikes emerge. The amplitude of the spike is about 1Ud, and the time width is equal to the preset dead time. They should not occur according to the normal control logic, so they are called "unexpected output levels".

The reason for the unexpected output levels are presented based on the analyses on the switching modes conversions for hybrid clamped multilevel inverters. The effect of the PWM carrier waveforms on the switching modes conversions are discussed in terms of triangle carrier construction. A novel AHPWM (Alternative Hybrid PWM) method for hybrid

clamped multilevel inverters is proposed. It can completely eliminate all the unexpected output levels.

II.   ANALYSES ON THE SWITCHING MODES CONVERSIONS

Seen from TABLE 1, the highest output level (2Ud) and the lowest one (-2Ud) respectively occur only with one kind of switching mode, and the middle output levels 1Ud, 0Ud, -1Ud with two kinds of switching modes respectively. TABLE 2 lists all the switching modes conversions for a hybrid clamped five-level inverter topology.

From TABLE 2, it can be seen that there are two cases of switching modes conversions: (1) Switching modes conversions with the different output levels; (2) Switching modes conversions with the same output levels. Taking the topology of a hybrid clamped multilevel inverter as an example these two cases of switching modes conversions will be analyzed as follows.

A.   Switching modes conversions with the different output levels

According to the switching rules of the hybrid clamped multilevel inverter topology, the switching modes conversions with the different output levels only occur between two conversions with the adjacent output levels, as shown in TABLE 2, that is to say, 1111 ↔ 0111 (+2Ud ↔ +1Ud), 0011 ↔ 0111 (0Ud ↔ +1Ud), 0011 ↔ 0001(0Ud ↔ -1Ud), 0001 ↔ 0000 (-1Ud ↔ -2Ud), etc.. In this case, only one pair of complementary devices will be turned on or off, i.e. one is turned from on to off, the other is turned from off to on. Switching modes conversions cause the output levels conversions, and during the dead time between two levels conversions no unexpected output levels emerge.

B.   Switching modes conversions with the same output levels

In the case of switching modes conversions with the same output levels, two pairs of complementary devices will be turned on or off, i.e. among the upper four devices one is turned from on to off, the other is turned from off to on, as well the lower four devices. In TABLE 2, switching modes conversions of No.5, 6, 11, 12, 17, 18 all belong to this case, in which the unexpected output levels will emerge. The magnitudes of these unexpected output levels are determined by the switching mode and the load current direction.

The current flows from the inverter bridge to the load is set as the positive current, and the current flows from the load to the inverter bridge is set as the negative one.

Figure 4 shows the change of the current path during the switching modes conversion from "1001" to "0011" with the output level 0Ud and the negative load current. The output level during the dead time is changed to +1Ud.

In the switching mode"1001", Sa1, Sa4, Sa4' and Sa3' are ON, but the load current actually flows through Sa4' and Sa3' instead of Sa1 and Sa4 because of the negative load current direction, as shown in Figure 4 (a). During the modes conversion from "1001" to "0011", Sa1 and Sa3' should be turned off, Sa3 and Sa1' should be turned on, that is to say,

978-1-4244-4782-4/10 $26.00 © 2010 IEEE

TABLE II. SWITCHING MODES CONVERSIONS FOR A HYBRID CLAMPED FIVE-LEVEL INVERTER

No.	Switching modes conversions	Output level conversions	No.	Switching modes conversions	Output level conversions
1	1111→0111	2Ud → Ud	12	1001→0011	0 → 0
2	1111→1011	2Ud → Ud	13	0011→0001	0 → -Ud
3	0111→1111	Ud → 2Ud	14	1001→1000	0 → -Ud
4	1011→1111	Ud → 2Ud	15	0001→0011	-Ud → 0
5	0111→1011	Ud → Ud	16	1000→1001	-Ud → 0
6	1011→0111	Ud → Ud	17	0001→1000	-Ud → -Ud
7	0111→0011	Ud → 0	18	1000→0001	-Ud → -Ud
8	1011→1001	Ud → 0	19	0001→0000	-Ud → -2Ud
9	0011→0111	0 → Ud	20	1000→0000	-Ud → -2Ud
10	1001→1011	0 → Ud	21	0000→0001	-2Ud → -Ud
11	0011→1001	0 → 0	22	0000→1000	-2Ud → -Ud

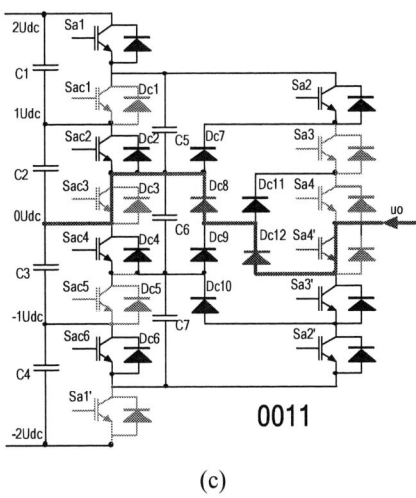

(c)

Figure 4. Change of current path during switching modes conversion (1001→0011) with the negative load current

(a) switching mode "1001"

(b) mode during the dead time

(c) switching mode "0011"

two pairs of complementary devices will be turned on or off.

At the beginning of the dead time, the turn-off signals of Sa1 and Sa3' come, thus Sa3' can be turned off immediately. The clamping diodes Dc8 and Dc12 are turned on and Dc10 is turned off, so the current path is changed, as shown in Figure 4 (b). On the other hand, the turn-off signals of Sac2, Sac4 and Sac6, which are the same as the turn-off signals of Sa1, come simultaneously. The actual load current flows through the anti-paralleled diodes of clamping IGBTs instead of clamping IGBTs themselves owing to the negative load current direction, so the anti-paralleled diode of Sac4 is still in current flow condition, thus the output level is changed to +1Ud, as shown in Figure 4 (b). At the end of the dead time, the turn-on signals of Sa3 and Sa1' come, Sa3 and Sa1' can be turned on immediately and the voltage of Sa1' is decreased to zero. On the other hand, Sac1, Sac3, Sac5 is turned on and the voltages of them are also decreased to zero, thus the voltages of the adjacent clamping IGBTs Sac2, Sac4, Sac6 are increased to Ud, so the anti-paralleled diode of Sac4 is turned off and Sac3 will be in current flow condition, as shown in Figure 4 (c). The current path is changed again, thus the output level is changed back to 0Ud. It is concluded that during the modes conversion from "1001" to "0011" with the negative load current, the output level will be changed from 0Ud to +1Ud, and then back to 0Ud, which means the emergence of output voltage spike during the dead time. The duration of voltage spike is equal to the width of the dead time.

Figure 5 shows the change of the current path during the switching modes conversion from "0011" to "1001" with the output level $0U_d$ and the positive load current. The output level in the dead time is changed to -1Ud.

In the mode "0011", Sa3, Sa4, Sa4' and Sa1' are ON, but the load current actually flows through Sa3 and Sa4 instead of Sa4' and Sa1' because of the positive load current direction, as

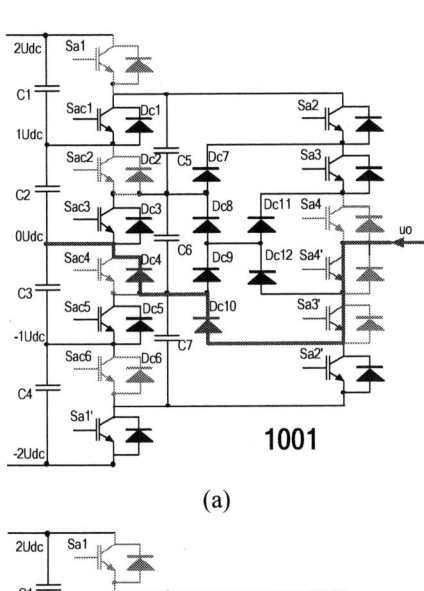

1001

(a)

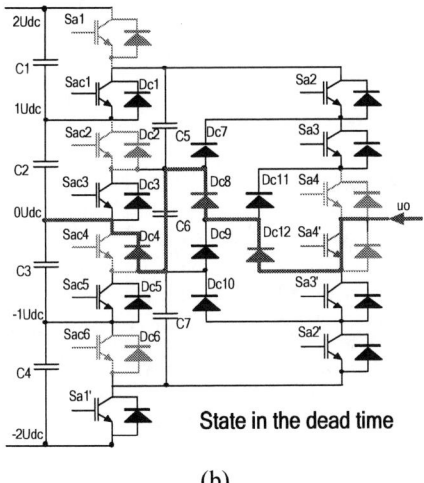

State in the dead time

(b)

978-1-4244-4782-4/10 $26.00 © 2010 IEEE   1536

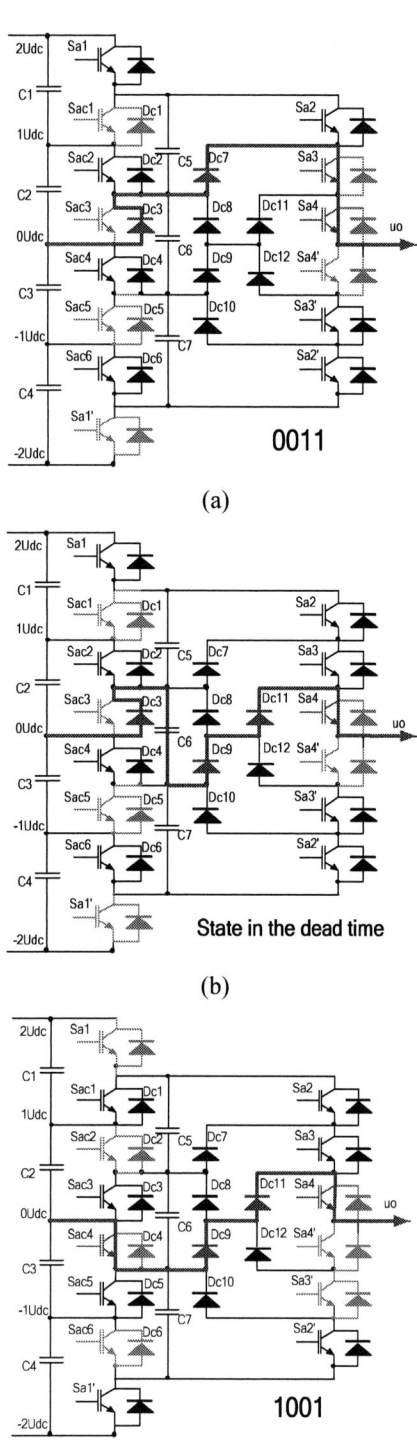

Figure 5. Change of current path during switching modes conversion
(0011→1001) with the positive load current

(a) switching mode "0011"

(b) mode during the dead time

(c) switching mode "1001"

shown in Figure 5 (a). During the modes conversion from "0011" to "1001", Sa3 and Sa1' should be turned off, Sa1 and Sa3' should be turned on, which means two pairs of complementary devices will also be turned on or off.

At the beginning of the dead time, the turn-off signals of Sa3 and Sa1' come, thus Sa3 can be turned off immediately. The clamping diodes Dc9 and Dc11 are turned on and Dc7 is turned off, so the current path is changed, as shown in Figure 5 (b). On the other hand, the turn-off signals of Sac1, Sac3 and Sac5, which are the same as the turn-off signal of Sa1', come simultaneously. The actual load current flows through the anti-paralleled diodes of clamping IGBTs instead of clamping IGBTs themselves owing to the positive load current direction, so the anti-paralleled diode of Sac3 is still in current flow condition, thus the output level is changed to -1Ud. At the end of the dead time, the turn-on signals of Sa1 and Sa3' come, Sa1 and Sa3' can be turned on immediately, but no load current flow through Sa3' owing to the positive current direction. The voltage of Sa1 is decreased to zero. Sac2, Sac4 and Sac6 is turned on and the voltages of them are also decreased to zero respectively, thus the voltages of the adjacent clamping IGBTs Sac1, Sac3, Sac5 are increased to Ud, so the anti-paralleled diode of Sac3 is turned off and Sac4 will be in current flow condition, as shown in Figure 5 (c). The current path is changed again, thus the output level is changed back to 0Ud. It is concluded that during the modes conversion from "0011" to "1001" with the positive load current, the output level will be changed from 0Ud to -1Ud, and then back to 0Ud, which means the emergence of output voltage spike during the dead time. The duration of voltage spike is equal to the width of the dead time.

Based on the above analyses on two typical switching modes conversions, it is concluded that as long as the switching modes conversions with the same output levels occur, the unexpected output levels will emerge during the dead time, and their magnitudes are determined by the switching modes and the load current directions.

III. EFFECTS OF THE CARRIER WAVEFORMS ON THE SWITCHING MODES CONVERSIONS

Seen from Figure 2, if all carrier waveforms of Sa1, Sa2, Sa3 and Sa4 are combined together, four intact carrier bands can be attained, so the carriers of PWM shown in Figure 2 can be also regarded as a PDPWM method. It can be noticed, however, the carrier waveforms of a certain switching device are positioned on two carrier bands alternatively, which is different from the general PDPWM method.

Figure 6 shows the carriers of a general PDPWM method for the hybrid clamped multilevel inverter. The initial phase angle of the triangle carrier waveform is zero, so it is called "M" PDPWM method. In Figure 6, the colors of the carrier waveforms and the corresponding PWM pulse waveforms for a certain switching device are identical and four main switching devices are corresponding to four colors respectively. Symbol "(1)", "(2)" and "(3)" in Figure 6 respectively identify three cases of the carrier waveforms intersecting the modulation waveform. In the case of "(1)", when the modulation waveform intersects the certain carrier band and the two adjacent intersecting points all locate in the

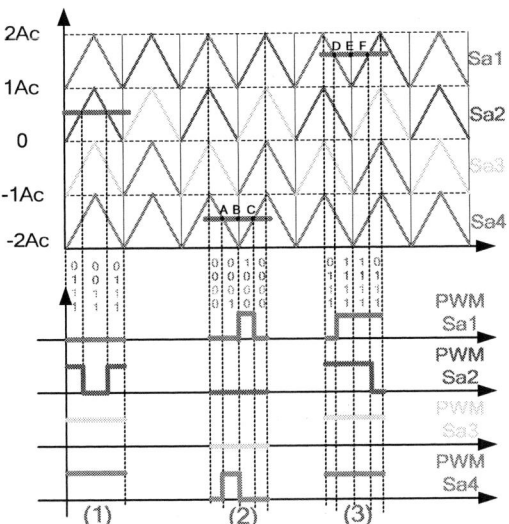

Figure 6. Carrier waveforms and the representative PWM pulse waveforms of the "M" PDPWM method

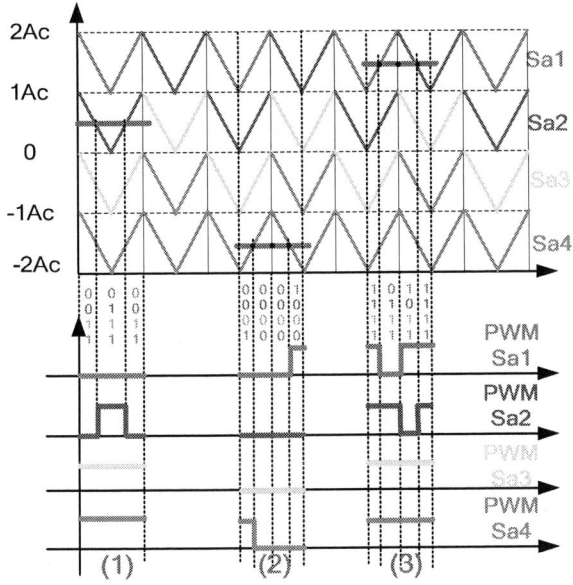

Figure 7. Carrier waveforms and the representative PWM pulse waveforms of the "W" PDPWM method

carrier waveforms of a certain switching device, only one switching device of the upper bridge leg is turned on or off during the switching modes conversion, consequently the arrange sequence of the carriers from top band to bottom band keeps unchanged during the corresponding carrier cycle, i.e. Sa1, Sa2, Sa3, Sa4 in turn. In the case of "(2)", when the modulation waveform intersects the second or the third or the fourth carrier band and the two adjacent intersecting points respectively locate in the carrier waveforms of two different switching devices, the switching modes conversion with the same output level occurs at the edge of the two kinds of carrier waveforms with different colors. Two switching devices of the upper bridge leg are turned on or off during the switching

modes conversion. For example, for the sector "AB", the arrange sequence of the carriers from top band to bottom band is Sa1, Sa2, Sa3, Sa4 in turn and the switching mode is "0001"; for the sector "BC", the arrange sequence of the carriers shifts rotationally, i.e. Sa2, Sa3, Sa4, Sa1 in turn and the switching mode is "1000". It can be seen from Figure 6, one more switching modes conversion (0001 → 1000) emerges in the point "B", the arrange sequence of the carriers shifts from "Sa1Sa2Sa3Sa4" to "Sa2Sa3Sa4Sa1". In this case, the unexpected output level will emerge, which is known from the above analysis on switching modes conversions. In the case of "(3)", when the modulation waveform intersects the first carrier band, though the two adjacent intersecting points respectively locate in the carrier waveforms of two different devices, no switching modes conversion occurs at the edge of the two kinds of carrier waveforms with different colors, which means no unexpected output level will emerge. For example, for the sector "DE", the arrange sequence of the carriers is Sa1, Sa2, Sa3, Sa4 in turn and the switching mode is "1111"; for the sector "EF", the arrange sequence of the carriers still shifts rotationally, i.e. Sa2, Sa3, Sa4, Sa1 in turn and yet the switching mode is still "1111". Though rotation shift of the arrange sequence of the carriers occurs, all switching devices of the upper bridge leg are ON, so the switching modes are unchanged.

According to the analyses on the switching modes conversions in PART II, it is known that once switching modes conversions with the same output levels occur, the unexpected output levels will emerge because of the dead time. So it is concluded that the causes of the dead time effect are: (1) two adjacent carrier cells (triangle waveform) is corresponding to two different switching devices; (2) the rotation shift of the arrange sequence of the carriers occurs at the edge of the two kinds of carrier waveforms corresponding to the different switching devices.

IV. PRINCIPLE OF THE ALTERNATIVE HYBRID PWM METHOD

As we known, the switching modes conversions with the same output levels cause the unexpected output levels, so the key to eliminate the dead-time effect is to avoid this kind of switching modes conversions.

Figure 7 shows the carriers of another PDPWM method. The initial phase angle of the triangle carrier waveform is 180 degrees, so it is called "W" PDPWM method.

Seen from Figure 6, it is known that when the modulation waveform intersects the first carrier band, even if the two adjacent intersecting points respectively locate in the carrier waveforms of two different devices, no switching modes conversion occurs at the edge of the two kinds of carrier waveforms, which means no unexpected output levels will emerge; when the modulation waveform intersects the second or the third or the fourth carrier band and the two adjacent intersecting points respectively locate in the carrier waveforms of two different devices, switching modes conversions with the same output levels will occur at the edge of the two kinds of carrier waveforms, which means unexpected output levels will emerge. In Figure 7, when the modulation waveform

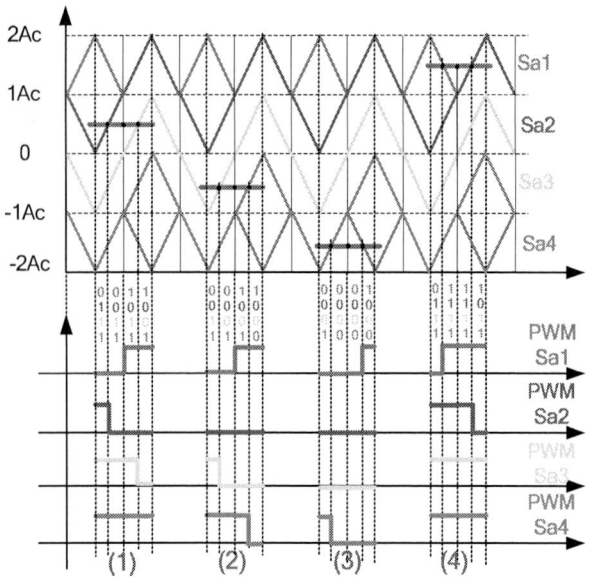

Figure 8. Carrier waveforms and the representative PWM pulse waveforms of the AHPWM method

intersects the fourth carrier band, even if the two adjacent intersecting points respectively locate in the carrier waveforms of two different devices, no switching modes conversion occurs at the edge of the two kinds of carrier waveforms, which means no unexpected output level will emerge; when the modulation waveform intersects the first or the second or the third carrier band and the two adjacent intersecting points respectively locate in the carrier waveforms of two different devices, switching modes conversions with the same output levels will occur at the edge of the two kinds of carrier waveforms, which means unexpected output levels will emerge. Obviously, if the "M" carriers are applied to the first carrier band and the "W" carriers to the fourth carrier band, the switching modes conversions causing the unexpected output levels can be avoided. On the other hand, in order to avoid the switching modes conversions causing the unexpected output levels for the second and third carrier bands, the carrier cells (triangle waveforms) in opposite phase alternatively are applied to these two carrier bands respectively, as shown in Figure 8, which is called AHPWM (Alternative Hybrid PWM) method.

Figure 8 shows four cases of the modulation waveform intersecting four carrier bands respectively. Case "(3)" is similar to the case "(2)" in Figure 7, and case "(4)" is similar to the case "(3)" in Figure 6, that is to say, no switching modes conversions occur at the edge of two kinds of carrier waveforms with different colors, so no unexpected output levels emerge. In the case of "(1)" and case of "(2)", applying the alternative hybrid carrier cells (triangle waveforms), the switching modes conversions at the edge of two kinds of carrier waveforms with different colors is corresponding to the different output levels respectively, so no unexpected output levels will emerge, which is known from the above analyses.

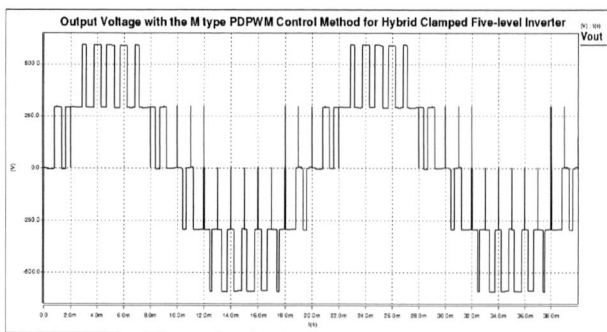

Figure 9. Simulated waveform of the output voltage for a hybrid clamped five-level inverter with the "M" PDPWM method

V. SIMULATED AND EXPERIMENTAL RESULTS

Figure 9 shows the simulated output voltage waveform for a hybrid clamped five-level inverter with the "M" PDPWM method. It can be seen some unexpected output levels emerge in the simulated waveform during the negative semi-cycle of modulation sine waveform, which is identical to the experimental waveforms, shown in Figure 3.

Figure 10 and Figure 11 respectively show the simulated and the experimental output voltage waveforms for a hybrid clamped five-level inverter with the "W" PDPWM method. It can be seen that some unexpected output levels emerge in the simulated waveform during the positive semi-cycle of modulation sine waveform, which is identical to the experimental waveforms, shown in Figure 11.

Figure 12 and Figure 13 respectively show the simulated and the experimental output voltage waveforms for a hybrid clamped five-level inverter with the proposed AHPWM method. It can be seen that no the unexpected output levels emerge, and the simulated waveform and the experimental one are identical.

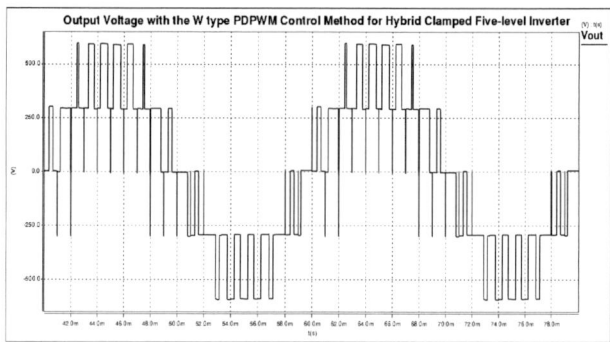

Figure 10. Simulated waveform of the output voltage for a hybrid clamped five-level inverter with the "W" PDPWM method

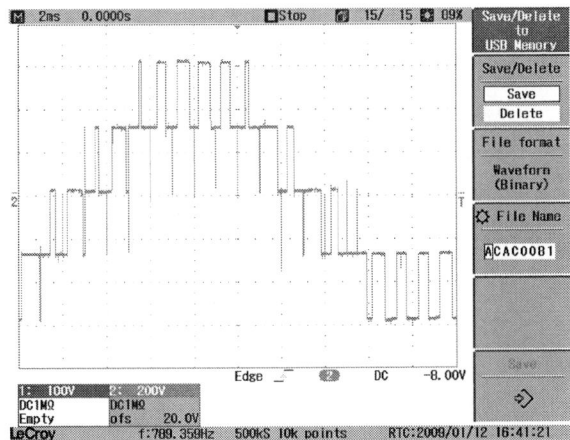

Figure 11. Experimental waveform of the output voltage for a hybrid clamped five-level inverter with the "W" PDPWM method

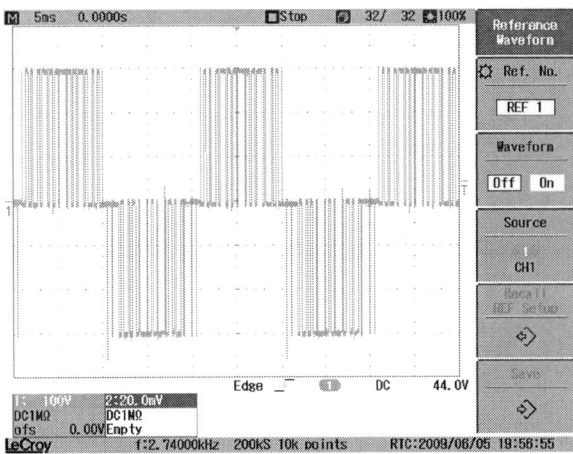

Figure 14. Experimental waveform of the output voltage for a hybrid clamped five-level inverter with the AHPWM method at ma=0.4

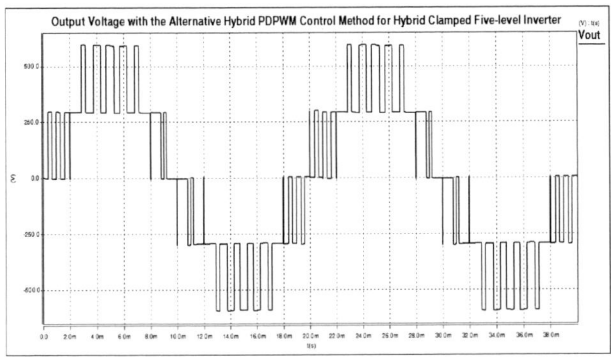

Figure 12. Simulated waveform of the output voltage for a hybrid clamped five-level inverter with the AHPWM method

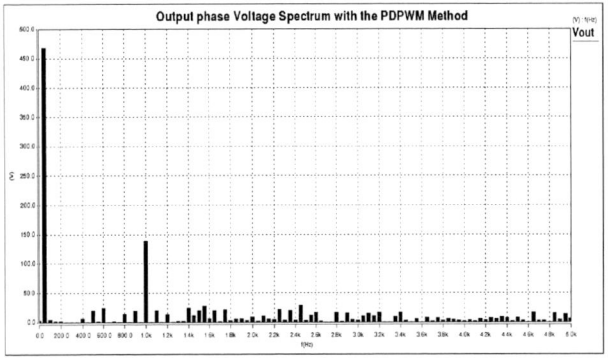

Figure 15. Simulated spectrum of the phase voltage for a hybrid clamped inverter with the PDPWM method

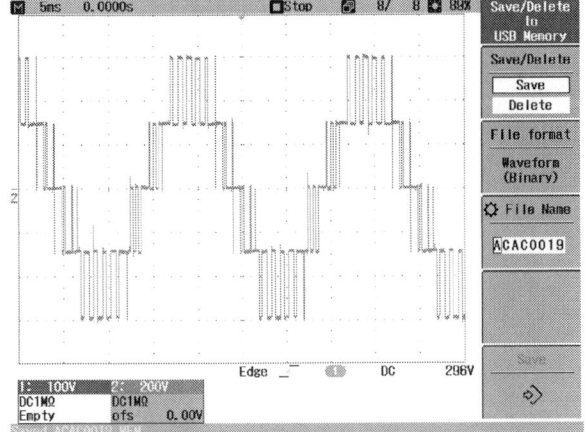

Figure 13. Experimental waveform of the output voltage for a hybrid clamped five-level inverter with the AHPWM method

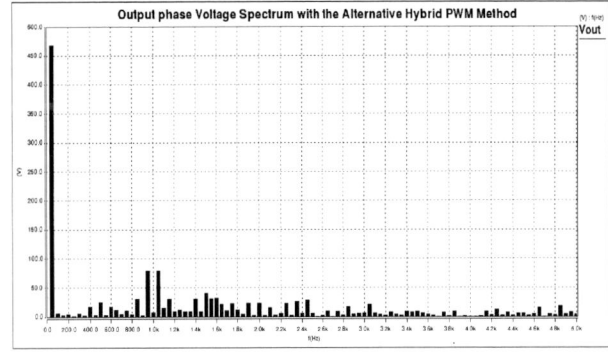

Figure 16. Simulated spectrum of the phase voltage for a hybrid clamped inverter with the AHPWM method

The Saber simulation parameters are: amplitude modulation index m_a=0.8, carrier frequency f_c=1 kHz, modulation frequency f_m=50 Hz, DC link voltage 1200V. The experiments are performed on an 18kW prototype of the hybrid clamped five-level inverter with the DC link voltage

1200V and the resistor load.

The experiment at the lower modulation index m_a=0.4 is also performed, and consequently no unexpected output levels emerge, as shown in Figure 14.

The output voltage spectrums with the traditional PDPWM method and the AHPWM method are simulated, as shown in Figure 15 and Figure 16 respectively. TABLE 3 lists the

TABLE III. CALCULATED RESULTS OF THD AND THE LOWER HARMONICS ENERGY P

PWM method		AHPWM	PDPWM
m_a=0.8	THD	35.95%	36.44%
	P	26.22%	30.86%
m_a=0.4	THD	61.90%	61.91%
	P	26.89%	61.22%

calculated results of THD and the lower harmonics energy with the proposed AHPWM method and the traditional PDPWM method at the modulation index m_a=0.8 and m_a=0.4 respectively. The variable P in TABLE 3 is calculated by

$$P = \sqrt{\sum_{n=2}^{21} V_n^2} \Big/ V_1 \qquad (1)$$

where V_1 is the fundamental amplitude, V_n is the harmonics amplitude.

Seen from TABLE 3, the THD value with the AHPWM method is less than that with the PDPWM method. The harmonics energy of lower orders with the AHPWM method is much less than that with the PDPWM method, especially in the low modulation index. The reason is that the harmonics energy transfers from the lower harmonics band to the higher harmonics band within broad modulation index range.

VI. CONCLUSIONS

The unexpected output levels emerge if applying the traditional PDPWM method to the hybrid clamped multilevel inverter, which is caused by the dead-time effect. The representative switching modes relevant to the dead-time effect are analyzed in terms of circuit principles. The reason for the unexpected output levels is that many switching modes conversions with the same output levels exist. The effect of the PWM carrier waveforms on the switching modes conversions is also discussed. A novel PWM control method called the AHPWM method is proposed for hybrid clamped multilevel inverters. It can eliminate the dead-time effect of the output voltage within broad modulation index range. On the other hand, the harmonics content and distribution of the output voltage spectrum can be changed by applying the AHPWM method, i.e. the harmonics energy mostly transfers from lower harmonics band to the higher one, especially at the low modulation index. The reduction of the energy of lower harmonics can simplify the design of output filter and reduce its size.

ACKNOWLEDGMENT

The authors would like to thank the support of Fuji Electric Advanced Technology Co., Ltd, particularly Dr. Yasushi Abe and Mr. Masaaki Wakou.

REFERENCES

[1] J.S.Lai and F.Z.Peng, "Multilevel converter-A new breed of power converters," IEEE Trans. Ind. Applicati. , vol. 32, pp.509-517.

[2] Rodriguez, J., Jih-Sheng Lai and Fang Zheng Peng, "Multilevel inverters: a survey of topologies, controls, and applications", IEEE Trans. Ind. Electron., vol.49, no.4, pp.724-738, Aug.2002.

[3] A. Nabae, I.Takahashi and H.Akagi, "A New Neutral-Point-Clamped PWM Inverter," IEEE Proc. of IAS'80, Vol.2 pp.761-766, 1980.

[4] Marchesoni, M., Tenca, P, "Diode-clamped multilevel converters: a practicable way to balance DC-link voltages", IEEE Trans. Ind. Electron., vol.49,no.4,pp.752-765,Aug.2002.

[5] X.Yuan, H.Stemmler and I.Barbi, "Investigation on the clamping voltage self-balancing of the three-level capacitor clamping inverter," in Proc. IEEE PESC, 1999, pp.491-496.

[6] Hongyan Wang, Man Lu, Yan Deng, Rongxiang Zhao and Xiangning He, "Relationship between flying capacitor multilevel inverter PWM methods and switching loss minimized PWM method for flying capacitor multilevel inverter", Power Electronics Specialists Conference, 2004. PESC 04. 2004 IEEE 35th Annual, vol.6, pp.4418-4422, Jun.2004.

[7] Song. S. G., Kang. F. S. and Park. S.-J., "Cascaded Multilevel Inverter Employing Three-Phase Transformers and Single DC Input," IEEE Trans. Ind. Electron., vol.56, no.6, pp.2005-2014, Jun. 2009.

[8] Chen, A. and Xiangning He, "Research on hybrid-clamped multilevel-inverter topologies", IEEE Trans. Ind. Electron., vol.53, no.6, pp.1898-1907, Dec.2006.

[9] Alian Chen and Xiangning He, "A hybrid clamped multilevel inverter topology with neutral point voltage balancing ability," Power Electronics Specialists Conference,2004, IEEE35th Annual, vol.5, 20-25, pp.3952-3956, Jun.2004.

[10] Jong-Lick Lin, "A new approach of dead-time compensation for PWM voltage inverters", IEEE Transactions on Circuit and System, vol.49, no.4, pp.476-483, Apr.2002.

[11] Seung-Gi Jeong and Min-Ho Park, "The analysis and compensation of dead-time effects in PWM inverters", IEEE Transaction on Power Electronics, vol.38, no.2, pp.108-114, Apr.1991.

[12] Kaitwanidvilai S., Khan-Ngern W. and Panarut M., "The impact of deadtime effect on unwanted harmonics conducted emission of PWM inverters", Environmental Electromagnetics, 2000. CEEM 2000. Proceedings. Asia-Pacific Conference on, pp.232-237,May.2000

[13] Dongsheng Zhou and Rouaud D.G., "Dead-time effect and compensations of three-level neutral point clamp inverters for high-performance drive applications", IEEE Transactions on Power Electronics, vol.14, no.4, pp.782-788, Jul.1999.

[14] Videt A., Le Moigne P., Idir N., Baudesson P. and Ecrabey, J., "A New Carrier-Based PWM for the Reduction of Common Mode Currents Applied to Neutral-Point-Clamped Inverters", Applied Power Electronics Conference, APEC 2007 - Twenty Second Annual IEEE, pp.1224-1230, Feb.2007

AUTHOR INDEX

A

Abdel-Rahman, Osama	2073
Abe, S.	30
Abu Qahouq, Jaber A.	19, 120, 1723, 1778, 1800
Acero, J.	92, 439, 1328
Agarwal, Pankaj	512
Agelidis, Vassilios G.	295
Aggeler, Daniel	1584
Agostinelli, Matteo	170
Agostini Junior, Eloi	1911
Aguilar, D.	605
Ahmad, Hani	1871
Ahmadi, Damoun	1038
Ahmed, S.	63, 881
Ahmed, Tarek	1825
Ahsanuzzaman, S.M.	980
Akhavan Fomani, Armin	132
Akin, Bilal	1990
Al Mamun, Mostafa	1261
Alahmad, Mahmoud	672
Alcalá, Janeth	1651
Alderman, Arnold	525
Alesi, Larry	849
Al-Hoor, Wisam	627, 1723
Alico, Jurgen	1113
Aller, J.M.	343, 1139
Almukhtar, Basil	1922
Alonso, J.M.	743
Alonso, Rafael	92, 439
Alou, P.	271, 723, 729, 781
Amin, Mahmoud M.N.	1640
Ang, Simon S.	750
Arias, Manuel	196
Arnet, Beat	474
Aroca, J.	1792
Arthur, Stephen	401, 1598
Asadi, Peyman	1578
Avestruz, Al-Thaddeus	444, 580, 2305
Ayana, Elias	1804
Ayyanar, Raja	2149
Azcona, R.	1300
Azongha, S.	216

B

Badstuebner, U.	773
Bae, Chae-Bong	112
Baek, Seunghun	1666
Baggu, Murali M.	2121
Bai, Sanzhong	1145
Baiju, M.R.	1963
Baker, Jonathan	143
Bakhshai, A.	149, 155, 768, 1209
Bakkaloglu, Bertan	1871
Balda, Juan Carlos	321
Ball, Arthur	533
Bao, Jianyu	1097
Bao, Weibing	1097
Barbaroux, Jean	248
Barbi, Ivo	550, 1911
Barlow, Fred	1108
Barrado, A.	1026, 1131, 1279
Barragán, L.A.	309, 439
Barreto, L.H.S.C.	837
Basu, Supratim	244
Batarseh, Issa	143, 627, 1723, 2073
Bates, John	474
Batschauer, Alessandro L.	909
Bazzi, Ali M.	256
Beaupre, Richard A.	1591
Beaupre, Richard A.	1603
Beechner, Troy	2174
Benavente, C.	729
Bendl, Jiri	895
Benfatto, I.	1622, 1810
Ben-Yaakov, Sam	928
Berzoy, Alberto	343
Bhattacharya, Subhashish	761, 1010, 1243, 1666
Bianco, A.	1287
Biela, J.	773, 1397, 1584, 1865
Bing, Zhonghui	336
Birdane, E.	1197
Boel, R.K.	1711
Boroyevich, D.	355, 408, 881, 1272, 1378, 1487, 1521
Brockerhoff, Philip	1970
Bucher, A.	557, 1763
Bueno, A.	1139
Bueno, Alexander	343
Bull, Chris	398
Burdío, J.M.	92, 309, 439, 1328
Burgos, R.	355, 408, 881

C

Cai, Jun	1018
Campa, L.	743

Canales, Francisco	550
Cao, Dong	1365
Cao, Lingling	920
Cao, Yue	968
Cárdenas, Víctor	1651
Carp, C.	1177
Carretero, Claudio	92
Casady, Jeff	1838
Celanovic, Ivan L.	961
Céspedes, Mauricio	2174
Ceyhan, Adil	1197
Cha, Hanju	1659
Chai, Jianyun	2104
Chan, Walker R.	961
Chan, Yick Po	571
Chang, Soon-Jyh	1043
Chang, Wei-Hsu	1727
Chang-Chien, Le-Ren	1043
Chapman, Patrick L.	2138, 2294
Cheah, Sze Kwan	1804
Chen, Baifeng	887, 1704
Chen, Chien Liang	619
Chen, Chingchi	676
Chen, Ching-Jan	1727
Chen, Dan	1727
Chen, Dong	818
Chen, Emil	1081
Chen, Fu-Zen	188
Chen, Guozhu	361, 1514
Chen, Henglin	642, 691
Chen, Jhih-Han	420
Chen, Jifeng	2091
Chen, Lihua	676, 1119, 1124
Chen, Min	935, 1204, 1340, 1674
Chen, Qianhong	920
Chen, Qing Su	948
Chen, W.	594, 1238, 1358
Chen, Yan	1534
Chen, Yang	1578
Chen, Yu	1435, 1441, 1464
Chen, Zheng	1572
Chen, Zhong	1448, 1471, 1616, 1627
Cheung, Chun	1081
Cheung, Victor Sui-pung	491
Chiang, T.-Y.	1413
Chinthavali, Madhu S.	1108
Chiu, Chen-Hua	1727
Chiu, Huang-Jen	948
Cho, B.-H.	1373
Cho, Un-Kwan	1561
Choi, Hangseok	36
Choi, Mun-Gi	1833

Choi, Seung-deog 1990
Choi, Sewan 1934
Choi, Sungjin 512
Choi, W.P. 2251
Choi, Woojin 466, 2166
Choi, Woo-Young 42, 1494
Chomat, Miroslav 895
Choo, Fook Hoong 2143
Chou, Welly 2057
Chowdhury, Badrul H. 2121
Chuang, S.-A. 1413
Chung, Bong-Gun 1698, 1833
Chung, Henry Shu-hung 491, 1214, 1904
Ci, Song 672
Clifford, Zachary 444
Cobos, J.A. 271, 723, 729, 781
Cochran, Travis 2154
Coelho, Enane Antônio Alves 2258
Cooke, Philip 183
Cooley, John J. 444, 2264, 2305
Corradini, L. 277
Corrêa, M.B.R. 239
Corzine, Keith 58, 452
Costabeber, A. 1287
Cox, Robert 1547
Crebier, Jean-Christophe 248, 2238
Crespo, M. 743
Cruz, C.M.T. 837
Cui, Xizhi 1610, 2042

D

da Câmara, Raphael A. 837
da Silva, Fábio Vincenzi Romualdo 2258
Das, Pritam 564, 1222
Davoudi, Ali 2138, 2294
Day, Jon 1243
de Britto, Jonas Reginaldo 2258
De Doncker, Rik 696
de Freitas, Luiz Carlos 2258
de M. Fernandes, Eisenhawer 1984
de Nie, Robert 1
De Novaes, Yales R. 550
Deboy, G. 1397
DeCarlo, Ray 480
Delgado, Eladio C. 1603
Delicado, Bernardo 1300
Deng, Yan 575, 1069, 1266
Deng, Zhe 2021
Deng, Zhiquan 1018
Dhople, Sairaj V. 2138, 2294
Dias, José A.A. 755

Diaz, D. ... 723, 729
Dick, Christian P. ... 696
Dickson, Andrew ... 702
Ding, Xiaodong ... 519
Ding, Yi ... 1555
Djabbari, Ali ... 1056
Domahidi, Alexander ... 1995
Dominguez-Garcia, Alejandro ... 256
Dong, Yan ... 79
Donlon, John F. ... 392
dos Santos Girio, J.A. ... 1306
dos Santos, Jr., E.C. ... 1191
dos Santos, Euzeli ... 755, 1183
Du, Chengrui ... 861
Du, Weijing ... 823, 1392
Du, Xiaoli ... 1732
Du, Yu ... 1145, 1666
Duerbaum, T. ... 557, 1763
Duryea, Timothy P. ... 686
Dutta, Sumit ... 761, 1666

E

Eckler, Kyle Roger ... 2202
Edrington, C.S. ... 216
Effler, Simon ... 315, 1087
Egan, Michael G. ... 787
Egelkraut, S. ... 231
Elasser, Ahmed ... 1598
El-Barbari, Said ... 1487
Elmes, John ... 143
Emadi, Ali ... 1957
Emanuel, Alexander E. ... 2096
Endredy, John ... 1172
Englehart, Amy ... 2305
Enjeti, Prasad ... 63
Erb, Dylan C. ... 2066
Eren, Suzan ... 149, 768
Ertl, H. ... 986

F

Fahimi, Babak ... 68, 1498, 2231
Falcones, Sixifo ... 2149
Fan, Haifeng ... 210
Fan, S.-Y. ... 1842
Fang, Xiong ... 1745
Farias, Valdeir José ... 2258
Fei, Wanmin ... 1034, 1732
Feng, Yupeng ... 915, 1093
Feng, Zhuomin ... 935
Ferdowsi, Mehdi ... 58, 452, 2111

Fernández, A. 196, 1313, 1792
Fernández, C. 1131, 1279
Fernández, Carlos 1300
Fernández, Cristina 1026
Ferrieux, J.-P. 1817
Filho, Faete 968
Firmansyah, E. 30
Fleming, F. 216
Foley, Raymond 525
Forsyth, A.J. 1306
Frey, D. 1817, 2238
Frey, L. 231
Friedli, Thomas 1527
Fu, Dianbo 940
Fu, Jizhen 702, 1482
Fu, P. 1622, 1810
Fujita, Atsushi 1825
Fukuda, Kenji 2030
Fukushima, Kentaro 289

G

Gacio, D. 743
Galperti, C. 2281
Gamboa, Gustavo 143
Gao, Feng 1555
Gao, G. 1622
Gao, G. 1810
Gao, Mingzhi 1204
Gao, Mingzhi 1674
Garces, Luis 1295
Garcia, J. 743
García, O. 271, 723, 729, 781
Gargoom, A. 162, 2132
Garrett, Jerome 1598
Gazel, Nicolas 1272
Ge, Baoming 1124
Ge, Qiongxuan 1419
Geng, Hua 2126
Gillmor, Colin 1384
Gimeno, A. 729
Glaser, John S. 401, 654
Gong, Jinwu 887, 1704
Gonzalez, M.C. 271, 781
Gonzalez-Llorente, Jesus 666, 1062, 2161, 2226
Gowda, Arun V. 1591, 1603
Graham, Jeff 2220
Green, R. 1568
Grogan, S.A.S. 873
Guépratte, K. 1817
Guerrero, Josep M. 380
Guo, Rong 1172

Guo, Suxuan	887, 1704
Guo, Zhiqiang	662
Gupta, Ranjan K.	901

H

Ha, Dong Sam	2154
Habetler, T.	343, 1139
Halton, Mark	1075, 1087, 2207
Hamilton, Christopher	143
Han, Baikhee	512
Han, Byung M.	303
Han, Jung Hee	1108
Han, Sangmin	2288
Han, Yunlong	1534
Hang, Lijun	2021
Haque, M.E.	162, 2132
Harada, Shinsuke	2030
Harada, Yosuke	289
Harbaugh, Mark	425
Harris, John H.	1048
Hartmann, L.V.	239
Hartmann, M.	986
Hartnett, Kevin J.	787
Haruni, A.M.O.	162, 2132
Hasan, Jaber	750
Hayes, John G.	787
He, Chao	361
He, Xiangning	575, 681, 801, 1069, 1266, 1454, 1534, 2080, 2300
He, Yanhui	915, 1093
He, Yingjie	1692
Hegarty, Tim	1056
Heldwein, Marcelo L.	909
Henze, C.P.	605
Herbert, Edward	1048
Herbsommer, Juan A.	398
Hewson, Christopher R.	2050
Hirose, Fumitoshi	1879
Ho, W.C.	994, 2251
Holmes, D.G.	873
Hong, Xiaoyuan	1238, 1358, 2021
Hosoda, Hiromi	1261
Hsu, G.-W.	1849
Hu, Haibing	519, 627, 1785, 2073
Hu, Qingcong	2314
Hu, Yuequan	203
Huang, Alex	761, 849, 1010, 1145, 1172, 1477, 1666, 1875, 2181
Huang, Hong	1770
Huang, Jing-Yi	1043
Huang, Xiucheng	433, 823, 1392
Huber, Laszlo	203
Huemer, Mario	170

Huh, Dong-Young 1885, 1949
Hui, Joanne 155
Hui, S.Y.R. 86, 594, 994, 1346, 2251
Hung, Chung-Wen 420
Husain, Iqbal 2007
Hutchens, Chris 1056
Hwu, K.I. 507, 710, 737, 1942
Hyeon, B.-C. 1373

I

Ide, Kozo 103
Inman, Daniel J. 2154
Ioinovici, Adrian 1214, 1904
Ishii, Kenichiro 74
Ishizuka, Yoichi 1879
Itoh, Jun-Ichi 1684
Izquierdo, D. 1300

J

Jacobina, C.B. 755, 1183, 1191, 1984
Jain, P.K. 14, 149, 155, 499, 544, 768, 1209, 1248, 1334, 2321
Jakobsen, Uffe 98
Jang, Minsoo 295
Jang, Sang-Ho 1833
Jang, Yungtaek 23
Jayakanthan, Gnanavel 794
Jeannin, P.-O. 248, 1817, 2238
Jeon, Yong-Seog 1949
Jeong, In Wha 1166
Jeong, Yu-Seok 303
Ji, Biao 1448, 1471
Ji, Feng 1448, 1471
Ji, Young-Hyok 2275
Jia, Liang 124
Jiang, Dong 408
Jiang, Wei 68
Jiménez, O. 309
Jimenez-Brea, Emil 666, 1062, 2161
Jing, Wei 1010
Jovanović, Milan M. 23, 203
Jung, Ha-Jin 1678
Jung, Sungyoon 2002
Jung, Yong-Chae 1885, 2275

K

Kanai, Takeo 1101
Kang, Sung-In 1885
Kang, Yong 1272, 1435, 1441, 1464
Kato, Koji 1684

Kazimierczuk, Marian K. 2212
Kelleher, Paul 1922
Kelley, Robin 1838
Kelly, Anthony 328, 1922, 2189
Keogh, Bernard 1384
Kerkman, Russ J. 634
Kesler, Metin 374
Khaligh, Alireza 1755, 2066, 2245
Khazarei, Mostafa 452, 58
Khoobroo, Amir 2231
Kim, Deuk-Soo 1678
Kim, Eun-Soo 1698, 1833, 1885, 1949
Kim, Gi-Taek 1678
Kim, Gyeong-Hun 2085
Kim, Hyun-Cheol 112
Kim, Jang-Mok 112
Kim, Jin-Tae 540
Kim, Joo-Hoon 1698, 1885, 1949
Kim, Jun-Gu 2275
Kim, Rae-Young 1890
Kim, Sang-Hyun 466, 2166
Kim, Sungmin 103
Kim, Tae-Hoon 466, 2166
Kim, Teahoon 512
Kim, Wook 466, 2166
Kim, Young-Gook 112
Kim, Young-Ho 2275
Kim, Young-Ju 2085
Kimball, Jonathan W. 1508, 2121, 2202
Kirlin, R. Lynn 1749
Kirtley, Jr., James L. 1547
Kisacikoglu, Mithat C. 458
Knight, Andy 2013
Koellner, Walter 1158
Kolar, J.W. 773, 986, 1397, 1527, 1584, 1865
Kong, Na 2154
Kong, Pengju 1424
Koo, Gwan-Bon 540
Krein, Philip T. 256
Krishnamurthy, Mahesh 1957
Kutkut, Nasser 627
Kwon, Bong-Hwan 1494
Kwon, Soon Kurl 1230
Kye, Moon-Ho 1698, 1833

L

Lai, Jih-Sheng 42, 387, 474, 619, 1056, 1494, 1890
Lai, Pengjie 1927
Lai, Rixin 355, 408
Lam, John 2321
Lamar, Diego G. 196

Lascu, Cristian	1749
Laughman, Christopher	1547
Lázaro, A.	1026, 1131, 1279
Leão, J.F. Araujo	239
Lee, Beomseok	2002
Lee, C.K.	86
Lee, Fred C.	79, 176, 533, 940, 1424, 1927
Lee, Hoi	686
Lee, Jae-Sam	1885, 1949
Lee, Jong-Hak	466, 2166
Lee, Jun-Young	303
Lee, Kwang-Ho	1698, 1949
Lee, Sangwon	1934
Lee, Ting-Peng	948
Lee, Tzung-Lin	380
Lee, Yuang-Shung	619
Leeb, Steven B.	444, 580, 1547, 2194, 2264, 2305
Lei, Qin	844, 854, 1002
Leslie, Scott	474
Li, Chushan	681, 2300
Li, Duo	935
Li, Hong	1740
Li, Hui	210, 223, 807
Li, Jian	176
Li, Jin	1521
Li, Jiping	2036
Li, Jun	1010
Li, Ming	1745
Li, Qiang	79, 533
Li, S.N.	594
Li, Weichen	1454, 2080
Li, Wuhua	801, 1069, 1454, 2021, 2080
Li, Xiao	861
Li, Yaohua	1419
Li, Yong	349
Li, Yongdong	1736, 2104
Liang, Xiaoguo	794
Liang, Zhigang	849, 1477
Liao, XiaoZhong	662
Lim, Michele	533
Lim, Sungkeun	1875
Lima, A.M.N.	239, 1984
Lin, Cheng-Tsung	420
Lin, D.Y.	1346
Lin, Fei	1740
Lin, Hai	818
Lin, Hung-Chih	2154
Lisi, Gianpaolo	1056
Litvinov, Alexander	474
Liu, Chih-Wen	420
Liu, Chui Pong	571
Liu, Congwei	1419

Liu, Jingbo	425
Liu, Jinjun	915, 1093, 1521, 1633, 1692, 2116
Liu, Kun	1674
Liu, Liming	223
Liu, Ting	2116
Liu, X.	86, 994, 2251
Liu, Yan-Fei	124, 702, 1482
Liu, Zeyuan	1018
Lo, Yu-Kang	948
Loh, Poh Chiang	1555, 2143
López del Cerro, F.J.	1300
Lopez, Osvaldo	398
Lorduy, Abad	1026
Losee, Peter	401, 1598
Lu, Kaiyuan	98
Lu, Ying	1785
Lu, Zhengyu	519, 1238, 1358, 1542, 2021, 2091
Lucena, Carlos	1026
Lucía, O.	92, 309, 439, 1328
Lukic, Srdjan	1145
Lukić, Zdravko	1, 315
Luo, Fang	1272
Luo, Yingpeng	1616, 1627

M

Ma, Chongguang	1340
Ma, Dongsheng	284, 813
Maciel, A.M.	1191
Maddaleno, Franco	1166
Mahdi, Abdulhussain E.	2207
Maksimović, Dragan	188, 277, 980
Mankani, A.D.	1622, 1810
Mantooth, H. Alan	321
Mao, Jingxin	1740
Mao, Xiaojing	1405
Mariéthoz, Sébastien	1995
Marsili, Stefano	170
Martin, Daniel	619
Marxgut, Christoph	1865
März, M.	231
Massoud, Ahmed	63
Matocha, Kevin S.	401
Matsuo, Hirofumi	1879
Mattavelli, P.	953, 1287, 2281
Mazumdar, Joy	1158
McGrath, B.P.	873
Melkebeek, J.A.A.	1711
Meng, Peipei	433, 642, 691
Metwally, M.K.	414
Meyer, Eric	124
Miaja, P.F.	715

Miftakhutdinov, Rais 1897
Milivojevic, Nikola 1957, 2245
Millan, Ignacio 92, 439, 1328
Miller, Greg 1081
Millner, Alan 1572
Min, Chen 1460, 1503
Ming, Zhengfeng 1919
Ming, Zheng-Feng 109
Mishima, Tomokazu 1230
Mo, Qiong 1674
Moghe, Rohit 1158
Mohammed, O.A. 1640
Mohan, Ned 901, 1804
Mohapatra, Krushna K. 901
Molina Cardozo, Diogenes D. 321
Moon, Sang-Cheol 540
Mooney, James 2207
Morari, Manfred 1995
Moschopoulos, Gerry 564, 829, 1222, 1320
Motto, Eric R. 392
Mourra, O. 1313, 1792
Mousavi, Ahmad 564, 1222
Muller, Sean 2194
Murai, H. 648
Muralidhar, Gautam 19
Mussa, Samir A. 909

N

Nakano, Shinya 74
Nakaoka, Mutsuo 1230, 1825
Nam, Kwanghee 2002
Narveson, Brian 525
Nasadoski, Jeffrey J. 401
Navarro, D. 309
Neely, Jason 480
Negnevitsky, M. 162, 2132
Neuman, Sabrina 2194
Neumeyer, C. 1810
Ng, W.M. 594, 1346
Ng, Wai Tung 132
Ngo, K.D.T. 533, 2036
Ngo, Phong 813
Nguyen, Do Hung 750
Nguyen, The-Van 2238
Ni, Guang-Zheng 109
Nilles, Gerald 2294
Ning, Puqi 1378
Ninomiya, T. 30, 289
Nishi, Mariko 1879
Nondahl, Thomas 425
Noquil, Jonathan 398

Norford, Les K. ... 1547
Norigoe, Isami ... 289
Núñez, Ciro ... 1651

O

Oh, J.S. ... 1810
Ohashi, Hiromichi ... 648, 1101, 2030
Oliveira, Jr., D.S. ... 837
Oliveira, Alexandre C. ... 1984
Oliver, J.A. ... 271, 723, 729, 781
O'Malley, Eamon ... 328, 1922, 2189
Ó'Mathúna, Cian ... 525
Omori, Hideki ... 1825
Onar, Omer C. ... 1755, 2066
Ongaro, F. ... 2281
Orabi, Mohamed ... 1778, 1800
Orietti, E. ... 953
Orji, Uzoma A. ... 1547
Ortega, F.J. ... 729
Ortiz-Rivera, Eduardo I. ... 666, 1062, 2161, 2226
Otsuki, Etsuo ... 74
Ozdemir, Engin ... 367, 374
Ozdemir, Sule ... 367
Ozpineci, Burak ... 458

P

Page, Sarah ... 2194
Pahlevaninezhad, Majid ... 149, 768
Pallo, Nathan A. ... 961
Pan, S. ... 499, 1248
Pang, H.M. ... 973, 1857
Parayandeh, Amir ... 980
Pardo, J.M. ... 729
Paris, James ... 580, 1547, 2194, 2305
Park, Jinseok ... 2181
Park, Jun-Ho ... 1885
Park, Minwon ... 2085
Park, Sung-Yeul ... 387
Parkhideh, Babak ... 1666
Parto, Parviz ... 1578
Pasquesoone, Gregory ... 2007
Pautsch, Adam G. ... 1591
Pawellek, A. ... 557
Pei, Xuejun ... 1435, 1441, 1464
Pei, Yunqing ... 1610, 2042, 2060
Pekarek, Steve ... 480
Peng, Fang Z. ... 818, 844, 854, 1002, 1119, 1124, 1365, 2288
Peng, Li ... 1435, 1441, 1464
Pepper, Michael ... 143
Perin, Arnaldo J. ... 909

Perreault, David J. 961
Pilawa-Podgurski, Robert C.N. 961
Pokryvailo, A. 1177
Pong, M.H. Bryan 973, 1857
Pong, Man Hay 571
Poon, Ngai Kit 571
Poucand, M. 1306
Praça, P.P. 837
Priewasser, Robert 170
Prodić, Aleksandar 1, 315, 980, 1113, 1256

Q

Qi, Tao 2220
Qian, Hao 474, 1056
Qian, Wei 2288
Qian, Zhaomin 1674
Qian, Zhaoming 642, 691, 818, 823, 935, 1002, 1204, 1340, 1392
Qian, Zhijun 2073
Qiu, Weihong 1081
Quesada, Isabel 1026

R

Radić, Aleksandar 1, 315
Rahimian, Mina M. 1990
Ramamurthy, Anand 1243
Rauch, M. 231
Ray, William F. 2050
Remscrim, Zachary 444, 1547, 2194
Ren, Xiaoyong 920
Ren, Zheng 1204, 1674
Restrepo, J. 343, 1139
Reutzel, Evan 1430
Rico-Secades, M. 743
Rinne, Karl 328, 1075, 1087, 1922, 2189
Ritenour, Andrew 1838
Rivas, Juan M. 654
Rocha, Nady 1183
Rodríguez, A. 715
Rodríguez, M. 715
Rodríguez-Valdez, Carlos D. 634
Rosas, Emanuel 1651
Royak, Semyon 425
Ruan, Xinbo 920, 1214, 1405
Rylko, Marek S. 787

S

Sabate, Juan 1487
Sadakata, Hideki 1825
Sagawa, Natsumi 2212

Saggini, S.	953, 1287, 2281
Saha, Bishwajit	1825
Salah Morsy, Ahmed	63
Salazar-Llinas, Andres	666, 1062, 2161, 2226
Salem, T.E.	1568
Salmon, John	2013
Samsi, Rohan	183
Sanders, Seth	1430
Sanz, M.	1279
Satoh, K.	392
Sayed, Khairy Fathy	1230
Scanlan, Tony	1075
Scapellati, C.	1177
Schantz, Christopher	1547, 2194
Scharrer, Martin	1075
Schletz, A.	231
Schmidt, Peter	425
Schreier, Ludek	895
Schulz, Martin	1970
Schutten, Michael	1598
Schweizer, Mario	1527
Sebastián, J.	196, 715
Seger, Eric	2264
Seidlitz, Steve	1804
Sekiya, Hiroo	2212
Sen, P.C.	702, 1482, 1719
Sepahvand, Hossein	58, 452
Sha, Deshang	662
Shao, Jianwen	601
Sharif, Hamid	672
Shaw, Steven R.	2194, 2264
Shen, Guoqiao	861
Shen, John	627
Shen, Weixiang	2143
Sheng, Honggang	1572
Sheng, Z.	1622, 1810
Sheridan, David	1838
Shi, Lei	1448, 1471
Shi, Wei	1785
Shih, Frank	948
Shim, Won-Sul	1678
Shinohe, Takashi	1101, 2030
Shiny, G.	1963
Shoyama, M.	30, 289
Shuai, Peng	550
Silva, C.E.A.	837
Simanjorang, Rejeki	648
Singh, Bhim	1976
Singh, Sanjeev	1976
Slepchenkov, Mikhail	1166
Slowey, John	525
Smedley, Keyue	47, 264, 1166

Smith, Chris 474
Smith, Greg 1056
Solovitz, Stephen A. 1591
Somani, Apurva 901
Somayajula, Deepak 2111
Song, Bo 2060
Song, Byeong-Mun 2085
Song, Z.Q. 1810
Spiazzi, G. 953
Stamenkovic, Igor 1957, 2245
Steimer, Peter K. 1865
Steiner, Reto 1865
Stephan, H. 1817
Steurer, M. 216
Stevanovic, Ljubisa D. 401, 1591, 1603
Straeussnigg, Dietmar 170
Stum, Zachary 1598
Su, Gui-Jia 1152
Su, Jen-Ta 420
Su, Y.-H. 1842
Su, Y.P. 86
Sugimura, Hisayuki 1230
Sul, Seung-Ki 103, 611, 1561
Sullivan, Charles R. 1048
Sumiyoshi, Shinichiro 1825
Sun, Jian 336, 2174, 2220
Sun, Jianjun 887, 1704
Sun, Julu 533, 1927
Sun, Pengwei 387, 474
Sun, Yi 176
Sung, Kyungmin 1101
Suzuki, K. 392
Szczesny, Paul 1295

T

Takahashi, T. 392
Takao, Kazuto 1101, 2030
Takeda, Takashi 648
Tamyurek, B. 1197
Tan, Kai 1419
Tan, Kuan Khoon 1555, 2143
Tanaka, Yasunori 1101
Tang, Lixin 1152
Tang, Yu 867
Tao, J. 1622, 1810
Tao, X.H. 594
Thomas, Brinda A. 588
Thompson, Chris 1243
Titiz, Furkan Kaan 696
Tjokrorahardjo, Andre 1352
Todd, R. 1306

Todeschini, Grazia 2096
Tolbert, Leon M. 458, 968, 1108
Toliyat, Hamid A. 1990
Tomioka, S. 30
Tomita, Koji 103
Tonicello, F. 1313, 1792
Torrico-Bascopé, R.P. 837
Tran, Manh Hung 248
Trowler, Derik 321
Trzynadlowski, Andrzej M. 1749
Tschirhart, Darryl J. 14, 544, 1334
Tseng, S.-Y. 1413, 1842, 1849
Tsukakoshi, Kenta 289

U

Ucar, Mehmet 367
Undeland, Tore M. 244
Urciuoli, D.P. 1568
Urriza, I. 309

V

Vafakhah, Behzad 2013
Vagnon, Eric 2238
Vaks, Nir 480
Valdivia, V. 1131, 1279
Vasić, M. 271, 723, 729
Veillette, Robert J. 2007
Vickery, Dan 444, 2305
Vieira, Jr., João Batista 2258
Viola, Julio C. 343
Visairo, H. 271, 781
Vodyakho, O. 216
Volfson, Oleg 138
Vu, Trung-Kien 1659
Vyncke, T.J. 1711

W

Wada, Keiji 1101
Waldron, Finbarr 525
Wang, Dong 124
Wang, Fred 355, 408, 881, 1272, 1378, 1487, 1572
Wang, Gangyao 761, 1666
Wang, Huai 491, 1904
Wang, Jin 676, 1038
Wang, Jun 1266
Wang, K.-C. 1413, 1849
Wang, Ke 1745
Wang, Kunrong 7
Wang, Laili 1610, 2042, 2060

Wang, Meng 794
Wang, Mingliang 2036
Wang, Peng 1555, 2143
Wang, Ruxi 1378
Wang, Shunqing 1627
Wang, Shuo 940, 1272
Wang, Siran 1097, 1238, 1358, 1542, 2021, 2091
Wang, Yen-Ching 380
Wang, Yousheng 818
Wang, Yue 915, 1093, 1745
Wang, Zhan 807
Wang, Zhaoan 1610, 1692, 2042, 2060
Wang, Zhengshi 575
Wang, Zhiqiang 361, 1514
Watson, Luke 2121, 2202
Wegner, Hagen 1384
Wei, Jukui 867
Wen, Jun 47
Wichakool, Warit 580, 1547
Wijeratne, Dunisha 829
Wilson, Jr., Thomas G. 183
Wolbank, T.M. 414
Won, Chung-Yuen 2275
Wood, R.A. 1568
Wu, Bin 1034, 1732, 2126
Wu, Chun-Hsun 1043
Wu, D. 1306
Wu, Guan-Hong 948
Wu, Haimeng 575
Wu, Jiande 575, 681, 2300
Wu, Jinlong 915, 1093
Wu, W.-C. 1842
Wu, Xinke 642, 691
Wu, Zhichao 223

X

Xiao, Xi 1736
Xie, Chuan 361, 1514
Xie, Huikai 2036
Xie, Shaojun 867
Xing, Lei 2174
Xing, Yan 1785
Xu, Biwen 1340
Xu, Chunchun 1295
Xu, Dehong 861
Xu, Dewei 2126
Xu, L. 1622
Xu, L.W. 1810
Xu, Ligang 920
Xue, Jianren 1785
Xue, Tao 519

Y

Yamada, Yusuke	1879
Yamaguchi, Hiroshi	648
Yamazaki, Mikio	648
Yan, W.	1346
Yang, Bing-Zhong	109
Yang, Binjian	1266
Yang, Bo	801
Yang, C.-M.	1849
Yang, Geng	2126
Yang, Jianyou	642
Yang, Joonhyun	512
Yang, Liyu	1172, 2181
Yang, Shuitao	844, 854, 1002
Yang, Xinyi	1340
Yang, Xu	1610, 2042, 2060
Yao, Kai	1405
Yao, Wei	1204, 1674
Yao, Wenxi	519, 2091
Yau, Y.T.	507, 710, 737, 1942
Yazdani, D.	1209
Ye, Shaoshi	1238, 1358
Ye, Zhihong	1405
Yim, Jung-Sik	1561
Yin, Zhenggang	1419
Ying, Yucheng	533
Yisheng, Yuan	1460, 1503
Yoo, Hyunjae	611
Yoon, Young-Doo	103
York, Ben	1890
Yoshihiura, Y.	392
Yoshino, Teruo	1261
You, Xiaojie	1740
Young, George	1384
Youssef, Mohamed	1778, 1800
Yu, Haidong	1498, 2025
Yu, In-Keun	2085
Yu, Wen Long	948
Yu, Wensong	387, 474, 1056
Yu, Wen-Song	42
Yuan, Wei	433, 823, 1392
Yuan, Xibo	2104

Z

Zane, Regan	2314
Zawodniok, Maciej	1508
Zeltser, Ilya	928
Zeng, Jie	761
Zha, Xiaoming	887, 1704
Zhang, Di	1487
Zhang, Guoxing	433

Zhang, Hui 1108
Zhang, Jianhui 1056
Zhang, Jing 1514
Zhang, Jiucai 672
Zhang, Jun 861
Zhang, Junming 433, 823, 1392
Zhang, Leqiang 1745
Zhang, Xin 1214
Zhang, Xuan 2116
Zhang, Yanli 1034
Zhang, Yi 284
Zhang, Yingqi 1719
Zhang, Zhe 935
Zhang, Zhiliang 702, 1482
Zhang, Zhongchao 1097
Zhao, April 132
Zhao, Guopeng 1610, 1633, 1745, 2042, 2060
Zhao, Jing 1266, 1534
Zhao, Rongxiang 1534
Zhao, Tiefu 761, 1666
Zhao, Yi 801, 1069, 1454, 2080
Zhao, Zhenyu 1256
Zhaoming, Qian 1460, 1503
Zheng, Cong 2245
Zheng, Feng 1919
Zheng, Jerry 349
Zheng, Sheng 691, 818
Zheng, Trillion Q. 1740
Zheng, Yuzhen 2080
Zhixin, Xu 1266
Zhong, W.X. 994
Zhou, Liang 47, 264
Zhou, Linyuan 2116
Zhou, Xia 2091
Zhou, Xiaohu 849, 1145, 1666
Zhou, Xin 1477
Zhu, Guangyong 7
Zhu, Haipeng 2143
Zhu, Hao 1736
Zhu, Yinyu 1616, 1627
Zou, Ke 676, 1038
Zou, Yunping 1692
Zumel, P. 1131, 1279